THE
ENCYCLOPEDIA
OF PHYSICS

THE ENCYCLOPEDIA OF PHYSICS

THIRD EDITION

EDITED BY

Robert M. Besançon

VNR VAN NOSTRAND REINHOLD
————————————— New York

Library of congress Catalog Card Number 84-13045
ISBN 0-442-00522-9 (pbk.)

Manufactured in the United States of America

Van Nostrand Reinhold
115 Fifth Avenue
New York, New York 10003

Van Nostrand Reinhold International Company Limited
11 New Fetter Lane
London EC4P 4EE, England

Van Nostrand Reinhold
480 Latrobe Street
Melbourne, Victoria 3000, Australia

Nelson Canada
1120 Birchmount Road
Scarborough, Ontario MIK 5G4, Canada

15   14   13   12   11   10   9   8   7   6   5   4   3   2   1

Library of Congress Cataloging in Publication Data

Main entry under title:

The Encyclopedia of physics.

   1. Physics—Dictionaries. I. Besançon, Robert M.
(Robert Martin)
QC5.E546 1984   530'.03'21   84-13045
ISBN 0-442-25778-3
ISBN 0-442-00522-9 (pbk.)

CONTRIBUTORS

JEROME L. ACKERMAN, Department of Chemistry, University of Cincinnati, Ohio. *Magnetic Resonance*.

KÊITSIRO AIZU, Hitachi Central Research Laboratory, Tokyo. *Ferroicity, Ferroelectricity, and Ferroelasticity*.

CHRISTOPHER W. ALLEN, Department of Chemistry, University of Vermont. *Chemistry*.

DOUGLAS L. ALLEN, Emery Industries, Cincinnati, Ohio. *Vapor Pressure and Evaporation*.

CHARLES L. ALLEY, Department of Electrical Engineering, University of Utah. *Modulation*.

ROBERT C. AMME, University of Denver, Colorado. *Ionization*.

DAVID L. ANDERSON, Department of Physics, Oberlin College, Ohio. *Electron*.

C. L. ANDREWS, Emeritus, Department of Physics, State University of New York at Albany. *Doppler Effect*.

ROBERT E. APFEL, Applied Mechanics, Yale University. *Cavitation*.

H. L. ARMSTRONG, Department of Physics, Queen's University, Kingston, Ontario. *States of Matter*.

ATAM P. ARYA, Department of Physics, West Virginia University. *Simple Machines*.

P. W. ATKINS, Physical Chemistry Laboratory, University of Oxford, England. *Physical Chemistry*.

GEORGE E. BACON, University of Sheffield, England. *Neutron Diffraction*.

D. C. BAIRD, Royal Military College of Canada, Kingston, Ontario. *Measurements, Principles of*.

RADU BALESCU, Faculty of Sciences, Université Libre de Bruxelles, Belgium. *Statistical Mechanics*.

WILLIAM BAND, Department of Physics, Washington State University. *Mathematical Physics* and *Quantum Theory*.

L. E. BARBROW, National Bureau of Standards, Washington, D.C. *Photometry*.

JAMES A. BARNES, National Bureau of Standards, Boulder, Colorado. *Atomic Clocks*.

C. J. BARTLESON, Research Laboratories, Eastman Kodak Company. *Color*.

ROBERT P. BAUMAN, Department of Physics, University of Alabama in Birmingham. *Absorption Spectra*.

WILLIAM L. BAUN, The Materials Laboratory, Wright Patterson Air Force Base, Ohio. *X-rays*.

JOSEPH J. BECKER, General Electric Research and Development Center, Schenectady, New York. *Magnetism*.

ALBERT C. BEER, Battelle-Columbus Laboratories, Ohio. *Hall Effect and Related Phenomena*.

BARRY A. BELL, Electrosystems Division, National Bureau of Standards. *Electrical Measurements* (with Forest K. Harris).

DAVID A. DELL, Emeritus, University of Hull, England. *Cybernetics*.

H. E. BENNETT, Michelson Laboratory, Naval Weapons Center, China Lake, California. *Reflection*.

REUBEN BENUMOF, Department of Applied Sciences, College of Staten Island, New York. *Alternating Currents*.

M. J. BERAN, School of Engineering, Tel Aviv University, Israel. *Coherence*.

LEO L. BERANEK, Acoustical Consultant, Winchester, Massachusetts. *Architectural Acoustics*.

ERIK BERGSTRAND, *Velocity of Light*.

ARTHUR I. BERMAN, Risø Library, Denmark. *Astronautics*.

ROBERT M. BESANÇON, Editor, The Encyclopedia of Physics, 515 Grand Avenue, Dayton, Ohio. *Physics*.

GEORGE L. BEYER, Retired, Eastman Kodak Company. *Molecular Weight*.

P. J. BILLING, *Induction Heating*.

EDWARD A. BIRGE, Department of Botany and Microbiology, Arizona State University. *Molecular Biology*.

CHARLES A. BITTMANN, Solid State Laboratory, Hewlett-Packard Company, Palo Alto, California. *Transistors*.

ALFRED K. BLACKADAR, Department of Meteorology, Pennsylvania State University. *Meteorology*.

JOHN P. BLEWETT, Brookhaven National Laboratory. *Accelerators, Linear*.

N. BLOEMBERGEN, Harvard University. *Light*.

WARREN B. BOAST, Emeritus, Department of Electrical Engineering, Iowa State University. *Potential*.

ANDREW H. BOBECK, AT&T Bell Laboratories. *Ferrimagnetism* (with W. H. von Aulock).

H. V. BOHM, Department of Physics, Wayne State University, Detroit, Michigan. *Fermi Surface* (with Norman Tepley and George Crabtree).

BRUCE BOLT, Department of Geology and Geophysics, University of California, Berkeley. *Seismology.*

JILL C. BONNER, Department of Physics, University of Rhode Island. *Antiferromagnetism.*

ROBERT M. BOYNTON, Department of Psychology, University of California, San Diego. *Vision and the Eye.*

G. E. BRIGGS, Emeritus, University of Cambridge, England. *Osmosis.*

FREDERICK C. BROCKHURST, Electrical Engineering Department, Virginia Polytechnic Institute and State University. *Motors, Electric* and *Electric Power Generation.*

STANLEY J. BRODSKY, Stanford Linear Accelerator Center, Stanford University. *Quantum Electrodynamics* (with Toichiro Kinoshita).

JAMES J. BROPHY, University of Utah. *Electronics.*

LAURIE M. BROWN, Department of Physics and Astronomy, Northwestern University, Evanston, Illinois. *Gauge Theories.*

STEPHEN G. BRUSH, Department of History and Institute for Physical Science and Technology, University of Maryland. *Kinetic Theory* and *Irreversibility.*

H. A. BUCHDAHL, Australian National University, Canberra. *Thermodynamics.*

DONALD G. BURKHARD, Department of Physics and Astronomy, University of Georgia. *Irradiance Calculations, Microwave Spectroscopy,* and *Solar Concentrator Design, Optics of.*

E. R. CAIANIELLO, Laboratorio di Cibernetica del Consiglio Nationale delle Ricerche, Naples, Italy. *Field Theory.*

ELTON J. CAIRNS, Lawrence Berkeley Laboratory and University of California, Berkeley. *Energy Storage, Electrochemical.*

EARL CALLEN, Department of Physics, American University, Washington, D.C. *Magnetostriction.*

G. S. CARGILL, III, IBM Thomas J. Watson Research Center, Yorktown Heights, New York. *Amorphous Metals.*

THEODORE G. CASTNER, Department of Physics and Astronomy, University of Rochester, New York. *Electron Spin.*

NICHOLAS CHAKO, Retired, Department of Mathematics, Queens College, City University of New York. *Aberrations.*

B. S. CHANDRASEKHAR, Department of Physics, Case Western Reserve University, Cleveland, Ohio. *Superconductivity.*

FRANK CHORLTON, Department of Mathematics, University of Aston in Birmingham, England. *Differential Equations in Physics.*

BRUCE P. CLAYMAN, Department of Physics, Simon Fraser University, Vancouver, British Columbia. *Interference and Interferometry.*

KENNETH J. CLOSE, Department of Physics, The Polytechnic of Central London, England. *Vacuum Techniques* (with John Yarwood).

IRA COCHIN, Department of Mechanical Engineering, New Jersey Institute of Technology, Newark. *Gyroscope* and *Inertial Guidance.*

E. RICHARD COHEN, Rockwell International Science Center, Thousand Oaks, California. *Constants, Fundamental.*

C. SHARP COOK, Department of Physics, University of Texas at El Paso. *Fallout.*

JOHN C. CORBIN, U.S. Air Force Aeronautical Systems Division, Wright-Patterson Air Force Base, Ohio. *Skin Effect.*

H. COTTON, University of Nottingham, England. *Optics, Geometrical.*

HERMAN V. COTTONY, H. V. Cottony Consulting Service, Bethesda, Maryland. *Antennas.*

GEORGE CRABTREE, Argonne National Laboratory, Illinois. *Fermi Surface* (with H. V. Bohm and Norman Tepley).

ROBERT G. CUNNINGHAM, Manufacturing Technology Division, Eastman Kodak Company. *Static Electricity* (with D. J. Montgomery).

R. H. DAVIS, Department of Physics, Florida State University. *Proton.*

L. WALLACE DEAN, III, Pratt and Whitney Aircraft Division of United Aircraft, East Hartford, Connecticut. *Physical Acoustics.*

PETER G. DEBRUNNER, Department of Physics, University of Illinois, Urbana. *Mössbauer Effect* (with Robert L. Ingalls).

BARBARA DECKER, *Volcanology* (with Robert W. Decker).

ROBERT W. DECKER, U.S. Geological Survey. *Volcanology* (with Barbara Decker).

JOHN DeSANTO, Department of Mathematics, Colorado School of Mines. *Ocean Acoustics.*

N. G. DESHPANDE, Department of Physics and Institute of Theoretical Science, University of Oregon. *Current Algebra* and *Electroweak Theory.*

R. E. DE WAMES, Advanced Technology Energy Systems Group, Canoga Park, California. *Spin Waves* (with T. Wolfram).

A. DINSDALE, Retired, British Ceramic Research Association, Stoke-on-Trent, England. *Viscosity.*

RICHARD H. DITTMAN, Department of Physics, University of Wisconsin-Milwaukee. *Heat.*

ROBERT H. DOREMUS, Materials Engineering Department, Rensselaer Polytechnic Institute, Troy, New York. *Crystallization.*

GLENN L. DOWNEY, Department of Mechanical Engineering and Engineering Mechanics, University of Nebraska-Lincoln. *Dynamics.*

G. DRESSELHAUS, Lincoln Laboratory, Massachusetts Institute of Technology. *Cyclotron Resonance (Diamagnetic Resonance).*

MILDRED S. DRESSELHAUS, Department of Electrical Engineering and Physics, Massachusetts Institute of Technology. *Semiconductors.*

H. G. DRICKAMER, School of Chemical Sciences, University of Illinois at Urbana-Champaign. *Pressure, Very High.*

ROBERT H. EATHER, Physics Department, Boston College, Chestnut Hill, Massachusetts. *Aurora.*

ERNST R. G. ECKERT, Emeritus, Department of Mechanical Engineering, University of Minnesota. *Heat Transfer.*

D. EDELSON, AT&T Bell Laboratories, Murray Hill, New Jersey. *Polar Molecules.*

JOHN A. EISELE, Space Systems Division, Naval Research Laboratory, Washington, D.C. *Tensors and Tensor Analysis* (with Robert M. Mason).

LEONARD EISNER, Norwalk State Technical College, Norwalk, Connecticut. *Radiation, Thermal.*

RAYMOND J. EMRICH, Department of Physics, Lehigh University. *Fluid Dynamics* and *Fluid Statics.*

DUANE D. ERWAY, Xerox Medical Systems, Pasadena, California. *Solar Energy Utilization* (with Abe Zarem).

HOWARD T. EVANS, JR., U.S. Geological Survey, Reston, Virginia. *Crystallography* and *Crystal Structure Analysis.*

A. G. FISCHER, Department of Electrical Engineering, University of Dortmund, West Germany. *Electroluminescence.*

GRANT R. FOWLES, Physics Department, University of Utah. *Schrödinger Equation.*

MARTIN M. FREUNDLICH. *Electron Microscope.*

SUSUMU FUKUDA, Department of Electronics, Kyoto University, Japan. *Photoelasticity.*

RICHARD M. FULLER, Department of Physics, Gustavus Adolphus College, St. Peter, Minnesota. *Density and Specific Gravity* (with Robert G. Fuller).

ROBERT G. FULLER, Department of Physics, University of Nebraska-Lincoln. *Density and Specific Gravity* (with Richard M. Fuller).

HAROLD P. FURTH, Department of Astrophysical Science, Princeton University. *Magnetic Field.*

T. H. GEBALLE, Department of Applied Physics, Stanford University. *Calorimetry* (with Frances Hellman).

BARRY A. GEORGE. *Electron Optics.*

H. A. GERSCH, School of Physics, Georgia Institute of Technology, Atlanta. *Boltzmann's Distribution.*

ANTHONY B. GIORDANO, Polytechnic Institute of Brooklyn. *Microwave Transmission.*

JOSHUA N. GOLDBERG, Syracuse University, Syracuse, New York, *Gravitation.*

PAUL GOLDHAMMER, Department of Physics, University of Kansas. *Nuclear Structure.*

R. H. GOOD, JR., Department of Physics, Pennsylvania State University. *Photon.*

JOHN B. GOODENOUGH, Inorganic Chemistry Laboratory, University of Oxford, England. *Diamagnetism.*

CHARLES D. GOODMAN, Department of Physics, Indiana University, Bloomington. *Critical Mass* and *Isospin.*

CLARK GOODMAN, 95 Antigua Court, Coranado, California. *Cross Section and Stopping Power.*

JOSEPH W. GOODMAN, Department of Electrical Engineering, Stanford University. *Fourier Analysis.*

S. J. GREGG (retired), Department of Chemistry, University of Exeter, England. *Adsorption and Absorption.*

CLARK B. GROSECLOSE, Lawrence Livermore Laboratory, University of California. *Positron* (with William W. Walker).

JOHN B. GRUBER, Departments of Physics and Chemistry, Portland State University, Oregon. *Rare Earths* (with Richard P. Leavitt).

VINCENT P. GWINN, Department of Chemistry, University of California, Irvine. *Neutron Activation Analysis.*

Y. M. GUPTA, Shock Dynamics Laboratory, Department of Physics, Washington State University. *Shock Waves.*

CECIL W. GUINN, Physicist-Consultant, U.S. Air Force Department of Defense (retired). *Bionics.*

WALTER J. HAMER, Washington, D.C., *Electrochemistry.*

A. O. HANSON, Department of Physics, University of Illinois, Urbana-Champaign. *Compton Effect.*

W. HAPPER, Department of Physics, Princeton University. *Optical Pumping.*

AKIRA HARASIMA, Emeritus, Tokyo Institute of Technology, Japan. *Surface Tension.*

FOREST K. HARRIS (retired), Electricity Division, National Bureau of Standards. *Electrical Measurements* (with Barry A. Bell).

ROLAND H. HARRISON, National Institute for Petroleum and Energy Research, Bartlesville, Oklahoma. *Gases: Thermodynamic Properties.*

R. W. HART, Applied Physics Laboratory, Johns Hopkins University. *Light Scattering.*

RYUKITI R. HASIGUTI, Faculty of Engineering, University of Tokyo, Japan. *Lattice Defects.*

SHERWOOD K. HAYNES, Emeritus, Department of Physics, Michigan State University. *Auger Effect.*

G. E. HAYTON, Department of Electronic Engineering, University of Hull, England. *Feedback* (with P. M. Taylor).

RAYMOND W. HAYWARD, National Bureau of Standards, Washington, D.C. *Lorentz Transformations, Parity.*

JOHN HEADING, Department of Applied Mathematics, University College of Wales, Aberystwyth. *Matrices.*

EUGENE HECHT, Department of Physics and Astronomy, Adelphi University, Garden City, New York. *Mechanics.*

FRANCES HELLMAN, Department of Applied Physics, Stanford University. *Calorimetry* (with T. H. Geballe).

ANTONY HEWISH, Cavendish Laboratory, Cambridge, England. *Pulsars.*

FRED A. HINCHEY, Department of Mathematics, Northeastern University, Boston. *Vectors in Physics.*

RUSSELL K. HOBBIE, Space Science Center, University of Minnesota. *Biomedical Instrumentation.*

JOHN F. HOGERTON, S.M. Stoller Corporation, New York. *Atomic Energy.*

CHARLES A. HOLT, Department of Electrical Engineering, Virginia Polytechnic Institute and State University. *Induced Electromotive Force.*

ROBERT E. HOPKINS, Optizon Corporation, Rochester, New York. *Lens.*

ROLF HOSEMANN, Gruppe Parakristallforschung, % Bundesanstalt für Materialorufung, Berlin. *Colloids, Thermodynamics of; Diffraction by Matter and Diffraction Gratings; Microparacrystals; Microparacrystals, Equilibrium State of; Paracrystals.*

KAREL HUJER, Emeritus, Department of Physics and Astronomy, University of Tennessee at Chattanooga. *History of Physics.*

MCALLISTER H. HULL, JR., Department of Physics, University of New Mexico. *Calculus of Physics.*

T. S. HUTCHISON, Royal Military College, Kingston, Ontario. *Ultrasonics* (with S. L. McBride).

ROBERT L. INGALLS, Department of Physics, University of Washington. *Mössbauer Effect* (with Peter Debrunner).

MAX JAMMER, Bar-Ilan University and Hebrew University, Israel. *Statics.*

J. V. JELLEY, Nuclear Physics Division, Atomic Energy Research Establishment, Harwell, England. *Čerenkov Radiation.*

R. J. JOENK, IBM Information Products Division, Boulder, Colorado. *Ferromagnetism* (with T. R. McGuire).

RUSSELL H. JOHNSEN, Department of Chemistry, Florida State University. *Elements, Chemical.*

FRANCIS S. JOHNSON, University of Texas at Dallas. *Space Physics.*

ROBERT A. JOHNSON, Materials Science Department, University of Virginia. *Irradiation, Displaced Atoms.*

JESS J. JOSEPHS, Department of Physics, Smith College, Northampton, Massachusetts. *Musical Sound.*

P. K. KABIR, Department of Physics, University of Virginia. *Weak Interactions.*

G. MICHAEL KALVIUS, Physik-Department Technische Universität, München, Federal Republic of Germany. *Magnetometry* (with H. J. Litterst).

WILLIAM M. KAULA, Department of Earth and Space Sciences, University of California, Los Angeles. *Geodesy.*

ROBERT W. KENNEY, Lawrence Berkeley Laboratory, University of California. *Bremsstrahlung and Photon Beams.*

D. W. KERST, Department of Physics, University of Wisconsin, Madison. *Betatron.*

WILLIAM F. KIEFFER (retired), Department of Chemistry, College of Wooster, Ohio. *Mole Concept.*

ALLEN L. KING, Emeritus, Department of Physics and Astronomy, Dartmouth College. *Refrigeration.*

GERALD W. KING, Department of Chemistry, McMaster University, Hamilton, Ontario. *Molecules and Molecular Structure.*

RUDOLPH KINGSLAKE (retired), Institute of Optics, University of Rochester, New York. *Refraction.*

R. H. KINGSTON, Lincoln Laboratory, Massachusetts Institute of Technology. *Laser.*

TOICHIRO KINOSHITA, Laboratory of Nuclear Studies, Cornell University. *Quantum Electrodynamics* (with Stanley Brodsky).

RANDALL D. KNIGHT, Department of Physics, Ohio State University. *Molecular Spectroscopy.*

HENRY J. KOSTKOWSKI, Spectroradiometry Consulting, Fairfax, Virginia. *Pyrometry, Optical.*

ALLAN D. KRAUS, U.S. Naval Postgraduate School. *Circuitry.*

REINOUT P. KROON (retired), University of Pennsylvania. *Dimensions.*

H. G. KUHN, Emeritus, Clarendon Laboratory, Oxford University. *Atomic Spectra.*

KAILASH KUMAR, Research School of Physical Sciences, Australian National University, Canberra. *Many-Body Problem.*

A. BARRY KUNZ, Department of Physics and Materials Research Laboratory, University of Illinois at Urbana-Champaign. *Solid State Theory.*

C. G. KUPER, Department of Physics, Technion-Israel Institute of Technology, Israel. *Polaron.*

DONALD W. KUPKE, Department of Biochemistry, School of Medicine, University of Virginia. *Centrifuge* (with Ralph A. Lowry and Houston G. Wood, III).

K. O. KUTSCHKE, Division of Chemistry, National Research Council of Canada, Ottawa. *Photochemistry.*

ROBERT T. LAGEMANN, Emeritus, Department of Physics, Vanderbilt University. *Wave Motion.*

HELMUT E. LANDSBERG, Institute for Physical Science and Technology, University of Maryland. *Geophysics.*

C. T. LANE. *Superfluidity.*

KENNETH R. LANG, Department of Physics, Tufts University. *Cosmology.*

PAUL LANGACKER, Department of Physics, University of Pennsylvania. *Grand Unified Theories.*

D. F. LAWDEN, Emeritus, University of Aston in Birmingham, England. *Mathematical Principles of Quantum Mechanics.*

RICHARD P. LEAVITT, Applied Physics Branch, Harry Diamond Laboratories, Adelphi, Maryland. *Rare Earths* (with John B. Gruber).

REUBEN LEE, Consulting Engineer. *Transformer.*

R. J. W. LE FÈVRE, Emeritus, School of Chemistry, Macquarie University, North Ryde, Australia. *Dipole Moments (Electrical and Magnetic).*

MARC D. LEVENSON, IBM Research Division, San Jose, California. *Kerr Effects.*

SUMNER LEVINE, Department of Materials Science and Engineering, State University of New York at Stony Brook. *Thermionics.*

DAVID R. LIDE, JR., Standard Reference Data, National Bureau of Standards. *Chemical Physics.*

INGOLF LINDAU, Electronics Laboratory, Stanford University. *Photoelectricity.*

ROBERT LINDSAY, Department of Physics, Trinity College, Hartford, Connecticut. *Resonance.*

RAPHAEL M. LITTAUER, Laboratory of Nuclear Studies, Cornell University. *Pulse Generation.*

F. JOCHEN LITTERST, Physik-Department Technische, Universität München, Federal Republic of Germany. *Magnetometry* (with G. Michael Kalvius).

MICHAEL G. LITTMAN, Department of Mechanical and Aerospace Engineering, Princeton University. *Zeeman and Stark Effects.*

A. L. LOEB, Department of Visual and Environmental Studies, Harvard University. *Heisenberg Uncertainty Principle.*

JOSEPH J. LOFERSKI, Division of Engineering, Brown University. *Photovoltaic Effect.*

EDWARD J. LOFGREN, Lawrence Berkeley Laboratory, University of California. *Cyclotron* and *Accelerators, Particle.*

RALPH R. LOWRY, School of Engineering and Applied Science, University of Virginia. *Centrifuge* (with Donald W. Kupke and Houston G. Wood, III).

ROBERT A. LUFBURROW, Department of Physics, St. Lawrence University. *Carnot Cycles and Carnot Engines.*

H. R. LUKENS, IRT Corporation, San Diego, California. *Radioactive Tracers.*

PAUL S. LYKOUDIS, School of Nuclear Engineering, Purdue University. *Magneto-Fluid-Mechanics.*

DAVID N. LYON, Department of Chemical Engineering, University of California, Berkeley. *Liquefaction of Gases.*

WILLIAM J. MACKNIGHT, Polymer Science and Engineering, University of Massachusetts. *Polymer Physics* (with Lawrence E. Nielson).

J. D. MACKENZIE, School of Engineering and Applied Science, University of California, Los Angeles. *Vitreous State.*

ALFRED U. MAC RAE, AT&T Bell Laboratories, Holmdel, New Jersey. *Electron Diffraction.*

S. M. MAHAJAN, Institute for Fusion Studies, University of Texas at Austin. *Plasmas.*

FRED C. MAIENSCHEIN, Oak Ridge National Laboratory. *Nuclear Radiation Shielding.*

B. W. MANGUM, Temperature and Pressure Division, Center for Basic Standards, National Bureau of Standards. *Temperature and Thermometry.*

STEPHEN P. MARAN, Laboratory for Astronomy and Solar Physics, NASA-Goddard Space Flight Center. *Radio Astronomy.*

D. MARCUSE, AT&T Bell Laboratories, Holmdel, New Jersey. *Fiber Optics.*

HUMPHREY J. MARIS, Physics Department, Brown University. *Phonons.*

JOHN W. MARTIN, Department of Metallurgy and Science of Materials, University of Oxford, England. *Metallurgy.*

ROBERT M. MASON, Consultant, CLEF, Peterborough, New Hampshire. *Tensors and Tensor Analysis* (with John A. Eisele).

P. T. MATTHEWS, Department of Applied Mathematics and Theoretical Physics, University of Cambridge, England. *Strong Interactions.*

R. D. MATTUCK, Physics Laboratory I, H. C. Ørsted Institute, University of Copenhagen. *Feynman Diagrams.*

JOSEPH E. MAYER, University of California, San Diego. *Liquid State.*

S. L. MCBRIDE, Royal Military College of Canada, Kingston, Ontario. *Ultrasonics* (with T. S. Hutchison).

DAVID E. MCCULLOUGH, Tektronix, Inc., Beaverton, Oregon. *Oscilloscopes.*

C. B. A. MCCUSKER, School of Physics, University of Sydney, Australia. *Quarks.*

D. K. MCDANIELS, Physics Department, University of Oregon. *Solar Energy Sources.*

RAYMOND H. MCFEE, Consultant, Laguna Hills, California. *Infrared Radiation.*

T. R. MCGUIRE, IBM Thomas J. Watson Research Center, Yorktown Heights, New York. *Ferromagnetism* (with R. J. Joenk).

A. E. E. MCKENZIE (deceased). *Optical Instruments* (with Nigel C. McKenzie and J. D. Walker).

NIGEL C. MCKENZIE. *Optical Instruments* (with A. E. E. McKenzie and J. D. Walker).

G. T. MEADEN, The Journal of Meteorology, Tornado and Storm Research Organization, Bradford-on-Avon, England. *Conductivity, Electrical.*

HOWARD C. MEL, Donner Laboratory, University of California, Berkeley. *Radiation, Ionizing, Basic Interactions* (with Paul Todd).

MAEL A. MELVIN, Emeritus, Department of Physics, Temple University. *Antiparticles.*

HAROLD METCALF, Physics Department, State University of New York, Stony Brook. *Biophysics.*

DIETRICH MEYERHOFER, RCA Laboratories, Princeton, New Jersey. *Tunneling.*

WOLFGANG E. MOECKEL (retired), NASA-Lewis Research Center, Cleveland. *Electric Propulsion.*

ORREN C. MOHLER, Emeritus, Department of Astronomy, University of Michigan. *Solar Physics.*

D. J. MONTGOMERY, College of Engineering, Michigan State University. *Static Electricity* (with Robert G. Cunningham).

KARL Z. MORGAN, School of Nuclear Engineering, Georgia Institute of Technology. *Health Physics.*

A. H. MORRISH, Department of Physics, University of Manitoba, Canada. *Paramagnetism.*

J. MORT, Webster Research Center, Xerox Corporation. *Photoconductivity.*

ERWIN W. MÜLLER, Emeritus (deceased), Department of Physics, Pennsylvania State University. *Field Emission* (with Tien T. Tsong).

RAYMOND L. MURRAY, Emeritus, Nuclear Engineering, North Carolina State University. *Nuclear Reactors.*

RAYMOND R. MYERS, Department of Chemistry, Kent State University (Ohio). *Rheology.*

NORMAN H. NACHTRIEB, Department of Chemistry, University of Chicago. *Diffusion in Liquids.*

GÉRARD NADEAU, Départment de Physique, Université Laval, Québec. *Elasticity.*

PAUL NELSON, Department of Mathematics, Texas Tech University. *Transport Theory* (with G. Milton Wing).

JACOB NEUBERGER, Department of Physics, Queens College, City University of New York. *Expansion, Thermal.*

LAWRENCE E. NIELSON, Redmond, Oregon. *Polymer Physics* (with William J. MacKnight).

JOHN S. NISBET, Ionosphere Research Laboratory, University of Pennsylvania. *Ionosphere.*

JOHN F. NOXON, Fritz Peak Observatory, Aeronomy Laboratory, NOAA, Boulder, Colorado. *Airglow.*

ALLEN NUSSBAUM, Department of Electrical Engineering, University of Minnesota. *Fourier Optics* and *Optics, Geometrical, Advanced.*

R. F. O'CONNELL, Department of Physics and Astronomy, Louisiana State University. *Faraday Effects.*

HARRY F. OLSON, late of RCA Laboratories. *Noise, Acoustical.*

JOHN M. OLSON, Institute of Biochemistry, Odense University, Denmark. *Photosynthesis.*

STEPHEN J. O'NEIL. *Servomechanisms.*

D. D. OSHEROFF, AT&T Bell Laboratories. *Cryogenics.*

RALPH T. OVERMAN. *Radioactivity.*

THORNTON PAGE, NASA Johnson Space Center, Houston, Texas. *Astrophysics.*

WILLIAM E. PARKINS, Consultant, Woodland Hills, California. *Energy Levels, Atomic* and *Work, Power and Energy.*

ROBERT PETERS, Speech and Hearing Sciences, University of North Carolina. *Hearing.*

NORMAN E. PHILLIPS, Department of Chemistry, University of California, Berkeley. *Heat Capacity.*

RICHARD A. PHILLIPS, Foster Grant Corporation. *Optics, Physical.*

JULIAN M. PIKE, National Center for Atmospheric Research, Boulder, Colorado. *Coriolis Effect.*

J. J. PINAJIAN, Oak Ridge National Laboratory. *Isotopes.*

MARTIN A. POMERANTZ, Bartol Research Foundation of The Franklin Institute, University of Delaware. *International Solar-Terrestrial Physics Programs.*

ALAN Y. POPE, Albuquerque, New Mexico. *Aerodynamics.*

G. M. POUND, Materials Science Department, Stanford University. *Condensation.*

R. D. PRESENT, Department of Physics and Astronomy, University of Tennessee. *Gas Laws* and *Intermolecular Forces.*

A. EDWARD PROFIO, Department of Chemical and Nuclear Engineering, University of California, Santa Barbara. *Nuclear Instruments.*

JOHN E. PRUSSING, Department of Aeronautical and Astronautical Engineering, University of Illinois at Urbana-Champaign. *Kepler's Laws of Planetary Motion.*

ERNEST RABINOWICZ, Department of Mechanical Engineering, Massachusetts Institute of Technology. *Friction.*

S. RAIMES, late of Imperial College, London. *Wave Mechanics.*

LAWRENCE L. RAUCH, Jet Propulsion Laboratory, California Institute of Technology. *Telemetry.*

FREDERICK REINES, Department of Physics, University of California, Irvine. *Neutrino.*

JOHN R. REITZ, Research Staff, Ford Motor Company. *Propagation of Electromagnetic Waves.*

D. C. REYNOLDS, Air Force Wright Aeronautical Laboratories, Avionics Laboratory, Wright-Patterson Air Force Base, Ohio. *Excitons.*

J. A. REYNOLDS, Culham Laboratory, United Kingdom Atomic Energy Authority, Abington, Oxfordshire. *Fusion Power* and *Laser Fusion.*

JAMES A. RICHARDS, JR., Agricultural and Technical College, State University of New York, Delhi. *Brownian Motion* and *Fermi Dirac Statistics and Fermions.*

WOLFGANG RINDLER, Department of Physics and Mathematics, University of Texas at Dallas. *Relativity.*

PETER H. ROGERS, School of Mechanical Engineering, Georgia Institute of Technology. *Electroacoustics.*

JACQUES E. ROMAIN, Centre de Recherches Routières, Belgium. *Time.*

DONALD M. ROSS, Propulsion Consulting Engineer. *Flight Propulsion Fundamentals* (with G. P. Sutton).

THOMAS D. ROSSING, Department of Physics, Northern Illinois University. *Acoustics.*

MILTON A. ROTHMAN (retired), Department of Physics, Trenton State College (New Jersey). *Conservation Laws and Symmetry.*

ROGERS D. RUSK, Emeritus, Department of Physics, Mount Holyoke College. *Nuclear Radiation.*

HAJIME SAKAI, Department of Physics and Astronomy, University of Massachusetts. *Spectroscopy.*

R. T. SANDERSON, Emeritus, Department of Chemistry, Arizona State University. *Bond, Chemical* and *Periodic Law and Periodic Table.*

ROBERT B. SCHAINKER, Electric Power Research Institute, Palo Alto, California. *Energy Storage, Thermal Mechanical.*

J. E. SCHIRBER, Sandia National Laboratories. *DeHaas-van Alphen Effect.*

H. M. SCHLICKE, Consulting Engineer, Milwaukee, Wisconsin. *Capacitance.*

PETER A. SCHROEDER, Physics Department, Michigan State University. *Thermoelectricity.*

WILLIAM T. SCOTT, Department of Physics, University of Nevada. *Electricity.*

GLEN T. SEABORG, Lawrence Berkeley Laboratory, University of California. *Transuranium Elements.*

ARTHUR H. SEIDMAN, Department of Electrical Engineering, Pratt Institute. *Diode (Semiconductor).*

J. M. H. LEVELT SENGERS, Institute for Basic Standards, National Bureau of Standards. *Compressibility, Gas.*

R. S. SHANKLAND, late of Case Western Reserve University. *Michelson-Morley Experiment.*

A. G. SHARKEY, JR., Pittsburgh Energy Technology Center. *Mass Spectrometry.*

WILLIAM F. SHEEHAN, Department of Chemistry, University of Santa Clara (California). *Chemical Kinetics.*

ERIC SHELDON, Department of Physics, University of Lowell (Massachusetts). *Neutron.*

HOWARD A. SHUGART, Department of Physics, University of California, Berkeley. *Atomic and Molecular Beams.*

WILLIAM A. SHURCLIFF, Harvard University, *Polarized Light.*

R. P. SHUTT, Brookhaven National Laboratory. *Spark and Bubble Chambers.*

W. A. SIBLEY, Oklahoma State University. *Color Centers.*

MIRIAM SIDRAN, Department of Natural Sciences, Baruch College/City University of New York. *Photography.*

LESTER S. SKAGGS, Cancer Therapy Institute, King Faisal Specialist Hospital, Riyadh, Saudi Arabia. *Medical Physics.*

MERRILL I. SKOLNIK, Radar Division, Naval Research Laboratory. *Radar.*

L. SLIFKIN, Department of Physics, University of North Carolina. *Diffusion in Solids.*

J. SMIDT, Laboratorium voor Technische Natuurkunde, Technische Hogeschool Delft, The Netherlands. *Relaxation.*

HOWARD M. SMITH, Eastman Kodak Company, Rochester, New York. *Holography.*

M. G. SMITH, Faculty of Computing, Management Science, Mathematics and Statistics, City of London Polytechnic. *Laplace Transforms.*

CHARLES P. SMYTH, Emeritus, Department of Chemistry, Princeton University. *Dielectric Theory.*

B. A. SOLDANO, Physics Department, Furman University. *Mass and Inertia.*

S. L. SOO, Department of Mechanical and Industrial Engineering, University of Illinois at Urbana-Champaign. *Equilibrium.*

DAVISON E. SOPER, Institute of Theoretical Science, University of Oregon, *Quantum Chromodynamics.*

J. W. T. SPINKS, University of Saskatchewan. *Radiation Chemistry* (with R. J. Woods).

M. T. SPRACKLING, Department of Physics, Queen Elizabeth College, University of London. *Mechanical Properties of Solids.*

J. C. SPROTT, Department of Physics, University of Wisconsin-Madison. *Electron Tubes.*

H. EUGENE STANLEY, Center for Polymer Studies and Department of Physics, Boston University. *Critical Phenomena.*

ROBERT L. STEARNS, Department of Physics, Vassar College. *Bose Einstein Statistics and Bosons.*

SAMUEL STEIN, Time and Frequency Division, National Bureau of Standards. *Frequency Standards.*

K. AA. STRAND, (formerly) U.S. Naval Observatory, Washington, D.C. *Astrometry.*

E. C. G. SUDARSHAN, Center for Particle Theory and Department of Physics, University of Texas at Austin. *Elementary Particles.*

G. P. SUTTON, Consulting Engineer, Danville, California. *Flight Propulsion Fundamentals* (with Donald M. Ross).

J. D. SWIFT, School of Physics, University of Bath, England. *Electrical Discharges in Gases.*

S. M. SZE, AT&T Bell Laboratories, Murray Hill, New Jersey. *Semiconductor Devices.*

P. M. TAYLOR, Department of Electronic Engineering, University of Hull, England. *Feedback.*

R. E. TAYLOR, Department of Anthropology, University of California, Riverside. *Radiocarbon Dating.*

NORMAN TEPLEY, Department of Physics, Oakland University, Rochester, Michigan. *Fermi Surface* (with H. V. Bohm and George Crabtree).

JAMES TERRELL, Los Alamos National Laboratory. *Fission.*

RUDOLPH E. THUN, Raytheon Company, Lexington, Massachusetts. *Thin Films.*

PAUL TODD, Program in Molecular and Cell Biology, Pennsylvania State University. *Radiation, Ionizing, Basic Interactions* (with Howard C. Mel).

RICHARD TOUSEY, Consultant, Space Science Division, Naval Research Laboratory, Washington. *Ultraviolet Radiation.*

LAWRENCE M. TRAFTON, Astronomy Department and McDonald Observatory, University of Texas at Austin. *Planetary Atmospheres.*

L. E. H. TRAINOR, Department of Physics, University of Toronto, Canada. *Theoretical Physics.*

MYRON TRIBUS, Center for Advanced Engineering Study, Massachusetts Institute of Technology. *Entropy.*

G. J. F. TROUP, Physics Department, Monash University, Victoria, Australia. *Maser.*

JOHN G. TRUMP, Department of Electrical Engineering, Massachusetts Institute of Technology. *Accelerator, Van de Graaff* and *High Voltage Research.*

N. W. TSCHOEGL, Department of Chemical Engineering, California Institute of Technology. *Viscoelasticity.*

TIEN T. TSONG, Department of Physics, Pennsylvania State University. *Field Emission* (with Erwin W. Müller).

W. N. UNERTL, Laboratory of Surface Science and Technology, University of Maine at Orono. *Surface Physics.*

EDGAR VILCHUR, Foundation for Hearing Aid Research, Woodstock, New York. *Reproduction of Sound.*

WILHELM H. VON AULOCK, AT&T Bell Laboratories, Whippany, New Jersey. *Ferrimagnetism* (with Andrew H. Bobeck).

JEARL WALKER, Department of Physics, Cleveland State University, Ohio. *Optical Instruments* (with A. E. E. McKenzie and N. C. McKenzie).

WILLIAM W. WALKER, Department of Physics, University of Alabama at Tuskaloosa. *Positron* (with B. Clark Groseclose).

FRANKLIN F. Y. WANG, Department of Materials Science and Engineering, State University of New York at Stony Brook. *Rectifiers.*

KENNETH M. WATSON, Marine Physical Laboratory of the Scripts Institution of Oceanography, University of California, San Diego. *Collisions of Particles.*

A. H. WEBER, Emeritus, Department of Physics, Saint Louis University. *Rotation-Curvilinear Motion.*

WALTER L. WEEKS, School of Electrical Engineering, Purdue University. *Electromagnetic Theory.*

VERNON G. WELSBY, late of the Department of Electronic and Electrical Engineering, University of Birmingham, England. *Inductance.*

H. L. WELSH, Emeritus, Department of Physics, University of Toronto, Canada. *Raman Effect and Raman Spectroscopy.*

CHARLES WERT, Department of Metallurgy and Mining Engineering, University of Illinois, Urbana. *Solid State Physics.*

N. REY WHETTEN, General Electric R&D Center, Signal Electronics Laboratory. *Secondary Emission.*

FREDERICK E. WHITE, Emeritus, Boston College, Chestnut Hill, Massachusetts. *Vibration.*

MILTON G. WHITE, late of Department of Physics, Princeton University. *Synchrotrons.*

DONALD J. WILLIAMS, Applied Physics Laboratory, Johns Hopkins University. *Magnetospheric Radiation Belts.*

FERD WILLIAMS, Physics Department, University of Delaware. *Luminescence.*

JOHN H. WILLS, Kennett Square, Pennsylvania. *Phase Rule.*

RICHARD C. WILLSON, Jet Propulsion Laboratory, California Institute of Technology. *Solar Total Irradiance and its Spectral Distribution.*

A. J. C. WILSON, Crystallographic Data Centre, University Chemical Laboratory, Cambridge, England. *X-ray Diffraction.*

G. MILTON WING, Los Alamos National Laboratory. *Transport Theory* (with Paul Nelson).

DENNIS E. WISNOSKY, Industrial Systems Group, GCA Corporation, Naperville, Illinois. *Computers.*

FRANK L. WOLF, Department of Mathematics, Carleton College. *Statistics.*

HUGH C. WOLFE, American Institute of Physics. *Symbols, Units, and Nomenclature.*

T. WOLFRAM, Physics Department, University of Missouri. *Spin Waves* (with R. E. DeWames).

HOUSTON G. WOOD, III, Department of Chemical and Aerospace Engineering, University of Viriginia. *Centrifuge* (with Ralph Lowry and Donald Kupke).

G. K. WOODGATE, St. Peter's College, University of Oxford. *Atomic Physics.*

R. J. WOODS, Department of Chemistry, University of Saskatchewan. *Radiation Chemistry* (with J. W. T. Spinks).

JOHN YARWOOD, (retired), Department of Physics, The Polytechnic of Central London. *Vacuum Techniques* (with Kenneth J. Close).

HSUAN YEH, Towne School of Civil and Mechanical Engineering, University of Pennsylvania. *Impulse and Momentum.*

A. M. ZAREM, Beverly Hills, California. *Solar Energy Utilization* (with Duane D. Erway).

ALEXANDER ZUCKER, Oak Ridge National Laboratory. *Nuclear Reactions.*

PREFACE TO THE THIRD EDITION

Welcome to physics, the science of relativity and gravitation, matter and energy, quarks and quanta, lasers and masers, and many other intriguing scientific areas.

Since the appearance of the second edition many new developments have become important, and this edition has been expanded to include as many as possible of these new ideas without neglecting older, but still valid, concepts.

In a major effort to strengthen coverage of the rapidly moving and vitally significant field of particle physics, four new articles were added: Gauge Theories, Quantum Chromodynamics, Electroweak Theory, and Grand Unified Theories. In addition, many revisions were made in second edition articles in this field, including Elementary Particles, Weak Interactions, Strong Interactions, Parity, Current Algebra, and Quantum Theory.

Similarly, to increase emphasis on the important role of physics in medicine and biology, articles were added on Molecular Biology and Biomedical Instrumentation, and second edition entries on Biophysics, Medical Physics, Health Physics, and Bionics were brought up-to-date. Many other fields of physics, such as optics and energy storage, have received additional emphasis.

The general plan of this edition follows that of the first two editions and is discussed in their prefaces which follow. A reader with a limited background in physics should find it worth while to start by reading the article on "Physics," and then reviewing the articles on major areas of physics mentioned in that article and listed as cross-references at the end of the article.

The major credit for this book must go to the contributors who made it possible. Thanks are also due to the staff at the Van Nostrand Reinhold Company, who suggested the third edition, did a large amount of editing, and made a book out of a collection of manuscripts. To my wife, Leigh, goes my deep appreciation for accomplishing the vast amount of record keeping and secretarial work needed, and for doing much of the checking of manuscripts and proof-reading.

ROBERT M. BESANÇON

Dayton, Ohio

PREFACE TO SECOND EDITION

This second edition of the *Encyclopedia of Physics* follows the same general plan as was used for the first edition; that is, each article is written so as to be of primary value to the type of reader who is most apt to look for the particular topic. There are articles on major areas of physics which are at a low technical level, so as to be of maximum value to the reader with little prior knowledge of physics. There are also articles on major divisions and subdivisions of these areas. In general, these latter start with an introduction intended to define the topic and describe the concepts involved. This is followed by more detailed and advanced treatment for the reader with a stronger background in physics.

To cover more of physics, the book has been considerably expanded, both by adding new articles and by including new material on topics in the first edition. Many of the articles have been completely rewritten, others received major changes, while others, particularly those on major areas of physics, required little or no change.

As in the first edition, the major credit for any success the book may achieve belongs to the authors, many of whom not only contributed a tremendous amount of time and effort in preparing articles, but made valuable suggestions for other parts of the book.

The editors at the Van Nostrand Reinhold Company contributed a great deal to the readability and accuracy of the book, and to my wife, Leigh, goes credit for much careful proofreading, for the preparation of the extensive index, and for the typing and detailed record-keeping required in assembling a book of this magnitude.

To all of these workers my heartfelt thanks are due.

ROBERT M. BESANÇON

Dayton, Ohio
April 10, 1974

PREFACE TO FIRST EDITION

THE AIM of this book is to provide in one volume concise and accurate information about physics. It should be of use to physicists who need information outside of their own special areas of interest, to teachers and librarians who must answer inquiries, to students who wish to add to their funds of knowledge, and to engineers and scientists who encounter physical concepts in pursuit of their professions. The book has been made possible by the thoughtful and generous cooperation of more than 300 authors, both in this country and abroad, who have unstintingly contributed their time, skill and knowledge. Their names and affiliations are shown immediately before this preface.

The most challenging problem for the editor was deciding which topics to include and which to leave out, since the space available was very limited compared with the vast amount of knowledge that could have been included. The approach used was to provide short introductory articles on physics, on the history of physics, on measurements, and on symbols, units and nomenclature, plus general articles on the major areas of physics: heat, light, mechanics, acoustics, etc. To these were added entries on divisions and subdivisions of the major areas; these are more detailed and pitched at somewhat higher technical levels than the broader, more general articles. Other topics lie on the interfaces between major areas of physics or are on subjects that include both physics and other disciplines. These include, among others, astrophysics, geophysics, biophysics, and mathematical biophysics. Finally, a few articles cover sciences that are so closely related to physics that the differences are frequently merely matters of emphasis.

Each article attempts to provide not just a definition of a term but an explanation of an area of physics. No attempt was made to hold all articles at the same technical level; on the contrary, the level for each entry was aimed at those readers who would be most apt to look for information on that specific topic. The contents of each article was left to the discretion of the author as the one most capable of making the proper selection. Some of the authors found it necessary to use mathematics, as is done in many books on physics. However, the reader with a limited mathematical background will find many articles with no mathematics at all, and others with very little, while the reader who is so inclined can sink his teeth into the more mathematical paragraphs.

Most of the authors have provided references to summary articles and books, and in addition, cross-references to other articles in this book have been added wherever it was felt that they might be of particular help to the reader. A few cross-references are shown by the use of small capitals in the body of the text (thus, MECHANICS); others are listed at the end of the article. The index should serve to locate particular topics that might not be subjects of complete articles.

I should like to extend my heartfelt thanks to the authors who contributed so much and to Mr. G. G. Hawley and Mr. H. Simonds of the Reinhold Publishing Corporation who invited me to compile and edit this book as one of the series of scientific and technical encyclopedias published by that company. I also owe a very great deal to Mrs. Alberta Gordon and her staff, who did much of the editing and proofreading, and to my wife, Leigh, who contributed the bulk of the tremendous amount of clerical work involved as well as adding a great deal of enthusiasm and inspiration.

ROBERT M. BESANÇON

Dayton, Ohio
November 1, 1965

THE
ENCYCLOPEDIA
OF PHYSICS

A

ABERRATION

1. Introduction* When light is considered as an electromagnetic process or disturbance, it is characterized as a vibration phenomenon taking place in time and space. It propagates from one region to another in the form of waves. Mathematically, such a vibratory motion is governed by certain second-order partial differential equations—wave equations—which are derived from Maxwell's equations of the electromagnetic field. The solutions of these equations are distinguished by the presence of certain quantities accompanying periodic motion in space and time, namely wavelength λ and frequency ν. They depend on the medium in which propagation of waves takes place. However, in the classical description of light propagation it was assumed that the process acts like a stream of particles; consequently such light phenomena are described as *rays*—the old corpuscular hypothesis. In this description, the periodicity of waves is lost, i.e., the wavelength λ does not have any direct effect on the paths of the light

Principle of Least Action for charged or uncharged particle motion, namely,

$$V(A, B) = \int_B^A \mu(x, y, z) \, ds, \qquad (1.1a)$$

$$S(A, B) = \int_A^B \sqrt{h - \varphi(x, y, z)} \, ds, \quad (1.1b)$$

where $\mu(x, y, z)$ is the refractive index of the medium and $\sqrt{h - \varphi}$ is the action function of mechanics, h being the constant of energy and φ the potential energy function. In both problems, the first variation along the paths from A to B must vanish.†

The differential equations satisfying the variational conditions are known as the *Euler-Lagrange equations* of light rays and particles, respectively.

The equations of ray paths in a medium of refractive index $\mu(x, y, z)$ and of relativistic (fast) electrons in an electromagnetic field E, H are:[3],[4]

Light Optics	Electron Optics

$$\frac{d}{ds}\left(\mu \frac{dx}{ds}\right) = \frac{\partial \mu}{\partial x} \qquad \frac{d}{ds}\left(\sqrt{\varphi(1 + \epsilon\varphi)} \frac{dy}{ds}\right) = -\frac{1}{2} \frac{1 + 2\epsilon\varphi}{\sqrt{\varphi(1 + \epsilon\varphi)}} E_x + \lambda_0\left(B_y \frac{dz}{ds} - B_z \frac{dy}{ds}\right)$$

$$\frac{d}{ds}\left(\mu \frac{dy}{ds}\right) = \frac{\partial \mu}{\partial y} \text{ (I)} \qquad \frac{d}{ds}\left(\sqrt{\varphi(1 + \epsilon\varphi)} \frac{dy}{ds}\right) = -\frac{1}{2} \frac{1 + 2\epsilon\varphi}{\sqrt{\varphi(1 + \epsilon\varphi)}} E_y + \lambda_0\left(B_z \frac{dx}{ds} - B_x \frac{dz}{ds}\right) \text{ (II)}$$

$$\frac{d}{ds}\left(\mu \frac{dz}{ds}\right) = \frac{\partial \mu}{\partial z} \qquad \frac{d}{ds}\left(\sqrt{\varphi(1 + \epsilon\varphi)} \frac{dz}{ds}\right) = -\frac{1}{2} \frac{1 + 2\epsilon\varphi}{\sqrt{\varphi(1 + \epsilon\varphi)}} E_z + \lambda_0\left(B_x \frac{dy}{ds} - B_y \frac{dx}{ds}\right)$$

$$\lambda_0 = \frac{e}{2m_0}, \qquad \epsilon = \frac{e}{2m_0 c^2}$$

rays through a medium. In analogy to the motion of particles, the propagation of light as rays must be governed by differential equations similar to the equations of particles. Indeed, the equations of light rays and of particles are derived from variational principles, the *Fermat Principle* for light propagation as rays and the

*This article gives a theoretical treatment of optical aberration. For an introductory discussion, see the articles entitled LENS and OPTICS, GEOMETRICAL. For astronomic aberrations, see VELOCITY OF LIGHT.

†The differential ds is the element of arc of the path of the rays or particles. *Fermat's Principle* is of least time, i.e., $\int_{t_0}^{t_1} c \, dt = 0$, from which (1.1a) is derived by writing $dt = (1/v) \, ds$. The corresponding statement in mechanics is Hamilton's Principle, i.e.,

$$\delta \int_{t_0}^{t_1} L \, dt = 0, \quad \text{from which (1.1b) is derived, pro-}$$

vided t does not enter explicitly into the Lagrangian L.

where ϵ is the relativistic correction factor, φ is the electric potential, and $E = -\text{grad } \varphi$. If $\epsilon = 0$, the electron optics equations reduce to the equation for electrons of nonrelativistic velocity.

A heuristic procedure for deriving the equations of geometric optics is to assume a waveform solution of the scalar wave equation $\Delta u - (1/v^2)U_{tt} = 0$, i.e.,

$$u(x, y, z, t) = A(x, y, z)$$
$$\cdot \exp[ik(V(x, y, z) - ct)] \quad (1.2)$$

Substituting (1.2) into the wave equation and afterwards letting $\lambda \to 0$ or the propagation constant $k \to \infty$, the remaining terms involve only the phase function V:

$$[\text{grad } V(x, y, z)]^2 = \frac{c^2}{v^2} = \mu^2(x, y, z) \quad (1.3)$$

where μ is the refractive index of the medium. This equation is called the *characteristic equation of Hamilton* or the *Eiconal Equation*. The solution of this equation will give all the information about the imaging properties of a medium with refractive index μ. Unfortunately, complete (closed) solutions of (1.3) are known for only a few cases.

The phase function $V = $ constant represents surfaces which propagate through the medium. At each point of the surface or wavefront a light ray intersects it normally. The amplitude on the surface Σ is given by $A(x, y, z)$, so each intersection point of a light ray with the phase function surface Σ becomes a source or disturbance which propagates from it in the form of a surface or wavefront, the *wavelets* of Huygens.

Equation (1.2) is equivalent to the propagation of light as rays, since grad V is proportional to a vector $\mathbf{s}^0$ which is normal to the wavefront. To account for the optical properties of the medium, the direction of the rays is $\mathbf{s} = (p, q, r^0)$, derived from $\mathbf{s}^0$ as follows. If α, β, γ are the geometric directions, so that $\mathbf{s}^0 = (\alpha, \beta, \gamma)$, then we have $\mathbf{s} = \mu\mathbf{s}^0$, with

$$p = \mu\alpha, \quad q = \mu\beta, \quad r^0 = \mu\gamma, \quad \mathbf{s} \cdot \mathbf{s}^0 = 1.$$

Therefore, we have grad $V = \mathbf{s}^0$, with

$$\text{grad } V \cdot \mathbf{s}^0 = \frac{dV}{ds} = \mu(\mathbf{s} \cdot \mathbf{s}^0) = \mu. \quad (1.4)$$

Furthermore, multiplying by the element of arc ds (along the rays), we obtain

$$\int_A^B (\text{grad } V \cdot \mathbf{s}^0) \, ds = \int_A^B \frac{dV}{ds} \, ds$$

$$= V(B) - V(A) = \int_A^B \mu \, ds \quad (1.5)$$

which is the Fermat formulation (1.1a).

The expression $V(B) - V(A)$ gives the optical distance of the path of a ray from a point $A(x, y, z)$ of the wavefront Σ_0, $V(A) = c_0$, to the corresponding point $B(x', y', z')$ on Σ', $V(B) = c_1$. Since the rays are normal (normal congruence) to the wavefronts, the latter are optically equidistant from each other—a two-parameter of *transversal* surfaces. They may be regarded as an optically parallel (transversal) family of surfaces, the rays constituting the geodesics of the optical medium. Now each point $P(x, y, z)$ on the wavefront $V(A) = C_0$ may be regarded as giving rise to a source, or disturbance of a wave at an instant t', with amplitude $A(x, y, z)$. At another instant t, the wavelet arrives at a point $B(x', y', z')$. The totality of all wavelets originating on Σ_0 will now form a surface wave Σ', which is the envelope of all the wavelets. The distance from Σ_0 to Σ' is expressed by the characteristic function (point) $V(x, y, z, x', y', z') = v'(t' - t)$, where v is the velocity of the wave in the medium. The transformation of Σ_0 to Σ', or the rays issuing from points on Σ_0 to the corresponding rays intersecting Σ', describes a finite contact transformation, and the directions (optical) (p_0, q_0) on Σ_0 and (p', q') on Σ' are given by the equations

$$p_0 = -\frac{\partial v}{\partial x_0}, \quad p' = -\frac{\partial v}{\partial x'}$$

$$q_0 = -\frac{\partial v}{\partial y_0}, \quad q' = -\frac{\partial v}{\partial y'}$$

$$r_0 = -\frac{\partial v}{\partial z_0}, \quad r' = -\frac{\partial v}{\partial z'}$$

where $r_0^2 = \mu_0^2 - p_0^2 - q_0^2$, etc., and $r = \sqrt{\mu^2 - p^2 - q^2}$ is the so-called Geometrical Optics Hamiltonian.

2. Aberrations A light ray originating at a point $P_0 = (x_0, y_0, z_0)$ on the object plane $z = z_0$, is imaged by the optical system at a point $P_1 = (x_1, y_1, z_1)$, which we assume to be on the Gaussian plane $z_1 = z_g$ of the system, and let the point $P_g = (x_g, y_g, z_g)$ be the Gaussian image of P_0. The two planes, $z = z_0$ and $z = z_g$ are conjugate planes.

We select in the image space a wave surface (front) Σ' which crosses the axis of the system at $O_a = (o, o, o_a)$. Let the ray intersect the surface Σ' at $P' = (x', y', z')$ as shown in Fig. 1. Now draw a spherical reference surface S of radius R equal to the distance $\overrightarrow{P_1 O_a}$. (Fig. 1).

The displacement vector $\overrightarrow{P_g P_1}$ measures the deviation of the actual image of P_0 from the corresponding (aberration free) Gaussian image. This deviation is called the *geometrical aberration of the ray*, or the *ray aberration*. On the other hand, the optical distance measured along the ray, namely, $\overrightarrow{P'P_g}$ is called the *aberration of the wave front*, or simply the *wave aberration*, which is positive if the reference surface is between the wave surface Σ' and the image plane $z = z_1$, negative otherwise.

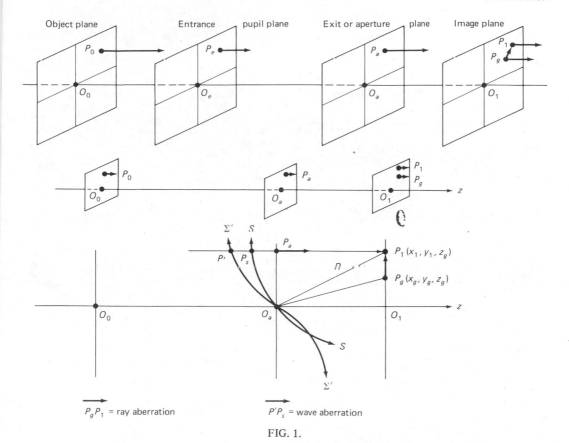

$P_g P_1$ = ray aberration $P' P_s$ = wave aberration

FIG. 1.

This wave aberration is expressed by the Hamilton characteristic function $V(x_0, y_0, z_0; x_1, y_1, z_1)$ or the difference of the characteristics at P_S and P', namely,

$$V(x_0, y_0, z_0; x_S, y_S, z_S)$$

$$- V(x_0, y_0, z_0; x', y', z'). \quad (2.1)$$

Consequently, we can say that any ray, or bundle of rays emerging from the object plane (space) will reappear as another (unique) ray or bundle crossing the image* plane (space). The latter are characterized by the coordinates and directions in the image space, which will depend on the corresponding quantities of the object ray or bundle and the nature of the optical medium. Mathematically, this is called a mapping of the object space into the image space by the optical system; the intervening medium thus determines the type of mapping or transformation function.†

*This is true except at singular points of the mapping function, i.e., at focal points, caustics, etc.

†The type of transformation determining the image ray from the object ray in terms of the characteristic function V is called *contact transformation*, which changes any surface in object space into any new surface in the image space.

Let $P_0 = (x_0, y_0, z_0)$ and $P_1 = (x_1, y_1, z_1)$ be the coordinates of the object ray and the image ray, respectively. We write this change of the object ray into the image ray in the form of a transformation or mapping given by the relations

$$\begin{aligned} x_1 &= x_1(x_0, y_0, p_0, q_0, z_0, z_1) \\ y_1 &= y_1(x_0, y_0, p_0, q_0, z_0, z_1) \end{aligned} \quad (2.1a)$$

$$\begin{aligned} p_1 &= p_1(x_0, y_0, p_0, q_0, z_0, z_1) \\ q_1 &= q_1(x_0, y_0, p_0, q_0, z_0, z_1) \end{aligned} \quad (2.1b)$$

Eliminating p_0, q_0 from the first of set equations (2.1a) with the aid of the second set (2.1b) (assuming the Jacobians do not vanish), say

$$\begin{aligned} p_0 &= p_0(x_0, y_0, p_1, q_1, z_0, z_1) \\ q_0 &= q_0(x_0, y_0, p_1, q_1, z_0, z_1) \end{aligned} \quad (2.2)$$

we rewrite equations (2.1a, b) as follows:

$$\begin{aligned} x_1 &= x_{1i}(x_0, y_0, p_1, q_1, z_0, z_1) \\ y_1 &= y_1(x_0, y_0, p_1, q_1, z_0, z_1) \\ p_0 &= p_{0i}(x_0, y_0, p_1, q_1, z_0, z_1) \\ q_0 &= q_0(x_0, y_0, p_1, q_1, z_0, z_1) \end{aligned} \quad (2.3)$$

By separating the linear terms in the arguments, (2.1a) becomes

$$x_{1i} = \sum_i a_{ij} x_{0j} + \sum_j b_{ij} p_{0j} + f_i(x_{0j}, p_{0j})$$

$$p_{ij} = \sum_i e_{ij} x_{0j} + \sum_j d_{ij} p_{0j} + g_i(x_{0j}, p_{0j})$$

$$(ij = 1, 2) \quad (2.4)$$

where $f_i(x_{0j}, p_{1i})$, $g_i(x_{0i}, p_{0j})$ are the deviations from an aberration-free image point. The latter is represented by the linear forms. The four functions f_i, g_i are called the *ray aberrations* of geometrical optics. Knowledge of them will determine the deviation of the image from the object produced by the optical system.

The constants a_{ij} etc. will depend also on the positions of the object and image planes. However, by properly choosing their positions and making the principal ray coincide with the optic axis of the instrument, the linear forms may be reduced to

$$x_{1i} = Mx_{0i}$$
$$p_i = Nx_{0i} + Lp_{0i} \qquad (i = 1, 2) \quad (2.4')$$

where M is the magnification factor, N is the reciprocal of the focal distance, and $L = M^{-1}$. If $f_i = g_i = 0$ (no aberration), (2.4) defines an *ideal* optical system.

Consequently, the main task in any optical imaging problem is to calculate the position x_i and the directions (optical) p_i in image space, usually on a fixed plane $z = z_i$ and most often on a Gaussian image plane, from the knowledge of the functions f_i, g_i. In practice, one determines x_i, p_i by tracing individual rays through the system, applying the laws of refraction or reflection (mirrors) or, in the case of isotropic media, one employs numerical integration. The labor of obtaining fairly good approximate values of f_i, g_i is long, but with modern computers one is able to reduce the time by several orders of magnitude, so that now it is possible to calculate the image position function to a high degree of accuracy, at least for moderate orders of aberration.

Theoretically, the image functions can also be derived if one of the characteristic functions of the optical system is known, for instance, the point characteristic V or the mixed characteristic W. They are found from the relations

$$p_i = +\frac{\partial V}{\partial x_i}, \qquad x_i = -\frac{\partial W}{\partial p_i} \qquad (i = 1, 2) \quad (2.5)$$

where $W = V - (x_1 p_1 + y_1 q_1)$, exception being made for telescopic systems, where the angular characteristic applies.

We have already mentioned the difficulties encountered in finding V or W in explicit form, except in a few optical systems. However, for theoretical discussions an analysis of aberrations and, in a number of practical cases; the characteristic functions are overall better suited for classifying aberrations.[3,4,5,6]

3. General Theory of Aberrations In the design and construction of image forming systems, one tries to minimize the effects of aberrations on the image quality,* or at least remove some of the harmful aberrations by compensating different aberrations of various orders.[4,5]

In general, the balancing of aberrations is a difficult problem. However, a general (formal) analysis of aberrations without specifying the characteristic properties of the optical system—except for geometric configurations (symmetry etc.) based on the wavefront concept of light propagation—will facilitate the discussion of characterizing and classifying the aberrations.

The fundamental functions needed to make a formal analysis of the image forming properties of an optical system are the characteristic functions of Hamilton: the *point characteristic*, or *eiconal*, $V(x_{0i}, x_i)$; two *mixed characteristics*,† $W(x_{0i}, p_i)$, $\overline{W}(x_{1i}, p_{0i})$;‡ and the *angular characteristic* $T(p_{0i}, p_{1i})$. These functions are related to each other through a so-called Legendre transformation. For instance, if the point characteristic $V(x_{0i}, x_i)$ is known, then the mixed characteristic $W(x_{0i}, p_i)$ is obtained by subtracting a term $x_1 p_1 + y_1 q_1$ from V, i.e.,

$$W(x_{0i}, p_i) = V(x_{0i}, x_i) - (x_1 p_1 + y_1 q_1) \quad (3.1)$$

where x_1, y_1 are expressed in terms of $(x_0, y_0; p_1, q_1)$.§ We then obtain the image coordinates on the image plane $z = z_1 = 0$ and the directions p_0, q_0 on the object plane $z = z_0 = 0$ from

$$x_1 = -\frac{\partial W}{\partial p_1}, \qquad p_0 = -\frac{\partial W}{\partial x_0}$$
$$y_1 = -\frac{\partial W}{\partial q_1}, \qquad q_0 = -\frac{\partial W}{\partial y_0}. \qquad (3.2)$$

The above relations, together with the relations $x_0 = \partial V/\partial p_0$, $x_1 = \partial V/\partial p_1$, given by the point characteristic, and the expression $V - W = x_1 p_1 + y_1 q_1$, constitute a Legendre transformation; the inverse of this transformation, i.e., W, is obtained from V by subjecting $x_1 p_1 + y_1 q_1$ (the image coordinates) to a Legendre transformation.

The other mixed characteristic (object characteristic) is simply $\overline{W}(x_1, y_1, z_1, p_0, q_0, z_0)$.

The angular characteristic, often used in telescopic reflecting systems, etc., is related to V

---

*In electrical terminology, "noise."

†The *image* and the *object* characteristics, respectively.

‡We have assumed the object and image planes to be conjugate planes in the Gaussian sense. In general, W will depend on the six parameters, and for the case where the image plane is in a homogeneous medium it will depend on five parameters: x_0, y_0, p_1, q_1, z_1.

§$\overline{W} = v + y_0 p_0 + y_0 q_0$.

as follows. The coordinates of the object and image planes are given by the expressions

$$x_0 = \frac{\partial T}{\partial p_0}, \quad x_1 = -\frac{\partial T}{\partial p_1}$$

$$y_0 = \frac{\partial T}{\partial q_0}, \quad y_1 = -\frac{\partial T}{\partial q_1}. \tag{3.4}$$

The relations above, together with those corresponding to V, also constitute a Legendre transformation between T and V, and $T = V + x_0 p_0 + y_0 p_0 - x_1 p_1 + y_1 p_1$.

In the analysis of wave or ray aberrations, it is more convenient to deal with the mixed characteristic $W(x_0, y_0, p_1, q_1; z_0, z_1)$, especially in the diffraction theory of optical systems, since it enters in the phase function of the solution of the wave equation discussed in section 1. Now we see that apart from the linear terms, the ray aberration functions $f_i(r_{0i}, p_i)$, $g_i(r_{0i}, p_i)$ are derived by simple differentiation of W or V or T, whichever is appropriate to the problem.

We have already mentioned the difficulties of obtaining explicit (closed form) expressions for V or W, therefore we express W in a series of homogeneous polynomials of various degrees (orders) in the four variables, say (x_0, y_0, p_1, q_1) for fixed object $(z = z_0)$ and image $(z = z_1)$ planes: *

$$W = W_0 + W_1 + W_2 + W_3 + W_4 + \cdots + W_N + \cdots$$

$$\tag{3.5}$$

where W_N is a homogeneous polynomial of degree N in the arguments. The leading term W_0 does not play any role, and W_1 represents free space. The quadratic polynomial W_2 may be written as

$$W_2 = \sum_{i,j=1}^{4} a_{ij} u_i u_j \tag{3.6}$$

where $u_1 = x_0$, $u_2 = y_0$, $u_3 = p_1$, $u_4 = q_1$. We write in full

$$W_2 = a_{11}x_0^2 + a_{12}x_0 y_0 + a_{22}y_0^2 + a_{13}x_0 p_1$$

$$+ a_{23}y_0 p_1 + a_{14}x_0 q_1 + a_{24}y_0 q_1$$

$$+ a_{33}p_1^2 + a_{34}p_1 q_1 + a_{44}q_1^2. \tag{3.7}$$

Therefore, from (3.2) the image coordinates are

$$x_1 = \frac{\partial W_2}{\partial p_1} = a_{13}x_0 + a_{23}y_0 + 2a_{33}p_1 + a_{34}q_1$$

$$y_1 = \frac{\partial W_2}{\partial q_1} = a_{14}x_0 + a_{24}y_0 + a_{34}p_1 + 2a_{44}q_1.$$

$$\tag{3.8}$$

*The W_k depend on z_0, z_1. However, being fixed quantities, they are absorbed by the constant coefficients: A_{ij}, etc.

We do not write the directions (p_0, q_0) of the ray at the object point (x_0, y_0). Equations (3.8) are the same linear terms we discussed in section 2. Therefore, W_2 represents what is known as the paraxial approximation of the characteristic function and of the image functions (x_1, y_1). In certain arrangements (special optical systems) mentioned in section 2, the coefficients a_{23}, a_{34}, and a_{14}, and even a_{33}, a_{44}, may be eliminated so that one obtains an ideal system. Consequently, one does not consider this approximation to W as an aberration.

The next approximation is the homogeneous polynomial of degree 3, W_3. Its expansion is given by

$$W_3 = \sum_{i,j,k=1}^{4} a_{ijk} u_i u_j u_k. \tag{3.9}$$

Further, W_4 is expanded as

$$W_4 = \sum_{i,j,k,l=1}^{4} a_{ijkl} u_i u_j u_k u_l. \tag{3.10}$$

There are altogether 20 terms in W_3. If we fix the object plane, $z = z_0$, then the homogeneous terms in $x_0 y_0$ (alone) are omitted, since they do not account for any displacement in the image plane. There are four such terms in W_3,

$$x_0^3, x_0^2 y_0, x_0 y_0^2, y_0^3$$

so the number of independent wavefront aberrations reduces to 16.

In W_4 there are altogether 35 terms, but by discarding the terms in x_0, y_0 alone, the number of independent wave aberrations diminishes to 30.† On the other hand, if the system has a plane of symmetry, say, either the x–z or y–z plane, i.e., rays lie on a meridian plane, then the number reduces further to 16 terms. Identifying $u_1 = x_0$, $u_2 = y_0$, $u_3 = p_1$, $u_4 = q_1$, the terms with the following coefficients a_{ijkl} vanish: ‡

$$a_{0031} = a_{0013} = a_{1003} = a_{0130} = a_{1021}$$

$$= a_{2011} = a_{1120} = a_{1102} = a_{3001}$$

$$= a_{2110} = a_{0310} = a_{1201}.$$

The number of wave aberration terms in W_N increases considerably with N. Therefore we shall limit the analysis to W_4, i.e., to third-order (ray) aberrations for optical systems possessing rotational symmetry, since in most if not all

†The number of individual terms in W_N is given by

$$N = \frac{(n+1)(n+2)(n+5)}{3!}.$$

‡The numbers appearing in the subscripts indicate the exponents of the variables associated with the indices. For example, a_{1030} corresponds to the term $a_{1030}x_0 p^3$, a_{1111} to $a_{1111}x_0 y_0 pq$; etc.

practical cases the instruments possess geometrical symmetry. The wave aberration function W then depends only on three parameters,

$$u_1 = x_0^2 + y_0^2, \qquad u_2 = p_1^2 + q_1^2,$$

$$u_3 = 2(x_0 p_1 + y_0 q_1), \qquad (3.11)$$

known as the three optical invariants of the system. In the expansion of W, the odd polynomials disappear, and we have

$$W = W_0 + W_2 + W_4 + \cdots.$$

The image functions are

$$x_1 = \frac{\partial W_2}{\partial p_1} = 2(a_3 x_0 + a_2 p_1),$$

$$y_1 = \frac{\partial W_2}{\partial q_1} = 2(a_3 y_0 + a_2 q_1). \qquad (3.12)$$

These expressions give an image free of aberration.

The W_4 term contribution is given by

$$W_4 = a_{11} u_1^2 + a_{22} u_2^2 + a_{33} u_3^2 + a_{12} u_1 u_2$$

$$+ a_{13} u_1 u_3 + a_{23} u_2 u_3. \qquad (3.13)$$

The image functions (deviations from the ideal or Gaussian image) are derived from the expressions

$$\delta x_{1i} = \frac{\partial W_4}{\partial p_{1i}} = \frac{\partial W_4}{\partial u_j} \frac{\partial u_j}{\partial p_{1i}},$$

$$\delta p_{0i} = -\frac{\partial W_4}{\partial x_{0i}} = -\frac{\partial W_4}{\partial u_j} \frac{\partial u_j}{\partial x_{0i}}$$

$$(i = 1, 2; j = 1, 2, 3) \qquad (3.14)$$

In (3.13) one notices the presence of only six aberration coefficients, a_{11}, a_{22}, ..., a_{23}. Of these a_{11} is associated with the object coordinates $x_{0i} = (x_0, y_0)$. It is not considered an aberration per se. To obtain the five types of third-order aberrations from (3.13) we assume without loss of generality rays restricted to points $(x_0, 0)$ on the object plane and, after differentiation, letting $p_1 = \rho \cos \varphi, q_1 = \rho \sin \varphi$, we arrange the right-hand side of (3.13) in descending powers of ρ. The x_{1i} will now depend on five aberration coefficients, excluding a_{11}, whereas p_{0i} will be free of a_{22}. This arrangement for δx_{1i} will determine the five geometrical "Seidel" aberrations of third order. The terms in $(\delta x_1, \delta y_1)$ depending on ρ^3 alone give the *spherical* or *aperture* aberration; the terms in ρ^2 give the *primary coma*; the linear terms in ρ are known as *astigmatism* and *curvature of field*; and the term independent of ρ produces *distortion* of the *pincushion* or *barrel* type.

For a detailed discussion, we refer the reader to the treatises in Refs. 3 and 5, and for electron optical systems to those listed in Ref. 4. However, in diffraction problems[2,5] the ex-

pansion of the wave aberration functions for rotational symmetric systems is usually given in terms of three new parameters σ, ρ, φ, which are related to the ray functions x_0, y_0, p_1, q_1 in object and image space (planes) as follows:

$$\sigma^2 = u_1 = x_0^2 + y_0^2, \qquad \rho^2 = u_2 = p_1^2 + q_1^2,$$

$$\sigma \rho \cos \varphi = u_3 = 2(x_0 p_1 + y_0 q_1). \quad (3.15)$$

The wave aberration function is written in the form

$$W(x_0, y_0, p_1, q_1, z_1) = \sum_{l, n, m} b_{lnm} \sigma^{2l+m} \cos^m \varphi$$

$$(3.16)$$

where $l, m = 0, 1, \ldots$, $n = 0, 1, 2$, and $n - m > 0$ and even. This expansion is the *standard* or *classical* expansion used by earlier investigators before the Zernike–Nijboer papers on diffraction theory of aberrations. The degree of the polynomial W_K is equal to the sum of the exponents of σ and ρ in the term $b_{lnm} \sigma^{2l+m} \rho^n \cos^m \varphi$.

For instance, if we let $2l + n + m = 2$, i.e., W_2, then the three terms of W_2 are obtained by letting $m = 0$, $l = 1$, $n = 0$; $m = 0$, $l = 0$, $n = 2$; $m = 1, l = 0, n = 1$.

A more convenient expansion of W is a Fourier-like expansion*

$$W = \sum_{l, n, m} b_{lnm} \sigma^{2l+m} \rho^n \cos m\varphi. \quad (3.17)$$

Here, a single term $b_{lnm} \sigma^{2l+m} \rho^n \cos m\varphi$ represents any of the homogeneous polynomials discussed above. For example, W_2 is obtained by letting $l = 1$, $n = m = 0$; $l = m = 0$, $n = 2$; and $l = 0$, $m = n = 1$. The W_2 polynomial is given by the expressions

$$W_2 = b_{100} \sigma^2 + b_{020} \rho^2 + b_{011} \sigma \rho \cos \varphi \quad (3.18)$$

which is the same as in the standard expansion. However, the Fourier form has the advantage of being an orthogonal function in φ, so it facilitates the integration over φ in the diffraction integrals.

On the other hand these expansions are superseded in their application to the diffraction theory of aberrations by the Zernike–Nijboer form of expansion, i.e.,

$$W = \sum_{l, n, m} b_{lnm} \sigma^{2l+m} Z_n^m(\rho) \cos m\varphi \quad (3.19)$$

where l, n, m have the same meaning as in the earlier expansions, but here $Z_n^m(\rho)$ is a polynominal of degree n in ρ and not a single term ρ^n as in the other expansions. Furthermore, the Zernike polynomials, Z_n^m are orthogonal over a circular plane region. In the next sections we shall discuss the advantage of this expansion

*For nonsymmetric systems one introduces another sum in sin $m\varphi$, i.e., a complete Fourier-series expansion.

over the other two, the standard and the Fourier developments.

In (3.20) below and in Tables 1 and 2 are listed the expansions of W_2, W_4, and W_6 in the three expansion forms[5] and the Zernike polynomials $Z_n{}^m$ up to $n = 6$, $m = 6$.

In each of them the order for a single aberration member is $N = 2l + n + m - 1$, and the number of types in a single aberration is $M = (1/\gamma)(N + 1)(N + 8)$. The degree of the polynomial W_K representing the wavefront function is $K = N + 1 = 2l + n + m$.

$$W_2 = a_1 \sigma^2 + a_2 \rho^2 + a_3 \sigma\rho \cos \varphi$$

$$W_4 = a_{11} \sigma^4 + a_{12} \sigma^2 \rho^2 + a_{22} \rho^4$$
$$+ a_{13} \sigma^3 \rho \cos \varphi + a_{23} \sigma\rho^3 \cos \varphi$$
$$+ a_{33} \sigma^2 \rho^2 \cos^2 \varphi$$

$$W_6 = a_{111} \sigma^6 + a_{112} \sigma^4 \rho^2 + a_{122} \sigma^2 \rho^4$$
$$+ a_{222} \rho^6 + a_{113} \sigma^5 \rho \cos \varphi$$
$$+ a_{123} \sigma^3 \rho^3 \cos \varphi + a_{133} \sigma^4 \varphi^2 \cos^2 \varphi$$
$$+ a_{223} \sigma\rho^5 \cos \varphi + a_{233} \sigma^2 \rho^4 \cos^2 \varphi$$
$$+ a_{333} \sigma^3 \rho^3 \cos^3 \varphi \qquad (3.20)$$

4. Classifications of Aberration for Rotational Symmetric Systems[2,4,5]

To classify the ordinary (ray) aberrations, we shall examine the different terms of the W_n of the wavefront aberration function W given by the three expansion forms u_1, u_2, u_3. Each W_n depends on the three optical invariants $u_1 = x_0{}^2 + y_0{}^2$, $u_2 = x_0 p + y_0 q$, $u_3 = p^2 + q^2$, or in σ, ρ, and φ given by the relations $\sigma^2 = u_1$, $\rho^2 = u_3$, and $\sigma\rho \cos \varphi = u_2$. The first three terms of W, i.e., W_2, W_4, W_6 are given in Tables 1 and 2, from which we can obtain the ordinary aberrations of first, third and fifth orders. A typical term of W_n is

TABLE 2.

$Z_0{}^0 = 1$	$Z_5{}^1 = 10\rho^5 - 12\rho^3 + 2\rho$
$Z_1{}^1 = \rho$	$Z_5{}^3 = 5\rho^5 - 4\rho^3$
	$Z_5{}^5 = \rho^5$
$Z_2{}^0 = 2\rho^2 - 1$	
$Z_2{}^2 = \rho^2$	$Z_6{}^0 = 20\rho^3 - 30\rho^4 + 12\rho^2 - 1$
	$Z_6{}^2 = 15\rho^6 - 20\rho^4 + 6\rho^2$
$Z_3{}^1 = 3\rho^3 - 2\rho$	$Z_6{}^4 = 6\rho^6 - 5\rho^4$
$Z_3{}^3 = \rho^3$	$Z_6{}^6 = \rho^6$
$Z_4{}^0 = 6\rho^4 - 6\rho^2 + 1$	
$Z_4{}^2 = 4\rho^4 - 3\rho^2$	
$Z_4{}^4 = \rho^4$	

Zernike–Nijboer development (new)

$$b_{lnm} \sigma^{2l+m} Z_n{}^m(\rho) \cos m\varphi \qquad (4.1a)$$

Fourier development

$$b_{lnm} \sigma^{2l+m} \rho^n \cos m\varphi \qquad (4.1b)$$

Standard (classical) development

$$a_{lnm} \sigma^{2l+m} \rho^n \cos^2 m\varphi \qquad (4.1c)$$

The order N of a single aberration (geometrical or ray) is $N = 2l + n + m - 1$, and the number of terms or types of aberration is $M = \frac{1}{8}(N + 1)(N + 7)$. The degree L of the polynomial W_n giving the order of the wave aberrations is $L = N + 1 = 2l + n + m$, where $l = 0, 1, 2, \ldots, n = 1, 2, \ldots, m = 0, 1, 2, \ldots$, and $n - m$ is even, making L even and equal to the sum of the powers of σ and ρ which appear in each individual member of the wave aberration function.

Consider first-order aberrations, i.e., W_2. Since $N = 1$, there are actually three terms in W_2. From Tables I and II they are

Zernike–Nijboer

TABLE 1.

n^m	0	1	2	3
		$N = 1$		
1		$\sigma\rho \cos \varphi (011)$		
2	$\rho^2 (020)$			
		$N = 3$		
1		$\sigma^3 \rho \cos \varphi (111)$		
2	$\sigma^2 \rho^2 (120)$		$\sigma^2 \rho^2 \cos^2 \varphi (022)$	
3		$\sigma\rho^3 \cos \varphi (031)$		
4	$\rho^4 (040)$			
		$N = 5$		
1		$\sigma^5 \rho \cos \varphi (211)$		
2	$\sigma^4 \rho^2 (220)$		$\sigma^4 \rho^2 \cos^2 \varphi (122)$	
3		$\sigma^3 \rho^3 \cos \varphi (131)$		$\sigma^3 \rho^3 \cos^3 \varphi (033)$
4	$\sigma^2 \rho^4 (140)$		$\sigma^4 \rho^2 \cos^2 \varphi (042)$	
5		$\sigma\rho^5 \cos \varphi (051)$		
6	$\rho^6 (060)$			

$$b_{100}\sigma^2 Z_0{}^0(\rho), \qquad b_{011}\sigma Z_1{}^1(\rho)\cos\varphi,$$

$$b_{020}Z_2{}^0(\rho) \qquad (4.2a)$$

Fourier

$$b_{100}\sigma^2, \qquad b_{011}\sigma\rho\cos\varphi, \qquad b_{020}\rho^2 \quad (4.2b)$$

Standard

$$a_{100}\sigma^2, \qquad a_{011}\sigma\rho\cos\varphi, \qquad a_{020}\rho^2. \quad (4.2c)$$

We shall ignore the term or terms which depend solely on σ. Therefore we have only two members, whose coefficients are b_{011} and b_{020}, etc. However, these two terms are not aberrations in the sense of representing a deformation of the image. They simply shift the focal point, the b_{020} term along the principal axis and the b_{011} in the image plane at $(x_1, 0, 0)$ from the origin $(0, 0, 0)$—the intersection of the principal ray with the image plane. Consequently, one does not consider the first-order terms as aberrations.

The next order aberrations come from W_4. These are called third-order aberrations, five in all according to the value of N (see above). Again we ignore terms solely in σ, since they do not influence the image pattern. The terms in W_4 are given below. From the tables, we have:

Zernike–Nijboer type

$$l = 0, \, n + m = 4$$

$$b_{040}Z_4{}^0(\rho) = b_{040}(6\varphi^4 + 6\rho^2 + 1)$$
$$m = 0, \, n = 4,$$

$$b_{031}\sigma Z_3{}^1(\rho)\cos\varphi = b_{031}\sigma(3\rho^3 - 2\varphi)\cos\varphi$$
$$m = 1, \, n = 3,$$

$$b_{022}\sigma^2 Z_2{}^2(\rho)\cos 2\varphi = b_{022}\sigma^2\rho^2\cos 2\varphi$$
$$m = 2, \, n = 2, \quad (4.3a)$$

$$l = 1, \, n + m = 2$$

$$b_{120}\sigma^2 Z_2{}^0(\rho) = b_{120}\sigma^2(2\rho^2 - 1)$$
$$m = 0, \, n = 2,$$

$$b_{111}\sigma^3 Z_1{}^1(\rho)\cos\varphi = b_{111}\sigma^3\rho\cos\varphi$$
$$m = 1, \, n = 1,$$

Fourier type

$$b_{040}\rho^4, \qquad b_{031}\sigma\rho^3\cos\varphi, \qquad b_{022}\sigma^2\varphi^2\cos 2\varphi,$$

$$b_{120}\sigma^2\rho^2, \qquad b_{111}\sigma^3\rho\cos\varphi \quad (4.3b)$$

Standard type

$$a_{040}\rho^4, \qquad a_{031}\sigma\rho^3\cos\varphi, \qquad a_{022}\sigma^2\rho^2\cos^2\varphi,$$

$$a_{120}\sigma^2\rho^2, \qquad a_{111}\sigma^3\rho\cos\varphi. \quad (4.3c)$$

From the tables we can write the individual members of W_6 from which we could obtain the individual fifth-order aberrations by differentiation.*

We can classify various types of aberrations from the subscripts appearing in the coefficients b_{lnm}. The aberrations associated with the coefficients $b_{0,2k,0}(k = 1, 2, \ldots)$ depend only on ρ. Such members produce spherical aberrations of order $2k - 1$; thus in W_4 the term of b_{040} is called spherical aberrations of third order. Terms whose coefficients are $b_{l,2k+1,1}$ produce *coma* of order k, where $l = 0, 1, 2, \ldots, k = 1, 2, \ldots$. For $l = 0$, $k = 0$ we have a simple distortion of the image and it is not considered an aberration. The term with coefficient b_{031} is the first-order coma or primary coma, with b_{051} secondary or second-order coma, with b_{071} third-order coma, etc. However, terms with coefficients $b_{l,2k+1,1}$ are also coma aberrations, where $l = 1, 2, \ldots, k = 1, 2, \ldots$. To distinguish those terms with $l = 0$ from the others with $l = 1, 2, \ldots$ we call the aberrations with $l = 0$ *pure*, so all terms with coefficients $b_{0,2k+1,1}$ are *pure* coma aberrations of order k, the others *mixed* comas. Terms of the wave aberration function W_n having for coefficients $m = 2$, $l = 0, 1, 2, \ldots$, $n = 2k$, $k = 1$, $2, \ldots$, produce *astigmatism* of order k. For $l = 0$, $k = 1$, $m = 2$, the aberration term b_{022} produces first-order or primary astigmatism; for $m = 2$, $k = 2$ secondary astigmatism; etc. When we take $m = 3$, etc., no particular name is given this type of aberration. The special terms of the type $b_{022}\sigma^2\rho^2\cos^2\varphi$ is called *wing aberration*, and for $m = 3$, $n = 1$, $l = 0, 1$, etc., *arrow*-type aberrations. For $m = 0$ and $l > 0$, the aberration wave function produces a curvature of the field, thus all the terms associated with $b_{l,2k,0}$ produce a *curvature of field* of order k, producing a change of focus depending on the field given by σ^{2l}.

In comparing the three expansions of W_4, the Z–N development contains additional terms of lower power in ρ. The additional terms displace the point (focal) to maximum illumination that is compensating (balancing) higher-order aberrations with the lower orders. This is an important advance over the standard expansion of the wave aberration functions, especially when the Z–N expansion is introduced in the phase function of the diffraction integral, not to mention the advantage of employing orthogonal functions over the domain of integration for evaluating the diffraction integrals; in most optical situations, the domain is circular.

5. **Diffraction Theory** The starting point of

$$\left.\begin{aligned}
*\delta x_1 &= \alpha\left[W_\rho\cos\varphi - W_\varphi\frac{\sin\varphi}{\rho}\right]\\
\delta y_1 &= \alpha\left[W_\rho\sin\varphi + W_\varphi\frac{\cos\varphi}{\rho}\right]
\end{aligned}\right\} \quad w = W(x_{0i}, \rho, \varphi).$$

α depends on the distance from the reference surface to (x_1, y_1) and the maximum value of ρ.

the modern theory of diffraction of optical imaging systems may be traced to the famous paper on the diffraction theory of the phase contrast method by Zernike.[1] He and his students made significant advances in the theory of diffraction. Nijboer[2] was able to calculate to an unprecedented degree of accuracy the intensity distribution of the field (contours) produced by an optical system affected by several types of aberration of various orders. Moreover, experiments were performed in Zernike's laboratory by another of his pupils, Nienhuis, to verify the theory. The laboratory results agreed to a high degree with Nijboer's calculations. Likewise, experiments with microwaves were carried out at McGill University by Bachynski and Bekefi[2] using lenses made of dielectric materials; the results were compared with calculations based on Nijboer's formulation of diffraction of aberrations. Again, comparison of calculations with experimental observations surprisingly shows a great likeness to the contours of the intensity distribution on the Gaussian plane. Further investigations have shown the usefulness and the advantages of the Zernike–Nijboer formulation of the diffraction of aberrations over the earlier formulations in other fields than optics.

The Zernike–Nijboer formulation of diffraction of aberrations rests on the expansions of the wave aberrations function in a series of orthogonal functions, known as Zernike polynomials, which are orthogonal over a circular region. Since the geometrical shapes of apertures or stops in lens systems, microscopes, cameras, or telescopes, are circular, this form of expansion is the most appropriate one for treating diffraction, scattering and other problems including problems of vibrating elastic systems. The advantage of introducing this kind of expansion in the diffraction integrals lies in the simplification achieved not only in the evaluation of the integrals, but also on account of the orthogonality of the polynomials over the domain of integration; the coefficients of the expansion do not mix with each other when the aberrations are small. The extension to large aberration was carried out by van Kampen.[2]

Since then, great advances have been made, both in theory and experimental observations leading to important applications in the improvement of optical instruments. This success has been paralleled in other fields of science such as in electron optical image formation, in microwave optics, and in the design of image formation apparatus employing high- or low-frequency waves (radio, infrared, ultraviolet, X-rays, and even acoustic waves). Significant contributions have been made in the design of microwave lenses (dielectric), microwave imaging instruments such as satellite cameras, mirrors for collecting (exploring) information on sources of electromagnetic radiation of different frequencies from distant objects such as planets, stars, and galaxies, as well as in the development of the new branch of optics

named *holography*.[7]* In holography, the theory of diffraction of aberrations plays an important role in the design of optical and other image forming systems to obtain faithful images of objects emitting or illuminated by different kinds of radiation—coherent, incoherent, or mixed. The vast usefulness of the diffraction theory of aberrations is accounted for in the similarity of the mathematical differential equations governing wave phenomena and the analogous solutions, which, subject to certain appropriate conditions, resemble each other in mathematical form.

We would like to refer also to some of the earlier significant contributions to the theory of diffraction of aberrations by Ignatowski, Fischer, Steward, Picht and more recently by Luneburg and others.[3]

The basis of the diffraction theory of aberrations of optical systems is founded on Kirchhoff's integral or a modified form of it, namely,

$$U(P) = -\frac{ikn}{2\pi} \iint_S \sqrt{K} U_0(Q)$$

$$\cdot \exp ik[W + (\mathbf{r} \cdot \mathbf{s})] \, dS \qquad (5.1)$$

where $U_0(Q)$ is the value of the field on the wave surface (front) S after passing through the optical system, usually located in the image space; n, K, and $\mathbf{r}$ are, respectively, the refractive index of the medium in image space, the Gaussian curvature of the wavefront, and the distance vector from the point Q, located on S, to the image point (observation point) P. The function W is Hamilton's mixed characteristic of the optical system which depends on the coordinate of the object and the optical direction of the ray at Q, namely the vector $\mathbf{s}$. Here the field represented by $U_0(Q)$ is the geometrical optics wave solution of the harmonic scalar equation or a component of the vector wave equation derived from Maxwell equations. Here U_0 is considered a scalar quantity, or a component of the vector field.

A more convenient form of (5.1) used frequently in actual problem is [3,4]†

$$u(P) = -\frac{ik}{2\pi} \iint_{D = p^2 + q^2 \leqslant n^2} g(p \cdot q)$$

$$\cdot e^{ik[W + (\mathbf{r} \cdot \mathbf{s})]} \, dp \, dq, \qquad (5.2)$$

where

$$g(p, q) = \sqrt{n^2 |\Delta|} \; U_0(p, q). \qquad (5.3)$$

The symbol Δ denotes the discriminant of the second differential form of S, i.e., $\Delta = LN - M^2$, with L, M, N given by the expressions

$$nL = W_{pp} - \frac{p^2 + r_0^2}{n^2 r_0^2} Z$$

---

*Recently acoustic waves have been used for wavefront reconstruction imaging.[9]

†$(\mathbf{r} \cdot \mathbf{s}) = xp + yq + z\sqrt{n^2 - p^2 - q^2}$.

$$nM = W_{pq} - \frac{pq}{n^2 r_0^3} Z \qquad (Z = pW_p + qW_q - W),$$

$$nN = W_{qq} - \frac{q^2 + r_0^2}{n^2 r_0^2} Z. \qquad (5.4)$$

where $\mathbf{s} = (p, q, r_0)$ are the optical direction cosines of a ray on Σ with $r_0^2 = \sqrt{n^2 - p^2 - q^2}$. A point $P(x, y, z)$ of Σ is given by the equations

$$x = -W_p + \lambda(p, q)p, \qquad y = -W_q + \lambda(p, q)q,$$

$$z = -\lambda(p, q)r_0 \qquad (5.5)$$

where $\lambda(p, q)$ is the distance from P to the image point.

In practice one takes as the surface of integration the aperture of the exit pupil. The integrals (5.1) and (5.2) form the basis of the diffraction theory of aberrations. They give all the information about the image field or the intensity distribution at any point in image space. They are known as *Debye–Picht–Luneburg Integrals*.

Evaluation of the integrals (5.1) or (5.2) is difficult even for simple imaging systems, on account of the presence of W, which is not known explicitly except for simple optical instruments. In addition, the standard expansion of W makes the analysis of the field distribution or equal illumination contours on the image plane difficult. This is simplified when W is expanded according to the Zernike–Nijboer[2] or Fourier procedures.

The problem of spherical aberrations of any order has been treated exhaustively by Boivin, not only for circular apertures, but for annular apertures as well. The calculation of the diffracted field in the presence of higher-order aberrations, including the more general problem of nonsymmetric optical systems, is given in reference 3.

All the methods discussed above are valid only for small aberrations. For large or moderately large aberrations, one must resort to asymptotic methods, which at present are sufficiently developed to include most of the interesting cases occurring in the theory of diffraction of optical systems. When these analytical methods are combined with the present progress in computational methods, the intensity distribution produced by an optical system can be calculated to any desired degree of accuracy.

There are many branches of science in which aberrations play an essential, though "negative," role, wherever image forming systems are involved or stability of dynamic systems is studied. In the theory of reconstruction of wavefronts,[6] spherical aberration and astigmatism have to be controlled in order to reconstruct the object structure accurately. In high-energy accelerators (linear, circular), especially in toroidal accelerators for achieving thermonuclear fusion (deuterium–tritium plasma ignition), both the accelerating field and the magnetic field are used to confine the beam of particles within a tubular region close to the axis (optical) of the torus, which is circular. Since the fields are not only functions of space but of time as well (periodic in time), the electron-optical analysis of the system will involve aberrations both in space (geometrical) and in time (stability). In addition, as the charged particles are accelerated the energy varies, and this variation introduces a continuous variation in the de Broglie wavelength of the particles, which introduces an additional kind of aberration known in optics as *chromatic aberration*.[7] Therefore, the problem here involves not only geometrical aberrations (stability), but also chromatic aberrations, which interact with each other, making the analysis very complex.

In microwave focusing[8] (spectroscopy, microscopy) and especially in the construction of giant mirrors and telescopes for exploring the sources of radio, ultraviolet, and X-rays in outer space, care is taken in the design to minimize the aberrations, in particular spherical aberrations and astigmatism, caused by temperature variations on the surface of these instruments.

Recently, sound and ultrasound waves[9] have been used in studying structural defects in solid materials, and in the exploration and surveying of the ocean floor and land masses. In addition, the recent development of apparatus used for imaging biological matter, in particular the internal organs of the human body,[10] is a striking example of the advances which have been made in extending the optical ideas of imaging systems to the whole wave spectrum, including elastic and sound waves on the one hand and charged particles on the other, thanks to the dual nature of particles and waves.[11]

NICHOLAS CHAKO

References

1. Zernike, F., *Physica*, **1**, 689 (1934).
2. Nijboer, B. R. A., "The Diffraction Theory of Aberrations," Groningen thesis, 1942. For the experimental part, see the thesis by Nienhuis, K., Groningen, 1948. For microwave experiments see: Bachynski, M. P. and Bekefi, G., *IRE Trans.*, **AP-4**, No. 3, 412 (1955). "Studies in Microwave Optics," *McGill Univ. Tech. Rept.*, **38** (1957).
3. Steward, G. C., "The Symmetrical Optical System," Cambridge, Cambridge Univ. Press, 1928.
   Picht, Johannes, "Optische Abbildung," Braunschweig, 1931.
   Luneburg, R. K., "Mathematical Theory of Optics," Providence, R.I., Brown University, 1944. Reproduced by the University of California Press, Berkeley and Los Angeles, 1964.
   Linfoot, E. H., "Recent Advances in Optics," London, Oxford Univ. Press, 1955.
   Born, M., and Wolf, E., "Principles of Optics," New York, Pergamon Press, 1959.
   Boivin, A., "Théorie et Calcul des Figures de Diffraction de Révolution," Paris, Les Presses de l'Université Laval, Québec and Gauthier-Villars, 1964.
   Maréchal, A., and Françon, M., Diffraction Struc-

ture des Images, Paris, Masson et Cie, Editeurs, 1970.

Françon, M., in "Handbuch der Physik," Vol. 24, Berlin, Springer, 1956.

Chako, Nicholas, "Contribution à la Théorie de la Diffraction," Centre d'Etudes Nucléaires de Saclay, CEA-R-3151, Saclay, France, 1969. "Etudes sur les Développements Asymptotiques des Intégrales Multiples de la Physique Mathématique" (large aberration diffraction), CEA-R-3263, Saclay, France, 1968.

4. de Broglie, L. "Optique Electronique et Corpusculaire," Paris, 1950.

Glasser, W. "Grundlagen der Elektronenoptik," Berlin, 1955.

Sturrock, P. "Static and Dynamic Electron Optics," Cambridge, 1955.

5. Buchdahl, H. "An Introduction to Hamiltonian Optics," Cambridge, 1970.

Herzberger, M., "Geometrical Optics," Academic Press, New York, 1958.

Pegis, R. J., "The Modern Development of Hamiltonian Optics," *Progress in Optics*, I, Ed. E. Wolf, Amsterdam, North Holland Publ. Co. (1961).

Focke, J., "High Order Aberration Theory," *Progress in Optics*, IV (1965).

"Handbuch der Physik," Vol. XXIX, Ed. S. Flügge, Berlin, Springer Verlag, 1967. Articles by Welford, Walter T., "Optical Calculations and Optical Instruments, An Introduction"; Marechal, Andre, "Methode de Calcul des Systemes Optiques"; Helmut, Max, "Theorie der Geomelrisch-Optischen Bildfehler."

Hawkes, P. W., "Quadrupole Optics, Electron Optical Properties of Orthogonal Systems," Springer Series, "Tracts in Modern Physics," Vol. 42, Berlin, 1966.

See also the Luneburg, Chako, Linfoot, and Born and Wolf titles listed under Ref. 3.

6. Gabor, D. Microscopy by reconstructed wavefronts, I, *Proc. Roy. Soc.* **A197**, 454 (1949); II, *Proc. Phys. Soc.* **64**, 449 (1951).

Stroke, G. S. "An Introduction to Coherent Optics and Holography," Academic Press, New York, 1966.

Okoshi, T. "Three-Dimensional Imaging, Techniques," Academic Press, New York, 1976.

7. Septier, A. "Focusing of Charged Particles," Vols. I and II, Academic Press, New York, 1967–68.

Baber, R. K., and Cosslett, V. E. (Eds.), "Advances in Optical and Electron Microscopy" (serial), Vol. I, Academic Press, New York, 1966.

8. Cornbleet, S. "Microwave Optics," Academic Press, New York, 1976.

Proc. Conf. Electron Microscopy, Kyoto, 1966.

9. Greguss, D. "Ultrasonic Imaging," Focal, New York, 1981.

Bergmann, P., et al., "Physics of Sound of the Sea," Parts I–IV, Gordon and Breach, New York, 1967–68.

Wade, Glen (Ed.), "Acoustic Holography" (serial), Vol. IV, Plenum Press, New York, 1972.

10. Coulon, M., et al., "The Physical Basis of Medical Imaging," Appleton-Century-Crofts, New York, 1981.

Herman, G. T., Image reconstruction from projections, in "Fundamental of Computerized Tomography," Academic Press, New York, 1980.

Gruning, J. R. *Medical Physics*, Proc. Intern. School of Physics "Enrico Fermi," 1981, Vol. 76, Varenna, Italy.

11. Synge, J. L. "Geometrical Mechanics and de Broglie Waves," Cambridge Univ. Press, 1954.

Cross-references: DIFFRACTION BY MATTER AND DIFFRACTION GRATINGS; LENS; OPTICAL INSTRUMENTS; OPTICS, GEOMETRICAL; OPTICS, PHYSICAL; VELOCITY OF LIGHT.

ABSORPTION SPECTRA

Spectroscopy is the measurement of the amount of light, or other radiant energy, transmitted, absorbed, or emitted by a sample of matter, as a function of the frequency, or wavelength, of the radiation. Absorption spectra provide information on individual atomic or molecular units and their immediate environment. Thus it is possible to determine what species are present in a sample and in what form they are present, including the sizes and shapes of individual molecules and often the environments of the molecules. The information obtained directly is the amount of absorption or emission, related to the quantity present and the probability of the radiative transition, and the frequencies absorbed or emitted, which are proportional to the separations of the energy levels in the sample. Spacings of energy levels are equal to Planck's constant times the frequency of the radiation, $\Delta E = h\nu$, and the frequency is related to the vacuum wavelength λ by $\lambda \nu = c$, where c is the speed of light in vacuum. In practice, energy information is often expressed as a vacuum wavenumber, σ(or $\tilde{\nu}$), which is the number of waves per unit of length, and hence the reciprocal of the wavelength; $\sigma = 1/\lambda = \nu/c$. Wavelengths are measured in meters or fractions of meters (e.g. nanometers, nm); wavenumbers are measured in waves per meter or, more often, in waves per centimeter, cm^{-1}.

Spectroscopic studies extend across the electromagnetic spectrum. Gamma rays and x-rays provide information about particles inside the nucleus and about inner-shell electrons of atoms. Ultraviolet and visible radiation interact primarily with electrons in the outer shells of atoms and molecules. Infrared and microwave frequencies produce molecular vibrations and rotations. Radio-frequency spectroscopy measures energies of reorientation of nuclear spin angular momenta in magnetic fields. The present article will consider only absorption of infrared, visible, and ultraviolet radiation by molecules, omitting Mössbauer and atomic absorption, fluorescence, and spin resonance techniques, as well as microwave and Raman spectroscopy.

Experimental Methods Conventional spectroscopy relies upon dispersion of radiation, or separation according to frequency, by a prism or grating monochromator. Most current instruments follow similar principles of design. In a Littrow mount, radiation passing through the entrance slit is collimated, or converted to approximately parallel rays, by a spherical or paraboloidal mirror. The collimated beam is passed through a prism and returned by a reflecting surface through the same prism, or is reflected by a plane grating, returning to the collimating mirror and from there to the exit slit, from which it is focused by an ellipsoidal mirror onto a small detector. The Czerny–Turner mount employs two spherical mirrors in place of the single collimating mirror. These are side by side, with the grating lying along the bisector of the line between them so that, except for the angular position of the grating, the arrangement resembles a Littrow system and its mirror image, sharing a common grating. An Ebert monochromator substitutes a single large spherical mirror for the two mirrors of the Czerny–Turner design. If a concave grating is substituted for the plane grating, no other lenses or mirrors are required. The slit, grating, and image point will lie on a large circle called the *Rowland circle*. Some instruments incorporate a large part of the circle; others include little more than the three critical points inside the case, shifting the circle with respect to the instrument, by rotating the grating, instead of moving components around the circle.

Prisms offer good dispersion properties for the visible and ultraviolet regions, but are generally inferior to gratings for the infrared region and for highest resolution spectra. Most gratings have been made by ruling an aluminum surface with a diamond tip, then casting replicas of the ruled surface in plastic. Gratings with lower levels of scattered light may now be produced holographically, both plane and concave. For very high resolution, a coarse grating can be employed in a very high order (an echelle grating) with cross dispersion by another element to separate the orders. Low resolution infrared spectra may be obtained with a wedge interference filter.

Some research instruments employ only a single beam passing through the sample and monochromator to the detector, but nearly all commercial instruments are now of double-beam design, with a reference beam serving to correct for solvent and atmospheric absorption and, especially, for variations in source intensity with frequency. These may be compared by an optical null system that automatically attenuates the reference beam to match the sample beam, or by an electronic comparison of beam intensities. This is increasingly being accomplished with digital electronics under microcomputer control. To avoid sample decomposition, ultraviolet spectrometers disperse the beam before the sample, whereas infrared spectrometers usually place the sample before the monochromator to avoid displacements of the image of the exit slit on the small detector surface.

A very different design concept employs interferometry to provide the coding by which frequency information is extracted. The beam is not dispersed, but the movement of a mirror with time superimposes an interference pattern on the beam that is different for each frequency. Taking the Fourier transform of the signal, with a computer, generates the spectrum as intensity vs. frequency. When applied to infrared spectroscopy the technique is commonly called FTIR. Interferometers have a substantial energy advantage over dispersive instruments. They can accept radiation from a larger aperture, and because the entire spectrum is examined at once, rather than measuring small frequency increments sequentially, there is an energy gain for thermal detectors that increases with the number of resolved frequency intervals in the spectrum. This second factor, often called *Fellget's advantage*, disappears for photon detectors because the noise increases with total radiation striking the detector surface.

The newest experimental approach takes advantage of tunable sources, eliminating the need for a monochromator or interferometer. First applied in microwave spectroscopy, the technique has now been extended to the infrared and visible regions with tunable dye lasers and diode lasers. Photoacoustic spectroscopy is a variation in which the absorption of radiation produces expansion of the absorber, and thus an acoustic signal.

Conventional sources for the ultraviolet region are hydrogen, deuterium, or xenon discharge lamps. For the visible and near infrared regions, tungsten filament lamps are satisfactory. Common infrared sources include a hollow cylinder of refractory oxides (Nernst glower), a rod of silicon carbide (Globar), alumina tubes, Nichrome wires, and, for the far infrared, silica jacketed mercury arcs.

The standard detector for the visible and ultraviolet region is the photomultiplier, which has almost entirely replaced photographic emulsions. For the near and medium infrared, solid state photoconductors (PbS, PbSe, Au-doped Ge, etc.) provide high sensitivity, especially if cooled. Triglycine sulfate, a pyroelectric detector, and the Golay pneumatic cell, a thermal detector, have advantages for the far infrared. Thermal detectors, including thermocouples and bolometers, are less sensitive than photon detectors, such as cooled photoconductors, for high frequencies but are usable over the entire spectrum.

Solid samples and strongly absorbing liquids are usually dissolved in a solvent. Water is transparent in the ultraviolet (to beyond 200 nm, or 50,000 cm^{-1}). Other suitable solvents, in order of decreasing range, include ethanol and ether, saturated hydrocarbons (e.g. isooctane and cyclohexane), methylene chloride, chloroform, and carbon tetrachloride. Absorption cells usually have a path length of 1 cm. There are no good transparent solvents in the infrared, but carbon tetrachloride is clear over a large region and carbon disulfide is transparent in most regions where carbon tetrachloride is not, for cells less than 1 mm thick. The thin spacings required, coupled with lack of durability of most infrared cell windows, makes maintenance of infrared cells a significant problem.

Sample cells for the visible region may be of glass or any other transparent material, but special glasses or, especially, quartz or fused silica are required for the ultraviolet. (CaF_2, NaCl, and other materials are also transparent but are not competitive because of price, workability, and durability disadvantages.) Infrared cells are typically made with NaCl windows (transparent to about 625 cm^{-1}) or other alkali halides (KBr, CsBr, CsI). For selected regions, quartz or silica, calcium fluoride, magnesium fluoride, and certain special glasses (e.g., arsenic compounds) are advantageous. Polyethylene and silica are transparent in the very far infrared. Rolled AgCl and KRS-5, a mixed thallium bromide–iodide, are impervious to water and transparent over a large part of the infrared.

Solid samples that cannot be dissolved or prepared in thin enough sections for transmission measurements may be immersed in a medium of similar refractive index (e.g., mulled in mineral oil or pressed in KBr or other soft material into a pellet) or examined by attenuated total internal reflectance. A half-rod or prism of transparent material with a refractive index greater than that of the sample is placed against the sample and the beam is totally internally reflected at the interface. As the sample index changes with frequency, because of absorption bands, the total reflectance is modified, giving a spectrum that resembles the absorption spectrum. The method shows high sensitivity, especially for thin films where the prism acts as a light pipe to give multiple internal reflections from the surface.

Structure Determinations To a good approximation, the electronic energy of a molecule may be found by ignoring the motions of nuclei; the energy obtained by solving the Schröedinger equation then contains an error, as a function of positions of nuclei, that is just the vibrational potential function. Similarly, the vibrational energy may be evaluated by ignoring rotations. This separation of the wave function is called the *Born–Oppenheimer approximation*. It allows first-order interpretation of spectra in different regions as if they were solely electronic, vibrational, or rotational in origin.

The primary theoretical tool of the molecular spectroscopist is group theory, applied to the symmetry properties of the system under study. Absorption of radiation normally requires a change in dipole moment of the absorber during the transition. The intensity of absorption is determined by the transition moment,

$\Psi_f{}^* M \Psi_i \, d\tau$, the integral over initial and

final state wave functions and the dipole moment operator. From symmetry arguments it can be shown that certain transitions must have zero transition moment (at least in the usual approximations, such as isolated harmonic oscillators or one-electron wave functions.) These conditions provide selection rules, that

predict which transitions may be active in absorption for various possible structures. Comparison of observed with predicted spectra then usually allows a determination of the actual structure of the molecule.

Prediction of intensities is much more difficult, depending more strongly on detailed models for the absorber, but successful correlation of intensities and structure can therefore provide significant additional information.

Analysis of low-frequency spectra of crystals relies on the symmetry of the unit cell. Vibrational motions, called *phonons*, travel through the crystal and may be analyzed according to the theory of transmission lines with discrete elements. The shapes of absorption bands are determined primarily by the densities of states. If phonon energy is plotted against phonon wavenumber in the crystal, the curves are flat for certain wavenumber values (often near the boundaries of the Brillouin zones), so the probability of absorption at these energy values is significantly increased. The energy determines the frequency of the external radiation that will be absorbed. Symmetry arguments in crystals are modifications of those for isolated molecules, involving lattice groups as well as point groups.

Selection rules may also be derived on the basis of conservation of angular momentum. A photon carries one unit ($\hbar$) of angular momentum, and thus adds this unit to the molecule upon absorption. Depending upon the nature of the molecular transition, this angular momentum will change the rotational state, or a degenerate vibrational state, or the orbital momentum of an electron. Spin angular momentum is seldom affected, except insofar as it is coupled with other motions. Because angular momentum follows vector addition rules, the angular momentum of the molecule can change in magnitude by ± 1 or 0 units, unless it is initially zero. Group theory provides additional insight into the axis along which the angular momentum can change.

Models of the electronic structures of molecules are usually based upon one-electron wave functions, called *orbitals*. Some examples include the united-atom and separated-atom models for diatomic molecules and their extension to polyatomics, the free-electron model for unsaturated and aromatic structures, and Hückel calculations for aromatics. In addition, theoretically based rules have been given by Walsh and by Woodward and Hoffman that predict which orbitals will be occupied and the effects on molecular shapes and reactivities. Ab initio calculations, employing self-consistent field approximations, have been carried out for small molecules, providing a theoretical basis for extrapolations to larger systems.

Qualitative Analysis Molecular vibrations can be described as superpositions of normal, or independent, modes of vibration. Each normal mode involves the entire molecule and has the symmetry of one of the symmetry species, or irreducible representations, of the point group that describes the ground-state symmetry of the molecule. Nevertheless, certain normal modes are affected primarily by specific functional groups within the molecule. For example, C—H stretching modes appear at a wavenumber near 3000 cm^{-1}, somewhat higher if the hydrogen is near an electron-rich group such as an aromatic ring, an oxygen, or a chlorine, and generally a little lower for aliphatic C—H. Stiffer bonds absorb at higher frequencies. N—H and O—H are near 3300–3700 and S—H around 2500 cm^{-1}; C=O is near 1650–1750, compared with C—O, closer to 1000–1200 cm^{-1}. Larger mass moves the vibration to lower frequencies; C—Cl stretching bands appear near 600–900 cm^{-1} and C—I lower yet. Vibrational absorptions that always tend to appear in the same place are called *characteristic frequencies*. By contrast, low-frequency vibrations, including modes involving primarily bending motions, and stretches that may couple with adjacent bonds, such as C—C in aliphatic chains, are much more variable in position, depending on the particular molecule in which they occur. Thus the low-frequency end of the spectrum is called the *fingerprint region*. Often the nature of a compound can be determined from the characteristic frequencies and the exact compound then identified by comparing the fingerprint region with spectra of known compounds.

One must be particularly careful in identifying spectra of gases and solids. Rotational structure can appear in the gas phase, especially for small molecules, that depends to some extent on the gas pressure. Spectacular misinterpretations of spectra have been reported when different rotational branches of a single absorption band have been assigned to different molecular absorbers. Often solids will give spectra very similar to solution spectra, but the spectrum may also vary markedly depending upon crystal structure and orientation of crystallites with respect to the beam. The halide pellet technique tends to introduce spurious hydroxyl bands and the high pressure or chemical reactions sometimes alter the spectrum significantly.

A first approximation to explanation of electronic absorption spectra, in the visible and ultraviolet, was provided by the model of *chromophores* and *auxochromes*. Certain structural fragments (chromophores) were recognized that cause absorption. For example, a single aromatic ring (benzene) absorbs weakly around 250 nm (40,000 cm^{-1}), as does a carbonyl or nitro group. Substituents (auxochromes) on the molecule shift the absorption, most often to lower frequencies. For example, methyl, chloro, hydroxy, methoxy, amine, and acetyl substituents on the benzene ring produce shifts, increasing in the order given, of the strong absorption band initially near 200 nm (49,000 cm^{-1}). Although the chromophore–auxochrome model was initially empirical, theoretical justifications of many of the absorption bands and frequency shifts are now possible.

Despite advances in theory underlying absorption spectra, most qualitative identifications rely on empirical correlation of observed spectra with the spectra of known compounds. In the ultraviolet and visible, relatively few bands are observed and these are often broad and sometimes featureless. Nevertheless, the positions offer significant information on possible absorbers, which can be supplemented by changing sample conditions. For example, a nonpolar molecule such as benzene will show little frequency shift with change of solvents; a polar compound such as a ketone will show an appreciable shift; and addition of acid to an amine will produce the amine hydrochloride, which has a totally different absorption spectrum. Extensive compilations of electronic spectra for comparison are now available.

Quantitative Measurements Except for beams of unusually great intensity, the absorption is determined by the Lambert–Bouguer law or Beer's law. The fractional intensity lost in any small increment of sample thickness is the same, and is proportional to the thickness increment and to the concentration of the absorbing species. This leads to the idealized law, for monochromatic radiation,

$$I = I_0 \, 10^{-abc}$$

where a is the absorptivity of the absorbing species, present at concentration c, in a cell of length b. In practice, however, deviations are often observed. If the absorptivity varies over the frequency spread of the beam, the more strongly absorbed frequencies will be depleted, and the absorptivity will therefore decrease with sample path length. Chemical and physical interactions in the sample may modify the nature of the absorbing species, causing the absorptivity to vary with concentration. Thus it is necessary to check each sample to determine whether Beer's law is obeyed before attempting a quantitative analysis. So long as the absorbance, $A = abc = -\log$ (transmittance), is additive for all absorbing species present, nonlinearities in the calibration curve of measured absorbance vs. concentration can be accommodated by graphical techniques or successive approximations.

Any number of components may be determined simultaneously by making measurements at the same number of frequencies, but accuracy will suffer unless several conditions can be satisfied. There should be significant differences of absorptivity of the various species at the frequencies selected. No absorptivity should change rapidly with frequency at any of the analytical frequencies. If an absolute background (I_0, or $A = 0$) line is not available, the reference background line must be drawn between fixed frequencies, rather than between similar features (e.g., transmittance maxima) of the curves.

Greater accuracy in a multicomponent analysis is achievable by making measurements at more frequencies than the number of unknowns. The analytical equations may be expressed in matrix form, $A = KC$, where A is a column matrix of measured absorbance values at the several analytical frequencies, C is a column matrix of the (unknown) concentrations of the species present, and K is a rectangular matrix of calibration values, one measured value (ab) at each analytical frequency for each species. Multiplication of each side on the left by the transpose of K gives $\tilde{K}A = A' = \tilde{K}KC = K'C$, where A' has the same dimension as C and K' is a square matrix. Solution of $A' = K'C$ gives the best values of the concentrations to a least squares approximation.

Quantitative analysis is usually much more difficult for solid and gaseous samples than for solutions. Solid samples are generally not spatially homogeneous, so Beer's law cannot be expected to apply. However, if the sample has been very finely divided before mulling in oil or pressing in a transparent matrix, the deviations may be small enough. It is often necessary to resort to some form of internal standard, a compound of known absorptivity added in known concentration to the sample, so that the amount of sample in the beam can be determined.

Gases are spatially homogeneous but because of rotational fine structure may appear inhomogeneous with respect to frequency. The beam typically averages transmittance over several narrow absorption bands and the gaps between, so the observed absorbance depends strongly on the widths of the absorption bands, which depend, in turn, on the pressures of the gases in the sample. The apparent absorptivity increases as the gas pressure increases (i.e., absorbance increases faster than pressure, for given cell length) because pressure broadening makes the sample more nearly homogeneous with respect to frequency over the spectrometer bandwidth.

Stray radiation, scattered within the monochromator and reaching the detector mixed with the signal at another frequency, also may cause quantitative errors. The problems are most severe near the ends of the spectrometer range, where the signal intensity is low.

ROBERT P. BAUMAN

References

L. J. Bellamy, "The Infra-red Spectra of Complex Molecules," Chapman and Hall, London, 1975.

N. B. Colthup, L. H. Daly, and S. E. Wiberley, "Introduction to Infrared and Raman Spectroscopy," 2nd Ed., Academic Press, New York, 1975.

J. G. Grasselli and W. M. Ritchey (Eds.), "CRC Atlas of Spectral Data and Physical Constants for Organic Compounds," 2nd Ed., 6 volumes, CRC Press, Cleveland, Ohio, 1975.

G. Herzberg, "Molecular Spectra and Molecular Structure," Van Nostrand Reinhold, New York, Vol. 1, "Spectra of Diatomic Molecules," 2nd Ed., 1950; Vol. 2, "Infrared and Raman Spectra," 1945;
Vol. 3, "Electronic Spectra of Polyatomic Molecules," 1967.

J. W. Robinson, "Handbook of Spectroscopy," Vol. II, CRC Press, Inc., Cleveland, Ohio, 1974.

R. M. Silverstein, G. C. Bassler, and T. C. Morrill, "Spectrometric Identification of Organic Compounds," 4th Ed., John Wiley & Sons, Inc., New York, 1981.

B. P. Straughan and S. Walker (Eds.), "Spectroscopy," Vols. 2 and 3, Chapman & Hall, London, 1976.

Cross-references: ATOMIC SPECTRA; COLOR; MOLECULAR SPECTROSCOPY; OPTICS, PHYSICAL; POLARIZED LIGHT; RADIATION, THERMAL: RAMAN EFFECT AND RAMAN SPECTROSCOPY; REFRACTION; SPECTROSCOPY; X-RAYS.

ACCELERATOR, PARTICLE

Particle accelerators are electromagnetic devices used to generate energetic beams of charged particles—electrons, protons or other ions. They are widely used in research in many fields of physical science and they have many practical applications in medicine, manufacturing, and engineering.

The earliest forerunners of particle accelerators were the gas discharge tubes and x-ray tubes of the late 1800s. They provided some of the early technological base for accelerator development, but they were special in purpose, very limited in voltage, and did not provide the motivation for accelerator development.

The need for particle accelerators became apparent in the 1920s after Rutherford had demonstrated the existence (1911) and the disintegration (1919) of the atomic nucleus using alpha particles from a radioactive substance to probe the structure of the atom. Clearly these experiments gave promise of a radical new understanding of the nature of matter. But just as clearly, particles from radioactive substances were not adequate for the task of exploring atomic and nuclear structure. Beams from such sources were very limited in intensity, poor in collimation, lacked control of energy, and were limited to β-rays (electrons) and α-particles (helium nuclei). Reviewing the need for particle accelerators and the technological base for developing them, Rutherford stated in a famous address in 1927:

It would be of great scientific interest if it were possible in laboratory experiments to have a supply of electrons and atoms of matter in general, of which the individual energy of motion is greater even than that of the α-particle. This would open up an extraordinarily interesting field of investigation which could not fail to give us information of great value, not only on the constitution and stability of atomic nuclei but in many other directions.....[B]ut it is obvious that many experimental difficulties will have to be surmounted before this can be realised.

Inventors and experimenters were already at work to overcome the difficulties with a variety of approaches: electrostatic devices, Tesla coils,

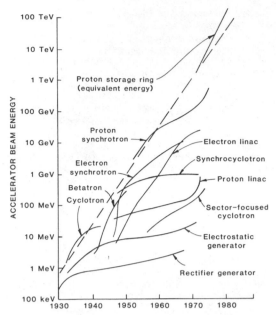

FIG. 1. The maximum energy of each type of accelerator is plotted against the year it was achieved. For colliding beams, the energy of an equivalent fixed target accelerator is plotted. Maximum energy has increased by about 8 orders of magnitude in 5 decades.

transformers, voltage multipliers, radio-frequency resonance acceleration, even atmospheric electricity. By the early 1930s several types of particle accelerators had been invented and successfully used in nuclear experiments.

The electric force is the only macroscopic force strong enough to accelerate particles. The magnetic force, while also strong, is exerted perpendicular to a line of motion, and thus may be used to change the direction of, but not to accelerate, particles. Applying these forces, many kinds of accelerators have been developed, but all are based on a few physical concepts. The emergence of these concepts will be used as a framework to discuss the principal types of accelerators.

Direct-Potential-Drop Accelerators The simplest particle accelerator concept entails a source of charged particles at one end of an insulating evacuated tube and a source of high voltage placed across the ends of the tube. The particles are accelerated from the source end of the tube to the target end by the electric field and gain kinetic energy. The energy of accelerated particles is universally expressed in electron-volts, eV, the energy gained by a particle bearing a charge equal to that of an electron accelerated through a potential difference of one volt. The energies that are of interest in nuclear studies are several hundred kiloelectron-volts, keV, and upwards without a presently preceived limit. The technological difficulties encoun-

tered in the realization of this concept are connected with generating the high voltage and with electrical breakdown both internally and externally in the accelerating tube.

J. D. Cockcroft and E. T. S. Walton, who were the first to achieve nuclear disintegration by electrically accelerated particles in 1932, used a voltage-multiplying rectifier circuit of four stages which effectively charged capacitors in parallel and discharged them in series to reach about 400 kV. They used a glass accelerating tube which was divided into two segments with an intermediate electrode at mid-potential. The ion source was a low-voltage discharge in hydrogen yielding a supply of protons. This type of accelerator has been continuously improved by increasing the number of power supply stages and the segments of the accelerating tube, increasing the frequency of the charging circuit, immersing the voltage supply and the accelerating tube in insulating fluid or pressurized gas to reduce breakdown and other variations and refinements. The *Cockroft–Walton accelerator* remains in use today in many applications up to a few MeV and as a preaccelerator typically operating at 750 keV to inject a beam into a higher-energy accelerator.

The other application of the simple concept of direct application of voltage to an accelerating tube that has had enduring success is the *Van de Graaff* or *electrostatic accelerator*. In this accelerator the potential is supplied electrostatically by an insulating belt transporting charges between ground and a large, usually spherical, high-voltage terminal. R. J. Van de Graaff demonstrated an electrostatic generator of this type in 1931 and the first application to nuclear studies was made by a group headed by M. A. Tuve at the Carnegie Institution in Washington in 1933 with a beam of 600 kV protons. This type of accelerator has also been subject to many improvements and variations. The charging belt and the accelerating tube may be vertical or horizontal. After the earliest examples these accelerators were invariably housed in pressure tanks to exploit the superior voltage holding properties of various gases at high pressures. Refinements in the design of segmented accelerating tubes and the use of shells at intermediate potentials between the grounded pressure tank and the high voltage terminal have been the main factors leading to reliable operation of Van de Graaff accelerators at over 10 MeV. More than twice that voltage is achieved in the tandem design in which negative ions are accelerated from ground potential to a positive terminal where they are stripped of electrons to form positive ions and then further accelerated to ground potential.

Both the Cockcroft–Walton and the Van de Graaff accelerators are characterized by good regulation, easy control of voltage, and excellent beam collimation. They also work with either sign of charged particle and a comparatively simple change of ion source may permit acceleration of partially or completely stripped nuclei of any element.

Resonance Acceleration The concept of resonance acceleration, that is, repeated acceleration by radio-frequency power at relatively low voltages to produce high-energy particles, made it possible to avoid the severe technological problem of electrical breakdown at high voltage. R. Wideroe in 1928 demonstrated the principle with a single tubular electrode supplied with radio-frequency power mounted between two grounded electrodes. Sodium ions, Na^+, were accelerated into the electrode while it was negative, passed through, and further accelerated at the other end when the potential of the tube became positive, thus attaining an energy corresponding to twice the applied radio-frequency voltage. In 1929 E. O. Lawrence elaborated the concept to include the effect of a magnetic field and invented the *cyclotron*. If, in a uniform magnetic field, ions are accelerated perpendicular to the field by a radio-frequency voltage applied to a reentrant electrode, then for appropriate values of the field and frequency the ions will pursue a circular path and return to the opening of the electrode when the voltage has reversed and the ions will again be accelerated. Continuing, the ions will describe a spiral path increasing in radius and energy as they enter and leave the accelerating electrode in resonance with the applied radio-frequency voltage, making it possible to reach energies corresponding to hundreds of times the applied voltage. An additional requirement is that the ions remain in the mid-plane of the cyclotron; that requirement, focusing, is met by a slight decrease of the magnetic field with radius. The great advantage of achieving high energies without the necessity of generating high voltages led to the rapid development of the cyclotron as an accelerator of protons, deuterons, and alpha particles of energies of several tens of MeV. It was the leading type of accelerator until the end of World War II and over a hundred were in operation in laboratories all over the world. As the requirements of research called for increasing energy the limitation of the classical cyclotron became apparent. The cyclotron resonance condition specifies a frequency of the accelerating voltage equal to the rotational frequency and proportional to the magnetic field and the charge to mass ratio of the ions, but as the energy of the ion increases the mass of the ion increases (relativistic effect), decreasing the rotational frequency and violating the resonance condition. This limits the number of turns that the ions will stay in phase with the accelerating voltage and limits the energy of the cyclotron to some tens of MeV.

The Induction Accelerator There were many attempts in the 1920s and 30s to devise an accelerator using the electric field induced by a changing magnetic flux (transformer action) to accelerate ions. The requirements of a magnetic field to hold ions in a circular orbit and to provide a changing flux linking the orbit to accelerate the ions were demanding, and success was not achieved until 1941 when D. Kerst demonstrated electron acceleration in his *betatron*. The betatron was immediately recognized as being very well adapted to accelerate electrons for high-energy x-ray production, and it was commercially developed and extensively used for that purpose. A typical energy was about 25 MeV. It was also developed as a research tool to an energy of about 300 MeV but was soon rendered obsolete by other developments.

More recently, special purpose *linear induction accelerators* have been developed to provide very intense, short bursts of electrons at energies of a few MeV.

The Principle of Phase Stability In 1944 and 1945 V. Veksler and E. McMillan, seeking a means of circumventing the energy limitation of the cyclotron, independently formulated the principle of phase stability. They pointed out that in the acceleration of charged particles by a radio-frequency field, particles in a certain phase band were stable; that is, if they had small errors of phase with respect to the accelerating field or of energy, the acceleration itself automatically tended to correct the error. This principle had far-reaching consequences; it effectively removed the energy limitation (except the economic one) of accelerators and it led to the development of several new designs.

In the *synchrocyclotron*, as the ions are accelerated and their mass increases, the frequency of the accelerating voltage is slowly decreased. The ions automatically remain in the proper phase and increase in energy to the limit imposed by the size of the magnet. Synchrocyclotrons were rapidly developed for protons, deuterons, and alpha particles to energies of hundreds of MeV. The magnets weighed several thousand tons and had pole diameters up to about 5 meters. The size and cost of the magnet became the limiting factor as the need for higher energies in nuclear research continued.

The principle of phase stability coupled with the cyclotron resonance condition provided the means around the limitation of the synchrocyclotron also. In the *proton synchrotron*, low-energy ions are injected into a ring-shaped magnet. Both the frequency of the accelerating voltage and the magnetic field are increased, holding the ions in a nearly constant radius orbit as they are accelerated. The weight and cost of a magnet for constant-radius orbits are very much less than those of one spiral orbits. As in the cyclotron, the ions are focused by a small decrease of field with radius. Proton synchrotrons of this type extended the practical energy limit to over 10 GeV (10,000 MeV).

Electron synchrotrons or just *synchrotrons* are based on the same principles, with one simplifying feature but a new limitation. The simplification is that since an electron attains nearly the constant velocity, c, at an energy of a few MeV, only the magnetic field need be varied if the electrons are introduced into the synchrotron with a few MeV energy. The limitation compared with the proton synchrotron is an energy limitation and is due to the radiative loss of energy of charged particles in circular orbits. This is called *synchrotron radiation* and is a special case of bremsstrahlung. This effect, which placed a practical energy limit of about 1 GeV on the first generation of synchrotrons, is not significant in proton accelerators because the energy loss varies inversely as the square of the mass of the particle.

Linear Accelerators The concept of the *linear accelerator*, that is the acceleration of charged particles along a linear path by radiofrequency fields, goes back to the 1920s, and early development was carried out in the 1930s. However, the necessary rf power technology was not available at that time and the cyclotron and the Van de Graaff accelerators were so successful that linear accelerator development languished. Radar and communications developments during World War II resulted in great advances in rf technology and specifically in the availability of high-power high-frequency tubes of several kinds. This led to the development of two kinds of linear accelerator in the immediate post-war years.

The first was the *electron linear accelerator* or *electron linac* by W. Hansen in 1947. In this accelerator a traveling rf wave is introduced into a waveguide which has been loaded with a series of washer-shaped irises to reduce the phase velocity of the wave to c. Electrons preaccelerated to about 2 MeV, where their velocity is $0.98c$, ride the crest of the advancing wave as, in analogy, a surfer rides a water wave. Typically the frequency is about 3000 MHz and the diameter of the waveguide accelerating tube is 8 cm. The maximum energy achieved with accelerators of this type has been 24 GeV at the SLAC 2-mile accelerator. In the lower energy range, about 100 MeV, hundreds of these accelerators have been built commercially and are used as x-ray sources both for therapy and radiography.

The second was the *proton linac* developed by L. Alvarez in 1948. In this accelerator, a standing rf wave is set up in a resonant tank in a mode in which the maximum electric field is along the axis of the tank. A series of "drift tubes" of appropriate length, shape, and spacing are distributed along the axis so that charged particles are accelerated by the rf field when they are between drift tubes, but are shielded from the field during the reverse half cycle. An initial energy at injection is necessary; in the modern proton linac this is typically a Cockcroft–Walton accelerator operating at 750 KeV. Proton linacs themselves are used as injectors into proton synchrotrons. For this purpose their energy may be 50 or 200 MeV. The highest-energy proton linac is an 800 MeV accelerator designed to exploit the high current capability of linacs.

Linear accelerators similar in principle to proton linacs may also be used to accelerate nuclei of any atom, including Uranium. Such heavy-ion linear accelerators (*hilacs*) are more complicated because they must provide for acceleration of particles of various charge-to-mass ratios corresponding to different charge states and different nuclei.

Sector-Focused Cyclotrons We have seen that the energy of the classical cyclotron is limited by violation of the resonance condition as the mass of the particle increases with energy. The synchrocyclotron provided a way around this difficulty but only at the expense of intensity because the magnets are pulsed. If the magnetic field of a cyclotron increased with radius, the resonance condition could be matched to the increasing particle mass; however, an increasing field defocuses and all the particles would be lost. L. H. Thomas in 1938 pointed out that a focusing force could be restored if the magnet poles were sectored, producing alternate regions of high field and low field even if the average field increased to match the resonance condition. This idea was not immediately exploited, but developments beginning in 1949 resulted in numerous variations of cyclotrons characterized by azimuthally varying magnetic fields and constant rotational frequency. Accelerators of this type have been built for protons to energies of about 600 MeV with currents of 150 μA, a factor of 100 greater than can be achieved with a synchrocyclotron. The sector focusing idea introduced a flexibility into design, making it possible also to build cyclotrons which could accelerate ions of different species and with variable energy. These developments have rendered both the classical cyclotron and the synchrocyclotron obsolescent.

Alternating-Gradient Focusing The principle of phase stability removed the energy limitation of the classical cyclotron and it made possible the design of accelerators using annular magnets, but the focusing requirement was still met as in the case of the cyclotron by introducing a negative gradient of the magnetic field with radius which gives a force restoring ions to the median plane of the magnet (vertical focusing in the usual orientation). This focusing force, which is relatively weak, determines the space required by the beam, hence the size and cost of the magnet. The focusing force cannot be increased simply by increasing the gradient of the magnetic field, because then the ions would not be confined in the radial direction (horizontal defocusing).

N. C. Christophilos in 1950 and E. D. Courant, M. S. Livingston, and H. S. Snyder in 1952 independently devised a new focusing scheme called *alternating-gradient* or *strong focusing*. If a magnet is divided into segments, alternating segments with vertical focusing and radial defocusing forces with segments having vertical defocusing forces and radial focusing forces the net effect will be focusing in both directions.

Strong focusing incorporated into proton synchrotron design reduced magnet aperture cross sections by a factor of ten or more, making it economically possible to design proton accelerators up to several hundred GeV. The largest of these at the Fermi National Accelerator Laboratory (1972) and at the European Organization for Nuclear Research, CERN, (1976) have annular magnet systems of 2 km major diameter and aperture cross sections of about 5 by 15 cm. The maximum proton energies of these accelerators are 500 and 450 GeV, respectively. The energy of electron synchrotrons incorporating strong focusing is still limited by radiative energy loss, but the advantage of small magnet cross section has made it possible to achieve energies of more than 10 GeV.

Colliding Beams In a collision between a particle and a stationary target nucleus, not all the kinetic energy is available to induce a reaction. Part of the energy, as required to conserve momentum, goes to the motion of both particles after the collision. For accelerated and target particles of equal mass and for energies where relativistic effects are small, the available energy is approximately one-half the energy of the accelerated particle. This is not a serious loss; however, for particles accelerated to higher energies, an increasing fraction goes to the energy of motion. For protons at relativistic energies striking target protons, the available center-of-mass energy is approximately $\sqrt{2E}$ GeV, where E is the energy of the incident particle in GeV. Thus the largest proton synchrotrons of about 500 GeV energy can deliver only about 30 GeV to a reaction. If two beams of particles of energy E traveling in opposite directions could be made to collide head on, an energy $2E$ would be available for reactions. The possibility that this obvious, but very difficult to achieve, objective might be realized was derived from suggestions made independently by D. W. Kerst, G. K. O'Neill and others in 1956. Because even the most intense accelerator beams are not adequate to give a useful interaction rate if two accelerator beams are simply pointed at each other, it is necessary to collect, store and recycle the accelerated particles. This is done by injecting the beam from an accelerator into an annular magnet with a constant magnetic field. If the magnetic field is very precise and if the pressure in the vacuum chamber very low ($<10^{-9}$ Torr), the beam may be made to circulate for many hours, even days. Such *storage rings* may be constructed in intersecting pairs with provisions for loading them in opposite directions with particles from an accelerator. At the beam intersections a small fraction of the particles interact and the noninteracting ones continue around for repeated chances to interact.

The ISR (Intersecting Storage Rings) at CERN provides for collisions of proton beams at 30 GeV, giving a total energy of 60 GeV. A single beam of 30 GeV on a target would give only about 7.75 GeV. A single storage ring may also be used to store two counter-rotating beams of particles of opposite sign. Thus PETRA at Hamburg and PEP at Stanford are single storage rings designed for electrons and positrons at about 20 GeV each. At CERN the 450 GeV proton synchrotron has been reconfigured with a complicated set of auxiliary rings to accelerate, store, and collide protons and antiprotons at 270 GeV, giving collisions of 540 GeV. To produce this collision energy with a single accelerator and a fixed target would require an energy of about 15 TeV (15×10^{12} eV)!

Uses of Accelerators While the demands of nuclear and particle physics research have been the strongest driving force in the development of new accelerators and the achievement of high energies, the applications of accelerators in other sciences and in industry has been widespread and the contributions very important. Usually the accelerators designed for practical applications operate at less than the maximum energy for their type, but they may often be required to meet other demands at the limit of technology—intensity, reliability, compact size, etc.

In medicine, accelerator-produced radioisotopes are routinely used to image internal structures and to monitor functions. Thousands of small compact electron linear accelerators are used in hospitals to generate penetrating x-rays for cancer therapy. Accelerated particles ranging from protons to silicon nuclei and secondary beams of neutrons and pions are also used for cancer therapy but on an experimental basis. In engineering and manufacturing, electron accelerators are used to generate penetrating x-rays to examine large structures; small ion accelerators are used to implant controlled impurities in the fabrication of semiconductor devices. Radioisotope tracers are used to study and monitor chemical reactions, wear, and other processes. Small accelerators are used to log oil wells and other bore holes by analysis of the characteristic radiation from various elements when excited by neutrons. Plastics with superior electrical and chemical properties are produced by curing organic polymers with electron beams. Extremely sensitive and nondestructive analysis can be accomplished by inducing characteristic x-ray emission by proton or alpha-particle beams from cyclotrons or electrostatic accelerators. Synchrotron radiation, electromagnetic radiation from energetic electrons confined to orbits by magnetic fields, a limiting factor in energy of electron accelerators, is an extremely useful source of intense, highly collimated radiation extending from the infra-red to x-rays. There are many applications of this radiation in chemistry, metallurgy, and biology.

The Future of Accelerators The course of accelerator development may be displayed in an interesting way in a plot first due to M. S. Livingston. The maximum energy achieved with each type of acclerator is plotted against the year it was achieved, see Fig. 1. It will be noted that, as each type of accelerator reaches or approaches a limiting energy, a new type appears. A linear envelope of these curves shows that maximum accelerator energy has increased by a factor of about 8 each decade for 50 years. There appears to be no letup in the demand for higher energies for research directed towards the ultimate structure of matter; yet the sizes of the largest accelerators are measured in kilometers and the cost in hundreds of millions of dollars. It seems likely then that further advances in accelerator performance will depend upon the emergence of new concepts to circumvent the limits of size and cost. Superconductivity is already coming into use to provide higher magnetic fields at lower power costs and will be exploited more fully. Strong electric fields are associated with intense laser beams; a way may be found to apply these fields to accelerate particles. The very strong magnetic fields associated with an intense electron beam may be useful to confine other particles. The collective effects of a swarm of particles may be used to transfer energy to other particles. Invention and development are continuing and there will be new concepts almost surely leading to new types of accelerators with performance going well beyond the present large accelerators.

EDWARD J. LOFGREN

References

Livingood, John Jacob, "Principles of Cyclic Particle Accelerators," Van Nostrand Reinhold, New York, 1961.

Livingston, M. Stanley, and Blewett, John P., "Particle Accelerators," McGraw-Hill Book Co., New York, 1962.

Hicks, J. W. (Ed.), Eighth International Conference on Cyclotrons and Their Applications, *IEEE Transactions*, NS-26(2), (1979).

Livingston, M. Stanley, Early History of Particle Accelerators, in "Advances in Electronics and Electron Physics," Vol. 50, pp. 1–88, Academic Press, New York, 1980.

Duggan, Jerome L., and Morgan, I. L., 1980 Conference on the Application of Accelerators in Research and Industry, *IEEE Transactions*, NS-28(2), (1980).

Newman, W. S. (Ed.), "11th International Conference on High-Energy Accelerators," Birkhäuser Verlag, Basel, 1980.

Placious, R. C. (Ed.), 1981 Particle Accelerator Conference, Accelerator Engineering and Technology, *IEEE Transactions*, NS-28(3), (1981).

Cross-references: ACCELERATORS, LINEAR; ACCELERATORS, VAN DE GRAAFF; BETATRON; CYCLOTRON; SYNCHROTRON.

ACCELERATORS, LINEAR

Linear accelerators (often abbreviated to "linacs") are used for acceleration of electrons, protons, and heavy ions. Electron linear accelerators have yielded electrons at energies above 20 GeV; proton linear accelerators have not yet reached energies above 800 MeV.

Although the term "linear accelerator" is occasionally used to describe systems in which particles are accelerated by electrostatic fields (Cockcroft-Walton or electrostatic accelerators), the term is generally used to apply to systems in which particles are accelerated along a linear path by application of rf fields. Only accelerators of this type will be discussed in this article.

The linear accelerator has the advantage that the accelerated beam is easily extracted for experimental use. In principle it is capable of producing well-focused beams of higher intensity than are available from circular machines of the synchrotron or synchrocyclotron type. It does, however, require very high power levels at frequencies where conversion equipment is relatively expensive. For a given final energy, a linear accelerator will usually be materially more expensive than a synchrotron. (For a general discussion of accelerators see ACCELERATORS, PARTICLE.)

Field Patterns Used in Linear Accelerators The rf fields used for acceleration are set up in a long cylindrical cavity whose axis is to be the axis of the accelerated beam. Hence for acceleration the field pattern must have a major electric field component parallel to the axis. This requirement is satisfied by the TM_{010} waveguide mode in which a paraxial electric field has its maximum strength at the axis and falls to zero at the cavity wall. Azimuthal magnetic fields lie in planes normal to the axis, have small values near the axis and increase to maximum values at the cavity walls. Usually the field pattern is maintained by coupling to these magnetic fields by loops or apertures excited by external power sources. Corresponding to the high rf magnetic field at the wall, paraxial currents flow in the walls and are responsible for a major fraction of the power loss in the system. When high electric fields are required on the axis to accelerate to high energy in reasonable distances, the wall currents are correspondingly high. For acceleration rates of 2 MeV/m, power losses in copper walls will be of the order of 50 kW/m.

Both standing wave and traveling wave patterns are used in linear accelerators. If traveling waves are used, as is the case in most *electron* machines, the phase velocity of the waves must be made equal to the velocity of the particles accelerated; as the particle velocity increases, the phase velocity also must increase. But phase velocities in simple waveguides always are greater than the velocity of light, and loading must be introduced to reduce the phase velocity to the desired value. This is accomplished by introduc-

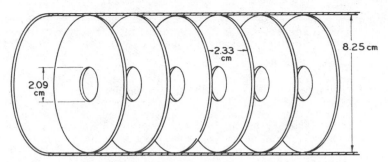

FIG. 1. Cutaway of iris-loaded waveguide for electron linear accelerator.

tion at intervals of washer-shaped irises, as shown in Fig. 1.

Standing wave patterns are used in *proton* linear accelerators. Cavities many meters in length are excited in the TM_{010} mode in which the axial field is uniform from one end of the cavity to the other. Protons which enter the cavity at a low injection velocity may arrive at a phase of the rf field at which they are immediately accelerated, but before they have traveled more than a few centimeters, the field will reverse and become decelerating. To protect the particles from the field in its decelerating phase, "drift tubes" are introduced, as shown in Fig. 2. These are pipes coaxial with the cavity and of such length that the particle is protected from the field during its reverse phase and emerges only after a complete rf cycle when the field again is accelerating. As the particles gain energy, the drift tubes are increased in length.

It would appear that the rather complicated drift-tube structure is conceptually and mechanically inferior to the rather simple iris-loaded traveling-wave system. It is adopted at the relatively low phase velocities required for protons in the range below about 200 MeV because the extreme loading required to reduce the phase velocity of the iris-loaded system to velocities below one-half of the velocity of light results in very high losses. From the point of view of rf power consumption, the drift-tube structure is much superior at low phase velocities.

Electron Linear Accelerators Electrons very rapidly approach the velocity of light (c) as they are accelerated. At 1 MeV an electron already has reached 94 per cent of its ultimate velocity. At energies higher than this satisfactory acceleration will be achieved if all sections of the accelerator are made to have phase velocities equal to c. This makes much easier the tasks of construction and of operation. For example, rf excitation of a section of the accelerator may fail and the whole machine will still be operative, although at a slightly lower final energy.

Losses per unit length in waveguides generally decrease as the square root of the rf wavelength for equal axial fields. Hence, where possible, it is desirable to operate at as high a frequency as possible. But, as wavelength is decreased the diameter of the structure and of the beam aperture decrease correspondingly. The highest frequency that gives convenient beam apertures and at which adequate power sources are available is in the 3000-MHz range. For reasons that are primarily historic, most electron linear accelerators in the United States are operated at a frequency of 2856 MHz.

Both the phase velocity and the group velocity in the guide are determined by the dimensions of the guide and the loading irises. The group velocity is fixed also by the capabilities of the rf power sources. Klystrons with outputs of the order of 20 MW have become standard; each klystron can excite a section of waveguide 3 m long to axial fields of 10 MV/m. The group velocity suitable for this operation is 1 per cent of the velocity of light. The dimensions indicated in Fig. 1 result in a phase velocity of c and a group velocity of $0.01c$ when the guide is excited at 2856 MHz.

Injection is from a conventional electron gun. In some cases a short "bunching section" pregroups the electrons around the peak of the accelerating wave. In this section, the phase velocity is matched to the electron velocity by suitable choices of dimensions.

The power levels required are so high as to preclude continuous operation. Typical operation is with two-microsecond (2-μsec) pulses repeated several hundred times per second. Of the 2-μsec pulse, the first half is required to build up the accelerating field.

Electron linear accelerators in the energy range below 100 MeV are widely used for x-ray production and are commercially available. Most of the pioneer work on electron linear accelerators was done at Stanford where a machine two miles long is in operation at over 20 GeV. At Orsay, France, a 1.3-GeV electron machine is in operation, and in Kharkov in the USSR a 2-GeV accelerator is also in operation.

Proton Linear Accelerators Because of the lower velocities of protons at million-electron-volt energies, proton linear accelerators suffer from several limitations from which the electron machines are free. Injectors for protons usually are Cockcroft-Walton voltage multiplier sets giving energies of 500 to 750 keV. At 750 keV the velocity of a proton is only $0.04c$. The accelerating field component at such low phase velocities varies strongly with radius at a rate that is approximately proportional to the square of the frequency. This effect sets an upper limit of about 200 MHz for the frequency of the accelerating field, and most proton linear accelerators are operated in the neighborhood of 200 MHz. A cavity resonant in the TM_{010} mode at that frequency will be about 90 cm in diameter.

Figure 2 is a schematic cross section through a 50-MeV proton linac formerly used at the Brookhaven National Laboratory as the injector for the 33-GeV synchrotron. Sections are shown at the injector end, at the region where the protons have an energy of about 10 MeV, and at the high-energy end. The over-all length of the machine is about 33 m. The drift tube shapes indicated have the purpose of keeping each section of the machine resonant to give a uniform accelerating field pattern and, at the same time, of holding the resistive losses in the walls of the drift tubes to levels as low as possible.

The principle of phase stability is operative in proton linacs whereas, at the extreme relativistic velocities of multi-MeV electrons, electron linacs do not enjoy phase stability and require extreme precision in axial dimensions. In the proton linac, the drift tube lengths increase at a rate corresponding to acceleration at a phase displaced $20°$ or $30°$ from the peak of the wave. The phenomenon of phase stability (see ACCELERATORS, PARTICLE) results in continual restoration to the correct phase of protons which enter the machine at phases in the neighborhood of the correct phase. Often prebunchers are used to collect a large fraction of the injected beam around the accelerating phase. These prebunchers have the same design as the modulating gap in a klystron and function in the same fashion.

At the stable phase the field across an accelerating gap is rising as the proton crosses the gap. As the proton enters the gap, the accelerating field has a focusing component, but as it enters the next drift tube it feels a larger defocusing field and the net effect is a strong defocusing. In early proton linacs this effect was overcome by the introduction of rudimentary grids at the downstream end of each gap. These grids give unsatisfactory performance because they intercept a large fraction of the beam and because their poor optical quality results in loss of many protons. With the advent of alternating gradient focusing, grids in linacs were largely abandoned and focusing is now accomplished by quadrupole magnets imbedded in the drift tubes. This has resulted in an increase in output current by two orders of magnitude to levels of the order of 100 mA.

As in electron machines, the high rf power level required forces operation at a relatively low duty cycle. Since, at this frequency, the time required to build up the field in the linac cavity is about 200 μsec, pulse lengths for research use are chosen to be several hundred microseconds. Duty cycles are rarely larger than 1 per cent.

The first proton linac was the 40-ft machine at the Lawrence Radiation Laboratory in which protons were accelerated to 32 MeV. This machine was made possible by the development during World War II of powerful rf sources for radar applications. It was a pulsed machine

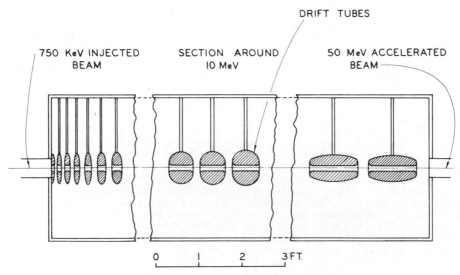

FIG. 2. Cross section through proton linac.

operating at peak currents in the microampere range.

In the early 1950s a daring attempt was made at Livermore to build a continuously operating prototype for a very-high-energy linear accelerator to accelerate protons to energies of hundreds of MeV where they could be used in production of fuels for reactors and weapons. The prototype was built and operated. In those days the highest frequency practical for continuous operation was 12 MHz and as a consequence the resonant cavity was 60 ft in diameter, so big that a railroad spur was run into the resonator to deliver heavy parts. The prototype was 60 ft long. Focusing was by solenoids inside the drift tubes—alternating gradient focusing had not yet been invented. After about two years of work it was discovered that there was considerably more natural reactor fuel than had been suspected and the project was abandoned. The project, called the MTA, was highly classified at the time but was declassified in 1957.

In the late 1950s, 50-MeV pulsed linacs operated at 200 MHz were built to serve as injectors for the proton synchrotrons at the Argonne National Laboratory (ANL), the Brookhaven National Laboratory (BNL) and the CERN International Laboratory in Geneva. These were in operation by 1961. In a few years the BNL linac was replaced by an updated 200-MeV machine which is still in operation in 1982. A similar linac was built at the newly formed Fermi National Laboratory in Batavia, Illinois to be the injector for the second stage of the 200–500-GeV proton synchrotron.

A 20-MeV proton linac is in operation at the KeK Laboratory in Japan where it is the injector for the 12 GeV synchrotron. This machine's KeK Laboratory in Japan where it is the injector for the 12 GeV synchrotron. This machine's structure has some valuable innovations that serve as protection against damage by earthquakes.

The first proton linac to be built by industry for industrial use (production of radiopharmaceuticals) came into operation in 1981 at 40 MeV. It was built by the New England Nuclear Corporation near Boston.

The highest energy reached in a proton linac by 1982 was achieved at Los Alamos where an 800-MeV "meson factory" has been in operation for about a decade. This interesting machine combines a 100-MeV drift-tube linac with a 100–800 MeV iris-loaded section. The iris-loaded structure is not as simple as were the early electron linacs; coupling down the linac is via a cavity on the side of the structure which is coupled to the cavities on each side of an iris. This is reported to improve stability and to decrease rf losses. At 100 MeV the proton velocity is approaching half of the velocity of light and the iris-loaded structure becomes as efficient as the drift-tube system.

Heavy-Ion Linear Accelerators Heavy ions such as C, N, O and Ne and higher masses can be accelerated in a structure similar to a proton linac but usually operated at a lower frequency. Multiply charged ions are injected at a few hundred electron volts and accelerated to energies of the order of 1 MeV per nucleon. The earliest heavy-ion machines resulted from a joint design study between Berkeley and Yale. Machines were built at both centers and proved very valuable in research on nuclear physics and nuclear chemistry. The Yale machine has been shut down but the Berkeley accelerator has undergone massive improvements. In 1982 the Berkeley "Superhilac" had its beam transported a considerable distance across the campus and then injected into the Bevatron, a large synchrotron where much higher energies are achieved.

Heavy-ion accelerators have been particularly valuable in production of transuranium elements. Their energies are continually being pushed to higher levels in the hope of production of elements in the predicted "islands of stability" far beyond the end of the presently collected periodic table of the elements.

A very ambitious heavy-ion accelerator was built during the 1970s at a new laboratory in Darmstadt. It combines several linear accelerator types in sequence, each peculiarly suitable for use in the energy range in question. In 1981 this machine—the UNILAC—was undergoing major rebuilding and modernization. When this is complete the machine's output can include heavy ions with energies of 14 MeV per nucleon.

New Developments Three new developments during the decade of the 1970s merit attention.

First is the "radio-frequency quadrupole" or RFQ. It was proposed in the Soviet Union during the years after 1971 that a structure could be designed in which the rf accelerating fields would not defocus the beam but could provide focusing as well as acceleration. The structure would consist mainly of four bars parallel to the particle orbits, one above, one below, and one on each side of the orbit. The surfaces of these bars facing the orbits would be lightly corrugated, the corrugations above and below facing each other, and the rising part of the side corrugations facing the low part of the upper and lower ones. There was considerable skepticism in the United States about this scheme, and it was not tested in the US until 1978. The tests were at Los Alamos where it was shown that the Russian predictions were completely correct and that the RFQ system has notable advantages at low energies over conventional drift-tube systems. These include possible injection at much lower energies than the conventional 750 keV—the Los Alamos system used 100-keV injection. Possible applications of the RFQ will be mentioned below.

Another step forward, applicable to the linac as well as in many other places is the evolution of rare earth permanent magnets. Pioneering work on these materials has been carried on in many places, mainly industrial, in the United States, Japan, and elsewhere. These materials would be used primarily in linacs in the form of quadrupoles. Permanent magnet quadrupoles for linac focusing were proposed and studied at Brookhaven in the 1950s, but the barium ferrite permanent magnet materials did not have sufficient remanent fields. The new materials have remanent fields a factor of ten higher—of the order of 1 Tesla. The problem of how to construct these materials into a powerful quadrupole was first solved in LBL; since then contributions have been made in a number of laboratories. The first working application of rare earth quadrupoles is in the New England Nuclear accelerator mentioned above.

Sporadic work has been in progress for twenty-five years on development of superconducting rf cavities. Early work in the High Energy Physics Laboratory at Stanford led to optimism that rf fields could be increased and losses drastically reduced in superconducting niobium cavities. But difficulties intervened. There have been few applications; notable has been the superconducting cavity in the Cornell synchrotron. More recently a superconducting heavy-ion linac is almost completed at Argonne. In 1981 most of the linac was assembled and in operation. Eventually it will accelerate heavy ions to energies of 25 MeV per nucleon.

Design Studies and Proposals *FMIT*. A major problem in the design of fusion reactors will be the choice of materials for the "first wall," the wall closest to the reaction area. This wall will be subjected to bombardment by about 10^{14} 15-MeV neutrons per second. Various projects using the D–T reaction have been undertaken but generally they produce neutron fluxes too low by two orders of magnitude. Finally it became evident that the only way to meet the parameters required was by use of a 30-MeV deuteron linac operating continuously with currents of the order of 100 milliamperes. With a liquid metal target this device could yield as much as 10^{16} neutrons per second. The original design was done at Brookhaven; later, the effort was joined by several other laboratories. Eventually the detailed design study was awarded to the Hanford Engineering organization, later to be joined by Los Alamos. This team has made important progress in the design effort. In 1982 it is not clear if or when construction will be approved.

SNQ. At Karlsruhe, a design study is in progress on a linac to serve as a spallation source. A 1.1-GeV proton linac with a current of 100 mA will yield neutron fluxes comparable with those from high-flux reactors. The Karlsruhe linac would have a 5–10% duty cycle. There is some optimism that this project may be approved.

Breeders. The basic idea behind the MTA project mentioned above has been revived. Advances in linac technology make it possible to build a much more reasonable machine than was the MTA. The parameters of such an accelerator, for use in production of plutonium from U^{238} or production of tritium, would be similar to those just described for the SNQ except that it would have a 100% duty cycle. Its energy and current might be higher than those of the SNQ by, perhaps, a factor of two. The design and use of such an accelerator have been discussed for some time at several centers, particularly at Brookhaven and at the Chalk River Laboratory of Atomic Energy of Canada, Ltd.

Medical Applications Electron linacs have been used for several decades in hospitals for radiography and for treatment of tumors. A favorite energy is around 50 MeV. They are produced commercially. Medical and industrial applications involve over 1000 electron linacs in the United States alone.

Neutrons produced by disintegration of deuterons accelerated in cyclotrons have been used for tumor irradiation since the 1930s. More recently protons accelerated in linacs have been used in neutron production. Most notable is the neutron beam generated at the Fermilab in Illinois by an extracted 67 MeV beam from the injector for the Fermilab synchrotron.

Proton beams from linacs and synchrotrons have proved useful in radiography; they appear, for example, to yield more sensitive detections of breast cancers than do the conventional X-ray beams.

Perhaps the most spectacular medical accelerator application involves the use of negative pions. These are unstable antiparticles. They can penetrate human flesh without doing much damage; then, at the end of their orbits they join with a stable particle in an explosive annihilation. Since the orbits have well defined ranges this is evidently a very desirable particle for tumor irradiation. During the late 1970s a design study was done at the Stanford University Hospital. One conclusion of the study was that the facility would be very expensive. A linac with four rooms for patient irradiation would cost some tens of millions of dollars. Hence construction has been postponed pending results from a pion study at Los Alamos. A pion beam has been derived from the Los Alamos "meson factory" and a number of patients have been treated. In the meantime the Los Alamos group has been attempting to design a facility that can be materially less expensive. Part of the Los Alamos enthusiasm for development of the RFQ has been for inclusion in a pion irradiation facility.

Theory Since linear currents have been brought to high levels—hundreds of milliamperes—a number of mysterious phenomena have been observed. In electron linacs spurious field modes have deflected beams into the accelerator walls. In proton linacs the beam "emittance"— its size and angular distribution—has increased for no immediately visible reason in the low-energy part of the machine.

In 1982 it can be said that intensive work using modern computers has resolved these mysteries. This comes as a result of studies in the United States, Europe, and the USSR. Consequently it is now possible to design linacs with confidence that the machine performance will be as predicted by the linac theory groups.

JOHN P. BLEWETT

References

Smith, Lloyd, Linear Accelerators in "Handbuch der Physik," Vol. 44 pp. 341–389, Springer-Verlag, Berlin.
Lapostolle, P. M., and Septier, A. L. (Eds.) "Linear Accelerators," North-Holland, Amsterdam, 1970.
By far the best references on linear accelerators are the *Proceedings* of a series of linear accelerator conferences begun in 1960 at Brookhaven. The eleventh of this series was held in October, 1981 at Santa Fe, New Mexico. The *Proceedings of the 1981 Linear Accelerator Conference* can be obtained from Los Alamos. It is Los Alamos report LA-9234-C.

Cross-references: ACCELERATOR, PARTICLE; ACCELERATORS, VAN DE GRAAF; BETATRON; CYCLOTRON; SYNCHROTRONS.

ACCELERATORS, VAN DE GRAAFF

The electrostatic particle accelerator originated by American physicist Robert Jemison Van de Graaff is widely used for nuclear structure research. These constant -potential accelerators make use of the electrostatic belt generator invented by Van de Graaff about 1930. They belong to the *direct* accelerator family in which the high voltage power is applied directly across the terminals of a highly evacuated multi-electrode tube. Electrified atoms or electrons from a source within the high-voltage terminal gain velocity and energy as they move along the tube axis to ground under the action of the applied electric field. As each particle emerges from the accelerator, it is moving with a kinetic energy equal to qV where q is the particle charge and V the generator voltage.

While a Rhodes Scholar at Oxford during 1927 and 1928, Van de Graaff selected the electrostatic approach to fulfill the need, much emphasized by Rutherford, for more copious sources of atomic particles comparable in energy to those spontaneously emitted from naturally radioactive materials (see ELECTROSTATICS). Subsequently, at Princeton University, Van de Graaff produced over one million volts between the spherical terminals of two small electrostatic belt generators of a new and surprisingly simple design; in 1931, he described the electrostatic belt generator principles, and their suitability for the bombardment of atomic nuclei, before the American Physical Society. The method was first applied to nuclear investigations at the Carnegie Institution of Washington in 1932. The early machines, insulated in atmospheric air, produced streams of light positive ions such as protons and deuterons homo-geneous in energy and with smooth control over the voltage range of the machine. General acceptance of the Van de Graaff accelerator as the precision instrument for experimental nuclear research followed rapidly, and its further development for this purpose has been continuous since that time. Greater compactness and higher voltage were attained by insulating the belt generator and tube with compressed gas; greater beam intensity came through improved ion source and acceleration tube technology. About 300 such accelerators were in use by 1960, producing particles and radiation with energies from 400 keV to 10 MeV. At that time, Van de Graaff accelerators for nuclear science incorporated the "tandem acceleration" principle described below. It opened the way to far higher particle energies by applying the tandem principles to multiply charged heavy atoms.

Van de Graaff accelerators can accelerate any electrified particle, including any of the 92 elements, electrons, and clumps of matter simulating micrometeorites. In addition to use in experimental nuclear physics with high-energy positive ions, Van de Graaff electron accelerators designed for voltages in the 1 to 5 MeV range are used to produce megavolt x-rays for the treatment of malignant disease and for the radiographic inspection of heavy opaque structures such as metal forgings, weldments, and rocket engines. Streams of electrons from such accelerators are also used for radiobiological and radiochemical research and for the treatment of skin malignancies. Radiation processing studies for such purposes as the sterilization of surgical materials, the cross-linking of polyethylene and other plastics, the deinfestation of grains, and increased shelf life of foods have often made use of Van de Graaff accelerators.

Van de Graaff Generator Operating Principles
Although a variety of electrostatic machines had been developed since the first frictionally excited generator of Otto von Guericke in the middle of the seventeenth century, all have been superseded by the Van de Graaff generator because of its greater voltage capability and comparative simplicity. The essential components of the generator, outlined in Fig. 1, include a well-rounded metal terminal supported by an insulating column and an endless insulating belt system which physically conveys electric charge from ground to the high voltage terminal.

Electric charge of the desired polarity is deposited on the moving belt surface by corona from a row of metal points at a controllable voltage with respect to the lower pulley toward which they are directed. In addition to overcoming friction and windage, the motor-driven belt does work in carrying this charge from ground to the terminal potential. Transfer of the charge from belt to terminal is accomplished by again presenting a row of points toward the electrified belt. This time the electric field of the surface-bound charge produces the gaseous ionization needed for conduction across

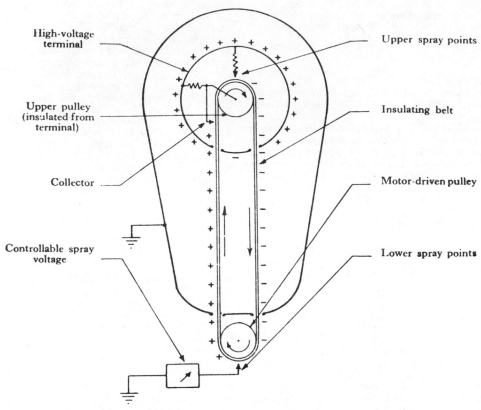

High-voltage terminal

Upper spray points

Upper pulley (insulated from terminal)

Insulating belt

Collector

Motor-driven pulley

Controllable spray voltage

Lower spray points

FIG. 1. Diagram of Van de Graaff electrostatic belt generator. Reproduced by permission of The Institute of Physics and The Physical Society from the article by R. J. Van de Graaff, J. G. Trump, and W. A. Buechner, *Reports on Progress in Physics*, **11**, 1 (1948).

the point-to-belt gap. Van de Graaff pointed out that these ionized charge-transfer processes remain independent of the terminal voltage if they are located in the field-free space within the hollow terminal or below the ground plane. The current of such an electrostatic generator is limited by the maximum charge density which can be insulated in the gaseous medium surrounding the belt and by the total area per second of charge-laden surface entering or leaving the terminal. To increase the current capability of the system, the return run of belt may be charged within the terminal in a similar manner but with the opposite polarity.

The potential, V, of the high voltage terminal of a Van de Graaff generator is determined by the amount and polarity of the accumulated charge on its insulated terminal. At any instant $V = Q/C$ where Q is the net positive or negative charge on the terminal and C is the capacitance of the terminal system to ground. Although the Van de Graaff generator is inherently a constant-current machine, it can be maintained steadily at the desired voltage by balancing the current arriving at the terminal against the total current delivered to the load. The load usually

includes the particle current through the accelerating tube, the current through resistors which divide the terminal voltage uniformly along the supporting column, and any corona from the terminal itself arising from the high electric field at its surface. By adjusting either the belt current or the load current, the terminal voltage may be maintained at any desired value up to the maximum which can be insulated. This maximum voltage depends only on the physical size and geometry of the terminal and on the electrical strength of dielectric medium surrounding it. An isolated metallic sphere would be the ideal terminal, but modifications are necessitated by the supporting column, belt, and tube.

The pair of generators built by Van de Graaff at Princeton in 1930 each had an aluminum spherical terminal 2 ft in diameter supported in air by a slender glass rod 7 ft long. A silk ribbon was employed as the insulating charge conveyor. The voltage insulated in atmospheric air between these two generators, one accumulating positive and the other negative charge, was more than twice any previously attained constant voltage.

About 5.5 million volts were insulated in air between two larger generators constructed by Van de Graaff in the early 1930s for nuclear research. This voltage required spherical terminals 15 ft in diameter supported on insulating tubular columns 25 ft high. This historic equipment, shown in an early sparking demonstration in Fig. 2, was used in a modified form for precision nuclear research at Massachuesetts Institute of Technology for nearly 20 years. It is now impressively installed at Boston's Museum of Science for daily demonstrations of high-voltage phenomena before large audiences and explanation of the underlying electrical principles.

The need for still higher constant voltages for nuclear investigations, and the desire for more compact apparatus, led to the use of high-pressure gases for the insulation of electrostatic accelerators. Today nearly all Van de Graaff accelerators operating at potentials in excess of one-half million volts are within a steel pressure tank and insulated in gases compressed to 10 to 25 atmospheres. Electronegative gases such as sulfur hexafluoride (SF_6) and "Freon" (CCl_2-F_2) are now increasingly used instead of mixtures of nitrogen and CO_2, since they insulate approximately the same voltages at one-third gas pressure.

Acceleration System The evacuated acceleration tube, the source of positive ions or electrons, and the target to which the energized particles are directed, constitute the particle accelerating system of the Van de Graaff accelerator. The insulating length of the evacuated acceleration tube is divided into many sections by metal disk-like electrodes, each with an axial opening for the passage of the particle beam. Each disk is mounted between annular rings of glass or porcelain to form a slender vacuum-tight accelerating column. The tube electrodes take their potential from the metallic members in the generator column along which the terminal voltage is divided by resistors. The charged particles, acted upon by the electric field be-

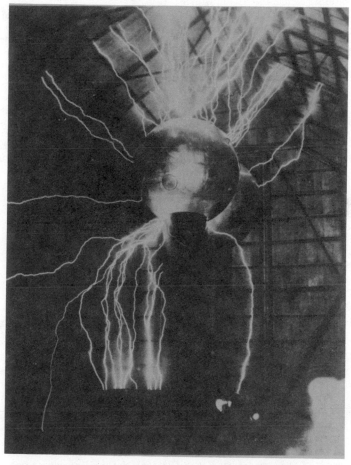

FIG. 2. 5.5-million volt Van de Graaff generator in sparking demonstration.

tween these electrodes, are progressively accelerated and focused as they move through the electric fields between the electrodes. At the remote end, the beam emerges as a collimated and directed stream of energetic particles.

Tandem Acceleration and Multiply Charged Ions Van de Graaff accelerators for nuclear science now reach higher particle energies with a given terminal voltage by switching the polarity of the accelerated particles. In the two-stage tandem diagramed in Fig. 3, negatively charged ions are produced at ground and then accelerated toward a high-voltage positive terminal. Within this terminal, the swiftly moving negative ions are stripped of electrons by passing through a thin gaseous region. The resultant positive ions continue through the tube under the second accelerating action of the positive terminal. A singly charged particle, such as a proton, thus arrives at the ground end of the system with an energy of $2qV$.

At sufficiently high energy, atoms of higher atomic number may be stripped of several or even of all their satellite electrons. An ion which lacks N electric charges during the second acceleration stage gains a total energy of $(N + 1)V$ in a two-stage tandem accelerator. Three-stage acceleration is secured by adding an additional in-line two-stage accelerator with a central negative terminal and using it to produce one stage of negative ion acceleration for injection into the second tandem. In 1967 the first three-stage tandem Van de Graaff, developed by the High Voltage Engineering Corporation and using terminals at 6 MV, was brought into use for nuclear research at the University of Pittsburgh. This was shortly followed by a three-stage tandem 7.5-MV terminal at the University of Washington in Seattle and in 1970 by two in-line 10-MV

"Emperor" tandem Van de Graaffs at the Brookhaven National Laboratory.

Although the light elements, hydrogen and helium, were almost exclusively used as atomic projectiles in nuclear structure physics until 1960, interest in heavier nuclei developed rapidly as higher energies became possible. It is estimated that, by applying tandem acceleration principles, a two-stage Van de Graaff accelerator with a 15-MV positive terminal can produce a beam of uranium ions with energies up to 400 MeV. In large part because of the more complete electron stripping attained at higher energies, three-stage acceleration could produce uranium ions with energies over 1000 MeV.

During the 1970s a number of two-stage Van de Graaff accelerators for national physics research became operational at university and national laboratories in several countries at controllable terminal voltages up to 16 MV. In England at Daresbury an SF_6-insulated, vertically mounted two-stage Van de Graaff designed for terminal voltages up to 30 MV reached the testing stage in 1981.

Medical Applications of Van de Graaff Accelerators Since x-rays are the form of electromagnetic energy, similar to light, produced by the sudden stopping of high-energy electrons, Van de Graaff accelerators are often used as x-ray sources for the treatment of malignant disease and for radiography. In this application, the high-voltage terminal is operated at negative polarity, the electrons are emitted from a tungsten source at the terminal end of the acceleration tube, and they are suddenly stopped after traversing the length of the tube by striking a water-cooled metal target, usually of tungsten or gold.

A 2 million volt x-ray generator of this type,

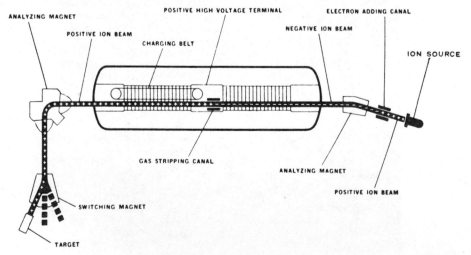

FIG. 3. Diagram of two-stage tandem Van de Graaff accelerator. Reproduced from the article by R. J. Van de Graaff in *Nuclear Instruments and Methods*, **8**, 195–202 (1960), by permission of the North-Holland Publishing Co.

in which a gold target is bombarded with 300 μA of electrons, yields an x-ray intensity of 100 r/min measured 1 meter from the target in the electron direction. The quality of this radiation is closely similar in its physical properties to that of the gamma rays from radium or from the radioactive isotope cobalt 60. To equal this x-ray intensity would require over 4000 curies of cobalt 60 or 6000 grams of radium. This Van de Graaff accelerator for therapy is housed in a steel tank 3 ft in diameter and 6 ft long and is insulated by a mixture of nitrogen and CO_2 at 300 psi.

JOHN G. TRUMP

References

Van de Graaff, R. J., Trump, J. G., and Buechner, W. W., "Electrostatic Generators for the Acceleration of Charged Particles," *Rept. Progr. Phys.,* **11**, 1 (1948).

Van de Graaff, R. J., "Tandem Electrostatic Accelerators." *Nuclear Instr. and Methods,* **8**, 195–202 (1960).

Wittkower, A. B., Rose, P. H., Bastide, R. P., and Brooks, N. B., "Injection of Intense Neutral Beams into a Tandem Accelerator," *Rev. Sci. Instr.* **35**, 1–11 (January 1964).

Wright, K. A., Proimos, B. S., and Trump, John G., "Physical Aspects of Two Million Volt X-ray Therapy," *Surg. Clin. North Am.,* **39**, 1–12 (June 1959).

Livingston, M. Stanley, and Blewett, J. P., "Particle Accelerators," Ch. 3, New York, McGraw-Hill Book Co., 1962.

Trump, J. G., "New Developments in High Voltage Technology," *IEEE Trans. Nuclear Sci.,* NS-**14**, No. 3, 113–119 (1967).

Goldie, C. H., "High current Multimegavolt Ion Accelerators," Proc. 3rd Intern. Conf. on Electrostatic Technology, Oak Ridge National Laboratory, April 13–16, pp. 254–257 (1981).

Cross-references: ACCELERATOR, PARTICLE; ACCELERATORS, LINEAR; CYCLOTRON; HIGH VOLTAGE RESEARCH; STATIC ELECTRICITY; SYNCHROTRON.

ACOUSTICS

The word *sound* is used to describe two different things: (1) an auditory sensation in the ear; (2) the disturbance in a medium, which can cause this sensation. (Making this distinction answers the age-old question, "If a tree falls in a forest and no one is there to hear it, does it make a sound?")

The science of sound, which is called *acoustics*, has become a broad interdisciplinary field encompassing many academic disciplines—physics, engineering, psychology, speech, audiology, music, architecture, physiology, and others. Among the branches of acoustics are physical acoustics, architectural acoustics, psychoacoustics, musical acoustics, electroacoustics, noise control, shock and vibration, underwater acoustics, speech, physiological acoustics, and bioacoustics.

Physical acoustics deals with the production, propagation, and detection of mechanical waves in continuous media. Of particular interest are the radiation, reflection, refraction, diffraction, attenuation, and scattering of longitudinal waves. Much attention has been given to acoustic waves of high intensity (nonlinear acoustics) and those of high frequency (ultrasonics).

Two interesting acoustical effects, whose understanding requires the use of quantum mechanics, are the propagation of *phonons* (lattice vibrations of very short wavelength) in solids and sound propagation in liquid helium. At least five different kinds of sound have been identified in liquid helium at very low temperatures, where it becomes a superfluid. Acoustic signals generated by high-energy particles in water appear to be promising as a means for detecting high-energy protons, muons, and neutrinos.

Architectural acoustics deals with sound in buildings, and in particular the efficient distribution of desirable sound and the exclusion or reduction of undesirable sound. It is usually this branch of acoustics which comes to mind when the layperson hears mention of the term acoustics.

In most auditoriums, the intensity of the reflected sound exceeds the intensity of the direct sound reaching most listeners. Thus the character of the perceived sound is very much dependent on the nature of the reflected sounds and especially their temporal and spatial distributions. The reflected sound which reaches the listener within about 50 milliseconds of the direct sound is often called the *early* sound. Later-arriving reflections make up the *reverberant* sound. One of the parameters characterizing an auditorium is the *reverberation time*, usually defined as the time required for the reverberant level to decrease 60 dB after the sound source ceases.

Psychoacoustics deals with the perception of sound. Loudness, pitch, timbre, and duration are attributes used to describe sound. These attributes depend in a rather complex way on measurable quantities such as sound pressure, frequency, spectrum of partials, duration, and envelope. The relationship of the subjective attributes of sound to physical quantities is the central problem of psychoacoustics.

Musical acoustics considers special problems connected with the production, transmission, and perception of musical sound. Of considerable interest is the physics of musical instruments, including studies of modes of vibration and feedback processes which sustain oscillations in wind instruments.

Musical acoustics overlaps several other branches of acoustics, such as architectural acoustics (concert halls and music listening rooms), psychoacoustics (perception of loudness, pitch, and timbre), speech communication (singing), and electroacoustics (reproduction of music).

Electroacoustics deals mainly with transducers, such as microphones and loudspeakers that convert sound to electrical signals and vice versa. Sometimes the amplification, recording, and reproduction of sound are included as well. The multi-billion dollar audio and entertainment industries depend upon and apply the principles of electroacoustics, and so it is familiar to the consumer.

Unwanted sound or *noise* has been receiving increasing recognition as one of our critical environmental pollution problems. Like air and water pollution, noise pollution increases with population density; in our urban areas it is a serious threat to our quality of life. Noise-induced hearing loss is a serious health problem for millions of people employed in noisy environments. Finding technical solutions to many of our environmental noise problems is central to the branch of acoustics called *noise control*.

Vibration is a term that describes oscillation in a mechanical system. *Shock* is a rather loosely defined aspect of vibration wherein the excitation is sudden, severe, and nonperiodic. The branch of acoustics referred to as *shock and vibration* includes theoretical and experimental studies of both deterministic and random vibration. Methods of measuring shock and vibration have received considerable attention, as have methods for predicting their effects on physical structures.

Underwater acoustics deals with the propagation of sound in water, especially in sea water. The special interest in this field results from two important applications: underwater communication and sonar (SOund NAvigation and Ranging). Slight changes in sound velocity due to temperature gradients in the oceans result in refraction of sound, the creation of special channels and shadow zones, and other effects of importance to nautical and naval personnel.

Acoustical studies of *speech communication* constitute an important branch of acoustics. The production, analysis, and synthesis of speech have been popular areas of research for acousticians, and their work has been greatly facilitated by the availability of the digital computer.

Physiological acoustics is mainly concerned with the auditory system, how it responds to sound, and evoked responses to sound. New insight into the mechanics of the ear have resulted from probing with laser light and the Mössbauer effect, and this has contributed to some new mathematical models of how the cochlea or inner ear functions.

Bioacoustics deals with the interaction of sound waves with biological tissues in humans and animals. Much recent research in this area is concerned with the use of high-frequency *ultrasound* in medical diagnosis and treatment.

The human auditory system responds to sounds with frequencies ranging from about 20 to 20,000 Hz. Below this audible range is *infrasound* and above it is *ultrasound*. It is possible to generate sound waves in air with frequencies of hundreds of megahertz, and in solids frequencies up to 2.5 THz (2.5×10^{12} Hz) have been generated.

The strength of a sound field is measured by its mean square pressure p expressed as a *sound pressure level* L_p in decibels. Decibels are logarithmic units that compare a sound pressure to a reference pressure, usually 20 micropascals or 2×10^{-5} N/m^2, which is near the threshold of hearing for a healthy young person with normal hearing. Prolonged exposure to sound levels above 85 dB can damage the hearing mechanism, and at levels of 130 dB damage can occur almost instantaneously.

The *loudness* of a sound is a subjective quality that depends mainly upon the sound pressure, but to a lesser extent on the frequency, spectrum, and duration of a sound as well. In an effort to obtain a quantity proportional to the loudness sensation, a loudness scale was developed in which the unit is called the *sone*. One sone is defined as the loudness of a 1000-Hz tone at a sound pressure level of 40 dB. The ear shows a marked decrease in sensitivity at frequencies below 200 Hz, and this decrease is most pronounced at low levels.

Various methods are available for expressing the loudness of complex sounds from their sound pressure levels in octave or third-octave bands. Environmental noise levels are usually measured with a sound level meter using the A-weighting network, and these A-weighted levels correlate reasonably well with subjective loudness. For some types of noise, such as that of a jet aircraft, a scale of perceived noise level may be preferable.

Acoustic disturbances can usually be regarded as small perturbations to an ambient state (which in a fluid may be described by the pressure, density, and fluid velocity). The linear approximation (sometimes called the acoustic approximation) considers only first-order changes in these variables, and assumes that their time averages are zero. In a sufficiently intense sound field, however, the time average of one or more of these variables may differ from its ambient value. This is an example of *nonlinear* behavior. Steepening of wavefronts in intense sound waves is another.

THOMAS D. ROSSING

References

Books

Beranek, L. L., "Acoustics," New York, McGraw-Hill Book Co., 1954.

Beranek, L. L., "Noise and Vibration Control," New York, McGraw-Hill Book Co., 1971.

Kinsler, E. E., A. R. Frey, A. B. Coppens, and J. V. Sanders, "Fundamentals of Acoustics," 3rd ed., New York, John Wiley & Sons, 1982.

Kuttruff, H., "Room Acoustics," London, Applied Science, 1973.
Lindsay, R. B., "Acoustics: Historical and Philosophical Developments," Stroudsburg, PA, Dowden, Hutchinson & Ross, 1973.
Morse, P. M., and K. U. Ingard, "Theoretical Acoustics," New York, McGraw-Hill, 1968.
Pierce, A. D., "Acoustics: An Introduction to its Physical Principles and Applications," New York, McGraw-Hill, 1981.
Rossing, T. D., "The Science of Sound," Reading, MA, Addison-Wesley, 1982.

Periodicals

Journal of the Acoustical Society of America (1929–).
Acustica (1951–).
Proceedings of the International Congresses on Acoustics.
Journal of the Audio Engineering Society (1953–).
Journal of Sound and Vibration (1964–).
Applied Acoustics (1968–).

Cross-references: ARCHITECTURAL ACOUSTICS; ELECTROACOUSTICS; HEARING; MUSICAL SOUND; NOISE, ACOUSTICAL; PHONONS; PHYSICAL ACOUSTICS; REPRODUCTION OF SOUND; SONAR; ULTRASONICS; VIBRATION; WAVE MOTION.

ADSORPTION AND ABSORPTION

When a porous solid such as charcoal is exposed, in a closed space, to a gas such as ammonia, the pressure of the gas diminishes and the weight of the solid increases; this is an example of the adsorption of a gas by a solid. It is termed physical adsorption because the forces bringing it about are the "van der Waals" forces of attraction which act between the molecules of the gas and the atoms or ions comprising the solid. It is now known that all solids, whether porous or nonporous, will adsorb all gases physically, whereas the phenomenon of *chemisorption* is specific in nature. Thus hydrogen is chemisorbed by transition metals such as nickel or iron but not by oxides such as alumina.

In physical adsorption the amount of gas taken up per gram of solid depends on the temperature T, the pressure p, and the nature of both the gas and the solid: $n = f(p, T, \text{gas, solid})$. For a given gas (the "adsorbate") adsorbed on a particular solid (the "adsorbent") at a fixed temperature, the amount adsorbed depends only on the pressure of the gas, and the relationship between n and p, viz., $n = f(p)_{T, \text{gas, solid}}$, is termed the *adsorption isotherm*. For vapors, the alternative form $n = f(p/p^0)_{T, \text{gas, solid}}$ is preferable (p^0 = saturation vapor of the adsorbate at temperature T).

Adsorption isotherms may be classified into the five types of the Brunauer–Emmett–Teller (BET) classification. The basic isotherm, the Type II, is concave to the p/p^0 axis at low relative pressures, then passes through a point of inflection (situated usually between ~0.05 and $\sim0.30p^0$) to become convex to the p/p^0 axis at higher relative pressures; the point of inflection corresponds to the *monolayer capacity* n_m of the adsorbent, i.e., the amount of adsorbate which can be accommodated in a completed single molecular layer ("monolayer") on the surface of the solid; the convex branch corresponds to the building up of a multimolecular layer ("multilayer"). From n_m it is possible in principle to calculate the specific surface A (= surface area per gram) of the solid, by assigning a value to a_m, the area occupied per molecule of adsorbate in the monolayer. In practice, however, very few adsorbates other than nitrogen (at its b.p., 77 K) are found to be suitable, and the "BET-nitrogen" method is now the standard procedure for estimation of the specific surface of finely divided or porous solids. (When $A <$ ~1 m^{2}g^{-1}, krypton, though less reliable, has to be used).

If the solid contains mesopores (pores having a width between tens and hundreds of Å) the isotherm is similar to Type II except that the convex branch is replaced by a hysteresis loop in which the curve for desorption lies above that for adsorption, (Type IV isotherm); frequently, with xerogels for example, the isotherm bends over to become effectively horizontal as saturation pressure is approached. By application of the Kelvin equation to points on the loop (usually the desorption branch) it is possible to calculate the pore size distribution of the adsorbent, but owing to the arbitrary nature of the assumptions that have to be made, including oversimplified pore models, the detailed interpretation of the results is uncertain. Even so, the method (again with nitrogen as adsorbate) is valuable for comparative purposes, and is indeed virtually the only method available for the lower end of the mesopore range.

Since adsorption is exothermic, the amount adsorbed at a given pressure must fall as temperature increases. The differential molar heat of adsorption $\dot{q}$–i.e., the limit of the ratio $(\delta Q/\delta n)$, where δQ = heat evolved for an increment δn in the amount adsorbed—is a function of the amount adsorbed. In general, $\dot{q}$ decreases gradually from an initial high value (e.g., $\sim2q_L$) as n increases over most of the monolayer range, then rises to a low maximum (e.g., $\sim1.3q_L$) as n approaches n_m, and finally falls to a nearly constant value close to q_L in the multilayer region (q_L = molar heat of condensation). The initial fall is ascribed to surface heterogeneity (adsorption occurring preferentially on the high-energy sites), and the low maximum is explained in terms of the attractive interaction between adsorbed molecules in the monolayer as they become closely packed together.

The *net heat of absorption* $(\dot{q} - q_L)$ determines the shape of the isotherm in the monolayer region: the greater the value of $(\dot{q} - q_L)$ the sharper is the "knee" of the isotherm; and if $(\dot{q} - q_L) \approx 0$, the isotherm actually becomes convex to the p/p^0 axis, giving a Type III or a Type V isotherm. (The latter, like the Type IV, bends over as p/p^0 approaches unity).

In micropores (i.e., pores of width $\leqslant 15$ Å), the attractive fields from neighboring walls overlap so that the net heat of adsorption, and with it the adsorption at a given relative pressure, is enhanced; consequently the isotherm rises steeply from the origin. Thus a solid containing micropores plus mesopores will give a Type IV, and one with micropores plus a large external surface will yield a Type II isotherm, but with a much sharpened knee in each case. With a wholly microporous solid, where mesopores are absent and the external surface is negligible, the point of inflection vanishes and the isotherm soon reaches a plateau and remains horizontal for the rest of its course (Type I isotherm). The mechanism of adsorption in micropores is still a matter of controversy and reliable methods for the estimation of micropore size distribution are still lacking.

The detailed course of the isotherms of all types, and of the corresponding curves of $\dot{q}$ against n, vary considerably with the mode of preparation and subsequent treatment of the solid, which influence both the porosity of the solid and the structure of its surface. However, in recent years some success has been achieved in preparing standardized samples of a few substances, notably silica, γ-alumina, and graphitized carbon blacks.

Since physical adsorption results from van der Waals forces, the greater the condensability of the gas or vapor as measured by its boiling point or its critical temperature, the greater is the amount of gas or vapor adsorbed at a given pressure. Thus at room temperature and atmospheric pressure, the "permanent gases" such as hydrogen or nitrogen are only slightly adsorbed even on a good adsorbent such as charcoal, while carbon dioxide is more adsorbed, and benzene and carbon tetrachloride are strongly adsorbed. At very low temperatures the adsorption is correspondingly greater, so that nitrogen at its boiling point of $-195°C$ has an adsorption, on a given solid, comparable with that of benzene at $25°C$. For the adsorption to be readily measurable, however, the solid needs to have a relatively large area—a completed monolayer of nitrogen, 1 square meter in extent, weighs only 0.3 mg, for example—so that adsorption phenomena may escape notice unless the solid is "highly disperse," i.e., has an area exceeding several square meters per gram.

Chemisorption results from valency forces—from the sharing of electrons between the adsorbate molecule and the adsorbent—so that, in effect, a surface chemical compound is formed. Chemisorption is characterized by a high heat of adsorption (of the order of many tens of kilojoules per mole, in contrast to the 20–30 kJ mol^{-1} of physical adsorption) and by difficulty of reversal: to desorb a chemisorbed gas in a reasonable time requires a temperature much higher than that at which the chemisorption occurred. Even so the adsorbate may be released in a chemically changed form; thus carbon monoxide chemisorbed on zinc oxide at room temperature is desorbed as carbon *dioxide* at 300°C.

Chemisorption is an essential primary step in heterogeneous catalysis. At least one of the reactants must be chemisorbed on the surface of the catalyst, and each of its molecules then forms, on the surface, a "transition complex" with a chemisorbed molecule of the second reactant B, or with a molecule of B which hits it directly from the gas phase.

Physical adsorption is an extremely widespread phenomenon, frequently unwanted. The adsorption of water vapor by chemicals, by textiles, by building materials and by glass is frequently troublesome and can only be avoided by taking extreme precautions; sometimes, however, the adsorption of water may be beneficial, and it plays an important role, for instance, in the hygiene of clothing.

Adsorption, whether physical or chemical, also reduces the adhesion, and therefore the friction, between solids; gases can accordingly act as lubricants. In addition, adsorption diminishes the tensile strength of brittle solids; the breaking stress of glass when exposed to nearly saturated water vapor is four times less than when exposed to a vacuum. The mechanism is a matter of controversy, but it is probably connected with the fact that adsorption reduces the free surface energy (in the thermodynamic sense) of the solid. Adsorption also causes a small (a fraction of 1 per cent) expansion of the solid, but the swelling pressure set up—i.e., the pressure which would have to be exerted on the solid to prevent expansion—is very high and may reach many atmospheres. Stresses set up in structures made up of porous solids, such as cement and mortar, when they take up or lose vapors, particularly water, may be so great as to cause cracking.

Absorption is said to occur when the molecules of the gas or vapor actually penetrate into the solid phase itself, so that a solid solution is formed; hydrogen is absorbed by iron at elevated temperature in this way, and many synthetic polymers absorb water vapor; benzene vapor is extensively taken up by rubber and water vapor by gelatin. Extensive swelling occurs and if the solid is mechanically weak, the absorption may continue until the system acquires the consistency of a jelly, or even the fluidity of a sol. An absorption isotherm (analogous to the *adsorption* isotherm discussed earlier) can be determined, but is generally complicated and is best handled theoretically as a branch of solution thermodynamics.

In *adsorption from solution*, when a solid having appreciable surface area (say ~ 1 square meter per gram) is shaken up with a solution of substance A in solvent B, both A *and* B are adsorbed, but to different relative extents. This manifests itself in a change in the composition, e.g., a change Δx_A in the mole fraction of A in the solution. The problem is thus more complicated than in the adsorption of gases, and the measured isotherm—the curve of Δx_A against x_A—is not susceptible to any simple theoretical treatment. In a *dilute* solution of A, however, A is always relatively more adsorbed than the solvent B; and if A is colored, the resulting diminution in the concentration of A in the solution will be readily detected by eye or colorimetrically.

S. J. GREGG

References

Gregg, S. J., and Sing, K. S. W., "Adsorption, Surface Area and Porosity," 2nd Ed., London, Academic Press, 1982.

Steele, W. A., "The Interaction of Gases with Solid Surfaces," Oxford, Pergamon, 1974.

Flood, E. A. (Ed.), "The Gas–Solid Interface," 2 Vols., London, Arnold, 1967.

Parfitt, G. D., and Rochester, C. M. (Eds.), "Adsorption from Solution at the Gas–Solid Interface," London, Academic Press 1983.

Cross-references: INTERMOLECULAR FORCES, SURFACE PHYSICS, VAPOR PRESSURE AND EVAPORATION.

AERODYNAMICS

Aerodynamics is the science of the flow of air and/or of the motion of bodies through air. It is usually directed at achieving flow or flight with the maximum efficiency. Aerodynamics is a branch of aeromechanics; the other main branch is *aerostatics* (lift of balloons, etc.). In popular usage aerodynamics differs from *Gasdynamics* in that the latter considers other gases and products of combustion (and combustion); from *aerophysics*, which implies substantial molecular changes in the gas; and from *hydrodynamics*, which implies employing a medium of density approximating that of useful bodies in it, and not infrequently a sharp limit to its extent (i.e., a water surface).

Aerodynamics is conveniently divided into low and high speed regimes. (The latter is in the articles on COMPRESSIBILITY, FLUID DYNAMICS, and SHOCK WAVES; here low-speed aerodynamics is discussed.)

The many facets of aerodynamics include: (1) aerodynamic performance, (2) aerodynamic design, (3) aerodynamic loads, (4) aerodynamic structures, (5) aero-elasticity, (6) aerodynamic heating, (7) aerodynamic compressibility, and (8) aerodynamic research for all of the above.

The computation of aerodynamic effects is based on four laws (given as adapted for fluids):

(1) *Newton's Second Law*: "A force applied to a fluid results in an equal but opposite reaction which in turn causes a rate of change of momentum in the fluid."

(2) *The Equation of State*: (also called the Gas Law): "The product of pressure and volume of a gas, divided by its absolute temperature, is a constant."

(3) *The Continuity Equation*: "The mass that passes a station in a duct, or in a natural tube bounded by streamlines in a given time, must equal that passing a second station in the same time."

(4) *The Energy Equation*: "The total energy in a mass of air remains constant unless heat or work is added or subtracted."

The above relations, written algebraically and limited or combined, yield a vast array of equations used to calculate the practical problems of aerodynamics. One of the *equations* thus derived is due to Bernoulli and is widely used in low speed aerodynamics. It states that in free flow the sum of the static and dynamic pressures is a constant. Static pressure is that pressure which is equal in all directions; dynamic pressure is the pressure rise realized by bringing the fluid to rest. Bernoulli's equation states that as air speeds up, its pressure falls, thus explaining the "lifting suction" as the airstream traverses the curved upper surface of a wing.

The overwhelmingly important *law* of low speed aerodynamics is that due to Newton (1, above). Thus a helicopter gets a lifting force by giving air a downward momentum. The wing of a flying airplane is always at an angle such that it deflects air downward. Birds fly by pushing air downward. Propellers and jet engines make a forward force ("thrust") by giving air a rearward momentum. For some aircraft in some flight modes, *both* the engine *and* the wing are used to provide downward momentum. In the above statements, it is much more accurate to use the expression "downward (or rearward) momentum," rather than "downward (or rearward) velocity," since less dense air at high altitudes produces smaller forces for comparable velocity changes. Forcing the air downward does not occur instantaneously; it takes place over the lifting surface. Indeed, the motion reaches ahead of the airplane so that some downward velocity occurs before the airplane arrives. Thus, airplanes (helicopters, birds, etc.) are always "flying uphill" which is another way of saying it takes a force to fly even when the air is considered frictionless. This force is used to push air *forward*, and is called "drag due to lift." It *increases* as an airplane *slows down* or *flies higher*, and it may be reduced (for low-speed aircraft) by *increasing* the wing span. Supersonic aircraft, operating under the same laws, but in a different manner, still have a drag due to lift which increases as above, but is reduced designwise by *reducing* the span; hence the new airplanes whose wings crank back for high-speed flight. (The "uphill" concept has an analogy in the rolling resistance of a wheel. The weight a wheel carries deflects the surface on which it rests so that it sits in a "gully." Either way it rolls, the path is uphill).

A second important phenomenon (but not a "law" in the sense that it cannot be circumvented) is the manner in which air flows over say, an airplane surface. Away from the surface some free-stream velocity exists. As the surface is approached the local velocity becomes less than freestream and finally becomes zero at the surface. The zone in which the air is appreciably slower is called the *boundary layer*. Slowing the air reduces its momentum, and by Newton's law (above), a force is produced which acts in a direction to slow the airplane. Like forward force needed to produce lift, this *frictional drag* force must have a forward force supplied from somewhere to balance it. For high-speed aircraft a heating occurs in the boundary layer which requires additional force. Skin friction drag is *decreased* by reducing the amount of surface the air scrapes against (i.e., making the aircraft smaller) and by *flying slower* or *higher*—essentially the opposite of the actions which reduce the drag due to lift. Thus the science of aerodynamics seeks the most efficacious melding of the two types of losses. Factors which must be added include providing space for fuel and people, and enough wing to yield a reasonable landing speed. The drag due to lift and the drag due to friction are balanced by the forward thrust provided by the propeller or jet engine (which also operates by Newton's law). However, this is not enough. The distribution of lift on wing and tail must be so located that the aircraft is aerodynamically balanced. Like a child's swing, upon being disturbed it should tend to return to its original ("trimmed") condition.

In the above paragraphs the concern has been for aerodynamic efficiency through optimum design for optimum performance. After these have been achieved (by studies of previous designs and tests in *wind tunnels*) the aerodynamicist provides aerodynamic loads to which an aero-structural engineer must design. Aero-structural design is one of the most challenging of all design problems as the loads must be carried by minimum weight. Aeroplanes carry no "factor of safety" (sometimes called "factor of ignorance"). The aerodynamic-loads engineer furnishes the maximum air loads the aircraft is ever expected to see; the structure is designed to withstand these loads without being permanently bent. If it ever sees a greater load it will be bent or destroyed; there is about a 50 per cent difference between maximum no-permanent set load and catastrophic destruction, depending on the material of which the aircraft is constructed.

In an effort to keep aerodynamic structures light, they often become flexible, and, in turn, become susceptible to flutter, a motion similar to that of a flag in a wind. The motion gets worse with speed. It is the job of the aero-elasticity engineer to assure that flutter will not occur, usually through a redistribution of inter-nal weights (fuel, etc.) and, rarely, through strengthening the structure.

At the higher speeds, the performance, loads, structures and elasticity problems are greatly worsened by the aerodynamic heating which occurs. This is discussed in the article on COMPRESSIBILITY.

Aerodynamics is not only concerned with aircraft. The wind loads on signs, buildings, trees; the aerodynamics drag of autos, boats, trains; the air pollution from smoke stacks of factories and ships; the evaporation of open water; the blowing of sand and snow; and the internal losses of air-conditioning ducts—all deserve and are getting scrutiny by aerodynamicists. The forces on all are proportional to the rate of change of momentum they give, or is given by the air; and all have friction drag in their boundary layers. Aerodynamic research scientists seek to further understand and improve the air flow involved in each.

Aerostatics Aerostatics is the science of making things (balloons, zeppelins, etc.) statically buoyant in the air.

The basic principle of aerostatics is due to Archimedes. "A body immersed in a fluid [or gas] is buoyed up by a force equal to the weight of the fluid [or gas] displaced." Thus for buoyancy the weight of structure plus the weight of the contained gas must equal the weight of the air displaced.

Wind Tunnels Wind tunnels are devices which provide an airstream of known and steady conditions in which models requiring aerodynamic study are tested. The essential elements of a tunnel are:

(1) A drive system consisting of either a compressor for continuous operation or a tank of compressed air for intermittent operation;

(2) a test section in which models are held and their orientation is changed.

(3) instrumentation capable of reading force, pressure, temperature, and optical effects produced by the model;

(4) an air efflux system consisting of free exit to the atmosphere or to a vacuum tank, or a tunnel returning the air to the compressor.

There are approximately 500 wind tunnels in the country, ranging in test section size from 1 in. $\times$ 1 in. to 80 $\times$ 120 ft, with speeds from 50 to 7000 mph.

ALAN POPE

References

Perkins, Courtland, D., and Hage, Robert E., "Airplane Performance, Stability, and Control," New York, John Wiley & Sons, Inc., 1949.

Kuethe, A. M., and Shetzer, J. D., "Foundations of Aerodynamics," New York, John Wiley & Sons, Inc., 1959.

Pope, Alan, and Harper, John J., "Low-Speed Wind Tunnel Testing," New York, John Wiley & Sons, 1960.

Cross-references: ASTRONAUTICS,   DYNAMICS,
INERTIAL   GUIDANCE,   MECHANICS,   SHOCK
WAVES.

AIRGLOW

The aurora has been known to mankind since antiquity (see AURORA). The intensity, color, and motion of an active auroral display invariably create a vivid and memorable impression upon those fortunate enough to observe it. The aurora is a luminescence of air at high altitude which results from the penetration of energetic charged particles (principally electrons) into the atmosphere. The existence of the Earth's magnetic field restricts the entry of these particles to high latitudes; thus only those living in the far North or South enjoy regular entertainment by this magnificent natural phenomenon.

There is another way by which the upper atmosphere becomes luminescent; this arises from the release of chemical energy stored up during the day by the absorption of sunlight. This luminescence is known simply as the *airglow*, although in earlier times it was rather ponderously dubbed "the light of the night sky." The sequence of events which lead to airglow may be understood by the following example.

Over the range of wavelength (or colors) to which the eye is sensitive the atmosphere is highly transparent. This is of course no accident but a natural result of evolution, as is the further fact that the human eye is most sensitive in the yellow-green ($\sim$5500 Å) just where the sun radiates light most strongly. But deep in the ultraviolet (below 1800 Å) oxygen molecules cease to be transparent and absorb light strongly; in this process the molecule is broken up into two oxygen atoms. Most of the sunlight which thus dissociates molecular oxygen is absorbed between 100 and 150 km above the Earth's surface. At such altitudes an atom makes a collision with another atom relatively infrequently, perhaps 100 times per second. At the surface of the Earth collisions occur at a rate of about a billion per second. But even if one oxygen atom encounters another they will not recombine to reform a molecule unless a third particle is also present to satisfy laws requiring that energy and momentum be conserved. The frequency of such triple collisions is of course much less than the frequency of simple two-body collisions. Thus it is that above 100 km a newly released oxygen atom could expect to live for months before it joins up with another in a triple collision to reform a molecule. But at lower altitudes the lifetime is shorter. So the oxygen atoms created by sunlight above 100 km drift downward until they arrive at the region between 90 and 100 km where their lifetime is short enough that they there recombine to form molecules rather than continue their downward drift. The recombination releases just the amount of energy originally required to split the molecule and so,

in principle, this energy can appear in the form of light emission at any wavelength longer than about 1800 Å. In most cases the energy actually goes into heating the atmosphere but when it does come out as light one has a component of the airglow. In effect, what has happened is that a small fraction of the incoming sunlight has been trapped in the atmosphere and later released as longer wavelength light. One may think of this as a phosphorescence with a time delay imposed both by the need for oxygen atoms to drift downward to altitudes where they can recombine and by the recombination lifetime of many days even at this latter altitude.

This sort of airglow process includes not only dissociation and recombination of neutral molecules but also the case in which a molecule or atom is ionized by very short solar wavelength; here the neutral particle is stripped of one of its electrons, leaving a heavy positively charged ion behind. Eventually the free electron will meet and recombine with an ion, and in this process there also can be emission of light. It is the cloud of temporarily liberated electrons which reflect radio waves back to Earth (see IONOSPHERE).

Not all of the airglow comes from such recombination processes, although they are the principal source of the night airglow. During the daytime atoms and molecules in the upper atmosphere can absorb sunlight at wavelengths too long to dissociate or ionize. This energy is almost instantaneously re-emitted either at the same wavelength or at a longer wavelength; it can be considered as a fluorescent component in the day airglow. Yet another type of day airglow comes from the process in which a fast moving electron, newly formed by ionization, collides with an atom or molecule and transfers energy to it; the excited atom or molecule can then emit light.

From the Earth's surface one can only determine the total airglow intensity coming from above. To find out the altitude from which the glow arises one must measure either by using a rocket which passes through the emission layer or by viewing the horizon from a satellite. Such measurements show that nearly all of the airglow comes from the altitude region from about 50 km to 400 km. There is also a powerful means by which one can, in some cases, determine the emission altitude of airglow from the ground. This is to make measurements during the twilight period after sunset or before sunrise when the upper atmosphere is still illuminated by the sun even though it is dark at the surface. After sunset the dayglow emissions will persist until the Earth's shadow reaches the altitude from which the glow arises. The time at which the glow fades away will indicate what this altitude is.

Despite all this furious chemical activity in the upper atmosphere the net result is only a rather feeble glow, particularly at night. All the various recombination processes together produce a night airglow which is not immediately obvious even far from cities. One way to see that there *is* such a glow is to notice that one's hand looks dark against the night sky in some direction where there are very few stars, on a moonless night far from cities. Another way is to look from a jet aircraft just above the horizon at night (if the cabin lights are out). A faint greenish glow can be seen near the horizon; this is airglow from about 95 km altitude which appears brighter because one is looking nearly tangentially through the atmosphere. The green glow comes from oxygen atoms which have been excited by the energy released in the recombination of two other atoms. As it happens, it is this same radiation from oxygen atoms which gives aurora its yellow-green color, but in that case the atoms have been excited by an energetic stream of electrons injected into the atmosphere from above (see AURORA).

The day airglow is of course not detectable to the eye owing to the enormously more intense visible sunlight scattered in the lower atmosphere. It can be seen and photographed from above and astronauts have done just this.

Scientists have found that a great deal can be learned about the upper atmosphere by studying the airglow using sensitive devices (spectrometers) which isolate one or another of the many different wavelengths present in the airglow emission spectrum. Even the day airglow can be detected and measured from the Earth's surface with the help of special instruments although this is best done using spectrometers carried aboard satellites. One can identify the atom or molecule responsible for airglow emission at a certain wavelength. Then one can use the result of laboratory studies on how recombination processes lead to light emission to infer what particular atoms or ions must be recombining in order to produce that particular emission. It is even possible to use the airglow as a thermometer and find out the temperature of the atmosphere at the altitude from which a particular airglow emission arises. One way this is done is to use the Doppler effect; an atomic emission line in the airglow actually possesses a finite width in wavelength (about 1/100 Å) determined by the velocity of the emitting atoms. This velocity is a measure of the atmospheric temperature. Sometimes the temperature is seen to oscillate up and down over a period of an hour or less by as much as 50°C; when this happens one knows that waves are moving up in the atmosphere causing a periodic compression and heating. These waves generally start in the lower atmosphere, often from weather disturbances; they grow in amplitude as they rise through the atmosphere and so produce large oscillations in temperature when they arrive at, say, 100 km altitude. Their importance lies in the fact that they carry energy from the lower atmosphere to high altitudes. At the altitude where the wave dissipation occurs the character of the upper atmosphere is modified by the wave energy deposited there.

While much continues to be learned about our atmosphere using ground based studies of the airglow, there is today even more being done from space. An example is a study of the ozone layer using infrared airglow from the oxygen molecule. Dangerous ultraviolet light from the sun (just below 3000 Å) is blocked by the ozone layer and when it is absorbed the ozone is dissociated into an oxygen atom and an oxygen molecule. The molecule is excited and radiates near 13,000 Å; a measurement of this airglow radiation allows one to determine the density of the parent ozone molecules and to study how and why the ozone varies in both space and time.

JOHN F. NOXON

References

McCormac, B. (Ed.), "Physics and Chemistry of Atmospheres," Dordrecht, Holland, D. Riedel, 1973.

Cross-references: AURORA; GEOPHYSICS; IONIZATION; IONOSPHERE; LUMINESCENCE; MAGNETIC FIELD; SOLAR PHYSICS.

ALTERNATING CURRENTS

Definition of Alternating Current An alternating current is a periodic function of the time, the function being such that the average value is zero. A special case of an alternating current is shown in Fig. 1. The square wave is clearly periodic, and in any one cycle, the area under the curve above the horizontal axis is equal to the area below the horizontal axis. If the two areas are not equal, the current may be described as an alternating current superposed on a direct current, provided the resultant current varies in a cyclic manner.

In general, any alternating current may be considered to be the sum of a Fourier series of sinusoidal waves. For example, the square wave shown in Fig. 1 may be written as

$$\frac{4A}{\pi} \left(\sin 2\pi ft + \frac{1}{3} \sin 6\pi ft + \frac{1}{5} \sin 10\pi ft + \cdots \right)$$

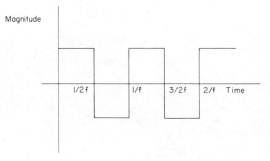

FIG. 1. A square wave alternating current.

where A is the amplitude of the square wave, f is the frequency in hertz (cycles per second), and t is the time in seconds. Since any alternating current may be expressed as the sum of a series of sinusoidal terms, the remainder of this article will be devoted to a discussion of sinusoidal voltages and currents.

Root-mean-square Value The equation for an alternating current i may be written as

$$i = I_m \sin(\omega t - \delta) \tag{1}$$

where I_m is the maximum or peak value of the current, ω is 2π times the frequency f in hertz (Hz), and δ is a phase angle. A graph of the current i is shown in Fig. 2. Since the positive and negative loops are mirror images, the average value of the current over a complete cycle is zero. The latter statement is valid for all alternating currents and, hence, gives no information about a particular alternating current.

A useful way of stating the magnitude of an alternating current is to give its effective or root-mean-square value. The term root-mean-square is derived from the idea of taking the square root of an average square of the current. Thus, by definition, the effective value I_e of the current i given by Eq. (1) is

$$I_e = \sqrt{\frac{\omega}{2\pi} \int_0^{2\pi/\omega} I_m{}^2 \sin^2(\omega t - \delta)\, dt} \tag{2}$$

where $2\pi/\omega$ is the time for one cycle. In effect, the quantity under the square root sign is the sum of the squares of the currents during one cycle divided by the time for one cycle. The value of I_e may be found by performing the integration. The result is

$$I_e = \frac{I_m}{\sqrt{2}} = 0.707 I_m \tag{3}$$

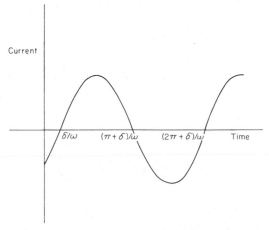

FIG. 2. A sinusoidal alternating current.

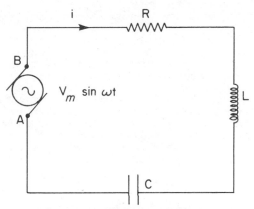

FIG. 3. An ac series circuit.

Clearly, the effective value of a sinusoidal alternating current is 70.7 per cent of the maximum or peak value. Similarly the effective value of a sinusoidal voltage is 70.7 per cent of the maximum or peak value.

Alternating-current Series Circuit A simple alternating-current series circuit is shown in Fig. 3. At the instant considered in the diagram, the current is in the direction shown. The circuit consists of a generator connected in series to a pure resistance R, a pure inductance L, and a capacitance C. It is important to understand the relationship of the current to the potential difference across each element. In each case, the best starting point is a basic definition. According to Ohm's law for a pure resistance,

$$R = \frac{V_R}{i} \tag{4}$$

where V_R is the voltage drop across the resistance. The inductance L of a coil is given by

$$L = \frac{N\phi}{i} \tag{5}$$

where N is the number of turns and ϕ is the magnetic flux passing through one of the turns of the coil as a result of the current i. Writing Eq. (5) as

$$Li = N\phi$$

and then differentiating both sides, we obtain

$$L\frac{di}{dt} = N\frac{d\phi}{dt} \tag{6}$$

According to Faraday's law, the right-hand side of Eq. (6) is the magnitude of the induced emf. The left-hand side, $L\,di/dt$, therefore, is the voltage drop V_L across the inductance L. Finally, the capacitance C is by definition

$$C = \frac{q}{V_C} \tag{7}$$

where q is the instantaneous charge on the positive plate and V_C is the drop in potential in going from the positive plate to the negative plate.

The relation between the impressed voltage $V_m \sin \omega t$ and the instantaneous current i follows from Kirchhoff's law that the sum of the differences in potential in going around a complete circuit must be zero. At the instant shown in Fig. 3, there is a potential rise $V_m \sin \omega t$ in going from A to B and there are potential drops, V_R, V_L, and V_C in traversing the rest of the circuit. According to Kirchhoff's law

$$V_m \sin \omega t - V_R - V_L - V_C = 0$$

$$V_m \sin \omega t - Ri - L\frac{di}{dt} - \frac{q}{C} = 0$$

Since the current is the rate of flow of charge

$$i = \frac{dq}{dt} \qquad (8)$$

It is now possible to express the current i as a function of t. The result neglecting initial transient effects is

$$i = \frac{V_m \sin(\omega t - \delta)}{\sqrt{R^2 + (\omega L - 1/\omega C)^2}} \qquad (9)$$

where $\tan \delta = (\omega L - 1/\omega C)/R$. The phase angle δ is the angle by which the current i lags behind the impressed voltage.

The maximum or peak value of i is

$$I_m = \frac{V_m}{\sqrt{R^2 + (\omega L - 1/\omega C)^2}}$$

If both sides of this equation are divided by $\sqrt{2}$, we obtain

$$\frac{I_m}{\sqrt{2}} = \frac{V_m/\sqrt{2}}{\sqrt{R^2 + (\omega L - 1/\omega C)^2}}$$

$$I_e = \frac{V_e}{\sqrt{R^2 + (\omega L - 1/\omega C)^2}} \qquad (10)$$

Equation (10) states the relation between the effective value of the current and the effective value of the impressed voltage.

Impedance and Reactance The denominator of Eq. (10) may be defined as the impedance Z of the circuit. We may therefore write

$$I_e = \frac{V_e}{Z} \qquad (11)$$

Equation (11) is similar in form to Ohm's law, Eq. (4). The impedance Z is the square root of the sum of the squares of two terms. The first is the resistance R, and the second is $\omega L - 1/\omega C$. The latter is called the reactance. The quantity ωL is the inductive reactance whereas $1/\omega C$ is capacitative reactance. If the inductive reac-

tance is greater than the capacitative reactance, the phase angle δ is positive and the current lags behind the voltage. If the inductive reactance is less than the capacitative reactance, the current leads the voltage.

Vector Diagram The current and the various voltages may be related in a meaningful way by means of a vector diagram. Equation (10) may be rewritten as follows:

$$V_e = \sqrt{R^2 I_e^2 + (\omega L I_e - I_e/\omega C)^2} \qquad (12)$$

Equation (12) implies that V_e is the resultant of a vector RI_e at right angles to a vector $\omega L I_e - I_e/\omega C$. This is shown in Fig. 4. The current I_e is drawn along the horizontal axis, and the effective voltage drop across the resistance, RI_e, is also drawn along this axis. The effective voltage drop across the coil is $\omega L I_e$, and this potential difference is drawn along the positive vertical axis. Finally, the effective potential drop across the capacitor is $I_e/\omega C$, and this vector is drawn along the negative vertical axis. The resultant of the three vectors is V_e, in agreement with Eq. (12).

Resonance If the capacitance C can be varied, the current I_e will be a function of C in accordance with Eq. (10). When

$$\omega L = \frac{1}{\omega C} \qquad (13)$$

the effective current will be a maximum. The circuit is then said to be in resonance. Actually, the inductance L or the angular frequency ω may be varied instead of the capacitance C. The circuit will be in resonance whenever Eq. (13) holds. At resonance, the impedance Z is equal to R, and the circuit, under such circumstances, acts as though it contains resistance only. The process of obtaining resonance is called tuning the circuit.

Average Power The potential difference V across an ac generator at any instant is the work required to transfer a unit charge from the

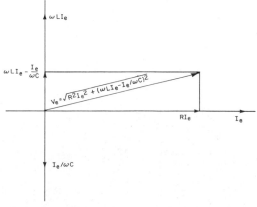

FIG. 4. A vector diagram for an ac series circuit.

negative to the positive terminal. The work done in transferring a charge dq is consequently $V\,dq$, and the work done per unit time is

$$P = \frac{V\,dq}{dt} \qquad (14)$$

where P is, by definition, the instantaneous power and dt is the time interval to transfer the charge dq. Since

$$i = \frac{dq}{dt}$$

Eq. (14) may be written

$$P = Vi$$

The instantaneous power is the product of the instantaneous voltage and current.

When alternating current circuits are considered, the average power $\overline{P}$ rather than the instantaneous power is of interest. By definition

$$\overline{P} = \frac{\omega}{2\pi} \int_0^{2\pi/\omega} Vi\,dt \qquad (15)$$

where, as before, $2\pi/\omega$ is the time for a complete cycle. The average power may be evaluated by making the following substitutions in Eq. (15):

$$V = V_{\mathrm{m}} \sin \omega t$$

$$i = I_{\mathrm{m}} \sin (\omega t - \delta)$$

The result is

$$\overline{P} = \tfrac{1}{2} V_{\mathrm{m}} I_{\mathrm{m}} \cos \delta \qquad (16)$$

Equation (16) may be rewritten

$$\overline{P} = \frac{V_{\mathrm{m}}}{\sqrt{2}} \frac{I}{\sqrt{2}} \cos \delta$$

$$\overline{P} = V_e I_e \cos \delta \qquad (17)$$

Evidently, the average power is the effective voltage times the effective current multiplied by the cosine of the phase angle. In this connection, $\cos \delta$ is called the power factor. Equation (17) may be interpreted to mean that only the component of V_e in phase with I_e contributes to the average power. The other component may be said to be wattless. Since

$$V_e = I_e Z$$

and

$$\cos \delta = \frac{R}{Z}$$

Eq. (17) may be written as follows:

$$\overline{P} = I_e^2 R \qquad (18)$$

From the latter form, it may be concluded that the average power is the average rate at which heat is developed in the circuit. Equation (18) also shows that a direct current having a value I_e would produce the same heating effect as an alternating current having an effective value I_e.

The Complex-number Method In the foregoing, an alternating-current series circuit was discussed by representing voltages as vectors in the real plane. For more complicated circuits, this method is too clumsy. It is much more convenient to deal with vectors analytically by utilizing the j-operator. By definition,

$$j = \sqrt{-1}$$

When a real number is multiplied by j, it becomes an imaginary number. In other words, a point on the real axis is rotated through $90°$ so that it becomes a point on the imaginary axis. The "complex" impedance of a series circuit may thus be written

$$Z = R + j\left(\omega L - \frac{1}{\omega C}\right) \qquad (19)$$

since the reactance may be considered to be at right angles to the resistance. When several impedances are connected in series, the total complex impedance is

$$Z = Z_1 + Z_2 + Z_3 + \cdots \qquad (20)$$

and, when several impedances are connected in parallel, the total impedance is given by

$$\frac{1}{Z} = \frac{1}{Z_1} + \frac{1}{Z_2} + \frac{1}{Z_3} + \cdots \qquad (21)$$

The effective voltage V across the generator may be considered to be a vector along the real axis. The effective current I furnished by the generator is therefore

$$I = \frac{V}{Z} \qquad (22)$$

By solving Eq. (22), the magnitude of I and the phase relation between I and V may be found. Although new mathematical techniques are needed, the saving of time usually justifies the use of the complex number method of handling complicated ac circuits.

Transfer Function From a more advanced point of view, we may consider the voltage V in Eq. (22) as the source and the current I as the response. If we set $j\omega$ equal to a new variable s, called the complex frequency, then we may refer to $1/Z(j\omega) = 1/Z(s)$ as the transfer function. For a complicated network, the transfer function may generally be expressed as the ratio of two polynomials in s. Thus

$$\frac{1}{Z(s)} = \frac{N(s)}{D(s)}. \qquad (23)$$

The zeros of the transfer function are those values of s for which $N(s) = 0$, and the poles of the network are those values of s for which $D(s) = 0$. At a zero, the network does not respond to the source. At a pole, the network may oscillate at a natural frequency given by the imaginary part of s. For a sinusoidal input, the amplitude of the response may be very large at a pole. Network analysis is facilitated through the use of the transfer function. Basically, the transfer function expresses all the characteristics of the network.

REUBEN BENUMOF

References

Benumof, Reuben, "Concepts in Electricity and Magnetism," Ch. 14, New York, Holt, Rinehart, & Winston, 1961.

Benumof, Reuben, "Concepts in Physics," 2nd Ed., Ch. 14, Englewood Cliffs, NJ: Prentice-Hall, Inc., 1972.

Kirwin, G. J., and Grodzinsky, S. E., "Basic Circuit Analysis," Ch. 5, Boston, Mass., Houghton Mifflin Co., 1980.

Bobrow, Leonard S., "Elementary Circuit Analysis," Ch. 8-10, New York, Holt, Rinehart & Winston, 1981.

Hayt, W. H., and Kemmerly, J. E., "Engineering Circuit Analysis," Ch. 8-11, New York, McGraw-Hill Book Co., 1978.

Cross-references: CAPACITANCE; CIRCUITRY; CONDUCTIVITY, ELECTRICAL; ELECTRICAL MEASUREMENTS; ELECTRON; INDUCED ELECTROMOTIVE FORCE; POTENTIAL.

AMORPHOUS METALS

Amorphous metals are materials with good electrical and thermal conductivity, with lustrous appearance, and with other common metallic properties, but with atomic arrangements which are not periodically ordered as in more familiar, crystalline metallic solids. *Noncrystalline, amorphous*, and *vitreous* are equivalent terms used to describe solids in which atoms are not periodically arranged and which in fact lack any sort of long-range order. The term *glass* has often been reserved for amorphous solids formed by continuous solidification of a liquid, but metallic glasses are now commonly considered to include amorphous metals produced in a variety of ways, including evaporation, sputtering, and electro- and chemical deposition, as well as cooling from the liquid state.

Widespread interest in metallic glasses has developed since 1960, when it was first demonstrated that metallic glasses could be formed by very rapidly cooling, or quenching, some molten metallic alloys. Liquid quenching techniques, illustrated in Fig. 1, have been developed for producing metallic glasses as continuous ribbons with widths of several centimeters and at speeds in excess of kilometers per minute.

Research on amorphous metals has resulted from both scientific and technological interest in these materials. They provide opportunities for investigating effects of structural disorder on the basic physical properties of metallic solids, and they offer promise for achieving new and useful combinations of properties for technological ends. Many of the unique properties of metallic glasses are thought to result from the remarkable isotropy and homogeneity of most of these materials on all scales greater than a few atom diameters.

Formation Metallic glasses, like other amorphous solids, are always unstable with respect to some crystalline phase or phase mixture. However, glasses can sometimes be formed from the liquid state because liquids become progressively more viscous and atomic mobility within the liquids becomes more limited as the temperature is reduced. If crystallization does not intercede, many liquids undergo large continuous changes in viscosity in a narrow temperature interval somewhere below their equilibrium freezing temperature T_m. The large increase in viscosity on cooling through this interval causes the supercooled liquid to become a rigid glass below a "glass transition" temperature T_g at which atomic mobility is severely reduced and crystallization is thereby suppressed. Although glasses are ultimately unstable against crystallization, they can be retained almost indefinitely at temperatures well below T_g.

The high fluidity of metallic liquids for temperatures greater than T_g, in contrast to more familiar glass formers like SiO_2, greatly facilitates crystallization as the liquids are cooled below T_m; but for temperatures less than T_g the high viscosity and very limited atomic mobility make formation of stable crystalline nuclei and growth of these nuclei very unlikely. Thus the critical temperature range in which rapid crystallization is liable to occur at appreciable rates is from T_m to T_g, and the less time spent in traversing this temperature interval, the

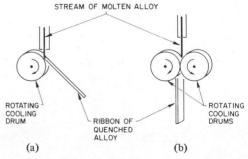

FIG. 1. Continuous quenching devices for making metallic glasses: (a) melt spinning and (b) roller quenching.

higher the probability of successful glass formation. The rate of crystallization R_c is expected to peak in this interval.

Many alloy systems and alloy compositions have been discovered which can be quenched to form a glass with cooling rates of 10^5–10^7 °K/sec. These alloys involve compositions for which T_m is suppressed with respect to melting points of nearby crystalline phases, and compositions of high glass forming tendency are those with the lowest T_m values. Two devices used for preparing metallic glasses are illustrated in Fig. 1. Rapid quenching is accomplished by spreading the molten alloy on a metal surface which serves as a heat sink. This speeds the cooling by reducing the thickness across which heat must diffuse and by ensuring that the alloy remains in good thermal contact with the heat sink during the critical cooling period. Metallic glass ribbons produced in this way generally have thicknesses of less than 50 μm.

Several classes of metallic glasses which have been made by liquid quenching are listed in Table 1, and the list continues to grow. The most widely studied class consists of the transition metal–metalloid alloys with about 20 atomic percent metalloid. Electrodeposition and chemical or electroless deposition have also been used to produce several amorphous alloys of the transition metal–metalloid type.

An even larger range of materials has been prepared as amorphous solids by vapor quenching methods involving deposition by evaporation or sputtering onto a cooled substrate. Here the problem is to obtain metastable, noncrystalline atomic arrangements at sufficiently low temperatures to make spontaneous crystallization very improbable. Some nearly pure metals have been prepared as amorphous thin films on substrates cooled to 4°K. Nominally pure films of cobalt crystallize at about 50°K, and intentional addition of impurities, for example a few percent of silicon, greatly increases the crystallization temperature of such amorphous films. Amorphous metallic alloys which can be retained to temperatures well above room temperature have also been prepared by vapor quenching, including alloy systems and compositions for which glass formation by liquid quenching has been unsuccessful.

Structure X-ray, electron, and neutron scattering experiments, together with electron microscopy, have provided evidence for nonperiodic atomic arrangements in the amorphous alloys discussed above. These materials produce diffuse diffraction patterns, with a few broad,

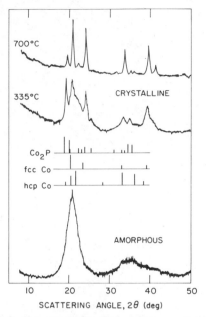

FIG. 2. X-ray scattering patterns for amorphous and polyphase, crystalline $Co_{80}P_{20}$.

overlapping peaks. Part of the diffraction pattern for an amorphous Co-P alloy with about 20 atomic percent phosphorus is shown in the lower section of Fig. 2. Heating this alloy changed the scattering pattern to that shown in the center of the figure. Appearance of sharp maxima indicates that the alloy began to crystallize to the expected crystalline phases of Co_2P and Co. Further heating to 700°C produced the sharper scattering pattern shown at the top of the figure, indicating improved perfection of the crystalline materials. Diffraction patterns of amorphous alloys look more like those of liquid metals than like diffraction patterns of crystalline materials. Electron micrographs of most amorphous alloys are featureless compared with those of crystalline solids.

The diffuse scattering patterns produced by amorphous solids can be used to calculate radial distribution functions, $RDF(r)$, which describe correlations among atomic positions in the materials. The methods of analysis are the same as those used for liquid alloys. Examples for two amorphous metal–metalloid alloys are shown in Fig. 3. Maxima in $RDF(r)$ occur for r-values corresponding to frequently occurring

TABLE 1. CLASSES AND EXAMPLES OF METALLIC GLASSES.

Transition metal–metalloid alloys	$Fe_{80}P_{13}C_7$
Late transition metal–early transition metal alloys	$Ni_{60}Nb_{40}$
Transition metal–rare earth metal alloys	$Co_{20}Gd_{80}$
Simple metal alloys	$Ca_{65}Al_{35}$

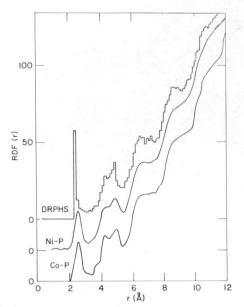

FIG. 3. Radial distribution function for amorphous transition metal–metalloid alloys $Ni_{76}P_{24}$ and $Co_{78}P_{22}$, and for a dense random packing of hard spheres (DRPHS) structural model. [From G. S. Cargill III, *AIP Conf. Proc.* **24** 138 (1975).]

interatomic separations in the amorphous alloys. RDFs illustrate several differences between atomic arrangements in amorphous and crystalline alloys. Long-range structural periodicity is absent in the amorphous alloys. There are only weak correlations between atomic positions separated by more than four or five atom diameters. There is no unique nearest neighbor distance in these amorphous alloys. Widths of the nearest neighbor maximum in the RDFs are typically between 0.4 Å and 0.5 Å full width at half maximum (FWHM). Radial distribution functions for crystalline Co or Ni would have much narrower nearest neighbor maxima, with widths less than 0.2 Å FWHM. Crystallization of these alloys increases their density by less than one percent. The RDFs and density measurements together provide statistical descriptions of atomic arrangements in amorphous alloys and serve as critical tests for three-dimensional structural models.

The most extensively studied models for atomic arrangements in metallic glasses are based on the nonperiodic "random" packing of spheres and are called *dense random packing of hard spheres models*. The RDF for a physically constructed dense random packing of several thousand single-sized steel spheres is shown in Fig. 3, together with the experimentally determined RDFs for two transition metal-metalloid alloys. The agreement is impressive, although the model takes no account of the alloy nature of the experimentally studied

materials. Efforts to improve dense random packing models for binary alloys have used computers to generate dense random packings with two sizes of spheres, chosen to represent the two elements of the binary alloy. These structures have been relaxed by allowing those small displacements of atom sites which reduce the energy, calculated by assuming that the atoms interact with one another by pairwise forces. Although such models have been fairly successful in reproducing experimentally observed distribution functions for most metal-metalloid and rare earth metal–transition metal alloys, the more general applicability of dense random packing as a structural model for metallic glasses has not yet been established.

Electrical Resistivity Resistivities of amorphous metallic alloys are typically between 100 and 200 $\mu\Omega$-cm, values which are higher than resistivities of most crystalline metals and alloys but are similar in magnitude to electrical resistivities of liquid metals. Although the reversible temperature dependence of electrical resistivity for amorphous alloys is less than for most crystalline metals, when the alloys crystallize, their resistivities decrease abruptly and irreversibly, as shown for example in Fig. 4.

The major features of the reversible resistivity behavior can be explained in terms of a model originally developed for liquid metals. The basic features of this model are that the electrical resistivity arises from scattering of conduction electrons by the amorphous structure and that the structure-dependent scattering can be evaluated from experimentally measured X-ray, electron, or neutron scattering patterns of the material.

A number of amorphous alloys become superconductors at sufficiently low temperatures. Much early work on vapor quenched amorphous alloys was stimulated by the discovery that amorphous bismuth was metallic and became superconducting at 6°K, although crystalline bismuth is a non-superconducting semimetal.

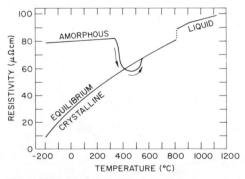

FIG. 4. Electrical resistivity for a $Pd_{80}Si_{20}$ alloy in amorphous, equilibrium crystalline, and liquid forms. [From P. Duwez, *Trans. Am. Soc. Metals* **60**, 607 (1967).]

However, subsequent research has failed to discover any amorphous superconductors with transition temperatures above 10°K.

Magnetic Properties Metallic glasses are as diverse in their magnetic behaviors as are crystalline metals and alloys, ranging from strong ferromagnetism to weak diamagnetism, including ferrimagnetism and spin glass type magnetism. The largest amount of research has been carried out on ferromagnetic transistion metal-metalloid (TM–M) alloys and on ferrimagnetic transition metal–rare earth metal (TM–RE) alloys. The magnetically ordered amorphous TM–M systems are well described as aligned ferromagnets, as illustrated schematically in Fig. 5(a), where metalloid atoms are not shown. Some TM–RE systems are ferrimagnets, with RE moments pointing opposite to the TM moments. For alloys with gadolinium as the rare earth element, the moments of TM atoms (small circles in Fig. 5(b)) are closely antiparallel to those of the RE atoms (large circles), but for other rare earths, notably terbium and dysprosium, the RE moments are strongly coupled to local easy directions, giving a canted ferrimagnetic arrangement like that shown in Fig. 5(c).

It is usual to think of glasses and amorphous solids as being macroscopically isotropic, and indeed these materials are isotropic with respect to most of their physical properties. Magnetic anisotropies for most amorphous transition metal–metalloid alloys are very small and their effects on coercivity and ease of magnetic saturation can be largely eliminated by suitable annealing treatments, which relax internal strains and allow some local rearrangements of atom positions. Coercivities for suitable annealed TM–M alloys can thus be as small as 10^{-2} Oe. Annealing in the presence of applied magnetic fields can also be used to induce macroscopic magnetic anisotropies in the amorphous TM–M alloys. Low coercivity and high permeability are very desirable properties for a variety of applications, including low-loss transformer cores, presently one of the most widely pursued applications for metallic glasses.

The behavior of magnetic domains in amorphous Gd–Co alloys is often dominated by large perpendicular easy axis magnetic anisotropies of 10^4–10^5 erg/cm^3 induced in the sputtered films by appropriate deposition conditions. This class of amorphous alloys has been of special interest because of its potential usefulness in magnetic memory devices, for which the magnetic anisotropy and ease of magnetic domain wall movement are essential.

Other basic aspects of magnetic behavior of amorphous alloys which have been studied are the overall temperature dependence of magnetization, the characteristic magnetic excitations, including spin waves, and the effects of structural and chemical disorder on phase transition phenomena, including sharpness of the ferromagnetic–paramagnetic transition and the critical exponents associated with this transition.

Mechanical Properties Many mechanical properties of metallic glasses have been investigated, including elastic constants, yield and fracture strengths, modes of plastic deformation, the dependence of these properties on thermal history and measurement temperature. Most of these studies have employed quenched-from-the-liquid glasses.

Metallic glasses, like most polycrystalline solids, are mechanically isotropic, so only two constants, for example the bulk modulus and the shear modulus, are needed to characterize their elastic properties. Elastic stiffness constants of the glasses are generally smaller than those of the same materials in crystalline form, which are usually mixtures of crystalline phases. The bulk moduli differ only by 5–10%, but the shear modulus μ and Young's modulus E for glasses are lower by 30–50% than for corresponding polycrystalline solids. The lower resistance to shear deformation has been explained in terms of nonuniform atomic displacements which can occur in the glasses but which are restricted by symmetry constraints in the crystalline forms of these materials.

Metallic glasses are typically very strong, having observed yield strengths $\sigma_y \approx E/50$ as high as 370 kg/mm^2 (640×10^3 psi) for Fe$_{80}$B$_{20}$, similar to that of polycrystalline, cold drawn, high carbon steel. However, unlike high strength polycrystalline alloys, which are inherently brittle, most metallic glasses are microscopically ductile and can sustain appreciable local plastic deformation, particularly in compression or shear. Although there have been several proposals for using metallic glass wires or ribbons to form reinforced composites with plastic, ceramic, or metallic matrices, no commercially important applications of this sort have yet emerged.

Plastic deformation in metallic glasses for temperatures much lower than the glass temperature T_g is inhomogeneous, occurring in localized shear bands along directions of maximum resolved shear stress. In contrast to the usual work hardening which accompanies plastic deformation in crystalline solids, the localized shear bands of metallic glasses are manifestations of "work softening." The glassy material is weakened as a result of the local disruption of

(a)   (b)   (c)

FIG. 5. Atomic magnetic moments in amorphous magnets: (a) ferromagnetic as in Fe$_{80}$B$_{20}$, (b) ferrimagnetic as in Gd$_{30}$Co$_{70}$, and (c) canted ferrimagnetic as in Tb$_{33}$Fe$_{67}$.

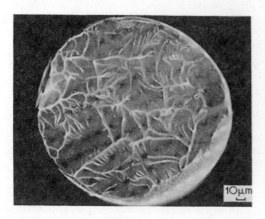

FIG. 6. Fracture surface of a glassy Pd–Cu–Si wire. [From L. A. Davis and S. Kavesh, *J. Mat. Sci.* **10**, 453 (1975).]

short-range order or the local creation of excess free volume by the previous plastic deformation. Tensile failure of metallic glass wires or ribbons occurs in the localized shear band, with characteristic veinlike fracture surface features like those shown in Fig. 6, which are suggestive of local melting. For temperatures closer to the glass transition temperature, deformation takes place by homogeneous viscous creep, with the strain rate proportional to the applied stress.

The ductility of many metallic glasses is lost through irreversible embrittlement when the alloys are annealed, even though the annealing temperatures are too low to cause any detectable crystallization. Most metal–metalloid alloys show this behavior, and those containing phosphorus are particularly susceptible. Some experiments suggest that the embrittlement is caused by fine-scale phase separation or segregation involving metalloid elements.

G. S. CARGILL III

References

P. Chaudhari, B. C. Giessen, and D. Turnbull, "Metallic Glasses," *Scientific American* **242**, 98–117 (April 1980).

Güntherodt, H. J. and Beck, H. (Eds.), "Glassy Metals I, Ionic Structure, Electronic Transport, and Crystallization," Topics in Applied Physics, Vol. 46, Springer-Verlag, New York, 1981.

Masumoto, T. and Suzuki, K. (Eds.), "Proceedings of the Fourth International Conference on Rapidly Quenched Metals," Japan Inst. of Metals, Sendai, 1982.

Gilman, J. J. and Leamy, H. J. (Eds.), "Metallic Glasses," American Society for Metals, Metal Park, Ohio, 1978.

ANTENNAS

Communication systems characteristically consist of cascaded networks, each network designed to carry out some operation on the energy conveying the information. In radio communication, transmitting systems' antennas are the networks serving to transfer the signal energy from final circuit network to space. In receiving systems, on the other hand, antennas serve to transfer the signal energy from space to the input circuit networks. In circuits the flow of energy is restricted to one of two directions. The effectiveness of the transfer of energy between the antenna and its adjoining circuit is, therefore, determined solely by the terminal impedance of the antenna and that of the adjoining circuit network. The terminal impedance of a circuit network can generally be designed for operation over a relatively wide frequency band. On the other hand, the terminal impedance of an antenna may, in some cases, vary outside acceptable limits, thus limiting the useful frequency bandwidth of the antenna. The terminal impedance of an antenna and its variation with frequency is, therefore, one of the important parameters describing the performance of the antenna.

The relationship between an antenna and space, however, is more complex. The distribution of the radiated energy varies with direction in space, giving rise to the directive properties of the antenna. The energy radiated by the antenna is in the form of electric and magnetic fields. These are vector quantities which, at a distance from the antenna, are at right angles to each other and to the direction of propagation. The planes in which these vectors are located, and whether they are stationary or rotate with time, determine the polarization of the radiated field. The performance of an antenna can, therefore, be fully described only by specifying several parameters, such as radiation pattern, power gain,[1,2] directive gain, and polarization. In discussing antenna properties, it is convenient to consider the antenna as a radiating rather than as a receiving network. The antennas, however, are passive and linear networks and, therefore, subject to the law of reciprocity.[3] Therefore, the performance of an antenna, in terms of radiation pattern, gain, or polarization, is the same whether the antenna radiates or absorbs radiation.

Except for the region in the immediate vicinity of the antenna, termed the *near-field region* of the antenna, the radiated electromagnetic energy propagates radially away from the antenna, and the radiation intensity* varies inversely as the square of the distance from the antenna. This attenuation with distance is the *propagation loss.* In radio communication systems it is the principal component of the system's loss and must be compensated for by increasing the power of the transmitter, the gains of the transmitting and receiving antennas, and the sensitivity of the receiver. In describing antenna performance, it is customary to disregard the propagation loss and to represent the distribution of the radiated power as a function of the two direction angles only. Such distribution is commonly represented graphically in the form of a radiation pattern. Radiation patterns can take a variety of forms. One form is a polar diagram with radial displacement proportional to the field strength or to the radiation intensity in that direction. Another form is a rectangular diagram with abscissa representing one of the directional angles, and the ordinate representing the radiated field intensity. The intensity may be represented linearly as power or logarithmically in decibels.[4] For representing the directive properties of an antenna in all directions, contours of equal radiation intensity may be plotted, with the two direction angles as abcissas and ordinates, respectively.

The directive properties of an antenna also lead to the concept of antenna *gain.* The *directive gain* of an antenna in a specified direction is the radiation intensity in that direction compared to what it would be if the total radiated power were distributed equally in all directions. Besides the directive gain there is also the concept of *power gain.* The latter differs in that the total input power rather than the radiated power is used as the reference. The power gain for an antenna is always smaller than the directive gain by the factor of *radiation efficiency.* IEEE Standards Publications 145 and 149 (Refs 1 and 2) should be consulted for exact terminology. For some applications, such as point-to-point communication, high values of antenna gain are desired because such antennas concentrate the available power, thus effectively increasing it. Similarly, in receiving applications, such antennas are more responsive to radiation arriving from the desired direction and, at the same time, reduce the response to signals arriving from other (possibly interfering) directions. Conversely, for other applications such as broadcasting, broadly directional antennas with wide coverage may be desired.

The gain of an antenna is dependent principally

upon the size of the antenna, expressed in wavelengths. The larger the antenna, the greater is likely to be the gain. The values of gain for different antennas range from 1.5 for an electrically small dipole to values hundreds and thousands of times greater. In practice, antenna gains are usually expressed logarithmically, in decibels. For the low-frequency end of the ratio spectrum (15 kHz to 3 MHz) antennas, although large physically, are relatively small in terms of wavelengths. Therefore, the directive gains of these antennas seldom exceed 3 (4.8 dB). The radiation efficiencies of low-frequency antennas are usually very low. As a result, their power gains are significantly lower than their directive gains and may be negative when expressed in decibels. In the high-frequency band (3–30 MHz), which is used principally for long-distance communication, antenna gains of 10–100 (10–20 dB) are frequently encountered. At microwave frequencies, where the wavelengths are a fraction of a meter, gains of several hundred and even thousand times (20 to over 30 dB), are common.

When an antenna has one or more of its dimensions significantly larger than a wavelength, its radiation pattern is likely to have more than one maximum. The radiation pattern in such cases is said to have a *lobe* structure. That part of the radiation pattern which encompasses the direction of the largest maximum and the radiation immediately to each side of it is referred to as the *main lobe.* The radiation about the minor maxima is referred to as the *secondary* or *side lobes.* One of the common goals in antenna design is the reduction in the levels of secondary lobes. These may, at times, be a source of interference to other transmissions.

In common with light, radio waves consist of electric and magnetic fields at right angles to each other and to the direction of propagation. The orientation of these fields, specifically the electric field, determines the polarization of the wave. Thus, if the electric field vector is parallel to the ground, the radio wave is termed *horizontally polarized.* Although the polarization of the energy radiated by an antenna, in general, varies with direction, an antenna is usually designated to be horizontally (or vertically, or circularly, etc.) polarized, depending on the polarization of its radiation in the direction of the main lobe maximum.

The importance of polarization in radio engineering lies principally in the different reflective properties of the ground for horizontally and vertically polarized waves. Different radio services are served best by different polarizations. Antennas for use in the low-frequency end of the radio spectrum, 15 kHz through about 3 MHz, are almost invariably vertically polarized. This includes the AM broadcast band (535–1605 kHz). In the high-frequency band, 3–30 MHz, both horizontal and vertical polarizations are employed. For television broadcast service

---

*In this discussion, radiation intensity has the dimensions of power flow per unit area, normally, watts per square meter. Electric field strength, on the other hand, is in volts per meter. Definitions of these and other terms can be found in Reference 1.

in the United States and many other countries (but not in the United Kingdom), horizontal polarization is in use. In the United States horizontal polarization is also standard for FM broadcasting service. However, to accommodate reception in private automobiles, FM broadcast stations are now permitted to add a vertically polarized component to the transmission, but it may not exceed the horizontally polarized component.

Recent years have witnessed a very rapid growth in the importance of communication using satellite relays. Since the antennas used in this service are highly directive, and the directions to satellites are significantly above the horizon, ground reflections play no part in such communications. For this reason various organizations involved in satellite relay communication make use of horizontal, vertical, and circular polarizations, depending on local considerations. One current practice is to use circular polarization in both directions: left-hand circular polarization for the up-link (ground-to-satellite) and right-hand circular polarization for the down-link. At the time of this writing (December 1981) a growing practice is to reuse frequencies. By this is meant the use of the same frequency for two channels in the same direction; one channel uses right-hand circular polarization, the other, left-hand circular polarization. Isolation of 27 dB (power ratio of 500:1) between the two channels is being reported using this practice.

The types and varieties of antennas encountered in practice are very numerous. Each type has some advantage over the others for some specific applications. Some of the more important and frequently encountered requirements are those for operating bandwidth, high radiation efficiency, a specified degree of directivity,

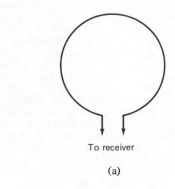

(a)

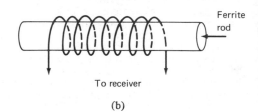

To receiver

(b)

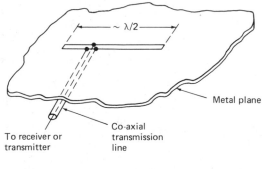

(c)

FIG. 2. Some examples of elementary magnetic radiators: (a) single-turn loop antenna; (b) ferrite rod antenna; (c) half-wavelength slot antenna.

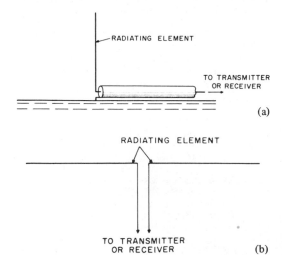

FIG. 1. Two types of elementary radiators: (a) Monopole over ground; (b) dipole.

whether high or low, polarization, etc. By no means the least of the requirements is that of economy; it is a poor engineering practice to overdesign the antenna system. Two fundamental types of antennas encountered in practice are illustrated in Figs. 1 and 2. Fig. 1 shows two of the electric radiators, a monopole and a dipole. A monopole, shown in Fig. 1(a), in one form or another, is used almost exclusively throughout the low-frequency end of the radio spectrum. It is vertically polarized. The dipole, shown in Fig. 1(b), is more versatile since it can be oriented to give either horizontal or vertical polarization. It is frequently used as an elementary radiator in large, array-type antennas. Basically, a monopole is a special case of a dipole where one-half of the dipole is replaced by its electromagnetic "image" reflected by the ground. An electric dipole may be considered to be one of the two most fundamental types of antenna. In its most elementary form of two point charges it is used to derive the mathematical expressions for the radiated fields. The other fundamental type is the magnetic radiator shown in Fig. 2. Fig. 2(a) illustrates the simplest form of a magnetic radiator, a single-turn loop. In the early days of broadcast radio the loop antenna, in a multiturn form, was sometimes used with home receivers. Because of its small size and portable nature it is still used for this purpose, especially for portable receivers, but in the form of a compact, ferrite-loaded coil, such that shown in Fig. 2(b). In industrial applications a loop antenna is sometimes used in a modified form for FM transmitting antennas. For microwave applications the magnetic radiator finds use in the form of a "slot" antenna. An elementary form of the slot antenna is shown in Fig. 2(c). It consists of a narrow, half-wavelength-long slot in a sheet of metal and is excited at two points across the slot by a transmission line. For proper impedance match to the transmission line, the points of excitation are usually well off the midpoint of the slot. Slot antennas usually find their application in large, array-type antennas; in such cases the slot antennas are often excited by waveguides passing beneath the slots.

Departing from the elementary radiators, Figs. 3, 4, and 5 show three antenna types employing wire conductors which have widespread use in the lower end of the useful radio spectrum, up to somewhere between 30 and 300 MHz. Figure 3 is a diagrammatic representation of a *Yagi antenna*. It consists of a single driven element—a half-wavelength dipole with a single parasitic

element to the rear which acts as a reflector and several parasitic elements in the forward direction, known as directors. Parasitic elements are so termed because they derive their excitation by an electromagnetic coupling to driven elements. The reflector element is slightly longer than its resonance at the operating frequency, which gives it a positive reactance. The directors, on the other hand, are shorter than their resonant length, which gives them a negative reactance; this causes them to enhance the radiation in the forward direction. The Yagi antenna is compact, with a moderately high directive gain. However, it has a very narrow operating bandwidth.

An antenna resembling the Yagi antenna physically, but with very different performance characteristics is the *log-periodic antenna*, illustrated in Fig. 4. Like the Yagi antenna, it consists of a tandem array of dipole elements; however, all the elements are driven, being connected to the transmission line running axially through the array. At any one frequency, only three or four of the elements are radiating, those whose lengths are near resonance at that frequency. At the lowest operating frequency for which the antenna is designed, the longest elements, those at the rear end of the antenna, are in operation. As the frequency of operation is increased, the "active region" moves forward along the antenna, and at the highest design frequency only the forward elements are in operation. In contrast to the Yagi antenna, the log-periodic antenna has only a moderate directivity, but it can be designed for a very wide frequency operating band. This last characteristic has made the log-periodic design popular for consumer applications for TV reception; there the VHF band alone extends from 54 to 216 MHz, a 4 to 1 spread.

Another antenna employing wire conductors but very different in design, is the *rhombic*

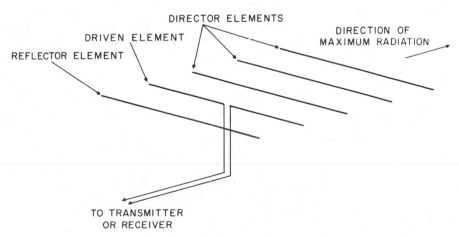

FIG. 3. A five-element Yagi antenna.

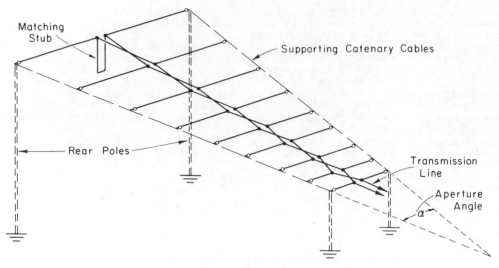

FIG. 4. Horizontally polarized log-periodic antenna.

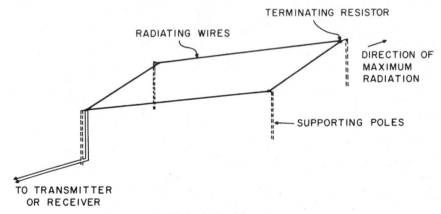

FIG. 5. A rhombic antenna.

antenna, illustrated in Fig. 5. It consists of two wires in rhombic configuration; each side of the rhombus is, characteristically, two to five wavelengths long. Its principal use is for long-distance point-to-point communication in the HF (3–30 MHz) band. Its construction is simple and trouble-free; however, it does require a large site. Its design and construction is described in detail by Harper.[5]

An antenna type which may be considered as transitory between the lower frequency, wire-conductor types and the microwave, reflector-type antennas is the *corner-reflector antenna*, illustrated in Fig. 6. The active element shown in the illustration is a half-wavelength dipole; however, collinear arrays of dipoles can be and have been employed. The key element of

this antenna type is the reflector. This consists of two rectangular reflecting surfaces joined at the axis, forming a corner. In operational antennas the metal reflector surfaces are usually replaced by wire screens. The width of the corner reflector must be at least one-half wavelength. In practice, the lengths and the widths are usually one to two wavelengths. The greatest usefulness for this antenna type is in the VHF (30–300 MHz) band. The gains and radiation patterns of the corner-reflector antenna, as functions of length, width, and aperture angle, have been measured in detail at the National Bureau of Standards and reported by Cottony and Wilson.[6,7]

Figure 7 shows two antenna types, a *horn antenna* and a *paraboloid reflector antenna*.

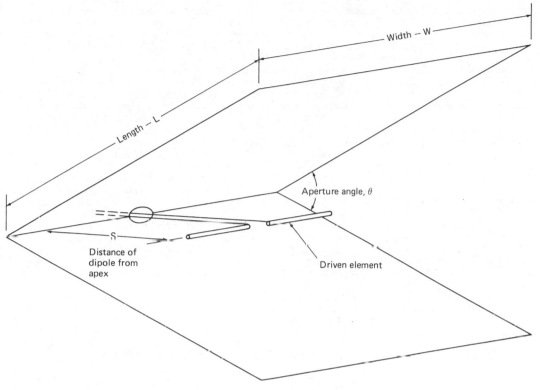

FIG. 6. Diagram of a corner-reflector antenna showing the nomenclature of its parameters.

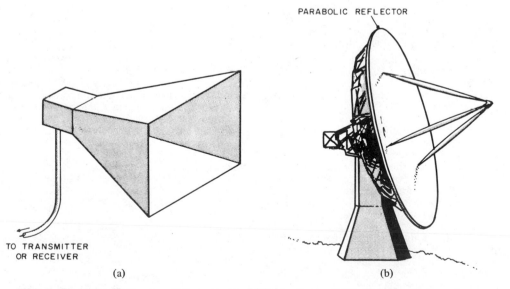

FIG. 7. Examples of microwave-type antennas: (a) horn antenna. (b) paraboloid-reflector antenna.

These are characteristic of the antennas used for microwave frequencies. The horn antenna, Fig. 7(a), is used generally where moderate directivity suffices, but a high front-to-back ratio of directivity is desired. The paraboloid antenna, on the other hand, is used for high-gain applications; it is a quasi-optical device. It is the type widely used for ground stations for satellite communication. For this application it is beginning to some extent to enter the consumer market.

The statement in the introductory paragraph regarding the importance of the terminal impedance of the antenna understates one of the fundamental antenna problems: the relationship between the size of the antenna, expressed in wavelengths, and the properties of the antenna, both its terminal impedance and its directivity. As the size of the antenna is reduced below one-half wavelength, the resistive component of the terminal impedance rapidly diminishes while the reactance increases. As a result, either the operating bandwidth or the radiation efficiency, or both, are rapidly reduced. Numerous papers have been published in an attempt to quantify this relationship, the most recent being that by Hansen.[8] Hansen's paper also lists numerous references to earlier work. It is appropriate to note, however, that the mathematical operations employed by Hansen implicitly limit the validity of the final results to dipoles.

The relationship between the size of the antenna and its power and directive gain is easier to grasp. In the case of quasi-optical antennas, such as horn and paraboloid reflector antennas, (Fig. 7), the opening of the antenna is known as the *aperture*, a term borrowed from optics. The gains of such antennas are directly proportional to the areas of the apertures in wavelengths modified by the illumination taper. The concept of aperture is carried directly to the broadside arrays of dipoles or other elementary radiators. In the case of end-fire arrays and other antennas which depend upon length for directivity, such as Yagi antennas, the concept of aperture loses its applicability. Instead, *equivalent aperture* is sometimes used. The gains of such antennas are roughly proportional to the square root of the antenna lengths expressed in wavelengths.

The preceding discussion of the relationship between the size of the antenna and its gain is correct insofar as the practical applications are concerned. To be complete, however, one should mention the concept of *supergain* or *superdirectivity*. Woodward and Lawson[9] demonstrated that, using arrays of dipoles, it is in theory possible to realize much higher directivities than those predicted by the aperture concept. To realize such directivities, special excitation of elements must be employed. The excitation is such that, while some elements radiate power, others absorb it, feeding it back to the elements that radiate. The gain then becomes a function of the number of elements rather than of the area over which they are arrayed. Uzkov[10] showed that, for broadside arrays, the theoretical gains are directly proportional to the number of elements and, for end-fire arrays, to the square of the number of elements. Bloch, Medhurst, and Pool[11] presented a comprehensive survey of supergain and superdirectivity. To realize even a modest degree of supergain, one should be able to control the excitation currents, both in magnitude and phase, with a very high order of precision. Supergain and superdirectivity are, therefore, of theoretical rather than practical interest.

For a more complete listing and discussion of antenna types the reader is referred to texts on antennas such as the "Antenna Engineering Handbook."[12] For additional discussion of the principles underlying antennas, texts by Kraus[13] and Schelkunoff and Friis[14] are suggested.

H. V. COTTONY

References

1. IEEE Standard No. 145-1983, "Definitions of Terms for Antennas," The Institute of Electrical and Electronics Engineers, Inc., New York, 1983.
2. IEEE Standard No. 149-1979, "Test Procedures for Antennas," The Institute of Electrical and Electronics Engineers, Inc., New York, 1979.
3. Jordan, E. C. "Electromagnetic Waves and Radiating Systems," pp. 327–328, Prentice-Hall, Inc., Englewood Cliffs, N.J., 1950.
4. Schelkunoff, S. A. "Electromagnetic Waves," pp. 25–26, Van Nostrand Reinhold, New York, 1943.
5. Harper, A. E. "Rhombic Antenna Design," D. Van Nostrand Co., Inc., New York, 1941.
6. Cottony, H. V., and Wilson, A. C., "Gains of finite-size corner-reflector antennas," *IRE Trans. on Antennas and Propagation*, AP-6(4), 366–369 (October 1958).
7. Wilson, A. C., and Cottony, H. V., "Radiation patterns of finite-size corner reflector antennas," *IRE Trans. on Antennas and Propagation*, AP-8, 144–157 (March 1960).
8. Hansen, R. C. "Fundamental limitations in antennas," *Proc. IEEE*, 69(2), 170–182 (February 1981).
9. Woodward, P. M., and Lawson, J. D., "The theoretical precision with which an arbitrary radiation pattern may be obtained with a source of finite size," *Journal IEE*, 95(Part III), 362–370 (Sept. 1948).
10. Uzkov, A. I., "An approach to the problem of optimum directive antennae design," *Comptes Rendus (Doklady) de l'Académie des Sciences de l'URSS*, 53(1), 35–38 (1946).
11. Bloch, A., Medhurst, R. G., and Pool, S. D., "A new approach to the design of super-directive aerial arrays," *Proc. IEE*, 100(Part III, No. 67), 303–314 (Sept. 1953).

12. Jasik, H. (Ed.), "Antenna Engineering Handbook," McGraw-Hill Book Co., New York, 1961.
13. Kraus, J. D., "Antennas," McGraw-Hill Book Co., New York, 1950.
14. Schelkunoff, S. A. and Friis, H. T., "Antennas: Theory and Practice," John Wiley & Sons, New York, 1952.

Cross-references: ALTERNATING CURRENT; CIRCUITRY; PROPAGATION OF ELECTROMAGNETIC WAVES; RESONANCE; WAVE MOTION.

ANTIFERROMAGNETISM

Antiferromagnetism is the most common form of magnetic order. It is found in the majority of inorganic compounds of the transition metals, rare earths and actinide elements; it is also found in Cr, Mn, Pt, Pd, and rare earth metals and alloys, although the situation is rather more complicated in the case of metals, and the discussion here will focus mainly on insulators. However, during the last decade great interest has arisen in organic conductors, of which TTF–TCNQ is a typical example. These systems are strongly anisotropic in lattice character, i.e., are quasi-one-dimensional, and their magnetic character is antiferromagnetic. A striking experimental development since the mid-1960s has been the synthesis of magnetic materials which have an effective dimensionality D less than three. Such "low-dimensional" magnets are chemically engineered by inserting large diamagnetic organic "spacer" molecules to minimize the magnetic interactions in one or two directions in the crystal. Low-dimensional antiferromagnets show striking differences in behavior from the familiar three-dimensional antiferromagnets, such as MnF_2, because quantum effects (fluctuations) are particularly pronounced in such systems.

The principal feature of antiferromagnetism is the spontaneous antiparallel alignment of electron spins on neighboring magnetic ions, which takes over from the paramagnetic state (where spin–spin interactions are essentially negligible) as the temperature is lowered. The critical temperature of the antiferromagnetic phase transition is called the Néel temperature (T_N). The strength of the ordering interaction is characterized by the magnitude of T_N, which typically ranges from below 1 K to above room temperature. Some antiferromagnets whose properties have been studied in detail are listed along with their Néel temperatures: (3D) NiO, 520 K; Cr_2O_3, 310 K; MnF_2, 67.4 K; (2D) K_2NiF_4, 97 K; (1D CuCl$_2 \cdot 2(pyr)_2$, (CPC) 1.1 K; [(CH$_3$)$_4$N] Mn Cl$_3$ (TMMC), 0.8 K. (*Note:* (pyr) denotes large organic pyridine complexes.)

The ordering interaction between neighboring metal ionic spins in an insulator is called *superexchange*, since it takes place via an intervening anion, O, F, S, Cl, etc. (cf. direct exchange between adjacent ions as encountered in FERROMAGNETISM). Superexchange results from charge transfer (see BOND, CHEMICAL), and it is best illustrated by a typical example, say MnO.

Ground and Excited States of $(MnOMn)^{++}$

$$Mn^{++}O^{--}Mn^{++} \qquad Mn^{+(+-)}O^-Mn^{++}$$

ground state   excited state

In the ground state, the purely ionic configuration, there is no interaction between metal ions. If, however, one of the two bonding electrons of O^{--} is transferred to the Mn^{++} at left, there will be strong Hund's rule coupling within that ion, and also the unpaired electron on O^- can couple with the Mn^{++} at right. Since the two bonding electrons on O^{--} have opposite spins, the overall interaction will appear as antiparallel exchange coupling (i.e. antiferromagnetic coupling) between the two Mn ions.

The archetypal form of the Hamiltonian function (the energy) which describes the magnetic properties of the exchange interaction is the Heisenberg exchange form

$$\mathcal{H} = \sum J_{ij} S_i \cdot S_j,$$

where S_i and S_j are the spin angular momentum vectors (operators) of a pair of nearest-neighbor ions. (Interactions between more distant neighbors are much weaker and will largely be ignored in what follows.) J_{ij} is the exchange integral (exchange constant) between the spins S_i and S_j, and is a measure of the strength of the magnetic interaction. For relatively simple systems, J may be calculated quantum-mechanically from first principles. However, in general, J is best regarded as a phenomenological constant, to be inferred from experiment. The Heisenberg Hamiltonian can describe either ferromagnetism or antiferromagnetism. J is negative for ferromagnets, since then parallel spin alignment is energetically favored. Antiferromagnetism is characterized by J positive, which favors antiparallel order.

While the antiferromagnetic exchange interaction produces antiparallel alignment of the spins, their direction with respect to the crystalline axes is a consequence of the magnetic anisotropy. There are three origins of anisotropy: (1) dipole–dipole interactions among the array of ionic moments which gives anisotropy in all but cubic symmetry; (2) Stark-effect interactions of each single ion with the local crystalline electric fields; and (3) anisotropic exchange, a result of spin-orbit coupling and super-exchange between excited orbital states. The latter two mechanisms are important for non-S-state ions, especially in crystals of low symmetry. Crystal field effects are represented by including a term of the type $D \sum_i (S_i^z)^2$ in the Hamiltonian, where S_i^z is the z-component of the spin angular momentum (S_i) of the ith ion. This term gives rise to spin-anisotropy such that when $D < 0$ easy-axis anisotropy is favored, and when $D < 0$ easy-plane anisotropy is favored. For anisotropic exchange in real systems the Hamiltonian may effectively be written:

$$\mathcal{H} = \sum_{ij} (J_{ij}{}^z S_i{}^z S_j{}^z + J_{ij}{}^x S_i{}^x S_j{}^x + J_{ij}{}^y S_i{}^y S_j{}^y).$$

If $J^z > J^x \simeq J^y$ uniaxial anisotropy is present, and when $J^z < J^x \simeq J^y$ easy-plane anisotropy is present. The extreme limit of uniaxial anisotropy occurs when $J^x = J^y = 0$, and is called the *Ising model* of magnetism. The extreme limit of easy-plane anisotropy is when $J^z = 0$, and this is called the *XY model*. When $J^z = J^x = J^y$, the *isotropic Heisenberg exchange model* is recovered.

The effects of an applied magnetic field on the antiferromagnet are represented by the addition to the Hamiltonian of a Zeeman energy term $-g\mu_B H \sum S_i{}^z$, where g is the Landé g-factor and μ_B is the Bohr magneton. The phase diagram of an Ising antiferromagnet as shown in Fig. 1. In the H–T plane, the antiferromagnetically ordered phase is enclosed by a second-order phase boundary. Note that antiparallel order is destroyed both by increasing temperature and by increasing magnetic field. The critical field, H_c, is reached when the Zeeman energy tending to align a pair of spins parallel balances the antiferromagnetic exchange energy. At $T = 0$, antiferromagnetic order exists for $H < H_c$, while paramagnetic disorder occurs for $H > H_c$. H_c is therefore a special point. For the pure Ising model, to which Fig. 1 applies, nonphysical effects occur at H_c. There is a nonzero entropy at $T = 0$ and $H = H_c$, which can be calculated exactly in 1D and 2D, and approximately in 3D. This implies a violation of the third law of thermodynamics and reflects the somewhat artificial, or classical, character of the Ising model. Inclusion of terms such that J^x or $J^y \neq 0$ eliminates this defect.

Also shown in Fig. 1 is a different kind of field, namely a staggered field, H_s, which reverses sign at every successive spin site. This is the so-called *ordering field* for an antiferromagnet. For example, an antiferromagnet in a staggered field shows phase behavior similar to a ferromagnet in a direct (ordinary) applied field. It might seem that a staggered field is an artificial concept, not realizable in the laboratory, which is generally true. However, in the case of the complex, six-sublattice-structured antiferromagnet dysprosium aluminum garnet (DAG), application of a direct field in general results in the appearance, in addition, of an "internal" staggered field.

The obvious magnetic parameter for experimental measurement is the susceptibility, denoted χ, which is effectively a measure of the response of the spins to an external applied field. In general, there are two susceptibilities

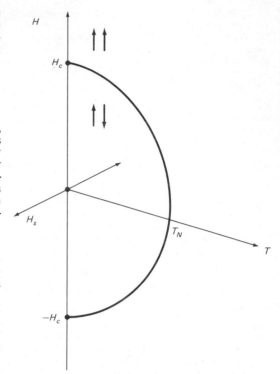

FIG. 1. Phase diagram (schematic) of an ideal classical antiferromagnet. T is temperature, H is direct field, and H_s is staggered field. The arrows suggest the predominant pattern of spin ordering.

to measure in the case of a single crystal sample. One is the susceptibility measured with the field parallel to the easy-axis (easy-plane), called $\chi_\parallel$, and the other, $\chi_\perp$, measured with the field in a direction perpendicular to the easy-axis (easy-plane). In a truly spin-isotropic, Heisenberg, system $\chi_\parallel$ and $\chi_\perp$ are equivalent above T_n. This is not true in general, however. When any type of anisotropy is present $\chi_\parallel$ and $\chi_\perp$ are different. Mean-field theory is a simple, approximate calculational approach applicable to both ferromagnets and antiferromagnets. Further, it is believed to describe phase behavior correctly for magnetic systems in dimensions $D \geqslant 4$, Fig. 2 shows the susceptibilities of an antiferromagnet, in the limit of a vanishing field, in the mean-field approximation. Both $\chi_\parallel$ and $\chi_\perp$ vanish at high temperatures where spin–spin interactions become negligible. $\chi_\parallel$ then exhibits a maximum at the transition temperature T_N and subsequently vanishes as $T \to 0$. $\chi_\perp$ by contrast, goes to a nonzero value as $T \to 0$. There is a vanishing total magnetization as the antiparallel alignment of the spins becomes increasingly complete.

It is interesting to observe how Fig. 2 is modified for real antiferromagnets with $D < 4$. This

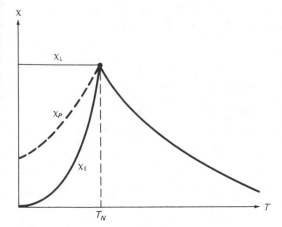

FIG. 2. The susceptibility (schematic) of an antiferromagnet in zero applied field, as a function of temperature, in the mean-field approximation, $\chi_\parallel$ and $\chi_\perp$ are as explained in text. χ_p is the powder susceptibility which is the weighted average of $\chi_\parallel$ and $\chi_\perp$.

is now well known due to exact and accurate approximate calculations in 1D, 2D, and 3D, and is shown schematically in Fig. 3. Both susceptibilities pass through a rounded maximum T_{max}, which is *not* the critical temperature T_N, as the temperature is lowered. In zero-field T_N is manifested as a vertical tangent in both $\chi_\parallel$ and $\chi_\perp$ on the low-temperature side of T_{max}. The difference $\Delta = T_{max} - T_N$ is a measure of dimensionality, increasing as D decreases. Typically, Δ is about 8% of T_{max} or less for 3D

antiferromagnets, 50–70% for 2D antiferromagnets, and 90% or more for good 1D antiferromagnets. Strictly, $T_N = 0$ for an ideal 1D antiferromagnet with short-range interactions. However, weak residual 3D interactions allow 3D ordering to take place at appropriately low temperatures.

Antiferromagnets show very rich and complex phase behavior when competing interactions are present, and/or as a function of the degree of anisotropy. A special class of Ising-like antiferromagnets are called *metamagnets*, and their H–T phase diagram is shown in Fig. 4, which resembles Fig. 1, except now a high-field portion of the antiferro–paramagnetic phase boundary has become first order. The point at which the boundary changes character, T^*, is a special kind of critical point called a *tricritical* point. At a normal critical point, or along a line of critical points, divergences occur in the thermodynamic properties which are characterized by so-called *critical exponents*. The critical exponents at a tricritical point are markedly different from those at an ordinary second-order critical point. Figure 5, which is an extended version of Fig. 4 including one additional parameter, the staggered field H_s, illustrates the phase behavior of a metamagnet in a more illuminating way. The first-order boundary of Fig. 4 now appears as a line of triple points where three first order

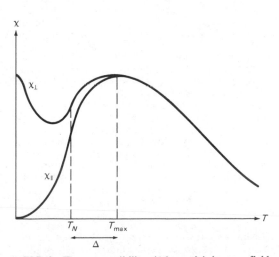

FIG. 3. The susceptibility (schematic) in zero field as a function of temperature according to non-mean-field theories. The difference between T_N and T_{max} is a function of dimensionality. T_N corresponds to a vertical tangent in $\chi_\parallel$ and $\chi_\perp$.

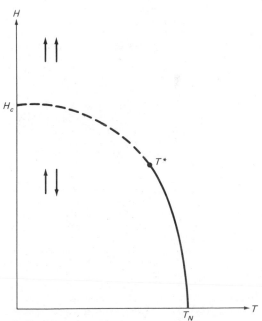

FIG. 4. The H–T phase diagram (schematic) of an Ising-like metamagnet. The hatched portion of the phase boundary is first order and T^* denotes the tricritical point.

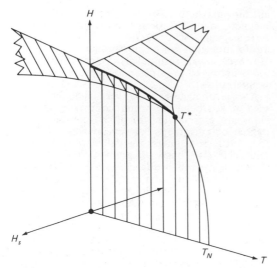

FIG. 5. The extended phase diagram of a metamagnet in terms of parameters T, H, and staggered field H_s. The shaded areas denote first-order two-phase coexistence surfaces.

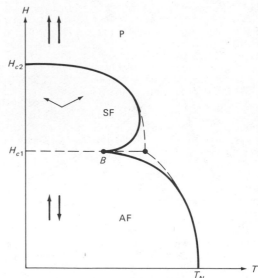

FIG. 6. H–T phase diagram of a spin–flop antiferromagnet in a field parallel to the easy axis. The low-field phase region has antiparallel aligned spins along the easy axis. The high-field phase region has "flopped" spins, aligned at an angle to the field. The boundary between the phases is first order. B is the bicritical point. The dashed lines illustrate the phase boundaries in the neighborhood of the bicritical point in the mean-field approximation.

phase boundary surfaces meet. The tricritical point T^* is the point at which three critical phase boundary lines meet. Examples of metamagnetic systems are ferromagnetic layers with antiferromagnetic interlayer coupling, and systems with nearest-neighbor antiferromagnetic coupling and weak next-nearest-neighbor ferromagnetic coupling. The phenomenon of metamagnetism was first observed experimentally in $FeCl_2$, which is an example of the layer-type system. DAG is also a well known metamagnet.

Another interesting class of antiferromagnets, characterized by rather small anisotropy, consists of the *spin-flop* antiferromagnets. The H–T phase diagram of a spin-flop system is shown in Fig. 6. Two distinct ordered phases appear. In addition to the regular antiferromagnetic phase, a new phase appears at high fields. In this phase, in a classifical vector-spin picture, the spins tend to align themselves *perpendicular* to the applied field on crossing the boundary line, which is first order, ehnce the name spin–flop. This system has two critical fields at $T = 0$, H_1 (the spin-flop field) and H_2. Two lines of critical points meet the first-order line at a special point B. B is a second example of a special type of critical point (these special points are generally called *multicritical points*). B is called a *bicritical point*; from Fig. 6 it may be observed that two critical lines meet at B. When a staggered field is included, the phase structure has several interesting features, but is too complicated to describe here. It should be noted that early calculations of mean-field type showed the three phase boundaries meeting at an angle. Modern calculations of non-mean-field

type have indicated that the two second-order lines meet the first-order line tangentially. This effect is difficult to observe in 3D antiferromagnets, but is very prominent in quasi-1D antiferromagnets on account of the low dimensionality. The effect has been experimentally confirmed in both 3D and 1D antiferromagnets. Examples are MnF_2 and TMMC, respectively.

An interesting recent development concerns phase transitions in 3D antiferromagnets of basic lattice structure fcc (or bcc). The low-temperature ordered phases of these multisublattice magnetic systems can be fairly complex in terms of spin-ordering arrangements, characterized by doublings of the magnetic unit cell relative to the crystal unit cell as the system goes from a paramagnetic state into an ordered phase as the temperature is lowered. Examples are: Type I antiferromagnets such as UO_2; Type II antiferromagnets such as TbSb, MnO, NiO; and Type III antiferromagnets such as K_2IrCl_6. An order parameter of unusual complexity is required to describe such systems, and experimental testing continues of calculations based on group theoretical predictions of Landau and Lifshitz combined with the currently very popular technique of renormalization group analysis.

Other current areas of investigation where

antiferromagnetic exchange interactions are important are Ising systems showing Lifshitz points, multiphase points, and a sequence of modulated phases. This theoretical work has been stimulated by recent magnetization and neutron-scattering experiments on rare-earth pnictides, particularly CeSb.

Spin-glasses are under intensive investigation at this time. They are commonly regarded as systems with random ferromagnetic and antiferromagnetic interactions, resulting ultimately from a modulated, long-range RKKY interaction. Spin-glasses exhibit many curious experimental features, but despite much theoretical effort no clear and definitive microscopic explanation has yet emerged. Systems which have features in common with random spin-glasses, but where enough regularities remain to render the system exactly solvable in a number of cases, are called *spin-frustrated systems*. A typical example is the Ising model on a triangular lattice with nearest-neighbor interactions. The condition for antiparallel spin alignment cannot be satisfied around a triangular unit of the lattice. Hence the origin of the term spin frustration.

Spin dynamics is another area where new theoretical breakthroughs are being made, and where antiferromagnets play a major role. Approximate analytic and computer-based classical calculations are now available in 3D. By contrast, in 1D spin $\frac{1}{2}$ antiferromagnets, many striking quantum mechanical features have been observed, enhanced by the low dimensionality. These developments have led to advances in nuclear magnetic resonance experimental techniques, with corresponding insight into the details of charge distributions around ions. In addition to susceptibility and magnetic resonance, an important experimental technique in antiferromagnetism is neutron scattering, which reveals details of the spin structure array and the effective spin dimensionality D, and probes the nature of the energy excitations.

JILL C. BONNER

References

1. Mattis, D. C., "The Theory of Magnetism I: Statics and Dynamics," Springer-Verlag, Berlin–Heidelberg–New York, 1981.
2. Bonner, J. C., and Müller, G., "Low Dimensional Magnetism," Oxford University Press (forthcoming).
3. de Jongh, L. J., and Miedema, A. R., "Experiments on Simple Model Systems," *Advances in Physics*, **23**, 1 (1974). L. J. de Jongh, "Some Recent Experiments on Quasi 1- or 2-Dimensional Magnetic Systems," in "Recent Developments in Condensed Matter Physics," J. T. Devreese (Ed.), Vol. 1, p. 343, Plenum, New York, 1981.
4. Nagle, J. F., and Bonner, J. C., "Phase Transitions—Beyond the Simple Ising Model," *Annual Review of Physical Chemistry*, **27**, 291 (1976).
5. Stryjewski, E., and Giordano, N., "Metamagnetism," *Advances in Physics*, **26**, 487 (1977).

Cross-references: BOND, CHEMICAL; CRITICAL PHENOMENA; FERRIMAGNETISM; FERROMAGNETISM; MAGNETISM; MAGNETOMETRY; PARAMAGNETISM.

ANTIPARTICLES

I. Overview: Particles and Quanta. Classification by Statistical Behavior and by Behavior with Respect to the Four Fundamental Interactions One of the great discoveries of modern physics is that for every type of elementary entity of matter and radiation—"particle"— there exists a corresponding conjugate type of entity—"antiparticle." In the antiparticle certain of the particle-defining properties are identical— "conjugation-invariant"—and others—"conjugation-reversing"—are reversed in sign. The reversed sign in a conjugation-reversing property allows one to maintain a conservation law for that property in the dramatic processes of *pair creation* and *pair annihilation* in which an antiparticle is observed to appear and disappear together with the particle to which it is conjugate. In those cases where all the conjugation-reversing properties occur with zero values, the antiparticle is identical with the particle. The progressive recognition of the existence of antiparticles was initiated by Dirac's relativistic anti-electron theory in 1931, and by Anderson's independent experimental discovery of the anti-electron (positron) in 1932.

A simple descriptive and inductive introduction to the fact of existence of antiparticles, and conceptualizations of it, is given in the article entitled ANTIPARTICLES in the first edition of this encyclopedia, and in the articles in the present edition on FIELD THEORY and ELEMENTARY PARTICLES. An enumeration of known particles and antiparticles up to 1983 will be found in the latter. Here we attempt to convey the most general theoretical setting in which the existence of antiparticles, and symmetry of conjugation between particles and antiparticles, may be established. At first a brief overview will be given and then a more detailed systematic discussion. Some of the technical terms used are defined in the sequel.

A major classification of all particle types is according to their statistics or "social behavior." This is based on symmetry or antisymmetry of the mathematical function describing an identical multiparticle state (i.e., a state with two or more particles of the same kind) under interchange in any pair. Those particle types for which there is symmetry under intrapair exchange are *bosons;* those for which there is antisymmetry are *fermions*, which fall into the two subtypes: *quarks* and *leptons*.

Cutting across the boson-fermion classification

of particle types according to their *statistical behavior* are other classifications based on their *interaction behavior* or "charges." This refers to their participation or neutrality in the four fundamental physical interactions found in nature: hadronic, electromagnetic, weak, and gravitational. Formerly these were considered to be distinct, as once upon a time, the magnetic interaction and the electrical interaction were considered to be distinct only to be united later in the one electromagnetic interaction. Similarly, at the levels of energy of experiments in which they have been observed until now, the four interactions are distinguished by markedly different intrinsic strengths (coupling constants), roughly in the ratio $1:10^{-2}:10^{-14}:10^{-39}$, as well as different symmetries. Only the first three have hitherto played an important role in quantum particle physics. Particles subject to hadronic interactions are called *hadrons*, the elementary ones, out of which all others are composed being called *quarks*. Composite fermionic hadrons are called *baryons*; all other hadrons are *mesons*. It is found that baryons all carry a nonvanishing *baric charge B* which is very nearly, perhaps absolutely, *conserved*; i.e., up to a very high probability—perhaps certainty—*the total baric charge of the component entities after a reaction equals the total baric charge of the component entities before a reaction.* At both ends of the reaction the total baric charge is computed by simple addition. The baric charge of the elementary hadrons—the quarks—out of which all other hadrons are constructed is $\frac{1}{3}$, and the baric charge of antiquarks is $-\frac{1}{3}$. The only fundamental hadrons directly observed in nature are the composite baryons (quark triplets) with baric charge 1 and the mesons (quark-antiquark pairs) with baric charge 0. The leptons also have baric charge 0. The mesons, however, in contrast to the leptons, do interact hadronically, which is to be expected from their dipolar structure with respect to baric charge. The leptons are elementary, have no quark content and therefore no hadronic interactions. The leptons in turn display what seem to be other types of conserved charges. There appear to be three types, tau-leptic charge (for tauon and its neutrino), mu-leptic charge (for muon and its neutrino) and e-leptic charge (for electron and its neutrino). Table 1 diagrams these classifications.

The most fundamental theory that we have of source particles and their interactions is Quantum Field Theory (QFT). The central concept in the classical ancestor of this theory is that of the *force field*, in which the effects of action by a source at any one point are propagated continuously to other points via an intermediary influence, i.e., not by "action at a distance" but by local action from point to point. The quantum interaction field retains some of these features, but in modified form.

The quantum modification of the classical field concept of propagation of influence is a picture of emission and absorption of intermediary entities or excitations—the interaction-quanta—which are exchanged between source-particles and which can also exist free. The conception of these intermediary excitations first appeared explicitly and with incisive success for phenomenological description and prediction in Einstein's 1905 work on light quanta or photons (for which, together with his "work in

TABLE 1. CLASSIFICATION OF PARTICLE TYPES ACCORDING TO SYMMETRY UNDER PERMUTATION OF IDENTICAL PARTICLES (STATISTICS)

		Bosons	Fermions
Hadrons	Baric Charge 0	Mesons	
	Baric Charge ≠ 0		Baryons (nucleons, hyperons, baryon resonances, . . .)
Leptons	Baric and Leptic Charge 0	Photon / Weak-Interaction Boson (?)	
	Leptic Charge ≠ 0		Tau-Leptons (taus and tau-neutrines) Mu-Leptons (muons and mu-neutrinos) e-Leptons (electrons and e-neutrinos)
		Graviton (?)	

theoretical physics," he was awarded the Nobel prize in 1921, with no mention of the relativity theories!). The similar intermediary quanta that have been established since that time in theory for the other fundamental interactions, are interpreted as excitations of fields whose potential is distributed over spacetime. Thus we now speak of *gluon fields*, and *intermediate vector boson fields*, along with the photon field.

In QFT, source-particles are also interpreted as excitations of the modes of a field—the appropriate source-particle—field. Upon interacting, a given source field and a given interaction-quanta field can excite or de-excite each other. The excitations which make up our world of substance "sources" are traditionally recognized as systems of one or more "particles;" this corresponds to the fact that whenever we have a "pure" excitation of the source field, i.e., such as occurs in one and only one field-mode, it always comprises an exact whole number of elemental excitations identical with each other. (Similarly for "quanta" in an interaction field.) This exact whole-number multiplicity is already a sufficient basis for calling the theory "quantum." For such an elementary excitation, we shall reserve the term "particle" in the case of source fields ("quantum" in the case of interaction fields) or, more technically, "one-particle" (one-quantum) state. The term "excitation," though not in common technical use, is really more appropriate than "particle" or "quantum" because "excitation" connotes aspects of wave or vibrational patterns, as well as the wave property that all elementary excitations of a given type are indistinguishable from each other.

In summary: The hadronic interaction—also called "strong" or "nuclear" interaction—acts fundamentally between the members of one of the two basic families of *elementary source-particles*. The members of this basic hadronically interacting family have been named "quarks." The other basic family of source-particles, which have no hadronic interactions, are called "leptons." The term "source-particles" is used here for the members of these two families to contrast them with other fundamental massive and radiation entities which transmit interactions; these interaction-transmitting entities we shall designate *"interaction-quanta."* Among these are the "photon" and "weak-interaction vector bosons" and the "gluons" which transmit the hadronic interaction between both quarks and leptons. Tables 1 and 2 diagram these classifications.

Within the two particle families—quarks and leptons—there are many finer distinctions. In our subsequent discussion we shall see that any specific type of source-particle has attributes that come from its belonging to a corresponding space-time symmetry type, one of the "irreducible representations" of the symmetry group of spacetime. Among these attributes, to which different values are assigned in the different spacetime symmetry types, are *mass* and *spin*. Besides being of a given spacetime or *external symmetry type*, each source-particle also is of an *internal symmetry type*—one of those which serve as representations of an internal symmetry group. Corresponding type attributes are *charges* of various kinds—hadronic, electric, and weak, which play a role in the interaction of the various types of particles with the corresponding forces. Interaction-quanta possess similar symmetry attributes.

The electromagnetic interaction is perhaps the simplest. Its internal symmetry group is U_1, the group of all phase factors, i.e., complex numbers of unit magnitude; the corresponding attribute is electric charge. The electromagnetic interaction operates only between electrically charged or magnetically active particles, exchanging energy and momentum. It does not, however, transfer electric charge, i.e., it is itself neutral. Its intrinsic strength, 1/137, is labeled α_e.

The situation is quite different with the hadronic and weak interactions. The hadronic force, and likewise the weak force—in contrast to the long-range electromagnetic and gravitational forces—both are "short-range" forces. This means that the region of appreciable manifestation of the hadronic force is limited to a very small volume around the source particles, basically a quark, from which the force emanates. This force acts only on other quarks or quark structures, but not on leptons.

Both quarks and their intermediary hadronic interactions—"gluon interactions"—are classified according to the symmetry types of an exact $SU_3{}^c$ group, the indicated internal symmetry group of the hadronic interactions. The superscript c stands for "color," a label which takes on the three primary color values R (red), G (green), and B (blue), each symbolic of a particular kind of hadronic charge. Antiquarks come with the complementary three "anticolors" or anticharges. Because of the exact symmetry, the intrinsic strength of all the color charges is the same.

Besides the color charge classification of quarks there are two other type classifications found empirically that are not given by the exact $SU_3{}^c$ group, but which provide an approximate symmetry grouping. There are six "flavors" divided into three families." The three families, each comprising two flavors, are: (1) up (u), down (d); (2) charm (c), strange (s); (3) top (t), bottom (b).

The quark color classification corresponds to the first nontrivial symmetry-type, *triplet* (3) of $SU_3{}^c$. Its threefoldness is represented by R, G, and B. There is a corresponding *antitriplet* ($\bar{3}$) with the three complementary anticolors $\bar{R}$, $\bar{G}$, and $\bar{B}$ for the antiquarks. Thus, combination of three quarks having each of the three colors makes a "white" particle (baryon), or combination of a quark of a certain color and its antiquark makes another type of "white" particle

TABLE 2. CLASSIFICATION OF PARTICLE-TYPE ACCORDING TO SYMMETRY UNDER PERMUTATION OF IDENTICAL PARTICLES (STATISTICS) AND THEIR ROLE AS INTERMEDIARY QUANTA OR PARTICLES IN THE STANDARD MODEL

$$(SU_3)_{\text{strong}} \times (SU_2 \times U_1)_{\text{weak}},$$

WHERE WITHIN EACH WEAK MULTIPLET EXCESS ELECTRIC CHARGE ≡ WEAK CHARGE, I.E., $Q - Q_{av} \equiv R_3$. The question marks in parentheses indicate "not yet verified experimentally."

Particles presently considered to be composite bound states of the more elementary QUARKS and LEPTONS are written in lower-case type.

	BOSONS (intermediary quanta and $q\bar{q}$ structures)	FERMIONS (matter particles)	
HADRONICALLY INTERACTING Baric Charge $B \neq 0$		QUARKS q_r^α ($\alpha = R, G, B$) ($r = u, d, c, s; t, b$) (baryons hyperons, baryon resonances, ...)	
Baric Charge $B = 0$	HADRONIC (STRONG) INTERMEDIARY QUANTA — Mesons GLUONS G_β^α ($\alpha, \beta = R, G, B$)		
		QUARKS	**LEPTONS**
		D UP and DOWN QUARK: u_D, d_D D STRANGE and CHARM QUARK: s_D, c_D D TOP (?) and BOTTOM QUARK: t_D, b_D	D ELECTRON: e_D^+ D MUON: $\mu\bar{D}$ D TAUON: $\tau\bar{D}$
ELECTROWEAKLY INTERACTING $SU_2 \times U_1$ Singlets Weak Charge $R_3 = 0$	ELECTRO-WEAK INTERMEDIARY QUANTA — PHOTON A NEUTRAL WEAK-INTERACTION BOSON (?): Z^0		
SU_2 Doublets Weak Charge $R_3 \neq 0$	CHARGED WEAK-INTERACTION BOSONS (?): W^+ W^-	L UP-DOWN QUARKS: $\binom{u}{d} L$ L STRANGE-CHARM QUARKS: $\binom{c}{s} L$ L TOP (?)-BOTTOM QUARKS: $\binom{t}{b} L$	L NEUTRINO e — e^-: $\binom{\nu_e}{e^-} L$ L NEUTRINO μ — μ^-: $\binom{\nu_\mu}{\mu^-} L$ L NEUTRINO τ — τ^-: $\binom{\nu_\tau}{\tau^-} L$
GRAVITATIONALLY INTERACTING	GRAVITON (?): Γ		

(meson). Strong interactions between the various quarks depend only on their colors and are governed by an exact unbroken $SU_3{}^c$ gauge group, implying eight massless gluons, which mediate the quark interactions. All structures are governed by the following fundamental rule: *While various flavored quarks can be found mixed in various proportions, only "white" color singlets are found in independent particle entities.*

It is this rule which summarizes in a formal manner the phenomenon of "quark confinement" which one requires in the current theory of strong interactions that has become known as "quantum chromodynamics." As already stated, the hadronic force does not operate on the members of the other set of elementary source-particles found in nature, the leptons; in other words, leptons have no hadronic (or gluon) charges. Also, leptons, unlike quarks, are actually observed in a state of freedom—a fact which suggests that the fundamental hadronic interquark (gluon) interaction is necessary for the "total confinement" which seems to be operating for quarks. The phrase "total confinement" signifies that only 3-quark and quark-antiquark composites have been observed— their quark content being inferred, however, from compelling evidence. This suggestion has been incorporated in the quantum chromodynamic field theory, where the fundamental hadronic force is comprised of the octet of gluon interactions. The mediating gluons are endowed with color and are therefore self-interacting, thereby producing antiscreening between source particles. This property, if it could be proved to increase without limit with distance, would lead to total confinement.

Residually, the hadronic force acts also between the 3-quark composites which we recognize as nucleons, neutrons and protons, which constitute units out of which nuclei of atoms are constructed. This force is, however, relatively (only relatively!) small compared to the interquark force at the same distances. It is analogous to the Van der Waals forces between electrically neutral atoms.

II. Antiparticle Existence Principle: Symmetry, Conjugation As we have already indicated, it is one of the great theoretical and empirical discoveries of modern physics that there is a general *duality* in nature in that to every particle type there exists a mutually annihilating "antiparticle" type with certain *exactly* identical and certain other *exactly* reversed defining properties. The exact duality of particle and antiparticle serves to classify the two sets, that of exactly identical and that of exactly reversing properties. Among the exactly identical properties are the mass and spin, and among the exactly reversing properties are all the fundamental charges. The members of both sets will be called *type-observables* and the following discussion will be concerned with specifying them.

Both sets of type-observables are associated with exact symmetries of the particle types and their interactions. What a "symmetry," or "symmetry operator," is will be defined more precisely later; here we shall designate the aspects of nature to which it applies, and indicate the meaning of the symmetry concept, by the following statement: Nature lends itself to a classification into two kinds of aspects, that of *systems* and that of *natural processes*. A natural process (e.g., motion in an arbitrary electromagnetic field) is generally considered to be a more fundamental feature than a particular system (e.g., the earth), since the process is taken to correspond with a *law of nature* whereas a particular system corresponds to the more special features which we label *side conditions*. (There is a blurring of the distinction in microphysics where the properties—to a certain extent unexplained—of isolated microphysical systems appear to be instances of laws of nature rather than of side conditions; this goes, of course, with the fact that these isolated microphysical systems are found in vast numbers of identical replicas: "elementary particles.") Symmetry operators are applicable to either side conditions or laws of nature or, alternatively, to either systems or natural processes: A symmetry operator of a system or process is a possible change which leaves certain overall relevant characteristics of the system or process unaltered (a possible change is one that can be actually, or in imagination, carried out on the system or process). If we are looking at nature as a whole, and are concerned principally about processes, the relevant characteristics to be preserved are the *dynamics* of transitions between arbitrary initial conditions and the possible final conditions in all *spontaneous natural processes*. In classical physics, this dynamics is given by *differential equations* and it is the form of these and certain parameters appearing in the equations which are to be preserved; in the truer-to-nature quantum physics the dynamics may be expressed even more directly by an array (S-matrix) of *transition probabilities* each of which is to be preserved individually. The additional invariant parameters here are such quantities as the mass and spin of the system (see next paragraph).

The symmetry operators which preserve the dynamics are, first of all, those which correspond to the *homogeneity* (invariance under translations) and *isotropy* (invariance under space-space and space-time rotations) of space-time in regions small enough so that the effects of gravitation may be neglected, i.e., over regions small compared with the reciprocal of the local value of the acceleration of gravity—in units in which the speed of light is 1 and all speeds are dimensionless ratios to this speed. (For all moderate gravitational fields these are very large regions, e.g., for an acceleration equal to the earth's surface gravity the linear dimension of the limiting region is $\approx 10^{18}$ cm ≈ 1

light year.) This set of homogeneity and isotropy operators characterizes the special relativity theory of spacetime and is called the
Poincaré group. Thus, besides the fundamental
conjugation-reversing interaction charges, the
principal symmetry-related observables which
define particles and antiparticles are those
which result from the continuous spacetime
symmetries which express homogeneity and
isotropy: The invariance of natural evolution
processes under an arbitrary spacetime translation (homogeneity) implies existence of conserved four-component energy-momentum P;
the invariance under an arbitrary spacetime rotation (isotropy) implies existence of conserved
six-component angular momentum $J \sim (\vec{M}, \vec{\pi})$.
The conjugation-invariant type-observables here
are mass m (with which may be associated
phenomenologically, as an imaginary part, lifetime τ) and magnitude of *spin*—intrinsic angular
momentum—s. Both mass m and spin-magnitude
s are spacetime scalars or "invariants," which
are algebraic functions of P and J; by saying
that they are invariants, we mean that they are
left unchanged by any spacetime translation or
rotation. (Actually, as we shall see, it is natural
to require that they be invariants—but not
necessarily functions—of *every symmetry operation*.) Because two translations *commute*
(give the same result if performed in alternate
order) there is no theoretical restriction on the
possible values of the mass, though empirically
such restrictions exist. Because rotations are
noncommuting in a characteristic manner, the
theoretically allowed values of spin are integral
or half-integral multiples of an elementary unit,
as observed. In short, the theory of (linear
operator) representations of the relativity symmetry group (Poincaré group) prescribes
clearly the *existence* of the particle-characteriz-
ing properties of spin and mass but only partially
explains their *values*: (1) The specific *range* of
observed spin values (all half-integral or integral multiples of a fundamental unit) is fully
explained; the *allocations* to specific types of
particles are partly explained by the general
spin-statistics theorem (see following). (2) The
specific range of observed masses is not fundamentally interpreted; however, a partial interpretation of the *mass differences* among the
strongly interacting particles in a multiplet and
supermultiplet in terms of an *approximate
algebraic internal symmetry group, SU_3*, does
exist.

The occurrence of the duplicity of particle
types with the same conjugation-invariant properties (like mass and spin) but opposite conjugation-reversing properties (like electric and baric
charge) has been much discussed. In general each
member of the double is called the "antiparticle"
of the other, and the basic fact that all particle
types may be ranged in doubles (allowing for
cases of identity between particle and antiparti-

cle types) may be called the "*antiparticle
existence principle.*" Later we shall discuss the
derivation of this principle as a theorem in the
context of other more general principles. We
shall review the general basis that has been
proposed for it in quantum field theory—which
adds the principle of *locality* to general algebraic
quantum theory—and in S-matrix theory—which
is a stripped-down and occasionally extrapolated
version of quantum field theory.

The validity of the antiparticle principle or
theorem means that there exists an operator $\hat{\Theta}$
which divides all particle type-observables into
two sets and thereby ranges existing particle
and antiparticle types in one-to-one correspondence. We call this operator *particle conjugation*
and we may restate the antiparticle existence
principle: *There exists a particle conjugation
operator $\hat{\Theta}$, and a division of symmetry-related
observables ("type-observables") in two sets,
such that $\hat{\Theta}$ acting on the type-observables of
any existing elementary entity of matter and
radiation—"particle"—gives the type-observables
for an existing conjugate entity—"antiparticle"
—with the first set of observables identical—
"conjugation-invariant"—and second set reversed—"conjugation-reversing"—in sign. Fur-
ther, $\hat{\Theta}$ is an "involution": acting on the type-
observables of antiparticles it restores the values
for the corresponding particles. Thus the operator $\hat{\Theta}$ and the identity, together, make up a
two-member "particle-conjugation group," with
the conjugation-invariant and conjugation-reversing type-observables belonging respectively
to the even and odd symmetry types ("irreducible representations") of the group.*

What "acting on" an observable means formally will be defined later.

It is to be emphasized that the *exact doubling*,
and no more than doubling, as between particle
and antiparticle types, is in contrast to the
approximate higher multiplicities found in
nature. Such are the pion triplet and N*
quadruplet of electric charge values, which
respectively fit into the larger multiplicities of
the meson charge-hypercharge octet and the
baryon charge-hypercharge decuplet; these are
ascribed to representations of higher algebraic
groups of approximate symmetry, which will
be discussed at a later point.

The conjugation exists not only for free
particles and antiparticles but also for *interacting* systems of particles and antiparticles.
This larger content of the antiparticle existence
principle may be stated as follows: *The conjugation $\hat{\Theta}$ acting on any type-observable of
any possible system of particles and antiparticles gives the observable of a corresponding
possible system of antiparticles and particles.*

Again, in the following we shall discuss the
derivation of this extended principle as a
theorem—in the usual interpretation of $\hat{\Theta}$ known

as the "C$\mathscr{P}$T theorem"—in quantum field theory and S-matrix theory.

III. Superselection and the Associated Charges; Conjugation Uniqueness Principle; Partial Conjugation Operators

Besides the antiparticle *existence* principle there appears to be another principle relating to the *uniqueness* of particle-antiparticle conjugates which, as far as we know, has not been emphasized in the literature. This may be called *the principle of unique superselective conjugation*. We now describe it.

Among the type-observables—the basic observable properties used to define particle types—it is an empirical fact that there are some which are *simultaneously sharply measurable with all other observables*, and yet are not trivial "constant observables"—they are not trivial in that they exhibit different values for different systems. These "superselection observables" which we discuss in detail later are principally the "*superselection charges*" baric B, electric Q, mu-leptic $L\mu$, electron-leptic L_e, and tau-leptic L_T (the latter three are included with some reservations in that their empirical basis is not as firm as that for B and Q). B, Q, and (L_e, $L\mu$, L_T) are related respectively to the hadronic, electromagnetic, and weak interactions. Besides these, by virtue of the Poincaré-group space-time symmetry there exists another superselection observable; this is the *valence* index $(-1)^{2s_3}$ [for massless particles $(-1)^{2\lambda}$] defined by the *observable component* s_3 of spin [or, for massless particles, the *helicity* λ] defined in the following. The empirical *evidence* for the superselection nature of all these quantities is the total absence of interference terms between systems with different values of any of the charges or of the valence index. (The superselection of the valence index can be correlated theoretically with the fact that the continuous mathematical function describing half-integral spin [helicity] particles requires *two* whole rotations to repeat, in contrast with the continuous function describing integral spin [helicity] particles, which is periodic with respect to a single whole rotation about a point; thus *no coherent linear combination* of the two mathematical functions—i.e., with a definite complex-number phase on one function with respect to the other—is consistent, and physically this means: *no interference phenomena*.)

Besides spin and mass, the numerical values of all the "*superselectables*" (superselection observables) are the principal quantities needed to specify a particle type and to distinguish antiparticle from particle. Under conjugation the valence index remains unaltered: The antiparticle of an integral (half-integral) spin particle also has integral (half-integral) spin. That this must be so is a clear consequence of the fact that s_3 or λ remains integral or half-integral under conjugation; even with s_3 or λ reversed in sign, $(-1)^{2s_3}$ or $(-1)^{2\lambda}$ remains unchanged. By contrast, under conjugation the superselec-

tive charges are all reversed. Except in those "totally neutral" cases where these charge values are all zero, the particle type can be distinguished uniquely, under all circumstances, from its antiparticle type by a reversal of sign in all charge values. How is it that when more than one charge occurs with a nonzero value, they all reverse *together* for the antiparticle type? Thus, for example, there is a unique antiproton, negative electrically ($Q = -1$) and with negative baric number ($B = -1$); there does not exist a negative baryon ($Q = -1$, $B = +1$) or a positive antibaryon with protonic mass and spin. The striking general fact that there is exact doubling—with whatever occurring nonzero B, Q, L_e or $L\mu$, and L_T reversing simultaneously—seems to be of special significance.

In our discussion we asserted that the spin [helicity], and therefore the valence, remains unchanged under conjugation while the superselective charges undergo the required reversal. A priori there were two possibilities, of which this was one. The other would have been that there exist mutually annihilating particle-antiparticle conjugates which have the spin (helicity) integral for one and half-integral for the other; the two members of such a doublet would exhibit the same, or equal opposite, baric, electric, or leptic charge but opposite valence. Because of conservation of angular momentum, upon annihilation an odd number of half-integral spin particles would appear. Such a situation seems to be in marked contrast to what we observe in nature where, for example, any antimeson and its meson (both hadrons with zero baric charge) have the same integral spin, whereas an antibaryon and its baryon have the same half-integral spin.

The principle involved may be stated as follows: *For reasons as yet unknown, there exist the three fundamental quantum-relevant interactions: hadronic, electromagnetic, weak, with associated superselective charges: baric B, electric Q, tau-leptic L_T, mu-leptic $L\mu$ and e-leptic L_e. The principle of unique superselective conjugation—an empirically verifiable (or, in principle refutable) assertion—states: The multiplication of particle types by reversals of sign of superselective charges is an exact doubling; i.e., there is a simultaneous reversal of sign of all superselective charges, whereas the type-observables associated with the Poincaré group—mass, spin, and valence—remain unchanged. This allows us to speak of a "unique" antiparticle to every particle. Alternatively described: The particle conjugation operator $\hat{\Theta}$ is inique and, together with the identity operator, defines a unique particle conjugation group such that all superselective charges below to its odd representation.* Only the situation we have described under the heading of the "principle of unique superselective conjugation" appears to occur in nature. The behavior of the valence can be inferred from the fact that the conjugation is a symmetry operation which leaves all Poincaré-group type-observables unchanged.

The uniqueness of conjugation—the simultaneous reversal of sign of all the superselective charges—requires a more subtle explanation. It is customary to interpret the rigorous conjugation operator to be what is commonly called $C\mathcal{P}T$; here C is the electric charge-conjugation operator found to be an exact symmetry in quantum electrodynamics, as also are $\mathcal{P}$ the parity of space reflection operator, and T "time-inversion" (actually "motion reversal") which unlike the first two is an *antilinear* operator (defined formally in the following). By taking the product only of the first two, one defines a linear operator Θ' commonly known as $C^{\mathcal{P}}$ (it has also been called "coparity" by the writer). To our present knowledge Θ' is an exact symmetry for all processes with the sole exception of the weak decays of one single type of particle—the kaon; in these decays invariance under Θ' appears to be broken at the level $\sim 10^{-3}$ [the number 10^{-3} refers to the ratio: Rate $(K^0_L \rightarrow 2\pi)$/Rate $(K^0_L \rightarrow 3\pi)$].

Following the survey here, and an introductory theoretical orientation, we propose a reason why the principle of unique superselective conjugation must hold: If the universal conjugation operator can be identified with Θ — a symmetry which though nonobservable is linear—and we also assume the von Neumann property of the algebra of all observables, then the principle follows. In the standard interpretation of the conjugation with another operator not Θ', but

$$\hat{\Theta} = C\mathcal{P}T \equiv \Theta$$

all the homogeneity operators, energy-momentum $\underline{P}$, are conjugation-invariant and all the isotropy operators, space-space and space-time angular momentum $\underline{J}$, are conjugation-reversing and so likewise is the (related) observable component s_3 of the spin. These quantities are not type-observables but rather "state-observables" and we shall discuss them in the following. Their behavior under conjugation is not so clearly indicated empirically as that of the type-observables. For example, with Θ' as a conjugation operator, the $\vec{P}$ components, rather than the $\underline{M}$ components, are conjugation-reversing.

Following the discussion of Θ' we give an account of some of the conventional formalism which has been successful in deriving the antiparticle existence principle and the conjugation principle (Θ interpreted as $C\mathcal{P}T$) as theorems in a general context of relativistic quantum theory. There then follows a brief discussion of the special and more stringent form which the general conjugation operator takes for the basic interactions: By combining Θ with parity $\mathcal{P}$ and time-inversion T, or Θ' with $\mathcal{P}$, one defines the operator $\Theta\mathcal{P}T \equiv \Theta'\mathcal{P} \equiv C$, known traditionally as "charge conjugation," which is a kind of "purely internal and unitary conjugation operator." Except for weak-interaction systems and processes, it is again a symmetry.

To understand all this in detail it is necessary first to clarify what we mean by "particles." The fundamental entities of matter and radiation constituting the physical world are associated with observables which are: (I) *quantum-mechanical*; (II) *symmetry-governed*; (III) *invariant* or *covariant* under the governing symmetry. (I) By the quantum-mechanical aspect we mean that a key role is played by probabilistic—more explicitly: *statistically deterministic*—concepts rather than detailed deterministic concepts as in classical physics. This is because of the fundamental fact of nature that any two observables A and B are *not necessarily compatible*, i.e., starting in a given situation and *measuring* the two observables *in alternation*, they do not each repeat their original measure values; instead, the measurements give for each a distribution of *measure-value probabilities* characteristic for that observable when preceded by a measurement of the other. (II) By the symmetry-governing aspect we mean that it is of central importance to classify observables according to their behavior under physically relevant *groups of symmetries*; each individual symmetry is by definition a transformation ("change") which (a) leaves measure-value *probabilities unchanged*, (b) is evolution-independent, and (c) leaves the spacetime type-observables, mass, spin, and valence, unchanged. (III) By the invariance-covariance aspect we mean that among all possible observables—and *relative to a given degree of generality of observation procedures*—there is a division into two classes—the type-observables which are *invariant* with respect to the groups of symmetries, i.e., go into themselves when transformed by any symmetry, and the state-observables which are *covariant*, i.e., mix nontrivially with other members of a set of observables when transformed.

A more explicit discussion follows.

IV. Observables and Their Algebras. Symmetries (I) The modern algebraic version of the quantum-mechanical concepts is streamlined in comparison with the earlier formulations in that the concept of a given state space has receded to the background; with the greater economy and generality of algebraic quantum theory, many results may be demonstrated more elegantly.

(Ia) We consider all directly or indirectly measurable properties which we call *observables* A. The observables constitute an algebra with a certain formal structure. We first remark that observables in quantum mechanics are not all mutually sharp or compatible: In any given situation, keeping all other conditions constant, if we measure an observable A repeatedly we get consistently the same measure value. The same is not necessarily the case if any two observables A and B are measured in alternation repeatedly, keeping all other conditions constant; we find a distribution of values for each rather than a single value. We then say: "A and B are not compatible." Two observables are compatible if in a given system, prepared appropriately and repeatedly in the same way, each observable always exhibits one characteristic definite measure number ("quantum number," "eigenvalue") upon consecutive repeated measurements appropriate to each observable in either order. Constant observables, i.e., those which exhibit only one measure value under *all* circumstances with *all* systems (multiples of the "identity"), are of course compatible with all others. If there is also a nonconstant observable—i.e., in experiments with various systems it yields differing measure numbers—which is compatible with all other observables, we call it a *superselection observable*. We shall discuss some examples later, among them electric charge Q.

In the following, let the complex conjugate of a scalar number, e.g., λ, be represented by a superior bar, $\bar{\lambda}$. Consider an infinite-dimensional linear complex vector space $\mathcal{H}$ with an inner product (ϕ, ψ), i.e.,

$$(\phi, \psi + \theta) = (\phi, \psi) + (\phi, \theta)$$

$$(\phi, \lambda\psi) = \lambda(\phi, \psi) \quad (\lambda\phi, \psi) = \bar{\lambda}(\phi, \psi)$$

$$(\phi, \psi) = (\overline{\psi, \phi})$$

for all vectors ϕ and ψ; $\mathcal{H}$ is assumed complete in the norm $\| \phi \|$ defined by $| \phi, \phi |^{1/2} \equiv \| \phi \|$, and has a countable basis (every Cauchy sequence of vectors has a limit vector in the space and a complete orthonormal basis can be introduced), i.e., $\mathcal{H}$ is a Hilbert space. The unit vectors of physical interest are only those lying in certain subspaces of $\mathcal{H}$ defined by certain values of the superselectables. These unit vectors, up to an arbitrary phase factor, represent states. Each observable is represented formally by a linear *self-adjoint operator* or "matrix" A acting on the vectors of $\mathcal{H}$. The eigenvalue λ of A for an eigenstate ϕ, i.e., such that $A\phi = \lambda\phi$, is the measure value found for the observable corresponding to A when measured in the state ϕ. The role of self-adjointness is to ensure that all such eigenvalues are bounded and real. What "self-adjoint" means formally and what further properties are to be expected of A is clear from the following discussion.

Besides their action on the vectors of $\mathcal{H}$, one also has the effect of one operator in $\mathcal{H}$ acting on another to give a resultant operator, and in the case of the operators representing two observables, this product of operators represents the successive measurement of the two observables. Two compatible observables are then represented by two commuting operators corresponding to the fact that the two successive measurements give the same result in either order. For brevity the operator representing an observable will also be called an observable. A general operator A is said to be *bounded* or to have a *norm* if there exists a positive number b such that

$$\| A\phi \| \leqslant b \| \phi \|. \quad \text{all } \phi \text{ in } \mathcal{H}$$

The smallest number b with this property is the *norm of A* and is denoted $\| A \|$. It follows that

$$\| A + B \| \leqslant \| A \| + \| B \| \quad \| AB \| \leqslant \| A \| \| B \|$$

$$\| \lambda A \| = | \lambda | \| A \|$$

We now turn to the characterization of the algebra of observables in $\mathcal{H}$. What the algebra formalizes are (1) the property that the numerical values obtained as the result of any measurement are (a) real and (b) bounded; (2) the property that not only is every observable or algebraic function of observables compatible with all superselectables, but that every operator compatible with all superselectables is an observable or algebraic function of observables; in short compatibility with superselectables characterizes the algebra of observables.

The formal structure which satisfies the requirements of these two sets of properties is that of a *von Neumann algebra*, a type of subalgebra $\mathfrak{A} \equiv \{A\}$ of the algebra of all linear bounded (or, equivalently, continuous) operators in $\mathcal{H}$, which subalgebra satisfies two conditions:

1. $\mathfrak{A}$ *IS A C* ALGEBRA.* A C^* algebra involves (a) an *involution* applied to (b) *normed operators.*

a. *Involutive application:* There is a * or adjoint application of $\mathfrak{A}$ into itself such that, for any A in $\mathfrak{A}$, A^*—the adjoint of A—has the properties

$$(A+B)^* = A^* + B^*, \quad (\lambda A)^* = \bar{\lambda} A^*,$$

$$(AB)^* = B^* A^*, \quad A^{**} = A.$$

An operator for which $A^* = A$ is said to be *self-adjoint.*

b. *Normed operators:* In a C^* algebra we have that all operators are normed and that their norms satisfy

$$\|A^*\| = \|A\|, \quad \|A^* A\| = \|A\|^2, \quad \text{(all A).}$$

An operator H which satisfies the equation between inner products

$$(\psi, H\phi) = (H\psi, \phi)$$

is called hermitian. It is easy to verify that every eigenvalue of a hermitian operator is real:

$$(\phi, H\phi) = (\phi, \lambda\phi) = \lambda(\phi, \phi) = (H\phi, \phi) = \bar{\lambda}(\phi, \phi).$$

Likewise for every operator which is self-adjoint. Every self-adjoint operator is hermitian, but unbounded hermitian operators are not self-adjoint.

(It can be proved conversely that a hermitian operator which is everywhere self-adjoint is necessarily bounded.)

2. $\mathfrak{A}$ *IS EQUAL TO ITS DOUBLE-COMMU-TANT $\mathfrak{A}''$.* The latter is the set of all linear operators in $\mathcal{H}$ which commute with the first commutant which, in turn, is the set of all linear operators in $\mathcal{H}$ which commute with $\mathfrak{A}$.

The algebra of observables is a von Neumann algebra based on self-adjoint operators where simultaneous observability of observables means that the self-adjoint operators, which represent the observables, commute. Since $\mathfrak{A}$ contains all observables and algebraic functions of them, condition (2) means here that a linear operator which is a nonmember of the observable algebra cannot commute with all members of the commutant $\mathfrak{A}'$ of $\mathfrak{A}$. Since $\mathfrak{A}'$ contains, besides the constant operators, only the set of all superselectables and functions of these, the assumption that $\mathfrak{A}$ is a von Neumann algebra requires such a linear nonmember of the observable algebra not to commute with one or more of these superselectables. We shall see in the sequel that this is one of the key points in rationalizing the occurrence of *unique* superselective conjugation as described above.

The possible measure numbers of the observables are represented by the *spectrum of eigenvalues a_i* of A. For simplicity suppose these spectra are discrete. By the *spectral theorem* of Hilbert space theory, A is a linear combination with real coefficients of bounded self-adjoint operators called *projections* (or "projectors") P^A, having the idempotent property $(P^A)^2 = P^A$, with P^A corresponding to the eigenvalue λ. In particular, for describing various specific conditions in which a given system is prepared, one is interested in the subset $\mathcal{S}$ of self-adjoint operators with all eigenvalues a_i nonnegative and such that the sum of all a_i—the "trace"—is equal to unity. We call these unit-trace nonnegative self-adjoint operators "density matrices." These are in one-to-one correspondence with the "states," both pure and mixed, of the earlier less-streamlined formulation of quantum mechanics. (It should be noted, however, that the modern C^*-algebraic quantum theory recognizes much more general "states" than can be represented by density matrices.) Even this subset, consisting of nonnegative self-adjoint operators of trace 1, however, is too large for physics when superselection observables are present. $\mathcal{H}$ is then divided up into mutual totally incoherent subspaces corresponding to the different eigenvalues of the superselection observables; the operators representing *physical observables* act only within each subspace but not between them.

(Ib) *Probabilities*. The concept of probability gives physical meaning to the projection P_χ^A corresponding to an eigenvalue λ. Suppose a system is prepared repeatedly in the same manner, so that a given density matrix M describes the state of the system, and we measure the observable A each time. On the one hand, the occurrence

frequency or *probability measure* for the eigenvalue λ of A in the state M is $|\dim P_\chi^A| a_\lambda$ where a_λ is the coefficient of P_χ^A in the spectral-theoretic direct sum expansion of M—and this probability measure is given by Trace (MP_χ^A). On the other hand, the eigenvalue λ itself occurs as the coefficient of P_χ^A in the expansion of A. Thus if—having prepared the system each time in the same way—we repeatedly measure the observable A, the expectation value of A over a large number of similar trials is Trace (MA).

(II) *Symmetries*. A symmetry is a mapping of the set $\mathcal{S}$ of all density matrices onto $\mathcal{S}$ which

   (a) preserves probabilities,
   (b) is evolution-independent,
   (c) leaves mass, spin, and valence index invariant.

A symmetry may or may not also be an observable.

(IIa) Just as in the case of complex numbers in the complex plane, each of the operators in state space may be viewed in an active role as operating on other objects or in a passive role as one of the objects being operated on. A particular kind of operation on an operator A by another operator G is called "transformation of A by G" or a similarity mapping of A by G: GAG^{-1}. (This gives the operator in the G-transformed state space which corresponds to A in the original space, i.e., it accomplishes the corresponding operation upon corresponding states.) If under transformation by G, A goes into itself: $GAG^{-1} = A$, then A is said to be *invariant under G*; more generally, if for each member of a set $\{A_i\}$, GA_iG^{-1} gives a member $A_i' = A_i$ of the set, A_i and all its partners in $\{A_i\}$ are said to be *covariant under G*. If the entire set $\mathfrak{A}$ of all observables is transformed onto itself by G, then we say G performs a similarity mapping of $\mathfrak{A}$.

Condition (IIa) says that a symmetry is a probability-conserving similarity mapping of $\mathcal{S}$. Probabilities are given by the coefficients in the "convex" (i.e., with real positive coefficients) linear combinations into which density matrices can be decomposed according to the spectral theorem. Thus the preservation of probabilities is equivalent to the preservation of convex linear combinations or projectors. There are two kinds of operators which, acting as transforming operators, preserve convex linear combinations of projectors.

1. A *unitary* operator U is *linear*

$$U(\alpha\phi + \beta\psi) = \alpha U\phi + \beta U\psi$$

and *isometric*, meaning there is equality in absolute value of both sides of

$$(U\phi, U\psi) = (\phi, \psi).$$

2. An *antiunitary* operator Θ differs only in that it is *antilinear isometric*, i.e.,

$$\Theta(\alpha\phi + \beta\psi) = \bar{\alpha}\Theta\phi + \bar{\beta}\Theta\psi \, ; (\Theta\phi, \Theta\psi) = (\overline{\phi, \psi}).$$

The second condition implies that the antiunitary transform of an operator A is given by

$$A' = (\Theta^{-1} A\Theta)^*, \text{ and therefore } (AB)' = B'A',$$

or there is a reversal of order of operator products if Θ is antiunitary.

It is clear that the transformation $\hat{\Theta} A \hat{\Theta}^{-1}$, where $\hat{\Theta}$ is a unitary or antiunitary operator, preserves convex linear combinations of observables. When we limit A to the subalgebra $\mathcal{S}$ of all density matrices in $\mathcal{Q}$, the converse is also true. By the general Wigner-Kadison theorem (Wigner, 1931–1959; Bargmann, 1964; Kadison, 1951) every convex combination-preserving one-to-one map of $\mathcal{S}$ on $\mathcal{S}$ is either unitary or antiunitary. Moreover this mapping operator is unique up to a phase factor. (Wigner's formulation of the theorem was in terms of preservation of absolute scalar products of states rather than of convex linear combinations of density matrices).

We should qualify that, in the presence of superselectables, the symmetry mapping may be from a subalgebra $\mathcal{S}_1$ of density matrices onto some other coherent subalgebra $\mathcal{S}_2$ isomorphic with $\mathcal{S}_1$, but labeled by different values for the superselective observables. This is the case when the symmetry operator Θ' is not a member of the observable algebra. If, in addition, Θ' is linear, we can draw an important conclusion. As we have already emphasized, in this case—because the algebra of observables is a von Neumann algebra—the linear nonobservable symmetry operator Θ' cannot commute with all superselective charges. As the notation has already indicated $CP \equiv \Theta'$ is such an operator; unlike $\Theta \equiv C\mathcal{P}T$, Θ' is linear—$\mathcal{P}$ is linear and C (see end of article), as dictated for instance by quantum electrodynamics, is linear. Also, unlike $\mathcal{P}$ or C separately, Θ' is a symmetry operator in all cases, with the apparent exception of the neutral kaon decays. Further, except in the small subspace of totally neutral particle states, Θ' is not a member of the observable algebra: There is no preparation or measurement procedure, or combination of such procedures, corresponding to turning particles into antiparticles. Thus, overlooking for the moment the difficulty with the neutral kaon decays, Θ' is the natural candidate for a nonobservable linear symmetry operator representing particle conjugation, which cannot commute with all the superselective charges. That Θ' must actually anticommute with the superselective charges, e.g., letting Q be the relevant charge:

$$\Theta'Q = -Q\Theta'$$

follows from the fact that all probabilities, including the square of the expected value of the charge, must be preserved. That this happens simultaneously with all superselective charges (principle of "unique superselective conjugation") becomes evident when we recognize the empirical fact that there exist particle types for each case of only one charge differing from zero—all other charges having zero values:

	B	Q	L_μ	L_e	L_τ
baryon, otherwise neutral: Λ^0	1	0	0	0	0
electrically charged hadron, otherwise neutral: π^+	0	1	0	0	0
mu-lepton, otherwise neutral: ν_μ	0	0	1	0	0
e-lepton, otherwise neutral: ν_e	0	0	0	1	0
tau-lepton, otherwise neutral: ν_τ	0	0	0	0	1

Considering each of these systems in turn, we draw the conclusion that each of the superselective charges, when occurring singly, anticommutes with the symmetry operator Θ', so that, for consistency, when several occur together with nonzero values, they must all reverse together. An example is the proton p ($B = 1, Q = 1$) which, under conjugation of the decay $\Lambda^0 \to p^+ + \pi^-$ has to give the unique antiparticle $\bar{p}$ ($B = -1, Q = -1$).

We note that to be able to draw the conclusion that a superselective charge is reversed by a nonobservable symmetry—"conjugation"—it was necessary for that superselective charge to occur with an eigenvalue unequal to zero, while all others have zero eigenvalues (this is what we mean by "otherwise neutral"). The argument will, however, not go through if we have not established the conjugation invariance of that one remaining superselective observable, which is not a charge, i.e., the valence $(-1)^{2s_3}$ [or $(-1)^{2\lambda}$ for the massless case]. This conjugation invariance is assured, however, because the valence must always be $+1$ or -1 depending upon the integer or half-odd integer value of the spin; but, by condition (c), a symmetry leaves the spin and therefore the valence index invariant.

It is striking that the only violations of Θ' which prevent it from being taken as a universal conjugation under which interactions are invariant are in the neutral kaon decays. The neutral kaon case is precisely the one in which there exists no superselective charge which by its reversal defines the difference between particle and antiparticle. Only the approximately conserved "strangeness" or "hypercharge" (see following) is available as a possible charge to discriminate between K and $\bar{K}$.

(IIb) We note with emphasis the importance of condition (IIb). Though every unitary or antiunitary operator preserves probabilities, it is not necessarily a symmetry. The requirement of evolution independence is what brings the enormous number of unitary and antiunitary operators in $\mathcal{H}$ down to the physically significant subset of symmetry operators. Evolution independence may be described as a "generalized time independence." The evolution matrix—known as the "S-operator" or "S-matrix"—is an operator which takes any one of a set of *initial asymptotic* states to any one of a set of *final asymptotic* states, where the states are characterized by the values of certain observables and where "initial" and "final" refer to appropriate boundary conditions. The initial and final set can be the same and be a complete set spanning the entire state space. (In the case of the "collision states" appropriate to scattering problems, this condition is known as *asymptotic completeness*, and it goes with unitarity, $S^* = S^{-1}$, of the S-matrix). The evolution matrix is a replacement of a "moment-to-moment" time-displacement operator by an overall or global "before-and-after" operator. The need for such a replacement is evident when we reflect on the fact that the passage of "time" is relative to the observer, according to the special relativity principle which governs nature—at least locally. In fundamental processes, only the before-and-after relations described by the S-matrix have an observer-invariant significance, and *the formal expression of the property of evolution independence, which every symmetry G must satisfy* may be stated: *When the symmetry is performed first and followed by evolution, it gives the same result as when evolution is performed first and then the symmetry.*

For a unitary symmetry U this condition takes the form

$$SU = US$$

which may be read either way as: $SUS^{-1} = U$, "the evolution matrix leaves every unitary symmetry invariant"; or $USU^{-1} = S$, "every unitary symmetry leaves the evolution matrix invariant."

For an antiunitary symmetry Θ, because initial and final states are interchanged under the action of Θ, the evolution-independence condition has to be stated as

$$S\Theta = \Theta S^*$$

where S^*, the adjoint of S, takes final to initial states; this equation is easily checked by letting the right and left hand sides operate respectively on a final state f. The results of the sequence of operations on the right and left can be symbolized as follows (primes indicate Θ-transformed states and f indicates the result of S operating on i):

$$(f' \leftarrow i', i' \leftarrow f) = (f' \leftarrow i, i \leftarrow f).$$

The relation for S and Θ may then be rewritten

$$S^* = \Theta^{-1} S\Theta$$

and, referring back to the expression given earlier for the antiunitary transform of an operator, we see that, even though it is not a simple similarity invariance, this relation can be expressed by a statement parallel to that for the unitary case: "every antiunitary symmetry leaves the evolu-

tion matrix invariant." It is clear that for finding the S-matrix, which is central in elucidating fundamental processes, the knowledge of its symmetries is invaluable.

The symmetries of any system form one or more groups. In general there are discrete groups of symmetries (e.g., the permutations of identical particles in a multiparticle system), and continuous groups of symmetries. In the latter case the *generators* are of particular importance. Let G be any symmetry operator differing infinitesimally from the identity operator I. Then the "business end" of this operator defines a generator $\mathcal{G}$: $G = I + \epsilon \mathcal{G}$. In this case the transformation by G of any operator A goes—to the first order in ϵ—into A plus the commutator $[\mathcal{G}, A]$ of $\mathcal{G}$ with A:

$$GAG^{-1} \simeq A + \epsilon[\mathcal{G}, A] \qquad [\mathcal{G}, A] \equiv \mathcal{G}A - A\mathcal{G}.$$

The invariance of A under all members of a group $\{G\}$ requires then that the commutator of A with all generators of $\{G\}$ be zero. In particular the S-operator must commute with the generators of any symmetry group.

The S-operator governing the evolution of fundamental processes, is a particular though important case. Its structure, as well as the definition of particle types which undergo the fundamental processes, is elucidated by considering the invariant and covariant bedfellows of the S operator. Quite generally, in its passive role as an operand, any operator will be found to have certain transformation or covariance properties under all the generators of a given group of state space transformations. This characteristic transformation behavior, as well as the relation structure of operators with each other, plays a key role in deciding whether an operator may be interpreted as representing a physical quantity having that transformation behavior. The generators themselves are usually merely covariants of the group, but certain polynomials or functions of them may be invariant; a function of generators which is invariant—i.e., commutes with all generators—is called a "Casimir operator" or "Casimir invariant" of the group.

(IIc) This condition is roughly equivalent to the one otherwise expressed by the statement that a symmetry operation should leave the subspace of one-particle states invariant. We have already seen its usefulness in our argument concerning the invariance of the valence index under conjugation.

(III) As we have already indicated, by the "invariance-covariance" aspect of the fundamental physical entities we mean that among all possible observables there is a (heuristic) division in two classes:

I. A subset which we call "invariant" or "particle type-defining";

II. all other observables which we call "covariant" or "state-defining."

In our formulation both subsets will be characterized by labels associated with symmetry groups. The first set of observables—used to define fundamental *particles*—are certain *external and internal Casimir invariants*, so called because they are associated with external and internal symmetry groups governing the fundamental systems. In contrast the second set of observables—used to characterize the *states* in which fundamental particles are found—are *covariants* of these same external and internal symmetry groups. We shall see how this very valuable distinction is nevertheless relative to the degree of generality of the phenomena being considered.

The external symmetry group is the *proper spacetime symmetry group*, i.e., the group of all rigid translations and rotations in spacetime (Poincaré group). The covariants here are physical representatives of the four translation generators—the components of the energy-momentum P—and of the six rotation generators—the space-space and space-time components of the angular momentum $J \sim (\vec{M}, \vec{\mathfrak{M}})$. The two principal Casimir invariants here define the principal external properties of elementary particles: Their magnitudes give the numerical values of the first two of these properties, mass m and spin s. These numerical values label the possible "symmetry types" or *irreducible representations* ("reps") of the proper spacetime symmetry group.

We have discussed the general doubling in Nature of particle types for which the conjugation-invariant observables are identical, whereas the values of all other type-defining observables are exactly opposite for the two members of the double; and we have introduced the mapping operator $\hat{\Theta}$ which acts on observables. Formally, $\hat{\Theta}$ "acting on" an observable means transforming according to $\hat{\Theta}(\ )\hat{\Theta}^{-1}$. The superselective charges: B, Q, L and ℓ, and λ are included among the type-defining conjugation-reversing observables, and give $-B, -Q, -L, -\ell, -\lambda$ under this transformation.

What about the behavior of state-characterizing observables under $\hat{\Theta}$? These also may be sorted out once and for all into two sets, conjugation-invariant and conjugation-reversing. From now on we use the generic symbol Σ to represent all conjugation-invariant observables, and the symbol R to represent all conjugation-reversing observables, whether type or state. Thus, $\hat{\Theta}$ is an operator which maps the subalgebra of symmetry-related observables onto itself according to

$$\hat{\Theta} \, \Sigma \, \hat{\Theta}^{-1} = \Sigma \qquad \hat{\Theta} \, R \, \hat{\Theta}^{-1} = -R.$$

Closely related to the existence of antiparticles—almost but not quite as firmly established empirically—is the further fundamental fact that not only may all *types* be arranged in matched particle-antiparticle pairs by $\hat{\Theta}$, but also that $\hat{\Theta}$ is *represented in $\mathcal{H}$ by an antiunitary symmetry* Θ. We call this *the conjugation principle*. That $\hat{\Theta}$ should be represented ("implemented") by an *operator* Θ, unitary or antiunitary, in $\mathcal{H}$ is a purely mathematical consequence of $\hat{\Theta}$ mapping onto itself the subalgebra of density matrices (Wigner-Kadison). The nontrivial physical point of the conjugation principle—a key point, commonly referred to as "the $C\mathcal{P}T$ Theorem"—is that Θ is not just an operator but is a *symmetry* operator. In other words the evolution operator S, and the Poincaré invariants, mass spin, and valance (and therefore the dynamics of all fundamental processes), are invariant under Θ, i.e., $\hat{\Theta}$, *which we have defined as a mapping of symmetries on symmetries, itself gives rise to an antiunitary symmetry.* More explicitly, the conjugation principle may be stated: *The conjugation operator $\hat{\Theta}$ is represented in state space $\mathcal{H}$ by* (1) *an evolution-independent mapping Θ of states in $\mathcal{H}$ onto other states in $\mathcal{H}$, such that:* (2) *transition probabilities are preserved, with initial states being mapped on final and final on initial;* and (3) *mass, spin, and valence are preserved.*

$$S\Theta - \Theta S = 0 \tag{1}$$

$$(\psi_{\Sigma', R'}, \psi_{\Sigma, R}) \longrightarrow (\Theta \psi_{\Sigma', R'}, \Theta \psi_{\Sigma, R})$$
$$= (\psi_{\Sigma, -R}, \psi_{\Sigma', -R'}) \tag{2}$$

$$\Theta m^2 - m^2 \Theta = 0 \qquad \Theta s(s+1) - s(s+1)\Theta = 0 \tag{3}$$

We emphasize that the conjugate of the original initial (final) state plays the role of the final (initial) state in the conjugate process, of which the rate is identical to that of the original process. Again we shall discuss later how the conjugation principle may be derived as a theorem in the context of general theory.

V. Conjugation-Invariant and Conjugation-Reversing Observables The mass squared is the squared magnitude of the energy-momentum four-vector (Table 3). The spin comes in as follows: In general there is defined a four-component polarization operator W_α made up by the outer product of the linear momentum operators and the angular momentum operators. (The time component $W_0 = \vec{P} \cdot \vec{M}$ is the *longitudinal polarization*.) There are two cases generally recognized to be of physical interest, $m^2 > 0$ and $m^2 = 0$. In the massive case, W_α yields as the measure of its invariant magnitude the spin s according to the formula $W^2 = m^2 s(s+1)$.

By a general spin-statistics theorem of relativistic quantum theory, the spin of a one-boson system is integral and of a one-fermion system half-integral.

In the case of zero-mass particles, W splits into two parts, each consisting of a pair of components. The pair of components perpendicular to the momentum define, by the sum of their squares, the "continuous spin" invariant r^2 where r is a real nonnegative number. The remaining components, longitudinal and time-like, are equal in this case, and they yield another invariant, the projective index $\sigma = 0$ or $\frac{1}{2}$ or valence index $+1$ or -1, which specifies the integral or half-integral character of the forward angular momentum (i.e., the projection along the direction of motion). For $r > 0$ this forward angular momentum takes on an infinite discrete ladder of values, while for $r = 0$ the ladder breaks up into its individual segments (the representation matrix reduces to the direct sum of its diagonal entries). In these $r = 0$ cases the forward angular momentum—now an invariant—takes on a unique integral or half-integral value, the helicity, λ, and the valence index ceases to be an independent invariant—just as it happens in the massive case. All these relations and cases are summarized in Table 2.

In nature only cases where the continuous invariant r is equal to zero seem to occur. As we have indicated, the two sets—particle-defining observables and state-defining observables—are associated respectively with two sets of labels of mathematical groups—groups which describe the symmetries of natural systems. We have discussed one of these sets: Each fundamental entity belongs (in its mathematical description) to an irreducible representation—"rep"—of the symmetry groups governing all physical systems, and the type-characterizing observables are associated with the Casimir operators whose eigenvalues are the labels specifying the rep. What are the other labels with which the state-characterizing properties are associated? They are the labels for the *rows* of the reps and do not have the same invariant significance as the rep labels since various basis systems of states may be chosen to span the vector space constituted by each rep. The type-observables have relatively stable values, defining the identity of the particle, so that it may be recognized as the same at the end as at the beginning of a process; whereas the state observables are more labile, having values which can vary between the beginning and end of the many processes in which the identity—e.g., m and s—of the particle is unaltered. For example, a rep of the Poincaré group comprises a vector space of states, vector space consisting, e.g., of states of all momenta and longitudinal polarizations associated with a particle of given mass and spin. Alternatively, a basis of angular momentum

TABLE 3. GENEALOGY OF THE EXTERNAL-PROPERTY INVARIANTS
MASS m AND SPIN s OR HELICITY λ

	Energy-Momentum	Total Angular Momentum	Polarization Vector
	Four Components of a Spacetime Four-vector P_0, P_1, P_2, P_3	Three Space-space Components of a Spacetime Six-vector $J_{23} \equiv M_1\ J_{31} \equiv M_2\ J_{12} \equiv M_3$; and Three Spacetime Components $J_{10} \equiv \mathfrak{M}_1\ J_{20} \equiv \mathfrak{M}_2\ J_{30} \equiv \mathfrak{M}_3$	Four Components of a Spacetime Four-vector (Axial) $W_0 \equiv P_1 J_{23} + P_2 J_{31} + P_3 J_{12}$ $W_1 \equiv P_0 J_{23} + P_2 J_{30} + P_3 J_{02}$ $W_2 \equiv P_0 J_{31} + P_3 J_{10} + P_1 J_{03}$ $W_3 \equiv P_0 J_{12} + P_1 J_{20} + P_2 J_{01}$
Timelike Energy Momentum. Standard form taken when state is limited to the subspace of states in which the center of mass is at rest.	$P_0^2 - \sum\limits_{k=1}^{3} P_k^2 \equiv m^2$ $P_0 = m$ $P_1 = 0$ $P_2 = 0$ $P_3 = 0$	$M_1 = S_1$ $M_2 = S_2$ $M_3 = S_3$	$\sum\limits_{k=1}^{3} W_k^2 - W_0^2 = m^2\, s(s+1)$ $W_0 = 0$ $W_1 = mS_1$ $W_2 = mS_2$ $W_3 = mS_3$
Lightlike Energy Momentum. Standard form taken when state space is limited to the subspace of states in which the momentum is in the x_3 direction.	$m = 0$ $P_0 = p$ $P_1 = 0$ $P_2 = 0$ $P_3 = p$	$M_1 - \mathfrak{M}_2 = T_1$ $M_2 + \mathfrak{M}_1 = T_2$ $M_3 \quad\ \ = \mathcal{H}_3$	$\sum\limits_{k=1} W_k^2 - W_0^2 = r^2$ $W_0 = W_3 = p\mathcal{H}_3$ $W_1 = pT_1$ $W_2 = pT_2$ In case $r = 0$ $W_a \equiv \lambda\, P_a$ $\lambda = \underset{\sim}{M} \cdot \underset{\sim}{P}/|\underset{\sim}{P}|$

might be chosen. The state-characterizing observables include then energy-momentum P, and angular momentum J, or W, depending on the basis chosen.

We have in the Poincaré group an example of how the very valuable distinction between particle-defining labels and state-defining labels is relative to the degree of generality of the observations. Suppose that we limit the allowed changes of observation systems (or "frames of reference") to space rotations, so that the external symmetry group reduces to the rotation subgroup of the Poincaré group. Then the fourth component—the energy P_0—of P, which in the full group is merely a covariant, becomes an invariant, $P_0 = (\vec{P}^2 + m^2)^{1/2}$, corresponding to the Casimir operator $\vec{P}^2$ of the space rotation group. This example illustrates the general proposition that as the scope of observation is increased, it becomes more and more natural to view different particles merely as different states associated with one supersystem.

We turn now to the internal properties of particles. As already remarked, Fermionic hadrons are called *baryons*, bosonic hadrons are called *mesons*; nature permits us to define *baric* charge B equal to ± 1 for the former and 0 for the latter in such a way that the total baric charge is conserved in all processes. For some unknown reason baric charge is related to spin or statistics.

So far only fermionic leptons of two types—electron or e-type and muon or mu-type—are known, but the existence of bosonic leptons involved in weak interactions has been conjectured. Again, in accord with exact conservation in all processes, nature permits us to define: (1) *electron leptic charge* ℓ, equal to ± 1 for electrons and e-neutrinos, and 0 for muons and mu-neutrinos; (2) *mu leptic charge* L, equal to ± 1 for muons and mu-neutrinos, and 0 for electrons and e-neutrinos. (The universality of the muon- and electron-leptic charge labeling—based on the phenomena observed in weak interactions—is not quite certain.)

Intermediate between hadronic and weak interactions in strength is the electromagnetic interaction. Here nature permits the definition of a conserved *electric charge* Q equal to ± 1 and 0 for various particles, hadronic and leptonic, as the case may be.

Another way to regard the existence of these internal charges, in which the aspect of conservation is derivative rather than primary, is in terms of symmetries. In this way too we are led to four independent universally observable internal properties; baric, electric, muon- and electron-leptic charge B, Q, L, and ℓ, defined by symmetries associated with the behavior of particles under the three fundamental quantum-relevant interactions: strong, electromagnetic, weak. Each of the types of charge may be associated with an SU_1 ("special"—i.e., uni-

modular—"unitary" group on one object), i.e., groups of *evolution-independent* transformations having the form $e^{iB\phi}$, $e^{iQ\phi}$, $e^{iL\phi}$, $e^{i\ell\phi}$, respectively (ϕ is an arbitrary parameter ranging over all elements of each group). In these cases B, Q, L, and ℓ are the sole generators—indeed Casimir operators of their respective SU_1 groups.

With the specification that they are generators of symmetries—i.e., evolution-independent—it comes about that, as in the case of energy-momentum and angular momentum, the sum total value of each of the internal charges is conserved in any process. In the case of mass and spin of a composite system, the addition is vectorial rather than simply algebraic, and in the case of mass, there may also be a contribution due to internal energy changes. In contrast, the charges have values which are simply additive in a composite system. As already mentioned, the sign reversal of the values of the charges for antiparticles is essential to maintaining the conservation laws in the processes of *pair creation* and *pair annihilation* in which antiparticles are observed appearing and disappearing with their conjugate particles. Only those particles for which the values of all charges are zero, e.g., the photon γ, the neutral pion π^0, and the η, ρ, ω, and ϕ mesons are identical with their antiparticles; these are called "self-conjugate particles." Any number of these may occur together with pair creation or annihilation. Examples are

$$e^- + e^+ \to 2\gamma \qquad e^- + e^+ \to 3\gamma$$

depending upon whether the spins of the electrons are antiparallel or parallel.

As indicated earlier there is an external property or type observable, the *parity operator* $\mathscr{P}$ associated with invariance of non-weak evolution processes with respect to the discrete operation of space inversion (reflection). When one extends the proper Poincaré group to include space inversion, the symmetry types and the corresponding states double. Naively one says that $\mathscr{P}^2 = 1$, and that the parity of a one-particle state or one-antiparticle state is $+1$ or -1 according as the state function is preserved or reversed in sign when the space coordinates upon which the state function depends are all reversed. (Properly speaking the term "parity" as used here means "the eigenvalue of the parity operator when applied to the state," and the state must therefore be an eigenstate of $\mathscr{P}$.) Actually the result of a double inversion can be asserted to equal the identity only up to a phase factor—a complex number of modulus 1, namely $\mathscr{P}^2 = \exp(i2\alpha_R)$, where α_R is some arbitrary phase angle which may be different for the states associated with each superselective family—i.e., for each set of particle types with given values of the charges, electric, baric, muleptic, electron-leptic, and hypercharge Y (see later discussion): $\{Q \ B, L, \ell; Y\} \equiv \{R\}$. Examples of such families are (1) the self-conjugate or totally neutral ($Q = B = L = \ell = 0$) "non-strange" ($Y = 0$) mesons; (2) all the nonstrange mesons of positive electric charge; etc. With $\mathcal{P}^2 = \exp\,(i2\alpha_R)$ the two possibilities for the effect of $\mathcal{P}$ are multiplication by $\exp\,(i\alpha_R)$ and $-\exp\,(i\alpha_R)$. Within any given family, by renormalizing, $\mathcal{P} \to \exp\,(i\alpha_R)\,\mathcal{P}$, one can reduce to the two alternatives $+1$ and -1. See PARITY.

More explicitly, we describe the situation as follows. Within any *superselective sector*, i.e., a superselectively separated subspace of state space, we assume the "principle of maximum coherence" to hold: *any two physical states ϕ and ψ can cohere into a single physical state, with the relative phase angle of ψ with respect to ϕ in the coherent superposition $\phi + \psi$ unambiguously determined.* Observationally the phase angle β appears in the interference term $2\,|\,\phi, \psi)\,|\cos\beta$ in the scalar product

$$(\phi + \psi, \phi + \psi) = (\phi, \phi) + (\psi, \psi)$$
$$+\, 2\,|\,(\phi, \psi)\,|\cos\beta.$$

If, under reflection, ϕ and ψ behave the same ($\phi + \psi \to \phi + \psi$) the interference term remains unchanged, and we say the *relative parity* of ϕ and ψ is $+1$; if they behave oppositely ($\phi + \psi \to \phi - \psi \equiv \beta \to \beta + \pi$) the interference term changes sign, and ϕ and ψ are said to be of relative parity -1. Relative parity, being a special case of relative phase factor, has the transitivity property, i.e.,

$\mathcal{P}(A$ with respect to $B) \times \mathcal{P}(B$ with respect to $C)$

$= \mathcal{P}(A$ with respect to $C)$.

In the coherent superposition case where ψ has relative parity -1 with respect to ϕ, $\phi + \psi$ is not a parity eigenstate though it is a physical state. For simplicity and brevity we shall refer to all physical states with definite charges and Poincaré group indices, which are also parity eigenstates, simply as particles, even though from some points of view they are composite: n-particle states, particle-antiparticle states, etc.

The vacuum state (i.e., with no particles present) has the same values, $\{R\} = 0$, for the superselective charges as the totally neutral particles. Thus the relative parity of any totally neutral particle with respect to the conventionally chosen parity of the vacuum, $\mathcal{P} = +1$, is empirically determined. This is called the *intrinsic parity* of the particle. Because of the extraordinary empirical fact that the set of all fermions is coextensive with that of all leptons or baryons, as exhibited by the equivalence: half integral spin $\rightleftharpoons$ either B or L or $\ell \neq 0$, the totally neutral particles occur only among the bosons and only for these is the intrinsic parity unambiguously determined. Another way of obtaining the same result follows from the fermion side of the above empirical equivalence. The

reason why a boson can have an unambiguous parity (relative to the conventionally chosen parity of the vacuum, $\mathcal{P} = +1$) is that the continuous function describing it obeys the "Columbus principle," i.e., that a $360°$ movement of the observing apparatus around an axis restores the original description. Because the description of a one-fermion system does not obey this principle, the intrinsic parity of a one-fermion system is not uniquely defined.

A fermion intrinsic parity is, however, often assigned conventionally, as for the nucleon family where it is taken to be $+1$. More generally, as a consequence of the isospin and SU_3 groupings of particle types with respect to hadronic interactions (see later discussion), it is natural to assign those in the SU_3 multiplet the same intrinsic parity. It is important to emphasize however that the choice of relative parity between two superselectively separated particle types (e.g., neutron and proton) is only conventional. This is because any two physical states ϕ and ψ lying in different superselective sectors can not cohere into a physical state, and the behavior of the relative sign of ϕ with respect to ψ under a space inversion cannot be deduced since there is no interference to be observed. It is true however that within each superselective sector a choice, albeit conventional, of the intrinsic parity of one particle, A, determines that of any other particle, B, according to the rule derivable from the transitivity property

(Intrinsic Parity)$_B$

$=$ (Relative Parity)$_{B/A}$ (Intrinsic Parity)$_A$

By its nature as a phase factor—the eigenvalue of a discrete symmetry operation rather than of a generator of a continuous symmetry operation—the parity of a composite system (product state) is the product of the parities of the parts. Thus given the parities of the parts the parity of the composite is determined, but there are cases where the converse is not true: other than in totally neutral cases, particle states are prevented from combining into states (coherent superpositions) with the antiparticle states by the different values—of one or more superselective charges. This has the consequence that the intrinsic parity of the antiparticle with respect to that of the particle is not defined. In the case of a system made up of a particle and its own antiparticle (which is therefore always equivalent to a totally neutral boson and consequently in the same superselective sector as the vacuum) the intrinsic parity of the *pair* is however defined. In the case where the pair consists of a boson and its antiboson the intrinsic parity of the system is $+1$. (If the boson itself is totally neutral then of course it and the antiboson must have the same parity, either $+1$ or -1; but if it is not totally neutral it can be assigned $\mathcal{P} = e^{i\alpha}$, and then the antiboson must be assigned $\mathcal{P} = e^{-i\alpha}$; α can be arbitrary, e.g., $\alpha = \pi/2$ so that boson and antiboson could be assigned opposite parity.) The intrinsic parity of a given fermion-antifermion pair (constituting, therefore, a one-boson system) is, however, always -1. (If the conventional assignment of $\mathcal{P} = +1$ is made for the nucleon, then $\mathcal{P} = -1$ for the antinucleon; but if $\mathcal{P} = e^{i\phi}$ for the nucleon, then $\mathcal{P} = e^{i(\pi-\phi)}$ for the antinucleon—and ϕ can be arbitrary, e.g., $\phi = \pi/2$ so that both nucleon and antinucleon have the same parity, i.

The general result that $\mathcal{P} = +1$ for a boson particle-antiparticle pair and $\mathcal{P} = -1$ for a fermion particle-antiparticle pair is derivable theoretically by applying the assumption of analyticity, in the sense of complex function theory, to the description of scattering processes. This is done in the manner indicated later, in connection with the derivation of the antiparticle existence theorem, by analytic continuation of the scattering amplitude for the process

$$a + b \rightarrow a + c$$

to a region where it corresponds to the process

$$b \rightarrow \bar{a} + a + c.$$

It is confirmed observationally by selection rules which hold in various pair annihilation processes.

We also have some observables associated with approximate internal symmetries in processes in which only the strong interactions are important: The hypercharge Y, isospin I, and isoparity G are observables characterizing certain isomultiplets of nearly coincident mass but differing electric charge into which the mesons or baryons group. These isomultiplets are of the internal symmetry group SU_2. The members of each isomultiplet have the same baric charge and differ by consecutive values of the electric charge Q. The hypercharge Y is the sum of the lowest and the highest electric charges in the isomultiplet. The isospin magnitude I is a number which measures the multiplicity of the isomultiplet as $2I + 1$. The full significance of Y comes in a further grouping of these SU_2 isomultiplets into hypermultiplets belonging to the higher symmetry group SU_3 which has been found very successful. The isomultiplets within each hypermultiplet still have the same baric charge B but differ by consecutive values of the hypercharge Y. Further suggested higher internal symmetries have been found partially successful.

In the hadronically relevant isomultiplets (associated with an SU_2 group) the generator I_3 is an observable which is conjugation-reversing. The hypercharge Y which, under purely hadronic interactions, is a generator of an SU_1 symmetry group, is again a conjugation-reversing observable, i.e., behaves like a "charge." Both of these results are confirmed as fol-

lows. There are three generators I_1, I_2, I_3 for the isospin group SU_2. This group has identically the same Lie algebra—and therefore the same theory—as spatial angular momentum. By convention the observable component—represented by a diagonal matrix—is taken to be I_3. The ladder of diagonal elements has of course the three properties found with angular momentum: (1) It is equal-runged; (2) it is symmetrical about zero—thereby specifying integral or half-integral units of isospin as eigenvalues; (3) it is finite, i.e., begins with $(I_3)_{min} = -I$ and ends with $(I_3)_{max} = I$. The operational significance of the observability of I_3, i.e.,

$$I_3 = Q + \text{const}$$

plus the definition of Y as $Q_{max} + Q_{min}$, then leads to the Gell-Mann-Nishijima relation

$$Q = I_3 + Y/2.$$

Clearly, along with Q, I_3 and Y are conjugation-reversing.

The invariant isospin I of an isomultiplet is, like spin s to which it is analogous, conjugation-invariant. Under appropriate conditions there is also an *isoparity* G which, like space parity $\mathcal{P}$, is conjugation-invariant. G is defined unambiguously only for hadronically interacting systems for which $B = Y = 0$ (and of course $L = \ell = 0$), i.e., for hadronic systems which, "neutral" in every other respect, need not be electrically neutral. All such systems can be connected directly or through a sequence of intermediate systems to a basic number of pions; the latter are stable against hadronic decays. The isoparity G equals $+1$ or -1 according to whether the basic number of pions is even or odd. Individual pions then have odd isoparity and, as its name and the rule indicate, isoparity is multiplicative for a composite system. Among mesons, G can be defined only for the $Y = 0$ "nonstrange" isomultiplets for which I is integer because $Q_{max} = -Q_{min} = I$ cannot be half-integral insofar as has been established for any actual particle. (In practice such isomultiplets are either isosinglets or isotriplets since in none of the well established meson systems does I exceed 1.) There are 17 such hadronically unstable mesons for which G is very well known; in every case it is found to be conserved in the purely hadronic decays of these particles. (*Review of Particle Properties*, April 1973). G is violated in other interactions as, for instance, in the decay of the hadronically stable lightest η meson (550 Mev) into three pions by virtual electromagnetic interactions.

In the generalization to SU_3, with the grouping of isomultiplets to hypermultiplets, B remains an invariant while Y which was (a non-Casimir operator) invariant for SU_2 becomes a (covariant) generator. The electric charge Q which was originally the unique generator—and therefore, trivially, the Casimir invariant—of an SU_1 group, and became a displaced generator of the isospin SU_2 group through the Gell-Mann-Nishijima relation, becomes a regular generator in the SU_3 group.

To sum up then, the set of type-defining observables includes (in the massive case) mass m, spin s, parity $\mathcal{P}$ (with certain qualifications in fermion systems), and also electric, baric, muon- and electron-leptic charges (Q, B, L, ℓ) and (with certain qualifications) hypercharge Y, isoparity I, isoparity G, and SU_3 rep labels p and q. In the massless cases occurring in nature the type-characterizing observables comprise besides helicity λ, parity $\mathcal{P}$, isoparity G, and muon- and electron-leptic charges L and ℓ (the latter three with qualifications). The values taken by these observables classify all existing entities into particle types.

VI. Derivation of Antiparticle Existence and Conjugation Principles from Relativistic General Quantum Theory; Other Results We turn now to a discussion of proposed theoretical derivations of the antiparticle principle and the conjugation principle from very general theory with a minimum of supplementary features.

Successful theories in physics are often built in stages: (1) a *general theory* with room for further structure (thermodynamics, Maxwell's *E-M* theory without constitutive relations, etc.); (?) more particular theories filling out the structure. The primary general theory of elementary particles and antiparticles is *relativistic general quantum theory*, and the two more particular theories which fill out, in alternative ways, what is to a large extent the same structure are: (1) *general quantum field theory* and (2) its somewhat stripped-down version known as *S-matrix theory*. (In taking these theories to be adequate we are trusting that the problems of making them applicable universally, i.e., also to infinite range interactions—zero mass quanta—as in electrodynamics, can be resolved by appropriate procedures or reformulations, e.g., algebraic quantum theory. It is to be noted that difficulties associated with the possible occurrence of zero mass particles are common to both the S-matrix theory and the field theory.)

The formal ingredients of relativistic general quantum theory, which is the common basis of the two more particular theories, are:

(I) Relativistic elements expressing: (A) continuous *spacetime symmetry or invariance* of natural processes under the Poincaré (inhomogeneous Lorentz) group of translations, rotations, and uniform velocity shifts; (B) operational *possibility of distinguishing before and after* in natural processes (distinction between forward and backward light cones). (C) More specific than any of the preceding postulated elements is the assumption that no zero mass particles occur.

(II) Quantum mechanical elements expressing:

(A) the merely *statistical determination* in nature; (B) *the disturbance by the act of observation of physical systems*—which is related to mere statistical determination; (C) within the context of mere statistical determination, the *causal independence* of well separated experiments; (D) *analyticity of the collision or S-matrix elements* in its arguments.

(III) A somewhat more specific quantum mechanical assumption than the above refers to the operational *impossibility of counting, as distinct, particles of the same kind*.

(I) We have already discussed in some detail the manner in which the existence of spacetime symmetry allows us to define particle types and states.

(IIA) This general quantum mechanical principle takes the form of assuming the existence of a Hilbert space of *asymptotically free 1-particle momentum eigenstates*. A scalar product of these states with themselves and with each other is defined. The one-particle states may be added. When added they obey the *superposition principle*: Any linear combination (except for certain superselection restrictions) is itself a state. Conversely any given first state may be considered as a linear combination of a complete set of states built up beginning with any other second state. The probability of finding the second state immediately after establishing the first state is given by the square of the *probability amplitude* which is itself given by the scalar product of the second state with the first.

(IIB) Takes the form of recognizing that arbitrarily chosen observables are not necessarily compatible with each other. A *complete set of compatible observables* (C.S.C.O.) is any set needed to define the states uniquely, and the individual state is labeled by the set of values which the C.S.C.O. takes on for it. The states are represented as rays in a Hilbert space, the observables as linear operators. Compatibility (noncompatibility) of two observables is represented by commutativity (noncommutativity) of the corresponding two operators. In many instances two such operators anticommute, and this is how quantum theory describes the reversals of sign of certain properties under Θ as indicated above. Thus we have

$$\Theta Q + Q\Theta = 0 \quad \text{or} \quad \Theta Q\Theta^{-1} = -Q$$

$$\Theta B + B\Theta = 0 \quad \text{or} \quad \Theta B\Theta^{-1} = -B, \text{ etc.}$$

where the similarity transformation on the left of the second equation on each line represents the effect of conjugation on the corresponding charge operator.

(IIC) Besides the scalar product and the addition of one-particle states, one may also consider taking ordered outer products of one-particle states with the objective of representing multiparticle states. This is straightforward in the quantum field theory. In a somewhat old-fashioned but picturesque manner the quantum field may be described as a linear superposition of creation and annihilation operators. The field with its constituent operators and associated excitation states may be described as somewhat analogous to an idealized piano constructed with an infinitude of key-hammers (creation operators) and individual pedals (destruction operators) associated with each possible tone (excitation state); this tone can be excited to various degrees (corresponding to the number of particles in that state). All nonempty states are produced by the creation operators, acting on the *vacuum state*; the latter corresponds to the totally quiescent piano. In this formalism, in the simple case of noninteraction the basic states are *n-particle states, n* having all possible values. A specific *n*-particle state $|n_1, n_2, \cdots> \equiv |n>$ is characterized by the set of *occupation numbers* $n_1, n_2, \cdots (n_1 + n_2 + \cdots = n)$ specifying the numbers of particles having quantum numbers $q_1, q_2, \cdots$. The state $|0, 0, \cdots> = |vac>$ is called the *vacuum state*, and assumption (IC) excluding zero mass particles implies that there is a *mass-gap above the vacuum state*.

In the quantum field theory the assumption that corresponds to causal independence of well separated regions is that of locality, i.e., that field operators at spacelike-separated points commute. This, together with the assumption (IC) that zero mass particles do not occur, suffices to derive the existence and uniqueness of the set of *stationary collision states* in which arbitrary numbers of beams of stable particles collide and produce final stable particles as interaction products. These collision states define a unique *evolution-governing collision matrix* (*S*-matrix) whose matrix elements are the probability amplitudes for transitions between the asymptotic initial and final states at $t = -\infty$ and $t = +\infty$ (assumption (IB) tells us that we can distinguish between these "in" and "out" states). The collision matrix has the *"cluster decomposition"* or *"connectedness" property*; i.e., when particles interact the interaction (amplitude) is a sum of subinteractions (subamplitudes) by clusters, of particles two at a time, three at a time, etc., each cluster accompanied by a remaining set of noninteracting particles. The contribution to the *S*-matrix subamplitude for each noninteracting particle involves a Dirac δ-function expressing the constancy of momentum and energy of that particle, whereas the interacting clusters contribute combinations of amplitudes which are non-δ-functions with momentum-energy variables as arguments.

In the *S*-matrix approach, which seeks to circumvent the assumption of a local field, the existence of the *S*-matrix is postulated directly, with the cluster decomposition property as that one of its properties which expresses the causal independence of well separated experiments. This leads (Froissart and Taylor, 1967) to the representation of multiparticle or collision states by ordered outer products of one-particle states, such that these are given by the standard creation operator formalism of the field theory.

If in addition one more assumption is made: *asymptotic completeness*—that there are no additional states besides those that are superpositions of collision states—the quantum field theory also derives the *unitarity*, $SS^+ = S^+S = 1$, and certain *analyticity properties of the collision matrix*. The unitarity expresses the fact that the total probability summed over all processes is unity—the conservation of probability. Unitarity and analyticity have to be assumed ad hoc in the *S*-matrix approach, as the price of avoiding the locality assumption of the field theory.

The dependent variables of which the *S*-matrix elements are functions are prescribed by the "relativity" or "Lorentz-Poincaré invariance" of the theory (IA). This requires that the laws of nature be invariant under spacetime-interval preserving transformations. Here it means that any *S*-matrix element is a scalar under these transformations, or that the amplitude combinations corresponding to interacting clusters are functions of spacetime-invariant variables only, i.e., scalar products of the four-momenta of the interacting particles (or their sums and differences).

(IID) This assumption is that of *analyticity of the probability amplitudes* as functions of the invariant variables. As we have remarked, it is derivable—in a complicated way—from locality in the field theory, but it has to be introduced ad hoc in the sparser short-circuiting *S*-matrix approach. The assumption is sometimes called "causality," but it is probably more restrictive mathematically than what is implied by the causality concept.

(III) leads to a condition on the *statistics* associated with a given particle type. *Identical multiparticle states*, i.e., states containing two or more particles of the same kind, occur in nature only with *symmetry or antisymmetry* (of the mathematical function describing the multiparticle state) under interchange in any pair. This symmetry difference gives a major classification of all particle types; this is usually described as a classification according to "statistics" or "social behavior," since those with interchange symmetry—called *bosons*—tend to aggregate in the same momentum state, whereas those with antisymmetry—called *fermions*—tend to exclude each other from the same state.

From these general ingredients, either in the field theory version (Streater and Wightman,

1964; Jost, 1965), or in the stripped-down S-matrix version subject to a subtle proviso concerning analytic continuations (Olive, 1964; Lu and Olive, 1966; Froissart and Taylor, 1967), one can derive a number of remarkable general results:

1. *Hermitian analyticity.* Hermitian conjugate amplitudes are associated with the same analytic functions.

2. *Existence of antiparticles.* The argument by traditional field theory is detailed later. The argument as presented by the S-matrix protagonists is interesting and worth sketching. It depends on matching singularity structures of different subamplitudes when the cluster-structure is inserted in the unitarity relation $SS^* = 1$. This leads, e.g., to a relation between the amplitude A_{44} for the complete cluster interaction

$$A + B + C + D \rightarrow E + F + G + H$$

and the amplitudes A_{23}, A_{32} for the sequential interactions

$$A + B \rightarrow E + F + J \qquad C + D + J \rightarrow G + H.$$

The bubble diagrams and the equation below express this relation

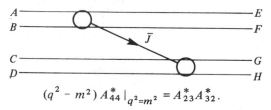

$$(q^2 - m^2) A_{44}^* \big|_{q^2 = m^2} = A_{23}^* A_{32}^*.$$

Left and right sides of this equation are analytic functions of their variables and one can continue the equation analytically to a different part of the physical region for A_{44}. In the new part of the physical region the continued pole represents the two successive interactions

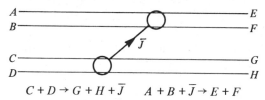

$$C + D \rightarrow G + H + \bar{J} \qquad A + B + \bar{J} \rightarrow E + F$$

The new particle $\bar{J}$ has the same mass as J but opposite internal charges (since they obey a rule of additivity). Thus the antiparticle principle is proved and becomes the *antiparticle theorem* (provided the analytic continuation brings us back to the "right side" of the physical region singularity of A_{44}^*).

3. *Crossing.* Since, under analytic continuation, residues also continue (provided there exists a path of analytic continuation in the complete amplitude joining the two points and

lying within the mass shell section—sufficient condition) the argument also demonstrates the property known as "crossing," i.e., that the amplitude for a process in which a given initial particle disappears without issue is the same function as for a process, otherwise identical, except that the corresponding antiparticle appears without antecedent:

$$\text{Ampl}\,(A + \cdots + N + J \rightarrow A' + \cdots + N')$$
$$\sim \text{Ampl}\,(A + \cdots + N \rightarrow \bar{J} + A' + \cdots N').$$

4. *$C\mathcal{P}T$ or Conjugation Theorem.* In the S-matrix approach to the $C\mathcal{P}T$ theorem, similar considerations of residues of single particle poles are used as in discussing the crossing theorem. The proviso, made in all the S-matrix arguments, concerning existence of "right paths" joining particle and antiparticle poles are related to the two basic facts used in the field theory proofs of the conjugation theorem:

a) vacuum expectation values of products of field operators are boundary values of analytic functions;

b) the spacetime inversion $x \rightarrow -x$ can be connected with the identity transformation through complex Lorentz transformations.

With the help of these two facts one can prove the existence of an antiunitary Θ, acting on fields, which has the properties of a symmetry. For example, for a scalar field $\Phi(x)$ the two basic facts a) and b) lead to the identities (ψ_0 is the vacuum state):

$$[\psi_0, \Phi(x_1) \cdots \Phi(x_n)\, \psi_0]$$
$$= [\psi_0, \Phi(-x_n) \cdots \Phi(-x_1)\psi_0]$$

for the vacuum expectation values of the products of field operators at any number of spacetime points $x_1, \cdots x_n$. Because of the reversal of order of operators, these identities are equivalent to the existence of an antiunitary Θ satisfying

$$\Theta\Phi(x)\Theta^{-1} = \bar{\Phi}(-x).$$

But then it follows that Θ maps stationary collision in-states onto stationary out-states of the corresponding antiparticles with the same momenta and opposite spin projection.

$$\Theta\psi_{\text{in}}(p_1 \cdots p_k) = \psi_{\text{out}}(p_1' \cdots p_k').$$

The symbol p_j represents the momentum p_j and spin of the jth particle of the colliding beams, and p_j' stands for the same momentum but opposite spin projection for the antiparticle. As an immediate consequence, we have the equation between the S-matrix element for the particle reaction $p_1 \cdots p_k \rightarrow p_1 \cdots p_m$ and the S-matrix element for the antiparticle reaction

$$\underline{p}_1' \cdots \underline{p}_m' \to p_1' \cdots p_k':$$

$$[\psi_{\text{out}}(\underline{p}_1, \cdots \underline{p}_m), \psi_{\text{in}}(p_1 \cdots p_k)]$$

$$= [\Theta\psi_{\text{out}}(\underline{p}_1 \cdots \underline{p}_m), \Theta\psi_{\text{in}}(p_1 \cdots p_k)]$$

$$= \overline{[\psi_{\text{in}}(\underline{p}_1' \cdots \underline{p}_m), \psi_{\text{out}}(p_1' \cdots p_k')]}$$

$$= [\psi_{\text{out}}(p_1' \cdots p_k'), \psi_{\text{in}}(\underline{p}_1' \cdots \underline{p}_m')]$$

5. *Spin-statistics theorem.* It is found in nature, so far without any demonstrated exception, that bosons have integral spin and fermions half-integral spin. This is the spin-statistics relation. It has been derived in field theory with various degrees of generality. (See Streater and Wightman, 1964, for a very general proof.) In the S-matrix approach (Lu and Olive, 1966), again provided certain plausible features are contained in the singularity structure, the relation between spin and statistics is established. It is found that connectedness—together with Lorentz invariance, unitarity, and analyticity—implies that the connected parts for processes involving particles with the wrong relation between spin and statistics vanish, so that such particles are unobservable.

Besides the behavior under Θ of the S-matrix and the type- and state-characterizing observables, we may be interested in the behavior under Θ of fields, and interactions between fields, which are also operators in quantum field theory. For all but the weak interactions, an additional symmetry over and above the universal Θ symmetry holds. This is what is traditionally known as "charge conjugation" and denoted by the symbol C. As we have already remarked, it corresponds to $\Theta \mathscr{P} \text{T}$ where $\mathscr{P}$ is the (unitary) space reflection operator, and T is the (antiunitary) time-inversion (more properly "motion-reversal") operator. Besides interchanging initial with final states, T reverses the signs of the three space-space components of $\underset{\sim}{J}$ (the angular momentum $\vec{M}$) and of the three space components of linear momentum $\vec{P}$; $\mathscr{P}$ reverses the signs of the latter and of the three spacetime components of J, the centroidal moments $\vec{\underset{\sim}{\mathfrak{M}}}$. The net result of combining T with Θ then is to undo the antiunitarity, restoring the original order of before and after, and also to undo the conjugations of all external state observables. Thus $\Theta \mathscr{P} \text{T} \equiv C$ is a kind of "internal conjugation operator."

VII. Charge Conjugation in Lagrangian Quantum Field Theory C-conjugation ("charge conjugation") plays an important role in the history and elementary discussions of fundamental particles. We here review its traditional treatment in the Lagrangian form of quantum field theory—a theory which was originally inspired by the analysis of oscillations and waves in classical physics where fields are simply numerical functions of space and time. The quantum version of the theory requires that the fields be operators. Also, the theory, necessarily a many-particle theory, is best described at first for the noninteracting case when we can take the basic states to be n-particle states, characterized by the set of *occupation numbers* $n_1, n_2, \cdots (n_1 + n_2 + \cdots = n)$ specifying the numbers of particles having quantum numbers $q_1, q_2, q_3, \cdots$ corresponding to a complete set of commuting observables (C.S.C.O.): $Q_1, Q_2, Q_3, \cdots$. The state $|0, 0, 0, 0, \cdots >$ with all $n_i = 0$ is the *vacuum state*. As in any vector space, there is also the *zero vector* 0. The single-step destruction operator a_α for particles, and creation operator b_α^* for antiparticles associated with the αth quantum number are introduced in the usual way (see FIELD THEORY) such that they satisfy the standard commutation (anticommutation) rules if the particles satisfy Einstein-Bose (Fermi-Dirac) statistics. One then introduces the *field operator*, which is a linear expansion in the a_α and b_α^* (all α), and which formally satisfies certain field equations, the latter usually being chosen on the basis of a classical analogy or on the basis of relativistic covariance and general agreement with experiment. All of the consequences of interest then follow from the definition of the field operator and from that of the C-conjugation operator:

$$Ca_\alpha C^{-1} = \eta_c b_\alpha \qquad Ca_\alpha^* C^{-1} = \overline{\eta}_c \, b_\alpha^*$$

$$Cb_\alpha^* C^{-1} = \eta_c a_\alpha^* \qquad Cb_\alpha C^{-1} = \overline{\eta}_c \, a_\alpha.$$

Here η_c is a phase constant ($|\eta_c| = 1$) which is nonmeasurable and can be chosen +1 by convention. If particle and antiparticle are identical then $b_\alpha = a_\alpha$, and η_c becomes measurable and equal to ± 1, two physically distinct cases.

For a specific theory, particularly if it is given in Lagrangian form in which case each of the symmetry observables is defined by a symmetry operation on the Lagrangian, these observables can be written as bilinear functionals in the field operators.

As an example, we consider the case of a scalar (spin zero) charged field $\Phi(x)$ associated with a mass m, and satisfying the Klein-Gordon equation

$$(\Box + m^2)\Phi(x) = 0$$

For the C.S.C.O. we choose the linear momentum and the (not-independent) energy. Denoting the destruction operator for the particle of momentum $\underset{\sim}{k}$ and energy $k_0 = \sqrt{\underset{\sim}{k}^2 + m^2}$ by a_k, and the creation operator for the corresponding antiparticle by b_k^*, the field operator is given by

$$\Phi(x) = \frac{1}{\sqrt{2(2\pi)^3}} \int \frac{d^3 k}{k_0} \left(a_k e^{-ikx} + b_k^* e^{ikx}\right)$$

An explicit representation for C is

$$C = \exp\left[\frac{\pi i}{2} \sum_k (a_k^* - \overline{\eta}_c\, b_k^*)(a_k - \eta_c b_k)\right]$$

and for the total energy-momentum operator, obtained for instance from the Lagrangian for the Klein-Gordon equation above, we have

$$P_\mu = \int \frac{d^3k}{k_0}\, k_\mu (a_k^* a_k + b_k^* b_k)$$

It is now possible by direct calculation to establish whether P_μ (or any other observable) commutes or anticommutes with C. In this way we can derive for all the free physical fields which have been considered applicable to nature, the results stated earlier in this article concerning preservation and reversal of signs of the fundamental quantum numbers.

It is possible in a similar way to examine the validity of invariance under C-conjugation of the *interactions* between fields. For instance, the electromagnetic interaction between the electron-field current j_μ and the photon-field potential A_μ is $j_\mu A_\mu$. The two results found for the separate fields by the methods described in the foregoing.

$$Cj_\mu C^{-1} = -j_\mu \qquad CA_\mu C^{-1} = -A_\mu$$

then guarantee that

$$Cj_\mu A_\mu C^{-1} = j_\mu A_\mu$$

Similarly, C-invariance holds for the accepted forms of strong interactions.

C-invariance *does not hold* for the V-A (polar vector minus axial vector),four-fermion type of weak interaction occurring, e.g., in β-radioactivity. As we have seen, however, for quite general interactions with *proper orthochronous* spacetime symmetry, provided they are *local*, there is automatically invariance under the combined operation $C\mathcal{P}T \equiv \Theta$ ($C\mathcal{P}T$ theorem). "Proper orthochronous" spacetime symmetry of the interactions means that they are invariant under rotations, translations in space and time, and shifts to uniformly moving frames: "locality," in practice, means that the interactions consist of a linear combination of products of the fields and finite-order derivatives of the fields. Equality of mass and lifetime, antiequality of charge and magnetic moment, and (with certain restrictions) conjugacy of decay schemes for a particle and its antiparticle all follow from $C\mathcal{P}T$ invariance alone. And this rests only on proper spacetime symmetry, general principles of quantum theory, and the locality requirement (or, in the S-matrix rendition, the requirements of connectedness and analyticity).

M. A. MELVIN

References

Dirac, P. A. M., "Quantized Singularities in the Electromagnetic Field," *Proc. Roy. Soc. London, Ser. A*, 133, 60 (1931).

Wigner, E. P., "Gruppentheorie" (Germany, Frederick Vieweg und Sohn, Braunschweig, 1931, pp. 251–254; "Group Theory," (New York, Academic Press, 1959, pp. 233–236.

Kadison, R., "Isometries of Operator Algebras," *Ann. Math.*, 54, 325 (1951).

Wick, G. C., Wightman, A. S., and Wigner, E. P., "The Intrinsic Parity of Elementary Particles," Phys. Rev., 88, 101 (1952).

Bargmann, V., *J. Math. Phys.*, 5, 862 (1964).

Wolfenstein, L., and Ravenhall, D. G., "Some Consequences of Invariance under Charge Conjugation," *Phys. Rev.*, 88, 279 (1952).

Lee, T. D., and Yang, C. N., "Elementary Particles and Weak Interactions," Office of Technical Services, Department of Commerce, Washington, D.C., 1957.

Melvin, M. A., "Elementary Particles and Symmetry Principles," *Rev. Mod. Phys.*, 32, 477 (1960).

Melvin, M. A., "Remarks on the Infinite Spin Case for Zero Mass Particles," *Particles and Nuclei*, 1, 34 (1970).

Jost, R., "TCP-Invarianz der Streumatrix und interpolierende Felder," *Helv. Phys. Acta*, 36, 77 (1963).

Jost, R., "General Theory of Quantized Fields," Amer. Math. Soc. Publications, 1963.

Olive, D., "Exploration of S-Matrix Theory," *Phys. Rev.*, 135, B 745 (1964).

Streater, R. F. and Wightman, A. S., "PCT Spin and Statistics and All That," New York, Benjamin, 1964.

Ekstein, H., "Rigorous Symmetries of Elementary Particles," *Ergebnisse der exakten Naturwissenschaften*, 37 (Berlin, Springer, 1965).

Lu, E. Y. C., and Olive, D. I., "Spin and Statistics in S-Matrix Theory," *Nuovo Cimento*, 45, 205 (1966).

Froissart, M., and Taylor, J. R., "Cluster Decomposition and the Spin-Statistics Theorem in S-Matrix Theory," *Phys. Rev.*, 153, 1636 (1967).

Wightman, A. S., "What is the Point of So-Called 'Axiomatic Field Theory?' " *Physics Today*, p. 53 (September 1969).

Armenteros, R., and French, B., "Antinucleon-Nucleon Interactions," in "High Energy Physics," Vol. IV, New York, Academic Press, 1969.

Lasinski, Thomas A. et al., "Review of Particle Properties," Rev. Med. Phys., 45, Supplement, 51 (Apr 1973).

Shapiro, I. S., "The Physics of Nuclear-Antinuclear Systems," *Physics Reports* 35, 129 (1978).

Perl, M. L., "The Tau Lepton," *Annual Reviews of Nuclear and Particle Science*, 30, 299 (1980).

Berko, S. and Pendleton, H. N., "Positronium," *Annual Reviews of Nuclear and Particle Science* 30, 543 (1080).

Fry, J. N., Olive, K. A., and Turner, M. S., "Evolution of Cosmological Baryon Asymmetries I. The Role of Gauge Bosons" and ". . . II. The Role of Higgs Bosons," *Phys. Rev.* D 22, 2953 and 2977. These papers contain a summary and detailed further analysis of the suggestions made by earlier investigators to account—on the basis of standard cosmology and the SU_5 Gauge Unified Theory of quarks and leptons (see GRAND UNIFIED THEORIES)—for the: a) matter content ($10^{-9 \pm 1}$ baryons per photon), b) matter-antimatter asymmetry (practically no naturally occurring and antibaryons) in the universe.

Cross-references: BOSE-EINSTEIN STATISTICS AND BOSONS; CONSERVATION LAWS AND SYMMETRY; ELECTRON; ELEMENTARY PARTICLES; FERMI-DIRAC STATISTICS AND FERMIONS; FIELD THEORY; GRAND UNIFIED THEORIES; PARITY; POSITRON; QUANTUM THEORY; RELATIVITY.

ARCHITECTURAL ACOUSTICS

Although the practice of architectural acoustics involves a wide variety of special problems and techniques, the basic reasons for acoustical design are simply:

(a) to provide a satisfactory acoustical environment, not too noisy and often not too quiet, for people at work and relaxation;

(b) to provide good hearing conditions for speech; and

(c) to provide a pleasant acoustical environment for listening to music.

Designing for Satisfactory Acoustical Environment Each acoustical situation must be treated as a system comprised of three parts; source, transmission path, and listener. When the properties of the source are known, the transmission path can be modified to attenuate the sound to suit the listener's needs.

Sources. Noise sources are specified in terms of the total acoustical power radiated in each of a number (generally between 8 and 25) of contiguous frequency bands.[1,2] A standard set of ten bands is listed in Table 1.

Because of the wide range of sound powers encountered in practice, it is customary to express them in a logarithmic form. Thus we speak

TABLE 1. STANDARD OCTAVE FREQUENCY BANDS

Lower and Upper Frequency Limits of Each Band (Hz)		Geometric Mean Frequency of Each Band (Hz)
22.1–	44.2	31.5
44.2–	88.5	63
88.5–	177	125
177 –	354	250
354 –	707	500
707 –	1,414	1,000
1,414 –	2,828	2,000
2,828 –	5.655	4,000
5,655 –	11,310	8,000
11,310 –	22,620	16,000

of the strength of a sound source in terms of *sound power level*, W, in decibels, defined by item 1 in Table 2.

We note that sound power W is expressed in watts.

A listener does not experience the total sound power from a source, since it radiates in all directions, but rather the proportion that arrives at his ear. Thus we speak of *sound intensity*, I, as the sound power passing through a small area at the point of observation. The units are watts per square centimeter or per square meter. *Sound intensity level*, L_I, in decibels, is defined by item 2 in Table 2.

Until recently there has been no commercially available instrument for measuring sound intensity, so it has usually been determined indirectly from the mean-square sound pressure, p^2, i.e., the time average of the square of the instantaneous sound pressure in the acoustic wave. This quantity can be determined readily with a pressure microphone. The relation is given by:

$$I = p^2 / \rho c \text{ watts/m}^2 \qquad (1)$$

where, ρ is the density of air (or other gas) in kilograms per cubic meter and c is the speed of sound in air in meters per second. *Sound pressure level*, L_p, in decibels, is defined by item 3 in Table 2.

Instruments and techniques for the measurement of sound pressure levels are widely available.[2] Typical measured values of sound power levels for many sources are given in references 1 to 3.

Paths. Sound may travel from a source to a receiver by many paths, some in the air (outdoors or in a room), some through walls, and some along solid structures. In the latter two cases, the sound is radiated into the air from the vibrations of the surfaces.

Outdoors, the relation between the sound pressure level measured at distance r from a source and the sound power level of the source is given by,

$$L_{p\theta} \doteq L_w + DI_\theta - 20 \log_{10} r - 11 \text{ dB} \qquad (2)$$

where it is assumed that the source is near a hard-ground plane at a distance r in meters from the receiver (also near the plane) and that the source produces different sound intensities in different directions, θ, as described by a *directivity index*, DI_θ (see reference 2). If the source radiates sound equally in all directions, then $DI = 0$. In practice, sources generally have directivity indexes in the range of 0 to 12 dB in the direction of maximum radiation. At large distances, r, there will be losses in the air itself at frequencies above 1500 Hz. Also, wind, temperature gradients, and air turbulence may reduce or augment the value of $L_{p\theta}$ determined from Eq. (2).

TABLE 2

Decibel Scale	Abbreviation	Reference Quantity	Definition	
Sound power level	L_W	$W_{ref} = 10^{-12}$ watt	$10 \log_{10} \dfrac{W}{W_{ref}}$	dB
Sound intensity level	L_I	$I_{ref} = 10^{-12}$ watt/m^2 $= 10^{-16}$ watt/cm^2	$10 \log_{10} \dfrac{I}{I_{ref}}$	dB
Sound pressure level	L_p	$p_{ref} = 20$ micropascal $= 0.0002$ microbar	$10 \log_{10} \dfrac{p^2}{p^2_{ref}}$ $= 20 \log_{10} \dfrac{p}{p_{ref}}$ dB	

In a room, the sound pressure level produced by a nondirective source is given by,

$$L_p \doteq L_W + 10 \log_{10} \left(\frac{1}{4\pi r^2} + \frac{4}{R} \right) \text{dB} \quad (3)$$

where r is the distance between the receiver and the microphone and R is the *room constant* in square meters. $R = S\bar{\alpha}$, where S is the total area of all wall, ceiling and floor surfaces in square meters and $\bar{\alpha}$ is the average absorption coefficient for the whole room (see Reference 2). Typical values of R are found in Fig. 1.

In practical cases, we are often interested in the sound pressure level produced in a room separated by a partition (wall) from a room in which the source is located. We assign a *transmission loss, TL*, in decibels to the intervening wall. Curves of transmission loss versus frequency for several different building structures are given in Fig. 2 and reference 4. The equation relating

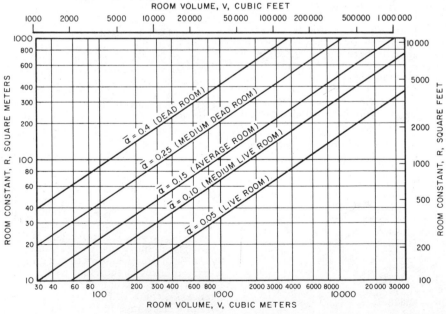

FIG. 1. Approximate value of room constant R for five categories of rooms ranging from "live" to "dead." Metric units referenced at bottom and left, English units top and right. The Greek letter $\bar{\alpha}$ indicates the percentage of the energy that is removed from a sound wave when it reflects from an "average" surface of the room. It is called the average sound absorption coefficient.

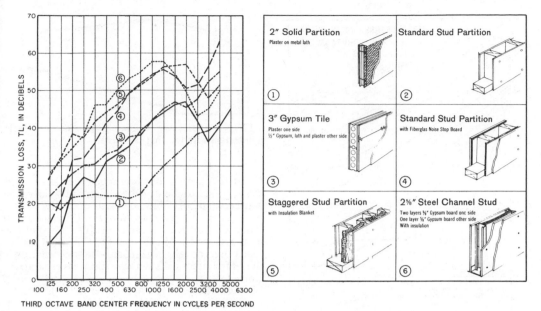

THIRD OCTAVE BAND CENTER FREQUENCY IN CYCLES PER SECOND

FIG. 2. Transmission loss, TL, of six typical building structures.

L_p to the L_W of the source* is,

$$L_{p_2} = L_{W_1} - TL + 10 \log_{10} \left(\frac{S_w}{R_1} \right)$$

$$+ 10 \log_{10} \left(\frac{1}{S_w} + \frac{4}{R_2} \right) \text{dB} \qquad (4)$$

where L_{p_2} is the sound pressure level in the second room produced by a source in the first room; TL is the transmission loss of the common wall; S_w is the area of the wall in square meters, and R_1 and R_2 are the room constants for the first and second rooms respectively, in square meters. It is assumed that the sound pressure level is measured near the common wall; in the center of the room it may be 3 to 5 dB lower.

Technique. It is apparent from Eq. (4) that the techniques for noise reduction indoors are threefold. First, make every effort to reduce the sound power radiated by the source, i.e., use quiet ventilating fans, quiet typewriters, quiet factory machinery, and so forth. Enclose noisy machinery in separate rooms or in enclosures. Mount vibrating machinery on resilient pads or springs. Second, provide walls with suitably high transmission losses between rooms. For example, between adjoining apartments, walls 4 to 6 of Fig. 2 are usually satisfactory, while walls 1 to 3 are not. On the other hand, walls 2 and 3 would be satisfactory between rooms of the same

*When the only sound path between the two rooms is through the common wall.

apartment, while wall 1 would not. Finally, increase the room constants by adding sound-absorbing materials to either or both rooms, e.g., carpets and draperies, or acoustical materials on ceiling or walls or both. The sound-absorbing efficiencies of various materials are given in references 2, 3, and 5.

It is of great importance to observe that when a wall is placed between two rooms or when an enclosure is built around a noisy machine, the structure must be hermetically sealed, or, if airflow is necessary, it must be conducted in and out of the enclosure through suitable silencers. A hole even as small in diameter as a pencil can render an otherwise satisfactory wall or enclosure acoustically inadequate.

In cases of very high noise levels where it is acoustically impractical to quiet or isolate the machine, then ear plugs, ear cushions, or both, must be worn by personnel exposed to the noise.

Criteria for Design. Acceptable noise levels in rooms of various types in each of eight octave frequency bands are shown by Fig. 3 and Table 3.

Auditoriums for Speech Three goals must be met in the design of auditoriums for speech. First, the ambient noise levels must be sufficiently low (see Table 3). Second, speech must be loud enough in all parts of the room so that faint syllables can be heard in the presence of normal audience noise. This second goal is achieved in small auditoriums (under about 500 seats) by proper shaping of the front part of the hall so that the speaker's voice is directed uniformly to all parts of the hall. In large halls (over 500 seats), electronic amplification of

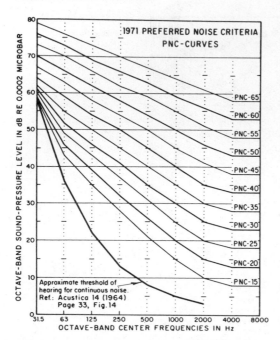

FIG. 3. Preferred Noise Criteria (PNC) curves for various types of building spaces given in Table 3. Measurements are made with an octave band filter and the readings in each band should not exceed that shown on the appropriate PNC curve.

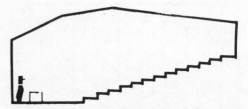

FIG. 4. Satisfactory ceiling shape for a speech auditorium with less than 500 seats.

(That is to say, a loud sound should take about 1.5 seconds to die down to inaudibility after its source is cut off abruptly. This quantity, *called mid-frequency reverberation time*, is measured with full audience present at 500 to 1000 Hz and averaged.) Music of the Classical period (early Beethoven, Mozart and Haydn) was composed for larger halls with medium reverberation times (about 1.8 sec). On the other hand, music of the Romantic period (after 1850), was in general, composed for fairly large halls with long reverberation times (about 2.0 sec). Today, halls must not only accomodate a musical repertoire extending over centuries, but often they must seat so large an audience that they become an entirely new type of space in which to perform music.

In the development of the design of a hall, the acoustics dictate the cubic volume and strongly influence the orientation of every sound-reflecting surface, the interior materials, and even the seating.

Concert hall and opera house design is complex,[6,7] but some guiding principles stand out. The seating capacity should be low, below 2200 if possible. The ceiling should have an average height of 45 ft, if there are no balconies, or 55 ft with balconies, measured above the floor beneath the main floor seats. The hall should be narrow, or other means such as suspended panels should be provided for producing early sound reflections at listener's positions. Finishes for the interior should primarily be plaster on lath. Not

speech is usually necessary. Third, the reverberation in the auditorium should be sufficiently low that speech is distinct. In auditoriums where there is no sound system, this requirement means that either the ceiling should have an average height of less than 30 ft above the main floor, assuming that the seats are upholstered and that there are no large floor areas without seats. If the ceiling height is over 30 ft, sound-absorbing materials will have to be added to the walls, and perhaps to the rear ceiling to control the reverberation. A satisfactory shape of a 500-seat auditorium for unamplified speech is shown in Fig. 4.

Auditoriums for Music There appears to be no single, ideal architectural solution for the acoustical design of a hall for music. Successful acoustics have been achieved with rectangular, fan or wedge, horseshoe, and even asymmetrical, plans. But though this is true, the many attributes of musical-architectural acoustics are so closely interrelated that if a hall is to be successful, the architect must solve all requirements simultaneously.

The music of each era of the past was composed for a different acoustical environment. Music of the Baroque period (Bach and earlier), except for organ music, was composed for small halls with relatively short reverberation times.

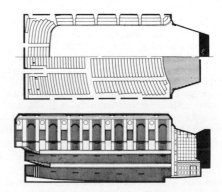

FIG. 5. Drawings of Symphony Hall, Boston, Mass.

TABLE 3   RECOMMENDED CATEGORY CLASSIFICATION AND SUGGESTED
NOISE CRITERIA RANGE FOR STEADY BACKGROUND NOISE AS HEARD IN
VARIOUS INDOOR FUNCTIONAL ACTIVITY AREAS.

Type of Space (and acoustical requirements)	PNC Curve*	Approximate L_A, dBA
Concert halls, opera houses, and recital halls (for listening to faint musical sounds)	10 to 20	21 to 30
Broadcast and recording studios (distant microphone pickup used)	10 to 20	21 to 30
Large auditoriums, large drama theaters, and churches (for excellent listening conditions)	Not to exceed 20	Not to exceed 30
Broadcast, television, and recording studios (close microphone pickup only)	Not to exceed 25	Not to exceed 34
Small auditoriums, small theaters, small churches, music rehearsal rooms, large meeting and conference rooms (for good listening), or executive offices and conference rooms for 50 people (no amplification)	Not to exceed 35	Not to exceed 42
Bedrooms, sleeping quarters, hospitals, residences, apartments, hotels, motels, etc. (for sleeping, resting, relaxing)	25 to 40	34 to 47
Private or semiprivate offices, small conference rooms, classrooms, libraries, etc. (for good listening conditions)	30 to 40	38 to 47
Living rooms and similar spaces in dwellings (for conversing or listening to radio and TV)	30 to 40	38 to 47
Large offices, reception areas, retail shops and stores, cafeterias, restaurants, etc. (for moderately good listening conditions)	35 to 45	42 to 52
Lobbies, laboratory work spaces, drafting and engineering rooms, general secretarial areas (for fair listening conditions)	40 to 50	47 to 56
Light maintenance shops, office and computer equipment rooms, kitchens, and laundries (for moderately fair listening conditions)	45 to 55	52 to 61
Shops, garages, power-plant control rooms, etc. (for just acceptable speech and telephone communication). Levels above PNC-60 are not recommended for any office or communication situation	50 to 60	56 to 66
For work spaces where speech or telephone communication is not required, but where there must be no risk of hearing damage	60 to 75	66 to 80

*See Reference 2 for other rating curves.

over 20 percent of them should be wood if the strength of the bass tone is to be preserved. Irregularities on all the surfaces should be provided to produce diffusion and blending of the sound. Above all, avoid echo, noise and tonal distortion. Finally, the orchestra enclosure should provide sectional balance in the orchestra and permit the musicians to hear each other.

Boston Symphony Hall, one of the world's best-liked concert halls, is rectangular, as shown in Fig. 5, and meets the general requirements listed above. Its mid-frequency reverberation time, with full audience, is 1.8 sec.

LEO L. BERANEK

References

1. Peterson, A. P. G., and Gross, E. E., Jr., "Handbook of Noise Measurement," Gen Rad Co., West Concord, Mass. (Ninth Ed.) 1980.
2. Beranek, L. L., Ed., "Noise and Vibration Control," New York, McGraw-Hill Book Co., 1971.

3. Harris, C. M., Ed., "Handbook of Noise Control," New York, McGraw-Hill Book Co. (2nd Ed.) 1979.
4. "Performance Data–Architectural Acoustical Materials," Acoustical Materials Association, New York, N.Y., published annually (A.I.A. No. 39–B).
5. "Solutions to Noise Control Problems in the Construction of Houses, Apartments, Motels, and Hotels," Owens-Corning Fiberglas Corp., Toledo, Ohio, 1963 (A.I.A. No. 39–E).
6. Beranek, L. L., "Music, Acoustics, and Architecture," New York, John Wiley & Sons, Inc., 1962.
7. Furrer, W., "Room and Building Acoustics" (translated by E. R. Robinson and P. Lord), London, Butterworths, 1964. W. Furrer and A. Lauber, "Raum- und Bauakustik-Laermabwehr" (3rd Ed.) Birhauser Verlag, Basel, Switzerland, 1972.
8. Kinsler, L. E., and Frey, A. R., "Fundamentals of Acoustics," New York, John Wiley & Sons, Inc., 2nd Ed., 1962.
9. Beranek, L. L., "Acoustics," New York, McGraw-Hill Book Co., 1949.

Cross-references: ACOUSTICS; HEARING; MUSICAL SOUND; NOISE, ACOUSTICAL; PHYSICAL ACOUSTICS; RESONANCE; VIBRATION.

ASTROMETRY

Astrometry deals with the space-time behavior of celestial bodies and therefore belongs to the classical field of astronomical studies. It is often referred to as fundamental, positional or observational astronomy.

Early astrometric investigations were directed mainly toward establishing a suitable frame of reference for the determination of the complex motions of the planets, while the studies of the positions and motions of the individual stars as well as the various stellar systems gradually developed as improved precision of observations made it possible to discover and observe these motions.

The fundamental, and perhaps most difficult, problem of astrometry is the establishment of a reference system against which the motions of the celestial bodies can be measured.

The principal planes involved in the spherical coordinate systems usually used in astrometry are the equator, defined by the rotation of the earth on its axis, and the ecliptic, defined by the revolution of the earth around the sun. The positions of both these planes vary continuously in a most complicated manner due to gravitational forces and couples between earth and the moon, the sun, and the principal planets. Such motions of the reference planes are reflected in the positions of the stars referred to them.

The motions of these planes cannot be derived entirely from theory alone, but must be deduced from observed changes in the positions of the stars which, in turn, are also in motion. This complication has forced the construction of the astronomical coordinate system to proceed by

a series of successive approximations which are still in progress. Initially, the sun, and planets were observed against the "fixed stars." From these observations came the first approximations of the motions of the solar system by the laws of dynamics and of the effect of the changing orientation of the earth's axis of rotation (precession) upon the positions of the stars. Successive repetitions of the observational process have gradually improved our knowledge of these and other motions affecting the fundamental planes of the coordinate system, each improvement resulting in an increase in our knowledge of the positions and motions of the stars.

Observational programs for the improvement of the celestial coordinate system are long and tedious and must be conducted with meticulous care. They make use of highly developed instruments and observing techniques which, in combination with adopted theories of the rotation of the earth and its motion around the sun, enable the positions of the equator and equinox to be derived anew and the positions of the stars to be related to them. Each such program is an independent effort to reconstruct the celestial coordinate system. Meridian circles have generally been used for this kind of work. The results of such programs are said to be fundamental and are usually published in the form of star catalogs.

From time to time, when sufficient fundamentally observed catalogues have accumulated, they are combined with similar earlier material to form a *Fundamental Star Catalog*. This catalog is usually regarded as the best representation that may be had of the celestial coordinate system at the time of its publication; the right ascensions and declinations of the stars in the catalog define the system for the equinox and epoch chosen for the catalog. The proper motions in combination with the adopted values of the constant of precession permit the system to be referred to equinoxes and equators at other epochs.

The latest and most precise of the fundamental catalogs is designated the *FK4* and was published by the Astronomischen Rechen-Instituts, Heidelberg, Germany in 1963. The catalog contains the positions (right ascension and declination) and the changes with time (precession and proper motion) of 1535 stars. These data were compiled from nearly 200 star catalogs containing observations over a span of 110 years. Its successor, the *FK5*, will appear in 1984, with data compiled from more than 250 catalogs of approximately 3 million observations. Two other fundamental catalogs that have been extensively used are the *GC, Albany General Catalogue of 33342 Stars* and the *N30, Catalog of 5268 Stars*.

The coordinate system provided by the positions and motions of the stars in a fundamental catalog serves as a reference system for the measurement of other star positions and proper motions which must be carried out for a variety of problems originating in the study of stellar motions, in geodesy, in the determination of time, in space research and others.

With the exception of the *GC*, which contains all the stars brighter than the 7th magnitude, fundamental catalogs do not contain a complete list of all stars down to a certain magnitude as, for example, the survey catalogs do. The prototype of the survey catalogs for star positions is the *Bonner Durchmusterung* which contains the positions of 320000 stars to a limiting magnitude of 9.5 and north of declination $-2°$. Although the observations for the catalog were made in the middle of the past century, the catalog and the charts made from it have been an extremely useful tool for astronomers for identification of star fields. The survey was later extended to the south celestial pole by the Bonn, Cape and Cordoba Observatories.

Positions of the fainter stars on a fundamental system are obtained by a close coordination between visual and photographic programs. The positions of a selected number of moderately bright stars (7th to 9th magnitude) are related to the fundamental system by meridian circle observations. These stars are then used as a position reference for the photographic observations of the fainter stars, thus tying them to the fundamental system.

An example of this procedure is the large astrometric project initiated toward the end of the nineteenth century and carried out by international cooperation.

The fundamental system adopted for this undertaking was embodied in the *FC* (*Fundamental-Catalog für die Zonen-Beobachtungen am Nördlichen Himmel*) developed by Auwers. The visual program, designated the *AGK* (*Astronomische Gesellschaft Katalog*), was carried out through the collaboration of 12 northern hemisphere observatories and resulted in the determination of the positions with respect to the *FC* of 144128 stars to the limiting magnitude of 9 and north of $-2°$ declination. The extension of the visual work into the southern skies was gradually carried out by other observatories. The adjunct photographic program known as the *Carte du Ciel* or the *Astrographic Catalogue* called for observations down to approximately the 11th magnitude covering the entire sky by $2° \times 2°$ fields. Originating in 1887, the program was completed by 1970 and involved the participation of 18 different observatories. The positions in the catalogs are given in the form of rectangular coordinates as measured on the plates, but by means of auxiliary tables, these coordinates can be translated into right ascension and declination. Each field was photographed a second time with a longer exposure with a limiting magnitude of 14 to be used for the purpose of star charts.

Several other catalogs of photographically derived positions have been published. Among the catalogs of this nature may be mentioned the *AGK2* (*Zweiter Katalog der Astronomischen Gesellschaft*) and the Yale and Cape photographic catalogs.

The *AGK2* was related to the fundamental system represented by the *FK3* (*Dritter Fundamentalkatalog des Berliner Astronomischen Jahrbuchs*) through the use of simultaneous visual observations of about 13000 moderately bright stars in making the plate reductions. The *AGK2* plates were taken at the Bonn and Bergedorf Observatories and covered the sky in $5° \times 5°$ overlapping fields from $-2°$ to the north pole. The resulting catalog contains the positions of over 180000 stars for the mean epoch of 1930. Between 1956 and 1959 a second photographic series of observations of these stars was carried out at the Bergedorf Observatory. The measured positions on the plates were reduced to the system of the *FK4* by use of the positions of some 21000 reference stars observed simultaneously through an international cooperative program involving 12 meridian circles in the northern hemisphere. A comparison of the plate results at the two epochs gives rather accurate proper motions with respect to the fundamental system for the entire 180000 stars. The majority of these stars are brighter than the 9th magnitude. A good many, however, are as faint as the 11.5 photographic magnitude. The catalog is named the *AGK3*. Except for gaps between $+85°$ to $+60°$, $+50°$ to $+30°$, and $-50°$ to $-60°$, the Yale Zone Catalog covers the sky from $+90°$ to $-90°$ declinations, while the Cape catalogs provide a complete coverage from $-30°$ declination to the south pole. Both series of catalogs were taken by zones of declination by use of wide-angle cameras (from $5° \times 5°$ to $10° \times 14°$) and were reduced to a fundamental system (not always the same one) by use of contemporary meridian circle observations. The mean epochs of the positions in these catalogs range from the early 1930's to the late 1940's. The stars in these catalogs are similar in magnitude range to those in the *AGK2* and *AGK3*. A catalog especially prepared for geodesy and other computer oriented research based on accurate satellite orbits is the Smithsonian Astrophysical Observatory Catalog (SAO). It is derived from selected visual and photographic catalogs, contains positions and proper motions of 259,000 stars, and covers the entire sky in zones of 10-degrees-wide declinations. It has been published in book form and as a set of star charts, but like the other catalogs previously mentioned, it is also available in machine readable form from the Centre de Données Stellaires at the Observatoire de Strasbourg, and from NASA-Greenbelt, Maryland. A program of observing 20000 stars in the Southern Hemisphere with meridian circles is now nearing completion. Known as the *SRS* (*Southern Reference Star*) *Program*, it has been carried out through international cooperation with northern observatories (U. S. Naval Observatory, Pulkovo Observatory, and Hamburg-Bergedorf) participating with stations in Argentina, Chile, and Australia, respectively. The positions and proper motions of these stars in the fundamental system are intended for use in obtaining new positions and proper motions of some 200000 stars in the Southern Hemisphere observed photographically.

An important source for obtaining proper motions of the fainter stars is a combination of early photographic plates with recent ones taken with the same telescope. The proper motions derived in this way are relative proper motions and require further reductions for transformation into absolute proper motions in a fundamental system. This procedure has been followed in several extensive programs aimed at solving such problems as determining the solar motion, and deriving secular parallaxes and galactic rotation.

Proper motions for tens of thousands of stars have been obtained by this method while radial velocities for a lesser number of stars have been determined from the spectroscopic application of the Doppler principle. Besides the proper motion and radial velocity, the distance of a star is needed to determine its motion in space. For stars beyond 100 parsecs from the solar system it becomes increasingly difficult to obtain all three factors involved, and often knowledge of stellar motion is either based on proper motions or radial velocities alone. However, by various statistical devices substantial information about stellar motions has been obtained.

On the basis of these studies, the sun's velocity has been determined to be about 20 km/sec towards a point in space not far from Vega, although the amount and direction of the motion varies depending upon the chosen group of stars.

As a result of the sun's motion through space, the stars show a parallactic or secular shift which can be used to determine their distances. Because of the individual motions of the stars, this method is applicable only to groups of stars with the assumption that their individual motions are random. By means of the secular parallax method, general ideas of the distances of stars up to 1000 parsecs have been obtained.

From statistical studies of proper motions and radial velocities, it was found in 1927 that the stars in our galaxy are moving in orbits not greatly inclined to the galactic equator. The observations are consistent with the assumption that the principal force governing the motions is gravitational with the center of mass near the galactic center. The period of rotation at the sun's distance from the center is 2×10^8 years.

There are, at the present time, two programs in progress which will attempt to establish absolute stellar proper motions using the distant galaxies as a reference frame, the assumption being made that those objects do not show any systematic rotation with respect to the local inertial frame of rest.

A large number of proper motion studies of galactic clusters has been carried out in order to establish membership of the individual stars in the field. Because of the high internal precision required for this work, these studies have been confined primarily to long-focus telescopes, with plates taken over time intervals of 50 or more years.

Several surveys of the sky for stars with high proper motion, largely with the aim of finding absolutely faint stars, have been carried out over the past several decades. Two such surveys are one with the 48-inch Schmidt telescope at Mt. Palomar and the other with the 13-inch telescope at Lowell Observatory. They cover 80 percent of the sky to a limiting magnitude of 21 and 17 respectfully. In surveys of this magnitude, it is essential that telescopes of not too large focal length be used to limit the number of plates needed to cover the sky and that the "moving" stars be found by rapid scanning of the plates. These surveys have drastically increased the known number of white-dwarf, sub-dwarf, and faint red-dwarf stars, which, at the present time attract much interest among astronomers.

An important area within astrometry is the determination of the distances of individual stars. Because of the extremely small quantities to be measured, the ultimate in precision is required. The geometric method of measuring distances is based upon the surveyor's principle: the object is observed from both ends of a base line. In determining the trigonometric parallax of a star, the semimajor axis of the earth's orbit is used as the base line. Reliable individual distances have been measured in this way for several thousands of stars within 30 parsecs of the solar system. The majority of the photographic plates in these determinations were obtained with long-focus refracting type telescopes, and the plates measured on manually operated measuring machines. The discovery of the large number of intrinsically faint stars in the solar neighborhood with apparent magnitudes beyond the practical limits of the refracting type telescope demanded the development of a reflecting-type telescope with greater light gathering power, and with a highly stable optical system. Such a telescope, named an astrometeric reflector, has been in operation since 1964. In combination with a high precision automatic measuring machine, parallaxes have been achieved with a precision of 2 milliarcseconds, thereby obtaining reliable distance determinations to 150 parsecs by this method. The importance of stellar distance determination is realized from the fact that the distance of a star must be known before its intrinsic luminosity and its rate of energy generation can be determined.

Studies of the stars in the solar neighborhood have revealed that the majority of them are components of double and multiple systems. Since the motions of the stars within a system are governed by their mutual gravitational attraction, it is possible to determine their masses by use of Kepler's third law (see KEPLER'S LAWS OF PLANETARY MOTION), whenever their orbital motions and the parallax of the system become known. This is the only direct way masses of the stars can be determined.

Routine observations of the motions in binary systems began about 150 years ago. Originally all observations were carried out visually. Although this method continues to be used for close pairs, it has been largely replaced by a more accurate photographic method for wider pairs.

Various searches for double stars have produced some 75000 visual binary systems, but for only a small fraction (approximately 100) of these systems are data available for determining the individual masses with an accuracy of 30 per cent or better. These masses range from about 0.08 solar mass for a star 3000 times less luminous than the sun to 6 times the sun's mass for a star 100 times more luminous than the sun. Larger masses, as high as 50 to 100 times the solar mass, are found among the very close binaries such as eclipsing and spectroscopic binaries. Although these objects cannot be resolved into individual components, their orbital motions can be determined from the periodic variation in light and radial velocity (observed Doppler shift).

Stellar masses smaller than the value of 0.08 quoted above have been discovered in recent years by intensive photographic studies of nearby single stars and components in double stars. These studies, representing the ultimate in accuracy in photographic astrometry, have revealed unseen companions of such small masses that, according to theoretical estimates, they are either stars so small that they will never burn nuclear fuel or planets of the size of Jupiter.

Aside from demands for such utilitarian purposes as navigation, geodesy and space research, astronomers themselves are making heavy demands for substantial gains in quality and quantity in astrometric observations, extended to fainter and fainter stars.

The discovery of the intrinsically faint stars in the solar neighborhood has demanded an extensive parallax program with an entirely new telescope of special design.

Positions and space velocities on a large scale of the individual stars and of stellar systems within our galaxy are essential to understand its dynamics and evolution as well as the physical properties and evolution of the individual stars which populate it.

To accomplish this, new instrumental and analytical techniques are currently being introduced which take advantage of the latest technological developments in automation.

Within recent years measurements with a precision of a few milliarcseconds have been achieved in astrometry in both the optical and radio spectrum by means of interferometry. In the optical spectrum, speckle interferometry of close binaries has achieved standard errors of 5 milliarcseconds in their relative positions, while long baseline optical interferometry of the order of 50 meters is capable of very high resolution of one milliarcsecond. Similar accuracies are expected in transcontinental radio interferometry. High precision optical and radio position measurements are currently being carried out of radio sources and their optical counterparts to improve the relationship between optical and radio reference frames, with the possibility of establishing an inertial reference frame from the measurements of the cosmological distant quasars.

Astrometry from space observations will be a reality within the next few years. Among these projects can be mentioned the astrometric satellite by the European Space Agency (ESA), named *Hipparcos*. With this satellite, in a geostationary orbit, it will be possible to observe some 10,000 stars during a 2–3 year mission. Positions, yearly proper motions, and parallaxes with an accuracy of two milliarcseconds are projected. Similar accuracies are expected in the astrometric programs planned with the *Space Telescope* of NASA, which is scheduled to be in orbit by 1986.

K. AA. STRAND

References

Eichhorn, H., "Astronomy of Positions," New York, F. Ungar Publisher, 1974.

Mueller, I. I., "Spherical and Practical Astronomy," New York, F. Ungar Publisher, 1969.

Prochazka, F. V., and Tucker, R. H. (Eds.), "Modern Astrometry," International Astronomical Union Colloquium No. 48, Vienna, Austria, University Observatory Vienna, 1978.

Kraus, J. D., "Radio Astronomy," New York, McGraw-Hill Book Company, 1966.

Abell, G. O., "Exploration of the Universe" (3rd Ed.), New York, Holt, Rinehart, & Winston, 1975.

Taff, L. G., "Computational Spherical Astronomy," New York, John Wiley & Sons, 1981.

Strand, K. Aa., "Basic Astronomical Data," Chicago, University of Chicago Press, 1963.

Cross-references: ASTRONAUTICS; ASTROPHYSICS; COSMOLOGY; DOPPLER EFFECTS; GRAVITATION; INTERFERENCE AND INTERFEROMETRY; KEPLER'S LAWS OF PLANETARY MOTION; PULSAR; ROTATION–CURVILINEAR MOTION; SPACE PHYSICS.

ASTRONAUTICS*

Future historians may record that the age of space flight marked a turning point in modern times. Its physical principles—the laws of motion of celestial bodies—marked the turning point of medieval times. From Copernicus to Kepler to Galileo to Newton, the Aristotelian myth of a man-oriented universe succumbed to the conception of a detached mechanically oriented universe, operating through laws which were a synthesis of the new knowledge gained in the formerly separate domains of terrestrial and celestial mechanics. Mankind never quite recovered from that detachment.

Weight and Weightlessness In Newtonian mechanics, weight is understood to mean the force that an object exerts upon its support. This would depend on two factors: the strength of gravity at the object's location (things weigh less on the moon) and, as Newton called it, the quantity of matter in a body (its "mass"). At any given location, where gravity is fixed, mass can be measured relative to a standard by noting the extension of a spring to which it and the standard are successively attached. Alternatively, the unknown and standard may be hung at opposite ends of a rod and the balance point noted. However, by an entirely separate experiment, mass can also be measured by noting the resistance of the object to a fixed force applied horizontally on a frictionless table. The measured acceleration provides the required basis of comparison with the standard. Needless to say, all objects measure identical accelerations when freely falling in the vertical force of gravity. This merely means that, unlike the arbitrary force we apply horizontally in the experiment above, gravity has the property of adjusting itself in just the right amount, raising or lowering its applied force, to maintain the acceleration constant.

It was well known that objects appear to increase or decrease their weight (alter the extension of the spring) if the reference frame in which the measurement takes place accelerates up or down. As gravity did not really change, however, most people were inclined to draw a distinction between *weight* defined as mg, where m is the mass and g is the local gravity field, and the *appearance of weight*, the force of an object on its support as measured by the spring's extension. One way to avoid the difficulty has been to speak of an *effective* g, which takes into consideration the frame's acceleration. For example, at the equator of the earth, we measure, say by timing the oscillation of a pendulum, the effective g, some 0.34 per cent less than the g produced by the mass of earth beneath our feet.

*John Wiley & Sons, Publishers, has kindly permitted extensive use of copyright material from "Physical Principles of Astronautics" by Arthur I. Berman, who is also the author of this article.

If the earth were rotating with a period of an hour and a half instead of 24 hours, our centripetal acceleration at the equator would cause the effective g to vanish completely, our scales would not register, objects would be unsupported, and for all practical purposes they and we would be weightless.

Formally, we could state that any accelerating frame produces a local gravitational field g_{acc} that is equal and opposite to the acceleration. Thus, a rotating frame generates a centrifugal g_{acc} opposing the centripetal acceleration. We have at any point

$$g_{eff} = g + g_{acc} \tag{1}$$

where g is the field produced by matter alone (e.g., the earth). By identical reasoning, an object in orbit, whether falling freely in a curved or in a straight path, will carry a reference frame in which g_{eff} is zero, for its acceleration will always exactly equal the local g by the definition of the phrase, "freely falling."

This concept was placed on a firm footing by Einstein who maintained that Eq. (1) is reasonable not only in mechanics but in all areas of physics including electromagnetic phenomena. We arrive at the inevitable conclusion that we cannot distinguish by any physical experiment between an apparent g accountable to an accelerating frame and a "real" g derived from a local accumulation of mass. This central postulate of the General Theory of Relativity also unified the two separate conceptions of mass. An object resting on a platform that is accelerating toward it will resist the acceleration in an amount depending on its inertia. It presses against the platform with a force equal to that it would have if placed at rest on the surface of a planet with local field equal and opposite to the acceleration of the frame.

General Principles of Central Force Motion The gravitational force between point masses is inverse square, written

$$mg = - \frac{\gamma m'm}{r^2} \hat{r} \tag{2}$$

where the center of coordinates from which the unit vector $\hat{r}$ is described lies in m', one of the masses. Thus, the force on m is directed $-\hat{r}$, toward m' and is proportional to $1/r^2$ with γ the constant of proportionality. The quantity g is the force on m *divided by* m (or normalized force) for which the name "gravitational field of m'" is reserved. Of course, if m were in the field of a collection of mass points, or even in a continuous distribution of mass, the summated or integrated g at the location of m would no longer be an inverse square function with respect to any coordinate center. However, in one special case, the inverse square functional form would be preserved: if the source mass were symmetrically distributed about the coordinate center. This would be the case if the source were a spherical shell or solid sphere, of density constant or a function only of r. The sun and earth can be regarded, at least to a first approximation, as sources of inverse square gravitational fields.

There are some important general statements we can make about the motion of an object placed with arbitrary position and velocity in a centrally directed force field, i.e., a field such as the one described, which depends only on distance from a central point (regardless of whether or not the dependence is inverse square). As the force has only a radial and no angular components, it cannot exert a torque about an axis through the center. This means that the initial angular momentum is conserved. Now angular momentum is a vector quantity and therefore is conserved both in direction and magnitude. It is defined by $\mathbf{r} \times \mathbf{p}$, where $\mathbf{r}$ is the position vector to the mass of momentum $\mathbf{p}$. The direction of the angular momentum vector is thus perpendicular to the plane containing $\mathbf{r}$ and $\mathbf{p}$. As this direction is permanent, so also must be the plane. The planar motion of the object can be expressed in polar coordinates, so that by writing $\mathbf{r} = r\hat{\mathbf{r}}$ and $\mathbf{p} = m(\dot{r}\hat{\mathbf{r}} + r\dot{\phi}\hat{\phi})$, we find the specific angular momentum (angular momentum per unit mass) called h, to be

$$h = r^2 \dot{\phi} \qquad (3)$$

This too then must be a constant of the motion.

Consider now the rate at which area is swept out by the radius vector, dS/dt. We recall from analytic geometry that $dS = \frac{1}{2}r^2 \dot{\phi}$. Thus

$$\frac{dS}{dt} = \frac{h}{2} \qquad (4)$$

so that this is a constant of the motion as well. On integration, we conclude that the size of a sector that is swept out is proportional to the time required to sweep it out. In the case of a closed orbit, the total area S would then be related to the specific angular momentum as

$$S = \frac{hT}{2} \qquad (5)$$

This sector area-time relationship is Kepler's second law of planetary motion which was induced from Tycho Brahe's observation of Mars without prior knowledge of gravity and its central character.

The Laws of Kepler   Kepler stated two other laws of planetary motion: The orbits of all the planets about the sun are ellipses (a radical departure from the circles of Copernicus), and the squares of their periods are proportional to the cubes of their mean distance from the sun, this mean being the semimajor axis of their ellipses. The third law pertained to the one characteristic common to all the planets: the sun. Taken together, the three laws led Newton to the concept of gravitational force and its inverse-square form.

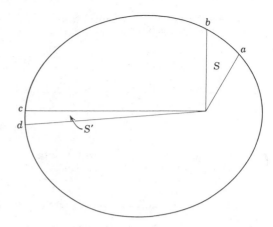

FIG. 1. Kepler's second law. The sector area S swept out is proportional to the time required for the planet to move from a to b. Thus, if $t_{cd} = t_{ab}$, then $S' = S$.

By applying Newton's law of motion $\mathbf{F} = m\mathbf{a}$, a relationship between $\mathbf{a}$, the second derivative of the position vector, expressed in polar form, and $\mathbf{F}/m$ or $\mathbf{g}$, as given by Eq. (2), leads to the familiar conic solution for the trajectory of an object in an inverse square field.

$$\frac{1}{r} = \frac{\gamma m'}{h^2} + A \cos(\phi - \phi_0) \qquad (6)$$

where A and ϕ_0 are constants. A rotation of axis will eliminate ϕ_0, thereby aligning the coordinate axis with the conic's major axis. Also, by expressing the general conic, an ellipse or hyperbola, in terms of the usual parameters of semimajor axis a and eccentricity ϵ, we can relate the geometric parameters to the gravitational-dynamical constants, viz:

$$h = [\gamma m' a(1 - \epsilon^2)]^{1/2} \qquad (7)$$

and

$$\frac{1}{r} = \frac{\gamma m'}{h^2}(1 + \epsilon \cos \phi) \qquad (8)$$

Note that by substituting Eq. (7) into Eq. (5) and expressing the area of an ellipse as $S = \pi a^2 (1 - \epsilon^2)^{1/2}$ we arrive at Kepler's third law,

$$T = \frac{2\pi}{(\gamma m')^{1/2}} a^{3/2} \qquad (9)$$

The *energy* of the orbiting object can be calculated with ease by evaluating it at an extremal point, say the nearest point to the gravitational source, called pericenter or perifocus. As the energy is constant, it is immaterial where the calculation is made. Here the velocity has only an angular component so that the kinetic energy for a unit orbiting mass is $\frac{1}{2}v^2 = \frac{1}{2}r^2\dot{\phi}^2$. The potential energy at pericenter is $-(\gamma m'/r_{pe})$ where

r_{pe} is the distance of the unit mass from m', the focal point. Here $\phi = 0$ so that by Eq. (8),

$$\frac{1}{r_{pe}} = \frac{\gamma m'}{h^2} (1 + \epsilon) \qquad (10)$$

On substituting Eq. (7), we find the total kinetic and potential energy to be

$$E = -\frac{\gamma m'}{2a} \qquad (11)$$

Our conclusion: All objects in orbit with the same major axes have identical periods and identical energies per unit mass. Knowledge of E is invaluable in determining an object's speed when its distance from the source is known, and vice versa.

In the event that the orbiting object's mass is not negligibly small compared with that of the gravitational source, one must take note that the combined center of mass, from which the acceleration is described, no longer may be assumed to lie in the center of the gravitational source. This complicates our equations somewhat, for the accelerating force still is expressed relative to the center of the source (if spherical). The adjustment that results, when center of mass coordinates are transformed to relative coordinates in the expression for acceleration, requires our equations to take the form $\gamma(m' + m)$ wherever formerly $\gamma m'$ appeared.

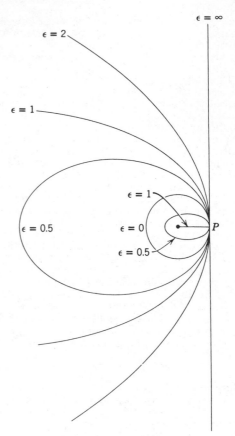

FIG. 3. Orbits of differing eccentricities and major axes which pass through a common point. Higher speeds correspond to higher energies and longer major axes. For an ellipse the eccentricity ϵ is the ratio of the distance to the focus (gravitational center) from the point of symmetry of the curve to the semimajor axis of the ellipse. For a hyperbola it is the ratio of the distance to the focus from the point of symmetry of the hyperbolic pair to the semimajor axis.

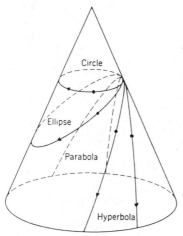

FIG. 2. Conics formed by the intersection of planes inclined at various angles with the axis and sides of a cone. Ellipses and hyperbolas are general curves in astronautics, i.e., general solutions of the two-body problem, while the circle and parabola are special cases. An object in circular orbit is at constant distance from the gravitational source and therefore has constant potential energy; it moves at constant speed and therefore has constant kinetic energy. An object in a parabola is in a transition curve between an ellipse and hyperbola; its major axis, therefore, is infinite and its total energy zero.

Disturbances in the Central Field The earth, of course, is spherical to only a first approximation. More accurately, it is an ellipsoid of revolution about a minor axis—an oblate spheroid. Still more accurately, it appears to be slightly pear-shaped and, in addition, its figure is distorted by continuous local variations. The spheroidal figure, nevertheless, accounts for nearly all the anomalous effects on satellite orbits. For one thing, the gravitational force on the satellite is no longer centrally directed; the excessive mass in the equatorial plane produces a force on the satellite directed out of its orbital plane. The resultant torque causes the direction of the angular momentum vector to change; i.e., the plane containing the satellite's ellipse turns. The plane turns continuously about the polar axis maintaining its angle with the axis and with the

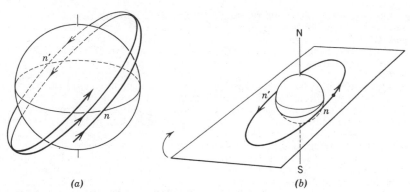

(a)   (b)

FIG. 4. The orbit of an earth satellite. The earth's equatorial bulge causes retrograde motion of the points of intersection n and n' of the orbit and equatorial plane. This can alternatively be interpreted as a retrograde motion, about the north-south axis, of the plane containing the closed orbit. The plane moves in the direction shown by the arrow in (b), maintaining a constant angle with the axis.

equatorial plane constant. The turning rate is greatest for low orbits and small angles of inclination with the equator. For polar satellites, the plane remains fixed. A separate effect of this equatorial bulge perturbative force is the slow turning of the ellipse's major axis *within* the orbital plane. This effect vanishes at an inclination of 63.4°; the major axis turns backward at inclinations above this angle and forward below.

Astronautical Examples From the preceding material some interesting problems in astronautics can be solved readily. Suppose (Fig. 5) an astronaut releases himself at a speed δv of 10 m/sec from a satellite in a small circular orbit about the earth. What would be his velocity and that of his satellite when at their maximum separation if his initial thrust is (a) the same as that of the satellite in its orbit, (b) toward the earth's center, and (c) normal to the orbital plane? Also, what would be the maximum separation between astronaut and satellite and where would it occur?

Writing E in terms of both the kinetic-potential energy sum at any point in orbit, and its constant value given by Eq. (11), we find that

the speed at the astronaut's perigee in (a) can be written

$$v^2 = \frac{\gamma m'}{r_0}\left(2 - \frac{r_0}{a}\right) \qquad (12)$$

where r_0 is the satellite's orbital radius and a the astronaut's semimajor axis. If

$$v_c^2 = \frac{\gamma m'}{r_0} \qquad (13)$$

where v_c is the satellite's circular speed, found by letting $a = r_0$ in (12). We have all the information needed to find the distance of separation between ellipse and circle as

$$2a - 2r_0 \simeq \frac{4r_0\,\delta v}{v_c}$$

$$= \frac{4(6.38)\,(10^3\text{ km})\,(10\text{ m/sec})}{7.9\,(10^3)\text{ m/sec}}$$

$$= 32\text{ km}. \qquad (14)$$

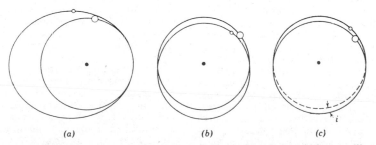

(a)   (b)   (c)

FIG. 5. Orbits of an astronaut released from a circularly orbiting satellite (a) in the direction of satellite motion, (b) toward the center of attraction, and (c) normal to the orbital plane. In the last case the astronaut moves in a new orbital plane inclined to that of the satellite by angle i.

By applying angular momentum conservation—that is, the apogee speed times the focal distance at apogee equals the perigee speed times the focal distance there (or simply Eq. (7))—we find that the astronaut will have a relative speed at apogee of 30 m/sec.

In (b), the astronaut will orbit in an ellipse in the plane of the satellite's circle that intersects the circle twice. As his injection velocity is radial, his specific angular momentum is that of the satellite. This, together with Eq. (12), where now v at launch is not the linear sum of δv and v_c but the square root of the sum of the squares of these terms, leads directly to the orbital parameters a and ϵ. We find the latter as $\delta v/v_c$ and the orbital separation at apogee as 8 km with the speed there relative to the satellite as 10 m/sec, the same as at launch.

The astronaut moves out of the satellite's orbital plane in (c) and into a new plane inclined at angle $i = \delta v/v_c$. The maximum separation between orbits now becomes $ir_0 = 8$ km. But the astronaut never departs more than 20 m from his initial distance from earth, so his speed is always close to that of the satellite.

Rocket Propulsion A rocket operates by the simple principle that if a small part of its total mass is ejected at high speed, the remaining mass will receive an impulse driving it in the opposite direction at a moderate speed. As δm_e, the propellant, leaves at speed v_e with respect to the rocket, the remaining rocket mass m receives a boost in speed δv such that

$$\delta m_e v_e = m \, \delta v \qquad (15)$$

If additional equal propellant mass is ejected at the same speed, the boost in rocket speed is slightly greater than before as the rocket mass has been slightly depleted by the prior ejection. Indeed, if the residual rocket mass eventually were minuscule, its boost in speed could reach an enormous value. The integrated effect of these nonlinear boosts is found as

$$v_t - v_0 = v_e \log_e \frac{m_0}{m_t} \qquad (16)$$

where v_0 and m_0 are the rocket speed and mass at some arbitrary initial time and v_t and m_t are the same quantities at some time t later.

From these simple considerations, it is apparent that the highest rocket velocities are attained if we could increase the propellant speed as well as the mass ratio m_0/m_t. The mass ratio can be maximized by obvious methods such as choosing a high-density propellant which cuts the tankage requirement or avoiding unnecessarily complicated apparatus for ejecting propellant at high speed. A nuclear rocket, for example, may perform well in its ability to eject propellant an order of magnitude higher in velocity than conventional chemical rockets; nevertheless, the penalty required in reactor weight and shielding severely limits its effectiveness.

Specific impulse is one performance characteristic which applies to the propellant's ability to be ejected at high speed regardless of the weight penalty required to do this. It is the impulse produced per mass of propellant ejected, or $m \delta v/\delta m_e$, or, by Eq. (15), simply v_e. In engineering usage, it is impulse per *weight* of propellant ejected, or v_e/g_e where g_e is the acceleration of gravity at the earth's surface. Its units are seconds, and it can be interpreted as the thrust produced by a rocket per weight of propellant ejected per second. By itself, thrust is of little importance unless it is sustained for a significant time by a large backup of propellant tankage. It is here that the mass-ratio term in Eq. (16) would play an important role in any evaluation of a rocket's true performance.

It is important to realize how sensitive is the energy imparted to a moving vehicle to a given boost in speed. If an object, say a rocket, moving at some speed v gets a boost in the same direction in an amount δv, then its increase in kinetic energy is $v\delta v + \frac{1}{2}(\delta v)^2$. Thus, if a rocket receives a small boost in speed, say from turning on its engine for a short time, the rocket can thereby receive a very *large increase* in kinetic energy provided it was already moving fast before getting the boost.

This principle is made use of in the firing of a multistage rocket. Suppose, for example, that we have a two-stage rocket of similar components: The "burnout" speed of each stage is the same, lifting each to the same apogee. We compare the effect of two firing programs: In the first case, firing of the upper stage is delayed until its apogee as booster payload is reached, but in the second case its firing takes place close to the ground immediately following booster burnout. Assuming that the force of gravity is constant at all heights, air drag is negligible, and the burnout time is small compared with the flight time, the apogee of the payload of the second case will be twice as high as in the first.

Transfer Orbits If one wishes to leave one orbit and enter another by rocket, an optimum path is generally chosen to minimize the total propellant required. Nevertheless, this should not be done at the expense of unduly long flight times, complicated guidance equipment, or high acceleration stresses. These would require unprofitable weight expenditures which would offset the frugality in propellant tankage.

Let us examine a simple but recurring example of a transfer problem, that of leaving a reference frame or space platform in one circular orbit and entering another larger one concentric with the first. If the transfer path were radial or near radial (a so-called ballistic orbit) then one would have to launch at a large angle to the direction of motion of the frame, accomplished only by a velocity component opposed to the frame's motion. On reaching the outer platform, a soft landing can be made only by a substantial rocket velocity boost tangent to the orbit.

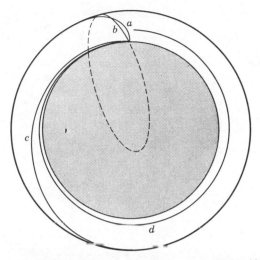

FIG. 6. Four launch trajectories into a satellite orbit about a planet. (a) If that planet has an atmosphere, the rocket may ascend in a "synergic" trajectory from the planetary surface to the final orbit, i.e., it cuts through the denser portions in an initially vertical path and gradually bends over into a horizontal path during burnout. (b) If there is no atmosphere it may ascend from the ground in a ballistic ellipse. This same ascent path may be chosen if the departure is from a parking orbit or "space platform" close to ground level. A far better choice would be (c) the Hohmann ellipse, with pericenter at the planet's surface and apocenter at the satellite orbit. Burnout time is assumed short in both this and the ballistic case. (d) A vehicle such as an ion rocket, which can sustain a microthrust for a very long time, cannot be launched from the ground but only from a parking orbit. It will spiral out to the desired altitude with few or many turns about the planet, depending on the magnitude of the thrust relative to that of the gravitational force.

Clearly, the total propellant expenditure would be far greater than one alternative of launching the rocket in the direction of motion of the first frame with just sufficient speed to reach the outer circle, timed so that the outer frame will meet the spacecraft. The transfer orbit will be an ellipse contangent with both circles. The outer frame will be moving much faster of course at the contact point as the major axis of its orbit is much greater [see Eq. (11)], but the difference in speed is not nearly as pronounced as for the ballistic transfer case. A differential speed increment at contact completes the maneuver.

The return trip, from an outer to inner circle, is made by following the second half of this co-tangent ellipse, named the Hohmann transfer orbit after the German engineer who discovered its optimal property with regard to propellant expenditure. In the return case, the spacecraft is launched in opposition to the outer platform's motion. This removes kinetic energy and forces the spacecraft to fall in closer to the attractive center in order to make cotangent contact with the inner circle. The total propellant expenditure from the outer to the inner platform is the same as for the original journey.

An interesting question arises if one wishes to leave a frame for an outer orbit when it initially is in an elliptical orbit rather than a circle. Should we depart from apocenter where we are furthest from the gravitational source and closest to our destination? Or should we depart instead from some other point in the ellipse?

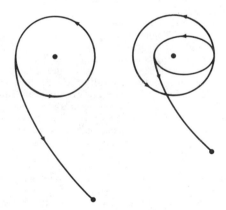

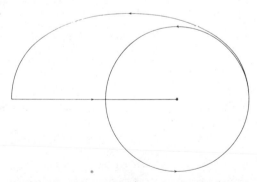

FIG. 7. Reaching the center of attraction from a circular orbit by a double-thrust maneuver. The rocket enters an eccentric ellipse by a forward thrust, and at apocenter reverses thrust to cancel the small orbital speed there, enabling it to fall radially into the source. By this maneuver radioactive waste could be brought to the sun at up to 40% saving in propellant.

FIG. 8. Alternative ways to escape from a circular orbit to "infinity." *Left:* The rocket simply blasts off into a parabola or hyperbola. *Right:* The rocket *reverses thrust* to fall into a close-in ellipse and then at the *pericenter* or *near apsis* of the ellipse (perigee if orbiting about the earth) it blasts off into a parabola or hyperbola. Despite the loss of circular orbital energy in maneuver (b), it is preferred to (a) provided the energy remaining in the hyperbola after completely leaving the gravitational field exceeds the escape speed at the radius of the circular orbit. (If the arrows were reversed the figures would represent alternative maneuvers for *entering* a circular orbit from infinity.)

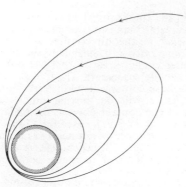

FIG. 9. Atmospheric-entry braking "ellipses," a trajectory controlled by gravity and the reverse force of atmospheric drag. Drag is concentrated at the pericenter, where air density is high. This acts to decrease the major axis and eccentricity of the ellipse, so that the orbit gradually approaches a circle. This analysis is somewhat idealized, however, and in an actual case the pericenter lowers slightly and the major axis turns in each successive cycle.

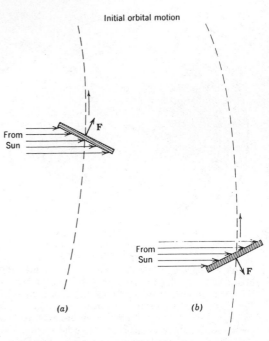

FIG. 10. Orientation of a solar sail for a voyage from earth to (a) a superior, and (b) an inferior planet. F is the vector sum of the forces due to the incident and reflected beams. When F is aligned as in (b) it constitutes a drag and causes the sail to spiral into the sun.

Paradoxically, our best launch point is at pericenter, for here the largest possible amount of energy will be transferred to the spacecraft for a given expenditure of propellant. This was noted in the previous section. A given thrust applied for a given time interval will do more work on the spacecraft when it is moving fast, as at pericenter, for it covers a greater distance during the interval. This advantage offsets the undesirability of being at a lower potential energy point at pericenter.

Powered Trajectories In the usual operation of a solid- or liquid-propelled rocket, the propellant is depleted in a time negligibly small compared with the total flight time. The trajectory analysis may generally be considered as that of a free orbit subject to burnout initial conditions as in the discussion above. If, however, the propellant ejection is sustained over long periods, as in an ion-propelled rocket, the trajectory analysis is necessarily complicated, for, in addition to the varying gravitational force, the vehicle, of slowly diminishing mass, is subject to a thrust which may be changing both in direction and magnitude. Even one of the simplest thrust programs, a constant thrust in the direction of motion, requires an electronic computer analysis in order to obtain the position and velocity at future times (see ELECTRIC PROPULSION).

The continuous-thrust trajectory is a spiral with many advantages over the orbital ellipses. First, the lower sustained thrust precludes the high-acceleration stresses associated with rapid-burning chemical rockets. Much of the structural weight usually needed to withstand these stresses can be replaced by propellant. Also, flights to the extremities of a gravitational re-

gion may take a shorter time in a spiral trajectory. In a long Hohmann ellipse, for example, most of the journey is made at very low speed. In a powered spiral, on the other hand, the spacecraft could be made to move fast, for the thrust, though small, is integrated over many months.

The spiral concept is ideal for rockets where very high ejection velocities are feasible by using electromagnetic or electrostatic particle accelerators, but only at the expense of a low propellant flow rate and relatively heavy power-generating equipment. However, the propellant reserve, and thrust, could then last the required long time. Such an ion rocket with its very low thrust-to-weight ratio could hardly be expected to take off from the ground, and could only take off from an orbital platform. In the vacuum of space, the ion beam meets its ideal environment.

ARTHUR I. BERMAN

References

Berman, Arthur I., "The Physical Principles of Astronautics," New York, John Wiley & Sons, 1961.

Moulton, Forest Ray, "An Introduction to Celestial Mechanics," New York, The Macmillan Co., 1914.

Danby, J. M. A., "Fundamentals of Celestial Mechanics," New York, The Macmillan Co., 1962.

Sterne, Theodore E., "An Introduction to Celestial Mechanics," New York, Interscience Publishers, 1960.

Berman, Arthur I., "Space Flight," New York, Doubleday Book Co., 1979.

Cross-references: AERODYNAMICS; DYNAMICS; ELECTRIC PROPULSION; FLIGHT PROPULSION FUNDAMENTALS; GRAVITATION; INERTIAL GUIDANCE; KEPLER'S LAWS OF PLANETARY MOTION; MASS AND INERTIA; ROTATION–CURVILINEAR MOTION; WORK, POWER, AND ENERGY.

ASTROPHYSICS

Starting with the advent of photography and the study of stellar spectra in the second half of the nineteenth century, astrophysics now includes optical and radio observations of planets, stars, clusters, interstellar material, galaxies and clusters of galaxies, and their interpretation. Radiation from these external sources provides information on the direction of the source, its velocity, composition, temperature and other physical conditions, including magnetic fields, density, degree of ionization, and turbulence. The term "astrophysics" is generally understood to include all these aspects except the measurement of direction (ASTROMETRY–positions of stars in the sky and changes due to parallax and proper motion), and the orbits of planets, asteroids and comets (celestial mechanics). Because of its proximity, the sun can be studied in more detail than other stars; its structure and its influence on the nearby planets and comets are the concern of SOLAR PHYSICS and are closely related to geophysics and stellar astrophysics. Study of the motions of stars in pairs, groups, clusters, associations, and galaxies is the overlap of celestial mechanics with astrophysics, and the study of the distribution and patterns of motion of the distant galaxies is the overlap with "COSMOLOGY."

Astrophysical studies of *planets* began in the 1930s with measurement of their spectra, showing the composition of the atmospheres of Venus, Mars, Jupiter, Saturn, and Neptune. These spectra showed bands of CO_2, CO, H_2, CH_4, NH_3, and other, more complex molecules. The atmospheres of Jupiter, Saturn, Uranus, and Neptune seemed to have about the same atomic composition as the Sun, and theoretical models of Jupiter showed that its core must consist of metallic hydrogen under high pressure in the interior. Starting in 1959, the space age brought spacecraft missions to Mars, Venus, Mercury, Jupiter, and Saturn, with landings on Venus and Mars (as well as Earth's Moon), and measurements *in situ* of their atmospheric composition, density, and temperature. Venus probes in 1970–75 found the most extreme conditions; surface pressure and density about

100 times the Earth's, temperatures of 750 K, and possible seas of acetic acid. The high temperature is attributed to atmospheric gases, mostly CO_2, that let sunlight in but prevent the outward, long-wavelength thermal radiation from leaving the atmosphere.

High-speed jet-stream motions were detected in the atmospheres of Venus, Mars, Jupiter, and Saturn. On Venus, they produce changing cloud patterns, on Mars they produce global dust storms, and on Jupiter and Saturn they produce vortices that show as small white spots, but bypass the giant red spot on Jupiter (explanation still unknown). By far the most dramatic Voyager (1981) photos are those of Saturn's rings (Fig. 1). Earth-based observations showed none of the detail of the hundreds of single rings, some of them entwined by the gravitational effects of nearby small moons. Less impressive rings were also found around Jupiter, and Earth-based observations showed rings around Uranus. These rings are related to stages in the origin and evolution of the solar system, probably the remnants of proto-planet clouds of gas and dust that formed the giant planets in much the same way as the solar nebula formed the Sun and planets. From all this, the new science of planetology developed among astrophysicists, geophysicists, meteorologists, and geochemists.

Comets have been observed and recorded by men for several thousand years. Starting in the 1920s, astrophysicists observed their spectra, which showed that as a comet approaches the Sun, sunlight boils off various molecules to form a coma around the nucleus, then two long tails, pushed out away from the Sun by radiation pressure and by the solar wind–outward moving electrons and ions ejected by the Sun. The dust tail, showing in reflected sunlight, is formed of dust released from the comet nucleus as frozen gases boil off; the ion tail is formed from ionized gases that show an emission-line spectrum. These two tails deviate from one another because the solar wind blows in a slightly different direction from solar radiation.

The Oort cloud of comet nuclei (named after Jan Oort, a Dutch astrophysicist) is believed to be the outer fringe of the solar nebula condensed from gas and dust to form nuclei revolving about the Sun at distances larger than 50 Astronomical Units (1 A.U. = 93 million miles, the distance from Sun to Earth). These small bits of frozen gas and dust represent the early composition of the solar nebula. Every now and then, one gets deflected from its circular orbit to a highly elliptical orbit, falling toward the Sun, sweeping around at perihelion close to the Sun with large coma and long tails, then receding from the Sun out to aphelion, losing coma and tails, then falling back in again. Halley's Comet is the archetype of these long-period comets, returning to the Sun every 75 years. In 1985–86 it will reappear. European and Soviet spacecraft are expected to rendezvous with Halley's Comet then in an effort to

FIG. 1. Saturn's Rings; NASA JPL color photo.

photograph the size of its nucleus and to sample the gas and dust boiling off. Astrophysicists want to determine what the frozen gases in the nucleus are; spectra of the coma show only molecules that have been dissociated by sunlight. The nature of these gases, and the included dust, may help with calculations of how the primordial solar nebula cooled and contracted about 5 billion years ago. The planets' sizes, masses, densities, rotations, distances from the Sun, satellites and rings, and the thousands of asteroids between Mars' and Jupiter's orbits, provide the rest of the data needed for a theory of the origin and evolution of the solar system. A recent discovery adds the fact that Pluto has a small moon, whose orbit provides the first firm measure of Pluto's mass, much smaller than expected.

The *instruments* used by astrophysicists have multiplied during the past two decades, and their resolving power (needed, for instance, to detect Pluto's small moon close to that planet) has increased enormously. Telescopes for the 1980s (see reference) include the 2.4-meter (95-inch) Space Telescope (Fig. 2), the Very Large Array of radio receivers near Soccorro, N.M., and the orbiting Einstein x-ray telescope launched in 1978. The European Space Agency is supplying Space Lab with a cooled infrared telescope to be flown on NASA's Space Shuttle. Space Telescope and VLA can resolve objects as close as 0.02 arc-sec apart, and detect objects 100 times fainter than previously possible. Very long baseline interferometry is achieved with radio dishes thousands of miles apart; the slightly different arrival times of radio waves from the same object provides resolution of 0.001 arc-sec or better.

Optical telescopes, both in orbit and ground-based, are fitted with spectrographs that spread the incoming light into a spectrum of colors from short wavelength to long, using a finely ruled grating or an echelle to resolve wavelengths as close as 1/100,000 of their size. This high spectrographic resolving power allows very small Doppler shifts to be measured, and the spectral lines of different atomic isotopes to be detected.

Needed for orbiting telescopes and spectrographs, electronic detectors have been developed to substitute for photographic emulsion, and are now being used in ground-based observatories as well. The most successful is the charge-coupled device (CCD). Arrays of 800 minute CCDs, 15 microns on a side, turn out digital information on each pixel (picture ele-

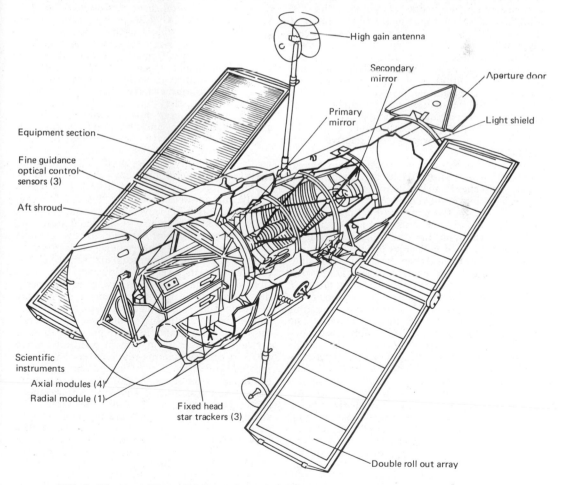

FIG. 2. Diagram of Space Telescope; Longair & Warner, "Sci Res with ST" NASA CP-2111.

ment) in an image or a spectrum. These digitized data can be fed directly into a computer for processing. Such use of electronics and computers has changed the observer's role in astrophysics. Instead of spending long cold nights at the eyepiece of a telescope in an unheated dome and developing his plates or film in a photographic darkroom, he now prepares computer programs in advance, picks the object he wants on a TV screen in the comfortable control room, and waits during a suitable "integration time" for his processed results to be printed out or graphed by the computer.

Early studies of *stellar spectra* revealed differences due primarily to surface temperature and described by the sequence of spectral types ranging from "O" (30 000 K or more) through "B," "A," "F," "G," and "K" to "M" (2000 to 3000 K). The type of a spectrum is set by relative intensities of lines and bands due to ions, atoms, and molecules. The earth's atmosphere limits the wavelength region observable from terrestrial observatories since ozone and other constituents are opaque at wavelengths $\lambda <$ 3000Å, and water vapor bands block much of the region $1.2\mu < \lambda < 8\mu$. The ionized layers block radio waves longer than 20 meters. Nevertheless, thousands of spectrum lines, mostly in absorption, have been identified in the range $3000 < \lambda < 12000$Å, and the pattern of lines within one spectral type has been found to vary with a second parameter, the "luminosity class" designated by roman numerals "I" (highly luminous "super giants") through "II", "III", "IV" to "V" ("dwarfs" of relatively low luminosity).

The continuum between spectral lines has an intensity distribution with wavelength that roughly matches Planck's theoretical distribution for a blackbody, $B(\lambda)\,d\lambda = 2hc^2\lambda^{-5}(e^{hc/\lambda kT} - 1)^{-1}$ where T is the temperature that accounts approximately for the ionization and excitation of atoms producing the star's line spectrum, or the dissociation of molecules producing bands—that is, for the spectral type. In so far as the lines can be ignored, the color of a star is approximately that of a blackbody of temperature T and the total ("bolometric") luminosity is given by $L_b = 4\pi R^2 \sigma T^4$, where R is the radius of the star and σ is Stefan's constant. The "apparent brightness" of a star (observed optical flux) depends upon its distance, D, and its luminosity in the wavelength region observed— approximately $4000 < \lambda < 6500$Å for visual observations and $3700 < \lambda < 5000$Å for photographic observations through glass optics. Recently the introduction of photoelectric equipment has allowed more accurate measurements in smaller wavelength regions. Standard measures are designated U (ultraviolet), B (blue), V (visual), I (infrared), etc., and are usually expressed in magnitudes, an inverse, logarithmic scale. An increase of 5 magnitudes corresponds to a decrease in brightness by a factor 100. By successive approximations in such measure-

ments of many stars, it has been possible to correct for the effects of interstellar absorption and limited wavelength range, as well as for the inverse-square law $(1/D^2)$, to obtain total luminosities and colors. The luminosities are often expressed in "suns", that is, multiples of the sun's luminosity (about 4×10^{33} ergs/sec). Analogous measurements at radio frequencies are expressed as the flux in watts/square meter/cycle/second. Spectrophotometric measurements from rockets and artificial satellites above the earth's atmosphere have been made in the far ultraviolet, and it is to be expected that the intensity distribution, $I(\lambda)$, will soon be observed for all wavelengths from 10^{-4} Å to several km.

X-ray observations started in 1962, using ionization chambers and Geiger counters mounted on rockets and sent up for a few minutes above the earth's atmosphere. As the rocket rotated, the x-ray detectors scanned part of the sky, but the angular resolution was at first poor (about 3°). Still, these early observations showed several discrete sources (Sco X-1, Sco X-2, Tau X-1, Cyg X-1, etc.) named for the constellation in which each is located.

Later orbiting x-ray observatories (Uhuru, HEAO, Einstein Observatory) have detectors with much higher space resolution; they can locate x-ray sources to within a few arc-minutes. Many of these sources have been identified with optical objects; they are many light-years distant, and their energy output is very high. Some are star-like in size, and thought to be the remnants of old supernovae—possibly neutron stars; others are whole galaxies, and our Milky Way has a faint x-ray background. The measured flux is given in counts/cm²-sec, each count representing a quantum of about 1 Å wavelength, or about 10^4 eV, or 1.6×10^{-8} erg.

Gamma rays of wavelength about 10^{-3} Å (energy 10 MeV) can also be detected above the atmosphere using scintillation counters—crystals that emit a flash of light when penetrated by a gamma ray. Such detectors are sensitive, but not accurate in direction. However, the directions of gamma-ray sources have been accurately determined by using three or more gamma-ray detectors timing gamma-ray pulses or "bursts." These times, accurate to microseconds, determine direction from the differences in time of arrival at spacecraft in known positions. Since 1975, scores of gamma-ray bursters have been located, and there is a diffuse gamma-ray background all over the sky, indicating many more distant sources. Solar flares emit gamma rays, and other stars probably do the same. The strong gamma-ray sources are probably distant galaxies, but bursters seem to be in our Milky Way Galaxy, and are thought to be neutron stars or black holes with matter falling into them to produce the bursts. At least six are located at the centers of globular clusters.

The major gap in observed spectra, $I(\lambda)$, is from $\lambda = 0.08$ to 2.5 mm, between the far infrared and short radio waves. In the far infrared, measurements from high-flying aircraft and spacecraft show many star-like objects and a few galaxies that are strong emitters between λ = 0.02 and 0.08 mm. In 1983 NASA's Infrared Astronomical Satellite (IRAS), cooled by liquid helium to 16 K, surveyed the whole sky in the band 0.02 to 0.3 mm, and discovered thousands of far-infrared sources.

The *masses of stars* are determined from motions of double stars, ranging from widely separated visual binaries whose relative motions can be photographed, to close spectroscopic and eclipsing binaries with orbits calculated from variations in radial velocity (observed Doppler shift). Among some 50 pairs, masses of individual stars are found from 0.08 to 20 solar masses, and there is less definite evidence of others as low as 0.03 and as high as 50 or 100. (One solar mass is 2×10^{33} gm.) Over most of this range the luminosity L is proportional to M^3.

Astrophysical theory has achieved considerable success in explaining the spectra of stars by theoretical models of the atmospheres, involving the surface temperature, surface gravity, abundances of chemical elements, turbulence, rotation, and magnetic fields. The strengths of absorption lines are found to fit a "curve of growth," the relation between measured line strengths and the strengths predicted by quantum theory for unit abundance of the one ion, atom, or molecule involved. The theory of stellar interiors further relates mass M, luminosity L, and radius R with chemical abundances, the opacity of the material, and the generation of energy by nuclear reactions. More spectacularly, it has explained stellar evolution in terms of changes due to nuclear reactions.

The theoretical models of *stellar interiors* are based on stability and two modes of transferring energy outward to the surface: radiative transfer and convective transfer. Radiative transfer, by repeated emission, absorption, and emission of light, implies a temperature gradient dependent on the opacity and on the flow of radiative energy, or L. After calculating the opacity of gaseous stellar material (about 60 per cent hydrogen, 35 per cent helium, and 5 per cent heavier elements, by weight) at various temperatures and densities, the astrophysicist can compute the temperature, density and pressure in shells at various depths inside a star, starting with a definite radius and surface temperature, and adding up the shell masses to get the total mass. At some level, the temperature and density are sufficiently high for nuclear reactions to take place, the simplest being conversion of hydrogen to helium generating 7×10^{18} ergs of energy per gram. The rate of energy generation over the whole inner core must match L. Calculations for convective transfer follow a similar pattern but depend upon matching the adiabatic gas law for over-all stability. A combined model may have a convective core surrounded by a radiative shell and that surrounded by an outer convective shell.

The successful fitting of nuclear energy generation into such gas-sphere models of stars by 1940 led to the idea of *stellar evolution*, the conversion of hydrogen to helium in its core causing a star to age. Direct evidence of this aging was first obtained from clusters of stars. The stars in one cluster, relatively close together in space, are assumed to have been formed at the same time, and the pattern of stellar characteristics differs from one cluster to another in a systematic way. The pattern is easily recognized on a Hertzsprung-Russell ("H-R") diagram of log L vs spectral type (or color) on which the vast majority of stars appear near a diagonal line, the "main sequence" (Fig. 3).

Several other classes of stars can be distinguished by location on the H-R diagram ("red giants," "white dwarfs," etc.) and theories of stellar evolution account for a change in location along an "evolutionary track" for any one star. The rate of such change will vary in general; for instance, the large-mass, high-L blue stars are expected to exhaust their hydrogen in a few million years, whereas yellow and red dwarfs remain for billions of years on the main sequence. H-R diagrams for clusters confirm the aging (also the theoretical star models and the assumption of cluster origin) and provide evidence of the age of each cluster.

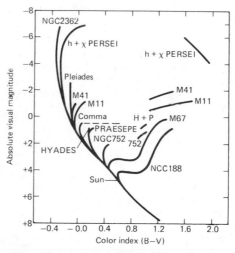

FIG. 3. This is the *H-R* diagram for galactic clusters used by Allan R. Sandage to estimate their relative ages. The turnoff from the age-zero main sequence (on which the sun is located) is lowest for the oldest clusters, NGC₁88 and M67. The stars in the uppermost sequences, such as the red giants of the Perseus Double Cluster, are intrinsically some 10,000 or more times brighter than the sun. (From Thorton Page and Lou W. Page, eds. *The Evolution of Stars*, in *Sky & Telescope Library of Astronomy*. Copyright © 1967 by Macmillan Publishing Co.)

A large part of astrophysical research is devoted to filling in the details of stellar evolution and the variety of nuclear reactions involved. For example, after its hydrogen is exhausted, the core of a giant star contracts and heats up to a billion degrees; then helium combines to form carbon, providing a new intense source of energy, and gas is probably blown off the star. Later, a small white dwarf remains. The explosion driving off a large fraction of a star's mass probably accounts for *supernovae*, which are seen every 100 years or so in our Milky Way and in other galaxies. One that was seen to blow up in A.D. 1054 now has a large expanding cloud of ionized gas around it—the Crab Nebula in Taurus. Near the center is a peculiar small star thought to be the core of the supernova, possibly a *neutron star* with density (of pure neutrons) about 10^{15} gm/cm^3. Such supernova remnants (SMRs) are found in other galaxies such as the Large Magellanic Cloud (LMC).

In 1968 this was confirmed when astronomers in England discovered a rapidly pulsating radio source, the first of about 300 *pulsars* now known, one of which is near the center of the Crab Nebula. By 1970 it was fairly well agreed that pulsars are rapidly rotating neutron stars with strong magnetic fields that interact with surrounding gas to emit a radio pulse on each rotation. This pulse timing is accurately periodic—1.33730 sec for the first pulsar discovered —but the pulses measured in different radio frequencies (wavelengths) are out of step, an effect due to the ionized gas along the line of sight. In several cases, the period of the pulses (star rotation period) has been found to be increasing by 10^{-6} sec or so each year, showing the "braking" action of the surrounding gas. (see PULSARS).

The *formation of a star* starts with gravitational contraction of a large cloud ("nebula") of interstellar gas and dust, including matter ejected from previous generations of giant stars. The recycling of material back and forth from nebulae to stars involves changes in composition—the abundances of helium and heavy elements increasing with time. Since 1954, astrophysicists have therefore been concerned with *nucleogenesis,* the creation of the chemical elements in stars (or in an early stage of the evolving universe). Differences in composition of stars in various locations are now interpreted as evidence of past star making.

The *formation of the solar system* (sun, planets, asteroids, meteors, and comets) is one case of star formation studied in great detail by astrophysicists, geologists and chemists. Radioactive dating of minerals in the earth, moon, and meteorites places this event about five billion years ago when a slowly rotating nebula contracted, forming earth and planets, but losing a good deal of its mass in the process. Fractionation of chemical elements and compounds during the condensation is linked with astrophysical interpretation of chemical analyses of meteorites, and lunar and terrestrial minerals.

Interstellar material in the form of bright nebulae has been known since telescopes were first used. During the first two decades of this century, evidence was collected showing less obvious clouds of dust and interstellar gas in the plane of the Milky Way, based on the obscuring and color effects of dust and the spectral absorption lines of gas (primarily sodium and ionized calcium). In 1945 the polarization of starlight caused by the interstellar dust was discovered, and in 1950 the radio telescope added the emission by interstellar atomic hydrogen at 21-cm wavelength. This interstellar medium is now known to extend in a thin, flat slab centered in the Milky Way. When highly luminous blue stars are in or near it, the gas is ionized by ultraviolet radiation, and the resulting electrons produce emission lines of hydrogen, oxygen, helium and other elements by recombination or by electron excitation. In addition to such "H II regions," the dimensions of which depend on the temperature and luminosity of the exciting star and the density of the medium, astrophysicists have studied more complex nebulae in which the material density varies from one place to another. The interstellar medium is often denser near young clusters or individual blue stars, as expected from the theory of star formation

Since 1963, radio astronomers have detected over 50 different kinds of molecules in the intersteller medium, mostly in dark clouds of dust where light from stars cannot penetrate. These molecules consist of the most common atoms, H, C. N, and O; two include sulfur (S), and one silicon (Si). Absorption lines of CH and CN were discovered in optical spectra of stars about 1940, and H_2 in far-ultraviolet spectra taken from a rocket above the atmosphere in 1969. The study of these molecules is the new subject of *astrochemistry*, and involves such questions as how the interstellar molecules are formed, what other ones should be there, and the conditions (density, temperature, turbulent motions) of the gas. Preliminary estimates are that most of the 8 simple diatomic molecules may be formed by collisions in space; the more complex ones, such as H_2CO_2, CH_3CN, and NH_3CO, are probably formed on grains of interstellar dust in clouds where the concentration of hydrogen molecules (H_2) is 10^6/cm^3 or higher. These molecules leave the dust grains after combination, but other molecules stick to the dust until strong light "boils" them off. Hence dust grains can deplete the interstellar gas in large dark nebulae—another process going on in our Milky Way.

The molecular absorption lines are at slightly different wavelengths when heavy *isotopes* ^{13}C or ^{18}O are involved instead of ^{12}C or ^{16}O, and the line strengths show the relative abundances, or "isotope ratios," which are related to the source of the interstellar gas. Relative strengths of different spectral lines from the same molecule show that the excitation temperature in instellar clouds is very low—about 3 K, and the absolute strengths of the spectral lines show that the density of, for instance, CO is 20 molecules/cm^3. Two molecules, OH and H_2O, show maser interaction; i.e., these molecules are excited so as to "pump" energy into a few selected radio wavelengths, which may come out of the gas cloud in a single direction.

The whole *Milky Way system* of stars, nebulae and interstellar gas and dust is assumed to be in dynamic equilibrium; that is, the mass distribution can be calculated from individual motions under the gravitational attraction of the whole galaxy, another important part of modern astrophysics. Since 1920 the dimensions of the galaxy have been determined from distances of the large bright globular clusters and from the distances of nearer stars. The resulting model, a flat disk with a high-density nucleus (total mass about 10^{11} suns) also fits the average motions of stars within a few thousand light years' distance from the sun, and the radial motions of cold atomic hydrogen out to 50 000 light-years determined by Doppler shifts in the 21-cm radio emission line. The stellar motions are derived from statistics of Doppler shifts and changes in direction, allowing for random individual motions differing from the general circulation.

Most of the stars in the Milky Way share in a circular velocity, v_c, around the center of the galaxy, and the mass distribution is inferred from the observed change of v_c with distance from the center. The globular clusters and many other stars appear to move in orbits at high inclination to the Milky Way plane, forming a "halo" around the center having little or no angular momentum. It thus appears that there are two populations in the galaxy: "Population I" stars, nebulae, gas and dust in the outer parts of a thin rotating disk, and "Population II" stars in the nonrotating halo, probably formed at an earlier time. A large fraction of Population I in the disk is in the form of nonluminous interstellar dust and gas, although this interstellar material is only 10 to 15 percent of the total mass of the system. During the 1970s, after the "mass discrepancy" was recognized, several astrophysicists assumed the presence of much more nonluminous material in the halo. They found some evidence of such "dark haloes" in other galaxies (see below), and calculated that as much as half the mass of a galaxy is in its dark halo. Efforts have been largely successful in confirming this assumption for the Milky Way system.

There is a weak *magnetic field* between the disk stars which probably lines up the interstellar dust particles like small magnetic needles, and this pattern accounts for the polarization of starlight passing through. The field also deflects moving ions and electrons, and may account for the energy of very high-speed *cosmic rays*. Lower-speed cosmic rays are ejected by solar flares, and other stars probably do the same. It is possible that these ejected ions are accelerated to enormous speeds by the uneven magnetic field in the Milky Way system. Cosmic rays are studied above the atmosphere to determine the relative abundance of different ions, their energy spectrum, and source. Their energy flux is about the same as starlight, and they probably affect the temperature of the interstellar gas. They certainly produce x-rays and gamma rays by collisions with atoms and molecules.

The many *other galaxies* well outside our own are found to include some (classed as "spirals") very similar to our Milky Way galaxy in structure and internal motions. Others are strikingly different (classed as "ellipticals"), probably due to different conditions of formation. All the techniques of astrophysics are being applied to the study of these objects: measurement of their sizes, luminosities, colors and masses, their proportion of interstellar material, the formation and evolution of their stars, etc. These physical characteristics are fairly well correlated with morphological type; the spirals are similar to our Milky Way, but the ellipticals have almost no interstellar material, and are much more massive for their size and luminosity than the spirals.

The *evolution of galaxies* is due to the ageing of their member stars. This is a slow process, and no change can be observed during a man's lifetime, except for supernova explosions, and light variations in the nuclei of a few "active galaxies." But as we look to more distant (fainter) galaxies, we are looking back in time, seeing them at a younger age. This is because faint galaxies are hundreds of millions of lightyears away, and the light we see now left them hundreds of millions of years ago. By 1986, it may be possible to look back more than 5 billion years, using the new Space Telescope, launched into orbit then by Space Shuttle. Astrophysicists have traced galactic evolution backwards and expect that elliptical galaxies were very luminous 10 billion years ago, when their blue supergiant stars had just formed. They are eager to confirm this, and also to observe young spiral galaxies with Space Telescope.

The *mass* of a single galaxy can be measured from the rotational velocity, v_c, near the rim and the radius, R. Then $M = v_c^2 R/G$, where G is the gravitational constant, and $v_c = v/\sin i$, where v is measured from the relative Doppler shift (rim to center), and i is the inclination of the disk to the line of sight. However, R is poorly determined because galaxies do not have sharp rims. Moreover, the observed change of v_c with distance from the nucleus r does not always follow the expected form of a rapid rise to a maximum and then a decrease ($v_c \propto 1/\sqrt{r}$) after r includes most of the galaxy's mass. During the 1970s many such velocity curves were measured for spiral galaxies, and several showed no peak; that is, v_c remained at its peak value out to r for the faintest rim stars observable. This is considered evidence of a spherical dark

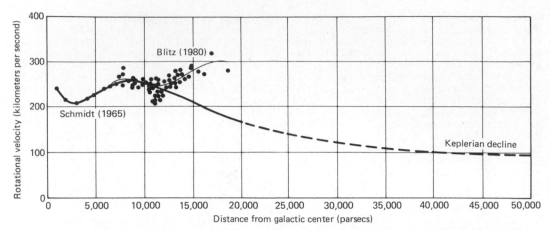

FIG. 4. Rotation curve graphs the circular velocity of matter in rotation about the center of the Milky Way. Here two such curves are drawn. A rotation curve plotted in 1965 by Maarten Schmidt of the Hale Observatories (*solid black line*) shows a circular velocity that declines toward the limit of the visible galaxy at 20,000 parsecs. If all the mass in the galaxy lay inside that limit, a test mass placed farther out would rotate at lower speed, in approximate obedience to a law first formulated (for the motion of the planets) by Johannes Kepler (*broken line*). Data analyzed by Leo Blitz and his colleagues at the University of California at Berkeley now yield a rotation curve (*light solid line*) that rises toward a value of 300 kilometers per second at 20,000 parsecs. The rise of the newer curve implies unseen mass in great quantity outside the visible limit of the galaxy. Each point in the newer data is the circular velocity of a cloud of hydrogen atoms. (From Bart J. Bok, "The Milky Way Galaxy," *Scientific American*, March 1981. Copyright © 1981 by Scientific American, Inc. All rights reserved.)

halo—nonluminous material extending almost to the edge of the luminous disk. Fig. 4 shows the velocity curve for the Milky Way system.

Another method can be used to measure the masses of the many galaxies in pairs, one orbiting around the other. In such a pair, $M_1 + M_2 = (v_1 - v_2)^2 S/G$, where S is the separation. However, the inclination of the orbit cannot be measured, and an average value of M must be obtained from statistics of many pairs. This method shows that the average spiral has $M_S = 4 \times 10^{10}$ suns $= 8 \times 10^{43}$ gm, while the average elliptical has 30 times this mass. The luminosity of either type is 10^{10} suns, on the average, so a spiral has about four times as much mass per watt of light output as the sun does, while an elliptical has over 100 times as much.

Except for the dark haloes, the mass of most galaxies seems to be concentrated in their nuclei, and several of these nuclei appear to be exploding. The most dramatic is the irregular galaxy, Messier 82, with jets coming out of the nucleus (Fig. 5). Doppler shifts in these jets show that they are moving outward at about 1000 km/sec. A few dozen *Seyfert galaxies* have active nuclei that vary in brightness and show strong, broad emission lines indicating explosive conditions. Both jets and Seyfert nuclei, as well as the superluminous quasistellar objects (QUASARS) are thought to be powered by black holes at the center. The power comes from mass falling into the black hole at rates of up to 1 solar mass (2×10^{33} gm) per year. Such black holes are estimated to have

mass M_H, about 10^8 suns within a radius $R_H = 2GM_H/c^2 = 3 \times 10^{13}$ cm $= 3 \times 10^{-5}$ lightyear. (This "Schwarzschild radius" R_H comes from setting the velocity of escape v_e equal to the velocity of light c in the equation $\frac{1}{2}mv_e^2 = GmM/R$.) With such a small radius (0.01 lightday), the black hole in the nucleus can vary rapidly in brightness as gas falls into it.

Galaxies are observed in increasing numbers at larger and larger distances, roughly in uniform distribution, but with marked clustering. In 1981, a large void was found, about 1 billion lightyears from us in the direction of the constellation Bootes, where astrophysicists can detect no galaxies in a region 300 million light years across. There is some evidence that compact clusters contain a preponderance of ellipticals; the total number of galaxies in the Coma Cluster is about 500, of which only a dozen are spirals. The total mass of such clusters can be measured by the relative velocities of galaxies near the edge, and when this was done (first in 1932), astrophysicists were surprised to find that the cluster mass is 5 to 10 times larger than the sum of all galaxy masses in it. This *mass discrepancy* means that we see only 10 to 20 percent of the material in a large region of space—possibly in the rest of the observed universe, as well. The "missing mass" may be a thin intergalactic gas or discrete, nonluminous objects, or due to the fact that many galaxies have dark haloes and masses 10 times as large as the "average mass" used. Another possibility was brought out in 1980, when physicists

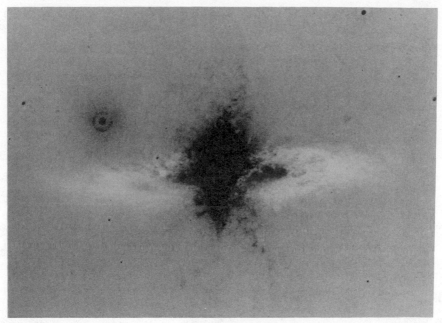

FIG. 5. The irregular galaxy Messier 82, photographed in red light of hydrogen with the 200-inch Palomar reflector. The enormous puff of luminous hydrogen gas escaping from the nucleus of this system stretches over 10,000 light years above and below the disk of the galaxy, which is about 20,000 light years across. (Palomar Observatory Photograph.)

found evidence that neutrinos have a small mass. Before then, neutrinos were considered to have a rest mass of zero. Nuclear reactions in the Sun's core are spewing out neutrinos at the rate of 10^{38} per second, and more luminous stars do so at even higher rates. Over the past 15 billion years, all space must have been filled with neutrinos, although their number levelled off as the density increased and neutrino pairing began to destroy them as fast as they are being created. Light as they are, the large numbers of neutrinos add up to a mass somewhat larger than all the visible galaxies, and provide a good part of the "missing mass" in the universe.

The spectra of distant galaxies all show large red shifts which, if interpreted as Doppler shifts, indicate a recessional velocity proportional to distance (Hubble's Law). In fact, the red shift is used to measure all distances (D) larger than about 40 million light-years by $D = v_R/H$, where H = 17 km/sec/million light years. Starting in 1958, radio astronomers began to survey the whole sky for faint radio sources, and found that many galaxies are strong radio emitters. The most puzzling were faint, starlike (quasi-stellar) objects, soon named *quasars*. In 1963 these were found to have very large red shifts, therefore very far away (up to 5 billion light-years) and about 100 times more luminous than the average galaxy. This large energy output is probably an enormous explosion, lasting but a brief fraction of a galaxy's life. At these large distances, we see now the quasars as they were 5 billion years ago, and it seems likely that many galaxies exploded at that time (see QUASARS).

Here astrophysics leads into COSMOLOGY, based on general RELATIVITY. Observations of the short-wave (3-mm) radio emission coming from all over the sky (isotropic) seems to confirm the "big-bang" cosmological model of the universe, and to disprove the steady-state theory which assumes continuous creation of matter. The *isotropic background radiation* with $I(\lambda)$ matching a black body at 3 K, is the remnant of the very hot explosion about 20 billion years ago that sent the galaxies moving outward to the great distances where we see them today. Astrophysical evidence for the missing mass in clusters of galaxies shows that the average density of matter in the universe is probably higher than 10^{-28} gm/cm^3 so that the space curvature predicted by general relativity fits the numbers of galaxies we count at different distances. For higher density of matter, the predicted curvature is larger, and at the extreme, space "wraps itself around" a large collapsed mass so that no light or radio waves can get in or out. This leaves a *"black hole"* which exerts gravitational attraction on distant masses, but cannot be seen—a possible explanation of the missing mass in clusters of galaxies.

THORNTON PAGE

References

Abell, G. O., "Exploration of the Universe," 2nd Ed., New York, Holt, Rinehart and Winston, 1969.

Pasachoff, J., and Kutner, M., "University Astronomy," Philadelphia, Saunders, 1978.

Page, T., and Page, L. W., "The Evolution of Stars," New York, Macmillan, 1968.

Page, T., and Page, L. W., "Stars and Clouds of the Milky Way," New York, Macmillan, 1968.

Bok, B., and Bok, P., "The Milky Way," Cambridge, Mass., Harvard Univ. Press, 1981.

Page, T., and Page, L. W., "Beyond the Milky Way," New York, Macmillan, 1969.

Page, T., and Page, L. W., "Space Science and Astronomy," New York, Macmillan, 1976.

Burbidge, G., and Hewitt, A., "Telescopes for the 1980s," Palo Alto, Annual Reviews, 1981.

Longair, M. S., and Warner, J. W., "Scientific Research with the Space Telescope," NASA CP-2111, Washington, US Govt. Printing Office, 1979.

Burbidge, Burbidge, Fowler, and Hoyle, "Synthesis of the Elements in Stars," *Rev. Mod. Phys.*, 29, 547 (1957).

Menzel, D. H., Whipple, F., and deVaucouleurs, G., "Survey of the Universe," Englewood Cliffs, N.J., Prentice-Hall, 1971.

Cross-references: ASTROMETRY, COSMOLOGY, GRAVITATION, DOPPLER EFFECT, PULSARS, QUASARS, RELATIVITY, SOLAR ENERGY SOURCES, SOLAR PHYSICS.

ATOMIC AND MOLECULAR BEAMS

This field of research utilizes a collision-free stream of neutral atoms or molecules as they traverse a vacuum chamber. With 10^{-6} mm Hg pressure in a vacuum chamber, air molecules at room temperature travel on the average about 300 meters between collisions and move with an average speed of about 500 m/sec. Between collisions these molecules are essentially "free" and unperturbed by molecules of the residual gas or by atoms in the walls of the apparatus. The mathematical description of such isolated systems is much less complicated than for denser gases, liquids, or solids containing interacting particles.

Since 1911, when Dunoyer proved that a stream of neutral atoms would remain collimated in a vacuum, atomic- and molecular-beam research has become one of the most versatile, precise, and sensitive techniques for studying the properties of isolated atomic systems and interactions between such systems. Numerous fundamental discoveries in beam research have contributed to the present understanding of physical laws. The earliest experiments (1920) sought the molecular velocity distribution, which is important in the kinetic theory of gases. Atomic diameters (cross sections), van der Waals' interaction potentials, and polymer vapor composition were obtained after later refinements of technique.

The Stern-Gerlach experiment (1924) demonstrated the validity of space quantization of angular momentum and established the electron spin as 1/2. This historic work placed quantum mechanics on a firmer foundation and initiated beam investigations of atomic and nuclear electromagnetic properties. Although many low-precision results appeared in subsequent years, high-precision spectroscopy began in 1937 with the introduction of the magnetic-resonance method by Rabi. In this method, transitions between quantum states separated by an energy $h\nu$ are induced by a radio-frequency field of frequency ν (h is Planck's constant). From the Heisenberg uncertainty principle, the width of the rf resonance is small, owing to the long lifetimes of the beam quantum states and to the ability to irradiate the beam with radio frequency for as long as a few milliseconds along its path.

Precision atomic-beam measurements have contributed to many theoretical and practical developments. The deuteron quadrupole moment (1939) pointed to the necessity of a tensor interaction in nuclear forces. The anomalous electron moment (1949) and the Lamb shift (1950) in the atomic-hydrogen fine structure were resolved by quantum electrodynamics. Nuclear spins (I) as well as magnetic-dipole (μ) and electric-quadrupole (Q) moments have been important in providing test information for the shell model (1949) and collective model (1953) of the nucleus. Atomic hyperfine-structure constants (dipole, a; quadrupole, b; and octupole, c), which describe the interaction between the electrons and the nucleus, as well as the numerous constants required to describe a molecule and its internal interactions, have contributed to the theory of atomic and molecular structure. The cesium "clock" or frequency standard (1952) represents a widespread practical application of beam technology by using the hyperfine-structure transition at 9192.631770 MHz (Ephemeris time) to regulate a quartz-crystal oscillator. Other frequency standards such as the thallium clock, ammonia maser (1954), and hydrogen maser (1960) also employ beam techniques for quantum-state selection. Very recent, high-energy nuclear accelerators have been equipped with atomic-beam sources to produce polarized protons.

Among nonresonant experiments, beams impinging on solid surfaces produce information on the wave nature of particles, work functions, and accomodation coefficients. Charge-exchange cross sections and interaction potentials are obtained from experiments with crossed beams, one neutral and one charged. Chemical reaction kinetics in isolated systems are studied in crossed beams of two reactants. Precision optical studies combine beam and laser techniques.

Four individuals have received Nobel Prizes for beam research: Otto Stern (1943) "for his contribution to the development of the molecular-ray method and for his discovery of the

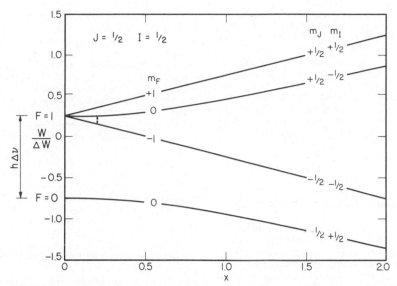

FIG. 1. Hyperfine-structure energies (ordinate) of an atom with $J = 1/2$ and $I = 1/2$ in an external magnetic field (abscissa).

magnetic moment of the proton"; I. I. Rabi (1944) "for his application of the resonance method to the measurement of the magnetic properties of atomic nuclei"; P. Kusch (1955) "for his precision determination of the magnetic moment of the electron"; and W. E. Lamb (1955) "for his discoveries concerning the fine structure of the hydrogen spectrum."

A discussion of the energy levels of a simple atom and of one particular apparatus will illustrate the magnetic-resonance technique. An isolated atom in an external field, H, has energy states which are calculable from the Hamiltonian

$$\mathcal{H}/h \text{ (Hz)} = a\mathbf{I} \cdot \mathbf{J} + b \text{ (quadrupole operator)}$$

$$- g_J(\mu_0/h)\mathbf{J} \cdot \mathbf{H} - g_I(\mu_0/h)\mathbf{I} \cdot \mathbf{H}.$$

where some symbols have been defined previously, $g_J = \mu_J/(J\mu_0)$, $g_I = \mu_I/(I\mu_0)$, and μ_0 is the magnitude of the Bohr magneton. The first two terms represent the dipole and quadrupole hyperfine-structure interactions between the electrons and nucleus; the last two terms express the interaction of the electron and nuclear magnetic moments with the external magnetic field. For the simple case of $I = 1/2, J = 1/2, b =$

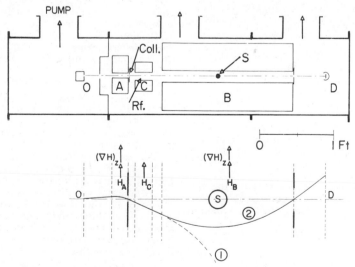

FIG. 2. Schematic of an atomic-beam, magnetic resonance apparatus.

0, as in the ground electronic state of atomic hydrogen, the energy levels that arise from different "relative orientations" of the nuclear (I) and electronic (J) spins are shown in Fig. 1 as a function of the magnetic-field parameter $X = [(-g_J + g_I)\mu_0 H]/\Delta W$. Here ΔW is the hyperfine-structure separation at $H = 0$. The levels are labeled by either the low-field quantum numbers (F, m_F) or by the high-field numbers (m_J, m_I), where $F = |I \pm J|$, and the m's are the projections of F, I, or J along the field direction.

Of the many magnetic or electric resonance apparatuses, one specialized type has proved valuable for measuring atomic properties of both stable and radioactive isotopes. In Fig. 2, the "oven" or source, O, may take one of many forms—a microwave discharge to dissociate gaseous diatomic molecules, a closed tantalum crucible (with an exit slit) heated by electron bombardment, or one of many other devices for evaporating atoms. The atoms pass between the poles of three separate electromagnets (denoted A, C, and B, successively, from oven to detector). The inhomogeneous A and B magnets have eccentric cylindrical pole tips which produce a field gradient, $\partial H/\partial Z$. In this field, an atom experiences a force $F = \mu_{eff}(\partial H/\partial Z)$, where $\mu_{eff}(= -\partial W/\partial H)$ is the negative slope of an energy level in Fig. 1. Within the homogeneous C field, a superimposed rf field induces state changes. Thus an atom which remains in a single state [(1, 0) for example] is deflected similarly by the strong A and B magnets and follows trajectory 1 in Fig. 2. The stopwire, S, shields the detector from fast atoms and atoms with small deflections. If, in the C-field region, a transition occurs that causes the high-field slope ($-\mu_{eff}$) to change sign [e.g., $(1, 0) \rightarrow (1, -1)$], the A and B deflections are opposite, and the atom follows trajectory 2 to the detector D. A resonance is observed as an increase in beam intensity at the detector. Values of the constants in the Hamiltonian are deduced from the observed resonant frequencies of the atoms in known magnetic fields. Some detection methods in frequent use are:

(a) Deposition on a surface with subsequent assay by radioactive counting, neutron activation, or optical means (earliest detector).

(b) Ionization of alkali atoms on a hot tungsten wire, and measurement of the resulting ion current.

(c) Electron-bombardment ionization with subsequent mass analysis to discriminate against background gas ions. The beam ions are frequently counted by using electron multiplier tubes.

(d) Other detectors employing the principles of radiometers, pressure manometers, thermopiles, bolometers, laser fluorescence, and changes in space charge.

HOWARD A. SHUGART

References

W. J. Childs, "Hyperfine and Zeeman Studies of Metastable Atomic States by Atomic-Beam Magnetic-Resonance," in *Case Studies in Atomic Physics*, Vol. 3, No. 4, North-Holland, Amsterdam, 1973.

English, T. C., and Zorn, J. C., "Molecular Beam Spectroscopy," in "Methods of Experimental Physics," vol. 3, 2nd edition, Academic Press Inc., New York, 1973.

Estermann, I., Ed., "Recent Research in Molecular Beams," a collection of papers dedicated to Otto Stern, on the occasion of his seventieth birthday, New York, Academic Press, 1959.

M. A. D. Fluendy and K. P. Lawley, "Chemical Applications of Molecular Beam Scattering," Chapman and Hall, London, 1973.

Kopfermann, H., "Nuclear Moments," English translation by E. E. Schneider, New York, Academic Press, 1958.

Kusch, P., and Hughes, V. W., in Flügge, S., Ed., "Handbuch der Physik," Vol. 37/1, Berlin, Springer-Verlag, 1959.

Nierenberg, W. A., in *Ann. Rev. Nucl. Sci.*, 7, 349 (1957).

Ramsey, N. F., "Molecular Beams," London Oxford University Press, 1956 (reprinted in 1963).

Smith, K. F., "Molecular Beams," London, Methuen and Company, 1955.

Cross-references: COLLISIONS OF PARTICLES, CROSS SECTION AND STOPPING POWER, MAGNETIC RESONANCE, CHEMICAL KINETICS.

ATOMIC CLOCKS

Time Standards The concept of time has at least two distinct aspects: (1) duration (or time interval), and (2) date. Ordinary clocks can be used for either aspect, while stopwatches measure duration only. The concept of date includes not only the calendar day but finer subdivisions as well. For example, a date might be 10:27 AM EST, November 13, 1982. Both duration and date use the same set of units; that is days, hours, minutes, and seconds can refer to either date or duration.

Almost all clocks have three main parts: (1) A pendulum or other periodic phenomenon to control the clock rate; (2) a counting element to count the cycles of the periodic element just mentioned; and (3) a display mechanism to display the current count (i.e., time). Atomic clocks use natural resonances in atoms or molecules for the periodic part. It is possible to find natural resonances which are very insensitive to environmental conditions as well as being insensitive to the detailed construction of the clock. This allows independent laboratories to fabricate atomic clocks which agree in rate with other comparable clocks to within a few parts in ten to the 14th power.

The Early Atomic Clocks In 1949, Harold Lyons at the U.S. National Bureau of Standards announced the operation of the world's first atomic clock. This clock was based on microwave absorption in ammonia and was stable to about one part in a hundred million, which was roughly comparable to the best clocks of that time (pendulum and quartz).

In the 1950s and early '60s, scientists and engineers investigated various atomic and molecular resonances as well as various means of extracting the desired information to run a clock. In the mid 1940's, I. I. Rabi suggested using cesium in an atomic beam as a frequency standard. With innovations due to N. Ramsey, the essential design of the cesium beam clock was complete as we know it today, with, of course, numerous small refinements.

In the mid-'50s, C. H. Townes and others at Columbia University demonstrated the ammonia MASER (Microwave Amplification by Stimulated Emission of Radiation). In the early '60s, N. Ramsey and others developed the hydrogen maser. This maser is unique in that the hydrogen maser stores hydrogen atoms in a quartz bulb for periods of a second or so. This relatively long interaction time results in a very sharply defined resonance (i.e., a high Q). The hydrogen maser has demonstrated the highest frequency stability of any atomically controlled device for sampling periods out to a few hours. Even today, hydrogen masers are used in special applications where this high stability is needed. Because the hydrogen atoms collide with the walls of the quartz storage bulb, the hydrogen resonance is shifted. Various wall coatings have been used in attempts to avoid this and other problems, but so far the hydrogen maser has not been able to compete with the cesium beam device for absolute frequency accuracy.

Also in the '50s, scientists and engineers developed the rubidium gas cell. This device optically pumps rubidium atoms (contained in a sealed cell) to a higher energy state. A microwave signal relaxes the atoms to the lowest hyperfine level in the ground state provided this microwave signal is suitably close to the natural rubidium resonance. The resonance condition is sensed typically by monitoring the light from the optical pumping as a function of the microwave frequency. A buffer gas is added to the cell containing the rubidium vapor to reduce the effects of collisions with the cell walls. The commercial versions of rubidium devices are less expensive and less stable than cesium clocks, and typically display a systematic frequency drift. Still there are several thousand commercial rubidium cells in use today.

The Cesium Beam Atomic Clock In the mid-'50s commercial cesium beam clocks became available for the first time. The early versions were accurate and stable to about one part in

ten to the tenth. Although expensive, several standards laboratories used them.

In a joint experiment between the US Naval Observatory (USNO) and the National Physical Laboratory (NPL) at Teddington, England, the frequency of the appropriate transition in cesium was measured in terms of the standard of the second as then defined (the Ephemeris Second). This experiment involved NPL's laboratory cesium frequency standard and USNO's determination of Ephemeris time. The experiment lasted about three years and resulted in a frequency value for cesium of 9,192,631,770 Hz. The accuracy was thought to be about one or two parts in ten to the ninth, due mostly to the difficulties in measuring Ephemeris time, not the cesium resonance.

In the '60s and '70s competing firms developed smaller and more advanced cesium beam models. Although the basic design remained about the same, the beam tube was made smaller and the electronics were reduced greatly in size with the advances in solid-state electronics. The performance of cesium clocks was close to the theoretical limits imposed by the granularity of the beam (i.e., "shot-noise"), and design limitations such as size, power, and reliability.

Several primary standards laboratories built laboratory versions of the cesium beam standard. The basic philosophy was to build a device which could allow the laboratory to measure any physical parameter which might possibly influence the frequency. Experiments were performed to test agreement of theory and performance of the standard. The laboratory then made a final accounting of uncertainties caused by such things as imperfect knowledge of the magnetic fields within the beam tube and many others. The resulting figure is an accuracy estimate of that laboratory's frequency standard. Historically, the individual accuracy estimates of the few primary frequency standards in existence have normally (but not always) been in accord with the measured differences between standards. Any disparity never has been really large.

On Friday, October 13, 1967, the General Conference of Weights and Measures (CGPM) announced a new definition of the second:

> The second is the duration of 9,192,631,770 periods of the radiation corresponding to the transition between the two hyperfine levels of the ground state of the cesium-133 atom. [13th General Conference of Weights and Measures (1967), Resolution 1.]

International Time Scales Historically, time has always been closely tied to the position of the sun in the sky. An ordinary sundial, for example, indicates apparent solar time. Keppler's laws allowed scientists to correct apparent solar time for the earth's elliptic orbit and the inclination of the earth's spin axis to the ecliptic plane (the plane of the earth's orbit). This correction is called the "Equation of Time," and can often be found engraved on sundials. These corrections amount to as much as plus or minus 15 minutes during the year. This corrected time is often called UT0, Universal Time, "zeroth" approximation.

Astronomers discovered that UT0 measurements at different observatories were not in agreement. This disagreement amounted to about ±30 milliseconds and was traced to the fact that the earth is not precisely fixed to its axis of rotation. That is, the pole wanders around within a circle of about 15 meters in diameter. Of course, the natural response was to correct time for the wandering of the pole. This new time was called UT1.

In the 1930s, scientists discovered that UT1 had periodic variations. Although corrections were applied to UT1 to get UT2, this latter time scale does not indicate the actual variations in the earth's rotation rate nor is it significantly more uniform than UT1. Today, one seldom hears of UT2.

Back near the turn of the century, Simon Newcombe, at the US Naval Observatory, computed a table of future positions of the sun, moon, and some of the principal planets for several years in advance. The computations were based on the best theories available. Such a table is called an "Ephemeris." Newcombe discovered, however, that the various celestial objects systematically departed from their predicted positions. He noticed that if the *time* were somehow in error, then the tables would be in good agreement for all of the objects. Newcombe correctly recognized that the earth's rate of rotation was not constant.

The natural response was to use the Ephemeris backwards. That is, when the sun reached its predicted position, the *time* was the value listed in the Ephemeris. Conceptually, this kind of time, called Ephemeris Time, is based on the orbital motion of the earth, and should be more uniform in its rate than the earth's rotation, since the earth can undergo geometrical changes that alter its rotational rate.

The definition of the second, the unit of measure in the International System (S.I.) of units, was defined as the fraction 1/86,400 of the mean solar day (24 X 60 X 60 = 86,400), prior to 1956. In 1956 the second was redefined as the fraction 1/31,556,925.9747 of the tropical year 1900. Since 1967, the definition of the second has been in terms of cesium, as noted above.

The definition of the International Atomic Time Scale (TAI) incorporates the definition of the second. The formal definition reads:

International Atomic Time (TAI) is the time reference coordinate established by the Bureau International de l'Heure (BIH) on the basis of the readings of atomic clocks functioning in various establishments conforming to the definition of the second, the time unit of the International System of Units. [14th CGPM (1971), Resolution 1.]

International Atomic Time is a coordinate time scale. The reference elevation is mean sea level and frequency corrections are applied in transferring time measurements between laboratories. The largest gravitational "blue-shift" is 1. 8 parts in ten to the 13th for the National Bureau of Standards (elevation 1.7 km). Accurate and precise comparisons of time scales on an international basis now require relativistic corrections.

In spite of the importance of TAI, many operations require a time reference tied to earth position (i.e., UT1). Unfortunately, the earth is a poor clock compared to atomic clocks. Still, earth time is quite important. For example, small boat owners often use sextants for celestial navigation. For these purposes (and others), pure atomic time is not useful since it is not related to earth position. Beginning in 1972, nations agreed to a compromise time scale between the needs for atomic time and earth time. The compromise time scale, called Coordinated Universal Time (UTC), runs at precisely the same rate as TAI, but offset an integral number of seconds. When UT1 accumulates nearly an entire second error relative to UTC, the UTC clocks are reset one complete second, called a "leap second."

The rate of rotation of the earth is variable at a few parts in ten to the eighth. (Atomic clocks are almost a million times less erratic.) Unlike leap years, leap seconds are not predictable years in advance. Since 1972, eleven leap seconds have been added (Fall, 1982) to UTC; bringing the total time difference to 21 seconds. (TAI was set to agree with earth time January 1, 1958.)

The official time almost everywhere in the world is based on UTC, with an integral number of hours difference for the appropriate time zone.

Who Needs It? Many of the uses of precise time are related to distance measurements. In fact, a new definition of the meter is expected to be in terms of the distance traversed by a light wave in a specified fraction of a second. Thus, the cesium beam will be the standard for *both* the second and the meter. The speed of light then becomes a defined constant rather than a subject for measurement.

At this point, there are probably a few thousand commercial cesium beam devices in operation. They can be found in television stations, telephone control systems, NASA tracking stations, scientific laboratories, standards laboratories, and military systems, to name a few. The Department of Defense is currently developing a satellite navigation system, called the Global Positioning System (GPS), which is dependent on state-of-the-art timing systems. Atomic clocks are essential to the success of the GPS program.

Future Clocks For over 25 years now, the cesium beam frequency standard has surpassed all potential rivals. The accuracies specified by the national standards laboratories have steadily dropped to finer and finer tolerances. Today's standards (1982) quote accuracies of a very few parts in ten to the 14th. Still there are new potential rivals, as well as ideas to improve the cesium beam standard.

The leading candidate in several laboratories is trapped ions. The basic idea is to take a number of ions of some particular element and hold them in a Penning (or rf) trap. The trap can be arranged to cause only insignificant frequency shifts in the ions, and can hold the same set of ions for extended periods of time. This allows long interaction times with the interrogating signal and hence provides very narrow resonance linewidths.

A modification of the trapped ion method cools the ions to temperatures below one Kelvin. The cooling is accomplished by irradiating trapped ions with a laser set slightly below an optical transition frequency. The motion of warm ions will Doppler shift the laser radiation into resonance with the ion, and the ion will absorb a photon. At some time later, the ion will reradiate the photon in a random direction. The ions thus absorb photons with an energy bias below the ion's natural resonance frequency but reradiate photons with no such bias. The energy imbalance is supplied by the kinetic energy of the ions—thus cooling them.

In present atomic frequency standards, the motion of the atoms cause the primary limitations on improved accuracy. Although a lot of effort has been expended on attempts to cancel first order Doppler shifts, significant difficulties remain. Second order Doppler shifts are also significant. Laser cooling may solve both of these problems.

JAMES A. BARNES

References

Ashby, N., and Allan, D. W., "Practical Implications of Relativity for a Global Coordinate Time Scale," *Radio Science*, **14**, 649 (1979).

Helwig, H., Evenson, K. M., and Wineland, D. J., "Time, Frequency, and Physical Measurement," *Physics Today*, **31**, 23 Dec. (1978).

Markowitz, W., Hall, R., Essen, L., and Parry, J., "Frequency of Cesium in Terms of Ephemeris Time," *Phys. Rev. Lett.*, **1**, 105–107 (Aug. 1958).

D. J. Wineland, "Laser Cooling of Atoms," *Phys. Rev. A*, **20**, 1521 (1979).

Cross-references: ATOMIC AND MOLECULAR BEAMS, ATOMIC PHYSICS, ATOMIC SPECTRA, DOPPLER EFFECT, FREQUENCY STANDARDS, KEPPLER'S LAWS OF PLANETARY MOTION, LASERS, MASER, MICROWAVE SPECTROSCOPY, MICROWAVE TRANSMISSION, OPTICAL PUMPING, QUANTUM THEORY, RELATIVITY.

ATOMIC ENERGY

The terms "atomic energy" and "nuclear energy" are used interchangeably in the contemporary literature to mean energy that originates within the atomic nucleus. Events that release atomic energy involve basic changes in nuclear structure and result in the formation of one or more different nuclides, which may be isotopes of the original atom or altogether different elements. The release of atomic energy is thus a more fundamental process than the release of chemical energy, which merely involves a regrouping of intact atoms into different molecular forms.

To date, three basic atomic energy mechanisms have been exploited in practical applications: (1) the fission of certain heavy nuclides; (2) the fusion of certain light nuclides; and (3) the process of radioactive decay. These will be discussed in the order listed.

Fission In fission, a heavy nuclide splits into two lighter and predominantly unstable nuclides, commonly referred to as fission products, with the accompanying emission of several neutrons and the release of approximately 200 MeV of energy. Nuclides that readily undergo fission on interaction with low-energy or "slow" neutrons (< 0.5 eV) are referred to as fissile materials. There are three primary fissile materials:

(1) Uranium 235, which is a natural constituent of the uranium element and accounts for 0.71 per cent by weight of that element as found in nature.

(2) Plutonium 239, formed by neutron irradiation of uranium 238.

(3) Uranium 233, formed by neutron irradiation of thorium 232.

Uranium 238, which does not undergo fission on interaction with slow neutrons, does so on interaction with high-energy or "fast" neutrons (> 0.1 MeV). Table 1 lists representative fission energy distributions for the four nuclides cited.

A useful rule of thumb is that an energy release of 200 MeV per fissioning atom corresponds to an output of approximately one megawatt-day of thermal energy per gram of fissioned matter.

Practical applications of fission are based on the principle of a self-sustaining fission chain reaction, i.e., a reaction in which a neutron emitted by atom A triggers the fission of atom B, and one from atom B triggers the fission of atom C, and so on. For this to be achieved requires the assembly of a "critical mass" of fissile material, i.e., an amount sufficient to reduce the probability of neutron losses to a threshold value. The amount required depends on a number of factors, notably the concentration of the fissile material used and the composition and geometry of the reaction system.

There are two basic application concepts. One is the essentially instantaneous fission of a mass of highly concentrated fissile material in such a way as to generate an explosive force. This, of course, is what occurs in atomic weapons. Atomic explosives are also of interest in connection with peaceful uses such as large-scale excavation projects.

TABLE 1. ENERGY DISTRIBUTION IN FISSION

Type of Energy	Quantity of Energy (MeV)			
	Slow Fission			Fast Fission
	^{235}U	^{239}Pu	^{233}U	^{238}U
Kinetic energy of fission products	165	172	163	163
Kinetic energy of neutrons emitted	5	6	5	5
Instantaneous emission of gamma rays	8	7	7	7
Beta emission during fission product decay	9	9	9	9
Gamma emission during fission product decay	7	7	7	7
Total[a]:	194	201	191	191

[a]Exclusive of nonrecoverable energy associated with neutrino emission during fission product decay. All numbers are rounded to the nearest integer. It should be mentioned that 8 or 9 MeV of additional energy become available in a nuclear reactor as the result of neutron capture and subsequent gamma-decay phenomena.

The other application concept is that of the controlled and gradual fission of an atomic fuel in a nuclear reactor, which may be designed for one or more of the following principal purposes:

(1) To provide fluxes or beams of neutrons for experimental purposes. This category of use includes research and materials-testing reactors.

(2) To produce materials by neutron irradiation. Examples are reactors used primarily to produce plutonium for atomic weapon stockpiles or for the production of various radioisotopes for use in science and industry.

(3) To supply energy in the form of heat for such applications as the generation of electric power, the propulsion of ships or space vehicles, or the production of process steam.

The first demonstration of a fission chain reaction was achieved by E. Fermi and co-workers on December 2, 1942 when the world's first nuclear reactor (Chicago Pile No. 1) was successfully operated in a converted squash court beneath Stagg Field at the University of Chicago.

The most important application of fission promises to be in the electric power field. The basis for this expectation is that, if exploited efficiently, known and inferred deposits of atomic fuels represent a potential energy reserve many times larger than that of the fossil fuels (coal, oil and natural gas) on which the world's electric energy economy largely depends at present.

By the end of 1981 approximately 150,000 electrical megawatts of atomic power capacity were in operation, under construction, or planned for construction in the United States.

Fusion Fusion is a general term for reactions in which the nuclei of light elements combine to form heavier and more tightly bound nuclei with the simultaneous release of large amounts of energy. In order for this to occur the interacting nuclei must be brought sufficiently close together to permit short-range nuclear forces to become operative. This means that one or both nuclei must be accelerated ("heated") to velocities sufficient to overcome the strong electrostatic repulsion that exists between particles having the same electrical charge. The velocities required correspond to particle "temperatures" of the order of tens or hundreds of millions of degrees, which in turn correspond to particle energies of thousands or tens of thousands of electron volts. The term "thermonuclear" reactions is reserved for fusion reactions in which both nuclei are traveling at high velocity (as distinct from reactions between an accelerated projectile particle and a static target nucleus, as in particle accelerator experiments).

The only practical application of thermonuclear reactions developed to date is in thermonuclear weapons (so-called "hydrogen bombs") in which the energy released by a charge of fissile material serves to create the conditions required to bring about the reaction of "fusionable" materials. The first test of a thermonuclear weapon, which was the first demonstration of a man-made thermonuclear reaction, took place on October 31, 1952 at a U.S. testing site in the Marshall Islands. Peaceful uses of thermonuclear explosives are being studied and have the advantage, relative to straight fission-based explosives, that problems of radioactive contamination are greatly reduced. This reflects the fact that the nuclides formed by fusion are stable and hence, apart from neutron activation effects, the formation of radioactive substances is limited to the fission component of the explosive.

Research has been in progress for two decades on techniques for controlling the fusion process as a means of supplying energy for electric power generation. The thermonuclear reactions of primary interest in this context are the deuterium-tritium reaction:

$$D + T \rightarrow {}^4He + n + 17.6 \text{ MeV}$$

and the deuterium-deuterium reactions:

$$D + D \begin{cases} \nearrow {}^{3}\text{He} + n + 3.2 \text{ MeV} \\ \searrow T + p + 4.0 \text{ MeV} \end{cases}$$

Deuterium is a stable isotope of hydrogen with a natural abundance of 0.0015 per cent. Tritium is an unstable hydrogen isotope with a radioactive half-life of 12.3 years and is produced from lithium 6 by the neutron-alpha reaction. The latter thus represents a relatively expensive "fuel" for thermonuclear reactions; however, the ignition temperature of the deuterium-tritium reaction is roughly an order of magnitude lower than that of the deuterium-deuterium reactions and the energy release is greater.

In most controlled fusion systems as presently conceived, the fuel is in the form of an ionized gas, or "plasma," confined by magnetic pressure within a high-vacuum apparatus. In effect, the plasma is held in a "magnetic bottle," thereby preventing fuel particles from dissipating heat in collisions with the physical walls of the apparatus. The objective is to achieve a situation in which an adequately hot plasma of adequate density can be magnetically confined for a long enough interval of time for the desired reaction to take place. One approach is to constrict and confine a high-current discharge of fuel ions and hold the resulting dense plasma in confinement while its temperature is raised by adiabatic compression or other methods. Another approach is to accelerate fuel ions to high energies and then trap them in a magnetic field, maintaining confinement long enough for a dense plasma to accumulate.

In experiments in various experimental devices, the time-temperature-density multiple has steadily been increased; however, there appears to be no conclusive evidence that true thermonuclear conditions have been achieved in any laboratory. Beyond laboratory demonstration of controlled fusion per se lies the problem of demonstrating that devices can be designed to produce more power than they consume and beyond that lies the problem of demonstrating the economic feasibility of practical thermonuclear power plants.

The chief incentive for thermonuclear power development is the promise of a virtually inexhaustible energy source, assuming the ultimate use of deuterium as the primary fuel.

Radioactive Decay As radioactive atoms undergo decay by alpha, beta or gamma emission, heat is generated by the interaction of the radiation with surrounding matter. Devices that utilize this heat to produce electricity are known as isotopic power generators. Research has been conducted on a range of such devices for specialized applications requiring from fractions of a watt to tens of watts of electricity. Thermoelectric or thermionic techniques are used to convert the heat to electricity. At present, isotopic power generators are being used on an experimental basis in a number of applications such as navigational satellites, automatic weather stations and coastal light buoys, all of which require a compact power source that can operate unattended for sustained periods (months or years). In the case of space applications, alpha-emitting radionuclides such as plutonium 238 or curium 244 are mainly used as the fuel. In terrestrial applications, the principal fuel used to date is strontium 90, a beta emitter.

JOHN F. HOGERTON

References

Hogerton, John F., "The Atomic Energy Deskbook," New York, Van Nostrand Reinhold, 1963.
Glasstone, Samuel, "Sourcebook on Atomic Energy," New York, Van Nostrand Reinhold, 1958.

Cross-references: FISSION, FUSION, ISOTOPES, NUCLEAR REACTIONS, NUCLEAR REACTORS, NUCLEONICS, RADIOACTIVITY.

ATOMIC PHYSICS

Throughout the twentieth century the study of atoms, their internal structure and their interaction with electromagnetic radiation and with other particles has played an important part in attempts to understand the physical world. The subject of atomic physics has been the testing ground of fundamental theories and at the same time the source of many applications.

The concept of an atom as an indivisible particle, the ultimate building block of bulk matter, is attributed to the ancient Greeks. During the nineteenth century such atoms, of which it was recognized that there had to be many different kinds, were identified with the chemical elements of the periodic table. But the era of modern atomic physics may be said to have begun with two discoveries, that of the electron and that of the nuclear atom. The idea of the indivisible atom was to be abandoned.

The first discovery was made in 1897 by J. J. Thomson, who was able to isolate an electrically charged particle, an electron, and to measure the ratio of its charge e to its mass m. Following his work it came to be believed that electrons were constituents of atoms. It was a little later that R. A. Millikan measured directly the electron charge itself: this charge e was the same for all electrons (by convention, it is taken to be negative). The two results taken together led to the conclusion, already guessed at, that the mass of the electron is only about 1/2000 of the mass of a hydrogen atom, the lightest element in the periodic table.

The second discovery was made by Sir Ernest Rutherford in 1911. At his instigation Geiger and Marsden carried out experiments in which alpha particles (which are positively charged) were scattered from thin foils. The result, very surprising at the time, that many alpha particles were scattered through large angles was explained quantitatively by Rutherford. He concluded that the mass of an atom of the foil is concentrated in a nucleus whose radius is of the order of 10^{-14} m, very much smaller than the radius of an atom, 10^{-10} m, and that the nucleus carries a positive charge Ze where Z, an integer, is the atomic number appropriate to the ordering of chemical elements in the periodic table. The model of a neutral atom, then, is that it consists of a small massive nucleus of positive charge Ze surrounded by Z electrons, each of negative charge e, filling a much larger volume. Electrostatic forces bind the whole thing together.

However, the nuclear atom, resembling a miniature solar system, would not be stable according to the laws of classical physics because the orbiting electrons, being charged, would emit electromagnetic radiation in a continuous spectrum of frequencies, lose energy and spiral into the nucleus. This does not happen: atoms are stable. It is true that they can emit radiation, but only when they are given enough energy, for example when free atoms are bombarded by electrons in a light source such as an electric arc. The emitted spectrum consists not of a continuous frequency distribution but rather of discrete frequencies characteristic of a particular atom. Spectroscopists working in the nineteenth century had studied these "spectral lines," many of which appear as light in the visible region, and had measured their frequencies with considerable precision. So more than one difficulty was raised by the model of the nuclear atom, and the problem was tackled in 1913 by Niels Bohr. He had at his disposal the concept of a "quantum" of energy of radiation which had arisen in Planck's treatment of blackbody radiation and in Einstein's theory of the photoelectric effect. A quantum of radiation has energy $E = h\nu$ where ν is the frequency of the radiation and h is a universal constant known as Planck's constant, having the value 6.626×10^{-34} J Hz^{-1}. Bohr developed a set of rules which required the angular momentum of the orbiting electron in hydrogen to be "quantized," that is, to have only integral values in units of $h/2\pi$. It followed that the energy of the atom was also quantized with the consequence that the frequency of emitted radiation had to satisfy the relation $h\nu = E_m - E_n$ where E_m and E_n are the discrete allowed energies of the mth and nth "energy levels" of the atom. Bohr's calculation of the allowed frequencies in the spectrum of atomic hydrogen, based on the mechanics of his model, agreed with precise spectroscopic measurements. This was a brilliant achievement.

Yet Bohr's calculations only worked well for hydrogen, the one-electron atom. Further, there was still no adequate explanation for the stability of the lowest energy level nor for the rate at which radiation would take place when an atom decayed from a higher level to a lower one. Even more difficult would be the attempt to predict in detail what would happen in an electron-atom or atom-atom collision. Whereas the idea had already been grasped that a light wave, through its quantum nature, can have the feature of a particle, namely localization in space, what was missing was the equally extraordinary idea that a particle, say an electron, can have the feature of a wave—a spreading-out in space. The latter idea was put forward by Louis de Broglie in 1924: he attributed to a particle of momentum p a wavelength $\lambda = h/p$. Planck's constant appears again in this relation and its size is crucial: electrons bound in atoms have momentum such that their de Broglie wavelength λ is of the order of the size of the atom. These electrons are only localized to the extent that they are somewhere inside the atom, which is now visualized as a nucleus surrounded by an electron charge cloud having no well defined boundary.

These quantum ideas were put into mathematical form in 1925. Schrödinger's wave mechanics of matter was soon generalized in the formal theory of quantum mechanics by Born, Heisenberg, Dirac and others. Quantum mechanics, which includes the famous Uncertainty Principle, is the basis for all calculations of physical problems in which the magnitude of Planck's constant cannot be considered to be negligible, and by this criterion the theory is certainly needed to describe atomic physics. The birth of quantum mechanics has been a revolution in the theory of physics, and this revolution has had in its turn a profound influence on philosophical thinking.

The hope that one now understands the principles of atomic physics is one thing, but the task of making quantitative predictions is another, for it involves making detailed calculations which are especially difficult in many-electron atoms. Fortunately, approximation methods can be used in treating the forces which govern atomic structure, at least in simple atoms with few electrons. The largest forces are the electrostatic attraction between nucleus and electrons, and the electrostatic repulsion between the electrons themselves. These are dealt with first. Electrons are found to have, in addition to their charge, an intrinsic angular momentum (called electron spin) together with an associated magnetic moment. The interaction between this magnetic moment and the magnetic field which an electron experiences within an atom is a small perturbation in our hierarchy of approximations. This effect leads to a "fine-structure" splitting of the energy levels of an atom. Even smaller is a "hyperfine-structure" splitting which arises from the interaction between the electron and the magnetic moment with which the nucleus is endowed (the nucleus, too, has the property of spin angular momentum).

On the theoretical side the aim has generally been to calculate the energies of excitation available to an atom in terms of all the interactions mentioned above, along with other properties such as the average lifetime of an excited atom before it decays to a lower energy state, emitting electromagnetic radiation (typical lifetimes are of the order of 10^{-8} s). In all this the role of the modern computer has been important, especially for many-electron atoms whose complexity would otherwise present intractable problems. On the experimental side there have been two main ways of approaching the study of atomic structure. One is by measuring the frequencies and intensities of spectral lines (spectroscopy) and the other is by causing collisions between atoms and particles—electrons or other atoms, for example. Whereas optical spectroscopy had been a traditional experimental discipline since the middle of the nineteenth century and collision physics began in the 1920s, both experimental approaches enjoyed the benefits of the enormous technical advances made during and after the Second World War.

The ever increasing precision of spectroscopic methods has had important consequences. One of these is exemplified by the Lamb-Rutherford experiment (1950) which is justly famous not only because it was a tour-de-force of experimental technique but also because it played a central part in an effort to develop a new and deeper understanding of physics—always the primary goal. Lamb adapted the methods of radio-frequency spectroscopy to measure the small separation between two energy levels in the lowest excited state of atomic hydrogen, a separation subsequently called the "Lamb shift." Even the sophisticated relativistic quantum mechanics developed by Dirac had not predicted a Lamb shift. The theory of quantum electrodynamics (also initiated by Dirac in 1927), in which electromagnetic fields were taken to be quantized, was needed to explain the Lamb shift, and it was Lamb's precise experiment which stimulated efforts to make accurate calculations from the theory. Quantum electrodynamics is regarded as a striking success because of its ability to achieve precise agreement (parts in 10^6) between theory and experiment in the interaction between radiation and matter. The theory is also the forerunner of more general quantum-field theories needed in the attempt to understand elementary particle physics.

The particular radio-frequency method to which Lamb's experiment is related is called atomic-beam magnetic resonance. This has been developed by I. I. Rabi and his co-workers, beginning in 1938, and has been very fruitful. The method ensures, in an especially elegant way, that the frequency of the peak of a spectral line can be measured accurately because the frequency bandwidth of the line is made very narrow. Moreover, the atoms, which are made to travel as a directed beam through a vacuum, do not collide with each other and so may be regarded as completely unperturbed by their neighbors: they interact only with the applied radio-frequency field. By this method a large number of measurements, particularly of hyperfine-structure splittings, have been made in nearly all the elements. Such work provided a body of data on nuclear spins and nuclear moments at a time when nuclear physicists could make good use of it in constructing models to describe the nucleus; the precise measurement of hyperfine structures also led to a better quantitative understanding of the behavior of the atomic electrons when they are close to the nucleus.

Technological developments soon followed. The hyperfine structure of the lowest energy level of cesium could be measured so precisely and reproducibly that a cesium atomic-beam apparatus was adopted as a time standard, that is, the second is now defined in terms of the frequency of an atomic clock:

$$\Delta\nu(\text{Cs}) = 9\ 192\ 631\ 770\ \text{Hz}.$$

During this period of prolific activity in the 1950s the maser was invented. The first maser was made to work by Townes and his colleagues in 1954 and subsequently many other workers contributed to the development of various kinds of maser. This device (the word is an acronym for *m*icrowave *a*mplification by *s*timulated *e*mission of *r*adiation) relies for its operation on the maintenance, in a dynamic equilibrium, of a population inversion in an assembly of atoms: this means that, given two energy levels of an atom, there are more atoms in the upper level than in the lower. In this way it is possible to make an oscillator tuned to the frequency of an atomic spectral line. A particular version, the hydrogen maser developed by Ramsey, has led to a measurement of the hyperfine structure in the lowest level of hydrogen which is utterly remarkable for its precision:

$$\Delta\nu(\text{H}) = 1\ 420\ 405\ 751.7662 \pm 0.0030\ \text{Hz}.$$

An optical analog of a maser, called a laser, was first made to work in 1960. It is not an overstatement to say that the laser, in its various forms, has brought about a revolution in optical technique. For example, the use of a dye laser, which is a tunable, nearly monochromatic, powerful light source, is now widespread in spectroscopic work. Many new experiments requiring high resolution and high power have been performed on atoms and molecules and many more are likely to be performed during the 1980s. Just one example of new work is that on very highly excited levels of neutral atoms. These so-called Rydberg atoms may be regarded as being very large in the sense that the outermost electron has a large "orbit" of the order of 10^{-6} m radius, and they are also fragile because only a tiny perturbation is required to ionize them.

The study of controlled bombardment of atoms (or molecules) by other particles, in particular electron-atom, atom-atom and ion-atom collisions, has had a development parallel to that of spectroscopy. The experiments have to be most carefully performed because, in obtaining quantitative results on the behavior of a variety of atoms undergoing collisions, it is necessary to distinguish one of several possible processes from another. For example, electrons used as bombarding particles may be scattered by atoms, they may excite them, they may ionize them and they may be captured by them. Reliable results on the probability of such events have to be achieved over a wide range of bombarding energies with good resolution. Only then can they be used both as a severe test of the quantum-mechanical description of atoms and as a means of understanding the collision processes occurring in more complicated environments.

Naturally, refinement of experimental technique has brought progress in many different directions. For example, as a result of the pioneering work of Dehmelt on ions trapped in electromagnetic fields it has become possible to study a single, isolated ion trapped for several hours on end—a remarkable feat. Some modern work has overlapped other fields: tests of fundamental symmetry principles and conservation laws, often regarded primarily as the concern of high-energy particle physicists, have been conducted by means of ingenious experiments on atoms; and close to the interests of nuclear physicists has been the study of "exotic atoms," those containing unstable particles playing the role of nucleus or electron, for example positronium, muonium, muonic and kaonic atoms.

Atomic physics is an applied science in the sense that in other branches of physics it is necessary to know about atoms. Obviously the astrophysicist needs to know about lifetimes and structures of excited states of atoms because he is trying to learn about a source of radiation (a star) which is in no way under his control! Similarly the study of atmospheres, both planetary and stellar, relies heavily on the results of collision physics and of spectroscopy conducted in the laboratory, as does work on plasmas associated with nuclear fusion. Solid-state physicists clearly must understand the atoms which are the components of their assemblies of interacting systems. None of these other fields can be thought of as divorced from atomic physics itself.

G. K. WOODGATE

References

Enge, H. A., Wehr, M. R., Richards, J. A., "Introduction to Atomic Physics," Addison-Wesley, 1972.

Born, M., "Atomic Physics," London, Blackie & Son, 1969.

Woodgate, G. K., "Elementary Atomic Structure," Oxford Univ. Press, 1980.

Massey, H., "Atomic and Molecular Collisions," London, Taylor & Francis Ltd., 1979.

November 1981 issue of *Physics Today*, American Institute of Physics, 1981.

Fortson, E. N., Wilets, L., "Parity Nonconservation in Atoms: Status of Theory and Experiment," in "Advances in Atomic and Molecular Physics," Vol. 16, p. 319, Academic Press, 1980.

Cross-references: ATOMIC ENERGY; COLLISIONS OF PARTICLES; CONSERVATION LAWS AND SYMMETRY; ELECTROMAGNETIC THEORY; ELECTRON; ELECTRON SPIN; ELEMENTS, CHEMICAL; LASER; MASER; MOLECULES AND MOLECULAR STRUCTURE; NUCLEAR STRUCTURE; PERIODIC LAW AND PERIODIC TABLE; QUANTUM ELECTRODYNAMICS; QUANTUM MECHANICS; SPECTROSCOPY.

ATOMIC SPECTRA

Fundamental Facts Light from electric discharges in gases shows *line spectra* due to free atoms excited by electron collisions. Noble gases and metal vapors produce almost pure atomic spectra, while discharges in molecular gases show both molecular *band spectra* and atomic line spectra. Some spectral lines can also be observed in absorption when white light is made to pass through the gas into a spectroscope. Under high spectroscopic resolution, all lines are found to have nonzero width. This is due to random motion (DOPPLER EFFECT) and disturbing influences of neighboring atoms, molecules, ions or electrons (pressure broadening); but even after allowance for these effects, a spectral line has a definite, generally very small, width due to radiation damping (natural width). Precision measurement of wavelengths or resolution of very fine structures requires light sources giving narrow lines—discharges at low gas- and current-density and low temperature—or even atomic beams at right angles to the line of sight. *Continuous* atomic spectra are generally weak under laboratory conditions; in emission, they are due to recombination of an electron with a positive ion; in absorption, to the reverse process of *photoionization*.

The term atomic spectra includes positive ions, with the following terminology: spectrum of Na, *arc spectrum*, NaI; of Na^+, Na^{++}, $\cdots$: first, second, $\cdots$ *spark spectrum*, or NaII, NaIII, $\cdots$. Spectra of highly ionized or *stripped* atoms are important in astrophysics and occur in high-temperature plasmas. Systems with the same number of electrons, such as Na, Mg^+, Al^{++} show marked similarities and are called *iso-electronic* sequences.

In the ultraviolet, visible or infrared, the spectroscope, in the form of a grating or interferometer, measures primarily the wavelength (λ) of the spectral lines. It is generally expressed in angstrom units (Å) defined as 10^{-8} cm (10 nm) or, by recent international convention, the fraction $1/6056.12525$ of the wavelength of a line of the isotope 86 of krypton, in air under standard conditions. The wave number ($\tilde{\nu}$ or σ), the reciprocal of the wavelength in vacuo, is measured in cm^{-1} or kayser (K), or in millikayser (mK). The frequency ν is derived by multiplying by c, the velocity of light in vacuo; in the range of microwaves and radio frequencies, ν is measured directly (1 mK = 29.9793 MHz).

In contrast to frequencies, intensities of lines are strongly dependent on experimental conditions, and special experiments are required for deriving quantities expressing the strength of a line as a characteristic constant of the atom. This can be defined in various forms; the f-value is a number giving the ratio of the absorptive or dispersive power of the line to that of the classical, harmonic electron oscillator of the same frequency; the transition probability or Einstein A-value is the probability, per second, of an excited atom emitting a light quantum.

Hydrogen-like Spectra The spectra of atoms containing one electron only (H, He^+, $Be^{++} \cdots$) are very simple if the fine structure is disregarded; they form the basis of the classification and theory of atomic spectra. Balmer's empirical discovery of a numerical relationship between the wavelengths of the visible hydrogen lines led to a formula expressing the wave numbers of all hydrogen-like spectra by one constant R, the charge number Z (=1 for H; =2 for He^+, $\cdots$) and two integral numbers, n, $n' > n$:

$$\tilde{\nu} = Z^2 R(1/n^2 - 1/n'^2) = T_n - T_{n'} \qquad (1)$$

A *series* arising from a sequence of values n' is characterized by regularly decreasing spacings and intensities of the lines towards increasing wave numbers. Substitution of $n = 1, 2$ and 3 in Eq. (1) with $Z = 1$ gives the Lyman, Balmer and Paschen series, in the ultraviolet, visible and near infrared respectively. The wavelengths of the Balmer lines, H_α, H_β and H_γ ($n' = 3, 4, 5$) are 6562.8, 4861.3 and 4340.5 Å. The Lyman α line, the *resonance line* of hydrogen, has the wavelength 1215.7Å.

The relation of Eq. (1) can be derived theoretically by applying nonrelativistic quantum theory to a model consisting of a point electron of mass m and charge $-e$ and a fixed point nucleus of charge Ze. In the *Bohr-Sommerfeld* theory, this is done by imposing quantum conditions on the classical orbit of the electron; in the more rigorous Schrödinger theory, by solving the wave equation with the assumption of constant energy E. Provided $E < 0$ (bound state), it assumes discrete values, those of the stationary states of motion or the *eigenvalues* of the wave equation. Emission and absorption arise from transitions between two *energy levels* E_n, $E_{n'}$, with the frequency of the light given by

$$\nu_{n,n'} = (E_{n'} - E_n)/h \qquad (2)$$

where h is Planck's constant. Equation (1) is a special case of Eq. (2), with $E_n = -hZ^2Rc/n^2$. Allowance for the motion of the nucleus of finite mass M causes R to differ slightly for different M; it is given by $R_\infty/(1 + m/M)$ where $R_\infty = 109737.3$ cm^{-1}.

The solution of the SCHRÖDINGER EQUATION for a mass point in space leads to 3 quantum numbers. In polar coordinates, with a force derived from a central potential $V(r)$, the quantum numbers n_r, μ and m give the numbers of nodes of the wave function in the range of the coordinates r, ϑ and φ. Introducing the *azimuthal quantum number* $l = |m| + \mu$ and the *principal quantum number* $n = n_r + l + 1$, we find the set n, l, m to have the following meaning: the z-component of the angular momentum is $L_z = m\hbar$, where $\hbar = h/2\pi$, the square of its absolute value is $|L|^2 = l(l + 1)\hbar^2$, and the energy E depends on n and l only. For the special case of the Coulomb field $V \sim 1/r$, E depends on n alone: $E_n = $ constant$/n^2$. An energy level E_n has to be considered as consisting of a number g of *states* of different l and m. This situation is described as *degeneracy*, and g is the *statistical weight* of the level. The degeneracy in m is due to the central symmetry of the force field and occurs in all atoms in the absence of external fields. The degeneracy in l is peculiar to the Coulomb field in nonrelativistic treatment.

Alkali-like Spectra The spectra of the alkali atoms and their isoelectronic ions (Li, Na, $\cdots$, $Be^+, Mg^+, \cdots$) show lines arranged in series similar to those of hydrogen. Their wave numbers can be represented by empirical relations which are generalizations of Eq. (1). Series of term values T_n can be defined in such a way that $T_n \to 0$ for $n \to \infty$, and the observed wave numbers are equal to term differences $T_n - T_{n'}$ (Ritz combination principle). In contrast to Eq. (1), however, there are several series of terms, so that apart from n, a second index number l has to be introduced. The term values $T_{n,l}$ can then be identified with quantized values of $-E/ch$ and the index numbers n and l with quantum numbers, as implied by the letters chosen, if we assume that in these atoms one electron moves in a central force field different from a Coulomb field. This *valency* or *optical* electron has to be imagined as more loosely bound than the others and moving in the field of electrostatic attraction by the *core* consisting of the nucleus of charge Ze and the remaining $Z-1$ electrons. At large distances from the core, the field is like that caused by a single charge e as in hydrogen. At smaller distances, the optical electron penetrates into the electron cloud of the core and experiences an increased attraction. This picture leads to a qualitative understanding of the term diagrams and spectra of alkali atoms as exemplified for Na in Fig. 1, where terms with $l = 0, 1, 2, 3, 4, \cdots$ are conventionally described as $S, P, D, F, G, \cdots$ terms. For any given n, an energy level is the further below that of hydrogen (the term value the larger) the smaller l becomes, because a smaller angular momentum decreases the centrifugal force and brings the electron closer to the core. Emission or absorption of radiation according to Eq. (2) does not occur for all pairs of levels but is subject to the *selection rule* $\Delta l = \pm 1$. Transitions from P levels to the lowest S level form the *principal series*. Its lines can also be observed in absorption since the lower level is the ground level; the first member is the well-known yellow *resonance* line. Transitions from D or higher S levels to the lowest P level form the *diffuse* and *sharp* series, those from F levels to the lowest D level, the Bergmann or *fundamental* series.

A feature that cannot be explained by this model is the *doublet* structure, a doubling of all except the S levels. It is due to the fact that the electron possesses a *spin*, i.e., an intrinsic angular momentum $\mathbf{S}$ of fixed absolute value given by $|\mathbf{S}|^2 = s(s + 1)\hbar^2$, where $s = \frac{1}{2}$, and connected

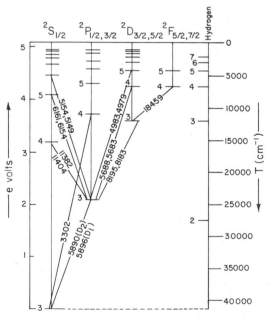

FIG. 1. Term diagram of Na. Approximate wavelengths in angstroms. The doublet splitting of the terms is not shown.

with a magnetic dipole moment μ. The interaction of μ with the magnetic field due to the orbital motion, the *spin-orbit coupling*, causes any one energy level of given n and l (except for $l = 0$) to split into two levels. They are characterized by a new quantum number j associated with the total angular momentum resulting from vector addition of $\mathbf{L}$ and $\mathbf{S}$. It can have the two values $j = l \pm s = l \pm \frac{1}{2}$. Optical transitions are subject to the selection rule $\Delta j = \pm 1$ or 0. The width of the doublet splitting increases with core penetration and thus with decreasing l; it is most prominent in P terms. It decreases rapidly with increasing n and increases from Li to Cs. In the term symbol, j is written as suffix and the doublet character is indicated by superscript 2. For example, two transitions forming the yellow resonance doublet of Na are written $3^2S_{1/2}$ – $3^2P_{1/2}$ and $3^2S_{1/2}$ – $3^2P_{3/2}$. The absolute value of n (= 3 in this case) can be deduced by comparison with hydrogen (Fig. 1).

Since for all term values $T_n \to 0$ for $n \to \infty$, the extrapolation of a series $T_n - T_{n'}$ for $n' \to \infty$ gives the term value T_n. If this is the ground term, chT_n is the ionization energy of the atom. A convenient conversion formula is 8066 cm^{-1} = 1 eV.

Under high resolution, hydrogen-like spectra are found to have a rather complex structure known as *fine structure* (the same name is often applied to the much wider doublet or multiplet structures); it can be explained by a relativistic velocity dependence of the electron mass removing the degeneracy of states of different l, and by magnetic spin-orbit coupling causing a doublet splitting. In fact, these two effects are related since the spin itself is relativistic in origin. The theory gives the result that a hydrogen level of given n depends on j only, so that, e.g., $n = 2, l = 0, j = \frac{1}{2}$ should coincide with $n = 2, l = 1, j = \frac{1}{2}$. In fact, such terms show a small difference known as *Lamb shift*. Its existence can be explained by quantum electrodynamics.

A more complete description of the state of an alkali atom has to treat the core as a dynamical system of many particles. This cannot be done rigorously, but as a useful zero-order approximation, quantum numbers n and l can be assigned to individual electrons; in extension of the symbolism used before, values $l = 0, 1, 2 \cdots$ for individual electrons are described by small letters $s, p, d \cdots$, and the number of electrons of the same n and l, called *equivalent* electrons, by a superscript. Thus two 2p-electrons ($n = 2, l = 1$) are said to form the *configuration* $2p^2$. The number of equivalent electrons in one atom is limited by the *Pauli principle* to $2(2l + 1)$, i.e., to 2 for s-electrons, 6 for p-electrons and 10 for d-electrons. When this limiting number is completed, a closed *subshell* is said to be formed. Thus the 10 electrons of the atom Ne form the three closed subshells $1s^2$, $2s^2$, and $2p^6$, and the entire atom contains the two closed shells $n = 1$ ($1s^2$) and $n = 2$ ($2s^2 2p^6$). The same applies to the 10 electrons forming the core of the atom of Na.

In a complete subshell or shell the orbital angular momenta of the electrons cancel one another out, as do the spins of the individual electrons; the total charge distribution has spherical symmetry. In an alkali atom the quantum numbers l and s of the outer electron therefore determine the orbital- and spin-angular momentum of the entire electron structure of the atom. The ground level of Na, e.g., can be described either as an s-level or an S-level, since $L = l = 0$.

The classical concept of the core as a *rigid* charge distribution is expressed by the assumption that the quantum numbers of the core electrons remain constant during the excitation of the outer electron.

Other Simple Spectra Helium, and also Li$^+$, Be^{++}, $\cdots$, have two electrons, and the atoms in the second column of the periodic table Be, Mg (also B$^+$, Al$^+$, $\cdots$) have two electrons outside closed shells. The analysis of the spectra leads to two systems of terms, *singlets* and *triplets*, with only weak intercombinations; the S terms are single also in the triplet system. These facts can be formally described, in analogy to alkali spectra, by vector addition of the two spins to form the two possible resultant spin quantum numbers $S = 0$ and 1. The first alternative produces singlets ($j = l$), the second triplets ($j = l - 1, l, l + 1$) unless $l = 0$. This implies a strong interaction forming the resultant S of the two spins and a weaker interaction forming j. The latter interaction is the same magnetic spin-orbit interaction that causes doublet splitting in alkali atoms, but the former is, in a less obvious way, due to the *electrostatic repulsion* between the two electrons. In all the terms concerned, the symbol for the configuration shows only *one* electron to be excited, e.g., $1s2p$ in He, but owing to the identity of the electrons, it is not possible to attribute the excited state $2p$ to one particular electron. The situation is analogous to that of two identical, coupled, linear oscillators showing two normal modes of vibration, each involving both oscillators in a *symmetrical* and *antisymmetrical* way. Application of Pauli's principle to the wave mechanical description of this two-electron system leads to two energy states, one with parallel and one with anti-parallel spins, $S = 1$ and 0 respectively. Elements in the subgroup (Zn, Cd, Hg) show similar singlet and triplet spectra; in the heavier elements, however, the magnetic spin-orbit interaction is no longer weak compared with the electrostatic repulsion, and the division into singlet and triplet terms has a very restricted meaning.

In the elements of the third column, B, Al, Ga, In, Tl, the single electron outside a closed subshell produces doublet spectra, but in contrast to the alkali spectra, the ground term is a P term.

Complex Spectra The assumption that the quantum numbers n and l of only one of the electrons outside closed shells change in emission or absorption of a spectral line forms a good approximation for most of the strongest spectra in the optical range, from near infrared to near ultraviolet. Somewhat arbitrarily but conveniently one can define complex spectra as those in which more than one electron outside closed subshells assumes a value of $l > 0$. One then has to consider the interaction of more than two angular momentum vectors of orbits and spins, and a simple interpretation in terms of a vector model is meaningful only in certain limiting conditions. Very often, especially in low-lying levels, a description in terms of the Russell Saunders coupling scheme (L, S coupling) is possible because the electrostatic interaction predominates over the magnetic spin-orbit interaction. As a result, we can define *terms*, each of which is characterized by a set of values L, S (or a multiplicity $2S + 1$), and each term is split into *levels*, each characterized by a value of j, the highest of which is equal to $L + S$. The classification and terminology are obvious generalizations of those for two-electron spectra. For configurations of 3; 4; 5 electrons the possible multiplicities are respectively: doublets and quartets; singlets, triplets and quintets; doublets, quartets and sextets, etc. The strongest lines arise from transitions between terms of the same multiplicity. Such line multiplets are often recognizable by their characteristic groupings of the lines and their intensity ratios. The level spacings are governed by the *interval rule*: they are in the ratio of the j values, e.g., the levels of $^4D_{7/2,5/2,3/2,1/2}$ have spacings in the ratio 7:5:3:1.

Other, often much more complex forms of coupling occur, especially in the higher levels, and L and S then lose their meaning. One important property of any level which always remains well defined is the *parity*, and the *Laporte rule* states that even terms combine only with odd terms and vice versa. If the configuration is defined, a level is *even* or odd if Σl is even or *odd*. As one proceeds to higher levels or to heavier elements the concept of the configuration becomes less distinct. A given energy level may still be said to belong to a certain configuration, but a more accurate description often requires the inclusion of one or more other configurations of the same parity in the form of a *perturbation*. In other cases the *configuration interaction* is so strong that even the lowest approximation has to be based on the concept of mixed configurations.

Hyperfine Structure (hfs) and Isotope Shift These structures are usually of the order of fractions of 1 cm^{-1} and generally require interferometric methods for their study. Hyperfine structure is primarily due to the magnetic interaction of the nuclear magnetic moment μ_N with the field produced by spins and orbital motions of the electrons. A level of given j splits into hyperfine levels, each characterized by a quantum number F, where F can assume the values

$j+I$, $j+I-1$, $\cdots$, $|j-I|$. The nuclear spin I is a characteristic property of each nucleus and has integral or half-integral values for even or odd values of the atomic number. The structure of hyperfine multiplets is similar to that of fine structure multiplets, with F, I, j taking the place of j, S, L. However, there is often also an electrostatic interaction between the electrons and the nucleus if the nuclear charge distribution has no spherical symmetry. Deviations from the interval rule in hyperfine multiplets have led to the discovery of such nuclear deformations described mainly by the *quadrupole moment* Q. Hyperfine structures in ground states can be measured very accurately by methods of atomic beam resonance. Hyperfine structure studies lead to values of I, approximate values of μ_N and Q, and accurate values for the ratio of the two latter for different isotopes.

For different isotopes of an element the spectral lines, or the centers of gravity of their hyperfine multiplets, are often displaced against each other; this *isotope shift* has two quite different causes. The atomic nucleus, though heavy, does not stay at rest during the motion of the electrons, and its mass has a small influence on the energy levels, thus causing a *mass shift* between lines of different isotopes. For atoms with only *one* electron such as H or He$^+$, the effect is fully described by a factor $1/(1 + m/M)$ in the term values. For other spectra this factor, often referred to as normal mass shift, is at best a rough approximation; the influence of the other electrons is rarely negligible and not easy to calculate. All mass shifts, however, depend on M in a regular, monotonous way for any given spectral line, and are generally much *smaller* for heavier elements. These facts often allow them to be separated from the more important *field shifts* (or volume shifts) which markedly *increase* with atomic weight in the list of elements, a fact pointing to the nonzero size of the nucleus as a cause. Within the volume occupied by the nuclear charge, the electrostatic attraction acting on the electron is much smaller than that caused by a positive point charge. This raises the energy levels compared with their fictitious values for a point nucleus. For any two isotopes, the difference in neutron number and the resulting difference in nuclear size or shape leads to a difference of the *energy levels*. In the transition to another term, a *line shift* between the two isotopes is then observed, equal to the *difference* of the level differences. The field shift of levels is particularly large for s-electrons, ($l = 0$), on account of their high probability density near the center, and the observed line shift is therefore large in lines involving transitions of one or two electrons from their s-states. Studies of field shifts have provided valuable information on nuclear structure, especially the shapes and sizes of charge distribution and their dependence on the number of neutrons.

In the spectroscopy of hyperfine structures and isotope shifts, a limiting factor is often the Doppler width of spectral lines caused by the random motion of the atoms. Cooling of discharge tubes and the use of atomic beams for absorption or emission of light have often greatly reduced this effect but rarely quite eliminated it. The recent development of *tunable lasers*, in the form of *dye lasers*, has opened up new possibilities by providing the spectroscopist with a highly monochromatic source of continuously adjustable frequency and very high intensity. The possibility of selective and powerful excitation offers obvious advantages for emission spectroscopy, but the less obvious advantages for absorption spectroscopy are best explained by two examples. The absorption spectrum is scanned by means of the laser whose spectral width determines the spectroscopic resolving power. If, instead of the transmitted beam, the fluorescent light is focused on the detector, the recorded signal is proportional to the beam intensity, and the great radiation density of the laser beam allows the study of *very weak* absorption lines. In this way absorption lines due to *two-photon transitions* have been observed. These can occur between two atomic levels of equal parity, of energy E and E', at exactly half the frequency corresponding to the level difference,

$$\nu = \tfrac{1}{2}(E' - E)/h$$

with a probability proportional to the *square* of the light intensity, in accordance with the concept of the simultaneous absorption of two photons. In the case of Na (see Fig. 1). Laser light of wavelength 6022.3 Å has been found to cause transitions from the ground state $3\,^2S_{1/2}$ to $5\,^2S_{1/2}$ which were detected by fluorescence of wavelength 6161 Å ($5\,^2S_{1/2}$ – $3\,^2P_{3/2}$). The *virtual* energy level introduced by the theory of this process half-way between E and E' is then slightly below $3\,^2P_{3/2}$. Its closeness to this level accounts in the theory for the comparative ease of observing the two-photon absorption. If the atomic vapor is exposed to two laser beams traveling in opposite directions, e.g., by being placed inside the laser cavity, the spectrum shows a very sharp, virtually *Doppler-free* peak caused by two photons of opposite direction, flanked by much weaker wings of full Doppler width due to photon pairs of the same direction.

Of probably wider application is the technique of crossing a laser beam with an *atomic beam* at right angles. In contrast to earlier work it is now possible to use atomic beams of much lower density and therefore of much higher collimation, since the probability of an atom getting excited during its passage through the laser beam is close to 1. Such methods of Doppler-free spectroscopy are proving to be of great value in work on isotope shift and hyperfine structure.

Magnetic and Electric Effects The effects caused by magnetic fields play a great part in research on atomic spectra, particularly in magnetic resonance methods. An external magnetic field removes the degeneracy due to the spherical symmetry of atomic force fields and causes the energy to depend on the magnetic quantum number m. This leads to the formation of the *Lorentz triplet* or *normal Zeeman effect* in singlet spectra and to the more complex structures of *anomalous Zeeman effects* in multiplet lines. Zeeman effects in hyperfine structures are especially important for the determination of nuclear spins. The *Stark effect,* due to electric fields, is of somewhat less importance. It causes the energy to depend on $|m|$ only, thus not removing the degeneracy completely. While magnetic splittings are proportional to the field strength, Stark splittings are generally proportional to the square of the electric field.

Direct Measurement of Hyperfine Structures. The optical spectroscopy of hyperfine structures aims at the determination of a very small quantity (a level difference in an atom) by means of measuring the difference between two much larger quantities (wavenumbers or frequencies). If, instead of these, the small level difference can be measured directly, as a single, low frequency, one can expect a gain in accuracy on general grounds; more specifically, since the Doppler width is proportional to the frequency, such direct measurements are virtually *Doppler-free*. Many hyperfine structures, and also some other small, spectroscopic effects such as *Lamb-shifts*, have been measured in this way to an accuracy beyond that of optical methods. It has to be remembered that *isotope shifts* can *not* be measured by such direct methods because they are not due to a level difference in *one* atom.

Radiofrequency Methods. All hyperfine levels and magnetic sublevels within a given configuration have the same parity, and electric dipole transitions between them are forbidden. An electromagnetic wave can, however, act through its *magnetic* field on the spin and orbital motions and their precession about the direction of an external field B. Whenever the frequency of the radiation field equals the Larmor frequency ($1/h$ times the spacing of the energy levels), transitions $m \rightleftarrows m \pm 1$ can occur, with equal probability in both directions, and are described as *magnetic resonances*. A change of m implies a change of the component of angular momentum and magnetic moment along the direction of B. This field is described as weak, intermediate, or strong according to its effect as compared with the hyperfine structure splitting. Each resonance point is found by scanning B at constant frequency. In weak fields, transitions between the magnetic levels of each hyperfine level are measured separately and allow the nuclear spin I to be determined; measurements in intermediate fields involve all values of

F of the term and yield an accurate value of the hyperfine splitting for $B = 0$. The various techniques differ mainly in their methods of *detecting* resonances.

In the *atomic beam resonance* method (see ATOMIC AND MOLECULAR BEAMS) the beam passes through two strong, magnetic fields having *gradients* of *opposite* directions, one before and the other after passage through the resonance region with a *homogeneous* field; the fields are all in the same direction. Though the quantization is different in weak and strong fields, the total component of the angular momenta of the atom remains the same during its passage ($m_F = m_j + m_I$), unless resonance occurs, and the two deflections cancel out. A resulting deflection can thus be used for the *detection* of resonance.

In the methods of *double resonance* and *optical pumping*, changes of intensity or polarization of light in emission or absorption are used for the detection of resonances. These phenomena depend on the *population* of sublevels of different m within a large assembly of atoms. Since the *Boltzmann* factor is close to 1, all these levels have normally the same population, and resonance transitions between them have statistically no effect. Absorption of polarized light in a resonance line, however, causes an anomalous population in the excited state, as shown by the polarization of the fluorescent light. When a radiofrequency field is applied, resonances in the excited state tend to restore normal population and can thus be detected by the resulting *depolarization* of the fluorescence. The method is known as *double resonance*. If *circularly* polarized light is used in the fluorescence experiment, the optical transition always increases (or decreases) the value of m to $m + 1$ (or $m - 1$), according to the sense of polarization and produces a state of partial *orientation* in the *excited* state. The reemission, in accordance with the selection rules, then shifts the population in the *ground* state in the same sense. A state of partial orientation of the angular momenta and the magnetic moments has thus been produced solely by the action of light, a process known as *optical pumping*. Such a state can be remarkably long-lived; it can easily be monitored by the strength of absorption in the resonance line. Resonances in the *ground* state can then be produced and detected in the same way.

Level Crossing and Quantum Beats. A group of methods for measuring hyperfine structures and radiative lifetimes of excited states is based on the properties of atomic systems in or close to a condition of *degeneracy*. A plot of the energy levels of different m, of any hyperfine state, as function of B shows $2F + 1$ lines crossing one another at $B = 0$. Some of the levels arising from different values of F and differing m also cross at *intermediate* fields. Any such crossing point represents a *degenerate* system; a transition from this to the ground state "corresponds" to a classical, *isotropic* oscillator. It responds to a light wave by oscillating in the direction of the electric vector, emitting light waves of the appropriate direction and polarization. Slow changes of B from 0 or from its value at other crossing points remove the degeneracy and cause precessional motions of the oscillator, thereby changing or destroying the polarization of the fluorescence. This change takes place when the frequency of precession $(E_m - E_{m\pm1})/h$ is equal to the reciprocal of the *radiative lifetime* of the excited state which can thus be measured. For zero-field crossing, these phenomena, known as the *Hanle effect*, are often used for measuring radiative lifetimes. More information is gained from crossings at intermediate fields. The exact value of B for any crossing point can be determined by the very distinct change of polarization, and this allows the width of the field-free hyperfine splitting to be derived; this is known as the method of *level crossing*.

In the emission of light from two closely adjacent levels excited by the same source, the atom passes through a stage during the radiative lifetime τ when it can be described by a mixture of two time-dependent functions. This causes a light emission with an intensity modulation superimposed on the exponential radiative decay. The emission from different atoms having different velocities is *not* optically coherent but contains common *Fourier* components of the *beat frequency*. In excitation by pulses of duration $\ll\tau$ these *beats* can be synchronized and thus made observable, e.g., on an oscilloscope; they are known as *quantum beats*. This *time-resolved* spectroscopy has been applied to different kinds of close level structures, also in *beam-foil* spectroscopy where the foil produces the sudden excitation and the beam provides the time resolution.

Far Ultraviolet and X-ray Spectra Atomic spectra of frequencies about 1000 times those of optical spectra have long been known as *characteristic* x-ray spectra (see ENERGY LEVELS). Though generally observed in condensed matter, they are characteristic of the constituent atoms, only slightly modified by chemical bonds or crystal structure. They are due to the inner electrons of the atom which are very tightly bound by the strong attraction of the nuclear charge, only partly screened by the other electrons of the same or lower n. The energy of the inner electrons depends primarily on n, as in hydrogen-like spectra, and in x-ray terminology one classifies electrons and shells by the value of n as K-, L-, M, $\cdots$ electrons and shells for $n = 1, 2, 3, \cdots$. Since the inner shells are filled in atoms of not too small Z, the absorption spectrum shows mainly continuous bands, due to the removal of an electron from one of these shells. The photoionization energies for electrons of $n = 1, 2, \cdots$ mark the low-frequency

edges of the K-, L- $\cdots$ bands. The state of an atom with a "hole" in one of the inner shells, caused by x-ray absorption or electron bombardment, represents a highly excited state of the atomic ion. It leads to emission of the x-ray line spectra by transition of an electron from a higher shell to the vacant state of the inner shell.

Spectra in the gap between x-rays and the near ultraviolet have recently received increasing attention, partly owing to their importance to astrophysics. Work in this *far* and *extreme* ultraviolet requires vacuum spectrographs, usually with gratings used at grazing incidence. *Emission* spectra of highly ionized atoms (*stripped* atoms) can be observed in the solar corona and in stellar nebulae, and in the laboratory in condensed sparks and high-temperature plasmas. Their strongest lines are in the extreme ultraviolet; e.g., in work ranging down to about 20 Å, all the spectra of the isoelectronic sequence from NaI ($Z = 11$) to Cu XIX ($Z = 29$) are known. The study of *absorption* spectra of neutral atoms has been extended into the extreme ultraviolet, partly by the use of the continuous background radiation from synchrotrons. Such spectra are mainly due to two kinds of processes: either the excitation of one of the electrons in closed shells, e.g., of one of the electrons $1s$, $2s$, or $2p$ in Na, or the simultaneous excitation of two electrons. A simple example of the second type is the absorption series $1s^2\ {}^1S - 2snp\ {}^1P$ in He between 206 and 165 Å. In both types of absorption spectra, many lines show a peculiar, strongly asymmetric profile known as *Fano profile*. It can be explained by configuration interaction in the following way. The same energy as that of the two-electron excitation state $2snp\ {}^1P$ can be reached from the normal atom by removal of a single electron with excess kinetic energy, i.e., the absorption line falls in the range of the continuous absorption extending towards higher frequencies from the limit $n = \infty$ of the absorption series $1s^2\ {}^1S - 1snp\ {}^1P$. Configuration interaction between the discrete state and the continuous state of the same energy and parity causes the anomalous line profile by a kind of interference effect. The absorption of light within this line leads either to re-emission of the same frequency or to nonradiative transition to the ionized state, a process known as *autoionization*. In x-ray spectroscopy, the analogous process had been detected by the appearance of fast, free electrons and is known as *Auger* effect (see AUGER EFFECT). The process of autoionization is closely related to the inverse process of *resonance capture* of electrons by ions.

Particle accelerators normally used in nuclear structure research have been successfully applied to the optical spectroscopy of multiply ionized atoms. In a technique known as *beam-foil method*, the high-speed ion beam is made to pass through a very thin foil (usually of carbon) and the spectrum is observed at different distances from the foil. The rate of decrease of intensity allows life times of excited states to be determined. In spite of difficulties due to cascading processes numerous transition probabilities could be measured in this way. Modulation structures superimposed on the exponential decay are sometimes observed and are known as *quantum beats*. They are due to finer level structures and can be understood in the way indicated above.

H. G. KUHN

References

Kuhn, H. G., "Atomic Spectra," 2nd ed., London, Longmans, and New York, Academic Press, 1971.

Woodgate, G. K., "Elementary Atomic Structure," Oxford Univ. Press, 1980.

Hanle, W., and Kleinpoppen, H., "Progress in Atomic Spectroscopy," Vol. A, Plenum Press, New York and London, 1978; Vol. B, 1979.

Corney, A., "Atomic and Laser Spectroscopy," Oxford Univ. Press, 1977.

Berry, H. G., "Beam Foil Spectroscopy," *Rep. Progr. Phys.* **40,** 155 (1977).

Cross-references: ATOMIC AND MOLECULAR BEAMS, ATOMIC PHYSICS, AUGER EFFECT, ELECTRON SPIN, ENERGY LEVELS, MOLECULAR SPECTROSCOPY, OPTICAL PUMPING, RAMAN EFFECT AND RAMAN SPECTROCOPY, SCHRÖDINGER EQUATION, SPECTROSCOPY, X-RAYS, ZEEMAN AND STARK EFFECTS.

AUGER EFFECT

Definition and History The Auger effect is the filling of an electronic vacancy in the atom by one electron from a less tightly bound state, with the simultaneous emission not of a photon but of a second electron from another less tightly bound state.

Following experiments by Barkla (1909) and Sadler (1917) in which the number of characteristic K x-rays emitted by material absorbing higher-energy x-rays appeared to be substantially less than the number of x-rays absorbed in the K shell, Kossel (1923) suggested that the remaining vacancies might be filled by a radiationless transfer of the excess energy to an emitted electron. This interpretation was reiterated by Barkla and Dallas (1924), who observed an increase in the number of electrons emitted when x-rays were absorbed. Wilson (1923) had observed in a cloud chamber, simultaneous ejection of two electrons from the same atom. It remained for Auger (1925, 1926) to make systematic investigations of this phenomenon in argon. The effect has since been called the Auger effect, and the ejected electrons have been called Auger electrons.

Principal Features Auger showed that:

(1) The photoelectron and its Auger electron arise at the same point.

(2) The Auger-electron track length is independent of the wavelength of the primary x-rays, but the photoelectron track length increases with x-ray energy.

(3) The direction of ejection of the Auger electron is independent of that of the photoelectron.

(4) Not all photoelectron tracks show a coincident Auger track.

Filling of vacancies by the Auger effect can occur for any vacancy for which there are two electrons in the atom sufficiently less tightly bound that a net positive energy is available for the ejected Auger electron. Because photon emission is more easily detected and has played such an important role in the development of quantum theory, it is not generally realized that Auger emission is much more probable. Only for vacancies in the K shell in atoms with atomic number above 32 and in vacancies in the outer two electron states of an atom does photon emission dominate. The Auger effect also occurs after capture of a negative meson by an atom. As the meson changes energy levels in approaching the nucleus, the energy released may be either emitted as a photon or transferred directly to an electron which is emitted as a fairly high-energy Auger electron (keV for hydrogen, MeV for heavy elements). Finally we note that each Auger process increases the positive ionization of the atom by changing one initial vacancy into two final vacancies.

Energy Spectra of Auger Electrons Auger spectroscopy is the measurement of the number, energy, and intensity of lines present. The spectrum of Auger electrons resulting from a given vacancy is more complex than the corresponding photon spectrum. The energy of the Auger electron resulting from the filling of a vacancy V of energy $E(V)$ by production of vacancies X_i and Y_j of energies $E(X_i)$ and $E(Y_j)$ is $E(V - X_iY_j) = E(V) - E(X_i) - E(Y_j) - \Delta Ex_iy_j$, where ΔEx_iy_j can be interpreted as either the increase in binding energy of the Y_j electron due to an X_i vacancy, or vice versa. Experimental energy determination in Auger spectra then consists of determining ΔEx_iy_j for each transition. Exact calculation of the number of possible Auger transitions, their energies, and their relative probabilities necessitates the use of a relativistic intermediate-coupling theory. In the above notation, X and Y refer to the total quantum number of a group of levels, and i and j to the individual substates within the group.

At low resolution the Auger spectrum from a vacancy with initial total quantum number n always consists of at least three well separated groups which can be characterized by vacancies as $n \rightarrow (n + 1)(n + 1), n \rightarrow (n + 1)(n + m)$, and $n \rightarrow (n + m)(n + m)$, where $(n + m)$ represents all final vacancies with total quantum numbers greater than or equal to $(n + 2)$. Thus for an initial K vacancy we have K-LL, K-LX, and K-XY groups where X and Y stand for all vacancies with quantum numbers equal to or greater than $n + 2$ (in this case 3). Similarly, for an L_3 primary vacancy we have L_3-MM, L_3-MX, and L_3-XY, and similarly for L_1 and L_2. But for the L shell the groups do not appear well separated because the L_1 and L_2 groups overlap each other and the L_3 groups.

When a higher resolution is used each band is seen to be composed of numerous lines of which many are ordinary lines resulting from a single initial vacancy and others are satellites resulting from multiple initial vacancies, multiple Auger processes, and other complex phenomena.

Complete interpretation of the ordinary lines necessitates relativistic-intermediate coupling which is a combination of L-S coupling (small binding energies) and j-j coupling (very large binding energies). As an example of these interpretations we show in Table 1 the designations for the K-LL group.

Auger spectra also enable the relative intensities of the lines or groups of lines to be determined. At high resolution the K-LL intensities of all the ordinary lines can be accurately measured. No theoretical calculation gives the correct relative intensities for all values of atomic number but relativistic j-j coupling gives quite good agreement above 80. K-LX and K-XY lines are less well resolved and theories are less developed. For L-MM spectra the experimental intensities are much less precise due to the much lower energies and the overlapping of the bands. Nonrelativistic L-S and j-j coupling theoretical intensities have been calculated, and crude agreement is obtained with the former for low atomic numbers and with the latter for high atomic numbers. Very little good data exist for

L-MN and L-NN spectra. Considerable medium resolution data exist for M and N Auger spectra.

While the global K-Auger intensity is easy to obtain by integrating under the spectrum, the total L_1, L_2, and L_3 intensities are difficult to obtain because of the overlap of the three bands.

Relative Probability of Auger and X-ray Emission The evaluation of the relative probability of x-ray and Auger emission for different initial vacancies, and the determination of the relative intensities of various Auger lines constitute one of the most important aspects of Auger effect research.

The fluorescence yield, ω_i, for any initial vacancy i is defined as the fraction of vacancies filled by emission of photons. The Auger yield is defined correspondingly as the fraction filled by emission of Auger electrons. The Auger yield is divided into two parts, one (denoted by a_i) which transfers the vacancy to a level with a higher total quantum number, and the other (denoted by f_{ij}) which transfers the vacancy to a lower-energy vacancy with the same total quantum number. The latter process is called the Coster-Kronig effect.

For the K shell the following equation holds

$$1 = \omega_K + a_K.$$

Although in principle $\omega_K(Z)$ should be readily calculable, the number of (frequently relativistic) electron wave functions which must be known for each Z, and the number of permutations and combinations of these functions which must be handled in order to calculate the individual probability of every line and, by summing, the total K-Auger probability per unit time, present a formidable problem even with the aid of a sophisticated computer. The K fluorescence yield is therefore normally

TABLE 1

L-S Coupling (light elements)		Intermediate Coupling	j-j Coupling (heavy elements)
$K - 2s2s$	$^1S_0$	K-L_1L_1 $^1S_0$	K-L_1L_1
	$^1P_1$	K-L_1L_2 $^1P_1$	
$K - 2s2p$		K-L_1L_2 $^3P_0$	K-L_1L_2
	$^3P_{0,1,2}$	K-L_1L_3 $^3P_2$	
		K-L_1L_3 $^3P_2$	K-L_1L_3
	$^1S_0$	K-L_2L_2 $^1S_0$	K-L_2L_2
	$^1D_2$	K-L_2L_3 $^1D_2$	
$K - 2p2p$		K-L_2L_3 $^3P_1$ parity forbidden	K-L_2L_3
	$^3P_{0,1,2}$	K-L_3L_3 $^3P_0$	
		K-L_3L_3 $^3P_2$	K-L_3L_3

For K-Auger electrons, measurements and comparison with theory are extensive.

found by fitting theoretical expressions for ω_K containing empirical constants to the experimental values. For $20 < z < 55$, ω_K is given to a few per cent by

$$\omega_K = (1 + 7.8 \times 10^5 \, Z^{-4})^{-1}$$

and somewhat less accurately for $10 < z < 20$. For $z < 10$ the experimental errors are large because of solid state and molecular effects, and the theoretical estimates may also be considerably in error.

For $Z > 50$ it is better to estimate a_K since relativity has an important effect on a_K but not on ω_K which is very close to unity. The following equation probably predicts a_K to 10 to 15 per cent and hence ω_K to a per cent or better for $z > 50$:

$$a_K = [1 - (1 + 7.8 \times 10^5)^{-1}] \, (1 + 3.5 \times 10^{-5} \, Z^2$$
$$+ 3 \times 10^{-8} \, Z^4).$$

For the L shell in the j-j coupling limit the following equations hold:

$$1 = \omega_3 + a_3$$
$$1 = \omega_2 + a_2 + f_{23}$$
$$1 = \omega_1 + a_1 + f_{12} + f_{13}.$$

In addition, the average L-fluorescence yield $\overline{\omega}_L$ for an atom with an L vacancy, having probability n_1, n_2, and n_3 of being in each of the three subshells, is

$$\overline{\omega}_L = 1 - \overline{a}_L = n_1(\omega_1 + f_{12}\omega_2 + f_{13}\omega_3$$
$$+ f_{12}f_{23}\omega_3) + n_2(\omega_2 + f_{23}\omega_3) + n_3\omega_3.$$

The values of the nine L-shell constants are much less well known than the two constants for the K shell. Below atomic number 50 the ω_i are all less than 6 per cent and appear to be less than 50 per cent for all elements. ω_3 is given reasonably accurately by

$$\omega_3 = (1 + 0.82 \times 10^8 \, Z^{-4})^{-1} \qquad Z > 65$$
$$\omega_3 = (1 + 1.08 \times 10^8 \, Z^{-4})^{-1} \qquad Z > 50.$$

ω_2 is much less accurately known but the agreement between theory and experiment is fairly good. The knowledge of ω_1 is relatively poor. The Coster-Kronig yields go through sudden changes in value as certain transitions become energetically possible or impossible. For example, $L_2 - L_3 \, M$ transitions are forbidden for Z from 50 to 73. Experimental values for f_{13} and f_{23} have precisions ranging from 5 to 20 per cent, while f_{12} has only a precision of about 30 per cent. Agreement with theory is not very good. The $\overline{\omega}_L$ and $\overline{a}_L$ clearly depend on the type of excitation.

Although initial interest in the Auger effect arose through creation of vacancies by ejection of photoelectrons by x-rays, there has been a strong recent upsurge in interest in several other types of experimental work which involve Auger spectra closely:

(1) Vacancies created in radioactive decay.
(2) Vacancies created in electron-ion and ion-ion collisions in gases.
(3) Auger electrons emitted from the surface layers of solid targets in ultrahigh vacuum when the surface is bombarded by electrons up to 3 keV and sometimes by ions.

When radioactive decay occurs, internal conversion, shake-off, accompanying beta-decay, and electron capture all create vacancies. In fact, the only method of studying the relative probability of orbital capture in the various shells and subshells of a nuclide involves the study of the x-ray or Auger spectrum of the product nuclide. The interpretation of these spectra necessitates the knowledge of the constants discussed above.

The importance of the Auger effect for radioactive nuclei, as well as the almost 100 per cent probability of Auger and Coster-Kronig emission for the M and higher levels, is indicated by measurement of the total charge accumulated by certain radioactive nuclei. For example, a vacancy produced in xenon 131m by internal conversion gives rise in some cases to as many as 21 Auger processes, leaving a xenon ion with a charge of $-22e$.

The highest-resolution Auger spectra are those from vacancies created in gaseous atoms or molecules at low pressure by bombardment by electrons or other ions. Such spectra are particularly rich in lines (satellites) arising from multiple vacancies, and their interpretation has contributed to our understanding of the processes which take place in such collisions.

Since about 1967 there has been great interest in the study of surface impurities by means of the Auger spectra of solid surfaces in ultrahigh vacuum bombarded by low energy, up to 3 keV, electrons. The experiments are usually done by modifying a Low Energy Electron Diffraction (LEED) apparatus to provide it with a retarding grid. By varying the retarding voltage, an integral spectrum of the scattered and secondary electrons is obtained. By applying a small alternating voltage to this grid and tuning the detector amplifier to twice the frequency, a very sharp differential line spectrum of scattered and Auger electrons is obtained. Impurity detection of better than $\frac{1}{50}$ of a monolayer is claimed in some cases. Although it can be shown that a deflection-type spectrometer with an alternating voltage on the detector has in principle a better signal-to-noise ratio than the retarding grid, the combination of the latter with electron diffraction studies of the same surface constitutes an extremely powerful tool for the study of surface physics and chemistry.

SHERWOOD K. HAYNES

References

Burhop, E. H., "The Auger Effect and Other Radiationless Processes," Cambridge, The University Press, 1952.

Listengarten, M. A., "The Auger Effect" (a Review), *Bull. Acad. Sci. USSR*, **24** (9), 1050 (1960).

Burhop, E. H. S., and Asaad, W. H., "The Auger Effect," in "Advances in Atomic and Molecular Physics," New York, Academic Press 1972.

Chang, C. C., "Auger Electron Spectroscopy," *Surface Sci.*, **25**, 53–79 (1971).

Sevier, K., "Low Energy Electron Spectroscopy," New York, John Wiley & Sons Inc., 1972.

Haynes, S. K., "Phenomenological Systematics of L-Auger Spectra," Summary of the IUPAP Conference on Inner Shell Ionization Phenomena, Atlanta, Georgia, April 1972.

Cross-references: ATOMIC PHYSICS, ISOTOPES PHOTOELECTRICITY, X-RAYS.

AURORA

The earliest auroral descriptions date back to the Old Testament and early Chinese chronologies. Up until the Renaissance period, there was little scientific understanding and most reports involved superstitious, prognostic, or religious interpretations. The metaphor "aurora borealis" seems to have been first used by Galileo, and many famous 17th–19th Century scientists (including Halley, Descartes, Celsius, Dalton, Biot, and Ångström) speculated on the nature of the phenomenon. However, little real progress in understanding was made until the era of polar exploration resulted in systematic data gathering from high latitude regions.

The general term "aurora" (*aurora borealis* or northern lights in the northern hemisphere, *aurora australis* or southern lights in the southern hemisphere) refers to the luminous emissions commonly seen in the night sky in polar regions. The aurora is much brighter than airglow (see AIRGLOW), which is a uniform, subvisual luminosity and is worldwide in occurrence.

Auroral forms typically take the shape of arcs or bands that align along parallels of geomagnetic latitude (see Fig. 1), and often exhibit a vertical striated structure (rays) that align along the local direction of the earth's magnetic field (see Fig. 2). The height of the aurora may be as low as ~80 km and as high as 300–500 km, but most auroras occur around 110 km. Because the extent of auroras commonly exceed the limited field of view from a ground station, many of the auroral forms reported from the ground (arcs, bands, coronas, etc.) are only apparent shapes, distorted by perspective effects.

The aurora occurs in two oval-shaped (nearly circular) bands encircling the north and south geomagnetic poles, locating about 25° of latitude from these poles on the night side of the earth, and about 12° from the poles on the dayside (see Fig. 3). This instantaneous auroral distribution is called the "auroral oval." As the aurora is typically brighter near local midnight, the region of highest probability of seeing an aurora is the projection of the midnight sector of the auroral oval onto the rotating earth; this is a circular region about 25° latitude from the geomagnetic poles, called the "auroral zone." During periods of geomagnetic storms (see GEOPHYSICS) the auroral oval shifts to lower latitudes (see Fig. 4) and at rare times is seen over the tropics.

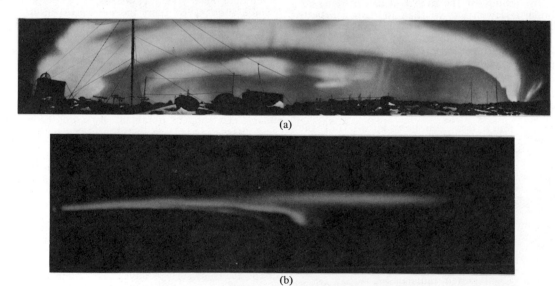

(a)

(b)

FIG. 1. (a) Wide angle photograph showing typical appearance of an auroral arc as seen from the ground. (b) Pictures of an auroral arc as seen from space (*Spacelab*) (R. H. Eather).

FIG. 2. An illustration of the variety of auroral structures (NASA).

Auroras are observed to occur simultaneously in the northern and southern hemispheres, showing remarkably similar shapes and radiances at opposite ends of the same magnetic field line. Such auroras are said to be magnetically conjugate (see Fig. 5).

The auroral light arises from spectral emissions by atmospheric gases in atomic, molecular and ionized states. The most important auroral emissions, and approximate relative radiances, are listed in Table 1. The strongest emission is the yellow-green line of atomic oxygen.

The total integrated radiance of the aurora may vary over many orders of magnitude, and is often classified on a scale of 1 to 4 called the International Brightness Coefficient (see Table 2).

The auroral light is produced by bombardment of the upper atmosphere by energetic electrons and protons in the energy range of about one-tenth to tens of keV, with electrons normally being the dominant energy source. The energy of the particles determines how far they penetrate into the atmosphere, and so determines auroral height. Because atmospheric composition changes significantly with height above the turbopause, the relative importance of various spectral features (Table 1), and hence auroral color, also are functions of height. If auroral radiance does not exceed the color

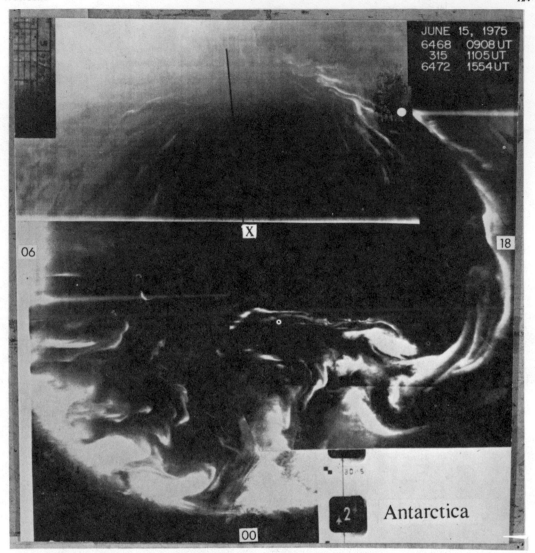

JUNE 15, 1975
6468   0908 UT
315   1105 UT
6472   1554 UT

06

18

X

00

Antarctica

FIG. 3. A composite of satellite pictures taken over Antarctica near midwinter. The geomagnetic south pole is marked with an X, and the geographic south pole with ●. Local time is indicated around the edges of the picture. The active bright nightside aurora is illustrative of a intense substorm; the dayside aurora is much less intense, and shows a common fanning out of structures from near midday (S. I. Akasafu).

threshold of the eye (∼IBC 2, see Table 2), the aurora appears whitish in color. When this threshold is exceeded (scotopic vision), the strong auroral emissions in the blue, green and red regions of the spectrum (Table 1) can combine to give a wide range of hues. Typically, however, bright auroras appear yellowish-green (557.7OI), sometimes with blue- (N_2^+) or red (N_2, O_2^+) lower borders, and sometimes with red upper regions (630.0/636.4 OI). The auroras seen at low latitudes during large magnetic storms are often excited by lower energy electrons, giving high red auroras.

The energetic electrons (and protons) that generate the aurora have their primary source in the solar wind (see MAGNETOSPHERIC RADIATION BELTS and Fig. 6). (A secondary probable source of auroral particles is the earth's upper ionosphere.) Details of the mechanism whereby solar wind particles gain entry to the magnetosphere are not clear but probably involve merging of the earth's magnetic field with the interplanetary magnetic field imbedded in the solar wind. It is likely that particles entering the magnetosphere through the dayside cusp regions have fairly direct access to the dayside

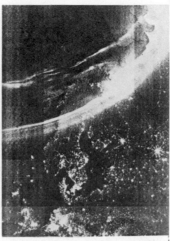

FIG. 4. Satellite pictures showing the location of the auroral oval over western Europe for different levels of magnetic activity. For quiet conditions (left panel) the aurora locates north of the northern coast of Scandinavia. For weak magnetic activity (center panel), the aurora is seen over central Scandinavia. During moderate disturbances, auroras are seen from southern Scandinavia (right panel) (Air Force Geophysics Lab, R. H. Eather).

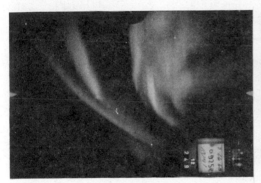

FIG. 5. Photographs taken from aircraft flying under the same magnetic field line in the northern hemisphere (over Alaska) and the southern hemisphere (between New Zealand and Antarctica). The pictures are mirror images of each other (auroral conjugacy) (T. Neil Davis).

TABLE 1. Main Auroral Emissions in Visible Region.

Emission	Spectral Region (nm)	Approx. Height	Relative Radiance*
OI	557.7 yellow-green	110 km	1.
	630.0/636.4 red	175 km	$.1 \rightarrow 10$
N_2+ 1st neg	391.4 violet	110 km	.6
	427.8 blue	110 km	.2
N_2 1st pos	red	90–100 km	.5
N_2 2nd pos	violet	90–100 km	1.
O_2+ 1st neg	yellow-red	90–100 km	.1
H Balmer	486.1 H, 656.3 H	120 km	.1

*Radiance refers to an emitting surface and is not strictly the correct term in this context. Auroral "radiance" is actually an integrated column emission rate, and is often quoted in units of Rayleighs, where $1 R = 10^{10}$ photons m^{-2} (column) sec.

TABLE 2. Auroral Radiance* Classification.

International Brightness Coefficient	557.7 OI Column Emission Rate (kR)*	Description of Brightness	Approximate Luminance (ft Lamberts)
1	1	Milky Way	10^{-4}
2	10	thin moonlit cirrus	10^{-3}
3	100	moonlit cumulus	10^{-2}
4	1000	provides ground illumination similar to full moonlight	10^{-1}

*See note at bottom of Table 1.
NOTE: Approximate photographic exposure required (f1.4 lens, ASA 400) is 1–2 secs for IBC 3 aurora.

ionosphere and generate dayside aurora. These precipitating particles are typically low energy (0.1–1 keV) and generate weak, reddish aurora (often subvisual) at heights of ≳150 km. The particles generating the more intense nighttime aurora have higher energies (~1–10s keV) and a source region in the plasma sheet. The acceleration mechanism required to energize the solar wind particles to keV energies have not been definitively identified. Some energization prob-ably occurs by magnetospheric electric fields that drive particles from the flanks of the magnetosphere to the plasma sheet region, and additional (possibly primary) energization by field aligned electric fields has been discovered at low altitudes (10–15,000 km) above the auroral ovals.

The auroral oval displays considerable dynamic behavior on a time scale of the order of hours (Fig. 7). Periodically, intense auroral

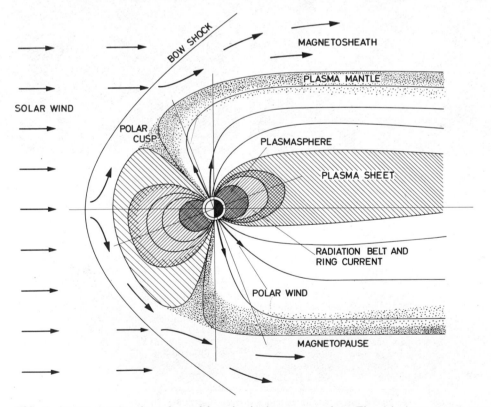

FIG. 6. Sketch showing the main particle region in the magnetosphere. The nighttime aurora occurs at the base of magnetic field lines that thread the plasma sheet. Dayside auroras occur at the base of the polar cusp field line (G. Paschman).

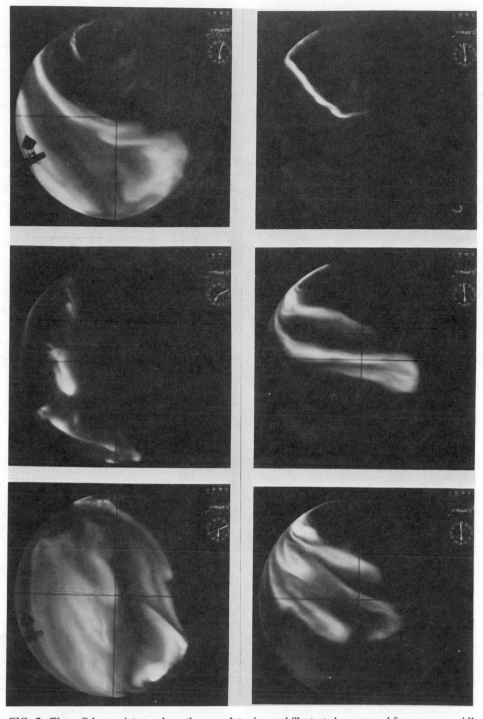

FIG. 7. These fisheye pictures show the complete sky, and illustrate how auroral forms may rapidly change during a substorm. This sequence shows changes over a 10 min. period of substorm development (R. H. Eather).

brightenings appear in the midnight region, followed by a characteristic poleward expansion in the midnight sector, westward travelling surges in the evening sector, pulsating forms in the morning sector, and equatorward shifts of dayside aurora. This sequence of events, lasting $\frac{1}{2}$–1 hour, is called an auroral substorm. The frequency and intensity of substorms increases during worldwide magnetic storms. The triggering process that initiates this substorm sequence is not yet identified. The uncertainties concern whether it is an internal (to the magnetosphere) relaxation process that suddenly releases stored energy in the tail, or whether the process is externally driven by fluctuations in the solar wind.

There are many phenomena associated with the aurora other than the visible light. The auroral ovals carry large electric currents (the auroral electrojet) that represent closure paths for large-scale magnetospheric current systems. These currents have strong magnetic effects at ground level. Ionization in the aurora results in radio-wave reflections at lower frequencies and absorption and scintillations at higher frequencies. Electromagnetic emissions from the aurora include x-ray, ultraviolet and infrared, vlf, radiowave, and ulf emissions. Auroral heating leads to a wide variety of chemical changes.

The aurora is the only visible manifestation of the dynamic electromagnetic environment around the earth. This facilitates coordinated experiments between ground-based and satellite based detectors. It is such coordinated experiments, between ground station arrays and multiple satellites both inside and outside the magnetosphere, that will eventually result in definitive answers to three very important unanswered questions—what is the entry mechanism(s) for solar wind particles into the magnetosphere; what is the acceleration mechanism(s) inside the magnetosphere; and what triggers and drives the substorm process?

ROBERT H. EATHER

References

Chamberlain, J. W., "Physics of the Aurora and Airglow," Academic Press, New York and London, 1961.

Eather, R. H., "Majestic Lights—The Aurora in Science, History and the Arts," Amer. Geophys. Union, Washington, D.C., 1980.

Omholt, A., "The Optical Aurora," Springer-Veralg, Berlin-Heidelberg, 1970.

Vallance Jones, A., "Aurora," D. Reidel, Dordrecht-Holland, 1974.

Cross-references: AIRGLOW, ELECTRICAL DISCHARGE IN GASES, GEOPHYSICS, IONOSPHERE, MAGNETIC FIELD, MAGNETOSPHERIC RADIATION BELTS, PLASMAS, SOLAR PHYSICS, SPACE PHYSICS.

B

BETATRON

A betatron is a particle accelerator which uses a sustained induced voltage to accelerate charged particles to full energy during the whole period of acceleration of the particle. Since this method of acceleration seemed most applicable to electrons, the name *betatron* was used to indicate that it was the agency for producing high-speed electrons. The accelerating action in a betatron is similar to the action of an electrical *transformer* in which a high-voltage winding of many turns is used. In a transformer, the voltage can be stepped up from a primary voltage V_1 to the secondary voltage V_2,

$$V_2 = V_1 N_2 / N_1$$

where N_2 is the large number of turns of the secondary coil and N_1 is the small number of turns in the primary winding. For example, diagnostic x-ray transformers producing high voltage, such as 100 000 volts, have very many turns of fine wire and consequently raise the primary voltage by a large factor.

The accelerating or voltage generating structure of a betatron is really a transformer; particle guiding, or focusing, magnets are arranged around the transformer core where a secondary winding might be put. A toroidal vacuum vessel is placed between the poles of the focusing magnets so that electrons can travel hundreds of thousands of times around the core. Each time the electron circulates around the core it acquires an energy equivalent to the voltage which would have been induced in one turn of wire at that instant. The very long path requires magnetic and electrostatic field errors to be small, particularly when the particles are starting at low speed and are perturbed by small magnitude field errors.

In order to guide the electrons the focusing magnet can be such that its magnetic field decreases with increasing radius. Then the lines of force going from pole to pole bulge outwardly across the orbital plane. This provides forces, which are always perpendicular to the field line, which have a component directed back toward the orbital plane in case the particle strays away from this plane. This bulging field is also used for vertical focusing cyclotrons. However, it is necessary that the magnetic field decrease less rapidly than $1/r$, where r is the radius of the orbit. If this latter requirement is met, the radial focusing will be insured; the required centripetal force to hold the particle in the circle going around the core decreases as $1/r$. Consequently, if the magnetic force decreases less rapidly than this, it will be too strong at large radii to permit the orbit to remain circulating at a large radius, and it will be too weak at small radii to maintain the particle circulating at a small radius. The particle thus will oscillate about the so-called *equilibrium orbit radius* when the magnetic force has the right value to supply the required centripetal force. The axial or radial oscillations of the particle about the equilibrium orbit are called betatron oscillations, and this name appears in the scientific literature referring to particle motions and focusing in other accelerators.

It is possible to use "strong focusing" magnets with much more rapid variation of magnetic field with radius to provide axial and radial focusing. In this case a succession of strong focusing and defocusing magnets must be used which alternately focus vertical (axial) and radial motion. The net result is a focusing action, just as for optical lenses. Such magnets are called *alternate gradient focusing magnets*, and their value is that they can limit the amplitude of oscillation of a particle about the equilibrium orbit to a very small size. This makes possible very large circumference high energy acclerators without immense magnetic guide field spaces. For example, the magnets for the large synchrotrons at CERN and at the Fermi National Accelerator Laboratory providing energies of several hundred GeV are such alternate gradient magnets; but in these accelerators the acceleration is not done by a continuously rising betatron flux within the orbit. Instead, small transformer cores which operate at radio frequencies link the orbit so that whenever the particle returns to the vicinity of the core the flux is rising in the direction to accelerate the particle. Thus the extremely large iron core of a betatron is not needed.

The usual betatron has a focusing magnetic field which rises proportionally with the increase in the transformer's magnetic flux within the orbit. Thus the guiding field increases proportionately with the momentum gained by

FIG. 1. The 320 MeV betatron at the University of Illinois near completion. The six guide field magnets placed around the central leg of the big accelerating core would form a synchrotron if a radiofrequency accelerating cavity were used for acceleration instead of the transformer. 5 watts of x-rays were produced by this accelerator. The light radiation loss from the revolving 320 MeV electrons was nine percent, which had to be made up by extra flux driven through the core. The original 2-MeV typewriter-size betatron, if made in the image of this 320 MeV betatron and biased, would be the size of a match box.

the transformer action, and it provides sufficient magnetic force to hold the particle at a constant radius.

The first betatron of this type produced 2 MeV and radiation equivalent to 2 grams of radium. It is now in the Smithsonian Museum in Washington, D.C. This accelerator is the size of a typewriter. The largest betatron (see Fig. 1) could generate beams of 320 MeV. The x-rays and electrons from it were used for experiments to produce mesons and numerous nuclear disintegrations. At such energies circulating electrons radiate an important fraction of their energy as light. This is a limitation on the use of betatron action for very high energy electrons, but synchrotrons are better able to make up this energy loss. The most commonly used betatrons are for 25 to 35 MeV. These provide x-rays of maximum penetration in iron for industrial radiography, and they provide x-rays and electrons with optimum depth dose characteristics for x-ray or electron beam therapy of the human body.

The intensity of radiation from the medical or industrial betatrons is of the order of 100 to 200 roentgens/min at a meter from the target for x-rays. With the extracted electron beam, the ionization doses would depend on how widely the electrons are spread at the point of treatment, but comparable doses are obtained. The large 320-MeV betatron at the University of Illinois, which is shown in the figure, produced intensities of the order of 20 000 roentgens/min at a meter or other terms, of the order of 5 watts.

There are possibilities for increasing the intensity of radiation from betatrons. One is the fixed-field alternating-gradient betatron (FFAG betatron). In this case, the focusing field is constant in time, and the particle orbit can be caused to spiral either outwardly or inwardly with increasing energy to the high field region of the direct current magnet or permanent magnet poles. The electrons can thus be injected and accepted by the guide field continuously; and with a transformer accelerating core with a sinusoidal flux variation a beam of x-rays could be produced about 20 percent of the time.

Betatrons with time-varying focusing fields just give one pulse every cycle of the transformer core, and the pulses are only of the order of a microsecond duration. Because of the large duty factor available with FFAG betatrons, it should be possible to achieve intensities of 10 000 watts of electrons. Although FFAG combination betatrons and synchrotrons have been made, full advantage has not been taken of a large duty factor achievable by incorporating a full-size transformer core within the betatron orbit.

Another possibility for increased intensity is operation at a higher frequency or loading each cycle with more electrons. Medical and industrial betatrons are filled to approximately their space-charge limit at the injection energy of about 60 000 volts. But as the energy increases the space-charge limit rises because the beam's rising current magnetic pinching of itself counteracts its space-charge repulsion. At relativistic energies the limit on current is very high—hence injection at higher energy allows much more current to be held in the guide field. Some attempts at holding high currents in large aperture guide fields have been made.

In the case of the conventional betatron with constant orbit radius, the relation between the strength of guiding field and total flux change within the orbit can be found as follows:

The momentum of the particle in the orbit

$$P = \frac{e}{c} BR$$

while the rate of change of momentum

$$\frac{dp}{dt} = eE$$

where e is the charge of the electron, c is the velocity of light, B is the magnetic field in gauss, R is the radius in centimeters, and E is the electric field in electrostatic units.

$$2\pi RE = \frac{1}{c} \frac{d\Phi}{dt}$$

is the electrostatic volts per turn, where Φ is the flux linking the orbit. Combining the last two equations

$$\frac{dp}{dt} = \frac{e}{2\pi RC} \frac{d\Phi}{dt}.$$

Thus, after a lapse of time

$$p = \frac{e}{2\pi RC} (\Phi_2 - \Phi_1).$$

Combining this with the first equation yields

$$2\pi R^2 B = \Phi_2 - \Phi_1.$$

Thus the flux change within the orbit during acceleration is twice as big as the flux would be if the flux density B were uniform within the orbit. Therefore, the transformer core must be adjusted so the proper excess flux provided within the orbit meets the conditions of this last relation if the orbit is to be at a constant radius, the assumption made in the above derivation.

The flux condition allows the core of the transformer to be biased with Φ_1, reversed compared with Φ_2. Thus the iron in the transformer can be started at $-16\,000$ gauss and reversed to $+16\,000$ gauss while the orbit field B goes from zero to its maximum value. Biased betatrons are thus much smaller than unbiased betatrons.

D. W. KERST

Cross-references: ACCELERATOR, PARTICLE; ACCELERATORS, LINEAR; ACCELERATORS, VAN DE GRAAFF; CYCLOTRON; SYNCHROTRON.

BIOMEDICAL INSTRUMENTATION

This article is restricted to examples of physical measurements on humans which are useful in the diagnosis and management of disease. It ignores measurements of physiologic variables for research purposes, measurements in other animal species or in plants (for crop management, for example), and measurements based primarily on chemical processes. It also excludes devices to augment patient function, such as artificial organs, limbs, and other prostheses.

Many measurements involve the use of a transducer to convert some physical variable into another variable more easily sensed by the observer. Often the transduction process involves an electrical signal as an intermediate stage, because of the ease with which such signals can be processed (amplified, filtered, etc.). An electrical intermediary is not always required: blood pressure is measured with a cuff and stethoscope, and x-rays blacken a photographic film.

Measurable mechanical quantities which may be useful are displacement, velocity, acceleration, force or pressure. Pressure within the eye is measured (tonometry) by determining the force required to flatten a defined area of the cornea. The blood pressure maximum (systole) and minimum (diastole) can be measured with the familiar cuff and stethoscope. Measurements of blood pressure as a function of time are usually made with catheters inserted in an artery. These either have a small pressure transducer at their tip or contain fluid which transmits the pressure to an external transducer. Recently, schemes have been developed which measure the blood pressure changes noninvasively by applying a time-varying external pressure which unloads the arterial wall (Wesseling; Yamakoshi).

The classical instrument for measuring respiratory flow is the water spirometer. A bell jar, balanced by a counterweight and closed at its lower, open end by a water seal or bellows, moves up and down as the patient breathes into the space it encloses. Movement of the bell jar can be related to volume changes in the lung by using the gas law. Flow rate is the time derivative of the lung volume. Pneumotachometers measure flow directly. Some pneumotachometers use a small turbine in the air stream; others measure the pressure drop across a section which offers a fixed resistance to the air flow. Still other models measure the convective cooling of a hot wire by the flowing air.

Thermal and dye dilution techniques are used to measure cardiac output. For thermal dilution, catheters are threaded through a vein into the right atrium and the pulmonary artery. A known amount of cool liquid injected into the right atrium warms up as it passes through the right ventricle into the pulmonary artery. The time constant for warming can be related to the flow rate from the heart. Cardiac output can be determined by a similar mathematical analysis when a dye or a radioactive isotope spreads after it is injected.

The most obvious application of acoustics is in the diagnosis of hearing disorders. Audiograms show the threshold of sensitivity of the ear to tone bursts of different frequencies. Measurements of the acoustic compliance of the ear drum as external air pressure is changed (tympanogram) can be used to distinguish different problems in the middle ear. *Evoked response audiometry* or *auditory brainstem response* records small electrical signals from the scalp due to audible stimuli. Because these signals have an amplitude of about 100 nV and are masked by other signals of about 50 μV, the signal must be extracted by averaging over many stimuli. The amplitude and time delay of the various components of this signal provide diagnostic information (Thornton).

Ultrasound is used extensively as a diagnostic tool. The basic pulse-echo technique detects discontinuities in acoustic impedance at anatomical features and locates them by the time required for the echo to return to the transducer. Two-dimensional maps are made by recording the direction in which the transducer points along with the echo time. By pointing the ultrasound beam at the heart, motion of the heart wall and valve leaflets can be recorded as a function of time. The Doppler effect is also exploited in ultrasonic diagnostic to measure the velocity of the reflecting surface. Detecting only moving interfaces improves the signal-to-noise ratio for monitoring fetal heart or lung movement. The magnitude of the frequency shift when ultrasound is scattered from red cells is used to measure blood flow. Details of various ultrasound techniques can be found in the references by Wells.

One frequently measures directly the electrical potential difference between two points on the body. Various kinds of measurements are shown in Table 1. These potentials arise because the interior of a cell is normally at a potential of 70–90 mV less than the outside. As a nerve cell conducts or a muscle cell prepares to contract, the cell depolarizes and the interior becomes more positive than the outside. The resulting currents flowing in the body give rise to potential differences at the surface. (These currents also produce magnetic fields which can be measured with SQUID magnetometers (Wikswo; Geselowitz)).

The measurement of these potentials requires great care in the design of the electrodes. It is necessary to minimize battery effects (polariza-

TABLE 1. **Measurements of Electrical Potential Difference Between Points on the Body.**

Measurement	Source or Stimulus	Measurement	Amplitude	Frequency Response
Electroencephalogram	nerve activity in the brain	electrodes on scalp	10–100 μV	0–100 Hz
Electrocardiogram	muscle activity in the heart	surface electrodes on the limbs and chest	1 mV	0–100 Hz
Electromyogram	electrical activity in skeletal muscles during contraction; motor and sensory nerve conduction	needle electrodes surface electrodes	1 mV 50 μV	10 Hz–3 kHz
Electroretinogram	flash of light periodically moving pattern	cornea to lid	200–500 μV 10 μV	1–600 Hz 0.5–100 Hz
Electrooculogram	eye movement	skin on nose and temples, or corners of eye	1 μV	0–10 Hz
Visual evoked response	flash of light; changing pattern	scalp electrodes over visual (occipital) cortex	1 μV	1–600 Hz
Auditory brainstem response	sound stimulus	scalp	100 nV	50 Hz–2 kHz

tion) and spurious signals which can be generated in the outer layers of the epidermis (Webster; Plonsey).

The principles of geometrical optics are used in ophthalmology to examine the eye and determine errors of refraction. The ophthalmoscope works on the principle that an object at the focus of a lens produces a beam of parallel rays. If both the patient and examiner have normal vision and the patient's retina is illuminated, the parallel beam of light emerging from the patient's eye will be focused on the retina of the examiner. A laser, used as a flying-spot scanner, has been used to produce raster images of the retina (Webb). Total internal reflection has made possible fiber optic endoscopes which are quite flexible and which can be used to examine the patient's colon, stomach, lungs, urinary bladder, etc., with much less discomfort than with the old rigid-tube instruments. Moiré patterns have been used to record body surface contours.

The most widespread use of lasers in medicine has been for photocoagulation, which is outside the scope of this article. Speckle patterns have been used for refracting the eye. Holographic techniques have been suggested for the diagnosis and quantification of glaucoma.

The use of x-rays, nuclear radioactivity, and ultrasound for diagnosing and treating patients is called *medical physics* in the United States. Conventional radiography measures anatomical features by using a point source to cast a shadow of the patient on a sheet of film. X-ray photons of 60–80 keV are used so that the photoelectric effect, which depends strongly on atomic number, will help separate bone from soft tissue. It is often necessary to introduce contrast agents of high atomic number into the esophagus, stomach, bowel, kidneys, blood vessels, or hepatobiliary tree, in order to visualize the anatomical features.

Tomography (Greek *tomos*: a cut or section) has been used to blur all but one plane within the body by pivoting both the camera and the film around a point in the plane of interest. Computed tomography produces genuine sections by reconstructing a function $f(x, y)$ from

a series of projections $F(x) = \int f(x, y)\, dy$ taken

at many angles through the body (Fig. 1). In transmission tomography, the attenuation of a pencil beam of x-rays is measured, and the function $f(x, y)$ is the x-ray attenuation coefficient (Pullan). Emission tomography measures the γ-rays emitted by a radioactive pharmaceutical agent distributed along a line through the body, and $f(x, y)$ is the concentration of the radioactive substance.

Ultrasonic tomography is also being developed. For ultrasound $f(x, y)$ can be the reciprocal of the velocity (if $F(x)$ corresponds to the propagation delay of a pulse) or $f(x, y)$ can be the ultrasonic attenuation coefficient. The analysis is complicated by the fact that ray optics is not a good approximation to the propagation of the ultrasound.

Digital radiography uses electronic detectors similar to those used in computed tomography to record a two-dimensional projected image similar to that of a conventional x-ray. In addition to providing greater dynamic range than film, this technique allows images to be made by digital subtraction. For example, a recording without an intravenous contrast agent, sub-

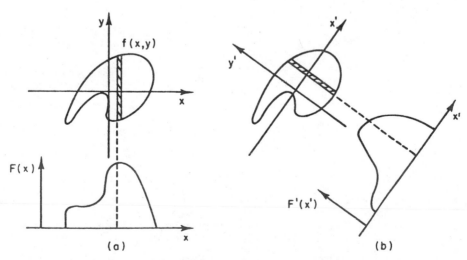

FIG. 1. Computed tomography reconstructs a function $f(x, y)$ from a series of projections $F(x)$, $F'(x')$, etc. (From Hobbie; reproduced by permission of John Wiley & Sons, Inc.)

tracted from one with the agent, provides an image of the vascular system. Similar images in the past usually required an intra-arterial injection of larger amounts of the contrast material, with greater discomfort and risk for the patient.

A single scan of a limb with a monoenergetic photon source and a scintillation detector allows quantitative determination of bone mineral content. The source for this small, desk-top machine is a radioactive nucleus which emits a gamma ray photon (Cameron, 1963).

X-ray fluorescence is used to measure the concentration of heavy metals such as lead or mercury in the body. When the x-ray fluorescence lines are favorably situated this method is sensitive to 10–20 ppm (Bloch).

Nuclear medicine measures the γ-ray emission of radioactive pharmaceuticals taken up in various organs of the body. Detectors range from a small unit placed over the organ of interest to a large disk of scintillator viewed by many photomultiplier tubes, which can record a plane-projected image of the body. Emission tomography, mentioned above, uses an array of detectors surrounding the patient (Brownell). X-ray and nuclear medicine techniques provide complementary information. X-ray pictures typically have resolutions of 1 mm and measure anatomical features; the resolution of a nuclear medicine image may be a factor of 10 worse, but since the radioactive pharmaceutical has been selectively taken up by some organ, the nuclear medicine image shows the physiological function of different parts of the organ (MacIntyre).

Neutron activation analysis is used to measure specific elements in the body. A neutron source such as ^{252}Cf irradiates the region of interest. Some of the nuclei in the body are made radioactive; the excited nuclei emit γ-rays of characteristic energy, the intensity of which is proportional to the concentration of the element being measured.

Nuclear magnetic resonance measures the absorption and emission of radiofrequency photons as a nuclear spin magnetic moment changes orientation in an external magnetic field. The photon frequency is proportional to both the external magnetic field and the nuclear magnetic moment. The strength of the signal is proportional to the concentration of the nuclei for which the apparatus is tuned. Studies are most often made of protons (hydrogen), although other atoms such as ^{31}P also possess a nuclear magnetic moment. Pulsed NMR also allows measurement of relaxation times, during which the excess nuclear orientation energy is exchanged with the surrounding lattice or with neighboring nuclei. Since the NMR resonance frequency is proportional to the external magnetic field, it is possible to image either hydrogen concentrations or relaxation times by varying the magnetic field across the patient. The reconstruction techniques are similar to those in computed tomography (Brownell; Lai).

NMR is also used to measure blood flow. The measurement relies on the fact that once the nuclei have been aligned, the RF resonance field misaligns them even as it produces a measurable signal. If the fluid is flowing, new aligned nuclei are brought into the sensitive volume at a rate proportional to the flow, and the signal is increased (Battocletti).

An article this short can only survey the field of biomedical instrumentation. The references which follow will provide the reader with more detailed information about any topic mentioned. Topics which do not have specific references above are discussed in the books by Cameron (1978), Cobbold, and Webster.

RUSSELL K. HOBBIE

References

Battocletti, J. H., Halbach, R. E., and Salles-Cunha, S. X., "The NMR blood flow meter—Theory and history," *Med. Phys.* 8:435–443 (1981).

Bloch, P., and Shapiro, I. M., "An x-ray fluorescence technique to measure the mercury burden of dentists in vivo," *Med. Phys.* 8:308–311 (1981).

Brownell, G. L., Budinger, T. F., Lauterbur, P. C., and McGeer, P. L., "Positron tomography and nuclear magnetic resonance imaging," *Science* 215:619–626 (1982).

Cameron, J. R., and Skofronick, J. G., "Medical Physics," New York, Wiley, 1978.

Cameron, J. R., and Sorenson, J., "Measurement of bone mineral in vivo: An improved method," *Science* 142:230–232 (1963).

Cobbold, R. S., "Transducers for Biomedical Measurements: Principles and Applications," New York, Wiley, 1974.

Geselowitz, D. B., "Magnetocardiography: an overview," *IEEE Trans. Biomed. Eng.* BME-26:497–504 (1979).

Hobbie, R. K., "Intermediate Physics for Medicine and Biology," New York, Wiley, 1978.

Lai, C.-M. and Lauterbur, P. C., "True three-dimensional image reconstruction by nuclear magnetic resonance zeugmatography," *Phys. Med. Biol.* 26: 851–856 (1981).

MacIntyre, W. J., several chapters in J. R. Greening, ed., "Proceedings of the International School of Physics 'Enrico Fermi,' Course LXXVI," Amsterdam, North Holland, 1981, pp. 163–232.

Plonsey, R., "Bioelectric Phenomena," New York, McGraw-Hill, 1969.

Pullan, B. R., several chapters in J. R. Greening, ed., "Proceedings of the International School of Physics 'Enrico Fermi,' Course LXXVI," Amsterdam, North Holland, 1981, pp. 7–65.

Thornton, A. R. D., several chapters in J. R. Greening, ed., "Proceedings of the International School of Physics 'Enrico Fermi,' Course LXXVI," Amsterdam, North Holland, 1981, pp. 345–397.

Webb, R. H., and Hughes, G. W., "Scanning laser ophthalmoscope," *IEEE Trans. Biomedical Engineering* BME-28:488–492 (1981).

Webster, J. G., ed., "Medical Instrumentation Application and Design," Boston, Houghton-Mifflin, 1978.

Wells, P. N. T., "Biomedical Ultrasonics," London, Academic Press, 1977.

Wells, P. N. T., several chapters in J. R. Greening, ed., "Proceedings of the International School of Physics 'Enrico Fermi,' Course LXXVI," Amsterdam, North Holland, 1981, pp. 398–440.

Wesseling, K. H., "Physics of the cardiovascular system," in J. R. Greening, ed., "Proceedings of the International School of Physics 'Enrico Fermi,' Course LXXVI," Amsterdam, North Holland, 1981, pp. 66–162.

Wikswo, J. P., "Noninvasive magnetic detection of cardiac mechanical activity: (1) Theory, (2) Experiments," *Med. Phys.* 7:297–306 and 307–314.

Yamakoshi, K.-I., et al., "Indirect measurement of instantaneous arterial blood pressure in the human finger by the vascular unloading technique," *IEEE Trans Biomedical Engineering* **BME-27**:150 (1980).

Cross-references: BIOPHYSICS; HEALTH PHYSICS; MEDICAL PHYSICS; MOLECULAR ENGINEERING.

BIONICS

Bionics is defined as the branch of knowledge pertaining to the functioning of living systems and the development of nonliving systems which function in a manner characteristic of, or resembling, living systems. The definition of bionics infers the use of scientific skills and techniques from biological, physical, mathematical, and applied sciences in carrying out research in which: (1) the functions of chosen biological components and systems are studied and analyzed to determine underlying principles and processes that may lead to methods for improving physical components or systems, and (2) the theories and techniques of chemistry, physics, and mathematics are applied to advance our knowledge of the principles upon which these functions are based.

Bionics research depends on the acceptance of certain postulates. These postulates are of two types—one essentially operational, the other essentially technical. The important operational postulates are: (1) Common experience shows that biological systems perform operations that no nonbiological system can now perform efficiently, e.g., operations such as pattern recognition and identification, discrimination, and learning; (2) biological components perform such functions as detection, filtering, and information transfer more efficiently and with greater certainty over broader bandwidths than do present nonbiological components; (3) intensive study, analysis, and application of the principles that make superior biological performance possible can lead to nonbiological systems that equal or may, in some cases, exceed biological systems capabilities. The technical postulates are: (1) The functional advantages of biological systems are implied in the unique methods for information transfer, memory storage, and retrieval, united with unique ways of correlating and integrating data from many sensors or sensor systems. These unique methods depend mainly on: (a) many converging and diverging information transfer channels and many connections between channels, and (b) the special properties of biological components at the points where these channels are interconnected. (2) The superior capabilities of biological components or particular elements of the components (e.g., receptor cells or nerve muscle junctions) are derived from specific ways of interconnection and probably as well from specific molecular properties. (3) The relevant data describing biological systems that will represent major improvements over existing physical and engineering hardware are analyzed and studied. (4) The present rapid advances in microminiaturization techniques suggest that, for the first time, the possibility exists for the development of physical components or systems that incorporate these superior biological principles and processes.

The definition of bionics suggests the methodology of procedure. We design, grow, or in some way obtain nonbiological systems that function in a way "resembling" living systems. The physical component simulates the biological way of doing or carrying out its function. To obtain this objective, we first choose the biological components that perform the desired function; second, we compile the descriptive biological data; third, we translate these data into engineering terms; and fourth, we apply the translated data for the physical simulation of the function.

This process requires application of relevant mathematics to describe clearly, and as rigorously as possible, the biological function by some mathematical theory or model. This process may also require various techniques from the physical sciences to arrive at the necessary data defining the biological function of interest.

While the bionics research procedure requires the description, mathematically, of the function to be performed by the nonbiological system, it may also require mathematical and physical descriptions of the properties of the materials used to construct the nonbiological analog.

Gaining enough data to describe the given biological function may require study of the biological component or one of its elements at the molecular level. Similarly, solid-state or molecular techniques may be required to construct the physical system that is the appropriate simulation of the desired biological function.

Successful bionics activity may be modified by several factors. These include the complexity of the biological functions; the kind, quality, and quantity of data available to describe the functions; and the existence of relevant mathematical and physical techniques essential to the simulation of the functions.

In terms of the bionics objectives, the scope of the work must include research on the following components and/or systems: (a) receptors or sensors; (b) receptor systems—including central interconnections and interconnections among receptor systems; (c) central nervous system networks and the interconnections among parts of the nerve network; (d) effectors and actuators; (e) effector systems—including the feed-back and feed-forward connectors to and from the central processor; (f) the integrated system made up of sensors and their input channels, the central correlating, control, and computing networks, and the channels from the central system to the effectors and actuators.

This complete research program will extend over a long period of time. However, progress has been made. Data have been acquired, analyzed, and put into engineering terms to provide a set of specifications for the functional properties of several types of neurons. The neuron is presumed to be that physiological component underlying observed psychological parameters such as learning and adaptation. Therefore, the construction of a network of artificial neurons should, to some extent, simulate these observed behavioral parameters. In contrast, data have been acquired and analyzed from the field of experimental psychology on functions such as learning. These data can be translated into engineering and can suggest types of components which can simulate the learning function directly. This example illustrates a procedure which is common in bionics research. One can start at the operational level with an observed set of functional parameters and attempt to synthesize the class of mechanisms that could simulate these functions, or one can start with the biological component which is presumed to give rise to these functions and attempt to simulate this component. The choice of one method over the other depends greatly on the assurance one has of the validity of the data at one level or the other.

Applications, using the former (functional simulation) type of component, have been successfully made to flight control systems, variable geometry jet engine control, radar detection and search problems, and various industrial process control problems such as nondestructive materials inspection techniques. The functional simulation type of component has been successfully realized in both microcomputer and LSI (large scale integrated circuit) forms.

At the biological component level of simulation a somewhat different approach has been initiated, mostly by the work of Dr. A. H. Klopf (see references).

Heretofore goals for adaptive networks were presumed specified at the systems level or at most at the subsystem level; rarely at the individual component level. Dr. Klopf's work has been exploring the possibilities of goal seeking systems realized with goal seeking components.

The goals for the components may be of several types, but the emphasis has been on designing the components so as to maximize certain parameters associated with signal propagation. There is evidence that similar processes may occur in physiological neurons. A preliminary extensive assessment (circa 1980) has been made. A few of the relevant papers are listed in the references under Applications.

CECIL W. GWINN

References

General

"Cybernetic Problems in Bionics," London, Gordon and Breach, 1968.

Beer, S., "Decision and Control," New York, John Wiley & Sons, 1966.

Buckley, W. (Ed.), "Modern Systems Research for the Behavioral Scientist," Chicago, Aldine, 1968.

Sommerhoff, G., "Logic of the Living Brain," New York, Wiley & Sons, 1975.

Applications

Barron, R. L., and Gwinn, C. W., "Applications of Self Organizing Control to Aeronautical and Industrial Systems," Design Engineering Conference ASME, New York, April 1971.

Mucciardi, A. W., "Self Organizing Probability State Variable Parameter Search Algorithms for Systems that must Avoid High-Penalty Operating Regions," *IEEE Transactions on Systems, Man, and Cybernetics*, **SMC-4** (No. 4) (July 1974).

Zeger, A. and Burgess, L., "Adaptive Techniques for Radar Control," General Adaptronics Corp. under contract F33615-72C-1842, with U.S. Air Force Avionics Laboratory, AFAL-TR-394, Nov. 1973.

Mucciardi, A., Shankar, R., et al., "Adaptive Nonlinear Signal Processing for Characterization of Ultrasonic NDE Waveforms," Adaptronics Inc., under contract with the U.S. Air Force Materials Laboratory, F33615-74-C-5122, AFML-TR-75-24, Feb. 1975 and AFML-TR-76-44, April 1976.

Hampel, D., "LSI Electronically Programmable Arrays (Configurable Polynomial Arrays)," RCA Corp. under contract with U.S. Air Force Avionics Laboratory, AFAL-TR-76-228.

Klopf, A. H., "Brain Function and Adaptive Systems—A Heterostatic Theory," Air Force Cambridge Research Laboratories AFCRL-72-0164, Bedford, MA, 1972.

Barto, A. G., and Sutton, R. S., "Goal Seeking Components for Adaptive Intelligence: An Initial Assessment," University of Massachusetts under contract to U.S. Air Force Wright Aeronautical Laboratories F33615-77-C-1191, AFWAL-TR-81-070, April 1981.

Klopf, A. H., "The Hedonistic Neuron: A Theory of Memory, Learning and Intelligence," Washington, D.C., Hemisphere Publishing Corp., 1982.

Cross-references: BIOPHYSICS, CYBERNETICS, FEEDBACK.

BIOPHYSICS

Biophysics is a very broad field with poorly defined boundaries. It encompasses all applications of the principles of physical science to the functioning of living organisms. It has considerable overlap with areas of biomedicine, bioengineering, biochemistry, and physiology, as well as strong connections to space medicine, sports medicine, fluid dynamics, psychology of perception, molecular biology, membrane physiology and other related topics. This article will provide a very brief summary of some selected topics; it will neither cover all areas nor discuss those chosen in great depth.

In a strict sense, biophysics is as old as science itself. In Western intellectual history, those who made significant contributions to physical science also had important roles in the biological sciences. Aristotle wrote extensively (often incorrectly) about human and animal functioning, Newton and Galileo studied many aspects of human physiology, and René Descartes made significant contributions to our understanding. Thomas Young, a physician who described optical interference and translated the Rosetta Stone, provided a theory of color vision that has a significant remnant in modern theories of this most difficult subject. Galvani's discovery of bioelectricity was another important event in the history of biophysics.

More modern progress has resulted from surprising and unrelated discoveries. The work of Becquerel and Roentgen provided X-ray methods that produced an incredible amount of new knowledge, even up to the application of the diffraction technique of Bragg for the analysis of the structure of DNA. Arrhenius and other chemists provided quantitative descriptions of the behavior of ions in solution. Coupled with advances in fluid dynamics, these discoveries induced further progress in understanding the behavior of biological fluids. The organic chemists provided information about metabolism and other processes leading to understanding of amino acids and proteins and modern achievements such as Sanger's synthesis of insulin.

Contemporary biophysics contains very many diverse areas of investigation. Some of these are extensions of older subjects while others represent completely new areas of investigation. A small number of topics from each of these categories is discussed below.

One of the oldest subjects in biophysics is the study of energy expenditure, conservation, and conversion in the performance of physiological functions. Ordinary mechanics can be used to calculate the optimum performance expected from human athletes in a variety of events. The results of these studies compare favorably with Olympic records, thereby supporting the hypothesis that mechanical action for men and machines follow the same rules. For example, equating the potential energy at the top of a jumper's trajectory h to the work done by the leg muscles in accelerating the body upward by distance s gives $Fs = mgh$ where F is the average muscular force. For humans, s is about $\frac{1}{2}$ meter and the maximum force F is about mg, suggesting that we can raise our center of gravity about $\frac{1}{2}$ meter by jumping. This is supported by experience. In another example, we find that the peak power output of a human athlete is about 3600 W whether running at maximum speed (10 m/sec), throwing a 1 kg football 50 meters or a 7 kg shotput 16 meters, or broad jumping 3.5 meters. Each of these calculations is based on certain idealized but sensible models of the body, in accordance with the practice of physicists trying to understand nature.

The food requirement, energy and oxygen consumption, and metabolism changes for these activities are now well documented. Studies of animal motion are now showing that there are many different ways to use the metabolic energy available from the food supply to move about, conserve water, escape from predators, etc.

Measurements and studies of energy consumption in the ordinary processes of life have provided considerable knowledge about physiological functions as well as the basis for a wide variety of diagnostic tools and methods for the practice of modern medicine. The power required to keep warm, maintain respiration and circulation, and provide necessary chemical gradients constitute the basal metabolism rate (BMR). Control of the flow of this energy between storage areas and external reservoirs is the subject of a major field of study. Elementary aspects of feedback control are apparent in some processes; others are so very complicated that they haven't been unravelled. There are many pathological circumstances that can be traced to improper or inadequate control of energy exchange.

On the microscopic scale, many aspects of energy transfer are still not known. We know that mitochondria are the power plants of cells, converting the energy of metabolism into a phosphor bond in ATP, the basic fuel for life. The ATP transports stored energy for muscle contraction, Na^+ expulsion, etc., but many details of the process are not understood. Activity and progress in the study of energy in biological progress spans the range from how horses run to the phosphorylization of ATP.

Another very important subject in biophysics is sensory perception. The discussion in this article will deal primarily with vision and hearing because both the stimuli and the responses are more amenable to quantitative description than are taste, smell, and kinesthesis (touch). The sciences of optics and acoustics are very highly developed and as such allow careful study of animal response to light and sound.

We begin with the optics of the human eye. The $f/4$ (approximately) system has a resolution limited to a few μm on the retina by diffraction, and this is remarkably close to the size of the receptor cells (this near agreement is startling). Measurements of visual acuity reveal that our angular resolution of detail is indeed equal to this 10^{-4} radian limit. But the naive notion of point-to-point transmission of images from eye to brain that would allow connection of these two facts is shattered by the additional information that 10^8 cells 2 μm in diameter cover the retinal surface, but there are only 10^6 individual neurons in the optic nerve. This 100:1 reduction is the most elementary of many pieces of information that demand a more sophisticated description of human vision.

The Nobel prize winning work of H. K. Hartline (1967) and of D. Hubel and T. Weisel (1981) have shown that lateral inhibition (LI) among receptor cells plays an indispensible role in our visual process. LI arises when the response of a particular receptor cell to a fixed stimulus is reduced by the stimulation of an adjacent cell. It provides a mechanism for enhancing contrast and increasing sensitivity to edges and boundaries. Furthermore, there are higher levels of LI that result in considerable data processing, leading many scientists to the feeling that the retina is an extension of the brain.

It has long been known that there are many ways nerves can form junctions (synaptic junctions) and that some of them are inhibitory. That is, if two axons feed into a single ganglion cell, it is possible that the ganglion's response to a fixed stimulus from one of the axons could be reduced by a stimulus from the other axon. In this case we say that the other axon has an inhibitory effect.

In order to understand how inhibition affects perception, we consider an array of receptors as shown in Fig. 1. Suppose that each cell in the array could inhibit the response of its nearest neighbor, but only by some fixed fraction of the original cell's response. If the array is subject to a nonuniform stimulus that has a maximum located at one of the cells, that cell will exert a stronger inhibitory effect on its neighbors than they will exert on it, resulting in an enhancement of its relative strength. In the numerical example of Fig. 1, we assume each cell can inhibit the response of its neighbors by 25%. The cell whose uninhibited response would be 80 is reduced by 25% of 50 (= 12.5) and 25% of 100 (= 25) so that its net response is only 42.5 (similarly for the other cells). We notice that the center cell that had enjoyed only a 25% higher level of response to distinguish it as being at the maximum before inhibitory effects is now elevated to a 40% higher level over its neighbor. Inhibition can act in this way to sharpen sensory perceptions.

Resolution of detail in a biological optical system is not as simple as a discussion of ideal components with well understood physical limits. Indeed, the system is subject to all the irregularities of biological growth as well as the effects of the scattering of light from impurities in ocular fluids. The problem of information retrieval from severely distorted images by a system of limited resolution is a fertile area for physical study.

The visual system provides much more information than simply the image in the field of view: one of its most striking capabilities is color

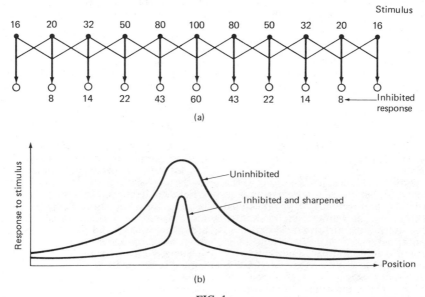

(a)

(b)

FIG. 1.

perception. Although different wavelengths of light are perceived as different colors, it has been conclusively demonstrated that different color perceptions can be achieved with the same wavelength, and the same color perception with different wavelengths. There is an extraordinarily large body of data on color vision, but no single, clear, simple theory has emerged. The role of LI in the process is probably as important as it is in detail perception.

Our hearing process, particularly tonal discrimination, is also enhanced by LI. The Nobel prize winning work of G. von Bekesy (1961) showed that tonal discrimination over a large region of the spectrum of audible sound stems from excitation of spatially distinct regions of the basilar membrane, a thin sliver of tissue curled up in the cochlea in the inner ear. But tones having small frequency differences can excite overlapping regions of the membrane, resulting in limits to our ability to distinguish between slightly different tones. The detection of vibration of the basilar membrane is also subject to LI, much like image detection, in a system that functions to enhance discrimination. The ability of trained musicians to sense off-key notes to a precision of less than 1% attests to the capability of the system.

Another area of major interest in biophysics is transport. This includes the gross transport of fluids through the circulatory system for the purpose of distributing heat and nutrients, as well as the microscopic transport of molecules through membranes and fluids. The most obvious example of fluid transport in human physiology is blood flow. Classical fluid dynamics can be used to describe flow when friction can be ignored ($P + \rho g h + \rho v^2/2 = $ constant—Bernoulli's equation) and also when viscous friction plays a dominant role ($Q = \pi R^4 \Delta P/8\eta L$—Poiseuille's equation). Each of these equations describes ideal fluids (continuous, homogeneous, incompressible, etc.) flowing under ideal conditions (smooth straight pipe with rigid walls, no turbulence, steady flow, etc.). Application of these models to human circulation must be done with the greatest of care because blood contains both rigid and deformable bodies of various sizes, flows in vessels with elastic walls that bend and divide, and is pumped into an elastic aorta by a cyclic heart that produces pulsatile flow.

Ventricular contraction produces a rapid injection of blood into the aorta. Since resistance to flow in the peripheral circulatory system is rather high, the elastic walls of the aorta expand to accommodate the extra volume: the kinetic energy of the blood is briefly stored as potential energy in the aortic walls and then slowly returned as the collapsing aorta squeezes the blood through the rest of the system. The flow is very fast at first (Reynolds number $\cong 2000$) and then slows as the dominant pumping comes from the aortic walls. A good electrical analogy is a capacitor ripple filter on an ac current source. Classical fluid dynamics may apply instantaneously to some parts of this sequence of events, but the boundary conditions are changing so rapidly that quantitatively accurate descriptions are not available.

Flow in the peripheral arteries is somewhat simpler, but branches and elastic walls still provide difficulties. The simple relationship between radius R and pressure difference ΔP for an elastic tube of wall thickness t ($R/R_0 = 1/(1 - x)$, $x = R_0 \Delta P/yt < 1$) that might be used in Poiseuille's equations is inapplicable because the Young's modulus Y is not necessarily constant—arterial walls contain both elastin and collagen, resulting in a nonlinear stress-strain relation.

Flow in the smallest vessels is complicated by the presence of blood cells. An average viscosity is no longer a tenable description, and very complicated conditions can arise. Some capillary vessels are so narrow that the cells must deform to squeeze through them. Sometimes a few red cells can join into larger objects called rouleaux, further complicating the problem by producing a rather large distribution of cell sizes. In some circumstances there is a tendency for the cells to concentrate in the central core of flow leaving a layer of liquid adjacent to the vessel walls, and thus producing a *decrease* in the apparent viscosity (Fahreus-Lindquist effect).

For these and other reasons, a physicist approaches the problem of blood flow with some trepidation. The models must be carefully thought out, applied in the proper domains, and the results must be interpreted with the proper limitations in mind.

The other meaning of transport in biophysics applies to movement of molecules, ions, and dissolved gases across membranes and fluid boundaries. This problem has very many facets, only some of which will be described here. To begin, the transport of solutes across concentration gradients in the usual kinetic sense (osmosis, diffusion, etc.) is very important in many circumstances. There are several distinct phenomena involved in various examples, but they can most readily be divided into the two categories of kinetic and carrier-mediated diffusion. Kinetic diffusion is the usual topic described by the random walk process. The techniques of statistical analysis provide a quantitative description called Fick's laws, whose equations are familiar to many physicists. Carrier-mediated diffusion is not significantly different in some concentration domains, but the transported substance may reach concentrations that saturate the capabilities of the carriers, resulting in a nonlinearity that doesn't appear in Fick's equations. Nevertheless, a differential equation similar to that of catalyzed chemical reactions describes the process.

One of the most important areas of current study is transport through membranes. This process may be suitably described by kinetic methods for substances soluble in the membrane's lipids, but this is usually not the case. In the more usual case, transport occurs through pores or channels that are either large or comparably sized with the solute molecule, and different conditions prevail in these cases. Sometimes channel transport is carrier mediated, sometimes the walls of the channel or the nature of the solute molecule dictate the dominant process of transport, and sometimes the properties of the channel depend on time or on the chemical environment. Transport against a concentration gradient, called active transport, is most well-known for the "sodium pump" that maintains a lower Na^+ concentration inside electrically active cells (nerve and muscle). The details of this process are not yet completely understood.

The uptake of nutrients, the exchange of O_2 and CO_2 in the lungs, the response to chemical transmitters of nervous information, and many other processes involve membrane transport. Each of them has special characteristics and mitigating circumstances that are amenable to a physicist's model-making approach to nature. A great deal of understanding of many of these processes has been achieved by the application of physical methods to biological problems; many others are under active investigation.

This article has presented a brief summary of some restricted aspects of three major topics in biophysics. The interested reader is referred to the references which follow and the work cited in them for further information on these topics as well as discussions on very many others.

HAROLD METCALF

References

Ackerman, Ellis, and Williams, "Biophysical Science," Prentice-Hall, Englewood Cliffs, New Jersey, 1979.

Hobbie, R. K., "Intermediate Physics for Medicine and Biology, John Wiley & Sons, New York, 1978.

Metcalf, H., "Topics in Classical Biophysics," Prentice-Hall, Englewood Cliffs, New Jersey, 1980.

Ruch and Fulton, "Medical Physiology and Biophysics," Saunders, Philadelphia, 1960.

Selkurt, "Physiology," Little, Brown and Co., Boston, MA, 1976.

Cross-references: BIOMEDICAL INSTRUMENTATION, BIONICS, HEALTH PHYSICS, MEDICAL PHYSICS, MOLECULAR BIOLOGY, X-RAYS.

BOLTZMANN'S DISTRIBUTION

Boltzmann considered a gas of identical molecules, able to exchange energy upon colliding, but otherwise independent of each other. Such a system, in which the interactions between the molecules between collisions is neglected, is applicable to real gases which are sufficiently rarefied, for in that case the molecules are almost always sufficiently far apart for the interaction force between them to be negligible. An individual molecule of such a gas moves freely between collisions, but experiences abrupt random changes in velocity in each collision, so that no exact statements can be made concerning its state at a particular time. However, when the gas comes to equilibrium at some fixed temperature, predictions can be made concerning the average fraction of molecules which are in a given single state, or equivalently, the fraction of time spent by one molecule in that state. (The concept of a single state will be described below.) Since these average fractions are equivalent to probabilities, one can say that predictions can be made for the probability distribution for a molecule over its possible single states. To describe this distribution, let the set of energies available to each molecule be denoted by ϵ_j. If l and m stand for two single states of a molecule and ϵ_l and ϵ_m the energies of the molecule when it is in these states, then the ratio of the probability P_l of finding a molecule in the single state l to the probability P_m of finding a molecule in the single state m is

$$\frac{P_l}{P_m} = \frac{\exp(-\epsilon_l/kT)}{\exp(-\epsilon_m/kT)} \tag{1}$$

This distribution of molecules over different states is called *Boltzmann's distribution*, after the founder of statistical mechanics, Ludwig Boltzmann (1844–1906). Equation (1) indicates that the ratio of probabilities for a molecule to be in two selected states depends only on the energies of these two states and on the thermal energy kT, which is proportional to the equilibrium temperature T. When T is measured in kelvin ($T = 273.16$ K at the triple point of water) then the universal constant k, called *Boltzmann's constant*, has the value 1.38041×10^{-23} joules/kelvin.

Using the fact that the sum of the probability P_l over all accessible states for a molecule must be unity leads to an alternative form of Eq. (1):

$$P_l = \frac{\exp(-\epsilon_l/kT)}{\sum_j \exp(-\epsilon_j/kT)}. \tag{2}$$

Equation (2) indicates that the probability for a molecule to be in a single state l varies exponentially with the negative of the energy in the state, divided by kT. The exponential $\exp(-\epsilon_l/kT)$ is called the *Boltzmann factor*. The sum of the Boltzmann factor over all single states for the molecule which appears in the denominator of Eq. (2) is called the *partition function* for the molecule. This alternative form of Boltzmann's distribution indicates that the probability of finding the molecule in a single state l is smaller, the larger is the energy ϵ_l, and that the rate of decrease of probability with increasing energy is faster, the smaller is the temperature T.

Boltzmann's original derivation of the probability distribution described by Eqs. (1) and (2) was restricted to the molecules of a gas. Subsequently, Gibbs (Josiah Willard Gibbs, 1839–1903) recast the reasoning to show that the Boltzmann distribution holds not only for a molecule, but also for an arbitrary system in thermal equilibrium with a much larger system, the reservoir, with which it can only exchange energy. The combination of system plus reservoir forms a closed system, insulated from all external influences. With this greatly extended applicability, the Boltzmann distribution is of vast utility in analyzing the equilibrium behavior of a wide range of both classical and quantum systems. In such applications, it is essential to understand the meaning of the concept of single state used above. This is simpler to describe for quantum systems, and will lead to a somewhat different but related description for classical systems, which can be regarded as limiting cases of quantum systems.

Quantum mechanics asserts that a complete description of a physical system is given in terms of probability amplitudes and provides a method for calculating these amplitudes. The probability amplitudes describe the state of the system, for using them one can predict the probabilities for the result of measurement of a set of compatible dynamical variables of the system, such as the three Cartesian components of momentum for a free particle, p_x, p_y, and p_z, or alternatively, its three coordinates x, y, and z. Special importance is attached to states, called *eigenstates*, for which measurements of particular dynamical variables yield one unique result, rather than a range of possibilities. The special values of the dynamical variable corresponding to each eigenstate are called the *eigenvalues* for that dynamical variable. For our present purposes we are interested in the eigenvalues of the energy for a system, corresponding to the ϵ_l in Eqs. (1) and (2), and its corresponding eigenstate. Confined systems, like a gas of molecules in a container have a discrete set of energy eigenvalues. It may be that a particular one of these discrete energy eigenvalues ϵ_l corresponds to not one, but rather several linearly independent probability amplitudes, each of which describes the system with the same energy ϵ_l. Each of these linearly independent probability amplitudes represents what we previously called a single state. One says that such an energy level ϵ_l is *degenerate*, and a g_l-fold degenerate energy level is one which is equally well described by g_l linearly independent eigenstates (single states). As an example, take the physical system to be a single structureless particle of mass m freely moving in a container, which is a rectangular box with length a, breadth b, and height c. The energy eigenvalues for this system are

$$\epsilon_{n_x, n_y, n_z} = \frac{h^2}{8m}\left(\frac{n_x^2}{a^2} + \frac{n_y^2}{b^2} + \frac{n_z^2}{c^2}\right) \qquad (3)$$

where h is Planck's constant, $h = 6.6256 \times 10^{-34}$ joule-seconds, and n_x, n_y, and n_z are independent integers, positive or zero, which are called *quantum numbers*. The probability amplitude, or energy eigenstate ψ is the following function of particle coordinates x, y, and z:

$$\psi_{n_x, n_y, n_z}(x, y, z) = \left(\frac{8}{abc}\right)^{1/2} \sin\frac{n_x \pi x}{a}$$

$$\cdot \sin\frac{n_y \pi y}{b}\sin\frac{n_z \pi z}{c}. \qquad (4)$$

In the case that no two dimensions a, b, c of the box are in a ratio of integers, the energy levels corresponding to various sets of values of the three quantum numbers are all different, with one and only one probability amplitude ψ associated with each. Energy levels of this type are *nondegenerate*. If, however, any pair of a, b, c are in a ratio of integers, there will occur certain values of the energy ϵ_{n_x, n_y, n_z} corresponding to two or more distinct sets of values of the three quantum numbers n_x, n_y, and n_z, and to two or more independent probability amplitudes ψ. This is the case of a degenerate energy level. For example, if the container is a cubical box with $a = b = c$, most of the energy levels will be degenerate, since all eigenstates with $n^2_x + n^2_y + n^2_z = q$, a fixed integer, have the same energy, $\epsilon_q = h^2 q / 8ma^2$. Say $q = 30$, then any of the six assignments of integers 1, 5, and 2 to n_x, n_y, and n_z yield the same energy, but correspond to different amplitudes ψ. This example illustrates the connection between degeneracy and geometrical symmetry—for the cubical box the directions x, y, and z are all equivalent.

It may often be more suitable to express the Boltzmann distribution in terms of probabilities $P(\epsilon_l)$ for a system to have a given energy ϵ_l, rather than P_l, the probability for the system to be in a single state. In that event, the ratio of probabilities for finding the system in two energy levels will include the degeneracy factors g for each level, and Eq. (1) will become:

$$\frac{P(\epsilon_l)}{P(\epsilon_m)} = \frac{g_l \exp(-\epsilon_l/kT)}{g_m \exp(-\epsilon_m/kT)}. \qquad (5)$$

For example, if the difference in the two energies $\epsilon_l - \epsilon_m$ is much less than the thermal energy kT, the probability ratio will just be the ratio of the degeneracy factors for these two energy levels.

Atoms and molecules are properly described in terms of quantum mechanics, but a description in terms of classical mechanics may sometimes be a useful approximation. We now consider the concept of *single state* in classical mechanics. Here the state of a single particle is completely described in terms of its position coordinates x, y, and z, and corresponding momenta p_x, p_y, and p_z. The specification is complete, since the laws of classical mechanics are such that knowledge of position and momentum at any one time permits prediction of these variables at any other time. Specification of the state of the particle is then equivalent to specifying a point in the six-dimensional Cartesian space spanned by three coordinate axes x, y, z and three momentum axes p_x, p_y, p_z. As the position and momentum of the particle change with time, its representative point moves through this space, which is called the *phase space*. The particle has three degrees of freedom, and is represented in a six-dimensional phase space. In general, a system with f degrees of freedom has its state described by f coordinates and f conjugate momenta, and is represented by a point in this $2f$-dimensional phase space.

Connection between the classical description of a state with its continuously changing coordinates and momenta and the quantum mechanical description in terms of discrete states for dynamical variables labelled by quantum numbers is made by considering the classical limit of quantum systems. In this limit, the discrete eigenvalues of dynamical variables change by only a very slight fractional amount with small changes in the quantum numbers which determine them. For example, the classical limit for the particle in a box, with quantized energies given by Eq. (3) is achieved for large energies, or large quantum numbers n_x, n_y, and n_z, when energy levels become very closely spaced. Consider the case of a cubical container with volume a^3 and energies $\epsilon_n = h^2 n^2/8ma^2$, with $n^2 = n^2_x + n^2_y + n^2_z$. For large values of the quantum numbers n_x, n_y, and n_z, small changes in these numbers will produce a change Δn in n, and a corresponding change $\Delta\epsilon_n$ in energy, such

p and $p + \Delta p$ is just $(2a/h)^3$ times the volume in n_x, n_y, n_z space, or $a^3 4\pi p^2 \Delta p/h^3$. One now sees that this number of states, which is the phase space volume $a^3 4\pi p^2 \Delta p$ divided by h^3, could also have been obtained by dividing the six-dimensional phase space into cells of size h^3, then counting the number of cells which lie in that region of phase space which corresponds to the spatial volume a^3 and the region between p and $p + \Delta p$ in the momentum space. This result implies that each quantum state for the particle occupies a volume h^3 in the classical phase space. It is an example of a general result: for a system with f degrees of freedom, each of its quantum states occupies a volume h^f in the $2f$-dimensional phase space in the classical limit. We may remark that the finite size h for a cell in the phase space x, p_x of a particle with one degree of freedom corresponds to Heisenberg's uncertainty principle, according to which the position and momentum of a particle cannot be more exactly defined than is consistent with the relation $\Delta x \Delta p_x \approx h$ between products of their uncertainties. In view of this relation, it would be meaningless to make a finer division of the phase space, as it is impossible to decide by experiment in which of these cells a particle lies.

We can now apply Boltzmann's distribution law to find the momentum distribution for a free particle in a cubical container in thermal equilibrium at temperature T, in the classical limit. The particle has a (kinetic) energy $\epsilon = (p_x^2 + p_y^2 + p_z^2)/2m$. Suppose we want the probability $P(p_x, p_y, p_z)\,dp_x dp_y dp_z$ to find its momenta in the range between p_x and $p_x + dp_x$, p_y and dp_y, p_z and dp_z. Corresponding to this specification are a number of quantum states equal to $a^3 dp_x dp_y dp_z/h^3$. Multiplying this number of states (degeneracy factor) by the Boltzmann factor $e^{-\epsilon/kT}$ and dividing by the integral over all states gives the desired probability as

$$P(p_x, p_y, p_z)\,dp_x dp_y dp_z = \frac{\exp\left(-\dfrac{p_x^2 + p_y^2 + p_z^2}{2mkT}\right) dp_x dp_y dp_z}{\displaystyle\iiint_{-\infty}^{\infty} \exp\left(-\dfrac{p_x^2 + p_y^2 + p_z^2}{2mkT}\right) dp_x dp_y dp_z} \tag{6}$$

that $\Delta\epsilon_n/\epsilon_n = 2\Delta n/n$ is smaller, the larger is n. Then the number of quantum states for which n lies between n and $n + \Delta n$ will be the volume V of one octant of a spherical shell in n_x, n_y, n_z space with radius n and thickness Δn, or $V = 4\pi n^2 \Delta n/8$. In this classical limit, we expect the particle energy to depend on momentum as $\epsilon_n = p_n^2/2m$, with $p^2 = p_x^2 + p_y^2 + p_z^2$. Comparing the two expressions for the energy gives $p_n = nh/2a$. Then the number of quantum states for which the particle momentum lies between

Equation (6) is called *Maxwell's distribution*, after James Clerk Maxwell (1831–79), who obtained it in 1860 before Boltzmann's more general derivation in 1871. We may note that this probability distribution is the product of three independent factors, each of which defines the probability distribution for a separate component p_x, p_y, or p_z of the momentum. The probability distribution for momenta given by Eq. (6) is applicable to the translational motion of the molecules of a gas, and in fact does not depend at all on the type or strength of the interactions between the molecules in the classical limit.

For systems of weakly interacting indistinguishable constituent elements such as atoms, molecules, electrons, and photons, the Boltzmann distribution properly describes how such elements are distributed among their own individual states only when the average number of elements or particles in every single state is considerably less than unity. Physically this case corresponds to the gas of particles being sufficiently rarefield. When this condition is not satisfied, quantum mechanical symmetry requirements on probability amplitudes for the system under exchange or any two identical particles lead to two new distribution laws. Particles with integral intrinsic spin, like photons, are described by the *Bose-Einstein distribution*. This distribution is applicable, for example, to the treatment of electromagnetic radiation, described as photons, in a hollow enclosure which has come to thermal equilibrium. It leads to the Planck radiation law for the density of radiant energy from a black body. Particles with half-integral intrinsic spin, like protons, neutrons, and electrons, are described by the *Fermi-Dirac distribution*. This distribution is applicable to the treatment of conduction electrons in metals and to dense stellar interiors.

An interesting physical system which involves both the Boltzmann and the Bose-Einstein or Planck distributions consists of a rarefield gas of atoms which absorb or emit electromagnetic radiation. The radiation may be considered as a gas of noninteracting photons. When thermal equilibrium is established between the gas of atoms and the gas of photons, the atoms of the rarefield gas will be distributed among their individual energy states according to the Boltzmann distribution, while the photons are distributed among their individual energy states according to the Bose-Einstein or Planck distribution. The physical consistency of this result was first proved by Einstein in 1917, for slowly moving atoms, and generalized by Dirac in 1927 to the case of atoms with arbitrary speeds.

H. A. GERSCH

References

Wannier, G. H., "Statistical Physics," John Wiley & Sons, New York, 1966.

Reif, F., "Statistical and Thermal Physics," McGraw-Hill, New York, 1965.

Kittel, C., and Kroemer, H., "Thermal Physics," W. H. Freeman and Co., San Francisco, 1980.

Krylov, N. S., "Works on the Foundation of Statistical Physics," Princeton Press, Princeton, 1979. (Appendix by Ya. G. Sinai.)

Cross-references: BOSE-EINSTEIN STATISTICS; FERMI-DIRAC STATISTICS; STATISTICAL MECHANICS; RADIATION, THERMAL.

BOND, CHEMICAL

Every atom consists of a relatively compact nucleus bearing positive charge, surrounded by a cloud of electrons in sufficient number exactly to balance the nuclear charge by their negative charge. Being thus electrically neutral, atoms might be expected to exert among themselves only the negligibly weak gravitational forces predictable for such low mass. However, there are two principal means by which more significant interatomic attractions can occur, and these are largely responsible for the existence and properties of chemical substances.

One is the result of the fact that the electrons surround the nucleus in an easily deformable cloud. Although opposite charges tend to distribute themselves if possible so that their centers coincide, when two atoms come near to one another, the repulsion between the two negatively charged electronic clouds results in mutually induced distortions of the clouds so that the charge centers no longer coincide, creating atomic dipoles. Although these dipoles oscillate rapidly, their net effect is to create an electrostatic attraction between the atoms. Such attractions are known as *van der Waals forces*. Although such forces become very significant as exerted between molecules consisting of many atoms, they are so slight per atom pair that at ordinary temperatures the attractions among small molecules are usually completely overcome by the disruptive forces of kinetic energy, making them gaseous. They are insufficient to hold atoms or small molecules together in a condensed state, liquid or solid, unless most of the kinetic energy has been removed by cooling. However, large molecules do cluster together at ordinary temperatures as a consequence of these intermolecular attractions, commonly so strongly that decomposition occurs before the temperature is high enough for melting or vaporization.

In contrast, much stronger per atom pair is the interatomic force known as the *chemical bond*. This depends on the electronic structure of each atom, described by quantum theory as based on occupancy by electrons of certain energy regions of limited number and availability. In atoms of only six of the chemical elements, helium, neon, argon, krypton, xenon, and radon, this electronic configuration, with two outermost electrons in helium, eight in all the others, results in a sufficiently thorough covering up of the nuclear charge that no region of the atom remains within which an electron from another atom might be stably accommodated. In atoms of all the other chemical elements, all of which have fewer than eight outermost electrons, however, there remain regions, called *orbitals*, within which an electron from elsewhere would, despite repulsions by the other electrons, experience a net electrostatic attraction between itself and the effective positive charge of the nucleus. Each such orbital can accommodate two electrons only, and there are four outermost orbitals in all atoms beyond helium in atomic number. Vacancies exist wherever the number of outermost electrons is less than eight, this being the maximum that the outermost shell can hold in any isolated atom. (For more on the effective nuclear charge that can be sensed in such vacancies, see PERIODIC TABLE AND PERIODIC LAW.) The presence of outermost vacancies provides the possibility that outer electrons from another atom may be accommodated, these electrons now being held by both nuclei. Thus the atoms are held together by simultaneous attraction of both nuclei for the same electrons. All chemical bonding is essentially of this nature.

There are four ways by which vacancies may be utilized to form chemical bonds:

(1) If the atoms have more outermost vacancies than electrons, then the outermost electrons of each can spread out into the vacancies of the others, resulting in what is called *delocalized* or *metallic bonding*.

(2) If each atom provides one outermost orbital that is half-filled, the single electron of each can be accommodated within the vacancy of the other so that a pair of electrons is shared between the two atoms, the atoms being held together because both nuclei are attracted to the same bonding electrons. This is called a *covalent bond*. The shared electrons are essentially concentrated within the internuclear region, and thus localized. If additional half-filled orbitals are available on each atom, multiple bonds between the same two atoms may form.

(3) If one atom has an outermost vacant orbital and the other an outermost filled orbital, or electron pair, this pair may be accommodated within the vacant orbital to provide electron-pair sharing, termed *coordinate covalence*.

(4) If each atom has the requisites for forming a covalent bond but one attracts electrons much more strongly than the other, the former may acquire essentially all the bonding electrons, thus gaining a unit negative charge at the expense of the other atom left with unit positive charge. The electrostatic attraction between the opposite charges will hold the atoms together in what is commonly called an *ionic bond*. The existence of truly ionic bonds is in serious question, most so-called ionic bonds being partly covalent and therefore more accurately described as *polar covalent*.

All substances derive their special physical and chemical properties from the nature of the atoms that compose them and the way in which these atoms are attached to one another. An understanding of chemical bonds is therefore at the very heart of understanding chemistry. All substances except the six elements mentioned above normally exist in the form of atoms bonded together. Therefore all chemical reaction involves the breaking of existing bonds and the formation of new ones. At ordinary temperatures the usually dominant tendency is for bonds to be broken in favor of forming new, stronger bonds. Therefore a knowledge of bond strength and the origin of this strength is also essential to understanding chemistry.

Chemical bonds occur between atoms of all the chemical elements except the six previously mentioned, and in all chemical compounds. If the atoms linked together are all alike, the substances is a *chemical element*, but if the substance contains bonds between unlike atoms (different elements), it is a *chemical compound*.

The following qualities of chemical bonds are of special interest: number of bonds an atom can form, multiplicity, bond length, bond angle, and bond strength. Under most circumstances, atoms tend to form the maximum number of bonds of which they are capable. They can form one covalent bond for each outermost half-filled orbital they can supply. The number of single bonds or their equivalent in multiple bonds that an atom can form is usually called its *valence*. A two-electron bond is called *single*, a four-electron bond *double*, and a six-electron bond *triple*. Bond length is important because bond energy is greater at shorter internuclear distances. It is measured experimentally as the distance between the two nuclei. When an atom forms more than one bond, the angle between bonds is important in determining the geometry of the molecule. In general, electron locations in the exterior of an atom consist of bonds or lone pairs. These locations tend to follow the laws of like charge repulsions and be separated as far from one another as possible. If all outermost electrons are in two locations, these must be at opposite sides of the atom, and if in bonds, the bond angle is 180°. If all electrons are in three locations, these are at the corners of an equilateral triangle planar with the nucleus, and any bond angles are 120°. If all the electrons are in four locations, these are at the corners of a regular tetrahedron, corresponding to bond angles close to 109°28′. Five bonds usually assume a trigonal bipyramid structure, two trigonal pyramids having a common base. Six bonds most commonly are directed toward the corners of a regular octahedron. Most molecular structure or geometry can be satisfactorily rationalized through this simple type of reasoning. As for bond strength, to be discussed in detail presently, chemical bonds exhibit a wide range of strength, but even the weakest are recognized as bonds if they are sufficiently stable to allow experimental studies of the combined atoms. In general, chemical bonds are far stronger, per atom pair, than the van der Waals forces described previously, and quite able to persist at ordinary temperatures despite the kinetic energy which threatens their disruption. Although all bonds tend to break at sufficiently elevated temperatures, some persist even at several thousand degrees C.

Chemical bonds may join atoms together into finite molecules or into indefinitely extensive three-dimensional arrays of atoms in solids, called *nonmolecular*. Molecular substances themselves may become solid, if sufficient kinetic energy is removed, through van der Waals forces which hold the molecules together in crystalline array. In molecular substances, each molecule contains the same whole number of each kind of atom, and thus can be assigned a molecular formula which indicates its exact composition but does not necessarily reveal the exact arrangement of the atoms in the molecule. For nonmolecular substances, all of which are solid, the composition is commonly accurately representable by definite chemical formulas, but many examples exist wherein the composition is slightly variable and the formulas cannot represent the composition accurately in exactly whole numbers of atoms. These differences are indicated by calling the exact molecular compositions or nonmolecular compositions *stoichiometric* and the inexact compositions of certain nonmolecular compounds *nonstoichiometric*. Only in the solid state, and then only for certain combinations of elements, is nonstoichiometry possible in what are called *pure compounds*. Most naturally occurring substances are mixtures of different chemical substances, to which assignment of specific chemical formulas would be meaningless and misleading.

Covalent bonds involve bonding electrons localized within the internuclear region, but bonding electrons spread out if they can. Their delocalization in metals, not possible in nonmetals where there are no extra vacancies into which the electrons can spread out, removes the limitation of valence imposed by the number of available orbitals. Regardless of the number of normal valence electrons, the atoms of metals tend most commonly to pack together as closely as is possible for like sized spheres, which gives each interior atom 12 closest neighbors, or in other close packing wherein each interior atom has 8 direct neighbors and 6 more just a little farther away. The special properties of the metallic state, good conductivity of heat and electricity, malleability, and ductility, are the consequences of the delocalization of electrons, allowing them to flow among the atoms without affecting the bonding, and allowing the rearrangement of atoms without the breaking of specific localized bonds. Atoms of metal may usefully be regarded as positive ions held together by the negative electron glue (bonding electrons) that fills the interstitial space.

The exact description of chemical bonds or the molecules or nonmolecular solids which they hold together has been the unrealized, and probably unrealizable, goal of theoretical chemists applying quantum mechanics for more than half a century. (See PHYSICAL CHEMISTRY.) The immense complexity of mathematical calculations involved requires the extensive use of computers even for approximations, and so far defies efforts to obtain exact results. One cause lies in the difficulty inherent in problems involving, as do atoms, many-particle interactions. Another, with respect to chemical bonds, lies in the fact that the energies of interactions among atoms are generally very much smaller than the total energies of these atoms. For example, the total energy of a molecule of potassium bromide, KBr, is more than 8 million kJ per mole but differs from that of the separate atoms by only about 370 kJ. To evaluate the bond energy as the difference between the total energies of the separated atoms and the molecule is to determine accurately a small difference between very large numbers, in practice impossible.

Nevertheless by far the major effort has been directed toward quantum mechanical applications, even though these require numerous simplifying assumptions and approximations. Two ways of looking at a covalent bond that have received most attention are: (1) the atomic orbital, or valence bond approach, according to which electron sharing is visualized as permitted through the overlap of bonding orbitals on each atom so that they occupy a region in common within which the bonding electrons are attracted to both nuclei; and (2) the molecular orbital approach, according to which, ideally, all the electrons of a molecule belong to all the nuclei, occupying molecular orbitals which are the equivalent of atomic orbitals in atoms. Wave equations for these molecular orbitals are usually obtained as a *linear combination of atomic orbitals* (LCAO). For a diatomic molecule, coalescence of two bonding atomic orbitals is pictured as forming two new molecular orbitals, one concentrating the electrons within the internuclear region and called a *bonding orbital*, and the other dispersing them away from the internuclear region and called an *antibonding orbital*. Each electron within a bonding orbital serves to hold the atom together and each electron within an antibonding orbital cancels the bonding effect of one bonding electron.

It is the molecular orbital approach which receives most current attention, and numerous molecular properties are studied in this way. However, not surprisingly, far less complicated, more empirical studies have led to a more practical understanding of chemical bonds. In particular, a simple concept of covalence allows a practical method of calculating bond energies from atomic properties which produce accurate results for most molecular species and for many nonmolecular solids as well. Thus it allows an understanding of the origins of the heats of formation and reaction compiled by thermo-chemists, and the prediction of the strengths of bonds and the direction of chemical reactions.

The concept is based largely on that of electro-negativity, which is a measure of the attractive force between a bonding electron and the effective nuclear charge acting over the distance of the covalent radius. Atoms of different elements differ in their electronegativity, or ability to attract bonding electrons. Consequently the only chemical bonds in which the bonding electrons are evenly shared are between like atoms. In all compounds, the bonds involve uneven sharing of electrons, which results in acquisition of a partial charge by each atom. The initially more electronegative atoms acquire more than half share of the bonding electrons, leaving the initially less electronegative atoms with a partial positive charge and themselves acquiring partial negative charge. This uneven sharing results in an equalization of electronegativity within the compound, since atoms initially low in attraction for electrons become partially positive, thus increasing the attraction for electrons, while atoms initially high in attraction for electrons become less so as the result of acquiring partial negative charge. Thus each bond between unlike atoms is polar covalent, meaning that electrons are unevenly shared. The energy of such a bond is considered to consist of two contributions. One is the nonpolar covalent energy E_c which would result if the two electrons were evenly shared at the observed bond length. This is easily evaluated as the geometric mean of the two homonuclear bond energies, corrected for any difference between the sum of the nonpolar covalent radii and the actual bond length. The other is the ionic energy E_i which would result if one atom monopolized these two electrons creating a positive and a negative ion. The ionic energy replaces a part of the total possible covalent energy, but the ionic contribution is always larger than the part of the covalent energy which it replaces, so that polarity always strengthens the bond. For this reason, chemical change tends to occur in the direction of forming more polar bonds, and the earth's matter exists primarily as compounds having polar bonds rather than pure elements held together by nonpolar bonds.

A polar covalent bond is thus pictured as a blend of a covalent and an ionic contribution, both based on hypothetical extremes. The blending coefficients are easily evaluated from the partial charges, the ionic coefficient t_i being half the difference between the two charges, and the sum of the ionic coefficient and the covalent coefficient t_c being 1.00. In turn the partial charges are evaluated from the initial electronegativities. The electronegativity in the compound is evaluated as the geometric mean of all the initial atomic electronegativities. The change in electronegativity corresponding to

acquisition of unit charge is a constant times the square root of the atomic electronegativity. The partial charge is the ratio of the actual change in electronegativity in forming the compound to the change corresponding to unit charge.

The energy E of a polar covalent bond, in kJ per mole, when bond length and radii are expressed in picometers, is

$$E = t_c E_c + t_i E_i = t_c R_c (E_{AA} E_{BB})^{1/2}/R_0$$
$$+ 138909 t_i / R_0$$

where R_c is the sum of the nonpolar covalent radii and R_0 the bond length.

The homonuclear single covalent bond energies (E_{AA} and E_{BB}) are determined directly or indirectly from experiment. Here a very important anomaly is recognized. Normally, as the electronegativity increases and the radius decreases from left to right across the major groups, one expects the homonuclear single bond energy to increase steadily also. From M1 to M4 (e.g., Li–C) it does. Beyond M4, however, there are two changes. The outermost shell now acquires its first lone pair of electrons, having five to be accommodated within only four orbitals. At the same time, the homonuclear bond energy decreases. It is assumed that the lone pair somehow weakens the bond, so the effect is called the *lone pair bond weakening effect* (LPBWE). The effect is very large in nitrogen, oxygen, and fluorine, and significant but much smaller in the heavier elements of these groups. It is halved when the atom forms a double bond and eliminated when a triple bond is formed. Similar reduction of this weakening is also observed in certain single bonds. Although not yet thoroughly understood, the LPBWE provides quantitative explanation of many common and important chemical facts, such as the existence of nitrogen, oxygen, and carbon dioxide as gases instead of solids like phosphorus, sulfur, and silicon dioxide.

The bond energy equation given below, with appropriate modification for solids, has been successfully applied to thousands of different bonds in hundreds of compounds, including most of the common functional types of organic molecules, with results agreeing usually within 1–2% of the experimental values.

The bond energy obtained in this way is the *contributing bond energy* (CBE), defined as that part of the total atomization energy which this bond provides. Only for diatomic molecules is this the same as the *bond dissociation energy* (BDE). When breaking a bond liberates a fragment—*free radical*—which consists of more than one atom, the liberated bonding electron may either decrease or increase the strength of the remaining bonds. If it decreases the remaining bond strength, extra energy must be absorbed when the original bond is broken, making it that much more difficult to break and increasing the bond dissociation energy. If breaking the bond increases the remaining bond strength, energy so liberated will permit the original bond to be broken that much more readily, decreasing the bond dissociation energy. This energy of re-adjustment may be termed the *reorganizational energy* of the radical. It is applicable whenever that particular is liberated by a bond dissociation. For example, all bonds to the phenyl group, C_6H_5-, require about 50 kJ per mole more to break than would be expected from the contributing bond energy, this being the reorganizational energy of the phenyl radical. Such values can be very useful in organic chemistry, especially in understanding free radical reactions.

Finally, no discussion of bonds, however brief, should omit mention of the special bonding exhibited in compounds of hydrogen. Since an atom of hydrogen has only one electron, used in its bonding, the nucleus remains quite un-protected when the hydrogen bears partial positive charge. If a small atom on another molecule bears partial negative charge and a lone pair of outer electrons, a substantial attraction can occur between the positive hydrogen of one molecule and this electron pair, bridging the two molecules together. Commonly termed a *hydrogen bond* but better called a *protonic bridge*, this kind of bond is only 5–10% as strong as a typical covalent bond but nevertheless very widespread and important. It is an important factor, for example, in deter-mining the structure of liquid and solid water and of proteins. Hydrogen can also form *hydridic bridges*, in which the hydrogen atom is bonded to two other atoms simultaneously through what is sometimes called a *three-center bond*, corresponding to half a bond to each other atom. In a protonic bridge, a positive hydrogen bridges two negatively charged atoms each having a pair of electrons. In a hydridic bridge, a negative hydrogen with a pair of elec-trons bridges two positively charged atoms each having a vacant orbital. Intermediate examples of bridging by hydrogen are also observed.

Understanding chemical bonding remains the most important and fundamental problem in theoretical chemistry. Continuing evolution of these concepts is to be expected into the indefi-nite future.

R. T. SANDERSON

References

McWheeny, R., "Coulson's Valence," Oxford Univer-sity Press, New York, 1980.
Sanderson, R. T., "Chemical Bonds and Bond Energy," Academic Press, New York, 1st Ed. 1971; 2nd Ed. 1976.
Sanderson, R. T., "Polar Covalence," Academic Press, New York, 1983.

Cross-references: CHEMISTRY; ELECTRONS, ELE-MENTS, CHEMICAL; PERIODIC LAW AND PERI-ODIC TABLE; PHYSICAL CHEMISTRY.

BOSE-EINSTEIN STATISTICS AND BOSONS

Bose-Einstein statistics is a type of quantum statistics concerned with the distribution of particles of a particular kind among various al-lowed energy values taking into account the quantization of the energy values. Quantum statistics is a branch of STATISTICAL ME-CHANICS which treats the average or statistical properties of a system composed of a large num-ber of particles using standard mathematical techniques and the properties of the constituent particles. It is different from classical statistical mechanics only in that the particles of the sys-tem are described quantum mechanically.

Let us consider a system of N non-interacting particles. Three different distributions of the particles among the various energy levels are possible depending upon the assumptions that are made about the particles. If it is assumed that each arrangement or distribution which conserves energy is equally probable and also that the particles are distinguishable, and if each permutation of particles among the possible levels is counted as a different distribution, one obtains an average for the relative number of particles in the various levels known as the Maxwell-Boltzmann distribution. If the particles are treated as indistinguishable and only the number of different combinations of particles is counted, the Bose-Einstein distribution is ob-tained. A third distribution, known as the Fermi-Dirac distribution, results if, in addition to indistinguishability, it is required that the particles obey the Pauli exclusion principle which permits no more than one electron in each quantum state.

These three distributions may be expressed mathematically as follows where $n(\epsilon)$ gives the number of particles per energy level at energy ϵ when the particles are in thermal equilibrium at temperature T:

(1) $n(\epsilon) = \dfrac{1}{e^{(\epsilon-\epsilon_0)/kT}}$ Maxwell-Boltzmann

(2) $n(\epsilon) = \dfrac{1}{e^{(\epsilon-\epsilon_0)/kT} - 1}$ Bose-Einstein

(3) $n(\epsilon) = \dfrac{1}{e^{(\epsilon-\epsilon_0)/kT} + 1}$ Fermi-Dirac

where k is the Boltzmann constant and ϵ_0 is related to the number of particles present and depends on the temperature in such a way that for energies large compared to kT (so that the probability of occupation for a level becomes considerably less than unity), all three distributions reduce to the Maxwell-Boltzmann distribution.

The appropriate form of statistics to apply to an assembly of particles can also be discussed in terms of the symmetry properties of the wave functions describing the particles. Two classes of wave function ψ (a solution of the SCHRÖDINGER EQUATION for two or more identical particles) result from interchanging all the coordinates, both spatial and spin, in the wave function. It should be noted that this symmetry class does not change as a function of time. The wave functions for particles obeying Fermi-Dirac statistics (fermions) are antisymmetric, while those for bosons (Bose-Einstein statistics) are symmetric. Therefore, for a system of bosons, if all the coordinates of any pair of identical particles are interchanged in the wave function, the new wave function will be identical with the original.

Photons, all mesons (except the mu meson which is really a lepton) and all nuclei of even mass number are bosons, while nucleons (i.e., neutrons and protons), quarks, all nuclei of odd mass number and all leptons, such as electrons, neutrinos and the muon are fermions. All known bosons have angular momentum $nh/2\pi$, where n is an integer or zero and h is Planck's constant. The statistics of some nuclei have been determined experimentally by the observation of the relative intensities of successive lines in the band spectra of homonuclear, diatomic molecules.

One application of Bose-Einstein statistics to a physical situation is the treatment of a "photon gas." It is possible to obtain the Planck distribution law for blackbody radiation by treating the electromagnetic radiation inside an enclosure at constant temperature as a gas of particles of zero rest mass which obey the Bose-Einstein distribution law. This treatment provides an interesting example of the wave-particle duality found in nature, since it is in marked contrast to the original derivation which was based on the wave nature of electromagnetic radiation.

Another interesting application is the qualitative explanation of the superfluid properties of liquid helium which occur below the so called lambda point of 2.186 K. At atmospheric pressure ^{4}He condenses into a liquid at 4.3 K. If the pressure is then reduced the liquid boils until the temperature reaches the lambda point, where the boiling immediately stops. The lack of boiling is caused by a very large increase in the thermal conductivity, which becomes essentially infinite, so that all parts of the liquid are at the same temperature. Below the lambda point, liquid ^{4}He seems to behave as if it had nearly zero viscosity when the viscosity is measured by passing the liquid through a fine capillary, but when the viscosity is measured by observing the drag on parallel plates moved through the fluid it is not zero. To explain these peculiar properties, F. London[4] suggested that liquid ^{4}He, below the lambda point (often called He II) is composed of two interpenetrating fluids, a normal and a "superfluid." The superfluid is composed of those molecules in the ground state or lowest energy state whereas the normal fluid would consist of the remaining molecules. The finite viscosity is due to the normal component. As the temperature is reduced toward absolute zero the number of molecules in the superfluid component increases until at absolute zero one would expect to have only a superfluid. Since ^{4}He is a boson and the Pauli exclusion principle does not apply, there is no limit to the number of molecules which can exist in the lowest energy state. Thus this completely different type of "condensation" caused by the quantum mechanical properties of the molecules not only provides a qualitative explanation of the strange superfluid nature of He II but also provides one of the few examples of a macroscopic system for which the mechanical behavior cannot be explained by classical mechanics. It is significant to note that no superfluid behavior has been observed for ^{3}He which has spin $\frac{1}{2}$ and obeys Fermi-Dirac statistics and thus is unable to condense into the lowest energy state even at temperatures as low as a few thousandths of a degree Kelvin.

ROBERT L. STEARNS

References

1. King, A. L., "Thermophysics," San Francisco, W. H. Freeman Co., 1966.
2. Mandl, F., "Statistical Physics," New York, John Wiley & Sons, 1971.
3. Landau, L. D., and Lifshitz, I. M., "Statistical Physics," Reading, Mass., Addison-Wesley Pub. Co., 1958.
4. London, F., *Phys. Rev.* 54:947 (1938).

Cross-references: FERMI-DIRAC STATISTICS AND FERMIONS, PHOTON, SCHRÖDINGER EQUATION, STATISTICAL MECHANICS, SUPERFLUIDITY.

BREMSSTRAHLUNG AND PHOTON BEAMS*

An electron can suffer a very large acceleration in passing through the Coulomb field of a nucleus, and in this interaction the radiant energy (photons) lost by the electron is called bremsstrahlung,[1] (bremsstrahlung† sometimes designates the interaction itself). If an electron

*This work was supported by the U.S. Atomic Energy Commission.

†"Bremsstrahlung"–German; bremsen, to brake and Strahlung, radiation.

whose total energy $E_0 \gtrsim 800/Z$ MeV traverses matter of atomic number Z, the electron loses energy chiefly by bremsstrahlung. This case is considered here.

Bremsstrahlung in the coulomb fields of the *atomic electrons* is adequately included by replacing Z^2 in the formulas by $Z(Z + 1)$. For $Z \lesssim 5$, more complicated correction is required.[1]

Protons and heavier particles radiate relatively little because of their large masses (radiation rate is proportional to the square of the acceleration, inversely proportional to the square of the mass). If a very energetic electron traverses one radiation length (X_0) of any matter, bremsstrahlung reduces the electron's energy to $1/e$ of its incident value on the average. Some examples are:

Element	Air	C	Al	Fe	Cu	W	Pb
Radiation length X_0 cm	29800	20	9.1	1.7	1.42	0.32	0.51

The energy dependence of radiation loss per centimeter by an electron of energy E_0, traversing matter of density n atoms/cm^3 is given by $dE/dx = -nE_0\phi_{rad}$ where ϕ_{rad} is given by the curves in Fig. 1.

A beam of energetic electrons incident upon a radiator produces a bremsstrahlung beam that is directed sharply forward. Photon angular distributions for typical "thick" tungsten targets are shown in Fig. 2. Curves for other heavy elements are similar if all radiator thicknesses are measured in units of the radiation length. In such thick radiators, the incident electrons scatter appreciably, as well as radiate, making any

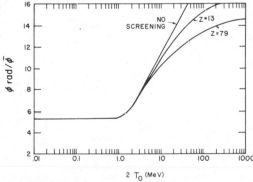

FIG. 1. Dependence of the total radiation cross section $\phi_{rad} = (1/E_0)\int_0^{E_0} k\,d\sigma$ on the initial electron kinetic energy, E_0. The parameter $\bar{\phi} = Z^2 r_0^2/137$.[1]

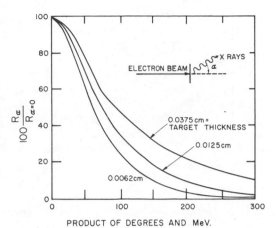

FIG. 2. Theoretical bremsstrahlung angular distributions from thick tungsten targets for relativistic energies. These data are obtained from the Natl. Bur. Std. Handbook, 55. R_α is defined as the fraction of the total incident electron kinetic energy that is radiated per steradian at the angle α.[1]

The angle α is measured with respect to the incident electron's direction. Since the electron may scatter before it radiates, $\alpha \neq \theta$ where α is shown in Fig. 3.

observed photon distribution actually an average over electron scattering angles of the basic bremsstrahlung distribution. The basic bremsstrahlung angular distribution has a zero at $\alpha = 0$, which is quite different from the curves of Fig. 2. The basic spectral shape is a weak function of photon angle and, in thick radiators, electron scattering modifies this shape slightly (Fig. 3). Examples of thick radiator spectra are shown in Fig. 4 for various incident electron energies. The bremsstrahlung spectra depend upon screening of the nuclear coulomb field by atomic electrons through the parameter $\gamma = 51k/[E_0(E_0 - k)Z^{1/3}]$, where k is the photon energy in million electron volts. For complete screening ($\gamma \approx 0$), the thick radiator spectrum is given by

$$\frac{d\sigma_b}{dk} = \frac{4Z^2 r_0^2}{137k}\left\{\left[1 + \left(\frac{E}{E_0}\right)^2 - \frac{2}{3}\frac{E}{E_0}\right]\right.$$
$$\left. \cdot \ln(183Z^{-1/3}) + \frac{1}{9}\frac{E}{E_0}\right\} \text{cm}^2/\text{MeV}$$

where E is the final electron total energy in million electron volts and r_0 is 2.82×10^{-13} cm. For no screening ($\gamma \gg 1$),

$$\frac{d\sigma_b}{dk} = \frac{4Z^2 r_0^2}{137k}\left[1 + \left(\frac{E}{E_0}\right)^2 - \frac{2}{3}\frac{E}{E_0}\right]$$
$$\cdot \left[\ln\frac{2E_0 E}{0.51k} - \frac{1}{2}\right] \text{cm}^2/\text{MeV}$$

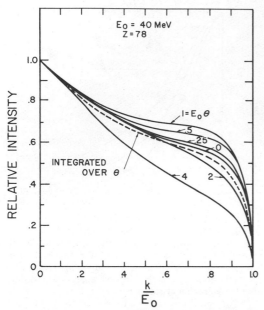

FIG. 3. Dependence of the spectral shape (Schiff's calculation) on the photon emission angle, θ. k is the photon energy in MeV. These curves are from reference 1. E_0 is the incident electron energy in MeV.

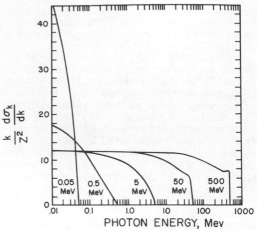

FIG. 4. Dependence of the Born-approximation absolute cross section (integrated over photon directions) on the photon and electron energy. These curves are from reference 1.

Intermediate screening ($2 < \gamma < 15$) leads to much more complicated formulas.[1]

A remark concerning formulas is in order. Generally, expressions for a given cross section are very different depending upon whether the electron energy is small or very large, upon whether the screening is zero, intermediate, or complete, and upon whether one is dealing with the most usual electron-nucleus collisions or with purely electron-electron collisions. Most calculations have been done in Born approximation. The reader is referred to Koch and Motz[1] for an excellent review article on the subject.

The absolute number of bremsstrahlung photons in the photon energy interval dk radiated by a single electron of energy E_0 traversing a radiator of thickness dt and n atoms/cm^3 is given by $(d\sigma_b/dk)n\, dt\, dk$, where $d\sigma/dk$ can be found from Fig. 4.

It must be noted that photon-electron showers begin developing in approximately one radiation length, and these formulas and curves apply only to the basic bremsstrahlung interaction itself or to radiators somewhat thinner than one radiation length.

Conventional bremsstrahlung beams are partially polarized only from extremely thin radiators ($< 10^{-3}$ radiation lengths) because the angular region of polarization is sharply peaked about the angle $\theta = m_0 c^2/E_0$. Electron scattering in the radiator broadens the peak and shifts the maximum to larger angles. Polarization is

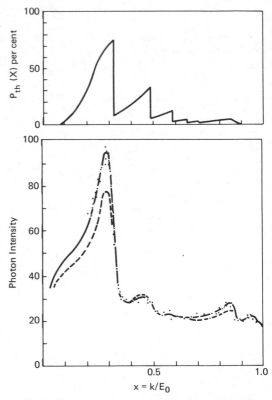

FIG. 5. Photon intensity and photon polarization from a diamond target; experimental spectra compared with the averaged intensity and theoretical polarization for photon angle equal to 50 mrad and electron energy $E_0 = 4.8$ GeV. The solid line represents Hartree potential, the dashed line shows exponential potential.

defined by

$$P(\theta, E_0, k) = \frac{d\sigma_\perp(\theta, E_0, k) - d\sigma_\parallel(\theta, E_0, k)}{d\sigma_\perp(\theta, E_0, k) + d\sigma_\parallel(\theta, E_0, k)}$$

where an electron of energy E_0 radiates a photon of energy k at angle θ. $\perp$ and $\parallel$ directions are with respect to the plane defined by the incident electron and the radiated photon. When the electron is relativistic before and after the radiation, the electric vector is most probably in the $\perp$ direction. Polarization in conventional beams is difficult to observe because thin, low-yield radiators are required. Practical thick-target bremsstrahlung shows no polarization effects whatever. One usually deals with this unpolarized bremsstrahlung and therefore averages over all possible states of polarization of the incident photons.

It is clear that if one makes use of polarized photons when investigating electromagnetic interactions, additional information on spin and angular momentum states can be obtained. The need for this additional information is so compelling that special techniques are frequently employed at the highest-energy electron accelerators (e.g., SLAC, DESY*) to generate polar-

*SLAC—Stanford Linear Accelerator Center, Stanford, California; DESY—Deutsches Elektronen-Synchrotron, Hamburg, Germany.

ized photons. Unfortunately, yields are low by normal intensity standards, but not unusably so.

Coherent bremsstrahlung from an electron incident upon a properly oriented single crystal (e.g., diamond) is discussed in a definitive review article by Palazzi.[2] This effect depends upon the interactions being coherent from the scattering centers in a given crystal plane. A typical spectral shape (Fig. 5) and polarization (Fig. 5a) prove to be extremely useful even though the spectrum is not monochromatic and free from background. Techniques for orienting the crystal radiator are given by Luckey and Schwitters.[3] Typical examples of high-energy polarized photon beams and their applications to physics are found in Ballam et al.,[4] Bingham et al.,[5] and in Bologna et al.[6]

Polarized photon beams from high-energy accelerators are frequently generated through the inverse Compton effect. Nearly 100 per cent polarization can be achieved with reasonable spectral shapes, but photon yields are somewhat lower than from the coherent bremsstrahlung process discussed above. Linearly polarized photons from a high-power pulsed laser are directed into a nearly head-on collision with a high-energy electron beam. Those photons that are back-scattered through approximately 180 degrees carry a large fraction of the electron's

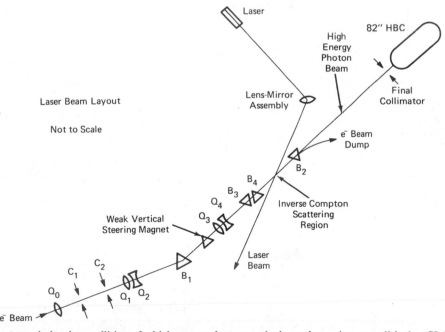

FIG. 6. A nearly-head-on collision of a high energy electron and a laser photon is accomplished at SLAC in the manner shown schematically here. The electron beam employed is part of the central beam facility at SLAC. Angles are exaggerated and there is no scale. Practical considerations eliminated the possibility of crossing the two beams at angles much less than 3 mrad. The high energy photon beam finally enters the 82″ hydrogen bubble chamber where its interactions with hydrogen are studied.

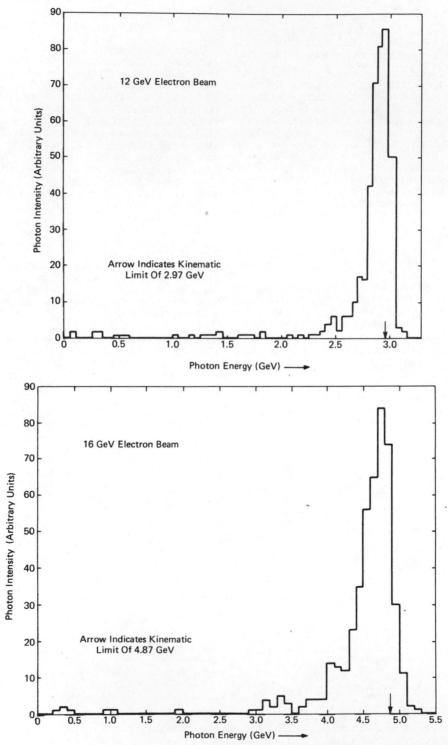

FIGS. 7 and 8. Energy spectra obtained from operation with 12 and 16 GeV electron beams, giving 2.97 and 4.87 GeV scattered photons respectively. These spectra were obtained by measuring e^+e^- pairs produced in the bubble chamber along the known beam line.

kinetic energy and retain the full polarization of the original laser photons. Several hundred photons of $\approx$ 95 per cent polarization and 5 GeV energy are routinely obtained from each pulse of a two-joule ruby laser in conjunction with the linear accelerator electron beam at SLAC.[7] (See Fig. 6.)

Kinematics of these "laser beams" are given by

$$k = \frac{E_0(1 - a)}{1 + a(\Gamma\theta)^2} \qquad \theta \ll 1$$

where k = scattered photon energy in the lab, E_0 = electron energy in the lab, $a = [1 + (4\Gamma k_i/m)]^{-1}$, m = electron mass (0.5 MeV), and $\Gamma = E/m$. k_i = laser photon energy (ruby laser light has $k_i = 1.786 \times 10^{-6}$ MeV) and θ = lab angle (in radians) of scattered photon. (Electron beam has $\theta = 0$ rad.) For example, if $E_0 = 20.000$ MeV, k_i = ruby laser energy, then $k \sim 7070$ MeV. At photon energies in the several thousand MeV range (several GeV range), θ must be restricted to $\approx 10^{-5}$ radians to restrict the lower spectral limit of the photons to approximately 90 per cent of the maximum photon energy. Typical spectra are shown in Figs. 7 and 8.

Measurement of the photon flux in an accelerator bremsstrahlung beam is required in order to make quantitative determinations of cross sections and to normalize observations. An instrument called a "quantameter"[8] is used quite successfully for this purpose and is accurate to the order of 1 per cent. It basically provides sufficient matter (copper plates) to contain the entire electron photon shower volume generated by the incident bremsstrahlung beam as well as to sample and integrate the intensity of the showers over their entire extent.

"Inner bremsstrahlung" is an interesting example of true bremsstrahlung. In beta decay interactions and in orbital electron capture, one sees, on the average, a low-intensity photon continuum, the quantum limit of which is equal to the transition energy of the interaction. These photons are bremsstrahlen emitted by the electron in its transition to the final state.

The Feynman approach to theoretical treatment of the bremsstrahlung process is detailed by Williams.[9] A definitive review article on bremsstrahlung, especially from gaseous targets, has been written by Blumenthal and Gould.[10] They include the closely allied topic of synchrotron radiation and Compton scattering as well as some other interesting radiative effects. They provide an excellent and quite current bibliography.

A recent paper by Chahine[11] has shown the importance of the long radiative tail to wide angle hard bremsstrahlung in very inelastic scattering of mu mesons. Its bibliography is useful in further study of this aspect of bremsstrahlung.

A process called hard gluon bremsstrahlung has been newly observed (1979) in the head-on collisions of beams of high energy electrons and positrons.[12] At energies near 15 GeV for each colliding particle, jets of hadrons (collimated groups of strongly interacting particles, e.g., pions, protons, etc.) are sometimes produced. The usual two-jet structure has been seen to be accompanied occasionally by a third less energetic jet of hadrons. A model and theory for interactions of high energy particles, quantum chromodynamics (QCD), furnishes a quantitative explanation of the observed three-jet structures in terms of a fundamental (?) particle (quark) emitting a field particle (gluon) in the scattering process. Gluons in the interaction, as well as quarks, have the property of materializing in the laboratory as jets. The least energetic jet is usually interpreted as the materialization of a gluon emitted with relatively lower energy than the parent quark. In analogy with a charged particle (usually an electron) emitting a corresponding field particle (a photon), this latter process, properly termed bremsstrahlung, has given its name to the process of gluon emission by a quark. An excellent review article by Duinker[12] includes this subject as well as references to the current literature.

Hadronic jets have also been observed in the basic process of scattering of neutrinos by nucleons.[13] The hadronic jet structure occasionally arising from these interactions is, again, quantitatively explained in QCD only by invoking the idea of hard gluon bremsstrahlung.

These latter two processes are found in collisions of particles at the highest accelerator energies currently available. As energies increase with the advent of newer and larger accelerators, gluon bremsstrahlung will likely remain an important part of the scattering process.

ROBERT W. KENNEY

References

1. Koch, H. W., and Motz, J. W., Rev. Mod. Phys. 31, 920, 1959. Extensive survey of formulas and excellent presentation of curves for numerical calculation.
2. Palazzi, G. D., Rev. Mod. Phys., 40, 611 (1968).
3. Luckey, D., and Schwitters, R. F., Nucl. Inst. Meth., 81, 164 (1970).
4. Ballam et al., Phys. Rev. Letters, 24, 1364 (1970); 24, 960 (1970); 23, 498, 817 (F) (1970).
5. Bingham et al., Phys. Rev., Letters, 24, 955 (1970).
6. Bologna et al., Nuovo Cimento, 42A, 844 (1966).
7. Ballam et al., Phys. Rev. Letters, 23, 499 (1969); Sinclair, C. K. et al., IEEE Trans. Nucl. Sci., 16, 1065 (1969).
8. Yount, D., Nucl. Ins. Meth., 52, 1 (1967).
9. "An Introduction to Elementary Particles," 2nd Ed., W. S. C. Williams, New York, Academic Press, 1971 (page 313 ff.).
10. Blumenthal, G. R., and Gould, R. J., Rev. Mod. Phys., 42, 237 (1970).

11. Chahine, C., *Phys. Rev. Lett.* 47, 1374 (1981).
12. Duinker, P., *Rev. Mod. Phys.* 54, 325 (1982).
13. Ballagh, H. C., et al., *Phys. Rev. Lett.* 47, 556 (1981).

Cross-references: ATOMIC AND MOLECULAR BEAMS, COLLISIONS OF PARTICLES, QUANTUM CHROMODYNAMICS, QUARKS.

BROWNIAN MOTION

Brownian motion is the randomly agitated behavior of colloidal particles suspended in a fluid. The phenomenon is named for its discoverer, Robert Brown, an English botanist. In 1828 he observed the "perpetual dance" of microscopic pollen grains suspended in water. Initially, this effect was interpreted as being due to the motions of living matter, but it was later found that any tiny particles in suspension exhibit Brownian motion.

In 1888, M. Gouy attributed the motion to the bombardment of the visible particles by invisible thermally excited molecules of the suspension. In 1900, F. M. Exner expressed the view that the kinetic energy of the visible particles must equal that of the surrounding suspension particles, and he attempted to estimate molecular velocities on this basis.

In a series of papers published from 1905 to 1908, Einstein[1] successfully incorporated the suspended particles into the molecular-kinetic theory of heat. He treated the suspended particles as being in every way identical to the suspending molecules except for the vast difference of their size. He set forth several relationships which were capable of experimental verification and he invited experimentalists to "solve" the problem.

Several workers undertook this task. The most notable of these was Perrin.[2] Perrin's special success was due to his technique of preparing particles to suspend which were of uniform and known size. The uniformity was achieved by fractional centrifuging, and the size was established by noting that they could be coagulated into "chains" whose length could be measured and whose "links" could be counted. The microscopic observation of these uniform particles enabled Perrin and his students to verify the Einstein results and to make four independent measurements of Avogadro's number. These results not only established our understanding of Brownian motion, but they also silenced the last critics of the atomic view of matter.

Probably the simplest example of Perrin's experiments was his test of the Law of Atmospheres. If we assume that the air is at rest and has the same temperature from ground level upward, it can be shown that the pressure (and concentration) of the air falls off exponentially with increasing altitude. For particles of mass m and density ρ suspended in a medium of density ρ' at absolute temperature T, the ratio of the particle concentrations n_1 to n_2 at heights h_1 and h_2 is given by

$$\frac{n_1}{n_2} = exp \left[- \frac{mg(\rho - \rho')N_0(h_1 - h_2)}{\rho RT} \right]$$

where N_0 is Avogadro's number, g is the acceleration of gravity, and R is the universal gas constant. Although the concentration of air varies slowly with height, the concentration of the relatively heavy particles varied significantly over a height change of a few millimeters. By observing the concentration variation as a function of height, all quantities in the given equation were known except Avogadro's number which could therefore be determined.

JAMES A. RICHARDS, JR.

References

1. Einstein, Albert, "Investigation of the Theory of the Brownian Movement," A. D. Cowper, translator, New York, Dover Publications, 1956.
2. Perrin, Jean, "Atoms," D. L. Hammick, translator, London, Constable, 1923.

Cross-references: ATOMIC PHYSICS, HEAT.

C

CALCULUS OF PHYSICS

To label a topic as the "calculus of physics" is not intended to imply the establishment of some new type of mathematics, but rather that a point of view different from that comfortable to the professional mathematician is to be employed in its discussion. Concepts are introduced for the immediacy of their application to the description of physical phenomena, and a heuristic approach is used to introduce them. We shall not hesitate to ignore interesting but uncommon exceptions to our statements and shall make use of pictorial representations and special cases to illustrate our points.

Functions Since many of the processes of physics are continuing, with the state of things at a given instant developing smoothly out of the state of things in the previous instant, the means of describing these processes compactly is with the help of continuous functions. In some sense, any way of naming the members of a set of objects when given a member of another set comprises a functional relationship. Thus if one is given the set of numbers

$$x = 1, 2, 3, 4 \cdots$$

and the relation

$$y = x^3, \tag{1}$$

then one immediately knows that

$$y = 1, 8, 27, 64, \cdots.$$

Eq. (1) is one way of representing the function. Another would be to tabulate x and y side by side:

x	1	2	3	4
y	1	8	27	64

and another is to draw a graph of y against x, as in Fig. 1. The points contain exactly the same information as the table; the curve contains much more information, but not as much as Eq. (1). Thus it would be a hypothesis to say from the table alone that to $x = \frac{1}{2}$ there corresponded $y = \frac{1}{8}$. The curve (if it could be read accurately enough) would allow one to make that assignment, but would not allow one to conclude that $x = 5$ corresponds to $y = 125$. Equation (1) contains all of this information, and more. We understand from it that to *any*

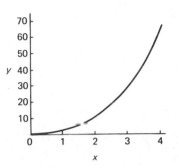

FIG. 1

value of x one may obtain y by multiplying x by itself and the result by x again. In this case, x is the independent variable, y the dependent variable. The concept of continuity is contained in the idea that between any two values of x another can be found; and the concept of continuous function, that to all such values of x a value of y can be assigned according to the prescription in Eq. (1).

Our example is one of the simplest types of function: an algebraic function. Functions of several variables may be considered: functions of complex variables, trigonometric functions, exponential functions, etc. Because of their importance in representing physical processes, let us consider the trigonometric functions a little more closely.

The trigonometric functions are functions of the variable θ (or x or y, or any other symbol you choose), which need not be an angle in the narrow sense, although in physics applications we shall insist that the variable be dimensionless as an angle is. (The distinction between *units* in which the size of a quantity is expressed relative to some standard and dimensions, which are fundamental attributes of a quantity in terms of mass, length, and time, will not hold us here. Suffice to say that, e.g., a *second* is a unit for the dimension time, and the ratio of an arc length to a radius of the arc, which may be expressed in the unit *radian*, is dimensionless.) The tables of trigonometric functions provide a discrete representation, and the familiar graphs are even more useful in visualizing their properties. Figure 2 is a graph of the function $y = \sin \theta$ and Fig. 3 of $y = \cos \theta$. The horizontal scale of each of Figs. 2, 3 should be thought of

155

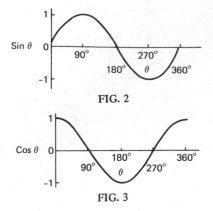

FIG. 2

FIG. 3

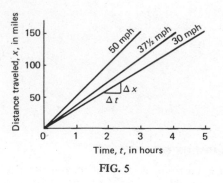

FIG. 5

as extending to the left and right indefinitely with the curves repeating the behavior as shown every 360°. The vertical scales for the sine and cosine functions need be no larger than shown, since their curves oscillate between plus and minus one.

These are examples of continuous functions of a single variable (θ in these illustrations). They are *periodic* functions, i.e., they repeat their values periodically as the independent variable continuously changes. It is the property of periodicity which makes these functions suitable for representing certain physical phenomena.

In Fig. 4 we represent two more useful functions for physics applications: the exponential curve, $y = e^x$ and its inverse, $y = e^{-x} = 1/e^x$.

As a matter of notational convenience, one frequently replaces y by $f(x)$, to be read as "function of x." As a rule, continuous functions can be represented by sums of *algebraic* functions of the independent variable. For example:

$$\sin \theta = \theta - \frac{\theta^3}{3!} + \frac{\theta^5}{5!} - \frac{\theta^7}{7!} + \cdots + \cdots$$

$$\cos \theta = 1 - \frac{\theta^2}{2!} + \frac{\theta^4}{4!} - \frac{\theta^6}{6!} + \cdots + \cdots$$

where θ must be expressed as a dimensionless ratio; i.e., in radians rather than degrees, and

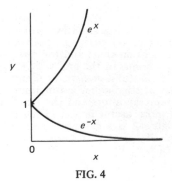

FIG. 4

$$e^x = 1 + x + \frac{x^2}{2!} + \frac{x^3}{3!} + \cdots + \cdots$$

$$e^{-x} = 1 - x + \frac{x^2}{2!} - \frac{x^3}{3!} + \cdots + \cdots .$$

Derivative We can now introduce a few of the concepts of the branch of mathematics which deals with functions and their properties, the calculus. To be specific, we should like to mention the *derivative* and the *integral* of a function. The reader will know something of these already: e.g., the speed of an automobile is the time derivative of its position. This is shown in Fig. 5. The slope of each line represents the speed with which the automobile traveled from New Haven to Boston: 50 mph, $37\frac{1}{2}$ mph, 30 mph. The speed, v, is given by $\Delta x/\Delta t$, where Δt is the time required to traverse the distance Δx. From the graph, one sees that the slope is also the tangent of the angle made by the line with the horizontal axis. These are general properties of the derivative.

The question arises as to what happens if the line is not straight. Then one defines the derivative in the same way, but expects its value to change from point to point along the curve. At each point, one draws the tangent line and calculates its slope. This is the derivative. The abstract definition of derivative is based on the notion of *limit*. In mathematical language, if a function approaches a fixed value as close as one pleases while the independent variable approaches a given value arbitrarily, the function is said to approach a limit as the parameter approaches its value. The notation is $\lim_{x \to a} f(x) = f(a)$, to be read as "the limit of $f(x)$ as x approaches a is $f(a)$." For simple algebraic functions the concept is rather obvious: If $f(x) = ax + bx^2$, then $\lim_{x \to 2} f(x) = 2a + 4b$. The reader should be warned, however, that the situation is not always so obvious. For example, let $f(x) = (ax + bx^2)/cx$. As $x \to 0$, both numerator and denominator approach 0, so $\lim_{x \to 0} f(x)$ seems to be 0/0, which is indeterminate. However, the limit is actually finite; namely, it is a/c. The derivative has been heuristically defined as $\Delta f(x)/\Delta x$, but this has signifi-

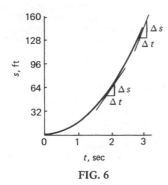

FIG. 6

cance only so long as the ratio remains determinate as $\Delta x \to 0$. Formally, one says

$$\frac{df(x)}{dx} = \lim_{x_2 \to x_1} \frac{f(x_2) - f(x_1)}{x_2 - x_1}, \text{ or}$$

$$\frac{df(x)}{dx} = \lim_{\Delta x \to 0} \frac{\Delta f(x)}{\Delta x}$$

For all the cases we shall be interested in discussing, the limit exists, is equal to the derivative, and is equivalent to the tangent to a curve of $f(x)$ vs x at the point in question. This is illustrated in Fig. 6, which is a graph of $s = 16t^2$. At 2 sec, the tangent has a slope of 64 ft/sec, and at 3 sec it is 96 ft/sec. Since the speed is changing with time, we can graph it and find its rate of change or derivative, as shown in Fig. 7. The slope is seen to be a constant: 32 ft/sec². What is its meaning? This is the *acceleration* of the object which is moving according to the graph in Fig. 6. Thus the acceleration is defined as the time derivative of the speed. For an object moving with constant speed, the acceleration is zero—as it should be to conform with common sense.

While speed and acceleration are among the most familiar examples of derivatives, we should note that the derivative of a function need not be taken with respect to time. If f is a function of an arbitrary variable, x, then the derivative is equal to $\Delta f(x)/\Delta x$, where Δf and Δx are measured on the tangent drawn at point x. The notation, which we shall have occasion to employ, is df/dx for the derivative, and

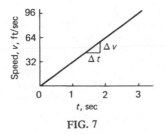

FIG. 7

d^2f/dx^2 for the derivative of the derivative (this would be the acceleration if $f(x)$ were distance and x were time). The reader will imagine that more derivatives may be taken, and wonder if the process is limitless. For some functions, e.g., e^x, there is no highest derivative to be taken. For others, e.g., x^3, all derivatives above a given one are zero—the zero values begin with the fourth derivative in this case.

Examples of derivatives which can be used in calculations are $da/dx = 0$, where a is a constant, $dx/dx = 1$, $dx^2/dx = 2x$, $dx^n/dx = nx^{n-1}$ where n is any number, and $d(ax^n)/dx = nax^{n-1}$, where a is a constant.

$$\frac{de^x}{dx} = e^x, \frac{de^{-x}}{dx} = -e^{-x},$$

$$\frac{d \sin \theta}{d\theta} = \cos \theta, \frac{d \cos \theta}{d\theta} = -\sin \theta.$$

Integral The other important operation of the calculus is *integration*. It may be simply defined as the inverse of taking the derivative, although such a definition has only limited usefulness—mainly it lulls the unwary into thinking he may know something of the process. An operational definition, lacking elegance, may be more nearly indicative of the true nature of the integral: It is a function so constructed that its derivative yields the function whose integral was to be found. From our examples of derivatives, the curve of Fig. 7 may be written as

$$dx/dt = 32t, \tag{2}$$

and of Fig. 6 as

$$x = 16t^2. \tag{3}$$

We have been at some pains to show that $dx/dt = 32t$, so by our definition of integral, x is the integral of dx/dt. The notation for integral is shown in Eq. (4):

$$x = \int (dx/dt)\, dt. \tag{4}$$

Although this is rather a special case, it contains a number of interesting features. If it were a legitimate operation to "multiply" dx/dt by dt, the expected product would be dx, and Eq. (4) would become

$$x = \int dx, \tag{5}$$

which somehow looks like an identity. In fact, if we think of dx replaced by Δx and $\int$ by "sum of," then

$$x = (\text{"sum of"}) \Delta x$$

is pretty obvious. Pictorially, we may think of

TABLE 1

t Interval	0 to 1 sec	0 to 2 sec	0 to 3 sec
Area	16	64	144

the integral of a function as the area under a curve giving the graphical representation of the function. In the general case, the notation reads

$$\int f(x)\,dx,$$

where the dx performs some of the functions in the derivative notation: It identifies the independent variable and implies how the operation is to be carried out.

To give an example, let us return to Fig. 7 and calculate the area under the curve. Table 1 contains the results for the area up to 1 sec, up to 2 sec, etc. It is clear that the numbers in the "area" row can be obtained from Eq. (3) by evaluating it for $t = 1, 2, 3$ sec, respectively.

What of the integral as area when the curve is not as simple as our example? Even the case graphed in Fig. (6) appears to be beyond the definition. We are rescued from this dilemma by recalling that we deal with continuous functions, so we may employ as small a Δx as we please. Thus the integration becomes the summation of many areas whose bases are Δx and heights the values of $f(x)$ at the point in question. Some error will remain: The lined areas shown in Fig. 8 will not always cancel as they must for exact total area calculation. That is, our calculation of the area of the strip as $f(x) \cdot \Delta x$ omits the piece with vertical shading lines and incorrectly includes the piece with horizontal shading. However, as Δx becomes smaller, these two pieces will come nearer and nearer to canceling for each strip. The penalty for increased accuracy is increasing the number of strip areas to obtain and sum.

In fact, for the simpler functions it is possible to obtain the integral without adding areas: The means we have already used of looking for a function whose *derivative* is the function in hand is one method of doing so. Of course, with the advent of high-speed digital computers, the task of adding up many little strips to calculate the integral numerically is reduced to preparing a program to control the computer—and the program will work for any function which can be tabulated.

Sample integral formulas which can be used are

$$\int dx = x + c, \quad \int x\,dx = \frac{1}{2}x^2 + c,$$

$$\int ax\,dx = \frac{ax^2}{2}, \quad \int x^n\,dx = \frac{1}{n+1}x^{n+1} + c,$$

$$(n \neq -1)$$

$$\int \cos\theta\,d\theta = -\sin\theta + c, \quad \int \sin\theta\,d\theta = \cos\theta + c,$$

$$\int e^x\,dx = e^x + c, \quad \int e^{-x}\,dx = -e^{-x} + c,$$

where c is a constant which cannot be determined in the integration.

MCALLISTER H. HULL, JR.

References

1. Hull, M. H., "The Calculus of Physics," New York, W. A. Benjamin, 1969.
2. Goldstein, L. S., Lay, D. C., and Schneider, D. I., "Calculus and its Applications," New Jersey, Prentice Hall, 1980.
3. Thomas, G. B., and Finney, R. L., "Calculus and Analytic Geometry," Massachusetts, Addison-Wesley, 1980.
4. Goodman, A. W., and Saff, E. B., "Calculus: Concepts and Calculations," New York, Macmillan, 1981.

Cross-references: DIFFERENTIAL EQUATIONS IN PHYSICS, MATHEMATICAL PRINCIPLES OF QUANTUM MECHANICS.

CALORIMETRY

Calorimetry is the science of measuring the quantity of heat absorbed or evolved by matter when it undergoes a change in its chemical or physical state. The apparatus in which the measurement is performed is a *calorimeter*, and the experimenter is frequently referred to as a *calorimetrist*.

When matter is involved in a chemical or physical process, its total energy content is usually altered. The difference in energy between its initial and final states, ΔE, must be transferred to, or from, the environment of the

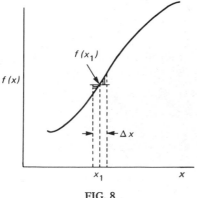

FIG. 8

system. This energy exchange between the system and its environment is in the form of heat or work or both. In calorimetry, the energy exchanged as heat is quantitatively evaluated. The heat absorbed by the system, q, is related to the work done by the system on its environment, ω, and the increase in internal (total) energy of the system, ΔE, by the thermodynamic relationship

$$q = \Delta E + \omega. \tag{1}$$

When calorimetric measurements are performed at constant pressure and only pressure-volume work is involved, q is equal to the increase in heat content or enthalpy ΔH. Most calorimetric measurements are performed under these conditions, but when other conditions are imposed, appropriate consideration must be made in the thermodynamic treatment of the data.

The process selected for calorimetric study may be a simple change in the physical state of matter, such as a change in temperature of the material, or it may consist of a series of complex chemical reactions such as are encountered in the combustion of many fuels. In fact, nearly any process involving a chemical or physical change in matter might well become a necessary subject for calorimetric investigation.

Calorimetric determinations of energy changes are essential in many theoretical and practical problems. Heat capacity or specific heat data are vital to the design of heat exchange equipment. The thermal properties of steam and certain metals are major considerations in the design of modern boilers and turbines. The heats of combustion of fuels are essential in rocket, engine, and gas turbine design. The heat liberated by chemical reactions must be considered in the development of chemical process equipment. Often the equilibrium constant required to determine directions and extent of chemical reactions is most conveniently obtained by a simple calculation from the free energy change ΔF. For a great many processes, numerical values of ΔF can be obtained from the change in heat content ΔH, and the entropies of the participating substances S, using the thermodynamic relationship

$$\Delta F = \Delta H - T\Delta S, \tag{2}$$

where T is the absolute temperature. The entropies of the individual substances can generally be evaluated from heat capacity measurements that extend to very low temperatures. In addition to these primarily practical design considerations, calorimetric measurements can provide information about the microscopic parameters of a material, such as its structure or the energy levels of its electrons and atoms. This information in turn can provide insight into the fundamental interactions in the material.

The design and constructional details of calorimeters vary widely because of the diversified nature of the processes suitable for calorimetric study. However, the basic principles are general, and their consideration constitutes a common requirement in practically all designs. Suitable devices and procedures for three essential measurements are usually required, but one or two can sometimes be omitted by operating under certain restrictions. The measurements are: (1) the temperature of the calorimeter and its contents, (2) the quantity of energy that is added to the calorimeter from an external source, and (3) the quantity of heat that is exchanged between the calorimeter and its environment.

Most calorimetric operations involve a temperature change, since the heat liberated (or absorbed) during the process is stored in the calorimeter and its contents by virtue of their combined heat capacity. Thermocouples, thermopiles, and resistance thermometers are commonly used for temperature measurements. The quantity of energy liberated or absorbed in a calorimetric process is most commonly evaluated in terms of electrical energy. This is done by three similar methods. (1) In an exothermic process where heat is liberated, the calorimeter is cooled to the original temperature; the temperature rise is then duplicated using an electrical resistance heater. (2) The heat absorbed in an endothermic process is supplied by an electrical heater at such a rate as to keep the temperature constant. (3) In heat capacity measurements an electrical heater supplies known amounts of energy to the sample. The resultant temperature change is then monitored. Electrical energy and temperature can be measured very accurately by modern methods.

The quantity of heat exchanged between the calorimeter and the environment is a more difficult problem. When two adjacent bodies (such as a calorimeter and its environment) are not at exactly the same temperature, heat is transferred from the warmer to the cooler body. This transfer is made by three major processes: (1) gaseous convection, (2) radiation, and (3) conduction. Gaseous convection can be completely avoided by evacuating the space between the calorimeter vessel and its environment. When evacuation is impractical, convection can be minimized by suitable geometrical considerations in the design of the calorimeter. It is very important to avoid or at least minimize convection, since the heat transported is a complex function of the temperature difference and an accurate evaluation is impossible. For small temperature differences, radiation is usually not a serious problem at low temperatures but is a major contributor to heat exchange at elevated temperatures. Heat exchange by radiation can be limited to a few percent of the blackbody (maximum) values by the use of suitable reflecting surfaces on the outside of the calorimeter and on the adjacent environment. Conduction by air or other gases is also usually minimized by evacuating as much as possible of the space between the calorimeter proper and its environment. Conduction in solid materials, used for supporting the calorimeter and for electrical leads is optimized by proper choice of materials and geometrical design.

In the absence of convection and for small temperature differences, the heat transferred, Q, is predominantly via conduction and is essentially proportional to the temperature difference ΔT and time t, in accordance with Newton's law of cooling,

$$Q = k\Delta Tt. \qquad (3)$$

There are two approaches chosen currently. For large samples, this heat exchange is minimized by reducing the quantities on the right-hand side of Eq. (3). The constant k is a measure of the thermal link to the environment and is minimized by design considerations. Adiabatic calorimeters maintain the calorimeter and the environment at the same temperature so that the temperature difference ΔT equals zero. In this approach, corrections are made when the heat transferred is not exactly zero. For smaller samples, the connections necessary for measuring temperature and supplying electrical energy are a source of significant heat exchange and the corrections become rather large. The second approach therefore, instead of minimizing the heat transferred, actually uses the heat exchange to measure the heat capacity. The rate of change of the temperature of the sample $d\Delta T/dt$ times the heat capacity of the sample, C, is equal to the rate of heat exchange:

$$\frac{dQ}{dt} = - \frac{Cd\Delta T}{dt}. \qquad (4)$$

By accurately measuring the quantities on the right hand side of Eq. (3), Eq. (4) may be used to obtain C.

In calorimeters containing liquids, there is a possibility of a fourth mechanism for transporting heat. This method involves the transport of matter from the calorimeter and its subsequent condensation on the surrounding surfaces. The effect can be avoided by keeping the environment warmer than the liquid or by completely enclosing the liquid. However, even in a completely enclosed system the possibility of vaporization into the space above the liquid with increasing temperature must be considered for volatile liquids.

There are many different varieties of calorimeters, each being particularly suited for a specific type of measurement. Some general features of several representative types are discussed below.

Low temperature calorimetry, used down to the temperatures available with liquid and solid hydrogen, ~10 K, has become an important source of heat capacity data for the evaluation of entropies of substances from measurements extending from near the absolute zero to room temperature or slightly above. This information

may, in particular, be used by chemists to predict chemical reactions. The calorimetric vessel consists of a vacuum-tight metal container in good thermal contact with an electrical resistance heater and a thermocouple or resistance thermometer. The sample under study is sealed in the container along with a small amount of gaseous helium. The helium aids in attaining thermal equilibrium at low temperatures because of its high thermal conductivity. The calorimetric vessel is suspended in an evacuated chamber by some material, such as a strong thread, having low thermal conductivity. This chamber is often within a massive copper block which provides a uniform and stable thermal environment. The temperature of the protective block is kept at a temperature near that of the calorimetric vessel. The heat exchanged is evaluated by observing the temperature difference ΔT as a function of time and applying Eq. (3) in an integrated form. The constant k is evaluated by observing the change in temperature of the calorimeter vessel and its contents under equilibrium conditions. During this rating period the temperature change is due entirely to heat exchanged with the environment. Some calorimetrists use the adiabatic principle and maintain the temperature of a protective shield as near as possible to that of the calorimeter. This procedure results in the elimination of heat exchange corrections but is not entirely free from objections. Although low-temperature calorimeters are used chiefly for heat capacity determinations, heats of transition, heats of fusion, and heats of vaporization are also measured.

At very low temperatures, calorimetry is used to measure energies associated with the ordering of the magnetic moments of nuclei, transitions into the superconducting state of metallic elements and compounds, and other phenomena which require only small amounts, or quanta, of energy to be activated. A decade decrease in temperature means a decade decrease in the size of the energy quanta which can be studied, and in this sense the range from 0.1 to 0.01 K covers a range equivalent to that between 1000 K and 100 K (i.e., from far above, to far below room temperature at 300 K). Temperatures down to 0.3 K are achieved by reducing the pressure over a bath of the helium isotope of mass 3, a relatively simple process. Using a more complex process, a helium dilution refrigerator produces millikelvin temperatures and thus heat capacities can be measured in the decade around 0.01 K. One method which has been used to measure the heat capacity of small samples such as thin films weighing less than 1 mg (10^{-4}–10^{-6} moles) at low temperatures is the so-called *relaxation method*. The sample is heated by an electrical heater to a small ΔT above its environment. The electrical leads provide a previously measured heat link to the environment, so that when the heater is turned off the temperature of the sample will relax back to that of the environment with a time constant proportional to the heat capacity of the sample. Here, as in all heat capacity measurements, the heat capacity of thermometers, heaters, and electrical leads is lumped together with that of the sample. To isolate the heat capacity of the sample, therefore, all of these addenda must be accurately known and subtracted from the total.

An interesting type of measurement is time-dependent heat capacity. In some materials, it is possible to put energy into only the electrons or only the ion lattice because the time needed for the electrons and the lattice to exchange energy and reach equilibrium is very long (up to seconds). Measuring the heat capacity as a function of the time probes both the equilibration processes and the electron and/or lattice heat capacities.

The *dropping method* is the most common of the accurate high-temperature procedures for measuring heat contents. This apparatus consists of a carefully regulated furnace and a suitable calorimeter, such as a Bunsen ice calorimeter, operating near room temperature. The sample under investigation is sealed inside a container that will not undergo chemical reaction at the highest temperature of the measurements. The sample and container are thermally equilibrated with the furnace and then dropped into the calorimeter. The empty container is afterwards studied in an identical manner and the difference in the two measurements gives the heat content of the sample relative to the room temperature reference. Heat capacities are derived from a series of such measurements as a function of temperature and the thermodynamic relationship

$$C_p = \frac{\partial (H)}{(\partial T)_p} = \left[\frac{(\partial H - H_0)}{\partial T} \right]_p, \qquad (5)$$

where C_p is the heat capacity at constant pressure, H the heat content, H_0 the heat content at the reference temperature, and T the absolute temperature.

The *Bunsen ice calorimeter* is an example of an isothermal calorimeter that is operated at a fixed temperature. The calorimeter is usually surrounded by ice, making it also adiabatic and thus free from heat exchange. Bunsen's design makes use of the very large difference between the specific volume of ice and water. The calorimeter contains a closed chamber which is full of ice and water. A pool of mercury is maintained in the bottom of the chamber, and as the ice melts, additional mercury enters and keeps the chamber full. The calorimeter has a universal calibration in the form of energy per unit mass of mercury. In early versions, the quantity of ice melted was used as a measure of the heat liberated in the calorimeter. By replacing the ice with other suitable substances, the restriction of operating at one fixed temperature can be removed.

Quantitative measurements of the heat liberated (or absorbed) during the solution of a solid or of another liquid by a solvent are performed in *solution calorimeters*. Heats of solution, dilution, and mixing are common determinations of this type. In addition to participating in the process under investigation, the solvent is used as a means of attaining uniform temperature and composition throughout the calorimeter. This feature necessitates stirring, which is usually accomplished with mechanically or magnetically driven stirrers. Sometimes the calorimeter itself is rotated. Regardless of the method used, the quantity of heat introduced by the stirring must be determined either directly or indirectly and a suitable correction must be applied. Another feature characteristic of solution calorimeters is the method of adding the sample. Either it must be equilibrated with the solvent in the calorimeter, or its heat content relative to the calorimeter temperature must be determined. A common method for solids is immersing a capsule containing the sample in the solvent and breaking it at the desired time.

The heat of combustion of fuels and similar materials is usually measured by *bomb calorimetry*. The solid or liquid sample is contained in a bomb (pressure vessel) containing excess oxygen or other suitable gas under pressure. The bomb is immersed in a calorimeter containing a liquid, usually water. The reaction is initiated by igniting the sample with a measured amount of electrical energy, and the heat evolved is measured in terms of the temperature rise of the calorimeter. Electrical energy is usually used to duplicate the temperature rise and thus evaluate the heat liberated. However, sometimes a standard sample of a substance having a known heat of combustion, such as benzoic acid, is used to calibrate the apparatus. In bomb calorimetry, corrections to standard conditions must be applied (Washburn corrections) since the system is under pressure and because solutions are usually formed.

There are many other important types of calorimeters, such as flow calorimeters, microcalorimeters, flame calorimeters, etc. Nearly any process can be studied by the investigator who is ingenious enough to devise the appropriate apparatus and who has the resources and patience to undertake an extensive project. Although calorimetric measurements are in general time-consuming and tedious, they are essential for a fundamental and practical understanding of many important chemical and physical processes.

T. H. GEBALLE
F. HELLMAN

Cross-references: ENTROPY, HEAT, HEAT CAPACITY, HEAT TRANSFER, THERMODYNAMICS.

CAPACITANCE

Definition and Fundamental (Quasi) Static Properties If a constant voltage V[V = volts] is applied between two conductors insulated from each other, electrical charges Q [As = coulomb] are so distributed that the conductors form equipotentials. The measure for the charges stored is the capacitance C [F = farad = 10^6 μF = 10^9 mμF = 10^{12} pF] of the capacitor so formed.

$$Q = C \cdot V \qquad (1)$$

(Q in coulombs, C in farads, V in volts.) It is often more convenient to express this storing capacity in terms of energy

$$E = (\tfrac{1}{2}) V^2 C \qquad (2)$$

(E in watts.) C is defined by

$$C = \frac{1}{V} \int i \, dt \quad \text{or} \quad C = I \left| \frac{dv}{dt} \right. \qquad (3)$$

For capacitor discharge (E_0 = starting voltage),

$$e_c/E_0 = \epsilon^{-t/\tau} \qquad (4a)$$

and for capacitor charge (E_b = battery voltage)

$$e_c/E_b = 1 - \epsilon^{-t/\tau} \qquad (4b)$$

with the time constant τ

$$\tau = CR \qquad (5)$$

where R is the resistor through which the capacitor is being (dis)charged.

For sinusoidal excitation of angular frequency ω, the reactance of the lossless capacitor is

$$V[\text{V}]/I[\text{A}] = (-)jX[\Omega] = 1/j\omega C[\Omega] \qquad (6)$$

If, in electrical circuits, capacitors are connected in parallel, their capacitances add

$$C = \sum_{k=1}^{n} C_k \qquad (7a)$$

If capacitors are connected in series, their elastances (the reciprocal of capacitance, S) add

$$S = \sum_{k=1}^{n} S_k \qquad (7b)$$

Losses in the dielectric may be expressed by a complex relative dielectric constant

$$\epsilon = \epsilon' - j\epsilon'' \qquad (8)$$

where $\epsilon'/\epsilon'' = Q_e$ determines the dielectric quality factor. For $Q_e > 10$, the loss resistance of the capacitor is given by

$$(1/\omega C)/r_s = Q_e = R_p/(1/\omega C) \qquad (9)$$

where r_s is the equivalent series and R_p is the corresponding parallel loss resistance. $Q_e = 1/DF$

(DF = dissipation factor). The power factor is related to DF by

$$PF = DF \sqrt{1/(1 + DF^2)} \qquad (10)$$

The loss factor = $(DF) \cdot \epsilon$ is proportional to the energy loss/cycle/voltage^2/volume.

Capacitors are used for: (1) frequency determining or selective networks [LC circuits and filters; cf. Eq. (6)]; (2) energy storage [Eq. (2)], for instance, the capacitor being slowly charged and quickly discharged [Eqs. (9) and (10)] in a short burst of energy; and (3) integrators and differentiators [in conjunction with R; cf. Eq. (3)].

Geometry *Uniform Fields.* For a uniform field as, for instance, given between two closely spaced parallel metallic plates (area A in square meters, distance l in meters) and disregarding edge effects

$$C[\text{F}] = \epsilon_0 \epsilon A/l \qquad (11)$$

with $\epsilon_0 \epsilon$ = dielectric constant of free space = $(36\pi \times 10^9)^{-1}$ [F/m] and ϵ = the relative dielectric constant (dimensionless) of the material between the plates.

Discontinuity in Uniform Fields. If, in the above case, the dielectric consists of two sheets of different materials with ϵ_1 (having thickness l_1) and ϵ_2 (having thickness l_2)

$$\frac{E_1}{E_2} = \frac{\epsilon_2}{\epsilon_1} \qquad (12)$$

where E_n is electric field strength = V_n/l_n. (13)

Equation (12) is of great practical significance if one of the ϵ's is very high, since then the sheet with the low ϵ carries nearly all voltage (for this reason, for example, higher-ϵ ceramic capacitors have to have fired-on electrodes).

Nonuniform Fields. The most common capacitance with nonuniform fields is the coaxial capacitor (inside diameter d, outside diameter D). Its capacitance is

$$C'[\text{pF/m}] = 55.6 \epsilon / \ln (D/d) \qquad (14)$$

Extreme cases of nonuniformity, often causing corona, exist on the sharp edges of plate capacitors. Remedy: For field equalization, deform plates to follow equipotential lines of half potential in a capacitive field with twice the spacing of the original, flat plates (Rogowski profile).

Dielectrics The dielectric "constant" is often not constant but a function of crystal orientation (anisotropy), temperature, voltage, and frequency (dispersion).

The objective of developing a good fixed capacitor is to have the largest capacity in the smallest possible volume for a given operating voltage. Ideally, the capacitance is not to change with voltage, temperature, time, mechanical stress, humidity, and frequency, and (in most cases) is to have a minimum of losses. The greatest capacitance can be achieved by maximizing ε (Case a) and A (Case b), and minimizing l (Case c) [cf. Eq. (11)].

Typical for *Case (a)* are *ceramic* capacitors made in discoidal or tubular form (and now recently also as coaxially laminated capacitors). There are four classes of ceramic dielectrics:

(1) Semiconducting, so-called layerized, ceramics with dielectric constants above 10^5. These can be used only for very low voltages (transistor circuits), are quite lossy, and have a strong dispersion of ε in the megacycle range.

(2) High-ε' dielectrics (mostly barium titanates) with ε' in the order of 6000. These are quite temperature- and voltage-sensitive (nonlinearity and hysteresis) and are used as guaranteed-minimum-value capacitors (GMV).

(3) So-called stable dielectric capacitors with an ε' of 2000 or, if doped with rare-earth materials, with an ε' of 3000 to 4000. These are much less dependent on temperature and applied dc voltage.

(4) Linear, high-Q (in the order of several thousand) temperature-compensating capacitors made with a prescribed (P positive, N negative, or NPO) temperature coefficient of the capacity for incorporation in temperature-stable tuned circuits (compensation of the temperature coefficient of the inductance). The ε of such materials lies between 10 and 100.

Case (b) (large A) is exemplified best by stacked plates [silvered mica (for military use; excellent Q, temperature coefficient about –100 ppm) or ceramic (monolithic)] or rolled dielectric strips [polystyrene (excellent Q; commercial use; also about –100 ppm T.C.); "Mylar"; oil-impregnated paper; "Teflon" etc.].

Case (c) (small l) is represented by polarized capacitors (to make them unpolarized, two capacitors are connected in series in polarity opposition, usually in the same housing), and it includes the older, larger, and cheaper types like the aluminum foil electrolytics. The newer, more costly, but much smaller, types (having much less leakage current) are tantalum oxide capacitors. Ta_2O_5 stands continuously the extraordinary field strength of 3×10^6 V/cm with an ε' of 25, l being measured in angstroms. The Q is about 100. For microminiaturization, silicon monoxide or dioxide or tantalum oxide films of very small l are utilized.

Rating The reliability of a capacitor is predominantly determined by the dielectric and the seal of the housing. One has to distinguish between failure value and withstand value. The failure value of dielectric strength is the voltage at which the material fails and is conventionally given as the average failure voltage. In contrast, the withstand value is a voltage below which no failure can be expected.

Deterioration of capacitors with time (aging) can be greatly reduced by systematic "physics

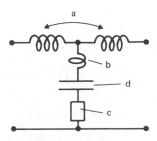

FIG. 1. High-frequency behavior of capacitors.

of failure" investigations. Typical failure mechanisms are, for instance, precorona discharge in adsorbed air layers, or silver migration.

Non-ideal Behavior at Higher Frequencies Equation (6) presumes ideal conditions. An actual capacitor, particularly if considered over many decades of frequency, and more so, if used as a shunting element across lines, is much more aptly describable as a three-terminal network. Figure 1 marshals the four key deviations from the ideal behavior:

a. At very high frequencies, inductive input-output coupling may override the shunting effect of the capacitor. Remedy: use feed-through capacitors where input and output leads are separated by a shield.

b. Again, at high frequencies, unless as a remedy a feed-through configuration is selected, a series L in the shunt branch results in the capacitor behaving as an inductor above the resonance frequency thus determined ($\omega = (LC)^{-1/2}$).

c. If the capacitive reactance at high frequencies becomes very small, the resulting transfer impedance may be determined by the series loss resistance. A typical case is a tantalytic capacitor behaving this way. Remedy: It must be paralleled by a smaller capacitor of less high frequency losses.

d. At very high frequencies, let us say 100 MHz and above, even ceramic feed-through capacitors start to resonate internally (transmission line effect) rendering them useless above certain frequencies. Remedy: See, for instance, bibliography 1.

Nonlinear Capacitance The dielectric of highly nonlinear capacitors is the depletion layer formed at the p-n junction by application of proper bias. These back-biased diodes have a reasonable Q and are widely used as nonlinear reactances in parametric amplifiers for VHF and higher frequencies and for varactor tuning in TV receivers. Nonlinear ceramics are less suitable for this purpose because of their high losses and great temperature dependency.

"Parasitic" Capacitances Capacitances play a significant role in voltage limiters used extensively for interference control. If one employs sparkgap limiters, one has the advantage of a small shunt capacity. Thus, with $R = Z_0$ (characteristic impedance of line), the time constant is small and the bandwidth of the line is large—as it often needs to be. But sparkgaps have disadvantages like slow response time, possibility of sustaining arcs, etc. Solid state voltage limiters (single-crystalline Si Zeners or polycrystalline metal-oxide varistors) do not have such drawbacks, but have large C's which may be reduced (a) or exploited (b). To maintain large bandwidth, a small series capacitance is to be added (a). On the other hand (b), metal-oxide varistors have variable resistance, very high below the knee voltage and very high above it, and in parallel a considerable capacitance. Their dielectric constant is on the order of 3000 and is not bias sensitive and much less temperature dependent than corresponding ceramics. Feedthrough capacitors made of such material are not yet commercially available, but they are very small integrated interference suppression devices that filter at low voltages and limit at high voltages.

H. M. SCHLICKE

Reference

Schlicke, H. M. "Electromagnetic compatibility," Marcel Dekker, New York, 1982.

Cross-references: CIRCUITRY, DIELECTRIC THEORY, POTENTIAL.

CARNOT CYCLES AND CARNOT ENGINES

Background Steam engines were first built to do heavy, tedious jobs such as pumping water out of deep mines. Not only could they pull harder than a team of horses, but they did not get tired. These engines were built by rule of thumb by practical men, for at that time there was no thermodynamics to guide the design. Still engines could be rated by how much water was pumped for a bushel of coal burned. For a century the quality of these engines improved, and the obvious question was whether there was any limit to this improvement. Carnot's work addressed that question. He set out to devise a heat engine against which all imaginable heat engines could be compared.

Description The engine Carnot devised is often represented, as shown in Fig. 1, by a device which uses an ideal gas as its working substance. A quantity of gas is confined in a cylinder with a wall so well insulated that no heat can flow through it. The cylinder's heat-conducting base rests on the first reservoir, whose constant temperature is T_h, and the gas assumes this temperature. A weighted insulating piston holds the pressure of the gas at P_1 at which pressure its volume is V_1. The gas is then said to be in thermodynamic state 1 characterized by P_1, V_1, and T_h. Little by little the weight on the piston

is now set aside until the pressure is reduced to P_2 and the volume is expanded to V_2. The gas is now in state 2 characterized by P_2, V_2, and T_h. The transition from one thermodynamic state to another is called a thermodynamic process. A process is called reversible if done so slowly that no temperature differences arise within the gas and if the piston moves without friction. The cylinder base is kept at temperature T_h, so during this process the temperature of the gas remains at T_h; that is, it is an isothermal process. When this is shown on a pressure-volume diagram, the process appears as a portion of the T_h isotherm. To hold the temperature constant, some heat energy Q_h must flow from the reservoir into the gas. In expanding against the weight on the piston, the gas does work W_{12} which is represented on the P-V diagram by the crosshatched area. The cylinder is next moved to an insulated pad where the pressure is further decreased by setting aside more weights, and the gas again expands. No heat energy flows into the gas from the outside during this expansion, and this is called an adiabatic process. The temperature decreases and when it reaches T_c, the temperature of the second reservoir, the process is stopped. The gas is now in state 3 characterized by P_3, V_3, and T_c. During this process the gas does work W_{23} against the load. The cylinder is then moved to the second reservoir where enough weights are slowly replaced to bring the gas to state 4, the point on the T_c isotherm from which state 1 can be reached by an adiabatic process. During this isothermal compression, heat energy Q_c flows from the gas into the reservoir, and the piston does work W_{34} on the gas. To show that this work is done on the gas while previously the work was done on the piston, the appropriate portion of the cross-hatched area has been removed. The cylinder is finally placed on an insulated pad, the remaining weights are slowly added, and the gas returns adiabatically to state 1. Again the fact that the piston does work W_{41} on the gas, is shown by the removal of the cross-hatched area. One Carnot cycle is now completed.

Definition and Characteristics A Carnot cycle is any reversible cyclic thermodynamic operation composed of four processes which are alternately isothermal and adiabatic. (The working substance need not be an ideal gas, but traditionally this is used in discussions.) Since no natural process is strictly reversible, the Carnot cycle is an idealization.

Although heat energy has entered and left it and work has been done on it, the gas undergoes no detectable physical changes for having passed through the Carnot cycle. Heat energy Q_h was removed from the hotter reservoir, and a smaller amount Q_c flowed into the cooler reservoir so that the heat energy budget of the gas increased by $Q = Q_h - Q_c$. The net work done by the gas on the piston is $W = W_{12} +$

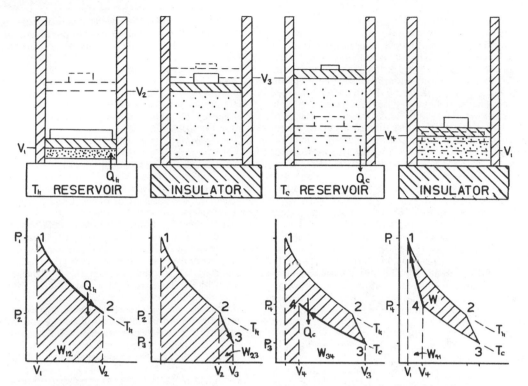

FIG. 1. The steps of a Carnot cycle and the corresponding pressure-volume diagrams. The piston location at the start of each process is shown in solid lines; at the end, in dashed lines. The pressure on the gas can be estimated by the area of the weight shown on top of the piston. The dots indicate that the molecules of the ideal gas are close together when the volume is small and are farther apart when the volume is large.

$W_{23} + W_{34} + W_{41}$, and the mechanical energy budget of the gas decreased by W which is represented on the P-V diagram by the area enclosed by the phase lines. The first law of thermodynamics requires that $W - Q = 0$ so no energy residue is left in the gas. If the cycle is traversed as described, heat energy Q_h is removed from the higher-temperature reservoir. Part of this remains in the form of heat energy Q_c as it flows into the cooler reservoir, and part of it is converted to mechanical energy as the work W done on the piston. A Carnot cycle operated in this direction is called a Carnot engine. If the direction of operation were reversed, the cycle would be called a Carnot refrigerator. In such a device mechanical energy, the work W done by the piston, is converted into heat energy which combines with the heat energy Q_c which flows from the cooler reservoir into the gas. All this heat energy Q_h flows out of the gas into the hotter reservoir.

The efficiency η of any engine is the fraction of the heat energy input Q_h which is converted into mechanical energy W; that is, $\eta = W/Q_h$. The following properties of Carnot engines are derived in many textbooks of thermodynamics.

(1) The efficiency of a Carnot cycle depends only on the temperatures of the two reservoirs. When these temperatures are measured on the absolute scale in Kelvin, then $\eta = 1 - (T_c/T_h)$.

(2) No heat engine operating in cycles between two reservoirs can have a greater efficiency than a Carnot engine operating between those reservoirs.

The Carnot engine is thus a standard against which other heat engines can be compared. Historically it was the source of a number of ideas that are now basic to the study of thermodynamics.

Although the Carnot cycle is the most efficient heat engine, it does not produce any power P. The requirement of reversibility means that a very long time t is required to complete even one cycle. Thus the power (defined as $P = W/t$) is zero. To speed up the cycle a temperature difference between the working substance and reservoirs can be introduced. Under this condition the power efficiency η' is found[1] to be $\eta' = 1 - (T_c/T_h)^{1/2}$. The efficiency of real heat engines is described rather well by this equation.

History of the Carnot Cycle In 1824 Sadi Carnot[1a] (1796–1832) analyzed a heat engine assuming that heat can perform mechanical work in falling from a higher temperature to a lower just as water can do work falling from a higher level to a lower, and assuming that no heat would be lost just as no water was lost. (His work preceded by more than 20 years the theory of Joule and Helmholtz on the mechanical equivalence of heat.) In his study he proposed an ideal heat engine that operated in a continuous cycle and was reversible. He then showed that it is impossible in a cyclic operation to obtain work from a single constant-temperature heat source and that no more work can be obtained from any process than is required to reverse it.

His ideas escaped notice until 1834 when Clapeyron[2] recognized their merit, suggested some of the details of the device described above, and plotted its behavior on a P-V diagram. Again the ideas were neglected until William Thomson (later Lord Kelvin) learned of Carnot's work through Clapeyron's memoir. In 1848, Thomson described[3] how a Carnot engine could be used to define a temperature scale that was absolute in the sense that it did not depend on what thermometric substance was used. It was based on a series of Carnot engines, each of which did the same amount of work. This was the first important idea drawn from a study of the Carnot cycle. In 1850 Clausius[4], who learned of Carnot's ideas through Thomson and Clapeyron, showed how Carnot's assumption (no loss of heat) could be reconciled with the newer views of Joule and Helmholtz (which now form the basis of the first law of thermodynamics). It was only required that the engine exhaust less heat energy, by the amount of the work done, than it accepted. Thomson independently reached the same conclusion[5] in 1851.

In 1854 Clausius[6] in his study of the Carnot cycle identified the physical property he later named "entropy." This was the second important idea drawn from study of the Carnot cycle. In 1877 Boltzmann[7] took the principle of Clausius that real processes evolve naturally toward states of higher entropy (which is the second law of thermodynamics) as a basic point in his theory and thus had no need to consider the Carnot cycle from which that principle was derived. It is now common practice to discuss thermodynamics axiomatically rather than historically so that the Carnot cycle no longer plays the important role it once did. The most complete discussions are found in the older volumes,[8] but some recent books do describe the cycle in detail.[9] An historical account[10] with an elementary presentation of the theory and a review[11] of Carnot's work with excerpts from his original paper have been published.

ROBERT A. LUFBURROW

References

1. Curzon, F. L., and Ahlborn, B., *Amer. Journ. Physics*, **43**, 22–24 (Jan. 1975).

1a. Carnot, S., "Réflexions sur la Puissance Motrice du Feu," Paris, Bachelier, 1824. Reprinted (together with Refs. 2 and 4 below) by Dover Publications, New York.
2. Clapeyron, E., reprinted in *Ann. Physik* [2] **59**, 446–451, 566–586 (1843). See Ref. 1a.
3. Thomson, W., "Mathematical and Physical Papers," Vol. I, pp. 100–106, Cambridge Univ. Press, 1882.
4. Clausius, R., *Ann. Physik* [2] **79**, 368–397, 500–524 (1850). See Ref. 1a.
5. Thomson, W., *Ann. Physik* [2] **79**, 174–316 (1850).
6. Clausius, R., *Ann. Physik* [2] **93**, 481–506 (1854); **125**, 390 (1865).
7. Boltzmann, L., "Lectures on Gas Theory" (translated by S. G. Brush) Berkeley, Univ. California Press, 1964.
8. Britwistle, G., "The Principles of Thermodynamics," Cambridge Univ. Press, 1925. Preston, T., "The Theory of Heat," Third edition, London, The Macmillan Co., 1919.
9. Shortley, G. H., and Williams, D. E., "Principles of College Physics," Englewood Cliffs, N.J., Prentice-Hall, 1959.
Zemansky, M. W., "Heat and Thermodynamics," 4th Ed., New York, McGraw-Hill Book Co., 1957.
10. Sandfort, J. F., "Heat Engines," Garden City, N.Y., Doubleday & Co., 1962.
11. Wilson, S. S., *Sci. Am.* (August 1981, 134–145).

Cross-references: HEAT, THERMODYNAMICS.

CAVITATION

Cavitation is defined as the formation of one or more pockets of gas (or *cavities*) in a liquid.* Here the word *formation* can refer, in a general sense, both to the expansion of a newly created cavity or of a preexisting one to a size where macroscopic effects can be observed. The cavity's gas content refers to the liquid's vapor, some other gas, or combinations thereof. Sometimes these cavities are referred to as *bubbles* or *voids*, depending on the relative amount and type of gas.

Cavitation usually occurs in response to a reduction of the pressure sufficiently below the vapor pressure or the gas saturation pressure of the liquid or to the elevation of the temperature above the boiling point, although chemical, electrical, and radiation-induced phenomena can be important.

Although liquids, by definition, cannot sustain shearing forces without flow, they can sustain uniform tension. The tensile strength of a liquid refers to the limiting tension, or negative pressure, that a liquid can sustain before cavitation occurs. Ultimate cavitation *thresholds* can be reached only in the absence of weak spots or cavitation *nuclei* (usually gas or vapor pockets) which exist on solid surfaces of container walls and suspended impurities, or which

*The first two paragraphs are from Ref. 3 and are reprinted with the permission of Academic Press.

are generated by the interaction of ionizing radiation with the tensed liquid (see also BUBBLE CHAMBERS).

Because of the prevalence of these weak spots (typical tap water has in excess of 10,000 particles per cubic centimeter), the ultimate tensile strength of a liquid, theorized to range from 150 to 1500 atmospheres for most materials that are liquid at room temperature, has rarely been measured. Commonly measured tensile strengths measured by static or dynamic means are typically two to three orders of magnitude lower (approaching the vapor pressure of the liquid). Static tensile stresses can occur in the laboratory, in a liquid filled capillary that is spun at sufficient speed, and also in nature, it is presumed, in the xylem paths of tall trees, as water transpiring from the leaves is drawn up from the tree's roots.

Dynamic tensile stresses are produced hydrodynamically and acoustically. Hydrodynamic cavitation results from the local reduction of pressure produced by the motion of a body in a liquid (Bernoulli effect), as would be found in the vicinity of a rotating propeller of a ship or near the surface of a fast submarine traveling at an insufficient depth for the hydrostatic pressure to suppress the flow-induced tension. Acoustically generated cavitation occurs when a sufficiently energetic acoustic wave produces periods of high negative pressure exceeding the ambient hydrostatic pressure, as can happen near the surface of a high-power acoustic transmitter used for long range communication in the ocean (see SONAR).

In both hydrodynamic and acoustic cavitation, the size of cavitation nuclei and the degree they are wetted by the liquid will dictate the cavitation threshold (which is usually very low in the hydrodynamic case); the number of cavitation nuclei will dictate the amount of cavitational activity. And in both hydrodynamic and acoustic cavitation, the duration of the period of tension will determine how big the bubbles grow before the bubble will collapse under the positive ambient hydrostatic pressure. Inside a collapsing bubble the pressure and temperature will rise, with the severity of the collapse increasing with maximum bubble size and with the scarcity of trapped permanent gas, which tends to cushion the collapse. Collapsing bubbles can generate in the surrounding liquid shock waves and hydrodynamic jetting. Such effects in the vicinity of a solid surface can lead to severe erosion, as is commonly found on screw propellers of ships.

In addition to these effects, a number of interesting phenomena can occur in acoustic cavitation. Temperatures inside a collapsing bubble in fairly outgassed liquid can reach thousands of degrees Celsius and pressures can reach thousands of atmospheres. Light can be produced (called *sonoluminescence*) and sonochemical effects may occur due to free radical formation inside the bubble. These transient phenomena can be attributed to the concentration of energy which occurs when bubbles grow to at least twice their original size in one or no more than a few acoustic cycles. For this to occur, peak acoustic pressures must be at least an atmosphere greater in magnitude than the hydrostatic pressure, and the acoustic frequency must be sufficiently low (as must be the liquid's viscosity) so as to allow for adequate bubble growth during the tension part of the cycle.

There is also a relatively more stable and periodic type of acoustic cavitation that occurs when a bubble filled with relatively inert gas (often air, in practical circumstances) oscillates with an amplitude that is small compared to the bubble size. It may slowly grow by a process called *rectified gas diffusion* when the acoustic pressure is only a fraction of an atmosphere. If the bubble reaches a resonant size (for example 6500 Hz for a 1-millimeter-diameter air bubble in water at an ambient pressure of one atmosphere), it will go through much larger oscillations, producing acoustic emission in the liquid not only at the driving frequency but also at its harmonics ($2f, 3f, \ldots$), subharmonics ($f/2, f/3, \ldots$), and ultraharmonics ($2f/3, 3f/4, \ldots$). The bubble may then become unstable and transient, eventually breaking up into smaller bubbles. The flow-induced effects induced in the liquid by this cavitational activity are used in a variety of practical applications, from removing embedded dirt (as with ultrasonic cleaners) to improving the efficiency of chemical reactions.

ROBERT E. APFEL

References

Knapp, R. T., Daily, J. W., and Hammitt, F. G., "Cavitation," New York, McGraw-Hill Book Co., 1970.
Neppiras, E. A., "Acoustic Cavitation," *Physics Reports* 61, 160–251 (1980).
Apfel, R. E., "Acoustic Cavitation," in Edmonds, P., Ed., "Ultrasonics," Vol. 19, pp. 356–413, New York, Academic Press, 1982.

Cross-references: ACOUSTICS, ULTRASONICS, LIQUID STATE.

CENTRIFUGE

The centrifuge is a device consisting of a rotating container in which substances of different size, shape, or density are separated by the centrifugal field developed along the radial direction. In most cases the centrifuge is used with fluids for producing sedimentation of macromolecules of molecular weight $M > 10^4$ grams/mole or daltons.[1,2] It has been used extensively also in the separation of gaseous components and for determining the strength of materials as well as a number of other applications.[3,4] Centrifugal fields as large as 10^9 times gravity have been achieved; however, for most purposes the fields are of the order of 10^4–10^5 times gravity. The effective centrifugal force F on a particle of mass m and density ρ in a fluid of density ρ', is expressed by $F = m(\rho - \rho')\,\omega^2 r/\rho$, where ω is the angular velocity and r is the radial distance of the particle from the axis of rotation. For a spherical particle, F is opposed by the frictional force of the fluid on the particle according to Stokes' law, $6\pi\eta a v$, where η is the coefficient of viscosity, a the radius and v the speed of the particle (provided that the Reynolds number, $2av\rho/\eta$, is less than unity and wall effects are neglected). Thus, the average rate of sedimentation v of spherical particles is described by

$$\tfrac{4}{3}\pi a^3 (\rho - \rho')\,\omega^2 r = 6\pi\eta a v = f v \qquad (1)$$

where f is the frictional coefficient. Nonspherical particles can be handled by the introduction of appropriate shape factors. Since the measured quantity in sedimentation is $v = dr/dt$ and the quantities ω, r, ρ, ρ', and η are also measurable, a and hence m can be determined. When m is expressed in terms of the molecular weight M, the buoyancy of the particles as a function of concentration C (mass/volume) is given by $(\partial\rho/\partial C)_\mu$, where ρ is now the density of the solution; this derivative is virtually constant for dilute macromolecules in a given solvent of much smaller molecular species. (In the general case for multicomponent solvents, such as when salts are added to reduce the effect of charge on the particles, the chemical potential μ of the solvent mixture is held constant rather than its composition as C changes.[5]) In dilute solutions the molar frictional coefficient is RT/D where D is the diffusion coefficient, which is measured independently. Thus, the equation for calculating M from the rate of sedimentation, in the absence of substantial charge effects, is

$$M = \frac{RT(dr/dt)}{D(\partial\rho/\partial C)_\mu \omega^2 r}. \qquad (2)$$

The rate of sedimentation in a unit field $(dr/dt)/\omega^2 r$ is called the *sedimentation coefficient* s, which has the dimension of time. The coefficient s is often expressed in Svedbergs (10^{-13} sec), a unit named in honor of T. Svedberg, the pioneer in analytical centrifugation. The value of s or the Svedberg is very important in characterizing a substance in solution.

As noted, the net transport of particles through a solvent medium in a centrifugal or gravitational field is opposed by the tendency to restore the previously distributed condition through the process of diffusion. Thus BROWNIAN MOTION from thermal agitation prevents sedimentable particles from settling out if v, although finite, is sufficiently small. The average displacement X of a particle in time τ owing to Brownian motion is $\overline{X^2} = 2D\tau$, and for a sphere of radius a the average velocity v_τ over a period τ is expressed by

$$v_\tau = \frac{\sqrt{\overline{X^2}}}{\tau} = \left(\frac{RT}{N}\,\frac{1}{3\pi\eta a\tau}\right)^{1/2}$$

where N is the Avogadro number. As v_τ becomes larger than v, particles will no longer settle out; i.e., net transport becomes zero at a sufficiently small centrifugal field so that an equilibrium distribution of the particles along the column of solution in the radial direction will be reached. When this equilibrium state is achieved between the tendency to sediment and to back diffuse, the concentrations at any two levels in the column can be related to M by

$$\ln\frac{C_2\gamma_2}{C_1\gamma_1} = \frac{M(\partial\rho/\partial C)_\mu \omega^2 (r_2{}^2 - r_1{}^2)}{2RT} \qquad (3)$$

where the numerical subscripts refer to two arbitrary positions in the radial direction and γ is the activity coefficient. In dilute solutions, as is usually the case in analytical work, $\gamma \to 1$ and the practical equation for determining the weight-average molecular weight M_w in sector-shaped cells (to avoid wall effects) is given by

$$M_w \approx \frac{2RT}{(\partial\rho/\partial C)_\mu \omega^2 (r_2{}^2 - r_1{}^2)}\,\frac{C_2 - C_1}{C_0} \qquad (4)$$

where C_0 is the initial concentration of the solution before sedimentation is initiated. The concentrations at radial distances r_1, r_2 are normally determined (by various optical means) at the meniscus and at the bottom of the solution column, respectively. Since charge effects may be large with sedimenting macromolecules, suppression of this effect is usually carried out by raising the ionic strength (via a neutral salt) if the system cannot be studied at the isoelectric point. In Eqs. (2)–(4) accurate densities as a function of C must be available from which to calculate $(\partial\rho/\partial C)_\mu$;[6] the chemical potential μ of the solvent is maintained constant for this purpose by first dialyzing to equilibrium the sedimentable species against the chosen solvent via a semipermeable membrane.[5]

Preparative Centrifuges For the purification and concentration of substances in solution or suspension in a fluid, the centrifugal field is made high enough so that appreciable separation of each material in a mixture is realized. Often one substance is sedimented to the bottom of the centrifuge cell (i.e., toward the periphery of the rotor), thereby concentrated as a pellet or simply enriched relative to the slower sedimenting substances which are present. Repeated centrifugation of the redissolved pellet material then further purifies it from the slower moving impurities. Conversely, the pelleted material may be the contaminant which is freed from the medium containing the desired substance. Centrifuges for these purposes are called *preparative centrifuges*; the variety, convenience, and sophistication of preparative instruments for a wide range of purposes in both industry and research is great indeed. Equations (1) and (2) are used for estimating the required rotor speed, size and style of rotor, and the time needed for achieving the desired degree of differential sedimentation of mixtures. These equations hold strictly only when the sedimentation is radial and so that no turbulence or remixing occurs. Moreover, the temperature must be uniform throughout the sedimentable column so that convection does not occur. The force generating thermal convection is roughly proportional to the density gradient along the column times the centrifugal field. Since the latter is often large, the temperature gradient must be small. In high-speed centrifugation the rotor is usually spinning in a good vacuum ($<10^{-5}$ torr) in a temperature-controlled instrument, so that heating and thermal gradients are avoided. In general, biomacromolecules, including enzymes, are not inactivated by the increase in pressure along the radial path in the fluid at high rotor speeds. Ordinarily, these pressures amount at most to a few hundred atmospheres, which are insufficient to irreversibly alter the structure of these macromolecules.

A major development in preparative centrifugation is toward the separation of substances along a density gradient. When a column of fluid is prepared such that the density increases smoothly along the sedimenting path and then a thin layer of solutes or suspended particles to be separated is applied to the top of the column, the solutes tend to segregate into distinct bands during the centrifugation. For the preparation of the density gradient, nearly inert, denser materials, such as sucrose, cesium chloride, or osmotically inert silica sols are often used and methods exist for constructing these gradients in the fluid column. This technique has become a great boon in several kinds of research. In some methods the mixture of test materials moves down the column until each material has reached a level where it becomes isodense with the layer of fluid of a particular density (isopycnic centrifugation). In this way each test material exhibiting a different density in solution concentrates as a thin band at different levels along the column because the density gradient, made up of much smaller molecules or stable sols, is approximately invariant relative to the transport of the macromolecules or larger entities in the centrifugal field. The isolation attempts with live cells, viruses, nucleic acids and proteins have met with remarkable success by the use of the isopycnic method.[7] For example, the clear separation of complementary strands of nucleic acid and the parent double-stranded molecule have been achieved owing to the fact that each complementary strand, while of the same size, has a slightly different density in solution or partial volume, $(\partial V/\partial m)_\mu$.[8] This quantity is contained in the buoyancy term of Eqs. (2), (3), and (4), wherein

$$\left(\frac{\partial\rho}{\partial C}\right)_\mu = \left[1 - \rho'\left(\frac{\partial V}{\partial m}\right)_\mu\right]$$

Analytical Centrifuges When a centrifuge is designed for analytical work it is called an *ultracentrifuge*.[1] The ultracentrifuge is employed in one of two general methods. The first method makes use of Eqs. (1) and (2), whereas the second is based on Eqs. (3) and (4); both methods are applied extensively. In the first method, called *velocity sedimentation*, comparatively large centrifugal fields (e.g., 2×10^5 times gravity) are generated in order to produce an easily measured rate of sedimentation v of a molecular species. The value of v is usually determined by optical means and s is computed. The latter is normalized to the rate (per unit field) under standard conditions, such as at $20°$ in a medium corrected to the density and viscosity of water ($s_{20,w}$). Because s depends on the concentration of the sedimenting molecules its value is further reduced to that at vanishing concentration ($s^0_{20,w}$) by extrapolating s at finite concentrations to that at $C = 0$. In this way different macromolecules can be compared on the basis of their rates of sedimentation. The slopes of the regression lines, s versus C, reflect on the shape and other properties of the sedimenting molecules and in their interactions with other chemical species present.[2] The high centrifugal field quickly produces a small density gradient along the radial path which stabilizes the liquid column. As a result, highly precise control of the temperature and speed of the rotor is not mandatory, although it is desirable. The time requirements of velocity sedimentation is comparatively small (1–2 hours). Also, if the solution contains more than one sedimenting species, the characteristic value of s for each can be measured by observing certain precautions.[2] Thus, the method is useful for comparing the effects of treatments on the product output and for following the degree of purification during isolation procedures. Historically, velocity sedimentation serves as a criterion for the degree of homogeneity of a purified macromolecular species. Furthermore, the self association and interaction of species in a mixture often can be determined. These and other aspects yielding important information have been studied both theoretically and experimentally on a broad front.[2,9,10] The velocity method, however, requires knowledge of D or another measurement reflecting f, which are determined less accurately than is s. Hence, the value of M is more uncertain than by the second method (below).

The second method, known as *equilibrium sedimentation*, is on more firm theoretical grounds, being based upon equilibrium thermodynamics. Also, the value of M, not being dependent upon a separately determined function of f, is more reliable. The equilibrium method permits an analysis of the various factors contributing to non-ideal behavior of solutions from which much valuable information can be gained about the total system. Further, the equilibrium method has been employed successfully in resolving equilibrium constants for self-associating and other interacting macromolecules in mixed systems.[9,10,11] In general, high purity of the sedimenting species is required. Also, this method usually entails relatively small rotor speeds in order that the test molecules do not sediment to the bottom but rather distribute at equilibrium along the column of solution as a function of the radial distance. Thus, much more time is needed to achieve the relevant data than by the velocity method. However, various procedures have been devised to shorten the time requirement. Molecules as small as $M \sim 10^3$ daltons as well as particles of $M > 10^7$ daltons have been studied successfully, the latter using fields of only a few gravities by use of a magnetically suspended rotor in order to reduce rotor precession effects to a safe level.[12] For larger species, the gravitational field of the earth can be used.[13]

Gas Centrifuge The gas centrifuge has been used for removing fine particles suspended in gases, for the separation of gaseous mixtures, and for the separation of isotopes, with the last of these being one of the most important large scale applications. The first suggestion of using a centrifugal field to separate isotopes was in 1919,[14] and the first successful demonstration was made in 1934 by J. W. Beams at the University of Virginia with the isotopes of chlorine.[15] In recent times, large scale gas centrifuge plants have been built for separating the isotopes of uranium to produce uranium enriched in the fissionable isotope U^{235} for fuel in light water power reactors. These plants employ tens or hundreds of thousands of centrifuges depending on the size and peripheral speed of the rotor.

When a centrifugal field is applied to an ideal gas, it sets up a pressure gradient

$$\frac{dp}{dr} = \frac{Mp}{RT} \omega^2 r \qquad (5)$$

where p is the pressure, M the molecular weight of the gas, R the gas constant and T the absolute temperature. For the case of an isothermal centrifuge, Eq. (5) is readily integrated to yield

$$p(r) = p(0) \exp (M\omega^2 r^2 / 2RT) \qquad (6)$$

which gives the pressure $p(r)$ at any radial position r in terms of the pressure at the axis $p(0)$. For a mixture of two ideal gases of molecular weights M_1 and M_2, each gas would have a pressure governed by Eq. (6) and the ratio of the two equations gives the radial separation under equilibrium conditions (i.e., no gas circulation). An equilibrium separation factor between the two gases is therefore given by

$$\alpha_0 = \frac{x_1(0)}{x_2(0)} \bigg/ \frac{x_1(a)}{x_2(a)}$$

$$= \exp [(M_2 - M_1) \omega^2 r^2 / 2RT] \qquad (7)$$

where x_1 and x_2 are the concentrations of species 1 and 2, respectively. The details of the theory of isotope separation by the gas centrifuge have been considered by numerous authors,[16-19] and the separative performance depends upon the internal circulation of the process gas which has been analyzed in detail.[20]

The operation of a gas centrifuge is illustrated in Fig. 1. The rotor is suspended at the bottom by a low friction bearing and at the top by a frictionless magnetic bearing. In the case of the separation of uranium isotopes, uranium hexafluoride (UF_6) is introduced into the spinning rotor from the stationary central post and removed from stationary pipes called scoops located at either end of the rotor. In practice gas centrifuges are spun in a vacuum at very high peripheral speeds, 400 m/s or greater, so that the process gas is compressed into a thin stratified layer adjacent to the cylindrical wall of the rotor. If we consider the process gas to be a binary mixture of the two isotopic species $U^{235}F_6$ and $U^{238}F_6$, then the heavier molecules containing U^{238} will tend to be concentrated near the cylinder wall and the lighter molecules containing U^{235} will tend to be concentrated near the axis.

In addition to removing mass, the stationary scoop at the bottom of the rotor induces a countercurrent flow by removing angular momentum and pumping the gas radially inward, forcing it to travel up near the axis and down along the cylinder wall. In order to prevent an opposing circulation, the scoop at the top is shielded by a baffle which rotates with the rotor and has holes which allow the gas to enter the scoop chamber and be removed through the scoop. A similar countercurrent flow can be induced by heating the bottom of the rotor and cooling the top of the rotor, which is analogous to the high altitude winds driven by thermal gradients in the earth's stratified atmosphere. This countercurrent flow produces a net transport of heavy isotopes to the bottom of the rotor and a net transport of light isotopes to the top of the rotor, which establishes a concentration gradient in the axial direction. Therefore, the gas removed by the top scoop is enriched in U^{235} (product) and the gas removed by the bottom scoop is depleted in U^{235} (tails). Cascades are formed by centrifuges connected in series to obtain the desired enrichment and in parallel to obtain the desired throughput.

The maximum theoretical separative capacity of a gas centrifuge has been shown to be[16,18]

$$\delta U(\max) = \frac{\pi Z \rho D}{2} \left(\frac{\Delta M \omega^2 a^2}{2RT} \right)^2 \qquad (8)$$

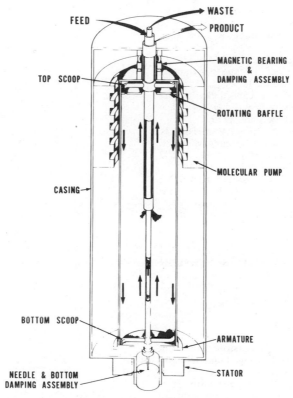

FIG. 1. Gas centrifuge.

where Z is the length of the rotor, ΔM is the difference in molecular weights, and δU is the separative capacity in moles per unit time. Therefore, long, high speed rotors are desirable features of the gas centrifuge process. The peripheral speed is limited by the strength of the rotor material. As in the case of a rod that is rotated, the rotor will have certain natural frequencies which will be determined by the materials of construction, the rotor length to diameter ratio, and the damping of the suspension systems. Centrifuges which operate at rotational frequencies which are below the lowest natural flexural frequency of the rotor are called *subcritical centrifuges*, and those which operate at rotational frequencies above the first flexural natural frequency are called *supercritical centrifuges*.

DONALD W. KUPKE
RALPH A. LOWRY
HOUSTON G. WOOD, III

References

1. Svedberg, T., and Pederson, K. O., "The Ultracentrifuge," Clarendon Press, Oxford, 1940.
2. Schachman, H. K., "Ultracentrifugation in Biochemistry," Academic Press, New York, 1959.
3. Beams, J. W., *Physics Today*, **12**, 20 (1959).
4. Lavanchy, A. C., Keith, F. W., Jr., "Encyclopedia of Chemical Technology" (R. E. Kirk and D. F. Othmer, Eds.), 3rd. Ed., Vol. 5, p. 194, Wiley (1979).
5. Casassa, E. F., and Eisenberg, H., *J. Phys. Chem.*, **65**, 427 (1961); *Advan. Protein Chem.*, **19**, 287 (1964).
6. Kupke, D. W., in "Physical Principles and Techniques in Protein Chemistry," Part C (S. J. Leach, Ed.), pp. 1–75, Academic Press, New York, 1973.
7. Meselson, M., Stahl, F. W., and Vinograd, J., *Proc. Nat. Acad. Sci., U. S.*, **43**, 581 (1957); Wolff, D. A., *Methods in Cell Biology*, **10**, 85 (1975).
8. Meselson, M., and Stahl, F. W., *Proc. Nat. Acad. Sci., U. S.*, **44**, 671 (1958).
9. Williams, J. W., Van Holde, K. E., Baldwin, R. L., and Fujita, H., *Chem. Revs.*, **58** (1958).
10. Fujita, H., "Mathematical Theory of Sedimentation Analysis," Academic Press, New York, 1962.
11. Williams, J. W., "Ultracentrifugation of Macromolecules," Academic Press, New York, 1972.
12. Weber, F. N., Jr., Elton, R. M., Kim, H. G., Rose, R. D., Steere, R. L., and Kupke, D. W., *Science*, **140**, 1090 (1963).

13. Weber, F. N., Jr., Kupke, D. W., and Beams, J. W., *Science*, **139**, 837 (1963).
14. Lindemann, F. A., and Aston, F. W., *Phil. Mag.*, **37**, 523 (1919).
15. Beams, J. W., *Rev. Mod. Phys.*, **10**, 245 (1938).
16. Cohen, K., "Theory of Isotope Separation," McGraw-Hill, New York, 1951.
17. Soubbaramayer, "Topics in Applied Physics: Uranium Enrichment" (S. Villani, Ed.), Vol. 35, p. 183, Springer-Verlag, New York, 1979.
18. Hoglund, R. L., Shacter, J., and Von Halle, E., "Encyclopedia of Chemical Technology" (R. E. Kirk and D. F. Othmer, Eds.), 3rd ed., Vol. 7, p. 639, John Wiley & Sons, New York, 1979.
19. Benedict, M., Pigford, L. H., and Levi, H. W., "Nuclear Chemical Engineering," 2nd Ed., McGraw-Hill, New York, 1981.
20. Wood, H. G., and Morton, J. B., *J. Fluid Mech.*, **101**, 1 (1980).

Cross-references: BROWNIAN MOTION, ISOTOPES, MOLECULAR WEIGHT, ROTATION–CURVILINEAR MOTION.

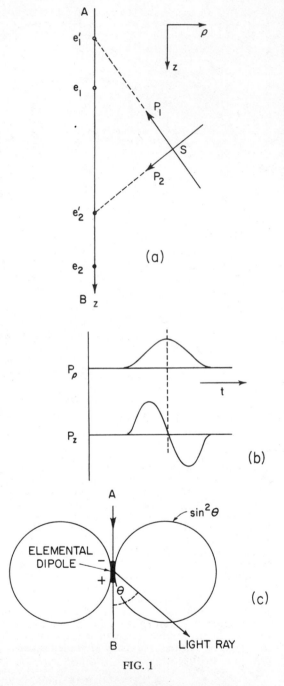

FIG. 1

ČERENKOV RADIATION

This is a feeble radiation in the visible spectrum, which occurs when a fast charged particle traverses a dielectric medium at a velocity exceeding the velocity of light in the medium. It is thus a shock-wave phenomenon, the optical analog of the "supersonic bang." The radiation arises from the local and transient polarization of the medium close to the track of the particle. Consider, Fig. 1(a), an arbitrary element S of the medium to one side of the track AB of a fast electron, the track defining the z-axis. At a particular instant of time, when the electron is at say e_1, the local polarization vector P_1 will be directed along $S\,e_1{}'$, to a point $e_1{}'$ slightly behind e_1, owing to the retarded fields. As the particle goes by, the vector P_2 will turn over and, when the electron reaches e_2, will be directed to a point $e_2{}'$. The variation of P with time, may be resolved into radial and axial components P_ρ and P_z, as shown in Fig. 1(b). Owing to cylindrical symmetry, this polarization, viewed at a point distant from the particle, appears as an elementary dipole lying along the axis z, Fig. 1(c). As the particle plunges through the medium, radiation arises from the coherent growth and decay of this sequence of elementary dipoles. Two essential features of the radiation become at once apparent. First, since it is only the P_z component which is important, the field variation, Fig. 1(b), is that of a double δ-function. Thus, from Fourier analysis, if the circular frequency is ω, we will expect a spectrum of the form $\omega \cdot d\omega$, i.e., radiation which is bluer than that from an equi-energy spectrum. Secondly, since the radiating element is an axial dipole, the angular distribution, for this element alone, will be of the form $\sin^2\theta$, Fig. 1(c). It is impor-

tant to realize that the radiation arises from the medium itself, not directly from the particle. Since the medium is stationary, the intensity and angular distributions do not contain the relativistic factor (mc^2/E); in this respect, it is essentially different from Bremsstrahlung or synchrotron radiation.

The description above applies only to one

element along the track. The most characteristic feature of Čerenkov radiation, its coherence, is at once apparent when we now consider an extended region of track. In Fig. 2(a) it is easily seen that there is only one angle θ at which it is possible to obtain a coherent wave front. If the velocity of the particle is v ($=\beta c$, where c is the velocity of light in vacuo), and n is the refractive index of the medium, the particle travels a distance AB in a time Δt, given by $AB = \beta c \cdot \Delta t$; in the same time the radiation, emitted at A, travels a distance $AC = (c/n)\Delta t$, from which we obtain the Čerenkov relation:

$$\cos \theta = (1/\beta n) \qquad (1)$$

From Eq. (1) it is at once evident that there is a threshold velocity given by $\beta = (1/n)$, below which no radiation takes place. At ultrarelativistic velocities, as $\beta \to 1$, the Čerenkov angle θ tends to a maximum value $\theta(\max) = \cos^{-1}(1/n)$. The polarization vectors E and H of the radiation which, owing to symmetry, takes place over the surface of a cone, are shown in Fig. 2(b).

The radiation yield, from the theory of Frank and Tamm, is

$$\frac{dW}{dz} = \frac{e^2}{c^2} \int_{\beta n > 1} \left[1 - \frac{1}{\beta^2 n^2} \right] \omega \cdot d\omega, \text{ergs/cm path}$$

$$(2a)$$

or

$$\frac{dN}{dz} = 2\pi \left(\frac{e^2}{hc} \right) \cdot \left[\frac{1}{\lambda_2} - \frac{1}{\lambda_1} \right] \cdot \sin^2 \theta$$

$$\text{photons/cm path} \quad (2b)$$

between wavelength limits λ_1 and λ_2 (in cm). The spectral distribution is $(dW/d\omega) \propto \omega$ or $(dW/d\lambda) \propto \lambda^{-3}$, expressed as energy per unit circular frequency or per unit wavelength, respectively. The radiation has, therefore, a continuous spectrum toward the blue and ultraviolet. There is no radiation in the x-ray region, for which $n < 1$. For example, in the case of a fast electron in water, $n = 1.33$, we find from Eq. (2b), that when $\beta \to 1$ and $\theta(\max) = 41°$, the yield (dN/dz) is ~ 200 photons/cm, between λ_1 and λ_2 of 3500 and 5500Å, respectively,

Before elaborating on the many applications of the effect, we will first trace a few of the theoretical developments following the original classical theory of Frank and Tamm,[2] which applied to isotropic transparent media, and which neglected quantum effects. A quantum treatment of the effect, which shortly followed,[3] revealed results which deviated slightly from those obtained by the classical treatment. These deviations in the threshold condition and in the angular and intensity distributions of the emitted light were, however, extremely small, as one might have expected, since $\hbar\omega \ll E$, where E is the kinetic energy of the particle; e.g., $\hbar\omega$ would be ~ 3 eV for blue light, while E would be typically $\sim$MeV.

There followed, in the forties and fifties, a spate of theoretical papers, mostly in Russian journals, extending the general theory to many specialized cases. These are too numerous to discuss in detail, but some of them will be mentioned in passing. If the medium is anisotropic (e.g., a crystal), or optically active, the threshold condition, angular distribution, intensity

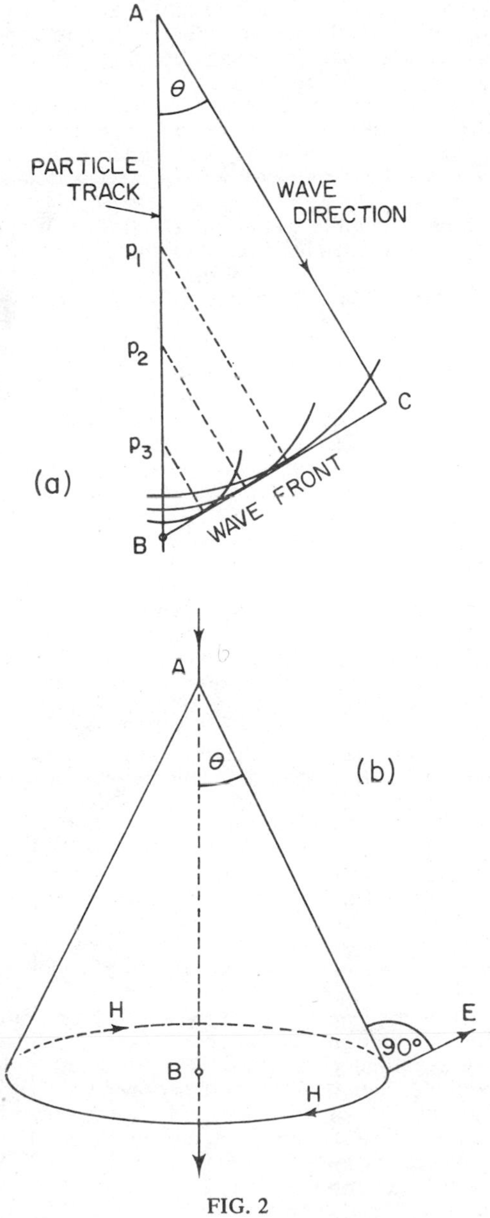

FIG. 2

distribution, and especially the polarization characteristics of the radiation, depend in a complex way on the relationship of the direction of the path of the particle relative to the orientation of the crystal axes (anisotropic media), or the axis of gyration (optically active media).

Much work was also devoted to cases in which the fast-moving single-point charge is replaced by one which also carries a magnetic moment, and to cases in which the moving particle is either an electric or magnetic dipole; continuing further, work was also carried out on the radiation caused by the motion of multipoles and oscillators through the medium. In the latter case, below the normal Čerenkov threshold, a straightforward Doppler effect is observed, while above the threshold, a "complex" Doppler effect occurs, but not truly related to Čerenkov radiation.

Another case of some interest, one which occurs mainly in the UV part of the spectrum, is the region of anomalous dispersion, where one meets a "complex" index of refraction; in this case the Čerenkov spectrum deviates from the smooth λ^{-3} law, above, and is broken up into bands which occur in those regions where the real part of the dielectric constant $\epsilon > \beta^{-2}$.

Most of these theoretical problems mentioned so far, apply to transparent media and to the optical region of the spectrum.

It was early realized, however, that Čerenkov radiation could also be expected to occur in ferrites (both isotropic and anisotropic), having a permeability μ, under the same conditions of threshold, i.e., $n = \beta^{-1}$, where n in this case is $(\epsilon\mu)^{1/2}$. This can occur in the microwave and radio-frequency regions of the spectrum. In these cases, referring back to Fig. 1, it is only necessary that the distance between a point S and the track of the particle shall be $<\lambda$, so that, in the microwave region of the spectrum, the particle need not pass through the "body" of the medium, but can travel down a tunnel, eliminating the otherwise accompanying ionization losses. Some enhancement of the radiation, poor in this case, because of the λ^{-3} relation, can however be achieved, if it is possible to "bunch" the electrons. Loaded waveguides are also potential media for Čerenkov radiation.

We mention briefly three other associated effects, that radiation, in limited bands, can occur in a plasma, that there are various diffraction and scattering effects,[4] and that the phenomenon of Čerenkov radiation has a certain relationship with that of transition radiation.[5]

Much of what we have discussed above is of mostly academic interest as far as applications are concerned, though these topics lay the foundations for the many applications.

While a summary of the results of these theoretical studies will be found in Ref. 6, the most comprehensive review of these aspects of the phenomenon will be found in the work of Bolotovskii.[7]

At this stage let us discuss the situation in which the medium is a gas rather than a liquid or solid. Take air for instance. In this case, we have $n = 1.000293$, $\theta_{max} = 1.3°$, E_{min}(electrons) = 21 MeV, and $dN/dl = 0.3$ photons cm^{-1} path.

These energy yields, both in gases and even in solids and liquids, are extremely small, relative to the ionization energy losses, which amount to ~ 2 MeV gm^{-1} cm^{-2} for relativistic particles. However, while the energy yields in Čerenkov radiation are small compared with the ionization losses, the emitted photons, having energies ~ 2–3 eV, are so numerous, and light detectors such as photomultipliers so efficient, that it is possible to utilize the effect in many branches of high-energy nuclear and cosmic-ray physics.

Let us now turn to the experimental and application aspects of the phenomenon of Čerenkov radiation. While the early experiments carried out by Čerenkov himself were based on relatively strong radioactive sources and used the photographic technique, the first detection of *single* charged particles in a liquid is attributed to the writer,[8] as is the first detection of the Čerenkov light from single cosmic-ray particles in air,[9] in conjunction with a colleague.

While most of the theoretical work on the topic has been carried out in the USSR, it would appear that the West has contributed by far the larger fraction of the work on the experimental side.

The phenomenon has found considerable application in the fields of high-energy and nuclear physics, and in cosmic-ray research. The unique directional and threshold properties of the radiation may be used in a number of different ways. For example, by velocity selection, it is possible to distinguish between particles of different mass having the same energy, and it is also possible to measure particle velocities directly, by measuring θ. Other examples may be cited: The e^2 dependence, Eq. (2) above, has been used to determine the charge spectrum of the primary cosmic rays, and transparent lead-loaded glasses have been developed as total absorption spectrometers for high-energy γ-rays.

Reference 6 contains a review of some of the early types of Čerenkov detector, while Refs. 10 and 11 cover somewhat later developments.

More recently, developments have taken place in very sophisticated detectors, especially ring-imaging systems based on gaseous media which have proved to be especially valuable in high-energy particle physics.[12,13]

Čerenkov radiation in the atmosphere produced by cosmic-ray showers from ultra-high-energy primary particles[14] has become a most valuable tool in high-energy γ-ray astronomy. This in fact provides the only tool, and it is ground-based, for searching for point sources of γ-rays in the energy region 10^{11}–10^{13} eV.[15,16]

A further application of Čerenkov radiation is the search for neutrinos of cosmic origin in seawater, in a deep ocean trench off the island of Hawaii, a project known as DUMAND (Deep Underwater Muon and Neutrino Detector),[17] the volume of water involved in this detector being ~1 km³. A somewhat similar type of experiment is likewise under investiagation in attempt to observe the decay of the proton.[18]

Three further aspects of the phenomenon will also be made in passing: (1) the application of the effect to produce extremely weak standard light sources for testing astronomical equipment;[19] (2) the interpretation of a particular type of radio emission from sunspots;[20] and (3) the inverse Čerenkov effect,[21] proposed as a way of accelerating charged particles.

By contrast, there have been two applications in biological areas. One of these is that flashes of light observed by astronauts involved in the NASA Apollo program during periods of translunar flight have been attributed, partially or possibly totally, to Čerenkov radiation produced by primary cosmic rays of relatively high nuclear charge traversing the retinas of the eyes of the astronauts on these missions.[22]

And finally we mention that there is evidence[23] that Čerenkov radiation may be a contributing or even dominant effect in the repair mechanism by photoreactivation of the DNA in the bacterium *Esherichia coli* subsequent to damage by ionizing radiation.

J. V. JELLEY

References

1. Čerenkov, P. A., *C. R. Acad. Sci (USSR)* **2**, 451 (1934).
2. Frank, I. M., and Tamm, I., *C. R. Acad. Sci. (USSR)* **14**, 109 (1937).
3. Ginsburg, V. L., *Zh. Fiz. SSSR* **2**, 441 (1940).
4. Dedrick, K. G., *Phys. Rev.* **87**, 891 (1952).
5. Frank, I. M., and Ginsburg, V., *Zh. Fiz. SSSR*, 9(5), 353 (1945).
6. Jelley, J. V., "Čerenkov Radiation and its Applications," Pergamon Press, London, 1958.
7. Bolotvskii, B. M., *Usp. Fiz. Nauk* 75(2), 295 (1961).
8. Jelley, J. V., *Proc. Phys. Soc.* **A64**, 82 (1951).
9. Barclay, F. R., and Jelley, J. V., *Nuovo Cimento*, Series 10, 2(1), 27 (1955).
10. Hutchinson, G. W., *Prog. Nucl. Phys.* **8**, 195 (1960).
11. Fabjan, C. W., and Fischer, H. G., *Rep. Prog. Phys.* 43(8), 1003 (1980).
12. Gilmore, R. S., Malos, J., Bardsley, D. J., Lovett, F. A., Melat, J. P., Tapper, R. J., Giddings, D. I., Lintern, L., Morris, J. A. G., Sharp, P. H., and Wroath, P. D., *Nuc. Instrum. and Methods*, **157**, 507 (1978).
13. Ekelot, T., Sequinot, J., Tocqueville, J., and Ypsilantis, T. *Physica Scripta* **23**, 718 (1981).
14. Jelley, J. V., *Phil. Trans. Roy. Soc. London.* **A301**, 611 (1981).
15. Porter, N. A., and Weekes, T. C., Smithsonian Astrophys. Obs. Spec. Rep. No. 381 (1978).
16. Turver, K. E., and Weekes, T. C., *Phil. Trans. Roy. Soc. London* **A301**, 615 (1981).
17. Roberts, A., Ed., "Proc. 1978 DUMAND Summer Workshop," DUMAND Scripps. Inst. Oceanogr., 1978.
18. News Section, "Search and Discovery," *Physics Today* **33**, 17 (1980).
19. Anderson, W., and Belcher, E. H., *Brit. J. Appl. Phys.* 5(2), 53 (1954).
20. Marshall, L., *Astrophys. J.* **124**, 469 (1956).
21. Veksler, V. I., *CERN Symposium, Geneva*, **1**, 80 (1956).
22. Fazio, G. G., Jelley, J. V., and Charman, W. N. *Nature* **228**, 260 (1970).
23. Moss, S. H., and Smith, K. C., *Int. J. Radiat. Biol.* 38(3), 323 (1980).

Cross-references: COHERENCE, DIELECTRIC THEORY, DIPOLE MOMENTS, FOURIER ANALYSIS.

CHEMICAL KINETICS

Besides offering useful rate equations to describe the speeds of chemical reactions, chemical kinetics attempts to describe exactly how each reaction occurs. It does so in terms of one or more elementary steps, which are reactions having no observable intermediate chemical species. The ultimate goal is a theory interrelating energy, structure, and time for these single chemical events. Many of the ideas, developed since 1850 when the first quantitative rate study was made, have been extended by analogy to explain electron-hole processes in semiconductors, various solid-state processes, and thermonuclear reactions.

Reaction rates depend on the nature of the reactants, temperature, pressure, kind and intensity of radiation, nature of catalyst or solvent, and many other factors. The extent of a reaction can be followed by withdrawal of samples for early chemical analysis. It is more common, however, to analyze the main reaction mixture continuously and nondestructively by spectroscopic means or by observing physical properties like density, electrical conductivity, optical activity, dielectric constant, and so on.

The rate v of a reaction $aA + bB \rightarrow eE + fF$ is best related to the rate of change of the concentrations [A] and [B] of reactants by rewriting the equation as $0 = eE + fF - aA - bB$. If the change is an elementary step, its rate is

$$v = -\frac{1}{a}\frac{d[A]}{dt} = -\frac{1}{b}\frac{d[B]}{dt} = +\frac{1}{e}\frac{d[E]}{dt} =$$

$$+ \frac{1}{f}\frac{d[F]}{dt} = k[A]^a[B]^b$$

Here k is a rate constant independent of concentration. If the step is not elementary, its rate is often presumed to be expressible as proportional to certain empirically observed powers $m, n, \cdots$ of the concentrations of species present in the reaction vessel. The values of m, n, $\cdots$ need not be integers and cannot be predicted from the balanced chemical equation if the step is not elementary. The over-all order of a reaction is the sum of these exponents, and the order with respect to a particular species is its own exponent. The order of a reaction can be determined in several ways. In the method of initial rates, concentrations of all but one reactant are held constant, if possible at great values, and v is observed at the start of reaction for several values of [A]. The order with respect to A is the slope of a graph of log v vs. log [A].

The first major reaction rate theory was founded on the kinetic theory and classical mechanics. It still is very useful when reactants approach each other in an attractive potential field (e.g., ions), when the distribution of energies is nonequilibrium (e.g., electrical discharge in gases), or when the molecules involved are tremendous in size. In general, however, the collision theory suffers from its inability to satisfactorily predict effective cross sections (molecular sizes for reaction) or the effect of isotopic substitution.

Most modern theories suppose the existence of an undetectable transition state of high energy and fleeting existence. The configuration of this activated complex lies, as it were, atop the mountain pass of lowest height between energy-valleys of reactants and products. It is mechanically stable to all vibrations except the one that describes the progress of reaction over the saddle-point at the pass. This one motion is assigned a very low (sometimes imaginary) frequency, but otherwise the activated complex is just another molecule. It is supposed to be in dynamic equilibrium with reactants and its free energy can be calculated from its partition function by the usual methods of STATISTICAL MECHANICS. The rate constant k_n then takes the form $(kT/h) \exp(-\Delta A/kT)$ where k is the gas constant per molecule, T is the absolute temperature, h is Planck's constant, and ΔA is the increase in free energy on going from reactants to activated state. This resembles the well-known relation $k_n = s \exp(-E/kT)$ discovered empirically in 1889 by Arrhenius. In it, s and E are approximately independent of T, and E is called the activation energy.

A recent theory of promise treats reactants as a wave packet that gradually spreads in time. The rate of reaction is taken to be the probability that the packet will be found in a configuration that is indentified with products.

Thermal decomposition of an initially pure gas is seemingly a simple change, yet its order often changes gradually from first to second as the pressure falls. Moreover, there is always the question why like molecules do not all decay at once. The answers lie in understanding the mechanism, which is presently taken to be collision with any other molecule M ($0 = A^* + M - M - A$ with rate constant k_2') to yield an energized molecule A^* that may suffer a stabilizing collision ($0 = A + M - M - A^*$, with k_2) or internal change ($0 = A\ddagger - A^*$, with k_1) that leads to the activated state $A\ddagger$. The reaction is called unimolecular because the activated complex $A\ddagger$ contains only one reactant molecule. The rate of decomposition of A at any instant is $v = k_2'[M][A] - k_2[M][A^*]$, and the rate of decomposition of A^* is $v^* = k_1[A^*] - k_2'[M][A] + k_2[M][A^*]$. Since the v's are time derivatives, these are simultaneous differential equations.

It is generally impossible to solve the simultaneous differential equations that describe a mechanism. The least restrictive and most useful simplification is generally the steady-state approximation, wherein the concentration of a species of low concentration is assumed to reach an effectively constant value after a certain reasonable time (induction period) has passed. If [A*] reaches a steady-state concentration, $v^* = 0$ and the v^* differential equation becomes an algebraic one for [A*]. Moreover, the rate of decomposition of A then is $v = k_1[A^*] = k_1 k_2'[M][A](k_1 + k_2[M])^{-1}$. This rate equation becomes second order if $k_2[M] \ll k_1$; this occurs at low pressure or when A^* changes rapidly into $A\ddagger$. If A is a simple molecule of few atoms, the activation energy easily becomes effective in one bond to cause decay. On the other hand, the order becomes first if $k_2[M] \gg k_1$; this corresponds to high pressure or an A with many degrees of freedom to accommodate the activation energy.

A generally more restrictive way to simplify the mathematics of a sequence of reactions is to assume that one step is so much slower than the others that it alone limits the rate. All steps besides the rate-limiting one are assumed to be at equilibrium in this approximation. If, in unimolecular decomposition, the rate-limiting step is $A^* \to A\ddagger$, then the rate is $k_1[A^*]$. The reaction is then first order in A because $v = 0$ for the equilibrium $A + M = M + A^*$. If, however, the rate-limiting step is $A + M \to A^* + M$, then the rate is $k_2'[M][A]$.

Bimolecular reactions, wherein the activated state consists of two reactant species, are very common. The rates of the fastest of these are limited by the rates of diffusion of reactants and have rate constants of the order of 10^{10} liter mole^{-1} sec^{-1} in aqueous solution. Typical examples are the aqueous neutralizations $NH_4^+ + H_2O \to NH_3 + H_3O^+$ and $H_3O^+ + F^- \to HF + H_2O$. Typical bimolecular gaseous reactions are the linear chain reactions $X + H_2 \to HX + H$ and $H + X_2 \to HX + X$ where X is H, D, Cl, Br, or I. A nice way to initiate these reactions of H_2 and X_2 is by a photon: $X_2 + \text{photon} \to X + X$. Thermal dissociation of X_2 is also sufficient to start reaction.

Carbon compounds undergo many reactions, but most of them can be classified into a few types. Nucleophilic substitution, wherein a basic reactant replaces another initially on C by a net reaction $X + RY \to XR + Y$, may be first order in RY alone or in both X and RY. If first order in just RY, the mechanism is labeled S_N1 and the rate-limiting step is conceived as production of the active carbonium ion R^+ by the process $RY \rightarrow R^+ + Y^-$. If second order (S_N2), the rate-limiting step is considered to be production of the bimolecular activated complex $X \cdots R \cdots Y$. Elimination reactions typically yield a double bond with loss of part of the organic reactant RCH_2CH_2Y. If first order in organic reactant (type E_1), the rate-limiting step is said to be production of the carbonium ion $RCH_2CH_2^+$, which then swiftly eliminates H^+ to become $RCH = CH_2$. If first order in both base and organic reactant (type E_2), the rate-limiting step is thought of as production of a bimolecular complex which eliminates H^+ and Y^- almost simultaneously. A fifth class of organic reaction (S_Ni) describes how an electrophilic reagent (e.g., NO_2^+ in mixed HNO_3 and H_2SO_4 or Br^+ in Br_2 with $FeBr_3$) may attack an aromatic ring like that in benzene to form a positively charged intermediate that soon loses H^+ to a base in the solution.

A catalyst is a species that changes the rate of a reaction and yet is regenerated by that reaction so that it seems to be unchanged in the net reaction. Catalysts do not affect the equilibrium state but they do lower the activation energy and sometimes may provide a needed steric arrangement. The most general mechanism of catalysis is $A + C_1 \rightarrow D + C_2$ followed by $B + C_2 \rightarrow E + C_1$ to give the net change $A + B \rightarrow D + E$. Many so-called catalysts of industry need regeneration ($C_2 \rightarrow C_1$) by a reaction other than that catalyzed. For example, silica-alumina cracking catalysts used in making gasoline must be cyclically burned free of carbon deposited during cracking. A catalyzed reaction is almost always first order in catalyst concentration (or surface area) and usually has an order that is less than the true order by unity.

Enzymes are biological catalysts. Many act by the well-known Michaelis-Menten mechanism $E + S \rightleftharpoons C \rightarrow P + E$, where enzyme E attacks substrate S with a rate $k_A[E][S]$ to form a complex C that may yield products P at a rate $k_P[C]$ or may revert to S with rate $k_R[C]$. The rate of disappearance of S is $v_S = k_A[E][S] - k_R[C]$ and the rate of appearance of C or disappearance of E is $v_E = k_A[E][S] - k_R[C] - k_P[C]$. The total concentration of enzyme in the system is $[E]_0 = [C] + [E]$. In the steady state, $v_E = 0 = k_A([E]_0 - [C])[S] - (k_R + k_P)[C]$ so that

$$v_S = k_P[C] = \frac{k_P k_A [S][E]_0}{(k_R + k_P) + k_A[S]}.$$

The rate of disappearance of S is always first order in total enzyme but may change from first to zero order in S as [S] increases. The maximum rate at which E can act occurs when [S] is great and $v_S = k_P[E]_0$.

Much industrial research is devoted to finding catalysts which speed (or slow) specific desired (or undesired) reactions. Academic research on rates tends, however, to search for mechanisms and to propose models useful for prediction. Very fast reactions are of particular interest and they offer several kinds of challenge to experimenters. If a decrease in temperature does not affect the mechanism, a very fast reaction can perhaps be slowed to fit the response time of the analytical devices in use. Typically for analysis in real time one uses visible and ultraviolet spectrometers or a mass spectrometer. If a reaction is fast, the reactants must be mixed in a time very short relative to the time of the reaction; for times less than a millisecond, reactants are generally mixed before reaction, which is later initiated by a pulse of pressure or radiation (e.g., flash photolysis) or introduction of a catalyst. Or, instead of working in real time, a reaction may be maintained in a steady state (e.g., a flame) with concentrations noted as a function of position, with of course some direct link between position and time. For reactions that occur in a time of the order of 10^{-3} sec, typical experimental methods are: nuclear magnetic resonance, sudden changes in pressure or temperature, flash photolysis, shock waves, and various methods of controlling flow. For reactions done in 10^{-6} sec, one has: electric field displacement, ultrasonic methods, electron paramagnetic resonance, dielectric relaxation, and pulsed radiolysis. A few special techniques (e.g., fluorescence quenching) are useful at 10^{-9} sec. In liquid solution, the rate of a fast reaction may be independent of the nature of the reactants and become, instead, a function of liquid viscosity, with a temperature coefficient appropriate to viscosity's change with temperature. Rate of reaction is thus controlled by diffusion; for this, the activation energy of the chemical change seldom exceeds about 20 kJ per mole.

Very fast gaseous reactions and their mechanisms are often studied by crossing molecular beams in a vacuum. A heated reservoir with small hole allows molecules to effuse (ca. 10^{-7} torr). A beam is defined by slits or orifices, and a selection of velocities is then made by rotating sectors in the beam, which at the start contains the usual Boltzmann distribution of velocities. For easier detection, a much more intense beam of high-energy molecules can be made by fluid flow through a supersonic nozzle, but extremely rapid pumping is needed to maintain the necessary vacuum. The simplest collision in crossed molecular beams is "elastic," with momentum and translation energy conserved by whole molecules. Inelastic collision modifies their internal energies. Reactive scattering of beam by beam involves a transfer of atoms during an encounter. To link the results of such scattering to a bulk reaction rate, one must average properly over angles, energies, and so on. While typical reactive collisons in gases occur at distances of the order of (or less than) 100 Å, certain other reactions involving a preliminary transfer of an electron may occur at distances of the order of 200 Å. These last are rather poetically called "harpooning."

WILLIAM F. SHEEHAN

References

Most physical chemistry textbooks contain introductions to chemical kinetics. Some of their many authors are: R. A. Alberty and F. Daniels, G. M. Barrow, G. W. Castellan, S. Glasstone, I. N. Levine, E. A. Moelwyn-Hughes, W. J. Moore, R. M. Rosenberg, and W. F. Sheehan.

Ausloos, P. (Ed.), "Kinetics of Ion-Molecule Reactions," New York, Plenum Press, 1979.

Avery, H. E., "Basic Reaction Kinetics and Mechanisms," London, The Macmillan Press, 1974.

Benson, S. W., "The Foundations of Chemical Kinetics," New York, McGraw-Hill Book Co., 1960.

Bradley, J. N., "Fast Reactions," Oxford, Clarendon Press, 1975.

Erdey-Grúz, T., "Kinetics of Electrode Processes," New York, Wiley-Interscience, 1972.

Glasstone, S., Laidler, K. J., and Eyring, H., "The Theory of Rate Processes," New York, McGraw-Hill Book Co., 1941.

Hinshelwood, C. N., "The Kinetics of Chemical Change," London, Oxford University Press, 1940.

Mulcahy, M. F. R., "Gas Kinetics," New York, John Wiley & Sons, 1973.

Roberts, D. V., "Enzyme Kinetics," London, Cambridge Univ. Press, 1977.

Slater, N. B., "Theory of Unimolecular Reactions," Ithaca, NY, Cornell Univ. Press, 1959.

Wilkinson, F., "Chemical Kinetics and Reaction Mechanism," New York, Van Nostrand Reinhold Co., 1980.

There are several timely review articles in various volumes of "Advances in Chemical Physics," I. Prigogine, Ed., London and New York, John Wiley & Sons, 1958.

Values of rate constants for specific reactions are listed in National Bureau of Standards Circular 510, its supplements, and in NBS Monograph 34. For critically evaluated rate constants and mechanisms, see S. W. Benson and H. E. O'Neal, "Kinetic Data on Gas Phase Unimolecular Reactions," NSRDS-NBS 21 (Feb. 1970), Washington, D.C., U.S. Government Printing Office.

Cross-references: CHEMISTRY, PHYSICAL CHEMISTRY, STATISTICAL MECHANICS.

CHEMICAL PHYSICS

Chemical Physics is a term used to describe a rather broad set of research activities that fall on the borderline between the traditional disciplines of chemistry and physics. As with all interdisciplinary endeavors, it is difficult to arrive at a precise definition of chemical physics that is generally acceptable to the practioners of the field and that remains meaningful over an extended period of time. In particular, the distinction between chemical physics and physical chemistry is rather arbitrary, depending more on the background and training of the individual conducting the work than on the nature of the work itself. Many universities have a curriculum option that requires formal training in both chemistry and physics departments. The graduates of such programs tend to think of themselves as chemical physicists; others, who have a primarily chemistry background, may identify themselves as physical chemists even though their research is very similar.

Chemical physics did not emerge as an identifiable research field until well into the twentieth century. The atomistic approach to chemistry was introduced by Dalton and others at the start of the nineteenth century, and important contributions were made by physicists such as Avogadro. However, the lack of a detailed model for the atom and molecule made it impossible to relate the microscopic structure of matter to macroscopic behavior. Thus, physics and chemistry pursued their own evolutionary development through most of the nineteenth century, with rather limited interaction.

These paths began to intersect at the turn of the century: J. Willard Gibbs, a professor of mathematical physics, developed a powerful tool for chemists in the form of statistical mechanics; Peter Debye created detailed physical models that explained the properties of electrolyte solutions in terms of forces between electrically charged particles; William and Lawrence Bragg developed x-ray techniques for determining the structure of crystals, which opened the chemistry of solids to detailed investigation for the first time; Niels Bohr recognized that the theory which explained atomic spectra should also permit a physical explanation of the periodic table of the elements. With the advent of quantum mechanics in the 1925–30 period, physics and chemistry were drawn so closely together that there was no longer a definable boundary. Quantum mechanics provided for the first time a quantitative theory capable of explaining the forces between atoms that lead to the formation of molecules. During a remarkably prolific period the foundations were laid for a detailed understanding of molecular structure, the kinetics of chemical reactions, the relation of thermodynamic properties to structure, the interaction of light with matter, and many other subjects of long-standing importance in chemistry.

Another important factor was the development by physicists of instruments that allowed chemical phenomena to be probed in new ways. Infrared and mass spectrometers were used successfully by chemists in the 1930s, and the postwar period saw the introduction of more and more sophisticated instrumentation—microwave, nuclear magnetic resonance, electron spin resonance, and Mössbauer spectrometers, to name only a few. More recently, the laser has opened many new avenues for chemical research. In a pattern which has been repeated many times, a new instrumental technique is developed by physicists, applied first to problems of current interest in physics (usually involving atoms or simple molecules), and then adapted by chemists to study larger molecules and more complex chemical phenomena.

The birth of chemical physics as an identifiable research field can perhaps be dated from the establishment of *The Journal of Chemical Physics* in 1933. The driving force for this event was the rapid increase in research papers, resulting from the new theoretical and experimental tools described above, which could not be accommodated by the traditional journals of either physics or chemistry. More than any other single measure, the contents of this journal have defined the scope of chemical physics. More recent journals and publication series using the term *chemical physics* in the title have tended to adopt the same scope and emphasis as *The Journal of Chemical Physics*.

It is therefore instructive to analyze the contents of *The Journal of Chemical Physics*. The articles appearing in the journal are currently sorted into five categories, which are listed below along with the approximate distribution of papers in early 1982:

Spectroscopy and light scattering	35%
Molecular interactions and reactions, scattering, photochemistry	22%
Statistical mechanics and thermodynamics	20%
Quantum chemistry, theoretical electronic and molecular structure	18%
Polymers, surfaces, and general chemical physics	5%

Titles of papers on spectroscopy show emphasis on topics such as laser-induced fluorescence, double resonance, and other techniques that permit measurement of the fine details of molecular structure. Studies of energy distribution in molecules, ionization processes, and reaction kinetics as determined by molecular beam experiments are prevalent in the second category. The application of statistical mechanics to the understanding of phase transitions and critical phenomena is now an active topic. Papers in quantum chemistry range from diatomic molecules to complex aromatics but stress the calculation of the full range of molecular properties, not only energy levels and stability. The relatively small number of papers classified as surface physics is somewhat misleading, in view of the growing interest of chemical physicists in this area of research. The explanation probably lies in the existence of a number of competing journals.

Another measure of the topics currently emphasized in chemical physics research is provided by a 1981 survey of members of the Division of Chemical Physics of the American Physical Society. This division has a membership of about 2,800, which gives a rough idea of the number of scientists in the United States who regard themselves as chemical physicists. The areas of special interest were ranked as follows in the survey:

Spectroscopy and light scattering
Molecular interactions and reactions
Photochemistry
Surfaces
Statistical mechanics and thermodynamics
Scattering processes
Polymers

Here the importance of surface science is more evident.

In recent years the overlap between chemical physics and solid state (condensed matter) physics has become stronger. A number of ideas derived from the chemist's traditional approach to bonding have been introduced into solid state physics. Likewise, the distinction between atomic physics, which in the past focused on interactions between electrons, photons, and atoms, and the part of chemical physics that deals with molecular interactions and reactions has become blurred. There is also a trend toward a closer relation between chemical physics and biophysics.

In summary, chemical physics is a research area of somewhat diffuse and fluid boundaries which is characterized by use of the theoretical and experimental tools from physics to further the understanding of phenomena of interest to chemistry. It bridges the gap between two historically different approaches to natural science, whose distinction is gradually disappearing.

DAVID R. LIDE, JR.

References

The Journal of Chemical Physics (American Institute of Physics, New York).

Chemical Physics (North Holland Publishing Company, Amsterdam).

Chemical Physics Letters (North Holland Publishing Company, Amsterdam).

Advances in Chemical Physics (John Wiley & Sons, New York).

Cross-references: CHEMISTRY; MATHEMATICAL PHYSICS; PHYSICAL CHEMISTRY; PHYSICS; THEORETICAL PHYSICS.

CHEMISTRY

Chemistry is the branch of natural science that is concerned with the nature, composition, and structure of matter. The chemist attempts to provide a consistent model of the macroscopic behavior of materials in molecular terms. Because of the focus on the behavior of matter, chemistry is in a central position in the array of natural sciences and has strong interdisciplinary ties to both physics and biology. In fact the core of conceptual models of chemistry is familiar to the physicist and includes primarily quantum mechanics and thermodynamics.

The drive to understand matter at the molecular level has lead the chemist to follow the age old goal of transformation of matter both at the molecular and, more recently, the atomic level. In this field man is able to synthesize new combinations of elements, substances and materials. Actually, chemists have created in the laboratory several chemical elements (43, 61, 87, and 93 to 105 inclusive) that apparently are not normally present in or on our planet. Even a cursory examination of the concept of isomerism (the existence of two or more different compounds that have identical compositions) yields the conclusion that literally trillions and trillions of different compounds of carbon could be synthesized. In fact, there are possible at least 62×10^{12} different compounds (called isomers) all having an identical composition indicated by the formula $C_{40}H_{82}$. This specific combination of atoms is just one of millions of other possible combinations, and some of these would have thousands of isomers. The current chemical literature contains documented evidence for either the existence, or the synthesis by man, of well over a million different compounds of carbon. And carbon is only one of the 105 known chemical elements.

As with most areas of science, the explosive growth of chemical knowledge in the twentieth century, particularly since the Second World War, has lead to the development of several subdisciplines of chemistry. The four main areas which are generally recognized are analytical, inorganic, organic, and physical chemistry. Interfacial areas bridging the natural sciences range from medicinal and biochemistry to geochemistry, astrochemistry, theoretical chemistry, and chemical physics.

A unique and important factor associated with chemistry is the existence of a strong industrial connection. The chemical industry is one of the major components in the economy of any developed nation. The continually increasing need for commodities such as structural materials, fabrics, fertilizers, pesticides, and pharmaceuticals will enhance the chemical industry. The consumption of sulfuric acid, for instance, has for several decades been considered a significant indication of general industrial activity. This strong academic/industrial link is shown in the development of various subdisciplines such as polymer and industrial chemistry.

Historical Background The history of alchemy is a fascinating record of man's earliest investigations of matter. However, the mixture of rational and mystic approaches hampered significant progress. Consequently, it was not until the rationalist eighteenth century that the serious origins of chemistry developed. Robert Boyle and Antoine Lavoisier, two of the earliest proponents of exact quantitative experimentation, exerted profound influences on late eighteenth and early nineteenth century chemistry. Their techniques were mainly physical measurements.

Around the middle of the nineteenth century, Friedrich Kekulé and Archibald Couper independently proposed a system for writing graphic formulas for chemical compounds. Their concepts were based apparently on the notion that physical forces held together the atoms in compounds. About the same time, general acceptance was finally accorded Amedeo Avogadro's hypothesis (proposed about 50 years earlier) which stated that equal volumes of gases at the same temperature and pressure contain equal numbers of molecules. The importance of the establishment of a direct link between observable and molecular scale behavior contained in Avogadro's hypothesis cannot be overestimated.

In addition to this interest in conceptual models, considerable activity revolved about phenomenological models. These studies culminated in the development of the periodic table of the elements by Mendeleev. In this scheme regularities in the chemical and physical behavior are organized and provided a powerful predictive tool for the discovery of new elements, compounds, and physical properties.

During the period 1860-1920, the first great period of synthesis of new compounds occurred. Both organic and to a lesser degree inorganic chemistry flourished, and hundreds of compounds were prepared, correctly analyzed and identified, and logically classified as to structure and reactivity. And yet, throughout this period, chemists had no knowledge whatsoever about the structure and composition of atoms. Also they had a rather shallow conception of chemical bonding and molecular geometry. The growth of organic chemistry during the nineteenth century is an epic example of the productivity of sound inductive and deductive reasoning.

As was the case with physics, the quantum revolution in the early decades of the twentieth century had a profound effect on the direction and development of modern chemistry. Although the influence of quantum theory is pervasive throughout chemistry there are three particularity significant cases which bear mentioning. The organization of elements in the periodic table is directly correlated to electronic configurations hence establishing electronic structure as the basis of chemical behavior. Due to the pioneering efforts of Lewis, Pauling, and others, quantum theory provides a basis for both a qualitative and quantitative understanding of the chemical bond. The explosive growth of various forms of atomic and molecular spectroscopy which accompanied the development of quantum theory has added a significant battery of methodologies which are available for modern chemical research. These techniques have proven invaluable both in the measurement of amounts of materials present in a given sample and in the elucidation of molecular structure. Types of spectroscopic measurements include: atomic absorption/emission, ultraviolet, photoelectron, fluorescence, infrared, microwave, nuclear magnetic resonance, electron spin resonance, nuclear quadrupole resonance, and Mössbauer. The last mentioned technique is an interesting example of the rapidity of information transfer between physics and chemistry in that within a few years after Mössbauer's discovery of the effect, it had become a significant tool for investigation of the geometric and electronic structure of iron and tin compounds. Other, nonspectroscopic, physical methods are also of importance. In particular, one can cite mass spectrometry, which is of value in the determination of molecular structure and the study of isotope masses and diffraction methods (x-ray and electron) for the determination of geometrical structure.

During and following World War II there was considerable investigation of the transuranic elements and of technetium, francium, and promethium in relation to their production and use. All these radioactive elements were produced by extraordinary nuclear reactions, whereas all ordinary chemical reactions involve only the electrons outside the nucleus. Ordinary chemical reactions are, in a sense, extranuclear.

The utilization of chromatographic techniques such as column, paper, thin-layer, vapor phase (gas), and high performance liquid chromatography has allowed chemists to effect the separation of the components of complex mixtures. The use of ion exchange in the resolution of ionic mixtures and in the removal of ions from solutions has been extended during the past twenty years.

The applications of the molecular orbital and ligand-field concepts have been highly successful in the description of the bonding in and geometry of complex molecules. These concepts have firm physical bases.

Chemists, like most scientists, take a special delight in doing experiments that are considered to be theoretically impossible; consequently the discovery in 1962 by Bartlett of a stable compound of xenon, $Xe^+(PtF_6)^-$, was an epic event. Since the discovery of the noble gases during the period 1894–1900, most chemists had assumed or believed that these elements were chemically inert. Therefore, Bartlett's discovery prompted much activity to produce other noble gas compounds.

The importance of chemistry in modern biology, and especially in medicine, has been reemphasized by knowledge of the role of the nucleic acids, such as DNA and RNA, in the genetic scheme. The increased use of chemotherapy in medicine is common knowledge.

Analytical Chemistry The qualitative and quantitative determination of elemental composition of matter resides in the branch of chemistry called analytical chemistry. Any means of determining molecular structure is often called an analytical technique. Until relatively recently most analyses were performed by using specific chemical reactions and techniques in liquid solutions. Recent advances in spectroscopy and other physical techniques have yielded a variety of instruments that greatly facilitate chemical analyses. The ability of the analytical chemist to measure trace quantities of materials has played a central role in the development of awareness of the hazards of environment pollution.

Biochemistry Investigations of the chemical phenomena in, and the constituent compounds of, living organisms are performed mainly by biochemists. Because every chemical reaction in any living organism involves compounds of carbon, biochemistry is in part the application of organic chemistry to investigations of vital systems. However, physical and analytical and more recently inorganic chemistry are essential to biochemistry.

Inorganic Chemistry This branch of chemistry is mainly that of all forms of noncarbon compounds. Although its potential scope is huge, it has attracted, until recently, much less attention than have organic and physical chemistry. The availability of sophisticated physical methods have made detailed knowledge of inorganic materials possible. The very breath of this subdiscipline has lead to the existence of several interdisciplinary activities. The most important of these are organometallic chemistry, in which one studies carbon derivatives of the elements, and bioinorganic chemistry, which focuses on the role of inorganic species, especially metal ions, in biological systems.

Organic Chemistry This division of chemistry is essentially that of the compounds of carbon. Originally organic chemistry was confined to materials in or from living organisms, probably because nearly every compound either isolated from or produced by a living organism is a compound of carbon. Although there are more compounds of hydrogen than of any other element, the compounds of carbon are next in line. The property of catenation (ability of identical atoms to bond together) is exhibited most extensively by carbon. The hundreds of different carbon-atom skeletons of the thousands of known organic compounds attest this fact. All foods, nearly every fabric, every ordinary commercial fuel, and almost all pharmaceuticals are organic in the sense that they contain compounds of carbon. The great tradition of syntheses of new compounds has developed to an art form in modern organic chemistry. One now can prepare structures ranging from highly symmetrical geometric forms to huge, complex naturally occurring molecules such as vitamin B-12.

Physical Chemistry The quantitative measurements of the properties and behavior of the elements and their compounds are the major concern of the physical chemist. Nearly every technique and concept has been adopted from physics. The development of new chemical concepts follows logical consideration of quantitative data. The major branches of physical chemistry are spectroscopy, nuclear chemistry, kinetics, thermodynamics, quantum and statistical mechanics, and solution and surface chemistry. Physical organic and inorganic chemistry have gained prominence during the past 25 years. The availability of modern high speed computers has allowed for the application of the methods of quantum chemistry to molecules of increasing size and complexity. Calculations are usually divided into ab initio and semiempirical methods, the latter having a higher degree of presupposition and/or parameters derived from experimental data.

There are certain discernable trends in the current direction of chemical research which are worth noting. The first of these is the increasing tendency of interdisciplinary activities between subdisciplines. The most prominent of these include bioinorganic, bioorganic, physical inorganic, physical organic, and organometallic chemistry. On a larger scale, joint programs involving chemistry, physics, and engineering may be found in the rapidly increasing number of material sciences programs. One should also note the pervasive utilization of analytical instrumentation in virtually all areas ranging from synthetic organic to experimental physical chemistry. In similar fashion, methods traditionally associated with physical chemistry, such as X-ray diffraction and molecular orbital calculations, are now extensively used by inorganic and organic chemists. There has also been a strong interest in chemistry related to industrial processes giving rise to continued growth not only of polymer chemistry but also the study of catalysis and surface chemistry.

CHRISTOPHER W. ALLEN

References

A readable introduction to chemistry may be found in:
  Dickerson, R. E., Gray, H. B., and Haight, G. P., Jr., "Chemical Principles", 3rd Ed., Benjamin/Cummings, Menlo Park, CA, 1979.
Brief surveys of current activity in various subdisciplines of chemistry may be found in *Chemical and Engineering News:* 58(33), 30 (1980) (analytical); 59(9), 28 (1981) (biochemistry); 59(46), 42 (1981) (inorganic); 58(11), 34 (1980) (organic); 58(22), 20 (1980) (physical); 58(50), 30 (1980) (polymer).

Cross-references: BOND, CHEMICAL; ELEMENTS, CHEMICAL; ISOTOPES; MOLECULAR WEIGHT; MOLECULES; PHYSICAL CHEMISTRY; PHYSICS; SPECTROSCOPY.

CIRCUITRY

Basic Concepts An electric circuit may be defined as a path or group of interconnected paths capable of carrying an electric current or an arrangement of one or more closed paths to accommodate an electron flow. Electrons bear *electric charges*, and the unit of electric charge is the coulomb, defined as the quantity of charge possessed by 6.24×10^{18} electrons, and the electrical *through variable* is the *current*. The unit of current is the ampere, defined as the flow of one coulomb (6.24×10^{18} electrons) per second (1 ampere = 1 coulomb/sec).

The passage of a current through a *circuit element* yields a change of potential designated as the *voltage drop* or *voltage* (in volts) and convention dictates that in passive elements such as resistance R, inductance L, capacitance C, short circuits and open circuits, the current flow *through* the element is the direction of voltage drop or voltage *across* the element. In source elements such as ideal voltage sources and ideal current sources, the current flow is in the direction of the voltage rise across the source. The use of the words *through* with regard to current and *across* with regard to voltage may be observed.

Three types of equations are used in the analysis and synthesis of electrical circuits. These are the continuity, compatability and elemental equations. A continuity equation is a relationship among through variables, and in the electrical case this is the *Kirchhoff Current Law:* The algebraic sum of the currents entering a point (node) in an electrical circuit (network) must be zero ($\sum i = 0$). A compatability equation is a relationship among across variables, and in the electrical case, this is the *Kirchhoff Voltage Law:* The algebraic sum of the voltage drops around a closed path (loop) in an electrical circuit (network) must be zero ($\sum \Delta v = 0$). It is to be noted that in applying the Kirchhoff Laws, currents flowing *away* from a point are treated as a negative current flowing *toward* the point and negative voltage *drops* within a path are treated as voltage *rises*.

Elemental Equations The third type of equation is the elemental equation which is a relationship between the through and across variables for a particular element. For the cases of R, L and C, the elemental equations are

$$v = Ri; \qquad i = v/R = Gv \qquad (1)$$

which is known as *Ohm's Law* and where it is to be noted that $G = 1/R$,

$$v = L \frac{di}{dt}; \qquad i = \frac{1}{L} \int i \, dt \qquad (2)$$

and

$$i = C \frac{dv}{dt}; \quad v = \frac{1}{C} \int v\, dt. \quad (3)$$

Equations (1), (2), and (3) demand that the elements be *linear*, that is, they possess an output (either i or v) that is in direct proportion to the input (either i, v, di/dt or dv/dt) which shows that R, L, and C may be viewed as simple proportionality constants. Equations (1), (2), and (3) may also be derived experimentally or by an interpretation of electromagnetic field phenomena.

For the case of the short circuit

$$i = i; \quad \Delta v = 0 \quad (4)$$

which states that a short circuit passes all of the current with no voltage drop and for the open circuit

$$i = 0; \quad \Delta v = v \quad (5)$$

which states that an open circuit possesses the impressed voltage but passes no current.

Circuit Connections Circuit elements may be connected in *series* in which they are connected in such a manner as to provide a single path for the flow of current or in *parallel* in which they are connected so as to provide a division of current among the elements. It is apparent that in series circuits, the Kirchhoff Voltage Law must prevail and in parallel circuits, the Kirchhoff Current Law must apply to the element connection points. Equivalent single values of R, L, and C may be determined (R_{eq}, L_{eq} and C_{eq}) where for the series circuit

$$R_{eq} = \sum_{i=1}^{m} R_i; \quad L_{eq} = \sum_{j=1}^{n} L_j; \quad C_{eq} = \frac{1}{\sum_{k=1}^{p} \frac{1}{C_k}} \quad (6)$$

and for the parallel circuit

$$R_{eq} = \frac{1}{\sum_{i=1}^{m} \frac{1}{R_i}}; \quad L_{eq} = \frac{1}{\sum_{j=1}^{n} \frac{1}{L_j}}; \quad C_{eq} = \sum_{k=1}^{p} C_k. \quad (7)$$

It is useful to define a quantity called the *impedance to exponentials* or merely the *impedance* as the ratio of an exponential voltage, $v = V e^{st}$ to an exponential current $i = I e^{st}$. With the *admittance to exponentials* taken as the reciprocal of impedance, Eqs. (1), (2), and (3) can be used to show that, for R, L, and C,

$$Z_R(s) = R; \quad Y_R(s) = 1/R = G \quad (8)$$

$$Z_L(s) = Ls; \quad Y_L(s) = 1/Ls \quad (9)$$

$$Z_C(s) = 1/Cs; \quad Y_C(s) = Cs. \quad (10)$$

Then if a *direct current (d-c) voltage* $v = v(s = 0)$ is applied $Z_R(0) = R$, $Z_L(0) = 0$, $Z_C(0) = \infty$ and $Y_R(0) = 1/R = G$, $Y_L(0) = \infty$ and $Y_C(0) = 0$. If an *alternating current (a-c) voltage* is applied, Euler's equations can be used to show that $s = j\omega$ ($j = \sqrt{-1}$) and $Z_R(j\omega) = R$, $Z_L(j\omega) = j\omega L = jX_L$, $Z_C(j\omega) = -j(1/\omega C) = -jX_C$, $Y_R(j\omega) = 1/R = G$, $Y_L(j\omega) = -j(1/\omega L)$ and $Y_C(j\omega) = j\omega C$. In the foregoing X_L and X_C are respectively called the *inductive reactance* and the *capacitive reactance*.

The Kirchhoff Laws can be employed to yield the familiar voltage and current divider relationships. For n impedances in series with an applied voltage $V_{in}(s)$, the voltage drop *across* the ith impedance is given by the *voltage divider relationship*

$$\Delta V_i(s) = \frac{Z_i(s)}{\sum_{k=i}^{n} Z_k(s)} V_{in}(s),$$

$$k = 1, 2, 3, \ldots, i, \ldots, n. \quad (11)$$

In similar fashion, for n admittances in parallel (recall that $Y(s) = 1/Z(s)$) with an applied current $I_{in}(s)$, the current *through* the ith admittance is given by the *current divider relationship*

$$I_i(s) = \frac{Y_i(s)}{\sum_{k=1}^{n} Y_k(s)} I_{in}(s),$$

$$k = 1, 2, 3, \ldots, i, \ldots, n. \quad (12)$$

Circuit Analysis Electrical circuit or network analysis involves the determination of unknown currents or voltage drops when the circuit is specified and subjected to a forcing function. For a particular quantity of interest such as the current through a series connection of circuit elements when subjected to a voltage forcing function, the response is composed of a transient and a steady state response. For example, consider the connection of a resistor, capacitor, inductor and a switch all in series with a voltage source $v_{in}(t)$. Kirchhoff's Voltage Law can be employed to yield the integro-differential equation

$$L \frac{di}{dt} + Ri + \frac{1}{C} \int i\, dt = v_{in}(t) \quad (13)$$

where it is presumed that the switch closes at $t = 0$. The *total response* is given by

$$i(t) = i_c(t) + i_p(t)$$

where $i_c(t)$ is the *complementary function* (also termed the *natural* or *transient response*) and $i_p(t)$ is the *particular integral* (known also as the *forced* or *steady state response*).

The complementary function derives from the assumption of an exponential solution, $i(t) = Ie^{st}$ and its substitution into the *homogeneous* integro-differential equation yields

$$L \frac{di_c}{dt} + Ri_c + \frac{1}{C} \int i_c \, dt = 0$$

which leads to the factored and algebraically adjusted auxiliary equation

$$L \frac{s^2 + \dfrac{R}{L} s + \dfrac{1}{LC}}{s} Ie^{st} = 0.$$

Here it is observed that because $I = 0$ or $e^{st} = 0$ will yield trivial solutions

$$s^2 + \frac{R}{L} s + \frac{1}{LC} = 0.$$

With $\alpha = R/2L$ and $\omega_n^2 = 1/LC$, this equation becomes the simple quadratic

$$s^2 + 2\alpha s + \omega_n^2 = 0$$

with two solutions for the exponent s in the assumed $i_c(t) = Ie^{st}$,

$$s_1, s_2 = -\alpha \pm \sqrt{\alpha^2 - \omega_n^2}.$$

The form of the response is dictated by the discriminant $\alpha^2 - \omega_n^2 = \omega_d^2$. If $\alpha > \omega_n$, then ω_d is positive and s_1 and s_2 are both negative, real, and distinct. The complementary solution will be a pair of exponentials, each with an arbitrary constant of integration

$$i_c(t) = I_1 e^{(-\alpha + \omega_d)t} + I_2 e^{(-\alpha - \omega_d)t} \quad (14)$$

which is the *overdamped* case.

If $\alpha = \omega_n$, then s_1 and s_2 are negative, real and equal. In this case $s_1 = s_2 = s = -\alpha$ and the solution will be given by

$$i_c(t) = I_1 e^{-\alpha t} + I_2 t e^{-\alpha t}. \quad (15)$$

This is the *critically damped* case.

If $\alpha < \omega_n$, then $\alpha^2 - \omega_n^2$ is negative and $s_1 = -\alpha + j\omega_d$ and $s_2 - \alpha - j\omega_d$, a pair of complex conjugate roots with negative real parts. A damped oscillation occurs

$$i_c(t) = e^{-\alpha t}(I_1 \cos \omega_d t + I_2 \sin \omega_d t) \quad (16a)$$

which may be written in terms of a sinusoidal function and a phase angle

$$i_c(t) = Ie^{-\alpha t} \cos (\omega_d t + \sigma) \quad (16b)$$

or

$$i_c(t) = Ie^{-\alpha t} \sin (\omega_d t + \phi). \quad (16c)$$

Any of the oscillating forms of (16) describe the *underdamped* case.

The particular integral of Eq. (13) may be found by standard methods such as undetermined coefficients or variation of parameters. The complete general response is then the sum of $i_c(t)$ and $i_p(t)$ and the arbitrary constants in $i_c(t)$ are evaluated after this sum is taken from two different initial conditions to form the particular response.

Other Methods Other methods for analysis of the complete response are in frequent use. One of these is the method of *Fourier analysis*, which is useful when the circuit is subjected to a periodic input. The input function is represented as a *Fourier series*. Because the circuit is linear, the principle of *superposition* (the total response to the total stimulus is equal to the sum of individual responses due to individual stimuli acting alone) permits the response to be evaluated from a term-by-term solution summation from the individual terms of the input Fourier series.

Fourier analysis may be extended beyond Fourier series to first the *Fourier integral*, then to the *Fourier transform*, and finally to the *Laplace transform*. This permits the entire analysis to be conducted algebraically in the complex frequency domain instead of on a differential equation basis in the time domain via the Laplace transform of $f(t)$:

$$\mathcal{L}[f(t)] = F(s) = \int_0^\infty f(t) e^{-st} \, dt \quad (17)$$

where $s = \alpha + j\omega$, a complex frequency.

Finally, by making use of the idealization of the *unit impulse* defined as a pulse of unbounded magnitude acting over an infinitesimal time period so that its behavior is not defined by its magnitude but by its behavior under integration

$$\int_{0^-}^{0^+} \delta(t) \, dt = 1 \quad (18)$$

one can employ the *Faltung* or *convolution integral* (or simply the convolution of two functions)

$$r(t) = \int_0^t f_1(t - \tau) r_2(\tau) \, d\tau \quad (19)$$

where $f_1(t)$ is the actual input function, $r_2(t)$ is the circuit response to the unit impulse function, $r(t)$ is the actual circuit response, and τ is a dummy variable of integration. This procedure is known as *Borel's Theorem*.

Steady State a-c and d-c Analysis The concepts of circuit impedance and admittance permit a direct analysis of the steady state condition. The simplest possible case is that of the *ladder network* in which every element current or voltage can be determined from any or all of the following: the two Kirchhoff Laws, voltage or current dividers, and systematic combination of impedances (admittances) in series or parallel.

When the network possesses loops in the graph theoretical sense, the method of *loop (mesh) analysis* to yield loop currents or *node analysis* to yield node-to-datum voltages can be used. Both of these methods involve systematic application of one of the Kirchhoff Laws to each loop or node (loops and nodes are evident from the *topology* of the circuit)—the voltage law for loop analysis and the current law for node analysis. The inevitable result is a set of n simultaneous, linear (and linearly independent) algebraic loop or node equations which are easily solved by several methods.

For certain networks, the *superposition theorem* (discussed above) or the network theorems of *Thévenin* (also *Helmholtz*) and *Norton* may be employed to advantage. The Thévenin and Norton Theorems permit calculation of the performance of a portion of the circuit (possibly a single circuit element) from the terminal properties of the rest of the network.

Nonlinear Circuits and Circuits with Time-Varying Elements Analysis of circuits with nonlinear and time-varying elements poses a problem that is substantially more difficult than the analysis of their linear counterparts. For nonlinear elements such as the *pn* junction diode or tunnel diode (resistors), the varactor diode (capacitor), and the common inductor driven to flux saturation by large currents, analysis can be conducted using perturbation techniques or by successive small signal approximations.

Examples of time-varying elements are the potentiometer or rheostat (resistor) and the moving plate capacitor used in communications systems for tuning. Analysis of circuits with such elements requires a valid mathematical model of the element variation. For example, for a resistor which is time varying, Eq. (13) becomes

$$L \frac{di}{dt} + R(t)i + \frac{1}{C} \int i \, dt = v_{in}(t) \quad (20)$$

and the degree of analysis difficulty will depend on the form of $R(t)$.

Microelectronics and Microelectronic Circuitry Microelectronics, as the prefix *micro* implies, pertains to the area of technology that is associated with the fabrication of and operation of circuits, devices and systems (and subsystems) utilizing extremely small components. Four technologies are in widespread use:

Thin-film hybrid circuits consist of conductor, dielectric, and resistor materials which are deposited on glass, ceramic, or crystalline substrates by evaporation, sputtering or electroplating. The form of the deposition is obtained by patterning through photoengraving, or by printing through a mask. In this technology,

semiconductors, monolithic circuits, capacitors, and inductors can also be added consistent with the device function.

Thick-film hybrid circuits consist of conductor, dielectric, and resistor patterns which are screen printed on ceramic substrates. Miniature components such as monolithic integrated circuits, semiconductor devices, capacitors, inductors, and transformers are then added after the basic circuitry patterns are formed.

Monolithic bipolar circuits consist of bipolar junction and monopolar junction field-effect transistors, Schottky diodes, junction diodes, resistors and capacitors formed by oxidation, diffusion, ion-implantation, chemical vapor deposition and photoengraving on and perhaps within a single crystal semiconductor substrate. Interconnection is accomplished through the use of thin film circuitry.

Monolithic metal-oxide-semiconductors (MOS) consist of MOS field-effect transistors, capacitors, conductors and resistors which are formed by oxidation, diffusion, photoengraving, chemical vapor deposition, thin-film deposition, and ion-implantation on and within a single-crystal semiconductor or substrate.

ALLAN D. KRAUS

Cross-references: ALTERNATING CURRENTS; CAPACITANCE; CONDUCTIVITY, ELECTRICAL; ELECTRICITY; FEEDBACK; INDUCTANCE; LASER; MASER; SEMICONDUCTORS; SEMICONDUCTOR DEVICES.

COHERENCE

Basic Definitions The term coherence as it is used in electromagnetic radiation studies is best explained by a discussion of Young's interference experiment. Referring to Fig. 1, we consider a self-luminous radiating source S (like a mercury arc lamp) placed a distance l_1 from an opaque screen A. In the screen A two pinhole openings P_1 and P_2 are made a distance d apart. The radiation passing through the pinholes impinges upon and is recorded upon a photographic plate B a distance l_2 from screen A.

If l_1 is a very large distance from A (say $l_1 \gg bd/\bar{\lambda}$, $\bar{\lambda}$ being the average radiation wavelength) and the spectral width of the radiation is made very narrow by filtering, then the fringes observed on B in the neighborhood of O will be very sharp, and we say that the radiation fields impinging *upon* P_1 and P_2 are very coherent. This is in accord with our usual notions about the radiation from a point source. If, on the other hand, we use the same source and l_1 is taken to be small (say of the order of b) and the dimension b is large compared to d then fringes will not be observed on B, and we say that the radiation fields impinging *upon* P_1 and P_2 are incoherent. Again this is to be expected

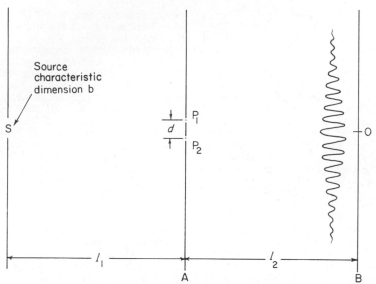

FIG. 1. Young's interference experiment.

since in this case we observe the radiation just as it emerges from the lamp. As the distance l_1 is increased from the order of b, faint fringes will begin to appear upon B. As l_1 increases, the fringes will become progressively stronger until they become very pronounced when $l_1 \gg bd/\bar{\lambda}$. The intermediate states, when the fringes are present but not necessarily very strong, are termed states of partial coherence. This experiment may be performed using a variety of sources. The experiment, as we have emphasized, measures the coherence of the radiation when it reaches P_1 and P_2; it does not measure the coherence of the source.

A quantitative measure of the fringe strength called the visibility $\mho$, was given by Michelson.[7] It is defined as

$$\mho = \frac{I_{\max} - I_{\min}}{I_{\max} + I_{\min}} \qquad (1)$$

where $I_{\max}$ is the maximum intensity recorded in the vicinity of O and $I_{\min}$ is the minimum intensity. $\mho$ varies between 0 and 1 and, roughly, we term radiation fields at two points coherent when $\mho = 1$ (strong fringes) and incoherent (no fringes) when $\mho = 0$.

In the above introduction, d was held fixed but it is most important to realize that the fringe visibility at O is a function of the spacing d and that $\mho$ may be close to 1 for one spacing of d and close to zero for another spacing of d. For example, if we fix l_1, we will generally decrease the visibility of fringes at O by making d larger (more precisely it will decrease and then increase in a succession of oscillations with the successive peaks being reduced in magnitude.) The visibility may thus be viewed as a measure of the correlation (coherence) between the radiation at P_1 and the radiation at P_2. To give a more precise definition of coherence, Wolf[9] considered the correlation function $\Gamma(P_1, P_2, \tau)$, commonly termed the mutual coherence function, defined as

$$\Gamma(P_1, P_2, \tau) = \langle V(P_1, t + \tau) V^*(P_2, t) \rangle \quad (2)$$

where $V(P_1, t)$ is the radiation field at P_1 and time $t + \tau$ (in a scalar approximation), V^* is the radiation field at P_2 at time t and the brackets $\langle \rangle$ denote a time average. For convenience Wolf used a complex notation (V^* is the complex conjugate of V), but this need not concern us here.

It can be shown that the magnitude of the normalized form of $\Gamma(P_1, P_2, \tau)$, $\gamma(P_1, P_2, \tau)$, where

$$\gamma(P_1, P_2, \tau) = \frac{\Gamma(P_1, P_2, \tau)}{\sqrt{\Gamma(P_1, P_1, 0)\Gamma(P_2, P_2, 0)}} \quad (3)$$

is equal to the visibility $\mho$ when the radiation is quasi-monochromatic (narrow spectral width) and all path lengths in the problem are small compared to $c/\Delta v$ (where c is the velocity of light and Δv is a characteristic spectral spread). In this case it is appropriate to fix τ at some value τ_0 usually taken to be zero.

The definition given by Eq. (2) was intended to consider polychromatic fields in addition to the quasi-monochromatic fields so often studied using the visibility $\mho$. Beran and Parrent[1] have shown that the full function $\Gamma(P_1, P_2, \tau)$ is, in principle, measurable by an extension of the techniques used in a Young's interference experiment. Modern discussions of coherence now center on the calculation and measurement of the mutual coherence function $\Gamma(P_1, P_2, \tau)$. For calculation it is convenient to note that Wolf[9] has shown that $\Gamma(P_1, P_2, \tau)$ satisfies the pair of wave equations

$$\nabla_i{}^2 \Gamma(P_1, P_2, \tau) = \frac{1}{c^2} \frac{\partial^2 \Gamma(P_1, P_2, \tau)}{\partial \tau^2} \quad (i = 1, 2) \quad (4)$$

The vector properties of the electromagnetic field may easily be introduced into the mutual coherence function. In general one must consider the tensor

$$\mathcal{E}_{ij}(P_1, P_2, \tau) = \langle E_i(P_1, t + \tau) E_j{}^*(P_2, t) \rangle \quad (5)$$

where $E_k(P_s, t)$ is the kth component of the electric field at the space point P_s and time t. In general, this function has proved most useful in the study of polarization effects for fields in which the radiation has a principal direction of propagation. The form of $\mathcal{E}_{ij}(P_1, P_2, \tau)$ has, however, been derived for the radiation in a black body cavity.

The above considerations have been for radiation fields that are stationary in time. That is, the statistical parameters, as opposed to the detailed structure of the radiation field, are independent of the absolute scale of time. For fields in which this is not true we must introduce the concept of an ensemble average (similar to that used in statistical mechanics). The mutual coherence function $\mathcal{E}_{ij}(P_1, P_2, \tau)$ must be replaced by the function $\mathcal{E}_{ij}{}^E(P_1, t_1; P_2, t_2)$ defined as

$$\mathcal{E}_{ij}{}^E(P_1, t_1; P_2, t_2) = \overline{E_i(P_1, t_1) E_j(P_2, t_2)} \quad (6)$$

where the overbar denotes an average over an ensemble of systems.

Higher-order Coherence Functions and Quantum Aspects As the concept of coherence grew and the statistical formulation of the theory came more into the fore it was realized that a two-point moment like $\mathcal{E}_{ij}(P_1, P_2, \tau)$ or $\mathcal{E}_{ij}{}^E(P_1, t_1; P_2, t_2)$ was inadequate to completely describe the radiation field. Two fields could have the same mutual coherence function and yet differ in the statistical content of the field. To completely describe the field, it was necessary to consider higher-order moments like $L_{ijkl}(P_1, t_1; P_2, t_2; P_3, t_3; P_4, t_4)$ defined as

$$L_{ijkl}(P_1, t_1; P_2, t_2; P_3, t_3; P_4, t_4) =$$
$$\overline{E_i(P_1, t_1) E_j(P_2, t_2) E_k(P_3, t_3) E_t(P_4, t_4)} \quad (7)$$

and, in fact, for a complete description it was necessary to consider the probability density function $P[E_i(P_1, t_1)]$ defined roughly as the probability of the occurrence of a particular realization of the field.

There are a number of problems that require consideration of higher-order moments. These moments are necessary in intensity interferom-

etry, the study of laser radiation and the study of the radiation from turbulent gases. The measurement of the contracted fourth-order moment

$$R_{ik}(P_1, P_2, 0) = \langle |E_i(P_1, t)|^2 |E_k(P_2, t)|^2 \rangle \quad (8)$$

(called intensity interferometry) has received considerable attention since it entailed consideration of the quantum aspects of the electromagnetic field. This moment may be thought of as the correlation of instantaneous intensities

$$I_i(P_1, t) \equiv |E_i(P_1, t)|^2$$

at two points and was originally studied as an alternate method to the measurement of $\Gamma(P_1, P_2, 0)$ for determining the angular diameter of visible and radio stars.[4,5] In measuring $\Gamma(P_1, P_2, 0)$, we may use the Young's interference experiment described above; the quantum nature of the field rarely explicitly enters since averaging times are usually long enough to permit a classical analysis. To measure R_{ik}, however, we need to correlate the two signals $I_i(P_1, t)$ and $I_k(P_2, t)$ which are recorded using photomultipliers or coincidence counters. Since the relationship between the impinging electric field and the ejected photoelectron is a statistical one, classical considerations did not suffice for a deep understanding of the problem or for consideration of the very important signal-to-noise problems resulting from inadequate averaging times.

A quantum field formalism for coherence theory has been studied by a number of authors. References 6 and 14 summarize and reference most of the basic work. Optical correlation phenomena have been studied extensively and particular consideration has been given to the fundamental differences between laser light and thermal light. Laser models have been developed using coherence concepts. In addition it is possible to study the basic interaction between light and matter from a coherence point of view. For example, the Kramers-Heisenberg dispersion formula may be generalized to include the effects of partially coherent incident electromagnetic fields, and spontaneous radiation may be studied when the initial atomic states are correlated.

Applications (Measurement of the Angular Diameter of Stars) Coherence theory is of great use in the treatment of imaging and mapping problems. It is especially convenient for treating problems involving the effects of the turbulent atmosphere on resolution and acoustic scattering in the ocean.[2,10,11] Radiometry has also been studied using coherence theory.[12,13] To present the reader with a definite example of the use of the coherence theory formalism, we will conclude this brief discussion with an outline of the measurement of the angular diameter of visible stars.

Let us suppose that the source in Fig. 1 is a visible star so that l_1 is many light-years. There is no telescope big enough to resolve any star, and to make direct measurements of the angular diameter of a star, Michelson introduced the use of interference experiments to essentially give one a bigger effective aperture.

For purposes of visible star measurement we can replace the star of diameter D by a circular disk of diameter D lying in a plane parallel to A. The radiation leaving the star is assumed to be incoherent so that for all points on the disk we may take $\Gamma(P_1, P_2, 0) = c\delta(P_1 - P_2)$. Using Eq. (4) this allows us to solve for $\Gamma(P_1, P_2, 0)$ on the earth (screen A) if we filter the starlight to insure the validity of the quasi-monochromatic approximation. We find

$$\Gamma(P_1, P_2, 0) = \text{const.} \frac{J_1\left[\dfrac{\bar{k}dD/2}{l_1}\right]}{\dfrac{\bar{k}dD/2}{l_1}} \quad (10)$$

where $\bar{k}$ is the average wave number of the light and J_1 is a first-order Bessel function.

If we define the angular diameter of the star as $\theta = D/l_1$, we see that J_1 equals zero when $\theta = 1.22\bar{\lambda}/d$, where $\bar{\lambda} = 2\pi/\bar{k}$. When J_1 equals zero, there are no fringes on the screen B. Hence to find the angular diameter of the star, we need only increase d from zero, when there will be high contrast fringes, to the separation d when there are no fringes. Putting this latter value into our expression for θ gives us the angular diameter of the star. The effects of turbulence may be taken into account in this problem if Eq. (4) is generalized to include a variable index of refraction.

Basic References For a fundamental treatment of coherence theory we refer the reader to the following basic texts: Born and Wolf,[2] O'Neill,[8] Beran and Parrent,[1] Klauder and Sudarsham,[6] Françon[15], Marathay.[16]

MARK J. BERAN

References

1. Beran, M., and Parrent, G., Jr., "Theory of Partial Coherence," Englewood Cliffs, N.J., Prentice-Hall, 1964.
2. Beran, M., *Radio Science*, 10, 15, (1975).
3. Born, M., and Wolf, E., "Principles of Optics," London, Pergamon Press, 1970.
4. Hanbury Brown, R., and Twiss, R., *Proc. Royal Soc. London Ser. A*, 242, 300 (1957).
5. Hanbury Brown, R., and Twiss, R., *Proc. Royal Soc. London Ser. A*, 243, 291 (1957).
6. Klauder, J., and Sudarshan, E., "Fundamentals of Quantum Optics," New York, W. J. Benjamin, 1968.
7. Michelson, A., *Phil. Mag.* (5), 30, 1 (1890).
8. O'Neill, E., "Introduction to Statistical Optics," Reading, Mass., Addison-Wesley, 1963.
9. Wolf, E., *Proc. Royal Soc. London Ser. A*, 230, 246 (1955).
10. Ishimaru, A., "Wave Propagation and Scattering in Random Media," Vols. I and II, New York, Academic Press, 1978.
11. Flatlé, S. (Ed.), "Sound Transmission Through a Fluctuating Ocean," Cambridge, U.K., Cambridge Univ. Press, 1979.
12. Marathay, A. S., *Optica Acta* 23, 785 (1976).
13. Walther, A., *J. Opt. Soc. Am.* 58, 1256 (1968).
14. Arecchi, F. T., Bonifacio, R., and Scully, M. O. (Eds.), "Coherence in Spectroscopy and Modern Physics," New York, Plenum Press, 1978.
15. Françon, M., "Optical Image Formation and Processing," New York, Academic Press, 1979.
16. Marathay, A. S., "Elements of Optical Coherence Theory," New York, John Wiley & Sons, 1982.

Cross-references: INTERFERENCE AND INTERFEROMETRY, LASER, MASER.

COLLISIONS OF PARTICLES

Introduction Scattering experiments provide the principal technique by which physicists attempt to understand the structure and interactions of matter on a microscopic scale. Scattering theory provides the basis of analyzing and interpreting scattering experiments. Other applications are the scattering of acoustic waves by sonar systems and electromagnetic waves by radar systems. Detailed descriptions of scattering theory are available in several references.[1,2,3,4]

A description of the development of scattering theory may be divided into several topics. The oldest and simplest branch of scattering theory is that of potential scattering, or scattering of two particles which interact through a local potential.[1] Potential scattering was studied extensively in the first two decades following the development of quantum mechanics in the analysis of elastic scattering of particles by atoms and of nucleon-nucleon scattering. The latter topic, in particular, led to the introduction of an elaborate theory of scattering by noncentral interactions.[5] The development of nuclear physics, with the observation of resonance reactions, indicated the need for more general descriptions of scattering. The resulting theory of resonance reactions[6,7] has leaned only rather lightly on the details of the Schödinger equation. Quantum field theory was developed to describe electromagnetic phenomena.[8] Of major importance in the development of scattering theory was the introduction of renormalization techniques into FIELD THEORY.[9]

It might be claimed that modern scattering theory began with the integral equation formulation of Lippmann and Schwinger[10] and the introduction of S-matrix theory by Heisenberg[11] and others.[12] This work has stimulated much of the development of theoretical physics in the last two decades. Of particular significance are the clarification of the study of rearrangement collisions and the development of the so-called dispersion techniques and current S-matrix formalism.

The Scattering Cross Section The properties of scattering interactions are usually expressed most conveniently in terms of the *scattering cross section*. To define this term, we consider the following scattering experiment: A beam of particles (called *beam particles*) is directed on a scatterer consisting of *target particles*. As a result of collisions between beam and target particles, there are particles which emerge from the reactions (called *reaction products*) and these are detected in *particle detectors*. To describe this quantitatively, we suppose that the scatterer contains N_t target particles and that this is uniformly illuminated by a flux F_B (expressed as the number of beam particles per unit area per unit time arriving at the target) of beam particles. We suppose also that the scatterer is sufficiently small that the beam is negligibly attenuated in passing through it. Then, if there are δN_s scattering interactions per unit time which lead to detected particles, we define the scattering cross section $\delta\sigma$ as

$$\delta\sigma = \frac{\delta N_s}{N_t F_B} \qquad (1)$$

In the limit that the detectors subtend very small solid angles, as seen from the target, we define the differential scattering cross section $d\sigma$. When a single detector subtending a solid angle $\delta\Omega$ is used to define $\delta\sigma$, we may define the cross section per unit solid angle as

$$\frac{d\sigma}{d\Omega} = \lim_{\delta\Omega \to 0} \frac{\delta\sigma}{\delta\Omega} \qquad (2)$$

The total scattering cross section σ is obtained by summing $\delta\sigma$ over all scattering events:

$$\sigma = \sum \delta\sigma \qquad (3)$$

(Eq. (3) does not exist for scattering by a coulomb force).

The scattering cross section may be expressed in terms of the square of the magnitude of a scattering amplitude (or S-matrix, or $\mathcal{T}$ = matrix element) and is completely described as a function of the momenta and internal states of the particles in the initial and final states. Thus, for the two particles prior to collision[13] we may take the momenta p_1 and p_2 and the internal state quantum numbers s_1 and s_2 as variables. (For example, s_1 and s_2 may describe spin orientation, isotopic spin, etc. For colliding molecules these variables will describe vibrational, rotational and electronic states). We may suppose there to be μ particles in the final state following the collision and specify this state by the momenta and internal variables $k_1 \ldots k_\mu, s_1' \ldots s_\mu'$. The scattering cross section may be expressed in terms of these variables. Because of symmetries, the number of variables required

to describe $\delta\sigma$ may ordinarily be reduced. The most commonly encountered of these symmetries are: (1) energy and momentum conservation; (2) rotational invariance; (3) the Lorentz invariance of the scattering cross section $\delta\sigma$.

Energy and momentum conservation imply that $\delta\sigma$ contains the Lorentz invariant factor

$$\delta(\mathbf{P}' - \mathbf{P})\delta(E' - E),$$

where $\mathbf{P}$ and E are the respective initial total momentum and energy and the primed variables refer to the final state. The quantities δN_s and N_t in Eq. (1) represent integer numbers of particles and are thus Lorentz invariants. The beam flux F_B is also invariant under Lorentz transformations along the beam axis parallel to p_1. For such transformations $\delta\sigma$ is therefore invariant. It is, in fact, customary to define $\delta\sigma$ to be invariant under an arbitrary Lorentz transformation. This will be the case if F_B in (1) is replaced by F_I, the value of F_B in the *laboratory* frame of reference (for which $p_2 = 0$).

The particle detectors for which Eq. (1) is defined will ordinarily record events for a restricted domain of the final spins and momenta $s' \ldots k_\mu$. From this and its Lorentz invariance we infer that $\delta\sigma$ is of the form

$$\delta\sigma = \sum_{s_1' \ldots s'} \int \frac{d^3k_1}{\epsilon_{k1}} \cdots \frac{d^3k_\mu}{\epsilon_{k\mu}} (\mathbf{P}' - \mathbf{P})\delta(E' - E)I.$$

$$(4)$$

Here ϵ_{k_1} is the energy of a particle with momentum k_1 and integration is over the domain of detection. Since d^3k_1/ϵ_1 etc. are Lorentz invariant, we see that I must be a Lorentz invariant function of the momenta and spin variables.

We have noted that in the *laboratory* frame of reference the target particles are initially at rest, so $p_2 = 0$. In the center-of-mass, or *barycentric*, frame of reference the total momentum is zero, so $p_2 = -p_1$.

Potential Scattering We briefly illustrate the discussion of the preceding section with the example of nonrelativistic scattering by a local central potential $V(r)$. The SCHRÖDINGER EQUATION for scattering in the barycentric coordinate system is[14]

$$[\nabla_r^2 + \kappa^2 - v(r)]\psi_\kappa^+(\mathbf{r}) = 0 \qquad (5)$$

Here $\hbar K$ is the momentum of particle "1" in the barycentric system and $v(r) = (2M_r/\hbar^2)V(r)$, with M_r the reduced mass of the two particles. In the limit of large separation $\mathbf{r}$ between the particles the wavefunction ψ_κ^+ has the asymptotic form [our notation is such that we represent a unit vector in the direction of $\mathbf{K}$ by $\hat{\mathbf{K}}$]

$$\psi_\kappa^+(\mathbf{r}) \to (2\pi)^{-3/2} \left[e^{i\boldsymbol{\kappa}\cdot\mathbf{r}} + \frac{e^{i\kappa r}}{r} f(\hat{\mathbf{k}}\cdot\hat{\mathbf{r}}) \right] \quad (6)$$

Here $f(\hat{\mathbf{k}}\cdot\hat{\mathbf{r}})$ is the scattering amplitude for scattering particle "1" from the direction $\hat{\boldsymbol{\kappa}}$ into the direction $\hat{\mathbf{r}}$. The corresponding cross section per unit solid angle is

$$\frac{d\sigma}{d\Omega} = |f(\hat{\mathbf{k}}\cdot\hat{\mathbf{r}})|^2 \quad (7)$$

The wave function $\psi_{\boldsymbol{\kappa}}^+$ may be expanded into partial waves as follows:

$$\psi_\kappa^+(\mathbf{r}) = \sum_{l=0}^{\infty} \frac{(2l+1)}{4\pi\kappa r} P_l(\hat{\mathbf{k}}\cdot\hat{\mathbf{r}}) i^l e^{i\delta l} w_l(\kappa;r) \quad (8)$$

Here P_l is the Legendre polynomial of order l, δ_l is the scattering phase shift [see Eq. (10) below], and $w_l(\kappa;r)$ satisfies the differential equation

$$\left[\frac{d^2}{dr^2} + \kappa^2 - \frac{l(l+1)}{r^2} - v(r) \right] w_l = 0 \quad (9)$$

This is to be integrated subject to the condition that w_l is regular at $r = 0$. For large r, w_l has the asymptotic form

$$w_l(\kappa;r) \to \sqrt{\frac{2}{\pi}} \sin\left(\kappa r - \frac{\pi l}{2} + \delta_l \right) \quad (10)$$

It is Eq. (10) which permits the determination of the phase shift δ_l. The quantity

$$S_l(\kappa) = \exp\left[2i\delta_l(\kappa) \right] \quad (11)$$

is an eigenvalue of Heisenberg's S-matrix.[11]

For scattering by noncentral forces, the potential $V(\mathbf{r}, S_1, S_2)$ is a function of $\mathbf{r}$ (and sometimes the orbital angular momentum operator) and the spin operators S_1 and S_2 of the two colliding particles (if either has no spin, we consider its spin operator to be zero). Spin eigenfunctions $u(\nu_1, \nu_2)$ may be introduced as depending on the orientations ν_1 and ν_2. Then the wave function $\psi_{\kappa,\nu_1,\nu_2}^+$ is to be labeled with the initial spin orientations ν_1 and ν_2. The asymptotic form corresponding to Eq. (6) is

$$\psi_{\kappa,\nu_1,\nu_2}^+ \to (2\pi)^{-3/2} \left[e^{i\boldsymbol{\kappa}\cdot\mathbf{r}} u(\nu_1,\nu_2) \right.$$
$$\left. + \frac{e^{i\kappa r}}{r} \sum_{\nu_1',\nu_2'} \langle \nu_1', \nu_2' | f(\hat{\mathbf{k}},\hat{\mathbf{r}}) | \nu_1, \nu_2 \rangle u(\nu_1', \nu_2') \right]$$
$$(12)$$

Here $\langle \nu_1', \nu_2' | f(\hat{\mathbf{k}},\hat{\mathbf{r}}) | \nu_1 \nu_2 \rangle$ is the scattering amplitude for scattering to a final spin orientation ν_1, ν_2'. The cross section per unit solid angle is in this case

$$\frac{d\sigma}{d\Omega} = |\langle \nu_1', \nu_2' | f(\hat{\mathbf{k}},\hat{\mathbf{r}}) | \nu_1, \nu_2 \rangle|^2 \quad (13)$$

For an unpolarized initial state, corresponding to a uniform mixture of the $(2S_1 + 1)(2S_2 + 1)$ spin states, the cross section for scattering particle "1" into the direction $\hat{\mathbf{r}}$ with any spin orientation is

$$\frac{d\bar\sigma}{d\Omega} = \frac{1}{(2S_1 + 1)(2S_2 + 1)}$$
$$\times \sum_{\nu_1',\nu_2'} \sum_{\nu_1,\nu_2} |\langle \nu_1', \nu_2' | f | \nu_1, \nu_2 \rangle|^2 \quad (14)$$

where the sums extend over all spin orientations.

Following scattering by noncentral forces, the particles will in general have preferred spin orientations, or be polarized. When, for example, particle "1" has spin one-half with a spin operator σ_1, we define its polarization vector $\mathbf{P}(\nu_1, \nu_2)$ by the equation

$$\mathbf{P}(\nu_1, \nu_2) = \left\{ \sum_{\nu_1'',\nu_1',\nu_2'} [\langle \nu_1'', \nu_2' | f | \nu_1, \nu_2 \rangle]^* \right.$$

$$\times \langle \nu_1'' | \sigma_1 | \nu_1' \rangle \langle \nu_1', \nu_2' | f | \nu_1, \nu_2 \rangle \Big\}$$

$$\times \left\{ \sum_{\nu_1',\nu_2'} |\langle \nu_1', \nu_2' | f | \nu_1, \nu_2 \rangle|^2 \right\}^{-1}$$
$$(15)$$

For an unpolarized initial state, the polarization is

$$\bar{\mathbf{P}} = \frac{1}{(2S_1 + 1)(2S_2 + 1)} \sum_{\nu_1,\nu_2} \mathbf{P}(\nu_1, \nu_2) \quad (16)$$

The study of polarization following scattering has provided an important tool for analyzing nuclear and elementary particle reaction.[15,16] In particular, the role of noncentral interactions in nucleon-nucleon scattering has been studied in great detail.[17]

Formal Scattering Theory To describe a general scattering reaction Lippmann and Schwinger[10,18] introduced a scattering matrix $\mathcal{T}_{ba}$ to describe scattering from an initial state χ_a to a final state χ_b.[18]

$$\mathcal{T}_{ba} = (\chi_b, V\psi_a^+) \quad (17)$$

where κ_a^+ is the steady-state wave function for the event and V is the scattering interaction. Since momentum is conserved for an isolated scattering, we may write

$$\mathcal{T}_{ba} = \delta(\mathbf{P}_b - \mathbf{P}_a) T_{ba} \quad (18)$$

where $\mathbf{P}_a$ and $\mathbf{P}_b$ are the total momenta of the particles in the initial and final states, respectively, and T_{ba} is defined only for states b and a corresponding to $\mathbf{P}_b = \mathbf{P}_a$.

The scattering cross section $\delta\sigma$ [Eq. (1)] is expressed in terms of T_{ba} as

$$\delta\sigma = \frac{(2\pi)^4}{v_{rel}} \sum_b \delta(\mathbf{P}_b - \mathbf{P}_a)\delta)(E_b - E_a)|T_{ba}|^2$$

(19)

Here v_{rel} is the relative velocity of beam and target particles, E_b and E_a are the respective total energies of the particles in states b and a, and the sum on b extends over those states which lead to the reaction products striking the detectors and thus to register an event. We emphasize that Eq. (19) is Lorentz invariant and conforms to the general structure of Eq. (4).

The Heisenberg S-matrix[11] is given by the expression

$$S_{ba} = \delta_{ba} - 2\pi i \delta(E_b - E_a)\, \mathcal{T}_{ba} \qquad (20)$$

where δ_{ba} is a Dirac δ-function. The S-matrix is unitary, so

$$\sum_b S_{cb}^\dagger S_{ba} = \delta_{ca} \qquad (21)$$

On substituting Eq. (20) into this, we obtain the equivalent expression of unitarity

$$i[\mathcal{T}_{ca} - \mathcal{T}_{ca}^\dagger] = 2\pi \sum_b \mathcal{T}_{cb}^\dagger \delta(E_b - E_a)\mathcal{T}_{ba}$$

(22)

which is defined only for states c and a on the same energy shell (corresponding to $E_c = E_a$).

The fundamental problem of scattering theory is to determine the $\mathcal{T}$-matrix on the energy shell (or, equivalently, the S-matrix). The first step in doing this is to make use of general symmetry principles (such as Lorentz invariance) to limit the functional forms allowed. Following this a dynamical principle is needed. Such dynamical principles (reviewed in Chapters 5 and 10 of Ref. 2) have been proposed in a great variety of forms including integral equations, variational principles, and conditions of functional analyticity.

Scattering from Composite Systems Scattering from systems composed of two or more particles is generally very complex. This is in part due to the occurrence of sequential interactions, in part due to the dynamics of the scattering system, and in part due to the possibility of rearrangement phenomena.

Description of sequential interactions can be given in terms of the multiple scattering and optical model equations.[19]

Rearrangement collisions (i.e., collisions in which bound particles rearrange themselves) have been studied extensively following the development of formal scattering theory. Much of the early work[20] was stimulated by the observance of apparent paradoxes. Later work has tended to be directed toward specific applications. The formulation of Fesbach, for example, has led to a variety of applications in nuclear and atomic physics.[21] The eikonal approximation has been used in the description of rearrangement collisions of slow ions, atoms, and molecules.[22]

The careful description of three-body scattering given by Fadeev[23] has led to active study of this and related phenomena.[24]

Variational principles have also been developed for application to several body collisions.[25]

Field Theory Quantum field theory was originally developed to describe electromagnetic phenomena. It was applied in a promising context during the 1930s to β-decay and to the meson theory of nuclear forces. The great optimism following the development of renormalization theory[9] faded for want of adequate mathematical techniques for handling strong interactions. Semiphenomenological calculations of Chew and others[26] gave useful insight into the strong interaction phenomena. Heisenberg suggested in 1946[11] that a proper quantum theory of scattering would deal with only observable quantities such as the S-matrix and should not require off-the-energy-shell matrix elements of such quantities as $\mathcal{T}$ [Eq. (18)]. Considerable impetus for this point of view has been given by the development of dispersion theory, following early suggestions of Wigner and others.[27] The first attempt at a systematic formulation of a dispersion relation within the context of quantum field theory was made by Gell-Mann, Goldberger, and Thirring.[28] Further development followed applications of formal scattering theory to quantum field theory.[29] The development of the Mandelstam representation[30] provided an important step toward obtaining a "dynamical principle." A further important step was the proposal by Chew and Frautschi[31] and Blankenbecler and Goldberger,[32] who suggested that the only sinularities of the S-matrix are those required by the unitarity condition [Eq. (22)] and that families of particles should be associated with Regge Trajectories.[33]

Considerable impetus was given to the application of quantum field theories by the development of gauge theories of weak interactions, following a model of Yang and Mills.[34,35] Quantum chromodynamics, an $SU(3)$ gauge theory, has provided insight into the "strong" particle interaction scattering phenomena.[36]

KENNETH M. WATSON

References

1. The early development of scattering theory is well described in the classic work of Mott, N. F., and Massey, H. S. W., "The Theory of Atomic Collisions," Oxford, Clarendon Press, 1933.
2. Goldberger, M. L., and Watson, K. M., "Collision Theory," Huntington, N.Y., Krieger, 1964.
3. Newton, R. G., "Scattering Theory of Waves and Particles," New York, McGraw-Hill, 1966.
4. Joachain, C. J., "Quantum Collision Theory," Amsterdam, North-Holland, 1975.

5. Rarita, W., and Schwinger, J., *Phys. Rev.* **59**, 436 (1941); Christian, R. S., and Hart, E. W., *Phys. Rev.* **77**, 441 (1950); Christian, R. S., and Noyes, H. P., *Phys. Rev.* **79**, 85 (1950).

6. Breit, G., and Wigner, E. P., *Phys. Rev.* **49**, 519, 642 (1936).

7. Wigner, E. P., *Phys. Rev.* **70**, 15, 606 (1946); Wigner, E. P., and Eisenbud, L., *Phys. Rev.* **72**, 29 (1947); Sachs, R. G., "Nuclear Physics," Reading, Mass., Addison-Wesley Publishing Co., 1953.

8. See, for example, Mandl, F., "Introduction to Quantum Field Theory," New York, Interscience Publishers, 1959. The older work is admirably described in G. Wentzel, "Quantum Theory of Fields," New York, Interscience Publishers, 1949.

9. Feynman, R. P., *Phys. Rev.* **76**, 749 (1949); Dyson, F. J., *Phys. Rev.* **75**, 486 (1949); Tomonaga, S., *Progr. Theoret. Phys.* (Kyoto) **1**, 27 (1946); Schwinger, J., *Phys. Rev.*, **74**, 1439 (1948).

10. Lippmann, B., and Schwinger, J., *Phys. Rev.*, **79**, 469 (1950).

11. Heisenberg, W., *Z. Naturforsch.* **1**, 608 (1946).

12. Wheeler, J. A., *Phys. Rev.*, **51**, 1107 (1937); Moller, C., *Kgl. Danske Videnskab. Selskab, Mat. Fys. Medd.* **23**, 1 (1948).

13. The case that more than two particles collide is important for the discussion of chemical reactions in gases and liquids. This is discussed, for example, in Ch. 5 and Appendix B of Ref. 2.

14. The notation used here follows that of Ref. 2.

15. Wolfenstein, L., and Ashkin, J., *Phys. Rev.* **85**, 947 (1952); Simon, A., and Welton, T., *Phys. Rev.* **93**, 1435 (1954); Wolfenstein, L., *Am. Rev. Nucl. Sci.* **6**, 43 (1956).

16. A comprehensive account of the theory is given in Ch. 7 of Ref. 2.

17. Moravcsik, M. J., and Noyes, H. P., "Theories of Nucleon-Nucleon Elastic Scattering," *Ann. Rev. Nucl. Sci.* **11**, 95 (1961).

18. Gell-Mann, M., and Goldberger, M. L., *Phys. Rev.* **91**, 398 (1953).

19. Watson, K. M., *Phys. Rev.* **89**, 575 (1953); **105**, 1388 (1957); Francis, N. C., and Watson, K. M., *Phys. Rev.* **92**, 291 (1953); see also Chapter 11 of Ref. 2.

20. Foldy, L., and Tobocman, W., *Phys. Rev.* **105**, 1099 (1957); Epstein, S., *Phys. Rev.* **106**, 598 (1957); Lippmann, B., *Phys. Rev.* **102**, 264 (1956); Brening, W., and Haag, R., *Forschr. Physik.* **7**, 183 (1959); Cook, J., *J. Math. Phys.* **6**, 82 (1957). A somewhat more flexible interpretation has been made in Chapter 4 of Ref. 2.

21. Feshbach, H., *Ann. Phys.* (New York) **5**, 357 (1958); ibid. **19**, 287 (1962); Hahn, Y., and Spruch, L., *Phys. Rev.* **153**, 1159 (1967); Hahn, Y., *Phys. Rev.*, **C1**, 12 (1970); Chen, J. C. Y., *Phys. Rev.* **156**, 159 (1967); Weller, H., and Roberson, N., *Rev. Mod. Phys.* **52**, 699 (1980).

22. Chen, J. C. Y., and Watson, K. M., *Phys. Rev.* **174**, 152 (1968); ibid. **188**, 236 (1969); Hatton, G., Chen, J., Ishihara, T., and Watson, K., *Phys. Rev.* **A12**, 1281 (1975).

23. Fadeev, L. D., *Zh. Eksperim. i Teor. Fiz.* **39**, 1459 (1960); English trans., *Soviet Phys. JETP*, **12**, 1014 (1961).

24. Weinberg, S., *Phys. Rev.* **133**, B232 (1964): Newton, R. G., *Nuovo Cimento*, **24**, 400 (1963); Watson, K. M., and Nuttall, J., "Topics in Several Particle Dynamics," San Francisco, Holden-Day, 1967.

25. Schwartz, C., *Phys. Rev.* **124**, 1468 (1961); O'Malley, T. F., Sprach, L., and Rosenberg, L., *J. Math. Phys.* **2**, 491 (1961), and earlier references; Sugar, R., and Blankenbecler, R., *Phys. Rev.* **136**, B472 (1964).

26. This subject is reviewed by Wick, G. C., *Rev. Mod. Phys.* **27**, 339 (1955).

27. The history of this subject is reviewed in Ch. 10 of Ref. 2.

28. Gell-Mann, M., Goldberger, M. L., and Thirring, W., *Phys. Rev.*, **95**, 1612 (1954).

29. Lehmann, H., Symanzik, K., and Zimmerman, W., *Nuovo Cimento*, **1**, 205 (1955).

30. Mandelstam, S., *Phys. Rev.*, **112**, 1344 (1955); **115**, 1741, 1759 (1959).

31. Chew, G. F., and Frautschi, S. C., *Phys. Rev. Letters* **8**, 41 (1962).

32. Blankenbecler, R., and Goldberger, M. L., *Phys. Rev.* **126**, 766 (1962).

33. Reviews of the S-matrix theory of scattering can be found in Chew, G. F., "S-Matrix Theory of Strong Interactions," New York, W. A. Benjamin, 1961. Omnes, R., and Froissart, M., "Mandelstam Theory and Regge Poles," New York, W. A. Benjamin, 1963.

34. Weinberg, S., *Rev. Mod. Phys.* **46**, 255 (1974); ibid. **52**, 515 (1980).

35. Taylor, J. C., "Gauge Theories of Weak Interactions," Cambridge, U.K., Cambridge Univ. Press, 1976.

36. Marciano, W., and Pagels, H., *Phys. Rep.* **36C** (Nov. 3, 1978).

Cross-references: ATOMIC AND MOLECULAR BEAMS, CONSERVATION LAWS AND SYMMETRY, CROSS SECTIONS AND STOPPING POWER, ELEMENTARY PARTICLES, FIELD THEORY, SCHRÖDINGER EQUATION, GAUGE THEORIES.

COLLOIDS, THERMODYNAMICS OF

All colloids can be physically described by agglomerates of microparacrystals (mpc's). Their size depends on the paracrystalline lattice distortions (see MICROPARACRYSTALS and MICROPARACRYSTALS, EQUILIBRIUM STATE OF). Within each microparacrystal the distortion energy U at the boundaries reaches such high values that the netplanes are destroyed because the angles of the valence bonds are overstrained. Figure 1 shows schematically an atom and its valence bond in unstrained position; hence $r = 0$ has a minimum potential ΔU. ΔU increases with the square of the deviation r from the undistorted position. Averaging over all atoms one obtains for ΔU a value proportional to $\overline{r^2}$. In the case of microparacrystals this value is proportional to the number n of netplanes (see

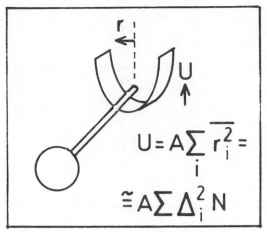

FIG. 1. Atom with one of its valence bonds to a neighboring atom and its potential U which depends on the value r by straining the valence angle.

MICROPARACRYSTALS, EQUILIBRIUM STATE OF), hence $\Delta U = \sum_i A r_i^2 = 2AN^2 \int_0^{N/2} n\Delta_1^2 = A_0 N^4 g^2/4$, where $\Delta_1^2 = g^2 \bar{d}^2$ is the distance variance between adjacent netplanes: $\Delta U \sim A_0 g^2 N$. The free energy ΔG of a cubic microparacrystal now contains, in addition to the well known terms of the surface free energy σ and the volume enthalpy ΔG_v (which are proportional to N^2 and N^3, respectively), a third

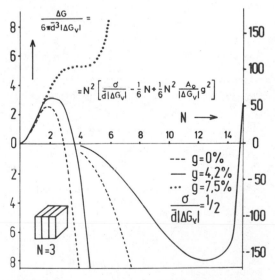

FIG. 2. Free enthalpy of solid microparacrystals $(0 < g < 7.5\%)$, crystals $(g = 0\%)$, and liquid microparacrystals $(g \sim 7.5\%)$. For details see text.

term proportion to N^4:

$$\Delta G = \bar{d}^3 \left(\frac{N^2 \sigma}{\bar{d}} + \frac{1}{6}N^3 \Delta G_v + \frac{1}{6}N^4 A_0 g^2 \right).$$

Here $\bar{d}$ is the mean netplane distance. For $g = 0$ the well known dependence of ΔG on N is plotted as the dashed line in Fig. 2. When $g = 4.2\%$, the size of the critical nuclei increases by 10% and at $N = 13$ there is a stoppage of growth (solid line in Fig. 2). In the limiting case, $g = 7.5\%$ (dotted line in Fig. 2), the critical nuclei have the same sizes as the mean number $\bar{N}$ of the equilibrium state. Now one is concerned with microparacrystals in a melt. The thermodynamics of solid colloids can be understood as having g-values smaller than those of the limiting case.[1]

ROLF HOSEMANN

Reference

1. Hosemann, R., *Colloid & Polymer Sci.* **260,** 864–870 (1982).

Cross-references: PARACRYSTALS; MICROPARACRYSTALS; MICROPARACRYSTALS, EQUILIBRIUM STATE OF.

COLOR

Definitions

1. *Color (n)* The noun *color* is used in different ways by different people to convey something about what we see when we look at an object. It has been used to describe differences among spectral power distributions of objects, differences among chemical constitutions of colorants, and conditions for color measurement by colorimetry, in addition to the perceived appearances of objects or media. There is, then, no single, universal, definition of color. The Commission Internationale de l'Éclairage (CIE) recommends two definitions: *perceived color* and *psychophysical color*. CIE perceived color is "an aspect of visual perception by which an observer may distinguish differences between two fields of view of the same size, shape, and structure, such as may be caused by differences in the spectral composition of the radiation concerned in the observation." CIE psychophysical color is said to be "a characteristic of a visible radiation by which an observer may distinguish differences between two fields of view of the same size, shape, and structure, such as may be caused by differences in the spectral composition of the radiation concerned in the observation."

2. *Color Perception* An attribute of vision consisting of *chromatic* and *achromatic* components. Chromatic components of color perceptions allow organisms to distinguish among *hues*, such as those called red, green, yellow,

blue, etc., and to distinguish among their *saturations* or degrees of chromatic content. Achromatic components of color perceptions refer to intensive aspects of the perceptions such as those commonly designated by the words white, gray, black, bright, dim, light, or dark.

3. *Color Stimulus* The agent by which an organism is stimulated to see color; color stimuli commonly consist of radiant energy called *light*. In general, then, color stimuli consist of visible radiation entering the eye and producing the sensation of color, either chromatic or achromatic.

4. *Color Measurement* The practice or methodology of attempting to measure the relationships between color stimulation and color perception; one form of *psychophysics*, the science of determining correspondences between stimuli and responses. A special case of color measurement is called *colorimetry*, which deals mainly with the specification of stimuli that match in color appearance.

Perception of Color All color perceptions have at least three attributes: hue, saturation or chromatic content, and brightness. When a color stimulus is perceived as part of an array of other stimuli, it may also have other attributes derived from the perceived relationships of its basic attributes to those of the other stimuli in the field of view. These are called relative hue, relative chromatic content (or saturation), and relative brightness (or lightness). These and other derivative attributes of color perceptions are conscious awarenesses that arise in the brain as a result of stimulation of the organism and signal processing carried out within the organism's nervous system.

The mechanism of human color vision consists of an optical system for collecting and forming images of light, neural networks that detect and respond to light and that transmit encoded signals, and neural cells in the brain that elaborate and interpret the signals. The visual mechanism then includes the eye, the neural pathways, and the brain.

The lens of the eye forms an image of light on the *retina* at the rear of the eyeball. The retina contains six kinds of cells called *neurons*. One class of neurons consists of photodetectors, which are capable of responding to light or changes in light. They are called *receptors*. There are about 10^8 *rod* receptors and 8×10^6 *cone* receptors, so named because of their respective shapes. Each receptor contains one of four kinds of photolabile *pigments*, the stereochemical forms of which can be altered by absorption of *photons* (light quanta). The distribution of rods and cones varies over the retina; there are many more cones than rods near the visual axis. Rods are useful for night vision but do not contribute significantly to color vision or acuity. There are three classes of cones, distinguished by their different spectral sensitivities to light. One class of cones (ρ) is most sensitive to long wavelengths of light, with a maximum

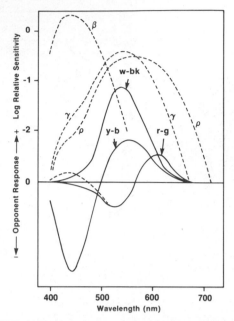

FIG. 1. Estimated ρ, γ, β cone receptor systems sensitivities (upper dashed curves) and sensitivities of achromatic (w-bk) and chromatic (r-g and y-b) neural processes (solid curves), both as functions of wavelength.

sensitivity in the region of 560 nm. A second class of cones (γ) is most sensitive to middle wavelengths of light and has a peak sensitivity around 530 nm. The third class of cones (β) is most sensitive to light of short wavelengths near 420 nm. Estimations of the distributions of spectral sensitivities of ρ, γ, and β cones are illustrated by dashed curves in Fig. 1.

Discrimination among color stimuli of different spectral powers occurs as a result of the different spectral sensitivities of the cones. However, regardless of differences in spectral power, when two color stimuli excite the cones equally, the stimuli cannot be distinguished as different; that is, when the rate of photon absorption is the same for each stimulus, the stimuli appear identical. This *principle of univariance* is the basis for *metameric* color matching, where two spectrally different stimuli can be made to match in color by properly adjusting their relative radiances.

The remaining neurons of the retina combine, elaborate, and encode electrophysiological signals set up in response to absorption of photons by the receptors. Many receptors may be interconnected, and their signals converge on higher order (i.e., more proximal) neurons. The axons of the most proximal neurons of the retina (ganglion cells) form the *optic nerve*, a sheath of about 10^6 fibers that transmit signals to the *lateral geniculate nucleus* of the thalamus (the next highest level of the visual mechanism). The encoded signals that leave the eye through the optic nerve are of two kinds: spectrally *opponent* and spectrally *nonopponent* (most are also spatially opponent). Spectrally opponent cells signal to different extents and with different polarities as a function of wavelength of stimulation, whereas spectrally nonopponent cells signal to different extents with the same polarity as a function of wavelength. The spectrally nonopponent signals (A) are thought to be the result of additive combinations of signals from ρ and γ cone networks. The spectrally opponent signals (C_1 and C_2) are thought to consist of differences among classes of cone signals. The neural signals are conventionally represented mathematically as:

$$A = a_{11}\rho + a_{12}\gamma \qquad (1)$$

$$C_1 = a_{21}\rho - a_{22}\gamma \qquad (2)$$

$$C_2 = a_{31}\rho + a_{32}\gamma - a_{33}\beta \qquad (3)$$

although there are a number of variations on these expressions.

A represents a neural signal analogous to achromatic perceptions: white to black or bright to dim (sometimes symbolized *w-bk*). C_1 represents a neural signal corresponding to redness (for positive values) or greenness (negative values); it is therefore sometimes symbolized *r-g*. C_2 represents a neural signal corresponding to yellowness (positive values) or blueness (negative values); it is therefore sometimes symbolized *y-b*. When $C_1 = C_2 = 0$, only A remains for any suprathreshold stimulus and so it is seen as achromatic (white, gray, black, etc.). The spectral distributions of sensitivities for the A, C_1, and C_2 neural mechanisms depend upon those of ρ, γ, and β and upon the interconnections in signal processing; there are also differences between threshold and suprathreshold processing conditions. Estimated spectral sensitivities of the neural functions are shown as the solid curves in Fig. 1. The viewing or adaptation condition represented by that figure is a physiological neutral state such as that associated with dark adaptation or, for the dashed extension of the *r-g* curve, suprathreshold daylight adaptation.

Figure 1 corresponds to observers with normal color vision. Deviations from normal color vision are thought to arise when one or more of the cone classes is missing or reduced in effectiveness and when neural processing differs from that subsumed for people with normal color vision. These deviations are of three kinds: anomalous trichromatic vision, dichromatic vision, and monochromatic vision. Their names derive from the minimum number of color stimuli required by an observer to match colors satisfactorily. A person with normal trichromatic vision can match the hues of all stimuli with additive combinations of three appropriately selected lights. Those with anomalous trichromatic vision also require three lights but make different matches from those of normal observers. Dichromatic observers require only two lights, and monochromatic observers are satisfied with a single light.

The wavelengths at which the opponent response functions of Fig. 1 are zero identify the spectral stimuli that elicit *unitary hues* for normal observers under the viewing conditions represented in the graph. The suprathreshold *r-g* curve is zero at about 475 nm. That leaves only the *b* signal at that wavelength, so it corresponds to the spectral stimulus that elicits a blue color response with no trace of any other hue (for the viewing condition in question). The *y-b* curve crosses the zero line at about 505 nm, leaving only *g*, so that would be the wavelength for a unitary green blue. The *r-g* curve again becomes zero at about 580 nm, so that wavelength corresponds to a unitary yellow hue. Unitary red hue can be produced only by a combination of short- and long-wavelength stimuli for the viewing conditions illustrated. The hues corresponding to all other stimuli contain proportions of two hues. These other perceptions are therefore known as *binary hues*. All hues are either unitary or binary. That is, they may appear red or green (but not both at the same time), yellow or blue (but not both at the same time), yellow-red, yellow-green, blue-red, or blue-green, and in various degrees of lightness or darkness. Even the color normally called brown can be described as a dark yellow-red color. Table 1 lists several relationships between attributes of perceived colors and conditions of the A, C_1, and C_2 neural signals of Eqs. (1)–(3).

Additional elaboration and analysis of neural signals takes place in the thalamus and occipital cortex of the brain, but the general form of color-coded signals seems to be the same as those leaving the retina. These signals are organized into *receptive fields* representing the combined responses of many receptors and interneurons. In this way, stimulation in one area of the retina influences signal responses in adjoining and nearby areas. These interactions give rise to contrast and induction effects. When, for example, a gray paper is viewed against a white background, it appears darker than when it is seen on a dark background. The higher level of activity elicited by the white background inhibits response to the gray paper, making it appear darker. Chromatic induction or contrast may also occur to alter perceived hues and saturations. These kinds of interactions make it difficult or impossible to predict color appearances from measurements or specifications of only single focal stimuli; all stimuli in the field of view must be considered and their interactions must be taken into account.

The quantitative nature of these interactions is not well known. Signal processing is both complex and nonlinear in its effect. For example, although the visual mechanism can respond over a range of luminances (spectrally

TABLE 1. EXAMPLE OF THE WAY IN WHICH ATTRIBUTES OF COLOR PERCEPTION COULD BE RELATED TO NEURAL SIGNALS OF THE VISUAL MECHANISM.

Let:*

$$A = a_{11}\rho + a_{12}\gamma \qquad = w\text{-}bk$$

$$C_1 = a_{21}\rho - a_{22}\gamma \qquad = r\text{-}g$$

$$C_2 = a_{31}\rho + a_{32}\gamma - a_{33}\beta = y\text{-}b$$

Attribute of Color Perception	Symbol	Condition
Achromaticness	A	$C_1 = C_2 = 0, \quad A > 0$
Chromaticness	K	$(C_1 + C_2) > 0$
Saturation (relative chromaticness)	S	$0 \leqslant \dfrac{C_1 + C_2}{C_1 + C_2 + A} \leqslant 1$
Brightness	B	$C_1 + C_2 + A > 0$
Lightness (relative brightness)	L	$0 \leqslant \dfrac{C_1 + C_2 + A}{(C_1 + C_2 + A)_w} \leqslant 1$
Constant hue	H_i	$C_1/C_2 = \text{constant}$
Unitary hues	H_r	$C_1 > 0, \quad C_2 = 0$
	H_g	$C_1 < 0, \quad C_2 = 0$
	H_y	$C_1 = 0, \quad C_2 > 0$
	H_b	$C_1 = 0, \quad C_2 < 0$

*Where + stands for some unspecified form of combination and w refers to a white reference.

weighted radiances) of about 10^{12}, neural signals are greatly compressed and nonlinearly related to input powers (a range of about 10^2 in the receptors and 10^1 in other neurons). This compression requires that to be effective, the visual mechanism must adjust or adapt the sensitivities of its component processes to suit the level and quality of stimulation provided it. The process of adaptation permits us to see a "white" paper as white by moonlight or bright sunlight and by yellowish incandescent lamp or blue sky illumination.

Color Stimuli Color stimuli have extents (spatial and temporal), intensities, and qualities. Spatial extent (size and location) can be specified by length and its derivative measures or by angular subtense and location in the visual field. Temporal extent can be specified by time and its derivative measures. Intensity and quality are specified by spectral power or radiance concentration.

Normal color stimuli can be sources of direct radiation (lights) or of indirect radiation (e.g., reflecting or transmitting objects). Light can be produced by valence-shell electronic excitations from higher levels back to the ground state. In heated solids, the emission spectra of lights are characteristic of the temperature of the material. In gases, the emission spectra are characteristic of the molecular or atomic structure of the material. In either case, the stimulus function $\Psi(\lambda)$ represents the radiance of the light source: $\Psi(\lambda) = L_e(\lambda)$.

Light can also cause low-energy valence-shell electronic excitations in atoms. Spectral selectivity is produced when molecules selectively absorb certain frequencies of light because of these electronic transitions between quantum electronic states. Both organic and inorganic compounds absorb light frequencies by transition of the molecule from the ground state to an excited state, but the physical phenomena that cause the transition may differ in two cases. In either instance, absorption of a particular frequency of light by excitation of electrons whose energy difference corresponds to that frequency removes light of that wavelength from the spectrum of irradiation. Selective alteration of light can also occur by other physical phenomena such as polarization and Rayleigh scattering and by fluoresence and photophorescence. The light which is not absorbed (scattered or re-emitted) is reflected or transmitted. In these cases, the color stimulus function is designated by the product of the irradiance function and the reflectance function $[\Psi(\lambda) = \rho(\lambda)E(\lambda)]$ or the transmittance function $[\Psi(\lambda) = \tau(\lambda) \cdot E(\lambda)]$. In the general case, the product of the relative radiance of a medium with the irradiance of the light source defines the color stimulus function $[\Psi(\lambda) = \beta_e(\lambda)E(\lambda)]$. When $L_e(\lambda)$ is expressed in $W \cdot m^{-2} \cdot sr^{-1}$ or $E(\lambda)$ in $W \cdot m^{-2}$, then $\Psi(\lambda)$ specifies both intensity and quality of the color stimulus. When those values are expressed relative to the corresponding value for some standard of reference (e.g., the Lambert reflector or a specified thickness of air) then $\Psi(\lambda)$ specifies only the quality of the spectral stimulus.

Color Measurement Color measurements may be classed in three broad categories: *color matches*, *color differences*, and *color appearances*. In all three, an atempt is made to relate $\Psi(\lambda)$ or its derivative expression to some invariant criterion of color perception.

The criterion for color matching (upon which colorimetry is based) is the invariance of perceptual identities. The amounts of three reference stimuli or *primaries* that are required to produce the same color appearance as that of some sample are specified as a color match by *color mixture*. Because of the principle of univariance combined with the fact that there are but three kinds of cone receptors, a minimum of three primaries can be used to match the colors of all stimuli, provided that the primaries are perceptually *independent* (i.e., none of them can be matched in color by any mixture of the other two). Certain conventions have been adopted internationally to simplify the practice of colorimetry and enhance communication. The spatial extents of stimuli are restricted to about $2°$ and $10°$ diameter in visual subtense centered on the visual axis. Certain geometric conditions are also specified for $\rho(\lambda)E(\lambda)$ and $\tau(\lambda)E(\lambda)$. In order to avoid dealing with the nonlinear complexities of the visual mechanism, only color matches (conditions of equal quantum catches by the receptors) are addressed by CIE conventions adopted in 1931. These colorimetric conventions lead to specifications of the conditions for color matches in terms of the amounts of standard primaries required by a standard observer; they do not specify color perceptions or differences among them. Color matches for stimuli subtending $1-2°$ diameter have been found to hold over a range of luminances from about $1-2$ cd $\cdot$ m^{-2} to around $1,000-2,000$ cd $\cdot$ m^{-2} and over changes in chromatic adaptation among daylight sources or Planckian radiators whose spectral radiance functions are not greatly different from that of average daylight. Over these ranges, the nominal process of color matching is *linear* (stimulus radiances can be increased or decreased by any positive factor and the match will continue) and *persistent* (the match holds over changes in sensitivity occasioned by variations in chromatic adaptation). Therefore, any set of color mixture primaries that is a linear transform of some appropriate set will serve as a standard. The CIE has selected a set for the *CIE 1931 Standard Colorimetric Observer* (for $2°$) and a second set for the *CIE 1964 Supplementary Standard Colorimetric Observer* (for $10°$) and has specified the amounts required of them for the standard observer to match the colors of spectral components of an equi-energy spectrum. These amounts are symbolized as $\bar{x}(\lambda)$, $\bar{y}(\lambda)$, and $\bar{z}(\lambda)$ [or $\bar{x}_{10}(\lambda)$, $\bar{y}_{10}(\lambda)$, $\bar{z}_{10}(\lambda)$ for the CIE 1964 observer]. The color mixture coefficients at each wavelength ($\bar{x}_\lambda$, $\bar{y}_\lambda$, $\bar{z}_\lambda$) represent the tristimulus values of the components of the equi-energy spectrum and are illustrated in Fig. 2. When these color mixture coefficients are used, the condition of color match for two stimuli $\Psi_1(\lambda)$ and $\Psi_2(\lambda)$ can be specified as:

$$k \int \Psi_1(\lambda)\bar{x}(\lambda)\,d\lambda = k \int \Psi_2(\lambda)\bar{x}(\lambda)\,d\lambda$$

$$(4)$$

$$k \int \Psi_1(\lambda)\bar{y}(\lambda)\,d\lambda = k \int \Psi_2(\lambda)\bar{y}(\lambda)\,d\lambda$$

$$(5)$$

$$k \int \Psi_1(\lambda)\bar{z}(\lambda)\,d\lambda = k \int \Psi_2(\lambda)\bar{z}(\lambda)\,d\lambda$$

$$(6)$$

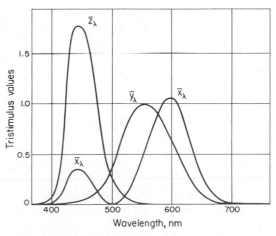

FIG. 2. CIE 1931 Standard Colorimeteric Observer's color mixture functions.

or similarly for the CIE 1964 observer [using $\bar{x}_{10}(\lambda)$, $\bar{y}_{10}(\lambda)$, $\bar{z}_{10}(\lambda)$]. The limits of integration are usually 380 nm to 780 nm. The integrals of Eqs. (4)–(6) are called *tristimulus values* and symbolized X, Y, and Z. The scale factor k may have any convenient value but most often is assigned one of two values: $k = 680$ lumens · W^{-1} (which yields tristimulus values that are consistent with luminance since $\bar{y}(\lambda) = V(\lambda)$, the CIE 1924 luminous efficiency function) or $k = 100 \left[\int E(\lambda)\bar{y}(\lambda)\,d\lambda \right]^{-1}$ (which yields tristimulus values consistent with luminance factor).

If $\Psi_1(\lambda) \equiv \Psi_2(\lambda)$ (the spectral distributions of both stimuli are identical), then Eqs. (4)–(6) will be satisfied for all observers under all illuminants; such matches are sometimes said to be *isomeric*. If Eqs. (4)–(6) are satisfied but $\Psi_1(\lambda) \neq \Psi_2(\lambda)$, then the match is a *metameric* one that may not satisfy observers whose color mixture functions differ from those of the CIE standard observer, and the match may not hold for the standard observer under a different illuminant.

It is often convenient to plot the *relative* amounts of tristimulus values. They are called *chromaticity coordinates* and are derived from the tristimulus values as follows:

$$x = \frac{X}{X + Y + Z} \qquad (7)$$

$$y = \frac{Y}{X + Y + Z} \qquad (8)$$

$$z = \frac{Z}{X + Y + Z}. \qquad (9)$$

Since $x + y + z = 1$, only two are needed for a complete specification of chromaticity; x and y are conventionally used. When the chromaticity coordinates of the components of the equienergy spectrum are plotted in a diagram of y versus x, and the chromaticity coordinates for the extremes of wavelength integration are joined together, the roughly "horseshoe" area of Fig. 3 results. The diagram is called a *chromaticity diagram*, and the area depicted contains chromaticities corresponding to all color stimuli.

When a normal trichromatic observer matches the color of any one color stimulus many times, a distribution of chromaticities results representing the variability of determination. Figure 4 illustrates bivariate normal ellipses representing such variability for 25 different color stimuli; the ellipses are plotted as approximately $\frac{10}{3}$ the size of the threshold of perceptibility. The ellipses differ in size and orientation. If a threshold color difference is assumed to be perceptually equal everywhere in color space, then the chromaticity diagram of Fig. 4 must be judged

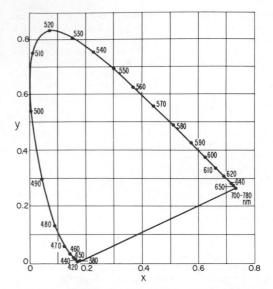

FIG. 3. CIE 1931 chromaticity diagram with coordinates y versus x. Locus of chromaticities corresponding to spectral color stimuli is shown by the curve, along which wavelengths are indicated, and the locus of nonspectral mixtures of additive combinations of spectral stimuli having wavelengths of 380 and 780 nm is shown by the line connecting the ends of the spectrum locus.

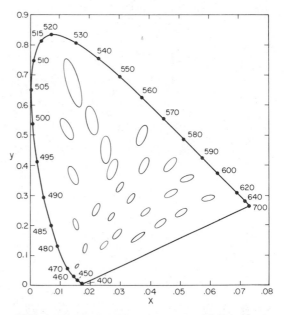

FIG. 4. Bivariate normal ellipses representing $\frac{10}{3}$ of the threshold variability of color matching for 25 color stimuli. [Based on data reported by MacAdam, D. L., "Visual sensitivities to color differences in daylight," *J. Opt. Soc. Am.*, **32**, 247–274 (1942).]

to be nonuniform. A uniform chromaticity diagram would be one in which the distributions of color matches are represented as circles of constant diameter throughout the graph. Straight lines in such a diagram would represent *geodesics*, lines depicting the relative sizes of perceived color differences. The (constant) distances corresponding to threshold differences in color might be taken to represent the additive unit of such a color metric.

Many attempts have been made to derive such a *uniform chromaticity diagram* or scale (*UCS*). None have been completely successful. In 1976, the CIE provisionally recommended two transformations of CIE 1931 tristimulus values as approximations to a uniform color metric: *CIE 1976 L*u*v** and *CIE 1976 L*a*b** diagrams. They are related to the tristimulus values X, Y, Z as follows:

$$L^* = 116(Y/Y_w)^{1/3} - 16 \qquad (10)$$

$$u^* = 13L^*(u' - u'_w) \qquad (11)$$

$$v^* = 13L^*(v' - v'_w) \qquad (12)$$

where

$$u' = (4X)(X + 15Y + 3Z)^{-1}$$

$$v' = (9Y)(X + 15Y + 3Z)^{-1}$$

and

$$L^* = 116(Y/Y_w)^{1/3} - 16 \qquad (13)$$

$$a^* = 500[(X/X_w)^{1/3} - (Y/Y_w)^{1/3}] \qquad (14)$$

$$b^* = 200[(Y/Y_w)^{1/3} - (Z/Z_w)^{1/3}]. \qquad (15)$$

The subscript w refers to a "white" achromatic reference stimulus. The coordinates u', v' form a supplementary chromaticity diagram since they can be defined for a plane of constant luminance; the $L^*a^*b^*$ metric does not have planes of constant luminance and, therefore, has no associated chromaticity diagram. Color differences (constrained to small or near-threshold in size) are specified in the two color spaces of Eqs. (10)–(15) as the Euclidean distances:

$$\Delta E_{uv} = [(\Delta L^*)^2 + (\Delta u^*)^2 + (\Delta v^*)^2]^{1/2} \qquad (16)$$

and

$$\Delta E_{ab} = [(\Delta L^*)^2 + (\Delta a^*)^2 + (\Delta b^*)^2]^{1/2}. \qquad (17)$$

CIE has not yet recommended a method for specifying large differences in color or magnitudes of color appearances. In practice, any of a number of *color order systems* are used for such specification. A color order system is a systematic structure, usually represented by material

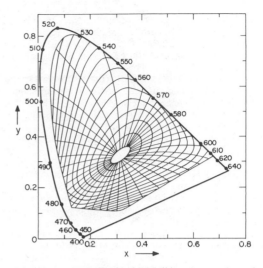

FIG. 5. Loci of constant Munsell hue (approximately radial lines) and chroma (approximately concentric circles) in a CIE 1931 chromaticity diagram for samples having luminance factors of 0.1977 (a Munsell value of 5).

standards, that attempts to array color stimuli according to some plan. The *Optical Society of America uniform color order system* arrays color stimuli approximately according to geodesics of suprathreshold color differences. The *Swedish natural color system* arrays color stimuli according to the magnitudes of their relative hues, saturations, and lightnesses. The *Munsell notation* arrays samples according to a combination of their hues, chromas (a form of relative chromatic content), and values (a function of lightness), and their differences in color. Figure 5 illustrates the chromaticities of the Munsell notation array for color stimuli having luminance factors of 0.1977. The approximately radial curves represent constant hues, and the approximately concentric circles correspond to constant chromas. These relationships permit specification of differences or magnitudes in terms of CIE tristimulus values. However, all current systems for specifying near-threshold and suprathreshold color differences are inexact to various degrees and, as with conventions for specifying color matches, they should be constrained to certain conditions to be useful.

C. J. BARTLESON

References

Billmeyer, F. W., Jr., and Saltzman, M., "Principles of Color Technology," 2nd Ed., New York, Interscience, 1981.
Boynton, R. M., "Human Color Vision," New York, Holt, Rinehart and Winston, 1979.
CIE [Commission Internationale de l'Éclairage],

"Colorimetry," CIE Publication No. 15, Bureau Central de la CIE, Paris, 1971.

Grum, F., and Bartleson, C. J. (Eds.), "Optical Radiation Measurements, Vol. 2. Color Measurement," New York, Academic Press, 1980.

Grum, F., and Bartleson, C. J. (Eds.), "Optical Radiation Measurements, Vol. 5. Visual Measurement," New York, Academic Press, 1984.

Hunt, R. W. G., "The Reproduction of Colour," 3rd Ed., London, Fountain Press, 1975.

Hurvich, L. M., "Color Vision," Sunderland, Mass., Sinauer Associates, 1981.

Jameson, D., and Hurvich, L. M. (Eds.), "Handbook of Sensory Physiology, Vol. VII/4. Visual Psychophysics," Berlin, Springer-Verlag, 1972.

Judd, D. B., and Wyszecki, G., "Color in Business, Science and Industry," 3rd Ed., New York, John Wiley & Sons, 1975.

Wyszecki, G., and Stiles, W. G., "Color Science," 2nd Ed., New York, John Wiley & Sons, 1982.

Cross-references: LENS; LIGHT; OPTICS, GEOMETRICAL; OPTICS, PHYSICAL; PHOTOMETRY; PHOTON; REFLECTION; REFRACTION.

COLOR CENTERS

Color center is a generic term first coined to characterize the entity responsible for the colored appearance of alkali halide crystals after exposure to x-rays. Today this term is broadly used to describe the microscopic defects that are responsible for the optical property changes throughout the ultraviolet, visible, and infrared regions of the spectrum that occur in irradiated or chemically treated materials. Although impurities may associate with color centers to produce optical absorption or emission bands, color centers are generally defined as vacancies, interstitials, or clusters of these types of defects. These defects can be produced in transparent insulator-type crystals. Recent studies have emphasized both oxide and fluoride materials, but the great wealth of experimental information and tradition still resides within the alkali halide crystal series. For a number of years it was thought that color center research would only yield prototype information for more complex materials with application potential. In the last few years, however, it has been acknowledged that color center research is important in such diverse applied projects as nuclear waste disposal and tunable infrared *F*-center lasers.

The *F* center is the best characterized color center. This center is formed when a negative ion is displaced from a lattice site into an interstitial position in the lattice or to the surface of the crystal. The displacement of this negative ion results in a potential well formed by the next neighbor positive ions that surround the negative ion site. This potential well traps an electron and forms the *F* center—a negative ion vacancy with a trapped electron for charge compensation. Both the "particle-in-the-box" and "hydrogenic" models can be used as crude approximations to determine the energy levels for the *F* center. The former model predicts that the absorption energy of *F* centers is inversely proportional to the square of the lattice constant (potential well size). This approximation is surprisingly accurate in predicting the absorption energies of *F* centers in alkali halides and is helpful for other materials such as MgO, MgF_2, and CaF_2. However, much more sophisticated theoretical treatments are necessary to understand the width of the absorption bands, the temperature dependence of these bands and the lifetimes of the transitions. Table 1 lists the energy for the maximum of the *F*-center absorption in a number of materials.

Although the negative ion vacancy is a "good" electron trap only about 80% of the electron charge is centered within the vacancy. This means that the electron spends most of its time in the potential well interacting with next neighbor positive ions, but for a significant amount of time the electron interacts with other neighbors as well. Since the motion of the surrounding ions continually changes the dimensions of the potential well for the electron, the *F* band absorption is very broad. Of course, as the temperature of a sample is lowered and the ion motion is decreased the width of absorption band is also decreased. At very low temperature (less than -200°C) ion motion is almost entirely the result of zero-point vibrations and the width of the absorption remains constant.

TABLE 1. COLOR CENTER ABSORPTION IN SELECTED MATERIALS.

	Wavelength of the Maximum of the Absorption Bands (nm) at 300K		
	F	F_2 (M)	F_3 (R)
LiF	250	447	310
			380
NaF	340	499	450
KCl	560	822	680
			740
KI	660	1010	810
			905
RbBr	680	957	805
			860
MgF_2	260	322	300
		355	
		370	
$RbMgF_3$	290	387	300
	340		
$KMgF_3$	270	446	396

When an F center absorbs a photon of the appropriate energy the electron is promoted to an excited state. For device applications how this electron returns to the ground state is most important. For example, the electron can return to the ground state by interacting with neighboring ions to give up its excess energy as heat to the crystal lattice. This is referred to as a *nonradiative* transition and can be highly temperature dependent. The electron can also pass to the ground state through the emission of a photon (luminescence). This emission is especially important for phosphor applications or for laser action. The emission band is broad for the same reason that the absorption is broad. Since in every instance some energy is shared with the lattice neighbors while the electron is in the excited state, there is an energy difference between the absorption and the emission. This is known as the *Stokes shift*. The emission for most alkali halide F centers occurs in the near infrared where tunable lasers are sorely needed. Such lasers have been constructed using F centers and impurity perturbed F-cluster centers. Long-term stability at room temperature is the most serious deficiency of these small solid state lasers.

In some materials the excited state of the F center is close to the conduction band in energy. Thus, at higher temperatures it is possible under light excitation for the excited state electron to move into the conduction band. This results in ionization of the F center, and photoconductivity is observed. The last method of deexcitation of the excited electron which will be considered is energy transfer. It is possible when other defects with similar energy levels are nearby for energy to be transferred from one defect to another. This mechanism has been most thoroughly studied for impurities, but pertains to color centers as well when the concentration of such centers is high. Clearly, the understanding of optical properties of materials is necessary for the development of practical applications. Because of the ease of growth and preparation of alkali halide crystals and the atomic simplicity of these materials, numerous experimental and theoretical advances have occurred through the study of color centers. The great number of high technology techniques which have been developed for detailed investigation of color center phenomena are now being applied to more complex materials such as quartz.

Although the F center is the best known and understood color center, a number of defects have been observed and characterized. A partial list of these defects is given below. Table 1 lists the absorption energy for these defects in several types of crystals:

1. The F center. In a monovalent anion material such as KCl or MgF_2 the F center consists of an electron trapped at a negative ion vacancy (charge compensation is essentially complete). In the case of a divalent anion material such as MgO, the F center consists of two electrons trapped at the negative ion vacancy (charge compensation is again essentially complete). When charge compensation is not complete, e.g., one electron trapped at a divalent anion site, the center is referred to as an F^+ center, indicating a net positive charge at this site. F centers may also be trapped by impurities. When this occurs the notation F_A center is utilized.

2. F-center aggregates. In many instances F centers aggregate to form clusters. The simplest such cluster is the F_2 or M center, which consists of two F centers in neighboring positions. This center resembles a hydrogen molecule and its optical properties have been approximated using such a model. When the two negative ion vacancies share only one electron the center is referred to as an F_2^+ center. This center can be approximated as an H_2^+ molecule. Because of the stability and optical properties of these centers, a number of infrared laser system using them have been constructed. When three F centers occupy next neighbor positions the complex is given the notation F_3 or R center. As with the F_2 centers, because of the planar nature of these defects nonequivalent dipoles exist and more than one absorption band is associated with these defects.

3. Hole traps. In addition to electron centers, certain defects can trap holes and give rise to optical absorption and emission. The two best studied such centers are the X_2^- or V_K center and the so-called H center. The X_2^- center is formed in highly ionic crystals such as the alkali halides and alkaline earth flourides. This defect has not been detected generally in oxide materials. The V_K center consists of a hole trapped between two negative halide ions. It may be pictured as a X_2^- molecule that occupies two normal halide sites in a crystal in which X^- denotes the negative ion halide. The H center consists of a hole that is shared by four negative ion halide ions occupying three normal halide ion lattice sites in a crystal. The H center forms an interstitial defect. Since the halide is interstitial, a negative-ion vacancy must exist elsewhere in the crystal. H centers are stable only at low temperature. As the temperature of a material is increased, the H centers aggregate to form interstitial clusters or dislocation loops.

There are a number of ways to produce color centers. The most generally used methods involve irradiation with x-rays, electrons, protons, or neutrons; the use of chemical heat treatment; or the passage of a dc electrical current through the sample while it is at a high temperature. The latter two treatments generate mostly F-type centers without introducing interstitials. When crystals are irradiated both interstitials and vacancies are formed. In highly ionic materials photochemical production of these defects occurs during irradiation. Vacancy-interstitial pairs are produced with as little as

20 eV energy expenditure. The production of electron-hole pairs in the crystal is sufficient to form X_2^- centers which then degenerate into F centers and H centers. In the photochemical process one 1 MeV electron can produce as many as 10,000 F centers. For those materials in which X_2^- centers are not easily formed, such as oxide crystals, this type of damage mechanism is negligible. Instead the impinging radiation has the same effect as in metal crystals. Interstitial-vacancy pairs are produced when the impinging particle has sufficient energy to displace an ion into an interstitial position. In MgO the required energy is about 60 eV so that electrons with a minimum energy of 330 KeV are required to produce ionic damage. These materials are much more difficult to damage by irradiation than are the more ionic halides. Typically about one stable F center is produced for every ten 1-MeV electrons.

The thermal, electrical, mechanical, and optical properties of most crystals are changed considerably when color centers are present. Only a few parts per million of interstitial type color centers can change the hardness of a material appreciably. Because of these changes, the study of color centers in those materials with potential for practical applications is particularly important. Oxide materials such as MgO, Al_2O_3, and quartz have been studied as have most flouride systems. New work is progressing on wide band gap semiconductors and even more complicated mixed crystals.

W. A. SIBLEY

References

Books

Schulman, J. H. and Compton, W. D., "Color Centers in solids," London, Pergamon Press, 1962.
Fowler, W. B. (Ed.), "Physics of Color Centers," New York, Academic Press, 1968.
Markham, J. J., "*F*-Centers in Alkali Halides," Supplement No. 8, *Solid State Physics*, (F. Seitz and D. Turnbull, Eds.), New York, Academic Press, 1966.
Crawford, J. H. and Slifkin, L. M. (Eds.), "Point Defects in Solids," Vol. 1, p. 201, New York, Plenum Press, 1972.
Henderson, B., and Wertz, J. E., "Defects in the Alkaline Earth Oxides," London, Taylor and Francis Ltd., 1977.
Stoneham, A. M., "Theory of Defects in Solids," Oxford, Clarendon Press, 1975.

Review Articles

Sibley, W. A., and Pooley, D., "Radiation Studies of Materials Using Color Centers," in "Treatise on Materials Science and Technology," Vol. 5, (H. Herman, Ed.), p. 45. New York, Academic Press, 1979.
Mollenauer, L. F., "Color Center Lasers," in "Methods of Experimental Physics," Vol. 15B, (E. L. Tang, Ed.) p. 1, New York, Academic Press, 1979.
Litfin, G. and Welling, H., "Color Center Lasers,"
in "Laser Advances," (B. S. Wherrett, Ed.) p. 39, New York, John Wiley & Sons, 1980.

Other Articles

Kabler, M. N. and Williams, R. T., *Phys. Rev.* B18, 1948 (1978).
Catlow, C. R. A., Diller, K. M., and Hobbs, L. W., *Phil. Mag.* 42, 123 (1980).

Cross-references: EXCITONS; LATTICE DEFECTS; LUMINESCENCE; PHOTOCONDUCTIVITY; RADIATION, IONIZING, BASIC INTERACTIONS.

COMPRESSIBILITY, GAS

(a) The *compressibility* of a gas is defined as the rate of volume decrease with increasing pressure, per unit volume of the gas. The compressibility depends not only on the state of the gas, but also on the conditions under which the compression is achieved. Thus, if the temperature is kept constant during compression, the compressibility so defined is called the isothermal compressibility β_T:

$$\beta_T = -\frac{1}{V}\left(\frac{\partial V}{\partial P}\right)_T = \frac{1}{\rho}\left(\frac{\partial \rho}{\partial P}\right)_T. \qquad (1)$$

If the compression is carried out reversibly without heat exchange with the surroundings, the adiabatic compressibility at constant entropy, β_S, is obtained:

$$\beta_S = -\frac{1}{V}\left(\frac{\partial V}{\partial P}\right)_S = \frac{1}{\rho}\left(\frac{\partial \rho}{\partial P}\right)_S. \qquad (2)$$

The two compressibilities are related by

$$\beta_S/\beta_T = C_V/C_P. \qquad (3)$$

Here P is the pressure, V the volume, ρ the density, T the temperature, S the entropy, and C_V, C_P specific heats at constant volume and pressure, respectively.

The *compressibility factor* of a gas is the ratio PV/RT (cf. GAS LAWS). This name is not well chosen since the value of the compressibility factor is no indication of the compressibility.

(b) The *experimental behavior* of the compressibility as a function of pressure and temperature is as follows: dilute gases obey the laws of Boyle and Gay-Lussac, $PV=RT$, to a good approximation. The compressibilities β_T and β_S of a dilute gas are then given by

$$\beta_T = \frac{1}{P}; \; \beta_S = \frac{(C_V/C_P)}{P}. \qquad (4)$$

Compressed gases, however, show large deviations from the behavior predicted by Eq. (4). This is demonstrated in Fig. 1, where the isothermal compressibility of argon, divided by the corresponding value for a perfect gas at the

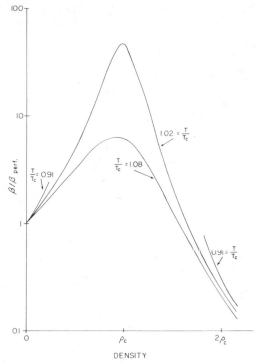

FIG. 1. The ratio $\beta/\beta_{perf.}$ of the isothermal compressibility of argon to that of a perfect gas at the same density, as a function of the density, at 0.91, 1.02, and 1.08 times the critical temperature. The critical density is indicated by ρ_c.

force of impact increase as well, giving rise to an extra increase of pressure. Therefore $\beta_S < \beta_T$. The actual magnitude of the temperature rise depends on the internal state of the molecules; the more internal degrees of freedom available, the more energy can be taken up inside the molecule and the smaller the temperature rise on adiabatic compression. If molecules have many internal degrees of freedom, the difference between β_T and β_S, just like the difference between C_P and C_V, becomes small.

Molecular theory also shows that if the molecules have negligible volume and do not interact with each other, the gas follows the perfect gas laws of Boyle and Gay-Lussac so that Eq. (4) holds for the compressibilities β_T and β_S. However, in real gases the molecular volume is not negligible. Consequently, in states of high compression little free space is left to the molecules and thus the real gas and the liquid have low compressibilities as compared to the perfect gas (Fig. 1). On the other hand, the mutual attraction molecules experience as they approach each other makes a real gas easier to compress than a perfect gas. This explains the initial rise of $\beta_T/\beta_{perf.}$ at temperatures not too far above critical.

(d) *Experimental values* for the compressibility could be obtained in principle by measuring the pressure increase on a small volume decrement. In practice a measurable pressure increase is obtained only in regions of low compressibility, i.e., in the dense gas. Usually, compressibilities are determined in indirect ways. Thus, the isothermal compressibility, being proportional to $(\partial V/\partial P)_T$, can be deduced from experimental PVT data if these data are sufficiently accurate and densely spaced. For obtaining the adiabatic compressibility from PVT data, some additional information is needed, for instance SPECIFIC HEAT data in the perfect gas limit. For reviews of experimental methods for determining PVT relations and deriving thermodynamic properties see Refs. 1a, b.

The adiabatic compressibility is readily obtained from speed-of-sound data through the relation

$$v^2 = \frac{1}{\rho \beta_S} \qquad (5)$$

same density, Eq. (4), is pictured as a function of density for various temperatures. At all temperatures the compressibility at high densities falls to a small fraction of the value for a perfect gas. As the critical temperature is approached from above, a large maximum occurs in β_T at densities near critical. The isothermal compressibility diverges strongly at the critical point (just like C_P, see Ref. 1d) and is infinite everywhere in the two-phase region, where the pressure does not rise on isothermal compression. The adiabatic compressibility, however, diverges only weakly (like C_V, see Ref. 1d) when the critical point is approached, and it is finite in the two-phase region.

(c) Simple notions taken from *molecular theory* can be used to explain the general features of the compressibility in its temperature and density dependence. The pressure of the gas is caused by the impact of the molecules on the wall. If the volume is decreased at constant temperature, the average molecular speed and force of impact remain constant, but the number of collisions per unit area increases and thus the pressure rises. If the gas is compressed adiabatically, the heat of compression cannot flow off, thus the average molecular speed and

Eq. (5) is valid only when compressions and expansions of the sound wave are truly reversible and adiabatic, i.e., if the frequency is low and the amplitude small. Experimental techniques for determining the speed of sound are discussed in Ref. 1c.

The isothermal compressibility, through the fluctuation theorem of statistical mechanics, is related to the density fluctuations in the system.[2,3] These density fluctuations are responsible for the scattering of light. Thus, the isothermal compressibility, in principle, could be directly obtained from the intensity of scattered light.[2,3] In gases, sufficient intensity is obtained only in regions of large density fluctuations, i.e., near the critical point. Experimental methods and results are discussed in Refs. 1d, 3.

(e) *Theoretical predictions* for the isothermal compressibility are obtained from evaluation of the statistical-mechanical partition function or the radial distribution function. In either case, a model for the intermolecular potential is required. An approximate calculation of the partition function valid at low densities leads to the virial expansion;[2] this expansion expresses PV/RT in a power series in density, the coefficients being related to the interactions of two, three, etc. particles; the compressibility then follows straightforwardly.

The virial expansion is not useful for dense gases, because convergence is slow and higher virials are hard to calculate. For dense systems, methods have been developed for evaluation of the radial distribution function $g(r)$, which is the ratio of the average density of molecules at a distance r from any given molecule to the average density in the gas. The compressibility is related to $g(r)$ by the fluctuation theorem

$$kT \beta_T = 1/\rho + \int [g(r) - 1] \, 4\pi r^2 \, dr \qquad (6)$$

Integral equations for $g(r)$, relating it to the molecular potential, its derivatives, and higher-order distribution functions, have been developed. They can be solved after approximations about the form of these higher-order distribution functions are made. These solutions are hard to obtain for any but spherical interactions, and invariably suffer from internal inconsistency. For systems of hard spheres, good solutions are available that have been verified by comparison with results from computer simulation.

The most promising route toward predicting the equation of state of dense fluids of nonspherical molecules has been the approximate evaluation of the partition function by the so-called *perturbation method*. Here it is assumed that the partition function is known for a reference system, which may be a system of hard spheres. The departure of the real interaction from that in the reference system is then treated as a perturbation in a high-temperature expansion of the partition function. Considerable progress has been made in recent years with the prediction of properties of systems of molecules with nonspherical hard cores, "soft" repulsions, dipolar and quadrupolar interactions, etc. Validation of the results of the perturbation method is increasingly performed by comparison with the results of simulation of molecular systems on high-speed computers. Ref. 4 gives a complete survey of and references to the more successful integral equation and perturbation methods.

Many semiempirical equations of state with varying degrees of theoretical foundation have been developed and can be used with Eq. (1) to calculate the compressibility. Van der Waals' equation, a two-parameter equation which gives a qualitative picture of the PVT relation of a gas and of the gas-liquid transition, is an example. For surveys of useful semiempirical equations see Refs. 1b, 2 and 4.

J. M. H. LEVELT SENGERS

References

1. Vodar, B., and Le Neindre, B. (Eds.), "Experimental Thermodynamics" (I.U.P.A.C.): Volume II: "Experimental Thermodynamics of Non-Reacting Fluids," New York, Plenum Press; London, Butterworths, 1973/74. The following articles are particularly pertinent.
   a. Trappeniers, N. J., and Wassenaar, T., "*PVT* relationships in gases at high pressure and moderate or low temperatures."
   b. McCarty, R. D., "Determination of thermodynamic properties from the experimental *PVT* relationship."
   c. Van Dael, W., "Measurement of the velocity of sound and its relation to the other thermodynamic properties."
   d. Levelt Sengers, J. M. H., "Thermodynamical properties near the critical state."
2. Hirschfelder, J. O., Curtiss, C. G., and Bird, R. B., "Molecular Theory of Gases and Liquids," New York, John Wiley & Sons, 1964.
3. McIntyre, D. M., and Sengers, J. V., "Studies of fluids by light scattering," in "Physics of Simple Liquids," (H. N. V. Temperley, J. S. Rowlinson, and G. S. Rushbrooke, eds.), Ch. 11, p. 449, Amsterdam, North Holland Publishing Co., 1968.
4. Boublík, T., Nezbeda, I., and Hlavatý, "Statistical Thermodynamics of Simple Liquids and Their Mixtures," Amsterdam, Elsevier Publ. Co., 1980.

Cross-references: DENSITY AND SPECIFIC GRAVITY, GAS LAWS, KINETIC THEORY.

COMPTON EFFECT

Introduction The Compton effect refers to the collision of a photon and a free electron in which the electron recoils and a photon of longer wavelength is emitted as indicated in Fig. 1. It is one of the most important processes by which x-rays and γ-rays interact with matter and is also one which is accurately calculable theoretically.

A discussion of the effect is found in most textbooks on atomic physics. Particularly complete presentations have been made by Evans.[1,2]

History Barkla and others (1908) made many observations on the scattering of x-rays by different materials. The diffuse scattering was interpreted qualitatively by J. J. Thomson in terms of the interaction of electromagnetic waves with electrons which he had shown to be a constituent of all atoms. As more experiments were carried out with light elements, it was established by J. A. Gray (1920) that the diffusely scattered x-rays were less penetrating. This implied that the scattered radiation had a longer wavelength than the incident radiation. This could not be reconciled with Thomson's theory which represented x-rays as continuous electromagnetic waves with wavelengths unchanged by scattering.

The effect which now bears his name was established quantitatively by Arthur Holly Compton (1923) when he published careful spectroscopic measurements of x-rays scattered at various angles by light elements. He found that x-rays scattered at larger angles had systematically larger wavelengths. In searching for an explanation of the data, he discovered that the observations were accounted for by considering the scattering as a collision between a single photon and a single electron in which energy and momentum are conserved.

The important place which the effect occupies in the development of physics lies in his interpretation of the effect in terms of the newly emerging quantum theory. The essential duality of waves and particles was demonstrated in an especially clear way, since the collision conserved energy and momentum while both the incident and scattered x-rays revealed wave-like properties by their scattering from a crystal. In recognition for this contribution, Compton was awarded the Nobel Prize in 1927.

A complete theory for the effect was worked out in 1928 by Klein and Nishina using Dirac's relativistic theory of the electron. The calculation was one of the brilliant successes of the Dirac theory. It represents quantitatively, within the experimental uncertainties, all phenomena associated with the scattering of photons by electrons for energies up to several billion electron volts. Because of the confidence with which photon interaction with electrons can be interpreted, the Compton effect has been important in the analysis of the energy and the polarization of gamma rays from many sources.

Kinematics The relations between the energies and directions of the incident and scattered photons and the recoil electron are determined by the conservation of energy and of the components of momentum parallel and at right angles to the incident beam. In the usual case, where the electron is initially at rest and the energy and momentum of the incident photon are $h\nu$ and $(h\nu/c)$, the equations are:

$$h\nu = h\nu' + T \tag{1}$$

$$\frac{h\nu}{c} = \frac{h\nu'}{c} \cos \theta + p \cos \phi \tag{2}$$

$$0 = \frac{h\nu'}{c} \sin \theta - p \sin \phi \tag{3}$$

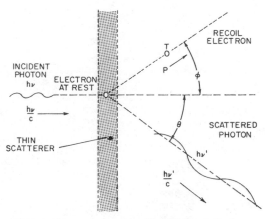

FIG. 1. Diagram showing the initial and final energies and momenta for Compton scattering.

where c is the velocity of light, h is Planck's constant, and the angles are those indicated in Fig. 1. The relativistic relation between the kinetic energy T of the recoiling electron and its momentum p is

$$pc = \sqrt{T(T + 2mc^2)} \tag{4}$$

where m is the mass of the electron. These equations can be combined to obtain relations which are useful in the interpretation of data. The Compton shift is

$$\lambda' - \lambda = \frac{c}{\nu'} - \frac{c}{\nu} = \frac{h}{mc}(1 - \cos \theta) \tag{5}$$

This relation was first found experimentally by Compton, who noted that the shift in wavelength ($\lambda' - \lambda$) depended on the angle, but not on the wavelength, of the incident photon. The quantity (h/mc), which is the shift at $90°$, is called the Compton wavelength of the electron and is one of the useful constants (2.4262×10^{-10} cm).

$$h\nu' = \frac{mc^2}{1 - \cos \theta + \dfrac{mc^2}{h\nu}} \tag{6}$$

In this form, the energy of the scattered photon is seen to vary from that of the incident photon at $0°$ to less than $(mc^2/2)$ at $180°$. At high energies the angle θ for which $h\nu'$ is $(h\nu/2)$ is approximately $2(mc^2/h\nu)$ radians.

The kinetic energy of the recoiling electron is

$$T = \frac{h\nu(1 - \cos \theta)}{(1 - \cos \theta) + \dfrac{mc^2}{h\nu}} \tag{7}$$

The relation between the scattering angles of the electron and photon is

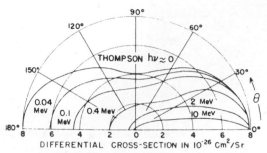

FIG. 2. Differential cross section for photons scattered at angles, θ, for a number of incident energies.

$$\cot \phi = \left(1 + \frac{h\nu}{mc^2}\right)\left(\frac{1 - \cos \theta}{\sin \theta}\right) \qquad (8)$$

Graphs of these kinematic relations and of the scattering cross section are given by Evans[2] and by Nelms[3].

Scattering of Unpolarized Radiation The differential cross section for the scattering of unpolarized radiation at an angle θ is given by the Klein and Nishina equation.

$$\frac{d\sigma}{d\Omega} = \frac{r_0^2}{2}\left(\frac{\nu'}{\nu}\right)^2\left(\frac{\nu}{\nu'} + \frac{\nu'}{\nu} - \sin^2 \theta\right) \qquad (9)$$

where r_0 is the electron radius $= e^2/mc^2 = 2.8177 \times 10^{-13}$ cm, and ν' is obtained from Eq. (6). The cross section is shown as a function of θ for several energies in Fig. 2. The classical Thomson cross section $r_0^2 (1 + \cos^2\theta)/2$ can be seen to hold for low energies where $\nu' \approx \nu$.

The total cross section obtained by integrating this cross section over angle is important in the attenuation of well-defined beams in passing through a material. The relative importance of Compton scattering as compared to the photoelectric effect and pair production is illustrated for aluminum in Fig. 3 where the attenuation coefficient α is shown as a function of energy. The fraction of the photons surviving without an interaction upon passing through x g/cm^2 of aluminum is $e^{-\alpha x}$. The Compton effect is the major one between 0.5 and 2 MeV. Extensive tables and graphs for other elements are available.[2,4]

In detectors whose response is proportional to the energy deposited by the recoil electrons, the distribution of electron energies associated with a photon of known energy is of interest. The distribution is given by the relation

$$\frac{d\sigma}{dT} = \frac{\pi r_0^2 mc^2}{(h\nu)^2}\left\{2 + \left(\frac{T}{h\nu - T}\right)^2\right.$$

$$\left.\left[\frac{(mc^2)^2}{(h\nu)^2} + \frac{h\nu - T}{h\nu} - \frac{2mc^2(h\nu - T)}{h\nu T}\right]\right\}$$

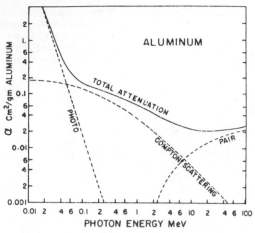

FIG. 3. The attenuation coefficients, α, for the absorption of photons in aluminum as a function of energy. The broken lines represent the separate contributions of the photoelectric effect, the Compton effect, and pair production to the absorption.

where T varies from 0 to $T_{\max} = 2(h\nu)^2/(2h\nu + mc^2)$. A number of these distributions are shown in Fig. 4.

Scattering of Plane Polarized Radiation The differential cross section for the scattering of plane polarized radiation by unoriented electrons was also derived by Klein and Nishina. It represents the probability that a photon, passing through a target containing one electron per square centimeter, will be scattered at an angle θ into a solid angle $d\Omega$ in a plane making an angle η with respect to the plane containing the electric vector of the incident wave.

$$\frac{d\sigma}{d\Omega} = \frac{r_0^2}{2}\left(\frac{\nu'}{\nu}\right)^2\left(\frac{\nu}{\nu'} + \frac{\nu'}{\nu} - 2\sin^2 \theta \cos^2 \eta\right) \quad (10)$$

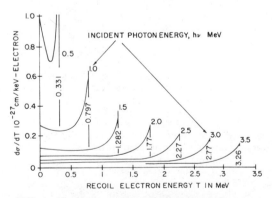

FIG. 4. The energy distribution of the Compton recoil electrons for several values of the incident photon energy $h\nu$. Based on figure in "Compton Effect" by R. D. Evans in "Handbuch der Physik," Vol. XXXIV, pp. 234–298, 1958, J. Fluge, Ed., by permission of Springer-Verlag, publishers.[2]

The cross section has its maximum value for $\eta = 90°$, indicating that the photon and electron tend to be scattered at right angles to the electric vector of the incident radiation.

This dependence is the basis of several instruments for determining the polarization of photons. For example, it was used by Wu and Shaknov[5] to establish the crossed polarization of the two photons emitted upon the annihilation of a positron electron pair; by Metzger and Deutsch[6] to measure the polarization of nuclear gamma rays; and by Motz[7] to study the polarization of bremsstrahlung.

Scattering of Circularly Polarized Radiation
The scattering of circularly polarized photons by electrons with spins aligned in the direction of the incident photon is represented by

$$\frac{d\sigma}{d\Omega} = r_0^2 \left(\frac{\nu'}{\nu}\right)^2 \left[\left(\frac{\nu}{\nu'} + \frac{\nu'}{\nu} - \sin^2 \theta \right) \right.$$
$$\left. \pm \left(\frac{\nu}{\nu'} - \frac{\nu'}{\nu} \right) \cos \theta \right] \quad (11)$$

The first term is the usual Klein-Nishina formula for unpolarized radiation. The + sign for the additional term applies to right circularly polarized photons. The ratio of the second term to the first is a measure of the sensitivity of the scattering as a detector of circularly polarized radiation and is shown in Fig. 5.

In practice, the only source of polarized electrons has been magnetized iron where 2 of the 26 electron spins can be reversed upon changing its magnetization. Although the change in the absorption or scattering is usually only a few

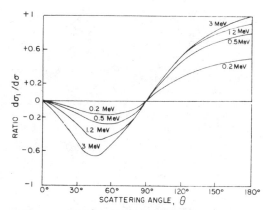

FIG. 5. The ratio of the partial cross section dependent on the spin orientation of the electrons to the average cross section as represented by the second and first terms of Eq. (10). Based on figure in "Compton Effect" by R. D. Evans in *Handbuch der Physik*, Vol. XXXIV, pp. 234–298, 1958, J. Fluge, Ed., by permission of Springer-Verlag, publishers.[2]

per cent, this is often sufficient to get accurate and reliable measurements of circular polarization.

Cross sections for some practical arrangements and discussions of earlier work are presented by Tolhoek.[8] Applications to the determination of the helicities of photons, electrons, and neutrinos in confirming the two-component theory of the neutrino are reviewed in considerable detail by L. Grodzins.[9]

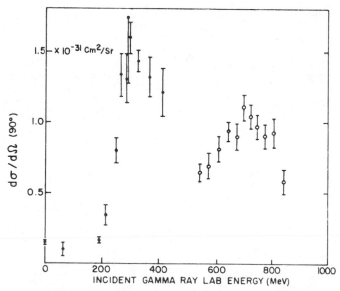

FIG. 6. The differential cross section for the scattering of high-energy photons by protons at 90° in the center of mass system. Based on figure in Steining, Loh and Deutsch, *Phys. Rev. Letters*, **10**, 536 (1963).[10]

Scattering from Moving Electrons Compton backscattering of low-energy photons by high-energy electrons can produce photons with energies comparable with those of the electrons. Milburn[10] and Arutyunian et al.[11] pointed out that backward scattering of an intense polarized laser beam by a beam of high energy electrons could produce a useful beam of monoenergetic photons of intermediate energy. The maximum energy of the backscattered photons is given by

$$hv' = hv\frac{4(E/mc^2)^2}{1 + 4hvE/(mc^2)^2} \qquad (12)$$

where E is the total energy of the electron. For photons from a ruby laser ($\lambda = 6943$ Å, $hv = 1.79$ eV) the photons backscattered from 6-GeV electrons will have energies of 848 MeV. Photons from a ruby laser interacting with the electron beam at the Stanford Linear Accelerator Center produced monoenergetic polarized photons of 1.44, 2.8, and 4.7 GeV for the study of photon interactions in a hydrogen bubble chamber.[12] Higher-intensity facilities for producing intermediate-energy photons for photonuclear research has been proposed by Italian and U.S. groups.[14] These facilities would use photons from high-intensity lasers interacting with 1.5 GeV and 2.5 GeV electron beams in storage rings.

Proton and Deuteron Compton Effect Particle-like scattering of high-energy photons by protons and deuterons has been observed and has been referred to as the proton and deuteron Compton effect. The kinematic equations are identical to those for electrons except that the mass is that of the proton or deuteron.

Although the cross sections are smaller than that for electrons, by the square of the ratio of the masses, the scattering is easily distinguished by the characteristically higher energy of the radiation at large angles. At energies above the pion threshold, the cross section is dominated by pion nucleon resonances. The experimental cross sections for the scattering by protons, as presented by Steining, Loh, and Deutsch,[15] are shown in Fig. 6. Some experimental results and calculations on the coherent scattering from deuterium are described by Jones, Gerber, Hanson, and Wattenberg.[16]

Measurements at photon energies up to 7 and 16 GeV and comparisons with theoretical predictions of the vector dominance model have been reported by groups from Hamburg[17] and Stanford.[18]

A. O. HANSON

References

1. Evans, R. D., "The Atomic Nucleus," Chapter 23, New York, McGraw-Hill Book Co., 1955.
2. Evans, R. D., "Compton Effect," In Flugge, S., Ed., "The Encyclopedia of Physics," Vol. 34, pp. 234–298, Berlin, Springer-Verlag, 1958.
3. Nelms, A. T., "Graphs of the Compton Energy-Angle Relationship and the Klein-Nishina Formula from 10 keV to 500 MeV," *Natl. Bur. Std. Circ.*, **542** (1953).
4. White, G. R., "X-Ray Attenuation Coefficients form 10 keV to 100 MeV," *Natl. Bur. Std. Rept.*, **1003** (1952).
5. Wu, C. S., and Shaknov, I., "Angular Correlation of Scattered Annihilation Radiation," *Phys. Rev.*, **77**, 136 (1950).
6. Metzger, F., and Deutsch, M., *Phys. Rev.*, **78**, 551 (1950).
7. Motz, J. W., "Bremsstrahlung Polarization Measurements for 1 MeV Electrons," *Phys. Rev.*, **104**, 557 (1956).
8. Tolhoek, H. A., "Electron Polarization Theory and Experiment," *Rev. Mod. Phys.*, **28**, 277 (1956).
9. Grodzins, L., "Measurement of Helicity," in Frisch, O. R., Ed., "Progress in Nuclear Physics," New York, Pergamon Press, 1959.
10. Milburn, R. H., *Phys. Rev. Letters* **10**, 75 (1963).
11. Arutyunian, F. R., and Tumanian, V. A., *Phys. Letters* **4**, 176 (1963).
12. Ballam, J., et al., *Phys. Rev. Letters* **23**, 498 (1969).
13. Federici, L., et al., *Lettere al Nuovo Cimento*, **27**, 339 (1980).
14. Sandorfi, A. M., et al., IEEE NS **30**, 3083 (1983).
15. Steining, R. F., Loh, E., and Deutsch, M., "The Elastic Scattering of Gamma Rays by Protons," *Phys. Rev. Letters*, **10**, 536 (1963).
16. Jones, R. S., Gerber, H. J., Hanson, A. O., and Wattenberg, A., "Deuteron Compton Effect," *Phys. Rev.*, **128**, 1357 (1962).
17. Buschhorn, G., Criegee, L., Franke, G., Heide, P., Kotthaus, R., Poelz, G., Timm, U., Vogel, G., Wegener, K., Werner, H., and Zimmerman, W., Proton Compton Scattering between 2.2 and 7 GeV," *Phys. Lett.*, **37B**, 207 (1971).
18. Boyarski, A. M., Coward, D. H., Ecklund, S., Richter, B., Sherden, D., Siemann, R., and Sinclair, C., "Forward Compton Scattering from Hydrogen and Deuterium at 8 and 16 GeV," *Phys. Rev. Lett.*, **26**, 1600 (1971).

Cross-references: COLLISIONS OF PARTICLES, CONSERVATION LAWS AND SYMMETRY, ELEMENTARY PARTICLES, QUANTUM THEORY, X-RAYS.

COMPUTERS

Introduction A computer may be defined as a device capable of solving problems by accepting data (input), performing prescribed operations on the data (processing), and providing the results of these operations (output). The basic distinction between electronic calculators, which also fit this definition, and computers is that the latter provide speed in performing complex operations, virtually infinite program and data storage, and the ability to interact with the environment (including other computers) on a real time basis.

Computers may be hydraulic, mechanical, electromechanical, or electronic devices. They are broadly classified as being *digital, analog,* or *hybrid,* i.e., analog and digital linked together. The first two types differ fundamentally in the manner in which data are stored and operated upon. Analog computers operate on continuous variables, in the form of voltage or current (the electrical analog of a number or physical quantity), that represent continuous data. Digital computers operate on discrete numerical data represented by a series of binary digits. The data and instructions are stored internally and are indistinguishable in memory. Digital computers perform calculations by adding binary numbers according to instructions derived from coding an arithmetical expression or series of expressions (an algorithm) that represent the problem to be solved. Even though the problem may include differential or trigonometric expressions, each is reduced to simple addition or subtraction (addition of the complement) since addition is the only direct operation a digital computer can perform. Instructions, which are written sequentially, are executed sequentially (one at a time). Calculations usually culminate in a numerical display or graphical representation of the results. Because of the sequential nature of the digital process, the amount of time spent in solution is proportional to the problem complexity. Analog computers, on the other hand, are composed of elements which perform summation, integration, multiplication and differentiation directly. Rather than a serially coded algorithm, the instructions are in the form of basic modifications of the analog computer's circuitry by a user-wired patch board panel. There is no internal data or program storage as such. Instructions are executed effectively si-multaneously (in parallel) while the solution is displayed in graphical form in continuous fashion. Since operations are performed in parallel, increased problem complexity demands more computer components, not more time. Solution time is therefore a function of the time characteristics of the problem, not the machine. The speed of the analog computer, therefore, is orders of magnitude faster than the digital computer in solving complex problems that involve calculus. This speed is paid for by a precision of results that is orders of magnitude less than for a digital computer. Hybrid computers attempt to combine the best features of both types i.e., the speed of the analog with the precision of a digital. This is brought about through a digital "front-end" which is used to set up and check out the analog computer and its program. The actual computation time is equivalent to that of a pure analog machine, but the total time (including set-up time) is significantly reduced. Indeed, problems may be solved that would be cost prohibitive on a digital computer and, for all practical purposes, impossible to set up on an analog computer.

As described below, in physics all three types of computers are utilized. Digital computer applications range from theoretical calculations to pure data collection and experiment control. Applications of analog computers involve primarily solutions of differential equations which serve as models of physical systems. Hybrid systems come into play for extremely complex simulation. Examples include determination of tactical envelopes for missiles, optimization of control settings in a nuclear power plant, and development of chemical kinetic models. Whatever the branch of physics, one can be sure that computers today play some significant role.

Computers of the Past The abacus has been

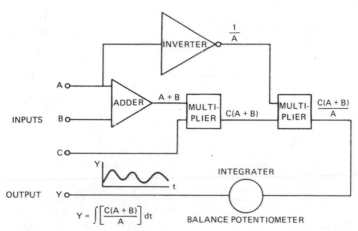

FIG. 1. Typical analog computer configuration to solve the equation for y. Requires: 1 adder, 1 inverter, 2 multipliers, 1 integrator.

called the earliest computing machine. For more than 3000 years after the recording of its invention, this was the sole device available to aid in arithmetic calculations. Finally, in the year 1642, a Frenchman, Blaisé Pascal, invented the first mechanical adding machine. His stylus operated "arithmetic machine" had the ability to handle carry overs from one column to the next. Nearly 200 more years elapsed before C. X. Thomas, also a Frenchman, working on a concept proposed by the German, Gottfried Wilhelm Leibniz, built an "arithmometer" in 1820 which, besides addition and subtraction, could perform multiplication and division using the concept of repeated addition and subtraction.

While this was going on, an independent development in the weaving industry saw Frenchmen Jacques de Vaucanson, in 1741, and Joseph Marie Jacquard, in 1804, use holes punched, first in metal drums, later in punched cards, as the control or programmer for textile looms. On a more esoteric plane, Englishman Charles Babbage, in 1823, began construction of his "difference engine." This device, based on the fact that an equation of degree n will have a constant nth difference, was used to make calculations for trigonometric and logarithmic tables. Babbage succeeded in constructing a machine to solve a second degree equation. But his "analytical engine," which had a memory, control, an arithmetic unit, and an input/output section could not be built at that time with sufficient precision to produce reliable results. Earliest analog computers of the type designed by Vannevar Bush and others in the 1930's actually followed these same mechanical principles until electronics began to take over in the late forties. This activity was culminated perhaps by the Maddida, which appeared in 1951. Analog computers since that time have differed mainly in the application of advanced modular electronics and in linkages with digital computers. Developments of calculators after Pascal and Thomas are not reliably documented. Two American mechanical adders, which spawned present day companies, are worthy of mention, however. These include the Felt "macaroni box" made in 1885 by a firm which later became the Victor Comptometer Corporation, and the "listing accountant" made by a forerunner of the Burroughs Corporation.

The real push behind the development of modern digital computers came about because of U. S. Government requirements, first to count people and tabulate corresponding data; later, during World War II, for developing artillery trajectory tables and performing calculations for the Manhattan project. In preparation for the 1890 census, it became clear that, using available counting and sorting techniques, it would be nearly time to perform the 1900 census before 1890 figures could be determined.

Herman Hollerith, who had worked for the census office and was a mechanical engineer, solved the problem by devising a punched card, which could contain all pertinent data, and a series of machines for punching, counting and sorting. Using this equipment, census figures were published in less than three years with unheard-of accuracy. Hollerith revised his punched cards in 1894. They contained 80 columns, each with holes for 0 through 9; this is exactly the format in predominant use today. Hollerith's efforts eventually led to the formation of the International Business Machines Corporation, the world's largest manufacturer of computers. Tremendous quantities of work have been accomplished and are still possible using punch card techniques. However, this methodology is practical only for sorting, counting and selecting in a limited number of ways.

Electronic and electrical techniques were required to make practical, accurate computations of the vast quantities of data to solve problems and prove theories which today serve as the cornerstone of modern science. These machines are based upon the foundations laid by George Boole, who pioneered in the field of symbolic logic, and Allan Turing, who hypothesized a universal computer. George Boole's algebra provides a mechanism for representing logic in mathematical symbols and rules for calculating the truth or falsity of statements. Digital computers carry out these operations an infinite number of times. Allan Turing's paper, "Computable Numbers," in 1937, described a hypothetical Turing machine that can solve any type of mathematical problem which can be reduced to coding in a given set of commands within the memory capacity of the machine.

First application of these principles resulted in the Bell Telephone Laboratories relay computer in 1939. Five years later, in 1944, the second significant relay computer, which was also the first general purpose digital machine, was constructed at Harvard by Professor Howard Aiken, with funds provided partially by IBM. Known as the Mark I, this machine was really a huge electro-mechanical calculator. The Electronic Numerical Integrator and Calculator, "ENIAC," the first electronic computer, was developed in 1946 by J. P. Eckert and J. W. Mauchly of the Moore School of Engineering in Philadelphia. Funds were provided by the U. S. Army with the promise that the machine would be suitable for calculating ballistic tables. This machine contained 18,000 vacuum tubes which replaced the former mechanical relays as switching elements. As might be expected, it was huge, weighing over 30 tons, and terribly unreliable. Nevertheless, it existed until 1955 after over 80,000 hours of operation had been logged.

Development proceeded thereafter at a much more rapid rate. IBM continued work, producing the Selective Sequence Electronic Calculator

(SSEC) in 1947. The Moore School developed a second machine known as the Electronic Discrete Variable Computer (or EDVAC) which became operational in 1952. EDVAC is accepted as the first stored program computer. Unlike earlier machines which were programmed at least partially by setting switches or using patch boards, this machine and all others which were to follow, stored instructions and data in identical fashion. EDVAC used acoustical delay lines, which were simply columns of mercury through which data passed at the speed of sound, as the main memory. This type of storage has given way to magnetic core memory and semiconductor memory in computers of today.

In 1946, John Von Neumann, a mathematician at the Institute of Advanced Study (IAS), Princeton, New Jersey, presented a paper entitled "Preliminary Discussion of the Logical Design of an Electronic Computing Instrument." This paper, which was prepared jointly under a contract with the U. S. Army, suggested the principles under which all digital computers which followed would be built. This included internal program storage, relocatable instructions, memory addressing, conditional transfer, parallel arithmetic, internal number base conversion, synchronous internal timing, simultaneous computing while doing input/output, and magnetic tape for external storage. An IAS computer, employing most of these concepts, actually went into operation in the early fifties. All machines which followed used virtually the same principles. UNIVAC I (Universal Automatic Computer) was the first commercially available digital computer (circa 1950). This machine was a direct descendent of ENIAC and EDVAC, having been built by Remington Rand following acquisition of the Echert-Mauchly Computer Corporation. Eventually, 48 UNIVAC I's were built making Remington Rand the number one computer manufacturer until International Business Machines (IBM) began in earnest in 1954 with the introduction of its 700 line. This machine and its successor, the 650, made IBM the number one computer manufacturer in the world, a position which it holds today.

Digital Computer Hardware As shown in Figs. 2 and 3, general-purpose digital computers are comprised of four basic components: an arithmetic or computing unit, a high speed internal storage unit, input-output devices, and a control unit. Data and instructions are indistinguishable within the computer; both are represented by binary patterns of semiconductor or magnetic core states. Each binary unit is called a "bit". A fixed number of bits is referred to as a computer "word". A word is, generally, the smallest directly addressable whole unit of memory. The number of bits in the word determines the accuracy of the basic machine and its cost. Small word machines are employed in real world applications to control

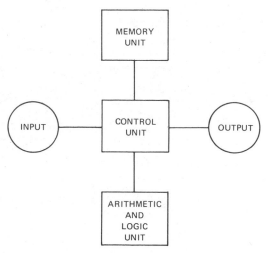

FIG. 2. Basic configuration of a digital computer. Problems are of the form

Input A = 3.0, B = 4.0
Output C = SQRT (A * A + B * B)

processes and make basic decisions. These small machines are referred to as minicomputers. Large word computers serve as theoretical number crunchers, often with smaller machines preprocessing input data or controlling output devices. Computer word size ranges from eight bits, or a single word accuracy of one part in 256, to sixty bits for an accuracy of one part in 2^{60} or a maximum positive decimal number of about 1 152 921 504 000 000 000.

In setting (programming or coding) instruction sequences, the user arranges for the control unit to determine, through an instruction decoder, whether the instruction itself refers to some memory location where data resides or calls for some basic arithmetic or logical operation to be performed. These operations are performed between data registers. Registers always provided include: an accumulator, a program counter, a memory address indicator, and an overflow indicator. Operations themselves may modify the stored instructions by performing calculations on them, branch to a new set of instructions, and/or cause the machine to interact with the outside world. Response from this interaction may cause the machine to "decide" what operations to perform next depending upon the options built into the program. The computer can make no decision of its own accord; it is simply directed to "conclude" which one of a given set of alternatives best fits a given situation. This being the case, it is important to remember that computers don't solve problems, people that program them do. The advantage possessed by the computer is speed of operation. Even the slowest machines can perform an addition of two numbers in less

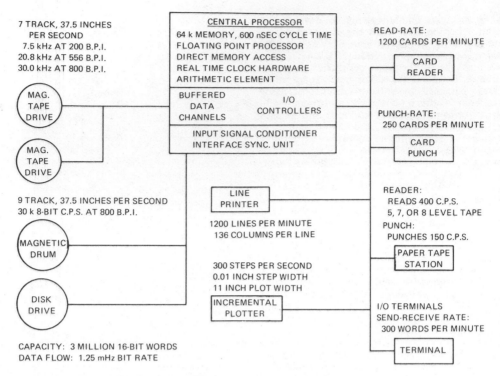

FIG. 3. Typical 3rd generation, medium size computer.

than 3 microseconds. Using special circuitry, multiplication of any two numbers takes less than ten microseconds. These arithmetic functions, plus logical operations like AND, OR, XOR, repeated thousands of times in one second enable the machine to use iterative techniques to do all operations necessary for calculating mathematical tables, as well as to keep up with the real world in monitoring and control applications where data at rates of up to 100,000 characters per second are encountered.

Figure 3 shows the configuration of a typical third generation, medium size machine (first introduced after 1968, and employing medium scale integrated circuitry with a multiprogramming operating system). Its purpose is to show the diversity of peripheral gear which is tied to a computer and the speed at which this equipment operates. The numbers shown are representative and by no means show the highest capacities available. This configuration will support a dedicated operation (one user at a time) or several users simultaneously in a multiprogramming or time shared mode. These latter terms refer to the ability of the system to run more than one user program simultaneously. The difference between time sharing and multiprogramming, as the terms are usually used, is that several users are connected simultaneously to the system in time sharing, while more than one program is run at the same time under

multiprogramming. This assumes, of course, that no single program requires total system resources for more than a small time period. Hardware controllers have been optimized to take advantage of this usual case and permit virtually simultaneous user access to the central processor.

Digital Computer Software Computer *software* refers to all programs which direct machine activities. This software can be generally divided into three classes: an operating or executive system; language assemblers, translators and compilers; user applications programs, mathematical program libraries, and data analysis packages. The first class generally goes under acronyms similar to Disk Operating System (DOS), Mass Storage Operating System (MSOS), Real Time Executive (RTE), UNIX, CPIM, or other such names. It is the function of this master software to schedule users and users' programs, allocate processor and peripheral resources, perform internal "housekeeping," and handle emergencies. It is the operating system that permits time sharing and multiprogramming to be carried out. All but the most basic systems offer this software. The second software class operates within or is subject to the OS. It includes such things as FORTRAN (Formula Translator) compilers, BASIC (Beginners All-Purpose Symbolic Instruction Code) interpreters, a system editor, and an assembler.

These and other conversion programs reside within the machine and are used to change English language statements or data to machine bit patterns which are decoded internally. Standards have been devised for FORTRAN, BASIC, ALGOL, COBOL, PL/1 and other languages so that a program written on one manufacturer's machine will (almost) run on another. Assembly languages, however, are machine independent and in no way compatible. The final class of software includes those programs designed and written to accomplish a specific task. For example: find the roots of a polynomial expression of a certain type; or given a set of data describing the elongation versus temperature and load of an alloy specimen, as well as known constants, calculate stress and strain values at the break

point. On small dedicated computer systems a single application package may be the only software loaded other than input/output drivers which communicate with peripheral devices. This is true of microprocessors, experiment controllers, front-end machines and, in general, all computers that have less than 8000 words of central memory. Software clearing houses that supply computer programs worldwide for exclusive manufacturers, as well as for general use, have gone into business to meet the demand for application programs.

Scientific Applications The evolution of the digital computer and modern scientific knowledge are closely linked. We have seen that development of the ENIAC was the direct result of the need to perform millions of calculations

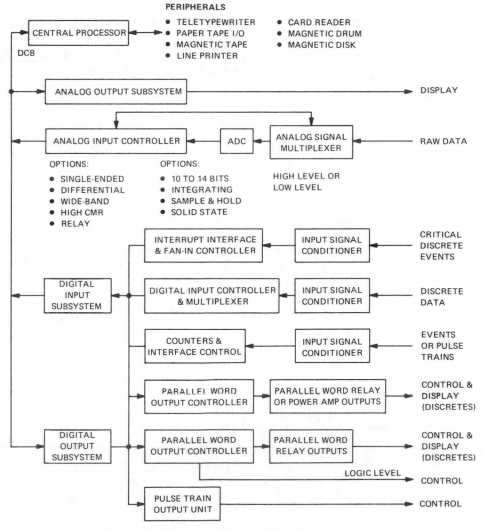

FIG. 4. Typical data acquisition and analysis configuration.

on the Manhattan Project that would have taken years during a time when years were not available. A second example is the fact that no attempt was made to find a rigorous solution of the "3-Body Problem" of celestial mechanics until the advent of the modern digital computer. Today these servants of scientists and engineers are found in their smaller versions in more physics laboratories than not. Theoretical data or that produced by passive data acquisition devices is then transmitted to larger systems for further processing. Where theoretical studies are involved, large systems perform computations that would simply not be practically possible otherwise. Examples of typical applications or user software now in use to solve particular scientific problems include programs to: calculate Cartesian coordinates for all atoms in a molecule, compute numerical eigenvalues and matrix elements for the quantum mechanical radial equation, and determine the coulomb lattice energy of an ionic crystal. There is also a complete system of programs for quantitative theoretical chemistry. In the area of direct data acquisition and instrument control, small computer systems of the configuration shown in Fig. 4 are most often employed. These systems are built as part of electron spin and nuclear magnetic resonance spectrometers, all types of optical spectrophotometers, real time Fourier analyzers, and x-ray spectrometers, to name only a few. They are used to control and acquire data in applications ranging from remote seismic stations to laboratory spark chamber monitoring. In physics education they find uses ranging from teaching basic principles to performing semiautomatic pattern recognition for bubble chamber film data analysis.

Personal Computers The latest entry into the computer industry is the personal computer (PC). In the past eight years, the personal computer has moved from the province of hobbyists to take a place in business offices and private homes. Thousands of people are purchasing personal computers each week, not only low-end microcomputers, but also high-end systems such as the IBM PC. They have proven to be incredibly popular and several hundred thousand of them were sold in only the past three years. Further, for every person who buys a personal computer, two or three more are considering a purchase. This has been attributed to both the videogame explosion and the public's recognition of the power and capability of the microcomputer. These computers are reasonably inexpensive, and designed to allow the average person to learn about the computer and use it to solve everyday problems.

The personal computer is a genuine computer which has most of the features of the big mainframe computers. As computers became smaller, they also became faster, and today processing time is measured in microseconds and nanoseconds (billionths of a second). For example, the IBM PC is four times faster than the IBM 360. It can be equipped with enough capacity to handle the accounting and inventory control tasks of most small businesses. It can also perform computations for engineers and scientists, and it can be used to keep track of home finances, budget management, investment analysis, and many, many other applications ranging from educational to recreational to business functions.

Figure 5 shows the basic configuration of a personal microcomputer. Like other computers it consists of the CPU and the system board that connects the CPU to other devices for input and output (I/O) and storage. There are three basic types of memories: ROM (Read-Only Memory) which is not user-modifiable and which contains programs you would never want to change; RAM (Random Access Memory), in which data can be either written or read and which can be thought of as the microprocessor's work space; the mass storage device, which al-

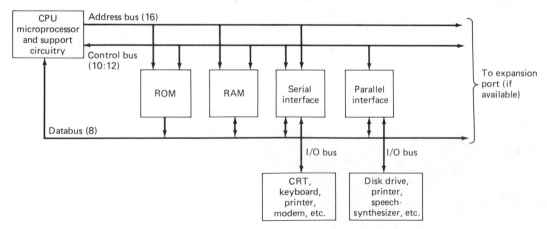

FIG. 5. Block diagram of a basic personal microcomputer system.

lows the storing of great quantities of information and programs on magnetic disk or tape outside the computer itself. Parallel and serial interfaces connect the microcomputer to other devices such as printers, keyboards, etc. Early computers had only serial ports that received and returned data serially, or, one bit at a time. Processing time in computers was vastly improved by the development of parallel ports which handle and move large amounts of data simultaneously. The process utilizes electrical conductors (printed circuit boards) that will carry multiple electrical impulses around the innards of the computer. These multiple impulse carriers are known as busses, and several used together are called bus systems.

Although the industry is yet dominated by 8-bit microprocessors, all new systems are 16-bit. Within a few years they will have replaced the 8 bit processors as the brains of medium-priced personal computers. The power of the 16-bit microprocessor—with its ability to address over 1 million characters of computer memory—is needed to run software that is more flexible and easier to use than the software commonly used today. Microprocessor chips (such as Intel Corporation's 8088 and Motorola's MC68000) are already turning up in the more advanced personal computers and manufacturers of the Corvus Concept and the IBM Personal Computer, among others, are putting that power to good use. Further, during 1984 the first 32-bit biped PCs will be introduced. This will put on a desk top computer power of limitless dimensions to most people.

Future Developments As we have shown, the computer industry only recently celebrated its silver anniversary, yet there is almost no area of our life which it has not affected. In terms of dollars, the hardware, software and manpower investment is well over 100 billion. Despite these impressive figures, which point to success, many serious shortcomings in computer technology and its applications are evident. Most experts agree that hardware advances at the component level will slow down. Peripheral devices such as line printers, card readers, etc., appear to be approaching their capacity limits. These devices must give way to new technologies or be reduced in cost to the extent that they can proliferate infinitely. It seems more reasonable that they will give way to more direct user involvement through input/output terminals and direct data transfer. The direction of central and mass memory supports this contention since costs, speed, and density continue to improve drastically. This trend will continue for some time with mechanical devices giving way to new approaches employing chemical and basic molecular phenomena. Real future advances will occur as a result of changes in arithmetic and control elements and in the software systems which drive computers. Micro-processors made possible by large scale integrated circuitry (LSI)

will allow "intelligence" to be built into all computer peripherals and make possible special purpose modules to be employed for handling a single arithmetic or control function. This will permit many more operations to be performed in parallel, thus decreasing time to solution. The low cost of these modules will cause them to begin to appear almost everywhere. Few scientific instruments will be built where they are not employed as control elements or to provide at least intermediate results directly. Software development will respond to make use of this distributed internal processing and permit better machine and manpower utilization. More emphasis will be placed on high level languages similar to PL/I and on the development of applications packages which will handle a given problem from start to finish. Less distinction will be made between systems and language routines. A single tool will be available to accomplish a given objective. Problems like code conversions between units will be handled by hardware. Modular programming techniques and top-down software management will harness software development making distributed intelligence and application of automatic data processing a reality in all fields.

DENNIS E. WISNOSKY

References

Weik, Martin H., "Standard Dictionary of Computers and Information Processing," New York, Handen Book Co., 1969.

Davis, Gordon B., "An Introduction to Electronic Computers," New York, McGraw-Hill Book Co., 1965.

Bernstein, Jeremy, "The Analytical Engine: Computers, Past, Present, and Future," New York, Random House Inc., rev. ed. 1981.

Boole, George, "An Investigation of the Laws of Thought," New York, Dover Publications, 1954.

Quantum Chemistry Program Exchange, Indiana University, Bloomington, Indiana 47401.

Korn, Granizo Arthur, "Minicomputers for Engineers and Scientists," New York, McGraw-Hill Book Co., 1973.

Perone, Sam P. and Jones, David O., "Digital Computers in Scientific Instrumentation," New York, McGraw-Hill Book Co., 1973.

"Digital Products and Applications," Digital Equipment Corporation, Maynard, Massachusetts, 1971.

"CDC System 17 Computer Systems Applications Guide," Control Data Corporation, Minneapolis, Minnesota, 1973.

Petrocelli, O. R., Ed., "The Best Computer Papers of 1971," Princeton, Auerbach Publishers Inc., 1972.

Frederick, Franz, "Guide to Microcomputers," Washington, D.C., Association for Educational Communications and Technology, 1980.

Goldstein, Larry and Goldstein, Martin "IBM Personal Computer: An Introduction to Programming and Applications," Robert J. Brady Co., Prentice-Hall

Publishing and Communications Company, Bowie, Maryland, 1982.

Dertouzos, Michael L. and Moses, Joel, "The Computer Age: a Twenty-year View," MIT, 1979.

"All About Personal Computers," Debran, N.J., Datapro Research, 1982.

Mark, Frank, "Discovering Computers," London: Stonehenge, 1981.

Cross-references: ELECTRONICS; MEASUREMENTS, PRINCIPLES OF; CYBERNETICS; SEMICONDUCTOR DEVICES.

CONDENSATION

The condensation of a vapor to form a liquid, amorphous phase, or crystal generally occurs by the mechanism of nucleation and subsequent growth. *Nucleation* is a thermally activated process which leads to a stable fragment of the condensed phase. In the absence of surfaces of certain condensed phases, reactive foreign molecules, or other potent catalysts to the nucleation process, it is usually the slower step and occurs at an appreciable rate only under conditions considerably removed from equilibrium. For the usual case where nucleation catalysts are present, the process is characterized as heterogeneous nucleation, but if there are no such catalysts at all, it is called *homogeneous nucleation*.

In principle, statistical thermodynamics would appear to offer the most attractive approach to nucleation rate theory. For example, Band[1] and Hill[2] have given formal treatments for the equilibrium concentration of clusters of molecules in a vapor. However, the internal partition functions have thus far eluded quantitative evaluation. Further, in the case of metallic systems, there is at present little knowledge of the electronic energy states in clusters containing only a few atoms.

Accordingly, even to the present day, most treatments follow that of Volmer[3] and coworkers, who evaluated the free energy of formation of clusters by ascribing macroscopic thermodynamic properties to them. Thus the free energy of a droplet is described as the sum of a surface term (area times surface tension) and a volume term (volume times the negative bulk free energy change). An attractive feature of this approach is that it permits ready visualization of the origin of the free energy barrier to nucleation in terms of the maximum in the above sum as a function of size. However, a very unattractive aspect is that the calculated size of the critical nucleus, i.e., the cluster size at the top of the free energy barrier, is only about 100 molecules, which leads one to doubt the applicability of macroscopic concepts in the present examples. Nevertheless, following standard methods,[4] this spherical-drop model leads directly to a rather simple expression for the rate of homogeneous nucleation of droplets from supersaturated vapor.

The remarkable agreement[3,5] of this macroscopic theory with observations of the critical supersaturations for appreciable nucleation rate of various liquids in cloud chambers stood for many years as the basis of our knowledge of nucleation. For example, referring to the data of C.T.R. Wilson[6] and of C. F. Powell[7] who reported a "fog limit"* for homogeneous nucleation of water droplets at a supersaturation ratio of about 5.0 at 275 K,† agreement with the macroscopic theory is excellent.[8] However, in recent years it has been pointed out[4,5,8,9,10] that the external partition functions for free translation and rotation had been neglected in the macroscopic theory. When these contributions are included, the theory predicts a critical ‡ supersaturation ratio of about 3.0 for water vapor at 275 K, in poor quantitative agreement with Powell's observations. This situation stimulated a great deal of experimental and theoretical work in the field. On the experimental side, two new techniques were developed to measure critical supersaturation ratios for homogeneous nucleation: (1) the diffusion cloud chamber[11,12,13] and (2) the supersonic nozzle method.[14,15] Work is still in progress, but it appears that virtually all substances, with the possible exception of argon[16] and ammonia,[17] follow the original macroscopic theory which ignores the contributions from free translation and rotation. On the theoretical side, much effort has been directed toward ascertaining the magnitude of another correction to the macroscopic theory, which tends to counterbalance the effects of free translation and rotation. This is termed the *replacement partition function*, and it describes the free energy due to the six internal degrees of freedom the isolated cluster does not have because it is not a part of the bulk phase. Efforts to calculate this quantity by classical phase integral methods[18,19,20] have not been entirely successful because of the great difficulty in defining an embedded cluster in a liquid. In fact, some treatments yield large[18,19] replacement partition functions while others give small ones.[20] However, the calculation is somewhat easier for crystals, and the result[20] indicates that the replacement partition function is not large and hence does not appreciably offset the contributions from free translation and rotation in the case of homogeneous nucleation of crystallites from the vapor. Unfortunately, there are very few experimental data for this case, but the theoretical result has been used in astrophysics calculations.[21] In summary, one might con-

---

*Denoting production of many droplets, i.e., of the order of 10^7 cm^{-3}.

†Meaning the ratio of actual to equilibrium partial pressure of vapor.

‡Critical for a given, usually high, nucleation rate.

clude at the present time that the good agreement between the original macroscopic theory and experiments on liquids is fortuitous and due to (i) a large replacement partition function or (ii) the circumstance that the macroscopic surface tension overestimates the droplet entropy and underestimates the potential energy.[22,23]

One might think that the above issues could be settled by computer calculations using fairly realistic potential functions to calculate the cluster free energies, and much work has been done in this area.[22,24] In one of the more elaborate efforts,[25] Monte Carlo methods were used to calculate the surface tension and chemical potential of bulk Lennard-Jones argon, and these results were applied to estimate the free energy of isolated L-J argon droplets. The resulting free energies of formation were then compared with the actual Monte Carlo free energies of isolated liquid L-J argon clusters. It was found that the contributions from free translation and rotation were required to describe the Monte Carlo cluster data. This means that, unlike the case of physical experiments on real liquids, the original macroscopic theory must be modified by the contributions from free translation and rotation in order to describe L-J argon. However, a major shortcoming of this and other computer calculations is that pairwise potentials were used. This is a serious difficulty in view of the importance of three-body potentials in describing surface properties.

On the other hand, molecular dynamics computer simulations on highly supersaturated L-J vapor have been useful in predicting spinodal decomposition as another condensation mechanism, which is an alternative to nucleation and growth of droplets. Spinodal decomposition has long been recognized[26] as a mechanism of phase separation in binary solid systems in which there is no symmetry change and for which the free energy-composition curve is continuous. At high supersaturations the initial phase is unstable for all infinitesimal density variations having a wavelength greater than some "critical" wavelength. The molecular dynamics work on L-J vapor[27] exhibited the characteristic interconnected structure of vapor and liquid, and the process was consistent with the recognized laws for spinodal decomposition. There are at present few, if any, physical experiments to corroborate this prediction.

A considerable amount of work has been done on the heterogeneous nucleation of metal crystals from thermal vapor beams onto substrates.[5,28,29] Many theoretical approaches follow a macroscopic treatment similar to that outlined above. In view of the high supersaturation ratios, typically 10^6–10^{33}, and resulting small critical nucleus sizes (1 to 5 atoms) involved in this case, an "atomistic" theory[30,31] has been introduced to replace the older macroscopic theory. However, quantitative agreement with observed supersaturations for appreciable nucleation rate is not good for either the old or the new theory. Nevertheless it has been established that in most cases nucleation occurs by the processes of adsorption, surface diffusion, and statistical fluctuation to form the crystalline nuclei. Also, particularly for metals on ionic and semiconductor substrates, many nucleation rates have been quantitatively measured by means of both kinematic and *in situ* transmission electron microscopy. Indeed, within the past 15 years there has been a remarkable increase in the degree of sophistication of such experiments. Ultra high vacua (of the order of 10^{-10} Torr) leading to much greater cleanliness are now common and the characterization of both substrate and deposited crystallites has been greatly enhanced by the application of LEED, Auger spectroscopy, mass spectrometry, transmission electron diffraction and a host of other new techniques. However, interpretation of the rate data is often obscured by ignorance of the actual (often defect) sites for nucleation on the substrate surface. Interestingly enough, it has been possible to demonstrate and measure the mobility of small (10–20 atom) clusters,[32] e.g., of solid gold on the (100) surface of alkali halide crystals. In connection with the above experiments, selected-zone dark-field electron microscopy (SZDF) has been developed[33] to determine the crystal structure, orientation, and degree of perfection (absence of twinning) of each of the myriad of isolated microcrystallites on a substrate surface. SZDF employs annular objective-lens apertures of different geometries to select only a well-defined number of Debye-Scherrer diffraction rings. Thus it is possible to map, within one image, all specimen areas of random azimuthal orientation that diffract into the selected range of Bragg angles.

Historically, field emission microscopy has yielded interesting information on nucleation of metals in deposition from the vapor onto clean tungsten field emitter tips.[34,35] Many metals, e.g., copper and gold, form critical adsorbed coverages of several monatomic layers before nucleation of three-dimensional crystals occurs. This is the Stranski-Krastonow mechanism[36] for deposition of crystals from the vapor and may be expected for situations in which the binding energy of the adatoms to the substrate is high. In some cases the initial monolayers are thought to be pseudomorphic* with the substrate.[37] Another important mechanism for cases of high binding energy is the monolayer-by-monolayer overgrowth (MO) or Frank-van der Merwe mode,[38] which also may occur with pseudomorphism in the initial monolayers. In some cases the monolayers are initiated by nucleation of 2-dimensional discs.[39] Misfit dis-

---

*Of exactly the same crystal structure and lattice parameter.

locations to accommodate the strain are thought to form at the substrate-deposit interface as the deposit thickens. Ultimately, the deposit becomes noncoherent with the substrate and attains its natural crystal structure. The misfit dislocations even facilitate phase transformation in cases where the bulk deposit and substrate are of different crystal structure. Examples of the MO mechanism are SnSe on (001) SnTe and Pt on (001) Au.[39] Recent studies[39] have provided much quantitative information, such as Burgers vector, spacing and mechanism of origin, on misfit dislocations. Also, they have revealed ordered structures of monolayer or sub-monolayer thickness, e.g., Pb on the (111) Ag surface. This may have a relevance to the preparation of epitaxial thin films.

Growth is the process by which the stable nuclei continue to grow and thereby consume the supersaturated vapor. In general, several mechanisms are involved in the growth process, and some of these are thermally activated. However, the free energies of activation are usually low, and hence most growth processes proceed at an appreciable rate even under conditions close to equilibrium where the gross evaporation flux almost equals the gross condensation flux. The first step in growth from the vapor is thought to be adsorption of the impingent molecule. The bulk of both experimental and theoretical work indicates that the impingent atoms or molecules are in most cases thermally accommodated and adsorbed at the surface before being either re-evaporated or integrated into the liquid or crystalline structure.

In the case of liquids, it is thought that the molecular mobility is sufficiently high that the adsorbed molecules are taken almost immediately into the liquid structure. However, the situation is quite different for crystals, whose surfaces are still most conveniently visualized in terms of the original "atomic building block" model of Kossel[40] and Stranski.[41] Thus, in the case of certain surfaces of high index, the kinks in the steps of the atomically rough surface provide ready sinks for adsorbed molecules. In fact, experiment shows that such planes grow so rapidly that they quickly eliminate themselves from the crystal growth form, leaving the smoother surfaces of low index. These closely packed planes contain no steps and kinks to serve as sinks for the admolecules diffusing on the surface. Accordingly, for a perfect crystal, it is thought that growth can proceed only by nucleation of new monomolecular layers, whose edges provide the sinks, and their lateral propagation. A typical supersaturation ratio for appreciable growth by this mechanism is of the order of 1.5 for molecular substances, and there is a large amount of theoretical and experimental evidence[42] for the general occurrence of this type of growth from the vapor at high supersaturation.

The fact that real crystals do indeed grow at much lower supersaturation ratios, of the order 1.01, continued to present a theoretical problem for many years. Then in 1949 Burton, Cabrera, and Frank[43] showed that certain emergent dislocations of the screw orientation* must provide a source of monomolecular steps for growth at low supersaturations. Further, they demonstrated that the resultant growth form on the close-packed surface, the growth spiral, cannot exterminate itself as do other types of steps or ledges. In the usual case, crystal growth by this mechanism is thought to be controlled by surface diffusion of the admolecules. Experimental verification of these predictions is now voluminous.[44]

Inasmuch as crystal growth from the vapor is linked to the motion of monatomic steps on the surface, it becomes of great interest to describe the step dynamics arising from surface diffusion gradients. Most of the studies thus far have been concerned with the closely related reverse process of crystal evaporation, and the theory of step dynamics has been extended to cover stationary-state evaporation[5] with the steps emanating from crystal edges or from dislocations and the bunching of evaporation steps under transient conditions.[45,46] Recently the evaporation rate from (100) KCl single crystal surfaces was measured as a function of undersaturation under stationary-state conditions,[47] and remarkable agreement with theory was found. In general the evaporation coefficient† will lie between $\frac{1}{3}$ and 1 and the catalytic effect of dislocations in providing additional monatomic steps disappears as the undersaturation ratio rises above 0.9. Similarly, in recent electron microscope work on vacuum evaporated (100) NaCl surfaces,[48] the transient fluctuations in monomolecular step spacing caused by a sudden change in evaporation temperature were quantitatively described by the theory.

Most of the theoretical studies of growth in deposition on substrates have not been in terms of the fundamental microscopic mechanisms, namely, the screw dislocation spiral mechanism and the monatomic disc nucleation mechanism. Perhaps this is because of the experimental difficulty in observing growth rates of the small crystallites on substrates. Rather, the main thrust of the theoretical work on growth thus far has had as its goal prediction of the crystallite number density and size distribution. This is of course a complicated problem affected by nucleation, depletion of the adatom concentration about growing clusters, surface dif-

---

*Or dislocations with a component of the Burgers vector perpendicular to the surface.

†Ratio of actual to equilibrium gross evaporation flux.

fusion fields, competition between crystallites for the available adatoms and the effect of capillarity[†] in retarding crystallite growth. Most of the studies have confined themselves to the case of incomplete coverage of the substrate. There are many treatments of these phenomena, but one useful approach is numerical integration of the divergence of cluster current in size space with consideration of capillarity in the boundary conditions.[49] The results give a reasonably good description of the observed time evolution of cluster size distribution.[‡]

In the interests of brevity, the complex and interesting effects relating to diffusion in the vapor,[5] adsorption of impurities,[5] chemical reaction,[42] and dissipation of the heat of condensation[8] have been omitted from the above discussion. The subject of crystal growth morphologies is, of course, huge and beyond the scope of the present article.[5,42]

G. M. POUND

References

1. Band, W., "Quantum Statistics," New York, Van Nostrand Reinhold, 1955.
2. Hill, T. L., "Statistical Mechanics," New York, McGraw-Hill Book Co., 1956.
3. Volmer, M., "Kinetik der Phasenbildung," Dresden and Leipzig, Steinkopff, 1939.
4. Frenkel, J., "Kinetic Theory of Liquids," London, Oxford Univ. Press, 1946.
5. Hirth, J. P., and Pound, G. M., "Condensation and Evaporation, Nucleation and Growth Kinetics," Oxford, Pergamon Press, 1963.
6. Wilson, C. T. R., *Phil. Trans. Roy. Soc. London* **192**, 403; **193**, 289 (1899).
7. Powell, C. F., *Proc. Roy. Soc. London* **199**, 553 (1928).
8. Feder, J., Russell, K. C., Lothe, J., & Pound, G. M., *Adv. Phys.* **15**(57), 111 (1966).
9. Dunning, W. J., "Nucleation," (A. C. Zettlemoyer, Ed.) p. 1, New York, Marcel Dekker, 1969.
10. Lothe, J., and Pound, G. M., "Nucleation," (A. C. Zettlemoyer, Ed.) p. 109, New York, Marcel Dekker, 1969.
11. Franck, J. P. and Hertz, H. G., *Z. Physik* **143**, 559 (1956).
12. Katz, J. L., *J. Chem. Phys.* **52**, 4733 (1970).
13. Katz, J. L., Mirabel, P., Scoppa, C. J., and Virkler, T. L., *J. Chem. Phys.* **65**, 382 (1976).
14. Wegener, P. P., and Parlange, Jean-Yves, *Naturwissenschaften* **57**, 525 (1970).

15. Wegener, P. P., and Wu, B. J. C., p. 325 in "Nucleation Phenomena," Vol. 7, Adv. Colloid Interface Sci. (A. C. Zettlemoyer, Ed.), New York, Elsevier, (1977).
16. Hoare, M. R., Pal. P., and Wegener, P. P., *J. Colloid Interface Sci.* **75**, 126 (1980).
17. Dawson, P. B., Willson, E. J., Hill, P. G., and Russell, K. C., *J. Chem. Phys.* **51**, 5389 (1969).
18. Reiss, H., p. 1 in Ref. 15.
19. Kikuchi, R., p. 67 in Ref. 15.
20. Nishioka, K., and Pound, G. M., p. 205 in Ref. 15.
21. Czyzak, S. J., Hirth, J. P., and Tabak, R. G., "The Formation and Properties of Grains in the Interstellar Medium," Vistas in Astronomy 1982 (in press).
22. Abraham, F. F., "Homogeneous Nucleation Theory," Supplement 1 to Advances in Theoretical Chemistry, New York, Academic Press, 1974.
23. Binder, K., and Kalos, M. H., *J. Statistical Physics* **22**, 363 (1980).
24. Hoare, M. R., *Advances in Chem. Phys.* **40** (1979).
25. Miyazaki, J., Pound, G. M., Abraham, F. F., and Barker, J. A., *J. Chem. Phys.* **67**, 3851 (1977).
26. Cahn, J. W., *J. Chem. Phys.* **42**, 93 (1965).
27. Mruzik, M. R., Abraham, F. F., and Pound, G. M., *J. Chem. Phys.* **69**, 3462 (1978).
28. Voorhoeve, R. J. H., "Molecular Beam Deposition of Solids on Surfaces: Ultra Thin Films," Treatise on Solid State Chemistry, (N. B. Hannay, Ed.) New York, Plenum, 1976.
29. Matthews, J. W. (Ed.), "Epitaxial Growth," Parts A & B, New York, Academic Press, 1975.
30. Walton, D., *J. Chem. Phys.* **37**, 1282 (1962).
31. Rhodin, T. N., p. 31 in "Proceedings of a Conference on Single Crystal Films at Bluebell, Penna." (M. H. Francombe and H. Sato, Eds.), Oxford, Pergamon, 1964.
32. Metois, J. J., Zanghi, J. C., Erre, R., and Kern, R., *Thin Solid Films* **22**, 331 (1974).
33. Poppa, H., p. 215 in Ref. 29, Part A.
34. Jones, J. P., *Proc. Roy. Soc. (London)* **284A**, 469 (1965).
35. Gretz, R. D., and Pound, G. M., *Applied Phys. Letters* **11**, 67 (1967).
36. Stranski, I. N., and Krastonow, L., *Akad. Wiss. Deut. Math. Nat. Kl.* **146**, 797 (1938).
37. Bauer, E., and Poppa, H., *Thin Solid Films* **12**, 167 (1972).
38. Frank, F. C., and van der Merwe, J. H., *Proc. Roy. Soc.* **A200**, 125 (1949).
39. Honjo, G., and Yagi, K., "Studies of Epitaxial Growth of Thin Films by In Situ Electron Microscopy," p. 197 in "Current Topics in Materials Science, Vol. 6," (E. Kaldis, Ed.), Amsterdam, North Holland, 1980.
40. Kossel, W., *Nachr. Akad. Wiss. Goettingen Math. Phys. Kl.* **1**, 135 (1927).
41. Stranski, I. N., *Z. Phys. Chem.* **136**, 259 (1928).
42. Strickland-Constable, R. F., "Kinetics and Mechanism of Crystallization," London, Academic Press, 1968.
43. Burton, W. K., Cabrera, N., and Frank, F. C., *Phil. Trans. Roy. Soc. London* **A243**, 299 (1950).

[†]Surface tension.
[‡]Recently (unpublished work) continuous and uniform thick films of amorphous metals and alloys have been prepared by deposition from vapor beams onto very cold substrates. Here the mechanism is one of random impingement and adherence of the immobile atoms to the surface.

44. Dekeyser, W., and Amelinckx, S., "Les Disloca-
    tions et la Croissance des Cristaux," Paris, Masson,
    1956.
45. Mullins, W. W., and Hirth, J. P., *J. Phys. Chem.
    Solids* 24, 1391 (1963).
46. Surek, T., Pound, G. M., and Hirth, J. P., *Surface
    Sci.* 41, 77 (1974).
47. Nordine, P. C., and Gilles, P. W., *J. Chem. Phys.*
    74, 5242 (1981).
48. Bethge, H., Hoeche, H., Katzer, D., Keller, W. K.,
    Bennema, P., and van der Hoek, B., *J. Crystal
    Growth* 48, 9 (1980).
49. Robertson, D., and Pound, G. M., *J. Crystal
    Growth* 19, 269 (1973).

Cross-references: CRYSTAL STRUCTURE ANAL-
YSIS, CRYSTALLIZATION, CRYSTALLOGRAPHY,
FIELD EMISSION, STATES OF MATTER, VAPOR
PRESSURE AND EVAPORATION.

CONDUCTIVITY, ELECTRICAL

The electrical conductivity of a substance is an
intrinsic property denoting the ability with
which electric charge can flow through the substance. The meaning of a definite conductivity
is most commonly associated with solids, although electrical conduction also occurs in
liquids, electrolytes, and ionized gases. Electrons are the usual charge carriers in solids, but
ionic conduction can be important for some
materials, such as the alkali halides and compounds of the KAg_4I_5 class, while proton conduction has been demonstrated for ice.

A suitable definition for the electrical conductivity of an isotropic material is provided by
Ohm's law. This is the statement that the
direct-current density **J** within a conductor is
proportional to the dc electric field **E**. At a
given temperature and pressure, the constant of
proportionality is the electrical conductivity σ,
thus:

$$\mathbf{J} = \sigma \mathbf{E}$$

If the material is *anisotropic*, the magnitude of
J depends not only on the magnitude of **E** but
on its direction as well. **J** and **E** are then non-
parallel for some orientations of the material,
and σ is a tensor of the second rank. Since the
ability of a material to conduct electricity is influenced by the mechanisms resisting the flow
of charge, it is also helpful to work in terms of
the *specific electrical resistivity* ρ, which is the
reciprocal of σ.

If a voltage V between the ends of a conductor of length ℓ and uniform cross-sectional area
A maintains a current I through the conductor,
Ohm's law may be set in a practical form by
combining the relations:

$$\mathbf{J} = \sigma \mathbf{E}, \quad \mathbf{E} = \frac{V}{\ell}, \quad \mathbf{J} = \frac{I}{A}, \quad \sigma = \frac{1}{\rho}.$$

Hence, $I/A = (1/\rho)(V/\ell)$ or $V = I(\rho\ell/A)$, so that
$V = IR$. The ratio $\rho\ell/A$ is called R, the *electrical
resistance* of the conductor. It is a property of
a particular sample because it involves the dimensions A and ℓ, whereas ρ and σ are intrinsic properties of the constituent material.
With V in volts and I in amperes, R is measured
in ohms. The electrical resistivity $\rho = RA/\ell$ is
then given in ohm-meters if A is in square
meters and ℓ in meters, while σ is given in
(ohm-meters)$^{-1}$ or mhos per meter.

The major factors determining the magnitude of the electrical conductivity for a material
are the conduction electron, or ion, density, and
the nature of the interatomic forces (which
decide the mobility of the charge carriers). The
actual current flow also depends on the size of
the electric field. Part of the energy carried by
the current is inevitably consumed as Joule
heat, but there are many commercial devices in
which such heat or light conversion is put to
good use (electric fires, toasters, cookers, light
filaments, fuses, etc.). The rate per unit volume
at which energy is converted is I^2R or VI, in
units of watts or joules per second. Where it is
required to add resistance to an electrical circuit,
resistors made of carbon, graphite, or metallic
alloys are often selected. When wires of low resistance are needed, copper is the most com-
mon material ($\rho \sim 1.7 \times 10^{-8}$ Ωm) although at
ordinary temperatures silver is the best conduc-
tor ($\rho \sim 1.6 \times 10^{-8}$ Ωm).

In alternating current circuits, the conductivity or resistivity depends on the frequency
of the applied electric field. Deviations from
the dc value are not appreciable at low frequencies but may become significant for microwave
or higher frequencies.

Gases can conduct electricity if they are
ionized. Practical applications include discharge
tubes, electronic vacuum tubes, and the arc
discharge. Natural ionization of the atmosphere
results from cosmic rays or radioactive sources
in the ground. The conductivity of the atmosphere is quite low at ground level but it increases rapidly with altitude up to 50 km because of the greater cosmic radiation and the
lower density of scattering centers.

Many liquids or solutions known as electrolytes (besides a few solids) can be decomposed
by an electric current into charged particles
called anions and cations. Such processes involve a transfer of matter through the conductor. The conductance of a solution is defined as
the current flowing per unit charge applied to
the electrodes immersed in the solution and
per unit concentration of electrolyte between
the electrodes. It is dependent on the number
and mobilities of the ions in the solution. In
geophysics there is an important field of activity involving electrical conductivity measurements of the surface layers of the crusts of
the earth and the moon. But in laboratory
physics a principal area of activity lies in

solid-state physics, and the remainder of this article will be devoted to such work.

Solids may be classified in various ways (according to their binding, ductility, crystalline or amorphous nature, etc.) but a particularly convenient one considers their conduction properties. Two distinctive qualities are of interest: (i) the magnitude of σ or ρ at a suitable comparison temperature (say, room temperature), and (ii) the temperature variation of σ or ρ. Three broad classes of solids may thus be characterized:

(1) *Metals* having high conductivities (when pure) with specific resistivities at room temperature lying in the range 1.6×10^{-8} Ωm (for silver) to 140×10^{-8} Ωm (for manganese and plutonium). Less pure metals and alloys may have resistivities up to 1000 times bigger than these, while very pure metals at liquid helium temperatures may have resistivities 10^5 times smaller. In general, the resistivities of metals increase with temperature. Also the effect of adding small amounts of impurities is that of adding a temperature-independent contribution to the resistivity (*Matthiessen's rule*). If ρ_T is the thermal resistivity and ρ_0 is the impurity contribution, the total resistivity $\rho = \rho_T + \rho_0$.

(2) *Semiconductors* which have much lower conductivities than do metals, with resistivities in the range 10^{-5} to 10^5 Ωm. In contrast to metals, their resistivities decrease with rising temperature and very rapidly with the addition of impurities.

(3) *Insulators* whose electrical conductivities are lower still, with resistivities ranging from 10^6 to 10^{16} Ωm. The feeble conductivity is little affected by impurity additions, but it improves rapidly as the temperature is raised.

The basic differences between these classes can be understood in terms of atomic and quantum-mechanical principles which explain the varying degree of availability of free electrons or mobile ions for conduction purposes. The main features are outlined below. An additional distinct class, that of superconductivity, which is a spectacular quantum situation in which the low-temperature state of a number of metals and a few semiconductors is one of zero resistance, is not dealt with here (see SUPERCONDUCTIVITY).

Metals are characterized by their high density of free conduction electrons which transport negative charge $(-e)$ through the interstices of the crystal lattice composed of positive ions. The origin of the conduction electrons is some or all of the valence electrons from the previously neutral atoms. In the absence of an electric field the conduction electrons have high-speed random motions $(\sim 10^6$ ms$^{-1})$, and there is no directed charge flow in any particular direction. But when a field is applied, they acquire a steady net drift, of much lower speed than their kinetic speeds, in exactly the opposite direction to the field. Part of the kinetic energy

gained in this way is just as steadily returned to the lattice as Joule heat via collision processes with it. The current density $\mathbf{J}$ is then $-ne\mathbf{v}$ for an electron density of n per unit volume and drift velocity $\mathbf{v}$. The mean-free path between collisions commonly exceeds 100 interatomic spacings at room temperature, or even 10^7 spacings in pure metals at low temperatures. An assumption that the collisions are elastic permits the use of a relaxation time τ, at least under the conditions pertaining at high temperatures for thermal scattering or at low temperatures for impurity scattering. τ is a quantity inversely related to the probability per unit time of an electron undergoing a collision. A basic equation of the form $\rho = m/(ne^2\tau)$, in which m is closely related to the electron mass, can be derived without undue difficulty, so that the major barrier to calculating the magnitude of ρ or σ of a metal revolves about understanding and evaluating τ and its temperature dependence.

A conceptual hurdle here is how can an electron proceed more than a few atomic spacings without being scattered by the massive, closely spaced lattice ions? The reason is that the lattice is the source of a periodic electrostatic field and that the electron waves are modulated by a function having the same period. It can then be shown that such waves are propagated with no loss of energy if the lattice is perfectly periodic. In practice, the lattice potential is never perfect, for it is disturbed by both thermal vibrations and impurity atoms or physical defects. The thermal vibrations are quantized with discrete energy values called *phonons*. For temperatures exceeding the Debye temperature, the phonon density and hence the electron-phonon scattering and the resistivity increase almost in proportion to T. At very low temperatures, $\rho_T \propto T^5$. This is directly related to the T^3 variation in the Debye phonon spectrum which gives a T^3 specific heat at low temperatures. Obtaining realistic estimates of the resistivity magnitudes for metals other than the alkali metals remains a matter of considerable complexity. A useful semiempirical equation is the Grüneisen-Bloch relation because it facilitates the analysis and discussion of experimental data. It represents the variation of the thermal resistivity of a wide selection of metals rather well. In some multivalent metals, but more importantly in semiconductors, electrons behave in a way which can be described by the displacement of positive charge carriers called *holes*. Experimental results are sometimes discussed as if the current arises from the flow of electrons and holes.

Magnetic metals have additional resistive effects due to scattering from localized-spin assemblies. An important field is *dynamic cooperative phenomena* using the divergence of $d\rho/dT$ at magnetic critical points as a tool. Another active field is the *Mott transition* whereby certain materials can be switched from a metallic to an insulating condition using small changes of pressure, temperature, or electron-to-atom ratio. Resistivity is also used to study atomic order-disorder and crystal phase transitions, and in the study of defect production and migration. Its temperature dependence is used for *thermometry* and its strain dependence for strain gauges. Its magnetic-field dependence is the basis of the major field of *magnetoresistivity*. When used together with the related Hall effect, valuable information is provided on the effective sign, number, and mobility of the charge carriers in semiconductors, while in metal single crystals certain details of the FERMI SURFACE can be deduced.

Insulators and *pure semiconductors* have no free electrons available at 0 K for conduction. Diamond, silicon, and germanium are typical examples. All their four valence electrons are fully occupied in forming chemical bonds in the solid. Raising the temperature energizes, and frees for conduction duties, a small fraction of these electrons. The empty energy states left behind (holes) also aid in conduction. The fraction of carriers is $\sim$1 in 10^9 for Si and Ge at room temperature; the number varies approximately as $T^{3/2} \exp(-\Delta E/2kT)$ where ΔE is the energy to excite a bound electron and k is the Boltzmann constant. ΔE is $\sim$5.2, 1.2, and 0.75 eV for diamond, silicon, and germanium respectively. The chief difference between the behavior of diamond (a typical insulator) and pure Si and Ge (typical semiconductors) is the greater ease with which temperature can induce conduction in the latter. The conduction of semiconductors, but not of insulators, is readily improved by adding certain impurities. This has the effect of introducing electrons (*n*-type) or of producing holes (*p*-type). Such materials are termed *extrinsic*, or *impurity*, *semiconductors*. Their great practical application is in transistors and diodes. (See SEMICONDUCTORS and SOLID-STATE PHYSICS.) Conduction in many semiconductors and insulators is also increased by the photoelectric action of incident light radiation (see PHOTOCONDUCTIVITY).

G. T. MEADEN

References

Blatt, F., "Physics of Electronic Conduction in Solids," New York, McGraw-Hill, 1968.

Kittel, C., "Introduction to Solid State Physics," Fifth Edition, New York, Wiley, 1976.

Meaden, G. T., "Electrical Resistance of Metals," New York, Plenum, 1965; London, Iliffe, 1966.

Meaden, G. T., "Conduction Electron Scattering and the Resistance of the Magnetic Elements," *Contemporary Physics*, **12**, 313–337 (1971).

Ziman, J. M., "Electrons and Phonons," Oxford, Clarendon Press, 1960.

Cross-references: ELECTRICITY, FERMI SURFACE, HALL EFFECT AND RELATED PHENOMENA, PHOTOCONDUCTIVITY, SEMICONDUCTORS, SOLID STATE PHYSICS, SUPERCONDUCTIVITY.

CONSERVATION LAWS AND SYMMETRY

Among the most basic of the laws of nature are the conservation laws. A conservation law is a statement saying that in a given physical system under specified conditions, there is a certain measurable quantity that never changes regardless of the actions which go on within the system. One of the tasks of physics is to determine which properties of a given system are actually conserved during the course of specific types of interactions.

In classical (pre-quantum and pre-relativity) physics the following conservation laws were known:

(1) *Conservation of Mass.* In a closed system the total mass is constant.

(2) *Conservation of Energy.* In a closed system the total amount of energy is constant. (In relativistic physics these two laws are identical due to the equivalence of mass and energy.)

(3) *Conservation of Momentum.* The total momentum of a system is constant if there is no outside force acting on the system. (The momentum of an object is defined as the mass multiplied by the velocity; the total momentum of a system is the vector sum of all the individual momenta of the parts.) This means that the internal forces within the system have no effect on the total momentum.

(4) *Conservation of Angular Momentum.* The total angular momentum of a system is constant if there is no torque acting on the system from without. (The angular momentum of an object relative to a point O is its momentum multiplied by the perpendicular distance between its line of travel and the point O.)

Historically these laws arose out of a philosophical belief that the universe was created with a definite amount of motion which remained unchanged following the original creation. As a result of the attempt to clarify what kind of "motion" was conserved, the concepts of momentum and kinetic energy were developed. As early as the seventeenth century Huygens recognized that both momentum and kinetic energy were conserved in the collisions of elastic balls. During the nineteenth century the existence of various "forms" of energy was recognized, and the more general law of conservation of energy arose out of measurements involving reactions in which energy was transformed from one form to another (e.g., mechanical, thermal, electrical). The measurements showed that within certain limits of accuracy, the total amount of energy in a closed system was unchanged by any of the reactions tested.

From the modern point of view, it is not necessary to make measurements involving large-scale systems, for the macroscopic behavior of matter results from the interactions between relatively few types of elementary particles. Therefore it is sufficient to investigate the conservation laws as they apply to the basic interactions between fundamental particles.

At present only four fundamental types of interactions have been recognized: the gravitational, the weak nuclear, the electromagnetic, and the strong nuclear force. Each of these interactions individually obeys the classical conservation laws. As a result those laws must be obeyed in any kind of action involving interactions between particles. This rule, of course, applies to every activity in the universe.

For example, when we compress a spring, the potential energy of the spring is increased. The modern picture visualizes the energy as stored in the electric fields between the atoms of the spring as they are pushed closer together. Thus, the spring's potential energy is ultimately of an electrical nature.

The development of a conservation law is seen to depend on a combination of theoretical concept and experimental measurement. The scientist forms in his mind an abstract concept of a physical quantity such as energy which can be measured by a given set of operations. Measurements then show that (within limitation of error) this quantity is conserved under a given set of conditions. Modern measurements have been able to verify the conservation laws to very high degrees of accuracy. Conservation of energy has been verified to within 1 part out of 10^{15}, using the Mössbauer effect. (See Reference 1.)

An important function of the conservation laws is that they allow us to make many predictions about the behavior of a system without going into the mechanical details of what happens during the course of a reaction. They give us a direct connection between the state of the system before the reaction and its state after the reaction. In particular we can say that any action which violates one of the conservation laws must be forbidden. For example, many problems involving rotational or orbital motion are solved very simply by noting that the motion must be such that the angular momentum of the system remains constant. No further information concerning the forces or accelerations are required.

With the development of the Hamiltonian method of solving physical problems, and particularly with the growth of importance of quantum mechanics, it has become clear that the conservation laws are closely connected with the concept of symmetry in nature. This is based upon the fact that the interaction between two or more objects can be described in terms of a potential energy function (more precisely a mathematical function called the Hamiltonian of the system). If the potential energy of the system is known for any position of these objects in space, then we can predict the future motion of the objects in the system.

A detailed solution of the equations of motion will describe the position and velocity of each particle in the system at any time during their interaction. However, certain general predictions can be made without going through the complete solution of the problem, if there exist certain symmetries of space and time. The following examples illustrate the various geometrical or space-time symmetries encountered in classical physics.

(1) If the potential energy function does not depend explicitly on one of the space coordinates, then the component of momentum associated with that coordinate never changes—it is a constant of the motion, and thus obeys a conservation law. Particular situations most frequently encountered are as follows:

(a) An object moves in a three-dimensional space where its potential energy is a constant. That is, the expression describing the potential does not explicitly contain the coordinates x, y, or z, so it does not make any difference where the origin of the coordinate system is located. This means that the description of the system is invariant with respect to a translation of the origin of the coordinate system in any direction. As a result of this symmetry the momentum of the object in all three dimensions is constant. In technical terms, conservation of linear momentum is associated with translational symmetry (or homogeneity) of space.

(b) An object moves in a world which is flat, so that the force of gravity is in the vertical (z) direction. The potential energy depends on the height of the object above the ground, but does not depend on its location in the horizontal plane. That is, the description of the system is invariant with respect to a translation of the coordinate system in the x-y plane. Since there is symmetry in the x-y plane, the object's momentum is conserved as far as motion in that plane is concerned, but is not conserved in the z direction.

(c) Two spherical bodies interact in such a way that the potential energy depends only on the distance between the two bodies. This interaction has spherical symmetry, and the system is invariant with respect to a rotation of the coordinate system about any axis; i.e., it is isotropic. In spherical coordinates there are two angle variables, so there are two components of angular momentum to be conserved. As a result the two bodies orbit around their common center of mass in such a way that the magnitude of the total angular momentum is constant, while the plane of the orbit in space never changes. In other words, conservation of angular momentum is due to the isotropy of space.

(2) If the interaction between two objects

does not depend explicitly on the time coordinate, then the actions which take place do not depend on when we start measuring time. That is, the properties of the system are invariant with respect to a translation of the origin along the time axis. As a result of this symmetry it is found that the total energy of the system is conserved. In other words, conservation of energy is associated with a symmetry in the time dimension.

Use of a four-dimensional coordinate system in accordance with Einstein's principle of relativity allows us to combine both space and time symmetries into a single space-time symmetry. With this scheme the three dimensions of space and the one dimension of time make up a single four-dimensional space. Analogously, energy is regarded as the fourth component of a four-dimensional vector whose first three components are the three components of momentum. The symmetries associated with translation and rotations in this space-time continuum are called Poincaré symmetries.

With the rise in importance of elementary particle physics, a new type of symmetry has proven very valuable. These are "internal symmetries"—symmetries involving the internal properties of particles. The general philosophy underlying the study of elementary particle interactions is that anything can happen as long as it is not expressly forbidden by a law of nature. Among elementary particles there are a vast number of conceivable reactions that might take place. However, most of these reactions are forbidden by "selection rules," which are essentially conservation laws. For example, the total electric charge during any reaction cannot change. This rule immediately forbids such reactions as the conversion of a neutron into a proton plus a neutrino. An electron must also be created to balance the charge.

The study of symmetries and conservation laws is especially important in elementary particle physics because the exact nature of the strong and weak nuclear interactions is not known, so one cannot make detailed predictions concerning the results of reactions involving these forces. However, a knowledge of symmetry principles gives one a great amount of general information concerning these reactions, so one can estimate the probability of each reaction taking place.

While a law such as conservation of energy is true for all interactions, a number of the internal symmetries lead to conservation laws that do not apply to all of the four fundamental interactions. Such symmetries are therefore called "approximate" or "broken" symmetries. The implication of the term is that undistorted nature would be completely symmetrical, but that the presence of certain forces leads to an asymmetry, or breaking of the symmetry.

One important class of conservation laws has

to do with the constancy of certain essential numbers during the course of particle reactions. These "number laws" are as follows:

(1) *Conservation of Electric Charge.* If P is the number of positive charges in a system, and N is the number of negative charges, then $Q = P - N$ is the net number of charges. The charge number Q is unchanged by any reaction. For example, the creation of a positive charge must always be accompanied by the formation of an equal negative charge (e.g., an electron-positron pair is created by a high-energy photon). Conservation of electric charge is associated with a symmetry property of Maxwell's equations known as gauge invariance, which states that the absolute value of the electric potential (as opposed to the relative value) plays no part in physical processes. In quantum field theory conservation of electric charge is connected with the fact that the properties of a system of particles do not depend on the phase of the wave function describing the system.

(2) *Baryon Conservation.* Baryons are a class of elementary particles including the proton, the neutron, and several heavier particles such as the lambda, the sigma (plus, minus, and neutral), and the omega (minus). Baryons are particles that interact with the strong nuclear force. Each baryon is given a baryon number 1, each corresponding antibaryon is given a baryon number -1, while the light particles (photons, electrons, neutrinos, muons, and mesons) are given baryon number 0. The total baryon number in a given reaction is found by algebraically adding up the baryon numbers of the particles entering into the reaction. During any reaction among particles the baryon number cannot change. This rule ensures that a proton cannot change into an electron, even though a neutron can change into a proton. Similarly, to create an antiproton in a reaction, one must simultaneously create a proton or other baryon. Baryon conservation ensures the stability of the proton against decaying into a particle of smaller mass. Both conservation of charge and baryon conservation are absolute selection rules.

(3) *Lepton Conservation.* Leptons are a class of light particles that include electrons, neutrinos, and muons, as well as their antiparticles: the positrons, antineutrinos, and antimuon. Each lepton is assigned a lepton number +1, while each antilepton has a lepton number -1. All other particles have lepton number zero. In any reaction the algebraic sum of lepton numbers is conserved. This rule determines the course of beta decay, muon decay, and other reactions governed by the weak interaction.

(4) *Isospin Conservation.* Since the strong nuclear force acting between two neutrons is found to be the same as the force acting between two protons, as well as between a neutron and a proton, it is found useful to consider the neutron and proton as two states of the same

particle (the *nucleon*). These two states are considered to differ only by the different positions of a vector property called the *isospin* (or isotopic spin). This concept arises by analogy from the fact that two electrons in an atom can exist in a state of spin "up" and spin "down." These two states are indistinguishable in the absence of an external magnetic field because of symmetry of space with respect to rotations around an arbitrary axis. Similarly, a proton is a nucleon with isospin "up" and the neutron is a nucleon with isospin "down." In particle physics, whenever a system can exist in a discrete state, characterized by a definite quantum number, there exists a property (in this case isospin) that is conserved. In the absence of electromagnetic interactions there is no difference between the two isospin states because of symmetry with respect to rotation in "isospin space." Electromagnetic interactions make a difference because of the charge on the proton. Isospin conservation implies equality of p p, n-n, and n-p forces, except for the effect of the electromagnetic force. Thus isospin conservation is only an approximate symmetry.

(5) *Strangeness Conservation. Strangeness* is a property of elementary particles found useful to classify hyperons (particles more massive than nucleons) into families. Each particle is assigned a strangeness quantum number S which is related to the electric charge Q, the isospin number T, and the baryon number B by the formula $Q = T + (S + B)/2$. ($T = \frac{1}{2}$ for a proton and $-\frac{1}{2}$ for a neutron; other particles may have $T = 0$ or 1, depending on the type.) Strangeness is conserved in reactions involving the strong interaction. The selection rules resulting from strangeness conservation are very important in explaining why some reactions take place much more slowly than others.

A very important set of conservation laws is related to symmetries involving parity (P), charge conjugation (C), and time reversal (T). Parity is a property that is important in the quantum-mechanical description of a particle or system of particles. It relates to the symmetry of the wave function that represents the system. If the wave function is unchanged when the coordinates (x, y, z) are replaced by ($-x$, $-y$, $-z$) then the system has a parity of +1. If the wave function has its sign changed from positive to negative (or vice versa) when the coordinates are reversed, then the system is said to have a parity of -1. During a reaction in which parity is conserved, the total parity number does not change.

Changing the coordinates (x, y, z) into ($-x$, $-y$, $-z$) converts a right-handed coordinate system into a left-handed coordinate system. In terms of symmetry, the meaning of conservation of parity is that in any situation where parity is conserved, the description of the reaction will not be changed if the word "left" is changed to the word "right" and vice versa. This means that

such reactions can provide no clue that will distinguish between the directions right and left.

Prior to 1956 it was believed that all reactions in nature obeyed the law of conservation of parity, so that there was no fundamental distinction between left and right in nature. However, in a famous paper by C. N. Yang and T. D. Lee it was pointed out that in reactions involving the weak interaction, parity was not conserved, and that experiments could be devised that would absolutely distinguish between right and left. This was the first example of a situation where a spatial symmetry was found to be broken by one of the fundamental interactions.

The principle of charge conjugation symmetry states that if each particle in a given system is replaced by its corresponding antiparticle, then nobody will be able to tell the difference. For example, if in a hydrogen atom the proton is replaced by an antiproton and the electron is replaced by a positron, then this antimatter atom will behave exactly like an ordinary atom, if observed by people also made of antimatter. In an antimatter universe the laws of nature could not be distinguished from the laws of an ordinary matter universe.

However, it turns out that there are certain types of reactions where this rule does not hold, and these are just the types of reactions where conservation of parity breaks down. For example, consider a piece of radioactive material emitting electrons by beta decay. The radioactive nuclei are lined up in a magnetic field which is produced by electrons traveling clockwise in a coil of wire, as seen by an observer looking down on the coil. Because of the asymmetry of the radioactive nuclei, most of the emitted electrons travel in the downward direction. If the same experiment were done with similar nuclei composed of antiparticles and the magnetic field were produced by positron current rather than an electron current, then the emitted positrons would be found to travel in the upward, rather than in the downward, direction. Interchanging each particle with its antiparticle has produced a change in the experiment.

However, the symmetry of the situation can be restored if we interchange the words "right" and "left" in the description of the experiment at the same time that we exchange each particle with its antiparticle. In the above experiment, this is equivalent to replacing the word "clockwise" with "counterclockwise." When this is done, the positrons are emitted in the downward direction, just as the electrons in the original experiment. The laws of nature are thus found to be invariant to the simultaneous application of charge conjugation and mirror inversion.

Time reversal invariance describes the fact that in reactions between elementary particles, it does not make any difference if the direction of the time coordinate is reversed. Since all reactions are invariant to simultaneous application of mirror inversion, charge conjugation, and time reversal, the combination of all three is called *CPT* symmetry and is considered to be a very fundamental symmetry of nature.

A new type of space-time symmetry has been proposed to explain the results of certain high-energy scattering experiments. Called "scale symmetry," it pertains to the rescaling or "dilation" of the space-time coordinates of a system without changing the physics of the system. (See Reference 5.) Other symmetries, such as chirality, are of a highly abstract nature, but aid the theorist in his effort to bring order into the vast array of possible elementary particle reactions.

It is a temptation to say that "nature likes symmetries" in order to prove the theoretical necessity of a conservation law. However it must be realized that only human beings can like anything. The search for symmetries in nature leads to experiments that test the theory. While a symmetry idea may suggest a conservation law, the conservation law must be tested by experiment to see if nature really behaves that way in a given situation.

MILTON A. ROTHMAN

References

1. Rothman, M. A., "Discovering the Natural Laws: The Experimental Basis of Physics," New York, Doubleday & Co., Inc., 1972.
2. Rothman, M. A., "The Laws of Physics," New York, Basic Books, Inc., 1963.
3. Swartz, C. E., "The Fundamental Particles," Reading, Mass., Addison-Wesley Publishing Co., 1965.
4. Sakurai, J. J., "Invariance Principles and Elementary Particles," Princeton, N.J., Princeton University Press, 1964.
5. Jackiw, R., "Introducing Scale Symmetry," *Physics Today* (January, 1972).

Cross-references: ANTIPARTICLES; ELECTRO-WEAK THEORY; ELEMENTARY PARTICLES; GAUGE THEORIES; GRAND UNIFICATION THEORY; IMPULSE AND MOMENTUM; IRRE-VERSIBILITY; PARITY; POTENTIAL; QUANTUM CHROMODYNAMICS; QUARKS; ROTATION—CURVILINEAR MOTION; WEAK INTERACTIONS; WORK, POWER AND ENERGY.

CONSTANTS, FUNDAMENTAL*

Perhaps the basic concept of modern physical theory is that there are fundamental entities underlying the structure of the universe, that the atoms and molecules which make up the

*Preparation of this article supported in part by National Bureau of Standards (US), Grant NB81-NADA2087.

universe are the same and that they obey the same physical laws in the farthest galaxy as they do on earth. One of the goals of science then, is to determine the relationships that exist among these entities so that the set of independent fundamental quantities may be reduced to a minimum. Thus, 19th Century chemistry led the way to the understanding of the vast complexity of matter in terms of less than a hundred "irreducible" chemical elements. In the first half of the 20th Century physics expanded the realm of chemistry by the discovery of isotopes and increased the number of atomic elements to perhaps a thousand, but at the same time reduced the number of fundamental entities to three—the proton, the neutron and the electron—out of which the thousand different isotopes and billions of different molecules are built. Physics is now undergoing a similar simultaneous expansion and reduction. The electron has been joined by the muon, tauon, and neutrinos (and their anti-particles) and the proton and neutron are now associated with pions, kaons, and an array of "strange" and "charmed" mesons and baryons. On the other hand there are emerging theories of a more elementary particle—the quark—out of which the observed particles are "constructed," and the "grand unification," which will describe gravitation, electromagnetism, and the strong and weak nuclear forces as manifestations of a single entity.

It is this universality that makes it meaningful to attempt to measure the properties of these fundamental entities with all possible precision. Although the numerical values of such quantities as the mass and electric charge of a proton, or the velocity of propagation of electromagnetic waves are expected to be constant in time and space, our knowledge of those values is variable and in general changes with each new measurement. There is no valid evidence however, at the present time, that these quantities are not indeed constant. It may be that the values of the fundamental constants are changing over time scales comparable to the age of the universe and that this so-called "age" is not the indication of a time of beginning but is instead the time constant describing the rate of change of an ever-existing, unending universe. Should such a model indeed prove correct, it would not in principle change the concept of fundamental, invariable constants of nature. Although it would fundamentally alter our theoretical understanding, and change the models one uses for the description of the universe, these changes would still be described in terms of fixed parameters, which are then the fundamental constants.

Fundamental Units The existence of fixed, constant values associated with atomic and molecular physics has important implications for metrology and the establishment of units and standards. By the end of the nineteenth century it had already been suggested that basic physical processes could be used to establish standards for units of measurement. Since atomic processes are uniquely reproducible and universally available they can be better standards than arbitrary artifacts such as a metre bar or a prototype kilogram. Artifacts can change or be damaged. Does the build-up of a surface film or the absorption of atmospheric gases on the prototype kilogram change the size of the international unit of mass? How can we know? On the other hand the mass of a proton, or of a carbon atom in its ground electronic state, is invariant, the same everywhere and at all times. Similarly, one may consider the Bohr radius or the Compton wavelength of an electron as a fundamental atomic unit of length. Because of the limitations of measurement, the use of such microscopically small standards of length would sacrifice accuracy in the comparison of the standard with macroscopic lengths. The optical interferometer, however, allows one to compare lengths in terms of the wavelength of a monochromatic light source. Therefore in 1960 the platinum-iridium prototype metre bar in use since 1889 was replaced by the definition of the metre as the length equal to 1650763.73 wavelengths in vacuum of the electromagnetic radiation arising from the optical transition $2p_{10}$-$5d_5$ in the isotope of krypton of mass 86.

The unit of time (the second) is ordinarily defined as $1/86400$ of a day. As the available precision of measurement and the demands of science and technology increased, this definition proved inadequate, since the period between two successive noons changes throughout the year due to the nonuniform speed of the earth in its orbit around the sun. The specification was made more precise, and the "day" was defined for time-keeping purposes to be the mean length of the day averaged over a full year. However, this definition also proved inadequate because the rotation of the earth on its axis is not constant: as a result of tidal forces between the earth and the moon and of physical changes in the distribution of mass over the earth's surface during the year, the rate of rotation is decreasing by 0.03% per million years and varying seasonally by as much as a part in 10^7. In 1960 the international General Conference on Weights and Measures therefore adopted a definition of the second as "the fraction 1/31556925.9747 of the tropical year for 1900 January 0 at 12 noon ephemeris time." This definition directly demonstrates one of the problems associated with nonatomic standards. The year (the time between successive vernal equinoxes) is not a constant so that the mean motion of the earth in its orbit must be specified at a specific instant. With the development of atomic clocks and the ability to make accurate and precise measurements of frequency it was clear that such clocks could provide convenient and reproducible standards for the measurement of time. Following the recommendation of the General Conference in 1964 an atomic definition of the second was internationally adopted in 1967: "the second is the duration of 9192631770 periods of the radiation corresponding to the transition between the two hyperfine levels of the ground state of the cesium-133 atom" ($F = 4$, $m_F = 0 \leftrightarrow F = 3$, $m_F = 0$).

In the decade between 1972 and 1982 intercomparisons of the wavelengths of many absorption-stabilized gas lasers demonstrated their reproducibility at the level of parts in 10^{10} with stability of parts in 10^{11}. During the same period it became clear that the ^{86}Kr transition defining the metre could not provide an unambiguous standard of wavelength, because of its inherent spectral profile and lack of spectral purity, beyond the order of several parts in 10^9. At the same time, nonlinear optical techniques were being developed that provided efficient microwave harmonic generation. These, coupled with frequency synthesis developments and the inherent monochromaticity and stability of absorption-stabilized lasers, made possible the direct comparison of the frequency of optical transitions ($\lambda \sim 10^{-4}$ cm) with that of the cesium clock ($\lambda = 3.26$ cm). When both the frequency and wavelength of a wave are known its velocity is simply $c = \lambda f$. In this way the velocity of light was found to be $c = 299792458 \pm 1.2$ m/s. The precision of this result is limited not by the measurements themselves but by the inherent uncertainties associated with the krypton lamp. Such a situation implies that the wavelength standard should no longer be krypton but should instead be a much more reproducible and monochromatic laser. If, however, one can transfer frequency (or wavelength) measurements from the optical region to the microwave region and hence know the wavelength of the radiation used to define the duration of the second, or the frequency of the radiation used to define the length of the metre, is it necessary to have two separate standards? Should one not use a single atomic transition to define both the unit of length and the unit of time? In essence, why not define the metre in terms of the second and an accepted value for the velocity of light? This is exactly what has now been done and in October 1983 the metre was redefined by adopting $c = 299792458$ m/s as an exact quantity.

The unit of temperature is related in a less direct manner to atomic processes, but since the liquid or gaseous state of matter is a result of intermolecular forces, the realization of the international temperature scale is in principle also atomic, defined by fundamental constants, and reproducible without reference to a specific arbitrary artifact, by chosing the triple point of water to have an assigned temperature of 273.16 degrees Kelvin, exactly.

Only the unit of mass is defined on the basis of a man-made object; the kilogram is nothing more than the mass of the prototype platinum-iridium kilogram kept in the laboratory of the

Bureau International des Poids et Mesures in Paris.

These units and standards, which define the international metric system of units ("Système International" or SI), are not without significance to those English-speaking nations who have not yet adopted the metric system. There is no independent English system; the foot is internationally defined as 0.3048 m and the pound is exactly 0.45359237 kg (see Table 1).

Universal Constants In view of the attempts to replace arbitrary or man-made standards of measurement with standards based on atomic phenomena it is appropriate to ask whether there is a truly universal set of fundamental units of length, mass and time. The so-called Planck units are based on a recognition that the existence of the universe itself provides a basis for constructing such quantities. We may postulate that the universe contains matter and energy. The Newtonian constant of gravitation, G, is the fundamental quantity defining the interaction between matter in the large—without bringing into question its composition or internal structure. Similarly, Planck's constant (divided by 2π), $\hbar$, and the speed of electromagnetic radiation, c, are the fundamental quantities associated with light and electromagnetic energy. A more fundamental basis can indeed be found. In the special theory of relativity the constant c appears as a factor which unifies the description of space and time. Its numerical value depends on the units in terms of which one measures length and time. It has the dimensions of a velocity (length/time) but its identification with the velocity of electromagnetic radiation is a consequence of the theory and the fact that the photon has zero rest-mass. In the general theory of relativity the fundamental constant is G/c^4, determining the extent to which the energy density produces a curvature of space. The fundamental constant of the

Planck radiation law (describing the distribution of electromagnetic energy and how it changes with temperature) is hc. In fact, the thermodynamics of black-body radiation requires, solely on the basis of dimensional analysis, that the constant hc must exist independently of quantum theory. From these three constants one may construct a quantity with the dimensions of a mass, $m^* = (\hbar c/G)^{1/2} = 21.8$ μg, a quantity with the dimensions of length $l^* = (\hbar G/c^3)^{1/2} = 1.62 \times 10^{-35}$ m, and a quantity with the dimensions of time, $t^* = l^*/c = (\hbar G/c^5)^{1/2} = 5.38 \times 10^{-44}$ s. The physical significance of these quantities is not clear. The length l^* and the time t^* may be related to an ultimate quantization of space and time and may represent inherent limits of measurement, so that it would be impossible to distinguish two points which are closer together in space or time as separate points. The same kind of interpretation cannot be given to m^*, an object of that mass is within the range of size visible to the eye, certainly easily seen in a microscope of even low magnification, and would contain the order of 10^{17} atoms! It remains for future theorists to provide a meaning to these fundamental quantities.

Experimental Determinations The only test one has of validity of physical theory is the agreement between the results of measurements and the predictions of the theory. Discrepancies between theory and experiment may be due to experimental errors or to misinterpretation of the experimental results, or they can demonstrate the need for revision or rejection of the theory and its replacement by a new theory. The difficulties inherent in measuring physical constants to an accuracy of a few parts per million are great. Direct measurements of the mass of the electron or of the electronic charge are not as accurate as measurements which determine instead various combinations of these quantities. Whereas R. A. Millikan was able to measure the elementary charge on an electron in 1912 to one part in a few thousand (and this experiment can hardly be improved upon today), our current knowledge of the electronic charge, with an accuracy of approximately 1 part in 500,000, comes from combining the ratio of the Faraday constant to the Avogadro constant with measurements of the Sommerfeld fine structure constant, the gyromagnetic ratio of the proton, the Josephson effect, the realization of the SI ohm, the magnetic moment of the proton relative to the Bohr magneton, and a half-dozen or so other measurements which affect the final result to a greater or lesser degree. In fact, the present knowledge of the numerical values of all of the so-called fundamental constants of physics come from such indirect measurements. Of these physical constants, only the universal gravitational constant G is measured independently of the others. In the first place, no theoretical relation is known which relates G to the other physical constants,

TABLE 1. METRIC EQUIVALENTS.

1 inch	0.0254 m
	2.54 cm
1 foot	0.3048 m
1 mile	1609.344 m
1 pound (avdp)	0.45359237 kg
	453.59237 g
1 ounce (avdp)	28.349523125 g
1 pound (troy)	373.2417216 g
1 ounce (troy)	31.1034768 g
1 gallon (231 in^3)	3.785411784 L*
1 quart	0.946352946 L
1 ounce (fl)	29.5735295625 cm^3

NOTE These numbers are exact by definition. For most calculations they may be rounded off considerably.

The legal definition of length is in terms of the yard. 1 yard = 0.9144 m.

*L $\equiv$ litre = 1 dm^3 = 1000 cm^3.

and in the second place, the accuracy with which G is known is at least an order of magnitude poorer than the accuracy of the atomic constants.

With the growth in our knowledge of natural laws and of the technical means of making precise physical and chemical measurements, an increasing number of relationships have been discovered between the fundamental constants of physics and chemistry. The situation regarding our knowledge of these constants is a spiderweb of interconnected data which can be visualized as a bridge truss made up of elastic members, or, more simply, an interconnected network of springs, in which the length of each member represents the experimentally measured relationship between constants, and the stiffness is a measure of the accuracy of this measurement. The problem is to determine the positions of the nodes of this network. One recognizes that the alteration of any one member will produce an effect which will be transmitted throughout the entire structure. This will be true whether we change the length of a member, or its stiffness, or remove it entirely.

In order to determine the values of the physical constants from such an overdetermined set of data, it has become common to use the method of least squares. This can be considered as equivalent to the problem of minimizing the stored potential energy in our multidimensional network. The fundamental requirement, however, is basically one of establishing a method of analysis which is consistent and independent of the choice of variables used to describe the situation. The method of least squares not only does this but also provides a procedure which yields "best" estimates for the values of the constants in the sense that these estimates are the most accurate. For these reasons, least squares adjustment has been used for all of the significant determinations of the values of the fundamental physical constants over the past 35 years.

Numerical Values The numerical values given in Tables 2 and 3 are based on the most accurate and consistent measurements available as of June 1982. Significant measurements are the frequency-voltage relation in a superconducting Josephson junction which determines the ratio $2e/h$; the Faraday constant, N_Ae; the ratio of the magnetic moment of the proton to the magnetic moment of the electron μ_p/μ_e, or to the magnetic moment of the muon, μ_p/μ_μ; the proton magnetic moment measured in units of the Bohr magneton μ_p/μ_B, or in units of the nuclear magneton $\mu_p/\mu_N = (m_p/m_e)\mu_p/\mu_B$; the quantized Hall effect in semiconductors which yields e^2/h; and the Avogadro constant N_A. The spectroscopy of the energy levels of one-electron atoms (positronium, muonium, hydrogen, and deuterium) and of two-electron helium provide the basis for determining the Ryberg constant and the Sommerfeld fine structure constant. An accurate determination of the fine structure constant may also be obtained from the observed anomalous magnetic amount of the electron and (with less precision) from the anomalous magnetic moment of the muon. On the other hand, the agreement of the observed magnetic moment of the muon with that predicted by the theory of Quantum Electrodynamics (QED) and the value of α derived from the electron magnetic moment provides perhaps the most accurate verification of QED, and may be considered as providing a probe of the elec-

TABLE 2. DEFINED VALUES AND EQUIVALENTS.

Meter (m)	the length of path traveled by light in vacuum during a time of 1/299792458 of a second.
Kilogram (kg)	mass of the international kilogram
Second (s)	9192631770 cycles of the radiation of the hyperfine transition $F = 4$, $m_F = 0$ to $F = 3$, $m_F = 0$ of the ground state of the atom ^{133}Cs.
Degree Kelvin (K)	In the thermodynamic scale, 273.16 K = triple point of water $t(°C) = T(K) - 273.15$ (freezing point of water, $0.0000 \pm 0.0002°C$)
Unified atomic mass unit (u)	$\frac{1}{12}$ the mass of an atom of the ^{12}C nuclide
Standard acceleration of gravity (g_n)	9.80665 m/sec^2 980.665 cm/sec^2
Normal atmosphere (atm)	101325 N/m^2 1013250 dyne/cm^2
Thermochemical calorie (cal$_{th}$)	4.184 J 4.184 × 10^7 erg

TABLE 3. GENERAL PHYSICAL CONSTANTS[a]

Constant	Symbol	Value
Speed of light	c	$299792458 \text{ m} \cdot \text{s}^{-1}$
Gravitational constant	G	$6.673 \times 10^{-11} \text{ N} \cdot \text{m}^2 \cdot \text{kg}^{-2}$
Elementary charge	e	$1.60218 \times 10^{-19} \text{ C}$
Avogadro constant	N_A	$6.0221 \times 10^{23} \text{ mol}^{-1}$
Mass unit	u	$1.66054 \times 10^{-27} \text{ kg}$
Electron mass	m_e	$9.1094 \times 10^{-31} \text{ kg}$
		$5.4858 \times 10^{-4} \text{ u}$
Proton mass	m_p	$1.67263 \times 10^{-27} \text{ kg}$
		1.00727647 u
Neutron mass	m_n	$1.67493 \times 10^{-27} \text{ kg}$
		1.0086649 u
Faraday constant	F	$96485 \text{ C} \cdot \text{mol}^{-1}$
Planck constant	h	$6.6261 \times 10^{-34} \text{ J} \cdot \text{s}$
	$\hbar = h/2\pi$	$1.05458 \times 10^{-34} \text{ J} \cdot \text{s}$
Fine structure constant	α	0.00729735
	α^{-1}	137.0360
Josephson Frequency	$2e/h$	$483594 \text{ GHz/V}_{BI}$
		483597 GHz/V
Magnetic flux quantum	Φ_0	$2.06785 \times 10^{-15} \text{ T} \cdot \text{m}^2$
Rydberg constant	R_∞	$10973731.5 \text{ m}^{-1}$
Bohr radius	a_0	$0.529177 \times 10^{-10} \text{ m}$
Compton wavelength	$\lambda_c = h/m_e c$	$2.42631 \times 10^{-12} \text{ m}$
Electron radius	r_e	$2.81794 \times 10^{-15} \text{ m}$
Thomson cross section	$8\pi r_e^2/3$	$6.65246 \times 10^{-29} \text{ m}^2$
Gyromagnetic ratio of protons in H_2O	γ_p'	$26751.5 \times 10^4 \text{ s}^{-1} \text{ T}^{-1}$
	$\gamma_p'/2\pi$	$42.5763 \text{ MHz} \cdot \text{T}^{-1}$
Bohr magneton	μ_B	$9.27406 \times 10^{-24} \text{ J} \cdot \text{T}^{-1}$
Nuclear magneton	μ_N	$5.05081 \times 10^{-27} \text{ J} \cdot \text{T}^{-1}$
Proton magnetic moment in H_2O	μ_p'/μ_B	1.5209931×10^{-3}
	μ_p'/μ_N	2.792776
Free proton magnetic moment	μ_p/μ_B	1.5210322×10^{-3}
	μ_p/μ_N	2.792848
	μ_p	$1.410615 \times 10^{-27} \text{ J} \cdot \text{T}^{-1}$
First radiation constant	$8\pi hc$	$4.9925 \times 10^{-24} \text{ J} \cdot \text{m}$
	$2\pi hc^2$	$3.7418 \times 10^{-16} \text{ J} \cdot \text{m}^2 \text{ s}^{-1}$
Second radiation constant	hc/k	$0.014388 \text{ m} \cdot \text{K}$
	$N_A hc$	$0.1196266 \text{ J} \cdot \text{m}^1 \cdot \text{mol}^{-1}$
Gas constant	R	$8.3145 \text{ J} \cdot \text{K}^{-1} \cdot \text{mol}^{-1}$
Boltzmann constant	k	$1.38066 \text{ J} \cdot \text{K}^{-1}$

[a]Based on an analysis of the data available June 1982. The numerical values are expected to be accurate to within one or two units in the last digit given. All quantities are in SI (Systeme International) units:

C	= coulomb	N	= newton
G	= gauss	T	= tesla
Hz	= hertz (cycles per second)	u	= atomic mass unit
J	= joule	W	= watt
K	= kelvin (degrees)	Wb	= weber

tromagnetic interaction to distances as small as 10^{-20} m.

E. RICHARD COHEN

References

1. Cohen, E. R., Crowe, K. M., and DuMond, J. W. M. "The Fundamental Constants of Physics," New York, Interscience Publishers, 1957.
2. Rossini, Frederick D. "Fundamental Measures and Constants for Science and Technology," Cleveland, Ohio, CRC Press, 1974.
3. Taylor, B. N., Parker, W. H., and Langenberg, D. N. "The Fundamental Constants and Quantum Electrodynamics," New York, Academic Press, 1969.

Cross-references: ELEMENTARY PARTICLES; FREQUENCY STANDARDS; GRAVITATION; MEASUREMENTS, PRINCIPLES OF; PHOTOMETRY; PHOTON; QUANTUM THEORY; QUARKS; RELATIVITY; SPECTROSCOPY; SYMBOLS, UNITS AND NOMENCLATURE IN PHYSICS; TIME.

CORIOLIS EFFECT

A marksman fires his rifle due north. In the absence of wind, he might well expect it to travel in a straight line and land due north of him. But will it? The physicist would, in general, answer no on the basis that the earth is rotating and is not, therefore, an inertial frame of reference (see ROTATION—CIRCULAR MOTION). G. G. Coriolis first analyzed this effect in 1844, and he is acknowledged in its name. For large artillery projectiles this effect may be significant, but for hand carried weapons it is usually negligible. For instance, a typical .22-caliber rifle bullet might be horizontally deflected 0.2 meter in traveling one kilometer.

That a projectile will normally follow a path which is curved in the horizontal leads conversely to the idea that to follow a straight path over the rotating earth requires the application of a sidewise force. Even though this appears to violate Newton's first law (see DYNAMICS), it is perfectly true in a non-inertial system—hence, the reason that this force is sometimes called "fictitious."

A body which is moving with constant speed in a straight line in an inertial system is not accelerating and is subject to no net force. An observer in a rotating coordinate system will, however, observe the same object to follow a curved path. The observer may treat this apparent deflection from a straight line as an acceleration which is always perpendicular to the path of the object. It is called the Coriolis acceleration. The apparent force applied to the body to cause the deflection from a straight path into a curve is called the Coriolis force.

The true acceleration of a body moving with constant speed in a straight line in a rotating coordinate system is equal in magnitude to the apparent acceleration just described. An expression for it may be derived simply and quite rigorously, but not generally, for the case of a body moving with constant radial speed in a rotating system. It consists of two distinct components both easily evaluated for the case mentioned. One component arises from the change in direction of the radial velocity of the body, the other from its change in tangential velocity due to changing distance from the center of rotation (see ROTATION—CIRCULAR MOTION).

Suppose the rotating system has a constant angular velocity ω, and the body moves radially with constant speed v. It is initially at distance r_1 from the center of rotation with velocity $\bar{v}_1$. After a small time interval Δt, it is at distance r_2 with velocity $\bar{v}_2$ (see Fig. 1). The two velocities $\bar{v}_1$ and $\bar{v}_2$ are related since the final velocity $\bar{v}_2$ is the vector sum of the initial velocity $\bar{v}_1$ and the change in velocity $\Delta \bar{v}$. Refer to Fig. 2 for a vector diagram of the preceeding statement (see STATICS). If we consider instantaneous values, Δt approaches zero, the angle $\omega \Delta t$ approaches zero, and the chord Δv approaches its arc in length. We may then use the well known angle-arc relationship (arc length) = (radius) (angle in radians), and write

$$\Delta v = v \omega \, \Delta t, \text{ or } \Delta v / \Delta t = \omega v$$

Now $\Delta v / \Delta t$ is the acceleration component (a_1) resulting from the change in direction of $\bar{v}$.

The tangential velocity of a body equals the product of its angular velocity and its radius of rotation. Equating changes in these quantities yields

$$\Delta v_t = \omega (r_2 - r_1) = \omega \, \Delta r$$

Or, dividing by Δt

$$\Delta v_t / \Delta t = \omega \, \Delta r / \Delta t$$

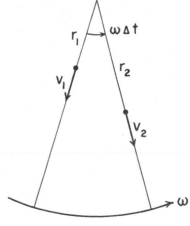

FIG. 1

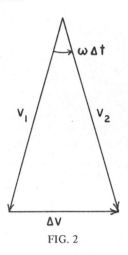

FIG. 2

Now $\Delta v_t/\Delta t$ is the acceleration component (a_2) arising from changing tangential velocity, and $\Delta r/\Delta t$ is the radial speed v. Hence a_2 also equals ωv.

Since both a_1 and a_2 lie in the same direction, being perpendicular to the radial velocity and to the right in the figure, their magnitudes may be added together with the sum equaling the magnitude of the Coriolis acceleration a_c

$$a_1 + a_2 = a_c = 2\omega v$$

Using Newton's second law $F = ma$, the Coriolis force is

$$F_c = 2m\omega v$$

The earth is not, of course, a rotating plane. The Coriolis acceleration reaches a maximum at the poles and vanishes at the equator. If v_h denotes horizontal velocity on the earth's surface, the resulting Coriolis acceleration may be evaluated by reference to Fig. 3 where ϕ is the lati-

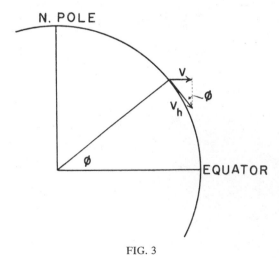

FIG. 3

tude and v is the true radial speed with respect to the earth's axis.

$$v/v_h = \sin \phi$$

$$v = v_h \sin \phi$$

and the Coriolis acceleration becomes

$$a_c = 2\omega v_h \sin \phi$$

and the Coriolis force is

$$F_c = 2m\omega v_h \sin \phi$$

The Coriolis effect applies to any object moving on the surface of the earth, and a more general treatment will show it to be completely independent of the direction of motion. The quantity $2\omega \sin \phi$ is commonly known as the Coriolis parameter. Since the earth rotates 2π radians in 24 hours, or at a rate of 7.27×10^{-5} radians/sec, the parameter is quite small, and equals exactly 10^{-4} sec^{-1} at about $43\frac{1}{2}°$ latitude. This small value for the Coriolis parameter means that in everyday life its effects are small and go largely unnoticed. For instance, the Coriolis force on an automobile driving at turnpike speeds might typically be five newtons. The acceleration can cause considerable deflection of long range artillery, however, and appropriate corrections must be made.

The Coriolis effect plays a large role in the great mass movements of the oceans and atmosphere. In the northern hemisphere, the apparent deflection is always to the right of the direction of motion. For example, air drawn toward a center of low pressure is deflected to the right and eventually flows *around* the low pressure area in a counterclockwise motion. This motion characterizes frontal storms typical of temperate climates. If the pressure force and Coriolis force are equal, the resulting wind velocity is said to be geostrophic (i.e., "turned by the earth"). Above one kilometer, the winds are closely geostrophic.

Though it has been only a little over a century since the first analysis of this effect, the Coriolis force has through the centuries influenced man's environment through its control of the motions of winds and waters, and hence the distribution of the sun's heat over the earth. Ancient man, not aware of the motion of his planet, was nevertheless profoundly influenced by it.

JULIAN M. PIKE

References

Byers, Horace Robert, "General Meteorology," fourth edition, New York, McGraw-Hill Book Co., 1974.

Coriolis, G. G., "Traite de la Mecanique de Corps Solides," Paris, 1844.

Fowles, Grant R., "Analytical Mechanics," third edition, New York, Holt, Rinehart and Winston, 1977.

Stephenson, Reginald J., "Mechanics and Properties

of Matter," third edition, New York, John Wiley & Sons, Inc., 1969.

Cross-references: DYNAMICS; MECHANICS; ROTATION–CURVILINEAR MOTION.

COSMIC RAYS

Cosmic rays are high energy subatomic particles which populate the interstellar space within galaxies, and to some extent between galaxies. The primary cosmic rays are atomic nuclei and electrons which have been accelerated by the action of electric and magnetic fields in a hierarchy of astrophysical processes. Only a few of these processes have been clearly identified. The primary particles produce secondary cosmic rays in collisions with other forms of matter.

The kinetic energy of these particles is typically 1–10 GeV; they have far more energy than the alpha, beta, and gamma rays emitted by radioactive atoms. Collisions between cosmic rays and target nuclei release enough energy to create mesons, both directly and via decay. Most of the cosmic rays which reach the earth's surface are in fact muons from the decay of charged pions. The pions are produced high in the atmosphere by collisions between primary cosmic rays and the nuclei of oxygen and nitrogen atoms.

Investigations of the unexpected behavior of these muons, and of other cosmic-ray phenomena, stimulated the development of large machines for producing "artificial cosmic rays," and evolved directly into the discipline now called elementary particle physics. Many of the experimental devices that are used in this new field were originally developed to study cosmic rays. They include Geiger-Muller and proportional counters, and devices for delineating tracks, descendents of the Wilson cloud chamber. Positrons, muons, charged pions, kaons, the first hyperons—all were discovered using cosmic rays.

Because the cosmic-ray energy spectrum extends, roughly as an inverse power law, to energies much greater than accelerating machines can duplicate, a good deal of activity still centers on the use of cosmic rays for studying elementary particle physics. With the advent of proton-antiproton colliders achieving TeV center-of-mass energies, this kind of work will be restricted henceforward to cosmic-ray energies greater than 10^{15} eV. The intensity of such cosmic rays is so small that it will be difficult indeed to do worthwhile experiments. At the present time much cosmic ray work is devoted to a second application, obtaining information about the astrophysical setting in which cosmic rays are produced and propagate.

Cosmic rays provide the only directly accessible sample of matter from outside the solar system. Hence the chemical composition (nuclear charge spectrum) of this material has been studied extensively, using detectors carried to great heights by means of balloons, rockets and spacecraft. The charge spectrum of "normal" (solar system) material, accounted for by nucleosynthesis in the big bang followed by processing in stars, is characterized by strong contrasts. The odd-Z elements are much less abundant than those with even Z, the light elements Li, Be, and B are practically absent, and so forth. The cosmic-ray charge spectrum shows much less contrast. The differences can largely be accounted for by spallation of primary heavy nuclei in collisions with interstellar hydrogen. The path length required to explain the anomalous composition of the nuclei arriving at the earth (the so-called "escape length") is about 7 g/cm^2 below 1 GeV per nucleon, decreasing to less than 1 g/cm^2 above a few hundred GeV per nucleon. The abundances of the primary elements (those which are actually accelerated) show a suggestive correlation with the first ionization potential of the atoms. Studies carried out by measuring artificial etch-pits in meteoritic crystals have shown that the cosmic-ray charge spectrum extends beyond Fe at least to Z about 80, the vicinity of Pt and Pb in the periodic table. These "ultra-heavy" nuclei are especially interesting probes since their interaction mean free paths are much shorter than the escape length. The conditions they require in order to be synthesized are also more restrictive than for lighter nuclei. Large counter systems and large sheets of etchable plastic are now being used to study the ultra-heavy component.

Measurements have also been made of isotopic abundances in cosmic rays. In case of primary constituents it is expected that the results will provide especially clear evidence regarding the varieties of nuclear processing that the cosmic-ray material has been through prior to its acceleration. One of the first findings is that the abundance ratio ^{22}Ne/^{20}Ne (extrapolated to the source) is significantly greater than for solar system material. However in other cases (^{26}Mg/^{24}Mg and ^{30}Si/^{28}Si) the abundances are not dissimilar.

The stable isotopes of secondary constituents provide information on the escape length and its energy dependence. The unstable isotopes provide a clock for measuring the average time spent in traveling from the sources to the solar system. By combining the lifetime with results on the path length one can obtain the density of the scattering material, averaged over those parts of the galaxy through which the cosmic rays have propagated. Results of this kind have come mainly from the isotope ^{10}Be (half-life 1.6×10^6 years), and they are limited thus far to energies less than 1 GeV per nucleon. They indicate that the lifetime is at least 10^7 years. This value, which is supported by data on the electron and positron energy spectra, described below, is unexpectedly large. It implies a mean density of scattering material equal to 0.3 g/cm^3, substantially less than the accepted value for interstellar matter in the galactic disk. These results indicate that cosmic rays spend much of their lifetime in regions of lower than average density, perhaps outside the disk.

Attempts to detect nuclei of antimatter in cosmic rays have shown that for charge numbers $Z \geqslant 3$ their abundance is less than 10^{-4}. Antiprotons have been found in two different energy bands. They appear to be secondaries produced in the sources or in passage through the interstellar medium. If the anomalously high intensity in the lower of these bands is confirmed it may indicate that cosmic-ray protons have a different life history than the heavier nuclei.

About 1% of the cosmic rays that strike the earth are electrons. High energy electrons have difficulty in propagating very far in the galaxy because they lose energy in producing synchrotron radiation under the influence of the magnetic fields that are present. On the other hand, the fact that they do produce this radiation has made it possible to show that cosmic-ray electrons (and presumably nuclei) are present in large numbers in supernova remnants such as the Crab Nebula, and in curious structures (jets and radio lobes) associated with unusually active external galaxies. Observationally, synchrotron radiation is distinguished by its nonthermal spectrum and high degree of polarization. From the intensity of this radiation one can deduce the amount of power that must be supplied to electrons in these sources. It proves to be a significant fraction of the visible power, indicating that the mechanism by which cosmic rays are accelerated is remarkably efficient.

If the electrons were a secondary component, produced by nuclear collisions, half of them would be positively charged. Measurements show that positrons make up only one-tenth of the total, so most of the electrons must be directly accelerated. Since the positrons are secondary and are subject to the life-shortening synchrotron process, measurements of their energy spectrum afford an alternative approach to determining the cosmic-ray lifetime.

Cosmic γ-rays and neutrinos with energies less than 10 or 20 MeV are a mixed bag. Some of these γ-rays come from nuclear reactions in stellar atmospheres and some from $e^{\pm}$ annihilation, others are thermal radiation from accretion on neutron stars and black holes; some of these neutrinos come from nuclear reactions inside stars. But γ-rays and neutrinos with higher energies are true cosmic rays, albeit secondary.

High energy neutrinos produced locally, by cosmic rays striking the earth, have been observed in deep mines. Initial steps have been taken toward construction of DUMAND, a deep underwater muon and neutrino detector. Having a target mass of order 10^9 tons (target volume ~ 1 km^3), it will be capable of detecting high energy neutrinos from distant events such as supernova explosions.

Work on cosmic γ-rays is further advanced. One of the original objectives was to determine the spatial distribution of cosmic-ray flux in this galaxy, using the γ-rays from decay of π^0 mesons produced by collisions of nuclei with interstellar gas and dust. Progress has been made, but this is proving to be more difficult than expected because of competition from γ-rays produced by cosmic-ray electrons. Two mechanisms are involved, bremsstrahlung and collisions with low energy photons (microwave through infrared to visible). It is found that the disk-shaped region in which electrons are present is considerably thicker than the disk containing most of the gas and dust of the galaxy. With their limited angular resolution, the instruments used to date respond mainly to the electron-initiated γ-rays. However these γ-rays are themselves a source of important astrophysical information.

The most abundant cosmic rays, having energies on the order of 1–10 GeV, are strongly affected by magnetic fields within the solar system. The first proof that the bulk of cosmic rays are charged particles rather than γ-rays was provided by the latitude effect, a reduction in cosmic-ray intensity at low latitudes resulting from the shielding effect of the earth's dipole field. Experiments detecting an east-west asymmetry in the intensity of obliquely incident cosmic rays proved, on the basis of geomagnetic theory, that primary cosmic rays have predominantly a positive charge.

The weaker but more extensive fields in interplanetary space show their influence in more subtle ways, through time variations correlated with solar activity. The effects range from an 11-year periodicity in the local cosmic-ray intensity to sudden intensity modulations associated with individual solar flares. The equipment used to study these variations ranges, in turn, from networks of earth-based monitors to satellites and interplanetary probes. Both the short term and long term variations are understood in terms of an interaction between cosmic rays and the "solar wind." The solar wind, an extension of the sun's corona, consists of outward-streaming plasma carrying with it a magnetic field that is partly regular, resulting in convection, and partly turbulent, resulting in diffusion. At the boundary where the solar wind meets the interstellar medium some particle acceleration seems to take place, making that region an important plasma physics laboratory. Particle acceleration also occurs from time to time near the sun's surface, in flares. Many cosmic-ray investigations are aimed primarily at deriving information about the sun and the interplanetary medium.

It is important for use of the ^{14}C radiocarbon dating technique in archaeology to know whether the intensity of cosmic rays reaching the earth has varied appreciably over the past 10^4 years. This unstable isotope, generated in the earth's atmosphere mainly through the reaction $^{14}N(n, p)^{14}C$, is one of several produced by cosmic-ray bombardment. By comparing ^{14}C dates with those determined by counting tree rings it has been found that the apparent production rate of ^{14}C has varied by some 20% over the past 3000 years. However, carbon participates in complex exchanges between the atmosphere, biosphere, and hydrosphere. It is not clear what relative importance should be given, therefore, to variations in the earth's magnetic field, in solar activity, or in climate, when interpreting these data. Extensive measurements of cosmogenic radionuclides in meteorites indicate that the intensity of cosmic radiation in interplanetary space has been constant within experimental error (about a factor of 2) for the past 100 million years.

Below 1 GeV per nucleon the energy spectra of the various cosmic-ray constituents are increasingly affected by the solar wind. Below 100 MeV per nucleon, corrections for the residual solar modulation, even for quiet-sun data, are so great that the shapes of the spectra in interstellar space are uncertain. For at least 3 decades above 1 GeV per nucleon the energy spectra have approximately inverse power-law form. The exponent has about the same value, 2.6, for all primary nuclei. The spectra of elements that are wholly or predominantly secondary, such as Li, Be, B, are somewhat steeper. These differences are successfully explained by models that properly take into account fragmentation during propagation from the source or sources to the solar system. By inverting the calculation one obtains "source abundances," for comparison with solar system material, for example. It is found that for self-consistency in these results the path length should decrease with increasing energy, as remarked above.

Above 10^3 GeV per nucleon the evidence from "direct" experiments, those in which the primary nuclei are identified by measuring their charge, is still scanty. Such experiments must be carried out above the atmosphere using balloons or spacecraft, but the particle energies must be measured calorimetrically, using thick absorbers. The two requirements combine to limit the product of area, time and solid angle, and hence the total count for a given intensity. Ground based experiments can have much larger sensitive areas and can use the atmosphere as an absorber, but they must rely on indirect evidence for identifying the primary particles.

The ground-based experiments are of two types, those which detect unaccompanied high energy muons, and those which detect air showers. The muons arise from the decay of charged pions (and kaons) produced when the primary cosmic rays collide with air nuclei, high in the atmosphere. In the same collisions, neutral pions are produced which give rise, by their decay, to photons, which in turn produce electromagnetic cascades via pair production and bremsstrahlung. The colliding nuclei retain half of their energy, on the average, so this process

repeats for several generations. At high enough energies many of the charged pions collide before decaying, because of relativistic time dilation, further augmenting the hadronic cascade. The composite cascade, consisting of hadrons, muons, and electrons (plus γ-rays) is called an air shower.

The cosmic radiation observed at ground level consists largely of muons from very small air showers whose electron and hadron components have been absorbed in the upper atmosphere. Above 10^{11} eV primary energy the electromagnetic cascades can be detected at ground level by means of atmospheric Čerenkov light collected with large mirrors. This light is very faint, so the observations must be made on clear, moonless nights. The same technique applies, of course, to showers produced by γ-rays. Cosmic γ-rays from several pulsars have been detected in this way.

When the primary energy reaches about 10^{14} eV the electrons reaching mountain altitude are numerous enough to be detected at distances of order 100 m from the shower core, or axis. This marks the beginning of a new experimental regime in which large counter assemblies are used to make detailed measurements, preferably on several components, of individual showers. The equipment used includes ionization calorimeters for studying the high energy hadrons, counters deep underground for detecting "hard" (high energy) muons, counters with smaller amounts of shielding to measure "soft" (~ 1 GeV) muons, unshielded counters for registering the total number of particles at ground level (most of which are electrons), and detectors of atmospheric Čerenkov light accumulated over the growth and decay of the shower above ground level. The simpler and less costly of these detectors are deployed in large numbers, forming arrays with collecting areas ranging from a small fraction of a square kilometer to several tens of square kilometers. Using electronic coincidence circuitry, the array counters select the events to be measured. Commonly they are given the added task of determining the trajectory of each selected shower: the core location from the intensity pattern, the direction from the timing pattern. Aside from the trajectory requirements, which determine the geometrical factor of the experiment, the selection depends primarily on the shower energy being greater than some threshold. The actual energies of the individual selected showers are calculated after the fact, in most cases using data from additional detectors over and above those which produced the trigger.

Large air shower arrays, operating continuously for many years, have measured the cosmic-ray energy spectrum over a range extending from about 10^{17} eV to more than 10^{20} eV per particle, while smaller arrays, operating at high mountain altitudes, have covered the range from 10^{14} to 10^{17} eV. This spectrum can be extended to lower energies by adding together the low energy spectra described above for all values of Z. The result exhibits several features. The (integral) slope, which at 10^{12} eV has become 1.7, a trifle greater than at lower energies, first decreases to ~1.5 and then, above the so-called "knee" at a few times 10^{15} eV, increases rather abruptly to ~2.0. After this rapid fall has continued for 3 decades the spectrum again flattens, having an "ankle" at ~10^{19} eV, above which the slope is only ~1.6.

Air shower experiments also explore the directional characteristics of high energy cosmic rays. At low energies, disregarding effects of the geomagnetic field and solar wind, cosmic rays are remarkably isotropic in direction. This is understood as resulting from repeated magnetic scatterings as the cosmic rays diffuse through the interstellar medium. Up to energies corresponding to the knee, the magnitude and direction of the small observed anisotropy (<0.1%) is explained by the motion of the solar system relative to its surroundings (the Compton-Getting effect) and a tendency of cosmic rays to stream outward from the denser parts of the galaxy in the direction of the local interstellar magnetic field. At energies corresponding to the knee the anisotropy begins to increase. The direction of the maximum intensity begins to change, becoming perpendicular to the galactic disk at about 10^{17} eV. At that point the anisotropy is several percent. As the energy increases further, so does the anisotropy, but no clear pattern emerges until the highest energies are reached. Above 5×10^{19} eV, where the Larmor radius of a proton in a regular magnetic field of reasonable strength (3 microgauss) is greater than the size of the galaxy (20 kiloparsec), the arrival directions are predominantly perpendicular to the plane of the galactic disk. It is generally believed that these cosmic rays must be extragalactic. But they cannot come from the extreme depths of space. Because of the Doppler effect, photons of the 2.7° microwave background radiation will appear to a 10^{20} eV proton to be 100 MeV γ-rays capable of producing photopions. The resulting energy loss sets a limit of about 30 megaparsec on the range of such protons. Photodisintegration imposes a similar limit on the range of heavier nuclei.

The identity of the particles that produce large air showers must be deduced from secondary features of the showers, such as the "elongation" (atmospheric depth of maximum development) or the ratio of muons to electrons. But these secondary features also depend on the character of elementary particle collisions at energies where direct evidence is unobtainable. The extreme proposals can be ruled out by rather general arguments: practically none of the large air showers are initiated by relativistic dust grains, and few if any by neutrinos or γ-rays. Just as at low energies, most of the very high energy cosmic rays are atomic nuclei.

The preponderance of indirect evidence on the composition of these nuclei (for equal energy) indicates that from 10^{12} eV to the knee of the energy spectrum the average mass increases. It then appears to decrease markedly over the next 1.5 to 2 decades. After that it seems to remain constant. At the knee, where the average mass seems to be greatest, the fraction of Fe nuclei may be 50–100%; at the highest energies the fraction of protons is at least 50%, probably greater.

Seventy years have passed since the balloon flights by Victor Hess which established the existence of cosmic rays. Except for small contributions from solar activity, there is still no clear evidence of where or how they are produced. It is becoming more and more difficult to account for the nonsolar cosmic rays reaching the earth by means of a single source or class of sources. There is much to be said in favor of an association between supernovae and the bulk of galactic cosmic rays, those with energies not much higher than 10^{15} eV, but even in that domain alternative models have some support. The possibility of accounting for the knee in terms of a "pulsar bump" made up of particles enriched in heavy nuclei cannot yet be ruled out. It is very probable but not absolutely certain that the highest energy particles are extragalactic. Does the ankle at 10^{19} eV mark a transition from predominantly galactic to extragalactic? Alternatively, if the knee marks the onset of enhanced leakage from the galaxy, one might expect the extragalactic component to become dominant within a couple of decades, by a few times 10^{17} eV, but there the spectrum seems to be smooth.

In the highest energy region, at least, the study of cosmic rays remains a frontier, both in physics and in astrophysics.

JOHN LINSLEY

References

Progress in Cosmic Ray Physics, **1–3**, and subsequent volumes entitled *Progress in Elementary Particle* and *Cosmic Ray Physics*, Amsterdam, North Holland Publishing Co.

Ginzberg, V. L., and Syrovatskii, S. I., "The Origin of Cosmic Rays," New York, Pergamon Press, 1964.

Hayakawa, S., "Cosmic Ray Physics," New York, Wiley-Interscience, 1969.

Greisen, K., "The Physics of Cosmic X-ray, γ-ray, and Particle Sources," New York, Gordon and Breach, 1971.

Hillas, A. M., "Cosmic Rays," New York, Pergamon Press, 1972.

Origin of Cosmic Rays, Dordrecht, The Netherlands, Reidel, 1981.

Conference Papers, 17th International Cosmic Ray Conference, **1–14**, Dordrecht, The Netherlands, Reidel, 1982.

Cross-references: ASTROMETRY; ASTROPHYSICS; DOPPLER EFFECT; ELEMENTS, CHEMICAL; RELATIVITY; SOLAR PHYSICS; SPACE PHYSICS.

COSMOLOGY

The Size and Shape of the Universe Although philosophical speculations about the origin, structure, and evolution of the Universe have continued since the time of the ancient Greeks, the observational basis for scientific cosmology has only recently been obtained. For nearly two centuries, the universe was, for example, thought to be limited to a finite, static system of stars which make up our Milky Way Galaxy. Observations supporting a finite stellar Universe with an observable edge were presented by William Herschel in the late eighteenth century and reaffirmed by Jacobus Kapteyn in the early 20th Century. By assuming that all stars shine with roughly the same intrinsic brightness, they could use counts of the number of faint stars in different directions to determine the structure of the stellar system. In this way both Herschel and Kapteyn concluded that the stars are concentrated in a flattened disk with the Sun at the center and with the greatest extent in the plane of the Milky Way. Herschel was unable to provide a distance scale for his picture, however, and it remained for Kapteyn to specify that the edges of the stellar Universe are located at about 1,500 pc in the direction perpendicular to the galactic plane and at about 8 times that distance in the direction of the plane. (One parsec, or pc, is equal to ten thousand billion kilometers, or 10^{13} km.)

One of the first objections to the Sun-centered Universe was provided by Harlow Shapley in the early 1920s. His observations indicated that the globular clusters outline an immense galactic system that is nearly a factor of ten larger than the visible stellar system. According to this scenario, the Kapteyn Universe is simply a local star cloud located in the outer regions of the grander Milky Way structure. Although Shapley did not realize it at the time, interstellar dust absorbs the light of distant stars, and for this reason the star counts vastly underestimated the scale of the stellar system. This was not all, for it was soon realized that the entire system of stars is rotating like a gigantic pinwheel about a remote, massive center. For instance, the investigations of Bertil Lindblad and Jan Oort, in 1925 and 1927 respectively, demonstrated that the Sun is located about 10,000 pc from the galactic center, and that the Sun revolves around this center at a very rapid speed of about 250 kilometers per second. In addition the enormously massive center, which contains a mass equivalent to about 100 billion stars, gravitationally controls the motion of the Milky Way, producing differential rotation in which the more remote stars move at slower speeds. In any event, Harlow Shapley's arguments for a dramatically larger size for the Milky Way were correct. The known Universe had to be centered far from the Sun, and its boundary had to be increased by a factor of ten. As it turned out however, Shapley was decidedly incorrect in arguing that the so-called spiral nebulae belong to the Milky Way. When it was realized that the spiral nebulae are actually extragalactic "island universes" containing literally billions of stars, the known size of our Universe had to be increased by factors of hundreds and then thousands, and even tens of thousands.

Although the once-fashionable island universe theory of spiral nebulae went into decline after the spectroscopists demonstrated the gaseous nature of certain diffuse nebulae, the extragalactic interpretation of the spirals revived its popularity after Vesto M. Slipher serendipitously discovered their large radial velocities. In the early twentieth century there was a widespread belief that the spiral nebulae are primitive solar systems in the making, and Percival Lowell asked Slipher to use the instruments at the Lowell Observatory to investigate their expected rotations. As a result, in 1914 Slipher "accidentally" discovered their astonishingly high velocities which sometimes exceeded 1,000 kilometers per second. Because it was hard to believe that objects with such enormous speeds could long remain a part of the Milky Way system, these observations proved to be a major impetus for reviving the island universe theory. In other words, the velocities of the spiral nebulae exceeded the escape velocity of the Milky Way, and they were therefore probably extragalactic objects which lie outside the gravitational control of our Galaxy.

This was nevertheless, a controversial conclusion, for it seemed to contradict Adrian van Maanen's measurements of the proper motions of spiral nebulae. If these objects lie well outside our Galaxy, then their angular motions correspond to absurdly high linear speeds approaching the velocity of light. Van Maanen's observations were simply incorrect, however, and it was eventually shown that the spiral nebulae lie at the enormous distances suggested by their high radial velocities. The great debate over the extragalactic nature of spiral nebulae was, in fact, finally settled in 1925 when Edwin Hubble announced the discovery of Cepheid variable stars in the Andromeda nebula (M 31) and the great spiral in Triangulum (M 33). The observed period and apparent brightness of these stars could be combined with the period-luminosity relation to infer a distance of about 480,000 pc for these two nebulae. This meant that they are considerably more distant than the most remote parts of our own Milky Way Galaxy. In fact, we now know that all of the spiral nebulae are remote galaxies, or island universes of stars, which are scattered throughout extragalactic space at distances as large as 10 billion pc.

Hubble next embarked on a survey of the realm of the nebulae. Just as the counts of stars as a function of brightness and direction in the sky had been used to show that the Milky Way is a finite disklike system, nebula counts could be used to explore how the nebulae, or galaxies, are distributed in space. As Hubble probed deeper into space to fainter and fainter magnitudes, the number of galaxies increased proportionally just as would be expected for a uniform distribution of galaxies in ordinary flat Euclidean space. Expressed algebraically, the number $N(m)$ of nebulae per unit solid angle brighter than visual magnitude m was:

$$N(m) = 1.43 \times 10^{-5} \times 10^{0.6m}$$
galaxies per steradian.   (1)

(Astronomers use a logarithmic scale of brightness, or magnitudes, in which larger apparent magnitudes refer to apparently fainter objects.)

Although some galaxies are aggregated into groups or clusters, Hubble found that they are uniformly distributed when viewed on very large scales. That is, the counts of faint galaxies in different directions and to different depths in the same direction indicated that there is a homogeneous and isotropic distribution of galaxies. This meant that we are not in a privileged position in the Universe, and that the Universe has no preferred shape or center. The distribution of galaxies would appear the same, apart from small scale fluctuations, when studied from any galaxy. In addition, there was no sign of the number of galaxies thinning out as Hubble looked further into the depths of space. There simply was no indication of an edge of the Universe. In fact, observations with the most powerful modern telescopes have reaffirmed Hubble's basic conclusions about the large scale structure of the Universe. It has a nearly homogeneous distribution of mass with no visible edge and no preferred center.

The Redshift-Distance Relation As early as 1914 Vesto M. Slipher had published the radial velocities for fifteen spiral nebulae or galaxies, noting that most of them are moving away from the Earth at velocities which exceed those of any other known astronomical object. According to Slipher, this indicated "a general fleeing from us or the Milky Way." To be precise, Slipher actually measured the difference between the wavelengths λ_0 of the spectral lines of the galaxies and the wavelengths λ_L of the same lines observed in terrestrial laboratories. This difference specifies the redshift z through the relation

$$z = \frac{\lambda_0 - \lambda_L}{\lambda_L}.   (2)$$

The most straightforward interpretation of the redshift is in terms of the Doppler effect due to the recession of the galaxies. According to this interpretation, the galaxies are moving away from the Earth with radial velocities V_r along the line of sight given by

$$V_r = c \times z \quad \text{for} \quad V_r \ll c   (3)$$

where $c = 2.99792456 \times 10^5$ kilometers per second is the velocity of light, and it is assumed

that V_r is much smaller than c. In any event, seven years after Slipher's pioneering work, C. Wirtz looked for correlations between these velocities and other observable properties of the galaxies. Wirtz found that, when suitable averages of the available data were taken, "an approximate linear dependence of velocity and apparent magnitude is visible. . . . The dependence of the magnitudes indicates that the spiral nebulae nearest to us have a lower outward velocity than the distant ones." Thus, both Slipher and Wirtz were remarkably close to "discovering" the expanding Universe; but it remained for Edwin Hubble to provide the data which conclusively indicated that the most distant galaxies are rushing away from us the fastest.

It actually was not until 1929 that Hubble provided better quantitative data for the distances of the spiral nebulae and mentioned somewhat incidentally the linear correlation of distance D and redshift. Because he was primarily interested in determining the solar motion with respect to the distant nebulae, he did not specifically state the now-famous Hubble law

$$cz = V_r = H_0 \times D.   (4)$$

(The parameter H_0 is now called Hubble's constant.) Hubble's first formulation of this redshift-distance relation referred only to relatively nearby galaxies (velocities less than 1,000 kilometers per second and distances less than 2 megaparsecs or 2 million parsecs), but he initiated a program to explore the relation to the largest distances possible. By the early 1930s Milton Humason had extended the spectrographic observations to the limits of the 100-inch telescope at Mount Wilson, and together Humason and Hubble found the redshift-distance relation to hold for velocities as large as 20,000 kilometers per second and for distances as large as 32 megaparsecs. This was done by reformulating Hubble's law in the form:

$$\log (cz) = 0.2m + B,   (5)$$

where m is the apparent magnitude of the object and the constant B depends on Hubble's constant H_0 and the intrinsic brightness of the galaxy. Because of the impossibility of measuring the distances to the weaker nebulae, Hubble and Humason assumed that all nebulae had the same intrinsic brightness and inferred a velocity-distance relation from the log redshift-apparent magnitude relation given in Eq. (5). They were also able to derive the value of Hubble's constant $H_0 = 558$ kilometers per second per megaparsec. This work was then amplified and extended by Humason, Nicholas Mayall, and Allan Sandage. In 1956 they published the log redshift-apparent magnitude relation for 474 extragalactic nebulae, using data obtained during a 25-year period at the Mount Wilson, Palomar, and Lick Observatories. Within the accuracies of the data, the relation is linear and the slope has a value of 0.2, which is the expected value for a homogeneous, isotropic expanding Universe. Moreover, by this time Walter Baade had shown that the period-luminosity relation for Cepheids in spiral arms is quite different from that for Cepheids in galactic nuclei or in globular clusters. When this difference was taken into account, the extragalactic distance scale was revised, and Hubble's constant was revised downward to about 180 kilometers per second per megaparsec. (During the past 15 years Allan Sandage and his colleagues at the Mount Wilson and Palomar Observatories have published at least five successively smaller values for Hubble's constant, the most recent value being about 50 kilometers per second per megaparsec.)

The log redshift-apparent magnitude relation was extended much deeper into space and further into the past when radio galaxies and quasi-stellar objects, or quasars, were discovered. Although normal galaxies like the spiral nebulae emit most of their radiation at optical wavelengths in the visible part of the spectrum, the radio galaxies and quasars emit tremendous amounts of energy at radio wavelengths. In fact, these objects were first discovered by radio astronomers who measured their angular sizes and determined their positions in the sky using interferometric techniques. These positions were then used to locate optical counterparts whose redshifts could be measured. Thus, for example, the first known radio galaxy, Cygnus A, was identified with an extragalactic object in 1954. Its redshift was 0.057, at the time one of the largest known. Nearly a decade later the first two quasars, 3C 273 and 3C 48, were identified with blue objects with starlike images and redshifts of 0.16 and 0.37, respectively. In retrospect, the designation quasi-stellar object appears to this writer a misnomer, however, for most quasars are comparable to galaxies in overall extent, and their starlike images are actually due to their enormous distances.

These were mind boggling discoveries, for the quasars had to be receding from our Galaxy at speeds which are a substantial fraction of the velocity of light, and both radio galaxies and the quasars had to be intrinsically much brighter than the nearby spiral galaxies. Although Nobel prize winner Martin Ryle once reasoned that the discrete radio sources are stars located within the Milky Way, and Geoffrey Burbidge once argued that quasars do not participate in the expansion of the Universe, there is now a general consensus that both of these superluminous objects are the active nuclei of young galaxies whose remote distances are specified by Hubble's law. Even their unusually bright radiation seems to be explained by a central energy source involving a massive black hole. At any rate, the extragalactic radio objects have approximately the same intensity over a wide range of wavelengths, in contrast to the spectrum of the normal galaxies. As a result, the radio galaxies

and quasars can be used to probe the remotest regions of the Universe. In fact, quasars with redshifts as large as 3.5 have been discovered, corresponding to a distance of about 10 billion pc. These quasars are so far away that they emitted the radiation we now detect when the Universe was less than 20% of its present age. Moreover, their radial velocities are so large that astronomers must use the relativistic expression

$$V_r = c \times \left[\frac{(z+1)^2 - 1}{(z+1)^2 + 1} \right] \qquad (6)$$

to infer a radial velocity V_r from the measured redshift z. As illustrated in Fig. 1, the quasars and radio galaxies nevertheless exhibit a log redshift-apparent magnitude relation which is consistent, within the errors, with that expected for a homogeneous, isotropic expanding Universe. Unhappily, the great dispersion in the observed data makes it impossible to determine the details of the expansion. The data suggest instead that the distant quasi-stellar objects are about a factor of 10 brighter than the nearer radio galaxies, which are in turn about a factor of 10 brighter than the very nearby normal galaxies. Because the light with which we view distant objects left them some time ago, this change in brightness suggests that younger extragalactic objects are brighter than older ones.

The Expanding Universe: Origin and Evolution
The most obvious and simplest interpretation of the observed galaxy redshifts is that they are due to the Doppler effect of receding galaxies, and that the Universe is therefore expanding. Nevertheless, there was initially quite a different explanation, for the Universe was still believed to be limited to the Milky Way system of stars. Around 1917 theoreticians were trying to find a solution to Einstein's gravitational theory which would keep this finite, static system from collapsing under its own weight. Willem de Sitter showed that there is one solution in which the redshifts of galaxies increase with the square of their distance, but in this case the finite Universe remains static and the galaxies do not move apart. That is, the redshift-distance relation becomes a mathematical curiosity of the General Theory of Relativity, and has nothing to do with an expanding Universe. Curiously enough, Hubble at first attributed his linear redshift-distance relation to this so-called de Sitter effect. Most astronomers, including Hubble, eventually accepted the expanding Universe, whose observed properties can be explained by simple Newtonian gravitation, but additional astronomical evidence was at first required to "prove" that the Universe is expanding.

Indirect evidence for an expanding Universe is provided by the fact that the sky is dark at night. If a static Universe is composed of a uniform and infinite distribution of stars, the night sky should be about as bright as the Sun. This so-called Olbers' paradox arises from the

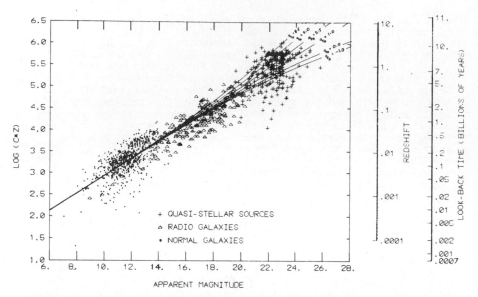

FIG. 1. The author's composite Hubble diagram, in which are plotted 663 normal galaxies, 230 radio galaxies, and 265 quasi-stellar objects. The vertical axis at right gives the look-back time in billions of years, calculated on the assumption of a Hubble constant of 50 kilometers per second per magaparsec. The solid lines denote the theoretical log redshift-apparent magnitude relations for a homogeneous isotropic Universe with different values of the deceleration parameter q_0. Because of the wide range in the intrinsic properties of these extragalactic objects, the observations provide no definitive statement about the value of q_0 and the fate of the expanding Universe.

fact that every line of sight in a static infinite Universe will eventually intersect the surface of a star, and the sky should therefore be as bright as a typical star's surface. This paradox is resolved in an expanding Universe in which the redshift moves the starlight of distant galaxies out of the visible optical wavelengths and into the larger invisible wavelengths. In fact, in order to avoid a theoretically bright night sky, astronomers have been compelled to conclude that the radiation from energetic young galaxies has to be greatly redshifted, and that an expanding Universe is therefore required.

If we accept the notion of an expanding Universe, the redshift-distance relation provides a method for studying the past history of the Universe, while also providing a time scale for creation in an evolving Universe. For instance, the light we detect from radio galaxies and quasars was emitted during past epochs, or higher redshifts, when the Universe was comparatively young. The fact that these objects are more abundant at higher redshifts provides support for the idea that we live in an evolving Universe that changes with time. In fact, a backwards extrapolation of the presently observed expansion indicates that there must have been a time in the past when all of the matter in the Universe was highly compressed and extremely hot. This was the time of the big bang which gave rise to the expanding Universe.

For example, in a Universe which expands forever in flat Euclidean space, the big bang occurred at a time

$$t_0 = \frac{2}{3H_0} = 13 \text{ billion years ago,} \qquad (7)$$

where Hubble's constant $H_0 = 50$ kilometers per second per megaparsec. The astrophysically determined ages of the oldest stars agree, within a factor of two, with this estimated time since the big bang. This agreement therefore provides additional evidence for an expanding Universe which originated in a big bang.

During the early stages of the expanding Universe, thermal radiation provided the dominant form of mass-energy, and the expansion was therefore gravitationally controlled by the radiation. As shown by George Gamow in the 1940s, the temperature $T(t)$ at time t since the big bang is, under these conditions, given by

$$T(t) = \left[\frac{c^2}{32\pi Gat^2} \right]^{1/4} \approx \frac{10^{10}}{t^{1/2}} \text{ °K.} \qquad (8)$$

Here, c is the velocity of light, G is the Newtonian constant of gravitation, a is the radiation constant, and in the numerical approximation the time t is in seconds. As Gamow realized, the temperatures during the first few minutes are hot enough for protons and neutrons to combine to form helium nuclei. The amount of helium synthesized is about 27% by mass, and this abundance agrees with that observed in various spots in our Galaxy, in nearby galaxies and in quasars. Moreover, the observed abundance of helium in our Galaxy far exceeds the amount which would be synthesized in all of the stars in our Galaxy over its entire lifetime. On the other hand, the observed abundance of the heavier elements in our Galaxy shows appreciable variation, and the much lower abundances can be synthesized within stars. In other words, the observed abundance of helium cannot be produced within the stars, and even if it was there would be an excessive abundance of the heavier elements. We can only conclude that the observed helium originated in the early stages of the expanding Universe before galaxies and stars were formed. As a matter of fact, detailed calculations indicate that the amount of helium produced depends upon the "frozen-in" proton-neutron ratio, and that this ratio is very sensitive to the rate of expansion. If this rate differed by even 50% from the accepted value, the helium abundance would disagree with that observed.

Direct astronomical evidence for the expanding Universe was provided in 1965 when Arno Penzias and Robert Wilson accidentally detected the $3°K$ microwave background radiation that proved to be the relic radiation of the big bang. When testing the microwave (or short wavelength radio wave) receiving system intended to be used for measurements of the high-latitude continuum radiation of our Galaxy, Penzias and Wilson found a few degrees of unexpected noise temperature. Because the excess noise showed no sidereal, solar, or directional variation, they first supposed that is arose in the receiving antenna itself, but they concluded that there remains an external noise contribution with an antenna temperature of $3.5 \pm 1.0°K$ at the wavelength of 7.3 centimeters. In a companion letter in the same issue of the *Astrophysical Journal*, Robert Dicke and his colleagues explained the excess $3°K$ radiation as the residual temperature of the primeval explosion that initiated the expansion of the Universe. This concept had nevertheless already been developed by George Gamow, Ralph Alpher, and Robert Herman in the late 1940s, when they showed that a present temperature of about $5°K$ results when the 10 billion degree primeval fireball has been cooled by expansion in the enlarging cavity of the world. In any event, if the cosmic noise discovered by Penzias and Wilson is the relic fireball radiation, it should have a thermal spectrum. Because of the interactions of matter and radiation, the primeval radiation will very rapidly reach thermodynamic equilibrium, and the resulting thermal spectrum will be preserved for all time as the Universe expands and cools. The expected thermal spectrum was promptly confirmed, and the microwave background was additionally shown to be extremely isotropic. In fact, the most precise measurement of the large-scale isotropy of the Universe now comes from the microwave background, which is isotropic to better than 0.1% on angular scales greater than one minute of arc. (An anisotropy of a few millidegrees has been attributed to either the motion of our Galaxy and its neighbors relative to the background radiation at a velocity of 500 kilometers per second, or to a Universe which is anisotropic at the 0.1% level.) At any rate, the high energy density, thermal spectrum, and isotropy of the $3°K$ microwave background cannot be explained by the radiation of conventional astronomical objects, and it must be attributed to the relic radiation of a denser, hotter epoch. This discovery therefore provides unambiguous evidence for the singular origin of the expanding Universe, while also providing an important constraint on the radiation temperature throughout the evolution of the Universe.

The radiation temperature $T_r(t)$, the radiation energy density $\rho_r(t)$, and the matter density $\rho_m(t)$ at any time t can, for example, be inferred using the boundary conditions

$$T_r(t_0) = 3 \text{ degrees Kelvin}$$

$$\rho_r(t_0) = 4 \times 10^{-34} \text{ grams per cubic centimeter}$$

$$\rho_m(t_0) = 1 \times 10^{-31} \text{ grams per cubic centimeter}$$

$$(9)$$

where the time $t_0 = 13$ billion years and the mass density is for galaxies. For a Universe of radius $R(t)$ at time t we have

$$\rho_m(t) \propto [R(t)]^{-3}$$

$$\rho_r(t) \propto [R(t)]^{-4} \qquad (10)$$

$$T_r(t) \propto [R(t)]^{-1}$$

so that

$$\frac{\rho_m(t) T_r(t)}{\rho_r(t)} = \text{constant for all times.} \quad (11)$$

We are presently in a matter dominated era in which ρ_m exceeds ρ_r. At the decoupling time these two forms of energy density were equal, and before that time radiation dominated matter.

Fate of the Universe What is the fate of the Universe? Will it continue to expand forever, or will it stop its expansion in the future and begin to contract? The answer to this profound question is found by considering the condition for which the outward motion of a remote galaxy is just balanced by the inward gravitational pull of all of the other galaxies in the Universe. This critical condition occurs when the velocity of the remote galaxy is just equal to the escape velocity of the rest of the Universe. When Hubble's Law is used to specify the galaxy's velocity and the galaxy's kinetic energy is placed equal to the gravitational potential energy resulting from the gravitational tug of all the other galaxies, we obtain a critical mass density:

$$\rho_c = \frac{3H_0{}^2}{8\pi G} = 4.7 \times 10^{-30}$$

grams per cubic centimeter,   (12)

where G is the Newtonian gravitational constant and the numerical value corresponds to $H_0 = 50$ kilometers per second per megaparsec.

If the present mass density $\rho_m(t_0)$ of the Universe exceeds ρ_c then the present expansion will eventually reverse, the galaxies and stars being destroyed and finally engulfed in a hot, dense fireball analogous to the one which initiated the present expansion. In other words the Universe will eventually stop its expansion, contract, and undergo another big bang. Alternatively, if the present mass density is less than or equal to the critical value, the Universe will continue to expand forever. The galaxies will disperse and fade into a black void. Will the expanding Universe end with a whimper or a bang? Edwin Hubble answered this question in 1926 when he essentially showed that the mass density of galaxies is less than the critical density, and that the Universe will therefore continue to expand forever.

Of course, we now have greater confidence in the conclusion that the expansion will continue indefinitely. Astronomers have reaffirmed the fact that the mean density of matter in the luminous visible parts of galaxies falls short of the critical value by a factor of about 30. Dark or invisible matter has been inferred from dynamical considerations as well as from radio and X-ray measurements of galactic halos and the intergalactic medium in clusters of galaxies. Nevertheless, when these contributions have been added, the mass density of the Universe is at least 10 times too small to halt its expansion. This has been confirmed by satellite measurements of the abundance of interstellar deuterium, which was presumably created in the big-bang explosion. These measurements indicate that the present mass density of the Universe in both visible and invisible forms is $\rho_m(t_0) \lesssim 5 \times 10^{-31}$ grams per cubic centimeter. Theoreticians will continue to create esoteric theories and observational astronomers will continue to collect uncertain data; but the available evidence indicates that the conclusions will not change. The evolving Universe will continue to expand forever, and the observable properties of this expansion will probably continue to be explainable in terms of Newton's theory of gravity and flat Euclidean space.

KENNETH R. LANG

References

Harrison, E. R., "Cosmology: The Science of the Universe," Cambridge, U.K., Cambridge Univ. Press, 1981.

Hubble, E., "The Realm of the Nebulae," New Haven, Yale Univ. Press, 1936; New York, Dover Pub., 1958.

Lang, K. R., "Astrophysical Formulae," New York, Springer-Verlag, 1980.

Lang, K. R., and Gingerich, O., "A Source Book in Astronomy and Astrophysics, 1900–1975," Cambridge, MA, Harvard Univ. Press, 1979.

Sandage, A. et al. (Eds.), "Galaxies and the Universe," Chicago, Univ. of Chicago Press, 1975.

Sciama, D. W., "Modern Cosmology," Cambridge, U.K., Cambridge Univ. Press, 1971.

Silk, J., "The Big Bang: The Creation and Evolution of the Universe," San Francisco, W. H. Freeman, 1980.

Weinberg, S., "Gravitation and Cosmology: Principles and Applications of the General Theory of Relativity," New York, Wiley, 1972.

Weinberg, S., "The First Three Minutes: A Modern View of the Origin of the Universe," New York, Basic Books, 1977.

Cross-references: ASTROMETRY; ASTROPHYSICS; DOPPLER EFFECT; ELEMENTS, CHEMICAL; RELATIVITY; SOLAR PHYSICS; SPACE PHYSICS.

CRITICAL MASS*

The mass of fissionable material required to produce a self-sustaining sequence of fission reactions in a system (a reactor, for example) is the *critical mass* for that system.

The chain of reactions will be self-sustaining if, on the average, the neutrons released in each fission event initiate one new fission event. The system is said to be *critical* when that condition exists.

Neutrons released from fissioning nuclei may escape from the system; they may be captured in non-fissioning reactions, or they may produce new fissions. The critical mass depends on the relative probabilities of these processes and on the average number of neutrons released per fission. Evaluation of these probabilities is the concern of criticality calculations which are important in the design of neutron chain reactors.

The escape probability becomes larger for smaller systems, inasmuch as the ratio of surface to volume increases as a system is made smaller. Thus, there is a *critical size* below which the chain reaction in a given system cannot be made self-sustaining. The concept of critical size is often discussed along with critical mass.

Neutrons colliding with non-fissionable nuclei in the system may be absorbed and thus lost to the chain reaction. In fact, not every neutron absorption by a fissionable nucleus results in a fission. Non-fission absorption must be taken into account in calculating the critical mass. For example, a system containing pure ^{235}U can be made to have a low critical mass. If the same

*Research sponsored by the USAEC under contract with Union Carbide Corporation.

configuration were loaded with a sufficient quantity of natural uranium (0.0057 per cent ^{234}U, 0.72 per cent ^{235}U, and 99.27 per cent ^{238}U) to contain the same total amount of ^{235}U, it would not be critical because at certain energies the ^{238}U readily absorbs neutrons without fissioning.

The probability that a neutron striking a fissionable nucleus will cause it to fission depends on the fission cross section (see FISSION) which in turn depends on the energy of the neutron, increasing as the neutron energy gets lower. Thus the addition of a moderator, that is, a material which takes up energy from the neutrons without absorbing them, will lower the critical mass of a system. Water and carbon are good moderators.

The critical mass also depends on the average number of neutrons released per fission. This number changes slightly with neutron energy. For ^{235}U it is about 2.45 for thermal neutrons and about 2.65 for 1 MeV neutrons. The numbers are slightly higher for ^{233}U.

A complete criticality calculation must take into account the fission cross section as a function of neutron energy and the average neutron yield per fission as a function of neutron energy. Also to be considered are the geometrical distributions in the system of the fissionable nuclei, the absorbing nuclei, and the moderator, and how the neutrons scatter from them. Furthermore, the configuration of reflecting material outside of the fuel volume has a marked influence on the critical mass. A complete calculation would construct the spatial and energy distributions of neutrons in the system through the use of a mathematical procedure that models the history of neutrons from their release to their capture or escape. Neutron diffusion theory or transport theory is usually used for such calculations. Actual calculations use idealizations and produce approximate results. Often, criticality experiments are required to verify results.

Finally, as example of critical masses, a sphere 32 cm in diameter containing ^{235}U dissolved in water has a critical mass of about 2.1 kg. The same sphere with ^{233}U has a critical mass of about 1.1 kg.[1] The Oak Ridge National Laboratory graphite reactor was loaded with 31 tons of natural uranium. This contains about 203 kg of ^{235}U.

A thorough discussion of neutron chain reactors is given in reference 2.

CHARLES D. GOODMAN

References

1. Callihan, A. D., Morfitt, J. W., and Thomas, J. T., *Proc. Intern. Conf. Peaceful Uses At. Energy, Geneva,* **5**, 145 (1956).
2. Weinberg, A. M., and Wigner, E. P., "The Physical Theory of Neutron Chain Reactors," Chicago, University of Chicago Press, 1958.

Cross-references: CROSS SECTIONS AND STOPPING POWER; FISSION; NUCLEAR REACTIONS; NUCLEAR REACTORS.

CRITICAL PHENOMENA

Introduction The field of phase transitions and critical phenomena has seen extraordinary developments within the past few years,[1] and in this brief account we shall attempt to provide an overview of some of these, concentrating our discussion on the concepts of scaling and universality.

We shall organize the introduction to critical phenomena about three simple questions: (1) "What happens?", (2) "Why study?", and (3) "What do we actually do?"

"What Happens?" What happens near the critical point is easily explained by means of the example of a simple magnet. The most striking macroscopic property of a magnet is that it is in fact "magnetized"—it can, for example, pick up thumb tacks! Suppose we measure the number of thumbtacks or magnetization as we heat the magnet at a uniform rate. As we heat, the tacks fall off one by one, but the rate of falloff suddenly diverges to infinity at a certain temperature (see Fig. 1) and there are no thumbtacks left for temperatures higher than this temperature. We call this temperature the *critical temperature*, T_c, and we call phenomena associated with the critical point *critical phenomena*.

Examples of other critical phenomena are the singularities in two important *"response functions"*:

(1) the constant-field specific heat (see Fig. 2(a))

$$C_H \equiv \text{response in heat content to a change in temperature}$$

$$\equiv T \left(\frac{\partial S}{\partial T} \right)_H, \qquad (1)$$

where here $S = S(T, H)$ denotes the entropy of the system, and

(2) the isothermal susceptibility (see Fig. 2(b))

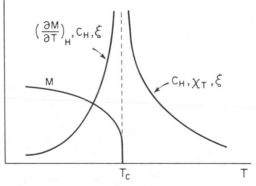

FIG. 1. Behavior (schematic) of some common physical quantities near the critical point.

$\chi_T \equiv$ response of magnetization to a
change in magnetic field

$$= \left(\frac{\partial M}{\partial H}\right)_T, \qquad (2)$$

where $M = M(T, H)$ is the magnetization and H is the magnetic field.

Thermodynamic functions are *macroscopic* in nature. What is happening at a *microscopic* level to account for the anomalies in the thermodynamic functions? Simply said, the motion of increasing numbers of particles is becoming correlated; e.g., the correlation function,

$$C_2(T, H, \underset{\sim}{r}) \equiv \langle s_0 s_{\underset{\sim}{r}} \rangle - \langle s_0 \rangle \langle s_{\underset{\sim}{r}} \rangle, \qquad (3)$$

which describes the degree of correlation among the constituent magnetic moments s_r of our magnet, is becoming long-range. Here the angular brackets denote thermal averages.

In fact, the "moments" of the correlation function,

$$\mu_a(T, H) \equiv \int |\underset{\sim}{r}|^a C_2(T, H, \underset{\sim}{r}) \, d\underset{\sim}{r}, \qquad (4)$$

are found to diverge for all positive a and even for a limited range of negative a (roughly $a \gtrsim -2$)! In particular, the zero-field isothermal susceptibility is directly proportional to the zeroth moment

$$\chi(T, H = 0) \propto \mu_0 (T, H = 0) = \int C_2(T, 0, \underset{\sim}{r}) d\underset{\sim}{r}, \qquad (5)$$

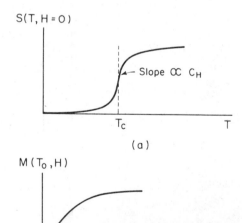

(a)

(b)

FIG. 2. Definitions of two response functions, the constant-field specific heat $C_H(T, H = 0)$, and the isothermal susceptibility evaluated in zero field $\chi_T(T, H = 0)$. Both functions are singular at the critical point of a simple magnet.

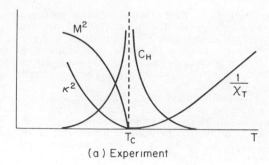

(a) Experiment

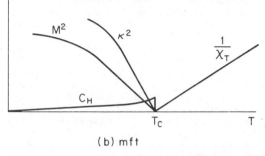

(b) mft

FIG. 3. Behavior (schematic) of common physical quantities near the critical point, (a), compared with the predictions of the mean field theory, (b). Here $\kappa \equiv 1/\xi$.

so the divergence of the susceptibility can be seen to be directly related to the increase in range of $C_2(T, H, \underset{\sim}{r})$.

A particularly useful measure of the range of the correlation function is the correlation length, $\xi(T, H)$, defined by the equation

$$\xi^2(T, H) = \mu_2(T, H)/\mu_0(T, H) \qquad (6)$$

This quantity can be measured experimentally, e.g., by light scattering, and it is found to diverge near the critical point (see Fig. 1).

"Why Study?" There are, of course, many answers to the question "why study critical phenomena," but certainly the simplest answer is that there exist rather striking discrepancies between the predictions of closed-form theories and our findings in the real world. For example, in Fig. 3 we compare, in a qualitative fashion, the experimentally observed temperature dependences of the functions discussed above with the predictions of the *mean field theory* (mft) of cooperative phenomena. The mean field theory corresponds to a model in which each magnetic moment interacts with all other magnetic moments in the entire system with an *equal* exchange interaction. Such a model is unlikely to be a realistic model for the task of describing the effects noted above, and indeed we know that interaction energies in a system fall off with distance between the magnetic moments. Thus one reason for studying critical

phenomena is that it provides a testing ground for theories that concern the microscopic interactions in matter.

A second reason for the burgeoning interest in critical phenomena is that a wide variety of seemingly disparate physical systems are found to behave quite similarly near their "critical points." For example, phenomena analogous to those shown in Fig. 1 are found in a liquid-gas system (see LIQUID STATE), in a ferroelectric, in a superfluid, in a superconductor, and in a binary alloy or mixture. Accordingly, workers from a variety of disciplines have been drawn together by the problem of critical phenomena, adding to the interdisciplinary flavor of the field.

"What Do We Actually Do?" The answer to this question may be given on many different levels, but basically it is that we simply want to understand on a microscopic level what forces exist between the constituent particles in matter, and how they "contrive" to produce the anomalies discussed above and sketched in Fig. 1. As a first step in this direction, it is useful to provide a quantitative measure of the phenomena.

It is found that in almost all systems the limiting behavior of the functions of interest is well described by a simple power law—so that if one plots one's data on log-log paper, the data fall on a straight line sufficiently near the critical point (see Fig. 4). It is important to emphasize that the power law behavior sets in for some systems only quite near the critical point, and when we write

$$f(x) \sim x^\theta, \qquad (7)$$

meaning "$f(x)$ varies as x to the θ power for x near zero," we mean, formally, that

$$\theta \equiv \lim_{x \to 0} \frac{\log f(x)}{\log x}. \qquad (8)$$

There is in general a different exponent for each function and each path of approach to the critical point. The thermodynamic functions considered above all concern approaches to the critical point ($T = T_c$, $H = 0$) in which $H = 0$

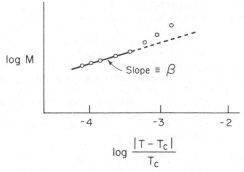

FIG. 4. Sketch of definition of the critical-point exponent β $[M \sim |\tau|^\beta]$ to illustrate the general definition (8).

and $\tau \equiv T - T_c \to 0$. Accordingly, we define the three exponents α, β, and γ in Table 1 to describe, respectively, the behavior of the specific heat, the spontaneous magnetization, and the isothermal susceptibility along this path.

The second column gives a typical range of experimental values for the exponents, while the third column gives the values predicted by the mean field theory. The reader would do well to verify from inspection of Fig. 3 that the mean field exponents are correctly listed. That the agreement is far from perfect is consistent with the discrepancies between Figs. 3(a) and 3(b) (experiment and mft).

Theorists have endeavored to calculate the exponents for as many models as possible. For example, the fourth column shows the exponents calculated for the two-dimensional Ising model, for which many of the zero-field properties are known exactly. We see that these exponents agree no better with experiment than do the mean field exponents—in fact, they err in the opposite direction, with the experimental numbers lying in between the mft and two-dimensional Ising predictions.

We shall subsequently argue that the reason that the two-dimensional Ising model disagrees with experiments on three-dimensional systems is that the system dimensionality plays a crucial

TABLE 1. DEFINITIONS AND TYPICAL VALUES OF SELECTED
CRITICAL-POINT EXPONENTS FOR A SIMPLE MAGNET.

Exponent	Definition	Experiment	Mean Field Theory (mft)	$d = 2$ Ising	$d = 3$ Ising
α', α	$C_H \sim (-\tau)^{-\alpha'}$ $\sim \tau^{-\alpha}$	−0.1 to 0.2	0	0	$\simeq 1/8$
β	$M \sim (-\tau)^\beta$	0.2 to 0.4	1/2	1/8	$\simeq 5/16$
γ', γ	$\chi_T \sim (-\tau)^{-\gamma'}$ $\sim \tau^{-\gamma}$	1.1 to 1.5	1	7/4	$\simeq 5/4$
$\alpha' + 2\beta + \gamma'$		$\simeq 2$	2	2	$\simeq 2$

role in determining critical-point exponents. Indeed, the fifth column, showing the results of numerical approximation procedures for the three-dimensional Ising model, is seen to agree rather better with the experimental results.

Scaling Hypothesis One could continue discussing exponents for other systems, for other functions, or for other paths of approach to the critical point. Until a few years ago this discussion would serve to summarize most of the research activity in critical phenomena. However, we are sooner or later going to seek a deeper understanding of the exponents, and this understanding is just beginning to arise. It is coming not from the exact solutions of model systems—for which the complexity of the derivation all but totally obscures any physical insights concerning the magnitude of the exponent obtained—but rather from an altogether different approach.

This approach began historically with the introduction of rigorous relations among the critical-point exponents—these relations took the form of inequalities, and generally involved three exponents. The inequality involving α', β, and γ' is simply

$$\alpha' + 2\beta + \gamma' \geqslant 2, \qquad (9)$$

and is generally called the Rushbrooke inequality.[2] That model systems appeared to satisfy Eq. (9) (and most of the other "rigorous" inequalities) as *equalities* (Table 1), and that most experimental systems were not inconsistent with the possibility that (9) is an equality, did not go unnoticed. However, all attempts to rigorously prove (9) as an equality have been singularly unsuccessful.

It is in the finest tradition of theoretical physics that when one cannot solve the original problem, one seeks to replace it with a simpler problem that one can solve. In this instance, the "breakthrough" occurred in 1965—two years after Rushbrooke proposed the Rushbrooke inequality—by many investigators working independently.[3] Their work generally goes by the name of the *"homogeneity"* or *"scaling"* hypothesis, and is perhaps most easily formulated in terms of a class of functions called *generalized homogeneous functions* (GHFs). A func-

tion $f(x, y, z, \cdots)$ is a GHF if we can find functions $g_x(\lambda)$, $g_y(\lambda)$, $\cdots$, $g_f(\lambda)$ such that for all positive λ,

$$f[g_x(\lambda)x, g_y(\lambda)y, \cdots] = g_f(\lambda)f[x, y, \cdots], \quad (10)$$

where the functions $g_i(\lambda)$ are arbitrary except that they possess inverses. It is elementary to show[3] that (10) is equivalent to the statement that there exist numbers $a_x, a_y, \cdots, a_f$ such that

$$f(\lambda^{a_x} x, \lambda^{a_y} y, \cdots) = \lambda^{a_f} f(x, y, \cdots). \quad (11)$$

The scaling hypothesis is just that—a hypothesis. It involves GHFs, and because of the properties of GHFs, it can be made about a variety of functions. One can make a scaling hypothesis about three different classes of functions: thermodynamic functions (TF), static correlation functions (SCF), and dynamic correlation functions (DCF).

TF hypothesis: Close to the critical point $\tau = H = 0$, the singular part of the Gibbs potential per spin $G(\tau, H)$ is "asymptotically" a GHF.

SCF hypothesis: Close to the critical point and for large $|r|$, the static correlation function $C_2(\tau, H, r)$ is a GHF.

DCF hypothesis: Close to the critical point and for large $|r|$ and t, the dynamic correlation function $C_2(\tau, \tilde{H}, r, t)$ is a GHF.

One can show that these statements are not entirely independent of one another and that in fact:

DCF hypothesis $\Rightarrow$ SCF hypothesis

$\Rightarrow$ TF hypothesis.

The TF hypothesis predicts that Eq. (9) holds as an equality, and thus is consistent with the numbers shown in Table 1. However, it also makes other predictions, not anticipated before. For example, it predicts that all data taken near the critical point can be made to "collapse" onto a single curve providing the data are plotted in the correct units. The single curve is called a "scaling function," and examples of scaling functions are shown in Fig. 5 and defined in Table 2. The scaling function has just

TABLE 2. SUMMARY OF SCALING FUNCTIONS PREDICTED BY THE SCALING HYPOTHESIS FOR THE M-H-T EQUATION OF STATE. HERE δ IS DEFINED BY $H \sim M^\delta$. SKETCHES OF THE THREE FUNCTIONS ARE SHOWN IN FIG. 5

Function	Dependent Variable	Independent Variable
$H_\tau = \mathcal{J}_{\mathrm{sgn}\,\tau}^{(1)}(M_\tau)$	$H_\tau \equiv H/|\tau|^{\beta\delta}$	$M_\tau \equiv M/|\tau|^\beta$
$H_M = \mathcal{J}_{\mathrm{sgn}\,M}^{(2)}(\tau_M)$	$H_M \equiv H/|M|^\delta$	$\tau_M \equiv \tau/|M|^{1/\beta}$
$M_H = \mathcal{J}_{\mathrm{sgn}\,H}^{(3)}(\tau_H)$	$M_H \equiv M/|H|^{1/\delta}$	$\tau_H \equiv \tau/|H|^{1/\beta\delta}$

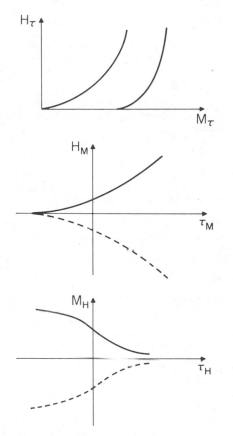

FIG. 5. Sketches of the three possible "scaling functions" for interpretation of MHT equation of state data near the critical point. A dashed line indicates that data are not taken for this "branch" of the scaling function due to MH symmetry.

recently been calculated directly for the Heisenberg model, and agreement with experimental data on magnetic systems is striking.[4]

Universality Hypothesis The universality hypothesis has arisen from attempts to answer the question "On what features of a model do critical properties (exponents, scaling functions, etc.) depend?" Universality states that the only properties are the dimensionality d and the "symmetry" n, where n actually denotes that dimensionality of the "order parameter" or spin space.

To make this concept more precise, let us introduce a general Hamiltonian which encompasses, as special cases, a very large proportion of the models that exhibit critical phenomena. This Hamiltonian utilizes the concept of classical spin vectors, situated on sites $\underset{\sim}{r}$ and $\underset{\sim}{r}'$ in a d-dimensional lattice, and interacting with energy parameters $J_{r-r'}{}^\alpha$ that depend upon the vector $(\underset{\sim}{r} - \underset{\sim}{r}')$. The important thing is that the spin vectors are taken to be unit vectors span-

ning a n dimensional space, i.e.,

$$\underset{\sim}{S} \equiv (S^1, S^2, \cdots, S^n), \qquad (12a)$$

with

$$(S^1)^2 + (S^2)^2 + \cdots (S^n)^2 = 1. \qquad (12b)$$

Thus the Hamiltonian under consideration is[5]

$$\mathcal{H} = - \sum_{\underset{\sim}{r},\underset{\sim}{r}'} \sum_{\alpha=1}^{n} J_{\underset{\sim}{r}-\underset{\sim}{r}'}{}^{(\alpha)} S_{\underset{\sim}{r}}{}^\alpha S_{\underset{\sim}{r}'}{}^\alpha .(13a)$$

Note that when $n = 1$, this Hamiltonian reduces to the Ising model, since the spins become simply one-dimensional "sticks" capable of assuming the two discrete orientations $+1$ (up) and -1 (down).

For $n = 2$, the Hamiltonian (13a) describes a system of two-dimensional vectors, and is generally called the XY of "plane rotator" model. It has also been called the Vaks-Larkin model, because Vaks and Larkin have considered it a lattice model for the superfluid transition in a Bose fluid.

For $n = 3$, (13a) describes the general anisotropic Heisenberg model, which has proved particularly useful in describing a variety of magnetic materials near their critical points.

For $n \to \infty$, (13a) reduces to the spherical model.[6]

A study that has recently been done concerns which parameters in the general model Hamiltonian (13a) are important for determining the values of critical-point exponents, and which are "irrelevant." To this end, one considers

(1) the lattice dimensionality, d;

(2) the spin dimensionality, n,

(3) the "lattice anistropy" or "nonuniformity of the interaction"–i.e., the dependence of $J_{r-r'}{}^\alpha$ upon the direction of $\underset{\sim}{r} - \underset{\sim}{r}'$.[7] There are many important materials for which the interaction strength in one crystal direction is different than in another direction, so this is not "of merely academic interest."

(4) the "spin space anistropy," or dependence of $J_{\underset{\sim}{r}-\underset{\sim}{r}'}{}^\alpha$ upon α. Probably no material is *perfectly* isotropic and there are certainly some materials for which the anistropy is believed to play a very important role.[8]

(5) the range of interaction, or dependence of $J_{r-r'}{}^\alpha$ upon $\underset{\sim}{r} - r'$. Again, there are probably no materials for which the commonly made assumption that only nearest neighbors of one another interact, and that all other pairs of spins are completely coupled, is valid.[9]

(6) the spin quantum number, S.[10] The Hamiltonian (13a) was for classical spins, and corresponds to the $S = 1/2$ Ising model and the $S \to \infty$ limits of the other systems. Materials in nature have a wide range of spin quantum number S, and accordingly we must consider what happens to these models for general quantum number S.

The conclusion of this work is that the only features of Eq. (13a) that are important are lattice dimensionality d and the symmetry of the ground state (which can be measured by an "effective dimensionality" of the spin). This conclusion is sometimes called "*the universality hypothesis*," and it has the implication—if believed—that one need only consider *two* parameters, d and n.[11,12] Hence (13a) may be replaced by[5]

$$\mathcal{H}_U(d, D) \equiv -J \sum_{\underset{\sim}{r}, \delta} \sum_{\alpha=1}^{n} S_{\underset{\sim}{r}}^{\alpha} S_{\underset{\sim}{r}+\delta}^{\alpha} \quad (13b)$$

where here δ denotes a vector between nearest-neighbor pairs of lattice sites.

Conclusion and Outlook   In this all-too-brief introduction to the subject of critical phenomena we have perforce concentrated our discussion on a single type of system, namely a simple magnet. One principal attraction, as noted above, is that a great variety of physical systems appear to behave quite similarly near their critical points. Indeed, the role of cooperative phenomena (and the proposed role of phase transitions) in certain biological systems has long been a subject of fascination for many, and it may be that the concepts that are eventually uncovered in our attempts to understand physical systems near their critical points will prove useful in elucidating biological behavior.[13-16]

H. Eugene Stanley

References

1. Stanley, H. E., "Introduction to Phase Transitions and Critical Phenomena," London and New York, Oxford University Press, 1971. Second Edition 1983.
2. Rushbrooke, G. S., *J. Chem. Phys.*, 39, 842 (1963).
3. See, e.g., the recent GHF approach of A. Hankey and H. E. Stanley, *Phys. Rev.*, **B6**, 3515 (1972); a comprehensive list of references to all earlier work is found therein.
4. Milošević, S., and Stanley, H. E., *Phys. Rev.*, **B5** 2526 (1972); **B6**, 986 (1972); **B6**, 1002 (1972).
5. Stanley, H. E., *Phys. Rev. Letters*, 20, 589 (1968).
6. Stanley, H. E., *Phys. Rev.*, 176, 718 (1968).
7. Paul, G., and Stanley, H. E., *Phys. Rev.*, **B5**, 2578 (1972).
8. Jasnow, D., and Wortis, M., *Phys. Rev.*, 176, 739 (1968).
9. Paul, G., and Stanley, H. E., *Phys. Rev.*, **B5**, 3715 (1972).
10. Lee, M. H., and Stanley, H. E., *Phys. Rev.*, **B4**, 1613 (1971).
11. Kadanoff, L. P., in "Proceedings of the Varenna Summer School on Critical Phenomena," M. S. Green, Ed. London and New York, Academic Press, 1972.
12. Griffiths, R. B., *Phys. Rev. Letters*, 24, 1479 (1970).
13. Herzfeld, J., and Stanley, H. E., *J. Mol. Biol.* 82, 231 (1974).
14. Stanley, H. E., in "Proceedings of the Enrico Fermi Summer School on Phase Transitions," K. A. Müller, Ed., London and New York, Academic Press, 1973.
15. Stanley, H. E., Ed. "Cooperative Phenomena near Phase Transitions: A Bibliography with Selected Readings," Cambridge, Mass., MIT Press, 1973.
16. Stanley, H. E., Ed., "Biomedical Physics and Biomaterials Science," Cambridge, Mass., MIT Press, 1972.

Cross-references: ANTIFERROMAGNETISM, COMPRESSIBILITY, FERRIMAGNETISM, FERROELECTRICITY, GAS LAWS, GASES: THERMODYNAMIC PROPERTIES, HEAT, HEAT CAPACITY, LIGHT SCATTERING, LIQUID STATE, MANY BODY PROBLEM, MATHEMATICAL BIOPHYSICS, POLYMER PHYSICS, SUPERCONDUCTIVITY, SUPERFLUIDITY.

CROSS SECTION AND STOPPING POWER

Cross section, σ, is a conceptual quantity widely used in physics, particularly in nuclear physics, to represent the probability of collision between particles. For example, if a beam of neutrons (n) is incident from the left on a nucleus (A), a certain fraction of the neutrons will be removed from the beam by interaction with A.

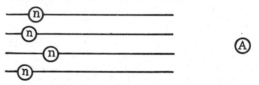

By definition, σ is the fraction of neutrons, contained in 1 cm^2 of beam, that interact with A.

For a thin layer of material of thickness dx containing N nuclei/cm^3 the number of nuclei/cm^2 is $N\,dx$. If the flux of incident neutrons is ϕ/cm^2, the fractional decrease in traversing the thin layer will be:

$$- (d\phi/\phi) = N\sigma \, dx \quad (1)$$

or for a finite thickness x and incident flux ϕ_0:

$$\phi = \phi_0 e^{-N\sigma x} \quad (2)$$

For a given type of nucleus there are, in general, a number of possible interactions, a, b, c, $\cdots$. The total cross section σ_t is the sum of the cross sections for the individual interactions:

$$\sigma_t = \sigma_a + \sigma_b + \sigma_c \cdots \text{etc.} \quad (3)$$

A convenient unit for nuclear cross sections is the *barn*, an area of 10^{-24} cm^2, which is approximately equal to the cross-sectional area of medium weight nuclei.

The *stopping power, Sp,* of a material for an incident particle is the quantity $-dT/dx$, i.e., the energy loss per unit length of path, generally expressed in ergs per centimeter. This type of attenuation is used primarily in considerations of the passage of heavy charged particles, such as protons, deuterons and alpha particles, through matter. For example, for a particle of any spin having a rest mass $M(\geqslant m_0$, the rest mass of an electron), charge ze, and velocity $V(=\beta c)$, the energy loss (as excitation and ionization) per element of path $-dT/dx$ to a homogeneous medium containing N atoms/cm^3, each of atomic number Z, is given by:

$$-\frac{dT}{dx} = \frac{4\pi e^4 z^2}{m_0 V^2} NZ \left[\ln \frac{2m_0 V^2}{I} - \ln(1-\beta^2) - \beta^2 \right]$$

$$(4)$$

where I is the mean atomic excitation potential calculable from the Thomas-Fermi electron distribution function to be $I = kZ \approx 11.5\ Z$ electron volts.

The *relative stopping power S* is the inverse ratio of the length of a material to the length of a standard substance having equivalent stopping power (usually referred to aluminum as $S_0 = 1$) at 15°C and 76 cm pressure:

$$S = \frac{(dT/dx)_1}{(dT/dx)_0} = \frac{N_1 B_1}{N_0 B_0} = \frac{\rho_1 B_1 A_0}{\rho_0 B_0 A_1}$$

$$(5)$$

where ρ is the density, A the atomic weight, and $B = Z \ln 2m_0 V^2/I$ from Eq. (4).

CLARK GOODMAN

Cross-references: COLLISIONS OF PARTICLES, NUCLEAR REACTIONS, NUCLEAR REACTORS, RADIOACTIVITY.

CRYOGENICS

Cryogenics is the science of very low temperature refrigeration and the study of phenomena which occur at these temperatures. In a general sense, cryogenics can be divided into two parts: the methods used to obtain and maintain very low temperatures, and the special techniques associated with measuring physical properties, particularly temperature, in the low temperature environment. Cryogenic temperatures are usually defined as those below about 90 K, (90 Kelvin) the liquification point of oxygen, although in the laboratory temperatures as low as 0.005 K (5 mK) are becoming commonplace.

Applications Industrial cryogenic applications include the shipment and storage of liquified gases such as nitrogen, oxygen, and natural gas fuel, and the use of these fluids through evaporation, particularly liquid nitrogen, as refrigerants. Liquid oxygen at 90 K is used extensively in the steel industry and in providing hospitals with oxygen gas for breathing. In addition, liquid oxygen and liquid hydrogen are used as rocket fuels in the space program. Liquid nitrogen, which is relatively inert, quite inexpensive ($\approx$ \$0.15/liquid liter), and has a large latent heat at 77 K of 1.6×10^5 joules/liter, is used as a powerful refrigerant in a variety of applications including the frozen foods industry, the biological storage of tissue and live cells, the refrigeration of sensitive detectors of electromagnetic radiation, and in the production and storage of colder liquid cryogens such as hydrogen at 20.3 K and helium at 4.2 K. Perhaps the largest use for liquid nitrogen, however, is in the storage and transfer of nitrogen to provide clean and relatively inert atmospheres in the chemical and semiconductor industries.

Liquid ^{4}He at 4.2 K is at present used almost exclusively in laboratory applications, but may soon find its way into hospitals and even into the business world thanks to new technologies being developed based on superconductivity in metals. Superconductivity is a phenomena in certain metals where below a transition temperature T_c the electrical resistance of the metal vanishes completely, and electrical currents can flow with no dissipation of energy. For most elemental superconductors, T_c lies in the range of 1 K to about 20 K. If a magnetic solenoid wound of superconducting wire is energized and its leads are then joined together, the current source can be removed and the magnetic field will remain as the current continues to flow through the solenoid. Magnetic fields as high as 120,000 Gauss can be produced with such solenoids, which decay no more than one part in 10^8 per day. Less stable superconducting magnets have produced fields as high as 165,000 Oe. A conventional room temperature magnet able to produce such a high field would dissipate many millions of watts of electrical energy in maintaining the field, and this energy as heat would then have to be removed or the magnet would melt. Applications for superconducting magnets include uses in nuclear magnetic resonance (NMR) studies and in some large electric motors. NMR spectrometers capable of observing resonances in a whole human body are being built utilizing superconducting magnets which can produce three-dimensional images of the subject. Such devices may soon replace x-ray machines as diagnostic medical tools.

In addition to the absence of electrical resistivity, superconductors possess a variety of other exotic properties. For example, the current flowing in a closed superconducting loop is quantized in units of $h/2e$, where h is Planck's constant and e is the absolute value of the charge on an electron. *Superconducting quantum interference devices* (SQUIDs) are able to use this fact and the tunneling properties of electrons in superconducting weak links to measure minute magnetic fields. For example, the fields produced by electrical activity in the human brain, at the level of 10^{-12} Gauss are being studied with SQUID devices in a number of low temperature laboratories. Other applications of superconducting quantum tunneling have resulted in computer logic elements smaller and faster than conventional semiconductor devices. Such elements may ultimately be utilized to produce super powerful computers operating at 4.2 K.

Cryogenics serves three major purposes in the physics laboratory. First, many physical properties are temperature dependent, and can be more easily understood if studied as functions of temperature rather than at a single temperature. Second, thermal energies often mask important features which, when discovered and understood, allow a far deeper understanding of physical behavior to be obtained. One example, of such a feature was the discovery by de Haas and Van Alphen in 1930 of oscillations in the magnetic susceptibility of bismuth at low temperatures in a high magnetic field. This observation later provided the key to understanding complicated electronic properties of many metals. Thirdly, phenomena such as superconductivity can only exist at low temperatures. Superfluidity in liquid ^{4}He below 2.17 K and ^{3}He below .0027 K are other examples of ordered states which can only exist at very low temperatures. In these states the viscosity of the fluids vanish and they are able to flow without dissipating energy: a true case of the frictionless pulley. In the case of liquid ^{3}He the fluid becomes anisotropic, exhibiting liquid-crystal-like textures and many unique magnetic properties. In all superfluids heat can propagate as a wave, as opposed to the usual diffusive heat flow, and in the case of one of the superfluid phases of ^{3}He, this heat wave carries magnetization with it.

In other low temperature studies, physicists have been able to form crystalline lattices of free electrons on the surface of superfluid ^{4}He and measure their mechanical properties. In additional low temperature studies of two-dimensional electron gases, such as those found in MOS-FET devices, a quantized Hall effect is observed when the samples are placed in a very intense magnetic field. This unusual quantum effect is providing a new and better standard for the unit of electrical resistance.

Refrigeration Most refrigeration processes used to obtain very low temperatures involve the change of state of some fluid refrigerant. The most simple refrigerators involve the liquification of a gas under elevated pressure with the latent heat of condensation being removed to the outside world, followed by evaporation at a much lower pressure and temperature. Such processes will operate only between their critical temperatures and their triple point temperatures. Oxygen, for example, has a critical point at 50 atmospheres pressure and 154 K. Its triple point is at 54.4 K. To provide continuous cool-

ing from room temperature such refrigerators must operate in series with other refrigerators using working fluids with higher temperature critical points, the hottest being above room temperature.

By far the most widely used refrigeration process involves the expansion of a gas under elevated pressure through a porous plug, the Joule-Thompson process. In such an expansion an ideal noninteracting gas will neither warm nor cool; however, with real gases the expansion alters the average potential energy between the gas particles, and as a consequence the temperature of the gas is changed. If the kinetic energy of the gas particles is sufficiently low (at low temperatures) so that the atoms or molecules feel mainly the attractive part of the interatomic (intermolecular) forces during collisions, the expansion will result in cooling; however, if the temperature is sufficiently high so that the gas particles feel mainly the repulsive short range interactions, the expansion will result in warming. In the limit of hard sphere gas particles with no attractive interactions and an infinite repulsive interaction, the volume excluded by the particles themselves causes the pressure to drop further during the expansion than it would for an ideal gas. As a result of this excess drop in pressure, more mechanical energy is expended in forcing the gas though the plug than can be recovered from the expanding gas on the low pressure side. This difference in energy results in heating of the gas. Not surprisingly, the maximum temperature at which cooling will occur, the inversion temperature, is related to the boiling temperatures of the liquified gases, since both temperatures depend on the interatomic forces. Helium (^{4}He), for instance, boils at atmospheric pressure at 4.2 K and has an inversion temperature of 51 K. Hydrogen boils at 20.4 K and has an inversion temperature of 205 K. Nitrogen and oxygen boil at 77.3 K and 90.2 K respectively, and have inversion temperatures of 621 K and 893 K. Since helium and hydrogen have inversion temperatures below room temperature, they cannot be used for this type of cooling process directly, but must be precooled by a prior refrigeration process to well below room temperature. Joule-Thompson expansion is the process most generally used in the liquification of oxygen and nitrogen, however, and it is the final refrigeration process used in the liquification of helium.

If rather than allowing a gas to expand through a porous plug as in the Joule-Thompson process, one allows the gas to expand against a piston or other moving wall (such as the vane of a gas turbine), any gas will cool at any starting temperature. This occurs because on the average when the gas particles recoil off the moving surface in an expanding volume they have lower kinetic energies than they had prior to their collisions with the moving wall. From a macroscopic viewpoint the gas is doing work against the moving piston, $P\delta V$. Expansion engines using either moving pistons or turbines are almost always employed in the liquification of helium, although a three-stage Joule-Thompson process to liquify helium using nitrogen, hydrogen and helium is possible. The disadvantage of expansion engines is that they involve closely fitting moving parts which require very high gas purity. Ice and frozen air are constant hazards in such machines.

Liquid ^{3}He and ^{4}He hold a special place in the field of cryogenics, largely because they are the only stable materials available to man which remain in a fluid state as their temperature is lowered arbitrarily close to absolute zero. (Both can be solidified near absolute zero, however, by applying pressures on the order of 30 atmospheres.) This fact not only makes ^{3}He and ^{4}He unique materials insofar as low temperature physics is concerned, it also allows them to be useful refrigerants to exceedingly low temperatures.

Liquid ^{3}He and ^{4}He have boiling temperatures at atmospheric pressure of 3.2 K and 4.2 K, respectively. By pumping away the vapor above the liquid phase, however, one can lower the boiling points to about 0.25 K for ^{3}He and 0.7 K for ^{4}He. Below these temperatures the vapor pressures above the liquids, which are decreasing exponentially with decreasing temperature, are insufficient to allow significant amounts of liquid to be evaporated, and the cooling capacities of ^{3}He and ^{4}He pumped cryostats drop rapidly to zero.

In the mid-1960s a novel refrigeration device called the ^{3}He-^{4}He dilution refrigerator was developed which overcame the problems of the vanishingly small vapor pressures of ^{3}He and ^{4}He at low temperatures. This device virtually revolutionized low temperature physics during the 1970s, and one can now purchase a dilution refrigerator capable of continuous refrigeration down to 0.004 K (4 mK) for about $100,000.

The key to the operation of the dilution refrigerator is the finite solubility of ^{3}He in superfluid ^{4}He. As ^{3}He is added to liquid ^{4}He at low temperatures (below about 0.2 K), the ^{3}He mixes with the ^{4}He to a maximum concentration of about 6%. In the mixing or *dilution* process the distance between the ^{3}He atoms increases and in the process the ^{3}He absorbs heat. The superfluid ^{4}He has very low specific heat and thermal conductivity at these temperatures, and in fact, the ^{3}He atoms do not scatter off the ^{4}He atoms in the superfluid. At these temperatures superfluid ^{4}He is often referred to as a *dense vacuum*. In a crude sense, one can therefore consider the mixing of ^{3}He in superfluid ^{4}He to be a form of evaporation. However, the density of ^{3}He in the ^{4}He is so high and the temperature is so low that this simple model breaks down. In fact, the heat absorbed by a quantity of ^{3}He mixing with superfluid ^{4}He is proportional to T^2. A moderately powerful dilution refrigerator has a cooling capacity at 10 mK of only a few micro-watts.

The ^{3}He which is diluted into the ^{4}He-rich fluid is withdrawn from it by connecting the region where the mixing occurs, the *mixing chamber*, to a *still* by a long capillary tube. The still temperature is regulated at about 0.8 K, where liquid ^{3}He has a much higher vapor pressure than liquid ^{4}He. Thus a pump which removes the vapor above the liquid in the still distills the ^{3}He out of the ^{4}He-rich phase. The ^{3}He removed from the still is pressurized to about 0.1 atmosphere and is reliquified in a pumped ^{4}He bath. It is then returned to the mixing chamber through a series of heat exchangers which transfer heat from the warm returning liquid to the cold dilute ^{3}He rising up to the still.

There is one additional refrigeration process involving ^{3}He called *Pomeranchuk* or *compressional cooling*. It relies on the unusual property that the liquid phase of He^3 below 0.3 K is more highly ordered than the solid phase. This is due to the random orientations of the ^{3}He nuclear spins in the solid, which give it a substantial disorder even at low temperatures. If one solidifies liquid ^{3}He below 0.3 K (about 34 atmospheres at 0.02 K), as solid ^{3}He forms it absorbs heat from the surrounding liquid and the system cools. A carefully designed Pomeranchuk cell precooled with a dilution refrigerator to about 25 mK can reach about 0.8 mK for short periods of time and can maintain 1.5 mK for several hours. Lower temperatures are not feasible because at 1.0 mK the nuclear spins in the solid order antiferromagnetically, and the difference in the degree of order between the liquid and solid drops rapidly to zero. Although not an important cooling process for external samples, this self-cooling of ^{3}He has been crucial in experiments designed to study the unusual superfluid phases of liquid ^{3}He below 2.7 mK and the nature of nuclear magnetic ordering in solid ^{3}He below 1.0 mK.

In many ways the most powerful and by far the most common technique now used to obtain temperatures below those available with dilution refrigerators is that of *adiabatic demagnetization*. In this technique a collection of spins, either electronic or nuclear, is partially ordered by the application of a large magnetic field. The magnetic moments associated with the spins tend to align along the direction of the magnetic field. The system is precooled, usually with a dilution refrigerator, until some significant fraction of the spins are aligned (the net polarization varies as $1/T$ for small polarizations). Then the magnetic refrigerant is thermally isolated from the precooling stage, and the magnetic field is lowered. As the field decreases, thermal excitations flip the aligned spins against the magnetic field and in the process thermal energy is absorbed. If the heat capacity of the magnetic coolant totally dominates the heat capacity of the sample being cooled, the ratio of the temperature to the sum of the external and effective internal magnetic fields stays constant. For nuclear coolants such as copper, the internal fields, which arise largely from the magnetic moments associated with the spins, are only a few Gauss. If the external field is lowered from 80,000 Gauss to 80 Gauss, the temperature of the copper nuclear spin system will drop by a factor of 1000. If the starting temperature is 10 mK, the final nuclear temperature will be 10 μK. Electronic moments and hyperfine enhanced nuclear moments may be polarized more easily at higher temperatures due to their larger magnetic dipole moments, but due to higher internal fields, they do not get as cold. The lowest temperature one can reach with even a very dilute paramagnetic salt such as cerium magnesium nitrate is about 2 mK.

Although one can achieve nuclear spin temperatures well below 1 μK in copper nuclear demagnetization cryostats, heat leaks have not allowed external samples to be cooled to this very low temperature. Thermal conductivities vanish as the temperature approaches absolute zero, and even heat leaks as small as 10^{-9} watts become important in determining sample temperatures. External metallic samples have not been cooled much below 50 μK with copper demagnetization refrigerators, and ^{3}He, which is harder to cool, has not been below about 120 μK.

Thermometry There are a wide variety of thermometry schemes for measuring cryogenic temperatures. Most are calibrated against certain easily reproducible fixed points whose absolute temperatures are established by some thermodynamic means such as the ideal gas law thermometer. Down to 90 K the accepted international thermometry scale is based on the electrical resistivity of platinum wire, which is nearly proportional to temperature in this range due to electron-phonon interactions. The vapor pressures of certain gases in equilibrium with their liquid phases provide highly reproducible thermometry scales over limited temperature intervals. In particular, the vapor pressures of ^{4}He from 5 K to 1 K and ^{3}He from about 3 K to 0.5 K are internationally accepted temperature scales. Thermocouple devices, particularly gold doped with iron vs chromel, can provide useful thermometry down to 1 K and below, although their use in the laboratory has waned in recent years.

By far the most convenient and prevalent secondary thermometry schemes are those based on the temperature dependent resistance of some stable device. Pure metal wires are commonly employed down to liquid nitrogen temperatures. Commercially available carbon resistors are the cheapest and most prevalent thermometers, being useful from close to room temperature down to temperatures as low as 10 mK. These thermometers do not tend to be very reproducible, however, and for high precision work doped germanium resistors are generally used down to about 50 mK. Virtually anything whose resistance is temperature dependent in a reproducible way can serve as a secondary resistance thermometer.

Below about 20 mK a variety of Curie law thermometers are generally used. These rely on the temperature dependence of the polarization of either electronic or nuclear magnetic moments in a magnetic field. Detection of the polarization can be either through a mutual inductance measurement, a static magnetization measurement, nuclear magnetic resonance, or a measurement of the anisotropy of gamma radiation emitted from polarized radioactive nuclei. Nuclear magnetic resonance of platinum wire or powder is the thermometer of choice below about 1 mK, and the lowest temperatures measured this way are somewhat below 50 μK.

Certain other thermometry schemes are also occasionally used at ultralow temperatures, such as those based on the temperature dependent dielectric constant of certain materials (a scheme virtually immune to even very large magnetic fields), the thermal noise power (Johnson noise) of a suitable resistor, and the ^{3}He melting pressure. In the range from about 100 mK to just above 1 mK a simple ^{3}He melting pressure thermometer can be produced with a resolution of somewhat better than .0001 mK or 100 nano Kelvin.

The major difficulty associated with thermometry schemes below 0.3 K is the absence of suitable calibration or fixed points whose temperatures have been established by thermodynamic means. A series of superconducting transition temperatures has been established by the U.S. National Bureau of Standards recently, but the lowest is at about 15 mK. Ultimately the superfluid ^{3}He transition temperatures at various sample pressures should provide a highly reproducible temperature scale from 1.0 mK to 2.7 mK, and almost all provisional temperature scales in this range are currently being referenced to superfluid ^{3}He transition temperatures.

The engineering aspect of cryogenics is itself a very challenging area. The properties of materials near absolute zero are generally quite different from those exhibited at room temperature. In particular, thermal contraction of materials being cooled to 4.2 K can amount to changes in length as large as 1% or as small as 0.1%. Materials which fit together well at room temperature may bind or leak badly at low temperatures. Some become too brittle to be of any use, while the mechanical properties of others improve dramatically. The problems of providing adequate thermal contact between the different parts of a cryostat and adequate heat exchange between warm and cold cryogenic fluids are enormous. Often the low temperature physicist or cryogenic engineer must exhibit considerable resourcefulness in utilizing the unusual properties of materials at low temperatures to overcome technical problems. Considerable progress is being made, however, and the facility with which one can now produce and maintain very low temperature environments is vastly expanding the applications of the cryogenic science.

D. D. OSHEROFF

References

Wilks, J., "The Properties of Liquid and Solid Helium," London, Oxford Univ. Press, 1967.

White, G. K., "Experimental Techniques in Low Temperature Physics," second edition, London, Oxford Univ. Press, 1968.

Lounasmaa, O. V., "Experimental Principles and Methods Below 1 K," New York, Academic Press, 1974.

Bennemann, K. H., and Ketterson, J. B., "The Physics of Liquid and Solid Helium Part II," New York, John Wiley & Sons, 1978.

Cross-references: CONSERVATION LAWS AND SYMMETRY, DE HAAS-VAN ALPHEN EFFECT, ENTROPY, HEAT TRANSFER, LIQUEFACTION OF GASES, HEAT, SUPERCONDUCTIVITY, SUPERFLUIDITY.

CRYSTAL STRUCTURE ANALYSIS

Diffraction of radiation by the ordered, periodic arrangement of atoms in a crystal yields mainly information about: (1) the nature of the radiation (spectrum); (2) the geometry of the crystal (repeat unit and symmetry); and (3) the distribution of the scattering material in the repeat unit of the crystal. In the last category lies the powerful technique of *crystal structure analysis* as revealed primarily by x-ray diffraction, and also by neutron and electron diffraction. With x-rays the electron density distribution is determined, and from this rather precise information about atomic locations (± 0.005 Å), as well as thermal motions, polarization effects, and disorder phenomena, is obtained. Neutrons yield information about positions and thermal motions of atomic nuclei, and also about distributions of magnetic and spin vectors in the crystal.

Experimental Aspects When atomic properties are of primary importance, the radiation used is monochromatic, usually generated as characteristic x-radiation from a specified target material in a Coolidge tube (typically $CuK\alpha$, $\lambda = 1.5418$ Å, or $MoK\alpha$, $\lambda = 0.7107$ Å). The diffraction effects are observed by various photographic methods, or by means of a goniometer or diffractometer fitted with a pulse counter.

The geometry of a crystal can be determined by routine methods, in which the diffraction effects observed are interpreted in terms of the reciprocal lattice (see CRYSTALLOGRAPHY). Various single-crystal film cameras are arranged to show separately the x-ray spectra produced by as many gratings in the crystal as possible (each spectrum corresponds to a node in the reciprocal lattice), and if the x-rays are monochromatic, each spectrum (or "reflection") appears as a spot-image of the crystal on the photographic film, or a sharp peak on the diffractometer trace. Within the limitations of Friedel's Law ($|F(hkl)|^2 = |F(\bar{h}\bar{k}\bar{l})|^2$, so that a center of symmetry is always present in the diffraction effects when the scattering is purely elastic), such observations show the crystal system, point symmetry, and dimensions of the unit cell of the crystal. Further information about glide and screw symmetries may lead to a unique assignment to one of the 230 space groups (such as the common $P2_1/c$, $Pcab$ or $P2_1 2_1 2_1$), but may leave a 2- to 4-fold ambiguity. The space group assignment and unit cell dimensions place severe, often crucial restrictions on the atomic arrangement in the crystal, and are a necessary prerequisite for any crystal analysis.

The *intensities* of the spectra, or reflections, form the basis of the determination of the atomic arrangement. The determination (using x-rays) is based on the development of the electron density distribution in the unit cell by means of the relationship:

$$\rho(x,y,z) = \frac{1}{V}\sum_h \sum_k \sum_l F(hkl)\, e^{-2\pi i(hx+ky+lz)} \quad (1)$$

in which V is the volume of the unit cell. The structure amplitudes $F(hkl)$ are obtained experimentally from

$$|F(hkl)|^2 = \frac{K^2}{ALp} I(hkl) \quad (2)$$

where A is an absorption factor, L is a geometric correction (Lorentz) factor and p is the polarization factor; $I(hkl)$ is the total intensity in the reflection peak (integrated intensity). The scale factor K is not usually determined, and so the observed data set consists of the relative structure amplitudes $|F(\text{obs})| = |F|/K$, each indexed and identified by the Miller indices h, k, and l.

Though the number of terms in (1) is theoretically infinite, the actual number is always limited by the wavelength and the angular range available. With x-rays (but not with neutrons) the average $|F|$ values decrease rapidly with Bragg angle, so that often no intensities can be measured at higher angles. Within these limitations, an effort is generally made to collect all of the independent relative amplitudes that can be produced by a crystal. Such a data set may contain about 1000 (for simple inorganic structures) or as many as 30 000 (for proteins) observations.

The Crystal Structure Problem A computation of the electron density distribution in the unit cell by Eq. (1) will clearly provide all the crystal structure information that can be extracted from the measured data. Unfortunately, there is one link missing between this desired result and the experimental observations, namely, the relative phases of the terms in Eq. (1). These phase angles cannot be determined experimentally and must be derived indirectly. This situation constitutes the notorious "phase problem" of crystal structure analysis, a major stumbling block in the procedure for the first 50 years following the discovery by Laue of x-ray diffraction in 1912, and one which strongly affected the character of crystal structure research during all that time. In the 1960s powerful theoretical arguments showed how the phase information was contained in and could be extracted from the intensity distributions within the data set itself. Thus, the chasm of the phase problem was finally bridged, so that there is no longer any limit on the complexity of structures that can be determined, and crystal structure analysis has become a standard and exceedingly powerful tool in the service of chemistry and physics.

Trial and Error Methods Formerly, an empirical trial structure model served as a starting point of an iterative procedure that was used to try to converge on a true set of phases for equation (1). From such a model, theoretical amplitudes and phases can be directly calculated by:

$$F(hkl) = \sum_n t_n f_n e^{2\pi i(hx_n + ky_n + lz_n)} \quad (3)$$

in which f_n is the form (scattering) factor of the atom n (a tabulated function of the Bragg angle), t_n is a thermal parameter of the form $\exp\left[-(2\pi \bar{u}_n \sin \theta)^2/\lambda^2\right]$ ($\bar{u}_n$ is the root-mean-square amplitude of vibration of atom n), and x_n, y_n, and z_n are the coordinates of the atom n. F is thus a complex quantity

$$F(hkl) = A(hkl) + iB(hkl) \quad (4)$$

where $A = F(hkl) \cos \alpha$ and $B = F(hkl) \sin \alpha$. In general, the phase angle α may have any value, but if the origin of coordinates is a center of symmetry, F must be an even function and therefore $B = 0$. In that case the phase problem reduces to a choice of sign for each term in the equation

$$\rho(x, y, z) =$$

$$\frac{1}{V} \sum_h \sum_k \sum_l \pm |F(hkl)| \cos 2\pi(hx + ky + lz) \quad (5)$$

This condition enormously simplifies the phase problem.

The quality of a given model is judged by comparison of $|F(\text{obs})|$ with $|F(\text{calc})|$ obtained from equation (3). The traditional figure of merit commonly used in crystal structure analysis is a relative first moment coefficient

$$R = \frac{\sum \||F(\text{obs})| - |F(\text{calc})|\|}{\sum |F(\text{obs})|} \quad (6)$$

known as the "R factor," "reliability index," etc. For analytical purposes a so-called "weighted R factor" is also used:

$$R_w \text{ or } R_2 = \frac{\sum w \left(|F(\text{obs})| - |F(\text{calc})|\right)^2}{\sum w |F(\text{obs})|^2} \quad (7)$$

where $\sqrt{w}$ is the weight of the observation $|F(\text{obs})|$. In structure analysis where the observations often outnumber the parameters by more than 10 to 1, the two R indices follow each other closely.

If R for the initial model is < 0.50, the model may be at least partly correct, and is then tested by forming an electron density synthesis from Eq. (1) or (5), using only $F(\text{obs})$ terms whose phases are strongly indicated by Eq. (3). This map will show the starting partial model plus other features in the structure not included in that model (additional atoms, shifts in assumed atom positions, etc.), from which an improved model can be derived. This model is then used to calculate new values of $F(\text{calc})$, leading favorably to substantially reduced R, and thus initiating an iterative process of structure refinement.

Vector Maps Considerable information about the structure can be gained from another type of Fourier synthesis that does not depend on knowledge of phase angles, the so-called "Patterson function":

$$P(x, y, z) =$$

$$\frac{1}{V} \sum_h \sum_k \sum_l |F(hkl)|^2 \cos 2\pi(hx + ky + lz) \quad (8)$$

It was shown by A. L. Patterson in 1935 that this function contains maxima at the points in crystal space which represent interatomic vectors in the crystal structure standing at the origin, and that the height of these maxima is proportional to the product of the electron densities of the two atoms forming these vectors. Thus, if there are n atoms in the unit cell, there will be $n(n - 1)$ peaks in the Patterson map, the remaining n peaks being superposed at the origin representing self-vectors for each atom. The Patterson map is a convolution of the electron density over every point in space, or essentially every atom in the structure, and many highly ingenious techniques have been devised to effect a deconvolution. Nevertheless, when n becomes very large, interpretation becomes impossible. The Patterson map and its various convolutions are called generally "vector maps."

Obviously, the task of interpretation of a vector map is greatly simplified if there is one or a very few relatively heavy atoms in the structure (with large f in Eq. (3)), because these will be identified easily in the mapping, and will tend to dominate the distribution of phases in Eq. (1). This is the basis of the so-called "heavy-atom method." Where it is applicable, it can avoid much of the labor involved with more direct methods, and it is still of primary importance in protein structure analysis.

Statistical Methods of Phase Determination D. Sayre in 1952 first noted that when large F terms are concerned, there is a strong tendency for

$$S[F(h_1 \pm h_2, k_1 \pm k_2, l_1 \pm l_2)] \approx$$

$$S[F(h_1 k_1 l_1)] \cdot S[F(h_2 k_2 l_2)] \quad (9)$$

where S signifies $+1$ or -1 according to the centrosymmetric phase of F. This relationship, which depends on the required positivity of the function (1), is the root of the most powerful phase-determining procedures in modern use, notably the "symbolic addition procedure" evolved by Karle and Karle (see ref. 6). These techniques, consisting of the analysis of vector combinations of the strongest F terms (usually normalized to emphasize the trigonometric part of equation (3)), have been extended to non-centrosymmetric cases with considerable general success, so that the phase problem may now confidently be said to be for the most part overcome.

Refinement of the Crystal Structure Once a satisfactory model structure has been evolved, defined by a given set of parameters (which may amount to hundreds or even thousands), the refinement of these parameters is carried out by differential methods. The least squares analysis of the $F(\text{obs})$ data set (in which a function based on Eq. (7) is minimized) is the most easily adapted to computers and is most widely used. Methods based on minimization of $\Delta\rho$ (using "difference maps" calculated by Eq. (1) with ΔF as coefficients instead of F) are also commonly used. Convergence when $R \sim 0.15$ is considered satisfactory (with standard error of bond lengths ~ 0.03 Å), but as the quality of data measurement and techniques of refinement have improved, R often nowadays reaches 0.05 or lower (standard error of bond lengths < 0.005 Å).

Crystal Structure Data The most important information yielded by crystal structure analysis, namely, the configuration of atoms in molecules and the solid state, and the distances between the atoms, forms a vast body of critical data on the properties of substances which is expanding at a rapid rate. Compilations of these and of unit cell geometry data are published with a lag usually of 10 years or more, so that reference to the journal literature is usually necessary to find reliable and recent data concerning a particular substance. Reference to original papers is also very desirable to answer the important question of the reliability and distribution of errors in a given structure determination.

Data compilations and treatments of techniques of crystal structure analysis are given in a list of selected references below.

HOWARD T. EVANS, JR.

References

1. Bacon, G. E., "Neutron Diffraction," 3rd ed., Oxford, Clarendon Press, 1975.
2. Buerger, M. J., "Vector Space," New York, John Wiley & Sons, Inc., 1959.
3. Buerger, M. J., "Crystal-Structure Analysis," New York, John Wiley & Sons, Inc., 1960.
4. Donnay, J. D. H., and Ondik, H. M., eds., "Crystal Data," 3rd ed., 4 vols., Nat. Bur. of Stand. and Jt. Comm. on Power Diffr. Stand., 1973–78 (compilation of unit cell data, references to structure determinations).
5. Ewald, P. P., and Hermann, C., eds., "Strukturbericht," vol. 1–7, Leipzig, Akademische Verlagsgesellschaft M.B.H., 1931–43 (detailed compilation of crystal structure determinations, 1913–1939).
6. Karle, J., and Karle, I. L., *Acta Cryst.*, **21**, 849 (1966) (symbolic addition procedure).
7. Lipson, H., and Cochran, W., "The Determination of Crystal Structures," 3rd ed., Ithaca, N.Y., Cornell Univ. Press, 1966.
8. Stout, G. H., and Jensen, L. H., "X-ray Structure Determination," London, The Macmillan Co., 1968.
9. Warren, B. E., "X-Ray Diffraction," Reading, Mass., Addison-Wesley Publ. Co., 1969.
10. Wilson, A. J. C., and Pearson, W. B., eds. "Structure Reports," vol. 8–47, Utrecht, N.V.A. Oosthock's Vitgenvers Mij. 1956–1980 (continuation of "Strukturbericht" (ref. 5), covering literature 1940–1976).
11. Woolfson, M. M. "Direct Methods in Crystallography," Oxford, Clarendon Press, 1961.
12. Wyckoff, R. W. G., "Crystal Structures," 2nd ed., vol. 1–6, New York, John Wiley & Sons, Inc. 1963–1969. (compilation of results of crystal structure determinations).

Cross-references: CRYSTALLIZATION, CRYSTALLOGRAPHY, DIFFRACTION BY MATTER AND DIFFRACTION GRATINGS, LATTICE DEFECTS, NEUTRON DIFFRACTION, PARACRYSTALS, X-RAY DIFFRACTION, X-RAYS.

CRYSTALLIZATION

The forms of natural crystals have been studied by mineralogists for many years and have been classified by symmetry, interfacial angles, perfection of shape, and more detailed criteria. These forms are often related to the molecular structure and growth of the crystals. The equilibrium shape of a crystal is that for which the surface energy is a minimum. Since atomic planes of densest packing usually have the lowest surface energy, these planes predominate in the surface facets of equilibrium crystals, resulting in a correspondence between the atomic structure and shape of the crystal. However, natural and even synthetic crystals rarely have the equilibrium shape, because for crystals larger than about 10 μ in dimension, the differences in surface energy between faces are too small to transport enough material over the distances required. Therefore the morphology of crystals is usually determined by the rate of crystal growth, rather than by the equilibrium shape.

A crystal is bounded by those faces whose rate of growth is slowest, since fast-growing faces grow out of existence. Close-packed planes frequently grow most slowly, so even when kinetic factors control the crystal shape there is usually a relation between the faces of a crystal and its molecular structure.

Unusual crystalline morphologies result from particular conditions. Dendritic (treelike) shapes result when crystals grow with high driving force, or with rapid transfer of heat.

Since crystallization proceeds by propagation of the nucleus into the parent phase, the surface separating these phases is the site of incorporation of molecules into the crystal. Thus the structure of this interface between phases is critical in determining the mechanism of growth. A crystal-vapor interface becomes molecularly "rough" above a critical temperature, so that molecules can be incorporated anywhere on it. For low-index faces this critical temperature is close to the melting temperature, and at lower temperatures the surface is molecularly smooth at equilibrium. If there is a monomolecular step on the surface, incorporation of molecules into the crystal will occur preferentially at the step. Such a step contains kinks or jogs, which are the final sites for incorporation. Thus the progress of a molecule from the vapor into the crystal is: (1) transport through the vapor to the crystal surface, (2) adsorption onto the crystal surface, (3) movement on the surface to a step (surface diffusion), (4) adsorption onto a step, (5) transport along the step to a kink, and (6) incorporation at the kink. Steps (2), (4), and (6) can involve reorientation and desolvation of the molecules. The rates at which crystals grow can be controlled by any one or several of these steps. In growth from the vapor it is usually step 3 involving surface diffusion that controls the rate of growth of the crystal.

If the crystal surface is molecularly perfect, a pillbox of material must be nucleated on it to create a step, which then grows to another perfect surface. Under these circumstances, continued surface nucleation is required, and growth occurs only below a certain undercooling. However, experimentally, crystals often grow at much smaller undercoolings than this calculated one, so that a continuous source

of steps must exist. For this source F. C. Frank postulated a screw dislocation in the crystal that emerges at the crystal surface. This emergent dislocation provides a step pinned at one end, so that as it propagates it winds up into spiral and is always available for incorporation of molecules. Many spirals have been observed on crystal surfaces; one is shown in Fig. 1. The surface nucleation and screw dislocation mechanisms for crystal growth were definitely confirmed by the elegent experiments of G. W. Sears on the growth of perfect metallic filaments (whiskers), metallic platelets, and para-toluidine crystals.[10]

Impurity molecules can modify the morphology and growth rate of crystals by their effects on the relative surface energies and growth rates of different crystal faces. These molecules can poison growth on certain planes by adsorption at kinks in steps on these planes, slowing the growth of these steps. Impurities can also change the rate of adsorption and surface diffusion of incorporating molecules. The rate of pillbox nucleation can be increased by the lowering of surface tension by impurity adsorption. Therefore impurities can produce different crystal habits and either faster or slower growth rates.

Crystal growth from dilute solution is similar to growth from the vapor, but impurity effects are more marked and morphologies more varied. The growth rate is often controlled by the rate of diffusion of solute up to the crystal surface rather than interfacial processes, especially for crystals larger than a few microns in size. Convection in the liquid solution can change crystal morphologies and growth rates and make compositional and size control more difficult.[11]

The rate of crystallization from a pure liquid, often called *solidification* or *freezing*, is usually controlled by the rate at which the heat of fusion is removed from the interface. Only for slowly growing crystals in viscous melts do interface processes control the rate of crystallization. Nevertheless the structure of the liquid-solid interface can influence crystal morphologies. Many solid-liquid interfaces are "rough," but others appear to be similar to vapor-crystal and solution-crystal interfaces, as evidenced by growth facets in such materials as silicon, germanium, P_2O_5, and many organics. Growth in concentrated solutions (e.g., metallic alloys) is influenced by convective instabilities in the melt as well as instabilities of the interface (dendritic growth).

Crystal growth and liquid-crystal interfaces are being simulated in computers.[12]

To make crystals for laboratory and industrial use a great variety of techniques have been used. Growth by either condensation or chemical reaction from the vapor phase can give crystals with high purity and special structures and form. For large-scale industrial use, this method is too costly, although it is valuable for certain

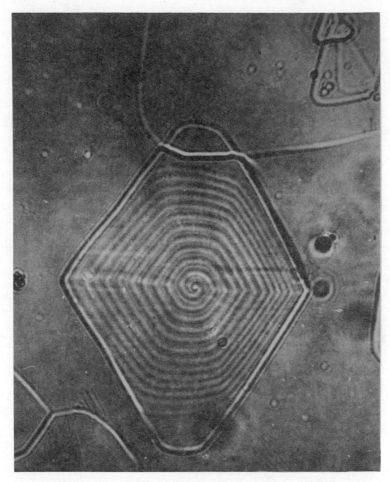

FIG. 1. Growth spiral on a paraffin crystal, observed by C. M. Heck. From Doremus, Roberts and Turnbull "Growth and Perfection of Crystals," by permission of John Wiley & Sons, Inc., New York.[4]

special applications. Luminescent crystals of zinc and cadmium sulfides are grown from vapor for industrial use. Metallic crystals with few impurities and defects, and in the special forms of thin films or whiskers, can readily be grown from vapor. Other crystals made in this way are silicon, germanium, iodine, selenium, phosphorus, and a variety of organic crystals. The study of the growth of ice crystals from water vapor has special importance in meteorology.[13]

The most common method of growing metallic and semiconducting crystals is by solidification of their melts. Special techniques have been developed to grow single crystals of these materials and many others. In the *Czochralski method*, a seed crystal is touched to the melt, and the crystal is "pulled" from it by slowly withdrawing the seed. In the *Bridgman technique*, the melt is slowly moved through a temperature gradient in a furnace, so that crystallization starts at one point in the melt and propagates through it relatively slowly. In the *Vernueil method*, powder is added to the molten surface of a crystal so that a crucible of other material is not needed. This method is used for materials with high melting temperatures, such as alumina, spinels, rutile, mullite, ferrites, and yttrium-iron garnet. A new method of growing sapphire crystals from the melt, called *edge-defined film-fed growth*, has been announced by Tyco Laboratories. The crystal is pulled rapidly from a die, to which molten alumina is transported as a film. Solid crystals are often purified by zone melting, in which a molten zone is moved through the crystal. Segregation of impurities into the melt purifies the crystal.

Precipitation from liquid solution is a common method of growing crystals. Ionic salts are grown from aqueous solutions both industrially and in the laboratory. Sugar is crystallized from water solution. Other organic crystals, including polymers, are grown from a variety of solvents. Quartz crystals are grown from aqueous solution at elevated temperatures and pressures (hydrothermal growth). Various crystals have been grown from more exotic solvents, for example: garnets, titanates, and ferrites from molten salts (fluxes); tin, iron, and phosphorus from mercury; and diamond from a molten metal under pressure.

Crystallization from the solid phase is also possible. Growth of grain size in a single-phase solid, called *recrystallization*, is often used to improve the properties of polycrystalline materials, particularly metals. Crystalline compounds can be made from high-melting-point materials by pressing together mixtures of their powders and diffusing them together at high temperature (sintering). Crystals can be grown from a solid solution. This type of precipitation is frequently used to improve the properties of metals, for example, to harden them.

R. H. DOREMUS

References

1. Holden, A., and Singer, P., "Crystals and Crystal Growing," Garden City, New York, Doubleday and Co., 1960. A simple, nonmathematical discussion of crystal growth and structure.
2. Strickland-Constable, R. F., "Kinetics and Mechanism of Crystallization," London and New York, Academic Press, 1968. Emphasis on nucleation, and crystallization from the vapor.
3. Laudise, R. A., "The Growth of Single Crystals," Englewood Cliffs, N.J., Prentice-Hall, 1970. A guide for persons wanting to grow crystals, with accent on practical methods together with some background material.

Papers from Symposia

4. Doremus, R. H., Roberts, B. W., and Turnbull, D. (Eds.), "Growth and Perfection of Crystals," New York, John Wiley & Sons, 1958.
5. Frank, F. C., Mullin, J. B., and Peiser, H. S. (Eds.), "Crystal Growth in 1968," *J. Crystal Growth*, 3 and 4 (1968).
6. Laudise, R. A., Mullin, J. B., and Mutaftschiev, B., "Crystal Growth 1971," Amsterdam, North-Holland, 1972.

Special Techniques

7. Bockris, J. O., and Razumney, G. A., "Fundamental Aspects of Electrocrystallization," New York, Plenum Press, 1967.
8. Henisch, H. K., "Crystal Growth in Gels," University Park, PA, The Pennsylvania State Univ. Press, 1970.

Review Article

9. Parker, R. L., "Crystal Growth Mechanisms: Energetics, Kinetics, and Transport," in "Solid State Physics," Vol. 25, New York, Academic Press, 1970, p. 152.

Research Papers

10. Sears, G. W., *J. Chem. Phys.* 24, 868 (1956) and other references by the same author.
11. Pimputkar, S. M., and Ostrach, S., *J. Cryst. Growth* 55, 614 (1981). Convective effects in crystals grown from the melt.
12. Gilmer, G. H., "Computer Models of Crystal Growth," *Science* 208, 355 (1980).
13. Kuroda, T., and Laemann, R., "Growth Kinetics of Ice from the Vapor Phase and its Growth Forms," *J. Cryst. Growth* 56, 189 (1982).

Cross-references: CONDENSATION, CRYSTAL STRUCTURE ANALYSIS, CRYSTALLOGRAPHY, LATTICE DEFECTS, PARACRYSTALS, VAPOR PRESSURE AND EVAPORATION.

CRYSTALLOGRAPHY

Crystallography is the science of the geometric properties of matter in the ordered solid state. When atoms or molecules condense into a solid phase from a liquid or gaseous phase, the lowest energy state is achieved if they become arranged in as regular a way as possible, usually by forming a small basic unit of structure which is repeated indefinitely in three dimensions throughout the solid to form a *crystal*. The geometric properties of this unit and its manner of regular repetition are highly characteristic of the substance in question, and constitute an exceedingly useful subject of study in connection with any field of science involving the solid state. Occasionally, no extended, regular repetition of structure is present in the solid phase, but this glassy state has many properties of a liquid and lies outside the realm of crystallography. In other cases, extended order may occur in one direction only (as in fibres) or in two directions (as in some clays), but by far the most common condition of the solid state is full three-dimensional order, and it is with this type of order that crystallography is primarily concerned.

The familiar outward manifestation of the three-dimensional order of the atomic structure of the solid is the polyhedral shape commonly exhibited by crystals. These remarkable shapes were admired for centuries (see, for example, Albrecht Dürer's engraving "Melancholia," 1514), but the underlying principle governing them was first discovered by Steno in 1669. This principle is expressed as the Law of the Constancy of Interfacial Angles, according to which the dihedral angles between the faces of all crystals of a given substance remain unchanged regardless of how the relative sizes and shapes of the faces may vary. René Just Haüy in the late eighteenth century was the first to present a systematic account of the characterization of substances by the measurement of interfacial angles, that is, crystallography, and thus establish it as a science. Haüy was a mineralogist, and through his influence crystallography was subsequently developed and applied by workers mainly in mineralogy, and to a small extent in chemistry. The link between external form and internal structure was dramatically completed by M. von Laue's discovery of x-ray diffraction in 1912, and from then on crystallography was rapidly developed and advanced in physics laboratories, and later more and more in chemistry.

A crystal may or may not exhibit external faces, but if it does, these may be studied in terms of their distribution and development, which constitutes the *morphology* of the crystal, by special techniques of *crystallometry*, usually making use of an instrument that reflects beams of light from the crystal faces into a telescope, the *two-circle goniometer*. If the crystal has no faces, its internal geometric properties may be studied by its interaction with radiation, by the methods of *optical crystallography* if refraction of infrared, visible or ultraviolet light is involved, or by *x-ray crystallography* if diffraction of x-rays (also neutrons or electrons) is studied.

The crystal can be defined completely (except for chance irregularities and defects) in terms of the arrangement of the atoms within a finite unit of volume called the *unit cell* (whose size is usually of the order of 10Å on an edge) and the way this unit is repeated in three dimensions to fill up the volume of the crystal. The shape and dimensions of the unit cell provide parameters characteristic of the substance and constitute the first primary geometrical property of crystals. The most general unit cell (triclinic case) is a parallelopiped which can be defined by six constants, three edge lengths (a, b and c) and three interedge angles (α, β and γ). This unit cell is repeated by *translation*, a shift along each of the cell edges by an integral number of edge lengths. If the unit cell and its contents are represented by a point in space, the crystal consists of a regular array of such points called a *lattice*, in which each point is related to every other by an integral number of vectorial translations corresponding to the unit cell edges. The lattice should be distinguished from the *crystal structure*, which refers to the arrangement of atoms within the unit cell, although the term "lattice" is sometimes loosely used in reference to the structure.

The atoms within the unit cell may be related to each other by a number of geometric operations called *symmetry*, and this phenomenon constitutes the second chief geometric property of crystals. The unit cell must embrace all of the different types of atoms related by symmetry that are not related by simple translation. On the other hand, the symmetry operations which apply to one unit cell are also operated on by the lattice translations, so that these symmetry operations must also apply to the entire crystal. Thus, the morphology of the crystal and all its other properties must obey this symmetry. The detection and definition of the symmetry also serve to characterize the substance and are equally as important in crystallography as the measurement of the lattice parameters.

The way in which symmetry operations can interact consistently with each other is strictly limited by the geometry of coincidence and can be rigorously analyzed by the mathematical methods of group theory, both as to what symmetry operations are possible and how they may be combined. The problem is usually approached by constraining all symmetry operations to pass through a single point in space, but special restrictions are introduced by the requirement that this point must be consistent with any point in the crystal lattice, that is, the symmetry groups must be consistent with the translational operations of the lattice. An important symmetry operation is the axis of rotation by which any motif is reproduced by a rotation around an axis of $360/n$ degrees, where n is the order of the axis; n successive operations then superimpose the object on itself. In crystals, because of the requirements of the lattice, n can only have the values 1, 2, 3, 4 and 6. A 5-fold axis is not possible, for example, for the same reason that it is not possible to fit regular pentagons into a regular two-dimensional pattern which will fill all space. Further, there are only 11 ways to combine the 5 axes together at a point; these are called the 11 axial point groups. These form a convenient basis for classifying all crystals into 6 *crystal systems*, which, while not strictly rational in their definition, provide a fundamental link between the symmetry of the crystal and its dimensional properties. Reference axes are generally chosen parallel to lattice translation directions, of course, but further, they are customarily taken parallel to prominent symmetry axes. Four of the axial point groups, for example, have a single 3-fold axis or 6-fold axis, with or without a number of 2-fold axes at right angles to them. In all these, the reference c axis is customarily set parallel to the unique 3- or 6-fold axis, and the other two axes a_1, and a_2, which are equivalent by symmetry, are taken normal to the c axis along lattice directions 120° apart, coincident with the 2-fold axes if present. These groups are all included in the hexagonal system.

The axial symmetry operations are operations of the first kind, that is, they reproduce left-hand motifs as left-hand motifs. Other symmetry elements of the second kind, that is, which reproduce left-hand motifs as right-hand motifs, are the center of symmetry and the mirror plane of symmetry. When these operations are added to the axial groups, 32 *point groups* are produced.

Referring to the lattice, the introduction of symmetry gives rise to a number of lattice groups in which various rational relationships

TABLE 1. THE SIX CRYSTAL SYSTEMS

System	Independent Lattice Parameters	Axial Relationships	Number of Symmetry Groups			
			Axial Point Groups	Point Groups	Bravais Lattices	Space Groups
Triclinic	$a, b, c, \alpha, \beta, \gamma$	none	1	2	1	2
Monoclinic	a, b, c, β	$\alpha = \gamma = 90°$	1	3	2	13
Orthorhombic	a, b, c	$\alpha = \beta = \gamma = 90°$	1	3	4	59
Tetragonal	a, c	$b = a$ $\alpha = \beta = \gamma = 90°$	2	7	2	68
Hexagonal	a, c	$b = a$ $\alpha = \beta = 90°$ $\gamma = 120°$	4	12	2	52
Cubic (isometric)	a	$b = c = a$ $\alpha = \beta = \gamma = 90°$	2	5	3	36

exist between the lattice parameters. One important result of this interaction is the appearance of *centered lattices*, in which the lattice unit cell chosen according to the rules used to set up the 6 crystal systems contains additional lattice points on body or face diagonals. There are 14 such "Bravais lattices." The symmetry groups so far mentioned can in favorable circumstances all be detected from the external morphology of the crystal.

When the combinations of lattice translations and symmetry operations, that is, the symmetry properties of the crystal structure, are analyzed, new symmetry operations are evolved (screw axes and glide planes) and each of the point groups contains many such combinations, adding up to a total of 230 *space groups*. These are detected by diffraction methods. Table 1 summarizes the relations between the crystal systems and the symmetry groups.

A fundamental concept of great importance to all aspects of crystallography is that of the *reciprocal lattice*. If the unit vectors of the direct lattice are a, b, and c, then a reciprocal lattice exists whose unit vectors are a* = b × c, b* = c × a, c* = a × b, with lengths given by $a^* = (1/V)\ bc \sin \alpha$, $b^* = (1/V)\ ca \sin \beta$, $c^* = (1/V)\ ab \sin \lambda$; the reciprocal unit cell volume $V^* = 1/V$. Because the Bragg diffraction angle 2θ is a function of the reciprocal of spacings in the direct lattice ($2 \sin \theta = \lambda/d$), the various orders of diffraction are associated with nodes in the reciprocal lattice. These nodes are designated by their integral coordinates in the lattice h, k, and l, known as *Miller indices*. Each node in the reciprocal lattice corresponds to a set of net planes in the direct lattice, and vice versa; the duality is complete. Crystal faces, which are parallel to the direct lattice net planes are therefore designated by Miller indices. This duality of direct and reciprocal space is exactly parallel to that represented by the Fourier transform:

$$f(x,y,z) = \iiint g(h,k,l)e^{-2\pi i(hx+ky+lz)}\ dxdydz$$

where h, k, and l may now be continuously variable (nonintegral). The potential $f(x, y, z)$ at any point x, y, z in direct space (e.g., electron density) is thus a synthesis of all the reciprocal potential $g(h, k, l)$ (e.g., x-ray scattering) in reciprocal space (and vice versa). In a crystal, the reciprocal potential is sampled at the reciprocal lattice points (vanishing for nonintegral h, k, l) and the above integral becomes a triple summation. The concept is used also in the interpretation of electron momentum space in a crystal, in which the Brillouin zones are defined by the reciprocal lattice (see FERMI SURFACE).

HOWARD T. EVANS, JR.

References

deJong, W. F., "General Crystallography," San Francisco, W. H. Freeman and Co., 1959.

Phillips, F. C., "An introduction to Crystallography," New York, Longmans, Green and Co., 1946.

Terpstra, P., and Codd, L. W., "Crystallometry," New York, Academic Press, 1961.

Buerger, M. J., "Elementary Crystallography," New York, John Wiley & Sons, Inc., 1956.

Hilton, H., "Mathematical Crystallography and the Theory of Groups of Movements," Oxford, Clarendon Press, 1903; New York, Dover Publications, Inc., 1963.

Cross-references: CRYSTALLIZATION, DIFFRACTION BY MATTER AND DIFFRACTION GRATINGS, CRYSTAL STRUCTURE ANALYSIS, LATTICE DEFECTS, X-RAY DIFFRACTION.

CURRENT ALGEBRA

Current algebra is a study of hadronic matter, (i.e., strongly interacting particles like protons, neutrons, and π mesons), through their electromagnetic and weak interaction properties. This study leads to a degree of unification of the three different forms of elementary particle interactions: strong, weak, and electromagnetic. The electromagnetic four current density $J_{e.m.}{}^{\mu}(x) \equiv (\rho(x), \vec{J}(x))$, with $\vec{J}$ and ρ respectively the densities of electric current and charge, is long familiar from Maxwell's equations. The conservation of this current, of course, leads to charge conservation in nuclear reactions. Mathematically this is expressed by saying $Q = \int d^3x \rho(x)$ is a time-independent constant. In 1958 it was established that the weak decays of hadrons are also described by currents. This led to the highly successful (V-A) theory of weak interactions, proposed by Marshak and Sudarshan and independently by Feynman and Gell-Mann. (Here V is a vector current and A is an axial vector current.) Soon after the successful proposal of SU(3) symmetry for classification of hadrons (suggested by Gell-Mann and Ne'eman in 1961) Gell-Mann laid down the foundations of current algebra. In attempting to interpret in precise terms the notion of symmetry violated by the strong interactions, Gell-Mann suggested that SU(3) symmetry operators be identified with charges associated with the weak and electromagnetic currents. In particular he emphasized the equal time commutation rules of these charges, and suggested that they may remain unchanged even in the presence of SU(3) breaking interactions, showing how this leads to a precise notion of universality of weak and electromagnetic interactions.

In the SU(3) scheme, one can define eight vector currents, with the charges corresponding to these operators being the generators of the group. These charges obey the equal time commutator algebra.

$$[F_i, F_j] = if_{ijk}F_k \qquad i, j, k = 1 \cdots 8 \quad (1)$$

where f_{ijk} are antisymmetric structure constants of the group SU(3). One may also define an octet of axial currents, and the charges associated with these currents, $F_i{}^5$, obey the algebra

$$[F_i, F_j{}^5] = if_{ijk}F_k{}^5 \qquad i, j, k = 1 \cdots 8. \quad (2)$$

Gell-Mann further postulated that axial generators among themselves obey a similar algebra,

$$[F_i{}^5, F_j{}^5] = if_{ijk}F_k \qquad i, j, k = 1 \cdots 8. \quad (3)$$

Such an algebra leads to an abstract group SU(3) X SU(3). Although such a group is in fact not a symmetry group of particle states, nevertheless the generators at equal time are assumed to obey the commutator algebra. Thus the axial charges are not constants of motion, nor are all the vector charges constants of motion. The charges that play a role in electromagnetic and weak interactions are respectively

$$Q = F_3 + F_8/\sqrt{3} \quad (4)$$

and

$$J^{\pm} = \cos\theta \, (F_1 \pm i F_2) + \sin\theta \, (F_4 \pm i F_5)$$
$$-\cos\theta \, (F_1{}^5 \pm i F_2{}^5) - \sin\theta \, (F_4{}^5 \pm i F_5{}^5). \quad (5)$$

Here θ is a parameter called the Cabibbo angle, and experimentally $\theta \cong 15°$.

Gell-Mann suggested that the current algebra be used in the same way as the familiar quantum condition, i.e., $[q, p] = i\hbar$, to derive sum rules as in atomic physics. His proposal remained dormant until use was made of another concept called the partial conservation of axial current hypothesis (PCAC). This postulate connects the pseudoscalar pion field to the divergence of the axial current. This hypothesis was used to obtain a value for the weak decay of the π meson from the known strong interaction pion-nucleon coupling constant (Goldberger-Treiman relation) and is known to be accurate to about 10 percent.

Among the successful applications of current algebra are prediction of weak decay rates of hyperons [e.g., $\Sigma^- \to n + e^- + \bar{\nu}$]. Such decays involve vector and axial parts; the vector parts are predicted to be just Clebsch-Gordon coefficients of the group SU(3), while the axial parts are also predicted in terms of two unknowns (f and d type couplings). As for the meson decay rates, the $K \to \pi e \nu$ decay is predicted accurately, although in meson systems symmetry-breaking corrections have to be made in general before predictions can be compared.

Application of current algebra leads to *sum rules* between different observables, as was the case in atomic physics. To illustrate the use of commutators in atomic physics, we give a derivation of the Thomas-Reiche-Kuhn sum rule. One starts with the fundamental quantum-mechanical assumption

$$[\vec{\epsilon} \cdot \vec{q}_i, \; \vec{\epsilon} \cdot \vec{p}_j] = i\hbar\delta_{ij} \quad (6)$$

where i and j label different electrons and $\vec{\epsilon}$ is an arbitrary unit vector. Assuming that forces are velocity-independent, we may write

$$p_i = im \, [H, q_i] \, \hbar \quad (7)$$

Using Eqs. (1) and (2) we may take the expectation value between the ground state $|0\rangle$ of an atom with Z electrons.

$$\sum_{ij} \langle 0 | [\vec{\epsilon} \cdot \vec{q}_i, [H, \vec{\epsilon} \cdot \vec{q}_j]] | 0 \rangle = \frac{Z}{m}\hbar^2 \quad (8)$$

Now inserting a complete set of intermediate states one gets

$$\sum_n |\langle 0| \sum_{i=1}^{Z} \vec{\epsilon} \cdot \vec{q}_i \, | n \rangle|^2 (E_n - E_o) = \frac{Z\hbar^2}{2m} \quad (9)$$

The usefulness of the sum rule arises from the fact that the left-hand side of the equation involves matrix elements of dipole operators that are observable in the transition spectrum of an atom. Very similar methods can be used with current algebra to obtain sum rules.

A sum rule obtained by Adler and Weisberger for the axial vector coupling in β-decay is among the triumphs of the axial-charge algebra. Use was made of the commutation relation

$$[F_1^5 + iF_2^5, F_1^5 - iF_2^5] = 2 F_3 \qquad (10)$$

to derive

$$1 = \left[\frac{G_A}{G_V}\right]^2 + \frac{F_\pi}{\pi} \int_\mu^\infty \frac{d\nu\sqrt{\nu^2 - \mu^2}}{\nu^2} [\sigma_{\pi^- p}(\nu)$$

$$- \sigma_{\pi^+ p}(\nu)] \qquad (11)$$

Here G_A is the axial vector coupling, G_V is the vector coupling, F_π is the pion decay constant, ν is the laboratory pion energy, and σ is the pion-nucleon total cross-section. G_A calculated from this relation is in good agreement with experiment. Other results include: *Adler's neutrino sum rule*, which relates neutrino to antineutrino scattering on protons at high energy; *Cabibbo-Radicati sum rule*, which relates anomalous magnetic moments of nucleons to cross-sections for absorption of isovector photons on nucleons; and *photoproduction sum rules*. These sum rules, wherever tested, are in good agreement with experiment.

Another class of applications leads to derivation of *low-energy theorems*. These theorems especially relate to processes where pions are emitted. It was shown that the Adler-Weisberger sum rule could be obtained as a low-energy theorem relating (G_A/G_V) to s-wave pion-nucleon scattering lengths. Other theorems deal with leptonic decays of K-mesons. A theorem due to Callan and Treiman for the decay $K \to \pi e \nu$ states that the two form factors that enter this decay are related to the K and π decay constants.

$$f + (m_K^2) + f - (m_K^2) = F_K/F_\pi \qquad (12)$$

Other theorems due to Weinberg relate form factors in the $K \to \pi \pi e \nu$ decay and obtain $\pi \pi \to \pi \pi$ scattering lengths. Low-energy theorems that combine PCAC and current algebra are not tested as well by experiment and some corrections may have to be made.

Other applications of the current algebra lead to a value of the electromagnetic mass difference of $\pi^+ - \pi^0$ mesons. Current algebra has been further extended by assuming that the currents are proportional to vector fields, like ρ, ω, K^* mesons. This hypothesis, called the current-field identity, has led to predictions for a large number of strong interaction decay rates of mesons. Such applications, however, involve a large number of free parameters. Nevertheless this enables calculations to be made where no method existed before.

In recent years tremendous progress has been made in understanding the strong interactions. Quarks have emerged as the basic constituents of hadrons, and their interactions are described by a gauge theory called *quantum chromodynamics*. In such a theory current algebra is exact. The great utility of current algebra arises from the smallness of quark masses, especially the "up" and "down" quarks. The notion of PCAC can then be made very precise. Thus all the basic hypotheses of current algebra, which were arrived at mostly through intuition, are now on very firm ground. The problem of how quarks form hadrons has not yet been solved in quantitative detail. Current algebra continues to provide a strong tool for the study of hadronic properties.

N. G. DESHPANDE

References

1. Adler, S. L., and Dashen, R. F., "Current Algebras and Applications to Particle Physics," New York, Benjamin, 1968.
2. Bernstein, J., "Elementary Particles and Their Currents," San Francisco, W. H. Freeman, 1968.
3. Marciano, W., and Pagels, H., "Quantum Chromodynamics," *Physics Reports* **36C**(3), 137–276 (1978).

Cross-references: CONSERVATION LAWS AND SYMMETRY, ELECTROMAGNETIC THEORY, ELECTROWEAK THEORY, ELEMENTARY PARTICLES, GAUGE THEORIES, GRAND UNIFIED THEORY, PARITY, QUANTUM CHROMODYNAMICS, QUANTUM THEORY, QUARKS, STRONG INTERACTIONS, WEAK INTERACTIONS.

CYBERNETICS

Cybernetics was defined by Norbert Wiener in 1948[1] as "control and communication in the animal and the machine." Communication clearly implies the communication of *information*; and purposeful control requires information about the current position and about the end being sought. Thus cybernetics has a strong "information" connotation and may be equated to "information handling." Note that it is information *handling*, rather than communication, for two reasons: first, the information may be used to modify the system (rather than merely passing through it), and second, it may be necessary to marshal information from various sources and process it (e.g., by a computer) before using it to effect control to a desired end. Wiener's definition can then be rewritten as "the handling of information in animals and machines in order to achieve desired effects." This includes the idea of "goal-seeking" systems which adapt themselves, to a greater or less extent, to the requirement of moving from the current situation to the desired goal. The least extent of this is found in the constant-parameter closed-loop system, the design of which may nonetheless involve considerable sophistication in order to achieve optimum performance (see FEEDBACK, SERVO-MECHANISM). A further component in the idea of cybernetics is that the system in question should be stochastic, i.e., that it should show random fluctuations, usually about a defined average trend. Wiener was concerned with this aspect and contributed to the theories both of the frequency spectra of stochastic time series and of the filtering of stochastic signals in order to get the best estimate of the underlying trend. The three elements in cybernetics, information, spectra of stochastic processes, and filtering of stochastic signals, can be summarized as follows.

Information Control systems respond to signals which carry information. For example, when a call is made for an elevator car in a modern multiple installation, the computer-like controller takes account of all calls which have not yet been answered and the relative positions and directions of travel of all cars before deciding which car shall answer each call. This is an example of information *handling* rather than communication. Another example is that in the CAD/CAM (Computer-Aided Design and Computer-Aided Manufacture) organization of an engineering plant, information fed to a computer during design of components may be used to control the machine tools which make the components. An objective definition of information, independent of the observer and of the medium in which the information exists, was provided by Shannon in his mathematical theory of communication.[2] (Information is intimately related to communication, first because it is what is communicated and second because it can be observed only when it is communicated to the observer.) In human terms the information possessed by an individual may be thought of as a weighted sum of probabilities (many of which may be practical certainties). This accords with Shannon's measure of information in physical communication systems, although this measure was derived from entirely different and specific considerations. Shannon's measure of the information conveyed by each unit signal is numerically equal to the negative of the entropy *H*, where

$$H = - \sum_{i=1}^{n} p_i \log p_i.$$

Here, p_i is the probability of the ith state of the receiver after receiving the signal, e.g., the printing of the ith character out of a set of n in a teletype system. This assumes that there is a finite number n of possible states. See Shannon's paper for a derivation of the number of possible states of a communication channel of given

signal-to-noise ratio and the theorem that any waveform can be represented by a number of discrete signals equal to two per cycle of the maximum Fourier frequency (the *sampling theorem**).

Spectrum of a Stochastic Process The frequency spectrum corresponding to a specified time function can be found by Fourier transform, but this is not applicable to a stochastic time function of which the amplitude at any given time can only be stated as a probability. In that case one proceeds via the autocorrelation function $R_{xx}(t_1, t_2)$ which is defined[4] as the expectation (in the statistical sense) that the variable x has the values x_1, x_2 at times t_1, t_2. If the stochastic function is stationary one can replace t_1 and t_2 by the difference $\tau = t_2 - t_1$ and average over all times:

$$R_{xx}(\tau) = \lim_{\tau \to \infty} \frac{1}{2T} \int_{-T}^{T} f(t) f(t - \tau) \, dt.$$

Having determined the autocorrelation function for all values of τ one obtains the power (squared amplitude) spectrum of frequencies via the Wiener-Khintchine transform,

$$W(f) = \int_{-\infty}^{\infty} R_{xx}(\tau) \cos 2\pi f\tau \, d\tau$$

Wiener Filters The classic idea of filtering is to separate frequency components, but Wiener posed a more fundamental question: Given the past history of a stochastic time series, how can one make a least-squared-error estimate of its value at any time? An exact answer requires the solution of an integral equation, which is not always feasible. Moreover, if "any time" includes the future, this implies prediction of the future, which is not consistent with a physically realizable filter. One method of avoiding prediction is to introduce delay, leading to the *infinite-lag* filter. By shifting to the frequency domain it is possible to design a filter of characteristic

$$Y(f) = \frac{W(f)}{W(f) + N(f)} \, e^{-j\beta 2\pi f}$$

where $W(f)$ is the power spectrum of the signal (the underlying trend) and $N(f)$ that of the noise (the superimposed fluctuation). The exponential term represents a delay which tends to infinity as the filter is elaborated to make $Y(f)$ exact, and this form of filter is physically realizable. In order to obtain a physically realizable filter without delay (the *zero-lag* filter) one must return to the time domain and factor the

---

*Shannon assumed a low-pass frequency characteristic, but bandpass characteristics have subsequently been treated.[3]

filter function into two components, having zero value for $t < 0$ and $t > 0$, respectively. The former corresponds to a physically realizable filter since it has no output before its input. This factor can be translated into the frequency domain to combine with a function of the signal and noise spectra to specify the optimum zero-lag filter. This is inevitably less effective than the infinite-lag filter, but delay cannot be tolerated in closed-loop systems.*

Adaptive and optimizing control can be applied to chemical process plant to maintain specified conditions of temperature, pressure, flow etc., in spite of changes in external circumstances which may include changes in the characteristics of input materials. One can go further and, for example, make a fractional distillation plant optimize the output of any desired fraction.

The first generation of industrial robots were not adaptive but merely followed a fixed cycle of operations, which was programmed by "leading" the robot through the cycle which it could then repeat exactly as often as required. But present development is concerned with achieving adaptive behavior by adding sensors to the robot, e.g., tactile sensors[5] or solid-state television cameras to provide vision.[6] For example, an arc welder may be made to follow the joint to be welded, in spite of variations in positions of component parts; or a mobile robot can avoid obstacles.[7]

The application of cybernetics to robots is very close to *artificial intelligence*, i.e., making machines behave as though they were intelligent living creatures. This was first demonstrated in the form of "mechanical animals,"[8] but now includes such things as chess-playing computers.

The recent tendency is to apply the term "cybernetic" to more abstract systems which exhibit feedback plus adaptation and often some degree of stochastic behavior. There is then a relationship between "cybernetics" and "general systems theory" which is discussed in a group of papers.[9] One application of this type is to education,[10] where there may be various forms of feedback from taught to teachers, and a random element is introduced by the varied abilities and motivation of students. Other applications have been to government policy,[11] power systems,[12] chemical analysis,[13] and the development and manufacture of silicon chips.[14] The point in the last is that there are two coupled loops, for process and for product development, and only the use of an *adaptive* controller overall can ensure stability. Some of these topics are often grouped under the heading of *economic cybernetics* while others fall within *biocybernetics*.[15] The whole field of cybernetics is reviewed from time to time in a "progress" volume.[16]

D. A. BELL

References

1. Wiener, N., "Cybernetics," New York, Wiley, 1948.
2. Shannon, C. E., "A Mathematical Theory of Communication," *Bell Syst. Tech. J.* **27**, 379–423 and 623–656 (1948).
3. Linden, D. A., "A Discussion of Sampling Theorems," *Proc. I.R.E.*, **47**, 1219–1266 (1959).
4. Korn, G. A., and Korn, T. M., "Mathematical Handbook for Scientists and Engineers," New York, McGraw Hill, 1961.
5. Anon., "Robots Feel their Way into Tight Situations," *New Scientist* **87**, 591 (21st August 1980).
6. Marsh, P., "Robots See the Light," *New Scientist* **86**, 238–240 (12th June 1980).
7. Belenkov, V. D., Gusov, S. V., Zotov, Yu. K., Rushanskiy, V. I., Timofeyev, A. V., Frolov, V. M., and Yakubovich, V. A., "Adaptive System for Control of Autonomous Mobile Robot," *Eng. Cyb.* (*U.S.A.*) pp. 37–45, translation of *Tekh. Kibern.* (*U.S.S.R.*) no. 6, pp. 52–63 (Nov.–Dec. 1978).
8. Walter, W. G., "An Electromechanical 'Animal,'" *Discovery* **11**, 90 (March 1950).
9. Various authors, "Cybernetics and General Systems—A Unitary Science?" *Kybernetes* **8**, 7–15; 17–23; 25–32; 33–37; 39–43; and 45–49 (1979).
10. Landa, L. N., "Cybernetic Methods in Education," *Educ. Technology* **17**, 7–13 (1977).
11. Berlin, V. N., and Weiss, R. G., "The Role of Evaluation Systems in the Government Policy and Programme Change," *Proc. Int. Conf. on Cybernetics & Society, Washington D.C. 19–21 Sept. 1977*, New York, IEEE, 1977.
12. Sukhanov, O. A. and Kristov, Kh. K. "Synthesis of Stable Cybernetic Models for Investigating the Dynamics of Electrical Systems," *Power Engineering* (*U.S.A.*) **15**, 20–25 (1977), translation of *Izv. Akad. Nauk SSSR Energy and Transp.* (*USSR*) **15**, 22–27 (1977).
13. Stepanenko, V. E., "Group Identification of Substances based on Cybernetic Models," *J. Anal. Chem.* **35**, 404 (1980).
14. Garte, D. "Cybernetic Model for the Process of Product Development of Solid-state Circuits considering the Semiconductor Process Development," *Nachrichten Tech. Elektron.* **30**, 312–317 (1980).
15. Nalecz, M., "Some Problems in Modern Biocybernetics and Biomedical Engineering," *Proc. 2nd. Int. Symposium on the Theory and Practice of Robots and Manipulators, Warsaw 14–17 Sept. 1976*, Amsterdam, Elsevier, 1977.
16. "Progress in Cybernetics and Systems Research," edited by R. Trappl and others. Vols. 1–5, published by Halsted Press, New York; vol. 6 and

*Those who want to explore the mathematics in detail are referred to Wiener's book, "The Extrapolation, Interpolation and Smoothing of Stationary Time Series," New York, Wiley, 1949. It is advisable to start with a paper by Levinson which is reprinted as Appendix C in the book.

subsequent by Hemisphere Publications, Washington, D.C.; vol. 9, 1981.

Introductory and General Books.

Rothman, Milton A. "The Cybernetic Revolution: Thought and Control in Man and Machine," New York, Franklin Watts, 1972.

George, F. H., "Foundations of Cybernetics," London, Gordon & Breach, 1977.

Pask, Gordon, "Cybernetics of Human Learning and Performance," New York, Crane-Russak, 1975.

Trappl, Robert (Ed.), "Cybernetics: A Source Book," Washington, D.C., Hemisphere Publications, 1982.

Cross-references: BIONICS, FEEDBACK, COMPUTERS.

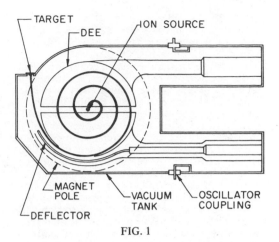

FIG. 1

CYCLOTRON*

The cyclotron is an accelerator of ions widely used to study the nucleus, to produce radioactive substances, and to study the interactions of ionizing radiation with living systems and with inert matter. It is equally important as the first of a class—*Magnetic Resonance Accelerators*—which includes the various kinds of synchrotron (see SYNCHROTRON) as well as synchrocyclotrons and sector focused cyclotrons. The essential feature of this type of accelerator is that acceleration of charged particles to high energies is achieved by a successive application of small accelerations in synchronization with the rotational period of the particles in a magnetic field. The condition for synchronization is simple and can be derived as follows: A charged particle moving perpendicularly to the lines of force in a magnetic field will describe a circle which is defined by the equilibrium between the Lorentz force $F_Q = eBv$ and the centrifugal force $F_c = mv^2/r$. Equating these, one may solve for the rotational frequency of the particles which is set equal to the frequency of the accelerating field. This is the *Cyclotron Resonance Condition:*

$$f_a = f_0 = \frac{eB}{2\pi m} \qquad (1)$$

where

f_a = frequency of accelerating field
f_0 = rotational frequency
e = charge of ion
m = mass of ion
B = magnetic field strength.

The important fact is that the rotational frequency is independent of the energy of the particle and depends only on quantities which are (approximately) constant. In 1929 the possibility of using this relationship as the basis for an accelerator occurred to Ernest O. Lawrence, who like many other physicists at the time had

been inspired by Rutherford's success in disintegrating atoms with alpha particles from natural sources to seek a means of producing a controlled beam of high energy particles. The practicability of the idea was demonstrated, and most of the essential features of the Cyclotron were developed by Lawrence, M. Stanley Livingston, N. E. Edlefsen, and others during the next few years.

Figure 1 is a schematic diagram showing the principle components of a cyclotron. The dees are two hollow semicircular electrodes in a vacuum tank located between the poles of an electromagnet which provides an approximately uniform magnetic field over the entire region. The dees are part of an electrical resonant circuit which may be excited by an oscillator whose frequency is adjusted to the rotational frequency given by Eq. (1). Ions are produced by an electric discharge in a source located at the center. They are drawn from the source and accelerated into a dee while it is negative, they follow a semicircular path in the (electrostatic) field free interior of the dee and again arrive at the gap between the dees where, by that time, the voltages are reversed in sign and they are accelerated again. The ions describe semicircles of increasing radius as their velocity and energy increase as a result of repeated accelerations. When they reach the maximum radius of the dee, they enter a channel between a septum in one of the dees and the deflector. The deflector is charged negatively and draws the particles out where they may strike a target in the target chamber or they may travel some distance as a beam outside the cyclotron before they are used.

The kinetic energy of the accelerated particles is given by:

$$T = \frac{1}{2} \frac{B^2 R^2 e^2}{m} \qquad (2)$$

where T is the kinetic energy and R is the radius of ion path at point of extraction. For protons, Eq. (1) reduces to $f = 1.52B$ MHz and Eq. (2) to $T = 0.484 B^2 R^2$ MeV with B in kilogauss and R in meters. The usual values of B are from 15 to 22 kilogauss.

In addition to the resonance condition, a successful cyclotron requires that the orbits be stable, i.e., they must remain in the median plane and at the appropriate radius. The first is achieved by introducing in the magnetic field a small negative gradient with respect to radius. The field lines are then bowed as shown in Fig. 2, and the Lorentz force on a particle off the median plane has a vertical focusing component. Radial stability results from the fact that the orbit of the particle is an equilibrium orbit with the inward Lorentz force predominating at radii larger than the equilibrium orbit and the centrifugal force predominating at smaller radii. Ions which are displaced either vertically or radially then execute oscillations about the equilibrium orbit. If the magnetic field is described by the index,

$$n = - \frac{r}{B} \frac{\partial B}{\partial r} \qquad (3)$$

where r is the radius, then it can be shown that for $0 < n < 1$ stable oscillations occur with frequencies:

$$f_z = f_0 \sqrt{n}$$
$$f_r = f_0 \sqrt{1 - n} \qquad (4)$$

where f_z is the frequency of vertical oscillations and f_r is the frequency of radial oscillations.

The negative gradient in the magnetic field results, however, in the situation where the rotational frequency, Eq. (1), is not exactly the same at all radii. In addition, it must be noted that the mass in Eq. (1) is the relativistic mass, $m = m_0 + T/c^2$ and increases with energy. The result of these two discrepancies is that the rotational frequency of the ion decreases as it is accelerated and there is an accumulated phase lag between the ion and the accelerating field which when it approaches π radians, results in

no further acceleration. The energy limit of the conventional cyclotron imposed by this phase error can be shown to be proportional to the square root of the accelerating potential and to be about 30 MeV for protons with a dee-to-dee potential of 200 kV. It has not been possible to reach the theoretical maximum energy in practice, and for reasons made clear in the next sections, the incentive to do so has disappeared. The maximum energy which has been attained with protons is 22 MeV and that required about 500 kV on the dees. Currents in cyclotrons are usually of the order of 100 μA but up to 1 mA has been attained. The most commonly used ions are protons, deuterons, and alpha particles, although heavier ions such as carbon, nitrogen and oxygen ionized to +3 or +4 have also been accelerated.

A possibility of achieving higher energies with cyclotrons was opened up in 1945 when V. Veksler and E. M. McMillan independently pointed out the phase stable characteristic of the cyclotron resonance condition [Eq. (1)] which may be explained if the equation is rewritten in terms of particle energy:

$$f_0 = \frac{Be\,c^2}{2\pi(E_0 + T)} \qquad (5)$$

where E_0 is the rest energy of the particle.

Consider a particle rotating in a cyclotron at the resonant frequency and crossing the accelerating gap at a phase such that it gains no energy and that a later arrival causes it to lose energy. If this particle is perturbed by an excess of energy, f_0 decreases and the particle loses energy. If the particle is perturbed in phase so that it arrives at the accelerating gap too early, it gains energy, f_0 decreases, and the phase slips back. Perturbations in energy or phase thus result in oscillations about the equilibrium phase. Under these conditions, if the accelerating frequency of a cyclotron is slowly decreased, the ions will execute stable oscillations about that phase which will give sufficient energy gain so that the radius and energy are matched as the orbits expand. This is the *Principle of Phase Stability* as applied to the Synchrocyclotron. It completely removes the energy limitation of the cyclotron previously discussed. This principle was immediately exploited and synchrocyclotrons (also sometimes called Frequency Modulated or FM Cyclotrons, and in the U.S.S.R., Phasotrons) have been built which give protons up to 1000 MeV. The only limit is the economic one due to the large size of the magnet.

The important structural difference between a synchrocyclotron and a conventional cyclotron is in the provision for a variable frequency. This is accomplished by placing a variable capacitor in the resonant dee circuit. Rotary blade capacitors have been in common use for this purpose, but in more recent designs, vibrating blade capacitors have been preferred. The required frequency swings are about two to one,

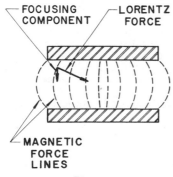

FOCUSING COMPONENT — LORENTZ FORCE

MAGNETIC FORCE LINES

FIG. 2

and the usual modulation frequencies are about 60 to 100 Hz. The ions are accelerated in pulses as the accelerating frequency sweeps through its modulation cycle in contrast to the continuous acceleration in a conventional cyclotron. The result is that average currents in synchrocyclotrons are about 1 per cent of cyclotron currents, thus removal of the energy limit has been accomplished at the expense of a current limitation.

Another method of circumventing the energy limit of the cyclotron was proposed by L. H. Thomas in 1938, seven years before the principle of phase stability was enunciated. In the Thomas proposal, the average magnetic field increases with radius so that the resonance condition may be exactly matched by a constant accelerating frequency as the ion gains energy. The axial focusing force is supplied by an azimuthal variation of the magnetic field which may be obtained by using sectored magnet poles, Fig. 3a. The ion orbits are then no longer circular, and the radial component of velocity interacting with the azimuthal component of magnetic field produces an axial focusing force. This is the "edge focusing" which occurs when an ion crosses a fringe field obliquely, and it has long been used in mass spectrometers and other devices.

This idea was well in advance of the theory and practice of the cyclotron art at the time and was not immediately exploited. Development beginning in 1949 and extending to recent years has resulted in a whole subclass of cyclotrons characterized by a fixed rotational frequency and focusing forces derived from spatial variation in the magnetic field. For example, if the sectors are spiral shaped as in Fig. 3b, additional focusing forces of alternating gradient type are developed. These cyclotrons are variously called Sector Focused, Isochronous, and AVF (azimuthally varying field) cyclotrons. They have energies well beyond the energy limit

of the conventional cyclotron and, at the same time, are capable of high average currents because they operate at a constant frequency. Provision of auxiliary magnet coils on the pole tips, to trim the field shape over a range of values of average field, and adjustable frequency oscillators, to provide for different ions and a variation in maximum energy, have made the modern cyclotron of this type very flexible.

In a further application of the Thomas principle, the magnet assembly is made up of individual sectors with field-free spaces between them. Fixed-frequency accelerating cavities are located between the sectors. This design variation of the sector-focused cyclotron makes it economically possible to go to higher energies, but at a sacrifice of some of the features permitting easy variation of energy and ion species. Cyclotrons may also be used two or even three in tandem. The extracted beam from a cyclotron designed for optimum acceleration of certain ions in a low energy range is used to provide an intense beam for injection into a larger cyclotron designed for a higher energy. Superconducting magnets have been used in some of the most recent cyclotron designs to produce magnetic fields more than twice that of conventional magnets, resulting in more compact installations and saving in magnet power.

The flexibility of cyclotron design, resulting from the application of the Thomas principle and concurrent technological developments, has rendered the classical cyclotron and synchrocyclotron obsolescent. A compilation of cyclotrons made in 1978 in connection with the Eighth International Conference on Cyclotrons and their Applications (see last reference) lists a total of 108 cyclotrons world-wide, in operation or under construction, of which 14 are synchrocyclotrons and only 4 classical. The remainder are variations of the sector-focused design. Two of these use superconducting magnets.

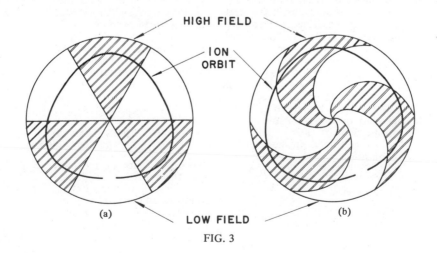

FIG. 3

TABLE 1

	CYCLOTRON 60-inch Cyclotron, University of Washington, Seattle	SYNCHROCYCLOTRON 184-inch Cyclotron Lawrence Berkeley Laboratory, Berkeley	SECTOR FOCUSED CYCLOTRON 88-inch Cyclotron, Lawrence Berkeley Laboratory, Berkeley	SECTOR FOCUSED CYCLOTRON (Two in tandem) SIN, Swiss Institute for Nuclear Research
First Operation	1951	1946 rebuilt 1957	1962	1974
Magnet	1.52 m diameter, 19 kG max. field, 197,000 kg	4.8 m diameter, 23.4 kG max. field, 3,900,000 kg	2.24 m diameter, 3 spiral sectors 20 kG max. field, 272,000 kg	9.2 m diameter, 8 separated sectors, 21 kG max. field, 2,000,000 kg
RF	Fixed frequency, 11.6 MHz, two dees, 250 kV	Variable frequency, 18–36 MHz for protons, Modulation frequency 64 Hz, single dee, 11 kV	Fixed frequency, adjustable 5.5–16.5 MHz for various particles, single dee, 75 kV	Fixed frequency, 50 MHz, 4 cavities, 600^{kV}/cavity
Beam	Protons 11 MeV, deuterons 21 MeV, alpha particles 42 MeV, 150 μA max. current	Protons 740 MeV, deuterons 460 MeV, alpha particles 910 MeV, 1 μA max. current.	Protons 60 MeV, alpha particles 140 MeV, $^{12}C^{4+}$ 193 MeV $^{16}O^{6+}$ 315 MeV 3 mA max. current.	Protons 588 MeV, 110 μA design current.

Source of data, "Eighth International Conference on Cyclotrons and Their Applications," *IEEE Transactions* NS-26, Number 2, 1979.

Table 1 gives a comparison of the salient design features and performance of typical examples of a classical cyclotron, a synchrocyclotron, a sector focused cyclotron designed for high intensity and variable ion and energy, and a sector focused cyclotron designed for high intensity and energy. The latter, giving copious secondary beams of pi mesons, are sometimes called meson factories.

The early and continuing impetus for cyclotron development has come from its use in research in nuclear physics and chemistry and particle physics. But applications in medicine where cyclotrons have provided radioisotopes for diagnosis and neutron and charged particle beams for therapy have been not far behind. Some of the other important applications have been trace elements analysis by activation of a sample and characterization of the resultant activity; solid state studies by creation of irradiation damage and by implantation of impurities; and production of many different radioisotopes for use as tracers in many fields of science and technology.

EDWARD J. LOFGREN

References

Livingston, M. Stanley, and Blewett, John P., "Particle Accelerators," New York, McGraw-Hill Book Co., 1962.
Kolomensky, A. A., and Lebedev, A. N., "Theory of Cyclic Accelerators," New York, John Wiley & Sons, 1966.
Burgerjon, J. J., and Strathdee, A. (Eds.), "Cyclotrons—1972" (Proceedings of the Sixth International Cyclotron Conference), New York, American Institute of Physics, 1972.
Joho, W. (Ed.), "Seventh International Conference on Cyclotrons and their Applications," Basel and Stuttgart, Birkhäuser Verlag, 1975.
Hicks, J. W., (Ed.), "Eighth International Conference on Cyclotrons and their Applications," *IEEE Transactions* NS-26, Number 2, 1979.

Cross-references: ACCELERATORS, LINEAR; ACCELERATOR, PARTICLE; BETATRON; SYNCHROTRON; ACCELERATOR, VAN DE GRAAFF.

CYCLOTRON RESONANCE (DIAMAGNETIC RESONANCE)†

The term cyclotron resonance is used to designate the resonant coupling of electromagnetic power into a system of charged particles undergoing periodic orbital motion in a uniform static magnetic field. The frequency of the electric

†Support of U.S. Air Force is acknowledged.

field at resonance is simply related to the orbital frequency of the electron in the magnetic field. The effect has been observed and studied extensively in gases and in solids.

One important application of the cyclotron resonance principle is made in the acceleration of charged particles, as in a cyclotron. In a uniform magnetic field, H, a charged particle of mass, m_c, undergoes orbital motion with an angular velocity

$$\omega_c = \frac{eH}{m_c c},\qquad (1)$$

in which e is the charge and c the velocity of light. Energy from the electromagnetic fields, i.e., from the alternating electric and magnetic fields, is transferred into kinetic energy of the particle, and the radius of the particle orbit is increased with no change in angular velocity. Particle acceleration takes place *in vacuo* in order to prevent energy transfer to the gas by means of collisions.

In solids, cyclotron resonance has been successfully applied to studies of electronic energy band structure. The perfectly periodic array of atoms in an ideal solid scatters electrons coherently. An electron experiencing such coherent scattering can be described by the same equations of motion as the free electron, except that the free electron mass is replaced by an effective mass, m^*. Incoherent scattering from crystalline imperfections causes electronic collisions which limit the number of completed electron orbits, thus giving rise to a frequency bandwidth for the cyclotron resonance absorption. The observation of cyclotron resonance requires that the charged particle execute about one complete cyclotron orbit without collisions, or $\omega_c \tau \gtrsim 1$, in which the collision time, τ, is the mean time between incoherent scatterings. A long collision time is achieved by using samples of the highest possible purity and lattice perfection and by cooling to very low temperature (usually liquid He temperature, 4 K) to eliminate the thermal motion of the atoms. The condition for cyclotron resonance can also be satisfied by increasing ω_c through the use of high magnetic fields, e.g., 100-kilogauss static fields are currently available which for free electrons results in $\omega_c \approx 1.5 \times 10^{12}$ rad/sec or an electromagnetic wave length of about 1 mm.

Electrons moving in the periodic lattice of a solid occupy energy levels which are specified by the wave vector quantum number, $\mathbf{k}$, or by the crystal momentum, $\hbar \mathbf{k}$. Since the number of electrons is very large, the wave vectors assume an almost continuous range of values. A knowledge of the functional form of the dependence of the energy on wave vector is necessary for a complete description of the behavior of electrons in solids. The simplest form of the relation between energy and wave vector valid for energy bands in cubic crystals is

$$E(\mathbf{k}) = \frac{\hbar^2 k^2}{2m^*}.\qquad (2)$$

In this case, the constant energy surfaces in wave vector space are spheres, and the cyclotron mass of Eq. (1) is just the effective mass, m^*. For energy extrema located at general points in wave vector space, Eq. (2) becomes

$$E(\mathbf{k}) = \frac{\hbar^2}{2}\left(\frac{k_x^2}{m_x} + \frac{k_y^2}{m_y} + \frac{k_z^2}{m_z}\right)\qquad (3)$$

in which the extremal point is taken as the origin, and m_x, m_y, and m_z are three components of an effective mass tensor. This generalization is also necessary in describing the energy bands for crystals with symmetry lower than cubic.

An expression for the cyclotron effective mass which is valid for an electron orbiting on a constant energy surface of energy E for an arbitrary $E(\mathbf{k})$ is

$$m_c(E, k_H) = 2\pi\hbar^2 \left(\frac{\partial A}{\partial E}\right)_{k_H}\qquad (4)$$

in which k_H is the wave vector component parallel to the magnetic field, A is the area of the electron orbit in wave vector space, and $(\partial A/\partial E)_{k_H}$ is the derivative of this area with respect to energy evaluated at constant k_H. For spherical or ellipsoidal constant energy surfaces, the cyclotron mass is independent of both energy and k_H, and for these two simple cases, m_c is given, respectively, by $m_c = m^*$, and

$$\left(\frac{1}{m_c}\right)^2 = \frac{\alpha^2}{m_y m_z} + \frac{\beta^2}{m_x m_z} + \frac{\gamma^2}{m_x m_y}\qquad (5)$$

in which α, β, γ are the direction cosines of the magnetic field with respect to the axes of the ellipsoidal constant energy surface.

For solids which have relatively low carrier density (e.g., insulators, semiconductors, and semimetals) the electronic states which are important in the transport properties are located near energy band extrema. For nondegenerate extremal points in wave vector space, $E(\mathbf{k})$ can be expanded in a Taylor's expansion. The leading term of such an expansion would be given by Eq. (3). For degenerate points (positions where two or more levels have the same energy), a simple generalization of a Taylor's expansion must be used. In solids with relatively high carrier density (e.g., metals), the transport properties are determined by electronic states which are far from the energy extrema and the $E(\mathbf{k})$ relation is not adequately described by a Taylor's expansion.

Cyclotron resonance experiments have been particularly successful in the quantitative determination of the band parameters of the semiconductors silicon and germanium. The successful application of this technique in these semiconductors is attributed to the high quality of the available material, and to the complete classification of the possible forms of the theoretical band structure model. Since in these materials the intrinsic carrier concentration is extremely small at low temperatures, electrons are optically excited out of filled valence levels in the crystal in order to produce sufficient carriers to obtain a measurable signal. Resonances are observed both for the excited electrons and for the holes left behind in the empty levels in the valence band.

In metals, the high carrier density requires modification of the conventional cyclotron resonance experiment. Two important consequences of this high carrier density are the non-uniform penetration of the electromagnetic field in the skin depth and the inapplicability of the simple effective mass theory to describe the electronic states. To overcome the problem of the small electromagnetic penetration depth, the geometrical arrangement suggested by Azbel and Kaner is used. The static magnetic field is applied in the plane of a flat sample, so that the electrons near the surface can be accelerated by the electromagnetic fields, and the orbits described by a cyclotron radius which is large compared with the skin depth. In this way, whenever the applied frequency is a multiple of the cyclotron frequency, a resonant condition is satisfied. This type of cyclotron resonance experiment yields an effective mass at the Fermi energy given by Eq. (4), which is, in general, dependent on the wave vector component parallel to the magnetic field. The interpretation of these experiments is not simple but when coupled with experiments which measure the shape of the Fermi surface, such as DE HAAS-VAN ALPHEN EXPERIMENTS, a fairly complete determination of the electronic band structure is possible. These techniques have been successfully applied in the study of copper.

Cyclotron resonance in ionic crystals allows the measurement of polaron effects. The POLARON denotes the charge carrier together with its local lattice distortion. Cyclotron resonance observed in AgBr has been interpreted as a polaron orbiting in the applied magnetic field.

G. DRESSELHAUS

References

Kittel, C., "Introduction to Solid State Physics," sec. ed. p. 371, New York, John Wiley & Sons, Inc., 1956.

Lax, B., and Mavroides, J. G., "Solid State Physics," Vol. XI, p. 261, New York, Academic Press Inc., 1960.

Cross-references: CYCLOTRON, DE HAAS-VAN ALPHEN EFFECT, DIAMAGNETISM, FERMI SURFACE, RESONANCE.

D

DE HAAS-VAN ALPHEN EFFECT*

The de Haas-van Alphen effect is the periodic oscillation of the magnetization with inverse magnetic field. These oscillations were discovered in 1930 in Bi at very low temperature by W. J. de Haas and P. M. van Alphen. For some time the effect was thought to be unique to Bi but it was observed in zinc in 1947 and since then in a very large number of metals and intermetallic compounds. Related magneto-oscillatory behavior in the electrical resistance has even been observed recently in organic metals.

Peierls first explained the effect in 1933 as a consequence of the quantitization of the motion of the conduction electrons in a magnetic field. In zero field, electrons have a nearly continuous range of energies up to a maximum called the Fermi energy E_F. In momentum or wave number space, all states are occupied up to the Fermi surface. The Fermi surface is thus the boundary between filled and unfilled states in momentum space.

When a magnetic field H_z is applied, only discrete energy levels E_n are available for a given k_z, and the separation between these levels increases with field. If the energy surfaces are ellipsoidal so that E is proportional to k^2, E_n is exactly soluble from the Schrödinger equation. Typically the energy surfaces are much more complex but an approximate approach due to Onsager can be used. By application of the Bohr-Sommerfield quantization condition to orbits normal to the applied field direction, Onsager showed that the area A enclosed in an orbit satisfies

$$A = 2\pi(n + \gamma)\, eH/\hbar c \qquad (1)$$

where n is an integer, γ is a constant near $\frac{1}{2}$ (exactly $\frac{1}{2}$ for quadratic energy surfaces), $\hbar$ is Planck's constant divided by 2π, and c is the velocity of light. For a free-electron metal ($E \propto k^2$) at $T = 0$, the electrons are contained in a sphere in k-space of maximum energy E_F, while in a field H_z they are found in a series of cylinders with axes parallel to H_z as shown in Fig. 1. As the field magnitude increases the

*This work performed at Sandia National Laboratories supported by the U.S. Department of Energy under Contract Number DE-AC04-76DP00789.

cylinders grow in size, and at a critical field given by

$$1/H = 2\pi(n + \gamma)e/hcA_0 \qquad (2)$$

the nth cylinder becomes too large to be contained in the surface. Here A_0 is the maximum cross-sectional area of the sphere. This gives rise to oscillations in the free energy F periodic in $1/H$. This same result occurs for more irregular shaped Fermi surfaces, with the oscillations corresponding to *extremal* cross sections.

The effect is usually observable only below 4 K because the Fermi surface is blurred out over an energy range $\sim kT$ at finite temperatures T. If we define an effective carrier mass m^* by

$$m^* = \hbar^2/2\pi(dA/dE)_{k_z} \qquad (3)$$

then the oscillatory term in the free energy F is given[1] by

$$F \propto TH^{3/2} \exp\left[-2\pi^2 k(T + X)m^*c/eH\hbar\right]$$
$$\cdot \cos\left[c\hbar A_0/eH) + (\pi/4) - \delta\right] \qquad (4)$$

for the case when the argument of the exponential is much greater than unity. Here k is the Boltzmann constant, δ is a phase factor, and X is a term that takes into account the impurity collision broadening of the energy levels. By varying the temperature as the field is held constant m^* can be determined, and then by varying field at constant temperature the scattering factor X can be measured.

Historically the de Haas-van Alphen effect was measured using torsion balance techniques, usually at slowly varying fields considerably below 40 kG or using pulsed field techniques to a maximum of $\sim$200 kG. In the former method the torque exerted on a sample as the field is changed is measured with the advantages of high sensitivity and accuracy. The disadvantages are insensitivity to spherical segments of the Fermi surface and practical limitation to moderate fields, which make large cross sections difficult or impossible to observe. The pulsed field technique employs 100–200 kG fields with rise times of the order of 10 msec. These high fields permit observation of very high de Haas-van Alphen frequencies but accuracy is limited by the necessity of making measurements of the susceptibility and field magnitude in a few msec.

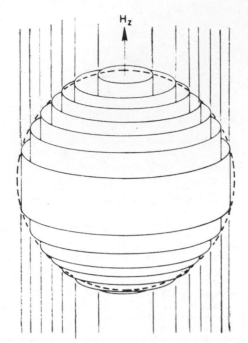

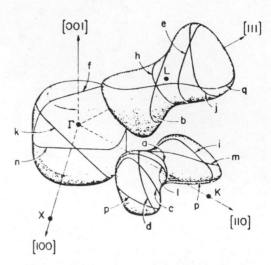

FIG. 2. Fermi surface of thorium as calculated by Gupta and Loucks.

FIG. 1. Quantitized orbits for a free-electron metal in a magnetic field H_z. The dashed line is the boundary of the Fermi surface.

In the past decade, the bulk of the detailed de Hass-van Alphen measurements have employed what is known as the *field modulation technique*.[2] Here a small balanced pickup coil detects the oscillations in the magnetization of the sample as the applied field is slowly varied

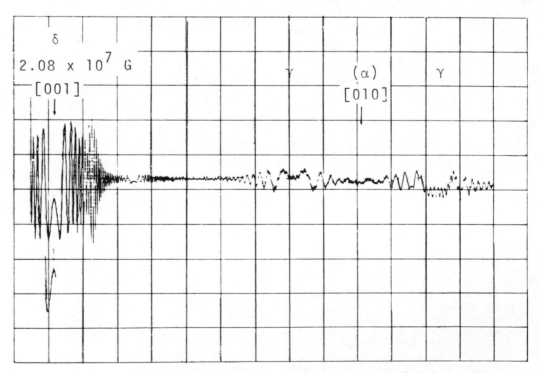

FIG. 3. De Haas-van Alphen oscillations obtained by rotation of the magnetic field about a single crystal of α-U at ~100 kG, 9 kbar and 1 K.

while being modulated sinusoidally. The technique lends itself to use with superconducting coils, which can now be obtained in the 100–150 kG range. The flexibility to vary the modulation amplitude and direction when coupled with detection of various harmonics of the modulation allows spectrometer action in sorting out complicated de Haas-van Alphen spectra. Fourier analysis of the spectra is also routinely employed. These techniques have resulted in an enormous amount of very accurate, and in many cases complete, experimental mappings of the Fermi surfaces of a great many metals. With high-quality, long-mean-free-path crystals, the limitation on the accuracy of the cross-sectional area data is reduced to the limitations in counting the oscillations and determining the absolute value of the magnetic field (which can be monitored by *in situ* nuclear magnetic resonance).

The calculated Fermi surface of thorium is shown in Fig. 2 and all of the extremal orbits shown on each of the three sheets have been observed. The detailed angular dependence of the sheets can be determined by rotation of the magnetic field (at constant magnitude) with respect to the crystallographic axes as shown in Fig. 3 for α-U. These data were taken at 1 K at 100 kG at a pressure of 9 kbar ($\sim$9000 atmospheres) in order to avoid low-temperature phase changes below 43 K. This is an indication of the versatility of present-day experimental de Haas-van Alphen techniques.

De Haas-van Alphen data have been important in the development of band calculations of electronic structure of metals. Because of its great sensitivity and the fact that it is nearly a dc experiment, the field modulation technique also lends itself to measurements of the pressure and strain dependence of the Fermi surface. This type of experiment provides critical and direct tests for the theoretical model descriptions which can be calculated easily at various interatomic spacings.

J. E. SCHIRBER

References

1. Lifshitz, I. M., and Kosevich, A. M., *J. Exper. Theor. Phys.* **29**, 730 (1955).
2. Stark, R. W., and Windmiller, L. R., *Cryogenics* **8**, 272 (1968).
3. Boyle, D. J., and Gold, A. V., *Phys. Rev. Lett.* **22**, 461 (1969).
4. Schirber, J. E., and Arko, A. J., *Phys. Rev.* **B21**, 2175 (1980).

Cross-references: FERMI SURFACE, MAGNETISM.

DENSITY AND SPECIFIC GRAVITY

Definitions The amount of a substance that occupies a given amount of space is an intrinsic property of the substance. This intrinsic prop-

erty of substances is called *density*. Thus, density is a measure of the amount of matter that occupies a given amount of space. Precisely defined, density is the mass of a substance per unit volume. In the international system of units (Système International) the density of a substance is given in kilograms per cubic metre (kg/m^3). In the centimetre/gram/second (cgs) system the density of water is 1 g per cubic centimetre. It is possible to remove the system of units from the numerical value of density by defining a property called *specific gravity*. The specific gravity is the ratio of the density of the given substance to the density of a standard substance. It has been customary to define the specific gravity of a substance as the ratio of the density of the substance to the density of water. As a result, since the density of water in the cgs system is 1 g/cm^3, the specific gravity of an object has the same numerical value as its density in cgs units.

Density is the ratio of two independent extrinsic properties of a given amount of a substance, its mass divided by its volume. In common useage the concept of density is frequently confused with the weight of an object. In addition it is not uncommon to find weight and volume confused in common discussions of objects. The common confusion among weight, volume, and density is displayed in the child's trick question, "Which is heavier—a ton of bricks or a ton of feathers?" When trying to convey scientific information about the density of an object to the public, care must be taken to explain this intrinsic property of matter. It may be best to contrast density to the daily experiences of heft or heaviness.

Density is an intrinsic property of an object, i.e., it will not vary from one part of a homogeneous object to another, but it does depend upon the volume of the object. Hence, density is a function of those variables that can change the volume of an object. In general, the volume of an object is a function of both temperature and pressure. This functional dependence of volume on temperature and pressure is most striking for gaseous materials (see GAS LAWS) but temperature is also an important consideration for all three states of matter (see EXPANSION, THERMAL).

The concept of density as mass per unit volume seems to imply a continuous medium. According to the modern atomic theory of matter (see ATOMIC PHYSICS) all matter is made up of discrete entities called *atoms*. Hence, the density of a substance is a microscopic measure of how tightly the atoms of the substance are packed together. If a solid sphere model is adopted for an atom, then the density of individual atoms of various elements can be calculated and compared to the densities of normal materials (see the table at the end of this article). Similarly according to the nuclear model of atoms (see ATOMIC PHYSICS), an atom consists of a massive nucleus situated in the middle of an electronic cloud. Once again, if a solid spherical model is adopted for a nucleus, then one can take typical nuclear properties and calculate nuclear densities to compare with the densities of other kinds of matter (see the table at the end of this article).

Solids Since the compressibility of solids is very small, their density is essentially independent of pressure and only shows a small temperature dependence. The direct measurement method of density involves measurement of both mass and volume under the same conditions. For example the mass of 1 cubic meter of aluminum is 2700 kg. The hydrostatic method of measuring densities is based on the application of Archimedes' principle, which states that the buoyant force on an object immersed in a fluid is proportional to the mass of the volume of the displaced fluid. The hydrostatic method for measuring the density of a solid weighs the solid object in air and in water and uses the loss of weight in water as a way to determine the density of the object. If the solid has a weight in air W_a and a weight in water W_w then the density of the solid ρ_s, neglecting the buoyancy of air, is given by

$$\rho_s = W_a \rho_w / (W_a - W_w) \, (\text{kg/m}^3)$$

where ρ_w is the density of water.

The determination of the density of a solid by the hydrostatic method can be corrected for the buoyant force of the air. Then the density of the solid is given by

$$\rho_s = W_a (\rho_w - \rho_a)/(W_a - W_w) + \rho_a$$

where ρ_a is the density of air.

Sources of error in the hydrostatic method are largely due to surface tension effects and trapped air on the surface of the solid. As a rule of thumb a minimum volume of 5 cm³ is required for an accuracy of one part in 10^4 in the corrected density.[1]

For solids less dense than water, a sinker (denser than water) is used. The sinker is weighed in air and in water as described above and then weighed with the unknown. The density of the unknown sample ρ_u is then found in terms of:

W_{ua} = weight of unknown in air;

W_{sa} = weight of sinker in air

W_{usw} = weight of unknown and sinker in water;

W_{sw} = weight of sinker in water

$$\rho_u = W_{ua} \rho_w / (W_{ua} + W_{sa} - W_{usw})$$
$$- (W_{sa} - W_{sw}))$$

$$\rho_u = W_{ua} \rho_w / ((W_{ua} - (W_{usw} - W_{sw}))$$
$$\cdot (\text{kg/m}^3)$$

where ρ_w is the density of water.

Liquids A standard object and the hydrostatic method can be used to find the density of a liquid. This process for determining the density of a liquid calls for measuring the weight of the object in air W_a and in water, W_w, and in the liquid, W_l, then the density of the liquid is given by

$$\rho_l = (W_a - W_l) \rho_w / (W_a - W_w) \, (\text{kg/m}^3).$$

Accurate density measurements must minimize temperature effects and are usually made at specified temperatures (e.g., 15°C).

A mechanical system for the rapid measurement of the density of liquids with extremely high accuracy has been developed.[2] This system measures the period of oscillation of a U-tube containing the liquid of unknown density. The period of oscillation is directly related to the density of the liquid in the U-tube.

The operation of this system is based on the physics of a damped harmonic oscillator. The instrument provides an external electromagnetic force to balance the damping forces on the oscillating U-tube. While this force balance is maintained the resonant oscillations of the system are established. The period of the resonance oscillations is then measured and related to the density of the liquid as follows:

The total mass of the vibrating objects is given by

$$M = M_t + V\rho$$

where M_t is the mass of the empty U-tube, V is the volume of the tube, and ρ is the density of the liquid in the tube.

The period at resonance is given by

$$T = 2\pi (M/k)^{1/2} = 2\pi ((M_t + V\rho)/k)^{1/2}$$

where k is the restoring force constant of the oscillator.

The density of the liquid can be computed from the period of oscillation and the values of the instrument constants. In practice the instrument constants are determined by measuring the periods of oscillations of fluids of known density such as dry air and pure water.[3]

The commercially available instrument is designed to measure total time for a fixed number of oscillations in time ranges from 0.7 to 480 seconds.* During the time of measurement the temperature of the sample must be controlled to within $\pm 10^{-2}$°C for absolute errors in the density of the fluid less than 10^{-3} kg/m³. In practice this temperature control is difficult and leads to the use of small sample volumes, i.e., seven milliliter volumes are the most common. The instrument specifications call for a maximum precision of 1.5×10^{-3} kg/m³ for a measuring range of $0-3 \times 10^3$ kg/m³ when density differences are measured. In comparing

*The commercial instrument is manufactured by the A. Paar Company of Graz, Austria and is marketed by the Mettler Instrument Corp. in the U.S.A.

the measurements using this mechanical oscillation instrument to other sources of density measurement of standard sodium chloride solutions, the precision of this instrument was found to be better than 5 parts in a million for all cases.

Gases The densities of a gas are strongly dependent on both temperature and pressure. It is the custom to give density values at the standard temperature and pressure (STP) values of 0°C and 1 atmosphere, respectively. Measurements under other conditions may be reduced to STP values by using the appropriate equation of state. In many cases, the ideal gas law can be used to give the STP density ρ from measurements at any pressure and temperature as follows:

$$\rho = M/V = MP_0(1 + 0.00366(T - T_0))/PV$$

where M is the mass of the gas, V is the volume of gas at a pressure P and temperature $T(K)$, and P_0 and T_0 are the standard values of pressure and temperature. The experimental method used is the direct measurement of a known volume V of a gas at ambient temperature and pressure. A standard flask of air is evacuated, then filled with the gas to ambient pressure, and its mass is measured.

Applications Since density is an intrinsic property of a substance and is, in general, a unique value for each substance, density can be used as a way to distinguish one substance from another. In systems where the force on objects is proportional to mass, as in a gravitational field (See gravity) differences among the densities of fluids can be used to separate fluids from one another.

Typical Values of Densities The materials we commonly encounter have densities that vary only over a few orders of magnitude from air with a density of 1.3 kg/m³ to gold with a density of 19×10^3. However, as shown in the Table of Typical Densities, in the whole universe the densities of matter vary over 45 orders of magnitude.

ROBERT G. FULLER
RICHARD M. FULLER

TABLE OF TYPICAL DENSITIES

Substance	Density (kg/m^3)
Smooth density of galactic material throughout the universe	2×10^{-28}
Mean density of interstellar gas	3×10^{-21}
Mean density of the moon	3.3×10^3
Mean planet densities	
Saturn	0.70×10^3
Jupiter	1.33×10^3
Mars	3.93×10^3
Venus	5.24×10^3
Earth	5.515×10^3
Typical densities of earth materials	
Air	1.293
Teak wood	0.58×10^3
Alcohol	0.80×10^3
Ice (0°C)	0.917×10^3
Water (0°C)	0.999841×10^3
Water (4°C)	0.999973×10^3
Water (20°C)	0.998203×10^3
Aluminum	2.70×10^3
Copper	8.96×10^3
Silver	10.5×10^3
Mercury	13.6×10^3
Gold	19.3×10^3
Platinum	21.5×10^3
Mean densities of atoms (using solid sphere model)	
Hydrogen	4.4×10^1
Oxygen	2.0×10^3
Uranium	1.7×10^5
Densities of stars	
The sun	1.41×10^3
White dwarf	1×10^9
Neutron star	2×10^{17}
Density of nuclear matter (using solid sphere model)	2×10^{17}

References

1. Marton, L., "Classical Methods," Volume 1 of "Methods of Experimental Physics" (I. Estermann, Ed.), Academic Press, New York, 1959.
2. Stabinger, H., Kratky, O., and Leopold, H., *Montash. Chem* 98, 436 (1967).
3. Elder, J. P., Volume 61 of "Methods of Enzymology" (C. H. W. Hirs and S. N. Timasheff, Eds.), Academic Press, New York, 1979.
4. Fuller, H. Q., Fuller, R. M., and Fuller, R. G., "Physics Including Human Applications," Harper and Row, New York, 1978. See pp. 182 and 690-691.
5. Anderson, H. L. (Ed.), "Physics Vade Mecum," American Institute of Physics, New York, 1981.
6. "Handbook of Chemistry and Physics," 64th Edition, The Chemical Rubber Company, Boca Raton, FL, 1982.

Cross-references: ATOMIC PHYISCS; COMPRESSIBILITY, GAS; EXPANSION, THERMAL; GAS LAWS; MASS AND INERTIA.

DIAMAGNETISM

Magnetic susceptibility is defined as $\chi_m = M/H$, where M is the magnetic moment per gram (gram susceptibility) or per mole (mole susceptibility) that is induced by an external magnetic field strength H. If $M > 0$, the susceptibility is *paramagnetic*; if $M < 0$, the susceptibility is *diamagnetic*. Whereas most magnetic phenomena, including PARAMAGNETISM, are manifestations of ELECTRON SPIN, diamagnetism reflects electron angular momentum.

If an external field strength H is applied to a conductor so as to change the number of lines of flux that thread through it, there is induced in the conductor an electric current whose associated magnetic field opposes the change (*Faraday's Law of Induction* and *Lenz's Law*). In most conductors the current I that is thus induced is rapidly dissipated as heat through the I^2R loss, where R is the electrical resistance. These currents are known as *eddy currents*, and they are of great practical interest in ac applications. However, such transients do not influence the dc measurement of χ_m. There are three other classes of electron-momentum change induced by H that are not dissipated: electron currents in SUPERCONDUCTORS, where the resistance is $R = 0$; currents of atomic dimension induced in atoms or molecules or the atomic "core" electrons of solids; and microscopic conduction-electron helical currents having quantized helical radii.

In a superconductor, switching on of an H induces eddy currents that permanently shield the inside of the conductor from penetration by the magnetic-field lines. Therefore the superconductor is an ideal diamagnet, except for a small skin depth at the surface. If a superconductor is cooled through the normal-conducting $\rightleftharpoons$superconducting transition temperature in the presence of H and after the eddy currents induced in the normal-conducting state have been dissipated, the field lines are rapidly expelled from the superconductor (*Meissner effect*). This proves that the ideally diamagnetic state is thermodynamically stable.

An external field H superposes on the motion of atomic or molecular electrons (or the atomic core electrons in solids) a common circular motion about H of angular frequency $\omega_L = eH/2mc$, where e/m is the electronic charge-to-mass ratio (*Larmor's theorem*). This atomic current produces an atomic moment that is proportional to the square of the distance of a classical electron from the nucleus, $r_i \sim 1\text{Å}$. Therefore the diamagnetic contribution from electrons localized about an atomic nucleus is

$$\chi_m{}^{core} = - (Ne^2/6mc^2) \sum_i \overline{r_i{}^2},$$

where N is the number of atoms per gram (or mole).

In addition to macroscopic eddy currents, conduction electrons tend to move in microscopic helical paths in the presence of an H. The contribution to χ_m from this helical motion is a purely quantum mechanical effect. The radii of the H-induced helical paths are quantized, which leads to a "bunching" of the energy levels within an energy band of conducting states, and at large H these "bunches" can be resolved. They are known as *Landau levels* because Landau[1] first presented the quantum mechanical theory of conduction-electron diamagnetism, which for single parabolic energy bands gives

$$\chi_m{}^{cond} = - \frac{2}{3} \mu_B{}^2 N(E_F),$$

where μ_B is the *Bohr magneton* and $N(E_F)$ is the density of energy levels at the *Fermi energy* E_F. Since $N(E_F)$ oscillates with H as successive Landau levels pass through E_F, $\chi_m{}^{cond}$ shows oscillations in large H (DE HAAS-VAN ALPHEN EFFECT). Transitions between Landau levels, which are split by an energy $\hbar\omega_p = eH/m^*c$, may be induced by an electromagnetic field of angular frequency ω_p. This gives rise to resonance power absorption as ω passes through ω_p (*cyclotron or diamagnetic resonance*). These two effects are used to map out the contours in momentum space of the Fermi energies in metals.

JOHN B. GOODENOUGH

Reference

1. Landau, L. D., *Z. Physik*, **64**, 629 (1930).

Cross-references: DE HAAS-VAN ALPHEN EFFECT; MAGNETISM; SUPERCONDUCTIVITY.

DIELECTRIC THEORY

A dielectric is a material having electrical conductivity low in comparison to that of a metal. It is characterized by its dielectric constant and dielectric loss, both of which are functions of frequency and temperature. The dielectric constant is the ratio of the strength of an electric field in a vacuum to that in the dielectric for the same distribution of charge. It may also be defined and measured as the ratio of the capacitance C of an electrical condenser filled with the dielectric to the capacitance C_0 of the evacuated condenser:

$$\epsilon = C/C_0$$

The increase in the capacitance of the condenser is due to the polarization of the dielectric material by the applied electric field. Since the dielectric constant is not a constant, it is frequently called the "dielectric permittivity." The relative permittivity or dielectric constant is the ratio ϵ/ϵ_0, where ϵ_0 is the permittivity or dielectric constant of free space. In the mks system of units, the dielectric constant of free space is 8.854×10^{-12} farad/m, while in the esu system the relative and the absolute dielectric constants are the same. The relative dielectric constant, which is dimensionless, is the one commonly used. When variation of the dielectric constant with frequency may occur, the symbol is commonly primed. When a condenser is charged with an alternating current, loss may occur because of dissipation of part of the energy as heat. In vector notation, the angle δ between the vector for the amplitude of the charging current and that for the amplitude of the total current is the loss angle, and the loss tangent, or dissipation factor, is

$$\tan \delta = \frac{\text{Loss current}}{\text{Charging current}} = \frac{\epsilon''}{\epsilon'}$$

where ϵ'' is the loss factor, or dielectric loss, of the dielectric in the condenser and ϵ' is the measured dielectric constant of the material.

At low frequencies of the alternating field, the dielectric loss is normally zero and ϵ' is indistinguishable from the dielectric constant ϵ_{dc} measured with a static field. Debye has shown that

$$\frac{\epsilon_{dc} - 1}{\epsilon_{dc} + 2} = \frac{4\pi N_1}{3}\left(\alpha_0 + \frac{\mu^2}{3kT}\right) \qquad (1)$$

where N_1 is the number of molecules or ions per cubic centimeter; α_0 is the molecular or ionic polarizability, i.e., the dipole moment induced per molecule or ion by unit electric field (1 esu = 300 volts/cm); μ is the permanent dipole moment possessed by the molecule; k is the molecular gas constant, 1.38×10^{-16}, and T is the absolute temperature. An electric dipole is a pair of electric charges, equal in size, opposite in sign, and very close together. The dipole moment is the product of one of the two charges by the distance between them.

In Eq. (1) $\mu^2/3kT$ is the average component in the direction of the field of the permanent dipole moment of the molecule. In order that this average contribution should exist, the molecules must be able to rotate into equilibrium with the field. When the frequency of the alternating electric field used in the measurement is so high that dipolar molecules cannot respond to it, the second term on the right of the above equation decreases to zero and we have what may be termed the optical dielectric constant ϵ_∞, defined by the expression

$$\frac{\epsilon_\infty - 1}{\epsilon_\infty + 2} = \frac{4\pi N_1}{3}\alpha_0 \qquad (2)$$

ϵ_∞ differs from n^2, the square of the optical refractive index for visible light, only by the small amount due to infrared absorption and to the small dependence of n on frequency, as given by dispersion formulas. It is usually not a bad approximation to use $\epsilon_\infty = n^2$. The general Maxwell relation $\epsilon' = n^2$ holds when ϵ' and n are measured at the same frequency. The Debye equation may be written in the form

$$\frac{\epsilon_{dc} - 1}{\epsilon_{dc} + 2} - \frac{\epsilon_\infty - 1}{\epsilon_\infty + 2} = \frac{4\pi N_1}{9kT}\mu^2 \qquad (3)$$

A much better representation of the dielectric behavior of polar liquids is given by the Onsager equation

$$\frac{\epsilon_{dc} - 1}{\epsilon_{dc} + 2} - \frac{\epsilon_\infty - 1}{\epsilon_\infty + 2}$$
$$= \frac{3\epsilon_{dc}(\epsilon_\infty + 2)}{(2\epsilon_{dc} + \epsilon_\infty)(\epsilon_{dc} + 2)}\frac{4\pi N_1 \mu^2}{9kT} \qquad (4)$$

Kirkwood developed an equation differing from that of Onsager (Eq. 4) in that μ^2 is multiplied by a correlation parameter g in an attempt to account for the hindrance of the rotational orientation of dipolar molecules by their neighbors. The departure of the value of g from unity is a measure of the hindrance to molecular rotation by short-range intermolecular forces. Positive deviations of g from unity occur when short-range hindering torques favor parallel orientation of neighboring dipolar molecules, while negative deviations occur when these torques favor antiparallel orientation. Empirically determined values of g are not far from unity for normal or unassociated liquids, but may depart considerably from unity for abnormal or associated liquids. When $g = 1$, the Kirkwood and the Onsager equations are identical. Both equations contain the approximation involved in treating the dipolar molecules as spheres. Johari has been successful in treating the water in certain clathrate hydrates, in which he finds the dipole moment of H_2O increased by nearly 30% over its vapor phase value.

An equation derived by Debye for the change of dielectric constant or permittivity $\Delta\epsilon$ produced by application of an electric field of intensity E to a liquid or gas

$$\Delta\epsilon = \frac{-4\pi N_1 \mu^4 E^2}{45(kT)^3}\frac{(\epsilon + 2)^4}{81} \qquad (5)$$

showed that any departure from linear dependence of polarization upon the intensity of the applied field was too small to be evident at the field intensities (<100 volts cm^{-1}) normally used in measurement of ϵ, but at higher applied fields in the kilowatt region and, especially with molecules of large dipole moments, departure from linearity was sufficient to give significant values of $\Delta\epsilon$. The important work of A. H. Piekara and his coworkers in this area has been reviewed and discussed by Davies (Hill, Vaughan, Price, and Davies, 1969) and more recently by Davies and by Piekara himself (Neel, L., Ed.). The Onsager treatment has also been introduced to change the factor $(\epsilon + 2)^4/81$ in Eq. (5). Several simple liquids, such as ethyl ether, chloroform, and chlorobenzene, the molecules of which have moderate-sized dipole moments, were found to show the predicted decreases in dielectric constant when a strong electric field was applied. However, in the case of the more polar nitrobenzene and 1,2- and 1,3-nitrotoluene, positive values of $\Delta\epsilon$ were observed, changing to negative on dilution with a nonpolar solvent, which presumably reduced or destroyed pairwise orientation of the dipolar molecules. These and more or less similar changes in dielectric constant or permittivity produced by application of a high electric field are called *nonlinear dielectric* or *saturation effects*, although the degree of saturation attainable before dielectric breakdown occurs is normally small. An exception cited by Davies is a dioxan solution of poly(γ-benzyl-L-glutamate), which has an enormous molecular dipole moment of about 2000 D. At a field intensity of about 40 kV cm⁻¹, the dielectric constant of the solution is reduced by an amount $\Delta\epsilon = -0.10$ to equal that of the pure solvent, which means that virtually complete dielectric saturation has occurred. Liszi, Mészáros, and Ruff have developed an equation for the field dependence of the dielectric constant or permittivity of liquids taking into account a field dependence of the Kirkwood correlation parameter g. This equation predicts both normal and anomalous dielectric saturation and the structure making and breaking effects of the field.

Anomalous dielectric dispersion occurs when the frequency of the field is so high that the molecules do not have time to attain equilibrium with it. One may then use a complex dielectric constant

$$\epsilon^* = \epsilon' - i\epsilon'' \qquad (6)$$

where $j = \sqrt{-1}$. Debye's theory of dielectric behavior gives

$$\epsilon^* = \epsilon_\infty + \frac{\epsilon_{dc} - \epsilon_\infty}{1 + j\omega\tau} \qquad (7)$$

where ω is the angular frequency (2π times the number of cycles per second) and τ is the dielectric relaxation time. Dielectric relaxation is the decay with time of the polarization when the applied field is removed. The relaxation time is the time in which the polarization is reduced to $1/e$ times its value at the instant the field is removed, e being the natural logarithmic base.

Combination of the two equations for the complex dielectric constant and separation of real and imaginary parts gives

$$\epsilon' = \epsilon_\infty + \frac{\epsilon_{dc} - \epsilon_\infty}{1 + \omega^2\tau^2} \qquad (8)$$

$$\epsilon'' = \frac{(\epsilon_{dc} - \epsilon_\infty)\omega\tau}{1 + \omega^2\tau^2} \qquad (9)$$

These equations require that the dielectric constant decrease from the static to the optical dielectric constant with increasing frequency, while the dielectric loss changes from zero to a maximum value ϵ''_m and back to zero. These changes are the phenomenon of anomalous dielectric dispersion. From the above equations,

it follows that

$$\epsilon_m'' = (\epsilon_{dc} - \epsilon_\infty)/2 \qquad (10)$$

and that the corresponding values of ω and ϵ' are

$$\omega_m = 1/\tau \qquad (11)$$

and

$$\epsilon_m' = (\epsilon_{dc} + \epsilon_\infty)/2 \qquad (12)$$

The symmetrical loss-frequency curve predicted by this simple theory is commonly observed for simple substances, but its maximum is usually lower and broader because of the existence of more than one relaxation time. Various functions have been proposed to represent the distribution of relaxation times. A convenient representation of dielectric behavior is obtained, according to the method of Cole and Cole, by writing the complex dielectric constant as

$$\epsilon^* = \epsilon_\infty + \frac{\epsilon_{dc} - \epsilon_\infty}{1 + (j\omega\tau_0)^{1-\alpha}} \qquad (13)$$

where τ_0 is the most probable relaxation time and α is an empirical constant with a value between 0 and 1, usually less than 0.2. When the values of ϵ'' are plotted as ordinates against those of ϵ' as abscissas, a semicircular arc is obtained intersecting the abscissa axis at $\epsilon' = \epsilon_\infty$ and $\epsilon' = \epsilon_{dc}$. The center of the circle of which this arc is a part lies below the abscissa axis, and the diameter of the circle drawn through the center from the intersection at ϵ_∞ makes an angle $\alpha\pi/2$ with the abscissa axis. When α is zero, the diameter lies in the abscissa axis, there is but one relaxation time, and the behavior of the material conforms to the simple Debye theory. When, as may arise from intramolecular rotation, a substance has more than one relaxation mechanism, or, when the material is a mixture, the observed loss-frequency curve is the resultant of two or more different curves and, therefore, departs from the simple Debye or Cole-Cole curve.

The behavior of a good many materials which have been found, mostly at low temperatures, to depart from the symmetrical arc given by Eq. (12) has been well represented by the empirical equation of Cole and Davidson, in which the exponent $1 - \alpha$ in Eq. (13) is replaced by β. The corresponding curve is a skewed arc and the empirical parameter β, which has values between 0 and 1, measures the degree of skewness. Glarum has accounted for skewed-arc behavior in terms of the diffusion of lattice defects. Anderson and Ullman have treated the reorientation probability of a molecule as a function of the free volume, which fluctuates as the result of random thermal motion. If the rate of free-volume fluctuation is slow compared with that of molecular reorientation, the Cole-Cole plot is symmetrical, but flatter, the greater the dependence of the relaxation rate on the free volume. When the rate of change of free volume is much greater than that of reorientation, all of the molecules have the same environment and a single relaxation time should be observed. For an intermediate situation a skewed-arc plot is to be expected. These and other approaches to the problems of dielectric behavior have been described by Hill, Vaughan, Price, and Davies. Nee and Zwanzig have formulated a theory of dielectric relaxation involving dielectric friction on the rotating dipole, which leads to a frequency-dependent relaxation time. Frequency-dependent dynamic viscosity has been used by Johari and Smyth to explain an apparent wide distribution of the relaxation times of supercooled solutions of rigid polar molecules as well as the seemingly low dielectric relaxation times found for polar molecules at high frequencies or high viscosities. The complexity of molecular behavior in condensed matter is such that no molecular theory of dielectrics is completely satisfactory at the present time.

If the dielectric material is not a perfect dielectric, and has a specific dc conductance k' (ohms^{-1} cm^{-1}), there is an additional dielectric loss

$$\epsilon_{dc}'' = \frac{3.6 \times 10^{12} \pi k'}{\omega} \qquad (14)$$

The effective specific conductance is given by

$$k' = \frac{1}{4\pi} \frac{(\epsilon_{dc} - \epsilon_\infty)\omega^2 \tau}{1 + \omega^2 \tau^2} \qquad (15)$$

It is evident from this equation that k' increases with ω, approaching a limiting value, k_∞, the infinite-frequency conductivity, which is attained when 1 can be neglected in comparison with $\omega^2 \tau^2$, so that

$$k_\infty = \frac{\epsilon_{dc} - \epsilon_\infty}{4\pi\tau} \qquad (16)$$

In a heterogeneous material, interfacial polarization may arise from the accumulation of charge at the interfaces between phases. This occurs only when two phases differ considerably from each other in dielectric constant and conductivity. It is usually observed only at very low frequencies, but, if one phase has a much higher conductivity than the other, the effect may increase the measured dielectric constant and loss at frequencies as high as those of the radio region. This so-called Maxwell-Wagner effect depends on the form and distribution of the phases as well as upon their real dielectric constants and conductances. Each type of form and distribution requires special treatment. For a commercial rubber, for example, the observed loss may be

$$\epsilon''(\text{observed}) = \epsilon_{dc}'' + \epsilon''(\text{Maxwell-Wagner})$$
$$+ \epsilon''(\text{Debye}) \qquad (17)$$

CHARLES P. SMYTH

References

Böttcher, C. J. F., "Theory of Electric Polarization," Second edition, revised by O. C. Van Belle, P. Bordewijk and A. Rip, Vol. I, New York, Elsevier, 1973.

Debye, P., "Polar Molecules," reprinted by Dover, New York, 1945.

Fröhlich, H., "Theory of Dielectrics," Second edition, London, Oxford University Press, 1958.

Hill, N. E., Vaughan, W. E., Price, A. H., and Davies, M., "Dielectric Properties and Molecular Behavior," New York, Van Nostrand Reinhold Co., 1969.

Johari, G. P. and Smyth, C. P., *J. Am. Chem. Soc.* **91**, 5168 (1969); *J. Chem. Phys.* **56**, 4411 (1972).

Johari, G. P., *J. Chem. Phys.* **74**, 1326 (1981).

Kirkwood, J. G., *J. Chem. Phys.* **7**, 911 (1939).

Liszi, J., Mészáros, L., and Ruff, I., *J. Chem. Phys.* **74**, 6896 (1981).

Nee, T. W. and Zwanzig, R., *J. Chem. Phys.* **52**, 6353 (1970).

Neel, L. (Ed.), "Nonlinear Behavior of Molecules, Atoms and Ions in Electric, Magnetic or Electromagnetic Fields" (Proc. Soc. Chim. Phys.), Amsterdam, Elsevier Scientific Publishing Company, 1979, pp. 301–353.

Onsager, L., *J. Am. Chem. Soc.* **58**, 1486 (1936).

Scaife, B. K. P. (Compiler), "Complex Permittivity," London, The English Universities Press Ltd., 1971.

Smyth, C. P., "Dielectric Behavior and Structure," New York, McGraw-Hill Book Co., 1955; *Ann. Rev. Phys. Chem.* **17**, 433–456 (1966).

Smyth, C. P., "Molecular Interactions" (Ratajczak, H. and Orville-Thomas, W. J., Eds.), Vol. 2, Chap. 7, John Wiley and Sons, Ltd., London, 1981.

Cross-references: CAPACITANCE, DIPOLE MOMENTS, POLAR MOLECULES, REFRACTION, RELAXATION.

DIFFERENTIAL EQUATIONS IN PHYSICS

Because of their high frequency of occurrence and importance in the physical sciences it is fitting to introduce some of the more common differential equations that arise and to use these as a basis for discussion and development.

(i) When a given mass of a radioactive substance disintegrates it is well known that if at any time t the mass remaining is m, then the rate of decay of mass, $-dm/dt$, is proportional to the amount remaining. This implies that

$$-\frac{dm}{dt} = km \qquad (k = \text{positive constant}). \qquad (1)$$

(ii) Suppose that an alternating voltage $E \cos \omega t$ is applied to an electrical circuit consisting of a resistance R, inductance L, and capacitance C in series connection, t specifying time. If i denotes the current flowing in the circuit at time t and q the charge on the capacitance plate into which it flows, then $i = dq/dt$ and so the potential differences across the three components are respectively Ri, $L\,di/dt$, q/C. Equating their sum to the applied voltage gives

$$R\frac{dq}{dt} + L\frac{d^2q}{dt^2} + \frac{q}{C} = E\cos\omega t. \qquad (2)$$

(iii) When heat is conducted along a metal bar the temperature $u(x, t)$ at time t at a distance x from one end is known to satisfy

$$\frac{\partial u}{\partial t} = k\frac{\partial^2 u}{\partial x^2}, \qquad (3)$$

in the case when there are no radiation losses. For homogeneous material and constant cross section, k is a positive constant.

(iv) If $y(x, t)$ denotes the lateral displacement at time t of a point distant x from one of the fixed ends of a string set in vibration, then it can be shown that

$$\frac{\partial^2 y}{\partial t^2} = c^2\frac{\partial^2 y}{\partial x^2}, \qquad (4)$$

where c is a constant for a uniform string of constant cross section.

The equations (1) through (4) are examples of *differential equations*. In each case it is seen that a relation exists between a quantity or function whose value is sought, i.e., the *dependent variable*, one or more *independent variables*, and the derivatives of the dependent variables with respect to the independent ones. Equations (1), (2) are examples of *ordinary differential equations* and they involve only total derivatives as there is but one independent variable. Equations (3), (4), however, involve partial derivatives and are called *partial differential equations*. Partial differential equations involve two or more independent variables. In both cases the *order* of the differential equation is that of the highest-order derivative it contains. (1) is of first order, but (2), (3), (4) are all of second order.

In many physical problems which are formulated as differential equations auxiliary conditions are imposed upon the dependent variable and possibly also on certain of its derivatives. These conditions compounded with the partial differential equation constitute a *boundary value problem*. If, as often happens, such conditions are prescribed at $t = 0$, where t is an independent variable specifying time, the compound problem is called an *initial value problem*.

(1) Ordinary Differential Equations We now consider the ordinary differential equation of order n

$$a_n\frac{d^n y}{dx^n} + a_{n-1}\frac{d^{n-1}y}{dx^{n-1}} + \cdots + a_1\frac{dy}{dx} + a_0 y = f(x)$$

or

$$L(D)y = f(x) \qquad (5)$$

where $L(D) \equiv a_n D^n + a_{n-1}D^{n-1} + \cdots + a_1 D + a_0$ and $D^r \equiv d^r/dx^r$. The coefficients $a_0, a_1, \cdots, a_n$ may be either constants or functions of x only. Note that y and all its derivatives in (5) occur only to the power unity and that there are no products of these quantities. Such an equation is said to be *linear*. Suppose that (5) is satisfied by the particular value $y = Y(x)$ so that $L(D)Y(x) = f(x)$. Then $Y(x)$ is called a *particular integral* of (5). Further suppose that the equation $L(D)y = 0$ has n linearly independent solutions $y_1(x), y_2(x), \cdots, y_n(x)$. Then it is easily seen that

$$L(D)(A_1 y_1 + A_2 y_2 + \cdots + A_n y_n) = 0,$$

where $A_1, \cdots, A_n$ are arbitrary constants, and so $y = A_1 y_1 + \cdots + A_n y_n$ satisfies the homogeneous equation $L(D)y = 0$. This solution of the homogeneous equation is called the *complementary function* of the nonhomogeneous equation (5). The number of constants in it is equal to the order of the differential equation. Since

$$L(D)\{Y(x) + A_1 y_1 + A_2 y_2 + \cdots + A_n y_n\}$$
$$= L(D)Y + 0 = f(x),$$

Eq. (5) is satisfied by

$$y = Y(x) + A_1 y_1(x) + \cdots + A_n y_n(x). \qquad (6)$$

This form can be shown to be the most general form of solution to (5) and it is called the *complete primitive*. Thus the task of finding the complete primitive to (5) consists of finding the complementary function, obtained by taking $f(x) = 0$, and adding to this a particular integral.

For the special case when the coefficients $a_0, a_1, \cdots, a_n$ in (5) are all constants—and this assumption will be made from now on—the general solution to the equation $L(D)y = 0$ is obtained by making the trial substitution $y = e^{mx}$, where m is an undetermined constant. We find

$$a_n m^n + \cdots + a_1 m + a_0 = 0. \qquad (7)$$

Let $m = m_1, m_2, \cdots, m_n$ be the n roots of the *auxiliary equation* (7). Then $L(D)y = 0$ has the n solutions $e^{m_1 x}, e^{m_2 x}, \cdots, e^{m_n x}$ and it can be shown that its most general solution is

$$y = A_1 e^{m_1 x} + A_2 e^{m_2 x} + \cdots + A_n e^{m_n x}, \qquad (8)$$

where $A_1, \cdots, A_n$ are constants and the m's are all distinct. As an example the differential equation (1) is a first-order homogeneous equation, and by putting $m = e^{\lambda t}$ we find $\lambda = -k$. Thus the general solution is $m = Ae^{-kt}$, involving but one arbitrary constant. If it is given that $m = m_0$ when $t = 0$, we find $A = m_0$ and so the solution of this initial value problem is $m = m_0 e^{-kt}$.

By way of further example, consider the second-order homogeneous equation

$$\frac{d^2 y}{dx^2} + (a + b)\frac{dy}{dx} + ab\,y = 0$$

$(a, b$ unequal constants$)$.

Putting $y = e^{mx}$ gives the auxiliary equation $m^2 + (a + b)m + ab = 0$ having roots $m = -a, -b$. Hence the general solution is

$$y = Ae^{-ax} + Be^{-bx} \qquad (a \neq b).$$

This form is suggested immediately if the differential equation is written in the operator form

$$(D + a)(D + b)y = 0 \qquad (a \neq b),$$

since $y = Ae^{-ax}$ and $y = Be^{-bx}$ satisfy $(D + a)y = 0$ and $(D + b)y = 0$ respectively. In the case when $a = b$, it can easily be verified that xe^{-ax} is a second solution to $(D + a)^2 y = 0$ and so its general solution is

$$y = Ae^{-ax} + Bxe^{-ax}.$$

For the general solution of (5), it follows that the main task is the finding of a particular integral. Various techniques are available. Here we describe two methods using suitable examples.

The first is called the method of *variation of parameters*. Suppose it is required to solve for $y(t)$ the differential equation

$$\frac{d^2 y}{dt^2} + \omega^2 y = \cos \omega t. \qquad (9)$$

This is the resonance equation and we observe that when $R = 0$ and $\omega^2 = 1/LC$, Eq. (2) is of this form. The complementary function, found from the auxiliary equation $m^2 + \omega^2 = 0$ is clearly $y = ae^{i\omega t} + be^{-i\omega t}$. This may be more simply expressed in the form $y = A \cos \omega t + B \sin \omega t$ using the relations $e^{\pm i\omega t} = \cos \omega t \pm i \sin \omega t$, the constants A, B being given by $A = a + b$, $B = i(a - b)$. To obtain a particular integral of (9) take

$$y(t) = A(t) \cos \omega t + B(t) \sin \omega t; \qquad (10)$$

then

$$\dot{y}(t) = -\omega A(t) \sin \omega t + \omega B(t) \cos \omega t, \qquad (11)$$

provided A and B are so chosen that

$$0 = \dot{A}(t) \cos \omega t + \dot{B}(t) \sin \omega t. \qquad (12)$$

Differentiating $\dot{y}$ in (11),

$$\ddot{y}(t) = -\omega^2 A(t) \cos \omega t - \omega^2 B(t) \sin \omega t, \qquad (13)$$

provided A and B are so chosen that

$$\cos \omega t = -\omega \dot{A}(t) \sin \omega t + \omega \dot{B}(t) \cos \omega t. \qquad (14)$$

Solving equations (12), (14) for $\dot{A}$ and $\dot{B}$ we find

$$\dot{A}(t) = -\frac{1}{\omega} \sin \omega t \cos \omega t, \quad \dot{B}(t) = \frac{1}{\omega} \cos^2 \omega t$$

$$= \frac{1}{2\omega}(1 + \cos 2\omega t)$$

and so integrating these,

$$A(t) = -\frac{1}{2\omega^2} \sin^2 \omega t,$$

$$B(t) = \frac{1}{2\omega}\left(t + \frac{1}{2\omega} \sin 2\omega t\right).$$

Thus on substituting these into (10) and simplifying we obtain the particular integral $y = (t/2\omega) \sin \omega t$. Hence the complete primitive is

$$y = A \cos \omega t + B \sin \omega t + \frac{t}{2\omega} \sin \omega t$$

$(A, B$ constants$)$.

As a second method we introduce the *Laplace transform*. This device is eminently suited to solving initial value problems and it is consequently favored by electrical and control engineers. The Laplace transform of $f(t)$ is denoted either by $\mathcal{L}\{f(t)\}$ or by $\bar{f}(s)$ and it is defined to be

$$\mathcal{L}\{f(t)\} = \bar{f}(s) = \int_0^\infty e^{-st} f(t)\, dt.$$

By elementary integration one can compile the following useful table of Laplace transforms for different forms of $f(t)$:

$f(t)$	1	e^{at}	$\cos at$	$\sin at$	t^n
$\bar{f}(s)$	$\dfrac{1}{s}$	$\dfrac{1}{s-a}$	$\dfrac{s}{s^2+a^2}$	$\dfrac{a}{s^2+a^2}$	$\dfrac{n!}{s^{n+1}}$

Here a is constant and $n = 1, 2, 3, \cdots$. Also it is easy to show by integration by parts that

$$\mathcal{L}\left\{\frac{df(t)}{dt}\right\} = s\,\bar{f}(s) - f(0),$$

$$\mathcal{L}\left\{\frac{d^2 f(t)}{dt^2}\right\} = s^2\bar{f}(s) - s\,f(0) - \dot{f}(0).$$

Let us use this method to solve again equation (9), subject to the initial conditions $y(0) = 0$, $\dot{y}(0) = 0$. Let $\bar{y}(s) = \mathcal{L}\{y(t)\}$. Then, using the initial conditions, $\mathcal{L}\{dy/dt\} = s\,\bar{y}(s)$, $\mathcal{L}\{d^2 y/dt^2\} = s^2\bar{y}(s)$. Also the table shows that $\mathcal{L}\{\cos \omega t\} = s/(s^2 + \omega^2)$. Hence on taking the Laplace transform of both sides of (9),

$$(s^2 + \omega^2)\bar{y}(s) = s/(s^2 + \omega^2)$$

and so

$$\bar{y}(s) = s/(s^2 + \omega^2)^2.$$

Now

$$\frac{s}{s^2 + \omega^2} = \int_0^\infty e^{-st} \cos \omega t\, dt,$$

and so differentiating both sides of this partially with respect to ω,

$$-\frac{2\omega s}{(s^2 + \omega^2)^2} = \int_0^\infty \frac{\partial}{\partial \omega} (e^{-st} \cos \omega t) \, dt,$$

$$= \int_0^\infty (-t \, e^{-st} \sin \omega t) \, dt.$$

Dividing through by -2ω, we have

$$\frac{s}{(s^2 + \omega^2)^2} = \int_0^\infty e^{-st} \left(\frac{t}{2\omega} \sin \omega t \right) dt.$$

Thus

$$\bar{y}(s) = \mathcal{L} \left\{ \frac{t}{2\omega} \sin \omega t \right\},$$

and so

$$y(t) = \frac{t}{2\omega} \sin \omega t.$$

(2) Partial Differential Equations The simplest methods of solution are based on the technique of *separation of variables* and on the use of an *integral transform* such as the Laplace transform. These are illustrated by means of examples.

Suppose it is desired to solve the one-dimensional heat conduction equation (3) holding along a uniform rod of length ℓ, being given that when $t = 0$, $u = u_0 x$ and, for all $t \geqslant 0$, $\partial u / \partial x = 0$ at $x = 0$ and $x = \ell$ (i.e., both ends are thermally lagged). Making the trial solution $u = X(x) T(t)$, we find

$$X''(x)/X(x) = T'(t)/kT(t).$$

As the left side of this equation is a function of x only and the right one of t only, each is constant. The physical nature of the problem implies that as t increases, u decreases for any particular value of x. Hence T decreases as t increases and so $T' < 0$. Thus we take the constant to be negative, say $-m^2$, and obtain the ordinary differential equations

$$\begin{cases} X''(x) + m^2 X(x) = 0, \\ T'(t) + m^2 kT(t) = 0. \end{cases}$$

These have the general solutions

$$\begin{cases} X(x) = A \cos mx + B \sin mx, \\ T(t) = C \exp (-m^2 kt). \end{cases}$$

Thus, writing $E = AC$, $F = BC$, a solution for $u(x, t)$ is

$$u(x, t) = XT$$

$$= \exp (-m^2 kt)(E \cos mx + F \sin mx).$$

Since, for all t, $\partial u / \partial x = 0$ when $x = 0$ and when $x = \ell$,

$$F \, m \, \exp (-m^2 kt) = 0,$$
$$m \, \exp (-m^2 kt)(-E \sin m\ell + F \cos m\ell) = 0.$$

The first of these equations implies that $F = 0$ and the second that $m\ell = n\pi$, where n is an integer. Thus a solution satisfying the end conditions is

$$u(x, t) = E \exp (-n^2 \pi^2 kt/\ell^2) \cos (n\pi x/\ell),$$

where n is an integer. A more general solution may be obtained by superposition (as (3) is linear) in the form

$$u(x, t) = \sum_{n=0}^\infty E_n \exp (-n^2 \pi^2 kt/\ell^2) \cos (n\pi x/\ell)$$

Using the initial condition $u(x, 0) = u_0 x$, the last form leads to the half-range Fourier series representation

$$\sum_{n=0}^\infty E_n \cos (n\pi x/\ell) = u_0 x \quad \text{for} \quad 0 < x < \ell.$$

Determining the coefficients by the usual method,

$$E_0 = \frac{1}{\ell} \int_0^\ell u_0 x \, dx = \frac{1}{2} u_0 \ell,$$

and for $n = 1, 2, 3, \cdots$,

$$E_n = \frac{2}{\ell} \int_0^\ell u_0 x \cos \left(\frac{n\pi x}{\ell} \right) dx$$

$$= -\frac{2u_0 \ell}{n^2 \pi^2} [1 - (-1)^n].$$

Thus the solution to the boundary value problem is

$$u(x, t) = \frac{1}{2} u_0 \ell$$

$$-\frac{4u_0 \ell}{\pi^2} \sum_{n=1}^\infty \frac{\exp \{-(2n - 1)^2 \, k\pi^2 t/\ell^2\}}{(2n - 1)^2}$$

$$\cdot \cos \frac{(2n - 1)\pi x}{\ell},$$

where $t \geqslant 0$ and $0 \leqslant x \leqslant \ell$. It can be shown that the solution is unique.

To illustrate the Laplace transform technique, consider the boundary value problem of solving for $f(x, y)$ the partial differential equation

$$\frac{\partial^2 f}{\partial x \partial y} = f(x, y)$$

in the region $x \geqslant 0$, $y \geqslant 0$ and subject to the boundary conditions

$$\begin{cases} f(x, 0) = 0, \\ f(0, y) = a \ (= \text{const.}). \end{cases}$$

The first boundary condition suggests taking the Laplace transform with respect to y. To this end write

$$\bar{f}(x, s) = \int_0^\infty e^{-sy} f(x, y) \, dy.$$

Then

$$\mathcal{L}\{\partial f/\partial y\} = s \bar{f}(x, s) - f(x, 0) = s \bar{f}(x, s),$$

and

$$\mathcal{L}\left\{\frac{\partial^2 f}{\partial x \partial y}\right\} = \frac{\partial}{\partial x} \mathcal{L}\left\{\frac{\partial f}{\partial y}\right\} = s \frac{d\bar{f}(x, s)}{dx}.$$

Thus the partial differential equation transfor: into

$$\frac{d\bar{f}(x, s)}{dx} = \frac{1}{s} \bar{f}(x, s)$$

and this ordinary differential equation has the general solution

$$\bar{f}(x, s) = \bar{f}(0, s) \exp (x/s).$$

Since $f(0, y) = a$, $\bar{f}(0, s) = a/s$ and so

$$\bar{f}(x, s) = \frac{a}{s} \exp\left(\frac{x}{s}\right)$$

$$= a\left[\frac{1}{s} + \frac{x}{s^2} + \frac{x^2}{2! \, s^3} + \frac{x^3}{3! \, s^4} + \cdots\right].$$

From the table, $\mathcal{L}\{y^n\} = n!/s^{n+1}$ for positive i tegral n, and so

$$f(x, y) = a\left[1 + xy + \frac{x^2 y^2}{(2!)^2} + \frac{x^3 y^3}{(3!)^2} + \cdots\right].$$

In terms of the modified Bessel function this may also be written as $f(x, y) = aI_0(2\sqrt{xy})$.

FRANK CHORLTON

References

1. Carslaw, H. S., and Jaeger, J. C., "Operational Methods in Applied Mathematics," London, Oxford University Press, 1947.
2. Chorlton, F., "Boundary Value Problems in Physics and Engineering," New York, Van Nostrand Reinhold, 1969.
3. Courant, R., and Hilbert, D., "Methods of Applied Mathematics," New York, Interscience, 1962.
4. Churchill, R. V., "Fourier Series and Boundary Value Problems," New York, McGraw-Hill, 1941.
5. Forsyth, A. R., "Differential Equations," London, Macmillan, 1921.
6. Ince, E. L., "Ordinary Differential Equations," New York, Dover, 1956.
7. Piaggio, H. T. H., "Differential Equations," London, Bell, 1942.
8. Sneddon, I. N., "Elements of Partial Differential Equations," New York, McGraw-Hill, 1957.

Cross-references: CALCULUS IN PHYSICS, FOURIER ANALYSIS, LAPLACE TRANSFORM, MATHEMATICAL PHYSICS, MATHEMATICAL PRINCIPLES OF QUANTUM MECHANICS.

DIFFRACTION BY MATTER AND DIFFRACTION GRATINGS

According to the principle of Huygens (1629–1695) each point in the space which is touched by a wave gives rise to a spherical secondary wave, which again produces tertiary waves, and so on. Every wave interferes with the next one and quite generally gives rise to diffraction phenomena (see OPTICS, PHYSICAL). Such phenomena in the case of visible light first were observed by F. M. Grimaldi (1618–1663) and mathematically explained by J. Fresnel (1788–1827). G. R. Kirchhoff (1824–1887) gave the first exact mathematical solution of the scalar wave differential equation in terms of a boundary integral. If both the primary and the diffracted rays are parallel (e.g., small source, large distances between diffracting sample and source and detector), we have the experimental conditions of Fraunhofer (1787–1826). With two lenses L_1 and L_2, a collimator pinhole P in the focal plane of L_1, and a photographic plate F in the focal plane of L_2, this condition is fulfilled even for short distances (Fig. 1). Monochromatic light is produced by a Hg lamp with a Schott filter S and an aqueous solution of $CuSO_4$. If s_0 and s are unit vectors in the direction of the primary and the diffracted beam, and λ is the wavelength of the source, then the diffracted intensity I is proportional to

$$I(\mathbf{b}) \cong f_e^2 \, f_\theta^2 \, |R|^2 ; \, R(\mathbf{b}) = F(\rho) \qquad (1)$$

where

$$\mathbf{b} = \frac{\mathbf{s} - \mathbf{s}_0}{\lambda} \qquad (2)$$

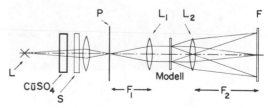

FIG. 1. Equipment for Fraunhofer diffraction.

and

$$F = \int e^{-2\pi i(\mathbf{bx})}\, dv_x \qquad (3)$$

is the symbol of the Fourier transform, $\mathbf{x}$ is a vector in physical space, $\rho^2(\mathbf{x})$ is the transparency of the object M at the end point of the vector $\mathbf{x}$, which lies in the plane of the object, dv_x is a surface element of the object, f_θ^2 and f_e^2 are explained in Table 1 and $\mathbf{b}$ is given by Eq. (2). Figure 2(a) shows an object $\rho(\mathbf{x})$ in the form of a parallelogram with the edge vectors $\mathbf{L_1}$ and $\mathbf{L_2}$, and Fig. 2(b) give its Fraunhofer pattern:

$$|R|^2 = |S|^2 ;$$

$$S(\mathbf{b}) = |\, \mathbf{L_1} \wedge \mathbf{L_2}\,| \frac{\sin \pi(\mathbf{bL_1})}{\pi(\mathbf{bL_1})} \cdot \frac{\sin \pi(\mathbf{bL_2})}{\pi(\mathbf{bL_2})} \qquad (4)$$

Fraunhofer used assemblies of N parallel oriented metal wires with an intermediate distance d and the distance a (from center to center), in the direction s_1. Hence

$$\rho(x_1) = \begin{cases} 1 \text{ for } na - \dfrac{d}{2} \le x_1 \le na + \dfrac{d}{2} \\ 0 \text{ for all other } x_1 \end{cases} \qquad (5)$$

and

$$I(b_1) = \frac{1}{a}\, f^2 \widehat{Z |S|^2} \qquad (6)$$

$$f = \frac{\sin \pi b_1 d}{\pi b_1}; \quad Z = \frac{1}{a} \sum_n^\infty P(b_1 - n/a);$$

$$S = \frac{\sin \pi b_1 Na}{\pi b_1} \qquad (7)$$

TABLE 1. FACTORS f_e^2 AND f_θ^2 FOR DIFFERENT RADIATIONS

		f_e^2	f_θ^2	
1	Visible light	$1/\lambda^2$	$\dfrac{1 + \cos^2 2\theta}{2}$	2θ scattering angle
2	X-rays	$\left(\dfrac{e^2}{m_0 c^2}\right)^2$	$\dfrac{1 + \cos^2 2\theta}{2}$	e electric charge of an electron m_0 rest mass of an electron
3	Electrons	$\dfrac{m_0 e^2 \lambda^2}{2h^2}$	$1/\sin^4 \theta$	c velocity of light h Planck's constant
4	Neutrons	Cross section	Polarization factor	λ (de Broglie) wavelength

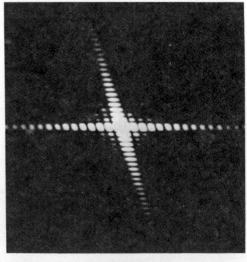

(a)   (b)

FIG. 2. (a) Parallelogram as diaphragm. (b) Fraunhofer pattern. Secondary maxima of the shape factor S^2 (Eq. (4)).

f is the Fourier transform of a single slit, S that of the shape of the whole lattice (length Na). Z is the "reciprocal" lattice point function (lattice factor) of the centers of the wire, since $P(b_1 - 0)$ is a normalized point function at $b_1 = 0$. The symbol of convolution $\frown$ is defined for both functions $G_1(b)$ and $G_2(b)$ in Fourier space, and $g_1(x)$ and $g_2(x)$ in physical space by

$$\widehat{G_1(b)G_2}(b) = \int G_1(c)G_2(b - c)dv_c \quad (8)$$

$$\widehat{g_1(x)g_2}(x) = \int g_1(y)g_2(x - y)dv_y \quad (9)$$

In the one-dimensional case of Eq. (8) c is to be replaced by the scalar quantity c_1 and dv_c by dc_1. Two-dimensional gratings are of high interest, since their Fraunhofer pattern gives all the information we need to understand the more complicated structural theories of three-dimensional matter. This will be shown below.

The technique of preparing such models is quite easy: in the examples of Figs. 4, 5, 6, 7, 8, 10 and 12, the objects were painted with india ink on paper 15×15 inches and then photographed on fine-grained films 0.5×0.5 inch with steep gradation characteristics (for instance Peruline film F 10). The black ink points now become transparent on a black background. In the models of Figs. 3, 9 and 11, steel balls of 2 to 3 mm diameter were placed into the focal plane of a lens system and photographed with a linear reduction $1:15$ on the same film material mentioned above.

By defocusing the system, one obtains diaphragms, where the single balls do not touch each other (Figs. 3 and 9, but not Fig. 11). This is quite advantageous, since the width of the atom-factor f^2 [see Eq. (6)] is larger and more reflections are visible.

Fig. 3(a) shows a model of a two-dimensional ideal periodic "point lattice," and Fig. 3(b) represents its Fourier transform.

$$\rho(x) = \sum_r^\infty P(x - x_r); \; x_r = p_1 a_1 + p_2 a_2 \quad (10)$$

$$R^2 \sim Z(b) = \frac{1}{|a_1 \wedge a_2|} \sum_h^\infty P(b - b_h);$$

$$b_h = h_1 A_1 + h_2 A_2 \quad (11)$$

where a_1 and a_2 are the vectors of a lattice cell of the model, and A_1 and A_2 are those of the "reciprocal" lattice cell (in b-space)

$$A_1 = \frac{a_2 \wedge a_3}{a_1(a_2 \wedge a_3)}; A_2 = \frac{a_3 \wedge a_1}{a_2(a_3 \wedge a_1)} \quad (12)$$

a_3 is an arbitrary vector orthogonal to a_1 and a_2, and p_1, p_2, h_1 and h_2 are integers. Since the lattice of Fig. 3(a) is bounded if one multiplies $\rho(x)$ by the shape function $s(x)$ of the lattice (which is 1 inside the lattice, and zero out of it) the p-summation can be extended to infinity. From the convolution theorem one obtains generally

$$Fg_1g_2 = \widehat{G_1 G_2}; F\widehat{g_1g_2} = G_1 G_2 \quad (13)$$

Hence the Fourier transform of the bounded lattice $\rho(x) \cdot s(x)$ is given again by the convolution product of Eq. (6), where $S(b)$ is the Fourier transform of $s(x)$ and $Z(b)$ is the Fourier transform of Eq. (11). $f^2 = 1$ applies only for "pointlike" atoms, otherwise f is the Fourier transform of the shape of each "point."

M. v. Laue in 1911 prepared an article "Wellenoptik" for the "Encyclopedie der Math. Wissen-

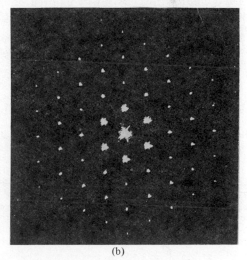

| (a) | (b) |

FIG. 3. (a) Steel balls in a crystalline lattice. (b) Bragg reflections with shape factor S^2.

schaften" and found that Eq. (11) can be easily developed to the Fourier transform of three-dimensional point lattice, where a_3 is a third lattice vector, non-coplanar to a_1 and a_2. Then b_h in Eq. (11) must be replaced by

$b_h = h_1 A_1 + h_2 A_2 + h_3 A_3;$

$$A_3 = \frac{a_1 \wedge a_2}{a_3(a_1 \wedge a_2)} \quad (14)$$

Together with W. Friedrich and P. Knipping using x-rays of a wavelength of the same order of magnitude as that of the atomic distances, M. v. Laue (1912) found three-dimensional diffraction effects in single crystals. $f(b)$ then is the atom form amplitude

$$f(\mathbf{b}) = F(\rho_0) \quad (15)$$

where $\rho_0(x)$ is the electron density distribution of one atom, whose center lies at $x = 0$.

C. J. Davisson and L. H. Germer (1927) observed the same diffraction phenomena using electron beams. $\rho(x)$ must then be understood as density distribution of both the electrons (negative) and protons (positive). If single diffraction processes occur, Eq. (6) remains unchanged.

D. P. Mitchell, P. N. Powers, H. v. Halban and P. Preiswerk (1936) found the same diffraction phenomena using thermal neutrons. In this case $\rho(x)$ is the density distribution of the nuclei, each one weighted by the mean of the square root of its respective cross section. The proportional factors f_e^2, f_θ^2 of Eq. (1) for the different radiations are given in Table 1.

The vector b defined by the integral of Eq. (3) expands the three-dimensional Fourier space and is connected with the unit vectors s, s_0 of the diffracted and primary beam by Eq. (2). This is the construction of Ewald (1914). For a fixed λ and s_0 all values $R(b)$ are in reflection positions, which lie on a sphere with radius $1/\lambda$ and the center at $b_0 = s_0/\lambda$.

P. Debye (1915) found that the molecules in the gaseous state give rise to diffraction patterns depending on the structure of the single molecules, without any intermolecular interferences.

In 1927, F. Zernike and I. A. Prins discussed quantitatively diffraction phenomena of liquids and laid down the fundamentals of the structure analysis of amorphous matter. Moreover, P. Debye (1913) and I. Waller (1927) found that real crystals never have a periodic lattice similar to that of Fig. 3(a), since the atoms show thermal oscillations around their ideal positions. All these different phenomena can be studied quite easily with the help of two-dimensional statistical models and their Fraunhofer patterns, since Eq. (1) holds for all diffraction phenomena. In Fig. 4(a) we have the "frozen" structure of a point lattice with thermal oscillations. They are quite anisotropic and occur only in the horizontal direction. If $H(x)$ is the frequency of the

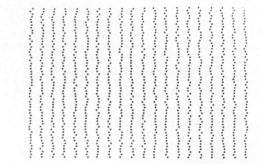

(a)

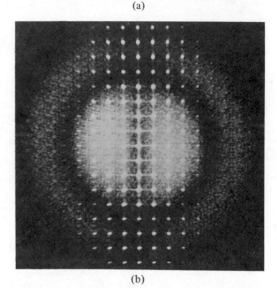

(b)

FIG. 4. (a) Linear thermal oscillations without correlations. (b) Debye factor and thermodiffuse background.

center of an atom being at the distance x from its ideal position and if the atoms oscillate independently from each other, then R^2 is given by

$$R^2(b) = Nf^2(1 - D^2) + \frac{1}{v_r} f^2 D^2 \widehat{Z|S|^2};$$

$$D(\mathbf{b}) = F(H) \quad (16)$$

(v_r = volume of a lattice cell). In Fig. 4(b) can be clearly recognized the first term of Eq. (16) as a diffuse background. It has a structure in the horizontal direction. The "Bragg reflections" are weakened by the "Debye-Waller factor" D^2, the more, the stronger is the diffuse background $(1 - D^2)$. In nature there exist correlations between the different oscillations and "elastic waves" with an "acoustical" and "optical" frequency creep through the lattice.

Fig. 4(a) gives a single undamped longitudinal and sinusoidal wave with a horizontal wave vector b_t and a wavelength λ_t and amplitude a_t

$$\lambda_t = \frac{1}{|b_t|} = 8a; \quad |a_t| = \frac{1}{4}a \qquad (17)$$

According to the theories of M. v. Laue (1927) and Laval (1941) in Fig. 5(b) at the reciprocal lattice points

$$\mathbf{b} = \mathbf{b}_h + m\mathbf{b}_t \qquad (18)$$

"Extra Laue spots" ("Laval spots") occur. According to conventional theories, $m = 1$ is called one-phonon scattering, $m = 2$ is two-phonon scattering and so on; and it is said that the frequency ν_0 of the incident radiation is changed into $\nu_0 + m\nu_t$ ("inelastic scattering," ν_t-frequency of the phonons).

In Fig. 6(a) we have again a horizontal wave with the same wave vector [Eq. (17)], which now is transverse. From Laue's theory it follows, that the intensity of a Laue spot of the mth order is proportional to the square of the Bessel function

$$I_m(2\pi(\mathbf{ba}_t))$$

where m is the order of the Bessel function and a_t is the amplitude vector of the elastic wave. Hence in Fig. 6(b), strong Laval spots occur only in the vertical $\mathbf{b}_h$ direction; in Fig. 5(b), in the horizontal $\mathbf{b}_h$ direction. Since $I_1 (I_2)$ has its maximum at $\mathbf{ba}_t = 0.25\ (0.5)$, in Figs. 5(b) and 6(b), the Laval spot $m = 1$ at the reflection h_1,

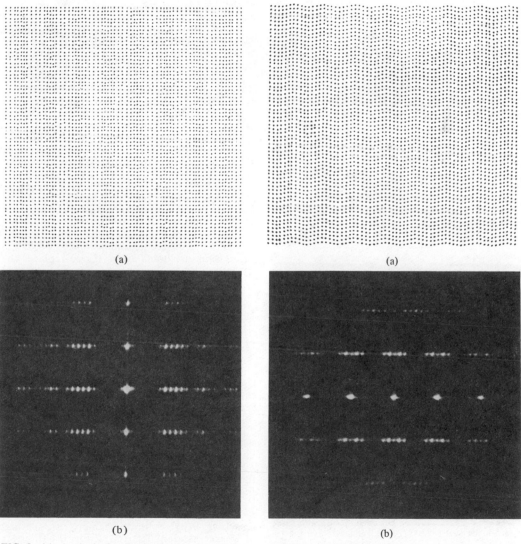

(a)   (a)

(b)   (b)

FIG. 5. (a) A single longitudinal wave. (b) Extra Laue spots.

FIG. 6. (a) A single transversal wave. (b) Extra Laue spots.

$h_2 = 1.0$ (0.1) is much stronger and in the 2.0 (0.2) reflection, much weaker, than the Laval spot $m = 2$.

In nature, one never finds such sharp Laval spots, but a completely smooth "thermodiffuse" scattering, and it is said that a "white" spectrum of undamped elastic waves exists.

In Fig. 7(a) strongly damped transverse waves according to Eq. (17) consisting only of two maxima and minima are introduced. Now in Fig. 7(b) quite diffuse spots appear. Hence, in nature, such damped waves could also occur, and if they had a spectral distribution, they could give rise to the same observable thermo-

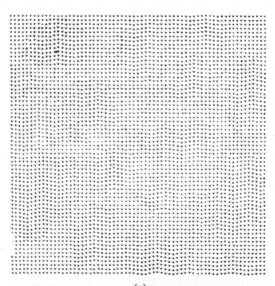

(a)

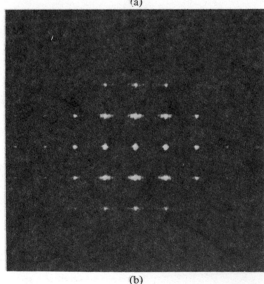

(b)

FIG. 7. (a) Twelve damped transversal waves. (b) Diffuse extra Laue spots.

diffuse scattering. According to Debye's theory of heat capacity, every wave has a different amplitude a_t, following a Boltzmann statistic. Then in certain regions of a single crystal, the atoms oscillate statistically with different amplitudes than in others. As a result, the lattice cells exhibit different sizes and paracrystalline distortions occur.

Close to the melting point, the amplitudes a_t of the elastic waves become so large that the electron clouds of the atoms suffer large deformations, which damp these waves more and more.

Another physical reason for paracrystalline distortions below the melting point exists if the "motives" in the single lattice cells have different shapes. Such phenomena have been observed in the "macrolattices" of natural and synthetic high polymers.

Figure 8(a) shows a paracrystalline model, where both coordination statistics H_k are horizontal line functions. Figure 8(b) represents its Fraunhofer pattern and Fig. 8(c) the x-ray small-angle pattern of the β-keratin of the quill of a sea gull (Bear and Rugo (1951)). The cell edge a_2 in the vertical direction parallel to the fiber axis has a constant length of 185Å but statistically changes its direction (with respect to the macroscopic fiber axis) within ±5°, while the orthogonal edge length a_1 has an average value of 34Å and changes its length statistically within ±2.5Å. As a result of the van der Waals forces, which allow a variation of the length a_1 orthogonal to the fiber axis, the homopolar forces along a_2 have freedom to change their direction of the molecular chain, within one paracrystal. This paracrystal itself, therefore, shows a flexible character in atomic dimensions.

Equation (6) holds again, if one replaces the crystalline lattice factor Z of Eqs. (7) and (11) by the paracrystalline lattice factor

$$Z(\mathbf{b}) = \frac{1}{v_r} \prod_{k=1}^{3} Re \, \frac{1 + F_k(\mathbf{b})}{1 - F_k(\mathbf{b})} \qquad (19)$$

$$F_k(\mathbf{b}) = F(H_k) \qquad (20)$$

H_k is a so-called coordination statistic. $H_k(\mathbf{x})$ is the "a priori" probability of finding a cell edge vector $a_k = \mathbf{x}$ in a certain paracrystalline lattice cell.

For the same reason, in mixed crystals consisting of atoms of different sizes, paracrystalline distortions can also occur.

Figure 3(a) showed a model of steel balls of the same size, building up a crystalline lattice. Figure 3(b) represents its Fourier transform. In Fig. 9(a) steel balls of different sizes built up a paracrystalline lattice, whose Fraunhofer pattern [Fig. 9(b)] exhibits characteristic features of a paracrystalline lattice [cf. Fig. 8(b)].

In Fig. 10(a) another model of a mixed crystal is given without paracrystalline lattice distor-

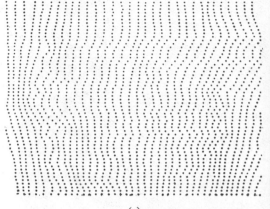

(a)

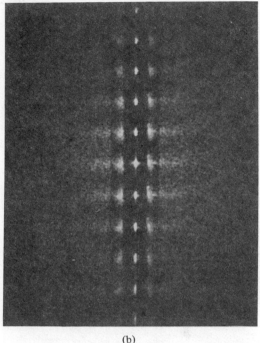

(b)

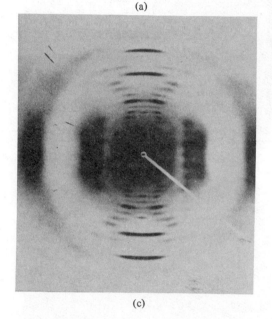

(c)

FIG. 8. (a) Monoparacrystal with linear horizontal coordination statistics. (b) Its Fraunhofer pattern. The reflections (o, h_2) are crystalline, all others are more or less diffuse. (c) Small-angle x-ray pattern from β-feather-keratin (Bear-Rugo, 1951).

tions. Now the two kinds of atoms are not distributed totally randomly to the lattice points. Hence the Fraunhofer pattern of Fig. 10(b) shows a diffuse background, which is not given by

$$N(\overline{f^2} - \overline{f}^2) \qquad (21)$$

and which presents "diffuse walls" in the vertical direction at $h_1 = \pm\frac{1}{2}$ of a width $\delta b_1 \sim \frac{1}{8}A_1$. This means that in the horizontal direction, rows of about 8 atoms show a kind of "superstructure" (*Nahordnung*, "cooperative order"). This "superstructure" has a psychological background: The technician, painting the models row by row in the horizontal direction was anxious to choose thick and thin atoms quite arbitrarily. However, he did not use a Monte Carlo Method, but tried as arbitrarily as possible

to draw statistical sequences. Unfortunately, after he painted a thick atom he tended to choose a thin one, etc. Unconsciously, he introduced "cooperative forces."

Similar diffuse walls are observed in ferroelectric $NaNO_2$ above the Curie point: In this case, rows of about six $NaNO_2$ molecules have parallel oriented polar axes, and have built up microferroelectric domains in a statistically paraelectric matrix (M. Canut and R. Hosemann, 1964).

In Fig. 11(a), similar to Fig. 9(a), steel balls of different sizes which now built up a structure of single small paracrystallites were used. The Fraunhofer pattern, Fig. 11(b), shows the typical features of a liquid or melt or "amorphous" solid. Hosemann and Lemm (1964) have proved, that in molten gold and lead such paracrystals can be found with average diameters of 12Å to 40Å.

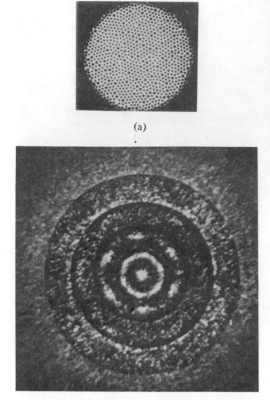

(a)

(b)

FIG. 9. (a) Mixed single paracrystal (steel balls). (b) Intensity function (Eq. (6) and (19)).

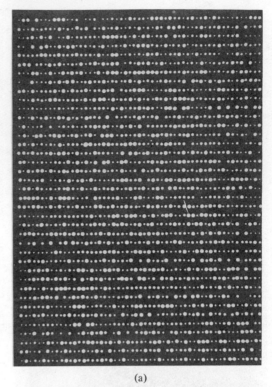

(a)

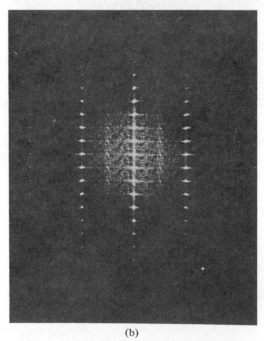

(b)

FIG. 10. (a) Mixed single crystal with "cooperative forces." (b) Background with diffuse walls (*Nahordnung*).

After having completed the step from crystals to amorphous matter, Fig. 12(a) gives an example of a special gas and Fig. 12(b) at larger b-values, the gas-interferences. Lord Rayleigh (1842–1919) proved that here phase relations between the single scattering centers are destroyed in the average as a consequence of their irregular positions. Since in Fig. 12 only 5200 centers were used, "ghosts" remain in the Fraunhofer pattern. Besides this fluctuation the intensity at large b-values is given by the shape of the single points. Moreover in Fig. 12(a) every 40 points cluster together. The clusters are arranged in a crystalline lattice. Hence, at small angles in Fig. 12(b), "Bragg-reflections" whose intensity is proportional to the squared transform Eq. (15) do occur. ρ_0 is now a forty-point function different for each cluster, and in Eq. (6), f must be replaced by the average $\bar{f}$. The diffuse background of Fig. 12(b), which now is the statistical fluctuation of the density distribution ρ_0 is given again by Eq. (21). If we replace the word "cluster" by zircon atom and "point" by electron, Fig. 12(a) gives an instantaneous picture of the electron configuration in a zirconium crystal. The Bragg reflections give in this case information about Schrödinger's wave functions for electrons

$$\overline{\rho_0(x)} = \psi\psi*$$

and the fluctuation term of Eq. (21) gives some information about the structure of a single electron. In reality, Compton processes disturb the

(a)

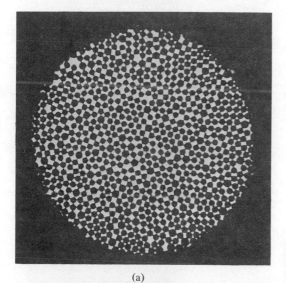

(a)

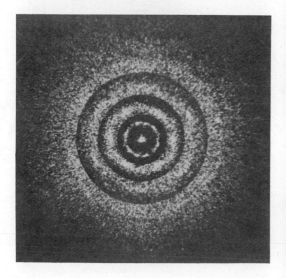

(b)

FIG. 11. (a) Polymicroparacrystalline assembly of steel balls. (b) Intensity function of "amorphous" matter.

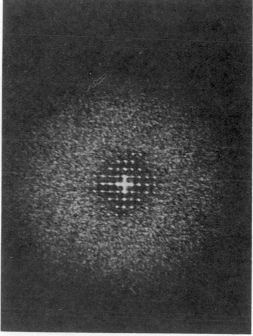

(b)

FIG. 12. (a) Zirconium crystal with discrete electrons. (b) Rayleigh scattering of the electrons.

diffraction. There is some hope for further detailed studies.

ROLF HOSEMANN

Cross-references: ABERRATION, DIFFRACTION THEORY OF; CRYSTALLOGRAPHY; ELECTRON DIFFRACTION; NEUTRON DIFFRACTION; OPTICS, PHYSICAL; PARACRYSTAL POLYMER PHYSICS; SCHRÖDINGER EQUATION; X-RAY DIFFRACTION.

DIFFUSION IN LIQUIDS

Diffusion, in a macroscopic sense, is a universal process that leads to the eliminating of concentration gradients in gases, solids, or liquids. At the molecular level, it arises because atoms or molecules undergo small, essentially random displacive movements as a result of their thermal energy. It is a property having great fundamental importance, because all useful theoretical efforts to understand the dynamical behavior of liquids at the molecular level must lead to agreement with accurate macroscopic measurements. It is one of four transport properties of matter: viscosity, diffusion, thermal conductivity, and (in electrically conducting media) electrical conductivity. Each of these transport properties measures the flux of some quantity in a gradient: *viscosity*, the flow of momentum in a velocity gradient; *diffusion*, the flow of mass in a concentration gradient; *thermal conductivity*, the flow of heat in a temperature gradient; *electrical conductivity*, the flow of electric charge in a gradient of electrical potential. Since their formal mathematical structures are the same, it is not surprising that there is often a quantitative correlation among them.

The diffusive motions of individual molecules may be likened to a kind of aimless three-dimensional *random walk*. If we limit our attention to the displacement of a single particle in a given direction from its position at some arbitrary zero time, the *probability* that it will be found at a distance, $\pm x$, from its origin after a time, t, is given by:

$$P(x, t) = \frac{1}{2(\pi D t)^{1/2}} \exp(-x^2/4Dt) \qquad (1)$$

In this equation, the parameter D, called the *diffusion coefficient*, is a measure of the average rate with which the displacement of the particle occurs.

If, instead of attempting to follow the random motion of an individual particle, we introduce a large number of particles C_0 at a point within the system, their concentration $C(x, t)$ will vary with time and distance in a given direction according to:

$$C(x, t) = \frac{C_0}{2(\pi D t)^{1/2}} \exp(-x^2/4Dt) \qquad (2)$$

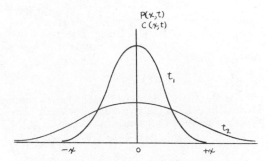

FIG. 1. Concentration or probability profiles for one-dimensional diffusion for times t_1 and t_2.

Both equations have the form of the well-known Gaussian error curve, which gives the distribution of random errors to be expected in a large set of measurements. Figure 1 illustrates these equations, showing the probability of finding a single particle at a distance from its origin (or the concentration distribution of a finite quantity of a substance) after times, t_1 and t_2.

Equation (2) is a particular solution of a pair of more general differential equations known as Fick's laws. Consider a plane of unit cross-sectional area in a system, across which a concentration gradient, $\partial C/\partial x$, exists. There will be a net flux of matter, J_x, through the plane from the region of higher to lower concentration given by Fick's first law:

$$J_x = -D \frac{\partial C}{\partial x} \qquad (3)$$

In principle, it is possible to determine the coefficient of diffusion on the basis of Eq. (3) by measuring the net quantity of matter that crosses through unit area of the plane in unit time. This proves to be difficult in liquids or solids, however, and an additional complication is that the concentration gradient $\partial C/\partial x$ is not constant but decreases with time.

For this reason, it is more feasible to measure the accumulation of matter in a small volume element after a measured period of time. We may consider such a volume element to be bounded by parallel planes of unit area, separated from one another by a distance dx, as shown in Fig. 2. The rate of change of concentration within the volume element is expressed by Fick's second law:

$$\frac{d}{dx}\left[J_x - \left(J_x + \frac{\partial J}{\partial x} \cdot dx \right) \right] = \frac{dC}{dt} = D \frac{\partial^2 C}{\partial x^2}$$

$$(4)$$

This is the fundamental equation, upon which all experimental studies of diffusion depend.

One of the most widely used of the absolute

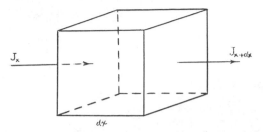

FIG. 2. Diffusive flux across parallel planes.

methods for measuring diffusion coefficients of liquids is the "open capillary" technique, devised by Anderson and Saddington.[1] A capillary tube, usually less than 1 mm in diameter and 2–3 cm in length, is filled with an isotopically labeled substance and immersed in a thermostatted bath of the unlabeled liquid. Interdiffusion of the labeled and unlabeled molecules occurs across the open end of the capillary, and at the end of the experiment, the average concentration of labeled substance remaining in the capillary is determined by suitable radiochemical or mass spectrometric techniques. The relation between the diffusion coefficient and the initial and final concentration of labeled substance in the capillary is given by a series solution of Fick's second law:

$$\frac{C_{av}}{C_0} = \frac{8}{(\pi)^2} \sum_{n=0}^{\infty} \frac{1}{(2n+1)^2}$$

$$\exp\left(-(2n+1)^2 \pi^2 Dt/4L^2\right) \quad (5)$$

In this equation, for which L and t are, respectively, the capillary length and time, the series converges rapidly and as a rule may be terminated after the first or second term with negligible error.

Other methods have been developed for the measurement of diffusion coefficients, usually limited to the liquids or solutions of a particular kind. One of the most accurate, limited to electrolytic solutions,[2] is based upon the measurement of the change in the electrical conductance as a function of time in a direction transverse to a concentration gradient in the solution. Such a method leads to the mean diffusion coefficient of ion pairs of a 1:1 salt, rather than the individual ionic diffusion coefficients if the concentration gradient is initially established between the electrolyte and pure water.

The diffusion coefficients of electroreducible or electrooxidizable ions in polar solvents may be determined by the method of polarography[3] or related techniques. Here the principle is the establishment of an ion-depleted zone of the solution next to the cathode or anode of a polarographic cell, and the measurement of the current carried by the appropriate ions as they diffuse through the depleted zone to the electrode. The Ilkovic equation gives the relation between the diffusion current and parameters of the system when the electrode is a flowing liquid metal, such as mercury:

$$i_t = 4\left(\frac{7\pi}{3}\right)^{1/2} nFD^{1/2} C\left(\frac{3m}{4\pi\rho}\right)^{2/3} t^{1/6}$$

$$(6)$$

where n is the number of equivalents involved in the electrode reaction,

F is the Faraday,

C is the concentration of reducible or oxidizable ions in the bulk solution,

m is the mass flow rate of metal (mercury),

ρ is the density of metal, and

D is the ionic diffusion coefficient.

Optical methods[4] are applicable to solutions that are transparent, and have been used extensively to measure the diffusion coefficients of polymers, proteins, and other biologically important compounds. They depend upon the variation of the index of refraction of a solution with the concentration of the dissolved substance. Schlieren images of a slit or interferometric fringes of a collimated beam of light that is deviated by refraction as it passes through a concentration gradient of the solution permit measurement of the time-dependent displacement of the solute.

A completely different method for the measurement of diffusion coefficients in the liquid state is based upon nuclear magnetic resonance (NMR). An assembly of atomic nuclei which have been excited to some nonequilibrium spin distribution will return to thermal equilibrium by a mechanism that involves the coupling of the nuclear spins with their local molecular environment. The rate of the return to equilibrium is characterized by a *relaxation time* $(T_1,$ the spin-lattice relaxation time), and is governed by the variations in the local magnetic field at nuclei which are induced by the translational motions of neighboring molecules. In favorable circumstances the spin-lattice relaxation time may be obtained from the width of the NMR resonance line. Bloembergen, Purcell, and Pound[2] developed the basic theory which relates the spin-lattice relaxation time to the diffusion coefficient of molecules, and made the first measurements on water and a series of hydrocarbons to test its validity. A more direct determination of the self-diffusion coefficient of liquids is provided by NMR spin-echo experiments. These techniques employ rf pulses and either steady or pulsed magnetic field gradients to determine the rate of decay of the induced nuclear spin polarization. Spin echo measurements of the self-diffusion coefficients of liquid lithium and liquid sodium, as well as of H_2O and D_2O, are in excellent agreement with the results of isotope labeled diffusion experiments.[5]

A great deal of information[6] about the mechanisms of molecular motion is contained in the NMR signal and its time evolution. The processes of nuclear spin relaxation involve coupling to magnetic fields and electric field gradients (when $I \geqslant 1$) both internal and external to a molecule, and are modulated by their thermally driven fluctuations. Various theoretical models[7] have been devised to infer the details of translational and rotational motions of molecules from several characteristic correlation times that are observable by spin echo NMR methods. (A correlation time is the average time required for a molecule to lose the memory of its orientation in a prepared spin state.) It is often possible, for example, to deduce the rates with which molecules of low symmetry rotate about their axes, to ascertain whether the angular motion is through a small or large angle, and to characterize such motions by rotational diffusion coefficients or equivalently, correlation times.

Efforts to relate observable transport properties of a system of motions at the molecular level are based upon the so-called *fluctuation dissipation theorem*, which links the linear transport coefficients to time correlation functions of locally fluctuating dynamical variables (e.g., force, position, velocity, momentum). For translational self-diffusion, the self-diffusion coefficient is given[8] by the velocity autocorrelation function:

$$D = \frac{1}{3} \int_0^\infty dt \, \langle v(0) \cdot v(t) \rangle \qquad (7)$$

where $\langle v(0) \cdot v(t) \rangle$ is the average of the particle velocity over a short time Δt, which is nevertheless long compared with the period between "collisions." Similar time correlation functions are defined for other transport properties, such as viscosity and electrical conductivity, and for molecular rotation. The correlation times for molecular rotation are:

$$\tau_{1,2} = \int_0^\infty dt \, \langle Y_{lm} \Omega(0) \cdot Y_{lm} \Omega(t) \rangle \qquad (8)$$

and

$$\tau_J = \int_0^\infty dt \, \frac{\langle J(0) \cdot J(t) \rangle}{\langle J(J+1) \rangle} \qquad (9)$$

where, in Eq. (8), $l = 1$ for dielectric relaxation and $l = 2$ for nuclear spin relaxation by intramolecular dipolar or quadrupolar interactions, and in Eq. (9), $J(t)$ is the rotational angular momentum operator that governs nuclear

relaxation by spin-rotation interactions. The relative magnitudes of τ_1 and τ_2 distinguish between small and large step angular motions, and τ_J depends upon the moments of inertia of the molecule.

A completely different approach to the study of the dynamics of atom motion in liquids is based upon slow neutron scattering. The angular and energy dependence of the cross section for inelastic scattering of neutrons by the nuclei of atoms in a liquid metal, for example, may be used to deduce the Van Hove space-time correlation function.[5] This function contains information about the diffusive motions of nuclei, and experiments on liquid metals indicate that a transition from the gas-like free motion of particles to solid-like diffusive displacements occurs on a time scale of 10^{-13} sec.

Studies of the temperature dependence of the diffusion coefficient of liquids generally lead to an empirical relationship of the form:

$$D = D_0 \exp(-(Q/RT)) \qquad (10)$$

where D_0 and Q are experimental parameters that are essentially temperature-independent and characterize the diffusion process in the system at hand. Because of the exponential form of Eq. (10), the mechanism of diffusion in liquids is often assumed to be an activated process, by analogy with other kinetic processes to which the *absolute reaction rate theory* has been applied with much success. According to this theory, the rate-limiting step of a kinetic process is determined by the frequency with which atoms or molecules acquire sufficient energy through thermal fluctuations to surmount an energy barrier identified by the parameter Q. It has not yet proved possible to make accurate *a priori* calculations of D_0 and Q on the basis of the activated state theory, however, and the exponential form of Eq. (10) is no proof that the mechanism of atom transport in liquids is a thermally activated process in any simple sense.

In recent years high speed computers have made possible the simulation of self-diffusion experiments by molecular dynamics calculations. In such experiments the classical equations of the motion are solved by difference methods for a set of particles of given mass and assumed pair interaction potential at certain temperatures and particle densities. The time intervals are on the order of 10^{-14} seconds, and D is calculated from the Einstein equation with the mean square displacements $\langle R^2 \rangle$:

$$\langle R^2 \rangle_{av} = 6Dt. \qquad (11)$$

Calculations have been made for (noninteracting) hard spheres, liquid argon,[9] liquid sodium metal,[10] and molten sodium chloride,[11] and the agreement is generally very good. They show clearly that diffusion occurs by small steps of less than molecular diameter. This rules out the once favored "hole theory" of transport and corroborates other evidence that diffusion is a highly cooperative process, involving the participation of many particles. This, together with other evidence[12] based upon the small pressure-dependence of self-diffusion in liquid mercury, effectively disproves all models that appeal to the activated state theory of transport processes for the elementary diffusive step as a single particle jump over an intermolecular distance. The concept of an activated state also loses its meaning when the number of particles needed to define it becomes large and ill-defined. The characterization of the exponential temperature-dependence of self-diffusion by an "activation energy" should therefore be regarded as an experimental parameter, and not as a barrier that a single molecule must surmount. The same must be said of other pseudo-thermodynamic quantities (enthalpy, entropy, and free energy) that have been adduced to the transport properties of liquids.

NORMAN H. NACHTRIEB

References

1. Anderson, J. S., and Saddington, K. J., *J. Chem. Soc.* 381 (1949).
2. Harned and Owen "The Physical Chemistry of Electrolytic Solutions," 2nd Ed., p. 595, Reinhold, New York, 1950.
3. Kolthoff, I. M., and Lingane, J. J., "Polarography," Vol. I, p. 40, Interscience Publishers, New York, 1952.
4. Geddes, A. L., and Pontius, R. B., in "Physical Methods of Organic Chemistry," Vol. I, Part II, pp. 895–1005, Interscience Publishers, New York, 1960.
5. Murday, J. S., and Cotts, R. M., *J. Chem. Phys.* **53**, 4274 (1970).
6. Jonas, J., *Science*, **216**, 1179–1184 (1982).
7. Boden, N., in "Nuclear Magnetic Resonance," Vol. 1, p. 121, The Chemical Society, Burlington House, London, 1972).
8. Zwanzig, R., *Ann. Rev. Phys. Chem.* **16**, 67–102 (1965).
9. Rahman, A., *Phys. Rev.* **136**, A405 (1964).
10. Takeuchi, S., Tanaka, M., Fukui, Y., Watabe, M., and Hasegawa, M., in "The Properties of Liquid Metals," Proc. 2nd Int. Conf. Tokyo, Japan (1972), p. 143, Taylor & Francis Ltd., London, 1973.
11. Woodcock, L. V., in "Advances in Molten Salt Chemistry," (Braunstein, J., Mamantov, G., and Smith, G. P., Eds.), Vol. 3, p. 40, Plenum Press, New York, 1975.
12. Nachtrieb, N. H., and Petit, J., *J. Chem. Phys.* **24**, 746 (1956).

Cross-references: BROWNIAN MOTION, DIFFUSION IN SOLIDS, KINETIC THEORY, LIQUID STATE, MAGNETIC RESONANCE, RELAXATION.

DIFFUSION IN SOLIDS

The term "diffusion" refers to the random motion, generally activated by local fluctuations of thermal energy, of particles through a medium. The particles with which we shall be concerned are atoms and molecules; the medium can be various types of solids. Of special interest will be crystalline solids—these include all metals, most ionic substances, and many covalent ones—in which the atoms occupy periodic and well-defined sites.

The migrating particles may themselves be uniformly distributed constituents of the host solid; this is called self-diffusion. When the diffusing system contains chemical inhomogeneities or when a foreign substance diffuses in from the surface, we speak of chemical or of impurity diffusion. Diffusion processes are technologically important in the oxidation and tarnishing of metals, where one reactant must migrate through the layer of reaction product, and in the annealing of deformed or radiation-damaged materials. Self-diffusion is an essential step in the photographic process in silver halides, and impurity diffusion is widely used in the fabrication of semiconductor devices such as transistors. Many metals, such as steel and duralumin, are hardened by solid-state precipitation and reaction, in which diffusion plays a dominant role. It is also significant in the powder metallurgy technique of fabrication of parts from high-melting metals.

If, in the medium, there are variations in the concentration of the migrating atoms—perhaps chemically different atoms or, in the case of self-diffusion, radioactive tracer isotopes—then there occurs a net drift of the diffusing species from regions of high concentration to those of lower concentration. This flow takes place even though each individual atom may migrate completely at random. It is a statistical result of the fact that if there are more atoms per unit volume of, say, A to the left of a given plane than to the right, then even with random, nondirected motion, more A atoms will cross the plane from the left than from the right. We can define the flux of A as the net excess of A atoms crossing a plane of unit area in unit time. Experimentally, this flux is found to depend on the chemical natures of the medium and the diffusing species. If the medium is isotropic or is a crystal of cubic symmetry, the flux is along the direction of the concentration gradient. Moreover, if the system is not too thermodynamically nonideal, the flux is proportional to the concentration gradient; the constant of proportionality is called the diffusion coefficient, D. Thus, we write the flux $J = -D \, dc/dx$, where the negative sign indicates that the net flow is toward the region of lower concentration. This statement is known as Fick's law. If lengths are measured in centimeters and time in seconds, D is in units of square centimeters per second. It follows from

Fick's law that at any given point in the medium, the concentration of the diffusing entity A will change with time at a rate governed by the variation of the flux with distance [$\partial c/\partial t = \partial/\partial x(D\partial c/\partial x)$; for the particularly simple case where D is independent of distance, as in self-diffusion, then $\partial c/\partial t = D\partial^2 c/\partial x^2$]. Also, from the theory of random flights, it can be shown that the root-mean-square displacement of atoms resulting from diffusion for a time t increases as the square root of the product Dt; $R_{rms} = (6Dt)^{1/2}$.

It is observed that for any given system, D increases rapidly with increasing temperature, almost invariably following the Arrhenius relation $D = D_0 \exp(-H/RT)$. Here, D_0 and H are positive constants for a given system and R is the universal gas constant. The parameter H is called the activation energy and generally increases as the melting point of the host crystal increases. Typically, in crystals which melt at 400 to 500°C, H is about 20 000 to 25 000 cal/mole, or 1 eV/atom, for self-diffusion or diffusion of substitutionally dissolved impurities. In crystals which melt near 1000°C, such as the noble metals, the activation energy is approximately 2 eV/atom. The value of the parameter D_0 is usually in the range 0.01 to 100 cm²/sec. It is interesting that for a large number of metals and simple ionic crystals, the diffusion coefficients for self-diffusion and for most impurities lie near to 10^{-8} cm²/sec at temperatures approaching the melting point. Thus, after diffusing for one day at such a temperature, the value of R_{rms} is about 1 mm, rather a large distance when compared to the spacing between atoms in a crystal.

Because of the three-dimensional regularity of atomic positions in a crystalline solid, the unit step in diffusion must be the jump of an atom from one site to a neighboring, crystallographically equivalent site. Large-scale diffusion is the result of random superposition of many such jumps, all of the same length λ but distributed among the various jump directions allowed by the crystal. It is readily shown that the relation between the macroscopic diffusion coefficient D and the microscopic atomic jump frequency Γ is $D = 1/6\lambda^2\Gamma$. This equation may be compared with that given above for R_{rms} by noting that for random jumps $R_{rms} = \lambda(\Gamma t)^{1/2}$. Now λ will depend on the details of the mechanism of diffusion but it must be of the order of the interatomic spacing, about 3 $\times$ 10^{-8} cm. Then the typical high-temperature diffusion coefficient of 10^{-8} cm²/sec requires each atom to make 10^7 to 10^8 jumps each second.

In most crystals the atoms are rather densely packed; thus the means whereby such a high frequency of jumps can be accomplished is not obvious. Conceptually, the simplest possibility is the simultaneous exchange of sites between two atoms, but this is ruled out because of the excessive activation energy that would be required to push aside the mutual neighbors of the pair. A dramatic demonstration that diffusion must proceed by a mechanism which allows independent motion of individual atoms is the Kirkendall effect. Two mutually soluble specimens of differing composition, say A and B, are welded together with inert markers imbedded at the interface. Subsequent interdiffusion of A and B results in a drift of the markers relative to the ends of the specimen, indicating that more atoms have left one side of the couple than have entered it from the other. Clearly a pair exchange mechanism cannot be operative here.

Extensive evidence is now available that in most cases of self-diffusion or of substitutionally dissolved impurities, migration proceeds as a result of the presence and mobility of vacant lattice sites. These vacancies exist in the crystal in thermodynamic equilibrium, at concentrations which increase with temperature as exp $(-H_f/RT)$, where H_f is the energy required to form a vacancy (about 1 eV in the noble metals). At temperatures near the melting point, the fraction of sites vacant is typically 0.01 to 0.1 per cent. Vacancies move by the jumping of adjacent atoms, at a rate which varies as exp $(-H_m/RT)$. H_m is the activation energy for the migration process. The average jump frequency of an atom must then be the product of the jump frequency of a vacancy and the fraction of atomic sites that are vacant. Comparing the temperature dependence of the diffusion coefficient with that of these two factors, it follows that H must equal the sum of H_m and H_f. Quantitative experimental verification of this equality in a number of substances has firmly established the vacancy mechanism for diffusion in such crystals.

Other mechanisms, however, are also known to operate. For example, linear or planar defects within the crystal can provide paths for easier and more rapid diffusion. Thus, along boundaries between crystal grains, on the external surfaces, and along the linear defects known as crystal dislocations, the atomic regularity is interrupted and binding energies are correspondingly decreased. As a result, the activation energy for diffusion is locally smaller and the diffusion coefficients are therefore larger; one speaks of a "short-circuiting" effect. Along dislocations and internal boundaries, the diffusion activation energy is typically only about half that for diffusion within the bulk crystal, and short-circuiting contributions become especially important at temperatures below about half of the melting point. Diffusion along external surfaces appears to depend strongly on the nature of the ambient atmosphere, and the details of the process are not well understood at present.

Another mechanism which is found in some systems involves the diffusing atom (or ion) migrating from one interstitial position (i.e., squeezed in between proper atom sites) to another. This process is not surprising in those cases in which a very small impurity ion is already dissolved interstitially, such as for carbon in iron; its occurrence is perhaps unexpected, however, in cases where the migrating ion is present primarily substitutionally (i.e., in normal atomic sites of the crystal). Thus, in the silver halides, self-diffusion of silver and the diffusion of several cationic solutes proceed primarily via the interstices of the lattice; perhaps the "softness" of the silver halides encourages such a mechanism. Also, in the loosely packed crystals of germanium and silicon, some solutes (such as copper or gold) diffuse interstitially, although they are dissolved mainly substitutionally. And more recently, it has come to be appreciated that in many polyvalent metals such as lead and tin, the noble metals and their divalent neighbors diffuse by some sort of interstitial process. In all such cases the diffusion coefficients at high temperatures are much larger, by several factors of ten, than would be expected for ordinary substitutional diffusion by means of vacancies.

Within the last decade, numerical procedures for calculating the energies of formation and migration of point defects have been developed to such a point that the interpretation of experimental data can often proceed with considerable confidence, especially in the case of diffusion and ionic conductivity in such relatively simple materials as halides and oxides. Thus, experimental deviations from the simple Arrhenius relation can be analyzed in terms of the dependence on temperature (through the thermal expansion of the lattice) of the parameters D_0 and H, and in terms of the operation of more than one mechanism of atom migration. In other cases, in which no unambiguous ion transport mechanism could otherwise be identified, the results of these calculations now serve as an invaluable guide. The most successful of these schemes has been the HADES code developed at AERE Harwell, in which point defect parameters are calculated from interatomic potentials which, in turn, are obtained by fitting the known dielectric and lattice mechanical properties of the substance. Examples of the use and versatility of this procedure may be found in the volume of *Journal de Physique* cited in the references.

Another recent development has been the rapid growth of interest in those ionic materials in which one component displays an unusually high mobility, giving rise to diffusivities and ionic conductivities in the crystalline solid that are comparable to those typically found only in liquids. These substances are referred to as fast ionic conductors, "superionic" conductors, or solid electrolytes—this last term reflecting their potential application in electrolytic cells.

Apart from the technological interest in these fast ion conductors, there is also a great scientific challenge in understanding the microscopic mechanisms and interactions involved. In such substances, one sublattice retains the normal crystalline order, but the other one invariably displays an extraordinarily high density of point defects—often in the range of several percent—which gives rise to the high mobility of the corresponding ionic component. For example, in α-AgI and $RbAg_4I_5$, the anion sublattice is relatively perfect, while the silver ion distribution is highly disordered. In many lead and alkaline earth halides, on the other hand, it is the anion sublattice which becomes disordered. In the several beta-alumina-type materials, blocks of perfectly ordered ions are separated by parallel planes containing a disordered distribution of mobile, monovalent cations.

One of the interesting questions posed by such materials is the origin, extent, configuration, and temperature-dependence of the disorder in just one of the sublattices. Another problem involves the very low activation energies of migration of the mobile ions and defects. Still a third point of interest hinges on the nature of cooperative interactions between the mobile ions. This is especially fascinating, since fast ion conductors are known in which the mobile ions can move either in all three dimensions (as in α-AgI), in only two dimensions (as in the conducting planes of the beta-aluminas), or along only one-dimensional channels (as in the mineral hollandite). It seems likely that these substances may well yield as rich a reward in scientific understanding as they promise to do in technical applications.

LAWRENCE SLIFKIN

References

The first reference describes experimental techniques for determining diffusion coefficients; those following are recent brief reviews in order of increasing sophistication or date of publication. References to more detailed discussions are given in these.

Tomizuka, C. T., in Lark-Horovitz, K., and Johnson, V., Eds., "Methods of Experimental Physics," Vol. 6A, p. 364, New York, Academic Press, 1959.

Girifalco, L. A., "Atomic Migration in Crystals," New York, Blaisdell, 1964.

Shewmon, P. G., "Diffusion in Solids," New York, McGraw-Hill Book Co., 1963.

Lazarus, D., "Diffusion in Metals," in Seitz, F., and Turnbull, D., Eds., *Solid State Phys.*, **10**, 71 (1960).

Peterson, N. L., "Diffusion in Metals," in Seitz, F., and Turnbull, D., Eds., *Solid State Phys.*, **22**, 409 (1968).

Neumann, G., and Neumann, G. M., "Surface Self-Diffusion of Metals," in Wöhlbier, F. H., Ed., Diffusion Monograph Series, No. 1; Solothurn (Switzerland), The Diffusion Information Center, 1972.

Various chapters in: Crawford, J. H., Jr., and Slifkin, L., Eds. "Point Defects in Solids," New York, Plenum Press, 1972 and 1973.

Adda, Y., and Philibert, J., "La Diffusion dans les Solides," Paris, Presses Universitaires de France, 1966.

Nowick, A. and Burton, J., Eds., "Diffusion in Solids: Recent Developments," New York, Academic Press, 1975.

Examples of the calculation of atomic jump parameters may be found in "Lattice Defects in Ionic Crystals," *J. Physique*, 41 Colloq. No. 6 (1980).

For recent experimental and theoretical research on fast ion conductors, see *Solid State Ionics* 5 (1981), and other issues of this journal.

Murch, G. E., "Atomic Diffusion Theory in Highly Defective Solids," Aedermannsdorf (Switzerland), Trans Tech, 1980.

Cross-references: CRYSTALLIZATION, CRYSTALLOGRAPHY, DIFFUSION IN LIQUIDS, LATTICE DEFECTS, SOLID-STATE PHYSICS, SOLID-STATE THEORY.

DIMENSIONS

When describing natural phenomena, certain physical attributes or characteristics, such as mass or length, are distinguished. To these are assigned numerical *magnitudes* by prescribed measuring procedures, in which selected *scales* are employed. The length of a Foucault pendulum might be measured with a meter stick and found to equal 64 *units* of measurement, namely 64 meters. The length is called a physical *quantity*.

Customarily, certain quantities, such as length and time, are chosen to be primary or fundamental. Other quantities, like velocity, can then be expressed in terms of these. They are termed secondary or derived quantities. When the meter and the second are taken as units of length and time, the magnitude of a velocity can be expressed in meters per second. The exponent of the power of any primary quantity is called the *dimension* of the secondary quantity in that primary quantity. The dimensions of a secondary quantity are then written in terms of the primary quantities from which they are constituted and the powers to which they are raised. Thus the dimensions of a velocity v, here written as $[v]$, with length L and time T as primary dimensions, are LT^{-1}.

Most physical statements, such as those relating to the conservation of energy or mass, or to equilibrium of forces, can be considered as accounting statements, certifying that there is to be neither a gain nor a loss of a certain quantity. The terms in the equations, in which such statements are expressed, then must all refer to the same quantity: The equations must be *dimensionally homogeneous*.

Consider Bernoulli's equation for steady, non-viscous, incompressible flow in a gravity field. It may be written as

$$p/\rho + v^2/2 + gz = \text{const.}$$

where p is the pressure, ρ the mass density, v the fluid velocity, g the acceleration of gravity, and z the vertical distance measured from some arbitrary level. When mass M, length L, and time T are selected as primary quantities, the dimensions of a force can be derived from Newton's second law as that of mass times acceleration: $[F] = MLT^{-2}$. Therefore the dimensions of pressure, being force per unit area, are $[p] = ML^{-1}T^{-2}$. Further

$$[\rho] = ML^{-3}, \quad [v] = LT^{-1}, \quad [g] = LT^{-2}, \quad [z] = L.$$

All terms in Bernoulli's equation have the dimensions L^2T^{-2}, which are those of the square of a velocity, but also of energy per unit mass. Dimensional homogeneity can be used as a check against errors.

Dimensional Analysis Mathematical relations, with which one attempts to describe nature's order, cannot depend on arbitrary units of measurement (meter, second, etc.). These relations must be expressible in dimensionless form by means of dimensionless parameters. Dimensional analysis, making it possible to find such parameters and to say something about the relation between them, is useful particularly in dealing with complex problems not amenable to direct analysis. It provides the rationale for all model experiments.

To perform dimensional analysis, one must have an insight into the quantities which are relevant to the problem in question. Typically, one lists these relevant quantities and selects the primary quantities in which their dimensions are to be expressed.

For example, assume that a homogeneous body of arbitrary shape and having a characteristic dimension ℓ, having been kept at uniform temperature, is suddenly subjected, over a portion of its surface, to a temperature which is greater by the amount θ_0. The remainder of the surface is to be insulated. What can be said about the temperature rise θ at any point, designated by some radius vector $\mathbf{r}$, after a time t has elapsed?

Heat conduction being a geometric phenomenon, the heat-absorbing capacity per unit volume is judged to be relevant. With the specific heat c of the material being defined as the heat energy which raises the temperature of a unit mass by one degree, and designating the mass density by ρ, the heat capacity per unit volume is $c\rho$. Another relevant quantity must be the thermal conductivity k of the material, defined as the heat transmitted per unit time per unit cross section per unit temperature gradient.

For this problem heat energy $[E]$, length $[L]$, time $[T]$, and temperature $[\theta]$ are con-

TABLE 1.

Quantities	Exponents of Dimensions			
	E	L	T	θ
ℓ		1		
$\mathbf{r}$		1		
θ_0				1
θ				1
$c\rho$	1	-3		-1
k	1	-1	-1	-1
t			1	

venient primary quantities. The quantities judged to be significant, with their exponents in terms of these primary dimensions are listed in Table 1. The number of dimensionless parameters to be obtained is determined by Buckingham's PI theorem. Formal ways to determine them are given in references 1 through 5. In this case, inspection shows that there are three dimensionless parameters, which can be written as:

$$\pi_1 = \theta/\theta_0, \quad \pi_2 = \mathbf{r}/\ell, \quad \pi_3 = kt/c\rho\ell^2$$

The result can be stated as: $\theta/\theta_0 = f(\mathbf{r}/\ell, kt/c\rho\ell^2)$. The temperature rise θ at any point and at any time is proportional to θ_0. The time t needed to attain a temperature rise which is a specified fraction of θ_0 is proportional to the square of the characteristic dimension ℓ (the size of the body). The material is fully described by the quantity $k/c\rho$, known as the thermal diffusivity.

Particularly in fluid mechanics and heat transfer, some dimensionless parameters have gained general significance. The Mach number is the ratio of local fluid velocity to local velocity of sound. The Reynolds number is $v\ell\rho/\mu$, where v is a characteristic velocity, ℓ a characteristic dimension, ρ the fluid density, and μ the viscosity of the fluid.

Dimensions and Physical Constants Physical constants appear when relations are established among quantities which are already dimensionally connected. For example, Newton's second law when written as $F = md\mathbf{v}/dt$ fixes the dimensions of force as $[F] = MLT^{-2}$. But Newton's law of gravitation also relates a force (that of gravitation) with masses and their distance. With MLT dimensions taken to be independent, the gravitational law is written $F = Gm_1m_2/r^2$ where G is a dimensional gravitational constant. It is possible to write the gravitational law $F = m_1m_2/r^2$, but then M, L, T can no longer be independent. For example, the mass dimensions could be expressed in terms of L and T: $[M] = L^3T^{-2}$. Since physical constants are dimensional, they can be expected to be constant

TABLE 2. SOME QUANTITIES AND THEIR TYPICAL DIMENSIONS

Physical Quantity	MLT	Primary Quantities $MLT\theta$	$MLTQ$
Velocity	LT^{-1}		
Acceleration	LT^{-2}		
Force	MLT^{-2}		
Energy, Torque	ML^2T^{-2}		
Momentum	MLT^{-1}		
Density	ML^{-3}		
Viscosity	$ML^{-1}T^{-1}$		
Heat, Energy		ML^2T^{-2}	
Specific Heat		$L^2T^{-2}\theta^{-1}$	
Conductivity		$MLT^{-3}\theta^{-1}$	
Entropy		$ML^2T^{-2}\theta^{-1}$	
Electric Current			$T^{-1}Q$
Electric Potential			$ML^2T^{-2}Q^{-1}$
Impedance			$ML^2T^{-1}Q^{-2}$
Capacitance			$M^{-1}L^{-2}T^2Q^2$

only when the operational methods with which the fundamental units are defined are properly correlated. They then take on the character of conversion factors. The velocity of light, having the dimensions LT^{-1}, can be universally invariant only when the operational definitions of L and T make it so, as can be done by making use of radiation phenomena.

In thermal problems E, L, T, and θ can be selected as primary quantities. But, as $[E] = ML^2T^{-2}$, obviously M, L, T, θ can also be chosen where convenient. When physical constants are relevant to a particular problem with the dimensions selected, they must be listed among the relevant quantities. When the gas law $p = \rho R\theta$ applies to the problem, the gas constant R must thus be listed. Alternatively, R can be taken to be dimensionless, making $[\theta] = L^2T^{-2}$. Then θ would no longer appear as a primary quantity.

In electricity, a variety of primary quantities can be used, based on the manner in which the electrostatic and electromagnetic laws are stated. Often M, L, T, and Q (electric charge) are employed.

REINOUT P. KROON

References

1. Buckingham, E., "Dimensional Analysis," *Philosophical Magazine*, **48**, 141 (1924).
2. Bridgman, P. W., "Dimensional Analysis," Cambridge, Mass., Harvard University Press, 1931.
3. Huntley, H. E., "Dimensional Analysis," London, MacDonald and Co., Ltd., 1958.
4. Sedov, L. I., "Similarity and Dimensional Methods in Mechanics," New York, Academic Press, 1959.
5. Kline, S. J., "Similitude and Approximation Theory," New York, McGraw-Hill, 1965.
6. Kroon, R. P., "Dimensions," *J. Franklin Inst.*, **292** (July 1971).

Cross-references: CONSTANTS, FUNDAMENTAL; SYMBOLS, UNITS, AND NOMENCLATURE IN PHYSICS.

DIODE (SEMICONDUCTOR)

There exists a class of two-terminal devices which have the property of permitting current to flow with practically no resistance in one direction and offer nearly infinite resistance to current flowing in the opposite direction. These devices are called *diodes*. The applications of diodes to electronic circuits are numerous. To mention a few, they include rectification of alternating current to a unidirectional current, detection of radio waves, and gating circuits used in digital computers.

The basic materials utilized for making semiconductor diodes are germanium (Ge) and silicon (Si). These elements are included in column IV of the periodic table (see PERIODIC LAW AND PERIODIC TABLE). Both Ge and Si are *tetravalent* elements, i.e., they have 4 valence electrons. Elements in their pure state are said to be *intrinsic*.

Each element under column III of the periodic table has 3 valence electrons and is referred to as *trivalent*. Examples of trivalent elements include indium (In) and gallium (Ga). Elements in column V of the table have 5 valence electrons and are called *pentavalent*. Arsenic (As) and antimony (Sb) are examples of pentavalent elements.

The process of introducing one of the elements from column III or V into intrinsic Ge or Si is called *doping*. The doped material becomes impure or *extrinsic*. If a trivalent impurity is introduced in Ge or Si (trivalent elements have one less valence electron than Ge or Si) holes are created and the material is said to be *p*-type. Introduction of a pentavalent impurity (pentavalent elements have one more valence electron than Ge or Si) creates free electrons and the material is *n*-type.

Because of thermal effects, free electrons and holes are always being produced in Ge and Si (intrinsic generation of electron-hole pairs). Consequently, there will be some electrons in the *p*-type material and some holes in the *n*-type material. These carriers are referred to as *minority* carriers. Electrons in *n*-type material and holes in *p*-type material are termed *majority* carriers. The most widely used diode is the *p-n junction diode*. Imagine a single crystal of Ge (or Si) doped so half the material is *p*-type and the other half, *n*-type. The internal boundary between the two extrinsic regions is a *p-n* junction, and the resulting device is a junction diode (Fig. 1). The electrical symbol for the junction diode is illustrated in Fig. 2.

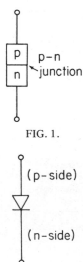

FIG. 1.

FIG. 2.

What are the characteristics of the *p-n* junction? To answer this question, three possible conditions are considered. Referring to Fig. 3, these are:

(1) *Unbiased:* *p-* and *n*-sides are connected by a wire.

(2) *Reverse biased:* the *p*-side is connected to the negative terminal of battery E, and the *n*-side connected to the positive terminal.

(3) *Forward biased:* the *p*-side goes to the positive terminal of E, and the *n*-side to the negative terminal.

Simple energy diagrams for the three conditions are shown in Fig. 3 for the electron. Similar diagrams can be generated for holes. When the diode is unbiased, no net flow of electrons takes place across the junction. Assuming that some electrons on the *n*-side have sufficient energy to overcome the potential hill, electrons on the *p*-side (minority carriers) "slide down" the hill making the net current flow zero. For the reverse biased case, the potential hill is raised and only the few minority carriers from the *p*-side "slide down." This results in a minute reverse saturation current. When the diode is forward biased, the potential hill is lowered. This enables electrons to climb over the hill and current flow occurs. The same considerations apply to holes. In fact, the total diode current is equal to the sum of the electrons and holes flowing across the junction.

The characteristic curve of a semiconductor diode is shown in Fig. 4. An equation for this curve, called the *rectifier equation*, is expressed as:

$$I = I_s(e^{-11600\,E/T} - 1)$$

where

I = diode current, amperes

I_s = reverse saturated current (which is temperature dependent), amperes

E = diode biasing voltage (+ E for forward bias; $-E$ for reverse bias), volts

T = absolute temperature ($^\circ C + 273^\circ$), degrees Kelvin.

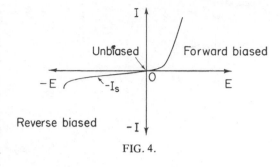

FIG. 4.

At room temperature (300 K) and $E > 0.1$ volt:

$$I \cong I_s e^{39\,E}$$

When E is more negative than 0.1 volt:

$$I \cong -I_s$$

An example of a simple rectifier employing a *p-n* junction diode is given in Fig. 5. During the positive half-cycle (0° to 180°) of the ac sinusoidal waveform v_s, the diode is forward-biased and conducts. The voltage v_L across load resistor R_L is therefore nearly identical to that of v_s for the positive half-cycle. For the negative half-cycle (180° to 360°) the diode is reverse-biased and does not conduct. No current flows in R_L, and $v_L = 0$ during the negative half-cycle. Because the diode conducts for only one-half cycle, the circuit of Fig. 5 is called a *half-wave rectifier*.

The waveform of v_L is only unidirectional. To obtain steady dc, like that from a battery, a filter is required. An example of an elementary filter is a large-valued capacitor placed across the load resistor.

The circuit of Fig. 5 can also be used as a detector of amplitude-modulated (AM) radio waves. Figure 6(a) illustrates the components of an AM wave. If this is applied to the input of Fig. 5, the wave is rectified and the output ap-

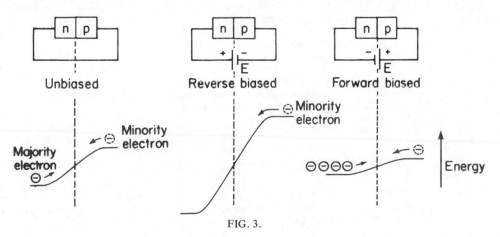

FIG. 3.

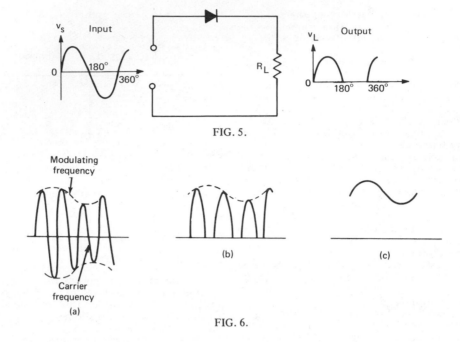

FIG. 5.

FIG. 6.

pears as shown in Fig. 6(b). Placing a small-valued capacitor across R_L filters out the carrier frequency and the desired modulating signal is obtained (Fig. 6(c)).

Besides the *p-n* junction diode, there are a number of other diode types which find use in specialized applications. These include the following diodes:

Gunn: used for the generation of microwave power.

Hot carrier (*Schottky* barrier): used for fast switching of waveforms, such as found in computers.

IMPATT: used for the generation of microwave power.

Injection: used for the generation of laser frequencies.

Light-emitting (LED): used for alpha-numeric displays.

Varactor: a reverse-biased junction diode that behaves like a variable capacitor, as a function of the applied voltage across the device.

Zener (*avalanche*): a reverse-biased junction diode that exhibits a dc voltage which is nearly independent of a specified range of current flowing in the device. The *Zener* diode finds wide use as a voltage reference in regulated power supplies.

ARTHUR H. SEIDMAN

References

Ghaznavi, C., and Seidman, A. H., "Electronic Circuit Analysis," New York, The Macmillan Company, 1972.

Hunter, L. P., "Handbook of Semiconductor Electronics," 3rd Ed., New York, McGraw-Hill Book Company, 1970.

Kaufman, M., and Seidman, A. H., "Handbook for Electronics Engineering Technicians," New York, McGraw-Hill Book Company, 1976.

Millman, J., "Microelectronics," New York, McGraw-Hill Book Company, 1979.

Seidman, A. H., and Waintraub, J. L., "Electronics: Devices, Discrete and Integrated Circuits," Columbus, Ohio, Charles E. Merrill Publishing Company, 1977.

Streetman, B. G., "Solid State Electronic Devices," 2nd Ed., Englewood Cliffs, N.J., Prentice-Hall, Inc., 1980.

Cross-references: ENERGY LEVELS, POTENTIAL, SEMICONDUCTOR, SEMICONDUCTOR DEVICES, SOLID-STATE PHYSICS, SOLID-STATE THEORY, TRANSISTOR.

DIPOLE MOMENTS (ELECTRICAL AND MAGNETIC)

Uncharged molecules can be classified as nonpolar or polar dependently on whether, in the absence of an electric field, the centers of gravity of their constituent positive and negative charges are coincident or not. A body containing two opposite charges, $\pm Q$, separated by a distance d, is characterized by an electric dipole moment, $Q\mathbf{d} = \mu$; μ is a vector quantity, expressed conveniently in debye (D) units: 1 debye $= 10^{-18}$ esu $= 3.33 \times 10^{-30}$ [coul m].

In the presence of an applied field a normally nonpolar molecule becomes dipolar by induction, i.e., by deformation of its electronic and atomic arrangements: $\mathbf{m} = (\alpha_e + \alpha_a)\mathbf{E}$, where the coefficients of proportionality α_e and α_a are the electronic and atomic polarizabilities, respectively. In the general case of an anisotropically polarizable molecule, α_e and α_a are tensors, the components of which may be evaluated from observations of electric birefringence (Kerr effect), the depolarization of (Rayleigh) scattered light, refractive index dispersion, etc. An estimate of the mean of the three principal polarizabilities is given by $3R/4\pi N$, where R is a molecular refraction by the Lorenz-Lorentz formula; when R is extrapolated to infinite wavelength the mean polarizability obtained refers to the electronic deformations alone. Polarizabilities are expressed in volume units (cubic centimeters) (N = the Avogadro number).

A field $\mathbf{E}$ exercises a torque on an electric dipole μ, tending to align it in the field direction in opposition to the randomness caused by thermal agitation. In a large assembly of molecules, therefore, a statistical and temperature-dependent equilibrium is achieved which corresponds to a slight excess of molecules having their permanent dipoles oriented antiparallel to the field so that the average moment $\overline{m}$ of one molecule is apparently proportional to the field intensity, i.e., an orientation polarizability α_o is exhibited.

The electric dipole moment per unit volume of a dielectric material is the polarization vector $\mathbf{P}$, understandable in magnitude as the charge density bound at the electrodes by a polarized dielectric. Based on the arguments of Mossotti (1850) and Clausius (1879), the polarization per mole is related to the dielectric constant ϵ by $M(\epsilon - 1)/d(\epsilon + 2) = 4\pi N\alpha/3$, where M/d is the molecular volume, N is the Avogadro number, and α is the over-all polarizability. Debye (1912) showed α_o to be $\mu^2/3kT$ (k = Boltzmann's constant, T = absolute temperature) so that $\alpha = \alpha_e + \alpha_a + \alpha_o$, and the total polarization per mole $_TP$ is the sum of the electronic, atomic, and orientation polarizations: $_TP = {}_EP + {}_AP + {}_OP$. A possible fourth polarization mechanism, the blocking or trapping of migrating charge carriers in a dielectric, although ignored in the classical molecular theory, may also contribute to the apparent ϵ of solid or macromolecule-containing systems.

The commonest method for the determination of dipole moments involves the dispersion of ϵ: $_TP$ is measured at radio and optical wavelengths (the second of these is a molecular refraction since the square of the index of refraction of a nonabsorbing, nonmagnetic material equals the dielectric constant at the same frequency), then approximately $R = {}_EP + {}_AP$, and $\mu^2 = 9kT(_TP - R)/4\pi N$. Although strictly valid only for gaseous dielectrics the Mossotti-Clausius-Debye equations have proved applica-

ble also to solutes in nonpolar solvents, and by using alligation formulas, values of $_TP$ for a dissolved species can be obtained at infinite dilution; such estimates are usually close to, but not identical with, the true $_TP$'s directly observed on the vaporized solutes. Over the past thirty years, much effort has been devoted to theoretical or empirical treatments of "solvent effects." Since distortion polarizations are almost invariant with temperature, the temperature dependence of $_TP$ follows as $(_TP)_T = A + B/T$; the constants A and B, when fitted to experimental data by least squares, give $A = {}_EP + {}_AP$ and $B = 4\pi N\mu^2/9k$, whence $\mu = 0.012812B^{0.5}$ esu; results for about 350 gases are listed by Marryott and Buckley.

Practical details concerned with the measurement of dielectric constants, and other properties, necessary for the deduction of μ's of solutes or vapors, are described fully in the books (cited below) by Le Fèvre, Smith, and Smyth, wherein also references are made to other, but less simple, techniques by which dipole moments can be determined (e.g., Stark splitting in microwave spectra of gases at low pressures, the dielectric losses or power factors of dilute solutions, molecular beam studies, etc.); the first two of these are useful since they can detect very small moments which the ordinary dielectric constant methods cannot reveal accurately; the third technique—involving the deviation undergone by a thin ribbon of gaseous molecules in passing through an intense nonhomogeneous electric field—is applicable to substances, such as metal salts, which through insolubility or low volatility would be otherwise unexaminable.

By the end of 1961, some 7000 dipole moment values for more than 6000 substances had been recorded (see McClellan's Tables); they fall mostly in the range 0 to 5 debyes.

Chemical interest is largely due to the relationships between polarity and molecular structure. Monatomic molecules, diatomic molecules of the type AA, and centrosymmetric polyatomic molecules, are nonpolar; a linear triatomic molecule ABA is nonpolar, but if bent or constructed as AAB it is polar; pyramidal tetratomic molecules AB_3 are polar, etc. A more quantitative approach supposes that characteristic polarities are associated with covalent chemical bonds, e.g., that two bonds, having "bond moments" μ_1 and μ_2, mutually inclined at θ^0, produce a resultant of $(\mu_1{}^2 + \mu_2{}^2 + 2\mu_1\mu_2 \cos \theta)^{0.5}$. On this basis, bond moments deduced from the resultant moments of molecules with known structures, often permit the discovery or testing of stereo specifications of further molecules. However, caution is necessary since bond moments are not independent of bond environments, but may be modified by induced moments—determined by the fields of neighboring polar bonds or centers and the (anisotropic) polarizabilities of the bonds under consideration—or by other internal electronic effects (c.f. resonance, mesomerism, hybridization, etc.). Completely successful calculations of dipole moments from *a priori* theory have yet to be made.

Some concepts developed for electrostatic fields have magnetic counterparts; thus in place of polarization $\mathbf{P}$ there is magnetization $\mathbf{I}$, the magnetic dipole moment per unit volume caused in a material by an externally applied field $\mathbf{H}$; internally the magnetic flux density (the magnetic induction) is $\mathbf{B} = \mathbf{H} + 4\pi\mathbf{I}$; the ratio $\mathbf{I}/\mathbf{H}$ is the volume susceptibility κ ordinarily measured. Individual magnetic monopoles are not known to exist in nature, but movements and spins of electrons in atoms and molecules—if viewed classically as direct currents flowing in closed circuits—can create fields identical with those expected from magnetic dipoles having moments dimensionally equivalent to products of pole strengths and distances. The elementary magnetic moment is the "Bohr magneton," 9.273×10^{-21} [erg gauss^{-1}], assumed to be the magnetic moment of an electron "spinning" on its own axis. Atoms may possess orbital moments (due to mechanical angular movements of electrons) and spin moments (one for each electron). Magnetic moments can be induced or permanent. A unit volume containing ν particles each of magnetizability α_m, subjected to a field $\mathbf{H}$, displays a magnetization $\mathbf{I} = \nu\alpha_m\mathbf{H} = \nu\overline{m}$, where $\overline{m}$ is the average magnetic dipole moment per particle; thus $\kappa = \nu\alpha_m$, and the molar susceptibility $\chi = \kappa V = N\alpha_m$ (where V is the molar volume and N the Avogadro number); α_m can be split into $\alpha_i + \alpha_p$, to correspond with the contributions to $\mathbf{I}$ made by the induced and permanent moments respectively.

An electron in an orbit of radius r represents a current loop; application of a magnetic field H perpendicularly to the loop plane will induce a voltage tending to create a field opposing that applied; the effect will be manifest as an apparent induced moment antiparallel to $\mathbf{H}$ and—by classical calculations—of the value $-e^2r^2\mathbf{H}/6mc^2$ (here e is the electronic charge, m is the electronic mass, c is the velocity of light); hence $\alpha_i = -e^2r^2/6mc^2$, and for a monatomic substance with spherical atoms the molar diamagnetic susceptibility $\chi = -(Ne^2/6mc^2)\sum nr_i^2$, where r_i^2 is the mean value of r^2 for the ith electron and the sum is taken over n electrons. The χ's observed for the inert gases, the C atoms in diamond, the Cl atoms in Cl_2, etc. have agreed with reasonable magnitudes of $\sum r^2$. Pascal (1910) showed diamagnetic susceptibility to be an "additive-constitutive" property, so that the χ's of polyatomic molecules can be approximately predicted by summing "atom" and "bond" susceptibilities in numbers and kinds appropriate to the molecular structure under consideration. The diamagnetic susceptibility of an individual molecule is a tensor

quantity; χ/N by experiment is an average of three principal magnetic susceptibilities directed along three mutually perpendicular principal axes of magnetic susceptibility; these can be investigated through torsional movements of crystals in magnetic fields (Krishnan's method) or from magnetic birefringence measurements (Cotton-Mouton effects) in conjunction with data for χ_{mean} secured with a Gouy balance.

A permanent magnetic dipole will experience a torque in a magnetic field. Langevin (1905) showed that the mean moment $\overline{m}$ of a gaseous molecule in the field direction (provided that H is not too large) is $\overline{m} = (m_p^2/3kT)\mathbf{H}$, where m_p is the actual moment of each molecule; therefore the molar paramagnetic susceptibility χ_p is $N\overline{m}/H = N\alpha_p = Nm_p^2/3kT$. In practice χ_p is extracted from the observed χ by treating this as the algebraic sum of a negative diamagnetic susceptibility (estimated from Pascal's constants) and a positive paramagnetic susceptibility; thus m_p follows as $(3kT\chi_p/N)^{0.5}$ [erg gauss^{-1}] or as $2.84(T\chi_p)^{0.5}$ [Bohr magnetons]. Molar diamagnetic susceptibilities are independent of temperature, while molar paramagnetic susceptibilities in general vary as $1/T$ or $1/(T - T_c)$. The small paramagnetisms of alkali metals, Cu, Ag, etc., or of certain salts (e.g., $KMnO_4$ or $K_2Cr_2O_7$), attributable respectively to uncompensated spins of conduction electrons, or to uncompensated paramagnetisms of complex ions, are temperature invariant.

Normally any atom or molecule with unpaired electrons shows paramagnetism and possesses a magnetic moment. Magnetic properties can therefore provide important information on valency states in free radicals, molecules containing first period elements with unpaired p electrons, transition elements having unpaired d electrons, lanthanides with unpaired $4f$ and actinides with unpaired $5f$ electrons. Theoretical expressions exist to calculate paramagnetic moments in terms of atomic structures and spin and orbital angular momenta of unpaired electrons. Simple examples are the ions of transition metals where, if n is the number of unpaired electrons, m_p is approximately predicted as $[n(n + 2)]^{0.5}$ Bohr magnetons (for a full discussion of such relations see Nyholm's review). Determination of m_p thus gives n, which is often of value in deciding the three-dimensional arrangements and bond types involved in molecules, especially those built around a central metal atom.

For most substances χ is independent of field strength, but a few paramagnetic compounds can, below the characteristic temperature T_c (see above), show "ferromagnetism" due to spontaneous parallel alignments of spins of atomic magnets. Materials which are ferromagnetic at ordinary temperatures (e.g., soft iron) have nonlinear magnetization-field characteristics, develop large magnetizations in weak fields, rapidly approach saturation conditions, exhibit hysteresis, etc.; whole domains about 0.01 mm in diameter and magnetically saturated are thought to be undergoing orientation during such processes. Ferromagnetism—and related phenomena such as "antiferromagnetism" and "ferrimagnetism"—have at present few applications in chemistry; in electronics (e.g., ferrites in antennas and in magnetic tape), they are frequently important.

R. J. W. LE FÈVRE

References

Debye, P., "Polar Molecules," New York, The Chemical Catalog Co., Inc., 1929.

Hippel, A. R. von, "Dielectrics and Waves," New York, J. Wiley & Sons, Inc., 1954.

Le Fèvre, R. J. W., "Dipole Moments," Third edition, London, Methuen and Co., Ltd., 1953.

Le Fèvre, R. J. W., "Molecular Refractivity and Polarizability" in "Advances in Physical Organic Chemistry," London and New York, Academic Press, 1965.

Le Fèvre, R. J. W., "Polarization and Polarizability in Chemistry," Rev. Pure and Applied Chem. (Aust.), 20, 67 (1970).

Le Fèvre, C. G., and Le Fèvre, R. J. W., in "Physical Methods of Chemistry," Vol. I, Weissberger, A. and Rossiter, B., Eds., New York, Wiley-Interscience, 1972 (part IIIc, pp. 399–452).

McClellan, A. L., "Tables of Experimental Dipole Moments," San Francisco and London, Freeman and Co., 1963.

Maryott, A. A., and Buckley, F., "Table of Dielectric Constants and Electric Dipole Moments of Substances in the Gaseous State," Natl. Bur. Std. Circ., 537 (1953).

Nyholm, R. S., Quart. Rev. London, 7, 377 (1953).

Selwood, P. W., "Magnetochemistry," New York, Interscience Publishing, 1956.

Smith, J. W., "Electric Dipole Moments," London, Butterworth's Scientific Publications, 1955.

Smyth, C. P., "Dielectric Behavior and Structure," New York, Toronto, London, McGraw-Hill Book Co., Inc., 1955.

Van Vleck, J. H., "The Theory of Electric and Magnetic Susceptibilities," Oxford, Clarendon Press, 1932.

Cross-references: BOND, CHEMICAL; DIELECTRIC THEORY; FERROMAGNETISM; MAGNETISM.

DOPPLER EFFECT

The wave effect by which astronomers measure the radial velocities of galaxies, and policemen determine the speeds of approaching automobiles, was in spite of its simplicity not discovered until the nineteenth century. In 1842, Christian Doppler predicted that the frequencies of received waves were dependent on the motion of the source or observer *relative to the propagating medium*. His predictions were promptly checked for sound waves by placing the source or observer on one of the newly developed railroad trains.

In his original article on the special theory of relativity (see RELATIVITY), Einstein[1] developed the expression for the Doppler shift of light waves which was dependent upon the velocity of the source *relative to the observer*. From the photon hypothesis for light, Schrödinger[2,3] obtained the same results. Thus, the Doppler effect provides one of the illustrations of the equivalence of the wave and particle descriptions of light.

Classroom demonstrations of the Doppler effect for water waves are made in shallow glass-bottom ripple tanks. Instead of giving the vibrating source a constant velocity, one lets the sheet of water as medium flow continuously by the source.

The circles of Fig. 1 are snapshots of the crests of a water wave or compressions in a sound wave observed when the source is moving at constant velocity v to the right relative to the medium. Points 1 and 2 are positions of the source one and two periods after passing O. The largest circular crest originated at O, the next at 1 and the smallest at 2. A crest is about to leave point 3 at the time the snapshot is taken. If the position P of the observer is a large distance from the source compared to the distance the source moves in one period, then with good approximation we may assume that two successive crests are moving in the same direction as they pass P. If v is the velocity of the source in the direction OA, the source moves a distance vT in one period T. In one period, the source comes closer to P by the amount $vT \cos \theta$, where θ is the angle between the direction of the velocity of the source and the line from the

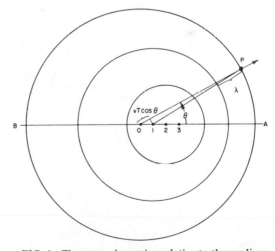

FIG. 1. The source is moving relative to the medium along the line BA. The circles represent crest of the wave at an instant. (Andrews, C. L., "Optics of the Electro-Magnetic Spectrum", Englewood Cliffs, N.J., Prentice-Hall, Inc., 1960).

source to the observer at P. Now λ_0 is the wavelength and ν_0 the frequency when the source is at rest; λ is the observed wavelength and ν the observed frequency when the source is in motion. Because of the motion of the source, the wavelength received at P is reduced by $\nu T \cos \theta$.

$$\lambda = \lambda_0 - \nu T \cos \theta$$

If c is the velocity of the wave, $T = \lambda_0/c$ and $\lambda = \lambda_0[1-(\nu/c) \cos \theta]$, but $\lambda = c/\nu$ and $\lambda_0 = c/\nu_0$. Therefore,

$$\frac{\nu}{\nu_0} = \frac{1}{1 - \dfrac{\nu}{c} \cos \theta} \tag{1}$$

when the *source is in motion relative to the medium*.

In Fig. 2 the source is at rest, but the observer at P has a velocity v with respect to the medium. The velocity of the wave relative to the observer is equal to the vector sum of the velocity of the wave relative to the medium and the velocity of the medium relative to the observer. In Fig. 2, c is the velocity of the wave and v the velocity of the observer relative to the medium. Let λ_0 be the wavelength, ν_0 the frequency of the source, and ν the frequency received by the moving observer. The radial velocity of the wave relative to the observer is $c + v \cos \theta$ so that

$$\nu \lambda_0 = c + v \cos \theta$$

For an observer at rest $\nu_0 = c/\lambda_0$. Substituting for λ_0, we obtain

$$\frac{\nu}{\nu_0} = 1 + \frac{v}{c} \cos \theta \tag{2}$$

when the *observer is in motion relative to the medium*.

By a postulate of relativity, the velocity of light is the same relative to all observers. The theory of relativity yields the frequency

$$\frac{\nu}{\nu_0} = \frac{1 + \dfrac{v}{c} \cos \theta_0}{\sqrt{1 - \dfrac{v^2}{c^2}}} \tag{3}$$

in which $v \cos \theta_0$ is the component of the velocity of the source toward the observer. The angle θ_0 is measured in the source system. If θ is the angle measured in the observer's system, then

$$\cos \theta_0 = \frac{\dfrac{v}{c} - \cos \theta}{\dfrac{v}{c} \cos \theta - 1} \tag{4}$$

Figure 3 is a graphical plot of ν/ν_0 against v/c for the radial motion in the three cases we have treated. (1) The linear relation is that for the observer in motion relative to the medium that propagates sound or other mechanical waves.

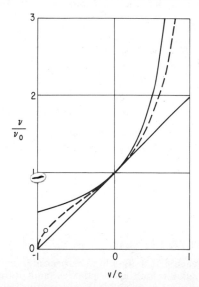

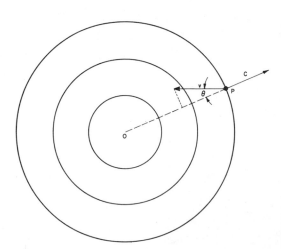

FIG. 2. The source is at rest at point O and the observer at point P is moving with velocity v relative to the medium. (Andrews, C. L., "Optics of the Electro-Magnetic Spectrum," Englewood Cliffs, N.J., Prentice-Hall, Inc., 1960).

FIG. 3. Graphical plots of the ratio of the observed frequency to the frequency at the source against the ratio of radial velocity to the velocity of the wave for three cases: (1) sound waves from a moving source, (2) sound waves to a moving receiver, (3) electromagnetic waves. The circle represents the red shift of light received from the most distant galaxies observed.

(2) The other solid curve is for the source of sound in motion. (3) The broken curve represents the Doppler effect for electromagnetic waves such as x-rays, light, and radio waves.

By comparing several spectral lines of elements observed in a star with a laboratory spectrum of the same elements, astronomers use the Doppler effect to measure the radial components of velocity of astronomical bodies toward or away from the earth. Spectra of the edges of the sun's disk are measured to determine the velocities toward and away from the earth. The radial velocities of the principal stars of our galaxy have been recorded. The spectral lines of some of the stars are doublets which periodically come together and separate again indicating that the light comes from two stars revolving about a common center of gravity (see ASTROMETRY).

In the expanding universe, the radial velocities of other galaxies away from our galaxy are proportional to their distances from the observer. Thus, the Doppler red shift provides a means of determining the dimensions of the observed universe. In 1964, some of the most intense radio sources (see RADIO ASTRONOMY) were located with high precision by observing these sources when the moon passed in front of them. With this knowledge of position, the same sources were located with a light telescope.[4] The measured red shift was surprisingly high. The sources were not stars as previously thought but the most distant galaxies known. In a period of eighteen years the Doppler red shifts were measured for more than a thousand quasi-stellar objects. Quasar OQ172 in the Catalogue of Quasi-Stellar Objects[5] had the largest red shift. If this shift were due solely to the Doppler effect, the astronomers had to conclude that this quasar was moving away from us with 91% of the speed of light. The Doppler frequency and velocity of this source are indicated by a circle on the broken curve of Fig. 3.

If a microwave beam is reflected from a moving microwave mirror, such as a person, an automobile or a man-made satellite, the image of the primary source may be considered as another source moving with twice the velocity of the mirror. Since the speed is small compared with the speed of light, the squared terms of Eq. (3) may be neglected. Thus

$$\nu = \nu_0 \left(1 + \frac{2v}{c} \cos\theta\right)$$

Direct frequency measurements cannot be made to enough significant figures to distinguish ν from ν_0. However, if the two frequencies are combined they give beats or the difference frequency

$$\Delta\nu = \nu_0 \frac{2v}{c} \cos\theta$$

Since the beat frequency is porportional to the radial velocity, a frequency meter may be calibrated in miles per hour. The precision of such a speed detector depends upon the frequency of the source being so stable that it varies less than the Doppler frequency shift during the time that the wave travels from the source to the mirror and back. The same phenomena of beats between the direct wave from the source and the wave reflected from a moving mirror may be observed with light. If one of the mirrors of Michelson's interferometer is moved at constant speed, the frequency with which dark bands pass the cross hair is the difference in frequency of the two waves.

The numerator of Eq. (3) contains a term for the radial component of velocity. However, the second-order term in the denominator is independent of direction. Thus, as v/c approaches unity, one may expect to detect a *tangential Doppler effect*. Ives and Stilwell[6] have measured the predicted value of the Doppler shift in frequency due to a stream of radiating molecules for which v/c was 10^{-2}. This experiment was a direct proof of time dilatation (see RELATIVITY) for the transverse case. In order to separate the tangential from the radial effect, Ives and Stilwell produced a sharply collimated beam of molecules. In order that θ be precisely $90°$, they set a mirror accurately normal to the line of observation and altered the line of observation until nearly the same wavelengths were given by direct and reflected light.

At relativistic speeds the transverse Doppler effect will be an important consideration in the final determination of the nature and positions of the quasars.[7]

C. L. ANDREWS

References

1. Einstein, A., *Ann. Physik*, **17**, 891 (1905).
2. Schrödinger, E., *Physik. Z.*, **23**, 301 (1922).
3. Michels, W. C., *Am. J. Phys.*, **15**, 449 (1947).
4. Burbage, G., and Burbage, M., "Quasi-Stellar Objects," San Francisco, W. H. Freeman Co., 1967.
5. Hewitt, A., and Burbidge, G., "Catalogue of Quasi-Stellar Objects," *Astrophysical J. Supplement* **43**, 57, 14-A1 (1980); Shipman, H. L., "Black Holes, Quasars and the Universe," Boston, Houghton Mifflin, 1980.
6. Ives, H. E., and Stilwell, A. R., *J. Opt. Soc. Am.* **31**, 369 (1941).
7. Gordon, K. J., *Am. J. Phys.* **48**, 514 (1980).

Cross-references: ASTROMETRY, RADAR, RADIO ASTRONOMY, RELATIVITY, WAVE MOTION.

DYNAMICS

Introduction *Dynamics*, a branch of mechanics, is often defined in two basic parts. First, *kinematics* is the study of the motion of bodies with no consideration of what has caused the motion, and second, *kinetics* is the study which relates the action of forces on the bodies to the resulting motion.

This order of presentation is often utilized in textbooks and other treatises on the subject. With various interpretations of the word *bodies*, the subject has also been divided into the areas of particle dynamics, rigid body dynamics, and fluid dynamics. As needs have developed, various specializations have been created and new theories have been formulated. These special areas include mechanical vibrations, flight dynamics, space dynamics, gas dynamics, magnetohydrodynamics, dynamic systems,[1] and relativistic dynamics, to mention a few.

The present state of knowledge recognizes all motion as relative, since no fixed reference is known for finding the absolute motion of any body. The earth, which is frequently used as a frame of reference is rotating about its own axis and is also revolving about the sun. The solar system, which consists of the sun and its planets, is a minute part of the Milky Way galaxy that is known to be revolving in space[2]. Beyond this, there is limited knowledge of the nature of motion that exists.

In the field of astronomy, measurements are evaluated in a coordinate system that is located relative to the *fixed* stars. These stars are located at such a vast distance from the earth that they appear as points of light that are almost motionless in space. In this frame of reference, the motions of celestial bodies are described with extremely great precision, and the motions of bodies within the solar system can be predicted accurately over periods of hundreds of years.

Some applied areas of dynamics, exemplified by the space exploration program, also require the degree of extreme accuracy that is possible with a celestial frame of reference. In many other areas, this extreme accuracy is not essential and measurements based upon this frame of reference would be tedious and impractical. In such cases, motion may often be adequately described in a coordinate system located relative to the earth.[3]

Kinematics A particle is a body having dimensions that are small, relative to other dimensions of the system, so that its motion may be considered equivalent to the motion of a point at its mass center with rotational effects neglected. Thus, particles may be either small or large. In the solar system, it would be possible to assume the earth to be a particle but, in a terrestrial system, this assumption could be totally unjustifiable.

A rigid body is a group of particles having unvarying external and internal configuration. The size of the rigid body would be appreciable in comparison to the other dimensions of the system, so that the rotational effect would have to be considered.

A fluid body is a group of particles with varying external and or internal configuration. The analysis of this type of system will not be considered in this article. (See FLUID DYNAMICS.)

The kinematic analysis of the motion of a particle may be approached through the establishment of a position VECTOR r, directed from the origin of a specified fixed coordinate system to the point representing the position of the particle, to give

$$r = x\,i + y\,j + z\,k \qquad (1)$$

where i, j, and k are unit vectors along the x, y, and z axes, respectively. Differentiating Eq. (1) with respect to time yields a velocity equation of the form

$$v = \frac{dr}{dt} = \frac{dx}{dt}i + \frac{dy}{dt}j + \frac{dz}{dt}k \qquad (2)$$

where the instantaneous velocity, v, of the particle at any position on its path is the instantaneous time rate of change of displacement. A second differentiation with respect to time yields the acceleration of the particle in the form

$$a = \frac{dv}{dt} = \frac{d^2 r}{dt^2} = \frac{d^2 x}{dt^2}i + \frac{d^2 y}{dt^2}j + \frac{d^2 z}{dt^2}k \qquad (3)$$

where the instantaneous acceleration, a, of the particle at any position on its path is the instantaneous time rate of change of velocity.

If x, y, and z are scalar functions of time, then Eqs. (1), (2), and (3) may be used to trace the path of the particle and determine the velocity and acceleration at any instant. These equations may be easily adapted to the cases of rectilinear translation and curvilinear translation of a particle in plane motion.

The kinematic analysis of a rigid body moving in a plane often involves the trace of two points, which may be called A and B, that are located on the body. These points move with the body and remain a fixed distance apart. Two coordinate systems may be used to define the position of the body in the plane. An X-Y coordinate system, with origin O, is a fixed reference, and an x-y coordinate system, with origin o located at A, is attached to the body so that it moves and rotates with the body. In the fixed reference system, a position vector R is directed from O to point A on the body and a second position vector ρ is directed from O to point B. In the moving coordinate system, a vector r is directed from A to B. An equation relating the position of the two points may be written as

$$\rho = R + r \qquad (4)$$

Using the I, J, K unit vectors for the X-Y-Z coordinate system and the i, j, k unit vectors for the x-y-z coordinate system, Eq. (4) may be rewritten as

$$\rho = X\,I + Y\,J + x\,i + y\,j \qquad (5)$$

Differentiating Eq. (5) with respect to time yields the velocity of B in the form

$$v_B = \frac{d\rho}{dt} = \frac{dX}{dt} I + \frac{dY}{dt} J$$

$$+ \frac{dx}{dt} i + x \frac{di}{dt} + \frac{dy}{dt} j + y \frac{dj}{dt} \quad (6)$$

Noting that $dx/dt = dy/dt = 0$, Eq. (6) simplifies to

$$v_B = v_A + v_{B/A} \quad (7)$$

where v_B and v_A are the velocities of points B and A, respectively, and $v_{B/A}$ is the velocity of point B relative to point A.

A second differentiation of Eq. (5) with respect to time yields the acceleration of B in the form

$$a_B = a_A + a_{B/A} \quad (8)$$

where a_B and a_A are the accelerations of points B and A, respectively, and $a_{B/A}$ is the acceleration of point B relative to point A.

A second important case of rigid body motion exists when point B is not attached to the same body as point A but is moving along a constrained path on this body. For the analysis of this motion, it is convenient to designate the fixed reference as body 1, the body to which point A is attached as body 2, and the body to which point B is attached as body 3. The same general arrangements of coordinate systems are used but in this case, the fixed X-Y coordinate system may be considered as attached to body 1 while the moving x-y coordinate system is attached to body 2. In the general case, vector **r** within the x-y coordinate system is varying in both magnitude and direction. Differentiating the position expression

$$\rho = R + r \quad (9)$$

once with respect to time and simplifying yields

$$v_{B3} = v_{A2} + v_{B2/A2} + v_{B3/2} \quad (10)$$

where v_{B3} is the velocity of point B on body 3, v_{A2} is the velocity of point A on body 2, $v_{B2/A2}$ is the velocity of point B on body 2 relative to point A on body 2, and $v_{B3/2}$ is the velocity of point B on body 3 relative to body 2.

A second differentiation with respect to time gives

$$a_{B3} = a_{A2} + a_{B2/A2} + a_{B3/2} + 2\omega_2 \times v_{B3/2}$$

$$(11)$$

where a_{B3} is the acceleration of point B on body 3, a_{A2} is the acceleration of point A on body 2, $a_{B2/A2}$ is the acceleration of point B on body 2 relative to point A on body 2, $a_{B3/2}$ is the acceleration of point B on body 3 relative to body 2, ω_2 is the angular velocity of

body 2, and $v_{B3/2}$ is the velocity of point B on body 3 relative to body 2. The term $2\omega_2 \times v_{B3/2}$ is often referred to as the CORIOLIS component of acceleration.

Equations 7 and 8 may be adapted to the case of rotation of a rigid body about a fixed axis at point A by considering point A to be fixed. Thus, v_A and a_A are both zero and

$$v_B = v_{B/A} \quad (12)$$

$$a_B = a_{B/A} \quad (13)$$

For a body rotating about a fixed axis, analysis of the rotational motion yields

$$\frac{d\theta}{dt} = \omega \quad (14)$$

$$\frac{d\omega}{dt} = \alpha \quad (15)$$

where θ is the angular displacement in radians, ω is the angular velocity in radians per second, and α is the angular acceleration in radians per second per second. It should be noted that time may be expressed in other units.

Kinetics *Newton's Laws of Motion* The laws of Newton are based upon the motion of a particle relative to a fixed frame of reference in which the particle can be made completely free of all outside influences. Under such a condition, the particle at rest will remain at rest and a particle in motion will continue to move at a constant velocity. This ideal frame of reference is often referred to as a Newtonian or as an inertial frame of reference.[4]

Since it is not possible to actually establish the Newtonian frame of reference, Newton's laws of motion are used in a celestial or a terrestrial frame of reference. The gyroscopic instruments and the stabilized platforms represent attempts to achieve a fixed or stabilized frame of reference for aircraft or space vehicles.

In modern terminology, Newton's laws of motion for a particle may be interpreted as

(1) A particle tends to remain at rest or continues to move at a constant velocity if there is no unbalanced force acting upon it.

(2) An unbalanced force acting on a particle will produce a time rate of change of momentum, $d(mv)/dt$, which, at any instant, will be proportional to the force and will be in the same direction as the force.

(3) The forces that exist between two contacting particles are equal in magnitude, are opposite in direction, and are collinear.[5]

The concept of the first law is the fundamental principle used for the analysis of forces acting on stationary particles and also for the analysis of forces acting on particles moving with a constant velocity. The concept has been expanded to include rigid bodies and fluid bodies.

The second law is the foundation of the analysis of forces acting on particles moving with accelerations and, again it has been extended to include rigid bodies which involve rotary motion and to include fluid bodies.

The third law is fundamental to the force analysis of interconnecting systems of particles under both static and dynamic conditions. It has also been extended to include simple contact between any pair of bodies and, with some modification, to include any type of interaction between bodies.

Force and Acceleration In general, Newton's second law is stated as

$$\sum \mathbf{F} = k \frac{d(m\mathbf{v})}{dt} \qquad (16)$$

where $\sum \mathbf{F}$ is the net unbalanced force on the particle; k is a constant of proportionality, consistent with the units used, that is determined experimentally; m is the mass of the particle; and $\mathbf{v}$ is the instantaneous velocity of the particle.

For a particle that is not shedding or accumulating mass, Newton's second law reduces to

$$\sum \mathbf{F} = km\mathbf{a} \qquad (17)$$

where $\sum \mathbf{F}$, k, and m are as previously defined and $\mathbf{a}$ is the instantaneous acceleration of the particle. By the proper choice of units in Eq. (16) and (17), the constants of proportionality can be made equal to unity.

For a rigid body in plane motion, both translational and rotational acceleration must be considered. By extending the concept of Newton's second law, it may be stated that

$$\sum \mathbf{F} = k_1 m\mathbf{a} \qquad (18)$$

$$\sum \mathbf{T} = k_2 \bar{I}\alpha \qquad (19)$$

where $\sum \mathbf{F}$ and $\sum \mathbf{T}$ are the unbalanced force and unbalanced torque, respectively, acting on the body; k_1 and k_2 are constants of proportionally, consistent with the units used, that are determined experimentally; m is the mass of the body; $\bar{I}$ is the mass moment of inertia of the body about a centroidal axis that is perpendicular to the plane of the motion; $\mathbf{a}$ is the linear acceleration of the center of mass of the body; and α is the angular acceleration of the body. Again, with proper choice of units, k_1 and k_2 can be made equal to unity.

The concepts and equations presented herein can be applied for the solution of a wide variety of problems which involve systems of particles, systems of bodies, or combinations thereof. The usefulness of this approach may be further broadened by the introduction of the closely related concepts of energy and impulse—momentum.

It should be noted that, in Newtonian mechanics, the fundamental property of the particle is an unvarying mass and time is absolute. It should also be noted that when the speed of the particle approaches the speed of light, this theory becomes inaccurate in compariosn to a theory based upon a more exact mathematical model attained through the application of the principles of RELATIVITY.[6]

GLENN L. DOWNEY

References

1. Cannon, Robert H., Jr., "Dynamics of Physical Systems," New York, McGraw-Hill, 1967.
2. Robertson, H. P., "The Universe," *Sci. Am.*, **195** (3), 73–81 (September 1956).
3. Kane, Thomas R., "Dynamics," New York, Holt, Rinehart and Winston, Inc., 1968.
4. Goodman, L. E., and Warner, W. H., "Dynamics," Belmont, California, Wadsworth, 1964.
5. Smith, G. M., and Downey, G. L., "Advanced Engineering Dynamics," 2nd ed., Scranton, Pa., International Textbook, 1968.
6. Synge, J. L., and Griffith, B. A., "Principles of Mechanics," 3rd ed., New York, McGraw-Hill, 1959.
7. Goldstein, Herbert, "Classical Mechanics," 2nd Ed., Reading, MA, Addison-Wesley, 1980.
8. Shigley, Joseph E., and Uicker, John J., "Theory of Machines and Mechanisms," New York, McGraw-Hill, 1980.

Cross-references: ASTRONAUTICS, CORIOLIS EFFECT, FLUID DYNAMICS, IMPULSE AND MOMENTUM, MECHANICS, STATISTICS.

E

ELASTICITY*

Elasticity is the part of mechanics dealing with deformations that vanish entirely once the forces that have caused them are removed. Most solid bodies behave elastically for sufficiently small deformations, and we will be concerned here with the infinitesimal theory of elasticity. Also we will consider only isotropic bodies, that is, bodies whose elastic properties are the same in all directions.

The fundamental quantities in elasticity are second-order tensors, or dyadics: the deformation is represented by the *strain dyadic*, and the internal forces are represented by the *stress dyadic*. The physical constitution of the deformable body determines the relation between the strain dyadic and the stress dyadic, which relation is, in the infinitesimal theory, assumed to be linear and homogeneous. While for anisotropic bodies this relation may involve as much as 21 independent constants, in the case of isotropic bodies, the number of elastic constants is reduced to two.

Let $s(r)$ be the displacement vector, due to the deformation, of a particle that before the deformation was situated at point P having r as position vector with respect to some arbitrary origin. A neighboring point Q, whose position vector was $r + dr$ before the deformation, will suffer a displacement $s(r + dr)$ which will differ from $s(r)$ by the quantity

$$ds = dr \cdot \nabla s$$

The hypothesis of small deformations means that ds, the change in the displacement vector when we go from P to the neighboring point Q, is very small compared to dr, the position vector of Q relative to P. Consequently, the scalar components of the dyadic ∇s are all very small compared to unity. The geometrical meaning of the dyadic ∇s is obtained by separating it into its symmetric part $S = \frac{1}{2} (\nabla s + s\nabla)$ and its antisymmetric part $R = -\frac{1}{2} 1 \times (\nabla \times s)$, where 1 is the unity dyadic. The antisymmetric part is interpreted as follows: if at some point M the symmetric part vanishes, then we have for the neighborhood of M the relation

$$ds = dr \cdot R_M = \omega_M \times dr$$

*See MECHANICAL PROPERTIES OF SOLIDS for a less mathematical introduction to elasticity.

where $\omega_M = \frac{1}{2} (\nabla \times s)_M$ is an infinitesimal vector. This means that the neighborhood of point M undergoes an infinitesimal rigid rotation, without any change in shape or size. Consequently, the deformation is represented by the symmetric part S, which is called the *strain dyadic*.

In a Cartesian orthonormal basis, in which we have $r = \sum_{i=1}^{3} x_i a_i$, we write $s = \sum_{i=1}^{3} s_i a_i$, and obtain

$$S = \sum_{i, j=1} a_i a_j S_{ij}$$

where $S_{ij} = \frac{1}{2} \left[\frac{\partial}{\partial x_i} s_j + \frac{\partial}{\partial x_j} s_i \right]$. The diagonal components S_{11}, S_{22}, and S_{33} are the coefficients of linear extension in the directions a_1, a_2, and a_3, respectively, while the non diagonal components $S_{12} = S_{21}$, $S_{13} = S_{31}$, and $S_{23} = S_{32}$ are called shear strains. For instance, $2S_{12}$ is the change in the angle of the dihedron formed by the planes that before the deformation were respectively normal to the directions a_1 and a_2. The shear strains are not essential for the complete representation of a deformation since they can be made to vanish by expressing S in the basis of its principal axes.

If an infinitesimal element of the body occupies the volume dV before the deformation and the volume dV' after, the relative increase of volume, or volumetric dilatation, is given by

$$\frac{dV' - dV}{dV} = S_{11} + S_{22} + S_{33} = |S| = \nabla \cdot s$$

The forces applied to a finite deformable body are either body forces acting on every volume element dV and represented by the notation $dV F = dV \rho K$, where F is the force per unit volume, K is the force per unit mass, and ρ is the density, or surface forces acting on every element dS of the bounding surface and represented by $dS T$, where T is the surface stress, or surface force per unit area. The effect of these applied forces is transmitted throughout the body, so that through any surface element inside the body, there is a force exerted by the matter on one side of the element upon the matter on the other side. Such forces are called internal stresses and are defined as follows: let dS be a surface element completely inside the body, and let us choose arbitrarily the positive sense of the normal n to this surface element; this defines for dS a positive side, the one containing n, and a negative side. Then T_n, the stress vector on the positive side of dS is defined as a vector such that $dS T_n$ is the surface force on the positive side of dS—i.e., the resultant of all the forces exerted through dS by the matter on the positive side of dS upon the matter on the negative side. In general there is a normal component $T_n \cdot nn$, which is a pressure or a traction depending upon whether the sign of $T_n \cdot n$ is negative or positive, and a tangent component $n \times T_n \times n$ called the shear stress. The value of stress vector T_n depends upon the orientation of the normal n, so that we can characterize the state of stress at a point by defining the *stress dyadic* T through the relation

$$T_n = n \cdot T$$

The mechanical equilibrium conditions applied to an arbitrary volume V, bounded by the closed surface S, and completely inside the deformable body give

$$\int_V dV \mathbf{F} + \int_S dS \, n \cdot \mathbf{T} = 0$$

and

$$\int_V dV \mathbf{r} \times \mathbf{F} + \int_S dS \mathbf{r} \times (n \cdot \mathbf{T}) = 0$$

By the use of the divergence theorem, the first condition gives the equation

$$\nabla \cdot \mathbf{T} + \mathbf{F} = 0$$

at any point inside the body, and the second condition implies that T is a symmetric dyadic. On the external surface of the body, we have usually to fulfill the boundary condition

$$n \cdot \mathbf{T} = \mathbf{T}$$

where T is the applied external force per unit area. Other boundary conditions can also be met, such that the value of the displacement be prescribed.

For infinitesimal deformations, we assume that the relation between strain and stress is expressed by Hooke's law: the deformation is proportional to the applied force. For isotropic bodies, this linear relation is

$$\mathbf{S} = \frac{1}{E} [(1 + \nu)\mathbf{T} - \nu |\mathbf{T}| \mathbf{1}]$$

where E is Young's modulus and ν is Poisson's ratio. These two elastic constants can be defined by considering the stretching of a cylindrical bar by normal traction forces uniformly distributed on the end sections; then we have

Young's modulus =

$$\frac{\text{Normal traction force/unit cross sectional area}}{\text{Relative longitudinal extension}}$$

and

$$\text{Poisson's ratio} = \frac{\text{Relative lateral contraction}}{\text{Relative longitudinal extension}}$$

We can also write

$$\mathbf{T} = 2\mu \mathbf{S} + \lambda |\mathbf{S}| \mathbf{1}$$

where $\mu = E/2(1 + \nu)$ and $\lambda = \nu E/(1 + \nu)(1 - 2\nu)$ are Lamé's constants. μ is the rigidity modulus, the only constant necessary when the volumetric dilatation vanishes everywhere.

Substituting the preceding relation into the equilibrium equations, we transform them into

$$2\mu \nabla \cdot \mathbf{S} + \lambda \nabla |\mathbf{S}| + \mathbf{F} = 0 \text{ inside the body}$$

and

$$2\mu n \cdot \mathbf{S} + \lambda n |\mathbf{S}| = \mathbf{T} \text{ on the bounding surface.}$$

These vector relations are not sufficient for the complete determination of the symmetric dyadic S. To insure that a solution of the above equations corresponds to a possible displacement vector s, we must be able to integrate the relation

$$\mathbf{S} = \tfrac{1}{2} (\nabla \mathbf{s} + \mathbf{s}\nabla)$$

i.e., from a given expression for S, obtain the value of s. From the vanishing of the curl of a gradient, it is easily seen that this integrability condition, also called the compatibility equation, is

$$\nabla \times \mathbf{S} \times \nabla = 0$$

By elimination of the vector products, we obtain the equivalent form

$$\nabla \nabla \cdot \mathbf{S} + \nabla \cdot \mathbf{S}\nabla - \nabla \nabla |\mathbf{S}| - \nabla \cdot \nabla \mathbf{S} = 0$$

Using the stress-strain relation and the equilibrium conditions, we obtain the Beltrami-Michell form of the compatibility equation:

$$\nabla \cdot \nabla \mathbf{T} + \frac{1}{1 + \nu} \nabla \nabla |\mathbf{T}| = - \frac{\nu}{1 - \nu} \nabla \cdot \mathbf{F}\mathbf{1}$$
$$- (\nabla \mathbf{F} + \mathbf{F}\nabla)$$

Finally, by expressing the strain dyadic in terms of the displacement vector, we obtain Navier's form of the equilibrium equations:

$$\mu \nabla \cdot \nabla \mathbf{s} + (\lambda + \mu) \nabla \nabla \cdot \mathbf{s} + \mathbf{F} = 0 \text{ inside the body}$$

and

$$\lambda n \nabla \cdot \mathbf{s} + 2\mu n \cdot \nabla \mathbf{s} + \mu n \times (\nabla \times \mathbf{s}) = \mathbf{T}$$
$$\text{on the bounding surface.}$$

Dealing here directly with the displacement vector, there is no need of considering the compatibility equation.

The propagation equation for elastic disturbances is obtained by adding the inertia force to the body force. We get then

$$\mu\nabla \cdot \nabla s + (\lambda + \mu)\nabla\nabla \cdot s + \rho K = \rho \frac{\partial^2}{\partial t^2} s$$

inside the body.

The stress-strain relation and the boundary conditions are not affected, but we generally have to take into account initial conditions.

The energy density u, or energy per unit volume, is given by

$$u = \frac{1}{2} S : T + \frac{1}{2}\rho \frac{\partial s}{\partial t} \cdot \frac{\partial s}{\partial t}$$

where the first term is potential, or strain energy, and the second term is kinetic energy. The energy flux density vector

$$S = -\frac{\partial s}{\partial t} \cdot T$$

is a vector such that $dSn \cdot S$ gives the quantity of energy that flows per unit time through the surface element dS in the positive direction of n, the normal to dS. At any point the energy continuity equation

$$\frac{\partial u}{\partial t} + \nabla \cdot S - \rho \frac{\partial s}{\partial t} \cdot K = 0$$

expresses the conservation of mechanical energy.

Gérard Nadeau

References

Godfrey, D. E. R., "Theoretical Elasticity and Plasticity," London, Thames and Hudson Co., 1959.

Green, A. E., and Zerna, W., "Theoretical Elasticity," New York, Oxford University Press, 1954.

Jaunzemis, W., "Continuum Mechanics," New York, Macmillan, 1967.

Lai, W. M., Rubin, D., and Krempl, E., "Introduction to Continuum Mechanics," New York, Pergamon Press, 1978.

Nadeau, G., "Introduction to Elasticity," New York, Holt, Rinehart and Winston, Inc., 1964.

Pearson, C. E., "Theoretical Elasticity," Cambridge, Mass., Harvard University Press, 1959.

Sokolnikoff, I. S., "Mathematical Theory of Elasticity," New York, McGraw-Hill Book Co., Inc., 1956.

Cross-references: MECHANICAL PROPERTIES OF SOLIDS, POLYMER PHYSICS, VECTOR PHYSICS, VISCOELASTICITY.

ELECTRICAL POWER GENERATION

About 98% of the electric power produced in this country is by three-phase generators. It is transmitted and distributed this way. Advantages of three-phase generators lie in economy of apparatus, lower transmission losses, inherent starting torque for polyphase motors, and constant running torque for balanced loading. A generator is built with axial slots for armature coils in a stationary hollow cylindrical iron core called the *stator*. The windings are placed in the slots so that when carrying current they produce a chosen even number of alternate magnetic poles. The coils over each magnetic pole are grouped in three equal bands to give a three-phase balanced system of terminal voltages.

An inner rotor has coils which carry direct current to give the same number of alternate magnetic poles as on the stator. Rotor current strength is controlled by a rheostat or voltage from a dc generator. Voltages are produced in the stator windings by flux cutting as the rotor magnetic flux sweeps by them, currents flow when the generator terminals are connected to a three-phase load impedance. The three-phase stator line voltages are equal in magnitude and 120 electrical degrees apart in time sequence. So also are the line currents for a balanced three-phase load. Generator voltages are of the order of 12,000 to 30,000 volts for large machines.

Generator frequency is the product of the pairs of magnetic poles and the speed in revolutions per second. At 60 Hz (cycles per sec), a two-pole generator runs at 3600 rpm and a six-pole generator at 1200 rpm. The maximum speed of 3600 rpm has been increasingly adopted even for very large machines because high speed means decreased size and weight for a given kilowatt rating and better steam-turbine performance. Waterwheels and water turbines show best characteristics at much lower speeds—roughly a range of 100 to 600 rpm. A frequency of 60 Hz prevails in this country for public utility power generation. Because of weight and space limitation, 400 Hz is found in the aircraft industry. Europe is basically on 50 Hz.

In the large central station steam power plants, single generators (*units*) reach, or go somewhat beyond, 1,250,000 kW in rating. Some units have ratings up to 1.5 million kilowatts.

Direct-current generators are built with their dc magnetic poles in the stator. Armature conductors in the rotor have ac voltages induced in them as they are rotated; the same principle of flux cutting holds as before. An automatic mechanical switching device, called a *commutator*, is placed on the shaft. It carries fixed brushes, and with its many insulated copper bars connected to the armature coils, it inverts every other alternation of the voltage at the two armature terminals. It is the commutator that requires the rotor to be the armature so that coils and their switching arrangement always move exactly together. Direct-current generators are generally limited to several thousand kilowatts and their application lies mainly in industrial plants.

The majority of central station generating plants in the USA are steam turbine generators. Coal, oil, or gas is used to heat water to around $1000°F$ at pressures of 2000–3500 psi. This superheated steam is successively expanded through high pressure, medium pressure, and low pressure turbines, all mounted on a common shaft with the generator. Steam is extracted at perhaps five to seven points through the turbines. This extracted steam is used to reheat the water from the condenser in the boiler feedwater heaters. The effect is to "square-off" the Rankine cycle and bring the system efficiency closer to Carnot efficiency. Actual generating plant efficiencies approach 40%.

Nuclear power plants utilize the same Rankine steam cycle as do fossil fuel plants, but, the heat is provided by nuclear fission. Temperatures and pressures are not as high as in fossil fuel plants, and therefore operating efficiencies are not quite as high. Nuclear power generation has suffered some growing pains that have served to increase the wariness of the populace. Such wariness has slowed the growth of nuclear power.

Other methods of generating electricity include gasoline- and diesel-driven generator sets, gas turbine-driven generator sets, fuel cells, solar cells, wind driven generators, magneto-hydrodynamic (MHD), generators, and geothermal generators. The engine- and gas-driven generators are normally used as backup sytems or peaking systems for high demand periods. They are also used for prime power in remote, isolated installations. Fuel cells and solar cells are being used in very small power demand applications such as microwave relay sites and spacecraft. Both have yet to demonstrate the efficiencies necessary for economic feasibility. Wind driven generators up to a few hundred kw are being developed to relieve the burden on fossil fuels. Relability and economic feasibility have not yet been demonstrated. MHD generators received much attention during the 1970s, but many technological challenges have not yet been met. MHD units have had fair success in Britain when combined with conventional steam turbine-generator plants to raise the overall plant efficiency.

The limited efficiency of steam turbines imposed by the thermodynamic properties of steam has stimulated the development of methods to convert heat directly into electricity. The MHD generator is one in which a thermally ionized gas is forced at high temperature, pressure, and velocity through a duct situated in a transverse magnetic field. An induced voltage appears in the third mutually perpendicular direction (the Hall effect), and this voltage may be tapped by electrodes within the duct (see MAGNETOFLUID-MECHANICS).

When the exhaust gas from the MHD generator is used to heat steam for a conventional generator, a larger portion of the thermal spectrum is utilized and the system efficiency may be raised from the present 40% to possibly 50 or 55%. Heat for the system may come from the use of fossil fuel, nuclear reactors, or as expected in the future, fusion reactors.

Geothermal production of electric power uses natural steam obtained from the earth through steam wells and piped to turbines. Italy produces about one-third of a million kW in this manner and New Zealand has slightly less installed capacity of this kind. The only US installation is about 175,000 kW on the West Coast. Geothermal power is limited. Temperatures and pressures are low, but there is a lower capital investment and absence of fuel cost. A serious problem lies in elimination of contamination in the steam. These systems remain in prototype stages.

Probable Future Trends Considerable work is being done on boiler materials to increase the operating temperatures and pressures of steam plants. Ceramics look promising. Nuclear fusion research will provide the answers in clean nuclear power. Fluidized bed combustion techniques will reduce fossil fuel emissions.

FREDERICK C. BROCKHURST

References

US Department of Energy Reports, "Electric Power Supply and Demand for the Contiguous United States."

Mablekos, Van E., "Electric Machine Theory for Power Engineers," New York, Harper & Row Publishers, 1980.

Cross-references: ELECTRICITY; ENERGY STORAGE, ELECTROCHEMICAL; ENERGY STORAGE, THERMAL-MECHANICAL; HALL EFFECT AND RELATED PHENOMENA; MAGNETO-FLUID-MECHANICS; NUCLEAR REACTORS; PHOTO-ELECTRICITY; THERMOELECTRICITY; WORK, POWER AND ENERGY.

ELECTRIC PROPULSION

Electric propulsion is a form of rocket propulsion in which electric power, generated on board the propelled vehicle, is used to eject propellant rearward at high velocity to produce thrust. Electric propulsion systems can be considered to be made up of two major components: (1) the *electric power generation system*, which converts power from a basic power source (such as a nuclear reactor or the sun) into electric power, and (2) *the thruster*, which uses this electric power to produce thrust by ejecting the propellant.

The primary potential advantage of electric

rockets over chemical rockets (or hypothetical solid-core nuclear rockets) is that much higher propellant ejection velocities can be attained. Higher ejection velocities, in accordance with Newton's law, produce higher thrust per unit mass of propellant, so that the total mass of propellant needed for space missions can be greatly reduced. The mass of required electric power generation equipment is appreciable, however, so that some of the saving in propellant mass is offset by the mass of the power generation system. The net mass saving possible using electric propulsion therefore depends strongly on the performance parameters of the system.

One of the most important of these performance parameters is the propulsion-system specific mass α, which is defined as

$$\alpha = \frac{m_{ps}}{P_j} \quad \frac{kg}{kW} \tag{1}$$

where m_{ps} is the total propulsion system mass (in kilograms) and P_j is the jet power produced (in kilowatts). If this parameter is less than about 30 kg/kW, electric propulsion systems can be employed to advantage over nuclear or chemical rockets for many unmanned interplanetary exploration missions. For such missions, typical required power levels range from about 25 kilowatts to several hundred kilowatts, to propel vehicles having initial mass in earth orbit in the range of 1000 to 10 000 kg.

If α is less than about 5 kg/kW, mission studies indicate that electric propulsion is superior to nuclear rockets, with regard to required initial weight and trip time, for manned expeditions to the near planets.[1] For these missions, power levels of several megawatts would be needed for vehicle weights (in orbit) of the range of 100 000 to 1 000 000 kg.

In other possible applications, such as providing small amounts of thrust for attitude control or orbit control of satellites, the specific mass is less important, since the required electric power is small compared to that used by the other on-board equipment.

Most of the mass of an electric propulsion system resides in the electric power generation system; however, the performance of the other major component, the thruster, is of equal importance in determining the over-all specific mass. The most important parameter for the thruster is the efficiency η with which the electric power is converted into jet power. If this efficiency is low, the required electric power, and therefore the power-plant mass, is correspondingly high.

Another important parameter for the thruster (as for all rockets) is the specific impulse I. This parameter is defined as the thrust F produced per unit weight flow of propellant:

$$I = \frac{F}{\dot{m}_p g_0} \quad sec \tag{2}$$

where $\dot{m}_p$ is the mass flow rate of propellant and g_0 is the acceleration of gravity at the earth's surface (9.8 m/sec²) which relates mass to weight. The relation of thrust, specific impulse, and propellant ejection velocity is

$$F = \dot{m}_p g_0 I = \dot{m}_p v_j \quad newtons \tag{3}$$

where v_j is the mean propellant ejection velocity (more commonly called *effective jet velocity*). The first and last terms in Eq. (3) express Newton's law that force is equal to the time rate of change of momentum. The last two terms show that specific impulse is directly proportional to effective jet velocity.

The *jet power* is the time rate of change of jet kinetic energy, or

$$P_j = \tfrac{1}{2}\dot{m}_p v_j^2 = \tfrac{1}{2}Fv_j = \tfrac{1}{2}g_0 IF \quad newton\text{-}m/sec$$

or, in kilowatts,

$$P_j = \frac{g_0 IF}{2000} \quad kW \tag{4}$$

For constant thrust and jet velocity, the total propellant mass m_p needed for a mission can be written [from Eq. (3)] as

$$m_p = \frac{Ft}{g_0 I} \quad kg \tag{5}$$

where t is the total propulsion time and Ft is the total impulse required for the mission. From Eq. (1) and (4), the propulsion system mass can be written:

$$m_{ps} = \alpha P_j = \frac{\alpha g_0 IF}{2000} \quad kg \tag{6}$$

These equations show that, although propellant mass can be reduced indefinitely by increasing the specific impulse [Eq. (5)], the power required (and therefore the power-plant mass) is increased when this is done [Eq. (6)]. It is, therefore, desirable to use that value of specific impulse for which the *sum* of the masses of propellant and propulsion system is lowest. This optimum specific impulse will yield the least total mass for the mission, or the highest payload mass for a given total mass. For lunar and interplanetary missions and for specific mass likely to be obtained, calculations show that the optimum specific impulses range from about 1500 to 15 000 seconds (corresponding to jet velocities of about 15 to 150 km/sec). These specific impulses compare with values of about 450 seconds that are typical for high-energy chemical rockets and about 900 seconds that may be possible with solid-core nuclear rockets.

Another characteristic feature of electric propulsion systems is the very low thrust generated in comparison with chemical or nuclear rockets. This can be seen from Eq. (6) which can be written:

$$\frac{F}{m_{ps}g_0} = \frac{2000}{\alpha I g_0{}^2} \qquad (7)$$

For a specific mass α of 10 kg/kW, and a specific impulse I of 5000 seconds, Eq. (7) yields a thrust-to-weight ratio of about 4×10^{-4}. This very low value results partly from the higher specific impulse typical of electric rockets, but mostly from the specific mass, which is of the order of 1000 or more times higher than that obtainable with solid-core nuclear rockets or chemical rockets. The low thrust-weight ratio means that electric propulsion systems cannot be used for launching from planetary surfaces. They are best suited for propelling vehicles between orbits about the planets or between orbits about the earth and the moon.

Because the thrust-weight ratio is so low, electric rockets must operate for much longer periods of time (of the order of 1000 times longer) than chemical or nuclear rockets to produce the same total impulse. Typically, for interplanetary missions to the near and far planets, these required operating times range from many months to several years. The removal of limitations on jet velocity, therefore, is obtained at the expense of greatly increased propulsion system mass and required operating lifetime.

Power Generation Systems The need for low specific mass dominates the selection of suitable methods for generating electric power for primary propulsion of space vehicles. The requirement that power be generated with very little consumption of mass dictates that either nuclear or solar energy must be used as the basic energy source.

Among the possible methods of converting this energy into electric power, the most direct are photovoltaic solar cells. Considerable progress has been made in reducing the thickness, and hence the weight of photovoltaic solar cells;[2] eventual achievement of a specific mass near 5 kg/kW appears possible. A lightweight radioisotope cell, in the range of 1 kg/kW, has been proposed and analyzed[3] but not yet demonstrated. This cell is basically a very high-voltage, low-current device, which converts a fraction of the kinetic energy of the isotope decay particles directly into electric power.

Somewhat less direct in energy conversion are systems that use thermionic cells

$$\left(\begin{matrix} \text{nuclear} \rightarrow \\ \text{solar} \rightarrow \end{matrix} \text{heat} \rightarrow \text{electricity} \right)$$

A nuclear reactor or solar concentrator is used to heat a suitable material (such as tungsten) to temperatures high enough to produce thermal emission of electrons. These electrons traverse a gap to a cooled collector electrode, thereby producing electric power at a potential of the order of 1 volt. Many thousands of these thermionic cells must be connected in series-parallel combinations to achieve the required power levels

and voltages. Also, to produce useful power densities, emitter temperatures must be in the range 1500 to 2000 K. Conversion efficiencies (heat into electric power) of 15 to 30 per cent are possible. The remaining 70 to 85 per cent of the thermal power must be radiated into space. The collector electrodes, where this waste heat appears, must be adequately cooled by a heat-transfer fluid that is pumped past the collector to pick up the waste heat and carry it to a radiator. In order that the radiator be of adequately low size and weight, it must operate at temperatures of about 900 K or higher. Analyses for a complete nuclear thermionic system yield specific masses of the order of 4 to 10 kg/kW, but numerous severe performance, design, and engineering problems remain to be solved before such systems can be developed to mission status [4]

Still more indirect, in the conversion of energy, are the turboelectric systems

$$\left(\begin{matrix} \text{solar} \rightarrow \\ \text{nuclear} \rightarrow \end{matrix} \text{heat} \rightarrow \text{mechanical} \rightarrow \text{electric} \right)$$

For these, as well as the thermionic systems, the nuclear reactor appears to be a better basic energy source than the sun, because it provides a more compact and versatile system, suitable for operation in shaded regions and at any distance from the sun.

A nuclear turboelectric system for electric propulsion (as illustrated in Fig. 1) is basically a lightweight adaptation to space conditions of small ground-based nuclear power stations.[5] The chief differences result (as for the thermionic systems) from the lack of means other than radiation to eliminate the waste heat resulting from inescapable conversion inefficiencies. To produce specific mass below 10 kg/kW, the waste-heat radiator must operate at temperatures above 900 K, which in turn requires that the nuclear reactor operate at temperatures in excess of 1200 K.

The most suitable working fluid, at these temperatures, is potassium, if a liquid-vapor thermodynamic cycle (Rankine cycle) is used. In a single-loop version of this cycle, the liquid metal is vaporized in the nuclear reactor; the resulting vapor drives the turbine, which in turn drives the generator to produce electric power. The vapor passes from the turbine through the radiator, where it is recondensed, and the liquid is then recirculated through the reactor. A major problem is to develop materials with adequate corrosion resistance during long periods of high-temperature operation with alkali liquid metals. As illustrated in Fig. 1, the radiator is the largest and heaviest part of the system.

A possible alternative to the turboelectric system is an MHD (magnetohydrodynamic) generator, which replaces the turbogenerator with a duct through which a hot, electrically conducting fluid is passed. The duct is embedded in a

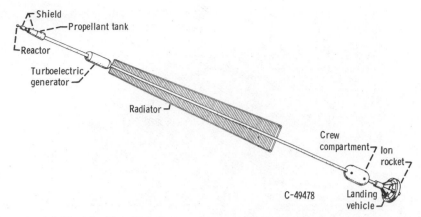

FIG. 1. Conceptual design of space vehicle for manned Mars mission. Nuclear turboelectric propulsion system.

strong magnetic field which must be produced by superconducting coils to minimize the power consumption. The most suitable fluid seems to be a noble gas seeded with cesium to make it electrically conducting.[6] When this fluid, heated by the nuclear reactor, is forced through the magnetic field, electric currents are induced, and power can be extracted by electrodes embedded in the duct. Such a system can tolerate higher temperatures than turbines, and should be particularly attractive for required power levels in the multi-megawatt range.

Thrusters A large number of methods are possible to eject propellant by use of electric power. These are generally divided into three categories: (1) *electrostatic thrusters*, in which atoms (or heavier particles) are electrically charged and then accelerated rearward by means of an electrostatic field; (2) plasma thrusters, in which the propellant is made into an electrically conducting gas and accelerated rearward by application of electromagnetic forces; (3) electrothermal thrusters, which use the electric power to heat the propellant, and then accelerate it rearward by thermal expansion through a nozzle.

Electrostatic thrusters that accelerate atomic ions (ion rockets) have received the most research and development attention, as a result of early demonstrations of good efficiencies in the range of specific impulses needed for major space missions. Typical of these ion rockets is an electron-bombardment thruster (such as that shown in Fig. 2) which uses mercury vapor as propellant.[7] The propellant atoms are ionized by collision with electrons emitted by the cathode and attracted toward the anode. A weak axial magnetic field is maintained in the ionization chamber to make the electrons spiral around on their way to the anode, thereby increasing their path length and their probability of colliding with propellant atoms. The resulting positive ions are extracted through a screen grid by means of an accelerating grid that is main-

tained at the proper voltage difference (usually several thousand volts) to produce the desired ejection velocity (specific impulse). A second electron emitter (not shown) is placed adjacent to the ion beam, downstream of the accelerator, to neutralize both the ion space charge and the net current leaving the thruster. Experimental efficiencies in converting electric power into jet power range from 60 to 80 per cent at specific impulses in the range 2500 to 9000 seconds. Thrusters in sizes up to 150 cm in diameter, with jet powers near 180 kW have been successfully operated.[7]

Other ion thrusters, using contact ionization of cesium atoms on hot tungsten to produce the ions (rather than electron bombardment), have achieved somewhat lower performance. In these thrusters, cesium vapor is passed through porous tungsten, which must be heated to about 1500 K to evaporate enough cesium ions from the ionizer surface. The high work function of tungsten and the low ionization potential of cesium make these two substances the most promising for contact ionization thrusters.

Atomic-ion thrusters tend to become less efficient at low ejection velocities (low specific impulse), because a certain fixed amount of energy is needed to ionize the propellant atoms. As the ejection velocity decreases, the jet power approaches the power required for ionization, and the efficiency decreases. A possible way to increase the efficiency is to increase the mass of each charged particle so that its kinetic energy, at a given jet velocity, is higher. This approach leads to use of colloidal particles in place of atomic ions. Because of the much higher mass per unit charge, voltages in the hundreds of kilovolts are needed to produce the desired jet velocities.

Although the efficiencies attainable with electrostatic thrusters are high, there remains a limitation which, although not crucial, is undesirable, namely, a low thrust (or power) per unit beam area, due to limitations on ion beam cur-

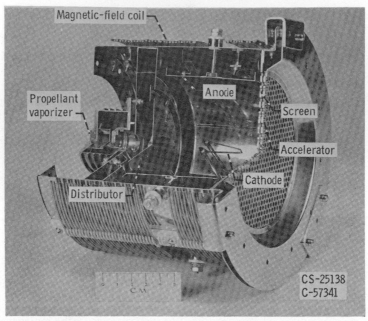

FIG. 2. Cutaway photograph of electron-bombardment ion thruster. With about 1 kW of power, this unit produces a thrust of 0.025 Newtons grams at a specific impulse of 5000 sec.

rent density. These limitations result from two sources: (1) space charge and (2) accelerator electrode erosion. The space-charge limited current is determined by the accelerating voltage and the distance between accelerator electrode and ion source. The voltage, in turn, is approximately fixed by the desired specific impulse, and the accelerator spacing is limited by electrical breakdown and thermal warping. The erosion limitation appears to be even more restrictive on thrust per unit area than space charge. As the current density is increased, there is greater impingement of ions on the accelerator electrode. For an accelerator lifetime of the order of 1 year, estimates indicate a limit for thrust per unit beam area of about 2 newtons (0.2 kg) per square meter (about 50 kW/m^2) at a specific impulse of 5000 seconds. Higher values are allowable as specific impulse increases.

Plasma thrusters, which operate on the principle of accelerating an electrically conducting gas (plasma) are not subject to the space charge limitation, and require no accelerator grid. Consequently, a higher thrust per unit area with adequate lifetime may be achievable. A variety of plasma thruster types have been investigated during the past decade,[8] but so far the efficiencies have been much lower than those of ion thrusters. Sizable effort has been devoted to the magnetoplasmadynamic (MPD) arc jet, which consists of a coaxial discharge between a central cathode and a surrounding anode in the presence of a magnetic field.[9]

Electrothermal thrusters are primarily of two types—the electric-arc jet and the electrically powered hydrogen heater (also called resistojet). The hydrogen heater is limited to specific impulses less than about 1000 seconds, because of the limitation on the wall temperature of the heater. High efficiencies, however, have been achieved.[10] The arc jet, which heats the propellant by means of continuous electric discharge as the propellant flows by, can achieve somewhat higher specific impulses (up to about 2000 seconds), but the efficiency is generally less than 50 per cent, due to losses involved in dissociation and ionization of the propellant atoms, and losses to the walls of the arc chamber and nozzle. Because of the lower specific impulse range, electrothermal thrusters are not useful for interplanetary missions, but may be used for more limited applications such as satellite orientation control and orbit correction.

History and Status The possibility of reducing propellant consumption by ejecting the propellant electrically at high velocities was recognized by early space flight and rocket pioneers, such as Goddard[11] and Oberth,[12] but the practical feasibility of such propulsion systems was not demonstrated. With the advent of nuclear-electric power and large rockets during and after World War II, more interest in electric propulsion was aroused, and between 1946 and 1956, a number of preliminary analyses of nuclear and solar electric systems were published.[13,14]

Comparative studies of the applicability of electric, nuclear, and chemical propulsion to future space missions, together with an engineering study of large electric power systems for space use were completed in 1957.[15] These and similar studies led to the initiation of major research programs in electric propulsion and power generation at U.S. government and industrial laboratories. Low-thrust trajectory studies and mission analyses[1,16,17] have further clarified the role of electric propulsion in future space missions.

The use of cesium-tungsten contact ionization for electrostatic thrusters was proposed by Stuhlinger,[14] and early experimental work in the United States, beginning in 1957, was concentrated on this approach.[18-20] However, the invention and development of the electron bombardment ion thruster[21] showed that it could achieve higher efficiencies with less sensitivity to fabrication techniques and materials. In 1964, the first successful space flight test of ion thrusters was accomplished with the launching of SERT-I (Space Electric Rocket Test I).[22] One cesium-tungsten and one electron bombardment thruster were launched into a 20-minute ballistic space trajectory. The cesium-tungsten thruster developed a high-voltage breakdown, but the electron bombardment thruster operated successfully. The test demonstrated that ion beam neutralization in space was no problem, and that the thrust level was the same as in ground test vacuum facilities. After further research and development of the electron bombardment thruster, a long-duration satellite orbital test (SERT II) was launched in February 1970 (Fig. 3).[23] A one-year operating period was planned. One of the thrusters operated continuously for five months and the other for three months before high-voltage short circuits occurred. Data analysis indicated that ion-beam erosion of the accelerator grid probably produced metal chips which shorted across to the screen grid. A long period of partial shading of the spacecraft occurred from 1972 through 1978, so that further testing was not possible. Almost continuous sun exposure again occurred for several periods from 1979 to May 1981. Early in 1979, the short circuit of one of the thrusters was cleared by maneuvering the spacecraft with the attitude control rockets. Thereafter, this thruster was operated normally during sun exposure periods until the mercury propellant was exhausted. Many startup and component performance tests were accomplished during these periods.[24] These remarkably successful tests of the long-time survivability of a solar-electric spacecraft and its thruster system provided evidence that solar electric propulsion was feasible for use in planetary missions.

Following the initial tests of SERT II (1971), further development of thrusters and solar-cell arrays was carried out by NASA. A Solar Electric Propulsion System (SEPS) program was initiated to produce a complete, reliable mission-ready system. Initial application was to be a Halley's Comet rendezvous mission in 1985, but budgetary restraints eliminated this program and other missions for SEPS.[25]

The performance parameters achieved for a

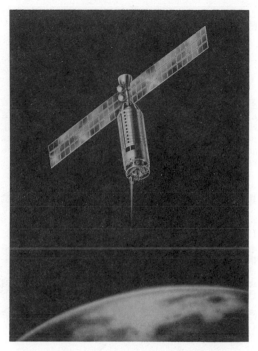

FIG. 3. Artist's drawing of SERT-II Spacecraft.

FIG. 4. Artist's representation of SEP (Solar Electric Propulsion System) Spacecraft. 25 KW power level.

25 killowatt version of SEPS are as follows (Fig. 4):

Solar Array Mass: 375 kg
Power Processing Subsystem Mass: 375 kg
Thruster system mass: 312 kg
Thruster system efficiency: 60%

Thus the overall specific mass was about 40 kg/kW. Operating the thrusters directly from the solar array (to eliminate power processing mass) is considered feasible in future systems so that specific mass below 30 kg/kW should be readily achievable.

Other research and development programs with more modest intended applications (attitude control, station-keeping and orbit control of satellites) have been carried out for many years in Europe[26] and Japan,[27] as well as the United States.[28,29] These programs have studied a variety of electric thruster types from pulsed plasma ejectors to smaller ion thrusters. In addition, research is underway to improve the performance of magneto-plasmadynamic (MPD) thrusters to make them more competitive with ion propulsion. Some research is also continuing on nuclear-thermionic power generation for electric propulsion.[29]

W. E. MOECKEL

References

1. Moeckel, W. E., "Comparison of Advanced Propulsion Concepts for Deep Space Exploration," *J. Spacecraft and Rockets*, 9 (12), 863–868 (December, 1972).

2. Rappaport, P., "Photovoltaic Power," *J. Spacecraft and Rockets*, 4 (7), 838–841 (July 1967).

3. Mickelsen, W. R., and Low, C. A., Jr., "Potentials of Radioisotope Electrostatic Propulsion," *Astronautics Aerospace Eng.*, 1 (9), 52–57 (October 1963).

4. Becker, R. A., "Thermionic Space Power Systems Review," *J. Spacecraft and Rockets*, 4 (7), 847–851 (July 1967).

5. Zipkin, Morris A., "Alkali-Metal Rankine-Cycle Power Systems for Electric Propulsion," *J. Spacecraft and Rockets*, 4 (7) 852–858 (July 1967).

6. Nichols, Lester, D., "Comparison of Brayton and Rankine Cycle Magnetogasdynamic Space Power Generation Systems," NASA TN D-5085, 1969.

7. Richley, Edward A., and Kerslake, William R., "Bombardment Thruster Investigations at the Lewis Research Center," *J. Spacecraft and Rockets*, 6 (3), 289–295 (March 1969).

8. Seikel, George R., "Generation of Thrust—Electromagnetic Thrusters," NASA SP-11, 1962, pp. 171–176.

9. Connolly, D. J., Sovie, R. J., Michels, C. J., and Burkhart, J. A., "Low Environmental Pressure MPD Arc Tests," *AIAA J.*, 6 (7), 1271–1276 (July 1968).

10. Jack, John R., "NASA Research on Resistance-Heated Hydrogen Jets," "Advanced Propulsion Concepts," Vol. I, Gordon and Breach Science Pub., Inc., 1963 (pp. 75–89).

11. Lehman, Milton, "This High Man," New York, Farrar, Straus, and Co., 1963.

12. Oberth, H., "Wege zur Raumschiffahrt," Munich and Berlin, Verlag von Oldenbourg, 1929 (reprinted by Edwards Bros. Inc., 1945).

13. Shepherd, L. R., and Cleaver, A. V., "The Atomic Rocket," Pt. I, *J. Brit. Interplanet. Soc.*, 7, 185–189 (1948); Pt. II, ibid., 7, 234–241 (1948); Pt. III, ibid., 8, 23–37 (1949); Pt. IV, ibid., 8, 59–70 (1949).

14. Stuhlinger, E., "Electrical Propulsion System for Space Ships with Nuclear Power Source," *J. Astronautics*, 2 (4), 149–152 (1955); 3 (1), 11–14 (1956); 3 (2), 33–36 (1956).

15. Moeckel, W. E., Baldwin, L. V., English, R. E., Lubarsky, B., and Maslen, S. H., "Satellite and Space Propulsion Systems," NASA TN D-285, 1960. (Unclassified versions of material presented at NACA Flight Propulsion Conference, November 22, 1957.)

16. Irving, J. H., and Blum, E. K., "Comparative Performance of Ballistic and Low-Thrust Vehicles for Flight to Mars," *Vistas Astron.*, 2, 191–218 (1959).

17. Sauer, C. G., and Melbourne, W. G., "Optimum Earth-to-Mars Trip Trajectories Using Low-Thrust, Power-Limited Propulsion Systems," Rep. TR 32-376, Jet Prop. Lab., C.I.T., 1963.

18. Forrester, A. T., and Spenser, R. C.: "Cesium-Ion Propulsion," *Astronautics*, 4 (10), 34–35 (October 1959).

19. Childs, J. H., "Design of Ion Rockets and Test Facilities," Paper 59-103, Inst. Aero. Sci., Inc., 1959.

20. Brewer, G. R., Etter, J. R., and Anderson, J. R., "Design and Performance of Small Model Ion Engines," Paper 1125-60, ARS, 1960.

21. Kaufman, Harold R., "An Ion Rocket with an Electric Bombardment Ion Source," NASA TN D-585, 1961.

22. Cybulski, R. J., Shellbauer, D. M., Lovell R. R., Domino, E. J., and Kutnik, J. J., "Results from SERT-I Ion Rocket Flight Test," NASA TN D-2718, March 1965.

23. Kerslake, W. R., Goldman, R. G., and Neiberding, W. C., "SERT-II: Mission, Thruster, and In-Flight Measurements," *J. Spacecraft and Rockets*, 8 (3), 223–224 (March 1971).

24. Kerslake, W. R., "SERT II Thrusters—Still Ticking After Eleven Years," AIAA paper 81-1539, AIAA/SAE/ASME 17th Joint Propulsion Conference, 1981.

25. Austin, R. E. and Kesteu, W., "Solar Electric Propulsion Systems (SEPS) Program Plans and Systems Development," AIAA paper 79-2119, 14th International Electric Propulsion Conference, 1979.

26. Loeb, H. W., et al., "European Electric Propulsion Activities," AIAA Paper 79-2120, 14th International Electric Propulsion Conference, 1979.

27. Azuma, H., et al., "Experimental Plan for Electron Bombardment Ion Thruster on Engineering Test Satellite III," AIAA paper 81-0662, 15th International Electric Propulsion Conference, 1981.

28. Vondra, R. T., "U.S. Air Force Programs in Electric Propulsion," AIAA paper 79-2123, 14th

International Electric Propulsion Conference, 1979.

29. Hudson, W. R., "NASA Electric Propulsion Program," AIAA paper 79-2118, 14th International Electric Propulsion Conference, 1979.

Bibliography

Angrist, S. W., "Direct Energy Conversion," Boston, MA, Allyn and Bacon, Inc., 1965.

Jahn, Robert G., "Physics of Electric Propulsion," New York, McGraw-Hill Book Co., 1968.

Stuhlinger, Ernst, "Ion Propulsion for Space Flight," New York, McGraw-Hill Book Co., 1964.

Journal of Spacecraft and Rockets, Space Power Issue, Vol. 4, No. 7, July, 1967.

Cross-references: ASTRONAUTICS, PHYSICS OF; DYNAMICS; FLIGHT PROPULSION FUNDAMENTALS; IMPULSE AND MOMENTUM; MAGNETO-MECHANICS; PHOTOELECTRICITY; PLASMA.

ELECTRICAL DISCHARGES IN GASES

Motion of Slow Electrons in Gases Suppose that a swarm of electrons traverses a gas in which a uniform electric field X exists. In general the distribution of energy among the electrons will depend on the distance x which they have traveled in the field. However, provided x is sufficiently large, the energy distribution attains a steady value independent of x. In this steady state, the average rate of supply of energy to an electron from the field is equal to the average rate of loss of energy in collisions with gas molecules.

Many important quantities in this subject are related to $eX\lambda$, the energy gained by an electron of charge e in traveling the mean distance λ between two successive collisions with gas molecules. Since λ is inversely proportional to the gas density, the above quantity can be expressed in the form X/P_0 where P_0 is the gas pressure reduced to some standard temperature.

The mean energy of an electron in the swarm, $\bar{\epsilon}$, is a function only of X/P_0 for a particular gas. Figure 1 shows the form of this variation for a monatomic gas (He) and a diatomic gas (N_2). Here $k = \bar{\epsilon}/\bar{\epsilon}_g$, where $\bar{\epsilon}_g$ is the mean kinetic energy of a gas molecule at $15°C$ (0.037 eV). It is seen that the mean electron energy greatly exceeds the mean energy of a gas molecule even when X/P_0 is small. This is due to the inefficient energy exchange in collisions between electrons and gas molecules. If the collisions are elastic, it is readily shown that f, the mean fractional energy lost by an electron in a collision, is $\sim 2m/M$ where m is electron mass and M is molecular mass. Clearly $f_{el} \ll 1$. At a given X/P_0, $\bar{\epsilon}$ is generally lower in polyatomic than in monatomic gases. Owing to the possibility of inelastic collisions involving vibrational or rotational excitation of the molecule, $f \gg 2m/M$ in the former case. In the latter case, only electronic excitation of the atom can occur and this requires much higher energies in general.

In addition to their random motion, the electrons must obviously possess a superimposed drift motion in the direction of the applied field. Figure 2 shows the variation of the drift velocity W_e with X/P_0; normally W_e is small compared to the mean random speed of the electrons.

Ionization by Electron Collision When the energy of an electron exceeds a certain critical value ϵ_i, ionization can occur at a collision with a gas molecule. As X/P_0, and hence $\bar{\epsilon}$, is increased, an increasing fraction of electrons in the swarm will have energies exceeding ϵ_i. The size of the electron swarm will then increase with the distance x traveled in the field direction.

This growth is most conveniently studied under conditions where $\bar{\epsilon}$ is kept constant. This can be done by releasing electrons from the cathode of a plane-parallel system and varying the electrode gap d and potential difference V in such a way that the electric field $X(=V/d)$ is fixed. It is then found that the electron current at the anode, i, increases exponentially with d or V. That is

$$i = i_0 \exp(\eta V) \qquad (1)$$

where i_0 is the electron current released from the cathode and η is the electron ionization coefficient: this is defined as the average number of ionizing collisions made by an electron in moving through a 1-volt potential difference.

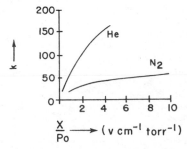

FIG. 1. Mean electron energy as a function of X/P_0 for He and N_2.

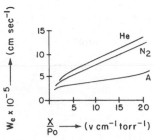

FIG. 2. Electron drift velocity as a function of X/P_0 for He, A and N_2.

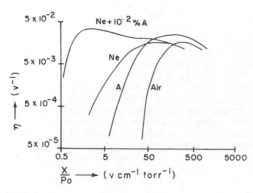

FIG. 3. Electron ionization coefficient as a function of X/P_0 for air, Ne, A, and Ne + 10^{-2} per cent A.

η is also a function only of X/P_0 for a given gas (Fig. 3).

It is important to note that the curve of η against X/P_0 passes through a maximum. The decrease in η at low X/P_0 is due to the increasing importance of excitation compared with ionization as X/P_0 decreases; since the excitation energy losses are larger in polyatomic than in monatomic gases, as remarked earlier, the decrease in η occurs more rapidly in air than in neon. The decrease in η at high X/P_0 (300 volts cm^{-1} torr^{-1}), where excitation losses are comparatively unimportant, is due to the fact that an increasingly large fraction of the energy supplied from the field is used in maintaining the kinetic energy of the swarm.

The curve for the gas mixture Ne + 10^{-2} per cent A is of great interest. Since the excitation potential of the most important metastable state of Ne (16.5 volts) exceeds the ionization potential of A (15.8 volts), the process Ne* + A → Ne + A$^+$ + e can occur. This reaction has a very high probability, of the order unity per collision, and causes a great increase in η at low X/P_0 above the value for pure neon, since the effective excitation energy losses are now considerably reduced. The double maximum in the curve of η vs X/P_0 arises from the fact that the direct and indirect ionization processes have their maximum efficiencies at different X/P_0 values (~70 and 2 volts cm^{-1} torr^{-1}, respectively).

Secondary Ionization Processes It is found that Eq. (1) no longer holds at larger values of V; i now increases more rapidly leading ultimately to spark breakdown. This is due to the occurrence of secondary ionization processes, in addition to ionization by collision between electrons and gas molecules. In general, the most important secondary process is the release of electrons from the cathode surface. If various simplifying assumptions are made, it can be shown that the ionization current is now given by:

$$i = \frac{i_0 \exp{(\eta V)}}{1 - \gamma[\exp{(\eta V)} - 1]} \tag{2}$$

where γ is a generalized secondary ionization coefficient. This is defined as the probability of a secondary electron being released from the cathode per positive ion arriving at the cathode. Included in γ are contributions to the secondary emission arising from radiation quanta and metastable molecules.

Since γ depends largely on the mean energies of the electrons and ions, it is, like η, a function only of X/P_0 though the function now depends on the nature of the cathode as well as on the gas.

Spark Breakdown It is clear from the above equation that the ionization current tends to become very large as the potential difference across the gap approaches the value V_s given by:

$$\eta V_s = \log\left(1 + \frac{1}{\gamma}\right) \tag{3}$$

This is the condition for spark breakdown and can be best explained in the following manner.

Suppose that a primary electron current i_0 is released from the cathode when $V = V_s$. The electron current reaching the anode is then i_0 exp (ηV_s). Hence, the positive ion current reaching the cathode due to the current i_0 is i_0 [exp $(\eta V_s) - 1$]. This will give rise to a secondary electron current of value γi_0 [exp $(\eta V_s) - 1$]. If V_s is given by Eq. (3), then γ [exp $(\eta V_s) - 1$] = 1 and the secondary current is equal to the original primary current i_0. Hence it is clear that the process can continue even if the initiating current ceases. When V is less than V_s, however, the discharge current i is proportional to i_0 [Eq. (2)]. Thus $i = 0$ when $i_0 = 0$. It follows that $V = V_s$ marks the transition from a non-self-maintained to a self-maintained discharge. V_s is best defined as the potential difference required to maintain a small discharge current i when the primary current $i_0 = 0$. V_s is independent of i provided this is sufficiently small to avoid space charge distortion of the field.

Since η and γ are both functions only of X/P_0 and $X = V_s/d_s$ at breakdown, it follows from Eq. (3) that

$$V_s = F(P_0 d_s) \tag{4}$$

Thus, for a given gas and cathode material, the breakdown potential between large plane-parallel electrodes depends only on the product of the reduced gas pressure and electrode separation. This result, which is known as Paschen's law, has been confirmed experimentally over a wide range of P_0 and d_s.

The variation of V_s with $P_0 d_s$ for a number of gases and cathode materials is shown in Fig. 4. It should be noted that the curves all exhibit a minimum; this corresponds to the maximum in the curve of η vs X/P_0. It will be seen that the rise of V_s at high values of $P_0 d_s$ is most marked in air, less in pure Ne, and less still in the Ne + A mixture. This is readily understood by reference

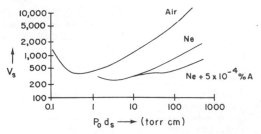

FIG. 4. Breakdown potential as a function of the product of the reduced gas pressure and electrode separation, $P_0 d_s$ for air, Ne, and Ne + 0.0005 per cent A with an iron cathode.

to the decrease in η at low X/P_0 $(=V_s/P_0 d_s)$ in these gases (Fig. 3).

Time Lag of Spark Breakdown If a potential difference $\geqslant V_s$ is suddenly applied to a discharge gap, a finite time elapses before the initial current i_0 has increased to a self-maintained discharge current $\sim 10^{-7}$ ampere/cm^2. This time lag consists of two parts. First of all, there is a statistical lag which arises from the fact that the primary and secondary ionization processes are both subject to statistical fluctuations. Thus, although $V > V_s$ where $\gamma\,[\exp(\eta V_s) - 1] = 1$ implies that on the average one electron leaving the cathode will give rise to one secondary electron, this may not happen in any particular case. Clearly the mean statistical lag t_s will decrease as the initial current is increased and it may be shown that

$$t_s = \frac{1}{PN_0} \qquad (5)$$

where N_0 is the number of primary electrons leaving the cathode per second and P is the probability that any particular electron leads to breakdown. The latter quantity is zero at the sparking threshold V_s but increases rapidly for $V > V_s$. $P \simeq 1$ provided $V > 1.25\ V_s$.

The second component of the total time lag is the formative lag t_F. This can be regarded as the time that must elapse after the appearance of a suitable initiatory electron before the various ionization processes generate a self-maintained current of any given magnitude. This current can be chosen arbitrarily to specify breakdown of the gap and is generally taken to be $\sim 10^{-7}$ ampere/cm^2. Clearly t_F will depend on the relative importance of the various secondary mechanisms mentioned earlier; positive ion transit times are typically $\sim 10^{-6}$ second, while the time lags involved in the contribution of radiation quanta and metastable molecules to γ are $\sim 10^{-8}$ and 10^{-3} second, respectively. The observed variation of t_F with overvoltage ΔV $(=V - V_s)$ for various fixed values of X/P_0 in H$_2$ is shown in Fig. 5. Comparison with theory enables an estimate to be made of k, the relative contribution of photons at the cathode to the

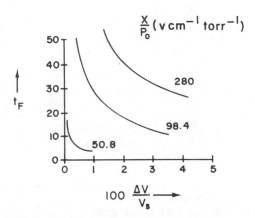

FIG. 5. Variation of formative time lag with overvoltage for various fixed values of X/P_0 in H$_2$ with copper electrodes.

total γ. This ranges from 0.75 at $X/P_0 = 50$ to 0.50 at $X/P_0 = 300$ volts cm^{-1} torr^{-1}.

Glow Discharge We have seen that any small current i can be maintained even in the absence of initiatory electrons when the potential difference between the electrodes reaches a value V_s given by Eq. (3). V is only independent of i when the latter is less than $\sim 1\ \mu A$. At higher currents the space charge concentration becomes sufficient to cause X and hence η to vary across the gap, and ηV_s in Eq. (3) must be replaced by $\int \eta\ dV$. The static V-i characteristic is normally negative since the field redistribution produced by the space charge effects increases the over-all ionization efficiency. Once breakdown has taken place, the current increases to a value determined by the impedance of the voltage supply.

If the current density is sufficiently small (< 0.1 ampere/cm^2), the cathode is not heated to a high enough temperature for thermionic emission to be a significant factor in the maintenance of the discharge. This regime is termed a glow discharge and the field variation across the gap in a long cylindrical tube is indicated in Fig. 6. We can distinguish five main regions here:

(1) The cathode fall, in which the field decreases from a high value at the cathode to approximately zero.

(2) The negative glow, in which ionization and excitation are due largely to fast electrons arriving from the cathode fall. The length of this region is normally controlled by the distance traveled by the electrons before their energy is reduced below the minimum required for excitation.

(3) Faraday dark space. In many cases the ionization in the negative glow is so intense that the electron current here exceeds the total discharge current. A region is therefore required where electrons are lost by diffusion and not replenished by ionization; usually $X \leqslant 0$ here.

(4) The positive column, where X has a small constant value such that the corresponding elec-

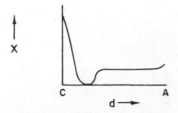

FIG. 6. Variation of axial field with distance from the cathode for a glow discharge in a long cylindrical tube.

tron energy distribution gives an ionization rate which just balances the loss of electrons and ions by radial diffusion to the walls.

(5) The anode fall, where X again increases.

Regions (1) and (2) are the most important regions of the discharge; the primary and secondary ionization by which the discharge is maintained take place here. The fall of potential across region (1), usually termed the cathode fall (V_c) is clearly an important parameter of the discharge.

It should be noted that the section extending from the negative glow to the anode has only a small field strength and small resultant space charge with $| n_i - n_e | \ll n_e$, where n_i and n_e are the ion and electron concentrations. This region is generally called a plasma. In many cases, the electrons here have a random motion which is large compared to their drift motion in the field direction. Our earlier discussion on electron swarms is valid here. On the other hand, the regions which occur near the cathode and the walls have a high field strength and resultant space charge with $n_i \gg n_e$. The electrons and ions behave here as a beam rather than as a swarm.

Ambipolar Diffusion The radial diffusion of ions and electrons to the wall in the plasma region (4), above, does not occur at the same rate as when only one type of carrier is present. Clearly, the electrons will tend to diffuse to the walls much more rapidly than the ions leaving an excess of positive charge. A space charge field is set up which retards electrons and accelerates positive ions so that their effective diffusion rates are equalized. This process can be described in terms of the ambipolar diffusion coefficient D_a which is given approximately by:

$$D_a \dot{=} D_i \left[1 + \frac{T_e}{T_i} \right] \qquad (6)$$

where D_i is the normal ion diffusion coefficient and T_e and T_i are the effective electron and ion temperatures, respectively.

Cathode Fall When the current is sufficiently small (< 10 mA for a cathode of area ~ 1 cm^2), the discharge does not occupy the entire cathode area. The current density in the covered portion j_n is approximately constant, and the cathode fall of potential V_c is nearly independent of current and pressure. This is termed the normal cathode fall. The abnormal cathode fall occurs when $i > j_n S$, where S is the total cathode area. V_c now increases with current.

Arc and High Current Discharges If i is increased sufficiently, a stage will eventually be reached where the cathode temperature is high enough for thermionic emission to be important. V_c now decreases with further increase in i (Fig. 7), and we are in the region of the arc discharge. The transition current clearly depends on the rate of loss of heat from the cathode and only has a definite value when the surface is uniform. In some arc discharges (e.g., Hg), the emission mechanism is probably not thermionic; these are not fully understood however.

The current density is much higher in the arc than in the glow discharge. The charged particle density is typically in the range 10^{14} to 10^{18} electrons cm^{-3} in the core of the arc when the pressure is approximately atmospheric. Because of the very high frequency of collisions between the electrons, positive ions, and neutral molecules, thermal equilibrium is often established for all the groups of particles present in the arc positive column. The temperature at the axis of the arc is typically in the range 4 000 to 10 000 K. At high pressures the temperature diminishes laterally very quickly whereas at low pressures it remains constant over a large portion of the cross section.

We have assumed hitherto that the current is always sufficiently low for the magnetic field produced by the current to play an unimportant role in the discharge mechanism. At high currents this is no longer true, and the interaction of the self-magnetic field of the discharge and the current produces forces on the ionized gas comparable to the other forces acting. The required currents increase with the gas pressure p; at normal temperatures, $i > 10^3$ amperes and $p < 1$ torr are required. The force due to the magnetic field tends to constrict the discharge, and a column so constricted is said to be pinched. This pinch effect offers a possible method of confining the hot gas to a channel remote from the

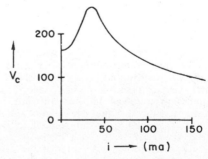

FIG. 7. Variation of cathode fall of potential with current for discharge in A at 30 torr pressure with spherical tungsten electrodes, 1.8 mm diameter.

walls of the containing vessel. However, a major obstacle to the achievement of a steady pinched discharge is the inherent instability of such a channel to lateral perturbations. This causes the pinched column to leave the axis of the containing tube and take up a helical path in contact with the walls. It is, however, possible that a suitable arrangement of magnetic fields may help to stabilize the discharge, leading to the prospect of continuous operation of a pinched discharge.

J. D. SWIFT

References

Craggs, J. D., and Meek, J. M., "Electrical Breakdown of Gases," London, Oxford University Press, 1953.

Loeb, L. B., "Basic Processes of Gaseous Electronics," Berkeley, Cal., University of California Press, 1955.

Jones, F. Llewellyn, "Ionisation and Breakdown in Gases," London, Methuen & Co., Ltd., 1957.

Acton, J. R., and Swift, J. D., "Cold Cathode Discharge Tubes," London, Heywood and Co., Ltd., 1963.

von Engel, A., "Ionized Gases," London, Oxford University Press, 1965.

Somerville, J. M., "The Electric Arc," London, Methuen & Co., Ltd., 1959.

Nasser, E., "Fundamentals of Gaseous Ionization and Plasma Electronics," New York, John Wiley & Sons, 1971.

Llewellyn-Jones, F., "The Glow Discharge and an Introduction to Plasma Physics," London, Methuen & Co., Ltd., 1966.

Raether, H., "Electron Avalanches and Breakdown in Gases," London, Thornton Butterworth, 1964.

Hoyaux, M., "Arc Physics," Berlin, Springer-Verlag, 1968.

Brown, S. C., "Introduction to Electrical Discharges in Gases," New York, John Wiley & Sons, 1966.

Franklin, R. N., "Plasma Phenomena in Gas Discharges," Clarendon, Oxford Univ. Press, 1979.

Howatson, A. M., "An Introduction to Gas Discharges," London, Pergamon, 1976.

Meek, J. M., and Craggs, J. D., (Eds.), "Electrical Breakdown of Gases," New York, John Wiley & Sons, 1978.

Cross-references: ELECTRICITY, IONIZATION, POTENTIAL, STATIC ELECTRICITY.

ELECTRICAL MEASUREMENTS

In an electrical measurement, one is concerned with the evaluation of an electrical quantity—resistance, capacitance, inductance, charge, current, voltage—or of a quantity that depends on some combination of them. The measurement means may be a ratio device, such as a potentiometer or bridge in which similar quantities are compared; or it may be an electromechanical system in which a force or torque is developed; or the heating effect of an electric current may be utilized; or some effect associated with quantum dynamics may be employed.

The basis of any meaningful measurement of an electrical quantity must ultimately be the *national reference standards* maintained by the National Bureau of Standards (NBS), and assigned by means of *absolute* measurements in which certain electrical quantities are determined in terms of appropriate mechanical quantities. The electrical units are related to the metric system of mechanical units in such a way that the units of power and energy are identical in both systems. In 1960 the name Système International (SI),[1] now in use throughout the world, was assigned to the measurement system based on the *meter, kilogram, second, ampere*.

Because four independent relations tie the six electrical quantities together, only two absolute measurements are needed to fix the national reference standards. Historically, these have been an *ohm* and an *ampere* determination.[2] In the ohm determination, a resistance is compared to the reactance at a known frequency of an inductor or capacitor whose magnitude is calculated from its measured dimensions together with the assigned magnetic constant ($4\pi \times 10^{-7}$ H/m in SI units), or in the case of a calculable capacitor, the corresponding electric constant derived from the magnetic constant and the speed of light in vacuum. Thus the *ohm* is assigned in terms of the *meter* and the *second*. In the ampere determination, a current carried by two coils is evaluated in terms of their measured dimensions, the magnetic constant, and the force with which they interact. This force is opposed by gravity acting on a known mass. The *ampere* is assigned in terms of the *meter, kilogram*, and *second*. The measured current is passed through a known resistor, and the *volt* is assigned using Ohm's law.

The National Reference Standards, in terms of which the *legal* electrical units are maintained, are groups of stable resistors, capacitors, and standard cells (Weston saturated cells) whose values have been assigned by *absolute* measurements. At NBS the reference standard of resistance is a group of 1-Ω resistors, fully annealed and mounted strain-free out of contact with the air, in sealed containers. The reference standard of capacitance is a group of 10 pF fused-silica-dielectric capacitors whose values are assigned in terms of the calculable capacitor used in the ohm determination. The reference standard of voltage is a group of standard cells maintained at a constant temperature. Their values are periodically reassigned by means of a "Josephson experiment" to be described below. The individuals within each reference group are intercompared routinely to detect any drift with respect to the group mean. Representatives of the group are compared with the national standards maintained by the national laboratories of other countries to detect any differences that could develop in their "as-maintained" units. These international comparisons are carried out by the Bureau International des Poids et Mesures (BIPM) in Sèvres, France. Stability of the national reference standards has also been monitored in recent years by experiments which relate an electrical unit directly to some atomic constant—a natural invariant. In one such experiment, the *ampere* is related to *proton gyromagnetic ratio.*[3] A characteristic magnetic moment and spin are associated with the proton—it behaves both as a magnet and a gyroscope—tending to align itself in a magnetic field and to precess about the field direction with an angular velocity proportional to field strength if its alignment is disturbed. In fact, proton precession frequency has become a widely used method of measuring magnetic field strength. Thus the repeatability of proton precession frequency in the field of a dimensionally stable solenoid, excited by a current established in terms of the *legal ohm* and *volt*, is a measure of the stability of the *legal ampere*. In another experiment use is made of the voltage appearing across a Josephson junction (made of two superconductors separated by a barrier), irradiated with microwave power and simultaneously biased with direct current.[4] The voltage across the junction increases with increasing bias current, in discrete quantum jumps related to the frequency of the microwave irradiation. This relation is $Nh\nu = 2eV_N$, where N is step number, h is Planck's constant, ν is the microwave frequency, e is elementary electron charge, and V_N is the junction voltage. Thus the stability of the *legal volt* is monitored in terms of frequency and the ratio of Planck's constant to electron charge by comparing the Josephson voltage to that of the cells in the national reference group.

The *legal volt* was reassigned on January 1, 1969, based on recent *ampere* determinations. This assignment was such that the number stating the emf of a standard cell was increased by 8.6 μV over its previous (1948–68) assignment. Other national laboratories also made adjustments such that all national reference standards of emf (including the standard at BIPM) were in agreement at that time. It was determined through Josephson volt experiments that the emf of the cells in the national reference group was slowly drifting (as much as 0.3 μV/year), and it was decided that the emf of these cells would be periodically reassigned to keep the *legal* volt at a constant value.[5] Thus, since July 1, 1972 the *legal* volt has been maintained in terms of an assigned frequency/voltage ratio through the Josephson volt experiment. Recent *ohm* determinations,[6] based on a calculable capacitor, indicate that the 1948 assignment of the national reference standard of resistance (USA) was correct within a microohm, and the value of the *legal ohm* continues to be maintained by a group of 1-Ω resistors on the basis of their 1948 assignment.

The *voltage divider* is the basic element of many measurement networks. In general it consists of a group of series-connected resistors (or impedors using resistance, inductance, capacitance, or some combination of them). Its operating principle is the following: when the series circuit is tapped at an intermediate point but no current is drawn from the tap, the ratio of the voltage between the tap point and a divider terminal to the voltage between terminals equals the ratio of tapped resistance (or impedance) to the total divider resistance (or impedance). Modern dc potentiometers are dividers that may achieve an accuracy of a few parts-per-million (ppm) in comparing direct voltages and, with appropriate range-extending dividers, can be used to measure voltages to 1500 V or more. Factors limiting accuracy in high-voltage dividers may be the heating effect of power dissipated in the divider and leakage currents across its insulating structure. Standard cells may be intercompared to a few parts in 10^8, using special potentiometers designed to minimize parasitic voltages. Direct currents can be measured to a few ppm by comparing the voltage drop the current produces in a known resistance, with a known reference voltage.

Two voltage dividers may be connected in parallel to the same source to form a *bridge*. Equality of divider ratios—indicated by zero potential difference at their tap points—permits accurate comparison of impedances and is relatively insensitive to minor variations in the level of supply voltage. For dc resistance measurement, the Wheatstone network is used. One of its two dividers provides a known ratio, and the other includes the unknown and a known resistor with which it is compared. In this simple form the Wheatstone bridge is used for resistors (usually greater than 1 Ω) which have only two terminals, i.e. whose potential and current connections coincide. For 4-terminal resistors, whose potential and current terminals are separate, the Kelvin bridge is used, with two dividers having known identical ratios in addition to the divider that incorporates the current circuits of the unknown and reference resistor. Using a direct substitution method with one of these bridges, nominally equal resistors can be compared at the ppm level or better, in the range 10^{-4}–10^4 Ω.

The more general impedance bridge is a 4-arm network, similar to the Wheatstone bridge—or a more complicated network that can be reduced to an equivalent 4-arm array by appropriate delta-wye transformations[7]—in which, by proper choice of components, inductances may be intercompared or measured in terms of capacitance and resistance, or capacitances may be intercompared. The accuracy of such bridges is usually limited by the stability of ratios and reference components. Additional limits may be imposed by coupling between bridge elements or to nearby objects or ground. Such coupling may be an ambient magnetic field inducing a parasitic emf in a bridge element or in an open loop between elements, or by capacitance or leakage between elements or to a source or to ground. Such effects can be reduced or eliminated by choice and arrangement of components and by the use of shields maintained at appropriate potentials to eliminate the effect on bridge balance of stray capacitance and leakage.

The *inductive voltage divider*, used extensively as a bridge element at power and audio-frequencies, is in principle a multi-decade, selectable tap-point auto-transformer whose decades may be wound on separate magnetic cores. With proper construction this type of divider can be operated at ppm or better accuracy.

Completely shielded bridges whose ratio arms are closely-coupled secondary windings of a transformer with a high-permeability core, (known as *transformer-ratio-arm bridges*) are capable of high accuracy and are widely used for capacitance measurements. For example, stable 3-terminal 10-pF capacitors can be routinely compared to one part in 10^8 in such a bridge. The voltage ratio supplied by the coupled secondaries may be 1/1, and is rarely more than 10/1. The conjugate of the transformer ratio-arm bridge is based on the *current comparator* and has much greater ratio flexibility.[8] In this arrangement, two currents are compared by equality of their opposed, uniformly distributed linkages with a common core, following Ampere's circuital law,

$$\oint H dl = \sum I,$$

i.e., the line integral of magnetizing force along a closed loop is determined by the sum of the currents passing through the loop. The balance point, corresponding to zero flux density in the high-permeability core, is sensed by zero induced voltage in a detector winding around the core. There is generally an auxiliary winding into which a current can be injected to compensate the difference in ampere-turns of the ratio windings, and so bring the system to balance. The whole arrangement must be shielded from external magnetic fields. Current ratios as high as 1000/1 are feasible, with correspondingly high voltage ratios impressed on the elements being compared. This range of ratios makes possible a relatively simple bridge for the evaluation of high-voltage power cables, insulating structures, and energy-storage capacitors. DC versions of the current comparator have been adapted to use in potentiometers and bridges for the comparison of resistance standards and for the evaluation of current-carrying shunts.

Superconducting Quantum Interference Devices (*SQUIDs*) are very sensitive magnetic flux sensors. A SQUID comprising a superconductor formed into a loop interrupted by a Josephson junction can be used as a flux detector in a current comparator for the measurement of cryogenic resistance ratios.[9] Comparator sensitivity can amount to 0.5 nA-turns. Such an arrangement must be magnetically shielded and is inherently limited by the Johnson noise of the resistors at their operating temperature.

In any measurement network (such as a bridge or potentiometer) using voltage divider or current comparator techniques, a sensitive *detector* must be included to show the absence of current (or voltage difference) in an appropriate network branch. In a dc network the simplest detector is the D'Arsonval galvanometer, consisting of a coil of fine wire suspended by conducting filaments in a radial magnetic field. A light-beam reflected from a mirror attached to the moving system indicates rotation arising from current in the coil. The light beam may be focused on a scale for direct observation, or may be shared by differentially connected photocells whose output is supplied to a second galvanometer, producing an amplified deflection. If the galvanometer is isolated from mechanical disturbances, unbalance detection at the nanovolt level is possible in circuits having resistance up to several hundred ohms.

Electronic instruments which have been developed for low-level dc signal detection are more convenient, more rugged, and less susceptible to mechanical disturbances than is a D'Arsonval galvanometer. However, the measuring circuits of electronic detectors must use considerable filtering, shielding, and guarding to minimize electrical interference and noise. The galvanometer is an extremely efficient low-pass filter, and (when operated to make optimum use of its design characteristics) is still the most sensitive low-level dc detector. Electronic detectors generally use either a mechanical or transistor chopper driven by an oscillator whose frequency is chosen to avoid the local power frequency and its harmonics. This modulation converts the dc input signal to ac which is then amplified, demodulated, and displayed on an indicating instrument, fed to a recording device, or subjected to further processing.

Adjustable-frequency amplifier detectors are used with ac bridges and basically incorporate a low-noise preamplifier followed by a high-gain amplifier around which is a tunable feedback loop whose circuit has zero transmission at the selected frequency, so that the negative feedback circuit controls the overall transfer function of the detector to eliminate passage of signals at other than the chosen frequency. The display element which accepts the detector output may be a cathode-ray oscillograph (CRO) or a rectifier-type moving-coil instrument. Frequently a phase-selective element is interposed between the amplifier and display elements so that in-phase and quadrature bridge adjustments are independently displayed. If a CRO is used, the unbalance bridge signal may be impressed on the vertical deflection plates and a phase-adjustable signal derived from the bridge supply impressed on the horizontal plates. The result is an elliptical screen pattern (Lissajou figure), the slope of whose major axis (after appropriate phase adjustment) represents one of the quadrature unbalance signals, and whose minor axis represents the other. Balance is indicated when the ellipse is collapsed to a straight line and the line is brought to the horizontal. Nanovolt sensitivity is possible also in some types of ac detectors.

Indicating instruments are used to measure current, voltage, phase, and power. With special circuit arrangements they can be used for other electrical quantities or for nonelectrical quantities for which transducers can be devised that will convert the measurand into an electrical signal. Indicating instruments fall into two categories, *analog* and *digital*. *Analog* instruments generally have a pivoted rotating element driven by the electrical signal, the turning motion being opposed by springs so that equilibrium position of the system under driving and restoring torques is indicated on a scale by a pointer attached to the moving element, or by an optical system employing a mirror. The accuracy of such a system is limited by how well the pointer position can be read, generally a percent or so, a tenth percent in the best of pointer/scale instruments, and a hundredth percent for some mirror/light-beam readouts. Some analog instruments also have front-end electronics which can improve sensitivity and input impedance of electro-mechanically-based instruments. *Digital* instruments convert an electrical signal into a digital output which may be either displayed as a numeric readout or further processed in a computer. The accuracy of such a system is limited by its internal reference and by its critical network components such as the signal conditioner, sample-and-hold amplifier, and analog-to-digital converter. Additional limitations in many cases may be offset, gain, nonlinearity, quantization "noise," or other sources of error attributable to the analog-to-digital converter. Even though a digital instrument may have a 5-digit numeric display, its overall accuracy may be the equivalent of only 3 or 4 digits (0.1 or 0.01%), depending on the critical network components. A calibration service for precision analog-to-digital and digital-to-analog converters has been available since 1981 at NBS based on a reference digital-to-analog converter standard developed for this purpose.[10]

DC analog instruments usually have a permanent-magnet moving-coil system whose operation is identical with that of the D'Arsonval galvanometer, i.e., a response proportional to the average value of current in the moving coil.[11] In a milliammeter, the coil may directly carry the current to be measured; in an ammeter the bulk of the current is carried by a shunt; in a voltmeter a series resistance limits the current and the scale is marked in terms of the voltage drop between the instrument terminals.

An ac instrument requires an arrangement for which the direction of torque does not reverse with the direction of current at the instrument terminals. This requirement may be provided by four rectifier elements arranged in a square with the input across one diagonal and a D'Arsonval meter across the other, the rectifier arrangement being such that the current in the dc meter is the same for either direction of input current. Response is to *average* value of the rectified current, but because one is usually concerned with *effective* (rms) value, the scale is marked in terms of rms value for an assumed sine-wave input. Thus, a waveform error is present for a nonsinusoidal input. *Rectifier* instruments of this type are used as voltmeters and milliammeters in the power and audiofrequency range.

Thermocouple instruments use the heating effect of current in a fine wire or thin-walled tube to which a thermocouple is attached. (These structures and related devices are referred to as *thermoelements*, and they are also employed in thermal voltage and current converters as transfer elements.) In the analog-type instrument the thermocouple output goes to a low-range dc millivoltmeter. Since temperature rise of the heater elements is proportional to the square of the current, there is no waveform error in the indication of effective (rms) value. This type of instrument may be used for current measurement from dc to rf range (200 MHz or more with some constructions). As a voltmeter, thermal converter-based instruments are usually restricted to audio frequencies, but if a multiplier having low distributed capacitance is used, the range may be extended to a megahertz or more without serious error.

In *moving-iron* instruments a soft-iron piece forms the moving element. It is immersed in the field of a coil that carries the current to be measured, and its motion is such as to increase the inductance of the system with increasing current. Energy storage, and hence torque, depend on current squared, and there is no waveform error for rms indication. By suitably shaping fixed and movable iron pieces, the scale can be made nearly linear over the upper 80% of its range. In other arrangements, part of the scale can be compressed and a small portion of total range expanded to cover much of the scale. This scheme is useful in a voltmeter that monitors a nearly constant voltage.

Electrodynamic instruments have a moving coil immersed in the field of a fixed-coil system. The interaction of their fields produces a torque proportional to the square of the current for a series (voltmeter) connection, or to the product of the currents in a shunt (ammeter) connection, so that the scale indication is of effective (rms) value as in a moving-iron instrument. However, eddy currents can be much less, and electrodynamic instruments are generally useful over an extended low-frequency range. Also, since there is no magnetic polarization (present on dc in a soft-iron instrument), the dc response can be error-free if the coil arrangement is astatic or is shielded from ambient magnetic fields. Electrodynamic instruments are also used as *wattmeters* to measure power at frequencies up to 1 KHz. The moving coil with a series resistance is connected across the supply as the voltage circuit of the wattmeter. The fixed coils are connected to carry the load current. At any instant, the driving torque is proportional to the instantaneous product of line voltage and load current; the moving coil, by reason of its inertia, takes up a position that represents average power over a cycle.

A *digital voltmeter* (DVM) compares a direct voltage input to an internally generated reference voltage. This comparison may be done by successive approximations in an automatic self-balancing potentiometer; the reference voltage may be used to generate a ramp function that opens a counting gate when the ramp equals the input signal and closes the gate when the ramp voltage is zero, resulting in a voltage-to-time conversion; a ramp voltage may be generated at a rate proportional to the input until it equals a reference value, returned to zero, and repeated. The number of repetitions in a fixed time (proportional to the input voltage) is counted and displayed, the time interval being chosen so that the number of pulses makes the meter direct-reading. In a dual-slope instrument, the input signal may be made to charge a capacitor at a proportional rate, switch to a reference voltage of opposite polarity, and discharge the capacitor at a constant rate, the discharge time being used to gate clock pulses to a counter for voltage-time con-

version. For ac measurements the DVM is preceded by an ac/dc signal converter. If this converter is a rectifier followed by a filter, to obtain an average value of the rectified signal, the display of an equivalent rms value (for an assumed sine-wave input) will be subject to error for other than sinusoidal waveforms. Alternatively, thermoelements can be employed to provide accurate measurement of the true rms value of ac voltages, irrespective of wave shape. Typically, the thermoelement output is part of a feedback network which provides a dc output voltage equal to (or proportional to) the true rms value of the ac input voltage. The thermoelement may be switched alternately to the ac input and then to the dc output as shown schematically in Fig. 1. With either input applied to the thermoelement, the output emf of the thermocouple is balanced against an opposing emf (produced automatically by the feedback loop). Equivalency between the input rms ac voltage and the dc output voltage is thus obtained by virtue of detecting a null voltage under both input conditions.

The reference voltage of a DVM is usually based on the reverse breakdown of a Zener diode, and (depending on selection, aging, and compensation of temperature coefficient) may achieve 5-place accuracy. In fact, portable direct voltage standards of this type can, with temperature control, achieve short-time accuracy of a ppm and stability of 2ppm/year or better.[12] Such a voltage standard can replace a standard cell for many applications, although an inherently higher noise level is to be expected.

The *ac-dc transfer* function is an important one, since the basic standards of resistance and emf, and the potentiometer techniques for accurately measuring current and voltage, are available only on dc. Electrodynamic instruments have been used as ac-dc transfer standards in the range below a kilohertz.[13] Thermoelements with appropriate shunt and series resistors are used over an extended frequency

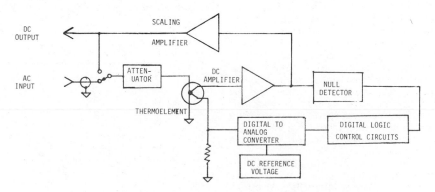

FIG. 1. Automatic thermoelement-based rms ac voltmeter.

range, and are both portable and convenient to use.

A recent electrical measuring instrument development is that of the *transient digitizer* or *waveform recorder*. These kinds of instruments have grown out of a need for measurements with the speed of a CRO and the accuracy of a DVM. The waveform recorder typically consists of a wide-band amplifier, high-speed sample-and-hold and analog-to-digital converter circuits, and a temporary (or buffer storage) memory. Similar to most oscilloscopes, the position and duration of the captured segment of the input waveform can be selected. Both digital and reconstructed analog output interfaces are also incorporated. Some of the best waveform recorders claim 20–100 MHz sampling rates with 0.25–0.1% resolution[14] Accuracy specification for these instruments is a subject that is receiving considerable attention and one in which physical as well as performance standards must be developed.

Instrument transformers are used at power frequencies to extend current and voltage measurement capability beyond the nominal input range of indicating instruments. *Current* transformers rated at 20 kA are used in some installations, as are *voltage* transformers rated at 350 kV. Low-range current transformers of special construction are capable of accurate operation throughout the audio-frequency range. Instrument transformers, consisting of a primary and secondary winding coupled by a magnetic core, are designed to accurately reproduce the primary current (or voltage) on a reduced scale in the secondary circuit. They are quite different from power transformers in details of design and operation, although their basic operating principle is the same. The usual current transformer, designed for use with a 5-A ammeter and to supply the current circuit of a wattmeter or watthourmeter, is capable of delivering only the small amount of power required by these instruments. Current transformers operate under nearly short-circuit conditions with their magnetic core at low flux density. The usual voltage transformer, designed for use with a 120-V voltmeter and the voltage circuit of a wattmeter or watthourmeter, also has a low power rating, and operates under nearly open-circuit condition with its magnetic core approaching saturation. Instrument transformers of good design may be expected to have errors of only a few hundredths percent when operated within their design limits. Performance of a current transformer may be evaluated in terms of a *current comparator*, whose magnetic core operates at zero flux density and is therefore free from the major sources of error that arise from core excitation in a transformer.

In power measurements and ac impedance measurements in general, audio *phase angle* is an important electrical quantity. Although analog phase-angle meters using precision bridges and rotating coil indicators are still in use (with accuracies on the order of 0.5%), digital phase-angle meters are rapidly displacing them, where resolutions of 0.01 degree and 0.05 degree accuracy are possible.[15] As with a DVM however, there are certain limitations with respect to input signal levels, zero offset, and full-scale (gain) corrections, as well as nonlinearities which must be taken into account to achieve the best accuracy possible with digital phase meters. A precision audio-frequency phase-angle calibration standard has been developed at NBS, which uses digital waveform synthesis techniques to produce a reference and variable pair of sinusoidal waveforms whose relative phase displacement can be accurately controlled to within 0.005 degree.

The *induction watthourmeter*, for measurement of electrical energy, is probably the most familiar and widely used of all electrical instruments. More than 9×10^7 are in continuous use to meter electrical energy consumed in the USA, representing in excess of 30 billion dollars per year. The essential features of an induction watthourmeter are indicated schematically in Fig. 2. The voltage winding has many turns of fine wire and carries a current that is almost in quadrature with the line voltage; special lag arrangements ensure that the magnetic field of the voltage pole V lags the line voltage by 90 degrees. The current coils (a few turns of coarse wire) carry the load current, and the field of the current poles C–C is in phase with the load current. The eddy currents produced in the disk by the field from the current poles are in quadrature with this field, and hence have a component in phase with the field from the voltage pole, corresponding to the in-phase (power) component of the load current. The reaction of this component of current-pole-induced eddy currents in the disk, with the voltage-pole field, produces a driving torque on the disk. Similarly, eddy currents in the disk, induced by the voltage-pole, are in phase with the field associated with the power component of the load current in the current-poles; this also produces a driving torque proportional to power through the meter. A braking torque proportional to disk speed

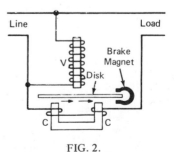

FIG. 2.

results from interaction of the brake magnet field with the eddy currents it induces in the moving disk. As a result, disk speed is proportional to load power, and the number of disk revolutions (read from a counter geared to the disk) is a measure of the total energy consumed by the load.

Electric power and energy can also be measured by processing and combining current and voltage signals electrically. This can be accomplished either by some form of analog multiplication, e.g., using a time-division multiplier,[16] or by sampling the instantaneous currents and voltages periodically, i.e., digitizing the samples so obtained and computing the power from the digital values obtained. In general, higher accuracies can be obtained than with the methods previously described, but the instrumentation required is quite complex. A portable system for measuring ac voltage, current, power, and energy,[17] based on a thermal converter which compares a current derived from the quantity being measured with an equivalent direct current, has been found to have an accuracy of 20 ppm in an international comparison of power measurements at unity and 50% power factor.

FOREST K. HARRIS
BARRY A. BELL

References

1. "The International System of Units (SI)," NBS Special Publication 330, U.S. Govt. Printing Office, Washington, D.C., 1971.
2. "Precision Measurement and Calibration—Electricity (Low Frequency)" NBS Special Publication 300 (Vol. 3), U.S. Govt. Printing Office, Washington, D.C. 1968.
3. Driscoll, R. L., and Olsen, P. T., "Application of Nuclear Resonance to the Monitoring of Electrical Standards" (Precision Measurements and Fundamental Constants), NBS Special Publication 343, 117–121, U.S. Govt. Printing Office, Washington, D.C., 1971.
4. Harris, F. K., Fowler, H. A., and Olsen, P. T., "Accurate Hamon-pair Potentiometer for Josephson Frequency/Voltage Measurements," *Metrologia* **6**, 134–142 (1970).
5. Field, B. F. et al., "Volt Maintenance at NBS via 2e/h: A New Definition of the NBS Volt," *Metrologia* **11**, 155–166 (1973).
6. Cutkosky, R. D., "New Measurement of the Absolute Farad and Ohm," *Trans. IEEE* **IM-23**, 305–309 (1974).
7. Hague, B., and Foord, T. R., "AC Bridge Methods," Pitman, New York, 1971.
8. Kusters, N. L., "Precise Measurement of Current Ratios," *Trans. IEEE* **IM-13**, 197–209 (1974).
9. Gallop, J. C., and Petley, B. W., "SQUIDs and their Applications—A Review Article," *Jour. Phys.—E (Sci. Inst.)* **9**, 417–429 (1976). Dzuiba, R. F., and Sullivan, D. B., "Cryogenic DC Comparators and their Applications," *Trans. IEEE* **MAG-11**, 716–719 (1975).
10. Souders, T. M., Flach, D. R., and Bell, B. A., "A Calibration Service for Analog-to-Digital and Digital-to-Analog Converters," NBS Tech Note 1145, U.S. Govt. Printing Office, Washington, D.C., 1981.
11. Harris, F. K., "Electrical Measurements," Wiley, New York, 1952.
12. Spreadbury, P. J., and Everhart, T. E., "Ultra-Stable Portable Voltage Sources," IEE Conference Publication 174, pp. 117–120, 1979.
13. Hermach, F. L., "AC-DC Comparators for Audio-frequency Current and Voltage Measurements of high Accuracy," *Trans. IEEE* **IM-25**, 489–494 (1976).
14. Lawton, R., "Proceedings of the Waveform Seminar," NBS Special Publication, U.S. Govt. Printing Office, Washington, D.C., 1982.
15. Turgel, R. S., Oldham, N. M., Stenbekken, G. N., and Kibalo, T. H., "NBS Phase Angle Calibration Standard," NBS Tech Note 1144, U.S. Govt. Printing Office, Washington, D.C., 1981.
16. Tomoto, M., Sugiyama, T., and Yamaguchi, K., "An Electronic Multiplier for Accurate Power Measurements," *Trans. IEEE* **IM-17**, 245–251 (1968). Turgel, R. S., "Digital Wattmeter using a Sampling Method," *Trans. IEEE* **IM-23**, 337–341 (1974).
17. Schuster, G., "Thermal Instrument for Measurement of Voltage, Current, Power, and Energy at Power Frequencies," *Trans. IEEE* **IM-29**, 153–157 (1980). McAuliff, R., Lentner, K. J., Moore, W. J. M., and Schuster, G., "An International Comparison of Power Measurements at 120 V, 5 A, 60 Hz," *Trans. IEEE* **IM-27**, 445–449 (1978).

Cross-references: ALTERNATING CURRENTS; CAPACITANCE; ELECTRICITY; INDUCTANCE; MAGNETOMETRY; MEASUREMENTS, PRINCIPLES OF; SEMICONDUCTOR DEVICES; SYMBOLS, UNITS, AND NOMENCLATURE.

ELECTRICITY

Electricity is the material interaction that arises from electromagnetic forces between *electric charges* at rest and in motion. The properties of charge that are responsible for the main features of electrical phenomena are:

1. *Superposition.* The effects of a group of charges are the sum of the independent effects of each one, both in producing fields of force and in being acted on by them. The principle of superposition gives significance to the total amount of charge q or Q on an object or in a system, and also to the concepts of *volume charge density* ρ, *surface charge density* σ, and *linear charge density* λ. The *current density* vector **J** gives the direction and magnitude of the rate per unit area at which charge moves through a surface perpendicular to its velocity. The total *current* through a surface **S** is then

$$I = \int_S \mathbf{J} \cdot d\mathbf{S} \text{ taken over the surface.}$$

2. *Conservation*. Positive and negative charges have opposite effects in the superposition sums, and tend to be present in nearly equal numbers in most circumstances so that electric neutrality is widely prevalent in matter. The algebraic sum of the charges present in any closed system is strictly constant, a fact which is represented by the differential equation

$$\nabla \cdot \mathbf{J} + \partial\rho/\partial t = 0. \quad (1)$$

A current of positive charges in the positive direction is equivalent in the absence of magnetic fields to a current of negative charges in the opposite direction.

3. *Quantization*. In ordinary matter, not subject to high energy interactions (300 MeV or more), negative charge is carried exclusively by *electrons*, and positive charge by protons, including those in nuclei. Electrons and protons have quantum-mechanical properties:

(a) Charge is quantized so that each electron has the same charge $-e$ and each proton has $+e$, where $e = 1.6092 \ldots \times 10^{-19}$ coulomb. This value is so small that the quantization of charge is not noticeable in macroscopic electromagnetic effects. Nevertheless a quantity of charge q may be defined in principle as an integer multiple of $\pm e$.

(b) Wherever situated, electrons and protons have available distinct quantized values or *levels of energy*, and each may generally be considered to occupy a single level at a time.

(c) The wave properties of electrons mean that on a microscopic scale electrons can not be localized at points of space but are represented by a quantum-mechanical charge density. These properties allow electrons to move freely through condensed matter provided they can be excited into appropriate energy levels.

4. *Mobility*. Charge can move through solid matter in electron waves, through liquids in the form of ions in solution, and through gases as both electrons and ions. Without external forcing, charge mobility always tends to neutralize the containing matter.

5. *Force*. A small piece of matter or particle with a charge q moving at a speed v in some frame of reference will experience a force given by the *Lorentz force law*:

$$\mathbf{F} = q(\mathbf{E} + \mathbf{v} \times \mathbf{B}) \quad (2)$$

where the vectors $\mathbf{E}(x, y, z)$ and $\mathbf{B}(x, y, z)$ are, respectively, the *electric* and *magnetic field vectors* at the position (x, y, z) of q, and are determined by the magnitudes, positions, and velocities of all the other charges in the neighborhood. The magnetic force $q\mathbf{v} \times \mathbf{B}$ is responsible for the motor and generator effects (see ELECTRIC MOTORS, ELECTRIC POWER GENERATION). It is readily seen that $\mathbf{E}$ is an ordinary or polar vector, whereas $\mathbf{B}$ is a pseudovector or axial vector (see VECTORS). For each vector there is defined the *flux* through an arbitrary surface $\mathbf{S}$ by

$$\Phi_\mathbf{E} = \int_S \mathbf{E} \cdot d\mathbf{S}, \quad \Phi_\mathbf{B} = \int_S \mathbf{B} \cdot d\mathbf{S}. \quad (3)$$

The variations of $\mathbf{E}$ and $\mathbf{B}$ in a region may be represented by *lines of force* which are tangent at every point to the respective vector at that point.

6. *Potential*. When the charges responsible for $\mathbf{E}$ are stationary, and in cases where they move so that the currents are steady, the electric field is conservative. That is, $\mathbf{E}(x, y, z)$ is derivable from a single-valued *scalar potential* $V(x, y, z)$:

$$\mathbf{E}(x, y, z) = -\nabla V(x, y, z). \quad (4a)$$

Equivalently, V can be found from $\mathbf{E}$ in a region by integration along an arbitrary line in space:

$$V(x, y, z) = -\int_{P_0}^{P} \mathbf{E} \cdot d\mathbf{l} \quad (4b)$$

where P is the point (x, y, z) and P_0 is an arbitrary reference point. $V(x, y, z)$ is the potential energy per unit charge of a small test charge placed at P, considering its energy to be zero at P_0. A line of electric force can be seen to proceed always from higher to lower potential and can never cross or come back to itself. The variation of potential in a region can be described by constructing a set of *equipotential surfaces*, which are everywhere perpendicular to the electric lines of force. (See Fig. 1.)

7. *Coulomb's Inverse Square Law*. When a charge is at rest or moving at a speed much lower than the speed of light c, it produces a field $\mathbf{E}$ at a distant point P given by *Coulomb's Law*:

$$E = \frac{q\hat{\mathbf{r}}}{4\pi\epsilon_0 r^2} \quad (5)$$

where $\hat{\mathbf{r}}$ is the unit vector from q to P and r is the distance. The constant ϵ_0 is found to have the value $\epsilon_0 = 8.845 \ldots \times 10^{-12}$ in SI units. The field produced at P by many q_k's at different places is given by a vector sum of (5) over all the q_k's.

From *Coulomb's Law* may be deduced the *Gauss Flux Law* for the electric flux through an arbitrary closed surface:

$$\oint \mathbf{E} \cdot d\mathbf{S} = \sum_i q_i/\epsilon_0 \quad (6)$$

where the flux is calculated through a closed arbitrary surface and the q_i are the charges within the surface and no others. The *Maxwell*

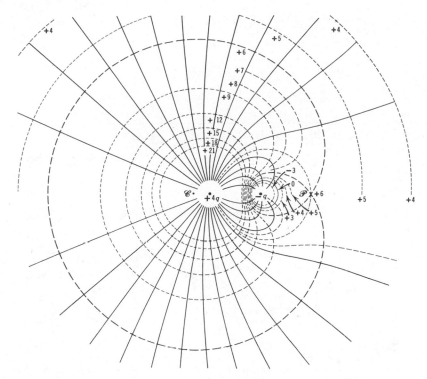

FIG. 1. Equipotentials and lines of force for $+4q$ and $-q$. The light dashed line through P separates flux that passes from $+4q$ to $-q$ from flux that passes from $+4q$ to infinity. (From Scott, p. 52; originally from Garrett, Milan Wayne, "Electricity and Magnetism," Ypsilanti, MI, University Litho-Printers, 1941, p. 48, Fig. 4, 7.)

differential-equation form of Gauss' Law is

$$\nabla \cdot E = \rho/\epsilon_0. \qquad (7)$$

A *tube of flux* in free space, the sides of which are composed of flux lines, contains the same flux across any section; the flux passes from positive charge at one end of the tube to negative charge at the other. Gauss' Law can be used to find the electric field in situations of high symmetry. The intensity of the field outside an infinite uniform plane distribution of charge density σ is $E = \sigma/2\epsilon_0$ regardless of distance; outside a uniform line of charge density λ it is $E = \lambda/2\pi\epsilon_0 r$, where r is the cylindrical radius coordinate; and outside a spherically symmetric distribution of total charge Q it is $E = Q/4\pi\epsilon_0 r^2$, with r being the spherical radius coordinate. Just outside a conductor with charges at rest, at a point where the surface density is σ, E is normal to the surface with intensity $E = \sigma/\epsilon_0$.

In the general case in which all charges can be treated as volume distributions and are at rest or slowly moving, the scalar potential can be found from Coulomb's Law by

$$V(x, y, z) = \frac{1}{4\pi\epsilon_0} \int \frac{\rho\, d\tau}{r} \qquad (8)$$

where r is the distance from the volume element $d\tau$ to the point (x, y, z).

8. *The Poisson and Laplace Equations.* From (4) and (7) we find *Poisson's Equation*:

$$\nabla^2 V(x, y, z) = \rho/\epsilon_0 \qquad (9)$$

which in charge-free space becomes *Laplace's Equation*:

$$\nabla^2 V(x, y, z) = 0. \qquad (10)$$

A unique solution of (9) exists for a set of fixed conductors with each conductor having either a given value of V or a given value of total charge Q, so that the fields between the conductors and the charge distributions on them can be determined.

9. *Capacitance.* For a bundle of flux leaving a portion of the surface of one conductor and ending on another, the ratio of the charge at the positive end to the potential difference between the two conductors is the *capacitance* of that pair of conductor areas. In general, the charges on the conductors are linear functions of the potentials:

$$Q_j = \sum_{k=1}^{n} C_{jk} V_k \qquad (11)$$

where the C_{jk} are coefficients of capacitance. When only two conductors are present, and $Q_1 - Q_2 = Q$, we call the pair a capacitor and set

$$Q = CV \qquad (12)$$

where V is the potential difference between the two, C is the capacitance, and Q is the charge on the positively charged conductor.

10. *Energy in Electrostatic Systems.* The energy in a capacitor is

$$U = \tfrac{1}{2}CV^2 = \tfrac{1}{2}Q^2/C \qquad (13)$$

and for a set of n conductors is

$$U = \tfrac{1}{2}\sum_{i=1}^{n}\sum_{j=1}^{n} C_{ij}V_iV_j \qquad (14)$$

$$= \tfrac{1}{2}\sum_{i=1}^{n} Q_iV_i$$

which can be converted into a volume integral over the intervening space

$$U = \tfrac{1}{2}\epsilon_0 \int E^2 \, d\tau. \qquad (15)$$

11. *The Ampere-Biot-Savart Law.* A charge q moving at a nonrelativistic velocity **v** creates a magnetic field **B** at a distant point P given by

$$\mathbf{B} = \frac{\mu_0 q}{4\pi}\,\frac{\mathbf{v}\times\hat{\mathbf{r}}}{r^2} \qquad (16)$$

where $\hat{\mathbf{r}}$ is the unit vector from the charge q to the point P and r is the distance. The constant μ_0 is given the value $4\pi \times 10^{-7}$ in SI units. An element of current of strength I and length $d\mathbf{l}$ makes a field $d\mathbf{B}$ given by replacing $q\mathbf{v}$ by $I d\mathbf{l}$, so that for a finite length l of current we have

$$\mathbf{B} = \frac{\mu_0 I}{4\pi}\int_l \frac{d\mathbf{l}\times\hat{\mathbf{r}}}{r^2}. \qquad (17)$$

For a closed loop of current, it can be shown from (17) that

$$\mathbf{B} = -\frac{\mu_0 i}{\pi}\,\boldsymbol{\nabla}\Omega \equiv -\mu_0\boldsymbol{\nabla}V_m \qquad (18)$$

where Ω is the solid angle subtended by the current loop at the point where **B** is sought. The quantity V_m is the occasionally used multiple-valued magnetic scalar potential. Equations (17) and (18) can be used to obtain the magnetic fields of a number of shapes of wires and coils, and when applied to microscopic current loops can account for the phenomena of magnetism.

If **B** in (16) is integrated around an arbitrary close loop l' as in Fig. 2, it can be shown that

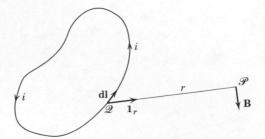

FIG. 2. Ampere-Biot-Savart law, Eq. (17).

$$\oint_{l'} \mathbf{B}\cdot d\mathbf{l}' = \mu_0\epsilon_0\, d\Phi_E/dt. \qquad (19)$$

The quantity $\epsilon_0\, d\Phi_E/dt$ is called the *displacement current*, where Φ_E is the electric flux from q through l'. For a finite segment of current using (17) we find

$$\oint_{l'} \mathbf{B}\cdot d\mathbf{l}' = \frac{\mu_0 I}{4\pi}\,\Delta\Omega \qquad (20)$$

where $\Delta\Omega$ is the difference between the solid angles subtended by l' at the ends of the current segment. (See Fig. 3.) $\Delta\Omega$ is 4π for a closed loop. For a combination of closed current loops I_k and moving or variable charge elements q_j, we have *Ampere's generalized circuital law*:

$$\oint_{l'} \mathbf{B}\cdot d\mathbf{l}' = \mu_0\sum_k I_k + \mu_0\epsilon_0\, d\Phi_E/dt \qquad (21)$$

where the sum is over all the current loops that thread through l' and Φ_E is the total electric flux produced by the net charges in the region. Segments of current that are not closed, as in the case of a capacitor being charged, may be counted either in the first term as currents *or* in the second as changing charges at the ends of current segments.

The corresponding Maxwell differential equa-

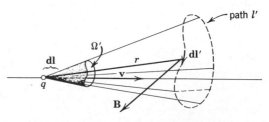

FIG. 3. Calculation of $\oint \mathbf{B}\cdot d\mathbf{l}'$ around a closed path in the neighborhood of a moving charge q subtending a solid angle Ω'.

tion is

$$\nabla \times \mathbf{B} = \mu_0 (\mathbf{J} + \epsilon_0 \partial \mathbf{E}/\partial t) \qquad (22)$$

12. *Vector potential.* From (17) it also follows that **B** can be found from a *vector potential* **A**:

$$\mathbf{B} = \nabla \times \mathbf{A}; \qquad \mathbf{A} = \mu_0 \sum_k \frac{I_k}{4\pi} \int \frac{d\mathbf{l}}{r^2} \qquad (23)$$

where the sum is over all the closed or open loops into which the current sources of **B** can be analyzed. The flux through an arbitrary loop l' can be rewritten in two ways:

$$\Phi_B = \int_S \mathbf{B} \cdot d\mathbf{S} = \oint_{l'} \mathbf{A} \cdot d\mathbf{l}' \qquad (24)$$

From (23) the third Maxwell equation follows:

$$\nabla \cdot \mathbf{B} = 0 \qquad (25)$$

which asserts that the flux through any closed surface is zero, $\oint \mathbf{B} \cdot d\mathbf{S} = 0$. The lines of **B** have no beginnings or ends; they generally circulate indefinitely without closing.

13. *Faraday's Law of Induction.* For any closed loop in space, whether moving or stationary, *Faraday's Law* asserts that

$$\mathcal{E} = \int_l \mathbf{E} \cdot d\mathbf{l} = -d\Phi_B/dt = -d\left[\oint_l \mathbf{A} \cdot d\mathbf{l}\right]\Big/dt \qquad (26)$$

where ϵ is the electromotive force (emf) around the loop l and Φ_B is the magnetic flux through l. The loop may be stationary or moving. From (26) it follows that the electric field of magnetic origin is not conservative and is given by $\mathbf{E} = -\partial \mathbf{A}/\partial t$, so that the complete expression for electric field in terms of potentials is

$$\mathbf{E} = -\nabla V - \partial \mathbf{A}/\partial t. \qquad (27)$$

The corresponding Maxwell equation is

$$\nabla \times \mathbf{E} = -\partial \mathbf{B}/\partial t. \qquad (28)$$

From (17) and (26) it follows that any circuit with a changing current will induce an emf in any neighboring circuit, or in itself. In the latter case, the negative sign in (26) guarantees that the induced emf will oppose the current changes producing it (*Lenz' Law*). We have

$$\mathcal{E}_{jk} = -M_{jk} \, dI_k/dt \qquad (29)$$

for the emf in circuit j produced by a changing current in circuit k, where M_{jk} is the *mutual inductance* and obeys the rule $M_{jk} = M_{kj}$. Self-

inductance L is defined by the analogous relation

$$\mathcal{E} = -L \, dI/dt. \qquad (30)$$

It may be shown that the magnetic energy in a collection of circuits is

$$U = \tfrac{1}{2} \sum_k L_k I_k^2 + \tfrac{1}{2} \sum_{k,j} M_{kj} I_k I_j \qquad (31)$$

which can be written as an integral over the volumes of all the wires

$$U = \tfrac{1}{2} \int_{\text{wires}} \mathbf{J} \cdot \mathbf{A} \, d\tau \qquad (32)$$

and over all space as

$$U = \frac{1}{2\mu_0} \int_{\text{space}} B^2 \, d\tau. \qquad (33)$$

14. *Relative Motion.* From (2) it is easy to see that if the source of **B** moves at $-\mathbf{v}$ and q remains at rest, the moving source must create an electric field at the position of q:

$$\mathbf{E}' = \mathbf{v} \times \mathbf{B}. \qquad (34)$$

From (16) it is easy to see that if a collection of charges which creates a field **E** at a point is all set into motion at a common speed **v**, the moving source of **E** creates a **B**' according to

$$\mathbf{B}' = \mu_0 \epsilon_0 \mathbf{v} \times \mathbf{E}. \qquad (35)$$

Motion of Charges in Matter

1. *Solids.* (a) Levels for the electrons in solids that take part in binding lie in continuous bands separated by finite gaps. Because electrons are fermions with spin $\tfrac{1}{2}$, each level can hold at most two electrons; at equilibrium each band is filled up from the bottom. The average energy of a group of electrons added to or removed from a solid is called the *Fermi level* (F.l.) and lies essentially at the top of the filled group of levels. A change in the electric potential of a region in the piece of matter by ΔV will change the F.l. by $-e\Delta V$.

(b) A conductor has its F.l. below the top of a partly filled band called the conduction band. When a gradient $\nabla V = -\mathbf{E}$ is established in the conductor, the gradient in F.l. causes electrons to jump freely into slightly higher levels corresponding to motion in the direction of $-\mathbf{E}$, thus producing a current density

$$\mathbf{J} = \sigma_c \mathbf{E} \qquad (36)$$

where σ_c is the conductivity. Eq. (34) is a form of *Ohm's Law.* Applied to a wire of length l and cross-section S, (36) becomes

$$V = \int_0^l \mathbf{E} \cdot d\mathbf{l} = I \int_0^l \frac{dl}{\sigma S} = IR \qquad (37)$$

where R is the resistance and V is the potential drop along the wire. The power dissipated into electron transitions at crystal imperfections is $IV = I^2R$. This power can be proved (Poynting theorem) to enter the segment of wire through the sides, not the ends.

No current in a conductor implies $\mathbf{J} = \sigma\mathbf{E} = 0$ in the interior of the conductor, which becomes an equipotential volume. By (7), ρ also becomes zero in the interior; any net charge on the conductor must reside on its surface. Application of Gauss' law to a surface in the material surrounding an empty cavity in the conductor shows that there can be no charge on the inside surface and no field in the cavity. The potential in such a cavity must therefore be the same as the potential of the conductor found by use of (8), and will be called the *cavity potential*.

(c) A thin wire carrying steady current must have $\mathbf{J}$ and $\mathbf{E}$ within it parallel to it sides. The surface charges of its distributed capacitance must move with $\mathbf{J}$. They are locally responsible for a normal component of $\mathbf{E}$ outside the wire, and over the whole circuit are the main sources of the field both inside and outside the conductors.

(d) The energy difference per unit charge between the F.l. and a point in a field-free cavity in the conductor is called the *work function* ϕ of the solid. When two conductors of the same cavity potential and differing ϕ's are placed in electrical contact, electrons will transfer toward the lower of the two F.l.'s until the F.l.'s come to equality with a difference $\Delta V = \Delta\phi$ between the two potentials, which is the *contact potential difference*. (See Fig. 4.)

(e) When a temperature gradient exists in a conductor, the varying extent of thermal electron jumps will tend to cause a current. Ohm's law becomes

$$\mathbf{J} = \sigma(-\boldsymbol{\nabla}V + S^*\boldsymbol{\nabla}T) \qquad (38)$$

where S^*, the *entropy transport per unit charge*, is responsible for thermoelectricity. (See THERMOELECTRICITY.)

(f) An insulator has all the levels of one band filled and a substantial gap below the next, empty band. The presence of a small field $\mathbf{E}$ will shift electrons only within molecular distances, creating dipoles in the phenomenon of dielectric polarization. (See DIELECTRIC THEORY.) A sufficiently large $\mathbf{E}$ may transfer some electrons to the empty band and provide them enough energy of motion to cause dielectric breakdown.

(g) A semiconductor is an insulator with a narrow interband gap, which contains either donor levels from atoms with excess electrons not covalently bound, or acceptor levels requiring the addition of electrons to satisfy their covalent bonds. In the former case (n-type) the donated electrons are readily lifted into conductor levels, above the gap, and in the latter (p-type) the residual "holes" become transferred down to the filled band and function as positive carriers. (See SEMICONDUCTORS.)

2. *Aqueous liquids* are like insulators in having no accessible levels corresponding to macroscopic electron transfer. However, electrons can transfer between molecules or atoms, and the resulting positive and negative ions are free to move under the influence of an electric field. The conductivity of a positive or negative ionic species is expressed by

$$v_\pm = \pm u_\pm E \quad \text{and} \quad J_\pm = neZ_\pm v_\pm \qquad (39)$$

where $v_\pm$ is the *ionic velocity*, $u_\pm$ the *mobility*, E the local field intensity, Z the *valence*, n the *ionic density*, and $J_\pm$ the partial current density due to one positive or negative species.

In a liquid with a particular ionic constitution there is always a highest occupied electron level and a lowest empty one. If a metallic electrode is inserted, equilibrium will generally be established between one of these levels and the Fermi level in the metal; the electron transfer between phases often results in deposition of neutral atoms on the metal or removal from the metal. The F.l. of the electrode thus takes up a definite position in relation to the cavity potential

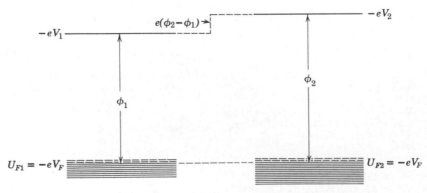

Fig. 4. Energy levels for adjacent metals in contact when equilibrium has been established (Fermi levels U_F coincident).

of the solution. A *cell* is an arrangement with two electrodes in which a solution has two different constitutions at the electrodes that are capable of reacting by electron transfer but cannot react directly with each other because of physical separation. The two electrode F.l.'s came to different positions in relation to the solution cavity potential, the difference being the electomotive force (emf) $\mathcal{E}$ of the cell.

When current passes through a cell, the suppressed reaction can proceed via electron transfer through the outer circuit. Owing to internal resistance r in the cell, the external potential or F.l. difference becomes $V = \mathcal{E} - Ir$. The amount of reaction at either electrode is such that one mole of reactant of valence Z will be deposited or dissolved when the charge transferred $Q = It = N_A Ze$, an expression of Faraday's laws of electrochemistry. The transition between two or more types of current carriers in the solution and one type only at the electrode is accomplished by the occurrence of diffusion processes.

3. In *gases*, electrons can move freely between molecules once knocked loose by molecular collision or by energetic charged particles. Local fields will accelerate electrons and ions in between collisions. A field strength of about 3×10^6 v/m is enough to initiate a breakdown in air at S.T.P. by allowing electrons to gain enough energy between collisions to ionize more molecules. Somewhat lesser fields can render a gas partially conducting. In a highly ionized condition, the gas becomes a plasma (PLASMA).

Electrostatic phenomena of charge transfer make use of gaseous conduction, of charge redistribution in conductors under the influence of fields, and of forces exerted on the charges that reside on the surfaces of conductors. For the production of fields by charges moving at relativistic velocities, see PROPAGATION OF ELECTROMAGNETIC WAVES.

WILLIAM T. SCOTT

References

Scott, W. T., "The Physics of Electricity and Magnetism," Second Edition, New York, John Wiley, 1966; Huntington, N.Y., Robert E. Krieger, 1976.

Reitz, J. R., Milford, F. J., and Christy, R. W., "Foundations of Electromagnetic Theory," Third Edition, Reading, Mass., Addison-Wesley, 1979.

Lorrain, P., and Corson, D., "Electromagnetic Fields and Waves," Second Edition, San Francisco, W. H. Freeman, 1970.

Moore, A. D. (Ed.), "Electrostatics and Its Applications," New York, John Wiley, 1973.

Portis, A. M., "Electromagnetic Fields: Sources and Media," New York, John Wiley, 1978.

Cross-references: CAPACITANCE; CIRCUITRY; CONDUCTIVITY, ELECTRICAL; DIELECTRIC THEORY; ELECTRIC POWER GENERATION; ELECTRICAL DISCHARGES IN GASES; ELECTRICAL MEASUREMENTS; ELECTROCHEMISTRY; ELECTROMAGNETIC THEORY; ELECTRON; FERROMAGNETISM; HIGH VOLTAGE RESEARCH; INDUCED ELECTROMOTIVE FORCE; INDUCTANCE; MOTORS, ELECTRIC; POTENTIAL; SEMICONDUCTOR; STATIC ELECTRICITY.

ELECTROACOUSTICS

Introduction Electroacoustics is concerned with the transduction of acoustical to electrical energy and vice versa. Devices which convert acoustical signals into electrical signals are referred to as *microphones* or *hydrophones* depending on whether the acoustic medium is air or water. Devices which convert electrical signals into acoustical waves are referred to as *loudspeakers* (or *earphones*) in air and *projectors* in water.

Transduction Mechanisms *Piezoelectricity* Certain crystals produce charge on their surfaces when strained and conversely become strained when placed in an electric field. Important piezoelectric crystals include quartz, ADP, lithium sulfate, rochelle salt, and tourmaline. Lithium sulfate and tourmaline are *volume expanders*, that is, their volume changes when subjected to an electric field in the proper direction. Such crystals can detect hydrostatic pressure directly. Crystals which are not volume expanders must have one or more surfaces shielded from the pressure field in order to convert the pressure to a uniaxial strain which can be detected. Tourmaline is relatively insensitive and used primarily in blast gauges. Quartz is used principally in high-Q ultrasonic transducers.

Certain ceramics such as lead zirconate titanate (PZT), barium titanate, and lead metaniobate become piezoelectric when polarized. They exhibit relatively high electromechanical coupling, are capable of producing very large forces, and are used extensively as sources and receivers for underwater sound. PZT and barium titanate have only a small volume sensitivity hence must have one or more surfaces shielded in order to detect sound efficiently. Piezoelectric ceramics have extraordinarily high dielectric coefficients and hence high capacitance and are thus capable of driving long cables without preamplifiers.

Recently, it has been discovered that certain polymers, notably polyvinylidene fluoride, are piezoelectric when stretched. Such piezoelectric polymers are finding use in directional microphones and ultrasonic hydrophones.

Magnetostriction Some ferromagnetic materials become strained when subjected to a magnetic field. The effect is quadratic in the field so a bias field or dc current is required for linear operation. Important magnetostrictive metals and alloys include nickel and permendur. At one time, magnetostrictive transducers were used extensively in active sonars but have now been largely replaced by ceramic transducers. Magnetostrictive transducers are rugged and reliable but inefficient and configurationally awkward. Recently, it has been discovered that certain rare earth iron alloys such as terbium-dysprosium-iron possess extremely large magnetostrictions (as much as 100 times that of nickel). They have relatively low eddy current losses but require large bias fields, are fragile, and have yet to find significant applications. Metallic glasses have also recently been considered for magnetostrictive transducers.

Electrodynamics Electrodynamic transducers exploit the forces produced on a current-carrying conductor in a magnetic field and, conversely, the currents produced by a conductor moving in a magnetic field. Direct-radiation moving-coil transducers dominate the loudspeaker field. Prototypes of high power underwater projectors have been constructed using superconducting magnets. Electrodynamic microphones, particularly the directional ribbon microphones, are also common.

Electrostatics Electrostatic sources utilize the force of attraction between charged capacitor plates. The force is independent of the sign of the voltage so a bias voltage is necessary for linear operation. The forces are relatively weak so a large area is needed to obtain significant acoustic output. The effect is reciprocal, with the change in the separation of the plates (i.e., the capacitance) produced by an incident acoustic pressure generating a voltage. The impedance of a condenser microphone, however, is high, so a preamplifier located close to the sensor is required. Condenser microphones are very flat and extremely sensitive. The change in capacitance induced by an acoustic field can also be detected by making the capacitor a part of a bridge circuit, or alternatively part of an oscillator circuit. The acoustic signal will then appear as either an amplitude or frequency modulation of some ac carrier. The charge storage properties of electrets have been exploited to produce electrostatic microphones which do not require a bias voltage.

Magnetism Magnetic transducers utilize the force of attraction between magnetic poles and, reciprocally, the voltages produced when the reluctance of a magnetic circuit is changed. Magnetic speakers are used extensively in telephone receivers.

Hydraulics Nonreversible, low frequency, high power underwater projectors can be constructed utilizing hydraulic forces acting to move large pistons. Electroacoustic transduction is achieved by modulating the hydraulic pressure with a spool valve actuated by an electrostrictive (PZT) stack.

Fiber Optics An acoustic field acting on an optical fiber will change the optical path length by changing the length and index of refraction of the fiber. Extremely sensitive hydrophones and microphones can be made by using a fiber exposed to an acoustic field as one leg of an optical interferometer. Path length changes of the order of 10^{-6} optical wavelengths can be detected. The principal advantages of such sensors are their configurational flexibility, their sensitivity, and their suitability for use with fiber optic cables. Fiber optic sensors which utilize amplitude modulation of the light (microbend transducers) are also being developed.

Parametric Transducers The nonlinear interaction of sound waves can be used to produce highly directional sound sources with no side lobes and small physical aperture. In spite of their inherent inefficiency, substantial source levels can be achieved and such "parametric sonars" have found a number of underwater applications. Parametric receivers have also been investigated but practical applications have yet to be found.

Carbon Microphones Carbon microphones utilize a change in electrical resistance with pressure and are used extensively in telephones.

Sensitivty and Source Level A microphone or hydrophone is characterized by its free-field voltage sensitivity M, which is defined as the ratio of the output voltage E to the free-field amplitude of an incident plane acoustic wave. That is, for an incident wave which *in the absence of the transducer* is given by

$$P = P_0 \cos (\mathbf{k} \cdot \mathbf{R} - \omega t), \qquad (1)$$

M is defined by

$$M = E/P_0. \qquad (2)$$

In general M will be a function of frequency and the orientation of the transducer with respect to the wave vector k (i.e., the direction of incidence of the wave). Thus, for a given frequency, M is proportional to the directivity of the transducer. It is usually desirable for a microphone or hydrophone to have a flat (i.e., frequency independent) free field voltage sensitivity over the broadest possible range of frequencies to assure fidelity of the output electrical signal.

A loudspeaker or projector is characterized in a similar manner by its transmitting current response S which is defined as the ratio of the acoustic source level to the driving current I. In the farfield of a transducer the acoustic pressure is a spherical wave which can be expressed as

$$P(R) = P_s(\theta, \phi) (R_0/R) \cos (kR - \omega t) \qquad (3)$$

where θ and ϕ are elevation and azimuth angles and R_0 an arbitrary reference distance (usually 1 meter). $P_s(\theta, \phi)$ is defined as the source level. Thus S is given by

$$S = P_s(\theta, \phi)/I \qquad (4)$$

which is a function of θ and ϕ and the frequency ω. For high fidelity sound reproduction S should be as flat as possible over the broadest possible bandwidth. For some purposes, however, such as ultrasonic cleaning or long range underwater acoustic propagation, fidelity is unnecessary and high-Q resonant transducers are employed to produce high intensity sound over a narrow bandwidth.

Reciprocity Most conventional transducers are *reversible*, that is, they can be used as either sources or receivers of sound (a carbon microphone and a fiber optic hydrophone are examples of transducers which are *not* reversible). A transducer is said to be *linear* if the input and output variables are linearly proportional (hot-wire microphones and unbiased magnetostrictive transducers are examples of *nonlinear* transducers). A transducer is said to be *passive* if the only source of energy is the input electrical or acoustical signal (a microphone with a built in preamplifier and a parametric receiver are examples of *nonpassive* transducers). Most transducers which are linear, passive and reversible exhibit a remarkable property called *reciprocity*. For a *reciprocal* transducer of any kind (moving coil, piezoelectric, magnetostrictive, electrostatic, magnetic etc.) the ratio of the free field voltage sensitivity to the transmitting current response is equal to the reciprocity factor J, which is independent of the geometry and construction of the transducer. That is:

$$\frac{M(\omega, \theta, \phi)}{S(\omega, \theta, \phi)} = J(\omega) = \frac{4\pi R_0}{\rho_0 \omega} \qquad (5)$$

where ρ_0 is the density of the medium and R_0 is the reference distance used in defining the source level. Equation (5) has a number of useful consequences: (1) the receiving and transmitting beam patterns of a reciprocal transducer are identical, (2) a transducer cannot be simultaneously flat as a receiver and transmitter since S has an additional factor of ω, and (3) Eq. (5) provides the basis for the three transducer reciprocity calibration technique whereby an absolute calibration of a hydrophone or microphone can be obtained from purely electrical measurements.

Canonical Equations and Electroacoustic Coupling Simple acoustic transducers can be characterized by the following canonical equations:

$$E = Z_e I + T_{em} V \qquad (6)$$

$$F = T_{me} I + Z_m V \qquad (7)$$

where V is the velocity of the radiating or receiving surface, F is the total force acting on the surface (including acoustic reaction forces), Z_e is the blocked ($V = 0$) electrical impedance, Z_m is the open circuit mechanical impedance and T_{em} and T_{me} are the electromechanical coupling coefficients. For reciprocal transducers $T_{em} = \pm T_{me}$. For example, for a moving coil transducer where the "motor" is a coil in a radial magnetic field B,

$$T_{em} = -T_{me} = BL \qquad (8)$$

where L is the length of the wire in the coil and the electrical impedance Z_e is largely inductive. For a piston transducer with a piezoelectric "motor"

$$T_{me} = T_{em} = -id_{33}/(\epsilon^T s \omega) \qquad (9)$$

where d_{33} is the piezoelectric strain coefficient, s is the compliance, ϵ^T is the permittivity at constant stress, and the electrical impedance Z_e is largely capacitive.

If a piston transducer is placed in an acoustic field such that the average pressure over the surface of the piston is P_B then $F = P_B A$, where A is the area of the piston, and for a receiver $I = 0$, so

$$E = (T_{em} A/Z_m) P_B. \qquad (10)$$

If the transducer is small compared with an acoustic wave length $P_B \approx P_0$ (in general $P_B = D P_0$, where D is the diffraction constant) and the free field voltage sensitivity is given by

$$M = T_{em} A/Z_m. \qquad (11)$$

From Eq. (5) the transmitting current response is

$$S = \frac{\rho_0 \omega T_{em} A}{4\pi R_0 Z_m}. \qquad (12)$$

From these simple considerations a number of principles of practical transducer design can be deduced. The mechanical impedance Z_m is in general given by

$$Z_m = \frac{K_m}{i\omega} + i\omega M + R_m \qquad (13)$$

where K_m is an effective spring constant, M the mass, and R_m the mechanical resistance. For a piezoelectric transducer (Eq. (9)) T_{em} is inversely proportional to frequency, hence from Eqs. (10) and (11) we see that a piezoelectric transducer will have a flat receiving sensitivity below resonance (i.e., where its behavior is controlled by stiffness). On the other hand a moving coil microphone must have a resistive mechanical impedance to have a flat response. From Eq. (12) we derive the fundamental tenet of loudspeaker design, that a moving coil loudspeaker will have a flat transmitting current response above resonance (i.e., where it is mass controlled). Accordingly, moving coil loudspeakers are designed to have the lowest possible resonant frequency (by means of a high compliance, since the output is inversely proportional to the mass) and piezoelectric hydrophones are designed to have the highest possible resonant frequency.

An interesting and important consequence of electromechanical coupling is the effect of the motion of the transducer on the electrical impedance. In the absence of external forces (including radiation reactance) from Eqs. (6) and (7)

$$E = \left(Z_e - \frac{T_{em}T_{me}}{Z_m}\right) I. \qquad (14)$$

That is, the electrical impedance has a "motional" component given by $T_{em}T_{me}/Z_m$. The motional component can be quite significant near resonance where Z_m is small. This effect is the basis of crystal controlled oscillators.

Radiation Impedance An oscillating surface produces a reaction force F_R on its surface given by

$$F_R = -Z_R V \qquad (15)$$

where Z_R is the radiation impedance. We can thus rewrite Eq. (7) as

$$F_{ext} = T_{em} I + (Z_R + Z_m) V \qquad (16)$$

where F_{ext} now includes only external forces. For an acoustically small baffled circular piston of radius a,

$$Z_R = \pi a^4 \rho_0 \omega^2 / 2c - i(8/3)\omega\rho_0 a^3. \qquad (17)$$

The radiation impedance thus has a masslike reactance with an equivalent "radiation mass" of $(8/3)\rho_0 a^3$ and a small resistive component proportional to ω^2 responsible for the radiated power. A transducer will thus have a lower resonant frequency when operated underwater than when operated in air or vacuum. The total radiated power of the piston transducer is given by

$$\pi = R_e Z_R |V|^2 = (\pi a^4 \rho_0 \omega^2 / 2c) V^2. \qquad (18)$$

Most transducers are displacement limited so for a direct radiating transducer V in Eq. (18) is limited. To obtain the most output power the piston should have the largest possible surface area consistent with keeping the transducer omnidirectional (the transducer will become directional when $a \gtrsim \lambda$). This is easy to do in air but difficult in water, since it is hard to make pistons which are both lightweight and stiff enough to hold their shape in water. Alternatively, the driver can be placed at the apex of a horn. For a conical horn the fluid velocity at end of the horn (where the radius is a_e) will be reduced to $V(a/a_e)$ but the radiating piston will now have an effective radius of a_e so the radiated power will increase by a factor of $(a_e/a)^2$. For high power operation at a single frequency the driver can be placed at the end of a quarter wave resonator.

Directivity It is often desirable for transducers to be directional. Directional sound sources are needed in diagnostic and therapeutic medical ultrasonics, for acoustic depth sounders and to reduce the power requirements and reverberation in active sonars, etc. Direc-

tional microphones are useful to reduce unwanted noise (e.g., to pick up the voice of a speaker and not the audience), directional hydrophones or hydrophone arrays increase signal-to-noise and aid in target localization. One way to achieve directionality is to make the radiating surface large. A baffled circular piston has a directivity given by

$$D_\theta = 2J_1 (ka \sin \theta)/ka \sin \theta \qquad (19)$$

D_θ equals unity for $\theta = 0$ and $\frac{1}{2}$ when $ka \sin \theta = 2.2$. For small values of ka, D_θ is near unity for all angles.

Some transducers respond to the gradient of the acoustic pressure rather than pressure, for example, the ribbon microphone which works by detecting the motion of a thin conducting strip orthogonal to a magnetic field. Such transducers have a directivity which is dipole in nature, i.e.,

$$D_\theta = \cos \theta. \qquad (20)$$

Note that since the force in this case is proportional not to P_0 but to kP_0, a ribbon microphone (which like a moving coil microphone is electrodynamic) will have flat receiving sensitivity when its impedance is mass controlled. By combining a dipole receiver with a monopole receiver one obtains a unidirectional cardioid receiver with

$$D_\theta = (1 + \cos \theta) \qquad (21)$$

P. H. ROGERS

References

Hunt, F. V., "Electroacoustics," Cambridge, MA, Harvard Univ. Press, and New York, John Wiley & Sons, 1954.

Bobber, R. J., "Underwater Electroacoustic Measurements," U.S. Gov. Printing Office, Washington, D.C., 1969.

Bouyoucos, "Hydroacoustic Transduction," *J. Acoust. Soc. Am.* **57**, 1341 (1975).

Meeks, S. W., and Timme, R. W., "Rare Earth Iron Magnetostrictive Underwater Sound Transducer," *J. Acoust. Soc. Am.* **62**, 1158 (1977).

Bacaro, J. A., Dardy, H. D., and Carome, E. F., "Fiber Optic Hydrophone," *J. Acoust. Soc. Am.* **62**, 1302 (1977).

Moffett, M. B., and Mellon, R. M., "Model for Parametric Acoustic Sources," *J. Acoust. Soc. Am.* **61**, 325 (1977).

Ricketts, D., "Electroacoustic Sensitivity of Piezoelectric Polymer Cylinders," *J. Acoust. Soc. Am.* **68**, 1025 (1980).

Sessler, G. M., and West, J. E., "Applications," in "Electrets" (G. M. Sessler, Ed.), New York, Springer-Verlag, 1980.

Bobber, R. J., "New Types of Transducer," in "Underwater Acoustics and Signal Processing (L. Bjorno, Ed.), Dordrecht, Holland, D. Riedel Publishing Company, 1981.

Cross-references: ACOUSTICS, CRYSTALLOGRA-
PHY, DIELECTRIC THEORY, LASER, MAGNE-
TISM, RESONANCE, SOLID-STATE PHYSICS, UL-
TRASONICS.

ELECTROCHEMISTRY

Electrochemistry is that branch of science which deals with the interconversion of chemical and electrical energies, i.e., with chemical changes produced by electricity as in electrolysis or with the production of electricity by chemical action as in electric cells or batteries. The science of electrochemistry began about the turn of the eighteenth century. In 1796 Alessandro Volta observed that an electric current was produced if unlike metals separated by paper or hide moistened with water or a salt solution were brought into contact. Volta used the sensation of pain to detect the electric current. His observation was similar to that observed ten years earlier by Luigi Galvani who noted that a frog's leg could be made to twitch if copper and iron, attached respectively to a nerve and a muscle, were brought into contact.

In his original design Volta stacked couples of unlike metals one upon another in order to increase the intensity of the current. This arrangement became known as the "voltaic pile." He studied many metallic combinations and was able to arrange the metals in an "electromotive series" in which each metal was positive when connected to the one below it in the series. Volta's pile was the precursor of modern batteries (see BATTERIES).

In 1800 William Nicholson and Anthony Carlisle decomposed water into hydrogen and oxygen by an electric current supplied by a voltaic pile. Whereas Volta had produced electricity from chemical action these experimenters reversed the process and utilized electricity to produce chemical changes. In 1807 Sir Humphry Davy discovered two new elements, potassium and sodium, by the electrolysis of the respective solid hydroxides, utilizing a voltaic pile as the source of electric power. These electrolytic processes were the forerunners of the many industrial electrolytic processes used today to obtain aluminum, chlorine, hydrogen, or oxygen, for example, or in the electroplating of metals such as silver or chromium.

Since in the interconversion of electrical and chemical energies, electrical energy flows to or from the system in which chemical changes take place, it is essential that the system be, in large part, conducting or consist of electrical conductors. These are of two general types—electronic and electrolytic—though some materials exhibit both types of conduction. Metals are the most common electronic conductors. Typical electrolytic conductors are molten salts and aqueous solutions of acids, bases, and salts. Owing, in general, to the low solubility and ionization of solutes in nonaqueous solvents, ethyl alcohol for example, nonaqueous solutions usually have much lower electrical conductivity than aqueous systems and as a consequence find less use in electrochemical applications. Molten salts, on the other hand, are ionic fluids, freely mix, and have high electrical conductivity but are restricted to electrochemical uses at high temperatures.

A current of electricity in an electronic conductor is due to a stream of electrons, particles of subatomic size, and the current causes no net transfer of matter. The flow is, therefore, in a direction contrary to what is conventionally known as the "direction of the current." In electrolytic conductors, the carriers are charged particles of atomic or molecular size called *ions*, and under a potential gradient, a transfer of matter occurs.

An electrolytic solution contains an equivalent quantity of positively and negatively charged ions whereby electroneutrality prevails. Under a potential gradient, the positive and negative ions move in opposite directions with their own characteristic velocities and each accordingly carries a different fraction of the total current through any one solution. Each fraction is referred to as the ionic transference number. Furthermore, the velocity increases with temperature causing a corresponding increase in electrolytic conductivity. This characteristic is opposite to that observed for most electronic conductors which show less conductivity as their temperature is increased.

The concept that charged particles are responsible for the transport of electric charges through electrolytic solutions was accepted early in the history of electrochemistry. The existence of ions was first postulated by Michael Faraday in 1834; he called negative ions "anions" and positive ones "cations." In 1853, Hittorf showed that ions move with different velocities and exist as separate entities and not momentarily as believed by Faraday. In 1887, Svante Arrhenius postulated that solute molecules dissociated spontaneously into *free ions* having no influence on each other. However, it is known that ions are subject to coulombic forces, and only at infinite dilution do ions behave ideally, i.e., independently of other ions in the solution. Ionization is influenced by the nature of the solvent and solute, the ion size, and solute-solvent interaction. The dielectric constant and viscosity of the solvent play dominant roles in conductivity. The higher the dielectric constant, the less are the electrostatic forces between ions and the greater is the conductivity. The higher the viscosity of the solvent, the greater are the frictional forces between ions and solvent molecules and the lower is the electrolytic conductivity.

In 1923 Debye and Hückel presented a theory which took into account the effect of coulombic forces between ions. They introduced the concept of the ion atmosphere, or continuous charge distribution, which is a continuous function of r, the radial distance from a central or reference ion rather than a discrete or discontinuous charge distribution. The ion atmosphere acts electrostatically somewhat like a sphere of charge $-\epsilon$ at some average distance from a central or reference ion of charge $+\epsilon$, with the value of the average distance approximating that of the ionic radii of ionic crystals. This interionic attraction leads to two effects on the electrolytic conductivity. Under a potential gradient, an ion moves in a certain direction. However, the ion cloud, being of opposite sign will tend to move in the opposite direction, and because of its attraction for the central ion, will have a retarding effect on the ion velocity and thereby lead to a lowering in the electrolytic conductivity. On the other hand, the central ion will tend to pull the ion cloud with it to a new location. The ion atmosphere will adjust to its new location in time, but not instantaneously, and the delay results in a dissymmetry in the potential field around the ion. This also causes a lowering in the conductance of the solution. These effects become more pronounced as the concentration of the solution is increased; for dilute solutions, below about 0.1 molal, the equivalent conductance decreases with the square root of the concentration. For more concentrated solutions, the relation between conductivity and concentration is much more complex and depends more specifically on individual solute properties.

Interionic attraction in dilute solutions also leads to an effective ionic concentration or activity which is less than the stoichiometric value. The *activity* of an ion species is its thermodynamic concentration, i.e., the ion concentration corrected for the deviation from ideal behavior. For dilute solutions the activity of ions is less than one, for concentrated solutions it may be greater than one. It is the ionic activity that is used in expressing the variation of electrode potentials, and other electrochemical phenomena, with composition.

When electricity passes through a circuit consisting of both types of electrical conductors, a chemical reaction always occurs at their interface. These reactions are electrochemical. When electrons flow from the electrolytic conductor, oxidation occurs at the interface while reduction occurs if electrons flow in the opposite direction. These electronic-electrolytic interfaces are referred to as *electrodes*; those at which oxidation occurs are known as *anodes* and those at which reduction occurs, as *cathodes*. An anode is also defined as that electrode by which "conventional" current enters an electrolytic solution, a cathode as that electrode by which "conventional" current leaves. Positive ions, for example, ions of hydrogen and the metals, are called *cations* while negative ions, for example, acid radicals and ions of nonmetals are called *anions*.

In 1833, Michael Faraday enunciated two laws of electrolysis which give the relation between chemical changes and the product of the current and time, i.e., the total charge (coulombs) passed through a solution. These laws are: (1) the amount of chemical change, e.g., chemical decomposition, dissolution, deposition, oxidation, or reduction, produced by an electric current is directly proportional to the quantity of electricity passed through the solution; (2) the amounts of different substances decomposed, dissolved, deposited, oxidized, or reduced are proportional to their chemical equivalent weights. A chemical equivalent weight of an element or a radical is given by the atomic or molecular weight of the element or radical divided by its valence; the valence used depends on the electrochemical reaction involved. The electric charge on an ion is equal to the electronic charge or some integral multiple of it. Accordingly, a univalent negative ion has a charge equal in magnitude and of the same sign as a single electron, and its chemical equivalent weight is equal to its atomic weight, if an element, or to its molecular weight, if a radical. A trivalent ion has $+3$ or -3 electronic charges, depending on whether it is a positive or a negative trivalent ion. For trivalent ions, then, the equivalent weight would be equal to its atomic weight, if an element, or to its molecular weight, if a radical, divided by three.

The quantity of electricity required to produce a gram-equivalent weight of chemical change is known as the *faraday*. A faraday corresponds, then, to an *Avogadro number of charges*. The most accurate determination of the faraday has been made by a silver-perchloric acid coulometer in which the amount of silver electrolytically dissolved in an aqueous solution of perchloric acid is measured. This method gives 96487 coulombs (or ampere-seconds) per gram-equivalent for the faraday on the unified C^{12} scale of atomic weights.

The *electrochemical equivalent* or, preferably, the *coulomb equivalent* of an element or radical is that weight in grams which is equivalent to one coulomb of electricity and is given by the gram-equivalent weight divided by the faraday (96487 coulombs per gram-equivalent); for example, the electrochemical equivalent of silver is given by $107.870/96487$ or 0.00111797 g/coulomb where 107.870 is the atomic weight of silver based on the unified C^{12} scale. The electrochemical equivalents of other elements may be calculated in like fashion.

In electrolysis and in any electric cell or battery, there is an electromotive force (emf) or voltage across the terminals. This emf is expressed in the practical unit, the volt, which is equal to the electromagnetic unit in the meter-kilogram-second system. In any one cell, the emf is the sum of the potentials of the two electrodes and of any liquid-junction potentials that may be present. Neither of the individual electrode potentials can be evaluated without reference to a chosen reference electrode of assigned value. For this purpose, the hydrogen electrode has been universally adopted and is arbitrarily assigned a zero potential for all temperatures when the hydrogen ion is at unit activity and the hydrogen gas is at atmospheric pressure. A hydrogen electrode consists of a stream of hydrogen gas bubbling over platinized platinum or gold foil and immersed in a solution containing hydrogen ions; the electrochemical reaction is: $1/2H_2 (gas) = H^+(solution) + \epsilon$, where ϵ represents the electron. The potential of the hydrogen electrode, E_H, as a function of hydrogen-ion concentration and hydrogen-gas pressure is given by

$$E_H = E_H{}^0 - (RT/nF) \ln (a_{H^+}/p_{H_2}{}^{1/2})$$
$$= E_H{}^0 - (RT/nF) \ln (c_{H^+}f_{H^+}/p_{H_2}{}^{1/2}).$$

where $E_H{}^0$ is the standard quantity assigned a value of zero, R is the gas constant, T the absolute temperature, n the number of equivalents, F the faraday, p_{H_2} the pressure of hydrogen, and a_{H^+}, c_{H^+} and f_{H^+}, respectively, the activity, concentration, and activity coefficient of hydrogen ions. When a_{H^+} and $p_{H_2}{}^{1/2}$ equal one, $E_H = E_H{}^0$. For very dilute solutions below 0.01 molal f_{H^+} may be taken as unity without appreciable error.

The standard potentials, E^0, of other electrodes are obtained by direct or indirect comparison with the hydrogen electrode. Values thus obtained at 25°C for some typical elements are listed in Table 1.

The reducing power of the elements decreases on going down the column. These values are for the ions at unit activity, and reversible or thermodynamic values as a function of metal or radical concentration are given by equations similar to the one above. For the general reaction: $M = M^{n'} + n\epsilon$, the potential is given by $E_M = E_M{}^0 - (RT/nF) \ln a_M{}^{n+}$.

In electrolysis, at very low current densities, the potentials of the electrodes approximate in magnitude their reversible values and deviate somewhat from these values because of an IR drop in the solution and possible concentration polarization (the concentration at the electrode surface may differ from that in the bulk of the solution). Also for high current densities, especially for the generation of gases such as hydrogen, oxygen or chlorine, the voltage required exceeds the reversible voltage; the excess voltage is known as overvoltage, or overpotential for a single electrode, and arises from energy barriers at the electrode. Overpotential, in general, increases logarithmically with an increase in current density.

The electrochemistry of the solid state is similar in many ways to that of aqueous solutions. The basic problems of ionic crystals can be treated in many cases by analogy to aqueous solutions. Differences do occur. In silver halides, for example, practically only cations are mobile, while in lead chloride and bromide only anionic conduction occurs. Mixed conductance, in that both anions and cations migrate, is exhibited by alkali halides and lead iodide. Other solids at temperatures below about 600°C, cuprous oxide for example, show only electronic migration, while above this temperature they exhibit in addition some cationic conduction. Ionic conduction in the solid state is attributed superordinately to lattice defects. Even for pure substances lattice defects which may contribute to the electrical conductivity may be due to interstitial cations, cation vacancies, interstitial anions, anion vacancies, excess electrons, or electron holes. Conduction occurs, for instance, when a cation in one interstitial site jumps to an adjacent interstitial site. Lattice defects in pure substances may be artificially produced by trace doping with an unlike substance of proper characteristics. When, for example, silicon, a semiconductor with four valence electrons, is doped with a five-valence-electron element, such as antimony, arsenic, or phosphorus, loosely bound electrons are introduced into the lattice of silicon which are free to migrate leading to electrical conduction. Such a semiconductor is known as an n-type. A p-type semiconductor is produced if silicon is doped with a three-valence-electron element, such as boron, indium, or gallium, whereby a hole in

TABLE 1. SOME STANDARD ELECTRODE POTENTIALS AT 25°C

Electrode	Potential (V)	Electrode	Potential (V)
$Li = Li^+ + \epsilon$	−3.045	$Cu = Cu^{++} + 2\epsilon$	+0.337
$Ca = Ca^{++} + 2\epsilon$	−2.87	$Cu = Cu^+ + 1\epsilon$	+0.521
$Na = Na^+ + \epsilon$	−2.714	$2I^- = I_2 + 2\epsilon$	+0.536
$Mg = Mg^{++} +2\epsilon$	−2.37	$2Hg = Hg_2{}^{++} + 2\epsilon$	+0.789
$Al = Al^{+++} + 3\epsilon$	−1.66	$Ag = Ag^+ + \epsilon$	+0.799
$Mn = Mn^{++} + 2\epsilon$	−1.18	$Pd = Pd^{++} + 2\epsilon$	+0.987
$Zn = Zn^{++} + 2\epsilon$	−0.763	$Pt = Pt^{++} + 2\epsilon$	+1.20
$Fe = Fe^{++} + 2\epsilon$	−0.440	$2Cl^- = Cl_2 + 2\epsilon$	+1.36
$Ni = Ni^{++} + 2\epsilon$	−0.250	$Au = Au^+ + \epsilon$	+1.68
$H_2 = 2H^+ + 2\epsilon$	0.000	$2F^- = F_2 + 2\epsilon$	+2.87

the silicon lattice is produced which can move about and impart conductivity.

The passage of electricity through gases is sometimes included under electrochemistry. However, in electrical discharges in gases, the principles are entirely different from what they are in the electrolysis of electrolytic solutions. Whereas in the latter, ionic dissociation occurs spontaneously as a result of forces between solvent and solute and without the application of an external field, for gases relatively high voltages must be applied to accelerate the electrons from the electrode to a velocity at which they can ionize the gas molecules they strike. In this case, the resulting chemical reaction taking place between ions, free radicals, and molecules occurs in the gas phase and not at the electrodes as in the electrolysis of solutions. Studies of the electrical conduction of gases, accordingly, are generally considered under the physics of gases.

In addition to the above topics, it is frequently customary to include under electrochemistry: (1) processes for which the net reaction is physical transfer, e.g., concentration cells; (2) electrokinetic phenomena, e.g., electrophoresis, electroosmosis, streaming potential; (3) properties of electrolytic solutions if determined by electrochemical or other means, e.g., activity coefficients and hydrogen-ion concentration; (4) processes in which electrical energy is first converted to heat which in turn causes a chemical reaction to occur that would not do so spontaneously at ordinary temperature. The first three are frequently considered a portion of physical chemistry, and the last one is a part of electrothermics or electrometallurgy.

Electrochemistry finds wide application. In addition to industrial electrolytic processes, electroplating, and the manufacture and use of batteries already mentioned, the principles of electrochemistry are used in chemical analysis, e.g., polarography, and electrometric or conductometric titrations; in chemical synthesis, e.g., dyestuffs, fertilizers, plastics, insecticides; in biology and medicine, e.g., electrophoretic separation of proteins, membrane potentials; in metallurgy, e.g., corrosion prevention, electrorefining; and in electricity, e.g., electrolytic rectifiers, electrolytic capacitors, Josephson junctions.

WALTER J. HAMER

References

Dubpernell, George, and Westbrook, J. H. (Eds.), "Selected Topics in the History of Electrochemistry," Proceedings Volume 78-6, Princeton, N.J., The Electrochemical Society, 1978.

Bard, A. J., (Ed.), "Encyclopedia of Electrochemistry of the Elements," Vol. 14, New York, Marcel Dekker Inc., 1980.

Bard, A. J., and Faulker, L. R., "Electrochemical Methods: Fundamentals and Applications," New York, John Wiley & Sons, 1980.

Rand, D. A. J., Power, G. P., and Ritchie, I. M. (Eds.), "Progress in Electrochemistry," New York, Elsevier Scientific Publishing Co., 1981.

Cross-references: BATTERIES; CONDUCTIVITY, ELECTRICAL; DIELECTRIC THEORY; ELECTRICAL DISCHARGES IN GASES; IONIZATION; MOLECULAR WEIGHT; POTENTIAL; VISCOSITY.

ELECTROLUMINESCENCE

Electroluminescence is a process which generates light in crystals by conversion of energy supplied by electric contacts, in the absence of incandescence, cathodo- or photoluminescence.

It occurs in several forms. The first observation of the presently most important form, "radiative recombination in p-n junctions," was made in 1907 by Round, then more thoroughly by Lossev from 1923 on, when point electrodes were placed on certain silicon carbide crystals and current passed through them. Explanation and improvement of this effect became possible only after the development of modern solid-state science since 1947.

If minority carriers are injected into a semiconductor, i.e., electrons are injected into a p-type material, or "holes" into n-type material, they recombine with the majority carriers, either directly via the bandgap, or through exciton states, or via impurity levels within the bandgap, thereby emitting the recombination energy as photons. Part of the recombinations occur nonradiatively, producing only heat.

Exploitation of the effect was strongly dependent on progress in compound semiconductor crystal preparation and solid-state electronics, since crystal perfection (absence of defects) is of prime importance. At present, single-crystalline dome-shaped p-n diodes made of gallium arsenide, GaAs, a "III-V compound" (from groups III and V of the periodic system), yield the highest efficiencies (40 percent of the electrical power input converted into optical power output) in the near-infrared, and diodes made of gallium phosphide, GaP, doped with oxygen and zinc, yield red and infrared light, efficiency peaked at 10 percent. GaP diodes doped with nitrogen emit green light, with .5 percent efficiency. Very important are alloys such as $In_x Ga_{1-x}P$, $Al_x Ga_{1-x}As$, and especially $GaAs_x P_{1-x}$ (red-emitting), where the color of luminescence can be changed by changing the composition. For high efficiency, it is important that the material have a "direct" bandgap, allow-

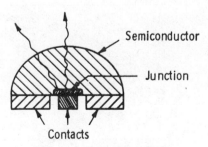

FIG. 1. Light-emitting diode.

ing electron-hole recombination without phonon participation. Wavelengths of light emitted by III-V crystals range from 6300 Å to 30 μ.

An important phenomenon, "injection laser action," was discovered in GaAs diodes in 1962. The crystal faces at the ends of a p-n junction are made optically parallel so as to form a Fabry-Perot optical cavity. Beyond a certain injection current density, (the "threshold current") the individual recombination processes no longer occur randomly and independently of each other, but in phase, so that a near-parallel beam of coherent light (~9000 Å) of enormous intensity (10^7 W/cm^2 in pulsed operation) is emitted. The efficiency has been improved by using graded bandgap $Al_xGa_{1-x}As$ heterojunctions and special doping profiles so that the lasing region near the p-n junction acts as a "light pipe," preventing light straying out sideways. AlAs has the same lattice constant as GaAs, yet higher bandgap. Therefore, these junctions are free of strain-generated imperfections. The current threshold for lasing is now reduced to 200 A/cm^2 at room temperature, allowing continuous operation at low total power, to prevent heating.

These coherent or incoherent electroluminescent p-n diodes are small point sources used for pilot lights, alpha-numeric readouts, optoelectronic processing, ranging systems, direct-sight communication, and as IR-lamps for night vision devices.

Another kind of electroluminescence, discovered by Destriau in 1936, uses inexpensive powders consisting of small particles of essentially copper-doped zinc sulfide, ZnS, a II-VI compound, embedded in an insulating resin and formed into a large flat plate capacitor with one plate transparent (e.g., SnO$_2$-coated glass). If an ac voltage is applied, light is emitted (blue, green, red, depending on the exact material composition) twice per cycle, with brightnesses up to thousands of foot-lamberts. Brightness increases linearly with drive frequency, and exponentially with voltage, until saturation occurs. The efficiency is about 1 percent but it decreases with increasing brightness.

Microscopic examination of the interior of an individual particle reveals that the light is emitted

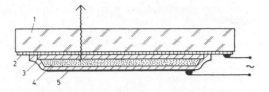

FIG. 3. AC-driven thin-film electroluminescent device. (1) Glass substrate; (2) transp. cond. In$_2$O$_3$-Sn front electrode; (3) insul. film of high permittivity; (4) ZnS:Mn EL film, ca. 500 nm thick; (5) black back electrode.

inhomogeneously, in the form of two sets of comet-like striations which light up alternatingly, each set once per cycle. These comets coincide with long, thin conducting copper sulfide precipitates which form along crystal imperfections. The applied field relaxes in these needles and concentrates at the tips, so that electrons and holes are alternatingly field-emitted into the surrounding insulating luminescent ZnS. The holes are trapped there until they recombine with the more mobile electrons, emitting the typical luminescent spectra. The local brightness can be as high as 10^5 foot-lamberts.

Instead of these powder particles embedded in an insulating resin one also uses contiguous, vacuum-evaporated or sputtered or epitaxially grown ZnS:Mn films sandwiched between similarly prepared inorganic insulator films (Fig. 3). With 150 volt, 5000 cps applied, yellow light emission (brightness up to 3000 fL) takes place twice per cycle in this ZnS:Mn film, due to multiple avalanche breakdown in close-spaced channels. The current is capacitively limited by the insulating films. The manganese centers are excited by impact with field-accelerated avalanche electrons. The optical contrast of these transparent film structures is increased by using black back electrodes which absorb the ambient light so that the addressed segments stand out bright against the black background even in sunlight. The operational longevity of these film structures is

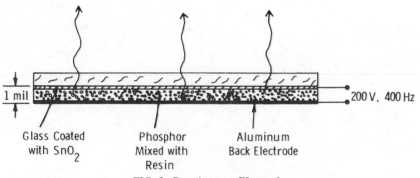

1 mil

200 V, 400 Hz

Glass Coated with SnO$_2$ Phosphor Mixed with Resin   Aluminum Back Electrode

FIG. 2. Destriau-type EL panel.

very good, probably owing to the absence of copper doping.

Low voltage dc-driven-electroluminescence of polycrystalline films of ZnS on glass, doped with copper and manganese, has also been achieved. The mechanism involves high-field-aided hole injection. The operation with DC simplifies the addressing of multi-element display panels. Longevity is still not satisfactory. However, operation with low-voltage DC instead of high-voltage AC would be very desirable to simplify the addressing of multi-element displays.

Simpler to produce in large areas are DC-EL layers consisting of ZnS:Mn particles coated with conducting copper sulfide. They are held together by a minimum of organic binder so that there are many random interparticle contacts. During a "forming" process, the copper sulfide diffuses away from the ZnS particle surfaces that get the warmest, creating a high voltage drop there, with the concomitant appearance of electroluminescence owing to avalanche injection at conducting spikes. The copper-sulfide-coated particles before and behind the light-emitting ones act as current-limiting series resistors to prevent local breakdown. This is a requirement for all DC-EL devices, just as the capacitive current limitation is a requirement for all AC-EL devices. By using rare earth doping instead of manganese, and CaS host material instead of ZnS, blue, green and red emission has been achieved also (Fig. 4).

The latest advance is a combination of the most desirable features such as low DC voltage operation, and high contrast in bright ambient due to black back electrodes. This requires film structures, not powder layers. It consists of an n-type conducting, luminescent film of II-VI material on a transparent conducting front glass substrate, backed by a resistive

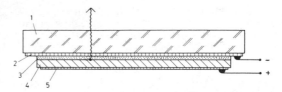

FIG. 5. DC-driven film EL cell. (1) Glass substrate; (2) transp. cond. front electrode; (3) n-type conducting II-VI material film; (4) black resistive layer, high work function; (5) metallic back electrode.

p-type cermet layer which provides multiple hole-injecting contacts, current limitation, and black background. Such EL area sources are of great interest for use in luminous dials, instrument faces, alphanumeric computer readouts, and ultimately in flat image display panels (Fig. 5).

A. G. FISCHER

References

Fischer, A. G., "Electroluminescence in II-VI Compounds," in "Luminescence of Inorganic Solids," P. Goldberg, Ed., New York, Academic Press, 1966 (pp. 541–602).

Dean, P. J., "Junction Electroluminescence," in "Applied Solid State Science," R. Wolfe and C. J. Kriessman, Eds., New York, Academic Press, 1969 (pp. 1–151).

Bergh, A. A., and Dean, P. J., *Proc. IEEE*, **60**, 156–223 (1972) (LEDs).

Cross-references: LIGHT; LUMINESCENCE; OPTICS, GEOMETRICAL; OPTICS, PHYSICAL; VISION AND THE EYE; COLOR.

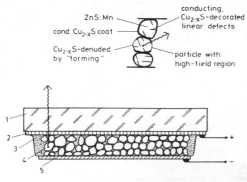

FIG. 4. DC-driven powder EL layer. (1) Glass substrate; (2) cond. transp. front electrode; (3) sparsely applied organic binder; (4) densely packed small ZnS:Mn,Cu powder particles, $Cu_{2-x}S$ coated; (5) aluminum or silver paint back electrode.

ELECTROMAGNETIC THEORY

The task of electromagnetic theory is to account for the effects of electrical charges in various states of motion. Although historically electromagnetic theory was developed from Coulomb's celebrated law, it is at present more economic to develop it differently.[11,12] The macroscopic effects are described with remarkable accuracy by the following set of equations (rationalized mks system of units)

$$\mathbf{F} = q\mathbf{E} + q\mathbf{v} \times \mathbf{B} \qquad (1)$$

$$\nabla \cdot \mathbf{J} + \frac{\partial \rho}{\partial t} = 0 \qquad (2)$$

$$\nabla \times \mathbf{H} = \frac{\partial \mathbf{D}}{\partial t} + \mathbf{J} \qquad (3)$$

$$\nabla \times \mathbf{E} = -\frac{\partial \mathbf{B}}{\partial t} \qquad (4)$$

$$\mathbf{D} = f_1(\mathbf{E}) \qquad (5)$$

$$\mathbf{B} = f_2(\mathbf{H}) \qquad (6)$$

$$\mathbf{J} = f_3(\mathbf{E}, \mathbf{H}) \qquad (7)$$

provided the functional relationships indicated in Eqs. (5), (6), and (7) are known explicitly. With these equations and the laws of mechanics, classical electromagnetic theory becomes essentially a branch of applied mathematics.

Equation (1), sometimes known as the Lorentz force equation, defines the field quantities, $\mathbf{E}$, the electric field intensity, and $\mathbf{B}$, the magnetic induction, in terms of an observable, the force $\mathbf{F}$ on a charge q. In Eq. (1), $\mathbf{v}$ is the velocity of the charge relative to the observer. Equation (2) is a statement of the law of conservation of electric charge in terms of the charge density ρ and the total current density $\mathbf{J}$. Equation (3) is the differential form of Ampère's law,

$$\oint_{\mathbf{c} \text{ of } \mathbf{s}} \mathbf{H} \cdot d\mathbf{l} = \iint_{\mathbf{s}} \mathbf{J} \cdot d\mathbf{S} = I$$

which relates the magnetic field intensity $\mathbf{H}$ to the current, but includes also the displacement current density term $\partial \mathbf{D}/\partial t$, which was added by Maxwell to make the law applicable to time-varying fields. The term $\mathbf{J}$ represents the total current density. Equation (4) is the differential form of Faraday's law of electromagnetic induction. Equations (5), (6), and (7) are functional relationships, for the most part determined experimentally, by means of which the effects of different materials are accounted for. Mathematically, these equations are employed to reduce Eq. (3) and (4) to a pair of equations in only two unknowns. In free space, Eq. (5), (6) and (7) take their simplest form, respectively, $\mathbf{D} = \epsilon_0 \mathbf{E}$, $\mathbf{B} = \mu_0 \mathbf{H}$, $\mathbf{J} = 0$ (or $J = J_s$, a source current independent of $\mathbf{E}$ and $\mathbf{H}$), where ϵ_0 and μ_0 are constants whose value depends on the system of units (in the mks system $\epsilon_0 = 8.854 \times 10^{-12}$ farad/meter, $\mu_0 = 4\pi \times 10^{-7}$ henry/meter). Since matter itself is a relatively dilute collection of charged particles, it is always theoretically possible to define terms so that the theory is a description of the effects and interactions of charges in free space, with consequently no essential distinction between $\mathbf{D}$ and $\mathbf{E}$ or between $\mathbf{B}$ and $\mathbf{H}$, as indicated above. In practice however effects of materials are usually best handled in another way.[3,8,9,10,11,12] Dielectric polarization effects are accounted for by making the $\mathbf{D}$ vector include the electric dipole moment density $\mathbf{P}$, $\mathbf{D} = \epsilon_0 \mathbf{E} + \mathbf{P}$, and then introducing a material constant, the permittivity ϵ, such that $\mathbf{D} = \epsilon \mathbf{E}$. The relative permittivity of a dielectric material is then equal to one plus the electric susceptibility.

Magnetic polarization effects are handled similarly by defining the field vector $\mathbf{B}$ so that it includes the magnetic dipole moment density $\mathbf{M}$, $\mathbf{B} = \mu_0(\mathbf{H} + \mathbf{M})$. The material permeability is then introduced so that it depends upon the magnetic susceptibility analogously, and $\mathbf{B} = \mu \mathbf{H}$. Effects of conductors are represented by a material conductivity σ, such that $\mathbf{J_c} = \sigma \mathbf{E}$. With these simple forms for Eq. (5), (6) and (7), Eq. (3) and (4) take on the useful form

$$\nabla \times \mathbf{H} = \epsilon \frac{\partial \mathbf{E}}{\partial t} + \sigma \mathbf{E} + \mathbf{J_1} \qquad (8)$$

$$\nabla \times \mathbf{E} = -\mu \frac{\partial \mathbf{H}}{\partial t} \qquad (9)$$

provided μ and ϵ are constant in time. The term $\mathbf{J_1}$ here includes currents arising from charges in free space plus any (source) currents which are independent of $\mathbf{E}$ and $\mathbf{H}$. If there are no free charges in the region, $\mathbf{J_1}$ includes only the source currents; these latter are known, so Eq. (8) and (9) may be solved for $\mathbf{E}$ and $\mathbf{H}$. Since the equations are partial differential equations, boundary conditions over closed surfaces are required for unique solutions. Boundary conditions on the field quantities, which must hold at any boundary between two regions, may be derived from these equations. The conditions are: across a boundary (a) tangential $\mathbf{E}$ must be continuous, (b) tangential $\mathbf{H}$ must be continuous, (c) normal $\mathbf{D}$ and normal $\mathbf{B}$ must be continuous. Idealizations of material properties are sometimes helpful. For example, a perfect conductor has no non-static fields inside it, and at its surface, tangential $\mathbf{E}$ and normal $\mathbf{B}$ are zero, tangential $\mathbf{H}$ is equal and perpendicular to any surface current density, and normal $\mathbf{D}$ is equal to any surface charge density.

Two additional equations, especially useful in static problems, may be deduced from Eq. (2), (3) and (4):

$$\nabla \cdot \mathbf{D} = \rho \qquad (10)$$

$$\nabla \cdot \mathbf{B} = 0 \qquad (11)$$

Solutions to the field equations are most readily obtained by imposing a restriction on the time dependence. If the fields are assumed to be independent of time (static), then Eq. (3) and (4) or (8) and (9) decouple. One of the equations becomes $\nabla \times \mathbf{E} = 0$. This means that $\mathbf{E}$ is irrotational and may be represented by a scalar potential function ϕ, $\mathbf{E} = -\nabla \phi$. Combining this with Eq. (10) gives the fundamental equation of electrostatics,

$$\nabla^2 \phi = -\rho/\epsilon \qquad (12)$$

Poisson's equation. This equation for the electrostatic potential is solved by the standard methods of partial differential equations. The boundary conditions on the potential may be found from the boundary conditions on the fields.[11],[12] In practice, it is frequently necessary to solve for the potential and electric field in a restricted region in which the charge density is zero, but the potential at the boundary is held at some particular value(s). The problem then is to solve Laplace's equation, $\nabla^2 \phi = 0$, subject to the stated boundary conditions. The standard techniques for solving boundary value problems are employed. However, if the region of interest is partially open, known analytical techniques are sometimes inadequate to solve the problem. In two-dimensional problems of such a difficult type, the method of conformal transformations (conjugate functions) is often helpful.[10],[12] The widespread availability of high speed digital computers having large memories has made it feasible to obtain acceptable solutions to a myriad of problems which previously could not be handled. The most useful general methods are called the *finite difference method* and the *finite element method*.[1],[2] The finite difference method essentially determines the potential at a finite number of points (the grid) by means of a numerical solution of the appropriate differential equation. The finite element method divides the region into element spaces and finds a solution for each element space via a matrix solution of an integral equation obtained from an energy or a variational constraint.

The main applications of electrostatic theory are in (a) the theory of material properties, (b) the calculation of charged particle trajectories in electron guns, deflection systems, and accelerators (here in conjunction with magnetostatic theory), (c) the calculation of circuit component values, such as capacitance, and (d) the determination of potentials, potential gradients, and induced currents in connection with insulation breakdowns and safety questions.[14]

Magnetostatic theory is developed from Eq. (11) and (8). Since **B** is divergenceless, it can be represented by the curl of a vector **A**, which is known as the magnetic vector potential. Equation (8) can usually be written in terms of this potential as follows:

$$\nabla^2 \mathbf{A} = -\mu \mathbf{J} \qquad (13)$$

Taken one rectangular component at a time, this equation is of the same form as Poisson's equation [Eq. (12)] and may be solved in the same way. The boundary conditions on **A** may be found from those on **B** and **H**. In regions with no current, Eq. (8) becomes $\nabla \times \mathbf{H} = 0$ so that **H** may be represented by a scalar potential function $\mathbf{H} = -\nabla\phi_m$. In such regions then, in view of Eq. (11), the magnetic scalar potential, ϕ_m, must satisfy Laplace's equation

$$\nabla^2 \phi_m = 0 \qquad (14)$$

provided $\nabla\mu = 0$ in the region. The techniques and solutions of electrostatics are applicable to

many magnetostatic problems. Unfortunately, however, in practice many of the systems designed to establish a given magnetic field incorporate ferromagnetic materials. For such materials, the magnetic susceptibility (and hence the permeability) is not independent of the field intensity and the field equations become nonlinear. Present mathematical techniques for handling nonlinear problems are severely limited. Practical magnetostatic problems are, therefore, frequently solved by some approximation. One of the simplest and most useful approximations is a representation by a magnetic circuit.[8],[10],[12] Series and parallel branches of the magnetic circuit may be recognized, and the techniques of linear and nonlinear circuit analysis can be applied to obtain a solution. When more accurate solutions are required, digital computational techniques of the types mentioned above for electrostatic problems are employed widely.[1],[2]

Magnetostatic theory is applicable to a myriad of magnetic devices, including deflection systems, motors, generators, relays, magnetic pickup devices, permanent magnets, memories, transducers, coils, magnetic containment and suspension systems. Although, historically, the need for particular solutions arose before sound computational techniques were available, earlier empirical techniques are being supplemented and replaced by numerical solutions.

Energy is required to establish electric and magnetic fields, and such energy is associated with the fields. The field energy in a given volume may be computed in most cases from a volume integral of one or both of the following energy density expressions $W_e = \frac{1}{2}\epsilon E^2$, $W_m = \frac{1}{2}\mu H^2$, respectively the electrostatic and magnetostatic values.

When the fields are time varying, Eq. (8) and (9) are coupled and must be solved simultaneously. Almost invariably, a potential function such as a vector potential or a Hertz potential is introduced.[11],[12] For example, Eq. (11) implies that **B** may be replaced by a vector potential such that $\mathbf{B} = \nabla \times \mathbf{A}$. Equation (9) then suggests the convenience of thinking of the electric field as being the sum of two parts:

$$\mathbf{E} = -\nabla\phi - \frac{\partial \mathbf{A}}{\partial t}. \qquad (15)$$

Then **H** and **E** may be replaced in Eq. (8), and with the condition on **A**, $\nabla \cdot \mathbf{A} = \mu\epsilon\partial\phi/\partial t$, the following equations may be obtained for **A** and ϕ (σ assumed zero here)

$$\nabla^2 \mathbf{A} - \mu\epsilon \frac{\partial^2 \mathbf{A}}{\partial t^2} = -\mu \mathbf{J} \qquad (16)$$

$$\nabla^2 \phi - \mu\epsilon \frac{\partial^2 \phi}{\partial t^2} = -\rho/\epsilon \qquad (17)$$

That is, both the vector potential **A** and the scalar potential ϕ satisfy a differential equation known as the inhomogeneous wave equation.

Because of their simplicity and practical importance, solutions for those sources and fields which simply oscillate at a single frequency have been studied extensively.[7,11,12] In this case, the time is eliminated as an independent variable, as if by a transform operation. (In fact, transform methods are often the best means of obtaining transient field solutions.) In the equations, the time derivatives are replaced by frequency multipliers so that the resulting equations are functions of the space variables only. The vector potential may then be found by standard techniques of partial differential equations and boundary value problems. Having A, the field quantity B is found from $B = \nabla \times A$ and E is found from Eq. (8). In practice, a theorem which can be derived from the field equations, called the reciprocity theorem,[7,12] is often helpful. The theorem relates the fields E_a and E_b produced respectively by a pair of current distributions J_a and J_b. The theorem is

$$\iiint E_a \cdot J_b dv = \iiint E_b \cdot J_a dv \qquad (18)$$

For example, if J_b is selected to be a point current at point P, directed along x (represented mathematically by a Dirac delta function), then Equation (18), $E_{ax}(P) = \iiint E_b \cdot J_a dv$, gives a formula for the computation of the field due to J_a which is equivalent to a superposition integral involving a Greens function.

Perhaps the most fundamental problem of electromagnetic theory is the determination of the fields of a point charge, at rest, in oscillation, or in some general state of motion. For a point charge q, at rest in free space, the solution may be obtained by solving Eq. (12) in spherical coordinates. With the point charge at the origin, symmetry conditions may be employed to eliminate the angular variation, and the remaining differential equation in r can be solved subject to Eq. (10) to give $\phi_G = q/4\pi\epsilon_0 r$ for the potential associated with the point charge. A superposition integral

$$\phi = \iiint \frac{\rho \, dv}{4\pi\epsilon_0 r} \qquad (19)$$

may then be employed to find the potentials associated with more complicated distributions. The field of an oscillating dipole, which is equivalent to a point alternating current, is also of great interest. This solution may be obtained from Eq. (16) (single frequency version). If the point current is directed along z, the z-component of the vector potential may be found by a procedure similar to that employed for a point charge. The final result is

$$A_{zG} = \frac{I\Delta z}{4\pi\mu_0 r} \cos \omega(t - \sqrt{\mu_0\epsilon_0}r) \qquad (20)$$

where $I\Delta z$, the current moment, is equal to $\omega q\Delta z$, the maximum dipole moment of the oscillating dipole. The factor $(t - \sqrt{\mu_0\epsilon_0}r)$ exhibits the time delay required for the effects of the oscillating charges to propagate to distant points. The electric and magnetic fields may be computed from Eq. (20) as indicated above. The magnetic field strength produced by an oscillating dipole (point current) is, for example, in the spherical coordinate system (r, θ, φ)

$$H_\varphi = \frac{I\Delta z}{4\pi} \sin \theta \left[\frac{\cos \omega(t - \sqrt{\mu_0\epsilon_0}r)}{r^2} \right.$$
$$\left. - \frac{\omega\sqrt{\mu_0\epsilon_0}}{r} \sin \omega(t - \sqrt{\mu_0\epsilon_0}r) \right]$$

This form, like Eq. (20), shows that the crests and valleys of the field oscillations are propagated in spherical waves at the speed of light $v = (\mu_0\epsilon_0)^{-1/2}$. The solution for a point current may be employed in an integral similar to Eq. (19) to find the vector potential of a more complicated distribution of current. Such solutions may also be employed to find the radiation patterns and input impedances of antennas.[13] In practice, however, the distribution of currents on antenna structures is not known a priori, i.e., the antenna currents are not truly source currents. High speed digital computers have made it possible to employ a numerical technique related to the finite element method, known as the method of moments,[5] for the determination of the currents on antennas and scattering bodies of rather arbitrary shapes, given the boundary and source conditions. Once these currents have been determined, the fields and impedances are then calculated as indicated above.

The potentials and fields produced by a charge moving in an arbitrary way may also be obtained.[6,11] The results may be found in Stratton.[11,pp.475-476]

In regions free of source currents and charges, the fields and potential satisfy the homogeneous wave equation [for example Eq. (16) with $J = 0$]. Then one of the simpler solutions which can be obtained is that of the plane electromagnetic wave. With appropriate orientation of the rectangular coordinate system, the solutions show that plane waves may progress along z, with components as follows:

$$E_x = E_0 \cos \omega(t - \sqrt{\mu_0\epsilon_0}z)$$

$$H_y = E_0 \sqrt{\frac{\epsilon_0}{\mu_0}} \cos \omega(t - \sqrt{\mu_0\epsilon_0}z)$$

where E_0 is an arbitrary constant amplitude. Note that E, H and the direction of propagation are all perpendicular to one another. The Poynting vector, $S = E \times H$, points in the direction of propagation. Moreover, the power carried through a closed surface by an electromagnetic field may be computed from a surface integral of the Poynting vector.

With single frequency fields in source free regions, both H and E can be represented by vector potentials,[11,12] $H_1 = \nabla \times A_1$, $E_2 = \nabla \times A_2$, and moreover the coordinate systems may be oriented so that A_1 and A_2 each have a single component.[12] In cylindrical systems, this single component is commonly along z. H_1 is then transverse to z (TM) and the set of fields, E_1, H_1, derivable from A_1, are called TM fields. E_2 is likewise transverse to z and the set of fields, E_2, H_2, derivable from A_2, are called TE fields. This procedure is particularly helpful in problems involving transmission lines and WAVEGUIDES and is developed in detail in Weeks.[12, Ch. 4-6]

Some of the most interesting and fundamental problems of electromagnetic theory are concerned with the scattering and diffraction of electromagnetic waves.[6,7,10-12] For example, exact solutions are available for the scattering by cylinders and spheres, as well as an infinitely long slit. Approximate solutions are available for many other shapes. The methods are the analytical and numerical ones outlined above, supplemented by generalizations of the principles of Huygens and Babinet.

Another topic of wide interest is the nature of fields in ionized gases or plasmas. The applications range from ionospheric propagation to microwave devices to nuclear apparatus to magneto-hydrodynamics to satellite re-entry problems. The simplest theory for these effects is developed from Eq. (3) and (4) (single frequency version) by separating the ion current term $J_e = \rho v$ from J, and employing Newton's law to eliminate v in favor of E, H and whatever mechanical constraints are applicable[3,11,12] (see PLASMAS).

Effects peculiar to charges moving with very high velocities have not been included in this discussion (see RELATIVITY THEORY). Quantum effects are also discussed elsewhere (see QUANTUM ELECTRODYNAMICS and QUANTUM THEORY).

W. L. WEEKS

References

1. Chari, M. K. V., and Silvester, P. P., "Finite Elements in Electric and Magnetic Field Problems," New York, John Wiley & Sons, 1980.
2. Finlayson, B. A., "The Method of Weighted Residuals and Variational Principles," New York, Academic Press, 1972.
3. Good, R. H., and Nelson, T. J., "Classical Theory of Electric and Magnetic Fields," New York, Academic Press, 1971.
4. Hayt, W. H., "Engineering Electromagnetics," New York, McGraw-Hill Book Company, 1981.
5. Harrington, R. F., "Field Computation by Moment Methods," New York, Macmillan, 1968.
6. Jackson, J. D., "Classical Electrodynamics," New York, John Wiley & Sons, 1962.
7. Kong, J. A., "Theory of Electromagnetic Waves," New York, John Wiley & Sons, 1975.
8. Panofsky, W., and Phillips, M., "Classical Electricity and Magnetism," Reading, Mass., Addison-Wesley, 1955.
9. Peck, E. R., "Electricity and Magnetism," New York, McGraw-Hill Book Co., 1953.
10. Smythe, W. R., "Static and Dynamic Electricity," Second edition, New York, McGraw-Hill Book Co., 1950.
11. Stratton, J. A., "Electromagnetic Theory," New York, McGraw-Hill Book Co., 1941.
12. Weeks, W. L., "Electromagnetic Theory for Engineering Applications," New York, John Wiley & Sons, 1964.
13. Weeks, W. L., "Antenna Engineering," New York, McGraw-Hill Book Co., 1968.
14. Weeks, W. L., "Transmission and Distribution of Electric Energy," New York, Harper & Row, 1981.

Cross-references: ELECTRICITY, PLASMAS, POTENTIAL, PROPAGATION OF ELECTROMAGNETIC WAVES, QUANTUM ELECTRODYNAMICS, QUANTUM THEORY, RELATIVITY, STATIC ELECTRICITY.

ELECTRON

The electron is the smallest known electrically charged particle. Its existence and characteristics were inferred from many experiments clustered in and around the last decade of the nineteenth century. In the 1830's, Faraday had tentatively suggested that his experiments in ELECTROCHEMISTRY could be interpreted in terms of a small unit of charge attached to ions. This notion of individual "atoms of charge" was somewhat eclipsed, however, by the enormous success of Maxwell's theory of electromagnetism, which was generally interpreted, by 1880, as favoring a view that electrical phenomena were due to continuous charge distributions and motions. G. Johnstone Stoney, in 1874, and Helmholtz, in 1881, had suggested again an atomic interpretation of electricity, but it was not until the brilliant experiments of Perrin, J. J. Thomson, Zeeman, and others in the 1890's that the concept of the electron received firm experimental foundation. Later experiments and theory (Millikan, Bohr, etc.) established the constancy of the electronic charge and interwove the concept of an electron of definite charge and mass into the basic structure of the atom.

The Cathode Ray Controversy After the discovery of the cathode ray in high-vacuum discharge tubes by Plücker in 1858, there developed, with the experiments of Goldstein,

Crookes, Hertz, Lenard, and Schuster, a controversy over the nature of the rays. A predominately German school held that the rays were a peculiar form of electromagnetic rays. The British physicists thought they were negatively charged particles. The controversy provides a classic "case history" of the typical scientific controversy in which two quite different models both explain most, but not all, of the observable facts. The proponents of each model designed ingenious experiments, but in some cases were so trapped in their preconceptions that they badly misinterpreted their observations. The Germans were especially impressed by the fact that the rays could go through thin foils—something no known particles could do. The British were firm in pointing out that the rays could be deflected by magnetic fields—something not possible with electromagnetic waves. Hertz, in what he thought was a crucial experiment, was unable to detect deflection of the rays by electric fields, but this very phenomenon was demonstrated by J. J. Thomson and made the basis for his conclusive experiments that the rays had velocities less than that of light. Thomson showed, further, that if one assumed that the rays were composed of particles, then the particles had the same ratio of charge to mass regardless of the cathode material or the nature of the residual gas. Perrin's classic experiment, meanwhile, proved that the rays did indeed convey negative charge. In the decade between 1896 and 1906, Thomson and others showed that negatively charged particles from sources other than cathode rays had the same ratio of charge to mass: the negative particles emitted by hot filaments in the Edison effect, the beta rays emitted by some radioactive materials, and the negative particles emitted in the photoelectric effect that had so ironically been discovered by Heinrich Hertz in his great experiment which demonstrated the electromagnetic rays predicted by Maxwell's equations.

Thomson's Determination of, e/m In 1897 Thomson devised an apparatus in which he could deflect a beam of cathode rays with a magnetic field of induction B and also with an electric field of strength E. If the fields are perpendicular to each other, and to the original path of the beam, and if they occupy the same region, then (with proper polarities and magnitudes of fields) the electric force on the beam can equal the magnetic force, so that the beam hits the same point on a fluorescent screen as when no fields are applied. If e is the charge of a given particle, m its mass, and v its velocity, $v = E/B$. Thus, velocities of typical cathode ray beams could be measured. If the magnetic field is used alone, and the radius of curvature R of the beam is measured, then one can equate centripetal and magnetic field forces: $mv^2/R = Bev$, and then deduce $e/m = v/BR$. With v known from the previous experiment, e/m can be calculated. Thomson's early values were not

very precise, but later experiments of a similar type gave values close to 1.76×10^{11} coulomb/kg. More recent evaluations, drawing on measurements of many kinds, give $e/m = 1.7588 \times 10^{11}$ coulombs/kg as a 1982 value calculated from values for e and m given under CONSTANTS, FUNDAMENTAL, in this book.

The Zeeman Effect In 1896 Zeeman discovered the broadening of spectral lines when a light source was in a strong magnetic field. Experimental refinements of Zeeman and others, and theoretical work by Lorentz and Zeeman, permitted the interpretation of this effect as due to the influence of the magnetic field on oscillating or orbiting negatively charged particles within the light-emitting or absorbing atoms. From the spectroscopic data, the ratio of charge to mass of these hypothetical particles, and the sign of their charge, could be shown to be equal to that of cathode rays. The Zeeman effect thus provided the first experimental evidence that the negative particles emitted by atoms when heated (Edison effect) or subject to high fields and/or ionic bombardment (cathode rays) or bombarded by short-wavelength light (photoelectric effect) were, indeed, actual constituents of the atoms and were probably responsible for the emission and absorption of light.

The Charge on the Electron In the decade following 1897, many different methods were evolved for determination of ionic charges. Some methods depended upon measuring the total charge of a number of ions used as nuclei for cloud droplet formation. Other methods were more indirect—experiments, for example, which, combined with the kinetic theory of gases, could give crude values for avogadro's number, N (see MOLE CONCEPT). By dividing the Faraday constant (the charge carried in electrolysis by ions formed from one gram-atom of a univalent element) by N, one could determine the average charge per ion. Similarly, the constants in Planck's theory of blackbody radiation, when evaluated experimentally, could provide a numerical value for N, as could certain experiments in radioactivity. All such methods gave values of N of the order of 6×10^{23}, and hence 1.6×10^{-19} coulomb for the ionic charge. None of these methods measured individual charges; strictly speaking, the value for the ionic charge could be thought of only as an average value.

Millikan's experiments with single oil drops, beginning in 1906, provided a method for measuring extremely small charges with precision. He was able to show that the charge on his drops was *always* ne, with $e = 1.60 \times 10^{-19}$ coulomb (modern value) and n a positive or negative integer.

He observed the motions of very small charged oil drops in uniform vertical electric fields. The drops were so small that they moved with constant velocity (except for Brownian fluctuations) for a given force. The force in each case was due to gravity acting on the mass of the drop and to the electric field (if any) acting on the charge, q, on the drop. The charge on a given drop could be changed by shining x-rays upon it. Using Stokes' Law, in a form modified to correct for the fact that the drops were *not* large in comparison to the inhomogenieties of the surrounding air, and the velocity of a drop in free (gravitational)fall, Millikan could infer the diameter and mass of a given drop, and then calculate its charge. The charge q always equaled ne. (See reference 1 or 2 for experimental details.) A few other physicists, in similar experiments, thought they had detected electric charges smaller than Millikan's e, but their experimental techniques were probably faulty.

Millikan's experiment did not prove, of course, that the charge on the cathode ray, beta ray, photoelectric, or Zeeman particle was e. But if we call all such particles electrons, and assume that they have $e/m = 1.76 \times 10^{11}$ coulomb/kg, and $e = 1.60 \times 10^{-19}$ coulomb (and hence $m = 9.1 \times 10^{-31}$ kg), we find that they fit very well into Bohr's theory of the hydrogen atom and successive, more comprehensive atomic theories, into Richardson's equations for thermionic emission, into Fermi's theory of beta decay, and so on. In other words, a whole web of modern theory and experiment defines the electron. (The best current value of e (June, 1982) = 1.6028 $\times 10^{-19}$ coulomb (see CONSTANTS, FUNDAMENTAL).

The Wave Nature of the Electron In 1924, L. DeBroglie suggested that the behavior of electrons within atoms could be better understood if it were assumed that the motion of an electron depends upon some sort of accompanying wave, the length of which would be h/p (h = Planck's constant and p the momentum of the electron). This suggestion led to the development of QUANTUM MECHANICS by Schrödinger, Heisenberg, and others. The concept of electron waves provided an explanation for experiments on reflection of electron beams by metallic crystals, carried out from 1921 onward by Davisson and others, and provided an impetus for the experiments of G. P. Thomson on the diffraction of electron beams by thin films (see ELECTRON DIFFRACTION).

Other Characteristics of Electrons In applying quantum mechanics to certain problems in atomic spectroscopy, in 1925 and 1926, Pauli, and Goudsmit and Uhlenbeck found that electrons must possess angular momentum of amount $\pm \frac{1}{2}(h/2\pi)$. Dirac's work on a generalized quantum theory of the electron showed that it possessed a related magnetic dipole moment of magnitude $eh/4\pi mc$ (see ELECTRON SPIN). The ratio of the dipole moment to the angular momentum (e/mc) is larger than can be accounted for in classical terms with any homogeneous wholly negative model. The concept of electronic dipole magnetic moment is essential not only in spectroscopy but in theories of ferromagnetism (see MAGNETISM).

One may speak of the "classical radius of the electron," $a = e^2/mc^2$, derived by setting the self-energy of the coulomb field of a charge e contained at a radius a equal to the relativistic rest energy, mc^2 of the electron. This $a = 2.82 \times 10^{-13}$ cm, comfortably smaller than any atom, but larger than the usual estimates of sizes of protons and neutrons.

Positive Electrons Dirac's paper in 1928 could be interpreted as predicting the existence of electrons that are positive. But until such particles were found experimentally by C. D. Anderson in 1932 in cloud chamber pictures of cosmic ray particle tracks, most physicists preferred other interpretations of Dirac's paper. Positive electrons, or POSITRONS are now known (1) to occur as decay products from certain radioactive isotopes, (2) to be produced (paired with a negative electron) in certain interactions of high-energy gamma rays with intense electric fields near nuclei, and (3) to be the product of certain decays of certain mesons. In principle, positrons could form anti-atoms with nuclei made from anti-protons and anti-neutrons, but in practice almost all positrons produced in ordinary matter quickly meet their end by annihilating themselves together with some hapless negative electron. The end product of a positron-electron annihilation is a pair of gamma rays.

Recent Theoretical Developments The relativistic quantum mechanical theory of the electron, in its earlier forms, led to embarrassing predictions of infinite electronic mass and charge. Schwinger and others have developed methods for coping with these infinities, so the theory, in general, now satisfactorily agrees with observations. Further developments are not unlikely. (See QUANTUM ELECTRODYNAMICS).

Experiments and theory, first in radioactive decay, and more recently in elementary particle physics, have made it clear that electrons can be thought of as members of a class of particles called "leptons." These particles, which include neutrinos (see NEUTRINO) and muons (see ELEMENTARY PARTICLES) interact with each other and with other particles in so-called "weak interactions," apart from whatever other reactions they may share, such as gravitational, magnetic, and electrical interactions.

Applications of Electrons Aside from their inherent usefulness in physical theories of magnetic, electrical, optical, and mechanical properties of matter, electrons either in beams or in conductors can be made to do all sorts of useful things. Cathode ray oscilloscopes, electron microscopes, image converters, certain memory devices for computers, television picture tubes, and most "radio tubes" depend upon beams of electrons controlled by electric or magnetic fields (see ELECTRON OPTICS). In ordinary metallic conductors, electricity is carried primarily by electrons. The behavior of electrons in SEMICONDUCTORS and in superconductors (see SUPERCONDUCTIVITY) has in recent years been the basis both of intense theoretical interest and of interesting and useful devices.

DAVID L. ANDERSON

References

1. Millikan, R. A., "The Electron," edited with an introduction by J. W. M. DuMond, Pheonix, Science Series, PSS523, University of Chicago Press, 1963.
2. Anderson, D. L., "The Discovery of the Electron," New York, Arno Press, 1981.
3. Shankland, R. S., "Atomic and Nuclear Physics," Second edition, New York, The Macmillian Company, 1960.
4. Condon, E. U., and Odishaw, H., "Handbook of Physics," p. 7-169, New York, McGraw-Hill Book Co., Inc. 1958.
5. Borowitz, S., and Bornstein, L. A., "A Contemporary View of Physics," Chapter 12, New York, McGraw-Hill Book Co., 1968.
6. An excellent (although somewhat disputatious) discussion of the electron will be found in "Evolution of the Concept of the Elementary Charge," by L. Marton and C. Marton in *Advances in Electronics and Electron Physics*, 50, 449-472 (1980).

Cross-references: ELECTROCHEMISTRY, ELECTRON DIFFRACTION, ELECTRON OPTICS, ELECTRON SPIN, ELEMENTARY PARTICLES, MAGNETISM, MOLE CONCEPT, NEUTRINO, PHOTOELECTRICITY, POSITRON, QUANTUM ELECTRODYNAMICS, QUANTUM MECHANICS, SUPERCONDUCTIVITY, ZEEMAN AND STARK EFFECTS.

ELECTRON DIFFRACTION

The discovery of electron diffraction independently by C. J. Davisson and L. H. Germer (1927) and G. P. Thomson (1927) verified L. de Broglie's earlier hypothesis (1924) that matter exhibits both corpuscular and wavelike characteristics. This hypothesis served as a stimulus for the formal development of quantum mechanics by E. Schrödinger, M. Born, W. Heisenberg, and others. Following this momentous discovery, which eventually resulted in the award of a Nobel Prize to Davisson and Thomson, electron diffraction was immediately utilized as a tool for the study of the structure of matter.

Electron, x-ray and neutron diffraction are all used for structure studies. Electron diffraction is used particularly for those structural studies that involve small numbers of atoms. This is due to the strong interaction of electrons with matter. Thus the principal area of application of electron diffraction is for the study of thin films, surfaces, gases, crystalline defects, and small samples.

The different energy ranges that were used in the Davisson-Germer and Thomson experiments provide a natural division for a description of the types of equipment, areas of application,

and analytical techniques that have evolved since 1927. These classic experiments were performed with electrons having energies in the vicinity of 150 eV and 15 keV, respectively. De Broglie's relationship $\lambda = h/mv$, where h is Planck's constant and λ is the wavelength associated with a mass m traveling with a group velocity v, reduces to $\lambda = \sqrt{150/V}$ for electrons in the nonrelativistic limit, where V is the accelerating voltage and λ is expressed in Ångstroms. The two experiments thus used electrons having a wavelength of approximately 1 and 0.1 Å. The longer wavelength is comparable to the spacing between atoms in crystals.

While 50–100 keV electrons, which are used in commercial electron diffraction instruments, penetrate to a depth of about 10^3 Å into a crystal, 150 eV electrons penetrate only about 10 Å. Since the higher-energy electrons are capable of passing through the several layers of adsorbed foreign material that normally are present on the surface of a crystal, surface cleanness and therefore the vacuum requirements for 100 keV electron diffraction are not as stringent as those for low-energy electron diffraction. This fact, in addition to the relative ease in focusing intense high-voltage beams, and individual interests, resulted in the wide application of high-energy electron diffraction for structure studies. A typical instrument of this type operates at about 50 keV, has provisions for producing and focusing the electrons, contains specimen manipulators, means to record the diffraction patterns and typically is contained in a chamber capable of being evacuated to 10^{-6} torr. Diffraction patterns are obtained either by transmission of the beam through very thin specimens or by working at grazing incidence and reflection.

One common use of high-energy electron diffraction by transmission or reflection is the study of films on amorphous, polycrystalline, and single crystal substrates. This includes films that have been formed by the oxidation or corrosion of a surface as well as those formed by the deposition of material on a substrate. In many instances these films consist of crystallites having an orientation that is related to the structure and orientation of the substrate material.

The transmission ELECTRON MICROSCOPE has been used widely for the study of atomic arrangements at structural imperfections, such as dislocations and stacking faults. Image contrast is obtained by local differences in the intensities of diffracted beams. In addition, many electron microscopes are constructed in such a way that it is possible to obtain the diffraction pattern associated with the material in the area being studied. Considerable information on the crystallography of domain structures, order-disorder phenomena, and phase transitions has been obtained by this selective area diffraction technique.

Electron diffraction in this high-energy range

is also useful for the determination of the atomic arrangements, bond distances, bond angles, and mean square atomic vibrational amplitudes in gaseous molecules.

Low-energy electron diffraction was used sparingly until approximately 1960. Improved diffraction equipment, which enabled the direct display of the diffraction pattern on a fluorescent screen by accelerating the diffracted electrons after they had passed through grids, and the commercial availability of ultrahigh (10^{-10} torr) vacuum equipment, resulted in a resurgence of interest in this field. The structure of clean surfaces, the arrangements of foreign atoms on these surfaces at a monolayer or less coverage, and many aspects of the initial stages in the oriented overgrowth of thin films have been studied with electrons having energies in the range of 2 to 10^3 eV. It has been revealed that the atomic arrangement at the clean surfaces of semiconductors such as germanium and silicon is quite unlike that found in the bulk of these materials. At low coverages, foreign atoms are normally adsorbed in structures that have a symmetry and dimensions that are simply related to the orientation of the substrate plane. A multitude of such structures has been found on semiconductors and metals. Their atomic array is dependent on many parameters, such as the amount of adsorbed material, the temperature, orientation and cleanness of the substrate. This same equipment is also used to determine the presence of impurity atoms on the surface by the technique of Auger spectroscopy. The energy of the secondary electrons is characteristic of the surface atoms and can be used to make a quantitative identification of surface impurities, which can assume an important role in affecting surface structures.

While electron diffraction techniques are widely used, interpretation of much of the data can be challenging. This is due to the strong interaction of the incident and scattered electrons with the substrate atoms. It is often necessary to apply dynamical scattering theory to explain observations. While interpretation of results has been very successful at the high energies of 100 keV, it has been less successful in the low energy regime of 100 eV. Here, the dependence of the scattered beams' intensity on incident electron energy is complex and is indicative of large and energy dependent scattering cross sections, multiple scattering, strong absorption of incident and scattered electrons, and effects that are substrate dependent. However, considerable progress has been made in recent years in the understanding of these interactions and in the development of computational tools. This has led to the interpretation of data to obtain surface atom spacings, surface atom vibrational amplitudes and relatively simple surface structures. This has provided an improved understanding of chemical bonding at surfaces. Other important and successful applications of low-energy electron diffraction have been made in studying crystal surface defects such as steps, islands, and domain structures. It is expected that work of this type will result in detailed understanding of surface phenomena such as bonding, catalysis, and corrosion.

ALFRED U. MAC RAE

References

Cowley, John M., "Diffraction Physics," Amsterdam, North-Holland Publishing Co., 1975.

Amelinckx, S., Gevers, R., Van Landuyt, J., "Diffraction and Imaging Techniques in Material Science," Amsterdam, North-Holland Publishing Co., 1978.

Van Hove, M. A., and Tong, S. Y., "Surface Crystallography by LEED," New York, Springer-Verlag, 1979.

Cross-references: DIFFRACTION BY MATTER AND DIFFRACTION GRATINGS, ELECTRON, ELECTRON MICROSCOPE, ELECTRON OPTICS.

ELECTRON MICROSCOPE

Introduction The standard light microscope uses light in the visible range to produce a magnified image. The electron microscope (EM) uses electrons for magnification. Since electrons are easily absorbed in air at atmospheric pressure the EM has to operate in a vacuum. The imaging electrons are usually accelerated to between 30 and 100 kV. The EM magnifies in two or three stages by means of electromagnetic or electrostatic lenses.

Resolution In 1978 Ernst Abbé and, independently, H. von Helmholtz proved that in the light microscope the resolution is limited by the wavelength of the illuminating light. No matter how perfect and free of aberrations the optical system is, the image of a geometrical point is not a point but a disc, the "Airy disc." Regardless of any further magnification, two separate points cannot be resolved as separate unless their centers are the distance d apart, whereby d is the radius of the Airy disc referred to the object plane:

$$d = \frac{0.5\,\lambda}{n \sin \alpha} \qquad (1)$$

where

λ = wavelength of the illuminating light
n = index of refraction of medium between object and lens (in air, $n = 1$)
α = aperture of lens, i.e., half-angle of collected light beam
$n \sin \alpha = NA$ = numerical aperture of lens (in air, $NA_{max} = 0.95$).

The wavelength of green light for which the eye is most sensitive is:

$$\lambda_{green} \simeq 500 \text{ nm}* \quad (5000 \text{ Å}).$$

The resolution of the light microscope is, therefore:

$$d \simeq 250 \text{ nm} \quad (2500 \text{ Å}).$$

Even if we go to the extreme of using the ultraviolet line of mercury ($\lambda = 253.7$ nm) oil immersion optics ($NA = 1.4$), quartz lenses and microphotography, the best resolution obtainable with the light microscope is still:

$$d \simeq 100 \text{ nm} \quad (1000 \text{ Å})$$

The electron microscope makes use of the wave properties of the moving electron. Its "de Broglie" wavelength is:

$$\lambda = \frac{h}{mv} = \frac{1.23}{V^{1/2}} \lfloor \text{nm} \rfloor \quad (2a)$$

for nonrelativistic electrons (below 100 kV) and

$$\lambda = \frac{1.23}{(V + 10^{-6} V^2)^{1/2}} \text{ [nm]} \quad (2b)$$

for relativistic electrons, where

h = Planck's constant
m = mass of electron
v = velocity of electron
V = accelerating voltage in volts.

For electrons of $V = 50$ kV the wavelength is

$$\lambda_{50kV} = 0.00535 \text{ nm}$$

Objective lenses for electrons, unlike those for light optics, cannot be made free of spherical aberrations. They have to operate with numerical apertures that are 500 to 1000 times lower. The best theoretical resolution for magnetic lenses working with electrons of 50 kV energy is:[1]

$$\ddot{\delta} = 0.21 \text{ nm}$$

The best practical resolution is between 0.5 and 1 nm (5-10 Å).

Early History The first electron microscope was developed by M. Knoll and E. Ruska at the Technical University of Berlin early in 1931.[2] Ruska obtained the first magnification in two stages by means of electromagnetic lenses on 7 April 1931.[3] The EM celebrated, therefore, its 50th anniversary on that date in 1981. The greatest magnification obtained with this first somewhat crude instrument was 17X. Knoll and Ruska improved the electron microscope

*1 nanometer (nm) = 10^{-9} meter (m) = 1 millimicron (mμ) = 10 angstrom units (Å).

step by step. They added a condenser lens and built an iron shield with a narrow center gap around their magnetic lens. Ruska, working from 1932 by himself, equipped the magnetic lenses with narrow pole pieces and was able to demonstrate in 1933 a resolution of 50 nm better than the best resolution obtainable with the light microscope (magnification 12,000X).[4] Figure 1 shows a functional diagram of his supermicroscope. Figure 2 shows the details of one of his lenses.[†]

The first commercially available EM was developed by Ruska together with von Borries at Siemens & Halske, Berlin, in 1939. The first American commercial EM came out in 1941 by RCA.

Parallel to the development of the magnetic electron microscope went the development of an electrostatic one. In 1931, Brüche and Johannson of the Research Institute of A.E.G. in Berlin imaged the emitting surface of the cathode with an electrostatic immersion objective. In 1932, they employed unipotential or Einzellenses.

Electron Lenses *Magnetic Lenses.* A charged particle entering a uniform magnetic field moving parallel to the lines of force will not be deflected. It will move in a straight line. Moving perpendicular to them it will describe a circle. The radius of this circle, the "cyclotron radius," is:

$$\rho = \frac{m}{e} \frac{v}{B} \quad (3)$$

where

m = mass of particle
e = charge of particle
v = velocity
B = magnetic field intensity (gauss).

The circle described by an electron is

$$\rho_e = 3.372 \frac{\sqrt{V}}{B} \text{ [cm]} \quad (4a)$$

for nonrelativistic electrons,

$$\rho_e = \frac{1}{B} \sqrt{11.3 V + 1.11 \times 10^{-5} V^2} \text{ [cm]} \quad (4b)$$

for relativistic electrons,

where V is in volts and B is in gauss.

The time it takes a particle to describe a cyclotron circle is:

$$\tau = \frac{2\pi\rho}{v} = 2\pi \frac{m}{e} \frac{1}{B} \text{ [sec]} \quad (5)$$

[†]The two first EM were destroyed during WW II. Ruska has recently constructed exact replicas of them which were shown at various EM Congresses in Europe and are scheduled to be permanently exhibited at Deutsches Museum, Munich.

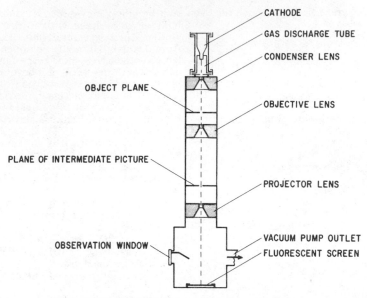

FIG. 1. First Supermicroscope (Ruska 1934, Ref. 4).

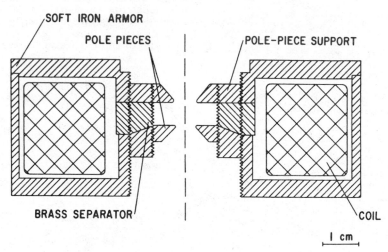

FIG. 2. Magnetic Lens (Ruska 1934, Ref. 4).

A charged particle entering a uniform magnetic field at an angle will describe a cycloid. While its velocity component normal to the field lines causes it to describe a circle, the velocity component parallel to the field lines remains unchanged. Since the time required to describe a circle is independent of the normal velocity, all circles, large or small, are traversed in the same time interval. In an electron beam with low divergence where, therefore, the velocities parallel to the magnetic field (v_z) are identical, all electrons leaving one point P on the axis will meet downstream at another

point P′ at a distance

$$d = \frac{v_z}{\tau} \qquad (6)$$

with angles to the axis identical to those they had at point P (see Fig. 3).

A "long" magnetic lens, having a uniform magnetic field extending from the object to the image point, will form an upright picture of the object with an image to object ratio of 1 to 1.

A "thin" magnetic lens is a lens with a magnetic field short compared to the object-to-image distance. A "weak" thin lens is a lens

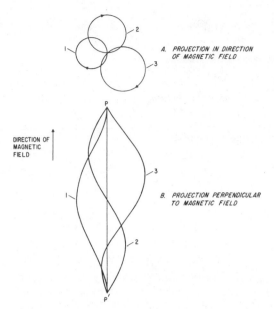

A. PROJECTION IN DIRECTION OF MAGNETIC FIELD

DIRECTION OF MAGNETIC FIELD

B. PROJECTION PERPENDICULAR TO MAGNETIC FIELD

FIG. 3. Cycloids, described by electrons in uniform magnetic field.

with the focal length large compared to the axial length of the magnetic field. The refractive power, i.e., the reciprocal of the focal length, of such a lens is determined by:

$$\frac{1}{f} = \frac{0.022}{V} \int_{-\infty}^{+\infty} B_z{}^2 dz \ [\text{cm}^{-1}] \qquad (7)$$

It images according to the general optical equation:

$$\frac{1}{f} = \frac{1}{a} + \frac{1}{b} \qquad (8)$$

where f is the focal length, a is the object distance and b is the image distance. The picture is turned around from the position of the object. The angle it is turned is determined by:

$$\theta_1 = \frac{0.149}{\sqrt{V}} \int_{-\infty}^{+\infty} B_z dz \ [\text{radians}] \qquad (9)$$

The objective and projector lenses of the electron microscope require extremely short focal lengths in order to obtain high magnifications without going to extremely long microscopes. These lenses have, therefore, pole pieces which limit the extent of the magnetic field, both in the axial and radial dimensions. The treatment of strong, thin magnetic lenses can be found in references 5 and 6.

Electrostatic Lenses *Immersion lenses* consist of two apertures or two coaxial cylinders at different potentials. They are important as lenses for television or oscillograph cathode-ray tubes. The lenses used in electrostatic transmission electron microscopes are usually *unipotential* or *Einzel-lenses*. They have the same potential on either side of the lens. They consist of three apertures: the two outer apertures are at ground or anode potential, and the center electrode can have either a positive or negative potential. Regardless of whether a positive or a negative potential is applied, the lens will always be convergent. Figure 4(a) shows the equipotential lines of such a lens; Fig. 4(b) shows the focal length vs V_L/V_0 for this lens, where V_L is the voltage applied to the center electrode and V_0 the cathode potential. Figure 5 shows the design of a typical electrostatic lens.

The Standard Transmission Microscope (TEM) The best known electron microscope is the magnetic transmission microscope where the image is formed by electrons which have passed through the specimen. It is composed of the following major sections: electron gun, condenser lens, specimen chamber, objective lens, projector lens, viewing and photographing chamber. It has the following ancillary equipment: (1) power supplies for high voltage, cathode heater voltage, and focusing currents; (2) vacuum systems.

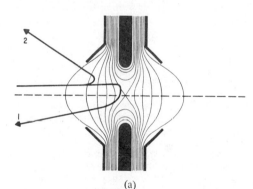

(a)

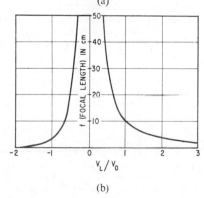

(b)

FIG. 4. Unipotential or Einzel-lens: (a) equipotential lines (Ref. 7); (b) focal length vs V_L/V_0 (Ref. 8).

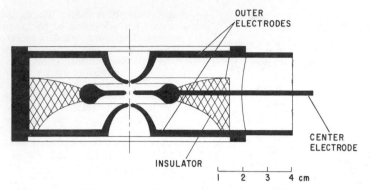

FIG. 5. Typical electrostatic lens (see ref. 9).

The *electron gun* generates the electron beam which illuminates the object. It has to provide the required electron density within a certain limited divergence. It consists of the cathode, a hairpin tungsten filament, enclosed in a cup-shaped electrode at cathode or a more negative potential called the grid, shield or Wehnelt cylinder. The fine emitting point of the cathode must be accurately aligned with the aperture of the grid. The negative grid bias against the cathode determines the beam current that passes through the grid. The "crossover point" which is located near the grid aperture has the smallest beam diameter inside the gun. While the anode facing the grid is at ground potential, the cathode is maintained at a high negative potential, usually between 50 and 100 kV.

The *condenser lens* or lenses increase the electron density reaching the specimen by concentrating the beam. An aperture in the condenser lens of 0.25 to 0.5 mm diameter reduces the amount of stray electrons reaching the specimen.

The *objective lens* provides the first magnification. It is a strong, thin lens with a high refractive power. It has, therefore, narrow precision pole pieces. An aperture of 25 to 100μ diameter is usually inserted in the gap of this lens to limit the beam divergence. The specimen is mounted either close above or inside the magnetic gap.

The *projector* or *image lens* selects a small portion of the intermediate image produced by the objective lens and magnifies it again. A third electron-optical magnification is in many cases produced by a second projector lens. The magnetic lenses are usually wirewound coils. In some cases permanent magnets are used with or without additional coils for fine adjustment.

The *specimen chamber* is located above the objective lens. It can be opened to the outside for inserting new specimens. It is pumped out before reconnecting it to the main column. Adjustments by means of micrometer screws permit shifting the specimen in the object plane in order to locate accurately the area of interest. In many cases a number of different specimens can be moved under the electron beam by rotating or sliding specimen holders. Some models have special facilities to keep the specimen at certain high or low temperatures.

The *viewing chamber* at the bottom of the microscope column contains the fine-grain fluorescent screen that can be observed through glass windows either directly or through a telescope. If a permanent record is desired, the fluorescent screen is moved aside and a photographic plate inside a plate holder is exposed. The electrons produce a latent image directly. Each plate can be removed from the vacuum separately. A film pack may be used instead of plates. Rollfilm is used when a whole series of pictures are to be taken.

In some cases when an extremely high magnification, for instance 100 000 times, is desired, it may pay to magnify electron-optically to a somewhat lower magnification and to add a final photographic enlargement later; the adjustment of the electron microscope is then much easier, since the field of vision is so much larger.

The *vacuum system*, generally maintaining a vacuum of 10^{-4} to 10^{-5} torr (1 torr = 1 mm Hg), consists usually of an oil diffusion pump backed by a mechanical forepump. A second mechanical pump is used to purge the specimen chamber and the photographic chamber before they are opened to the main vacuum column. A water or liquid-nitrogen-cooled baffle is used to reduce backstreaming of oil vapor into the chamber.

While this type of vacuum system may be satisfactory for a great number of applications, it proved to be unsatisfactory for more demanding investigations. It is impossible to prevent completely the backstreaming of diffusion pump oil into the system. The electron beam interacts with the oil molecules, causing hydrocarbon "varnish" to be deposited on the specimen and on critical apertures. This limits the exposure time and requires frequent cleaning of apertures. Heating or cooling the specimen stage will improve the situation, but the vacuum may still not be satisfactory, especially in cases where film depositions are studied.

When a dry vacuum is desired, getter ion pumps or turbomolecular pumps are used, often in conjunction with titanium sublimation pumps. To bring the system down to the pressure where these pumps take over, rotary oil pumps or liquid-nitrogen-cooled zeolite molecular sieve traps are used.

Scientists who converted their system from oil diffusion pumps to dry pumps report a longer cathode life, reduced exposure time, sharper pictures, and practically the elimination of the varnish problem. They report a reduction in varnish build-up from .5 to 1 nm sec^{-1} down to .1 to .5 nm min^{-1}.*)

To reduce the contamination from elastomer o-rings, they are sometimes replaced in critical locations, for instance in the specimen chamber, by metal gaskets—usually oxygen-free copper rings. Viton A o-rings are used for less critical seals. These o-rings have to be outgassed for many hours in vacuum at a temperature of 150°C before they can be installed in the system. Photographic plates must be outgassed too before they can be placed in the chamber.

Rotating or sliding motion which is usually accomplished by rods fed through O-ring seals or through greased glands can in critical cases be done by means of metal bellows.

Power Supplies. The voltage regulation required for the power supplies of the *electrostatic* microscope is not very critical. As long as the lens voltage and cathode potential maintain the same linear ratio—very often they are identical—a good image is obtained.

The refractive power of the *magnetic* lens depends directly on the square of the magnetic flux, which is proportional to the lens current—if the magnetic circuit is not saturated—and varies inversely with the cathode potential [see Eq. (7)]. It is, therefore, paramount that all power supplies be extremely well-regulated.

Image Formation. In the light microscope the image is formed by the difference in the absorption of light in the various sections of the object. There is very little absorption of electrons in the TEM. The image is formed essentially by the scattering of electrons during their passage through the specimen. Those sections that are denser or thicker or are composed of heavier atoms will scatter more electrons and will, therefore, appear darker in the final image on the fluorescent screen. They will be lighter on the photographic plate.

Three types of electrons emerge after transit through the specimen: unscattered electrons, elastically scattered electrons, and inelastically

scattered electrons. The *unscattered* electrons are those that did not interact with the atoms of the specimen. They traversed it without deviating from their trajectory. The *elastically scattered* electrons interacted with the nuclei of atoms in their path. The interaction is especially strong with heavy nuclei. The number of elastically scattered electrons is proportional to the $\frac{4}{3}$ power of Z, the atomic number of the atoms. The elastically scattered electrons did not lose kinetic energy but changed their direction significantly. They are removed, to a large extent, by the limiting aperture in the gap of the objective lens. *Inelastically scattered* electrons did interact with the electrons of the atoms in their path. They emerge within a narrow angle but have given up some of their energy. Their number is proportional to the $\frac{1}{3}$ power of Z. Inelastically scattered electrons cause chromatic aberrations. They are focused in a plane other than the image plane of the unscattered electrons. They degrade the image to some extent.

Picture Enhancement. The microscope picture can be improved by means independent of the microscope. It can be intensified by using closed-circuit television. In this case it is projected from the fluorescent screen onto the face of a television pick-up tube and is, after amplification, displayed on the face of a television picture tube. This can be accomplished, for instance, by using a fiber-optic window at the end of the viewing chamber. Such a window makes it possible also to take contact photographs of the fluorescent screen without bringing the photographic material into the vacuum. It is also feasible to feed the fluorescent image directly into a high-voltage image intensifier. The picture contrast and resolution can be improved using data correlation techniques (see, for example, Ref. 10). It can also be sharpened considerably by means of holographic technology by using optical deburring filters.[11]

Other Types of Transmission Electron Microscopes *Electrostatic EM.* Its development is as old as the development of the magnetic electron microscope.[12] It uses mostly unipotential lenses. Electrostatic transmission electron microscopes are used where a high resolution is not required. Since electrostatic lenses do not require highly regulated power supplies, they are in general less costly than magnetic microscopes.

Million Volt Electron Microscope. The first extra high voltage microscope went into operation in 1960. It was built by Dupouy and Perrier at the Laboratoire d'Optique Electronique, Toulouse, France.[13] It was designed for 1.5 MV operation. The first 1 MV microscope in the United States started operation in 1967 at the U. S. Steel Corp. Research Center at Monroeville, Pennsylvania.

The ever increasing number of these rather expensive instruments indicates that they are able to provide information that is not obtainable in any other way. The MV microscope has

*The build-up is measured by observing the decrease in the radius of a small hole, approximately 1 μ in diameter, in a carbon film deposited on a thin substrate.

the advantage of a higher resolution due to the shorter wavelength of the electrons at the higher voltage. It has a five to eight times greater penetration power than the conventional transmission microscope. It permits, therefore, the investigation of specimens of greater thickness. Biological specimens can be examined that could not be penetrated and that had to be sectioned before. Bulk properties of various materials, for instance of crystalline materials, can be investigated that may be different from those of thin foils. Materials can be studied that could not be sliced thin enough or that would be affected or contaminated by the slicing process. Another advantage is that, at the higher voltage, the electrons have a lower cross section for inelastic scattering. This leads to lower radiation damage in materials such as polymers that are easily damaged at voltages between 15 and 100 kV. The reduced inelastic scattering leads also to reduced chromatic aberration.

Scanning Electron Microscope (SEM) While in the standard transmission electron microscope the total area under observation is irradiated at the same time by the electron beam, only a single element of this area is irradiated at any one time by the "probe" in the scanning electron microscope (SEM). This probe is a very fine electron beam that has been demagnified by magnetic lenses. It scans the specimen in a television-type raster. The finer the probe, the higher is the resolution of the microscope. The signal originating from each element while it is under irradiation modulates the electron beam of a display tube. The beam of the display tube is deflected in synchronism with the probe scan.

A system of this type without probe demagnification, and therefore, with low resolution was first used by M. Knoll in 1935.[14] A demagnified probe was used by M. von Ardenne in 1938. The SEM has been developed since that time into an extremely useful tool through the work of many researchers. Many models are now commercially available.

The SEM can operate in a number of different modes that will supply different information about the specimen. The SEM can make use of:

(a) Secondary electrons from the surface of a thick specimen.
(b) Backscattered electrons from the surface.
(c) Transmitted electrons.
(d) Cathodo-luminescence and x-rays.

The SEM operates in its most common form at a voltage between 5 kV and 50 kV. It has a tungsten hairpin cathode and two demagnifying lenses that are built like the lenses of the standard transmission microscope but have a short image distance compared to a long object distance (see Fig. 6). The final lens has built inside its inner cylinder two sets of vertical and horizontal scanning coils. One set near the top of the lens deflects the beam away from the axis,

while the second set deflects it in the opposite direction towards the axis, so that the beam crosses the axis in the exit aperture of the lens. Stigmators are used to correct imperfections of the lens.

The specimen is held several millimeters below the exit aperture of the final lens at a tilting angle between 30° and 60°. The distance has to be great enough to prevent the magnetic field of the lens from interfering with the measurement. Backscattered electrons are collected by a ring electrode mounted below the exit aperture of the lens. Secondary electrons are collected by a collector mounted on the side, facing the specimen. The collector is at a high enough positive potential to attract the secondary electrons that leave the specimen with an energy of a few volts. Inside the collector box is a scintillation crystal e.g., europium activated calcium fluoride [CaFe(Eu)] at a still higher positive potential. The light flashes produced by the electrons impinging on the crystal are conducted through a light pipe to a photomultiplier tube located outside the vacuum chamber. The signal from the photomultiplier is amplified and used to modulate the beam of the display cathode ray tube. The display tube has, usually, a long-persistent phosphor screen. The micro-

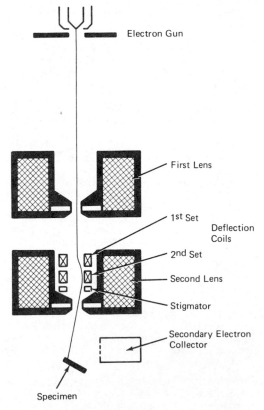

FIG. 6. Typical scanning electron microscope (SEM).

graph is taken photographically from the face of the display tube. It is possible to display different signals side by side that are obtained from different types of collectors. Storage tubes may be used in place of ordinary display tubes. The signal can also be stored on magnetic tape or on a magnetic drum and can be fed to a computer for image enhancement. The magnification of the SEM is determined by the ratio of the size of the raster on the display tube to the size of the raster on the specimen. The number of scanning lines per frame can be varied between 250 and 1000 lines or more. The time per scan can be varied from a fraction of a second, for observation, to several minutes for recording. Since the probe currents are very small, on the order of 10^{-12} to 10^{-10} A, long exposure times are necessary in order to obtain a good signal-to-noise ratio. The exposure time is limited by the stability of the voltage supplies and the difficulty of eliminating mechanical vibrations completely.

The SEM has a very small beam-convergence angle—between .005 and .01 radian. This means that it has a great depth of focus, much greater than that of optical or standard transmission

microscopes. This accounts for the three-dimensional qualities of its pictures.

In the secondary emission mode the contrast in the picture is generated by a change in the composition of the specimen, causing a change in the secondary emission ratio. Where the composition of the specimen surface is so uniform that the secondary emission ratio varies very little, the contrast may be produced by the topography of the surface. The reason for this is that the secondary emission ratio increases as the angle of incidence of the primary electron increases.

The resolution of the SEM described above is limited to 10 to 25 nm (100 to 250 Å). This limitation is determined by the size of the electron probe. This in turn is determined by the diameter of the crossover near the aperture of the grid, by the demagnification of the electron optical system, and by the increase in spot size due to the range in kinetic energy of the electrons emerging from the cathode. The hot tungsten cathode has a low emission density even when operated at the highest temperature compatible with a useful life (approx. 2900°K). Owing to its high temperature, the emitted elec-

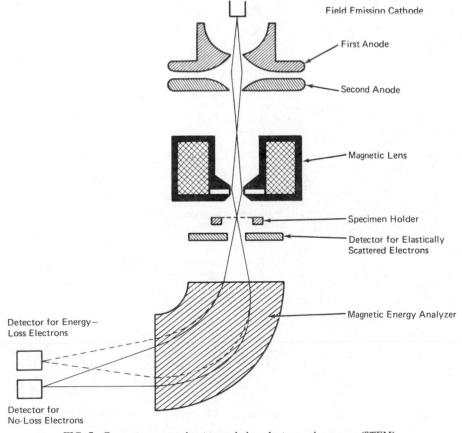

FIG. 7. Crewe-type scanning transmission electron microscope (STEM).

trons have a great range in kinetic energy (approximately 0.9 V). A different cathode at a lower temperature and higher emission density would permit a smaller limiting aperture for the same optical system and the same probe current. The lanthanum hexaboride [LaB_6] cathode introduced by A. N. Broers[15] is a great improvement over the tungsten filament.

Scanning Transmission Microscope (STEM). Another major improvement was achieved by A. V. Crewe of the University of Chicago by introducing a field emission electron source.[16] The field emission technique was invented in 1937 by E. W. Müller (see FIELD EMISSION). In Crewe's STEM (see Fig. 7) the field emission cathode, which requires a vacuum of 10^{-9} to 10^{-10} torr, consists of a fine tungsten point. A negative potential of 3 kV is applied to the cathode against a spherical first anode. A second anode at ground potential is at a potential of 30 kV against the cathode. The two anodes together form an immersion lens. Though the tunsten emitting point has a diameter of 100 nm the electrons seem to emanate from a virtual electron source that has a diameter of only 3 to 10 nm (30 to 100 Å) due to the very strong electric field close to the cathode. Emission currents of several microamperes are obtainable.

Since the tungsten tip is at room temperature, the range of kinetic energy of the emitted electrons is less than 0.2 V. With a demagnifying magnetic lens, a probe diameter of 0.5 nm (5 Å) is possible with a probe current of 10^{-11} to 10^{-10} A.

Crewe collects separately the three types of electrons that emerge after passing through the specimen, namely: unscattered, elastically scattered, and inelastically scattered electrons. A ring electrode mounted underneath the specimen collects essentially all elastically scattered electrons that have changed their direction significantly. The opening in the ring permits the unscattered and the inelastically scattered electrons to pass. A magnetic— or electrostatic—analyzer separates the two types. In a homogeneous magnetic field perpendicular to the electron path, the inelastically scattered electrons will describe a circle of smaller radius than the unscattered electrons due to their lower energy. Two different detectors can then collect the two types separately. Since the number of elastically scattered electrons increases much faster than the number of inelastically scattered electrons with increasing atomic number, the signals can be used to determine the atomic number of the atoms at the spot under investigation.

MARTIN M. FREUNDLICH

References

1. Ruska, E., "Fifth International Congress for Electron Microscopy," New York, Academic Press, 1962.

2. Freundlich, M. M., "Origin of the Electron Microscope," *Science*, **142** (3589), 185–188 (1963).

3. Ruska, E., "Die frühe Entwicklung der Elektronenlinsen und der Elektronenmikroskopie," *Acta Historica Leopoldina* #12, 1979, Deutsche Akademie der Naturforscher Leopoldina, Halle/Saale, East-Germany. See on page 32 a copy of the page of Ruska's notebook for this date.

4. Ruska, E., "Uber Fortschritte im Bau und in der Leistung des Magnetischen Elektronenmikroskops," *Z. f. Physik*, **87** (9 & 10), 580–602 (1934).

5. Ruska, E., *Arch. Elektrotechn.*, **38**, 102–130 (1944).

6. Hall, C. E., "Introduction to Electron Microscopy," Chapter 5, New York, McGraw-Hill Book Co., 1953.

7. Mahl, H., and Pendzich, A., *Z. Tech. Physik*, **24**, 38–42 (1943).

8. Johannson, H., and Scherzer, O., *Z. f. Physik*, **80**, 183–202 (1933).

9. Mahl, H., *Jahrb. AEG Forsch.*, **7**, 43–56 (1940).

10. Andrews, H. C., Tescher, A. G., and Kruger, R. P., "Image Processing by Digital Computers," *IEEE Spectrum*, **9** (7), 20–32 (July 1972).

11. Stroke, G. W., "Sharpening Images by Holography," *New Scientist* 23 Sept. 1971; "Optical Computing," *IEEE Spectrum*, **9** (12), 24–41 (December 1972).

12. Brüche, E., "Elektronenmikroskop," *Naturw.*, **20**, 49 (1932).
    Brüche, E., and Johannson, H., "Elektronenoptik und Elektronenmikroskop," *Naturw.*, **20**, 353–358 (1932).

13. Dupouy, G., and Perrier, F., *J. Microscopie*, **1**, 167–192 (1962).

14. Knoll, M., "Aufladepotential und Sekundäremission Elektronenbestrahlter Körper," *Z. Techn. Physik*, **16** (11), 467–475 (1935); "Aenderung der Sekundären Elektronenemission von Isolatoren und Halbleitern durch Elektronenbestrahlung," *Naturw.*, **24** (22), 345 (1936).

15. Broers, A. N., *Rev. Sci. Instr.*, **40**, 1040 (1969).

16. Crewe, A. V., "Scanning Electron Microscopes: Is High Resolution Possible?" *Science*, **154** (3750), 729–738 (1966).
    Crewe, A. V., "A High Resolution Scanning Electron Microscope," *Sci. Amer.*, **224** (4), 26–35 (1971).

Cross-references: ELECTRON OPTICS; FIELD EMISSION; LENS; OPTICS, GEOMETRICAL.

ELECTRON OPTICS

The invention of wave mechanics in 1926 by Heisenberg and Schrödinger saw a revolution in physics. It became apparent that the ideas of classical dynamics could be formally replaced when a stream of particles was considered. According to the well-known De Broglie hypothesis, a wavelength λ can be assigned to any material particle such that

$$\lambda = h/mv \qquad (1)$$

where h is Planck's constant, m is the particle mass, and v is the particle velocity.

As one of the many consequences of these ideas, the new science of electron optics emerged. In the same year, Busch demonstrated that the action of a short axially symmetrical magnetic field on a beam of electrons was similar to that of a glass lens on light. Terms used before then in optics found their use in describing electron devices. Electron "lenses" and "mirrors" having "focal lengths" and "resolutions and aberrations" were described. This analogy has proved useful in the study of the behavior of electrons in electronic valves, magnetrons and klystrons, traveling wave tubes, cathode ray tubes, and electron microscopes, to name only a few devices. The original concept of the electron microscope was evolved by direct analogy with the light microscope.

An electron which has fallen through a potential V has a kinetic energy $\Gamma = \frac{1}{2}mv^2 = eV$. Hence $v = (2eV)^{1/2}/m^{1/2}$, and using Eq. (1), we obtain the useful formula expressing electron wavelength in terms of volts

$$\lambda = \left(\frac{150}{V}\right)^{1/2} \qquad (2)$$

where λ is in angstroms.

In light optics, the least resolvable distance S between two objects is given by the Abbe expression $S = \lambda/(2n \sin i)$, where n is the refractive index of the material between object and lens and $2i$ is the angle subtended at the lens by the object; $n \sin i$ is called the numerical aperture. In the case of white light of equivalent wavelength $\lambda = 5600$ Å,

$S_{(\text{minimum, light})} = 1800$ Å, i.e., 1.8×10^{-5} cm.

Using the same expression and Eq. (2), which gives λ for an electron = 0.04 Å for $V = 100$ kV,

$$S_{(\text{minimum, electrons})} \simeq 0.04 \text{ Å}$$

i.e., small fractions of an angstrom can be resolved. However, the theory does not take into account the fact that lenses are imperfect. Spherical aberrations drastically affect the situation. In addition, certain diffraction defects and scattering impose limitations. These facts and others make it impossible to obtain resolutions near the theoretical maximum. Instead, 2.0 Å is a better theoretical value. Practical limiting resolutions obtained are of the order of 2.5 Å to 3.0 Å.

Source of Electrons The basic source of electrons in electron tubes is the heated filament or disk. The well-known equation $J = AT^2 \exp(-\phi/kT)$ describes the emission, where J is the current density, T is the temperature, ϕ is the work function of the emitting material, A is a constant, and k is Boltzmann's constant. J is enhanced by increasing T or reducing ϕ. Typical cathode materials used are tungsten, tantalum, and the oxides of barium and strontium.

A voltage is applied to accelerate the emitted electrons, and the current density is given by $J = AT^2 \exp(-\phi/kT)$ if they are removed to the anode. However, many electrons stay near the cathode and repel other emitted electrons back to the cathode. Most devices operate in this "space-charge limited" way. For any electrode configuration, the current density is then $J = GV$ where V is the applied voltage and G is a constant for the particular electrode configuration. G is roughly equivalent to electrical conductance and is called the "perveance." The constant is fundamentally important. The higher G is in value, the greater is the efficiency of the beam system.

In simple designs, the axial portion of the cathode is overloaded, producing excessive emission and cathode burn-out. To avoid this situation, carefully shaped electrodes must be used, but the acceptable design depends upon the application. A bent hairpin point cathode is used in, for example, the electron microscope. It produces low perveance but high emission density. Since G is low, the field at the cathode can be high, and this tends to reduce ϕ, giving emission at lower temperature. The electron guns for klystrons, traveling wave tubes, and metallurgical applications such as vacuum melting and welding require higher efficiency and current density. Hence they are high-perveance guns.

Beam Control Electrostatic and magnetic fields control the motion of an electron according to the following equation:

$$\overline{F} = m\frac{dv}{dt} = e\,(\overline{E} + \overline{v} \times \overline{B}) \qquad (3)$$

where

$\overline{F}$ = force
$\overline{E}$ = electrostatic field
B = magnetic field
$\overline{v}$ = electron velocity
e = electronic charge

The electron must travel in a vacuum if it is not to be scattered and lose its kinetic energy by collision with relatively massive gas molecules. Nuclear particles such as protons and neutrons are each about 1837 times larger in mass than the electron.

Electrostatic Electron Lenses If in Eq. (3), $\overline{B} = 0$, then $\overline{F} = e\overline{E}$. The equation says that an electron in a field $\overline{E}$ experiences a force $\overline{F}$ in the direction of the field. Thus in the system shown in Fig. 1 where there is a voltage V_1 on cylinder 1 and V_2 on cylinder 2, and where $V_2 > V_1$, the field and the path of an electron are as shown. The electron moving from left to

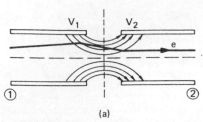

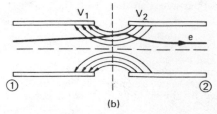

FIG. 1. Electrostatic focusing of electron beam by cylinder lenses ① and ②. (a), voltage $V_2 > V_1$; (b), voltage $V_2 < V_1$.

right increases its velocity and is deflected towards the axis. After passing the median, the force is away from the axis, but since velocity is increased, the electron spends less time in this part of the field. Therefore, deflection away from the axis is less than it was towards it, and there is a net convergence. If $V_2 < V_1$, as in Fig. 1 (b), then for an electron beam traveling from left to right the lens will still be convergent because maximum deflection will occur after the electrons have slowed down. All such lenses are convergent for $\overline{E} = 0$ on both sides of the lens.

Magnetic Electron Lenses If in Eq. (3) $\overline{E} = 0$, then $\overline{F} = e\overline{v} \times \overline{B}$. This says that the force $\overline{F} = e|\overline{v}||\overline{B}| \sin \theta \, \overline{\epsilon}$ where $|\overline{v}|$ and $|\overline{B}|$ are the numerical values of the vectors and θ is the angle between them; $\overline{\epsilon}$ is a unit vector perpendicular to both, and indicates the direction of the force. For $\overline{F}$ to be greater than 0, $\overline{v}$ must be greater than 0. The force constrains the electron to move in a circle of radius $\rho = (\overline{v} \sin \theta)/(\overline{B} \, e/m)$. At the same time it moves in a perpendicular direction with velocity $\overline{v} \cos \theta$, and therefore, it traces the path of a helix and returns to the axis in a time $T = 2\pi/(\overline{B} \, e/m)$ which is independent of both $\overline{v}$ and θ. The net effect is that all electrons are focused by $\overline{B}$ to produce an image of the source from which they diverge.

The simple magnetic lens consists of a short coil of wire contained in a surrounding shield of magnetic material. A small gap in the case material concentrates the escaping field when the coil is energized. The electron microscope (shown in Fig. 2) illustrates the use of magnetic lenses. The coils are wound in opposite sense to cancel the spiral distortion of the electron beam which is introduced by the individual lenses.

Electrostatic Deflection If in Fig. 3(a) a voltage V_d exists between the plates $x_1 \, x_2$ and an electron enters along the axis with a constant velocity $v = (2 \, Ve/m)^{1/2}$, then because of the field $\overline{E}_y$ caused by V_d, a transverse force $\overline{F}_y$ causes a motion in this direction to be impinged on the electron, i.e., $m \, (d^2y/dt^2) = V_d e/d$, from which $dy/dt = (V_d/d) \, (e/m) \, t$ and $y = \frac{1}{2} \, (V_d/d) \, (e/m) \, t^2$. The force acts for time $t = \ell/v$; thus deflection $y = \frac{1}{2} \, (V_d/d) \, (e/m) \, (\ell/v)^2 = \frac{1}{4} \, (V_d/d) \, (\ell^2/V)$. The additional deflection, after F has ceased to act on the electron, is $y^1 = (V_d/V) \, (\ell L/2d)$.

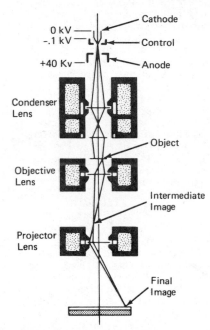

FIG. 2. Schematic diagram illustrating the general arrangement of a transmission electron microscope.

Magnetic Deflection See Fig. 3(b). If $\overline{B}$ is a sharply defined field, then the deflection on leaving the field is given by $y = \ell^2/2\rho$. Total deflection $D = L \tan \alpha + y$. To a good approximation

$$D = \frac{\overline{B}L\ell}{v} \frac{e}{m} \left(1 + \frac{1}{2L}\right).$$

It should be emphasized that these considerations are only approximate. In practice, factors such as the inability to sharply define a field edge in space cause added complications.

Cathode Ray Tubes A device worthy of brief consideration is the cathode ray display tube. Its basic components are as follows: an electron gun, an acceleration and focus system, a deflection, and a display system which is normally a phosphor screen. Depending on the type of tube the applied voltage causes electrostatic or magnetic deflection. In the oscilloscope tube electrostatic deflection is used while in the large display tube, such as the television

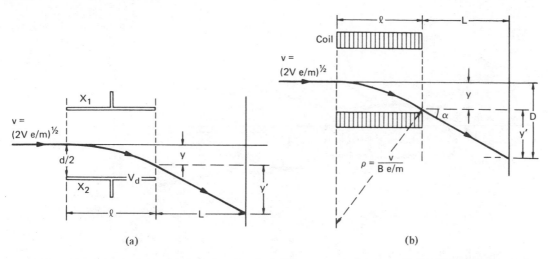

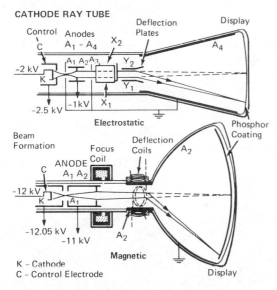

FIG. 3. (a) Deflection of electron beam by electric field between parallel plates X_1X_2. (b) Deflection of electron beam by magnetic field. Electron path is an arc of a circle only while it is influenced by field. After leaving the field the path is linear.

CATHODE RAY TUBE

FIG. 4. Schematic diagram of cathode ray tubes.

tube, magnetic deflection is used because of the large scan angles which are required. The tubes are illustrated in Fig. 4.

Advanced Oscilloscope Tubes The instrument tube shown in Fig. 4 is of low cost and has limited frequency performance to 5 to 10 MHz. If voltage on A_4 is increased, then it becomes increasingly difficult to deflect the beam in such a tube. In recent years performance of the comparatively low-cost tube has been increased to greater than 50 MHz by the introduction of post-deflection acceleration systems, the most successful of which is shown in the tube illustrated in Fig. 5. Here a grid of fine wires placed over the end of the gun

assembly shields the deflection area from the effect of the high acceleration field required for high brightness and high writing speed. The electrons pass through the grid and those which collide with it cause low energy emission of secondary electrons which can be collected readily by placing a small positive voltage on a neighbouring electrode such as A_4, and therefore they do not cause background illumination of the image produced by the primary beam.

Table 1 illustrates the operating features and performance of the tube. The display envelope is commonly rectangular and an external magnetic coil is needed to align the beam axes with those of the display face. Plate capacitances are reduced by the use of side pins. Nevertheless these capacitances and transit time phenomena contribute to the main limitations for use at higher frequencies. However, other tubes operating at frequencies greater than 500 MHz have been developed.

Color Television Tube In the television tube the beam must be deflected to trace each part of the screen in sequence while it is simultaneously intensity modulated. The resulting picture is a mosaic of dark and light elements. Three primary red, green and blue pictures must be displayed in exact registration to obtain a a high-quality color picture. Color mixing to produce the wide range of colors required is obtained subjectively and depends on the relative brightnesses of the three primary light yields at a particular point on the screen. There are numerous ways of achieving the color picture, but the "shadow mask tube" has been developed to provide the most economical method. There are three primary guns in the tube, each tilted to converge to a central point near the screen. A single de-

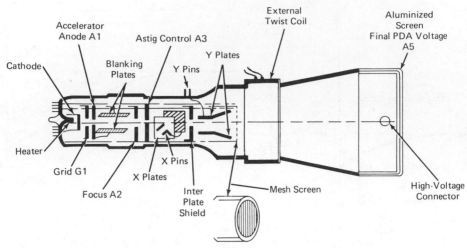

FIG. 5. Advanced Instrument Cathode Ray Tube. Mesh screen, held near mean plate potential, prevents PDA field from reducing deflection plate sensitivity. Note other features such as side pin connection to plates to reduce capacitance; also the blanking arrangement for blanking the beam during time-base flyback. (The method has advantages over using grid G1.) See Table 1 for other information.

TABLE 1. PERFORMANCE OF MESH PDA TUBE (See Reference 5)

Beam deflection plate potential	1250 V[a]	1500 V
First acceleration plate potential	1250 V	1500 V
Focusing plate potential	160–80 V	200–100 V
PDA voltage	12.5 KV	12.5 KV
Typical screen current	10 μA	10 μA
Line width 10 μA	0.36 mm	0.32 mm
Linearity	2 per cent	2 per cent
Y sensitivity	4.8 Vcm^{-1}	5.8 Vcm^{-1}
X sensitivity (time base)	9 Vcm^{-1}	11 Vcm^{-1}
Y scan	80 mm	
X scan	100 mm	
Tube length	350 mm	

[a]Voltages w.r.t. cathode.

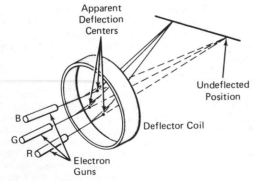

FIG. 6. Simultaneous deflection of three electron beams in shadow mask color tube.

flection coil is used to scan the beams in registration (as shown in Fig. 6. Additional deflection yokes carry currents derived from the main deflection waveform to introduce convergence corrections as needed to maintain registration over the complete screen.

The screen must emit red light when bombarded by electrons from the red gun, green from the green gun, and blue from the blue gun. To enable this to occur a perforated metal mask is positioned about 12 mm from the screen and the beams in fact converge to its central point rather than that of the screen. Electrons must pass through the mask holes to reach the screen as shown in Fig. 7. The beams appear to originate from three apparent deflection centers in the deflection field. Thus, they

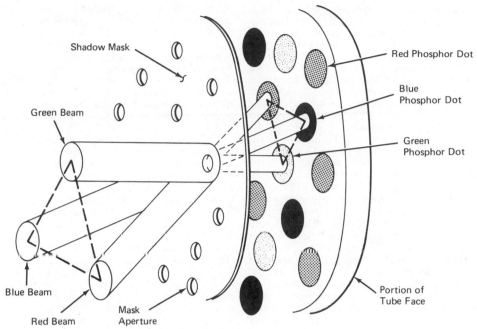

Shadow Mask

Green Beam

Red Phosphor Dot

Blue
Phosphor Dot

Green
Phosphor Dot

Blue Beam

Red Beam

Mask
Aperture

Portion of
Tube Face

FIG. 7. Operation of shadow mask in color television tube.

will strike the face in three points of a triangle when passing through a particular hole. Red, green, or blue emitting phosphor "dots" are deposited at these points as appropriate. The plate is called a shadow mask because when one gun is switched on, the unwanted dots are not energized since they lie in its shadow. Typical mask holes are 0.5 mm in diameter and about 0.7 mm apart. The total number of holes is about 400 000 for a 25-inch tube.

During manufacture the mask itself is used to process the screen. The phosphor is deposited with a photosensitive lacquer and a UV light source is placed at the appropriate deflection center. The screen is exposed through the mask, developed, and the process repeated for the other two colors. The deflection centers move slightly towards the screen for large deflection angles, and a correction lens has to be used to compensate during exposure. The accurate positioning of the million or more phosphor dots is achieved in this way.

BARRY A. GEORGE

References

1. Klemperer, O., "Electron Optics," Cambridge, The University Press, 1953 (for a rigorous treatment of electron optics).
2. Bakish, R., "Introduction to Electron Beam Technology," New York, John Wiley & Sons, 1962 (for the practical applications of electron optics).
3. Bakish, R., Ed., "Electron and Ion Beams in Science and Technology," New York, John Wiley & Sons, 1965.
4. Pierce, J. R., "Theory and Design of Electron Beams, New York, Van Nostrand Reinhold, 1949.
5. "Performance data relates to modern German CRT," AEG Telefunken D14-131, Table 1.

Cross-references: ELECTRON; ELECTRON MICROSCOPE; ELECTRON TUBES; OPTICS, GEOMETRICAL; OSCILLOSCOPE; THERMIONICS.

ELECTRON SPIN

The electron, as one of the stable fundamental particles (a lepton) of physics, has a small rest mass m and a classical unit of charge $-e$, but also possesses the attribute of spin, a strictly quantum mechanical property. Associated with this spin is an angular momentum of magnitude $\hbar/2$ ($\hbar$ is Planck's constant) and a magnetic moment given approximately by $e\hbar/2mc$, where $e\hbar/2mc = \mu_B$ is known as the Bohr magneton ($\mu_B = 0.9273 \times 10^{-20}$ ergs/gauss). On the other hand, orbital angular momentum, such as for electrons in atoms, is quantized in integral units of $\hbar$. Thus, the electron spin quantum number of $\frac{1}{2}$ is different than that for orbital quantum numbers. Furthermore, the *electron spin gyromagnetic ratio* (the spin magnetic moment divided by the spin angular momentum) is twice the classical value for orbital angular momentum. This feature plays an important role in the Zeeman effect of optical spectra of atoms. Electron spin and its magnetic moment are predicted by the relativistically invariant Dirac equation. Whereas quantum mechanical wave functions for orbital motion are invariant under 2π rotations, electron spin wave functions are not invariant under 2π rotations, but instead change sign. The vector components of the electron spin operator can be conveniently described in terms of the 2×2 Pauli spin matrices. These two-component quantities are termed *spinors*.

The first direct experimental evidence for quantized electron spin S (having a z-component $S_z = m_s\hbar$, with $m_s = \pm\frac{1}{2}$) was obtained from the Stern-Gerlach experiment in which a collimated beam of neutral silver atoms was split into just two separate beams with opposite deflections when the silver atoms passed through an inhomogeneous magnetic field. This experimental approach developed into the molecular beam approach, which was so fruitfully pursued by Rabi, Kusch, and colleagues in the study of free atoms and molecules. These workers first used the magnetic resonance technique to change atoms from one spin orientation (with respect to a space quantization axis) to another, thus altering the populations to be deflected by the inhomogeneous field. Another important ramification of electron spin was in the Zeeman effect of optical spectra of atoms in an external magnetic field. Uhlenbeck and Goudsmit in 1925 proposed the concept of spin in order to explain the spectroscopic fine structure of atoms. The different gyromagnetic ratios for orbital and spin motion produce a state-dependent Zeeman splitting ($\Delta E = g\mu_B H$), where the g-factor is given by the famous Landé result which depends on the spin (S), orbital (L), and total (J) angular momentum of the atom. This result explained the anomalous Zeeman effect.

Early studies of spin-dependent effects in solids and liquids were made utilizing susceptibility and magnetization techniques, techniques determining the bulk magnetic moment (including orbital contributions) of a macroscopic sample. In 1945 Zavoisky reported the first observation of *electron paramagnetic resonance* [EPR, sometimes also called *electron spin resonance* (ESR)] in a solid. *Nuclear magnetic resonance* (NMR) was reported independently a year later by two groups. These developments utilizing the magnetic resonance approach, a much more sensitive and accurate method for detecting the presence of spins, represented a natural extension of the magnetic resonance technique first used in molecular beam methods.

The detection of magnetic resonance, described in quantum mechanical terms, represents the absorption of power in a radio-frequency (RF) coil or microwave cavity in the form of photons of frequency ν and energy $2\pi\hbar\nu$ such that $2\pi\hbar\nu = g\mu_B H$. $g\mu_B H$ is the Zeeman splitting resulting from the local magnetic field H. H may contain a contribution from other sources (i.e., hyperfine and exchange interactions) besides the external field. For $H = 3000$ gauss, $\nu \simeq 9 \times 10^9$ Hz for g $\simeq 2$, a frequency conveniently in the microwave range, suggesting microwave spectroscopy methods for EPR. A block diagram of a typical EPR microwave spectrometer is shown in Fig. 1. Rapid advances in microwave- and radio-frequency techniques before and during World War II greatly benefitted the rapid growth of both EPR and NMR.

The power of the EPR technique results not only from the enhanced sensitivity, but from the microscopic information obtained from the EPR spectra about the environment of the unpaired electron spins. This microscopic information consists of g-shifts [$\Delta g = g - g_{f.e.}$ (f.e. = free electron), $g_{f.e.} = 2(1 + \alpha/2\pi - \cdots) \simeq 2.0023$ since the fine structure constant $\alpha \simeq 1/137$] resulting from the spin-orbit and orbital Zeeman interactions, fine structure splittings (for spin $S \geq 1$ only) resulting from crystalline fields, hyperfine spectra resulting from electron spin–nuclear spin interactions, line widths and shapes resulting from various spin-spin interactions, the motion and diffusion of spins, and spin-lattice relaxation (spin-phonon interactions). Pulsed EPR (the spin-echo technique) and dynamic (field swept adiabatic fast passage) measurements have yielded information on spin-spin interactions, spin diffusion, and spin-lattice relaxation. The detailed microscopic information obtained from the spin Hamiltonian parameters tells us much about the symmetry properties and spatial extent of the electron (hole) wave function of a particular paramagnetic species. Figure 2 shows the EPR spectrum of atomic hydrogen situated at a cubic interstitial site in CaF_2 for $H \parallel [100]$. The principal splitting into two patterns results from the interaction of the unpaired electron with its hydrogen nucleus with spin $I = \frac{1}{2}$ (two hyperfine fields for $m_I = \pm\frac{1}{2}$). The two symmetric patterns of 9 lines each (intensity ratios $1:8:28:56:70:56:28:8:1$) result from the interaction with eight equivalent ^{19}F ($I = \frac{1}{2}$) nuclei. The interaction with the eight ^{19}F neighbors is termed the superhyperfine or ligand hyperfine structure. For each ^{19}F nucleus the hyperfine interaction has axial symmetry and is characterized by tensor components $A_\parallel$ and $A_\perp$. For other external field orientations the eight ^{19}F nuclei are not all equivalent and additional splittings result. A detailed theory of the embedded H-atom wave function, taking account of orthogonalization to the core electrons of the neighboring ^{19}F ions is required to explain the hyperfine tensor components.

A vast amount of work has been done on transition metal ions and rare earth ions in a wide variety of hosts, including ionic crystals, covalent crystals, and certain metals. Here we limit our discussion to S-state ($L = 0$) ions such as Fe^{3+} or Mn^{2+} [$(3d)^5$ configuration for free

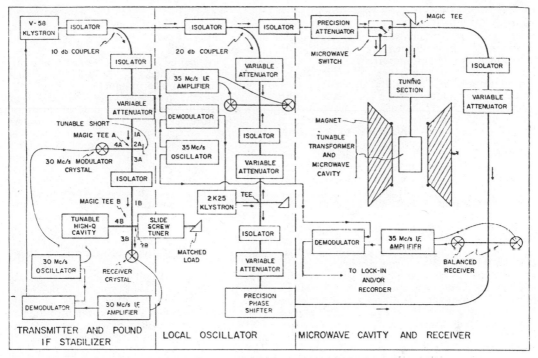

FIG. 1. Block diagram of a high-sensitivity superheterodyne X-band EPR spectrometer. [After Holton, W. C., and Blum, H., *Phys. Rev.* **125**, 89 (1962).]

ion]. For the $S = \frac{5}{2}$ ground state ions the crystal field can play an important role in determining the EPR spectrum. In the strong crystal field case zero field spin splittings are larger than typical Zeeman energies. Even for an iostropic g-value and no hyperfine interaction the EPR spectrum can be very anisotropic and depends critically on the symmetry and strength of the local crystal field. Certain rare earth ions in appropriate symmetry hosts have been utilized

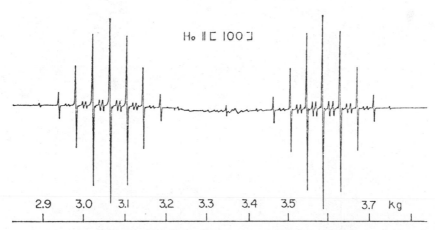

FIG. 2. EPR absorption derivative spectrum of CaF_2:H for H∥ [100]. The sharp line at 3.35 kG is a g-marker while the doublets between the major lines of the two g line patterns are forbidden transitions. [After Hall, J. L., and Schumacher, R. T., *Phys. Rev.* **127**, 1892 (1962).]

to make a 3- or 4-level maser (microwave amplification by stimulated emission of radiation). Such solid state microwave amplifiers, first proposed by Bloembergen, have been of considerable technological importance.

It frequently happens that the paramagnetic ion of interest has an orbital degeneracy in a particular crystalline field. This leads to the interesting Jahn-Teller effect. A static or dynamic distortion (the latter represents a tunnelling between equivalent static distortions) occurs, lowering the total energy of paramagnetic ion plus environs. This distortion can have a profound effect on the EPR spectrum. Analysis of these spectra yields information on the symmetry and magnitude of the distortion.

Non-S-state ions or impurities frequently have large anisotropic g-shifts resulting from the spin-orbit interaction which depend critically on the position of excited orbital states. Analysis of the g-tensor thereby provides useful information on the excited state energies of an impurity ion or defect. One example is the O_2^- molecular ion, frequently observed as a substitutional impurity in ionic crystals. Consider an O_2^- ion, oriented along a [110] axis, replacing a halogen ion in an alkali halide. The axially symmetric free O_2^- would have a g-tensor with components $g_{\parallel}$ and $g_{\perp}$ for **H** parallel and perpendicular to the molecular axis, respectively. In the alkali halide the $2p\pi_g$ orbitals are split by both the spin-orbit interaction and the larger crystalline field producing a correction to $g_{\parallel}$ and a splitting of $g_{\perp}$ into two components $g_{[1\bar{1}0]}$ and $g_{[001]}$. The sign of $g_{[1\bar{1}0]} - g_{[001]}$ determines whether the unpaired electron in the $2p\pi_g$ orbital is along the [1$\bar{1}$0] or [001] axes. Experimental results show

the unpaired $2p\pi_g$ orbital is aligned along the [1$\bar{1}$0] axis for potassium and rubidium halides, but is along the [001] axis in the sodium halides.

EPR line shapes in solids often are inhomogeneously broadened by residual hyperfine interactions with nearby nuclei. In these circumstances the unresolved lines observed in EPR spectra do not allow the determination of the hyperfine tensors of these residual hyperfine interactions. Feher developed an ingenious electron-nuclear double resonance technique (ENDOR) that permits the resolution of these otherwise unresolved hyperfine interactions. The idea depends on the much narrower line widths of the NMR transitions. An EPR transition is partially saturated at fixed field while a RF source is swept through the NMR transitions of the nuclei coupled through the residual hyperfine interactions with the unpaired electron (hole). Strong saturation of these NMR transitions alters the populations of the electron (hole) spin energy levels and alters the magnitude of the EPR signal. The EPR transitions are sensitive probes of weak NMR transitions from a small number of nuclei. ENDOR has been utilized to map the complicated oscillatory wave function of shallow donors in silicon and accurately determine the position in momentum space of the conduction band minima. Another interesting ENDOR example in Fig. 3 shows part of the ENDOR spectrum of the F-center and M-center (excited triplet $S = 1$ state) in KCl at $T \sim 90$ K. (a) gives the spectrum of the first shell of K-nuclei for the F-center without optical excitation. (b) shows additional ENDOR transitions from the

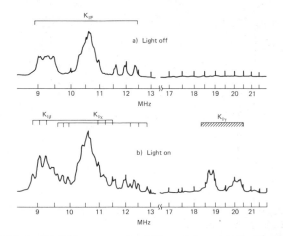

FIG. 3. ENDOR spectra of a KCl crystal containing F-centers and M-centers. B_0 = 3345 gauss $\parallel$[100], T = 90 K. (a) Without light excitation; first shell (K-nuclei) of the F-center; (b) during 3650-Å light excitation, additional ENDOR lines of K-nuclei of M-centers in the triplet state. [After Seidel, H., *Phys. Letters* 7, 27 (1963).]

nearest neighbor shells of K-nuclei of the excited $S = 1$ state of the M-center after irradiation with 3650 Å light. The ENDOR technique provided direct experimental proof of the models of these centers—the F-center consisting of an electron trapped at a halogen ion vacancy and the M-center composed of two adjacent F-centers.

Another important type of double resonance technique is the optical detection of magnetic resonance (ODMR). The optical-RF double resonance technique, first proposed by Brossel and Kastler, was demonstrated by Brossel and Bitter and has been extensively used in studying excited states in gases. The ODMR approach is particularly useful when the excited state population is very small and direct observation of the excited state EPR spectrum would be impossible. The sensitivity is enhanced by detecting optical photons from the fluorescing excited states while the magnetic field is swept through the excited state EPR spectrum. Absorption of microwave photons induces changes in the optical photon emission spectrum. The principle is similar to the ENDOR technique in that the sensitivity is enhanced by detecting higher energy photons. Figure 4 shows the ODMR spectrum of a deep defect in silicon observed at $T = 1.7$ K for two cases—with and without the application of a compressional

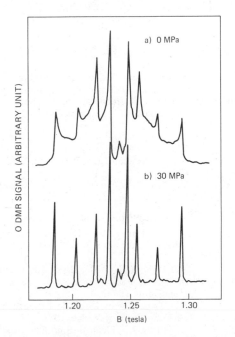

FIG. 4. ODMR spectrum the 0.97-eV luminescence in neutron-irradiated silicon. B ∥ [011], $T = 1.7$ K, $\nu_{EPR} = 35.0$ GHz. (a) Zeros stress spectrum; (b) Spectrum with 30 MPa compressional uniaxial stress, **T** ∥ [01̄1]. [After Lee, K. M., et al., *Phys. Rev. Lett.* 48, 37 (1982).]

stress. This data is valuable in testing various theoretical models of these defects.

Many other variations and applications of EPR should be mentioned. These include spin echoes, a novel, powerful technique developed by Hahn for NMR, but which has also been useful in EPR. In concentrated spin systems in crystalline solids the isotropic Heisenberg exchange interaction, in addition to smaller anisotropic exchange interactions, exist between pairs of neighboring spins. These systems frequently show spontaneous magnetization (below a critical temperature), long range spin order, and ferro-, ferri-, or antiferromagnetic behavior. The spin modes of these systems can be studied using *ferromagnetic resonance* (FMR) or *antiferromagnetic resonance* (AFMR) techniques. The *magnetic resonance of conduction electrons* (CESR) has not only been studied in certain high purity metals and heavily doped semiconductors, but also in metals containing magnetic impurities. EPR has given useful information on unpaired spins in amorphous semiconductors, on lunar samples, on low-dimensional systems (1 and 2 dimensions), on systems subjected to hydrostatic pressure or uniaxial stress, and on a wide range of unpaired spins in the form of molecular free radicals. Furthermore, spin systems have afforded an ideal system to study phenomena in nonequilibrium statistical mechanics.

For several decades EPR has been utilized as a powerful tool, which has been broadly applied to chemical and biophysical applications. While it is impossible to address the many important applications of the EPR technique to chemistry and biology, a few illustrative examples should be mentioned. Paramagnetic molecules or ions in solution undergo translation and a rotational tumbling motion. In these cases the EPR spectra, line widths, and shapes are strongly affected by the magnitude of the anisotropic terms (anisotropic g-shifts and hyperfine interactions) and the correlation times characteristic of the molecular tumbling rates. The EPR spectra, along with associated theory, yield useful information on the molecular rotational tumbling of the paramagnetic molecules. The EPR of numerous free radicals, produced as intermediates in chemical reactions or by radiation damage, is another topic studied in chemistry and biology. One most important example of EPR radiation damage studies is of the individual amino acids and the DNA helix. However, many biologically important molecules are not paramagnetic. A vast field of growing importance has arisen, termed spin labeling, in which an appropriate paramagnetic molecule is added to a given system to act as a "neutral" (noninterfering) probe of the diamagnetic biological molecule. Certain parts of important proteins, such as the heme group (with Fe^{3+}) and other porphyrins, have been extensively studied with EPR and ENDOR techniques. Figure 5 shows the ENDOR spectra of

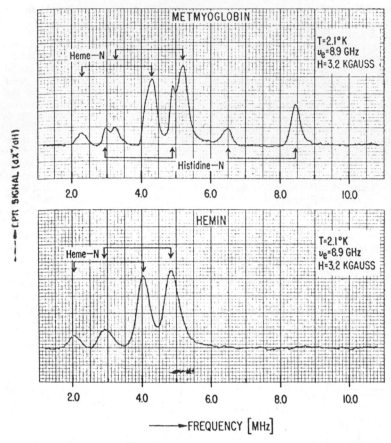

FIG. 5. ENDOR spectra due to ^{14}N interactions in hemoglobin. Top trace: Normal human methemoglobin A, conc. 6 mM in 0.1M phosphate buffer, pH = 7; Bottom trace: Hemoglobin M$_{Hyde\ Park}$. Washed whole oxygenated red cells were used. Note the absence of the high frequency peaks which had been assigned to the proximal histidine. [After Scholes, C. P., Isaacson, R. A., and Feher, G., *Biochim. Biophys. Acta* **263**, 448 (1972).]

normal human methemoglobin A and also that of the mutant hemoglobin M$_{Hyde\ Park}$. The differences in the two spectra give information on the binding of histidine nitrogens to the iron and on the nature of the mutant. Along with NMR, EPR, ENDOR, and ODMR have all become valuable tools utilizing spins in the study of biological systems.

THEODORE G. CASTNER

References

Electron Spin

Stern, O., and Gerlach, W. *Ann. Physik.* **74**, 673 (1924); *Z. Physik* **41**, 563 (1927); **8**, 110; **9**, 349 (1922).

Uhlenbeck, G. H., and Goudsmit, S. A., *Physica* **5**, 266 (1925); *Nature* **117**, 264 (1926).

Van Vleck, J. H., "Electric and Magnetic Susceptibilities," Oxford, Clarendon Press, 1932.

Ramsey, N. E., "Molecular Beams," Oxford, Clarendon Press, 1956.

Dirac, P. A. M., "The Principles of Quantum Mechanics," 4th Ed., Oxford, Clarendon Press, 1958, Ch. 11.

Electron Paramagnetic Resonance and Applications

Abragam, A., and Bleaney, B., "Electron Paramagnetic Resonance of Transition Ions," Oxford, Clarendon Press, 1970.

Pake, G. E., and Estle, T. L., "The Physical Principles of Electron Paramagnetic Resonance," Reading, Mass., W. A. Benjamin, 1973.

Slichter, C. P., "Principles of Magnetic Resonance," 2nd Ed., New York, Springer-Verlag, 1978.

Wertz, J., and Bolton, J., "Electron Spin Resonance: Elementary Theory and Practical Applications," New York, McGraw-Hill, 1972.

Carrington, A., and McLachlan, A. D., "Introduction to Magnetic Resonance with Applications to Chemistry and Chemical Physics," New York, Harper & Row, 1967.

Poole, D. P., Jr., "Electron Spin Resonance: A Comprehensive Treatise in Experimental Techniques," New York, Wiley-Interscience, 1967.

Manenkov, A. A., and Orbach, R. (Eds.), "Spin-Lattice Relaxation in Ionic Solids," New York, Harper & Row, 1966.

Geschwind, S. (Ed.), "Electron Paramagnetic Resonance," New York, Plenum Press, 1972.

Symonds, M., "Chemical and Biochemical Aspects of Electron-Spin Resonance Spectroscopy," New York, John Wiley & Sons, 1978.

Berliner, L. J. (Ed.), "Spin Labeling: Theory and Applications," Vols. I and II, New York, Academic Press, 1976.

Berliner, L. J., and Berliner, J. (Eds.), "Biological Magnetic Resonance," Vols. I and II, New York, Plenum Press, 1978.

Clark, R. H. (Ed.), "Triplet State ODMR Spectroscopy: Techniques and Applications to Biophysical Systems," New York, Wiley-Interscience, 1982.

Cross-references: ELECTRON, MAGNETIC RESONANCE, ZEEMAN AND STARK EFFECTS.

ELECTRON TUBES

Although electron tubes have been replaced with solid state electronic devices in most applications, tubes will retain their importance in applications involving high voltage, high power, high frequency, and visual display.

The basic electron tube consists of an evacuated glass envelope containing a *cathode* that emits electrons by thermionic emission and a positive *anode* that collects these electrons. The tube is evacuated so that electrons can travel without colliding with gas molecules. Other forms of electron tubes operate in a high pressure regime where gas ionization is important (i.e., thyratrons, ignitrons, voltage regulator tubes, etc.) but high pressure tubes will not be discussed. In addition, one or more transparent wire *grids* are typically inserted between the cathode and anode to control the flow of electrons.

Diodes The simplest such tube is called a *diode* (two electrodes) and is shown schematically in Fig. 1. If the anode (or *plate*) voltage is positive relative to the cathode, the electrons that boil off the cathode will be drawn to the anode and collected. Hence an electrical current flows from anode to cathode (opposite to the electron flow) in a diode. On the other hand, if the anode is negative relative to the cathode, the electrons are repelled, and no current flows.

The relationship between current and voltage in the vacuum diode is shown in Fig. 2. For small positive voltages the electric field that draws electrons away from the cathode is partially shielded by the cloud of electrons that surrounds the cathode. This is called the *space-charge-limited region*, and the current in am-

peres is given approximately by *Child's law:*

$$I = 2.33 \times 10^{-6} \, A V^{3/2} / d^2$$

where A is the area of the cathode in square meters, d is the separation of the cathode and anode in meters, and V is the voltage difference between the anode and cathode. For large positive voltages the electric field is strong enough to collect all the electrons emitted by the cathode, and the current is independent of voltage but depends strongly on the absolute temperature T as described by *Richardson's equation:*

$$I = 1.3 \times 10^6 \, A T^2 e^{-e\phi/kT}$$

where k is *Boltzmann's constant* (1.38×10^{-23} joules/K), ϕ is the work function of the cathode (typically a few volts depending on the material), e is the electronic charge (1.6×10^{-19} coulombs), and A is the area of the cathode in square meters. This is called the *emission-limited region*.

Triodes If one inserts a transparent conducting grid between the cathode and anode, as indicated schematically in Fig. 3, the device is called a *vacuum triode* (three electrodes). If the grid is made positive relative to the cathode, but not as positive as the plate, it will accelerate the electrons and increase the cathode current, provided the tube is operating in the space-charge-limited region. Some of the electrons are collected by the grid, but if the grid is transparent, many of the electrons will pass through the grid and will be collected by the even more positive plate. A large grid current is usually undesirable because it requires that the source connected to the grid provide power, and this power must be dissipated by the grid, which in extreme cases may overheat the grid. Consequently, the vacuum triode is normally operated with its grid negative relative to its cathode. In such a case, it is energetically impossible for electrons to reach the grid, and so the grid current is always zero. On the other hand, if the grid is sufficiently close to the cathode, and not too negative, some of the electrons feel the electric field from the positive plate and pass through the grid and are collected by the plate. In this way, a small voltage applied to the grid can be used to control the flow of current be-

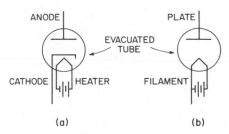

FIG. 1. Vacuum diodes: (a) heated cathode; (b) heated filament.

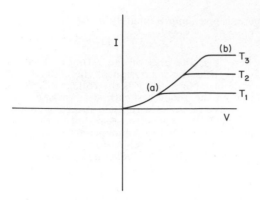

FIG. 2. I vs V characteristics of a vacuum diode for various cathode temperatures: (a) space-charge-limited region; (b) emission-limited region ($T_3 > T_2 > T_1$).

tween the plate and the cathode. The important property of the vacuum tube is that the source which controls the grid voltage supplies no power, since the grid current is zero, and yet it can significantly alter the power delivered by a source connected between the plate and cathode, and thus can be used as an amplifier.

The incremental change in plate current i_p can be calculated from the incremental change in grid-to-cathode voltage v_{GC} and plate-to-cathode voltage v_{PC} according to

$$i_p = g_m v_{GC} + v_{PC}/r_p$$

where g_m is called the *grid-plate transconductance* and r_p is called the *plate resistance*. The plate-to-cathode of a vacuum tube can thus be considered as a voltage source of value $-\mu v_{GC}$ (where $\mu = g_m r_p$) in series with a resistance r_p. The fact that the *amplification factor* μ is typically 100–1000 means that a vacuum tube can be used in a circuit to provide voltage amplification. Since the input resistance of a vacuum tube circuit usually exceeds 1 megohm whereas the plate resistance r_p is much smaller (kilohms), a vacuum tube amplifier provides a very large power gain.

Tetrodes In order to raise the amplification factor and reduce the grid-to-plate capacitance for high frequency operation, a second grid, called the *screen grid*, is often placed between the control grid and plate. This grid is usually held at a constant voltage, positive relative to the cathode, but negative relative to the plate. Such a tube is called a *tetrode* and has a plate current that is approximately independent of the plate-to-cathode voltage.

Pentodes One drawback of the tetrode is that, whenever the plate becomes negative relative to the screen, secondary electrons are knocked off the plate by the incident primary electrons. These secondary electrons are attracted to the screen. This causes an undesirably high screen current and can cause the plate current to reverse direction. To eliminate this effect, a third grid, called the *suppressor grid*, is inserted between the screen and plate. Such a tube is called a *pentode*. The suppressor grid is usually held at a constant voltage near that of the cathode and, in fact, is often connected internally to the cathode.

Cathode Ray Tubes When the electrons strike the plate of a vacuum tube, their kinetic energy is normally transferred to heat, resulting in a plate dissipation power limit that should not be exceeded without risking damage to the tube. However, it is also possible to have the electrons strike a fluorescent screen, in which case some of the electron energy is transformed to visible light. If the electrons are focused into a narrow beam by the use of apertures, and deflected in two dimensions by means of deflection plates or magnetic fields, a visible spot can be produced and moved rapidly across the face of the tube to produce a visual image. Such a tube is called a *cathode ray tube* (*CRT*) and is shown schematically in Fig. 4. Cathode ray tubes find application as visual display devices in instruments such as oscilloscopes, television receivers, and computer displays.

Klystrons Above a few hundred MHz, the plate voltage and plate current exhibit a significant phase difference generally due to electron transit time effects, and conventional tube designs become ineffective. At microwave frequencies, special tubes are used which employ velocity-modulation of an electron beam. An external magnetic field may be used to confine the electron beam. In the *klystron* the electron beam is modulated by traveling through a pair of closely spaced grids connected across a resonant cavity. The *reflex klystron* shown in Fig. 5 is used as an oscillator. It consists of a single

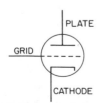

FIG. 3. Vacuum triode.

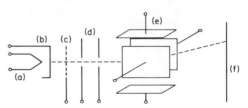

FIG. 4. Cathode ray tube: (a) filament; (b) cathode; (c) control grid; (d) focusing electrodes; (e) deflection plates; (f) fluorescent screen.

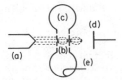

FIG. 5. Reflex klystron: (a) filament; (b) grid space showing bunched electrons; (c) resonant cavity; (d) reflector; (e) coupling loop.

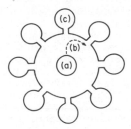

FIG. 7. Magnetron: (a) cathode; (b) path of electron in magnetic field; (c) resonant structure.

cavity followed by a drift space and an electrode known as a *reflector*. The streaming bunches of electrons are reflected back into the grid/cavity area in the proper phase to sustain oscillation. The *amplifier klystron* shown in Fig. 6 consists of many cavities in series, with the input applied to the first cavity and the output taken from the last. Power is applied or extracted from these cavities usually by means of a coupling loop.

Traveling Wave Tubes In a *traveling wave tube* (*TWT*) the electron beam is focused and confined by an external solenoidal magnetic field or by a periodic permanent magnet structure in a relatively long cylindrical cavity. The electron beam is bunched by means of an rf current applied to a helical coil which surrounds the beam. An electromagnetic wave is thus produced which travels down the tube at a velocity slower than the velocity of the electrons in the beam. The electrons are thus able to transfer some of their energy to the wave, resulting in an amplified rf current in the helix. Because no resonant cavities are employed, the TWT can be used as a broadband amplifier or tunable oscillator at frequencies above about 1000 MHz.

Magnetrons In the *magnetron*-type tube, electrons are emitted from a filament and move radially outward in two dimensions toward a coaxial anode as shown in Fig. 7. However, an external magnetic field along the axis of the cylinder causes the electrons to move in roughly circular orbits around the cathode. The anode consists of a periodic sequence of resonant cavities. The electrons are bunched by the electric field across the entrance to these cavities and thus give up energy to the cavity field at the resonant

frequency. This energy is extracted from the cavities by coupling loops or a waveguide. Magnetrons are used as fixed-frequency oscillators in high power cw service (such as microwave ovens) and short pulse radars.

Light-Sensitive Tubes Electron tubes can also be used as sensitive detectors of light and other electromagnetic radiation in the near infrared and ultraviolet. A common form of such a detector is the *photomultiplier* (*PM*) tube. As shown in Fig. 8, such a device is a vacuum tube consisting of a cold, photosensitive cathode that emits electrons when struck by light. The electrons are attracted to a nearby positive electrode (called a *dynode*), where they typically release 3–6 secondary electrons which are attracted to the next, even more positive dynode, and the whole process repeats through many stages, producing a large current at the anode. Such a device is so sensitive that it can detect a single photon of light, in which case the output consists of a negative voltage pulse with a size determined by the capacitance of the measuring instrument.

An alternative and more compact configuration is the *channeltron electron multiplier* (*CEM*) in which a long, thin, evacuated glass tube is coated on the inside with a low-work-function conducting material that takes the

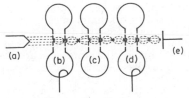

FIG. 6. Klystron amplifier: (a) filament; (b) input cavity; (c) intermediate cavity; (d) output cavity; (e) anode.

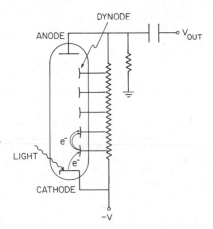

FIG. 8. Photomultiplier tube.

place of the individual dynodes in the PM tube. A high voltage is applied between the ends of the tube, and an avalanche of electrons is formed whenever light is incident on the more negative end of the coating. By combining many of the characteristics of the photomultiplier tube and the cathode ray tube, various types of image-sensitive tubes, such as the *image orthicon* and *vidicon* tube have been created. These tubes are used in television cameras.

J. C. SPROTT

References

Cobine, J. D., "Gaseous Conductors," New York, Dover Publications, Inc., 1958.

Langford-Smith (Ed.), "Radiotron Designers Handbook," 4th Ed., RCA, 1960.

Hutter, R. G. E., "Beam and Wave Electronics in Microwave Tubes," New York, Van Nostrand Reinhold, 1961.

Gewartowski, J. W., and Watson, H. A., "Principles of Electron Tubes," New York, Van Nostrand Reinhold, 1966.

Anderson, L. W., and Beeman, W. W., "Electric Circuits and Modern Electronics," New York, Holt, Rinehart, and Winston, 1973.

Brophy, J. J., "Basic Electronics for Scientists," New York, McGraw-Hill, 1977.

Sprott, J. C., "Introduction to Modern Electronics," New York, Wiley, 1981.

Cross-references: CIRCUITRY, DIODE (SEMICONDUCTOR), POTENTIAL, THERMIONIC EMISSION, TRANSISTOR.

ELECTRONICS

The field of electronics comprises the design, analysis, and application of electric circuits and the devices which control electric currents in such circuits. Since electric currents arise from the flow of electrons in conductors and the properties of control devices depend upon the motion of electrons in them, electric circuits, electronic devices, and their combinations depend upon the properties of the fundamental unit of electric charge, the electron. The age of electronics is reckoned from the discovery that the electron current in a vacuum triode can be controlled by electric signals applied to one of the triode's three terminals, the grid. Invention of the transistor in 1948 expanded the possibilities of electronic circuits to such an extent that semiconductor devices have to a large extent supplanted vacuum tubes, except for special applications. The expansion of semiconductor technology to integrated circuits now extends electronic techniques to nearly all aspects of contemporary society, ranging from communications and entertainment to information processing for business and military, and to measurement and control in manufacturing and science.

Understanding of electronic circuits implies knowledge of the currents in all parts of the circuit. An electric current accompanies the drift motion of free electrons in a conductor in response to an electric field, and the energy required to effect successive electron accelerations between collisions within the solid is called the electric potential difference, or *voltage* (in honor of the early Italian worker in electricity, Alessandro Volta). The *resistance* of a conductor to electron flow involves both material parameters and geometrical factors so that, for example, the resistance of a long thin wire is larger than that of a short thick wire of the same material. An electric circuit component having a specific value of resistance is called a *resistor*; according to *Ohm's Law* (after Georg Simon Ohm), the constant of proportionality between the current I, carried by a resistor, and the potential difference V is just the resistance:

$$V = RI.$$

This relation is a fundamental to all electronic circuit analysis and design.

The energy source that sustains steady currents in circuits is often a chemical battery. Calculating the currents in various branches of a network of resistors and batteries can be achieved by successive applications of Ohm's Law, but is greatly facilitated by *Kirchhoff's Rules* (after Gustav Kirchhoff). These are: (1) the algebraic sum of all currents at every junction in a network is equal to zero; and (2) the sum of voltages around any complete loop of a network is equal to zero. The first rule is essentially a statement of the conservation of electric charge, and the second is a consequence of the conservation of energy. It often proves possible to replace all or a portion of a network by an equivalent circuit in order to simplify the analysis. Thus, for example, the *Thévenin equivalent circuit* (named after M. L. Thévenin) which can represent the properties of an entire network is simply a single voltage source in series with a single resistor. Thus, very complicated systems can be analyzed and designed.

The currents in practical electronic circuits are time dependent, and the simplest time-varying current alternates direction sinusoidally. Since even the most complex current or voltage waveforms may be represented by a Fourier series consisting of harmonically related sine waves, ac circuit analysis is reduced to simple sinusoidal currents and voltages. Capacitors and inductors impede the varying currents in ac circuits basically because of the energy that must be stored and dissipated in the electric field between the electrodes of a capacitor and in the magnetic field associated with the current in an inductor. This effect is conveniently expressed in complex number notation by writing Ohm's Law describing, for example, an ac circuit consisting of a resistor, capacitor and inductor in series as follows:

$$V = [R + j(\omega L - 1/\omega C)]I$$

where $j = \sqrt{-1}$, ω is the angular frequency, and the current and voltage are sinusoidal. The quantity in brackets is called the *complex impedance* of the circuit, and its imaginary part is termed the *reactance*. If the frequency is such that the inductive reactance equals the capacitative reactance ($\omega L = 1/\omega C$) the circuit is said to be *in resonance*; the current is a maximum and is in phase with the voltage. An important and familiar application of resonance is to select different channels in radio and TV receivers by adjusting the circuit to resonance at the given channel frequency with a variable capacitor.

The key elements in all electronic circuits are the active devices which can control strong currents between two terminals of the device in response to weak electric signals applied to a third, thus amplifying input signals almost without bound. Semiconductor devices, most notably the *bipolar transistor* and the *field effect transistor*, are the principal active components in modern electronic systems, but the great versatility of semiconductor materials, principally silicon, has spawned an amazing variety of other semiconductor devices. Of particular significance is the field effect transistor design which is fabricated with a metal control electrode deposited upon the insulating oxide layer of a small silicon crystal carrying the two output terminals; the descriptive terminology of this metal-oxide-silicon structure is MOSFET. Voltage signals applied to the metal electrode change the conductivity of the underlying silicon and thus the current between the output terminals, leading to amplification of the input signal. Ingenious fabrication techniques can produce such structures having dimensions of the order of 10 micrometers, so that many hundreds or thousands of transistors can be placed on the same single crystal. Suitable metallic interconnections between devices yield an entire electronic circuit completely contained within one solid wafer of silicon. Extensions of this integrated circuit technology lead to VLSI, very large-scale integration, in which networks of several hundred thousand transistors are included in a silicon chip a few millimeters square and less than a millimeter thick. Thus, an extremely complex electronic circuit can be produced in a practical size, and, because of the extreme miniaturization, require minimal electric power. Integrated circuit designs based upon complementary symmetry, that is, which use transistors that employ electronic conduction in combination with those that employ hole conduction, so-called CMOS technology, are particularly appropriate because of very low power consumption. On the other hand, bipolar integrated circuits tend to have high operating speeds, which is important in extensive systems.

Electronic circuits, be they composed of discrete transistors or integrated circuits, operate in two distinct modes. In one case, the output signal is a linear function of the input signal and covers the range from zero to a saturation value determined by device or circuit parameters. Digital circuits, on the other hand, operate such that the output is either zero or some finite voltage and no intermediate values are important. These two techniques are complimentary, but, although linear circuits take precedence historically, the world of electronics is becoming ever more digital. Turning first to linear amplifiers, the gain a of the two types of single transistor amplifier is given by

$$a = \frac{V_o}{V_i} = \begin{cases} -G_m R_L & \text{(FET)} \\ -\dfrac{h_{fe}}{h_{ie}} R_L & \text{(Bipolar)} \end{cases}$$

where V_o is the output signal, V_i is the input signal, G_m, h_{fe}, h_{ie} are device parameters, and R_L is a resistor carrying the output current. Clearly, the output is a linear function of the input; the minus sign indicates that the output signal polarity is inverted with respect to the input. Gains exceeding 100 are easily possible in a single stage and greater values are attained simply by cascading several stages. Practical amplifiers can be designed to amplify minute dc signals and others, particularly those employing resonant circuits, to amplify high frequency signals up to 50 gigahertz.

The performance of amplifiers is enhanced in several respects by returning a fraction βV_o, of the output voltage to the input, a process called *feedback*. The output signal of a feedback amplifier is then $V_o = a(V_i + \beta V_o)$, or

$$V_o = \frac{aV_i}{1 - a\beta} \cong \frac{1}{\beta} V_i \quad \text{if} \quad -a\beta \gg 1.$$

The inequality means that the polarity of the feedback signal is negative. It does not prove difficult to achieve sufficient gain for the inequality to hold, so that, as this expression shows, the amplifier gain depends only upon the feedback network and not at all upon transistor parameters. This achieves greater stability and linearity. If, on the other hand, β is some nonlinear factor, the output signal can be a complicated function of the input. In this case, the feedback amplifier may be looked upon as operating upon the input to produce a modified output signal. Such operational feedback amplifiers were first used in analog computers to add, multiply, integrate or differentiate input signals. Opamps have subsequently found broad application and in integrated circuit form are widely used as complete amplifiers in preference to systems built of discrete transistors.

Positive feedback in an amplifier, together with the condition

$$a\beta = 1$$

suggests an infinite output signal, according to the above expression, which in practice means that an output is obtained in the absence of an input signal, that is, the circuit oscillates. A resonant circuit feedback network assures that the equality is satisfied only at the resonant

and accuracy of digital circuits were first realized in digital computers but now find application in all fields of electronics. Digital circuits tend to be much more complex than linear circuits, however, so it remained for the advent of integrated circuit technology to make digital electronics viable.

Notation and logic in the binary number system are analogous to the more familiar decimal system. The binary number 10110, for example, stands for increasing powers of two, or

$$10110 \; : \; 1 \times (2^4) + 0 \times (2^3) + 1 \times (2^2) + 1 \times (2^1) + 0 \times (2^0)$$

$$: \quad 16 \quad + \quad 0 \quad + \quad 4 \quad + \quad 2 \quad + \quad 0$$

$$: \quad 22$$

frequency, and such oscillators are secondary time standards as well as generators of high frequency signals for radio and TV broadcasting, microwave transmission, satellite communications, etc. In this connection, the high frequency carrier signal radiated from the sending to the receiving station is modulated with the audio, video, or data signal by varying the carrier amplitude, frequency, or amplitude and frequency in combination. At the receiving end, the signal is amplified and demodulated to recover the initial signal. Widespread radio and television communication networks represented the first major application of electronics and have subsequently been joined by many diverse control and processing applications. In fact, the versatility of electronics is so great that it is common practice to develop electric signals analogous to phenomena of interest so that all processing may be carried out electronically.

Very strong positive feedback may cause an amplifier to oscillate between zero output and a finite saturation value, becoming, in effect, a digital oscillator. One form of such a multivibrator circuit can be caused to remain in one or the other state until triggered into the opposite state by an appropriate external signal. That is, the circuit remembers. Very large arrays of such binary circuits make up the memory devices needed in digital computers. The integrated circuit form of a semiconductor random-access memory, or RAM, can store as many as 262,144 bits (see below) of information in a single silicon chip.

In digital electronics, voltage signals are used to represent quantities in the binary number system. Since the binary number system has only two digits, 0 and 1, these can be represented by signal waveforms which have only two values, zero and some finite saturation value set by circuit or transistor parameters. Digital circuits are, therefore, inherently more reliable than linear circuits, which must handle a continuous range of signals. Furthermore, digital signals can be amplified indefinitely and stored accurately. The advantages of reliability

That is, the binary number 10110 represents the same quantity as the decimal number 22. Voltage pulses in digital electronics represent digital numbers in two ways. The series representation has regularly spaced voltage pulses corresponding to increasing powers of 2 appearing in time sequence beginning with 2^0. In the alternative method, parallel representation, voltage pulses for the increasing powers of 2^0 appear simultaneously on separate transmission paths. In either case, each digit, be it a 0 or a 1, is termed a *bit*, a contraction for *binary digit*. Parallel representation is extensively used in digital circuits because all bits of a number are available in a time equal to the voltage pulse for one bit. Conversely, a time equal to the number of bits in the number is needed for serial representation, but, since only one signal path is required, this method is conventionally used for long-distance transmission. Digital transistor circuits are considered to operate upon the pulse waveforms representing digital numbers according to the logic of Boolean algebra. Thus, for example, a single transistor amplifier with two inputs, A and B, performs the logical NOT-AND, or NAND, operation, since only when both A and B inputs are present is the output equal to zero (note the signal inversion in a single-stage amplifier). A NOT-OR, or NOR, logic gate is equally possible, and it is interesting to note that, according to Boolean algebra, all possible logic operations (among the simpler are addition and multiplication) can be achieved by suitable arrays of NAND gates alone, or NOR gates alone. This is important for design and maintenance of extensive logic arrays.

The synergism of digital electronics and integrated circuits reaches its ultimate in sophistication in the microprocessor, which is essentially an entire digital computer contained in a single silicon chip. The microprocessor combines small size and low cost with sophisticated logic processing that permits a seemingly unending parade of important and ingenious applications. A digital computer is a complex array of logic gates organized into five main parts: input, output, memory, arithmetic, and control units. The digital circuits are designed to carry out logic calculations of all kinds and, therefore, the computer is furnished a set of specific instructions, or a *program*, pertaining to any desired calculation or process. The input and output units present digital numbers and the digital words of the program to the computer and subsequently retrieve results. The memory stores each data number and program word at a specific address in memory until it is needed during execution of the program. The arithmetic unit contains the logic gates that perform the logic operations called for by the program, and the function of the control unit is to interpret each instruction and set the circuits of the computer accordingly. The control unit also regulates the basic speed at which the computer operates. Because the functions of the arithmetic unit and the control unit are central to performance of a microprocessor, they are often referred to as the *central processing unit*, or CPU, and fabricated together on one silicon chip. In this case, peripheral chips for memory and input/output operations must be associated with the microprocessor CPU chip. This permits flexibility in design, but complete 5-unit, single-chip microcomputers are also widely used.

After the program and data are placed in memory and the microprocessor started, the control unit reads the first instruction, prepares the circuits accordingly and causes the appropriate number to be read from memory as specified by the address in the instruction. After completion of the indicated operation, the result is returned to memory and the control unit passes on to the next instruction. The CPU proceeds sequentially through the program until the last instruction is reached and the end result transferred to the output unit. It proves possible to write programs in such a way that at a certain instruction the CPU can jump to one or the other of two different program continuations, depending upon the value of an intermediate result. In this fashion, the microprocessor adapts to its activity in a way that cannot be predicted by the human programmer. Note also that by simply changing the program, a microprocessor can undertake any task, which is an extremely powerful feature. On the other hand, often a dedicated microprocessor has a specific program for its one task embodied in a read-only-memory, or ROM, which is fixed during fabrication of the integrated circuit. In either case, the digital words of the program are various arrays of the digits 0 and 1 and this makes it extremely difficult to prepare a program in machine language. Many ingenious schemes have been devised which permit programming in stylized algebra or rudimentary English (or other human language, of course); in this case, a special, previously prepared program translates the high-level language program into machine language.

The power and flexibility of microprocessors is engendering an electronic revolution equivalent to that in communications permitted by linear circuits decades ago. In manufacturing and process control, the logic adapts to the job at hand and may respond by driving a robot to a given task. The business office of the present, and future, contains word processors that correct spelling, type at hundreds of words per minute, and edit text effortlessly; data storage and retrieval systems to describe the instantaneous state of the business; and electronic mail that eliminates memos and filing. Personal computers balance bank accounts, will soon permit electronic banking, and also play intelligent games. Smart home appliances talk and listen, and automobiles are made more fuel efficient and less polluting through adaptive microprocessor control. Even communication is revolutionized as computers communicate and digital techniques replace linear circuits. This extends to phonograph and TV records as well. Truly it may be said that electronics supplements human intellectual activities not unlike the way the steam engine supplemented human muscle.

There is also a branch of electronics, more properly termed *physical electronics*, which studies physical phenomena of electrons per se. Very often this focuses on the motion and control of electrons in a vacuum and from such studies have developed the familiar cathode ray tube used in TVs and computer terminals; the electron microscope and its several derivatives which provide enormous visual magnifications; photomultiplier tubes and image intensifiers which detect and display nuclear particles, star images, and TV shows. Another aspect treats features of electrons in solids, particularly semiconductors, transistors, and the host of other useful semiconductor devices such as solar batteries, photocells, and light emitting diodes. Even the large nuclear particle accelerators, kilometers in circumference, can be considered part of this branch of science.

JAMES J. BROPHY

References

Brophy, J. J., "Basic Electronics for Scientists," 4th Ed., New York, McGraw-Hill Book Co., 1982.

Horowitz, P., and Winfield, H., "The Art of Electronics," New York, Cambridge Univ. Press, 1980.

Abelson, P. H., and Dorfman, M. (Eds.), "Computers and Electronics," *Science* 215 (Feb. 1982).

Cross-references: ANTENNAS, CAPACITANCE, CIRCUITRY, DIELECTRIC THEORY, DIODE (SEMICONDUCTOR), ELECTRICITY, ELECTROMAGNETIC THEORY, INDUCTANCE, MICROWAVE TRANSMISSION, OSCILLOSCOPES, RECTIFIERS, SEMICONDUCTOR DEVICES, SEMICONDUCTORS, SERVOMECHANISMS, THERMIONICS, TRANSISTOR.

ELECTROWEAK THEORY*

By electroweak theory is meant the unified field theory that describes both weak and electromagnetic interactions. The development of a unified electroweak theory is certainly the most dramatic achievement in theoretical physics to occur in the second half of this century. It puts weak interactions on the same sound theoretical footing as quantum electrodynamics. Many theorists have contributed to this development, which culminated in the works of Glashow, Weinberg and Salam,[1] who were jointly awarded the 1979 Nobel Prize in physics. Some of the important ideas that contributed to this development are the theory of beta decay formulated by Fermi and parity violation suggested by Lee and Yang and incorporated into the immensely successful V-A theory of weak interactions by Sudarshan and Marshak. At the same time ideas of gauge invariance were applied[2] to weak interaction by Schwinger, Bludman, and Glashow. Weinberg and Salam then went one step further and wrote a theory that is renormalizable, i.e., all higher order corrections are finite, no mean feat for a quantum field theory! The theory had to await the development of the quark model of hadrons for its completion.

To understand the basic content of the theory, a simple beta decay of neutron into proton, electron and antineutrino can be expressed by the reaction $n \rightarrow p + e^- + \bar{\nu}$. The theory is much better behaved if this reaction is divided into two reactions

$$n \rightarrow p + W^-, \qquad W^- \rightarrow e^- + \bar{\nu}. \qquad (1)$$

The W particle is called an *intermediate W boson* because it is an intermediary in this weak process. The W particle is very heavy, and reactions in Eq. (1) occur only "virtually." The strength of the W interaction (or the coupling constant) is comparable to the electromagnetic interactions, but the large mass of the W makes the effective strength of the interaction both short ranged and weak. (The effective strength is approximately (coupling constant)^2/mass of $W)^2$). We now know that neutron and proton are respectively made of (udd) and (uud) quarks, where the u quark has charge $\frac{2}{3} e$ and the d quark has charge $-\frac{1}{3} e$. The transitions in Eq. (1) then arise at the quark level from the process

$$d \rightarrow u + W^-, \qquad W^- \rightarrow e^- + \nu. \qquad (2)$$

Reactions in Eq. (2) can be viewed as transitions between two doublets (u, d) and (ν, e^-). These doublets are referred to as *weak isospin doublets*. The crucial idea on how to construct a

*This work is supported in part by a grant from the Department of Energy under contract No. DE-AT06-76ER70004.

gauge theory for objects that are doublets (or higher multiplets of some group) was provided by Yang and Mills.[3] It is a generalization of the field concept that underlies quantum electrodynamics. In the latter, particles have scalar charges and do not transform to new states in an interaction; the carrier of the force, the electromagnetic field, is itself neutral. In this generalization the carriers of the force (like $W^\pm$) themselves can carry this more complicated charge and can have self-interactions. The field carriers would be massless like photons in the symmetry limit, so a new mechanism to make some of them massive had to be discovered. We begin the discussion of the theory in the so-called *symmetry limit*, a limit that could be realized at very high energies (energy larger than 100 GeV). In this limit the gauge group is $SU(2) \times U(1)$. The group $SU(2)$ is similar to the isospin group and the left-handed quarks and leptons form doublets under this group. $U(1)$ is like electromagnetic interactions, and the charge associated with this group is called *hypercharge* (Y). There is a fundamental relation between the charge Q, the third component of weak isospin I_3, and hypercharge Y:

$$Q = I_3 + Y \qquad (3)$$

The assignment of quantum numbers for quarks and leptons are:

Left-handed doublets $\begin{cases} u & \nu_e & I_3 = \frac{1}{2} \\ d & e & I_3 = -\frac{1}{2} \end{cases}$

Hypercharge $\qquad Y = \frac{1}{6} - \frac{1}{2}$

$$(4)$$

There are interaction strengths g and g' associated with the groups $SU(2)$ and $U(1)$ respectively. If a and b are members of any isodoublet in Eq. (4), the basic processes that occur with strength g are

$$a \rightleftarrows b + W^+$$
$$b \rightleftarrows a + W^-$$
$$a \rightleftarrows a + W^0 \qquad (5)$$
$$b \rightleftarrows b + W^0$$

and those with strength g' are

$$a \rightleftarrows a + B^0$$
$$b \rightleftarrows b + B^0 \qquad (6)$$

where $W^\pm$, W^0 form a triplet and are carriers of the weak $SU(2)$ force, while B^0 is the carrier of the $U(1)$ force. Former process occur with the strength g, and the latter with g' times the hypercharge.

This picture needs modification at low energies. The mass of the W should show up, and the photon should remain massless and interact with matter in the usual manner. How is this accomplished? This is where Weinberg and Salam introduced the idea of *spontaneous breaking of symmetry*. It is caused by a new iso-doublet—The *Higgs field*. It has the following remarkable property: its minimum of energy occurs not when the field strength is zero, but when it takes some finite value. This in turn implies a preferred direction in weak isospace, and this breaks symmetry. The effect of the destruction of symmetry is as follows:

(1) The hypercharge field B and the third component of the W triplet, W^0, mix to form new states

$$A = \cos \theta_W\, B - \sin \theta_W\, W^0$$

$$Z = \sin \theta_W\, B + \cos \theta_W\, W^0. \qquad (7)$$

Here A is the usual photon, and Z a new massive state.

(2) The Higgs fields interact with $W^\pm$ and Z, making them massive. A prediction of the theory is that

$$M_Z{}^2 = M_W{}^2 \sec^2 \theta_W \qquad (8)$$

where M_Z and M_W are masses of the Z and W boson, respectively.

(3) The coupling of the field A to quarks and leptons is identical to the coupling in quantum electrodynamics.

(4) The bosons $W^\pm$ carry charge, and their interactions lead to V-A theory of beta decay. (The name V-A derives from *Vector-Axial vector* structure of currents in this theory.)

(5) The gauge boson Z leads to new interactions that involve "neutral currents" (i.e., processes like $u \to u + Z$, with $Z \to \nu + \bar{\nu}$). This is a new, completely unexpected prediction. The test of this prediction had to await the experimental development of high energy neutrino beams. This prediction has now been verified and found to yield agreement with the experiment to the level of 5–10%. The Z boson interactions depend on the value of the mixing angle θ_W. Several different experiments lead to the value $\sin^2 \theta_W = 0.23 \pm 0.02$.

So much for the description of electroweak theory. The most important experimental verification, the observations of $W^\pm$ and Z bosons has been announced. Their masses were predicted to be $M_W \approx 80$ GeV and $M_Z \approx 90$ GeV. In the summer of 1983 the discovery of both W^+ and Z bosons was reported by C. Rubia using the proton-antiproton collider at CERN. The masses are in excellent agreement with the theoretical prediction. The observation of the Higgs particle is much more difficult to carry out experimentally, but is nonetheless an important test of the theory.

Where do new quarks and leptons fit into the scheme? They have to enter as doublets. Two more lepton doublets have been identified (μ^-, ν_μ) and (τ^-, ν_τ), while two quark doublets are also expected to occur, (c, s) and (t, b).

Except for the t quark, the others have been seen. No one has the slightest idea of why these additional doublets occur in nature. This problem is referred to as the *Generation Puzzle*. A refinement of the theory suggests that the actual doublets involve d', s' and b' which are mixtures of d, s, and b. This mixing, called *Cabibbo mixing*, permits decays that alter strangeness like $s \to u + W^-$ to occur with a small probability.

Is the theory complete? Most researchers suspect that strong, weak and electromagnetic forces should ultimately be united into one unified theory (called *grand unified theory*). Many attempts to do so have been made, leading to new predictions which would signal this kind of theoretical structure. Some theories predict the decay of the proton[4] although with a large but measurable lifetime, and others predict additional Z bosons.[5] The next few years should prove an exciting period in physics where many of these new ideas will be tested.

N. G. Deshpande

References

1. Glashow, S. L., *Nucl. Phys.* **22**, 579 (1961); Weinberg, S., *Phys. Rev. Letts.* **19**, 1264 (1967); and Salam, A., in "Elementary Particle Theory" (N. Svartholm, Ed.), Stockholm, Almquist and Wikscll, 1968.
2. Schwinger, J., *Ann Phys.* (*N.Y.*) **2**, 407 (1957); Bludman, S., *Nuovo Cimento* **9**, 433 (1958); Salam and Ward, *Phys. Lett.* **13**, 168 (1964).
3. Yang, C. N., and Mills, R. L., *Phys. Rev.* **96**, 191 (1954).
4. Pati, J. C., and Salam, A., *Phys. Rev.* **D10**, 275 (1974); George, H., and Glashow, S. L., *Phys. Rev. Lett.* **32**, 438 (1974).
5. Deshpande, N. G., and Iskandar, D., *Nucl. Phys.* **B167**, 223 (1980).

Cross-references: CONSERVATION LAWS AND SYMMETRY, CURRENT ALGEBRA, ELEMENTARY PARTICLES, GAUGE THEORY, GRAND UNIFIED THEORIES, QUANTUM CHROMODYNAMICS, QUANTUM ELECTRODYNAMICS, QUARKS, STRONG INTERACTIONS, WEAK INTERACTIONS.

ELEMENTARY PARTICLES

The search for the elementary constituents of matter is as old as physics itself, but any quantitative attempt at such a theory had to await the experimental discoveries of this century. Such a search is prompted by two considerations: the identification of the basic building blocks of nature and the hope that their laws of interaction would be essentially simple.

The atom had to yield its claim to indivisibility when it was found that electrons were constituents of all atoms; moreover, the electrons from various species of atoms were identical. Light, with its particle properties, seemed another universal entity connected with matter, since the photons (light quanta) which were emitted in atomic transitions appeared identical apart from their momenta. The *electron* and the *photon* were the first two elementary particles to be discovered, and a quantitative theory of the emission and absorption of photons by the electrons in an atom was possible only after the invention of quantum mechanics. The corresponding picture of the atom regarded the electrons in an atom as being subject to electrostatic attraction of the positively charged nucleus (and the mutual repulsion of other electrons), the photons having only a transistory existence being either emitted or absorbed in the transitions between the atomic states. The search for the structure of matter now became a search for the constituents of the nucleus.

A quantum theory of the nucleus (or rather nuclei) was made possible by the discovery of the *proton* and the *neutron*. The nuclear interaction which was responsible for holding the nucleus together (against the disruptive electrostatic repulsion of the protons) was found to be of an entirely new kind, much stronger than the electric interaction at short distances but decreasing very much more rapidly with distance. The various complex nuclei differ in the number of protons and neutrons they contain.

By that time, the theory of the interaction between electrons and photons had developed to the point where the electrostatic repulsion or attraction between electrically charged particles could be understood in terms of the exchange of photons between them. In the lowest nontrivial approximation, it gave the Coulomb law for small velocities. The basic interaction was the emission and absorption of "virtual" photons by charged particles. A similar mechanism could be invoked to explain the short-range nuclear interaction, and to picture the nuclear interaction as due to the exchange of particles which have nonzero masses which are a fraction of nuclear mass. (The approximation procedure used for deducing the static Coulomb force from the electron-photon interaction is no longer valid here; and the nuclear force has a rather complicated form. However, these theoretical considerations did predict the existence of a set of three particles called *pions*, which have since been discovered.)

Another kind of particle and another kind of interaction were discovered from a detailed study of beta radioactivity in which electrons with a continuous spectrum of energies are emitted by an unstable nucleus. The corresponding interactions could be viewed as due to the virtual transmutation of a neutron into a proton, an electron, and a new neutral particle of vanishing mass called the *neutrino*. The theory provided such a successful systematization of beta decay data for several nuclei that the existence of the neutrino was "well established" more than twenty years before its experimental discovery. The beta decay interaction was very weak even compared to the electron-photon interaction.

Meanwhile, the electron was found to have a positively charged counterpart called the *positron;* the electron and positron could annihilate each other, with the emission of light quanta. The theory of the electron did in fact "predict" the existence of such a particle. It has, since then, been found that the existence of such "opposite" particles (*antiparticles*) is a much more general phenomenon (see below and also see ANTIPARTICLES).

Our present catalog of elementary particles and decay modes contains many more entries. These particles fall into families: the *photon family*, the *lepton family*, and *meson family*, and the *baryon family*, as well as a number of postulated particles like *quarks* and *gluons*. Most of these particles are unstable and decay within a time which is often very small by normal standards but which is many orders of magnitude larger than the time required for any of these particles to traverse a typical nuclear dimension. There is a wide variety of reactions between them, but they could be understood in terms of three basic interactions— the *strong* (or *nuclear*) *interactions*, the *electromagnetic interactions*, and the *weak interactions*. The nuclear forces and the interaction between pions and nucleons belong to the first type; the electron-electron and electron-photon interactions to the second; and the beta decay interaction to the third. The present theoretical framework enables us to handle more or less quantitatively the electromagnetic and weak interactions and certain aspects of all interactions. Despite this, it is possible to understand many aspects of strong (as well as the other) interactions in terms of conservation laws and invariance principles (see CONSERVATION LAWS AND SYMMETRY).

The classical conservation laws of energy, momentum, and angular momentum are valid in the relativistic quantum theory of elementary particles also. The particles may possess intrinsic angular momentum or *spin*, which, expressed in natural units $h/2\pi$ of angular momentum, is restricted to an integer or a half-odd integer. Angular momentum conservation holds only when this spin angular momentum is included. But one finds that to every particle there corresponds an antiparticle with the same mass, same spin, and same lifetime. (In the case of the photon and the neutral pion, they are their own antiparticles; they are strictly neutral particles.) The particle and antiparticle have equal and opposite electric charges, and the antiparticle of the antiparticle is the original particle. *Conservation of electric charge* is another familiar (although nonclassical) conservation law satisfied by all known interactions.

In addition to additive conservation laws which arise from continuous symmetries, there is a set of "multiplicative" conservation laws which are associated with discrete symmetries. It is possible to examine the invariance of the physical laws under space inversion, i.e., using a left-handed coordinate system or vice versa; if the statement of the law is unaffected by this interchange, it is possible to show that a quantum number having the two values ±1 can be assigned to classify the quantum-mechanical states so that a state with the label +1 will not change to one with the label -1 due to any interaction. This quantum number is called *parity*. Just as particles may possess intrinsic angular momentum (spin), particles may also have intrinsic parity. Table 1 lists the particles (and their corresponding antiparticles) with their respective additive quantum numbers, intrinsic parities, and lifetimes. General principles of relativistic quantum theory imply that antiparticles of integral spin particles have the same parity as the particles; for half-odd-integral spin particles the antiparticle has the parity opposite that of the particle. All experimental checks are in accordance with this prediction.

In addition to invariance under space inversion, we may consider *particle conjugation* (replacement of particles by antiparticles) *invariance* and *time reversal invariance*, or combinations of these transformations. It turns out that strong and electromagnetic interactions are invariant under each of these three transformations (and hence any product of these), but weak interactions are invariant only under combined inversion (product of particle conjugation and space inversion) and under time reversal. It can be shown that all interactions are invariant under the product of the three transformations of space inversion, particle conjugation, and time reversal if some very general principles of the relativistic quantum theory of these particles are valid.

Even the statement that weak interactions are invariant under combined inversion has turned out not to be strictly true. In the decay of the neutral K meson we should have expected a short-lived particle (even under combined inversion) called K_1^0 and a longer-lived particle (odd under combined inversion) called K_2^0, provided combined inversion were strictly valid. We do observe such short- and long-lived components, but we also expect that the long-lived component K_2^0 cannot decay into two pions. Yet experimentally we find a small amount of decay into two pions. This violation of combined inversion (and, hence, of time reversal invariance) is only two-tenths of a percent, but it is definitely present. Thus, none of the discrete symmetries (except the product of the three) seems to be strictly valid.

One notices that the various particles belonging to a family have the same spin and the same values of the additive quantum numbers except the electric charge. The *photon* has a universal interaction with all charged particles; it has been found possible to connect the conservation of electric charge and this universal interaction structure, on the one hand, to the vanishing mass and unit spin of the photon, on the other. The electron and muon partake of both electromagnetic and weak interactions, but do not exhibit any strong interaction. In fact the muon family appears to be simply a duplicate of the electron family except for a change in the unit of mass. These light particles are collectively known as *leptons*. This fascinating puzzle has now been made more intriguing by the discovery of the τ-lepton with possibly its own neutrino. Why the repetition of this pattern, this recurrence?

For a long time people had assumed that "the" *neutrino* had zero mass like the photon. In weak interaction physics since the left-handed and right-handed fermions behave differently (see V-A chiral interactions, below) it is not necessary that the mass be zero. Contemporary thinking puts neutrino masses not necessarily zero, not even diagonal in the species: so that we could have neutrino oscillations in which one species slowly transmutes itself into another in a reversible manner very much like the oscillations of a pair of weakly coupled pendulums. No definite conclusions can be stated about neutrino masses at the present time except for certain numerical limits.

The *meson family* consists of eight members which fall into a triplet of *pions*, a singlet *eta*, a doublet of *kaons*, and a doublet of *antikaons*. They are all pseudoscalar (spin zero and odd parity) and exhibit strong interactions. The charged particles are of course coupled to the photon, but even the neutral members can participate in electromagnetic interaction by virtue of the large probability of virtual dissociation into charged particles. They participate in a variety of weak interactions including the nuclear beta decay interaction.

It is found that the kaons, the *hyperons* (baryons other than the neutron and proton), and their antiparticles, collectively known as *strange particles*, can decay by weak interactions not involving leptons or photons with a lifetime which is large compared to the natural periods appropriate to strong interactions. On the other hand, these particles are produced copiously in high-energy nuclear collisions. These two circumstances can be understood in terms of the existence of another additive quantum number, called *hypercharge*, which is conserved in strong and electromagnetic interactions but violated in weak interactions.

The meson-baryon system exhibits further regularities as far as strong interactions are concerned. The neutron and the proton have

TABLE 1. CATALOG OF ELEMENTARY PARTICLES

Particle	Family[a]	Spin	Mass (MeV)	Lifetime[b] (sec)	Antiparticle	Parity[c]	Charge	Hyper-charge	Baryon Number	Lepton Number	Charm
Photon, γ	Photon	1	0	Stable	Photon, γ	−	0	0	0	0	0
Electron neutrino, ν_e	Lepton	$\frac{1}{2}$	0	Stable	Antielectron neutrino, $\bar{\nu}_e$	Undefined	0	0	0	1	0
Electron, e^-		$\frac{1}{2}$	0.51100	Stable	Positron, e^+	+	−1	0	0	1	0
Muon neutrino, ν_μ		$\frac{1}{2}$	0	Stable	Antimuon neutrino, $\bar{\nu}_\mu$	Undefined	0	0	0	1	0
Muon, μ^-		$\frac{1}{2}$	105.659	2.20×10^{-6}	Positive muon, μ^+	+	−1	0	0	1	0
Tau, τ^-		$\frac{1}{2}$	1784	$<2.3 \times 10^{-12}$	Positive tau, τ^+	+	−1	0	0	1	0
Neutral pion, π^0	Meson	0	134.96	0.83×10^{-16}	Neutral pion, $\bar{\pi}^0$	−	0	0	0	0	0
Positive pion, π^+		0	135.56	2.60×10^{-8}	Negative pion, π^-	−	+1	0	0	0	0
Eta, η		0	549	7×10^{-19}	Eta	−	0	0	0	0	0
Neutral kaon, K^0		0	497.7	0.86×10^{-10}	Neutral antikaon, $\bar{K}^0$	−	0	+1	0	0	0
Positive kaon, K^+		0	493.67	1.24×10^{-8}	Negative antikaon, $\bar{K}^-$	−	+1	+1	0	0	0
Charmed D^+		0	1868	2.5×10^{-13}	Charmed D^-	−	+1	0	0	0	+1
Charmed D^0		0	1863	3.5×10^{-13}	Charmed $\bar{D}^0$	−	0	0	0	0	+1
Proton, p	Bayron	$\frac{1}{2}$	938.28	Stable	Antiproton, $\bar{p}$	+	+1	+1	1	0	0
Neutron, n		$\frac{1}{2}$	939.57	$0.93 \times 10^{+3}$	Antineutron, $\bar{n}$	+	0	+1	1	0	0
Lambda, Λ		$\frac{1}{2}$	1115.6	2.6×10^{-10}	Antilambda, $\bar{\Lambda}$	+	0	0	1	0	0
Positive sigma, Σ^+		$\frac{1}{2}$	1189.4	0.8×10^{-10}	Negative antisigma, $\bar{\Sigma}^-$	+	+1	0	1	0	0
Neutral sigma, Σ^0		$\frac{1}{2}$	1192.5	5.8×10^{-20}	Neutral antisigma, $\bar{\Sigma}^0$	+	0	0	1	0	0
Negative sigma, Σ^-		$\frac{1}{2}$	1197.3	1.48×10^{-10}	Positive antisigma, $\bar{\Sigma}^+$	+	−1	0	1	0	0
Neutral xi, Ξ^0		$\frac{1}{2}$	1315	2.9×10^{-10}	Neutral antixi, $\bar{\Xi}^0$	+	0	−1	1	0	0
Negative xi, Ξ^-		$\frac{1}{2}$	1321	1.6×10^{-10}	Positive antixi, $\bar{\Xi}^+$	+	−1	−1	1	0	0
Omega, Ω^-		$\frac{1}{2}$	1672	0.8×10^{-10}	Antiomega, $\bar{\Omega}^+$	+	−1	−2	1	0	0
Charmed Λ^c		$\frac{1}{2}$	2273	7×10^{-13}	Charmed $\bar{\Lambda}^c$	+	+1	0	1	0	+1

[a]Electron, muon, and tau families are collectively known as the *lepton family*. The proton and neutron are both nucleons; other members of the baryon family are the hyperons.

[b]The neutral kaon has a long-lived component K_2^0 and a short-lived component, which are quantum mechanical superpositions of the neutral kaon and the neutral antikaon.

[c]Electron, muon, tau, proton, neutron, and lambda parities are defined by convention. Antifermions have opposite parity from fermions, Antibosons have same parity as bosons.

very nearly the same mass and similar nuclear interactions, although their electromagnetic properties are quite different. The three pions have different electric charges, but again they have approximately equal masses and similar nuclear interactions. This kind of multiplet structure is evident for other strongly interacting particles: the *kaons* form a doublet, the *sigma hyperons* form a triplet, the *xi hyperons* form a doublet, and the *lambda hyperon* remains a singlet. In view of the relative weakness of the electromagnetic interaction, it is tempting to ascribe all deviations from exact equality of the masses to the indirect action of the electromagnetic interaction. In this framework, it is possible to consider the members of a multiplet to be different states of the same particle corresponding to the values of a new quantum number. What is remarkable is that if one takes this point of view, it is possible to show that the strong interactions exhibit a remarkable invariance under a group of continuous transformations which may be viewed as the group of rotations in a fictitious three-dimensional space (or more correctly as the special unitary group $SU(2)$ of transformations on two variables). The transformations act as follows: the singlet is unchanged, the doublet components transform like the components of a spinor, and the triplet components transform like the components of a vector. This property of strong interactions is called *charge independence*, and the corresponding conserved dynamical variable (with three components) is called the *isotopic spin*. It then turns out that hypercharge conservation is a consequence of isotopic spin conservation and electric charge conservation. While the conservation of isotopic spin is violated by the electromagnetic (and weak) interactions, the charge independence of nuclear interactions is still expected to be satisfied to within a few per cent and experimental tests confirm this. Since the symmetry associated with invariance under isospin transformations is not directly related to space-time properties, one often refers to it as an *internal symmetry*.

One might now raise the question: Which of these particles are basic constituents of matter? For the case of the atom, say the simplest of them all, the hydrogen atom, it seems easy to say that it is a composite system made up of an electron and proton bound together by an electrostatic force. However, this answer is not completely satisfactory, since the electrostatic force itself is due to the exchange of light quanta, and in the process of atomic transitions photons are emitted or absorbed. Yet we do not include them as constituents of the atom. In beta radioactivity, electrons and neutrinos emerge from the nucleus, yet the nucleus is not pictured as containing either of these varieties of particles but rather as made up of protons and neutrons. The beta electron and neutrino are rather assumed to be created at the moment of emission. With the mesons taking part in strong interactions, however, such distinctions are no longer obvious, and the question of whether a particle is elementary or is composed of several other particles cannot be answered except perhaps within the context of a more quantitative but limited model. A point of view that has gained some acceptance is that *none* of these particles are elementary and that each is a composite of several particles.

This view, while by no means inevitable or even well-established, is a possible picture, because in the realm of elementary particles we can not only add particles together to construct a composite system, we can also "subtract" particles by adding antiparticles. The claim that particles A and B go to make up the particle C is different to distinguish from the claim that particles $\bar{B}$ (antiparticle to B) and C go to make up the particle A. Further, particles play a dual role. On the one hand, they are constituents of a composite system; on the other hand, they are the objects which are exchanged to generate forces between the constituents. In any case, in view of the very large number of entries in Table 1, it is not desirable to accept all of them as the ultimate constituents of matter.

This is even more forcefully brought to our attention by the recent discovery of a very large number of ultra-short-lived particles. They appear as sharp resonances in multiparticle systems. Since these "resonances" disintegrate within a short time (even on the nuclear scale!), it is difficult to view them as elementary particles, but they seem to play an important role in interaction phenomena and are produced as often as the more stable (and familiar) mesons and baryons included in Table 1. It appears at the present time that they ought to be included on more or less the same footing. A list of the better established resonances is given in Table 2 along with the mesons and baryons from Table 1. Since an unstable particle lives only for a very short time its energy and consequently its mass cannot be sharp, and from elementary quantum mechanical considerations we should expect this "width" in the mass of a resonance to be inversely proportional to its lifetime. Since the width is what is measured experimentally the width (rather than the lifetime) is usually quoted in connection with resonances.

Since these particles are coupled in the strong interactions, one would expect them to occur in isospin multiplets. This is in fact observed. It turns out that since strong interactions are invariant under particle conjugation and are charge independent, we could define a multiplicative quantum number called *G-parity* which has definite values ±1 for mesons and meson resonances. These values are also included in Table 2.

With these resonances included among the "elementary particles" we have a situation

TABLE 2. SOME STRONGLY INTERACTING PARTICLES AND RESONANCES

Particle or Resonance	Spin	Mass (MeV)[a]	Width (MeV)[b]	Parity	Electric Charge	Hyper-charge	Isotopic Spin	G-parity
Pion, π	0	138.5	0	−	0, +1, −1	0	1	−
Kaon, K	0	495.8	0	−	0, +1	+1	$\frac{1}{2}$	Undefined
Antikaon, $\bar{K}$	0	495.8	0	−	0, −1	−1	$\frac{1}{2}$	Undefined
Eta, η	0	548.8	0	−	0	0	0	+
Rho resonance, ρ	1	776	158	−	0, +1, −1	0	1	+
Kaon resonance, K^*	1	893	50	−	0, +1	+1	$\frac{1}{2}$	Undefined
Antikaon resonance, $\bar{K}^*$	1	893	50	−	0, −1	−1	$\frac{1}{2}$	Undefined
Phi resonance, ϕ	1	1020	4	−	0	0	0	Undefined
Omega resonance, ω	1	782	10	−	0	0	0	−
J/ψ	1	3097	0.063	−	0	0	0	−
Nucleon, N	$\frac{1}{2}$	938.9	0	+	0, +1	+1	$\frac{1}{2}$	Undefined
Lambda, Λ	$\frac{1}{2}$	1115.6	0	+	0	0	0	Undefined
Sigma, Σ	$\frac{1}{2}$	1193.4	0	+	0, +1, −1	0	1	Undefined
Xi, Ξ	$\frac{1}{2}$	1318.4	0	+	0, −1	−1	$\frac{1}{2}$	Undefined
Nucleon resonance, N^*	$\frac{3}{2}$	1236	120	+	0, +2, +1, −1	+1	$\frac{3}{2}$	Undefined
Y resonance, Y^*	$\frac{3}{2}$	1385	36	+	0, +1, −1	0	1	Undefined
Xi resonance, Ξ^*	$\frac{3}{2}$	1530	7	+	0, −1	−1	$\frac{1}{2}$	Undefined
Omega minus resonance, Ω	$\frac{3}{2}$	1672	0		−1	−2	0	Undefined

[a]The average mass of the members of the isotopic multiplet is tabulated.

[b]Since the unstable particles of Table 1 live "practically forever" on the nuclear time scale, the corresponding widths are several orders of magnitude smaller than 1 MeV; these are quoted here as 0.

somewhat parallel to atomic spectroscopy *before* the discovery of quantum mechanics. The catalogue of the strongly interacting particles (collectively known as *hadrons*) now contains well over a hundred entries, and it would be difficult to consider a hundred plus "elementary" constituents. Yet how are we to select the genuine subset of elementary constituents? We have already remarked about the picture in which *every* hadron is a composite system. We should then look for regularities among them, including groupings into families, multiplets, etc., as well as for systematic relations between masses, spins, multiplet sizes, etc.

There are also practical questions regarding the identification and interpretation of resonances. When a number of reaction channels are open a resonance may not be easily visible as a pronounced peaking in crosssection or mass plot. Fortunately resonances like the ground-state hadrons seem to fall into multiplets, and this together with other systematics aid us in identifying resonances: observation is intimately tied in to the theoretical framework employed.

One notes also that the meson and baryon multiplets seem to fall into further super-multiplets. Following the analogy of the isospin group, we may now ask what internal symmetry group is responsible for this interaction. We must also remember that whatever is

responsible for the violation of this higher symmetry must itself be a part of the strong interaction. A scheme in which invariance under the special unitary group on three variables $SU(3)$ holds approximately has been successfull in correlating and predicting the spectrum of particles and their interactions. The isospin group $SU(2)$ is a subgroup of this unitary group. Just as for charge independence, no *basic* reason has been found for the origin of this "unitary symmetry." Still other symmetry groups, even wider than $SU(3)$ and generally incorporating it, and which are even more significantly violated, are being studied. It appears that the complete understanding of these higher internal symmetries would involve not only their origin, but also the origin of their violation.

The hadron multiplets appear to have other regularities. We can discern, by analogy with atomic physics, subfamilies consisting of a lowest-spin "excited states." The states listed in Table 2 may be viewed as the ground states. If we plot the masses squared versus the spin for several of these subfamilies, we find them to lie approximately on straight lines with a universal slope of about 1 $(GeV)^{-2}$. These may be thought of as the generalization of the bound state energy versus spin relation for potentials to the domain of resonances, and thus, as orbital excitations; and in this context they are known as *Regge families*.

The ground states themselves may be understood in terms of a generalization of the "internal symmetry" to include a spin aspect also, so that instead of $SU(3)$ we consider $SU(6)$. The spin $\frac{1}{2}$ even-parity baryon octet and spin $\frac{3}{2}$ even-parity baryon decuplet resonances together then form a single 56-dimensional representation of $SU(6)$. The nine spin 0 odd-parity mesons and the nine spin 1 odd-parity vector meson resonances form a mixture of the 35-dimensional and the 1-dimensional representation of $SU(6)$. We can combine this $SU(6)$ structure together with the orbital excitations mentioned above to bring about a phenomenological $SU(6) \times O(3)$ classification for hadrons.

It is very tempting to think of this $SU(6) \times O(3)$ structure as pointing to a substructure of the hadrons in terms of 3 hypothetical entities with spin $\frac{1}{2}$ and even parity which transform as a 3×2-dimensional representation of this group called *quarks*. The simplest baryon resonances are then to be viewed as three-quark compounds and the meson resonances as quark-anti-quark compounds. The spin, parity, and $SU(3)$ quantum numbers of most hadrons are consistent with this picture and some quantitative understanding of the resonance masses and decay parameters can be obtained within this framework. sometimes referred to as the *quark model*. It should be pointed out that so far no quarks have been discovered and physicists are not sure if they even expect them to exist as ordinary particles.

The baryons and baryon resonances identified as three-quark compounds correspond to a completely symmetric spin-internal symmetry wave function. The spin $\frac{1}{2}$ and spin $\frac{3}{2}$ baryons may be viewed as realizing a third rank symmetric tensor realization. But quarks, being spinorial entities, should behave as Fermions and hence should furnish a totally antisymmetric third rank tensor. It is therefore appropriate to endow quarks with an as yet unidentified property called *color* with *at least* three linearly independent states. If for economy we use only three colors, all known hadrons are color singlets: one may require "color confinement" as an essential ingredient of hadron structure.

What about the binding of the quarks and more generally of the quark-quark and quark-anti-quark forces? We can expect them to be quite strong since at low energies and momentum transfers the hadrons act as indissoluble. But we need a special feature for these forces: unlike usual forces in atomic and nuclear physics these forces must be "confining," and hence *increase with distance*. In contrast, in many high momentum transfer reactions which probe small distances these quark binding forces are relatively weak. It turns out that these features result from an interaction structure invented by Yang and Mills on the basis of gauge invariance

generalizing the electromagnetic coupling to charged particle fields to more general noncommutative groups. Identifying the color of quarks as the realization of an $SU(3)$ color group leads to the introduction of an eightfold vector meson field of zero mass coupled to the quarks as well as self-coupled, which leads to forces with precisely the properties required. The quanta of these octets of fields are called *gluons*. They are neutral with regard to isospin and its generalization, which in contrast to color is called *flavor*. The fundamental strong interaction is then the gluon interaction. The hadron-hadron interaction by contrast is the residual interaction between quark compounds, somewhat analogous to the van der Waal's forces between atoms. No wonder nuclear forces are so complicated!

At the present time most experts seem to believe that the quantum theory of coupled colored quarks and their Yang-Mills gluons is the theory of strong interactions. It is called *quantum chromodynamics* (QCD). The difficulties of principle that plagued the theory of photons and electrons, *quantum electrodynamics* (QED), are still there; but in addition we have new difficulties in calculation. As if in compensation we have many more phenomena to which QCD can be applied. Nonperturbative methods and semiphenomenological methods have been developed to cope with these difficulties. Particularly important have been the observation that the effective coupling strength is energy dependent and in QCD it *decreases* as the momentum transfer increases, leading to "asymptotic freedom." The precise manner in which this takes place is described by the renormalization group equations.

The elementary particle scene has been livened by the discovery of extremely narrow resonances and their identification as quark-antiquark bound states. This has led to the identification of a new flavor quantum number called *charm*. Together with charge and strangeness this suggests that the internal symmetry (flavor) group is $SU(4)$ rather than $SU(3)$ or $SU(2)$. It is only the breaking of the symmetry by the large mass differences between the various quarks that tended to hide these symmetries. The discovery of charm has also provided a means for accounting for the absence of strangeness-changing neutral hadronic currents in weak interactions.

Strong interactions are thus described in terms of a color triplet of four flavors of quarks which are called u, d, s, c (for *up, down, strange,* and *charm*). It is a question of some interest whether this is the end. It is generally accepted that one of the very narrow resonances should be identified as a new quark pair. If this is called *b* (for *beauty* or *bottom*) and a sixth called *t* (for *truth* or *top*) is also postulated, we would have these pairs: $u, d; s, c; t, b$. This would be then very similar to the leptons, where we seem to have $e, \nu_e; \mu, \nu_\mu; \tau, \nu_\tau$: three families of quark pairs and three of lepton pairs. This family recurrence suggests even more strongly that the leptons and quarks are intimately connected; and that we have not exhausted levels of possible unification. What was an electron-muon puzzle is now a full-fledged "family" affair!

As particles get organized into multiplets, so do interactions exhibit systematic properties. During the last century Maxwell unified electric and magnetic interactions into the electromagnetic theory. With the reinterpretation of the coupling of the charge flow to the electromagnetic field to be implemented by a gauge invariant interaction in quantum field theory we got quantum electrodynamics (QED). In the middle of this century, Sudarshan and Marshak started from Fermi's theory of beta decay of the neutron to unify weak interactions into a chiral ("handed") V-A interaction. In this theory for the first time we identified the left-handed fermions to be the only ones coupled; in contrast to electromagnetism, in which the left and right chiral components were both coupled but the neutrino was not. The adaptation of this theory to strange and charmed particles has now been carried out satisfactorily.

Is it possible to weld together the Maxwell theory of electromagnetism and the Sudarshan-Marshak theory of chiral weak interactions into a unified electroweak theory? Clearly the weak interactions are weak and if seen as due to charged counterparts to the photon these intermediate vector mesons should be about 40 times the proton mass. It turns out that Glashow, Weinberg, and Salam constructed precisely such a theory by starting with a Yang-Mills theory of a group $SU(2) \times U(1)$ involving the left-handed doublet comprising the electron and neutrino on the one hand and the right-handed electron on the other. This symmetry must break spontaneously and in the process generate substantial masses for the vector mesons. Such a theory is "renormalizable," so that higher order field theory corrections can be calculated in an unambiguous manner. It is not fully unified since there are two distinct coupling constants which can be related to the ratio of the charged to neutral vector bosons. It predicts a level of neutral current interactions in agreement with experiments. The mechanism of generation of masses involves the use of certain auxiliary Higgs fields which also serve to give the electron its mass. Once the electron-neutrino interactions are formulated we could use it for the two other lepton pairs and the three quark pairs.

The existence of *three* families rather than two throws some light on the puzzling small time reversal violation as "natural," in that the complex phases appearing in a mass matrix coupling left- and right-handed objects cannot all be "defined away" as soon as three families appear.

If we really pursue the goal of unification and are emboldened by the close similarity between quarks and leptons attempt to put them together as a grand multiplet, we would be led to a *grand unified theory* (GUT), or rather several theories. Symmetry is broken in these theories so that at ordinary energies they do not appear.

Particularly simple forms of GUT are based on the groups $SU(5)$ and $O(10)$; these theories do predict a quark-lepton transition induced by a very heavy boson. This would lead to the decay of the proton, possibly into a positron and neutral pion or other such modes. The lifetime for such a decay would depend on the heavy boson mass, varying roughly as the fourth power of the mass. Using renormalization group equations and the observed ratio of the couplings one can estimate the mass and hence the proton lifetime. This yields a lifetime on the order $10^{31 \pm 2}$ years. While this means an extremely small transition rate there are a great number of protons in the world, and detection, if they decay, is imminent.

We should also remark that the particles so far discovered all have either finite mass (Class I particles or *bradyons*), or zero mass (Class II particles or *luxons*). It is an interesting question to ask if particles of imaginary mass (Class III particles or *tachyons*) can and do exist. It used to be thought that such particles could not exist, since their existence would violate the principle of relativity, but now we know that such is not the case. If hadronic tachyons exist, the quantum theory of tachyons predicts that they will show up as fixed resonances in momentum transfer. Leptonic tachyons (and photonlike tachyons) could probably be best detected in astronomical phenomena. If they exist they would provide for a substantial pressure in the interior of hot stars and thus provide a balancing force to alleviate gravitational collapse. Whether the concept of particles of imaginary mass, which is the relativistic counterpart of geometric size used in simpler nonrelativistic physics, is useful in elementary particle physics is not yet clear.

One of the remarkable new developments is the close relationship between cosmology and elementary particle physics. This goes beyond the application of standard methods of nuclear synthesis and stellar astrophysics in dealing with the change in physical laws under cosmological conditions. The spontaneous symmetry breaking that makes weak interactions weaker than electromagnetism no longer operates when energy density and temperatures are very high. Similarly, masses and thresholds are dependent on the presence of radiation and other particles in high concentration. Finally the baryon to photon abundance ratio can be related to the violation of time reversal invariance.

Despite the breathtaking unifications we are

still short of two unifications. One is the unification of gravitation with particles interactions. While the emphasis on geometry characteristic of general relativistic gravitation theory is now permeating particle physics via gauge theories, the combination of gravitation with particle theories is still far from satisfactorily envisaged, much less carried out.

The other unification is of bosons and fermions by a supersymmetry. Such theories are now being actively pursued, since they seem to have some effect of diluting the infinities in quantum field theory. Supersymmetry is reluctant to "break" in a simple manner, yet break it must if experiment is our guide: bosons and fermions are not degenerate in their masses.

To sum up: we have a large number of "elementary" particles which are not so elementary. We have a number of theories which aid us in understanding the regularities observed in their spectrum and in their interactions. We have discovered a large number of invariance groups and the extent to which they break. And we have found that coupling to the flow of electric charge and its noncommutative generalizations as well as the flow of energy and momentum lead to a class of gauge theories which may underlie all the physical interactions. However, we still do not understand the multiplicity of particles and interactions. The past decade has been a time of great discoveries. Particle physics promise to be a rich area of physics in the foreseeable future. Perhaps yet another level of discovery awaits us in our search for the constitution of matter.

E. C. G. SUDARSHAN

References

Marshak, R. E., and Sudarshan, E. C. G., "Introduction to Elementary Particle Physics," New York, Intersicence Publishers, 1962.

Bernstein, J., "Elementary Particles and Thier Currents," San Francisco, W. H. Freeman, 1968.

Marshak, R. E. Riazzuddin, and Ryan, C. P., "Theory of Weak Interactions in Particle Physics," New York, John Wiley, 1969.

Streater, R. F., and Wightman, A. S., "TCP Theorem and All That," W. A. Benjamin, New York, 1964.

Weinberg, S., "The First Three Minutes," Basic Books Inc., New York, 1977.

Particle Data Group, "Review of Particle Properties," Particle Properties Data Booklet, April 1980, CERN, Geneva.

Cross-references: ANTIPARTICLES, CONSERVATION LAWS AND SYMMETRY, CURRENT ALGEBRA, ELECTROWEAK THEORY, GAUGE THEORIES, GRAND UNIFIED THEORY, QUANTUM CHROMODYNAMICS, QUANTUM ELECTRODYNAMICS, QUARKS, STRONG INTERACTIONS, WEAK INTERACTIONS.

ELEMENTS, CHEMICAL

The idea of a basic simplicity underlying the bewildering complexity of nature has always been a conceptual thread woven into man's view of the world. The Greek philosophers of antiquity were among the first to record their speculations, and we are to this day influenced in an unconscious way by their thoughts, concerning the elements, which they supposed to be the ultimate components of matter and chemical change. Thus we speak of "man's battle with the elements" and "the raging elements" in unconscious reflection of the ideas of Thales, Anaximenes, Heraclitus and Empedocles of the fifth century B.C. who believed that all matter was made of one or more of the elemental substances: earth, air, fire and water. These ideas did not prove particularly fruitful in advancing our understanding of the nature of matter and chemical change. Nevertheless, it was not until van Helmont (1648) that they were challenged on a rational basis.

In 1662, Robert Boyle, in the *Sceptical Chymist* gave a reasonably clear definition of a chemical element with an operational basis which we can accept today. "I mean by elements, . . . certain Primitive and Simple, or perfectly unmingled bodies; which not being made of any other bodies, or mingled bodies, are the Ingredients of which all those called perfectly mixt bodies are immediately compounded, and into which they are ultimately resolved." He gave no list of elements, however, this being left to Lavoisier who published a naturally incomplete, but remarkably accurate list in 1789, in his justly famous, *Traité de Chimie*.

The Definition of "Element" In modern language, Boyle can be paraphrased in the following way: an *element* is a chemical species that cannot, by ordinary chemical manipulation be decomposed into a number of simpler chemical species. It is the entity that survives intact the infinite variety of transformations that a sample of matter can be caused to undergo. Every *compound* is composed of two or more of these species and can be decomposed into them by suitable chemical procedures. This definition provides an operational means of identifying an element in terms of laboratory procedure, and by it, any species that defies decompositional efforts must be classified as an element. Such an assignment must of course be somewhat tentative, since in a number of cases, substances that have stubbornly resisted decomposition and therefore carried the classification, have ultimately yielded as new techniques developed.

With the recent growth in detailed knowledge of atomic structure, it became possible to define an element in terms of the submicroscopic structure of matter. Such a definition is relieved of the ambiguities mentioned above. Thus an element is a sample of matter that consists of only one kind of atom, the atoms being identified in terms of their atomic number, or nuclear charge. Each element is composed of atoms having characteristic nuclear charge and an equivalent number of electrons. This nuclear charge can be determined by the charged-particle scattering technique first employed by Rutherford or by the simpler and more precise method of Moseley which relates the frequency of the characteristic x-rays produced by the element upon electron bombardment to the nuclear charge z by means of the equation

$$w = R(z - b)^2$$

where w is the wave number of the x-ray, z is the nuclear charge, and b and R are constants.

This definition removes the ambiguities created by such observations as the decomposition of elemental molecular hydrogen or nitrogen by high temperatures into atomic species, or the decomposition of the rare gases into charged species in an electric discharge. This type of decomposition, which does not effect the underlying nuclear structure is thus excluded from the operational definition originating with Boyle.

Numbers and Kinds of Elements The number of substances recognized as chemical elements has steadily increased since the publication of Lavoisier's list which included about thirty of the true elements. Today it is recognized that there are some 90 naturally occuring elements, the exact number depending upon the level of abundance considered limiting. There are also about 15 artificial ones with this latter number possibly increasing as the techniques of nuclear science improve. The artificial elements include those with atomic numbers above 92 as well as promethium and technetium. Technetium is also observed in stars, and in that sense is naturally occuring. (See table in the article ISOTOPES).

The 105 presently known elements represent distinct chemical species differing by integral units of positive charge (corresponding to the charge of the proton) beginning with element number 1, hydrogen, and progressing through element number 105. Species having nuclear charges between 1 and 105 have all been identified, so it can be said that the list of elements is complete, except for the possibility of adding new ones with atomic numbers greater than 105. Considerable effort, both experimental and theoretical, is currently being expended in the effort to find super-heavy elements of unusual stability. With the discovery of element 103, the actinide series was completed. Both 104 and 105 are quite unstable; however, islands of stability are predicted to occur between atomic numbers 114 and 126 and again between 164 and 194.

Most of the elements exhibit variations in mass, due to the varying numbers of neutrons present in their nuclei. Atoms having the same nuclear charge but differing in mass number or atomic weight are referred to as ISOTOPES. If all of these are considered, then there are approximately 1000 different atomic species represented in the list of chemical elements. Only 18 of the elements existing in nature exhibit a single mass number, and some, tin being a good example, have as many as ten naturally occurring stable species differing only in their neutron number or mass. All of the elements exhibit a variety of mass modifications which are artificially produced, but these are unstable or radioactive. These artificially produced species, of course, make up the bulk of the previously mentioned 1000 different entities. The *naturally occuring* radioactive isotopes, in addition to those beyond bismuth, include carbon 14, chlorine 36, vanadium 50, potassium 40, rubidium 87, indium 115, lanthanum 138, neodymium 144, samarium 147, lutetium 176, tantalum 180, rhenium 187, and platinum 190.

The Natural Distribution of Elements The relative abundance of the elements is quite different for the earth's crust from what it is thought to be for the universe as a whole. These terrestrial abundances are listed in Table 1 from which it can be seen that only thirteen elements comprise over 98 per cent of the earth's crust, the oceans and the atmosphere. It will be noted that many of the common and important elements of commerce are not included in this list, but rather belong to the remaining 2 per cent of the earth's crust. Copper, lead, and nitrogen are especially conspicuous by their absence.

If we turn our attention now to cosmic abundances, the list has quite a different make up (see Table 2). This list of course is known with considerably less accuracy since it is our attempt to guess at the relative abundance of the chemical elements in the entire universe: our galaxy, all the other galaxies, and the vast, but

TABLE 1. THE THIRTEEN MOST ABUNDANT ELEMENTS ON THE EARTH'S SURFACE

Element	Abundance (%)
Oxygen	49.52
Silicon	25.75
Aluminum	7.51
Iron	4.70
Calcium	3.39
Sodium	2.64
Potassium	2.40
Magnesium	1.94
Chlorine	1.88
Hydrogen	0.88
Titanium	0.58
Phosphorus	0.120
Carbon	0.087

TABLE 2. THE THIRTEEN MOST ABUNDANT
ELEMENTS IN THE UNIVERSE

Element	Relative Abundance[a]
Hydrogen	3.5×10^8
Helium	3.5×10^7
Oxygen	2.2×10^5
Nitrogen	1.6×10^5
Carbon	8×10^4
Neon	2.4×10^4
Iron	1.8×10^4
Silicon	1×10^4
Magnesium	9×10^3
Sulfur	3.5×10^3
Nickel	1.3×10^3
Aluminum	8.8×10^2
Calcium	6.7×10^2

[a]These abundances are relative to silicon taken as
1×10^4.

not entirely empty spaces in between. This information has largely been gathered by spectroscopic studies of the light emitted by the luminous bodies in these galaxies and by careful
analysis of the samples of off-planet material
(meteorites) that constantly shower the earth,
as well as lunar-samples.

It can readily be seen that this list is quite different from the terrestrial abundance list. For
the universe as a whole, the elements hydrogen
and helium far outrank all others, while on the
earth, hydrogen is only tenth in abundance and
helium doesn't even appear on the list. Oxygen
remains high, and nitrogen, absent from the
terrestrial list, is the fourth most abundant element when cosmic abundances are considered.
In Fig. 1, some of the interesting variations in
abundance are displayed. The relative abundance is plotted against mass number.

It will be noted that elements with atomic
weights that are multiples of four and two are

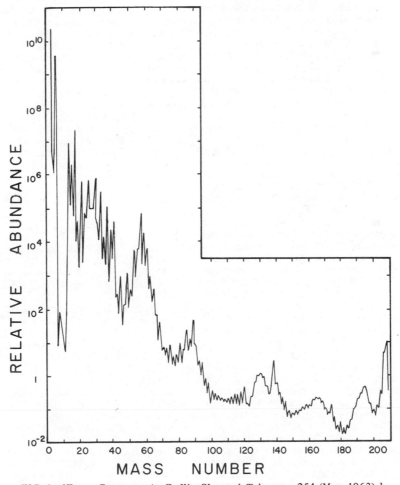

FIG. 1. [From Cameron, A. G. W., *Sky and Telescope*, 254 (May, 1963).]

more abundant than nearby elements with relatively similar atomic weights. Examination shows that those elements having *proton or neutron* numbers 8, 50, 82, and 126 also exhibit maxima. These numbers are so-called magic numbers confering an especially high degree of nuclear stability.

The Origins of the Elements The origin of the elements and their eventual appearance in the observed abundances in the earth has been the subject of intense research and speculation for many years. The currently accepted ideas involve a series of events, some dating back to the "beginning" and others continuing to occur in appropriate locations throughout the universe even today, which result ultimately in the production of all the known naturally occuring elements.

The first part of the story, one of remarkably short duration, involves a modified "big bang" of the type postulated by Gamow. At the beginning of our universe all matter was thought to exist in a highly condensed sphere of neutrons (the *ylem*). This giant nucleus underwent a cataclysmic explosion and with a half-life of around eleven minutes the neutrons were converted to protons and electrons. Because of the very high photon density in the ylem about 25% of the resulting expanding cloud was converted into helium by condensation of the hydrogen nuclei, concomitantly filling the universe with black body radiation as the cloud cooled. The existence of such radiation, corresponding to a temperature of 2.7°K has been confirmed experimentally.

The remainder of the story is spread over a period of from 10 to 15 billions of years and involves the cycle of stellar evolution familiar to the astronomer. The original cloud of interstellar matter suffered local condensation into billions of stars. As these stars consume their hydrogen by a fusion process they undergo a series of changes in internal conditions acting as the factories of the chemical elements. At first only helium is formed, but as the hydrogen is consumed and helium formed the temperature of the core falls, allowing it to contract by gravitational attraction. This contraction offsets the cooling trend and the core becomes even hotter—heating the outer hydrogen mantle and causing it to expand and cool, producing the so-called *red giant* star. Continued contraction of the core raises its temperature until helium fusion is possible, producing the lighter elements such as carbon, oxygen, neon, magnesium, and silicon. Essential to the production of these lighter elements is the possibility of utilizing the extremely short-lived product of the first condensation of two helium nuclei, ^{8}Be, ($t_{1/2}$ = 2×10^{-16} s). Fortunately the density of matter in the red giant cores is so great that the probability of subsequent reaction of beryllium with helium to produce the lighter elements is very high.

The heavier elements, up to and including iron, are generally the result of carbon fusion in very massive stars with interior temperatures of the order of one billion degrees. This produces a "soup" of nuclei, nucleons, alpha-particles, and photons in which reaction between the lighter particles and heavier nuclei leads to elements as heavy as iron.

To form elements up to the mass of bismuth the s- (or slow) process is invoked. This involves the equilibrium addition of neutrons to the intermediate nuclei with intervening beta-decay. This growth of heavier and heavier nuclides is arrested by the intervention of a number of very short-lived species. The trans-bismuth elements require an environment of even higher pressure and neutron abundance, as found in the cores of supernovae.

Eventually stars eject matter containing this array of elements into interstellar space. The Earth, the Sun, and other planets are presumably the result of accretion of this cosmic dust, in a whirling cloud which centrifuges the heavier nuclides outward, leaving the helium and hydrogen to coalesce to form our central star.

Synthetic Elements The first of the synthetic elements was made in 1937 by the bombardment of molybdenum with deuterons. This produced technetium, filling a long recognized gap in the periodic table for element number 43. The other gap, at number 61, promethium, was filled in 1947 when ^{149}Pm was isolated in small yield from the products of fission of ^{235}U.

Since that time 13 additional synthetic elements have been made, all lying beyond uranium in the periodic table. These are formed by a variety of processes involving neutron capture or bombardment with heavy ions.

RUSSELL H. JOHNSEN

References

1. Chaisson, E., "Cosmic Dawn—The Origins of Life and Matter," Boston, Little, Brown and Co., 1981.
2. Choppin, G. R., and Johnsen, R. H., "Introductory Chemistry," Reading, Mass., Addison-Wesley Inc., 1972.
3. Johnsen, R. H., and Grunwald, E., "Atoms, Molecules and Chemical Change," Third Ed., Englewood Cliffs, N.J., Prentice-Hall, Inc., 1971.
4. Pagel, B. E. J., and Edmunds, M. G., "Abundances in Stellar Population and the Interstellar Medium in Galaxies," *Ann. Rev. Astron. Astrophys.* **19**, 77 (1981).
5. Wannier, P. G., "Nuclear Abundances and Evolution of the Interstellar Medium," *Ann. Rev. Astron. Astrophys.* **18**, 399 (1980).
6. Keller, O. K., Jr., Burnett, J. L., Carlsen, T. A., and Nestor, C. W., Jr., "Predicted Properties of the Super Heavy Elements I., Elements 113 and 114," *J. Phys. Chem.* **74**, 1127 (1970).
7. Seaborg, G. T., "From Mendeleev to Mendelevium and Beyond," *Chemistry* **43**, 6 (1970).

8. "Physics and Chemistry of the Earth, Vol. 2: Origin and Distribution of the Elements," Proceedings of the 2nd Paris Symposium (L. H. Ahrens, Ed.), Pergamon Press, Oxford, 1977.

Cross-references: COSMOLOGY, ELECTRON, ISOTOPES, NEUTRON, PERIODIC LAW AND PERIODIC TABLE, PROTON, RADIOACTIVITY.

ENERGY LEVELS, ATOMIC

The term "energy level" is used in referring to discrete amounts of energy which atoms and molecules can have with respect to their electron or nuclear structure. The concept of permissible discrete energy levels was first introduced by Planck in explaining the physical basis for the spectral distribution of blackbody radiation. A second related principle due to Planck was that the emission and absorption of radiation are associated with transitions between these energy levels, the energy thereby lost or gained being equal to the energy, $h\nu$, of the quantum of radiation. Here h is Planck's constant and ν is the frequency of the radiation.

The first application of energy levels in the electron structure of atoms to explain optical spectra was made by Bohr. The original Bohr atom had as its basis that the only allowable states of an atom were those in which the electronic angular momentum was an integral multiple of $h/2\pi$. Circular orbits suggested by Bohr were extended by Sommerfeld to include the quantization of momentum in elliptic orbits, and to provide an improved explanation of optical spectra.

These early concepts were modified by the development of the theory of wave mechanics, in which it was shown that the allowable "stationary" states for the electrons in an atom must represent solutions of the Schrödinger wave equation. These solutions are conveniently represented by a set of "quantum numbers" for each electron. On this basis the electron structure of an atom containing any number of electrons can be built up. Two further concepts that are essential to this picture, however, are electron spin proposed by Uhlenbeck and Goudsmit, and the exclusion principle due to Pauli. In addition to the angular momentum of the electron in its orbit, each electron possesses angular momentum due to spin about an axis. The Pauli exclusion principle specifies that no two electrons in an atom can exist in the same quantum state, corresponding to the same set of quantum numbers.

Each electron in an atom can be characterized by four quantum numbers, n, l, m_l, m_s. The energy of an electron depends principally upon the positive integer n, and larger values of n correspond to larger electronic orbits. The quantum number l possesses physical significance in terms of angular momentum in the orbit, and is constrained to have the values of zero or positive integers less than n. The number, m_l, represents the component of l along a given axis, and must take on the values of zero or positive and negative integers whose absolute values are less than or equal to the value of l. The quantum number, m_s, can be $+\frac{1}{2}$ or $-\frac{1}{2}$, and represents the component of the spin along the axis.

Within a given atom, electrons having the same value of the principal quantum number, n, form a definite group or "shell." Those electrons possessing the same value of l for a given value of n are in the same subgroup or "subshell." The possible number of electrons in a shell or subshell depends upon the possible values of m_l and m_s. Whenever a subshell is filled, the total angular momentum of the electrons involved is zero. Electrons outside of filled subshells contribute additional angular momentum which is summed vectorially and assigned the numbers J, L and S. Here J is representative of the total angular momentum, L the orbital angular momentum, and S the spin angular momentum.

An atom is stable only when it exists in the state for which the quantum numbers of its electrons give the lowest total energy. The energy of the atom may be increased to a higher level by having an electron "excited" to another state represented by a different set of allowed quantum numbers. Transitions back again to the "ground" state will be accompanied by the emission of radiation. Wave mechanics indicates, however, that only certain transitions from one quantum state to another can be probable. These "selection rules" specify that $\Delta L = \pm 1$ and that $\Delta J = 0$ or ± 1.

Following the principles just given, an electron energy-level diagram can be constructed for the excited states of the atoms of any particular isotope. Discrete energy levels will exist for the allowed quantum states of excited electrons. The spectrum of radiation which can be emitted from the isotope will be determined by the energy differences between these states, where the transitions involved are allowed by selection rules. Figure 1 illustrates such an energy-level diagram for sodium.

This diagram indicates that very little difference in energy level results from changes in the electron spin orientation. These orientations correspond to different values of J for the same value of L. No attempt is made to show a separation in the diagram for the cases of $L = 2$ and $L = 3$. Such closely spaced energy levels due to the effect of the coupling of the electron spin give rise to "fine structure" in spectra.

Many spectral lines, when examined with high resolving power instruments, are found to exhibit a still finer structure of several lines very close together. This is termed "hyperfine structure" and has been found to be due to two causes. One is the isotope effect in which atoms

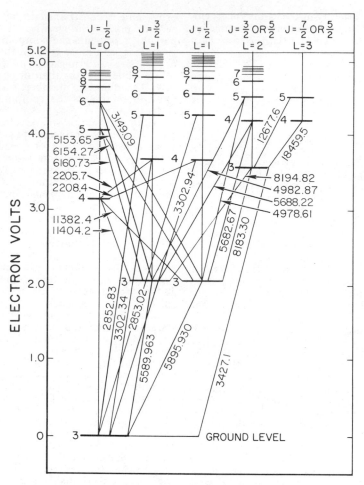

FIG. 1. Energy-level diagram for sodium. The large numbers are wave
lengths in A of radiation emitted or absorbed during the indicated transitions. Principal quantum numbers are shown as integers.

of different isotopes of the same element
possess slightly different excited electron energy
levels. The other cause of hyperfine structure
has been determined to be due to the fact that
the atomic nucleus also possesses angular momentum, which is vectorially added to the electronic angular momentum and quantized. Differences in the resultant states of the atom
correspond to very small differences in the
energy levels and hence in the observed
spectrum.

Another source of structure in spectra results
when the atoms emitting or absorbing the radiation are in a magnetic or an electric field. In a
field, space quantization in the direction of the
field takes place. The values of the magnetic
moment or electric moment of the atom associated with the various possible components of
angular momentum as quantitized in the field

direction result in different energy levels. This
splitting of levels is referred to as the Zeeman
effect in the case of an applied magnetic field,
and the Stark effect in the case of an applied
electric field (see ZEEMAN AND STARK EF-
FECTS). The amount of splitting increases with
the intensity of the superposed field. In addition, the spectral structure can vary due to a
tendency for the orbital momentum, L, and
the spin momentum, S, to become uncoupled
and undergo space quantization independently
in high fields. In weak fields, the space quantization is determined from the total angular momentum J.

The splitting of energy levels in electric or
magnetic fields has become much more important in another type of phenomenon referred to as magnetic resonance (see MAGNETIC
RESONANCE). Two examples are electron spin

resonance (esr) and nuclear magnetic resonance (nmr). If a single level is split into its Zeeman components by a magnetic field H, electromagnetic radiation of frequency ν will induce transitions between the Zeeman levels if the following condition is satisfied:

$$h\nu = g\mu H$$

Here g is energy level "splitting factor" and μ is the appropriate "magneton" (Bohr magneton μ_B for esr and nuclear magneton μ_N for nmr). (See reference 1.) For magnetic field strengths conveniently available in the laboratory, the resonance frequencies are in the short radio or microwave regions.

Any electron of an atom may be excited to some higher allowed energy level by absorption of the amount of energy specified by the difference in the energy levels involved. By absorption of a sufficient amount of energy, any electron can be removed from an atom, resulting in ionization. It is not necessary to consider only the outermost loosely bound electrons. When electrons from inner shells are excited or removed, the process of returning to the "ground" or lowest energy state involves the emission of "characteristic" x-rays. They are "characteristic" in that the x-ray spectrum produced is typical of the particular atom producing the radiation. Atoms of higher atomic number and transitions involving electrons in innermost shells produce higher-energy radiation. Absorption, as well as emission, of characteristic x-ray radiation is observed between allowed electronic energy levels. For further information on atomic spectra and energy levels, see reference 2.

In addition to the energy levels associated with particular types of atoms, wave mechanics shows that discrete quantum states and energy levels are associated with molecular structure. New energy levels arise from vibration and rotation of molecules. To illustrate the allowable vibrational states for a diatomic molecule, consider Fig. 2, which shows the mutual force, F, and the potential energy, V, plotted as a function of the separation, r, between the two atoms. Possible energy levels are indicated by the dotted lines. Even the lowest state corresponds to an energy greater than V_0, where the force between the atoms would be zero, and hence has some associated kinetic energy.

The rotation of a molecule has a quantitized angular momentum which can be vectorially combined with that of the electrons. Transitions from one level to another usually involve a change in the electronic state as well as in the rotational state. Since energy differences due to allowed changes in rotational motion are very small compared to the energy differences in electronic states or vibrational states, the effect of transitions in rotational states is to produce bands of very closely spaced frequencies in the emission and absorption spectra. Such "rota-

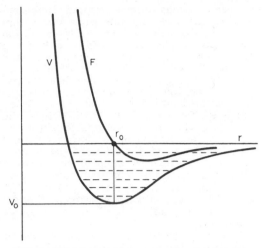

FIG. 2. Mutual potential energy V and force F as a function of atomic separation in a diatomic molecule. F is repulsive when positive and attractive when negative.

tional" bands are observed in molecular spectroscopy, depending upon what other transitions may be simultaneously involved, in the ultraviolet region, the visible, the infrared, and even the microwave region. See reference 3 for further details on molecular energy levels and associated spectra.

An atom may not only combine with others to form a molecule, but may be one of a large

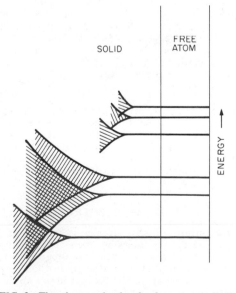

FIG. 3. The electron levels of a free atom split into bands when the atom enters a solid.

number of atoms forming a crystal. Here solutions to the wave equation show that within the solid, the individual energy levels of the free atom broaden into bands of overlapping levels. This is illustrated in Fig. 3. The bulk electrical and optical properties of solids are determined by the nature of these energy-level bands. Metals have the highest occupied energy band unfilled, while satisfying the exclusion principle that no two electrons occupy the same state. This permits electrons to gain energy under the application of an electric field and flow freely through the solid. Insulating crystals have the highest occupied energy band filled and appreciably separated from the next higher band. Semiconductors represent an intermediate situation where electrons can be injected into an unfilled band to contribute to conduction.

If the solid crystal is not completely regular but contains imperfections or impurity atoms, specific electron energy levels will be associated with these sites. This condition is responsible for luminescence and phosphorescence in certain solids. Additional information on energy bands in solids is given in reference 1.

WILLIAM E. PARKINS

References

1. Kittel, Charles, "Introduction to Solid State Physics," Fifth edition, New York, John Wiley & Sons Inc., 1976.
2. Herzberg, Gerhard, and Spinks, J. W. T., "Atomic Spectra and Atomic Structure," Second edition, New York, Dover Publications, 1944.
   Condon, E. U., and Odishaw, Hugh, "Handbook of Physics," Second edition, New York, McGraw-Hill Book Co., 1967.
3. Herzberg, Gerhard, "Molecular Spectra and Molecular Structure," Vol. 1, "Spectra of Diatomic Molecules," Second edition, New York, D. Van Nostrand Co., 1950, and Vol. 3, "Electronic Structure of Polyatomic Molecules," New York, D. Van Nostrand Co., 1966.

Cross-references: ATOMIC PHYSICS, ATOMIC SPECTRA, ELECTRON SPIN, MAGNETIC RESONANCE, MOLECULES AND MOLECULAR STRUCTURE, NUCLEAR STRUCTURE, SCHRÖDINGER EQUATION, SEMICONDUCTORS, SPECTROSCOPY, X-RAYS, ZEEMAN AND STARK EFFECTS.

ENERGY STORAGE, ELECTROCHEMICAL

Introduction Energy storage is important in alleviating discrepancies between energy supply and energy demand. These discrepancies may be either temporal or spatial or both. Energy storage is already being used in many applications, varying in size from a few milliwatt-hours (e.g., electric watches and calculators)

to many gigawatt-hours (pumped hydroelectric storage). Energy may be stored in many ways, including storage as potential energy in the earth's gravitational field (raising a weight), kinetic energy (a spinning flywheel), thermal energy (heating a substance), chemical energy (carrying out a reversible chemical reaction), or electrochemical energy (carrying out a reversible electrochemical reaction). This article deals with the storage of energy by means of electrochemical reactions which can be carried out in both forward and reverse directions. The devices which provide this capability are called rechargeable (or secondary*) electrochemical cells. Cells may be connected to series and/or parallel to form arrays called batteries.†,‡

There are many current applications for electrochemical energy storage devices, varying from aerospace (batteries for satellites' onboard power, recharged by solar cells; starting batteries for aircraft) to military (communications equipment, torpedoes, submarines) to industrial (electric forklift trucks, other electric vehicles) to consumer (battery-powered appliances, automobile batteries). Future applications that are providing incentive for current research and development include energy storage for electric utility systems (megawatt hours), for solar and wind-powered electric systems (kilowatt hours to megawatt hours), and for electric vehicles (tens of kilowatt hours).

The characteristics of rechargeable batteries that are important in many applications include the specific energy (energy stored per unit battery mass, Wh/kg), the specific power (power delivered per unit battery mass, W/kg), the cycle life (number of discharge-charge cycles before failure), the lifetime (time to failure), and the cost (usually expressed as $/kWh of energy storage capability). Of course, it is desirable to have batteries of high performance, long life, and low cost. The scientific implications of these goals are shown in Table 1.

There are a number of fundamental characteristics that enable the achievement of the goals of Table 1. The desired characteristics for each of the three essential parts of a cell (the negative electrode, the positive electrode, and electrolyte) are listed in Table 2. A large electronegativity difference between the reactants, and small equivalent weights yield a high theoretical specific energy for a cell:

$$\text{Theoretical specific energy} = \frac{-\Delta G}{\sum \gamma_i M_i}$$

*Primary electrochemical cells are based upon reactions that cannot be reversed; the cells are discarded after a single discharge.
†Often the term battery is misused in place of cell.
‡See ENERGY STORAGE, THERMAL-MECHANICAL for other types of energy storage systems.

TABLE 1. GENERAL GOALS FOR BATTERIES.

Goal	Implication
High specific energy	High cell voltage
	Low equivalent weight
	High utilization of active material
High specific power	High cell voltage
	Rapid reactions
	Rapid mass transport
	Low internal resistance
Long life	Reversible reactions
	Low corrosion rates
	Negligible rates of side reactions
Low cost	Inexpensive, plentiful materials
	Simple manufacturing processes
	High efficiency
	Little or no maintenance

where γ_i is the number of moles of reactant i, M_i is the molecular weight of reactant i, and ΔG is the Gibbs free energy change for the overall cell reaction. The theoretical specific energy is a useful guide in the selection of cells for applications in which high specific energy is important (e.g., mobile uses). For cells with solid reactants, it can be expected that 20–25% of the theoretical specific energy will be obtained from well-designed cells. A plot of the theoretical specific energy for a number of cells is shown in Fig. 1.

A careful examination of Fig. 1 reveals that the cells with the highest theoretical specific energy generally involve reactants that are not stable in contact with an aqueous electrolyte (Li, Na), whereas those of lower specific energy are compatible with aqueous electrolytes (Pb, Cd, Fe, NiOOH, etc.). This feature results in the subdivision of rechargeable cells into two major categories: those with aqueous electrolytes, which operate at ambient temperature, and those with nonaqueous electrolytes, most of

which operate at high temperatures (for reasons given below).

Ambient-Temperature Rechargeable Cells Rechargeable cells that operate at ambient temperature in almost all cases make use of aqueous electrolytes. The use of aqueous electrolytes limits the choice of electrode materials to those that do not react with the water in the electrolyte at an unacceptably high rate. This consideration eliminates the alkali metals (e.g., Li, Na, K) and the alkaline earth metals (e.g., Mg, Ca, Ba) from consideration, even though they have low electronegativities and low equivalent weights, and other desirable properties called for in Table 2. Materials that are sufficiently stable with aqueous electrolytes, have acceptably low electronegativities, take part in electrochemical reactions that can be reversed, and have other desirable properties as negative electrodes include Pb, Cd, Fe, and Zn. Elements that have high electronegativities are found in groups V, VI, and VII of the periodic chart of the elements. These have been used both as elements (e.g., Cl_2, Br_2) and in compounds (e.g., metal oxides such as PbO_2, NiOOH, AgO) in positive electrodes.

The lists of negative and positive electrode reactants given above can be used to select pairs of electrodes to make cells, along with suitable electrolytes. Aqueous electrolytes are usually strongly acidic (e.g., H_2SO_4) or strongly alkaline (e.g., KOH) in order to maximize the electrolytic conductivity, which contributes to achieving a low cell resistance and therefore a high specific power capability. The cells that are currently available, or are under development include $Pb/H_2SO_4/PbO_2$, $Cd/KOH/NiOOH$, $Fe/KOH/NiOOH$, $Zn/KOH/NiOOH$, $Cd/KOH/AgO$, $Zn/KOH/AgO$, $Zn/ZnBr_2\text{-}KCl/Br_2$, $Zn/ZnCl_2/Cl_2$.

The Pb/PbO_2 cell is the most widespread and least expensive of all those currently available. Its most common use is for starting, lighting, and ignition purposes in automobiles. The automobile battery stores about 1 kWh. With about 10^8 automobiles in the U.S., nearly 10^8 kWh of energy is stored in these batteries. For electric

TABLE 2. DESIRABLE CHARACTERISTICS OF SYSTEMS FOR HIGH-PERFORMANCE ELECTROCHEMICAL CELLS.

Characteristic	Negative Electrode Reactant	Positive Electrode Reactant	Electrolyte
Electronegativity	low (~ 1)	high (> 1.5)	–
Equivalent weight (g/g-equiv.)	low (< 30)	low (< 30)	low* (< 30)
Conductivity (ohm^{-1} cm^{-1})	high ($> 10^4$)	high ($> 10^4$)	high (> 1)
Electrochemical reaction rate (i_0, A/cm^2)	high ($> 10^3$)	high ($> 10^3$)	high ($> 10^3$)
Solubility in electrolyte (mol %)	low (< 0.1)	low (< 0.1)	–
Mass transport rate (equiv./(sec-cm^2))	high ($> 10^4$)	high ($> 10^4$)	high ($> 10^4$)

*A more important criterion for the electrolyte is low density.

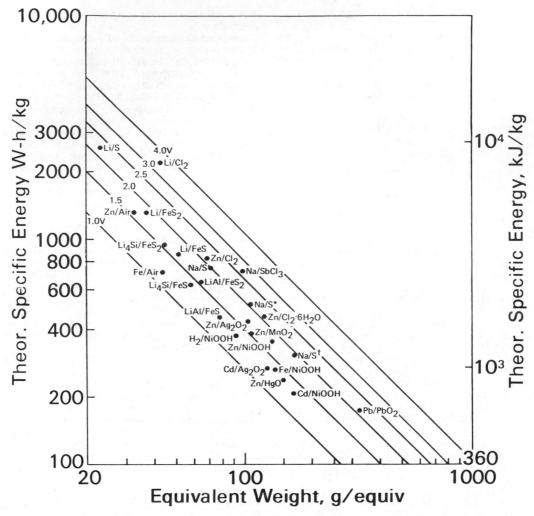

FIG. 1. Theoretical specific energy of candidate cells. Lines are for cell voltages from 1 volt to 4 volts, as labeled.

vehicle applications, the specific energy of Pb/PbO_2 at 25–40 Wh/kg is too low—70 Wh/kg is needed for an urban driving range of 150 km (a common goal). For electric utility energy storage, the cost of Pb/PbO_2 ($\sim$\$100/kWh) is too high, and the cycle life is too low.

The Cd/NiOOH cell, with its very long cycle life and ruggedness has found various military and aerospace applications that do not require storage of much energy (there is a limited supply of Cd, and it has a high cost). A significant consumer market for use in small battery-powered appliances has also developed.

The Fe/NiOOH cell, also known as the Edison cell, has disappeared from the marketplace, largely because of its high cost. There are now

efforts to develop a lighter-weight, lower-cost Fe/NiOOH cell for electric vehicle propulsion, but its modest specific energy of 40–50 Wh/kg, and its low efficiency ($\sim$60%) limit its applicability for electric automobiles or stationary energy storage.

The Zn/NiOOH cell has a very attractive specific energy of 70–75 Wh/kg, but the short cycle life of the Zn electrode ($<$300 cycles) has prevented it from achieving wide application. Research and development efforts aimed at improving the Zn electrode continue.

The Cd/AgO and Zn/AgO cells offer higher specific energy than their NiOOH counterparts, but the high cost of Ag limits them to military, aerospace, and specialty applications usually requiring the storage of small amounts of energy.

In recent years, Zn/halogen systems with flowing electrolyte-plus-reactant streams have been under development. These systems may be of acceptably low cost (less than $100/kWh) because they rely heavily on such inexpensive materials as carbon and polymers. The specific energies are in an intermediate range (about 60 Wh/kg for Zn/Br_2 and 70 Wh/kg for Zn/Cl_2). The safe storage of halogens is an important issue for these systems. Chlorine is stored as an icelike material $Cl_2 \cdot 8H_2O$ at temperatures below 9°C. Bromine is stored as a chemical complex formed from quaternary ammonium organic salts. This complex is a dense, oil-like liquid which spontaneously separates from the electrolyte. These Zn/halogen flow batteries are more complex than other batteries, and probably will not be acceptable for mobile applications if alternative batteries of similar specific energy are available. Stationary applications are good possibilities, especially if efficiencies above 70% are achieved.

There has been a great deal of research devoted to nonaqueous ambient-temperature rechargeable cells, especially those using Li electrodes, with the hope of achieving 100–200 Wh/kg. This has been a difficult research area because all the electrolytes investigated have been thermodynamically unstable in contact with lithium. Consequently, the lithium electrode has not been capable of being recharged very many times before failure, and there have been some explosions resulting from rapid reactions of lithium with the electrolyte. In addition, only a few candidate positive-electrode materials have shown the ability to be recharged repeatedly. One of these is TiS_2, a layered crystal that permits the entry of Li atoms between its crystal planes. Research continues on rechargeable ambient-temperature lithium cells, but none are available yet.

High-Temperature Systems It has been found that the alkali metals, which promise high specific energy, can be used more readily at elevated temperatures than at ambient temperature. This is the case because elevated-temperature operation permits the use of stable molten-salt or ceramic electrolytes which have high conductivities for alkali metal cations. For various chemical reasons, molten salt electrolytes are used with the lithium electrode, and ceramic electrolytes are used with the sodium electrode. The positive electrodes are sulfur or compounds containing sulfur. These cells operate at 350–475°C, and large batteries would be kept at operating temperature by means of efficient thermal insulation to contain the heat produced by the battery in its normal operation.

An example of the molten-salt cells is the $Li_4Si/LiCl-KCl/FeS_2$ cell. The compound Li_4Si is used instead of lithium because it is a lightweight solid, whereas Li is liquid at tempera-

tures above 180°C, and is therefore difficult to control in the cell. The same is true for solid FeS_2 as compared to sulfur, which melts at 118°C. The LiCl-KCl eutectic electrolyte melts at 352°C, and the cell is operated at about 450°C. The theoretical specific energy of this cell is 960 Wh/kg, as shown in Fig. 1. Practical values as high as 180 Wh/kg have been reported for single laboratory cells. Only a few small laboratory batteries have been built, many large cells and some multi-kWh batteries of LiAl/ LiCl-KCl/FeS cells have been demonstrated. This system is quite similar to $Li_4Si/LiCl-KCl/$ FeS_2, but it has a much lower specific energy (about 100 Wh/kg vs. 180–200 Wh/kg).

Molten-salt cells may find application for stationary energy storage, and for electric vehicles with high duty cycles, so that thermal cycling to ambient temperature is only infrequently required. In addition to high specific energy, these cells have inexpensive reactants. If the high-temperature corrosion problems can be overcome, molten-salt cells may be attractive.

The ceramic-electrolyte cells are best exemplified by the $Na/Na_2O \cdot 8Al_2O_3/S$ cell. The electrolyte is called beta alumina, and can conduct sodium ions in the crystallographic planes perpendicular to the C-axis. The purity and crystal size of the electrolyte are critical to achieving long life and high performance. This cell operates at 350°C, and uses liquid Na and liquid S as the reactants. A graphite felt material serves to collect the current in the nonconductive sulfur. The electrolyte is used in tube form, with one reactant inside, the other outside.

The ceramic electrolyte is made from materials which are basically inexpensive, but purity, quality control, and manufacturing conditions have made the process extremely expensive, and a major topic for continued development. The electrolyte should be strong, very dense (>97% of theoretical), of low resistivity (<10 Ω-cm), have controlled crystal size and structure and be highly pure.

Thousands of single cells of various sizes up to 100–200 Ah have been tested, as well as several multi-kWh batteries, and one 75–100 kWh battery. With additional development of the electrolyte, work on sulfur corrosion, and cost reduction, the Na/S cell could find applications similar to those mentioned above for Li_4Si/FeS_2.

Summary and Outlook The several rechargeable cells discussed above are listed with some of their important characteristics in Table 3. It is clear from this table that none of the systems at present meets the goals for the major energy storage applications of electric automobiles (>70 Wh/kg, >100 W/kg, >300 cycles, <$100/ kWh), or energy storage for electric utilities (>70% efficiency, 80 kWh/m² floor space, >2000 cycles, 10 years life, <$50/kWh). There are other applications that may be filled by the

TABLE 3. STATUS OF SOME RECHARGEABLE BATTERIES.

Cell	Voltage	Theor. Wh/kg	Actual Wh/kg	Actual W/kg	Cycle Life	Cost $/kWh	Comments
Ambient Temperature							
$Pb/H_2SO_4/PbO_2$	2.1	175	25–41	50–100	300–800	100	available now
$Cd/KOH/NiOOH$	1.3	202	40–50	200+	500–2000	>100	available, expensive
$Fe/KOH/NiOOH$	1.4	267	40–50	100	500+	>100	batteries on test, low efficiency
$Zn/KOH/NiOOH$	1.7	326	60–75	150	100–300	>100	cells and modules, short Zn life
$Cd/KOH/AgO$	1.4–1.1	257	50–65	200+	500	>>100	available, very expensive
$Zn/KOH/AgO$	1.8–1.5	432	80–110	200+	100	>>100	available, very expensive
$Zn/ZnBr_2\text{-}KCl/Br_2$	1.8	323†	60	75	400+	**	batteries on test, little life data, projected 35$/kWh
$Zn/ZnCl_2/Cl_2 \cdot 8H_2O$	2.1	405	66	70	1400*	**	complex system, batteries on test
High Temperature							
$LiAl/LiCl\text{-}KCl/FeS$	1.3	458	60–100	60–100	300+	**	small batteries on test
$Li_4Si/LiCl\text{-}KCl/FeS_2$	1.8, 1.3	944	120–180	100+	700	**	single cell status
$Na/Na_2O \cdot 8Al_2O_3/S$	2.0	758	100–160	60–130	250–1500	**	S corrosion problems, expensive ceramic, batteries on test

*1 kWh system, with electrolyte maintenance.
**Too early to predict; not available now.
†Includes weight of complexing agent.

newer systems, once developed. Batteries are already ubiquitous for storage of small amounts of energy (less than a few kWh); they may expand their applicability to many MWh.

ELTON J. CAIRNS

References

Cairns, E. J., in "Materials for Advanced Batteries" Murphy, D. W., and Broadhead, J., Eds.), Plenum Publishing Corporation, New York, 1982.

Birk, J. R., *EPRI Journal*, p. 6 (October, 1981).

Douglas, D. L., and Birk, J. R., *Annual Review of Energy* **5**, 61 (1980).

Kalhammer, F. R., *Scientific American* **241** (6), 56 (December, 1979).

Cairns, E. J., in "Energy and Chemistry" Thompson, R., Ed.), Royal Society of Chemistry, London, 1981.

Cairns, E. J., and Hietbrink, E. H., in "Comprehensive Treatise of Electrochemistry," Vol. 3 (Bockris, J. O'M., Conway, B. E., Yeager, E., and White, R. E., Eds.), Plenum Publishing Corporation, New York, 1981, p. 421.

E. J. Cairns, ibid., p. 341.

E. J. Cairns, in "Proceedings of the 3rd International Symposium on Molten Salts" (Mamantov, G., Blander, M., and Smith, G. P., Eds.), The Electrochemical Society, Pennington, New Jersey, 1981, p. 138.

Cross-references: ELECTROCHEMISTRY; ENERGY STORAGE, THERMAL-MECHANICAL; MOLE CONCEPT; WORK, POWER, AND ENERGY.

ENERGY STORAGE, THERMAL-MECHANICAL

Introduction* Energy storage systems serve to lower the cost of delivering energy in a suitable form to a user at the required time. For example, in electric utility systems the cost to produce energy can change by as much as a factor of five over a twenty-four hour period. By storing relatively cheap nighttime energy, the cost for supplying daytime energy can thus be significantly reduced.

Thermal energy storage systems rely on storing energy in the form of heat; whereas mechanical energy storage systems rely on storing energy in the form of potential and/or kinetic energy. Other forms of energy storage systems use chemical energy (e.g., batteries), electrical energy (e.g., capacitors) and electromagnetic energy (e.g., superconducting electromagnets).

Physical Principles—Mechanical Energy Storage Mechanical energy may be stored in a number of forms:

- Potential energy
- Kinetic energy
- Elastic (or strain) energy
- Compressed gas energy

The international system of units (SI) denotes the (J) as the unit to be used for energy. It is equivalent to $m^2 \cdot kg \cdot s^{-2}$ or $N \cdot m$, where m is

*See also ENERGY STORAGE, ELECTROCHEMICAL.

meters, kg is kilograms, s is seconds, and N is Newtons.

Potential energy is the energy associated with the position of an object in a gravitational field and is equal to $Mg\Delta h$, where M is the mass of the object in kg, g is the gravitational constant (at the earth's surface $g = 9.8 \text{ ms}^{-2}$), and Δh is the vertical distance (m) between two points within the gravitation field.

Kinetic energy can be stored in an object moving in a straight line where the stored energy equals $MV^2/2$ and V is the object's velocity (m s^{-1}). Kinetic energy can also be stored in a rotating object, with stored energy being equal to $I\omega^2/2$, where I is the object's *moment of inertia* (kg $\cdot$ m^2) about its axis of rotation and ω is it's angular velocity (s^{-1}). A wheel-shaped object in the form of a flat disc or radius R (m), thickness T (m), and density ρ (kg/m^3) has a moment of inertia about its center of $I = T\pi\rho R^4/2$, which equals $MR^2/2$.

Elastic (or strain) energy can be stored in the stretching or compressing of a solid object, where the stored energy equals $K(\Delta X)^2/2$. K is the object's spring constant (kg s^{-2}) and ΔX is the change of the object's length (m). The above can be generalized to two or three dimensions for solids and liquids.

Compressional gas energy can be stored in a gas that has been transformed from one pressure (p) and volume (V) to another pressure and volume. The expression for this stored energy is $\int_{V_2}^{V_1} p\,dV$, which represents the mathematical integral (or continuous summation) of the parameter p from volume V_1 to volume V_2. Pressure is represented by Newtons per square meter (N m^{-2}) and volume is represented by cubic meters (m^3).

Physical Principles—Thermal Energy Storage
Thermal energy may be stored in two forms:

- Sensible energy
- Latent energy

Sensible energy is stored when a substance at one physical state (e.g., the solid state) undergoes a temperature change. The expression for this stored energy is $CV(T_2 - T_1)$, where C is the *specific heat capacity* of a unit volume of the substance in J/kg $\cdot$ $^\circ$C, where $^\circ$C is temperature in degrees Centigrade; V is the volume of the substance (m^3); and T is temperature ($^\circ$C).

Latent energy is stored when a substance undergoes a change in physical state at constant temperature (e.g., from liquid state to gaseous state). The amount of stored energy depends on the type of substance, its mass, and the type of state change it undergoes. If the change of state occurs at the substance's melting point, the *heat of fusion* (J/kg) is the stored energy. If the change of state occurs at the substance's boiling

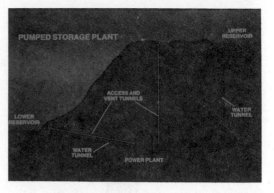

FIG. 1. Schematic diagram of pumped hydroelectric energy storage power plant.

point, the *heat of vaporization* (J/kg) is the stored energy.

Pumped Hydroelectric Energy Storage
Pumped hydroelectric plants (see Fig. 1) rely on the principle of potential energy storage and have been in use for over 50 years. The total power capacity (energy delivered per unit of time) all of such plants in the United States is currently approximately 13,000 megawatts (MW). A typical plant is rated at 1000 MW. The plant power rating is determined by the potential energy of the water stored in an upper reservoir (in relation to the vertical distance to the lower reservoir) divided by the time it takes for the water to flow through the water piping and pump-turbine generator separating the two reservoirs. These power plants have proved to be an economically attractive way for electric utilities to store inexpensive evening energy (used to pump the water into the upper reservoir) and thus provide daytime energy at a lower cost by allowing the stored upper reservoir water to flow through a turbine generator into the lower reservoir. Although these types of power plants are very efficient (approximately 72%), future construction efforts have been hampered by their large cost, diminished topographical opportunities, and their possibly negative environmental impact as perceived by nearby citizenry.

A good example of a pumped storage plant is the Northfield Mountain (Massachusetts) facility (1000 MW). The water stored in the upper reservoir can produce power for a nominal six hours; thus the stored energy is 6000 MWh which is equal to 2.16×10^{13} J. For comparison purposes the amount of energy stored in this one facility is approximately 20% of the storage capacity of all the automotive batteries in the United States.

Flywheel Energy Storage Flywheel energy storage devices rely on the principle of kinetic energy storage and have likely been used by man longer than any other type of energy storage device. They have smoothed or leveled man's energy sources in such historic applications as the potter's wheel, the grain mill, and the water wheel. More recently, flywheel technology has been applied in the design of the steam engine, automobile, pile driver, and gyroscope. The use of the potter's wheel permitting a craftsman with a heavy wheel to make uniformly shaped pots while he supplied leg energy in unequal amounts and at irregular times, probably dates back 5,000 years.

As the flywheel spins, a tensile stress or load is created in its structural material. The amount of stress increases as the square of the flywheel angular velocity. The ultimate strength of the flywheel material to withstand this stress limits the amount of energy that can be stored. The relationship between the strength σ (N/m^2) of the material, the stored energy E, the material mass M, and the density ρ (kg/m^3) is $\sigma = KE\rho/M$, where K is a constant that is greater than or equal to 1 and is determined from the geometry of the flywheel. The energy stored in a flywheel per unit volume V is $E/V = E\rho/M = \sigma/K$. For a flywheel with $K = 2$ (associated with a nontapered wheel rim) and $\sigma = 2 \times 10^8$ N/m^2 (a value corresponding to steel), the energy storage per unit volume is 10^8 $N/m^2 = 10^8$ J/m^3.

Compressed Air Energy Storage Compressed air energy storage (CAES) devices have recently become commercialized for electric utility applications. They incorporate modified state-of-the-art combustion turbines and site-specific underground reservoirs to store off-peak energy for later use during peak demand periods (see Fig. 2). The CAES power plants operate in different modes during different time periods. During an off-peak period, relatively inexpensive energy from coal-fired and/or nuclear baseload units powers the motor to compress air, which is then stored in an underground reservoir. During the subsequent peak period, the compressed air is withdrawn from storage and mixed with fuel. The combustion gases created by burning this mixture are expanded through turbines connected to the generator motor to produce electric power. Because the turbine is not required to drive a compressor during the peak time period (which it must do in a conventional combustion turbine system) a CAES plant reduces the use of petroleum- or gas-based fuels by more than 60%.

The underground caverns used to store the compressed air can be excavated in three types of geologic formations: salt domes, hard rock, and aquifers. Approximately three-fourths of the continental United States has suitable geology for compressed air caverns.

A 290 MW compressed air plant using underground salt-dome caverns is the first such plant in the world and is currently operating in Huntorf, Federal Republic of Germany, providing peak-leveling duty for the German utility Nordwestdeutsche Kraftwerke Ag. (Fig. 3). This power plant was commissioned in 1978 and has a storage capacity lasting 4 hours. The first USA application of this storage technology is currently under consideration by a number of USA utilities.

FIG. 3. Turbomachinery of the world's first compressed air energy storage power plant at Huntorf, Federal Republic of Germany.

COMPRESSED AIR STORAGE CYCLE

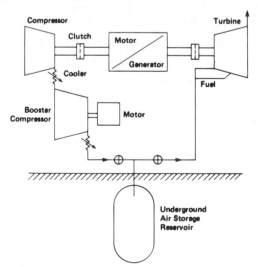

FIG. 2. Schematic diagram of compressed air energy storage power plant.

TABLE 1. EXAMPLES OF SENSIBLE HEAT AND
LATENT HEAT MATERIALS.

Material	Sensible Heat,* Specific Heat Capacity $(J/kg°C)$	Latent Heat	
		Heat of Fusion (J/kg)	Heat of Vaporization (J/kg)
Water, H_2O	4.18	3.34×10^5	2.26×10^6
Mercury	1.39×10^2	1.18×10^4	2.95×10^5
Lead	1.28×10^2	2.29×10^4	–
Glauber's salt	1.93×10^3	2.51×10^5	–

*At 1 atm., $20°C$

Sensible Heat Energy Storage Sensible heat thermal energy systems have been used by man for thousands of years to keep dwellings at relatively constant temperature. Thick adobe walls on a dwelling store heat during the day and keep the inhabitants warm at night. This and other ways to store solar energy utilizing sensible heat systems have recently been rediscovered because of increased interest in reducing both fuel consumption and associated heating and cooling costs.

Other ways to store heat often used in the past include the use of large stones, ceramic bricks, and even cast iron blocks near an early evening fire, which transmit their heat as radiant energy to warm the user after the fire goes out. More recently, in steam-electric power plants, the principle of sensible heat has been used to store energy from steam made available during low electricity demand periods. The steam is used to raise the temperature and pressure of water contained in tanks. When steam is needed during high electricity demand periods, a valve is opened on the upper portion of the tank to allow the pressure to decrease and produce the needed steam, which then lowers the temperature of the water. Although the input and output of this process is steam, the storage of the input energy is accomplished by the sensible heat of the water. The specific heat capacity of water as well as some other substances is provided in Table 1.

Latent Heat Energy Storage Latent heat thermal energy storage systems have not been as widely used as sensible heat storage systems, although latent heat systems usually have a greater energy storage potential than sensible heat systems. This is due to the high energy storage density per unit volume which can theoretically be achieved with latent heat systems, as shown in Table 1. Latent heat systems also have the advantage of operating over a narrow temperature range, which limits the design specification requirements to be met by the storage device materials, energy input, energy output, and control system components.

However, other technical issues have proved to be severe enough to hamper the widespread use of latent heat storage systems. These include the low thermal conductivity of candidate materials and their volumetric decrease during solidification, both of which inhibit the rate of heat transport during the energy input and output time periods. Also, many candidate materials do not completely solidify at their freezing point; rather, portions of the materials

TABLE 2. ENERGY STORAGE DENSITIES.

Energy Storage Technology	Energy Per Unit Mass (J/kg)	Energy Per Unit Volume (J/m^3)	Comments
Pumped hydroelectric	3×10^3	3×10^6	300 m hydraulic head
Flywheel	2×10^5	2×10^8	Fiber-epoxy composite material
Compressed air	6×10^5	10^7	Plant similar to Huntorf
Sensible head			
magnesite brick	4×10^5	2×10^8	
oil	10^5	10^8	
water	4×10^5	4×10^8	In liquid phase at 1 atm, $100°C$
Latent heat, salts	10^5	3×10^8	

stay in a supercooled liquid state below their theoretical freezing point. This is due to the slow nature of the propagation of solid crystals in a liquid at its freezing point.

One of the more promising materials which exhibits the above problems to a lesser extent than other materials is sodium sulfate decahydrate (Glauber salt, $Na_2SO_4 \cdot 10H_2O$). It has been successfully tested in two solar homes and has been used in heat pumps and air conditioning systems.

Summary A comparison of some of the thermal-mechanical energy storage systems available today is presented in Table 2. Each of the storage systems described above has inherent characteristics that determine its size, weight, and associated cost. Taking these factors together with user needs and the commercial maturity of the technology, each storage system has its preferred application. Some operate in a mode where the form of the input and output energy is the same as that which is stored (e.g., ceramic brick thermal storage systems) whereas others operate in a mode where the input and output energy is different from that which is stored (e.g., pumped hydroelectric energy storage systems). Thus, the comparison shown in Table 1 only gives a portion of the many factors that need to be considered when choosing the "best" storage device for a particular energy leveling or smoothing application.

Robert B. Schainker

References

1. Schmidt, Frank W., and Willmott, John, "Thermal Energy Storage and Regeneration," New York, McGraw-Hill Book Company, 1981.
2. Silverman, Joseph (Ed.), "Energy Storage: A Vital Element in Mankind's Quest for Survival and Progress, "Transactions of the First International Assembly held at Dubrovnik, Yugoslavia, 27 May– 1 June 1979, Conference jointly sponsored by the National Academy of Sciences (USA) and the Council of Academies of Science (Yugoslavia), Paragon Press, 1980.
3. "Energy Storage User Needs and Technology Applications," An Engineering Foundation Conference, sponsored by the Electric Power Research Institute and Energy Research and Development Administration, Asylomar, Pacific Grove, California, 8–13 February 1976, National Technical Information Service, CONF-760212, April 1977.
4. Hassenzahl, William V. (Ed.), "Mechanical, Thermal and Chemical Storage of Energy," Stroudsburg, PA, Hutchinson Ross Publishing Co., 1981.
5. Kalhammer, Fritz R., "Energy Storage Systems," *Scientific American*, **241** (6), 56–65 (December, 1979).
6. "An Assessment of Energy Storage Systems Suitable for Use by Electric Utilities," Electric Power Research Institute Research Project 225, Report EM-264, Vol. 1–3, July 1976.
7. "Preliminary Design Study of Underground Pumped Hydro and Compressed Air Energy Storage in Hard Rock," Electric Power Research Institute Research Project 1081, Report EM-1589, Vol. 1–13, 1981.

Cross-references: COMPRESSIBILITY, GAS; ELASTICITY; HEAT; HEAT TRANSFER; MECHANICS; SOLAR ENERGY UTILIZATION; VAPOR PRESSURE AND EVAPORATION; WORK, POWER, AND ENERGY.

ENTROPY

The word "entropy" was coined from the Greek by Rudolf Clausius in 1865 to mean "transformation." He defined the differential change in entropy dS by:

$$dS = dQ/T \quad \text{(reversible)} \tag{1}$$

where dQ is the element of heat added reversibly to a system at absolute temperature T. Since the values of dQ and T may be defined *operationally*, there is no need, in the classical view of thermodynamics, to explain entropy in molecular terms. As Lewis and Randall have written: "Thermodynamics exhibits no curiosity; certain things are poured into its hopper, certain others emerge according to the well known laws of the machine; no cognizance being taken of the mechanism of the process or the nature and character of the various molecular species concerned."

In 1948, Claude Shannon[3] demonstrated that the change in the function defined by

$$S_I = -k \sum_i p_i \log p_i \tag{2}$$

uniquely measures the amount of information in any message (p_i = probability that the receiver assigns to the receipt of the ith possible). The sum is over all possible messages. Here is a simple example of the meaning of entropy. Suppose you are asked to guess which number from 1 to 100 has been secretly written on a piece of paper. Taking each p_i equal to 1/100, it is found that in the above equation, $S_I = k \log (100)$. Letting k equal 1 and taking the logarithm to the base 2 gives $S = 6.67$ "bits." This means if each question has only two possible answers (i.e., "yes" or "no"), it will take between 6 and 7 questions to find the number. The questioner begins by asking, "Is the number between 1 and 50?" Depending on the answer, he continues by reducing the possible numbers by a factor of 2 with each question. Since $2^6 = 64$ and $2^7 = 128$, the questioner will surely find the number in 7 questions and 1/3 of the time he will find it in 6 questions. Entropy may therefore be said to measure the state of ignorance of a person relative to a well-defined question, if the person only knows a probability distribution. It measures the expected number of questions he will have to ask in order to go from his state of partial knowledge to a state in which he knows everything about the well-defined question. According to the Shannon derivation,[3] S_I represents the amount of information in a message telling the actual number written on the paper (for someone for whom $p_i = 1/100$). The interpretation of entropy is used in information theory in the design of codes and in the analysis of information transmission systems.

These two entropies would seem to be different and to be connected only through having the same name. However, they have been shown to be identical when applied to the analysis of how to interpret the messages an observer obtains from instruments which only can measure the average properties of quantum systems.[1] Consider the question: "In what quantum state is this system?" Of course one can never say in which quantum state a system resides but rather can only give a probability for the system being in a particular state. The probability distribution for the states must be consistent with the observer's knowledge. The information theory principle of maximum entropy says to choose a set of probabilities which agrees with the available data and maximizes the entropy, for, in accordance with the meaning of entropy, this is the most noncommittal view. A state of equilibrium is, by the definition given by Gibbs,[4] a state of maximum entropy. From the information theory point of view, it is a state in which all of the random motions which can take place are, in fact, taking place so that the observer knows as little about the systems as it is possible to know beyond his knowledge of the "constants of the motion." If we describe a system by giving only the pressure, temperature and volume (or other gross properties), we omit many details. Entropy measures how much more there is to be said before the quantum state is specified.

As an illustration, consider a closed system of particles. From quantum mechanics, we know the system is in some state, i, with energy ϵ_i. If we make an observation of c, the best we can do is infer that it represents the expectation energy, $\langle \epsilon \rangle$, for this is a repeatable quantity associated with the motion. To generate the appropriate probability distribution, we maximize S_I defined in Eq. (2), with the following constraints on the p_i:

$$\sum p_i = 1 \text{ (the system is in some state)} \quad (3)$$

$$\sum p_{i \in i} = \langle \epsilon \rangle \text{ (the system has an expectation energy)} \quad (4)$$

Maximization of the entropy subject to the two equations given leads to the probability distribution

$$p_i = e^{-\psi - \beta \epsilon_i} \quad (5)$$

The resultant entropy, $S_{I,\max}$, is, by Gibbs[1] definition, the equilibrium entropy. The probability distribution is known as the Boltzmann distribution. It is easy to demonstrate that the parameter β is equal to $1/kT$ where k is the Boltzmann constant and T is the absolute temperature. If the observer is limited to macroscopic observations concerned with the energy of a body, the above derivation leads to laws which connect the observations to one another. That is, the quantities ψ, β, $\langle \epsilon \rangle$, S and various combinations are related to one another by equations from which p_i and ϵ_i have been eliminated. These relations are known as the "Laws of Classical Thermostatics".

For example, since, from Eq. (4)

$$d \langle \epsilon \rangle = \sum_i \epsilon_i dp_i + \sum_i p_i d\epsilon_i$$

it is clear that the changes in the energy of a body may be divided into two classes: (a) those of the type $\sum \epsilon_i dp_i$ which *necessarily* change the probabilities and therefore the entropy (non-isentropic) and (b) those in which $\sum_i \epsilon_i dp_i = 0$ (i.e., isentropic). This division into two classes of energy exchange gives rise to the concepts of "heat" and "work." For a detailed account of the derivation see reference 2.

The information theory definition of entropy thus is shown to contain, as a special case, the entropy of Clausius. The two ways of approaching entropy are seen to be consistent with one another. In the classical case concepts such as heat, work, equilibrium, temperature and state are taken as "operationally defined." From the information theory perspective, these concepts are *derived* as necessary constructs in any consistent theory of molecular phenomena. As demonstrated in (1), (2) and (5) the concept of heat has to be invented to describe energy exchanges which take place on a scale too fine to be observed directly whereas the work concept describes energy transfers which are associated with changes in macroscopic, observable coordinates.

Entropy, as a logical device for generating probability distributions, has been applied in reliability engineering, decision theory, and the theory of steady-state irreversible processes. (See list of references with reference 5.) The generalized approach to entropy usage was initiated by Jaynes.[6] Applications of the entropy principle in engineering are given in reference 7. A comprehensive review of the method of entropy maximization in many fields is given in reference 8.

M. TRIBUS

References

1. Tribus, M., "Information Theory as the Basis for Thermostatics and Thermodynamics," *J. Appl. Mech.*, B (March 1961).

2. Tribus, M., "Thermostatics and Thermodynamics," New York, Van Nostrand/Reinhold, 1961.
3. Shannon, Claude, "A Mathematical Theory of Communication," *Bell System Tech. J.* (July and October 1948).
4. Gibbs, J. W., "Collected Works," Vol. I, p. 56, New Haven, Conn., Yale University Press.
5. Tribus, M., and Evans, R., "The Probability Foundations of Thermodynamics," *Appl. Mech. Rev.*, 765 (October 1963).
6. Jaynes, E. T., "Information Theory and Statistical Mechanics," *Phys. Rev.*, **106**, 620; **108**, 171 (1957).
7. Tribus, M., "Rational Descriptions, Decisions and Designs," Oxford, Pergamon Press, 1969.
8. Levine, R., and Tribus, M., "The Maximum Entropy Formalism," Cambridge, MA, MIT Press, 1978.

Cross-references: BOLTZMANN'S DISTRIBUTION LAW, IRREVERSIBILITY, PHYSICAL CHEMISTRY, STATISTICAL MECHANICS, THERMODYNAMICS.

EQUILIBRIUM

In the elementary sense of the macroscopic (visible to the naked eye) system, equilibrium is obtained if the system does not tend to undergo any further change of its own accord. Any further change must be produced by external means.

Mechanical and Electromagnetic Systems Equilibrium in mechanical and/or electromagnetic systems is reached when the vectorial summation of generalized forces applied to the system is equal to zero. In any potential field, that is, gravitational or electric potential or magnetic vector potential, force can be expressed as gradient of potential (magnetic force however, is a curl of a vector potential). The potential energy therefore has an extremum at the equilibrium configuration. For example, a system such as a mass suspended by a string against the gravitational force (or its weight) is at mechanical equilibrium if the tensile force in the string is equal to the weight of the mass it supports. The d'Alembert principle further states that the condition for equilibrium of a system is that the virtual work of the applied forces vanishes.

Thermodynamic Systems When a hot body and a cold body are brought into physical contact, they tend to achieve the same warmth after a long time. These two bodies are then said to be at thermal equilibrium with each other. The zeroth law of thermodynamics (R. H. Fowler) states that two bodies individually at equilibrium with a third are at equilibrium with each other. This led to the comparison of the states of thermal equilibrium of two bodies in terms of a third body called a thermometer. The temperature scale is a measure of state of thermal equilibrium, and two systems at thermal equilibrium must have the same temperature (see THERMODYNAMICS).

Generalization of equilibrium consideration by the second law of thermodynamics specifies that the state of thermodynamic equilibrium of a system is characterized by the attainment of the maximum of its ENTROPY. Thermodynamic coordinates are defined in terms of equilibrium states.

Equilibrium between two phases of a system is reached when there is no net transfer of mass or energy between the phases. Phase equilibrium is determined by the equality of the Gibbs functions (also called free enthalpy, free energy, or chemical potential) of the phases in addition to equality of their temperatures and stresses (such as pressure and/or field intensities—intensive properties). Equilibrium of first-order phase change requires continuity of slope or first derivative of the Gibbs function with respect to an intensive property and is generalized as the Clapeyron relation. Second- and higher-order phase changes are given by the condition of continuity of curvature or second derivative of the Gibbs function and so on.

Chemical or nuclear equilibrium of a reactive system is reached when there is no net transfer of mass and/or energy between the components of a system. At chemical or nuclear equilibrium, the Gibbs function of the reactants and the products must be equal according to stoichiometric proportions, in addition to uniformity in temperature and stresses. Chemical equilibrium is summarized in the form of the Law of Mass Action. The trend for the displacement from an equilibrium state is specified by Le Châtelier's principle.

Thermodynamic equilibrium is reached when the condition of mechanical, electromagnetic, thermal, phase, and chemical and nuclear equilibrium is reached.

Stability of Equilibrium A process or change of state carried out on a system such that it is always near a state of equilibrium is called a quasi-stationary equilibrium process. This requires that the process be carried out slowly. If a mechanical system is initially at the equilibrium position with zero initial velocity, then the system will continue at equilibrium indefinitely. An equilibrium position is said to be stable if a small disturbance of the system from equilibrium results only in small, bounded motion about the rest position. The equilibrium is unstable if an infinitesimal displacement produces unbounded motion. In the gravitational field, a marble at rest in the bottom of a bowl is in stable equilibrium, but an egg standing on its end is in unstable equilibrium. When motion can occur about an equilibrium position without disturbing the equilibrium, the system is in neutral (or labile, or indifferent) equilibrium, an example being a marble resting on a perfectly flat plane normal to the direction of gravity. It is readily seen that stable equilibrium is the case when the extremum of potential is a minimum.

When dealing with general thermodynamic systems, the fact that entropy tends to a maximum in the trend toward equilibrium of a natural process generalizes the above mechanical consideration with respect to stability. An equilibrium state can be characterized as a stable equilibrium when the entropy is a maximum; neutral equilibrium when displacement from one equilibrium state to another does not involve changing entropy; and unstable equilibrium when entropy is a minimum. Any slight disturbance from an unstable equilibrium state of a system will lead to transition to another state of equilibrium.

Statistical Equilibrium In the microscopic sense, that is, treating systems in terms of elemental particles such as molecules, atoms, and other material or quasi-particles (such as photons in radiation, phonons in solids and liquids, rotons in liquids), equilibrium states are recognized as the most probable states. An equilibrium state of a system is therefore defined in terms of most probable distributions of its elements among microscopic states which may be defined in terms of energy states. In this sense, statistical equilibrium is a condition for macroscopic equilibrium and an equilibrium state of a system is one of its extremal states. In the methods of STATISTICAL MECHANICS, the probability of distribution is expressed in terms of the density of distributions in the phase space. Based on the Liouville theorem, if a system is in statistical equilibrium, the number of the elements in a given state must be constant in time; which is to say that the density of distribution at a given location in phase space does not change with time. For an isolated system, the distribution is represented by a microcanonical ensemble. At equilibrium, no phase point can cross over a surface of constant energy, and the density of distribution is preserved. In this case individual molecules of a system can be represented by phase points. Any part of an isolated system in statistical equilibrium can be represented by a canonical ensemble. A subsystem of a large system in thermal equilibrium also behaves like the average system of a canonical ensemble. A system and a constant temperature bath together can be considered as an isolated system. A phase point in a canonical ensemble can represent a large number of molecules, thus accounting for strong interactions. A canonical ensemble is characterized by its temperature and is therefore pertinent to the concept of thermal equilibrium. When applied to equilibrium of systems involving mass exchange, such as a chemical system, we have a "particle bath" in addition to a constant temperature bath. The pertinent representation for equilibrium including mass exchange as well as energy exchange is known as a grand canonical ensemble, which accounts for the chemical potentials of its elements.

When applied to a system with a large number of elements, the distributions are measured by thermodynamic probability (W); the most probable distribution is such that W is a maximum. This optimal principle is consistent with the condition of maximum entropy (S) given in the above. The Boltzmann hypothesis states that $S = k \ln W$, where k is the Boltzmann constant.

Depending on the specifications of W, namely, those of Maxwell-Boltzmann (for low concentration of distinguishable particles, weak interaction and high temperature, such as a dilute perfect gas), Fermi-Dirac (for elemental particles with antisymmetric wave functions at high concentrations of indistinguishable particles and low temperatures, such as electrons in metal), or Einstein-Bose (for elemental particles with symmetric wave functions, such as He^4 at high concentration of indistinguishable particles and low temperature), equilibrium distributions take different forms (see BOSE-EINSTEIN STATISTICS and FERMI-DIRAC STATISTICS). The Maxwellian speed distribution in a dilute perfect gas is a distribution based on Maxwell-Boltzman statistics.

As a consequence of molecular considerations, when two systems are connected for transfer of mass without significant transfer of energy, such as two containers at different temperatures connected by a capillary tube, we have the relation of thermal transpiration.

Trend toward Equilibrium The mechanism by which equilibrium is attained can only be visualized in terms of microscopic theories. In the kinetic sense, equilibrium is reached in a gas when collisions among molecules redistribute the velocities (or kinetic energies) of each molecule until a Maxwellian distribution is reached for the whole bulk. In the case of the trend toward equilibrium for two solid bodies brought into physical contact, we visualize the transfer of energy by means of free electrons and phonons (lattice vibrations).

The Boltzmann H-theorem generalizes the condition that with a state of a system represented by its distribution function f, a quantity H, defined as the statistical average of $\ln f$, approaches a minimum when equilibrium is reached. This conforms with the Boltzmann hypothesis of distribution in the above in that $S = -kH$ accounts for equilibrium as a consequence of collisions which change the distribution toward that of equilibrium conditions. In terms of quantum states and their probabilities, it is also readily shown that any value of S below the equilibrium value will increase in time. Recursion of an original state for a fraction of one gmol of molecules may take a time period much longer than the estimated life of the universe of 10^{10} years.

Consideration of perturbation from an equilibrium state leads to methods for dealing with rate processes and methods of irreversible thermodynamics in general.

Fluctuation from Equilibrium A necessary consequence of the random nature of elemental particles in a body is that the property of such a body is not at every instant equal to its average value but fluctuates about this average. A precise meaning of equilibrium can only be attained from consideration of the nature of such fluctuations. In the above, we have repeatedly considered a "large" number of particles. It is important to know how large a number is "large." When considering fluctuation of energy from an average value in an isolated system, the ratio of the two is given to be proportional to $1/\sqrt{N}$, where N is the total number of elements in the system. This is also the magnitude of the fluctuation of number of particles in a system involving transformation of phases and chemical and nuclear species. An equilibrium state is one at which the longtime mean magnitude of fluctuation from the average state is independent of time and this magnitude has reached a minimum value.

Large perturbation from a given state of fluctuation leads to a relaxation process toward a state of equilibrium. The relaxation time, for instance, measures the deviation from quasistationary equilibrium of a process which is carried out at a finite rate.

S. L. Soo

References

Goldstein, H., "Classical Mechanics," Cambridge, Mass., Addison-Wesley Publishing Co., Inc., 1956.

Soo, S. L., "Analytical Thermodynamics," Englewood Cliffs, N.J., Prentice-Hall, Inc., 1962.

Mayer, J. E. and Mayer, M. C., "Statistical Mechanics," New York, John Wiley & Sons, 1977, Chapter 6.

Cross-references: BOSE-EINSTEIN STATISTICS, ENTROPY, FERMI-DIRAC STATISTICS, IRREVERSIBILITY, STATICS, THERMODYNAMICS, VECTOR ANALYSIS.

EXCITONS

Intrinsic Excitons The exciton is a quantum of electronic excitation produced in a periodic structure such as an insulating or semiconducting solid. This quantum of energy has motion and the motion is characterized by a wave vector. Frenkel[1] was the first to treat the theory of optical absorption in a solid as a quantum process consisting of atomic excitations. The excitation process implies that the excited electron does not leave the cell from which it was excited. In his attempt to gain insight into the transformation of light into heat in solids he was able to explain the transformation by first-order perturbation theory of a system of N atoms having one valence electron per atom with the following properties:

1. The coupling between different atoms in a crystal is small compared with the forces holding the electron within the separate atoms.
2. The Born-Oppenheimer approximation is valid.
3. The total wave function is a product of one-electron functions.

The Frenkel exciton is a tight-binding description of an electron and hole bound at a single site such that their separate identities are not lost. This model is applicable to insulating crystals. In the case of semiconductors, nonequilibrium electrons and holes are bound in excitons at low temperatures by Coulomb attraction. Semiconducting crystals are characterized by large dielectric constants and small effective masses, therefore the electrons and holes may be treated in a good approximation as completely independent particles, despite the Coulomb interaction. This results from the fact that the dielectric constant reduces the Coulomb interaction between the hole and electron to the extent that it produces a weakly bound pair of particles which still retain much of their free character. The exciton forms because it represents a state of slightly lower energy than the unbound hole-electron. The effective mass theory used to describe such weakly bound particles was developed by Wannier.[2] These weakly bound excitons are most appropriately described when the band structure of the solid is considered. The intrinsic fundamental-gap exciton in semiconductors is a hydrogenically bound hole-electron pair, the hole being derived from the upper valence band and the electron from the lowest conduction band. It is a normal mode of the crystal created by an optical excitation wave, and its wave functions are analogous to those of the Block wave states of free electrons and holes. When most semiconductors are optically excited at low temperatures, it is the intrinsic excitons that are excited. The energies of the ground and excited states of the exciton lie below the band-gap energy of the semiconductor. Hence, the exciton structure must first be determined in order to determine the band-gap energy. The exciton binding energy can be determined from spectral analysis of its hydrogenic ground and excited state transitions (this also gives central-cell corrections). Precise band-gap energies can be determined by adding the exciton energy to the experimentally measured photon energy of the ground state transition.

Both direct and indirect exciton formation occurs in semiconductors, depending on the band structure. The former is characterstic of many of the II-VI and III-V compounds, and the latter is characteristic of germanium and silicon. For indirect optical transitions, momentum is conserved by the emission or absorption of phonons. The detailed nature of the band structure of semiconductors with degenerate or simple valence bands is elucidated by understanding the intrinsic-exciton structure of these semiconductors.

High Density Electron-Hole Pairs Exciton complexes analogous to H_2, H_2^+ and H^- were first described by Lampert.[3] The exciton in real crystals may be considered as a boson since it consists of an electron and a hole, each being a fermion, bound together by Coulomb interaction. Two single excitons exhibit an attractive covalent interaction which results in the formation of an excitonic molecule if the electron and hole spins are antiparallel. The binding energy of such a molecule has been calculated by Akimoto, et al.[4] They showed that exciton molecules should be stable for any electron-to-hole mass ratio. Therefore, at high excitation intensities and if the mean time for the formation of exciton molecules is less than the radiative lifetime of single excitons, then the exciton molecule system is produced.

At high excitation intensities two states of the excited system have been proposed: (1) the excitonic-molecule gas, (2) the electron-hole droplet. Calculations of the ground state energies of the excitonic-molecule system and the electron-hole droplet system have been successful in determining which system is active in the particular material being studied. It was recognized that the band structure of the material would have a major impact on the ground state of the system.

With high exciting intensities the density of hole-electron pairs is increased and excitons are formed at a higher rate. At high concentrations the interaction among excitons becomes important, and when a certain threshold density of excitons is reached liquid droplets are formed. These collective droplets consist of nonequilibrium electrons and holes and, when electron-hole recombination occurs specific radiation is emitted. A luminescence line appears at a slightly lower energy than that of the free exciton in both Si and Ge which has been identified as light from an electron-hole liquid.

Motion of Excitons The motion of excitons has been dramatically demonstrated by Gourley et al.[5] in stressed Si. The crystal was stressed along the $\langle 100 \rangle$ direction by a rounded ($R \approx 5$ cm) steel plunger. The indirect band edge is lowered in energy by strain, as a result the droplets, free excitons, and excitonic molecules are accelerated in a strain gradient toward the point of maximum shear stress. The image of the photoluminescence at 15 K, produced by laser excitation at the side of the crystal was displayed by vidicon detection and is shown in Fig. 1. The parabolic strain well provides both spatial confinement and reduced electron-hole liquid binding energy. This experimental arrangement allowed them to achieve large concentrations of free excitons with low-level steady state excitation. The high density of free excitons allowed the production of excitonic molecules in appreciable numbers. In examining the spatial extent of the excitonic gas in a parabolic potential well, they treated the exciton gas as an ideal gas. The equipartition theorem $\langle V \rangle = \alpha \langle r^2 \rangle = \frac{3}{2} KT$ predicts that the radial extent of the gas is $\approx T^{1/2}$. They examined the spatial profile by slowly scanning a sharply focused image of the crystal across a narrow slit masking the Ge photodetector. In the temperature range 5–22 K, the $T^{1/2}$ radial profile was reasonably well obeyed.

Spatial Resonance Dispersion For those states in a crystal where the photon wave vector and the exciton wave vector are essentially equal, the energy denominator for exciton photon mixing is small and the mixing becomes large. These states are not to be considered as pure photon states or pure exciton states but rather mixed states. Such a mixed state has been called a *polariton*. When there is a dispersion of the dielectric constant, spatial dispersion has been invoked to explain certain optical effects of crystals. It was originally thought that it would introduce only small corrections to such things as the index of refraction until Pekar[6] demonstrated that, if there was more than one energy transport mechanism, as in the case of excitons, this was not true. Spatial dispersion addresses the possibility that two different kinds of waves of the same energy and same polarization can exist in a crystal differing only in wave vector. The one with an anomalously large wave vector is an anomalous wave. In the treatment of dispersion by exciton theory, Pekar showed that if the normal modes of the system were allowed to depend on the wave vector, a much higher order equation for the index of refraction would result. The new solutions occur whenever there is any curvature of the ordinary exciton band in the region of large exciton-photon coupling. These results apply to the Lorentz model as well as to quantum-mechanical models whenever there is a dependence of frequency on wave number.

Bound Excitons The intrinsic exciton may bind to various impurities, defects, and complexes and the subsequent decay from the bound state yields information concerning the center to which it was bound. Bound exciton complexes are extrinsic properties of materials. These complexes are observed as sharp-line optical transitions in both photoluminescence and absorption. The binding energy of the exciton to the impurity or defect is generally weak compared to the free exciton binding energy. The resulting complex is molecular-like (analogous to the hydrogen molecule or molecule-ion) and has spectral properties which are analogous to those of simple diatomic molecules. The emission or absorption energies of these bound exciton transitions are always below those of the corresponding free exciton transitions, due to the molecular binding energy.

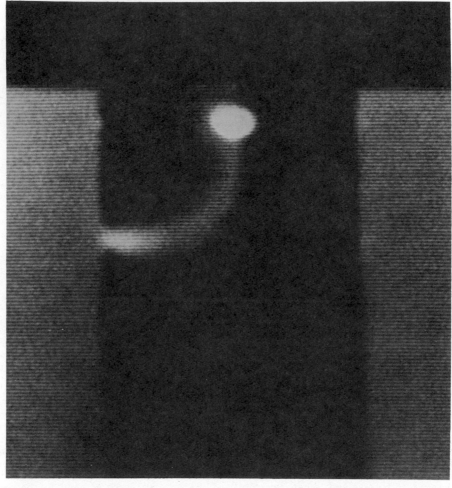

FIG. 1. Exciton luminescence emanating from a 1.5 × 1.5 × 4 mm crystal of ultrapure Si at $T \approx 10$ K. The top crystal surface, accurately flattened and polished, is pressed with a spherical contacting surface of a steel plunger cut from a ball bearing with radius $R = 3.8$ cm. Excitons produced at the left crystal surface by a CW Ar laser are drawn into a parabolic potential well corresponding to a shear strain maximum beneath the plunger (Gourley et al., Ref. 5).

Bound excitons were first reported by Haynes[7] in the indirect semiconductor silicon. He found that when group V elements were added to silicon sharp photoluminescent lines were produced and these lines were displaced in energy in a regular way. The binding energies of exciton complexes produced by adding different group V donors were described by the linear relation:

$$E = 0.1 E_i$$

where E is the binding energy of the exciton and E_i is the ionization energy of the donor. The small differences in ionization energies for different effective mass chemical donors result from central cell corrections. A similar relationship was found when the group III acceptors were added to silicon. A modified linear relationship has been found for donors and acceptors in compound semiconductors.

The sharp spectral lines of bound exciton complexes can be very intense (large oscillator strength). The line intensities will, in general, depend on the concentrations of impurities and/or defects present in the sample.

The theory of "impurity" or defect absorption intensities in semiconductors has been studied by Rashba.[8] He found that, if the absorption transition occurs at $k = 0$ and if the discrete level associated with the impurity approaches the conduction band, the intensity of the absorption line increases. The explanation offered for this intensity behavior is that the optical excitation is not localized in the impurity but encompasses a number of neighboring lattice points of the host crystal. Hence, in the absorption process, light is absorbed by the entire region of the crystal consisting of the impurity and its surroundings.

In an attack on the particular problem of excitons which are weakly bound to localized "impurities," Rashba and Gurgenishvili[9] derived the following relation between the oscillator strength of the bound exciton F_d and the oscillator strength of the intrinsic excitons f_{ex}, using the effective-mass approximation

$$F_d = (E_0/|E|)^{3/2} f_{ex},$$

where $E_0 = (2\hbar^2/m)(\pi/\Omega_0)^{2/3}$. E is the binding energy of the exciton to the impurity, m is the effective mass of the intrinsic exciton and Ω_0 is the volume of the unit cell.

It has been shown in some materials that F_d exceeds f_{ex} by more than four orders of magnitude. An inspection of the above relation reveals that, as the intrinsic exciton becomes more tightly bound to the associated center, the oscillator strength, and hence the intensity of the exciton complex line, should decrease as $(1/|E|)^{3/2}$.

In magnetic fields bound excitons have unique Zeeman spectral characteristics, from which it is possible to identify the types of centers to which the free excitons are bound. Bound exciton spectroscopy is a very powerful analytical tool for the study and identification of impurities and defects in semiconductor materials. It has been employed rather extensively over the last few years for the characterization of materials.

Multibound Excitons Sharp photoluminescent lines have been observed at energies less than the energy of the line associated with an exciton bound to a neutral donor in silicon, germanium and silicon carbide. Similar lines have also been observed which are associated with acceptors in silicon and gallium arsenide. The energies and widths of these lines were such that they could not be explained in terms of any recombination mechanism involving just a single exciton bound to a neutral shallow impurity center. A model involving a multiexciton complex bound to a donor (acceptor) was invoked in which each line was associated with radiative recombinations of an exciton in the bound multiexciton complex. The behavior of these lines in the presence of magnetic and stress fields helped to establish the viability of the bound multiexciton complex model.

Interaction of Excitons with Other Systems The interaction of excitons with phonons to form Stokes and anti-Stokes transitions are commonly observed in crystals. Interaction with the L_0 phonon is strong in most crystals. The exciton-bound-phonon quasiparticle is one form of interaction that has been observed in several materials. In the case of the quasiparticle the energy separating the parent transition and its L_0-phonon sideband is less than the L_0-phonon energy $h\omega_0$ by as much as 30% in some materials. The quasiparticle consists of the exciton-phonon bound state, having impurity modes associated with dielectric effects of the center rather than local modes associated with the mass defects of the substituents.

Resonant Raman scattering occurs with excitons as intermediate states, spin-flip scattering from bound excitons is observed and the production of excitons by two photon processes provides added information about a specific material. Exciton mechanisms are active in lasing transitions in many materials.

The study of free and bound excitons has contributed appreciably to the understanding of the basic intrinsic and extrinsic properties of many materials.

DONALD C. REYNOLDS

References

1. Frenkel, J., *Phys. Rev.* 37, 17 (1931).
2. Wannier, G. H., *Phys. Rev.* 52, 191 (1937).
3. Lampert, M. A., *Phys. Rev. Lett.* 1, 450 (1958).
4. Akimoto, O., and Hanamura, E., *J. Phys. Soc. Japan* 33, 1537 (1972).
5. Gourley, P. L., and Wolfe, J. P., *Phys. Rev. Lett.* 40, 526 (1978); *Phys. Rev.* 20B, 3319 (1979).
6. Pekar, S. I., *Sov. Phys.–JETP* 6, 785 (1958) [English transl: *Sov. Phys.–Solid State Phys.* 4, 953 (1962)].
7. Haynes, J. R., *Phys. Rev. Lett.* 4, 361 (1960).
8. Rashba, E. I., *Opt. Spektrosk.* 2, 508 (1957).
9. Rashba, E. I., and Gurgenishvili, G. E., *Fiz. Tverd. Tela* 4, 1029 (1962) [Engl. transl: *Sov. Phys.–Solid State* 4, 759 (1962)].

The following books, and references therein, present a broad coverage of the field of excitons:

Dexter, D. L., and Knox, R. S., "Excitons," New York, Interscience, 1965.
Knox, R. S., "Theory of Excitons," Suppl. 5 of *Solid State Physics*, New York, Academic Press, 1963.
Davydov, A. S., "The Theory of Molecular Excitons," New York, McGraw-Hill Book Company, Inc., 1962.
Reynolds, D. C., and Collins, T. C., "Excitons, Their Properties and Uses," New York, Academic Press, 1981.

Cross-references: PHONONS, PHOTON, SOLID-STATE PHYSICS, SOLID-STATE THEORY.

EXPANSION, THERMAL

Definition and General Remarks All substances change their shapes as a consequence of undergoing changes in temperature. A measure of this change is the thermal coefficient of expansion. In most cases the result is an increase in the length, area, or volume of a sample. The effect is by no means negligible but for moderate changes in the temperature a first-order correction suffices. The prevention of thermal expansion requires very large mechanical stresses. For example, a compressive stress of order 5×10^8 dynes/cm^2 is necessary to prevent a steel bar from expanding when the temperature is increased by 20°C. For solids in the form of a thin rod or cable, this change is confined (to first order) to a change in length, and a linear coefficient of expansion is thus defined by

$$\alpha_0 = \frac{1}{L_0}\left(\frac{\partial L}{\partial t}\right)_P \tag{1}$$

where L_0 is the length of the specimen at 0°C and the subscript P implies that the pressure is kept constant. Correspondingly, for fluids and for solids of arbitrary shape, one defines a cubical or volume coefficient of expansion, β_0, by the relation

$$\beta_0 = \frac{1}{V_0}\left(\frac{\partial V}{\partial t}\right)_P \tag{2}$$

with V_0 being the volume at the reference temperature (usually chosen to be 0°C). It may readily be shown that $\beta_0 = 3\alpha_0$. Thus for solids one usually tabulates values of α_0 while, of course, only β has meaning for fluids. Whereas β_0 is simply 1/273 for all ideal gases (this follows from the equation $PV = nRT$), there exists a wide variation for β values among liquids. Table 1 gives α and β values for several substances. The negative value of β for water below 4°C is anomalous and is caused by the comparatively open lattice structure of ice. In the case of nonisotropic crystals, the coefficient of linear expansion differs for different directions in the crystals and may even have opposite signs along different directions as is the case for $CaCO_3$.

Thermodynamic Relationships The cubical coefficient of expansion plays an important role in relating the molar specific heat at constant pressure, C_P, (which is usually measured directly in the laboratory) to the molar specific

heat at constant volume (which is most often obtained from theory). This relationship, based solely on the laws of thermodynamics is[1]

$$C_P - C_V = (\beta^2 VT)/X \tag{3}$$

where T is the absolute temperature and X the compressibility defined by

$$X = -\frac{1}{V}\left(\frac{\partial V}{\partial P}\right)_T \tag{4}$$

To obtain this result we write the first law of thermodynamics (see THERMODYNAMICS):

$$\delta Q = dU + P\,dV \tag{5}$$

Since

$$C_V = \left(\frac{\delta Q}{\delta T}\right)_V = \left(\frac{\partial U}{\partial T}\right)_V \tag{6}$$

$$dU = \left(\frac{\partial U}{\partial V}\right)_T dV + \left(\frac{\partial U}{\partial T}\right)_V dT \tag{7}$$

and

$$dV = \left(\frac{\partial V}{\partial P}\right)_T dP + \left(\frac{\partial V}{\partial T}\right)_P dT \tag{8}$$

we obtain

$$\delta Q = \left[\left(\frac{\partial U}{\partial V}\right)_T + P\right]\left(\frac{\partial V}{\partial P}\right)_T dP$$
$$+ \left\{C_V + \left[\left(\frac{\partial U}{\partial V}\right)_T + P\right]\left(\frac{\partial V}{\partial T}\right)_P\right\} dT \tag{9}$$

Thus

$$C_P = \left(\frac{\delta Q}{\delta T}\right)_P = C_V + \left[\left(\frac{\partial V}{\partial T}\right)_P\right]\left[P + \left(\frac{\partial U}{\partial V}\right)_T\right] \tag{10}$$

By the second law of thermodynamics, we have

$$dS = \frac{dU}{T} + \frac{P}{T}\,dV \tag{11}$$

On using Eq. (7) again this becomes

$$dS = \frac{1}{T}\left[\left(\frac{\partial U}{\partial V}\right)_T + P\right] dV + \frac{1}{T}\left(\frac{\partial U}{\partial T}\right)_V dT \tag{12}$$

Since

$$dS = \left(\frac{\partial S}{\partial V}\right)_T dV + \left(\frac{\partial S}{\partial T}\right)_V dT \tag{13}$$

one obtains on comparing Eqs. (12) and (13)

$$\left(\frac{\partial S}{\partial V}\right)_T = \frac{1}{T}\left[P + \left(\frac{\partial U}{\partial V}\right)_T\right] \tag{14}$$

TABLE 1. LINEAR COEFFICIENTS OF EXPANSION, α, FOR SOME SOLIDS AND CUBICAL COEFFICIENTS OF EXPANSION, β, FOR SOME LIQUIDS AT ROOM TEMPERATURE

	α (/C°)	β (/C°)
Aluminum	25.0×10^{-6}	
Copper	16.8×10^{-6}	
Nickel	12.8×10^{-6}	
Sodium	77.0×10^{-6}	
Mercury		18.2×10^{-5}
Glycerin		48.5×10^{-5}
Water (0–4°C)		-3.2×10^{-5}

$$\left(\frac{\partial S}{\partial T}\right)_V = \frac{1}{T}\left(\frac{\partial U}{\partial T}\right)_V \qquad (15)$$

Since

$$\frac{\partial}{\partial T}\left(\frac{\partial S}{\partial V}\right)_T = \frac{\partial}{\partial V}\left(\frac{\partial S}{\partial T}\right)_V \qquad (16)$$

we get from Eqs. (14) and (15)

$$\frac{1}{T}\frac{\partial^2 U}{\partial V \partial T} = -\frac{1}{T^2}\left[P + \left(\frac{\partial U}{\partial V}\right)_T\right]$$

$$+ \frac{1}{T}\left[\left(\frac{\partial P}{\partial T}\right)_V + \frac{\partial^2 U}{\partial T \partial V}\right] \qquad (17)$$

or

$$\left(\frac{\partial U}{\partial V}\right)_T = T\left(\frac{\partial P}{\partial T}\right)_V - P \qquad (18)$$

Inserting Eq. (18) into Eq. (10), yields

$$C_P - C_V = T\left(\frac{\partial V}{\partial T}\right)_P\left(\frac{\partial P}{\partial T}\right)_V \qquad (19)$$

Now

$$\left(\frac{\partial P}{\partial T}\right)_V\left(\frac{\partial T}{\partial V}\right)_P\left(\frac{\partial V}{\partial P}\right)_T = -1 \qquad (20)$$

Thus on replacing $(\partial P/\partial T)_V$, the final form of Eq. (19) becomes

$$C_P - C_V = -\frac{T\left(\frac{\partial V}{\partial T}\right)_P^2 V^2}{\left(\frac{\partial V}{\partial P}\right)_T V^2} \qquad (21)$$

or Eq. (3) when the definition of compressibility [Eq. (4)] is employed.

Thus for a substance where β and X are experimentally known C_V may be established from a measurement of C_P.

Grüneisen Relation[2] Grüneisen introduced the parameter $\gamma = \beta V/XC_V$ and on the basis of simple models reached the conclusion that γ is independent of temperature. This implies that the thermal expansion coefficient is proportional to the specific heat and has the same type of temperature dependence. This is true for many substances and has, in fact, been employed as a means for predicting values for C_V at low temperatures. To illustrate the physical basis of the Grüneisen relation, we will work with a crystal model of N oscillators of identical frequency, ν, and each having equilibrium energy, ϵ. In the region of $h\nu \ll kT$, the free energy is given by:

$$F = N\left(\epsilon + 3kT \log \frac{h\nu}{kT}\right) \qquad (22)$$

Since this must be a minimum at equilibrium, we obtain:

$$N\frac{d\epsilon}{dV} = -3NkT\frac{d(\log \nu)}{dV} \qquad (23)$$

To obtain the thermal expansion coefficient, one expands $d\epsilon/dV$:

$$\frac{d\epsilon}{dV} = \left(\frac{d\epsilon}{dV}\right)_{V=V_0} + (V - V_0)\left(\frac{d^2\epsilon}{dV^2}\right)_{V=V_0} \qquad (24)$$

But the compressibility is given by

$$\frac{1}{X_0} = NV_0\left(\frac{d^2\epsilon}{dV^2}\right)_{V=V_0} \qquad (25)$$

and thus

$$\frac{V - V_0}{V_0} = -3NkTX_0\frac{d(\log \nu)}{dV} \qquad (26)$$

Differentiating with respect to T and realizing that $3NkT$ is the thermal energy, we obtain:

$$\frac{\alpha V_0}{C_V X_0} = -\frac{d(\log \nu)}{d(\log V)} \qquad (27)$$

More exact crystal models yield values for

$$\gamma_i = -\frac{d(\log \nu_i)}{d(\log V)}$$

where ν_i is the ith frequency of a set of normal modes of vibration.[3] These and other refinements give rise to modifications of the simple Grüneisen theory.

Source of Thermal Expansion The dynamical basis for thermal expansion is the presence of an anharmonic component for the interaction potential. A qualitative argument for this is based on the property of typical potential energy of atoms in a lattice.[4] Figure 1 represents such a curve and it will be observed that the curve is not symmetrical about r_0, the equilibrium distance between atoms. As the internal energy E increases with an increase in temperature, the average value of r shifts to larger values. Clearly if the potential curve were perfectly parabolic about r_0, we would not have thermal expansion. This argument can be made qualitative as follows.[5] Taking x as the displacement of a lattice atom from its equilibrium neighbor separation, the potential energy has the form

$$V(x) = cx^2 - gx^3 \qquad (28)$$

Then $\bar{x}$, the average displacement using the Boltzmann distribution function becomes

$$\bar{x} = \frac{\displaystyle\int_{-\infty}^{\infty} xe^{-V(x)/kT}\, dx}{\displaystyle\int_{-\infty}^{\infty} e^{-V(x)/kT}\, dx} \qquad (29)$$

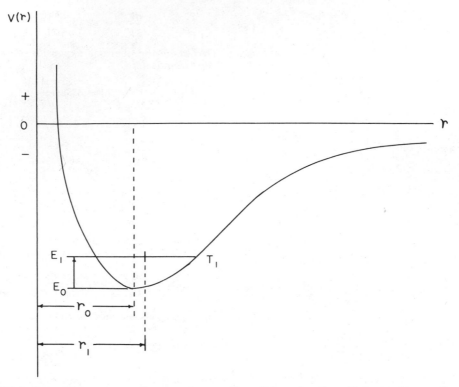

FIG. 1. Potential energy curve for atoms in crystalline solid as a function of interatomic separation.

Assuming that the anharmonic term is much less than the harmonic contribution, we expand $V(x)$ to yield

$$\int x e^{-V/kT}\, dx = \int e^{-cx^2/kT} \left[x + \frac{gx^4}{kT} \right] dx \quad (30)$$

and

$$\int e^{-V/kT} = \int e^{-cx^2/kT}\, dx \quad (31)$$

Both integrals are readily evaluated to give

$$\bar{\bar{x}} = \frac{3kTg}{4c^2} \quad (32)$$

or a constant temperature coefficient for thermal expansion. This simple derivation may be amplified to include specific interaction forces for the atoms of a lattice.[6]

Thermal Expansion and Curie Temperature The Curie-Weiss law for both ferroelectrics and ferromagnets may be shown to be connected with the thermal expansion coefficient.[7] Consider a region where the dielectric constant, ε, is large. If we let N be the cell density and A a constant, the Clausius-Mosotti formula is

$$\frac{\varepsilon - 1}{\varepsilon + 2} = AN \quad (33)$$

Differentiating with respect to T, the temperature, and making use of the assumption that $\varepsilon \gg 1$, one finds that

$$\frac{3d\varepsilon}{\varepsilon^2 dT} = \frac{1}{N}\frac{dN}{dT} = -\beta \quad (34)$$

On integrating between T and θ, the Curie temperature, one obtains

$$\varepsilon = \frac{3/\beta}{T - \theta} \quad (35)$$

which is a typical Weiss law and indicates the Curie constant is of the order of the reciprocal of the thermal expansion. The special electronic structure of ferromagnetic materials also gives rise to anomalous thermal expansion coefficients in the transition region. For some materials β values are depressed and for others β values increase more rapidly with temperature. Both magnetostriction and the variation of the energy of magnetization with the atomic size account for the anomalous behavior of different substances.

JACOB NEUBERGER

References

1. Any text in thermodynamics such as:
   Zemansky, M. W., "Heat and Thermodynamics," New York, McGraw-Hill Book Company, Inc., 1951.
2. Mott, N. F., and Jones, H., "The Theory of the Properties of Metals and Alloys," New York, Dover Publications, 1936.
3. Arenstein, M., Hatcher, R. D., and Neuberger, J., "Equation of State of Certain Ideal Lattices," *Phys. Rev.*, **13**, No. 5, 2087–2093 (1963).
4. Halliday, D., and Resnick, R., "Physics," Third edition Part I, New York, John Wiley & Sons, 1977 (p. 506).
5. Kittel, C., "Introduction to Solid State Physics," New York, John Wiley & Sons, 1976.
6. Peierls, R. E., "Quantum Theory of Solids," New York, Oxford Press, 1955. (Reprinted in 1964.)
7. Dekker, A., "Solid State Physics," Englewood Cliffs, N.J., Prentice Hall, 1957; Sinnott, M., "The Solid State for Engineers," New York, John Wiley & Sons, 1958.

Cross-references: BOLTZMANN'S DISTRIBUTION LAW; COMPRESSIBILITY, GAS; DIELECTRIC THEORY; HEAT; HEAT CAPACITY; MAGNETISM; THERMODYNAMICS.

F

FALLOUT

The term fallout generally has been used to refer to particulate matter that is thrown into the atmosphere by a nuclear process of short time duration. Primary examples are nuclear weapon debris and effluents from a nuclear reactor excursion. The name fallout is applied both to matter that is aloft and to matter that has been deposited on the surface of the earth. Depending on the conditions of formation, this material ranges in texture from an aerosol to granules of considerable size.

The topographic distribution of fallout is divided into three categories called local (or close-in), tropospheric (or intermediate), and stratospheric (or world wide) fallout. No distinct boundaries exist between these categories. The distinction between local and tropospheric fallout is a function of distance from source to point of deposit, while the primary distinction between tropospheric and stratospheric fallout is the place of injection of the debris into the atmosphere, above or below the tropopause. Whether radioactive debris from a nuclear weapon becomes tropospheric or stratospheric fallout depends on yield, height, and latitude of burst (the height of the tropopause is a function of latitude).

Because air acts as a viscous medium, a drag force is developed to oppose the gravitational force that acts on airborne particulate matter. This makes the velocity of fall dependent on particle size. The larger particles (diameters greater than about $20\mu m$) have a higher rate of settling and create local fallout. Smaller particles injected below the tropopause are carried by prevailing winds over large regions of the surface of the earth and create the tropospheric fallout. Tropospheric fallout particles larger than about 0.1-μm diameter continually mix through the circulating air mass that is in contact with the surface of the earth and gradually settle to the ground, or are washed down by rain or snow. Many smaller particles form nuclei for raindrops. Parts of the tropospheric fallout many remain in the atmosphere for a month or more, long enough to circle the earth several times. The mean residence time above the tropopause of stratospheric fallout is from 5 to 30 months, during which time it completely encircles the earth. It gradually returns through the tropopause, primarily in certain regions where mixing between the two layers is more probable.

The exact characteristics of the radiation associated with fallout depend on the nature of the nuclear processes from which its radioactivity originates. Generally these radioactive nuclides are fission products formed from the fissioning of uranium or plutonium, but, under appropriate circumstances, considerable quantities of radioactivity can be formed through nuclear reactions induced by neutrons that are produced by the weapon or reactor. The radiation problems associated with local fallout are usually those of high-intensity gamma-ray radiation fields resulting from the relatively large quantities of radioactive material that fall back to earth within a few tens of miles from the point of origin. The important radioactive materials consist in this case of short-lived fission products and neutron-induced radioactive nuclides. The hazards of worldwide fallout come more from the problems of the long-lived radionuclides, such as ^{134}Cs, ^{137}Cs, and ^{90}Sr, that can enter the human food chain and ultimately be absorbed by the body.

For a nuclear weapon burst in air, all materials in the fireball are vaporized. Condensation of fission products and other bomb materials is then governed by the saturation vapor pressures of the most abundant constituents. Primary debris can combine with naturally ocurring aerosols, and almost all of the fallout becomes tropospheric or stratospheric. If the weapon detonation takes place within a few hundred feet of (either above or below) a land or water surface, large quantities of surface materials are drawn up or thrown into the air above the place of detonation. Condensation of radioactive nuclides in this material then leads to considerable quantities of local fallout, but some of the radioactivity still goes into tropospheric and stratospheric fallout. If the burst occurs sufficiently far underground, the surface is not broken and no fallout results.

The considerable significance of radioactive fallout became apparent on July 24, 1946, when a nuclear explosive of about the same size as the Hiroshima and Nagasaki weapons was detonated 30 meters below the surface of the lagoon at Bikini atoll. This explosion produced a column one-half mile in diameter consisting of about

FIG. 1. Nuclear explosion at Bikini Atoll, July 24, 1946.

a million tons of water, as well as a base surge, as illustrated in Fig. 1. Instead of the radioactive debris being lifted by the explosion into the troposphere and stratosphere, as had been the situation for prior nuclear detonations, almost all the radioactivity produced by this underwater explosion was trapped in the water, spread out in a radioactive cloud, and fell as rain. The radioactive material was strongly absorbed by the painted surfaces of a group of ships which had been assembled as a target fleet. It also settled into joints, ventilating systems, almost everywhere. It clung so tenaciously that the paint sometimes had to be removed to reduce significantly the amount of radioactivity. The results were completely unexpected, since most planners had predicted a stronger shock wave to hit the ships than had been experienced from the earlier air detonation, but they greatly underestimated the amount of radioactive contamination that would occur. As a result of this test a large research effort was begun to try to understand the nature of fallout from nuclear detonations and the procedures needed to be followed to decontaminate the regions on which the distributed radioactive material was deposited.

Some type of radioactive fallout has been associated with every near-surface detonation of a nuclear explosive. On one occasion the radiation from fallout debris endangered human health. This was the detonation on March 1, 1954 at Bikini atoll, of a thermonuclear device with a yield equal to that of about 17 megatons (MT) of trinitrotoluene (TNT). The yield was greater than expected and an unexpected shift in wind direction caused significant quantities of radioactive fallout to be deposited over a region extending several hundred miles to the east, over the inhabited atolls of Ailingnae, Rongelap, Rongerik and Utirik and on the

Japanese fishing vessel *Lucky Dragon*. The people of these atolls have continued since that time to receive medical examinations on a regular basis. From the information obtained a reasonably thorough knowledge has accrued on the effects of fallout radiation on the health of one small group of individuals.

The aerodynamic principles governing deposition of fallout are the same regardless of the origin of the material which is thrown into the air. Therefore the deposition of any airborne matter should follow the same pattern as equivalent material produced by nuclear explosives. Two reasonably contemporary but very different nonradioactive phenomena which can be used for such comparisons have been the deposition of ash from the eruptions of Mount St. Helens and the deposition of acid rain in the northeastern United States and eastern Canada.

The eruption on May 18, 1980 of Mount St. Helens produced huge quantities of observable airborne particulate matter. This eruption has been determined to have released an energy equivalent to the detonation of about 35 MT of TNT, comparable to the largest nuclear explosions ever detonated. The basic difference between these two types of explosions is that the Mount St. Helens eruption was strictly mechanical, carrying with it several cubic kilometers of debris, considerably more than would be thrown into the air by even a very large nuclear explosion. However, the size distribution of the individual ash particles thrown into the air by the eruption was similar to particle-size distributions of airborne material from surface detonations of high-yield nuclear explosives. Also, the pattern of deposition of the ash was very similar to the expected deposition patterns for fallout from the surface detonation of an 8-MT nuclear weapon in the vicinity of Portland, Oregon, as predicted in 1959 during hearings held by the Joint Committee on Atomic Energy of the Congress of the United States.

The exact pattern of deposition of fallout depends on weather conditions. The eruption of May 18 happened to occur during a period when weather conditions were similar to the conditions assumed for the 1959 study. On the other hand, the eruption of May 25 occurred during a rainy period. Not as many ash particles were thrown aloft during the May 25 eruption as during the May 18 eruption but the number was still more than usually expected from a surface detonation of a nuclear weapon. On May 25 rain mixed with the ash to form mud, which fell locally and not uniformly. However the fallout pattern of this ash was similar to that predicted by Storebø for "rainout" of debris from nuclear explosives, again making the event worthy of study as a possible predictor of effects which could follow the detonation of a nuclear weapon. (See VOLCANOLOGY.)

The series of events which result in acid rain are at the other extreme of the fallout picture for, instead of consisting of particulate matter of finite size, much of the material released from chimneys, smokestacks and motor-vehicle exhausts is composed of individual molecules or groups of molecules of the oxides of sulfur and nitrogen. In the atmosphere these substances go through chemical reactions to form sulfates and nitrates, which are acidic and which dissolve in water and fall as precipitation at distances of 100 to 1000 kilometers downwind from their original source. The mechanical result is a type of fallout not unlike that from an air-burst of a nuclear weapon, during which essentially all matter is vaporized. Although the particulate matter cannot be physically observed, as in the case of the fall of ash from Mount St. Helens, the chemical effects of the acid rain can be observed, for example, in the increased acidity of the lakes of the Adirondack Mountains, where the water of some lakes has become too acidic to support fish life. Similar nuclear-weapon fallout would be observed simply as an increase in radioactivity in the water.

C. SHARP COOK

References

Brunner, H., and Pretre, S. (Eds.), "Radiological Protection of the Public in a Nuclear Mass Disaster," Proceedings of symposium at Interlaken, Switzerland, 26 May–1 June, 1968, Bern, Bundesamt für Zivilschutz, 1968.

Cook, C. S., "Initial and Residual Ionizing Radiations from Nuclear Weapons," in Attix and Tochilin (Eds.), "Radiation Dosimetry," Vol. III, New York, Academic Press, 1969, pp. 361–399.

Freiling, E. C. (Ed.), "Radionuclides in the Environment," Washington, D.C., American Chemical Society, 1970.

Danielson, E. F., "Trajectories of the Mount St. Helens Eruption Plume," *Science*, **211**, 819–820 (1981).

Storebø, P. B., "Prediction of Massive Wash-out of Nuclear Bomb Debris," *Health Physics* **11**, 1203–1211 (1965).

Babich, H., Davis, B. L., and Statzky, G. "Acid Precipitation: Causes and Consequences," *Environment* **22**(4), 6–13 (1980).

Cross-references: ATOMIC ENERGY, FISSION, FUSION, ISOTOPES, NUCLEAR REACTIONS, RADIOACTIVITY, VOLCANOLOGY.

FARADAY EFFECTS*

In 1845, Michael Faraday discovered the first magnetooptical effect[1] when he observed the rotation of the plane of polarization of light as a result of its passage through lead borate glass in a direction parallel to an applied magnetic field B. This is known as the *Faraday effect* or *Faraday rotation* and was important historically because it provided the first concrete evidence for a connection between magnetism and light. Since then many other magnetooptical effects have been investigated—notably cyclotron resonance, magnetic dichroism, and the Voigt and Hall effects—their common linkage being their dependence on various components of the dielectric or conductivity tensor, as well as on the magnetic permeability tensor in the case of magnetic materials. Faraday rotation has now been shown to be a general property of matter and has been observed in a variety of solids (especially semiconductors), liquids, and gases, over a wide range of frequencies.[2] It is often a very useful technique[3,4] for the determination of various quantities such as effective mass m, collision frequency ν, and mobility μ. In fact, not only does it complement cyclotron resonance determinations of m but it is especially useful in determining electron and hole effective masses in solids in cases where cyclotron resonance is unobservable.[5] The latter circumstance occurs when $\nu \gg \omega$, where ω is the angular frequency of the radiation, and corresponds to large damping (since $\tau = \nu^{-1}$, where τ is the damping or relaxation time). Also, since cyclotron resonance occurs when $\omega = \omega_c$, where $\omega_c = (eB/mc)$ is the cyclotron frequency with e being the magnitude of the charge, it is clear that for a typical m value of $10^{-1} m_0$ (where m_0 is the free electron mass) and a maximum B value of 100 kG we have resonance at $\omega = 1.77 \times 10^{13}$ s^{-1}, corresponding to a wavelength $\lambda = 326$ μm, i.e., for practical purposes sharp cyclotron resonances typically do not occur for wavelengths smaller than infrared, and in many cases actually no smaller

*Research for this article was partially supported by the Department of Energy, Division of Materials Science, under Contract No. DE-AS05-79ER10459.

than microwave. In other words, Faraday rotation may be measured over a far wider range of frequencies and in a far wider range of materials than cyclotron resonance. However, it has the disadvantage of measuring an average effective mass in cases where the effective mass is anisotropic. It is also used to deduce interstellar magnetic field values as well as providing support for the conclusion that there is little or no antimatter in the galaxy. We use cgs units in this article.

A linearly polarized wave can be decomposed into two waves of opposite *circular polarization*, called *right* and *left* (RCP and LCP), which propagate independently. In general, optical rotation occurs in a medium when its refractive indices for right- and left-circularly polarized radiation, n_+ and n_-, are unequal so that the phase velocities c/n_+ and c/n_- are also unequal. In the case of *natural* optical activity, this arises from asymmetry among the atomic layers whereas *Faraday* rotation arises from the anisotropy produced by the magnetic field. The former (which will not concern us here) disappears on reflection back through a sample whereas the latter is doubled. In the case of an absorbing medium, the difference in absorption coefficients for the two components (referred to as *dichroism*) causes the emerging beam to be elliptically polarized.

The Faraday rotation θ is defined to be one-half the phase angle change between the RCP and LCP waves and corresponds to the amount of rotation of the major axis of the transmitted polarization ellipse.[3] For radiation of frequency ω propagating a distance d through the medium along the direction of the magnetic field, the rotation is given by

$$\theta = \frac{\omega d}{2c}(n_+ - n_-) = \frac{\pi d}{\lambda}(n_+ - n_-), \quad (1)$$

where λ is the wavelength in vacuum. A theoretical evaluation of $n_\pm$ in the case of nonmagnetic materials (magnetic materials will be discussed below) starts with the relation

$$\epsilon_\pm = (n_\pm + ik_\pm)^2, \quad (2)$$

where k, the imaginary part of the complex refractive index, arises from absorption and $\epsilon_\pm$ is related to the components of the dielectric tensor ϵ_{ij} ($i, j = x, y, z$) by

$$\epsilon_\pm = \epsilon_{xx} \pm i\epsilon_{xy}. \quad (3)$$

Thus the problem of calculating θ is reduced to a calculation of ϵ_{ij}. In general, rotation arises due to interaction of the radiation with either free or bound charge carriers. There are five basic frequencies to be considered: the wave frequency ω, the plasma frequency $\omega_p = (4\pi Ne^2/m)^{1/2}$, where N is the number of charge carries per unit volume, the collision frequency ν, the cyclotron frequency ω_c, and ω_0, which refers to either the natural frequency of the bound charges (classical model) or the frequency separation of spectral lines (quantum mechanical model). In a solid there are contributions to θ from both free and bound electrons and also the nuclei but, as a general rule, at optical and infrared frequencies, the dominant contribution to θ is from the free electrons. For low photon frequencies only transitions within the same band (intraband) are of importance in a semiconductor but as we approach the optical region the band to band (interband) effects must be included. A *classical* calculation of ϵ_{ij} is based on the Boltzmann equation or, more frequently although less rigorously, on the Drude model, which assumes that all charge carriers act independently. The Drude model result is often referred to as the *cold-plasma limit* since it corresponds to the result obtained by use of the Boltzmann equation in the limit of extreme degeneracy (zero temperature).

Consider now the case of electrons moving freely ($\omega_0 = 0$) in a crystal-lattice background—while being cognizant of the fact that, except for a change in the sign of θ, similar results hold for positive carriers such as, for example, holes in the valence band of semiconductors. Then the Drude model leads to[3]

$$\epsilon_\pm = \epsilon_l - [\omega_p{}^2/\omega(\omega \pm \omega_c + i\nu)], \quad (4)$$

where $\epsilon_l = n^2$ is the (real) dielectric constant of the lattice. It follows that, if ω is much larger than the other three frequencies, the Drude model leads to

$$\theta = d\omega_c\omega_p{}^2/2cn\omega^2. \quad (5)$$

Thus θ is proportional to the magnetic field B and also to m^{-2}, making clear how a measurement of θ can determine the effective mass m. This formula is also used to determine galactic magnetic fields from observations, over a range of frequencies, of the Faraday rotation associated with polarized radio waves from such objects as pulsars.[6] Estimates of primordial magnetic field values have also been made from observations of the θ of a distant extragalactic radio source.[7] Furthermore, a measurement of the Faraday rotation of radio waves emitted by artificial satellites and transmitted through the ionosphere can be used to measure the electron density along the path.[8] Turning to the question of how much antimatter there is in the galaxy, we note that positrons and electrons cause rotations in opposite directions. But polarized light traversing the interstellar medium does suffer Faraday rotation, demonstrating that there are not comparable numbers of electrons and positrons. In fact, when these results (which in essence give the difference in the number of electrons and positrons) are combined with dispersion measures (which depend on the sum of the number of electrons and positrons), it is found that the number of positrons is negligibly small.[9]

In the case where ω_c is much greater than the other three frequencies (high-field, low-frequency approximation),

$$\theta = -d\omega_p^2/2cn\omega_c, \qquad (6)$$

i.e., θ is now negative, and it is proportional to B^{-1} and independent of ν again and also ω. Also if $\omega_p \ll \omega$ then a zero in θ occurs[10] at a photon frequency $\omega = (\omega_c^2 + \nu^2)^{1/2}$, which can lead to a determination of ν. If $\nu \gg \omega$, $\omega_c \gg (\omega_p/n)$ then we get the *low-frequency Faraday rotation*

$$\begin{aligned} \theta &= -d\omega_c\,\omega_p^2/2cn\nu^2 \\ &= -2\pi d\sigma_0\mu B/c^2 n, \end{aligned} \qquad (7)$$

where $\mu = e\tau/m$ is the carrier mobility and $\sigma_0 = ne^2\tau/m$ is the static conductivity. Since the latter expression for θ does not contain m or ν explicitly, one can use known values of σ_0 and the other parameters to deduce the mobility of charge carriers from a measurement of θ.

In strong magnetic fields, account must be taken of the fact that electron energies in a magnetic field are confined to discrete *Landau levels* and, in the case of interband transitions, this gives rise to oscillatory effects in the Faraday rotation, while there is evidence for a contribution also from exciton transitions. In general, there are other complications[3] which are sometimes of importance. For example, the collision frequency ν can be frequency- and magnetic-field dependent. Also, in polar semiconductors there is a contribution from optical lattice vibrations to the dielectric tensor.

In the case of thin samples, multiple reflections can play an important role with an attendant increase in the complexity of the analysis.[11,12] An example of where such multiple reflections play a role[13] is the *two-dimensional electron space-charge layer* (the motion being quantized in one direction whose effective width is negligible compared to the wavelength of the transmitted radiation), which is formed in various modern microelectronics systems, such as at the semiconductor surface in a metal-oxide-semiconductor (MOS) system. There is also a contribution to θ due to boundary effects in the transmission of radiation through different material. In the case of metals, the usual method of observing θ in transmission is not convenient except for very thin films because metals are very good reflectors in the visible and infrared regions. As an alternative, θ is measured on reflection (the *polar reflection Faraday effect*) and gives information on the electron band structure of nonferromagnetic metals.

The advent of high-intensity laser radiation has motivated the inclusion of various nonlinear terms into the laws of optics. In particular, the Faraday effect is intensity dependent, especially in a strong magnetic field.[14] A closely related phenomenon is the *inverse Faraday effect*, i.e., the magnetization of the medium by intense polarized radiation,[15] which has been suggested as the basis of a nondemolition optical quantum counting measurement.[16]

Faraday rotation has also been used as a diagnostic tool to study and measure the large magnetic fields which are produced both in controlled thermonuclear fusion plasma and in laser-produced plasma,[17] as well as being one of the first phenomena to be studied by the megagauss magnetic fields which are being increasingly produced in many laboratories.[18]

All of the effects discussed so far depend on the dielectric tensor and arise from the interaction of the charge carriers with the electric field of the electromagnetic wave and, in addition, the spatial dependence of the electric field is generally neglected (electric dipole approximation). However, for magnetic materials the Faraday rotation depends on the magnetic permeability tensor. In ferromagnetic metals very large rotations occur which are proportional to the net magnetization and not to the external magnetic field.[19,20] Thus the rotation per cm in a magnetic field of 10^4 gauss is of the order of 2 degrees in quartz, 10^2 degrees in aluminum and 1.3×10^5 degrees in iron. The large value of the latter rotation arises from a spin-orbit interaction: the magnetic moment of an electron, due to its spin, interacts with the magnetic field which arises by virtue of its motion through the electric field created by the nuclei and all the other electrons in the absence of radiation. This phenomenon is often called the *ferromagnetic Faraday effect* and refers to the transmitted beam, whereas effects associated with reflection from a ferromagnetic material are called *polar Kerr magnetooptic effects*. In a certain sense the spin-orbit interaction can be looked on as the effect of a large internal magnetic field acting on the electrons. Also, the role of the external magnetic field is peripheral in that it serves only to magnetize the sample in a certain direction. It should also be mentioned that there is also a contribution from spin-orbit effects in nonferromagnetic material, albeit small compared to the situation for ferromagnetics.

Absorption in the metallic ferromagnets is very large except in the case of very thin films. On the other hand, ferrimagnetic substances are particularly good magnetooptical materials because they combine the low absorption of a good insulator with high permeability. For example, yttrium iron garnet (YIG) is transparent in the optical region and also gives rise to a large Faraday rotation which makes it an excellent material for the observation of magnetic domains. In fact, measurements of θ can be used to measure several macroscopic magnetic properties of thin rare earth garnet films which are used for magnetic bubble devices.[21] Ferrimagnetic materials are used extensively in microwave technology,[22] their importance stemming from the fact that they can be used to make Faraday isolators which permits a signal to be transmitted with low attenuation in one direction but causes the reflected signal to be highly attenuated. Thus, for example, this permits the decoupling of an oscillator from a measuring system. Similar devices have also been used in laser systems.[23] At higher frequencies it turns out that antiferromagnetic materials perform better.

R. F. O'CONNELL

References

1. Barr, E. S., "Men and Milestones in Optics V: Michael Faraday," *Appl. Optics* **6**, 631 (1967).
2. Palik, E. D., and Henvis, B. W., "A Bibliography of Magneto-Optics of Solids," *Appl. Optics* **6**, 603 (1967).
3. Palik, E. D., and Furdyna, J. K., "Infrared and Microwave Magnetoplasma Effects in Semiconductors," *Rep. Prog. Phys.* **33**, 1193 (1970).
4. Piller, H., "Faraday Rotation," in Willardson and Beer (Eds.), "Semiconductors and Semimetals," Vol. 8, Academic Press, New York, 1972, pp. 103–179.
5. Lax, B., "Resonance Spectroscopy of Solids and Plasmas," *J. Mag. and Mag. Materials* **11**, 1 (1979).
6. Manchester, R. N., and Taylor, J. H., "Pulsars," W. H. Freeman and Co., San Francisco, 1977.
7. Shapiro, S. L., and Wasserman, I., "Massive Neutrinos, Helium Production, and the Primordial Magnetic Field," *Nature* **289**, 657 (1981).
8. Ratcliffe, J. A., "An Introduction to the Ionosphere and Magnetosphere," Cambridge Univ. Press, Cambridge, U.K., 1972, pp. 196–198.
9. G. Steigman, "Observational Tests of Antimatter Cosmologies," *Ann. Rev. Astron. Astrophys.* **14**, 339 (1976).
10. O'Connell, R. F., and Wallace, G. L., "Null Faraday Rotation—A Clean Method for Determination of Relaxation Times and Effective Masses in MIS and Other Systems," *Solid State Commun.* **38**, 429 (1981).
11. Donovan, B., and Medcalf, T., "The Inclusion of Multiple Reflections in the Theory of the Faraday Effect in Semiconductors," *Brit. J. Appl. Phys.* **15**, 1139 (1964).
12. O'Connell, R. F., and Wallace, G. L., "Multiple Reflections in the Theory of the Faraday Effect," *Phys. Lett.* **86A**, 283 (1981).
13. O'Connell, R. F., and Wallace, G., "Ellipticity and Faraday Rotation due to a Two-Dimensional Electron Gas in a Metal-Oxide-Semiconductor (MOS) System," *Phys. Rev.* **B26**, 2231 (1982).
14. Manakov, N. L., Ovsiannikov, V. D., and Kielich, S., "Nonlinear Variations in the Faraday Effect caused in Atomic Systems by a strong Magnetic Field," *Phys. Rev.* **A21**, 1589 (1980).
15. van der Ziel, J. P., Pershan, P. S., and Malmstrem, L. D., *Phys. Rev. Lett.* **15** 190 (1965).
16. Braginskii, V. B., and Khalili, F. Ya., "Optico-Magnetic Effects in Nondestructive Quantum Counting," *Sov. Phys.—JETP* **51**, 859 (1980).
17. Luhmann, N. C., Jr., "Instrumentation and Techniques for Plasma Diagnostics: An Overview," and Vernon, D., "Submillimeter Interferometry of High-Density Plasmas," in Button (Ed.), "Infrared and Millimeter Waves," Vol. 2, Academic Press, New York, 1979, pp. 1–135; Stamper, J. A., McLean, E. A., and Ripin, B. H., "Studies of Spontaneous Magnetic Fields in Laser-Produced Plasmas by Faraday Rotation," *Phys. Rev. Lett.* **40**, 1177 (1978).
18. Fowler, C. M., Caird, R. S., Garn, W. B., Erickson, D. J., and Freeman, B. L., "High Field Faraday Rotation of Some Zn(VI) Compounds," *Journal of Less-Common Metals* **62**, 397 (1978).
19. Argyres, P. N., "Theory of the Faraday and Kerr Effects in Ferromagnetics," *Phys. Rev.* **97**, 334 (1955).
20. Bennett, H. S., and Stern, E. A., "Faraday Effect in Solids," *Phys. Rev.* **137**, A448 (1965).
21. Tanner, B. H., "Magneto-Optical Experiments on Rare Earth Garnet Films," *Am. J. Phys.* **48**, 59 (1980).
22. Button, K. J., and Hartwick, T. S., "Microwave Devices," in Rado and Suhl (Eds.), "Magnetism," Vol. I, Academic Press, New York, 1963, pp. 621–666.
23. Wang, S., Shah, M., and Crow, J., "Studies of the Use of Gyrotropic and Anisotropic Materials for Mode Conversion in Thin Film Optical Wave Guide Application," *J. Appl. Phys.* **43**, 1861 (1972).

Cross-references: HALL EFFECT AND RELATED PHENOMENA, KERR EFFECT, LIGHT, MAGNETISM, POLARIZED LIGHT, PROPAGATION OF ELECTROMAGNETIC WAVES, SEMICONDUCTORS.

FEEDBACK

The concept of feedback lies at the heart of modern systems theory and control engineering. The term itself seems to have been used for the first time in a technical sense in 1920,[1] and refers to the return or feedback of system output signals to the inputs in order to improve or change the behavior of the system. A very simple example is the control of a heating system by using a bimetallic strip (Fig. 1). The system output is heat, the input the voltage across the heating element. When the desired temperature (the *set-point*) is reached the heat causes the bimetallic strip to deform sufficiently to switch off the heating element. As the temperature falls back below the set-point the strip returns to its original shape, contact is reestablished, and the heater switches on again. A second illustrative example may be found in the flyball governor (attributable to Sir James Watt) and used to control steam engines (Fig. 2). As more steam is fed to the engine the shaft

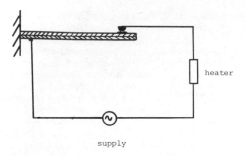

FIG. 1. Simple bimetallic strip heating control.

accelerates and the centrifugal forces generated are used to reduce the amount of steam made available, notice here the importance of *negative* feedback—as the output exceeds the set-point the input signal is decreased and vice-versa. If the governor had been incorrectly installed so that the steam supply increased with increasing velocity (*positive* feedback) the resulting instability would have destroyed the steam source or engine or both. The governor acts as a regulator. If the load is increased the speed of the shaft will initially decrease but with negative feedback the governor will then increase the steam supply pressure to restore the speed close to its nominal value. This illustrates one of the most important uses of feedback, namely to reduce the sensitivity of the controlled variable to disturbances made to the system. Even earlier examples of such regulation are mechanical clocks and lift tenters designed to regulate the grinding of corn. A detailed discussion of the mechanisms used in centrifugal governors and of the early history (up to 1930) of control engineering may be found in Ref. 1.

The true birth of feedback control in the analytical as well as the practical sense can perhaps be pinpointed to the publication of Nyquist's seminal paper of 1932,[2] which arose

from an attack on the problem of feedback amplifier stability. The three succeeding decades saw the emergence of what is now considered the classical theory of control as a coherent discipline which held and continues to hold a pivotal position in a world of increasing technical complexity and automation. The classical results, including the stability criteria and design techniques of Nyquist, Bode, and Nichols, are concerned primarily with linear systems having a single control input and a single output. The interested reader can do no better than consult Refs. 3, 4, or 5 or any other of the excellent introductory textbooks available on the subject. The historical perspective has been discussed by Macfarlane.[6]

From the early 1960s system complexity increased exponentially and various problems arose which were not amenable to the classical methods of feedback design. Many practical systems have not one, but many inputs and outputs, and these exhibit varying degrees of interaction with a signal to one input causing responses on several outputs. For example, the adjustment of the wing flaps of an aeroplane will cause a change in the direction of flight, but will also cause the aircraft to bank, and a change in the set-point of one of the propulsion units of a jet foil will cause changes in the roll, pitch, and yaw motions of the vessel. Implementation of classical single-loop controllers on such systems frequently leads to stability problems traceable to this interaction, and hence Rosenbrock, Macfarlane, and others[7,8] produced extensions of the classical techniques to allow the design of *multi-input, multi-output* (MIMO) *feedback systems*. Unlike the earlier methods these are in no sense "pencil and paper" techniques; they usually require the use of computer graphics terminals and extensive, interactive, computer-aided design programs. They resemble earlier work, however, in that they present information concerning system dynamics in a graphical form easily interpreted by the practiced engineer.

Most analysis and design techniques require that the system to be controlled be linear, and that a linear controller be used. However, many practical systems are highly nonlinear, and the design of a nonlinear controller, such as the bimetallic strip in the heating control, can lead to a particularly elegant solution to a control problem. Analysis and design of *nonlinear feedback systems* can be difficult. However, the stability theorems of Liapunov[9] have provided a basis for the development of techniques such as the cricle criteria[8] which allow analysis and design of a restricted set of nonlinear feedback systems using traditional frequency domain techniques for both single-input-single-output and multivariable systems. Although closed loop stability can be guaranteed by such methods, design tends to be conservative, and in practice there has been some success in the

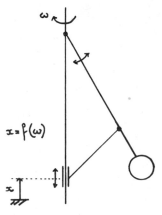

$x = f(\omega)$

FIG. 2. Simple governor.

application of approximate methods such as the describing function[10] whereby a nonlinear element is replaced by a set of equivalent linear elements for a set of specified input signal amplitudes.

Another area of increasing importance is that of *adaptive* or *"self tuning" feedback*. This has been an identifiable topic in the literature of control engineering for at least the last two decades, although the majority of practical applications are of relatively recent date. Adaptive techniques have been mainly concerned with classically structured feedback systems, the word *adaptive* signifying that the controller "learns" about the controlled process while at the same time controlling its behavior. It is thus expected that control will improve with the passage of time and further that the control will be robust in the presence of minor variations in the parameters of the controlled process and its environment. In crude terms the controller may be described as one which "learns from its own mistakes" and may thus be regarded as having some degree of intelligence. To date the majority of applications have been in nonlinear, stochastic systems. A good overview of both theory and application may be found in Ref. 11.

The work described above is concerned with the design of feedback systems; the hardware used to implement the controllers clearly varies from application to application, but it is of interest to note that virtually every area has been affected by the availability of inexpensive microprocessors. The availability of such computers as control elements allows much greater flexibility in the design of controllers. The use of a digital controller however causes additional delays around the feedback loop because of finite sampling rates, computation times, and digital-to-analog conversion times. Designs are generally undertaken either directly on the discrete systems model[12] or via a design of a controller for a continuous system.[13] Sampling rate, wordlength, and computer algorithm need to be carefully chosen for successful design.

Currently one of the most exciting applications of feedback is in the rapidly growing area of *robotics*. One method of increasing the industrial applicability of robotic devices is to give them the capability of sensing their environment via simple vision and touch systems. The robot gripper is frequently a useful location for such sensors. Signals from the sensor are processed to extract the relevant information, which is then passed to the robot control system, which in turn uses it to determine an optimal control strategy and thus obtain a degree of closed loop feedback.[14]

The impetus to understand and design feedback systems has come from all branches of engineering, in particular electronic and mechanical engineering. However, the concept of feedback has much wider scope than this. In particular, biological systems have very successfully used feedback for millions of years. Attempts have also been made to model ecological and economic systems in order to gain further understanding of these complex feedback systems and to be able to apply appropriate control to modify their behavior.

G. E. HAYTON
P. M. TAYLOR

References

1. Bennett, S., "A History of Control Engineering 1800–1930," Stevenage, U.K. and New York, Peter Peregrinus, 1979.
2. Nyquist, H., "Regeneration Theory," *Bell Syst. Tech. J.* 11, 126–147 (1932).
3. D'Azzo, J. D., and Houpis, C. H., "Feedback Control Systems Analysis and Synthesis," New York, McGraw-Hill,
4. Di Stefano, J. J., Stubberud, A. R., and Williams, I. J., "Theory and Problems of Feedback and Control Systems," Schaums Outline Series, New York, McGraw-Hill, 1967.
5. Zadeh, L. A., and Desoer, C. A., "Linear System Theory: A State Space Approach," New York, McGraw-Hill, 1963.
6. Macfarlane, A. G. J. (Ed.), "Frequency Response Methods in Control Systems," New York, I.E.E.E. Press, 1979.
7. Rosenbrock, H. H., "Computer Aided Control System Design," New York, Academic Press, 1974.
8. Patel, R., and Munro, N., "Multivariable System Theory and Design," Elmsford, N.Y., Pergamon, 1981.
9. Gibson, J. E., "Nonlinear Automatic Control," New York, McGraw-Hill, 1963.
10. Atherton, D. P., "Nonlinear Control Engineering," New York, Van Nostrand Reinhold, 1975.
11. Harris, C. J., and Billings, S. A., (Ed.), "Self-Timing and Adaptive Control," Peter Peregrinus, Stevenage, U.K. and New York, 1881.
12. Kuo, B. C., "Digital Control Systems," New York, Holt-Saunders, 1980.
13. Katz, P., "Digital Control using Microprocessors," Englewood Cliffs, N.J., Prentice-Hall International, 1981.
14. "Proceedings of the First International Conference on Robot Vision and Sensory Controls, April 1–3 1981, Stratford-upon-Avon, U.K.," I.F.S. Conferences, Kempston, Bedford, England, 1981.

Cross-references: BIONICS, CIRCUITRY, CYBERNETICS, MECHANICS.

FERMI-DIRAC STATISTICS AND FERMIONS

Solid metals are good conductors of heat and electricity because about one electron per atom is free to migrate through the volume of the conductor. These electrons were once thought to behave like gas molecules which obey Maxwell-Boltzmann statistics in which the number of particles at higher energies falls off exponentially according to a relation of the form

$$n_{E,T} = \frac{1}{e^{E/kT}}$$

where E is the energy, k the Boltzmann Constant and T the absolute temperature. This electron gas theory was qualitatively useful in explaining many metallic properties, but it was never quantitatively successful. One notable failure was its prediction that electrons should contribute to the specific heats of metals.

Bohr had shown that the electron in hydrogen is not free to assume any energy, but is restricted to certain permitted energies called quantum states or energy levels. When this quantum view of atomic electron structure was extended to more complex atoms, it was found that electrons obey the Pauli exclusion principle—only two electrons in any one atom having oppositely directed spin can occupy the same energy state. Thus in an atom with many electrons, no more than two can have the lowest permitted energy, no more than two may have the next higher permitted energy, etc. An unexcited atom with all its electrons in their lowest possible energy states includes many electrons whose energy is well above the energy of the lowest two. The old electron gas theory of metals recognized that the inner electrons associated with each atom were quantized but assumed that the electrons that were not bound to particular atoms were entirely free to migrate through the metal with no *a priori* restrictions on their energy. Fermi-Dirac statistics describes the behavior of the electron gas under the assumption that *all* electrons within the conductor have their energies quantized and obey the Pauli principle. This new viewpoint leads to a distribution of electron energies according to a relation of the form

$$n_{E,T} = \frac{1}{e^{(E-E_i)/kT} + 1}$$

where the new symbol, E_i, is a critical energy characteristic of the metal more fully described below. If the metal is at high temperature, this function approaches the Maxwell-Boltzmann distribution. We can see this by noting that if $T \to \infty$, the exponent of e approaches zero regardless of E. Thus, in both cases, the number of electrons of each energy tends to become uniform. The high-temperature electrons have so many states available to them that quantum restrictions make little difference. If we let the temperature approach absolute zero, the difference between these distributions becomes extreme. If T is very small, the Maxwell-Boltzmann distribution is strongly dependent on E with most particles having low E and few having high E. Indeed for $T = 0$, the number of particles with $E \neq 0$ becomes zero—in a Maxwell-Boltzmann gas all particles come to rest at absolute zero. A Fermi-Dirac gas behaves very differently at absolute zero. The exponent of e is plus or minus infinity depending upon whether E is greater or less than E_i. The exponential term is either infinity or zero. The denominator is either infinity or one. All energy states below E_i are filled whereas all those above E_i are empty. Thus, consistent with the assumptions, at absolute zero the electrons do not crowd into one state of zero energy but are uniformly distributed among those states which are below the critical energy E_i called the Fermi energy or the Fermi level. Fermi energies depend on the kind of metals but they are of the order of several electron volts. Thus, even at absolute zero, some electrons have energies which would be typical of a Maxwell-Boltzmann electron gas only if that gas were at several thousand degrees.

The contrast may be dramatized by the following analogy. If grains of sand are spilled on an open floor, they will spread out so they are only one deep and each has zero potential energy. If the grains are poured into a drinking straw, the straw will fill to a certain height and some grains will have considerable potential energy.

Heating a metal from absolute zero to room temperature adds only .025 eV to the average energy of its particles. Since the electrons already have a *much* greater average energy, heating a metal has but a slight effect on the energy distribution of the electrons. This accounts for the fact that electrons make a negligible contribution to the specific heats of metals, and it also explains why metals must be glowing hot before electrons acquire enough additional energy to escape from the metal surface as in the filaments of radio tubes. Since the quantum view of electrons in a metal provides both a qualitative and quantitative picture of many metallic properties, we know metallic electrons are quantized Fermi particles rather than unquantized Maxwell particles. The application of Fermi-Dirac statistics to semi-conductors accounts for their special properties as demonstrated by transistors.

From the standpoint of wave mechanics, all particles which are confined in any way are quantized. Those whose spin is integral have symmetric wave functions and do not obey the Pauli principle. If they are so numerous that they must be treated statistically, they are called *bosons* and are described by Bose-Einstein statistics. Photons are the most common bosons. Those particles whose spins are odd multiples of $\frac{1}{2}$ have antisymmetric wave functions and obey the Pauli principle. They are called *fermions* and obey Fermi-Dirac statistics. Although electrons are the most common example, protons, neutrons, and μ-mesons are all fermions with spin $\frac{1}{2}$. At high temperatures, the quantum nature of both bosons and fermions becomes insignificant and both obey the classical statistics of Maxwell-Boltzmann. The technique of deriving these distributions is called statistical mechanics.

To convey the over-all method of STATISTICAL MECHANICS, we note that it is a probability theory in which the basic technique is to compute the number of possible ways in which a system can arrange itself subject to restrictions as to the number and total energy of the particles. These ways are all assumed equally likely. (There are 52 factorial, 52!, different arrangements which might result from the shuffling of a deck of playing cards. Each is equally likely.) Then, depending on the nature of the particles, bosons or fermions, the number of distinguishable ways is computed. (In the game of bridge, there are many fewer deals $52!/(13!)^4$, than there are shuffles because the order in which a player receives his cards does not change his "hand.") The probability of any particular distinguishable distribution is proportional to the number of ways in which it can be achieved. (If we flip a coin five times, there are $2^5 = 32$ orders in which the coin can fall. Of these, there are ten ways to get two heads and only one way to get five heads. We therefore find getting two heads ten times more probable than getting five heads.) The actual expected distribution is the one which can be achieved in the largest number of ways.

JAMES A. RICHARDS, JR.

Reference

Leighton, Robert B., "Principles of Modern Physics," New York, McGraw-Hill Book Co., 1959.

Cross-references: BOLTZMANN'S DISTRIBUTION LAW, BOSE-EINSTEIN STATISTICS AND BOSONS, ELECTRON SPIN, STATISTICAL MECHANICS.

FERMI SURFACE

The Fermi surface of a metal, semi-metal, or semiconductor is that surface of constant energy in momentum space which separates the energy states which are filled with free or quasi-free electrons from those which are unfilled. [Momentum space is defined in terms of three orthogonal axes, the components of the momentum vector, p_x, p_y, and p_z (or alternatively, the components of the wave vector, k_x, k_y, and k_z: $\mathbf{p} = \hbar\mathbf{k}$, where $\hbar$ = Planck's constant divided by 2π). The components of momentum of an electron at a given instant of time may be thought of as the coordinates of a point in momentum space which then moves about as various forces act on the electron.] The Fermi

surface exists simply because the electrons obey Fermi-Dirac statistics.

Consider first an elementary model of a metal consisting of a lattice of fixed positive ions immersed in a sea of conduction electrons which are free to move through the lattice. Every direction of electron motion is equally probable. Since the electrons fill the available quantized energy states starting with the lowest, a three-dimensional picture in momentum coordinates will show a spherical distribution of electron momenta and, hence, will yield a spherical Fermi surface. In this free electron model, no account has been taken of the interaction between the fixed positive ions and the electrons; indeed the only restriction on the movement or "freedom" of the electrons is the physical confines of the metal itself.

A short derivation starting with the Schrödinger equation shows that the total energy of an electron (and thus also its kinetic energy) is given by

$$E = \hbar^2 k^2/2m = p^2/2m$$

where m is the mass of the electron. A plot of E against k is then a parabola, as shown in Fig. 1(a). The Cartesian components of those values of k which are possible solutions to the Schrödinger equation are $k_i = 2\pi n_i/L$, where the n_i's are integers and L is a physical dimension of the metal. Since for each energy value so defined there are actually two states (one for an electron with spin up, one with spin down), it can be shown that the density of energy states available to the electrons is

$$g(E) = \frac{(2m)^{3/2}}{2\pi^2\hbar^3} E^{1/2}$$

where $g(E)\, dE$ is the number of states in the energy range E to $E + dE$. Then $n(E)$, the number of electrons per unit volume occupying energy states in this energy range, is

$$n(E)\, dE = g(E)f(E)\, dE$$

where $f(E) = \{\exp[(E - E_f)/bT] + 1\}^{-1}$, a function characteristic of particles which obey Fermi-Dirac statistics. In this expression, T is the absolute temperature, b is Boltzmann's constant, and E_f is a parameter depending on the number of electrons involved and indeed turns out to be the Fermi energy. E_f can be evaluated by integrating $n(E)\, dE$ from $E = 0$ to $E = \infty$ and recognizing that the integral is equal to N, the total number of electrons per unit volume. The result (at $T = 0$ K) is

$$E_f = \frac{\pi^2\hbar^2}{2m}\left(\frac{3N}{\pi}\right)^{2/3}$$

At $T = 0$ K, for $E < E_f$, $f(E) = 1$, while for $E > E_f$, $f(E) = 0$. Physically this means that the probability of a state below the Fermi level

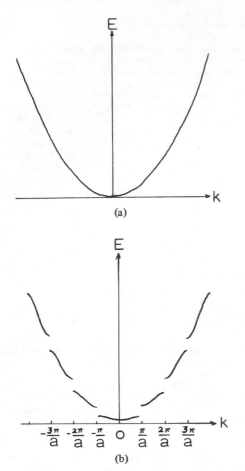

FIG. 1. (a) Energy plotted against wave number for the free electron model. (b) Energy plotted against wave number for the "quasi-free" electron model, showing energy discontinuities at Brillouin zone boundaries.

being occupied is one; whereas for states with $E > E_f$ the occupancy probability drops abruptly to zero. For temperatures greater than absolute zero, the occupancy probability drops smoothly from 1 to 0 in a range of energy of width approximately equal to bT. This shell of partially filled states gives rise to the following definition: The Fermi level is the energy level at which the probability of a state being filled is just equal to one half.

A numerical evaluation of the Fermi energy for a simple metal having one or two conduction electrons per atom yields a value of approximately 10^{-11} erg, or a few electron volts. The equivalent temperature, E_f/b, is several tens of thousands of degrees Kelvin. Thus, except in extraordinary circumstances, when dealing with metals, $bT \ll E_f$; i.e., the energy range of partially filled states is small, and the Fermi

surface is well defined by the statement above. It must, however, be noted that this is not necessarily true for semiconductors where the number of free electrons per unit volume may be very much smaller.

The foregoing treatment gives a qualitative insight into the physics of metals and, under some circumstances, semi-metals and semiconductors. A more detailed analysis requires that the effects of the ions in the lattice be recognized. This can be accomplished by introducing the periodic potential due to the lattice through which the electrons must move. Then the electrons are no longer "free," but, depending on the strength and character of the potentials and the approximations used in solving the Schrödinger equation, act as "quasi-free" particles. Another approach is the "tight-binding approximation"; occasionally a combination of the two approaches is used. In any case, introduction of lattice effects changes the characteristics of the model; the total energy and kinetic energy of an electron are no longer equivalent. The periodic lattice can be described conveniently in terms of Brillouin zones, each of which is large enough (in momentum space) to accommodate two electrons per atom. The Brillouin zone boundaries appear to the electrons as Bragg reflection planes or energy discontinuities, resulting in an energy versus wave number plot as shown in Fig. 1(b).

For many metals, the "nearly free" electron description corresponds quite closely to the physical situation. The Fermi surface remains nearly spherical in shape. However, it may now be intersected by several Brillouin zone boundaries which break the surface into a number of separate sheets. It becomes useful to describe the Fermi surface in terms not only of zones or sheets filled with electrons, but also of zones or sheets of holes, that is, momentum space volumes which are empty of electrons. A conceptually simple method of constructing these successive sheets, often also referred to as "first zone," "second zone," etc., was demonstrated by Harrison.[1] An example of such a construction is shown in Fig. 2. This construction works quite well, for example, for aluminum which has three valence electrons per atom. Experiments, and indeed more elegant theoretical calculations, show that the fourth zone is totally unoccupied and that the third zone monster is not multiple-connected in the manner shown. The recipe for constructing these figures, some of which may even be pleasing to art connoisseurs, cannot be developed in the limited space of this article but will be found in the references.[1,2]

The intense research effort of the last 15 years on the Fermi surfaces of metals and semimetals originated, to a great extent, with Pippard's ingenious deductions, based on anomalous skin-effect experiments, concerning the Fermi surface of copper.[3] Prior to Pippard's

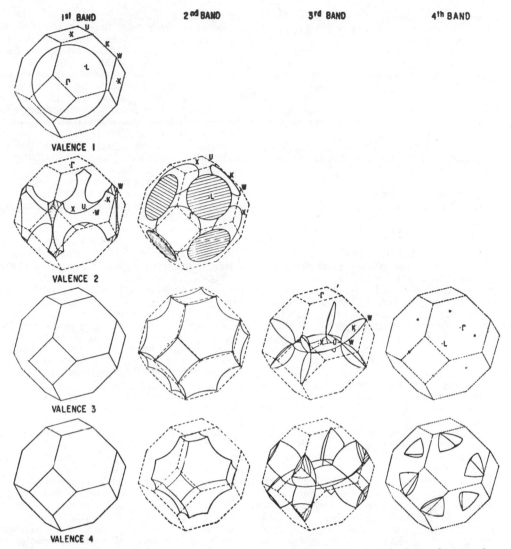

FIG. 2. Fermi surfaces in several zones or bands, for face-centered cubic metals having various numbers of "quasi-free" electrons per atom, as constructed by Harrison.[1]

work, it was taken for granted that in copper, with one quasi-free electron per atom, the first Brillouin zone would be only half filled and, hence, would have a nearly spherical Fermi surface. His work suggested that a series of eight necks pull out and touch the Brillouin zone boundaries in the [111] crystallographic directions. This shape has now been confirmed and precisely mapped, not only for copper but also for silver and gold.

A variety of experimental techniques has been developed, capable of yielding both overlapping and complementary information concerning Fermi surfaces of metals. Some of these techniques are described briefly:

(1) The DE HAAS-VAN ALPHEN EFFECT[4] is an oscillatory behavior of the magnetic susceptibility (or more generally, oscillatory behavior of any electronic property) due to the quantization of magnetic flux through an electron orbit in units of hc/e (where h = Planck's constant, c = speed of light, and e = electron charge). Measurements of the periods of the oscillations can be directly related to extremal cross-sectional areas of the Fermi surface.

(2) Cyclotron resonance[5] refers to oscillations in the magnetic field dependence of high-frequency surface impedance arising when the electron's cyclotron frequency is an integral multiple of the rf frequency. The periods of

these oscillations measure dA/dE, the rate of change of cross-sectional area with energy at extremal cross sections.

(3) Magnetoresistance[6] refers to extra resistance in a metal in the presence of a magnetic field. This magnetoresistance may be changed by changing the direction of the magnetic field and/or its magnitude. This direction and field dependence gives information about Fermi surface topology, particularly regions of contact with Brillouin zone boundaries.

(4) The magnetoacoustic effect[7] refers to oscillations in the magnetic field dependence of ultrasonic attenuation. The oscillations occur when dimensions of cyclotron orbits of electrons in extremal states on the Fermi surface are equal to integral multiples of the ultrasonic wavelength so that measurements of these oscillations give extremal linear dimensions of the Fermi surface.

(5) The anomalous skin effect[3] occurs when the electron mean free path is long compared to the rf skin depth. In this circumstance, electrons moving nearly parallel to the sample surface will dominate the conductivity. A measurement of surface impedance then gives an integral of the radius of curvature over that part of the Fermi surface containing the effective electrons.

(6) Positron annihilation in metals[8] gives directly the number of electron states in various cross sections of the Fermi surface. Because a positron impinging upon a metal is quickly thermalized, when it annihilates with an electron the resulting gamma rays must carry off the annihilated electron's momentum. Thus gamma ray angular correlation measurements can be related to the distribution of electron momenta.

(7) The Gantmakher effect[9] (or radio frequency size effect) occurs in thin samples when a magnetic field parallel to the sample surface is adjusted so that extremal electron orbits just fit in the sample thickness. RF energy can then be transported across the thickness of the sample by the electrons. Observation of these Gantmakher resonances, like the magnetoacoustic resonances, gives extremal linear dimensions of the Fermi surface.

These techniques have been used primarily to measure extremal properties of the Fermi surface geometry. Considerable effort has also been applied to the measurement of nonextremal properties[10,11] which can do much to elucidate Fermi surface geometry. The precision and applicability of each of these techniques is dependent on the material under investigation. Among pertinent factors are the number of quasi-free electrons per atom, crystallographic structure, magnetic properties, purity, and practicality of sample preparation.

The fundamental interest in the measurement and calculation of Fermi surface properties derives from their central role in understanding metallic behavior. Because the Pauli exclusion principle prevents more than one electron from occupying a given quantum state, the electrons in most of the filled states below the Fermi energy are "frozen," that is, they cannot be excited to a higher energy state by an external electric or magnetic field because the higher states are already occupied. Only the relatively small number of electrons within about bT of the Fermi energy can respond to external influences, and these "Fermi surface" electrons determine all the basic metallic properties, e.g., electrical conductivity, magnetic susceptibility, heat capacity, catalytic activity, thermoelectric effects, etc. Thus the experimental characterization and theoretical understanding of Fermi surface properties is one of the major activities in solid state physics.

Extensive application of the experimental techniques described above has resulted in a great deal of Fermi surface information about a large number of metals.[12] Nearly all the elemental metals (except those with severe sample preparation problems) have been investigated,[13] and many are extremely well characterized.[14] The complicated Fermi surface geometries that occur in the transition metals can be conveniently described by any of several parametrization schemes, the most successful of which is based on the Koringa-Kohn-Rostoker (KKR) method of energy band calculation.[15] This technique allows even the most intricate Fermi surfaces to be accurately described by a small number of physically meaningful parameters (usually seven or less), so that the anisotropy of various fundamental quantities over the Fermi surface can be studied and correlated in a simple way. These parametrization schemes have been used in detailed studies of the variation over the Fermi surface of the electronic Fermi velocity,[15] scattering lifetime due to impurities and defects,[16] the electron-phonon interaction,[17] and superconducting energy gap.

Paralleling this experimental work has been a great deal of theoretical effort devoted to first-principles calculation of energy bands and Fermi surface properties.[18,19] Like the experimental work, theoretical treatments of the Fermi surfaces of elements have been highly developed,[20] to the extent that agreement with experiment is usually within 10% and often much better.

Increasingly, both theory and experiment are turning away from the elements to alloys[21,22] and intermetallic compounds,[23] where a much greater range of unusual metallic behavior is to be found. Of particular interest is the behavior arising from various many-body effects: superconductivity from the electron-phonon interaction, spin fluctuations and itinerant magnetism from exchange interactions, and electron localization and local moment formation from electron correlation effects. Experimentally these many-body effects reduce the Fermi velocity of electrons at the Fermi surface and can be studied through effective mass measurements in the de Haas-van Alphen effect[23] and cyclotron resonance. The theoretical description of these many-body effects is less well developed, but is receiving increasing attention.[24] Understanding these many-body effects and their influence on Fermi surface properties and metallic behavior is likely to be a major theme of both theory and experiment in the coming years.

H. V. BOHM
NORMAN TEPLEY
GEORGE CRABTREE

References

1. Harrison, W. A., *Phys. Rev.*, **118**, 1190 (1960).
2. Ziman, J. M., "Electrons in Metals; A Short Guide to the Fermi Surface," London, Taylor and Francis, 1963.
3. Pippard, A. B., *Phil. Trans. Roy. Soc. London Ser. A*, **250**, 323 (1957).
4. Shoenberg, D., in "Proceedings of the Ninth International Conference on Low Temperature Physics," J. G. Daunt, D. O. Edwards, F. J. Milford, and M. Yaqub editors, New York, Plenum Press, 1965 (p. 665).
5. Kip, A. F., in "The Fermi Surface," W. A. Harrison, and M. B. Webb editors, New York, John Wiley & Sons, 1960 (p. 146).
6. Pippard, A. B., "The Dynamics of Conduction Electrons," New York, Gordon and Breach, 1965 (p. 90).
7. Tepley, N., *Proc. I.E.E.E.*, **53**, 1586 (1965).
8. Stewart, A. T., in "Positron Annihilation," A. T. Stewart and L. O. Roellig editors, New York, Academic Press, 1967 (p. 17).
9. Gantmakher, V. F., *Zh. Eksperim, i Teor. Fiz.*, **43**, 345 (1962). (English Transl.: *Soviet Physics JETP*, **16**, 247 (1962).)
10. Dooley, J. W., and Tepley, N., *Phys. Rev.*, **187**, 781 (1969).
11. Henrich, V. E., *Phys. Rev. Letters*, **26**, 891 (1971).
12. Cracknell, A. P. and Wong, K. C., "The Fermi Surface," Oxford, Clarendon Press, 1973.
13. Young, R. C., *Rep. Prog. Phys.* **40**, 1123 (1977).
14. See, for example, Karim, D. P., Ketterson, J. B., and Crabtree, G. W., *J. Low Temp. Phys.* **30**, 389 (1978) and Dye, D. H., Campbell, S. A., Ketterson, J. B., and Vuillemin, J. J., *Phys. Rev.* **B23**, 462 (1981).
15. Crabtree, G. W., Dye, D. H., Karim, D. P., and Ketterson, J. B., *J. Magnetism and Magnetic Materials* **11**, 236 (1979).
16. "Proceedings of the International Conference on Electron Lifetimes in Metals." (D. H. Lowndes and F. M. Meuller, Eds.), *Phys. Cond. Matter* 19, 1–423 (1975).
17. Crabtree, G. W., Dye, D. H., Karim, D. P. Koelling, D. D., and Ketterson, J. B., *Phys Rev. Letters* **42**, 390 (1979).
18. Harrison, W. A., "Pseudopotentials in the Theory of Metals," New York, W. A. Benjamin, 1966.
19. Koelling, D. D., *Rep. Prog. Phys.* **44**, 139 (1981).
20. Mackintosh, A. R., and Andersen, O. K., in "Electrons at the Fermi Surface," (M. Springford, Ed.), Cambridge, U.K., Cambridge Univ. Press, 1980, p. 149.
21. Saito, Y., and Maezawa, K., in "Proceedings of Twelfth International Conference on Low Temperature Physics," (E. Kamda, Ed.), Kyoto, Academic Press of Japan, 1971, p. 583.
22. Coleridge, P. T., in "Electrons at the Fermi Surface," (M. Springford, Ed.), Cambridge, U.K., Cambridge Univ. Press, 1980, p 321.
23. Crabtree, G. W., Johanson, W. R., Campbell, S. A., Dye, D. H., Karim, D. P., and Ketterson, J. B., "Proceedings of the International Conference on Physics of Transition Metals," (P. Rhodes, Ed.), *Inst. Phys. Conf. Ser.* **55**, 79 (1981).
24. Wilkins, J. W., in "Electrons at the Fermi Surface," (M. Springford, Ed.), Cambridge, U.K., Cambridge Univ. Press, 1980, p. 46.

Cross-references: CYCLOTRON RESONANCE; DE HAAS–VAN ALPHEN EFFECT; ENERGY LEVELS; SOLID STATE PHYSICS; SOLID STATE THEORY; TRANSPORT THEORY.

FERRIMAGNETISM

Snoek's publication (1946) of his wartime work on ferrites established the existence of new ceramic magnetic materials capable of combining the resistivity of a good insulator (10^{12} ohm-cm) with high permeability. (see MAGNETISM.) In 1948, Néel introduced the term ferrimagnetism to describe the novel magnetic properties of these materials. A simple ferrite is composed of two interpenetrating FERROMAGNETIC sublattices with magnetizations $M_a(T)$ and $M_b(T)$ which decrease with increasing temperature and vanish at the Curie point, T_c. In a ferromagnetic material, the resulting saturation magnetization, M, would be $M_a + M_b$; however, in a ferrite, strong antiferromagnetic interaction between sublattices results in antiparallel alignment, and $M = M_a - M_b$. In general $M_a(T) \neq M_b(T)$, and the material behaves in most respects like a ferromagnet, exhibiting domains, a hysteresis loop, and saturation of the magnetization at relatively low applied magnetic fields. Practical values for saturation magnetization and Curie temperature range from 250 to 5000 oersteds and from 100 to 600°C.

Ferrimagnetic materials have spinel, garnet, and hexagonal structures. A typical spinel ferrite is $NiFe_2O_4$. Other ferrites may be obtained by substituting magnetic (Co, Ni, Mn) or nonmagnetic (Al, Zn, Cu) ions for some of the Ni or Fe ions, e.g., $Ni_{1-y}Co_yAl_xFe_{2-x}O_4$, where x and y may be varied to modify M and T_c. Yttrium iron garnet (YIG), $Y_3Fe_5O_{12}$, is the classical ferrimagnetic garnet which combines very low magnetic loss with high resistivity. Substitution of magnetic RARE EARTH ions (Gd, Yb, Ho, etc.) for Y and of nonmagnetic ions (Ga, Al) for some of the Fe ions leads to many different ferrite compositions with a wide range of M and magnetic loss. The rare earth ions form a third magnetic sublattice with attendant magnetization M_c antiparallel to the resultant magnetization $M_{a,b}$ of the two Fe sublattices. Since M_c and $M_{a,b}$ exhibit different variations with temperature, the net magnetization may vanish twice, at T_c and at an intermediate temperature called the compensation point, T_{comp}, where $M_c = M_{a,b}$.

A typical hexagonal ferrite is $BaFe_{12}O_{19}$. Again, other magnetic ions such as Mn, Co, and Ni may be introduced to produce wide variations in M and T_c. Hexagonal ferrites are characterized by large anisotropy fields with an axis of symmetry which may be either a direction of hard (planar ferrites) or easy (uniaxial ferrites) magnetization.

To distinguish among major fields of applications, ferrites can be separated into five groups: soft, square-loop, hard, microwave, and single-crystal ferrites.

Soft ferrites have a slender, S-shaped hysteresis loop with low remanence and low coercive force permitting easy magnetization and demagnetization with little magnetic loss. Mn-Zn and Ni-Zn ferrites with spinel structure exhibit these properties and permit adjustment of M, and permeability, μ_i, over a wide range of values through variations in composition. Ni ferrite has $\mu_i = 15$ and $M = 3000$ G, whereas Ni-Zn ferrite may have as much as $M = 5000$ G combined with a permeability of several thousand. Mn-Zn ferrites have values of $\mu_i = 500$ to 5000 depending on composition. These ferrites are uniquely suited to low-loss inductor and transformer cores for radio, television, and carrier telephony.

Square-loop ferrites are materials exhibiting an almost rectangular hysteresis loop with two distinct states of remanence and with a coercive force of a few Oe. All practical square-loop ferrites have a spinel structure. The Mg-Mn (Zn) system has retained its preeminent position in computer memory applications two decades after its discovery in 1951. More recently, Li-Ni ferrites and more complex systems containing Li, Mn, and Al have become competitive in applications requiring stability and fast switching over a wide range of temperatures.

Hard ferrites are characterized by hexagonal structure, a hysteresis loop enclosing a large area, and a coercive force of several thousand Oe. These ferrites can store a significant amount of magnetic energy, and have found widespread application as permanent magnets in hi-fi loudspeakers, small motors, generators, measuring instruments, etc.

Microwave ferrites have garnet, spinel, or hexagonal crystal structure, and very low electric and magnetic loss factors. In general, the required M increases with the frequency, f, of application. Substituted and pure garnets, Mg-Mn-Al ferrites and Mg-Mn ferrites are used at the lower part of the microwave spectrum where $M = 200$ to 3000 G is adequate. In the millimeterwave region, $f = 30$ to 100 GHz, one uses Ni-Zn ferrites ($M = 5000$ G) and hexagonal ferrites of various compositions.

All microwave ferrite devices such as isolators, circulators, switches, phase shifters, limiters, parametric amplifiers, and harmonic generators are based on interactions of rf signals with the ferrite magnetization. Aligning M with an external biasing magnetic field, H_{dc}, and applying a microwave signal in an orthogonal direction leads to strong interaction and gyromagnetic resonance. On a microscopic scale, this is explained as application of a torque to the unpaired ELECTRON SPINS of the magnetic ions which causes them to precess at the rf frequency much like so many spinning tops. The precessional motion has a microwave RESONANCE frequency f_r dependent upon H_{dc} and the gyromagnetic splitting factor g_{eff}. In ferrimagnets with spinel structure, g_{eff} is related to the g-factors of the sublattices as follows:

$$g_{eff} = M/[(M_a/g_a) - (M_b/g_b)]$$

On a macroscopic scale, this interaction modifies the rf magnetic field in a manner which is described by introducing an antisymmetric permeability tensor $[\mu]$ whose complex components depend on M, H_{dc}, and frequency. When the frequency approaches f_r, one observes a resonance absorption line whose width, ΔH, is determined by the magnetic loss of the material. Values for ΔH cover a range from <1 oersted for single-crystal YIG to >1000 oersteds for some polycrystalline Ni-ferrites. The interaction of rf fields and electron spins becomes a maximum if the rf field is circularly polarized in the same sense as the precessional motion of the spins. Circular polarization in the opposite direction produces almost no interaction and no gyromagnetic resonance. This permits design of nonreciprocal ferrite devices. At high levels of microwave power, nonlinear coupling between microwave signal and precessional spin motion causes the parametric excitation of higher order modes of spin motion (magnetostatic modes and spin waves). This effect has been exploited in limiters and parametric amplifiers.

Single-crystal ferrites of practical importance are rare-earth garnets grown in a flux of molten lead oxide. Some of these are optically transparent permitting direct observation of magnetic domains. Interaction of infrared and visible light with the electron spins is called the magneto-optic effect. It permits electronic modulation of a beam of light which propagates through a single-crystal garnet. Devices of this type are of great potential interest in the rapidly developing laser technology.

Magnetic Bubbles Single-crystal, rare-earth garnet sheets have been grown on a substrate with a preferred direction of magnetization perpendicular to the plate. In these plates, tiny round magnetic domains called *bubbles* can be formed by an applied magnetic field. These bubbles can be propagated, erased, and manipulated to perform binary functions in computers including logic, memory, counting, and switching.

Bubbles are cylindrical magnetic domains whose magnetization is reverse to that in the remainder of the thin magnetic layer in which they are present. If the magnetic layer is a garnet then the process of bubble formation can be observed with a polarizing microscope as an external *bias* field applied perpendicular to the surface is slowly increased until bubbles develop from isolated island strip domains. This process is illustrated in Fig. 1. These single domain configurations will only occur if the magnetic material has a uniaxial anisotropy with the easy axis of magnetization perpendicular to the surface.

An external bias field is a necessary condition to support isolated bubble domains with the allowable ranging of the bias field approximately one-tenth the saturation magnetic moment, about 50 Oe for the typical garnet film supporting 2-μm-diameter bubbles. The upper limit of the bias field is set by bubble collapse, the lower limit by a distortion instability in the bubble shape leading to the conversion of a bubble into a long meandering strip domain. This instability occurs at the domain strip width at which the total energy is independent of the strip length. These rules are applicable for materials in which the uniaxial anisotropy field H_k is much greater than $4\pi M_s$ (M_s is the saturation magnetization in gauss). The ratio $H_k/4\pi M_s$ is defined as the material q and is, in fact, a measure of the "stiffness" of the magnetization. Experience has shown that for most useful bubble materials $q > 3$.

A material length $l = \sigma_w/4\pi M_s^2$ characterizes materials for the bubble domain size they support. Here l is in centimeters and σ_w is the domain wall energy in ergs/cm^2. The optimum magnetic film thickness $h = 4l$ results in the smallest stable domain diameter $d = \pi l$ at a bias field $H_A = (0.3) 4\pi M_s$.

The interplay of the material parameters σ_w and M_s in determining bubble stability is shown in Fig. 2 and 3. Note that the bias field acting directly on the bubble domain in the classical sense and the energy of domain wall both act to reduce the size of a bubble and are opposed by the stray magnetostatic fields which arise from the surface magnetic charges and which attempt to increase the bubble size. The expression for the stray field H_D shown in Fig. 2 is an approximation. Since the total energy ξ_T for the bubble has a minimum with $\partial \xi_T/\partial r = 0$ and $\partial^2 \xi_T/\partial r^2 > 0$ the bubble is statically stable.

For a film of thickness $h = 4l$ the nominal bubble diameter d, i.e., the diameter when the bias field is set in the middle of the stable range,

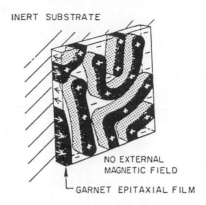

INERT SUBSTRATE

NO EXTERNAL
MAGNETIC FIELD
└ GARNET EPITAXIAL FILM

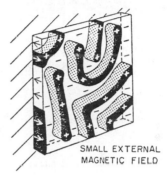

SMALL EXTERNAL
MAGNETIC FIELD

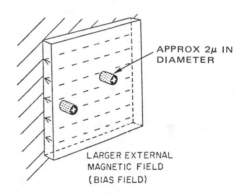

APPROX 2μ IN
DIAMETER

LARGER EXTERNAL
MAGNETIC FIELD
(BIAS FIELD)

BUBBLE FORMATION

FIG. 1. The garnet epitaxial film is grown on a nonmagnetic substrate. The serpentine nature of the magnetic domains in garnet films arises from a preferred "easy" axis of magnetization perpendicular to the film surface. With no external magnetic field, the magnetic domains are arranged such that the sample is magnetically neutral, half pointing up and half pointing down. As a small external magnetic field is applied, the domains, whose polarity is opposite to that of the field, shrink. If the external magnetic field is further increased, the stripes shrink into cylinders—the bubbles. The bubbles are free to move throughout the film and can be viewed as tiny magnets afloat in a magnetic field sea of the opposite polarity.

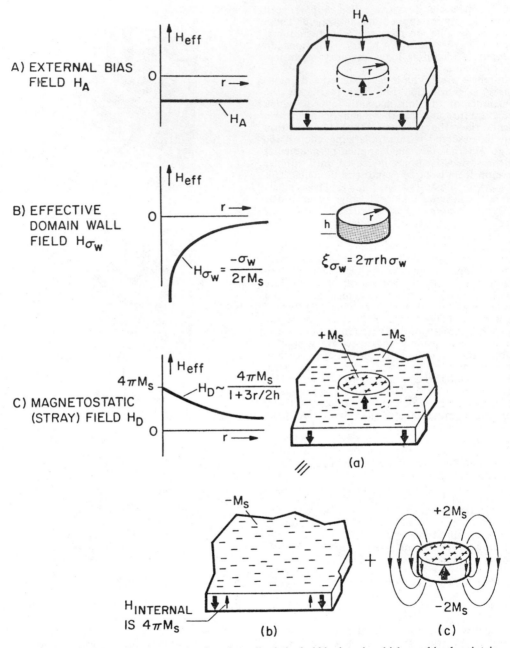

A) EXTERNAL BIAS FIELD H_A

B) EFFECTIVE DOMAIN WALL FIELD H_{σ_W}

$$H_{\sigma_W} = \frac{-\sigma_W}{2rM_S}$$

$$\xi_{\sigma_W} = 2\pi r h \sigma_W$$

C) MAGNETOSTATIC (STRAY) FIELD H_D

$$H_D \sim \frac{4\pi M_S}{1 + 3r/2h}$$

$+M_S$ $-M_S$

(a)

$-M_S$

$+2M_S$

$H_{INTERNAL}$ IS $4\pi M_S$

$-2M_S$

(b)   (c)

FIG. 2. The fields effective on the domain wall of the bubble domain which combined maintain static stability are indicated. The bias field H_A acts to diminish the volume of the bubble by exerting a direct force on the domain wall directed radially inward. An effective domain wall field results from a reaction by the domain wall energy density σ_W to reduce the wall area to zero. The magnetostatic (stray) field H_D operates to equalize the overall magnetic surface charge by an increase in the bubble volume. By resolving (a) into (b) and (c) the r-dependent part of the field is seen to be that of a cylindrical dipole magnet.

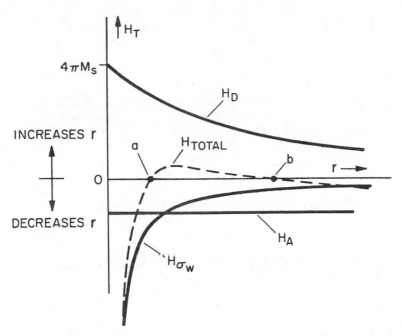

FIG. 3. There are two values of the bubble radius for which H_T, the sum of the fields illustrated in Fig. 2, is zero. Condition a is radially unstable and b is stable, a conclusion that can be reasoned from this figure.

is predicted by theory to be $8l$. To store the greatest amount of information in a given area is equivalent to packing the greatest number of bubbles into that area, thus the highest bit density will be realized in a material which supports the smallest-diameter bubbles.

Bubble propagation circuits such as T-bar, chevron, etc., provide a traveling local perturbation of the bias field. A bubble will move in a small magnetic bias field gradient $\partial H_z/\partial x$ at a velocity $v_x = \mu_w(2r\partial H_z/\partial x - 8H_c/\pi)/2$ where μ_w is the planar wall mobility in cm/s – Oe, H_c is the domain wall coercivity in oersted, and r is the bubble radius. At higher values of the applied field gradient, often reached in device operation, the velocity reaches a maximum value v_p which in turn limits the maximum data rate of a bubble device. A theoretical expression for this limiting velocity is $v_p = 24\gamma A/h\sqrt{k_u}$ where γ is the gyromagnetic ratio, A the exchange constant, K_u the uniaxial anisotropy constant ($H_k = 2K_u/M_s$), and h the magnetic film thickness. For most bubble film compositions in which the bubble thickness equals the bubble diameter, v_p is in the range 2000–3000 cm/s.

It was an unexpected discovery that bubbles with differing static and dynamic properties can exist in the same garnet film. Different domain wall states give rise to these unusual properties. The simplest bubble wall structure is a *Bloch wall*. In a Bloch wall the magnetiza-

tion rotates in a plane parallel to the wall as we move radially across the wall of the bubble. Rotation within the wall can be either clockwise (cw) or counterclockwise (ccw). In a hard bubble the domain wall is divided into cw and ccw Bloch segments linked to one another by Néel segments. In a Néel wall the magnetization rotates cw or ccw in a plane perpendicular to the wall. Néel and Bloch segments, really twists in the bubble wall, repel each other through the exchange energy which operates to keep "adjacent" spins parallel. Thus the static properties of hard bubbles differ from those of normal bubbles. Examples of wall configurations are shown in Fig. 4; bubbles (a) and (b) have normal static properties. Bubble (c) is the only bubble that moves parallel to a field gradient, i.e., "normal" in the dynamic sense.

Three techniques can be used to suppress hard bubbles. They are multilayer garnet films, ion implantation, and a very thin ($\sim$300 Å) Permalloy film directly on the garnet layer. Ion implantation is normally used. Figure 5 illustrates the effect of a hydrogen ion implant into the outer 0.5 μm surface of a nominally 6-μm-thick $(YGdTm)_3(GaFe)_5O_{12}$ garnet film on suppressing hard bubbles. The material parameters of $(YLuSm)_3(CaGeFe)_5O_{12}$ garnet films are less sensitive to temperature variation and are used exclusively in devices with bubble diameters 3 μm or less. Garnet films can be tailored to support 0.5-μm bubbles, which im-

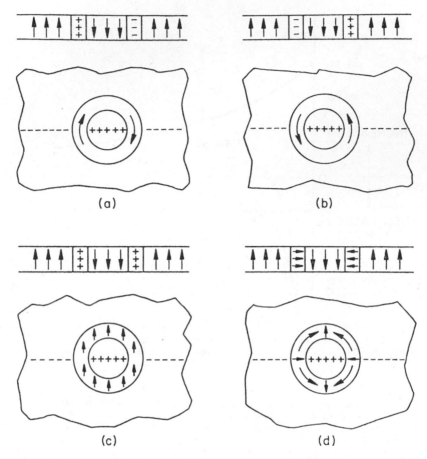

FIG. 4. (a) A simple Bloch wall. (b) a ccw sense for the Bloch wall. (c) A wall structure partially (a) and (b). (d) A more complex example of wall structure.

plies that a storage density of 2×10^7 bits/cm^2 can be realized with the garnet material system. Garnet films are grown epitaxially on polished (111) wafers of $Gd_3Ga_5O_{12}$.

Bubble operations used in chip designs are propagation, generation, detection, transfer, replication, annihilation, and swap. It is beyond the scope of this section to discuss all of these in detail; however, we will dwell on the most basic operation, that of propagation. Conductor propagation was the first method used to move bubbles but was replaced by feed access which features inductive power field to produce bubble motion via patterned permalloy features.

Field access propagation by T-Bar permalloy features is illustrated in Fig. 6. The Permalloy pattern is generally not in direct contact with the bubble supporting garnet material to guard against local spontaneous domain nucleation. Bubbles move to the right with a cw rotation of the in-plane field and to the left with a ccw rotation. The drive field should be imag-

ined as a constant amplitude rotating field in the plane of the figure whereas the bias field and the easy axis are normal to this plane. It can be seen that the bubble moves from a feature to the like feature one period away as the drive field rotates by 360°.

Propagation elements can be combined with bubble generators, detectors, etc. to build functional chips. Many chip organizations are possible. For example, Fig. 7 illustrates a 70 kbit "endless loop" shift-register architecture.

Bubble memory chips are packaged in an assembly of two mutually orthogonal wire-wound solenoids which are driven to provide a rotating in-plane drive field, and a permanent magnet which provides the bias field necessary to maintain stable bubbles. Because the field of the permanent magnet tracks the temperature variation of the bias field, the temperature range of satisfactory operation is wide. Data retention is completely nonvolatile.

Magnetic bubbles bridge the capacity–data

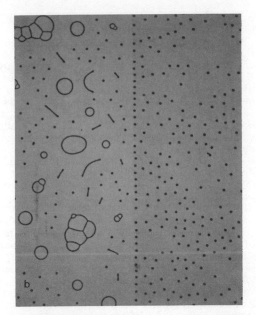

FIG. 5. The effectiveness of a hydrogen ion implant in suppressing hard bubbles is evident in this photograph. Only normal bubbles are found in the implanted area to the right, while a diverse collection of hard bubbles and strips are found to the left.

FIG. 7. Example of a 70-kbit bubble chip, approximate size 5 mm by 6 mm, processed at the Reading Division of Western Electric Research Department.

retrieval time gap left vacant by magnetic core and semiconductor devices on one side and the electromechanical magnetic tape and disk on the other. Improvements in bubble materials, circuit processing, and device design have advanced bubble technology to where it is used in applications requiring 10^6–10^8 bits and retrieval times less than 0.005 s. Bubble chips with 10 million storage cells are currently under development. Magnetic bubbles utilize remanent magnetization to provide data storage. How-

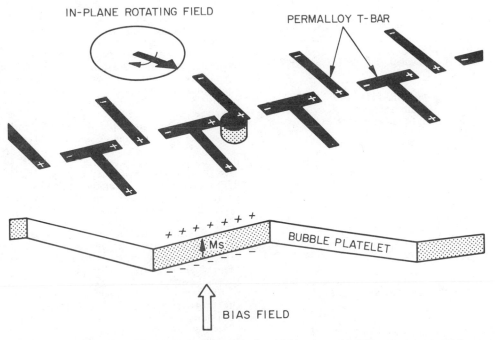

FIG. 6. The Permalloy T-Bar pattern (Permalloy is a highly permeable NiFe alloy) becomes magnetized in the presence of the drive field H_{xy}. Magnetic bubbles couple to precise positions on the track and move when H_{xy} is rotated.

ever, unlike magnetic disk or tape storage the operation is entirely nonmechanical.

For more information the reader is referred to a number of articles and books on the subject of magnetic bubbles that have recently appeared.

W. H. VON AULOCK
A. H. BOBECK

References

General

Standley, K. J., "Oxide Magnetic Materials," Oxford, Clarendon Press, 1962.

Smit, J., and Wijn, H. P. J., "Ferrites," New York, John Wiley & Sons, Inc., 1959.

Lax, B., and Button, K. J., "Microwave Ferrites and Ferrimagnetics," New York, McGraw-Hill Book Co., Inc., 1962.

von Aulock, W. H., "Handbook of Microwave Ferrite Materials," New York, Academic Press, 1965.

von Aulock, W. H., and Fay, C. E., "Linear Ferrite Devices for Microwave Applications," New York, Academic Press, 1968.

Snelling, E. C., "Soft Ferrites," London, Iliffe Books, Ltd., 1969.

Helszajn, J., "Principles of Microwave Ferrite Engineering," London, Wiley Interscience, 1969.

Magnetic Bubbles

1. Bobeck, A. H., *Bell Syst. Tech. J.* **46**, 1901 (October 1967).
2. O'Dell, T. H., "Magnetic Bubbles," London, MacMillan, 1974.
3. Bobeck, A. H., and Della Torre, E., "Magnetic Bubbles," Amsterdam, The Netherlands, North-Holland Publishing, 1975.
4. Chang, H., "Magnetic Bubble Technology," New York, IEEE Press, 1975.
5. Bobeck, A. H., and Scovil, H. E. D., *Scientific American,* p. 78, June 1971.
6. Bobeck, A. H., Bonyhard, P. I., and Geusic, J. E., *Proc. IEEE* **63**, 1176 (August 1975).
7. Nielsen, J. W., Licht, S. J., Brandle, C. D., *IEEE Trans. Magnetics* **MAG-10**, 474 (1974).
8. Tabor, W. J., Bobeck, A. H., Vella-Coleiro, G. P., and Rosencwaig, A., "A new type of cylindrical magnetic domain (hard bubble)," *AIP Conf. Proc.*, **10**, 442–457 (1972).
9. Slonczewski, Malozemoff, J. C., and Voegeli, O., "Statics and Dynamics of Bubble Containing Bloch Lines," *AIP Conf. Proc.* **10**, 458–477 (1972).
10. Eschenfelder, A. H., "Magnetic Bubble Technology," New York, Springer-Verlag, 1980.

Cross-references: FERROMAGNETISM, MAGNETISM, RESONANCE, TRANSFORMER.

FERROICITY, FERROELECTRICITY, AND FERROELASTICITY

Originally the concept *ferroic* was a unification and generalization of *ferromagnetic*, *ferroelec-tric*, and *ferroelastic,* but here only nonmagnetic crystals are considered.

When a given phase (phase I) has a structure that can be regarded as a slight distortion of another real or imaginary structure (structure II), it is called a *ferroic phase.* (*Distortion,* unlike *modification,* implies lowering in symmetry.) If structure II cannot be regarded as a slight distortion of any other structure, it is called the *prototype* of phase I. If there really exists a phase whose structure is an equisymmetric slight modification of the prototype, it is called the *prototypic phase.* The interatomic distances in the prototypic phase, like those in every phase, vary continuously with temperature (and pressure, etc.), while those in the prototype are fixed. The prototype is the zero point for all the order parameters, all the lattice vibration coordinates, and all the components of electric polarization vector and mechanical strain tensor. If we agree that the prototype is the same as the prototypic phase at a certain temperature T_p, then the prototypic phase at any other temperature is not the zero point. For example, let the point group of the prototype be $4mm$; the z component of spontaneous polarization vector and the (x, x) and (z, z) components of spontaneous strain tensor of the prototypic phase are zero at T_p and nonzero at any other temperature. In some cases, no prototypic phase exists really (owing to melting, dehydration, or the like); still it is possible to imagine the prototype.

The term "a ferroic phase" is often abbreviated to "a ferroic," and "a prototypic phase" to "a prototypic," omitting "phase."

If one phase is a slight distortion of a second phase, and this phase II is a slight distortion of a third phase, and this phase III cannot be regarded as a slight distortion of any structure, then phase III, not phase II, is recognized as the prototypic of phase I. Both phase I and phase II are ferroics derived from phase III. Phase III is the common prototypic of phases I and II. No phase can be both ferroic and prototypic. The relationship between ferroic and prototpyic phases is not simply the relationship between lower- and higher-symmetry phases.

The structure of a ferroic can be conceived to result from modulating the prototype in a wave. Let α denote the quotient ($\geqq 1$) of the wavelength by the unit-cell length of the prototype. According to whether α is integral or deviates slightly from an integer, the ferroic is said to be *commensurate* or *incommensurate.* (The deviation from an integer is not due to thermal expansion.) Only if it is commensurate is the ferroic three-dimensionally periodic (as the prototype is). We will consider below only commensurate ferroics.

As the *cell multiplicity* of a ferroic we refer to the quotient (integral) of the number of molecules in the primitive unit cell of the ferroic by that of the prototype. If the customary unit cell of the ferroic or prototype is not primitive, it needs to be rendered into a primitive unit cell.

A ferroic has a definite number of *situations*. Situation is a kind of state. The structure of the ferroic can be specified by assigning to every atom a displacement vector from its position in the prototype. When the ferroic is in a situation, the displacement vectors of all the atoms are uniquely determinate. The displacement vectors in one situation are not all equal to the displacement vectors in another situation. The latter displacement vectors (or the latter situation) can be obtained by performing a certain operation of the space group of the prototype upon the former displacement vectors (or the former situation). The mentioned operation is not uniquely determinate; there are many operations each changing the former situation to the latter. By performing every operation of the space group of the prototype upon one situation, all situations can be obtained. The number of situations is finite. Regardless of the situation the ferroic is in, it has the same free energy.

Different situations are regarded as the same *oriate* (short for *ori*entational *state*) if they can be changed to each other by pure translations. The number of situations belonging to one oriate equals the number of situations belonging to any other oriate. This number equals the cell multiplicity. The number of all situations of the ferroic, therefore, equals the product of the number of oriates and the cell multiplicity. Furthermore, the number of oriates equals the quotient of the order of the point group of the prototype by that of the ferroic.

Figure 1 shows a model ferroic. Points A, B, $\cdots$, F have $z = c_0/2$. Their images across the xy plane as a mirror are A$'$, B$'$, $\cdots$, F$'$. Points A, B, C, D, A$'$, B$'$, C$'$, D$'$ are the vertices of a primitive unit cell of the prototype. (Thus c_0 is a lattice constant of the prototype.) *White* atoms lie at the vertices. The unit cell, moreover, contains two *black* atoms which lie at $z = \pm d/2$ $(0 < d < c_0/2)$. The ferroic results from the displacement of every black atom by the drawn vector which is perpendicular to the z axis. (For simplicity it is assumed that no white

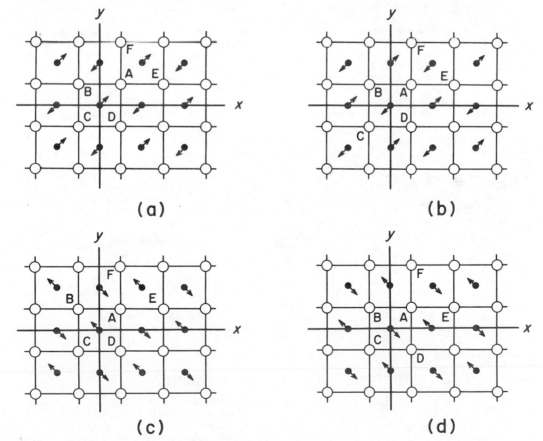

FIG. 1. A model ferroic; (a) situation δ_1, oriate $\mathcal{O}_1$; (b) situation δ_2, oriate $\mathcal{O}_1$; (c) situation δ_3, oriate $\mathcal{O}_2$; (d) situation δ_4, oriate $\mathcal{O}_2$.

atom is displaced.) Obviously, B, D, E, F, B', D', E', F' are the vertices of a primitive unit cell of the ferroic. Hence the cell multiplicity is 2. The ferroic has, in all, four situations which are represented by $\delta_1, \delta_2, \delta_3, \delta_4$. The translation along the x axis by length a_0 (= AB), obviously, changes δ_1 to δ_2, and δ_3 to δ_4. Thus δ_1 and δ_2 are regarded as the same oriate, $\mho_1$; δ_3 and $_4$ as the same oriate, $\mho_2$. Obviously, no pure translation changes δ_1 to δ_3 (the $\pi/2$ rotation about the z axis does so). Thus δ_1 and δ_3 are regarded as different oriates. In other words, $\mho_1$ differs from $\mho_2$. The number of situations 4, equals the product of the number of oriates, 2, and the cell multiplicity, 2.

A specimen of a ferroic often consists of several *domains* (or *terrains*); each domain (or terrain), as a whole, is in one oriate (or situation), and the oriates (or situations) of any two adjacent domains (or terrains) are different. Domain corresponds to oriate, and terrain to situation. But the number of domains (or terrains) in the specimen has no connection with the number of oriates (or situations) of the ferroic; the former can be any integer between 1 and ∞. Each domain comprises a number of terrains; this number may be any integer between 1 and ∞, having no connection with the number of situations per oriate, i.e., the cell multiplicity.

When at least two of the oriates of a ferroic are unequal with respect to (the direction of) the spontaneous polarization vector, the ferroic is said to be *ferroelectric* and those two oriates are said to be *ferroelectrically related*. (Two oriates that are equal with respect to the spontaneous polarization vector are said to be *ferroelectrically unrelated*.) According to whether all the oriates, or not all but some, are ferroelectrically related, the ferroic is said to be *fully* or *partially* (or a full or partial) *ferroelectric*. Analogously, when at least two of the oriates of a ferroic are unequal with respect to the spontaneous strain tensor, the ferroic is said to be *ferroelastic* and those two oriates are said to be *ferroelastically related*. According to whether all the oriates, or not all but some, are ferroelasticity related, the ferroic is said to be *fully* or *partially* (or a full or partial) *ferroelastic*. The oriates of a partial ferroelectric (or ferroelastic) can be divided into subsets such that any two oriates belonging to the same subset are ferroelectrically (or ferroelastically) unrelated and any two oriates belonging to different subsets are related. The number of oriates in one subset equals that in another subset. Hence this number is a divisor of N, the number of all oriates of the ferroic. It indicates the *degree of partiality*. In particular, if it is 1 the ferroic is fully ferroelectric (or ferroelastic); if N, not ferroelectric (or ferroelastic). Any ferroic with a prime N, especially with $N = 3$, is neither partially ferroelectric nor partially ferroelastic.

Since the spontaneous polarization vector and spontaneous strain sensor are invariant under every translation, any two situations belonging to the same oriate are ferroelectrically and ferroelastically unrelated.

A specimen of a full ferroelectric (or ferroelastic) can easily be changed from a multidomain state to a unidomain state, and from a unidomain state to another unidomain state, by the application of an electric field (or a mechanical stress). On the other hand, a specimen of a partial ferroelectric (or ferroelastic) can hardly be affected in this fashion. Even for a full ferroelectric (ferroelastic), if its cell multiplicity is greater than 1, a specimen of it can hardly be changed from a multiterrain state to a uniterrain state, and from a uniterrain state to another uniterrain state, by the application of an electric field (or a mechanical stress).

If a ferroic is both ferroelectric and ferroelastic, the ferroelectricity and ferroelasticity are not always coupled. Only when the condition is satisfied that ferroelectrically related oriates are all ferroelastically related and vice versa are the ferroelectricity and ferroelasticity completely coupled; i.e., a turning of the spontaneous polarization vector by a field is necessarily accompanied by a turning of the spontaneous strain tensor and, conversely, a turning of spontaneous strain tensor by a stress is necessarily accompanied by a turning of spontaneous polarization vector. Any ferroic that is both fully ferroelectric and fully ferroelastic satisfies the above condition.

Two oriates that can be changed to each other by a field (or stress) are not always ferroelectrically (or ferroelastically) related. If two oriates are unequal with respect to electric susceptibility tensor or elastic compliance tensor or piezoelectric modulus tensor, it is never impossible to change the oriates to each other by a field or a stress or a combination of both, respectively.

KÊITSIRO AIZU

References

Aizu, K., *Phys. Rev. B* **2**, 754 (1970).
Aizu, K., *J. Phys. Soc. Jpn.* **44**, 334 and 683 (1978).

Cross-references: CRYSTALLOGRAPHY, DIELECTRIC THEORY, MECHANICAL PROPERTIES OF SOLIDS.

FERROMAGNETISM

Ferromagnetism is an example of cooperative phenomena in solids. It is characterized by a spontaneous macroscopic magnetization M (magnetic moment per unit volume) in the absence of an applied magnetic field at temperatures below a critical value known as the Curie temperature, T_C. This property is exhibited by the transition metals, Fe, Co, and Ni; the rare earth metals, Gd, Tb, Dy, Ho, Er, and Tm; and by a variety of alloys, compounds, and solid solutions involving the transition, rare earth, and actinide elements. Ferromagnetic Curie temperatures range from a fraction of a degree to hundreds of degrees Kelvin.

Cooperative magnetic behavior results from the exchange interaction between electrons, which is qualitatively described as follows. Electrostatic coulomb repulsion between like electric charges acts to keep two electrons apart, a separation which is also favored by the Pauli exclusion principle if the electrons have parallel spins. Thus if two electrons are farther apart when their spins are parallel than they would be if their spins were antiparallel, the parallel state will have lower mutual electrostatic energy. However, the kinetic energies increase if the electrons are separated, and consideration of this energy may lead to lower total energy for antiparallel spins. In other words, the exchange interaction is electrostatic in nature, but is modified by details of kinetic energy and the exclusion principle, and is highly dependent on the spatial distribution of the electrons. The exchange energy between two electrons, though electrostatic, is usually expressed in the mathematically equivalent form, $-2Js_1 \cdot s_2$, where J is the quantum mechanical exchange integral, related to the overlap of the charge distributions 1, 2; and s_1 and s_2 are the spin angular momentum vectors of the two electrons. The value of J is usually expressed in energy units, such as ergs, when the angular momentum vectors are assigned nondimensional values, i.e., spin quantum numbers. When J is positive, parallel spins represent a lower energy state than antiparallel spins. Since each electron has a magnetic moment proportional to its spin angular momentum, a state of parallel spins corresponds to a state of parallel magnetic moments.

The principal effect of the exchange interaction is embodied in the empirical Hund's rules, which describe the combination of electron spins in an atom to form the atomic spin. The principal interaction between magnetic atoms, i.e., between atoms with magnetic moments as a consequence of spin angular momenta, is thought to be of exchange character also, and there has been considerable success in describing magnetic properties by assuming this interaction to have the form,

$$\mathcal{H} = -\sum_{i,j} J_{ij} \mathbf{S}_i \cdot \mathbf{S}_j.$$

Here $\mathbf{S}_i$ and $\mathbf{S}_j$ are the spin angular momentum vectors of atoms i and j; the exchange integral, J_{ij}, may vary for different pairs and is usually regarded as a phenomenological parameter to be evaluated by means of experimental data.

When J_{ij} is positive, parallel ordering or alignment of the atomic spins in a common direction is favored, and so there is a large spontaneous magnetization even in the absence of an applied field. For negative J_{ij}, antiparallel spins result in lower energy (see ANTIFERROMAGNETISM, FERRIMAGNETISM).

Maximum ordering obtains at the absolute zero of temperature where the randomizing effect of thermal agitation disappears. The initial decrease of the magnetization as the temperature is increased from zero is well represented by a superposition of wavelike disturbances known as spin waves. At the Curie temperature the magnetic ordering is destroyed by thermal agitation and the spontaneous magnetization is zero (Fig. 1). Above the Curie temperature a ferromagnetic material behaves paramagnetically and has a net magnetization only in the presence of an applied field (see MAGNETISM and PARAMAGNETISM).

At temperatures below T_C a sample of ferromagnetic material is usually divided into small regions called domains, which vary in size and shape with a typical dimension from 0.1 to 1000 μm. Within each domain the magnetization is uniform and has the maximum or saturation value, M_S, characteristic of the temperature of the material (Fig. 1), but the direction of alignment of the individual moments in each domain changes from one domain to the next. The magnetization of the sample, which is the resultant of the magnetization of all the domains, may be much less than the saturation value (of a single domain), or it may even be zero in a completely demagnetized state.

The magnetization of a sample increases when a magnetic field, H, is applied. The value of H required to saturate the magnetization may be as small as 0.01 oersted or as large as several thousand oersteds, depending on the material.

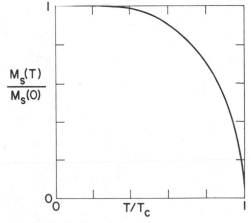

FIG. 1. Typical variation of spontaneous domain magnetization, M_S, as a function of temperature, T.

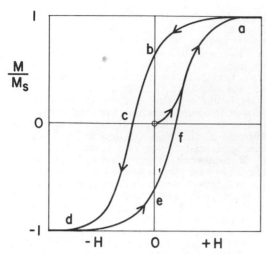

FIG. 2. Typical variation of sample magnetization, M, as a function of applied field strength, H; the magnetization curve depends on the magnetic history of the sample.

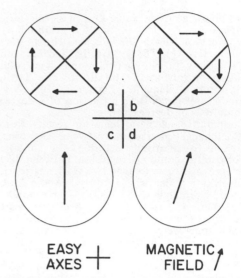

FIG. 3. Schematic representation of the change of domain structure and magnetization with applied field.

The magnetization process for a previously demagnetized material is represented by the 0-to-a portion of the magnetization curve in Fig. 2. When the applied field is subsequently removed, the magnetization exhibits the phenomenon of hysteresis, or lagging behind, tracing the path a-b and retaining a finite value, called the remanence, when the field is zero. The coercive force is the value of the field which must be applied opposite to the direction of the magnetic moment to trace the path b-c and reduce the magnetization to zero. Further increase of the field in this direction, followed by its removal, traces the path c-d-e. Repetition of the cycle then traces the curve e-f-a-b-c-d-e-, etc. Permanent magnets are characterized by a large coercive force; i.e., a large reverse field (typically thousands of oersteds) is necessary to destroy the magnetization However, in soft magnetic materials the coercive force may be less than one oersted.

Domains exist in a ferromagnetic material because their formation results in a lower total energy for the sample than it would have if the entire sample were a single domain. Total alignment of the magnetic moments is favored by the exchange forces, but these are usually short range forces acting between an atom and its neighbors. The dipole-dipole forces, although weaker, are long range and, alone, would orient the atomic moments like bar magnets, north pole to south pole, in closed chains to minimize the external field of the magnet (see DIPOLE MOMENT and MAGNETISM). One additional factor is required for the formation of domains. This factor is anisotropy, a result of the crystal structure of most solid materials. The structure of a crystal is not the same in all

directions; consequently its physical properties depend on direction. It is easier to magnetize a magnetic material in some directions, called easy axes, than in other directions. When the effect of anisotropy is superimposed on the effects of exchange and dipolar coupling, the ordered atomic moments break up into segments or domains, so that in each domain the magnetization is uniform and lies along or near one of the easy axes. There is a large change in the direction of magnetization from one domain to the next, with the reorientation occurring gradually (on an atomic scale) in a narrow transition region known as the domain wall.

The initial part of the magnetization process, 0-a in Fig. 2, is represented schematically in Fig. 3. In the demagnetized state the domains are arranged to minimize the external field due to the magnetization and to give zero net moment for the sample (Fig. 3a). In low fields the domain boundaries move so that the domains with magnetization direction near the direction of the applied field grow in size while other domains are depleted (Fig. 3b). As the field strength is increased, the domain boundaries are swept out of the sample (Fig. 3c), and finally the (single) domain magnetization rotates into the direction of the applied field until saturation is reached (Fig. 3d). When the field is removed, domains re-form in varied orientation along the easy axes, but there is a preference for domains whose magnetization vectors lie in the easy directions nearest the direction in which the field was applied, and consequently the sample as a whole has a magnetic moment (remanent magnetization).

In some thin magnetic films the domains are

cylindrical. Both the cylinder axis and the magnetization are perpendicular to the plane of the film, whereas in the domains shown in Fig. 3 the magnetization is in the plane of the diagram. Cylindrical domains—called bubbles because of their appearance and mobility at the surface—are caused when perpendicular anisotropy built into the film during preparation is stronger than the demagnetizing effect of the planar structure. Bubble domains are useful for data storage and control applications.

Amorphous materials have no crystalline order. They are often referred to as glasslike and have properties similar to those of liquids. Despite the total atomic disorder, the magnetic properties of amorphous alloys are sharply defined: Domain patterns are observed, including bubble domains; sharp Curie points are detected; and large permeabilities exist, almost as large as that of Permalloy.

Except when rare earth atoms are present, the exchange interactions in amorphous alloys are positive and lead to ferromagnetism. Using a variety of elements over a wide range of composition, one can obtain great flexibility of magnetic properties. When rare earth atoms are added, negative exchange interactions are induced, allowing further control of the magnetization and anisotropy energy to help stabilize domain formation and other factors affecting magnetic properties.

Ferromagnetism is also characterized by many secondary effects. Among the more important are the specific heat of the ferromagnetic state, which peaks at the Curie temperature, confirming that a phase transition has occurred; the magnetoresistance, which in metals can change the electrical resistance by as much as five percent, depending on the direction of magnetization; and the magnetostriction, which causes the sample dimensions to change during magnetization. These secondary effects give additional insight into the nature of the ferromagnetism and are useful in technical applications.

R. J. JOENK
T. R. MCGUIRE

References

Chen, C. W., "Magnetism and Metallurgy of Soft Magnetic Materials" ("Selected Topics in Solid State Physics," Vol. XV), Amsterdam, North-Holland Publishing Co., 1977.

Chikazumi, S., and Charap, S. H., "Physics of Magnetism," Melbourne, FL, Krieger Publishing Co., 1978.

Coey, J. M. D., "Amorphous Magnetic Order," *Journal of Applied Physics* 49, 1648–1652 (1978).

Craik, D. J., "Structure and Properties of Magnetic Materials," London, Pion Ltd., 1971.

Malozemoff, A. P., and Slonczewski, J. C., "Magnetic Domain Walls in Bubble Materials," New York, Academic Press, 1979.

Morrish, A. H., "The Physical Principles of Magnetism," Melbourne, FL, Krieger Publishing Co., 1980.

Vonsovskii, S. V., "Magnetism," 2 Vols., New York, John Wiley & Sons, Inc., 1974.

Cross-references: DIPOLE MOMENTS, FERRIMAGNETISM, MAGNETISM, MAGNETOSTRICTION, PARAMAGNETISM.

FEYNMAN DIAGRAMS

I. Introduction Feynman diagrams provide one of the most powerful methods known for finding the physical properties of systems of interacting particles, i.e., particles interacting with an external field and/or with each other. Since all physical systems—solids, liquids, gases, molecules, plasmas, atoms, nuclei, elementary particles—are composed of interacting particles, the Feynman diagram method is now used extensively in all branches of modern physics. In this article, the examples are from solid-state physics; the reader is referred to the References for examples from other branches. A very elementary account of the subject is found in Ref. 1.

The idea behind Feynman diagrams is this: In order to find the important physical properties of a system of interacting particles, it is not necessary to know the detailed behavior of the particles, but only their *average* behavior. The quantities which describe this average behavior are called "propagators" or "Green's functions," and physical properties may be calculated directly from them. For example, the *single-particle propagator* is defined thus: we put an extra particle into the system at point r_1 at time t_1, and let it interact with the external field and with the other particles for a while (i.e., we let it "propagate" through the system). Then the single-particle propagator is the probability (or in quantum systems, the probability *amplitude*) that at later time t_2, the particle will be observed at another point, r_2. From the single-particle propagator, we can calculate directly the energy and lifetime of certain excited states of the system, the ground state energy, etc. Similarly, there are the two-particle propagator, the no-particle propagator or "vacuum amplitude"), etc., which yield other physical properties (see Ref. 1).

Feynman diagrams give us a method of calculating propagators by means of pictures. We shall first show how this is done in a simple classical example from "liquid-state" physics, i.e., the drunken man propagator, then give a couple of illustrations from solid-state physics.

II. Classical Example: The Drunken Man Propagator A man who has had too much to drink (Fig. 1), leaves a party at point 1 and on the way to his home at point 2, he can stop off at one or more bars—Alice's Bar (A), Bardot Bar (B), Club 6 Bar (C), · · · etc. We ask for the

FIG. 1. Propagation of drunken man.

probability, $P(1, 2)$, that he gets home. This probability, which is just the propagator here (with time omitted for simplicity), is the sum of the probabilities for all the different ways he can propagate from 1 to 2 interacting with the various bars. Assuming, for simplicity, that the various processes involved are independent, this is just $P_0(1, 2)$ (= probability that he will go "freely" from 1 to 2, i.e., without stopping at any bar), plus $P_0(1, A) \times P(A) \times P_0(A, 2)$ (= probability he will go freely from 1 to bar A, times the probability that he stops off at A for a drink, times the probability that he then proceeds freely to 2), plus $P_0(1, B) \times P(B) \times P_0(B, 2)$ (= probability for route 1–bar B–2) plus etc. $\cdots$ plus $P_0(1, A) \times P(A) \times P_0(A, A) \times P(A) \times P_0(A, 2)$ (= probability for route 1–A–A–2) plus etc. $\cdots$. This gives us an infinite series for the propagator:

$$P(1, 2) = P_0(1, 2) + P_0(1, A)P(A)P_0(A, 2)$$
$$+ P_0(1, B)P(B)P(B, 2) + \cdots$$
$$+ P_0(1, A)P(A)P_0(A, A)P(A)P_0(A, 2) + \cdots . \quad (1)$$

Now this series is a complicated thing to look at. To make it easier to read, we draw a "picture dictionary" to associate diagrams with the various probabilities as shown in Table 1. Using this dictionary, series (1) can be drawn as in Fig. 2(a). Since, by Table 1, each diagram element stands for a factor, Fig. 2(a) is completely equivalent to series (1). However, Fig. 2(a) has the great advantage that it also reveals the physical meaning of the series, giving us a "map" which helps us to keep track of all the sequences of interactions the drunken man can have in going from 1 to 2.

The series may be evaluated approximately by selecting the most important terms in it and summing them. This is called "*partial summation.*" For example, suppose the man's favorite bar is Alice's bar, so that $P(A)$ is large and all other $P(x)$'s are small. Then Fig. 2(a) becomes approximately Fig. 2(b). Assuming for simplicity that all $P_0(r, s) = c$ (a constant) we have, using Table 1:

$$P(1, 2) \approx P_0(1, 2) + P_0(1, A)P(A)P_0(A, 2) + \cdots$$
$$= c + c^2 P(A) + c^3 P^2(A) + \cdots$$
$$= c [1 + cP(A) + c^2 P^2(A) + \cdots]$$
$$= c/[1 - cP(A)] \quad (2)$$

where we have used the fact that the expression in brackets is a geometric series. This same technique is used in the quantum case.

III. Single Electron Propagating in a Crystal
The example here is just like the previous one, except that instead of a propagating drunken man interacting with various bars, we have a propagating electron interacting with various ions in a crystal. A crystal consists of a set of positively charged ions arranged so they form a regular lattice, as in Fig. 3. An electron interacts with these ions by means of the Coulomb force. The single-particle propagator here is the sum of the quantum-mechanical probability amplitudes for all the possible ways the electron can propagate from point r_1 in the crystal, at time t_1, to point r_2 at time t_2, interacting with the various ions on the way. These are: (1) freely, without interaction; (2) freely from r_1, t_1 (= "1" for short) to the ion at r_A at time t_A, interaction with this ion, then free propagation from the ion to point 2; (3) from 1 to ion

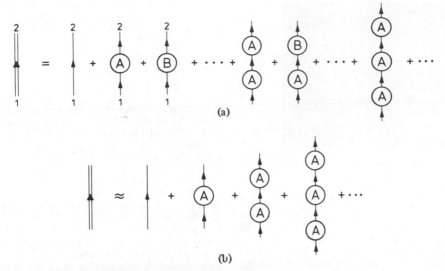

(a)

(b)

FIG. 2. (a) Feynman diagram series for drunken man propagator, or for electron propagating in crystal. (b) Approximate series ("partial sum") for drunken man propagator.

TABLE 1. FEYNMAN DIAGRAM DICTIONARY
FOR DRUNKEN MAN PROPAGATOR

Word	Picture	Meaning
$P(1, 2)$		Probability of propagation from 1 to 2.
$P_0(r, s)$		Probability of free propagation from r to s.
$P(X)$		Probability of stopping off at bar X for a drink.

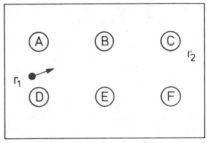

FIG. 3. Electron propagating among ions A, B, C ⋯ in crystal.

B, interaction at B, then from B to 2, etc. Or we could have the routes 1–A–A–2, 1–A–B–2, etc. We can now use the dictionary in Table 1 to translate this into diagrams, provided

the following changes are made: change "probability" to "probability amplitude," and change the meaning of the circle with an X to "probability amplitude for an interaction with the ion at X." When this is done, the series for the propagator can be translated immediately into exactly the same diagrams as in the drunken man case! That is, Fig. 2(a) is also the propagator for an electron in a crystal, provided that

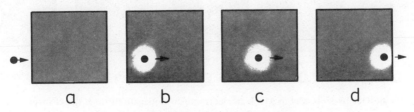

FIG. 4. Electron propagating in electron gas.

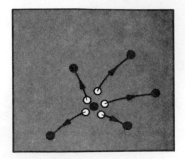

FIG. 5. "Hole" picture of empty space around electron in electron gas.

FIG. 7. Feynman diagram for interaction between two electrons.

we just use a quantum dictionary to translate the lines and circles into functions. The series can be partially summed, and from the resulting propagator we obtain immediately the energy of the electron moving in the field of the ions.

IV. Electron Propagating in an Electron Gas
We now look at the problem of many interacting particles, taking as an example a large number of electrons interacting with each other by means of the Coulomb force. It is assumed that there is a uniform, fixed positive charge "background" present which keeps the whole system electrically neutral. This system is called the "electron gas" and is used as a simple model for electrons in a metal. The propagation of an added electron through this system is shown in Fig. 4. Figure 4(a) shows the uniform charge distribution in the undisturbed system, with an extra electron entering from the left. In 4(b) the extra electron has entered, repels other electrons away from it (Coulomb repulsion between like charges), so we get an "empty space" near the extra electron and repelled electrons further away. The extra electron surrounded by the empty space is called the "*quasi electron*" and it propagates through the system as shown in Fig. 4(c), (d). It is convenient to view the empty space around the extra electron as composed of "holes" in the electron gas. That is, the Coulomb repulsion "lifts out" electrons from the electron gas in the immediate vicinity of the extra electron, thus creating "holes" in

the charge distribution of the gas, and puts these lifted-out electrons down again further away. This picture is shown in Fig. 5.

Using this picture, we can decompose the propagator into the sum of the probability amplitudes for all possible ways that the extra electron can propagate through the system interacting with the other electrons. The simplest way is free propagation without interaction. Another way is shown in the "movie" in Fig. 6. Figure 6(a) shows the extra electron entering the electron gas. In 6(b) we see the extra electron interacting with a nearby electron in the gas, creating a hole nearby and a lifted-out electron further away. The extra electron, the hole, and the lifted-out electron then propagate freely through the gas as in 6(c). In 6(d), the extra electron interacts with the lifted-out electron, causing it to fall back into the hole, thus annihilating both the lifted-out electron and the hole. Figure 6(e) shows the extra electron propagating out of the system. (It should be noted that unlike the drunken man case, the processes shown in Fig. 6 are not physical but rather "virtual" or "quasi-physical" processes, since they do not conserve energy and they may violate the Pauli exclusion principle.)

Let us represent the Coulomb interaction between an electron at point r and one at r' by a wiggly line as in Fig. 7. Then the sequence of processes in Fig. 6 may be represented by the Feynman diagram shown in Fig. 8(a) or 8(b). Note that hole lines are drawn as electron lines with a direction opposite to the direction of increasing time. By analyzing the other possible processes which can take place, we find the series for the electron propagator shown in Fig. 9. Again, these diagrams may be evaluated by writing the appropriate factor for each free electron and hole propagator and each interaction, and carrying out a partial sum. The result-

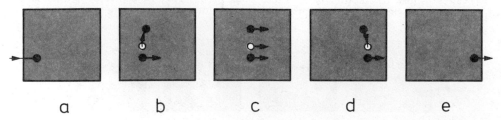

FIG. 6. One possible way in which an electron can propagate through the electron gas.

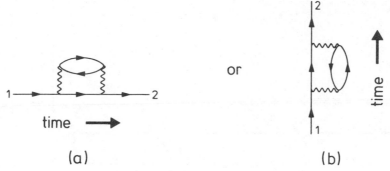

FIG. 8. Feynman diagram for the sequence in Fig. 6.

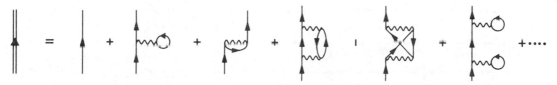

FIG. 9. Feynman diagram series for electron propagator in the electron gas.

ant expression for the propagator yields directly the energy and lifetime of the quasi electron.

For a more extensive and detailed account of Feynman diagrams, see Ref. 1.

R. D. MATTUCK

References

1. Mattuck, R. D., "A Guide to Feynman Diagrams in the Many-Body Problem," London, McGraw-Hill, 1967. (Most elementary account, examples from solid-state and nuclear physics.)
2. Fetter, A. L., and Walecka, J. D., "Quantum Theory of Many-Particle Systems," New York, McGraw-Hill, 1971. (Advanced, examples from solid state and nuclear physics.)
3. Mandl, F., "Introduction to Quantum Field Theory," New York, Interscience Publishers, 1959. (Elementary account, examples from elementary particle physics.)
4. Rickayzen, G., "Green's Functions and Condensed Matter," Academic Press, London, New York, 1980. (Intermediate, examples mostly from solid state physics.)
5. Bjorken, J. D., and Drell, S. D., "Relativistic Quantum Mechanics" and "Relativistic Quantum Fields," New York, London, McGraw-Hill, 1964.

Cross-references: ELECTRON, ELEMENTARY PARTICLES, GRAND UNIFIED THEORIES, MANY-BODY PROBLEM, PHOTON, QUANTUM CHROMODYNAMICS, QUANTUM ELECTRODYNAMICS.

FIBER OPTICS

As the name suggests, fiber optics deals with the transmission of light through thin strands of transparent dielectric materials such as glass or plastic. The general subject of fiber optics can be subdivided into three broad classes: fiber bundles for illumination and image transmission, fiber optic sensors, and optical fiber telecommunications. However, the physical principles involved in light transmission through optical fibers are the same in all three areas of applications.

Light Transmission Through Optical Fibers
The simplest kind of optical fiber consists of an inner circular cylinder of a homogeneous dielectric material, the *core*, that is surrounded by another homogeneous dielectric material of lower refractive index, the *cladding*. Because the refractive index profile of core and cladding resembles a step, this structure is called *step-index fiber*. A light ray traveling along a straight line in the core suffers total internal reflection if it impinges on the core-cladding boundary at a sufficiently small angle. Rays traveling at angles larger than a critical angle are only partially reflected and lose all their energy after only a few repeated reflections. Thus, the fiber is able to trap light within a cone defined by the critical angle. (See Fig. 1).

More complicated optical fibers consist of a core whose refractive index is inhomogeneous, changing continuously from a relatively high value on axis to a lower value at the core-cladding interface. Rays in such a graded-index

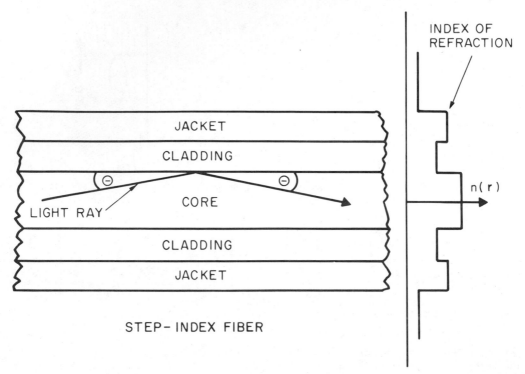

Fig. 1.

fiber travel on curved paths and remain trapped inside the core if their trajectory does not encounter the core-cladding interface (Fig. 2).

Light propagation in optical fibers can also be described in terms of wave optics. Instead of rays traveling at different angles we now speak of *modes*. A mode is defined as an electromagnetic wave that travels through a uniform fiber without changing the shape of its transverse field distribution. Depending on their core diameters and refractive index distributions, fibers can support varying numbers of modes. A fiber with a large diameter and/or large index difference can support many modes. On the other hand, it is possible to design a fiber with sufficiently small core diameter and/or small index difference so that it can support only one mode in each of two possible polarizations. Each mode can be represented by bundles of rays traveling at angles characteristic of the mode in question.

In step-index fibers, the distance and hence the travel time of light depends on the angle at which a ray propagates. Or, expressed in the language of wave optics, the travel time is different for each mode. This undesirable property of step-index fibers has the consequence that pulses, whose energy is spread over many rays traveling along different paths or on different modes, arrive at the fiber end spread out over a time interval corresponding to the difference

in arrival time of the slowest and fastest ray or mode. In graded index fibers, this *intermodal dispersion* can be compensated partially. The travel time of light depends not only on its geometrical distance but also on the index of refraction of the medium. Light moves more slowly in regions with high refractive index and faster in a low-index medium. Since the light trajectory in a graded-index fiber samples high-index regions near the fiber axis and lower-index regions farther from the axis, averaging takes place. It is possible to design graded-index fibers that very nearly compensate the travel time differences along all possible rays or for all possible modes. Such optimized fibers have a refractive index distribution that is nearly (but not exactly) parabolic as a function of the radial coordinate. Even though no perfect compensation is possible, a very substantial improvement over intermodal dispersion in step-index fibers can be achieved.

Fibers that support only a single mode (*single-mode* or *monomode fibers*) do not suffer from intermodal dispersion. However, a different source of pulse spreading is operative in all fibers: *chromatic dispersion*. This means that light of slightly different wavelength propagates at slightly different velocities. Light sources used to excite optical fibers have finite spectral width. But even if the unmodulated source were strictly monochromatic, the pro-

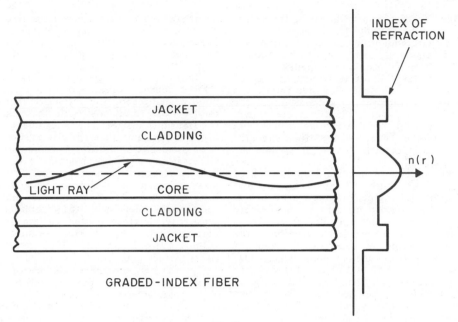

Fig. 2.

cess of modulation introduces new spectral components. Thus, chromatic dispersion is operative in all fibers. Fortunately, one of the most important materials for making optical fibers—fused silica—has the property that dispersion vanishes to first order at a wavelength near 1.3 μm. If operated at the minimum dispersion wavelength, extremely short pulses (psec) can be transmitted over long lengths (tens of km).

In addition to dispersion, the performance of an optical fiber is characterized by the power loss that a light signal suffers when traveling in it. The loss parameter depends mostly on the materials out of which the fiber is made. Fibers made of plastics have high losses and are useful only for light transmission over a few meters. Fibers made of fused silica have achieved extremely low losses indeed. The loss spectrum of fused silica has its lowest value near 1.5 μm. Losses as low as 0.2 dB/km have been measured in fused silica fibers. This means that one-half of the original input light power still arrives at the end of a fiber of 15 km length. For longer wavelengths the loss increases sharply because the light can excite vibrations of the atoms of fused silica. At shorter wavelengths the loss increases sharply because the light induces electronic transitions. In silica fibers, the window of high transparency extends roughly from 0.5 μm to 1.8 μm.

Fiber Fabrication Even though plastic fibers are cheap to fabricate, their high losses preclude their use in most telecommunications applications. For this reason we limit the discussion to a few methods for making glass fibers. A simple method consists in placing a solid glass rod of relatively higher refractive index into a glass tube of lower refractive index. The end of this rod-in-tube assembly is heated to the softening point and is pulled into a step-index fiber. The diameter of the resulting fiber depends on the diameter of the starting tube, the drawing temperature, and the drawing speed. More sophisticated methods use an arrangement of concentric crucibles filled with different molten glasses. The core and cladding glasses combine at the tips of the crucibles and are drawn into step-index fibers.

Low-loss fibers for telecommunications applications must be made of very pure materials. Contamination of the pure glass with metal ions or with water can increase the losses dramatically. In some chemical vapor deposition methods, glass is deposited on the outside of a mandrel by converting silicon tetrachloride and dopant vapors into doped silicon dioxide. In the modified chemical vapor deposition (MCVD) method a mixture of silicon tetrachloride and oxygen is passed through the inside of a heated silica tube. Added to this mixture are dopant gases such as germanium tetrachloride or phosphorus trichloride. Germanium or phosphorus increases the refractive index of fused silica, whereas the addition of boron or fluorine to silica glass decreases its refractive index. The chlorine compounds are converted to oxides and are deposited on the hot inner surface of the silica tube, where they fuse to form layers of doped silica glass. The rate of deposition is controlled by the concentration of the vapors inside the tube, by the temperature, and by the speed with which the heating torch is passed along its outside. By careful control of the ratio of silicon tetrachloride and dopant materials, desired refractive index distributions can be achieved. After the deposition step the tube is heated until it collapses to form a solid rod—the *preform*—out of which the fibers are drawn. Fibers for optical communications are protected against scratches (which weaken their strength) by a plastic jacket which is applied while the fiber is drawn.

Fiber Bundles An important application of fiber optics is the use of fiber bundles for illuminating or viewing inaccessible places such as body cavities (*endoscope*). Fiber bundles are formed by combining individual fibers. If no attention is paid to the relative position of the fibers at either end, we speak of an incoherent bundle. Fibers in incoherent bundles may have diameters between 50 to 200 μm. If the relative positions between all fibers are carefully maintained, a coherent bundle results. Incoherent bundles are cheaper to manufacture and are used for guiding light to illuminate inaccessible places. To make it coherent, the fiber bundle is first wound as a closed loop of the desired thickness and circumference. Next, the fibers of this closed-loop bundle are fused tightly to each other at one point of the circumference before they are cut at that point. Finally, the ends are polished. The fibers at both ends of the resulting linear bundle are now restrained to maintain their relative position but are free to flex in between. A coherent bundle can be used to transmit images. The image of an object is focused by lenses onto one end of the bundle. Each picture element is transmitted along one of the fibers. The resulting image can be viewed at the far end of the bundle through an eyepiece with a resolution that depends on the density and diameter of the fibers. A coherent bundle may contain from 5,000 to 50,000 fibers with diameters between 5 and 25 μm and may be several meters long. Flexible, coherent fiber bundles of sufficiently small diameter are used to view the inside of the stomach or other body cavities. Illumination is provided either by shining light through a subsection of the bundle or by a separate bundle that need not be coherent. Tapered coherent fiber bundles can be used to magnify or demagnify images. If desired, images can be scrambled by scrambling the relative positions of the fibers in the bundle.

Very short, coherent fiber bundles—*fiber optic plates*—are used for special applications. A fiber optic plate inserted into the face of a cathode ray (picture) tube can enhance image brightness by capturing more light from a fluorescent coating on its inside surface. Fiber optics plates are also used to connect successive stages in image intensifiers or infrared image converters.

An image can be transmitted through a single fiber with a parabolic refractive index profile. Short sections of such fibers act as tiny graded-index lenses. Inside a parabolic-index fiber an image focused onto one of its ends is refocused periodically along its length. The quality of the transmitted image depends on the quality of the index profile and the length of the fiber.

Fiber Sensors Because of their great length, minute changes of the refractive index along the fiber can result in sizeable phase changes of the transmitted light. The index change may be induced by many different effects such as magnetic fields, temperature, pressure, stress etc. Observation of the propagation properties of light in response to such external disturbances make fibers ideal as sensors. Pressure induced refractive index changes due to underwater sound make fibers useful as hydrophones. Their sensitivity to magnetic fields (Faraday effect) permits their use as current sensors on high voltage transmission lines; their sensitivity to stress enables their use as strain gauges.

A particularly interesting application of fiber optics is their use as rotation sensors. In principle, rotation can be sensed with any device that allows a comparison of the time of flight of light in the two opposite directions along a closed, rotating loop. Fibers have the advantage that the light path along the loop can be made very long by using fiber coils to magnify the effect. Rotation rate sensors for ships and aircraft are being developed.

Optical Fiber Telecommunications Because of their low losses, optical fibers are ideally suited as waveguides for the transmission of messages modulated on light signals. The discussion of the use of fibers as sensing elements may lead to the suspicion that optical fiber communications might be plagued by interference from electromagnetic fields, thermal changes, or sensitivity to vibrations. Fortunately, quite the opposite is true. The ability to sense changes in their environment is due to phase changes imparted by the fiber to a light signal passing through it. However, for communications applications the phase of the light field is not being sensed, so that one of the advantages of fiber communications is its insensitivity to electromagnetic interference. If fiber optics are used improperly, sensitivity to vibrations could be a problem (*modal noise*) but can be minimized by proper design of the system.

Typically, an optical fiber communications system consists of the following components: a modulated light source, the fiber waveguide, and a detector. For use with multimode fibers the light source can be either a cheap, reliable light emitting diode (LED) or a more sophisticated and expensive injection laser. Since single-mode fibers cannot be excited efficiently with an LED, a laser must be used. This is the reason why, in spite of their higher dispersion, multimode fiber systems coexist with single mode fiber systems.

Both LEDs and lasers work basically on the

same principle. Electrons and holes are injected into the junction region of a semiconductor diode where they combine to emit a light photon. In an LED the photons escape from the junction as soon as they are generated. In an injection laser each photon is reflected back and forth between the mirrors of a resonant cavity and is forced to stimulate the emission of more photons. Consequently, the light of an LED is completely incoherent while the output of a good laser is highly coherent. Lasers and LEDs are pulse modulated by pulsing their electrical feed current.

With the optimum refractive index profile and operating at the wavelength of minimum chromatic dispersion (near 1.3 μm) multimode fibers can transmit pulses at rates of 100 Mbits/sec over distances of several kilometers. With single-mode fibers, pulse transmission rates of several Gbits/sec can be achieved.

Several optical fibers are usually combined into fiber cables of various constructions. Each fiber in the cable provides one independent transmission path for light pulses that themselves may carry thousands of telephone conversations or tens of television programs. Some cable designs first combine several fibers into a ribbon and then incorporate several fiber ribbons into one cable. Other designs support loosely fitting individual fibers in plastic tubes several of which are incorporated into a cable. In all cases plastic or metallic support members are included in the cables for added strength. Metallic conductors are sometimes used for the purpose of powering repeaters along the route of the cable.

Optical fibers can be spliced. In ribbon cables all fibers in one ribbon are usually spliced simultaneously. In some methods, splicing is accomplished by sliding the ends of the fibers into plastic sleeves and bonding them into place. It is also possible to heat the ends of two fibers until they join to form a fusion splice.

Optical fiber cables are used to connect local exchanges and for long distance transmission between cities, and they are being considered as submarine cables.

In applications where the length of the optical fibers does not exceed a few kilometers (for use between local exchanges for example) no repeaters are needed. But for applications between cities or for submarine cables, repeaters must be used at distances from 5 to 20 (or even more) kilometers. At each repeater the light signal is first converted to an electrical signal which is used to regenerate the pulses before they are launched into the next section of the optical communications link. Amplification and pulse regeneration of the light signal itself, even though possible in principle, it not yet contemplated for any practical fiber optic communication system.

Light detectors are semiconductor devices which absorb photons and convert them into electron-hole pairs which cause a current to flow in an external circuit. Thus, detectors and light sources complement each other in their mode of operation.

In time, optical fibers may well replace most of the metallic conductors that are now being used in the telephone plant.

D. MARCUSE

References

1. Kapany, N. S., "Fiber Optics," Academic Press, New York, 1967.
2. Giallorenzi, T. G., "Fiber Optic Sensor," *Optics and Laser Technology*, **13**(2), 73–78 (April 1981).
3. Miller, S. E., and Chynoweth, A. G. (Eds.), "Optical Fiber Telecommunications," Academic Press, New York, 1979.

Cross-references: BIOMEDICAL INSTRUMENTATION; DIELECTRIC THEORY; OPTICS, GEOMETRICAL; REFLECTION; REFRACTION.

FIELD EMISSION

Field emission of electrically charged particles occurs when a sufficient high electric field is applied to the surface of a conductor. Specifically, field emission of electrons from cold metals into a vacuum is a basic physical effect comparable to thermionic, photoelectric, or secondary emission. Field electron emission is also termed *cold emission* or *autoelectronic emission*. Field emission of electrons from metals at room temperature requires an electric field of an order of magnitude of 3×10^7 volts/cm; this can be obtained by applying a few thousand volts to a sharply curved cathode, which may be either a fine wire, a sharp edge, or a needle tip. From such cathodes, field emission was first observed by R. W. Wood in 1897, and technical application in high-voltage rectifiers and x-ray tubes was attempted by J. E. Lilienfeld in the early 1920s. He failed because of the inadequate vacuum techniques available at that time. The quantum mechanical theory of field emission was given by R. H. Fowler and L. W. Nordheim in 1928, which agreed with the current-voltage relationship measured by R. A. Millikan and C. C. Lauritsen, but experimental work to further verify the predictions advanced only with the possibility of controlling highly perfect emitter tips in the field emission microscope by E. W. Müller in 1937. Subsequently field emission microscopy became an established research technique for surface phenomena connected with adsorption. By operating a point emitter at a positive potential in the presence of gas, Müller discovered field ionization (1951) and developed the low-temperature field ion microscope (1956) which surpassed all other microscopic devices with its capability of showing the individual atoms as the building blocks of the crystal lattice of the metal specimen. In the last three decades, technical application of field emission has also been successful with the development of powerful flash x-ray tubes (W. P. Dyke, 1955). The extremely large current densities of up to 10^8 amperes/cm^2 make a field emission cathode very attractive for mm wave tubes, cathode ray tubes, and electron microscopes, but the sensitivity to contamination and cathode sputtering are detrimental to stability and long lifetime.

Field emission can be explained with the concepts of quantum mechanics. The conduction electrons of a metal are moving in a potential trough, from which they can escape ordinarily only by addition of thermal energy (thermionic emission), by an energy transfer from photons (photoelectric emission), or by collision with other energetic particles (secondary emission). If the barrier of the trough is narrowed by the application of an external electric field to be comparable with the wavelength of the electrons inside the metal, then a small amplitude of the electron wave will penetrate outside the barrier. Thus there is a finite probability that the electron will tunnel through the potential barrier, even if its kinetic energy is insufficient to go over it. According to the Fowler-Nordheim theory of field emission, the current density J (in amperes/cm^2) as a function of field strength F (volts/cm) and of the work function ϕ (electron-volts) is approximately given by

$$J = 1.55 \times 10^{-6} \frac{F^2}{\phi} e^{-\frac{6.85 \times 10^7 \phi^{3/2}}{F}} .$$

A more refined theory takes into account the effect of the image force on the electron, which reduces the exponent of the above equation by a factor slightly smaller than unity and which depends upon $\phi\sqrt{F}$. The field required for a given current density is thus reduced by some 10–20%. The temperature dependence of field emission is found to be very small below about 1000 K (Fig. 1). Considerable increase in emission is observed when both the temperature and the field are high. This effect is called *T-F emission*.

Field emission of positive ions can occur when an adsorbed layer is desorbed from the emitter surface by a positive electric field (*field desorption*). At a field of a few volts per angstrom, lattice atoms can also be evaporated in the form of multiply charged ions (*field evaporation*). These ion currents cannot be sustained for a long period of time, since the adsorption layer or the emitter surface will soon be consumed. Continuous field ion emission is obtained by operating the positive emitter in a gas of low pressure. Gas molecules attracted to the tip surface by polarization effect can be adsorbed on the apex of surface atoms (*field adsorption*), or can be ionized when their valence electrons tunnel into the metal when the molecules are

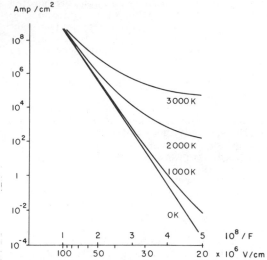

FIG. 1. Current density of field emission for a tungsten cathode plotted as a function of reciprocal field strength, for various temperatures.

a few angstroms above the surface (*field ionization*). Ion currents are small (less than 10^{-9} amp) because of the limited supply of the gas molecules. The fields needed for ionization are very high, about 2.2×10^8 volts/cm for hydrogen and 4.5×10^8 volts/cm for helium.

Field electron emission has been successfully applied to develop a powerful flash x-ray tube (W. P. Dyke), a high resolution scanning electron microscope (A. V. Crewe), and a bright point electron source for high resolution electron lithography. Field ion emission from liquid metals is now actively pursued as a continuous high brightness point ion source with a potential application in high resolution ion lithography (R. L. Seliger). The continuous miniaturization of electronic devices makes these techniques quite attractive because of their small sizes and high brightness.

Although its technical application is not yet widespread, field emission has been a subject of intensive investigations, and the field electron and the field ion microscopes have become productive tools of basic research in the study of solid surfaces. The specimen of the field emission microscope is a needle-shaped field emitter with a hemispherical tip of a radius of some 10^{-5} to 10^{-4} cm, arranged in a vacuum tube opposite a fluorescent screen. With a few thousand volts applied, field emitted electrons move away from the tip in a radial direction, displaying on the screen an enlarged projection image of the distribution of electron emission at the emitter. The magnification of this microscope is approximately equal to the ratio of screen distance to tip radius and can exceed one million. The lateral resolution is limited to 25 Å by a random tangential velocity component of the electrons due to the Fermi distribution inside the metal and by diffraction due to the de Broglie wavelength. As even traces of adsorption layers change the work function and, thereby, the emission, the presence of such layers can be readily detected on the screen. The imaged tip cap usually represents a single crystal, so that the behavior of adsorption layers can be studied in its dependence on crystallographic orientation of the substrate. As little as 10^{-3} of a monolayer of oxygen is clearly discernible on many metals. Because of the small temperature dependence of field emission, such layers, their surface migration, adsorption and desorption rates, and the activation energies of these processes can be measured in a wide temperature range. This is done by immersing the entire microscope into a cryogenic bath and by heating the emitter to any desired temperature up to near its melting point. Most studies have been done with the high-melting metals, W, Re, Ta, Mo, Nb, Ir, Pt, Rh, Pd, V, Ni, Fe, Ti, Cu, and various alloys and with adsorption layers of H_2, N_2, O_2, CO, CH_4, and the noble gases. Detailed patterns of

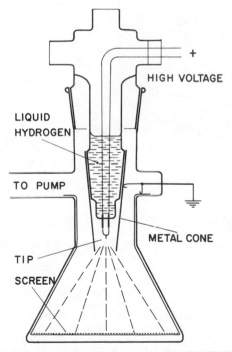

FIG. 2. Schematic diagrams of an early field ion microscope. The hemispherical cap of the tip, having a typical radius of 500 Å, is radially projected onto the fluorescent screen with the help of helium ions. Modern FIM uses a stainless steel chamber, and is equipped with a channel plate for image intensification. Otherwise, the construction is essentially the same.

individual organic molecules, such as phthalocyanine or similar aromatic compounds, cannot yet be fully explained. The interpretation of thermal desorption experiments is often difficult because of the occurrence of surface migration, while the well-defined field strength obtained in field desorption experiments cannot be fully utilized because of the complex polarization conditions at the various adsorption sites.

The field ion microscope (Fig. 2) was developed to overcome the limited resolution of the field electron microscope. In contrast to the case of electrons, the undesirable tangential component of the motion of the imaging ions can be reduced by lowering the emitter temperature. The short de Broglie wavelength of ions is also an advantage. Usually operated with helium or neon as the imaging gas and the tip cooled down to 20 to 80 K, a spatial resolution of ~2.5 Å can be achieved. Thus most atoms on the lattice can be resolved (Fig. 3). The specimen surface developed by field evaporation at low temperature, is atomically perfect and free from contamination. Helium atoms are ionized above field adsorbed atoms, and are then accelerated toward the screen assembly. Nowadays, the screen assembly consists of a microchannel plate and a fluorescent screen. The microchannel plate converts the ion image into an electron image and amplifies the image intensity by a factor of about 100 to 1000. Each protruding surface atom is imaged by 10^3 to 10^4 helium ions/sec. The surface stays atomically clean as all impurity gases, having a lower ionization potential than helium, are ionized in the low field region in space before they can reach the tip. Image stability requires that the evaporation field of a sample be higher than the ionization field of the imaging gas. Use of helium is thus limited to refractory metals. For soft metals such as Au, Cu, Fe, Ni, Al, etc. neon, argon or hydrogen may be used with a slightly inferior resolution. Many metals yield to the mechanical field stress $F^2/8\pi$ produced by the image field, which amounts to about 1000 kg/mm² at the helium best image field of 4.5×10^8 volts/cm.

The field ion images usually do not tell the chemical identity of the imaged atoms. This can be done by combining the field ion microscope with a time-of-flight mass spectrometer with a single atom detection sensitivity. This new instrument (Fig. 4), the atom-probe field ion microscope, allows the observer to select one atom by tilting the tip until its image falls onto a small probe hole in the screen. By superimposing a 10-nanosecond high voltage pulse to the dc imaging voltage the selected atom is field evaporated and travels through the probe hole and the time of flight tube to reach the particle detector, which is a Chevron channel plate. The flight time can be measured either with an oscilloscope or with a set of electronic timers. From

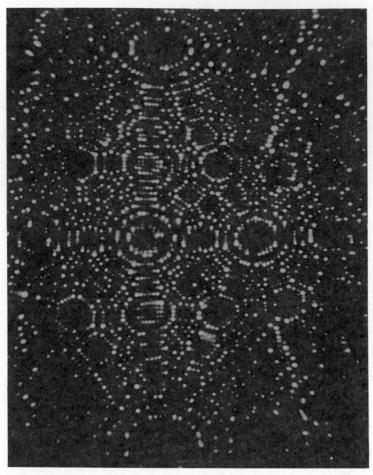

FIG. 3. Field ion micrograph of a fully ordered equiatomic platinum-cobalt alloy crystal, showing the individual atoms as single dots. Original magnification on the 5-inch screen was 1.5 million. In the image, only Pt atoms are visible.

the flight time, the mass to charge ratio and the chemical identity of the atom are identified. The energy focused time of flight atom probe is capable of separating isotopes of an element to the root of the mass lines (Fig. 5). The nanosecond pulsed field evaporation can also be done with laser pulses. High resistivity materials such as high purity silicon can only be studied with the pulsed laser atom-probe. A slightly different version of the atom-probe is the imaging atom-probe. It is essentially an ordinary FIM, but uses a screen assembly with single ion detecting sensitivity, or the Chevron channel plate. Although the mass resolution is severely limited by the short flight path, there is the advantage of finding the spatial distribution of field desorbed atoms from its desorption image. In fact by activating the chevron to detect only a certain ionic species by a time gating technique, the spatial distribution of a selected species can be displayed. A pulsed laser imaging atom-probe is shown in Fig. 6.

The atom-probe and the field ion microscope have been widely applied to study fundamental problems in metallurgy, materials science, and surface science. Single vacancies produced by quenching, cold working, or particle irradiation can be counted and localized. Interstitials resulting from impurities can be seen and their thermally activated migration to the tip surface can be measured. The structures of dislocations and the matching of the lattices along grain boundaries can be studied in atomic detail. Surface diffusion of single atoms and simple atomic clusters on atomically perfect surfaces, interaction between adsorbed atoms, and adsorbed layer superstructure formation can be studied quantitatively. The charge distribution of an adsorbed atom as manifest in the dipole moment and polarizability can be measured. The long-

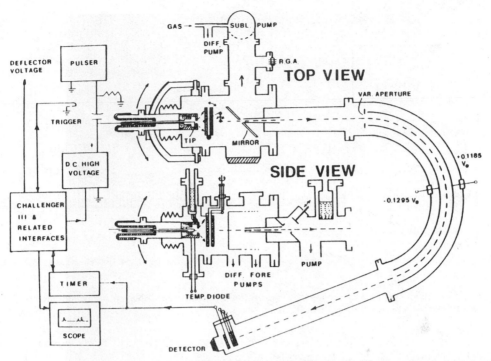

FIG. 4. The energy-focused time of flight atom-probe. The tip orientation can be adjusted with the help of an external gimbal mount and a metal bellows seal. The image can be seen through the 45° inclined mirror. By flipping the channel plate assembly away, the pulsed field evaporated atom can travel through the 163° spherical condenser plates, and be detected. The flight time is displayed in the oscilloscope, as well as recorded in an electronic timer.

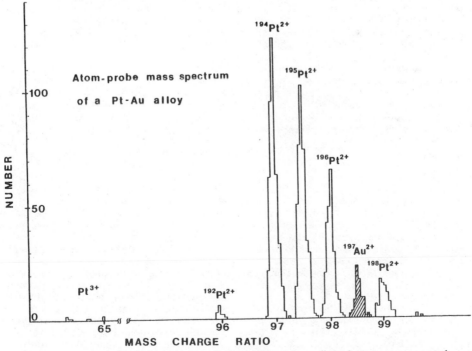

FIG. 5. An atom-probe mass spectrum of a Pt-Au alloy where all the isotopes are separated.

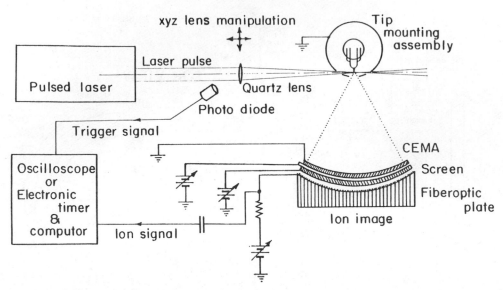

FIG. 6. A pulsed-laser imaging atom-probe. Desorption images can be displayed, and simultaneously chemical species of desorbed atoms identified from the flight times.

range and short-range order parameters in alloys can be investigated. The composition of surface layers in surface segregation of alloys can be measured with the atom-probe with true atomic layer depth resolution. The composition of precipitates, segregation of impurities to the grain boundaries, and cluster formation in alloys can be studied in the atom-probe. The atom-probe, an analytical tool of ultimate sensitivity, has also revealed the phenomena of field adsorption, thus provides new insights into the mechanism of the field ion image formation. The atom-probe has also been used to study gas adsorption, field evaporation, and exchange diffusion of surface atoms. As scientists are always interested in understanding physical phenomena through microscopic theories, atom-probe and field ion microscopy have a lot to contribute in the future.

ERWIN W. MÜLLER
TIEN T. TSONG

References

Good, R. H., Jr., and Müller, E. W., "Field Emission," in "Handbuch der Physik," Second Edition, Vol. XXI, pp. 176–231, Berlin, Springer-Verlag, 1956.

Gomer, R., "Field Emission and Field Ionization," Cambridge, Mass., Harvard Univ. Press, 1961.

Müller, E. W., and Tsong, T. T., "Field-Ion Microscopy, Principles and Applications," New York, London, Amsterdam, Elsevier, 1969.

Swanson, L. W., and Bell, A. E., "Recent Advances in Field Electron Microscopy of Metals," *Adv. Electron. Electron Physics* 32, 193–309 (1973).

Müller, E. W., and Tsong, T. T., "Field Ion Microscopy, Field Ionization and Field Evaporation," *Prog. Surface Sci.* 4, part I (1973).

Gadzuk, J. W., and Plummer, E. W. "Field Emission Energy Distribution (FED)," *Rev. Mod. Phys.* 45, 487–548 (1973).

Panitz, J. A., "Imaging Atom-probe Mass Spectroscopy," *Progr. Surface Sci.* 8, 219–262 (1973).

Tsong, T. T., "Quantitative Investigations of Atomic Processes on Metal Surfaces at Atomic Resolution," *Progr. Surface Sci.* 10, 165–248 (1980).

Kellogg, G. L., and Tsong, T. T., "Pulsed-Laser Atom-probe Field Ion Microscopy," *J. Appl. Phys.* 51 1184 (1980).

Cross-references: ADSORPTION AND ABSORPTION, CRYSTALLOGRAPHY, ELECTRON MICROSCOPE, LATTICE DEFECTS, POTENTIAL, SOLID-STATE PHYSICS, SOLID-STATE THEORY, THERMIONICS.

FIELD THEORY

The description of the physical world has evolved profoundly through the ages, at times because of, at other times being the cause of, sweeping changes in our philosophical, mathematical and experimental knowledge. Greek geometry concerned itself essentially with properties of "objects as such," a triangle or a cube being studied, for example, without any thought of their spatial environment; the Ptolemaic system enhanced this view into a clockmaker's dream, where celestial bodies parade around the earth, rigidly driven in circular motions. Only with Descartes' analytical geometry did objects become "portions of space" and the properties of space itself the main object of study; with

Galileo and Newton, a correct science of dynamics was born, which permits the prevision of an amazing number of mechanical phenomena in that space from a few first principles.

Field theory studies the phenomena of the physical world as due to interactions which propagate through space; the "geometrical emptiness," which is the space of mathematics, becomes the medium into and through which actions take place or, even more drastically, a structure which is itself determined by the properties of matter, as in general relativity (which will not be discussed in this article).

Suppose two bodies interact in space, e.g., the sun and the earth with Newton's law, or two electric charges with Coulomb's law. Two pictures of this situation are equally possible and correct. One is that this interaction cannot be conceived if *both* bodies are not there and that we should study primarily its effects without looking for a detailed mechanism for its propagation from one body to the other; this is the description of "action at a distance", in which *forces* are the main concepts and space is a vacuum into which bodies follow trajectories determined by the forces acting upon them. The other picture consists of imagining that each body, whether alone or not, modifies the structure of the space which surrounds it, geometrically or because in each point of that space there is now potentially a force, which becomes active if another body occupies that point, but should be conceived as existing there in any case; the main objective here is to study how these "fields of force" are created in space by material objects and how they propagate; this is the point of view of "action with contact," which finds its full development in field theory.

Mathematically, a field is characterized by assigning to each point of space a quantity which is *intrinsically* associated with it; a temperature, for instance, or a velocity, or a tensor or a spinor of arbitrary rank. "Intrinsic" means that if we change our frame of observation, this quantity does *not* change; supposing, e.g., that our field is that of the velocities at a given instant of all the points of a moving fluid, if we rotate our coordinate system we shall observe different values for the components of those velocities, just because we, not the velocities, have changed position. It is therefore essential that, together with the specification of the field quantities, their transformation laws also be assigned under changes of the reference frame; these laws are indicated by the description of the field quantity as a "scalar" (which does not change), a "vector" (which changes with the same law as the coordinates), a "tensor," etc.; the complete specification of all such possible laws is a standard chapter of group theory.

Physically, we have to account for the creation or the existence of the field, by describing the field quantities as generated by "sources," such as positive or negative charges for the electro-magnetic field, or the sources and sinks of hydrodynamics. Moreover, we have to describe in which way the values of the field quantities change when the point at which they are considered, or the time, is changed. In the absence of discontinuities, for instance in vacuum, one expects these values to differ by infinitesimal amounts if the corresponding points are infinitesimally close, in some way which is typical of the field considered; in other words, that the rates of change of the field quantities with respect to the space coordinates and time be connected by relations which specify both how these changes can occur compatibly with the geometrical properties of space, and how they are related to the sources. Group theory determines all the possible forms which are permissible for these relations, which take the name of *field equations*; each field theory is characterized by a special set of field equations, which are clearly *partial differential equations*.

RELATIVITY and QUANTUM THEORY have played a great rôle in the development of field theory; we shall briefly discuss, later, their influence both in the explanation of new physical phenomena and in the mathematical formulation of the theory.

Field theory has taken an entirely new shape with the so-called second quantization, which has led to several modern developments, of which some embody faithfully the concepts outlined thus far and others instead represent new views in natural philosophy; this is still a matter of controversy at present, and it is yet unpredictable whether a reasonably lasting description of nature will come out of such attempts or whether a new drastic turn in human thought will be necessary before we can hope to understand the fundamental laws of physics. Be that as it may, the ideas and the computational techniques of field theory have proved already of invaluable help in the description of many phenomena, from particle physics to superconductivity.

It is convenient to examine first the theories in which the field quantities are ordinary functions of space and time points, regardless of whether they have a direct physical meaning (as with the velocities of hydrodynamics and the electromagnetic forces) or not (as with the wave function which obeys a Schrödinger equation). This comprises of course most of classical and modern physics; mathematics permits again, however, a tremendous conceptual simplification. In the study of continuous media or fields one is, most often, interested in one of the following classes of phenomena:

(1) Phenomena which consist of the propagation of some action; the medium itself is not transported from one place to another; typical is the propagation of waves, whether they be seismic, fluid or electromagnetic;

(2) Phenomena in which there is transport or diffusion of a quantity in a medium: of heat in

a wall, of solute in a solvent, of neutrons in a pile;

(3) Equilibrium phenomena: deformations of strained elastic bodies, electro- or magnetostatic fields as determined by charges and boundaries.

Each class is ruled by essentially one type of equation. Let $\Delta = \partial^2/\partial x^2 + \partial^2/\partial y^2 + \partial^2/\partial z^2$ denote the Laplace operator; $\phi = \phi(x, y, z, t)$ the field quantity; F some function of x, y, z, t, ϕ and, at most, of the first-order derivatives of ϕ; v a velocity; and D a diffusion constant. The corresponding equations can be brought into the standard forms:

$$\Delta\phi - \frac{1}{v^2}\frac{\partial^2\phi}{\partial t^2} = F \text{ (hyperbolic partial differential}$$

$$\text{equation)} \quad (1)$$

$$\Delta\phi - \frac{1}{D}\frac{\partial\phi}{\partial t} = F \text{ (parabolic partial differential}$$

$$\text{equation)} \quad (2)$$

$$\Delta\phi = F \text{ (elliptic partial differential equation)}$$
$$(3)$$

If the field quantity has more than one component, one may deduce for each of its components an equation which is essentially of the same type, although it may be difficult or impossible to obtain an independent equation for each component.

This classification of physical phenomena according to the type of equation to which their study can be reduced is of the greatest importance: Eqs. (1), (2) and (3) are called in fact "the equations of mathematical physics"; more specifically, Eq. (1) is also called the wave equation, Eq. (2) the heat equation, and Eq. (3) the Laplace or potential equation. The study of their mathematical properties gives complete information on all the physical phenomena which they describe.

The equations of quantum mechanics can also be brought, at least formally, into the form of Eq. (1) or (2); the intervention of complex quantities modifies the situation somewhat, in a way which we cannot discuss here.

Electromagnetic phenomena fall typically into the category of Eq. (1): each of the components of the electric field $\mathbf{E} \equiv (E_x, E_y, E_z)$ and of the magnetic field $\mathbf{H} \equiv (H_x, H_y, H_z)$ satisfies, in vacuum, Eq. (1), with $F = 0$; the connections between $\mathbf{E}$ and $\mathbf{H}$ are given by the Maxwell equations, which characterize completely the theory, and lead in vacuum to the result just mentioned. When $\mathbf{E}$ or $\mathbf{H}$ does not vary with time, Eq. (1) reduces to Eq. (3), thus yielding electro- or magnetostatics.

The electromagnetic field, i.e., the vectors $\mathbf{E}$ and $\mathbf{H}$, generated by a distribution of moving charges or currents confined within a limited volume has a part which becomes dominant at a large distance from that volume, because it decreases only with the inverse of that distance (instead of the inverse-square law of static fields); this part constitutes the *radiation* field, which is responsible for the transmission of energy and signals (the radiated energy is, of course, supplied by the mechanism which drives the generating charges or currents). This is easy to understand: the energy radiated through a large sphere around the source is proportional to the area of the sphere times the square of E; it vanishes therefore with increasing radius for all but the radiative component, for which it stays constant: energy is actually removed from the source and radiated away to all distances. The study of radiation is a most important part of the theory, both macroscopically (telecommunications, radar) and microscopically (atoms, nuclei, elementary particles).

Relativity and quantum mechanics have extended and modified profoundly the classical picture presented so far. The very concepts of space and time change with special relativity: events which are simultaneous for an observer are not such when seen by another observer in uniform motion with respect to the first, because time and space are mixed together by the Lorentz transformations which relate the reference frames associated with the two observers. As a consequence, the laws of nature can retain their universal validity only if they are formulated in the same form by any such observer, i.e., if their form is not altered by a Lorentz transformation—technically speaking, if they are "Lorentz covariant." This requirement becomes a stringent dogma; it suffices to determine, with the help of group theory, the possible equations for any conceivable relativistic field theory; it is of invaluable help, when computations are made, in checking or correcting them.

The nonrelativistic Schrödinger equation for the wave function of a particle is, but for the appearance of complex quantities, of the type (2) described before: this is not acceptable in a relativistic world, because time and space are not treated alike. One needs either an equation which contains only second-order derivatives, or one with only first-order derivatives; for a free particle, this leads either to the Klein-Gordon equation, which is of type (1), or to the Dirac equation, which contains linearly only the first-order derivatives of the wave function, but has a mathematical structure which necessarily assigns special physical properties to the particles described by it. It was one of the greatest triumphs theoretical physics ever witnessed, to discover that such properties are actually displayed by all particles which obey the Dirac equation: spin, and the existence for each Dirac particle of a corresponding *antiparticle*, i.e., a particle having the opposite mechanical and electrical properties. (See article on ANTIPARTICLES.)

The requirement of relativistic covariance has thus led to fundamental physical discoveries; for each particle obeying the Dirac equation, the corresponding antiparticle has been experimentally found in nature; electron and positron, proton and antiproton, neutron and antineutron, etc. What is more, the theory predicts that a particle-antiparticle pair can be created in a collision phenomenon, if sufficient energy is available, or can annihilate itself, giving away its energy in the form of electromagnetic radiation or other particles.

The classical theory allowed only for the electromagnetic radiation emitted by moving charges or currents; the creation or absorption of particles in collision phenomena, as well as the creation or annihilation of pairs, were outside its scope and possibilities. A new formulation of the theory was needed, which could account consistently for all such phenomena, handling situations in which particles can be created and destroyed in any numbers. The formalism devised for this purpose is that of quantum field theory.

The basic idea is to describe each type of particle by means of a field which is not any more an ordinary numerical function of space and time, but an "operator," i.e., a quantity which changes the number of particles existing in any given state of the system. If the field operator is known, one can then evaluate the probability of a given state (so many particles, with determined energies and momenta) changing into an equally determined, different state. If the particles do not interact among themselves or with other particles, no change is possible; if there is interaction, the field operator has a structure which can cause such transitions. The field equations appear to be essentially the same as those of the classical Maxwell, Klein-Gordon, Dirac theories, etc.; their structure is however fundamentally different, because they now must be equivalent to infinite sets of ordinary equations, which couple states with different and ever-increasing numbers of particles.

Fields which are associated with particles obeying Bose-Einstein statistics (of which any number can be found in any given state) have radically different mathematical properties from fields associated with particles obeying Fermi-Dirac statistics (of which at most one can be found in any given state); examples of the first are photons (the massless neutral quanta of the electromagnetic field), pions (massive particles, with or without electric charge, which are believed to be responsible for nuclear forces), etc.; examples of the second are electrons and positrons, protons and antiprotons, etc.

The passage from numerical fields to operator fields is called "second quantization"; quantum field theory deals with operator fields.

Striking successes have been met with this approach. From a quantitative point of view, they are confined mostly to electrodynamics, where very small deviations from the values predicted by the non-quantized theory, which were observed in the measurement of the magnetic moment of the electron and in the so-called

Lamb shift, were accounted for with amazing accuracy by quantum field theory. Qualitatively, the new conceptual framework has proved extremely useful in understanding elementary phenomena, especially with the help of the diagrams devised by R. P. Feynman, which give a simple intuitive picture of collision and radiation processes involving elementary particles. Very little has been achieved quantitatively, though, for theories other than electrodynamics, because of the tremendous mathematical difficulties which arise as soon as the simplest approximation techniques are not applicable because the interaction is too strong; nevertheless, these ideas have proved greatly helpful in many ways, in combination with general principles of symmetry, Lorentz invariance and causality.

A beautiful consequence of this conception, which assumes that particles are the quanta of a field (as photons were recognized by Einstein to be the quanta of the electromagnetic field) was the discovery of H. Yukawa, that whenever such quanta have a mass different from zero, the force they create between two bodies which interact by exchanging such quanta with each other must be an exponentially decreasing function of distance; this force can become of a coulombian type only if the mass of the quanta vanishes. Thus, the Coulomb force can be explained as due to the exchange of photons among electric charges, the nuclear forces (which have typically short ranges) as due to the exchange of massive particles among nucleons. Exchanges of this nature are not observable in the laboratory, because this runs against Heisenberg's indeterminacy principle; if enough energy is supplied, however, such quanta can actually break loose and do appear as the particles created in collision processes.

The mathematical difficulties encountered in quantum field theory are many, and there is as yet lack of agreement as to the best way to circumvent some of them. Besides mathematical complexity, which prevents all but the simplest calculations, there are many unsolved problems of mathematical rigor and apparent inconsistencies which can be removed only by delicate analyses. Typical of the latter is the fact that unsophisticated calculations give infinite values for masses and charges of interacting particles, and a painstaking analysis is required to retrieve from them the significant physical values; this is the so-called renormalization procedure, which copes with infinities which partly are already present in the classical theory (such as the infinite electromagnetic contribution to the mass of a point-like charged particle, when computed from Maxwell's equations) and partly originate from the new formalism (which permits, for instance, pair creation).

It is not yet certain whether such difficulties are due to the lack of adequate mathematical techniques or are the expression of a fundamental inadequacy of the theory to describe ultimate laws of nature. For this reason, while, on the one hand, the attention of some theoreticians has been directed to perfecting the mathematical foundations of quantum field theory (giving rise to axiomatic field theory and to more rigorous methods of obtaining and studying the quantum field equations, etc.), on the other hand, most physicists have been trying new avenues, such as the S-matrix theory, dispersion relations, the so-called Regge poles, etc.; these approaches have certainly led to very useful results, but they leave altogether at least as many doubts as hopes.

Whatever may be the future prospects of field theory as the correct means for describing the fundamental laws of nature, its tremendous usefulness in providing a conceptual framework, in inspiring new ideas, and in suggesting computational techniques has been overwhelmingly demonstrated in the last decades. It has now found a new, very fertile ground of application in the study of systems containing a very large number of particles, where it has already provided a reasonably good quantitative understanding of superconductivity and superfluidity, and promises many other results of interest in the study of solids and liquids.

E. R. CAIANIELLO

Cross-references: ANTIPARTICLES, ELEMENTARY PARTICLES, QUANTUM THEORY, RELATIVITY, SCHRÖDINGER EQUATION, VECTOR PHYSICS.

FISSION*

Nuclear fission is the breakup of a heavy nucleus, such as that of uranium, into two medium-weight nuclei, with the release of a considerable quantity of energy. Also produced are a few neutrons, some gamma rays, and a number of beta-particles (electrons) from the radioactive decay of the two fragments. Fission occurs spontaneously in some cases, or may be induced by bombardment of the fissionable material with neutrons, protons, or other particles.

Discovery of Fission Although fission was not discovered until 1939, it had been realized, ever since Einstein published his theory of relativity in 1905, that there was a theoretical possibility of releasing tremendous energy from matter.

Fission is now known to have been first produced by Enrico Fermi and his co-workers in 1934, when they irradiated many elements, including uranium, with the newly discovered neutrons. They found a number of different β-activities to be produced from uranium, but believed that these were due to neutron cap-

*Work performed under the auspices of the U.S. Dept. of Energy.

ture. Later radiochemical work indicated that some of the new activities were from elements chemically similar to the much lighter elements Ba, La, etc.

Fission remained unrecognized until O. Hahn and F. Strassmann, German radiochemists, showed by very careful work that these products were not merely chemically similar to lighter elements, but *were* lighter elements. They published their startling results in the January 6, 1939 issue of *Naturwissenschaften*. On January 16, Lise Meitner and O. Frisch sent in to *Nature* (from Stockholm and Copenhagen) a paper in which they named the new process "fission," predicted that the fragments should have large kinetic energies, and explained the process in terms of a liquid-drop model. On the same date, Frisch sent in to *Nature* another paper in which he reported that he had observed the large electrical pulses from fission fragments in an ionization chamber.

Niels Bohr in the meantime had taken the news of the discovery of fission, and the prediction of large energies, to a conference on physics in Washington, D.C. During the conference physicists in a number of laboratories independently verified the tremendous kinetic energies of fission fragments, unaware as yet of Frisch's results.

During the months following the discovery of fission, there was feverish activity in laboratories around the world. It was soon discovered that neutrons were being produced by fission, and that almost all of the fission of U was taking place in the relatively rare isotope, ^{235}U. In that same year (1939), Bohr and Wheeler published their theory of fission, based on the liquid-drop model, which is still basic to modern fission theory.

Development of Atomic Energy On the date of publication of the Bohr-Wheeler paper, September 1, 1939, Germany invaded Poland, the Second World War was underway, and fission suddenly had a new importance. It was realized by many that a chain reaction was possible for fission, with the neutrons from each fission producing more fissions, resulting in the release of very large amounts of energy.

The fission process results in the conversion of 0.09 per cent of the mass of the original nucleus into kinetic energy. This amounts to about 200 MeV per fission, or 3.20×10^{-11} joules. The fission of 1 kilogram of ^{235}U thus releases a total energy equal to 8.21×10^{13} joules, or 2.28×10^7 kilowatt-hours. This is roughly equal to the daily output of a large hydroelectric power plant such as Hoover Dam, and very much greater than the energy released in chemical reactions. One kg of ^{235}U is equivalent in energy release to the burning of 3.45×10^6 kg of coal (C) by 9.20×10^6 kg of oxygen. This chemical process releases 7.2×10^{-11} of the mass as heat energy, more than 10 million times smaller than for fission.

The fission of 1 kg of ^{235}U is also equivalent

in energy released to the detonation of 19.6×10^6 kg (19.6 kilotons) of high explosive. A kiloton, 1000 metric tons, is conventionally taken to be equivalent to 10^{12} calories, or 4.186×10^{12} joules; a megaton is 1000 times larger.

Thus it was known to many people in 1939 that it might well be possible to produce the destructive effect of many thousands of tons of high explosive with a single bomb containing a relatively small amount of fissionable material. It seemed probable that Germany would press ahead with this development. Aware of this danger, scientists in the rest of the world largely ceased publishing fission results by 1940.

Work on fission was continued quietly at an increasing rate. In June 1942, the Manhattan Project (under U.S. Army direction) was set underway in the United States, with the objective of producing nuclear weapons, if possible. The first chain reaction was produced on December 2, 1942, under the direction of Enrico Fermi, who had arrived from Fascist Italy in January 1939. Fermi and his co-workers had piled up blocks of ordinary uranium and extra-pure graphite (carbon) to produce a nuclear reactor, under Stagg Field Stadium of the University of Chicago. The carbon was used to slow down fission neutrons and thus increase the likelihood of fission. Cadmium rods inserted in the reactor (at that time called a "pile") were used to control the chain reaction by capturing a certain fraction of the neutrons.

Thus, by December 1942, a fission chain reaction had been achieved. In the following years, research vital to the Manhattan Project was carried on at many laboratories. In order to produce the fissionable material for nuclear weapons, two tremendous industrial plants were set up beginning in 1943, at Oak Ridge, Tennessee, and Hanford, Washington.

The Oak Ridge plant was for the purpose of separating the more fissionable isotope, ^{235}U, from the much more common isotope, ^{238}U, which accounted for 99.3 per cent of the mass of ordinary uranium metal. The most successful separation process, which is still in use at Oak Ridge, was that of gaseous diffusion. Uranium in the form of a gas, uranium hexafluoride, is passed through a long series of porous barriers. The lighter isotope ^{235}U can diffuse more readily than ^{238}U, and the result of the process is enriched uranium, the ^{235}U content having been increased from 1 part in 140 to around 95 per cent.

The Hanford plant consists of giant nuclear reactors to produce a new element, plutonium, from ordinary uranium by neutron capture. It was thought, and eventually proved, that plutonium should be fissionable by slow neutrons in the same way as ^{235}U. Capture of a neutron by ^{238}U produces the heavier isotope ^{239}U, which then decays by beta-emission to the new element ^{239}Np (neptunium), and then by another beta-decay to ^{239}Pu.

The work of designing and building nuclear weapons was carried out at a laboratory set up at Los Alamos, New Mexico. This laboratory, under the direction of J. Robert Oppenheimer, began its work in early 1943. Before the end of the war, a good fraction of the world's most eminent nuclear physicists had come to work at Los Alamos, including Bohr, Fermi, Frisch, and many British scientists.

One of their most basic problems was to find a way to assemble the fissionable components of a weapon rapidly enough to produce a powerful chain reaction, lasting less than one μsec. The individual masses of enriched uranium or plutonium had to be of such a size and shape as not to be capable of a chain reaction; i.e., they had to be of less than critical mass. If these pieces were not assembled at sufficient speed, the result would be only a minor explosion, or perhaps the melting of the device. Two approaches were tried. One involved firing one piece of fissionable material at another in a short "gun." The other was the implosion method, in which the fissionable material is assembled into a highly compressed mass by the explosion of a surrounding spherical shell of high explosive.

Both methods of achieving a nuclear explosion were ultimately successful, but it was not known whether either method would work until July 16, 1945, when the first "atomic bomb," a plutonium implosion device, was set off in a desert area near Alamogordo, New Mexico. On August 6 a uranium gun weapon was exploded over Hiroshima, Japan, and on August 9 a second plutonium implosion warhead was detonated over Nagasaki. Each weapon had a yield equivalent to about 20 kilotons of high explosive and caused tremendous destruction. On August 10 the Japanese first offered to surrender, and accepted Allied terms on August 15; the mobilization of the massive United States invasion force was called off. Germany, which had surrendered on April 8, 1945, was found to have made little progress toward nuclear weapons during the war.

Following the end of the war, the work on nuclear weapons and nuclear power in the United States were placed under the newly created (1946) Atomic Energy Commission. Similar agencies have since been set up in many countries. On August 29, 1949, the U.S.S.R. detonated its first nuclear device. During the next several years, enormous production plants for ^{235}U were built at Paducah, Kentucky, and Portsmouth, Ohio, as well as facilities for producing ^{239}Pu and hydrogen isotopes on the Savannah River in South Carolina. On November 1, 1952, the Los Alamos National Laboratory exploded the first thermonuclear device (hydrogen bomb), with a yield equivalent to many megatons of high explosive. Such a weapon uses nuclear fission to trigger nuclear fusion, in which light elements (such as the various isotopes of hydrogen) are combined at exceedingly high temperature to give heavier nuclei, with considerable release of energy. On August 12, 1953, the USSR set off its first thermonuclear weapon. In the years since then, Great Britain, France, China, and India have all developed their own nuclear weapons.

Nuclear reactors and nuclear power have been developed extensively since the war, in many countries. Nuclear-powered submarines, aircraft carriers, and other vessels have revolutionized naval strategy. Nuclear-propelled rockets (the "Rover" program at Los Alamos) have been developed for space exploration. The possible use of nuclear explosives for earth-moving, increasing gas-well yield, and other engineering purposes has been an extensive project ("Plowshare") of the Lawrence Livermore National Laboratory, as well as of Los Alamos. Nuclear power plants already account for a significant fraction of electrical power production.

Chain Reactions The basic feature which makes both nuclear reactors and nuclear weapons possible is the "chain reaction," in which each fission produces another fission, or several, by means of the several neutrons emitted from fission. If there is not enough fissionable material, or it is not arranged compactly enough, no chain reaction will be possible. The fission neutrons emitted in such a situation will have too great a chance of escaping from the fissionable material, or of being absorbed in non-fissionable material, to continue the chain of fissions should one fission occur.

There is thus a "critical mass," or minimum amount of fissionable material, necessary for a chain reaction in any given arrangement. For a spherical mass of metal in air, the critical mass of highly enriched (94 per cent) ^{235}U, for instance, is 52 kg; the critical mass for ^{233}U or ^{239}Pu metal is lower, about 16 kg. It is the capture of neutrons by ^{238}U which leads to a higher critical mass in the first case. The critical mass can be lowered by mixing fissionable material with graphite or other material as a "moderator" to slow down the neutrons, or by surrounding the fissionable material with a reflector to scatter neutrons back.

The smallest critical masses are achieved by water (or heavy water) solutions of fissionable material, since H and D are most effective in slowing down neutrons. The "Water-Boiler" thermal reactor at Los Alamos operated ordinarily with about 1100 grams of enriched uranium in water solution, and achieved criticality with less than 600 grams.

The energy of the neutrons producing most of the fissions in a reactor determines whether it is called "fast," "intermediate," or "thermal." Fast reactors use neutrons only slightly slowed down from the 2-MeV average energy with which they are emitted. Thermal reactors use neutrons slowed down, by collision with moderator atoms, almost to the velocities corresponding to thermal motion at the temperature of the reactor. At 294 K, about room tempera-

ture, the most probable velocity is 2200 meters/sec., which corresponds to a neutron energy of 0.0253 eV. Intermediate reactors use partially slowed down neutrons, with less than 100-keV energy.

In order for a chain reaction to continue, it is necessary that each fission produce, on the average, at least one more fission. The average number of fissions produced by the neutrons from one fission is called k, the criticality factor or multiplication constant. If k is less than 1.0 (the critical value), a chain reaction, even if begun with many fissions, will soon die out.

A reactor must have $k = 1.0$ during normal, steady, operation, and must have k greater than 1.0 during the start-up operation. These changes in criticality factor can be brought about by introducing or removing from the reactor control rods made of neutron-absorbing elements, such as cadmium or boron.

This regulation of reactor power is made much easier by the phenomenon of delayed neutrons, which are emitted from some fission products for a few seconds after the fission. In the case of thermal-neutron fission of ^{235}U, 0.7 per cent of the neutrons produced (0.017 out of 2.43 neutrons per fission) are delayed in this way. Only if k were increased above 1.007 in this case could the reactor power increase rapidly, since each fission would then, on the average, produce one more fission promptly. Reactors are of course designed to avoid this "prompt critical" condition, since this can lead to overheating and destruction of the reactor. In any case, reactors cannot possibly achieve the high values of the criticality factor needed for a true nuclear explosion.

A "breeder" reactor is one which produces as much nuclear fuel as it consumes, such as by using the capture of neutrons in ^{238}U to produce thermally fissionable ^{239}Pu. It is of course necessary to reprocess the fuel elements to recover the ^{239}Pu. Breeder reactors are now being developed for more efficient power production. It is of interest that the uranium contained in coal, which is discarded with the fly ash and flue gases, could produce more power than the coal if efficiently used.

The Fission Process The fission of a nucleus can be understood on the basis of the liquid-drop model, which was first discussed by Meitner and Frisch and later greatly extended by Bohr and Wheeler. This model explains many features of fission, but others are still not fully understood, and the complete theory of fission is still awaited.

On the basis of this model, the nucleus is assumed to be similar to a uniformly charged drop of incompressible liquid. It will then be normally spherical, kept in this shape of minimum energy by the effect of surface tension. However, the individual parts of the positive charge, actually protons, tend to repel each other and to lessen the effective surface tension. It has been calculated that the two effects will cancel each other out for a value of Z^2/A equal to about 45 (Z is the atomic number or the number of protons; A is the mass number or total number of nucleons in the nucleus). The ratio of Z^2/A to this critical value is called the fissionability parameter x.

For a fissionability parameter equal to 1.0, the nucleus should have no effective surface tension and no stability against distortion, so that it should promptly elongate to a point where the coulomb (electrostatic) forces can blow it apart. Such a nucleus would have no "fission barrier" and could not exist long. Known nuclides have lower values of the fissionability parameter; for ^{235}U the value of x is about 0.8. For such nuclei, it takes 5 or 6 MeV of energy to deform the nucleus to the point where coulomb forces can cause fission. However, such a fission barrier may be overcome by the capture of a neutron, which adds an excitation energy of 5 or 6 MeV.

For the even-Z, odd-N nuclides ^{233}U, ^{235}U, and ^{239}Pu (N is the number of neutrons in the nucleus), fission can be induced by the capture of even a low-energy neutron. Even-even target nuclides require more energy, so that ^{238}U is very unlikely to fission upon capture of a thermal neutron, but needs about 1 MeV additional neutron energy to overcome the fission barrier. Even-even *compound* nuclei (i.e., after neutron capture) are evidently more likely to undergo fission. Even-even nuclei are also much more likely to undergo spontaneous fission, in which an occasional nucleus manages to overcome the fission barrier without any added energy. Fis-

sion may also be induced by gamma rays of energy equal to the fission barrier or by energetic particles of many kinds such as protons or alpha particles.

The probability that particles passing through a target of fissionable material will induce fission is measured by the fission cross section. The "cross section" is the effective area of a nucleus for a given process. For a thin target of thickness l cm having n nuclei per cubic centimeter, each with fission cross section σ_f cm^2, the probability that a particle passing through the target will induce a fission is $nl\sigma_f$.

Figure 1 shows the fission cross section for 11 fissionable nuclei, as a function of the energy (in MeV) of the neutron inducing fission. For those nuclides (^{233}U, ^{235}U, ^{239}Pu) which lead to even-even compound nuclei, the cross sections are very high for low-energy neutrons; these are "fissile" nuclides. For the other target nuclides, the fission cross section is extremely small until the neutrons have energy in excess of a threshold energy. All the cross sections may be seen to increase noticeably at about 7 MeV. This increase is due to the added possibility of fission following low-energy neutron emission, which becomes energetically possible here.

The energy released when fission takes place amounts to about 200 MeV. This is the difference in mass between a heavy fissionable nucleus and the two medium-weight nuclei, plus neutrons, into which it breaks up. This energy release of 200 MeV amounts to 21.3 per cent of the mass of a single proton or neutron. Most

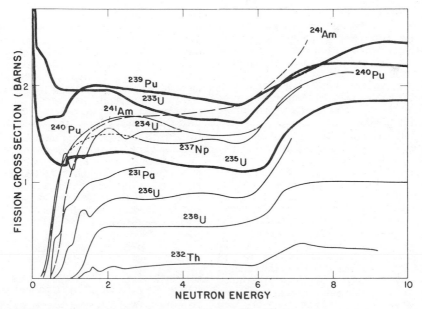

FIG. 1. Dependence of fission cross section (a "barn" is 10^{-24} cm^2) on inducing neutron energy (in MeV) for 11 different fissionable nuclei (reprinted with permission from R. L. Henkel, "Fission by Fast Neutrons," in J. B. Marion and J. L. Fowler, Eds., "Fast Neutron Physics, Part II," New York, Interscience Publishers, 1963).

of the energy appears in the form of kinetic energy of the two fission fragments, which are sent flying apart with an average total of 167 MeV in the case of thermal fission of ^{235}U.

The two fragments share the kinetic energy in the inverse ratio of their masses, since they have equal and opposite momenta. The mass division is usually asymmetrical, as may be seen in Fig. 2. The probability that neutron fission of ^{235}U will produce two given fragments of nearly equal mass (symmetrical fission) is about 600 times lower than that of the most probable case of asymmetrical division into light and heavy fragments. The initial fragments (before neutron emission) in this case have masses averaging 96 and 140 mass units. The lighter fragment has on the average about 99 MeV of kinetic energy, and the heavy fragment about 68 MeV. These energies correspond to initial velocities of 1.4 and 1.0 X 10^9 cm/sec, or 4.7 and 3.2 per cent of the speed of light, respectively. The total fragment energy is reasonably well approximated by $0.121 \, Z^2/A^{1/3}$ MeV.

Such fragment energies are due to the Coulomb interaction between the two fragments, and would be expected on the basis of any fission theory. The asymmetrical mass division, however, would not be predicted by liquid-drop theory and is evidently due to nuclear shell effects. The liquid-drop model would, in its simplest form, predict primarily symmetrical fission, which is rarely observed.

The fragments, as we have seen, carry away most of the 200-MeV energy release in the form of kinetic energy. The rest of the energy is released in the form of neutron energy, gamma rays, and beta decay. The prompt neutrons, which are emitted within 10^{-14} sec. following fission, will be discussed below. The fission fragments are strong sources of prompt gamma rays, emitting about 8 within a microsecond or less following fission. The gamma rays have a broad spectrum of energies, up to as much as 7 MeV, but they average about 1 MeV apiece, so that the total energy emitted in prompt gamma rays is about 8 MeV. This is more than simple theories would predict.

Immediately after the emission of neutrons and prompt gamma rays, the fragments (now called "fission products") begin the process of beta decay, which ultimately accounts for perhaps 22 MeV of fission energy release. The fission products have neutron-to-proton ratios which are nearly the same as that of the heavy nucleus from which they were formed (about 1.55). This is too many neutrons for stability in the fission-product mass region, where nonradioactive nuclei have neutron-proton ratios in the range 1.3 to 1.4. Since there is thus energy to be released by changing neutrons to protons, the result is a long sequence of beta decays averaging 3 or 4 for each fission product. Each beta decay results in the emission of a negative electron, a neutrino, and usually gamma rays,

and an increase of the charge of the nucleus by one unit. The neutrinos ultimately carry away about 10 MeV of the energy released, and for all practical purposes this energy is not detectable and is never seen again. The fission products are intensely radioactive immediately after fission, as the beta decays having the shortest half-life are completed, and gradually become less radioactive with the passage of time.

For a small fraction of the fission products it is energetically possible for a neutron to be emitted at some stage in the chain of beta decays. These delayed neutrons, amounting on the average to 0.017 per thermal-neutron fission of ^{235}U, are very useful in stabilizing the operation of nuclear reactors, as mentioned above. Because they are emitted immediately following beta decays, they appear to follow the same radioactive decay curves as some of the beta decays, and have varying half-lives, of the order of seconds.

In rarer cases—about one fission out of 400—an alpha particle (^{4}He) is emitted during the fission process, in addition to the two fragments. This process is usually called ternary fission, as distinguished from the usual binary fission. Because of the high energy (averaging 15 MeV) and direction (roughly perpendicular to fragment motion) of the alpha particle, it is probably formed at the same time as the two heavier fragments, and between them. In still rarer cases, other light charged particles such as nuclei of tritium (^{3}H) or of ^{6}He are seen. There seems even to be evidence that, in perhaps one fission out of 100 000, three fragments of roughly equal mass are formed.

In the vast majority of fissions, however, two fragments and several neutrons are the result. For low-energy fission it is clear that most, or perhaps all, of the neutrons are emitted by the fragments. The average number of neutrons per fission ($\bar{\nu}$) ranges from 2.0 to 4.0 for various nuclides. Fission at high energies, such as that induced by 100-MeV alpha particles, produces many more neutrons, most of which are emitted before fission.

In the typical case of thermal-neutron fission of ^{235}U, $\bar{\nu}$ is 2.43, including delayed neutrons. For individual fissions the number ν can vary from zero to 5 or 6, the standard deviation from the average being ±1.10. The prompt neutrons in this case have an average energy of 1.94 MeV. The neutron energy spectrum in this and other cases is well described by a Maxwellian distribution,

$$N(E) = (2/\sqrt{\pi T^3}) \sqrt{E} e^{-E/T}$$

in which $N(E)$ is the number of neutrons per unit energy E, and T is a parameter equal to two-thirds of the average energy $\bar{E}$. Such a spectrum is predicted from nuclear temperature theory.

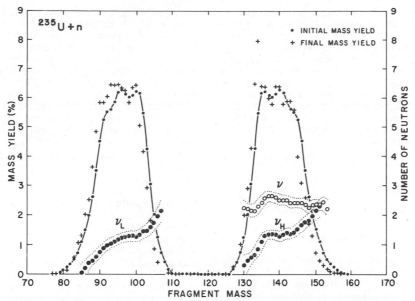

FIG. 2. Dependence of neutron yield on initial fragment mass for thermal-neutron fission of ^{235}U. Average numbers of neutrons emitted by light and heavy fragments are given the symbols ν_L and ν_H; the total from both fragments is ν. Standard deviations are indicated by dotted lines. Also shown are the initial and final mass yields [reprinted from J. Terrell, *Phys. Rev.*, 127, 880–904 (1962)].

Of the roughly 2-MeV average energy per neutron, about 0.75 MeV is contributed by the motion of the emitting fragments. There is evidence that the average neutron energy increases with increasing number of neutrons. This would be expected from considerations of nuclear temperature, which lead to the relation $\overline{E} = 0.75 + 0.65 \sqrt{\bar{\nu} + 1}$, in MeV, in agreement with experimental data.

The total kinetic energy carried away by the typical 2 or 3 neutrons per fission is thus about 5 MeV, of which 2 MeV is taken from fragment energy. The fragments after neutron emission thus have a total energy reduced (in the case of ^{235}U) from 167 to 165 MeV, and lower mass numbers.

The final fragment masses may be determined by radiochemical data on fission products. The final mass yields for ^{235}U are shown in Fig. 2, as crosses. The initial mass yields may be determined from simultaneous measurement of the velocities of the two fragments and are also shown in the Figure. The differences between initial and final mass yields are accounted for by emission of neutrons and may be used to determine the numbers of neutrons. The average numbers of neutrons emitted by individual fragments are shown (ν_L and ν_H) as functions of initial fragment mass; the total $\nu(= \nu_L + \nu_H)$ is also shown.

As may be seen in Fig. 2, the average number of neutrons emitted by a fission fragment depends strongly on the fragment mass; it is near

zero for the lightest of the light fragments, and also for the lightest of the heavy fragments, and rises to high values elsewhere. Thus the two fragments from a fission will usually emit quite different numbers of neutrons, which implies quite different initial excitation energies. This phenomenon came as quite a surprise when it was first reported by Fraser and Milton in 1954. It has since been found to be a common and perhaps universal feature of fission.

This sawtooth dependence of neutron number on fragment mass may be understood on the basis of varying stiffness of the fragments against deformation. The fragments having "magic" numbers (closed shells) of neutrons or protons would be expected to be stiff and to be exceptionally resistant to deformation and the consequent excitation. The magic number of neutrons, $N = 50$, will occur for fragments in the vicinity of mass 82, at the lower boundary of the light fragment peak. Similarly, two closed shells ($Z = 50$ and $N = 82$) occur at the lower edge of the heavy fragment peak, near mass 130. The low neutron yields seen at these masses may be quantitatively explained on the basis of the effects of closed nuclear shells on the deformation parameters. It is hoped that this fragment-deformation theory will be able to account quantitatively for the spectrum of fragment masses, since the mass yields and neutron yields tend to vanish near the same magic numbers.

Fission isomers have been of much interest recently. These are excited states of fissionable nuclides, produced by bombardment with energetic particles or nuclei, which exist for times of the order of 10^{-2} to 10^{-9} seconds before fissioning. Such quasi-stable states, lying a few MeV above the unexcited ground states, are believed to be "shape isomers" explainable by a double-humped fission barrier, for which there is other evidence in the intermediate structure of resonances in the fission cross sections. Such a fission barrier has a second minimum, at some deformation, in the energy required to deform the nucleus from the ground state. It is thought now that all fission barriers have such a double-humped structure, first explained by V. M. Strutinsky in terms of variation of shell structure with deformation.

JAMES TERRELL

References

Gindler, J. E., and Huizenga, J. R., "Nuclear Fission," Chapter 7 in "Nuclear Chemistry," (Ed. by L. Yaffe) Vol. II, New York, Academic Press, 1968.

Glasstone, S., and Sesonske, A., "Nuclear Reactor Engineering," Third Edition, New York, Van Nostrand Reinhold, 1980.

Glasstone, S., Ed., "The Effects of Nuclear Weapons," Third Edition, U.S. Government Printing Office, 1977.

Hyde, E. K., "The Nuclear Properties of the Heavy Elements III: Fission Phenomena," Englewood Cliffs, N.J., Prentice-Hall, Inc., 1964; New York, Dover Publications Inc., 1971.

I.A.E.A., "Physics and Chemistry of Fission," Proceedings of the Third I.A.E.A. Symposium, Vienna, I.A.E.A., 1973.

Keepin, G. R., "Physics of Nuclear Kinetics," Reading, Mass., Addison-Wesley Publishing Co., 1965.

Terrel, J., "Prompt Neutrons from Fission," in Proceedings of the I.A.E.A. Symposium on the Physics and Chemistry of Fission; I.A.E.A., Vienna, 1965.

Vandenbosch, R., and Huizenga, J. R., "Nuclear Fission," New York, Academic Press, 1973.

Cross-references: ATOMIC ENERGY, FUSION, NUCLEAR RADIATION, NUCLEAR REACTOR, NUCLEAR STRUCTURE.

FLIGHT PROPULSION FUNDAMENTALS

Propulsion is the act of changing, or maintaining, the motion of a vehicle flying in air or in space. This includes deceleration for reentering the earth's atmosphere with manned spacecraft. Fundamentally, the operating principles of propulsion devices, regardless of type or class, embody the basic laws of MECHANICS, THERMODYNAMICS, CHEMISTRY, ELECTRICITY, and other sciences. In propelling a vehicle, it is necessary to accelerate a mass, commonly called the *working fluid*, in a direction opposite to flight (or thrust force). Newton's basic laws of motion, the conservation of momentum (change

of momentum of the vehicle equals change of momentum of the working fluid), and the dictates of the perfect gas laws dominate the design and operation of the propulsion system, be it an aircraft engine with propeller, a turbojet engine, a solid rocket motor, a gas-turbine-driven rotor system in a helicopter, or any other propulsion device.

The advent of the "jet age" coincided with World War II and a wide variety of propulsion devices have been studied and developed since then. Table 1 lists the major types of propulsion devices and indicates their status. Note the three energy sources listed. In time, additional energy forms may prove feasible such as the laser beam, electromagnetic waves and proton or other particle beams transmitted to a flying receiver.

Fundamental Relations Assume an ideal engine inside of which the fluid receives energy and is heated and accelerated as shown schematically in Fig. 1. The acceleration of a mass flow of air or fluid $\dot{m}_a$ from an initial velocity v_0 (which equals the forward flight velocity of of the vehicle) to the jet velocity at the exit v_e will result in a net thrust F, which is equal to the mass flow rate multiplied by the velocity increment. Additional terms are added to this momentum relation in order to correct for the additional mass flow rate of the fuel or propellant ($\dot{m}_p$) (the fuel is carried in the vehicle) and for any difference in static pressure of the jet exit of the engine (p_e) and the atmospheric or ambient pressure p_0.

$$F = \dot{m}_a(v_e - v_0) + \dot{m}_p v_e + A_e(p_e - p_0)$$

The last term in this equation is called the pressure thrust; it is positive if $p_e > p_0$ (which occurs when there is incomplete expansion of the gases in the engine exit nozzle) or negative if $p_e < p_0$ (which occurs when there is overexpansion in the engine exit nozzle). For a rocket engine, the air mass flow $m_a = 0$, and the thrust is equal to the last two terms of the equation only. In the case of a propeller engine, there are really two different air flows, and the first term in the above equation is split into two separate terms: one flow which crosses the plane of the propeller and is accelerated by the propeller blades and a second smaller airflow which goes through the engine to furnish oxygen for combustion.

Consider a winged vehicle in equilibrium rectilinear flight in a two-dimensional (fixed plane) trajectory; assume all control forces, lateral forces and turning moments to be zero and the flight direction to be the same as the thrust direction. In the direction of the flight, the instantaneous vehicle mass m_v times the vehicle acceleration dv/dt has to equal the sum of all the forces, namely a component of the thrust F, the aerodynamic drag D, and a component of the gravitational attraction or the weight $m_v g$. The angles are as defined in Fig. 2.

TABLE 1. TYPES/STATUS OF FLIGHT PROPULSION DEVICES
(ELECTRIC PROPULSION EXCLUDED)

Propulsion Device	Energy Source*			Working Fluid
	Chemical	Nuclear	Solar	
Piston engine/propeller	D/P			Surrounding air
Turbojet	D/P	TFD		Surrounding air
Turbofan	D/P	TFD		Surrounding air
Turboprop	D/P			Surrounding air
Turbo–ramjet	TFD			Surrounding air
Ramjet	D/P			Surrounding air
Pulsejet	D/P			Surrounding air
Rocket	D/P	TFD		Stored propellant
Ducted rocket	TFD			Stored + surrounding air
Air turborocket	TFD			Stored + surrounding air
Solar heated rocket			TFD	Stored propellant
Photon rocket		TFND		Photon ejection (no stored propellant)
Solar sail			TFD	Photon absorption (no stored propellant)

*D/P: developed and/or considered practical; TFD: technical feasibility has been demonstrated, but development is incomplete; TFND; technical feasibility has not been demonstrated as yet.

$$m_v dv/dt = F - D - m_v g \sin \theta \qquad (2)$$

The vehicle mass m_v multiplied by the acceleration in a direction perpendicular to the flight path ($v \, d\theta/dt$) must equal the sum of all forces perpendicular to the flight direction; here, the lift force L must be considered.

$$m_v v \frac{d\theta}{dt} = L - m_v g \cos \theta \qquad (3)$$

The solution to these two equations results in the determination of a two-dimensional trajectory, maximum flight velocity, range, and other flight performance parameters. The actual solution is three-dimensional, and must usually be a numerical integration since m_v decreases with time, L and D vary with speed and altitude, and the direction of thrust is not the same as the flight direction; also, both the flight angle and the angle of attack are usually changing. For the case of a linear, simplified horizontal equilib-

rium flight, $\theta = 0$, $dv/dt = 0$, and thus Eq. (2) reduces to $F = D$. The vehicle mass m_v consists of the vehicle dry mass or final vehicle mass after expenditure of all propellant $(m_v)_f$ plus the propellant or fuel mass m_p. For steady fuel flow, $m_p = \dot{m}_p t$. For the case of gravity-free flight in a vacuum (true space environment), Eq. (1) can be rewritten for a rocket:

$$\frac{dv}{dt} = \frac{F}{m_v} = \frac{\dot{m}_p v_e}{(m_v)_f + \dot{m}_p t}$$

Integration gives

$$v_v = -v_e \ln \frac{m_v - m_p}{m_v} = v_e \ln \frac{(m_v)_i}{(m_v)_f}$$

Thus the maximum velocity attained by a rocket-space vehicle operating in a gravitationless vacuum is equal to the product of the average effective rocket exhaust velocity v_e and a logarithmic function of the initial vehicle mass $(m_v)_i = (m_v)_f + m_p$ (fully fueled vehicle) at start of the engine operation, divided by the

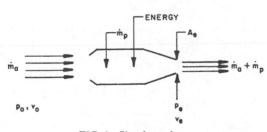

FIG. 1. Simple engine.

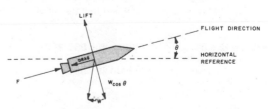

FIG. 2. Simple free body diagram of flying vehicle.

final vehicle mass (with all the fuel expanded) $(m_v)_f$ at the end of engine operation. This velocity v_v will be large when v_e is large, i.e., high energy is available from the propellant or the engine and when $(m_v)_i$ is small, i.e., when the dry mass of the vehicle (dry engine mass tanks, payload, or structure) is small and no unnecessary mass is designed into the vehicle. This means that m_p is large and the initial vehicle mass $(m_v)_i$ consists largely of propellant.

Two important measures of engine performance are *specific fuel consumption, sfc*, (for airbreathing engines) and *specific impulse I_s* (for rocket engines); one is the reciprocal of the other. Specific fuel consumption for turbojets, turbofans, and ramjets is expressed as pounds of fuel per hour per pound of thrust, and as pounds of fuel per hour per shaft horsepower for piston engines and turboprop engines. Specific impulse is usually expressed in seconds, a short designation for units of thrust (force) per units of propellant mass flow per second. Both of these parameters are an indication of engine design quality, the higher the I_s the better, and conversely for sfc.

By definition, specific impulse per second is:

$$I_s = \frac{F}{\dot{w}} = \frac{F}{\dot{m}g} \text{ (kg force/kg mass per second)}$$

with $\dot{w}$ being propellant flow rate per second. For a given I_s, the nozzle expansion ratio should be stated such as $1000 \rightarrow 14.7$ (sea-level specific impulse at 1000 psi chamber pressure), or $1000 \rightarrow 0.2$, indicating expansion at high altitude or space flight.

Some of the most significant types of engines are described briefly in the remainder of this section.

Rocket Engines These engines use both a fuel and oxidizing propellant and both are stored within the flying vehicle, making it in-

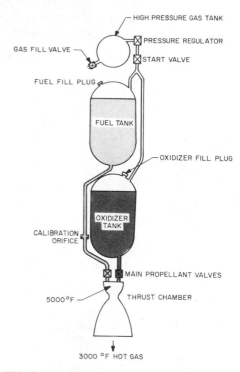

FIG. 3. Simplified schematic diagram of liquid propellant rocket engine with pressurized gas feed system and uncooled thrust chamber (reproduced by permission from McGraw-Hill Encyclopedia of Science and Technology, Vol. II, New York, McGraw-Hill Book Co.)

dependent from its surrounding fluid. Thus a rocket can operate in space, air, or under water. The supersonic nozzle jet exit velocity v_e of a rocket using ideal gas laws can be derived to be

TABLE 2. TYPICAL DATA FOR VARIOUS ROCKET ENGINES

Type	Typical Range of Thrusts (lbs)	Typical Range of Duration	Application
High thrust liquid propellant rocket	1 000 000 to 4 000 000 for each engine with several engines in a cluster	1 to 5 min	Booster and sustainer stages of large missiles and space vehicles
Large solid propellant rocket	100 000 to 3 000 000	1 to 3 min	Long range missiles, space boosters
Prepackaged storable liquid propellant	100 to 100 000	1 to 60 sec	Small missiles, lunar landing and takeoff missiles, lunar landing and takeoff
Jet assisted takeoff (solid propellant)	200 to 10 000	5 to 30 sec	Assist takeoff of airplanes
Space vehicle attitude control	1 to 150	0.01 to 10 sec/cycle; accumulate up to an hour	Control position, angle and orientation of spacecraft

$$v_e = \sqrt{\frac{2gkR}{(k-1)} \frac{T}{M} \left[1 - \left(\frac{p_e}{p_c}\right)^{(k-1)/k}\right]}$$

where

v_e = nozzle exit velocity
g = gravitational constant
k = ratio of specific heats of gas
R = universal gas constant
M = molecular weight of hot gas
T = absolute combustion temperature
p_e = nozzle exit gas pressure
p_c = combustion chamber pressure.

The exhaust velocity (or the specific impulse which is $I_s = v_e/g$) increases as the molecular weight M is decreased or as the combustion temperature T is increased. Because of the pressure ratio effect, there is actually a slight (10 to 20 per cent) increase in specific impulse as the altitude is increased (lower ambient pressure) or as chamber pressure is increased. Values of v_e or I_s calculated for a given propellant and engine from thermochemical and thermodynamic data are usually very close to actual performance (usually within 5 to 10 per cent), because rocket combustion efficiencies are high and nozzle losses are usually low. Schematic diagrams of several liquid and solid propellant systems are shown in Figs. 3 to 6, and some important applications are shown in Table 2.

In *liquid propellant rocket engines* the propellants are fed under pressure from tanks in

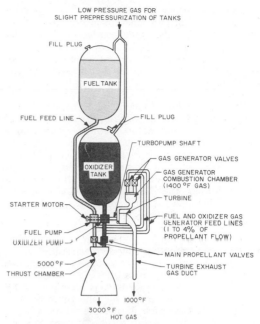

FIG. 4. Simplified schematic diagram of liquid propellant rocket engine with turbopump feed system and regeneratively cooled thrust chamber (reproduced by permission from McGraw-Hill Encyclopedia of Science and Technology, Vol. II, New York, McGraw-Hill Book Co.)

FIG. 5. F-1 rocket engine used in booster stage of advanced Saturn space vehicle. (Courtesy of Rocketdyne, A Division of North American Aviation, Inc.)

the vehicle into a thrust chamber where they are injected, mixed, and burned at high pressures and very high temperatures to form the gaseous reaction products, which in turn, are accelerated in a nozzle and ejected at high velocities. The feed system for transferring the propellants into the thrust chamber includes valves and controls.

The principal components of a *thrust chamber* (Figs. 3 and 4) are the *nozzle*, the *chamber*, and the *injector*. An injector introduces and meters the flow of the liquid propellants and also atomizes and mixes them in the correct proportions in such a manner that they can be readily vaporized and burned. In the combustion chamber, the burning of the liquid propellant takes place at high pressure, usually between 5 and 150 atmospheres.

The *gas pressure feed system* (Fig. 3) offers one of the simplest and most common means of transferring propellants by displacing them with a high-pressure gas which is fed into the tanks under a regulated pressure. In a *turbopump feed system*, the propellant is pressurized by means of pumps driven by one or more turbines (Fig. 4) which derive their power from the expansion of hot gases. A separate gas generator ordinarily produces these gases in the required quantities and at a temperature which will not hurt the turbine buckets (1200 to 2000°F).

Liquid Propellants. A *bipropellant rocket* unit has two separate propellants, a fuel and an oxidizer (such as liquid hydrogen and liquid oxygen), which are not mixed until they come in contact with each other in the combustion chamber. Most liquid propellant rockets have been of this bipropellant type. A *monopropellant* contains oxidizing agent and combustible matter in a single substance. It can be a mixture of compounds, such as hydrogen peroxide with liquid alcohol, or it may be a homogeneous chemical agent, such as hydrazine. Typical values of liquid and solid propellant characteristics are given in Table 3.

Solid Propellants. In a *solid propellant rocket engine* all the propellant is contained within the combustion chamber. The hardware includes, in addition to the combustion chamber nozzle, an igniter and provisions for mounting the rocket (Fig. 6). Solid propellants themselves usually have a plastic, cakelike appearance (specific gravity is approximately 1.6) and burn at high pressure (10 to 150 atmospheres) on their exposed surfaces to form hot exhaust gases which are ejected through the nozzle. The physical mass or body of the propellant is called the *grain*.

Processed (including curing) propellants fall into three general types: (1) double-base; (2) composite; and (3) composite double-base.

A double-base propellant forms a homogeneous cured propellant, usually a nitrocellulose-type of gunpowder dissolved in nitroglycerin, plus minor percentages of additives. Both the major ingredients are explosives, and both contribute to the functions of fuel, oxidizer, and binder.

A composite propellant forms a heterogeneous propellant grain with the oxidizer crystals and a powdered fuel (usually aluminum), held together in a matrix of synthetic rubber (or appropriate plastic) binder, such as polybutadiene. Normally, composite propellants are less hazardous to manufacture and handle than double-base propellants.

Composite double-base propellants are a combination of the double-base and composite propellants, usually comprising a crystalline oxidizer (ammonium perchlorate) and powdered aluminum fuel, held together in a matrix of nitrocellulose-nitroglycerin. The hazards of processing and handling this type of propellant are similar to those of double-base propellants.

Representative formulations for the three basic types of propellants are given in Table 4. In actual practice, each manufacturer of a propellant has a proprietary precise formulation and processing procedure. The exact percentages of ingredients, even for a given propellant, such

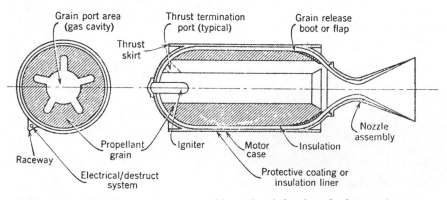

FIG. 6. Typical solid propellant motor with case-bonded grain and other components.

TABLE 3. TYPICAL PERFORMANCE OF SEVERAL CHEMICAL ROCKET PROPELLANTS.

Propellant	Theoretical Specific Impulse (1000 psi → 14.7)	Bulk Specific Gravity	Optimum Mixture Ratio (Oxidizer–Fuel)	Combustion Temperature, F	Molecular Weight of Exhaust Gas lb/mole	Burning Rate in/sec
Cryogenic Liquid						
Oxygen and kerosene	300	1.02	2.6	5800	22	
High Energy Liquid						
Fluorine and hydrogen	412	0.46	8.0	4700	9	
Fluorine and hydrazine	365	1.31	1.3	7300	18	
Oxygen and hydrogen	390	0.29	4.2	4500	8	
Storable Liquid						
Nitrogen tetroxide and hydrazine	292	1.21	1.3	5100	22	
Monopropellant (Liquid)						
Hydrogen peroxide	149	1.44		1365	34	
Hydrazine	245	1.0		1800	14.6	
Solid						
Double base solid	255	1.6		5300	24	.4 to .5
Composite solid	264	1.9		5500	24	.4 to .8
Composite double base	272	1.9		6600	21	.4 to .8

TABLE 4. REPRESENTATIVE SOLID PROPELLANT FORMULATIONS.*

Double Base		Composite		Composite Double Base	
Ingredient	Wt %	Ingredient	Wt %	Ingredient	Wt %
Nitrocellulose	51.5	Ammonium perchlorate	70.0	Ammonium perchlorate	20.4
Nitroglycerin	43.0	Aluminum powder	16.0	Aluminum powder	21.1
Diethylphthalate	3.2	Polybutadiene– acrylic acid– acrylonitrile	11.78	Nitrocellulose	21.9
Ethyl centralite	1.0	Epoxy curative	2.22	Nitroglycerin	29.0
Potassium sulfate	1.2			Triacetin	5.1
Carbon black	<1%			Stabilizers	2.5
Candelilla wax	<1%				

*Furnished by U.S. Air Force Rocket Propulsion Laboratory, Edwards, Calif.

as PBAN, will not only vary from one manufacturer to another, but also often varies from one motor application to the next.

Compared to liquid propellant, the solid propellant requires no pump or pressurization for the fuel tank, and hence is mechanically simpler. The combustion chamber with solid propellant is larger, especially for large rockets, and is frequently operated at higher pressure than for a liquid engine. Solid rockets are simpler, more storable, and are usually more immediately ready for use, but are generally lower in performance when compared to liquid propellant rockets.

In addition to producing a thrust force, solid and liquid propellant engines can be used also for producing auxiliary power and control torques to be applied to the vehicle. This latter is called *thrust vector control* and basically it is usually a mechanical means for altering the direction of the engine's thrust during flight.

Air-breathing Engines As shown by Figs. 7 to 10, there exists a variety of different types of air-breathing propulsion engines. The range of performance values shown in Table 5 for different types of air-breathing engines is representative and does not correspond to data for specific engines. Each engine is optimized for a specific flight operating condition of speed (Mach number) and altitude. For example the turbojet engine listed in Table 5 for a specific fuel consumption of 1 pound of thrust per pound of fuel flow per hour operating at a Mach number of $M = 0.8$, is a different engine from the one operating at $M = 3.0$ with a fuel consumption of 1.5; also it operates at different altitudes.

In an aircraft or missile engine, the atmospheric air is usually first compressed (by a mechanical rotary compressor or by a diffuser); then heated by burning with fuel; sometimes sent through a turbine (which provides the

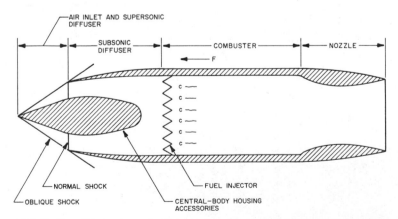

FIG. 7. Simplified schematic diagram of ram jet (reproduced by permission from H. H. Koelle, "Handbook of Astronautical Engineering," New York, McGraw-Hill Book Co., 1961).

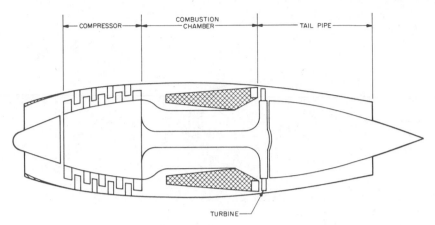

FIG. 8. Simplified schematic diagram of turbojet (reproduced by permission from H. H. Koelle, "Handbook of Astronautical Engineering," New York, McGraw-Hill Book Co., 1961).

rotary power for driving the compressor and accessories, such as electrical generators or hydraulic pumps); and then ejected at high velocities (usually supersonic) through a nozzle. The process in air-breathing engines approaches an ideal thermodynamic cycle (such as the Otto cycle for reciprocating engines or the Brayton cycle for a turbojet) which limits their maximum theoretical efficiency. The inlet duct serves to scoop up the desired air mass flow and to convert some of the kinetic energy of the flow into pressure, thus reducing the velocity. After heating the flow in a combustion chamber, the reverse process occurs adiabatically in the nozzle, where it is desired to attain a maximum exhaust velocity. The efficiency of energy conversion in the inlet duct, nozzle, combustors, compressors, and turbines is a very important factor, and becomes a predominant criterion at supersonic and hypersonic velocities, when compression is achieved usually by a series of oblique shock waves commencing at an inlet spike. To maintain good efficiency and the desired airflow, some diffusers and nozzles incorporate a variable wall contour or cross-section geometry. In general, air-breathing engines have been well developed to a high state of reliability and have given millions of hours of good service.

The available oxygen from the air limits the combustion process. For example, at constant flight speed, the thrust thus decreases with altitude (or oxygen density) and below a combustion pressure of approximately 3 psi, combustion is not easily sustained (flameout limit). Available high-temperature materials will set an upper limit to the maximum combustion temperature at approximately 1700 to 2400°F. At high speeds and high altitude, the ram-compression of the air causes its temperature to rise substantially, so that the amount of energy that can be added by combustion (without damaging turbine materials) is thus limited; also, special cooling provisions are required.

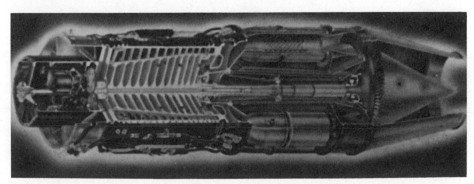

FIG. 9. Photo of J-47 turbojet engine. (Courtesy of General Electric Company.)

TABLE 5. TYPICAL DATA FOR AIR-BREATHING ENGINES.

Engine Type	Flight Speed (mph)	Altitude (feet)	Cruise Specific Fuel Consumption or Specific Thrust	Thrust to Weight or Power to Wt. Ratio	Typical Applications
Piston engine with propeller	0–400 mph	0–35 000	0.37 to 0.52 lb/hp-hr	1.0 to 2.0	Transport aircraft Small airplanes Helicopters Target drones
Turbojet	0–2000	0–100 000	0.8 lb/lb-hr at M = 0.8 1.5 lb/lb-hr at M = 3.0	0.8 to 5.0 (depending on afterburner)	Bomber Fighter Missiles Transport aircraft Target drones
Turbofan	0–600	0–40,000	0.5 lb/lb-hr at M = 0.1 0.65 lb/lb-hr at M = 0.9	3.5 to 5.5	Transport aircraft
Turboprop	0–450	0–35 000	0.8 lb/hp-hr at M = 0.6 0.25 lb/hp-hr at M = 0	2.0	Fast transport aircraft Small airplanes Helicopters
Ramjet	1200 to 4200	0–120 000	1.5 lb/lb-hr at M = 2 2.0 lb/lb-hr at M = 6	up to 20.0	Anti-aircraft missiles

FIG. 10. Sectional photo of CFU56-2 turbofan engine (Courtesy of General Electric Company).

The simplest air-breathing engine is a *ramjet* (Fig. 7). It does not produce static thrust (at zero flight speed, such as during takeoff) and thus needs a rocket engine or some other engine to bring it to its minimum operating speed.

The *piston engine with a propeller* was the very first engine to fly. It is the most economic engine for subsonic flight speed and is now used mostly in small airplanes. The hot gases do work against a piston (not a turbine), which in turn requires a crank mechanism to convert the reciprocating piston motion into shaft rotation. The use of a variable pitch propeller and superchargers for precompression of the inlet air further increases the economy.

The *turbojet* (Figs. 8 and 9) can be designed for a variety of speeds, altitudes and thrust ranges. It often includes special design features such as afterburners and water injection to increase the thrust.

The turbofan (Fig. 10) followed the turbojet in development and utilizes the same design principles. It is widely used in cargo and passenger airplanes. Its advantages over a turbojet include reduced fuel consumption, higher thrust to weight ratio and reduced noise—but it is limited to subsonic flight speeds.

The *fuel* used is usually a narrow-cut petroleum refinery hydrocarbon product, having the approximate formula of $CH_{1.95}$.

Photon Propulsion. Solar sail engines rely on the reflection of photons from the sun (radiation pressure at the distance of the earth from the sun is about 5×10^{-7} kg/m^2). Although this force is limited by being directed only "away" from the sun, solar "sailing" can provide low thrusts, and attitude control, turning a spaceship completely around in a few hours. No working fluid is carried in the vehicle.

For the vehicle to carry its own light source, a photon rocket engine of adequate thrust would necessitate energies and techniques far beyond present capabilities.

G. P. SUTTON
D. M. ROSS

References

1. Sutton, G. P., and Ross, D. M., "Rocket Propulsion Elements," Fourth edition, New York, J. Wiley & Sons, 1976.
2. Morgan, H. E., "Turbojet Fundamentals," Second edition, New York, McGraw-Hill Book Co., 1958.
3. "Janes' All the World's Aircraft, 1981–82," New York, Janes' Publishing Inc., 1982.
4. Constant, E. W., "Origins of the Turbojet Revolution," Baltimore, Johns Hopkins Univ. Press, 1980.

Cross-references: AERODYNAMICS, ASTRONAUTICS, CHEMISTRY, DYNAMICS, ELECTRIC PROPULSION, FLUID DYNAMICS, IMPULSE AND MOMENTUM, MECHANICS, STATICS, THERMODYNAMICS.

FLUID DYNAMICS

Fluid dynamics is study of the motion of matter in the gas, liquid, plastic, or plasma state. When restricted to flow of incompressible (i.e., constant density) fluids, it is called *hydrodynamics*; when dealing with electrically conducting fluids with magnetic fields present, it is called *magneto-fluid-dynamics*; when dealing with practical problems of air flow past airplane wings, through ventilating equipment, etc., it is called *aerodynamics*; when dealing with nonlinear materials such as wax and polymers, it is called *rheology*.

Fundamentally, two approaches to describing the motion of fluids are taken. The first, and most common, is continuum or field dynamics, the physical elements of which were stated in modern form by Stevin (about 1590), Torricelli and Pascal (1647), D. Bernoulli (1738), Euler (1755), and Stokes (1845). Clarification of the complex mathematical nature of the correct physical equations describing the deformation and flow of matter has involved and continues to involve the most astute mathematicians.

The second approach is kinetic theory and nonequilibrium statistical mechanics. In this approach, matter is not thought of as a continuum or field, but as consisting of discrete molecules moving in empty space (gas) or bound to one another by forces (condensed matter—liquids and solids). In either case, the molecules are in continuous motion, either colliding with other molecules or vibrating about a "home position." This approach, discussed even in antiquity, was formulated in more modern form by D. Bernoulli (1738), Clausius (1857), Maxwell and Boltzmann (1860s) and Gibbs (1902), and continues to be developed and applied (see KINETIC THEORY).

Continuum Dynamics In continuum dynamics, fluid properties—namely, velocity components u_1, u_2, u_3, density ρ, pressure p (more generally stress), temperature T, viscosity μ, conductivity, dielectric constant, electric field, electric current density, etc.—are assumed to be meaningful functions of three spatial variables x_1, x_2, x_3, and time t. The internal forces (pressure, stress), the velocity components, and the density are usually unknowns to be calculated, while weight, dielectric constant and some other properties are presumed given.

Measurement of these physical quantities in a moving field is important both for practical reasons and as the means for developing the working concepts of fluid dynamics. Examples of methods for measuring fluid velocity components u_i are given later in this article. For methods of measuring pressure, see FLUID STATICS.

When electromagnetic body forces are present, the situation is quite complicated, and we will in this article consider only the simpler situations where the only body force is weight. For information on situations in which electromagnetic forces exist, see MAGNETO-FLUID-MECHANICS.

In addition to internal forces (pressure, stress) and external body forces (for purposes of this article, weight), additional "fictitious" body forces appear in those cases where an accelerated frame of reference is used to describe the flow situation. Particularly common are *Coriolis* and *centrifugal* forces which arise in a frame of reference rotating relative to an inertial frame. (See CORIOLIS EFFECT, DYNAMICS.) A frame fixed to the earth has this property, but the fictitious forces can usually be ignored; the centrifugal force is combined with the gravitational force to produce the ordinary measured weight (see FLUID STATICS), and the Coriolis force is so small that it usually can be ignored. The Coriolis force is, however, extremely important in fluid dynamics of the atmosphere, which involves motion over global scales and is the basis for long range weather prediction.

A fluid is, by definition, a substance in which there are no shear stresses when *at rest*. All substances *in motion* have shear stresses acting in addition to the normal stresses. In motion, the normal stresses in a fluid are not equal in all directions, but the average of the normal stresses is employed and called *pressure* as an extension of the concept of pressure which arises in fluid statics. Many fluids, e.g., tar, waxes, oils, honey, bread dough, and many synthetic polymers, have both normal stress and shear stress components which depend on the recent deformation history as well as on the current rates of deformation (see RHEOLOGY).

The relations between the stress components and the current strain components and rate of strain components are extremely important in formulating the equations of fluid dynamics, elasticity, and rheology. The relations which exist for a particular substance determine, in fact, whether the flow of the substance is to be described within the field of fluid dynamics, elasticity, or rheology. In fluid dynamics, the stress-deformation relation is assumed to be a simple linear dependence of each stress component on the current rates-of-strain at the same point. Because the substance is also considered to be isotropic (no crystal axes), these linear relations can be shown mathematically to involve only two proportionality constants. The proportionality constants are called the *first viscosity coefficient* and the *second viscosity coefficient*; except in very unusual situations, the second viscosity coefficient is zero, so that it is common to consider the first viscosity coefficient as *the* viscosity μ. A substance which adequately follows this simple linear stress-deformation relation is called a *Newtonian substance* (see VISCOSITY).

With the linear relation between stress components and rate-of-strain components, the stress components can be expressed in terms of the pressure (i.e., the average of the normal stresses), the viscosity, and the spatial derivatives of the velocity components. Then the basic dynamical equation of continuous media—which equates the sum of all forces on an element to the element's mass times its acceleration—takes the following form, known as the *Navier-Stokes equations*:

$$\frac{\partial u_i}{\partial t} + u_j \frac{\partial u_i}{\partial x_j} = -\frac{1}{\rho} \frac{\partial p}{\partial x_i} + g_i$$

$$+ \frac{\mu}{\rho} \left(\frac{\partial^2 u_i}{\partial x_j \partial x_j} + \frac{1}{3} \frac{\partial}{\partial x_i} \frac{\partial u_j}{\partial x_j} \right). \quad (1)$$

The Navier-Stokes equations are components of a vector equation. The index notation used for subscripts is that an index i or j may be 1, 2, or 3, representing components of vectors along axes 1, 2, or 3, respectively, and in any term where the same index occurs twice, a sum of terms with that index taking successively the values 1, 2, and 3 is implied. μ is assumed to be a known and constant value characteristic of the fluid, and g_i are components of the known weight per unit mass, nominally 9.8 N kg^{-1} downward.

The left-hand side of Eq. (1) represents the acceleration component of the parcel of fluid at the point x_1, x_2, x_3 at time t, and the right-hand side is the force per unit volume divided by the mass per unit volume.

Two additional equations are needed to determine the five unknowns u_1, u_2, u_3, p, ρ, since the three equations of motion (with subscript i taking on the values 1, 2, and 3) are insufficient.

One of these needed equations is the so-called *continuity equation*,

$$\frac{\partial \rho}{\partial t} + \frac{\partial}{\partial x_j} (\rho \mu_j) = 0. \quad (2)$$

This equation, employing the same index notation, and based on the concept that ρ represents the quantity of matter as well as the inertial mass of the fluid, essentially states that matter flowing into a fixed volume is equal to the time rate of increase of the matter in that volume. It is a kinematical statement.

The fifth equation is, in various situations, some form of thermodynamic equation of state,

an example being the *adiabatic equation*,

$$\frac{p}{p_0} = \left(\frac{\rho}{\rho_0} \right)^\gamma \quad (3)$$

applying to an ideal gas. (γ, the ratio of heat capacities of the gas, is equal to 1.4 for air.) A variety of assumptions, usually involving the first and second laws of thermodynamics and thermodynamic equations of state of the substance, are needed in various situations.

Two further special assumptions are made in formulating situations dealt with in a large part of the fluid dynamics literature. One is the assumption that the viscosity coefficient μ is zero. The Navier-Stokes equations, then, with $\mu = 0$, are called the *Euler equations*. These equations are not really applicable to any fluid substance, but there are types of flows where important characteristics are illustrated. Such flows are said to be *ideal fluid* or *inviscid fluid* flows. The second special assumption is that the fluid is incompressible, i.e., has constant density.

These partial differential equations, even the Euler equations, are nonlinear and have no general solutions, even for the most restrictive boundary conditions. Particular solutions are carried out for various idealized flows.[1,2] Examples of particular solutions for selected geometrical boundaries are described later in this article.

Kinetic Theory and Nonequilibrium Statistical Mechanics In this approach, fluid properties are associated with averages of properties of microscopic entities. Density, for example is the average number of molecules per unit volume times the mass per molecule. While much of molecular theory in fluid dynamics aims to interpret processes already adequately described by the continuum approach, additional properties and processes are presented. The distribution of molecular velocities (i.e., how many molecules have any particular velocity), time dependent adjustments of internal molecular motions, and momentum and energy transfer processes at boundaries are examples.[3] See KINETIC THEORY and STATISTICAL MECHANICS.

Examples of Flows In Fig. 1, the special flow called Couette flow is indicated schematically. The flow is between parallel plates, lower plate at $y = 0$ at rest, upper plate at y_B

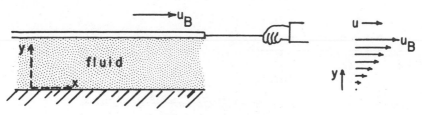

FIG. 1. Couette flow.

moving with constant speed u_B in the x direction. Stress throughout the fluid is constant, given by $P_{xy} = \mu(du/dy) = \mu(u_B/y_B)$. This is pure shear flow and experimentally is often considered to define and measure the viscosity coefficient μ assumed constant for the homogeneous fluid. The velocity profile appearing at the right in the figure shows by velocity arrows of different length at the various positions y how the velocity varies with position. Steady flow (no dependence of any quantity on time), constant pressure, constant density and laminar flow are additional assumptions for Couette flow. The flow is realized experimentally by confining the fluid in the narrow annulus between rotating concentric cylinders of nearly equal radius; the cylinders rotate at different speeds.

In Fig. 2, the special flow is in a pipe of uniform cross section, pressure is assumed to be constant across each cross section but to vary linearly with distance x along axis of pipe so $dp/dx = (p_1 - p_2)/L$. Pistons driving the flow are assumed to be infinitely far away, so that the flow velocity, parallel to pipe axis, has the same dependence upon y and z for all x. The velocity profile is parabolic in both the two-dimensional case (infinite parallel plates) and in the circular cross-section case. Mean flow velocity u_m and viscosity coefficient μ are assumed constant; the flow is assumed steady and laminar. For a circular cross-section pipe of radius a, at any distance r from the center, $u = 2u_m(1 - r^2/a^2)$, and the volume passing a cross section per second is $Q = \pi a^2 u_m = \pi a^4 (p_1 - p_2)/8\mu L$. Since these formulas do not apply near pipe entrances, caution in applying them to pipes of finite length is necessary even when the flow is steady and laminar. (See later discussion of turbulent and laminar flows.)

Other examples of idealized solutions are one-dimensional flow of an ideal gas through a normal shock wave, flow of an ideal gas without viscosity through a pipe of slowly changing cross section (wind tunnel), and one-dimensional finite waves in an ideal gas. Many other solutions involve making whatever approximations and assumptions are necessary to obtain descriptions of observed flows.

When motion of the fluid consists of only small fluctuations about a state of near-rest, the continuum equations are linearized by neglecting nonlinear terms and become the equations of acoustics. See ACOUSTICS. A large variety of fluid motions are described as sound waves; when the small-motion or acoustic description can be used the *principle of superposition* is valid. This powerful principle allows addition of simple simultaneous motions to represent a more complex motion, such as the sound reaching the audience from the instruments of a symphony orchestra. The superposition principle does not apply to large-scale (nonacoustical) motions, and the subject fluid dynamics (as distinct from acoustics) treats nonlinear flows, i.e., those which cannot be described as superpositions of other flows. The description of small motions in a small region of even a nonlinear flow is useful; at each place in the flow there is a "local sound speed."

Since sound waves travel with a sound speed relative to the fluid, waves moving in a moving fluid can sometimes be carried off in a direction opposite to the direction of sound travel. The flow where such a thing happens is called *supersonic*; the flow speed is greater than the sound speed at the spot where the flow is supersonic. Supersonic flow occurs around high-speed vehicles and missles, and in pipes when high pressure gas escapes through a constriction such as a partially open valve into a region of sufficiently lower pressure.

A steady supersonic flow always must pass through a *shock front* to slow down to subsonic flow again.[1,4] A shock front is a surface of discontinuity separating fluid of high pressure, high density, and high temperature from lower pressure, density, and temperature fluid. If the shock front is in motion, it is called a *shock wave*. The component of fluid velocity perpendicular to the shock front is also discontinuous, while the other components are continuous. See SHOCK WAVES.

The continuum description of flow fails to describe nearly all actual flows because actual flows when looked at carefully are *turbulent*. Turbulent flows have violent and erratic fluctuations of velocity and pressure which are not associated with any corresponding fluctuations of the boundaries containing or driving the fluid. Turbulence is generally considered to be the manifestation of the nonlinear nature of the fundamental equations. Under certain conditions as mentioned earlier in describing Couette and Poiseuille flows, nonturbulent or *laminar* flow exists. A common example is cigarette smoke rising from a cigarette held at rest; near

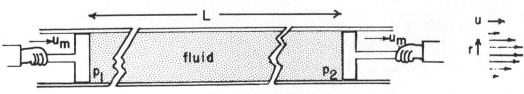

FIG. 2. Poiseuille flow.

the cigarette, the stream is smooth and straight, or laminar, and further up the flow breaks into turbulence.

Reynolds showed that Poiseuille flow in a pipe occurs when $\rho u_m a/\mu$ is smaller than 2000. The combination of variables is dimensionless and is called the *pipe Reynolds number*. Blood flow in capillaries is laminar, but water flow in household pipes is turbulent unless the flow is about that allowed by a leaky faucet or less. Other types of flow have Reynolds numbers characterizing transition from laminar to turbulent; for example a sphere falling in a fluid of viscosity μ obeys *Stokes' Law*

$$mg = 6\pi a\mu u$$

where u is the constant speed of fall, m is the sphere mass, a its radius, and g the weight per unit mass, but the law is obeyed only if the *sphere Reynolds number* $\rho u a/\mu$ is smaller than about 1.

Because of turbulence and viscosity, the very simple and useful *Bernoulli formula* is not valid; it can be derived as applying to a constant-density fluid with zero viscosity in laminar flow. However, under certain conditions, the formula applies approximately even when the flow is turbulent, predicting properties within 5 to 20 per cent of the observed values. The Bernoulli formula states

$$p + \rho gh + \tfrac{1}{2}\rho u^2 = \text{same constant in all places in the fluid}$$

where p is the fluid pressure, ρ is the fluid mass density (which must be treated as constant), u is the fluid speed, h is the vertical height above some convenient reference level, and g is the weight per unit mass.

When combined with the equation of continuity, the Bernoulli formula gives a simple description of the Venturi, used in the automobile carburetor (see Fig. 3). Continuity states $u_1 A_1 = u_2 A_2 = $ volume crossing any cross section of the pipe per second. $u_1/u_2 = A_2/A_1$ is small so u_1 is smaller than u_2. The Bernoulli formula states $p_1 + \tfrac{1}{2}\rho u_1^2 = p_2 + \tfrac{1}{2}\rho u_2^2$, and since u_1 is smaller than u_2, p_1 is larger than p_2. The atmospheric pressure p_1 pushes the liquid up into the lower pressure region p_2.

Another common situation described by the Bernoulli formula is the discharge of (constant

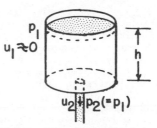

FIG. 4. Discharge through a small hole in a bucket.

density) fluid from a small hole. For the cylindrical bucket of water in Fig. 4, equate the sum of the quantities in the Bernoulli formula at the top surface to the sum at the hole: $p_1 + \rho gh + 0 = p_2 + 0 + \tfrac{1}{2}\rho u_2^2$. The pressure at both top and bottom is atmospheric pressure p_1; the speed at the top is approximately zero because the hole is considered to be very small. The predicted speed of the emerging water is therefore $(2gh)^{1/2}$ regardless of the size of the hole (as long as the hole is small).

A valuable instrument in the form of a probe for observing fluid speed is the Pitot tube; its operation is described by the Bernoulli formula. A glass or metal tube with an open end points into the flow, and the pressure difference Δp between the stagnant fluid in the tube and the moving fluid allows calculation of the fluid speed at the place where the tube tip is inserted by $u = (2\,\Delta p/\rho)^{1/2}$ where ρ is the fluid mass density. When observing air speed, the pressure difference Δp is easily measured by connecting the open ended tube via rubber hose to a glass U-tube water manometer. Errors as much as 50 per cent may easily occur in various practical situations, but order-of-magnitude measurements at least are usually possible.

The Pitot tube is the only *simple* instrument for measuring fluid velocity at a point. It is useful only in certain steady flows (no dependence of any quantity on time) and the degree of its accuracy is hard to determine in a given situation. *Absolute velocity measurement* is possible only by observing the velocity of tracer particles by optical methods.[5] Many other velocity measuring instruments are *calibrated*, either by comparison to tracer velocity measurements or by towing the instrument at a known speed through the fluid at rest. A tracer must be small, particularly if its density differs from that of the fluid, so that it moves exactly as the fluid surrounding it is moving. Usable tracers are liquid droplets or solid particulates in gases, and gas bubbles or solid particulates in liquids. Tracer dimensions are usually 1 to 100 micrometers, the smaller ones being required when the velocity being measured is changing spatially or temporally. Observation of the tracer is carried out either by photography with repetitive flash illumination (called *chronophotography*) or by Doppler methods.

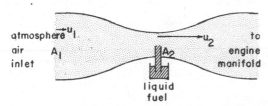

FIG. 3. Pipe flow with a constriction; carburetor employing Venturi.

The *laser Doppler velocimeter* has had many applications in recent years.[5] In this instrument light is scattered by tracers from two coherent laser beams which cross one another in a chosen small volume inside the moving fluid; the fluid carries micron-size tracer particles and the scattered light from them in the intersection volume is gathered by a lens on to a photomultiplier. If the direction of the velocity of the tracer is known relative to the bisector of the directions of the laser beams, the measured frequency of oscillation of the photomultiplier current gives the speed v of the tracer by means of the formula:

$$v_b = v \cos \theta = \frac{\nu_D \lambda_0}{2 \sin (\alpha/2)}$$

where θ is the angle between the bisector of the incoming laser beams and the tracer velocity $\mathbf{v}$, λ_0 is the wavelength of the light in the fluid medium, α is the angle between the incoming laser beams, and ν_D is the measured frequency of the photomultiplier signal, called the *Doppler frequency*. The quantity determined is the magnitude of one component of the velocity; the sense (sign) of the component is not measured but must be learned from other information. The electronic equipment to measure and display or record the Doppler frequency is rather elaborate, but available commercially. Fluid speed as a function of position is measured by moving the optical system mechanically about in the fluid field. The intersection volume is ordinarily small—much less than a cubic millimeter.

Measurement of fluid velocity at small scales and in small times is the most promising method for study of turbulence. The laser Doppler velocimeter can measure the speeds of successive tracer particles passing through a small measuring volume in times as close together as 1 ms, but the measurement is not very accurate. Another instrument which must be calibrated, the *hot-wire anemometer*, is used to assemble more accurate data collected in an air or water stream. The hot-wire anemometer is a probe consisting of two steel prongs supporting at their ends a tungsten wire of diameter 2–5 μm and length 1–3 mm carrying an electric current which would heat it white-hot except for the cooling produced by the fluid passing over it. Its resistance is related to its temperature, which in turn is related to the fluid velocity component perpendicular to the wire. Thus, after it is calibrated, its resistance as a function of time can be interpreted as a time record of the stream speed in which it is immersed. At larger scales than millimeters, data on the successive positions of injected dye spots or hydrogen bubbles in water are collected by cinematography using television recorders.

Measurement of other fluid variables—density, composition, and temperature—with precision, at small scales, and in rapid succession, is carried out by optical means. For example, intensity of Raman components of scattered laser light can be analyzed to determine temperature, density and even species concentration. Pressure at walls of channels or at surfaces of inserted objects is measured by "wall taps" or by miniature diaphragm or piezoelectric gages. A "wall tap" is a small hole at the point on the surface where the pressure is to be measured which connects with a pressure gage by a small channel filled with a static fluid.[5]

RAYMOND J. EMRICH

References

1. Landau, L. D., and Lifshitz, E. M., "Fluid Mechanics," London, Pergamon Press, 1959.
2. Lamb, H., "Hydrodynamics," First American Edition, New York, Dover Publications, 1945.
3. Hirschfelder, J., Curtiss, C. F., and Bird, R. B., "Molecular Theory of Gases and Liquids," New York, John Wiley & Sons, 1954.
4. Gaydon, A. G., and Hurle, I. R., "The Shock Tube in High-Temperature Chemical Physics," New York, Van Nostrand Reinhold, 1963.
5. Emrich, R. J. (Ed.), "Fluid Dynamics," Vol. 18 of series "Methods of Experimental Physics," New York, Academic Press, 1981.

Cross-references: COMPRESSIBILITY, GAS; FLUID STATICS; GAS LAWS; KINETIC THEORY; LIQUID STATE; MAGNETO-FLUID-MECHANICS; STATISTICAL MECHANICS.

FLUID STATICS

The definition of a fluid—in distinction to a solid—is a substance in which there are no shear stresses when it is at rest. A fluid is an infinitely slippery material. In reality, a substance is considered a fluid if shear stresses in it vanish so rapidly when motion ceases that, for times under consideration, it may be considered shear-free. Substances like tar and glass require very long times for shear stresses to relax, in comparison with fluids like water and gasoline. The sense that everyone has of the difference between a fluid and a solid is more clearly felt than is the nature of internal forces (stress) between the parts of the substance. Shear stresses are the internal forces opposing *sliding* of one part of a substance past the neighboring parts. All real fluids *in motion* do have shear stresses (see FLUID DYNAMICS), and solids also have shear stresses which help maintain their solid shape when weight and other forces act on them (see MECHANICAL PROPERTIES OF SOLIDS).

Historically the complexity of internal forces introduced great confusion (from our modern viewpoint) into the fundamental principles of mechanics. Principles applying to fluids at rest were clarified by Torricelli and by Pascal whose experiments in 1647 settled the question of the existence of an absolute vacuum and of weight of the air. Sir Isaac Newton devoted over one quarter of the famous *Principia* to analysis of fluids (Book II, 1713); his principles are more subtle in application to fluids, and Newton's contributions to fluid mechanics are seldom quoted today. Probably Euler was the first to give, in 1755, a clear idea of the complex relation of the internal forces (stress and body forces) and the motion of fluids and solids.

The concept of pressure, and methods of measuring pressure of gases, soon led to the discovery of Boyle's law (1662), and eventually to the clarification of the concept of temperature through its measurement by the gas thermometer.

Meaning of Pressure If we consider a surface within a substance we expect that material on one side will be pulling material on the other side with a force having components both normal and tangential to that surface. In a static fluid, however, the tangential component must be zero; otherwise the material would have a shear stress and would not be called a fluid. Pascal realized that at a given place in the fluid, if a surface with a different orientation is considered, the force per unit area on it is again perpendicular to that surface and has the same magnitude as the force per unit area on the original surface. (This is not true for a solid except for very special cases of loading.) The magnitude of the force per unit area for all these orientations at the same point is called the *pressure* at that point. The strange property of a force per unit area, which is a vector, having the same value at a point for all orientations seems to give it the property of a scalar. Actually, since the property of a tensor is to relate one vector to another vector, the stress tensor for a static fluid is expressed by the negative of the pressure times the unit tensor at all points. This expresses the force per unit area as perpendicular to whatever surface is considered at the point. (In general in a solid or in a moving fluid, the force per unit area is related to the vector giving the orientation of the surface by the stress tensor. The vectors have different directions, and all six components of the stress tensor have their own dependence on spatial position.) In a static fluid, only the magnitude of the force per unit area varies from point to point, and the pressure gradient ∇p can be shown to be equal to a force per unit volume in the direction in which the pressure decreases most rapidly.

Stress components are positive when the neighboring material is *pulling*, and negative when *pushing*: tension (+) and compression (−). Pressure is, however, treated as positive when pushing, since in gases tension never occurs and in liquids tension seldom occurs.

Body Forces and the Law of Fluid Statics Stresses on a sample of matter are not the only

forces acting—weight and electromagnetic forces can act also. Both these are forces per unit volume and are called *body forces*, in distinction to stresses which are *surface forces* or *contact forces*.

In a fluid the only forces acting (see, however, "surface tension," described below) are weight, electromagnetic forces, and pressure. For a volume element at rest the sum of these must be zero:

$$\rho\mathbf{g} + \mathbf{J} \times \mathbf{B} + \rho_{el}\mathbf{E} - \nabla p = 0. \qquad (1)$$

Here ρ is density (mass per unit volume), $\mathbf{g}$ is weight per unit mass, $\mathbf{J}$ is electric current per unit area perpendicular to $\mathbf{J}$, $\mathbf{B}$ is magnetic field, ρ_{el} is electric charge per unit volume, and $\mathbf{E}$ is electric field. In a dielectric liquid, ρ_{el} can include polarization charge and, in a magnetic liquid, $\mathbf{J}$ may include amperean currents representing magnetic polarization. For more information on the nature of the electromagnetic force, see MAGNETO-FLUID-MECHANICS.

Eq. (1) is the basic equation of fluid statics; it is valid at all interior points.

In all the foregoing, it is assumed that "rest" is reckoned relative to an inertial frame of reference. In a frame of reference translating or rotating relative to an inertial frame, additional "fictitious" body forces must be added to Eq. (1) (see CORIOLIS EFFECT, DYNAMICS).

An additional force called *surface tension* is evident at interfaces between liquids which do not mix (e.g., oil and water), between liquids and vapors, and between liquids and solids (see SURFACE TENSION). This force plays the role of a boundary condition that must be satisfied when Eq. (1) is integrated over an interface in a fluid statics problem. Across a curved interface, with no electric or magnetic field present, the following relation between the surface tension S, the radii of curvature R_1 and R_2, and the pressure jump $p_1 - p_2$ over the surface exists at each point on the surface:

$$p_1 - p_2 = S\left(\frac{1}{R_1} + \frac{1}{R_2}\right). \qquad (2)$$

The surface tension S has units of force per unit length in the surface, and is a joint property of the two fluids. It depends markedly on temperature and, for a liquid and its vapor, becomes zero at the critical temperature of the substance.

The surface tension still exists in a case of a plane interface (it is a force which will do work in a quasistatic deformation of the fluids), but no pressure difference exists across the plane interface. This follows from the nature of pressure as *normal* force per unit area. At a dielectric boundary, with electric or magnetic fields present, an additional contribution to the pressure discontinuity given by Eq. (2) exists. At a liquid-vapor boundary, the additional contribution is known as *electrostriction*.[1]

Equation (1) is not only the fundamental expression of the meaning of pressure; it is also the basis for the measurement of pressure. Before describing methods of pressure measurement, however, we will look at the meaning of pressure from a molecular viewpoint.

Molecular Models of Pressure[2] Pressure is nearly always used as one of the thermodynamic variables in equations of state. Our extensive knowledge of properties of substances, as compiled in thermodynamic tables, provides a wealth of data for testing molecular models of liquids and gases, as well as practical information for engineering design. The model employed in the kinetic theory of gases has been so successful that there is a tendency on the part of some to accept this molecular model's view of pressure as "real" and universally applicable. It is however restricted to gases, and models of liquids are less successful and their view of pressure is somewhat more complicated.

The kinetic theory of gases, rooted in ideas going back at least as far as Daniel Bernoulli (1738), pictures matter in gaseous form as consisting of isolated molecules flying about in a force-free vacuum and making occasional collisions with one another and with solid containing walls (see KINETIC THEORY). At a hypothetical small surface within the gas, there is nothing most of the time, but occasionally a molecule flies through the surface unimpeded. The model assumes such events to be random and uncorrelated, but to have a distribution of directions and times between occurrences. The model further assumes that the number of molecules flying through a surface per unit time and in one sense is proportional to the area considered, and that the number per unit area, per unit time, is the same for all orientations at any one position. This is the case, however, only for fluid statics—not fluid dynamics—that is, there must be no net transfer of mass through the area (no mean motion.)

If the position of the surface is at or near and parallel to one of the solid boundaries, from which the molecules are considered to rebound, the equal flux of molecules into the surface and back out is associated with an average force per unit area applied to the wall. This average force per unit area at the wall, which is equal to twice the normal component of momentum flux through the surface in one sense, is then considered to be the pressure at the wall, and the flux through a surface not at a wall for which there is likewise an equal flux in the other sense, is interpreted as the pressure within the gas. In this way, a meaning is given to "normal force per unit area" at any point within the gas, and if the density or velocity distribution of molecules varies from place to place, twice the flux of normal component of momentum in one sense through a hypothetical surface, namely the pressure, will vary from place to place as well. Since molecules actually have some size in the model, those passing through a hypothetical surface at a point may collide with other mol-

ecules before reaching a distant wall, and if the molecules with which they collide have in a certain measure the density and velocity distribution characteristic of the point, it is thought that the reverse flux of molecules may also have the local characteristics. (Again it is necessary, in making this statement, to limit ourselves to cases of no net flow of mass through the area.)

This justifies, to some extent, the application of an equation of state locally within a fluid with gradients of density, temperature, and pressure. It is apparent that only within volume elements larger than about one mean free path cubed can an equation of state have meaning in this model.

Liquid state models suppose the molecules to be in close contact with one another, as in a solid, but over a range of several molecular diameters the forces between molecules are such that long range order is not maintained as in a crystal. The picture of pressure in this model is quite similar to the picture of stress in a solid, and the absence of shear stresses in liquids at rest is actually only a qualified one in that the time for a shear deformation to relax is shorter than some time considered to be of interest. The substance is called a *glass* if the relaxation time is longer. (See LIQUID STATE.)

In statistical models of a monatomic substance such as argon, the pressure is defined[3] as a sum of two terms, one a *kinetic term* equivalent to the momentum flux of molecules through a hypothetical area in empty space as described above for the kinetic theory of gases, and a *collisional transfer of momentum term* representing the ever-present forces between adjacent pairs of molecules. The microscopic pressure tensor t_{ij} (i.e., the negative of the stress tensor) for isotropic molecules such as argon can be expressed as

$$t_{ij} = \sum_{\alpha} \left\{ \frac{p_{\alpha i} p_{\alpha j}}{m} + \frac{1}{2} \sum_{\beta} (q_{\beta j} - q_{\alpha j}) \frac{\partial U_{\alpha\beta}}{\partial q_{\alpha i}} \right\} \delta(\mathbf{q}_{\alpha} - \mathbf{x})$$

where the subscripts i and j take on the values 1, 2, 3 representing coordinate axes, and the subscripts α and β are numbers characterizing each of the molecules in the system so that the sums are over all the molecules and all pairs of molecules. The momentum components of the molecules are denoted by $p_{\alpha i}$, the mass of each by m, the separations of two molecules by $(q_{\beta j} - q_{\alpha j})$, and the forces between them by $(\partial U_{\alpha\beta}/\partial q_{\alpha i})$. The delta function $\delta(\mathbf{q}_{\alpha} - \mathbf{x})$ restricts the sum to only those molecules lying within a volume cell denoted by position $\mathbf{x}$ in space. The macroscopic pressure tensor is obtained when t_{ij} is averaged over the equilibrium statistical distribution of all coordinates $\mathbf{q}_\alpha$ and momenta $\mathbf{p}_\alpha$.

The above pressure tensor expression implies that the thermodynamic and mechanical variable pressure has the same meaning in a vapor and in a liquid; the first, kinetic term predominates in a vapor, and the second, collisional transfer of momentum term predominates in a condensed liquid, so that the expression provides a continuous pressure across a plane interface (when there is no net transfer of matter and static equilibrium prevails). Because of mathematical complications associated with the statistical mechanics of many body systems, the consequences of the model have not yet been fully worked out. The picture of pressure in the model is however clear.

Measurement of Pressure Two basic instruments are used to measure pressure, and all other pressure gages are calibrated by these.

U-Tube Manometer. Two columns of liquid partly filling a piece of glass tubing bent into a U-shape with hoses connecting to two reservoirs in which the pressure is to be measured are seen frequently in laboratories. The liquid typically has a density 10^3 times the density of the gas which is connecting the manometer to the reservoirs. The manometer equation

$$p_2 - p_1 = \rho g(z_1 - z_2) = \rho gh \qquad (3)$$

gives the pressure differences $p_2 - p_1$ between the reservoirs, where ρ is the mass density of the liquid in the manometer, g is the weight per unit mass (nominally 9.8 N kg^{-1}), and z_1 and z_2 are the vertical coordinates of the respective surfaces between liquid and gas on the two sides of the U-tube. Eq. (3) is derived by integration of Eq. (1) along a path through the liquid, under the assumptions that ρ is constant (homogeneous liquid) and that the liquid-gas surfaces are either plane or have equal curvatures.

If absolute pressure is to be measured in one reservoir, the pressure must be maintained at zero in the other, usually by a vacuum pump. When used to measure the ambient pressure in the atmosphere, the arrangement is called a *barometer*. When the pressure in some reservoir is measured *relative to the ambient atmosphere*, one speaks of the pressure difference as the "gage pressure" of the reservoir.

For accurate measurement, the liquid must not evaporate; a "low vapor pressure" liquid such as mercury or silicone oil is commonly used.

If the vertical height of either reservoir is appreciably different from the corresponding liquid-gas surface in the manometer, a correction for the variation in pressure with height in the gas must be made. An equation similar to Eq. (3) is used, but ρ is then the gas density.

A more likely correction will be necessary due to different diameters of the glass tubing on the two sides and consequent differences in the curvatures of the meniscuses. This latter correction is minimized by using very large diameter tubing. Tubing of 1 cm radius is needed when the liquid is mercury, and accurate measurement of h by interferometry is used.

Use of the manometer is so common that pressure units stated as h and the name of the liquid employed in the manometer have become common. Table 1 lists some of the bewildering variety of pressure units found in the literature. The value of each unit is given in terms of the SI unit Pa. The table permits pressure units to be converted among each other, as well as to the standard SI unit.

Piston and Cylinder Gage. The other absolute pressure measuring instrument uses a tight-fitting piston in a precision bore cylinder. It is the basic standard for measurements above about 100 kPa and even within the range of usefulness of the U-tube manometer it is used to calibrate other gages. Also called a *deadweight gage* and a *gage tester*, it is seldom seen outside a standards laboratory because its operation requires skill, care, patience, and conditions of cleanliness.

The vertical piston supports loads of known weights, and a liquid (usually oil) or a gas (nitrogen or air) in the cylinder below the piston,

TABLE 1. PRESSURE UNITS.*

Name of Unit	Value	Name of Unit	Value
1 kgf/cm^2	98.07 kPa	1 in. Hg	3.38 kPa
1 dyne/cm^2	0.1 Pa	1 in. H$_2$O	249 Pa
1 bar	100 kPa	1 mm Hg (Torr)	133.3 Pa
1 atm	101.3 kPa	1 mm H$_2$O	9.81 Pa
0 acoustic decibel (reference level)**	10 Pa (transducers) or 20 μPa (hearing)	1 lb/ft^2	47.9 Pa
		1 lb/in^2 (psi)	6.895 kPa

*The SI unit of pressure is the *pascal* (symbol Pa); 1 Pa = 1 N m^{-2}.

**Sound pressure level, in decibels, is 20 times the logarithm to the base 10 of the ratio of the sound pressure to the reference pressure. The reference pressure should be explicitly stated; if it is not, the levels noted in the table are often used. Also, unless otherwise explicitly stated, it is understood that the sound pressure is the *rms* pressure *change*.

communicating with a gage to be calibrated, is compressed by a hand pump until the piston and loads rise a short distance. As the liquid or gas slowly leaks out through the small piston and cylinder clearance, the pressure is maintained. The *effective area* of the piston and cylinder has been determined in principle by a metrological microscope in a standards bureau. Piston and cylinder gages are obtainable from commercial manufacturers, who supply the effective area by comparing gages with gages built and maintained by national standards bureaus.

With care, pressures can be measured with errors as small as 0.01% in ranges from 100 Pa to 100 MPa (1 millibar to 1000 bars).

Other Gages. The common gages with a pointer indicating on a circular dial contain a metal chamber which deforms with the applied pressure. By combining metallurgical art with mechanical art to translate the motion of the deforming part into rotation of a needle on a dial without backlash, manufacturers have been able to supply gages with a wide variety of operating ranges, compensated for temperature change, and retaining precision on the order of 0.1% of the range.

Mechanical distortion of a thin diaphragm held rigidly at its edges is another effect widely used for pressure measurement. Sensing of diaphragm deflection by capacitatively coupling it to a fixed electrode and conversion of the capacitance change to an electrical signal has been developed to form a gage of high sensitivity and reliability. Such gages are the best way of measuring pressure in the range below 10 Pa, and are often found on vacuum systems. Their output is assumed to be a linear function of pressure, so that, with points at the ends of a straight line on a graph determined at vacuum and at 10 Pa (by means of a piston and cylinder gage) to an accuracy of 0.1%, a pressure scale in the range 0–10 Pa is defined.

Minute diaphragms formed in silicon chips by integrated circuit etching techniques are mass produced and employ deflection sensors of diffused dopant in the diaphragm itself. The sensors are electrical resistance paths which change their resistance when stretched. While quite sensitive to temperature as well as to pressure, their small size, adaptability to electrical signal generation, and cheapness make them useful in processing equipment and in other systems. Their accuracy is less than that of the other gages described.

Many other devices capable of calibration are used for pressure measurement.[4] It is apparent that pressure cannot be measured in the interior of a fluid but only at a solid surface bounding the fluid. Sometimes a probe with a pressure sensitive element is inserted into a fluid to attempt a measurement within, but there is no assurance that the pressure is not changed by inserting the probe. A "contactless" pressure measurement is impossible.

Deductions from the General Principles of Fluid Statics We have stated the elementary basis of fluid statics along with the meanings of pressure and surface tension in Eqs. (1) and (2). Some traditional wisdom now derivable from these equations preceded their statement, such as "water always runs downhill," "water seeks its own level," and "a body is buoyed up by a force equal to the weight of the displaced fluid" (Archimedes' principle). Let us derive some general conclusions from the laws of statics, namely from Eqs. (1) and (2).

Pressure Variation with Depth. We have already, in discussing the manometer, established the variation of pressure with depth in lakes and in the oceans. This is Eq. (3), valid because the density of water in the ocean varies so little with depth, and because even at the greatest depths, the distance from the center of the earth has changed by such a small fraction that g may be considered practically constant.

Hydraulic and Pneumatic Actuating Systems. Provision for application of a force or forces at distant sites by means of a fluid in tubing connected to piston-and-cylinder arrangements is very common. Movement of the control piston increases the pressure in the fluid and the increase in pressure occurs almost immediately (transmitted at the speed of sound in the fluid) everywhere in the fluid. This uniform pressure increase is sometimes called *Pascal's law* and follows from Eq. (1). The aspect of such arrangements which causes the most technical trouble is that the fluid finds the smallest holes or cracks in the containing system and leaks out. Sealing of pistons, valves and other moving parts with gaskets introduces friction, so that the computed forces are not precisely found in practice. By choosing the size (area) of the piston-and-cylinder, the magnitude of force to be delivered can be chosen. An extremely large force is achieved by this means in the *hydraulic press*, which allows forces to be "magnified," and by means of mechanical valves (in a pump) to produce significant displacements of the large force.

Surface of Water at Rest Is Horizontal. The existence of sea level on the earth, combined with the detectable curvature of the sea surface over several kilometers, follows from the absence of all body forces in Eq. (1) except weight. Weight, as represented by the common **g**, is the resultant of the gravitational force in the nearly spherical earth and the local centrifugal force due to the earth's rotation. (Centrifugal force is the fictitious force introduced by describing pressure on the earth in a rotating frame of reference.) Due to the oblate shape of the earth, as well as due to the centrifugal force, **g** is not directed exactly toward the earth's center. The meaning of "vertical" is given by the direction of **g**. That the surface of a liquid is perpendicular to **g**, i.e., "horizontal," follows from the vector nature of Eq. (1) and the knowledge that the surface is one of constant pressure. Since **g** changes in direction as we consider various locations on the earth, the surface of constant pressure, i.e., sea level, also changes direction, giving the nearly spherical shape of the oceans' surfaces. So long as no solid body—one able to maintain shear stresses—is between water in a stream and water in the ocean, the water will move to form the static surface at atmospheric pressure which is the surface at sea level. Thus water runs down until it gets to the ocean, which is the final static equilibrium state.

Capillary Rise of Liquid. That sap rises to the top of the tree—well above sea level—is understood by Eq. (2) and the knowledge that the pressure is reduced below the atmospheric pressure by the curvature of the meniscus in the sap column.

Archimedes' Principle. That an immersed body is buoyed up by a force equal to the weight of the displaced fluid follows readily from Eq. (1), the assumption being that the only body force is weight. The pressure over the surface of the body varies with depth and is larger underneath than on top; the pressure at opposite sides is the same at the same depth, so the sideways forces pushing on the body are opposite and there is no net sideways force. Equation (1) with weight **g** as the only body force takes the form

$$\rho\mathbf{g} - \nabla p = 0, \qquad (4)$$

where both terms are forces per unit volume. We divide the immersed body into many imaginary vertical prisms, one of which is shown in Fig. 1, with cross sectional area A and height h. The net upward force on the top and bottom surfaces of the prism is ∇ph times A and is, according to Eq. (4), identical to the weight of fluid in the prism of cross section A and height h, namely, $\rho g h A$. This is true for each prism and we can add the results for all the prisms, thus deriving Archimedes' principle.

Criterion for Unstable Atmosphere. Air is compressible, and over heights of hundreds of meters the density of the atmosphere changes appreciably. Mixing of the air at levels up to about 400 m above the ground is important in predicting whether rain will occur and whether pollutants inserted near the ground will stay there or be diluted as unstable air mixes. The criterion for stability is that the temperature T increase with height z at a rate greater than

$$\frac{dT}{dz} = -\frac{g}{C_p}, \qquad (5)$$

the so-called *adiabatic lapse rate*. The heat capacity per unit mass at constant pressure, C_p, is a property of air: $1010 \text{ J kg}^{-1} \text{ K}^{-1}$. Equation (5) is derived from the basic equation of fluid statics, with weight the only body force, namely Eq. (4), in the form*

*The sense of the vector **g** is down, whereas z is positive upward.

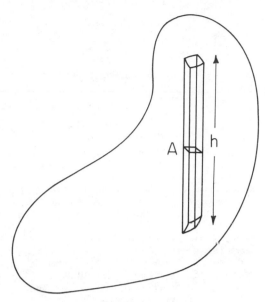

FIG. 1. Archimedes' principle.

$$\frac{dp}{dz} = -\rho g \qquad (4')$$

and the adiabatic equation of state for air

$$T = T_0 \left(\frac{p}{p_0}\right)^{(1-\gamma)/\gamma}, \qquad (6)$$

where γ is the ratio of heat capacities at constant pressure and at constant volume, and is a constant equal to 1.4 for air, and T_0 and p_0 are the temperature and pressure at any reference height z_0.

If a parcel of air at height z_0 should undergo a displacement to a different height $z_0 + dz$ it may find itself surrounded by air at a different temperature and density from its own temperature and density. If it has not changed its density by the correct amount, it will not be in equilibrium and will rise or fall further (unstable), or will be pushed back to its initial height (stable). The amount its pressure has changed is given by Eq. (4') and the amount its temperature has changed is then given by Eq. (6). The criterion for stability is

$$dT = \left(\frac{dT}{dz}\right)_0 dz = \left(\frac{dT}{dp}\right)_0 \left(\frac{dp}{dz}\right)_0 dz$$

$$= \left(\frac{dT}{dp}\right)_0 (-\rho_0 g)\, dz. \qquad (7)$$

Using Eq. (6) and the ideal gas equation of state in the form

$$\rho_0 = \frac{p_0}{C_p \left(\dfrac{1-\gamma}{\gamma}\right) T_0}, \qquad (8)$$

Equation (5) for the adiabatic lapse rate is obtained.

RAYMOND J. EMRICH

References

1. Landau, L. D., and Lifshitz, E. M., "Electrodynamics of Continuous Media," Oxford, Pergamon Press, 1960, pp. 64–69.
2. Hirschfelder, J. O., Curtiss, C. F., and Bird, R. B., "Molecular Theory of Gases and Liquids," New York, John Wiley & Sons, Inc., 1954, pp. 694 *et seq*.
3. Irving, J. H., and Kirkwood, John G., *J. Chem. Phys.* 18, 817 (1950).
4. Soloukhin, R. I., Curtis, C. W., and Emrich, R. J., "Measurement of Pressure," in "Methods of Experimental Physics–Fluid Dynamics" (R. J. Emrich, Ed.), Vol. 18, New York, Academic Press, 1981, pp. 499–515.

Cross-references: COMPRESSIBILITY, GAS; FLUID DYNAMICS; FLUID MECHANICS; GAS LAWS; KINETIC THEORY; LIQUID STATE.

FOURIER ANALYSIS

Fourier analysis is the mathematical representation of functions as linear combinations of sine, cosine, or complex exponential harmonic components. Such representations take their name from Jean Baptiste Fourier (1768–1830), French mathematician and physicist, who used them extensively in developing a mathematical theory of heat conduction. They continue to be widely used in mathematical physics and engineering, finding application in the study of such diverse subjects as diffraction, diffusion, image formation, spectroscopy, electrical networks, x-ray crystallography, and the theory of probability.

Fourier Series Let $f(x)$ be a periodic function with period L, i.e., $f(x)$ satisfies

$$f(x) = f(x - L) \qquad (1)$$

for all x. For a wide class of such functions, we can represent $f(x)$ by an infinite summation of sine and cosine harmonic components

$$f(x) = a_0 + \sum_{n=1}^{\infty} a_n \cos \frac{2\pi n x}{L} + \sum_{n=1}^{\infty} b_n \sin \frac{2\pi n x}{L}, \qquad (2)$$

known as the *Fourier series* representation of $f(x)$. If the left and right hand sides of Eq. (2) are multiplied by $\cos 2\pi m x/L$ or by $\sin 2\pi m x/L$, and the resulting equation is integrated with respect to x over one period L, the orthogonality of the harmonic sine and cosine components on this interval implies the following formulas for the *Fourier coefficients* a_0, a_n, b_n:

$$a_0 = \frac{1}{L} \int_{-L/2}^{L/2} f(x)\, dx$$

$$a_n = \frac{2}{L} \int_{-L/2}^{L/2} f(x) \cos \frac{2\pi n x}{L}\, dx \qquad n = 1, 2, \cdots$$

$$b_n = \frac{2}{L} \int_{-L/2}^{L/2} f(x) \sin \frac{2\pi n x}{L}\, dx \qquad n = 1, 2, \cdots$$

$$(3)$$

Other equivalent forms of the Fourier series can be obtained from Eq. (2). For example, $f(x)$ may be written

$$f(x) = a_0 + \sum_{n=1}^{\infty} C_n \cos \left(\frac{2\pi n x}{L} - \phi_n \right) \qquad (4)$$

where

$$C_n = \sqrt{a_n^2 + b_n^2}$$

$$\phi_n = \tan^{-1} \frac{b_n}{a_n}. \qquad (5)$$

Another widely used representation is the exponential (or complex) form of the Fourier series,

$$f(x) = \sum_{n=-\infty}^{\infty} \alpha_n e^{i(2\pi n x/L)} \qquad (6)$$

where

$$\alpha_n = \frac{1}{2}(a_n - i b_n) = \frac{1}{L} \int_{-L/2}^{L/2} f(x) e^{-i(2\pi n x/L)}\, dx.$$

$$(7)$$

For a real-valued $f(x)$, α_{+n} and α_{-n} are not independent, since $\alpha_{-n} = \alpha_{+n}{}^*$, where * indicates complex conjugate. However, when $f(x)$ is complex-valued, α_{+n} and α_{-n} are in general not related.

An important property of Fourier series, known as *Parseval's theorem*, states that the mean quadratic content of $f(x)$ (which often has the physical interpretation of average power) may be found by summing the squares of the Fourier coefficients. Thus, for the three representations used,

$$\frac{1}{L} \int_{-L/2}^{L/2} |f(x)|^2 \, dx = \begin{cases} a_0{}^2 + \sum_{n=1}^{\infty} \dfrac{a_n{}^2 + b_n{}^2}{2} \\[2ex] a_0{}^2 + \sum_{n=1}^{\infty} \dfrac{C_n{}^2}{2} \\[2ex] \sum_{n=-\infty}^{\infty} |\alpha_n|^2 \end{cases} \quad (8)$$

No discussion of Fourier series would be complete without some mention of the conditions under which the series actually converge to the functions they are to represent. Conditions which are both necessary and sufficient are not known. However, for the series to converge to $f(x)$ at each point of continuity, it is known to be a sufficient condition that $f(x)$ have only a finite number of maxima and minima in the interval $-L/2 \leqslant x \leqslant L/2$. Under this same condition, at points of discontinuity of $f(x)$ the series converges to the arithmetic mean of the values of $f(x)$ immediately to the left and right of the discontinuity.

Fourier Integrals Just as a periodic function can be represented by the sum of harmonic components in a Fourier series, so too an aperiodic function may be represented by an integral over a continuous spectrum of complex-exponential components. In this case we write

$$f(x) = \int_{-\infty}^{\infty} F(\nu) e^{i2\pi\nu x} \, d\nu \quad (9)$$

which is referred to as the *Fourier integral* representation of $f(x)$. Here $F(\nu)$, known as the *Fourier transform* (or Fourier spectrum) of $f(x)$, is given by

$$F(\nu) = \int_{-\infty}^{\infty} f(x) e^{-i2\pi\nu x} \, dx. \quad (10)$$

The variable ν is generally referred to as the *frequency* variable.

Note that a simple expansion of the exponential in Eq. (10) yields $F(\nu)$ in terms of sine and cosine integrals,

$$F(\nu) = \int_{-\infty}^{\infty} f(x) \cos 2\pi\nu x \, dx$$

$$-i \int_{-\infty}^{\infty} f(x) \sin 2\pi\nu x \, dx. \quad (11)$$

If $f(x)$ is an even function of x, the sine integral vanishes, leaving $F(\nu)$ expressed as a *Fourier cosine integral*,

$$F(\nu) = 2 \int_{0}^{\infty} f(x) \cos 2\pi\nu x \, dx. \quad (12)$$

Other forms of the Fourier integral and Fourier transform are often found in the literature. For example, frequency ν is often replaced by angular frequency $\omega = 2\pi\nu$, in which case Eqs. (9) and (10) become

$$f(x) = \frac{1}{2\pi} \int_{-\infty}^{\infty} F(\omega) e^{i\omega x} \, d\omega$$

$$F(\omega) = \int_{-\infty}^{\infty} f(x) e^{-i\omega x} \, dx. \quad (13)$$

Alternatively, the forms

$$f(x) = \frac{1}{\sqrt{2\pi}} \int_{-\infty}^{\infty} F(\omega) e^{i\omega x} \, d\omega$$

$$F(\omega) = \frac{1}{\sqrt{2\pi}} \int_{-\infty}^{\infty} f(x) e^{-i\omega x} \, dx \quad (14)$$

are sometimes found. Here we shall continue to use the forms of equations (9) and (10).

As in the case of Fourier series, the Fourier integral representation will converge to $f(x)$ only for a certain class of functions. Convergence at points of continuity is assured if $f(x)$ satisfies the following set of sufficient conditions:

   (i) $f(x)$ is absolutely integrable;
   (ii) $f(x)$ has only a finite number of maxima and minima in any finite interval; and
   (iii) $f(x)$ has no infinite discontinuities.
Under these conditions, at points of discontinuity of $f(x)$, the Fourier integral converges to the arithmetic mean of the values of $f(x)$ immediately to the left and right of the discontinuity. In general, any one of the above sufficient conditions can be slightly relaxed at the price of strengthening one or both of the additional conditions.

While the class of functions encompassed by the above conditions is wide, there do exist certain functions important in mathematical physics which are not included, such as $\sin x$, $\cos x$, and $f(x) = 1$. These may be included if the concept of the Dirac delta function, $\delta(\nu)$, is introduced, having the properties

$$\delta(\nu) = 0 \quad \nu \neq 0,$$

$$\int_{-\infty}^{\infty} \delta(\nu) \, d\nu = 1. \quad (15)$$

Strictly speaking, $\delta(\nu)$ is not a function in the usual mathematical sense. However, it can be

TABLE 1. FOURIER TRANSFORM PAIRS

$$f(x) = \exp(-\pi x^2) \qquad F(\nu) = \exp(-\pi \nu^2)$$

$$f(x) = \begin{cases} 1 & |x| \leqslant \frac{1}{2} \\ 0 & |x| > \frac{1}{2} \end{cases} \qquad F(\nu) = \frac{\sin \pi \nu}{\pi \nu}$$

$$f(x) = \begin{cases} 1 - |x| & |x| \leqslant 1 \\ 0 & |x| > 1 \end{cases} \qquad F(\nu) = \left[\frac{\sin \pi \nu}{\pi \nu} \right]^2$$

$$f(x) = 1 \qquad F(\nu) = \delta(\nu)$$

$$f(x) = \cos \pi x \qquad F(\nu) = \frac{1}{2}\delta(\nu - \frac{1}{2}) + \frac{1}{2}\delta(\nu + \frac{1}{2})$$

$$f(x) = \exp(i\pi x^2) \qquad F(\nu) = \exp\left[-i\pi(\nu^2 - \frac{1}{4})\right]$$

$$f(x) = \sum_{n=-\infty}^{\infty} \delta(x - n) \qquad F(\nu) = \sum_{m=-\infty}^{\infty} \delta(\nu - m)$$

treated rigorously using the theory of distributions.

In Table 1 are presented the Fourier transforms of a number of the more important functions encountered in mathematical physics and engineering. From the basic Fourier transform pairs listed in the table, many other pairs may be derived with the help of the following Fourier transform theorems.

Linearity: If $f(x) = a\ g(x) + b\ h(x)$, then $F(\nu) = a\ G(\nu) + b\ H(\nu)$, where $G(\nu)$ and $H(\nu)$ are the Fourier transforms of $g(x)$ and $h(x)$, respectively.

Similarity: If $f(x) = g(ax)$, then $F(\nu) = (1/|a|)G(\nu/a)$.

Shift: If $f(x) = g(x - a)$, then $F(\nu) = G(\nu) \exp(-i2\pi a\nu)$.

Convolution: If $f(x) = \displaystyle\int_{-\infty}^{\infty} g(\xi)h(x - \xi)\,d\xi$,

then $F(\nu) = G(\nu)H(\nu)$.

A relation entirely analogous to Parseval's theorem of Fourier series is *Plancherel's* theorem,

$$\int_{-\infty}^{\infty} |f(x)|^2\,dx = \int_{-\infty}^{\infty} |F(\nu)|^2\,d\nu. \quad (16)$$

In many physical applications this theorem leads to the interpretation of $|F(\nu)|^2$ as the energy spectrum of the function $f(x)$.

The Fourier integral representation (and the Fourier series representation) can be generalized to apply to functions of two or more independent variables. Most important are functions defined over a plane or over three-dimensional space. In the case of two independent variables we have

$$f(x,y) = \iint_{-\infty}^{\infty} F(\nu_X, \nu_Y)$$
$$\cdot \exp\left[i\,2\pi(\nu_X x + \nu_Y y)\right]\,d\nu_X d\nu_Y$$
$$F(\nu_X, \nu_Y) = \iint_{-\infty}^{\infty} f(x,y)$$
$$\cdot \exp\left[-i\,2\pi(\nu_X x + \nu_Y y)\right]\,dxdy. \quad (17)$$

For the most part, the properties of one-dimensional Fourier transforms carry over directly to the N-dimensional case. There is extra richness in the theory for the multidimensional case, however. For example, if $f(x,y)$ exhibits circular symmetry, i.e., is a function of only radius r in the plane, then Eqs. (17) can be reduced to the form of the so-called *Fourier-Bessel transform* (or Hankel transform of zero order),

$$f(r) = 2\pi \int_0^{\infty} \rho F(\rho)J_0(2\pi r\rho)\,d\rho$$

$$F(\rho) = 2\pi \int_0^{\infty} rf(r)J_0(2\pi r\rho)\,dr \quad (18)$$

where $r = (x^2 + y^2)^{1/2}$, $\rho = (\nu_X^2 + \nu_Y^2)^{1/2}$, and J_0 is a Bessel function of the first kind, zero order.

Discrete Fourier Transforms Of particular importance in the analysis of experimental data is the *discrete Fourier transform* (DFT), which is readily implemented on a digital computer. Let $f(x)$ be a function of finite duration (i.e., $f(x)$ is zero outside the range $0 < x < 2X$), and suppose that the bandwidth of $f(x)$ is limited to the approximate range $-B < \nu < B$. Let $f(x)$

be sampled with spacing $1/2B$ between samples, and define f_n to be equal to $(1/2B)f(n/2B)$. The number of such samples is $N = 4BX$. Then the DFT of the sampled data set f_n ($n = 0, 1, \ldots, N-1$) is defined by

$$F_m = \sum_{n=0}^{N-1} f_n \exp\left(-i2\pi \frac{mn}{N}\right) \qquad (19)$$

$$m = 0, 1, \ldots, N-1.$$

The corresponding inverse discrete Fourier transform relation is

$$f_n = \frac{1}{N} \sum_{m=0}^{N-1} F_m \exp\left(i2\pi \frac{mn}{N}\right). \qquad (20)$$

The relation between the resulting coefficients F_m calculated by Eq. (19) and the desired samples of the continuous Fourier transform $F(\nu)$ is relatively complex, but can be stated explicitly as

$$F_m = \sum_{k=-\infty}^{\infty} F[(m-kN)/2X].$$

In the event that there is little energy in the spectrum of $f(x)$ outside the interval $-B < \nu < B$, as supposed at the start, then to a good approximation F_m is equal to sampled values of the continuous spectrum $F(\nu)$. For indices in the range $m = 0, 1, \ldots, (N/2) - 1$, $F_m = F(m/2X)$ (these are the "positive-frequency" components of $f(x)$), while for indices in the range $m = (N/2) + 1, \ldots, N - 1$, $F_m = F((m - N)/2X)$ (these are the "negative frequency" components of $f(x)$).

The DFT can be computed very rapidly using a digital computer with the help of so-called *fast Fourier transform algorithms*. Note that a brute-force calculation of any single coefficient F_m of the DFT would require N complex multiply-and-adds, one for each term in the summation. (We refer to each complex multiply-and-add as an "operation.") Since there are N such coefficients, the total number of operations required would be N^2. When N is a power of 2, a reorganization of the order in which computations are performed allows the total number of operations to be reduced to $N \log_2 N$. The savings in computation time can be enormous for large N, and it can be said that the discovery of fast Fourier transform algorithms has made possible the widespread use of Fourier analysis in modern computational physics.

Applications The applications of Fourier analysis are so widespread throughout physics and engineering that to summarize them adequately in a brief space is nearly impossible. At best we can only outline some of the most important classes of applications, illustrating with a simple mention some of the particular branches of science that benefit from Fourier analysis in each case.

The applications of Fourier analysis can be conveniently divided into two major categories. Within the first category we find Fourier analysis used purely as an analytical tool, for the calculation and prediction of the results of experiments not yet performed, or for the explanation of experimental results already obtained. For example, in certain important cases the wave amplitude produced by a diffracting structure may be calculated using Fourier transforms, a consequence of the fact that the complete diffraction integrals may be reduced to Fourier transforms.

As a second example of this category, it might be desired to predict the image obtained when a certain object is viewed through an imaging system. The characteristics of such a system are often most easily specified in terms of its effects on spatial sine wave objects of various spatial frequencies. To predict the image obtained for a nonsinusoidal object, that object may be decomposed, by means of a Fourier integral, into sine wave (or complex exponential) components. The image is then found by synthesizing, again with a Fourier integral, the various sine wave image components, after their amplitudes and phases have been modified by the known properties of the imaging system. A similar approach is used in the analysis of linear, time-invariant electrical networks. It can also be used to calculate the diffusion of heat through a conducting plate when a known temperature distribution is impressed across one edge of the plate.

A second major category of applications utilizes Fourier analysis as an experimental tool, applied to experimental data to derive related physical information. For example, in Fourier spectroscopy the interferogram obtained from a scanned Michelson interferometer may be Fourier-analyzed to yield the spectral intensity distribution of the radiation source. In x-ray crystallography, the x-ray diffraction pattern produced by a crystal may be Fourier-analyzed to yield information about the crystal lattice structure. In radio astronomy the correlation coefficients produced when the signals collected by various elements of an array are cross-correlated may be Fourier-analyzed to yield a map of the radio brightness distribution of the sky. In all of these latter examples, Fourier analysis plays a fundamental role in transforming data which are readily collected by experiment into data which are more directly related to the physical properties of interest.

JOSEPH W. GOODMAN

References

1. Bracewell, R. N., "The Fourier Transform and Its Applications," New York, McGraw-Hill Book Co., 1965.

2. Goodman, J. W., "Introduction to Fourier Optics," New York, McGraw-Hill Book Co., 1968.
3. Papoulis, A., "Systems and Transforms with Applications in Optics," New York, McGraw-Hill Book Co., 1968.
4. Arsac, J., "Fourier Transforms and the Theory of Distributions," Englewood Cliffs, N.J., Prentice-Hall, Inc., 1966.
5. Titchmarsh, E. C., "Introduction to the Theory of Fourier Integrals," Oxford, Oxford University Press, 1937.
6. Gaskill, J. D., "Linear Systems, Fourier Transforms, and Optics," New York, John Wiley & Sons, 1978.
7. Brigham, E. O., "The Fast Fourier Transform," Englewood Cliffs, N.J., Prentice-Hall, Inc., 1974.
8. Papoulis, A., "Signal Analysis," New York, McGraw-Hill Book Co., 1977.

Cross-references: CALCULUS OF PHYSICS, DIFFERENTIAL EQUATIONS IN PHYSICS, FOURIER OPTICS.

FOURIER OPTICS

Diffraction Theory The articles OPTICS, GEOMETRICAL and OPTICS, GEOMETRICAL ADVANCED are based on the approximation that light is propagated as rays. For systems whose dimensions are of the magnitude of the wavelength of light, we must take into account its wave nature. A simple examle is the rectangular slit of Fig. 1, whose width w is about equal to the wavelength λ. Consider a plane wave striking this slit normally. Some energy from this wave will be absorbed and reradiated by the edges of the slit; the interaction of the initial and reradiated waves produces a change in direction, which we call *diffraction*.

The phenomenon of diffraction can be expressed quantitatively by using the fact that waves are specified in terms of sines or cosines; for mathematical convenience, we shall use the equivalent function exp (ikr), where r is the displacement along the wave. Since the wave is periodic, its amplitude at $r = 0$ and $r = \lambda$ is unchanged. Hence, exp $(ik\lambda) = 1$, $k\lambda = 2\pi$, or

$$k = 2\pi/\lambda \qquad (1)$$

gives the defining equation for the propagation constant k. To find the amplitude A that this diffraction wave has on a screen at some arbitrary distance to the right, we integrate over the area S of the slit, obtaining

$$A = C \int e^{ikr} \, dS \qquad (2)$$

where C is the constant involving the maximum strength of the wave and dimensionality factors. As Fig. 1 shows, $r = r_0 + x \sin \theta$; taking $dS = l\,dx$, and integrating from $-w/2$ to $w/2$, we obtain

$$A = Ce^{ikr_0} l \, \frac{\sin\,[(kw \sin \theta)/2]}{k \sin \theta}. \qquad (3)$$

But since we are interested in the intensity I, which is the square of the amplitude, we find that

$$\frac{I}{I_0} = \frac{\sin^2 \alpha}{\alpha^2} \qquad (4)$$

where $\alpha = \frac{1}{2}kw \sin \theta$, and I_0 incorporates all the constants.[1]

The function $(\sin^2 \alpha)/\alpha^2$ (which is the square of what is sometimes called the sampling function) is shown in Fig. 2; the alternate light and dark bands are the *Fraunhofer diffraction pattern* for a long, thin slit.

A similar calculation[1] for a circular aperture of radius R gives the expression

$$\frac{I}{I_0} = \left(\frac{2J_1(\rho)}{\rho}\right)^2 \qquad (5)$$

where $\rho = kR \sin \theta$ and J_1 is a well-known Bessel function. Tabulated values gives the plot of Fig. 3, which is qualitatively similar to Fig. 2.

The bright central spot, the *Airy disk*, contains about 84% of the energy which passes through the aperture. Its periphery corresponds to the first zero of $J_1(\rho)$, for which $\rho = 3.832$. The angle is then given by

$$\sin \theta = \rho/kR$$

or, approximately,

$$\theta = 3.83\lambda/2\pi R = 1.22\lambda/D. \qquad (6)$$

This formula indicates the angular resolving power of a lens of diameter D, showing that Fraunhofer diffraction imposes a fundamental limit. The same result applies to the problem of resolving two separate points; if their Airy disks overlap in the fashion indicated by Fig. 4, they are considered to be just distinguishable; this is the famous *Rayleigh criterion*.

It is possible to improve on this limit by putting more of the radiant energy into the central spot. This is done by *apodization* (from the Greek "without feet," which refers to the reduction of the side lobes in Figs. 2 or 3). The method involves placing a nonuniformly transmitting plate in front of the lens. For example, let a coating of the form

$$A(x) = \cos(\pi x/w)$$

be used with the slit of Fig. 1. This additional term inside the integral of (2) will lead to a transmission curve as shown in Fig. 5. The central bright line is now wider and more intense.

Spatial Filtering We next consider the diffraction process for a series of N slifts of width w with a uniform spacing d (Fig. 6); such a structure is called a *Ronchi diffraction grating*.

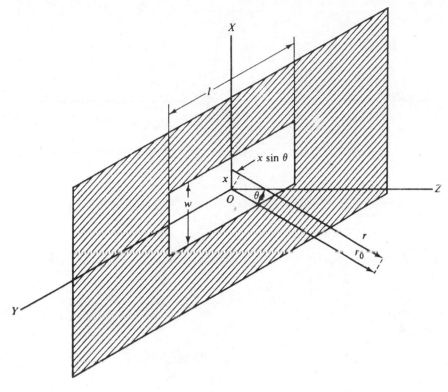

FIG. 1.

The integral in Eq. (2) is generalized to a sum of N integrals, one for each slit, and the result of the combined summation and integration is

$$\frac{I}{I_0} = \left(\frac{\sin \alpha}{\alpha}\right)^2 \left(\frac{\sin N\gamma}{N \sin \gamma}\right)^2 \quad (7)$$

where $\gamma = \frac{1}{2} kd \sin \theta$. The first term on the right is like Eq. (4), and comprises the *shape factor*, representing the contribution of the N identical slits to the pattern. The other term is the *grating factor*, and indicates the effect of repeating the slits. Figure 7 shows the shape factor as a

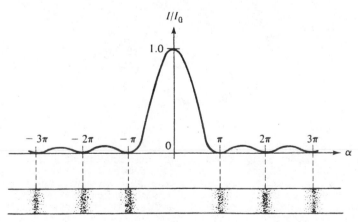

FIG. 2.

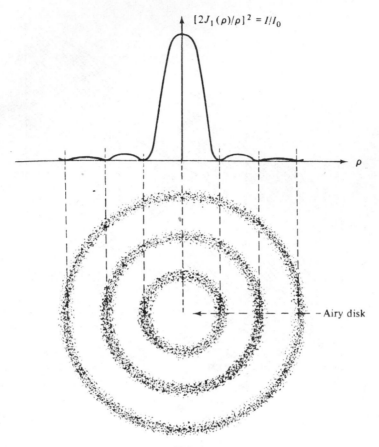

FIG. 3.

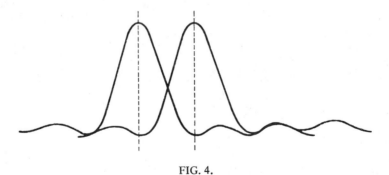

FIG. 4.

dotted envelope for the complete pattern; note how raising N from 5 to 20 (or higher) sharpens the bright lines forming the diffraction pattern. The pattern we show here is found in connection with other physical phenomena as well; electron and X-ray diffraction are examples (see discussion in DIFFRACTION BY MATTER AND DIFFRACTION GRATINGS).

The application we wish to consider involves the combination of a grating and a lens (Fig. 8a). There will be the usual image of the grating (Fig. 8b) at the place specified by the Gaussian lens equation (see OPTICS, GEOMETRICAL). In addition, the diffracted beams corresponding to $\gamma = 0, \pm\pi$ in Fig. 7, or to $\theta_0 = 0, \theta_1 = \lambda/d, \theta_{-1} = -\lambda/d$ (the zero and first order beams) are shown

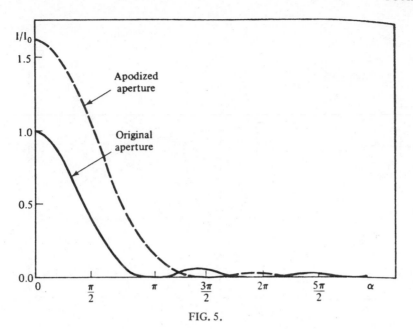

FIG. 5.

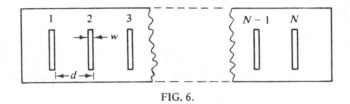

FIG. 6.

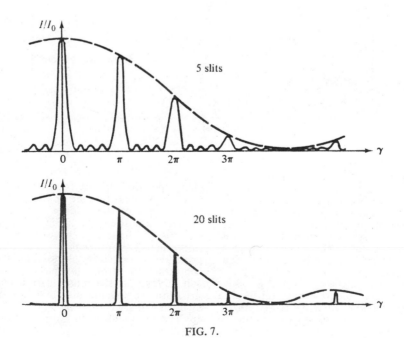

FIG. 7.

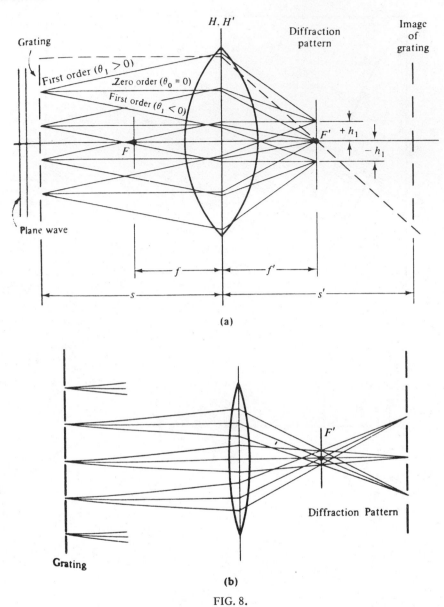

FIG. 8.

in detail in Fig. 8a. It is quite clear that the parallel zero-order beams will meet at the focal point F'. It can easily be shown by using the paraxial matrices of OPTICS, GEOMETRICAL, ADVANCED, that a set of rays which are parallel to themselves but not to the lens axis (such as the four for $\theta_1 > 0$) will meet on the plane through F' but off-axis.

Now we place an opaque aperture at F' which blocks all diffraction orders except for θ_0; this is the process of *spatial filtering*. If we think of the transmission properties of the grating as

being specified by the square wave of Fig. 9a, this is equivalent to removing the high frequency components of the Fourier series representing the wave and the result is Fig. 9b; i.e., the sharp corners of the image have been rounded or smoothed. This procedure could be applied in principle to satellite photographs, which are composed of strips; low-pass spatial filtering will remove the sharp boundaries. The converse situation—an opaque disk at F'—will then give high-pass spatial filtering (Fig. 9c), and one application could be to improve the readability

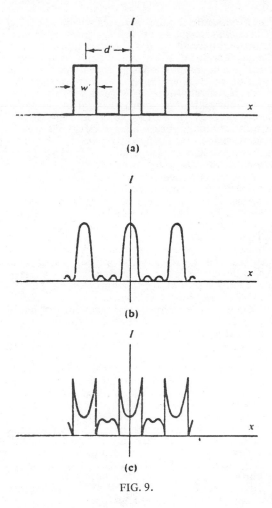

FIG. 9.

of illegible documents by sharpening edges. The details of evaluating the Fourier series are given by Garrard.[2]

Optical Transfer Functions The effect of a lens on light rays is the subject of geometrical optics (see associated articles) and the major problem is the inherent aberrations, which display themselves as blurry or distorted images. Light treated as waves brings in additional problems. The first of these is a direct consequence of Eq. (6), which indicates that the finite diameter of any optical system imposes a limit on its ability to resolve closely spaced objects. It is customary to use a test object in the form of alternate black and white bars; the resolution is expressed in terms of the largest number of line-pairs/mm that can be distinguished at the image plane. The other limitation is imposed by the ability to distinguish bars from spaces, i.e., the image contrast.

Both resolution and contrast are affected by the fact that when a light wave passes through a lens of variable thickness from center to edge, the time of transit varies with distance travelled in the glass, causing a shift in the phase of the wave from one part to another. This shift simply contributes an additional term in the exponent of the diffraction integral (2), where it has an effect analogous to the apodization mentioned above. It can be shown[1] with some elementary geometry and some very formidable algebra that the phase shift due to a spherical lens has a quadratic dependence on the x and y coordinates but that by positioning object and image so that they obey the Gaussian lens equation, this dependence reduces to a linear one. It will then be recognized that the diffraction integral (2), involving an exponential term of the form $\exp\left[(ik(x+y)\right]$, is generating a two-dimensional Fourier transform of the remaining factors inside that integral, these factors specifying the nature of the incoming wave. Thus, a lens is a device which performs a Fourier transform on an object, and also forms an image. We have already seen an example of this in Fig. 8; the Fourier transform of a periodic object, the grating, reduces to a discrete Fourier series, the result being located at the focal plane.

Fourier transform methods are used in science and engineering to solve the differential equations associated with transient phenomena; they give a way of computing the response of a circuit to an electrical impulse, for example. This response is identified as the admittance (the reciprocal of the impedance) and it is also called the transfer function. The optical equivalent of an impulse is a point source and the response is the *optical transfer function*. Like admittance or impedance, it will depend on frequency, but this is measured in line pairs/mm and is known as the *spatial frequency* ν_x. It is possible to calculate the dependence of the transfer function I on spatial frequency, with results as shown in Fig. 10. As already predicted, there is a maximum value ν_{x0} of ν_x beyond which the resolution vanishes. The curves shown represent the situation for a perfect lens; the presence of aberrations will

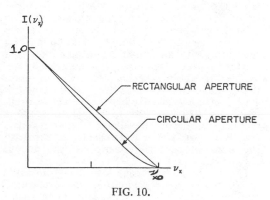

FIG. 10.

degrade these results. In addition, the transfer function can be shown to depend as well on the relative contrast of object and image and predict how this aspect of image quality decays with increasing spatial frequency.

ALLEN NUSSBAUM

References

1. Nussbaum, A., and Phillips, R. A., "Contemporary Optics for Engineers and Scientists," Prentice-Hall (1976).
2. Garrard, A., *Am. J. Phys.* **31**, 723 (1963).

Cross-references: DIFFRACTION BY MATTER AND DIFFRACTION GRATINGS; FOURIER ANALYSIS; LENS; OPTICS, GEOMETRICAL; OPTICS, GEOMETRICAL, ADVANCED; OPTICS, PHYSICAL.

FREQUENCY STANDARDS

Introduction A frequency standard is simply an oscillator whose characteristics of either frequency stability or accuracy satisfy certain subjective performance criteria. Such a device is approximately characterized by its nominal frequency and nominal amplitude of oscillation. But noise and systematic effects cause the actual frequency and amplitude of operation to differ from nominal. The level of the noise-induced frequency deviations is characterized by the quality called *frequency stability*, while the ability of the oscillator to achieve a predetermined frequency without calibration is called *accuracy*. Both quantities are usually stated on a fractional basis, $\Delta f / f_0$, where f_0 is the minimal operating frequency. Although oscillators cover the entire range of frequency stabilities smaller than 10^{-2}, for the purpose of this article frequency standards are considered to cover the range of stabilities smaller than 10^{-6}. Their size and weight can range anywhere from 10 cc and 50 gms to 1000 liters and 100 kilograms.

Nearly single frequency oscillation results from the use of a frequency selective element called a *resonator*. Two types of circuitry are used to produce a voltage output. An active oscillator is constructed by combining the resonator with an amplifier and a limiter in a feedback loop. However, it is sometimes desirable to obtain the characteristics of more than one resonator in a frequency standard. This is accomplished by using an active oscillator of the type just described to probe a second resonator. Feedback maintains the frequency of the active oscillator equal to the nominal center frequency of the second resonator.

The range of frequency stability from 10^{-6} to 10^{-9} is generally important for frequency control applications, such as maintaining broadcast transmitters and receivers within allocated frequency bands or providing signals for laboratory equipment such as counters and spectrum analyzers. The most commonly used device for such applications is the quartz crystal oscillator. Other applications require time-keeping in addition to frequency control. Precise navigation on earth, navigation in space very far from the earth, and certain digital communications applications require much greater performance than may be obtained with quartz crystal oscillators. These more stringent requirements are generally provided by atomic or molecular frequency standards called ATOMIC CLOCKS. In addition, scientists require a very precise and accurate frequency standards for applications such as precision spectroscopy and the measurement of relativistic effects. These special applications make use of a wide variety of frequency standards including those mentioned above, but also including such unusual devices as superconducting cavity stabilized oscillators, saturated absorption stabilized lasers, or even the Mössbauer effect.

Operating Principles For frequency standard applications, the most critical parameter of the resonator is its *relative linewidth*. The *loaded* Q of the resonator is the most commonly used figure of merit. It is equal to the total stored energy divided by the energy dissipated in one cycle of the oscillation and is inversely proportional to the linewidth. Quartz crystals have Q's as high as a few million. Superconducting cavities operating at microwave frequencies may have Q's as high as 10^{11}, and atomic resonances suitable for use in frequency standards today have Q's ranging from 10^7 to 10^{12}. A fundamental limitation on a frequency stability of the device is determined from the Q and signal-to-noise-ratio, S/N: The stability cannot be better than $(1/Q)(S/N)$.

There are many different types of noise sources. For quartz crystal oscillators, the fundamental sources are *thermal (Johnson) noise* which is associated with the dissipation of energy in real resistors and *shot noise* which is associated with the discrete nature of the charge carriers in the electrical circuitry. Since the frequency of the atomic standard is derived from the observation of a small number of atoms or molecules, frequency instabilities are observed which derive directly from the probabilistic nature of quantum mechanics.

Noise affects the performance in two distinct ways. First there is noise in the amplifiers which buffer or isolate the frequency standard from the user. Thermal or shot noise in these amplifiers adds to the output signal. The resultant frequency fluctuations have a low pass characteristic, i.e., they become smaller at higher frequencies. There are also noise sources within the feedback loop of the standard which do not simply add to the output signal, but perturb the frequency of the oscillator directly.

In this case a white voltage noise source produces white frequency fluctuations.

Accuracy is a property of atomic frequency standards. High accuracy is obtained by minimizing perturbations on the atomic system due to collisions with other atoms, interaction with electric and magnetic fields, and the effect of the velocity of the atoms themselves. For example, the definition of the second is 9,192,631,770 oscillations of the radiation from the cesium 133 atom in its ground state. The term *ground state* refers to the lowest energy state of an atom which is at rest in a field-free region of space. All present-day atomic frequency standards actually observe atoms in thermal equilibrium with their surroundings and therefore moving at velocities on the order of 100 m/s. From the point of view of the observer, the frequency emitted by such atoms is perturbed by two effects. The *first-order Doppler effect* is proportional to the velocity vector of the atom. It is analogous to the change in pitch of a train whistle passing a stationary observer. The *second-order Doppler effect*, so called because it is proportional to the square of the velocity, is a relativistic effect resulting from the property of time dilation. A radiating atom in motion behaves like a clock and according to the special theory of relativity, runs slow compared to a stationary clock.

One might think that the ideal frequency standard would combine the best characteristics described above. It would have as high a Q as possible, very small interactions, very small velocity for the atoms and a very high signal-to-noise ratio. However, these characteristics are inconsistent. For example, in an atomic standard, high signal-to-noise ratio generally means observing large numbers of atoms which in turn implies frequent atomic collisions and substantial interactions. The required compromises have resulted in the development of a wide variety of different frequency standards emphasizing different aspects of performance: accuracy, stability, size, weight, power, reliability, and cost.

Quartz Crystal Oscillators A quartz resonator is fabricated from a thin slice of material cut from a commerical or naturally grown quartz crystal. Since the material is piezoelectric, an acoustic vibration results in an electric potential between the two faces of the crystal. Coupling is provided by depositing thin film metal electrodes on the two faces of the crystal. To a first-order approximation the crystal acts like an electrical resonator composed of an inductor, capacitor, and resistor and is used by installing it in a standard oscillator circuit.

The center frequency of the resonance is determined from the dimensions of the resonator and is therefore inherently very sensitive to temperature. Frequency standard operation therefore requires some means of temperature compensation or control. At the 10^{-6} stability level, temperature compensation is the technique most often used. The circiut of the oscillator is modified by adding a second temperature sensitive element such as a thermistor which compensates for the temperature dependence of the resonator. Such devices can achieve frequency stability of 10^{-6} over wide temperature ranges and are both small and inexpensive, finding wide application in communications, radio, and television.

When higher precision is required, compensation is usually abandoned in favor of temperature control. The frequency of the resonator is a cubic function of the temperature, and there are two temperatures called the *turnover points* at which the frequency variation with temperature is zero. The crystal is placed in an oven and operated very near one of the turnover points. The stability of oscillators containing ovenized crystals varies between approximately 10^{-9} to 10^{-12} for one-second averages and the aging rate typically varies between 10^{-10} and 10^{-11} per day. Typical applications include high-quality laboratory instrumentation such as frequency counters and clocks for space vehicles which stress the small size, low power, and high reliability of quartz crystal oscillators.

Atomic Frequency Standards The change in state of an atom or molecule between two of its discrete energy levels is a resonance which may be used in a frequency standard. The frequency associated with the transition is the energy difference divided by Planck's constant. It is generally not convenient or practical to use most of the allowed transitions between energy states of an atom because the frequencies are usually too high for coherent signal processing techniques or because the transitions are extremely sensitive to either electric or magnetic fields. Consider a simple atom having a single unpaired electron such as hydrogen or any of the alkalai metals. The frequency corresponding to the transition of the electron between two of its discrete energy levels is typically in the ultraviolet or visible portion of the spectrum and therefore the atom must be used in its ground electronic state. If the electron has an asymmetric orbit, then it will have an associated magnetic moment which can interact with the magnetic moment of the nucleus. This interaction separates the ground state into several energy levels called *fine structure*. However, the transition frequencies are usually in the infrared, still too high for practical applications. If the electron is in a symmetric orbit then it has no orbital angular momentum but it still has a magnetic moment due to its spin. The interaction of this magnetic moment with the magnetic moment of a nucleus produces a set of energy levels called *hyperfine structure* which are typically at microwave frequencies. Transitions between such energy levels form the basis for all commercially available atomic clocks. Not until practical methods of frequency synthesis are developed for frequencies higher than 100 GHz will atomic clocks be used extensively in that range. Nevertheless, various laser techniques are used today at quite high frequencies as standards for atomic and molecular spectroscopy and length.

Conventional electronic or acoustic resonances are observed through the frequency selective nature of their power absorption. When coupled to an amplifier they can be used to provide frequency selective gain. The analogous situation holds for atomic resonances provided that an ensemble of atoms can be produced in one of the two energy states of the desired transition. In the case of hyperfine spectroscopy, the upper and lower energy states have very nearly equal populations, since their separation is very small compared to the thermal kinetic energy of the atoms. Microwave frequency standards generally employ either magnetic or optical state selection to create imbalance in the state populations.

Magnetic state selection depends upon the fact that the two hyperfine energy levels behave very differently in a strong magnetic field whose effect on the electron is larger than the electron's interaction with the nuclear magnetic moment. In an inhomogeneous magnetic field, the atoms feel a force proportional to the field gradient and the atoms in one state are deflected oppositely from the atoms in the other state. By intercepting one state while allowing the other to pass through an aperture, an atomic beam composed of atoms in the desired state is obtained.

Alternatively, optical state selection is accomplished by inducing a transition to a higher electronic energy level from which the atom decays spontaneously to the ground state. The light source must be narrower than the separation between the upper and lower ground-state hyperfine levels. Atoms induced to make transitions out of one hyperfine level to an excited state relax back to the two ground-state hyperfine levels in nearly equal proportions. Thus the hyperfine level to which the light source is tuned is depopulated.

The lifetime of the hyperfine transition is extremely long compared to the maximum length of time the atom could be observed in a practical spectrometer. However, without taking suitable precautions the observer will detect only a very broad line due to the Doppler shifts from many atoms traveling in random directions. Various techniques are used to eliminate the first-order Doppler effect. For example, in an atomic beam spectrometer the microwave radiation is applied to the atoms perpendicular to the mean velocity vector of the atomic beam and is reflected back upon itself. The net effect is to reduce the observed width of an atomic resonance to a value equal to the inverse of the observation time.

The operation of a cesium beam frequency standard illustrates all the aspects of the atomic beam frequency standard. The atomic transition used is the hyperfine transition in the ground state. This transition has a very small dependence on electric fields and is independent of magnetic field to first order at zero applied field. The atomic beam is formed by placing the cesium metal in an oven with a small exit hole. Cesium metal melts at 37°C, making it very easy to form an atomic beam of reasonable density. The desired hyperfine level is selected by passing the beam through the inhomogeneous magnetic field of a dipole magnet which deflects the desired atoms through a region containing a microwave cavity. A microwave field at the transition frequency of 9.2 GHz stimulates a change in atomic state which is analyzed by a second region of inhomogeneous magnetic field. The number of cesium atoms which make the transition between the two hyperfine levels is measured using a hot wire detector which ionizes the cesium atoms that successfully passed through the spectrometer. The microwave signal applied to the spectrometer is swept slowly back and forth across the resonance while the atom flux is measured. The signal is maximum when the frequency of the applied microwave field is equal to the frequency of the atomic transition. Electronic circuitry provides a correction signal which is used to maintain the average frequency of the microwave oscillator at the center of the atomic transition.

Commercial cesium standards have frequency stabilities which vary between six parts in 10^{13} and six parts in 10^{12} for observation periods of 100 s and the frequency stability improves inversely as the square root of the observation time. The best stabilities for cesium beam frequency standards generally vary between a few parts in 10^{14} and one part in 10^{13}. Laboratory cesium standards, called *primary cesium standards*, are constructed in such a way that the systematic effects which perturb the atomic resonance frequency may be measured very precisely. As a result, they are accurate and reproducible to a few parts in 10^{14}. Accuracy is limited principally by the ability experimentally to determine the residual first-order Doppler shift, although several other affects such as the relativistic time dilation, the effect of electric fields and the stability and homogeneity of the magnetic field also contribute to the inaccuracy. Fundamental advances in the accuracy of atomic frequency standards will probably come from devices which more nearly observe atoms at rest in free space.

SAMUEL R. STEIN

Cross-references: ATOMIC AND MOLECULAR BEAMS, ATOMIC CLOCKS, FEEDBACK, VIBRATION.

FRICTION

Introduction Friction is the resistance to motion which exists when a solid object is moved tangentially with respect to the surface of another which it touches, or when an attempt is made to produce such motion. Friction thus takes its place as one of the general systems of force which are considered in MECHANICS (others being GRAVITY, ELASTICITY, etc.). Unfortunately friction depends to a marked extent on the material properties of the contacting surfaces, and even more importantly, on any surface contaminants which may be present, so that it is very difficult to estimate values of the friction force theoretically, with an uncertainty of less than about 20 per cent. In many calculations in solid mechanics, this uncertainty in the friction constitutes the limiting factor in determining the accuracy of the over-all calculation.

The friction force arises from the fact that, when two solids are pressed together, bonding between their surface atoms occurs, and these bonds have to be broken before sliding can commence. Bonding of any considerable strength occurs only in places where the surface atoms come within range of each other's strong force-fields (i.e., closer than about 3×10^{-10} m); thus, when ordinary solids with appreciable roughness are used, bonding is confined to a few small patches (called junctions) over their interface, where the high spots (or asperities) of one material have made contact with asperities on the other material.

The energy used up in the friction process appears almost entirely in the form of heat. Generally this consists of a moderate temperature rise over the contacting bodies, and superposed on this there are higher "flash temperatures" at the junctions. At high sliding speeds, softening or even melting of the tips of the asperities may occur.

In addition to the bonding or "adhesion" effect, which is the principal cause of friction, there are four other mechanisms which use up energy during sliding; energy which must be supplied by the friction force. These mechanisms are:

(1) A roughness effect, caused by the interlocking of asperities and the need to lift one surface over the high spots of the other;

(2) A ploughing effect, whereby an asperity on a hard material can dig a groove in a softer material;

(3) A hysteresis effect, whereby there is elastic and plastic deformation of the material at or near the junctions, and not all the deformation energy is recoverable;

(4) An electrostatic effect (with electric insulators), where work must be done to separate electrically charged regions on the sliding surfaces.

In a great majority of applications these four mechanisms do not account for as much as 20 percent of the total resistance to sliding. The widely held belief that friction is due mainly to a roughness effect does not find experimental support. Cleaved mica (smooth to an atomic scale) shows very high friction.

Laws of Friction If a normal load L presses two surfaces together, then we may apply a tangential force up to some limiting value F_s, and the surfaces will remain at rest. Sliding occurs when the tangential force exceeds the static friction force F_s, and almost as soon as motion starts the tangential force takes on a characteristic value, F_k, and always acts in a direction opposite to the relative velocity of the surfaces. F_s is often some 30 per cent larger than F_k, but sometimes they are equal.

The ratio F_s/L is the static friction coefficient f_s(or μ_s), while the ratio F_k/L is the kinetic friction coefficient f_k. We may cite quite general statements, or "laws," involving these coefficients of friction. The three classical laws, dating back to the seventeenth and eighteenth centuries, are:

(1) The friction coefficient is independent of the load.

(2) The friction coefficient is independent of the contact area.

(3) The kinetic friction coefficient is independent of the sliding velocity.

More recently, it has been found possible to make another general statement about the friction coefficient:

(4) The friction coefficient is independent of the surface roughness.

We may summarise these laws in the comprehensive statement.

(5) The friction coefficient is essentially a material property of the contacting surfaces, and of the contaminants and other films at their interface.

Although in practice these "laws" are reasonably well obeyed, there are often systematic divergencies, some of which have important consequences. In the rare cases in which the friction coefficient varies with load, it is often because of some special effect (e.g., a surface coating which is broken up at heavy loads); however, materials with a high elastic limit (e.g., polymers) generally show a friction coefficient which goes down somewhat as the load is increased. The fourth law applies closely to the intermediate ranges of surface roughness, but it is found that very smooth surfaces have higher friction because they tend to seize, and very rough surfaces have higher friction because of asperity interlocking.

Velocity and Time Effects in Sliding Violations of the third friction law are important because they can lead to friction-induced oscillations. With some materials f_k is almost independent of velocity over a very wide range; however, with hard materials the friction generally goes down as the speed goes up (a so-

TABLE 1. COEFFICIENTS OF FRICTION

Materials	Surface Conditions	f_s	f_k
Metals on metals (e.g. steel on steel, copper on aluminum)	Carefully cleaned	0.4–2.0	0.3–1.0
	Unlubricated, but not cleaned	0.2–0.4	0.15–0.3
	Well lubricated	0.05–0.12	0.05–0.12
Nonmetals on nonmetals (e.g., leather on wood, rubber on concrete)	Unlubricated	0.4–0.9	0.3–0.8
	Well lubricated	0.1–0.2	0.1–0.15
Metals on nonmetals	Unlubricated	0.4–0.6	0.3–0.5
	Well lubricated	0.05–0.12	0.05–0.12

called negative characteristic), while soft materials show a negative characteristic at high sliding speeds and a positive characteristic at low sliding speeds. Often, f_k changes by about 10 per cent during a factor-of-ten change of velocity. In a sliding system which has elastic compliance, a negative characteristic introduces an instability, and at high speeds this takes the form of harmonic oscillations (e.g., a violin string), while at low speeds relaxation oscillations occur (e.g., a creaking door). The relaxation oscillations are usually referred to as stick-slip, because during a part of each cycle the surfaces are at rest. The severity of the stick-slip is enhanced because, while the surfaces are at rest, f_s increases somewhat with time of stick (about 10 per cent for every factor-of-ten increase in time of stick above 10^{-3} seconds).

Value of the Friction Coefficient In most situations the total real area of contact, or sum of all the asperities, is produced by plastic deformation of the asperities, and if we assume that strong bonds are formed joining the materials across the interface we find that the friction coefficient is given by

$$f = s/p$$

where s is the plastic shear stress, and p the plastic indentation hardness, of the softer of the two contacting materials. Since s and p are similar plastic strength parameters, their ratio tends to be constant, within the range 0.3 to 0.6, for a wide variety of materials. With some materials the real area of contact increases during sliding, and friction coefficients above 1.0 are then commonly observed. Especially bad in this connection are materials with a high ratio of surface energy to hardness, namely clean soft metals such as lead, aluminum and copper.

A lubricant acts by introducing a layer of lower shear stress s_1 at the interface, and lowers the friction coefficient accordingly, down to 0.05 in favorable cases. Many nominally clean materials also give low friction, as result of contamination during manufacture or handling. Slight differences in the degree of contamina-

tion can produce drastic differences in the friction.

Some typical friction values are given in Table 1.

Materials with Unusual Frictional Properties In this category we may place the hard nonmetals (diamond), which give low friction (~ 0.1); the elastomers (natural rubber) which give very high friction (~ 0.9); the layer-lattice substances (graphite, molybdenum disulfide, cadmium iodide) which give low friction (~ 0.1) and are used as solid film lubricants; the hexagonal close-packed metals (rhenium, magnesium, titanium, cobalt) which when clean give lower friction coefficients (~ 0.5) than do other metals; "Teflon," which adheres very poorly to other solids and accordingly gives very low friction (~ 0.05); and metals in group IVa of the periodic table (titanium, zirconium, hafnium) which cannot be well lubricated by any known liquid substances. Generally, the frictional anomalies of solids may be traced to peculiarities in their structure or surface properties.

Current Problems and Areas of Research Interest At one extreme there is concern with obtaining high and reproducible friction for traction purposes, as in the development of better brakes and clutches, and in the design of runway surfaces at airports which give high friction even in very wet weather. At the other extreme, work is being done to reduce friction in devices which run continuously and dissipate energy, such as automobile engines. Often, problems arise when new materials must be used. Thus the reduction in the use of asbestos because of health hazards and of cobalt because of restricted availablity has stimulated much research and development work. Perhaps the most persistent problem is that of controlling stick-slip, which is responsible for much of the noise pollution of our environment.

Related Fields of Interest Friction is considered one of a group of mechanical interaction phenomena, and research in the other members of the group, generally referred to as *tribology*, is of great interest to workers in the friction field. There is an extensive common literature. The other tribological phenomena are wear (the removal of surface material as result of mechanical action), lubrication (the properties of surface films which reduce friction and wear), adhesion (the tendency of solid objects pressed together to remain together), and electrical contact effects (the electrical resistance at the interface between contacting or sliding solids).

Other effects which are often classified as basic tribological effects are erosion (the wear produced by particles traveling in a fluid and impinging on a solid surface) and fretting (the friction and wear produced as result of vibration at the interface between contacting surfaces).

There has been a recent upsurge of interest in these fields, based on the realization that the various tribological phenomena have great practical importance. Thus, in a modern industrialized country excessive friction wastes about 0.5% of the Gross National Product, while the damage done by wear is about an order of magnitude greater.

ERNEST RABINOWICZ

References

Bowden, F. P., and Tabor, D., "Friction and Lubrication of Solids," Oxford, Clarendon Press, Part I, 1950, Part II, 1964.

Buckley, D. H., "Surface Effects in Adhesion, Friction, Wear and Lubrication," New York, Elsevier, 1981.

Ku, P. H. (Ed.), "Interdisciplinary Approach to Friction and Wear," NASA Report SP-181, Washington, D.C., National Aeronautics and Space Administration, 1968.

Rabinowicz, E., "Friction and Wear of Materials," New York, Wiley, 1965.

Suh, N. P., and Saka, N., "Fundamentals of Tribology," Cambridge, MA, MIT Press, 1980.

Szeri, A. S., "Tribology—Friction, Lubrication and Wear," New York, McGraw-Hill Book Co., 1980.

Cross-references: DYNAMICS, MECHANICS, STATICS.

FUSION POWER

Introduction When nuclei of certain light elements fuse together an excess binding energy is liberated. This is the energy source of stars, and it is the object of fusion research to find ways of using the energy for electrical power production on earth. Because of the abundance of suitable light elements the energy source is virtually unlimited.

In order to fuse, the reacting nuclei must have sufficient energy to overcome their coulomb repulsion, so nuclei with a high ratio of mass to charge heated to a high temperature are most suitable. The fusion reaction of interest is

$$D + T = {}^4He \ (3.5 \ Mev) + n \ (14.1 \ Mev) \quad (1)$$

where the fuel is deuterium (D) and tritium (T) and the reaction products are helium (the α particle, ${}^4He$) and a neutron (n); the numbers in parenthesis are their energies. The energy from fusion may be appreciated by noting that 36 m^3 of natural water contains 1 kg of deuterium, which on fusion with tritium produces an energy equivalent to 5×10^4 m^3 of gasoline. The deuterium is easily extracted from water, but the tritium, which does not occur naturally, is made by allowing the neutrons to combine with lithium, using the nuclear reactions

$$\begin{aligned} {}^6Li + n &= T + {}^4He + 4.8 \ Mev \\ {}^7Li + n &= T + {}^4He + n - 2.5 \ Mev. \end{aligned} \quad (2)$$

There is sufficient high grade lithium ore to supply the world's electricity requirements for thousands of years. The possibility of an abundant energy source combined with low potential radioactive hazard make fusion power attractive. Even allowing for the radioactive tritium, the hazard is much less than that for fission reactors, in that no actinides, iodine, cesium, strontium, or plutonium are produced.

Ultimately there is the possibility of fusion between deuterium nuclei alone using the reactions

$$\begin{aligned} D + D &= He^3 \ (0.82 \ MeV) + n \ (2.45 \ MeV) \\ D + D &= T \ (1.01 \ MeV) + H \ (3.02 \ MeV) \end{aligned} \quad (3)$$

thereby eliminating the need for lithium. However the energy produced is less, and the peak reaction rate for D/D fusion occurs at a temperature that is nearly ten times that for D/T fusion and is about ten times smaller. This results in a ratio of energy produced by the D/D fusion reactions to energy supplied as heat in the reactants of about 70, compared to about 1800 for D/T fusion reactions, making an overall energy gain more difficult to achieve. Thus attention is primarily directed first to achieving fusion power using D/T fuel. It is worth noting that D/D reactions are appreciable in a D/T fuelled reactor, increasing the average energy per reaction from the 17.6 MeV given in Eq. (1) to 22.4 MeV.

The D/T fusion cross section as a function of energy rises rapidly to a peak at 110 keV, equivalent to a temperature of 10^9 K. If the density is sufficient for a thermal distribution of particle velocities, the rapid rise in cross section results in the few particles with energies in the Maxwellian tail contributing most to the reactions and so the peak reaction rate is at a rather lower temperature, around 6×10^8 K, and conceptual fusion reactors are often designed to operate at a temperature of 10^8 K or so. At this temperature the fuel is fully ionized by collisions, so the electrons are not attached to any particular nucleus and electron charge density equals the nuclear charge density. This state of matter is called a *plasma*, and the problems of fusion power generation are mainly plasma physics combined with engineering, rather than nuclear problems.

Research into the feasibility of thermonuclear fusion power began in the mid-1950s, and for 25 years has been concentrated on the problems of heating and containing high temperature plasma. Progress in plasma physics has been sufficiently encouraging for a number of countries to start constructing large experiments to investigate plasma heated by its own thermonuclear fusion reactions: these will begin in the latter part of the 1980s. Successful operation of the experiments will demonstrate the scientific feasibility, and lead to research reactors to investigate the engineering and commercial potential of fusion power near the beginning of the next century.

The demonstration of the feasibility of fusion power is taking longer than fission because of the large characteristic dimensions required for thermal isolation of the plasma, and because a very high temperature is required to start the reactions. This means that comparatively expensive inflexible experiments are required to explore new ideas.

Fusion Reactor Parameters A fundamental problem of fusion power is thermally to isolate and confine the plasma while it is heated and the reactions occur. In stars the plasma is held by gravitational force, but this force is too weak to hold the small mass of plasma in a reactor. Also, practical electrostatic fields are too weak, even if Ernshaw's theorem (that there is no equilibrium for a system of charged particles in an electric field) can be overcome. Two approaches to solve the containment problem are being studied: one, which uses magnetic forces, is described in this article; the other, which uses inertial forces, is described in the article on LASER FUSION.

A magnetic field can exert pressure on plasma because the free electrons make the plasma a good electrical conductor. A magnetic field B external to the plasma causes an induced current in the plasma surface which interacts with the field to produce a pressure perpendicular to the magnetic field lines proportional to B^2; this follows from the fundamental laws of electromagnetics. The pressure of plasma that can be supported is usually appreciably lower than the applied magnetic field pressure because some magnetic field must be mixed with the plasma to make the overall system stable. The external magnetic field pressure must then balance the sum of the plasma pressure and the internal magnetic field pressure. A measure of the efficiency of using the confining magnetic field is β, defined by the ratio of plasma pressure divided by the external magnetic field pressure, i.e. $\beta \propto nT/B^2$, where n is the particle number density and T is the temperature.

The magnetic field also hinders the escape of

the charged particles produced by the thermonuclear reactions, provided the plasma dimensions exceed the particle cyclotron orbit diameter. This makes a self-sustained thermonuclear reaction possible because the energy from the particles can thermalize with fuel particles and balance power lost from the plasma as a whole. The minimum temperature for a self-sustained thermonuclear reaction is called the *ignition temperature*, and the plasma must be heated by external sources supplying power greater than the power loss up to the ignition temperature. At a minimum, the power loss is by bremsstrahlung photon radiation from free electrons deflected in coulomb collisions. With bremsstrahlung loss the ignition temperature is about 4×10^7 K. Much greater power loss may be produced by excitation radiation from electrons tightly bound to partially ionized heavy impurity atoms: only 0.1% of iron atoms will increase the ignition temperature to 7×10^7 K and $\geqslant 2.5\%$ of iron will prevent ignition completely. Since the fusion and radiation powers scale with density in the same way, the ignition temperatures are independent of density. It is therefore important to keep the plasma very pure.

Theoretically a uniform static magnetic field reduces the particle diffusion coefficient by about a factor $(\omega \tau_c)^2$, where τ_c is the collision time, ω ($=eB/m$) is the cyclotron frequency, and e/m is the charge-to-mass ratio of the particle. This is because the magnetic field restricts the trajectory of charged particles to small circular orbits so that the displacement at each collision is a cyclotron orbit diameter rather than a mean free path. The reduction of diffusion velocity is proportional to the square of the magnetic field intensity, and since e/m is larger for the electron than for ions, the electrons determine the overall plasma diffusion velocity; the ion density follows the electron density closely because very little charge imbalance can exist in a plasma. When the electron-ion collisions are simple two-body coulomb collisions, the diffusion is called *classical*. Fusion reactor parameters are such that $\omega \tau_c$ is sufficient for classical containment for the period of the fusion burn. However, in practice it is found that there are waves and instabilities in the plasma-magnetic field system and the particles act together in groups which effectively collide even under conditions where two-body coulomb collisions are negligible. The mechanism of these "cooperative phenomena" is not completely clear, but it appears that gradients of temperature, current, and density play a part. It has been argued that, at worst, the diffusion velocity in the absence of a magnetic field may be reduced by a factor $\omega \tau_c$ and not $(\omega \tau_c)^2$ as with classical diffusion. Such enhanced diffusion, with an appropriate numerical factor, called *Bohm diffusion*, scales as T/B and is independent of the collision frequency. Early experimental results indicated diffusion times of only a few times the Bohm rate, which is inadequate for a reactor, but in the last ten years the ratio has been increased to $10^3 - 10^4$, which is good enough.

Theoretically the thermal conduction should also be reduced by a factor of about $(\omega\tau_c)^2$ by the magnetic field, but, as with particle diffusion, much greater values are observed experimentally and often dominate the loss power. An increase of $\sim 100\times$ for electron thermal conduction and up to $\sim 10\times$ for ion thermal conduction are commonly observed. The loss is expressed in terms of an empirical energy containment time (τ) defined by the plasma energy density (nT) divided by the loss power. The fusion power balanced against the loss power gives the product $n\tau$ as a function of temperature. This has a minimum of

$$n\tau = 1.6 \times 10^{20} \text{ s/m}^3 \tag{4}$$

at about $T = 3 \times 10^8$ K, corresponding to another condition for ignition. Over a wide range of temperature near ignition the thermonuclear power is proportional to $n^2 T^2$, which together with the definitions of τ and β gives related ignition conditions $n T\tau \geqslant 4 \times 10^{28}$ or $\beta\tau > 3.3/B^2$ in the range of temperature of interest. For example, with $T = 1 \times 10^8$ K and $n = 2 \times 10^{20}$ m^{-3} an energy containment time of 2 seconds is required for ignition, which, with a typical value for the magnetic field of 5 Tesla, gives a requirement on β of 6.6%.

It is an important object of present research to understand the scaling and physical processes that determine τ and β and obtain ignition. The ignition condition given in Eq. (4) is similar to the condition for net power output from the reactor, derived by Lawson. If the efficiency of recycling the total power output to heat more plasma is 0.2 (probably optimistic), Lawson's condition gives $n\tau > 1 \times 10^{20}$ s/m^3 for a net power output.

Since the thermonuclear power density at the operating temperature is proportional to $n^2 T^2$, it follows from the definition of β that it is also proportional to $B^4\beta^2$. This shows that it is very important to operate at a high magnetic field and β. The limit on magnetic field is the strength of the magnet coils (which must support the magnetic pressure) or the loss of superconductivity. Typically, fusion reactors are designed for a magnetic field of between 5 and 10 Tesla. The limit on β is determined by equilibrium and stability of the plasma in the magnetic field which depending on the magnetic configuration of the system, may range from a few times 10^{-2} to a few times 10^{-1}.

The β, temperature and magnetic field determine the operating particle density of the fusion reactor to be near 10^{20} m^{-3}, which is about 10^{-5} of atmospheric particle density. After ignition some means of controlling the energy containment time is required to control the plasma temperature and therefore the thermonuclear burn, which could last hundreds of seconds. Too high a temperature may cause the β to exceed a stability limit, causing the plasma to expand and the burn to extinguish. Ideas to control the temperature are to vary the density, or inject particles, or adjust the loss by producing nonuniformities in the magnetic field, or change the magnetic field slightly to compress/decompress the plasma. Although the temperature of a thermonuclear fusion reactor is high, the low density means that the heat content is so low that there is no explosion hazard if the plasma were to expand to the wall of the containment vessel.

At the operating density and temperature, thermonuclear power density averaged over the plasma and blanket is considerably less than the power density in the core of a fission reactor. For instance, with $n = 2 \times 10^{20}$ m^{-3} and $T \approx 10^8$ K the total fusion power density is ~ 4 MW/m^3 compared to ~ 80 MW/m^3 core power density in a PWR fission reactor and ~ 500 MW/m^3 in a fast fission reactor. An important difference between fission and fusion reactors is the form in which the power appears. In fission the power is mostly thermal, whereas the power from D/T fusion is produced in two parts: 25% in α power and 75% in very fast neutrons. The neutrons pass out of the plasma and generate heat in a "blanket," which is a heat exchanger combined with a tritium breeder; it is coaxial with the plasma and has a thickness of ~ 0.5 m, fixed by the neutron mean free path. The minimum diameter of the tube containing the plasma (the "first wall") is determined by the thermal insulation produced by the magnetic field and the economic need to make it $\lesssim$ the thickness of the blanket and neutron shield. In practice the first wall diameter is in the region of 4 m.

The large minimum diameter of the first wall in fusion reactors also means that the first wall power load, which is proportional to the diameter, is an important limiting parameter in the reactor economy. The first wall conditions are more severe than for the pin clad in a fission reactor because of the following factors; a larger temperature gradient; a higher neutron energy (~ 10 times fast reactor value); thermal cycling due to pulsed burn of tens to hundreds of seconds; it must not liberate impurities; it must withstand the external atmospheric pressure; and it is more difficult to change because the complicated magnetic field coils and blanket system must be moved. At present, tentative fusion reactor designs are based on first wall total power loadings in the region of 1 MW/m^2 and allow for a new wall every 1 or 2 years. The crucial design base values for the first wall loading must await experimental measurements with high energy, high fluence neutrons to be produced in fusion engineering test facilities.

Although the size and cost of these facilities will be a considerable fraction of a full scale fusion reactor cost, a number are in the conceptual design stage, aiming at operation near the end of the 1990s.

The large characteristic dimension of the plasma and the limitation of power loading in the first wall may make the cost of the fusion nuclear island larger than that of the fission case. Recent estimates indicate a cost ratio of ~8, and, taking into account the cost of the conventional plant, the overall fusion cost has been estimated to be ~2 times that of a PWR. However, because so many of the factors are difficult to quantify, the costing of a fusion power station can at best be only very approximate.

Plasma Heating It can be shown simply from a balance of thermonuclear fusion power and power loss, allowing for a contribution from the hot α particles, that the power required to ignite the plasma is $\sim B^4 \beta_I^2 / 2$ MW/m^3 where β_I, the value of β at ignition, is $\sim 1.3 / \tau_e B^2$. For example, a thermonuclear fusion reactor with a plasma volume of 200 m^3, $B = 5$ T and β_I of 6% would require a pulse of ~56 MW heating power lasting for a few energy containment times (of ~1 second). With a heater efficiency of 15%, the power input to the heater would be ~380 MW. After ignition, the fusion burn may last for 100 seconds at a power output of 1 GW, so the total fusion energy produced is ~150 times the energy required to heat the plasma.

Many fusion reactor designs employ ohmic power to ionize and start heating the fuel charge. When the electron temperature (T_e) rises, the ohmic heating power falls because the resistivity is proportional to $T_e^{-3/2}$, and to ignite the plasma additional heating is required. Adiabatic compression heating, produced by suddenly increasing the confining field, is unsuitable because of the limited vacuum space left around the plasma which must be filled by expensively produced magnetic flux, and also because of engineering limits on the maximum field available. There are two methods being developed for the additional heating: neutral injection and the application of electromagnetic waves. At the moment which method is best is not clear.

The principle of neutral injection heating is to use a very high power beam of atoms which, being neutral, penetrates the magnetic field. When the beam collides with the plasma it is ionized and so is confined by the magnetic field, subsequently sharing its energy with the plasma. The neutral beam is made by passing positive ions from a high current ion source through a gas cell so that they pick up electrons from the gas atoms by charge exchange. Neutral injection heating has some fundamental limitations. First, the gas source current is limited by heating of the acceleration electrode grid to ~2 kA/m^2.

Second, to optimize the charge exchange efficiency in the gas cell, the beam energy should be chosen such that the velocity is approximately the orbital velocity of electrons in the atoms; against this the heating increases with energy so a compromise is chosen, typically ~160 keV for a deuterium beam with a corresponding efficiency of ~15%. Third, the plasma density must be such that the beam is absorbed in the plasma—too great a density causes absorption at the surface and too little density allows the beam to pass through the plasma. The density is probably rather lower than the density required for the thermonuclear burn, so further fuelling may be required, even allowing for fuel added by the beam. Last, because of the beam cross section, which is determined by the ion source current density limitations and beam divergence, a number of large holes are required in the vessel containing the plasma to allow the beams to enter. A significant fraction of fusion power and tritium production may be lost when neutrons escape through the holes, and, perhaps more important, neutron bombardment of the ion source will cause activation and maintenance difficulties. A way of overcoming the energy limitations which is at an early stage of development is to use a negative ion source. To produce the beam of neutral atoms, electrons can be removed from negative ions with a much higher efficiency (50–90%) than the charge exchange efficiency to positive ions. The efficiency is insensitive to the beam energy, enabling beam energies of 500 keV and a higher plasma density to be contemplated. At high beam energies, the time to thermalize the beam energy with the plasma may be important, during which a significant fraction of the fusion power may come from nonthermal particles.

Electromagnetic (RF) heating is at an earlier stage of development than neutral injection heating using positive ion sources. It has the advantages of higher efficiency and reduced size of the access holes in the blanket. There are three types of interest: electron cyclotron resonance heating (ECRH) which, typically, occurs at a frequency of 200 GHz (wavelength $\sim 1.5 \times 10^{-3}$ m); lower hybrid resonance heating, which occurs at a frequency around 3 GHz (wavelength $\sim 10^{-1}$ m); and ion cyclotron resonance heating (ICRH) with a frequency of about 65 MHz (wavelength ~5 m). Generation problems increase with frequency, whereas coupling the power to the plasma is best for ECRH.

The most successful method so far has been ICRH. This employs standard low frequency sources which have a high efficiency (around 90%) and overall, the efficiency of heating the plasma is between 20 and 60%. Input power densities of ~50 MW/m^2 have been investigated. The RF is launched from antennae near the plasma edge, producing a compression wave which propagates across the magnetic field towards the plasma center. The magnetic field is slightly nonuniform and the wave frequency is chosen to resonate with the ion cyclotron frequency at the plasma center so that absorption occurs. Originally it was intended to use absorption of the wave at the second harmonic because no absorption is possible in the fundamental mode. Although absorption at the second harmonic is expected to be strong near reactor temperatures it is rather weak in present experiments. Fortunately it was found that the presence of a few per cent concentration of another species (e.g., H or He) greatly increased the absorption efficiency because of the ion hybrid resonance and subsequently the majority species is heated by collisions. Potential difficulties with ICRH in a reactor are neutron bombardment and electrical breakdown of the insulator which support the antennae and power absorption near the wall which may liberate impurities. It may be possible to reduce the latter problem by focusing the waves toward the plasma center, but with ICRH it cannot be eliminated because the wavelength is appreciable compared to the plasma dimensions. A further problem is that the density must exceed $\sim 10^{19}$ m^{-3} for the waves to propagate, so that the antennae must be placed in the surface plasma.

In contrast, ECRH waves have no launching problems as long as the frequency exceeds the electron plasma frequency, which is proportional to the square root of the plasma density. The generation of powerful very high frequency fields has only recently been possible with the development of the gyrotron oscillator and the frequency condition should not seriously limit the plasma density. The waves are directed to the plasma via guides which are overmoded (approximately a few cm dimensions) to enable the high power densities to be carried. It is envisaged that ~ 200 0.3-MW gyrotons will be used. Advantages over ICRH are that because of the small wavelength the heating can be better focused at the plasma center; smaller holes are required in the blanket; and dog-leg waveguides can be used to avoid neutrons escaping. The overall efficiency is similar to ICRH, $\sim 50\%$. A crucial element of success will be the development of reliable high power sources.

Lower hybrid heating uses compression waves, like ICRH. The frequency, which is between the electron cyclotron frequency and the ion plasma frequency, is around a few GHZ, so dog-leg waveguides can be used, as with ECRH, avoiding neutrons escaping. The wave electric field must be polarized parallel to the plasma magnetic field for absorption. At the resonance, the wave becomes electrostatic and ions are heated directly. Difficulties are that the absorption depends on the plasma temperature, and the frequency of the waves must be changed as the heating proceeds; rather than move

radially, the wave spirals slowly to the plasma center, so that is spends a long time near the surface where it may heat the walls and liberate impurities; and lastly, like ICRH, the density must exceed a critical value for the wave to propagate. Powerful klystron (~ 1 MW/unit) sources are available to generate the waves, although development is required for long pulse operation. Low power experiments indicate an overall efficiency of $\sim 30\%$.

ECRH may have the potential for controlling certain plasma instabilities associated with adverse plasma current profiles as a function of radius. It is hoped to avoid the wrong current profile by using the heating to control the temperature and thereby the conductivity as a function of radius.

The currents are induced by a changing magnetic flux. Because of the limited magnetic flux swing that is possible in a transformer, fusion reactors that depend on plasma current may only be suitable for a pulsed burn. Pulsed systems are not desirable because the thermal and mechanical strains impose severe engineering limitations. A potential way of overcoming the problem is to use the heating method to drive the current continuously. This has been demonstrated in principle with lower hybrid heating, tangental neutral beam injection, and ECRH. Driven systems may have some economic penality in that about 10% of the reactor electrical output might be required to drive the current.

Magnetic Containment Although a uniform magnetic field greatly restricts diffusion perpendicular to the field lines, it has no effect on transport parallel to the field lines. To confine the plasma parallel to the field, the lines must be curved to form either a *toroid* or a *mirror*. A toroid is effectively a doughnut shape produced by bending a cylinder into a circle so that there are no ends through which the plasma can escape. The axis of the cylinder is called the *minor axis* and the axis about which it is bent is called the *major axis*. The magnetic field is produced by solenoidal coils placed around the bent cylinder. The magnetic mirror geometry is a straight cylinder with magnetic field again produced by coaxial solenoid coils. To restrict plasma flow parallel to the cylinder axis the magnetic field is increased near each end so that the field lines are curved near the region of increased field, creating a component of magnetic pressure which reflects particles moving axially back toward the main plasma. Unfortunately there is always a fraction of particles with sufficient energy directed parallel to the axis to overcome the magnetic pressure and escape. Furthermore, the fraction is maintained by collisions in the main body of the plasma. The great majority of magnetic confinement experiments are now toroidal, although in the US and USSR there is a "second line" of work on the mirror geometry. Advantages of mirror over toroid are better access, possibly better stability, and smaller unit size, but adequate containment of the axially moving particles remains a severe problem.

In a toroid, the magnetic field gradient is inward along the major radius because the solenoid coils must be more closely spaced on the inside of the cylinder, so concentrating the field. In this simple form, there would be loss of containment because the ions and electrons shift in opposite directions in the field gradient and the resultant charge separation causes an electric field to develop which, combined with the magnetic field, cause an outward drift of both ions and electrons. Containment can be restored by adding another magnetic field orthogonal to the first, looping the minor axis. The resultant magnetic field is then helical about the minor axis, and the angular displacement of the field line about the minor axis for one revolution about the major axis is known as the rotational transform. The effect is to allow particles moving parallel to the helical field lines to reduce the charge accumulation, thereby reducing the drift.

Three types of toroidal confinement geometry are being investigated: the stellarator, the tokamak, and the reversed field pinch. In a *stellarator* toroid, the helical transform field is generated by l pairs of helical coils outside the plasma, wound around the cylinder. Experimentally, values of $l = 2$ or 3 are investigated. The windings in each pair carry a positive and a negative current. A disadvantage of the stellarator field configuration is the fall-off of field from the coils, making a given field more expensive to produce and the maximum magnetic field in the plasma less than with a simple solenoid. Another problem is that, to accommodate the stellarator coil structure, a large ratio of major radius to minor radius is required which may result in an inconveniently large reactor size. Against this, there are the important potential advantages that the confining field is determined by currents in fixed conductors external to the plasma enabling the possibility of continuous operation, and containment and heating are possible without the necessity of currents in the plasma which may cause instabilities.

The second and most developed confinement system is the *tokamak* toroid; it originated in the USSR and its name derives from the Russian words for toroidal chamber and magnetic field. A simple solenoid, coaxial with the minor axis is used to produce the main or (toroidal) magnetic field (B_T), and the transform is made by a toroidal plasma current parallel to the minor axis instead of by external coils. This produces a weak component of magnetic field (the poloidal field, B_p) around the minor axis; typically $B_p/B_T \approx 0.1$. The current is induced by transformer action (either air-core or iron-core). A measure of the effectiveness of using the poloidal field to achieve confinement is the *poloidal beta*, defined by $\beta_p \propto nT/B_p^2$ which is ≈ 1. Since the current must not be allowed to fall to zero, the thermonuclear burn must be produced in a pulse lasting for the period of the transformer action.

The helical transform in the tokamak is not able to provide equilibrium because, unlike the stellarator, the current that produces the helical field is not flowing in a rigid conductor. Equilibrium is, however, easily achieved either by using a conducting wall (called a *flux-conserving tokamak*) so that the outward pressure compresses flux between the plasma and wall, or, more commonly, by applying a vertical magnetic field (B_v) which, acting with the plasma current I, produces a force $I \times B_v$ to balance the outward pressure force. With the latter method, B_v may be varied by a feedback circuit controlled by optical sensors of the plasma position. This method of maintaining equilibrium does, however, limit the plasma pressure, because of the condition that there should be no null field point in the plasma which would be a source of plasma loss. This gives a maximum $\beta_p \lesssim 1/2\epsilon$ where ϵ = plasma inverse aspect ratio = minor radius/major radius. For a typical ϵ of $\frac{1}{3}$, $\beta_p \lesssim \frac{3}{2}$.

A third type of toroidal confinement geometry is the *reversed field pinch*. It differs from the tokamak in that a shear condition for plasma stability is used. The condition is obtained by making $B_p \sim B_T$ with the field reversed near the plasma surface. As with the flux-conserving tokamak, equilibrium is obtained by using a conducting shell.

An important theoretical aspect of toroidal systems is the effect of the helical geometry on confinement. Particles moving along helical magnetic flux tubes experience a decrease in magnetic field as the flux loops around the outside of the minor axis, causing the tube to bulge. Thus, if a length of helical flux is unwound it would have a series of constrictions which act like a succession of magnetic mirrors. Thus particles with a low velocity parallel to the flux tend to congregate in the low magnetic field regions: these are called *trapped particles*. The rest of the particles, which have a high parallel velocity, pass unimpeded along the flux tube; these are called *passing particles*. Because of either trapping or collisions some particles cannot move around the helix to allow cancellation of the charge separation due to drifting in the magnetic field gradient. The noncancellation results in an outward drift, enhancing the diffusion rate above the classical value. Three regimes of enhanced diffusion (called *neoclassical*) have been identified, depending on the electron-ion collision rate ν. At low density the collision rate is so low that some particles can bounce back and forth many times between the mirrors. This is called the *banana diffusion regime*, which is named after the shape of the trajectory of the particle motion. The diffusion rate is $q^2 \epsilon^{-3/2}$ times the classical value where the safety factor $q = \epsilon B_T / B_p$. Typically, $q \sim 2$ and $\epsilon \sim \frac{1}{3}$ and the classical diffusion is increased by $\sim 20 \times$. When collisions within the mirror are much more frequent than the bounce frequency there are no trapped particles and charge cancellation is inhibited by collisions of the passing particles, increasing the classical diffusion by a factor $1 + q^2$ which is $\sim 5 \times$; this is called the Pfirsch-Schlüter regime. In the intermediate collisional regime the enhancement factor is v_e / Rqv where v_e is the electron thermal velocity. Since the classical diffusion is proportional to v, the enhanced rate is independent of v, producing a flat region of the graph of diffusion vs. v. Consequently this is called the plateau regime.

Other nonuniformities in the magnetic field may lead to other trapped particle effects. For example in stellarators, the multipole fields cause helical modulations which make the mean position of the bananas drift, leading to plots of the orbit positions that look like large bananas, and consequently called *super-bananas*.

A related effect of gradients on particle trapping is ripple caused when solenoidal cells are used. A ripple of a few per cent on the minor axis can cause serious loss, particularly for α particles because of their large cyclotron orbits.

Further theoretical predictions of trapped particles effects include the *bootstrap current* and the *trapped particle pinch effect*. The bootstrap current is caused by the radial diffusion velocity (v_r) of electrons in the banana regime. The diffusion velocity acting with the poloidal magnetic field produces a force proportional to $v_r B_p$ along the minor axis acts on the untrapped ions and electrons to produce an axial current. The current is predicted to be sufficient to provide a tokamak rotational transform up to a $\beta_p \approx 4\epsilon^{-1/2}$, offering a possible method for a continuously operating tokamak fusion reactor. The trapped particle pinch effect (also called the *Ware pinch effect*) is caused by the toroidal electric field used to drive the tokamak current. It makes the banana orbits drift towards the minor axis until the pinch force is balanced by the pressure gradient force.

Stability The temperature and density in a fusion reactor correspond to conditions in the banana neoclassical regime, and gives an $n\tau$ value easily satisfying the ignition and Lawson conditions. However the experimentally measured diffusion coefficients often exceed the neoclassical values. This is thought to be due to various instabilities in the plasma-magnetic field system. The instabilities are characterized by two mode numbers: m, corresponding to the number of perturbations in the minor cross section, and n, corresponding to the number of perturbations parallel to the plasma minor axis. The lowest surface mode, $m = 0$, is symmetrical about the minor axis and is due to

the poloidal magnetic field (B_p) produced by the plasma current when the toroidal magnetic field is comparatively small. The system is unstable because B_p increases as the plasma radius decreases and so a given initial inward perturbation grows. The resultant deformation, known as the *sausage instability*, may be stabilized by an axial field inside the plasma of magnitude $> B_p / \sqrt{2}$. The next surface mode, when $m = 1$, $n = 1$, is caused by increased pressure inside a given kink perturbation, where the field lines are close together, increasing the amplitude of the perturbation. It is stable when q at the plasma surface is greater than 1; this is the *Kruskal-Shafranov condition* and is basic to tokamak operation. In a tokamak higher mode stability can be achieved by suitably peaking the current profile toward the plasma axis. Kink modes may also be stabilized by image currents if a conducting wall is put close to the plasma.

Inside the plasma the interchange of plasma and tubes of magnetic flux can cause instability (called *internal pressure driven modes*). For stability, the average curvature of the lines of force must be convex towards the plasma, leading to the "minimum B" condition $\delta \int dl/B > 0$

taken along a flux tube. These internal modes are localized near surfaces at which the lines of force join on themselves when they pass around the torus, i.e., when $q = m/n$. Suydam showed that they can also be stabilized by changing the angle of the helical magnetic field as a function of distance from the minor axis by an amount depending on the pressure gradient. The sheared field acts like a basket weave that prevents interchange of plasma and flux.

However nonlinear effects can cause a more complicated situation. For example, the plasma current causes heating which changes the conductivity, so affecting the current and magnetic field distribution. Also the finite resistivity, which is a function of temperature, allows magnetic field diffusion such that oppositely directed field lines may join to form loops in the plasma. This is called *resistive instability* and the loops are called *magnetic islands*; resistive instabilities may limit the effectiveness of shear to stabilize internal modes. Further, various radial regions of the plasma may be connected by rapid parallel diffusion along wandering magnetic flux lines. These are called *ergodic regions*. Overall, it seems that while the plasma under some conditions may be grossly stable, mild nonlinear effects of instabilities within the plasma may tend to flatten the internal pressure profile, increasing the pressure gradient near the surface and enhancing diffusive loss processes.

Finally, a class of interchange instabilities called *ballooning modes* should be mentioned. These are local (high m, n) modes associated with the unfavorable curvature of magnetic field near the outer regions of the toroidal plasma. Ballooning may be imagined as like the unstable bulge at the weak point of an over-pumped bicycle inner tube. Calculations indicate that these modes may limit the maximum β in a tokamak to between 5 and 10%, although it is possible that nonlinear effects may allow the limit to be exceeded without disastrous loss.

It should be emphasized that plasma currents may be the source of many of these instabilities. Thus stellarators, in which the transform may be produced without plasma current, is likely to be most stable, followed by tokamaks and lastly by the reversed field pinch. The reversed field pinch operates with $q < 1$, relying on image currents in the conducting wall and magnetic shear for stability. Its cylindrical geometry necessitates a region of reversed magnetic field near the plasma surface. This automatically develops at an early unstable phase of the discharge, and is probably maintained by continuous mild resistive instabilities. The advantage of the reversed field pinch is its ability to operate at a much higher β than the tokamak; theoretically the maximum is 0.5.

Experiments In the last ten years world fusion research has mainly concentrated on the tokamak geometry showing continuous progress in extending the plasma parameters towards ignition with increased size of apparatus. Recent experiments have a major radius of ~1.3 m, a minor radius of ~0.5 m, $B_T \sim 5$ T and $I \sim 6 \times 10^5$ A, giving the record plasma parameters: $T \sim 7 \times 10^7$ K (at $n \sim 5 \times 10^{19}$ m^{-3}), $n\tau \sim 3 \times 10^{19}$ (at $n \sim 10^{21}$ m^{-3} and $T \sim 10^7$ K), $\beta \sim 4\%$ and $\tau \sim 10^{-1}$ s; these values were not all obtained at the same time, or in the same apparatus. In many respects the scaling of these quantities encourages the belief that a tokamak fusion reactor with reasonable dimensions may be physically feasible. A major milestone will be the control of plasma heated by thermonuclear fusion, which is an object of experiments at present being constructed.

However, although much sophisticated theory has been developed, many of the basic results continue to defy satisfactory explanation. The tokamak results give little support for neoclassical diffusion in that for years the electron thermal diffusion coefficient was observed to scale as $1/n$ and be independent of T and B. This corresponds to an energy confinement time scaling $\tau \propto na^2$ first discovered on the experiment Alcator. Recent larger experiments show the rather different scalings $\tau \propto na^2 T^{1/2}$ or $na^2 q^{3/4}$ at lower density and a peak of τ as a function of density at high density. Also no bootstrap current has been observed, the observed β as a function of heating saturates at high heating power, and the plasma is catastrophically unstable (disrupts) above an average density (the Murakami limit) which is proportional to the current density. The limits on

τ, β, and n may cause difficulties in achieving ignition or obtaining economical operation. However, they may be connected with the evolution of impurities at high power operation so that improvements can be expected from methods to improve plasma cleanliness.

Many of these effects seem to be connected with the presence of instabilities. Under many conditions ripples move along the plasma (Mirnov oscillations) and flat regions of the temperature profile have been observed which could be interpreted as the effect of magnetic islands. At the Murakami density limit the plasma may catastrophically disrupt. Although the plasma energy in a reactor is small (~1 MJ), there is concern that rapid release of magnetic energy may strain the system.

A number of tokamak experiments are aimed at finding ways of injecting fuel and keeping the plasma clean. Fuel injection is normally by puffing gas on the outside of the plasma, relying on an inverted density gradient or perhaps the trapped particle pinch effect to transport new fuel inward. Other experiments are to investigate the practicality of injecting small pellets of fuel. Experiments to cleanse the plasma involve scraping the outer layer of impure plasma into a special chamber where it is neutralized and pumped away. The scraping is achieved by diverting the magnetic field lines by using either special coils for a cooled deflector, called a *pumped limiter*.

The second type of toroidal system, stellarator, has received less support than tokamaks. This is probably because of the relative success of tokamaks in the early days when stellarators gave containment times of only a few Bohm times, probably due to a combination of impurities, field errors, and the heating method. However in the last decade stellarator progress has been encouraging. Small low temperature (~10^4 K) devices give classical containment, while larger devices ($T \sim 3 \times 10^6$ K) are beginning to give neoclassical results in the plateau regime, as expected. Recently ion temperatures of ~7×10^6 K, densities of 10^{20} m^{-3}, $\beta \sim 1\%$ and $n\tau \sim 4 \times 10^{18}$ have been reported with a radius of only 0.1 m and $B \sim 3$ T. In some experiments the plasma is heated by an axial current which causes an added transform; it is found that as the current is reduced, the energy confinement time increases. Experiments, where there is no current and the plasma is either heated by neutral beam injection or RF, show the best confinement time and have no MHD activity or major disruptions. This indicates that the plasma currents, essential for the tokamak configuration, are detrimental to containment and may be responsible for the density limitations. Theoretically $l = 2$ or 3 stellarator equilibria exist for β up to 5%, and there should be stability up to $\beta \sim 10\%$ when small ϵ (~0.05) and superimposed fields of different pitch are used. It appears that because of resistive instabilities, shear in the magnetic field may not be sufficiently maintained to suppress instabilities, and a minimum B magnetic configuration is required.

The origin of the reversed field pinch is the observation of a quiescent period that automatically developed in unstable plasma in the large Zeta toroidal experiment of the early 1960s. The parameters were major radius 1.5 m, minor radius 0.5 m and toroidal current 4×10^5 A, giving a plasma temperature $\sim 1.5 \times 10^6$ K, density $\sim 5 \times 10^{19}$ m^{-3}, $\beta \sim 0.1$ and energy containment time ~ 0.003 to 0.01 s. This was explained by subsequent low β theoretical analysis which showed that, in certain ranges of poloidal and toroidal magnetic field, a force-free shear-stabilized magnetic field configuration with reversed field near the surface is possible. In the 1970s the results were confirmed and experiments to extend them to higher temperature and β were started. A major objective is to determine whether the magnetic configuration can be maintained for the burn period of a fusion reactor either automatically by noncatastrophic small-scale instabilities or by the application of a reversed field at the plasma surface.

The only linear containment system with fusion reactor potential is the mirror. In its simple form it is unstable because interchange instabilities may develop where the magnetic field lines are concave towards the plasma. This was overcome by adding an extra field produced by current flowing in conductors placed outside of the plasma. This produces a complicated magnetic field configuration that is minimum β

stable; the curved field lines are, of course, only required near the ends when the system is long. Although the containment is grossly stable too many particles acquire enough parallel velocity to penetrate the mirrors. A possible way to overcome the problem that is being investigated is to heat the plasma in the mirror regions preferentially, so that the axial temperature gradient produces an electrostatic field directed to reduce the loss. Encouraging results of $T \sim 3 \times 10^6$ K, $n \sim 3 \times 10^{19}$ m^{-3}, $\beta \sim 0.1$-0.5, and $\tau \sim$ a few times 10^{-3}, have been achieved and work is proceeding on larger experiments to investigate the scaling.

Toward Reactors Although there are problems to be overcome in all confinement systems, the reactor potential of tokamak has received most attention, followed by stellarators, mirrors, and reversed field pinches.

Table 1 shows the parameters and objectives of large tokamak experiments on the path to a reactor. The first three will be in full operation by the mid 1980s. The largest, JET, has a D-shaped cross section which maximizes the β. The total cost will be $\sim \$450$ million. It will operate at high density, and may even achieve the ignition $n\tau$ product. In contrast, TFTR aims to operate at a lower density where nonthermal reactions rather than ignition are significant, possibly just achieving the Lawson $n\tau$ product for energy breakeven. As a physics milestone in fusion, the $n\tau$ value has been compared to criticality in fission development, although it should be emphasized that it is not so definitive.

The next three tokamak experiments listed

TABLE 1.

Experiment	Objective	R (m)	a (m)	B_T (T)	I_T (MA)	Pulse Time (s)
JET (Euratom)	Approach to ignition Maximize β Long pulse	3.0	0.3×2.1	3.4	4.8	20
TFTR (USA)	Approach to energy breakeven	2.7	0.9	5.2	2.5	1.6
JT60 (Japan)	Ignition parameters in hydrogen plasma Impurity control	3.0	1.0	4.5	2.7	5.0
TORUS II (France)	Superconducting coils Long pulse RF heating	2.2	0.7	4.5	1.7	—
T-15 (USSR)	Superconducting coils Approach to ignition	2.4	0.8	3.5	1–2	—
INTOR (IAEA)	Ignition Blanket design Fusion burn control Technology of electricity producing systems	5.2	1.3	5.5	6.4	100

are to investigate particular reactor requirements: impurity control, long pulse, RF heating, superconducting coils, etc. Lastly, INTOR is an ignition experiment to test crucial engineering aspects such as blanket design, materials, tritium handling, plasma burn control, and even limited electricity production. In 1982 it was in a conceptual design stage, the final design awaiting results on near-ignition conditions from the JET generation of experiments. It is hoped to be able to start construction of INTOR late in the 1990s. The subsequent stage of development will be a full scale prototype reactor. One of the major problems will be first wall erosion caused by the high energy neutrons, estimated to be ~100 metric tons of material per gigawatt (electric)-year.

It has been suggested that advantage could be taken of the high neutron energies by using the neutrons to produce fission power from U^{238}, or to breed plutonium from U^{238} which is ~10 times more efficient than in fast reactors because of the harder neutron spectrum. These schemes are known as *hybrid fusion-fission*. They have the advantage of lower energy breakeven $n\tau$ than pure fusion, but at the expense of complicated engineering construction and the loss of low radiation hazard compared to fission reactors.

The cost of magnetic confinement fusion research in 1982 was ~$330 million in Europe, $200 million in Japan, $460 million in the USA, and probably a similar amount in the USSR. These amounts are small compared to that spent on energy as a whole. Fusion power has a large number of unresolved problems which may only be investigated with large apparatus; the risk of failure is far from negligible, but the benefit of success would be great.

J. A. REYNOLDS

References

"Comments on Modern Physics, Part E," New York, Gordon and Breach.

Reports on the bi-annual conferences on Plasma Physics and Controlled Nuclear Fusion Research, Vienna, Austria, International Atomic Energy Agency.

Cross-references: BREMSSTRAHLUNG, FISSION, NUCLEAR REACTIONS, NUCLEAR REACTOR, PLASMAS, LASER FUSION.

G

GAS LAWS

The term "gas law" refers to the thermodynamic equation of state of a gas, which is an equation relating the pressure p, the volume V, the absolute temperature T, and the number of moles ν. The equation of state is a valid relation when and only when the gas is in a state of thermodynamic equilibrium; the pressure and temperature are then constant and uniform throughout the volume occupied by the gas.

Ideal or Perfect Gas The ideal or perfect gas is defined thermodynamically by the two conditions: (1) it obeys the equation of state: $pV = \nu RT$ where R is the gas constant per mole ($R = 8.3169 \times 10^7$ erg mole^{-1}°C^{-1}), and (2) the internal energy U is independent of pressure and volume and is a function only of the temperature ($(\partial U/\partial V)_T = 0$). The statistical-mechanical definition of an ideal gas is that it is a gas of noninteracting molecules, i.e., the molecules exert no appreciable forces of attraction or repulsion on each other. Since the notion of a finite "size" of a molecule connotes the existence of a repulsion which prevents two molecules from overlapping each other, the molecules of an ideal gas must be of negligible "size." The two thermodynamic properties can be deduced from the statistical mechanical definition.

The ideal gas equation: $pV = \nu RT$ embodies the experimental laws of Boyle, Charles and Gay-Lussac. It can be derived either from kinetic theory or from statistical mechanics. It is often written in the form: $p = nkT$ where n is the molecular number density and k is Boltzmann's constant ($k = 1.3804 \times 10^{-16}$ erg °C^{-1}). In the case of a mixture of inert, ideal gases, each gas obeys the equation: $p_i = n_i kT$ where p_i and n_i are, respectively, the partial pressure and partial density of the ith component gas. Boyle's law will not hold if the gases in the mixture react chemically since a change in p or V will in general change the value of ν.

Real or Imperfect Gas The ideal gas law is, of course, only an approximation which holds at temperatures sufficiently far above the critical temperature and at sufficiently low densities. The ordinary properties of bulk matter in the liquid and solid states require the existence of strong intermolecular repulsions which endow the molecules with a finite "size" and also require the existence of attractive forces to hold the molecules together. The equation of state of a real gas is therefore determined by the nature of the intermolecular forces. One of the earliest, simplest, and most useful equations is that of van der Waals

$$\left(p + \frac{a}{V^2}\right)(V - b) = RT \text{ (for 1 mole)} \quad (1)$$

where a and b are constants, determined empirically for each gas, which are related to the attractive and repulsive forces, respectively. This equation can be related theoretically, in first approximation, to a molecular model in which the molecules are represented by rigid elastic spheres that weakly attract each other. The van der Waals equation accounts qualitatively for the liquid-vapor phase transition. The constants a and b can be determined from critical point data.

Other equations of state for an imperfect gas have been proposed which are more accurate than the van der Waals equation, e.g., the equations of Dieterici, Berthelot, Beattie-Bridgeman, and Benedict-Webb-Rubin. These empirical equations are useful in treating the thermodynamic properties of gases at high densities. At low densities, the empirical equations have been superseded by *the virial equation of state*

$$\frac{pV}{RT} = 1 + \frac{B(T)}{V} + \frac{C(T)}{V^2} + \frac{D(T)}{V^3} + \cdots \text{ (for 1 mole)} \quad (2)$$

where $B(T)$, $C(T)$, and $D(T)$ depend on the nature of the gas and are called the second, third and fourth virial coefficients, respectively. The departures of a gas from ideality are represented in this case by a power series in the density. We may rewrite Eq. (2) as

$$p/kT = n + \hat{B}(T)n^2 + \hat{C}(T)n^3 + \hat{D}(T)n^4 + \cdots \quad (3)$$

where $\hat{B}, \hat{C}, \hat{D}$ are the virial coefficients referred to one molecule.

The basic experimental problem in this field is to measure the virial coefficients of different gases as functions of the temperature. The higher-order coefficients beyond $B(T)$ and $C(T)$

are very difficult to measure. The basic theoretical problem is to calculate the virial coefficients from an assumed form for the intermolecular potential energy and, ultimately, to derive the intermolecular potential from quantum mechanics.

Statistical Mechanics of the Imperfect Gas The derivation of the virial equation of state from the intermolecular potential involves several steps which may be summarized as follows:

lightest gases H_2 and He and especially at low temperatures, it is necessary to introduce quantum corrections in the equation of state.

Intermolecular Forces and the Equation of State In order to calculate the cluster integrals and virial coefficients, one must first choose a form for the intermolecular potential $u(r)$. In principle, the potential $u(r)$ is determined by quantum mechanics for any pair of molecules and could be found by solving the Schrödinger

---

Virial Equation   ←Helmholtz Function   ←Canonical Partition Function
                            ←Grand Potential (PV) ←Grand Partition Function

←Configuration Integral   ←Cluster Integrals   ←Intermolecular Potential

---

The two routes indicated are via the canonical and the grand ensembles. The most difficult step is the evaluation of the configuration integral:

$$Q_N = \int \cdots \int \exp \left[-\Phi(r_1, r_2, \cdots r_N)/kT \right]$$

$$d\tau_1 \, d\tau_2 \cdots d\tau_N \quad (4)$$

where Φ is the total intermolecular potential energy of the gas of N molecules. The proper way to evaluate Q_N was first sketched by Ursell and later carried through by Mayer who assumed central forces that were pairwise additive, i.e.,

$$\Phi = \sum_{i<j} u(r_{ij})$$

where $u(r_{ij})$ is the potential between molecules i and j. Neither of these assumptions is correct, but they appear to be good approximations, and with their aid, it is possible to evaluate Q_N rigorously in terms of the so-called cluster integrals, b_l, which are integrals over the coordinates of l molecules only. $B(T)$ is obtained directly from b_2, $C(T)$ is obtained from b_3 and b_2, and the lth virial coefficient requires evaluation of the cluster integrals up through b_l. Explicit formulas for $\hat{B}(T)$ and $\hat{C}(T)$ are:

$$\hat{B}(T) = -\frac{1}{2V} \iint f_{12} d\tau_1 \, d\tau_2 \quad (5)$$

$$\hat{C}(T) = -\frac{1}{3V} \iiint f_{12} f_{23} f_{13} \, d\tau_1 d\tau_2 d\tau_3 \quad (6)$$

where $f_{ij} \equiv \exp \left[-u(r_{ij})/kT \right] - 1$. Higher coefficients in the virial series are increasingly more difficult to evaluate.

The calculations just described are based on the classical Maxwell-Boltzmann statistics and are sufficiently accurate for most gases at ordinary temperatures. However, in the case of the

equation. In practice, this is virtually impossible, and quantum-mechanical calculations have been made only for the very simplest molecules. In the case of interactions between neutral, nonpolar, spherical molecules, e.g., noble-gas atoms, the quantum-theoretical interaction energy can be approximately decomposed into several parts, of which the two most important are the *dispersion energy* and the *valence repulsion energy*. The former corresponds to the van der Waals attraction and the latter to the van der Waals repulsion. The dispersion energy varies inversely with the sixth power of the distance. The valence-repulsion energy takes account of the short-range repulsion that sets in when the electron distributions of the two molecules begin to overlap, and it is associated with the Pauli exclusion principle. There is no simple, general form for the dependence of the valence repulsion potential on the distance: it is often empirically represented by $A e^{-ar}$ or by μr^{-n} where $n = 12$ is commonly used. In the case of molecules that possess permanent electric dipole or quadrupole moments, there are additional contributions to the van der Waals attraction but these are usually less important than the dispersion energy (H_2O is an exception).

In the absence of a complete quantum-mechanical expression for the intermolecular potential, it is necessary to approximate the potential by a semi-empirical formula, containing one or more adjustable constants, which is chosen on the grounds of physical plausibility and mathematical convenience. The semi-empirical force law is then used to calculate macroscopic properties that are known from experiment, and the parameters in the force law are adjusted to give the best agreement with experiment. Given the form of the intermolecular potential, it is possible to calculate not only the virial coefficients in the equation of state but also the kinetic-theory transport coefficients (i.e., the viscosity coefficient, the thermal conductivity, and the various diffusion coefficients of the gas) and the density, compressibility, and sublimation energy of the solid. A particular functional representation of the intermolecular potential can be considered satisfactory only if it is possible to secure agreement with all experimental data involving a particular pair of molecules with a single choice of the parameters that appear in the law of force.

The semiempirical law that is most frequently used to represent the interaction between nonpolar molecules is the *Lennard-Jones* (12, 6) *potential:*

$$u(r) = 4\epsilon[(\sigma/r)^{12} - (\sigma/r)^6] \qquad (7)$$

where ϵ and σ are parameters characterizing the particular pair of molecules. This simple two-parameter function, when inserted in Eq. (5), predicts a temperature variation of the second virial coefficient in good agreement with experiment. The same potential with slightly different values of ϵ and σ also explains the temperature variation of the viscosity coefficient over a substantial temperature range. Third virial coefficients calculated from Eq. (6) do not agree with experiment at low temperatures near the critical point. The large discrepancies have been attributed to three-body forces which invalidate the assumption of pairwise additivity. Calculations of virial coefficients and transport coefficients of gases and of equilibrium properties of the crystal lattices have also been made for other semiempirical potential functions, but the results are not very different from those found with the (12, 6) potential. Nevertheless, there is accumulating evidence both from experiment, e.g., atomic beam scattering, and from theory, e.g., quantum-mechanical calculations of the dispersion energy, that the (12, 6) potential has serious defects, and it has been replaced in recent work by more complicated multiparameter potential functions.

Further advances in this field will come, on the theoretical side, from a more detailed knowledge of the intermolecular forces, and on the experimental side, from more accurate ways of extracting virial coefficients from thermodynamic data. Current values of $C(T)$ are not only subject to experimental errors in the p, V, T measurements but also to substantial uncertainties incurred in fitting the data with polynomials in the density.

Dense Gases High-density gases cannot be conveniently represented by the virial equation of state because of the slow convergence of the virial series. Furthermore, the theoretical evaluation of the higher virial coefficients on the basis of any plausible molecular model would meet with great computational difficulties. Other approaches are therefore needed, e.g., the empirical equations of state already mentioned and the principle of corresponding states. In the latter method, one introduces the reduced, dimensionless variables: $p_r = p/p_c$, $V_r = V/V_c$, and $T_r = T/T_c$ where the subscript c refers to the critical point. The principle of corresponding states then asserts that all substances obey the same equation of state in terms of the reduced variables. The variables may also be reduced in terms of intermolecular potential parameters.

A promising theoretical approach to the equation of state of a dense gas or liquid is provided by the method of the radial distribution function $g(r, n, T)$. Because of intermolecular forces, the actual density at a small distance r from a given molecule is different from the bulk density n and is represented by $ng(r, n, T)$. Thus the radial distribution function measures the effect of intermolecular forces on the probability of finding two molecules close together. While it is difficult to determine $g(r, n, T)$ theoretically, it can be found experimentally from the diffraction pattern observed when x-rays are scattered by the fluid.

R. D. PRESENT

References

Cowling, T. G., "Molecules in Motion," London, Hutchinson & Co., Ltd., 1950 (for the general reader).

Hill, T. L., "Statistical Mechanics," Ch. 5 and 6, New York, McGraw-Hill Book Co., 1946 (statistical mechanics of imperfect gases and dense fluids).

Hirschfelder, J. O., Curtiss, C. F., and Bird, R. B., "Molecular Theory of Gases and Liquids," Chs. 3 and 4, New York, John Wiley & Sons, 1954 (covers all aspects of the subject and is the standard reference in this field).

Present, R. D., "Kinetic Theory of Gases," Ch. 6 and 12, New York, McGraw-Hill Book Co., 1958 (kinetic theory of the second virial coefficient; intermolecular forces).

Rushbrooke, G. S., "Introduction to Statistical Mechanics," Ch. 16, London, Oxford University Press, 1949 (good introduction to imperfect-gas theory).

Levelt Sengers, J. M. H., Klein, M., and Gallagher, J. S., "Pressure-Volume-Temperature Relationships of Gases, Virial Coefficients," AEDC-TR-71-39, 1971, and American Institute of Physics Handbook.

Cross-references: COMPRESSIBILITY, GAS; GASES: THERMODYNAMIC PROPERTIES; INTERMOLECULAR FORCES; KINETIC THEORY; THERMODYNAMICS.

GASES: THERMODYNAMIC PROPERTIES

Fundamental Principles The thermodynamic properties of a substance may be classified as either reference properties, energy functions, or derived properties.[1] The reference properties of a single-component system with their symbols and units are pressure, p, Pa or Nm^{-2}; volume, V, m^3; temperature, T, K; and entropy, S, JK^{-1} mol^{-1}. For a specific amount of a pure gas, it is necessary to specify only two of these reference properties to fix the state of the system and its properties. For mixtures of gases, the composition must also be specified to fix the system completely. The energy functions with their symbols and units are internal energy, U, J mol^{-1}; enthalpy, H, J mol^{-1}; Helmholtz energy, A, J mol^{-1}; and Gibbs energy, G, J mol^{-1}. These functions represent the energy available for performing useful work under various process conditions. Derived properties with their symbols and units include heat capacity, C, JK^{-1} mol^{-1}; fugacity, f, Pa or Nm^{-2}; compression factor, Z, unitless; and the Joule-Thomson coefficient, μ, K Pa^{-1} or Km^2 N^{-1}.

Properties are termed intensive if they are independent of the amount of the material. Examples are pressure and temperature. Properties such as volume and entropy, which are dependent on the amount of material, are termed extensive.

Absolute values may be determined for the reference properties, but the energy functions must be determined relative to an arbitrary zero reference point. The internal energy, U_0°, of the ideal gas at the absolute zero of temperature is generally taken as the zero reference point of the enthalpy and free energy functions. Other reference points include a zero value for the enthalpy of the ideal gas at the ice point, $H_{273.15}^\circ$, and another in which the sensible enthalpies are combined with chemical energies.[2] In the latter base, the value of $H_{298.15}^\circ$ is zero for the assigned reference elements so that the values of $H_{298.15}^\circ$ for the various compounds are equal to their heats of formation from the assigned reference elements.

Thermodynamic properties of gases are calculated for both the ideal gas and the real gas state. A gas is defined to be ideal if it follows the simple equation of state, $pV = RT$, for 1 mole of gas. Gases behave in this manner only at very low pressure, but the ideal gas state is a convenient reference state for the calculation of the thermodynamic properties. Thus, the thermodynamic standard state[3] is defined as the ideal gas at 1 atmosphere (= 101 325 Pa) pressure at each temperature, and it is denoted by a superscript degree mark as in H° and S°. The ideal gas properties have been calculated for many substances, but the real gas properties are known for relatively few substances.

Thermodynamic properties are used in the calculation of energy balances, reaction compositions at chemical equilibrium, reaction temperatures, and the work involved in the compression or expansion of gases in various systems.

Ideal Gas Properties The thermodynamic properties of an ideal gas such as heat capacity at constant pressure, C_p°, enthalpy function, $(H^\circ - H_0^\circ)/T$, entropy, S°, and Gibbs energy function, $(G^\circ - G_0^\circ)/T$, in units of JK^{-1} mol^{-1}, are calculated from theoretical equations and from an analysis of spectroscopic and molecular structure data.[4-7] These complex calculations are based upon the contributions from all of the energy states available to the molecule, such as translational, electronic, vibrational, and rotational. Contributions from excited electronic

states are important for diatomic molecules at higher temperatures but are entirely negligible for most polyatomic molecules. Vibrational energy levels are obtained from an analysis of infrared and Raman spectroscopic data by applying the principles of wave mechanics and group theory.[4,7,8] The interpretation of the spectra includes the assumption of a model for the molecule with parameters such as bond lengths, bond angles, and force constants. The parameters are varied within certain limits until the best agreement with observed spectra is obtained. Rotational energy levels are observed in infrared, Raman, and microwave spectra. The rotational energy includes not only the rotation of the molecule as a whole but also internal rotations by groups of atoms within the molecule and a pseudorotation in some ring molecules.[9]

For higher orders of accuracy in the calculation of the thermodynamic properties, additional contributions may be determined that are caused by the interaction between vibrational and rotational motions, centrifugal distortion of the molecule during rotation, and anharmonicity of the vibrations. Another contribution due to nuclear spin can be, and generally is, neglected for all molecules except H_2 and D_2 since it causes a detectable effect on measurable quantities only at very low temperatures. Adjustments are made for the presence of isotopes in some diatomic and polyatomic molecules.

The calculation of the thermodynamic properties of an ideal gas is based not only on theory but also on accurate experimental vapor heat capacity, heat of vaporization, and low-temperature calorimetric data.[9] The ideal gas heat capacity C_p° and entropy S° are derived from these data and are compared to theoretical values. When differences are found, the theoretical calculations are revised until there is good agreement. In this manner, new information is gained about the conformation of the molecule, its frequencies of vibration, etc. Thus, experimental data provide a firm base to test theoretical calculations and improve the calculation of all of the thermodynamic properties.

Theoretical calculations become increasingly complex as the molecular size increases. Thus, a method of increments has been devised to calculate the thermodynamic properties of large molecules based on an "anchor compound" for a given homologous series.[1,10]

Real Gas Properties The thermodynamic properties of real gases are determined primarily from experimental compressibility (pressure-volume-temperature) measurements (see COMPRESSIBILITY, GAS). All other properties are calculated either from equations of state or from a correlation of the individual experimental data points. In addition, the properties may be estimated from generalized correlations of the compression factor ($Z = pV/RT$). Here and in the equations that follow, V denotes molar volume in $m^3 \, mol^{-1}$ and R denotes the molar gas constant, in $JK^{-1} \, mol^{-1}$.

Experimental compressibility measurements have been made by a variety of methods[11,12] such as constant volume cells (Eucken and Meyer), variable volume cells (Beattie and Douslin, Michels, and Sage and Lacey), expansion systems with variable sample mass (Burnett), and differential systems (Whytlaw-Gray). The apparatus of Beattie and Douslin is used to measure both isometrics and isotherms up to 350°C and 400 atmospheres. The Michels apparatus may be used up to 3000 atmospheres but is limited to a temperature of 150°C. Sage and Lacey have made extensive measurements on both gas and liquid systems up to 240°C and 670 atmospheres.

Equations of state have been derived from compressibility data to represent the behavior of a gas over wide ranges of temperature and pressure.[11,13-14] Numerous equations have been published but one of the most important is the virial equation,

$$pV = RT(1 + B/V + C/V^2 + D/V^3 + \cdots)$$

$$(1)$$

It is quite important because the parameters B, C, D, etc., are related to the interactions between molecules according to the intermolecular potential energy theory. Other equations having wide applications are those of Beattie and Bridgeman and of Benedict, Webb, and Rubin (see GAS LAWS).

The energy functions, entropy, and heat capacity of a real gas are calculated as the sum of the ideal gas properties and a correction for the nonideality of the gas. The corrections for the nonideality of the gas are called difference or departure functions. For example, $S-S°$ is the entropy of the gas in the real state less that of the gas in the standard state at the same temperature. Theoretical equations needed for the computation of the thermodynamic properties have been derived in terms of pressure, volume, temperature, and the first derivatives $(\partial V/\partial T)_p$ or $(\partial p/\partial T)_v$.[13]

For example:

$$H - H° = \int_v^\infty [p - T(\partial p/\partial T)_v] \, dV + pV - RT$$

$$(2)$$

The heat capacity differences $C_p - C_p°$ and $C_v - C_v°$ are functions also of the second derivatives $(\partial^2 V/\partial T^2)_p$ or $(\partial^2 p/\partial T^2)_v$ depending upon whether the equations are written in terms of p and T, or V and T. The quantities appearing in Eq. (2) are usually evaluated from equations of state. However, the most accurate properties are those which are calculated from an analysis of isometric and isothermal p-V-T data.[15] The slopes of the isometrics $(\partial p/\partial T)_v$ are found by analytical, residual, and graphical techniques. The integrals as in Eq. (2) are integrated graphically or numerically.

Extensive correlations of data have been made to develop methods for estimating the properties of gases.[1,13] One method based on the theory of corresponding states presents the thermodynamic properties as a function of reduced temperature $(T_r = T/T_c)$, reduced pressure $(p_r = p/p_c)$, and the compressibility factor, Z_c. The subscript c refers to the critical state.

Gas Mixtures The thermodynamic properties of a mixture of gases may be calculated, but the procedures are only approximate unless compressibility data are available for the particular mixture.[11-13] Since few data for mixtures are available, the properties must be estimated from: (a) the equations of state of the pure gases assuming either additive volumes or additive pressures, (b) an equation of state for the mixture, or (c) generalized correlations of the compressibility factor based on pseudo-reduced conditions.[1]

ROLAND H. HARRISON

References

1. Hougen, O. A., Watson, K. M., and Ragatz, R. A., "Chemical Process Principles, Part Two: Thermodynamics," New York, John Wiley & Sons, Inc., 1959.
2. Guggenheim, E. A., "Thermodynamics," pp. 244–248, Amsterdam, North Holland Publishing Co., 1967.
3. "Selected Values of Properties of Hydrocarbons and Related Compounds," Thermodynamics Research Center Hydrocarbon Project, Texas A&M University, College Station, Texas, 3444 looseleaf sheets extant October 31, 1981.
4. Herzberg, G., "Molecular Spectra and Molecular Structure. II. Infrared and Raman Spectra of Polyatomic Molecules," New York, Van Nostrand Reinhold Co., 1945.
5. Stull, D. R., Westrum, E. F., Jr., and Sinke, G. C., "The Chemical Thermodynamics of Organic Compounds," New York, John Wiley & Sons, Inc., 1969.
6. Stull, D. R., and Prophet, H., "JANAF Thermochemical Tables," National Standard Reference Data Series, National Bureau of Standards (U.S.), 37, 1141 pp. (June 1971), U.S. Government Printing Office, Catalog No. C 13.48:37, Washington.
7. Knox, J. H., "Molecular Thermodynamics: An Introduction to Statistical Mechanics for Chemists," Revised Edition, New York, John Wiley & Sons, Inc., 1978.
8. Wilson, E. B., Jr., Decius, J. C., and Cross, P. C., "Molecular Vibrations—The Theory of Infrared and Raman Vibrational Spectra," New York, McGraw-Hill Book Co., Inc., 1955. Reprinted by Dover, New York, 1980.

9. McCullough, J. P., Pennington, R. E., Smith, J. C., Hossenlopp, I. A., and Waddington, G., "Thermodynamics of Cyclopentane, Methylcyclopentane, and 1,cis-3-Dimethylcyclopentane: Verification of the Concept of Pseudorotation," *J. Am. Chem. Soc.* **81**, 5880 (1959).

10. Scott, D. W., and McCullough, J. P., "The Chemical Thermodynamic Properties of Hydrocarbons and Related Substances," U.S. Bur. Mines Bull. 595, 1961.

11. Rowlinson, J. S., in Flugge, S. (Ed.), "Encyclopedia of Physics," Vol. XII, pp. 1–72, Berlin, Springer-Verlag, 1958.

12. Douslin, D. R., Harrison, R. H., and Moore, R. T., "Pressure-Volume-Temperature Relations in the System Methane-Tetrafluoromethane. I. Gas Densities and the Principle of Corresponding States," *J. Phys. Chem.* **71**, 3477–3488 (1967).

13. Beattie, J. A., and Stockmayer, W. H., in Taylor, H. S., and Glasstone, S. (Eds.), "Treatise on Physical Chemistry," Vol. 2, pp. 187–290, New York, Van Nostrand Reinhold Co., 1951; Beattie, J. A., *Chem. Rev.* **44**, 141–192 (1949).

14. Goodwin, R. D., and Haynes, W. M., "Thermophysical Properties of Propane from 85 to 700 K at Pressures to 70 MPa," National Bureau of Standards (U.S.), Monograph 170, 1982.

15. Harrison, R. H., and Douslin, D. R., "Derived Thermodynamic Properties of Ethylene," *J. Chem. Eng. Data* **22**(1), 24–30 (1977).

Cross-references: COMPRESSIBILITY, GAS; GAS LAWS; HEAT CAPACITY; KINETIC THEORY; THERMODYNAMICS.

GAUGE THEORIES

The principle of gauge invariance in classical electromagnetism expresses the postulate that electric and magnetic field strengths are measurable, and thus have objective meaning, while the scalar and vector potentials (from which the fields are usually derived) are neither measurable nor uniquely defined. For example, a static electric field is the negative gradient of a scalar potential function whose value is undefined up to a constant. Similarly, the magnetic field **B** is the curl of a vector potential, to which may be added any gradient of a scalar function without changing **B**, because the curl of a gradient vanishes identically. The importance of gauge invariance was first pointed out by Hermann Weyl in 1922. Its quantum mechanical significance was emphasized by Fritz London in 1927, namely: Gauge invariance fixes the form of the electromagnetic interaction and guarantees the atomicity of electric charge (without, however, fixing its value). Its application to relativistic field theories of elementary particles was made by Wolfgang Pauli and others.[1]

Quantum electrodynamics (QED) is the quantum theory of the electromagnetic field. It is the prototype of important quantum field theories that many believe describe all known fundamental interactions except gravitation; the theories of this type are called *gauge theories*. They are characterized by a principle that is a generalization of the gauge invariance principle of electromagnetism, namely: Gauge theories describe relativistic quantum fields that are invariant under a set of local symmetry transformations, i.e., operations of a continuous group (Lie group), whose elements are denoted by parameters that are functions of space and time. QED itself is a highly successful gauge theory, whose gauge group is Abelian, i.e., it has only mutually commuting elements.

A non-Abelian generalization of QED is the "standard" electroweak theory, which is a gauge theory that unifies the electromagnetic with the weak nuclear interactions that are responsible, e.g., for nuclear beta decay. Another non-Abelian gauge theory is quantum chromodynamics (QCD), designed to explain the binding of unobserved elementary particles called *quarks* into observed strongly interacting particles called *hadrons* (e.g., proton, neutron, pion). It postulates the exchange between quarks of eight massless "color gluons," analogous to the ordinary photon of electromagnetism, in order to account for the total confinement of quarks (whose nonintegral electric charges may not be observed) and to account for the strong nuclear force.[2] Non-Abelian gauge theories have also been proposed that unify all three of the above interactions (i.e., all fundamental interactions except gravity); they are called grand unification theories (GUTS).[3]

QED is formulated in terms of a relativistic four-vector field A^μ (μ = 0, 1, 2, 3), which is the quantum generalization of the electromagnetic four-potential. The principle of gauge invariance states that A^μ can be replaced by $A^\mu + \partial\Lambda/\partial x_\mu$, where Λ is any (relativistic) scalar function of space and time, without affecting any of the consequences of the theory. Among these consequences are:

- The appearance of massless photons (at all frequencies, spanning the entire electromagnetic spectrum)
- A unique form of electromagnetic interaction with charged particles
- The conservation of electric charge.

The last property follows from gauge invariance alone.

QED, proposed by P. A. M. Dirac in 1927, was plagued for two decades by so-called *divergences*, i.e., infinite values predicted for physical quantities that are actually finite. In the late 1940s, it was found that this problem could be circumvented by an ingenious calculational procedure called *renormalization*. Thus far, among quantum field theories applicable to elementary particles, only gauge theories have been found to be renormalizable.

The first gauge theory other than QED was formulated in 1954 by C. N. Yang and R. L. Mills. Among their motivations were:

- To find the consequences of assuming a law of conservation of *isotopic spin* (which was thought to be the strong-interaction analogue of electric charge)
- To have a principle that would select, from many possibilities, a unique form for the strong nuclear interaction.

The isotopic spin is a quantum-mechanical concept, having no classical analogue; it is represented by a vector operator whose components have the same commutation relations as the quantum mechanical angular momentum, hence the term *spin*. These operator components are the generators of a non-Abelian Lie group. In order to ensure that the strong-interaction Lagrangian be invariant under this local isospin group, a four-vector field $\mathbf{b}_\mu$ is seen to be necessary, having three components, and bearing the same relation to the isospin that the electromagnetic field bears to the electric charge. The field $\mathbf{b}_\mu$ (because of non-Abelian isospin gauge invariance) turns out to obey a nonlinear equation of motion.

In the case of electromagnetism, the interaction with an electric charge, described by a wave function $\psi(x)$, where $x = (\mathbf{x}, t)$, is obtained by replacing any gradient acting upon $\psi(x)$, by the so-called *covariant derivative* D^μ; that is,

$$\partial^\mu \rightarrow D^\mu = \partial^\mu - ie\,A^\mu,$$

where $\partial^\mu \equiv \partial/\partial x_\mu$ and A^μ is the electromagnetic four-potential. It follows from quantum mechanics that an arbitrary phase is allowed in the complex wave function $\psi(x)$. That is, the theory is invariant under the replacement

$$\psi(x) \rightarrow e^{i\alpha(x)}\,\psi(x).$$

To insure invariance under this transformation, terms in the Lagrangian must contain derivatives in the form D^μ, *not* ∂^μ, for only then:

$$D^\mu\{e^{i\alpha(x)}\,\psi(x)\} = e^{i\alpha(x)}D'^\mu\,\psi(x)$$

where

$$D'^\mu = \partial^\mu + i\partial^\mu\alpha(x) - ie\,A^\mu(x).$$

D'^μ is equivalent to D^μ because of the gauge invariance of $A^\mu(x)$, which allows the addition of the gradient of an arbitrary scalar function, i.e., $\partial^\mu\alpha(x)$.

Like the A^μ of QED, the isotopic spin invariant $\mathbf{b}_\mu$ field of Yang and Mills, whose gauge group is $SU(2)$, consists of massless "photons." However, since $SU(2)$ has three generators, while the Abelian group of electromagnetism has only one, there are three kinds of "photons." Similarly QCD, based upon the $SU(3)$ group that has eight generators, has eight kinds of "photons," called *gluons*. In all three cases (QED, Yang-Mills, and QCD) we deal with an "unbroken" gauge symmetry, characterized by a vacuum state which is itself gauge symmetric.

It is possible, however, to have a nonsymmetric vacuum state, leaving open the possibility that the quanta of the field (the "photons") are massive. This occurs as a result of a phenomenon known as *spontaneous symmetry breaking*. An example of such a field is the electroweak field theory of Weinberg, Salam, and Glashow.

LAURIE M. BROWN

References

1. See Pauli, W., "Relativistic Field Theories of Elementary Particles," *Rev. Mod. Phys.* **13**, 203–32, (1941).
2. So far, only one experimental group has claimed to observe free nonintegral charge.
3. For history and further descriptive materials, see Rosner, Jonathan L., *Am. J. Phys.* **48**, 90–103 (1980), and Yang, Chen Ning, *Annals. of N.Y. Acad. of Sci.* **294**, 86–97 (1977).

Cross-references: CONSERVATION LAWS AND SYMMETRY, ELECTROWEAK THEORY, ELEMENTARY PARTICLES, GRAND UNIFIED THEORY, ISOSPIN, QUANTUM CHROMODYNAMICS, QUANTUM ELECTRODYNAMICS, QUARKS.

GEODESY

Geodesy comprises the determination of the earth's external form and gravitational field, and the location of points with respect to earth-fixed reference systems. The earth's external form is customarily defined by the geoid: the equipotential of the earth's gravity field which most closely approximates the mean sea level.

The geoid is irregular in form, so that the mathematical representation thereof is necessarily an approximation. The most important approximation is an oblate ellipsoid of revolution, which is conventionally defined by its equatorial radius, a, and the flattening, f, equal to $(a - b)/a$, where b is the polar semidiameter. Location is conventionally expressed in coordinates referred to such an ellipsoid, in terms of the latitude ϕ, the angle between the normal to the ellipsoid and the equator; the longitude λ from the reference meridian, Greenwich; and altitude h above or below the ellipsoid.

If the ellipsoid is considered to be rotating with rate ω, and to be an equipotential for the combined effects (called gravity) of centrifugal and gravitational acceleration, additional parameters customarily required are γ_e, the acceleration of gravity at the equator; and m, the ratio of the centrifugal acceleration at the equator,

$\omega^2 a$, to γ_e. γ_e and m are connected to the total mass M contained in the ellipsoid by:

$$GM = a^2 \gamma_e [1 - f + 3m/2 - 15mf/14$$
$$+ O(f^3)] \quad (1)$$

where G is the constant of GRAVITATION $(6.67 \times 10^{-8} \text{cm}^3 \text{g}^{-1} \text{sec}^{-2})$. The customary formula for the acceleration of gravity γ at geodetic latitude ϕ is:

$$\gamma = \gamma_e [1 + (5m/2 - f - 17mf/14) \sin^2 \phi$$
$$+ (f^2/8 - 5mf/8) \sin^2 2\phi + O(f^3)] \quad (2)$$

The customary formula for the gravitational potential external to the ellipsoid is:

$$V = \frac{GM}{r} \left[1 - J_2 \left(\frac{a}{r} \right)^2 P_2 (\sin \phi) \right.$$
$$\left. - J_4 \left(\frac{a}{r} \right)^4 P_4 (\sin \phi) - O(f^3) \right] \quad (3)$$

In Eq. (3), V is written as positive, which is the convention of astronomy and geodesy, contrary to that of physics. In Eq. (3), P_2 and P_4 are Legendre polynomials, and

$$J_2 = 2f(1 - f/2)/3 - m(1 - 3m/2 - 2f/7)/3$$
$$+ O(f^3) \quad (4)$$

while J_4 is usually taken as a quantity determined observationally from satellite orbits.

The discrepancies in location of the actual geoid and a well-fitting ellipsoid are nearly always 10^{-5} or less of the radius vector, while the discrepancies in intensity of the gravitational acceleration from that of the standard ellipsoid are nearly always 10^{-4} or less of the total intensity. The mathematical representation of these discrepancies may either be in the form of spherical harmonic coefficients or in the form of mean values for areas; the former being preferable for effects on satellites in orbit, and the latter for use of terrestrial data. The potential theory dealing with the relationship between variations in the location of the geoid and the intensity of gravitational acceleration is known in geodesy as Stokes' theorem.

There are five principal categories of measurement in geodesy.

(1) Horizontal control comprises the determination of the horizontal components of position—latitude and longitude—starting from fixed values for a certain point. It includes measurement of distances over the ground by metal tapes or by pulsing or modulating radio or light signals, and measurement of angles about a vertical axis by theodolites. Over the land, the relative horizontal position of points is obtained either by triangulation—a system of overlapping triangles with nearly all angles measured, but only occasional distances measured; or by traverse—a series of measured distances at measured angles with respect to each other; or by trilateration—a system of overlapping triangles with all sides measured. Much of the land area of the world is covered by triangulation, which gives the difference in latitude and longitude between points in the same network with a relative error of about 10^{-5}.

(2) Vertical control comprises the determination of heights, which is performed separately from horizontal control because of irregularities in atmospheric refraction. The most accurate method, leveling, measures successive differences of elevation on vertical staffs by horizontal lines of sight taken at intermediate points over short distances (less than 150 meters) balanced so as to minimize differential refraction effect. The datum to which vertical control refers is mean sea level as determined by tide gages. The accuracy is such that the error in difference of elevation between points on the same principal network should be a few tens of centimeters or less. Current research emphasis in horizontal and vertical control is on the determination of temporal changes in areas of earthquake potential.

(3) Geodetic astronomy comprises the determination of the direction of the gravity vector and the direction of the north pole at a point on the ground. Astronomic longitude is the angle between the meridian of the gravity vector and the Greenwich meridian and is determined by measuring the time of intersection of a line of sight by a star. Astronomic latitude is the angle between the gravity vector and the equatorial plane, and is determined by measuring the maximum altitude attained by a star. In these types of astronomic observation, several stars are normally observed which are selected so as to minimize error due to atmospheric refraction. Astronomic azimuth is determined by the measurement of the horizontal angle between a target and Polaris or other reference star.

(4) Gravimetry comprises the determination of intensity of gravitational acceleration. Most gravimetric observations are made differentially, by determining the change, with change in location, of the tension on a spring supporting a constant mass. These measurements are connected through a system of reference stations to a few laboratory determinations of absolute acceleration of gravity. The relative accuracy of gravimetry is about $\pm.001$ cm/sec^2 on land and $\pm.005$ cm/sec^2 at sea. The principal difficulty in its geodetic application is irregular distribution of observations.

(5) Satellite tracking comprises the determination of the directions, ranges, or range rates of earth satellites (including the moon) from ground stations or other satellites, plus altitudes above the sea surface. These observations will be affected both by errors in positions of the station with respect to the earth's

TABLE 1. GEODETIC PARAMETERS.

Parameter	Current Estimate and Standard Deviation
Mean sidereal rotation rate, ω	$0.7292115085 \times 10^{-4} \text{ sec}^{-1}$
Gravitational constant $\times$ mass, GM	$39860047(\pm5) \times 10^7 \text{ m}^3/\text{sec}^2$
Equatorial radius, a	6378137 ± 2 meters
Oblateness, J_2	$108263(\pm0.5) \times 10^{-8}$

center of mass and by perturbations of the orbit by the earth's gravitational field; hence, in conjunction with a dynamical integration of the orbit, they are used to determine the position of tracking stations and the variations of the gravitational field. Radar altimetry from satellites is accurate to ±10 cm, and hence affected by oceanic temperature, salinity, and dynamics, as well as geoidal height and orbital perturbation. Satellites can also be used as elevated targets by simultaneous observations from several ground stations.

The principal practical application of geodesy is to provide a distribution of accurately measured points to which to refer mapping, navigation aids, engineering surveys, geophysical surveys, etc. The principal scientific interests in geodesy are the indications of the earth's internal structure from the variations of the gravity field and surface motions.

Numerical values of the leading geodetic parameters are given in Table 1.

WILLIAM M. KAULA

References

Bomford, G., "Geodesy," Fourth Edition, London, Oxford Univ. Press, 1980.

Moritz, H., "Geodetic Reference System 1980," *Bull. Geod.*, 54, 395–405 (1980).

Garland, G. O., "Introduction to Geophysics," Second Edition, Saunders, 1979.

Heiskanen, W. A., and Moritz, H., "Physical Geodesy," San Francisco, W. H. Freeman, 1967.

Committee on Geodesy, National Research Council, "Geodesy: Trends and Prospects," Washington, D.C., National Academy of Science, 1978.

Kaula, W. M., "Theory of Satellite Geodesy," Waltham, Mass., Blaisdell, 1966.

Cross-references: ASTROMETRY, ASTRONAUTICS, GEOPHYSICS, GRAVITATION, POTENTIAL, ROTATION—CURVILINEAR MOTION.

GEOPHYSICS

Geophysics is the physics of the earth and the space immediately surrounding it and the interactions between the earth and extraterrestrial forces and phenomena. It consists of a number of interlocking sciences dealing with physical properties, its interior, its atmosphere, oceans, glaciated areas, its age, motions and paroxysms, and the practical applications of the acquired knowledge. All of these sciences use the methods of physics for measurements and analysis. From observational material, often of an indirect nature, attempts are made to derive abstract models of states and processes through advanced mathematical concepts, but in many cases through statistical relations.

Geophysics is an ancient science. In its early stages it was developed by the Greeks who attempted to determine the shape and size of the earth (Eratosthenes, 275–194 B.C.). Among its most illustrious contributors have been Galileo Galilei (1564–1642), Sir Isaac Newton (1642–1727) who dealt with the motions of the earth and its gravitational field, Alexander von Humboldt (1769–1859) and Karl Friedrich Gauss (1777–1855) who observed and developed theories of the geomagnetic field; and Vilhelm Bjerknes (1862–1951) who laid the foundation for the hydrodynamic theories of the atmosphere and the oceans. During the current century its roster of distinguished scientists includes: L. Vegard (polar aurora), Sidney Chapman (aeronomy), Sir Harold Jeffreys and F. A. Vening-Meinesz (structure of the earth); Emil Wiechert, Boris Galitzin, Beno Gutenberg and J. B. Macelwane (seismology); H. U. Sverdrup (oceanography); C. G. Rossby and T. Bergeron (meteorology); A. Wegner, H. Hess and J. T. Wilson (tectonophysics).

Geophysics is the outstanding example of organized international scientific cooperation. In the 19th Century the International Polar Year (1882) served coordinated observations of the geophysical mysteries of the polar regions. Major milestone's in the development of the science were the International Geophysical Year (IGY) and the period of International Geophysical Cooperation (1957–1959), when 8000 scientists from 66 nations tackled a host of physical mysteries of our planet. The IGY started man's most spectacular ventures to date: The launching of artificial earth satellites and other space probes. It started the conquest of the icy wastes of the Antarctic continent where many geophysical secrets are locked, and through the Antarctic Treaty created the only weapon-free area in the world solely devoted to scientific research. The IGY took place in an interval of high solar activity. It was followed by the year of the quiet sun (IQSY, 1964–65) to cover the range of solar influences on the earth (see SOLAR-TERRESTRIAL PHYSICS PROGRAM). The Global Atmospheric Research Program (GARP) started with a number of observational experiments covering monsoons, tropical circulations, and a coordinated complete global coverage of weather phenomena (FIGI, 1980).

A brief survey of various subfields of geophysics follows.

Geochronology is study of the age of the earth and its various geological formations. Inadequate earlier methods of sedimentology have been replaced by the use of radioactive decay constants and isotope ratios. This has made it possible to date approximately all major geological eras. For the oldest rocks, the decay of ^{238}U to ^{206}Pb has led to ages of around 3×10^9 years.

Geodesy deals with the size and shape of the earth and its gravitational field. Because of the rotation of the earth, its lack of absolute rigidity, crustal mass distribution, and tidal forces, the shape is not quite spherical. The largest deviation from sphericity is the flattening at the poles. Historically this polar flattening has been determined by surface measurements and lunar observations, but since 1957 the most accurate determinations have been made by observations of the motions of artificial satellites. The current estimate of the flattening is 1:298.257. The sea-level surface of the earth, which is irregular, is referred to as the *geoid*, which represents an equipotential surface. The undulations of the geoid have been determined by surface gravity measurements and by utilization of data from artificial satellites. The value of gravity depends mainly on latitude because of the flattening and the variation of centrifugal force from pole to equator. The normal value of gravity at the earth's surface, in centimeters per second per second is represented by

$$\gamma = 978.0327(1 + 0.0053024 \sin^2 \varphi$$

$$- 0.0000058 \sin^2 2\varphi),$$

where φ is latitude. This formula is based on the Geodetic Reference System 1980, representing geodetic constants officially approved by the International Union of Geodesy and Geophysics in 1979.

Gravity measurements have shown that, in spite of large mass differences at the surface, the earth is nearly in isostatic equilibrium. Various crustal blocks act as if they floated in a dense subcrustal material. The undulations of the geoid do not exceed 100 meters. Approximate dimensional figures for the earth are: surface area 510×10^6 km^2; volume 1.083×10^9 km^3; average density 5.516 g/cm^3; mass 5.974×10^{27} grams; equatorial radius 6,378.137 km.

The deformations of the solid earth by tidal forces are an additional consideration. The twice daily occurring tides are observed by deflections of the vertical or variations of gravity. For the lunar tides the variations amount to 0.165 milligal, for the sun to 0.076 milligal (one milligal equals 10 micrometers per sec per sec). The maximal elevation of the geoid is 35.7 cm, the largest depression 17.8 cm, for the lunar effect; the total solar tide can reach 24.6 cm. The combined total at new and full moon is 78.1 cm.

Geomagnetism deals with one of the most important physical characteristics of the earth. Its magnetic field can be represented by a centered dipole, which at present is inclined at an angle of about 11° to the axis of rotation. Its field strength at the earth's surface is about 0.3 gauss at the equator and 0.6 gauss near the poles. There are small slow secular changes both in ionosphere direction and strength of the field. About $\frac{9}{10}$ of the field strength, designated as the main field, is assumed to be caused by internal forces, principally by differential rotation between the earth's core and its mantle. About $\frac{1}{10}$ of the field is externally produced by a ring current in the ionosphere. That part is often disturbed by electron and proton invasions from solar eruptions. Major flares cause magnetic storms lasting from a few minutes to several hours. In the polar zones these are often accompanied by auroras. In those regions field disturbances of 0.035 gauss have been observed. Small, short disturbances of the field are common. The earth's magnetic field, the *magnetosphere*, extends far beyond the atmosphere, especially on the side away from the sun, where the distance of the magnetopause is about 10 earth radii. The force of the magnetic field in the space surrounding the earth has created two zones where solar protons and electrons are trapped, the van Allen belts.

The magnetic field distribution at the earth's surface has been mapped, showing not only the total intensity but also inclination and declination. The latter is of importance for direction finding with magnetic compasses. The elements of the earth's magnetic field undergo slow secular changes and in geologic time intervals even reversals of polarity of the field have been observed. Paleomagnetisms, determined from the remanent magnetization from the time of formation of iron-bearing rocks, not only show rotations of the field but divergent motions of the continents.

Seismology deals with study of earthquakes. Most of them can be attributed to breaks in the earth's crust. A few may be subcrustal and caused by phase transformations. Some minor ones are associated with collapse of cavities and volcanic eruptions. All of them are characterized by the sudden release of elastic waves. These are longitudinal, transverse, Rayleigh or Love (surface) waves. They are recorded at seismographic stations, now often consisting of arrays of digitally recording sensors responding to various frequencies. Through analysis of the travel times of various wave groups from the point of origin and their amplitudes, the hypocenter and epicenter of an earthquake as well as its magnitude can be determined. Like other waves, earthquake waves can be reflected and refracted upon entering a different medium. The internal constitution of the earth has been derived primarily from seismic evidence. This reveals an upper and a lower crust, an upper and lower mantle, an outer and an inner core. The boundary between crust and mantle was first discovered by A. Mohorovičić in 1909. Its depth varies from about 5 km in some oceanic areas to 60 km in some continental regions. Seismic wave analysis also permits some statements on the strength of materials in the earth. The crust and upper mantle are rigid (lithosphere) with a more plastic layer below (asthenosphere). Present interpretations assume that the inner core (1370 km thick) is solid and the outer core (2100 km thick) is liquid. The scattering properties of various rock types in the crust reveal many inhomogeneities and seismic procedures, and are widely used in geophysical prospecting for mineral resources.

Careful records kept during the 20th Century, together with historical records, have yielded a fair picture of magnitude and distribution in time and space of earthquakes. Such seismicity studies permit risk estimates and hazard mitigation. The effort devoted to earthquake prediction has not yielded much in the way of progress toward that elusive goal. The principal approaches are geodetic observations of ground motions, interpretation of foreshocks, changes in wells, radon exhalations, and even animal behavior. Earthquakes under the oceans often result in seismic sea waves (tsunamis). A warning system for endangered coastal areas has been quite successful because of the slow travel times of these waves.

Geodynamics deals with the slow motions of crustal material. In 1912 Alfred Wegener presented a hypothesis of continental drift. This was fiercely contested until the 1960s. By then suboceanic cores had shown ocean-floor spreading and satellite geodetic observations yielded drifts of continental plates of several centimeters per year. This has led to the plate tectonics model which postulates the motion of lithospheric and upper mantle layers (100 km thick) over the less viscous asthenosphere, driven by convective subcrustal thermal circulations. There is evidence for different and separate convective systems under the continents and oceans. At the edges of the various plates form mountain systems, volcanic arcs, and earthquake zones. Extrusions of material from lower layers take place and some of the plates seem at their edges to submerge under others. Although problems and contradictions still remain with this model much research in this area promises clarifying discoveries.

Volcanology deals with some of the most spectacular phenomena in geophysics. They include explosive eruptions, lava flows, gaseous exhalations, magma intrusions, geysers, and hot springs. Since the last ice age it is estimated that about 1300 volcanoes have been active with over 5500 dated eruptions. Duration of eruptive activity and amounts of ejecta vary widely. In historical times the 7–12 April 1815 eruption of Tambora (Indonesia) was the greatest, with an estimate of 100 km^3 of tephra ejected. The 26–27 August 1883 eruption of Krakatoa (Indonesia) which killed 36,000 persons, threw an estimated 10^7 tons of dust into the air. This caused spectacular sunsets. The dust veil stayed in the stratosphere for several years. Some hypotheses of climatic fluctuations have been based on such veils. The dust from the much smaller eruption of Mt. St. Helens (Oregon) on May 18, 1980, after 123 years of dormancy, caused observable reduction in local maximum temperatures. Prediction of volcanic eruptions remains an uncertain art. It is based on precursory seismic activity, gaseous exhalations, and lava movements in or near the craters. Areas of present or ancient volcanic activity show abnormal geothermal gradients and are in some areas being tapped for electric energy production.

Meteorology and *aeronomy* are concerned with the physical state and the motions of the atmosphere, which is divided into a number of layers. The lowest is the *troposphere* with an average thickness of 7 to 8 km in polar regions and 13 km in the equatorial zone. Temperatures decrease to an interface, called the *tropopause*, where polar temperatures are about 218 K, at the equator about 193 K. The next higher layer is the *stratosphere*, where temperatures stay nearly isothermal in the first few kilometers and then increase again above about 20 km to a maximum of 270 K at 50 km where, above the stratopause, the *mesosphere* begins. The warming in the stratosphere is produced by absorption of the short wavelengths of solar radiation by photochemically produced ozone (O_3). The anthropogenic threat to the ozone layer by oxides of nitrogen and halomethanes has become a major environmental concern. In the mesosphere temperatures again decrease to 180 K at about 80 km, where notable dissociation of the permanent constituents of the atmosphere occurs. This is the region of rare noctilucent clouds; above this region lies the *thermosphere*. Here by vigorous interaction with the solar wind and cosmic rays many of the atmospheric constituents are in excited atomic states and ionized. There are several temporarily or permanently ionized layers in the thermosphere, often designated as *ionosphere:* The *D*-layer below 90 km, the *E*-layer between 90 and 180 km, and the *F*-layer above 160 km. The ionization undergoes a diurnal variation and large changes in response to the solar activity rhythms.

Most manifestations of weather take place in the troposphere. They are governed by the general atmospheric circulation, which is stimulated by the differential heating between tropical and polar zones. The resulting motions in the air are subject to the laws of fluid dynamics on a rotating sphere with friction. They are characterized by turbulence of varying time and space scale. Evaporation of water from the ocean and its transformation through the vapor state to droplets and ice crystals are important symptoms of the weather producing forces (see METEOROLOGY).

Hydrology studies the water cycle on the earth in detail. It includes the runoff from precipitation, the surface courses of water and their floods, the deposited forms of water as snow and ice, and the return of water to storage underground or the ocean. The study of ice caps, sea ice, and glaciers (cryosphere) is an important phase of this field and contributes to the understanding of the earth's heat balance. Stream flow has been successfully mathematically modelled but flash floods still present a major problem.

Oceanography has a broad overlap with both meteorology and hydrology. It includes the study of wind-driven waves and currents, and the storage of heat and its release to the atmosphere. Ocean currents are an important mechanism for the equalization of the temperature differences between low and high latitudes. Near some coast lines, wind-induced upwelling of highly nutrient deep ocean water is of major importance for the fishing industry. Temperature and salinity differences are additional causes for the dynamics of the ocean but most exchanges take place in a shallow mixed surficial layer, only about 100 meters thick, but occasionally reaching to 500 meters depth. Below that is a thermocline with high density cold water to the bottom. Exchange time of this abyssal water with higher layers is measured in centuries.

The tidal motions and their dependence upon configuration of ocean basins and coastal lines were among the earliest geophysical phenomena observed and analyzed by man. They were also the first to be predicted by computer. These predictions are now quite precise in timing and amplitude and are of great importance for ships with ever increasing draft.

Exploration geophysics uses physical methods derived from the study of earthquakes and magnetic and gravitational fields. These, using highly sensitive instruments, permit the discovery and mapping of the structure and inhomogeneities in the upper geological layers. These result in local anomalies and permit inferences about potential mineral resources. Magnetic anomalies, often associated with ore bodies, can be rapidly mapped by sensitive airborne magnetometers. Seismic techniques, using small explosions and a multiple array of sensors coupled to a computer use reflection and refraction of wave energy to map geological structures three-dimensionally. This technique is extensively used for petroleum exploration.

It can be readily applied also for mapping of off-shore stratigraphy. Radioactivity and geochemical procedures are auxiliary methods, often used in boreholes. Distortions of artificially created electric fields in the ground can indicate the presence of shallow ore bodies but are now more extensively employed for groundwater exploration.

H. E. LANDSBERG

References

Academic Press, *Advances in Geophysics*, Vols. 1–25, (H. E. Landsberg, H. E. Landsberg and J. van Mieghem, B. Saltzman, succeeding editors), New York, 1952–1983.

Aki, K., and Richards, R. G., "Quantitative Seismology," 2 Vols., W. H. Freeman and Co., San Francisco, 1980.

American Geophysical Union, *Reviews of Geophysics and Space Physics*, Vols. 1–22, Washington, D.C., 1963–1983.

Bird, J. M. (Ed.), "Plate Tectonics," American Geophysical Union, Washington, D.C., 1980.

Chovitz, B. H., "Modern Geodetic Earth Reference Models," *EOS* **62**, 65–67 (1981).

Dziewonski, A. M., and Boschi, E., "Physics of the Earth's Interior," Elsevier/North Holland Publishers, New York, 1980.

Eather, R., "Majestic Lights," American Geophysical Union, Washington, D.C., 1980.

Houghton, J. T., "The Physics of Atmospheres," Cambridge Univ. Press, New York, 1977.

Jacobs, J. A., "The Earth's Core," Academic Press, New York, 1975.

L'vovich, M. I. "World Water Resources and their Future," (English Translation, R. L. Nace, Ed.), American Geophysical Union, Washington, D.C., 1979.

McElhinny, M. W., "The Earth," Academic Press, New York, 1979.

Melchior, P., "The Tides of the Planet Earth," Pergamon Press, Oxford, 1978.

Ratcliffe, J. A., "An Introduction to the Ionosphere and Magnetosphere," Cambridge Univ. Press, New York, 1972.

Simkin, T., Siebert, L., McClennand, L., Bridge, D., Newhall, C., and Latter, J. H., "Volcanoes of the World," Hutchinson Ross Publishing Co., Stroudsburg, PA 1981.

Simpson, D. W., and Richards, P. G. (Eds.), "Earthquake Prediction," Maurice Ewing Series, Vol. 4, American Geophysical Union, Washington, D.C., 1981.

Telford, W. M., Geldart, L. P., Sheriff, R. E., and Keyes, D. A., "Applied Geophysics," Cambridge Univ. Press, New York, 1979.

Thurman, H. V., "Introduction to Oceanography," 2nd Ed., Charles Merrill Publishing Co., Columbus, OH, 1981.

Torge, W., "Geodesy," Walter de Gruyter, New York, 1980.

Cross-references: FLUID DYNAMICS, GEODESY, GRAVITATION, INTERNATIONAL SOLAR-TERRESTRIAL PHYSICS PROGRAM, MAGNETISM, METEOROLOGY, PLANETARY ATMOSPHERES, SEISMOLOGY, VOLCANOLOGY.

GRAND UNIFIED THEORIES

Introduction Much of our progress in understanding the basic interactions of matter has involved the development of unified theories, in which two or more apparently unrelated forces are realized to be different manifestations of a more fundamental underlying force. The first modern example was the unification of the electric and magnetic forces into the electromagnetic interaction by Maxwell and others in the 19th Century. Maxwell's equations not only reveal the intimate relation of the electric and magnetic fields, but also predict the existence of electromagnetic waves that can be emitted or absorbed by charged particles. We now know that the quanta of these waves are massless photons (γ) carrying an internal spin of one unit of Planck's constant $\hbar$. They are therefore referred to as *spin* -1 or *vector* (or *gauge*) *bosons*.

The next successful unification combined the electromagnetic force not with gravity (as had been unsuccessfully attempted by Einstein during much of his life) but rather with the weak force responsible for β decay. This electroweak theory of Glashow, Weinberg, and Salam (GWS), which reached its fruition in the early 1970s, combined the intermediate vector boson model of the weak interactions (Fig. 1a, b), in which the weak interactions are mediated by the exchange of massive electrically- charged vector bosons W^+ and W^-, with quantum electrodynamics, in which the electromagnetic force is mediated by the exchange of photons (see Fig. 1c). The result was a mathematically and physically consistent quantum field theory. Furthermore, the electroweak unification predicted the existence of a new electrically-neutral vector boson (the Z). The subsequent discovery of the weak neutral current interaction, mediated by the Z (Fig. 1d) along with the discovery of the charm quark (needed to correctly describe strongly interacting particles in the model) provided a dramatic verification of the electroweak unification.

Grand unified theories (GUTs), which are so far untested, go one step further and unify the electroweak interactions with the nuclear or strong force. (The strong interactions are thought to be due to the exchange of massless vector gluons (G) between quarks, which are believed to be the constituents of the proton, neutron, pion, and other strongly interacting particles. See Fig. 1e). In addition to unifying three of the four known interactions, GUTs have the attractive feature of regarding the quarks and the non-strongly interacting particles or leptons (electron, muon, tau, and neutrinos) as fundamentally similar.

Just as the electroweak unification predicted the existence of the Z boson, grand unified theories predict the existence of new, extremely massive vector bosons. In most GUTs, these new bosons can mediate the decay of the proton into lighter particles, such as a positron and neutron pion, as shown in Fig. 2. This is a remarkable prediction because it has generally been believed that the proton is absolutely stable. It is known experimentally that if the proton does decay, its average lifetime must be longer than 10^{30} yr, which corresponds to fewer than one decay per ton of matter per year! The strength of the new interactions can be predicted from the observed strength of the strong and electroweak interactions. Within the simplest grand unified theories, the proton lifetime is predicted to be $\tau_p \simeq 10^{29 \pm 2}$ yr, precisely the range that may be observable in new, dedicated experiments!

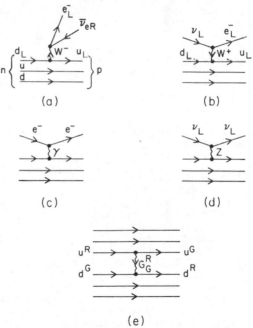

FIG. 1. (a) Beta decay $n \rightarrow pe^-\bar{\nu}$. (b) $\nu N \rightarrow e^-X$, where N is a nucleon (proton or neutron) and X represents any strongly interacting final state (e.g., $X = N, N\pi, N\pi\pi$). (a) and (b) occur via the exchange of charged weak bosons $W^{\pm}$. (c) The electromagnetic process $e^-N \rightarrow e^-X$. (d) The weak neutral current process $\nu N \rightarrow \nu X$. In (a)–(d), the quark color quantum number is not changed, but the flavor (u, d, etc.) is changed in charged current processes. (e) A strong interaction between quarks via the exchange of a gluon. Quark color (but not flavor) is changed in the transition.

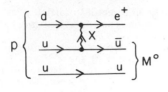

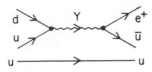

FIG. 2. Typical diagrams in which proton decay is mediated by the exchange of superheavy bosons carrying both flavor and color quantum numbers.

FIG. 3. (a) Vertex for fermion k to be transformed into fermion j upon emission or absorption of boson i. (b) Self-interactions between three or four bosons.

The same new interactions that lead to the instability of the proton may explain the long-standing puzzle of the apparent excess of matter over antimatter in the universe. Instead of having to postulate this excess as an asymmetric initial condition on the big bang, it is possible that it was generated dynamically by the new interactions in the first 10^{-35} seconds after the big bang when the universe was incredibly hot and dense! Of course, all matter will ultimately decay into a gas of electrons, positrons, neutrinos, and photons in a time $\gtrsim 10^{30}$ yr. if the universe does not collapse first!

In addition to their theoretical elegance and their fascinating predictions about the creation and decay of matter, grand unified theories yield explanations of the relative electric charges of quarks and leptons and of the value of the weak interaction angle, and they predict the existence of superheavy magnetic monopoles. These as well as difficulties of GUTs are detailed below. Particle physicists eagerly await the confirmation or refutation of GUTs in proton decay experiments.

The Standard Model There now exists a renormalizable (i.e. free from unmanageable divergences) field theory of the elementary particles and their interactions that is compatible with all known experimental facts. In this standard model the electroweak and strong interactions are gauge invariant, which means that the forms of the equations of motion are left invariant under a group of transformations of the fields into each other, which can be performed independently at different space-time points. Gauge invariance requires the existence of N vector or gauge bosons A_i, one for each of the N generators of the symmetry group. The amplitude for a fermion (lepton or quark) k to emit or absorb boson i and be transformed into fermion j is $gL^i{}_{jk}$, where g is the strength parameter or coupling constant of the theory and $L^i{}_{jk}$ is the jk element of the ith group generator (Fig. 3a). The forms of the self-inter-

actions between the gauge bosons are also prescribed by the gauge symmetry (Fig. 3b).

For example, the GWS electroweak theory is based on the gauge group $SU_2 \otimes U_1$. The SU_2 group implies three gauge bosons W^+, W^-, and W^0, with gauge coupling g, while the U_1 group involves one neutral boson B with coupling g'. The charged bosons W^+ and W^-, with masses $m_{W^\pm} \simeq 83$ GeV/c^2, mediate the ordinary charged-current weak interactions, the neutral boson $Z = W^0 \cos\theta_W - B \sin\theta_W$, with $m_Z \simeq 94$ GeV/c^2, is exchanged in neutral current processes, and the massless photon $\gamma = W^0 \sin\theta_W + B \cos\theta_W$ is responsible for electromagnetism. The weak angle is $\theta_W \equiv \tan^{-1} g'/g$ and the electric charge is $e = g \sin\theta_W$. Parity nonconservation in the weak interactions is introduced by assigning the left- and right-handed fermions to different representations of the $SU_2 \otimes U_1$ group. The left-handed up and down quarks (u_L, d_L) and the left-handed neutrino and electron (ν_{eL}, e_L) transform as doublets, which means that they are related by the symmetry operators and can be transformed into each other upon the emission or absorption of a W^+ or W^-. (The color quantum number is unchanged in these processes). The more massive left-handed particles (c_L, s_L), (t_L, b_L), ($\nu_{\mu L}$, μ_L^-) and ($\nu_{\tau L}$, τ_L^-) also transform as doublets. The right-handed fermions are unaffected by SU_2 transformations and do not couple to $W^\pm$ bosons (they do couple to the Z and γ).

The basic equations of gauge theories do not allow mass terms for the gauge bosons. The W^+, W^-, and Z masses are introduced by spontaneous symmetry breaking. This can occur if the ground state contains a nonzero expectation value of spin-0 Higgs fields (essentially a Bose condensation). The $W^\pm$ and Z bosons (but not the photon) couple to the Higgs field and develop a nonzero effective mass.

The strong interactions are described by quantum chromodynamics (QCD), which is a gauge theory based on the gauge group SU_3 with gauge coupling g_s. There are eight massless gauge bosons or gluons in QCD. Quarks are transformed from one color state (R = red, G = green, B = blue) into another upon emission or absorption of gluons, with the flavor quantum number (u, d, c, s, etc.) unchanged. The QCD couplings of left-and right-handed quarks are equal (parity is conserved), and there are no direct couplings of gluons to the SU_3 invariant leptons, which therefore do not participate in the strong interactions.

Despite the many successes of the standard model, most physicists believe that it is too arbitrary and complicated to be taken seriously as the fundamental theory of elementary particles. For example, the standard model involves the direct product of three different gauge groups with three distinct coupling constants ($\alpha_s = g_s^2/4\pi \simeq 0.25$, $\alpha_g \equiv g^2/4\pi \simeq 0.03$, $\alpha = e^2/4\pi \simeq 0.007$ when measured at laboratory energies). There is no fundamental explanation for the pattern of spontaneous symmetry breaking, for the breaking of parity in the weak but not the strong interactions, for the existence of repeated fermion families, or for the relation between the electric charges of the quarks and leptons. There are many arbitrary parameters, such as coupling constants, fermion masses, and weak mixing angles (19 in all if the neutrinos are massless). Finally, gravity is not incorporated in the theory.

Grand Unified Theories Grand unified theories constrain some of the arbitrary features of the standard $SU_3 \otimes SU_2 \otimes U_1$ model by embedding it in a simple (i.e., not a direct product) group G with a single coupling constant g_G. The simplest example is the SU_5 model of Georgi and Glashow. SU_5 implies 24 gauge bosons, of which 12 are those of the standard model. The other 12 are supermassive ($m > 10^{14}$ GeV/c^2) and carry both flavor and color quantum numbers. They are the X and Y bosons, with electric charges $\frac{4}{3}e$ and $\frac{1}{3}e$, each of which comes in three color states, and their antiparticles. Quarks, antiquarks, leptons, and antileptons are related by SU_5 transformations and are fundamentally similar. The left-handed fields (ν_{eL}, e_L^-, $\bar{d}_{\alpha L}$, $\alpha = R, G, B$) form a five-dimensional representation 5^*, which means that they can be transformed into each other upon emission or absorption of gauge bosons. Similarly, the fields (e_L^+, u_L^α, d_L^α, and $\bar{u}_{\alpha L}$) form a 10-dimensional representation (the right-handed fields are related by CPT). The heavier fermion families are assigned to additional 5^* and 10-dimensional representations.

Other grand unified theories are based on still larger groups, such as SO_{10} or E_6. They involve additional supermassive bosons emitted in transitions between the 5^* and 10, between different families, or between the ordinary and hypothetical new fermions.

Implications of Grand Unification (a) There is basically only one interaction. The complicated structure of interactions observed at low energies is due to the pattern of spontaneous symmetry breaking (gauge boson masses). Similarly, the quarks (q) and leptons (l) are closely related. This explains the simple relation $\frac{1}{2} q_u = -q_d = -\frac{1}{3} q_e = \frac{1}{3} e$ between quark and lepton charges (charge quantization).

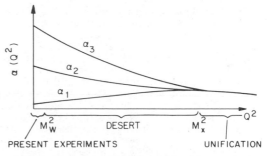

FIG. 4. Variation of the effective couplings $\alpha_3 = \alpha_s = g_s^2/4\pi$, $\alpha_2 = \alpha_g = g^2/4\pi$, and $\alpha_1 = \frac{5}{3}\alpha' = \frac{5}{3}g'^2/4\pi$ with momentum.

(b) Because of vacuum polarization and other higher order corrections, measured gauge couplings actually depend logarithmically on the typical squared momentum Q^2 in an experiment. The observed low energy couplings are predicted to approach a common value (up to known normalization factors) $\alpha_G = g_G^2/4\pi$ at momenta $Q^2 \gtrsim M_X^2$ for which spontaneous symmetry breaking can be ignored, as in Fig. 4. From the observed ratio of α_s/α at low energy and the theoretical Q^2 dependence, one can predict: (i) $M_X \simeq M_Y \simeq 2.4 \times 10^{14 \pm 0.5}$ GeV/c^2, which is twelve orders of magnitude larger than M_W and M_Z! (The threshold-free region between M_W and M_X is referred to as *the desert*.) (ii) The requirement that all three couplings meet at M_X yields the additional prediction: $\sin^2 \theta_W = \alpha/\alpha_g = 0.214 \pm 0.005$, which is in remarkable agreement with the experimental value 0.215 ± 0.012 obtained from neutral current reactions. Predictions (i) and (ii) hold for essentially all grand unified theories with no new mass scales in the desert and no exotic fermions or weak interactions.

(c) The X and Y bosons can mediate proton decay through diagrams such as those in Fig. 2. The average proton lifetime in years in the SU_5 model is predicted to be $\tau_p \simeq \lambda M_X^4/(\alpha_G^2 m_p^5) \simeq (0.8 - 13) \times 10^{29} (M_X/2.4 \times 10^{14}$ GeV/$c^2)^4 \simeq 3 \times 10^{29 \pm 2}$ yr, where the coefficient λ has been estimated using several phenomenological models of the proton and meson wave functions. This extremely long lifetime is due to the very short range ($R_X \sim \hbar/M_X c \simeq 10^{-28}$ cm) of the interaction. The predicted lifetime is comparable to current experimental limits $\tau_p^{\exp} \gtrsim 10^{30}-10^{31}$ yr and is within the sensitivity of several dedicated experiments now under construction. Most other theories give similar predictions for τ_p, although models with supersymmetry or large mixing effects can lead to longer lifetimes.

The dominant decay modes are $p \to e^+ M^0$ and $p \to \bar{\nu} M^+$, where M^0 represents neutral mesons (π^0, ρ^0, η, ω, $\pi^+ \pi^-$, ...) and $M^+ = \pi^+$, ρ^+, $\pi^+ \pi^0$, ... Bound neutrons can decay into $\bar{\nu} M^0$

and $e^+ M^-$. Different gauge groups (SU_5, SO_{10}, etc.) can be distinguished by their predictions for the ratio of $e^+ M^0$ to $\bar{\nu} M^+$ decays. All models with a desert between M_W and 10^{14} GeV/c^2 imply the selection rule $\Delta B = \Delta L$, where B and L are baryon and lepton number. ($B = +1$ or -1 for nucleons (p or n) or antinucleons, respectively. $L = +1$ for leptons (e^-, ν) and -1 for antileptons.) Theories with lower mass scales (which are disfavored by the $\sin^2 \theta_W$ test) can predict baryon number violating processes with different selection rules. For example, some models with particles with mass $\simeq 10^6$ GeV/c^2 predict that free neutrons can oscillate into antineutrons in a time scale $\tau_{n\bar{n}} \gtrsim 10^7$ sec and that two nucleons in a nucleus can annihilate in a time $\tau_{nuc} \gtrsim 10^{30}$ years.

(d) The baryon-number-violating interactions would have been very important in the first 10^{-35} sec after the big bang when the temperature kT was comparable to $M_X c^2$. Relics of that first instant may be observable today.

For example, grand unified theories may explain the baryon (matter-antimatter) asymmetry $n_B / n_\gamma \simeq 10^{-10 \pm 1}$, $n_{\bar{B}} / n_B \simeq 0$, observed in our part of the universe, where n_B, $n_{\bar{B}}$, and n_γ are the number densities of baryons (p, n), antibaryons ($\bar{p}$, $\bar{n}$), and blackbody photons. If baryon number were absolutely conserved and there is no large scale separation of baryons and antibaryons in the universe (no plausible separation mechanism has been found) then the net baryon number would have to be postulated as an asymmetric initial condition on the big bang. Alternatively, $n_B - n_{\bar{B}}$ may be generated dynamically in GUTs. The ingredients needed are (i) B violation (this alone is sufficient to erase or dilute any initial net baryon number), (ii) nonequilibrium of the $\Delta B \neq 0$ interactions, (otherwise $n_B = n_{\bar{B}}$), and (iii) CP and C violation (to distinguish baryons from antibaryons).

A possible specific mechanism is that superheavy spin-0 Higgs bosons and antibosons drop out of equilibrium at $kT \lesssim M_X c^2$. Their CP-violating decays into qq, $\bar{q}\bar{q}$, $\bar{q}l$, and ql can create a slight excess of quarks over antiquarks. These eventually form nucleons after most $q\bar{q}$ pairs annihilate. Detailed estimates are difficult, but the ratio $n_B / n_\gamma \simeq 10^{-10}$ appears quite plausible.

Grand unified theories also predict the existence of superheavy ($m \simeq 10^{16}$ GeV/c^2) magnetic monopoles, which would have been produced prolifically in the big bang. The monopole density depends sensitively on the nature and sequence of phase transitions (in the ground state spontaneous symmetry breaking pattern) in the early universe. Second-order transitions probably imply many orders of magnitude more monopoles than are allowed by limits on the total mass density of the present universe. Strongly first-order transitions may sufficiently suppress monopole production and may lead to a period of exponential expansion of the universe that could solve the cosmological horizon

and flatness problems. However, it is not clear that this expansion would terminate in an acceptable manner. Phase transitions and other aspects of GUTs may also be of relevance for galaxy formation.

(e) Almost all GUTs (with the exception of the SU_5 model) predict nonzero masses for the neutrinos. Estimates are very model dependent, typically ranging from $m_\nu = 10^{-9}$ eV/c^2 to 10^{+2} eV/c^2. Neutrino oscillations associated with mass $>10^{-5}$ eV/c^2 could explain the missing solar neutrinos. Neutrinos in the 10–100 eV/c^2 range could account for the missing mass in galactic clusters and would dominate the mass density of the universe.

Some of the simpler GUTs, such as the simplest SU_5 model, correctly predict the ratio of the b quark and τ lepton masses.

Outstanding Problems Despite their attractive features and successful predictions, grand unified theories have several serious problems and shortcomings. (a) Perhaps the most severe is the gauge hierarchy problem, which refers to the incredibly tiny ratio $(M_W / M_X)^2 \simeq 10^{-24}$. This quantity is not a natural feature of GUTs: in the usual models it must be put into the theory by hand by fine tuning parameters. Theories with composite Higgs bosons and models with an approximate low energy symmetry between bosons and fermions (supersymmetry) are promising approaches to this problem, but neither has been completely successful.

(b) Another problem is that GUTs have not shed much light on the fermion mass spectrum and mixings, the reason for heavy fermion families, or the origin of CP violation.

(c) Superheavy magnetic monopoles are a serious problem. Much work remains to be done on the dynamics of phase transitions in the early universe and on possible monopole suppression mechanisms.

(d) Finally, gravity has so far resisted all attempts at unification. Supergravity theories, in which spin-$\frac{3}{2}$ fermions are related to spin-2 gravitons, appear to be a promising direction.

PAUL LANGACKER

References

Introductory papers include:

1. Georgi, H., and Glashow, S. L., *Physics Today*, Sept., 1980, p. 30.
2. Weinberg, S., *Scientific American*, June 1981, p. 64.

Proton decay is reviewed in:

3. Goldhaber, M., Langacker, P., and Slansky, R., *Science* 210, 851 (1980).

Grand unified theories are reviewed in:

4. Langacker, P., *Phys Reports* 72, 185 (1981).

Cross-references: ELEMENTARY PARTICLES, GAUGE THEORIES, QUANTUM CHROMODYNAMICS, ELECTROWEAK THEORY, ANTIPARTICLES, COSMOLOGY, QUARKS.

GRAVITATION

Gravitation is the phenomenon characterized by the mutual attraction of any two physical bodies. This universal character of the gravitational force was first recognized by Sir Issac Newton who also gave its quantitative expression. For point masses or spherical bodies a simple expression results:

$$F = \frac{GM_1 M_2}{R^2} \qquad (1)$$

In addition to the masses M_1, M_2 of the two bodies and their distance apart R, the force depends only on a constant $G = 6.670 \times 10^{-8}$ dyne cm^2 which is independent of all properties of the particular bodies involved. The same force law describes the motion of the planets around the sun, of the moon around the earth, as well as the falling of an apple to the earth. A body moving under an inverse square law as given in Eq. (1) satisfies the three laws established by Kepler for the motion of the planets around the sun:

(1) The planets move in elliptical orbits with the sun at one focus (the general orbit is a conic section) (Fig. 1).

(2) The radius vector sweeps out equal areas in equal times.

(3) The square of the period of revolution is proportional to the cube of the semimajor axis: $a^3 = (2\pi)^{-2} GM_\odot T^2$. Here $M_\odot$ is the mass of the sun and T is the period of the planet. These results together with a detailed analysis of anomalies in the motion of the moon established the correctness of the Newtonian theory of gravitation (see KEPLER'S LAWS).

Recently, careful calculations have been carried out to determine the orbits of the artificial satellites which have been launched by the United States and the Soviet Union. These have required modifications in the force law Eq. (1) to take into account the deviation of the earth's figure from a sphere and the anistropy of the earth's density as well as the atmospheric drag.

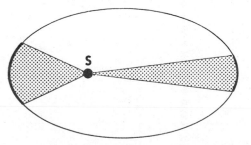

FIG. 1. An elliptical orbit for a planet around the sun. The shaded areas indicate equal areas swept out in equal times at different parts of the orbit. Clearly, the speed of the planet varies with its position in its orbit.

The success of the space program to date is an additional tribute to Newton's genius. Other calculations study powered space flight in order to examine possible orbits for exploration of the solar system. There is every reason to believe that Newton's gravitational theory is sufficiently accurate for this purpose. Einstein's modification of the theory, to be described below, will probably have little effect on our space program for some time to come.

The weight of a body of mass M on the earth is the force with which it is attracted to the center of the earth. On the surface of the earth the weight is given by

$$W = Mg$$

where the acceleration due to gravity is obtained from Eq. (1):

$$g = \frac{GM_E}{R_E^2} = 980 \text{ cm/sec}^2 = 32 \text{ ft/sec}^2$$

All freely falling bodies near the surface of the earth are accelerated at the same rate g. It is for this reason that Galileo found that both light and heavy objects take the same time to reach the ground when dropped from the Leaning Tower of Pisa.

An astronaut is said to be in a state of weightlessness when he is in orbit. Strictly speaking, he still has weight, for the earth's gravity still acts on him. Otherwise he would fly off into outer space. However, when in free fall, the local effects of the gravitational field are eliminated for the astronaut. Objects which are released fall together with him and hence remain in his vicinity, unlike the situation on the ground. Therefore, the organs of the body respond as though the gravitational field were absent and this gives the sensation of weightlessness. Conversely, we sense the earth's gravity and feel weight because we are supported by the earth's surface.

Gravitational Field According to Newtonian theory, the sun exerts the gravitational force directly on the earth without an intervening medium for transmitting that force. The behavior of such forces is called "action at a distance." To overcome the conceptual difficulty of a force acting directly over large distances, one assumes that a gravitational field fills all space. The force acting on any mass is determined by the gravitational field in its neighborhood. Thus, at the point P a distance R from the center of the earth, the gravitational field has the magnitude

$$\mathcal{G} = \frac{GM_E}{R^2}$$

and magnitude of the force on a mass M at P is simply $F = M\mathcal{G}$. Note that the field is to exist at P even in the absence of the mass M.

It is sometimes convenient to introduce the gravitational potential which determines the field through its gradient. For a spherical earth, it is defined as

$$\phi = - \frac{GM_E}{R}, \quad \mathscr{g} = - \text{ grad } \phi$$

In general ϕ will satisfy Poisson's equation

$$\frac{\partial^2 \phi}{\partial x^2} + \frac{\partial^2 \phi}{\partial y^2} + \frac{\partial^2 \phi}{\partial z^2} = 4\pi\rho \quad (2)$$

ρ is the density of matter. The potential energy of a mass M, in the field is simply expressed in terms of ϕ,

$$V = M\phi$$

This is equal to the work which must be done by an outside agent if he were to move the mass M to its location in the gravitational field from infinitely far away.

A body has enough speed to escape from the earth's attraction if its total energy, kinetic plus potential, is zero or greater. From the definition of potential energy above, the *escape velocity* at a distance R from the center of the earth is given by

$$v_{es}^2 = \frac{2GM_E}{R} \quad (3)$$

At the surface of the earth the $v_{es} = 11$ km/sec. It is interesting to note that when $R = R_s$, the Schwarzschild radius, the escape velocity is just equal to the velocity of light, $c(3 \times 10^5$ km/sec.):

$$R_s = \frac{2\,GM}{c^2} \quad (4)$$

The subscript E has been dropped from M because the *Schwarzschild radius R_s* is defined for any mass. For the earth, $R_s = 0.9$ cm while for the sun, $R_s = 3$ km. Now, according to special relativity, the speed of light is a limiting speed for matter and cannot be exceeded by any signal. Therefore, it is fortunate that the Schwarzschild radii of the sun and the earth are interior to their surfaces. Otherwise we would receive no energy from the sun and we could not explore the solar system.

Although one can introduce the gravitational field, it is an auxiliary concept in Newtonian theory for the field has no independent dynamical behavior as is true of the electromagnetic field (e.g., electromagnetic waves). At any time, the Newtonian gravitational field is determined by the configuration of masses at that instant and does not depend on previous history or state of motion. Thus, if the sun were to vanish, the gravitational force on the earth would immediately be removed. This property may be thought of in terms of an infinite velocity of propagation for the gravitational field. Letting the velocity of light become infinite in Maxwell's equations eliminates all independent dynamical behavior for the electromagnetic field. In that case there could be no radio or television. The special theory of relativity which is based on the velocity of light in vacuum being the maximum velocity for the transmission of energy, implies that Newton's theory requires modification.

Principle of Equivalence The mass of a body may be measured either by weighing $W = Mg$ (gravitational mass) or by observing its motion under a known applied force using Newton's second law of motion $F = MA$ (inertial mass). The equality of these two differently defined masses has been measured by V. B. Braginsky and V. I. Panov of the Soviet Union to an accuracy of 2×10^{-12} improving on earlier measurements by R. H. Dicke and R. V. Eötvös. It is this equality which distinguishes the gravitational force from all other forces in giving all bodies the same acceleration. The discussion of weightlessness pointed out that local effects of the gravitational field are eliminated for an observer in free fall precisely because all bodies fall at the same rate. It follows that the gravitational field measured by an observer will depend on his state of motion. In a sense there is an equivalence between a gravitational field down and an acceleration up for the observer. However, the equivalence is not complete, for real gravitational fields converge on their sources so that two particles released at the same time will drift closer together as they fall. On the other hand, acceleration fields have no effect on the separation of particles moving on parallel paths (Fig. 2). In a curved space, initially parallel geodesics—the "straight lines"—do not maintain a constant separation (e.g., great circles on a sphere). Thus, the gravitational field may have its explanation in the geometry of a curved space-time.

Red Shift According to the quantum theory, a photon of frequency ν has an energy $h\nu$ (h is Planck's constant), and by the relation $E = mc^2$,

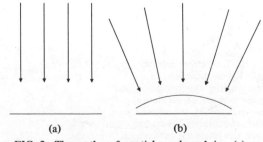

(a)	(b)

FIG. 2. The paths of particles released in: (a) an acceleration field (the acceleration is up, the apparent force is down); (b) a gravitational field showing convergence toward the source.

this quantum has a mass $m = h\nu/c^2$. To lift a mass m a height H requires expenditure of the energy mgH. Therefore, a photon emitted at the surface of the earth arrives at the height H with the energy

$$h\nu - (h\nu/c^2)\, gH = h\nu \left(1 - \frac{gH}{c^2}\right) = h\nu'$$

At the surface of the earth, the frequency shift amounts to

$$\frac{\Delta\nu}{\nu} = 1.1 \times 10^{-16}\, H(H \text{ in meters})$$

This shift was measured by Pound and Rebka, using the Mössbauer effect, in good agreement with the prediction. As standard clocks are determined by atomic transitions in freely falling atoms, it follows that if the same photon were emitted at the height H, it would be measured to have the frequency ν not ν'. Therefore, an observer at H must conclude that clocks at H run fast compared with the identical clocks on the ground in the ratio (ν/ν') (Fig. 3).

Einstein's Theory of Gravitation   Albert Einstein assumed that gravitation is a physical effect produced by the curvature of a four-dimensional space-time. The generalization of Newton's gravitational potential is the metric tensor $g_{\rho\sigma}$ in terms of which the four-dimensional distance, and hence the geometry of space-time, is determined:

$$ds^2 = \sum_{\rho,\,\sigma=1}^{4} g_{\rho\sigma}\, dx^\rho\, dx^\sigma$$

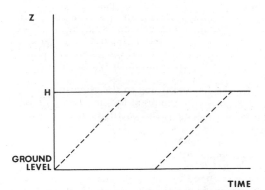

FIG. 3. Photons are emitted on the ground and are received at the height H. Between the two dotted lines representing the beginning and end of a pulse, the same number of oscillations, n, are received at H as are emitted at the ground level. Because of the red shift, the interval t' between oscillations at H is greater than the interval t between oscillations on the ground. Therefore, the time measured at H for the reception of the n oscillations is greater than the time required for their emission on the ground: $nt' > nt$. This result implies that clocks run faster at H than on the ground.

The curvature of space-time is defined in terms of a four-index tensor $R_{\nu\rho\sigma}{}^\mu$, the curvature tensor. The vanishing of the curvature tensor means that no real gravitational field is present. The field equations are ten linear combinations of the curvature components which are of the second order in the derivatives of the metric tensor and are a generalization of Poisson's equation [Eq. (2)]. Symbolically these equations are written

$$G^{\rho\sigma} = 8\pi\kappa\, T^{\rho\sigma}$$

where $T^{\rho\sigma}$ is a symmetric tensor which describes the distribution of matter and energy throughout space-time and $\kappa = G/c^2$. In a weak-field static approximation, these equations contain Newton's theory of gravitation with the Newtonian gravitational potential given by $2\phi/c^2 = g_{44} - 1$.

The metric tensor outside a static spherically symmetric mass distribution is given by the Schwarzschild solution:

$$ds^2 = \left(1 - \frac{2\kappa m}{r}\right) dt^2 - \left(1 - \frac{2\kappa m}{r}\right)^{-1} dr^2$$
$$- r^2 d\theta^2 - r^2 \sin^2\theta\, d\varphi^2$$

This geometry exhibits the red shift described above and in addition shows three other effects:

(1) The bending of a ray of light passing near the sun's edge by

$$\delta\theta = 1.75''.$$

(2) The precession of the perihelion of Mercury by

$$\delta\phi = 43''.03/\text{century}.$$

(3) The retardation of signals passing near the sun. For a radar pulse reflected from Mercury, this amounts to a maximum time delay

$$\Delta t = 1.6 \times 10^{-4} \text{ sec}.$$

Observations and experiments to check these predictions are still in progress.

Since one can see stars near the sun's edge only during an eclipse, the optical data on the bending of light have been slow and difficult to obtain and such measurements have poor reliability—about 10 to 25 per cent. However, using radio frequency measurements, a group including I. I. Shapiro has observed the angular positions of two sources, 3C279 and 3C273 which have an angular separation of about 10°. The latter source acts as the reference as 3C279 is occulted by the sun each year on October 8. Their results are in agreement with Einstein theory within 6 per cent. E. Fomalont and R. Sramek have carried out similar observation of three radio sources, the central one being occulted by the sun on April 11 of each year, which gives agreement within 2 per cent. Improvements in these observations are expected in coming years.

Shapiro has also reevaluated the optical data with regard to the solar system and has also taken new data using radar ranging. In both cases he finds agreement with the predicted value for the perihelion precession of Mercury within 3 per cent. If he combines the data, he reduces the error to 1 per cent.

As a new test of Einstein's theory of general relativity, Shapiro suggested measuring the retardation of radar echo signals from Mercury when the planet moves into a position of superior conjunction. The gravitational field of the sun, as represented by the Schwarzschild solution, not only produces a bending of the ray, but also affects the time of flight of the signal. Therefore, the time delay between the transmission of a radar pulse to Mercury and the reception of the reflected signal will depend not only on the relative positions of the earth and Mercury in their respective orbits, but also on whether the radar signals pass near the sun (Fig. 4). Current measurements give agreement within 2 per cent.

C. H. Brans and R. H. Dicke have proposed a "scalar-tensor" theory of gravitation in which the added scalar function determines the strength of the gravitational interaction. In this theory the gravitational constant G is no longer a universal constant, but depends on location and time. This effect has not been verified, but the theory also introduces small changes in the above predictions of the Einstein theory. Unfortunately, the experimental errors in the observations are such that one cannot distinguish between the Brans-Dicke and the Einstein theories. Nonetheless, the seeming arbitrariness of the scalar interaction makes the simpler Einstein theory preferable at the present time.

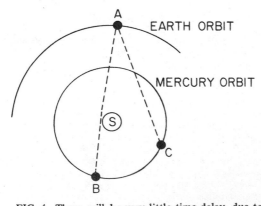

FIG. 4. There will be very little time delay, due to the sun's gravitational field, in the reception of the reflected radar signal when the earth-Mercury position is as in A-C compared to that when their position is as in A-B.

Gravitational Lens Effect A consequence of the bending of light by a massive object is that a distant galaxy may behave like a lens by bending light and other electromagnetic signals from even more distant objects.

Two quasars, labeled by their astronomical coordinates 0957 + 561 A, B, were observed to have identical red shifts, identical spectra, and very similar shapes. Further analysis indicated that they are two images of the same object produced by an intervening galaxy which acts as a gravitational lens. Qualitatively, if the earth, a galaxy, and a distant quasar are lined up, one can expect that signals passing near or through the outer limbs of the galaxy would give us two images displaced from the center while a central image might also be formed. Thus one may expect to see three images plus a central galaxy which might obscure the central image. As the galaxy moves off the earth-quasar line of sight, two of the images would move closer together, coalesce, and eventually would not be refocused to the earth if the galaxy is too far off the line of sight. Since, in fact, only two images are observed one infers that two of the images have merged and, indeed, it appears that the image of the galaxy itself is obscured by one of the images (see Fig. 5).

Gravitational Collapse. The gravitational force between any two masses is attractive. Therefore, given a quantity of matter, under action of gravity alone it will become as compact as possible. In the planets the compaction process is stopped by the electrical forces which act between atoms and molecules in close range. The pressure in the sun, however, is much too great to be supported by such solid body forces. This tremendous pressure is balanced primarily by the counter pressure of electromagnetic radiation which is produced by the nuclear processes at the sun's center. Stars in which the nuclear processes have ended undergo a further contraction which is stopped by the pressure of free electrons at the densities associated with white dwarfs. This pressure, which occurs because electrons obey the Pauli exclusion principle, is capable of supporting up to 1.4 solar masses within a volume of 10^{-4} to 10^{-8} of the solar volume. Objects which are more massive continue the crush. Neutrons become the most stable particles in the interior and the contraction is stopped by repulsive nuclear forces when a neutron occupies only about 10^{-39} cm^3, the nuclear volume. If the resulting neutron star is one solar mass, its radius is just 10 km and its volume 10^{-15} the sun's volume. Objects with more than about 1.2 solar masses cannot be stable as neutron stars. They continue to contract. Beyond this point the situation is confused by the abundance of exotic elementary particles, but there is no theoretical evidence that the contraction can be stopped.

One might have hoped that Einstein's theory of gravitation would contain a short-range re-

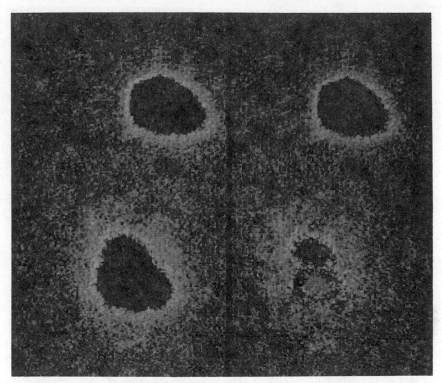

FIG. 5. This is a black and white print of a false color photograph. The display at the left was generated by a computer program that summed and color-coded many images of the twins. Elongation of both the northern twin and the southern one in the direction of four o'clock is an artifact of imperfect alignment of the telescope optics. The northern image was adjusted so that it had the same brightness as the southern one. Display at the right was made by subtracting the adjusted northern image (top) from the unadjusted one (bottom). What remains is an image of the galaxy one arc second north of the southern twin. (Courtesy of the Institute for Astronomy and Planetary Geoscience Data Processing Facility, University of Hawaii.)

pulsion which would stop this endless contraction. However, the opposite is the case. First of all, all forms of energy contribute to the attractive mass in general relativity, and secondly, the fact that matter determines the geometry means that there should be peculiarities in the space when an object is highly collapsed. There are several general theorems, particularly by R. Penrose and S. Hawking, whose general conclusion seems to be that as long as the energy density remains everywhere positive, collapse is inevitable. This does not mean that collapse actually occurs in nature. As a very massive star proceeds through the various stages indicated in the above paragraph, it may become unstable and throw off enough mass through an explosive process, such as a supernova, that it may settle down at a planetary size, or as a white dwarf, or as a neutron star. We have observational evidence for the existence of these objects. (A pulsar is thought to be a rapidly rotat-

ing neutron star.) So, not everything continues to collapse. But, there are many very massive stars and in the absence of more information it is unreasonable to rule out the possibility that some will indeed go through an indefinite collapse or that some may have already done so.

What physical effects result from the collapse? It was pointed out above, Eq. (4), that at the Schwarzschild radius the escape velocity from a point mass is the velocity of light. Thus, no signal can escape from a body which has collapsed below R_s. This result can in fact, be deduced from the Schwarzschild solution of the Einstein equations which is given above. As a result, our knowledge of events is limited at the Schwarzschild radius; the surface $r = R_s$ is an *absolute event horizon*. (Because we can receive no light or other signal from a source which has collapsed below its Schwarzschild radius, we call such objects *black holes*.)

Note that a neutron star of one solar mass has

a radius of 10 km while R_s = 3 km; a neutron star of 10 solar masses will have a radius of 30 km and R_s = 30 km. Thus, we have observational evidence for the existence of objects which are very nearly black holes. There is at present an active search for peculiar stellar motions which might indicate the presence of a black hole. On the other hand, if there should be a collapsed object in the nucleus of a galaxy, the very high concentration of stars near the black hole would produce a cusp in the surface 'brightness. So far there is no positive evidence, but the existence of black holes in the center of galaxies or as collapsed individual stars has not been ruled out.

Gravitational Waves Einstein's field equations require that the gravitational field have a finite velocity of propagation—the same as that for light. Therefore, the gravitational field has independent dynamical degrees of freedom which permit gravitational waves to exist in two states of polarization. These states are wholly transverse. That is, the waves act on matter only in planes which are orthogonal to the direction of propagation. In passing through matter, one state produces oscillations such that there is a compression followed by elongation along one axis and a corresponding elongation followed by compression along the perpendicular axis (Fig. 6). For a periodic wave this process repeats at the frequency of the wave. The other state of polarization has the same effect along axes rotated by 45°. This character for the modes is caused by the tensor nature of the potentials $g_{\rho\sigma}$ which limits the lowest order of gravitational waves to quadrupole radiation. A crude estimate of the energy radiated by the earth-sun system per year amounts to 10^{16} ergs (about 10^6 kWh). Radiating at this rate, the earth has lost about 10^{-15} of its available mechanical energy since its formation 5×10^9 years ago. Presumably there are stronger sources of gravitational waves available in the universe.

Experiments to detect gravitational radiation were begun in 1958 by J. Weber. For a detector he used an aluminum cylinder which is suspended in the earth's gravitational field. An in-

FIG. 7. One of the aluminum cylinders used by Professor Joseph Weber as a detector of gravitational waves. (Courtesy of Professor Weber.)

cident gravitational wave sets up transverse oscillations in the cylinder. These oscillations are transformed into electrical signals by piezoelectric crystals which are bonded to the surface of the cylinder (Fig. 7). However, the observations are made at room temperature and the thermal fluctuations produce displacement of the end faces of about 10^{-14} cm—an amount which, although only one-tenth the diameter of a nucleus, interferes with the expected gravitational effect. The gravity wave signal must be separated out from this background noise. To eliminate fluctuations due to earth tremors, trucks, or people dropping hammers, the whole apparatus is acoustically insulated from the ground.

The initial detection program used principally two identical cylinders, 153 cm long and 66 cm in diameter. One is located at the University of Maryland and the other at Argonne National Laboratory, 1000 km away. The electronic recording system is narrowly tuned to 1660 Hz which has an acoustic half-wavelength of 153 cm in aluminum. Thermal oscillations are randomly generated and one does not expect correlation between the outputs of two detectors, particularly if they are 1000 km apart. Therefore, Weber looks for coincidences in the output signals of the two detectors. The observation technique is to record each signal separately at its own location and at the same time to transmit the Argonne signal to Maryland where it can be compared directly with the Maryland signal. Coincidences of a certain pulse height are then marked (Fig. 8). A careful statistical analysis compares the coincidence rate due to random fluctuations with the observed rate. Weber concludes that there is a "significant coincidence rate of about one every two days."

Unfortunately other centers which have con-

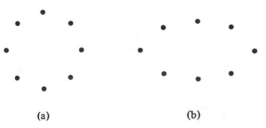

(a)   (b)

FIG. 6. (a) A circular arrangement of dust particles before a gravitational wave arrives. (b) The same particles after a passage of a wave consisting of one state of polarization. The second state of polarization would produce the same effect, rotated at 45°.

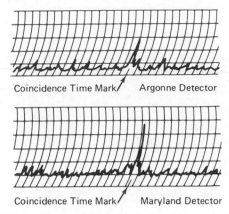

FIG. 8. Argonne National Laboratory and University of Maryland coincidence. (Courtesy of Professor Weber and *Physical Review Letters*.)

structed detectors for gravitational waves have been unable to corroborate Weber's observations. Therefore, at this time there is general agreement, among most workers in the field, that Weber's results cannot be accepted as evidence for the existence of gravitational waves. However, his claims have stimulated many other people to begin the search for gravitational radiation. Many improvements in the design of detectors have been proposed. The more obvious is to cool the detector with liquid helium to reduce thermal oscillations which mask the change in dimensions produced by an impinging gravitational wave. Several ingenious ideas have been proposed which could improve the sensitivity by as much as a factor of 10^5. Among these is the use of very large (several kilograms), carefully grown and annealed sapphire crystals instead of aluminum bars. Another is the use of a carefully designed microwave cavity whose frequency would shift by a very small amount when a wave impinges on it. A third design would use the shifting interference fringes of laser light when the distance between mirrors changes because a gravity wave is incident. Progress is slow because the technological problems are difficult. But there is confidence that in about ten years we will be monitoring the arrival of gravitational radiation inaugurating gravitational astronomy. When radio astronomy began, one did not know or understand the origin of the radiation. Now, in fact, radio observations rival and in some cases exceed the resolution of optical observations. These have shown structures which can not be seen optically. The same problems will arise with gravitational radiation; the interpretation of the observations will be much more difficult. But we may learn much about the universe which can not be discovered otherwise.

Although scientists are still waiting for the observation of gravitational waves, a stellar system has been observed which appears to exhibit the dynamical effect resulting from its emission of gravitational waves. A pulsar has been observed which is part of a double star system. The pulsar, identified as PSR 1913 + 16, has a precise spin rate of 16.94 revolutions per second. It is an object of 1.4 solar masses and has a diameter of about 20 kilometers. Its companion is about the same size, but there is no searchlight to indicate that it is also a pulsar.

Using equations of motion derived from Einstein's theory, the parameters describing the orbit have been determined. This is possible with high accuracy because the precise spin rate can be used as a clock. One found that the orbit is tilted 45° from the perpendicular plane to the line of sight, the separation of the two companions varies from 1.1 to 4.5 solar diameters, and that the periastron advances 4.2 degrees per year. Even more remarkable, however, the period of periastron passage is decreasing (Fig. 9). Over the six years of observation the period has decreased by 1 second. This means that the bodies have moved a little closer together, which indicates a loss of energy. This loss has been calculated to be that expected if the system PSE 1913 + 16 is radiating gravitationally in accordance with the Einstein theory.

Quantum Theory The gravitational interaction among the elementary particles is down by a factor of 10^{-40} from the electromagentic or strong nuclear interactions. Therefore, one cannot expect to observe quantum effects at the level of today's experiments. Nonetheless, because the gravitational field has two independent modes, the quantum field formalism requires that the gravitational field operators exist and satisfy appropriate commutation relations. However, the gravitational field equations are complicated by being nonlinear and by their covariance under arbitrary coordinate transformations. From its linear approximation, one expects that the quantized field will be a spin 2 boson field. However, the linear approximation may be deceptive because the gravitational field determines the geometry of spacetime. Quantization of the geometry may have the effect of undermining our current description of particle properties. Therefore, the labels we use to distinguish particle properties may be lost. Indeed, one can show that a black hole is characterized completely by its mass, angular momentum, and electric charge. Thus, when elementary particles are absorbed by a black hole, all of their characteristics, e.g., baryon number, flavor, and color, disappear. In this way, the gravitational field may be a link among all particles and all interactions.

As a result, a number of novel developments have occurred. One is an attempt by Roger Penrose to smear out space-time points by identifying them with the intersections of light rays. The light rays themselves are to be described by new mathematical objects called *twistors*. (Twistors have a similar relation to the con-

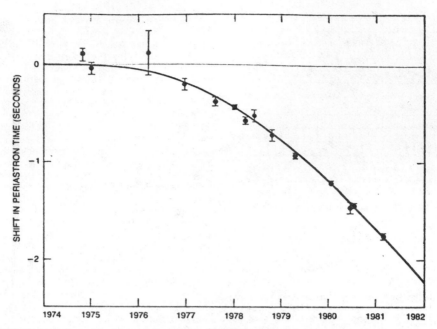

FIG. 9. Emission of gravitational radiation by PSR 1913 + 6 leads to an increasing deviation in the time of periastron passage compared with a hypothetical system whose orbital period remains constant. The solid curve corresponds to the deviation predicted by the general theory of relativity while the dots represent the measured deviation. The pulsar now reaches periastron more than a second earlier than it would if its period had remained constant since 1974. The data provide the strongest evidence now available for the existence of gravitational radiation. (Courtesy of J. H. Taylor, J. M. Weisberg, and *The Astrophysical Journal*.)

formal group in four dimensions as spinors have to the Lorentz transformations.) If one can construct a quantum theory of twistor fields, then space-time and its geometry would likewise be quantized. Another interesting suggestion is that through quantum gravity the distinction between Fermi-Dirac particles (half-odd-integral spin) and Bose-Einstein particles (integral spin) can be broken down. This work comes under the heading of "supergravity." While many interesting results have been obtained from studies of both of these new ideas, the results are still too preliminary to suggest that either is in the right direction.

On the other hand, Stephen Hawking has produced a striking result in a semiclassical treatment. He considers a quantized scalar field in a classical background metric—that of the Schwarzschild metric which has an absolute event horizon. Hawking assumes that in the infinite past the scalar field is in its lowest energy state—the vacuum state. He then shows that in the infinite future the scalar field is in a state corresponding to a black body temperature of

$$T_b = \frac{\hbar c^3}{8\pi GMk} \simeq 10^{-6} \, \frac{M_\odot}{M}.$$

An intuitive understanding of this process can be given as follows. The fluctuations of a quantized field produce particle-antiparticle pairs. If one particle of a pair should be produced with momentum toward the horizon, the other will be produced with momentum away from the horizon. There is a finite probability that one particle will fall across the horizon while its antiparticle moves away from the horizon and is observed. The net result is that black holes are unstable because of quantum fluctuations of the various quantized fields. However, the decay of black holes is a very slow process, the temperature of a $1M_\odot$ black hole being $T_\odot = 10^{-6}$ °K. It is estimated that any black hole created in the initial big bang smaller than 10^{15} grams would have radiated away all of its mass-energy by this time.

Cosmology One expects the gross structure of the universe to depend on global geometrical properties and the mean distribution of matter more than on the details of particle interactions. Among the various cosmological solutions of Einstein's equations are those models based on a homogeneous and isotropic matter distribution in which matter is streaming away from every point with a velocity dependent on its distance from that point. Such models agree with observations made by Hubble on the red shift in the spectra of distant galaxies. He interpreted this red shift as a Doppler effect due to a recessional velocity of the galaxy which is proportional to its distance:

$$v = Hr$$

with $H = 30$ km/sec/10^6 light-years and r measured in millions of light-years. This value of H leads to an age for the universe of about 10^{10} years.

The models based on Einstein's equations are evolutionary. This means that the rate of expansion of the universe, hence Hubble's constant H, depends on time. Bondi and his co-workers have suggested that the universe is in a steady state and that H is in fact independent of time. The steady-state theory says that the distribution of galaxies and the density of matter remain the same on the average. New matter is created in order to fill in the thinning density of matter as the universe maintains its expansion. One attempts to distinguish between the steady-state theories and the evolutionary models by counting the number of galaxies and the number of radio sources as a function of distance. These observations, though not definitive, suggest that the evolutionary models are correct.

There are two other observations which strengthen this conclusion. One of these is the recognition of the QSO (quasi-stellar objects) as extra-galactic objects. The QSO are very small, very luminous objects, and with very large red shifts. The simplest and most satisfactory interpretation of the observed red shifts is that it is the cosmological Doppler shift. With this interpretation, 3C273, the nearest, most luminous, and the first identified QSO, is at a distance of 1.6×10^9 light-years. This distance is so great that it is possible that an early state in the development of galaxies is being observed. On the other hand, the amount of energy being released is so great ($\sim 10^{49}$ ergs/sec) that a nuclear source for the energy is ruled out, and it is sug-

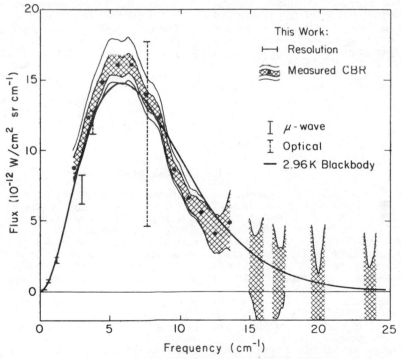

FIG. 10. Measured spectrum of the cosmic background radiation plotted as $\pm 1\sigma$ error limits. There are gaps in the data at the frequencies of strong atmospheric emission lines where the errors become very large. The shaded region includes contributions to the error which are uncorrelated across the spectrum. Additional solid lines are shown to represent the effect of changing the frequency-independent scale factor by $\pm 1\sigma$. The shaded region can be scaled up or down within these limits. The spectrum of the 2.96°K blackbody, which has the same integrated flux as the measured spectrum, and selected microwave and optical measurements of the CMB are also shown for comparison. (Courtesy of David P. Woody and Paul L. Richards and *The Astrophysical Journal*.)

gested that the energy comes from the gravitational contraction of 10^{10} $M_\odot$ down to its Schwarzschild radius of about 10^{10} km. However, it is equally possible that the QSO represent the opposite effect, the exploding of matter from a highly compacted state which is being thrust out. At present the models giving the energy through collapse seem to fit the observations. From the point of view of the steady-state cosmology, the QSO create serious problems. They are far away and there seems to be an overabundance of them at very large distances, hence a long time ago.

In 1949, George Gamow and his co-workers suggested that at an early stage in an evolutionary expanding universe, matter and radiation interacted very strongly and were in thermal equilibrium. As the universe expanded, the opportunity for interaction decreased, the radiation and matter were effectively decoupled. Matter cooled more rapidly and condensed into the galaxies and stars. The radiation maintained its thermal equilibrium distribution, its *blackbody* distribution, behaving like a gas which is undergoing an adiabatic expansion. Gamow estimated the temperature of this remnant of the primeval fireball at about 25°K. Fifteen years later, Dicke came to a comparable conclusion independently and set about to observe this radiation. In the meantime, Penzias and Wilson, while tracking down residual noise in a sensitive microwave radiometer, discovered an isotropic background radiation at a wavelength of 7.4 cm whose intensity corresponded to the emission of a blackbody at a temperature of 3.5°K. Subsequent measurements by many observers have verified the existence of this background radiation, but have reduced the temperature to 2.7°K.

The use of microwave radiometers from the ground was very successful in mapping the low frequency region. Unfortunately, because of atmospheric absorption and interference from local sources, this method of measurement could not be carried over the peak of the curve for black body radiation at 2.7 °K. However, recent measurements by Woody and Richards used a balloon-based spectrophotometer to measure the cosmic background radiation from the low frequency side well over the peak into the high frequency side of the black body curve. Within the errors of their measurements the total flux in the frequency range 2.4 – 13.5 cm^{-1} (λ^{-1}) agrees with the flux from a black body at a temperature of 2.96 °K with lower and upper limits of 2.88 and 3.09 °K respectively. These results together with those from earlier measurements are shown in Fig. 10.

The prediction and observation of the blackbody radiation and its interpretation as evidence of a primeval fireball give very strong support to the concept of an evolutionary universe, a universe which has expanded from a compact volume to its present size. However, it does not select a particular cosmological model from the many which are allowed by Einstein's gravitational theory, those permitted by the Brans-Dicke modification, or other possibilities which have not yet been constructed.

JOSHUA N. GOLDBERG

References

1. Bergmann, P. G., "The Riddle of Gravitation," New York, Charles Scribner's Sons, 1968.
2. Bonnor, W., "The Mystery of the Expanding Universe," New York, The Macmillan Co., 1964.
3. Chaffee, Frederic H., "The Discovery of a Gravitational Lens," *Scientific American* 243(5), 70 (November, 1980).
4. Davies, P. C. W., "The Search for Gravity Waves," Cambridge, U.K., Cambridge Univ. Press, 1980.
5. Gamow, G., "Gravity," Garden City, N.Y., Doubleday and Company, 1962.
6. Partridge, R. B., "The Primeval Fireball Today," *American Scientist* 57, 37 (1969).
7. Rindler, W., "Essential Relativity," New York, Van Nostrand Reinhold, 1969.
8. Sciama, D., "Modern Cosmology," Cambridge, U.K., Cambridge Univ. Press, 1971.
9. Weber, J., "General Relativity and Gravitational Waves," New York, Interscience Publishers, 1961.
10. Weisberg, J. M., Taylor, J. H., and Fowler, L. A., "Gravitational Waves from an Orbiting Pulsar," *Scientific American* 245(4), 74 (October 1981).

Cross-references: ASTROMETRY, ASTROPHYSICS, COSMOLOGY, POTENTIAL, RADIO ASTRONOMY, RELATIVITY, SOLAR PHYSICS.

GYROSCOPE*

The gyroscope consists of a flywheel or a sphere that is spinning (usually at high speed) about an axis. If this axis is free in space (such freedom may be provided by gimbals, by floating the spinning mass on a column of gas or fluid, or by suspension in a magnetic or an electrostatic field), the axis will remain parallel to its original position even though the gyroscope is mounted on a vehicle that translates and rotates in three dimensions. This property of the "free gyro" often referred to as "spatial memory" was used as an artificial horizon as early as 1744 by Serson. This permitted a ship's navigator to take readings with a sextant (measurement of the angular elevation of a star with respect to the horizon) when the horizon was obscured by darkness, fog, mist, etc. This early instrument has led to the modern vertical gyro.

In 1852, Leon Foucault built one of the first precise gyropscopes. Utilizing a flywheel 80 millimeters in diameter supported in near-frictionless gimbals, this instrument was sensitive

*Illustrations and text excerpted from Refs. 1 and 2.

enough to detect the rotation of the earth. This free gyro maintained a fixed orientation in space as the earth turned. The relative motion was observed through a microscope.

In 1896, the gyroscope saw its first application for guidance when Obry used it in a self-propelled torpedo. An *un*guided torpedo, under the influence of winds, currents, and ocean waves would *not* follow a prescribed course. The gyroscope, on the other hand, was pointed at the target before launching the torpedo, and by means of linkages connected to the spin axis support, it actuated the rudder of the torpedo, steering it along a straight course.

When the gyroscope is *not* free, that is, when the spin axis is forced to turn in space, the gyroscope develops a torque about an axis that is perpendicular to the plane containing the spin axis and the axis about which turning takes place. This property was utilized to find north by continually reorienting the spin axis until the gyroscope could *not* detect the *rotation* of the earth. This instrument, called the gyrocompass, was perfected in 1908 by Anschütz of Germany and by Elmer Sperry (1911) of the United States. This instrument used a wheel 0.1 to 0.2 meters in diameter, and was suspended in such a way that it tended to remain vertical. Since the vertical turned with the earth, the gyroscope following it produced a torque. This torque, in turn, was used to provide self-turning of the spin axis in a direction that would reduce the torque. In the final equilibrium position, the instrument pointed north. The gyrocompass is still widely used today.

The gyroscopic torque is also used to stabilize ships. Due to the motion of waves, the *un*stabilized ship rolls considerably. It is impractical to shift huge masses fast enough to counteract the irregular motion. On the other hand, a gyroscope develops torque instantly and with much less effort. This led to the development of the gyroscopic ship stabilizer. Utilizing a wheel 3 to 6 meters in diameter, and a hydraulic turning mechanism, angular rates are applied to the spin axis support producing tremendous gyroscopic torques upon the ship. In moderately rough seas, the gyro stabilizer can eliminate 70 to 80 per cent of the roll motion. The control signal to the turning mechanism is provided by a small "vertical gyro" which is very much like Serson's artificial horizon gyro. This essentially is a guidance system (like Obry's) using a small guidance gyro to provide the vertical and a large gyro to provide the "muscle."

If the motion of the gyro is free, and if a torque is applied to the structure containing the spinning mass, it will turn or precess about an axis that is perpendicular to the plane containing the torque axis and the spin axis. This is the converse of the gyroscopic torque phenomenon. By adjusting the amount of torque, the gyroscope provides a controlled rate. This property is utilized in an autopilot during a constant

rate of turn. The principle is also used for platform stabilization where the platform carries instruments which must bear specific orientations with respect to the earth. By carefully regulating the torque, the gyroscope and platform rotate with the earth without being in contact with the earth.

Gyroscopic Torque If the gyro is not free, but is forced tor turn, its spin axis is forced to change its orientation. Refer to Fig. 1(a). Let the initial position of the spin axis be represented by the vector ω_1 and the position after time Δt be the vector ω_2. The vector $\Delta\omega$ represents the change in angular velocity. It has a magnitude

$$\Delta\omega = \omega\Delta\phi \tag{1}$$

where $\Delta\phi$ is the angle turned during time Δt.

Likewise, the angular momentum vector $\mathbf{H}$ may be shown initially and after time Δt, as indicated in Fig. 1(b).

$$\mathbf{H} = I\omega \tag{2}$$

where I = mass moment of inertia of wheel about the spin axis

ω = angular velocity of wheel.

The change in magnitude of angular momentum is

$$\Delta H = I\Delta\omega \tag{3}$$

The torque required to change the angular momentum is T_i, and the equal and opposite torque T felt by the structure is

$$T = \frac{\Delta H}{\Delta t} \tag{4}$$

Apply Eqs. (1)–(3) to Eq. (4):

$$T = H\frac{\Delta\phi}{\Delta t} \tag{5}$$

In the limit, $\Delta\phi/\Delta t$ approaches $\dot\phi$. If the turn rate vector $\dot\phi$ makes an angle ψ with the plane containing the vectors $\mathbf{H}_1$ and $\mathbf{H}_2$, then the perpendicular component is

$$\dot\phi \sin \psi.$$

Applying the above concepts to Eq. (5), we have

$$T = H\dot\phi \sin \psi \tag{6}$$

as the magnitude of the torque. In vector notation, this becomes

$$\mathbf{T} = \mathbf{H} \times \dot{\boldsymbol{\phi}}. \tag{7}$$

In order to determine the sense of the torque vector $\mathbf{T}$, rotate the angular momentum vector $\mathbf{H}$ toward the input rate vector $\dot{\boldsymbol{\phi}}$ and apply the right-hand rule.

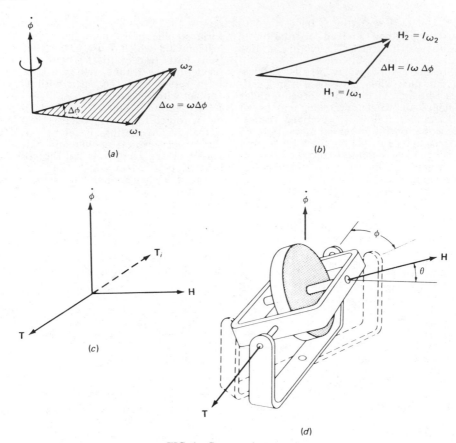

FIG. 1. Gyroscopic torque.

Gyro Drift The free gyro maintains its orientation in space, provided that there are no disturbances to the gyro. However, the presence of small torques produces a precession or *gyro drift*. Usually the magnitude and direction of the torque are unknown. Hence, gyro drift is usually unknown, and thus, constitutes an error.

One common source of this unknown or uncertain torque is the axial shift of the spin axis bearings. In a vertical gravity or acceleration field g, the torque T_d is given by

$$T_d = Mgb \tag{8}$$

where M = mass of gyro wheel
g = acceleration along the ϕ axis
b = axial shift.

The resulting gyro drift rate $\dot{\theta}_d$ is given by

$$\dot{\theta}_d = \frac{T_d}{H} = \frac{Mgb}{H}. \tag{9}$$

If the gyro wheel has most of its mass concentrated in its rim, then the angular momentum is approximately,

$$H = MR^2 \omega. \tag{10}$$

Then the gyro drift is given approximately by

$$\dot{\theta}_d = \frac{Mgb}{MR^2 \omega} = \frac{gb/R}{R\omega}. \tag{11}$$

In superprecise gyroscopes, the axial shift b is of the order of one millionth of the wheel radius, the wheel speed ω = 25,000 rpm = 2510 rad/sec, and the wheel radius R = 25 mm = 0.025 meters. Then,

$$\dot{\theta}_d = \frac{(9.8 \text{ m/s}^2) \, 10^{-6}}{(0.025 \text{ m}) (2510 \text{ rad/s})}$$

$$= 0.156 \times 10^{-6} \text{ rad/s}$$

$$= 0.0322°/\text{hr}.$$

A tolerance of one part in a million may seem extraordinary. However, as extreme as this seems, it is not good enough for typical applications in inertial guidance systems, where a drift rate of 0.01°/hr is required. For such applications, exceptional accuracy is demanded—possible only in ultrasophisticated instruments. One such device is the "ring laser gyro," which consists of three lasers (each about 0.1 meter long) arranged in a triangle. This instrument makes use of the linear momentum of a beam.

IRA COCHIN

References

1. Cochin, Ira, "Analysis and Design of Dynamic Systems," New York, Harper & Row, 1980.

2. Cochin, Ira, "Analysis and Design of the Gyroscope," New York, John Wiley & Sons, 1963.
3. Bulman, D. N., and Maunder, L. J., "Dynamics of the Dynatune Gyro," *Mech. Engr. Sci.*, **22**(3), 137–141, (June 1980).
4. Chow, Weng, and Hambenne, Jarel, "Multioscillator Laser Gyros," *IEEE Trans. Quantum Electron.*, QE-16(9), 918–935 (Sept. 1980).
5. McLeod, D. L. "Solid Rotor Electrostatic Gyro," IEEE Proc. Natl. Aerosp. Electron. Conf., vol. 3, pp. 1199–1205, May 1979.

Cross-references: ROTATION-CURVILINEAR MOTION: INERTIAL GUIDANCE; MASS AND INERTIA; MECHANICS.

H

HALL EFFECT AND RELATED PHENOMENA

If a current of particles bearing charges of a single sign and constrained to move in a given direction is subjected to a transverse magnetic field, a potential gradient will exist in a direction perpendicular to both the current and the magnetic field. This phenomenon is called the Hall effect, after E. H. Hall who discovered it in a metal in 1879. To commemorate the 100th anniversary of Hall's discovery, a symposium was held on November 13, 1979 at The Johns Hopkins University. It was at this university, while a graduate student under Henry Rowland in the Physics Department, that Edwin Herbert Hall did the experiments which led to his discovery of what is now universally known as the Hall effect. The published proceedings of the centennial symposium, entitled "The Hall Effect and Its Applications," occupy a total of 550 pages and illustrate the wealth of activity which has evolved from Hall's work. For example, studies of the Hall effect provide extremely useful techniques for obtaining information about the electronic properties of solids. As will be discussed in more detail, Hall data can be used to obtain concentrations and numerous other properties of the charge carriers. In addition, the Hall effect is the basis of a variety of specialized devices, in which the fundamental unit (referred to as a *Hall element* or, sometimes, as a *Hall generator*) is made of a material in which the voltage or, depending upon the application, the power produced by the Hall effect is especially large. More specific details are given later in the section entitled "Hall Effect Applications." Broadly stated, the utilization of the Hall effect yields a variety of miniature, light-weight, highly reliable devices which can efficiently be used in a number of applications such as: (1) replacement of the troublesome contacts commonly found in conventional circuitry; (2) measurement of position or velocity of rotating elements; (3) measurement of magnetic fields or magnetic field gradients by means of a simple probe and associated circuitry. The above applications are of especial importance in the guidance and sensor systems of satellites. Finally, Hall phenomena play an important role in plasmas in magnetic fields and in magneto-hydrodynamics. For example, it was pointed out in a paper given at the symposium mentioned above that the Hall currents in the aurora can be as large as one million amperes. In magnetohydrodynamic technology, Hall phenomena enter into the design of MHD electrical power generators.

Principles Involved The Hall effect is a manifestation of the force, usually known as the Lorentz force, which is exerted on a charged particle moving perpendicular to a magnetic field. The direction of the force is perpendicular to both the magnetic field and the velocity of the charge. If this sidewise thrust is not counteracted, the charged particle undergoes a deflection. The result of such motion is to create a charge unbalance and to produce a transverse electric field component, known as the *Hall field*. The force on the charged particle due to this Hall field tends to oppose the force resulting from the magnetic field. Mathematically, the forces and fields in question are represented by the following vector equation:

$$F = e\,[E + (1/c)v \times H] \qquad (1)$$

where F is the force, e the charge, and v the velocity of the particle, E is the electric field, and H is the magnetic field. A permeability of unity is assumed, and for the Gaussian system of units, c is the speed of light and F is in dynes. The use of the symbol H in the above equation, when more precisely the magnetic induction B is meant, is perhaps unfortunate. However, this notation was used almost universally in the early literature on Hall phenomena. In the Gaussian system the values of B and H are the same in nonmagnetic materials (a permeability of unity). As the equation indicates, the total force on the particle of charge e is the vector sum of that due to the electric field E and that resulting from the magnetic field. The latter enters via the *vector cross product* $v \times H$, which represents a vector perpendicular to both v and H and in the direction indicated by the right hand rule as v is rotated into H. The magnitude of the vector $v \times H$ is equal to the product of the individual magnitudes of v and H times the sine of the angle between them. Hence, only the component of v perpendicular to H contributes to the sidewise force.

The physics involved in the Hall effect is con-

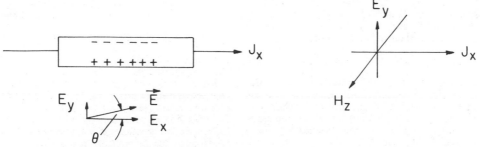

FIG. 1. Hall field due to action of magnetic field on positive charge carriers. For example shown, e, J_x, and H_z are positive and, therefore, v_x, E_x, and E_y are also. The resultant electric field and the Hall angle are shown at lower left. In the case of electrons, the Hall field is reversed for positive J_x, both v_x and e now being negative.

viently illustrated by considering a confined stream of free particles, each having a charge e and an initial velocity v_x. A magnetic field in the z direction produces initially a deflection of charges along the y direction. This charge unbalance creates an electric field E_y, and the process continues until the force on a moving charge due to the Hall field E_y counter-balances that due to the magnetic field so that further particles of the same velocity and charge* are no longer deflected. A pictorial representation of this is shown in Fig. 1. The magnitude of the Hall field follows at once from Eq. (1), with $\mathbf{F} = 0$ at steady state, namely

$$E_y = (1/c)v_x H_z = J_x H_z / nec \text{ (Gaussian}^\dagger \text{ units)}$$
$$(2)$$

In obtaining the last equality, the electric current density J_x was expressed in terms of the density of charge carriers n by the product nev_x. For *electronic* (i.e., as opposed to *ionic*) conduction in solids, the magnitude of e is

*If the particles have a *distribution* of velocities, then it is only those particles having a certain "average" velocity, which are undeflected. This point is expanded later.

†In the Gaussian system, mechanical quantities are in cgs units, electrical quantities in esu, and magnetic fields in gauss or oersteds. In the *practical* system, mechanical quantities are in cgs units, electrical quantities in volts and coulombs, and magnetic fields in gauss or oersteds. In all equations in this article, except for Eq. (1), conversion from the Gaussian system to the practical system is effected by replacing c by unity and replacing H by $H/10^8$, where these quantities explicitly occur. The practical units for R_H are cm^3/coulomb; those for mobility are cm^2/volt-sec. The corresponding units in the Gaussian system are cm^2-sec/esu and cm^2/statvolt-sec. If practical units are to be used in Eq. (1), to obtain the force in dynes it is necessary not only to make the changes specified above but also to include a factor of 10^7 on the right-hand side. This results from the additional procedure of converting charge in coulombs to esu and the field terms in volts/cm to statvolts/cm.

the electronic charge (4.8×10^{-10} esu in Gaussian units; 1.6×10^{-19} coulomb in practical units), and its sign is negative for transport by electrons and positive for transport by holes (deficit electrons). The *Hall coefficient* is defined by the ratio $E_y/J_x H_z$, namely

$$R_H \equiv E_y/J_x H_z = 1/nec \text{ (Gaussian units)} \quad (3)$$

Thus a very simple relation exists in the free-particle example between the Hall coefficient and the charge-carrier density. It is also seen that R_H is negative for conduction by electrons, positive for conduction by holes. Now the electric current density J_x exists by virtue of an applied electric field E_x. With the Hall field present, the *resultant* electric field in a solid‡ lies at some angle θ to the x axis. This angle is called the Hall angle, namely

$$\theta \equiv \tan^{-1} E_y/E_x \quad (4)$$

Thus the Hall effect may be described as a rotation of the electric field. At zero Hall field, the equipotential lines are perpendicular to $\mathbf{J}$, but when the Hall field appears, they are oblique, so that a Hall voltage exists across the specimen in a direction normal to the current. The rotation aspect is also brought out by considering the components of the *conductivity tensor*, which relate electric current densities and fields. For the boundary condition that $J_z = 0$, we may write

$$J_x = \sigma_{xx} E_x + \sigma_{xy} E_y$$
$$J_z = 0, H = H_z \quad (5)$$
$$J_y = -\sigma_{xy} E_x + \sigma_{xx} E_y$$

Equations (5) hold for media of sufficient symmetry§ that $\sigma_{xx} = \sigma_{yy}$, $\sigma_{yx} = -\sigma_{xy}$, and $\sigma_{xz} = \sigma_{yz} = \sigma_{zx} = \sigma_{zy} = 0$. The latter insures that E_z

‡Although the Hall effect can be discussed for any constrained electron gas, we shall, for simplicity, talk about a solid conductor.

§Isotropic media, e.g., or cubic systems with coordinate axes along cube axes.

vanish when J_z is zero. Since the boundary conditions for Hall effect require that J_y vanish, Eqs. (4) and (5) yield the following result for the off-diagonal elements σ_{xy}:

$$\sigma_{xy} = \sigma_{xx} \tan\theta \qquad (6)$$

The inverse of the conductivity tensor is the *resistivity* tensor, which relates **E** to **J**. A general definition of the Hall effect involves relating it to the antisymmetric* components of the resistivity tensor. This leads to the vector equation for the Hall field, namely

$$E_H = R_H \mathbf{H} \times \mathbf{J} \text{ (Gaussian units)} \qquad (7)$$

The identity in the first part of Eq. (3) is recognized as a special case of Eq. (7)

Application to Real Solids In a real solid the idealized free particle treatment no longer applies and must be replaced by a theory that takes into account the distribution of velocities and the interactions of the charge carriers with impurities, defects, and lattice thermal vibrations of the solid (i.e., the *scattering*), as well as the *band structure* of the solid. The latter consideration relates to the fact that the charge carriers in the solid are not free but exist in a potential energy field having the periodicity of the lattice. As a result of these constraints, only certain energy states, or *bands*, are allowed for the charge carriers. In addition, the relationship between the energy and velocity of the carriers is not the simple $\frac{1}{2}mv^2$ of the free electron, but is more complex. As an approximation, one frequently characterizes the charge carriers by an *effective* mass, m^*. Taking into account most of the complexities mentioned above, one still obtains an expression similar to Eq. (3), namely

$$R_H = r/nec \text{ (Gaussian units)} \qquad (8)$$

The *Hall coefficient factor*, r, actually depends on the nature of the scattering, the band structure, the magnetic field strength, and on the statistics characterizing the distribution of velocities of the carriers. Fortunately it depends weakly on these factors, and its value is usually within, say, 50 per cent of unity.

An important attribute of charge carriers is their mobility, i.e., their drift velocity per unit electric field. The conductivity mobility μ is related to the conductivity σ by

$$\mu = \sigma/ne \qquad (9)$$

It is also customary to define a *Hall mobility* μ_H for conduction by a single type of charge carrier by the relation

*These components are defined by the condition that $\rho_{ik} = -\rho_{ki}$, where the ρ_{ik} [cf. Eq. (5)] are defined by $E_i = \sum_{k=1}^{3} \rho_{ik} J_k$, $1 \equiv x$, $2 \equiv y$, $3 \equiv z$.

$$\mu_H H/c = \tan\theta, \text{ or } \mu_H = R_H \sigma c$$
$$\text{(Gaussian units)} \qquad (10)$$

With the use of Eqs. (8) and (9), it can be seen that the ratio of Hall and conductivity mobilities is precisely the Hall coefficient factor. It follows that the Hall angle is proportional to μ and H.

When transport by two or more kinds of charge carriers occurs, Eq. (8) is not applicable, and one must return to the general relations of Eq. (5). For conduction by electrons and holes (of respective densities n, p), the *weak-field* Hall coefficient can be written in the form

$$R_0 = -(r_e \mu_e^2 n - r_h \mu_h^2 p)/$$
$$[|e|c(\mu_e n + \mu_h p)^2] \qquad (11)$$

where $|e|, \mu_e$, etc., are positive. We note that the charge-carrier densities are now weighted by the mobilities μ_e, μ_h. There are also Hall coefficient factors for each carrier. For arbitrary H, the terms involve field-dependent factors.

Analysis of Hall-effect data is one of the most widely used techniques for studying conduction mechanisms in solids, especially semiconductors. For the single-carrier case, one readily obtains carrier concentrations and mobilities, and it is usually of interest to study these as functions of temperature. This can supply information on the predominant charge-carrier scattering mechanisms and on activation energies, i.e., the energies necessary to excite carriers from impurity levels into the conduction band. Where two or more carriers are present, the analysis is more complicated [cf. Eq. (11)], but much information can be obtained from studies of the temperature and magnetic field dependencies.

Unlike, for example, the magnetoresistance, the Hall effect is a first-order phenomenon. At weak magnetic fields it depends linearly on H, and it does not vanish in isotropic solids if all the carriers have essentially the same velocity or if the scattering is characterized by a relaxation time which is independent of the carrier energy. The Hall effect forms the basis of a number of devices used in isolating circuits, transducers, multipliers, converters, rectifiers, and gaussmeters (for measurement of magnetic fields). As mentioned earlier, the fundamental component of such devices is a slab of material (called a "Hall element" or a "Hall generator") possessing favorable Hall characteristics. Additional devices are mentioned in the last section.

Experimental Determination A number of techniques are available, the most direct being to measure, by means of a potentiometer or other high-impedance device, the Hall voltage V_H across a parallelepiped in a direction normal to both **H** and **J**. The Hall coefficient follows from Eq. (3) or (7), with attention to the footnote to Eq. (2):

$$R_H = 10^8 \, V_H t/IH, \text{ (practical units)} \qquad (12)$$

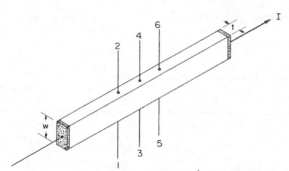

FIG. 2. Arrangement of contacts for measurement of Hall coefficient and related effects. Hall probes are 3 and 4. Probes 2 and 6 are for resistivity and magnetoresistance. Probes 1 and 5 allow a check of the uniformity of the specimen. For Hall effect and transverse magnetoresistance, the magnetic field is in the direction of the thickness t; for longitudinal magnetoresistance, it is along I. To avoid disturbances due to contact shorting, all probes should be at a distance of at least $2w$ from the end contacts.

where t is the thickness of the specimen and I is the total current. The arrangement is shown in Fig. 2. Since V_H may be the order of microvolts, extreme care must be taken to avoid extraneous voltages. An example is the misalignment voltage, caused by probes 3 and 4 not being on an equipotential plane when $H = 0$. This may be eliminated by taking measurements for opposite directions of H and taking half the difference, inasmuch as $V_H(H)$ is odd in H. Alternatively, one can adjust the position of the Hall probes for a null reading with $H = 0$. Other techniques involve use of resistances with sliding contacts suitably attached to the specimen. Other spurious voltages can arise from temperature gradients and resulting thermoelectric or thermomagnetic emf's (see following section). Some of these can be eliminated by taking appropriate averages among measurements for reversed polarities of I and direction of H. It is usually desirable to maintain good thermal contact between all points of the specimen and a constant-temperature bath. Although it is possible to analyze *adiabatic* Hall data (no heat flow to or from the specimen during measurement), *isothermal* data are preferred because of the simplicity of the equations. All of the relations in this article are for isothermal conditions. Errors in measurement can also result from a shorting of the Hall voltage, especially by the end contacts. With regard to the latter, the error is essentially negligible if the length-to-width ratio of the specimen is about 4 or more, and the Hall probes are near the center.

Related Effects A widely-studied galvanomagnetic effect is *transverse magnetoresistance*, usually written $\Delta\rho/\rho_0$. This phenomenon can be illustrated by the pictorial scheme in Fig. 1. If a charge carrier is deflected by the Lorentz force so as to traverse a longer path, it will contribute less to the conductivity, and there will be a positive magnetoresistance. It was noted, however, that if all charge carriers have the same velocity,

none will be deflected since the Hall field cancels the $\mathbf{v} \times \mathbf{H}$ force. This is the case in an isotropic metal, where the velocity of all the electrons is essentially the Fermi velocity. It is also true for electron scattering mechanisms described by a relaxation time which does not depend on energy. In these cases—assuming, of course, a single type of carrier—there is no magnetoresistance. If, however, there is a distribution of electron velocities—as in a semiconductor—then it is clear that only those electrons of a certain "average" velocity will be undeflected. The remaining carriers, having velocities either larger or smaller than the "average" will be deflected and will traverse longer paths, thus increasing the resistance of the conductor. A similar situation obtains if more than one type of carrier is present. It is also apparent that any mechanism which shorts out the Hall voltage—e.g., special geometry, shorting contacts, inhomogeneities in the specimen—will increase the magnetoresistance.

For the reasons discussed and the fact that $\Delta\rho/\rho_0$ varies as H^2 in weak fields, magnetoresistance is a second-order effect. It tends to saturate in strong magnetic fields, unless there is a disturbance of the Hall field as mentioned above. Magnetoresistance is even in H, and it is related to the *symmetric* components of the resistivity tensor, i.e., those for which $\rho_{ik} = \rho_{ki}$ [cf. footnote on p. 401]. It can be measured with the geometry shown in Fig. 2 by maintaining the current constant and determining the potential difference between probes 1 and 5 or 2 and 6 as a function of magnetic field. Magnetoresistance data can supply information on charge-carrier scattering and band structure. In the latter case, if anisotropy exists, it is useful to obtain data for different crystallographic directions. There is also a *longitudinal* magnetoresistance, measured when H is parallel to J. This effect vanishes in isotropic solids, and its presence indicates anisotropy in scattering or band structure, or inhomogeneities in the specimen.

By shorting the Hall field or by choosing a disk geometry so that such a field does not exist, one obtains a "magnetoresistance" (more strictly, a *magnetoconductivity*) which does not saturate. This is called the Corbino magnetoresistance or Corbino effect, after O. M. Corbino who studied circulating secondary currents in a "Corbino" disk carrying a primary radial electric current in a magnetic field.

There are a number of thermal effects in a magnetic field, which can produce transverse voltages or temperature gradients. These result from the velocity separation of charge carriers by the Lorentz force—the energetic ones going to one side, the slower ones going to the other. Temperature gradients are produced, and also electric fields. In the Righi-Leduc effect, a longitudinal temperature gradient produces a transverse temperature gradient (thermal analog of the Hall effect); in the Nernst effect, it produces a transverse electric field. In the Ettingshausen effect, a longitudinal electric current produces a transverse temperature gradient. This latter effect, if large, can lead to complications in the measurement of the Hall field since the potential probes and leads are seldom made of the same material as the specimen. Therefore the Ettingshausen temperature gradient can produce a thermoelectric voltage which adds to the Hall voltage.

Hall Effect Applications* The Hall effect finds application in many practical devices. These include a variety of contactless switches such as solid-state keyboards, plunger-operated switches, proximity switches, and vane switches. Other device applications include tachometers and various other counting systems, and the control of automobile ignition. For space application, such Hall effect devices as sensors, gaussmeters, magnetometers, brushless DC motors, and uses in connection with satellite magnetic-hysteresis damping and gravity-gradient stabilization at synchronous altitude are extremely important. Inasmuch as Hall elements can be made of silicon, their processing is compatible with silicon integrated-circuit technology and a number of devices utilizing Hall effect in conjunction with IC functions are now produced on silicon chips.

ALBERT C. BEER

References

Alfvén, H., and Fälthammar, C.-G., "Cosmical Electrodynamics," esp. pp. 180–185, Oxford, Clarendon Press, 1963.
Baynham, A. C., and Boardman, A. D., "Plasma Effects

in Semiconductors: Helicon and Alfvén Waves," esp. p. 30 ff., London, Taylor and Francis Ltd., 1971.
Beer, A. C., "Galvanomagnetic Effects in Semiconductors" (Supplement 4 to "Solid State Physics," F. Seitz and D. Turnbull, Eds.), New York, Academic press, 1963.
Chien, C. L., and Westgate, C. R. (Eds.), "The Hall Effect and Its Applications" (Proc. Commemorative Symposium on the Hall Effect and its Applications, Nov. 13, 1979), New York & London, Plenum Press, 1980.
Dunlap, W. Crawford, Jr., "An Introduction to Semiconductors," New York, John Wiley & Sons, Inc., 1957.
Fritzsche, Hellmut, "Galvanomagnetic and Thermomagnetic Effects," in "Methods of Experimental Physics" (L. Marton, Ed.), Vol. 6B, "Solid State Physics" (K. Lark-Horovitz and V. A. Johnson, Eds.), p. 145, New York, Academic Press, 1959.
Hall, E. H., *Amer. J. Math.*, **2**, 287 (1879); *Amer. J. Sci.*, **19**, 200 (1880).
Kuhrt, F., and Lippmann, H. J., "Hallgeneratoren" (in German), Berlin, Springer-Verlag, 1968.
Lindberg, O., *Proc. Inst. Radio Engrs.*, **40**, 1414 (1952).
Putley, E. H., "The Hall Effect and Related Phenomena" ("Semi-Conductor Monographs," C. A. Hogarth, Ed.), London, Butterworths, 1960.
Rosa, R. J., "Magnetohydrodynamic Energy Conversion," esp. p. 59 ff., New York, McGraw-Hill, 1968.
Seitz, Frederick, "The Modern Theory of Solids," New York, McGraw-Hill, 1940.
Shercliff, J. A., "A Textbook of Magnetohydrodynamics," esp. p. 18, Oxford, Pergamon Press, 1965.
Shockley, William, "Electrons and Holes in Semiconductors," New York, Van Nostrand Reinhold, 1950.
Swift-Hook, D. T., in "Direct Generation of Electricity" (K. H. Spring, Ed.), pp. 143–155, New York, Academic Press, 1965.

Cross-references: CONDUCTIVITY, ELECTRICAL; ELECTRIC POWER GENERATION; ELECTRICITY; MAGNETOMETRY; MAGNETO-FLUID-MECHANICS; POTENTIAL; SEMICONDUCTORS; THERMOELECTRICITY.

HEALTH PHYSICS

Health physics is the profession that is concerned solely with the protection of man from the damaging effects of radiation.[†] It attempts to understand the action of radiation on man and his environment, to establish appropriate limits for exposure to radiation, and to devise appropriate methods for detection, measurement and control of radiation exposure. Although this profession is relatively new, man's awareness and concern for the harmful effects of ionizing radiation are not of recent origin. Perhaps the earliest record of damage to man

---

*The information presented in this section can be found in the centennial symposium proceedings "The Hall Effect and Its Applications."

[†]The Health Physics Society in 1971 extended the scope of health physics to include also non-ionizing electromagnetic radiations.

from ionizing radiation dates back to about 1500 when the high incidence of lung diseases was recognized among the Schneeberg miners of Saxony and the Joachimsthal miners of Bohemia. In 1879 Herting and Hess performed the first autopsies on these miners and reported malignant growths in the lungs; however, the cause of these malignancies was not understood until after 1896 when Roentgen first announced his discovery of x-rays and Becquerel reported the discovery of radiation due to uranium. Even during the first year following the discovery of x- and γ-radiation, many things were learned about both the harmful and useful characteristics of this new source of energy. Grubbé, a manufacturer of Crookes tubes in Chicago, Illinois, was using his equipment to study the fluorescence of chemicals even before the public press on January 4, 1896, heralded Roentgen's discovery of x-rays. During January, 1896, he first noticed an erythema on the back of his hand and later the formation of a blister with skin desquamation and epilation. His hand was sufficiently painful that he sought medical aid on January 27, 1896. Realizing from first-hand experience the destructive power of x-rays Grubbé on January 29, 1896, treated a patient for carcinoma of the breast with his Crookes tube. Not only was this treatment significant because it was one of the earliest—if not the first—therapeutic uses of ionizing radiation, but it is noteworthy that he acted as one of the first health physicists when he used lead as a shield to protect the rest of the body of the patient. Even Becquerel and Madame Curie learned from first-hand experience the need for radiation protection when they received skin burns from the careless handling of radium (see RADIOACTIVITY).

S. Russ in 1915 made a comprehensive series of recommendations for radiation protection to the British Roentgen Society, and if these recommendations had been heeded, many of the early radiation fatalities might have been averted. It was not until 1928, when the International Commission on Radiological Protection was formed and published the first set of recommendations for radiation protection, that there began to be widespread interest and concern for this problem. In the following year the National Council on Radiation Protection was formed and it has set the standards in the United States.

Beginning at the time of the First World War and continuing until about 1930, there were many unfortunate exposures to radium. Some of these were the result of therapeutic injections of radium, the drinking of radium and radiothorium water, and occupational exposures of radium chemists. During this period, radium was considered to be a useful therapeutic agent, and as a result it was administered by physicians in the United States to hundreds of patients. In some cases it was taken as a general tonic, and in others it was given as a curative agent for hypertension, anemia, arthritis, and many other human ailments—even for insanity.

By far the most serious exposures were to young women engaged in the radium dial painting industry. Some of these women ingested relatively large quantities of radium as a consequence of tipping brushes with their lips as they applied radium paint on the dials of clocks and watches. The total number of radium dial painters and others who took radium by mouth or injection and, as a result, died with readily detectable symptoms of radiation damage is not known. The first recorded fatality due to radium-induced cancer, resulting from exposure in the radium dial industry, was in 1925, and since that time histories of over 1000 Ra cases have been recorded and studied in the United States.

In 1942–43 there was begun at the University of Chicago a program to explore the possibility of assembling a critical mass of natural uranium in such a way that a "pile" or NUCLEAR REACTOR could be operated for the production of the new element plutonium to be used in atomic weapons. A. H. Compton, the director, and his associates debated the wisdom of proceeding with this project because they realized that in a single large reactor there would be produced ionizing radiation equivalent to that from thousands of tons of radium. Yet all the radium that had been available to man throughout the world only amounted to about two pounds, and these men were acutely aware of the extensive suffering and death that had resulted from its misuse. They decided to call together a rather unique group of scientists to evaluate these health problems, to develop new instruments, equipment, and techniques and to establish radiation standards for the protection of nuclear energy workers. The group assembled was concerned with the *health* of the workers and consisted mostly of *physicists*; hence, they were called *health physicists*. E. O. Wollan was the leader and the other senior members were H. M. Parker, C. C. Gamertsfelder, K. Z. Morgan, R. R. Covcyou, J. C. Hart, L. A. Pardue, and O. G. Landsverk. Thus in 1942–43 health physics had its beginning at the University of Chicago. Although prior to this time many early pioneers such as S. Russ (England), L. S. Taylor (United States), G. Failla (United States), A. Mutscheller (Germany) and R. Sievert (Sweden)—to name only a few—had devoted considerable attention to the radiation protection problem, it was not until the advent of health physics that a professional group was organized with this as its sole objective.

As the nuclear energy programs expanded, large laboratories were established to carry on the program at Oak Ridge, Los Alamos, Hanford, Berkeley, Brookhaven, Savannah River, etc., and as the need for health physicists was recognized beyond nuclear energy programs and in private industry, hospitals, military organizations, utility power companies, state and federal agencies of public health, and colleges and universities, the profession of health physics grew and expanded very rapidly so that today (1982) it is estimated there are about 14 000 health physicists in the world. In 1956 the Health Physics Society was organized which now has a membership of about 5000. This society publishes the journal, *Health Physics*. In 1959 the American Board of Health Physics was formed for the certification of persons whose technical competence and judgment qualify them to be responsible for handling major problems of radiation exposure and/or contamination control. The International Radiation Protection Association was organized at the First International Congress on Radiation Protection in Rome on September 7, 1966, with K. Z. Morgan as its first president. In 1982 it included 28 affiliated societies with 10 000 members living in 70 countries.

There are three principal areas in which health physicists are employed—education and training, applied activities, and research—and these three areas will be discussed below.

In the early period, health physicists were scientists who had to develop their own competence during employment on the various atomic energy programs. In 1948 AEC Health Physics Fellowship programs were established at Oak Ridge National Laboratory in cooperation with Vanderbilt University and the University of Rochester and later with other national laboratories and universities. This program, over a twenty-five-year period in the U.S., became the principal source of senior health physicists. Unfortunately, during the period 1973–1976 the programs, which had expanded into some twenty universities, were phased out. Although most of the early health physicists began as physicists, there is today a need for health physicists with many different backgrounds, e.g., physics, biology, chemistry, mathematics, engineering. In the AEC Health Physics Fellowship program the student satisfied the usual Ph.D. requirements of courses, research, and thesis in one of these major departments, and at the same time, he met additional requirements such as special courses and summer work at one of the National Laboratories where he was given practical experience in health physics. Likewise, some of the graduate programs of the U.S. Public Health Service during this same period provided training in health physics. In spite of the various opportunities for education and training in health physics, the supply has not kept pace with the demand because of the expanding nuclear industry and the increasing uses of radiation in hospitals.

In addition to the above-mentioned graduate programs in health physics, there are education and training programs at all operating levels. The on-the-job programs are important because the success of health physics can be measured, to a considerable degree, in terms of how well plant managers, supervisors, scientists, engineers, technicians and operations personnel are made to realize their responsibility for protecting themselves and their associates from radiation damage. They must be ever aware that to some degree all radiation exposure is harmful and no

unnecessary exposure may be permitted unless it can be balanced by benefits of equal value. At the same time they must be made to respect ionizing radiation—not fear it.

The duties of the health physicist in applied operations are very diverse and differ considerably from place to place, depending upon the size and nature of the operation. For example, the health physicist in a reactor operation would have duties quite different from those of the health physicist associated with an accelerator program or the health physicist connected with a state public health organization charged with the survey of medical x-ray equipment. A few typical applied health physics activities may be summarized as follows:

(1) Aid in the selection of suitable locations for buildings in which radioactive materials are to be produced or used, and conduct pre-operation background surveys.

(2) Offer advice in the design of laboratories, hoods, remote control equipment, radiation shields, etc.

(3) Provide personnel monitoring meters for radiation dosimetry to all persons subject to radiation exposure and read these meters frequently; make thyroid counts, breath measurements, urine and feces analyses; check body with scanners and total body counters and conduct other tests to aid in estimating how much (if any) radioactive material is fixed in the body; and maintain accurate records of the accumulated dose from each type of ionizing radiation received by each individual for his protection and for the protection of the employer.

(4) Make frequent surveys of all accessible reactor areas, radioactive sources, x-ray equipment, high voltage accelerators, chemistry and physics laboratories, metallurgical shops, and other working areas where radiation exposure is possible.

(5) Advise scientists, supervisors and research directors of all radiation exposure hazards, of permissible working time in a given area, and of radiation protection measures and techniques (e.g., protective clothing, shields, remote control equipment) and aid supervision in the solution of new radiation problems as they develop.

(6) Make frequent surveys of all radioactive waste discharged beyond the area of immediate control, maintain accurate records of the level of this radioactivity in the air, water, soil, vegetation, milk, etc., and advise management of remedial measures as they are needed.

(7) Aid in all emergency operations where there are associated radiation hazards.

(8) Purchase and maintain in working order and in proper calibration suitable health physics survey and monitoring instruments which are used as aids in the protection of personnel from radiation damage.

(9) Prepare operations manuals on "Rules and Procedures Governing Radiation Exposure" and assist in preparation of Environmental Impact Statements required of nuclear power plants.

(10) Assist in radiation protection problems

related to civil defense, emergency planning, weapons fallout, space radiation, nuclear power plants, etc.

(11) Interface with members of the public in helping them to understand the risks of low level exposure and the desirability of keeping all exposures as low as reasonably achievable.

(12) Many health physicists work as consultants assisting radiation workers, atomic veterans, medical patients, etc. by litigation to receive compensation for alleged radiation damage.

Health physics research ranges from the applied and engineering programs to very basic studies. It is a working together of scientists of many disciplines—physicists, chemists, biologists, engineers, geologists, mathematicians—all studying the effects of ionizing radiation on man and on his environment. In these studies, they are working at all levels—nuclear, atomic, molecular, plasma, gas, solid, liquid, cell, animal, and the ecosystem. In this research program, radiation ecologists are studying the effects of low levels of radiation exposure on the environment. Some essential organisms in the environment are known to concentrate radioactive waste by a factor of 100 000 or more, and the health physicist must determine the importance of the indirect damage of ionizing radiation to man's environment as well as its direct effects. Internal dose studies are under way by researchers in health physics, studies which have led to the publication of the official handbooks on maximum permissible concentration of the various radionuclides in food, water, and air. These handbooks are issued by the National Council on Radiation Protection and the International Commission on Radiological Protection. They are under constant revision by the health physicist as more reliable and detailed information becomes available. Biologists in health physics are studying the uptake distribution and elimination of radionuclides which are taken into animals and man by the several modes of intake —ingestion, inhalation and skin penetration. Engineers in health physics are exploring and demonstrating new methods for the disposal of radioactive waste in deep wells thousands of feet below the earth's surface in salt mines and in a number of other geological formations. They are studying the seepage rates of radionuclides into various soil formations, its slow dissipation from packages of radioactive waste deposited on the ocean floor, and the dilution of airborne radioactive waste as it is discharged from stack of nuclear power plants under varying meterological conditions. The physicists in health physics are making basic studies of the various energy exchanges that take place in matter when it is exposed to ionizing radiation. This information aids in the development of better radiation detection systems and leads to an understanding of the true meaning and consequence of radiation exposure. When high energy radiation (in the MeV or keV region) strikes living matter, there are innumerable, complex energy exchanges that take place as the ENTROPY of the system increases. An understanding of the many low energy transitions is basic to a proper interpretation of the effects of ionizing radiation on man. Health physicists are working toward the ultimate goal of developing a coherent theory of radiation damage. Only when such a theory is available can they have complete understanding and confidence in the many extrapolations to man of the effects of ionizing radiation on animals. Such information is essential in developing reliable radiation protection standards and measures that are enforced by the applied health physicist. Many health physicists are working on programs to reduce unnecessary medical diagnostic exposure, which at present accounts for over 90 percent of exposure of the United States population.

The units used in health physics often result in some confusion. In the early period units such as the threshold erythema dose, roentgen, and rep were used, but they have been phased out and replaced by the rad and the rem, and in some cases these two units are now being replaced by the gray and the sievert. The rad corresponds to the absorption of 100 ergs per gram of medium and is the unit of absorbed dose. The rem is the unit of dose equivalent. We have the following relation:

Dose Equivalent (rem) = Absorbed Dose (rad) $\times Q \times N$ in which Q is a qualtiy factor that corrects for LET (linear energy transfer) or the density of ionization along the tracks of ionizing particles. Thus we have $Q = 1$ for x-, gamma, and beta radiations and $Q = 20$ for alpha radiation, recoil ions, and fast neutrons. The N is a biological correction factor which attempts to correct for such things as nonuniform distribution of radionuclide deposition in body organs, the radiosensitivity of the irradiated tissue and its essentialness to proper body function. It is common to set N equal to:

1 for x- and gamma radiation
1 for external exposure (radionuclide not inside body)
1 for radium when it is the parent radionuclide
1 for all tissue except bone
5 for bone if internally deposited radionuclide is not radium and the radiation is not x- or gamma.

The gray corresponds simply to 100 rads and the sievert to 100 rem.

The curie corresponds to an amount of a radionuclide which disintegrates at the rate of 3.7×10^{10} disintegrations per second. The becquerel has been introduced to replace the curie. It corresponds to a quantity of radionuclide which disintegrates at the rate of 1 disintegration per second. There is considerable opposition to the gray, sievert, and becquerel units—especially in the U.S., where they are seldom used.

There are many so-called maximum permissible dose levels, but the one most commonly used is the limit of 5 rem per year to the total body of a radiation worker. The present limits for internal dose in the U.S., upon which the present U.S. levels of maximum permissible body burden of radionuclides and their maximum permissible concentration in our water, air, and food are based, are set at 5 rem per year to total body, gonads, and red bone marrow; 30 rem per year to bone, thyroid, and skin; and 15 rem per year to all other body tissue. Various maximum permissible dose levels are in use for exposure to members of the public, but perhaps the most generally applied levels are those of the Environmental Protection Agency of the U.S. in reference to the uranium fuel cycle (i.e., nuclear power plants, reprocessing plants, etc.). These limits are 75 mrem per year to the thyroid and 25 mrem per year to all other body organs and the whole body.

The health physicist makes use of many types of instruments to determine the dose of ionizing radiation. For area survey measurements he uses geiger counters, ion chambers, proportional counters, and scintillation counters. Sometimes the instruments have sliding windows so that they can respond preferentially to alpha, beta, gamma, and low-energy x-rays. Neutron doses are measured with proportional counters, BF-3 counters, and threshold detectors (i.e., activation of radionuclides of indium, Pu-239, U-238, Np-237, S-32, etc.) The Ge-Li detectors (lithium drifted germanium detectors) are widely used to determine the energy of radionuclide emissions and thus identify them. When the radioactive material is contained inside the body, the body burden is determined by analyses of urine and feces and by total body and scan-counting, using large scintillators or Ge-Li detectors. Personnel monitoring used to be done by the use of pocket ion chambers and film badges. Today the pocket ion chambers are often replaced by direct-reading fiber dosimeters or small G-M counters which display the dose rate and the integrated dose. The film badges have for the most part been replaced by TLD's (thermoluminescent dosimeters) for beta, gamma, and x-ray monitoring. In the past, neutron personnel monitoring was done mostly by the use of thick nuclear emulsion photographic films. However, these are being replaced by polycarbonate foils which are read with optical instruments. This foil technique is a big improvement over the photographic film, in which fading of the tracks is a serious problem (the half-life of the tracks ranges from 3 weeks to 6 months depending on the type of emulsion and the relative humidity). For polycarbonate foils, track fading is almost nil and fast neutron detection sensitivity is about 1,000 times higher than that of photographic emulsions.

Only a relatively few health physicists are engaged in work with nonionizing radiations (UV, visible, microwave, rf, infrared, sonic, ultrasonic, and infrasonic), but here many new problems are being faced and there are uncertainties. For example, the maximum permissible occupational exposure limit of rf and microwave radiation is 1000 times higher in the U.S. than in the USSR (10 mW/cm^2 in the U.S. and 10 μW/cm^2 in the USSR).

There is disagreement and strong debate[3] among health physicists regarding the cancer risk from low-level exposure to ionizing radiation. The common estimates of general cancer risk range between 1×10^{-4} to about 8×10^{-3} lethal cancers per person rem. This author[4] believes one should not use a value smaller than 10^{-3} lethal cancers per person rem, or twice this value for total cancer incidence, until some important questions can be answered and some of the biases in the present studies can be evaluated.

Health physics continues to grow as a most interesting and challenging profession for scientists of many backgrounds. The success of these programs is attested by the fact that ionizing radiation with unparalleled potential for radiation hazards has expanded in its use and applications into almost every area of human endeavor and yet the nuclear energy industry has become one of the safest of all industries.

KARL Z. MORGAN

References

1. Morgan, K. Z., and Turner, J. E., "Principles of Radiation Protection," R. E. Krieger Pub. Co., 1973.
2. Morgan, K. Z., "Graduate Program for the Health Physicist in the United States," *Health Phys.* 11, 895 (1965).
3. Gofman, J. W., "Radiation and Human Health," Sierra Club Books, San Francisco, 1981.
4. Morgan, K. Z., "Hazards of Low-Level Radiation," in "Encyclopaedia Britannica," pp. 213–229, 1980.

Cross-references: MEDICAL PHYSICS, NUCLEAR INSTRUMENTS, NUCLEAR RADIATION, NUCLEAR REACTORS, RADIOACTIVITY, REACTOR SHIELDING.

HEARING

The role of the sense organ of hearing is to code acoustic disturbances into neural signals suitable for transmission to the brain. The study of this process necessarily involves anatomy and physiology of the ear, the nature of auditory pathways and central nervous system activity in hearing, properties of acoustic signals that elicit auditory responses, and observed phenomena of auditory behavior. These aspects serve to define and delineate areas for investigations of hearing and are the topics of discussion for this article.

In this approach to hearing, questions are asked about the structure of the system, how the system functions, and the relationships between inputs and outputs of the system. These three kinds of data—morphological, physiological, and psychological—need to be compared and correlated for a full understanding of hearing. It is important, however, that these three frames of reference be kept separate and not be confused. Although physiological functions may correspond in a general way to anatomical sequences, several physiological functions may occur in the same anatomical structure or a single function may require several anatomical units. In a similar manner, psychological functions cannot usually be identified with specific physiological functions, and it is recognized that the central nervous system, as well as the auditory system, is involved in any auditory response. The correlations between and knowledge about structure and functions are best developed for peripheral, rather than central, parts of the auditory system because the ear is more accessible for examination and study than are the more central parts of the auditory system.

Traditional theories of hearing have been largely concerned with pitch perception. A comprehensive theory does not exist that can account for the complexities of the processing of speech or music by the auditory system. However, there are a number of excellent models that depict various physiological, sensory, and perceptual functions of hearing. Much research

is concerned with empirical testing of these models, with results that are in the direction of a more coherent understanding of hearing. Investigators often endeavor to relate many phenomena of hearing to a particular model. An example is the critical-band model, a concept of a filterlike mechanism early in the auditory system that affects various auditory functions in a similar manner. Frequency analysis, masking, pattern perception, and changes in hearing with aging are but a few of the features of hearing that can be related to a critical-band filter model.

When the structure of the ear is examined, it is convenient to consider the external, middle, and inner ear separately. A cutaway drawing of the ear is shown in Fig. 1. From a functional point of view, the ear may be divided into an outer and inner part. The outer is concerned with the transformation of acoustic energy into mechanical energy and the inner with the transduction of mechanical energy into neural impulses. The auricle and external auditory meatus constitute the external ear. The meatus is an irregularly shaped tube approximately 27 mm long with a diameter of about 7 mm, terminated by the tympanic membrane. The outer ear serves as an acoustic resonator to produce a gain in acoustic energy of about 10 to 15 decibels in the frequency range from 2000 to 5000 Hz. This is a combined effect of the concha (the area at the entrance to the ear canal) and of the ear canal itself, each having a separate resonance. The

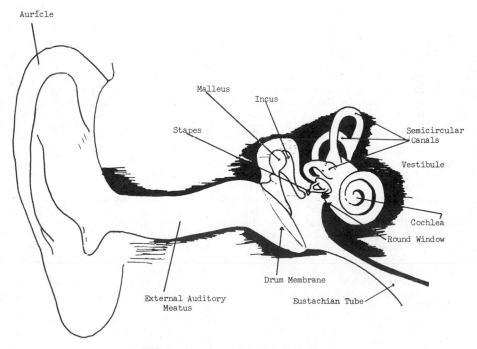

FIG. 1. A cutaway drawing of the ear.

former has a resonance at about 5000 Hz, the latter a resonance at slightly above 200 Hz. The ear drum is in a protected position at the end of the canal, and humidity and temperature conditions at the drum are relatively independent of those external to the ear.

The middle ear is an irregular, air-filled space in the petrous portion of the temporal bone. The three ossicles of the middle ear—the malleus, the incus, and the stapes—provide mechanical linkage between the tympanic membrane and the fenestra vestuli, an opening in the vestibule of the inner ear commonly known as the oval window. The handle of the malleus attaches to the tympanic membrane, and the footplate of the stapes attaches to the oval window. The important function of the middle ear is impedance matching and transformation of energy of air vibrations in the ear canal to fluid in the cochlea. This efficiency of sound transmission from one medium to another is accomplished primarily because the area of the tympanic membrane responding to vibrations of air is considerably larger than the area of the footplate of the stapes at the oval window to the inner ear. The lever action of the ossicles contributes to the efficiency of this transformation to a lesser extent than the area ratios of the two membranes. The amplification gain due to the middle ear is approximately 30 dB. The effectiveness of the middle ear action in increasing hearing sensitivity is evidenced in middle ear pathologies where the ossicular chain is disrupted. A hearing loss on the order of 25–40 dB occurs. The second function of the middle ear, that of protecting the inner ear from loud sounds, is accomplished by reflex action of the middle ear musculature, the stapedius, and the tensor tympani. The action of the muscles is to draw the stapes away from the oval window, retract the ear drum, and change ossicle vibrations in such a way as to decrease the transmitted pressure. The stapedius may be more responsive to accoustic stimulation than the tensor tympani. Latency of muscle contraction and possible muscle fatigue limit protection of the inner ear by these mechanisms. Middle ear air pressure is equalized by virtue of the eustachian tube which connects the middle ear and the nasopharynx. The pressure equalization is necessary for normal ear drum movement.

The inner ear is a system of cavities in the dense petrous portion of the temporal bone. One of the cavities is the cochlea, a bony labyrinth that is approximately 35 mm in length coiled around a central core for two and three-quarters turns. The spiral-shaped cochlea is divided into three ducts, two bony and one membranous. A cross section of cochlea showing the three ducts is shown in Fig. 2. The upper bony duct, the scala vestibuli and the lower bony duct, the scala tympani, are separated from each other by a membranous labyrinth, the cochlear duct. The cochlear duct is bound on top by Reissner's

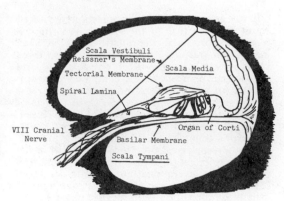

FIG. 2. Cross section of cochlea.

membrane and is bound below by the basilar membrane. The cochlear duct is filled with a viscous fluid called endolymph, and the duct is surrounded by a fluid called perilymph that has about twice the viscosity of water. The scala vestibuli and the scala tympani join at the apical end of the cochlea at a passage called the helicotrema. The scala tympani terminates at the basal end at the round window, a membrane-covered opening into the middle ear. The scala vestibuli is continuous with the vestibule; the oval window opens into the vestibule. Vibrations at the footplace of the stapes are transmitted into the fluid adjacent to the oval window. Vibration of the stapes and resultant disturbances in cochlear fluids results in movement of the basilar membrane. The cochlear duct contains the sensory receptors, specifically the organ of Corti, which lies upon the basilar membrane. There are about 12 000 hair cells; one end of each rests on the basilar membrane. The other ends of the hair cells are the cilia, very fine hairline processes, which make contact with the tectorial membrane, a membrane that overlaps the organ of Corti and that functionally behaves as if it were hinged at the cochlear wall. There are three rows of outer, and one row of inner, hair cells along most of the length of the basilar membrane. When vibrations are introduced into the inner ear and cause displacement of the basilar membrane a shearing action of the cilia occurs that results in neural activity. It is assumed that amplification occurs in the inner ear in that small pressures on the basilar membrane result in a shearing force of considerably greater magnitude that distorts the hair cells. The result is increased sensitivity of the hearing system. Physical properties of the cochlea are such that different frequencies tend to localize at different points along the basilar membrane. The basilar membrane is narrowest and stiffest at the basal end, and most lax and widest at the apical end of the cochlea. High-frequency sounds result in the greatest disturbances near the basal end, and low-frequency sounds tend to localize near the apical end. When the role of the cochlea in pitch and loudness analyses is considered, it is now realized that more is involved in pitch perception than the place of localization on the basilar membrane, although the particular neural fibers involved are probably relevant. Of particular interest in this regard is the phenomenon called low pitch, where there is no energy at the frequency that corresponds to the perceived pitch. Our impression that we are hearing a fundamental pitch for a person talking over a telephone is a good example. Telephone circuits do not transmit frequencies in the fundamental pitch range so there is, in fact, no energy in this range yet an experience of pitch is experienced by the listener. Loudness is probably related to the total number of neural impulses per unit time.

The auditory pathways provide for the neural impulses from the ear to be transmitted to the cerebral centers of the auditory cortex. Processing of the neural signals probably occurs at synaptic connections as well as in the cortex. The cell bodies of the receptor neurons are located in the spiral ganglion. Neurons of the auditory nerve make synaptic connections with the hair cells of the cochlea. Nerve fibers typically innervate many hair cells, and more than one nerve fiber may make a connection with the same hair cell. There is evidence to indicate that there are also descending neural pathways as well as ascending ones. The central nervous system may thus be involved in auditory processing at the cochlea. Spiral ganglion axons make synaptic connections with cells of the central nervous system at the cochlear nucleus. At this point, there is interconnection between the pathways for the two ears. Other synaptic stations between this point and the auditory cortex include the inferior colliculus and the medial geniculate body. A schematic depicting the auditory pathways is shown in Fig. 3. It can be observed that the major projection is to the contralateral (opposite) cerebral hemisphere for each ear. That is, the right ear projects primarily to the left hemisphere and the left ear to the right hemisphere. Some evidence has been presented that for right-handed people, the left cerebral hemisphere processes speech and the right hemisphere processes non-speech signals. Evidence from pathological auditory systems is of particular interest with respect to the auditory pathways. In fact, one excellent way to understand the auditory system is to study the system when it is in a breakdown or pathological state. An impaired cochlea, for example, may result in a better than normal response to small amplitude changes in a sound. A lesion of the VIIIth Nerve is frequently manifested by a rapid decrease in the ability to respond under sustained stimulation. The ability to process speech is markedly affected when there is an involvement of the lower central nervous system. Cortical involvement does not affect usual speech or pure tone inputs.

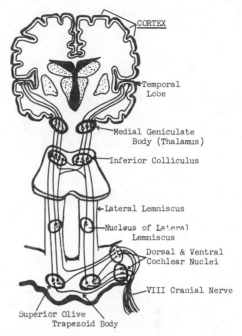

FIG. 3. Schematic depicting auditory pathways.

The stimulus for hearing is sound. Sound involves a disturbance in the air that is a forward and backward, rarefaction and compression, movement of air particles. The unit of force usually used in acoustics is the dyne. Sound pressure is frequently expressed in dynes per square centimeter. Intensities of sounds are usually measured on a decibel scale, a logarithmic ratio scale. The tremendous loudness range of the ear is exemplified by the fact that the most intense sound that can be tolerated is a million million times greater in intensity than a sound that is just audible. This is a range of approximately 120 dB. The frequency range of hearing is frequently given as 16 to 20 000 Hz. The ear is most sensitive in the middle frequency range of 1000 to 6000 Hz. In terms of discrimination of frequency and intensity, it is possible for about 1400 pitches and 280 intensity levels to be distinguished.

The truly phenomenal aspects of hearing can be observed in such behavior as localization of sounds, speech perception and particularly the understanding of one voice in the noisy environment of many, and the recognition of acoustic events that only last a few milliseconds. It is these and other behavioral phenomena that need to be accounted for in theories of hearing.

ROBERT W. PETERS

References

Books

Bekesy, Georg von, "Experiments in Hearing," New York, McGraw-Hill Book Co., 1960.

Rasmussen, Grant, Ed., and Windle, William F., "Neural Mechanisms of the Auditory and Vestibular Systems," Springfield, Charles C. Thomas, 1960.

Stevens, S. S., and Davis, Hallowell, "Hearing: Its Psychology and Physiology," New York, John Wiley & Sons, 1947.

Green, D. M., "An Introduction to Hearing," New Jersey, Lawrence Erlbaum, 1976.

Moore, B. C. J., "The Psychology of Hearing," Second Edition, New York, Academic Press, 1982.

Pickles, J. O., "An Introduction to the Physiology of Hearing," New York, Academic Press, 1982.

Schubert, E. D., "Hearing: Its Function and Dysfunction," in "Disorders of Human Communication, Vol. 1 (G. E. Arnold, F. Winckel, and B. D. Wyke, Eds.), Vienna and New York, Springer-Verlag, 1980.

Tobias, J. V. (Ed.), "Foundations of Modern Auditory Theory," New York, Academic Press, Volume I, 1970; Volume II, 1972.

Periodicals

The Journal of the Acoustic Society of America.

Cross-references: ACOUSTICS; ARCHITECTURAL ACOUSTICS; MUSICAL SOUND; NOISE, ACOUSTICAL; PHYSICAL ACOUSTICS; REPRODUCTION OF SOUND.

HEAT

Heat and Temperature *Heat* is (internal) energy in transit due to a decrease in temperature between the source from which the energy is coming and the sink toward which the energy is going. When heat is added to a system, the result may be either to raise the temperature or to cause a thermal transformation (e.g., melting, boiling, expansion, increase in resistance). Conversely, the removal of heat from a system may either lower the temperature or cause freezing, condensation, contraction, decrease in resistance, etc.

The concept of *temperature* begins with the physiological sensation of hotness and coldness. The hotter an object feels, the higher is the temperature, and vice versa. However, the sensation of temperature is easily influenced by the thermal condition of the skin and the thermal conductivity of the object being touched. Thus the concept of temperature must be quantified using inanimate instruments that reproducibly measure ranges of temperature over a continuous scale.

Many different types of thermometers are available, but the two most common are the colored-alcohol-in-glass and the coiled bimetallic strip. Historically the normal freezing and normal boiling points of pure water were chosen as the two fixed points of the principal temperature scales. In the English system of units the values of $32°F$ (F for *Fahrenheit*) and $212°F$ are assigned to the fixed points. In the practical metric system the values of $0°C$ (C for *Celsius*, previously *Centigrade*) and $100°C$ are assigned to the fixed points.

In the modern metric *Système International d'Unités* (abbreviated SI in all languages) temperature is measured in units of *Kelvin* (the word *degree* is dropped). The triple point of pure water is assigned the value of 273.16 K (corresponding to $0.01°C$) and a temperature interval of one Kelvin is the same as one degree Celsius.

When comparing the definitions of heat and temperature, it is possible to encounter a circular argument. For example, heat may be defined to be the flow of energy due to a temperature difference; and temperature may be defined as a property of an object which determines the direction of heat flow. The circularity in reasoning will be broken later in this article when heat will be shown under certain conditions to be equivalent to work, which can be independently defined and measured.

Measurement of Heat The *calorimetric* definition of heat was established by Joseph Black (1728–1799) during the eighteenth century. Equal masses of water, initially at different temperatures, were mixed in a container called a *calorimeter*, which was insulated from its surroundings. It was observed that the final temperature of the mixture was always exactly halfway between the two initial temperatures (if the effects of the calorimeter and the thermometer were negligible). The temperature changes ΔT of the hot and cold masses of water were equal and opposite,

$$\Delta T_h = -\Delta T_c.$$

If different masses of water were mixed, the temperature changes were observed to be inversely proportional to the respective masses,

$$\frac{\Delta T_h}{\Delta T_c} = -\frac{m_c}{m_h},$$

or

$$m_h \Delta T_h = -m_c \Delta T_c. \tag{1}$$

This experiment suggested thinking about something lost by the hot mass of water and gained in equal amount by the cold mass of water; or, in other words, that there was a transfer of *heat*. Until the middle of the 19th Century heat was assumed to be a material fluid that could be neither created nor destroyed, capable of flowing between atoms and molecules, and attracted by some materials and repelled by others. Antoine Lavoisier (1743–1794) gave the fluid principle of heat the name *caloric theory*.

Equation (1) was shown to be valid only when two samples of water were mixed at initially different temperatures, but failed to predict the correct temperature changes when water and another substance were mixed. Thus

Eq. (1) was modified to the general case:

$$C_A \, m_A \, \Delta T_A = -C_w \, m_w \, \Delta T_w, \qquad (2)$$

where A stands for a substance with a higher temperature mixed with water w, and C is a property of the material known as the *specific heat capacity* (usually shortened to *specific heat*). The left-hand term in Eq. (2) is the heat Q_A transferred to or from substance A, and the right-hand term is the corresponding amount of heat Q_w transferred from or to the water.

More accurate observations showed that the value of C for any particular substance was different and depended on the physical circumstances under which the final temperature was obtained. Gases, for example, showed a smaller value of C if the temperature changes occurred at constant volume than if the temperature changes occurred at constant pressure. Thus there needed to be a C_v for specific heat at constant volume and C_p for specific heat at constant pressure.

It was thought to be necessary to define a special unit in which calorimetric heat would be measured. Water was selected as the standard substance and C_p given the numerical value of unity. In the English system the *British thermal unit* (or Btu) is the heat needed to raise the temperature of the mass associated with one pound of water from $60°F$ to $61°F$ at one atmosphere of pressure.

In the pre-SI metric system the *calorie* was defined as the heat required to raise the temperature of a mass of one gram of water from $14.5°C$ to $15.5°C$ at one atmosphere of pressure. A large calorie or *kilocalorie* (*Calorie* written with a capital C) corresponds to 1000 ordinary calories and is used almost exclusively in dietetics. In the modern SI metric system there is no unique unit for heat. Instead heat is recognized as a form of energy and is measured in *joules*, as are all forms of energy.

Heat and Work The caloric theory was widely accepted toward the end of the 18th Century because of its simplicity and many successful applications. The first really conclusive evidence that heat could not be a material fluid was given by Benjamin Thompson (1753–1814), born in Woburn, Massachusetts, who left the United States because of royalist sympathies during the American Revolution and later became Count Rumford in Bavaria. In 1787, using a very delicate balance, he determined that there was no detectable difference in the weight of an object whether hot or cold, which raised doubts about the material nature of the caloric fluid.

In 1798 Rumford observed the temperature rise in brass chips produced during the boring of a cannon. The heat generated by friction between a moving boring tool and the cannon appeared to be *inexhaustible*. Obviously the caloric theory of heat could not be based on a material fluid that flowed without limitation. Rumford proposed instead that heat was a form of motion, but was unable to explain the nature of the motion.

The attack on the caloric theory continued in the 1840s, especially by the quantitative experiments of James Prescott Joule (1818–1889). In several different experiments Joule allowed falling objects to perform work on a sample of water that consequently experienced a rise in temperature. Heating one pound of water through $1°F$ he was able to determine that 1 Btu of heat was produced by 772 foot-pounds of work, which is sometimes called the *mechanical equivalent of heat*. In other words, energy could be converted from one form to another, an idea that would be incorporated into the formulation of the law of conservation of energy.

Recognition that heat is another form of energy means that the mechanical equivalent of heat is simply a conversion factor between different units of measurement. The currently accepted conversion factors are: 1 Btu is equivalent to 778.28 foot-pounds and 1 calorie ($15°C$) is equivalent to 4.1858 joules.

In order to discuss the modern definition of heat based on Joule's experiments it is necessary to specify the physical properties of the walls surrounding the system being investigated. When an *adiabatic* wall is used to separate two systems A and B, the macroscopic properties of one system are found to be unaffected by those of the other. For example, if a thick asbestos or styrofoam wall separates a gas from a wire, the properties of the gas (pressure, volume, temperature) have no influence on the properties of the wire (tension, length, resistivity). In other words, an adiabatic wall is a perfect insulator to heat.

If, however, the two systems are separated by a *diathermic* wall, such as a thin sheet of copper, the properties of the two systems will change spontaneously, until an equilibrium state of the combined system is reached, because a diathermic wall is a perfect conductor of heat. The two systems are then said to be in *thermal equilibrium*, which means they have the same temperature. Thus, temperature may also be defined as the property that determines whether or not two systems are in thermal equilibrium.

Consider next the concept of work. If there exists a macroscopic force F, undergoing a macroscopic displacement ds in a direction parallel to the force, the work dW is defined as Fds. Each simple system has its own expression for work, such as $-PdV$ for a hydrostatic system, Tdl for a wire, $\mu_0 HdM$ for a magnetic system, etc. Simple calculations show that the work in a finite process depends on the external processes (so-called *path*) used to change the system from its initial state to its final state. There is no such thing as the "work in a sys-

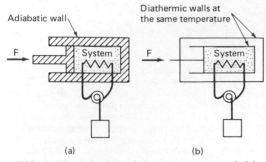

FIG. 1. The behavior of the system surrounded by adiabatic walls in (a) is the same as that whose diathermic walls and surrounding enclosure are at the same temperature in (b).

tem." The work is not a function of properties (so-called *coordinates*) of a system. In mathematical terms work is an inexact differential and the element of work may be written $đW$.

In Fig. 1(a) there are shown two ways in which work may be performed on a system surrounded by adiabatic walls. A force may cause the piston to move some distance; and a falling object may rotate an electric generator that provides a current in a resistor embedded in the system and included as part of the system. Consider two arbitrarily chosen states of the system, i and f (initial and final). It is possible using the apparatus in Fig. 1(a) to proceed from i to f along many different paths. One process might be to move the piston in and then maintain the current until the final state is achieved. Such a process is shown by path I in Fig. 2. Another process might be to perform the electrical work first and the work of compression second, as shown by path II in Fig. 2. The *total* work in proceeding from state i to state f is the same for each path. In other words, adiabatic work W_{ad} is independent of the path and depends only on the initial and final states of a system.

Interestingly enough, it seems that experiments of this type may never have been performed. There was a rapid and universal acceptance of the law of conservation of energy following the work of Joule and especially Hermann von Helmholtz (1821–1894). It is an accepted fact, indeed a law of nature known as *the first law of thermodynamics* (in a restricted case), that adiabatic work is independent of path. This statement is established beyond any reasonable doubt, however, because all the consequences of the first law of thermodyanics have been so well verified by direct experimental evidence.

It is an experimental fact that the apparatus shown in Fig. 1(b), in which the adiabatic wall is replaced by two rigid diathermic walls, *whose temperatures are maintained equal at all times*, is in all respects equivalent to the apparatus

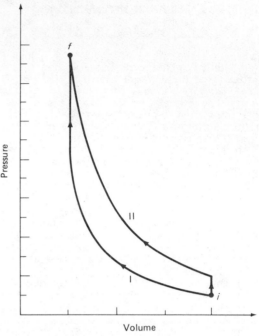

FIG. 2. The adiabatic work done on the system is the same whether going by path I or path II from the initial state i to the final state f.

with an adiabatic wall. This temperature equality is achieved experimentally by noting the readings of thermocouples whose junctions touch the inner wall enclosing the system and the outer wall surrounding the inner wall. When the thermocouple readings differ, energy is provided to either the inner wall or the outer wall.

Using this equipment, it is possible to measure the work W_{ad} when the system is brought from state i to state f adiabatically (by holding both diathermic walls at the same temperature). It is again observed that W_{ad} is independent of the path. The consequences of independence of path may now be derived. In mechanics the work needed to move an object slowly and without friction from one height to a higher elevation is independent of the path. Thus one is lead to the concept of gravitational potential energy. Similarly, since the adiabatic work in bringing the system in Fig. 1(b) from state i to state f is also independent of the path, then there must exist a function U of the coordinates of the system such that

$$\Delta U = U_f - U_i = W_{ad}. \tag{3}$$

The function U might justifiably be called the "adiabatic work function," but instead is called the *internal energy function*. Notice that if positive work is done on the system, its internal energy is increased, a sign convention universally used in physics. In some older discussions of the first law of thermodynamics there is a minus sign on the left in Eq. (3) and work is negative when performed on a system. The latter sign convention followed the engineering practice of defining work to be positive when a machine (the system) operates on its surroundings.

Suppose now that the inner wall of the apparatus in Fig. 1(b) is now deliberately set at a different temperature from the outer wall. While the temperature difference is maintained, the system is again brought from state i to state f and the work W is measured. This work would be found to be different from W_{ad}, that is, different from $U_f - U_i$. Since it is undesirable to give up the law of the conservation of energy, the conclusion is inescapable that the difference between W and W_{ad} (or W and $U_f - U_i$) is due to another form of energy that must have entered or left the system *by virtue of the temperature difference between the system and its surroundings*. This energy flow, which is internal energy moving between the system and its surroundings, is none other than the quantity that is called "heat."

We see that the old calorimetric definition and the modern thermodynamic definition of heat, as energy transferred by virtue of a temperature difference, are in agreement. However, any circular reasoning involving definitions of heat and temperature is now broken; heat can be independently defined by measurements in terms of work performed on a system surrounded by different walls. Denoting the heat that has entered or left the system again by the symbol Q, we get

$$Q = W_{ad} - W = U_f - U_i - W, \qquad (4)$$

which is the mathematical formulation of the *first law of thermodynamics*.

Rewriting Eq. (4) in the form

$$U_f - U_i = W + Q \qquad (5)$$

shows that W and Q are *methods of producing a change in internal energy*. W is the mechanical method of energy transfer, and Q is the non-mechanical or thermal method. Together they constitute the two thermodynamic methods in which U may be changed.

Work and heat are *processes* of energy transfer. When all flow is completed, the words "work' and "heat" have no longer any usefulness or meaning. It would be just as incorrect to refer to the "heat in a system," as it would be to speak of the "work in a system." The performance of work and the flow of heat are methods whereby the internal energy of a system is changed. As a result of these processes, the internal energy of the system has either increased, decreased, or remained constant. Once the processes of transferring energy are over,

we can only speak of the internal energy of the system. It is impossible to subdivide the internal energy into a mechanical and a thermal part, just as it is impossible to subdivide the water in a lake into parts resulting from (1) flow from a stream, (2) rain, (3) springs, etc.

RICHARD H. DITTMAN

Reference

Zemansky, Mark W., and Dittman, Richard H., "Heat and Thermodynamics," 6th Edition, New York, McGraw-Hill Book Company, 1981.

Cross-references: CALORIMETRY; HEAT CAPACITY; KINETIC THEORY; TEMPERATURE AND THERMOMETRY; THERMODYNAMICS; WORK, POWER, AND ENERGY.

HEAT CAPACITY

The heat capacity of any thermodynamic system is

$$C = \frac{\delta q}{dT}, \qquad (1)$$

where δq is the quantity of heat* required to produce the temperature increment dT. If the system consists of a single substance with definite chemical composition and physical state, C is proportional to the total quantity of matter in the system. The heat capacity of 1 gram is the specific heat, and the heat capacity of one mole is sometimes called the molar heat capacity. Commonly used units of heat capacity include cal mole^{-1} deg^{-1}, cal g^{-1} deg^{-1}, and J mole^{-1} deg^{-1}.

The heat capacity of a substance depends on the variables that determine its thermodynamic state—e.g., temperature, pressure, electric and magnetic field—and also on the constraints imposed during the absorption of heat. Here we limit consideration to a substance subject to a hydrostatic pressure P and to no other forces. In this case, the thermodynamic state is determined by two variables, usually taken to be temperature and either pressure or molar volume V. The usual conditions for which C is of interest are constant volume and constant pressure, and the symbols C_V and C_P are used for the heat capacities measured under these conditions.

The difference between C_P and C_V is related to other thermodynamic properties by the first law of thermodynamics, which, for processes of

---

*The symbol δ indicates a differential that is not exact, i.e., one for which a line integral is not determined by the initial and final points alone, but also depends on the path.

interest here, may be written

$$\delta q = dU + P\,dV = dH - V\,dP, \qquad (2)$$

where U is the energy and $H = U + PV$ is the enthalpy. (We write this and all subsequent expressions involving thermodynamic properties for one mole.) From Eqs. (1) and (2), we have

$$C_V = \left(\frac{\partial U}{\partial T}\right)_P \qquad (3)$$

and

$$C_P = \left(\frac{\partial U}{\partial T}\right)_P + P\left(\frac{\partial V}{\partial T}\right)_P = \left(\frac{\partial H}{\partial T}\right)_P. \qquad (4)$$

Subtraction of Eq. (3) from Eq. (4) and substitution of the mathematical relation

$$\left(\frac{\partial U}{\partial T}\right)_P = \left(\frac{\partial U}{\partial T}\right)_V + \left(\frac{\partial U}{\partial V}\right)_T \left(\frac{\partial V}{\partial T}\right)_P \qquad (5)$$

gives

$$C_P - C_V = \left\{\left(\frac{\partial U}{\partial V}\right)_T + P\right\}\left(\frac{\partial V}{\partial T}\right)_P. \qquad (6)$$

The quantity of heat required to produce a certain temperature increase dT is greater at constant pressure than at constant volume by the sum of the work of expansion in the constant pressure process, $P(\partial V/\partial T)_P\,dT$, and the increase in internal energy accompanying the expansion, $(\partial U/\partial V)_T (\partial V/\partial T)_P\,dT$.

An expression for $C_P - C_V$ that is more useful than Eq. (6) is

$$C_P - C_V = TV\alpha^2/\kappa, \qquad (7)$$

which can be obtained from Eq. (6) by introducing a relation based on the second law of thermodynamics.

$$\left(\frac{\partial U}{\partial V}\right)_T + P = T\left(\frac{\partial P}{\partial T}\right)_V,$$

the mathematical relation

$$\left(\frac{\partial P}{\partial T}\right)_V = -\left(\frac{\partial V}{\partial T}\right)_P \bigg/ \left(\frac{\partial V}{\partial P}\right)_T,$$

and the definition of the coefficients of thermal expansion α and isothermal compressibility κ,

$$\alpha = \frac{1}{V}\left(\frac{\partial V}{\partial T}\right)_P, \qquad (8)$$

and

$$\kappa = -\frac{1}{V}\left(\frac{\partial V}{\partial P}\right)_T. \qquad (9)$$

The subject of the following sections is the relationship of the heat capacities of ideal gases and solids to the energies associated with different degrees of freedom of the constituent particles. The heat capacity of a liquid is more complicated and is not considered here.

Ideal Gases An ideal gas obeys the equation of state, $PV = RT = NkT$. The gas constant R is 8.314 J mole^{-1} K^{-1}, N is Avogadro's number, and $k = R/N$ is Boltzmann's constant. By Eqs. (7), (8) and (9), $C_P - C_V = R$.

In the application of statistical mechanics to the calculation of C_P or C_V it is convenient to consider first the predictions based on classical principles. The pertinent result of classical statistical mechanics is the principle of equipartition of energy: any term in the energy of a particle proportional to the square of either a coordinate or a momentum contributes an average of $\frac{1}{2}kT$ to U. Since the molecules of an ideal gas have three translational degrees of freedom, each with an associated kinetic energy proportional to the square of a momentum, the expected translational contributions to the energy and heat capacity are $U = \frac{3}{2}NkT = \frac{3}{2}RT$ and, by Eq. (3), $C_V = \frac{3}{2}R$. For monatomic gases this is the only contribution to C_V and this value is in excellent agreement with experiment.

Diatomic molecules can rotate about each of two independent axes perpendicular to the internuclear axis, and the two atoms can vibrate with respect to each other along the internuclear axis. Each of the two rotational degrees of freedom has a kinetic energy proportional to the square of a momentum, and the vibrational degree of freedom has both a kinetic energy proportional to the square of a momentum and a potential energy proportional to the square of a coordinate (the vibration is approximately harmonic). In classical statistical mechanics, C_V is therefore expected to be $\frac{7}{2}R$, but for most diatomic gases at room temperature C_V is close to $\frac{5}{2}R$. This discrepancy, one of the historically important failures of classical theory, is resolved by proper consideration of quantum effects as suggested by Einstein. The allowed energy levels of a harmonic oscillator are not continuous as in classical mechanics, but are given by

$$\epsilon_n = nh\nu, \text{ with } n = 0, 1, 2, 3, \cdots,$$

where h is Planck's constant, ν is the natural frequency of the oscillator, and the energies ϵ_n are measured from the lowest level. For the average energy of a single oscillator, quantum statistical mechanics gives

$$\bar{\epsilon} = \frac{h\nu}{e^{h\nu/kT} - 1}.$$

Instead of the classical value k, the contribution to C_V is

$$\frac{d\bar{\epsilon}}{dT} = k\frac{\left(\frac{h\nu}{kT}\right)^2 e^{k\nu/kT}}{(e^{h\nu/kT} - 1)^2}. \qquad (10)$$

This contribution to C_V is negligible for $T \leqslant$ 0.1 $h\nu/k$ because most of the oscillators remain in the $n = 0$ level, but it increases rapidly near $T \approx \frac{1}{4}h\nu/k$ and approaches the classical value for $T \geqslant h\nu/k$. For many diatomic molecules, $h\nu/k$ is a few thousand degrees Kelvin or more, and a room-temperature C_V of $\frac{5}{2}R$ can be understood as the sum of the translational and rotational contributions. The details of the temperature dependence predicted by Eq. (10), including the approach of C_V to $\frac{7}{2}R$ at sufficiently high temperatures, have been verified experimentally for a number of gases.

In a few cases (H_2, HD and D_2), diatomic gases exist at temperatures for which kT is comparable to the spacing of the rotational energy levels, and quantum effects can be observed in the rotational contribution to C_V. For H_2, C_V drops below $\frac{5}{2}R$ as the temperature is reduced below about 300 K, and below 50 K it becomes equal to the $\frac{3}{2}R$ that is characteristic of translation. The translational energy levels of ordinary gases are so closely spaced that quantum effects are never important in the translational heat capacity.

The heat capacity of polyatomic gases can be treated by a straightforward generalization of the foregoing discussion. A molecule with n atoms has three translational degrees of freedom and, if it is nonlinear, three rotational and $3n - 6$ vibrational degrees of freedom. If it is linear, it has two rotational and $3n - 5$ vibrational degrees of freedom.

Solids For every solid there is a lattice heat capacity associated with vibrations of the atoms. If the interatomic forces are harmonic, the N atoms in one mole of a monatomic solid have $3N$ independent vibrational modes, and the lattice heat capacity is the sum of $3N$ terms given by Eq. (10). Since the spectrum of the $3N$ frequencies of a real solid is complicated and difficult to calculate, an approximation introduced by Debye is widely used. In the Debye model, the vibrational modes are sound waves in an elastic continuum; the boundaries of the solid determine the allowed wavelengths and the sound velocities then determine the frequencies. To limit the number of frequencies to $3N$, the spectrum is cut off at a maximum frequency $k\Theta_D/h$ where Θ_D, the Debye temperature, is determined by the sound velocities and is typically a few hundred degrees Kelvin. The cutoff corresponds to the fact that in a real crystal, the vibrations have a minimum wavelength comparable to the interatomic distance. The heat capacity is

$$C_V = \frac{12}{5}\pi^4 R \left(\frac{T}{\Theta_D}\right)^3 \qquad (11)$$

for $T \leqslant \Theta_D/20$, but it increases less rapidly at higher temperatures and approaches the classical limit 3R for $T > \Theta_D$. The predicted high-temperature limit is in agreement with the empirical rule of Dulong and Petit: for most monatomic solids C_V is approximately $3R$ at room temperature. However, this value is often exceeded at very high temperatures, partly as a consequence of anharmonicity in the interatomic forces. At low temperatures, Eq. (11) is in good agreement with experiment; C_V is found to be proportional to T^3, although often only at temperatures below $\Theta_D/100$, and Θ_D is given accurately by the sound velocities. This agreement is to be expected because at low temperatures only low-frequency vibrations contribute, and these are treated accurately in the Debye model. At intermediate temperatures, the agreement with experiment is only approximate.

Occasionally the lattice heat capacity of a molecular crystal is represented by the sum of a Debye heat capacity for the vibrations of the N molecules as units and the appropriate number of terms given by Eq. (10) for the intermolecular vibrations.

In metals, there is an electronic heat capacity related to the translational motion of the conduction electrons. The small mass and high density of the electrons make quantum effects important, and Sommerfeld showed that their heat capacity should be proportional to temperature for temperatures below about 10^4 K. This contribution is usually significant below a few degrees Kelvin, where the lattice heat capacity is small, and also at high temperatures, where it contributes to deviations from the rule of Dulong and Petit.

The heat capacities of a number of solids have bumps or peaks superimposed on the smoothly varying lattice and electronic heat capacities. These are usually called anomalies and two distinct types can be recognized. A Schottky anomaly is a smooth bump that arises from a set of energy levels for a single particle. The splitting of the rotational states of magnetic ions by electric fields, and of nuclei by electric or magnetic fields, are examples for which the associated anomalies occur in the ranges 10^{-1} to 10^3 K and 10^{-3} to 10^{-1} K, respectively. Lambda anomalies are sharp peaks produced by cooperative processes involving many particles in a transition from a low-temperature ordered state to a high-temperature disordered state. Examples are: the momentum ordering of ^{4}He atoms in liquid ^{4}He at 2.18 K and of electrons in superconductors at temperatures ranging from 10^{-1} to 10 K; the magnetic ordering of ferromagnets and antiferromagnets at temperatures from 10^{-3} to 10^3 K; the spatial ordering of different atoms of an alloy on a superlattice, e.g., β-brass at 750 K; and the rotational disorder in certain molecular crystals e.g., H_2 at 1.5 K.

NORMAN E. PHILLIPS

Cross-references: CALORIMETRY, GAS LAWS, HEAT, TEMPERATURE AND THERMOMETRY, THERMODYNAMICS.

HEAT TRANSFER

Establishment of thermodynamic equilibrium for a system consisting of a number of media requires that the temperature be locally uniform and time-wise constant. A departure from this condition causes a transfer of energy in the form of heat from locations with high temperature to locations with low temperature. Such an energy transfer occurs very frequently and is encountered in our everyday life as well as in many engineering applications or in scientific experiments. It has, therefore, been known for a long time. The fact that quantitative predictions have become possible in the recent past only is due to the situation that several mechanisms are usually involved and interrelated in such an energy transfer process. They are generally classified as conduction, convection and radiation.

Heat transfer by conduction is that process which transports heat in a medium from one location to another without involvement of any visible movement. It is generally the only or the dominating mode of heat transfer in a solid medium; however, it occurs also in liquids and gases. In such fluids, this energy transport is often augmented when parts of the fluid are in movement and carry energy along. This mechanism of heat transfer is classified as transfer by convection. All media can also release energy in the form of photons (electromagnetic waves). This energy travels in space essentially with the velocity of light until the photons are recaptured by some other atoms, causing in this way heat transfer by radiation. An example of this energy transport is the transfer of heat from the sun to the earth. The three modes of energy transfer mentioned above will be discussed consecutively in the following sections. However, it must be realized that they often occur simultaneously, so that in some cases the total energy transport will be the sum of the contributions of the individual mechanisms. In other cases, such a summation will not lead to the correct result when the individual transport mechanisms mutually interfere.

Conduction From a microscopic standpoint, thermal conduction refers to energy being handed down from one atom or molecule to the next one. In a liquid or gas, these particles change their position continuously even without visible movement and they transport energy also in this way. From a macroscopic or continuum viewpoint, thermal conduction is quantitatively described by Fourier's equation which states that the heat flux q per unit time and unit area through an area element arbitrarily located in the medium is proportional to the drop in temperature, $-\mathrm{grad}\ T$, per unit length in the direction normal to the area and to a transport property k characteristic of the medium and called thermal conductivity.

$$q = -k\ \mathrm{grad}\ T \qquad (1)$$

Predictions for the value of the thermal conductivity k can be made from considerations of the atomic structure. Accurate values, however, require experimentation in which the heat flux q and the temperature gradient, grad T, are measured and these values are inserted into Fourier's equation. Figure 1 presents thermal conductivity values for a number of media in a large temperature range. It can be recognized that metals have the largest conductivities and, among those, pure metals have larger values than alloys. Gases, on the other hand, have very low heat conductivity values. Electrically nonconducting solids and liquids are arranged in between. The low thermal conductivity of air is utilized in the development of thermally insulating materials. Such materials, like cork or glass fiber, consist of a solid substance with a very large number of small spaces filled by air. The thermal transport occurs then essentially through the air spaces, and the solid structure only supplies the framework which prevents convective currents. The range of thermal conductivities in Fig. 1 at ambient temperature extends through 5 powers of 10. This range is still small compared with the range for the electric conductivity of various substances where electric conductors have values which are larger by 25 powers of 10 than elec-

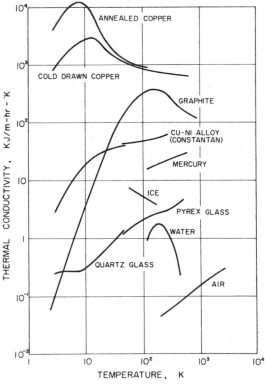

FIG. 1.

HEAT TRANSFER

tric insulators. As a consequence of this fact, it is much easier to channel electricity along a desired path than to do so with heat, a fact which accounts for the difficulty in accurate experimentation in the field of heat transfer.

Fourier's equation can be used together with a statement on energy conservation to derive a differential equation describing the temperature field in a medium. Fourier was the first one to develop this equation and to devise means for its solution. In vector notation, this equation is

$$\rho c \frac{\partial T}{\partial t} = \nabla (k \nabla T) \qquad (2)$$

where ρ is the density, c is the specific heat, t is time, ∇ is the Nabla operator. The temperature field in a substance can either change in time (unsteady state) or it can be independent of time (steady state $\partial T / \partial t = 0$). For a steady-state situation, the temperature field depends primarily on the geometry of the body involved and on the boundary conditions. The simplest case of a steady state temperature field is the one in a plane wall with temperatures which are uniform on each surface, however different at the two surfaces. The temperature in the wall then changes linearly in the direction of the surface normal as long as the variation of the thermal conductivity in the temperature range involved can be neglected. For an unsteady process, the capacity of the medium to store energy enters the energy conservation equation; correspondingly, the specific heat of the material and its density become factors for the conduction process, as well as the thermal conductivity. A combination of these properties, defined as the ratio of the thermal conductivity to the product of specific heat and density, called thermal diffusivity ($k/\rho c$), then determines how fast existing temperature differences in a medium equalize in time. It is found that metals and gases have thermal diffusivity values which are approximately equal in magnitude and considerably higher than thermal diffusivities of liquid and solid nonconductors. This means that temperature differences equalize much faster in metals and gases than in other substances.

Various other physical processes lead in their mathematical description to equations of the same form as Eq. (2), especially in its steady-state form. Such processes are, for instance, the conduction of electricity in a conductor or the shape of a thin membrane stretched over a curved boundary. This situation has led to the development of analogies (electric analogy, soap film analogy) to heat conduction processes which are useful because they often offer the advantage of simpler experimentation.

Convection It has been mentioned before that in fluids, energy is often transported by convection. In such a situation, conduction takes care of the transport of heat from one stream tube to another and is the dominating mode of transfer near solid walls. Convection transports heat along the stream lines and is dominating in the main body of the fluid where the velocities are large. In many situations, the flow is turbulent; this means that unsteady mixing motions are superimposed on the mean flow. These mixing motions contribute also to a transport of heat between stream tubes, a process which can be thought of as being described by an "effective" conductivity which often has values by several powers of ten larger than the actual conductivity of the fluid.

The movement of the fluid may be generated by means external to the heat transfer process, as by fans, blowers, or pumps. It may also be created by density differences connected with the heat transfer process itself. The first mode is called *forced convection*, the second one *natural* or *free convection*. Convective heat transfer may also be classified as heat transfer in duct flow or in external flow (over cylinders, spheres, air foils, or similar objects). In the second case, the heat transfer process is essentially concentrated in a thin fluid layer surrounding the object (boundary layer).

Of special interest in such heat transfer processes is the knowledge of the heat flux from the surface of a solid object exposed to the flow. This heat flux q_w per unit area and time is conventionally described by Newton's equation

$$q_w = h(T_w - T_f) \qquad (3)$$

where T_w is the surface temperature and T_f is a characteristic temperature in the fluid. This equation defining the heat transfer coefficient h is convenient because in many situations the heat flux is at least approximately proportional to the temperature difference $T_w - T_f$. Information on the heat transfer coefficients can be obtained by a solution of the Navier-Stokes equations describing the flow of a viscous fluid and the related energy equation, or they are found by experimentation. The availability of electronic computers has tremendously increased our ability to study heat transfer analytically at least for laminar flow, whereas in turbulent flow the bulk of our information is based on experiments. Such experimentation is made difficult by the large number of parameters involved. Dimensional analysis has therefore been applied to reduce the number of influencing parameters, and relations for convective heat transfer are correspondingly presented in modern handbooks as relations between dimensionless parameters. Such an analysis demonstrates, for instance, that heat transfer in forced flow can be described by a relation of the form

$$Nu = f(Re, Pr) \qquad (4)$$

in which the Nusselt number Nu is a dimensionless parameter hL/k, containing the heat transfer coefficient $h;$ the Reynolds number $Re = \rho(VL/\mu)$ describes essentially the nature of the flow; and the Prandtl number $Pr = c_p\mu/k$ can be considered as a dimensionless transport property characterizing the fluid involved. L and V are an arbitrarily selected characteristic length and velocity, respectively, ρ denotes the density, μ the viscosity, and c_p the specific heat of the fluid at constant pressure. Occasionally the Stanton number, $St = Nu/Re\,Pr$, is used instead of the Nusselt number as a dimensionless expression of the heat transfer coefficient. Equation (4) is based on the assumption that the thermodynamic and transport properties involved in the heat transfer process can be considered as constant. Larger variations of such properties are usually accounted for by additional terms in Eq. (4) expressing the ratio of the varying transport properties or of parameters (temperature, pressure) on which they depend. Fluids occurring in nature cover a very large range of Prandtl numbers. Liquid metals, for instance, have values of order 0.001 to 0.01. Gases have values between 0.6 and 1, and oils have values up to 10 000 and more. Some heat transfer relations for forced convection are presented in Table 1. Relations for other situations can be found in the various texts mentioned at the end of this section or in corresponding handbooks. With regard to the relations in Table 1 and other similar equations, it has to be kept in mind that they are valid over a restricted range of the independent parameters only.

Heat transfer by free convection is described by relations of the form

$$Nu = f(Gr, Pr) \qquad (5)$$

In this equation, the Grashof number

$$Gr = \frac{\rho^2 g\beta(T_w - T_f)L^3}{\mu^2}$$

(where g is the gravitational constant, β is the thermal expansion coefficient, T_f is the fluid temperature at a distance where it is not influenced by the heated or cooled object with surface temperature T_w) replaces the Reynolds number. Sometimes a dimensionless parameter called Rayleigh number $(Ra = Gr\,Pr)$ is used instead of the Grashof number. Equation (5) assumes again that the fluid properties involved are nearly constant. Examples for such relations are also contained in Table 1. Examples for free convection situations are the heat transfer from a heating register in the room. Free convection is also an important factor in the establishment of the temperature in the atmosphere. In the free convection relations of Table 1 and in Eq. (5), it has been assumed that the convection flows are generated by the gravitational field. Natural convection can also be created by other body forces, like centrifugal and Coriolis forces or electromagnetic forces.

TABLE 1   RELATIONS FOR CONVECTIVE HEAT TRANSFER

Forced Convection

Channel Flow
Flow through a tube:
  laminar $(Re < 3000)$

$$Nu = 3.65$$

  turbulent $(Re > 3000, Pr > 0.6)$

$$Nu = 0.116\,(Re^{2/3} - 125)\,Pr^{1/3}$$

$$\left(Nu = \frac{hD}{k}, Re = \frac{\overline{V}D}{\nu}, \right.$$

$$\left. D = \text{diameter}, \overline{V} = \text{mean velocity}\right)$$

External Flow
Flat plate parallel to flow:
  laminar $(1000 < Re < 500000, Pr > 0.6)$

$$Nu = 0.332\sqrt{Re}\,\sqrt[3]{Pr}$$

  turbulent $(Re > 500000, Pr > 0.6)$

$$Nu = \frac{0.0297\,Re^{4/5}\,Pr}{1 + 1.3\,Re^{-1/10}\,Pr^{-1/6}\,(Pr - 1)}$$

$$\left(Nu = \frac{hx}{k}, Re = \frac{\rho V_s x}{\mu}, x = \text{distance from leading edge},\right.$$

$$\left. V_s = \text{velocity outside boundary layer}\right)$$

Cylinder normal to flow: $1 < Re < 4000$

$$Nu = 0.43 + 0.48\sqrt{Re}$$

$$\left(Nu = \frac{hD}{k}, Re = \frac{\rho V_0 D}{\mu}, D = \text{diameter},\right.$$

$$\left. V_0 = \text{upstream velocity}\right)$$

Natural Convection

Vertical flat plate:
  laminar $(10^4 < Gr < 10^8)$

$$Nu = 0.508\,Gr^{1/4}\,Pr^{1/2}\,(0.952 + Pr)^{-1/4}$$

  turbulent $(Gr > 10^8)$

$$Nu = 0.0295\,Gr^{2/5}\,Pr^{7/5}\,(1 + 0.494\,Pr^{2/3})^{-2/5}$$

$$\left(Nu = \frac{hx}{k}, Gr = \frac{\rho^2 g\beta(T_w - T_f)x^3}{\mu^2}, x = \text{distance}\right.$$

from leading edge, T_w = wall temperature, T_f = fluid temperature at some distance from plate)
(All surfaces are assumed to have uniform temperature)

Space does not permit the discussion of other heat transfer processes, although such processes have found increasing attention in recent years. Especially large heat transfer coefficients are created by a boiling or condensation process. Boiling heat transfer is therefore used in applications which have to deal with very large heat fluxes like chambers and nozzles of rockets or the anodes in electric arc devices. Heat transfer is also often combined with mass transfer processes. This is, for instance, the case in evaporation devices. Heat transfer may also occur combined with chemical reactions as in processes involving gases at very high temperature where combustion, dissociation, or ionization occur.

Radiation Energy can also be transferred from one location to another within a medium or from one medium to another in the form of photons (electromagnetic waves). Usually a multiplicity of wavelengths λ is involved in such energy transfer.* In vacuum, all waves regardless of their wavelength move with the same speed 2.9977×10^8 m/sec. In various substances, the wave velocity c changes somewhat with wavelength, and the ratio of the wave velocity in vacuum to the velocity in a substance is equal to the optical refraction index. Air and generally all gases have refraction indices which differ from one only in the fourth decimal. Their wave velocity is, therefore, practically equal to that in vacuum.

Prévost's principle states that the amount of energy emitted by a volume element within a radiating substance is completely independent of its surroundings. Whether the volume element increases or decreases its temperature by the process of radiation depends on whether it absorbs more foreign radiation than it emits or vice versa. One talks about thermal radiation when the emission of photons is thermally excited, i.e., when the substance within the volume element is nearly in thermodynamic equilibrium. The discussion in this section will be restricted to thermal radiation. For such radiation, Kirchhoff was able to derive a number of relations by consideration of a system of media in thermodynamic equilibrium. If j_ν indicates the co-efficient of emission, i.e., the radiative flux at the frequency ν^* emitted per unit volume into a unit solid angle, and κ is the co-efficient of absorption at the same frequency, that is, the fraction of the intensity of a radiant beam which is absorbed per unit path length, then one of these relations states

$$c^2 \frac{j_\nu}{\kappa_\nu} = f(T, \nu) \qquad (6)$$

with c denoting the wave velocity. According to this relation, the combination of parameters on the left-hand side of the equation is a function of temperature T and frequency ν of the radiation only, but does not depend on the substance under consideration. Kirchhoff's law can also be expressed in parameters which refer to the interface of two media 1 and 2. It then takes the form

*Frequency and wavelength are used interchangeably. They are connected by the relation $\lambda\nu = c$.

$$c^2 \frac{i_\nu}{\alpha_\nu} = f(T, \nu) \qquad (7)$$

in which i_ν is the monochromatic intensity (flux per unit area and unit solid angle) of the radiative flux at frequency ν originating in medium 2 and traveling through the interface into medium 1 per unit solid angle and area normal to the direction of the radiant beam. α_ν is the monochromatic absorptance or absorptivity, i.e., that fraction of a radiant beam approaching the interface in the medium 1 in the opposite direction that is absorbed in medium 2. c is the wave velocity in medium 1. Kirchoff's law states that the combination of the parameters of the left-hand side of Eq. (7) is again a function of temperature and frequency only, but does not depend on the nature of the medium. A medium which absorbs all the radiation traveling into it through an interface ($\alpha_\nu = 1$) is called a blackbody. The intensity of radiation emitted by an arbitrary medium is, according to Eq. (7), in the following way related to the intensity of radiation $i_{b\nu}$ emitted by a blackbody at the same temperature and frequency

$$\frac{i_\nu}{\alpha_\nu} = i_{b\nu} \qquad (8)$$

From the consideration of a system in thermodynamic equilibrium, it is also easily shown that $i_{b\nu}$ is independent of direction and that the total monochromatic radiant flux emitted by a blackbody per unit interface area and unit time into all directions in space is equal to $\pi i_{b\nu}$.

The law describing the monochromatic intensity of radiation of a blackbody is given by Planck's equation

$$i_{b\nu} = \frac{2h\nu^3}{c^2(e^{h\nu/kT} - 1)} \qquad (9)$$

(where h is Planck's quantum constant and k is Boltzmann's constant). Experimentalists prefer to use the intensity $i_{b\lambda}$ per unit wavelength ($i_{b\lambda}d\lambda = i_{b\nu}d\nu$). Figure 2 presents the wavelength dependence of the intensity of blackbody monochromatic radiation for a number of temperatures. It may be observed that for each temperature, the intensity has a maximum at a certain wavelength and that this maximum shifts toward short wavelengths with increasing temperature (Wien's law). The wavelength λ is plotted on the abscissa in micrometers, μm, (1000 μm = 1mm). Our eye is sensitive to radiation in the range 0.4 to 0.7 μm (the dashed range). It may be observed that for temperatures with which we have largely to deal, the bulk of blackbody radiation is contained in the range of wavelengths larger than the visible ones (infrared range). This statement also holds for other media because Eq. (8) shows that no medium can have a monochromatic intensity which is higher than that of a blackbody. Only

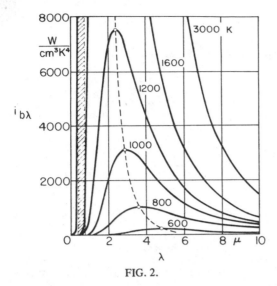

FIG. 2.

radiation coming from the sun has a major portion of the energy in the visible wavelength range (corresponding to a temperature of 6500 K).

The total energy flux q emitted per unit area and time from a blackbody into all directions in space can be obtained by integration of Eq. (9) over all frequencies and by multiplication of the result by π. For blackbody radiation into a vacuum (or with good approximation into a gas), the result is

$$q = \sigma T^4 \qquad (10)$$

The Stefan-Boltzmann constant σ has the value $5.67 \times 10^{-8}\,\mathrm{W/m^3 K^4}$

The following additional relation exists at an interface between two substances.

$$\rho_\nu + \alpha_\nu + \tau_\nu = 1 \qquad (11)$$

This equation describes that monochromatic radiant energy in a beam approaching in a medium 1 the interface with a medium 2 is found again either as radiation reflected back into the medium 1 or absorbed in the medium 2 or transmitted through the medium 2 into other media or back into medium 1. The monochromatic reflectance or reflectivity ρ_ν is the ratio of reflected to incident radiant energy, α_ν is the corresponding ratio for the absorbed, and τ_ν for the transmitted energy. The vast majority of solids and liquids absorb radiant energy over most wavelengths in the infrared range within a very thin layer adjacent to the interface (of order 1μ to 1 mm). In heat transfer calculations, it can therefore usually be assumed that the transmissivity of such substances is equal to zero and that reflectivity and absorptivity are connected to the temperature of the interface. One talks then often in a simplified manner

about radiative interchange between surfaces. Kirchhoff's law, Eq. (8), additionally connects the intensity of a beam emitted through the interface with the absorptivity and the intensity of a beam leaving a blackbody at the same temperature. Electromagnetic theory shows that electric nonconducting materials have in the infrared range generally high values of the absorptivity and correspondingly low values of the reflectivity. Metals (electric conductors), on the other hand, behave in the opposite way, having low absorptivity and high reflectivity values. This fact is utilized in aluminum-insulations and in vacuum thermos bottles. In the visible range, the appearance of surfaces on various materials to the eye already supplies information on approximate values of the reflectivity and absorptivity. The values of these parameters largely determine the reflective and absorption characteristics of surfaces for solar irradiation. A white surface, for instance, has a very low absorptivity and reflects the major part of solar irradiation. Gases behave differently with respect to radiation. They need fairly large layers in order to absorb the major part of incident radiation and radiate only in restricted wavelength ranges whereas solids and liquids have a more continuous spectrum. Diatomic gases (O_2, N_2) do not absorb energy with good approximation. Solar irradiation is absorbed in the atmosphere by carbon dioxide and water vapor. Values for reflectivities, absorptivities, and transmissivities of various substances are contained in the Handbook reference listed at the end of the ariticle.

Values for the radiation properties (ρ_ν, α_ν, τ_ν) together with the relations for blackbody radiation are the basis of calculations to determine heat exchange by radiation in a system with locally varying temperature. Calculations of such interchange are in general very involved, especially when substances with small absorption coefficients are involved. The formulation of such interchange leads to integral differential equations. The reader has in this connection to be referred to the books listed at the end of this section, and only a few relations for simple geometries will be presented here. Consider two area elements, dA_1 and dA_2, belonging to two blackbodies with the temperatures T_1 and T_2. The distance between the two area elements is s, and β_1 and β_2 are the angles between the two surface normals and the interconnecting line. The following equation then describes the net heat transfer dQ from area element dA_1 to area element dA_2 per unit time assuming that no radiation is absorbed or emitted in the space between the two surfaces.

$$dQ = \frac{\cos\beta_1 \cos\beta_2}{\pi s^2}\, dA_1\, dA_2\, \sigma(T_1{}^4 - T_2{}^4) \qquad (12)$$

If non-black surfaces are involved, then the process of radiant interchange is much more involved, since part of the incident radiation is now reflected from the surfaces and travels in this way back and forth until it is finally absorbed. Simple relations exist in this case for the radiative interchange between the surfaces of two concentric spheres or cylinders with areas A_1 and A_2 and temperatures T_1 and T_2. It is further assumed that both surfaces are emitting and reflecting in a perfectly diffuse way and that they are separated by a medium which does neither emit nor absorb radiation. The net monochromatic interchange $d\Phi_\nu$ between the two surfaces is then described by the equation

$$d\Phi_\nu = \frac{\pi}{\dfrac{1}{\alpha_{\nu 1}} + \dfrac{A_1}{A_2}\left(\dfrac{1}{\alpha_{\nu 2}} - 1\right)} dA_1(i_{b\nu 1} - i_{b\nu 2})$$

(13)

The monochromatic intensities $i_{b\nu 1}$ and $i_{b\nu 2}$ are calculated with Eq. (9). The relation changes when the outer cylinder reflects radiation mirror-like (specularly) to

$$d\Phi_\nu = \frac{\pi}{\dfrac{1}{\alpha_{\nu 1}} + \left(\dfrac{1}{\alpha_{\nu 2}} - 1\right)} dA_1(i_{b\nu 1} - i_{b\nu 2})$$

(14)

Both equations merge asymptotically into the same relation when the differences between the two radii become small. The corresponding relation then also holds for two parallel infinite planes. Equations (13) and (14) have to be integrated over all frequencies to obtain the net heat transfer between the two surfaces. The result is simple when the absorptances $\alpha_{\nu 1}$ and $\alpha_{\nu 2}$ are independent of frequency (gray surfaces). Equations (13) and (14) describe then the net heat transfer, when $i_{b\nu 1}$ and $i_{b\nu 2}$ are replaced by $(\sigma/\pi)T_1^4$ and $(\sigma/\pi)T_2^4$, respectively.

E. R. G. ECKERT

References

Eckert, E. R. G., and Drake, R. M., Jr., "Analysis of Heat and Mass Transfer," New York, McGraw-Hill, Inc., 1972.

Arpaci, V., "Conduction Heat Transfer," Reading, Mass., Addison–Wesley Publishing Co., 1966.

Kays, W. M., "Convective Heat and Mass Transfer," New York, McGraw-Hill Book Co., 1966.

Hottel, H. C., and Sarofim, A. F., "Radiative Transfer," New York, McGraw-Hill Book Co., 1967.

Patankar, S. V., "Numerical Heat Transfer and Fluid Flow," Washington, D.C., Hemisphere Publishing Corp., 1980.

"Handbook of Heat Transfer," Rohsenow, W. H., and Hartnett, J. P. (Eds.), New York, McGraw-Hill Book Co., 1973.

Cross-references: HEAT, HEAT CAPACITY, INFRARED RADIATION, PHOTON, REFLECTION, THERMAL RADIATION, THERMODYNAMICS.

HEISENBERG UNCERTAINTY PRINCIPLE

Classical physics is based on two assumptions that have been found experimentally to be untenable. The first of these is the existence of signals that can travel with infinite speed; the second is that the magnitude of the interaction between two systems can be reduced to arbitrarily small values. The realization that the speed of propagation of signals has a finite upper limit led to the development of relativity theory. The recognition of the existence of a finite quantum of action has been incorporated in quantum (wave) mechanics.

Quantum mechanics assigns a physical reality only to those variables whose value can, in principle, be experimentally determined. About the existence of phenomena or systems that cannot be experimentally observed, quantum mechanics is noncommittal. Questions regarding an isolated system are meaningless in quantum mechanics, for any observation made on such a system necessarily disturbs its isolation by at least one quantum of action. Heisenberg[5] observed that any measurement made on a system destroys some of the knowledge gained about that system through previous measurements. Any prediction about the future course of a system must be contingent on a knowledge about the measurements that will be made on that system, and is subject to uncertainties introduced by the measurements. Whereas one might speculate with some reliability about the future course of a system under the restriction that no more measurements will be performed on it, such speculations would be physically meaningless, as they could not be experimentally confirmed or denied.

In the broadest sense, then, the Heisenberg Uncertainty Principle states that the partitioning of the universe into observer (either a human observer, or a recording device such as a photographic plate) and observed is subject to a finite inaccuracy; one might say that the "knife" or "pencil" that makes the partition has a finite "thickness," h.

The concept of a monochromatic beam of radiation is not a difficult one to accept. Yet the experimental determination of the frequency of such a beam requires an infinite time interval; any finite portion chopped from the beam is shown by Fourier analysis to have a spectrum of finite width, hence not a single frequency at all. Quantum mechanics does not deny the existence of a monochromatic beam, but it does render the assignment of a definite frequency in a given time interval meaningless.

Fourier analysis shows that the specifications of the time interval, Δt, and of the spectral width, $\Delta \nu$, are reciprocally related:

$$\Delta \nu \, \Delta t \geqslant 1 \qquad (1)$$

This equation represents the uncertainty relation for classical waves: any attempt to specify the frequency at an instant of time results in a broadening of the frequency spectrum. This uncertainty relation applies to any wave, whether electromagnetic, acoustic, or otherwise.

Interference patterns observed when electron beams are reflected from crystalline surfaces[3] or transmitted through thin metallic films[7] indicate that these beams possess some wave characteristics. DeBroglie[2], by independent theoretical considerations (see WAVE MECHANICS), postulated the following relations between the dynamic variables, energy (E) and linear momentum (p) of the beam, on the one hand, and the wave variables, frequency (ν) and wavelength (λ), on the other:

$$E = h\nu \qquad (2)$$

$$p = h/\lambda \qquad (3)$$

Substitution of Eq. (2) in Eq. (1) gives the uncertainty relation between energy and time:

$$\Delta E \cdot \Delta t \geqslant h \qquad (4)$$

For a beam of free electrons of mass m, traveling in the x direction, $\Delta t = m \, \Delta x / p_x$, where p_x is the linear momentum and Δx is the distance traveled in the time interval Δt. Since $E = p_x^2 / 2m$, $\Delta E \cong p_x \, \Delta p_x / m$, so that $\Delta E \cdot \Delta t = \Delta p_x \cdot \Delta x$, and

$$\Delta p_x \cdot \Delta x \geqslant h \qquad (5)$$

The pairs of variables (E, t) and (p_x, x) are called canonically conjugate pairs of variables. In quantum mechanics, the operators corresponding to canonically conjugate variables do not commute (see QUANTUM THEORY). Heisenberg originally stated his uncertainty principle in the following form: the values of canonically conjugate variables of a given system can only be determined with a finite lower limit of accuracy.

Among the many important experimental phenomena illustrating the uncertainty principle in the COMPTON EFFECT. Here, a photon is scattered by an electron; the momentum of the photon is rendered uncertain as a result of its scattering by the electron and the electron is moved from its original position by the impact received from the photon. If we consider the photon and the electron as separate systems, then their interaction (the collision) introduces uncertainties for each system, given by Equation (5).

A general uncertainty relation follows from the postulates of WAVE MECHANICS. Consider two variables, a and b, of a system, whose operators are $\mathcal{C}$ and $\mathcal{B}$. The expectation values of a and b are called $\bar{a}$ and $\bar{b}$ respectively. The uncertainties in a and b can then be defined quantitatively as their respective rms deviations from their expectation values; it follows from the postulates that

$$(\Delta a)^2 \equiv \oint \Psi^* (\mathcal{C} - \bar{a})^2 \, \Psi \, dq$$

where Ψ is the normalized wave function of the system under observation, and $\oint \ldots dq$ indicates integration over all values of all coordinates. Similarly,

$$(\Delta b)^2 \equiv \oint \Psi^* (\mathcal{B} - \bar{b})^2 \, \Psi \, dq$$

If $\mathcal{C}$ and $\mathcal{B}$ are Hermitian,* $(\mathcal{C} - \bar{a})^2$ and $(\mathcal{B} - \bar{b})^2$ are also Hermitian. Therefore:

$$(\Delta a)^2 \cdot (\Delta b)^2 =$$

$$= \oint \Psi^* (\mathcal{C} - \bar{a})^2 \, \Psi \, dq \cdot \oint \Psi^* (\mathcal{B} - \bar{b})^2 \, \Psi \, dq$$

$$= \oint (\mathcal{C} - \bar{a}) \Psi (\mathcal{C} - \bar{a})^* \Psi^* \, dq$$

$$\cdot \oint (\mathcal{B} - \bar{b}) \Psi (\mathcal{B} - \bar{b})^* \Psi^* \, dq$$

$$= \oint |(\mathcal{C} - \bar{a}) \Psi|^2 \, dq \cdot \oint |(\mathcal{B} - \bar{b}) \Psi|^2 \, dq.$$

To put this product of two definite integrals in a more useful form, consider the function

$$f(q) = \lambda (\mathcal{C} - \bar{a}) \, \Psi + i (\mathcal{B} - \bar{b}) \, \Psi$$

where λ is real, and independent of coordinates, and $i \equiv \sqrt{-1}$. The function $|f(q)|^2$ must be non-negative:

$$\lambda^2 |(\mathcal{C} - \bar{a}) \Psi|^2 + i\lambda [(\mathcal{C}^* - \bar{a}) \Psi^* (\mathcal{B} - \bar{b}) \Psi +$$

$$- (\mathcal{C} - \bar{a}) \Psi (\mathcal{B}^* - \bar{b}) \Psi^*] + |(\mathcal{B} - \bar{b}) \Psi|^2 \geqslant 0$$

When the left-hand side of this inequality is integrated over all values of all coordinates:

---

*An operator $\mathcal{F}$ is Hermitian, if for any properly behaved functions u and v: $\oint u \mathcal{F} v \, dq = \oint v \mathcal{F}^* u \, dq$, where * indicates complex conjugation. Quantum mechanical operators usually are, or can be made to be, Hermitian.

$$\lambda^2(\Delta a)^2 + (\Delta b)^2 + i\lambda \left[\oint (\mathcal{C}^* - \overline{a})\Psi^*(\mathcal{B} - \overline{b})\Psi\, dq + \right.$$

$$\left. -\oint (\mathcal{C} - \overline{a})\Psi(\mathcal{B}^* - \overline{b})\Psi^*\, dq \right] \geqslant 0$$

Since λ is real, the left-hand side of this inequality becomes negative unless the discriminant becomes zero or negative:

$$4(\Delta a)^2(\Delta b)^2 \geqslant - \left[\oint (\mathcal{C}^* - \overline{a})\Psi^*(\mathcal{B} - \overline{b})\Psi\, dq + \right.$$

$$\left. -\oint (\mathcal{C} - \overline{a})\Psi(\mathcal{B}^* - \overline{b})\Psi^*\, dq \right]^2$$

The right-hand side of this inequality is reduced as follows (remember that Ψ is normalized):

$$\oint (\mathcal{C}^* - \overline{a})\Psi^*(\mathcal{B} - \overline{b})\Psi\, dq = \oint (\mathcal{C}\Psi)^*\mathcal{B}\Psi\, dq +$$

$$- \overline{a}\oint \Psi^*\mathcal{B}\Psi\, dq - \overline{b}\oint \Psi(\mathcal{C}\Psi)^*\, dq + \overline{a}\overline{b}$$

$$\oint (\mathcal{C} - \overline{a})\Psi(\mathcal{B}^* - \overline{b})\Psi^*\, dq = \oint (\mathcal{C}\Psi)(\mathcal{B}\Psi)^*\, dq +$$

$$- \overline{a}\oint \Psi(\mathcal{B}\Psi)^*\, dq - \overline{b}\oint \Psi^*(\mathcal{C}\Psi)\, dq + \overline{a}\overline{b}$$

Since $\quad \overline{b} = \oint \Psi^*\mathcal{B}\Psi\, dq = \oint \Psi(\mathcal{B}\Psi)^*\, dq, \quad$ and

$$\overline{a} = \oint \Psi^*\mathcal{C}\Psi\, dq = \oint \Psi(\mathcal{C}\Psi)^*\, dq,$$

$$4(\Delta a)^2(\Delta b)^2 \geqslant - \left[\oint (\mathcal{C}\Psi)^*(\mathcal{B}\Psi)\, dq + \right.$$

$$\left. - \oint (\mathcal{C}\Psi)(\mathcal{B}\Psi)^*\, dq \right]^2$$

Since $\mathcal{C}$ and $\mathcal{B}$ are Hermitian

$$\oint (\mathcal{C}\Psi)^*(\mathcal{B}\Psi)\, dq \equiv \oint (\mathcal{B}\Psi)(\mathcal{C}\Psi)^*\, dq =$$

$$= \oint \Psi^*\mathcal{C}\mathcal{B}\Psi\, dq,$$

$$\oint (\mathcal{C}\Psi)(\mathcal{B}\Psi)^*\, dq = \oint \Psi^*\mathcal{B}\mathcal{C}\Psi\, dq \quad (6)$$

$$\therefore\ 4(\Delta a)^2(\Delta b)^2 \geqslant - \left[\oint \Psi^*(\mathcal{C}\mathcal{B} - \mathcal{B}\mathcal{C})\Psi\, dq \right]^2$$

and $|\Delta a| \cdot |\Delta b| \geqslant \frac{1}{2}i\oint \Psi^*(\mathcal{C}\mathcal{B} - \mathcal{B}\mathcal{C})\Psi\, dq$

According to the postulates of wave mechanics, the operators for momentum and position

are given respectively by: $\mathcal{P} = -i\hbar\nabla$, where $\hbar = h/2\pi$, and $\mathcal{Q} = q$ (multiplication by q).

For linear motion in the x direction, $\mathcal{P}_x = -i\hbar\partial/\partial x$, $\mathcal{Q} = x$. Hence $(\mathcal{P}\mathcal{Q} - \mathcal{Q}\mathcal{P})\Psi = -i\hbar(\partial/\partial x)(x\Psi) + i\hbar x(\partial\Psi/\partial x) = -i\hbar\Psi$. When this expression is substituted into Eq. (6), it follows that:

$$|\Delta p_x|\,|\Delta x| \geqslant h/4\pi$$

which is in agreement with Eq. (5). Equation (4) can be similarly derived from the postulates of wave mechanics by setting $\mathcal{E} = (ih/2\pi)(\partial/\partial t)$, $\mathcal{T} = t$.

A. L. LOEB

References

1. Bohm, David, "Quantum Theory," Englewood Cliffs, N.J., Prentice-Hall, 1951.
2. De Broglie, L., *J. Phys. Ser.* 6, 7, 1 "Introduction to Wave Mechanics," London, Methuen, (1926).
3. Davisson, C. J., and Germer, L. H., *Phys. Rev.*, **30**, 705 (1927); *Proc. Natl. Acad. Sci. U.S.*, **14**, 317 (1928).
4. Harris, L., and Loeb, A. L., "Introduction to Wave Mechanics," New York, McGraw-Hill Book Co., 1963
5. Heisenberg, W., *Z. Physik*, **43**, 172; "The Physical Principles of the Quantum Theory," Chicago, University of Chicago Press, (1927).
6. Margenau, H., and Murphy, G. M., "The Mathematics of Physics and Chemistry," New York, Van Nostrand Reinhold, 1943.
7. Thomson, G. P., *Proc. Roy. Soc. London Ser. A*, **117**, 600 (1928); **119**, 651 (1928).

Cross references: COMPTON EFFECT, QUANTUM THEORY, WAVE MECHANICS.

HIGH-VOLTAGE RESEARCH

High-voltage research deals with phenomena evoked by high voltages and intense electric fields, with the behavior of dielectrics and electrical components under such electrical stress, and with the utilization of electrostatic fields and forces for various purposes of science and industry. In the laboratory, this electrical stress is produced by the presence of electric charge on the opposing surfaces of two electrodes between which a voltage is applied. Pulsed, alternating and constant voltages ranging from a few kilovolts to 20 MV have been used in such studies. In nature, air currents and water precipitation cause electric charge to become separated between cloud and earth or between clouds; the stressed region may reach electrical pressure differences in excess of 1000 MV. Lightning, a rapid high-current discharge, completes the breakdown of the over-stressed air and dissipates the accumulated electrical energy. Many aspects of this natural high-voltage phenomena have been the subject of investigation because of their scientific interest and the danger of lightning to life and to susceptible structures such as electric power systems.

In the industrial high-voltage laboratory, the direct and induced effects of lightning discharges on electrical apparatus are often simulated by high-voltage impulse generators. These use the Marx method of first slowly charging a number of condensers in parallel and then suddenly connecting them in series by spark-gap switches which at the same instant impress the multiplied voltage upon the test circuit. A typical voltage wave produced by such impulse generators rises to its peak value of several million volts in 1 μsec and then diminishes exponentially reaching half-voltage in 10 μsec.

Electric Field between High-voltage Electrodes The region between and around electrodes which have been charged by the application of a voltage V between them is occupied by the electric field of that charge. In an isolated system, the positive electrode has a deficiency of electrons exactly equal to the excess electrons on the negative electrode. The amount of electric charge Q on either electrode surface at any instant is given by $Q = CV$ where Q is in coulombs and C is the capacitance of the electrode system in farads. Energy is stored in the electrically stressed space between the electrodes; the amount of this electrical energy W can be expressed in terms of applied voltage or separated charge by $W = CV^2/2 = Q^2/2C$ joules.

The electric field intensity at any point is defined by the magnitude and direction of the force which would be experienced by a unit positive charge placed in the field at that point. Following Faraday, if we define a line of electric flux as a line drawn so that its direction is everywhere the direction of the force on a positive particle, and require that one line must originate on each unit positive charge and terminate on each unit negative charge, then the lines of flux will map out the electric field and the lines per unit area will be directly related to the electrical field intensity E. The electrostatic force acting on a particle with charge q placed in an electric field of intensity E volts per meter is given by $f = qE$ newtons. The electric field distribution depends upon the geometry of the electrodes but is affected by the presence of dielectric materials or of charged particles; a quantitative picture of the static or changing field picture is usually essential to high-voltage research.

Objectives of High-Voltage Research From antiquity to the present, the history of high-voltage research sparkles with many names well-known in electrical science—Thales of Miletus, von Guericke, Newton, Franklin whose kite experiment established the identity of natural lightning and electricity, Cavendish, Faraday, and Roentgen whose discovery of x-rays in 1895 marked the beginning of the atomic age. During the 60 years centered on the turn of the twentieth century, physicists studied the passage of electricity through gases at normal and reduced pressure, sought an understanding of long sparks and corona in atmospheric air, measured the conductivity and breakdown strength of liquid dielectrics, and examined the flashover of solid insulators in these media.

More recent research has been directed at the performance under high electrical stress of a wide range of solid, liquid, and gaseous dielectrics as well as vacuum-insulated systems. The solid materials vary from porcelains and glasses to hydrocarbon polymers and loaded epoxies. Superior gaseous insulation is now obtained by the combined use of such electronegative gas molecules as sulfur hexafluoride (SF_6) and Freon (CCl_2F_2) with elevated pressures sometimes exceeding 5 or 10 atmospheres. Solid insulator supports are indispensable in gas, liquid, and vacuum insulated systems and are studied in combination with these media.

Research in the high-voltage field may also be directed at testing and increasing the insulation strength of power equipment subject to lightning or switching surges. The increasing trend toward higher voltages for the transmission of electric power over long distances has directed research toward reduction of radio interference and power loss by corona and surface leakage from high-voltage transmission line conductors and their suspension insulators. The need to bring such power into urban areas has produced the requirement for ac power cables in which the center conductor is reliably insulated for hundreds of kilovolts above earth. The inherent efficiency and stability of dc power transmission have led to the development of more adequate high-voltage rectification and conversion apparatus. For these purposes, the low-pressure metallic vapor tube has reached the highest power levels though solid state devices offer much promise for high-voltage, high-power switching.

In the field of science and medicine, high-voltage research seeks improved methods of producing high constant voltages, of measuring and stabilizing such voltages, and of applying them to the acceleration of atomic ions and electrons to high energies. Such particle accelerators are needed for nuclear structure research; for the study of the properties of energetic atoms, electrons, x-rays, and neutrons; for the treatment of deep-seated tumors with energetic particles and radiation, and for the radiation processing of materials.

Industrial objectives of high-voltage research include the development of high-power, high-frequency tubes and their power supplies for radar and long-range radio communication, the ionization of particulate matter by corona and its electrostatic collection as in smoke and chemical precipitators, and the elimination of electrostatic hazards which arise in processes such as the transfer or mixing of volatile hydrocarbons, dusts, and explosive gases.

JOHN G. TRUMP

References

Graggs, J. D., and Meek, J. M., "High Voltage Laboratory Technique," Butterworths Scientific Publications, London, 1954.

Trump, J. G., "Electrostatic Sources of Electric Power," *Elec. Eng.*, **66**, No. 6, 525–534 (June 1947).

Trump, J. G., "New Developments in High Voltage Technology," *IEEE Trans. on Nuclear Science* **NS-14** (3), 113–119 (June 1967).

Cross-references: ACCELERATOR, VAN DE GRAAFF; ELECTRICITY; POTENTIAL; STATIC ELECTRICITY.

HISTORY OF PHYSICS

Any attempt to trace the beginning of man's ideas on the nature of the physical world brings us to anthropomorphic, mythological cosmology. It is impossible to trace any absolute adumbration of some scientific doctrine. If a hint could be suggested, then the Western world appears to have drawn more from Mesopotamian sources and traditions than from the Nile civilization. No archeological dust will reveal the inscrutable past. As for physics, in the limited space available we can start only with the era when specific names emerged that are associated with the earliest formation of scientific ideas, described at that time as philosophy. The Ionian period, that amazing flow of intellectual energy, paved the way for the rise of the fathers of Hellenic science. Physics, then closely linked with philosophy and mythological astronomy, played an auxiliary role in man's overwhelming concern in cosmography and cosmogony.

Although physics as an independent field originated in the age of Renaissance and was associated with such giants as Galileo, Kepler, and Newton, even in this brief sketch it is indispensable to trace the fountainhead of this science, which is in the cornerstone of Western civilization. The port of Miletus in Ionia, on the eastern coast of the Aegean sea, can be singled out as among the most significant birthplaces of physical science. Located in a favorable geographic position, its flourishing commerce since the second millenium before our era provided an excellent clearing house between two river civilizations, Egypt and Mesopotamia. Exposing people to divergent ideas and traditions, it created an atmosphere of open-mindedness where new, unrestrained ideas could flourish. Consequently, several most unique rationalizing minds had their roots in this town alone.

Of the galaxy of great pioneers of Ionian science, Thales of Miletus (624-565 B.C.) stands out as a symbol of the era. Sir James Jeans maintains that most of the major achievements of physical science of our age can be traced back to the stream of knowledge started by this Ionian intellectual giant in Miletus. Pythagoras, Democritus, Anaxagoras, Aristarchos of Samos, to mention a few, represent as one historian exclaims, that "miracle of ancient Greece" that prepared and shaped the climax of Plato and Aristotle, who for two millenia were to inspire and guide, for better or worse, the evolution of physics.

Liberated from the mythology of pre-Socratic time, these ancient astounding thinkers represent every school of philosophy from the extreme idealism of illusion and nonexistence of the world of sense perception, in the Pythagorean-Platonic sense, to the atomism and materialism of Democritus and Anaxagoras with their doctrine of the primacy of matter in a universe manipulated by accidental mechanism. It has been said that everyone by nature is a disciple of either Plato or Aristotle, and Raphael in his famous painting of the School of Athens on the wall of the Vatican Palace appropriately illustrates Plato pointing upward and Aristotle downward to the ground. Thus, the Alexandrian School in Hellenistic Egypt with its lighthouse, Pharos, as a symbol irradiated the glory of Hellenic science for centuries, nourishing Western civilization with Euclides' *Stoicheia* and Ptolemy's *Almagest* which, outside the Bible, were the most widespread literary sources in physical science relegated to antiquity.

With the rise of Christianity, followed by Islam half a millenium later, interest in the studies of natural science was temporarily paralyzed, to be later zealously renewed under the aegis of the theology of the new religion. A convert to Christianity, St. Augustine (A.D. 354–430), later Bishop of Hippo, was a pioneer in the realization that Plato's ideas on sense delusion conveniently responded to the devout Christian's search for the salvation of the soul. Like Socrates and Sophists about a millenium previously, so St. Augustine also turned his back on nature and advised, "Return to thyself. In the inner man dwells truth." Thus, for a thousand years the men who guided the thought of the Western world did not observe out-of-doors and learn from natural phenomena. Monasteries became leading establishments where sedentary monks pored over volumes of Plato and Aristotle. More than half a millenium after St. Augustine, St. Thomas Aquinas (A.D. 1227–1274) petrified Aristotelian peripatetic scholasticism into an authoritarian *Summa Theologica* that included all the answers man should know on the nature of the physical world. It must be mentioned that the science of Western Islam helped to shape the cosmological ideas of St. Thomas Aquinas and, through him, the whole thought of Catholic Europe. We should stress that Ptolemy's masterpiece, *Almagest*, first reached Europe in the Arabic language.

This marked the climax of the Middle Ages, described unjustly as "dark ages." Although theology was the queen of sciences, this period was not devoid of scientific activity because speculations in physics were constantly being nourished by the mystery of the *Primum Mobile*, a sphere beyond the fixed stars. It was this Prime Mover to which Aristotelians ascribed the first supernatural impulses or "impetus." How else could motion first have started? With the rise of Humanism, the rediscovery of Greek literature of antiquity revealed vast subjects dealing with fields other than theology. Invigorating new studies spread through Western Europe invading universities, not by-passing some monasteries. Bold, unusual views on the nature of the physical world and methods of investigation aroused the suspicion of watchful scholasticians but the age of Renaissance could not be diverted. Even the prominent ecclesiastic Oresme (1332–1382), Bishop of Lisieux, sustained the main interest by challenging the Aristotelian doctrine of the fixity of the earth. Nearly a century later this was continued by Cardinal Nicolas of Cusa who, like the Franciscan monk, Roger Bacon, in the thirteenth century also advocated experimentation in order to learn how the laws of nature operate. Very penetrating studies on the mystery of "violent" motion and inertia were accomplished long before their actual fruition with the appearance of Galileo and Newton.

Signs of a new era in physics were imminent with ideas of the universal genius, Leonardo da Vinci, maintaining like, Democritus and Anaximander that the whole universe conforms to unalterable mechanical laws. The coming dawn was evident when Nicolaus Copernicus (1473–1543) came to study in Bologna and Padua. Being more an ancient Greek philosopher in his use of geometry in support of the heliocentric system, Copernicus at least prepared the way of Galileo (1564–1642), who finally mobilized all known physics with his inventive experimentalism that adumbrated the first full stream of scientific revolution. Symptomatically, Galileo's first work was on motion, *De motu*, a subject of great concern through the Middle Ages, and his was the final challenging blow to the Aristotelian doctrine, when he verified that force primarily produces acceleration instead of mere movement.

As a sign of continuity, Newton was born (1642–1727) the year Galileo passed away. The trend of mechanism of physical phenomena as a consequence of mathematical determinism reached its portentous finalization with the Galilean-Newtonian revolution. Newtonian classical physics, associated with the world view of the majestic Newtonian universe, eternal and infinite, was formulated in three laws of motion, climaxed by the universal law of gravitation. This physics continued in its progressive refinement until it was confirmed by the triumphant mathematical discovery of the planet Neptune by J. C. Adams and Leverrier in 1845. It was then considered the final shape of knowledge man was in position to realize. Previously, Laplace produced an overwhelming impression on the entire century when, in 1798, he used in his *Origin of the World System* his deterministic equations in the formulation of his hypothesis on the origin of the solar system.

With advancing crystallization, Newtonian physics radiated with a galaxy of great names, inspired builders of the classical view of the physical world. Only a few principal milestones can be indicated, each a giant of his own in a panoramic view of the glorious century of promise of a scientific paradise. From Laplace at the beginning of the nineteenth century, the epic unfolds from Avogadro to Faraday, from Carnot and Joule to Kelvin and Helmholtz, attaining its pre-Einsteinian peak in Maxwellian equations formulating the electromagnetic theory of light. These equations were impressively described by Boltzmann, himself at the cradle of thermodynamics, when he quoted from Goethe's *Faust*: "Who was the god who wrote these lines?" Yet, these equations so brilliantly describing natural processes, pointed inevitably to a deterministic and mechanistic universe. Although rigorous mechanicism had flourished in the past in varying degrees, Newton must be regarded as the founder of the mechanistic world view even though he had difficulty in harmonizing mechanistic natural philosophy with his belief in a God who not only created the world but also constantly preserves it. This mechanization of the world picture systematically led to the conception of God as a retired engineer, and it was only another step to His complete exclusion. Therefore, the universe was ultimately knowable and predictable. This *Weltanschauung* of triumphant physics encouraged the rise of materialistic philosophy, that actually shaped the dialectic materialism of Marx and Engels, which became the official doctrine of the ruling communist state in the twentieth century.

Newtonian physics was not destined to remain the last form in the evolution of physics. When it appeared to reach its perfection, as some leading physicists advocated, the turn of the twentieth century again witnessed another tidal wave that changed the course of physics. By 1895, this second scientific revolution started with discoveries of the first magnitude, containing unfathomable consequences for the future. Becquerel's radioactivity, Roentgen's x-rays, J. J. Thomson's electron, Planck's quantum, Bohr's atom, Rutherford's nucleus, Einstein's relativity and equality of mass and energy, represent a revolution that will carry the exploring mind incomparably farther into the mysteries of the universe than Copernicus, Galileo, Kepler, or even Newton ever dreamed of. Heisenberg's principle of indeterminacy in the realm of microphysics not only shatters the once cherished corpuscular-kinetic determinism but points to a microcosmos much more complex than what Whitehead called "provincialism in time and space," and as valid in Newtonian mechanism. Combining this with the staggering discoveries in astrophysical macrocosmos, the Dopplerian red shift of external galaxies, quasars, neutron stars, and black holes, we confront a truly unprecedented era of future centuries that will bring about unimaginable amendments in the physics we know today and its subsequent world view.

As the end of the twentieth century looms ahead with all the problems of turbulent humanity combining utility with the catastrophic abuse of applied physical science, it is important to note in retrospect that the unparalleled advance of man's understanding of the physical world in the present century was not appropriately balanced by any corresponding increase in man's ethical enlightenment. Einstein's centenary, celebrated throughout the entire cultural world in 1979, more than justifies the inevitability of this comment in this encyclopedia. The genial author of the new, post-Newtonian concept of our universe never failed to express his concern in the fate of mankind alongside his world view of New Physics.

The Second Scientific Revolution, as historians like to call it, starting at the turn of this century, is not only intellectually longer in its three quarters of a century than were the three centuries since Newton, but far surpasses the length separating the world view of Newton from that of Aristotle, two thousand years ago. Even Laplace with his classical, mathematical physics belongs in a category that separates him from such wizards of the New Physics of our contemporary age as Einstein, Bohr, Planck, de Broglie, and Heisenberg. These conclusions are drawn because both Newton and Aristotle built world views from facts acquired by our sensory perceptions, as demanded by Victorian physics. Yet we can never perceive nor directly observe an electron, a nucleus, a quantum, or any ultimate particle of microphysics. We build their representation from a highly complex laboratory procedure which does not guarantee their actual reality. The great pioneer of New Physics, Ernst Mach, once stated that "Senses do not lie, but they do not tell us truth." This was supplemented a century later by J. M. Jauch, the late Director of the Institute of Theoretical Physics at the University of Geneva: "Bohr, by denying objective reality to properties whose simultaneous presence would require for their verification mutually exclusive physical situations, puts in question that very concept of reality which has been the cornerstone of all of physics as it has developed from the time of Galileo and Newton."

The dilemma arising from the Einsteinian world view was aptly described by Whitehead when he suggested our rooting in spatiotemporal provincialism. The Victorian physicist operated in what Reichenbach called the "world of middle dimensions," his Newtonian-Euclidian space between microcosmos and macrocosmos. In relativistic physics the rigid homogeneous Newtonian-Euclidian infinite space melts away in a constant spatiotemporal dislocation that depends on the presence of matter and a finite four-dimensional universe. Matter produces local distortions of space-time metric with its non-Euclidean geometry, which manifest themselves through force fields, inertia, and the "constant" of universal gravitation (G) that is subject to change with the progressive expansion of the universe.

In the microworld, Heisenberg's principle of indeterminacy denies the objective reality of sharply localized particles, so that the so-called particle blends into the background from which it should be distinct to deserve to be identified as a particle. The shocking problem of the inconstancy of the reality of such particles as the electron is well described by Oppenheimer:

If we ask whether the position of the electron remains the same, we must say "No;" if we ask whether the electron's position changes with time, we must say "No;" if we ask whether the electron is at rest, we must say "No;" if we ask whether it is in motion, we must say "No."

The Second Scientific Revolution has hardly begun to build its history with Einstein's unfinished dream of unified field theory and unfold a new universe from the immensity of extraordinary facts, whether observable or unobservable, pointing to an unprecedented magnitude of the changing new world of physics and its cosmology in the forthcoming centuries. It dwarfs all preceding world concepts of accommodating, enduring anthropomorphic views.

KAREL HUJER

References

1. Sarton, George, "A History of Science," 2 Vols., Cambridge, Mass., Harvard Univ. Press, 1959.
2. Singer, Charles, "A Short History of Scientific Ideas to 1900," London, Oxford Univ. Press, 1959.
3. Dijksterhuis, E. J., "The Mechanization of the World Picture," London, Oxford Univ. Press, 1969.
4. Capek, Milic, "Philosophical Impact of Contemporary Physics," D. Van Nostrand Co., 1961.
5. Frank, Philip, "Einstein, His Life and Time," New York, Alfred Knopf, 1947.
6. Dingle, Herbert, "The Sources of Eddington's Philosophy," Cambridge, U.K., Cambridge Univ. Press, 1954.
7. Koyré, Alexandre, "Des Révolutions des orbes célestes," Paris, Librairies Felix Alcan, 1934.

8. Heisenberg, Werner, "Physics and Philosophy," New York, Harper Torch Books, 1958.
9. de Broglie, Louis, "Physics and Microphysics," New York, Harper Torch Books, 1960.
10. D'Abro, A., "The Rise of New Physics: Decline of Mechanism," New York, Dover Publications, 1951.
11. Jauch, J. M., "Are Quanta Real;" Bloomington, IN, Indiana Univ. Press, 1971.
12. Oppenheimer, J. Robert, "Science and the Common Understanding," New York, Simon and Schuster, 1954.
13. Taylor, A.M., "Imagination and the Growth of Science," The Tallman Lectures, 1964–65, John Murrey Publications, 1966.
14. Manuel, F. E., "A Portrait of Isaac Newton," Cambridge, Mass., Harvard Univ. Press, 1969.
15. Calder, Nigel, "Violent Universe: An Eyewitness Account of the New Astronomy," New York, The Viking Press, 1969.

HOLOGRAPHY

In 1948, British scientist Dennis Gabor (1900–1979) proposed and demonstrated a new, two-step method of optical imagery. However, because of the extreme difficulty of implementing the method in those early days, it was not until the decade of the 1960s that the method became widely known and used. This renaissance of interest in the method was largely due to the general availability of the laser with the great temporal and spatial coherence of its light and to the efforts in the early 1960s of J. Upatnieks and E. N. Leith of the University of Michigan. As Gabor's method grew in significance, the genius of the idea was recognized and Gabor was awarded the Nobel Prize in physics in 1971.

The method proposed by Gabor is now known as *holography*. It is similar to photography in many respects, yet is fundamentally different. With photography, one generally records, by means of lens and film, the two-dimensional irradiance distribution in the image of an object. With holography, however, one records not the optically formed image of an object but the object wave itself. This wave is recorded (usually on photographic film) in such a way that a subsequent illumination of this record, called a *hologram*, reconstructs the original object wave. A visual observation of this reconstructed wavefront then yields a view of the object which is practically indiscernible from the original, including three-dimensional parallax effects.

Figure 1(a) shows schematically how a hologram is recorded. One starts with a single, monochromatic beam of light that has originated from a very small source. This single beam is split into two components, one of which is directed toward the object and the other to a suitable recording medium, most commonly a photographic emulsion. The component that is incident on the object is scattered by it, and this scattered radiation, now called the object wave, impinges on the recording medium. The wave that proceeds directly to the recording medium is called the *reference wave*. Since the object and reference waves originate from the same source, they are mutually coherent and form a stable interference pattern when they meet at the recording medium. The detailed record of this interference pattern constitutes the hologram.

When the hologram is illuminated with a beam similar to the original reference wave, it modulates the phase and/or amplitude of the illuminating wave in such a way that the transmitted wave divides into three separate components, *one of which exactly duplicates the original object wave*.

If the two interfering beams are traveling in substantially the same direction, the recording of the interference pattern is said to be a *Gabor hologram* or an *in-line* hologram. If the two interfering beams arrive at the recording medium from substantially different directions, the recording is a *Leith-Upatnieks* or *off-axis* hologram. If the two interfering beams are traveling in essentially opposite directions, the recorded hologram is said to be a *Lippmann* or *reflection* hologram, first invented by Y. N. Denisyuk.

Electromagnetic radiation is most commonly used, although acoustic radiation can be used. The most common electromagnetic radiation employed is light, but holograms have also been recorded successfully with electron beams, x-radiation, and microwaves.

Holograms can be classified by the way they diffract light. In an *amplitude hologram* the varying irradiance distribution of the interference pattern is recorded as a density variation of the recording medium. In this type of hologram the illuminating wave is always partially absorbed, i.e., the illuminating wave is *amplitude-modulated*. In a *phase hologram*, a *phase modulation* is imposed on the illuminating beam which in turn results in diffraction of the light. Phase modulation occurs when the optical path (thickness X index) varies with position. A phase hologram results from either relief-image or index variation, or both.

Either phase or amplitude holograms can be classified further as *Fresnel holograms* or as *Fraunhofer holograms*. Generally speaking, if the object is reasonably close to the recording medium, say just a few hologram or object diameters distant, the field at the hologram plane is the Fresnel diffraction pattern of the object. A hologram recorded in this manner is termed a *Fresnel hologram*.

If the object and hologram are separated by many object or hologram diameters, the field at the hologram due to the object alone is the Fraunhofer diffraction pattern of the object. A hologram recorded in this manner is termed a *Fraunhofer hologram*.

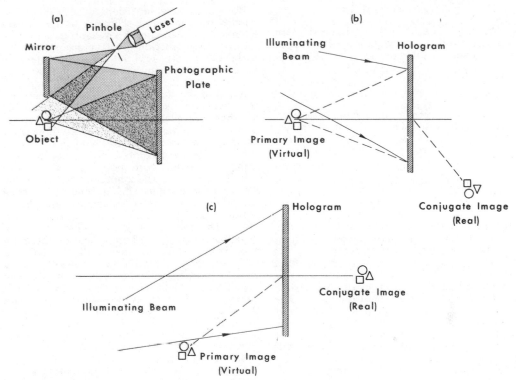

FIG. 1. A typical holographic arrangement. (a) Recording the hologram. (b) Reconstructing the primary object wave. (c) Reconstructing an undistorted conjugate wave.

Any of these hologram types may be recorded as either a *thick* or a *thin* hologram. A *thin hologram* is one for which the thickness of the recording medium is thin compared to the spacing between the recorded interference fringes. A *thick* or *volume hologram* is one in which the thickness of the recording medium is of the order of or greater than the spacing of the recorded fringes.

Conceptually, the simplest form of an off-axis hologram is one for which the object is just a single, infinitely distant point so that the object wave at the recording medium is a plane wave. If the reference wave is also plane, and incident on the recording medium at an angle to the object wave, the hologram will consist of a series of Young's interference fringes. These recorded fringes are equally spaced straight lines running perpendicular to the plane of incidence. Since the hologram consists of a series of alternating clear and opaque strips, it is in the form of a diffraction grating. When the hologram is illuminated with a plane wave, the transmitted light consists of a zero-order wave traveling in the direction of the illuminating wave, plus two first-order waves. The higher diffracted orders are generally missing or very weak since the irradiance distribution of a two-beam interfer-

ence pattern is sinusoidal. As long as the recording is essentially linear (irradiance proportional to final amplitude transmittance), the hologram will be a diffraction grating varying sinusoidally in amplitude transmittance, and only the first diffracted orders will be observe' One of these first-order waves will be traveling n the same direction as the object wave; this is the reconstructed wave.

To describe the recording of a hologram of a more complicated object, let O be a monochromatic wave from the object incident on the recording medium H, and let R be a wave coherent with O. The wave O contains information about the object, since the object has uniquely determined the amplitude and phase of O. The object can be thin and transmitting, such as a transparency, or it can be opaque and diffusely reflecting. Both the amplitude and phase of the wave O can be recorded with the aid of the reference beam R. The total field on H is $O + R$. A square-law recording medium, such as a photographic emulsion, will respond to the irradiance of the light $|O + R|^2$. Assume that after processing, the hologram possesses a certain complex amplitude transmittance $t(x)$ that can be expressed as a function of the exposure $E(x)$,

$$t(x) = f[E(x)]. \qquad (1)$$

The expansion of this function will yield a term linear in exposure, and by ignoring all terms except this one, one can write

$$t(x) = \beta E(x) = \beta \mid O + R \mid^2 \cdot T$$

$$= \beta T(\mid O \mid^2 + \mid R \mid^2 + OR^* + O^*R), \qquad (2)$$

where T is the exposure time, * indicates complex conjugate, and β is the constant coefficient of the linear term. When the hologram with this amplitude transmittance is illuminated with a wave C, the transmitted field at the hologram is

$$C \cdot t(x) = \beta T[C \mid O \mid^2 + C \mid R \mid^2$$

$$+ CR^*O + CRO^*]. \qquad (3)$$

If the illuminating wave C is sufficiently uniform so that CR^* is approximately constant across the hologram, the third term of Eq. (3) is $\beta TCR^*O = $ const. $\times O$. This term represents a wave identical with the object wave O. This wave has all of the properties of the original wave and can form an image of the object. The fact that this wave is separated from the rest can be seen most clearly by analogy with the diffraction-grating hologram described above. It can be shown that the other first-order diffracted wave corresponds to the conjugate image term $\beta TCRO^*$ of Eq. (3) and that the zero-order wave corresponds to the first two terms $\beta TC(\mid O \mid^2 + \mid R \mid^2)$. The object wave O is separated from the others and may be viewed independently.

Figure 1 illustrates the recording of a hologram and the subsequent reconstruction. In Fig. 1(a), the laser beam is first expanded and then divided by a mirror, which directs part of the beam directly onto the photographic plate; the rest of the light is reflected from the object. After processing, the hologram plate may be replaced in its original position [Fig. 1(b)] and the object removed. The light diffracted by the hologram forms, in part, the same wavefront that was originally scattered by the object. A viewer looking through the hologram will see an undistorted view of the object, just as if it were still present. Figure 2(a) shows a photomicrograph of an actual hologram, and 2(b), (c), and (d) photographs of three perspectives of the resulting image.

In addition to this virtual image, or *primary image*, a real, or *conjugate image* will be formed on the observer's side of the hologram. This image will appear unsharp and highly distorted, and it will also be inverted in depth, i.e., reversed front to back, as shown in Fig. 1(b). However, a distortion-free real image can be formed by changing the position of the illuminating beam so that all of the rays of the reference beam are reversed in direction. In this way, an undistorted, real, three-dimensional image of the object scene appears in front of the hologram as shown in Fig. 1(c).

Holograms may be recorded with diverging, parallel, or converging reference beams. If care is taken to maintain the recording geometry during reconstruction, it is even possible to form holograms with an arbitrary reference beam, the only requirement being that it be coherent with the object beam.

It is possible to produce color holograms by recording three separate holograms on a single photographic plate, each in a different color. Subsequent illumination with a three-color beam yields three separate wavefronts, one in each of the three colors representing the portion of the object corresponding to that color.

It is possible to make holograms that can be viewed in reflection. This is done by allowing the reference and object beams to enter the recording medium from opposite sides [Fig. 3(a)]. The fringes formed are planes lying approximately parallel to the plane of the hologram. When such a hologram is illuminated by a beam similar to the reference wave, a reflected wave is formed which exactly duplicates the object wave. The image is viewed in reflected light [Fig. 3(b)]. It is possible to illuminate this type of hologram with white light. The interference planes filter the light by acting as a $\lambda/2$ multilayer interference filter, in the same way as in Lippmann color photography.

Many applications have been proposed for holography. The most important of these are the ones that exploit the unique features of holography, rather than those that just do old tasks in a new way. The best example of this is holographic interferometry, whereby *arbitrary* wavefronts interfere with each other in such a way that interference bands are produced that depict only the *differences* between them. In practice the two wavefronts are nearly identical. There are several types of holographic interferometry.

For *single-exposure holographic interferometry*, a hologram of the object wave O is recorded in the usual way. The hologram is then placed in exactly the same position it occupied during exposure and it is illuminated with the reference wave and a slightly distorted object wave O'. The reference wave reconstructs the original object wave O, which interferes with the directly transmitted wave O'. This is a real-time technique, because when the wave O' changes, so do the interference bands. This technique can be used to measure surface deformations of all kinds, including thermal expansion and contraction, swelling caused by absorption, and any minute changes that might occur in an object.

Double-exposure holographic interferometry requires making two holograms on a single recording medium. One of the two holograms yields a primary image which constitutes the comparison wave, just as in the single-exposure case. The test wave is not the object itself, however, but a reconstructed wave from the changed object. Interference phenomena,

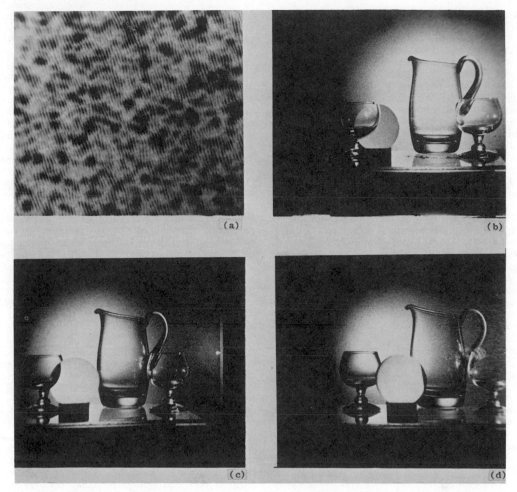

FIG. 2. A hologram of a diffusely illuminated object. (a) A highly magnified image of the hologram. (b), (c), (d) Three perspectives of the resulting image.

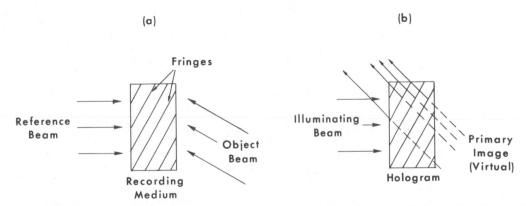

FIG. 3. Volume holograms that can be viewed in reflection. (a) Recording the hologram with object and reference waves incident in nearly opposite directions. (b) Reconstructing the primary wave in reflection.

caused by changes in optical path through the object between exposures, are produced when the doubly exposed hologram is illuminated. This technique is well suited to interferometric recording of transient phenomena, such as shock waves and fluid flow, when a pulsed laser is used as the source. All of the principles discussed thus far apply equally well to time-dependent events, and the very short pulse of light from a ruby laser can record the interference phenomena at a single instant of time. The very wide range of applicability of the method has been well demonstrated. This technique can also be extended to multiple exposures.

The idea of multiple-exposure interferometry can be extended to the limiting case of a continuum of exposures, resulting in what is called *time-average holographic interferometry*. This technique lends itself nicely to the problem of vibration analysis and may well be the best method yet devised for such analysis. The basis of the method is that since holography itself is an interferometric process, any instabilities of the interferometer cause fringe motion. Thus the hologram of a vibrating object is a record of the time-average irradiance distribution at the hologram plane. Since the amount of light flux diffracted from any region of the hologram depends on the fringe contrast, any object motion that causes the fringes to move during the exposure, causing a loss of contrast, will result in less diffracted flux from that region of the hologram. The strength of the reconstructed wave is therefore a function of the fringe motion during the exposure. If the object is vibrating in a normal mode, there will be standing waves of vibration on the surface, so that at the nodes the object motion will be very small or nonexistent. At the antinodes the vibration amplitude will be large. A hologram of such an object will then produce a bright image of the regions of the object for which little or no motion occurred during the exposure, whereas it will not produce images of antinodal points at all. This holographic method of vibration analysis has all of the advantages of all holographic interferometry. The method can be used regardless of the shape or complexity of the object; the vibration nodes can be examined in three dimensions, or at least from a variety of perspectives; and the method works regardless of whether the surface is optically smooth or diffusely reflecting.

One of the most striking aspects of the modern hologram is the three-dimensional image that it is capable of producing. This three-dimensional image indicates that there is a large amount of information contained in a single hologram—much more than is contained in a conventional photograph of the same size. Because of the many perspectives available, the hologram is well suited to display purposes. With a hologram, one can present all of the observable characteristics of a three-dimensional object clearly and concisely. Complicated molecular or anatomical structure can be simply presented with a single holographic image, with little chance of error or misinterpretation on the part of the viewer. Such a hologram could take the place of several conventional drawings or photographs. The use of holograms in textbooks would be a great aid to the student in many fields. Holograms made to be viewed with a small penlight and a colored filter have already been produced in large quantities and distributed in magazines and books.

Holographic microscopy is another important application. There are basically two distinct methods of holographic microscopy: (1) conventional holography with magnification achieved by changing the scale of the hologram, the illuminating wavelength, or the radius of curvature of the illuminating wavefront; by optically magnifying the holographic image; or by using any combination of these; or (2) holographically recording the optically magnified wavefront. For the first method, the lateral magnification is given by the formula

$$M = \frac{\mu m Z_c Z_R}{\mu Z_c (Z_R - Z_o) + m^2 Z_o Z_R},$$

where μ is the ratio of the illuminating and recording wavelengths, m is the factor by which the hologram has been scaled, and Z_c, Z_o, and Z_R are the radii of curvature of the illuminating, object, and reference wavefronts, respectively. Because aberrations are introduced when the wavelength is changed or the wavefront radii are changed, most of the current work in holographic microscopy involves optically magnifying the holographic image or using the second method, in which an optically magnified wavefront is recorded holographically. Holographic microscopy offers the following advantages: One can avoid the problems of limited depth of focus, off-axis observation, and the short working distance of the classical microscope. The hologram records a large volume of object space instantaneously. All of the usual image-processing techniques can be applied to the reconstructed wave, which may represent a large volume of space at an instant of time. No other method for doing this exists.

Recently, holograms used as optical elements (holographic optical elements, or HOEs) have become an important aspect of holography. An HOE is a generalized diffraction grating having many advantages over conventional optics (such as their light weight, thinness, and low cost), along with some disadvantages (such as requiring narrow bandwidth light). In principle, HOEs are able to perform the functions of conventional optical elements, such as beamsplitters, lenses, scanners, mirrors, and even spectral filters. HOEs can not only be designed with a computer, as can conventional optical elements, but they can also be generated with a computer, allowing inexpensive fabrication of such things as aspherics and off-axis elements.

HOWARD M. SMITH

References

1. Smith, H. M., "Principles of Holography," New York, Wiley-Interscience, 1969.
2. DeVelis, J. B., and Reynolds, G. O., "Theory and Applications of Holography," Reading, Mass., Addison-Wesley, 1967.
3. Collier, R. J., Burckhardt, C. B., and Lin, L. H., "Optical Holography," New York, Academic Press, 1971.
4. Vest, C. M., "Holographic Interferometry," New York, John Wiley & Sons, 1979.
5. Lee, T-C., and Tamura, P. N. (Eds.), Proc. SPIE 215, "Recent Advances in Holography," 1980.

Cross-references: COHERENCE; DIFFRACTION BY MATTER AND DIFFRACTION GRATINGS; INTERFERENCE AND INTERFEROMETRY; LASER; OPTICS, GEOMETRICAL; OPTICS, PHYSICAL; WAVE MOTION.

I

IMPULSE AND MOMENTUM

The concept of impulse and momentum derives directly from Newton's law of motion.

Consider first the case of *linear* impulse and *linear* momentum. Newton's law states that in the proper frame of reference, force is equal to the (time) rate of change of momentum, where momentum is defined as the product of mass and velocity. Consider a force F acting for a time Δt on a particle of mass m, thereby changing its velocity from v_1 to v_2. The rate of change of momentum in this case is the change of momentum divided by the time interval during which the change occurs. Thus, Newton's law states

$$F = \frac{(mv)_2 - (mv)_1}{\Delta t} \tag{1}$$

Now if we multiply each side of this equation by Δt, we obtain

$$F \Delta t = (mv)_2 - (mv)_1 \tag{2}$$

The left-hand side of this equation, $F \Delta t$, representing the product of a force and the time interval through which the force acts, is called "impulse." Thus, this equation states what is often known as the *Law of Impulse and Momentum*: Impulse is equal to the change of momentum.

It becomes apparent that a body will experience the same change of momentum irrespective of the separate values of F and Δt as long as their product $F \Delta t$ is the same. Thus, a large force acting briefly may have the same net effect as a smaller force acting longer, if the two impulses are the same. Going to the limit, we may consider an infinite force acting for an infinitesimal time such that their product remains a finite quantity. Under the action of such an impulse, a body will experience an instantaneous change of velocity. A common example is the change of velocity of a baseball as it is hit by a bat. For all practical purposes the change occurs instantaneously.

As a consequence of the Law of Impulse and Momentum described above, we find that if the total force is zero, so is the total impulse, and the momentum will remain unchanged. This is known as the *Conservation Law of Momentum*. It applies either to one particle or to a system of particles. Consider, for instance, the collision of two billiard balls, A and B. Taking each ball separately, the impulse on A by B is equal in magnitude but opposite in direction to the impulse on B by A. Thus, upon collision, the change of momentum of A is also equal and opposite to that of B. On the other hand, if we take *both* A and B as our system, then the two impulses are acting on the same system. Since the two are equal and opposite, they cancel out and the total impulse on this system due to collision is zero. The Conservation Law of Momentum then predicts that the total momentum of the system (which is the sum of momenta of all particles in the system) remains a constant no matter what goes on in the system as long as there are no external forces acting on the sytem.

The Conservation Law of Momentum forms one of the basic cornerstones in physics and engineering. Its application is all-pervading, from the motion of stars to the encounter and scattering of molecules, atoms and electrons.

We will now continue our discussion at a more precise level. First of all, in calculating the momentum mv, the mass m should be relativistic mass defined as

$$m = \frac{m_0}{\sqrt{1 - (v/c)^2}}$$

where m_0 is the rest mass, i.e., mass at zero velocity, and c is the velocity of light. It is seen that at velocities much less than the velocity of light, the relativistic mass and the rest mass are indistinguishable, and m may be considered constant. Secondly, both force and velocity are vector quantities, i.e., they have a magnitude and a direction and obey the parallelogram law of addition, and we shall use letters **F** and **v** to represent them. Finally, as the time interval Δt approaches zero as a limit, the rate of change of momentum during Δt becomes a derivative and Newton's law becomes:

$$\mathbf{F} = \frac{d}{dt}(m\mathbf{v}) \tag{3}$$

After integrating each side of this equation with respect to time t and taking the integration limits from $t = t_1$ to $t = t_2$, the result is:

$$\int_{t_1}^{t_2} \mathbf{F}\, dt = (m\mathbf{v})_2 - (m\mathbf{v})_1 \tag{4}$$

The integral on the left-hand side, analogous to the left-hand side of Eq. (2), is the "impulse," and we again reach the Law of Impulse and Momentum given previously. In evaluating this integral, we must know the variation of **F** as a function of time, i.e., $\mathbf{F} = \mathbf{F}(t)$. Furthermore, in applying this law to a system of particles, we need only to count those impulses that are caused by *external* forces acting on the particles due to sources outside the system. As has been illustrated by the previous example on two billiard balls, the *internal* forces that any two particles exert on each other will generally cancel out if both particles are included in the system.

A dramatic illustration of the above principle is the case of an exploding shell. Before the explosion, the shell will travel a parabolic path under the influence of the only external force, the gravitational force, if the forces due to air resistance are neglected. During explosion, all the forces acting on the exploding pieces due to explosion are internal forces and they cancel out if we include *all* pieces of the shell as our system. The conservation law of momentum then predicts that the center of gravity of the system will continue to travel along the same parabolic path even after explosion, until some new external force acts on the system (as, for example, when a piece of the shell hits an object).

Up to this point we have discussed the laws of linear impulse and linear momentum. Entirely similar laws hold for *angular* impulse and *angular* momentum. The angular momentum **L** of a particle about a point O is defined as the vector product of **r**, the radius vector from O to the particle, and $m\mathbf{v}$, the momentum vector of the particle. Thus,

$$\mathbf{L} = \mathbf{r} \times m\mathbf{v}$$

The moment of force or *torque* about O is defined as

$$\mathbf{T} = \mathbf{r} \times \mathbf{F}$$

If we take the cross product of **r** with each side of the expression for Newton's law, Eq. (3), we obtain

$$\mathbf{r} \times \mathbf{F} = \mathbf{T} = \mathbf{r} \times \frac{d}{dt}(m\mathbf{v})$$

The right-hand side can be identified to be just $d\mathbf{L}/dt$ on account of the vector identity:

$$\frac{d\mathbf{L}}{dt} = \frac{d}{dt}(\mathbf{r} \times m\mathbf{v}) = \mathbf{v} \times m\mathbf{v} + \mathbf{r} \times \frac{d}{dt}(m\mathbf{v})$$

where the first term on the right-hand side vanishes. Thus,

$$\mathbf{T} = \frac{d\mathbf{L}}{dt} \qquad (5)$$

We now integrate this equation with respect to t and take the integration limits from $t = t_1$ to $t = t_2$. The result is

$$\int_{t_1}^{t_2} \mathbf{T}\, dt = \mathbf{L}_2 - \mathbf{L}_1 \qquad (6)$$

Analogous to the linear impulse, we may call the integral on the left-hand side of the above equation the angular impulse. Thus this equation states: Angular impulse is equal to the change of angular momentum. In particular, if the total torque is zero, so is the total angular impulse, and the angular momentum will remain unchanged. This is known as the *Conservation Law of Angular Momentum*.

When applied to common problems, we take the axial component of the vector equations (5) or (6), i.e., along the direction perpendicular to the plane of motion. Along such a direction, the magnitudes of **T** and **L** assume the following values:

$$T = rF_t$$

$$L = mrv_t$$

where r again is the distance from the axis to the particle, F_t the tangential component of the force acting on the particle, and v_t the tangential velocity of the particle. Equation (5) then becomes a scalar equation.

$$T = rF_t = \frac{d}{dt}(mrv_t) \qquad (7)$$

If $T = 0$, then the angular momentum mrv_t remains constant.

Equation (7) is the basis of the operation of pumps and turbines. In the case of a turbine, the angular momentum of the working fluid is large at entrance and small or zero at exist. Hence the fluid imparts a torque to the shaft. In a pump or compressor the reverse is the case, i.e., the angular momentum of the working fluid is small at entrance but large at exit, hence the shaft supplies torque as well as energy to the fluid.

A remarkable example of the conservation law of angular momentum is the motion of planets around the sun. The attractive force between a planet and the sun is in the direction of the line connecting the two, hence it has no tangential component and contributes no torque. Equation (7) then concludes that rv_t must be a constant. This means that when the planet is at a larger distance from the sun, its tangential velocity must be smaller, and vice versa. This led Newton to explain the experimental finding of Kepler, namely, a planet sweeps equal sectorial areas in equal times. The same explanation applies to figure skaters and ballet dancers when they increase their spin speed by pulling in their arms.

HSUAN YEH

References

Yeh, Hsuan, and Abrams, Joel I., "Principles of Mechanics of Solids and Fluids," Vol. I, New York, McGraw-Hill Book Co., 1960.

Synge, John L., and Griffith, Byron A., "Principles of Mechanics," New York, McGraw-Hill Book Co., Third Edition, 1959.

Goldstein, Herbert, "Classical Mechanics," Reading, Mass., Addison-Wesley, 1950.

Feynman, Richard T., "Lectures on Physics," New York, Addison-Wesley Pub. Co., 1963.

Simon, Keith R., "Mechanics," 3rd. Ed., New York, Addison-Wesley Pub. Co., 1971.

Cross-reference: CONSERVATION LAWS AND SYMMETRY, DYNAMICS, MECHANICS, ROTATION—CIRCULAR MOTION, STATICS, VECTOR PHYSICS.

INDUCED ELECTROMOTIVE FORCE

Electromotive force and voltage drop are usually regarded as synonymous. When an electromotive force, or simply emf, is impressed on a closed metallic circuit, current results. The emf along a specified path C in space is defined as the work per unit charge W/q done by the electromagnetic fields on a small test charge moved along C. Since work is the line integral of force F, the work per unit charge is the line integral of the force per unit charge. Letting F/q denote the vector electromagnetic force per unit charge in newtons per coulomb, we have

$$\text{emf} = \int_C \frac{F}{q} \cdot d\mathbf{l} \text{ volts} \qquad (1)$$

The scalar product $(F/q) \cdot d\mathbf{l}$ is the product $(F/q) \cos \theta \, dl$, with θ denoting the angle between the vectors F/q and $d\mathbf{l}$.

The electric force per unit charge is the electric field intensity E (volts per meter) and the magnetic force per unit charge is $\mathbf{v} \times \mathbf{B}$, with $\mathbf{v}$ denoting the velocity of the test charge in meters per second and B denoting the magnetic flux density in webers per square meter. In terms of the smaller angle θ between $\mathbf{v}$ and B, the cross product $\mathbf{v} \times \mathbf{B}$ is a vector having magnitude $vB \sin \theta$; the direction of the vector $\mathbf{v} \times \mathbf{B}$ is normal to the plane of the vectors $\mathbf{v}$ and B, with the sense of that of the extended thumb of the right hand oriented so that its fingers curl through the angle θ from $\mathbf{v}$ toward B. As the total force per unit charge is $E + \mathbf{v} \times \mathbf{B}$, the emf in terms of the fields is

$$\text{emf} = \int_C (E + \mathbf{v} \times \mathbf{B}) \cdot d\mathbf{l} \qquad (2)$$

It might appear from Eq. (2) that the emf depends on the forward velocity with which the test charge is moved along the path C. However, this is not the case. If $\mathbf{v}$ and $d\mathbf{l}$ in Eq. (2) have the same direction, then the vectors $(\mathbf{v} \times \mathbf{B})$ and $d\mathbf{l}$ are normal, and their scalar product is zero. Consequently, only the component of $\mathbf{v}$ normal to $d\mathbf{l}$ can contribute to the emf. This component has value only if the differential path length $d\mathbf{l}$ has sideways motion. *Thus $\mathbf{v}$ in Eq. (2) represents the sideways motion, if any, of $d\mathbf{l}$.* The fields E and B of Eq. (2) may be functions of time as well as functions of the space coordinates. In addition, the velocity $\mathbf{v}$ of each differential path length $d\mathbf{l}$ may vary with time. However, Eq. (2) correctly expresses the emf, or voltage drop, along the path C as a function of time. That component of the emf consisting of the line integral of $\mathbf{v} \times \mathbf{B}$ is known as the *motional emf*, because it has value only when the path C is moving through a magnetic field, cutting lines of magnetic flux. For stationary paths there is no motional emf, and the voltage drop is simply the line integral of the electric field E.

For an emf to exist along a stationary path, it is necessary to have an electric field present. As electric charges are surrounded by electric fields, emfs are generated by devices that separate charge. A familiar example is the battery, which utilizes chemical forces to separate charge. Some other methods of separating charge are the heating of a thermocouple, the exposure of a photocell to incident light, and the rubbing together of different materials. Electric fields are also produced by time-changing magnetic fields, and this principle is extensively exploited, as is motional emf, to generate electric power. The remainder of this article is devoted to electromotive force induced by magnetic means.

A fundamental law of electromagnetism, often called the *Maxwell-Faraday law*, or the *first law of electromagnetic induction*, states that the line integral of the electric field intensity E around any closed path C equals $-\partial\phi/\partial t$, with ϕ representing the magnetic flux over any surface S having the closed path C as its contour. The positive side of the surface S and the direction of the line integral around the contour C are related by the right-hand rule; by this rule, the curled fingers are oriented so as to point around the loop in the direction of the integration and the extended thumb points out of the positive side of the surface S. The magnetic flux ϕ is the surface integral of the magnetic flux density B; that is,

$$\phi = \iint_S \mathbf{B} \cdot d\mathbf{S} \text{ webers} \qquad (3)$$

In Eq. (3) the vector differential surface $d\mathbf{S}$ has area dS and is directed normal to the plane of $d\mathbf{S}$ out of the positive side. The partial time derivative of ϕ is defined as

$$\frac{\partial \phi}{\partial t} = \iint_S \frac{\partial \mathbf{B}}{\partial t} \cdot d\mathbf{S} \text{ volts} \qquad (4)$$

and this is often referred to as the *magnetic current* through the surface S. For a moving surface S the limits of the surface integral of Eq. (4) are functions of time, but Eq. (4) still applies. It is important to understand that in evaluating $\partial \phi / \partial t$ over a surface that is moving in a region containing a magnetic field, *we treat the surface at the instant under consideration as though it were stationary*. The partial time derivative of ϕ is the time rate of increase of the flux over the surface S due only to the changing magnetic field $\mathbf{B}$; any increase in ϕ due to the motion of the surface in the $\mathbf{B}$-field is *not* included. The Maxwell-Faraday law is

$$\oint_C \mathbf{E} \cdot d\mathbf{l} = -\frac{\partial \phi}{\partial t} \qquad (5)$$

with ϕ being the magnetic flux in webers out of the positive side of *any* surface having the path C as its contour. The small circle on the integral sign indicates a closed path. We note from Eq. (5) that *an electric field must be present in any region containing a time-changing magnetic field*.

The application of Eq. (2) to a closed path C gives

$$\text{emf} = \oint_C \mathbf{E} \cdot d\mathbf{l} + \oint_C (\mathbf{v} \times \mathbf{B}) \cdot d\mathbf{l} \qquad (6)$$

Utilizing Eq. (5) enables us to write Eq. (6) in the form

$$\text{emf} = -\frac{\partial \phi}{\partial t} + \oint_C (\mathbf{v} \times \mathbf{B}) \cdot d\mathbf{l} \qquad (7)$$

Thus the emf around a closed path consists, in general, of two components. The component $-\partial \phi / \partial t$ is often referred to as the *variational emf* (or *transformer emf*), and the second component is, of course, the motional emf.

In Eq. (7) the relation $(\mathbf{v} \times \mathbf{B}) \cdot d\mathbf{l}$ can, by means of a common vector identity, be replaced with $-\mathbf{B} \cdot (\mathbf{v} \times d\mathbf{l}$ has magnitude $v\, dl$ and of $d\mathbf{l}$, the vector $\mathbf{v} \times d\mathbf{l}$ has magnitude $v\, dl$ and direction normal to the differential surface dS swept out by the moving length $d\mathbf{l}$ in the time dt. Letting B_n denote the component of $\mathbf{B}$ normal to this area, we note that $-\mathbf{B} \cdot (\mathbf{v} \times d\mathbf{l})$ becomes $-B_n v\, dl$, and Eq. (7) can be written

$$\text{emf} = -\left[\frac{\partial \phi}{\partial t} + \oint_C B_n v\, dl\right] \qquad (8)$$

Clearly the integral of $B_n v$ around the closed contour C, with v denoting the magnitude of the sideways velocity of each dl, is simply the time rate of increase of the magnetic flux over the surface bounded by C due to the path C cutting lines of magnetic flux. Hence, the complete expression in brackets is the time rate of increase of the magnetic flux ϕ, over any surface S bounded by the closed path C, *due to the changing magnetic field and also due to the moving path cutting through the magnetic field*. Equation (8) is often written

$$\text{emf} = -\frac{d\phi}{dt} \qquad (9)$$

It is important to note carefully the distinction between Eqs. (5) and (9). Equation (5) is only the variational emf, and Eq. (9) is the sum of the variational and motional emfs. In Eq. (5), the partial time derivative of the magnetic flux ϕ is the rate of change of the flux due only to the time-changing magnetic field; in Eq. (9), the total time derivative is the rate of change of the flux due to the time-changing $\mathbf{B}$-field and also to the path cutting through the magnetic field. Of course, if the closed path is not cutting lines of magnetic flux, then Eqs. (5) and (9) are equivalent. It is also important to note that $d\phi/dt$ in Eq. (9) does not necessarily mean the total time rate of change of the flux ϕ over the surface S. For example, the flux over a surface S bounded by the closed contour C of the left-hand electric circuit of Fig. 1 is changing when the coil is being unwound by the rotation of the cylinder. However, as $\mathbf{B}$ is static there is no variational emf, and since the conductors are not cutting flux lines, there is no motional emf. Consequently, $d\phi/dt$ as used in Eq. (9) is zero even though the flux is changing with time. Note that $d\phi/dt$ in Eq. (9) was *defined* as representing the bracketed expression of Eq. (8), and $d\phi/dt$ must not be more broadly interpreted.

In the applications of the equations which have been presented, we must refer all flux densities and movements to a single specified coordinate system. In particular, the velocities are with respect to this system and are not relative velocities between conductors and moving lines of flux. Of course the coordinate system is arbitrarily selected, and *the relative magnitudes of the variational and motional emfs depend upon the selection*. Let us consider two examples.

Example 1 An electric generator is shown in Fig. 2. The parallel stationary conductors separated a distance l have a stationary voltmeter connected between them. The electric circuit is completed through a moving conductor that is connected electrically by means of sliding taps. This conductor is at $y = 0$ at time $t = 0$ and moves to the right with constant velocity $\mathbf{v} = v\mathbf{a}_y$. The applied flux density $\mathbf{B}$, represented in Fig. 2 by dots, is $B_o \cos \beta y \cos \omega t\, \mathbf{a}_x$. Unit vectors in the directions of the respective coordi-

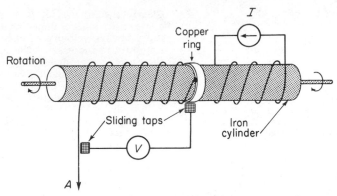

FIG. 1. The current generator produces a steady magnetic flux in the iron cylinder, which rotates as the wire is pulled at A.

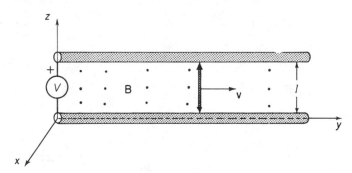

FIG. 2. Elementary electric generator.

nate axes are $\mathbf{a}_x$, $\mathbf{a}_y$, and $\mathbf{a}_z$. Find the instantaneous voltage across the voltmeter.

Solution. Let S denote the plane rectangular surface bounded by the closed electric circuit, with the positive side selected as the side facing the reader. The counterclockwise emf around the electric circuit is $-d\phi/dt$, with ϕ signifying the magnetic flux out of the positive side of S. As $d\mathbf{S} = l\,dy\,\mathbf{a}_x$, the scalar product $\mathbf{B} \cdot d\mathbf{S}$ is $B_o l \cos \beta y \cos \omega t\,dy$; integrating from $y = 0$ to $y = y_1$ gives

$$\phi = (B_o l/\beta) \sin \beta y_1 \cos \omega t \qquad (10)$$

with y_1 denoting the instantaneous y-position of the moving wire. The counterclockwise emf is found by replacing y_1 with vt and evaluating $-d\phi/dt$. The result is

$$\text{emf} = \omega B_o l/\beta \sin \beta vt \sin \omega t$$

$$- B_o l v \cos \beta vt \cos \omega t \qquad (11)$$

The variational (transformer) component is $-\partial\phi/\partial t$, which is determined with the aid of Eq. (10) to be $\omega B_o l/\beta \sin \beta y_1 \sin \omega t$, with $y_1 = vt$. This is the first component on the right side of Eq. (11). Note that y_1 was treated as constant when evaluating the partial time derivative of ϕ.

The motional emf is the line integral of $\mathbf{v} \times \mathbf{B}$ along the path of the moving conductor. As $\mathbf{v} \times \mathbf{B}$ is $-B_o v \cos \beta y_1 \cos \omega t\,\mathbf{a}_z$ and as $d\mathbf{l}$ is $dz\,\mathbf{a}_z$, we evaluate the integral of $-B_o v \cos \beta y_1 \cos \omega t\,dz$ from $z = 0$ to $z = l$, obtaining a motional emf of $-B_o l v \cos \beta y_1 \cos \omega t$. This component results from the cutting of lines of magnetic flux by the moving conductor.

If the voltmeter draws no current, there can be no electromagnetic force on the free electrons of the wires. Therefore, *the emf along the path of the metal conductors, including the moving conductor, is zero.* The total voltage of Eq. (11) appears across the voltmeter.

Example 2 Suppose the conductor with sliding taps in Fig. 2 is stationary ($\mathbf{v} = 0$) and located at $y = y_1$. Also suppose that the magnetic field $\mathbf{B}$ is produced by a system of steady currents in conductors (not shown in Fig. 2) that are moving with constant velocity $\mathbf{v} = v\mathbf{a}_y$. At time $t = 0$ the magnetic field $\mathbf{B}$ is $B_o \sin \beta y\,\mathbf{a}_x$. Determine the voltage across the voltmeter.

Solution. There is no motional emf because the conductors of Fig. 2 are stationary with respect to our selected coordinate system. However, the magnetic field at points fixed with respect to the coordinate system is changing with time, and hence there is a variational emf.

As the **B**-field at $t = 0$ is $B_O \sin \beta y \, \mathbf{a}_x$ and moving with velocity $v \, \mathbf{a}_y$, the **B**-field as a function of time is $B_O \sin [\beta(y - vt)] \, \mathbf{a}_x$. This is verified by noting that an observer at y_O at $t = 0$ moving in the y-direction with the velocity v of the moving current-carrying conductors, would have a y-coordinate of $y_O + vt$; hence according to the expression for **B**, he would observe a constant field. The magnetic current density is

$$\partial \mathbf{B}/\partial t = -\beta v B_O \cos \beta(y - vt) \, \mathbf{a}_x$$

The negative of the integral of this over the rectangular surface bounded by the electric circuit, with the positive side selected as the side facing the reader and with y limits of zero and y_1, gives the counter-clockwise emf. The result is

$$\text{emf} = B_O l v [\sin \beta(y_1 - vt) + \sin \beta v t]$$

This is the voltage across the meter.

CHARLES A. HOLT

References

Bewley, L. V., "Flux Linkages and Electromagnetic Induction," New York, Dover Publications, 1964.

Fano, R. M., Chu, L. J., and Adler, R. B. "Electromagnetic Fields, Energy, and Forces," New York, John Wiley and Sons, 1960.

Holt, C. A., "Introduction to Electromagnetic Fields and Waves, New York, John Wiley & Sons, 1963.

Moon, P., and Spencer, D. E., "Foundations of Electrodynamics," New York, Van Nostrand Reinhold, 1960.

Cross-references: ALTERNATING CURRENTS, CIRCUITRY, INDUCTANCE, INDUCTION HEATING, POTENTIAL.

INDUCTANCE

Inductance is a ratio of a magnetic flux Φ to an electric current i. The unit is the henry; 1 henry $\equiv$ 1 weber/ampere. The *mutual inductance* M of two circuits is defined as the ratio of the magnetic flux, linking with one circuit, to the current in the other

$$M_{12} = \frac{\Phi_2}{i_1}, \, M_{21} = \frac{\Phi_1}{i_2}$$

The *self-inductance* L of a single circuit is defined as the ratio of the flux linking the circuit to the current flowing in the circuit.

$$L = \frac{\Phi}{i}$$

In the absence of any magnetic material, M and L depend only on the geometry of the circuits concerned. The mutual inductance of two circuits 1 and 2 can be calculated from the Neumann equation*

$$M_{12} = M_{21} = \frac{\mu_0}{4\pi} \oint_1 \oint_2 \frac{\mathbf{dl}_1 \cdot \mathbf{dl}_2}{r}$$

where $\mathbf{dl}_1$ and $\mathbf{dl}_2$ are vector elements of length in circuits 1 and 2, respectively, and $\mathbf{r}$ is the distance between these two elements. Note that this expression is completely symmetrical with respect to the two circuits. The self-inductance is calculated from the same equation, where the elements $\mathbf{dl}_1$ and $\mathbf{dl}_2$ are now situated on the same circuit. The double integral is evaluated by first keeping the position of $\mathbf{dl}_2$ fixed and integrating $\mathbf{dl}_1/r$ around the circuit. The process is then repeated for all other elements such as $\mathbf{dl}_2$, and the results are summed.

The *external inductance* of a circuit is the part of its self-inductance which is due to flux lying outside the surface of the conductor while the *internal inductance* is the contribution from the magnetic flux within the conductor itself.

If the current, and therefore the magnetic flux associated with it, varies with time, an electromotive force (emf) will be induced in any circuit linked by the flux. This provides an alternative method of defining inductance in terms of the emf induced by a given rate of change of current.

$$\mathcal{E}_2 = -M \frac{di_1}{dt}$$

$$\mathcal{E}_1 = -M \frac{di_2}{dt}$$

$$\mathcal{E} = -L \frac{di}{dt}$$

The negative signs are used to imply that the direction of the emf is always such as to tend to oppose the change of current (Lenz's law). Note, however, that although the self-inductance must always be positive, the mutual inductance of two circuits may be either positive or negative. An inductance of 1 henry corresponds to an induced emf of 1 volt for a rate of change of current of 1 ampere/sec.

In order to maintain a current i, the induced emf $-L di/dt$ must be balanced by an equal and opposite applied voltage so that the total applied voltage is

$$v = Ri + L \frac{di}{dt}$$

and the power supplied to the circuit is

$$vi = Ri^2 + \frac{d}{dt} \left(\frac{1}{2} Li^2 \right)$$

*μ_0 is the permeability of free space; $\mu_0 = 4\pi \times 10^{-7}$ henrys/m.

where the resistance R is a measure of the power dissipated. This equation shows another way of interpreting inductance, namely as a measure of the amount of energy stored in the magnetic field when a given current flows. The stored energy is $\frac{1}{2}Li^2$ joules when L is measured in henrys and i in amperes.

Electric circuit theory is based on the use of sinusoidally varying currents and voltages. When the current varies sinusoidally with time, the rate of change of current has the same time waveform except for a phase shift of $\pi/2$ radians. This leads to the idea of a complex impedance, of which the real part is associated with the power lost from the circuit and the imaginary part is a reactance given by the ratio of the magnitude of the induced emf to that of the current. Since the phase of the applied voltage leads that of the current, the magnetically induced reactance is taken as being positive. We thus have

$$i = I \sin \omega t$$

$$v = Ri + L \frac{di}{dt}$$

The circuit impedance is Z where

$$Z = \frac{v}{i} = R + j\omega L$$

The self-inductance of a circuit can be regarded as a parameter which determines the inductive reactance presented to a sinusoidally varying current of given amplitude and frequency. In the same way, the mutual inductance determines the mutual reactance between two circuits.

Owing to magnetic hysteresis, the voltage and current time waveforms cannot both be sinusoidal if any magnetic material is present. Under these conditions, the reactance is defined as the ratio of the fundamental-frequency components of the voltage and current waveforms.

The sign of the mutual inductance can be specified in the following way. Suppose that the two circuits are connected in series. Then the total induced emf will be

$$-(L_1 + L_2 + 2M) \frac{di}{dt}$$

and the total reactance will therefore be

$$j\omega(L_1 + L_2 + 2M)$$

and will be greater than the sum of the reactances of the separate circuits when M is positive. A combination of two coils is said to be series aiding or series opposing, depending on whether they are connected so that M is positive or negative, respectively.

The *Q-factor* is used in describing the properties of an inductor. It is defined as

$$Q = \frac{\omega L}{R} = \tan \phi$$

where ϕ is the phase angle of the complex impedance Z. An alternative term is the *power-factor* $\cos \phi$, particularly when the emphasis is on power dissipated rather than on the damping of tuned circuits.

An impedance Z can be represented by an equivalent circuit consisting of a resistance R in series with an ideal *series inductance L*. An alternative is to start with the admittance $Y = 1/Z$ and to represent this by the parallel combination of a resistance R' and a *parallel inductance L'*.

$$R = \frac{R'}{1 + Q^2}$$

$$L = L\left(\frac{Q^2}{1 + Q^2}\right)$$

Note that, when Q is sufficiently large,

$$L' \simeq L$$

$$R' \simeq Q^2 R$$

Inductance Coil (Inductor). This is a device which is specially designed to possess inductance. The winding may have a ferromagnetic core composed of a dust-core material, metallic laminations or ferrite. An *air-cored* coil is one with no magnetic core.

Although it is the inductance which is of interest, the influence of the electric field often cannot be neglected and causes the effect known as self-capacitance. At relatively low frequencies, the inductor will behave as if an equivalent lumped capacitance were shunted across its terminals. This simple equivalent circuit fails as the frequency is approached at which the apparent lumped capacitance would resonate with the inductance of the coil. The problem then becomes one of electromagnetic wave propagation, and the concept of inductance is no longer relevant. The effect of a fixed parallel capacitance C is to reduce the Q-factor to Q_c, where

$$Q_c = Q\left[1 - \omega^2 LC\left(1 + \frac{1}{Q^2}\right)\right]$$

When Q is large, this becomes

$$Q_c \simeq Q(1 - \omega^2 LC)$$

The self-resonance effect is sometimes an advantage, e.g., in choke coils, where the object is to obtain a high impedance, irrespective of its phase angle. Generally however self-capacitance is an undesirable property, particularly in electrical networks in which the inductor forms part of a series resonant circuit.

The self-capacitance of a coil can be kept to a minimum by spacing the turns of the winding well apart and also, in the case of a multi-layer winding, by ensuring that wires which lie physically close together always belong to adjacent parts of the winding so that the potential difference between them is relatively small.

The inductance of some types of coil can be calculated directly, without recourse to the Neumann equation. An example is the *toroidal coil*, i.e., one consisting of a uniform winding around a ring-shaped former. In this case the magnetic flux exists in a well-defined magnetic circuit and the inductance is given, very nearly, by

$$L = \frac{\mu_0 N^2 A}{l}$$

where N is the number of turns in the winding and A and l are, respectively, the cross-sectional area and the length of the magnetic circuit. The same formula also holds for a long, thin *solenoidal coil*. Corrections must be applied unless both the diameter of each turn and the radial depth of the winding are small compared with the length of the magnetic circuit for the toroid or the length of the coil itself for the solenoid.

If the toroidal former is replaced by one made of a magnetic material having a relative permeability μ, the inductance will be increased by a factor μ. For coils of other geometrical shapes, where some of the flux linking the winding may lie partly or wholly outside the magnetic core, the presence of the latter will increase the inductance by a factor called the *effective permeability* μ_e. This can never exceed μ and may be considerably smaller than μ.

The relative permeability of a ferromagnetic material is not a constant but is a function of the instantaneous flux density. This increases power losses and causes distortion of the current and voltage time waveforms.

Power dissipation in a metallic core can be represented approximately by an equivalent series resistance R, where

$$\frac{R}{\mu_{ef} L} = c + h B_{max} + ef$$

and B_{max} is the peak flux density. The parameters c and h depend on the hysteresis properties of the core material. The parameter e is a measure of eddy-current losses in the core.

Ferrite materials are practically insulators; therefore eddy-current losses are negligible. For a ferrite, however, the relative permeability must be regarded as a complex quantity

$$\mu = \mu'(1 - j \tan \delta)$$

Both μ' and the dissipation coefficients $\tan \delta$ are functions of frequency.

The useful frequency range of coils with laminated cores is often limited by magnetic skin effect. This causes both the inductance and the Q-factor to start to fall as the depth of penetration of the electromagnetic field becomes comparable with the thickness of the laminations. The depth of penetration d is given by

$$\frac{2}{d^2} = \omega \mu_0 \mu \sigma$$

where σ is the conductivity of the core material.

The inductance of a coil with a magnetic core is reduced if there is a superimposed unidirectional polarizing flux, caused for example by a dc current in the winding. This may be an unwanted effect or it may be used as a means of controlling the inductance.

Transverse air gaps are often introduced into the magnetic circuit. There are several reasons for this. An air gap reduces the flux density for a given magnetizing force, thus reducing the inductance of a given coil. The power losses are, however, reduced at a greater rate than the inductance so that the power factor is improved. The performance of the inductor is made less dependent on the magnetic parameters of the core material. The waveform distortion due to hysteresis is also reduced. An air gap may be used to prevent saturation of the core by the polarizing flux when the coil is required to carry a dc current.

If the inductance of a coil, with a closed magnetic circuit of length l, is L, an air gap of length g will reduce the inductance to L_g where

$$L_g = L \cdot \left[\frac{\mu}{1 + \mu(g/l)} \right]$$

As the gap ratio g/l is increased, the inductance tends to become independent of μ, particularly when μ is large. Note however that the formula assumes that the presence of the air gap does not change the geometry of the flux distribution; an assumption which is generally only justified for relatively small gap ratios.

Dust-cored or ferrite-cored coils, for which the core eddy-current losses are small, are often wound with stranded wire. This is done to keep the eddy-current losses in the winding low by ensuring that the individual conductor strands have diameters which are small compared with the depth of penetration of the electromagnetic field into the material of which they are made. At very high frequencies, where this condition cannot be met, solid wire is often used and the diameter of the wire is increased to compensate for the fact that only part of its cross-sectional area is effective as a conductor.

V. G. WELSBY

References

Scott, W. T., "The Physics of Electricity and Magnetism," New York, John Wiley & Sons, 1959.
Plonsey, R., and Collins, R. E., "Principles and Appli-

cations of Electromagnetic Fields," New York, Mc-Graw-Hill Book Co., 1961.

Welsby, V. G., "The Theory and Design of Inductance Coils," London, MacDonald & Co., 2nd Edition, 1960.

Cross-references: INDUCED ELECTROMOTIVE FORCE, INDUCTION HEATING, TRANSFORMER.

INDUCTION HEATING

Induction heating is a technique for generating heat in electrically conductive articles by causing current to circulate in them by induction from an adjacent coil carrying alternating current.

The frequency of the current exciting the coil ranges from supply frequency, 60 Hz, up to typically 450 kHz. Except for 60-Hz applications, current may be obtained from rotary motor alternators, up to 10 kHz; magnetic multipliers using saturating cores, up to 540 Hz; valve oscillators, to over 1 MHz; and solid-state frequency changers up to 10 kHz. Coil-exciting voltages range from 400, 800, and 1 200 V in low-frequency systems (up to 10 kHz), to as much as 5 000 V for high-frequency valve oscillators.

Typical applications include general heating of metal parts, annealing, case and through hardening, melting in crucible-type furnaces, and heating of billets prior to forging.

In all cases of induction heating the fundamental principle is transformer action between the exciting coil and the workpiece which forms the secondary circuit. A typical arrangement is a solenoidal coil with a solid cylindrical workpiece. The currents induced in the charge flow in cylindrical paths centered on the longitudinal axis of the workpiece. Each of these paths has inductance and there is a consequent progressive decrease in current strength toward the center

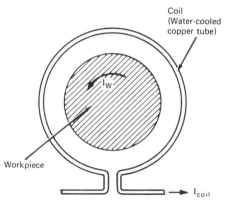

FIG. 1. Heating coil. I_w = current induced in material to be heated. I_{coil} = current through the water cooled coil.

TABLE 1. EFFECTIVE DEPTH OF CURRENT PENETRATION (in inches)

Frequency	50 Hz	1000 Hz	100 kHz
Copper	0.300	0.080	0.008
Aluminium	0.450	0.120	0.012
Brass	0.700	0.180	0.018
Steel	0.065	0.015	0.0015
Steel (above Curie)	3.200	0.750	0.075

of the workpiece. This effect increases with increase in frequency, and as the current at any depth from the surface determines the rate of heat generation at that depth, the frequency can be chosen to control the heat pattern.

The effective depth of current penetration is a function not only of frequency, but also magnetic permeability and specific resistivity. It can be expressed by the relation:

$$D = 1.98 \sqrt{\frac{\rho}{\mu \times f}}$$

where D = depth in inches, ρ = resistivity in $\mu\Omega$, μ = permeability, and f = frequency in Hz.

The very marked change in current depth in ferromagnetic material (steel) as it passes from its high permeability state through Curie temperature to its low permeability, higher resistivity state should be noted.

The exciting coil together with the workpiece can be represented by an inductor in parallel with a resistor. To achieve efficient operation when used with the power sources described it is necessary to compensate for the reactive KVA taken by the inductance by adding capacitance in parallel to bring the power factor close to unity. Once this is achieved the effective load resistance must then be matched to the source resistance to achieve maximum power transfer to the workpiece.

Induction heating makes possible power densities (KW/in.² of workpiece area) up to 100 KW/in.² compared with the maximum equivalent power density from an oxyacetylene flame of about 10 KW/in.².

P. J. BILLING

References

Simpson, P. G., "Induction Heating, New York, McGraw-Hill Book Co., 1960.

Tudbury, C. A. "Basics of Induction Heating," Vol. 1, 132 pp. and Vol. 2, 133 pp., New York, John Rider, 1960.

Vaughan, J. T., and Williamson, J. W., "Design of Induction Heating Coils for Cylindrical Nonmagnetic Load," *AIEE Transactions* 64, 587–592 (1945).

Vaughan, J. T., and Williamson, J. W., "Design of Induction Heating Coils for Cylindrical Magnetic Loads," *AIEE Transactions* 66, 887–892 (1947).

Baker, R. M., "Design and Calculation of Induction Heating Coils," *AIEE Transactions* 76, Part II, Apl. and Industry. 31–40 (1957).

"Heat Treating, Cleaning, and Finishing," in "Metals Handbook," 8th Ed., Vol. 2, p. 173, Metals Park, Ohio, American Society of Metals, 1962.

Cross-references: ALTERNATING CURRENTS; CONDUCTIVITY, ELECTRICAL; INDUCED ELECTROMOTIVE FORCE; INDUCTANCE.

INERTIAL GUIDANCE*

Inertial guidance is an on-board means to provide a steering control function that will hold a moving vehicle on a prescribed course, relying solely upon inertial measurements. *Inertial navigation* is an on-board means to determine the location of a vehicle relying solely upon inertial measurements. Hence, an inertial guidance system (IGS) employs an inertial navigation system (INS).

The steering function is formulated by comparing the location vector $\mathbf{Q}_{ins}$ (as determined by the INS) to the steering command vector $\mathbf{Q}_{sc}$. (See Fig. 1.) Note that solid lines represent signals while dotted lines represent physical or actual dynamics. In performing its task, the INS measures the vehicle acceleration vector $\ddot{\mathbf{Q}}$ and vehicle rate vector Ω. Employing a gyro-stabilized coordinate reference system, the INS computes the position or location vector $\mathbf{Q}_{ins}$. In actual practice, the vectors are handled as three scalar components, utilizing three channels each like that in Fig. 1.

Basic INS Essentially, the INS measures vehicle acceleration and integrates twice. See Fig. 2. Consider the X component of vehicle acceleration, $\ddot{X}$. The X accelerometer a_x measures this, producing an electrical signal. The signal is integrated twice, thus computing the X component of vehicle position. For a three dimensional system, there would be a similar channel each for the Y and Z components.

Stable Platform While Fig. 2 represents the basic INS, it is not practicable. It has several shortcomings—the lack of a coordinate system (implied in Fig. 1) being the primary one. In order to provide a coordinate system, it is necessary to physically orient the accelerometers appropriately. This can be accomplished by mounting the accelerometer on a stabilized platform (a surface whose orientation in space is maintained by physical means). For discussion purposes, let us employ a pendulum and a compass. (See Fig. 3.) For this hypothetical case, the X accelerometer would be level and pointed north.

The price paid for this particular form of stabilization is that the accelerometer rotates. (See Fig. 4). As a result, it senses the acceleration of constraint $\ddot{C}$ (Coriolis, centripital accelerations, etc.). This must be computed and then it must be subtracted from the accelerometer signal. (Return to Fig. 3.) Following this with an integration determines the vehicle velocity $\dot{X}$. Division by earth's radius R yields the angular rate $\dot{\lambda}$ (latitude rate about the earth), which upon integration provides the instantaneous latutude λ. (See Fig. 5.)

Schuler Tuned INS The simple pendulum used in Fig. 3 is unsatisfactory, since vehicle acceleration would cause the pendulum to seek a false vertical. Einstein's special theory of relativity states that it is impossible to distinguish between inertial reaction (acceleration) forces and mass attraction (gravitational) forces. Consequently, if a real pendulum were employed, the accelerometer would erroneously include a component of gravity. With an incorrect vertical, and an incorrect measure of acceleration, there would be no way to correct either.

Instead of a simple pendulum, a more sophisticated scheme is required. The pendulous effect is achieved by means of the accelerometer and a gyroscope. The same accelerometer will serve in two capacities (since they cannot be separated anyway). It will read the vehicle acceleration and act as part of the vertical seeking device. (See Fig. 6.) The vehicle rate $\dot{\lambda}$ is applied to the gyro, which precesses at that rate. After a time t, the platform turns through an angle λ which is the integral of $\dot{\lambda}$. Thus, when the system is operating properly, the platform is physically turned through the same angle that the vehicle has traveled around the earth. Consequently, the platform is still level. The INS is an analog of the actual system, maintaining a true vertical and measuring true acceleration. As such, the INS acts like a pendulum whose period T is given by

$$T = \frac{1}{2\pi}\sqrt{\frac{R}{g}}\,;\qquad(1)$$

whereupon, applying numerical values, the period becomes

$$T = 84 \text{ minutes.}\qquad(2)$$

This is referred to as a *Schuler pendulum* or a *Schuler tuned system*.

Complete INS The preceding section indicates one means by which the INS may be instrumented. Other schemes involve considera-

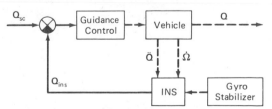

*Illustrations and text excerpted from Refs. 1 and 2.

FIG. 1. Inertial guidance system.

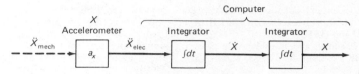

FIG. 2. Basic INS (X channel).

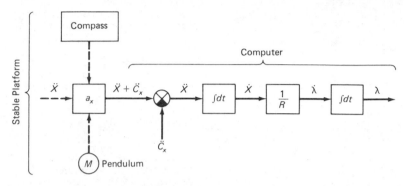

FIG. 3. INS stabilized with pendulum and compass.

tions such as: the shape of the earth, free gyros, platform fixed in space, platform affixed to the vehicle frame (strap down), free azimuth, slaved to arbitrary heading, etc. A universal model is shown in vector form in Fig. 7. The boxes marked G and T are universal transformations involving gravity and terrestrial coordinates, respectively.

A specific INS, one that is level and north-slaved, is shown in Fig. 8. Note that there are two Schuler loops, one each for the X and Y axes. The third axis, Z, is aligned along the

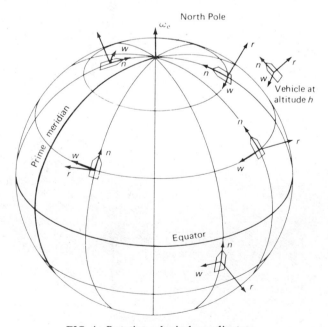

FIG. 4. Rotating spherical coordinates.

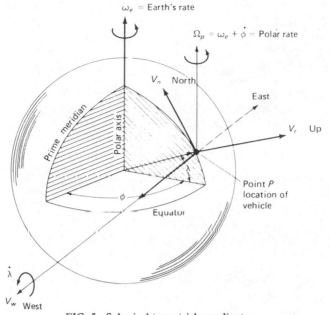

FIG. 5. Spherical terrestrial coordinates.

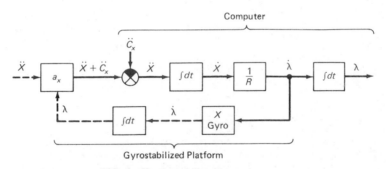

FIG. 6. X axis stabilized by gyroscope.

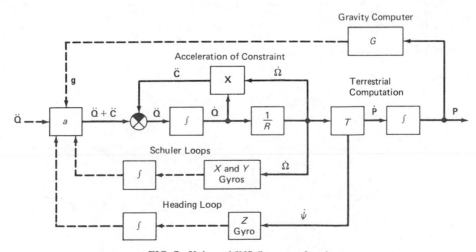

FIG. 7. Universal INS (in vector form).

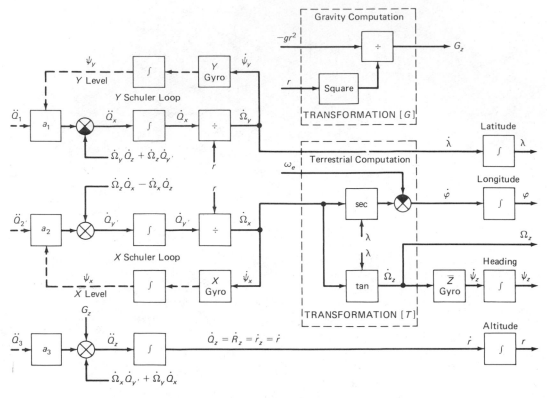

FIG. 8. A level north-slaved inertial navigation system.

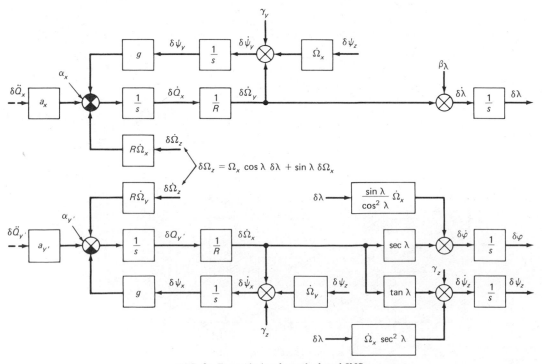

FIG. 9. Errors in level north slaved INS.

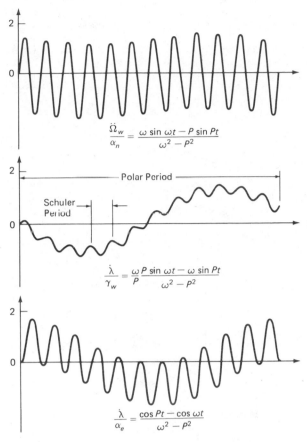

$$\frac{\ddot{\Omega}_w}{\alpha_n} = \frac{\omega \sin \omega t - P \sin Pt}{\omega^2 - P^2}$$

$$\frac{\dot{\lambda}}{\gamma_w} = \frac{\omega}{P} \frac{P \sin \omega t - \omega \sin Pt}{\omega^2 - P^2}$$

$$\frac{\dot{\lambda}}{\alpha_e} = \frac{\cos Pt - \cos \omega t}{\omega^2 - P^2}$$

FIG. 10. Typical errors in level north-slaved INS.

earth's radius vector (up). The accelerometer for this channel senses the full magnitude of gravity, requiring that this be computed accurately. The latitude channel λ has already been described. The longitude channel ϕ differs somewhat, since it requires some trigonometric manipulations. Refer to Fig. 5. A rate $\dot{\Omega}_x$, about a vector that is tangent to the earth and pointing north, is multiplied by sec λ in order to obtain the polar rate $\dot{\Omega}_p$, where,

$$\dot{\Omega}_p = \omega_e + \dot{\phi}. \qquad (3)$$

The polar rate includes the earth's rate of rotation ω_e, which must be subtracted in order to derive the longitude rate $\dot{\phi}$. Upon subsequent integration, this reveals the longitude ϕ. The azimuth Z gyro torquing rate $\dot{\Omega}_z$ is generated by multiplying $\dot{\Omega}_x$ by tan λ.

System Errors Upon examination of the INS in Fig. 8, it should be pointed out that many quantities are fed back, making the INS a multi-coupled system. Thus, error sources in various parts of the system will effect many other parts. Also, since these are fed back and cross coupled, the errors may either grow unbounded, or they may be self-limiting.

In order to quantitatively accomplish a complete error analysis, first perturb the system, as shown in Fig. 9. Approximate the error sources in the system as phantom inputs (step functions) by the following six quantities: accelerometer errors α_x and α_y; computer error β_λ; and gyro drift rates γ_x, γ_y, and γ_z. Assume that the vehicle travels at constant altitude ($\dot{R} = 0$) and at constant latitude ($\dot{\lambda} = 0$). Also assume that the Schuler loops limit the tilt errors to small values. The closed loop cross coupled analysis is treated with matrices,* and the resulting errors are shown in Fig. 10. Note the following:

ω = Schuler frequency whose period is given by Eq. (1), which is equal to 84 minutes at sea level.

P = Polar frequency $\dot{\Omega}_p$ given by Eq. (3), which is about 2 rev/day for a vehicle traveling east at the equator at about 1600 km/hr.

It should be noted that a gyro drift rate equal to 0.01°/hr results in maximum velocity

*See chapters 8 and 9 of Ref. 2.

errors of 2 km/hr; maximum latitude error of 2 km; maximum longitude error of 6 km; maximum platform tilt off the vertical of 2 arc-min; and maximum heading error of 6 arc-min. The most significant conclusion is that *no errors are unbounded*—a bonus provided by the multicoupling in the INS.

IRA COCHIN

References

1. Cochin, Ira, "Analysis and Design of Dynamic Systems," New York, Harper and Row, 1980.
2. Cochin, Ira, "Analysis and Synthesis of INS in Universal Terms," Ph.D. thesis Cooper Union, New York, 1969.
3. Bachman, K. L., "Ring Laser Gyro Navigator," *Navigation* 25(2), 142–152 (Summer 1977).
4. Daniel, H. L., and Hulslander, D. B., "Standard INS Program Status," *Navigation*, 27(1), 65–71, (Spring 1980).
5. Harrison, J. V., "Reliability and Accuracy Prediction for a Redundant Strapdown Navigator," *Guidance and Control*, 4(5), 523–529 (Sept. 1981).

Cross-references: ASTRONAUTICS, GYROSCOPE, MECHANICS, ROTATION-CURVILINEAR MOTION.

INFRARED RADIATION

The region of the electromagnetic spectrum between the wavelength limits 0.7 and 1000μm (7×10^{-5} and 1×10^{-1} cm) has become known as infrared radiation. The lower wavelength limit is set to coincide with the upper limit of the visible radiation region. Radiation of wavelength greater than 1000μm is generally thought of as the microwave spectrum. Both limits are arbitrary, and represent no change in characteristics as they are passed. Conventionally, the region between 0.7 and 1.5μm is called the *near infrared region*; that between 1.5 and 20μm, the *intermediate infrared region*; and that between 20 and 1000μm, the *far infrared region*.

For many applications, the location of infrared radiation in the spectrum is described by its wavelength in micrometers, μm, (1μm = 10^{-4} cm). In applications where the relative energy of the radiation is of interest, the *wave number*, σ, is used. The wave number is defined as the reciprocal of the wavelength, λ, in centimeters, and is expressed in units of cm^{-1} (called the kayser). This quantity is used more commonly than the frequency ν of the radiation, which is related to σ as follows:

$$\sigma = \nu/c \qquad (1)$$

where c is the velocity of light.

Infrared radiation is produced principally by the emission of solid and liquid materials as a result of thermal excitation and by the emission of molecules of gases. Thermal emission from solids is contained in a continuous spectrum, whose wavelength distribution is described by the relation

$$M_\lambda \, d\lambda = \frac{2\pi \, c^2 \, h\epsilon_\lambda}{\lambda^5} \frac{1}{e^{ch/\lambda kT} - 1} \, d\lambda \qquad (2)$$

where

M_λ = spectral radiant exitance of the solid into a hemisphere in the wavelength range from λ to $(\lambda + d\lambda)$

h = Planck's constant = 6.62×10^{-27} erg sec

ϵ_λ = spectral emittance

k = Boltzmann's constant = 1.38×10^{-16} erg/K

T = absolute temperature of the solid emitter, K.

The spectral emittance, ϵ_λ, is defined as the ratio of the emission at wavelength λ of the object to that of an ideal blackbody at the same temperature and wavelength. When ϵ_λ is unity, Eq. (2) becomes the Planck radiation equation for a blackbody. The distribution of radiant exitance with wavelength for blackbody radiators at different temperatures is shown in Fig. 1. It is apparent from the figure that blackbody radiation from emitters at temperatures below about 2000 K falls predominantly in the infrared region. An emitter which exhibits a constant value less than unity of spectral emit-

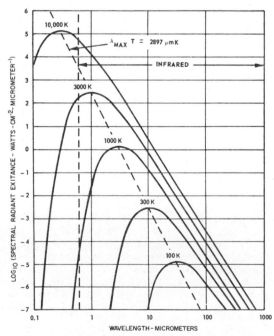

FIG. 1. Spectral radiant exitance of a blackbody at various temperatures.

tance at all wavelengths is called a *gray-body* radiator. Most solid radiators show a general decrease in spectral emittance with increasing wavelength in the infrared; however, over limited spectral ranges, many materials are approximately gray-body radiators. Radiators which approach the characteristics of ideal blackbodies can be made in the form of uniformly heated cavities. A relatively small aperture, through which the cavity can be observed, serves as the source of blackbody radiation.

Infrared radiation is also observed as emitted from excited molecules of gases. Many of the energy transitions which take place in gases excited thermally or electrically result in radiation emission in the infrared region. Gaseous emission differs in character from solid emission in that the former consists of discrete spectrum lines or bands, with significant discontinuities, while the latter shows a continuous distribution of energy throughout the spectrum. The predominant source of molecular radiation in the infrared is the result of vibration of the molecules in characteristic modes. Energy transitions between various states of molecular rotation also produce infrared radiation. Complex molecular gases radiate intricate spectra, which may be analyzed to give information of the nature of the molecules or of the composition of the gas.

The propagation of infrared radiation through various media is, in general, subject to absorption which varies with the wavelength of the radiation. Molecular vibration and rotation in gases, which are related to the emission of radiation, are also responsible for resonance absorption of energy. The gases in the atmosphere, for example, exhibit pronounced absorption throughout the infrared spectrum. The principal gases of the atmosphere, nitrogen and oxygen, do not absorb significantly in the infrared region. However, the lesser constituents, water vapor (H_2O), carbon dioxide (CO_2), and ozone (O_3), are responsible for strong absorption in the infrared. The absorption of radiation is so prevalent that those spectral bands in which relatively little absorption occurs are identified as *atmospheric windows*.

Solid and liquid materials show, as a rule, strong absorption in the infrared. There are, however, many solids which transmit well in broad regions of the infrared spectrum. Many materials, such as water and silica glasses, which show little absorption in the visible, are opaque to infrared radiation at wavelengths greater than a few microns. Many of the electrically insulating crystals, such as the alkali halides and the alkaline-earth halides, which transmit well in the visible, also are transparent to much of the near and intermediate infrared spectrum. Several of the semiconductor materials absorb strongly in the visible, but become transparent in the infrared beyond certain wavelengths characteristic of the semiconductor.

Detection of the presence, distribution and/or quantity of infrared radiation requires techniques which are, in part, unique to this spectral region. The frequency of the radiation is such that essentially optical methods may be used to collect, direct, and filter the radiation. Transmitting optical elements, including lenses and windows, must be made of suitable materials, which may or may not be transparent in the visible spectrum. Table 1 gives characteristics of several transmitting materials suitable for use in infrared optical systems. To avoid chromatic aberration, reflecting mirrors are commonly used in infrared optical systems to focus and deviate the radiation when broad spectral bands are observed. Filters for the infrared are designed and constructed like those for the visible, except for the choice of materials and, in the case of interference filters, the thickness of the layers.

The detector element for infrared represents the most unique component of the detection system. Photographic techniques can be used for the near infrared out to about $1.3 \mu m$. Photoemissive devices, comparable to the visible- and ultraviolet-sensitive photocells, are available with sensitivity also extending to about $1.3 \mu m$. The intermediate infrared region is most effectively detected by photoconductors. These elements, photosensitive semiconductors, are essentially photon detectors, which respond in proportion to the number of infrared photons in the spectral region of wavelength shorter than the cut-off wavelength. This cut-off wavelength corresponds to the minimum photon energy necessary to overcome the forbidden gap of the semiconductor. A number of sensitive photoconductors are available with spectral cutoff at various wavelengths in the infrared. Photoconductors are employed as resistive elements, as photovoltaic *p-n* diodes, or as photoelectromagnetic elements, according to the particular electrical advantage to be gained. All spectral regions from ultraviolet through visible, infrared, and microwaves, can be detected by an appropriately designed thermal element, which responds by being heated by the absorption of the incident radiation. In the infrared, thermal detectors take the forms of thermocouples, bolometers, and pneumatic devices. The thermal elements, in general, are not as sensitive or as rapidly responding as photoconductors in spectral regions where they both respond. However, the broad spectral response and uniform energy sensitivity characteristics make them highly useful. Two-dimensional image information is obtained in the infrared by one-dimensional scanning of a linear array of detectors, or by a two-dimensional array of detector elements which are scanned electronically or by electron beam. A "push-broom" scan arrangement of a row of detectors is commonly used for the preparation of infrared maps of the terrain. "Staring" sensors, with a

TABLE 1. INFRARED TRANSMITTING MATERIALS

Material	Useful Transmission Region (μm)	Refractive Index Near Transmission Peak	Special Characteristics
Optical glasses	0.3–2.7	1.48–1.70	Best for near infrared
Fused silica	0.2–3.5–4.5	1.43	Some types show absorption near 2.7μm
Arsenic trisulfide	0.6–12.0	2.4	A glass; subject to striations
Calcium aluminate	0.3–5.5	1.8	A glass; subject to attack by water
Sapphire	0.17–6.0	1.7	Single crystal, hard, refractory
Silicon	1.1–>20	3.4	Low density; opaque to visible
Germanium	1.8–>20	4.0	Opaque to visible
NaCl	0.2–15	1.52	Water soluble
KBr	0.21–27	1.54	Water soluble
LiF	0.11–6	1.35	Low solubility in water
CaF$_2$	0.13–9	1.41	Insoluble
Thallium bromide-iodide (KRS-5)	0.5–40	2.38	Fairly soft; cold flows
AgCl	0.4–25	2.0	Soft; cold flows
Irtran 1	0.5–9	1.3	Polycrystalline MgF$_2$
Irtran 2	0.4–14.5	2.2	Polycrystalline ZnS
Irtran 3	<0.4–11.5	1.3	Polycrystalline CaF$_2$
Irtran 4	0.5–22	2.4	Polycrystalline ZnSe
Irtran 5	<0.4–9.5	1.7	Polycrystalline MgO
Irtran 6	0.9–31	2.7	Polycrystalline CdTe

two-dimensional array of detectors, are used for imaging fields-of-view in which the object of interest may appear unpredictably. Imaging sensors may contain built-in amplification in the form of charge-coupled devices or electron beam scan with electron multiplication. Thermal-type response for infrared detection is also obtained by means of pyroelectric materials, such as triglycine sulfate, which change their surface charge with temperature. This type of detector shows wide spectral band response with short time constant at room temperature. Table

2 gives representative characteristics of several commonly used infrared detectors.

The most common application of infrared radiation is, of course, radiant heating. Solid radiators, such as hot tungsten filaments, alloy wires, and silicon carbide rods are employed extensively as sources of infrared to provide surface heating by radiation.

Infrared spectroscopy has become a powerful analytical tool in the chemistry laboratory. Organic molecules, in general, contain interatomic valence bonds which exhibit character-

TABLE 2. INFRARED DETECTORS

Detector (operating temperature)	Region (μm)	Specific Detectivity, D* at Peak (cmHz$^{1/2}$w^{-1})	Time Constant (sec)	Special Features
Si(295K)	Visible–1.0	2×10^{12}	5×10^{-6}	Photovoltaic crystal
PbS (295 K)	Visible–2.8	8×10^{10}	2×10^{-4}	Thin-film photoconductor
PbSe (195 K)	Visible–5.6	2×10^{10}	2×10^{-3}	Thin-film photoconductor
InSb (77 K)	1–5.6	10^{11}	$<2 \times 10^{-7}$	Photovoltaic crystal
(Hg · Cd)Te (77 K)	2–14	5×10^{9}	5×10^{-7}	Spectral cut-off varies with alloy composition
Ge (Hg doped) (25 K)	1–16	2×10^{10}	$<10^{-6}$	Photoconductor crystal
Ge (Cu doped) (5 K)	1–29	3×10^{10}	$<10^{-6}$	Photoconductor crystal
Ge (Zn doped) (5 K)	1–40	3×10^{10}	10^{-8}	Photoconductor crystal
Thermistor bolometer (295 K)	All	2×10^{8}	10^{-3}–10^{-2}	Flake of mixed oxides
Golay cell (255 K)	All	2×10^{9}	1.5×10^{-2}	Pneumatic
Thermocouple (295 K)	All	2×10^{8}	1.5×10^{-2}	Used in spectrometers
Pyroelectric (295 K)	All	3×10^{8}	2×10^{9}	High impedance

istic resonance frequencies which can be identified in the absorption spectrum of the material in gaseous form. Such information can be used to study the structure of complex molecules. It also serves in aiding the identification of the presence of known valence bonds in chemical analysis. Most absorption lines and bonds due to molecular vibrations fall in the frequency range 500 to 5000 cm^{-1} (wavelength range 2 to 20μm). A large quantity of data has been gathered on the detailed absorption spectra of many gaseous materials. The characteristic spectra of many organic molecules are such that identification of the presence of the molecules, as well as the presence of particular radicals within the molecules, can be readily observed. Petroleum chemistry, for example, has been greatly aided by the application of infrared spectroscopy to the identification of many of the complex constituents in petroleum products.

Observation of infrared absorption spectra is carried out by means of an infrared spectrophotometer, in which the transmission of monochromatic radiation by a gaseous sample in a cell is compared with that of a blank cell, while the wavelength of the radiation is scanned through the spectral range of interest. Prism dispersing elements are usually used in the infrared, rather than gratings, because of the difficulty with the latter of separating the several orders in the wide spectral range covered. Far infrared spectroscopy is complicated by the omnipresence of background and scattered radiation of shorter wavelength emitted inside the instrument at room temperature. Special techniques of filtering must be employed to eliminate the effects of the short-wavelength radiation.

Optical-electronic devices of many varieties have been designed to determine the direction of weakly radiating remote objects by means of detection of their infrared emission. Military applications have been found which have been made possible uniquely by this technique. Missiles can be guided to their target by infrared detection of the self-emission of heated segments of the target. Detailed maps of the earth's surface can be made from aircraft at night by observing the varying infrared emission of the ground. Personnel can be detected in total darkness by the infrared radiation they emit as warm objects. Space applications have included remote sensing of terrestrial surface characteristics for weather mapping, pollution detection and agricultural conditions detection from satellites. Nondestructive testing of electronic components, heat-processing plants, and thermal insulation effectiveness are now commonplace applications of infrared radiation detection.

Such devices require the detection of low-level radiation in the intermediate infrared region. Optical lenses or mirrors are used to collect the observed radiation and concentrate it onto the sensitive infrared detector. High-gain, low-noise electronic amplifiers must be provided to increase the weak signal from the detector to a level which can be used to operate controls or displays, as demanded by the application. Optical filtering is applied in order to restrict the observed spectral region to one in which the target is effectively detected, with a minimum of interference from radiation from its background. The wavelength of detection is such that angular resolution capability, as set by diffraction, is much greater with infrared devices than that of radar devices. Detection of targets at great distances through intervening atmosphere is more effective in the infrared than in the visible because of the much lower atmospheric scattering in the infrared.

Detailed discussions of the characteristics, detection and applications of infrared radiation may be found in the references.

R. H. MCFEE

References

Jamieson, J. A., McFee, R. H., Plass, G. N., Grube, R. H., and Richards, R. G., "Infrared Physics and Engineering," New York, McGraw-Hill Book Co., 1963.

Smith, R. A., Jones, T. E., and Chasmar, R. P., "The Detection and Measurement of Infrared Radiation," Fair Lawn, N.J., Oxford University Press, 1957.

Herzberg, G., "Infrared and Raman Spectra of Polyatomic Molecules," New York, Van Nostrand Reinhold, 1945.

Szymanski, H. A., and Alperts, N. A., "IR: Theory and Practice of Infrared Spectroscopy," Plenum Press, 1964.

Kruse, P. W., McGlauchlin, L. D., and McQuistan, R. B., "Elements of Infrared Technology," New York, John Wiley & Sons, 1962.

Hudson, R. D., Jr., "Infrared System Engineering," New York, John Wiley & Sons, 1969.

Wolfe, W. L., and Zissis, G. J. (Eds.), "The Infrared Handbook," Ann Arbor, MI, Environmental Research Institute of Michigan, 1978.

Keyes, R. J. (Ed.), "Optical and Infrared Detectors," Second Edition, Heidelberg, Springer-Verlag, 1980.

Martin, A. E., "Infrared Interferometric Spectrometers," Amsterdam, Elsevier Scientific Publishing Co., 1980.

Cross-references: ABSORPTION SPECTRA; LIGHT; RADIATION, THERMAL; SPECTROSCOPY.

INTERFERENCE AND INTERFEROMETRY

Interference is the term used to denote the physical effects of superimposing two or more waves of the same wavelength. *Interferometry* is the technique of measurement using these effects. Interference occurs whenever waves emanating from sources which have a constant phase relationship (i.e., which are coherent) are present at the same place. Depending on the phase difference between the waves, the interference may be constructive, so that the waves reinforce, or it may be destructive, so that the waves cancel each other. It may seem paradoxical that two waves—e.g., sound waves—can combine at one place to produce no effect—e.g., silence. However, the effects are easily observable under the proper conditions and are explicable by elementary theories.

Interference can occur for all wave phenomena. Vibrational waves such as sound or water waves and electromagnetic waves such as radio and light waves all behave similarly. Interference is a consequence of the *principle of superposition*. That principle states that the resultant effect of two or more waves—that is, the atomic displacements in the case of vibrational waves or the electric and magnetic fields in the case of electromagnetic waves—is simply the algebraic sum of the effects that the individual waves would produce alone. Superposition holds as long as the wave amplitudes are within the range of the normal, linear response of the propagating medium. It always holds when electromagnetic waves traverse a vacuum. Coherence of the sources is essential to the observation of interference (see COHERENCE). If the sources of the waves do not have a constant phase relationship, the superposition of the waves, while instantaneously producing a rapidly varying pattern of reinforcement and cancellation, will, on the average, produce no observable interference effects.

A simple illustration of interference is shown in Fig. 1. Concentric circles represent wavefronts moving out, at regular intervals (with wavelength λ), from the two sources S_1 and S_2. The dark areas represent the minima of the wave patterns and the light areas between them represent the maxima. The overlapping patterns show bands of reinforcement alternating with bands of cancellation. If the sources were light sources, a screen placed along the right edge of the figure would display alternating light (L) and dark (D) bands. Note that there is a light band at the center, equidistant from the two sources. Because the path of each wave from its source to that point is the same, there is no phase difference between them and constructive interference results. Away from the center, there is a path difference and hence a phase difference. When that path difference is an integral number of wavelengths, constructive interference results; when the path difference is an odd number of half-wavelengths (e.g., $\lambda/2$, $3\lambda/2$, $5\lambda/2$, . . .), there is destructive interference.

Figure 2 is a schematic representation of Thomas Young's classic experiment performed in 1802. His two sources were closely spaced pinholes which were both illuminated by light from a single pinhole (SL). Diffraction at this pinhole ensured that the two sources were coherent. Young placed a screen some distance away and observed a series of alternating bright and dark fringes. The fringes would vanish if either of the sources was covered; they would be replaced by a nearly uniform illumination of the screen. By this experiment, Young provided the first firm experimental evidence for the wave nature of light. The competing corpuscular theory of light, advocated by Isaac Newton a century earlier, still had many adherents. Young was also able to calculate the wavelengths of various colors of light; these were the first such measurements.

Theory A traveling wave can be represented mathematically by

$$M(x, t) = a \cos \left(2\pi \nu t - 2\pi \frac{x}{\lambda} + \alpha \right) \quad (1)$$

where $M(x, t)$ is the magnitude of the wave at distance x from the source and at time t, a is the amplitude, which is the maximum magnitude, ν is the frequency, λ is the wavelength, and α is the phase. The wave velocity $v = \lambda\nu$. The intensity of the wave, which is the observable quantity, is proportional to the square of the amplitude.

To calculate the effects of interference, one considers the sum of two waves of the same frequency and wavelength. Denoting the two

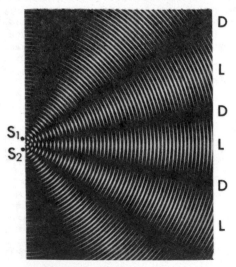

FIG. 1.

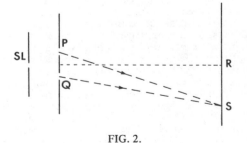

FIG. 2.

waves by subscripts 1 and 2, one obtains a resultant magnitude:

$$M = M_1 + M_2$$

$$= a_1 \cos (2\pi vt - 2\pi x_1 /\lambda + \alpha_1)$$

$$+ a_2 \cos (2\pi vt - 2\pi x_2 /\lambda + \alpha_2). \quad (2)$$

For simplicity, one considers the case of two sources of equal amplitude ($a_1 = a_2 = a$), which are in phase ($\alpha_1 = \alpha_2$). This situation is exemplified by Young's experiment.

Using the trigonometric relationship for the sum of the cosines of two angles, it is found that the amplitude A of the resultant wave is:

$$A = 2a \cos (\theta /2) \quad (3)$$

where $\theta = 2\pi(x_1 - x_2)/\lambda$ is the total phase difference between the waves at the point of interference. The quantity $x_1 - x_2$ is the difference in the paths from the sources to that point. The amplitude $A = \pm 2a$ when $\theta = 2m\pi$ ($m = 0, 1, 2, 3, \ldots$), so that $x_1 - x_2 = m\lambda$; $A = 0$ when $\theta = (2m + 1) \pi$, so that $x_1 - x_2 = (m + \frac{1}{2})\lambda$.

Because the intensity is proportional to A^2, the peak intensity is $4a^2$—four times that due to either wave acting alone and twice that due two incoherent sources (which would not interfere). Peaks in intensity occur whenever $A = \pm 2a$; i.e., whenever $x_1 - x_2 = m\lambda$, as stated above.

In Young's experiment (Fig. 2), the path difference $x_1 - x_2 = PS - QS$ may be easily calculated. Point S denotes the location of the first ($m = 1$) peak in intensity. Let $PQ = s$ and $RS = y$; let D be the distance of the screen from the plane of the slits. Then, using the Pythagorean theorem:

$$PS^2 - QS^2 = \left[D^2 + \left(y + \frac{s}{2}\right)^2\right]$$

$$- \left[D^2 + \left(y - \frac{s}{2}\right)^2\right] = 2ys$$

$$PS - QS = 2ys/(PS + QS).$$

For all cases of practical interest, $y \ll D$, so that $(PS + QS) \approx 2D$. At S, $PS - QS = x_1 - x_2 = \lambda$, so that

$$\lambda \approx ys/D. \quad (4)$$

Equation (4) provided Young with the basis for the first experimental determination of the wavelength of light. His results were remarkably accurate, yielding 420–700 nm as the range of visible wavelengths.

Practical problems made Young's experiment difficult to perform. For visible light, $\lambda \approx 500$ nm. If the separation of the sources $s = 1$ mm and the distance to the screen $D = 2$ m, the separation of the fringes $y \approx 1$ mm. Light sources available to Young were weak and slit separations of less than 1 mm were difficult to attain. This resulted in fringes that were weak and indistinct. The experimental results, while historically decisive, were not immediately convincing. However, over the following few years, refinements were made by Fresnel and others. These permitted greatly reduced separation of the sources which, in turn, increased the spacing and observability of the fringes. In addition, Fresnel produced a more complete theoretical formulation of interference.

The logical extension of Young's experiment to more than two slits yields a tremendously useful device, the diffraction grating. A grating consists of a large number of equally spaced rulings. (See DIFFRACTION BY MATTER AND DIFFRACTION GRATINGS.) Analysis of its properties, along the same lines which led to Eq. (4) above, predicts very narrow fringes. These occur in the same places as those produced by two slits whose separation is the same as that of the rulings of the grating. Narrow fringes permit the resolution of closely spaced spectral features. Many spectrometers and spectrophotometers used in the infrared, visible, and ultraviolet spectral ranges incorporate diffraction gratings as their central elements for spectral analysis.

Thin Film Interference An important application of the theory is in the analysis of interference which occurs in thin films of transparent media. The most common example of this is the interference of visible light in soap bubbles and in oil slicks, which produces a range of colors in reflection. These may be explained easily by the use of Eq. (3) and reference to Fig. 3. Here, light from the source reflects from the upper surface of the film at A and from the lower surface at B. The two coherent sources in this case are the two reflected images of the source. For light that is incident nearly perpendicular to the film, the geometrical difference between lengths of the two paths $x_1 - x_2 = 2d$. The optical path difference is $2dn$, where n is the index of refraction of the film (see REFRACTION). One additional complication is the phase reversal (corresponding to a phase shift of π) that occurs whenever light is reflected from a medium whose index of refraction is greater than the index of the medium in which the light is propagating. This is exactly the case at point A in Fig. 3; no such

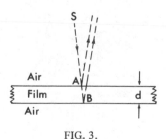

FIG. 3.

reversal occurs at point B since the index of refraction of air ($n = 1$) is less than that of the film. The condition for constructive interference remains a phase difference $\theta = 2m\pi$ [in Eq. (3)]; however, θ is now the sum of two terms:

$$\theta = \pi + 2\pi(2nd/\lambda). \qquad (6)$$

The first term is due to the phase reversal at A, the second to the optical path difference $2nd$.

Equation (6) has one surprising prediction: for very thin films, there is no light reflected. If $d/\lambda \ll 1$, $\theta \approx \pi$ so that $A \approx 0$ in Eq. (3). This effect is easily demonstrated by viewing in reflected light a film of soapy water held vertical in a loop of wire. As the fluid drains downward, and the thickness of the film decreases, the film appears to change color. The portion of the film at the top of the loop seems to disappear (no light is reflected from it) once its thickness is reduced to much less than the wavelength of light.

Equations (3) and (6) can be used to predict the apparent color of a thin film. For example, a film of soapy water whose thickness is 300 nm, and whose index of refraction $n = 1.40$ produces a phase difference in reflection of $\theta = \pi + 2\pi$ (840 nm/λ). Setting this equal to $2m\pi$ (for integer m) to obtain the conditions for constructive interference and solving for λ, it is found that only $m = 2$ gives a solution lying within the visible spectrum: $\lambda = 560$ nm. This corresponds to a yellow-green shade; if the film is viewed in reflection, it will appear that color.

Soap bubbles and oil slicks vary in thickness over their surfaces. Depending on the thickness, different spectral regions are enhanced by interference and brilliant bands of various colors are observed. Each continuous band corresponds to a region of constant film thickness.

A very common practical application of thin film interference is the "anti-reflection" coating used on lenses in optical instruments and cameras. The coating is a thin layer of transparent material whose thickness is carefully selected to minimize reflection (and thus maximize transmission) in the visible. The analysis is similar to that which led to Equation (6).

The Fabry-Perot interferometer is another application of thin film interference, one which leads directly to the topic of interferometry. It consists of two parallel plates of glass or quartz (see Fig. 4). Their inner surfaces are optically flat and partially silvered to enhance their reflectivity. Light from the source (S) is collimated by lens L_1 and is focused by lens L_2. The high reflectivity results in a large number of reflected images of the source combining to interfere in the focal plane (F) of lens L_2. As in the case of the diffraction grating, this produces very sharp interference fringes, permitting high resolution measurements. Optical filters based on the Fabry-Perot interferometer are used to produce nearly monochromatic light from polychromatic sources.

Interferometry Interferometers are instruments which utilize the principle of interference to make measurements. As in the case of the Fabry-Perot interferometer, a single wave is split into two or more waves which are recombined after they have traveled paths of different optical lengths. This introduces a phase difference, so that interference effects will generally be observed. The discussion here will focus mainly on several applications of the most widely used one, the Michelson interferometer.

The Michelson interferometer is shown schematically in Fig. 5. The discussion which follows will be in terms of visible light, but any other wave phenomenon will produce the same effects. Collimated light from source S is incident on the beamsplitter B which reflects half of the intensity to mirror M1 and transmits half to mirror M2. After reflection from the two mirrors, the two beams again encounter the beamsplitter. Half of each beam is now directed toward the observer or detector of O. Because the two beams at O originate at the same source, they will be coherent (as long as the difference in their optical paths does not exceed the coherence length of the light—typically ≈ 10 cm for conventional light sources). Thus they will interfere. Neglecting phase changes that may occur at the beamsplitter, one expects constructive interference if the optical path difference ($x_1 - x_2$) equals $m\lambda$ and destructive interference if it equals $(m + \frac{1}{2})\lambda$. If the mirrors M1

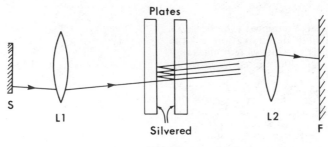

Plates

S   L1   Silvered   L2   F

FIG. 4.

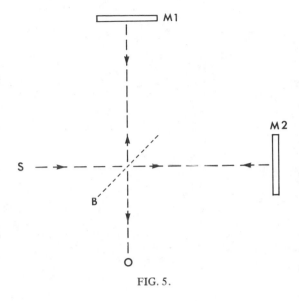

FIG. 5.

and M2 are aligned precisely perpendicular, analysis predicts the observation of a series of circular interference fringes at O. As the path difference is changed, say by moving mirror M1 toward the beamsplitter, the fringe pattern shifts: each time M1 is moved a distance $\lambda/2$, the optical path difference changes by λ, and the fringe pattern shifts by one fringe.

Among the applications of the Michelson interferometer that will be discussed here, there are two of historical significance. In 1892, Michelson and Benoit measured the primary standard of length, the International Prototype Meter which is kept in Paris, in units of the wavelength of the red line emitted by cadmium. By carefully counting fringes as one mirror was moved slowly, they determined that the standard meter is equivalent to 1,553,163.5 wavelengths of this line. Ultimately, it was decided by the International Commission of Weights and Measures (1960) to define the meter as exactly 1,650,763.74 wavelengths of the orangered line of krypton-86. This made the primary standard of length easily accessible to all scientists around the world, obviating the need to travel to Paris for a precise comparison of measured lengths with the standard. This line of krypton was chosen because it is extremely narrow and therefore has a very long coherence length. This permits the use of long optical path differences needed for measurements of objects of appreciable length.

The Michelson-Morley Experiment of 1887 was an ingenious attempt to measure the velocity of the earth in its orbital motion through the "luminiferous ether." Scientists up to that time had postulated the existence of the ether as the medium of propagation for light through the otherwise empty reaches of outer space. If one arm of a Michelson interferometer is ori-

ented along the direction of the earth's motion relative to the (fixed) ether, the other must be perpendicular to it. A calculation of the effective optical paths and transit times for light along the two paths through the ether is straightforward. It predicts a phase difference between the two beams, even if the geometrical path lengths are identically equal. It also predicts that when the interferometer is rotated 90° so that the roles of the two arms are interchanged, there should be a significant shift in the observed fringe pattern.

For the interferometer used in the Michelson-Morley experiment, this shift was predicted to be 0.4 of a fringe; the instrument could resolve a shift of about 0.01 fringe. Repeated measurements at differing times and in differing seasons never showed any shift upon rotation through 90°. This null result pointed to a fatal flaw in the classical theory of relativity. Only Einstein's theory of special relativity (1903), in which there is no ether and in which the speed of light is independent of the relative motion of source and observer, can provide a consistent explanation of the Michelson-Morley null result. (See RELATIVITY.)

Another application of the Michelson interferometer is in the measurement of the indices of refraction of gases. An evacuated cell with windows at each end is placed in one arm of the interferometer so that light passes through the windows. As gas is admitted slowly into the cell, the optical path in that arm is increased according to the laws of refraction. Thus the fringe pattern will shift due to the change in optical path. From the known length of the cell and the number of fringes that are observed to shift, the index of refraction of the gas may easily be found.

In the last 20 years the interferometer, in concert with digital computers, has revolutionized infrared spectroscopy. The light intensity detected at O in Fig. 5, as one mirror is moved steadily, is a sinusoidal function of the mirror displacement. The period of that function is proportional to the wavelength of the light. If the source is not monochromatic, but instead emits a range of wavelengths, the output signal at O is more complicated but is easily shown to be the Fourier transform of the spectral distribution of the source.

High-speed digital computers are capable of performing the transform in real time as the mirror is moved. Compared with conventional grating spectroscopy, the need for a computer is a disadvantage. However, this is heavily outweighed by the number of highly significant advantages in the areas of signal-to-noise ratio and spectral resolution. Such FTIR (Fourier transform infrared) spectrometers have gained rapid acceptance and application in several branches of physics and chemistry, including plasma diagnostics, astronomy, quantitative analysis, and solid state studies.

Two other applications of interference effects

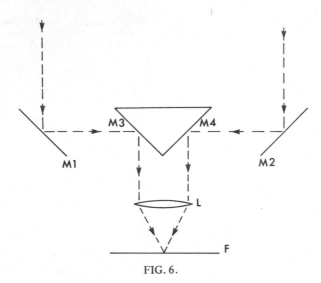

FIG. 6.

will complete this discussion of interferometry. The stellar interferometer constructed by Michelson (1920) can be understood as an example of the principle of Young's experiment. Here the distant star is the single source illuminating the device (see Fig. 6). However, the slits are replaced by two mirrors (M1 and M2), separated by several meters, which reflect the starlight into the objective lens (L) of an astronomical telescope. This produces a pattern of interference fringes in its focal plane (F). By varying the separation of the mirrors, the fringe pattern can be shifted. Analysis of the shifts permits the resolution of closely spaced binary stars or the measurement of the diameters of nearby red giant stars such as Betelgeuse and Antares. By this method, higher resolution than otherwise attainable with the same telescope can be realized.

Recent technological developments have permitted the extension of such stellar interferometry to the radio portion of the electromagnetic spectrum. In an exact analog of Michelson's stellar interferometer, signals from radiotelescopes separated by long distances can be combined to produce interference effects. This is termed very long baseline interferometry (VLBI). A Fourier transform converts the signals, which correspond to the fringes in Michelson's device, into an image of the astronomical object.

Originally, cables carried the signals to a central location for comparison. Subsequent development of highly stable and accurate atomic clocks permitted separations of the individual telescopes by thousands of kilometers. This has resulted in greatly improved performance: VLBI now yields angular resolution of 0.0001 arc-seconds, compared with 0.023 arc-seconds theoretically attainable with the 5 m optical telescope on Mt. Palomar. This is despite the tremendous difference in wavelength: resolution is proportional to λ, and radio wavelengths are typically a factor of 10^4 greater than those of light. The actual resolution of the VLBI system is close to that attainable theoretically because methods of eliminating the image-degrading effects of atmospheric turbulence have also been developed. This is not the case in optical astronomy, where such turbulence degrades the actual resolution of the Palomar instrument to about 1 arc-second. Thus, distant galaxies and quasars are being observed in far greater detail by very long baseline radio wave interferometry than by optical means.

BRUCE P. CLAYMAN

References

Born, M., and Wolf, E., "Principles of Optics," 5th Ed., Chs. 7 and 10, New York, Pergamon Press, 1975.

Cook, A. H., "Interference of Electromagnetic Waves," Oxford, Clarendon Press, 1971.

Jenkins, F. A., and White, H. A., "Fundamentals of Optics," Chs. 12, 13, 14, and 16, New York, McGraw-Hill, 1976.

Klein, M. V., "Optics," Chs. 5 and 6, New York, John Wiley & Sons, Inc., 1970.

Steel, W. H., "Interferometry," London, Cambridge Univ. Press, 1967.

Strong, J., "Concepts of Classical Optics," Chs. 8, 11, 12, and App. A, B, and F, San Francisco, W. H. Freeman and Co., 1958.

Cross-references: COHERENCE, DIFFRACTION BY MATTER AND DIFFRACTION GRATINGS, ELECTROMAGNETIC THEORY, LIGHT, MICHELSON-MORLEY EXPERIMENT, REFRACTION, RELATIVITY, WAVE MOTION.

INTERMOLECULAR FORCES

The terms "intermolecular forces" and "van der Waals forces" refer to the weak forces between molecules, and these forces are to be distinguished from the much stronger interatomic, intramolecular forces of chemical binding. The simplest example of an intermolecular force is provided by the interaction between two noble-gas atoms: The atoms are spherical and the force is central in this case.

Short-range Repulsion The volume of matter in the solid and liquid states is an extensive property, i.e., is proportional to the number of moles or molecules in the specimen, and this implies a molecular "size" or the existence of repulsive forces that prevent two molecules from occupying the same space at the same time. The very low compressibilities of solids and liquids indicate a very strong repulsion of two molecules when they begin to overlap each other. These self-evident conceptions were embodied in the nineteenth-century representation of molecules as little billiard balls. The origin of the "overlap forces" of repulsion was not explained until the advent of quantum mechanics and the Pauli exclusion principle. According to the latter, when two noble-gas atoms begin to overlap, the electrons tend to migrate from the crowded region in the middle to the far ends outside the nuclei, where they exert electrostatic forces that tend to pull the nuclei apart. This effect, which greatly exceeds the direct electrostatic repulsion of the partially shielded nuclei, cannot be represented by any simple analytic potential energy function. For reasons of mathematical convenience, however, an inverse-power law, especially r^{-12}, or a simple exponential function e^{-ar}, where r is the distance between the nuclei of the atoms, is often used to represent the potential energy of the overlap repulsion. The first calculation of an overlap repulsion (between H atoms) was made by Heitler and London in 1927.

Long-range Attraction When two noble-gas atoms are so far apart that there is negligible overlap of their charge clouds or wave packets, the atoms, although neutral and nonpolar, interact through a set of electrostatic multipole terms of which the most important is the lowest-order dipole-dipole term. An instantaneous electric dipole moment in one atom, which averages to zero, induces a dipole moment in the other atom which is proportional to the inducing moment and which interacts with it. It is readily seen that the interaction energy varies as r^{-6}, does not average to zero, and corresponds to an attraction between the atoms. Similarly an instantaneous electric quadrupole moment in one atom induces a dipole moment in the other which interacts with it to give an interaction energy, varying as r^{-8}, which is called the dipole-quadrupole term. These long-range multipole interactions, especially the dipole-dipole term, provide the explanation for the intermolecular attractive forces that are evidenced by imperfect-gas behavior, the Joule-Thomson effect, and the very existence of the liquid state. When the dipole-dipole term is calculated quantum-mechanically, the resulting attractive potential energy, varying as r^{-6}, is called the "dispersion potential." It was first treated by Wang (1927) and then by London (1930).

A Priori Calculations It has long been customary to assume that the intermolecular potential between two neutral, nonpolar molecules consists of a long-range dispersion attraction combined with a short-range overlap repulsion. The combined potential-energy curve for two noble-gas atoms then has a minimum at a distance r_m, and the potential vanishes at a distance σ which is referred to as the slow-collision diameter. However, it is questionable to decompose the interaction potential into parts, and in particular, to use the dispersion potential, which is an asymptotic expression, at distances as small as r_m. The coefficient of the dispersion potential for interactions between pairs of rare-gas atoms has been calculated with great accuracy in recent years, starting from the Schrödinger equation and using a small number of optical data. The a priori calculation of the complete potential energy curve, starting from Schrödinger's equation, is an extraordinarily difficult and laborious procedure even for the simplest atoms and even with the aid of modern electronic computers. Only since (1972) has it been possible to achieve sufficient accuracy to obtain the He–He potential in the neighborhood of its minimum.

Nonadditivity Effects When three or more rare-gas atoms are in proximity, the mutual energy of interaction cannot be expressed as the sum of potential energies for all the interacting pairs, except to a first approximation. The deviations from additivity are associated with many-body forces, and in recent years the three-body interaction has received particular attention. The asymptotic form of the nonadditive three-body interaction is called the triple-dipole potential; it arises from an instantaneous moment in one atom which induces moments in two other atoms, the induced moments then interacting to produce a three-body force. The quantum-mechanical calculation gives the result

$$u_{123} = \nu \, (r_{12} r_{23} r_{13})^{-3}$$

$$\cdot (1 + 3 \cos \theta_1 \cos \theta_2 \cos \theta_3)$$

for the triple-dipole potential, where r_{ij} are the sides and θ_i the interior angles of the triangular array and ν is a constant which can be calculated for rare-gas combinations. Little is known about three-body interactions in the region of overlap, and higher-order many-body forces have usually been neglected. Three-body forces appear to make a large contribution to the values of the third virial coefficients of the noble gases at temperatures close to the critical point.

Polar Molecules When the molecules possess permanent electric moments, additional electrostatic terms contribute to the intermolecular force. The dipole moments of simple polar molecules, such as LiH, can be calculated with considerable accuracy from the Schrödinger equation and compared with experimental values deduced from the temperature dependence of the dielectric constant. The potential energy of alignment of two polar molecules varies as r^{-6} and is usually smaller than the dispersion energy, but in the important case of H_2O at $20°C$ the alignment energy is four times as large as the dispersion energy. The nonspherical shape of polar molecules and the noncentral character of the electric force greatly complicate the mathematical analysis for most physical properties. Permanent electric quadrupole moments Q have also been calculated and measured, and their contributions to intermolecular forces through dipole-quadrupole and quadrupole-quadrupole interactions are important in special cases, e.g., for H_2O. Methods of measuring Q include molecular beam radiofrequency resonance, microwave pressure broadening, collision-induced absorption, nuclear magnetic resonance, and second dielectric virial coefficients.

Empirical Potential Functions Enough has been said to indicate that the nature of the intermolecular forces is qualitatively well understood but that accurate quantitative results cannot be obtained from wave mechanics except in a few special cases. The most detailed information about the actual magnitudes of intermolecular forces is obtained by the empirical procedure of (1) choosing an intermolecular force law, with two or more adjustable constants, on the grounds of physical plausibility and mathematical convenience, (2) selecting a macroscopic property that can be accurately measured and accurately calculated from the force law assumed, and (3) using the results of measurements over a wide range of experimental conditions to fix the values of the adjustable parameters in the force law. The most popular of the empirical potential functions has been the Lennard-Jones (12, 6) potential

$$u(r) = 4\epsilon \left[(\sigma/r)^{12} - (\sigma/r)^6 \right], \qquad \epsilon \equiv -u(r_m).$$

Here σ and ϵ are the adjustable parameters. The 6th-power term represents the dispersion forces and the 12th-power term, the repulsive forces. The parameter ϵ is the maximum energy of attraction, and σ is the value of r for which $u(r)$, the potential function, is zero. Although mathematically convenient and still widely used, this pair potential is now known to have serious defects (e.g., its behavior at small r and at large

r are both greatly in error) which preclude its use in accurate, realistic investigations. Other simple two-parameter and three-parameter potentials have been proposed but they are insufficiently flexible to be able to account for the results of increasingly accurate measurements of numerous physical properties extended over wide ranges of the experimental variables. As an example of the multiparameter potential functions which have been employed, we mention the so-called Morse–spline–van der Waals function which uses a Morse potential for the short-range interaction, a combination of r^{-6} and r^{-8} for the long-range interaction, and a spline interpolation formula to join the two segments. Even more elaborate functions are in current use, some containing as many as 12 adjustable parameters.

Experimental Methods The types of experimental measurements that can be used to derive information about intermolecular forces include the following: (1) thermodynamic properties of gases, (2) transport properties of gases, (3) equilibrium properties of solids and liquids, (4) molecular beam scattering cross sections, and (5) spectroscopic and x-ray diffraction data. Although much used in the past, the calculation of lattice energies, lattice distances, compressibilities, and elastic constants of solids from assumed pair potentials cannot be regarded as an accurate method of determining intermolecular pair potentials because of uncertainty with regard to the contributions from many-body interactions, i.e., from nonadditivity effects. The use of the radial distribution function $g(r)$ of a liquid, determined from x-ray and neutron diffraction experiments, in conjunction with the vapor pressure near the triple point, provides information about the pair potential, and one may also use $g(r)$ in combination with the isotopic separation factor in liquid-vapor equilibrium to test the repulsive region of the potential. The x-ray diffraction results for dense gases and liquids can be analyzed in terms of theories of dense fluids (e.g., Percus-Yevick) to deduce a numerical intermolecular potential. All of these liquid-state methods are subject to uncertainties with respect to nonadditivity effects.

Although the latter may be small, it is safer to deduce the pair potentials from phenomena that involve only binary interactions. In this category are the second virial coefficients $B(T)$, the viscosity coefficients $\eta(T)$, and other gas transport properties. The values of $B(T)$ extending from high to low temperatures do not provide a sensitive test for assumed forms of the pair potential; even the square-well and (12, 6) potentials give a good representation of $B(T)$ data. It is also possible to fit viscosity data over a wide temperature range with simple potential functions; former disagreements at high temperatures have recently been shown to have resulted from experimental errors. However, the parameters of the simple potentials that give the best fit to $B(T)$ measurements differ from those needed to fit the viscosity data. Other transport properties which provide useful information about the pair potential are the isotopic thermal-diffusion coefficient, the isotopic mutual-diffusion coefficient, and the thermal conductivity (of monatomic gases) for like-molecule interactions and the mutual-diffusion coefficient for unlike-molecule interactions.

The most direct method of investigating the force between two molecules is to measure the elastic-scattering differential and total cross sections at different energies. Molecular beams formed from neutralized accelerated ions may be used with stationary gas targets to determine the repulsive wall of the potential at high energies. In order to investigate the region of the potential near its minimum, it is necessary to employ energies close to thermal, and the motion of the target molecules then precludes the use of a single beam. Two molecular beams passing through velocity selectors and intersecting at right angles have been used but much greater intensity is available from crossing two supersonic nozzle beams. Differential elastic-scattering cross sections measured by the latter method, showing well-resolved rainbow structure and symmetry oscillations, have recently provided accurate potential functions for pairs of rare-gas atoms (Lee et al., 1972).

Molecular spectroscopy furnishes a precision tool for investigating the potential energy curves for different electronic states of diatomic molecules, but these interatomic potentials refer to intramolecular rather than intermolecular interactions and will not be further discussed. The pressure broadening of spectral lines and the phenomenon of collision-induced absorption, measured in the microwave and infrared regions of the spectrum, provide additional methods for studying intermolecular forces. These methods are not accurate but they have yielded useful information about molecular quadrupole moments. A recent and important spectroscopic method of determining the intermolecular potential near its minimum, which has been applied to Ar and Ne, is to measure the ultraviolet absorption bands that correspond to electronic transitions out of the ground state of the dimer (Ar_2 and Ne_2). The spectroscopic constants derived from the vibrational levels of the ground state provide accurate information about the bowl of the potential curve including the well depth (ϵ), the curvature at the minimum, and the anharmonicity (Tanaka and Yoshino, 1970, 1972). The best-known intermolecular potential, for the interaction between two Ar atoms, currently has an accuracy of about ±4 per cent in the well depth.

R. D. PRESENT

References

Hirschfelder, J. O., Curtiss, C. F., and Bird, R. B., "Molecular Theory of Gases and Liquids," New York, John Wiley & Sons, 1954.

Hirschfelder, J. O., Ed., "Intermolecular Forces," *Adv. Chem. Phys.*, **12** (1967).

"Intermolecular Forces," *Disc. Faraday Soc.*, **40** (1965).

Margenau, H., and Kestner, N. R., "Theory of Intermolecular Forces," Oxford, Pergamon Press, 1971.

Schlier, C., in Eyring, H., Ed., *Ann. Rev. Phys. Chem.*, **20**, 191–218 (1969).

Cross-references: CRITICAL PHENOMENA, GAS LAWS, LIQUID STATE, MOLECULAR SPECTROSCOPY, SOLID-STATE PHYSICS.

INTERNATIONAL SOLAR-TERRESTRIAL PHYSICS PROGRAMS

It has been recognized for some time that the sun, ultimate source of practically all the energy utilized on earth (the only current exception being the relatively small amount of power produced by nuclear fuels), influences many earthly phenomena. Aside from the obvious solar control of the weather through visible and invisible light—electromagnetic radiation—continuously emitted by the sun, other less well-understood effects occur. For example, the so-called *earth storm* (not to be confused with a weather disturbance in the troposphere near the ground) results from the arrival not only of radiation but also of matter—streams of electrically charged particles—spewed out by the sun following violent eruptions or flares in the chromosphere, just above the sun's visible surface (photosphere).

A multitude of geophysical effects arise from the interactions of these radiation and particle fluxes with the upper reaches of the earth's atmosphere, manifesting themselves as phenomena such as magnetic storms, radio disturbances, and auroral displays. The frequency of occurrence of these transitory happenings in the upper atmosphere waxes and wanes as the level of solar activity changes during the well-known sunspot cycle (approximately 11 years).

The phenomena that occur in the earth's upper atmosphere (i.e., the entire region above the troposphere) obviously recognize no international boundaries. Thus, it was natural that the earliest international scientific programs were organized by geophysicists. In 1625, Francis Bacon emphasized the need for "experiments in concert." One early effort, organized by Friedrich von Humboldt in early 19th Century, investigated the earth's magnetism by means of many simultaneous observations at widely different locations. Karl Friedrich Gauss subsequently founded the Göttingen Magnetic

Union in 1834 for studies of geomagnetic variations by a network of 44 magnetic observatories dispersed around the world.

IPY 1 The First International Polar Year (IPY) in 1882–83 was the original direct progenitor of the current mode of conducting international cooperative scientific enterprises on a global scale. Its program, confined to the Arctic and the approaches to the Antarctic, was limited to meteorological, geomagnetic, and auroral observations at 14 stations established by 11 participating countries. The most commonly cited result was the discovery of the oval-shaped auroral zone–a band with geomagnetic latitude roughly 65–70° where this phenomenon is most prevalent.

IPY 2 The first IPY was followed 50 years later by the Second International Polar Year in 1932–33, with 44 participating nations. Planned and coordinated by the International Meteorological Organization, the second IPY program included stations not only in the polar regions, but at lower latitudes as well, and investigated for the first time the role played by polar phenomena in the concert of worldwide effects. To the three types of observation conducted during the First International Polar Year was added a fourth one, the new field of ionospheric research which had been opened up by the development of radio during the intervening period.

IGY The International Geophysical Year (IGY) in 1957–58 was originally conceived as a third Polar Year, but it was generally agreed during the early planning phases that the terms of reference should be extended beyond a study of polar influence to include the whole earth and its far-reaching atmosphere. IGY was a huge success, and its impact upon the development of many fields of science was tremendous and permanent. It was truly distinctive in that it initiated the first extensive program of *in situ* investigations of the earth's tenuous envelope utilizing the rapidly developing new technology for transporting scientific instruments into space. Furthermore, it opened up Antarctica as a scientific treasure trove.

Responsibility for some part of the IGY program was shared among 67 nations operating 4,000 primary stations and an equal number of secondary posts. It is estimated that approximately 30,000 scientists, plus another 30,000 volunteer observers, participated. Although there is, of course, no satisfactory index of the productivity of a scientific enterprise, it is noteworthy that more than 3,000 articles on IGY results were published in scientific journals by United States workers alone. The vast wealth of data was deposited at the several World Data Centers which were established during IGY and which still continue to be utilized extensively.

Since the broad objective of IGY was to study every aspect of the earth as a planet, from the properties of the terrestrial interior to what we now know as an exceedingly complex region surrounding it, its program embraced many, sometimes overlapping, scientific disciplines. Thus, in addition to investigating the properties of the earth's interior (seismology, latitude and longitude, and gravimetry) and studying its surface characteristics (oceanography and glaciology), the program embraced observations in and beyond the earth's atmosphere, including investigations of sun-earth relationships (meterology, nuclear radiations, geomagnetism, ionosphere, aurora and airglow, cosmic rays, and solar activity). An additional discipline, rockets and satellites, was defined in terms of technique rather than scientific subject matter. As a consequence, the Space Age was spawned. And, to increase basic knowledge about the solar influences acting upon the earth, IGY was planned to cover a period at or near solar maximum. (See also GEOPHYSICS and SOLAR PHYSICS.)

As hoped, IGY did take place at a peak of sunspot activity. Propitiously, it was not only the maximum for that particular eleven-year cycle–but the level of solar activity in 1957–58 has probably not been matched at least since Galileo first observed sunspots in the early 17th Century.

IQSY The International Years of the Quiet Sun (IQSY) in 1965–66 was to be the sequel of IGY aimed at catching the myriad interrelated changing phenomena of sun, space, and earth at the sun's nadir of activity. But the later enterprise was not a small-scale repetition if its illustrious predecessor. On the contrary, in the fields that it embraced, the level of effort far exceeded that which it had been possible to attain only seven years earlier.

However, in a real sense, IGY had set the stage for IQSY. A new scientific discipline, *solar-terrestrial physics* (STP)–alternatively called *solar-terrestrial relationships* or *solar-terrestrial research* (STR)–was emerging to embrace all processes whereby diverse forms of energy generated by the sun influence the terrestrial environment, with the resulting complex interplay of physical-chemical processes in every element of the sun-earth system.

The purpose of IQSY was threefold. Some of the studies that were conducted were feasible, or are best undertaken, ony at a time of solar minimum. Others were concerned with observing in detail isolated solar events, uncomplicated by the superposition in time of a number of concurrent outbursts. Finally, some investigations provided data characteristic of solar minimum conditions as benchmarks for comparison with data observed previously during a most remarkable solar maximum.

Actually, at sunspot minimum solar outbursts do not cease completely, but when they do occur their effects can be observed under relatively "clean" conditions, free from confusing interferences. Hence, the entire sequence of events associated with a single outburst can be followed through all of its aspects. Furthermore, smaller effects, previously lost in the high-level background of activity, become discernible. Thus, many IQSY projects took advantage of this period to make observations with greatly increased resolution or "amplification."

The IQSY program was organized into eight disciplines, although in most cases there were overlapping interests, and no sharp boundary lines were drawn. These were: meteorology (see METEOROLGY), geomagnetism (see GEOPHYSICS), aurora (see AURORA AND AIRGLOW), airglow (see AURORA AND AIRGLOW), ionospheric physics and radio astronomy (see IONOSPHERE AND RADIO ASTRONOMY), the sun and the interplanetary medium (see SOLAR PHYSICS AND SPACE PHYSICS), cosmic rays and geomagnetically trapped radiation (see COSMIC RAYS AND MAGNETOSPHERIC RADIATION BELTS), and aeronomy (see PLANETARY ATMOSPHERES).

The sponsorship and supervision of the 65 worldwide networks of more than 2,000 IQSY stations were the responsibility of 71 participating committees, as they were called, established by the Academy of Science or an equivalent body in each country. Innumerable balloons, some 500 rockets, and about 140 satellites and space probes were launched during the two-year campaign. The IQSY bibliography lists 5,359 publications, which themselves formed the basis for an untold number of subsequent papers in the scientific literature.

IQSY was the final broadly based international scientific undertaking. It became evident that future cooperative ventures should be somewhat more sharply focused, and the mechanism for accomplishing this was the establishment of several permanent international committees arising from the temporary single purpose, national and international groups that had been formed specifically to organize IGY and IQSY. Each of the latter were dissolved after their missions, including the publication of their respective annals, had been completed.

IUCSTP The modus operandi of IQSY has continued to be employed for mounting imaginative interdisciplinary and international research projects. Thus, when in 1966 the IQSY program was drawing to an orderly conclusion and its international organization was closing down, a new and continuing body, the Inter-Union Commission on Solar-Terrestrial Physics (IUCSTP) was established by the International Council of Scientific Unions (ICSU) to plan and coordinate future programs in solar-terrestrial physics.

ISCU, from among the several interested unions among its membership, had established the bodies that were responsible for the international coordination of IGY and IQSY. IUCSTP was established on a permanent basis to plan and organize the international coordination of international projects of those research workers throughout the world who are interested in solar-terrestrial physics, and who wish to concentrate attention on some of the more important problems that still awaited solution. Several representative recent programs will be cited here.

MIDDLE ATMOSPHERE PROGRAM (MAP) The Middle Atmosphere Program (MAP) is an international cooperative enterprise developed under the auspices of the Scientific Committee on Solar-Terrestrial Physics of the International Council of Scientific Unions and the World Meteorological Organization. Scheduled to run from January 1982 through December 1985, its chief objective is a comprehensive understanding of the structure, chemistry, energetics, and dynamics of the middle atmosphere (defined for MAP purposes as the region extending upward from the tropopause, including the stratosphere and mesosphere, to the lower thermosphere, i.e., from about 10 to 100 km). Contemplated studies include ozone climatology; stratospheric composition; mesospheric composition; basic climatology of the middle atmosphere; planetary waves in the middle atmosphere; equatorial waves; tides, gravity waves, and turbulence; troposphere-stratosphere coupling; the influence of middle-atmospheric conditions on lower-level climate; aerosol formation and properties; solar radiation, especially in the ultraviolet; the effects of energetic particles and x-rays on the middle atmosphere; ion composition; and the electrodynamics of the middle atmosphere. These fourteen areas have been tagged observational or experimental MAP initiatives (MIs). The operational basis for MAP is an international plan for coordinated observations from spacecraft, rockets, balloons, aircraft, and ground-based facilities. This plan encourages cooperative data management, information exchanges on all appropriate time scales, and interaction between observers, modelers, and theoreticians.

THE INTERNATIONAL MAGNETOSPHERIC STUDY (IMS) The International Magnetospheric Study (IMS) is a highly coordinated, cooperative program designed to provide quantitative new information about certain key problems involving plasmas in the geomagnetic field. The IMS formally began in January 1976, and the primary data-collection period ran through 1979. A five-year follow-up program for data reduction and analysis was subsequently recommended by the IMS Steering Committee, and is in progress.

About 50 countries are participating in the IMS, and the four primary IMS spacecraft (ISEE-1, -2, -3 and GEOS-2) were successfully launched by the United States (NASA) and the European Space Agency (ESA). Japan and the Soviet Union also provided new spacecraft for the program. The key ingredients of the IMS were these dedicated new spacecraft and several sophisticated new ground-based instrument arrays, which greatly advanced international data-exchange standards by using simultaneous data collection, processing, and dissemination in near real time from central data facilities. Incoherent-scatter radars were operated at high latitudes, and a variety of ground-based, balloon, rocket, and aircraft experiments were conducted. While the IMS required a sizable number of specific new facilities, it also benefited greatly from the effective use of previously planned programs and from continued operation of certain older satellites and ground networks. The IMS Central Information Exchange (IMS CIE) Office was established in the World Data Center-A in Boulder, Colorado, and regional offices were set up in Paris, Moscow, and Tokyo; a monthly Newsletter transmitted schedules, plans for spacecraft and ground-based measurements, and general IMS news to several thousand scientists. NASA also established the Satellite Situation Center, which disseminated detailed trajectory data for all IMS spacecraft.

SOLAR MAXIMUM YEAR (SMY) During the next period of maximum solar activity following IQSY, programs were conducted in which several score observatories coordinated their observations with each other, and with spacecraft in particular. The NASA Solar Maximum Mission (SMM), carrying a new state-of-the-art set of instruments, ran from August 1979 through February 1981. Observing schedules were arranged by the SMY Steering Committee with special observational intervals being scheduled for the three activities subsumed under the SMY, namely the Flare Build-up Study (FBS), with emphasis on distinguishing the characteristics of localized solar active regions that subsequently produce flares from those that do not; Study of Energy Release in Flares (SERF), with emphasis on the flare process itself; and Study of Traveling Interplanetary Phenomena (STIP), with emphasis on the aftereffects of flares and other active phenomena. Coordination was effected by telex from several SMY control centers. Following the pattern of previous programs, the participants in the SMY engaged in a number of workshops to coordinate both their detailed planning and the analysis of their data sets.

CSTR In the United States, the Committee on Solar-Terrestrial Research (CSTR), formed in 1965 by the National Research Council's Geophysics Research Board, "looks after the health of this field" in our country, and represents the nation in International STR programs.

MARTIN A. POMERANTZ

References

1. "Annals of the IGY," New York, Pergamon Press, 1959.
2. "Annals of the IQSY," Cambridge, Massachusetts, M.I.T. Press, 1969.
3. Pomerantz, Martin A., "The IQSY and Solar-Terrestrial Research," *Proc. Nat. Acad. Sci.* **58**, 2136 (1967).
4. "Solar-Terrestrial Research for the 1980's," Washington, D.C., National Academy Press, 1981.
5. "The International Magnetospheric Study: Report of a Working Conference on Magnetospheric Theory," Washington, D.C., National Academy of Sciences, 1979.
6. Handbook for Middle Atmosphere Program, Scientific Committee on Solar-Terrestrial Physics (SCOSTEP), C. F. Sechrist, Jr., Ed., Urbana, IL 1981.

Cross-references: AIR GLOW, AURORA, COSMIC RAYS, GEOPHYSICS, IONIZATION, IONOSPHERE, PLANETARY ATMOSPHERES, MAGNETOSPHERIC RADIATION BELTS, RADIO ASTRONOMY, SOLAR PHYSICS.

IONIZATION

Ionization is the name given to any process by which a net electrical charge may be imparted to an atom or group of atoms. In the case of liquid solvents, molecules or ionic salts become dissociated to form positive and negative ions. This ionization process is known as *electrolysis*, and the name *electrolyte* is given to the solute or to the conducting solution. The study of electrolysis is embodied in the subject of *electrochemistry*. Of great interest in recent years has been the study of ionized gases. Rockets, hypersonic flight, and space physics have spurred investigations of plasmas, shock waves and high-temperature chemical processes arising in a variety of terrestrial and celestial phenomena. Indeed, ionized gases make up a major portion of all matter in the universe. Our chronic need for new energy sources has transformed the speculation of a controlled thermonuclear fusion reaction into one of the greatest research efforts in history. These and other considerations have induced a vigorous growth in the study of ionization phenomena.

Electrolytes The degree of ionization found in electrolytes is highly variable and depends upon the solute, the solvent, and the interaction between them. *Weak* electrolytes, such as many organic compounds, are solutes which are barely dissociated into ions except in the limit of infinite dilution. *Strong* electrolytes are highly dissociated at any concentration. Ions formed in solution may bear one or several electronic charges. The *electrochemical equivalent weight* is the atomic weight divided by the number of charges carried by the ion. If electrodes are placed in an electrolyte and a current flows in the external circuit, the ions with positive charges, called *cations* (cathode + ions), will migrate toward the negative electrode (cathode).

Those ions possessing a negative charge (*anions*) will migrate to the positive electrode (anode). The ions arriving at the cathode are neutralized by the acquisition of electrons; the atoms or molecules thus formed may then be evolved as a gas or retained as a deposit on the electrode. The cations are said to undergo *reduction*. Likewise, the anions experience a loss of electrons at the anode; this process is called *oxidation*. The quantity of electricity required to deposit one gram equivalent is called the faraday, in honor of Michael Faraday (1791–1867). Based on the physical scale of atomic weights, the faraday is numerically equal to 96 520 coulombs per equivalent.

As for any solid conductor of uniform cross sectional area A and length l, the electrical resistance R of an electrolyte is given by $l/\kappa A$. The conductivity κ is independent of geometrical shape and size, and bears the units (ohm cm)$^{-1}$. Of greater importance in the study of electrolytes is the *equivalent conductivity*:

$$\Lambda = \kappa/C$$

where C is the concentration of the solute in equivalents per cubic centimeter. Plots of Λ as a function of concentration show very different behavior for weak and strong electrolytes. The latter exhibit a limiting value of Λ as C diminishes to zero, while the weak electrolytes do not. Such behavior provides insight into the nature of the ions, their mobilities and their interactions with the solvent material.

Formation of Gaseous Ions Studies in 1895 by J. J. Thomson of the effects of newly discovered x-rays on gases marked the beginning of a series of experiments which established the existence of the electron and clarified many questions on the nature of atomic structure.

Just as in electrolytes, both positive and negative ions may exist in an ionized gas. In addition to the ions, the presence of free electrons may profoundly influence the character of the gas. Negative ions may be formed by the attachment of free electrons to a neutral atom or molecule, by the dissociation of a neutral molecule into positive and negative fragments, or by electron transfer upon collision of two neutral atoms or molecules. Positive ions may be formed by dissociation, charge transfer, neutral-particle or electron collisions, or by photoabsorption (the absorption of electromagnetic radiation). Still another mechanism for the formation of ions is the emission of a nuclear particle, such as beta decay. Several of these processes are discussed below.

Photoionization Photoabsorption leading to excitation and ionization is of interest because of its significance in astrophysics and geophysics. The ionosphere is constituted of molecular and atomic ions which result from the absorption of solar ultraviolet and x-radiation. The frequency ν of the electromagnetic radiation giving rise to ionization must satisfy the relation $h\nu \geq V$, where h is Planck's constant and V is the *ionization potential*. The latter is defined as the energy required to remove completely an electron from an atom or molecule in the ground state, leaving the resulting ion in its lowest state. Photons having energy less than V may be absorbed by atoms or molecules, giving rise to excitation of internal states or perhaps molecular dissociation, or both.

Many laboratory investigations of photoabsorption have been performed. One type of experiment requires the measurement of the absorption coefficient, α, of a photon beam:

$$I = I_0 e^{-ax}$$

in which I_0 is the initial intensity of the beam and I is the intensity after the beam has traversed a distance x in the absorbing gas. The absorption coefficient may then be studied as a function of photon energy (i.e., wavelength). It can be expressed in terms of a microscopic cross section for absorption or for ionization as a function of incident wavelength. This cross section curve for photoionization usually exhibits a sharp peak at the ionization threshold. Only a single, outermost electron is ejected from an atom which absorbs an ultraviolet photon. An x-ray photon generally will eject a more tightly bound electron from one of the atom's inner shells.

Other types of experiments utilize photoionization to study the deionization process for the ions thus formed. Of major importance are such processes as electron-ion recombination:

$$A^+ + e^- \rightarrow A$$

and ion-ion recombination if both charged species are ions.

Ionization by Heavy-particle Collisions By heavy particles is meant both atoms and molecules and their ions, ranging in mass from the hydrogen atomic ion (proton) to very heavy molecular systems of large atomic number. When two heavy particles collide with sufficient energy, one or more electrons may be ejected from either or both particles. In experimental work, the *target* molecules or atoms are in the form of a low-density gas having an energy corresponding to room temperature and usually negligible compared to the energy of the projectile particles. The latter are usually obtained through ionization of a selected gas in an ion source, and acceleration through a large electric potential difference E. Regardless of the mass of projectile particles, their kinetic energy will be equal numerically to E electron volts, if the potential difference E is in volts and if the particles carry but one elementary charge. One electron volt (eV) is equal to 1.6×10^{-12} erg.

If a beam thus formed with an intensity of B particles per second is incident on a target chamber of area A containing N target particles as a low-density gas, the electron ionization current i which is released is given by

$$i = BN\sigma_-/A \text{ electrons/ sec.}$$

This equation defines the effective ionization cross section σ_-. For energies at which multiple ionization is improbable, σ_- approaches the cross section for singly charged ions. Often the distinction is made between ionization of the target particle and the beam particle. Ionization of the latter is referred to as stripping.

In the quantitative description of heavy-particle collisions, one usually introduces the concepts of the (a) laboratory and (b) center-of-mass (CM) coordinate systems. The laboratory system is used to describe the motion of the particles as would be viewed by an observer standing at rest in the laboratory. The origin of the CM system moves with the center-of-mass of the two-particle system. If m and M are the masses of the projectile and target particles, the former moving with a velocity v much greater than the target velocity, the center-of-mass velocity V_c is

$$V_c = mv/(m + M) = \mu v/M$$

The latter relation defines μ, the reduced mass. The kinetic energy in the CM system is $\frac{1}{2}\mu v^2$, which is the projectile energy multiplied by $M/(M + m)$. For the case in which target and projectile are identical, the kinetic energy in the CM system is half that of the projectile. Using the law of conservation of linear momentum, one may show that the kinetic energy in the CM system is the maximum energy available for excitation and/or ionization.

Collisions in general are classified as (a) elastic in which no changes in internal states occur, and for which kinetic energy is conserved, and (b) inelastic, in which a part of the kinetic energy is converted to internal energy. A *superelastic* collision is one in which internal energy is transformed into kinetic energy. Because of the extreme complexity encountered in quantum-theoretical calculations of ionization cross sections for heavy-particle collisions, very little progress has been made in this important area of collision theory.

Ion-neutral collisions may give rise to free electrons, or simply *charge exchange* which, in its simplest form, is expressed by the equation.

$$A^+ + B \rightarrow A + B^+ + \Delta E$$

The neutral particle B has been ionized, but the electron has transferred to the incident ion, neutralizing it. The energy ΔE released in this process is the difference between the ionization potentials of the neutral particles A and B. For the case in which A and B are identical, $\Delta E = 0$ and the process is called *symmetric resonant* charge transfer. At low ion beam energies, this transfer proceeds with a large cross section.

Ionization by Electron Impact Of great importance in atomic physics is the ionization produced by collimated beams of electrons incident on heavy particles. Whether the target particles are atoms or molecules, the ionization is usually by the removal of single electrons from the outer most shell, as in photoionization. As a function of the incident electron energy, the ionization cross section rises rapidly from zero for energies just below the ionization potential and increases to a maximum value in the neighborhood of 50 to 100 eV; thereafter it decreases slowly and monotonically with increasing electron energy. Since an electron with a given energy travels at much higher speed than does a heavy projectile of the same energy, the electron collision induces a much more rapid perturbation on the target's orbital electrons. Thus, a larger ionization cross section at low energies is to be anticipated.

Much of the definitive work on electron impact ionization was performed in the 1930's by Tate, Smith, and Bleakney (cf. reference 4).

Collective Processes If, as in a glow discharge, a large number of charged particles are created, the collective interactions of these particles with each other and with external fields may permit the charged fluid to exhibit very unusual and distinct properties. An ionized gas possessing both positive and negative charges is called a *plasma* if the distance over which the gas can have an appreciable departure from charge neutrality is small compared to the dimensions of the gas. This distance is described by the *Debye-Hückel radius*, a quantity borrowed from the theory of strong electrolytes, which characterizes the decay of the shielded Coulomb potential surrounding the ionized particles of the fluid. If the charge neutrality in a plasma is disturbed in some manner, the electrons will be forced to oscillate about their equilibrium positions in simple harmonic motion with a frequency characterized by the electron density. Longitudinal oscillation of the ions and electrons as a whole constitutes another type of motion called *ion-acoustical* waves. *Hydro-magnetic* waves, which appear in the presence of a magnetic field are still another form of motion not observed in a nonionized medium. The description of such phenomena goes beyond the scope of the ionization process.

ROBERT C. AMME

References

1. Condon, E. U., and Odishaw, H., "Handbook of Physics," second edition, New York, McGraw-Hill Book Co., 1967.
2. Spitzer, L., "Physics of Fully Ionized Gases," Interscience Tracts on Physics and Astronomy, New York, Interscience Publishers, 1956.
3. Loeb, L. B., "Basic Processes of Gaseous Electronics," Berkeley, University of California Press, 1955.
4. McDaniel, E. W., "Collision Phenomena in Ionized Gases," New York, John Wiley & Sons, 1964.

5. Hasted, J. B., "Physics of Atomic Collisions," Washington D.C., Butterworth, 1964.

Cross-references: COLLISIONS OF PARTICLES, CROSS SECTIONS AND STOPPING POWER, ELECTRICAL DISCHARGES IN GASES, ELECTRO-CHEMISTRY, IONOSPHERE, MAGNETO-FLUID-MECHANICS, PLASMAS.

IONOSPHERE

The ionosphere is the region of the upper atmosphere in which the density of thermal electrons and ions is large enough to affect the propagation of radio waves. The ionosphere is produced by the effects of ionizing radiation from the sun, and high energy particles impinging on the neutral upper atmosphere. This produces photoelectrons and ions. These then interact with the neutral atmosphere and the resultant densities are controlled by the production and loss processes as well as horizontal and vertical transport. Ionospheres are thus a general feature of planetary atmospheres.

The ionosphere of the earth was divided up into vertical regions based on radio soundings which are sensitive to the gradient in the electron density; these regions also correspond to regions of different ion chemistry. Typical values of the densities in the earth's ionosphere are shown in Fig. 1.

The lowest region of the ionosphere was called the D layer to allow for up to three lower layers had they been discovered later. Ionization due to galactic cosmic rays does penetrate to the ground, and below the D region the conductivity decreases steadily. This ionization in the region of 30–50 km is detectable in in-situ observations and is often referred to as the C layer. In the lower D region between 50 and 80 km, although N_2^+ and O_2^+ are the major ions produced, the predominant positive ions are those of water clusters, $H_3O^+ \cdot [H_2O]_n$. During the day the negative charge is mainly carried by electrons, but at night negative ions become important. This is the only region of the ionosphere in which negative ions are numerically important.

In the upper D region the positive ions are mainly NO^+ and O_2^+, as they are in the E and F_1 regions above. The D region has very different behaviors during times when the sun is very active and when it is quiet. The reason for this was discovered by Nicolet and Aikin (1960) to be that when the sun is active the X-ray emission is enhanced by orders of magnitude and is the dominant production source. Under quiet conditions the Lymann alpha line of the sun's radiation, which varies little with solar activity, dominates. This line, which is of too low an energy to ionize any of the major constituents above these altitudes, penetrates to the D region and ionizes a very minor neutral constituent, nitric oxide.

The boundary between the E region and the D region is normally taken to be at around 90 km, or the lower boundary of penetration of ultraviolet radiation sufficiently energetic to ionize molecular oxygen. The production and loss processes in the E and F_1 regions are essentially identical in the region up to 120–140 km, though there is a continuous change in the portions of the extreme-ultraviolet (EUV) spectrum producing the ionization. X-rays between 30 and 100 Å, in particular, ionize N_2 and O_2 in the E region.

An important feature of the ionosphere in the region of 100 km is the formation, on occasion, of extremely thin layers of ionization known as "sporadic E." Direct sampling has shown the ions in these regions to be similar to the composition of the chondritic meteors. It is believed that these meteoritic ions are produced by charge exchange with ambient ions and are compressed into thin layers by the interaction of wind shears with the Earth's magnetic field. It is this type of sporadic E that is occasionaly responsible for long-distance television reception.

The F_2 layer, which is generally regarded as beginning at about 140–150 km, is responsible for by far the largest portion of the total ionization. In the lower regions the molecular ions NO^+ and O_2^+ predominate, but in the region of the peak and above O^+ becomes more important. In this region of the neutral atmosphere atomic oxygen is the major constituent and O^+ is the major ion produced. The most important loss process for this ion is charge exchange with O_2 and N_2, and this proceeds much faster at lower altitudes where the molecular densities are larger. Vertical transport is extremely important, with most ions being produced above the peak and recombining below the peak. The effect of neutral winds and electric fields on this transport is large and consequently the layer displays a very complex morphology with large variations in geographic and magnetic coordinates as well as altitude, time, and season. Other large fluctuations occur whose cause is not at present well understood. As shown in Fig. 1 there are large systematic changes in the F region between sunspot maximum and minimum following the solar cycle. These are due to the large variations in the neutral thermospheric densities as well as to the changes in the EUV ionizing radiation.

Above the region where O^+ predominates is the protonosphere. At the lower boundary the protons undergo charge exchange reactions with neutral atomic oxygen and neutral hydrogen charge exchanges with atomic oxygen ions. This results in a generally downward flux of protons when the F layer is decreasing and an upward flux when it is increasing. Because the atomic oxygen ions are much heavier than the electrons the equilibrium profile of the densities in which the fluxes produced by diffusion would be balanced by gravity are very different.

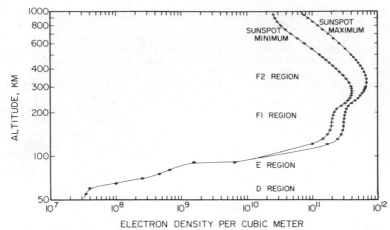

FIG. 1. Electron density at mid-latitudes at noon for high and low solar activity.

A vertical electric field is set up which keeps the two distributions the same. This electric field has a large effect on the protons in the region where O^+ predominates. In particular at high latitudes where the magnetic field lines are open to the solar wind, the protons are accelerated outward, leading to the escape of hydrogen. This is balanced by a poleward wind in the protonosphere.

The ionosphere of Mars and Venus, despite their lower atmospheres which are quite different from that of the earth, also have an O^+ "F_2 region" and an O_2^+ "F_1 region." The major difference is that, unlike the earth, the absence of a strong planetary magnetic field allows the solar wind to interact directly with the ionosphere, producing an extremely turbulent and variable upper boundary.

The thermal state of the ionosphere is quite interesting because the electrons, ions, and neutral particles all have different temperatures above about 200 km in the earth's ionosphere due to small collision frequencies between species, the very different energy inputs, and the importance of vertical transport. That part of the excess energy given to the photoelectrons produced in the ionization process which is not lost in inelastic collisions with the neutral gas is transmitted preferentially to the ambient electrons, which in turn heat the ambient ions. As a result, above about 200 km, in the daytime or when photoelectrons are entering the upper ionosphere from the conjugate hemisphere, the electrons are hotter than the ions, and the ions are hotter than the neutrals. Because of the comparative ease with which the electron and ion temperatures and densities can be measured remotely by radio waves these inputs provide a very valuable tool in studying the energy balance of both the ionosphere and the neutral atmosphere.

Many of the ions in the ionosphere are produced in excited states, or are excited by colli-

sions with higher energy particles. Some of this excess energy is radiated and results in aurora and the airglow radiations in the infrared, the visible, and the ultraviolet (see AURORA and AIRGLOW). The major source of visible illumination, for example, on a moonless night is from the 630 nm radiation from atomic oxygen in the F region of the ionosphere.

Electrical currents flowing in the ionosphere cause perturbations in the earth's magnetic field that can be measured at the ground. There are two main current systems. The "dynamo system" is produced by E region winds blowing the ions and electrons across the earth's magnetic field. The "solar wind" of plasma flowing outward from the sun produces a voltage across the polar caps resulting in field-aligned Birkeland currents entering the ionosphere at high latitudes. The resulting ionospheric currents are enhanced and can penetrate to low latitudes during disturbed times.

The ionosphere is important for ground-to-ground propagation because it can refract or reflect radio waves and thus makes communication beyond the horizon possible. For satellite-to-ground communication the combined effects of refraction and electron density irregularities produce scintillations. These can be a problem especially if the ray path crosses the equatorial region. For satellite navigation systems refractive errors have to be compensated for. Radio astronomy from the ground is limited to frequencies below the critical frequency of the F region.

JOHN S. NISBET

References

Al'pert, Ya. L., "Radio Propagation and the Ionosphere," 2nd Ed., Plenum, New York, 1973.
Banks, P. M., and Kockarts, G., "Aeronomy," Academic Press, New York, 1973.

Bauer, S. J., "Physics of Planetary Ionospheres," Springer-Verlag, Berlin and New York, 1973.

Nicolet, M., and Aikin, A. C., "The Formation of the D Region of the Ionosphere," *J. Geophys. Res.* **65**, 1469 (1960).

Nisbet, J., "On the Construction and Use of a Simple Ionospheric Model," *Radio Science* **6**, 437 (1971).

Cross-references: AIR GLOW, AURORA, INTERNATIONAL SOLAR-TERRESTRIAL PHYSICS PROGRAM, IONIZATION, PLANETARY ATMOSPHERES, MAGNETOSPHERIC RADIATION BELTS, PLASMAS, SPACE PHYSICS.

IRRADIANCE (ILLUMINANCE) CALCULATION

Introduction In geometrical optics the paraxial lens equation is applied at each surface to rough out a design. To determine the final image quality a designer may trace hundreds of nonparaxial rays to obtain a spot diagram in the image plane. Each spot is the point of intersection of a ray from a point on the object with the image plane, and the spot density is a measure of image spread and therefore of image quality.

A formula is available, however, for specifying the flux density, that is, the irradiance along each traced ray.[1] It is also applicable to ray acoustics and radiant heat transfer. For example, the formula enables one to vary a parameter, trace a new ray and immediately calculate the irradiance along that ray. To illustrate the ideas, we derive the formula here, referred to as the *flux equation*, for the special case of a two-dimensional geometry. This result is useful by itself since it gives some of the characteristics of the image formed by rays in the meridional plane of a three-dimensional geometry. The meridional plane is the plane formed by the object point (and image point) and the optical axis.

First, omitting mathematical details, we outline the procedure for calculating illuminance in three dimensions and then proceed in detail to the simpler two-dimensional geometry. As input to the flux equation in three dimensions one needs to know the principal radii of curvature of the wave front both before and after refraction. In any surface two curves may be formed at right angles to each other, one having a maximum curvature and the other a minimum curvature; they are called the *principal curvatures* of the surface. The radii of curvature are the reciprocals of the principal curvatures. The centers of the radii of curvature for a wave front emerging from a refracting surface define converging points for the wave front. They lie along the refracted ray vector and their loci, for a series of ray vectors, trace out two surfaces called *caustic surfaces*. In three dimensions, in optics, the principal curvatures after refraction may be obtained from the values of a pair of curvatures called the *parallel normal curvature* and the *perpendicular normal curvature* of the wave front. The parallel normal curvature is the curvature of the curve in the wave front whose tangent is a tangent of the surface and lies in the plane of incidence at the point of refraction. The perpendicular normal curvature of the wave front is the curvature of the curve in the wave front whose tangent is a tangent in the surface and is perpendicular to the plane of incidence at the point of refraction of a ray. There is an equation that relates the parallel normal curvature of the incident wave front to the parallel normal curvature of the refracted wave front. A second equation relates the corresponding perpendicular curvatures of the incident and refracted wave fronts. Two other equations relate the torsions of these respective curves. We call these the *generalized lens equations*. The two torsion equations can be shown to reduce to one, leaving three equations. From the generalized lens equations one can relate the principal curvatures of the wave front after refraction to the principal curvatures of the wave front before refraction, since the principal curvatures of a surface are related to the normal curvatures of the surface through an equation called *Euler's equation*. When the principal curvatures are thereby calculated one has the necessary data to evaluate the flux equation and also to locate points on the caustic surfaces associated with any ray, skew, or meriodional, in an optical system.

For the special case of meridional rays, the torsions of the normal curves are zero so the torsion equations do not appear. Also the normal curves and their associated curvatures become identical with the principal curves and their curvatures. Therefore, the two generalized lens equations become equations that directly relate the principal curvatures of the wave front before and after refraction. They also relate points on the caustic surface before refraction to the caustic surfaces after refraction; the caustic surfaces are therefore immediately determined.

The rather complicated three-dimensional problem described above is greatly simplified in two dimensions, as we shall see. Many of the key ideas are still involved, however, but can be made much clearer in two dimensions before proceeding to a full three-dimensional geometry. Figure 1 shows the three-dimensional caustic surfaces formed by collimated rays after refraction by a spherical surface. A hornlike surface is formed (tangential caustic) and a caustic spike (saggital caustic). The latter is associated with the radial convergence of rays through the optical axis because of the axial symmetry. If the source lies off the axis, the caustic spike also spreads out into a hornlike surface. In a strict two-dimensional geometry only the tangential caustic appears.

Two-Dimensional Geometry We now derive an equation that relates the curvature of a two-

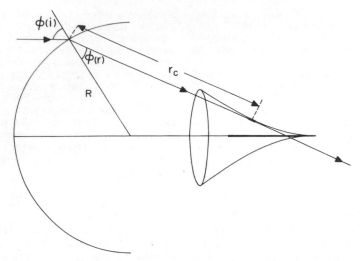

FIG. 1. Caustic cone and caustic spike along optical axis formed by incident collimated rays after refraction. In a 2-dimensional geometry only a slice of the cone in the plane of the figure is present.

dimensional wave front after refraction to the curvature of the wave front before refraction. This single equation is the two-dimensional analogue of the generalized lens equation for meridional rays in a three-dimensional geometry. Since the center of curvature of the wave front after refraction defines a point on the caustic, only one caustic is formed: a caustic curve, the tangential caustic. For small angles, with the optical axis, the two three-dimensional exact lens equations for meridional rays become the single paraxial lens equation. Likewise, the single two-dimensional generalized lens equation, the exact lens equation, becomes the par axial lens equation.

Definition of Curvature Let the position vector $x(s)$ of arbitrary origin define the equation of a refracting surface in parametric form. The parameter s may be arbitrary, but in most derivations is the arc length along the curve. I and J are cartesian unit vectors. Then

$$x(s) = Ix(s) + Jy(s), \qquad (1)$$

where $x(s)$ and $y(s)$ are arbitrary functions of s. The unit tangent vector to the curve is given by

$$t(s) = \frac{dx(s)}{ds}. \qquad (1a)$$

In turn the arc rate of change of the unit tangent vector is

$$\frac{dt(s)}{ds} = \kappa N(s) \qquad (1b)$$

where $N(s)$ is the unit normal to the curve, κ is the curvature, and $1/\kappa$ is the radius of curvature.

The arc rate of change of the normal vector is

$$\frac{dN(s)}{ds} = \kappa t(s). \qquad (1c)$$

Snell's Law in Vector Form Let a ray be incident upon the surface specified by $x(s)$. A position vector on the ray after refraction is given by

$$X = x(s) + rA(r)$$

where $A(r)$ is a unit vector along the refracted ray and r is a distance along the ray. $A(r)$ is related to $A(i)$, the unit vector along the incident ray, and N, the unit normal to the refracting surface, by Snell's law in vector form; that is

$$A(r) = \gamma A(i) + \Omega N. \qquad (2)$$

This equation merely expresses that fact that $A(r)$, $A(i)$, and N are coplanar, part of Snell's law, and that a vector $A(r)$ may be expressed in terms of two independent vectors, $A(i)$ and N. To determine γ, form the cross product of N with the above. Thus

$$A(r) \times N = \gamma A(i) \times N$$

so that $J \sin \phi(r) = J\gamma \sin \phi(i)$, where J is a unit vector perpendicular to the plane of the figure. Therefore, $\sin \phi(r) = \gamma \sin \phi(i)$. Snell's law as usually stated is

$$\sin \phi(r) = \frac{n(i)}{n(r)} \sin \phi(i),$$

where $\phi(i)$ is the angle of incidence and $\phi(r)$ is the angle of refraction. Therefore $\gamma = n(i)/n(r)$, the ratio of the index of refraction on the incident side to the refracted side.

To determine Ω, form the scalar product of $\mathbf{N}$ with Eq. (2). Then $\mathbf{A}(r) \cdot \mathbf{N} = \gamma \mathbf{A}(i) \cdot \mathbf{N} + \Omega$. Since $\mathbf{A}(r) \cdot \mathbf{N} = \cos \phi(r)$ and $\mathbf{A}(i) \cdot \mathbf{N} = \cos \phi(i)$ we obtain

$$\Omega = -\gamma \cos \phi(i) + \cos \phi(r).$$

For reflection $\phi(r) = \pi - \phi(i)$ since $\mathbf{N}$ is taken in the same directional sense as $\mathbf{A}(i)$, that is, it is positive on the emerging side of the refracting surface. Thus for reflection,

$$\gamma = 1 \quad \text{and} \quad \Omega = -2 \cos \phi(i).$$

With γ and Ω now known we form the complete differential of Eq. (2) and make use of the fact that the ray vector is perpendicular to the wave front. This is so because the wave front from the point source is spherical and is perpendicular to the ray vector which lies along the radius of the sphere. It is a general theorem in optics, known as the Malus-Dupin theorem,[2] that if the ray vector starts out perpendicular to the wave front it remains so after any number of refractions. Thus $\mathbf{A}(i)$, the unit ray vector before refraction, is also the unit normal to the wave front before refraction; likewise $\mathbf{A}(r)$, the unit ray vector after refraction is normal to the wave front after refraction. The complete differential of Eq. (2) is

$$\frac{d\mathbf{A}(r)}{ds(r)} ds(r) - \gamma \frac{d\mathbf{A}(i)}{ds(i)} ds(i)$$

$$- d\Omega \mathbf{N} - \Omega \frac{d\mathbf{N}}{ds} ds = 0,$$

(3)

where $ds(i)$, ds, $ds(r)$ are elements of length along the incident wave front, the refracting surface and the wave front after refraction. By Eq. (1c) we have

$$\frac{d\mathbf{A}(r)}{ds(r)} = \kappa(r) \mathbf{t}(r), \qquad \frac{d\mathbf{N}}{ds} = \kappa \mathbf{t},$$

$$\frac{d\mathbf{A}(i)}{ds(i)} = \kappa(i) \mathbf{t}(i),$$

(3a)

where $\kappa(r)$, $\kappa(i)$, and κ are the curvatures of the refracted wave front, the refracting surface, and the incident wave front, respectively, and $\mathbf{t}(r)$, $\mathbf{t}(i)$ and $\mathbf{t}$ are unit tangent vectors to the respective wave fronts.

If we form the vector cross product of $\mathbf{N}$ with Eq. (3) after inserting Eq. (3a) and use the fact that

$$\mathbf{t}(r) \times \mathbf{N} = \mathbf{J} \cos \phi(r), \qquad \mathbf{t} \times \mathbf{N} = \mathbf{J},$$

$$\mathbf{t}(i) \times \mathbf{N} = \mathbf{J} \cos \phi(i)$$

(3b)

where $\mathbf{J}$ is a unit vector normal to the figure, we have, after putting the coefficient of $\mathbf{J}$ in the resulting equation equal to zero,

$$\kappa(r) \cos \phi(r) ds(r) - \gamma \kappa(i)$$

$$\cdot \cos \phi(i) ds(i) - \Omega \kappa ds = 0.$$

(3c)

In order that flux be conserved (except for absorption) or that the wave front map continuously across the refracting surface, the following constraint must hold among $ds(i)$, ds, and $ds(r)$:

$$ds(i) = ds \cos \phi(i) \quad \text{and} \quad ds(r) = ds \cos \phi(r).$$

(4)

Putting Eq. (4) into Eq. (3b) yields

$$\kappa(r) = \gamma \kappa(i) \frac{\cos^2 \phi(i)}{\cos^2 \phi(r)} + \frac{\Omega \kappa}{\cos^2 \phi(r)}.$$ (4a)

In terms of wave front radii of curvature, $r(i)$ for incident, $r(r)$ after refraction,

$$\kappa(r) = \frac{1}{r(r)}, \qquad \kappa(i) = \frac{1}{r(i)}, \qquad \kappa = \frac{1}{R}$$

we have

$$\frac{1}{r(r)} = \frac{\gamma \cos^2 \phi(i)}{r(i) \cos^2 \phi(r)} + \frac{\Omega}{R \cos^2 \phi(r)}$$ (4b)

where $\gamma = n(i)/n(r)$ and $\Omega = -\gamma \cos \phi(i) + \cos \phi(r)$. The result applies to any surface. The corresponding mirror equation is

$$\frac{1}{r(r)} = \frac{1}{r(i)} + \frac{2}{R \cos \phi(i)}.$$ (4c)

Equation (4b) is the generalized lens equation, in that it relates the radius of curvature of the refracted wave front at the point of refraction to the radius of curvature of the incident wave front at the point of refraction and the radius of curvature R of the surface at the point of refraction. R may vary over the surface. The usual paraxial optics sign conventions apply if we begin by placing a minus sign in front of $r(i)$ in Eqs. (4b,c). When $\phi(r) \simeq \phi(i) \simeq 0$, Eq. (4b) becomes the paraxial equation

$$\frac{1}{r(r)} = \frac{\gamma}{r(i)} + \frac{\Omega}{R}, \qquad \Omega = -\gamma + 1.$$ (5a)

For reflection, Eq. (4c) becomes

$$\frac{1}{r(r)} = \frac{1}{r(i)} + \frac{2}{R},$$ (5b)

the mirror equation; $r(i)$ is negative in both cases, when the incident wave front is diverging from the left.

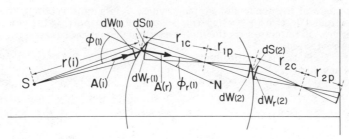

FIG. 2. Converging element of wave front of radius of curvature r_{2c} after refraction at surface 2. r_{2c} is also the distance along the refracted ray to a point on the caustic r_{2p} is the radius of curvature of the wave front incident upon surface 3.

Derivation of Flux Equation If the light originates from a point source of strength I_0, the element of flux dF emitted in angle $d\theta$ is

$$dF = I_0\, d\theta = I_0 \frac{dw(1)}{r(1)} = \sigma(1)\, dw(1) \quad (6)$$

where $\sigma(1) = I_0/r(1)$ is the flux density at position $r(1)$ and $dw(1)$ the element of wave front area. When the light originates from an element of length $ds(0)$ radiating in accordance with Lambert's law, then

$$\sigma(1) = B \cos \phi(0)\, ds(0)/r(1) \quad (7)$$

where B is the brightness and $\cos \phi(0) = \mathbf{A}(0) \cdot \mathbf{N}(0)$. Thus the flux incident upon $ds(1)$ is

$$E_{ds(0)-ds(1)} = \sigma(1) \cos \phi(1)\, ds(1). \quad (8a)$$

If the flux is refracted to $ds(2)$ then the flux density over $ds(2)$ is

$$E_{ds(0)-ds(2)} = \sigma(1)\rho_1 \cos \phi(1) \frac{ds(1)}{ds(2)}, \quad (8b)$$

where ρ_1 is the coefficient of absorption or reflection at $ds(1)$. Similarly if the flux

$$E_{ds(1)-ds(2)} \times ds(2)$$

incident upon $ds(2)$ is refracted to $ds(3)$ then the flux density over $ds(3)$ is

$$E_{ds(0)-ds(3)} = \sigma(1)\rho_1\rho_2 \cos \phi(1) \frac{ds(1)}{ds(2)} \frac{ds(2)}{ds(3)} \quad (8c)$$

and so on for any number of surfaces. Putting

$$ds(2) \cos \phi(2) = dw(2),$$
$$ds(3) \cos \phi(3) = dw(3), \quad (9)$$

which expresses continuity of flux across the interface, into Eq. (8c), one obtains

$$E_{ds(1)-ds(3)} = \sigma(1)\rho_1\rho_2 \cos \phi(1) \cos \phi(2)$$
$$\cdot \cos \phi(3) \frac{ds(1)}{dw(2)} \frac{ds(2)}{dw(3)}. \quad (10)$$

$\phi(1)$, $\phi(2)$, and $\phi(3)$, without subscripts, are angles of incidence at each surface; $\phi_r(1)$ and $\phi_r(2)$ are angles of refraction; $dw(1)$, $dw(2)$, and $dw(3)$, without subscripts, denote the elements of wave front on the incident side of the surfaces 1, 2 and 3; $dw_r(1)$ and $dw_r(2)$ denote the wave front as it leaves surfaces 1 and 2.

Likewise, using

$$\cos \phi_r(1)\, ds(1) = dw_r(1),$$
$$\cos \phi_r(2)\, ds(2) = dw_r(2), \quad (11)$$

in Eq. (10), we have

$$E_{ds(1)-ds(3)} = \sigma(1)\rho_1\rho_2$$
$$\cdot \frac{\cos \phi(1) \cos \phi(2) \cos \phi(3)}{\cos \phi_r(1) \cos \phi_r(2)}$$
$$\cdot \frac{dw_r(1)}{dw(2)} \frac{dw_r(2)}{dw(3)}. \quad (12)$$

The elements of wave front are related to the radii of curvature as follows:

$$dw_r(1) = r_{1c}\, d\theta_1, \qquad dw(2) = r_{1p}\, d\theta_1. \quad (13a)$$

$$dw_r(2) = r_{2c}\, d\theta_2, \qquad dw(3) = r_{2p}\, d\theta_2. \quad (13b)$$

See Fig. 3 for Eq. (13b). A similar figure applies to Eq. (13a) between surfaces 1 and 2. The radius of curvature of the wave front as it leaves

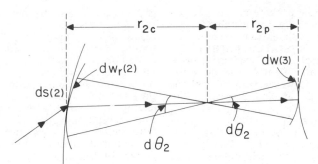

FIG. 3. Ray trace showing various quantities employed in the text.

surface 1 is denoted by r_{1c}. Its radius of curvature as it is incident upon surface 2 is denoted by r_{1p}. Likewise r_{2c} is the radius of curvature of the wave front as it leaves surface 2 at $ds(2)$ and r_{2p} is the radius of curvature of the wave front as it is incident at $ds(3)$. The orthogonality of the ray vector and the wave front is used in writing Eqs. (13,a,b). Putting Eqs. (13a,b) into Eq. (12) one obtains

$$E_{ds(0)-ds(3)} = \sigma(0)\rho_1\rho_2$$
$$\cdot \frac{\cos\phi(1)\cos\phi(2)\cos\phi(3)}{\cos\phi_r(1)\cos\phi_r(2)}$$
$$\cdot \frac{r_{1c}}{r_{1p}}\frac{r_{2c}}{r_{2p}} \qquad (14)$$

where $r_{1p} = r_{12} - r_{1c}$, $r_{2p} = r_{23} - r_{2c}$, with an obvious generalization to n surfaces. Equation (14) is the flux equation for a two-dimensional geometry, and r_{1c}, r_{2c} are the values of $r(r)$ at surfaces 1 and 2, respectively, obtained from Eq. (4b) for refraction and Eq. (4c) for reflection. This completes the derivation of the generalized lens equation and also the flux equation.

A word of caution. If the caustic curve from surface 1 intersects surface 2, then for the particular ray to that point on the caustic we have $r_{1p} = r_{12} - r_{1c} = 0$, since $r_{1p} = r_{12}$. Equation (14) then predicts a singular value for the flux density at $ds(3)$, for this ray, and very large values for the small pencil close to the ray. The singularity is not real, however, and is eliminated by writing

$$\frac{r_{1c}}{r_{1p}}\frac{r_{2c}}{r_{2p}} = r_{1c}\left(\frac{r_{2c}}{r_{1p}}\right)\frac{1}{r_{2p}}$$

and solving for r_{2c}/r_{1p} from Eq. (4b). In Eq. (4b) put $r(r) = r_{2c}$ and $r(i) = r_{1p}$, $\phi(i) = \phi(2)$ and $\phi(r) = \phi_r(2)$. Then

$$\frac{r_{2c}}{r_{1p}} = \frac{R_2\cos^2\phi_r(2)}{\gamma(2)R_2\cos^2\phi(2) + \Omega(2)r_{1p}}. \qquad (15)$$

If $r_{1p} \to 0$, then

$$\frac{r_{2c}}{r_{1p}} = \frac{\cos^2\phi_r(2)}{\gamma(2)\cos^2\phi(2)},$$

which is a finite result. If Eq. (4b) is always written as Eq. (15), this problem is automatically avoided.

DONALD G. BURKHARD

References

1. Burkhard, D. G., and Shealy, D. L., *Appl. Opt.* **20**, 897 (1981).
2. Born, M., and Wolf, E., "Principles of Optics," p. 131, Elmsford, New York, Pergamon, 1975.

Cross-references: LENS; OPTICS, GEOMETRICAL; OPTICS, GEOMETRICAL, ADVANCED; OPTICS, PHYSICAL.

IRRADIATION, DISPLACED ATOMS

The word *radiation* originally implied emanation of electromagnetic waves (photons, gamma rays), or alpha or beta rays (helium ions or electrons), and is used in that sense in the entries RADIATION, THERMAL and NUCLEAR RADIATION. It is now commonly accepted in the broader sense of emanation or bombardment by any elementary particle or ion. Indeed, the general subject area of the irradiation of solids is called *radiation effects*. The primary emphasis of the present entry is on irradiation which significantly alters the relative positions of the atoms in a solid, in contrast to cases in which the primary effect is a change in the state of ionization of the atoms (see RADIATION, IONIZING, BASIC INTERACTIONS; RADIATION CHEMISTRY; and COLOR CENTERS): Electron redistribution is sufficiently rapid in metals that ionization effects are not observable and hence the study of ionizing radiation normally is restricted to insulators and semiconductors.

The atoms in the majority of solids are arranged in very specific packing patterns which repeat three-dimensionally for distances large compared to atomic dimensions (see SOLID STATE PHYSICS and CRYSTALLOGRAPHY). This crystalline order is not usually evident: a piece of copper wire is quite flexible and does not look like a "crystal." However, truly amorphous solids (supercooled liquids and glasses) are uncommon, and the copper wire is in fact composed of many crystallites joined by grain boundaries, each "small" by human scale standards, but large enough so that the average copper atom is in a crystalline environment. Thus, the title of this entry refers to the decrease in this atomistic order caused by irradiation.

The simplest defects in the crystal lattice order of solids are *point defects* such as vacant atomic sites (vacancies) and extra atoms crammed into locations where they should not be (interstitials), and *line defects*, effectively wrinkles in the crystal structure (dislocations). Although there is still some controversy, it is generally acknowledged that self interstitials (i.e., not impurity interstitials) migrate freely in the lattice at very low temperatures ($< 100°K$), while vacancies typically might become thermally mobile at $100°C$. Self interstitials and vacancies are defined as the disturbance in the lattice order. Thus, interstitial migration does not involve a specific atom moving through the lattice, but the interstitial configuration, similar to a traveling wave, and the motion of a vacancy occurs by a neighboring atom hopping into the vacancy, leaving its site vacant.

The practical significance of the study of irradiation processes in solids is related, for example, to the degradation of mechanical properties of materials in nuclear reactors or electronic properties of semiconductors. To explain these gross effects, details of the structure and properties of solids must be understood, and so irradiation can be used as a probe to study many fundamentals of solid state physics and of defects in the crystalline order of solids.

As an energetic particle traverses a solid, it interacts with and transfers its energy to the crystal in a variety of ways. Much of the emphasis here will be on elastic collisions which impart kinetic energy to the lattice atoms. The probability of significant elastic interaction, called the scattering cross section, is sufficiently small under most circumstances that elastic collisions are spatially well separated, and can be treated theoretically as isolated events using nuclear collison theory. If, during a collision, a lattice atom recoils with kinetic energy above a threshold value, it is called a *primary knock-on*.

The theory for the creation of a primary knock-on is dependent on the irradiating particle. Photons, or gamma rays, first impart energy to electrons via the photoelectric effect or by Compton scattering, or transform to an electron and a positron by pair production (see PHOTOELECTRICITY, COMPTON EFFECT, and

POSITRON). These charged particles then travel through the lattice like incident electrons to create primary knock-ons. To reach threshold, photon energies of the order of 1 MeV are required for the two electron processes and of 5 MeV for pair production.

Electrons are so light compared to nuclei that they can only transfer a small fraction of their energy to a lattice atom in an elastic collision. For example, energies of the order of 1 MeV (and relativistic corrections in the theory) are required to reach threshold recoil energies in the 3d transition metals (e.g., Fe and Ni). High-voltage electron microscopes operate with electrons in this energy range and can be used both to create and observe the effects of radiation damage. With electron bombardment, the primary knock-on atoms normally do not have sufficient energy to create a significant number of additional knock-ons.

With ion bombardment, much of the incident particle energy can be transferred to a single lattice atom. At high energies, often in the tens of MeV range, there is a coulombic interaction between the incident ion and the lattice atom cores, and the scattering is approximated by Rutherford theory. Here the primary knock-on can have sufficient energy to act like an incident particle, which can lead to a cascade effect. However, most of the energy loss in this range is due to electronic excitation which, because it is less than 1 MeV, does not yield displaced atoms. As energies decrease, the interactions become more like hard-sphere collisions between atoms. The changeover from Rutherford to hard-sphere scattering occurs in the A-keV range, with A the atomic weight of the moving ion. Although most of the energy loss is in the Rutherford range, the majority of the displaced atoms are produced by hard-sphere scattering.

Neutrons can produce lattice atom displacements either by scattering processes which leave the atom unchanged or by nuclear reactions. For nuclear reactions, a lattice atom can absorb a thermal neutron and then undergo a radioactive transformation, such as $Fe^{58}(n, \gamma) Fe^{59}$ The iron atom recoils to conserve momentum as the γ is emitted, with energy, for example, somewhat less than a keV, but well above threshold to become a primary knock-on and act to produce a small cascade. Other reactions with thermal neutrons, such as (n, α), (n, β), and (n, f) (fission) also produce primary knock-ons.

Neutron scattering generally involves "fast" neutrons, i.e., with energies in the keV and MeV range. As in the thermal neutron case, lattice atoms generally receive kinetic energies well above the primary knock-on threshold. At the low energy end of neutron scattering, the interaction is primarily elastic, but as energies increase, lattice atom nuclei can become excited, leading to inelastic scattering.

The discussion so far has been concerned with the process by which a primary knock-on is created, and the concept of a threshold energy was presented as intuitively reasonable. It is similarly clear that, given the highly ordered crystal structure, the threshold energy must vary with direction, i.e., the threshold energy is a function of orientation. A minimum value of 25 eV is commonly accepted as typical, with 4.5 times the sublimation energy yielding a rough correlation with the experimental data. Minima tend to occur near high symmetry directions, with the sequence from most to least favorable being, for example, $\langle 110 \rangle$, $\langle 100 \rangle$, $\langle 111 \rangle$ in fcc copper and $\langle 100 \rangle$, $\langle 110 \rangle$, $\langle 111 \rangle$ in bcc iron.

If a knock-on atom only travels a short distance in the lattice, say several interatomic spacings, it momentarily becomes an interstitial, leaving a vacancy at its initial lattice site, but then this interstitial may spontaneously recombine with its vacancy to yield no net damage. As noted above, interstitials of this type move through the lattice as a disturbance in the order, not as a specific atom. The atom which returns to the knock-on site is generally not the atom initially there, so that this subthreshold event can, for instance, yield mixing of ordered alloys.

If the interstitial is somewhat further removed from the knock-on site, a stable interstitial-vacancy complex, called a *close Frankel pair*, is created. There is a binding energy between the two defects so that they preferentially recombine when one becomes mobile. At greater separations yet, the interstitial and vacancy do not interact, are called a *separated Frankel pair*, and migrate in a random path upon becoming mobile.

For primary knock-on energies above but not too much greater than threshold, the interstitial is formed at a distance from the knock-on site by a replacement chain process, which occurs most readily near high symmetry directions. The primary knock-on hits a neighbor almost head on, and, because they have the same mass, transfers most of its energy to that atom. The primary knock-on atom then remains in that neighboring position while the atom which has been there hits one of its neighbors, etc. While a highly energetic particle may have a long path in a solid, the majority of defects created in a cascade occur through this mechanism, often yielding a depleted zone surrounded by a shell rich in interstitials. This configuration, called a *displacement spike*, is raised to a very high temperature for a short duration of time because so much energy is deposited in a small region. The hot spot only lasts for perhaps 100 lattice vibrations before cooling below the melting temperature. Subthreshold collision chains similar to replacement chains are considered to be the primary means by which energy is rapidly removed from a displacement spike event. Experience with billiard balls would indicate that replacement chains or subthreshold collision chains would rapidly defocus, but the combined effects of the lattice structure and scattering theory more realistic than hard spheres indicates that conditions for focusing should be quite common.

Another focusing type of phenomenon is called *channeling*, whereby a particle (incident or knock-on) travels with little energy loss per unit length down an open channel in the crystal structure. Such channels are easily seen in ball-and-stick lattice models. Under careful laboratory conditions with properly oriented single crystals, incident particles which normally would be stopped can readily pass through a foil by such channeling.

Much of the interest in radiation damage theory is concerned with the estimation of the amount and spatial distribution of the disorder caused by a given irradiation. The ideas discussed above have been incorporated into complex formulations which give good estimates of the formation of damage but normally require numerical methods to solve. However, even if the disorder created was accurately known, direct comparison with experiment is extremely difficult because the defects are so small. Also, at all but very low temperatures, some if not all of the defects are mobile so that annealing takes place continuously. The irradiation drives the solid away from thermal equilibrium, but, if any kinetic pathways are available, the solid will adjust to thermodynamically lower its free energy.

The simplest annealing process is interstitial-vacancy recombination, called *annihilation*. The defects can also disappear at sinks such as surfaces, grain boundaries and dislocations, or clusters of a given type may form. These clusters can transform into dislocation loops, changing the effective sink concentration by orders of magnitude. The migrating defects can interact with alloying elements leading to solute segregation. The net effect of these, and many additional effects, is a continuous evolution of the microstructure of a solid under irradiation.

The initial defects created in a solid are too small to be observed directly, although strides have been made with field ion microscopy (see FIELD EMISSION). Electrical resistivity is probably the most standard experimental technique used to study the amount of disorder in a solid. A problem is that resistivity, and most other procedures, are not specific: vacancies and interstitials have about the same electron scattering strength, and, for example, a divacancy has about twice the scattering strength of a single vacancy. Sophisticated experimental methods such as positron annihilation spectroscopy in which different defects give different signals, are not without their own ambiguities.

The most significant practical effects in metals are related to mechanical properties. For example, an increase in the dislocation content produces a loss of ductility, dislocation climb leads to high-temperature creep, clustering of vacancies into voids yields bulk swelling, and segregation of impurities at grain boundaries can decrease fracture strength. The creation of damage in semiconductors (see SEMICONDUCTORS) is similar to that in metals, but here the electronic effects are most significant. The disorder gives rise to additional electronic states in the band gap which can act as donors, acceptors, or trapping sites, which in turn can completely alter the sensitive electronic properties built into a particular device.

R. A. JOHNSON

References

1. Dienes, G. J., and Vineyard, G. H., "Radiation Effects in Solids," New York, Interscience Publishers 1957.
2. Billington, D. S. (Ed.), "Radiation Damage in Solids," New York, Academic Press, 1962.
3. Peterson, N. L., and Harkness, S. D. (Eds.), "Radiation Damage in Metals," Metals Park, OH, American Society for Metals, 1976.
4. J. Gittus, "Irradiation Effects in Crystalline Solids," London, Applied Science Publishers, 1978.
5. Peterson, N. L., and Siegel, R. W. (Eds.), "Properties of Atomic Defects in Metals," New York, North Holland Publishing, 1978.

Cross-references: COLOR CENTERS; COMPTON EFFECT; CRYSTALLOGRAPHY; FIELD EMISSION; LATTICE DEFECTS; NUCLEAR RADIATION; PHOTOELECTRICITY; POSITRON; RADIATION CHEMISTRY; RADIATION, IONIZING, BASIC INTERACTIONS; RADIATION, THERMAL; SEMICONDUCTORS.

IRREVERSIBILITY

Physical systems commonly display a tendency to change spontaneously from one state to another, but not to change in the opposite direction. Examples are the tendency of heat to pass from regions of high temperature to regions of low temperature, the tendency of mechanical or electrical energy to be transformed into heat by friction or resistance, and the mixing or diffusion of different substances. While irreversibility appears to be an obvious feature of macroscopic natural phenomena, so much so that violations of this tendency are scarcely conceivable, it is not yet established whether it should be considered a general law applicable on both the atomic and the cosmological scales. Most physicists accept the "statistical" explanation of irreversibility, according to which complex systems with many degrees of freedom tend to spread out among more and more diverse states in the "phase space" of possible configurations (see STATISTICAL MECHANICS). Scientific discussions of this question go back to the time of Newton and continue up to the present day. A brief review of these earlier discussions is necessary in order to illustrate how the problem of irreversibility is related to our general view of the world. We shall be concerned here mainly with the qualitative aspects of irreversibility as distinct from quantitative theories of particular processes (see, e.g., the articles on DIFFUSION, HEAT TRANSFER, THERMOELECTRICITY, and VISCOSITY).

In the seventeenth century, the "clockwork universe" theory was popularized by the French philosopher René Descartes and the British scientist Robert Boyle. According to Descartes and Boyle, the physical world is like a perfect machine which, once created and set in motion by God, can run forever without any further need for divine intervention. This mechanistic view of nature dominated much scientific work (and conditioned many of the influences of physics on biology, psychology, philosophy, and political thought) up to the nineteenth century. Since its most striking successes were attained in the Newtonian theory of the solar system, it has often been attributed to Newton in such phrases as "the Newtonian world-machine." But Newton himself rejected it, pointing to the importance of irreversibility in phenomena such as the viscosity of fluids and the imperfect elasticity of solids; in his *Opticks* he stated that "motion is much more apt to be lost than got, and is always upon the decay." He also thought that mutual gravitational perturbations of planets in the solar system would accumulate over long periods of time, producing instabilities that could be corrected only by divine intervention.

In 1715 the German philosopher G. W. Leibniz attacked Newton's opinion, arguing that the suggestion that "God almighty needs to wind up his watch from time to time; otherwise it would cease to move" (as Leibniz put it) was a slur on God's ability to make a perfect world-machine. Newton retorted that the clockwork metaphor was too materialistic, and was likely to encourage the notion that God has no active role in the world at all. Newton's suspicions were well-founded; later improvements on his calculations of gravitational perturbations indicated that a solar system governed only by Newton's laws is stable, so that all changes in planetary orbits merely oscillate between fixed limits. Thus when Napoleon asked the French mathematician P. S. de Laplace why he had not mentioned God in his treatise on celestial mechanics, Laplace could reply: "Sir, I have no need of that hypothesis."

When Newton asserted that motion tends to be lost in processes such as friction or inelastic collisions, he was refusing to accept the suggestion (already well-known at that time) that it is simply transformed into invisible molecular motion, perceptible only as heat. The modern concept of irreversibility is different from Newton's because it involves a *dissipation* but not a *destruction* of energy. One might therefore expect that the modern theory of irreversibility could be established only after the law of conservation of energy had been accepted (around 1850); in a sense this is true, but it is somewhat misleading historically, as we shall see.

In 1852 the British physicist William Thomson (later known as Lord Kelvin) published a short paper, "On a Universal Tendency in Nature to the Dissipation of Mechanical Energy." The immediate stimulus for this pronouncement was Thomson's reflections on the consequences of Carnot's theory of steam engines, which he and the German physicist Rudolf Clausius had recently used as a basis for the second law of thermodynamics. In formulating the second law, Clausius and Thomson argued that it must be impossible to transfer heat from a cold body to a warm body without some kind of compensating process equivalent to the transfer of at least the same amount of heat in the opposite direction; or, as Thomson phrased it, it is impossible to obtain mechanical work by cooling a body below the temperature of its surroundings. But these "impossibility" or "impotency" statements of the second law were closely associated with the conviction that the forbidden processes were forbidden because they contravened a *natural tendency* for heat to flow from high temperatures to low. Thus, in the development of theories of energy dissipation, the simple flow of heat from hot to cold was the first irreversible process to be recognized; other processes were subsequently said to be irreversible, in part because they were equivalent or analogous to the equalization of temperature differences.

Although Thomson's explicit statement of the principle of dissipation of energy occurred in a thermodynamic context involving transformations of heat and mechanical work, it is evident from his own writings that he was equally concerned with an irreversible process that involved no change in the total quantity of heat: namely, the cooling of the earth from a hypothetical initial molten state to its present state (moderate surface temperature with residual internal heat), ultimately reaching the desolate cold of interplanetary space. This was the predicted "heat death" of the earth (and perhaps of the entire universe) that attracted the attention of popular science writers in the late nineteenth century, and was publicized further by the British astrophysicists J. H. Jeans and A. S. Eddington in the 1920s. But Thomson was not the first to discuss the cooling of the earth; indeed, this had been a favorite topic of debate among scientists in the eighteenth and early nineteenth centuries, and provided the occasion for several assertions that there is a natural tendency for heat to flow from hot to cold. Even the threat of a heat death as a fate common to all bodies in the universe had been hinted as early as 1777 by the French astronomer Jean-Sylvain Bailly. Fourier's theory of

heat conduction (whose development was motivated in part by the problem of terrestrial temperatures) was the first major physical theory in modern times to incorporate irreversibility in its basic postulates.

With the advent of thermodynamics in the 1850s, the concept of irreversibility could be extended to processes involving transformations among different forms of energy. Thomson's use of the word "dissipation" introduced a moral connotation based on the observation that whenever heat flows through a finite temperature difference a certain amount of mechanical work *could* be obtained by appropriate use of a heat engine; the maximum work is obtained when each temperature difference in the cycle of operations is made infinitesimally small so that the direction of heat flow could be reversed by a slight alteration of the temperatures (see CARNOT CYCLES AND CARNOT ENGINES). Whenever that is not done, heat flows irreversibly and the chance of doing mechanical work has been missed, hence the energy involved has been "dissipated."

In 1854, Clausius formulated his thermodynamic theory in terms of the "equivalence-value of a transformation," defined as the amount of heat transferred at a certain temperature divided by that temperature. For a cyclic process the total equivalence-value would be determined by an integral over the path of this ratio,

$$\int \frac{dQ}{T}.$$

For a reversible process the value of this integral would be zero, according to Clausius's statement of the second law; whereas for an irreversible process it would be positive. Since this particular mathematical expression proved to be useful in developing his theory, Clausius finally gave it the name *entropy* in 1865; more precisely, the *change* in entropy in a process is defined as

$$dS = dQ/T.$$

Thus entropy provided an indicator of irreversibility, and Clausius could state the second law in the generalized form, "the entropy of the world tends toward a maximum." It is in this form, or simply the phrase "entropy tends to increase," that the principle of irreversibility or principle of dissipation of energy is now ordinarily stated. Eddington asserted that this principle "holds the supreme position among the laws of Nature" because it determines the direction of time itself—"time's arrow."

Although one could simply accept the principle of irreversible entropy increase as a fundamental law of physics without further explanation (as is sometimes done in texts on thermodynamics), there is widespread sentiment among scientists that it must be possible to reduce the law to a more basic postulate about atomic behavior (or about the universe as a whole). This reduction seemed an urgent necessity for those physicists in the nineteenth century who believed that all properties of matter and energy were ultimately explicable by mechanical models, using Newton's laws of motion. Since the kinetic theory of gases was being developed during the same period when entropy was being introduced into thermodynamics, that theory offered a natural starting point. The following conclusions were reached as a result of analyses by Thomson, James Clerk Maxwell, and Ludwig Boltzmann (see KINETIC THEORY):

(1) It is reasonable to assume that the principle of irreversibility is *not* an absolute law of nature and that all processes involving individual atoms are perfectly reversible, since Newton's second law of motion is unchanged when $-t$ is substituted for t, and there is no direct evidence that interatomic forces are dissipative. In that case irreversibility can appear only when large numbers of atoms are involved, and is connected with the fact that observable macroscopic states (specified by parameters such as temperature and pressure) of a system correspond in general to enormously large collections of microscopic states, the latter being specified by values of the positions and velocities of all the atoms. In a typical irreversible process, a system passes from macroscopic state A corresponding to a small number of microscopic states, to another macroscopic state B corresponding to a much larger number of microscopic states. The process could be reversed only if there were some way to pick out the special microscopic states in B that had evolved from microscopic states in A, and then to reverse the directions of all atomic velocities.

(2) In this view irreversibility is a *statistical* property, which appears to be a general law of nature only because the probability of reversal (e.g., the probability that entropy will spontaneously decrease) is extremely small. There is some room for philosophical or methodological debate as to whether a statement that is always true in practice but *may* be wrong once in a billion years "in principle" (according to a theory of atomic behavior) should be considered a law of nature.

(3) The use of statistical terminology in explaining the principle of irreversibility nevertheless suggested that this principle may depend on *randomness at the atomic level*. In 1894, S. H. Burbury pointed out that the proof of Boltzmann's "H theorem" involved the assumption that two colliding molecules are uncorrelated *after* as well as *before* they collide, and argued that this assumption could be justified if the system were subjected to continuous random external disturbance. Boltzmann, and later Max Planck, agreed that such an assump-

tion could be used to derive irreversibility. Yet it was not until the advent of quantum mechanics in the 1920s that most scientists were ready to accept the idea that atomic behavior may involve an inherent element of randomness or indeterminacy. While the late-nineteenth-century debates on irreversibility thus foreshadowed the breakdown of determinism in physics, it has not yet been established that irreversibility should be attributed to atomic randomness.

The statistical theory developed by Maxwell, Thomson, and Boltzmann achieved a substantial clarification by describing irreversible processes in terms of the *mixing* of atoms with different properties; the flow of heat, or the transformation of mechanical energy into heat, can be seen as special cases in which the relevant atomic property is velocity. Thus entropy increases whenever one goes from an ordered state to a disordered state, and entropy itself can be regarded as a measure of disorder. That terminology also suggests that entropy is a measure of the amount of *information* (or rather *lack* of information) which we have about the system (see ENTROPY). An extreme interpretation based on this idea is that irreversibility is not an inherent property of physical systems but is merely a necessary feature of any human description of them.

There have been numerous attempts to construct or imagine devices that would permit a violation of the second law of thermodynamics, i.e., to accomplish "perpetual motion of the second kind" by arranging for heat to flow from cold to hot regions, or by extracting mechanical work from an isothermal heat reservoir. Brownian movement has sometimes been considered an example of the latter kind. Nevertheless there are no generally accepted instances of violations of irreversibility. On the other hand, scientists hesitate to accept the view of Clausius that irreversible entropy increase leads to a final state in which "no further change could evermore take place, and the universe would be in a state of unchanging death."

In addition to the statistical explanation of irreversibility, there have been attempts to discover violations of "time-reversal invariance" in the interactions of elementary particles (see PARITY) and to associate the "direction of time" with the expansion of the universe (see the article by Gal-Or cited below). The persistent philosophical interest in the nature of time ensures that debates about irreversibility will continue to enliven discussions of the foundations of physics for many years to come.

It is sometimes suggested that there is a conflict between biological evolution and the principle of irreversibility, since the former involves the transformation of simple forms of life into more complex ones while the latter involves the transformation from order to disorder. However, there is no conflict if it is recognized that the principle predicts an increase of entropy only in a closed system taken as a whole, and that the statistical nature of irreversibility leads one to expect local decreases in entropy which are compensated by larger increases elsewhere. In the case of biological evolution the system must include the sun; the entropy increase associated with the transfer of energy from the sun to the earth ($dS = dQ/T_{sun} - dQ/T_{earth}$) is enormous compared to the decrease in biological processes that utilize a small part of this energy.

As Boltzmann himself pointed out in a lecture in 1904, Darwin's theory of natural selection can be derived qualitatively from statistical thermodynamics. The equilibrium state of a molecular system is not simply the one of greatest entropy or disorder, but involves a balance between disorder and the forces that favor an ordered state; thus at sufficiently low temperatures water molecules will spontaneously form complex crystals. Such an ordered state, initially formed by chance, will be "selected" to survive under certain conditions. The chemical reactions that occur in living systems are quite consistent with thermodynamics, as Harold Blum showed in his classic monograph "Time's Arrow and Evolution (Princeton University Press, 1951).

During the past two decades several scientists have developed more detailed theories of the irreversible processes in biological systems. One of the best known is that of Ilya Prigogine and his collaborators. Their theory shows how a system driven far away from equilibrium may acquire a new kind of order, just as a fluid heated from below will develop a regular pattern of convection cells. Prigogine received the 1977 Nobel Prize in Chemistry for his theory of "dissipative structures," including its application to biology.

STEPHEN G. BRUSH

References

Brush, S. G., "Kinetic Theory," Vol. 2, "Irreversible processes," New York, Pergamon Press, 1966. (Includes reprints and translations of papers by Maxwell, Boltzmann, Thomson, Poincaré, and Zermelo.)

Brush, S. G., "The Development of the Kinetic Theory of Gases. VIII. Randomness and Irreversibility," *Archive for History of Exact Sciences* 12, 1 (1974), reprinted in "The Kind of Motion We Call Heat: A History of the Kinetic Theory of Gases in the 19th Century," Amsterdam, North-Holland, 1976, Chapter 14.

Brush, S. G., "Irreversibility and Indeterminism: Fourier to Heisenberg," *Journal of the History of Ideas* 37, 603 (1976), reprinted in "Statistical Physics and the Atomic Theory of Matter from Boyle and Newton to Landau and Onsager," Princeton, N.J., Princeton Univ. Press, 1983, Chapter 2.

Eddington, A. S., "The Nature of the Physical World," London, Cambridge Univ. Press, 1928, Chapter IV.

Fraser, J. T., Haber, F. C. and Müller, G. H. (Eds.), "The Study of Time," New York, Springer-Verlag, 1972.

Gal-Or, B., "The Crisis about the Origin of Irreversibility and Time Anisotropy," *Science* 176, 11 (1972).

Hinds, E. A., "Parity and Time-Reversal in Atoms," *American Scientist* 69, 430 (1981).

Kubrin, D., "Newton and the Cyclical Cosmos: Providence and the Mechanical Philosophy," *Journal of the History of Ideas* 28, 325 (1967).

Layzer, D., "The Arrow of Time," *Scientific American* 233 (6), 56 (December 1975).

Misra, B., and Prigogine, I., "Time, Probability, and Dynamics," in "Long-Time Prediction in Dynamics" (C. W. Horton et al., Eds.), New York, Wiley, 1982.

Park, David, "The Image of Eternity: Roots of Time in the Physical World," Amherst, Mass., Univ. Massachusetts Press, 1980.

Prigogine, I., Nicolis, G., and Babloyantz, A., "Thermodynamics of Evolution," *Physics Today* 25 (11), 23 (Nov. 1972), 25 (12), 38 (Dec. 1972).

Prigogine, I., "Time, Structure, and Fluctuations," *Science* 201, 777 (1978).

Reichenbach, H., "The Direction of Time," Berkeley, Univ. California Press, 1956.

Cross-references: CARNOT CYCLES AND CARNOT ENGINES, DIFFUSION, ENTROPY, HEAT TRANSFER, KINETIC THEORY, PARITY, STATISTICAL MECHANICS, THERMODYNAMICS, THERMOELECTRICITY, VISCOSITY.

ISOSPIN

Isospin (also called *isobaric spin* and *isotopic spin*) is an attribute ascribed to particles in a mathematical formalism introduced originally to simplify calculations involving the interactions among protons and neutrons through nuclear forces. In the isospin formalism, the neutron and proton are treated as though they were two different quantum-mechanical states of the same entity, the nucleon. A neutron is distinguished from a proton by the value of its isospin projection quantum number, t_z: the value $t_z = \frac{1}{2}$ is arbitrarily assigned to the neutron and $t_z = -\frac{1}{2}$ to the proton. (This convention is common in nuclear structure literature, but the opposite assignments for t_z are found in particle physics literature.)

The idea of using such a formalism for treating neutrons and protons was first suggested by Heisenberg,[1] who used the Pauli spin matrices to represent the attribute that we call isospin. Wigner[2] later named the concept *isotopic spin*. Subsequent authors pointed out that the attribute, when extended to nuclei, is relevant not to isotopes but to nuclei with the same mass, hence the name *isobaric spin* replaced *isotopic spin*. In recent literature the name has been shortened to *isospin*. In spite of the name, the concept has nothing to do with mechanical angular momentum. (See MATHEMATICAL PRINCIPLES OF QUANTUM MECHANICS for a discussion of quantum-mechanical angular momentum.)

The mathematical apparatus of the isospin formalism belongs to the general subject of group theory and is identical to that used for treating electron spin in atomic physics (see ELECTRON SPIN). Four quantum-mechanical operators, t_z, $t^+ \cdot t^-$, and t^2, constitute the basic elements of the formalism. The number t_z that distinguishes neutrons from protons is the eigenvalue of the operator t_z, analogous to the operator s_z that gives the projection of the electron spin on the quantum axis. The operator t^+ operating on the representation of a proton changes it into the representation of a neutron and t^- changes the representation of a neutron into that of a proton. The effects of these operators can be written symbolically as

$$t_z\, R(p) = \tfrac{1}{2} R(p); \qquad t_z\, R(n) = \tfrac{1}{2} R(n)$$

$$t^+ R(p) = R(n); \qquad t^- R(n) = R(p)$$

$$t^+ R(n) = 0; \qquad t^- R(p) = 0.$$

The letter R is used to indicate a representation of the entity specified in the parentheses. The representation can be a matrix, a wave function, a diagram, or any other representation appropriate to the method being employed in the calculation. The operator t^2 is related to the other operators by the equation $t^2 = \tfrac{1}{2}(t^+ t^- + t^- t^+) + t_z^2$. (Note that t^+ and t^- do not commute. Therefore $t^+ t^-$ does not equal $t^- t^+$.) For the case of the nucleon, where we have set up the defining equations to include only two states, t^2 has the eigenvalue $\tfrac{3}{4}$. The formalism may be applied also to other kinds of particles, and the generalized operator t^2 has eigenvalues of the form $t(t + 1)$, where t is either a half-integral or an integral number, and $2t + 1$ is the number of possible states. The value of t is called the *isospin of the particle*. Thus, it may be said that the nucleon is a particle of isospin $\tfrac{1}{2}$. From a general point of view, the relationships stated above may be viewed in terms of a vector operator t in a fictitious three-dimensional space where

$$t = t_x + t_y + t_z$$

$$t^+ = t_x + it_y$$

$$t^- = t_x - it_y.$$

This space, which we call *isospace*, is mathematically analogous to ordinary coordinate space but represents attributes of electrical charge rather than ordinary positions and lengths. If we represent an object by a vector $|R\rangle$ in isospace, we can determine the length of the vector by forming the scalar product $\langle Rt|tR\rangle$, which is equivalent in meaning to $\langle Rt^2 R\rangle$. If R is normalized so that $\langle R|R\rangle = 1$ and is an eigenfunction of t^2, we find that

$$\langle Rt^2 R\rangle = t(t + 1)$$

where t may be thought of as the length of the vector. The values that t may have are integers for bosons and half-integers for fermions.

The isospin representations are members of a rotational group in isospace. Starting with a vector $|R\rangle$ one can generate new orthogonal vectors in isospace through the application of the raising and lowering operators t^+ and t^-. The newly created vectors are characterized by the same eigenvalue of t^2 as the original vector but different eigenvalues of t_z. The operations may be pictured as rotations of the vector in isospace, and the set of vectors created in this way are the possible orientations of the vector $|R\rangle$ in isospace that can be generated through rotation alone. There are $2t + 1$ possible orientations. We say that $|R\rangle$ is a member of an *isospin multiplet* of multiplicity $2t + 1$.

The concept of isospin is especially useful whenever one wants to describe a set of particles differing in electrical charge but similar in other attributes. For example, positive, neutral, and negative pi mesons (or pions) may be described as an isospin triplet. To achieve multiplicity three, t must be 1 and the particle must be a boson.

At the time Heisenberg introduced this method of manipulating neutron and proton representations, it might well have looked like a contrived device of questionable utility. However, later in the 1930s, as evidence began to mount that the forces between nucleons were the same whether they were protons or neutrons, except for the additional electrical repulsion between protons, the concept of isospin gained in apparent utility. Eventually, the formalism evolved into something that could be applied to aggregate systems in which the constituents possessed the attribute of isospin, for example, to nuclei.

To gain insight into the physics underlying the concept of isospin one might note that part of the evidence for the charge independence of the nuclear force came from a study of the energy-level structure for mirror nuclei. Mirror nuclei are related to each other by an interchange of their neutron and proton numbers. For example, the nuclei ^{7}Li (3 protons and 4 neutrons) and ^{7}Be (4 protons and 3 neutrons) are a mirror pair and their energy levels accurately match each other except for an overall displacement. This correspondence suggests that it might be useful to view ^{7}Li and ^{7}Be as though one were related to the other as the neutron is related to the proton. The concept of total isospin in a many-nucleon system is defined in a way that is consistent with the single-nucleon formalism and gives meaning to statements of the type $T^+ R(^7\text{Be}) = R(^7\text{Li})$. The concept of total isospin together with a structure model can provide the relationship between the isospin of the nucleus and the arrangement of the constituent nucleons in the nucleus.

In the nuclear shell model, one pictures the nucleus as though it contains a set of single-particle quantum states that result from the sum of the interactions among the nucleons. The different energy levels of the nucleus are then pictured as different combinations of occupancy by the nucleons of the available single-particle states. The similarity of level structure for mirror nuclei suggests that similar sets of single-particle states exist for protons and neutrons.

In an isospin formalism each single-particle state occurs twice, once with $t_z = \frac{1}{2}$ and once with $t_z = -\frac{1}{2}$, and the states are occupied by nucleons. We can define a set of total isospin operators for the aggregate system as follows: The total isospin projection operator $\mathbf{T}_z$ is the sum of the individual nucleon isospin projection operators, $\mathbf{T}_z = \Sigma_i t_z$. To find its eigenvalue we need only to sum the isospin projection quantum numbers for all the occupied states. We find that the eigenvalue of $\mathbf{T}_z$ is $T_z = \frac{1}{2}(N - Z)$, where N is the number of neutrons and Z the number of protons in the nucleus. We can also define total operators corresponding to the other single-nucleon operators that we discussed previously:

$$\mathbf{T}^+ = \sum t^+; \qquad \mathbf{T}^- = \sum t^-;$$
$$\mathbf{T}^2 = \tfrac{1}{2}(\mathbf{T}^+\mathbf{T}^- + \mathbf{T}^-\mathbf{T}^+) + \mathbf{T}_z{}^2.$$

The operator $\mathbf{T}^+$ when applied to a representation of a many-nucleon system changes it to a representation with $T_z + 1$, if T_z is the original value. Similarly $\mathbf{T}^-$ changes the representation to a new one with $T_z - 1$. (Certain numerical factors have been omitted for simplicity. Refer to a textbook on quantum mechanics for a more complete treatment of raising and lowering operators.) The operator $\mathbf{T}^2$ has eigenvalues of the form $T(T + 1)$, where T is a number that is referred to as the *total isospin of the system*. (For a definition of eigenvalue, see MATHE-MATICAL PRINCIPLES OF QUANTUM ME-CHANICS). The value of T is always greater than or equal to the absolute value of T_z and is an integer for nuclei with an even number of nucleons and a half-integer for nuclei with an odd number of nucleons. An arbitrary representation of a many-nucleon system is not necessarily an eigenfunction of $\mathbf{T}^2$, but it can be reexpressed as a linear combination of representations that are eigenfunctions of $\mathbf{T}^2$.

Before the 1960s the concept of total isospin was used in formal classifications of shell model states, but it was not generally believed that the simple eigenfunctions of $\mathbf{T}^2$ would provide useful descriptions of real nuclear states except with respect to mirror nuclei and very light nuclei. It was thought that in a nucleus with a large number of protons the Coulomb force

would be strong enough to destroy the approximate symmetry between the proton and neutron states and isospin would not be a useful concept.

The discovery of isobaric analog states[3] gave isospin a new importance in nuclear structure physics. Analog states are energy levels of one nucleus that have a special relationship to energy levels in the neighboring nucleus with the same number of nucleons. If we compare the energy levels of a nucleus $A(N, Z)$ with N neutrons and Z protons, to the energy levels of a nucleus $B(N - 1, Z + 1)$, we find in B a set of levels at high excitation energies that match one-for-one the lowest few levels of A with respect to energy spacing and other properties.

In terms of isospin, analog states can be understood as states that differ only in their isospin projection T_z. That is, all quantum numbers except T_z of the state in nucleus A are the same as those of its analog in B. The ground state of a stable nucleus with more neutrons than protons is always characterized by $T = T_z = (N - Z)/2$. Its analog is characterized by the same value of T, but T_z is one unit lower, i.e., $T_z = T - 1$. That state may in turn have an analog in its neighboring isobar with $T_z = T - 2$, and so on until $T_z = -T$. States in mirror nuclei are simply analog states with T_z (for nucleus A) $= -T_z$ (for nucleus B).

The binding energy of an analog state differs from that of its parent state, and the energy difference is called the *Coulomb displacement energy*. The approximate value of this energy can be calculated in a rather simple way, if one assumes that the nuclear volume is filled with a uniform charge density. Since both parent and analog nuclei have the same number of nucleons, the volumes are the same. For the parent nucleus, the fraction $Z/(N + Z)$ of the nucleons are charged, while for the analog, $(Z + 1)/(N + Z)$ of the nucleons are charged. The value of the Coulomb displacement energy is approximately the difference in the electrostatic energy for spherical volumes of charge densities implied by those two charge fractions. This has the value of about 12 MeV for nuclei of about 100 atomic mass units, for example. In spite of large differences in absolute binding energies, the analog level spacings match those of the parent accurately for nuclei of all masses that have been studied. It thus appears that the similarity between neutron and proton single-particle states persists even when the Coulomb force produces a large change in the absolute binding energies of the states. The effect of the coulomb force is to introduce an approximately uniform shift in the energy levels of nuclei differing only in T_z.

Isospin considerations can be a guide to certain features of nuclear reactions. For example, a rule of conservation of isospin may be formulated by analogy with the principle of conservation of angular momentum. The rule states that if the initial system—the projectile plus the target—can be characterized by total isospin quantum numbers T and T_z, then the final system will also be characterized by the same values of T and T_z. The conservation of T_z is equivalent to the statement that the total number of protons and the total number of neutrons does not change in a nuclear reaction. The conservation of T implies restrictions on the rearrangements of the nucleons within the nuclei but is not a very strict rule, since it is based on neglect of the Coulomb force between protons. It has been shown, however, that processes that violate the rule are severely inhibited compared to processes that follow the rule. Some broader implications of isospin conservation have been pointed out by Adair.[4]

C. D. GOODMAN

References

1. Heisenberg, W., *Z. Physik* 77, 1–11 (1932).
2. Wigner, E., *Phys. Rev.* 51, 106–119 (1937).
3. Anderson, J. D., Wong, C., and McClure, J. W., *Phys. Rev.* 126, 2170–2173 (1962).
4. Adair, R. K., *Phys. Rev.* 87, 1041–1043 (1952).

For additional reading see: "Isospin in Nuclear Physics" (Wilkinson, D. H., Ed.), Amsterdam, North-Holland Publishing Co., 1969.

Cross-references: ELECTRON SPIN, ELEMENTARY PARTICLES, MANY-BODY PROBLEM, MATHEMATICAL PRINCIPLES OF QUANTUM MECHANICS, NEUTRON, NUCLEAR STRUCTURE, PROTON, QUANTUM THEORY, STRONG INTERACTIONS.

ISOTOPES

The word *isotopes*, stemming from the Greek word *isos* (same) and *topos* (place), refers to atoms of the same element which have differing masses. The term *isotope* was first proposed by Soddy (1913) to designate substances having different atomic weights and, at the same time, having chemical properties which were so closely allied as to make chemical methods of producing a separation ineffective. Hence, Soddy suggested that they were chemically identical, i.e., they occupied the same place in the periodic table. The term *nuclide* is, perhaps, a more accurate representation, since it is defined as a particular atomic species with an atomic number Z and a mass number A.

In 1905, Boltwood, studying the decay chains of uranium, identified a new element which he called *ionium* and which, when admixed with thorium, could not be separated by any of the known chemical processes. Similar behavior was observed by Soddy and others in other such mixtures. These studies depended on the detection and identification of minute amounts of a substance by observing the type of radiation and the half-life. Indeed, additional evidence came from the observation of the visible spectrum of a mixture of two of these substances and finding no new lines. Further, Boltwood, in his studies, noted the presence of lead in uranium minerals and suggested that lead might be the end product of the uranium series. As a result of the study of the relation of lead to uranium in a large number of minerals, this view was generally adopted. Soddy concluded that the end products of the uranium and thorium series should be lead with isotopic weights 206 and 208, respectively, whereas ordinary lead as found in nature has an atomic weight of 207.2. Soddy and Hyman reported (1914) that the atomic weight of lead as found in thorite, which consists mainly of thorium with 1–2% uranium and 0.4% lead, indeed had a slightly greater atomic weight than that of ordinary lead. It was not, however, until the development of the mass spectrograph (1919) that a detailed study of the occurrence and relative abundance of such species could be accurately made and it was not until the discovery of the neutron (1932) that a satisfactory explanation could be given.

Of the 81 naturally occurring stable elements, 22 elements are anisotopic, i.e., possess one stable isotope. The largest number of stable isotopes occurs in tin, which has ten. Since the atoms in multinuclidic elements are chemically identical and have the same number of protons in the nucleus, the varying masses are accounted for by the variable number of neutrons in the nucleus ($Z = 50$, $A = 112$, 114, 115, 116, 117, 118, 119, 120, 122, 124). Two of the elements below bismuth ($Z = 83$), technetium ($Z = 43$) and promethium ($Z = 61$), do not have any stable nuclides. Table 1 lists, in order of increasing atomic number Z (the charge or number of protons associated with the nucleus) the 106 known elements, and where available, the relative atomic weights. It should be noted that the conflict regarding the Soviet and United States claims to a discovery of elements 104 and 105 continues. The Berkeley workers have proposed the names of *Rutherfordium* for $Z = 104$ and *Hahnium* for $Z = 105$, while the Soviet workers have proposed *Kurchatovium* and *Nielsbohrium*. The International Union of Pure and Applied Chemistry has proposed *Unnilquadium* (Unq, $Z = 104$), *Unnilpentium* (Unp, $Z = 105$), and *Unnilhexium* (Unh, $Z = 106$). Popular usage, however, involves the direct use of $Z = 104$, 105, and 106.

Isotopes are divided into two groups: stable and radioactive (unstable). The total known isotopes number about 1950, of which 280 are stable and the balance are radioactive, having a transient existence ranging from millionths of a second to millions of years. Radioisotopes undergo transformation, or decay, emitting alpha, beta, gamma, or x-radiations during their

TABLE 1. TABLE OF MEAN RELATIVE ATOMIC WEIGHTS (1979)
BASED ON THE ATOMIC MASS OF $^{12}C = 12$.
(In Order of Atomic Number.)

Atomic Number	Name	Symbol	Atomic Weight	Atomic Number	Name	Symbol	Atomic Weight
1	Hydrogen	H	1.0079	55	Cesium	Cs	132.9054
2	Helium	He	4.00260	56	Barium	Ba	137.33
3	Lithium	Li	6.941*	57	Lanthanum	La	138.9055*
4	Beryllium	Be	9.01218	58	Cerium	Ce	140.12
5	Boron	B	10.81	59	Praseodymium	Pr	140.9077
6	Carbon	C	12.011	60	Neodymium	Nd	144.24*
7	Nitrogen	N	14.0067	61	Promethium	Pm	(145)
8	Oxygen	O	15.9994*	62	Samarium	Sm	150.36*
9	Fluorine	F	18.998403	63	Europium	Eu	151.96
10	Neon	Ne	20.179	64	Gadolinium	Gd	157.25*
11	Sodium	Na	22.98977	65	Terbium	Tb	158.9254
12	Magnesium	Mg	24.305	66	Dysprosium	Dy	162.50*
13	Aluminum	Al	26.98154	67	Holmium	Ho	164.9304
14	Silicon	Si	28.0855*	68	Erbium	Er	167.26*
15	Phosphorus	P	30.97376	69	Thulium	Tm	168.9342
16	Sulfur	S	32.06	70	Ytterbium	Yb	173.04*
17	Chlorine	Cl	35.453	71	Lutetium	Lu	174.967*
18	Argon	Ar	39.948	72	Hafnium	Hf	178.49*
19	Potassium	K	39.0983	73	Tantalum	Ta	180.9479
20	Calcium	Ca	40.08	74	Tungsten	W	183.85*
21	Scandium	Sc	44.9559		(Wolfram)		
22	Titanium	Ti	47.88*	75	Rhenium	Re	186.207
23	Vanadium	V	50.9415	76	Osmium	Os	190.2
24	Chromium	Cr	51.996	77	Iridium	Ir	192.22*
25	Manganese	Mn	54.9380	78	Platinum	Pt	195.08*
26	Iron	Fe	55.847*	79	Gold	Au	196.9665
27	Cobalt	Co	58.9332	80	Mercury	Hg	200.59*
28	Nickel	Ni	58.69	81	Thallium	Tl	204.383
29	Copper	Cu	63.546*	82	Lead	Pb	207.2
30	Zinc	Zn	65.38	83	Bismuth	Bi	208.9804
31	Gallium	Ga	69.72	84	Polonium	Po	(209)
32	Germanium	Ge	72.59*	85	Astatine	At	(210)
33	Arsenic	As	74.9216	86	Radon	Rn	(222)
34	Selenium	Se	78.96*	87	Francium	Fr	(223)
35	Bromine	Br	79.904	88	Radium	Ra	226.0254
36	Krypton	Kr	83.80	89	Actinium	Ac	227.0278
37	Rubidium	Rb	85.4678*	90	Thorium	Th	232.0381
38	Strontium	Sr	87.62	91	Protactinium	Pa	231.0359
39	Yttrium	Y	88.9059	92	Uranium	U	238.0289
40	Zirconium	Zr	91.22	93	Neptunium	Np	237.0482
41	Niobium	Nb	92.9064	94	Plutonium	Pu	(244)
42	Molybdenum	Mo	95.94	95	Americium	Am	(243)
43	Technetium	Tc	(98)	96	Curium	Cm	(247)
44	Ruthenium	Ru	101.07*	97	Berkelium	Bk	(247)
45	Rhodium	Rh	102.9055	98	Californium	Cf	(251)
46	Palladium	Pd	106.42	99	Einsteinium	Es	(252)
47	Silver	Ag	107.868	100	Fermium	Fm	(257)
48	Cadmium	Cd	112.41	101	Mendelevium	Md	(258)
49	Indium	In	114.82	102	Nobelium	No	(259)
50	Tin	Sn	118.69*	103	Lawrencium	Lr	(260)
51	Antimony	Sb	121.75*	104	(Unnilquadium)	(Unq)	(261)
52	Tellurium	Te	127.60*	105	(Unnilpentium)	(Unp)	(262)
53	Iodine	I	126.9045	106	(Unnilhexium)	(Unh)	(263)
54	Xenon	Xe	131.29*				

Values are those recommended by the Commission on Atomic Weights and Isotopic Abundances, Inorganic Chemistry Division, International Union of Pure and Applied Chemistry, in *Pure and Applied Chemistry*, 52, 2349–2384 (1980).
Values are considered to be ±1 in the last digit or ±3 in the last digit when followed by an asterisk (*).

return to a stable condition. Elements beyond bismuth ($Z = 83$) are radioactive. A clear graphic presentation of isotopes, both stable and radioactive, can be seen in the General Electric Knolls Atomic Power Laboratory "Chart of the Nuclides," now in its twelfth edition (revised to April 1977), where the nuclides are displayed by plotting nuclides of elements in increasing atomic number (Z) against the neutron number (N). (See RADIOACTIVITY).

Isotopes, both stable and radioactive, have continued to grow in importance to science and technology during the past 50 years. Since atoms can be characterized or marked by their radioactivity, or in some cases by a change in isotopic composition, the elements can be traced, a procedure of great value in physical and biological science, technology, and medical diagnosis. Further, radioisotopes emit particulate and/or electromagnetic radiations that can be used to probe into and through matter, affect it chemically and physically, produce heat and light, kill or alter microorganisms, and perform many other functions useful in today's complex industrial/technological society.

Stable Isotopes By 1900, physicists had found that positively charged particles formed by the passage of an electric discharge through an evacuated tube consisted of molecular ions of the gas present in the tube. Deflection of these positive rays by electric and magnetic fields offered a sensitive tool to study gaseous elements. By allowing the rays from a given element to fall on a photographic plate, a series of parabolic streaks were observed, each corresponding to a definite value of mass-to-charge ratio (m/e). Positive ray photographs of neon (atomic weight 20.2) obtained by Thomson exhibited a heavy neon line at mass 20 and a faint line at 22. In an effort to elucidate the situation, Thomson's assistant, Aston, passed neon gas through a porous pipe-clay tube repeatedly and was able to show a significant alteration in the atomic weight of the two extreme fractions. This alteration was reflected in changes in the relative brightness of the two lines in subsequent positive ray analyses.

Aston proceeded to redesign the positive ray apparatus so that the particles having the same mass were brought to a focus to produce a sharp line rather than a parabola; the resulting instrument was called a *mass spectrograph*. With this instrument, Aston was able to confirm the finding that neon exists in at least two forms (atomic weights of 20 and 22) and that the proportions appeared to be 10:1, giving an average atomic weight of 20.2 to neon. Aston next analyzed chlorine and also found that this gave two lines, corresponding to 35 and 37, and in time examined mercury, nitrogen, and the noble gases.

Independently Dempster, working at the University of Chicago, developed a *mass spectrometer* which he used to examine the isotopes of magnesium, lithium, potassium, calcium, and zinc.

Thus the pioneering work of Aston and Dempster in 1918–19 with the electromagnetic mass spectrometer unequivocally demonstrated the isotopic nature of the stable elements and is the historical starting point for separation and study of the isotopes of the elements. The electromagnetic separation of isotopes is relatively simple in principle (see MASS SPECTROMETRY).

Production of Stable Isotopes. There are a number of possible ways of separating isotopes using electromagnetic principles. However, the large-scale *electromagnetic mass separator*, known as a *calutron*, is the device now used predominantly. Within a tank maintained at high vacuum, ions of an element are produced by vaporization at high temperature, sometimes assisted by a chemical agent such as carbon tetrachloride (chlorination). The ions are accelerated by an electric potential and projected as a beam across a magnetic field. The trajectories of these ions are dependent on their masses; hence the path of the lighter ion has a greater curvature than that of the heavier ion. After traversing a circular path of 180–300°, the divergent particle paths are interrupted by catcher pockets, usually made of slots in graphite or copper water-cooled "receivers." The isotopes are then chemically recovered from the receiver pockets.

Only relatively small amounts of material can be separated in the calutron, since it separates the isotopes literally atom-by-atom. Nevertheless, ion currents up to one ampere can be maintained, allowing kilograms of material to be separated in a machine operating over a year's time. Virtually all of the isotopes of the elements have been separated in relatively high purity at Oak Ridge National Laboratory. The details on separated isotopes available and the procedures to be used in obtaining them are given in the ORNL Research Materials Catalogs.

Large-scale gaseous diffusion separation of ^{235}U is accomplished by diffusing uranium hexafluoride (UF_6) gas through a series of several thousand barriers. Because of the repetitive nature of the process, continuous diffusion through thousands of stages, or cascades, gaseous diffusion plants are extremely large industrial facilities.

More recently, large high-speed centrifuges have been developed for the separation of ^{235}U. Because the desired enrichment is not obtained in a single centrifuge, several machines are connected in a series or cascade. The gas centrifuge enrichment technology is economically superior to the gaseous diffusion enrichment process, requiring only 5% as much electrical energy.

Gaseous thermal diffusion is used to separate the noble gas isotopes (krypton, neon, and argon). Liquid thermal diffusion is currently used to separate the isotopes of chlorine, bromine, and sulfur.

Chemical exchange is used with such isotopes as hydrogen and nitrogen.

Cryogenic distillation is used to separate isotopes of carbon, oxygen, and nitrogen.

Electrolysis is used for deuterium separation and distillation and has been used for the enrichment of mercury isotopes.

Laser isotope separation has been used for ^{235}U separation and offers potential for the separation of other ions.

Other than the well-known uses of ^{235}U and ^{3}H in large-scale nuclear work, separated isotopes have been in the past primarily for fundamental scientific work, such as the measurement of nuclear reaction cross sections. There is now, however, a growing utilization of isotopic materials in all fields of fundamental research and as target materials for radioisotope production.

Radioisotopes Some radioisotopes occur in nature, e.g., uranium, radium, and thorium—ordinarily accompanied by their radioactive daughters (decay products). Radioisotopes that occur in nature have half-lives* greater than about 10^8 years or are the decay products of parent radioisotopes of such long-lived radioisotopes. These primordial radioisotopes were produced when the earth was formed and have not yet decayed away in the ensuing several billion years. Of the naturally occurring isotopes of the elements, roughly 280 are stable, and about 25 may be considered naturally radioactive. Some shorter-lived radioisotopes, such as 5530-year ^{14}C and 12.33-year ^{3}H are normally formed by cosmic ray interactions with atmospheric carbon and hydrogen. Irene Curie observed and identified the first artificially induced radioactivity in 1934 by irradiating targets of aluminum, magnesium, and boron with alpha particles from a 100-mCi† polonium source and noting that the targets continued to emit radiation after the alpha source was removed. This discovery offered the first chemical proof of artifical transmutation. After the introduction of the cyclotron and other particle accelerators, many elements were bombarded with deuterons and protons to produce hundreds of new radioisotopes, including the well-known ^{131}I, ^{32}P, and ^{14}C. Large-scale production of radioisotopes, however, did not come about until nuclear reactors were available after World War II to supply enormous amounts of neutrons. The number of artificially produced isotopes had reached 200 in 1937, and with the nuclear reactor as a source of neutrons in World War II, about 450

artificially radioactive isotopes were identified by 1944, and over 1650 by 1977. Each element has at least one radioactive isotope, and some have as many as 30.

The discovery of the first transuranic element, neptunium (Z = 93) in 1940 was followed by the identification of the other members of the actinide series (Z = 89-103), which are analogous to the lanthanide series or rare earths (Z = 57-71), thus completing the series. It is expected that element 104 and element 105 should demonstrate periodic characteristics of hafnium and tantalum, respectively. Indeed, predictions on both the chemical and nuclear basis have been made on the properties of other "superheavy" elements. For example, islands of stability of the elements are predicted around Z = 114, N = 184 and elements 117-120 lend themselves to reasonably detailed predictions of their macroscopic properties.

Radioisotopes are produced by disturbing a preferred neutron-proton ratio in the nuclei of elements. This is done by adding or removing neutrons, by adding or removing charged particles such as protons, or by a combination of both. Usually, a nuclear reactor is used as the source of neutrons; a cyclotron or other particle accelerator is used as the source of charged particles. The radionuclides formed by increasing the neutron-proton ratio generally decay (or transform) back to a stable configuration by having a neutron transform to a proton, with the emission of a negative electron (beta particles, β^-) and a neutrino (ν)—an almost undetectable uncharged particle of negligible mass (see NUCLEAR REACTIONS):

$$n \rightarrow p + \beta^- + \nu.$$

For those radionuclides resulting from a decrease in the neutron-proton ratio (i.e., neutron-deficient nuclei), the transformation again tends to reverse the cause of instability, and, where energetically possible, a proton in the nucleus is transformed into a neutron, with the emission of a positive electron (positron, β^+):

$$p \rightarrow n + \beta^+ + \nu.$$

In a competing process, neutron-deficient nuclei will regain stability by the capture of an orbital electron (EC). Indeed, it is the only β-decay mode possible for such nuclei when the decay energy (the mass difference between the decaying and product atom) is less than 2 mc^2:

$$p + e^- \rightarrow n + \nu.$$

The electronic vacancy produced in the K (or L, etc.) shell is filled by an electron from a less tightly bound state with the simultaneous emission of an x-ray (characteristic of the product element) or an electron produced by an

---

*A half-life ($T_{1/2}$) is defined as the time required for one half of an initially large number of radioactive atoms of a given species to decay.

†A curie (Ci) is defined as 3.70×10^{10} disintegrations/second.

internal photoelectric process (*Auger electrons*). These atomic rearrangements following electron capture, particularly in a heavy atom, may involve many x-ray emissions and Auger processes in successively higher shells. The adjustment can be quite extensive and includes such effects as *Coster-Kronig transitions* (Auger effect in the subshells). The *fluorescence yield* is defined as the fraction of vacancies filled by the emission of x-rays. The *Auger yield* is, in a similar fashion, defined as the fraction of vacancies filled by the emission of Auger electrons.

Many radioactive nuclei decay by two or more modes so that β^-, β^+, and EC decay with associated emissions and x-radiation are not uncommon. The *branching ratio* defines the relative amount of each mode of decay.

For heavy nuclei ($Z > 82$), the transformation to a more stable configuration usually takes place by the emission of an alpha particle (α or ^{4}He). As the nuclei become progressively heavier, the half-lives become shorter and the *fission process* becomes more dominant.

In alpha or beta decay processes the product nucleus may be left in either the ground state or, more frequently, in an excited state. A nucleus in an excited state may de-excite by the emission of electromagnetic radiation or photons (γ-radiation). Frequently the gamma transition does not proceed directly to the ground state, but rather may go in several steps involving intermediate excited states. The *angular correlation* between successive gamma rays depends on the multipole character of the radiations and on the spins of the intermediate states. An alternative to gamma-ray emission is the *internal conversion* process, an electromagnetic interaction between the nucleus and the orbital electrons. Thus, the transition between the two energy states of the nucleus is not evidenced by the emission of a photon. Instead the energy is imparted to an orbital electron which is ejected from the atom. The ratio of the internal conversion process to the rate of gamma emission is known as the internal conversion coefficient, α. The internal conversion process leaves the atom with a vacancy in one of the shells. The subsequent atomic rearrangement process is essentially identical to that following electron capture.

For some nuclei, only gamma radiation is emitted for the deexcitation from a metastable or isomeric state. Such decay is termed *isomeric transition* (IT) and is characterized by no change in mass number or atomic number. Here, too, the internal conversion process is a competing process.

Knowledge of the energies and intensities of the particulate radiation (α, β^-, β^+, and e^-) and the electromagnetic radiation (γ-rays and x-rays) serve to characterize a particular nucleus and as such is the principal means, along with the half-life, of identifying the radioisotope. Indeed, transuranic workers, using *Moseley's*

law, have used the coincidence between the alpha particles of the parent and the K x-rays of the daughter to simultaneously establish the parent-daughter genetic relationship and the atomic number independently of other nuclear or chemical information.

Production of Radioisotopes. The bulk of the artificially produced radioisotopes are made by neutron reactions in the high-volume neutron fluxes available in NUCLEAR REACTORS. Neutrons, having no charge, can easily penetrate the coulombic barriers of the nucleus. The atomic nuclei of the elements vary in their ability to capture thermal neutrons (i.e., neutrons slowed down to 2200 meters/second or ~ 0.025 eV) according to their *cross sections*, a term which expresses the probability of interaction of a neutron of a certain energy or velocity (see CROSS SECTIONS AND STOPPING POWER). When target materials are placed in the reactor and subjected to a flux of neutrons (ϕ, number of neutrons traversing a unit area per unit time), neutrons are captured in proportion to cross sections of the target element atoms present. Cross sections are expressed in *barns* (1 barn = 10^{-24} cm^2, which is approximately equal to the actual cross sectional area of a medium-weight nucleus). Certain materials, such as aluminum and graphite, have such small neutron capture cross sections that few neutrons are captured; others, such as cadmium, have such large cross sections that a thin foil will absorb almost all the thermal neutrons impinging upon it.

Radioisotopes are produced in a nuclear reactor by several different processes. Those processes that produce appreciable quantities of radioisotopes are described below.

(1) (n, γ) Process. In the (n, γ) process, which is most common, a neutron is captured by a target and simultaneously a photon is emitted. Since no change of atomic number Z occurs, the element remains the same as the target material. The (n, γ) reaction is primarily a thermal neutron reaction; cross sections for (n, γ) reactions vary from a few millibarns to many thousands of barns. For example:

$$^{23}\mathrm{Na}(n, \gamma)\,^{24}\mathrm{Na} \qquad (T_{1/2} = 15.02 \text{ h}).$$

The radioelement cannot be separated chemically unless a recoil collection is used. In the *Szilard-Chalmers process*, the recoil energy of the residual nucleus, resulting from the emission of the photon, is greater than the chemical binding energy of the nucleus in a compound.

Radioisotopes produced sometimes decay by beta emission (β^- or β^+) or electron capture to a radioactive daughter with a higher or lower atomic number. For example:

$$^{144}\mathrm{Sm}(n, \gamma)\,^{145}\mathrm{Sm}$$

$$^{145}\mathrm{Sm} \xrightarrow[\text{(340 d)}]{\text{EC}} {}^{145}\mathrm{Pm} \qquad (T_{1/2} = 17.7 \text{ y}).$$

The daughter can be chemically separated to obtain high-specific-activity* material.

With the availability of thermal neutron fluxes well in excess of $2 \times 10^{15} n/cm^2 \cdot$ sec, the preparation of millicurie amounts of radioisotopes by successive (n, γ) reactions has become feasible. For example:

$$^{64}Ni(n, \gamma)\,^{65}Ni \quad (T_{1/2} = 2.520 \text{ h})$$

$$^{65}Ni(n, \gamma)\,^{66}Ni \quad (T_{1/2} = 54.8 \text{ h}).$$

(2) (n, p) Process. In the (n, p) process, which requires neutrons of higher-than-thermal energies,† a neutron enters a target nucleus with sufficient energy to cause a proton to be released. The atomic number is reduced by 1, and the affected atom is transmuted into a different element, which can be separated chemically from the target material. Through chemical separation, high-specific-activity material can be obtained. For example:

$$^{32}S(n, p)\,^{32}P \quad (T_{1/2} = 14.28 \text{ d}).$$

Cross sections for such reactions, typically a few millibarns, are orders of magnitude less than those for (n, γ) reactions.

(3) (n, α) Process. The (n, α) process, like the (n, p) process, requires high-energy neutrons and typically has a cross section of a few millibarns. In the (n, α) process, a neutron of high energy enters a target atom and causes an alpha particle to be emitted. The atomic number of the target atom is reduced by 2, and a chemical separation yielding high-specific-activity material is possible. For example:

$$^{36}Cl(n, \alpha)\,^{33}P \quad (T_{1/2} = 25.3 \text{ d}).$$

(4) Fission. Under normal operating conditions, research reactors have 20–50% burnup of the fissile material. The asymmetric fission process yields fission products of mass ranging between $A = 72$ and $A = 162$. The maximum fission yields of 6.5% occur at $A = 95$ and $A = 138$. Since several isotopes of any one element are often produced, the isotopic purity will not be as high as that of radioisotopes produced by (n, p) and (n, α) reactions. This isotopic purity will depend somewhat upon the length of time that the uranium was exposed to neutrons and upon the elapsed time between removal from the reactor and the chemical separation. Fission products are routinely chemically separated and purified from high-level waste

*Specific activity is the amount of radioisotope per unit weight of the total element and is usually expressed as curies or millicuries/gram.

†A few exceptions are found among reactions with the light nuclei in cases where binding energy of a proton or particle is appreciably lower than that of a neutron: the reactions $^{10}B(n, p)\,^{10}Be$, $^{14}N(n, p)\,^{14}C$, $^{35}Cl(n, p)\,^{35}S$, $^{10}B(n, \alpha)\,^{7}Li$, and $^{6}Li(n, \alpha)\,^{3}H$ occur with thermal neutrons.

streams of DOE production facilities and represent an important source of such radioisotopes as ^{90}Sr (thermoelectric generator systems for terrestrial and underwater applications), ^{137}Cs (radiography, teletherapy, and large irradiation units), and ^{147}Pm (thermoelectric power generators).

The basic equation for radioisotope production is

$$A \xrightarrow[\sigma_A]{} B \xrightarrow[\sigma_B]{\lambda_B}$$

The target atom A captures neutrons to produce the product nuclide B, which in turn is transformed by decay or further neutron capture. The effective cross sections σ_A, σ_B and the decay rate constant λ_B enable the rate and equilibrium values for the transformation to be calculated for any particular irradiation conditions. The exact solution for the differential equation describing these rate processes for the number of atoms N of the product formed at time t in neutron flux ϕ is,

$$\lambda_B N_B = \frac{\lambda_B \phi \sigma_A N_A}{\lambda_B + \phi[\sigma_B - \sigma_A]}$$
$$\cdot [e^{-\phi\sigma_A t} - e^{-(\phi\sigma_B + \lambda_B)t}].$$

In most cases, one can neglect the burnup of the target atoms and the product radioisotope. In such cases, the above equation then reduces to:

$$\lambda_B N_B = N_{A_0} \phi \sigma_A (1 - e^{-\lambda_B t}).$$

Here N_{A_0} refers to the number of original target atoms at time zero.

For irradiations of sufficient length ($t \gg T_{1/2}$), and again neglecting burnup, the saturation factor $(1 - e^{-\lambda_B t})$ approaches 1 and the equation further reduces to

$$\lambda_B N_B = N_{A_0} \phi \sigma_A.$$

With the introduction of the *cyclotron* in the early 1930s, charged particle reactions started to play a significant role in the preparation of a large variety of neutron-deficient radioisotopes. Indeed, until the advent of nuclear reactors in the mid-1940s, proton-, deuteron-, and alpha-particle accelerators played the major role in supplying radioisotopes for medical, biological, and scientific research. During the period 1950–1970, the ORNL 86-Inch Cyclotron exploiting a 2.6-mA (2×10^{16} particles/sec) beam of 23-MeV protons, supplied the major portion of neutron-deficient radioisotopes. With the development of *compact cyclotrons*, the production and use of short-lived radioisotopes for medical purposes has gained impetus. Both hospitals and radiopharmaceutical and radiochemical organizations have found these cyclotrons extremely useful. Indeed, one organization is utilizing five such accelerators. Hospitals have found these accelerators particularly useful in producing short-lived radionuclides such as 20.38-m ^{11}C, 9.96-m ^{13}N, 122 s ^{15}O, and 109.8 m ^{18}F for on-site diagnostic procedures. *Heavy-ion accelerators*, developing beams of carbon, nitrogen, oxygen, neon, and argon, have opened additional areas for research. These devices, in conjunction with on-line mass separators, should make possible the production and identification of more of the 5000 or so theoretically possible nuclei. The Brookhaven Linac Isotope Producer (BLIP) and the Los Alamos Scientific Laboratory High Flux Meson Facility (LAMPF) offer the possibility of producing large amounts of neutron-deficient radioisotopes by high-energy proton-induced spallation reactions using 200-MeV and 750-MeV protons, respectively. Most recently, the completion of the Oak Ridge National Laboratory Holifield Heavy Ion Research Facility (HHIRF), a facility utilizing a 25-million-volt tandem electrostatic accelerator and the Oak Ridge Isochronous Cyclotron, makes it possible to accelerate ions up to the mass region of 160, i.e., rare earth elements, to the energies required to produce nuclear reactions, and extends the capability for creating nuclear species far from stability.

A method for preparing short-lived radio-isotopes off-site has found widespread use in nuclear medicine. The radioisotope generator employs a relatively long-lived parent which is sorbed onto an ion exchanger; the short-lived daughter may be eluted, as required, with a suitable agent. The most common such system, the ^{99}Mo–^{99m}Tc generator, yielding 6.02-h ^{99}Tc pertechnetate, has been established as a powerful tool for diagnostic scanning and has largely supplanted the use of ^{130}I. Other generators include the ^{87}Y–^{87m}Sr (2.83 h) and the ^{113}Sn–^{113m}In (99.5 m) systems.

Recent advances in *transuranic* isotope technology have made possible the production of large quantities of these radioisotopes. The original neutron irradiations to produce such nuclei were performed in the Oak Ridge and Hanford reactors in the 1940s, in the NRX reactor at Chalk River Laboratory in Canada in the late 1940s, and in the Materials Testing Reactor (MTR) in Idaho, in the 1950s. Both neutron irradiation and charge particle bombardment were used to produce and identify about 50 transplutonium nuclides covering the atomic number range up through $Z = 103$ and mass range up through 257. During the 1960–1970 period, the Savannah River reactors and the High Flux Isotopes Reactor (HFIR) at the Oak Ridge National Laboratory were used for large-scale production of an additional 50 transplutonium isotopes in milligram and gram quantities. Indeed, several grams of ^{252}Cf have been produced since 1966. Milligrams of materials such as ^{252}Cf and ^{241}Am represent excellent fission sources of neutrons presenting peak thermal neutron fluxes of $10^8 n/cm^2 \cdot$ sec (com-

pared to $10^5 n/cm^2 \cdot$ sec for a 1-g radium source mixed with beryllium). (See TRANSURANIUM ELEMENTS.)

Isotope Processing The techniques used for the processing of ultrahigh-purity chemicals are required for isotope work in recovering stable isotopes, preparing target materials, and separating and purifying radioisotopes. Practically every technique from traditional wet chemistry to ion exchange and chromatography is utilized, often with high-purity radioisotopes, at very low concentration levels (e.g., micrograms per liter). Sophisticated analytical methods (e.g., mass and radiation spectral analysis) are also required and make up a significant portion of the cost of isotope preparations.

The size of the radioisotope industry has not changed much during the past decade; about 100 private firms produce radioisotopes, radiopharmaceuticals, sealed sources, and equipment for medical, industrial, and scientific uses of radioisotopes. The magnitude of sales, however, has increased dramatically. Whereas in 1970 the U.S. estimated sales were more than $50 million, now, in 1982, a single firm has gross sales of substantially greater magnitude: New England Nuclear, a major supplier, reports net sales of $66 million for 1979 and $82 million for 1980. It is estimated that the annual sales of ^{99}Mo and associated generators well exceeds $50 million (perhaps as high as $100 million), ^{201}Ti exceeding $30 million, and ^{67}Ga exceeding $10 million. To place it in another perspective, approximately 225,000 thallium doses were administered to patients in the United States in 1980. Sales of stable isotopes at Oak Ridge National Laboratory exceed $4 million annually, that for radioisotopes is $6 million, and that for heavy elements (^{241}Am, ^{234}U, ^{229}Th, and others) exceeds $4 million. The overall market today is in the tens of billions of dollars.

It is estimated that 17 million in vivo and 100 thousand in vitro nuclear studies were made during 1980, a figure which corresponds to one out of every two or three hospitalized patients plus a large number of outpatients (275 thousand/day). There are about 22,000 individuals in the United States practicing nuclear medicine.

J. J. PINAJIAN

References

Nuclear and Radiochemistry

Barbier, Marcel, "Induced Radioactivity," Amsterdam, North-Holland Publishing Company, 1969.

Friedlander, G., Kennedy, J. W., Macias, E. S., and Miller, J. M., "Nuclear and Radiochemistry," Third Edition, New York, John Wiley and Sons, Inc., 1981.

Kocher, D. C., "Radioactive Decay Data Tables," Technical Information Center, Department of Energy Report, DOE/TIC-11026 (1981).

Lederer, C. M., and Shirley, V. S., "Table of Isotopes,"

Seventh Edition, New York, John Wiley and Sons, Inc., 1978.

Oak Ridge National Laboratory Research Materials Catalog, Oak Ridge, Tennessee, in press.

Seaborg, G. T., "Transuranium Elements: Products of Modern Alchemy," Stroudsburg, Dowden, Hutchinson, and Ross, 1978.

Subcommittee on Nuclear and Radiochemistry, Committees on Chemical Sciences, Assembly of Mathematical and Physical Sciences, National Research Council, "A Review of the Accomplishments and Promise of U.S. Transplutonium Research 1940–1980," Washington, D.C., National Academy of Sciences, 1982.

Mass Spectrometry

Roboz, J., "Introduction to Mass Spectrometry: Instrumentation and Techniques," New York, Interscience Publishers, 1968.

Atomic Weights

International Union of Pure and Applied Chemistry, Inorganic Division, Commission on Atomic Weights and Abundances, *Pure and Applied Chemistry* **52**, 2349–2384 (1980).

Stable Isotopes

Subcommittee on Nuclear and Radiochemistry, Committee on Chemical Sciences, Assembly of Mathematical and Physical Sciences, National Research Council, "Separated Isotopes: Vital Tools for Science and Medicine," Washington, D.C., National Academy Press, 1982.

Davis, W. C. et al., "Chemical Recovery and Refinement Procedures in the Electromagnetic Separation of Isotopes," Oak Ridge National Laboratory Report, ORNL-4583, August 1970.

Underwood, J. N., Love, L. O., Prater, W. K., and Scheitlin, F. M., "Calutron Experiments with Milligram Quantities of Charge Material," *Nucl. Instrum. Methods* **57**, 17–21 (1967).

Villani, S., "Isotope Separation," New York, American Nuclear Society, 1976.

Reactor Production of Radioisotopes

Aebersold, P. C., and Rupp, A. F., "Production of Short-lived Radioisotopes," in "Production and Use of Short-lived Radioisotopes from Reactors," Vol. 1, pp. 31–47, Vienna, International Atomic Energy Agency, 1962.

Binford, F. T., Cole, T. E., and Cramer, E. N., "The High Flux Isotope Reactor," Oak Ridge National Laboratory Report, ORNL-3572 (Rev. 2), May 1968.

Brookhaven National Laboratory, "Manual of Isotope Production Processes in Use at Brookhaven National Laboratory," BNL-864, August 1964.

Crandall, J. L., "The Savannah River High Flux Demonstration," Savannah River Laboratory Report, DP-999, June 1965.

Knoll, Peter, "The Technology of Isotope Production,"

Part I., "Irradiation Technology," (Zentralinstitut für Kernphysik Dresden) ZfK-RCH-1, December 1961 (in German); for English translation see Pinajian, J. J., Oak Ridge National Laboratory Report ORNL-tr-2400, November 1970.

Oak Ridge National Laboratory, "ORNL Radioisotope Procedures Manual," ORNL-3633, June 1964.

Pinajian, J. J., "Oak Ridge Research Reactor for Isotope Production," *Isotop. Radiat. Technol.* **1**, 130–36 (Winter 1963–64).

Rupp, A. F., and Binford, F. T., "Production of Radioisotopes," in "Nuclear Engineering Handbook" (H. Etherington, Ed.), Section 14, pp. 26–37, New York, McGraw-Hill Book Co., 1958.

Accelerator Production of Radioisotopes

Lange, J., and Münzel, H., "Estimation of Unknown Excitation Functions for (α, xn), (α, pxn), (d, xn), (d, pxn), and (p, xn) Reactions," Karlsruhe Nuclear Research Center Report, KFK-767, May 1968 (in German); for English translation see Pinajian, J. J., and Kern, L. H., Oak Ridge National Laboratory Report ORNL-tr-3020, October 1970.

Laughlin, J. S., Tilbury, R. S., and Dahl, J. R., "The Cyclotron: Source of Short-lived Radionuclides and Positron Emitters for Medicine" in "Progress in Atomic Medicine, Volume 3: Recent Advances in Nuclear Medicine" (J. H. Lawrence, Ed.), New York, Grune & Stratton, 1971.

Pinajian, J. J., "ORNL 86-Inch Cyclotron," in "Radioactive Pharmaceuticals," (G. A. Andrews, R. M. Kniseley, and H. N. Wagner, Eds.), Chapter 9, pp. 143–54, U. S. Atomic Energy Commission, 1966.

Rosen, L., Schillaci, M. E., Dropesky, B. J., and O'Brien, H. A., "Use of LAMPF for Isotope Production: Briefing to the AEC Division of Isotopes Development, December 15, 1970," Los Alamos Scientific Laboratory Report, LA-4587-MS, February 1972.

Stang, L. G., Jr., Hillman, M., and Lebowitz, E., "The Production of Radioisotopes by Spallation," Brookhaven National Laboratory Report, BNL-50195, August 1969.

Radiopharmaceuticals and Nuclear Medicine

Schneider, P. B., and Treves, S., "Nuclear Medicine in Clinical Practice," Amsterdam, Elsevier, North Holland Biomedical Press, 1978.

Spencer, R. P., "Radiopharmaceuticals: Structure, Activity, and Relationships," New York, Grune & Stratton, 1981.

Wagner, H. N., "Principles of Nuclear Medicine," Philadelphia, W. B. Saunders Co., 1969.

Cross-references: ATOMIC PHYSICS, CROSS SECTIONS AND STOPPING POWER, ELECTRON, NUCLEAR REACTIONS, NUCLEAR REACTORS, NUCLEAR STRUCTURE, PERIODIC LAW AND PERIODIC TABLE, PROTON, RADIOACTIVITY, TRANSURANIUM ELEMENTS.

K

KEPLER'S LAWS OF PLANETARY MOTION

The German astronomer and mathematician Johannes Kepler (1571–1630) worked briefly with the Danish astronomer Tycho Brahe, who gathered some of the most accurate observational data on planetary motion in the pre-telescope era. When Brahe died in 1601, Kepler inherited his data books and devoted many years of intensive effort to finding a mathematical description for the planetary motion described by the data. Kepler was successful in deriving three laws of planetary motion which led ultimately to our current understanding of the orbital motion of planets, moons, and comets, as well as man-made satellites and spacecraft. The first two laws were published in 1609, about the time Galileo was first making astronomical observations with his telescope; the third law did not appear until a decade later, in 1619.

Briefly stated, these laws are (see Fig. 1):

1. The orbit of each planet is an ellipse with the sun at one focus.
2. The line from the sun to a planet sweeps out equal areas inside the ellipse in equal lengths of time.
3. The squares of the orbital periods of the planets are proportional to the cubes of their mean distances from the sun.

1. The *first law* describes the geometrical shape of the orbit as an ellipse, which correctly

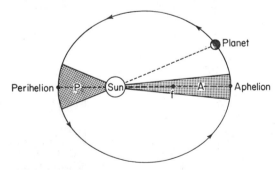

FIG. 1. Elliptical orbit of a planet with the sun at one focus. Equal areas P and A are swept out in equal lengths of time. (Taken from Second Edition article by Richard M. Sutton.)

accounts for the manner in which the distance from the sun to the planet changes as the planet travels along its orbit. As shown in Fig. 1, the perihelion is the point in the orbit at which the planet is closest to the sun; the aphelion is the point farthest from the sun.

Prior to Kepler's time, it was assumed that the planetary orbits were concentric circles with the sun at the center, because it was thought that only the perfect geometrical curve of the circle could describe the motion through the heavens of perfect celestial bodies. However, as the accuracy of the observational data increased, it became clear that the distance from the sun to each planet was not constant, as it would be for a circular orbit. Increasingly complicated geometrical constructions using several circles were proposed in an attempt to fit the data for an individual planet orbit. Kepler was the first person to recognize that an ellipse, rather than a circle, is the geometrical curve which describes the shape of a planet orbit in a simple and elegant manner.

2. The *second law* is a concise mathematical description of the observed fact that the rate at which the sun-planet line rotates through space (the angular velocity of the motion) increases as the planet moves closer to the sun and decreases as it moves farther from the sun. As shown in Fig. 1, the distance along the orbit traveled by the planet near perihelion in a given length of time (for example, one month) is greater than the distance travelled near aphelion during the same length of time. This phenomenon is described quantitatively by the statement that equal areas are swept out in equal lengths of time. The time required to sweep out the total area inside the ellipse is the orbit period of the planet.

3. The first two laws describe the motion of an individual planet. By contrast, the *third law* states the manner in which the motions of the various planets are related to each other. It states that the ratio formed by dividing the square of the orbit period of any planet by the cube of its mean distance from the sun is the same value for all planets in the solar system. The term *mean distance*, as used in the third law, is simply the average of the perihelion and aphelion distances. This mean distance is then half the distance between perihelion and aphelion, and is called the *semimajor axis* of the

ellipse. Another way of stating the third law is that the period of a planet is proportional to the 3/2 power of its mean distance from the sun.

Kepler, to his credit, formulated these three laws based entirely on empirical data, without the benefit of a fundamental theory which explained why planetary motion satisfied these laws. The missing ingredient was the concept of gravitational force, which was developed several decades later by Sir Isaac Newton. Newton, starting with Kepler's first law, deduced that a planet whould move in an elliptical orbit with the sun at one focus only if the force exerted on the planet by the sun was proportional to the inverse square of the distance between them. This is the so-called *inverse-square law* of gravitational force. Newton also showed that Kepler's second law was a consequence of the principle of conservation of angular momentum. The angular momentum is conserved because the gravitational force is a central force, that is, it acts along the line from the planet to the central body, the sun. The third law is a natural consequence of the inverse-square gravitational force field of the sun, which exists throughout the solar system. It is interesting to

note that the motion of any body orbiting the sun in a closed, periodic orbit is governed by this same force law and must satisfy Kepler's laws. Thus the motion of comets, asteroids, and interplanetary spacecraft, such as the Mariner spacecraft launched to Mars in 1969 and the Voyager spacecraft launched to Jupiter in 1977, are governed by these same laws.

One can easily verify Kepler's third law using modern astronomical data for the planetary orbits. If one plots the orbit periods of the planets against their mean distances from the sun on a log-log graph as shown in Fig. 2, a straight line of slope 3/2 connects the points, indicating that the orbit period T is proportional to the 3/2 power of the mean distance R. The unit of distance used in the figure is the astronomical unit (A.U.), which is approximately 150 million kilometers or 93 million miles.

In the few minutes it takes to plot the graph, one can discover what it took Kepler ten years to find! The data for the planets beyond Saturn (unknown to Kepler) also lie on this same straight line. It is interesting to also note that Halley's comet, which orbits the sun with a period of 76 years at a mean distance from the

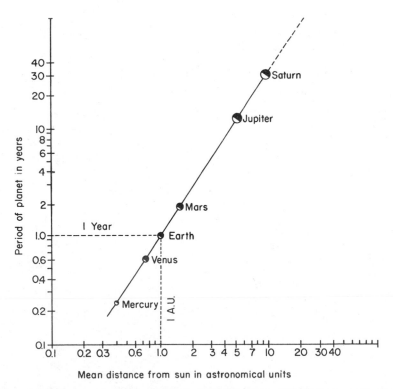

FIG. 2. Graph on log-log paper of periods of planets in years vs mean distance from sun in astronomical units (sun-earth distance is 1 A.U.). (Taken from second edition article by Richard M. Sutton.)

sun of 18 A.U. also falls on the same straight line.

The motion of Halley's comet, which will cross the earth's orbit in 1986, also provides a dramatic example of one consequence of Kepler's second law (the conservation of angular momentum). In order to satisfy this law, the ratio of the speed of the comet at perihelion to its speed at aphelion must equal the ratio of the aphelion to perihelion distances from the sun. Since the aphelion distance is 35.3 A.U. (outside of Neptune's orbit) and its perihelion distance is 0.6 A.U. (inside the earth's orbit at 1 A.U.), the speed of the comet at perihelion is nearly 60 times its speed at aphelion.

The magnitudes of the speeds of the planets are enormous by earth-based standards. Because the orbit of the earth is very nearly a circular orbit of radius 1 A.U., the distance travelled by the earth in one year is approximately 2π A.U. (the circumference of the circular orbit). Dividing this distance by the elapsed time of one year gives the value of the unit of speed called the EMOS, for *earth-mean-orbital-speed*. One EMOS is equal to about 30 kilometers per second or 67,000 miles per hour. Thus the reader, sitting motionless relative to his or her surround-ings, is actually traveling at this enormous speed around the sun. The EMOS is a convenient unit of speed used in mission planning for interplanetary spacecraft.

Kepler's laws also apply in the case of man-made or natural satellite orbital motion about a planet, such as the earth. In this application, the central body is the planet rather than the sun. If the satellite is close to the planet in relation to the distance from the sun to the satellite, the planet's gravitational force on the satellite is much greater than the gravitational force due to the sun. The satellite is then said to be *inside the sphere of influence of the planet*. The earth's moon and all artificial satellites closer to the earth are well within the earth's sphere of the influence because the sun is 400 times farther from the moon than is the earth. (Recall that the gravitational force is inversely proportional to the *square* of the distance).

Figure 3 illustrates on a log-log graph the periods in hours of satellites orbiting the earth plotted against their mean distances from the center of the earth, measured in units of earth radii. As in the case of the planetary orbits about the sun shown in Fig. 1, the graph is a straight line of slope 3/2, illustrating Kepler's

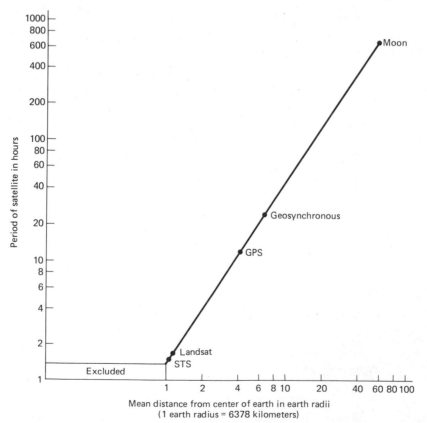

FIG. 3. Log-log graph of satellites of earth.

third law. The sphere of influence of the earth is located at approximately 150 earth radii from the center of the earth.

The lower left-hand region of the graph is excluded because no satellite can continually orbit the earth if its mean distance from the center of the earth is less than one earth radius, equal to 6378 kilometers, or approximately 4000 miles. In fact, the minimum mean distance is slightly greater than this value by a hundred kilometers or so because of the layer of atmosphere surrounding the earth. Satellites entering the atmosphere burn up due to the heat caused by atmospheric drag. However, if one for a moment imagines a fictitious satellite orbiting the earth at tree-top level and ignores the effect of atmospheric drag, the orbit period would be approximately 84 minutes or 1.4 hours, at an orbit speed of approximately 8 kilometers per second (about 18,000 miles per hour). This enormous speed is necessary in order to overcome the gravitational force attempting to pull the satellite down to the surface of the earth.

The satellites shown in Fig. 3 include the STS-1, the first orbital flight of the Space Transportation System, (the Space Shuttle), which was in orbit April 12–14, 1981. The Space Shuttle orbited the earth at an *altitude* of approximately 278 kilometers with an orbit period of 90 minutes. The altitude of a satellite is defined as the distance above the surface of the earth. The distance from the center of the earth is calculated by simply adding the altitude of the satellite to the radius of the earth. The Landsat satellites, which photograph the earth's surface, have an altitude of 920 km and a period of 1.7 hours.

Another satellite shown is a NAVSTAR GPS (Global Positioning System) satellite. A collection of these satellites are in orbits having 12-hour periods, and provide precision navigation information for ships and aircraft on or near the surface of the earth. Also shown is a geosynchronous satellite, which has an orbit period of one sidereal day, equal to 23 hours, 56 minutes, and 4 seconds (the period of the daily rotation of the earth relative to the distant stars). The orbital motion of geosynchronous satellites is synchronized with the daily rotation of the earth, causing these satellites to remain directly over fixed points on the surface of the earth at the equator. Geosynchronous satellites are used for radio, television, and other forms of communication over large distances on the earth. An example is the satellite WESTAR I, which is located above the equator at 99° west longitude. The earth's moon is also shown on the same straight line on the graph with its sidereal period of 27.3 days.

For a satellite in an elliptical orbit about the earth, the points in orbit nearest and farthest from the center of the earth are termed *perigee* and *apogee*, respectively. Data for satellite orbits is usually given in terms of perigee and apogee altitudes. To calculate the mean distance from the center of the earth for use in Fig. 3 to determine the orbit period, one simply adds the radius of the earth (6378 km) to the mean altitude (the average of the perigee and apogee altitudes). For example, a satellite having a perigee altitude of 1548 kilometers and an apogee altitude of 1800 kilometers has a mean altitude of 1674 km and a mean distance from the center of the earth of 8052 km, or about 1.3 earth radii. Referring to the graph of Fig. 3, this corresponds to an orbit period of 2 hours.

Other types of orbits besides ellipses are possible in an inverse-square gravitational force field, namely *parabolic* and *hyperbolic* orbits. These were entirely unknown to Kepler, who was only aware of closed, periodic orbits. Parabolic and hyperbolic orbits are open, in the sense that they do not close on themselves, but instead extend to infinity on either side of the central body. The motion along these orbits does not periodically repeat and Kepler's first and third laws do not apply in this case. However, Kepler's second law is still valid because angular momentum is conserved. Thus, Kepler derived a law which applies to orbits which he did not even know existed! Parabolic and hyperbolic orbits occur when the speed of the orbiting body is too great for the body to be captured by the central body. For this reason these types of orbits are also termed *escape orbits*.

Figure 4 shows an example of a hyperbolic orbit, the flyby of the planet Jupiter by the

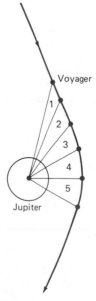

FIG. 4. Hyperbolic flyby of Jupiter by Voyager spacecraft. Numbered areas are equal and are swept out in equal lengths of time.

Voyager I spacecraft in March 1979. The spacecraft made a single high-speed passage by the planet and proceeded on to the next destination, the planet Saturn. The orbit is curved toward the planet due to the gravitational force of the planet pulling on the spacecraft.

Other examples of hyperbolic orbits are the orbits of the two Voyager spacecraft about the sun after their encounters with Jupiter in March and July of 1979. During their Jupiter flybys, the spacecraft were accelerated by the gravitational force of the massive planet to speeds which placed them on escape orbits from the solar system. After their subsequent encounters with the other planets, they will proceed out of the solar system into our galaxy never to return. As they escape the solar system on hyperbolic orbits, which were unknown to Kepler, their motion satisfies Kepler's second law, derived almost 400 years ago.

JOHN E. PRUSSING

References

1. Lodge, Sir Oliver, "Johann Kepler," in "The World of Mathematics," Vol. 1, (J. R. Newman, Ed.), Simon and Schuster, New York, 1956.
2. Sagan, Carl, "The Harmony of Worlds," in "Cosmos," Random House, Inc., New York, 1980.
3. Bate, R. R., Mueller, D. D., and White, J. E., "Fundamentals of Astrodynamics," Dover Publishing Co., New York, 1971.
4. Prussing, J. E., "The Mean Radius in Kepler's Third Law," *American Journal of Physics* 45 (12) (December 1977).

Cross-references: ASTRONAUTICS, DYNAMICS, GRAVITATION, ROTATION—CURVILINEAR MOTION, MECHANICS.

KERR EFFECTS

In electrooptics and nonlinear optics, a change in the index of refraction of a medium that is proportional to the square of an electric field or to the product of two electric fields is termed a *Kerr effect*.[1] In magnetooptics, various field-dependent changes in the polarization of a light beam reflected from the surface of a magnetized material are also called Kerr effects.[2] The details of the physics underlying each of these classes of phenomena depend upon the experimental conditions and the medium.

The electrooptic Kerr effect is intrinsically anisotropic; the index of refraction for light polarized along the direction of the field is altered by an amount different from that for light polarized perpendicular to the applied field. In a typical experiment a dc field is applied at 45° to the polarization of the incident light and the field-induced birefringence $\delta n_\parallel - \delta n_\perp$ causes the output beam to become elliptically polarized.[3] This change in polariza-

tion can be readily measured; the phase retardation due to the Kerr effect is proportional to a constant characteristic of the material and the length of the sample and inversely proportional to the wavelength.

In liquids and other optically isotropic media a dc Kerr constant is usually defined as:

$$B = \frac{\delta n_\parallel - \delta n_\perp}{\lambda |E|^2}$$

where λ is the wavelength in vacuum. The individual molecules of such media are often highly anisotropic with dipole moments μ oriented along a definite molecular axis as well as (frequency-dependent) dielectric polarizability tensors $\alpha_{ij}(\omega)$ with different values along each of three orthogonal directions. The overall isotropy of the medium results from the random orientations of the individual molecules. If the molecules are free to rotate, an applied electric field can perturb the initial random distribution of molecular orientations, causing more molecules to align parallel to the field than perpendicular to it. This realignment causes the majority of the birefringence detected in the Kerr effect.[1] Even when the molecular orientations are fixed, the applied field can perturb electronic energy levels. The change in energy of electronic transitions is called the *Stark effect*, and can often be detected as a shift or splitting of an absorption band. Since every absorption band causes dispersion in the index of refraction that extends to wavelengths beyond the absorption itself, the shift in energy levels implies a change in the index of refraction that contributes to the Kerr effect and is characterized by an electronic hyperpolarizability tensor γ_{ijkl}. The dc Kerr constant for a liquid composed of uniaxial molecules where the three components of the molecular polarizability tensor at zero frequency are $\alpha_{11}(0)$ and $\alpha_{22}(0) = \alpha_{33}(0) = \alpha_{11}(0) + \Delta\alpha(0)$ and the corresponding polarizabilities at optical frequencies are $\alpha_{11}(\omega)$ and $\alpha_{11}(\omega) + \Delta\alpha(\omega)$ and where the dipole moment is along α_{11} can be expressed as:

$$B = \frac{\pi N}{n\lambda} \left(\frac{n^2 + 2}{3}\right)^2 \left(\frac{\epsilon + 2}{3}\right)^2$$
$$\cdot \left[2 \langle \gamma_{xyyx} \rangle + \frac{\Delta\alpha(0)\,\Delta\alpha(\omega)}{15kT} + \frac{\mu^2\,\Delta\alpha(\omega)}{15k^2 T^2}\right].$$

In this equation N is the number density per unit volume, ϵ is the dc dielectric constant, $\langle \gamma_{xyyx} \rangle$ is the orientationally averaged molecular hyperpolarizability, k is Boltzmann's constant, and T is the absolute temperature. When $\mu \neq 0$ the term proportional to μ^2 usually dominates, while $\langle \gamma_{xyyx} \rangle$ is significant only for highly symmetric molecules such as CCl_4. The dc Kerr constant of nitrobenzene is 3.26×10^{-5} cm/statvolt^2 at $\lambda = 589$ nm, which is an order of magnitude larger than the constants for other simple molecules.[4] Nematic liquid crystals can show even larger effect, however.

In crystals, the polarization subscripts of the nonlinear susceptibility tensor must be referred to crystallographic axes. A quadratic field dependence of the index of refraction ellipsoid results from the Kerr effect, but is often called the *quadratic electrooptic effect*. It is smaller and less important technologically than the *linear electrooptic* or *Pockels effect*, which only occurs in noncentrosymmetric crystals.[5]

The optical Kerr effect occurs when the applied or orienting field is at optical frequency ω_1. The definition of the dc Kerr constant above can be modified to describe some effects expected in the optical case. The direction of a static dipole moment cannot reverse fast enough to follow a rapidly oscillating optical field so the third term in the equation for B vanishes. Also the dielectric constant ϵ must be replaced with the square of the index of refraction at the frequency of the applied field and the dc molecular anisotropy $\Delta\alpha(0)$ must be replaced by the optical frequency value $\Delta\alpha(\omega_1)$.[6] Additional complexities, however, result from the presence of resonances in the optical Kerr constant at a variety of optical frequencies.

The optical Kerr effect and its many variations are best discussed in terms of a third order nonlinear susceptibility tensor $\chi_{ijkl}^{(3)}(\omega_2, -\omega_1, \omega_1, -\omega_2)$ where the components of the dielectric polarization density is written as $P_i = \Sigma_j \chi_{ij} E_j + \Sigma_{kj} \chi_{ijk}^{(2)} E_j E_k + \Sigma_{jkl} \chi_{ijkl}^{(3)} E_j E_k E_l$.[7] In the Maker-Terhune convention, the frequency ω_2 is that of the optical probe, while the frequency ω_1 (which may be zero) describes the applied field. The subscripts relate to the polarization directions of the field, and each of the four subscripts is paired with the frequency at the corresponding position in the argument. Thus an applied field at frequency ω_1 in the x direction causes a change in the index of refraction in the direction parallel to the field:

$$\delta n_x = \delta n_\parallel$$

$$= \frac{24\pi}{n} \chi_{xxxx}^{(3)}(\omega_2, -\omega_1, \omega_1, -\omega_2)|E_x|^2,$$

while, perpendicular to the field,

$$\delta n_y = \delta n_\perp$$

$$= \frac{24\pi}{n} \chi_{yxxy}^{(3)}(\omega_2, -\omega_1, \omega_1, -\omega_2)|E_x|^2$$

in cgs units. These index of refraction changes cause self-focusing of high-power laser beams in most media.

Coherent beams crossing in a Kerr medium create an interference pattern of bright and dark fringes. The optical Kerr effect makes the index of refraction in the region of the dark fringes different from that in the bright fringes, creating a grating which can Bragg scatter other beams. The scattered beam contains information carried by the initial beams; the overall process has been widely studied and is generally termed *real-time holography*.

The tensor nature of the nonlinear susceptibility complicates the description of many technologically important applications of the optical Kerr effect. If the applied field at frequency ω_1 has projections $E_x(\omega_1)$ and $E_y(\omega_1)$ along the x and y axes, the field alters the polarization of a probe beam at frequency ω_2 in an otherwise optically isotropic medium. If the probing field is initially polarized along x, the birefringence of a sample of length l creates a component polarized along y with amplitude

$$E_y(\omega_2) = \frac{-i2\pi l}{n(\omega_2)\lambda_2} \{\chi_{yxyx}^{(3)}(\omega_2, -\omega_1,$$

$$\omega_1, -\omega_2)E_x^*(\omega_1)E_y(\omega_1)$$

$$+ \chi_{yyxx}^{(3)}(\omega_2, -\omega_1, \omega_1, -\omega_2)$$

$$\cdot E_x(\omega_1)E_y^*(\omega_1)\} E_x^*(\omega_2)$$

where $i = \sqrt{-1}$ and an asterisk denotes the complex conjugate.

Because the molecular reorientations can occur on a picosecond time scale, the optical Kerr effect can be used as a high-speed shutter. A cell filled with a Kerr liquid such as CS_2 is placed between crossed polarizers. A powerful 10-picosecond-long laser pulse, linearly polarized at $45°$ to the axes of the polarizers and incident from the side, can induce sufficient birefringence to permit rapidly occurring processes to be photographed.[8]

The nonlinear susceptibilities describing the optical Kerr effect show resonances and dispersion similar to the index of refraction, but often depending on more than one frequency. These effects have been exploited in "polarization spectroscopy," Raman induced Kerr effect spectroscopy, and other laser spectroscopy techniques.[9] These techniques benefit from the sensitivity with which polarization changes can be detected. In the polarization spectroscopy version of saturation spectroscopy, all the fields have the same frequency, but propagate in opposite directions through an absorbing vapor. When the laser frequency approaches that of a transition in the vapor, a field-dependent birefringence (and dichroism) can often be induced. In vapors at low density, the widths of the nonlinear resonances can be much less than that of the absorption itself.

In Raman induced Kerr effect spectroscopy, the difference of the input frequencies $\omega_1 - \omega_2$ must approach that of a Raman active vibration of the medium. Again a resonant Kerr effect occurs—accompanied by dichroism at exact resonance. The nonresonant part of the Kerr effect can be suppressed by using circularly polarized light for the applied field (i.e., $E_x(\omega_1) = \pm iE_y(\omega_1)$). Two-photon-absorption-induced birefringence occurs when the sum of the input frequencies approaches a two-photon transition. Dramatic polarization rotations occur when the applied field is circularly polarized.

The magnetooptic Kerr effects result from the different phase velocities of left and right circularly polarized light propagating along a magnetic field (see FARADAY EFFECT). This difference in phase velocity can be interpreted as a difference in the indices of refraction for the two polarization states. Fresnel's laws of reflection then imply that right and left circularly (or elliptically) polarized light will be reflected with different amplitudes and phases at the surface of a magnetized medium. Since linearly polarized light can be decomposed into circular (or elliptical) components, the difference in Fresnel coefficients implies that the reflected light will be elliptically polarized with major axis rotated from the initial polarization plane.[2]

If linearly polarized light is normally incident upon the surface, and if the magnetization is perpendicular to the surface, the rotation angle of the major axis of the ellipse is

$$\theta_k = -\operatorname{Im}\left\{\frac{n_+ - n_-}{n_+ n_- - 1}\right\}$$

and the ellipticity of the reflected light is

$$\epsilon_k = -\operatorname{Re}\left\{\frac{n_+ - n_-}{n_+ n_- - 1}\right\}$$

where $n_\pm = N_\pm - iK_\pm$ is the complex index of refraction for right (+) or left (−) circularly polarized light and Re and Im denote the real and imaginary parts, respectively. Both θ_k and ϵ_k are typically very small with θ_k less than 10 minutes of arc for the ferromagnetic metals.

If the light is not normally incident, there are three cases. (1) In the *polar Kerr effect*, the magnetization is normal to the surface, and the complex Fresnel reflection coefficients for *s* and *p* polarized incident light result in elliptically polarized reflected light. (2) In the *longitudinal Kerr effect*, the magnetization is parallel to the surface and lies within the plane of incidence. Again, linearly polarized incident light becomes elliptically polarized upon reflection. (3) In the *equatorial Kerr effect*, the magnetization is parallel to the surface and perpendicular to the plane of incidence. Light polarized along the magnetization (*s*-polarization) is *unaffected* by the magnetic field, while *p* polarized light has a reflection coefficient that contains a term *linear* in the magnetization. In absorbing materials the intensity of the reflected light will depend upon the sign of the magnetization. The equatorial Kerr effect has been an important tool for studying absorbing magnetic materials such as ferromagnetic metals.[10]

The polar Kerr effect has been exploited in magneto-optical memories for computer and video applications. A film of magnetic material can be magnetized with small domains having magnetic fields either up or down. An incident plane wave is reflected with polarization detectably rotated one way or the other by different domains. Up magnetization can be interpreted as a binary 1 while down magnetization would be binary 0. Such domains can be produced with dimensions as small as a wavelength of light by heating the film with a focused laser beam in a magnetizing field and allowing it to cool through a curie or compensation temperature. The calculated information density would be greater than 10^8 bits/cm^2.

MARC D. LEVENSON

References

1. Kerr, J., *Phil. Mag.* 50, 337, 446 (1875); 8, 85, 229 (1879); Beams, J. W., *Revs. Mod. Phys.* 4, 133 (1932).
2. Kerr, J., Rept. Brit. Assoc. Adv. Sci. P40 (1876); *Phil. Mag.* 3, 321 (1877); 5, 161 (1878).
3. McClung, F. J., and Hellwarth, R. W., *Appl. Optics Suppl.* 1, 103 (1962).
4. Gray, D. E. (Ed.), "American Institute of Physics Handbook," McGraw-Hill, New York, 1972, p. 6-232.
5. Kaminov, I. P., and Turner, E. H., *Appl. Optics* 5, 1612 (1966).
6. Mayer, G., and Gires, F., *C. R. Acad. Sci. (Paris)* 258, 2039 (1964).
7. Maker, P. D., and Terhune, R. W., *Phys. Rev.* 137, A801 (1965).
8. Dugay, M. A., and Mattick, A. T., *Appl. Opt.* 10, 2162 (1971).
9. Levenson, M. D., "Introduction to Nonlinear Laser Spectroscopy," New York, Academic Press, 1982, pp. 73ff, 139ff.
10. Freiser, M. J., *IEEE Trans. on Magnetics* MAG-4, 152 (1968).

Cross-references: FARADAY EFFECT, LASER, POLARIZED LIGHT, RAMAN EFFECT AND RAMAN SPECTRA, REFLECTION, REFRACTION.

KINETIC THEORY

The kinetic theory is a branch of THEORETICAL PHYSICS developed in the nineteenth century to explain and calculate the properties of fluids. It is most useful for studying the physical properties of gases, but it can also be applied to liquids, electrons in metals, and neutrons passing through solids. The word "kinetic" means "pertaining to motion," in this case the motion of molecules or subatomic particles.

Historical Development The first attempt to develop a kinetic theory of gases was made in 1738 by the Swiss mathematician Daniel Bernoulli. Bernoulli began with the idea that matter consists of tiny atoms moving about rapidly in all directions, which the Greek philosopher Democritus had presented, but was unable to prove. Bernoulli showed that the collisions of atoms against the walls of a container would produce a pressure which would be inversely proportional to the total volume of the container; he assumed that the space occupied by the atoms themselves is negligible compared to the total volume of the container and that the rest of the space is empty. He also found that the pressure would be directly proportional to the kinetic energy of motion of the atoms if the velocities of the atoms are changed while the volume is kept fixed. (The kinetic energy of an atom is half its mass multiplied by the square of its velocity.)

The British scientist Robert Boyle had already shown in 1662 that the pressure of air varies inversely as its volume if the temperature is held constant (see GAS LAWS). Thus Bernoulli's theory was able to explain a well-known fundamental property of air and other gases. It was also known that the pressure of a gas confined in a fixed volume increases with temperature. However, it was not until about 1800 that there was enough experimental evidence, and an accurate enough temperature scale, for Gay-Lussac (French) and others to establish a quantitative relation between pressure and temperature. This relation can be expressed by saying that pressure is proportional to the temperature measured from "absolute zero" (though it was not until later in the nineteenth century that the idea of absolute zero temperature was generally accepted). According to the kinetic theory, the absolute temperature is proportional to the kinetic energy of motion of molecules in a gas.

The kinetic theory was proposed again in the first half of the nineteenth century by two British scientists, John Herapath and J. J. Waterston. Neither of them was familiar with Bernoulli's theory, which had not made much impression on the world of science. Waterston obtained one important new result, which is now known as the "equipartition theorem": in a mixture of two or more different gases at the same temperature, the average kinetic energy of each kind of molecule will be the same. This means that heavy molecules will tend to move more slowly than light molecules, since when the mass of a molecule is greater, its velocity must be less in order to keep the kinetic energy the same.

In 1858, the German physicist Rudolf Clausius showed how the kinetic theory could be used to explain the rate of mixing of two gases and the rate of heat conduction. His work was extended by James Clerk Maxwell (British), who calculated the viscosity coefficient by the kinetic theory. He found that theoretically the VISCOSITY of a gas should be the same at different pressures, and should increase with temperature. This seemed to go against common sense, but later experiments by Maxwell himself and other physicists showed that the theory is correct. Soon afterward several scientists, starting with Josef Loschmidt (Austrian) in 1865, used the kinetic theory to calculate the diameter of an atom. At this time it began to appear that the atom is something that really exists in nature, since it can be measured, weighed, and counted, and is not merely a philosophical speculation. By then the atomic theory had already been accepted in chemistry as a basis for explaining chemical reactions, but it was the kinetic theory of gases that established the place of atoms in physics.

Starting from the foundations laid by Clausius and Maxwell, Ludwig Boltzmann (Austrian) and J. Willard Gibbs (American) worked out systematic methods for calculating all the properties of gases from kinetic theory (see STATISTICAL MECHANICS). Sydney Chapman (British) and David Enskog (Swedish) completed the theory, insofar as it pertains to the transport properties (diffusion, viscosity, and heat conduction) of gases at ordinary densities, although there are still some unsolved problems in the area of high-density gases and liquids, on the one hand, and rarefied (very low density) gases on the other. In the course of working out this theory, Chapman and Enskog discovered that it should be possible to separate the components of a mixture of a gas by making one side of the container hotter than the other. This effect—known as "thermal diffusion"—was soon afterwards established experimentally by Chapman and Dootson (British), thus confirming the prediction based on kinetic theory. (Thermal diffusion was used as one of the methods of separating isotopes of uranium during the development of the atomic bomb in World War II).

Although the kinetic theory was founded on the principles of classical Newtonian mechanics and led to some incorrect results because those principles are not valid on the molecular level, it is now generally agreed that the kinetic theory is valid for calculating the statistical properties of large numbers of molecules, provided that the properties of the individual molecules themselves are determined experimentally, or from the quantum theory. It is only when one tries to apply the kinetic theory to matter in extreme conditions (very low temperatures or very high densities) that he must take account of quantum-mechanical modifications of the statistical method itself (see QUANTUM THEORY and STATISTICAL MECHANICS).

Main Features of the Theory By assuming that the major part of the heat energy of a gas consists of kinetic energy of motion of the molecules, one finds that the average velocity of a molecule is several hundred meters per second under ordinary conditions. However, it is a fact of common observation that gases do not actually move as a whole at such speeds; a gas will eventually spread throughout any container in which it is placed, but it may be several seconds or minutes, for example, before chlorine gas generated at one end of a large laboratory is noticed at the other end. According to the kinetic theory, the reason for the relative slowness of gaseous diffusion, in contrast to the high average velocities of individual molecules, is that a molecule can travel on the average only a very short distance (its "mean free path") before it collides with another molecule and changes its direction of motion. In particular, at atmospheric pressure, if the molecular diameter is assumed to be about 0.00000001 cm (which is approximately true for most molecules), the mean free path would be approximately 0.00001 cm. At the same time, the average distance between neighboring molecules would be somewhat more than 0.0000001 cm, so that the fraction of the total volume occupied by the molecules themselves is less than 1 part in 1000. The average molecular velocity in air at 15°C is about 460 m/sec, so that a molecule will have about 4 600 000 000 collisions per second.

To simplify their calculations, Maxwell and Boltzmann made the following assumptions:

(1) Instead of trying to compute the exact path followed by every molecule, they assumed that, because of the enormous frequency of collisions, the velocities and postions of molecules in a gas are distributed at random over all possible values consistent with the known physical state of the gas. For example, it is assumed that the average total velocity is known, as is the temperature (which fixes the mean square velocity). If variations of temperature and density from one place to another can be ignored, then the molecular velocities can be described by a statistical distribution—the "Maxwell distribution"—which is similar to the normal "bell-shaped curve" or law of errors in statistics.

It should be noted that the effect of a large number of collisions is *not* to make all the velocities equal, but rather to produce a wide range of velocities from zero up to very large values—though the probability of large deviations from the average is quite small. The existence of this "spread" of molecular velocities has been verified directly by various experiments.

(2) The diameter of a molecule is so small, compared to the average distance between molecules, that simultaneous collisions of three or more molecules may be ignored. The validity of this assumption for low-density gases makes it possible to develop a very accurate theory of gas properties, since these properties can be related to the interactions of molecules taken two at a time, and the mathematical description of such two-particle interactions is relatively simple. The corresponding kinetic theory of dense gases and liquids, on the other hand, involves the solution of difficult many-particle problems, and reliable results have been obtained only within the last few years.

(3) In treating the collision of two molecules, one supposes that there is no correlation between their velocities before the collision. In the elementary kinetic theory, one assumes that each molecule has the average velocity characteristic of the region of the gas in which it has most recently undergone a collision. It thus "forgets" its past history every time it collides with another molecule. This is obviously not strictly true for each individual molecule, but it is a useful approximation for dealing with average properties of large number of molecules.

In modern physics research, it is usual to distinguish between "equilibrium theory" and "transport theory," both of which grew out of the elementary kinetic theory. Equilibrium theory is described in the article on STATISTICAL MECHANICS (see also EQUILIBRIUM); it is used to study such properties as the heat capacity (amount of heat needed to raise the temperature by a certain amount) and the compressibility (change in volume produced by a small change in pressure). Equilibrium theory also tries to explain the existence of phase transitions, such as the condensation of gases to liquids, or the appearance of magnetic ordering in solids. From the theoretical viewpoint, the calculation of equilibrium properties is simpler than that of transport properties such as viscosity, because one merely averages over all the possible states of the system (i.e., over all possible combinations of velocities and positions of the molecules) without having to worry about how one state follows another in time. TRANSPORT THEORY involves a detailed analysis of molecular collisions in order to determine how changes in the state of the system are related to external forces or nonuniform conditions imposed on it.

One of the most fruitful techniques in transport theory is the use of "Boltzmann's equation." This equation describes how the velocity distribution changes as a result of external forces and collisions between molecules. Unfortunately the equation is rather difficult to solve, because the term that expresses the effect of collisions on the velocity distribution is an integral over the values of the (unknown) velocity distribution itself for two colliding molecules. In order to calculate the transport properties it is necessary to resort to tedious computations with infinite series, except for certain artificial force laws (such as repulsive forces inversely proportional to the fifth power of the distance between two molecules) for which the integral can be simplified. In most cases, the results (as worked out by Chapman and Enskog) do not differ greatly from the ones obtained from the approximate elementary theory. However, the important phenomenon of thermal diffusion was discovered only because of a theoretical prediction by Enskog and Chapman; since the existence of this effect had not even been suggested by the earlier theories, this discovery must be regarded as one of the triumphs of mathematical analysis.

Irreversibility As indicated above, the kinetic theory assumes that the velocity of a molecule may depend on the conditions in the region where it has just suffered a collision, but is otherwise random—in other words, independent of its previous history. This assumption permits one to use the methods of probability theory even though, in classical mechanics, the actual motions of the molecules are regarded as completely determined by their initial configurations. As long as one uses the theory only to calculate properties of a gas that can actually be measured during a relatively short time, the assumption of randomness leads to no serious errors. However, it introduces an element of irreversibility which is inconsistent with the reversibility of the laws of classical mechanics. (A reversible process is one that can go equally well forwards or backwards, in contrast to an irreversible process, like scrambling an egg, which cannot be undone without a great expenditure of energy. The British physicist William Thomson, later Lord Kelvin, had pointed out the importance of irreversible processes in his "Principle of the Dissipation of Energy," in 1852.) The irreversible aspect of the kinetic theory is shown most clearly by Boltzmann's "H-theorem," which has led to a considerable amount of controversy about the foundations of kinetic theory. Boltzmann showed in 1872 that a certain quantity, later called H, which depends on the velocity distribution, must always decrease with time, unless the velocity distribution is Maxwell's distribution, in which case H remains constant. In the latter case, which corresponds to the equilibrium state, H is proportional to the negative of the entropy (see article on THERMODYNAMICS). Thus the H-theorem provides a molecular interpretation of the second law of thermodynamics or, in particular, the principle that the entropy of an isolated system must always increase or remain constant. Irreversible processes are those in which entropy increases. The entropy itself can be regarded as a measure of the degree of randomness or disorder of the gas, although it must be recognized that disorder really means just our own lack of knowledge about the details of molecular configurations. The equilibrium state represents the maximum possible disorder; the H-theorem implies that a gas which is initially in a nonequilibrium (partly ordered) state will eventually reach equilibrium and then stay there forever if it is not disturbed.

If the long-term consequences of the H-theorem were applicable to all matter in the universe, one might expect that the universe would eventually "run down"; although the total energy might always remain the same, no useful work could be done with this energy because all matter would be at the same temperature (see THERMODYNAMICS). This final state has been called the "heat death" of the universe.

The contradiction between the H-theorem and the laws of classical mechanics is shown by two famous criticisms of the kinetic theory, the "reversibility paradox" and the "recurrence paradox." The first paradox is based on the fact that Newton's laws of motion are unchanged if one reverses the time direction, so that it would seem to be impossible to deduce from these equations a theorem that predicts irreversible behavior. Kelvin discussed this paradox in 1874, and concluded that while any single sequence of molecular motions could be reversed, leading to an ordered state, the number of disordered states is so much greater than the number of ordered states that it is virtually impossible to stay in an ordered state for any period of time. Thus irreversibility is a statistical but not an absolute consequence of kinetic theory. Boltzmann gave a similar answer when the problem was pointed out to him by Loschmidt a few years later. The second paradox is based on a theorem of Henri Poincaré (French): if a mechanical system is enclosed in a finite volume, then after a sufficiently long time it will return as closely as one likes to its initial state. Hence H must return to its original value; if it has decreased during some period of time, it must increase during some other period. The time between successive recurrences of the same state for the molecules in 1 cc of air is much longer than the present age of the universe, so one does not have to worry about recurrences in any actual experiment. In his attempt to resolve the recurrence paradox, Boltzmann was finally led to a remarkable psycho-cosmological speculation: he suggested that the "direction of time" as perceived by an animate being is *determined* by the direction of irreversible processes in his environment and in his body. Thus when the time comes for a recurrence, entropy will decrease but subjective time will flow in the opposite direction; thus the law "entropy increases with time" is a tautology! This idea of alternating time-directions in cosmic history was further explored by H. Reichenbach (*The Direction of Time*, Berkeley, 1956) and has been proposed again in recent theories of the expanding (and contracting) universe.

Some other aspects of this problem and its connection with atomic randomness are discussed in the article on IRREVERSIBILITY.

Recent Developments Since World War II there has been a revival of interest in the "classical" kinetic theory of gases, based on the assumptions of Clausius, Maxwell and Boltzmann and ignoring quantum effects except insofar as these may determine the intermolecular force law. In part this interest is due to applications involving high-speed aerodynamics and plasma physics, in part to renewed attempts to construct reliable theories of liquids as well as dense gases. New methods for obtaining accurate solutions of the Boltzmann equation have been developed by H. Grad, C. L. Pekeris, E. Ikenberry and C. Truesdell, and many others. These solutions have been used to describe the behavior of gases in many circumstances more complicated than those treated in the nineteenth century (including the interactions of charged particles and magnetic fields). Problems such as the propagation and dispersion of sound waves have also been treated by G. E. Uhlenbeck and his collaborators.

In 1946, three general formulations of kinetic theory were published, by M. Born and H. S. Green, by J. G. Kirkwood, and by N. N. Bogoliubov. In each case the goal was to derive a generalized Boltzmann equation in a form that would be valid when simultaneous interactions among more than two molecules have to be taken into account, and thence to obtain solutions of the equation from which transport properties of dense gases and liquids could be calculated. In each formulation certain approximations had to be made in order to obtain practical results; because of the difficulty in estimating the error involved in these approximations, and the great complexity of the equations involved, there was no clear evidence that the results for properties such as the viscosity coefficient would be significantly more accurate than those obtained by Enskog from his modified kinetic theory for dense gases published in 1922. Eventually, in the early 1960s, attention was centered on the systematic derivation of series expansions for the transport coefficients in ascending powers of the density, together with attempts to calculate the first few terms in such series for special molecular models such as elastic spheres. In the meantime, an alternative and apparently more rigorous method for deriving theoretical expressions for transport coefficients, based on the "fluctuation-dissipation theorem" introduced in 1928 by H. Nyquist in electrical engineering problems, was developed by M. S. Green, H. Mori, and R. Kubo. This method had the heuristic advantage of bringing out clearly the connection between transport theory and the description of fluctuations in equilibrium statistical mechanics. Later it was proved that the Green-Mori-Kubo method gives results precisely equivalent to those that would be obtained from the Born-Green, Kirkwood, and Bogoliubov methods, and also those of yet another method developed by I. Prigogine, if in each case the calculations are done without approximation. Thus, just as in the case of quantum mechanics, several alternative approaches are equally valid in modern kinetic theory.

After intensive efforts to calculate terms in the density expansion of transport coefficients, it was finally discovered in 1965 that such a density expansion does not actually exist, for mathematical reasons associated with the persistence of weak correlations between colliding particles over very long times. The divergence of the expansion (and thus the inadequacy of the approximations on which most earlier theories had been based) was established almost simultaneously by J. R. Dorfman and E. G. D. Cohen, J. Weinstock, and R. Goldman and E. Frieman. The result has been a flurry of activity in kinetic theory, in which many of the intuitively plausible ideas about "relaxation" of initial states to steady nonequilibrium states, and destruction of correlations by intermolecular collisions, have been revised.

In spite of the impressive success of kinetic theory in solving numerous problems of both practical and theoretical interest, many scientists are not satisfied that the theory of nonequilibrium processes rests on a solid foundation comparable to that of equilibrium statistical mechanics. There is no single fundamental postulate analogous to the Gibbs canonical distribution from which everything can be deduced. Perhaps as a result of this gap in fundamental principles, there remains considerable disagreement about how to formulate some of the major outstanding problems which involve the connection between the atomic and the macroscopic levels, such as the nature of turbulence.

Other applications of the kinetic theory are discussed in the articles on AERODYNAMICS, BOLTZMANN'S DISTRIBUTION LAW, ELECTRICAL CONDUCTIVITY, LIQUID STATE, NUCLEAR REACTORS and PLASMA. See especially IRREVERSIBILITY.

STEPHEN G. BRUSH

References

For an elementary introduction, see Cowling, T. G., "Molecules in Motion," London, Hutchinson, 1950; reprinted by Harper Torchbooks, New York, 1960.

Diffusion and thermal diffusion: Furry, W. H., "On the Elementary Explanation of Diffusion Phenomena in Gases," *Am. J. Physics*, **16**, 63 (1948).

Comprehensive treatment of modern kinetic theory and its applications: Lifshitz, E. M., and Pitaevskii, L. P., "Physical Kinetics," New York, Pergamon Press, 1981; Berne, Bruce J. (Ed.), "Statistical Mechanics, Part B: Time-Dependent Processes," New York, Plenum Press, 1977, especially the chapter by J. R. Dorfman and H. Van Beijeren.

Recent developments and current research problems: Raveché, H. J. (Ed.), "Perspectives in Statistical Physics," Amsterdam, North-Holland, 1981, especially the chapters by J. R. Dorfman and R. Zwanzig.

History of kinetic theory: Brush, S. G. (Ed.), "Kinetic Theory," 3 Vols., New York, Pergamon Press, 1965-72; "The Kind of Motion We Call Heat: A History of the Kinetic Theory of Gases in the 19th Century," Amsterdam, North-Holland, 1976; "Statistical Physics and the Atomic Theory of Matter," Princeton, N.J., Princeton Univ. Press, 1983.

L

LAPLACE TRANSFORM

Introduction The Laplace transform $\bar{f}(s)$ of a function $f(t)$ is defined to be the integral

$$\bar{f}(s) = \int_0^\infty e^{-st} f(t)\,dt$$

if the integral exists. It will certainly exist if $f(t)$ is itself integrable between zero and an arbitrary upper limit, and if, for a large enough real value of k, $e^{-kt}f(t) \to 0$ as $t \to \infty$. Then if it exists for one complex value of s it exists also for all complex s of greater real part, and so the integral "transforms" the function $f(t)$ into a function $\bar{f}(s)$ defined on a half-plane of the complex Argand plane.[10,6]

Operational Methods The Laplace transform is perhaps the best known and most useful of a number of integral transforms, whose application is that province of applied mathematics usually called operational methods.

The essential idea is to exploit the analogy between certain differential and algebraic operators. Historically, the idea goes back to Leibnitz, and has been employed by a number of mathematicians including Laplace, Lagrange and Riemann.[2] It was treated extensively by Boole, and Oliver Heaviside developed it to the point where he could apply it, particularly to circuit theory in electricity. In his D-notation, given the ordinary differential equation with constant coefficients

$$a_0 \frac{d^n y}{dt^n} + a_1 \frac{d^{n-1} y}{dt^{n-1}} + \cdots + a_{n-1} \frac{dy}{dt}$$
$$+ a_n y = F(t),$$

we introduce the operator $P(D) = \sum_{k=0}^{n} a_k D^{n-k}$ and so write the equation as

$$P(D)y = F(t).$$

Then it is formally solved as

$$y = \frac{1}{P(D)} F(t),$$

and one is left to interpret this result. The key to this interpretation lies in the three operational relations

$$D \equiv \frac{d}{dt}, \quad D^{-1} = \int^t dt,$$

and

$$(1 - D)^{-1} = \sum_{k=0}^{\infty} D^k = \sum_{k=0}^{\infty} D^{-k-1}.$$

In these, D^{-1} is an indefinite integral and hence one of the major faults was that D and D^{-1} were not commutative, as $D^{-1}D - DD^{-1}$ would in general be a constant. Heaviside does not seem to have been clear which interpretation of $(1 - D)^{-1}$ to take, preferring the first but recognizing the difficulties of convergence implied. The method also fails to take account of the initial values under which the equation is to be solved.

Now it follows by successive integrations by parts that if $d^k y/dt^k$ has a Laplace transform it is

$$s^k \bar{y}(s) - \sum_{j=0}^{k-1} y_{k-j-1} s^j$$

where $\bar{y}$ is the Laplace transform of y, and $y_r = \lim_{t \to 0} d^r y/dt^r$. Thus the Laplace transform has the algebraic property of the D-operator, but the initial values are now included in the operation. If this is then applied to the differential equation above, the term to be interpreted becomes

$$\bar{y}(s) = \{\bar{F}(s) + Q(s)\}/P(s),$$

where $\bar{F}(s)$ is the Laplace transform of $F(t)$, and $Q(t)$ is a polynomial of degree $n - 1$ with coefficients depending on the constants y_r.

It can easily be verified that the Laplace transforms of $t^n e^{at}$, $e^{at}\cos bt$ and $e^{at}\sin bt$ are respectively $n!(s - a)^{-1-n}$, $(s - a)/\{(s - a)^2 + b^2\}$ and $b/\{(s - a)^2 + b^2\}$. If, then, the forcing function $F(t)$ is a finite sum of terms which are products of powers of t, exponentials, and trigonometric functions, the equation can be solved by resolving $\bar{y}(s)$ into partial fractions and interpreting separately the individual terms.

The Inversion Integral The justification for the procedure above involves an assumption that the relation between a function and its Laplace transform is unique. This would follow from a corresponding result for the Fourier transform, as we will show that the relationship for the Laplace transform can be deduced from the Fourier Integral Theorem. In fact, functions which differ on a set of points of zero measure will have the same Fourier or Laplace transform; but in most practical problems such an equivalence class of functions would not be distinguished in any case, so that effectively there is a uniqueness theorem. The Fourier integral theorem establishes, for a restricted class of functions, which vanish sufficiently strongly for large values of t, a unique reciprocal relation between a function $F(t)$ and its Fourier transform $\hat{F}(p)$,

$$(2\pi)^{1/2} \hat{F}(p) = \int_{-\infty}^{\infty} e^{ipt} F(t) dt,$$

$$(2\pi)^{1/2} F(t) = \lim_{X \to \infty} \int_{-X+ic}^{X+ic} e^{-ipt} \hat{F}(p) dp,$$

where c is chosen so that the integral is meaningful.

If we introduce the unit function $H(t)$, which is one for positive t, and zero for negative t, a function often associated with the name of Heaviside, and we also write $p = is$, $F(t) = (2\pi)^{1/2} f(t) H(t)$, then $\hat{F}$ (is) is the Laplace transform $\bar{f}(s)$. We therefore obtain the uniqueness theorem we desire, and in addition obtain an inversion formula for the Laplace transform, often known as the Bromwich integral

$$2\pi i f(t) H(t) = \lim_{X \to \infty} \int_{c-iX}^{c+iX} e^{st} \bar{f}(s) ds,$$

where c must be greater than the real part of any singularity of $\bar{f}(s)$.

We may use the method of residues to evaluate the inversion integral for a wide class of transforms,[6] and so with its use we can relax the restriction we had to place on the forcing function above.

An alternative form of the inversion formula can be given by[11]

$$f(t) = \lim_{n \to \infty} \frac{(-1)^n}{n!} \left(\frac{n}{t}\right)^{n+1} \bar{f}^{(n)} \left(\frac{n}{t}\right)$$

where $\bar{f}^{(n)}(s)$ is the nth derivative of $\bar{f}(s)$.

The Convolution Theorem We may also show that, for a suitably restricted class of functions, the Fourier transform has a convolution or Faltung theorem associated with it. This takes the form that if $\hat{F}(p)$ and $\hat{G}(p)$ are respectively the Fourier transforms of $F(t)$ and $G(t)$, then

$(2\pi)^{1/2} \hat{F}(p) \hat{G}(p)$ is the Fourier transform of

$$\int_{-\infty}^{\infty} F(u) G(t-u) du.$$

If we interpret this as a theorem for the Laplace transform we obtain the result that if $\bar{f}(s)$ and $\bar{g}(s)$ are respectively the Laplace transforms of $f(t)$ and $g(t)$, then $\bar{f}(s)\bar{g}(s)$ is the Laplace transform of the function

$$\int_{0}^{t} f(u) g(t-u) du.$$

This theorem has an immediate application to the solution of the differential equation we have introduced above, for we can assert that if $Y(t)$ is the function whose Laplace transform is $1/P(s)$, then $\bar{F}(s)/P(s)$ is the Laplace transform of

$$\int_{0}^{t} F(u) Y(t-u) du.$$

Now $P(s)$ is a polynomial in s, and so we can write

$$P(s) = a_0 \prod_{k=1}^{m} (s - \alpha_k)^{N_R}$$

where the α_k are in general complex, and $\sum_{k=1}^{m} N_k = n$. We can then resolve $1/P(s)$ into partial fractions

$$\sum_{k=1}^{m} \sum_{j=1}^{N_k} \frac{A_{kj}}{(s - \alpha_k)^j}$$

(see Refs. 5 and 6) and so

$$Y(t) = \sum_{k=1}^{m} \sum_{j=1}^{N_k} \frac{A_{kj}}{(j-1)!} t^{j-1} e^{\alpha_k t}.$$

This is equivalent to solving the differential equation by the method of variation of parameters.

The convolution theorem is also of great assistance in the solution of integral equations with difference kernels, i.e., with kernels of the form $K(t, u) = K(t - u)$. In particular we see that the Volterra equation

$$f(t) = g(t) + \lambda \int_{0}^{t} K(t - u) f(u) du$$

leads directly to

$$\bar{f}(s) = \bar{g}(s) / \{1 - \lambda \bar{K}(s)\}.$$

The corresponding Fredholm integral equation with a different kernel may also be solved using a transform. In this case, however, another integral equation is obtained which may be solved by the Wiener-Hopf technique.[6]

Other Properties of the Laplace Transform It can be proved quite easily[6] that within the half-plane in which it converges, $\bar{f}(s)$ is an analytic function of s, and so has derivatives of all orders with respect to s. It then follows that

$$\frac{d^k\bar{f}(s)}{ds^k} = (-1)^k \int_0^\infty e^{-st}t^k f(t)dt.$$

We may apply this relation to ordinary differential equations whose coefficients are polynomials in t. If we combine this formula with the relation for the transform of a derivative we have used before, the equation

$$\sum_{k=0}^{n}\sum_{j=0}^{mk} a_{kj}t^j \frac{d^{n-k}y}{dt^{n-k}} = F(t)$$

is transformed into

$$\sum_{k=0}^{n}\sum_{j=0}^{mk} (-1)^j a_{kj}\left\{\frac{d^j}{ds^j}(s^{n-k}\bar{y})\right.$$

$$\left. - \sum_{i=j}^{n-k-1} \frac{i!\,y_{n-k-i-1}}{(i-j)!}s^{i-j}\right\} = \bar{F}(s).$$

Hence an ordinary linear differential equation of order n whose coefficients are polynomials in t of maximum degree m, becomes an ordinary linear differential equation for the transform which is of order m and whose coefficients are polynomials of degree n. This is exactly the same situation as arises if we attempt to solve the original equation by a contour integral of Laplace type.[7]

In fact the Laplace transform is only a special case of this more general technique, which is in consequence rather more useful.

One advantage does, however, remain for the use of the Laplace transform in this context. It has asymptotic properties which can allow the deduction of certain properties of a function from its transform without explicitly inverting it. It is an immediate deduction from Watson's lemma[7] that if, for small t,

$$f(t) = t^\nu \sum_{n=0}^{\infty} a_n t^n, \qquad \nu > -1,$$

then for $|s|$ large

$$\bar{f}(s) \sim s^{-1-\nu} \sum_{n=0}^{\infty} \Gamma(n+\nu+1)a_n s^{-n}.$$

To deduce an expression for $f(t)$ for small t from the transform, a converse theorem is required and the conditions under which this is valid may be found in Ref. 4. The same text establishes conditions for the validity of another useful relation, which is that if a is the singularity of $\bar{f}(s)$ of greatest real part, and near it $\bar{f}(s)$ has an expansion

$$\bar{f}(s) = (s-a)^{-\nu-1}\sum_{n=0}^{\infty} A_n(s-a)^n,$$

then for large positive values of t

$$f(t) \sim t^\nu e^{at} \sum_{n=0}^{\infty} \frac{A_n}{\Gamma(\nu+1-n)t^n}.$$

Partial Differential Equations and Other Transforms Possibly the most useful application is to the solution of initial value problems for some of the linear partial differential equations of mathematical physics. This is exploited to the full in Ref. 1 and to a lesser extent in Refs. 5 and 6. In a number of these equations time derivatives appear with constant coefficients, so the use of the transform effectively reduces by one the number of independent variables. In particular, if there is only one other independent variable than the time, it produces an ordinary differential equation. The method has great flexibility since the asymptotic properties can often be used to extract information which conventional solutions in eigenvalue expansions conceal. It can be shown that if the Laplace transform of $f(t)H(t)$ is $\bar{f}(s)$, then the Laplace transform of $f(t-a)H(t-a)$ is $e^{-sa}\bar{f}(s)$. In problems involving wave propagation this can be exploited to analyze the solution into the components arising from multiple reflections.

It is clear from the way that the initial conditions appear in the Laplace transform of a derivative, that it can be used primarily for the solution of initial value problems, and hence for partial differential equations of hyperbolic or parabolic type only. In treating the solution of equations of elliptic type on infinite domains, in which the boundary conditions are given at both ends of the range, we require other integral transforms, the conditions at infinity being usually of a form which will ensure the convergence of the defining integral. The question of which transform to use is usually decided by the equation, the coordinate system, and the form in which the boundary conditions are given. It is convenient to illustrate some of the most commonly used transforms by reference to the Laplace equation $\nabla^2\phi = 0$.

In Cartesian coordinates this has the form

$$\frac{\partial^2\phi}{\partial x^2} + \frac{\partial^2\phi}{\partial y^2} + \frac{\partial^2\phi}{\partial z^2} = 0.$$

If the region in which a solution is sought extends over $(-\infty, \infty)$ in one coordinate, say x, and the boundary conditions are that $\phi \to 0$ as $|x| \to \infty$, we would use a Fourier transform, whose inversion formula and convolution integral have already been given. Like the Laplace transform, an asymptotic expansion of the transform is equivalent to the expansion of the function near the origin. The expansion of the transform near the singularity of least negative imaginary part gives the asymptotic expansion of the function near $x = \infty$, and of least positive imaginary part near $x = -\infty$.

If the range in x is only semi-infinite, we may use either a Fourier cosine transform, with the reciprocal relations

$$\bar{f}(s) = \left(\frac{2}{\pi}\right)^{1/2} \int_0^\infty f(x) \cos sx \, dx;$$

$$f(x) = \left(\frac{2}{\pi}\right)^{1/2} \int_0^\infty \bar{f}(s) \cos sx \, dx$$

or the Fourier sine transform with the reciprocal relations

$$\bar{f}(s) = \left(\frac{2}{\pi}\right)^{1/2} \int_0^\infty f(x) \sin sx \, dx;$$

$$f(x) = \left(\frac{2}{\pi}\right)^{1/2} \int_0^\infty \bar{f}(s) \sin sx \, dx.$$

Both transforms require the boundary condition $\phi(x) \to 0$ as $x \to \infty$, but the choice between them is decided by whether ϕ or $\partial\phi/\partial x$ is given on $x = 0$, for if $\phi(x) \to 0$ as $x \to \infty$, and the integrals exist,

$$\int_0^\infty \frac{\partial^2 \phi}{\partial x^2} \cos sx \, dx$$

$$= - \left(\frac{\partial \phi}{\partial x}\right)_{x=0} - s^2 \int_0^\infty \phi(x) \cos sx \, dx,$$

while

$$\int_0^\infty \frac{\partial^2 \phi}{\partial x^2} \sin sx \, dx$$

$$= s\phi_{x=0} - s^2 \int_0^\infty \phi(x) \sin sx \, dx.$$

Convolution theorems and the asymptotic properties can be deduced by noting that if ϕ is continued for $x < 0$ as an even function, the cosine transform is the Fourier transform of the continued function, and if ϕ is continued as an odd function we relate the sine transform to the Fourier.

In spherical polar coordinates the Laplace equation has solutions like $R^\nu S_\nu$ where S_ν is a spherical harmonic of degree ν. This suggests that to solve problems in conical regions of infinite extent we use the Mellin transform and its inverse

$$\bar{f}(s) = \int_0^\infty R^{s-1} f(R) \, dR;$$

$$f(R) = \lim_{X \to \infty} \frac{1}{2\pi i} \int_{c-iX}^{c+iX} R^{-s} \bar{f}(s) \, ds.$$

This can be derived from the Fourier transform by substituting $R = e^x$, which also gives the convolution theorem that $\bar{f}(s)\bar{g}(s)$ is the transform of

$$\int_0^\infty f(\rho) g\left(\frac{R}{\rho}\right) \frac{d\rho}{\rho}.$$

To apply this transform to the equation it is necessary, but not sufficient, that $R^s \phi$ vanish at infinity and at $R = 0$. These conditions will also dictate what choice of c in the inversion integral will be needed for convergence. The singularity of $\bar{f}(s)$ whose real part is less than c but nearest to it will dominate the behavior of $f(R)$ for small R, and the singularity with real part greater than c but closest to it will dominate the behavior of $f(R)$ for large R.

If the Laplace equation is expressed in terms of cylindrical polar coordinates (r, θ, z) it has solutions of the form $\Phi_\nu(r, z)e^{i\nu\theta}$, and to determine the form of Φ_ν we may use a Hankel transform of order ν, defined by

$$\bar{f}_\nu(s) = \int_0^\infty r f(r) J_\nu(sr) \, dr,$$

$$f(r) = \int_0^\infty s \bar{f}_\nu(s) J_\nu(sr) \, dr$$

where $J_\nu(sr)$ is the Bessel function of first kind and of order $\nu \geq 0$.

For integer order this can be derived from applying simultaneous Fourier transforms to the x and y variables and then converting to polar coordinates by $x = r \cos \theta$, $y = r \sin \theta$. For noninteger order it is better to derive the result directly using a modified version of the method needed for the Fourier theorem and exploiting the analogy between the Bessel and circular functions. There is no convenient convolution theorem, though convolution theorems can be deduced. Once again the behavior of $f(r)$ for small r is deducible from the values of $\bar{f}_\nu(s)$ for large s, and the properties of $f(r)$ for large values of r follow essentially from the singularity of $\bar{f}_\nu(s)$ nearest the real axis. Any of these transforms can be applied simultaneously with other ones.

All these transforms and also the Hilbert transform

$$\bar{f}(s) = \frac{P}{\pi} \int_{-\infty}^\infty \frac{f(t) \, dt}{t - s}, \quad f(t) = \frac{1}{\pi} P \int_{-\infty}^\infty \frac{\bar{f}(s) \, ds}{s - t}$$

where $P\int$ is a Cauchy principal value, are treated rigorously in Ref. 8.

An alternative to the Fourier transform, provided $f(t)$ vanishes strongly enough at $t = \pm\infty$, is the so-called two-sided Laplace transform

$$\int_{-\infty}^{\infty} e^{-st} f(t) dt,$$

which is really a complex Fourier transform. An extended account of this and its application can be found in Ref. 9. Very many other integral transforms have appeared in the literature, and tables of them with some theory can be found in Ref. 3. A more general transform which embraces a number of those above is described in Ref. 11.

More Recent Work The introduction of the concept of generalized functions has considerably increased the class of functions whose transforms can be utilized. Thus, for example, the Fourier transform of t^k exists as a generalized function and is $(-i)^k (2\pi)^{1/2} \delta^{(k)}(s)$, where $\delta^{(k)}(s)$ is the kth distribution derivative of the Dirac delta function. In particular the growth condition for a generalized Laplace transform is relaxed to become one that is bounded above by a polynomial in $|s|$.

A treatment of a number of these generalized integral transforms can be found in Ref. 12.

A number of authors have made use of the transforms of generalized functions, and most notably the Fourier transform, in the rapid development in the theory of Linear Partial Differential Equations which has taken place in the last twenty years.[13,14]

There is also quite a lot of interest in developing both new integral transforms, such as are published periodically in mathematical journals, and a general operational calculus not limited specifically to any particular transform.[15]

M. G. SMITH

References

1. Carslaw, H. S., and Jaeger, J. C., "Conduction of Heat in Solids," Oxford, Clarendon Press, 1959.
2. Davis, H. T., "The Theory of Linear Operators," Bloomington, The Principia Press, 1936.
3. Ditkin, V. A., and Prudnikov, A. P., "Integral Transforms and Operational Calculus," Oxford, Pergamon Press, 1965.
4. Doetsch, G., "Theorie und Anwendung der Laplace-transformation," Berlin, Springer, 1937.
5. Jeffreys, H., and Jeffreys, B. S., "Methods of Mathematical Physics," London, Cambridge University Press, 1956.
6. Smith, M. G., "Laplace Transform Theory," London, Van Nostrand Reinhold, 1966.
7. Spain, B., and Smith, M. G., "Functions of Mathematical Physics," London, Van Nostrand Reinhold, 1970.
8. Titchmarsh, E. C., "Introduction to the Theory of Fourier Integrals," London, Clarendon Press, 1948.
9. Van der Pol, B., and Bremmer, H., "Operational Calculus Based on the Two-sided Laplace Transform," London, Cambridge University Press, 1955.
10. Widder, D. V., "The Laplace Transform," Princeton, Princeton University Press, 1941.
11. Widder, D. V., and Hirschman, I. I., "The Convolution Transform," Princeton, Princeton University Press, 1955.
12. Zemanian, A. H., "Generalized Integral Transformations," New York, Interscience, 1968.
13. Treves, F., "Basic Linear Partial Differential Equations," New York, Academic Press, 1975.
14. Schechter, M., "Modern Methods in Partial Differential Equations," New York, McGraw-Hill, 1977.
15. Davies, B., "Integral Transforms and Their Applications," Berlin and New York, Springer-Verlag, 1977.

Cross-references: CALCULUS OF PHYSICS, FOURIER ANALYSIS, MATHEMATICAL PRINCIPLES OF QUANTUM MECHANICS.

LASER

"Laser" is an acronym for l(ight) a(mplification by) s(timulated) e(mission of) r(adiation). This device is identical in theory of operation to the MASER except that it operates at frequencies in the optical region of the electromagnetic spectrum, rather than in the microwave. Laser operation has been demonstrated at wavelengths from 1500 to over 1 000 000 Å or from 0.15 to 100μm. By common usage, these devices are all called lasers, although more descriptive terminology utilizes ultraviolet maser, optical maser, infrared maser, etc. Although the original microwave maser offers an extremely stable frequency source, its main use is as an amplifier with extremely low noise output. In contrast, the main significance of the laser is its ability to produce a single frequency at high intensity in the optical region, a feat heretofore impossible at these frequencies. Not only may the output be a single monochromatic wave, but the wave may be coherent, or in phase, over the whole surface of the radiator. In this mode of operation, the laser is actually an oscillator whose output depends upon the selective amplification of one of the single frequency modes of the resonant cavity containing the active laser medium.

Following the development of the microwave maser, Schawlow and Townes in 1958 proposed that optical maser action could be obtained by placing an active medium in an optical cavity. The medium would be a gas or solid which was excited electrically or by light in such a manner that any optical wave present would be amplified as it moved through the material. The cavity was proposed to be a Fabry-Perot resonator—two plane, parallel reflecting plates with a small transmission through which the radiation might escape. Upon excitation of the material, light will be emitted with a band of frequencies determined by the particular material. In addition, the direction of emission will be nominally random. In the presence of the cavity, some of the waves will escape after several back and forth reflections from the parallel plates, "walking off" the edge of the reflectors. Those waves which travel normal to the walls will remain in the cavity and be amplified provided they reinforce each other after each round-trip reflection at the two surfaces. This reinforcement or resonance is only satisfied if the spacing of the plates is an integral multiple of one-half the wavelength in the medium. Thus, after a short time, only that frequency which satisfies the resonant condition and those waves traveling normal to the reflector will build up to an appreciable intensity. The resultant light which is partially transmitted through one of the reflectors will thus be a single frequency or several discrete frequencies if there is more than one cavity resonance within the band of frequencies emitted by the laser material. In addition, the wave front will be in phase across the surface of the reflector since waves striking the surface at normal incidence are amplified most strongly. The resultant beam will then be diffraction limited, i.e., the beam will spread by an angle in radians given approximately by the ratio of the wavelength to the diameter of the beam. In actual practice, single-mode operation is obtained only under special conditions. Generally, several frequency modes are present due to the multiple resonances of the cavity and numerous "off-axis" modes are found which correspond to resonant waves which travel at small angles from the normal to reflectors. These waves "walk off" so slowly that they still are amplified appreciably. Refinements of the simple cavity proposed by Schawlow and Townes consist of concave reflectors which decrease the diffraction losses, or several parallel reflectors which limit the oscillation to a frequency common to each pair in the set.

The key to successful laser operation is of course the active medium which amplifies the wave. Qualitatively, a material which fluoresces or exhibits luminescence is an obvious candidate. In fluorescence, electrons are excited to an upper-energy state by short-wavelength light such as ultraviolet, while luminescence is produced by passing an electron current through the medium, such as in a gaseous discharge. In either process, stimulated emission can occur only if more electrons are produced in the upper-energy state than in the lower or terminal state for the radiating transition. In this case, an incident photon will stimulate further transitions and amplification will result. If the final state were more heavily populated, then the photon would cause more upward or absorbing transitions and the net effect would be absorption.

Solid-State Lasers The first optical maser was demonstrated by Maiman of Hughes Research Laboratories in 1960 using "ruby," which is single-crystal aluminum oxide "doped" with chromium impurities. By applying semitransparent reflective coatings on the ends of a rod about 2 inches long, he made the cavity and the crystal an integral unit. Then, exposure to an intense exciting light from a xenon flashtube was found to invert the population between the red-emitting level and the ground or lowest-energy state of the electrons. The result was a burst of intense red light emanating in a beam through the end reflectors. This was the first and is still one of the most powerful lasers. Advances in the art since that time have resulted in energies per pulse of the order of 1000 joules or watt-seconds. Peak powers are as high as 500 000 kW in short pulses of the order of 10^{-8} sec. Because of "off-axis" modes and multiple resonances, the output is not a single-frequency, single plane-wave mode, but generally consists of the order of 100 separate modes. The beam is still quite narrow, being the order of 1 milliradian or 0.05 degrees. As a comparison with conventional light sources, the energy radiated from 1 cm^2 of the brightest flash lamp is less than 10 kW and is distributed over the entire visible spectrum. In addition, the radiation is incoherent and is spread out uniformly in all angles from the source. Thus, the directivity and spectral purity of the laser source are many orders of magnitude superior to that of an incandescent source. The ruby laser suffers from a low efficiency, about 1 per cent, and except with elaborate cooling systems, only operates on a pulsed basis. Other crystalline or glass systems with impurity ions have been developed, which yield wavelengths from the ultraviolet to approximately $3\mu m$ wavelength in the infrared. Some, such as neodymium-doped yttrium aluminum garnet (YAG), operate in a continuous mode at the one-watt level while peak powers have reached values as high as 10^{14} W in pulses of the order of 10^{-12} sec. These ultrahigh powers are obtained in neodymium-doped glass systems using several stages of amplification and novel pulse-forming techniques.

Gas Lasers Historically, the next development came in 1961 when Javan, Bennett and Herriott demonstrated laser action in a gaseous discharge of helium and neon. Again, the parallel-plate reflector cavity was used but this time with a spacing of several feet. Later, concave mirrors were used to decrease the loss of energy out the sides of the cavity. This device operates continuously and delivers power at levels up to one watt. Pulsing the gas discharge yields peak powers as high as 100 W. The first laser radiated at 1.15 μm in the infrared, while further development with different gases has yielded outputs from the ultraviolet to 1000 μm

or 1 mm in the far infrared. In contrast to the ruby laser, the gaseous laser beam may be diffraction limited and the frequency is pure, i.e., oscillation may be limited to one mode. By careful design, the frequency may be stabilized to a few thousand cycles per second or approximately one part in 10^{13}. Although the original gas laser utilized electrical excitation of electronic transitions, later versions use vibrational transitions in molecules such as carbon dioxide, and the excitation mechanism may be electrical, chemical, or thermal. In the chemical laser, atomic species such as hydrogen and fluorine can be reacted to produce molecules in an excited vibrational state which in turn yields amplification or oscillation. Recent electrically excited lasers, particularly those using carbon dioxide at 10 μm, have been operated at atmospheric pressure using spark discharges or pre-ionization by voltages in the 100-kV range. The high pressure and the powerful electrical excitation result in peak powers in the 10 to 100-MW region. For continuous laser operation, the gas may be circulated rapidly to avoid excessive heating, and using an electrical discharge, powers from 1 to 10 kW have been obtained. An entirely new excitation process, essentially thermal, was announced by Gerry in 1970. In this, the gas dynamic laser, an appropriate fuel is burned to produce carbon dioxide and nitrogen at high temperature and pressure. When released through a nozzle into the optical resonator region, the gas cools rapidly in terms of its kinetic or translational energy, but the population of the vibrational energy levels of the carbon dioxide molecules becomes inverted since the lower level of the laser transition relaxes much more rapidly. In addition, the vibrationally excited nitrogen molecules are in near resonance with the upper laser state of the carbon dioxide and transfer energy with high efficiency to maintain the inversion. This type of laser has produced continuous powers as high as 60 kW.

Semiconductor Lasers The third main type of laser utilizes a solid material, in this case a semiconductor. Here the electron current flowing across a junction between p- and n-type material produces extra electrons in the conduction band. These radiate upon making a transition back to the valence band or lower-energy states. If the junction current is large enough, there will be more electrons near the edge of the conduction band than there are at the edge of the valence band and a population inversion may occur. To utilize this effect, the semiconductor crystal is polished with two parallel faces perpendicular to the junction plane. The amplified waves may then propagate along the plane of the junction and are reflected back and forth at the surfaces. The gain in the material is high enough so that the reflection at the semiconductor-air interface is sufficient to produce oscillation without special reflective coatings. The first such device used gallium arsenide and radiated at 8400Å or just beyond the visible region in the infrared. This laser was developed by groups at General Electric, International Business Machines, and Lincoln Laboratory in 1962. The efficiency is high, about 40 per cent, and the power source is low-voltage direct current. One shortcoming is the requirement of liquid nitrogen cooling (77 K) to maintain power output and efficiency. Powers as high as 3 W continuous have been produced. The cavity in this case is extremely small, the reflector spacing being less than a millimeter. As a result, it is fairly easy to limit the oscillation to one frequency mode although small irregularities in the junction prevent coherence over the full width of the narrow radiating junction strip. The compactness and efficiency of the semiconductor laser make it particularly attractive for systems use. Wavelengths as long as 30μm and as short as 6300Å have been generated using different semiconductors such as indium arsenide, indium phosphide, indium antimonide, or alloys such as gallium arsenide-phosphide. In addition, these lasers may be tuned over several percent of their nominal frequency of operation by varying the current flow through the device. The tuning results from the variation in temperature with current which in turn changes the index of refraction and the resultant resonant frequency of the cavity. Since the linewidth of the radiation is only about 1 MHz the tunable semiconductor laser is an excellent tool for high resolution spectroscopy.

Liquid Lasers Laser action may also be obtained in liquids using either a flash tube or another laser as the pump. Early versions used rare earths in an organic liquid, while more recently organic dyes have been found to be more efficient but require a separate laser for the exciting radiation. The dye laser has the special attraction that one laser may be tuned over a significant fraction of the visible spectrum by using a reflection grating as one of the cavity mirrors. Another type of liquid laser utilizes a different principle than those above, depending upon stimulated Raman scattering. Raman laser action was discovered by Woodbury in 1962 using a ruby laser and nitrobenzene. Here the laser excites the nitrobenzene, which in turn shows amplification at a frequency displaced from the ruby line by the vibrational frequency of the molecule. There is no true inverted population in this case. The incident photon is scattered by the molecule which absorbs an amount of energy determined by its vibrational energy. The molecule is left in an excited state and the scattered photon is frequency shifted by the energy loss. This process may be stimulated, since the rate at which the scattered photons are produced is proportional to the number of photons already present in the cavity at the scattering wavelength. As in the normal stimulated emission case, the frequency and phase of the output wave are identical with the wave which stimulates the scattering. The Raman laser normally operates using the Stokes line, or the wavelength corresponding to the loss of one vibrational quantum. Other modes of operation utilize the second or third Stokes lines corresponding to double or triple vibrational absorptions. Similarly, higher-order effects in the medium may produce a series of anti-Stokes lines which correspond to vibrational energy being added to the initial energy of the photons from the driving laser. The wavelength range of Raman lasers using different liquids is from the visible to the near infrared.

Q-Switching The high instantaneous powers quoted for ruby are obtained by using the "Q-switched" mode of laser operation. This technique, due to Hellwarth and McClung, uses a cavity resonator whose reflectivity or "Q" may be controlled externally. The laser, usually ruby, is first excited by the flash lamp while the cavity is in a state of low reflectivity and thus low feedback. As a result, the inverted population reaches an extreme value before oscillation occurs. At the peak of inversion, the reflectivity is "switched on," and the resultant high reflectivity produces an intense burst of energy which almost completely depopulates the high-energy states in a time of the order of 10^{-8} sec. The switching is accomplished either by a Kerr electro-optic shutter in the cavity or by rotating one of the mirrors so that it is lined up parallel with the opposite reflector at the optimum time during the flash lamp pulse.

Mode-Locked Pulses An alternative method for generating extremely short pulses utilizes the technique of mode-locking. Since a laser resonator has frequency modes equally spaced at a separation of $c/2L$, where L is the cavity length, oscillation can occur in any mode as long as it is within the natural emission linewidth of the laser transition. Many lasers oscillate in only one mode, since that with the highest gain takes over from any modes away from the line center. Now, if the cavity is modulated internally at a frequency equal to the mode spacing, all modes within the natural linewidth become coherently coupled and the result is a train of pulses at the modulation frequency with a pulse width roughly the inverse of the natural linewidth. An alternative way of looking at the process is to assume random noise pulses propagating back and forth between the cavity mirrors. Since the round-trip time is the same as the modulation period, a noise pulse which passes through the modulator at its maximum transmission will receive the most net gain in a round trip. Although the differential gain among differentially phased pulses is small, the cumulative effect after many round trips singles out the in-phase pulse train. The width of the pulse is determined by the amplifier bandwidth, which is the natural linewidth as mentioned above. After demonstration of this technique of pulse generation, it was discovered that a laser could be mode-locked by a saturable filter in the cavity, i.e., a material whose transmission loss decreased with increasing light intensity. In this case, a random pulse increases the transmission in the filter on each passage and produces its own transmission modulation. This so-called self-mode locking was soon discovered to exist in some lasers without the addition of a special saturable filter material. The measurement and discovery of these effects depends upon newly developed techniques of measuring short pulse lengths. In particular, the pulse train may be passed into a reflecting cell containing material whose fluorescence is proportional to the square of the light intensity. Upon reflection

TABLE 1. LASERS CLASSIFIED BY TYPE, WAVELENGTH, AND PUMPING SOURCE.

Type	Wavelength Range	Example	Pump	Comments
Solid (Insulating)	0.17–3.9 μm	$Al_2O_3:Cr$ Glass:Nd	optical	highest pulse powers
Gas	0.15 μm–1 mm	He–Ne Argon HF CO_2	electrical; gas dynamic; laser*	highest continuous powers
Semiconductor	0.6–30 μm	GaAs PbSnTe	electrical	electrically tunable over narrow bandwidth
Liquid	0.2–1.3 μm	organic dye	optical; laser	wide tuning range; picosecond pulses
Free electron	0.6 μm–3 mm	e^-	high-energy accelerator	requires relativistic electron beam

*Wavelengths from 50 to 500 μm usually require CO_2 laser pumping of an organic molecular gas.

back through the fluorescent material, the light output increases where one returning pulse passes an incoming pulse. In this manner, the physical width of the light pattern measures the pulse length. Using a laser material with a broad emission band has yielded pulse trains with widths of the order of 10^{-12} sec or a picosecond. This corresponds to physical lengths of the order of one millimeter.

Other Sources of Coherent Light   In addition to the lasers described above and summarized in Table 1, there are several other techniques for generating coherent infrared and optical radiation. Stimulated Raman scattering, the basis of the original liquid Raman laser, is now used to generate frequencies throughout the visible and infrared region. Harmonic generation, utilizing the nonlinear index of refraction of solids, liquids, and gases at high intensities, has produced radiation out to the extreme ultraviolet at wavelengths below 0.1 μm. Similarly, these same nonlinear materials may be used for parametric oscillators, which are tunable devices producing two output frequencies whose sum is equal to the driving laser frequency. An excellent review of laser principles and a description of these latter systems may be found in Ref. 1. The last entry in Table 1, the free electron laser, is included for completeness, although the radiation generation process is describable by relativistic electron dynamics and does not necessarily invoke the stimulated emission mechanism. The radiation in this case arises from the interaction of an electron beam near the velocity of light with a spatially periodic magnetic field. A description of this device and a detailed review of the complete laser field may be found in Ref. 2.

R. H. KINGSTON

References

1. Yariv, A., "Quantum Electronics," 2nd Ed., New York, John Wiley & Sons, Inc., 1975.
2. Weber, M. J., (Ed.), "Handbook of Laser Science and Technology," Vols. I and II, Boca Raton, CRC Press, Inc., 1982.

Cross-references: COHERENCE, LIGHT, MASER, OPTICAL PUMPING, RAMAN EFFECT AND RAMAN SPECTROSCOPY.

LASER FUSION

The object of laser fusion research is to use laser radiation to heat small pellets of deuterium/tritium mixture to produce the nuclear fusion reaction

$$D + T = {}^4He + n + 3 \times 10^{14} \text{ J/Kg}.$$

The reaction is exothermic with about one-fifth of the energy in the ^{4}He particles. Eventually it is hoped to find ways of using the excess energy for economic electricity generation. The idea was first reported in 1972.

To produce a significant reaction rate, a temperature of $\sim 10^8$ K is required and the corresponding pressure causes a rapid expansion of the deuterium-tritium fuel. The aim is to complete the nuclear fusion burn before the expansion can cool and extinguish the reaction. Since inertia limits the expansion velocity the approach is sometimes called *inertial confinement*, in contrast to magnetic confinement (see the article on FUSION POWER). Although other heating methods for inertial confinement (e.g., electron, light-ion, or heavy-ion beam bombardment) are being considered, laser heating is attracting most attention because focusing is easier, there is probably less undesirable core preheat, and it is possibly more economic.

The burn condition for a sphere of fuel of radius r, is that the number of particles burnt before significant expansion must be approximately equal to the total number of particles, i.e.,

$$n^2 \langle \sigma v \rangle t \approx n.$$

where t is the burn time ($\approx$ expansion timescale r/v), v is the ion thermal velocity, and σ is the fusion cross section. It happens that $\langle \sigma v \rangle / v$ is approximately independent of v in the temperature range important for fusion, so the condition is equivalent to $nr \approx$ constant or, since n is proportional to the final density ρ, $\rho r \approx$ constant. Numerically $\rho r > 30$ kg/m^2 is required, determined by the need to keep the dimensions greater than the ^{4}He particle range so that the ^{4}He energy can make up for heat radiated and thereby maintain the burn. The fusion energy is produced as a miniature thermonuclear explosion. It is estimated that the maximum energy of each explosion that can be handled is ~ 15 kg TNT, equivalent to $\sim 10^8$ J, and about 30 explosions per second would be required for a 1-GW(e) power station. The laser energy required to heat the fuel would be $\sim 10^6$ J. Each explosion would burn about 3×10^{-7} Kg of D/T mixture, corresponding to a small pellet of only $\sim 10^{-3}$ m radius. The burn time is $\sim 5 \times 10^{-10}$ s, and the pressure in the burning fuel would be $\sim 10^{12}$ atmospheres. The burn condition and the fuel radius correspond to $\rho \approx 3 \times 10^5$ kg/m^3 and a compression of ~ 1500 times the density of liquid D/T mixture. The feasibility of laser fusion depends on developing a way of getting this high compression.

The compression is produced by laser radiation, which exerts pressure on the pellet by heating and evaporating the surface material. In moving away, the ablated material carries momentum and, by Newton's Third Law, a corresponding pressure is exerted on the remaining material (the ablation pressure is, under the required conditions, much greater than the ponderomotive pressure of laser photons). However the simple application of an arbitrary pressure pulse of laser radiation to a sphere of fuel will not produce the required temperature and pressure. In plane geometry, conventional strong shock theory gives a compression of 4 because of preheating by the shock, and spherical convergence only increases the compression to ~30 because of the outward-going shock reflected at the fuel center. To obtain high compression, shocks must be avoided so that the compression is effectively isentropic and the temperature kept low until after the compression has been achieved. One method that has been proposed is continuously to increase the driving pressure while the implosion proceeds to that the shock and piston do not part, thereby avoiding shock preheating and the final compression limit due to the reflected shock from the center. This is not easy to achieve because of two factors: the high power required at the end of the implosion and the difficulty in getting the correct pressure variation in the short time available. These problems may be reduced by using a hollow shell target. A pressure p will accelerate a fuel shell of thickness δ to a final velocity $(2rp/\rho_s\delta)^{1/2}$, where ρ_s is the initial density, so the final pressure produced at the centre is a function of r/δ. This is further increased by convergence. For example, calculations show that with $r/\delta = 10$ the ratio of the final pressure to the driving pressure is ~600. The sensitivity to the pressure variation can be reduced by using a heavy "pusher" shell of another material which may be accelerated by a pressure more uniform in time. A further increase in energy may be obtained by surrounding the pusher with another massive shell with an intervening vacuum space. The massive shell is accelerated by the driving pressure and collides with the pusher/fuel shells which, by conservation of momentum (allowing for shocks in the shell thickness) move away with velocity $\sqrt{2}$ times the massive shell velocity. The use of massive shells to tailor the implosion velocity would reduce the overall efficiency because of the energy required to move the pusher/massive shells. Another important factor is the cost of complicated shell pellets.

The absorption of the laser radiation occurs at a critical radius where the radiation frequency equals the natural frequency of the electrons, called the *electron plasma frequency*, at a density $10^{15}/\lambda^2$ m^{-3}, where λ is the radiation wavelength in meters. About 20% of the incident radiation is absorbed. Heat is transported from the absorption to the ablation radius by electron thermal conduction; at high power the heat transport is restricted by the finite flow velocity of free streaming electrons, reducing the pressure available as a function of power. This limits the overall efficiency. Also, the faster electrons may have sufficient mean free path to penetrate the ablation radius to preheat fuel at smaller radii, so the resultant

pressure reduces the final compression. The preheating is wavelength dependent and is a severe limit to target design.

Another limit to the compression may be Rayleigh-Taylor instabilities which occur when the low density ablated material exerts acceleration pressure on the heavy shell. These become more restrictive with increasing ratio of instability wavelength (λ_R) to shell thickness: when $\lambda_R < \delta$ the ripples cause some turbulent heating of the shell; when $\lambda_R \sim \delta$ the shell is strongly turbulently mixed and preheated; and when $\lambda_R > \delta$ the instability disturbs the implosion symmetry. Although Rayleigh-Taylor instabilities cause heating and are a fairly severe limit on the symmetry of the implosion, it is possible that the problem may be reduced by damping due to viscosity and thermal conduction around the shell. Such damping occurs when the distance between the absorption and ablation fronts exceed about one-half the initial fuel radius.

Present experiments test laser fusion principles using glass microballoons of ~10^{-4} m in diameter and ~10^{-6} m wall thickness, containing D/T gas at a pressure of ~100 atmospheres. The laser pulse energy is a few kJ and power ~10^{13} W, produced in a number of beams (~10) arranged to exert pressure from all directions. In contrast to the ideal laser compression described above, the energy is delivered in a time short compared to the collapse time: so the thin, dense glass wall suddenly vaporizes and exerts pressure on the D/T gas; this is called an *exploding pusher*. A disadvantage of this method of operation is that considerable electron preheat can develop, but an advantage is that the preheat tends to inhibit Rayleigh-Taylor instabilities. The absorption of the laser light has been found to be complicated, involving such processes as inverse bremsstrahlung, stimulated Brillouin scattering, parametric instabilities, stimulated Raman scattering, and nonlinear density profile steepening. Compressions of a few times 100 are obtained and temperatures up to ~10^7 K. A fraction of up to 10^{-8} of the fuel is burnt, giving a neutron yield of ~10^{11} per pellet.

In conceptual laser fusion reactors, a reaction chamber of ~10 m diameter is envisaged. A heat exchanger/tritium breeder blanket surrounds the chamber in a way similar to magnetic confinement systems. The heat from the blanket would be used to generate electricity, some of which is used to drive the laser. The overall efficiency (the laser efficiency times the electrical generation efficiency) must exceed the inverse of the energy gain in the fusion reaction (~10^3) for net energy production. Since the generating efficiency is ~20%, the laser efficiency must $> ~0.5\%$, including coupling to the plasma. Short laser wavelengths favor good plasma coupling and reduce preheat, but the lasers are inefficient. A $\lambda = 1$ μm Nd glass laser has an efficiency of only 0.1%, whereas a $\lambda =$ 10 μm CO_2 laser has an efficiency of 1%. Attempts are being made to increase the coupling efficiency of Nd glass lasers by using short wavelength harmonics. However, glass lasers are probably unsuitable for the continuous high repetition rates required in a reactor, and so short wavelength gas lasers (e.g., using KrF) are being developed.

A proposal to increase the overall reactor efficiency is to use a liquid lithium wall in the reacting chamber, which will absorb the fusion energy pulse by evaporation and will be continuously replaced: this may give an order of magnitude increase in the power that can be handled. Among technical problems to be overcome in obtaining power from laser fusion are launching the pellets, laser beam or pellet steering, pumping gaseous debris so that it does not scatter the laser beam, production of cheap multilayer pellets ($<\sim 7\cent$/pellet, 1982 value) and disposal of ~ 100 metric tons/year of radioactive pellet debris.

It is hoped to demonstrate scientific breakeven in the mid-1980s by using multibeam lasers of energy $\sim 10^5$ J.

Support for inertial confinement research comes from its application in the areas of fusion power generation, the basic physics of matter at extreme densities and pressures. Indeed, confidence in the application to produce power derives from the knowledge that ignition may be achieved by compressional heating in thermonuclear weapons. The main countries interested in inertial confinement research are USA, USSR, Japan, UK, and France. In the last 10 years the US has spent $\sim$ $10 billion on laser fusion research; in 1982 the spend was $\sim$ $200 million, which was about 50% of the spend on the magnetic confinement approach to fusion power. However the US laser fusion spend is expected to decrease to $\sim$ $100 million in 1983, corresponding to about 25% of the spend on magnetic confinement.

J. A. REYNOLDS

References

Dolan, T. J., "Fusion Research," New York, Pergamon Press, 1982.

Cross-references: FUSION POWER, LASERS, NUCLEAR REACTIONS, PLASMAS.

LATTICE DEFECTS

The crystal consists of a regular array of atoms (or molecules) (see CRYSTALLOGRAPHY). But the perfect crystal, in which all atoms are on precisely defined lattice points, is nonexistent. Real crystals contain more or less *lattice defects* or *lattice imperfections*. Even a small amount of lattice defects can affect properties of crystals

remarkably, so that the understanding of the behavior of lattice defects is very important in modern solid-state physics.

Lattice defects are classified into three classes from the point of view of the dimensional extension. These are the *point defect*, the *line defect*, and the *plane defect*. Vacancies (atomic or ionic) and interstitials (atoms or ions) are the most elementary point defects. A dislocation is a line defect. Twin boundaries and extended dislocations are some examples of plane defects.

Dislocations Dislocations are lattice defects quite unique in that their movement produces the plastic deformation of crystals, and that mechanical properties of crystals are largely determined by the behavior of dislocations in crystals. The presence of dislocations, however, affects also various physical properties of crystals other than plastic and mechanical properties.

Figure 1 shows an edge dislocation in a simple cubic crystal, the dislocation line being perpendicular to the plane of the paper. The same atomic arrangement is repeated in the direction of the dislocation line. The edge dislocation is formed at the edge A of the inserted extra atomic half-plane AB. The edge dislocation A can also be formed by a partial slip of the crystal along a slip plane CAD by a shear stress shown by arrows in Fig. 1. The part CA of the slip plane has already slipped, but the part AD has not yet slipped. Therefore, the dislocation line A, which penetrates the plane of the paper perpendicularly at A, is defined as the boundary between the slipped and unslipped parts of the slip plane. When a dislocation sweeps the whole slip plane from C to D, the upper and lower half-crystals are sheared along the slip plane by an amount and in a direction shown in Fig. 1 by a vector b, which is called the Burgers vector. The slip produced by the movement of a dislocation is the elementary process of plastic deformation of crystals.

The Burgers vector of an edge dislocation is perpendicular to the line of dislocation as can be seen from Fig. 1. This is not the case in dislocations other than the edge dislocation. In the case of screw dislocation the Burgers vector is parallel to the line of dislocation. Atomic planes perpendicular to a screw dislocation form a continuous helicoid around the screw dislocation (AD) as shown in Fig. 2. In the case of a mixed dislocation, which consists of the edge component and the screw component, the Burgers vector and the dislocation line make an angle between 0° and 90°.

The stress and the strain around a dislocation can be accurately calculated by the theory of elasticity except in the region called the dislocation core which is within a few atomic distances of the center of a dislocation (broken circle in Fig. 1), as the crystal regions outside the core can be treated as an elastic continuum, because the strain there is sufficiently small. The following are some of the important results

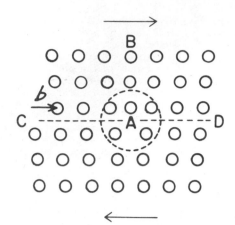

FIG. 1. An edge dislocation in a simple cubic crystal. A: center of a dislocation; AB: extra half-plane; CAD: slip plane; *b:* Burgers vector.

of the elasticity theory of dislocations. The stress around a dislocation decreases inversely proportionally to the distance from the dislocation axis. The total strain energy integrated over the entire crystal outside the core is proportional to the square of the magnitude of the Burgers vector, and is of the order of magnitude of μb^2 for a unit length of a dislocation, where μ is the shear modulus and b is the magnitude of the Burgers vector. Two or more dislocations interact with one another through the stress field. Two parallel screw dislocations attract each other if their helicoidal windings are in opposite directions (i.e., a left-handed screw and a right-handed screw), and repel each other if their windings are in the same direction. Two parallel edge dislocations with parallel Burgers vectors attract or repel in a somewhat complicated way (for details, see any textbook of dislocation theory, e.g., Refs. 1, 2, and 3).

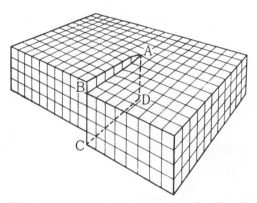

FIG. 2. A screw dislocation AD in a simple cubic crystal.

The stress, the strain, and the energy in a dislocation core can be computed if the discrete structure of the crystal is considered. The core energy of a dislocation in copper, for example, is of the order of 1 eV per dislocation length of an atomic spacing. This is considerably smaller than the elastic strain energy outside the core.

Another important property of dislocations which can be understood by considering the discrete and periodic structure of the crystal is the so-called Peierls potential. When an edge dislocation is located, e.g., at position A in Fig. 1, the atomic arrangement around the dislocation has a mirror symmetry so that the energy of the dislocation is at a minimum. When the dislocation has moved a distance b, the atomic arrangement around the dislocation resumes the same configuration as before and the energy is again at a minimum. Between these two positions with minimum energy the atomic configuration of the dislocation is not necessarily mirror symmetrical and the dislocation energy increases. Thus the potential energy of the dislocation changes periodically with the period b, when the dislocation moves along the slip plane. This potential barrier is called the Peierls potential. The Peierls potential is the intrinsic resistance to the movement of dislocations. It is relatively small in most metals, and is relatively large in such crystals as diamond, refractory oxides, and so on.

Incidentally, there are many extrinsic origins of resistance to the movement of dislocations, i.e., impurity atoms, point defects, plane defects, other dislocations, etc. These impurities and defects interact with the moving dislocations concerned, and act as obstacles to the movement of these dislocations. A larger resistance to the movement of dislocations means a larger resistance to the plastic deformation and accordingly a higher mechanical strength of crystals.

Properties of dislocations discussed so far are general ones and are not restricted to any specific crystal. But there are some properties which directly reflect specific crystal structures (e.g., extended dislocations) or specific physical characteristics of crystals (e.g., a space charge around a dislocation in semiconductors).

Now in discussing an extended dislocation, it would be more convenient to refer to a specific crystal structure, in which a dislocation is known to extend. The face-centered cubic (fcc) lattice gives a good example. In fcc metals a total dislocation, whose Burgers vector is a lattice vector $(a/2)\{110\}$, dissociates into two partial dislocations, whose Burgers vectors are of the type $(a/6)\{112\}$ and are smaller in magnitude than a lattice vector. An example of dissociative reaction is:

$$(a/2)[10\bar{1}] \to (a/6)[11\bar{2}] + (a/6)[2\bar{1}\bar{1}],$$

where a is the lattice constant. The passage of the first partial dislocation $(a/6)[11\bar{2}]$ along a

(111) slip plane in a perfect fcc lattice leaves behind it a fault of stacking of (111) atomic layers, which is a plane defect called a stacking fault. By the passage of the second partial dislocation $(a/6)[2\bar{1}\bar{1}]$ along the same slip plane, the faulted plane resumes its original perfect lattice. Thus between the two partial dislocation lines there is a strip of stacking fault. This dislocation configuration as a whole is called an extended dislocation. The width of the extended dislocation is determined by the balance of two forces, namely the repulsive force between two partial dislocations and the contractive force of the stacking fault. In copper and copper alloys, for example, the width is fairly large and is from 10 to 100 Å, while in aluminum, which has a large stacking-fault energy, the total dislocation practically does not extend. The extension of dislocations greatly affects the plastic and mechanical behaviors of crystals.

In elemental semiconductors with the diamond lattice, an edge dislocation has unpaired valence bonds or dangling bonds along the edge of the extra half-plane. In n-type germanium, for example, a dangling bond, which acts as an acceptor, traps an electron, so that an edge dislocation line is negatively charged. This negative line charge is shielded by a surrounding cylindrical positive space charge. This electrical structure of dislocations affects the electrical and other properties of semiconductors.

Dislocations are introduced into crystals by plastic deformation, and by precipitation of point defects such as vacancies or interstitials. They are also introduced during crystal growth to relax various internal stresses.

Point Defects Atomic or ionic vacancies and interstitial atoms or ions are the most elementary point defects. Di-vacancies, tri-vacancies, and higher clusters of vacancies, and di-interstitials, tri-interstitials, and higher clusters of interstitials can exist. Complexes of these point defects with impurity atoms are formed according to circumstances.

The most fundamental physical properties of point defects are the formation energy, the migration energy, the binding energy, and the atomic configuration around the point defect. These properties are widely different in different crystal structures and in different types of crystalline solids. In most metals and elemental semiconductors the formation energy of a single vacancy E_{fV} is smaller than the formation energy of a single interstitial E_{fI}, while the migration energy of a single vacancy E_{mV} is larger than the migration energy of a single interstitial E_{mI}. Typical examples of energy values are: $E_{fV} \approx 1$ eV and $E_{mV} \approx 1$ eV for copper; $E_{mV} \approx 0.2$ eV for p-type silicon and $E_{mV} \approx 0.3$ eV for n-type silicon; the binding energies between two single vacancies to form a di-vacancy are a few tenths of an eV in noble metals.

The relaxation of atoms around a single vacancy in most metals has a simple symmetry which is related to the symmetry of the crystal structure. But the atomic configuration around a single vacancy in covalent semiconductors is somewhat distorted because of the quantum-mechanical Jahn-Teller distortion. The atomic configuration of a single interstitial is rather complicated, because different configurations for an interstitial are possible in some cases, and because the configuration often shows a lower symmetry than the host crystal. In cubic metals, for example, the so-called split interstitial or dumbbell-form interstitial is the most stable one because of the lowest formation energy (for details, see any textbook of point defects, e.g., Refs. 1 and 4).

In ionic crystals one or more of the following point defects are considered according to respective circumstances, i.e., vacancies of positive ions and of negative ions, and interstitials of positive ions and of negative ions. Some of them form complexes with impurity ions. Point defects in alkali halides and similar ionic crystals have been extensively investigated in connection with their optical properties, and many color centers are well analyzed in terms of point defects (see COLOR CENTER; for details, see, e.g., Ref. 5). In alkali halides, for example, the F center is a negative ion vacancy with a trapped electron, the M center is a pair of negative ion vacancies with two trapped electrons, and the H center is an interstitial negative ion with a considerably distorted configuration, having captured a positive hole.

The migration of point defects results in the transport of materials in crystals, i.e., self-diffusion, mutual diffusion, ionic conduction, etc. In most metals and covalent semiconductors the self-diffusion and the substitutional impurity diffusion occur by means of the migration of vacancies. The diffusion and the ionic conduction in ionic crystals take place by means of the migration of ionic vacancies and/or ionic interstitials.

Point defects are produced in crystals in thermal equilibrium, their concentration being approximately equal to $\exp(-E_f/kT)$ at the absolute temperature T, if the formation energy is E_f. Here k is Boltzmann's constant. Point defects are also introduced into crystals in excess of thermal equilibrium by irradiation of energetic radiation or by plastic deformation. Excess point defects are annihilated by annealing, the kinetics of which is well investigated under various conditions (Ref. 4).

RYUKITI R. HASIGUTI

References

1. Hasiguti, R. R., "Crystal Lattice Defects," in "Solid State Physics," Ed. Kubo, R., and Nagamiya, T., New York, McGraw-Hill Book Co., Inc., 1969 (p. 719).

2. Friedel, J., "Dislocations," Oxford, Pergamon Press, 1964.
3. Nabarro, F. R. N. (Ed.), "Dislocations in Solids," in five volumes, Amsterdam, North-Holland Publishing Company, 1979 and 1980.
4. Leibfried, G., and Breuer, N., "Point Defects in Metals, I," Berlin, Springer-Verlag, 1978; Dederichs, P. H., Schroeder, K., and Zeller, R., "Point Defects in Metals, II," Berlin, Springer-Verlag, 1980.
5. Schulman, J. H., and Compton, W. D., "Color Centers in Solids," Oxford, Pergamon Press, 1962.

Cross-references: COLOR CENTERS, CRYSTALLIZATION, CRYSTALLOGRAPHY, METALLURGY, SEMICONDUCTOR, SOLID-STATE PHYSICS, SOLID-STATE THEORY.

LENS

A lens is any element that focuses light to form images. Many lenses are found in nature. Ice crystals, waves on the surface of water, and all the eyes of humans and animals are examples of lenses. These lenses have one or more curved surfaces and are made of a transparent material. Manufactured lenses are usually made out of glass, plastic or crystal material. The simplest lens consists of two ground and polished spherical surfaces. A line connecting the centers of the two spheres is called the optical axis of the lens. The lens is edged to form a cylindrical surface centered on the optical axis. The spherical surfaces may be convex or concave, resulting in lenses which are positive refracting or negative refracting. A positive lens collects the light from a distant object and focuses it to a real image. The negative lens disperses the light and causes it to diverge from a virtual image. Positive and negative elements are used in combinations to form optical lens systems. The optical axes of each of the lens elements usually coincide to form centered optical systems. Most optical systems are designed to be centered optical systems, but in manufacture the centering is seldom perfect, so the system will have various degrees of defective performance.

Spherical surfaces are usually used in optical systems because of stringent requirements on the manufacture of optical elements. In order to perform properly, a given surface in an optical system often has to coincide with the prescribed surface to within a few millionths of an inch. Such extreme tolerances can be achieved on spherical surfaces because spheres may be ground and polished with self-correcting techniques.

A few lenses have been made using non-spherical surfaces, but they usually have rotational symmetry around the optical axis. These surfaces are called rotationally symmetric aspheric surfaces. Aspheric surfaces of this type are difficult to generate so they are used infrequently.

Some lenses are made with cylindrical and toric surfaces. Spectacle lenses often have surfaces of this type. It is practical to use aspherics in spectacle lenses because the beam of light entering a person's eye is small in diameter. The performance requirements are therefore not great. Cylindrical or toric surfaces are seldom used in telescopes, or microscopes of high performance.

There are many types of glass used in optical lenses. Some glasses are more dispersive (see REFRACTION) than others. By combining positive and negative lenses of different glass it is possible to correct for chromatic aberrations.

Theory of the Lens Most of the performance of a lens or lens system may be understood by considering that light travels as rays in straight lines until it encounters a change of index of refraction. The light is then refracted according to Snell's law (see REFRACTION). Light is emitted from a point source of light in the object as a diverging beam of rays. A lens is able to collect these rays and refocus them to an image point (see Fig. 1).

With analytical geometry, one may derive equations for calculating the path of any ray as it passes through the optical system. The procedure is called ray tracing. The mathematical equations used to trace rays are long and complicated, and have to be computed with many significant figures. Prior to the use of modern digital computers, the design and analysis of lens systems was a long tedious job. An average lens design required many months of calculation. Today most of these calculations are done on large computers, and few people need to be concerned about being able to ray trace.

Paraxial Rays Paraxial rays pass through the center portion of the lens and the assumption is made that the object points are close to the optical axis. The ray-tracing equations for paraxial rays are simple, and by using the paraxial approximation, many useful theorems for lens performance may be worked out.

In the paraxial region, any optical system may be described by locating six cardinal points along the optical axis. Once these cardinal points are known, the position and size of the image of any object may be' computed from the following formulas (see Fig. 2).

$$m = \frac{\bar{y}_k}{\bar{y}_0} = -\frac{z'}{f} = -\frac{f'}{z} \qquad (1)$$

$$zz' = ff' \text{ and } \frac{f}{s} + \frac{f'}{s'} = 1 \qquad (2)$$

$$f/n_0 = f'/n_k \qquad (3)$$

$$P_1 N_1 = P_2 N_2 \qquad (4)$$

$$P_1 P_2 = N_1 N_2 \qquad (5)$$

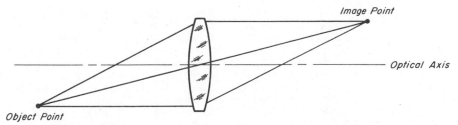

FIG. 1. A diagram showing how a lens collects diverging rays and focuses them at an image point.

$$F_1 N_1 = f' \qquad (6)$$

$$N_2 F_2 = f \qquad (7)$$

P_1 and P_2 are called the first and second principal points. N_1 and N_2 are called the first and second nodal points. F_1 and F_2 are the first and second focal points. f and f' are the front and back focal lengths. n_0 and n_k are the indices of refraction in the object and image space.

The following terms are commonly used in connection with lenses:

Field of View. The field of view usually refers to the half angle subtended by the object as seen from the first principal point P_1. For example, it would be $\tan^{-1}(\bar{y}_0/S)$ in Fig. 2. When specified for a lens, it usually refers to the maximum size of object which may be imaged by the lens. Optical designers tend to describe the field of view by its half angle, as shown in Fig. 2. Marketing firms often refer to the full field which is twice the half angle. If not clearly stated, confusion over this term may result.

Relative Aperture. This refers to the half angle of the cone of rays converging to the axial image point. The sine of this angle is often called the numerical aperture and is written NA. If NA is large, the lens collects a large cone of light and focuses it on the image. Another way to describe this NA is to use the term f-number. The f-number of a lens and the NA of a lens are related by the following equation

$$f\text{-number} = \frac{0.5}{NA}$$

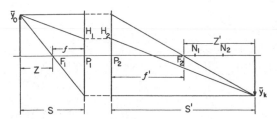

FIG. 2. Diagram showing the location of the six cardinal points in a lens system.

Aperture Stop. The aperture stop in a lens system is a diaphragm which determines the NA of the lens.

Lens Aberrations Lens designers attempt to combine elements and glass types to reduce the lens aberrations. All points in the object should be imaged as points in the image and should be located at or very near the position predicted by the paraxial rays. In lenses of large relative aperture and field of view, there are usually several residual aberrations that designers are unable to eliminate. There are the following types of aberrations:

(1) Spherical
(2) Coma
(3) Astigmatic
(4) Field curvature
(5) Distortion
(6) Axial chromatic aberrations
(7) Lateral chromatic aberrations.

These aberrations are corrected by using combinations of positive and negative elements. There are two general principles one may use as guide lines in correcting optical systems. (1) A closely spaced positive and negative lens with the aperture stop in contact may be corrected for spherical aberration, coma, axial and lateral chromatic aberration and distortion. (2) It is necessary to use positive and negative lenses with appreciable air space between them to correct field curvature and astigmatism.

There are many conflicting requirements in lens systems. Lenses of large relative aperture usually are designed to cover small fields of view. A large field of view normally dictates a small relative aperture. Wide fields and large aperture may be realized by using many elements or by compromising some of the image quality. For example, lenses of 140° total field working at $f/2$ are available, but they are complex and have large amounts of distortion. Periscopes allow one to look through a long pipe and see a wide field at high magnification, but there is always some residual chromatic aberration and field curvature left in the design.

Lenses are usually designed for specific applications, and the designer has made a careful balance between conflicting requirements. It is seldom that a lens designed for one application will perform optically in another. In the past there was a tendency to misuse lenses, because it was difficult to design a new system. Today it is easy to obtain a new design. It is, however, expensive to build a prototype of a new lens design. This is because the optical shop has to make the lenses without adequate tooling. It takes a skilled optician to make a prototype lens accurately enough to perform as designed. When production runs are large, modern tooling can ease the burden on the skilled optician and the cost per lens can be drastically reduced.

Lens Testing A designer specifies a set of curvatures, thicknesses and optical glasses for a nominal design. The work shop makes the lens to these specifications within certain tolerances. The final lens must then be tested to make sure that the over-all performance is as expected. It is rare indeed that a lens performs exactly as computed. It is then necessary to determine if the difference is negligible and, if not, what to do about it. First, one tests the complete assembled lens and attempts to predict the performance. By studying the defective image, it may be possible to locate the sources of error. The tests consist of studying the light distribution in the image of a point source. This is done on a lens bench or testing interferometer. Sometimes it is possible to locate the source of error by testing the over-all system, but usually the lens system is disassembled and all the lens surfaces, spaces and centering are checked separately.

The invention of the laser (1960), the astonishing progress in computing equipment (1950–1980), the development of small detectors, and the associated transistor technology have brought about major changes in lens testing. Interferometers are used more extensively, tolerances are held better in the optical shop and mounting techniques have improved. The trend now is to use better quality control at every step of the manufacturing process. The objective is to manufacture and mount the lenses to achieve a high yield of acceptable lenses. The disassembly of lenses as described above is seldom done beyond the prototype stage. In large production runs the rejected lenses are few and seldom are they repaired.

Lens Types There are several optical systems which may be classified as types. There is considerable overlap between the types, but there is some value in the following classification.

Microscope Objectives. Microscope objectives are used to magnify small objects. They are usually used with a microscope eyepiece. Microscope objectives range in focal lengths from 2 to 48 mm and are used at magnifications ranging from 1000 to 5 X. The high-power objectives are made up of many small elements. Some of the lenses are only a few millimeters in diameter. The lens making and mounting procedures for such small lenses are quite different from larger elements.

Telescope Objectives. Telescope objectives are used to view distant objects. Telescope objec-

tives have a wide variety of focal lengths and diameters. They are usually corrected precisely for spherical aberration, coma and axial chromatic aberration. Since telescope objectives cover small fields of view, astigmatism and field curvature usually are not corrected. Telescope lenses of large diameter (20 inches or more) become afflicted with chromatic aberration which cannot be corrected, so many of the large telescopes used by astronomers are mirrors instead of lenses.

Telescope objectives are used in binoculars, opera glasses, surveying instruments, gunsights, and many laboratory instruments.

Periscopes. Periscopes are used to enable one to look through a long tube. The submarine periscope is one well-known example, but there are many other types used in industrial and medical instruments. Gastroscopes and cystoscopes are examples of periscopes used in medical instruments. Periscopes are made up of a train of telescope objectives and eyepieces.

Camera Objectives. By far the largest class of optical lens systems would be classified as camera lenses. They are used to record images on films as in common landscape cameras, but today they are also used with many other types of image recording systems such as television image tubes, electrostatic plates, etc. The distinguishing features of camera lenses are wide field and large aperture. Usually the image is located on a flat image plane. Camera lenses range in complexity from a single meniscus lens to systems with more than ten elements. Camera lenses cover such a wide range of uses that one could claim all lenses to be a form of camera lens. For example, a long focal length lens used on a 35-mm camera may actually be very similar to a telescope objective.

Zoom Lenses. Zoom lenses are in common use on cameras and a wide variety of instruments. A zoom lens has a variable focal length, which is achieved by changing the spaces between elements. The lenses are positioned accurately with precision cams, which have become practical to manufacture since the introduction of numerically controlled machines. Zoom lenses first appeared on 8 mm and 16 mm movie and TV cameras. They were originally too large for the popular 35 mm still cameras. The latest designs and manufacturing methods have improved so much that the sizes are much reduced and 35 mm zoom lenses are now popular.

Photolithographic Lenses. Some of the most precise lenses made are used in the making of modern integrated circuits. These lenses are used to project images of precision masks onto silicon wafers. The devices on the silicon wafers have dimensions as small as one micrometer. The lenses have to form these minute images over large areas. Some of these lenses are capable of imaging several million independent image spots in a single exposure.

Satellite Lenses. Since the development of earth satellites, camera lenses have been made to photograph the cloud patterns covering the earth. These lenses are short focal length and wide field. Lenses of long focal length have been able to photograph lines less than 6 inches wide on the ground from a satellite in orbit 200 miles from the earth. Many spectacular pictures have been taken of planets and stars from satellite cameras.

Laser Scanning Lenses. The wide variety of gas and solid state diode lasers along with modern modulating systems has lead to many applications using laser scanners. The extremely intense small diameter beams from lasers may be reflected from oscillating mirrors to provide a rapidly scanning line of light similar to that generated in a cathode ray oscilloscope. The advantage is that the material to be scanned does not have to be in a vacuum. Nearly any material may be scanned with a laser beam. Laser scanners can be used to scan printed material and convert the light reflected into an electrical signal. The electrical signal may then be stored in a computer memory, and image processed. Finally, it can be read back through a similar laser scanner and be printed out on a light sensitive material of which there are now many kinds. Since these scanning lenses work with lasers the light is monochromatic, making it easier to design high quality lenses. Some of the scanning lenses can surpass the image quality of lithographic lenses because the light used is monochromatic.

Eyepieces. Eyepieces are quite clearly a distinctive class of optical lens system. Eyepieces are designed to match the sensitivity and physical requirements of the human eye. For example, the eye is sensitive in the visual part of the spectrum so eyepieces are designed for this range of wavelengths. An eyepiece must also be located with its aperture stop in an external position so that the observer's eye may be located within it. This requirement imposes serious limitations on eyepieces, and they are seldom useful in any other applications.

Magnifiers are essentially eyepieces except they are designed to view opaque material while an eyepiece is designed to view an aerial image formed by an objective.

Condensers. Condensers are used to collect and focus large amounts of light. They are found in projectors and substages of microscopes. A searchlight mirror is a form of a condenser. The numerical aperture of a condenser is usually very large, and for many applications, the image-forming properties are not important. Condensers are often made, therefore, with low-quality surfaces. Some condenser lenses are molded. Condensers are usually placed close to an intense light source which heats and cracks the lenses if made of glass. Condensers are often made out of quartz because of its ability to withstand heat. Aspheric condenser lenses are common in condensers. With some of the modern high-intensity light sources, it is necessary to correct for the image errors in order to obtain uniform illumination.

ROBERT E. HOPKINS

References

Hardy, A. C., and Perrin, F. H., "Principles of Optics," New York, McGraw-Hill Book Co., 1932.

Greenleaf, Allen, "Photographic Optics," New York, The Macmillan Co., 1950.

Conrady, A. E., "Applied Optics," London, Oxford University Press, 1929.

Cross-references: ABERRATIONS; MICROSCOPE; OPTICAL INSTRUMENTS; OPTICS, GEOMETRICAL; REFRACTION.

LIGHT

Light is a form of electromagnetic energy. It has a physical character similar to that of radiowaves. In order that the human eye may get the sensory perception of light, the electromagnetic waves entering the pupil should have a wavelength λ between 4000 and 7000Å (1Å = 10^{-8} cm). The wavelength of a wave is inversely proportional to its frequency ν. The product of the two quantities equals the velocity of propagation. For light in vacuum one has $c = \nu\lambda = 3 \times 10^{10}$ cm/sec. The frequency of light waves is therefore almost a billion times higher, their wavelength a billion times shorter, than the waves of standard radio broadcast bands. The perception of color depends on the distribution of the electromagnetic energy over the visible wavelengths. White light is a superposition of waves at many frequencies. It can be decomposed into its monochromatic spectral components by a prism or other spectral apparatus. The violet end of the spectrum is near 4000Å, the red end near 7000Å. Whereas light in its narrow definition should be confined to this relatively narrow portion of the electromagnetic spectrum, it is customary to extend the definition to the ultraviolet and infrared portion of the spectrum. One sometimes speaks loosely of ultraviolet and infrared "light," although electromagnetic waves at these frequencies are not detected by the eye. The human mind and hands have, however, devised a large variety of instruments by which such invisible radiation can be detected and measured. Photographic plates can be made sensitive to x-rays, with a wavelength shorter than the ultraviolet, or to the much longer wavelengths of the infrared. Geiger counters can detect electromagnetic radiation of very short wavelength (λ-rays and x-rays). Photoelectric cells are sensitive in the ultra-violet and visible portion of the spectrum. Photoconductivity can be used to detect infrared radiation. At still longer wavelengths, the microwaves and radiowaves are detected by diode detectors in appropriately arranged microwave and radioreceivers. All these types of radiation can also be converted into heat by absorption in a blackbody, i.e., a material that can absorb radiation at all wavelengths. The radiation can be felt as heat, if it is absorbed by the human skin.

The study of the human eye as a detector of light is the task of *physiological optics*. The impression of light is not necessarily always connected with the simultaneous presence of electromagnetic energy at the retina. We see "stars" from a heavy mechanical blow in the dark. The impression of light is retained for about 0.1 second after the light source is shut off. This fact is made use of in the movies to create the impression of motion by a series of still images. The eye is a detector with a relatively long response time. Photoelectric cells can react more than a million times faster. Color vision is also subject to physiological peculiarities which are quite complex (see COLOR and VISION AND THE EYE).

The property of light which is most immediately accessible to observation is its propagation along straight lines (shadows). If light rays pass from one medium to another, their direction is changed according to the law of REFRACTION. If the light in medium 1 propagates with a velocity v_1 and makes an angle ϑ_1 with the normal to the boundary between media 1 and 2, the direction ϑ_2 in medium 2, with a velocity of propagation v_2 is given by Snell's law, $\sin \vartheta_1 / \sin \vartheta_2 = v_1/v_2 = n$. The constant n is called the relative index of refraction of medium 2 with respect to medium 1. These three laws are the basis of *geometrical optics*. This branch of the science of light describes the paths of light rays, the formation of images by mirrors and lenses, the action of telescopes, microscopes, prisms and other optical instruments.

The wave character of light becomes apparent by more refined observations. The phenomena of diffraction, interference and polarization are the subjects of *physical optics*. Diffraction describes how waves are bent around obstacles. They represent corrections to and deviations from the laws of geometrical optics. These effects become pronounced only when the material has a characteristic dimension comparable to the wavelength of the wave. When light waves reach the same point along different paths, the resulting intensity may be smaller than that produced by each individual wave separately. The relative phases of the waves may be such that they interfere destructively, when the arrival of one wave with maximum positive deflection coincides with that of another wave with maximum negative deflection. Observations of light in crystals of calcite (iceland spar) first showed that there are two different modes of vibrations for each direction of propagation.

These are called the two transverse modes of polarization.

All phenomena of geometrical and physical optics are described consistently by Maxwell's equations of electromagnetic theory. Optical phenomena are, therefore, closely related to other electric and magnetic phenomena. Around 1900 the prevailing opinion was that the wave character of light was unambigously established and the nature of light well understood.

There was, however, a mathematical difficulty with the intensity of radiation of ultraviolet and higher frequencies. The photoelectric effect could also be interpreted only by considering light to have a quality of particles. The number of electrons emitted from a photosensitive surface is proportional to the intensity of the light. The energy of the individual electrons is, however, determined by the light frequency. This led to the postulate of light quanta with energy $h\nu$, where h is Planck's constant. This duality in nature, in which "wave-like" and "particle-like" properties are combined, is described without internal contradiction by quantum mechanics. The combined "particle and wave" character of light is revealed by the combination of properties of the light sources, the electromagnetic field describing the light waves, and the detectors.

The study of the interaction of light waves with matter in the sources and in the detectors is the subject of SPECTROSCOPY. This is a wide field which encompases atomic and molecular spectroscopy, parts of solid-state physics and photochemistry. The quantum theory was largely developed on the basis of spectroscopic data. A light quantum is emitted by an excited atom, molecule or other material system, when an electron in such a particle makes a transition or "quantum jump" from a state with higher energy to a state with lower energy. The energy difference between these states is equal to the quantum energy $h\nu$. Similarly, the absorption of light quanta is accompanied by an electronic transition from a state with a lower energy to a state with an energy higher by an amount $h\nu$. In this manner, the frequencies of spectral lines are characteristic for the electronic energy levels in each material. The frequency of the light may be said to correspond to the frequencies of the vibrating charges or oscillators, which are represented by the electrons.

Light sources are thus bodies with a sizeable population of electrons in excited states. This may be accomplished by raising the temperature of the material. The most important source of light is the sun. The moon and other planets are visible only because they reflect sunlight, just as all other objects on the earth which we can see by daylight, but not at night. The sun is a star. In stars, the temperature is maintained at a very high temperature by nuclear reactions.

Man-made light sources range from the primitive fire, candles, and oil and kerosene lamps to the electric light bulb, fluorescent gas-discharge tubes, arcs, etc. In early sources, the material particles of smoke or wick were heated by the chemical reaction of oxidation or burning; in the incandescent electric lamps, a wire is heated to a very high temperature by an electric current. There are so many energy levels in these luminous solid materials or gases at high pressures that the emitted light is white and contains essentially all frequencies. The higher the temperature, the more radiation is emitted and the higher the average frequency of radiation. It should be realized that most of the energy is emitted as invisible (infrared) radiation, even in the best incandescent lamps. Hot gases in flames may also emit sharp spectral lines characteristic of the atoms occurring in the flame. The yellow color which arises when kitchen salt is sprinkled in a flame is due to the characteristic yellow spectral line of sodium atoms.

In gas discharge tubes, atoms or molecules are excited by collisions with electrons in the ionized gas. The energy is provided by the generator which provides the voltage necessary to maintain the discharge current. An arc is a discharge in air or in a high-pressure vapor. Mercury and sodium discharges are used for street lighting. Fluorescent tubes for office and home lighting use a gas discharge with a substantial amount of ultraviolet components. This ultraviolet light excites electrons in fluorescent centers on the walls of the tube. The electrons drop immediately from the highly excited state to an intermediate state with a lower energy. From this state they finally drop down to the original ground energy level with the emission of visible light. Gas discharges at relatively low pressure may serve as spectroscopic sources to study the emission spectra of atoms, ions and molecules. From the relationship between the energy levels and the frequency of radiation, it follows that a material, when heated, can emit precisely those frequencies which it absorbs when it is in the lower energy level at low temperature.

All these light sources are incoherent in the sense that there is no phase relationship between the light waves emitted by the different atoms in the source. This is quite different from the property of the usual sources of electromagnetic radiation at lower frequencies. In the oscillator tubes of radio- or microwave transmitters, all electrons move and vibrate in step with each other. The analogy between light and low-frequency electromagnetic radiation raises the question, "Can coherent light sources be constructed?" Recently such coherent light sources have been developed. They are characterized by the emission of a highly directional, highly monochromatic light beam of high intensity. They are called LASERS because they are based on *l*ight *a*mplification by *s*timulated *e*mission of *r*adiation. In the conventional sources, all light is emitted spontaneously. In lasers, the original spontaneously emitted light forces the other excited atoms to emit their radiation in step, or coherently. If stimulated emission thus dominates the spontaneous emission, a laser results. This requires a high concentration of excited atoms and a sufficient feedback mechanism of light by mirrors. In its simplest form a laser consists of a gas discharge in a tube of suitably chosen dimensions and gas pressure between a set of parallel mirrors. Because the atoms in the laser source all act constructively in step, these sources provide a more efficient means to transmit light energy.

The high light intensities available in focused laser beams have led to the development of the branch of nonlinear optics. The optical properties of materials are different at high intensities, because the electronic oscillators are driven so hard that anharmonic properties become evident. A typical effect is the harmonic generation of light in which red laser light is converted into ultraviolet light at exactly twice the frequency, when the high-intensity beam traverses a suitable crystal such as quartz. It should be possible to duplicate at light frequencies all nonlinear effects known from the field of radio communications, such as modulation, demodulation, frequency mixing, etc. It is no longer correct to say that the propagation of a light wave is independent of the presence of other light waves. At high intensities, there is a noticeable interaction between light waves of different frequencies.

The combination of the laws of quantum mechanics and electromagnetic theory gives a consistent description of the generation, propagation and detection of light. Since these same laws also describe many other properties of matter such as electronic structure, chemical binding, electricity and magnetism, etc., it may be said that the nature of light is well understood. In this context, it is not necessary and not even desirable to pose the question, "What is it, precisely, that vibrates in a light wave in vacuum?" The electromagnetic fields acquire meaning only through their relationships with detectors and sources. Human knowledge or understanding is here used in the operational sense that a relatively simple framework of physical concepts and mathematical relationships exists, which gives an accurate description of the wide variety of optical phenomena at present accessible to observation or verification in experimental situations. The following references will introduce the reader to the vast literature of optics and spectroscopy.

N. BLOEMBERGEN

References

Whittaker, E. T., "A History of the Theories of Aether and Electricity," Vols. I and II, London, Nelson & Sons, 1952.

Born, M., and Wolf, E., "Principles of Optics," London and New York, Pergamon Press, 1959.

"Lasers and Light," in "Readings from Scientific American," San Francisco, Freeman, 1969.

Ditchburn, M., "Light," Third Edition, New York, Interscience Publishers, 1976.

Minnaert, M. G. J., "Light and Color in the Open Air," Ann Arbor, Mich., Dover Publications, 1953.

Cross-references: ELECTROMAGNETIC THEORY; INFRARED RADIATION; INTERFERENCE AND INTERFEROMETRY; LASER; OPTICS, GEOMETRICAL; OPTICS, PHYSICAL; PHOTOCONDUCTIVITY; PHOTOELECTRICITY; QUANTUM THEORY; REFLECTION; REFRACTION; SPECTROSCOPY; ULTRAVIOLET RADIATION; VISION AND THE EYE.

LIGHT SCATTERING

When a beam of light falls on a particle, part of this incident beam is diverted from its original path; that part which is diverted and not absorbed is *scattered*.

Light scattering is a familiar phenomenon. The colors of visible objects (other than light sources) are determined by the wavelengths which they scatter most effectively. Scattering by small particles was first studied experimentally by Tyndall (1869) in connection with the blue of the sky.

Classical physics is appropriate for the description of most light scattering phenomena. Thus, light scattering is explained in terms of the forces exerted by the electromagnetic field on the electronic charges which all matter contains. The oscillating electromagnetic field of the incident light exerts a periodic force on each electronic charge, causing it to execute harmonic motion at the frequency of the light wave. It is the fact that an oscillating charge radiates in all directions (except along the line of its motion) which accounts for the scattering. The intensity of the radiation scattered from a particle will be large in directions for which the radiation from the individual elements of the particle interferes constructively, and small in directions in which it interferes destructively.

For particles comparable in size to the light wavelength, the amount of energy scattered as well as the angular distribution of the intensity and polarization of the scattered light are influenced by the distribution of induced oscillating charge within each scatterer. Any correlation which may exist between the positions of the scatterers also affects the extent to which the radiation interferes constructively or destructively to make up the resultant scattered field. Thus, in principle, light scattering provides a tool for the investigation of surfaces as well as of the number, size, structure, and orientation of particles and their interactions.

The problem of relating the light scattering to these properties and vice versa has proved too difficult for exact solution in general, because it would be necessary to solve Maxwell's equations with the proper boundary conditions for each scatterer or scattering element. Many important cases have been solved, however, subject to certain approximations.[1,2,3] Widest success has been obtained for *single scattering*, i.e., for particles sufficiently dispersed that radiation scattered by any one particle can be considered to escape from the medium without being further scattered by other particles. Even for single scattering, however, exact solutions of Maxwell's equations have been obtained only for simple shapes such as spheres (Mie, 1908), spheroids, and cylinders. Thus, even for single scattering, it is often necessary to resort to approximation methods.

If the particles are less than about one-tenth of the light wavelength and if their index of refraction is near to that of their surroundings, only the induced electric dipole radiation is important. (The amplitude of the induced dipole moment is given by the polarizability of the scatterer times the amplitude of the electric vector of the light wave.) Lord Rayleigh (1871) explained Tyndall's principal results in terms of the intensity and polarization of the induced electric dipole radiation. This type of scattering has since become known as *Rayleigh scattering*.[1,2]

Rayleigh scattering is of particular importance. If the particles are dispersed at random (molecules of an ideal gas or widely dispersed macromolecules in an optically homogeneous solution), the individual particles may be regarded as independent sources. In this event, the total scattered intensity is merely the sum of the intensities scattered by the individual particles. The special case of isotropic particles and unpolarized light is both simple and illuminating. The Rayleigh formula is

$$\frac{\text{Intensity of scattered light}}{\text{Intensity of incident light}}$$

$$= \frac{8\pi^4 N\alpha^2 (1 + \cos^2 \theta)}{\lambda^4 r^2}$$

where N is the number of particles, α is their polarizability, θ is the angle of scattering, λ is the wavelength (in the surrounding medium), and r is the distance from the scattering system to the point of observation (with r much greater than any relevant dimension of the scattering system). Thus, Rayleigh scattering from independent particles is proportional to the number of particles. Moreover, if the dielectric constant of the surroundings is close to that of the scatterer, the polarizability is insensitive to the shape of the scatterer. Accordingly, the Rayleigh formula has been used to determine Avogadro's number and, when the total mass of scatterers is known, to determine molecular weights.

For electrically conducting scatterers (e.g., metals), the conduction electrons are relatively free so the incident electromagnetic wave induces relatively large amplitude oscillations of these electrons. For such particles, magnetic (as well as electric dipole) radiation may be important even for scatterers that are quite small relative to the light wavelength. Moreover, if the frequency of the incident light is close to the natural frequency of charge oscillation (*surface plasmon oscillation*), particularly large amplitude oscillations may be induced. At such frequencies, the polarizability may be enhanced by several orders of magnitude, making the particle an extraordinarily efficient and color-selective scatterer. The resonant frequency can be controlled, since it depends on the shape, the index of refraction, and the electrical conductivity of the scatterer as well as on the index of refraction of the surrounding medium. For example, gold and silver yield a wide range of striking colors, and their suspensions were used in the Middle Ages in stained and other colored glasses.

As we consider larger particles which begin to violate the criterion that their dimensions be very small compared with the wavelength, or the criterion that they are sufficiently dispersed that their positions are not correlated, the Rayleigh formula breaks down. This breakdown first appears at large scattering angles, (corresponding to large optical path differences) where the destructive interference is first significant, and quickly spreads to moderate and small angles. Nevertheless, for scattering angles near zero, the Rayleigh formula retains validity, since for zero scattering angle the radiation from all volume elements is essentially *in phase* regardless of particle positon or size (provided that the index of refraction of the scatter is close to that of its surroundings so that all relevant optical path differences are small). Thus, the Rayleigh formula, when properly applied, is useful over an extremely wide range of molecular weights (10^2–10^7).

For scattering from surfaces and dense media such as liquids in which there is local order over dimensions comparable to the light wavelength, individual scatterers cannot be treated as independent scatterers. Perhaps the most general formulation of the problem is in terms of the radial density function, which describes the correlation between the positions of scatterers, and thus determines the extent to which their radiation interferes constructively and destructively. For scattering by molecules of a solution, however, it is often preferable to relate the light scattering directly to thermodynamic properties of the medium. This may be accomplished through an ingenious approach due to Smoluchowski (1908) and Einstein (1910). It takes advantage of the fact that for molecules small compared with the wavelength, the scattered field may be regarded as made up of radiation from elements of volume small enough that each element may be considered an electric dipole source and yet large enough that the elements can be considered to be independent

of each other. If the index of refraction of every element were identical, the solution would be homogeneous and no scattering would result. But the index of refraction of an element will fluctuate according to the number of molecules it contains. The total scattering is found to be proportional to the mean square fluctuation in index of refraction, which is related to the thermodynamic properties of the solution through free energy. For crystalline solids and other media in which correlation extends over distances comparable to the wavelength, it is necessary to use more specialized techniques.

In many cases, light scattering is related to the composition and structure of the medium in a way similar to x-ray scattering. The criterion which must be satisfied is that the electromagnetic field within the scatterers should be closely approximated by the unperturbed incident field, just as in the x-ray case. Light scattering under this approximation is widely known as *Rayleigh-Gans*, or *Rayleigh-Debye*, scattering. It is applicable if the phase shift for radiation passing through a particle is not too different from the phase shift which would occur for radiation passing through the same distance in the surrounding medium. When this approximation is valid, the angular distribution of the scattered light is related, as in x-ray scattering, to a *form factor* which describes the structure of the individual scatterers and to the previously mentioned radial density function which describes their spatial correlation.

In general, each of the above approximate methods rests on approximating the dipole moment that is induced in the scatterers. When inadequate, these approximations can be very significantly improved by variational techniques. Variational principles for identical scatterers were described by Levine and Schwinger (1948), and have been developed and applied also to statistical distributions of nonidentical scatters by many investigators.[4]

To this point, the frequency of the scattered light has been regarded as identical to that of the incident light. Actually, as predicted by Brillouin (1914), frequency displacement and line broadening occur due to thermal and other motion of the scatters (via the DOPPLER EFFECT), and also due to variations in the directions or magnitudes of their polarizability tensors (e.g., due to chemical reaction). Highly monochromatic laser light and techniques of light-beating (correlation) spectroscopy extend the use of light scattering to the study of these kinetic phenomena, e.g., to measurement of velocities and properties of particles that influence their motion, such as diffusion constants.[3]

The scattered light also has a RAMAN SPECTRUM of relatively weak lines (or bands) first discovered by Raman (1928), originating from the light analog of the Compton effect, and explained by quantum theory.

R. W. HART

References

1. van deHulst, H. C., "Multiple Light Scattering: Tables, Formulas and Applications," Volumes 1 and 2, New York, Academic Press, 1980.
2. Young, A. T., "Rayleigh Scattering," *Physics Today* **35**, 42–48 (1982).
3. Degiorgio, V. (Ed.), "Light Scattering in Liquids and Macromolecular Solutions," New York, Plenum Publishing Co., 1980.
4. Feinstein, M. R., and Farrell, R. A., "Trial Functions in Variational Approximations to Long-Wavelength Scattering," (and references therein) *J. Opt. Soc. Am.* **72**, 223–231 (1982).

Cross-references: DOPPLER EFFECT; LIGHT; OPTICS, PHYSICAL; RAMAN EFFECT AND RAMAN SPECTROSCOPY.

LIQUEFACTION OF GASES

The liquefaction of all readily available gases has become a routine operation in industrial technology. Prominent among the reasons for converting a gas to a liquid are the net saving in the cost of storing or of transporting a normally gaseous material in liquid form, the convenience and flexibility of providing very low temperature refrigeration, in the form of a low-boiling liquid, to mulitple sites of modest or intermittent requirements, and the efficiency attainable in the separation of the components of a gaseous mixture by the partial liquefaction of the mixture, or its total liquefaction followed by rectification.

The transoceanic shipment of liquefied natural gas, the commercial distribution of liquid helium to scientific laboratories, and the production of pure oxygen and pure nitrogen from air are representative examples of the first, second, and third reasons, respectively. The first and last examples currently operate on scales such that thousands of tons of liquid are produced daily.

To produce a cold liquid product from gaseous raw material at ambient temperature requires a heat-pumping operation. Thermodynamic analysis gives the (unattainable) irreducible minimum work W_{min}, which must be expended in the heat pump operating in an environment at temperature T_0, to convert a unit mass of warm gas to liquid to be

$$W_{min} = (H_{liquid} - H_{gas}) - T_0(S_{liquid} - S_{gas})$$

where H_{liquid} and H_{gas} are the enthalpies and S_{liquid} and S_{gas} are the entropies per unit mass of liquid product and gaseous raw material, respectively. These thermodynamically reversible works of liquefaction are listed for various of the "permanent" gases in Table 1, which assumes that the starting material is gas at one atmosphere pressure and 300 K. In large-scale practical operations, the actual work requirement will range from ~3 times the minimum for a gas such as methane to ~15 times the minimum for helium.

To achieve the minimum thermodynamic work requirement for cooling and liquefying a stream of gas, an infinite sequence of perfectly efficient refrigerators operating at successively lower temperatures ranging from ambient to the boiling point of the material would be required. Various approximations to this theoretical ideal have been developed.

Cascade Process If the critical temperature of the gas which is to be liquefied lies well *above* the boiling point of some second fluid, whose critical temperature in turn lies above the boiling point of yet another fluid, and so on to some fluid that is condensable at ambient temperatures, then one can replace the infinite sequence of refrigerators of the thermodynamic ideal with this discrete series, or cascade, of liquid cooling baths.

Such an arrangement is shown schematically in Fig. 1. The raw material is compressed to the pressure necessary to condense it at the temperature of the final refrigerant bath. The resulting liquid is expanded through a throttle valve, and the vapor that boils off in the throttling process is recycled to conserve the refrigeration it represents. The penalty in increased work over the thermodynamic minimum arises from the small number of steps in the sequence with the attendant irreversible exchange of heat between the process gas and the much colder baths; from the throttling losses for the product liquid and the various cooling fluids; and from the imperfect efficiency of any real compressor.

For small-scale systems, the operational complexity and the equipment cost of a cascade are prohibitive. However, for large-scale operations, the economic performance of a cascade compares very favorably with any other process for the simple liquefaction of any gas for which a cascade of refrigerants can be found.

For large-scale operations such as the liquefaction of natural gas, one can realize significant savings in equipment costs if, rather than buying a smaller, separate compressor for each individual fluid in the cascade, one mixes the fluids together and buys a single, large compressor for the mixed refrigerants. This saving in equipment cost can more than compensate for the slightly lower thermodynamic efficiency

TABLE 1. MINIMUM WORK OF LIQUEFACTION OF VARIOUS GASES.

Substance	Boiling Point (K)	Work Required (kW-hr/kg)
Methane	111.7	0.320
Oxygen	90.2	0.176
Nitrogen	77.3	0.212
Hydrogen	20.4	3.26
Helium	4.22	1.90

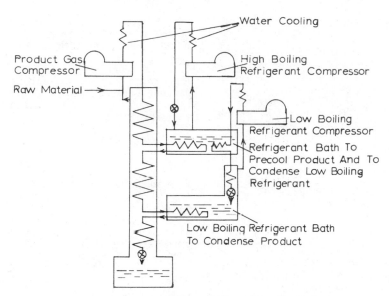

FIG. 1. Schematic arrangement for two-stage cascade.

(and higher power costs) that characterize such a system compared to a discrete cascade.

Such systems are represented schematically in Fig. 2. Commercially, they are frequently used for the production of liquefied natural gas (LNG) and are referred to variously as MCR (multicomponent refrigerant) or ARC (auto-refrigerated cascade) cycles. Cooling of the compressed mixture of refrigerants to ambient

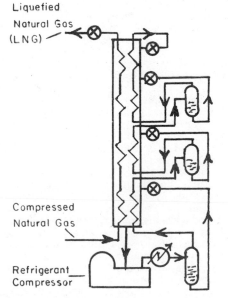

FIG. 2. Schematic arrangement for multicomponent refrigerant cascade.

temperature results in the liquefaction of a liquid rich in the highest boiling component of the refrigerant mixture. That liquid is separated from the residual gaseous phase and injected into the first exchanger to cool the compressed product stream and the residual gaseous refrigerant stream. That cooling condenses yet more of the refrigerant stream; this liquid is in turn separated and injected into the main exchanger, etc., until the final fraction of the refrigerant stream is throttled directly and returned to the exchanger as coolant.

Linde and Claude Processes A stream of cold gas, flowing countercurrent to the process stream in a heat exchanger that establishes perfect thermal equilibrium between the two streams at every point along their paths could substitute for the infinite sequence of refrigerators in the theoretical ideal system. The problem is just to produce the stream of cold gas (let alone to produce it with perfect efficiency) and to produce a refrigerator to extract the heat of vaporization from the product material at its boiling point.

Application of the first law of thermodynamics to the system shown in Fig. 3, consisting of a constant high-pressure source of fluid at P_1 that flows at constant rate through an insulated heat exchanger B, then through a throttling device C, and back through exchanger B, leaving the exchanger against some constant low pressure P_2 at the *same* temperature at which high-pressure fluid enters, gives

$$H_1 + Q = H_2 .$$

H_1 and H_2 are the enthalpies per unit mass of fluid entering and leaving the system, respec-

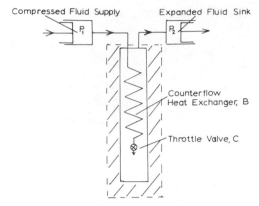

Compressed Fluid Supply   Expanded Fluid Sink

P_1 P_2

Counterflow
Heat Exchanger, B

Throttle Valve, C

FIG. 3. Adiabatic throttling flow system.

tively, and Q is the amount of heat *absorbed* by a unit mass of fluid in passing through the system. The initial and final kinetic and gravitational potential energies are assumed equal. H_1 will be smaller than H_2 for most gases at absolute temperatures less than 8–10 times the normal boiling point and for initial pressures of several hundred to a few thousand pounds per square inch. At higher temperatures, H_1 will be larger than H_2 so that Q is negative—heat is *liberated* within the exchanger-throttle valve system. If the thermal insulation of the exchanger-throttle valve system is perfect, then for $H_1 < H_2$, the exchanger-throttle valve system and the circulating gas itself will be continually cooled until some of the circulating gas accumulates within the system as liquid, and a new energy balance

$$H_1 = fH_{liquid} + (1 - f)H_2$$

is attained. The fraction f of the entering gas is withdrawn from the system as liquid product with enthalpy H_{liquid}, and the fraction $(1 - f)$ returns through the exchanger as its refrigerant.

Carl von Linde, in 1885, was the first to couple the relatively feeble ($40°C$ maximum for air expanded from 4000 psi to one atmosphere at room temperature) Joule-Thomson cooling, produced by throttling a high-pressure gas, and a regenerative heat exchanger to give the very simple system that, starting from room temperature, is capable of liquefying any gas except helium, hydrogen, or neon—albeit with poor efficiency. For helium, hydrogen, and neon, throttling at room temperature produces heating ($H_2 < H_1$). Sir James Dewar used a bath of liquid oxygen to precool compressed hydrogen to ~90 K where $H_2 > H_1$. The precooled hydrogen was then fed into a simple Linde liquefier, and Dewar first liquefied hydrogen in this way in 1898.

Kamerlingh Onnes used a bath of liquid hydrogen boiling under vacuum near its freezing point to precool compressed helium to

~14 K (where $H_2 > H_1$ for helium). The precooled helium was then fed into a simple Linde liquefier, and Onnes first liquefied helium in 1908.

For any real system, whose thermal insulation leaks q units of heat per unit mass of entering gas and whose exchanger permits gas to leave the system at $T_2' < T_1$, the energy balance becomes

$$H_1 + q = fH_{liquid} + (1 - f)$$
$$\cdot [H_2 + C_p(T_2' - T_1)].$$

Poor insulation (large q) and an inefficient exchanger (large $|T_2' - T_1|$) can easily reduce f to zero.

As the temperature of the gas entering the exchanger of a Linde liquefier approaches its critical temperature, the thermodynamic efficiency of the system as a refrigerator rises sharply. A simple Linde liquefier, coupled to any of several types of efficient auxiliary precooling refrigerators, forms the final stage of almost every large-scale liquefier in common use. Linde, himself, quickly modified his simple system to what, in essence, is a pair of simple liquefiers operating in cascade. The first (precooling) unit operates between the common initial high pressure and some intermediate pressure (chosen for optimum overall efficiency) rather than between the high pressure and one atmosphere. He also added a conventional ammonia refrigerator for precooling to further enhance efficiency.

It is possible to produce a stream of cold gas with relatively good efficiency by allowing compressed gas to expand in a reciprocating expansion engine, or in an expansion turbine. If the expansion engine is preceded by an efficient regenerative heat exchanger, relatively modest ratios of inlet pressure to exhaust pressure at the expansion engine can produce gas near its dew point. The thermodynamic efficiency of such expanders commonly approaches or exceeds 80%. A fraction of this cold exhaust gas can be used to refrigerate the feed to a simple Linde liquefier. Georges Claude in 1905 combined a reciprocating expansion engine with a simple Linde liquefier, as shown schematically in Fig. 4, to produce an air liquefier of improved efficiency which became known as the *Claude cycle*.

Peter Kepitza, in 1934, combined a precooling bath of liquid nitrogen in sequence with a reciprocating expansion engine to precool compressed helium feed for a final Linde stage and produced a Claude cycle which liquefied helium without the use of any auxiliary liquid hydrogen. Collins developed a similar machine which was produced commercially in relatively large numbers. Modifications in various sizes were developed and produced commercially. The machines typically use a cascade of two expansion engines in a Claude-cycle liquefier capable, when combined with liquid nitrogen precool-

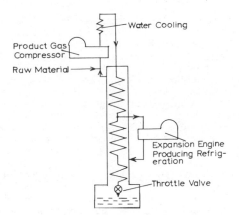

FIG. 4. Schematic arrangement for Claude cycle.

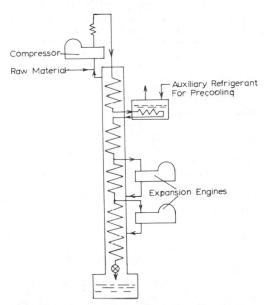

FIG. 5. Composite system for helium liquefaction.

ing, of producing up to 40 liters of liquid helium per hour. The schematic arrangement is shown in Fig. 5. The same schematic arrangement of a liquid precooling bath followed by expansion engines, all precooling the feed to a final Linde liquefier, describe in essence the large plants currently used for the large-scale liquefaction of hydrogen or helium.

DAVID N. LYON

References

Haselden, G. G., "Cryogenic Fundamentals," Ch. 2, New York, Academic Press, 1971.

Barron, Randall, "Cryogenic Systems," Ch. 3, New York, McGraw-Hill Book Co., 1966.

Vance, R. W., "Cryogenic Technology," Ch. 2, New York, John Wiley & Sons, 1964.

Collins, S. C., *Science* 116, 289 (1952).

Wenzel, L. A., *Advances in Cryogenic Engineering* 20, 90 (1975).

Gibson, B. M., *Advances in Cryogenic Engineering,* 25, 715 (1979).

Cross-references: CRYOGENICS; ENTROPY; GAS LAWS; GASES, THERMODYNAMIC PROPERTIES; LIQUID STATE; REFRIGERATION; STATES OF MATTER; THERMODYNAMICS.

LIQUID STATE

Liquid is the term used for a state of matter characterized by that of a pure substance above the temperature of melting and below the vaporization temperature, at any pressure between the triple point pressure and the critical pressure (see Fig. 1). The liquid state resembles the crystalline in the relatively low dependence of density on P and T, and resembles the gas state in the inability to support shear stresses (see reference to glasses below). Structurally the molecules are relatively close together but they lack long-range crystalline order. The mutual solubility of different liquids is also intermediate between the complete mutual solubility of most gases, and the relatively rare appreciable mutual solubility of pure crystalline compounds. Two liquids of similar molecules are usually soluble in all proportions, but very low solubility is sufficiently common to permit the demonstration of as many as seven separate liquid phases in equilibrium at one temperature and pressure (mercury, gallium, phosphorus, perfluoro-kerosene, water, aniline, and heptane at 50°C, 1 atm.

Stability Limits With the exception of helium and certain apparent exceptions discussed below, Fig. 1 gives a universal phase diagram for all pure compounds. The triple point of one P and one T is the single point at which all three phases, crystal, liquid, and gas, are in equilib-

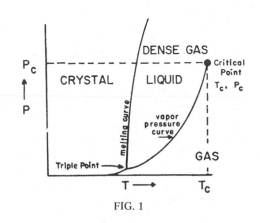

FIG. 1

rium. The triple point pressure is normally below atmospheric. Those substances, i.e., CO_2, $P_t = 3885$ mm, $T_t = -56.6°C$, for which it lies above, sublime without melting at atmospheric pressure.

From the triple point, the melting curve defines the equilibrium between crystal and liquid, usually rising with small but positive dT/dP, and presumably always with positive dT/dP at sufficiently high P values. The line is believed to extend infinitely without a critical point (it has been followed to $T \cong 16T_c$ for He, and calculations indicate that hard spheres would show a gas-crystal phase change). The gas-liquid equilibrium line, the vapor pressure curve, has dT/dP always positive and greater than the melting curve. The vapor pressure curve always ends at a critical point, $P = P_c$, $T = T_c$, above which the liquid and gas phase are no longer distinguishable. Since the liquid can be continuously converted into the gas phase without discontinuous change of properties by any path in the P-T diagram passing above the critical point, there is no definite boundary between liquid and gas.

The term *liquid* is commonly reserved for $T < T_c$, and "dense gas" is used for $T > T_c$. However, certain properties, such as the ability to dissolve solids, change rather abruptly at the critical density. In many respects, the dense gas resembles the low-temperature liquid of the same density more closely than it does the dilute gas.

The slope, dT/dP, of all phase equilibrium lines obeys the thermodynamic Clapeyron equation:

$$dT/dP = \Delta V/\Delta S = T\Delta V/\Delta H \qquad (1)$$

with ΔV, ΔS, and ΔH the differences, for the two phases, of volume, entropy, and heat content or enthalpy, respectively. The quantity ΔH is the heat absorbed in the phase change at constant P. Since always $S_{cr.} < S_{liq.} < S_{gas}$ and usually $V_{cr.} < V_{liq.} < V_{gas}$, one usually has $dT/dP > 0$; the relatively rare cases, including water, for which $V_{liq.} < V_{cr.}$ at low pressures leads to $dT/dP < 0$ for the melting curve near the triple point.

Figure 1 gives the P-T boundaries of the stable liquid phase. Clean liquids can readily be superheated or supercooled, and in vessels having walls to which the liquid adheres, they can be made to support negative pressures of several tens of atmospheres. Thus the properties of the metastable liquid can be investigated outside the limits shown in the diagram.

Two apparent exceptions to the universality of the phase diagram of Fig. 1 deserve mention. First, many of the more complicated molecules decompose at temperatures below melting or boiling, and the diagram is unobservable. Secondly, some liquids, notably glycerine and SiO_2 and many multicomponent solutions, supercool so readily that crystallization is difficult

to observe. In these cases, there is a continuous transition on cooling to a glass, which has the elastic properties of an isotropic solid. The structure of the glass is qualitatively that of the high-temperature liquid, lacking long-range order. Since glass and liquid are not sharply differentiated, the term *liquid* is sometimes used to include glasses, although common parlance reserves liquid for the state in which flow is relatively rapid.

Quantum Liquids The one real exception to the phase diagram of Fig. 1 is that of helium, Fig. 2. Both isotopes, ^{4}He and ^{3}He, have no triple point, the liquid is stable to 0 K below about 20 atm for ^{4}He and below about 30 atm for ^{3}He. The liquids have zero entropy at 0 K in both cases. This is also the only case in which isotopic mixtures form two liquid phases at equilibrium, the isotopic solution separating below 1 K. The isotope ^{4}He has itself two phases, He I above the dotted λ-line of the diagram, and He II with remarkable properties of superfluidity, second sound, etc., below the λ-line. The phase transition along the λ-line is second order; that is, whereas S and V are continuous, heat capacity and compressibility change discontinuously across the λ-line.

Although no completely satisfactory single theory of liquid helium has yet been formulated, one can say that most of the remarkable properties are qualitatively understood and are due to the predominance of quantum effects, including the difference in the statistics of the even and odd isotopes. Thus helium is the one example in nature of a quantum liquid, all other liquids showing only minor deviations from classical behavior.

Structure Considerable confusion in the description of liquid structure exists, due primarily to difficulties of precise formulation of verbal concepts. The geometric arrangement of any small number (say 10 to 12) of close lying molecules resembles the arrangement in the crystal, but the order rapidly disappears as larger groups are considered. Long-range order is lacking. The fact that numerical theories

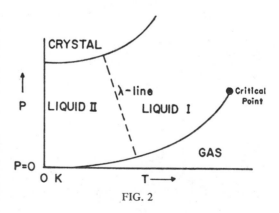

FIG. 2

based on a lattice or cell structure have some success is evidence only that most properties depend on the configuration of near neighbors alone. Insofar as the arrangement of nearest neighbors is describable in terms of that of the crystal, the structure of the normal liquid is probably characterized best by a somewhat closer spacing than the crystal of the same molecules, the reduced density arising from a considerable number of vacancies in the lattice; the coordination number or number of nearest neighbors is lower than in the crystal. The exception is water, in which the low coordination number, 4, of the crystal, is increased by interstitial molecules in the liquid, leading to a higher density of the liquid.

Structural descriptions of this nature usually lack the possibility of precise formulation. It is, however, possible to define for any disordered array of molecules in three-dimensional space an arrangement of contiguous cells, each containing one and only one molecule, the faces of the cells being the loci of the midpoints of neighboring molecules. The statistics of the fraction of cells with n faces and of the distances of the faces from the molecules would give the fraction of molecules having a given number of nearest neighbors and the distance distribution of these in a precisely defined manner. Neither present experimental information nor present theories lend themselves to analysis in such terms.

The only clearly defined manner of describing liquid structure in use at present involves the concept of a set of probability density functions, ρ_n, for ascending numbers, n, of molecules. The function ρ_n depends on the vector coordinates $r_1, r_2, \cdots, r_n$ of n molecules, and

$$\rho_n(r_1, r_2, \cdots, r_n)dr_1, \cdots, dr_n$$

is defined as being the probability that in the liquid of definite P and T, there will be at any instant of time, one molecule at each position, r_i, within the volume element, dr_i. For a fluid, unlike a perfect single crystal, $\rho_i(r)$ is a constant independent of r and equal to the number density: the number, ρ, of molecules per unit volume. The first significant member of the set is then the pair density function, $\rho_2(r_1, r_2)$, which depends only on the distance, $r = |r_1 - r_2|$, between the two molecules. At large distances $\rho_2(r \to \infty) = \rho^2$. This function can be found experimentally from the x-ray scattering intensities of the liquid (it is the three-dimensional Fourier transform of the scattering intensity at angle θ vs $(4\pi/\lambda)/\sin(\theta/2)$). A typical plot is shown in Fig. 3. The area under the ill-defined first peak integrated over $4\pi r^2 dr$ is the average number of nearest neighbors, and is of order 10 to 11 for normal liquids.

The quantity of dimensions of energy,

$$W_n(r_1, \cdots, r_n) = -kT \ln [\rho^{-n}\rho_n(r_1, \cdots, r_n)]$$

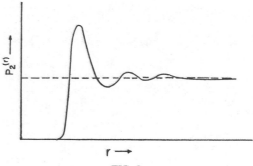

FIG. 3

can be shown to be the potential of average force of n molecules located at the positions $r_1, \cdots, r_n$. That is, if there are n molecules at these positions there will be some average force, $\bar{f}_{xi}$, along the x-coordinate of molecule i. This average is the sum of the direct force due to the other $n - 1$ plus the average of a fluctuating force due to the others, whose average position is affected by that of the n specified ones. This average force is

$$f_{xi} = -(\partial/\partial x_i)W_n(r_1, \cdots, r_n)$$

One frequently assumes that W_n is a sum of pair forces only,

$$W_n(r_1, \cdots, r_n) = \sum_{n \geqslant i > j \geqslant 1} \sum{}' W_2(r_{ij})$$

although this assumption is known to be only approximate. With this assumption, the pair average force potential, $W_2(r_{ij})$, can be computed as the solution of an integral equation, and the solutions agree quite well with the experimental curves.

The knowledge of the complete set of functions ρ_n plus that of the intermolecular forces would permit the computation of all equilibrium properties of the liquid, and indeed if the intermolecular forces are the sum of pair forces, only a knowledge of ρ_2 at all P, T values is necessary. An adequate, although numerically difficult, theory of the transport properties also exists, using the equilibrium functions, ρ_n. At present, only qualitative success is obtained in the completely a *priori* use of the equations.

Associated Liquids The description given above is adequate only for liquids composed of spherically symmetric molecules or molecules that are nearly so. These constitute the so-called normal liquids, which obey reasonably well the law of corresponding states, for which the entropy of vaporization at the boiling point has the Trouton's rule value of approximately 21 cal/deg. For molecules containing large dipole moments, or those forming mutual hydrogen bonds, the concept of the probability density functions must be extended to include angles or other internal degrees of freedom in the coordinates. Such inclusion is conceptually easy, but incredibly complicates the already difficult numerical evaluation of any equations. However, certain qualitative statements may be made.

Liquids composed of molecules with large dipole moments are frequently referred to as associated. Although in some instances relatively stable dimer or definite polymer units of relatively fixed orientation may exist, in many cases, notably water, it is extremely doubtful if an exact knowledge of the structure would reveal any distinguishable entities of associated molecules other than that of the whole liquid. In such cases, one would, however, expect that certain mutual angular orientations between neighboring molecules will be highly preferred, whereas in the dilute gas this will not be the case. The effect of this restriction on the internal coordinates will be to decrease the entropy of the liquid markedly compared to the gas. This effect is qualitatively the same as in association, and the properties of these liquids, particularly the high entropy of vaporization, will simulate those of a liquid composed of definite associated complexes.

JOSEPH E. MAYER

Cross-references: CRYSTALLOGRAPHY, DIPOLE MOMENTS, ENTROPY, SUPERFLUIDITY, SURFACE TENSION, THERMODYNAMICS, TRANSPORT THEORY, VAPOR PRESSURE AND EVAPORATION.

LORENTZ TRANSFORMATIONS

An event specified at a point $r \equiv (x, y, z)$ in ordinary space at an instant of time t defines an event described by the space-time coordinates (x, y, z, t). Such an event observed in a particular reference frame S may also be described in another reference frame S', in motion with a velocity v with respect to S. A Lorentz transformation relates the space-time coordinates in frame S' describing the event in frame S to the space-time coordinates in S. This Lorentz transformation forms the basic mathematical foundation for Einstein's *special theory of relativity* (see RELATIVITY).

An interval between two events may be described by an invariant measure of length, a metric, that is the same in all reference frames:

$$(\Delta s)^2 = (\Delta x_1)^2 + (\Delta x_2)^2 + (\Delta x_3)^2 + (\Delta x_4)^2$$
$$= (\Delta x_1')^2 + (\Delta x_2')^2 + (\Delta x_3')^2 + (\Delta x_4')^2$$

(1)

where we have set $(x_1, x_2, x_3, x_4) \equiv (x, y, z, ict)$ and where c is the velocity of light. A space-time with the above metric is called a *Minkowski space*. In the Einstein picture there is no invariant separation of space and time. Observation of an event from two different ref-

erence frames moving with a relative velocity v along the z axis (for simplicity) would be related by the Lorentz transformation:

$$x' = x$$
$$y' = y$$
$$z' = (1 - v^2/c^2)^{-1/2} (z - vt)$$
$$t' = (1 - v^2/c^2)^{-1/2} (t - vz/c^2).$$

(2)

One readily observes that these relations satisfy Eq. (1).

Every observer, related by a Lorentz transformation, observes the same physical process in all of space-time and conversely, any physical process must be Lorentz invariant. Such processes described by Newtonian mechanics, for example, are valid approximations when $v \ll c$ of a more general relativistic mechanics. Classical theories such as electromagnetism involving the velocity c are Lorentz invariant.

The Lorentz transformation of Eq. (2) is a subgroup of a general class of rotations in Minkowski space. We first consider ordinary rotations of the spatial coordinates leaving the x_4 coordinate unchanged. If we perform a rotation in the x_1-x_2 plane about the x_3 axis through an angle ω in a clockwise direction, a point described by (x_1, x_2, x_3, x_4) in the original coordinate system is described in the new (primed) coordinate system by

$$x_1' = x_1 \cos \omega + x_2 \sin \omega$$
$$x_2' = x_2 \cos \omega - x_1 \sin \omega$$
$$x_3' = x_3$$
$$x_4' = x_4.$$

(3)

These relations also satisfy Eq. (1). Such a transformation may be obtained by a unitary operator

$$\exp\left[-\omega\left(x_1 \frac{\partial}{\partial x_2} - x_2 \frac{\partial}{\partial x_1}\right)\right].$$

(4)

Noting that $\exp M = 1 + (M/1!) + (M^2/2!) + (M^3/3!) + \ldots$, and that

$$M^{2n}x_1 = (-1)^n \omega^{2n}x_1;$$
$$M^{2n}x_2 = (-1)^n \omega^{2n}x_2;$$
$$M^{2n+1}x_1 = (-1)^n \omega^{2n+1}x_2;$$
$$M^{2n+1}x_2 = (-1)^{n+1} \omega^{2n+1}x_1$$

one easily obtains the relations (2) that give the primed coordinates in frame S' in terms of the unprimed coordinates in S.

Similarly if we perform a "rotation" in the x_3-x_4 plane in Minkowski space through an imaginary angle $i\chi$ where $\tanh \chi = v/c$ such that the primed coordinate system moves along the x_3 axis with a velocity v, the primed coordinates may be obtained by use of the operator

$$\exp\left[-i\chi\left(x_3 \frac{\partial}{\partial x_4} - x_4 \frac{\partial}{\partial x_3}\right)\right].$$

The transformation obtained by this Lorentz transformation often called a Lorentz boost to distinguish it from the ordinary rotations yields

$$
\begin{aligned}
x_1' &= x_1 \\
x_2' &= x_2 \\
x_3' &= x_3 \cosh\chi + ix_4 \sinh\chi \\
x_4' &= x_4 \cosh\chi - ix_3 \sinh\chi.
\end{aligned}
\tag{7}
$$

Noting that $\cosh\chi = (1 - v^2/c^2)^{-1/2}$ and $\sinh\chi = (v/c)(1 - v^2/c^2)^{-1/2}$ one sees that expressions (7) are identical to those of (1) when expressed in the corresponding parameters.

One can express generally the Lorentz rotations and boosts of the coordinate frame in matrix form:

$$x_\mu' = a_{\mu\nu} x_\nu \tag{8}$$

with the orthogonality condition on the matrices

$$a_{\mu\nu} a_{\lambda\nu} = \delta_{\mu\lambda}. \tag{9}$$

A summation is to be made over repeated indices. The Lorentz transformations (8) correspond to a six-parameter homogeneous group.

However, events occurring in a coordinate system involve a physical process associated with a physical object in that coordinate system. An equivalent description may be obtained by "rotating" the physical system rather than the coordinate system, but in the inverse direction. A physical quantity may be expressed by a function involving the coordinates and may or may not have other intrinsic degrees of freedom not directly involving the space-time coordinates. Examples are scalars, spinors, polar four-vectors, antisymmetric tensors, etc. These other intrinsic degrees of freedom are designated as spin and they occur in the quantum mechanical description of elementary particles. The Lorentz transformation properties of these *wave functions* must be examined. We consider two examples, the transformation properties of a scalar and a polar four-vector wave function.

Consider a wave function in frame S' described by (1) that is considered to differ infinitesimally from S by an angle $\delta\omega$ in the x_1-x_2 plane. We make a Taylor expansion about the unprimed coordinate system where $x_1' = x_1 + \delta\omega x_2$ and $x_2' = x_2 - \delta\omega x_1$. The coordinates are expressed in a shorthand notation, i.e., $\phi(x) \equiv \phi(x_1, x_2, x_3, x_4)$.

$$\phi(x') = \phi(x) + \delta x_1 \frac{\partial\phi}{\partial x_1} + \delta x_2 \frac{\partial\phi}{\partial x_2},$$

consequently

$$\phi(x) = \left[1 + \delta\omega\left(x_1 \frac{\partial}{\partial x_2} - x_2 \frac{\partial}{\partial x_1}\right)\right]\phi(x').$$

The part within the parentheses is related to the orbital angular momentum operator L_{12} such that

$$\phi(x) = \left(1 + \frac{i}{\hbar}\delta\omega L_{12}\right)\phi(x'). \tag{10}$$

We may generally express a finite Lorentz rotation or boost in a particular x_μ-x_ν plane through an angle ω by the operator

$$\phi(x) = \exp\left(\frac{i}{\hbar}\omega L_{\mu\nu}\right)\phi(x') \tag{11}$$

which is obtained by exponentiation of (10)

The requirements of special relativity, that physical processes be independent of the reference coordinate frame, leads to rather profound conservation laws that may be obtained from variational principles involving infinitesimal variation of the Lagrangian function describing the system. The conservation of orbital angular momentum is a direct consequence of invariance under Lorentz transformations involving scalar particles.

A more complicated example is that of the four-vector field such as the electromagnetic vector potential $A_\mu(x)$ consisting of three spacelike components A_1, A_2, A_3 and a timelike component $A_4 = icV(x)$. The components may be represented by a column matrix:

$$A(x) = \begin{bmatrix} A_1(x) \\ A_2(x) \\ A_3(x) \\ A_4(x) \end{bmatrix}. \tag{12}$$

Omitting details too lengthy to treat here, we must introduce an operator analogous to $L_{\mu\nu}$ that rearranges the components of $A_\mu(x)$ into a new *local* arrangement $A_\mu'(x')$ referred to the transformed coordinate system. It should be emphasized that $L_{\mu\nu}$ involves the coordinates while the spin operator introduced here involves the components of $A(x)$.

These spin operators are defined in this Cartesian representation as

$$S_{23} = \hbar \begin{bmatrix} 0 & 0 & 0 & 0 \\ 0 & 0 & -i & 0 \\ 0 & i & 0 & 0 \\ 0 & 0 & 0 & 0 \end{bmatrix},$$

$$S_{31} = \hbar \begin{bmatrix} 0 & 0 & i & 0 \\ 0 & 0 & 0 & 0 \\ -i & 0 & 0 & 0 \\ 0 & 0 & 0 & 0 \end{bmatrix},$$

$$S_{12} = \hbar \begin{bmatrix} 0 & -i & 0 & 0 \\ i & 0 & 0 & 0 \\ 0 & 0 & 0 & 0 \\ 0 & 0 & 0 & 0 \end{bmatrix},$$

$$S_{14} = \hbar \begin{bmatrix} 0 & 0 & 0 & -i \\ 0 & 0 & 0 & 0 \\ 0 & 0 & 0 & 0 \\ i & 0 & 0 & 0 \end{bmatrix},$$

$$S_{24} = \hbar \begin{bmatrix} 0 & 0 & 0 & 0 \\ 0 & 0 & 0 & -i \\ 0 & 0 & 0 & 0 \\ 0 & i & 0 & 0 \end{bmatrix},$$

$$S_{34} = \hbar \begin{bmatrix} 0 & 0 & 0 & 0 \\ 0 & 0 & 0 & 0 \\ 0 & 0 & 0 & -i \\ 0 & 0 & i & 0 \end{bmatrix}.$$

The action of an operator similar to that of (11) involving these spin operators $S_{\mu\nu}$ may be obtained by expansion of the exponential

$$A'(x') = \exp\left(\frac{i}{\hbar} \omega S_{\mu\nu}\right) A(x).$$

Noting that, for example,

$$\left(\frac{i}{\hbar} \omega S_{12}\right)^{2n} = (-1)^n \omega^{2n} \begin{bmatrix} 1 & 0 & 0 & 0 \\ 0 & 1 & 0 & 0 \\ 0 & 0 & 0 & 0 \\ 0 & 0 & 0 & 0 \end{bmatrix}$$

and

$$\left(\frac{i}{\hbar} \omega S_{12}\right)^{2n+1} = (-1)^n \omega^{2n+1} \begin{bmatrix} 0 & 1 & 0 & 0 \\ -1 & 0 & 0 & 0 \\ 0 & 0 & 0 & 0 \\ 0 & 0 & 0 & 0 \end{bmatrix}.$$

After rearranging in trigonometric series and carrying out the matrix multiplication, one obtains expressions for the components of the four-vector field $A(x)$:

$$\begin{aligned} A_1'(x') &= A_1(x') \cos \omega + A_2(x') \sin \omega \\ A_2'(x') &= A_2(x') \cos \omega - A_1(x') \sin \omega \\ A_3'(x') &= A_3(x') \\ A_4'(x') &= A_4(x'). \end{aligned} \quad (14)$$

It is not obvious from the operation performed here that, not only are the components of the polar vector field rearranged by this operator, but also the dependence on the coordinates that are in the primed system. This feature would be made obvious in the transformation of the Lagrangian function involving these fields, which is beyond the scope of this article. This *local* dependence may be made global by inclusion of the operator L_{12} occurring in (11) that transforms the coordinates of the primed system to those of the unprimed system:

$$A'(x) = \exp\left[\frac{i}{\hbar} \omega(L_{12} + S_{12})\right] A(x) \quad (15)$$

so that we have

$$\begin{aligned} A_1'(x) &= A_1(x) \cos \omega + A_2(x) \sin \omega \\ A_2'(x) &= A_2(x) \cos \omega - A_1(x) \sin \omega \\ A_3'(x) &= A_3(x) \\ A_4'(x) &= A_4(x). \end{aligned} \quad (16)$$

The sum of $L_{\mu\nu}$ and $S_{\mu\nu}$ representing the orbital and spin angular momentum operators yield that for the total angular momentum $J_{\mu\nu}$.

The Lorentz rotation of a particle or field through an angle ω in the x_1-x_2 system is thus given by the operator

$$\exp\left(\frac{i}{\hbar} \omega J_{12}\right). \quad (17)$$

The corresponding Lorentz boost of a particle to velocity v along the X_3 direction is given by the operator.

$$\exp\left(\frac{i}{\hbar} \chi J_{34}\right) \quad (18)$$

where $\tanh \chi = v/c$. In this latter case the components of a four-vector field become

$$\begin{aligned} A_1'(x) &= A_1(x) \\ A_2'(x) &= A_2(x) \\ A_3'(x) &= A_3(x) \cosh \chi + iA_4(x) \sinh \chi \\ A_4'(x) &= A_4(x) \cosh \chi - iA_3(x) \sinh \chi \end{aligned} \quad (19)$$

where as before

$$\cosh \chi = (1 - v^2/c^2)^{-1/2} = E/mc^2$$

$$\sinh \chi = (v/c)(1 - v^2/c^2)^{-1/2} = p/me.$$

In these expressions, m is the rest mass of the particle, E is the total energy, and p is the momentum.

In any physical process involving particles or fields the invariance of the Lagrangian under infinitesimal homogeneous Lorentz transformations leads directly to the law of conservation of total angular momentum whenever the particle has an intrinsic spin. The orbital and spin angular momentum are not independently invariants in a dynamical process.

There are other Lorentz transformations allowable. The six-parameter homogeneous group (8) may be extended to the ten-parameter inhomogeneous group that includes the translations

$$x_\mu' = a_{\mu\nu} x_\nu + \epsilon_\mu. \tag{20}$$

The constant ϵ_μ represent a translation or shift of the coordinate x_μ. Such translations obey the invariant metric condition (1).

The corresponding Lorentz translation of a wave function is given by

$$\phi(x') = \exp\left(-\frac{i}{\hbar} \epsilon_\mu P_\mu\right)\phi(x) \tag{21}$$

where P_μ are the operators corresponding to the components of the four momenta $(p_1, p_2, p_3, (i/c)E)$.

$$P_\mu = \frac{\hbar}{i}\frac{\partial}{\partial x_\mu}.$$

Invariance of the Lagrangian function under infinitesimal Lorentz translations leads directly to the conservation of momentum and energy.

There are also the improper Lorentz transformations in which the space and/or the time coordinates are inverted. Space inversion corresponds to

$$x_i' = -x_i, \quad i = 1, 2, 3 \tag{22}$$

while time inversion corresponds to

$$x_4' = -x_4. \tag{23}$$

Both of these improper Lorentz transformations obey the invariance condition (1). They are discussed in detail in the article PARITY.

RAYMOND W. HAYWARD

References

Jackson J. D., "Classical Electrodynamics," Second Edition, New York, John Wiley & Sons, 1975.

Pauli, W., "Theory of Relativity," London, Pergamon Press, 1958.

Rindler, W., "Essential Relativity," Second Edition, New York, Springer-Verlag, 1977.

Sakurai, J. J., "Invariance Principles and Elementary Particles," Princeton, Princeton Univ. Press, 1964.

Cross-references: CONSERVATION LAWS AND SYMMETRY; ELECTROMAGNETIC THEORY; ELEMENTARY PARTICLES; MATRICES; MECHANICS; PARITY; TIME; VELOCITY OF LIGHT; WEAK INTERACTIONS.

LUMINESCENCE

Introduction Luminescence is the phenomenon of light emission in excess of thermal radiation. Excitation of the luminescent substance is prerequisite to the luminescent emission. Photoluminescence depends upon excitation by photons; cathodoluminescence, by cathode rays; electroluminescence, by an applied voltage; chemiluminescence, by utilization of the energy of a chemical reaction. Luminescent emission involves optical transitions between electronic states characteristic of the radiating substance. The phenomenon is essentially the emission spectroscopy of gases, liquids, and solids. The same basic processes may yield infrared or ultraviolet radiation in substances with suitable electronic energy states; therefore, such emission in excess of thermal radiation is also described as luminescence.

Luminescence can be distinguished from the Raman effect, Compton and Raleigh scattering and Čerenkov emission on the basis of the time delay between excitation and luminescent emission being long compared to the period of the radiation, λ/c, where λ is the wavelength and c is the velocity of light. The radiative lifetimes of the excited states vary from 10^{-10} to 10^{-1} sec depending on the identity of the luminescent substances whereas λ/c is approximately 10^{-14} sec for visible radiation. At ordinary densities of excitation, the spontaneous transition probability predominates so that the luminescent radiation is incoherent; under conditions of high densities of excitation in suitable luminescent substances, the induced transition probability may predominate, the emitted radiation is coherent, and laser action is attained.

The initial persistence of luminescent emission following the removal of excitation depends on the lifetime of the excited state. This emission decays exponentially and is often called fluorescence. In many substances, there is an additional component to the afterglow which decays more slowly and with more complex kinetics. This is called phosphorescence. For many inorganic crystals, the emission spectra for fluorescence and phosphorescence are the same; the difference in afterglow arises from electron traps from which thermal activation is prerequisite to emission. For organic molecules, the emission spectra for fluorescence and phosphorescence are often different: the former occurs from an excited singlet; the latter, from a triplet state.

Luminescence of Gases The simplest luminescent substances are monatomic gases. The electronic states are characteristic of the isolated atoms; therefore, the excitation and emission spectra depend only on the differences in energy of the stationary electronic states of the many-electron atom, and the spectral lines are broadened only by the lifetimes of the excited states or at higher pressures, by collisions. The transitions are to a good approximation one-electron transitions. Resonance fluorescence is photoluminescence in which the exciting radiation is the exact frequency or wavelength for the transition from the ground to the excited state and emission occurs with the same frequency. Resonance fluorescence is shown diagrammatically in Fig. 1 for low pressure alkali metal vapor. The well-known 2537Å line of mercury vapor is another example of resonance fluorescence. This emission can also be excited by electrons accelerated by 5 or more volts. The simplest case of sensitized fluorescence (photoluminescence in which absorption of the exciting radiation is by one substance and the excitation is transferred to another which emits radiation) occurs with mixtures of monatomic gases. For example, the characteristic fluorescence of thallium is observed when mixtures of Tl and Hg vapors are illuminated with the 2537Å radiation of Hg.

For diatomic and polyatomic gases, the energies of the electronic states are dependent on the interatomic distances of the molecule. This dependence is shown in Fig. 2 for the ground and excited states. For a diatomic molecule the coordinate R is the distance between the two atoms. For each electronic state, there is a series of vibrational levels which are also shown in Fig. 2. Optical transitions occur between individual vibrational levels of one electronic state and individual vibrational levels of another electronic state. These transitions occur in accordance with the Franck-Condon principle, i.e., with fixed nuclear coordinates, vertically

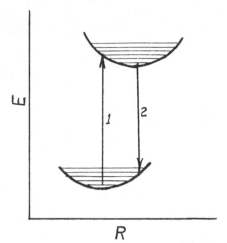

FIG. 2. Configuration coordinate model.

as shown in transitions 1 and 2 of Fig. 2. In most cases the emission will involve a smaller transition energy and occur at longer wavelength than the photo-excitation. This is referred to as Stokes' emission. In some cases, for example at high temperatures when the higher vibrational levels of the ground electronic state are populated thermally, anti-Stokes' emission is also observed. Additional structure in the photoexcitation and emission spectra arises from rotational states of the molecule. Iodine is a typical diatomic luminescent molecule excited by green light with visible emission at slightly longer wavelengths. Benzene and aniline are typical polyatomic molecules which are luminescent as vapors.

Luminescence of Organic Materials The electronic states of most organic luminescent materials in the liquid or solid phase, either as pure materials or as solutes in dilute concentration in inert solvents, are to a good approximation describable in terms of the electronic states of the free molecule in the gaseous phase. In other words, the intermolecular forces are weak compared to the intramolecular forces. The photoexcitation and luminescent emission spectra of these substances in condensed phases are similar to the spectra of the vapors. The intermolecular forces are, however, great enough to bring about broadening of absorption and emission lines and in some cases to bring about electronic energy transfer between molecules before intramolecular, vibrational relaxation with the accompanying Stokes' shift can occur. On the other hand, in a viscous or rigid medium collisional, nonradiative de-excitation is reduced.

Many of the organic luminescent materials are aromatic molecules related to dyes. The sodium salt of fluorescein in dilute aqueous solution is well known as an efficient fluorescent material. Other organic substances luminesce efficiently when dissolved in organic solvents. Terphenyl

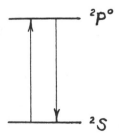

FIG. 1. Resonance fluorescence of atoms, e.g., Na.

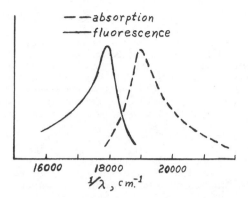

FIG. 3. Mirror symmetry of absorption and emission.

in xylene is a liquid β- and γ-ray scintillator with emission in the near ultraviolet. Some organic molecules luminesce most efficiently in a rigid medium. A solid solution of 1 per cent anthracene in napthalene is a scintillator with blue emission. For these solutions, energy is absorbed by the solvent molecules and transferred to the solute where the luminescent emission occurs. Crystals of some pure organic substances luminesce, particularly at low temperatures.

The fluorescent emission and the long-wavelength absorption of organic materials are often simply related as mirror images of each other. This is shown in Fig. 3 for rhodamine in ethanol and can be explained on the basis of the configuration coordinate model in Fig. 2. with the force constants for the two electronic states approximately equal. For organic molecules, the configuration coordinate R is interpreted as representing, schematically, all the intramolecular nuclear coordinates. In addition to the fluorescent emission, many organic substances exhibit phosphorescent emission. This arises from nonradiative relaxation from the excited singlet to the triplet state followed by radiative decay from the triplet to the ground singlet state, as illustrated in Fig. 4. Because

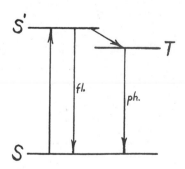

FIG. 4. Fluorescence and phosphorescence of organic molecules.

of the spin selection rule governing radiative transitions, the triplet has a long lifetime and the oscillator strength for direct excitation to the triplet is negligible. In suitable systems, polarized excitation and emission arising from the anisotropy of the organic molecules can be observed.

Chelates involving organic molecules as ligands bound to a metal atom or ion are a class of substances with members which luminesce. The fluorescence and phosphorescence of chlorophyll both *in vivo* and *in vitro* have been investigated for many years. On the other hand, rare earth chelates with organic ligands have been intensively investigated quite recently as lasers. The photoexcitation occurs in the broad absorption bands of the ligand; the energy is transferred to the localized $4f$ shell of the rare earth by a mechanism related to the transfer process occurring in the scintillators; luminescent line emission characteristic of the rare earth occurs as coherent radiation with high excitation intensities. A fluorinated Eu-acetonate dissolved in acetonitrile has been announced as a liquid laser operating near room temperature.

Luminescence of Inorganic Crystals Inorganic crystals which luminesce are often called phosphors. Their luminescence in most cases originates from impurities or imperfections. Exceptions include the luminescence of alkaline earth tungstates which is characteristic of the $WO_4{}^{2-}$ group perturbed by the crystal field, the luminescence of some rare earth salts, and radiative recombination of conduction electrons with valence band holes in semiconductors. The impurities and imperfections responsible for luminescence in inorganic crystals are of diverse atomic and molecular types whose characteristics depend on the structure of the defect and on the electronic structure of the pure crystal. In some cases, the electronic states involved in the luminescence can be described in terms of energy levels of the impurity ion perturbed by the crystal field; in other cases, in terms of the crystal band structure perturbed by the impurity. The existence of conduction bands in inorganic crystals, particularly in semiconducting crystals, introduces additional mechanisms for the excitation of luminescence and for phosphorescence. For example, suitable impurities can be excited by alternate capture of injected conduction electrons and valence band holes, thus providing one mechanism for electroluminescence; on the other hand, an excited luminescent impurity may lose an electron to another defect via the conduction band, and the thermal activation necessary for return to the luminescent impurity is responsible for phosphorescence.

The alkali halides are simple ionic crystals which become luminescent when doped with suitable impurities. Thallium substituted in dilute concentration at cation sites in potassium chloride has the absorption and emission shown

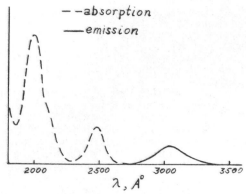

FIG. 5. Spectra of KCL:Tl.

in Fig. 5. The absorption bands involve the $^1S \rightarrow {}^3P^0,\ {}^1P^0$ transitions of the free ion perturbed by crystal interactions; the principal emission band, $^3P^0 \rightarrow {}^1S$, is similarly perturbed. The spectra can be understood qualitatively with the aid of Fig. 2, modified with a second excited state and with the configuration coordinate interpreted as symmetric displacement of the six nearest-neighbour Cl^- from the Tl^+. It is this interaction which is most dependent on the electronic state of the Tl^+ and is, therefore, largely responsible for the band widths and Stokes' shift.

Many inorganic crystals become luminescent when certain transition metal ions are dissolved in them. The luminescence involves intercombination transitions within the $3d$ shell; therefore, crystal field theory can be used to interpret the absorption and emission spectra. Divalent manganese is a common activator ion. Zn_2SiO_4, ZnS and $3Ca_3(PO_4)_2 \cdot CaF_2$ are important phosphors activated with Mn^{2+}. The last, activated also with Sb^{3+}, is the principal fluorescent lamp phosphor. The excitation at 2537Å from the Hg discharge occurs at the Sb^{3+}, whose energy level structure is similar to that of Tl^+; part of the energy is radiated in a blue band due to Sb^{3+} and part is transferred to the Mn^{2+} which is responsible for an orange emission band. It has been shown that the Sb^{3+} at a Ca^{2+} site is locally charge-compensated by an O^{2-} at a halide site and that the blue emission is in part an electron transfer transition. The ruby laser involves the luminescence of Cr^{3+} in Al_2O_3. Excitation occurs in a broad absorption band, and the system relaxes to another excited state from which emission occurs in a narrow band.

Rare earth ions, particularly trivalent, in solid solution in inorganic crystals and in glasses exhibit the emission characteristic of transitions in the $4f$ shell. For examples, samarium, europium, and terbium give visible emission; neodymium, infrared; and gadolinium, ultraviolet. Because of their narrow emission bands and stability, the rare earth activated phosphors are

of interest as lasers and as phosphors for fluorescent lamps and for color television. Crystal field theory can be used to explain the optical absorption and luminescent emission of the $4f$ transitions of rare earth ions in crystals and glasses. Phosphors doubly activated with rare earths, e.g., $YF_3:Yb$, Tm, have been found to be capable of large anti-Stokes' emission. This occurs by multiphoton infrared excitation and with visible emission.

The zinc sulfide phosphors, which are widely used as cathodoluminescent phosphors and well-known for their electroluminescence, are now recognized as large band gap, compound semiconductors. Two impurities or imperfections are essential to the luminescence of many of these phosphors: an activator which determines the emission spectrum and a coactivator which is essential for the emission but in most cases has no effect on the spectrum. Activator atoms such as Cu, Ag and Au substitute at Zn sites and perturb a series of electronic states upward from the valence band edge. In a neutral crystal containing only these activator impurities, the highest state is empty, i.e., it contains a positive hole and can accept an electron from the valence band; therefore, in semiconductor notation, the activator is an acceptor. In a similar way, coactivators such as Ga or In at Zn sites or Cl at S sites are donors in ZnS. The simultaneous introduction of both types of impurities results in electron transfer from donor to acceptor lowering the energy of the crystal and leaving both impurities charged. The coulomb attraction of the donor and acceptor leads to a departure from a random distribution over lattice sites and to pairing. The electronic states and some of the transitions of acceptors, donors and donor-acceptor pairs are shown in Fig. 6. The spectrum of ZnS:Cu, Ga is shown in Fig. 7. The longer-wavelength emission band involves the transition from the lowest donor state to highest acceptor state (transition 3) in approximately fifth nearest-neighbor pairs; the shorter-wavelength emission corresponds more nearly to transition 1 of Fig. 6. Luminescent emission from donor-acceptor pairs has been more clearly seen with gallium phosphide crystals. In addition to luminescence due to donors,

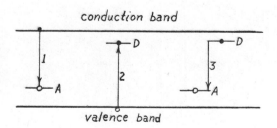

FIG. 6. Band model for acceptor, donor, and pair transitions.

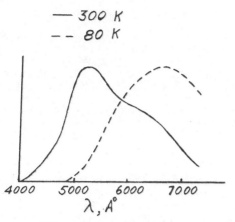

FIG. 7. Spectra of ZnS. Cu, Ga.

acceptors and their pairs, emission bands due to transition metals are well known for zinc sulfide as noted earlier. In zinc sulfide crystals, the donors which are unassociated with acceptors serve as electron traps and are responsible for long-persistent, temperature-dependent phosphorescence. Uncharged or isoelectronic dopants, such as N at phosphorous sites in GaP, have been shown to be efficient radiative recombination centers, particularly in indirect band gap semiconductors. The III-IV semiconductors with isoelectronic dopants or with donor-acceptor pairs are currently the principal electroluminescent light-emitting diodes.

Trends in Luminescence Research Both molecular and solid-state luminescence research have become increasingly sophisticated in experimental techniques and theoretical concepts; the materials investigated, more diverse and structured. Low temperatures, high electrical and magnetic fields, intense excitation and short times are among the conditions used to investigate luminescent phenomena. Luminescent excitation is most generally formulated theoretically as a collective excitation of a many-body problem encompassing the source of excitation and its probe, as well as the luminescent body itself. Inhomogeneously doped crystals, amorphous materials and man-made superlattices are now investigated as luminescent materials, as well as are homogeneous solutions, crystals and powders.

The configuration coordinate model shown in Fig. 2 for diatomic and polyatomic molecules is also applied to luminescent centers in the condensed phase. The configuration coordinate represents the local and/or normal atomic coordinates of the luminescent system. The model is based on the adiabatic approximation to the Born-Oppenheimer expansion for a system of many electrons and many nuclei, for which the electrons are assumed to move rapidly compared to the nuclei. However, luminescent excitation of these many-particle systems involves to some extent all the particles and thus the exact description of the excitation is a collective excitation.

Under some conditions of luminescent excitation (for example, very high excitation intensity), a more complete formulation is necessary, including the radiation field of the excitation, the detector of emission, and the source of any applied stresses to the luminescent body. Even at low intensities of excitation the coupling of the electromagnetic probe to transverse excitons must be included to describe excitonic luminescent transitions, that is, excitonic polaritons. Another example is the recent analyses of the effects of hydrostatic pressure on luminescent spectra in which a type of double adiabatic approximation is used, with the electrons, nuclei and pressure apparatus approximated respectively as fast, intermediate, and slow components of the total system.

The usual luminescent transitions depend on thermalization of the excited molecule or center among the vibrational levels illustrated in Fig. 2. Emission which occurs before thermalization is termed "hot" luminescence. It is well-known for molecular dopants such as O_2^- in alkali halides. This phenomenon is investigated more widely by nanosecond and picosecond time-resolved spectroscopy.

Hot electrons are responsible for the excitation of luminescence with high electrical fields, that is, high field electroluminescence. This phenomenon is investigated in thin film devices consisting of a phosphor sandwiched between insulators, all films prepared by evaporation or sputtering, and subjected to a.c. voltages. The mechanism is well-established for ZnS:Mn and ZnF_2:Mn as collision excitation of Mn^{2+} by hot electrons, and cross sections have been measured and calculated theoretically. These and rare earth doped components are used in thin film displays.

For systems exhibiting inhomogeneous spectral broadening, site selection spectroscopy is used to resolve components of the spectra. Ions or molecules at specific sites are selectively excited to fluorescence or phosphorescence with laser sources at low temperatures. Organic compounds in solutions, crystals, and amorphous materials are investigated by site selection spectroscopy. "Hole burning," that is, optical depletion of selective ground state molecules followed by optical probing of these molecules is another valuable tool for studying the origin of inhomogeneous broadening and also to investigate energy transfer.

FERD WILLIAMS

Reference

Pringsheim, P., "Fluorescence and Phosphorescence," New York, Interscience, 1949.
Curie, D., "Luminescence in Crystals" New York,

J. Wiley & Sons, 1963 (translated by G. F. J. Garlick).

Goldberg, P., Ed., "Luminescence of Inorganic Solids," New York, Academic Press, 1966.

Crosswhite, H. M., and Moos, H. J., Eds., "Optical Properties in Crystals," New York, Wiley-Interscience, 1967.

Lim, E. C., Ed., "Molecular Luminescence," New York, Benjamin, 1969.

Williams, F., Ed., "Proceedings of International Conference on Luminescence," Amsterdam, North Holland, 1970.

Birks, J. B., "Photophysics of Aromatic Molecules," New York, Wiley-Interscience, 1970.

Williams, F., Ed., "Luminescence of Crystals, Molecules, and Solutions," New York, Plenum, 1973.

Shionoya, S., Nagakura, S., and Sugano, S. (Eds.), "Proceedings of the 1975 International Conference on Luminescence," Amsterdam, North Holland, 1976.

DiBartolo, B. (Ed.), "Luminescence of Inorganic Solids," New York, Plenum Press, 1978.

Curie, D., Mattler, J., and Parrot, R. (Eds.), "Proceedings of the 1978 International Conference on Luminescence," Amsterdam, North Holland, 1979.

Williams, F. (Ed.), "Workshop on the Physics of Electroluminescence," *Journal of Luminescence* 23 (1,2) (1981).

Broser, I., Gumlich, H.-E., and Broser, R. (Eds.), "Proceedings of the 1981 International Conference on Luminescence," Amsterdam, North Holland, 1981.

Cross-references: COLOR CENTERS; CRYSTALLOGRAPHY; ENERGY LEVELS; LASERS; PHOTOCONDUCTIVITY; RADIATION, THERMAL; RESONANCE; SEMICONDUCTORS; SOLID-STATE PHYSICS; SOLID-STATE THEORY; SPECTROSCOPY.

M

MAGNETIC FIELD

Magnetic field is generated by the passage of electrical currents through conductors or by the circulation of microscopic charges within magnetic materials. The magnetic field exerts a mechanical stress on its sources.

Basic Equations To separate electromagnetic theory from the theory of the solid state, Maxwell's equations can be written in terms of the magnetic induction or flux-density $\mathbf{B}$ and the total current density $\mathbf{J}$:

$$\nabla \cdot \mathbf{B} = 0 \qquad (1)$$

$$\nabla \times \mathbf{B} = 4\pi \mathbf{J} \qquad (2)$$

where, for purposes of the present article, the displacement current $(1/c^2)\partial \mathbf{E}/\partial t$ is neglected relative to $4\pi \mathbf{J}$ (see ELECTROMAGNETIC THEORY and MAGNETISM).

The solution of Eqs. (1) and (2) is

$$\mathbf{B}(\mathbf{r}) = \int d^3 r_1 \frac{\mathbf{J}(\mathbf{r}_1) \times (\mathbf{r} - \mathbf{r}_1)}{|\mathbf{r} - \mathbf{r}_1|^3} \qquad (3)$$

where the integral includes all current-carriers, and $\mathbf{B}$ vanishes at infinity.

From Eq. (3) it follows, for example, that an infinite straight conductor [Fig. 1(a)] carrying a total axial current I_c gives rise to an azimuthally directed external magnetic induction of strength $\mathbf{B} = 2I_c/r$, where r is the distance from the conductor axis. The same is true for an axial current in any axisymmetric conductor, e.g., the toroidal conductor of Fig. 1(b). An infinitely long circular cylindrical conductor [Fig. 2(a)] carrying an azimuthal current density I_c' per unit length contains an axially directed magnetic induction of strength $\mathbf{B} = 4\pi I_c'$. The same expression holds for a straight cylindrical conductor of arbitrary cross section [Fig. 2(b)], or for an infinite plane current sheet. A circular cylinder like that of Fig. 2(a), but of finite length L and radius R, has a central magnetic induction of strength $\mathbf{B} = 4\pi I_c' L (L^2 + 4R^2)^{-1/2}$.

In terms of the vector potential $\mathbf{A}$,

$$\mathbf{B} = \nabla \times \mathbf{A} \qquad (4)$$

and the gauge $\nabla \cdot \mathbf{A} = 0$, one has

$$\nabla^2 \mathbf{A} = -4\pi \mathbf{J} \qquad (5)$$

and

$$\mathbf{A}(\mathbf{r}) = \int d^3 r_1 \frac{\mathbf{J}(\mathbf{r}_1)}{|\mathbf{r} - \mathbf{r}_1|} \qquad (6)$$

for $\mathbf{A}$ vanishing at infinity.

Magnetic lines of force are defined by $d\mathbf{r} \propto \mathbf{B}$

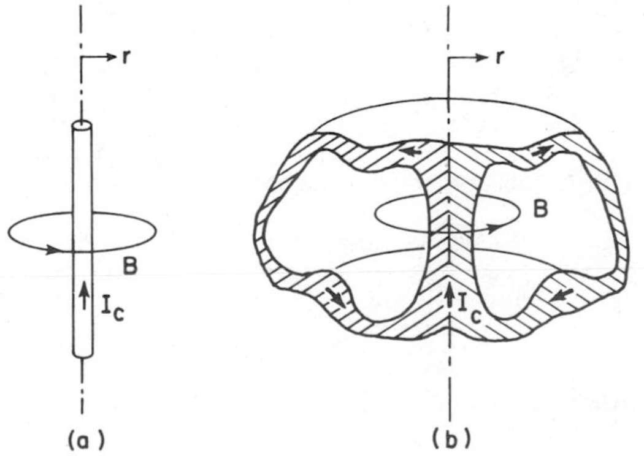

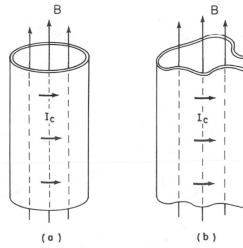

FIG. 1. Axisymmetric configurations with axially directed current I_c.

FIG. 2. Infinitely long cylindrical configurations with axially directed magnetic induction $\mathbf{B}$.

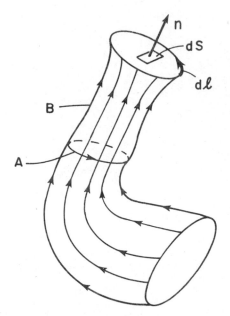

FIG. 3. Magnetic flux tube.

and are endless, by virtue of Eq. (1). The magnetic flux Φ through a surface S, bounded by a closed curve ℓ (Fig. 3), is given by

$$\Phi = \int dS\mathbf{B} \cdot \mathbf{n} = \oint d\ell \cdot \mathbf{A} \qquad (7)$$

where $\mathbf{n}$ is the normal to S. The flux tube defined by the field lines passing through ℓ contains constant flux, independent of S.

At a large distance from a localized current distribution at $\mathbf{r} = 0$, Eq. (6) gives the dipole potential

$$\mathbf{A} = \frac{\mathbf{m} \times \mathbf{r}}{r^3} \qquad (8)$$

where

$$\mathbf{m} = \frac{1}{2} \int d^3 r_1 \mathbf{r}_1 \times \mathbf{J}(\mathbf{r}_1) \qquad (9)$$

Equations in the Presence of Magnetic Materials A macroscopic current density $\mathbf{J}_M$ can be defined by local averaging of the microscopic current density $\mathbf{J}_m$ within magnetic materials

$$\mathbf{J}_M = \langle \mathbf{J}_m \rangle_{av} \qquad (10)$$

More conveniently, a magnetization vector $\mathbf{M}$ can be introduced, where

$$\mathbf{J}_M = \nabla \times \mathbf{M} \qquad (11)$$

In experiments, only $\mathbf{J}_M$ can be measured directly (via measurements on $\mathbf{B}$), and $\mathbf{M}$ is then uniquely derivable from Eq. (11) only with the added condition that it is to be a local state variable of the magnetic material (i.e., it is constant in uniform samples and constant fields). This condition follows automatically from the interpretation of $\mathbf{M}$ as a magnetic-moment density per unit volume:

$$\mathbf{M} = N\mathbf{m}_0 \qquad (12)$$

The theoretical molecular magnetic moment $\mathbf{m}_0$ (with number density N) is derived from $\mathbf{J}_m$ by evaluation of Eq. (9) over the molecular volume.

For macroscopic purposes, the total current density $\mathbf{J}$ of the preceding section is now specified by

$$\mathbf{J} = \mathbf{J}_c + \mathbf{J}_M \qquad (13)$$

The component $\mathbf{J}_c$ flows in conductors of resistivity η in accordance with Ohm's law:

$$\eta \mathbf{J}_c = \mathbf{E} \qquad (14)$$

The component $\mathbf{J}_M$ is derived from $\mathbf{M}$.

In the analysis of configurations involving magnetic materials, the magnetic field $\mathbf{H}$ is a convenient vector

$$\mathbf{H} = \mathbf{B} - 4\pi\mathbf{M} \qquad (15)$$

Then Eqs. (1) and (2) take the form

$$\nabla \cdot \mathbf{H} = -4\pi\nabla \cdot \mathbf{M} \qquad (16)$$

$$\nabla \times \mathbf{H} = 4\pi\mathbf{J}_c \qquad (17)$$

At the interface between two magnetic materials, Eqs. (1) and (17) imply continuity of the normal component of $\mathbf{B}$ and of the tangential component of $\mathbf{H}$.

Across a sheet-current of density I_c' per unit length, the tangential component transverse to $\mathbf{J}_c$ of both $\mathbf{H}$ and $\mathbf{B}$ undergoes an increment $4\pi I_c'$. The other components of $\mathbf{H}$ and $\mathbf{B}$ are unaffected.

The field patterns set up by a magnetized sphere are illustrated in Fig. 4. The magnetic induction (a) and magnetic field (b) are identical outside the sphere, but differ inside it, because of the magnetization (c). The same pattern of magnetic induction could be generated in the absence of magnetization by a surface current (d). The source of the magnetic field in Eq. (16), that is to say the quantity $\nabla \cdot \mathbf{M}$, is also referred to as the magnetic pole density. The north and south polar regions are indicated in (c).

For weakly magnetic materials, Eq. (15) can generally be written in terms of a scalar magnetic permeability μ

$$\mu\mathbf{H} = \mathbf{B} \qquad (18)$$

For ferromagnetic materials, one can still write

$$\mu\mathbf{H} = \mathbf{B} - 4\pi\mathbf{M}_0 \qquad (19)$$

where $\mathbf{M}_0$ is a permanent magnetization, but μ now depends on the time history as well as the

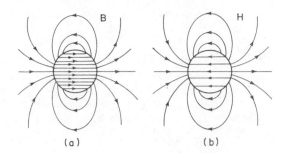

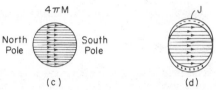

FIG. 4. Magnetic induction (a) and magnetic field (b) arising from sphere with magnetization pattern shown in (c) or with surface current pattern shown in (d).

magnitude of **H**. The typical relation between **B** and **H** for ferromagnetic materials is illustrated in Fig. 5.

When J_c is zero everywhere, one can define a scalar potential Ω, such that

$$\mathbf{H} = -\nabla\Omega \qquad (20)$$

$$\nabla^2\Omega = 4\pi\nabla \cdot \mathbf{M} \qquad (21)$$

If the boundary condition on Ω is simply that it vanish at infinity, the solution is

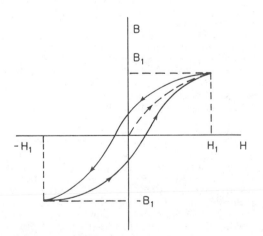

FIG. 5. As magnetic field **H** is initially raised to H_1, magnetic induction **B** rises to B_1. Cyclical pattern shown is typical of ferromagnetic materials: **B** saturates as **H** becomes large; **B** remains finite as **H** returns to null; **B** goes to $-B_1$ as **H** goes to $-H_1$. Double-valuedness of **B(H)** is known as hysteresis.

$$\Omega(\mathbf{r}) = -\int d^3r_1 \, \frac{\nabla_1 \cdot \mathbf{M}(\mathbf{r}_1)}{|\mathbf{r} - \mathbf{r}_1|} \qquad (22)$$

In the presence of current-carrying conductors Eqs. (20) and (21) still hold in the region where $J_c = 0$, but Eq. (17) now implies a multivalued potential

$$\oint d\boldsymbol{\ell} \cdot \mathbf{H} = \oint d\Omega = 4\pi I_c \qquad (23)$$

where the integral is taken around a loop enclosing the total conductor current I_c. To keep the potential single-valued, so that the solution of Eq. (22) remains valid, one may adopt the "magnetic-shell" approach: $\mathbf{J}_c$ is replaced with an equivalent **M**, in analogy with Eq. (11).

Magnetic Force and Energy From Maxwell's stress tensor, we find the volume force

$$\mathbf{f} = -\nabla\left(\frac{B^2}{8\pi}\right) + \frac{1}{4\pi}(\mathbf{B} \cdot \nabla)\mathbf{B} \qquad (24)$$

$$= \mathbf{J} \times \mathbf{B}$$

which agrees with the summation of the Lorentz forces on the moving charges composing **J**. The "magnetic pressure" against a current sheet bounding a region of finite **B** (as in the Meissner effect or ordinary skin effect) is thus $B^2/8\pi$, evaluated at the surface. The force and torque on a body localized in a nearly uniform field are

$$\mathbf{F} = (\mathbf{m} \cdot \nabla)\mathbf{B} \qquad (26)$$

$$\mathbf{N} = \mathbf{m} \times \mathbf{B} \qquad (27)$$

From the microscopic point of view underlying Eq. (2), the magnetic energy density is

$$w = \frac{B^2}{8\pi} \qquad (28)$$

In the presence of magnetic materials, one is more interested in the electrical input energy required to go from $\mathbf{B}_0$ to **B**, and this is given by

$$\Delta w = \frac{1}{4\pi}\int_{\mathbf{B}_0}^{\mathbf{B}} \mathbf{H} \cdot d\mathbf{B} \qquad (29)$$

For $\mathbf{H} = \mu\mathbf{B}$, with constant μ, this becomes

$$\Delta w = \frac{1}{8\pi\mu}(B^2 - B_0{}^2) \qquad (30)$$

The derivation of Eqs. (28) and (29) depends on the complete set of Maxwell's equations.

Units and Magnitudes The equations used here are based on the emu system. If the currents are expressed in amperes, they must be divided by 10 to give magnetic inductions in gauss or fields in oersteds. A magnetic induction of 10 kilogauss (one weber per square meter) can exert a maximum stress $B^2/8\pi$ of about 4 atm, and contains an energy density of 0.4 joules/cm^{-3}.

The strength of the earth's magnetic induction is about 0.2 gauss. For typical ferromagnetic materials, the maximum value of $4\pi M$ is 20 kilogauss; at this point of saturation, the permeability μ approaches unity. At magnetic fields below one gauss, permeabilities of 1000 or more can be reached. The strength of materials limits the magnetic induction obtainable nondestructively with laboratory electromagnets to peak values well below a million gauss.

HAROLD P. FURTH

References

Stratton, J. A., "Electromagnetic Theory," New York, McGraw-Hill Book Co., 1941.

Jackson, J. D., "Classical Electrodynamics," New York, John Wiley & Sons, 1962.

Cross-references: ELECTRICITY, ELECTROMAGNETIC THEORY, FERROMAGNETISM, FIELD THEORY, MAGNETISM.

MAGNETIC RESONANCE

The magnetic resonance phenomenon is the resonant interaction between an oscillating magnetic field and an orthogonal static magnetic field mediated by the presence of objects possessing both angular momentum and a magnetic moment. The objects in question are normally microscopic in character (molecules, atoms, atomic nuclei or subatomic particles: protons, neutrons, electrons, muons, etc.), although historically the effect was first demonstrated with magnetized iron rods.

In atoms and molecules, a magnetic moment may arise from the orbital motions of electrons, in the same manner as an electric current flowing in a loop of wire generates a dipole-like magnetic field. According to quantum theory, angular momentum for such orbital or rotational motions is restricted to values which are integral multiples of $\hbar (\hbar = h/2\pi$, where $h = 6.626176 \times 10^{-34}$ J s is the Planck constant). Thus, we may calculate the magnitude of the magnetic moment μ due to the orbital motion of an electron in the smallest ($n = 1$) orbit of the classical Bohr atom using Ampere's law (μ = current $\times$ area) and the fact that the angular momentum $L = mvr = nh$. The result,

$$\mu_B = \frac{e\hbar}{2m} = 9.274078 \times 10^{-24} \text{ J T}^{-1},$$

where e/m is the magnitude of the electron charge to mass ratio, is known as the Bohr magneton, and is a convenient unit in which to denote atomic and molecular magnetic moments.

The spin, or intrinsic angular momentum, of an elementary particle or a nucleus may take on values which are integral multiples of $\hbar/2$. The nuclear equivalent of the Bohr magneton, based on a classical model of the proton as a rotating sphere of charge is the nuclear magneton

$$\mu_N = 5.050824 \times 10^{-27} \text{ J T}^{-1}.$$

The actual intrinsic magnetic moments for the free electron and proton are respectively $-1.001160 \ \mu_B$ and $+2.792846 \ \mu_N$. The g factor is a dimensionless measure of magnetic moment (in units of $-\mu_B/2$) which is traditionally used for electron-associated moments.

A particle possessing both angular momentum $\mathbf{J}$ and a proportional magnetic moment $\boldsymbol{\mu} = \gamma \mathbf{J} = \gamma I \hbar$ will precess gyroscopically when placed in a magnetic field $\mathbf{B}_0$ in a manner similar to that of a spinning top precessing about the Earth's gravitational field. The proportionality constant γ is known as the *magnetogyric ratio*. The precession rate in radians sec^{-1}, the *Larmor frequency*, may be derived by a classical calculation to be

$$\omega_0 = \gamma B_0. \tag{1}$$

Coincidentally, an exact quantum calculation, using a Hamiltonian $\mathbf{H} = -\boldsymbol{\mu} \cdot \mathbf{B}_0$, yields an energy separation (Zeeman interaction) between adjacent energy levels of a particle with spin angular momentum quantum number I of $E_0 = \hbar \omega_0$. A total of $2I + 1$ states with energies $-\hbar \omega_0 I_z$ occur, I_z being the z component (magnetic) angular momentum quantum number and taking on the values $-I$, $-I + 1$, $\ldots, +I$.

This beautiful correspondence between the quantum and classical descriptions is made even stronger for isolated spins by the fact that a quantum dynamic calculation shows that the expectation values $\langle I_x \rangle$, $\langle I_y \rangle$, and $\langle I_z \rangle$ behave with time in precisely the same manner as the components of angular momentum of a wholly classical particle. Thus, we are entitled to formulate and visualize descriptions of magnetic resonance phenomena in terms of classically precessing vectors.

The vector description of magnetic resonance is due to Rabi. If a spin experiences simultaneously a static magnetic field and a field rotating in a plane perpendicular to the static field, resonance becomes possible. (In practice, linearly polarized transverse oscillatory fields are used; these may be decomposed into two counterrotating fields, one of which moves along with the precessing spin, the effect of the other being unimportant.) The detailed analysis of this situation is conveniently accomplished by a coordinate transformation to a frame of reference rotating synchronously with the rotating field (Fig. 1). By convention, the z-axis of the laboratory and rotating reference frames are collinear with the applied static

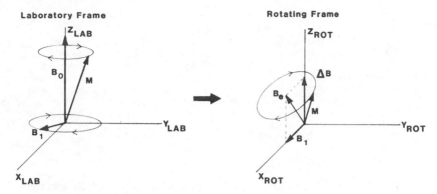

FIG. 1. Rotating coordinate transformation. The rotating reference frame rotates in synchronism with the applied $\mathbf{B}_1$ RF magnetic field. Within the rotating frame, the $\mathbf{B}_1$ field is static, the original $\mathbf{B}_0$ field is replaced by the resonance offset field $\Delta\mathbf{B}$, and the magnetization M precesses about the net field $\mathbf{B}_e$.

field $\mathbf{B}_0 = B_0 \mathbf{k}_{\text{LAB}}$. The rotating field amplitude is B_1, and the laboratory and rotating frame vectors are defined as

$$B_1 (\cos \omega t\; \mathbf{i}_{\text{LAB}} + \sin \omega t\; \mathbf{j}_{\text{LAB}})$$

and

$$B_1 \mathbf{k}_{\text{ROT}}$$

respectively, ω being the angular rotation frequency.

Thus, in the absence of a B_1 field a spin with true Larmor frequency ω_0 appears in the rotating frame to precess at $\omega_0 - \omega \equiv \Delta\omega$. In order to preserve the form of the dynamical equations upon transformation into the rotating frame, we adopt the idea of the spin experiencing an effective field given by $\Delta B = (\omega_0 - \omega)/\gamma$, which acts along $\mathbf{k}_{\text{ROT}}$.

Now the resonance effect becomes clear and simple to analyze. Upon application of a B_1 field, a net effective field $\mathbf{B}_e = B_1 \mathbf{i}_{\text{ROT}} + \Delta B \mathbf{k}_{\text{ROT}}$ results which is no longer parallel to the z-axis. The spin now precesses at rate γB_e about the effective field direction (often denoted by an effective field angle $\theta = \tan^{-1} [B_1/\Delta B]$) and may suffer large excursions in direction. When the frequency of the applied B_1 field is far above or below ω_0 [Fig. 2(b)] the effective field direction is essentially the same as that of the laboratory field, and no significant effects on the spin are observed. The maximum effect occurs when the frequency of the applied B_1 field exactly matches ω_0 [$\Delta B \to 0$, the reso-

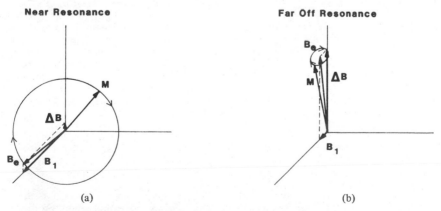

FIG. 2. Magnetic resonance—development of transverse magnetization. (a) On or near resonance ($|\Delta B| \lesssim B_1$) the magnetization M may be nutated from its thermal equilibrium position along z into the xy plane (90° pulse), inverted (180° pulse), and so forth. (b) When the applied RF is far from resonance ($|\Delta B| \gg B_1$) the magnetization is not significantly perturbed.

nance condition, Fig. 2(a)]. In this case it is possible to nutate the spin into the transverse direction and leave it there (a "90° pulse") or invert its orientation entirely (a "180° pulse," etc.).

The phenomenological description of the collective behavior of a system of weakly interacting spins was provided by a set of equations given by Bloch. Although they fail quantitatively in a number of specific instances (most notably for spin systems in solids or other strong-interaction situations), they offer an enormously useful and generally applicable conceptual framework:

$$\frac{dM_x}{dt} = \gamma (\mathbf{M} \times \mathbf{B})_x - \frac{M_x}{T_2}$$

$$\frac{dM_y}{dt} = \gamma (\mathbf{M} \times \mathbf{B})_y - \frac{M_y}{T_2} \qquad (2)$$

$$\frac{dM_z}{dt} = \gamma (\mathbf{M} \times \mathbf{B})_z - \frac{(M_z - M_0)}{T_1}.$$

Basically, the Bloch equations describe the time behavior of the components of the magnetization $\mathbf{M}$ (summation over the individual microscopic magnetic moments) in the rotating frame. M_0 is the magnitude of the magnetization in thermal equilibrium with the "lattice" (i.e., the remaining motational degrees of freedom of the material in which the spin system resides) given by the Curie law

$$M_0 = \frac{CB_0}{T}, \qquad (3)$$

C being the Curie constant and T the absolute temperature. In thermal equilibrium, the magnetization must be parallel to the static field (transverse components zero).

The cross-product terms in Eq. (2) represent the precessional or nutational behavior discussed above. The remaining terms are damping terms which describe the tendency of the system to return to thermal equilibrium. The time-dependent solution of the equations starting with an arbitrary $\mathbf{M}$ yields a z-component M_z which decays exponentially toward M_0 with a time constant T_1 (the longitudinal or "spin-lattice" relaxation time). The transverse components M_x and M_y decay exponentially to zero with time constant T_2 (transverse or "spin-spin" relaxation time). Longitudinal relaxation involves exchange of energy with the lattice, while transverse relaxation involves energy exchange between spins or dephasing of precessing components; hence the differentiation in relaxation times. It must always be true that $T_1 \geqslant T_2$. A low power $(B_1 \ll (\gamma^2 T_1 T_2)^{-1/2}$ steady-state solution gives a nonzero transverse magnetization parallel to B_e which may exhibit saturation (disappearance of the transverse component for sufficiently large values of B_1 near resonance).

The Bloch equations also suggest ways of experimentally observing magnetic resonance phenomena. The CW (continuous wave) technique is based on the steady-state solution. The static transverse component of $\mathbf{M}$ in the rotating frame corresponds in the laboratory frame to an oscillating transverse magnetization at the frequency of the applied field. It may be observed as a net power absorption by the sample or the appearance of a coupling between two orthogonal transverse radiofrequency (RF) coils containing the sample as field and applied RF are slowly brought into resonance according to Eq. (1) (either field or frequency may be swept). The transient response may be observed directly after a suitable (e.g., 90°) pulse has been applied. The resulting oscillatory decaying magnetization induces a voltage in a coil surrounding the sample (which may be the same coil used to apply the pulse). This signal is known as the *free induction decay* (FID), and contains components from all spins which were significantly affected by the pulse. The CW spectrum may be recovered by Fourier transformation of the FID (see FOURIER ANALYSIS).

The CW method is traditional for observation of nuclear magnetic resonance (NMR) and is still almost exclusively employed for electron spin resonance (ESR, or electron paramagnetic resonance, EPR). Over the past decade (1970s), pulsed, or Fourier transform, techniques have gradually supplanted CW in chemical NMR applications because of the higher potential sensitivity for a given measurement time (all of the spectral information is "captured" in one short time interval) and the convenience with which computer-based signal averaging may be adapted. (See Figs. 3–5.)

The great preponderance of magnetic resonance measurements are in NMR and ESR. Every chemical element has at least one observable isotope. Common examples are listed in Table 1. ESR measurements may be made on any substance with net unpaired electron spin density. Examples are organic free radicals, high-spin metal ions, conduction electrons in metals, some doped semiconductors, as well as systems with long-range collective interactions: ferromagnetic, antiferromagnetic, and ferrimagnetic materials. The spontaneous polarization of ferromagnetic systems allows the occurrence of NMR and ESR without an external B_0 field.

Several types of interactions affect NMR and ESR spectra. The presence of the B_0 field induces a circulation of electrons in materials which in turn gives rise to a small opposing field. This field-proportional shielding effect depends sensitively upon the chemical environment and orientation at the spin under observation. In NMR the effect is called the *chemical shift*, and is always reported with respect to a reference compound of the isotope in question;

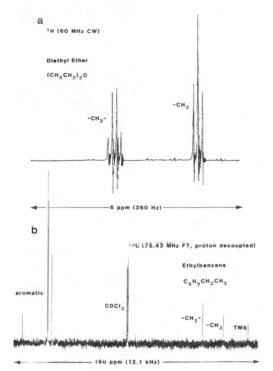

FIG. 3. Typical solution NMR spectra. (a) Proton 60 MHz (B_0 = 1.409 T) continuous wave spectrum of diethyl ether. Positions of spectral patterns are due to the chemical shift, line splittings within patterns are due to J couplings. Integral of each spectral pattern is proportional to the number of protons giving rise to that resonance. (b) ^{13}C 75.43 MHz (B_0 = 7.046 T) Fourier transform spectrum of ethylbenzene. Couplings to protons are removed by continuous irradiation of protons simultaneous with acqusition of carbon free induction decay. Carbon-carbon couplings do not appear because the low concentration of the isotope about 1% of ^{12}C) makes the occurrence of ^{13}C pairs rare. Solvent is deuterochloroform, $CDCl_3$. Chemical shift reference, tetramethylsilane (TMS), $(CH_3)_4Si$, has been added.

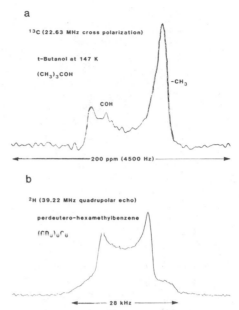

FIG. 4. Typical solid state Fourier transform NMR spectra. (a) ^{13}C spectrum of tertiary butanol, employing proton decoupling [see Fig. 3(b)]. The widths and shapes of the lines are due to the anisotropy of the chemical shift. (b) ^{2}H spectrum of deuterated hexamethylbenzene. The width and shape of this spectrum are due to the anisotropy of the quadrupole splitting interaction.

typical ranges of chemical shifts vary from 10 ppm (parts per million of field or frequency) for the three hydrogen isotopes to 250 ppm for carbon to several thousand ppm for the heavier elements or for shifts arising from conduction electrons in metals (these are called *Knight shifts*). This exquisite sensitivity to chemical effects has made NMR one of the most powerful and ubiquitous techniques of chemical analysis and research. Shieldings in ESR are reported as g values, which range from roughly 1.9 to 2.2. Shifts arising from bulk magnetic susceptibility are also observed.

Interactions also occur between all the magnetic dipoles in systems, such as between nuclei

of the same isotope (homonuclear dipole-dipole), between different isotoptes (heteronuclear dipole-dipole), between nuclei and unpaired electrons, or between electrons. These interactions may occur directly through space, or may be "conducted" through chemical bonds as a slight bias in spin polarization (*indirect* or *J coupling* when speaking of NMR, *hyperfine coupling* when speaking of couplings to nuclei in ESR spectra).

Couplings between molecular rotation-induced moments and nuclei or electrons can also occur (spin-rotation coupling). These are usually most apparent in their effect on relaxation (see next paragraph). Nuclei with $I > \frac{1}{2}$ have nonspherical charge distributions, and therefore interact with electric field gradients in materials. Quadrupole couplings can range from zero to small perturbations on NMR spectra to values which completely dominate the nuclear Zeeman energy. In the latter case, the quadrupole interaction can serve as the source of nuclear polarization rather than the B_0 field, making possible "zero-field" NMR (normally called nuclear quadrupole resonance, NQR).

All of these interactions may be time dependent due to atomic or molecular motions, or due to natural or experimentally induced mo-

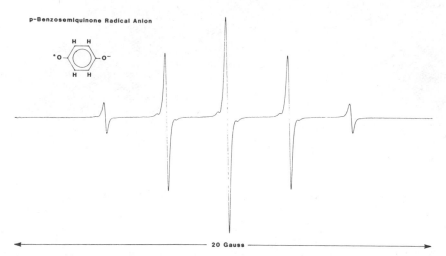

FIG. 5. Typical solution ESR spectrum. Most ESR spectra are obtained by a continuous wave method in the derivative mode. Spectrum of parabenzosemiquinone radical anion in alkaline ethanol. Line splitting (hyperfine coupling to the four protons) is 2.368 gauss, g-value is 2.005.

TABLE 1. MAGNETIC RESONANCE DATA FOR SELECTED PARTICLES.

Particle	Spin	Magnetic Moment in Units of μ_N[a]	Larmor Frequency in 2.3487-T Field, MHz	Electric Quadrupole Moment,[b] $\times 10^{-28}$ m^2	Natural Isotopic Abundance	Relative Receptivity per Particle[c]	Relative Receptivity at Natural Abundance[d]
e	$\frac{1}{2}$	1.7340593 μ_B	65,821.07	0	1.000	2.08×10^8	2.08×10^8
μ	$\frac{1}{2}$	1.7340706 μ_μ	318.33	0	0	32.3	0
n	$\frac{1}{2}$	-3.313670	68.51	0	0	0.322	0
^{1}H	$\frac{1}{2}$	4.873505	100.00	0	1.000	1.000	1.000
^{2}H	1	1.2125	15.35	2.73×10^{-3}	1.5×10^{-4}	9.65×10^{-3}	1.45×10^{-6}
^{3}H	$\frac{1}{2}$	5.1595	106.66	0	0	1.21	0
^{13}C	$\frac{1}{2}$	1.2166	25.15	0	1.11×10^{-2}	1.59×10^{-2}	1.76×10^{-4}
^{14}N	1	0.5706	7.22	0.016	0.996	1.01×10^{-3}	1.01×10^{-3}
^{15}N	$\frac{1}{2}$	-0.4900	10.13	0	3.7×10^{-3}	1.04×10^{-3}	3.85×10^{-6}
^{19}F	$\frac{1}{2}$	4.5509	94.08	0	1.000	0.833	0.833
^{23}Na	$\frac{3}{2}$	2.8610	26.45	0.12–0.15	1.000	9.25×10^{-2}	9.25×10^{-2}
^{29}Si	$\frac{1}{2}$	-0.9612	19.87	0	4.70×10^{-2}	7.85×10^{-3}	3.69×10^{-4}
^{31}P	$\frac{1}{2}$	1.9581	40.48	0	1.000	6.63×10^{-2}	6.63×10^{-2}
^{43}Ca	$\frac{7}{2}$	-1.4914	6.73	0.2 ± 0.1	1.45×10^{-3}	6.40×10^{-3}	9.28×10^{-6}
^{121}Sb	$\frac{5}{2}$	3.9537	23.93	-0.5 to -1.2	0.573	0.160	9.17×10^{-2}

[a]Magnitude of magnetic moment vector $\mu = \gamma \hbar [I(I+1)]^{1/2}$. Positron and muon moments in units of their respective magnetons. Values for all nuclei except ^{1}H are those observed in specific chemical compounds, uncorrected for shielding.

[b]Electric quadrupole moments are often known with only poor accuracy, due to certain assumptions which must be made in their experimental determination, and due to conflicting results from different experimental techniques. The table values represent a range from several sources.

[c]Approximate detection voltage signal to noise ratio, assuming equal B_0 fields, line shapes, relaxation times, measurement bandwidths, and total measurement noise, relative to protons; proportional to $\omega_0{}^3 I(I+1)$.

[d]Product of natural isotopic abundance and relative receptivity per particle.

tions of some of the magnetic dipoles in the system. Depending on the type of interactions, such motions can be the source of relaxation (e.g., T_1 and T_2 in the Bloch equations) between specific types of spins or between spins and the lattice.

Aside from the chemical and physical research and measurement, other applications of magnetic resonance include measurement of magnetic fields (proton magnetometers) and the use of ferrites (high RF resistivity ferrimagnetic ceramics of general formula $MOFe_2O_3$, M a divalent cation) as magnetic field-controllable microwave switches, phase shifters, attenuators, filters, circulators and isolators.

Most recently, magnetic resonance has been employed for noninvasive clinical imaging of the human body (zeugmatography). Lauterbur first demonstrated that the NMR spectrum of a compound with a single spectral line (e.g., protons in intracellular water) is the projection of the distribution of the compound in the object if a linear B_0 field gradient, rather than a uniform B_0 field, is applied across the object. Multiple projections in several directions may be used to reconstruct an image using algorithms similar to those employed in x-ray or emission computed tomography.

The formal description of magnetic resonance given above has been applied (by Feynman, Vernon, and Hellwarth) to the generalized two-level quantum system, and has proved extremely useful in understanding coherence phenomena in microwave, infrared and optical spectroscopies.

JEROME L. ACKERMAN

References

Carrington, A., and McLachlan, A. D., "Introduction to Magnetic Resonance with Applications to Chemistry and Chemical Physics," New York, Harper and Row, 1967.

Slichter, C. P., "Principles of Magnetic Resonance," Berlin, Springer-Verlag, 1978.

Abragam, A., "The Principles of Nuclear Magnetism," Oxford, Oxford Univ. Press, 1961.

Becker, E. D., "High Resolution NMR," New York, Academic Press, 1980.

Helszain, J., "Principles of Microwave Ferrite Engineering," London, Wiley-Interscience, 1969.

Kaufman, L., Crooks, L. E., and Margulis, A. R. (Eds.), "Nuclear Magnetic Resonance Imaging in Medicine," New York, Igaku-Shoin, 1981.

Lee, K., and Anderson, W. A., "Nuclear Spins, Moments and Magnetic Resonance Frequencies," in "The Handbook of Chemistry and Physics," (R. C. Weast, Ed.), Cleveland, The Chemical Rubber Company.

Steinfeld, J. I., "Molecules and Radiation," New York, Harper & Row, 1974.

Cross-references: COHERENCE, ELECTRON SPIN, FERRIMAGNETISM, FOURIER ANALYSIS, LASERS, MAGNETISM, MASERS, RESONANCE.

MAGNETISM

Magnetization Magnetic fields are produced both by macroscopic electric currents and by magnetized bodies. The first observed manifestations of magnetism were the forces between naturally occurring permanent magnets, and between these and the earth's field. North- and south-seeking poles could be identified. Poles were observed to be localized near the ends of long rods magnetized by contact with natural magnets or by a current-carrying coil. From the observed attraction and repulsion of unlike and like poles with an inverse square law came the concept of pole strength and the definition of the unit pole, that which acts on another in vacuum with a force of one dyne at a distance of one centimeter. The unit magnetic field, the oersted, could then be defined as that in which a unit pole experiences a force of one dyne. The magnetic moment of a long, uniformly magnetized rod of length l with a pole strength of m unit poles at each end is defined as ml, the largest couple that the sample can experience in a field of one oersted. The magnetization, M, is defined as the magnetic moment per unit volume, ml/al, where a is the cross-sectional area, and is thus also equal to the pole strength per unit area, m/a.

Magnetic Induction The induction, or flux density, B, is numerically equal to the field H in free space and is described as one line of flux per square centimeter for a field of one oersted. Its direction is that of the force on a unit north pole. If magnetic material is present, the flux density is equal to $H + 4\pi M$, since 4π lines of force emanate from the unit pole at each end of a dipole equivalent to a specimen of unit magnetization. Magnetic poles are observed to occur in pairs. Lines of B are continuous, i.e., div $B = 0$. If a material becomes strongly magnetized in a small field, the lines of flux can be considered to crowd into the material, leaving their original locations and reducing the field there. This is how magnetic shielding is accomplished. Changes of B within a coil induce voltages which can be measured and form the basis of galvanometer and fluxmeter measurement methods. For a coil of N turns of cross-sectional area a, in which the flux is changing at dB/dt gauss/sec, E in volts is given by

$$E = -10^{-8} Na \frac{dB}{dt}$$

Forces on magnetic bodies in field gradients are proportional to M.

Types of Magnetic Behavior In general a field H will produce a magnetization M in any material. If M is in the same direction as H, a sample will be attracted to regions of stronger field in a field gradient. It will be repelled if M is in the opposite sense. This experiment, as first performed by Faraday, is the basis for the broad classification of materials into paramagnetic, diamagnetic, and ferromagnetic. The susceptibility κ is defined as M/H. The force F_x on a small specimen of volume v in a field H_y and a field gradient dH_y/dx is

$$F_x = (\kappa_2 - \kappa_1) v H_y \frac{dH_y}{dx}$$

where κ_2 and κ_1 are the volume susceptibilities of the specimen and the surrounding medium, usually air.

For paramagnetic materials, κ is small and positive, usually between 1 and 1.001 at ordinary temperatures. These substances contain atoms or ions with at least one incomplete electron shell, giving them a non-zero atomic or ionic magnetic moment μ_a. Many salts of the iron-group and rare-earth metals are paramagnetic, as are the alkali metals, the platinum and palladium metals, carbon, oxygen, and various other elements. Antiferromagnetic substances also have small positive κ, as do ferromagnetics above their Curie temperatures. In the classical theory of paramagnetism, the orientations of the moments are considered to be initially thermally randomized in space. An applied field produces a net magnetic moment in its direction, as described by the classical Langevin function

$$\frac{M}{M_s} = \coth\left(\frac{\mu_a H}{kT}\right) - \frac{kT}{\mu_a H}$$

where k is the Boltzmann constant. M_s is the value of M attained for very large H/T. Under most conditions, only the initial portion of this curve is observed, with the corresponding constant κ. The conduction electrons at the top of the Fermi distribution in a metal can also give rise to a temperature-independent Pauli paramagnetism. The quantum-mechanical analogue of the Langevin function is called the Brillouin function (see PARAMAGNETISM).

For diamagnetic materials, κ is small and negative. Diamagnetism is a universal phenomenon but is often masked by paramagnetic or ferromagnetic effects. Net diamagnetic behavior is observed in a number of salts and metals, and in the rare gases, in which there is no net moment. The effect can be regarded as the operation of Lenz's law on an atomic scale (see DIAMAGNETISM).

Ferromagnetic materials show a value of M which may be of the order of 10^3 in small fields. Thus κ can be very large. It is common to describe their properties in terms of the permeability $\mu = B/H$. Since M saturates in ordinary fields, κ and μ are not constant. M is not necessarily in the same direction as H, so κ and μ are, in general, tensors. Furthermore, ferromagnetics generally exhibit hysteresis in the dependence of M on H, and the details are very structure-sensitive. Still another distinction is the rather abrupt disappearance of ferromagnetism at a characteristic temperature, the Curie temperature, T_c.

Atomic Magnetic Moments There are two possible sources for the moments of individual atoms. They are electron orbital motion and electron spin (see ELECTRON SPIN). In most ferromagnetic materials, most of the moment comes from spin rather than orbital motion, a fact that is revealed experimentally by gyromagnetic measurements (see FERROMAGNETISM) and by magnetic resonance experiments (see MAGNETIC RESONANCE). Orbital motions are quenched by the electric fields of the neighboring atoms in the crystal lattice. In the rare earth metals the unfilled shell is deep within the atom, orbital motion is not quenched, and the orbital contribution to the magnetic moment is observed. The unit of atomic moment is the Bohr magneton, μ_B, which is the moment associated with one electron spin, numerically equal to 0.9274×10^{-20} erg/oersted. The spin quantum number, S, is one-half the number of unpaired electrons. The moment per atom is $S g \mu_B$ where g is the gyromagnetic ratio, close to 2 for most materials. The moment in Bohr magnetons of an isolated atom or ion of the first transition series is equal to the number of unpaired d electrons, considering the first five electron spins to have one orientation and the next five the opposite (Hund's rule). The Ni^{++} ion, with eight d electrons, has the expected moment of $2\mu_B$ in ferrites, in which the ionic spacing is great enough so that the d levels are not disturbed (see FERRITES). In metallic nickel, however, the d levels overlap considerably, and the moment corresponds to only 0.6 μ_B per atom. Similarly the Bohr magneton numbers for metallic iron and cobalt are 2.2 and 1.7 respectively.

Ferromagnetism Ferromagnetism can only occur in a material containing atoms with net moments. Also, quantum-mechanical electrostatic "exchange" forces must be present, holding neighboring atomic moments parallel below the Curie temperature. These are much greater than the Lorentz force due to the average magnetization and are, in fact, equivalent to an effective field on the order of 10^6 oersteds. Such an effective "molecular field" was postulated in 1907 by Weiss in extending the Langevin theory of paramagnetism to include ferromagnetic behavior. The Langevin function predicts a temperature dependence of magnetization, for small M, of

$$M = \frac{CH}{T}$$

where C is a constant. The susceptibility is then C/T, which is Curie's law. Weiss pointed out that if the field H were augmented by an additional field NM proportional to the magnetization, the temperature dependence became

$$M = \frac{CH}{T - T_c}$$

where $T_c = NC$. This is the Curie-Weiss law, approximately obeyed by ferromagnetic substances above their Curie points. Below T_c, the presence of the molecular field produces an alignment of the atomic moments corresponding to the spontaneous magnetization M_s even when no external field is present. However, ferromagnetic samples can have any net externally measured value of magnetization, including zero, which seems to contradict this result. Weiss therefore postulated the existence of domains separated by boundaries. In each domain the atomic moments are parallel, the domain magnetizations having different orientations. The net external magnetization is then the vector sum of the domain magnetizations and can be varied by a rearrangement of the domain structure, which may happen in very small applied fields. This prediction has been completely verified by experiment. The motion of domain boundaries as observed under the microscope has been directly correlated with external changes in magnetization. Domain boundaries in iron are on the order of 1000 Å thick. Within a boundary, neighboring magnetic moments are not quite parallel. The change in orientation of the magnetization from one domain to another is distributed through the thickness of the boundary.

Within a domain, the magnetization will in general preferentially lie along some particular crystallographic direction. The energy difference between magnetization in the easiest and hardest direction may exceed 10^8 erg/cm^3. This anisotropy is described in an appropriate trigonometric series with coefficients K_i. Usually only a few terms are necessary. Often a material is described by a single K; this implies a uniaxial anisotropy energy of the form $K \sin^2 \theta$. The K_i may pass through zero and change sign with changing composition or temperature. Although such details cannot in general be predicted, the magnetocrystalline anisotropy will have the same over-all symmetry as the crystal structure. Anisotropy is best investigated in single crystals, by analysis of magnetization curves in various directions or from the relationship between the measured torque and the direction of the applied field (see FERROMAGNETISM). Dimensional changes are also associated with the position of the magnetization vector relative to the lattice (see MAGNETOSTRICTION).

There are two mechanisms available for changing the externally measured magnetization of a ferromagnetic material: domain boundary motion, and domain magnetization rotation. Broadly speaking, in magnetically soft materials, boundary motion accounts for most of the changes in low applied fields, leaving the magnetization in each domain in the easy direction nearest the applied field. Then rotation against anisotropy produces the remaining change in higher fields. In very low fields, boundary motion is practically reversible, but when boundaries move considerable distances, they experience a net drag from impurities and irregularities in the material, causing hysteresis in the dependence of B on H. There will in general be a remanence B_r, the flux density remaining after saturation when the field is reduced to zero, and a coercive force H_c, the reverse field required to reduce the flux density to zero. A loss associated with the irreversibility of magnetization changes also occurs in rotating fields. This loss becomes zero in very large fields, except in a few special cases.

Even in very slowly changing fields, a wall characteristically moves in jumps, each giving a sudden change in B. This irregularity has been known for a long time as the Barkhausen effect, and its physical origin is the irregularity of wall motion through various inhomogeneities in the material. Usually a very large number of these small jumps takes place. In special circumstances, however, the material may remain at B_r, until, in a sufficiently large field, a single wall will be nucleated and sweep all the way across the specimen, leaving it at B_r in the other direction. Such a material has only two stable states, $+B_r$ and $-B_r$, a useful behavior in some applications.

Direct microscopic observation of domains, e.g., by the Faraday effect or the magnetic Kerr effect, is an important research tool. Understanding and control of domain structures has progressed to the point that under appropriate conditions large numbers of tiny cylindrical domains can be deliberately produced and controllably moved in certain single-crystal materials, enabling the development of memory devices utilizing this ability.

It is also necessary to consider the behavior in rapidly varying fields, discussed below.

Antiferromagnetism Exchange forces can operate to hold neighboring moments antiparallel, rather than parallel. Materials whose magnetic moments are arranged in this way show no external permanent moment and are called antiferromagnetic. The sign of the exchange force may depend, among other things, on the atomic spacing. Metallic manganese, for example, is antiferromagnetic, while many alloys of manganese, in which the average Mn-Mn distance is greater, are ferromagnetic. In some antiferromagnetic compounds, the exchange interaction appears to be of a next-nearest-neighbor type, taking place through an intervening atom such as oxygen. This type of interaction is termed superexchange. Antiferromagnetic materials, having no net external moment, show small positive susceptibilities that reach a maximum at the temperature above which the exchange forces can no longer hold the moments aligned against thermal agitation. This temperature, T_N, the Néel temperature, corresponds to the Curie temperature of a ferromagnet. Magnetocrystalline anisotropy exists for antiferromagnets just as for ferromagnets (see ANTIFERROMAGNETISM).

Ferrimagnetism With more than one type of

magnetic ion present, in certain compounds, antiferromagnetic coupling may lead to a net external moment corresponding to a Bohr magneton number equal to the difference in ionic moments. Other more complicated cases occur. Ferrites, insulating oxides with the spinel structure, are important examples of this class of material, called ferrimagnetics (see FERRIMAGNETISM).

Exchange Anisotropy A ferromagnetic phase may be in exchange coupling with an antiferromagnetic phase, as in a cobalt particle covered with CoO. This leads to new phenomena, including non-vanishing high-field rotational hysteresis. Such a material cooled in a field through the Néel temperature, if $T_c > T_N$, may exhibit a hysteresis loop that is permanently displaced from the origin. This is equivalent to a unidirectional (not uniaxial) anisotropy and will appear in a torque curve as a $\sin \theta$ term. Ferromagnetic and antiferromagnetic regions in a single-phase alloy may also lead to these effects.

Other Configurations Atomic moments need not necessarily be either parallel or antiparallel. In a few materials they may be arranged in a triangular or spiral configuration. In some circumstances, an antiferromagnetic material may shift to a configuration having a large ferromagnetic moment in the appropriate combination of fields and temperatures (metamagnetism).

Rare Earths The rare earth elements have magnetic moments originating from unpaired electrons in the $4f$ shell. These electrons are close to the nucleus and are shielded by the $5s$ and $5p$ electrons. Thus direct exchange does not occur in the rare earths. However, several of them exhibit ferromagnetism at low temperatures, originating in indirect exchange via the three $5d$–$6s$ conduction electrons. The atomic moments can be large in the heavy rare earths, in which the spin and orbital moments add. In fact, Dy and Ho have a moment per unit volume, at low temperatures, half again as large as that of iron. The rare earths often exhibit complex magnetic ordering structures.

The rare earths form many solid solutions among themselves, in which the variation of moment and Curie temperature have been investigated. They also form many intermetallic compounds with other elements. Often a rare earth and another element will form several discrete binary compounds, sometimes showing extraordinary magnetic properties. A number of compounds, including RCo_5, have extremely high magnetocrystalline anisotropy. $TbFe_2$ and $DyFe_2$ exhibit the highest magnetostrictive strains known, on the order of 1%, thousands of times higher than values typical of other materials.

Amorphous Materials It is possible by rapid quenching from the molten state to prepare metallic samples whose atomic structure is amorphous. The atoms are not arranged in a crystal lattice but are randomly packed as in a glass. The saturation magnetizations and Curie temperatures of these materials, although generally somewhat less than those of their crystalline counterparts, are substantial. The demonstration of ferromagnetism in a glassy metallic structure is of great fundamental interest. Furthermore, in such a structure there is no macroscopic magnetic anisotropy. As a result, magnetization changes can take place readily in small fields. Thus these materials can show high permeabilities and narrow hysteresis loops, giving them potential usefulness in various devices.

Permanent Magnets A useful permanent magnet material should have as large a hysteresis loop as possible. In the early magnet steels, wall motion was made difficult by a heterogeneous alloy structure. A different approach is based on the theory that sufficiently small particles should find it energetically unfavorable to contain domain boundaries. The critical size is proportional to $K^{1/2}/M_s$. Reversal must then proceed by the difficult process of rotation against shape, strain-magnetostriction, or crystal anisotropy. Fine-particle (~ 1000Å) iron and iron-cobalt materials utilizing shape anisotropy have been developed. The Alnico permanent magnet alloys have very fine precipitate structures and are probably also best regarded as fine-particle materials. A magnetic oxide, $BaO \cdot 6Fe_2O_3$, utilizes magnetocrystalline anisotropy in fine-particle ($\sim 1 \ \mu m$) form. A new class of permanent magnet materials based on Co_5-(rare earth)$_2$ intermetallic compounds shows by far the highest permanent magnet properties of any material. These originate in the extremely high magnetocrystalline anisotropy of these materials.

Thin Films Since a surface atom's surroundings are different from those in the interior, the magnetization and Curie temperature of thin films should yield important information about the range of ferromagnetic interactions. Experimental difficulties, primarily with purity, have beclouded the subject to some extent, but it now appears that any surface layer on nickel having substantially different magnetic properties from the bulk cannot be more than a few Angstroms thick.

There have been many investigations of flux reversal in films, usually vapor-deposited on glass, which have been motivated by computer technology needs. Such films show a uniaxial anisotropy associated with fields present during deposition or sometimes with geometric effects such as the angle of incidence of the vapor beam.

Dynamic Behavior of Ferromagnetic Materials Changes in flux in a conductor induce emf's resulting in current flows whose fields tend to oppose the change in flux. For various time rates and geometries these can be calculated, leading to expressions for phase relationships and skin depth in conductors (see ELECTRO-MAGNETIC THEORY). These expressions have often been applied to magnetic materials at power frequencies by simply replacing H by B. This is in general not a good approximation and leads to erroneous results. It is more nearly correct to recognize that highly localized eddy currents around moving domain boundaries are the entire source of loss under these conditions. For a given dB/dt, the loss calculated in this way is much greater, decreasing to the classical value as the density of domain boundaries increases.

In bulk metals, domain wall velocities are usually determined by the damping associated with local eddy currents. In ferrites and thin films, other types of damping may predominate. These and many other aspects of the dynamic behavior of magnetic materials of all types have been investigated through resonance methods (see MAGNETIC RESONANCE).

Superparamagnetism For particles whose volume v is on the order of 10^{-18} cm^3 or less, the direction of the entire particle moment $M_s v$ may fluctuate thermally. An assembly of such particles will exhibit the Langevin function magnetization curve of a paramagnetic with the extremely large moment $M_s v$; thus it may be easily saturated with ordinary fields and temperatures. Such magnetization curves can be used to study particle sizes and size distributions.

Magnetic Bubbles A remarkable application of the principles of domain structure has been realized in the fabrication of computer memory components utilizing tiny cylindrical magnetic domains. In thin monocrystalline plates of a material such as a garnet or in thin amorphous films of certain alloys, having an appropriate combination of magnetization and anisotropy, it is possible to establish stable cylindrical magnetic domains passing completely through the material. These domains, universally referred to as *bubbles*, can be generated, erased, moved, and sensed by overlays of conducting strips and Permalloy guide patterns. Each bubble is a "bit" of information. The dimensions of these bubbles and the associated patterns are on the order of a few microns. The storage density and access times of these memories appear to fit them to an important range of applications. (See FERRIMAGNETISM.)

JOSEPH J. BECKER

References

Bozorth, R. M., "Ferromagnetism," New York, Van Nostrand Reinhold, 1951.

Kneller, E., "Ferromagnetismus," Berlin, Springer-Verlag, 1962.

Rado, G. T., and Suhl, H. (Eds.), "Magnetism," New York, Academic Press. Vol. I, 1963; Vol. IIA, 1965; Vol. IIB, 1966; Vol. III, 1963; Vol. IV, 1966.

Chikazumi, S., "Physics of Magnetism," New York, John Wiley & Sons, Inc., 1964.

Morrish, A. H., "The Physical Principles of Magnetism," New York, John Wiley & Sons, Inc., 1965.

Berkowitz, A. E., and Kneller, E. (Eds.), "Magnetism and Metallurgy," New York, Academic Press, 1969.

Nesbitt, E. A., and Wernick, J. H., "Rare Earth Permanent Magnets," New York, Academic Press, 1973.

Bobeck, A. H., Bonyhard, P. I., and Geusic, J. E., "Magnetic Bubbles—An Emerging New Memory Technology," *Proc. IEEE* **63**, 1176–1195 (1975).

Wohlfarth, E. P. (Ed.), "Ferromagnetic Materials," Vols. 1 and 2, New York, North-Holland Publishing Co., 1980.

Cross-references: AMORPHOUS METALS, ANTIFERROMAGNETISM, DIAMAGNETISM, FERRIMAGNETISM, FERROMAGNETISM, MAGNETIC FIELDS, MAGNETIC RESONANCE, MAGNETOMETRY, RARE EARTHS, THIN FILMS.

MAGNETO-FLUID-MECHANICS

Magneto-fluid-mechanics is the subject that deals with the mechanics of electrically conducting fluids (such as ionized gases and liquid metals) in the presence of electric and magnetic fields. Magneto-hydrodynamics is another name used extensively, but it suffers from the less general meaning of the words "hydro" and "dynamics." Other names used are: magneto-hydro-mechanics, magneto-gas-dynamics, magneto-plasma-dynamics, hydromagnetics, etc.

The fundamental assumptions underlying magneto-fluid-mechanics are those of continuous media. In this respect, magneto-fluid-mechanics is related to plasma physics (see PLASMAS) in the same way that ordinary fluid mechanics is related to the kinetic theory of gases. More specifically, such phenomenological coefficients as viscosity, thermal and electrical conductivities, mass diffusivities, dielectric constant, etc., are assumed to be known functions of the thermodynamic state, as derived from microscopic considerations or experiments.

From electromagnetic theory, we know that the "Maxwell stresses" give rise to a body force made up of the following components: electrostatic (applied on a free electric space charge); ponderomotive (the macroscopic summation of the elementary Lorentz forces applied on charged particles); electrostrictive (present when the dielectric constant is a function of mass density); a force due to an inhomogeneous electric field and its magnetic counterpart; and the magnetostrictive force. For any fluid the last two forces are negligibly small at normal temperatures, whereas the ones associated with the behavior of the dielectric constant, although normally small, are of the same order of magnitude as the buoyant forces under certain conditions. On the assumption that we deal with electrically neutral but ionized fluids, the only substantial force that remains is the ponderomotive force. Indeed, what is today called magneto-fluid-mechanics deals almost exclusively with this force.

Fundamental Equations The equations that govern magneto-fluid-mechanics are the following:

(1) Equation of conservation of mass, which is the same as in ordinary fluid mechanics.

(2) Equation of conservation of momentum, which is altered by the forces enumerated above. In particular, the ponderomotive force per unit volume is given by $\mathbf{J} \times \mathbf{B}$ where $\mathbf{J}$ is the vector current density and $\mathbf{B}$ the magnetic induction, both measured in the laboratory.

(3) Equation of energy conservation; the same as in ordinary fluid mechanics with the addition of the Joulean dissipation $\mathbf{E}' \cdot \mathbf{J}'$. The primes indicate that the electric field and current density are measured in a frame of reference moving with the fluid. In the nonrelativistic case and for zero space charge, we have $\mathbf{E}' = \mathbf{E} + \mathbf{q} \times \mathbf{B}$ and $\mathbf{J}' = \mathbf{J}$. The barycentric stream velocity is indicated by $\mathbf{q}$.

(4) Equation describing the thermodynamic state.

(5) Conservation of electric charge.

(6) Ampère's law.

(7) Faraday's law.

(8) Statement that the magnetic poles exist in pairs only.

(9) Ohm's phenomenological law.

(10) Constitutive equations linking the electric field with the displacement vector and the magnetic intensity field with the magnetic induction. Equations (1) to (3) are the conservation equations. Equations (5) to (8) are Maxwell's equations. For a large number of problems, the phenomenological coefficients of electrical and thermal conductivity, viscosity and the like are assumed to remain unaffected by the magnetic field. This implies that the collision frequencies among the particles are much higher than the cyclotron frequency associated with the property a charged particle has to rotate around a magnetic line under the influence of the Lorentz force. This means that the transfer of electric charge, mass, momentum, and energy is not realized in a preferential direction.

Physically, the magneto-fluid-mechanic system of the above equations is coupled in the following sense: A velocity field $\mathbf{q}$ cutting magnetic lines of flux $\mathbf{B}$ gives rise to an induced current whose magnitude is given by $\mathbf{J} = \sigma(\mathbf{q} \times \mathbf{B})$. At the same time the fluid feels an induced body force equal to $\mathbf{J} \times \mathbf{B}$. On the other hand, the electric currents induced by the motion create, according to Ampère's law, a magnetic field which distorts the original applied magnetic field. The basic mechanism of this distortion is the one created by the irreversibility introduced by the finite electrical conductivity, the same way that the distortion of the inviscid streamlines in ordinary fluid mechanics takes place by the action of viscosity.

Nondimensional Parameters and Some Important Theorems In order to study the nature of the solutions as they emerge from different problems, we shall form a number of nondimensional parameters that can be extracted from the different equations. The order of magnitude of the inertia force per unit volume is given by $\rho V^2/L$ where ρ is the mass density, V the velocity, and L a characteristic length; the order of magnitude of the ponderomotive force $\mathbf{J} \times \mathbf{B}$, after using Ohm's law, is equal to $\sigma B^2 V$. Also, the order of magnitude of the viscous force is: $\mu V/L^2$. The ratio of the typical inertia force over the viscous force is called the Reynolds number (Re) and, from the above, is found to be: $Re = \rho V L/\mu$. The ratio of the ponderomotive force over the inertia force is given by $\zeta = \sigma B^2 L/\rho V$. The ratio of the ponderomotive force over the viscous force is equal to $(Re) \cdot \zeta$ and is defined in the literature as the square of the "Hartmann number," denoted by M. We have $M = BL\sqrt{\sigma}/\sqrt{\mu}$.

The distortion of the magnetic field due to the hydrodynamic field can be studied best with the help of the following two equations:

$$\frac{d\Omega}{dt} = (\Omega \cdot \nabla)\mathbf{q} + \frac{\mu}{\rho} \nabla^2 \Omega$$

$$\frac{d\mathbf{H}}{dt} = (\mathbf{H} \cdot \nabla)\mathbf{q} + \frac{1}{\sigma\mu_e} \nabla^2 \mathbf{H}$$

In the above μ_e is the magnetic permeability and $\mathbf{H}$ the magnetic field intensity. The first equation describes the diffusion of vorticity $\Omega = \nabla \times \mathbf{q}$, whereas the second can be obtained by a combination of Ampère's and Ohm's laws after elimination of the electric field by using Faraday's law. In the ordinary fluid mechanic case, the streamlines obtained after solving the inviscid problem are distorted in regions of high vorticity through the mechanism of viscosity (last term in first equation). Similarly the magnetic field calculated in the case of ideal, nondissipative flow with $\sigma = \infty$ is distorted by the finite electrical conductivity (last term in second equation). The nondimensional number describing the influence of viscosity is the Reynolds number, and in perfect analogy as indicated by the above two equations, the magnetic field distortion is described by the number, $(Re)_m = \mu_e \sigma V L$ and is called the magnetic Reynolds number. When $(Re)_m$ is zero, the magnetic lines remain undisturbed, whereas in the limit $(Re)_m \to \infty$, the magnetic lines are frozen into the fluid in exactly the same way that vorticity is frozen according to Helmholtz's theorem. Mathematical similarities apart, the freezing of the magnetic lines with the motion is evident in the case of $\sigma \to \infty$ from the following physical considerations: An observer moving with the barycentric (stream) velocity in a medium of infinite electrical conductivity can measure only a zero electric field and hence he does not cut magnetic lines, which means that the magnetic lines must move along with his speed. From this argument it also follows that the total change in the magnetic flux through a given surface moving with the stream must be zero for an infinitely conducting medium. Finally, the remark should be made that except for stellar and interspace applications where the velocities and (especially) characteristic lengths are high, $(Re)_m$ is a small number. On the other hand, the assumption $(Re)_m = 0$ is a rather drastic one since it permits the uncoupling of Maxwell's equations from the conservation equations. In thermonuclear plasma physics, much of which is analyzed by MHD models, the approximation of frozen field lines is very useful, especially in dealing with phenomena which occur over brief time intervals, such as in MHD instabilities.

For the calculation of the ponderomotive force we can use Ampère's law ($\nabla \times H = J$) to find that $J \times B = (\nabla \times H) \times B$. Through regular vector operations, we can show that $(\nabla \times H) \times B = -\text{grad } (B^2/2\mu_e) + \text{div } (BB/\mu_e)$. One can identify the last term as representing a tension equal to $B^2/2\mu_e$ acting along the lines of force, whereas the first one corresponds to an equivalent hydrostatic pressure equal to $B^2/2\mu_e$. This term is frequently called "magnetic pressure" and in different problems is found to behave precisely as the static pressure does. Furthermore, one can show that if the magnetic lines are lengthened, the magnetic field intensity is increased.

Consider now the propagation of small disturbances in the form of acoustic waves for which the speed of sound for an ideal gas is proportional to $\sqrt{p/\rho}$. Now one can show through a linearization of the equations of conservation, assuming the presence of the magnetic pressure alone, that a small disturbance (for a gas of infinite electrical conductivity) will be propagated, in perfect analogy, with a speed equal to $\sqrt{B^2/\rho\mu_e}$. This is the so-called Alvén speed, and these waves are called magneto-fluid mechanic waves. Of interest also are combinations of several mechanisms of propagation which might include sound and gravitational waves.

Because of the property of the magnetic lines to increase their tension when lengthened, along with the additional ones of distortion and propagation of disturbances, their properties are presented in loose terms as resembling very much those of rubber bands.

Applications There are both astrophysical and terrestrial applications of magneto-fluid-mechanics. One of the earliest ones was perhaps suggested by Faraday, who thought to harness the river Thames with electrodes on its banks that would collect the induced electric current resulting from the flow of the river as it cuts the earth's magnetic field perpendicularly. Because of the small electrical conductivity of water, the small magnetic field of the earth and the small velocities, the interaction is too weak to be useful. However, in the laboratory, with a mutually perpendicular magnetic field, flow, and induced current density fields, a large interaction is possible when hot ionized gases are used in conjunction with strong magnetic fields. This area of research is called magneto-hydro-dynamic power generation, and its popularity emerges from the fact that mechanical energy can be converted to electrical without thermally stressed rotating parts. As a consequence, higher temperatures can be imparted to the working medium with better thermal efficiencies. This scheme, under development now, seems to be limited by losses due to heat transferred from the hot gas to the outside, corrosion of the electrodes, and Hall current losses. (When the gyrofrequency of the ionized particles is high compared to their collisional frequency, the particles drift in a direction parallel to the flow, and as a result, the current to be collected by the electrodes in the direction perpendicular to the flow, diminishes. The Hall effect can be turned to some advantage if it is designed to be substantial and if the current in the direction of flow is the one to be collected.)

One of the earliest astrophysical applications of magneto-fluid-mechanics lies in the area of solar physics and in particular the sunspots. Sunspots were seen and studied for the first time with the help of a telescope by Galileo about 1610. Three hundred years later, Hale discovered, through the Zeeman effect, that the magnetic field in the sunspots is very high (of the order of several thousand gauss). It was, however, only in the middle 1930s and in particular after the last world war that an explanation was sought in which the magnetic field was involved. At the writing of this article, there is no complete sunspot theory. However, the majority of workers in this area agree on the following rough picture. Because of mechanical equilibrium considerations, the pressure is the same at a given distance from the center of the sun in the sunspot proper or in the photosphere which is free of a magnetic field. This means that the magnetic pressure plus the static pressure in the sunspot region must balance the static pressure in the photosphere, a fact that implies that the static pressure in the sunspot region is smaller. If we picture the sunspot magnetic lines to be radial, the pressure gradient in this direction is independent of the magnetic field and balances exactly the gravitational force per unit volume ρg. Hence ρ is constant inside and outside the sunspot. Since the static pressure is proportional to density and temperature, the above arguments force us to accept a lower temperature inside the spot with a resulting darkening. The only question that rises is whether the order of magnitude of the magnetic pressure is enough for the effect to be significant. This seems to be so. If we assume a magnetic field of 1500 gauss (typical in a sunspot), the magnetic pressure is about 0.1 of an atmosphere which is the typical pressure in the photosphere.

An explanation for the bipolar nature of sunspots and the difference in the sign of their polarity has been offered. The differential rotation of the sun is invoked. The torroidal magnetic lines of the sun's field lying on its surface are twisted, since for very high electrical conductivity, they are frozen with the motion. As a result, the magnetic intensity is amplified and so is the magnetic pressure. Simple considerations based on the observed kinematics of the differential rotation establish the location in latitude with time where the intensities will be high enough to give rise to sunspot activity. The result compares favorably with observations. In fact, it can be shown that the sunspot activity migrates, time-wise, from the higher latitudes towards the equator as observations show. Because the twisted field is symmetric with respect to the equatorial plane, this model describes correctly the symmetry of the activity in the north and south hemispheres along with the fact that the polarity between two symmetric sunspot pairs is opposite in sign.

Efforts have been made to discover the mechanism for the generation and maintenance of cosmic magnetic fields, such as fields in stars, the earth, and galaxies. The most promising direction seems to lie in the so called "dynamo theories." Here, some general magnetic field is assumed (not necessarily strong), which upon interaction with the motion of a conducting medium (convective, or motion due to Coriolis forces), induces currents which reinforce the original magnetic field. As the magnetic field is reinforced, the ponderomotive force suppresses the motion until some kind of a steady state for both the motion and the magnetic field is reached.

Magneto-fluid-mechanics also studies problems related to magnetic confinement of plasmas and their stability to small disturbances. Consider for instance the so-called "pinch effect." Here, a strong current is passed through a cylindrical column made up of a plasma. The axial current filaments create an azimuthal magnetic field (the magnetic lines are then rings with the cylinder axis as the locus of their centers,) and as a result, a ponderomotive force is induced which compresses the plasma radially*. Through this confinement, it is hoped that temperatures of the order of 10^6 to 10^7 K will be created so that thermonuclear FUSION can take place. Such configurations are normally subject to instabilities. Consider, for instance, the case in which a small distortion in the form of a "kink" is formed in a cylindrical plasma

---

*Pinch-effect devices are also useful in metallurgy where molten metals can be confined away from solid boundaries in order to remain pure.

column, such that the rings in the concave side are pressed together, whereas the rings in the convex side are separated. As a result, the magnetic flux density (and hence the magnetic pressure) will be higher on the concave side resulting in a force tending to increase the concavity. We say that this configuration is unstable, since the force induced by the imposed disturbance acts in a destabilizing direction. Note that in this configuration, the center of curvature of the undistorted plasma boundary cross section falls inside the plasma. One can now create another example in which the curvature of the confining undistorted boundary of the plasma is opposite (the center of curvature falls in the vacuum) and show that the configuration will be stable. We can then state that a sufficient condition for stability is met when the magnetic lines are everywhere convex towards the plasma. If the magnetic lines induced by the currents going through the plasma are in an unstable configuration, externally imposed magnetic fields can be used in order to "stiffen" the configuration.

The "aurora borealis" can be explained in terms of the interaction of the solar wind (due to the continuous expansion of the solar corona with a velocity of about 500 km/sec) with the geomagnetic field. The inertia associated with this "wind" will penetrate the magnetic lines of the earth, only up to the point where the induced magnetic pressure is smaller than these inertial forces. The earth's magnetic field falls off with the inverse third power from the center of the earth. Knowing the mass density and the velocity of the solar wind, we can locate the remotest magnetic line from the earth that is strong enough to stop the penetration of the solar corpuscles. When this happens, these particles will glide along this magnetic line and eventually will come to the foot of this line at the surface of the earth. An elementary computation shows that the latitude of this line is the one where the "aurora borealis" is observed. (see AURORA AND AIRGLOW).

Convective motions can be effectively subdued by the presence of a magnetic field. Consider, for instance, the convection in a thin horizontal layer due to heating from below. Convective cells will be formed when the buoyant force is enough to counterbalance the viscous force of the motion. (These were formerly called Bénard cells because it was believed that Bénard had observed them. However, the name *convective cells* is preferred.) At the same time, balance of energy dictates that the heat convected upwards be equal to the heat conducted from the hot source at the bottom. The ratio of these two energies is called the "Rayleigh number," and for a given geometry, it must be higher than a critical value for the convective cells to appear. However, when a magnetic field is present, the ponderomotive force in general inhibits the motion and at the same time changes the geometry of the cell. The extent of this inhibition is given by the Hartmann number (defined earlier) so that the critical Rayleigh number is higher for higher Hartmann numbers. Available laboratory experimental results reconfirm the findings of this theory. On a cosmic scale, it has been hypothesized that the roll-like granulation in the sunspot penumbra is the result of the magneto-fluid-mechanic inhibition of the motion inside regular photospheric convective cells.

Magnetic fields are also known to inhibit the onset of turbulence. For instance, consider the flow of mercury in a channel. Experiments have shown that the flow can be laminar well above the critical Reynolds number of 2000 or so, if a coil is wrapped around the pipe thus creating an axial magnetic field. The small disturbances perpendicular to the direction of the main stream will be damped out through the action of the induced retarding ponderomotive force.

Many other cosmic scale phenomena seem to be explainable through magneto-fluid-mechanics. To list but a few, there are the solar flares and filaments, the spiral structure of some galaxies, the heating of the solar corona, explosion of magnetic stars and many others. Although order-of-magnitude analyses have been suggested to explain some of these phenomena, there are no complete self-consistent theories. Such theories seem to demand a simultaneous satisfaction of all the conservation and electromagnetic equations—a formidable, if ever possible, task. On the terrestial scale, many applications have been undertaken, and some of them are dependent upon technological development rather than fundamental physical understanding. To give a few more examples, in addition to those already mentioned, we list magneto-fluid-mechanic liquid metal pumps and flow meters, propulsion devices based on the acceleration of a neutral plasma through which a current and a normal magnetic field from the outside are supplied (an area called "plasma propulsion"), or a device in which positive ions (such as the ones easily produced by alkali metals) are accelerated with an electric field (ionic propulsion). Other examples are devices to reduce the heat transfer in reentry objects by using the decelerating action of a magnetic field carried by the vehicle or to use the ponderomotive force as a control force when needed for the navigation of space crafts.

PAUL S. LYKOUDIS

References

Ferraro, V. C. A., and Plumpton, C., "An Introduction to Magneto-Fluid-Dynamics," London, Oxford University Press, Second Edition, 1966.

Alfvén, H., and Fälthammar, C. G., "Cosmical Electro-Dynamics," Oxford, The Clarendon Press, 1963.

Bateman, G., "MHD Instabilities," Cambridge, Mass., MIT Press, 1978.

Cross-references: ASTROPHYSICS, AURORA AND AIRGLOW, CONSERVATION LAWS AND SYMMETRY, FLUID DYNAMICS, FLUID STATICS, HALL EFFECT AND RELATED PHENOMENA, IONIZATION, PLASMAS, SOLAR PHYSICS.

MAGNETOMETRY

Magnetometry is the art of determining accurately magnetic fields and the magnetic properties of matter. Both applications are of interest to pure science as well as to technology. The principles employed for measurements are based on the magnetostatic interaction between fields and moments, on voltages induced by flux changes, on the deflection of charge carriers in fields and on the precession of nuclear and electronic spins in a field.

Matter exposed to a magnetic field of strength H (as produced by a current-carrying solenoid) is magnetized. This phenomenon is described by the vector of magnetization $\mathbf{M} = N\langle\mathbf{m}\rangle$, where N is the number of atoms per unit volume having a mean dipole moment $\langle\mathbf{m}\rangle$. The magnetic induction (flux density) is then given by $\mathbf{B} = \mu_0(\mathbf{H} + \mathbf{M})$. One describes the magnetic response of a material by its susceptibility χ or its permeability μ_r using $\mathbf{M} = \chi\mathbf{H}$ and $\mathbf{B} = \mu_0\mu_r\mathbf{H}$. It follows that $\mu_r = 1 + \chi$. In magnetic materials the susceptibility χ is a function of temperature. Most often it will depend also on the magnitude and direction (relative to crystalline axes) of $\mathbf{H}$. Equations are given in SI units, which form a rational system. In all systems of units μ_r and χ are dimensionless numbers. μ_r is the same in SI and emu, but χ is smaller by $1/(4\pi)$ in emu. Note also that χ is related to the number of atomic dipoles per m^3 in SI and per cm^3 in emu.

Magnetometry is a still expanding field, mainly because of the many practical aspects of magnetism. New magnetometer designs appear constantly, but especially for absolute measurements the classical systems are still in use with only minor modifications.

Originally magnetometry developed from the interest in geomagnetism, with its importance for navigation. Nowadays exact measurements of variations in the earth's field are important for questions like the dynamics of the inner core of our planet and of the surface of the sun (via changes in the magnetosphere by sun spot activities).

The *classical method*, devised by Gauss, determines the horizontal intensity of the earth field H_- absolutely, with a relative accuracy better than 10^{-5}. Two measurements must be performed. First the torsional frequency of a small standard magnet having the magnetization M and being suspended by a torsion fiber is recorded. It is proportional to MH_-. Then a small magnetic needle is suspended by the same fiber and its deflection under the action of the standard magnet placed at a well defined position is measured. From this M/H_- can be determined. In modern systems the standard magnets are replaced by precision coils activated by exactly measured currents (*sine galvanometer*).

With the *earth inductor*, devised by Weber, one obtains the inclination of the earth field. The induced voltage in a rotating coil is sensitively measured. If the rotation axis is parallel to the field direction the induced voltage vanishes. Angles of a tenth of a minute of arc can be resolved. Portable systems are available as survey instruments, particularly for prospecting of iron ores and related uses.

Highly accurate field measurement (to 10^{-8}) are needed for the design of particle beam optical systems such as high energy accelerators or particle spectrometers. Similarly, such devices require a high temporal constancy of the field which can only be achieved by high gain feedback loops.

Rotating coil gaussmeters are modern descendants of the earth inductor. A small coil is wound on a nonmagnetic core and driven by a high speed electromotor. Such systems allow absolute measurements, but the probe averages over the volume covered by the coil. The mechanics of the electrical contacts can also give rise to problems.

A convenient, simple, rather pointlike field probe (which has to be calibrated) is the *Hall-effect gaussmeter*. A constant current is sent through a small conductor or semiconductor ($2 \times 2 \times 0.1$ mm^3). The field component perpendicular to the current flow will deflect the charge carriers via the Lorentz force and a voltage perpendicular to field and current is generated. It is proportional to the induction of the field. Most sensitive are semiconductor probes like InSb. Typically a Hall voltage of 5 mV is generated per T at 10 mA current. Such devices are now common laboratory equipment. They can be used over a fairly wide temperature range.

An even simpler and cheaper method for coarsely measuring or controlling fields in the laboratory is the measurement of the *magneto-resistance* of a semimetal wire or film or of the *forward diode bias* of commercial semiconductor diodes at liquid helium temperatures. They have moderate sensitivity and an overall nonlinear response.

A highly precise field measuring device is the *NMR magnetometer*. One determines the precession frequency of the nuclear magnetic moment of protons (or ^{7}Li) with a standard nuclear magnetic resonance circuit. The extreme sharpness of the resonance line (e.g., of protons in water) allows detection of variations in induction of 10^{-8} in 1 T. Measurements of broadening of the resonance line give information on the presence of small field gradients. An example is the field distortion caused by the presence of weak para- or diamagnetic impurities near the probe. The system may thus also be used for measurements of susceptibility. The sensor probe can be as small as 10^{-3} ml. NMR magnetometers lend themselves for telemetric readout and for incorporation into a field stabilizing circuit.

In so-called *volume averaging NMR magnetometers* the protons (e.g., of water or alcohol) are flowing through a tube in a strong B-field where their spins are longitudinally polarized. The fluid then flows through a region with unknown B', where the spins are turned. The remaining longitudinal spin polarization is monitored by standard NMR. The method can be used between 0.1 μT and 1 T with a typical relative accuracy of 10^{-7} around 1 mT.

A very sensitive device for measuring extremely weak fields is the *Zeeman magnetometer*. One of its applications is probing interplanetary fields in space. In the presence of a field the $^2S_{1/2}$ ground state of an alkali atom (e.g., Cs vapor) splits into a doublet. The population of the upper state is increased over the thermal equilibrium by optical pumping with circularly polarized light. The transition to the Zeeman ground state is forbidden. It can be stimulated by application of an rf field. This will bring the population ratio back to thermal equilibrium. The absorption of the pumping light in the vapor is dependent on the population ratio of the two Zeeman levels. By tracing the absorption maximum as a function of rf frequency one obtains the energy separation of the Zeeman levels and hence the acting field since the atomic moment of the alkali is precisely known. Sometimes a feedback circuit is used to keep the system at exact resonance. Its sharpness allows measurements down to 0.1 nT.

It should be mentioned that most information on stellar or interstellar fields comes from the observation of the Zeeman splitting of spectral lines of certain elements present in stars or interstellar matter. However, the width of spontaneously emitted optical lines is orders of magnitudes wider and thus the sensitivity is down.

Next we mention the magnetometer most widely employed in technical application, the *flux gate* (or *Förstersonde*). It consists of two soft magnetic cores in parallel orientation. They are driven to saturation by ac currents of fixed frequency applied to a primary coil wound around each core; the two windings are of opposite direction. Thus in a secondary sensing coil wound around both cores nominally no signal will be present. A superimposed, ambient dc field will produce a signal with twice the ac frequency in the sensing coil. The amplitude of this second harmonic is proportional to the dc field strength, its phase is related to the field direction. Modern systems use an additional field coil excited by the sensing coil signal in a feedback circuit. It compensates the ambient field and brings the sensing circuit back to zero.

Uses range from airborne survey of mineral deposits to minesweeping, submarine detection, treasure hunting (in archeology), and security checks. In these applications often the difference signal between two flux gates separated by a certain distance is monitored. Three devices mounted mutually perpendicular to each other are used to determine the vector components of a field. Magnetometers of this type were flown to the moon during the Apollo mission. The sensitivity of flux gates can be better than 10^{-5} A/m.

The most widespread use of magnetometry is the study of magnetic properties of matter. In pure science the quest is for the basic principles of magnetism and thus the electronic structure of matter. In the foreground of technological applications stands the design of new permanent magnets, of magnetic cores of transformers and inductances and recently mainly for magnetic storage and recording devices. These technological applications favor magnetically ordered materials. For example, new technologies appear on the horizon through the use of amorphous magnets. These types of measurements are less concerned with the magnitude and spatial distribution of fields but rather with the magnetic parameters of materials such as the susceptibility or the magnetization. A widely used instrument for such applications is the *vibrating sample* or *Foner magnetometer*. The specimen is vibrated perpendicular to a uniform magnetic field. Two signal coils wound in opposite direction and connected in series are placed around the sample with their axes parallel to the direction of motion. The dipole field of the specimen induces an ac signal in the coils. It is compared after amplification with the signal excited in a second pair of coils by a ferromagnetic reference sample (located usually near the vibrator) which is moved together with the unknown specimen. Both samples are mounted on a nonmagnetic shaft set into oscillations of ~80 Hz by a loudspeaker system. The magnetic field may either be generated by an electromagnet or superconducting coils. Careful design of pickup coils and phase-lock noise reduction allow the detection of changes in magnetization down to 10^{-6} A/m at fields of 10^6 A/m. The principle of the *vibrating coil magnetometer* is very similar. The sample stays fixed in a homogeneous magnetic field and the signal coils oscillate perpendicular to the field along the axis of the dipole moments of the sample. Demands on the homogeneity of the field are extremely high.

The other workhorse for studies of magnetic properties of matter is the *magnetic* (or *Faraday*) *balance*. Superimposed on the magnetizing uniform field is an inhomogeneous field. The dipole moments induced in a magnetic material experience a force parallel to the direction of the gradient of the external field. The sample is usually suspended on a thin fiber from one side of the arm of a microbalance mounted some distance above the field-producing magnets. In modern designs a feedback system is used which keeps the equilibrium by electromagnetic or electrostatic forces acting on the other side of the balance arm. The measurement is absolute. Sensitivity is about 10^{-6}–10^{-7} A/m in good systems. Fields are produced either by an electromagnet with inclined pole caps or by a system of superconducting coils.

Both the vibrating sample magnetometer and the magnetic balance allow the variation of sample temperature rather straightforwardly. Sample and vibrating rod (or suspending fiber) are mounted inside a small dewar system or oven. Complete systems are available commercially in highly advanced designs.

Another system which makes use of the force exerted on a magnetic material in a nonuniform field is the *pendulum magnetometer*. The sample is fixed at the upper end of a pendulum rod which is suspended in the middle. A counterweight is mounted on its lower end. For small amplitudes the force on the magnetic dipole acts as a restoring force and will thus change the frequency of oscillation. Measurements of pendulum frequency are made with and without field. The same basic principle is used in the *vibrating reed magnetometer*. The sample is attached to one end of the nonmagnetic, metallic reed (e.g., Au), while the other end is rigidly fixed. The reed is forced into oscillations by an inhomogeneous ac field superimposed on the uniform magnetizing dc field. The mechanical vibrations are converted by a piezoelectric transducer to an ac voltage which is proportional to the magnetization of the sample. Using look-in techniques one can resolve moments down to 10^{-13} A m^2.

The *classical astatic magnetometer* is still used to determine the magnetization of rod-shaped samples. Two small, identical magnets are mounted horizontally and antiparallel to each other at the ends of a long, nonmagnetic rod suspended vertically by a weak torsion fiber. This is the measuring system which is unaffected by the earth field. Two opposing coils are placed in the plane of rotation of one of the magnets. The aligned axes of the coils stand perpendicular to the axis of the magnet. Their fields cancel exactly at the position of the magnet. The rod shaped sample is inserted into one of the coils. The balance of fields is disturbed by its magnetization and the resulting torsional deflection can be read out optically. Moments down to 10^{-9} A m^2 can be determined absolutely.

Critical parameters for commercial magnets are the *coercive force* and the *remanence*. They are determined by recording the hysteresis loop (i.e., the B vs H curve for rising and decreasing field intensity) in a simple *inductance magnetometer*. The sample is inserted into a gap of the core of a toroidal solenoid. The current through the solenoid gives H. One obtains B by electronically integrating the voltage induced in a concentric pickup coil. Automatic systems of this type are common in materials testing laboratories. The sensitivity can be improved by first balancing the signals from two toroids with empty gaps. After inserting the sample into one gap the difference in outputs is measured.

For comparisons of the high field susceptibilities of ferromagnets the *orbiting sample magnetometer* is an advanced modern system. Several specimens are mounted on a disk which rotates with constant angular frequency of some 10 Hz. The uniform magnetizing field is directed along the axis of rotation. The samples pass successively by a sensing coil. Its output is stored in phase with the rotation. Digital signal averagers can be used to reduce noise.

In single-crystal samples there are directions along which the material is more easily magnetized. In general they coincide with the principal crystal symmetry directions. The free energy of a crystal thus contains a term dependent on the direction of magnetization relative to the crystallographic axes. It is called the *magnetocrystalline anisotropy* (energy) and is usually expressed in a set of parameters referred to as *anisotropy constants*. They yield basic information on the anisotropy of magnetic interaction of the atomic magnetic moments which in turn is often caused by the anisotropic electron distribution of the orbital ground state.

Magnetic anisotropy can be detected by all static methods for the measurement of magnetization if provision is made that the axis of the magnetizing field can be turned with respect to crystal orientation. Also used are ac bridges and ferro-, ferri-, or antiferromagnetic resonance. The most commonly employed system is, however, the *torque magnetometer*. The sample is suspended by a torsion wire between the poles of an electromagnet which can be moved around the sample. The magnetizing field tends to align the sample magnetization along an easy axis, and a torque is thus exerted on the sample, which is read out by optical or capacitive methods or by the variation in resistance of a set of strain gauge wires. A set of data is taken by varying the original orientation of the specimen relative to the external field.

In the *ripple field magnetometer* the magnetization of the sample is modulated by an ac ($\sim$100 Hz) "ripple" field superimposed parallel to the dc magnetizing field. A sensing coil oriented at 90 degrees to the field measures the perpendicular magnetization, which is proportional to the angular derivative of the anisotropy energy density with respect to a reference crystalline axis.

On a similar principle operate *rotating sample magnetometers*. The specimen (often a sphere) rotates slowly (0.02–0.1 Hz) within a dc field applied perpendicular to its axis of rotation. The variation in flux with rotation is sensed by search coils oriented parallel and perpendicular to the dc field. The anisotropy constants are derived from the two measured flux components.

Information similar to the magnetic anisotropy can be obtained by measuring the magnetostrictive changes in dimensions of single-crystal samples as a function of magnitude and direction of the applied field. X-ray techniques are usually too insensitive for this purpose, and strain gauges have found the widest use. Capacitive read out has also been reported.

A predecessor of this system is the *spinner magnetometer*. It rapidly rotates a ferromagnetic sample without external field. It is available as a survey instrument for anisotropy studies of rock samples.

Finally we discuss briefly a very modern system used for special magnetic measurements, the *superconducting quantum interference device* (SQUID) which is rapidly gaining importance. Basically a fluxmeter, it does not, however, measure the flux itself but rather, with extreme accuracy, minute variations in flux. In fact, sensitivity has reached about 10^{-5} of a single flux quantum ($\phi_0 = h/2e = 2.07 \times 10^{-15}$ Wb).

Applications of the SQUID in physics range from ac measurements of very small magnetic moments (e.g., magnetization in extremely weak fields) to a search for the elusive magnetic monopole. Measurements of volume susceptibilities down to 10^{-10} for 1 μg samples have been reported. The SQUID has also become the central tool for magnetomedical and -biological research. Examples are magnetocardiography and the study of fields generated by the action of the human brain.

The SQUID is based on two macroscopic quantum effects in superconductors. The first is *flux quantization:* The flux trapped inside a superconducting loop must be an integral multiple of ϕ_0. The second is the *dc Josephson effect:* A superconducting ring is interrupted at one point by a weak link. Examples are a thin ($\lesssim$ 1 mm) oxide layer or a point contact. The thickness of the insulating layer is such that electrons can tunnel through it. In a superconductor a current of Cooper pairs is flowing which requires no voltage across the barrier. Only when a critical current (which can be kept as low as some 10^{-5} A) is exceeded, a voltage proportional to the tunnel current appears at this so-called Josephson junction. The critical current drops rapidly when a field is applied perpendicular to the flow of Cooper pairs. After this field has increased so much that an additional flux quantum can be brought into the ring, the critical current jumps back to a higher value. These periodic discontinuities occur because the ring momentarily ceases to be superconducting, so that one quantum of flux can enter. In a wire loop, placed inside the ring, a voltage pulse is induced each time a flux jump occurs. By counting these pulses the total change of field through the ring can be calculated. For ~1 mm diameter a change in flux density of ~1 nT causes a jump.

A practical SQUID arrangement is the two-hole system: Inside an Nb cylinder a bore shaped like a dumbbell is drilled. The weak link is an Nb screw placed across the bar of the dumbbell. If the flux in one hole increases, an equal decrease of flux in the other hole must follow. The current through the Josephson junction depends on the flux difference between the holes. One uses an external search coil placed in the field the variations of which are to be measured. It is connected in series to a wire loop inside the first hole. This whole circuit works as a flux transformer. The second hole is inductively coupled to an LC circuit tuned at some 10 MHz. The appearance of a voltage drop across the junction due to flux brought into the first hole can be regarded as a change in inductance of the superconducting ring which mistunes the oscillator. The voltage across the circuit as a function of field acting on the search coil is saw tooth like with a period corresponding to trapping one flux quantum inside the two-hole ring. The sensing coil can be placed at a convenient distance from the SQUID. By giving it the appropriate shape one may also detect variations in the spatial derivatives of fields with extreme sensitivity. Furthermore, the SQUID has a fast response. Rates of 10^7 ϕ_0/s have been recorded.

F. Jochen Litterst
G. Michael Kalvius

References

1. Kohlrausch, F., "Praktische Physik" 22nd Ed., Vol. 2, p. 341, Stuttgart, B. G. Teubner, 1968.
2. Lark-Horovitz, K., and Johnson, V. A., "Methods of Experimental Physics," Vol. 6, "Solid State Physics," Part B, p. 171, New York, Academic Press, 1959.
3. Kalvius, G. M., and Tebble, R. S., "Experimental Magnetism," Vol. 1, John Wiley & Sons, Chichester, 1979.
4. Foner, S., "Vibrating Sample Magnetometer," Rev. Sci. Instr. 27 548 (1956).
5. Gallop, J. C., and Petley, B. W., "SQUIDs and Their Applications," J. Phys. E: Scient. Instr. 9, 417 (1976).
6. Primdahl, F., "The Fluxgate Magnetometer," J. Phys. E: Scient. Instr. 12, 241 (1979); 15, 221 (1982).
7. Pendlebury, J. M., et al., "Precision Field Averaging NMR Magnetometer . . . ," Rev. Sci. Instr. 50, 535 (1979).
8. Parsons, L. W., and Wiatr, Z. M., "Rubidium Vapour Magnetometer," Scient. Instr. 39, 292 (1962).
9. Romani, G. L., Williamson, S. J., and Kaufman, L., "Biomagnetic Instrumentation," Rev. Sci. Instr. 53, 1815 (1975).

Cross-references: GEOPHYSICS; HALL EFFECT AND RELATED PHENOMENA; MAGNETISM; MEASUREMENTS, PRINCIPLES OF; SUPERCONDUCTIVITY; ZEEMAN AND STARK EFFECTS.

MAGNETOSPHERIC RADIATION BELTS

The sun emits continuously a fully ionized gas, the *solar wind*, which flows radially outward throughout the solar system. The solar wind plasma is primarily made up of protons and electrons, and its properties although variable, have average values at earth orbit of bulk velocity $V \approx 350$ km/sec, number density $N \approx 5$ cm^{-3}, and temperature $T \approx 15$ eV. Because of its high conductivity, the solar wind carries with it an embedded magnetic field, which on average is parallel to the ecliptic plane and traces an Archimedean spiral back to the sun in this plane due to solar rotation. At the orbit of earth the interplanetary magnetic field is highly variable in direction and magnitude, having an average value of ~5 nT (10^{-5} gauss).

The interaction of the supersonic solar wind with the intrinsic dipole magnetic field of the earth forms a region, the *magnetosphere* (Fig. 1), whose boundary, the *magnetopause*, separates interplanetary and geophysical plasma and magnetic field environments.[1] Upstream of the magnetopause is a *bow shock* formed in the solar wind–magnetosphere interaction process. At the bow shock the solar wind becomes thermalized and subsonic and continues its flow around the magnetosphere as *magnetosheath plasma*, ultimately rejoining the undisturbed solar wind. The bow shock is of interest because of its collisionless character, and much work is presently being done to understand the nature and the development of the electric and magnetic field configurations required to establish a shock front in a collisionless medium.[2]

A rough estimate of the position of the dayside magnetopause is obtained by balancing the solar wind pressure against the geomagnetic field with the resistive pressure of the geomagnetic field itself:

$$\tfrac{1}{2}\rho V^2 = B^2/8\pi,$$

where ρ = solar wind mass density, V = solar wind velocity, and $B = 0.34/R^3$ gauss is the earth's field at the magnetic equator, with R = geocentric distance in units of earth radii. Use of the average solar wind values given above

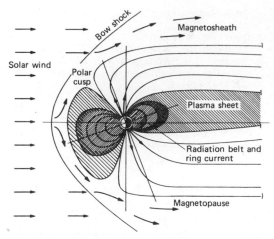

FIG. 1. An outline of the earth's magnetosphere. The lines represent magnetic field lines.

gives a dayside magnetopause distance of ~ 10.2 earth radii, as compared to an average observed distance of ~ 10.8 earth radii. More exact fluid and kinetic theory models give good agreement with the observed average latitudinal and longitudinal shape of the dayside magnetosphere.

In the antisolar direction, observations show that the earth's magnetic field is stretched out in an elongated *geomagnetic tail* (analogous to a cometary tail) to distances of several hundred earth radii. The geomagnetic tail field lines emanate from high geomagnetic latitudes from the vicinity of the auroral ovals to the geomagnetic pole. Topologically the geomagnetic tail consists of roughly oppositely directed field lines separated by a "neutral" sheet of nearly zero magnetic field. Surrounding the neutral sheet is a plasma of "hot" particles, the plasma sheet, having a temperature of 1–10 keV, a density of 0.01–1 particle/cm^3 and bulk flow velocity of a few tens to a few hundreds of km/sec. A definitive physical explanation of the extended geomagnetic tail has yet to be obtained.

Figure 1 is a schematic of the overall magnetospheric configuration. There are seasonal variations due to the $\sim 12°$ tilt of the earth's magnetic dipole axis relative to its spin axis. More important, variations in solar wind parameters cause large perturbations to the picture shown in Fig. 1. These perturbations are observed to have scale variations much larger than the $V^{-1/3}$ and $\rho^{-1/6}$ dependences predicted by the simple pressure balance discussed earlier. Therefore, physical mechanisms other than simple pressure balance are required to explain observed magnetospheric variations. For example, geomagnetic field lines are known to interconnect with interplanetary magnetic field lines (magnetic field component normal

to the magnetopause), causing a two-way transfer of particles and energy between the interplanetary medium and the solar wind. Neutral points (lines) are formed on the magnetopause and, by magnetic flux conservation, in the geomagnetic tail in the neutral sheet region. It is not known if sites of field line interconnection are responsible for large scale particle acceleration, but they are probably effective in altering the shape and size of the magnetosphere.

Electrostatic, induced, and polarization electric fields play a fundamental role in determining charged particle entry to and subsequent motion and acceleration within the magnetosphere. Much present work in magnetospheric physics is aimed at identifying these fields and deconvolving the subsequent currents responsible for sustaining the magnetospheric configuration and causing its variations.

The in situ phase of magnetospheric research began dramatically with the discovery by Van Allen and his colleagues[3] of a permanent, intense, trapped energetic particle population (the Van Allen radiation belts) residing in the geomagnetic field—a discovery made with data from the first United States satellite, Explorer I. Following this initial discovery, data obtained from the trapping regions showed the geomagnetic field to be at least at altitudes $\lesssim$ several earth radii, a very efficient and vast magnetic mirror machine.[4]

The most useful approach for describing the motion of charged particles in the earth's magnetic field has been the guiding center approximation and subsequent development of adiabatic invariant concepts.[5,6] In the guiding center approximation, the instantaneous position $\mathbf{r}$ of a particle moving in a magnetic field is broken down into its circular motion of radius $\boldsymbol{\rho}$, and the motion of the guiding center $\mathbf{R}$:

$$\mathbf{r} = \mathbf{R} + \boldsymbol{\rho}.$$

A general expression for the motion of the guiding center can be obtained by substituting the above into the equation of motion

$$m \frac{d^2\mathbf{r}}{dt^2} = m\mathbf{g} + \frac{e}{c}\frac{d\mathbf{r}}{dt} \times \mathbf{B} + \mathbf{E}e$$

where m = particle mass, e = electronic charge, c = velocity of light, $\mathbf{g}$ = acceleration of gravity, $\mathbf{B}$ = magnetic field, and $\mathbf{E}$ = electric field. This yields the nonrelativistic guiding center equation

$$\frac{d^2\mathbf{R}}{dt^2} = \mathbf{g} + \frac{e}{m}\left\{\mathbf{E} + \frac{1}{c}\frac{d\mathbf{R}}{dt} \times \mathbf{B}\right\} - \frac{\mu}{m}\nabla\mathbf{B}$$
$$+ O\left(\frac{\rho}{\alpha}\right)$$

where μ = particle's magnetic moment due to gyration, ρ = cyclotron radius, α = scale length over which the magnetic field changes appreciably, and $O(\rho/\alpha)$ = terms of order ρ/α. In this approximation the particle's motion in the earth's field is broken down into three components: gyration about a field line, bounce back and forth along the field line between mirror points, and a slow longitudinal drift around the earth. While these motions are not strictly separate from one another, the vast difference in the time scales associated with them makes such a separation possible and leads directly to the consideration of adiabatic invariants. These motions are illustrated in Fig. 2.

The adiabatic invariants may be considered constants of the particle's motion provided that magnetic field variations are small compared with the time and spatial scales associated with the particle's motion. The first of the adiabatic invariants is the magnetic moment generated by the particle as it gyrates around the field line,

$$\mu = mv_\perp^2/2B = mv^2 \sin^2 \alpha/2B = W_\perp/B$$

where α = angle between the field line and the velocity vector.

This leads directly to the mirror equation and definition of the particle's mirror point:

$$\mu = \frac{W \sin^2 \alpha}{B} = \text{constant}.$$

In a static field, W (the particle energy) is constant and

$$\sin^2 \alpha_1/B_1 = \sin^2 \alpha_2/B_2 = \text{constant}.$$

Using the earth's equatorial magnetic field as a reference, the mirror point B value on a given line of force is determined by the particle's pitch angle at the equator:

$$B_M = B_{eq}/\sin^2 \alpha_{eq}.$$

In a dipolelike field configuration, such as the earth's, having a minimum B value at the equator, particles will simply bounce back and forth between conjugate mirror points located

in the northern and southern hemispheres (cf., Fig. 2).

The second adiabatic invariant, obtained from the action integral and associated with the particle's bounce back and forth along a field line, is given by

$$J = 2 \int_M^{M^*} P_\parallel \, ds$$

where $P_\parallel = mv_\parallel$ is the particle momentum along the field line and the integral is taken along the field line between the two conjugate mirror points M and M^*.

Forces due to the gradient of the earth's magnetic field and field line curvature cause a longitudinal drift across field lines with electrons drifting eastward and protons drifting westward. In an ideal dipole field this effect produces a drift surface which is simply the figure of azimuthal revolution of a line of force. Associated with this drift motion is the third adiabatic invariant, the flux invariant

$$\Phi_M = B \, dS.$$

Φ_M, the magnetic flux linked in the drift orbit of the particle, is the weakest of the three invariants (μ, J, Φ_M) since it has associated with it the largest spatial and temporal scales. Therefore the conditions for adiabatic invariance are most easily violated for Φ_M.

In Table 1 we show for reference characteristic times associated with charged particle motion in the magnetosphere.

Radiation belt particles represent significant energy storage in the magnetosphere (2×10^{22} – 2×10^{23} ergs). Their gradient and curvature drifts establish a current encircling the earth, the ring current, which is responsible for worldwide depressions of the earth's surface magnetic field. During times of enhanced radiation belt intensities, particle energy densities significantly greater than the ambient magnetic field energy density are observed and can cause surface field variations up to several hundred nT. The bulk energy density of the ring current particles is contained within the energy range ~ 1–200 keV with a mean energy of ~ 85 keV. This high β plasma ($\frac{1}{2}\rho v^2 > B^2/8\pi$) decays primarily through charge exchange and ion-cyclotron wave generation.

Protons, helium, and oxygen together form the ring current but their relative contributions are unknown. Thus the ultimate source of radiation belt particles, the solar wind or the ionosphere, is still uncertain. It is expected that strong energy and spatial dependencies will be evident in the source mixture for the bulk of the radiation belt particles. The very high energy ($\gtrsim$ several tens of MeV) protons observed at low altitudes ($\lesssim 1.8$ earth radii)

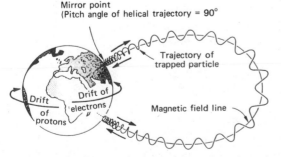

Mirror point
(Pitch angle of helical trajectory = 90°)

Trajectory of trapped particle

Magnetic field line

Drift of electrons

Drift of protons

FIG. 2. Illustration of the motion of a charged particle trapped in the earth's magnetic field.

TABLE 1. CHARACTERISTIC TIMES ASSOCIATED
WITH PARTICLES TRAPPED IN THE EARTH'S
MAGNETIC FIELD.

Particle	Energy (keV)	Gyroperiod ($\propto R^3$) (sec)	Bounce Period ($\propto R$) (sec)	Drift Period ($\propto 1/R$) (hr)
Electron	10	9.4×10^{-6}	0.64	36.7
	100	13×10^{-6}	0.23	4.1
	1000	80×10^{-6}	0.13	0.54
Proton	10	17×10^{-3}	27	36.7
	100	17×10^{-3}	8.6	3.4
	1000	17×10^{-3}	2.7	0.35

Values correspond to $\alpha_e = \pi/2$ and $R = 2R_e$.

are supplied by neutrons, generated in the atmosphere by cosmic rays, which leave the earth's atmosphere and decay in the geomagnetic field.

Radiation belt particles are accelerated to their final energies via $E \times B$ convection across field lines and betatron and Fermi acceleration processes due to slow diffusion across magnetic field lines under conservation of the first two adiabatic invariants. The relative importance of these mechanisms has not yet been estab-

lished. Radiation belt particles having a solar wind source obtain an initial heating in the geomagnetic tail where gradient and curvature drive across electric field equipotentials can increase particle energies by amounts up to several tens of keV. If the particles are from the ionosphere they are accelerated to energies of $\sim 1-10$ keV by electric fields parallel to magnetic field lines emanating from the auroral zones. In either case a radiation belt source, most likely the plasma sheet, can be formed at

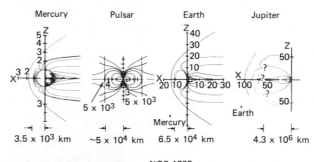

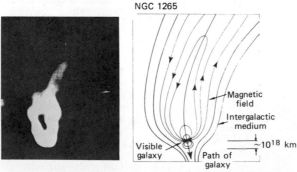

FIG. 3. Magnetospherelike systems are probably common throughout the universe, with a large range in scale sizes. For example, the subsolar magnetopause distance for Mercury is 3.5×10^3 km; for the radio galaxy NGC 1265, the analogous distance is roughly 10^{18} km.

altitudes $\gtrsim$ 6.5 earth radii which then can be accelerated as discussed above to form the trapped particle population.

The second large energy storage region in the magnetosphere is the extended geomagnetic tail (3×10^{22}–3×10^{23} ergs). The plasma sheet particles and the stretched geomagnetic field lines contribute roughly equal parts to this energy storage. The relationship between the plasma sheet, the aurora, and the radiation belts is an intimate one but not yet fully understood. As discussed above, the earthward portion of the plasma sheet is a likely source for the radiation belts, but the relative contributions of the solar wind and ionosphere are unknown. It is also likely that diffuse auroral forms are due to plasma sheet electrons scattered into the loss cone by electrostatic waves. On the other hand, discrete auroral forms are most probably caused by electric fields parallel to auroral magnetic field lines. It is not known whether these fields are due to very narrow electrostatic field geometries called *double layers* or the observed shear flows in the high altitude plasma ($\nabla \cdot E < 0$) coupled with ionospheric current continuity restrictions. There appears to be no requirement for anomolous resistivity mechanisms in auroral processes.

Associated with the parallel electric fields responsible for discrete auroral forms is an intense electromagnetic radiation in the 100–1000 kilohertz band, *auroral kilometric radiation*. The intensity of this radiation, normalized to a planetary radius distance scale, is comparable to Jupiter's emissions. In fact, integration over respective radiating solid angles may make the earth a radio source of the same order as Jupiter in total power output.

If we consider the earth's magnetosphere in a general sense as a rotating magnetized plasma, we find that such objects are plentiful throughout our solar system and perhaps the universe. This is not surprising, since most of the universe is filled with plasma and the basic interactions between plasmas, electric fields, and magnetic fields being uncovered in the earth's magnetosphere are present in the development of cosmic regions from small interstellar clouds to entire galaxies.

Interplanetary spacecraft have identified magnetospheres around Mercury, Saturn, and Jupiter.[7] Astronomers have detected similar structures around rotating neutron stars (pulsars) and radio wave-emitting galaxies. Figure 3 illustrates the scale sizes observed for these various magnetospherelike systems.

In all these cases it is evident that nature has been able to accelerate charged particles to very high energies. In the earth's magnetosphere there are at least four established methods of accelerating particles. The most general of these is magnetic field gradient and curvature drift across electric field equipotential surfaces. More specific mechanisms are betatron accelera-

tion, Fermi acceleration, and acceleration by electric fields parallel to magnetic field lines. It remains to be shown that field line interconnection can directly transfer energy from the magnetic field to charged particles (field line merging, reconnection) or if plasma turbulence effects are inportant as acceleration processes.

Magnetospheric systems, while similar, often have their own unique characteristics. For example, Jupiter and Saturn have moons in the heart of the charged particle populations which are effective absorbers creating distinctive features in their radiation belts. At Jupiter, the volcanic moon Io is a copius source of sulfur and oxygen, both of which have been detected at all energies throughout the Jovian magnetosphere. Jupiter's high spin rate (period = 9 hours 55 minutes 29.7 seconds) can produce effects to accelerate particles in addition to those found in the earth's magnetosphere. For example, low energy plasma corotating with Jupiter's magnetic field will exceed the Alfven speed and become supersonic well within the Jovian magnetosphere (30–40 Jovian radii). Even tiny Mercury, with neither atmosphere nor ionosphere, possesses a magnetosphere capable of accelerating large numbers of particles to high energies.

Work remains to be done to understand how the laws of physics operate in interacting magnetized plasma systems which display the range of boundary conditions seen throughout the solar system and in the universe.

DONALD J. WILLIAMS

References

1. For a detailed discussion of the magnetosphere and its interaction with the solar wind see, for example, Akasofu, S. I., and Chapman, S., "Solar Terrestrial Physics" 901 pp., London, Oxford Univ. Press, 1972, and Williams, D. J. (Ed.), "Physics of Solar Planetary Environments," 1038 pp., Washington, D.C., American Geophysical Union, 1976. For a brief summary of other magnetosphere systems see Stern, D. P., and Ness, N. F., "Planetary Magnetospheres," *Annual Review of Astronomy and Astrophysics* 20 (1982).

2. See collection of papers on recent bow shock results, *Journal of Geophysical Research* 86, 4319 (1981).

3. Van Allen, J. A., Ludwig, G. H., Ray, E. C., and McIlwain, C. E., "Observation of High Intensity Radiation by Satellites 1958 α and γ," *Jet Propulsion* 28, 588 (1958).

4. See for example, Roederer, G., "Dynamics of Geomagnetically Trapped Radiation," Heidelberg, Springer-Verlag, 1970, and Williams, D. J., "Charged particles Trapped in the Earth's Magnetic Field," *Advances in Geophysics* 15, 137 (1971).

5. Alfven, H., "Cosmical Electrodynamics," 1st Ed., London and New York, Oxford Univ. Press, 1950.

6. Spitzer, L., "Physics of Fully Ionized Gases," New York, Wiley-Interscience, 1956.
7. For Mariner-10 Mercury results see *Science* 185, 141 (1974).
   For Voyager-1 Jupiter results see *Science* 204, 945 (1979).
   For Voyager-2 Jupiter results see *Science* 206, 925 (1979).
   For Voyager-1 Saturn results see *Science* 212, 159 (1981).
   For Voyager-2 Saturn results see *Science* 215, 499 (1982).

Cross-references: ELECTRON, GEOPHYSICS, IONO-SPHERE, PLANETARY ATMOSPHERES, PROTON, SPACE PHYSICS.

MAGNETOSTRICTION

When a polycrystalline nickel sample is placed in a magnetic field, it contracts along the field direction by about 30 parts per million and elongates in the transverse direction by about half that amount. There is also a small volume change. Such changes in dimension of magnetic materials with variation of magnetic field strength or direction, are termed *magnetostriction*. They are measured by strain gages, optical dilatometers, capacitance variation, and x-ray analysis.

Below the Curie temperature, magnetostriction in weak fields is caused by domain rotation, becoming appreciable at fields near the knee of the *B-H* curve.

In saturating fields there is still a small linear dependence of magnetostriction on magnetic field strength, and above the magnetic ordering temperature magnetostriction is, except in rare instances, quadratic in magnetic field strength. Field strength dependent distortions in the saturated and paramagnetic regions, designated *forced magnetostriction*, are due to the paraprocess, the induction of a moment by the field.

The saturation magnetostriction of single crystals depends upon the direction of the (sublattice) magnetization, α, and the direction of measurement, β, with respect to the crystal axes. In a cubic crystal (with collinear sublattices), to lowest order,

$$\frac{\delta l}{l} = \lambda_0 + \frac{3}{2} \lambda_{100} [\alpha_1{}^2 \beta_1{}^2 + \alpha_2{}^2 \beta_2{}^2$$
$$+ \alpha_3{}^2 \beta_3{}^2 - \frac{1}{3}] + 3\lambda_{111} [\alpha_1 \alpha_2 \beta_1 \beta_2$$
$$+ \alpha_2 \alpha_3 \beta_2 \beta_3 + \alpha_3 \alpha_1 \beta_3 \beta_1] \quad (1)$$

Clark[1] gives higher order expressions for cubic and hexagonal symmetry.

For cubic crystals the fractional change in length along the field (and the magnetization) direction induced by a saturating magnetic field is, in principle, found by averaging Eq. (1) over directions. This gives the saturation magnetostriction

$$\lambda_s = \frac{2\lambda_{100} + 3\lambda_{111}}{5}. \quad (2)$$

The assumption underlying the averaging that yields Eq. (2) is that before the field is applied the material is unmagnetized and all polycrystal orientations are equally likely. This is in fact rarely the case; there is often both some remanent magnetization and some preferential orientation of crystallites. Measurements of $\lambda_\parallel - \lambda_\perp$, the difference in distortions parallel and perpendicular to the field direction, are more reproducible and significant, in that they are independent of the distortion in a fiducial "unmagnetized" state.

Magnetostriction coefficients vary greatly, depending upon the material, temperature, and magnetization state. For pure iron at room temperature, the saturation magnetostriction constants are $\lambda_{100} \sim 20 \times 10^{-6}$; $\lambda_{111} \sim -20 \times 10^{-6}$, while for alloys near 80Ni-20Fe (weight per cent) these constants are almost zero. The cobalt ion causes a large magnetostriction; for cobalt ferrite $\lambda_{100} \sim -500 \times 10^{-6}$ while for nickel ferrite $\lambda_{100} \sim -30 \times 10^{-6}$. The largest known magnetostriction is that of dysprosium metal.[2] As a magnetic field is rotated in the basal plane of this hexagonal crystal, there is a basal plane distortion of almost one per cent, at liquid nitrogen temperatures and below. At room temperature, $TbFe_2$ shows a magnetostriction 5 times larger than does any other material[3]; ($\lambda \sim 2 \times 10^{-3}$).

The source of magnetostriction is the dependence of magnetic energy on strain. Because the elastic energy is quadratic in strain while the magnetoelastic energy is linear in strain, the minimum free energy occurs at nonzero strain. For example, in a cubic crystal the equilibrium shear strain ϵ_{xy} is given by

$$\epsilon_{xy} = \frac{B_2(T, H)}{c_{44}} \alpha_x \alpha_y \quad (3)$$

Here c_{44} is the elastic constant, the α's are magnetization direction cosines, and $B_2(T, H)$ is a magnetoelastic coefficient representing the variation of magnetic energy (magnetic anisotropy, dipolar, anisotropic exchange) with strain.

Quantum mechanical calculations of the magnetoelastic coefficients are in a somewhat more satisfactory state in the case of nonconductors than for metals. Extensive calculations by Tsuya of the *B* coefficients of the spinels are reviewed by Kanamori.[4]

The temperature dependence (and "forced" field dependence) of the magnetostriction coefficients is due to statistical averaging as the individual spins fluctuate around the average magnetization direction α. For some materials this temperature dependence can be expressed entirely in terms of a known function of the (sublattice) magnetization. For ferrimagnets[5]

$$\lambda_i(T, H) = \sum_n \lambda_i{}^n(0)\, f_i[m_n(T, H)] \qquad (4)$$

That is, the magnetostriction coefficient $\lambda_i(T, H)$ is the sum over sublattices of temperature independent sublattice magnetostriction coefficients $[\lambda_1{}^n{}_1 (0) = B_2{}^n(0)/c_{44}]$ times a function f_i of the sublattice magnetization $M_n(T, H)/M_n(0)$. At sufficiently low temperatures this function reduces to

$$f[m_n(T, H)] = \left[\frac{M_n(T, H)}{M_n(0)}\right]^3 ; \ T \ll T_c \qquad (5)$$

for both λ_{100} and λ_{111}.

Clark[1], Bozorth[6] and Callen[7] give references.

EARL CALLEN

References

1. Clark, A. E., "Magnetostrictive Rare Earth-Fe₂ Compounds," in "Ferromagnetic Materials" (E. P. Wohlfarth, Ed.), New York, North Holland Publishing Co., 1980, Chapter 7, pp. 531–589.
2. Legvold, S., Alstad, J., Rhyne, J., *Phys. Rev. Letters* **10**, 509 (1963); Clark, A. E., Bozorth, R. M., and DeSavage, B., *Physics Letters*, **5**, 100 (1963).
3. Clark, A. E., Belson, H. S., AIP Conf. Proc. 5, 1498 (1972); Clark, A. E., Belson, H. S., Tomagawa, N., and Callen, E., Proc. Internat. Conf. on Magnetism, Moscow, August 1973.
4. Kanamori, J., "Magnetism" (Rado, G. T., and Suhl, H., Eds.), Vol. I, p. 127, New York, Academic Press, 1963.
5. Callen, E., Clark, A. E., DeSavage, B., Coleman, W., and Callen, H. B., *Phys. Rev.*, **130**, 1735 (1963).
6. Bozorth, R. M., *Ferromagnetism*, New York, Van Nostrand Reinhold, 1951.
7. Callen, E., *J. Appl. Phys.* **39**, 519 (1968).

Cross-references: FERRIMAGNETISM, FERROMAGNETISM, MAGNETISM.

MANY-BODY PROBLEM

Scope and Definition A large part of the experimental data of physics is concerned with natural objects which may be looked upon as being made up from smaller bodies. For example, we may think of the solar system as an object composed of the planets and the sun; ordinary matter, in solid, liquid or gaseous form, as composed of molecules and atoms; atoms and molecules themselves as made up from nuclei and electrons, the nuclei as composed from neutrons and protons, and so on. We shall call the composite object the system, and its constituents, the particles; and note that it seems most reasonable to suppose that the properties of the system can be explained on the basis of the law of interaction between the particles and the laws of dynamics. The latter may be classical or quantum mechanical according to the demands of the situation. At each level of refinement we refrain from asking about the internal structure of the particles. This is to achieve a natural simplicity of description; but still, at each such level, we have a rich variety of natural phenomena to explain.

The many-body theory is not concerned with any fundamental or complete explanation of nature. Its chief aim is to formulate schemes according to which calculations of certain physical quantities can be performed theoretically and the results can be compared with experimental measurements. It is inherent in its methods that the number of particles is considered as being large, and no attempt is made to find all the details of motions of the particles —a characteristic which distinguishes it from the so-called one-, two- or three-body problems.

The main approaches to the theory of quantum-mechanical many-body systems, such as nuclei, solids, and fluids, were worked out in a period of approximately ten years starting in the early 1950s. This led to a great deal of activity and attracted much attention. As a result, the term many-body problem has come to mean, almost exclusively, the theory of such systems at or near the absolute zero of temperature. The latter qualification serves to distinguish the many-body problem as such from the closely related field of STATISTICAL MECHANICS. (By convention many-body scattering theory is a separate subject). The new developments were based on the observation that when the number of particles is so large that it may be considered effectively infinite, then the system becomes very similar to that of interacting fields—except for the nature of interactions considered—and the general formal methods of quantum field theory and quantum electrodynamics may be used with advantage.

There is only one general theorem in many-body theory; it is known as Poincare's theorem. Roughly speaking, it states that any given initial state of a finite many-body system will be repeated provided one waits long enough. The quantum mechanical form of this theorem states that all observables in a finite system are almost periodic functions of time. This theorem has not had much practical use but has played an important role in discussions concerning the foundations of statistical mechanics.

The so-called many-body theory is mainly a collection of special approximate methods developed for particular problems. The chief common features of some of the methods, especially the ones connected with modern developments, will now be described.

Reduction to an Equivalent System of Noninteracting Particles The very fact that we can recognize some constituent particles leads us to believe that in the lowest approximation we may neglect their interactions. This approximation is already quite successful in derivation of perfect gas laws and electron theory of metals. A slightly different form of this assumption occurs in the case of atoms, which are treated as systems of non-interacting electrons moving in the field of force of the nucleus. For planetary systems a similar approximation is used.

The normal mode analysis of a lattice provides an example where a transformation of coordinates is used to achieve such a reduction. Instead of considering the coordinates of individual particles which interact with each other through harmonic forces, one considers certain linear combinations of displacements, the modes. In terms of the new variables there are no interactions and the solution is immediately obtained. This is an example of a *transformation* which introduces a *collective* description of the system.

Another type of situation occurs in nuclear theory, where it is found that a shell-model of the nucleus, built in analogy with the atomic shell-model, is very successful. The non-interacting particles of this model are called neutrons and protons, but interaction between them which must be used in this model is vastly different from that observed in two-body scattering experiments. As a first approximation one can completely ignore the mutual interaction and assume that the particles move in a common one-body potential. This circumstance suggests that what are called neutrons and protons in this model are not the same as the free ones but are only some *quasi-particles* which are appropriate to the model and happen to have many properties in common with actual particles. An analogous situation occurs in some solids where electrons as observed by means of cyclotron resonance experiments possess an effective mass different from the mass of free electrons. In fact, one may even say that many-body theories always deal with quasi-particles. That is true of most existing theories, but from such a point of view one loses sight of one of the basic motivations of many-body theory.

Effective Field Method This is one of the methods of taking into account the mutual interactions of the particles. One starts with a given motion of particles, e.g., from an approximation of the type described in the last paragraph, and calculates the field of force experienced by one of the particles under the influence of all the others. As a further refinement the field may be made *self-consistent*, that being the situation when the motion produced under the influence of the field is the same as that which generated it. But for the approximations made in the course of calculation, such as omission of the effects of correlations among the particles, a fully self-consistent theory would be a complete theory.

Examples are: Hartree-Fock theory, Fermi-Thomas approximations, Brueckner theory of nuclei, Wigner-Seitz cell model in solid state theory, and several others.

Density Functional Method This method is based upon a theorem of Hohenberg and Kohn, which states that if the ground state of a many-particle system is degenerate then the corresponding wave-function is a unique functional of the particle density. The theorem implies the existence of a universal functional of the external potential and particle density which is a minimum for the true particle density. Others have generalized the theorem to nonzero temperatures, nonlocal external potentials, relativistic systems, and spin-dependent systems. In the latter case a space and spin-dependent density is used. The universal functional of the theory is inferred from the basic many-body quantum mechanical description of the system. Given the functional, the ground-state energy and particle density are obtained from a variational principle involving the particle density alone. Appropriate approximations reproduce Thomas-Fermi theory and its known generalizations, but the density functional approach can go much further. Extensive applications to atomic, molecular and solid state problems have been made. (Unfortunately no comprehensive review of these works exists; we give references to the first basic papers.)

Collective Motion Theory In some phenomena, such as propagation of sound and plasma oscillations, it is clear that many particles are performing coordinated movements. To study such cases, one introduces some collective variables in addition to the usual ones, and the Hamiltonian is re-expressed in terms of these mixed variables. *Subsidiary conditions* have to be imposed upon this extended system of variables to preserve the original number of degrees of freedom. The collective variables should be such that there is no appreciable interaction between these and other degrees of freedom. When quantum mechanics is applicable, the collective motions are also excited in quanta which for all practical purposes may be treated as new (quasi) particles. The stability of collective motions is then expressed in terms of the lifetime of quasiparticles. Solid-state physics is particularly rich in exhibiting collective motions. Quasiparticles associated with some of them are: the phonons (sound, lattice vibration); the polarons (electron and its polarization field in dielectric); and the excitons (electron-hole excitations in insulators). Collective motions in nuclei can also be interpreted in a similar manner. Superconductivity and superfluidity are also examples of collective motion. The quasiparticles responsible for superconductivity are electron pairs with equal and opposite momenta and spin.

Use of Techniques of Field Theory With these techniques it is possible to obtain formal expressions which represent the effect of interparticle interactions to any order in perturbation theory.

By carrying out rearrangements and partial summations of terms in perturbation series it is possible to see that, as far as the motion inside the system is concerned, the relationship between the coordinates and momenta and the potential and kinetic energies is changed in such a way that it has to be described in terms of an effective mass and an effective interaction, which differ from the original quantities in a known way. In certain cases these effects can be calculated and are finite.

Brueckner's theory of nuclear matter is an example of this type. The effective mass is found to depend on the momentum of the particle inside the system, and the effective interaction, the so-called t- or K-matrix, is given by an integral equation involving the original interaction. A self-consistent calculation of the properties of the system (nuclei or atoms) can be based on this understanding.

Similar techniques can be used for studying collective motions. An example is the treatment of electron gas by Gell-Mann and Brueckner.

Perhaps the greatest advance has been made in the theory of superconductivity, where variational and canonical transformation methods have been used.

A combination of all these methods is needed to study the difficult problem of relationship between various excitations, i.e., the interaction between various quasiparticles of a many-body system. One of the most useful tools in these calculations is the representation of matrix elements by means of diagrams, first introduced by Feynman. Many of these methods were first developed in connection with the theory of interacting fields and they are usually employed in many-body problems for the limiting case of an infinite number of particles, but these restrictions are not essential; in fact, they are quite general methods for treating arbitrary quantum mechanical systems.

KAILASH KUMAR

References

ter Haar, D., "Introduction to the Physics of Many-Body Systems," New York, Interscience Publishers, 1958.

De Witt, B., "The Many-Body Problem," London, Methuen, 1959.

Thouless, D. J., "The Quantum Mechanics of Many-Body Systems," New York, Academic Press, 1961.

Fetters, A. L., and Walecka, J. D., "Quantum Theory of Many Particle Systems," New York, McGraw-Hill, 1971.

Hohenberg, P. and Kohn, W., "Inhomogeneous Electron Gas," *Phys. Rev.* **3B**, 864–871 (1964); Kohn, W., and Sham, L. J., "Self-Consistent Equations Including Exchange and Correlation Effects," *Phys. Rev.* **4A**, 1133–1138 (1965) (for density functional method).

Kumar, K., "Perturbation Theory and the Nuclear Many-Body Problem," Amsterdam, North-Holland Publishing Co., 1962.

Khilmi, G. F., "Qualitative Methods in Many-Body Problem," New York, Gordon and Breach, 1961 (for classical mechanics).

March, N. H., Young, W. H., Sampanthar, S., "The Many Body Problem in Quantum Mechanics," Cambridge, Cambridge University Press, 1967.

Ziman, J. M., "Elements of Advanced Quantum Theory," Cambridge, Cambridge University Press, 1969.

Cross-references: EXCITON, FEYNMAN DIAGRAMS, FIELD THEORY, KINETIC THEORY, NUCLEAR STRUCTURE, PHONON, PLASMAS, QUANTUM ELECTRODYNAMICS, QUANTUM THEORY, SOLID-STATE THEORY, STATISTICAL MECHANICS, SUPERCONDUCTIVITY, SUPERFLUIDITY.

MASER

The term "maser," coined by Townes and co-workers who pioneered this field, stands for m(icrowave) a(mplification by) s(timulated) e(mission of) r(adiation). "Microwave" has proved restrictive; stimulated emission amplifiers have operated in the UHF ($\sim$300 MHz), and at infrared, visible, and ultraviolet frequencies (see LASER). The principal advantage of the maser amplifier is its small intrinsic internal noise: the equivalent *noise input temperature* is but a few degrees Kelvin. The theoretical minimum noise input temperature is hf_s/k, where h is Planck's constant, k is Boltzmann's constant, and f_s is the signal frequency. This is 0.48 K at $f_s = 10$ Ghz (Giga Hertz) or 10×10^9 Hz. Maser oscillators can generate exceedingly monochromatic radiation, e.g., the ammonia maser has a short-term frequency stability of $\sim$5 parts in 10^{12}, and the atomic hydrogen maser has a short-term stability of better than 1 part in 10^{13}.

Because "quasi-optical" techniques are being employed increasingly in the millimeter and submillimeter regions, the distinction between LASER and maser in these regions is becoming eroded. Historically, maser oscillators used resonant systems of dimensions comparable to a cubic wavelength, (λ^3); laser oscillators used resonators with dimensions exceeding λ^3 by many orders of magnitude.

Stimulated Emission of Radiation Because its energy is quantized, a molecule (here a generic term) can exchange energy with the electromagnetic radiation field only in discrete amounts (quanta). The emission or absorption of a quantum (photon) is associated with a transition between molecular energy states. For two states, $|m>, |n>$ of energies W_m, W_n, $(W_m > W_n)$, the frequency f_{mn} of the radiation accompanying the (permitted) transition between them satisfies the Bohr condition

$$hf_{mn} = W_m - W_n. \tag{1}$$

A molecule in state $|n>$, exposed to radiation of frequency f_{mn} and energy density u, has a probability per unit time $u \times B_{nm}$ (B_{nm} is a constant) of absorbing a photon hf_{mn} and reaching state $|m>$. There is also a probability $u \times B_{mn}$ that a molecule in the upper state $|m>$ will *emit* a photon hf_{mn} and return to the lower state $|n>$. The upper state molecule is *stimulated* to emit radiation of frequency f_{mn} by the radiation field at this frequency. Stimulated emission, like absorption, is a process which is *phase coherent* with the incident radiation. Thermodynamical arguments by Einstein (1917) showed that

$$B_{nm} = B_{mn}. \tag{2}$$

A molecule in the upper energy state $|m>$ may also revert to the lower state $|n>$ by *spontaneously* emitting radiation of frequency f_{mn}. This spontaneous emission is a random process, which is phase incoherent with any incident radiation, and is therefore a source of noise in a maser.

The spontaneous emission probability A_{mn} is given by

$$A_{mn} = B_{mn} \times hf_{mn} \times \rho_f \tag{3}$$

where ρ_f is the number of wave modes per unit volume per unit frequency range open to radiation of frequency f_{mn}. Table 1 shows values of ρ_f under various conditions; c is the velocity of light, v_g is the group velocity of radiation.

In the microwave region (say, 1 to 100 GHz), $A_{mn} \ll B_{mn}$; spontaneous emission is therefore negligible except as a source of noise. However, maser spontaneous emission noise is usually exceeded by noise arising from losses in ancillary microwave circuit elements.

Molecular transitions are excited by either the electric or magnetic component of the radiation field, depending upon whether the change in molecular energy is primarily electric or magnetic in character. Each radiative transition has associated with it an effective oscillating electric or magnetic moment, usually dipolar. The probability B_{mn} given above depends directly on this dipole moment and inversely on the frequency spread (line width) δ of the transition.

TABLE 1

Environment	ρ_f
Enclosure large compared with the wavelength c/f_{mn}	$8\pi f_{mn}^2 / c^3$
Single mode resonant cavity, volume V, width of half-power response Δf	$2/(\Delta f \cdot \pi V)$
Waveguide, cross section A	$1/Av_g$

Conditions for Amplification Suppose radiation of frequency f_{mn} is incident on an assembly of molecules with an allowed transition at this frequency [Eq. (1)]. Let the number of molecules in the upper state $|m>$ be N_m, and in the lower state $|n>$ be N_n. If the incident radiation energy density is u, the power absorbed by the molecules will be

$$P_A = N_n u B_{mn} hf_{mn} \tag{4}$$

and the power emitted will be (see equation 2)

$$P_E = N_m u B_{mn} hf_{mn}. \tag{5}$$

Since at microwave frequencies spontaneous emission is negligible, the condition for amplification is

$$P_E > P_A; \text{ i.e., } N_m > N_n. \tag{6}$$

There must be an excess of molecules in the *upper* energy state of the transition associated with the signal frequency.

For thermal equilibrium at temperature T, Boltzmann statistics give

$$(N_m/N_n) = \exp\left(-(W_m - W_n)/kT\right]$$
$$= \exp\left(-hf_{mn}/kT\right) \simeq 1 - (hf_{mn}/kT) \tag{7}$$

at microwave frequencies, where $hf \ll kT$. Clearly a molecular system in thermal equilibrium is thus always absorptive. Equation (7) allows the definition of an "effective temperature" T_m for an emissive system; Eq. (6) and (7) show that T_m will be a "negative" temperature, and that $|T_m| \to 0$ for $(N_m/N_n) \to \infty$. Obtaining an emissive condition, obtaining a "negative temperature," and obtaining "population inversion" are thus synonymous. The excitation of a molecular assembly to an emissive condition is perhaps the crux of the maser problem. The schemes used depend on the conditions and on the molecular system. Discontinuous methods (pulse inversion, adiabatic fast passage) can be used, but the account here is confined to the principles of continuous methods. In a gas, actual separation of the upper-state molecules may be possible. For example, the upper-state molecules for the 23.87 GHz ammonia maser transition tend to increase their energy in a static electric field, while the lower-state molecules tend to decrease their energy (Quadratic Stark effect). In an inhomogeneous electric field, the wanted upper-state molecules will therefore drift to the low-field regions. An electrode system (with geometrical axial symmetry) which gives a low-field region along the symmetry axis will therefore confine the upper-state molecules in a beam along this axis while rejecting the lower-state ones.

In the atomic hydrogen maser, a state-selector *magnet* is used, with alternating north and south poles, again in an axial arrangement. The electron and nuclear spins can either be "parallel" ($F = 1$) or "antiparallel" ($F = 0$). The energy of the atoms with $F = 0$ decreases with the magnetic field as they go off axis, while that of the atoms with $F = 1$ increases. Hence the atoms with $F = 1$, those in the higher energy state, are focused by the magnet system along its axis, while the lower-state atoms are lost to the beam. The upper-state atoms enter a teflon-coated quartz bulb inside a microwave cavity resonant at the transition frequency. The teflon coating minimizes the chance of an atom emitting its energy because of a collision with the wall of the bulb.

Most masers operate on the multilevel excitation scheme, requiring an input of energy ("pumping") at some frequency other than the transition frequency; forms of energy other than electromagnetic may also be used. The principles of the scheme will be illustrated by reference to a molecule having 3 levels with energies $W_1 < W_2 < W_3$, such that all transitions between levels are allowed. (The transitions other than the signal transition need not radiate electromagnetically). In thermal equilibrium the number densities $(n_i)_e$ of the particles in the different states (i) will satisfy

$$(n_1)_e > (n_2)_e > (n_3)_e.$$

The frequencies f_{32}, f_{21}, f_{31} are defined from

$$f_{mn} = (W_m - W_n)/h.$$

Suppose now by some means, that the transition $1 \rightarrow 3$ is *saturated*, i.e., $n_1 \simeq n_3$. (This might be achieved by a sufficiently strong electromagnetic field at frequency f_{31} —known as the "pump" frequency). Under these conditions, it may happen either that $n_2 > n_1$, or that $n_3 > n_2$. In the first case, amplification will be possible at f_{21}; in the second case, at f_{32}, provided that the appropriate transition is electromagnetically radiative.

There are many variants of the simple scheme just described. The frequency f_{31} may lie in the optical region (OPTICAL PUMPING); the excitation may be by collision processes in a gas discharge; or more than three levels may be involved, and pump frequencies lower than the signal frequency can sometimes be used.

Maser Materials Maser action has been achieved in gases (e.g., ammonia, formaldehyde, hydrogen, rubidium vapor) and liquids (e.g., protons in water) but the most important maser materials are the solid-state ones, since these have a high concentration of active centers in a small space. Present emphasis is on the use of certain paramagnetic ions diluted in a host crystal lattice. Three-level excitation, or some variant, is usually employed.

PARAMAGNETISM is associated with ELECTRON SPIN. The directional quantization of angular momentum leads to the quantization of the energy of the ionic magnetic moments in a steady magnetic field. In general, the ground-state multiplet of these ions is split by the crystal field of the host lattice (Stark effect), and the levels are completely separated by steady magnetic field (Zeeman effect). When the steady magnetic field is applied at an angle to the major symmetry axis of the crystal field, and the resultant Zeeman splitting is comparable with the initial Stark splitting of the levels, the usually forbidden "leap-frog" transitions necessary for 3- or multiple-level excitation become allowed. In crystal fields of low symmetry, "leap-frog" transitions may be allowed at very low or even zero magnetic fields. Clearly, ions having three or more energy levels are wanted, and any processes competing with radiative processes—e.g., the interaction of the "spins" with the lattice—are usually required to be small. Spin-lattice interaction can usually be reduced by cooling the lattice to a low absolute temperature; and indeed most solid-state paramagnetic masers operate at liquid nitrogen (77 K) or liquid helium (4.2 K) temperatures. Some ions and host lattices with which maser action has been achieved are listed in Table 2.

A "spin-spin" interaction process, known as *cross-relaxation* must also be taken into account, since it may either aid or inhibit maser action. Cross relaxation is dependent on spin concentration, but not on temperature. Consequently, maser action may be achieved at comparatively high temperature (77 K) but not at low temperature (4.2 K) where the considerably longer spin-lattice relaxation time might be expected to give better maser action. Rearrangement of the level populations occurs because of single or multiple quantum transitions between the levels, in which energy is "almost" conserved on the microscopic scale, any differential being exchanged with the energy of the macroscopic spin system (total magnetic moment).

Amplifier Systems Maser amplifiers may be of either traveling-wave or resonant circuit (cavity) form. Their performances are expressed in terms of a molecular Q-factor, Q_m, defined over unit length for the traveling-wave maser and over the resonator volume for a cavity maser. At the signal frequency f_s,

$$Q_m = -2\pi f_s \times \frac{\text{Energy stored in the structure}}{\text{Power emitted by the molecules}}$$

$$(8)$$

since the Q's similarly defined for losses are positive.

TABLE 2

Ion	Effective Spin	Host Lattice
Cr^{3+}	$\frac{3}{2}$	Al_2O_3, alumina (ruby)
Cr^{3+}	$\frac{3}{2}$	TiO_2 rutile
Cr^{3+}	$\frac{3}{2}$	$Be_3Al_2(Si_6O_{18})$ emerald
Fe^{3+}	$\frac{5}{2}$	Al_2O_3, alumina
Fe^{3+}	$\frac{5}{2}$	TiO_2 rutile
Fe^{3+}	$\frac{5}{2}$	Al_2SiO_5 andalusite

For a magnetic dipole transition,

$$|Q_m| \propto \delta(N^* p_m^2 \eta)^{-1} \qquad (9)$$

where δ is the frequency width of the transition at half-intensity, N^* is the *excess* upper level population, p_m is the effective dipole moment for the transition, and η is the ratio of the magnetic energy coupled to the molecules to that stored in the microwave circuit.

Traveling-wave Maser The active maser material is placed in a waveguide carrying a pure traveling wave. The gain coefficient α_m is defined such that the power gain G for a length l of amplifier is given by

$$G = \exp(2\alpha_m l). \qquad (10)$$

It can be shown that

$$\alpha_m = (2\pi f s)/(|Q_m| v_g^{-1}) \qquad (11)$$

where v_g is the group velocity of radiation in the guide. Because p_m is typically of the order of a Bohr magneton, and the active centers are diluted, it is necessary to use *slow-wave structures* $(v_g \simeq c/100)$ in order to keep l to a reasonable value (a few centimeters). Suitable values of v_g are readily achieved by the resonant slowing obtained in periodic structures. Systems such as the Karp structure, comb structure, and meander line are favored, since these support waves with the magnetic field circularly polarized in a plane containing the direction of propagation and perpendicular to the plane of the periodic elements. A comb-structure traveling-wave maser is illustrated schematically in Fig. 1. The sense of circular polarization is reversed on crossing this plane and is opposite in any reflected wave to that in the forward wave. The nonreciprocal gyromagnetic properties of para- and ferrimagnetic materials may then be employed to obtain forward gain and reverse attenuation with these slow waveguides.

The noise input temperature T_{in} of a traveling wave maser is given approximately by

$$T_{in} \simeq |T_m| + T_1(|Q_m|/Q_e) \qquad (12)$$

where T_m is the effective negative temperature of the maser material, Q_m is the molecular Q (negative), Q_e is the similarly defined ohmic loss factor, and T_1 is the actual temperature of the waveguide (and contents). In this approximation, $|Q_m| \ll Q_e$. The bandwidth b_m of the amplifier is approximately equal to, but less than, δ.

The *Resonant circuit Maser* may be of either transmission (two-port) or reflection (one-port) type: only the reflection type is considered here, since it is superior in performance to the transmission type. A reflection cavity maser and necessary ancillary equipment are illustrated schematically in Fig. 2. Assuming that the unloaded resonant circuit (cavity) losses are negligible, the coupling to the external circuits will give rise to a Q-factor Q_e, say. The power gain G of the reflection cavity maser is then given by

$$G = (Q_e + |Q_m|)^2(Q_e - |Q_m|)^{-2} \qquad (13)$$

The bandwidth b_c depends on the gain in such a way that

$$G^{1/2}b_c \simeq 2|Q_m|/f_s(\text{for } G > 10, \text{ say})$$

The noise input temperature is given by Eq. (12) above, where now

$$Q_e^{-1} = Q_e^{-1} - |Q_m|^{-1}$$

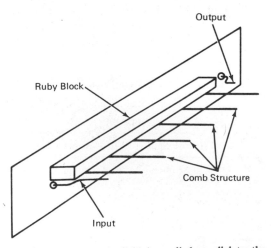

FIG. 1. A magnetic field is applied parallel to the "teeth" of the comb.

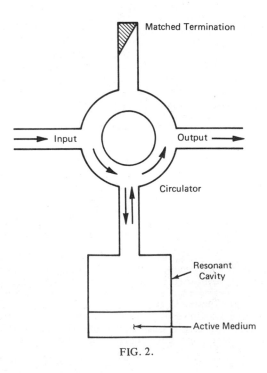

FIG. 2.

It is necessary to have some nonreciprocal device to separate the reflected amplified output from the input signal; the ferrite circulator is most commonly used. The bandwidth and gain stability of the cavity maser are inferior to that of the traveling-wave maser, but the cavity maser is more easily constructed. If three-level excitation is used, it is clear that any maser system must support both "pump" and signal frequencies.

Maser Oscillators Equation (13) indicates that if $|Q_m|$ is small enough, G becomes infinite; i.e., oscillation occurs when the stimulated emission is large enough to overcome all losses. The width of the signal emitted by a maser oscillator is very much less than δ, so that for narrow δ an extremely pure oscillation signal results, and a molecular transition which is relatively insensitive to external influences will thus give oscillations of high stability in frequency. The ammonia maser and the atomic hydrogen masers are two examples.

Two hydrogen maser oscillators have been compared against each other, and the relative frequency over several hours was stable to 1 part in 10^{14}. This stability has allowed the determination of the level separation of atomic hydrogen and its isotopes with greatly increased precision (e.g., 0.1 Hz in 1,420,405,751.768 Hz), and has thus allowed a further test of theoretical quantum electrodynamics, and given a more precise value for the fine structure constant. Hydrogen masers have also been flown in missiles and aircraft for the detection of small relativistic effects of the motion.

The Electron-Cyclotron Maser Relativistic free electrons gyrating in a static magnetic field undergo free-free transitions. If there is an energy-dependent level width, or an energy-dependent level spacing for the quantized free electron states, and a population inversion is also present, stimulated emission and hence amplification can occur.

The classical picture of cyclotron maser action can be obtained by considering the phases of electrons gyrating about the magnetic field. Each charge will radiate as an electric dipole; a multipolar contribution also occurs when relativistic effects are properly taken into account. Consider a system of monoenergetic electrons, initially distributed randomly in phase. If a phase bunching mechanism exists, coherent emission will take place. Because of the relativistic mass change, this bunching does in fact occur. Electrons absorbing radiation become more massive, and go back in phase; electrons emitting radiation become less "massive" and advance in phase. The ultimate phase distribution favors emission over absorption, thus increasing the intensity of the incoming electromagnetic wave.

From the quantum-mechanical viewpoint, a free electron in a uniform static magnetic field B is an anharmonic oscillator with quantized energy levels (neglecting spin)

$$W_n = mc^2 [1 + 2(n + \tfrac{1}{2})h\Omega_0/2\pi mc^2]^{1/2}$$
$$- mc^2 + p^2/2m$$

where Ω_0 is the rest electron gyrofrequency eB/m, and p is the unquantized momentum along the direction of B. Transitions between states $|n + 1\rangle$ and $|n\rangle$ occur at angular frequency $\omega_n = (1 - nh\Omega_0/2\pi mc^2)\Omega_0$ for $nh\Omega_0/2\pi \ll mc^2$; note that ω_n decreases as n increases. If a system has a greater population in the state $|n + 1\rangle$ than in $|n\rangle$, photons of angular frequency ω_n will induce more (downward) transitions $|n + 1\rangle \rightarrow |n\rangle$ than (upward) transitions $|n\rangle \rightarrow |n + 1\rangle$, because of the unequal level spacing; hence stimulated emission exceeds absorption. If the width of the level $|n + 1\rangle$ exceeds that of $|n\rangle$, a similar effect can occur. This situation can be obtained by causing the radiating atoms to suffer elastic phase-interrupting collisions (e.g., with neutral atoms) if the energy-dependent collision cross section is sufficiently strong.

In a typical device, the *gyrotron*, a solenoid creates an axially symmetric magnetic field about a gently tapering waveguide system, whose different sections act as interaction space (open or "quasi-optical" cavity) and output (and if necessary, input) apertures. The electrons are produced at a cathode with a large emitting surface, and accelerated towards a collector. The magnetic field increases in intensity from the cathode to the interaction space, which has an almost uniform magnetic field. In the nonuniform field region, the electron orbital velocity $v_\perp$ grows from the initial cathode orbital velocity according to $v_\perp^2/B =$ constant: the orbital energy is drawn from that of the longitudinal motion, and from the accelerating electrostatic field. The electrons deliver up the RF energy in the interaction space, then pass through a section of decreasing magnetic field to an extended surface collector. Such devices are capable of powers of many kilowatts in the millimeter and submillimeter wavelength regions.

Applications Maser amplifiers are now in use wherever the requirement for a very low noise amplifier outweighs the technological problems of cooling to low temperatures. They have been used in passive and active radioastronomical work, in satellite communications ("Project Echo") and as preamplifiers for microwave spectrometry. The "deep-space tracking" stations around the world use ruby maser preamplifiers for the reception of signals from planetary probes. The ammonia and the atomic hydrogen masers are being studied as frequency standards and have been used in a new accurate test of special relativity. Sources and amplifiers in the submillimeter, micron, and optical wavelength regions are being studied and developed (see LASER). Maser theory has been used to explain numerous atomic and molecular emission lines observed in radio astronomy ('celestial masers').

G. J. TROUP

References

Andronov, A. A., Flyagin, V. A., Gaponov, A. V., Gol'denberg, A. L., Petelin, M. I., Usov, V. G., and Yulpatov, V. K., "The Gyrotron: High-Power Source of Millimetre and Submillimetre Waves," *Infrared Physics* 18, 385 (1978).

Hershfield, J. L., and Granatstein, V. L., "The Electron Cyclotron Maser—An Historical Survey," *IEEE Transactions on Microwave Theory and Techniques* MTT-25(6), 522 (1977).

Microwave Journal Staff, "Low Noise Maser for Radio Astronomy," *Microwave J.* 21(3), 52 (1978).

Ramsey, N. F., "Hydrogen Maser Research," in "Fundamental and Applied Laser Physics" Proceedings of Esfahan Symposium (Feld, M. S., Javan, A., and Kurnit, N. A., Eds.), New York, John Wiley & Sons, 1973.

Weber, J., *Rev. Mod. Phys.* 31, 681 (1959).

Books

Cook, A. H., "Celestial Masers," Cambridge Monographs on Physics, Cambridge, U.K., Cambridge Univ. Press, 1977.

Ishii, T. Koryu, "Maser and Laser Engineering," Huntington, N.Y., Krieger, 1980.

Orton, J. W., Paxman, D. H., and Walling, J. C., "The Solid State Maser," London, Pergamon, 1970.

Siegman, A., "Microwave Solid State Masers," New York, McGraw-Hill, 1964.

Siegman, A., "An Introduction to Lasers and Masers," New York, McGraw-Hill, 1971.

Troup, G., "Masers and Lasers," 2nd Edition, London, Methuen and Co., 1963.

Weber, J. (Ed.), "Masers," International Science Review Series, New York, Gordon and Breach, 1967.

Cross-references: COHERENCE, ELECTRON SPIN, FERRIMAGNETISM, LASER, LIGHT, MICROWAVE SPECTROSCOPY, MICROWAVE TRANSMISSION, OPTICAL PUMPING, PARAMAGNETISM, QUANTUM THEORY, ZEEMAN AND STARK EFFECTS.

MASS AND INERTIA

Mass is, along with length and time, one of the three fundamental undefinables of Newtonian dynamics (the explanation of matter in motion). These three undefinables combine to form the important operational quantity called *force*. One in turn may consider force as a measure of the interaction of mass with its environment.

Force is defined in such a way that it is dependent on either one of two intrinsic properties of matter, called *gravitation* and *inertia*. First, all matter exerts an attractive force on all other matter (active gravitational mass) and is in turn attracted by all other matter (passive gravitional mass). Secondly, all matter resists any change in its motion (inertial mass).

The gravitational role of mass is quantified by Newton's universal law of gravitation,

$$F = -\frac{Gm_1 m_2}{r^2} \hat{u}_r \qquad (1)$$

wherein the force F of attraction (designated by the negative sign) between masses m_1 and m_2 is inversely proportional to the square of the distance r between their centers of mass, and directly proportional to the magnitudes of m_1 and m_2. F is called a *central force* because it acts in a straight line (the direction of which is given by a unit vector $\hat{u}_r$) which originates at the center of mass m_1 and terminates at the center of mass m_2.

This force is independent of the physical dimensions of m_1 and m_2. Its strength, compared with forces of other kinds such as that of electricity, is determined by the size of the gravitational constant G. In view of the high degree of accuracy over centuries of application inhering in Eq. (1), which does not require any specification as to which of the two masses is the active and which is the passive gravitational mass, no quantitative difference is expected between these two states of gravitational mass. Further, Newton's third law of action and reaction, to be considered below, implies their equality.

The inertial property of mass is described by Newton's three laws of motion. According to the first law, a body will continue in its state of rest or in a state of uniform motion (unchanging speed along a straight line) unless acted upon by a net external force. This means, for example, that the earth, instead of orbiting about the sun, would move in a straight line and thus escape from the solar system. The "escape" is prevented by the force of gravitational attraction between the sun and the earth. It also means that motion of mass can exist without force. It is through Newton's second and third laws however, that an operational definition of inertial mass is achieved, one based on the concept of force which can be a push or a pull (such as that inherent in the force called *weight*). The second law states that the net force F is equal to the change (or derivative) of momentum with time, momentum being the product of mass m and its velocity v:

$$F = \frac{d(mv)}{dt}. \qquad (2a)$$

Taking the derivative of the momentum with respect to time leads to a two-term expression for the force F:

$$F = m_i a + \frac{dm}{dt} v. \qquad (2b)$$

For most conventional situations the inertial mass is constant, so that the second term $(dm/dt)\mathbf{v}$ is equal to zero, thereby leaving $\mathbf{F} = m_i\mathbf{a}$. The acceleration $\mathbf{a}$, which is the derivative of velocity $\mathbf{v}$ with respect to time $(d\mathbf{v}/dt)$, represents a time-dependent increase or decrease in velocity of the mass or a change in its direction of motion such as that encountered in rotation. Velocity is speed in a specific direction, and is an example of a vector, namely it has magnitude (i.e., speed) and a specific direction. In the case of the unit vector $\hat{\mathbf{u}}_r$, the magnitude is one. Operationally, Eq. (2b) states that when the force $\mathbf{F}$ is fixed, mass and acceleration are inversely proportional to each other. The interdependence of the concepts of force and mass reflected by $\mathbf{F} = m_i\mathbf{a}$ emphasizes the lack of a unique definition of what inertial mass is. The latter, however, may not prove ultimately to be a drawback, for the following reason. At least three of the hierarchy of four forces, i.e., gravitation, weak, electromagnetism, and strong, contain empirical constants, such as the permittivity of free space ϵ_0, the gravitational constant G, or the Fermi coupling constant G_F. A delineation of the origin of these constants would provide a deeper insight into the nature of the interaction of mass with its environment, as well as a potentially richer understanding of the nature of mass itself.

For example, Einstein in banishing the concept of absolute velocity by focusing attention on the role of the relativity of the velocity term in Eq. (2b), enlarged the Newtonian operational view of the concept of inertial mass. In his theory of special relativity, the amount of mass associated with an object was shown to be a function of its relative velocity,

$$m = \gamma m_0, \qquad (3a)$$

a dependence quantified by the Lorentz transformation γ,

$$\gamma = \frac{1}{\sqrt{1 - (v^2/c^2)}}. \qquad (3b)$$

Here c, the speed of light in a vacuum, is considered independent of any motion of its source. The total mass of an object m was now considered to be the sum of a constant, intrinsic rest mass m_0 and an increment of variable mass depending upon its velocity relative to some reference frame. This latter effect negates defining mass as a measure of the quantity of matter in a body. Significantly, mass could now be shown to be equivalent to energy via the relationship $E = mc^2$. It is worth noting that this energy expression for mass does not reflect the fact that it is considerably easier for stable mass to be converted into radiant energy than the reverse. Significantly, the special relativistic Newtonian operational definition of mass has been found to be highly compatible with the

concept of electromagnetic fields as well as with the quantum of action. This success has culminated in the extraordinarily precise discipline of quantum electrodynamics.

It should not be concluded, however, that the concept of force is a unique prerequisite in Newtonian mechanics for the definition of inertial mass. Mach proposed an experiment for determining the relative inertial mass free of any consideration of force including that of weight. Two masses M and m on a frictionless surface are held together against a compressed spring between them. When released, they fly apart under the influence of the equal and opposite forces exerted by the spring. The second law states that the product of the mass and the magnitude of the acceleration for each object should be identical, or

$$mA = Ma. \qquad (4a)$$

If we replace M by one standard unit of mass, the kilogram, i.e., $M = 1$, then

$$m = \frac{a}{A}. \qquad (4b)$$

Mass m is therefore operationally defined as a ratio of two accelerations, i.e., that of the standard a and that of the object A; the ratio of the two accelerations is constant for a particular body and thereby quantifies an intrinsic property of matter. Its validity requires that Newton's third law hold, namely that for any force of action there must exist an equal and opposite force of reaction.

From the time of Galileo, there has existed reason to believe that the two intrinsic properties of mass might have a common origin. Since the mass of a falling object on earth is simultaneously inerting m_i and gravitating m_G, one can equate Eqs. (1) and (2b), where, on earth, M_e is the mass of the earth and r is the distance from the center of the earth to the falling object:

$$m_i\mathbf{a} = -\frac{GM_e m_G}{r^2}\hat{\mathbf{u}}_r. \qquad (5)$$

To a surprising degree of accuracy (about 1 part in 10^{12}) the acceleration for all falling bodies in the vicinity of the earth is equal to $-GM_e/r^2$, more commonly designated as $g \cong 9.8$ Newtons/ kilogram. It is the variation in r in the factor g and not a change in mass that causes the weight of a given object to be location dependent. This means that to the extent that all objects fall to the ground with the same acceleration, the ratio m_G/m_i, at fixed r, in Eq. (5) is constant so that m_i can be set equal to m_G by redefining the constant G. Einstein seized upon this apparent experimental equality of inertial and gravitational mass and made it absolute, thereby adopting the suggestion of Mach that inertial mass is simply the gravitational attraction of all the mass in the universe for the matter undergoing a change in its motion. This enabled him to reduce the number of fundamental undefinables of dynamics to two, namely length (three dimensions) and time. If we multiply the latter by the postulated invariant speed of light c, we obtain a fourth length or the fourth dimension. Thus, dynamics is reduced to four-dimensional geometry and the existence of mass is indirectly reflected by the warpage of this four-dimensional geometric structure. (The unusual behavior of black holes can be attributed to optical effects encountered under conditions of extreme warpage.) From a Newtonian standpoint, general relativity banishes absolute acceleration and quantifies the relativity of acceleration.

General relativity, in eliminating mass as a fundamental undefinable through the removal of any distinction between inertial and gravitational mass, redefines mass as a measure of the curvature of space-time. Despite many experimental confirmations involving matter and light; general relativity has resisted accommodation with electric charge, as well as with the related quantum of action h. Further, recent measurements of solar oblateness by Hill, if confirmed, would suggest that a small portion (~1.4%) of the advance of the perihelion of Mercury (43"/ century), until now a stunning, unique and exact prediction of general relativity, may have to be attributed to some other mass effect. It therefore remains an open question whether a description of mass based upon the elimination of all distinction between inertia and gravitation can be maintained.

B. A. SOLDANO

References

General Relativity

Klein, H. Arthur, "The New Gravitation: Key to Incredible Energies," Philadelphia and New York, J. B. Lippincott Company, 1971.

Sciama, P. W., "The Unity of the Universe," Garden City, N.Y., Anchor Books (Doubleday and Co.), 1959.

Lieber, L. R., and Lieber, H. G., "The Einstein Theory of Relativity," New York and Toronto, Rinehart and Company, Inc., 1945.

Black Holes

Taylor, J. G., "Black Holes," New York, Avon Books, 1973.

Penrose, Roger, "Black Holes," *Scientific American*, pp. 38–46, May 1972.

Ruffini, Remo, and Wheeler, John A., "Introduction to Black Holes," *Physics Today* 24(1), 30–41 (January 1971).

Newtonian Physics and Special Relativity

Hecht, E., "Physics in Perspective," Reading, Mass., Addison Wesley Publishing Company, 1980.

Eisenberg, R. M., and Lerner, L. S., "Physics: Foundations and Applications," Vol. I, New York, McGraw-Hill, 1981.

French, A. P., "Special Relativity," M.I.T. Introductory Physics Series, New York, W. W. Norton and Company, Inc., 1966.

Dicke, R. H., "The Solar Oblateness and the Gravitational Quadrupole Moment," *Astrophysical Journal* 59 1–24 (January 1970).

Cross-references: DYNAMICS, FRICTION, IMPULSE AND MOMENTUM, MECHANICS, RELATIVITY, STATICS.

MASS SPECTROMETRY

Mass spectrometry is based on observations of the behavior of positive rays by Thomson and Wien. In 1919, Aston demonstrated the existence of isotopes by introducing neon gas into a mass spectrograph. Prior to 1940, mass spectrographs and spectrometers were used primarily for isotopic studies in university laboratories. Analytical spectrometers became commercially available during the early years of World War II when their use for the rapid analysis of hydrocarbon mixtures was recognized.

Mass spectrometry provides information concerning the mass-to-charge ratio and the abundance of positive ions produced from gaseous species. There are several techniques for the production and measurement of the ions, and the design of an instrument is determined by its proposed application. The mass spectrograph, using a photographic plate for ion detection, had been used primarily for isotopic studies but later was used for the analysis of trace constituents in solids. The mass spectrometer uses an electrical detection and recording system giving a metered output that provides a more accurate measure of the abundance of the ions than the photographic plate provides. The mass spectrometer is used primarily for the quantitative analysis of gases, liquids, and a limited number of solids.

The five basic components of the instrument are the sample introduction system, the ion source, the mass analyzer, the ion detector, and the recorder. A sample pressure of approximately 5×10^{-5} torr is generally required for a satisfactory analysis. An elevated temperature inlet system or other means of converting the sample into a gaseous state is required for less volatile species. Direct insertion probes for introducing samples directly into the region of ionization are available on many instruments.

The most common methods of producing positive ions are electron impact, thermal ionization, spark, field emission desorption, and chemical ionization. The electron impact source is the most widely used. Positive ions are produced by removing one or more electrons from the molecules. Thermal ionization produces positive ions by vaporizing a material directly into the ion source from a filament coated with the sample. With a spark source, the material under investigation must be a conductor, or else suitable means must be provided for initiating and maintaining a spark. Ions produced in the spark are taken directly into the mass analyzer. In the field emission source, a high potential is applied between the sample— generally deposited on the tip of a tungsten wire—and another electrode. Ionic species representative of the sample are removed by a high-intensity electric field. Chemical ionization involves the reaction, directly in the ion source, of ions from a reactant gas such as methane with molecules of the sample.

The three most widely used types of mass spectrometers are (1) the single-focusing, magnetic deflection, (2) the double-focusing, and (3) the quadrupole. These three types of instruments differ primarily in the method used for mass separation. The single-focusing analyzer achieves direction but not velocity focusing of the ions. Ions of the same mass-to-charge ratio, having slightly different velocities resulting from different kinetic energies imparted in the ionization process, will not be focused simultaneously, thus producing a broadening of the peak. The resolution of commercially available instruments of this type is generally limited to about one part in 500. That, is, mass 499 can be separated from mass 500 with about a 10% valley. With double-focusing instruments, an electric sector and a magnetic analyzer are placed in tandem to produce both velocity and direction focusing of the ions. Several commercial models of the double-focusing design are available having resolutions in excess of one part in 50,000 with an electron impact source, and greater than one part in 3000 with a spark source. The double-focusing geometry is necessary with spark source operation because of the wide energy spread of ions produced in the spark. With both single and double focusing, the resolution varies directly with the radius of the analyzer tube and inversely with the width of the slits located in the ion source and ion collector regions. Sensitivity, the abundance of the ions collected per unit sample charge, varies inversely with slit width, and a compromise must be made between resolution and sensitivity. Combined gas chromatography-mass spectrometry with appropriate data processing units is a popular type of instrumentation. Separation of complex mixtures into individual components as well as analysis is accomplished by mating these two types of instrumentation. Recently, two mass analyzers have been used in tandem for the technique termed MS/MS. The first analyzer is followed by a reaction region and then by a second analyzer. Chemical ionization is commonly used for the source of positive ions. In this technique, a spectrum is obtained of ions at a single m/e in the initial spectrum.

A major advantage is the increased capability to detect individual organic compounds in complex mixtures.

The two types of ion detection and recording systems are the photographic plate and the electrical detector. The photographic method is commonly used with double-focusing instruments such as the Mattauch-Herzog design that focuses all ions simultaneously in one plane. The photographic plate records a complete spectrum (mass range $\sim 36:1$) in a time interval of a few seconds to 10 minutes. However, the response of the plate to the ion intensity is nonlinear and quantitative results are more difficult to obtain than by electrical detection. Electrical detection systems use an ion collector, amplifier (commonly an electron multiplier), and recorder.

Positive and negative ions and neutral species are produced by the electron bombardment of molecules. The mass spectrum of a compound is a record of the positive ions collected. Positive ions are produced by the removal of one or more electrons from the molecule and by the rupture of one or more bonds, fragmenting the molecule. While the majority of the positive ions are singly charged, doubly and triply charged ions are observed in many instances. Certain mass ions produced from organic molecules must be attributed to the rearrangement of hydrogen atoms during the ionization and fragmentation processes. Metastable ions, formed when ions decompose while traversing the path to the collector, are also frequently observed. Metastable ions generally appear at nonintegral mass units and produce broad, low-intensity peaks.

In the electron impact source, the electron energy is usually adjusted to 50–70 eV, which is considerably above the appearance potential for molecular and fragment ions. For simplification of a complex spectrum, the bombarding energy can be reduced to provide sufficient energy to ionize the molecule but not enough to rupture bonds, thus achieving a spectrum consisting primarily of molecular ions. Mass ions appearing in the normal mass spectrum correspond to the various atoms and combinations of atoms in the original molecule. The pattern of mass-ion intensities observed is independent of pressure. Differences in the patterns obtained for various compounds can be used as the basis for the analysis of complex mixtures. Quantitative analysis is based on the ion current varying linearly with the partial pressure of the gas.

Some of the common uses for mass spectrometry include analysis of petroleum products, identification of drugs, determination of the structure of organic molecules, determination of trace impurities in gases, residual vacuum studies and leak detection in high-vacuum systems, geological age determinations, tracer techniques with stable isotopes, determination of unstable ionic species in flames, identification of compounds separated by gas chromatography (combined gas chromatography-mass spectrometry), trace element analysis in metals and other solids, and microprobe studies of surfaces and of surface composition of various materials.

A. G. SHARKEY, JR.

References

1. Duckworth, Henry E., "Mass Spectroscopy," Cambridge, U.K., Cambridge Univ. Press, 1958.
2. Beynon, J. H., "Mass Spectrometry and is Applications to Organic Chemistry," Amsterdam, Elsevier Publishing Co., 1960.
3. Biemann, K., "Mass Spectrometry: Organic Chemical Applications," New York, McGraw-Hill Book Co., 1962.
4. McLafferty, F. W. (Ed.), "Mass Spectrometry of Organic Ions," New York, Academic Press, 1963.
5. Hill, H. C., "Introduction to Mass Spectrometry," 2nd Ed., London, Heyden and Son, Ltd., 1973.
6. Hamming, Mynard, C., and Foster, Norman G., "Interpretation of Mass Spectra," New York, Academic Press, 1972.
7. McLafferty, Fred W., "Tandem Mass Spectrometry," *Science* **214** (1981).
8. Cooks, Graham R., and Glish, Gary L., "Mass Spectrometry/Mass Spectrometry," *C & EN*, November 30, 1981.

Cross-references: FIELD EMISSION, IONIZATION, ISOTOPES, SPECTROSCOPY.

MATHEMATICAL PHYSICS

The term "mathematical physics" is almost synonymous with "THEORETICAL PHYSICS," but their difference is significant. It is like the difference between the descriptions of the electromagnetic field by Maxwell and by Faraday respectively. The theoretical (nonmathematical) description draws on analogies between elements of the field and familiar mechanical models—stretched strings, compressed fluids, vortex motion, etc.; the mathematical description made use of the abstract analytical properties of the elements of the field to set up a purely symbolic description without mechanical models. Classical theoretical physics was largely mathematical in content, but was nevertheless based on mechanical models in the spirit of Faraday's theory of the electromagnetic field. The atom and interactions between atoms were regarded as the "real," "external" objects in terms of which all physical phenomena could be explained. The mathematical formalism was merely a handy tool or language in terms of which to set up the explanation. The atoms themselves were not explained, but regarded as the fundamental "building blocks" of the physical world.

Einstein's theory of RELATIVITY is a magnificent historical example of mathematical physics we may cite to contrast with the classical atomic theory. Here mathematical abstractions, Minkowski space 4-vectors, Riemannian tensors, etc. were invented or adopted from the stock-in-trade of pure mathematicians, with analytical properties that were seen to match those of the data of experimental physics—velocities, forces, field variables, etc. Then the logical (i.e., mathematical) consequences of relations among these abstractions predicted new and unexpected relations among either already known or as yet undiscovered data of experimental physics. The construction of a self-consistent mathematical *description* of all physical phenomena, without the use of hypothetical building blocks of any kind, is the aim of mathematical physics as distinct from theoretical physics.

As a more recent example of this same concept, one may cite the deduction of conservation laws from generalized symmetry principles. The classically familiar conservation laws: energy, momentum, angular momentum, have for some time now been recognized as logical consequences of the homogeneity and isotropy of time and space. Interpreting these homogeneities and isotropy as meaning the invariance of physical laws under translation and rotation, one is led naturally to the theorem that each invariance principle corresponds to an appropriate conservation law. This theorem has yielded very significant discoveries in the study of elementary particles and nuclear structure where the most cunningly devised classical models have been not only fruitless but actually misleading. A non-technical account of this subject appears in Chapter 27 of R. K. Adair's text "Concepts in Physics," New York, Academic Press, 1969.

The activities of mathematical physicists have resulted in the invention of new mathematical abstractions some of which were at first rejected by pure mathematicians as illogical, only later to be granted a respectable status in the vocabulary of pure mathematics. Examples include Oliver Heaviside's operational calculus, J. Willard Gibbs' vector analysis, and P. A. M. Dirac's delta-function techniques. On the other hand, many branches of pure mathematics which initially had been regarded as so abstract as to be entirely "useless," have been found by mathematical physicists to serve as remarkably useful tools in describing physical phenomena. Examples include non-Euclidean geometry in the problems of COSMOLOGY; function space in modern QUANTUM MECHANICS; spinor analysis, or the theory of binary forms, in quantum FIELD THEORY. Again collaboration between mathematical physicists and mathematicians has in recent years resulted in the construction of new disciplines of great value, examples being group theory, operations analysis, the theory of random functions, information theory and CYBERNETICS. The names of many contemporary scientists are involved here, including Eugene P.

Wigner, John von Neumann, C. E. Shannon, Norbert Wiener and many others.

On closer examination it becomes difficult to distinguish clearly between mathematical physics and applied mathematics; very frequently the same individual may be responsible for discoveries in both areas. Classical examples of this may be cited: Isaac Newton, Laplace, Carl Friedrich Gauss, Henri Poincaré, David Hilbert, Ernst Mach, A. N. Whitehead. Evidently our attempt to define mathematical physics is degenerating into a simple catalog of items with only a vague hint of general characteristics common to all particulars. Physics has sometimes been defined as what physicists do, and one is expected to recognize the physicist without need for further definition than his own affirmation. Mathematical physics may then be defined as what physicists do with mathematics, or what mathematicians do with physics, or some superposition of the two. As the history of mathematical physics unfolds it becomes apparent that activity tends to cluster in a few fruitful directions at any one time. Current interests can be judged from the contents of the leading journals devoted to the subject; among these the reader should consult the *Journal of Mathematical Physics*, and the *Physical Review*, published by the American Institute of Physics; *The Proceedings of the Cambridge Philosophical Society; Comptes Rendus* (French Academy of Sciences); *Progress of Theoretical Physics* (Japan); *Nuovo Cimento* (Italy); *Indian Journal of Theoretical Physics; Zhurnal Eksperimental' noy i Teoreticheskoy Fiziki* (USSR) (in English Translation "JETP"); and other translations published by the American Institute of Physics. Probably the most popular fields in recent years have been in the wide application to solid-state physics and statistical mechanics of quantum field theoretical techniques introduced initially to deal with the phenomena of high energy physics—nuclear interactions, creation and destruction of particles, etc.

During the last decade the explosive growth in the availability of miniature computers and the wide accessibility of large computer facilities have had a revolutionary impact on the thinking of mathematical physicists, if not on mathematical physics itself. Exceptionally complicated problems that were easily proven "soluble in principle" have become soluble in practice, and this fact alone has freed creative thinkers to ask ever more elaborate and useful questions—e.g., finding the eigenvalues of very large matrices, solving nonlinear differential equations and adjusting parameters to make the solutions match experimental data, simulating the behavior of systems with many degrees of freedom* and the resulting understanding of

*A good example: Abraham, F. F., "Computer Simulation of Diffusion Problems . . . ," IBM Data Processing Division, and The Materials Science Department, Stanford University, Palo Alto, California, 1971.

shock formation in high-speed phenomena. For these purposes there are now several libraries of program subroutines useful in the study of literally thousands of unexplored problems. Computational projects that were never even contemplated because they would have taken decades to complete can now be undertaken and completed in hours or even minutes. The other side of the coin is: such masses of numerical data are generated that results can take years to digest and interpret. This difficulty is ameliorated in the simpler cases by the recently developed graphics display capability.

The most significant advance to date would seem to be the possibility of programming computers to perform abstract algebraic analysis without the use of purely numerical input. This may or may not help eliminate the one most serious weakness of computer science: computer programming is so absorbing an occupation that an expert can hardly be expected also to understand the historic significance, or even relevance, of the major problems at the frontiers of knowledge in mathematical physics. Perhaps the most serious error committed by schools of computer science has been to exempt their students from a solid course on mathematical physics. Physicists who wish to keep in touch with advances in computer science should watch the continuing series of annual volumes "Methods in Computational Physics" (Alder, Fernbach, Rotenberg, Eds.; Academic Press).

A philosophy of mathematical physics has gradually evolved with all this creative activity. For current thinking in this area, the reader is referred to two pertinent journals that have appeared in the past few years: *"Foundations of Physics,* and *International Journal of Theoretical Physics,* both from Plenum Press. Semipopular expositions are available in a number of recent texts in addition to the one by Adair, cited above: Kenneth W. Ford, "The World of Elementary Particles, 1967; F. A. Kaempffer, "The Elements of Physics," 1967; and Kenneth W. Ford, "Basic Physics," 1968; all from Blaisdell Publishing Co.

As for college level texts on mathematical physics, there are very few that are organized around physical concepts. We may refer to the classic series of volumes by Arnold Sommerfeld, and those by Slater and Frank, as prototypes. The standard modern work is the two-volume set by Morse and Feshbach, "Methods of Theoretical Physics," New York, McGraw-Hill, 1953. A less ambitious volume, William Band, "An Introduction to Mathematical Physics," New York, Van Nostrand Reinhold, 1959, served as an overall survey for advanced undergraduates. There is now a multitude of excellent texts whose major emphasis is the various mathematical techniques employed in theoretical physics, and we cite only three: Margenau and Murphy, "The Mathematics of Physics and Chemistry," New York, Van Nostrand Reinhold, 2 volumes, 1956 and 1964; George Arfken, "Mathematical

Methods for Physicists," New York, Academic Press, 1966; and James T. Cushing, "Applied Analytical Mathematics for Physical Scientists," New York, John Wiley & Sons, 1975.

WILLIAM BAND

Cross-references: COSMOLOGY, CYBERNETICS, FIELD THEORY, MATHEMATICAL PRINCIPLES OF QUANTUM MECHANICS, QUANTUM THEORY, RELATIVITY, THEORETICAL PHYSICS.

MATHEMATICAL PRINCIPLES OF QUANTUM MECHANICS

Classical Newtonian mechanics assumes that a physical system can be kept under continuous observation without thereby disturbing it. This is reasonable when the system is a planet or even a spinning top, but is unacceptable for microscopic systems such as an atom. To observe the motion of an electron, it is necessary to illuminate it with light of ultrashort wavelength (e.g., γ-rays); momentum is transferred from the radiation to the electron and the particle's velocity is therefore continually disturbed. The effect upon a system of observing it can never be determined exactly, and this means that the state of a system at any time can never be known with complete precision; this is *Heisenberg's uncertainty principle*. As a consequence, predictions regarding the behavior of microscopic systems have to be made on a probability basis and complete certainty can rarely be achieved. This limitation is accepted and is made one of the foundation stones upon which the theory of quantum mechanics is constructed.

Any physical quantity whose value is measured to determine the state of a physical system is called an *observable*. Thus, the coordinates of a particle, its velocity components, its energy, or its angular momentum components are all observables for the particle. A pair of observables of a system are *compatible* if the act of measuring either does not disturb the value of the other. The cartesian coordinates (x, y, z) of a particle are mutually compatible, but the x-component of its momentum p_x is incompatible with x (similarly, y, p_y are incompatible, etc.). $(p, q, \cdots w)$ constitute a *maximal set of compatible observables* if they are compatible in pairs and no observable is known which is compatible with every one of them. In quantum mechanics, the state of a system at an instant is fully specified by observing the values of a maximal set of compatible observables. Such a state is called a *pure state*. The act of observing a system in a pure state disturbs the system in a characteristic manner; the system is accordingly said to be *prepared* in the pure state by the observation. The cartesian coordinates (x, y, z) of a spinless (see below) particle

form a maximal set, and if the position of the particle is known, it is in a pure state; similarly, the momentum components (p_x, p_y, p_z) form a maximal set and a particle whose momentum is known is also in a (different) pure state.

A pure state S in which the observables of a maximal set S have known values is termed an *eigenstate* of S and the observable values are called their *eigenvalues* in the eigenstate. If $(p_0, q_0, \cdots, w_0)$ are the eigenvalues, the eigenstate is sometimes denoted by the symbol $|p_0, q_0, \cdots, w_0>$; in this article, pure states will be denoted more concisely by Greek letters. The eigenstates of S are represented by mutually orthogonal vectors, called the *base vectors*, defining a frame of rectangular axes F in an abstract *representation space*. The number n of such eigenstates is usually infinite, but for simplicity, it will first be assumed that each observable has only a finite number of eigenvalues and hence that the number n is finite; the representation space is then a straightforward generalization of ordinary space to n dimensions, with the additional requirement that the components of a vector will be permitted to take complex values. Any pure state of the system (not necessarily an eigenstate of S) is represented by a vector α in the representation space. If $(\alpha_1, \alpha_2, \cdots, \alpha_n)$ are the components of α with respect to F, and these components are arranged as a column matrix, this matrix is said to provide an S-representation of the state. The state, vector, and column matrix are all denoted by α.

The *scalar product* of two vectors α, β is denoted by (α, β) and is defined in terms of their components in F to be $\sum_{i=1}^{n} \alpha_i^* \beta_i$, where α_i^* is the complex conjugate of α_i. $\sqrt{(\alpha, \alpha)}$ is called the *norm* of α and corresponds to the length of an ordinary vector. All vectors representing pure states are taken to have unit norms and the S-eigenstates are accordingly represented in the S-representation by the columns $(1, 0, \cdots, 0)$, $(0, 1, \cdots, 0)$, etc.

Suppose a system is prepared in the state α at time t. The probability that the system can be observed in the state β at an instant immediately subsequent to t is taken to be $|(\alpha, \beta)|^2$. This event is termed a *transition* of the system from α to β. If β is identical with α, the probability of the transition is unity, as we expect. The probability of a transition from α into the eigenstate $(1, 0, \cdots, 0)$ is found to be $|\alpha_1|^2$. This provides a physical significance for the components of α in the frame F.

If the system is prepared in a state α on a number of occasions and an observable a is measured immediately after the preparation on each occasion, the values obtained will usually differ. If α is an eigenstate of a with eigenvalue a_1, then a will take this value with complete certainty. If, however, this is not the case, a will take all its possible eigenvalues a_i with associated probabilities p_i, and its *mean* or *expected value* $\bar{a}$ in the state α is given by $\bar{a} = \sum_i p_i a_i$. In the S-representation, an $n \times n$ matrix is associated with every observable a. This is said to represent a and is also denoted by a. Then, $\bar{a} = \alpha^\dagger a \alpha$, where α is the column matrix specifying the state and $\alpha^\dagger$ is its conjugate transpose. If a is a real observable, the matrix a is Hermitian (i.e., $a^\dagger = a$); this ensures that $\bar{a}$ is real. If a is one of the observables belonging to S, its matrix is diagonal, and the ith element in the principal diagonal is the eigenvalue a_i of a in the ith S-eigenstate; in this case, $\bar{a} = \sum_i \alpha_i^* \alpha_i a_i = \sum_i |\alpha_i|^2 a_i$, which is clearly correct, since $|\alpha_i|^2$ is the probability the system will be observed in the ith S-eigenstate.

A necessary and sufficient condition that the state α should be an eigenstate of a in which a takes the eigenvalue a_0 is that α should satisfy the matrix equation $a\alpha = a_0 \alpha$. a is said to be *sharp* in the state α.

Observables are often introduced as functions of other observables, e.g., the kinetic energy T of a particle of mass m is defined in terms of its momentum by $T = (p_x^2 + p_y^2 + p_z^2)/2m$. If a is defined in terms of $u, v, \cdots$ by the equation $a = \phi(u, v, \cdots)$, then this equation also defines the matrix representing a in terms of the matrices representing u, v, etc. There is, however, a proviso: wherever a product uv occurs in ϕ, this must be replaced by $\frac{1}{2}(uv + vu)$ before matrices are substituted, whenever the matrices u, v are such that $uv \neq vu$; this is called *symmetrization*. If $uv \neq vu$, we say that u, v do not *commute* and it is then found that the observables u, v are incompatible.

Thus far, the evolution of a system with time has not been considered. Suppose a system is prepared in a state $\alpha(0)$ at $t = 0$ and is not thereafter interfered with by further observation. Its state $\alpha(t)$ at a later time t is then determined by the *Schrödinger equation*,

$$H\alpha = \iota\hbar \frac{d\alpha}{dt}, \quad \iota = \sqrt{(-1)}$$

where $\hbar = h/2\pi$ (h is Planck's constant). Using the S-representation, α is the column matrix $(\alpha_1, \alpha_2, \cdots, \alpha_n)$ and H is an $n \times n$ matrix, characteristic of the system, called its *Hamiltonian*. H is Hermitian, and in the case of an isolated system, represents the *energy observable* of the system. An important property of H is that, if a is the matrix representing some observable and a and H commute, then a is a *constant* of the system; this means that, if a has a sharp value at one instant, it keeps this sharp value for all t, and otherwise, its probability distribution over its eigenvalues never changes.

If S' is a maximal set of compatible observables different from S, an alternative S'-

representation of the states and observables of a system can be developed. If α, α' are column matrices representing the same state in the two representations and a, a' are square matrices representing the same observable, the transformation equations relating the two representations take the forms

$$\alpha' = u\alpha, \quad a' = uau^{-1},$$

where u is a unitary $n \times n$ matrix (i.e., $u^{-1} = u^\dagger$) characteristic of the two representations. In particular, the Hamiltonian H transforms as an observable.

The type of representation we have been describing is called a *Schrödinger representation*. It is also possible, by rotating the frame in the representation space appropriately, to keep α constant as t increases. If this is done, the matrices $a, b, \cdots$ representing the observables of the system necessarily become functions of t and are not constants as assumed previously. This type of representation is called a *Heisenberg representation*. The Schrödinger equation above ceases to be valid and is replaced by equations of motion for the observable matrices $a, b, \cdots$ taking the form

$$\frac{da}{dt} = \frac{\iota}{\hbar}[H, a],$$

where $[H, a] = Ha - aH$ is called the *commutator* of H and a. If H and a commute, $[H, a] = 0$ and a is a constant of the system as already stated.

If the number of basic eigenstates of a representation is infinite but enumerable (i.e., they can be placed in a sequence $e_1, e_2, e_3, \cdots$), the matrices appearing in the theory will have an infinite number of rows. This creates convergence difficulties in most formulae (e.g., that for $\bar{a}$), but otherwise, the form of the theory is not affected. If, however, some of the observables upon which the representation is based have eigenvalues which are spread continuously over real intervals, the basic eigenstates will not be enumerable and the matrix-type representation must be abandoned. Such an observable is said to have a *continuous spectrum* of eigenvalues; examples are provided by the coordinates (x, y, z) of a particle. In these circumstances, the discrete sequence α_n of vector components is replaced by a function $\psi(p, q, \cdots)$ of the continuous eigenvalues $p, q, \cdots$ of the observables of the representation set S. ψ is called a *wave function*. If the system is in the state specified by ψ, the probability that if p, q, etc. are measured they will be found to have values lying in the intervals $(p, p + dp)$, $(q, q + dq)$, etc., respectively is $\psi^*\psi \, dp \, dq \cdots$. The representation space is now a function space called a *Hilbert space* in which each vector corresponds to a wave function and the scalar product of the vectors corresponding to the wave functions $\psi(p, q, \cdots)$, $\phi(p, q, \cdots)$ is defined by

$$(\psi, \phi) = \int \psi^* \phi \, dp \, dq \cdots,$$

the integration being over all possible eigenvalues of p, q, etc.

In this type of representation, observables are represented by *linear operators* which can operate upon the functions of the Hilbert space, transforming them into other functions belonging to the space. For example, if the system comprises a single particle and the representation being used is based on the particle's coordinates (x, y, z), its state is specified by a wave function $\psi(x, y, z)$ and its momentum components are represented by operators $p_x = (\hbar/\iota)(\partial/\partial x)$, $p_y = (\hbar/\iota)(\partial/\partial y)$, $p_z = (\hbar/\iota)(\partial/\partial z)$. This representation permits the immediate derivation of the important *commutation rules* for the coordinates and momenta, namely $[p_x, x] = \hbar/\iota$, $[p_x, y] = 0$, etc. Counterparts of all the results given earlier for a matrix representation can now be written down. Thus, if a is the operator representing some observable of a system which is in the state $\psi(p, q, \cdots)$, the expected value of a is given by

$$\bar{a} = \int \psi^* a \psi \, dp \, dq \cdots,$$

where a operates on the wave function ψ on its right. The necessary and sufficient condition for ψ to be an eigenstate of a with eigenvalue a_0 is $a\psi = a_0\psi$. Finally, the Schrödinger equation remains valid, but H is now an operator and α is replaced by the wave function ψ; for a single particle moving in a conservative field in which its potential energy is $V(x, y, z)$, employing the coordinate representation, the total energy $H = V + (p_x^2 + p_y^2 + p_z^2)/2m$, Schrödinger's equation becomes

$$-\frac{\hbar^2}{2m} \nabla^2 \psi + V\psi = \iota\hbar \frac{\partial \psi}{\partial t}.$$

The simplest example of an observable with a discrete spectrum of eigenvalues is the *angular momentum* of a system about a point O. If (M_x, M_y, M_z) are the three components of angular momentum and $M^2 = M_x^2 + M_y^2 + M_z^2$ is the square of its magnitude, the following relations hold between the matrices or operators representing these observables

$$[M_y, M_z] = \iota\hbar M_x, \quad [M_z, M_x] = \iota\hbar M_y,$$

$$[M_x, M_y] = \iota\hbar M_z.$$

These are the angular momentum *commutation rules*. M^2 commutes with each of the components. It follows that the components are mutually incompatible, but each is compatible with M^2. Simultaneous eigenstates of M^2 and M_z exist in which the eigenvalues of M^2 are

$\ell(\ell + 1)\hbar^2$, where $\ell = 0, \frac{1}{2}, 1, \frac{3}{2}, \cdots$, and of M_z are $m\hbar$, where $m = -\ell, -\ell + 1, -\ell + 2, \cdots, \ell - 1, \ell$.

Part of the angular momentum of a system is due to the orbital motion of its particles about O and the remainder is contributed by the intrinsic angular momentum or *spin* of the particles. Let (s_x, s_y, s_z) be the components of spin of a particle and s^2 the square of its magnitude; then, if the particle is a fundamental one (e.g., an electron, proton, or vector meson), s^2 will have only one eigenvalue $\ell(\ell + 1)\hbar^2$ and the particle is then said to have spin $\ell\hbar$. For an electron or a proton, $\ell = \frac{1}{2}$ and for a vector meson $\ell = 1$. The eigenvalues of a spin component are then $-\ell\hbar, (-\ell + 1)\hbar, \cdots, \ell\hbar$. Thus, any spin component of an electron has but two eigenvalues, $-\frac{1}{2}\hbar, \frac{1}{2}\hbar$; a spin component of a vector meson has three eigenvalues $-\hbar, 0, \hbar$.

If all aspects of the state of a fundamental particle except its spin are ignored, (s^2, s_z) constitute a maximal set of compatible observables. Employing a representation based on this set, the number of basic eigenstates for a particle of spin $\ell\hbar$ will be $(2\ell + 1)$ and the general spin state of the particle will be represented by a column matrix $(\alpha_\ell, \alpha_{\ell-1}, \alpha_{\ell-2}, \cdots, \alpha_{-\ell})$; for a particle in this state, $|\alpha_k|^2$ is the probability of measuring s_z to take the value $k\hbar$. In the special case of an electron or a proton, $\ell = \frac{1}{2}$ and its spin state is specified by a column $(\alpha_{1/2}, \alpha_{-1/2})$ called a *spinor*; in this representation, the three components of spin are represented by 2×2 matrices, thus:

$$s_x = \tfrac{1}{2}\hbar \begin{pmatrix} 0 & 1 \\ 1 & 0 \end{pmatrix}, \; s_y = \tfrac{1}{2}\hbar \begin{pmatrix} 0 & -\iota \\ \iota & 0 \end{pmatrix}, \; s_z = \tfrac{1}{2}\hbar \begin{pmatrix} 1 & 0 \\ 0 & -1 \end{pmatrix}.$$

The three 2×2 matrices appearing in these formulae are called the *Pauli matrices* and are denoted by $\sigma_x, \sigma_y, \sigma_z$.

In recent years, much thought has been given to the problem of constructing Schrödinger equations for the fundamental particles, which remain unchanged in form when subjected to the group of Lorentz transformations of special relativity theory. Dirac's equation for the electron is one such equation. The fields governed by these equations have themselves been treated as physical systems and quantized, thus leading to a quantum theory of fields. The symmetry of these fields under rotations and other transformations has been fully exploited by the application of group theory and the properties of Lie algebras. Details will be found in Ref. 10 of the bibliography below.

D. F. LAWDEN

References

1. Böhm, A., "Quantum Mechanics," Berlin, Springer-Verlag, 1979.
2. Cunningham, J., and Newing, R. A., "Quantum Mechanics," Edinburgh, Oliver and Boyd, 1967.

3. D'Espagnat, B., "Conceptual Foundations of Quantum Mechanics," Reading, Mass., W. A. Benjamin, 1976.
4. Jauch, J. M., "Foundations of Quantum Mechanics," Reading, Mass., Addison-Wesley, 1968.
5. Lawden, D. F., "Mathematical Principles of Quantum Mechanics," London, Methuen, 1967.
6. Matthews, P. T., "Introduction to Quantum Mechanics," New York, McGraw-Hill, 1974.
7. Merzbacher, E., "Quantum Mechanics," New York, Wiley, 1970.
8. Pauli, W., "General Principles of Quantum Mechanics," Berlin, Springer-Verlag, 1980.
9. Pilkuhn, H. M., "Relativistic Particle Physics," Berlin, Springer-Verlag, 1979.
10. Schweber, S. S., "Relativistic Quantum Field Theory," New York, Harper, 1961.

Cross-references: MATHEMATICAL PHYSICS, MATRICES, QUANTUM THEORY, SCHRÖDINGER EQUATION, THEORETICAL PHYSICS, WAVE MECHANICS.

MATRICES

Matrix notation and operations are introduced into theoretical physics so that algebraic equations and expressions in terms of rectangular arrays of numbers can be systematically handled.

An $m \times n$ matrix $A = (a_{ij})$ possesses m rows and n columns, having in double suffix notation the mn elements arranged to the form

$$A = \begin{pmatrix} a_{11} & a_{12} & \cdots & a_{1n} \\ \cdot & \cdot & \cdots & \cdot \\ a_{m1} & a_{m2} & \cdots & a_{mn} \end{pmatrix}.$$

The general element a_{ij} may be a complex number. If all elements are zero, A is the null matrix O or 0. When $n = 1$, the matrix is a column vector v. If $m = n$, A is square of order n; if all elements not on the leading diagonal a_{11}, $a_{22}, \cdots, a_{nn}$ are zero, the matrix is a diagonal matrix D, while D is the unit matrix I if

$$a_{11} = a_{22} = \cdots = a_{nn} = 1.$$

The sum or difference of two $m \times n$ matrices A and B is an $m \times n$ matrix $C = A \pm B$, where $c_{ij} = a_{ij} \pm b_{ij}$. The elements of αA are αa_{ij}.

The transpose of A is denoted by A^T; this is an $n \times m$ matrix whose ith row and jth column are identical respectively with the ith column and jth row of A. Hence v^T is a row matrix with m elements. For convenience, the column v is often printed as a row with braces, $\{v_1 \ v_2 \cdots v_m\}$.

The product $C = AB$ is only defined when the number of columns of A equals the number of rows of B; A and B are then conformable for multiplication. If A is $m \times n$ and B is $n \times p$, then C is $m \times p$, with

$$c_{ij} = \sum_{k=1}^{n} a_{ik} \, b_{kj}.$$

Generally, multiplication is not commutative, but it is always associative. The transpose of a product is given by $(ABC)^T = C^T B^T A^T$.

If A is square, then A is symmetric if $A = A^T$, while if $A = -A^T$ it is skew-symmetric. A quadratic form S_A in the n variables contained in the column $x = \{x_1 \ x_2 \cdots x_n\}$ may be written as $S_A = x^T A x$, where A is symmetric.

Let det $A \equiv |A|$ denote the determinant of the square matrix A. If det $A \neq 0$, A is non-singular. Then the definition of matrix multiplication ensures that

$$\det (AB) = \det A \det B$$

where A and B are square matrices of the same order.

The cofactor of a_{ij} in the square matrix A equals $(-1)^{i+j}$ times the determinant formed by crossing out the ith row and jth column in A. The sum of the n elements in any row (or column) multiplied respectively by their cofactors equals det A; the sum of the n elements in any row (or column) multiplied respectively by the cofactors of another row (or column) equals zero. We have

$$\det A^T = \det A; \qquad \det \alpha A = \alpha^n \det A.$$

The adjoint of A, denoted by adj A, is the transpose of the matrix formed when each element of A is replaced by its cofactor. We have

$$A \text{ adj } A = (\text{adj } A)A = (\det A)I$$

and $\det(\text{adj } A) = |A|^{n-1}$. The unique reciprocal or inverse of a non-singular matrix A is given by

$$A^{-1} = (\text{adj } A)/\det A.$$

This has the property that $AA^{-1} = A^{-1}A = I$. It follows that

$$(AB)^{-1} = B^{-1}A^{-1}$$

and

$$(A^{-1})^T = (A^T)^{-1}.$$

Linear equations relating n variables x_i to n variables y_i may be expressed as $x = Ay$; if det $A \neq 0$, the unique solution for the y_i in terms of the x_i is $y = A^{-1}x$.

The rank of an $m \times n$ matrix A is the order of the largest non-vanishing minor within A; of the m linear expressions Ax, the rank gives the number that are linearly independent. The m linear equations in n unknowns $Ax = d$, where d is a column with m elements, are consistent if the rank of A equals the rank of the augmented matrix $(A \ d)$.

If the m linear equations $Ax = d$ are inconsistent in the n unknowns $x_1, x_2, \cdots, x_n$, where

$n < m$ and rank $\mathbf{A} = n$, there the "best" solution for these n unknowns in the least squares sense is given by the normal equations

$$\mathbf{A}^T \mathbf{A} \mathbf{x} = \mathbf{A}^T \mathbf{d}$$

$\mathbf{A}^T \mathbf{A}$ being non-singular when rank $\mathbf{A} = n$.

An $n \times n$ matrix Λ is orthogonal if $\Lambda^T \Lambda = \mathbf{I}$, that is, if $\Lambda^{-1} = \Lambda^T$. Clearly, $|\Lambda| = \pm 1$. If $\mathbf{c}_i$ denotes the ith column of Λ, then $\mathbf{c}_i{}^T \mathbf{c}_i = 1$ and $\mathbf{c}_i{}^T \mathbf{c}_j = 0$ if $i \neq j$; similar results hold for the rows. The transformation $\mathbf{x}' = \Lambda \mathbf{x}$ represents a rotation of rectangular Cartesian axes in three dimensions; $|\Lambda| = +1$ if the right-handed character is preserved. The element λ_{ij} equals the cosine of the angle between the $x_i{}'$ and x_j axes. If $\mathbf{N}$ is skew-symmetric, then

$$(\mathbf{I} + \mathbf{N})^{-1}(\mathbf{I} - \mathbf{N})$$

is orthogonal.

Matrices with complex elements are manipulated according to the same rules. A square matrix $\mathbf{H}$ is Hermitian if $\mathbf{H}^{*T} = \mathbf{H}$, and skew-Hermitian if $\mathbf{H}^{*T} = -\mathbf{H}$, a star denoting the complex conjugate. A unitary matrix $\mathbf{U}$ satisfies $\mathbf{U}^{*T} = \mathbf{U}^{-1}$, the columns (and rows) enjoying the properties $\mathbf{c}_i{}^{*T} \mathbf{c}_j = 1$ if $i = j$ and 0 if $i \neq j$. If $\mathbf{N}$ is skew-Hermitian, then $(\mathbf{I} + \mathbf{N})^{-1}(\mathbf{I} - \mathbf{N})$ is unitary.

First- and second-order tensors, arising in many physical problems, may be expressed in matrix notation. If $\mathbf{x}' = \Lambda \mathbf{x}$ denotes a rotation of rectangular Cartesian axes, Λ being orthogonal, then $\mathbf{f}' = \Lambda \mathbf{f}$ and $\mathbf{F}' = \Lambda \mathbf{F} \Lambda^T$ define Cartesian tensors of orders 1 and 2 respectively. Evidently, if $\mathbf{u}$ and $\mathbf{v}$ are vectors or tensors of order 1, then $\mathbf{u}^T \mathbf{v}$ is the invariant scalar product, $\mathbf{F}\mathbf{u}$ is a tensor of order 1 and $\mathbf{u}\mathbf{v}^T$ is a tensor of order 2. For example, the vector product $\mathbf{u} \times \mathbf{v}$ may be written as $\mathbf{U}\mathbf{v}$, where

$$\mathbf{U} = \begin{pmatrix} 0 & -u_3 & u_2 \\ u_3 & 0 & -u_1 \\ -u_2 & u_1 & 0 \end{pmatrix}.$$

$\mathbf{U}$ is a tensor of order 2 if $\mathbf{u}$ is a vector; $\mathbf{U}^T$ is the dual of $\mathbf{u}$.

The differential operator (column)

$$\nabla \equiv \left\{ \frac{\partial}{\partial x_1}, \frac{\partial}{\partial x_2}, \frac{\partial}{\partial x_3} \right\}$$

is a vector or tensor of order 1; namely

$$\nabla = \Lambda \nabla.$$

Thus if ϕ is a scalar and $\mathbf{f}$ a vector,

grad $\phi \equiv \nabla \phi$ is a vector

div $\mathbf{f} \equiv \nabla^T \mathbf{f}$ is a scalar

curl $\mathbf{f} \equiv \nabla \mathbf{f}$ is a vector.

But if $\mathbf{x}' = \mathbf{A}\mathbf{x}$, where $\mathbf{A}$ is not orthogonal, $\mathbf{f}' = \mathbf{A}\mathbf{f}$ defines a contravariant vector, but $\mathbf{g}' = \mathbf{A}^{T-1} \mathbf{g}$ defines a covariant vector. The product $\mathbf{g}^T \mathbf{f}$ is now an invariant.

If $\mathbf{A}$ is square of order n, then the n homogeneous equations $\mathbf{A}\mathbf{k} = \lambda \mathbf{k}$ require

$$\det(\mathbf{A} - \lambda \mathbf{I}) = 0$$

for non-trivial solutions. This characteristic equation possesses n characteristic or latent roots; if they are all distinct, n corresponding characteristic or latent vectors exist. The vector $\mathbf{k}_i$ corresponding to the root λ_i may consist of the n cofactors of any row of $\mathbf{A} - \lambda_i \mathbf{I}$; at least one non-trivial row exists.

The Cayley-Hamilton theorem states that a square matrix $\mathbf{A}$ satisfies its own characteristic equation.

The following properties are important. If $\mathbf{A}$ is real and symmetric, and if λ_i and λ_j are distinct, then $\mathbf{k}_i{}^T \mathbf{k}_j = 0$; these two vectors are orthogonal. Again, if $\mathbf{A}$ is real and symmetric, the n values of λ are real, but if $\mathbf{A}$ is real and skew-symmetric, these n values are pure imaginary. For a real orthogonal matrix Λ, $|\lambda_i| = 1$ for all i. The characteristic roots of $\mathbf{A}^{-1}$ are $1/\lambda_i$, $\mathbf{k}_i$ still being the corresponding vectors. If $|\lambda_1|$ is the largest of the moduli of the n roots, then as $r \to \infty$, $\mathbf{A}^r \mathbf{x} \to \mathbf{k}_1$, where $\mathbf{x}$ is an arbitrary column.

If $\mathbf{A}$ is symmetric, n mutually orthogonal characteristic vectors may be found even if the roots are not all distinct. If each vector $\mathbf{k}_i$ is normalized, i.e., divided by $\sqrt{(\mathbf{k}_i{}^T \mathbf{k}_i)}$, then the matrix

$$\Lambda = (\mathbf{k}_1 \ \mathbf{k}_2 \ \cdots \ \mathbf{k}_n)$$

is orthogonal, and the product $\Lambda^T \mathbf{A} \Lambda$ equals $\mathbf{D}$, the diagonal matrix consisting of the n roots arranged down its leading diagonal in order. The matrix $\mathbf{A}$ is said to be diagonalized, and $\mathbf{D}$ is the canonical form of $\mathbf{A}$.

More generally, if $\mathbf{A}$ is a general square matrix of order n, then n independent vectors $\mathbf{k}_i$ may be found corresponding to the n roots if the latter are distinct. Then

$$\mathbf{T} = (\mathbf{k}_1 \ \mathbf{k}_2 \ \cdots \ \mathbf{k}_n)$$

transforms A into diagonal form, thus:

$$\mathbf{T}^{-1} \mathbf{A} \mathbf{T} = \mathbf{D}.$$

If some of the characteristic roots are identical, it may or may not be possible to find n corresponding independent columns (though it is always possible when $\mathbf{A}$ is symmetric). If it is possible, $\mathbf{T}$ is non-singular, and as before

$$\mathbf{T}^{-1} \mathbf{A} \mathbf{T} = \mathbf{D}.$$

If this is not possible, there still exists a distinct non-singular matrix $\mathbf{T}$ such that

$$\mathbf{T}^{-1} \mathbf{A} \mathbf{T} = \mathbf{C},$$

where $\mathbf{C}$, no longer diagonal, is the standard canonical form of $\mathbf{A}$, containing submatrices of the form

$$\begin{pmatrix} \lambda & 1 & 0 \\ 0 & \lambda & 1 \\ 0 & 0 & \lambda \end{pmatrix}$$

corresponding to the repeated roots.

In each case, respectively,

$$\mathbf{A}^n = \mathbf{T}\,\mathbf{D}^n\mathbf{T}^{-1} \quad \text{or} \quad \mathbf{T}\,\mathbf{C}^n\mathbf{T}^{-1}.$$

This enables us to define functions of square matrices whose canonical forms are diagonal, namely, if $\mathbf{A} = \mathbf{T}\,\mathbf{D}\,\mathbf{T}^{-1}$. If

$$f(x) = \sum_{r=0}^{\infty} a_r x^r$$

is convergent for $|x| < R$, then define

$$f(\mathbf{A}) = \sum_{r=0}^{\infty} a_r \mathbf{A}^r$$

$$= \mathbf{T}\left(\sum_{r=0}^{\infty} a_r \mathbf{D}^r \right) \mathbf{T}^{-1}$$

$$= \mathbf{T}\begin{pmatrix} f(\lambda_1) & 0 & \cdots \\ 0 & f(\lambda_2) & \cdots \\ \cdot & \cdot & \cdots \end{pmatrix}\mathbf{T}^{-1}$$

provided $|\lambda_1|, |\lambda_2|, \cdots, |\lambda_n| < R$. Thus, for example, if

$$\mathbf{J} = \begin{pmatrix} 0 & \theta \\ -\theta & 0 \end{pmatrix}$$

then

$$e^{\mathbf{J}} = \begin{pmatrix} \cos\theta & \sin\theta \\ -\sin\theta & \cos\theta \end{pmatrix}.$$

Similar remarks apply to Hermitian matrices $\mathbf{H}$. If $\mathbf{Hk} = \lambda\mathbf{k}$, all the n values of λ are real and n vectors can always be found such that

$$\mathbf{k}_i{}^{*T}\mathbf{k}_j = \delta_{ij}.$$

The unitary matrix $\mathbf{U} = (\mathbf{k}_1\ \mathbf{k}_2\ \cdots\ \mathbf{k}_n)$ transforms $\mathbf{H}$ into diagonal form.

Two quadratic forms $S_A = \mathbf{x}^T\mathbf{Ax}$, $S_B = \mathbf{x}^T\mathbf{Bx}$, where $\mathbf{A}$ and $\mathbf{B}$ are symmetric and of the same order, may be reduced simultaneously to sums of squares. The equations $\mathbf{Ak} = \lambda\mathbf{Bk}$ demand $\det(\mathbf{A} - \lambda\mathbf{B}) = 0$; this possesses n roots λ_i and n corresponding vectors $\mathbf{k}_i$. If

$$\mathbf{T} = (\mathbf{k}_1\ \mathbf{k}_2\ \cdots\ \mathbf{k}_n)$$

then $\mathbf{T}^T\mathbf{AT}$ and $\mathbf{T}^T\mathbf{BT}$ are both diagonal. The transformation $\mathbf{x} = \mathbf{Ty}$ yields the two sums of squares $S_A = \mathbf{y}^T(\mathbf{T}^T\mathbf{AT})\mathbf{y}$ and $S_B = \mathbf{y}^T(\mathbf{T}^T\mathbf{BT})\mathbf{y}$. In particular, if S_A is positive definite, $\mathbf{T}^T\mathbf{AT}$ will equal $\mathbf{I}$ if new columns $\bar{\mathbf{k}}_i$ are used in $\mathbf{T}$, where $\bar{\mathbf{k}}_i = \mathbf{k}_i/\sqrt{(\mathbf{k}_i{}^T\mathbf{Ak}_i)}$.

Necessary and sufficient conditions for the real quadratic form S_A to be positive definite for all real $\mathbf{x} \neq 0$ are that the n determinants

$$a_{11}, \quad \begin{vmatrix} a_{11} & a_{12} \\ a_{21} & a_{22} \end{vmatrix}, \quad \begin{vmatrix} a_{11} & a_{12} & a_{13} \\ a_{21} & a_{22} & a_{23} \\ a_{31} & a_{32} & a_{33} \end{vmatrix}, \ldots, \det \mathbf{A}$$

should be positive. This ensures that the n characteristic roots of $\mathbf{A}$ are all positive.

Finally, matrices may often usefully be partitioned employing matrices within matrices. Multiplication may still be performed provided each individual matrix product is permissible. For example,

$$\begin{pmatrix} a & \mathbf{b}^T \\ \mathbf{c} & \mathbf{D} \end{pmatrix}\begin{pmatrix} e & \mathbf{f}^T \\ \mathbf{g} & \mathbf{H} \end{pmatrix} = \begin{pmatrix} ae + \mathbf{b}^T\mathbf{g} & a\mathbf{f}^T + \mathbf{b}^T\mathbf{H} \\ \mathbf{c}e + \mathbf{Dg} & \mathbf{cf}^T + \mathbf{DH} \end{pmatrix}$$

where a, e are scalars, $\mathbf{b}$, $\mathbf{c}$, $\mathbf{f}$, $\mathbf{g}$ are 1×3 columns and $\mathbf{D}$, $\mathbf{H}$ are 3×3.

Applications *Differential Equations* If

$$d\mathbf{x}/dt + \mathbf{Ax} = \mathbf{f},$$

$\mathbf{A}$ being constant and

$$\mathbf{f} = \{f_1(t),\ f_2(t),\ \ldots,\ f_n(t)\},$$

then if $\mathbf{T}$ diagonalizes $\mathbf{A}$, $\mathbf{x} = \mathbf{Ty}$ yields n non-simultaneous equations $d\mathbf{y}/dt + \mathbf{Dy} = \mathbf{T}^{-1}\mathbf{f}$. If $\mathbf{y}_0(t)$ is a particular integral,

$$\mathbf{x}(t) = \mathbf{T}\begin{pmatrix} e^{-\lambda_1 t} \cdots & & 0 \\ & \cdot & \cdots & \cdot \\ 0 & & \cdots e^{-\lambda_n t} \end{pmatrix}$$

$$\times\ [\mathbf{T}^{-1}\mathbf{x}(0) - \mathbf{y}_0(0)] + \mathbf{Ty}_0(t).$$

Geometry. In three-dimensional Cartesian coordinates,

$$\mathbf{a}^T\mathbf{x} + d = 0$$

represents a plane, the perpendicular distance from $\mathbf{x}_1$ being

$$(\mathbf{a}^T\mathbf{x}_1 + d)/\sqrt{(\mathbf{a}^T\mathbf{a})}.$$

The equation $\mathbf{x}^T\mathbf{Ax} = d$ represents a central quadric. If Λ diagonalizes $\mathbf{A}$, the rotation

$$\mathbf{x} = \Lambda\mathbf{x}'$$

yields $\mathbf{x}'^T\mathbf{Dx}' = d$. The vectors $\mathbf{k}_1$, $\mathbf{k}_2$, $\mathbf{k}_3$ specify the three principal axes, of semi-lengths $\sqrt{(d/\lambda_i)}$ when $d/\lambda_i > 0$.

Dynamics. The rotational equations of motion of a rigid body with respect to moving axes fixed in the body and with the origin fixed in space or at the centre of mass are

$$\mathbf{g} = \mathbf{J}\dot{\omega} + \Omega\mathbf{J}\omega$$

where $\mathbf{g}$ = couple, ω = angular velocity, Ω^T = dual ω. $\mathbf{J}$ denotes the inertia tensor $-\Sigma m\mathbf{XX}$, where $\mathbf{X}^T$ = dual $\mathbf{x}$. Explicitly,

$$\mathbf{J} = \begin{pmatrix} A & -H & -G \\ -H & B & -F \\ -G & -F & C \end{pmatrix}.$$

The rotational kinetic energy is $\frac{1}{2}\omega^T\mathbf{J}\omega$. When principal axes of inertia are chosen, $\mathbf{J}$ is diagonal, yielding Euler's equations.

Small oscillations about a position of equilibrium are investigated by considering the second order approximations

$$K.E. = \dot{\mathbf{q}}^T\mathbf{A}\dot{\mathbf{q}}, \quad P.E. = \mathbf{q}^T\mathbf{B}\mathbf{q},$$

$\mathbf{q}$ containing n generalized coordinates measured from their equilibrium values. $\mathbf{A}$ and $\mathbf{B}$ are constant symmetric matrices. If the n roots of $\det(\mathbf{A} + \lambda\mathbf{B}) = 0$ are considered, and if $\mathbf{q} = \mathbf{Tx}$, where $\mathbf{T} = (\mathbf{k}_1 \ \mathbf{k}_2 \ \cdots \ \mathbf{k}_n)$ reduces $\mathbf{A}$ to the unit matrix $\mathbf{I}$, the equations of motion are

$$\ddot{x}_i + (1/\lambda_i)x_i = 0.$$

The elements of $\mathbf{x}$ are the *normal coordinates*; each individual solution x_i in terms of the q's is a *normal mode of* period $2\pi\sqrt{\lambda_i}$.

Electromagnetic Theory. Maxwell's 3×3 stress tensor in matrix notation is

$$\mathbf{T} = \frac{1}{2}[2\epsilon\mathbf{ee}^T + 2\mu\mathbf{hh}^T - \epsilon(\mathbf{e}^T\mathbf{e})\mathbf{I} - \mu(\mathbf{h}^T\mathbf{h})\mathbf{I}]$$

in mks rationalized units. The field exerts a force across an area element $\mathbf{n}\delta S$ equal to $\mathbf{Tn}\,\delta S$.

When electromagnetic waves are propagated in an ionized medium the equation

$$\text{curl curl } \mathbf{e} = k^2(\mathbf{I} + \mathbf{M})\mathbf{e}$$

arises, where

$$\mathbf{M} = -X(Y^2\mathbf{nn}^T + iY\mathbf{N} - \mathbf{I})/(Y^2 - 1)$$

in the usual notation with collisions neglected; here, $\mathbf{n}$ = unit vector directed along the external magnetic field, $\mathbf{N}_T$ = dual $\mathbf{n}$. These equations may be rearranged in terms of the matrix

$$\mathbf{f} = \{E_x, -E_y, Z_0 H_x, Z_0 H_y\}$$

giving $d\mathbf{f}/dz = -ik\mathbf{Tf}$, where $\mathbf{T}$ is a 4×4 matrix. If the characteristic roots $\lambda_i(z)$ of $\mathbf{T}$ are found, and if $\mathbf{R}$ diagonalizes $\mathbf{T}$, then the transformation $\mathbf{f} = \mathbf{Rg}$ yields

$$\frac{d\mathbf{g}}{dz} = -ik\mathbf{Dg} - \mathbf{R}^{-1}\frac{d\mathbf{R}}{dz}\mathbf{g}.$$

When $\mathbf{k}$ is large, approximate solutions are possible if the terms of $\mathbf{R}^{-1}(d\mathbf{R}/dz)\mathbf{g}$ arising from the non-diagonal elements of $\mathbf{R}^{-1}(d\mathbf{R}/dz)$ are much smaller in magnitude than the corresponding elements of $k\mathbf{Dg}$. Then

$$\frac{dg_j}{dz} \doteqdot -ikD_jg_j - \left(\mathbf{R}^{-1}\frac{d\mathbf{R}}{dz}\right)_{jj}g_j$$

and

$$g_j \doteqdot \exp\left(-ik\int D_j\,dz\right)$$

$$\cdot \exp\left[-\int\left(\mathbf{R}^{-1}\frac{d\mathbf{R}}{dz}\right)_{jj}dz\right]$$

giving rise to the characteristic waves propagating in the medium.

Special Relativity. If $\mathbf{x} = \{ict, x, y, z\}$ refers to an inertial frame S, and if a second parallel frame S' has uniform relative velocity U along Ox, the Lorentz transformation is $\mathbf{x}' = \Lambda_U\mathbf{x}$, where

$$\Lambda_U = \begin{pmatrix} \beta & -iU\beta/c & 0 & 0 \\ iU\beta/c & \beta & 0 & 0 \\ 0 & 0 & 1 & 1 \\ 0 & 0 & 0 & 1 \end{pmatrix}.$$

Λ is orthogonal, and $\beta = 1/\sqrt{(1 - U^2/c^2)}$. We have $\Lambda_V \Lambda_U = \Lambda_W$, where

$$W = (U + V)/(1 + UV/c^2).$$

For the general velocity $\mathbf{v}$ relating parallel frames,

$$\Lambda = \begin{pmatrix} \beta & -i\beta\mathbf{v}^T/c \\ i\beta\mathbf{v}/c & \mathbf{I} + (\beta - 1)\mathbf{vv}^T/v^2 \end{pmatrix}.$$

The operator $\square = \{\partial/ic\partial t, \partial/\partial x, \partial/\partial y, \partial/\partial z\}$ is a four vector satisfying $\square' = \Lambda\square$; so are the four-current $\mathbf{i}$ and the four-potential $\mathbf{b}$,

$$\mathbf{i} = \begin{pmatrix} ic\rho \\ \mathbf{j} \end{pmatrix}, \quad \mathbf{b} = \begin{pmatrix} i\phi/c \\ \mathbf{a} \end{pmatrix}$$

where $\mathbf{a}$ is the vector potential. They satisfy $\square^T\mathbf{i} = 0$ (conservation of charge), $\square^T\mathbf{b} = 0$ (the Lorentz relation). Maxwell's equations in mks units in free space take the form

$$\square^T\mathbf{F} = -\mathbf{i}^T/\epsilon_0 c^2$$

$$\square^T\mathbf{G} = 0$$

where

$$\mathbf{F} = \square\mathbf{b}^T - (\square\mathbf{b}^T)^T = \begin{pmatrix} 0 & i\mathbf{e}^T/c \\ -i\mathbf{e}/c & -\mu_0\mathbf{H} \end{pmatrix}$$

and

$$G = \begin{pmatrix} 0 & \mu_0 h^T \\ -\mu_0 h & -iE/c \end{pmatrix}$$

are tensors of order 2 under a Lorentz transformation. E^T and H^T are the respective 3×3 duals of e and h. All tensor equations are invariant in form in all frames of reference.

The tensor of order 2

$$T = \tfrac{1}{2}\epsilon_0 c^2 F\, F - \tfrac{1}{2}\mu_0^{-1} G\, G$$

$$= \tfrac{1}{2}\begin{pmatrix} \epsilon_0\, e^T e + \mu_0 h^T h & -2ie^T H/c \\ -2iEh/c & \epsilon_0\, ee^T + \epsilon_0\, EE + \mu_0\, hh^T + \\ & \mu_0\, HH \end{pmatrix}$$

contains the energy density, Poynting's vector, the momentum density and Maxwell's stress tensor in partitioned form.

Applications may likewise be made to circuit theory, to elasticity where 3×3 stress and strain tensors are defined, and to quantum mechanics, embracing, for example, matrix mechanics and the Dirac wave equation of the electron.

JOHN HEADING

References

Gourlay, A. R., and Watson, G. A., "Computational Methods for Matrix Eigenproblems," London and New York, John Wiley.

Heading, J., "Matrix Theory for Physicists," London, Longmans, Green & Co.

Heading, J., "Electromagnetic Theory and Special Relativity," Cambridge, University Tutorial Press.

Jeffreys, H., and Jeffreys, B., "Methods of Mathematical Physics," Cambridge, The University Press.

Liebeck, H., "Algebra for Scientists and Engineers," London and New York, John Wiley.

Perlis, S., "Theory of Matrices," Reading, Mass., Addison-Wesley.

Williams, I. P., "Matrices for Scientists," London, Hutchinson.

Cross-references: DIFFERENTIAL EQUATIONS IN PHYSICS, ELECTROMAGNETIC THEORY, LORENTZ TRANSFORMATIONS, MATHEMATICAL PRINCIPLES OF QUANTUM MECHANICS, QUANTUM THEORY.

MEASUREMENTS, PRINCIPLES OF

Measurement is the process of quantifying our experience of the world around us. It can be as simple as an elementary event-counting process (the number of automobiles passing a certain point per day or the number of β-particles entering a certain radiation detector in a certain time interval) but it is usually a more complex process involving comparison with some reference. For example, how many handspans wide is my desk? This is an elementary example, but even such a simple question reveals many of the essential and fundamental characteristics of measuring processes. First, the search for improved precision quickly makes desirable acceptance of a standard reference quantity. Second, the primitive process of comparison makes it clear that we can make statements about the measurement only within certain limits. (My desk is between 6 and 7 handspans wide). Our fundamental inability to make exact measurements leads to the concept of uncertainty, and a very extensive theory of the uncertainty of measurement exists.

Measurement Standards Standard, defined units of measurement have been in use for many thousands of years and the situation has been in continuous flux up to the present day. Every country maintains standards of measurement, not only of the basic quantities such as mass, length and time, which we normally think of in this context, but also of a large number of other items (such as the optical reflectivity of paper) which have been identified as important for trade and commerce. We shall restrict ourselves here to a few physical quantities which are important in scientific work. It will turn out that it may be a relatively simple matter to define a unit; it is usually a much more difficult problem to realize that unit in the laboratory so as to make possible the calibration of other instruments.

(a) *Mass.* The kilogram was defined in 1889 as the mass of a certain piece of metal (platinum-iridium) still preserved at the International Bureau of Weights and Measures near Paris, France. Copies of this prototype kilogram, compared with the original by carefully refined beam balance techniques, are kept in most countries to serve as that country's definition of a kilogram. The normal process of comparison using beam balances allows a precision for mass standards of around one part in 10^8, making mass standards substantially less precise than those of length and time.

(b) *Length.* The original 1889 definition of the meter (chosen to be one ten millionth part of the quadrant of the earth's surface at the longitude of Paris) was realized initially using engraved marks on a platinum-iridium bar. However, it very soon thereafter became clear, on the basis of the pioneering work of A. A. Michelson on optical interferometry, that a much much more precise, stable, and easily realizable standard was available using the wavelength of carefully selected spectrum lines. Despite Michelson's early suggestions to that effect it was not until 1960 that an international standard of length was defined in terms of the wavelength of a certain line in the spectrum of krypton-86. Recently, however, it has become clear that even the precision available from the krypton-86 line has been surpassed by precision in two other areas, measurements of the velocity of light and the standard of time. As a consequence it is now possible to define a unit of length in terms of the unit of time and the measurement of the velocity of electromagnetic waves. As accepted by the General Conference of Weights and Measures in 1983, the unit of length is the distance traveled in a time interval of 1/299,792,458 of a second by plane electromagnetic waves in a vacuum. Such a unit will be realizable in any laboratory with precise time standards, and will be translatable into actual distance measurements using standard techniques of interferometry.

(c) *Time.* For hundreds of years the most precise measurements of time were made by the astronomers, whose observations served to define the basic unit of time, the second, in terms of the axial rotation of the earth. When terrestrial clock systems achieved the precision required to show that the earth's rotation rate is not constant it became clear that replacement of the defined standard was necessary. Observations on the radiation frequency in atomic transitions can be made with very high precision and the present definition of the second (adopted in 1967) is the duration of 9,192,631,770 periods of the radiation associated with a certain transition in the cesium-113 atom. Cesium beam clocks now provide a realizable standard of time, not only for time-keeping, but also for practical purposes such as radio navigation systems and for experimental work like long baseline interferometry in radioastronomy.

(d) *Temperature.* The definition of temperature is based on the triple point of water, which was defined in 1954 to have a temperature of 273.16 K. The unit of temperature, the kelvin, is thus defined (since 1967) is as 1/273.16 of the temperature of the triple point of water. The realization of a temperature scale, particularly over wide temperature ranges, is a much more difficult matter. In practice the standards laboratories maintain a working scale called the International Practical Temperature Scale, 1968 (abbreviated IPTS-68). In this, thirteen fixed points have temperatures assigned to them and specified methods of measurement are used to provide intermediate temperatures for calibration purposes. Above the temperature of freezing gold (1337 K) optical pyrometry is used; between 903 K and 1337 K the standard measuring device is a thermocouple of platinum and an alloy of platinum with 10% rhodium; between 14 K and 903 K it is a platinum resistance thermometer. Standard temperatures are also available in the liquid helium temperature range between 0.5 K and 4.2 K using the vapor pressure of helium.

(e) *Electrical Quantities.* Once again we have a distinction between the fundamental definition of a standard quantity and its practical realization. In principle the fundamental electrical quantity is the unit of current (the ampere), which is defined in terms of the force between adjacent current-carrying conductors. In practice it is too difficult to implement this definition with sufficient precision, and the practical standards are those of potential difference and resistance. The unit of potential difference (the volt) is defined in terms of the Josephson effect, a phenomenon occurring in superconducting junctions, which provides an extremely sensitive, precise, and stable connection between potential difference and frequency. The practical realization of the volt is a bank of carefully preserved electrolytic cells which can be moved for international comparison and which are checked periodically for drift using a Josephson source. The standard of resistance (the ohm) is realized using a bank of 1 ohm resistors which can be compared internally and internationally.

Measurement Uncertainty As was mentioned earlier the primitive act of comparison between an object and a reference quantity leads to a value which is known only within a certain interval. No matter how sophisticated the measuring process we cannot evade this fundamental limitation; we can be confident about measurements only within a certain interval. The way in which we handle these uncertainties mathematically depends on the way in which our confidence varies along the scale.

(a) *Estimated Uncertainty.* If we are making our measurements by a personal, visual method, the outcome of the measuring process should be an interval, outside which we are certain the value does not lie. If, by careful examination of a scale, we feel confident that the value we seek does not lie below 24.6 and does not lie above 24.8, we can state that we are confident that our desired value lies inside the interval 24.6–24.8, although we can say nothing about its location within the interval. The interval is usually renamed 24.7 ± 0.1, and we call the quantity ± 0.1 the *uncertainty* of the reading. Frequently it is instructive to compare this uncertainty with the reading itself, and we call the ratio 0.1/24.7, usually expressed as a percentage, the *precision* of the measurement. This question of the uncertainty in reading a scale is only one contribution to our lack of exact knowledge of the value. There may be other effects, like calibration errors in the instrument, which affect all the readings in a similar way and constitute systematic errors in addition to reading uncertainty. To make satisfactory measurements we must always be aware of the presence of the reading uncertainty and also alert to the possible presence of systematic errors. These last must be identified and corrected if possible but, at the end of the whole process, it is important to express the reading in such a way that the quoted value provides a realistic appraisal of the complete range of uncertainty.

It is rarely sufficient to make such a measurement of a single quantity and we are more commonly faced with the problem of calculating some final quantity z as a function of a number of measured quantities x, y, etc., each of which has its own uncertainty Δx, Δy, etc. If

$$z = f(x, y, \ldots),$$

the value of Δz will be calculated from

$$\Delta z = (\partial f / \partial x) \Delta x + (\partial f / \partial y) \Delta y + \ldots.$$

If we were confident that the values of x and y lay within the measured intervals $x_0 \pm \Delta x$, $y_0 \pm \Delta y$, etc., then we can be equally confident that the value of z lies within the calculated interval $z_0 \pm \Delta z$, where $z_0 = f(x_0, y_0, \ldots)$. This general method of calculating uncertainties will be found to be useful for a wide range of functions.

Statistical Uncertainty Circumstances frequently do not permit the subjective estimation of an uncertainty interval as considered in the preceding section. Many measurement processes give results which are influenced by random fluctuations and we must resort to statistical treatment of the observations. Instead of identifying an interval within which we are confident our quantity lies we must be content with statements about probabilities. To make this possible we must build up experience of a fluctuating phenomenon by repeatedly making the measurement. We shall thereby obtain a sample of readings whose characteristics will give us as much information as is available from the system. This sample will have a certain frequency distribution along the scale of values and the uncertainty of the measuring process is related to the breadth of the distribution. One suitable measure of the breadth is the standard deviation of the sample, defined to be

$$S = \sqrt{\left[\sum (\bar{x} - x_i)^2 / N \right]}.$$

where $\bar{x}$ is the mean of the sample, x_i are the individual readings, and N is the number of readings in the sample. By making suitable assumptions about the basic distribution from which the sample was taken (often assumed to be Gaussian), we can now make numerical assertions about the sample. It turns out that any one reading has a 68% probability of falling within the interval $\bar{x} \pm S$ and a 95% probability of falling within $\bar{x} \pm 2S$. Rather than make assertions about single readings it is more useful to be able to make statements about probabilities for the sample mean. We calculate the standard deviation of the mean

$$S_m = S / \sqrt{N}$$

and we are then able to assert that the value we seek in the measuring process has a 68% chance of falling within the interval $\bar{x} \pm S_m$ and a 95%

chance of falling within $\bar{x} \pm 2S_m$. The mean and the standard deviation of the mean thus provide us with measures of probability which take us as far as randomly fluctuating phenomena allow.

The statistically determined interval serves as the measure of uncertainty for processes governed by random fluctuation so that, irrespective of the type of measurement, the outcome of the measuring process is an interval which has a certain probability of containing the value we seek. The measurements are now in suitable condition for the next step in a complete measuring process.

Systems and Models A measuring process is almost never a primitive process of simply comparing an object with a scale (unless we are satisfied simply to measure something with a ruler or read a temperature on a mercury-in-glass thermometer); the situation is almost invariably more complicated. Even if we are doing something as simple as comparing the weight of an object with that of a standard mass using a beam balance we have to make the assumption that the balance beam arms are of equal length. Our measuring process almost invariably involves some complete system and the result of our measuring process is dependent on the properties of the system. Since we can never know these exactly we are dependent on the set of assumptions we make about the system (like the equality of the balance beam arms), and this set of assumptions constitutes a model of the system. Our process of measurement almost invariably requires us, therefore, to consider the model of the system, and it is an integral part of all satisfactory measuring processes to check, not that the model is "correct" or "true," because all models are in principle oversimplified, but that the correspondence between the model and the system is good enough, at least at the level of precision under consideration.

Sometimes this process of testing the model of a measuring system will have been done for us if we buy an expensive piece of apparatus. If we buy a good quality slide-wire potentiometer from a reputable manufacturer it may be satisfactory to assume that the set of assumptions which constitute the model of the system (such as linearity of the slide wire resistance) has been adequately tested. But if we are making up our own measurement system, it is vital to include in the process adequate provision for testing the model on which our work with the system is predicated.

It is common to do the model testing graphically. For example a measurement of the resistance of an electrical component is incomplete without a study of the complete variation of the current through and the potential difference across the component. Only if the plotted values of V and i turn out to be compatible with a straight line can we assert that the resistance of the component is constant, with a value specified by the slope of the line. Usually the model of the system is more complicated than $V = iR$ and more complicated graphical methods of checking the model against the system are needed. In compensation, however, the value we seek as the objective of the experiment will usually be obtained from the graph, so that drawing the graph serves the dual purpose of checking the compatibility of the system and the model and of providing a computational procedure for obtaining the answer. Even when the measurements in an experiment are processed completely analytically (using a least squares procedure, for example, to fit a function to the observations), it is important not to forget the basically graphical nature of the process.

Measurement Systems The demands of modern experimenting and the availability of automatic data processing methods have recently revolutionized measuring methods. It is now common to think of a complete measuring system containing: (a) a conversion stage which may be desirable to convert the basic quantity under investigation into some other form more amenable to measurement; (b) a sensor, detector, or transducer stage to provide for conversion into an electrical signal; (c) a signal processing system to perform on the signal any necessary mathematical computations; (d) an output stage for display, storage, or telemetry of the information. Let us consider each of these stages in turn.

(a) *Conversion Stage.* This is not always present but is frequently necessary if it is impossible or inconvenient to work directly with the phenomenon under investigation. For example a gas thermometer bulb converts temperature into pressure, a moving coil ammeter converts currents into angles, a slide wire potentiometer converts values of potential difference into lengths, a prism, diffraction grating, or crystal converts the wavelength of electromagnetic radiation into angles, a digital voltmeter uses a ramp method to convert a potential difference into a time measurement, a mass spectrometer converts atomic masses into magnetic field values, etc.

(b) *Sensor, Detector, or Transducer Stage.* Almost invariably we wish to process our signal by electrical methods and so some process for converting the basic phenomenon into an electrical signal is required. Sensors are found in enormous variety depending on the type of physical phenomenon involved.

(i) Strain is measured using strain gauges, a small length of metallic or semiconducting material glued to the component under stress. Changes in length of the material are detected from the consequent alteration in electrical resistance.

(ii) Force or pressure transducers may rely on some elastic component (cantilever beam for force, membrane for pressure) to convert the force into a displacement which is then detected using strain gauges. Alternatively, a piezoelectric crystal may be used to obtain an electrical output directly.

(iii) Temperature sensors use a variety of phenomena. Gas thermometers use the change in pressure with temperature of a gas, often helium. Thermocouples produce an electrical emf directly but require careful calibration. From low temperatures up to 500 or 600 K the most commonly used junction materials are copper and constantan (an alloy of copper and nickel). At higher temperatures tungsten and tungsten alloys are used. Resistance thermometers use the change in electrical resistance with temperature of metals or semiconductors, the choice of material depending on the temperature range. Pure platinum gives high precision over a wide temperature range; germanium or ordinary carbon resistors are widely used at low temperatures; and thermistors, in which a sintered powder, generally of metallic oxides, provides rapid variation of resistance with temperature, can be made to suit a wide range of temperatures. The change in resistance which provides a measure of the temperature is commonly measured using some kind of bridge circuit. Temperatures too high for normal sensors are measured by pyrometers in which an absolute measure of the intensity of radiation at some fixed wavelength is interpreted using Planck's radiation equation to give values of temperature. Liquid-in-glass thermometers, although commonly used as an indicator of temperature, only rarely qualify for precise measurement and cannot be used as a transducer to supply an electrical output.

(iv) Optical sensors may be of several different types. The vacuum photodiode contains a surface, generally of some cesium compound, which reacts to illumination by emitting electrons which are collected by an anode. The photomultiplier tube uses successive stages of secondary electron emission in an avalanche process to provide amplification of the current. Solid state devices may be of the photoconductive type, in which the resistance of a semiconducting material such as selenium or cadmium sulfide changes with illumination, or the photovoltaic type in which a semiconducting junction produces its own emf in response to the light. The familiar "solar cell" is of this variety and usually contains a junction between silicon and a metal. Infrared detectors may be of the photoconductive or photovoltaic type. They normally use semiconducting materials or junctions involving compounds like indium antimonide or alloys containing such materials as lead, tin, and tellurium. For wavelengths which do not excite photoelectrons or for cases in which absolute measurements of intensity are required, a bolometer can be used. This device is constructed to absorb all the incident radiation, regardless of wavelength, and convert it into a temperature increase which can be measured using a thermocouple or resistance thermometer.

(v) Acoustic transducers take various forms depending on the frequency of the radiation. For ultrasonic applications at frequencies of tens of kilohertz up to megahertz a piezoelectric transducer, commonly of quartz, is used to convert the pressure fluctuation in the sound wave to an electrical signal. At the lower frequencies of the auditory region microphones can use the piezoelectric effect in materials like lead titanate ceramic or the electrostatic properties of electrets, materials possessing permanent electric polarization.

(vi) Transducers for magnetic field measurements frequently use the Hall effect, in which a current-carrying conductor exhibits a transverse potential difference when placed in a magnetic field. Hall effect probes make a sturdy component for many common measurements of magnetic field but for other applications higher sensitivity may be required. For example, airborne surveys of the earth's magnetic field are commonly carried out using a flux-gate magnetometer, which detects the out-of-balance signal produced by an external field in a carefully balanced magnetic circuit. Still higher sensitivity is available from the proton precession magnetometer. This relies on the relationship between the frequency of proton precession (often in a water sample) and the magnitude of the surrounding magnetic field, and can supply a sensitivity as low as 10^{-5} of the external field. Still higher sensitivity is available from the magnetic dependence of optical transitions in rubidium and cesium atoms in the form of the optically pumped rubidium or cesium magnetometer. For the very lowest values of magnetic field the Josephson effect in superconductors has extended the range of magnetic field measurements by many orders of magnitude. Such devices require cooling to liquid helium temperatures but make possible the measurement of magnetic field changes of the order of 10^{-12} tesla.

(vii) The various forms of ionizing radiation encountered in nuclear physics are detected by a large variety of techniques, of which some of the more common are described below.

Gas-filled tubes containing electrodes can be used under various conditions of pressure and potential difference to detect the ionization produced by fast particles or high-energy radiation photons. An ionization chamber can provide steady monitoring of particle flux densities and a similar gas-filled tube operated at higher potential constitutes a Geiger counter which provides a pulse of current for a single particle entering the counter. Scintillation counters detect individual photons of γ-rays or x-rays from the flash of visible light which is produced by passage through certain solid crystals or liquids. The solid materials are generally single crystals of organic materials such as anthracene or inorganic crystals of sodium iodide (doped with thallium). The light pulses emitted by the scintillation material are normally detected using photomultipliers. An enormous advantage of the use of a scintillation counter lies in the fact that the magnitude of the current pulse is closely proportional to the energy of the x-ray or γ-ray photon. These crystals thereby permit the analysis of such a beam into a spectrum of its energy components. X-ray and γ-ray beams can also be detected directly and analyzed using semiconducting crystals. Lithium-drifted germanium and silicon crystals, although requiring cooling to liquid air temperature, provide current pulses from individual x-ray or γ-ray photons which are accurately proportional to the photon energies.

Signal Processing Systems Signal processing procedures have been available since the introduction of electronic circuitry but were initially restricted to such elementary operations as amplification or heterodyning (the generation of a beat frequency to expedite the tuning and amplification of radio-frequency signals). Rather more sophisticated processing became available with the development of analog computers but was still limited to relatively simple arithmetic operations or integration and differentiation. The real revolution came with the development of digital data processing, and the availability of fast and powerful computers now makes it possible, in "real time," to perform almost any desired mathematical operation on the observations as they are made. If, as is frequently the case, the output of the sensor stage is in analog form, it is necessary to pass the signal through some form of analog-digital (a-d) converter. Examples of on-line signal processing procedures are given below.

(a) *Basic Arithmetic Operations.* Division, for example, can be used to calculate resistance directly as V/i or velocities from values of distance and time.

(b) *Statistical Operations.* These include calculation of such distribution parameters as mean, standard deviation, and correlation coefficient. Comparison with a prescribed function, linear or otherwise, can be carried out by least squares methods, or, in the absence of a specified function, a generalized polynomial fit can be obtained. In all cases progressive improvement of the accuracy of the calculations will result from continued revision as new observations become available.

(c) *Time Averaging.* This is a very powerful method to improve signal-to-noise ratios. If some phenomenon such as a resonance peak or a spectrum line is hidden in noise, we can carry out a process of repeated scans over the range of variable containing the signal. If we have some way of storing the information and averaging the results of the repeated scans, we shall find that the random noise will give, ultimately, an average of zero. The desired signal, on the other hand, will add positively on every scan and eventually appear free of the noise which formerly masked it.

(d) *Fourier Analysis.* It is very frequently desirable to analyze a time-varying phenomenon into a frequency spectrum by the method of

Fourier analysis. Such computation has recently been greatly facilitated by the development of the fast Fourier transform methods and is now commonly used in a wide range of applications. Typical is the Fourier transform spectrometer, in which optical spectra are processed in an interferometer. The resulting variation of fringe intensity with order of interference is analyzed to yield the frequency components of which the original beam was composed.

(e) *Optical Image Enhancement.* A two-dimensional picture can be converted into digital data by a scanning process, thereby becoming susceptible to computer processing. Photographs showing such defects as lack of contrast, out-of-focus, or smearing from camera movement can be analyzed to determine the precise nature of the defect and subsequently corrected to construct an improved image. Many of the most spectacular results of space exploration among the planets would not have been visible without various processes of image enhancement to emphasize detail in particular ways, and similar improvement is available for microscopic images of biological material. Computer processing can also be used to create "false-color" images of normally invisible phenomena such as infrared emission, accoustic signals, and others. These are now familiar in many applications, such as the thermographs which identify poorly insulated houses or human breast cancer and the satellite-based photography which is used to study earth resources such as crops.

(f) *Pulse-Height Analysis.* This is a process in which separate storage is provided for pulses of differing height. The resulting display can then give a direct picture of, for example, a spectrum of γ-ray energies from a scintillation counter or x-ray energies from a Li-drifted silicon counter. This very powerful technique can be used to record any phenomenon in which the quantity in which we are interested can be converted into pulses with height dependent on the original variable. For example, lifetime studies on excited atoms will use a time-to-pulse height converter to provide a direct picture of the decaying radiation.

(g) *Others.* Other examples of on-line signal processing too complex to be described here but of too great importance to omit include pattern recognition and the various procedures for clinical examination by tomography. In these a signal, derived usually from x-ray absorption but occasionally from other phenomena, is analyzed to provide a picture of a cross-section of the human body as an aid to diagnosis.

Display and Storage Systems The traditional methods of needle and scale are now less frequently encountered, except as indicators. For time-varying phenomena the strip-chart recorder remains useful at low rates of change and the cathode ray oscilloscope is absolutely irreplaceable for rapidly changing phenomena. The addition of digital processing and memory to the CRO has made it a uniquely powerful and versatile instrument for the display and study of oscillations or time-varying phenomena over a wide range of frequencies and times.

Digital data processing makes possible the presentation of output information using digital numerical displays, based often on neon tubes or light-emitting diodes. These offer the convenience of direct access to output information without need for further interpretation, e.g., a range scale on a laser range finder. Similar convenience is available from the two- or three-dimensional cathode-ray tube displays in which the output of computer processing of observations can be viewed directly in pictorial form.

For cases in which later use of the results is intended, digital storage methods using magnetic tape and disc or punched paper tape permit vast quantities of information to be stored and easily retrieved. If onward transmission of the information is required, the use of digital techniques makes possible the rapid and accurate transfer of observations over interplanetary distances.

D. C. BAIRD

References

Klein, H. A., "The World of Measurements," New York, Simon and Schuster, 1974.

Rossini, F. D., "Fundamental Measures and Constants for Science and Technology," Cleveland, CRC Press, 1974.

Baird, D. C., "Experimentation," Englewood Cliffs, N.J., Prentice-Hall, 1962.

"The International System of Units (SI)," Washington, D.C., National Bureau of Standards, and London, Her Majesty's Stationary Office.

Janossy, L., "Theory and Practice of the Evaluation of Measurements," Oxford, Oxford Univ. Press, 1965.

Plumb, H. H. (Ed.), "Temperature: Its Measurement and Control in Science and Industry," Instrument Society of America, 1972.

Quinn, T. J., "Temperature," Academic Press, New York, 1983.

Levi, L., "Applied Optics," New York, Wiley, 1980.

Keyes, R. J. (Ed.), "Optical and Infrared Detectors," Springer-Verlag, New York, 1981.

Peterson, A. P. G., "Handbook of Noise Measurement," Concord, Mass., GenRad Inc., 1980.

Zijlstra, H., "Experimental Methods in Magnetism," New York, Wiley, 1967.

Knoll, G. F., "Radiation Detection and Measurement," New York, Wiley, 1979.

Oppenheim, A. V., "Applications of Digital Signal Processing," Englewood Cliffs, N.J., Prentice-Hall, Inc., 1978.

Cross-references: ASTRONOMY; COSMIC RAYS; ELECTRICAL MEASUREMENTS; MAGNETOMETRY; NOISE, ACOUSTICAL; NUCLEAR INSTRUMENTS; OPTICAL INSTRUMENTS; PHOTOGRAPHY; PHOTOMETRY; TELEMETRY; THEORETICAL PHYSICS.

MECHANICAL PROPERTIES OF SOLIDS

When a material is in the solid phase, its constituent particles, which may be atoms, ions, or chemical molecules, vibrate about fixed equilibrium positions in which the interparticle force is zero. In most solids composed of small constituent particles, e.g., metals and ionic solids, these interparticle interactions produce an internal atomic or molecular arrangement which is regular and periodic in three dimensions over intervals which are large compared with the unit of periodicity. Such solids are called crystals.

Solids composed of larger units, e.g., polymers, can be crystalline, though the crystallinity is usually rather imperfect, or they can be amorphous.

When a solid is deformed by external forces, the constituent particles have their separations changed from the equilibrium values. The resultant of the interparticle forces acting on a particular particle is then no longer zero, but acts to restore the particle to its original position relative to its neighbors. When the solid is in equilibrium under the action of external forces, the interparticle (or internal) forces must be in equilibrium to give continuity of the material and must also be equal to the external forces, i.e., any element of the body must be in equilibrium. These internal forces are maintained as long as the external forces are applied. When the external forces are removed the internal forces restore to the constituent particles their original separations.

If, after unloading, the body returns exactly to its former size and shape its behavior is called perfectly elastic. If it retains completely its altered size and shape it is a perfectly plastic body. In general, the behavior of real bodies lies between these two extremes.

Stress and Strain Two types of forces may act on any element of a body: (a) surface forces, exerted by the surrounding material, which are proportional to the surface area of the element and (b) body forces, which are proportional to the volume of the element, e.g., gravitational forces. The effects of body forces are usually negligible compared with those resulting from surface forces.

For a body to be deformed and not merely accelerated when forces are applied to it the body must be in statical equilibrium under the action of the applied forces. The conditions of equilibrium are (a) there must be no unbalanced applied forces and (b) there must be no unbalanced applied couples. Further, the internal and external force equilibrium can be equated.

The effect produced in a given material by forces of given magnitudes depends on the size of the body to which they are applied, and hence, to enable a comparison to be made of the reaction to external loading of bodies of different size, the concept of *stress* is introduced.

The stress in an element is defined as force divided by area over which the force acts. It is described as a homogeneous stress if, for an element of fixed shape and orientation, the value is independent of the position of the element in the body. Usually the term stress is taken to mean stress at a point, and is the limiting value of force divided by area over which the force acts as the area tends to zero. If a force δF acts over a surface of area δA and makes an angle ϕ with the normal to the surface, the normal stress σ is

$$\sigma = \lim_{\delta A \to 0} \left(\frac{\delta F \cos \phi}{\delta A} \right)$$

and the tangential or shearing stress τ is

$$\tau = \lim_{\delta A \to 0} \left(\frac{\delta F \sin \phi}{\delta A} \right)$$

The stress is, of course, transmitted through the solid.

The change in the separation of the constituent particles of the solid produced by the applied forces is seen on the macroscopic scale as a change in the size and shape of the body. Since the deformation of different bodies of a given material subjected to a particular load is a function of the size of body, comparisons are made using the relative deformation, or strain, defined as

$$\text{strain} = \frac{\text{change in dimension}}{\text{original dimension}} .$$

A strain is homogeneous if, after deformation, lines of the body that were originally straight remain straight and lines that were originally parallel remain parallel.

The following strains are found to be convenient in describing the behavior of a body in various states of stress.

When a rod of unstretched length ℓ_0 has its length increased to ℓ by the application of external forces, the conventional, engineering, or nominal tensile strain ϵ is defined as

$$\epsilon = \frac{\ell - \ell_0}{\ell_0} .$$

Sometimes it is more convenient to use the true, natural, or logarithmic strain $\epsilon*$, defined as

$$\epsilon* = \sum \frac{\delta \ell}{\ell} = \log_e \left(\frac{\ell}{\ell_0} \right) .$$

Clearly

$$\epsilon* = \log_e (1 + \epsilon)$$

If, as the result of the application of a uniform hydrostatic force, the volume of a solid changes from V_0 to V, the bulk strain θ is defined as

$$\theta = \frac{V - V_0}{V_0} .$$

When a solid is sheared by the application of couples, the angle of shear is taken as a measure of the strain, in this instance a shear strain.

Elastic Behavior For very small strains ($< \sim 0.1$ per cent) the behavior of many solid materials is almost perfectly elastic. In this strain range a specimen will exhibit a linear relationship between the magnitude of the applied forces and the deformation produced. This relationship is known as the Hooke law and in terms of stress and strain it may be stated in the form

$$\text{stress} = \text{constant} \times \text{strain}.$$

The constant in this equation is called a modulus of elasticity. Each strain has a corresponding modulus of elasticity. These moduli are temperature-dependent, and in general, depend on the direction of measurement, but if elastically isotropic solids are considered, the value of a particular modulus is independent of the direction in which it is measured. A solid is effectively isotropic if it is composed of grains whose size is small compared with the smallest dimension of the solid and if the orientations of the grains are randomly distributed.

Consider a bar of uniform area of cross-section A and unstretched length ℓ_0 acted upon by forces F applied uniformly at the ends. If ℓ is the length when this load is applied, the stress σ is given by $\sigma = F/A$ and the strain ϵ is $\epsilon = (\ell - \ell_0)/\ell_0$. (Tensile stresses are counted positive.) When the Hooke law is obeyed, $F/A = E(\ell - \ell_0)/\ell_0$ where E is a constant for a given material at a given temperature and is known as the Young modulus of the material.

When a solid has a hydrostatic stress σ applied to it, the volume changes from V_0 to V so that if the Hooke law is obeyed

$$\sigma = \frac{K(V - V_0)}{V_0}$$

where K is a constant at a given temperature and is known as the bulk modulus of the material.

When a solid is deformed by couples producing a shear stress τ, the angle of shear γ is taken as a measure of the strain so that, if the Hooke law is obeyed,

$$\tau = G\gamma$$

where G is a constant at a given temperature and is known as the rigidity modulus for the material.

The axial deformation of a prismatic bar with unloaded prismatic surfaces is accompanied by a change in the cross-sectional area. Experiment shows that the ratio

$$\text{lateral strain/axial strain}$$

is a constant known as the Poisson ratio ν. For the small strains encountered in pure elastic be-havior the change in cross-sectional area is very small, so the difference in the stress calculated using the original area of cross-section and using that when the load is applied is negligible.

Elastic shear deformation, in contrast, takes place at constant volume. The elastic moduli are not independent and it can be shown that, for an isotropic solid,

$$E = 3K(1 - 2\nu)$$

and

$$G = \frac{E}{2(1 + \nu)}.$$

Plastic Behavior When a solid is deformed under an increasing stress, a stage is reached when the further deformation produced by a slight increase in stress, though still elastic, does not obey the Hooke law. The stress at which the departure from linearity of the stress-strain curve first occurs is called the proportional limit or elastic limit. If the stress is increased beyond the elastic limit, a value is reached at which permanent deformation occurs, i.e., the specimen does not recover completely its original size and shape on unloading. The stress at which permanent deformation is first detected has, for very many materials, a value characteristic of the material at that temperature and is called the yield stress. The corresponding point on the stress-strain curve is the yield point. For many materials the elastic limit and yield stress have almost the same value and are not readily distinguished. The deformation not recovered on unloading is called the permanent set and the specimen is said to have suffered plastic deformation.

Plastic Deformation of Simple Crystalline Solids, e.g., Metals and Ionic Solids. The simplest mechanical test that can be performed on a solid is the tension test, and measurements made during such tests are often used to characterize particular materials.

Many simple crystalline solids are ductile at temperatures greater than about 0.3 to 0.4 of the melting temperature in kelvins. The plastic deformation of such materials takes place at approximately constant volume, and hence, for a specimen tested in tension, the cross-sectional area decreases as extension proceeds. This change in cross-sectional area with strain necessitates a more careful definition of stress. Two definitions are in common use, namely, conventional stress σ_c, sometimes called the nominal or engineering stress, defined by

$$\sigma_c = \frac{\text{load}}{\text{original area of cross section}}$$

and true stress σ_t, defined by

$$\sigma_t = \frac{\text{load}}{\text{area of cross section under that load}}$$

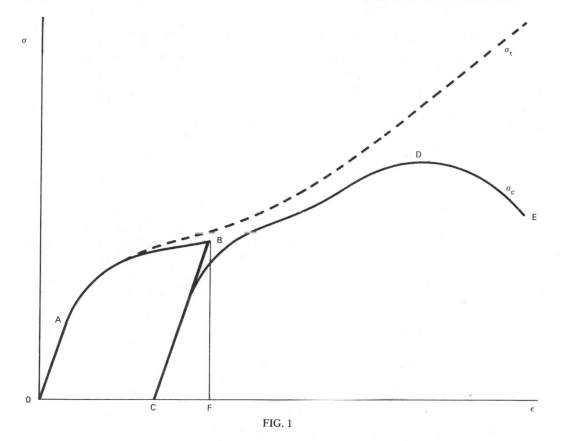

FIG. 1

When a tensile test is carried out on a fine-grained sample of a ductile material the conventional stress σ_c vs engineering strain ϵ graph has the form shown by the solid line in Fig. 1. The actual shape of the curve depends on many variables, e.g., purity of the material, temperature of testing, and rate of straining.

Over the region OA the graph is a straight line passing through the origin, the behavior is perfectly elastic, and the Hooke law is obeyed. When the stress exceeds that at A macroscopic permanent deformation occurs and the curve bends towards the strain axis; the stress at A is the yield stress σ_y and A is the yield point. However, the stress needed to produce further plastic deformation increases with strain and the material is said to work-harden or strain-harden. If the specimen is unloaded when B is reached, the unloading path is BC, which has almost the same slope as OA; the elastic properties of the material are little affected by plastic deformation. The elastic strain FC is recovered, but the material retains the plastic strain OC, which is the permanent set. The plastic strain becomes an increasing fraction of the total strain as the latter increases. When the test is continued the stress rises along CB (ignoring a small amount of hysteresis which is sometimes

observed), but before B is reached the curve bends towards the strain axis and then continues to rise as if unloading had not taken place. σ_c continues to rise until D is reached and then starts to fall. The load corresponding to D on the σ_c vs ϵ curve is the maximum load that the specimen can withstand in tension, and the value of σ_c corresponding to this load is called the ultimate stress or ultimate tensile strength σ_u.

For deformations represented by OD on the σ_c vs ϵ graph the extension is homogeneous, i.e., on the macroscopic scale the deformation is the same for all cross sections. At D, however, a neck forms in the specimen, all subsequent plastic deformation is restricted to this neck, and the load needed to produce further extension falls. The neck gets progressively narrower until fracture occurs at a strain corresponding to E.

When σ_t is plotted instead of σ_c, the curve has the form shown by the dotted line in Fig. 1. If, when the neck forms, σ_t is measured in the neck, σ_t continues to increase with strain up to fracture.

Mild steel and some other materials show a different behavior. The elastic range is terminated when the stress reaches a value known as

TABLE 1

Material	E (Nm^{-2})	K (Nm^{-2})	G (Nm^{-2})	ν	σ_y (Nm^{-2})	σ_u (Nm^{-2})
Al	71×10^9	75×10^9	26×10^9	0.33	26×10^6	60×10^6
Cu	130×10^9	138×10^9	46×10^9	0.34	40×10^6	160×10^6
Steel	210×10^9	168×10^9	83×10^9	0.28	0.4×10^9 (σ_{UYS}) 0.3×10^9 (σ_{LYS})	460×10^6

the upper yield stress σ_{UYS}. There is an abrupt partial unloading and macroscopic plastic deformation occurs locally in regions called Lüders bands. These bands spread along the specimen and the value of σ_c oscillates about a relatively constant value known as the lower yield stress σ_{LYS}. When the Lüders bands cover the whole specimen, further deformation is macroscopically homogeneous and σ_c rises as the material work-hardens.

The stress-strain curves usually plotted use the engineering strain, but it should be noted that if true stress is plotted against true strain the resulting curve is the same for both compression and tension tests.

Some typical values of elastic moduli and yield stresses are given in Table 1. These values refer to measurements on fine-grained wires at room temperature.

The Deformation of Solid Polymers Polymer molecules consist of very long chains of atoms, often containing short side groups at regular intervals. In many of the common polymers the linking between neighboring chains is weak. This type of polymer is rigid at low temperatures and soft and rubbery at high temperatures, the transition being reversible. Such long-chain polymers, whose properties are strongly temperature-dependent, are called *thermoplastics*.

Polymers in which there are frequent strong links between neighboring long chains are said to be *crosslinked*. Crosslinked polymers have properties that are rubberlike.

Crosslinking is also found in the thermosetting plastics, in which nonlinear structures are formed. When these materials polymerize, a process accelerated by raising the temperature, the monomers group themselves into a rigid framework that is not softened when the temperature is raised again. These materials show brittle behavior under an applied stress.

Thermoplastics can be either *amorphous* (or glassy), in which state the polymer chains are randomly oriented and intertwined, or *crystalline*, in which small regions of the structure exhibit a definite arrangement of the polymer chains.

At very low temperatures both amorphous and crystalline polymers show essentially brittle

behavior, having a Young's modulus of about 10^9–10^{10} N m^{-2} and breaking at a strain of about 5%. When tested at temperatures close to the melting point the deformation of both types of material is dominated by the sliding of polymer chains over each other, large irreversible strains are produced, and the behavior is termed *viscofluid*.

Between these extremes of behavior is an intermediate temperature range, the *glass transition range*, in which the mechanical behavior is strongly time dependent and is called *viscoelastic*. One manifestation of the glass transition is a fairly abrupt change in volume expansivity, and this can be used to define a glass transition temperature T_g. At temperatures above T_g the polymer chains have a certain freedom of movement relative to each other, whereas at temperatures well below T_g there is a complete locking of the polymer chains and their individual segments.

At temperatures up to T_g amorphous polymers deform elastically under tension until the so-called *yield stress* σ_y is reached, when the stress drops to a lower value σ_d (the *draw stress*) and a neck appears in the specimen (see Fig. 2). With further deformation this neck propagates along the specimen and the stress only rises again when the complete gauge length has been drawn down to a neck. In this process the polymer chains are oriented in the direction of the applied stress and the material becomes

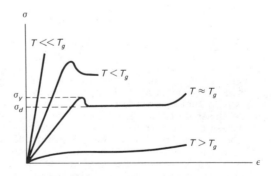

FIG. 2.

TABLE 2

Material	State at Room Temperature	E (MN m^{-2})	T_g (K)
Polyethylene	partially crystalline	70–280	153
Polyvinylchloride	amorphous/slightly crystalline	2500–3500	353
Polymethylmethacrylate	amorphous	2500–4000	380
Nylon 6	crystalline	2000–3000	323
Phenol formaldehyde resin	thermosetting glass	7000	–

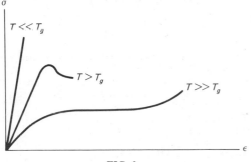

FIG. 3.

stronger. At temperatures above T_g large strains develop from the start of the test and no yield stress drop is observed. In both these regimes the strain can be recovered completely by heating the material at a temperature above T_g.

Crystalline polymers show rather similar curves (Fig. 3), but brittle behavior is observed for temperatures up to T_g. Above T_g a neck is produced in tensile deformation, but it results from recrystallization of the polymer chains in the direction of the stress, since the melting point of the aligned chains is higher than that of the unaligned chains. Consequently, no recovery of the strain takes place when the material is heated. At temperatures around T_g the propagation of the neck is usually terminated by flaws in the material, but well above T_g the neck propagates along the entire length of the specimen and the stress rises when the aligned polymer chains become strained.

Some data for common polymeric materials are given in Table 2.

M. T. SPRACKLING

References

Benham, P. P., "Elementary Mechanics of Solids," New York, Pergamon Press, 1965.

Calladine, C. R., "Engineering Plasticity," New York, Pergamon Press, 1969.

Hall, C., "Polymer Materials," London, Macmillan Press, 1981.

Honeycombe, R. W. K., "The Plastic Deformation of Metals," London, Edward Arnold, 1975.

Cross-references: ELASTICITY, POLYMER PHYSICS, SOLID-STATE PHYSICS, VISCOELASTICITY.

MECHANICS

"Give me matter and motion," proclaimed René Descartes, "and I will make the Universe." And what the renowned 17th-Century philosopher meant by that bold and somewhat cryptic remark was that at the very primal heart of things physical, there is *matter in motion*. Understanding the marveolus subtleties, the unity of that fundamental reality, is essential to knowing the Universe on any level. In its broadest sense *mechanics* is just that: *it is the study of the relative movement of objects* (actual and impending), or, if you will, *the study of motion and rest*, the latter merely being a special case of the former.

The subject developed historically along several different lines driven by very different practical concerns. For the most part, the descriptive aspects of the study evolved more successfully earlier on, followed only later by an effective explanatory capability. The result was an almost natural partition of mechanics into several broad subdivisions which are usually (though not universally) designated as kinematics, dynamics, and statics.

The description of every sort of motion, without regard to either the cause thereof or to the physical nature of that which is moving, is known as *kinematics*. Insofar as it deals with the changing locations of objects in space and time, it can be regarded as the geometry of motion.

By contrast, *dynamics* (sometimes referred to in part as *kinetics*) is the study relating motion and the changes therein with the corresponding causative interactions. Dynamics seeks to explain the motion described by kinematics.

As a special case of dynamics, *statics* relates specifically to the conditions governing constant relative motion (including "rest"). Historically the discipline evolved simply as the study of objects at rest and that is still the quintessence of the business; but insofar as we now know that absolute rest is a fiction, the purview of statics is specified more appropriately as that of objects in unchanging motion. Loosely speaking, statics deals with systems that can be imagined to be motionless (like the Brooklyn Bridge). It is the study of the balance of interactions operating on and within any material system which results in its effectively being at rest. Unlike dynamics, wherein time is central, the imagery of statics is quite independent of time—presumably nothing changes.

Kinematics The primary goal of kinematics is to provide a quantified description of motion, one in which the necessary concepts are inherently measurable. The formulation begins naturally enough with the familiar ideas of *space* and *time*. Even so, it is an illusion to think that we can satisfactorily define these most basic underlying concepts. Pragmatically we must content ourselves with measuring intervals of space and time with meter sticks and clocks, relying on intuition for conceptual meaning.

Clearly the faster an object moves, the farther it will travel in a given amount of time. That's the crucial insight that leads to a definition of *speed*, the measure of "how fast." The oldest surviving thoughts on the subject are those of Aristotle (384–322 B.C.) and, like his fellow Greeks of the era, he specified speed as the distance traversed in a given amount of time. And that's just the way it was framed for well over a thousand years—a thing traveled with a speed of "so many miles in so many hours." Nowadays we say almost, but not quite, the same thing defining *average speed* (v_{av}) *as the interval of distance traveled* (Δl) *divided by the interval of time* (Δt) *it took to do the traveling:* $v_{av} \equiv \Delta l/\Delta t$. However close the ancients got to the idea, they never actually carried through the division, apparently because of a hesitance to divide the "unlike" notions of space and time.

The scholars of the mid-1300s, especially at the University of Paris and Merton College, Oxford, dealt quite successfully with the idea of constant speed. But when they attempted to define the speed of an object at any given moment, the *instantaneous speed,* they failed. That was not surprising, for they lacked the mathematical imagery of motion, a calculus of change, something Newton would create just for the purpose centuries later. As the time interval Δt over which v_{av} is determined is made smaller and smaller, the ratio of $\Delta l/\Delta t$ approaches a value known as the instantaneous speed, or what we nowadays just call the speed (v). That limiting process actually defines the derivative and so in the notation of calculus, $v = dl/dt$: *speed is the time rate of change of distance*, the derivative of distance with respect to time.

If, instead of the distance traveled, we consider the *displacement* (**s**), which is the vector drawn from some origin to the moving body, we can express in a single concept both the speed and the direction of motion. Accordingly, *velocity* (**v**) *is the time rate of change of the displacement*, **v** = d**s**/dt, and the magnitude of the velocity is the speed.

Variations in motion are commonplace; change is the rule rather than the exception and the measure of this change is called *acceleration*. Aristotle considered the concept, hinting at it in his book *Physics*, but never quite grasping it clearly. His follower Strato (ca. 340–270 B.C.) seems to have been the first person to appreciate the real-world importance of the idea. He suggested that a body was accelerating when it traversed equal increments of distance in shorter and shorter times. During the 12th-Century revival of science, the alternative formulation was set forth that acceleration obtained when a body traveled greater and greater distances in successive equal intervals of time. Only later, in the 14th Century, did the realization begin to emerge that variations in the speed itself were the essence of the concept. Today we define *acceleration* (**a**) *as the time rate of change of velocity:* $\mathbf{a} \equiv d\mathbf{v}/dt$.

Along with the notions of displacement and time, velocity and acceleration complete the vernacular of kinematics. Constructing the *equation of motion* of the system (the expression of displacement as a function of time) becomes the primary task of the discipline.

Dynamics The explanations of dynamics go beyond the descriptions of kinematics, requiring additional concepts which reflect the physical nature of mover and moved. Perhaps the richest and, at the same time, most elusive of these notions is that of *mass*.

The idea that there had to be another measure of matter, in addition to the old standbys of weight and volume, was first proposed by the theologian Aegidius Romanus (ca. 1247–1316). In an effort to resolve some complex religio-philosophic questions concerning the Eucharist, he suggested that the true measure of a substance, the "how much" of a material entity, was its *quantity-of-matter*. Though in the end it had little or no influence on theology, the new insight, however undefined, found a welcome place in medieval dynamics. The Parisian physicist Jean Buridan (ca. 1300–1385) utilized the conception of quantity-of-matter in his highly influential *impetus theory*. By the 17th Century, the phrase *quantity-of-matter* and the term *mass* (already long in common unscientific usage) had become synonymous—Newton used them interchangeably.

Buridan is responsible for conceiving one of the most important dynamical ideas to come out of the Middle Ages. He began with a question which is essentially equivalent to this: "Given that both are traveling at the same speed, which would you rather get hit by, a firefly or a fire engine?" Obviously the firefly, but why? What aspect of motion, above and beyond just speed, is involved? He suggested that the essential measure of motion was proportional to both quantity-of-matter and speed, that is, it depended on their product. This new metaphor of motion would soon come to be known as the *quantity-of-motion* and ultimately it would be reinterpreted and renamed *momentum*. Momentum (mv) has proved to be one of the fundamental characteristics of the movement of all things.

Since the time of Aristotle, it had been widely assumed that the sustained motion of an object required the action of a sustained force. It was Galileo Galilei (1564-1642) who most convincingly challenged this seemingly reasonable view. Still, it should be pointed out that there were others before him who had thought, if somewhat tentatively, along similar lines. The Tuscan master performed a series of experiments which led him to conclude that an object once set in motion and left alone will continue in motion all by itself forever. That was the *law of inertia*, one of the first grand insights into the long hidden workings of the Universe. On a sizable planet like Earth, gravity and all sorts of friction conspire to obscure this all-important underlying principle—that is why it remained unrevealed for 2000 years. Had we been dwellers on a far smaller vessel in space, the natural tendency for things in motion to continue in motion would have been quite obvious. As it is, we are not, and it is not, and it took the genius of Galileo to see beyond the Scholastic fiction, the age-old error that the natural tendency of matter was to rest.

When Isaac Newton (1642-1727) came to codify motion in his masterpiece, the "Mathematical Principles of Natural Philosophy," he began with a series of definitions. The first was a rather unsatisfactory attempt at defining quantity-of-matter, while the second was a clear statement of quantity-of-motion framed as the product of mass and velocity. Newton then set out the three "axioms or laws of motion" which form the basis of dynamics (and statics) even to this day. The *first* of these was the law of inertia: *Every body continues in its state of rest, or of uniform motion in a straight line, except insofar as it is compelled to change that state by forces impressed upon it*. Force is the agent of change; it does not sustain motion, it changes it.

Newton's *second law* is a quantified recasting of the first law. Modernized somewhat in its language, it reads: *The rate of change of the quantity-of-motion (i.e., the momentum) of a body is equal to, and occurs in the same direction as, the net applied force*, $\mathbf{F} = d(m\mathbf{v})/dt$. From the first law we have that that which changes motion is force and now, more specifically, *force equals the change in the motion per unit time* (the measure of motion being the quantity-of-motion or the momentum).

If mass is taken to be constant, $\mathbf{F} = md\mathbf{v}/dt$ and so $\mathbf{F} = m\mathbf{a}$. This is the hallmark of Newton's theory and yet it does not appear anywhere in his work. In fact, it actually was introduced decades later by the Swiss mathematician Leonhard Euler. It is a real tribute to Newton's vision that his formulation in terms of momentum is in perfect accord with modern relativity theory, whereas $\mathbf{F} = m\mathbf{a}$ is not. Mass is a function of speed and therefore not constant in time—but that would not be shown until 1905 and Einstein.

Sir Isaac's third axiom, his *third law* of motion, completes the logical picture of force which is the very pillar of his dynamics. An isolated body follows the law of inertia. It cannot alter its own motion; that requires some outside intervention called *force*. And when two bodies, like billiard balls, interact, it's only reasonable to assume that both will be affected, both motions will be altered, *both will experience a force*. Leonardo de Vinci (1452-1519) had pointed the way long before. "An object offers as much resistance to the air," he wrote, "as the air does to the object. You may see that the beating of an eagle's wings against the air supports its heavy body in the highest and rarest atmosphere...." Whatever the source of inspiration, and others had grasped the essence of it, too, Newton provided the final link: "To every action there is always opposed an equal reaction: or the mutual actions of two bodies upon each other are always equal, and directed to contrary parts." *The interaction of two entities always occurs via an equal action-reaction pair: force and counterforce*. There is no such thing as a single force; force is a thing of pairs.

The second and third laws combine to reveal one of the guiding principles of modern physics, the *law of conservation of momentum*. If two objects interact, the forces acting on each will be equal and opposite and so, too, will be their resulting changes in momentum. In terms of the system as a whole, these paired opposite momentum changes cancel each other, leaving the net or combined momentum unaltered. *The total momentum of a system of interacting masses must remain unchanged provided that no net external force is applied*. As Newton put it, "the quantity-of-motion . . . suffers no change from the action of bodies among themselves." Amusingly, the idea had been speculatively anticipated by René Descartes (1596–1650), who wrote of the Creator: "He conserves continually . . . an equal quantity-of-motion" in the Universe.

To the kinematical ideas of time, displacement, velocity, and acceleration had been added the dynamical concepts of mass, momentum, and force, all united in the credo of the three laws. Brilliant physicists would spend the next two hundred years mathematically honing the fine edge of Newton's force-dynamics. And all the while another complementary piece to the scheme, another powerful vision, was slowly evolving—the concept of *energy*.

The necessity to formally quantify *work* came out of the practical needs of the engineers and scientists of the late 18th Century at the start of the Industrial Revolution. Work ≡ force X distance; work equals the force applied to an object multiplied by the distance through which it moves, a quantity easily measured with scales and meter sticks and just as easily bought and paid for. *Power*, the amount of work done per unit time, was also a practical measure dictated by the demands of the new machine age.

In "The Two New Sciences," Galileo had long before shown some grasp of the key idea. He talked about the physics of pile drivers and recognized that the weight of the hammer and the distance through which it fell determined its effectiveness—force and distance related in a crucial way.

Suppose we do work on a hammer, exert a force on it and cause it to accelerate, and then slam it down on a nail bringing it to rest, thereby doing work on the nail; we have work-motion-work. And if we recognize that work is the changer of energy, then the hammer apparently has some sort of *energy of motion.*

Interestingly, the underlying insight to all of this actually began to evolve roughly a hundred years before it reached maturity in the 19th Century. Christian Huygens (1629–1695) never cared much for Descartes's reliance on the idea of quantity-of-motion. Momentum to be really meaningful must be framed as a directional quantity, a vector, like force. A body at rest could explode into two pieces violently flying in opposite directions and yet the total momentum would remain zero throughout. That bothered Huygens, who suggested a different measure of the motion, one which would be independent of the direction of the velocity, one which would only vanish when all the motion actually ceased. He subsequently decided on the product of the mass and speed *squared*. Gottfried Leibniz (1646–1716), Newton's bitter rival, picked up the idea, calling mv^2 the *vis viva* or living force. The great and meaningless vis viva controversy between the followers of Descartes and Leibniz roared on for decades as each side claimed the fundamental notion.

It was not until the beginning of the 1800s that Thomas Young (1773–1829) shifted the imagery and spoke of mv^2 as *energy.* "Labour expended in producing any motion," he wrote, "is proportional . . . . to the energy which is obtained." Then Gaspard de Coriolis (1792–1843), using Newtonian mechanics, showed that the work done on a system was equivalent to a change in the quantity $\frac{1}{2}mv^2$. Lord Kelvin (1824–1907), years later, dubbed this the *kinetic energy.* Vis viva was forgotten, drowned out by the roar of the Industrial Revolution, and in its place stood its lookalike, kinetic energy.

Suppose that there is some sort of driving force constantly exerted on a body, like its weight. To move against that pull requires the application of a counterforce and the doing of work. The crucial point is that the force—be it elastic, electric, magnetic, gravitational, whatever—continues to act even after the displacement. Once let loose, that force will drive the body back, imparting kinetic energy in the process. Clearly it is possible to do work and not have it immediately appear as kinetic energy, and yet the potential for generating that energy is there. This retrievable stored energy, *energy by virtue of position in relation to a force*, is known as *potential energy*, a name given it by William Rankine (1820–1872).

When we consider the kinetic and potential energy of a body as a whole, it is understood that all of its atoms act together in an organized fashion. Alternatively, it is possible to impart motion to the individual constituent atoms which is disorganized, motion not of the body, but within the body. A pendulum swings until it ultimately comes to rest—organized kinetic energy is transformed into disorganized kinetic energy or, as it's called nowadays, *thermal energy*. The ubiquitous agent of that transformation is known as *friction.*

One of the great revelations of the previous century was the *law of conservation of energy* opened out to include both thermal and mechanical processes: *Energy can neither be created nor destroyed, but only transformed from one form to another.* Whatever energy is (and we have no satisfactory conceptual definition of that underlying quantity, although we know its various manifestations) it is conserved.

To the force-time-momentum imagery of the 18th Century was added the force-space-energy vision of the 19th Century. And then in the 20th Century Albert Einstein (1879–1955), questioning the very basic understanding of space and time, profoundly recast kinematics and dynamics in his *special theory of relativity.* Newtonian mechanics turned out to be the low-speed approximation of the new vision whose real significance becomes apparent only at speeds that are appreciable in comparison to the speed of light.

The theory builds from two basic postulates. The first is known as the *principle of relativity*, which states that *all the laws of physics are the same for nonaccelerating observers.* The second, the *principle of the constancy of the speed of light*, maintains that *light propagates in free space with a speed* ($c \simeq 300,000,000$ m/s) *that is independent of the motion of the source (and of the observer).* The speed of light in vacuum is absolute.

Among the many surprises provided by the new vision was the realization that rest, motion, simultaneity, time, length, and mass are not absolutes as had long been thought. Instead these fundamental quantities are relative, they depend on the motion of the observer.

The result, which Einstein himself thought was "the most important," was the *equivalence of mass and energy*. These two seemingly different concepts are actually manifestations of one single entity: *mass-energy, $E = mc^2$, the total energy E of an object equals its mass m multiplied by the speed of light squared.*

Statics The central problem that was never far from the surface in the early days of the development of kinematics and dynamics, was the motion of the heavens. Aristotle had woven his theory of motion into his cosmography to form a single fabric that would stand for two thousand years. By contrast, the motivation behind the development of *statics* was far more mundane and pragmatic. The ancients, who weighed out their goods on balances, who raised stone, hauled ships and pitched tents were all practitioners of statics, even before the formal body of knowledge evolved.

An object experiencing no change in its motion, *no acceleration*, is said to be in *equilibrium*. Hence it follows from Newton's second law that *translational equilibrium* ($a = 0$) *obtains when the sum of the forces acting on a body is zero,* $\Sigma \mathbf{F} = 0$. If the two teams in a tug of war pull equally hard in opposite directions, the net force is zero and the rope remains motionless. This is the first of the two conditions of equilibrium.

The simple equal-arm balance was already in widespread use well over 4000 years ago. It is not surprising then, that Aristotle, in his "Mechanica," attempted to analyze the important practical problem represented by balances, levers, and seesaws (which are just variations on one theme). When equal downward forces are applied to each end of the centrally pivoted rod, the system remains motionless, balanced horizontally. Any tendency to rotate clockwise is canceled by an equal tendency to rotate counterclockwise, and we say that the system is in *rotational equilibrium*. In fact, the word *equilibrium* derives from the Latin *aequus*, even or equal, and *libra*, a scale or balance.

It was the great Archimedes of Syracuse (287–212 B.C.) who, in his treatise "On Equilibrium," framed the *law of the lever* in a satisfactory way. *Unequal forces (of magnitudes F_1 and F_2) acting perpendicularly on a bar at unequal distances from the pivot (r_1 and r_2, respectively) balance each other provided that $F_1 r_1 = F_2 r_2$.* It is not enough to be concerned only with the sizes of the forces; the distances from the pivot at which they act are crucial, as well.

Da Vinci, aware that the human body achieved its mobility via a system of various kinds of levers, set himself to the study of these simple machines. He was among the first to recognize the significance of the idea of the *moment of a force*.

Envision a force $\mathbf{F}$ acting on a body which is pivoted at a point O: *The lever or moment-arm of the force $\mathbf{F}$ with respect to the axis passing through O is the perpendicular distance ($r_\perp$) drawn from O to the line of action of $\mathbf{F}$. The moment of the force about O is then defined as the product of the magnitude F and $r_\perp$.* Nowadays it is common practice to symbolize this quantity by the Greek letter *tau*, τ, and refer to it as *torque* (from the Latin, *torquere*, to twist). The law of the lever is then simply a requirement that the two opposing torques be equal. Formulated as a vector, torque becomes $\boldsymbol{\tau} = \mathbf{r} \times \mathbf{F}$.

In the case of a rigid body (viewed as a collection of interacting point masses) Newton's second and third laws lead to the *second condition of equilibrium: For a rigid body in equilibrium the sum of the torques about any point (due to all the externally applied forces acting on it) must be zero,* $\Sigma \tau = 0$.

The two conditions of equilibrium provide a basis for the analysis of the forces at work in all sorts of mechanical systems from trusses and bridges to the muscle-bone structure of the human body.

Mechanics, the tap root of physics, is the seminal discipline, whether it is general relativity on the cosmic scale, quantum mechanics in the micro-domain of the atom, or Newtonian mechanics in the macro-world of baseballs and ballistic missiles. "Give me matter and motion and I will make the Universe."

EUGENE HECHT

References

Hecht, E., "Physics in Perspective," Reading, Mass., Addison-Wesley Pub. Co., 1980.

Cajori, F., "A History of Physics," New York, Dover Pub., Inc., 1962.

Jammer, M., "Concepts of Mass," Cambridge, Mass., Harvard Univ. Press, 1961.

Toulmin, S., and Goodfield, J., "The Fabric of the Heavens," New York, Harper and Brothers, 1961.

Cross-references: DYNAMICS, ELASTICITY, GRAVITATION, MASS AND INERTIA, MECHANICAL PROPERTIES OF SOLIDS, STATICS.

MEDICAL PHYSICS

That physics has an important place in medicine can scarcely be denied. A physician's first move in examining a patient is to measure his temperature, count his pulse, listen to his heart sounds and take his blood pressure. Only much later does the physician get around to chemical and laboratory tests. Yet every hospital of any stature has a laboratory or a department of clinical chemistry. Laboratories of clinical physics are virtually nonexistent. While physics plays a large role in medical diagnosis and treatment, physicists have largely neglected the field.

Some of the earliest applications of the principles of physics to problems in medicine were in the fields of optics and sound. An early contributor was H. L. F. von Helmholtz, a physician as well as a physicist. His work in physiological optics and that on the sensations of tone are considered classics. Even earlier, J. L. M. Poiseuille, a French physician and physicist, seeking a better understanding of the flow of blood, studied the flow of water in rigid tubes. His work not only contributed to physiology but also established an important relation in the physics of viscous fluids. D'Arsonval, a French physicist, pioneered in the therapeutic use of high frequency electric currents and measuring instruments. Much earlier, that unusual artist, inventor and physicist, Leonardo da Vinci, had shown a keen interest in the fascinating mechanics of human locomotion.

With the intensive development of the sciences of physics and medicine in the latter part of the nineteenth century, the two drew further apart. This period also saw rapid development in the science of physiology which is concerned not only with chemical but also with physical processes in the body. Clinical physiology abounds with such concepts as the pressure-velocity relationships in the flow of blood, the mechanics of the cardiac cycle, the work of breathing, gas exchange in the lungs, voltage gradient in cellular membranes, and cable properties of nerves, to name but a few. These concepts have, of necessity, been worked out by scientists with training and experience in the basic biological and clinical procedures. Physicists have been inactive in the field and have made very little contribution to its development. But there is a growing awareness among physiologists of the importance of physical principles and the need for precise statement of physical law. An example of this conviction is Howell's Textbook of Physiology in which editions since the eighteenth have carried the title, "Physiology and Biophysics."[1]

A phenomenon of the mid-twentieth century has been the development of interdisciplinary fields of science. BIOPHYSICS combines the most fundamental of the biological and physical sciences. It has had an extremely rapid growth, with something like 30 to 40 university Departments of Biophysics in America alone. Its emphasis has been on the application of physical principles to all aspects of biology—cellular, botanical, zoological as well as clinical.

An even more recent phenomenon has been the development of biomedical engineering. Its basis has been the application of the tremendous developments in electronics to medical measurements and instrumentation. In fact, the field is frequently referred to as biomedical electronics. Such recent developments as vector electrocardiography, implantable pacemakers for the heart, and intensive-care physiological monitors and recorders are examples of the impact of electronics on medicine. While these fields border on medical physics, none are concerned primarily with the application of physical principles to clinical problems. Yet they compete so effectively with medical physics that it is difficult to delineate the boundaries of the latter.

The discovery of X-RAYS by Roentgen in 1895 had an immediate impact upon medicine. Within a few months, the new rays were used both diagnostically and therapeutically. Indirectly, their application set the stage for the development of medical physics. Therapeutic application of x-rays raised questions concerning their quality and quantity—both of which are important in accurate dosimetry. Evaluation of early successes and failures indicated the importance of the proper distribution of dose between neoplasm and normal tissue. The physician turned to the physicist for assistance. The late Otto Glasser was one of the early radiological physicists; he and Fricke in 1924 constructed an air wall ionization chamber for the measurement of radiation dose.[2] Their construction eliminated some of the nonlinear effects due to quality, i.e., photon energy distribution, in the evaluation of biological response. Other early workers in America were Edith Quimby and G. Failla. In England, L. H. Gray and W. V. Mayneord were active. In 1936, Gray proposed the Bragg-Gray formula for determining the absolute amount of energy delivered to a medium from ionization measurements.[3] The work of Fricke, Glasser, and Failla along with that of L. S. Taylor[4] and others contributed to the establishment in 1928 by the Second International Congress of Radiology of the roentgen as a unit of radiation dose based on the amount of ionization generated in a standard volume of air. The use of higher energies and ionizing radiations other than x-rays led during the 1950's to the abandonment of the roentgen as a unit of absorbed dose. Dissatisfaction with the roentgen was also due to a growing realization that biological response was more nearly related to the energy absorbed in a medium. The Bragg-Gray formula permitted the calculation of absorbed dose in a medium, and the work of J. S. Laughlin[5] established the dosimetry of high-energy radiations in energy units by calorimetric methods. While radiological physics is clearly a part of the broader discipline of medical physics, it included in the early days practically all that was organized of the later subject.

For many years there were no organizations of workers in the field of medical physics. In America, radiological physicists were associate members of the Radiological Society of North America, naturally dominated by radiologists. First in Britain (The Hospital Physicist's Association) and later in America specialty groups were organized. The American Association of Physicists in Medicine (AAPM) brings together those physicists working in hospitals and medical schools, and interested in an understanding of the physical side of medical problems. The membership has been largely drawn from those working in radiological physics, but an interest in all areas of the application of physics to medical problems is rapidly developing. The A.A.P.M. became affiliated with the American Institute of Physics in 1958 and is now showing the most rapid growth of its nine member organizations.[6] In 1973, the A.A.P.M. began publication of its own journal, *Medical Physics*. A further indication of the developing awareness of this field are the International Conferences on Medical Physics, which meet every three years (the Sixth met with the International Conference on Medical and Biological Engineering in Hamburg, Federal Republic of Germany in 1982).

One last word about a related field: Radiation protection was in the early days a part of radiological physics. In America, L. S. Taylor was active for many years at the Bureau of Standards in setting up guidelines for protection from radiation. During World War II, the Manhattan Project required large numbers of workers in the field of protection, and the term HEALTH PHYSICS was introduced. Since the war, the field has grown with the growth of the area of atomic energy. The Health Physics Society is a large and growing group with many local chapters and an international organization. The field seems, though, to be becoming more closely aligned with the area of public health than with clinical medicine.

When the first edition of the "Encyclopedia of Physics" was issued, the viability of medical physics as a profession was uncertain. It appeared that any breakthrough in the cure of cancer, eliminating radiation therapy as a modality of treatment, would also eliminate the livelihood of medical physicists. But much was already happening and the results that have changed the picture are now apparent. Electronics and nuclear physics have made many previously unmeasurable variables in medicine accessible to quantification. The introduction of the concept of modulation transfer function has made a science of the evaluation of diagnostic imaging quality. Nuclear medicine has had wide applications and attracted many physicists into the field. Ultrasound and computed tomography have had spectacular successes in improving medical diagnostics. And now nuclear magnetic resonance imaging is beginning to be applied. During the 1970s, with shrinking financial support for physics, many recent PhDs entered the field of medical physics as postdoctoral trainees. Physicists in general became more aware of the opportunities and the need to apply physical principles to the many problems in medicine. Many large physics accelerator installations developed programs for the treatment of cancer. Even the largest accelerator in America, Fermilab, has a large and continuing program for cancer treatment with high-energy neutrons. The award of a Nobel prize in Physiology and Medicine in 1977 to Roslyn S. Yalow[7] for the development of radioimmunoassay served notice to the world that physicists are contributing to the solution of problems in medicine. This was followed two years later by the award of Nobel prizes to a physicist, Allan M. Cormack,[8] and an engineer, Godfrey N. Hounsfield,[9] for the development of computed tomography. Certainly, one can now state that the new discipline of Medical Physics is healthy, growing and showing unmistakable signs of survival. And one can find few areas of physics where the challenge is greater, or the rewards more satisfying, than in making accurate evaluation of physical variables in the living human patient.

LESTER S. SKAGGS

References

1. Ruch, T. C., and Patton, H. D. (Eds.), "Physiology and Biophysics," 20th Ed., W. B. Saunders, Philadelphia, 1973.
2. Fricke, H., and Glasser, O., "Standardization of the Roentgen Ray Dose by Means of the Small Ionization Chambers," *Am. J. Roentgenol.* 13, 462 (1925).
3. Gray, L. H., "An Ionization Method for the Absolute Measurement of X-Ray Energy," *Proc. Roy. Soc., London Ser. A.*, 156, 578 (1936).
4. Taylor, L. S., and Singer, G., "An Improved Form of Standard Ionization Chamber," *J. Res., Natl. Bur. Std.* 5, 507 (1930).
5. Genna, S., and Laughlin, J. S., "Absolute Calibration of Cobalt-60 Gamma Ray Beam," *Radiology* 65, 394 (1955).
6. Porter, B. F., "AIP Member Societies Entering the 1980's," *Physics Today* 34, 27 (1981).
7. Yalow, R. S., "Radioimmunoassay: A Probe for the Fine Structure of Biologic Systems," *Science* 206, 1236 (1978).
8. Cormack, A. M., "Early Two-Dimensional Reconstruction and Recent Topics Stemming from It," *Science* 209, 1482 (1980).
9. Hounsfield, G. N., "Computed Medical Imaging," *Science* 210, 22 (1980).

Cross-references. BIOMEDICAL INSTRUMENTATION, BIOPHYSICS, HEALTH PHYSICS. MOLECULAR BIOLOGY, RADIOACTIVITY, X-RAYS.

METALLURGY

The metallurgical industry is one of the oldest of the arts, but one of the youngest of the subjects to be investigated systematically and considered analytically in the tradition of the pure sciences. It is only in comparatively recent times that any fundamental work has been carried out on metals and alloys, but there are now well-established and rapidly growing branches of science which are related to the metallurgical industry.

Extraction Metallurgy Extraction metallurgy, or the science of extracting metals from their ores, is broadly divided into two groups.

Ferrous. This branch is concerned with the production of *iron* (normally from iron ore, with coke and limestone in a blast furnace) and its subsequent refining into *steel*, by oxidizing the impurities either in an electric arc furnace by means of an appropriate slag on the surface or in a "converter," by blowing oxidizing gas through the molten iron. The most striking recent developments in this field have been the increasing use of pure gaseous oxygen in steel-making, and the increasing size of furnaces, with a resultant improvement in efficiency, rate of production, and quality of product. Over 500 million tonnes or 70% of the worlds's annual steel production is now made using oxygen converters.

Nonferrous. Some metals such as chromium, cobalt, and manganese, for example, are principally produced as alloying elements to improve the properties of steels. The nonferrous metals manufactured in greatest quantity include aluminum, copper, nickel, zinc, magnesium, lead, and tin, with titanium being an important newcomer in view of its low density, high melting point (1943 K) and resistance to corrosion. The precious metals, and the "refractory metals" of very high melting point (e.g., tungsten and molybdenum) are other important families.

Shaping of Metals This may be carried out in three main ways.

Casting. Most metals are initially cast into *ingots*, which may be subsequently forged to shape. The technique of *continuous casting* is increasingly used in this context to improve efficiency and to increase the rate of production. By 1990 it is expected that one-half of the world's steel output will be continuously cast. Many alloys are designed to be cast into their final shape by pouring the molten alloy into an appropriate mold. These may be sand molds if only a small number of objects are required, and very massive castings (e.g., over 10^6 Kg in mass) may also be produced in this way. A permanent mold, or die casting, is employed if large numbers of the object are required (particularly in alloys of low melting point, such as zinc-based alloys), and high dimensional accuracy can be achieved by these means. Cast iron is the cheapest metallic material. The microstructure of "gray cast iron" is shown in Fig. 1. It consists of flakes of graphite in a two-phase matrix of iron and iron carbide (Fe_3C). The brittleness of this material arises from the weakness of the graphite flakes, which act like cracks in the structure.

Forging. This entails shaping of the metal by rolling, pressing, hammering, etc., and may be carried out at high temperatures, when the

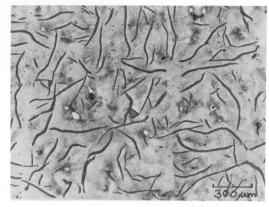

FIG. 1. The microstructure of gray cast iron: the black lines are flakes of graphite.

metal is soft (hot-working), or at lower temperatures (cold-working) where deformation leads to progressive hardening of the metal (work-hardening). In contrast with casting, forgings usually exhibit differing physical and mechanical properties in different directions, due to the directional nature of the shaping operation.

Much modern research in physical metallurgy is concerned with investigating the plastic flow and work-hardening behavior of metals and alloys. Metal crystals yield plastically at stresses several orders of magnitude lower than the theoretical value for the deformation of perfect crystals. This discrepancy is accounted for by the presence of linear imperfections known as "dislocation lines" within the crystals. Plastic flow takes place in metal crystals by "slip" or "glide" in definite crystallographic directions on certain crystal planes, due to the movement of dislocation lines under the applied stress. Dislocations multiply and entangle as deformation proceeds, thus making further flow increasingly difficult (work-hardening)—the density of dislocations rising from about $10^5 \, mm^{-2}$ in soft (annealed) metal to about $10^{10} \, mm^{-2}$ in work-hardened material. These and other types of crystal defect (such as *stacking faults*, which are planar in geometry) can be studied by x-ray diffraction and also by means of the electron microscope and the field-ion microscope (q.v.).

Powder Metallurgy. This is a method of shaping by pressing finely powdered metal into an appropriately shaped die. The "green compact" thereby produced is of low strength and is subsequently heated in an inert atmosphere ("sintered"); the pressing and sintering may be repeated until strong, dense products are obtained. The technology was first developed for metals which were of too high a melting point for conventional casting and forging methods, and tungsten lamp filaments were first produced by this means. Other refractory metals and hard metal-cutting alloys may thus be shaped, and some magnetic and other special alloys are prepared in this way by suitable blending of powders, which avoids any contamination that may be associated with the melting process. The pressing and sintering conditions may be arranged to leave some residual porosity in the structure of, for example, bronze bearing alloys. The pores are filled with oil, thus producing the so-called oil-less bearings which can operate without further lubrication.

Joining. The three important methods of joining metals are riveting, soldering or brazing (in which metal components are joined by means of a layer of alloy of lower melting point),

and welding (in which the metal itself is fused). *Weldability* is often the critical factor in the selection of an alloy for a given purpose, since the metallurgical changes produced by localized heating are often associated with the development of deleterious properties at, or adjacent to, the weld.

Alloy Constitution Phase equilibria in alloy systems are represented on *phase diagrams*, which represent the temperature ranges of phase stability as a function of composition. An example of such a diagram is given in Fig. 2; they are experimentally established by, e.g.,

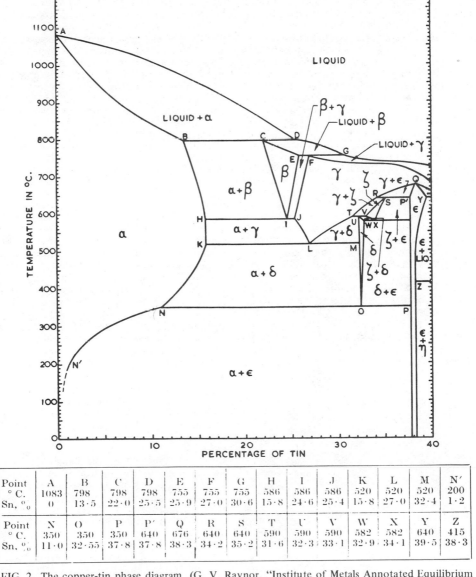

Point	A	B	C	D	E	F	G	H	I	J	K	L	M	N'
°C.	1083	798	798	798	755	755	755	586	586	586	520	520	520	200
Sn, %	0	13·5	22·0	25·5	25·9	27·0	30·6	15·8	24·6	25·4	15·8	27·0	32·4	1·2

Point	N	O	P	P'	Q	R	S	T	U	V	W	X	Y	Z
°C.	350	350	350	640	676	640	640	590	590	590	582	582	640	415
Sn, %	11·0	32·55	37·8	37·8	38·3	34·2	35·2	31·6	32·3	33·1	32·9	34·1	39·5	38·3

FIG. 2. The copper-tin phase diagram. (G. V. Raynor, "Institute of Metals Annotated Equilibrium Diagram," Series No. 2, 1944.)

thermal analysis, dilatometry, microscopical, and x-ray diffraction methods. Phase diagrams can also be calculated by computation of the Gibbs free energy minimum, if the thermodynamic functions of observable phases are determined by experiment. Diagrams such as that in Fig. 2 are invaluable in the interpretation of the structures of alloys observed under the microscope.

The microstructure of an alloy (and hence its properties) will be determined not only by its composition, but also by its thermal and mechanical history. Of particular importance is the metallurgical control of the mechanical properties of an alloy by *heat-treatment*, which affects the distribution of the phases present. Hardness, for example, will depend upon the state of deformation (i.e., the density of dislocations) and upon the composition of the alloy. Pure metal crystals can be hardened by other atoms in solid solution (solute hardening) as well as by finely dispersed particles of a hard second phase (precipitation, or dispersion hardening) which are effective in impeding the motion of dislocations when the crystal is stressed. Fig. 3 is an electron micrograph showing dislocations on the slip plane of a copper alloy crystal, and it illustrates how the presence of hard particles has caused local entanglement of the dislocations. The relationship between the microstructure and properties of metals and alloys is of fundamental importance and is a field of intense scientific activity.

Although many common alloys were not developed scientifically, a considerable theory of alloys is developing, springing from empirical rules and principles (notably those due to W. Hume-Rothery) which have generalized the facts and enabled predictions to be made. The early theories of the metallic state, due to Drude and Lorentz, and later to Sommerfeld, were developed and discussed by N. F. Mott and H. Jones in their book "The Theory of the Properties of Metals and Alloys." A great increase in our knowledge of transition metals and alloys has taken place, and some signs of general principles have begun to appear, although there is yet little theoretical knowledge enabling one to calculate properties or structures of alloys from fundamental principles. The propensity to form deleterious phases in nickel-based materials has been related empirically to the average number of electron vacancies (N_V) in a given alloy. Computer calculations, using a system known as PHACOMP, have identified the critical value of N_V above which such phases form.

The Effect of Environment Upon the Behavior of Metals *Low Temperature*. Some metals and alloys exhibit a spectacular change in mechanical behavior with decrease in temperature. Many metals of body-centered cubic crystal symmetry (e.g., iron and mild steel) which are tough and ductile at ordinary temperatures become completely brittle at subzero temperatures, the actual transition temperature depending upon the metallurgical condition of the alloy, the state of stress, and the rate of deformation. Some metals of hexagonal symmetry (e.g., zinc) exhibit this effect, but metals of face-centered cubic symmetry (e.g., copper) remain ductile to the lowest temperatures. This transition in behavior is clearly of critical importance in the selection of materials for low-temperature application.

High Temperature. Apart from problems of oxidation (discussed below), metals tend to deform under constant stress at elevated temperatures (the deformation is known as "creep"), and creep-resistant alloys are designed to provide strength at high temperatures. These are essentially alloys in a state of high thermodynamic stability, usually containing finely dispersed particles of a hard second phase which impede the movement of dislocations. Grain boundaries may be a source of weakness at elevated temperatures, and turbine blades in the form of alloy single crystals have recently been employed in engines of advanced design.

Fatigue. Metals break under oscillating stresses whose maximum value is smaller than that required to cause rupture in a static test, although many ferrous alloys show a "fatigue limit," or stress below which such fracture never occurs, however great the number of cycles of application. The phenomenon is associated with the nucleation of submicroscopic surface cracks in the fatigued component early in its life, which initially grow very slowly. Eventually a crack grows until the effective cross section of the piece is reduced to such a value that the applied stress cannot be supported, and rapid failure occurs. A typical fatigue fracture surface is shown in Fig. 4, in which two distinct zones are apparent. These correspond to the period of slow growth (left-hand side) and final failure, respectively.

Oxidation and Corrosion. With the exception of the "noble metals," which are intrinsically resistant to attack by the environment, metals

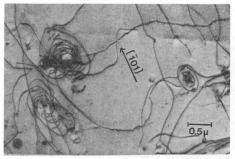

FIG. 3. A deformed copper alloy crystal containing hard particles. Electron micrograph showing interaction of dislocation lines with the second phase.

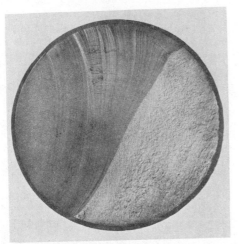

FIG. 4. A fatigue fracture surface upon a large steel shaft. (Courtesy of British Engine Insurance Ltd.)

in general owe their oxidation resistance, when they are heated in air, to the presence of impervious oxide films on their surfaces. Those which develop porous oxides (e.g., the refractory metals tungsten and molybdenum) oxidize very rapidly at high temperatures. Oxidation resistant alloys are designed to maintain a protective film under these conditions.

Corrosion occurs under conditions of high humidity or immersion in aqueous media. The phenomenon can be interpreted electrochemically—local anodes form at the region of metal dissolution, and local cathodes form where the electrons are discharged. "Galvanic corrosion" is encountered where dissimilar metals are in electrical contact under these conditions. Of particular importance is the *conjoint* action of stress and corrosion, where "stress corrosion" or (under fluctuating stresses) "corrosion fatigue" cracking may be encountered, in situations where no failure would occur under the action of the stress or the corrosive environment applied separately. Electrochemical principles are applied in the protection against corrosion.

Materials Technology The scientific principles which govern the behavior of metals are, of course, applicable to a wide range of other technologically important materials, such as polymers, ceramics, and glasses. In recent years many centers of metallurgical research both in industry and in universities have broadened their approach in this way and are often described as Departments of Materials Technology.

JOHN W. MARTIN

References

Metallurgical Data

The series of "Metals Handbooks," published by the American Society for Metals, Metals Park, Ohio.

Smithells, C. J., "Metals Reference Book," London, Thornton Butterworth Ltd., 1983.

Moffatt, W. G., "Handbook of Binary Phase Diagrams," Schenectady, N.Y., The General Electric Co., 1981.

General Reading

Street, A., and Alexander, W. O., "Metals in the Service of Man," London, Pelican, 1973.

Martin J. W., "Elementary Science of Metals," London, Wykeham Publications, 1969.

West, J. M., "Basic Corrosion and Oxidation," New York, Wiley, 1980.

Cross-references: CRYSTALLOGRAPHY, CRYSTAL STRUCTURE ANALYSIS, ELECTROCHEMISTRY, LATTICE DEFECTS, MECHANICAL PROPERTIES OF SOLIDS, SOLID-STATE PHYSICS.

METEOROLOGY

Meteorology is the study of the atmosphere. The word is derived from the classical Greek *meteoros* meaning "things lifted up." As early as 250 B.C. Aristotle wrote extensively about these topics in his treatise "Meteorologica." Meteorology, as practiced today, is broadly interdisciplinary, drawing extensively from the fields of hydrodynamics, thermodynamics, optics, chemistry, and mathematics. It has applications in industry, agriculture, transportation, economics, resource management, and many other human activities.

Almost every branch of physics and chemistry is in some way involved in atmospheric phenomena. This article is confined to brief overviews of modern methods of weather prediction, climatology, cloud physics (including attempts at weather modification), and boundary-layer processes.

Atmospheric Circulation and Weather Prediction The basic state of the atmosphere is described by seven variables: three components of velocity; temperature; pressure; density; and the water-vapor concentration (usually called the *mixing ratio*). The changes of these variables are governed by seven well known equations. The accelerations in the three principal directions are given by three differential equations based on Newton's second law. These were written in their hydrodynamic form by Euler about 200 years ago. Temperature changes are governed by the first law of thermodynamics. The other equations are the equation of state and the equations for the conservation of mass of the dry air and water vapor. The complete set of equations has been established for well over 100 years, but because of their complexity, their application to weather prediction has become a reality only during the last 25 years as modern computers have become available.

The numerical problems involved in solving these equations for the future values of weather elements are far from trivial. Essentially, the changes must be calculated for periods of a few minutes at a time based on the global distribution of the elements derived from observation or calculated during earlier time steps, and moved forward step by step. This method was first studied by L. F. Richardson in the early 1920s, but had to be abandoned because of inadequate computational capability. The problems were successfully confronted by John von Neumann and his colleagues in Princeton about 1950. In 1955 the U.S. Weather Bureau began numerical weather prediction on a twice-daily basis. Today, numerical weather prediction is being carried out routinely from greatly improved global models at several national and international centers. The use of models has brought about a considerable increase of predictive skill (measured by the degree to which accuracy exceeds that attainable using climatological averages). The skill achieved for two- to three-day forecasts by computer models is about that formerly attainable one day in advance by subjective methods. Demonstrable skill has been achieved out to ten or twelve days, but five days is a more realistic limit in temperate latitudes on a routine basis.

The use of circulation models in weather prediction requires upper-air observational input over at least the entire Northern Hemisphere (current models are now global in extent). The data are provided each 12 hours by several hundred soundings of pressure, temperature, humidity, and winds observed from balloons and telemetered to ground stations. The specifications for these data and their distribution are coordinated by the World Meteorological Organization (WMO). Data from several thousand surface stations and ships are also used.

Meteorological satellites have been employed extensively to supplement surface based data. Geostationary satellites, hovering at a height of 35,000 km above a fixed point on the equator provide quasi-hemispheric visible and infrared images every half hour. These are valuable for short-range prediction of thunderstorms, tornados, and other localized phenomena. They also provide important surveillance of tropical and extratropical cyclones in data-sparse areas, particularly in the southern hemisphere. Winds derived from satellite cloud-image motions have made important inputs to numerical prediction models. Also for this purpose, much effort has been devoted to developing methods of deriving temperature soundings remotely from polar-orbiting satellites. The inversion technique for reconstructing the temperature distribution from infrared fluxes observed at different wavelengths has been solved, but the technique can be used only in cloud-free regions.

Because of the global extent of the atmospheric circulation, much of the research is carried out in international programs. The most recent of these is the Global Atmospheric Research Program (GARP), planning and coordination of which is done jointly by the International Union of Geodesy and Geophysics (IUGG) and WMO. The program seeks to provide a better understanding of the large scale atmosphere, to define better the data requirements of atmospheric prediction models, and to exploit the fullest observational capabilities of modern technology.

Climatology Climate may be defined as the statistical summary of past weather at a fixed place. It is most often thought of as a running average using a fixed time interval, the length of which may be as little as a month in some conceptions or indefinitely large in others. Usually a period of about thirty years is preferred in order to filter out annual and shorter-term fluctuations while still allowing longer period variations to be studied. Strictly speaking, one cannot observe the climate of the present. However, the study of past climates and their changes provides a basis for projections into the present and the future, and such studies are receiving intensive effort and support at the present time.

Research in climatology may be subdivided into three main categories: (1) the reconstruction of past climates from weather observations and proxy data; (2) the understanding of the physical and mathematical bases of climate and its changes; and (3) assessment of the impact of climate on society. Contributions to the study of climate have been and are being made by scientists in a wide range of disciplines.

Instrumental records of weather go back for about two centuries at most. Therefore, the reconstruction of climates depends mainly on the analysis and interpretation of proxy data, such as tree rings, glacier ice cores, and ancient descriptions of the distribution of plants.

Geological evidence long ago indicated the occurrence of large variations of climate during the history of the earth. At least four major glaciations occurred in Europe and North America during the Pleistocene. Within the last ten years, a reasonably reliable chronology began to become available from isotope analyses of cores extracted from the deep ocean bottom. From these and other data is emerging an interesting pattern. The mean temperature of the earth declined significantly during Tertiary times and there may have been as many as 17 major glaciations during the last 1.7 million years. A period of about 100,000 years between major episodes is quite conspicuous, and the spectrum over the last 400,000 years also shows significant peaks near 41,000 and 22,000 years. The last glaciation reached its climax about 16,000 years ago and was followed by a warm extreme about 6,000 years ago. Since then the mean temperature has gradually declined, with numerous short-term fluctuations on many time scales.

Rapid strides have been made during the last

decade in understanding the physical and mathematical basis of climate. This understanding is greatly complicated by the large number of interactions that occur between different components of the atmosphere-ocean-earth system and by the response of such a complex system to perturbations imposed on it. As an example of such interactions or feedbacks, picture the effect of a temperature perturbation resulting possibly from some minor disturbance of the sun's atmosphere. An increase of surface temperature would increase evaporation from the ocean's surface, affecting in turn the mean cloudiness, snow cover, and probably the circulation pattern; these changes would in turn affect the distribution of absorption of solar radiation and thereby result in further changes of the temperature distribution. The complexity of the climate system has prompted the development of computer models capable of incorporating them. Such models are currently severely limited by the insufficient understanding of the physics and chemistry of some of these interactions, and also the difficulty of evaluating the covariances that appear when the atmospheric equations are averaged over long periods of time. One of the interesting suggestions emerging from mathematical and computational studies of climate models is the possibility that the system may spontaneously oscillate between two or more quite different states without any external forcing.

Many possibilities exist for both external and internal forcing of climate changes. Observations of solar energy output over the last 50 years have shown it to be remarkably steady; nevertheless, it is scientifically plausible that significant variations of this output rate may have occurred over longer time scales. It is also likely that the prevalence of suspended dust from volcanoes has varied widely. The impact of such dust on the terrestrial heat budget is not clear and may not have been great. It is becoming rather well accepted that on a time scale of tens of millions of years, continental displacements are effective causes of change; the major glaciations that occurred in India and Australia during Permian times, and probably also the Tertiary cooling, can be attributed to this cause. The presence of 22-, 41-, and 100-thousand-year periods in the Pleistocene chronology gives strong support to the astronomical theory proposed originally by Milankovitch. According to this theory, changes in the orbital elements of the earth caused by the Moon and planets give rise to variations of solar insolation with precisely these periods. The amplitude of these variations is rather small, and the mechanism by which they would produce the observed climatic response is not understood. However, on the basis of past relationships, it is possible to project that in the absence of anthropogenic influences, the mean temperature of the earth will decline over the next 60,000 years.

It is estimated that about half of the carbon dioxide given off from the burning of fossil fuels resides in the atmosphere and the rest is dissolved in the upper layers of the ocean. This atmospheric residue interacts with the infrared radiation in the atmosphere in such a way as to increase the surface temperature. At present the atmospheric content is increasing about 0.4% a year, and it is estimated that this increase, acting alone, would raise the mean temperature about 0.3 C over the next 50 years. The observational verification of carbon dioxide warming of the atmosphere has not been possible because of numerous other factors that cannot be controlled.

Cloud Physics and Weather Modification   The water vapor saturation required for the formation of clouds and precipitation normally comes from upward motion and accompanying adiabatic cooling of the air. Condensation actually begins at humidities slightly less than 100% on small, suspended hygroscopic nuclei derived from such sources as sea spray and combustion products. Because of the strong curvature of the surface of the growing droplets, a slight supersaturation is required for continued growth. Freezing nuclei are rare, and cloud droplets normally remain in the liquid phase at subfreezing temperatures down to -40 C.

Cloud droplets range in diameter from 1 to 50 μ depending mainly on the size of the original nucleus and the extent of coalescence with other droplets. An average raindrop contains about a million times the water mass of a single cloud droplet and cannot be formed in a reasonable time by the ordinary droplet growth mechanism. Two processes are believed to act naturally if the required circumstances are met. The first of these is coalescence of colliding droplets that fall at different speeds; such a process requires a broad spectrum of droplet sizes. The second mechanism is the Bergeron-Findeisen process, which requires the presence of a few ice crystals in a predominantly liquid cloud at subfreezing temperatures. The difference between the saturation vapor pressures over liquid and ice at low temperatures causes a rapid transfer of water from the liquid to the ice. Because the requirements for these processes are rather stringent, most clouds do not precipitate.

In principle it should be possible to trigger the first process by artificially injecting into the cloud liquid water masses, which fragment immediately into drops of a wide range of sizes. Each of these should then grow by coalescence and fracture into a large number of new seeds, effecting a chain reaction that can ultimately produce beneficial amounts of precipitation. The second process has been stimulated in natural clouds by seeding with freezing nuclei such as silver iodide or with "dry ice," which produces ice crystals by the sudden chilling of droplets.

Both of these methods have been employed

widely in rain-making operations and experiments. These techniques appear to have been effective in a few experiments, but the amounts of augmentation have usually been too small to demonstrate conclusively under statistically controlled conditions. In most (but probably not all) cases where there is sufficient liquid water in the cloud to produce a significant amount of precipitation and sufficient vertical motion to sustain it once it begins, the precipitation forms naturally. In a few conditions the artificially created latent heat of sublimation may provide enough additional buoyancy to trigger a large amount of additional cloud growth and precipitation.

Injection of silver iodide into cumulonimbus clouds has also been attempted for the purpose of preventing hail. By increasing the number of ice seeds that compete for the available cloud water it is hoped that one can substitute for a small number of large hail stones a vastly increased number of small stones that melt before reaching the ground.

Boundary Layer Meteorology The atmospheric boundary layer comprises the lowest kilometer (more or less), where the air properties are directly influenced by interactions with the surface of the earth. The layer is intensively studied for several reasons. Nearly all of the heat and water vapor that generate kinetic energy and precipitation pass through this layer and are controlled by processes near the earth's surface. The greater part of the kinetic energy that is generated by storms is dissipated within this layer. Finally, pollutants enter the atmosphere near the surface, and they are transported by the winds in this layer, undergo chemical transformations with other species, and are diffused by turbulence in varying degrees.

Boundary layer processes must be included in numerical prediction models. Without the inclusion of friction, storms tend to overintensify. It has also been found that longer period prediction of precipitation requires the inclusion of evaporation and the transfer of water vapor through the boundary layer into the free atmosphere.

Legislation has made it necessary to find and use working theories for relating ambient air quality to the industrial emissions that affect it. The pressure imposed by regulation authorities and the requirements of legislation have been a strong stimulus for research to develop better theories for transport, modification, and diffusion in natural environments. The problems are compounded by the fact that many industrial operations are situated near coastlines or in complex terrain that defy the application of simple theories. Moreover, the chemical transformations involve a large number of possible reactions under very low concentrations, often involving heterogeneous phases. Of particular importance is the conversion of SO_2 to sulfate and the precipitation of the resulting acidic ions (along with products derived from oxides of nitrogen) in what is commonly called *acid rain*. Natural rain is slightly acidic due to the presence of carbon dioxide. Rain downstream of pollutant sources sometimes reaches pH values of 4.0 or less.

Dilution of pollutant concentrations by turbulent diffusion has been intensively studied for more than 50 years, both in the laboratory and in field experiments. Turbulence in the atmospheric environment is more complicated than in the laboratory, and the challenge of these complications has attracted the attention of a number of prominent hydrodynamicists. During the last 25 years the emphasis has been directed mainly at observing the distribution of mean wind and turbulence on towers within the lowest 300 meters above the surface. The main efforts are now shifting toward computer modeling of turbulence throughout the entire boundary layer.

Research and Publications Research is meteorology is carried out in many countries and results are published in a variety of specialized journals. A partial list of these includes the *Journal of the Atmospheric Sciences, Journal of Climate and Applied Meteorology, Journal of Atmospheric and Terrestrial Physics, Quarterly Journal of the Royal Meteorological Society, Tellus, Izvestiya—Atmospheric and Oceanic Physics, Journal of Geophysical Research,* and *Atmospheric Environment.* A complete list of publications in this field can be found in *Meteorological and Geoastrophysical Abstracts* published by the American Meteorological Society.

ALFRED K. BLACKADAR

References

Anthes, R. A., Cahir, J. J., Fraser, A. B., and Panofsky, H. A., "The Atmosphere," 3rd Ed., Columbus, Ohio, Charles E. Merrill Publishing Company, 1981 (531 pp.).

Berger, A. (Ed.), "Climatic Variations and Variability: Facts and Theories," D. Reidel Publishing Co., 1981 (795 pp.).

Butcher, S. S., and Charlson, R. J., "An Introduction to Air Chemistry," New York, Academic Press, 1972 (241 pp.).

Byers, H. R., "Elements of Cloud Physics," Chicago, Univ. Chicago Press, 1965 (191 pp.).

Munn, R. E., "Descriptive Micrometeorology," New York, Academic Press, 1966 (245 pp.).

Pruppacher, H. R., and Klett, J. D., "Microphysics of Clouds and Precipitation," D. Reidel Publishing Co., 1978 (714 pp.).

Wallace, J. M., and Hobbs, P. V., "Atmospheric Science: An Introductory Survey," New York, Academic Press, 1977 (467 pp.).

Cross-references: AIR GLOW, COMPUTERS, PLANETARY ATMOSPHERES, TELEMETRY, TEMPERATURE AND THERMOMETRY.

MICHELSON-MORLEY EXPERIMENT*

Introduction The revival and development of the wave theory of light at the beginning of the nineteenth century, principally through the contributions of Young and Fresnel, posed a problem which proved to be of major interest for physics throughout the entire century. The question concerned the nature of the medium in which light is propagated. This medium was called the "aether" and an enormous amount of experimental and theoretical work was expended in efforts to determine its properties. On the experimental side, a long series of electrical and optical investigations were carried out attempting to measure the motion of the earth through the ether medium. For many years, the experimental precision permitted measurements only to the first power of the ratio of the speed of the earth in its orbit to the speed of light ($v/c \simeq 10^{-4}$), and these "first-order experiments" uniformly gave null results. It became the accepted view that the earth's motion through the ether could not be detected by laboratory experiments of this sensitivity. With the development of Maxwell's electromagnetic theory of light, and especially with its extensions by Lorentz in his electron theory, theoretical explanations for the null results obtained in the early ether drift experiments were provided. These results were in harmony with the Galilean-Newtonian principle of relativity in mechanics, which explains why the essential features of all uniform motions are independent of the frame of reference in which they are observed. In Maxwell's electromagnetic theory, however, the situation was altered when quantities of the second order in (v/c) were considered. According to the Maxwell theory, effects depending on (v/c)² should have been detectable in optical and electrical experiments. The presence of these "second-order effects" would indicate a preferred reference frame for the phenomena in which the ether would be at rest. At first, this feature of Maxwell's theory implying observable ether drift effects of the second order in (v/c) raised a purely hypothetical question, since the accuracy needed for such experiments was a part in a hundred million, and no experimental techniques then known could attain this sensitivity.

Michelson pondered this problem and it led him to invent the Michelson interferometer, which was capable of measurements of the required sensitivity, and to plan the ether drift experiment which he carried to completion in collaboration with Edward W. Morley at Cleveland in 1887. This famous optical interference experiment was devised to measure the motion of the earth through the ether medium by means of an extremely sensitive comparison of the velocity of light traveling in two mutually perpendicular directions. The experiment, when completed in 1887, gave a most convincing null result and proved to be the culmination of the long nineteenth century search for the ether. At that time, the definitive null result of the Michelson-Morley experiment was a most disconcerting finding for theoretical physics, and indeed for many years repetitions of this experiment and related ones were performed with the hope of finding positive experimental evidence for the earth's motion through the ether. These later experiments, however, have all been shown to be consistent with the original null result obtained by Michelson and Morley. In the years following 1887, their experiment led to extensive and revolutionary developments in theoretical physics, and proved to be a major incentive for the work of FitzGerald, Lorentz, Larmor, Poincaré, and others, leading finally in 1905 to the special theory of relativity of Albert Einstein.

The optical paths in the Michelson-Morley interferometer are shown in plan in Fig. 1. Light from a is divided into two coherent beams at the half-reflecting, half-transmitting rear surface of the optical flat b. These two beams travel at 90° to each other and are multiply reflected by two systems of mirrors d – e and d₁ – e₁. On returning to b part of the light from e – d is reflected into the telescope at f, and light from e₁ – d₁ is also transmitted to f. These two coherent beams of light produce interference fringes. These are formed in white

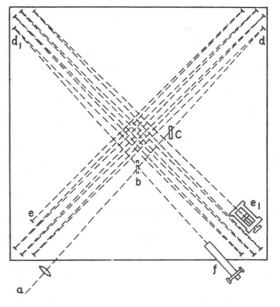

FIG. 1. Optical paths in the Michelson-Morley interferometer.

*The author passed away on March 5, 1982, after preparing this article.

only when the optical paths in both arms are exactly equal, a condition produced by moving the mirror at e_1 by a micrometer. c is an optical compensating plate. The effective optical path length of each arm of the apparatus was increased to 1100 cm by the repeated reflections from the mirror system.

Figure 2 is a perspective drawing of the Michelson-Morley interferometer showing the optical system mounted on a 5 foot square sandstone slab. The slab is supported on the annular wooden float, which in turn fitted into the annular cast-iron trough containing mercury which floated the apparatus. On the outside of this tank can be seen some of the numbers 1 to 16 used to locate the position of the interferometer in azimuth. The trough was mounted on a brick pier which in turn was supported by a special concrete base. The height of the apparatus was such that the telescope was at eye level to permit convenient observation of the fringes when the instrument was rotating in the mercury. While observations were being made, the optical parts were covered with a wooden box to reduce air currents and temperature fluctuations.

This arrangement permitted the interferometer to be continuously rotated in the horizontal plane so that observations of the interference fringes could be made at all azimuths with respect to the earth's orbital velocity through space. When set in motion, the interferometer would rotate slowly (about once in 6 minutes) for hours at a time. No starting and stopping was necessary, and the motion was so slow that accurate readings of fringe positions could easily be made while the apparatus rotated.

The experiment to observe "the relative motion of the earth and the luminiferous ether" for which this instrument was devised, was planned by Michelson and Morley as follows. When the interferometer is oriented as in Fig. 3 with the arm L_1 parallel to the direction of the earth's velocity v in space, the time required for light to travel from M to M_1 and return to M in its new position is,

$$t_{\parallel}(1) = \frac{L_1}{c-v} + \frac{L_1}{c+v} = \frac{2L_1}{c} \frac{1}{1-\beta^2} \quad \left(\beta = \frac{v}{c}\right)$$

The time for light to make the journey to and from the mirror M_2 in the other interferometer arm L_2 is,

$$t_{\perp}(1) = [\, 2L_2(1 + \tan^2\alpha)^{1/2}/c\,]$$

and since $\tan^2\alpha = v^2/(c^2 - v^2)$

$$t_{\perp}(1) = \frac{2L_2}{c} \frac{1}{(1-\beta^2)^{1/2}}.$$

When the interferometer is rotated through 90° in the horizontal plane so that the arm L_2 is parallel to v, the corresponding times are,

$$t_{\parallel}(2) = \frac{2L_2}{c} \frac{1}{1-\beta^2}$$

$$t_{\perp}(2) = \frac{2L_1}{c} \frac{1}{(1-\beta^2)^{1/2}}$$

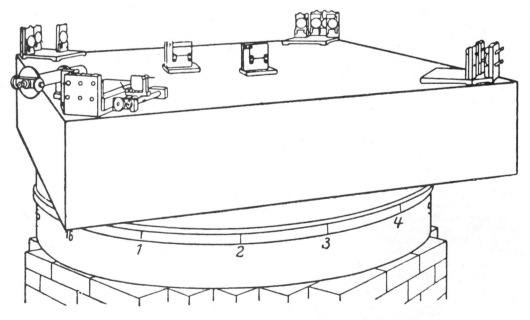

FIG. 2. Michelson-Morley interferometer used at Cleveland in 1887.

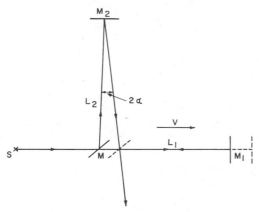

FIG. 3. The Michelson-Morley experiment.

Thus, the total phase shift (in time) between the two light beams expected on the ether theory for a rotation of the interferometer through 90° is,

$$\Delta t = \frac{2L_1}{c}\left[\frac{1}{1-\beta^2} - \frac{1}{(1-\beta^2)^{1/2}}\right]$$
$$+ \frac{2L_2}{c}\left[\frac{1}{1-\beta^2} - \frac{1}{(1-\beta^2)^{1/2}}\right]$$
$$= \frac{2(L_1+L_2)}{c}\left[\frac{1}{1-\beta^2} - \frac{1}{(1-\beta^2)^{1/2}}\right]$$

For equal interferometer arms, as used in this experiment,

$$L_1 = L_2 = L, \text{ and, since } \beta \ll 1,$$

$$\Delta t \simeq \frac{2L}{c}\beta^2$$

The observations give the positions of the fringes, rather than times, so the quantity of importance for the experiment is the change in optical path in the two arms of the interferometer.

$$A = c\,\Delta t = 2L(v/c)^2$$

This is the quantity of second order in (v/c) referred to above.

With the Michelson-Morley interferometer, the magnitude of the expected shift of the white-light interference pattern was 0.4 of a fringe as the instrument was rotated through an angle of 90° in the horizontal plane. Michelson and Morley felt completely confident that fringe shifts of this order of magnitude could be determined with high precision.

In July of 1887, Michelson and Morley were able to make their definitive observations. The experiments which gave their final measurements were conducted at noon and during the

evening of the days of July 8, 9, 11, 12 of 1887. Instead of the expected shift of 0.4 of a fringe they found "that if there is any displacement due to the relative motion of the earth and the luminiferous ether, this cannot be much greater than 0.01 of the distance between the fringes."

The result of the Michelson-Morley experiment has always been accepted as definitive and formed an essential base for the long train of theoretical developments that finally culminated in the special theory of relativity. The first important suggestion advanced to explain the null result of Michelson and Morley was G. F. Fitz-Gerald's hypothesis that the length of the interferometer is contracted in the direction of its motion through the ether by the exact amount necessary to compensate for the increased time needed by the light signal in its to-and-fro path. This contraction hypothesis was made quantitative by H. A. Lorentz in further development of his electron theory in which he introduced the formalism which has since been known as the "Lorentz transformation" for the analysis of relative motions.

H. Poincaré also contributed greatly to both the philosophical and mathematical developments of the theory. As early as 1899, he asserted that the result of Michelson and Morley should be generalized to a doctrine that absolute motion is in principle not detectable by laboratory experiments of any kind. Poincaré further elaborated his ideas in 1900 and in 1904 and gave to his generalization the name "the principle of relativity." He also completed the theory of Lorentz and it was he who named the essential transformation "the Lorentz transformation."

In 1905 Einstein published his famous paper on the "Electrodynamics of Moving Bodies" in which he developed the special theory of relativity from two postulates: (1) the principle of relativity was accepted as the impossibility of detecting uniform motion by laboratory experiments, and (2) the constancy of the speed of light was generalized to a postulate that light is always propagated in empty space with a velocity independent of the motion of the source. Both postulates have a close relationship to the Michelson-Morley experiment, which Einstein knew through his study of the work of Lorentz. Einstein's paper is generally considered as the definitive exposition of the special relativity principle, and the climax of the century-long developments which had begun with Young and Fresnel to explain the electrical and optical properties of moving bodies.

At all times the Michelson-Morley experiment continued to have great interest and was repeated many times throughout more than a half century. In 1904 Morley and Miller[4] showed that the Fitzgerald-Lorentz contraction is the same in several materials. All repetitions after 1887 failed to find the full expected "aetherdrift," although Dayton C. Miller's trials on Mount Wilson (1921–1926) gave a small effect, later shown to be due to temperature gradients.[5] The most certain null result was that obtained by Joos[6] using an interferometer built by Zeiss of Jena. Finally, experiments by Townes[7] with very sensitive laser techniques, gave definitive confirmation of Michelson and Morley's work.

In 1922 at the height of his fame for relativity, Einstein lectured widely in Japan. A reprint and discussion of these lectures has recently become available.[8] In his lecture on December 14, 1922 at Kyoto University, Einstein referred several times to the interferometer experiment, stating that he "had thought about the result even in his student days." In 1950 Einstein told the writer[1] that after 1905 he and Lorentz had discussed the Michelson-Morley experiment many times while he was working on the general theory of relativity. Today, both the experiment and the theories are among the prized achievements of Physics.

R. S. SHANKLAND

References

1. Shankland, R. S., "Conversations with Albert Einstein," *Amer. J. Phys.*, **31**, 47 (1963); **41**, 895 (1973); **43**, 464 (1975).
2. Shankland, R. S., "Michelson-Morley Experiment," *Amer. J. Phys.*, **32**, 16 (1964).
3. Shankland, R. S., "Michelson-Morley Experiment," *Sci. Amer.*, Nov. 1964, p. 107–114.
4. Morley, E. W., and Miller, D. C., *Phil. Mag.* **9**, 680 (1905).
5. Shankland, R. S., et al., *Rev. Mod. Physics* **27**, 167 (1955).
6. Joos, G., *Ann. Physik* **7**, 385 (1930); *Naturwiss*, **38**, 784 (1931).
7. Townes, C. H., *Phys. Rev. Letters* **1**, 342 (1958).
8. Ogawa, T., *Japanese Studies in the History of Science*, Nov. 18, 1979.

Cross-references: INTERFERENCE AND INTERFEROMETRY; LIGHT; OPTICS, GEOMETRICAL; OPTICS, PHYSICAL; RELATIVITY.

MICROPARACRYSTALS

Most noncrystalline matter consists of microparacrystals (mPCs). They can be identified by X-ray, neutron, and electron diffraction patterns. The reflections of microparacrystals show a characteristic broadening (see, for example, Figs. 8, 9, and 11 of the article DIFFRACTION BY MATTER AND DIFFRACTION GRATINGS. Plotting for instance the integral width $\delta\beta$ of the reflections ($h00$) against h we can calculate the number N of netplanes and the paracyrstalline distortion g by means of the following theoretical relationships:[1]

$$\delta\beta = \frac{1}{\bar{d}}\left[\frac{1}{N} + (\pi g h)^2\right]$$
$$g = \Delta_1/\bar{d} \tag{1}$$
$$\Delta_1 = (\overline{d^2} - \bar{d}^2)^{1/2},$$

where d is the distance between atoms of neighboring netplanes orthogonal to their surfaces and Δ_1 the variance of these distances d. Figure 1 shows a two-dimensional paracrystalline lattice, where 3% larger coins are mixed statistically with 97% smaller ones. At the center of the lattice there exist some crystalline-like domains. Near the boundaries the netplanes become more and more distorted until finally they are destroyed at some places. Similar phenomena exist in atomic dimensions. From statistical laws it is known that the distance variance Δ_N increases with $\sqrt{N}$ orthogonal to the surface between atoms of the first and $(N+1)$th netplanes. When Δ_N reaches $100\alpha^*\%$ of $\bar{d}$, then the valence bonds between atoms or molecules within the Nth netplane are strongly overstrained so that the netplane suffers a break. This leads to the so-called "α^*-relation:"

$$\Delta_N = \sqrt{N}\,\Delta_1 = \alpha^*\bar{d}, \quad \text{hence} \quad \sqrt{N} = \alpha^*/g$$
$$\alpha^* = 0.15 \pm 0.03. \tag{2}$$

In Fig. 2 is plotted $\sqrt{N}$ against $1/g$ for numerous different colloids. α^* always has values between 0.12 and 0.18. Another example is given by the steel-ball model (cf. the article DIFFRACTION BY MATTER AND DIFFRACTION GRATING, dotted line, Fig. 11). The α^*-relation was firstly published in 1967.[2] At that time its fundamental importance for the whole world of noncrystalline matter was not recognized. F. J. Baltá-Calleja[3] stimulated the publication of diagrams like Fig. 2. Nevertheless, the importance of the $\alpha^* =$ relation is rarely understood nowadays. On the right-hand side of Fig. 2, for instance, an example is given of the ammonia contact catalysts. Their microparacrystals are, on account of relation (2), very small compared with crystals and therefore build up a large thermostable "inner" surface of some 100 m^2 per 1 cm^3. This large surface is important in synthesizing ammonia (NH_4OH) from hydrogen (H_2), nitrogen (N_2), and water (H_2O) in a rational technical way:

$$3H_2 + N_2 + 2H_2O = 2NH_4OH.$$

In Figure 12 of the article on PARACRYSTALS it is shown that the microparacrystals of this catalyst consist of α-Fe lattices into which $FeAl_2O_4$ molecules are statistically imbedded and destroy the crystalline order of the Fe atoms. Here $g \sim 1\%$ and $\sqrt{N} \sim 15$. At temperatures higher than 400°C some $FeAl_2O_4$ molecules slowly begin to emigrate, the g-value therefore becomes smaller, and $\sqrt{N}$ reaches

FIG. 1. Two-dimensional model of a paracrystalline lattice with quadratic lattice cells. 97% smaller coins are mixed statistically with 3% larger ones.

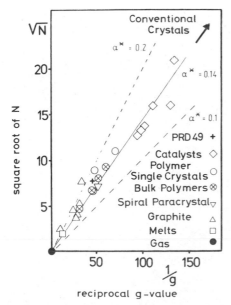

FIG. 2. The α*-law of colloidal systems. The relation of Eq. (2) gives the theoretical background for noncrystalline matter independent of its chemical composition.

values up to $\sqrt{N} = 20$. With higher annealing temperature the microparacrystals more nearly approach the crystalline state in the direction of the arrow until the catalyst loses its favorable properties. A new example of the existence of microparacrystals was found recently in the $NiAl_2O_3$ catalyst.[4] Adjacent to the catalyst in Fig. 2 are plotted mPCs ("single crystals" and bulk material) in polymers. Their g-values lie between 2 and 5%. The example for "crystalline" polymers is given in Fig. 9 of the article on PARACRYSTALS, where some microparacrystals can be recognized. They are linked together by long molecules and work therefore as knots of a three-dimensional network. During strain the chains glide along each other, each moment building up new microparacrystals which, step by step, become smaller until finally every microparacrystal is disintegrated into 30 smaller ones [Fig. 3(b)]. The most disordered microparacrystals are to be found in melts (left corner of Fig. 2). There they have values $g \sim 8\%$ and $\sqrt{N} \sim 2$. Figure 4 shows as an example a microparacrystal of molten Fe or Pb which builds up a cubic face-centered lattice similar to that below the melting point. In contrast to the solid state, the atoms have a high mobility and can move over to other nuclei building up icosahedra in some

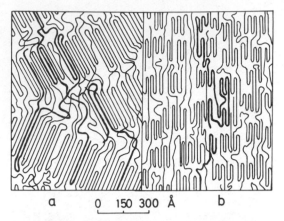

FIG. 3. Microparacrystals in polymers are the knots of a three-dimensional network: (a) before and (b) after stretching eight-fold.

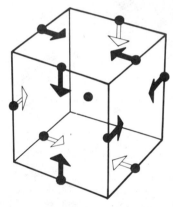

FIG. 4. Microparacrystals in molten iron, lead, and silver. The arrows indicate probable movements in tangential direction forming an icosahedron.

cases.[5] Equation (2) is not only an empirically detected relation but *the fundamental law* of colloid and surface science. It manifests an equilibrium state of microparacrystals which is unknown in conventional solid state physics of condensed matter. For details see the article MICROPARACRYSTALS, EQUILIBRIUM STATE OF.

ROLF HOSEMANN

References

1. Hosemann, R., *Ergeb. Ex. Nat. Wiss.* **24**, 142 (1951).
2. Hosemann, R., Lemm, K., Schönfeld, A., and Wilke, W., *Kolloid-Z. u. Z. Polymere* **216–217**, 103 (1967).
3. Baltá-Calleja, F. J., and Hosemann, R., *J. Appl. Cryst.* **13**, 521 (1980).
4. Wright, C. J., Windsor, C. G., and Puxley, D. C., Nat. Phys. Div. A.E.R.E. Harwell, Oxfordshire, U.K., MPD/NBS/189, Feb. 1982.
5. Steffen, B., and Hosemann, R., *Phys. Rev.* **B13**, 3232 (1976).

Cross-references: DIFFRACTION BY MATTER AND DIFFRACTION GRATINGS; MICROPARACRYSTALS, EQUILIBRIUM STATES OF; PARACRYSTALS.

MICROPARACRYSTALS, EQUILIBRIUM STATE OF

The physical meaning of the α^*-law is that the mean number $\bar{N}$ of netplanes in an ensemble of microparacrystals (the so called mPCs) depends only on the relative distance fluctuation g of atoms belonging to neighboring netplanes (see also the article MICROPARACRYSTALS. Here,

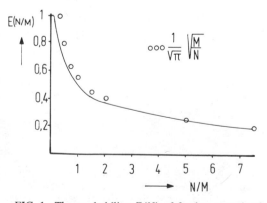

$$\circ\circ\circ \ \frac{1}{\sqrt{\pi}}\sqrt{\frac{M}{N}}$$

FIG. 1. The probability $E(N)$ of further growth of a microparacrystal consisting of N netplanes.

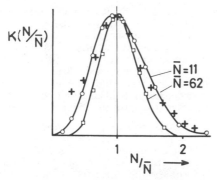

FIG. 2. Frequency distribution $K(N)$ of microparacrystals with $N + 1$ netplanes. Open circles and squares plot $K(N)$ by computer calculation; solid lines drawn by Maxwell approximation. Crosses indicate distribution as directly observed for $\bar{N} = 11$ microparacrystals in DuPont PRD 49 fibers.

FIG. 3. High-precision transmission-electron-microscopic picture of DuPont PRD 49 fibers.

from the probability $E(N)$, what amount of microparacrystals with $N - 1$ netplanes aggregate the next netplane (see Fig. 1), can be calculated by statistical mechanics. As in biology, where the expectation of surviving the next day decreases continuously day by day, here $E(N)$ begins with a plateau $E = 1$ which suddenly changes to a smooth decay. The frequency $K(N)$ of microparacrystals meeting $N + 1$ net-

planes is given by

$$K(N) = E(1)\,E(2)\,E(3)\ldots \qquad (1)$$
$$\cdot E(N - 1)\,(1 - E(N)).$$

$K(N)$ is plotted in Fig. 2 for $\bar{N} = 11$ (open circles) and $\bar{N} = 62$ (open squares) and can be approximated by Maxwellian functions.[2] The crosses in Fig. 2 show the $K(N)$ distribution directly observed[3] from high-precision transmission-electron-microscopic diagrams, with $\bar{N} = 11$ (Fig. 3). This agreement proves directly that the equilibrium state of microparacrystals is of utmost importance for the understanding of noncrystalline condensed matter.[4]

ROLF HOSEMANN

References

1. Hosemann, R., Schmidt, W., Lange, A., and Hentschel, M., *Colloid & Polymer Sci.* **259**, 1161 (1981).
2. Hosemann, R., *Colloid & Polymer Sci.* (in press).
3. Dobb, M. G., Hindeleh, A. M., Johnson, D. J., and Saville, B. P., *Nature* **253**, 189 (1975).
4. Hosemann, R., *Physica Scripta* (in press).

Cross-references: COLLOIDS, THERMODYNAMICS OF; MICROPARACRYSTALS; PARACRYSTALS.

MICROWAVE SPECTROSCOPY

Microwaves are electromagnetic waves which range in length from about 30 cm to a fraction of a millimeter or in frequency from 10^9 to 0.5×10^{12} cps. This corresponds to the rotational frequency range of a large class of molecules. Thus, microwave radiation passing through a gas can be absorbed when the rotating electric dipole moment of the molecule interacts with the electric vector of the radiation. Likewise, absorption can take place if the rotating magnetic moment of the molecule interacts with the magnetic vector of the radiation.

Most microwave spectroscopy is based on a study of transitions induced by interaction of the molecular electric dipole with the incident radiation.

A microwave spectrometer consists basically of a monochromatic microwave source (klystron), an absorption cell, and a detector. The absorption cell must transmit the microwave of interest and in the centimeter region may have cross-sectional dimensions of 1×4 cm and may be a few meters in length. Normally a metal strip is inserted along the length of the cell and is insulated from the cell. In this way, an auxiliary, spatially uniform electric field may be established in the absorption cell without affecting the microwaves. The Stark effect thereby produced splits the molecular energy levels into a series of levels and enables one to identify the transition.

The Hamiltonian for a rotating rigid asymmetric molecule including possible fine and hyperfine structure terms is given in Eq. (1). It is assumed, as is most commonly so, that the molecule is in a $^1\Sigma$ state, i.e., that there is no net electronic angular momentum and no net electron spin (singlet state). The Hamiltonian written is quite general and in many cases not all of the terms shown in Eq. (1) need be included to account for spectra observed under normal resolution.

A brief description will be given of each term in order that one may most simply understand the kinds of interactions which may occur and which may be pertinent to an understanding of the spectra of rotating molecules.

$$H = H_R + H_{dist.} + H_S + H_{ze} + H_Q$$
$$+ H_{zi} + H_D \quad (1)$$

1. H_R is the framework rotational kinetic energy and may be written

$$H_R = \frac{\hbar^2}{8\pi^2}\left[\frac{J_a{}^2}{I_a} + \frac{J_b{}^2}{I_b} + \frac{J_c{}^2}{I_c}\right]$$

where J_a, J_b, and J_c are the components of the total angular momentum in units of $\hbar$ referred to body-fixed principal axes. I_a, I_b, and I_c are the moments of inertia about the respective principal axes. The Hamiltonian may be written

$$H_R = AJ_a{}^2 + BJ_b{}^2 + CJ_c{}^2$$

or displaying the total angular momentum J

$$H_R = AJ^2 + (B - A)J_b{}^2 + (C - A)J_c{}^2$$

A, B and C are rotational constants with $A > B > C$. In this form, units can be chosen so as to give energy levels directly in megacycles per second. H_R describes a rigid symmetric top if I_a is equal to I_b. For a diatomic molecule, $I_a = I_b$ and $I_c \cong 0$. ($1/I_c$ becomes very large, and the rotational levels about the c axis are too far apart to become excited by microwaves). For the spherical rotor, $I_a = I_b = I_c$, but this implies no dipole moment and therefore no observable rotational spectrum.

Energy levels for the symmetric top may be determined by analytical methods,[1] by factorization methods[2], or by using the commutation properties of the angular momenta.[3] In Eulerian coordinates, the wave equation separates, and the wave function has the form

$$\psi_J = N(JKM)e^{iK\phi}e^{iM\psi}\Theta_{JKM}(\theta)$$

where Θ_{JKM} is the solution to the differential equation in the polar angle θ which results after separation of the simple terms in the azimuthal angle ϕ and the angle ψ which defines the direction of the line of nodes. The equation for Θ_{JKM} with appropriate change of variable becomes the equation for the Jacobi polynomials.

The solution ψ is characterized by three quantum numbers: J, the total angular momentum; K the component of angular momentum along the symmetry axis of the molecule; and M, the magnetic quantum number or projection of J along an arbitrary space axis. The energy does not depend on M in the absence of external electric or magnetic fields. The energy levels for the symmetric top have the form

$$E_{J,K} = BJ(J + 1) + (C - B)K^2$$

The rotational constants $A = B$ and C may typically range from 2000 to 300 000 MHz for presently observable spectra. A hertz is one cycle per second. A megahertz (MHz) is one million cycles per second.

Selection Rules In all but the accidentally symmetric top, the permanent dipole will be along the symmetry axis. In this case, the selection rules for absorption of radiation through rotation are:[4]

$$J \rightarrow J \pm 1, K \rightarrow K$$

For a component of the dipole moment perpendicular to the symmetry axis, the selection rules are

$$\Delta J = \pm 1, 0 \text{ and } \Delta K = \pm 1$$

In both cases, $\Delta M = \pm 1, 0$.

The wave functions of the asymmetric top are expressed as linear combinations of symmetric top functions. The energy remains diagonal in J but not in K. One must, therefore, arbitrarily label the energy levels and determine the selection rules. This will not be done here. In order to do this, however, one needs to know only the nonvanishing matrix elements of the three components of the dipole moment for the symmetric top given above.

The selection rule for diatomic molecules is simply $J \rightarrow J \pm 1, M \rightarrow M, M \pm 1$.

2. $H_{dist.}$ describes centrifugal stretching corrections to the energy levels which for an asymmetric molecule can be quite complicated. Corrections for a symmetric top molecule are easily derived, and the framework energy in this case including centrifugal stretching is given by

$$E_{J,K} = BJ(J + 1) + (C - B)K^2 - D_JJ^2(J + 1)^2$$
$$- D_{JK}J(J + 1)K^2 - D_KK^4$$

For a non-rigid diatomic molecule or linear polyatomic molecule, $K = 0$, and the energy is given by:

$$E_J = B_vJ(J + 1) - D_vJ^2(J + 1)^2$$

and since $J \rightarrow J + 1$, for absorption the line frequencies are:

$$\nu_r = 2B_v(J + 1) - 4D_v(J + 1)^3$$

where B_v is the "effective" spectral constant, $h(8\pi^2 I_v)$, for the particular vibrational state for which the rotational spectrum is observed, and where D_v is the centrifugal stretching constant for that state.

In terms of the constants B_e and D_e for the hypothetical vibrationless state, B_v and D_v for diatomic molecules are:

$$B_v = B_e - \alpha(v + \tfrac{1}{2})$$

$$D_v = D_e + \beta(v + \tfrac{1}{2})$$

where α and β are interaction constants which are very small in comparison with B and D, respectively, and where v is the vibrational quantum number. For linear polyatomic molecules, there is more than one vibrational mode, and the above equations must be written in the more general form

$$B_v = B_e + \sum_i \alpha_i \left(v_i + \frac{d_i}{2}\right)$$

$$D_v = D_e + \sum_i \beta_i \left(v_i + \frac{d_i}{2}\right)$$

where the summation is taken over all the fundamental modes of vibrations. The subscript i refers to the ith mode, and d_i represents the degeneracy of that mode.

In analysis of spectra, the variations of the stretching constants D_J and D_{JK} with vibrational state are customarily neglected. It is seldom possible to obtain sufficient data for the evaluation of these effects upon B and for determination of B_e even for the simpler symmetric tops.

Centrifugal stretching constants may typically range from 8.5 to 0.002 MHz or less.[6]

3. The third term H_S is the contribution to the Hamiltonian arising from the Stark effect and may be written as

$$H_S = \mu_e \cdot \mathcal{E}$$

where μ_e is the vector dipole moment and $\mathcal{E}$ is the external electric field.

If the dipole moment lies along the "c" body-fixed principal axis, H_S has the form

$$H_S = \mu_e(A_x{}^c \mathcal{E}_x + A_y{}^c \mathcal{E}_y + A_z{}^c \mathcal{E}_z)$$

where $A_x{}^c$, $A_y{}^c$, $A_z{}^c$ are the direction cosines of the c principal axis with space-fixed axes xyz. $\mathcal{E}_x$, $\mathcal{E}_y$, and $\mathcal{E}_z$ are the components of the electric field along the space-fixed axes. Additional terms will be added to this expression if the dipole moment has components along the remaining two principal axes. In order to obtain the contribution to the energy from this part of the Hamiltonian, one must evaluate matrix elements of the direction cosines with respect to symmetric top wave functions. Methods described in reference 5 enable one to do this.

In the case of a symmetric top, the dipole moment will have a component only along the c axis so that H_S consists of the single term $A_z{}^c \mathcal{E}_z$ when $\mathcal{E}_x = \mathcal{E}_y = 0$.

In this case, the energy associated with H_S is diagonal in all three quantum numbers JKM and has the form

$$E_S = - \mathcal{E}_z \mu_e \frac{MK}{J(J+1)}$$

where M, the "magnetic quantum number," measures the component of J along $\mathcal{E}_z$ and can take the values $M = J, J-1, \cdots, -J$. The selection rules for M are $M \to M$; $M \to M \pm 1$, depending on the polarization of the microwaves.

For asymmetric molecules, μ_e will in general have components along the A and B axes as well as C. The A and B components give rise to matrix elements off diagonal in J, K, and M. For a dipole moment of 1 debye and an electric field of 1 volt/cm, $\mu_e \mathcal{E}$ is 0.5 MHz.

4. H_{ze} is the contribution to the Hamiltonian due to the interaction of the external magnetic field with the magnetic moment which is created by rotation of the molecule.

We also include the interaction of the external magnetic field with the dipole moment of individual nuclei. For a molecule with two nuclear spins, I_1 and I_2, this may be written

$$H_{ze} = \sum_{j,k} \mu_n (J)_{jk} J_j H_k + \mu_n g_1 (I_1 \cdot H)$$
$$+ \mu_n g_2 (I_2 \cdot H)$$

where $g(J)$ is in general a tensor, J_j are components of J along axes to which H is referred, H_k are the components of the field H usually referred to the space-fixed axes, and g_1 and g_2 are called the nuclear magnetic g factors. The interaction between J and H is the same order as that between the nuclear spin and H. Therefore, we introduce the term μ_n so that g coefficients are of the order of unity. Thus for a field of one gauss, the quantity $\mu_n g H$ is 0.7 kHz.

For molecules with electronic angular momentum, μ_n in the first term is replaced by μ_0, the Bohr magneton which is 1836 times larger than μ_n. Thus, in this case $\mu_n g H$ is ~ 1.4 MHz for a field of one gauss. For a discussion of the problem of determining molecular g values, as well as magnetic susceptibility anisotropies and molecular quadrupole moments, see Ref. 5.

5. H_Q is the energy of interaction of the nuclear electric quadrupole with the gradient of the electric field produced by the electrons in the molecule at nucleus with spin I. For a nucleus on the axis of a symmetric top, the quadrupole operator is ordinarily considered to be of the form:

$$H_Q = \frac{-eQq}{2I(2I-1)(2J-1)(2J+3)} \left\{ 1 - \frac{3K^2}{J(J+1)} \right\}$$
$$\left[3(J+1)^2 + \frac{3}{2}(J \cdot I) - J^2 I^2 \right]$$

This operator yields only those matrix elements of the quadrupole interaction which are diagonal in J. The diagonal contributions are sufficient for most cases.

In the expression above, eQ is defined by

$$eQ = (I, I | \int \rho_n [3z_n^2 - r_n^2] d\tau_n | I, I)$$

where ρ_n is the nuclear charge density at a distance r_n from the center of charge of the nucleus and $d\tau$ is the differential volume element for the nuclear volume. z_n is the position coordinate along the direction of the nuclear spin I. The matrix element considered is that for which $M_I = I$.

The quantity q is defined as

$$q = \left[\frac{\partial^2 V}{\partial c^2} \right]_{(r_n = 0)}$$

where V is the electrostatic potential due to the electronic cloud and other nuclei surrounding the nucleus and c is the axis in the body-fixed system which is parallel to the symmetry axis of the molecule. The quantity eqQ varies from -1000 to 1000 MHz although the intermediate values are more common.[6] A tabulation of matrix elements for quadrupole interaction is given in Ref. 7. For a general discussion, see the book, Ref. 8.

6. H_{zi} represents the interaction between the magnetic field caused by rotation of the charged particles which make up the molecule and the nuclear magnetic moments of the nuclei. For the case of 2 nuclei, this takes the form

$$H_{zi} = \sum_{j, k} C(1)_{jk} J_j I_{1k} + \sum_{j, k} C(2)_{jk} J_j I_{2k}$$

$C(1)$ and $C(2)$ represent the internal magnetic moment tensors for the two nuclei. The C coefficients are of the order of 10^{-2} MHz.[9] J and I are pure numbers. This correction will therefore be unimportant for the large majority of molecules. For values of the coefficients as determined by molecular beam work, see Ref. 7 and 10.

7. H_D is the dipole interaction between the two nuclei which may be written in the form

$$H_D = \frac{g_1 g_2 \mu_n^2}{R^3} \left[I_1 \cdot I_2 - \frac{3(I_1 \cdot R)(I_2 \cdot R)}{R^2} \right]$$

where g_1 and g_2 are the nuclear gyromagnetic ratios of the nuclei, μ_n is the nuclear magneton and R is the distance between the two nuclei.

The operator which is usually used to represent this interaction is

$$H_D = \frac{g_1 g_2 \mu_n^2}{R^3} \cdot$$

$$\frac{[3(I_1 \cdot J)(I_2 \cdot J) + 3(I_2 \cdot J)(I_1 \cdot J) - 2I_1 \cdot I_2 J^2]}{(2J - 1)(2J + 3)}$$

This operator, like that given for the quadrupole interaction above, and usually quoted in the literature, will yield only those matrix elements which are diagonal in the quantum number J. The coefficient $g_1 g_2 \mu_n^2 / R^3$ may be of the order of a kilocycle. This correction is observed only in very rare cases.[9]

Matrix elements for all of the above-mentioned components of the Hamiltonian may be evaluated by the methods in references 11. The matrix elements themselves are too lengthy to be tabulated here.

There is an additional interaction which, for completeness, should be mentioned. The nuclear spins may interact with one another through mutual coupling with the surrounding electron cloud. This gives a correction of the form $C I_1 \cdot I_2$. The coefficient C may be larger than that in the dipole-dipole interaction term. In Tll, C has the value of 6.57 kHz.[7]

The preceding discussion emphasizes the interpretation of microwave absorption spectra. One is thereby led to a knowledge of the structure of the molecule, the value of nuclear spins, and various coupling constants. For a short review which emphasizes the experimental aspects of microwave spectroscopy see Ref. 12.

Microwave spectroscopy is also an effective technique for determination of barrier heights associated with internal rotation.[13,14,15] It may also yield the barrier to ring puckering and also the barrier to inversion, the earliest example of the latter being the inversion of nitrogen through the plane of the three hydrogens in ammonia; see Rudolph review.[23]

For a discussion of and references to microwave pressure broadening, line shape, and intensities see G. Birnbaum.[16] Among other things line width is related to the rate of energy transfer between molecules. The shape of a microwave line is broadened as pressure increases and this can camouflage fine structure.

We have not discussed electron spin resonance (ESR), also called electron paramagnetic resonance.[17] (See RESONANCE and MAGNETIC RESONANCE.) In this case transitions occur between energy levels created by unpaired electron spins in the presence of an external magnetic field. Absorptions of this type are observed in molecules, free atoms, radicals, and solids.

For a review of properties of high-temperature species as studied by microwave absorption spectroscopy see Ref. 18.

The role of transient effects in microwave spectroscopy is a new area of research. Transients may be achieved by switching the applied Stark field in a microwave spectrometer. The main usefulness of the technique appears to be in the determination of the rates of energy transfer between gas molecules. For details on this subject and many references to other reviews and papers including the same when applied to infrared laser spectroscopy, see the review by R. H. Schwendeman.[19]

Microwave spectroscopy has been extended to the submillimeter infrared region; however, this extension has been limited by the ability to generate harmonics and mix signals in point contact devices. With the advent of the laser, coherent sources have become available in the infrared region. This related subject has been reviewed by V. J. Corcoran.[20]

Although less common than the Stark effect technique, the Zeeman effect, that is, exposure of the molecules to very high magnetic fields (up to 30 kg) may also be used to obtain molecular information. The molecules investigated by this technique are in a "nonmagnetic" ground state. The Zeeman effect then is due to the very small molecular rotational magnetic moment which results from the rotation of the unequally distributed positive and negative charge. The main effect is first order in the magnetic quantum number M, but the slight nonlinear compression of the splitting pattern yields magnetic susceptibility anisotropies that permit calculation of the molecular quadrupole moment tensor. For the effect on the Hamiltonian and reference to original papers see the review by H. D. Rudolph.[23]

For further details and discussion of topics omitted, the reader is referred to the book[13] by W. Gordy and R. L. Cook. For other books see Refs. 6 and 21. For a review of microwave spectroscopy see the article by D. R. Lide, Ref. 22; also the reviews by H. D. Rudolph (1970), Morino and Hirota (1969), and Flygare (1967).[23]

The Hamiltonian required to explain the results of molecular beam electric resonance (MBER) experiments involving radio frequency transitions contains the same kind of interaction terms listed above. Therefore the literature on MBER may be referred to for further theoretical discussion. See, for example, English and Zorn, Ref. 22.

For a review of beam maser spectroscopy see Ref. 24.

For a comprehensive computation of microwave spectra including measured frequencies, assigned molecular species, assigned quantum number, and molecular constants determined from such data, the reader is referred to the multivolume work "Microwave Spectral Tables" prepared by personnel of the National Bureau of Standards.[25]

DONALD G. BURKHARD

References

1. Dennison, D. M., *Phys. Rev.*, **28**, 318 (1926); Reiche, F., and Rademacher, H., *Z. Physik*, **39**, 444 (1926), and **41**, 453 (1927).
2. Burkhard, D. G., *J. Mol. Spectry.*, **2**, 187 (1958); Shaffer, W. H., and Louck, J. D., *J. Mol. Spectry.*, **3**, 123 (1959).
3. Klein, O., *Z. Physik*, **58**, 730 (1929).
4. Dennison, D. M., *Rev. Mod. Phys.*, **3**, 280 (1931).
5. Flygare, W. H., and Benson, R. C., *Mol. Phys.*, **20**, 225 (1971).
6. Townes, C. H., and Schawlow, A. L., "Microwave Spectroscopy," New York, McGraw-Hill, 1955.
7. Stephenson, D. A., Dickinson, J. T., and Zorn, J. C., *J. Chem. Phys.*, **53**(4), 1529 (1970).
8. Lucken, E. A. C., "Nuclear Quadrupole Coupling Constants," New York, Academic Press, 1969.
9. Thaddeus, P., Krisher, L. C., and Loubser, J. M. N., *J. Chem. Phys.*, **40**, 257 (1964).
10. English, T. C., and Zorn, J. C., *J. Chem. Phys.*, **47**(10), 3896 (1967).
11. Condon, E. U., and Odabasi, Halis, "Atomic Structure," Cambridge, U.K., Cambridge Univ. Press, 1980; Landau, L. D., and Lifshitz, E. M., "Quantum Mechanics," New York, Pergamon, 1977.
12. Strandberg, M. W. P., "Microwave Spectroscopy," in "McGraw-Hill Encyclopedia of Science and Technology," New York, McGraw-Hill, 1977.
13. Gordy, W., and Cook, R. L., "Microwave Molecular Spectra," New York, Wiley, 1970.
14. Lin, C. C., and Swalen, J. D., *Rev. Mod. Phys.* **31**, 841 (1959).
15. Burkhard, D. G., *J. Opt. Soc. Am.* **50**, 1214 (1960).
16. Birnbaum, G., "Intermolecular Forces," in "Advances in Chemical Physics," Vol. 12, New York, Wiley, 1967.
17. Squires, T. L., "An Introduction to Electron Spin Resonance," New York, Academic Press, 1963; Alger, R. S., "Electron Paramagnetic Resonance: Techniques and Applications," New York, Interscience, 1968; Carrington, A., Levy, D. H., and Miller, T. A., "Electron Resonance of Gaseous Diatomic Molecules," in "Advances in Chemical Physics," Vol. 18, New York, Interscience, 1970.
18. Lovas, F., and Lide, D. R., *Adv. High Temp. Chem.* **3** (1972).
19. Schwendeman, R. H., *Ann. Rev. of Phys. Chem.* **29** (1978).
20. Corcoran, V. J., *App. Spectroscopy Revs.* **7** (1974).
21. Guillory, W. A., "Introduction to Molecular Structure and Spectroscopy," Boston, Allyn and Bacon, 1977. Wollrab, J. E., "Rotational Spectra and Molecular Structure," New York, Academic Press, 1967. Sugden, T. M., and Kenney, C. N., "Microwave Spectroscopy of Gases," New York, Van Nostrand Reinhold, 1965; Ingram, D. J. E., "Spectroscopy at Radio and Microwave Frequencies," 2nd Ed., New York, Plenum, 1967; Hedvig, P., and Zentai, G., "Microwave Study of Chemical Structures and Reactions," CRC Press, 1969; Svidziniskii, K. V., "Soviet Maser Research," New York, Plenum Press, 1964.
22. English, T., and Zorn, J. C., "Molecular Beam Spectroscopy," and Lide, D. R., "Microwave Spectroscopy," in "Methods of Experimental Physics," Vol. 3, 2nd Ed., New York, Academic Press, 1972.
23. *Ann. Rev. Phys. Chem.* **21** (1970); **20** (1969); and **18** (1967), resp.
24. Laine, D. C., *Repts. Prog. Phys.* **33**, 1001 (1970).
25. "Microwave Spectral Tables," Superintendent of Documents, U.S. Govt. Printing Office, Washington, D.C. 20402.

Cross-references: ATOMIC AND MOLECULAR BEAMS, MAGNETIC RESONANCE, MICROWAVE TRANSMISSION, SPECTROSCOPY, ZEEMAN AND STARK EFFECTS.

MICROWAVE TRANSMISSION

That portion of the electromagnetic spectrum adjacent to the far-infrared region is commonly referred to as the microwave region. It is bounded by wavelengths in the vicinity of 10 centimeters (10 cm) and 1 millimeter (1 mm). The longest wavelength of 10 cm corresponds to a frequency of 3×10^9 cycles per second (abbreviated 3000 megahertz or 3 kilomegahertz or 3 gigahertz or 3 GHz). The shortest wavelength of 1 mm corresponds to a frequency of 3×10^{11} cycles per second (abbreviated 300,000 megahertz or 300,000 MHz or 300 kilomegahertz or 300 gigahertz or 300 GHz).

The development of microwave transmission on a major scale was initiated in 1940 with the advent of the magnetron, an electronic generator of high-power microwaves. The magnetron spearheaded wartime radar at approximately 3 GHz and led to the utilization of waveguides for the efficient transmission of microwaves from the generator of the transmitting antenna and from the receiving antenna to the detector.

In essence, a waveguide is a hollow metal tube capable of propagating electromagnetic waves within its interior from its sending end to its receiving end. Unlike waves in space which usually propagate outward in all directions, waves in waveguides are fully confined while they propagate.

An electromagnetic wave is comprised of an electric field and a magnetic field. In free space, these fields are always perpendicular to one another and to the direction of wave propagation at any instant in time. However, when a wave travels through a waveguide, the confinement forces one of the fields, but never both, to have a component that is parallel to the direction of wave propagation.

In a waveguide, there are a number of possible field configurations. Each configuration is known as an operating *mode* and is determined by the operating frequency or wavelength and the lateral dimensions of the waveguide. There are two fundamental classes of modes that may propagate in a waveguide. In one class, the electric field is everywhere perpendicular, or transverse, to the direction of propagation and the magnetic field has a longitudinal component. It is referred to as the *transverse electric mode* or TE mode. In the other class, the magnetic field is transverse and the electric field has a longitudinal component. This configuration leads to the *transverse magnetic mode* or TM mode.

Propagation of either the TE mode or the TM mode is limited by the cross-sectional dimensions of the guide. For example, in a rectangular waveguide, the longest wavelength that can be propagated is equivalent to twice its width. Therefore, to transmit a 3 GHz signal in a rectangular waveguide, the width of the guide must be at least 5 cm. At 300 MHz, this width becomes 50 cm. Thus, waveguides are reasonable in size for the transmission of microwaves. Furthermore, when used with ferrites, thin metallic films and magnets, waveguide components can be designed to function as isolators, circulators, modulators, discriminators, or attenuators.

The propagation of electromagnetic waves in free space between transmitting and receiving antennas can be characterized in terms of ground waves, sky waves, and space waves. At microwave frequencies, ground waves attenuate completely within a few feet of travel, sky waves are influenced by the ionosphere and can penetrate through into outer space, and space waves travel through the atmosphere immediately above the surface of the earth. At microwave frequencies, space waves behave like light waves and travel in a direct line of sight. They follow many of the rules of optics. They can be reflected from smooth conducting surfaces and can be focused by reflectors or lenses.

If a space wave is radiated from a point antenna, the radiated energy spreads out like an ever-expanding sphere, and the amount of energy per square foot of wave front decreases inversely with the square of the distance from the antenna. The power that can be extracted from a wave front by a similar point antenna varies inversely with the square of the frequency. Thus, a point antenna receives power which is inversely proportional to both the square of the distance from the source and the square of the frequency. The ratio of the power received to the total power radiated is known as path attenuation.

When the receiving antenna is a parabola-shaped dish, power extracted from the wave front is greatly increased. The ratio of the power received by such an antenna to the power received by a theoretical point antenna is defined as *antenna gain*. The gain of a parabolic antenna increases with the antenna area and the operating frequency. Thus, for a given microwave transmission with fixed-sized antennas, the path attenuation increases with frequency, the antenna gain increases with frequency, and the overall result is that one tends to offset the other.

In radio broadcasting, the signal power radiates equally in all directions, and a receiving antenna picks up only a tiny fraction of the signal power. To overcome this low efficiency, the broadcast station must transmit a large amount of power. By contrast, a point-to-point microwave system radiates only a small amount of power, but it uses a directional transmitting antenna to concentrate power into a narrow beam directed toward the receiving antenna. Consequently, such systems are characterized by high efficiencies.

Because microwave transmission in free space follows essentially a straight line, reflectors are utilized to redirect a beam over or around an obstruction. The simplest and most common reflector system consists of a parabolic antenna mounted at ground level which focuses a beam on a reflector mounted at the top of a tower. This reflector inclined at 45° redirects the beam horizontally to a distint site where a similar "periscope" reflector system may be used to reflect the beam down to another ground level. If two sites are separated by a mountain, it may be necessary to use a large, flat surface reflector referred to as a "billboard" reflector. In a typical system, a billboard reflector might be located at a turn in a valley, effectively bending the beam to follow the valley. Many arrangements are possible which, in effect, resemble huge mirror systems.

Microwaves are ideally suited for communication systems where a broad frequency bandwidth of the order of several megacycles is required for the rapid transmission of signals which contain a large amount of information, such as, in television signals. Most of the major cities of the United States are serviced by microwave television links so that they can receive television programs which originate from other cities. These systems can also accommodate thousands of telephone channels.

In 1960, experiments were initiated aiming toward communicating over transoceanic distances via microwaves by utilizing balloons as reflectors. Echo I and Echo II were attempts in this direction as passive satellites. The first active repeater satellite (Telstar I) was launched in 1962 and resulted in live telecasts between Europe and the United States in addition to teleprint and other signals. Telstar II added more data to accent the value of satellite communications.

In 1962, Congress authorized the formation of Communications Satellite Corporation (Comsat). In 1964, Comsat took the lead in forming the International Telecommunications Satellite Organization (Intelsat) to coordinate international developments in the use of satellites. Over 100 nations are now members of this extraordinary effort. Today, all transoceanic "live" TV broadcasts and two-thirds of all transoceanic telephone and telegraph communications are via Intelsat satellites. They operate in the 12/14 GHz band in addition to the 4/6 GHz band.

Microwaves are broadly used for radar, navigation, and for the launching, guidance, and fusing of missiles. A typical defense project which uses microwave techniques is the DEW radar line which protects the United States from external enemy attacks.

The HAYSTACK facility which has been in operation since 1966 at Millstone Hill in Massachusetts is the first Western radar built for spacecraft tracking, space communications, and radar astronomy. Through radar astronomical techniques, the multipurpose HAYSTACK 120-foot paraboloid antenna reflector has greatly enhanced our knowledge of the galaxy and solar system. Another famous facility is the 210-foot GOLDSTONE antenna at the NASA Deep Space Institute (California).

ANTHONY B. GIORDANO

References

Nichols, E. J., and Tear, J. D., "Joining the Infrared and Electric Wave Spectra," *Astrophys. J.* 61, 17–37 (1923).
Carter, S. P., and Solomon, L., "Modern Microwaves," *Electronics* (June 24, 1960).
Southworth, G. C., "Survey and History of the Progress of the Microwave Art," *Proc. IRE* (May 1962).
Wheeler, G. J., "Introduction to Microwaves," Englewood Cliffs, New Jersey, Prentice-Hall, Inc., 1963.
Evans, J. V., and Hagfors, T. (Ed.), "Radio Astronomy," New York, McGraw-Hill, 1968.
Yeh, P., "Satellite Communications and Terrestrial Networks," Dedham, Massachusetts, Horizon House, Inc., 1977.
Topol, S., "Satellite Communications–History and Future," *Microwave Journal* (November 1978).
Cuccia, C. L., "Satellite Communications and the Information Decade," *Microwave Journal* (January 1982).

Cross-references: ANTENNAS, ELECTROMAGNETIC THEORY, MICROWAVE SPECTROSCOPY, PROPAGATION OF ELECTROMAGNETIC WAVES, RADAR.

MODULATION

Modulation is defined as the process, or the result of the process, whereby some characteristic of one wave is varied in accordance with some characteristic of another wave (ASA). Usually one of these waves is considered to be a carrier wave while the other is a modulating signal. The various types of modulation, such as amplitude, frequency, phase, pulse width, pulse time, and so on are designated in accordance with the parameter of the carrier which is being varied.

Amplitude modulation (AM) is easily accomplished and widely used. Inspection of Fig. 1 shows that the voltage of the amplitude modulated wave may be expressed by the following equation

$$v = V_c(1 + M \sin \omega_m t) \sin \omega_c t,$$

where V_c is the peak carrier voltage, ω_c and ω_m are the radian frequencies of the carrier and modulating signals, respectively, and t is time in seconds. The modulation index M may have values from zero to one. When the trigonometric identity $\sin a \sin b = \frac{1}{2} \cos(a - b) - \frac{1}{2} \cos(a + b)$ is used in the equation above, this equation

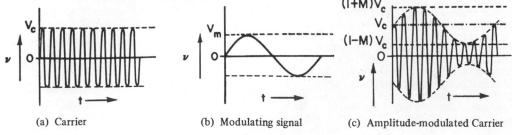

(a) Carrier   (b) Modulating signal   (c) Amplitude-modulated Carrier

FIG. 1. Amplitude modulation.

becomes

$$v = V_c \sin \omega_c t + \frac{M V_c}{2} \cos(\omega_c - \omega_m)t$$

$$- \frac{M V_c}{2} \cos(\omega_c + \omega_m)t$$

This equation shows that new frequencies, called side frequencies or side bands, are generated by the amplitude modulation process. These new frequencies are the sum and difference of the carrier and modulating frequencies.

Amplitude modulation is accomplished by mixing the carrier and modulating signals in a nonlinear device such as a vacuum tube or transistor amplifier operated in a nonlinear region of its characteristics. The nonlinear characteristic produces the new side-band frequencies. Frequency converters or translators and AM detectors are basically modulators. The various types of pulse modulation are actually special types of amplitude modulation.

A special type of amplitude modulation known as *pulse modulation* is commonly used in digital communication and other applications. In pulse modulation, the modulating signal abruptly changes the carrier amplitude from zero to some maximum amplitude V_m (or vice versa) as shown in Fig. 2(a).

Therefore, the modulation index is 1, or 100%, at all times. The side frequencies produced in the modulation process are determined by first using Fourier analysis to find the frequency

components of the rectangular modulating signal and then adding to the carrier frequency a pair of side frequencies for each of those components. The spacing of these components is $1/T$, which is the fundamental frequency, and the amplitude of these components vary as shown in Fig. 2(b). The envelope of this amplitude variation follows the familiar pattern of a $(\sin x)/x$ function. An infinite bandwidth would be required to either produce or reproduce a perfectly rectangular pulse, which of course is impossible to obtain. Bandwidths in the neighborhood of $2/t_d$, where t_d is the pulse duration in seconds, are commonly used to transmit a double-sideband pulse-modulated signal.

Frequency modulation (FM) is illustrated by Fig. 3. The frequency variation, or deviation, is proportional to the amplitude of the modulating signal. The voltage equation for a frequency modulated wave follows.

$$v = V_c \sin(\omega_c t + M_f \sin \omega_m t)$$

The modulation index M_f is the ratio of maximum carrier frequency deviation to the modulating frequency. This ratio is known as the deviation ratio and may vary from zero to values of the order of 1000. FM requires a broader transmission bandwidth than AM but may have superior noise and interference rejection capabilities. A large value of modulation index provides excellent interference rejection capability but requires a comparatively large bandwidth. The approximate bandwidth requirement for a fre-

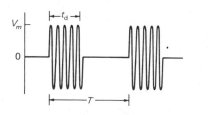

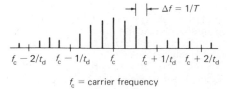

(a) Pulse-modulated carrier   (b) side frequencies generated.

FIG. 2. Pulse (amplitude) modulation.

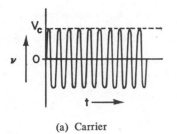

(a) Carrier

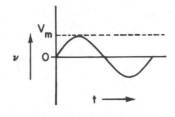

(b) Modulating signal

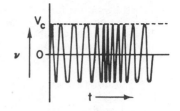

(c) Frequency-modulated Carrier

FIG. 3. Frequency modulation.

quency modulated wave may be obtained from the following relationship

Bandwidth = 2 (Modulating frequency) $(M_f + 1)$

The noise and interference characteristics of FM transmission are normally considered satisfactory when the modulation index or deviation ratio is five or greater.

Phase modulation is accomplished when the relative phase of the carrier is varied in accordance with the amplitude of the modulating signal. Since frequency is the time rate of change of phase, frequency modulation occurs when the phase modulating technique is used and vice versa. In fact, the equation given for a frequency-modulated wave is equally applicable for a phase-modulated wave. However, the phase-modulating technique results in a deviation ratio, or modulation index, which is independent of the modulating frequency, while the frequency modulating technique results in a deviation ratio which is inversely proportional to the modulating frequency, assuming invarient modulating voltage amplitude in each case.

The phase-modulating techniques can be used to produce frequency-modulated waves, providing the amplitude of the modulating voltage is inversely proportional to the modulating frequency. This inverse relationship can be obtained by including, in the modulator, a circuit which has a voltage transfer ratio inversely proportional to the frequency.

CHARLES L. ALLEY

References

Alley, C. L. and Atwood, K. W., "Electronic Engineering," Third Edition, New York, John Wiley & Sons, 1973.

Comer, David J., "Modern Electronic Circuit Design," Reading, Massachusetts, Addison Wesley, 1978.

DeFrance, J. J., "Communications Electronics Circuits," Second Edition, San Francisco, 1972.

Cross-references: MICROWAVE TRANSMISSION, PROPAGATION OF ELECTROMAGNETIC WAVES, PULSE GENERATION, RADAR, WAVE MOTION.

MOLE CONCEPT

The mole (derived from the Latin *moles* = heap or pile) is the chemist's measure of amount of pure substance. It is relevant to recognize that the familiar *molecule* is a diminutive (little mole). Formerly, the connotation of *mole* was a "gram molecular weight." Current usage tends more to use the term *mole* to mean an amount containing Avogadro's number of whatever units are being considered. Thus, we can have a mole of atoms, ions, radicals, electrons or quanta. This usage makes unnecessary such terms as "gram-atom," "gram-formula weight," etc.

A definition of the term is: *The mole is the amount of (pure) substance containing the same number of chemical units as there are atoms in exactly twelve grams of* ^{12}C. This definition involves the acceptance of two dictates—the scale of atomic masses and the magnitude of the gram. Both have been established by international agreement. Usage sometimes indicates a different mass unit, e.g., a "pound mole" or even a "ton mole"; substitution of "pound" or "ton" for "gram" in the above definition is implied.

All stoichiometry essentially is based on the evaluation of the number of moles of substance. The most common involves the measurement of mass. Thus 25.000 grams of H_2O will contain 25.000/18.015 moles of H_2O; 25.000 grams of sodium will contain 25.000/22.990 moles of Na (atomic and formula masses used to five significant figures). The convenient measurements on gases are pressure, volume and temperature. Use of the ideal gas law constant R allows direct calculation of the number of moles $n = (P \times V)/(R \times T)$. T is the absolute temperature; R must be chosen in units appropriate for P, V and T (e.g., $R = 0.0820$ liter atm mole^{-1} deg K^{-1}). It may be noted that acceptance of Avogadro's principle (equal volumes of gases under identical conditions contain equal numbers of molecules) is inherent in this calculation. So too are the approximations of the ideal gas law. Refined calculations can be made by using more correct equations of state.

Many chemical reactions are most conveniently carried out or measured in solution (e.g., by titration). The usual concentration convention is the *molar* solution. (Some chemists prefer to use the equivalent term *formal*). A 1.0 molar solution is one which contains one mole of solute per liter of solution. Thus the number of moles of solute in a sample will be

n = Volume (liters) $\times$ Molarity (moles/liter)

The amount of chemical reaction occurring at an electrode during an electrolysis can be expressed in moles simply as $n = q$ (coulombs)/$z\,\mathfrak{F}$ where z is the oxidation number (charge) of the ion and $\mathfrak{F}$ is the faraday constant, 96 487.0 coulombs/mole. Thus the *faraday* can be considered to be the charge on a mole of electrons. This affords one of the most accurate methods of evaluating the Avogadro number (6.0220 $\times 10^{23}$), since the value of the elementary charge is known with high precision.

Modern chemistry increasingly uses data at the atomic level for calculation at the molar level. Since the former often are expressed as quanta, appropriate conversion factors must involve the Avogadro number. Thus the *einstein* of energy is that associated with a mole of photons, or $E = Nh\nu$. Thus light of 2537Å wavelength will represent energy of

$$E = \frac{6.02 \times 10^{23}(\text{photons/mole}) \times 6.62 \times 10^{-27}(\text{erg-sec}) \times 3.000 \times 10^{10}(\text{cm/sec})}{2.537 \times 10^{-5}(\text{cm}) \times 4.184 \times 10^{7}(\text{erg/cal}) \times 10^{3}(\text{cal/kcal})}$$

$E = 113$ kcal/mole

If the SI system of units is used

$$E = \frac{6.022 \times 10^{23}(\text{mol}^{-1}) \times 6.626 \times 10^{-34}(\text{J} \cdot \text{s}) \times 3.000 \times 10^{8}(\text{ms}^{-1})}{2.537 \times 10^{-7}(\text{m})}$$

$= 4.740 \times 10^{5}(\text{J mol}^{-1})$

Another convenient conversion factor is 1 eV/particle = 23.05 kcal/mole.

The chemist's use of formulas and equations always implies reactions of moles of material, thus HCl(g) stands for one mole of hydrogen chloride in the gaseous state. Thermodynamic quantities are symbolized by capital letters standing for molar quantities, e.g., C_v (heat capacity at constant volume in cal mole^{-1} deg^{-1}), G (Gibbs function in cal/mole), etc. At times it is more convenient to convert an extensive property into an intensive expression. This is especially true in dealing with multicomponent systems. These are referred to as "partial molal quantities" and are given a symbol employing a bar over the letter. Thus the partial molal volume, $\overline{V}_1 = (\partial V/\partial n_1)$ is the rate of change of the total volume of a solution with the amount (number of moles) of component 1.

WILLIAM F. KIEFFER

References

Kieffer, W. F., "The Mole Concept in Chemistry," Ed. 2, New York, Van Nostrand Reinhold, 1973.

Lewis, G. N., and Randall, M., "Thermodynamics," Second edition, revised by Pitzer, K. S., and Brewer, L., New York, McGraw-Hill Book Co., 1961.

Cross-references: CHEMISTRY, ELECTROCHEMISTRY, GAS LAWS, MOLECULAR WEIGHT.

MOLECULAR BIOLOGY

Molecular biology, the study of biologically important molecules and their interactions, is the result of the progression of biology from the classical study of whole organisms to the more recent study of individual cells and their components. Its beginnings are usually dated from the announcement of the double-helical structure of DNA molecules made by Watson and Crick in 1953. During the early years of molecular biology, most attention was focused on bacteria and their viruses, since they were the most easily studied systems. For example, under appropriately controlled conditions, hundreds or thousands of liters of bacteria can be prepared in which every cell is essentially identical. Most of our fundamental knowledge about the ways in which cells synthesize and use their macromolecules was originally derived from the study of bacterial systems.

At the present time, however, the trend is in the opposite direction. A concerted effort is underway to apply the models developed for molecular biologic processes to multicellular organisms. Such organisms present a real challenge to biologists, since most complex organisms contain more than one kind of cell (the cells have "differentiated"), and the interactions between these groups of cells within an organism are carefully controlled. Moreover cells from multicellular organisms differ in fundamental ways from those of bacteria.

Examples of these differences can be seen by referring to Figs. 1A and 1B, which show transmission electron micrographs taken of thin sections of the two types of cells. Figure 1A is a bacterial cell which exhibits typical features such as a central cluster of DNA; basically featureless cytoplasm (the liquid portion of the cell) surrounded by a lipid bilayer (cell or unit membrane); and a rigid cell wall around the entire organism. This type of cell is considered

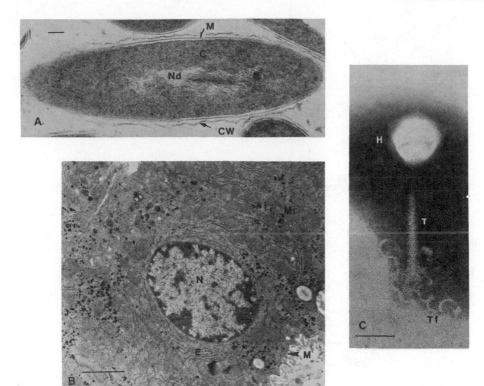

FIG. 1. Examples of biologic organization as seen in the transmission electron microscope. (A) A thin section of the prokaryote *Bacillus sphaericus* prepared by Dr. Elizabeth W. Davidson, Arizona State University. The length of the bar is 0.1 micrometer. (B) A thin section of a eukaryotic cell (from rat liver) prepared by Dr. Candice J. Coffin, Arizona State University. The length of the bar is 1.0 micrometer. (C) An intact bacterial virus, PBS1, negatively stained with potassium phosphotungstate by E. A. Birge. The length of the bar is 0.1 micrometer.

The labeled structures are: C, cytoplasm; CW, cell wall; E, endoplasmic reticulum; H, head; M, cell membrane; Mi, mitochrondria; N, nucleus; Nd, nucleoid; T, tail; Tf, tail fiber.

ancestral, in an evolutionary sense, to the type of cell shown in Fig. 1B, and is designated as prokaryotic. Figure 1B shows a eukaryotic cell with its typical nucleus surrounded by a unit membrane; cytoplasm containing energy-producing mitochondria and the membranous structures known as endoplasmic reticulum; and a unit membrane surrounding the entire cell. Animal cells do not have cell walls, although plant cells do. This type of cellular organization is typical not only of plants and animals but also the unicellular protozoa, fungi, and true algae.

Despite the differences noted above, there are basic similarities between prokaryotic and eukaryotic cells. The basic materials from which the cells are made are identical, as are many of the macromolecules within the cells. In both cases, the genetic material is deoxyribonucleic acid (DNA), which must be synthesized in a semiconservative manner (replicated) prior to each cell division. This process is facilitated by the double-stranded nature of the DNA molecule itself.

Both DNA and the related molecule ribonucleic acid (RNA) are polymers of nucleotide bases composed of certain nitrogenous bases (adenine, cytosine, guanine, thymine and uracil; abbreviated A, C, G, T, and U, respectively) coupled to a pentose (either deoxyribose or ribose) and then to phosphate, as shown in Fig. 2. The polymeric chain is formed by alternating pentoses and phosphates with the nitrogenous bases projecting to one side. RNA is generally single-stranded, incorporating the bases ACGU, while DNA is generally a double-stranded molecule incorporating the bases ACGT. The structure of DNA molecules is somewhat variable and dependent upon such factors as the temperature, salt concentration, and base composition, but a typical helical structure for DNA is shown in Fig. 3. The structure is formed by pairing bases from the parallel strands according to the following rule: A pairs with T and G pairs with C. When RNA is involved, U is substituted for T. During the replication process mentioned above, this base pairing is utilized to spontaneously line up the

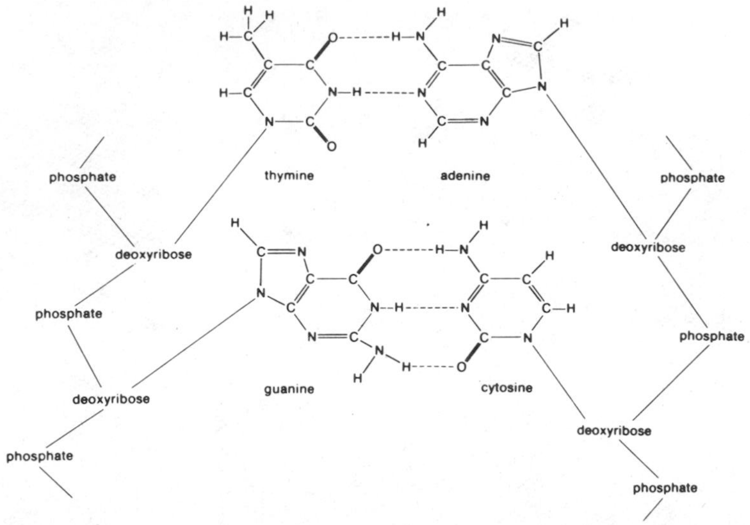

phosphate thymine adenine phosphate

deoxyribose deoxyribose

phosphate

guanine cytosine

phosphate

deoxyribose

deoxyribose

phosphate

phosphate

FIG. 2. Pairing of DNA strands to make a helix. In the diagram the backbones of the two DNA strands are shown along the right and left margins. The bases project into the space between and are held together by hydrogen bonds (dotted line). (Reproduced from Walter, W. G., McBee, R. H., and Temple, K. L., "Introduction to Microbiology," New York, D. Van Nostrand Company, 1973.)

precursor nucleotides so that a polymerase enzyme can join them together. Polymerases are extremely fast acting and may join as many as 250–1000 nucleotides per second under the appropriate conditions.

The genetic information for any cell is encoded in redundant form within its DNA base sequence due to the specificity of base pairing. This information is not, however, directly available for use. Instead an RNA copy of one of the two DNA strands (the "sense" strand) is made by an RNA polymerase following the usual base pairing rules in a process called transcription. Each coding region of the DNA is capable of producing specific RNA molecules whose functions are predetermined. Some are used as part of the subcellular structures called ribosomes and are designated rRNA. Others are used as highly specific carriers of the amino acids, the subunits of the polymers called proteins, and are designated transfer or tRNA. Still other molecules, the rarest class, contain the actual code which determines the sequence of amino acids used to construct a protein. These molecules are designated as messenger or mRNA molecules and tend to be unstable. They code for only one protein if isolated from a eukaryotic cell but may code linearly for as many as ten discrete proteins if isolated from a prokaryotic cell. To a first approximation, each region of the DNA coding for an rRNA, tRNA, or individual protein molecule represents a gene. Overlapping genes are rare but not unknown.

The next step in utilization of genetic infor-

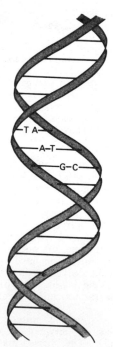

FIG. 3. The winding of a DNA double helix. A less magnified view of a DNA molecule than shown in Fig. 2, this diagram shows the B form structure. There are ten base pairs per turn of the helix, and each pair is rotated 36° with respect to the preceding pair. Reproduced from Walter et al.

mation is the translation of the genetic code into the appropriate sequence of amino acids. Each triplet of bases (codon) codes for a specific amino acid or punctuation signal, and all possible triplets are meaningful. Therefore ribosomes must always attach to mRNA at specific sites to ensure the proper "reading frame." During translation, codons on the mRNA are matched to corresponding anticodons on the tRNA by the usual base pairing rules to assure delivery of the correct amino acid.

In prokaryotes, which have no compartmentalization within their cytoplasm, translation occurs as soon as the mRNA is formed. In eukaryotes, however, the DNA and initial RNA transcripts are found in the nucleus, while the translation machinery is found in the cytoplasm. As a result the RNA must be exported from the nucleus. Before this can happen, the RNA must be processed to remove certain noncoding or intervening sequences which are present in most eukaryotic genes. The processed mRNA is passed through pores in the nuclear membrane into the cytoplasm. In either type of cell, translation occurs in a processive fashion on a complex of a single mRNA molecule and multiple ribosomes called a polysome. Although prokaryotic and eukaryotic ribosomes are similar in function, they are somewhat different in structure, with the prokaryotic ribosomes being smaller than their eukaryotic counterparts.

Interestingly certain organelles within eukaryotic cells, the chloroplasts (sites of photosynthesis) and the mitochondria, contain small DNA molecules coding for ribosomes of the prokaryotic type. It is now considered likely that at least chloroplasts are descended from primordial prokaryotic cells which colonized eukaryotic cells.

As protein molecules are produced, they fold spontaneously into three-dimensional configurations appropriate to their function. In the case of proteins intended for use outside the producing cell, they are generally produced in an inactive configuration which is altered by removal of a portion of the amino acid chain during transport through the cell membrane. As is true for most substances, movement of proteins across the cell membrane is an energy-requiring process.

All of the normal cellular processes mentioned above can be subverted by small obligate intracellular parasites called viruses. These entities represent the boundary between living and nonliving matter. When not in a cell, they have the appearance of complex crystals of protein and nucleic acid, as shown in Figure 1C. Viruses are generally rod-shaped or polyhedral structures consisting of a protein "coat" and a highly condensed nucleic acid molecule which may be RNA or DNA. Infection of a cell consists of movement of the nucleic acid across the cell membrane. In the case of prokaryotic cells, only the nucleic acid enters the cell. For eu-

karyotic cells the entire virus enters, and then the protein coat is removed.

Once inside the cell, viruses follow one of two patterns. In a "lytic" infection, the viral nucleic acid immediately uses the host cell enzymes to produce more viral nucleic acid and the proteins necessary to encase it. Other viral specific enzymes may be produced to facilitate this process by disrupting host cell functions not necessary for the production of viruses. New virus particles are produced not by division of a pre-existing entity, as is the case with cells, but rather by a sequential assembly process which shows remarkable similarities to formation of chemical crystals from a supersaturated solution. In most cases, the host cell is destroyed or "lysed" during release of the viral particles. This observation has given rise to an alternative name for bacterial viruses—bacteriophages or simply phages.

The second mode of viral infection is called a "temperate" infection. In this case the viral nucleic acid establishes a semipermanent relationship with the cell which preserves both virus and host cell. If the viral nucleic acid is RNA, it is converted to DNA. The DNA then sets up a stable association with the host DNA such that the host cell replicates the viral DNA at the same time as it replicates its own DNA. The viral DNA, instead of producing coat proteins and lytic factors, produces a protein repressor which acts to prevent the synthesis of mRNA molecules coding for the lytic functions. The quiescent viral DNA is now designated as a provirus or prophage, since the appropriate stimulus will cause it to revert to the lytic form and destroy its host cell. Not all viruses are capable of the temperate response, and those which are capable of the temperate response do not necessarily use it.

Proviral DNA has been observed to exist in two forms. It may integrate itself into the host cell DNA and become an actual physical part of the cell's genetic material. Alternatively it may exist as a small independent circular DNA molecule which replicates side-by-side with the host DNA. The latter form of DNA is called a plasmid and is more commonly found in prokaryotes.

The term plasmid actually encompasses a much larger group of DNA molecules than just viruses. Nonviral plasmids have been observed in most bacteria and many of the simpler eukaryotes such as yeast. These extra pieces of DNA are considered dispensable to the host cell even though they may, under the appropriate conditions, increase the cell's chances for survival. Examples of this include plasmids which make bacterial cells resistant to certain antibiotics or which allow them to break down certain complex molecules such as xylene for food.

Recent discoveries in molecular biology are having a profound effect on our understanding of molecular genetics and on the way in which

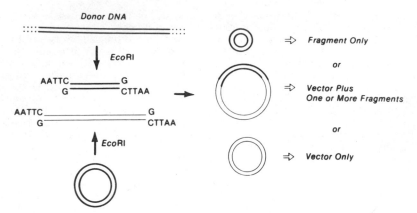

FIG. 4. A diagrammatic representation of gene splicing. The double circle at the bottom left of the diagram represents a double-stranded plasmid DNA molecule, while the double-stranded donor DNA at the top may be from any source. The *Eco*RI enzyme is the prototypical restriction endonuclease which always leaves identical single-stranded ends on the cut DNA. As shown on the right-hand side of the diagram, the single-stranded regions may pair so as to reform the original molecules or to form new constructs. (Adapted from Birge, E. A., "Bacterial and Bacteriophage Genetics: An Introduction," New York, Springer-Verlag, 1981.)

biologic problems can be solved. It is now known that while most portions of DNA molecules are stable over long periods of time, in both prokaryotes and eukaryotes certain small regions within the DNA molecules are naturally highly unstable. These unstable regions, called transposons, consist of special terminal elements with a wide variety of coding elements such as antibiotic resistance located between them. The terminal elements have the ability to cause the entire transposon to simultaneously replicate itself and insert itself into a new position on the same or a different DNA molecule. Transposons thus represent "jumping genes" and provide a way to move bits of DNA around in the cell in a more or less random fashion.

The greatest impact on modern biology has been made by combining the information presented above with the new techniques which have been developed for producing artificial rearrangement of DNA molecules—techniques known as gene splicing. The procedures all depend upon the action of certain restriction endonucleases produced by various bacteria. These enzymes attack all "foreign" DNA, molecules which have not been suitably modified by the addition of small substituents such as methyl groups at specific sites. They cleave the unmodified DNA at base sequence-specific sites to produce variably sized fragments. Since all DNA fragments produced by a given enzyme will have identical ends, it is a comparatively easy job to rejoin the fragments in new combinations and permutations in a manner such as shown in Fig. 4. When the spliced DNA is inserted into a cell and the cell is allowed to grow,

the result is a clone of cells all of which carry the particular DNA segment of interest.

Application of these techniques has led to a true biologic revolution. It is now possible to splice purified DNA into the middle of plasmids or transposons and to insert the spliced DNA into living cells. This is genetic engineering in the fullest sense of the term. The results that have been obtained from the process have included such oddities as bacteria which produce human insulin or growth hormone or animal cells which carry bacteriophage DNA. Constructs like these may some day permit us to understand precisely how cells regulate their internal processes as well as their interactions with neighboring cells. They certainly promise to revolutionize the study of biology.

As the preceding discussion indicates, it is apparent that a cell is constantly involved in many activities which require the movement of molecules. The extent of this feat becomes more apparent when a few size comparisons are made. A typical bacterial cell may be a rod approximately one micrometer in diameter and several micrometers in length. The DNA molecule of the same cell is a circular structure approximately one millimeter in length. In the case of a human cell, the diameter of the nucleus is about 10 micrometers and the 46 linear DNA molecules represent a length of about one meter. Clearly the DNA cannot exist as a random coil within the cell and still allow space for other activities.

Prokaryotic and eukaryotic cells have solved this problem in different ways. In prokaryotic cells the circular DNA molecule is formed into

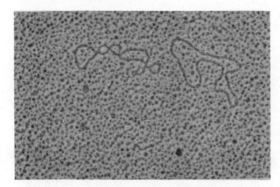

FIG. 5. Supercoiling of DNA. The two DNA molecules have been spread onto a thin plastic film; stained with uranyl acetate; and then shadowed with platinum and palladium. The molecule on the left has retained its supercoils, while the one on the right has one or more broken phosphodiester bonds and has therefore lost all of its superhelicity. (Electron micrograph by E. A. Birge.)

a series of loops, each of which is supercoiled by a family of enzymes called topoisomerases. These enzymes break either one or two strands of the DNA helix and can pass other strands of DNA through the nicked region. Therefore they also have the interesting property of being able to tie and untie knots within the DNA molecule. An example of the difference between supercoiled and "relaxed" DNA molecules can be seen in Fig. 5. The highly supercoiled DNA molecule comprises the nucleoid seen in Fig. 1A.

Eukaryotic cells literally coil their linear DNA around cylinders of protein called histones. There are four histone proteins, three of which are used to make cylinders and one of which covers the DNA which connects adjacent cylinders. The net effect is to take a single DNA molecule and coil it into a sort of "beads on a string" structure. Such a structure is called a chromosome and may exist in an extended state or in an even more condensed form during cell division.

Further information on packaging problems and many other topics discussed in this article can be obtained from the references.

EDWARD A. BIRGE

References

Adams, R. L. P., Burdon, R. H., Campbell, A. M., Leader, D. P., Smellie, R. M. S., "The Biochemistry of the Nucleic Acids," Ninth Edition. New York, Chapman and Hall, 1981.
Glover, D. M., "Genetic Engineering: Cloning DNA," New York, Chapman and Hall, 1980.
Primrose, S. B., Dimmock, N. J., "Introduction to Modern Virology," Second Edition, New York, John Wiley & Sons, 1980.
Wang, J. C. "DNA Topoisomerases," Scientific American 247(1), 94–109 (1982).
Watson, J. D., "Molecular Biology of the Gene," Menlo Park, California, W. A. Benjamin, Inc., 1976.

Cross-references: BIOMEDICAL INSTRUMENTATION, BIOPHYSICS, ELECTRON MICROSCOPE, MEDICAL PHYSICS, PHOTOSYNTHESIS.

MOLECULAR SPECTROSCOPY

Molecular spectroscopy encompasses the broad range of efforts to understand and utilize the interaction of gas-phase molecules with electromagnetic radiation. Spectroscopy is the basic tool for exploring the internal structure of molecules, and spectroscopic studies are of fundamental importance for understanding the microscopic properties of matter. Molecular spectroscopy is also of considerable practical value, since the spectrum of a molecule provides a characteristic "fingerprint" by which that molecule may be identified.

The spectrum of a molecule may be measured either by determining the wavelengths absorbed by the molecule (absorption spectroscopy) or by observing the wavelengths emitted by a sample of excited molecules (emission spectroscopy). In either case, the spectrum is far more complex than an atomic spectrum. Emission and absorption are found to occur from radiofrequency wavelengths through the infrared and visible regions of the spectrum and far into the ultraviolet. Under conditions of low resolution, the visible spectrum is observed to consist of numerous *bands*, hence the designation *band spectra*. Higher resolution demonstrates that each band is composed of numerous closely spaced lines. The microwave and infrared portions of the spectrum are much less congested and easier to analyze. Herzberg's three-volume work[1] on molecular spectroscopy is the major reference. The fact that Volumes I and II have been in print for over 30 years testifies to their enormous success. A more modern presentation is given by Steinfeld.[2]

Those wavelengths which are present in a molecular spectrum are governed by the law of quantum physics which states that a photon of frequency $f = (E_2 - E_1)/h$ is emitted or absorbed whenever the molecule undergoes a transition between energy levels E_1 and E_2. Here h is Planck's constant. Understanding the spectrum, then, is equivalent to understanding the energy levels of a molecule. All of the essential features of molecular spectra are present in diatomic molecules, to which the discussion below is restricted. The energy level structure of polyatomic molecules can be explained by extending the concepts developed for diatomic molecules. A recent compilation of data for diatomic molecules is given by Huber and Herzberg.[3]

It is customary in molecular spectroscopy to express frequencies in units of *wave numbers* (cm^{-1}). The wave number of a photon of frequency f is f/c, where c is the speed of light in vacuum. Since the frequency f and wavelength λ of an electromagnetic wave are related by $\lambda f = c$, it is seen that the wave number is the reciprocal of the wavelength. Using $f = (E_2 - E_1)/h$, the photon's wavenumber is $(E_2 - E_1)/hc$. The values E_1/hc and E_2/hc, which also have units of cm^{-1}, are known as the *term values* of the energy levels E_1 and E_2. Spectroscopists find it convenient to refer to energy levels by their term values, since a transition between two levels involves a photon whose wave number is given simply by the difference in the term values of the levels.

The electrons in a molecule are much lighter and move much more rapidly than do the nuclei. Consequently, it is possible, to a high level of accuracy, to separate the problems of electronic motion and nuclear motion. This procedure is called the Born-Oppenheimer approximation. Each electronic state is characterized by the value of its electronic angular momentum projected onto the internuclear axis. This component of the electronic angular momentum is conserved by virtue of the fact that a diatomic molecule is symmetric for rotations about the internuclear axis. In analogy with the atomic physics notation of $S, P, D, F, \ldots$, an electronic state is labeled Σ, Π, Δ, Φ, ... if its projected angular momentum, in units of $h/2\pi$, is 0, 1, 2, 3, ... The energy of an electronic state depends not only upon the electron configuration but also upon the internuclear separation R. Theorists are still challenged by the difficult problem of calculating accurate electronic energies. The transitions between different electronic states are responsible for visible and ultraviolet molecular spectra since, as in atoms, energy differences between the states are generally several electron volts. The appearance of bands rather than distinct lines is a consequence of the nuclear motion.

A physical model for the nuclear motion is obtained by considering the molecule to be a dumbbell which can vibrate along the internuclear axis as well as rotate end-over-end. The vibrational energy G and the rotational energy F must be added to the electronic energy T_e to give the total molecular energy: $T = T_e + G + F$. Nuclear vibration occurs in the potential well formed by the negative electronic binding energy, which is a function of R, and the positive energy due to Coulomb repulsion of the nuclear cores. The potential reaches a minimum at a particular value r_e of the internuclear separation. This is the equilibrium internuclear distance, and the nuclear separation oscillates about this equilibrium value. For small displacements from equilibrium, the vibrational motion can be approximated by that of a simple harmonic oscillator. The classical oscillation frequency for nuclei of mass M_1 and M_2 is given by

$$f_{\text{osc}} = \frac{1}{2\pi} \sqrt{\frac{k}{\mu}}$$

where k is the force constant and $\mu = M_1 M_2 / (M_1 + M_2)$ is the reduced mass. Solution of the Schrödinger wave equation for the quantum harmonic oscillator leads to energy levels given by

$$E_{\text{vib}}(v) = hf_{\text{osc}}(v + \tfrac{1}{2})$$

where $v = 0, 1, 2, 3, \ldots$ is the vibrational quantum number. Transforming to term values, by dividing by hc, the vibrational energy levels of a harmonic oscillator molecule are

$$G(v) = \omega(v + \tfrac{1}{2})$$

where ω is the vibrational frequency expressed in cm^{-1}.

Although the harmonic oscillator approximation displays the essential features of the vibrational motion, the actual potential in which the nuclei vibrate deviates rather sharply from a harmonic potential. A more complete expression for the vibrational energy can be developed as a power series in $(v + \tfrac{1}{2})$ and is given by

$$G(v) = \omega_e(v + \tfrac{1}{2}) - \omega_e x_e(v + \tfrac{1}{2})^2 + \omega_e y_e(v + \tfrac{1}{2})^3 + \ldots.$$

Here the subscript e refers to the equilibrium position, and the coefficients for each higher-order term ($\omega_e x_e$, $\omega_e y_e$, ...) become successively smaller. It is rarely necessary to go beyond the cubic term when analyzing experimental data. The energy difference between adjacent vibrational states is typically 0.1 electron volt, a factor of 100 less than the energy difference between electronic states.

A first approximation for the end-over-end rotational motion is to consider the molecule to be a rigid rotor. Quantizing the angular momentum leads to energy levels which are given by

$$E_{\text{rot}}(J) = \frac{h^2 J(J + 1)}{8\pi^2 I}$$

where I is the molecule's moment of inertia and $J = 0, 1, 2, 3, \ldots$ is the rotational angular momentum quantum number. Expressing these as term values, the rotational energy levels of a molecule are

$$F(J) = BJ(J + 1)$$

where

$$B = \frac{h}{8\pi^2 cI}$$

is called the rotational constant.

Again, the actual nuclear motion is more complex than this simple model. Centrifugal distortion of the molecule has the effect of introducing a term proportional to $J^2(J + 1)^2$. In addition, the rotational constant depends slightly upon the vibrational quantum number, since vibration changes the average moment of inertia. These considerations lead to a more general formula for the rotational energy, namely,

$$F(J) = B_v J(J + 1) - D_v J^2(J + 1)^2 + \ldots$$

where the subscript v indicates a dependence upon the vibrational quantum number. The separation between adjacent rotational energy levels is typically 0.001 electron volt.

Division of the molecular energy into electronic, vibrational, and rotational energies has been quite successful for understanding the primary features of molecular spectra. At the very highest levels of resolution, however, it is observed that each line in a band splits into several very closely spaced lines. This *fine structure* is a result of interactions, or couplings, between the various types of motion which, until now, have been considered separately. A typical example, known as Λ-doubling, is a coupling between the molecule's electronic and rotational motions. The existence of fine structure emphasizes the limitations of the Born-Oppenheimer approximation. After the various couplings are included, all known aspects of molecular spectra can be understood.

Each spectral line is a consequence of the absorption or emission of photons which occurs when molecules undergo transitions between two energy levels. Comparison of an observed spectrum with theoretical energy levels requires knowing which transitions are allowed and which are forbidden. Information of this sort is codified into *selection rules*. Most selection rules can be understood by considering the possible symmetries of a molecule.

Visible and ultraviolet spectra result from transitions between two different electronic states. The primary selection rule specifies that the molecule's angular momentum quantum number J can change only by $\Delta J = \pm 1$ or 0. Three groups of lines appear, each associated with a particular value of ΔJ, which are called P, Q, and R branches. Changes in the vibrational quantum number are not restricted by any selection rules. However, the Franck-Condon principle, which states that the internuclear separation cannot change during the emission or absorption of a photon, makes transitions between some pairs of vibrational levels far more likely than between other pairs. A band is formed from the combined P, Q, and R branches associated with a transition between a particular vibrational level of the upper electronic state and a particular vibrational level of the lower electronic state. Thirty or more rotational levels in the initial state of a room temperature gas can be populated, and each is allowed by the ΔJ selection rule to undergo a transition to three levels in the final state. Hence each band in a spectrum is comprised of 100 or more distinct lines. It is not surprising, then, that the analysis of molecular spectra made little progress before the advent of quantum physics.

Spectral lines in the infrared occur when a molecule undergoes a transition between two different vibrational levels within the same electronic state. The $\Delta J = \pm 1$ or 0 selection rule still holds, so P, Q, and R branches can again be identified (although the $\Delta J = 0$ Q-branch transitions are forbidden in Σ electronic states). If the vibrational motion were exactly that of a harmonic oscillator, a selection rule $\Delta v = \pm 1$ would apply to the vibrational quantum number. This rule is not rigorous since, as was noted, the harmonic oscillator model is not perfect. Nevertheless, $\Delta v = \pm 1$ transitions are usually the strongest.

Transitions between the rotational levels of a given vibrational level are characterized by frequencies in the far infrared and microwave regions of the spectrum. Only $\Delta J = \pm 1$ are possible for this case, so only a P and an R branch appear. Rotational spectra appear very simple and regular when compared to visible band spectra. Transitions between two rotational levels can often be detected by the absorption of microwaves. The high accuracy with which microwave frequencies can be measured allows rigorous tests of the theory of molecular structure. This aspect of molecular spectroscopy is discussed in a well-known text by Townes and Schawlow.[4]

Another important feature of a molecular spectrum is the intensities of the lines. Even casual observation of a spectrum reveals that some lines are quite intense while others are barely perceptible. This can be understood by noting that not all of the possible transitions from a particular initial state are equally probable. Each molecular transition is characterized, then, not only by a wavelength but also by a *transition probability*. The larger the transition probability, the more intense the spectral line. In emission spectroscopy, where the initial state is an excited state, the transition probabilities determine the average amount of time that a molecule spends in the excited state before emitting a photon. This quantity, known as the *lifetime* of the excited state, is typically 1×10^{-7} second. The measurement of excited state lifetimes is an important part of molecular spectroscopy.

Energy levels in polyatomic molecules may also be understood by a consideration of electronic, vibrational, and rotational motions. Instead of a single mode of vibration, an N-atom polyatomic molecule will have $3N - 6$ possible modes of vibration; $3N - 5$ for a linear molecule. Rotation in a polyatomic molecule is possible about three separate axes, although a useful simplification occurs for symmetric-top molecules, where two of the axes become equivalent. As a consequence of the additional vibrational and rotational motions, polyatomic molecules have many more energy levels than diatomic molecules. The spectrum which is actually observed is again dependent upon selection rules which govern the allowed transitions between energy levels. Even moderate-size molecules (3 or 4 atoms) exhibit spectra which are exceedingly complex, and full rotational resolution is, in general, obtainable only at the highest possible resolution. Spectra of larger molecules are usually analyzed only in terms of electronic and vibrational motions, rotational analysis not being possible. The study of polyatomic molecular spectra is aided considerably by the use of group theory.

Experimental molecular spectroscopy historically has been pursued with the use of grating spectrometers. While other techniques are now available, grating spectrometers continue to play a major role in contemporary research, especially for investigations of ultraviolet spectra. The wide spectral range of spectrometers, from the far infrared into the vacuum ultraviolet, makes them exceptionally versatile tools. Instruments range in size from portable tabletop models, with resolution of 1 part in 10^3, to room-size giants, with resolution of 1 part in 10^6. Spectrometers designed for use in the ultraviolet or far infrared are evacuated, to prevent atmospheric absorption, and utilize special optical materials.

Both absorption and emission spectra can be measured with a grating instrument. For absorption, light from a continuum source is passed through an absorption cell (sometimes the spectrometer itself) and then dispersed. Excitation spectra are obtained by dispersing the light from a discharge which contains the molecule of interest.

Photographic plates have long been the traditional means of detecting the dispersed light, but they are rapidly being replaced by photoelectric detectors, which are more sensitive as well as more compatible with computers.

Fourier transform spectroscopy[5] represents an alternative to grating spectroscopy. The basis for this technique extends back to Michelson, who noted that the output of an interferometer, as a function of time, is the Fourier transform of the light source spectrum, as a function of frequency. Practical realization, however, has awaited the appearance of high-speed, inexpensive computers. In a typical experiment, light from a continuum source passes through an absorption cell, containing the molecule of interest, and then through a Michelson interferometer. The interferometer's output intensity, as measured by an appropriate detector, is digitized as the mirror moves and then transferred to a computer, which calculates the Fourier transform of the data to produce the spectrum.

Resolution with a Fourier transform spectrometer can exceed that with a large grating spectrometer. Infrared spectroscopy has made the most use of this technique, with high resolution studies of, among many other molecules, H_2O, CO_2, CH_4, and C_2H_4. Fourier transform spectroscopy has been valuable as well for visible spectra, especially for mapping the complex spectrum of I_2, which extends throughout much of the visible region of the spectrum.[6] Because of the high absolute precision with which they were measured, I_2 wavelengths are now routinely used as standards in many spectroscopy experiments.

The recent development of lasers, especially tunable lasers, has awakened a new interest in molecular spectroscopy and motivated a rapid proliferation of new experimental techniques.[7] Although absorption spectroscopy with laser sources has been highly successful, particularly in the infrared, the most innovative techniques have utilized laser-induced fluorescence. Spectra are obtained in these experiments by detecting photons which are emitted from excited states (fluorescence) as the laser wavelength is varied. Excitation to the higher energy level occurs whenever the laser's wavelength coincides with a spectral line, enabling those molecules in the proper lower energy level to absorb laser photons and thus undergo the upward transition. Significant advantages of laser spectroscopy include high resolution, low background and noise, and exceptionally high sensitivity. Laser-induced fluorescence experiments have obtained spectra from molecules at densities as low as 10^4 cm^{-3}. This has been especially valuable for the study of free radicals (chemically unstable molecules) and molecular ions.

Laser spectroscopy techniques have been developed for both the near-infrared and far-infrared spectral regions. Tunable diode lasers and color center lasers now cover the entire near-infrared region from 1 to 30 μm. Wavelengths for transitions between different vibrational energy levels of a single electronic state typically fall within this range, and laser absorption spectroscopy, because of the narrow laser linewidths, provides good resolution of the rotational structure even for quite large molecules. Extensive investigations have been made for molecules such as NH_3, C_2H_2, CF_4, and SF_6.

Far-infrared lasers have wavelengths which are well-matched to rotational transitions in many molecules. These lasers, however, operate at fixed frequencies. Spectroscopy can nonetheless be performed by "tuning" the molecule. This is accomplished by applying a strong magnetic field which shifts the molecule's energy levels (Zeeman effect) until the transition frequency between the two levels matches the laser frequency. Laser magnetic resonance, as this procedure is called, is rapidly increasing in usage as more and more far-infrared laser lines are discovered.

Tunable laser radiation from dye lasers is available throughout the visible region of the spectrum. Nonlinear optical techniques, such as frequency doubling, extend the range of tunability to wavelengths as short as 200 nm. This is the wavelength region for band spectra associated with transitions between different electronic states, and both laser absorption and laser-induced fluorescence, as well as more exotic laser techniques, have been used to study a large number of molecules. In most laser spectroscopy experiments, the resolution is limited not by the laser's linewidth but rather by the Doppler width of the transition, a consequence of molecular motion. The limiting resolution of around 1 part in 10^6, often inadequate to resolve rotational details in polyatomic molecules, is poor when compared with the limit of 1 part in 10^9 or better which is set by the laser's linewidth. This has spurred considerable interest in techniques of Doppler-free spectroscopy. One such technique, which has been especially fruitful for ultra-high resolution spectroscopy, is optical-optical double resonance. It has been used for studies of molecules such as BaF_2, CaF_2, and NO_2. Other Doppler-free techniques have been applied to a variety of molecules.

Ultrahigh resolution alone is often inadequate for analysis of the highly complex spectra of polyatomic molecules. A new experimental technique, introduced to grapple with this problem, is laser spectroscopy of supersonic molecular beams.[8] Molecules of interest, often seeded into a rare gas, are forced at high pressure through an expansion nozzle into vacuum. The expansion process cools the molecules' vibrational and rotational motions down to temperatures of only a few degrees Kelvin, leaving the molecules in only a handful of the lowest energy levels. This greatly simplifies the spectrum and facilitates analysis. The small molecule NO_2 has a remarkably complex visible spectrum which for many years had stubbornly resisted attempts at analysis. New efforts with supersonic beams and Doppler-free laser spectroscopy have, however, finally succeeded in establishing a basis for understanding this molecule.

Recent advances in the spectroscopy of free radicals and molecular ions,[9] difficult to produce in large quantities, stem from the development of such high-sensitivity techniques as laser spectroscopy. The unpaired electron found in most of these species adds complexity as well as interest to their spectra. The number of radicals and ions studied, however, remains quite small in comparison with stable, neutral molecules. High resolution studies of several molecular ions have been performed by laser spectroscopy of ion beams. The infrared spectrum of the one-electron molecule HD^+ was measured with sufficient accuracy to test rigorously the foundations of molecular theory.

Another promising technique, which avoids problems due to the rapid recombination of molecular ions, is the use of laser spectroscopy to study ions which are stored in an ion trap.

Spectroscopic data for molecules are valuable far beyond the walls of the laboratory. Spectra are used in a wide range of applications, from the routine industrial analysis of chemicals to the identification of atmospheric pollutants. Perhaps the most important application in recent years of molecular spectroscopy has been in the field of interstellar chemistry.[10] Nearly 60 molecules in interstellar molecular clouds have been identified on the basis of their molecular spectrum, and the number continues to grow. While astronomers have observed and identified some visible spectral lines, most of the molecules have been detected by radio astronomers on the basis of microwave-emitting rotational transitions. Identification of these interstellar molecules has been possible only because of the extensive collection and tabulation of molecular spectroscopy data which has been going on for many years. In a few cases, suggestions by astronomers that some of the observed features were due to "exotic" molecules have prompted laboratory workers to produce and measure the spectra of these species. The interchange between astronomy and molecular spectroscopy has been beneficial for both sides, but with over 200 interstellar microwave lines still unidentified, much work remains to be done.

Both fundamental and applied spectroscopy have witnessed remarkable growth in the last decade. Increasing demand for spectroscopic data and continuing advances in technology will undoubtedly keep molecular spectroscopy vigorous for many years to come.

RANDALL D. KNIGHT

References

1. Herzberg, Gerhard F., "Molecular Spectra and Molecular Structure," Vol. I, "Spectra of Diatomic Molecules (2nd ed., 1950); Vol. II, "Infrared and Raman Spectra of Polyatomic Molecules" (1945); Vol. III, "Electronic Spectra and Electronic Structure of Polyatomic Molecules," (1966), New York, Van Nostrand Reinhold.
2. Steinfeld, Jeffrey I., "Molecules and Radiation Cambridge, Mass., MIT Press, 1978.
3. Huber, K. P., and Herzberg, G. F., "Molecular Spectra and Molecular Structure," Vol. IV, "Constants of Diatomic Molecules," New York, Van Nostrand Reinhold, 1979.
4. Townes, C. H., and Schawlow, A. L., "Microwave Spectroscopy," New York, McGraw-Hill, 1955.
5. Becker, E. D., and Farrar, T. C., "Fourier Transform Spectroscopy," *Science* 178, 361 (1972).
6. Gerstenkorn, S., and Luc, P., "Atlas du Spectre d'Absorption de la Molécule d'Iode (14 800–20 000 cm^{-1})," Paris, Editions du C.N.R.S.,

1978. A correction to the atlas is given in Gerstenkorn, S., and Luc, P., *Revue de Physique Appliquée* **14**, 791 (1979). Subtraction of 0.0056 cm^{-1} from all wavenumbers in the atlas results in an absolute accuracy of 0.002 cm^{-1} and a relative accuracy of 0.0007 cm^{-1}.

7. See papers in Hall, J. L., and Carlsten, J. L. (Eds.), "Laser Spectroscopy," Vol. III, Berlin, Springer-Verlag, 1977; and in Walther, H., and Rothe, K. W. (Eds.), "Laser Spectroscopy," Vol. IV, Berlin, Springer-Verlag, 1979.

8. Levy, Donald H., "Laser Spectroscopy of Cold Gas Phase Molecules," *Ann. Rev. Phys. Chem.* **31**, 197 (1980).

9. Saykally, R. J., and Woods, R. C., "High Resolution Spectroscopy of Molecular Ions," *Ann. Rev. Phys. Chem.* **32**, 403 (1981).

10. Green, Sheldon, "Interstellar Chemistry: Exotic Molecules in Space," *Ann. Rev. Phys. Chem.* **32**, 103 (1981).

Cross-references: ABSORPTION SPECTRA; ATOMIC SPECTRA; ENERGY LEVELS, ATOMIC; FOURIER ANALYSIS; LASER; RAMAN EFFECT AND RAMAN SPECTROSCOPY; SCHRÖDINGER EQUATION; SPECTROSCOPY.

MOLECULAR WEIGHT

The molecular weight of a chemical compound is the sum of the atomic weights of its constituent atoms. The molecule is the smallest weight of a substance which still retains all of its chemical properties. By convention, each atomic weight, and therefore molecular weights, are expressed relative to an arbitrary standard (see below). For example, the molecule of acetic acid, CH_3COOH, contains two atoms of carbon, four of hydrogen, and two of oxygen, so that its molecular weight is the sum of $2(12.01) + 4(1.01) + 2(16.00)$, which totals 60.06. This molecular weight value is clearly in arbitrary units, but a related quantity, the gram-molecular weight or mole, is the molecular weight expressed in grams. One mole of any compound has been found to contain 6.022×10^{23} molecules, and this number is called the Avogadro constant.

For many years, the standard used for atomic weights was the exact value 16 for the naturally occurring mixture of isotopes of oxygen. Another system of atomic weights, based on the value of 16 for the most abundant (99.8 per cent) oxygen isotope, came into use for comparisons involving single atoms or molecules where isotopic differences were important. A conference of the International Commission on Atomic Weights in 1961 adopted as the standard a value of exactly 12 for the carbon-12 isotope, and since then all atomic weights in use have been based on this standard.

The weights of molecules range from a value of about two for the hydrogen molecule to

several millions for some virus molecules and certain polymeric compounds. Molecular dimensions accordingly range from a diameter of about 4Å for the hydrogen molecule to several thousand angstroms—which has permitted viewing single large molecules in the electron microscope. Molecular sizes are generally much smaller and are not measured directly, but are deduced from x-ray diffraction studies of ordered groups of molecules in the crystalline state or from the physical properties such as hydrodynamic behavior of molecules in the gaseous or liquid state.

Many methods for determining molecular weights which are described below depend fundamentally on counting the number of molecules present in a given weight of sample. However, any usable sample contains a very large number of molecules: at least ten trillion of the largest known molecules are present in the smallest weight measurable on a sensitive balance. Therefore, an indirect count is made by measuring physical properties which are proportional to the large number of molecules present. A consequence of the large number of molecules sampled is the averaging of any variations in content of atomic isotopes in individual molecules, so that normal isotopic fluctuations lead to no measurable deviation of molecular weight values. Abnormally high concentrations of isotopes in radiation products may, however, produce altered molecular weights.

The term molecular weight is properly applied to compounds in which chemical bonding of all atoms holds the molecule together under normal conditions (see BOND, CHEMICAL). Thus, covalent compounds, as represented by many organic substances, usually are found to have the same molecular weight in the solid, liquid, and gaseous states. However, substances in which some bonds are highly polar may exist as un-ionized or even associated molecules in the gaseous state and in nonpolar solvents, but they may be ionized when dissolved in polar solvents. For example, ferric chloride exists in the gaseous state as $FeCl_3$ at high temperatures, as Fe_2Cl_6 at lower temperatures as well as in nonpolar solvents, but reverts to $FeCl_3$ in solvents of moderate polarity, and becomes ionic in water solutions—as chloride ions and hydrated ferric ions. Similarly, acetic acid and some other carboxylic acids associate as dimers in the vapor state and in solvents of low polarity, but exist as monomers with progressive ionization as the solvent polarity increases.

Truly ionic compounds, such as most salts, exist only as ions in the solid and dissolved states, so that the term molecule is not applicable and is not commonly used. Instead, the term, formula weight, is used; this denotes the sum of the atomic weights in the simplest formula representation of the compound. If a broad definition of a molecule as an aggregate of atoms held together by primary valence bonds is adopted, then salts in the crystalline state would appear to have a molecular weight which is essentially infinite and limited only by the size of the crystal, since each ion is surrounded by several ions of opposite polarity to which it is attached by ionic bonds of equal magnitude.

A further complication in the definition of molecular weights occurs with inorganic polymers, such as the polyphosphates and polysilicates, whose polymeric nature is clearly evident in both their crystal structure and their highly viscous behavior in the molten state. However, the magnitude of their molecular weights often cannot be found by conventional methods because they are either insoluble or react with solvents, with consequent degradation. These examples indicate that the molecular weight often depends on the conditions used for measurement and must be specified where compounds subject to association, dissociation or reaction are studied.

The history of the clarification of molecular weight concepts is of considerable interest, since this was so intimately related to other developments in chemical knowledge. Although Dalton had published a table of atomic weights in 1808, and by 1825 molecular formulas, derived from combining weights, were in use, many misconceptions of these formulas remained until about 1860. Then evidence from chemical reactions and from measurements of vapor densities firmly established the formulas of many inorganic and simple organic compounds as they are represented today. The vapor density method, based on Avogadro's hypothesis, was thus the first molecular weight method and continues to be useful for compounds that can be easily volatilized. It was not until 1881 that Raoult showed that the depression of freezing points was proportional to the molar concentration of solute. In 1884, van't Hoff related the osmotic pressure of solutions to the vapor pressure, boiling point, and freezing point behavior, and these methods were quickly put into use for determining molecular weights. The abnormal physical properties of salt solutions were explained in 1887 by the ionization theory of Arrhenius, and the very careful measurements of many of these properties furnished the strongest confirmation of the theory. While these measurements provided the most precise determinations of the extent of dissociation of weak electrolytes, they also contributed to the development of the Debye-Hückel theory for strong electrolytes.

Molecular Weight Distributions Most synthetic and many natural polymeric substances are mixtures of molecules having various chain lengths, and thus of different molecular weights—so-called polydisperse systems. In such cases, molecular weight values have an ambiguous meaning, and no single such value will completely represent a sample. Various techniques for measuring molecular weights, when applied to one of these materials, will produce values which often disagree by a factor of two or more. This disagreement arises from the different bases of the methods—for example, some methods yield so-called number-average molecular weights by determining the number-concentration of molecules in a sample, while other methods produce weight-average molecular weights which are related to the weight-concentrations of each species. Another common value is the viscosity-average molecular weight, which is related to the viscosity contribution of each species. Other bases are of importance for certain methods of study, and some of these are complex functions involving several averages. For some purposes, the determination of a single average molecular weight is sufficient for establishing relations between molecular weight and the behavior of polymers, but the type of molecular weight average must be so chosen as to have a close relation to the behavior property of interest. A more detailed knowledge of the constitution of a sample is sometimes required, particularly if several properties are to be considered, or if unusual forms of molecular weight distribution curve are present.

The problem of completely defining the molecular weight nature of polydisperse materials is most accurately solved by determining the frequency of occurrence of each molecular species and representing the results as a frequency distribution curve. Such a study is generally quite tedious, though there are a few methods which provide much of the required information in one experiment. The method currently most used for determining molecular weight distributions of polymers is size exclusion chromatography, which includes gel permeation chromatography and gel filtration. This involves measurement of the differences in extent of permeation of molecules of different sizes into pores of a solid or gel matrix. The distribution of molecular sizes found is converted into a distribution of molecular weights by calibration with standard polymer samples. The method is rapid and applicable to many polymer types. Alternatively, polymers can be separated by fractional precipitation or fractional solution into a series of fractions each of which contains a fairly narrow distribution of molecular weights. Each fraction can then be characterized by one of the methods described below to yield an average molecular weight. Finally, the molecular weight distribution curve can be constructed by summation of these results. While the curve derived is somewhat inexact, it is the best approach to samples which are not susceptible to analysis by the chromatographic methods. The ultracentrifuge is less commonly used for determining molecular weight distributions in a single experiment, partly because of high instrumentation costs and partly because of the complexity of methods needed to analyze the data.

Uses. Molecular weight measurements, in

conjunction with the law of combining proportions, have enabled the atomic weights of elements in compounds to be determined. When the atomic weights are known, molecular weight measurements permit the assignment of molecular formulas. Other applications to compounds of low molecular weight allow determination of the extent of ionization of weak electrolytes, and the extent of association of some uncharged compounds which aggregate. The study of molecular weights is becoming increasingly valuable in assessing the effects which various molecular species of a polymer sample have on the physical properties of the product. Through such knowledge, the synthetic process may be modified to improve the properties of polymers.

Methods of Measurement Many physical and certain chemical properties vary substantially with the molecular weight of compounds, and these properties are the bases of all molecular weight methods. The summary given in this section includes principally the methods which are most frequently used or have general applicability. The choice of the most suitable method for a given sample depends on its state (gas, liquid, or solid), the magnitude of the molecular weight and the accuracy required in its determination, as well as on the stability of the compound to physical or chemical treatment. Some mention of the applicability of the methods in these regards is given wherever possible.

Gases and Liquids Avogadro's hypothesis (1811) that equal volumes of different gases contain the same number of molecules under the same conditions made it possible to find how many times heavier a single molecule of one gas is than that of another. Thus, relative molecular weights of all gases could be established by comparing the weights of equal volumes of gases. The significance of the idea and utilization of this method were first clearly demonstrated by Dumas in 1827, but it was not until 1860 that the results were accepted by most scientists when Cannizzaro showed that a consistent system of atomic weights resulted. With the additional information from chemical experiments on the number of atoms of each kind present in each molecule, the relative weights of each atom were obtained. The assumption of the integral value, 16, for the atomic weight of oxygen (to give a value close to unity for the lightest element, hydrogen) then enabled molecular weights of all gaseous compounds to be determined. The method obviously can be applied to other molecules which normally occur in the liquid state but can be volatilized by heating. The Dumas and Victor Meyer methods are most used for molecular weight determinations with liquids in this way. These methods have been refined so that gas densities can now be determined with an accuracy of 0.02 per cent, and extremely small weights of material (about 1 μg) can be similarly studied with somewhat less accuracy. High temperatures up to 2000°C have been used to study substances which are volatilized only with difficulty, provided decomposition can be avoided.

Solids Measured by Colligative Methods It has been shown that nonvolatile molecules dissolved in a solvent affect several physical properties of the solvent in proportion to the number of solute molecules present per unit volume. Among these properties are a decrease of the vapor pressure of the solvent, a rise in its boiling point, a decrease in its freezing point, and the development of osmotic pressure when the solution is separated from the solvent by a semipermeable membrane. Properties such as these which are related to the number of molecules in a sample rather than to the type of molecule are called colligative properties. They are the basis for some of the most useful techniques for molecular weight determination. The magnitude of the effects and the ease of measurement differ greatly, so that certain of the colligative properties are preferred for this purpose. For example, an aqueous solution containing 0.2 gram of sucrose (molecular weight 342) in 100 ml has a vapor pressure 0.01 per cent less than that of the solvent, a boiling point 0.003°C greater, and a freezing point 0.011°C lower than the solvent, but will develop an osmotic pressure of 150 cm of water. Since the effects are related to the number-concentration of solute molecules, each method leads to a number average molecular weight if the sample consists of a mixture of molecules of different sizes. Accurate results with any of the techniques are obtained only when measurements at a series of concentrations are extrapolated to infinite dilution where the system becomes ideal, i.e. is not affected by interactions between molecules.

Direct vapor pressure measurements with a differential manometer are generally limited to the larger depressions produced by low molecular weight solutes, while refined techniques such as isothermal distillation require the most exact control of conditions. Isopiestic methods allow the comparison of the vapor pressure of solutions of an unknown with those containing a known substance, and several modifications have been used more than other vapor pressure methods. Ebulliometric techniques which depend on the elevation of the boiling point of a solvent are often used for solutes of low molecular weight and find some use for large molecules. Since boiling points are highly sensitive to the atmospheric pressure, it is either necessary to control pressure very precisely, or more commonly to measure the boiling points of both the solvent and solution simultaneously. Often a differential thermometer is employed to determine only the difference of the two temperatures, and these devices have been made so sensitive that molecular weights as large as 30 000 have sometimes been studied. Techniques involving the lowering of the freezing point of a solvent (cryoscopic methods) are much used for rapid approximate determinations of molecular weights in the identification of organic compounds. For this purpose a substance such as camphor, which is a good solvent for many organic compounds and has a large molar depression constant, is often chosen to magnify the difference in freezing point of the solvent and the solution of the unknown. Since freezing-point depressions are not sensitive to atmospheric pressure, they are easier to measure accurately than the methods described above, and much use has been made of them for precise studies of solutes having low molecular weights. The possibility of association or ionization of the solute must be considered with any of these methods, since these effects will greatly influence the result.

Osmotic pressures are so much larger than any other colligative property that they are most widely used for molecular weight measurements, particularly for long-chain polymers where the high sensitivity of the method is required. For accurate measurements, a membrane is required which permits the flow of solvent through its pores but completely holds back solute molecules. This condition is best satisfied where there are large differences in size of the solute and solvent molecules or of their affinity for the membrane. Membranes made from cellulose compounds are often successfully used for polymers which contain little material with molecular weights below about 10 000. Below this molecular weight the pore size of the satisfactory membranes is so small that solvent flow is very slow, and thus a very long time is required to reach constant osmotic pressure. In spite of this handicap, some of the most precise osmotic pressure measurements have been obtained with aqueous solutions of sucrose and similar small solutes by the use of membranes prepared by precipitating such materials as copper ferrocyanide in the pores of a solid support. The upper limit of molecular weights satisfactorily measured by osmometry is usually about 500 000, which is fixed by the lowest pressures that can be measured precisely and by the maximum concentrations of material which still give satisfactory extrapolations to infinite dilution. In comparing various colligative properties for the characterization of polymers, osmometry has the advantage that it is unaffected by the presence of impurities of very low molecular weight which will diffuse through membranes able to retain the polymer, whereas the other properties are greatly affected by the same impurities.

Modern instrumentation has provided commercial instruments utilizing several of these colligative properties for routine, accurate measurements in very short time and with small samples. This is true for boiling point, vapor pressure, and freezing point measurements of molecular weights up to several thousand, and for membrane osmotic pressure measurements of high molecular weight samples.

X-ray Diffraction X-ray diffraction analysis is a powerful method for determining exact molecular weight and structural characteristics of compounds in their crystalline state. However the method is complicated and slower than many techniques which provide molecular weights of accuracy sufficient for many purposes and so is usually employed only when the additional structural information is needed. The sample to be examined must have a high degree of crystalline order and is preferably a single crystal at least 0.1 mm in size; such samples are prepared fairly readily from many inorganic and non-polymeric organic compounds. Alternatively, crystalline powders of certain crystal types may provide suitable results. Diffraction patterns are then obtained by one of several methods, and the angular positions of the reflections are used to calculate the lattice spacings, and thus the size of the unit cell. This unit cell is the smallest volume unit which retains all geometrical features of the crystalline class, and it contains a small integral number of molecules. A rough estimate of the molecular weight of the compound is needed from a determination by an independent method in order to obtain this integral number. Finally, the resultant molecular volume is multiplied by the exact bulk density of the crystal and by the Avogadro number to yield the molecular weight (see X-RAY DIFFRACTION).

Light Scattering Measurements of the intensity of light scattered by dissolved molecules allow the determination of molecular weights. Most commonly the method is used for polymers above 10 000 units, though under optimum conditions molecular weights as low as 1000 have been determined. Since the intensity scattered by a given weight of dissolved material is directly proportional to the mass of each molecule, a weight-average value of the molecular weight is obtained for a polydisperse system. An average dimension of the molecule can also be obtained by a study of the angular variation of scattered light intensity, provided some dimension of the molecules exceeds a few hundred angstroms. The interaction between dissolved molecules substantially affects the intensity of scattered light so that extrapolation to infinite dilution of data collected at several polymer concentrations is required. The method has been so well developed in the last decade that it is now probably the most used method for determining absolute molecular weights of polymers. In addition, it provides information on sizes which is furnished by few other methods. The greatest problem encountered is in the removal of suspended large particles which otherwise would distort the angular scattering pattern of the solutions. This is rather easily accomplished by filtration in some cases, but it may be a formidable difficulty for particles which are highly solvated or are peptized by the molecules to be studied. Auxiliary information is required on the refractive index increment of the sample, i.e., the change in refractive index of the solvent produced by unit concentration of the sample. This information is supplied by a differential refractometer using the same wavelength of light as that employed in measurements of the intensity of scatter.

The Ultracentrifuge. The sedimentation of large molecules in a strong centrifugal field enables the determination of both average molecular weights and the distribution of molecular weights in certain systems. When a solution containing polymer or other large molecules is centrifuged at forces up to 250 000 times gravity, the molecules begin to settle, leaving pure solvent above a boundary which progressively moves toward the bottom of the cell. This boundary is a rather sharp gradient of concentrations for molecules of uniform size, such as globular proteins, but for polydisperse systems, the boundary is diffuse, the lowest molecular weights lagging behind the larger molecules. An optical system is provided for viewing this boundary, and a study as a function of the time of centrifuging yields the rate of sedimentation for the single component or for each of many components of a polydisperse system. These sedimentation rates may then be related to the corresponding molecular weights of the species present after the diffusion coefficients for each species are determined by independent experiments. Both the sedimentation and the diffusion rates are affected by interactions between molecules, so that each must be studied as a function of concentration and extrapolated to infinite dilution as is done for the colligative properties. The result of this detailed work is the distribution of molecular weights in the sample which is available by few other methods. At present, this method is only partly satisfactory for molecular weight determinations with linear polymers because of the large concentration dependence of the diffusion coefficients. Difficulties have been found in reliably extrapolating diffusion coefficients beyond the lowest polymer concentrations which are experimentally attainable at present.

A modification of the sedimentation method which avoids the study of diffusion constants is the sedimentation equilibrium method in which molecules are allowed to sediment in a much weaker field. Under these conditions, the sedimenting force is balanced by the force of diffusion, so that after times from a day to two weeks molecules of each size reach different equilibrium positions, and the optical measurement of the concentration of polymer at each point gives the molecular weight distribution directly. However, again extrapolation to infinite dilution must be used to overcome interaction effects. The chief difficulty here is the long time of centrifuging required, and the necessary stability of the apparatus during the period. A newer and somewhat faster technique, the

Archibald method, permits the determination of weight-average molecular weights of polymers by analysis of the concentration gradient near boundaries soon after sedimentation begins.

Chemical Analysis When reactive groups in a compound may be determined exactly and easily, this analysis may be used to determine the gram equivalent weight of the substance. This is the weight in grams which combines with or is equivalent to one gram-atomic weight of hydrogen. This equivalent weight may then be converted to the molecular weight by multiplying by the number of groups per molecule which reacted (provided they each are also equivalent to one hydrogen). If the number of reactive groups in the molecule is not known, then one of the physical methods for determining molecular weight must be used instead. The chemical method is convenient and often used for the identification of organic substances containing free carboxyl or amino groups which can readily be titrated, and for esters which can be saponified and determinations made of the amount of alkali consumed in this process. The equivalent weights of ionic substances containing, for example, halide or sulfate groups may also be determined by titration or by gravimetric analysis of insoluble compounds formed with reagents which act in a stoichiometric fashion. In the titration of acids, the "neutral equivalent" is the weight of material which combines with one equivalent of alkali, and a similar definition applies to the "saponification equivalent" of esters. If only one carboxyl or ester group is present in the molecule, these values equal the molecular weight of the compound.

In a similar way, if the terminal groups on polymer chains can be determined by a chemical reaction without affecting other groups in the molecule, the equivalent weight or molecular weight of the polymer may be obtained in certain cases. For polydisperse systems, a number-average value of the molecular weight is obtained because the process essentially counts the total number of groups per unit weight of sample. Since the method depends on the effect of a single group in a long chain, its sensitivity decreases as the molecular weight rises, and so is seldom applicable above molecular weights of 20 000. Particularly at high molecular weights, the method is very sensitive to small amounts of impurities which can react with the testing reagent, so that careful purification of samples is desired.

It is also important to know that impurities or competing mechanisms of polymerization do not lead to branching or other processes which may provide greater or fewer reactive groups per molecule. The analysis for end groups must be carried out under mild conditions which do not degrade the polymer, since this would also lead to lower molecular weight values than expected. Labeling of end groups either with radioactive isotopes or with heavy isotopes which can be analyzed with the mass spectrometer provides a rapid and convenient analysis for end groups. This labeling can be accomplished with a labeled initiator if this remains at the chain ends, or after polymerization is complete, by exchange of weakly bonded groups with similar groups in a labeled compound. Molecular weight determinations by end group analysis are often used for condensation polymers of lower molecular weights and are especially valuable in studying degradation processes in polymers.

GEORGE L. BEYER

References

Daniels, F., Williams, J. W., Bender, P., Alberty, R. A., Cornwell, C. D., Harriman, J. E., "Experimental Physical Chemistry," Seventh Edition, New York, McGraw-Hill Book Company, 1970.

Billmeyer, F. W., Jr., "Textbook of Polymer Science," New York, Wiley-Interscience, 1971.

Scholte, T. G., in "Polymer Molecular Weights," (P. E. Slade, Jr., Ed.), Part II, New York, Marcel Dekker, 1975.

Wells, A. F., "Structural Inorganic Chemistry," Fourth Edition, New York, Oxford Univ. Press, 1975.

Cross-references: ATOMIC PHYSICS; BOND, CHEMICAL; CENTRIFUGE; LIGHT SCATTERING; MOLECULES AND MOLECULAR STRUCTURE; OSMOSIS; POLYMER PHYSICS; VAPOR PRESSURE AND EVAPORATION; X-RAY DIFFRACTION.

MOLECULES AND MOLECULAR STRUCTURE

A molecule is a local assembly of atomic nuclei and electrons in a state of dynamic stability. The cohesive forces are electrostatic, but, in addition, relatively small electromagnetic interactions may occur between the spin and orbital motions of the electrons, especially in the neighborhood of heavy nuclei. The internuclear separations are of the order of 1 to 2×10^{-10} metres, and the energies required to dissociate a stable molecule into smaller fragments fall into the 1 to 5 eV range. The simplest diatomic species is the hydrogen molecule-ion H_2^+ with two nuclei and one electron. At the other extreme, the protein ribonuclease contains 1876 nuclei and 7396 electrons per molecule.

Historically, molecules were regarded as being formed by the association of individual atoms. This led to the concept of *valency*, i.e., the number of individual chemical bonds or linkages with which a particular atom can attach itself to other atoms. When the electronic theory of the atom was developed, these bonds were interpreted in terms of the behavior of the valence, or outer shell, electrons of the combining atoms. Each atom with a partly filled valence shell attempts to acquire a completed octet of outer electrons, either by electron

transfer, as in (a), to give an electrovalent bond, resulting from Coulombic attraction between the oppositely charged ions

$$ Na^+ \left[: \overset{..}{\underset{..}{Cl}} : \right]^- \qquad : \overset{..}{\underset{..}{Cl}} : \overset{..}{\underset{..}{Cl}} : \qquad R : \overset{..}{\underset{R}{N}} \overset{+}{:} \overset{..}{\underset{..}{O}} \overset{..}{:}{}^- $$

(a)   (b)   (R = CH₃)
                                             (c)

(W. Kossel, 1916); or by electron sharing, as in (b) and (c), to give a covalent bond (G. N. Lewis, 1916). In (b), each chlorine atom donates one electron to form a homopolar bond, which is written Cl—Cl where the bar denotes on this theory one single bond, or shared electron pair. In (c), the nitrogen-oxygen bond is formed by two electrons donated by only the nitrogen atom, giving a *semipolar*, or *coordinate-covalent* bond, which is written $R_3 N \rightarrow O$, and which is electrically polarized. Double or triple bonds result from the sharing of four or six electrons between adjacent atoms, as in ethylene (d) and acetylene (e) respectively.

$$ \overset{H}{\underset{H}{>}} C = C \overset{H}{\underset{H}{<}} \qquad H - C \equiv C - H $$

(d)   (e)

However, difficulties arise in describing the structures of many molecules in this fashion. For example, in benzene ($C_6 H_6$), a typical aromatic compound, the carbon nuclei form a plane regular hexagon, but the electrons can only be conventionally written as forming alternate single and double bonds between them. Furthermore, an electron cannot be identified as coming specifically from any of these bonds upon ionization. Such difficulties disappear in the quantum-mechanical theory of a polyatomic molecule, whose electronic wave function can be constructed from nonlocalized electron orbitals extending over all of the nuclei. The concept of valency is not basic to this theory, but is simply a convenient approximation by which the electron density distribution is partitioned in different regions in the molecule.

Molecular compounds consist of two or more stable species held together by weak forces. In *clathrates*, a gaseous substance such as SO_2, HCl, CO_2 or a rare gas is held in the crystal lattice of a solid, such as β-quinol, by van der Waals-London dispersion forces. The *gas hydrates*, e.g., $Cl_2 \cdot 6H_2O$, contain halogen molecules similarly trapped in ice-like structures. The hydrogen bond, with energy ~ 0.25 eV, is responsible not only for the high degree of molecular association in liquids such as water (O—H———O—H———) but also for such molecules as the formic acid dimer which contains two hydrogen bonds indicated by dashed lines. *Molecular complexes* vary greatly in their stability; in donor-acceptor complexes, electronic charge is transferred from the donor (e.g., NH_3) to the acceptor (e.g., BF_3), as in a semipolar bond. The $BF_3 \cdot NH_3$ complex has a binding energy with respect to dissociation into NH_3 and BF_3 of 1.8 eV. The bond here is relatively strong; the electron transfer can occur between the components in their electronic ground states. On the other hand, in weaker complexes such as $C_6H_6 - I_2$, with binding energy of about 0.06 eV, there is only a fractional transfer of charge from benzene to iodine. The actual ionic charge-transfer state lies at much higher energy than the ground state of the complex.

The discovery of $XePtF_6$ by Bartlett (1962) has been followed by the synthesis of many other rare gas compounds whose existence was not predicted by classical valency theories. Compounds such as XeF_2, XeF_4, XeF_6 and $XeOF_4$ are quite stable, the average $Xe-F$ bond energy in the square planar molecule XeF_4 being 1.4 eV.

A molecule X is characterized by:

(1) A *stoichiometric formula* $A_a B_b C_c \cdots$ where $a, b, c, \cdots$ are the numbers of atoms of elements A,B,C, $\cdots$ that it contains. The ratio $a : b : c : \cdots$ is found by chemical analysis for these elements. The absolute values of a, b, c, $\cdots$ are then fixed by determination of the *molecular weight* of X. For a volatile substance, the gas density of X and of a gas of known molecular weight are compared at the same temperature and pressure. The molecular weights are in the ratio of the gas densities, since *Avogadro's principle* states that equal volumes of gases at the same temperature and pressure contain the same numbers of molecules. For a nonvolatile substance, a known weight can be dissolved in a solvent, and the resultant lowering of vapor pressure, elevation of the boiling point, or depression of the freezing point of the solvent can be measured. Each of these properties depends upon the number of molecules of solute present, so the number of molecules per unit weight of X is found and, hence, the molecular weight. For substances of high molecular weight such as proteins (molecular weight $\sim 34\,000-200\text{-}000$) or polymers, the molecular weight is found from osmotic pressure measurements or the rate of sedimentation in a centrifuge. The molecular weight of a molecule in crystalline form is determined when the density of the crystal and the dimensions of the unit cell from x-ray analysis are both known. Finally, for stable volatile compounds, it is often possible to form

the ion X^+ and pass this through a mass spectrograph to determine the molecular weight.

(2) The spatial distribution of the nuclei in their mean equilibrium or "rest" positions. At an elementary level, this is described in geometrical language. For example, in carbon tetrachloride, CCl_4, the four chlorine nuclei are disposed at the corners of a regular tetrahedron, and the carbon nucleus is at the center. In the $[CoCl_4]^{2-}$ ion, the arrangement of the chlorine nuclei about the central metal nucleus is also tetrahedral, whereas in $[PdCl_4]^{2-}$ it is planar.

At a more sophisticated level, each molecule is classified under a *symmetry point group*. Most nonlinear molecules possess only 1, 2, 3, 4 or 6-fold rotation axes, and belong to one of the 32 crystallographic point groups. For example, the pyramidal ammonia molecule NH_3 has a threefold rotation axis C_3 through the nitrogen nucleus and three reflection planes σ_v intersecting at this axis, and belongs to the $C_{3v}(3m)$ point group. Tetrahedral molecules CX_4 belong to the $T_d(\bar{4}3m)$ point group. Linear diatomic and polyatomic molecules belong to either of the continuous point groups $D_{\infty h}$ or $C_{\infty v}$ according to whether a center of symmetry is present or not.

The symmetry classification does not define the geometry of a molecule completely. The values of certain *bond lengths or angles* must also be specified. In carbon tetrachloride, it is sufficient to give the $C-Cl$ distance (1.77×10^{-10} meters) since classification under the T_d point group implies that all four of these bonds have equal length and the angle between them is $109° \, 28'$. In ammonia, both the $N-H$ distance (1.015×10^{-10} meters) and the angle HNH ($107°$) must be specified. In general, the lower the molecular symmetry, the greater is the number of such independent parameters required to characterize the geometry. Information about the symmetry and internal dimensions of a molecule is obtained experimentally by SPECTROSCOPY, ELECTRON DIFFRACTION, NEUTRON DIFFRACTION, X-RAY DIFFRACTION, and MAGNETIC RESONANCE. (See these topics for details.) Nuclear magnetic resonance (NMR) is widely used to study molecular structure since it gives information about both the chemical environment of a given nucleus in a molecule and also the disposition of neighboring nuclei. While commonly employed on protons, its use is increasing for other nuclei with nonzero spin angular momentum.

(3) The *dynamical state* is defined by the values of certain observables associated with orbital and spin motions of the electrons and with vibration and rotation of the nuclei, and also by symmetry properties of the corresponding stationary-state wave functions. Except for cases when heavy nuclei are present, the total electron spin angular momentum of a molecule is separately conserved with magnitude $S\hbar$, and molecular states are classified as singlet, doublet,

triplet, $\cdots$ according to the value of the multiplicity $(2S + 1)$. This is shown by a prefix superscript to the term symbol, as in atoms.

The Born-Oppenheimer approximation permits the molecular Hamiltonian H to be separated into a component H_e that depends only on the coordinates of the electrons relative to the nuclei plus a component depending upon the nuclear coordinates, which in turn can be written as a sum $H_v + H_r$ of terms for vibrational and rotational motion of the nuclei (we may ignore translation here). The eigenfunctions Ψ of H may correspondingly be factorized as the product $\Psi_e \Psi_v \Psi_r$ of eigenfunctions of these three operators, and the eigenvalues E decomposed as the sum $E_e + E_v + E_r$. In general, we find $E_e > E_v > E_r$.

Electronic states of molecules are classified according to the symmetry properties of Ψ_e (which forms a basis for an irreducible representation of the molecular point group). Thus $^3B_{1u}$ is a term symbol for benzene (D_{6h} point group) that denotes a triplet electronic state whose wave function transforms like the B_{1u} representation of the group. In the case of diatomic and linear polyatomic molecules, the term symbol shows the magnitude of the conserved component of orbital electronic angular momentum $\Lambda \hbar$ about the axis, states being classified as Σ, Π, Δ, $\cdots$ according to $\Lambda = 0, 1, 2, \cdots$. The superscript + or - shows the behavior of Ψ_e for a linear molecule upon reflection in a plane containing the molecular axis; for centrosymmetric linear molecules ($D_{\infty h}$ point group) the subscript g or u shows the parity +1 or -1 respectively for Ψ_e with respect to inversion at the center.

The vibrational wavefunction Ψ_v can be approximated by a product of $3N - 6$ harmonic oscillator wave functions ψ_i, each a function of a normal displacement coordinate Q_i,

$$\Psi_v = \prod_{i=1}^{3N-6} \psi_i(Q_i)$$

The product is $(3N - 5)$, for a linear molecule; N is the number of nuclei. Each oscillatory mode can be excited with quanta $v_i = 0, 1, 2, \cdots$. When $v_i = 0$, ψ_i transforms like the totally symmetrical representation of the molecular point group; when $v_i = 1$, ψ_i transforms like Q_i. The symmetry of Ψ_v under the molecular group is found from the direct product for all the ψ_i. The vibrationless ground state with $v_1 = v_2 = \cdots = 0$ is always totally symmetrical.

Each rotational state is characterized by a value for the quantum number J, where $J(J + 1)$ $\hbar^2$ is the squared angular momentum for rotation of the nuclei (apart from spin). If I_a, I_b and I_c denote the moments about the principal axes of inertia of the molecule, then a spherical top has $I_a = I_b = I_c$; a molecule with two principal moments equal is either a prolate ($I_c = I_b > $

I_a) or an oblate ($I_c > I_b = I_a$) symmetric top; if $I_c > I_b > I_a$, the top is asymmetric. Symmetric top molecules have C_n symmetry axes with $n \geqslant 3$ and belong to point groups with degenerate representations. The component $K\hbar$ of rotational angular momentum about the top axis is conserved and the rotational levels are also characterized by the value of the quantum number $K = 0, 1, 2, \cdots J$. A symmetry classification is made for Ψ_r under the rotational subgroup of the molecular point group. Finally, each eigenstate is described as + or - according to the parity of Ψ under inversion in a space-fixed coordinate system.

(4) In order to distinguish between different electronic states Ψ_e of the same symmetry and spin multiplicity, a further classification is obtained by expanding Ψ_e as a product of n single-electron wave functions ϕ_i, each a function of the coordinates of one of the n electrons in the molecule.

$$\Psi_e = (n!)^{-1/2} \det |\phi_1(1)\phi_2(2)\phi_3(3) \cdots \phi_n(n)|$$

where $(n!)^{-1/2}$ is a normalization factor. Each of the molecular orbitals (MO's) ϕ_i is constructed to transform like an irreducible representation of the molecular point group and is usually formed by linear combination of atomic orbitals (LCAO) χ_i centered upon the individual nuclei

$$\phi_i = \sum_p C_{ip}\chi_p$$

The MO's are written in order of decreasing energy necessary to ionize the electrons which occupy them, and electrons are assigned to the MO's in accordance with the Pauli principle. For example, the electronic ground state of ammonia (C_{3v} point group) is written

$$(1a_1)^2(2a_1)^2(1e)^4(3a_1)^2 \qquad ^1A_1$$

where the superscripts show the distribution of the ten electrons among three MO's of a_1 symmetry and one of e symmetry, the electrons in the $(3a_1)$ orbital being most readily ionized. The symmetry of the resultant molecular wavefunction Ψ_e is found by taking direct products for each orbital occupied by an electron. Here Ψ_e belongs to the totally symmetrical representation (and is also singlet). Excited electronic states are obtained by promoting electrons into orbitals with higher energies, but the molecular symmetry in such states often differs from that in the ground state, as a result of changes in geometry.

In calculations of molecular properties, the MO's ϕ_i can be improved by variational methods which make them satisfy the *Hartree-Fock* equations. This gives *self-consistent field* (SCF) MO's, yielding a better wavefunction Ψ_e. However the latter is still, in practice, constructed from an incomplete set of basic functions. Further improvement is achieved by *configuration interaction* (CI), in which Ψ_e's of the same symmetry are allowed to mix in linear combination.

G. W. KING

References

Burdett, J. K., "Molecular Shapes," New York, John Wiley & Sons, 1980.

Drago, R. S., "Physical Methods in Chemistry," Philadelphia, W. B. Saunders Co., 1977.

Gillespie, R. J., "Molecular Geometry," London, Van Nostrand-Rheinhold, 1972.

King, G. W., "Spectroscopy and Molecular Structure," New York, Holt, Rinehart and Winston, Inc., 1964.

Levine, I. N., "Molecular Spectroscopy," New York, John Wiley & Sons, 1975.

Cross-references: BOND, CHEMICAL; ELECTRON DIFFRACTION; INTERMOLECULAR FORCES; MAGNETIC RESONANCE; MOLECULAR WEIGHT; NEUTRON DIFFRACTION; QUANTUM THEORY; SPECTROSCOPY; X-RAY DIFFRACTION.

MÖSSBAUER EFFECT

The Mössbauer effect is the phenomenon of recoilless resonance fluorescence of gamma rays from nuclei bound in solids. It was first discovered in 1958 and brought its discoverer, Rudolf L. Mössbauer, the Nobel prize for physics in 1961. The extreme sharpness of the recoilless gamma transitions and the relative ease and accuracy in observing small energy differences make the Mössbauer effect an important tool in chemistry, solid-state physics, nuclear physics, biophysics, metallurgy, and mineralogy.

Resonance fluorescence involves the excitation of a quantized system (the absorber) from its ground state (0) to an excited state (1) by absorption of a photon emitted from an identical system (the source) decaying from state (1) to (0). Not every nucleus has a suitable gamma transition; however, the Mössbauer effect has been observed in more than 60 different isotopes. The parameters characterizing the nuclear resonance process for some typical isotopes are illustrated in Fig. 1.

To conserve energy and momentum in the emission and absorption processes, each system, the source and absorber, must acquire a recoil energy R equal to $E^2/2Mc^2$, where E is the photon energy, M is the mass of the recoiling system and c is the speed of light. The energy available for the excitation of the absorber is thus reduced by $2R$, and resonance fluorescence can be achieved only if the missing energy $2R$ is not larger than the widths of the levels involved. Before 1958, it was thought that for all gamma transitions the width required to get overlap between the emission and the absorption line

was much larger than the natural width Γ, where Γ is related to the half-life $T_{1/2}$ of the excited nuclear level by the expression $\Gamma T_{1/2} = 4.55 \times 10^{-16}\,eV$ sec. In fact, techniques had been developed to compensate for the recoil energy loss by applying large Doppler shifts with an ultracentrifuge or through thermal motion. These methods necessarily broaden the intrinsically narrow lines thereby reducing the absorption cross section.

Mössbauer discovered that in some cases these difficulties may be removed by embedding the source and absorber nuclei in a crystal. Being part of a quantized vibrational system, these nuclei interact with the lattice by exchange of vibrational quanta or phonons only. If the characteristic phonon energy is large compared to the recoil energy R for a free nucleus, the probability for the emission of a gamma ray without a change in the vibrational state of the lattice is large. For such a zero phonon transition, the lattice as a whole adsorbs the recoil momentum and the recoil energy is negligibly small. At the same time, the emission and absorption lines achieve the natural width Γ.

For an atom bound by harmonic forces, the fraction f of events without recoil energy loss is given by $f = \exp(-4\pi^2\langle x^2\rangle/\lambda^2)$. Here $\langle x^2\rangle$ is the mean square displacement of the radiating atom taken along the direction of the photon with wavelength λ. In an environment of lower than cubic symmetry, $\langle x^2\rangle$, and therefore f, may be anisotropic. A large recoilless fraction may be obtained when $\langle x^2\rangle$ is small and λ large. The former condition implies small vibrational amplitude and thus low temperature, high vibrational frequency and large mass M, while the latter implies low photon energy, E. Both conditions imply small recoil energy R.

Recoilless transitions can also occur in amorphous substances like glasses and high-viscosity liquids. For the latter, the diffusive motion superimposed on the thermal vibration results in a broadening of the Mössbauer line.

For all Mössbauer isotopes, the nuclear half-life $T_{1/2}$, typically 10^{-8} second, is very long compared to the period of the lattice vibrations, typically 10^{-13} second. A conceivable first-order Doppler shift of the Mössbauer line due to the thermal motion will therefore average out to zero. The second-order Doppler effect, however, leads to an observable shift, sometimes called the temperature shift. The photons emitted by a source nucleus moving with a mean square velocity $\langle v_s^2\rangle$ are lower in energy by a fraction $\langle v_s^2\rangle/2c^2$ as compared to the photons emitted at rest. Similarly the transition energy of a vibrating absorber nucleus appears lower to the incident photon by a fraction $\langle v_a^2\rangle/2c^2$. In principle, the two shifts may be different whenever the source and absorber are of different composition and/or temperature.

Mössbauer performed his original experiment with ^{191}Ir at 88 K, obtaining a recoilless frac-

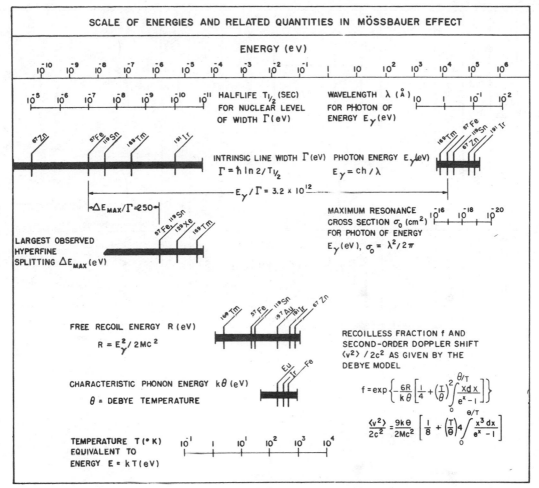

FIG. 1

tion of 1 per cent. Since the natural line width in ^{191}Ir, as in most other Mössbauer nuclides, is extremely narrow, Mössbauer was able to alter the degree of overlap between the emission and absorption lines by simply moving the source relative to the absorber at speeds v of the order of 1 mm/sec. Thus the gamma rays were slightly shifted in energy via the first-order Doppler effect by an amount $\Delta E = Ev/c$. By plotting the transmission through the absorber as a function of the relative source-absorber velocity, one thus obtains the characteristic Mössbauer velocity spectrum which exhibits the shape of the resonance curve. From such a plot, one can determine the recoilless fraction, the lifetime of the excited state and any possible energy differences between the emission and the absorption line.

With extreme care, it is possible to determine energy differences of the order of 1/1000 of the line width Γ. The latter typically varies with isotope from 10^{-10} to 10^{-15} times the actual gamma ray energy E. The Mössbauer effect therefore enables one to detect extremely small changes in this energy. One of the earliest applications of this great precision was the laboratory verification of the gravitational red shift by Pound and Rebka. According to Einstein's theory, photons have an apparent mass $m = E/c^2$. Thus if they fall toward the earth through a distance H, their energy increases by $\Delta E = mgH$, so that $\Delta E/E = gH/c^2 \cong 10^{-16}$ per meter. Using ^{57}Fe, which has a large recoilless fraction, and for which $\Gamma/E = 3 \times 10^{-13}$, the desired effect was observed when the photons were sent down the 22-meter tower at Harvard University.

It is well known from optical and high-frequency spectroscopy that a nucleus interacting with its environment through its charge distribution and magnetic moment can give rise to hyperfine shifts and splittings of the order of 10^{-9} eV to 10^{-5} eV. In Mössbauer experiments, such energy differences can readily be measured

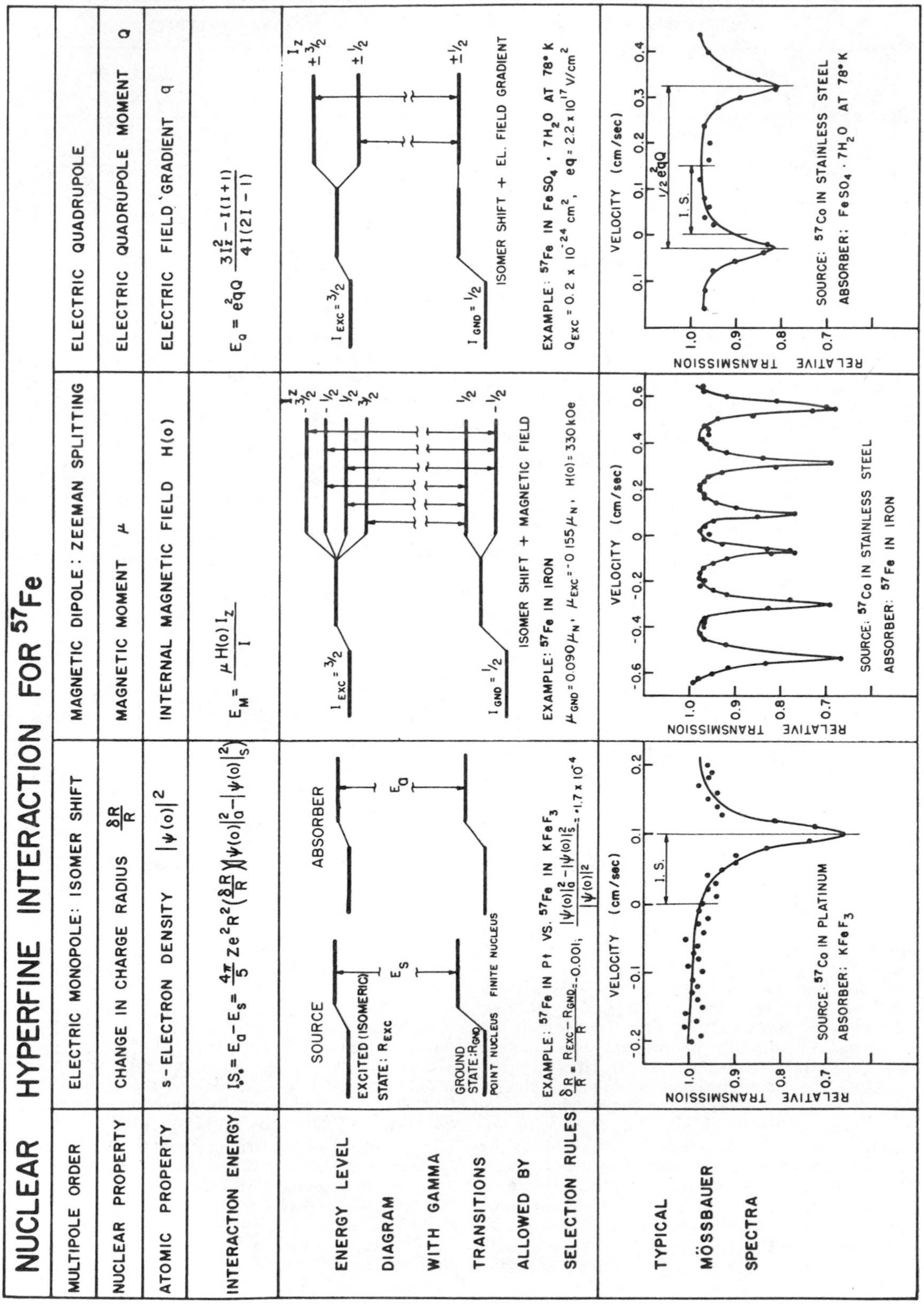

FIG. 2

NUCLEAR HYPERFINE INTERACTION FOR ^{57}Fe

	ELECTRIC MONOPOLE: ISOMER SHIFT	MAGNETIC DIPOLE: ZEEMAN SPLITTING	ELECTRIC QUADRUPOLE
MULTIPOLE ORDER	ELECTRIC MONOPOLE: ISOMER SHIFT	MAGNETIC DIPOLE: ZEEMAN SPLITTING	ELECTRIC QUADRUPOLE
NUCLEAR PROPERTY	CHANGE IN CHARGE RADIUS $\frac{\delta R}{R}$	MAGNETIC MOMENT μ	ELECTRIC QUADRUPOLE MOMENT Q
ATOMIC PROPERTY	s-ELECTRON DENSITY $\|\psi(o)\|^2$	INTERNAL MAGNETIC FIELD $H(o)$	ELECTRIC FIELD GRADIENT q
INTERACTION ENERGY	$I.S. = E_a - E_s = \frac{4\pi}{5} Ze^2 R^2 (\frac{\delta R}{R}) \{\|\psi(o)\|_a^2 - \|\psi(o)\|_s^2\}$	$E_M = \frac{\mu H(o) I_z}{I}$	$E_Q = \frac{e^2 qQ}{4} \cdot \frac{3I_z^2 - I(I+1)}{I(2I-1)}$
ENERGY LEVEL DIAGRAM WITH GAMMA TRANSITIONS ALLOWED BY SELECTION RULES	EXAMPLE: ^{57}Fe IN Pt vs. ^{57}Fe IN KFeF$_3$ $\frac{\delta R}{R} = \frac{R_{EXC} - R_{GND}}{R} = -0.001; \frac{\|\psi(o)\|_a^2 - \|\psi(o)\|_s^2}{\|\psi(o)\|^2} = -1.7 \times 10^{-4}$	ISOMER SHIFT + MAGNETIC FIELD EXAMPLE: ^{57}Fe IN IRON $\mu_{GND} = 0.090 \mu_N$, $\mu_{EXC} = -0.155 \mu_N$, $H(o) = 330 \text{kOe}$	ISOMER SHIFT + EL. FIELD GRADIENT EXAMPLE: ^{57}Fe IN FeSO$_4 \cdot$7H$_2$O AT 78°K $Q_{EXC} = 0.2 \times 10^{-24} \text{ cm}^2$, $eq = 2.2 \times 10^{17} \text{ V/cm}^2$
TYPICAL MÖSSBAUER SPECTRA	SOURCE: ^{57}Co IN PLATINUM ABSORBER: KFeF$_3$	SOURCE: ^{57}Co IN STAINLESS STEEL ABSORBER: ^{57}Fe IN IRON	SOURCE: ^{57}Co IN STAINLESS STEEL ABSORBER: FeSO$_4 \cdot$7H$_2$O AT 78°K

since the line width of the recoilless transitions is of the same order of magnitude. Perhaps, therefore, the most useful feature of the Mössbauer effect is that it may be used to obtain nuclear properties if the fields acting on the nucleus are known, and conversely, it is a powerful tool for probing solids once the various interactions are calibrated, i.e., the nuclear properties have been determined. Some representative results obtained with ^{57}Fe are illustrated in Fig. 2.

The most basic of these interactions is the effect of the finite nuclear size which, in general, is different for the ground state and the excited state. The electrostatic interaction of the nuclear charge with the s-electrons overlapping it raises the nuclear energy levels by an amount depending on the charge radii and s-electron density at the nucleus. Therefore under proper conditions, there appears a shift in the Mössbauer resonance, the isomer shift, which is proportional to $(\delta R/R)$ $\delta|\psi(0)|^2$, where $\delta R/R$ is the fractional change in the nuclear radius during the decay and $\delta|\psi(0)|^2$ is the difference in s-electron density between source and absorber. To determine the quantity $\delta R/R$, one compares the isomer shifts of two chemically simple absorbers, for which the s-electron density can be calculated. In the case of ^{57}Fe an isomer shift exists between compounds containing ferric ions, $Fe^{3+}(3d^5)$, and ferrous ions, $Fe^{2+}(3d^6)$. Although the number of s-electrons is the same for both ions, a detailed calculation shows that the shielding through the additional $3d$ electron changes the $3s$ density at the nucleus. For ^{129}I, the isomer shifts observed among different alkali iodides can be related quantitatively to the known transfer of $5p$ electrons to the ligands which affects the $5s$ density at the nucleus. Once calibrated, the isomer shift is a tool for measuring s-electron densities and is therefore of use in studying chemical bonding, energy bands in solids, and also in identifying charge states of a given atom.

One of the early successes of the Mössbauer effect was the observation of the completely resolved nuclear Zeeman splitting arising from the magnetic hyperfine interaction of ^{57}Fe in ferromagnetic iron. For this isotope, as well as for most other Mössbauer isotopes, the magnetic moment of the nuclear ground state is known from magnetic resonance experiments, and the calibration is therefore straightforward. Careful analysis of the velocity spectrum for magnetic samples is sufficient in general to reveal both the desired magnetic moment and internal magnetic field. The latter yields important information about the unpaired spin density at the nucleus, which in turn is related to the exchange interaction in crystals, molecular complexes, metals and alloys. For single crystals or magnetized samples, the intensities of the individual lines of the Mössbauer spectrum depend on the angle between the direction of the internal field and the emitted photon. From

a measurement of the intensity distribution, one therefore obtains the orientation of the internal magnetic field. The temperature dependence of the splitting can yield Néel and Curie temperatures and also relaxation times.

Whenever one of the nuclear levels possesses a quadrupole moment and an electric field gradient exists at the position of the nucleus, quadrupole splitting of the Mössbauer spectrum may be observed. If the quadrupole moment is known either for the ground state or for the excited state, then a Mössbauer measurement will readily yield the parameters of the field gradient tensor. Usually, however, the quadrupole moment is not known, and the field gradient tensor must be determined from other work or else calculated from first principles. This tensor exists whenever the symmetry of the surrounding charge distribution is lower than cubic, and it is generally specified by two independent parameters. This tensor is easiest to calculate for cases of axial symmetry, in which it is characterized by one parameter, the field gradient, q. For simple ionic systems, it is possible to estimate q with some degree of certainty, and thereby determine the quadrupole moment. Once this is done, the Mössbauer effect may be used to measure field gradient tensors in more complicated systems. Such measurements yield information about crystalline symmetries, crystalline field splittings, shielding due to closed shell electrons, relaxation phenomena and chemical bonding. In addition, with single crystals, a study of the relative intensity of the various lines of the resonance spectrum as a function of angle can yield information about the orientation of the crystalline field axes and, thus, the orientation of complexes in solids.

In cases where both magnetic and quadrupole splittings are present the spectrum depends markedly upon the relative orientation between the hyperfine magnetic field and the axes of the electric field gradient tensor. Paramagnetic complexes of lower than cubic symmetry and many magnetically ordered compounds are of this type. Careful quantitative analysis can then yield the magnitude as well as the relative orientation of the hyperfine interactions. There are also disordered systems such as alloys, amorphous solids, and especially spin-glasses, in which each atom has a slighty different environment. In such cases ^{57}Fe or other Mössbauer nuclides have been used to probe the distribution of isomer shifts, local magnetic fields, and/or electric field gradients. Moreover, the hyperfine interactions may be nonstationary, for instance as a result of spin fluctuations, diffusion, or other time dependent processes. Such systems have been successfully treated using dynamical models of the Mössbauer line shape.

This article has only covered the basic features of the Mössbauer effect and the phenomena which affect the Mössbauer velocity spectrum in a general way. The actual application of the effect is extremely far reaching, embracing not only almost all areas of physics but also the fields of chemistry, biology, geology, metallurgy, and engineering. The reader is advised to consult the references for more information.

R. INGALLS
P. DEBRUNNER

References

Mössbauer, R. L., *Science*, **137**, 731 (1962).

"Mössbauer Effect: Selected Reprints," New York, American Institute of Physics, 1963.

Stevens, J. G., and Stevens, V. E. (Eds.), "Mössbauer Effect Data Index," New York, IFI/Plenum, 1966–present.

Goldanskii, V. I., and Herber, R. H. (Eds.), "Chemical Applications to Mössbauer Spectroscopy," New York, Academic Press, 1964.

Greenwood, N. W., and Gibb, T. C., "Mössbauer Spectroscopy," London, Chapman and Hall, Ltd., 1971.

Gonser, U. (Ed.), "Mössbauer Spectroscopy," Berlin, Springer-Verlag, 1975.

Shenoy, G. K., and Wagner, F. E. (Eds.), "Mössbauer Isomer Shifts," Amsterdam, North-Holland, 1978.

Cohen, R. L., "Applications of Mössbauer Spectroscopy," Vol. 2, New York, Academic Press, 1980.

Gonser, U. (Ed.), "Mössbauer Spectroscopy II: The Exotic Side of the Method," Berlin, Springer-Verlag, 1981.

Cross-references: CONSERVATION LAWS AND SYMMETRY, DOPPLER EFFECT, ISOTOPES, LUMINESCENCE, PHONONS, RADIOACTIVITY, ZEEMAN AND STARK EFFECTS.

MOTORS, ELECTRIC

History Power conversion was discovered by M. Faraday in 1831; the commutator, by J. Henry, Pixii, and C. Wheatstone (1841); the electromagnetic field, by J. Brett (1840), Wheatstone and Cooke (1845), and W. von Siemens (1867); drum armatures, by Siemens, Pacinotti, and von Alteneck; ring armatures, by Gramme (1870); and disc armatures, by Desroziers (1885) and Fritsche (1890). Ring and disk types are now seldom used. Revolving magnetic fields (1885) and ac theory were discovered by G. Ferraris; polyphase motors and systems, by N. Tesla (1888); the squirrel-cage rotor, by C. S. Bradley (1889); and ac commutator motors, by R. Eickmeyer, E. Thomson, L. Atkinson, and others.

Principles These are explained by the laws of Ohm, Kirchoff, Lenz and Maxwell; specifically:

(1) Moving a conductor of length l across a magnetic flux field of density B with a velocity v generates in the conductor an electromotive force (emf) $e = vBl$ volts. In motors, e opposes the current i and decreases as load increases.

(2) The force on such a conductor equals

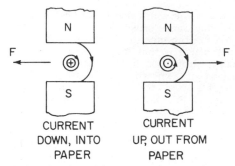

CURRENT DOWN, INTO PAPER   CURRENT UP, OUT FROM PAPER

FIG. 1. Direction of force due to current in a magnetic field.

$F = Bil$ newtons. Fig. 1 shows the directions of current and force.

(3) Magnetic structures tend to move to the position of minimum reluctance (maximum inductance) with force $F = dw/dx$, where w is stored magnetic energy and x is distance.

(4) The force between two coupled circuits is

$$F = \frac{i_1{}^2}{2} \frac{dL_1}{dx} + i_1 i_2 \frac{dM}{dx} + \frac{i_2{}^2}{2} \frac{dL_2}{dx}.$$

Frequently L_1 and/or L_2 are constant and they make no contribution to the force. Here i_1 and i_2 are currents, L_1 and L_2 are total self-inductances, and M is mutual inductance.

Symbols

B = flux density in Tesla
F = force in direction x in newtons
i, i_1, i_2 = instantaneous currents in amperes
I_a, I_2 = dc and effective currents in amperes
L_1, L_2, M = inductances in henrys
s = slip = (Syn. rpm – rpm)/Syn. rpm
v = velocity in meters per second
w = energy in joules
l, x = length, distance, in meters
ϕ_a = useful flux per pole in webers
ϕ_m = maximum flux per pole in webers

Motor Types Of many hundreds, the most used are:

(1) *Direct Current* (a) Series. The field coils are of heavy wire in series with the armature. The torque and current are high at low speeds, and low at high speeds. Torque and speed vary inversely as a square-law function. These motors will run away at light loads unless a speed-limiting device is used.

(b) Shunt. The field coils are of fine wire in parallel with the armature. The speed drops slightly, and the current and torque increase with load. Many small motors (up to 15 hp) use permanent magnets for the field, particularly where high acceleration rates are necessary.

(c) Compound. Both shunt and series fields are used in the same motor. Behavior is intermediate between (a) and (b). The armature current in all dc motors is described as $I_a = (V_T - E_a)/R_a$, where $E_a = p\phi_a Z_a S/(m_a \times 60)$; V_T = terminal voltage, E_a = counter electromotive force (cemf), R_a = armature resistance, m_a = number of paths, and Z_a = number of conductors, all for the armature. In addition, p = number of poles, S = rpm, ϕ_a = useful flux per pole. ϕ_a may be nearly constant or it may be a function of I_a.

Speed control is achieved by adjusting field current and/or armature terminal voltage, tapped series coils, or (more rarely) field reluctance or double commutator. To avoid damage from excessive currents during starting, either starting resistances or voltage controls are needed, except for very small motors.

(2) *Alternating Current* (a) Polyphase induction: These are usually three-phase, with phase windings distributed equally in slots around the periphery of the stator to produce alternate N and S poles. When fed with polyphase current a revolving field is set up which turns at $S = 120f/p$ rpm. This revolving field induces a counter emf $E = 2.22f\phi_m Zk_w$ volts per phase, where ϕ is flux per pole, k_w is the winding factor, and Z is the number of series turns.

The rotor is made of a number of short circuited conductors. If these conductors are open, or the rotor is running at synchronous speed S, each rotor phase behaves as an inductance and draws an exciting current lagging nearly $\frac{1}{4}$ cycle behind the emf. Shaft load reduces the speed from S to $S(1 - s)$, where s is slip. When referred to the stator mounted armature (primary), s causes the emf in each rotor phase to be sE (in volts). Each rotor phase sees current (in amperes):

$$I_2 = \frac{sE}{\sqrt{r_2{}^2 + (sx_2)^2}} = \frac{E}{\sqrt{\left(\dfrac{r_2}{s}\right)^2 + x_2{}^2}}$$

where r_2 is the resistance and x_2 is the rotor leakage reactance in ohms per phase.

The current I_2 produces the needed torque. It creates a new magnetomotive force (mmf) which turns at synchronous speed S in the same direction as the revolving stator field, and lags the stator field in space by the electrical angle $90° + \tan^{-1} (x_2/r_2)$. To balance this mmf the stator draws additional current sufficient to produce the load torque. Performance can be calculated (as for a transformer) if r_2/s is taken as the independent variable. Three-phase motors require less material than one- or two-phase motors.

(b) Single-phase induction. With the motor at rest there is no revolving field; the motor will start in either direction only if given a push. Rotation sets up an elliptical revolving field which turns synchronously at nonuniform velocity in the same direction as the rotor and pulsates between the limits ϕ_m and $\phi_m (1 - s)$. The stator current is the resultant of (stator + cross axis) magnetizing + load + loss currents. Analysis of this type of motor is less simple than that for a polyphase motor.

Starting is with a line switch, reduced voltage (auto-transformer or "compensator," wye-delta, series resistors, or chokes), wound rotors and resistors, part-windings, or more elaborate schemes. For single-phase motors, auxiliary start windings (split-phase or capacitor), or repulsion-start are used and are removed at approximately 65% of full speed. For very small, low-torque motors, shaded poles are used.

Speed control has historically been by reduced voltage, wound rotor, or elaborate degenerative feedback schemes. State-of-the-art is variable-voltage, variable-frequency electronic inverters. The motor speed tracks the applied frequency.

(c) Synchronous. Commonly these have a stationary phase-wound armature and revolving dc or permanent magnet field structure. When at full speed and synchronized, the revolving armature mmf stands still relative to the dc field. When the angle $(I_a, E_a) = 0$, the armature mmf poles stand midway between the field poles. When I_a lags E_a, the armature mmf assists the field; it opposes the field when I_a leads E_a. Armature current I_a adjusts to a value and time phase position such that the counter emf E_a and current I_a are correct to meet the existing load. Important modeling relations are $I_a = (V - E_a)/Z_a$; power converted = $mI_a E_a \cdot \cos (I_a, E_a)$; input = $mI_a V \cos (I_a, V)$; and armature copper loss = $mI_a{}^2 r_a$. Where E_a is a function of (I_a, I_f); angle (I_a, E_a) and E_a are in opposition; I_a increases with load; and m = number of phases.

When connected directly to the utility power system, synchronous motors are used where the rpm must be fixed, for power factor correction, regulating transmission line voltages, or speeds too low for good induction motor performance. They are not self starting unless special means are provided, usually squirrel-cage or phase windings in the pole faces. Precautions against high ac voltages in the dc field coils are needed when starting. The motors will carry some load with the dc field winding open. Small sizes (reluctance types) operate without field windings. Single-phase motors are less satisfactory because of lower efficiency, tendency to severe hunting, and problems in starting.

(d) ac Commutator. These are mostly single-phase series, repulsion, or combination of these types. Single-phase series motors are used in large quantities for portable tools, vacuum sweepers, garbage disposals, household appliances, etc. The ac series motor is similar to the dc motor except for a laminated field and precautions against low power factor and poor commutation. The runaway speed is high.

(e) *Brushless dc*. These are ac synchronous and induction motors which receive their power from electronic inverters. The inverters are fed from a dc source. By proper inverter control, the entire system responds the same as a dc motor. Uses include industrial drives, electric cars, railways, and aircraft control actuators.

(f) *Stepper Motors*. ac and dc motors specially designed for high torque and low inertia. Normally electronically controlled to discrete positions, or steps. These are used extensively for industrial positioners, computer line printers, robots, etc.

Probable Future Trends These include less expensive, higher energy product permanent magnets; electrical insulators with better heat conduction properties; and superconductors. Superconducting windings will require motors to be built of nonmagnetic materials, since the magnetic fields will exceed the saturation levels of magnetic materials. Rapid advances in power electronics will influence motor designs.

FREDERICK C. BROCKHURST

References

Slemon, G. R., and Straughen, A., "Electric Machines," Reading, Mass., Addison-Wesley Publishing Company, Inc., 1980.
Say, M. G., "Alternating Current Machines," New York, Halstead Press, John Wiley & Sons, Inc., 1976.

Cross-references: ALTERNATING CURRENTS, ELECTRICITY, INDUCED ELECTROMOTIVE FORCE, INDUCTANCE, MAGNETISM, TRANSFORMER.

MUSICAL SOUND

Musical sound may be characterized as an aural sensation caused by the rapid periodic motion of a sonorous body, while noise is due to nonperiodic motions. The above statement, originally made by Helmholtz, may be modified slightly so that the frequencies of vibration of the body fall into the limits of hearing: 20 to 20,000 Hz. This definition is not clear cut; there are some noises in the note of a harp (the twang) as well as a recognizable note in the sqeak of a shoe. In other cases it is even more difficult to make a distinction between music and noise. In some modern "electronic music" hisses and thumps are considered a part of the music. White noise is a complex sound whose frequency components are so closely spaced and so numerous that the sound ceases to have pitch. The average power per frequency of these components is approximately the same over the whole audible range, and the noise has a hissing sound similar to that one gets from FM radio that is tuned between stations. Pink noise has its lower frequency components relatively louder than the high frequency components and this is accomplished by keeping the average power the same in each octave (or in each $\frac{1}{3}$ octave) band from 20 to 20,000 Hz.

The attributes of musical sound and their subjective correlates are described briefly. The number of cycles per second, *frequency*, is a physical entity and may be measured objectively. *Pitch*, however, is a psychological phenomenon and needs a human subject to perceive it. In general, as the frequency of a single sinusoidal vibration of a sonorous body (pure tone) is raised, the pitch is higher. However, pitch and frequency do not bear a simple linear relationship. To define the relationship human subjects are used to construct a pitch scale so that one note can be judged to be two times the pitch of another and so on. The unit of pitch on this scale is called the *mel*, and a pitch of 1000 mels is arbitrarily assigned to a frequency of 1000 Hz. In general, it is observed that the pitch is slightly less than the frequency when the frequencies are higher than 1000 Hz, and slightly more than the frequency at frequencies less than 1000 Hz. Pitch also depends on loudness; for a 200 Hz tone if the loudness is increased, the pitch decreases, and the same happens for frequencies up to 1000 Hz. Between 1000 and 3000 Hz pitch is relatively independent of loudness, while above 4000 Hz, increasing the loudness raises the pitch. Small pitch changes with loudness do occur with complex tones and whether the pitch goes up or down with loudness seems to depend on the harmonic structure of the complex sound. A rapid variation in pitch when the variation occurs at the rate of from two to five times per second is called *vibrato*. The pitch variation in mels may be large or small but the rate at which the pitch is varied is rarely greater than five times per second. Violinists produce vibrato by sliding their fingers back and forth a minute distance on a stopped string. A variation in loudness occurring at the rate of two to five times a second is called *tremolo*. Singers often produce a combination of tremolo and vibrato to give added color to their renditions.

The ability to discriminate between pitches depends on other variables beside the acuity of the listener. The *just noticeable differences* (jnd's) also depend on the frequencies of the pure tones. Jnd's generally get larger as the frequency increases. Duration of a pure tone is also a requirement for recognition of pitch. Some pure tones of even 3 millisecond duration can still be recognized as having a definite pitch. The duration of time required to recognize pitch depends on the frequency and to some extent on the loudness of the tone. It takes longer to recognize the pitch of a low frequency pure tone than for one of a high frequency. In all cases if the time is too short one hears a click rather than a clear pitch. The pitch of a complex musical tone depends on the spectrum of the complex tone. If the complex tone is composed of the fundamental and exact overtones, the ear recognizes the pitch as that of the fundamental, even if the fundamental is weak or is *missing*. Manufacturers of small portable radios take advantage of this when they install small loudspeakers in these radios. These speakers are not capable of producing the fundamentals of certain low frequency musical sounds but the ear seems to fill in the "missing fundamental."

Like frequency, *intensity* is a physical entity defined as the amount of sound energy passing through unit area per second in a direction perpendicular to the area. It is proportional to the square of the sound pressure, the latter being the rms pressure over and above the constant mean atmospheric pressure. Since sound pressure is proportional to the amplitude of a longitudinal sound wave (see WAVE MOTION) and to the frequency of the wave, intensity since it depends on energy, is proportional to the square of the amplitude and to the square of the frequency. Sound intensity is measured in watts per second per square centimeter and, since the ear is so sensitive, a more usual unit is the microwatt per second per square centimeter. By way of example, a soft speaking voice produces an intensity of 0.1 micromicrowatt/cm^2 sec, while fifteen hundred bass voices singing fortissimo at a distance 1 cm away produce 40 watt/cm^2 sec. Because of such a large range of intensities, the *decibel scale* of intensity is normally used to designate intensity levels. An arbitrary level of 10^{-16} watts/cm^2 sec is taken as a standard for comparison at 1000 Hz. This level is very close to the threshold of audibility. At this frequency, other sound levels are compared by forming the logarithm of the ratio of the desired sound to this arbitrary one. Thus $\log I/10^{-16}$ is the number of bels a sound of intensity I has, compared to this level. Since this unit is inconveniently large, it has been subdivided into the decibel, one-tenth its size. Thus $10 \log I/10^{-16}$ equals the number of decibels (dB) the sound has. A few intensity decibel levels are listed:

	db
Quiet whisper	10
Ordinary conversation	60
Noisy factory	90
Thunder (loud)	110
Pain threshold	120

While intensity levels can be measured physically, loudness levels are subjective and need human subjects for their evaluation. The unit of loudness is the *phon*, and an arbitrary level of zero phons is the loudness of a 1000 Hz note which has an intensity level of 0 dB. Sounds of equal loudness, however, do not have the same intensity levels for different frequencies. From a series of experiments involving human subjects, Fletcher and Munson in 1933 constructed a set of equal-loudness contours for different frequencies of pure tones. These show that for quiet sounds (a level of 5 phons) the intensity

level at 1000 Hz is about 5 dB lower than an equally loud sound at 2000 Hz, for 30 Hz about 70 dB lower, and at 10,000 Hz about 20 dB lower. In general, as the intensity level increases, loudness levels tend to be more alike at all frequencies. This means that as a sound gets less intense at all frequencies, the ear tends to hear the higher and lower portions of sound less loudly than the middle positions. Some high fidelity systems incorporate circuitry that automatically boosts the high and low frequencies as the intensity level of the sound is decreased. This control is usually designated a *loudness control*.

At times it is necessary to have a scale of absolute perceived loudness. The unit of this perceived loudness is the *sone*. It is arrived at by a set of complicated procedures involving human subjects usually placed in a free field situation (e.g., anechoic chamber). One sone of loudness is defined as the sound pressure at the ear of 40 dB for a 1000 Hz pure tone. Two sones is perceived in these experiments as twice as loud, three sones three times, etc. In many situations, musical as well as noisy, the figures for sones of multiple sounds seem to add up in an arithmetical way; figures for db do not. Some consumer testing groups use the sone scale for evaluating the noise of various consumer products.

The entity which enables a person to recognize the difference between equally load tones of the same pitch coming from different musical instruments is called *timbre*, *quality*, or *tone color*. A simple fundamental law in acoustics states that the ear recognizes only those sounds due to simple harmonic motions (see VIBRATION) as pure tones. A tuning fork of frequency f, when struck, causes the air to vibrate in a manner which is very nearly simple harmonic. The sound that is heard does, in fact, give the impression that it is simple and produces a pure tone of a single pitch. If one now strikes simultaneously a series of tuning forks having frequencies f (the fundamental), $2f$, $3f$, $4f$, $5f$, etc. (overtones), the pitch heard is the same as that of the fork of frequency f except that the sound has a different timbre. The timbre of the sound of the series can be changed by altering the loudness of the individual forks from zero loudness to any given loudness. Another way to alter the timbre is to vary the time it takes for a composite sound to grow and to decay. A slow growth of an envelope, even though it contains the same frequencies, makes for a different timbre than one which has a rapid growth. The difference in timbre between a B-flat saxophone and an oboe is almost entirely due to the difference in growth or decay time.

A fundamental theorem discovered by the mathematician Fourier states that any complicated periodic vibration may be analyzed into a set of components which has simple harmonic vibrations of single frequencies. If this method of analysis is applied to the composite tones of musical instruments, it is seen that these tones consist of a fundamental plus a series of overtones, the intensity of overtones being different for instruments of differing timbre. Rise and decay times will also differ. The reverse of analysis is the synthesis of a musical sound. Helmholtz was able to synthesize sound by combining sets of oscillating tuning forks of various loudness to produce a single composite steady tone of a definite timbre. Modern synthesizers are more sophisticated. Electrical oscillators of the simple harmonic variety are combined electrically and then these electrical composite envelopes are electronically modified to produce differing rise and decay time. A transducer changes the electrical composite envelope into an acoustical one so that a sound of any desired timbre, rise and/or decay time can be produced. An alternative way to produce similar effects is to use an oscillation known as the *square wave*. When this oscillation is analyzed by the method of Fourier, it is shown to consist of a fundamental plus the odd harmonics or overtones. Another kind of oscillation, a *saw-tooth wave*, then analyzed, is shown to consist of the fundamental and all harmonics— even and odd. A square wave or a sawtooth wave produced by an appropriate electrical oscillator can be passed through an electrical filter which can attenuate any range of fre-

quencies of the original wave. This altered wave can later be transformed into the corresponding sound wave. In this way sounds having desired rise and decay times, plus the required fundamental and overtone structure, can be made as desired.

JESS J. JOSEPHS

References

Rayleigh, J. W. S., "The Theory of Sound," New York, Dover Publications, 1945.

Helmholtz, H., "On the Sensations of Tone," New York, Dover Publications, 1954.

Stephens, R. W., and Bate A. E., "Acoustics and Vibrational Physics," London, Edward Arnold, 1966.

Josephs, J. J., "The Physics of Musical Sound," New York, Van Nostrand Reinhold, 1967.

Winckel, F., "Music Sound and Sensation," New York, Dover, 1967.

Benade, A. H., "Fundamentals of Musical Acoustics," London, Oxford Univ. Press, 1976.

Backus, J., "The Acoustical Foundations of Music," New York, Norton, 1977.

Rossing, T. D., "The Science of Sound," Reading, MA, Addison-Wesley, 1982.

Cross-references: ACOUSTICS, ARCHITECTURAL ACOUSTICS, NOISE, FOURIER ANALYSIS, REPRODUCTION OF SOUND, VIBRATION, WAVE MOTION.

N

NEUTRINO

The neutrino is an elementary particle postulated by W. Pauli[1] in 1930 to explain the apparent non-conservation of energy and momentum in that class of nuclear radioactivity known as beta decay. A quantitative theory of beta decay incorporating the neutrino hypothesis was formulated by E. Fermi[2] in 1933 in analogy with the quantum theory of radiation and served to predict the nature of the neutrino and its extremely weak interaction with matter. According to the Pauli-Fermi ideas, the neutrino (in Italian, "little neutral one") is a particle of vanishingly small and possibly zero rest mass, no electrical charge, with spin 1/2, and the ability to carry energy and linear and angular momentum. Its interaction with matter is so weak that a 3-MeV antineutrino is predicted to be capable of penetrating an astronomical thickness of matter, e.g., 100 light-years of liquid hydrogen. In 1956, a group of Los Alamos physicists[3] succeeded in making a direct observation of the neutrino, $\widetilde{\nu}_e$, emitted from beta-decaying fission fragments produced in a powerful reactor at the Savannah River Plant operated by the du Pont Company for the U.S. Atomic Energy Commission. These investigators used giant liquid scintillation detectors to observe the inverse beta decay reaction

$$\widetilde{\nu}_e + p \rightarrow e^+ + n$$

where p is the target proton and e^+ and n are the product positron and neutron. The experiment consisted of observing the distinctive delayed coincidence between the prompt annihilation of the positron and the capture of the neutron by a cadmium isotope dissolved in the scintillator. In 1957, following the "overthrow" of parity conservation in weak interactions as a result of the work of Lee and Yang[4], the character of the neutrino was further elucidated. Two kinds of neutrinos were accepted: the neutrino, ν_e, produced in beta decay in association with positrons, and the antineutrino, $\widetilde{\nu}_e$, produced in beta decay in association with negative electrons. The neutrino emerged as completely polarized with the spin angular momentum parallel (antiparallel) to the linear momentum for the antineutrino (neutrino). A theory of weak interactions encompassing the neutrino, in which the

relativistically invariant forms, vector and axial vector, were found to be sufficient to account for most of the known characteristics of the weak interactions, was then formulated by Marshak and Sudarshan[5], and Feynman and Gell-Mann[6].

In 1962, an experiment at the Brookhaven National Laboratory by a Columbia-Brookhaven group,[7] using a heavily shielded 10-ton spark chamber array, showed that the neutrino most frequently associated with the decay of the Π meson differed from the neutrino produced in nuclear beta decay, thus enlarging the class of neutrinos to four; $\nu_e, \bar{\nu}_e, \bar{\nu}_\mu, \nu_\mu$. It now appears that any decay or inverse process involving an electron has associated with it an electron type neutrino while any such process involving a mu meson occurs in association with a mu-type neutrino. The discovery by Perl[8] (1975) of a heavy (1.9 times the proton mass) weakly interacting particle called a tau suggests that a third pair of neutrinos exists ($\nu_\tau, \bar{\nu}_\tau$).

An interaction of the neutrino which did not change the electrical charge of the target—a neutral current—was observed[9] at CERN (1973) in accordance with the theoretical prediction of Salam and Weinberg[10] (1967-72). The existence of the reactions $\nu_\mu + e^- \rightarrow \nu_\mu + e^-$ and $\nu_\mu + N \rightarrow \nu_\mu + N'$ supported the theoretical picture which resulted from unification of the weak and electromagnetic forces. More recent work[11] shows that, as expected, the neutral current reaction $\bar{\nu}_e + D \rightarrow \bar{\nu}_e + n + p$ also occurs. A further prediction of the theory is the existence of carriers of the weak interaction called intermediate bosons $W^\pm$, Z^0 having masses in the range some 90 times that of a proton. The existence of these particles was demonstrated at the CERN $\bar{p} > p$ colliding beam accelerator (1983).[12]

Recent experiments suggest that the neutrino may be more complicated than hitherto believed—in particular it may not have a vanishingly small rest mass. Theoretical models for such a possibility[13] advanced by Nakagawa et al. (1963) and Pontecorvo (1969) assume that the observed neutrinos are linear combinations of massive base states. Such combinations would give rise to observed states, e.g., ν_e, ν_μ, ν_τ, which differ in an energy-dependent manner with distance from a neutrino source. Such variations, called oscillations by analogy with the composite motion of coupled pendula, have been sought using reactor neutrinos $\bar{\nu}_e$ (1980) and higher energy neutrinos from accelerators, and hints of an effect[14] are being subjected to further tests.[15]

The implications of finite neutrino mass range from a mixing of neutrino types to consequences of an equally profound character in the formation of galaxies and the universe itself. A neutrino mass ~ 20 eV would result in a closed universe.

A search for ν_e from the sun[16] using a 400-ton perchloroethylene (C_2Cl_4) target located ~ 1 mile below the earth's surface to diminish the background due to cosmic rays has, thus far, yielded unexpectedly negative results. In this experiment the ^{37}Ar produced by solar neutrinos via the reaction $\nu_e + {}^{37}Cl \rightarrow {}^{37}Ar + e^-$ was sought by collecting the ^{37}Ar from the C_2Cl_4 container and counting in a tiny low background proportional counter. The limits set, a factor of ~ 3 below expectation, are difficult to reconcile with the current solar model.[17]

Because of the threshold for ^{37}Cl reaction (0.8 MeV) the solar ν_e signal is dependent on the details of the solar model. An experiment which is insensitive to the solar model is in process of development by Davis et al.[16] (1980). The reaction $\nu_e + {}^{71}Ga \rightarrow {}^{71}Ge + e^-$ has a threshold of 0.24 MeV and hence can be caused by ν_e from the essentially model independent $p + p$ reaction which is responsible for the bulk of the sun's energy:

$$p + p \rightarrow d + e^+ + \nu_e \quad (0.42 \text{ MeV endpoint}).$$

At this stage the experiment appears to be feasible, assuming the availability of the required 25 tons of gallium.

Two groups, one 2 miles underground in a South African gold mine,[15] and the other in the Kolar gold fields[20] at a slightly lesser depth, have detected high-energy muon neutrinos produced in the atmosphere by the interaction of cosmic ray primaries. A measure of the meager flux and weakness of the interaction is indicated by the small number of neutrino-induced events collected: the Indian, Japanese, British Kolar group recorded ~ 17 in two years of operation; the American, South African group detected ~ 100 with somewhat larger equipment and four years of operation.

Ambitious proposals are being formulated to study these "natural" neutrinos by means of giant (10^6–10^9 tons) water Čerenkov detectors to be located deep (~ 5 kilometers) in the ocean. As in the case of the gold mine experiments the neutrinos would be detected via product muons. Of special interest is the sensitivity of such an imagined detector to very energetic neutrinos ($\gtrsim 10^{10}$ eV). If current conjectures regarding energetic extraterrestrial sources are correct such a deep underwater muon and neutrino detector (DUMAND)[21] shows promise of observing them.

It is seen that neutrino physics encompasses low-energy reactions using fission reactors, work in the structure of the weak interaction using giant electronuclear machines, and the beginnings of various studies of neutrino sources such as the sun and high-energy cosmic rays[22] as well as the nature of the neutrino itself.

F. REINES

References

1. Pauli, W., Jr., "Rapports Septième Conseil Physique, Solvay, Bruxelles, 1933," Paris, Gautier-Villars, 1934.
2. Fermi, E., *Z. Physik* 88, 161 (1934).
3. Reines, F., and Cowan, C. L., Jr., *Phys. Rev.* 92, 830 (1953); Cowman, C. L., Jr., Reines, F., Harrison, F. B., Kruse, H. W., and McGuire, A. D., *Science* 124, 103 (1956); *Phys. Rev.* 117, 159 (1960).
4. Lee, T. D., and Yang, C. N., *Phys. Rev.* 105, 1671 (1957).
5. Marshak, R. E., and Sudarshan, E. C. G., *Phys. Rev.* 109, 1860 (1958); Proceedings of Padua-Venice Conference on Mesons and Newly Discovered Particles, Italy, September 1957.
6. Feynman, R. P., and Gell-Mann, M., *Phys. Rev.* 109, 193 (1958).
7. Danby, G., Gaillard, J. M., Goulianos, K., Lederman, L. M., Mistry, N., Schwartz, M., and Steinberger, J., *Phys. Rev. Lett.* 9, 36 (1962).
8. Perl, M. L., et al., *Phys. Rev. Lett.* 35, 1489 (1975).
9. Hasert, F. J., et al. *Phys. Lett.* 46B, 138 (1973).
10. Weinberg, S., *Phys. Rev. Lett.* 19, 1264 (1967); Salam, A., Proc. 8th Nobel Symposium (N. Svartholm, Ed.), 1969.
11. Pasierb, E., et al., *Phys. Rev. Lett.* 43, 96 (1979).
12. Armison, G. et al. (UAI Collaboration) Phys. Lett. 126B, 7 July 1983. See also *CERN Courier*, Nov. 1983 for a discussion of these important discoveries.
13. Nakagawa, M., Okonogi, H., Sakata, S., and Toyoda, A., *Prog. Theor. Phys.* 30, 727 (1963); Pontecorvo, B., *Zh. Eksp. Teor. Fiz.* 53, 1717 (1967).
14. Reines, F., Sobel, H. W., Pasierb, E., *Phys. Rev. Lett.* 45, 1307 (1980).
15. Mössbauer, R. L., Contribution to ν-82 Conference. Reines, F., *Nucl. Phys.*, A396, 469 (1983).
16. Davis, R., Jr., Rogers, L. C., and Radeka, V., *Bull. Am. Phys. Soc.* 16, 631 (1971).
17. Bahcall, J. N., and Sears, R. L., *Ann. Rev. Astron. Astrophys.* (1972).
18. The use of Gallium was suggested as a solar ν_e target by V. A. Kuzmin, *Soviet Physics—JETP* 22, 1051 (1966).
19. Reines, F., Kropp, W. R., Sobel, H. W., Gurr, H. S., Lathrop, J., Crouch, M. F., Sellschop, J. P. F., and Meyer, B. S. *Phys. Rev.* D4, 80 (1971).
20. Krishnaswamy, M. R., Menon, J. G. K., Narasimhan, F. S., Hinotani, K., Ito, N., Miyake, S., Os-

borne, J. L., Parsons, A. J., and Wolfendale, A. W., *Proc. Roy. Soc. London A* **323** (1971).

21. Several papers and references to DUMAND may be found in Vol. II of the 1981 Internat. Conf. on Neutrino Physics and Astrophysics, R. J. Cence, E. Ma, A. Roberts, Eds.

22. Further references and more extensive discussion may be found in the proceedings of the annual neutrino conferences.

Cross-references: ANTIPARTICLES, CONSERVATION LAWS AND SYMMETRY, ELEMENTARY PARTICLES, RADIOACTIVITY, WEAK INTERACTIONS.

NEUTRON

Significance The role of neutrons and their applications is of central importance to nuclear physics and, indeed, to science in general. Some indication of their wide-ranging significance in the current world is provided by the breadth of the topics covered in an International Conference on The Neutron and its Applications,[1] September 13–17, 1982, commemorating the fiftieth anniversary of the 1932 discovery of neutrons by (Sir) James Chadwick[2] in Cambridge (UK). The following sample of topics exemplifies the scope: Neutron sources; properties, lifetime, moments and structure of the neutron; ultracold neutrons; neutrons in astrophysics and neutron stars; nuclear interactions of the neutron; scattering, diffraction, and neutron optics; radiation; fission and fusion; reactor physics; transmutation, transfer, and transport theory; magnetism; molecular sciences, polymers, and materials sciences; molecular biology, biology, medical, and therapeutic applications; technological applications; theoretical and particle aspects; etc.

This recent overview follows surveys of neutron physics in previous international conferences that dealt with the interactions of neutrons with nuclei[3] and nuclear structure study with neutrons,[4,5] supplementing review articles,[6–8] and topical textbooks.[9–14]

The neutron is thus a fundamental particle of immense versatility, a constituent of all matter (except hydrogen, ^1H, whose nucleus consists solely of a proton) as one of the two principal particle species, collectively termed *nucleons*, that make up the nuclei of atoms. The protons that form the companion species have almost the same characteristic properties as the neutrons, with one important exception: they carry a positive electric charge, whereas neutrons are electrically neutral, i.e., uncharged, and therefore not susceptible to any electrostatic repulsive forces. Likewise, for this reason neutrons cannot attract (negatively charged) electrons to form neutronic "atoms," nor can they be deflected, focused, or detected directly through electric-field devices. But, being uncharged, they can and do approach, penetrate, and interact with nuclei much more readily than protons, since they have no Coulomb repulsion to overcome. Moreover, they can and do function as the "binding agents" within a nucleus to hold it relatively stably together despite the disruptive electrostatic forces that the constituent protons exert on each other. In this capacity they respond to the strong-interaction internucleon force (they are classified as *hadrons*, i.e., "the strong ones," from the Greek *hadrós*, "strong," a family of particles that includes not only the protons and other *baryons*, i.e., "the heavy ones," but also *mesons*, i.e., "the medium ones," when the latter are susceptible to strong interactions). This nuclear force is a powerful binding force that acts attractively between nucleons at short range, keeping them pinioned within the confines of a nucleus, whose diameter is on the order of 10^{-15} to 10^{-14} meters, about a million times smaller than the diameter of an atom.

The binding among neutrons and protons in a nucleus is mediated by the exchange of mesons: principally, $pi(\pi)$-mesons (or *pions*, the "primary" mesons), but also $rho(\rho)$-mesons, both of which species may carry a positive, negative, or zero electrical charge, and $omega(\omega)$-mesons, which exist only in the uncharged state. This meson-exchange model of the forces inside a nucleus also provides an explanation (indicated below) for the otherwise puzzling fact that the mean lifetime of many varieties of nuclei containing bound neutrons can greatly exceed the mean free-neutron lifetime. From measurements performed on large statistical aggregates of neutrons in the free state, their mean lifetime has been determined as $\tau = 925 \pm 11$ sec, which corresponds to a "half-life" of $T_{1/2} = 641 \pm 8$ sec, after which time interval on an average one-half of the aggregate will have decayed into products indicated by the following symbolic decay scheme:

$$n \rightarrow p^+ + e^- + \bar{\nu}.$$

This expresses the decay of any given neutron into a triad of particles comprising a proton ($p^+ = {}^1H$, the positively charged nucleus of the hydrogen atom), an electron ($e^- = \beta^-$, the particle identified as the constituent of β-rays in radioactivity) and an antineutrino ($\bar{\nu}$). This particular decay transition, termed "β-decay" in allusion to the presence of the β (electron) among the decay products, is an instance of a *weak* interaction. Thus the neutron in its *free* state participates in a *weak*-interaction decay, whereas in its *bound* state in a nucleus, it participates in a *strong*-interaction binding among the other nucleons. The heavy particles (n, p) featured in above β-decay transition are members of the baryon family; the light particles (e, $\bar{\nu}$) are *leptons*, wherein the presence of the antineutrino is indicative of a *weak* interaction. Of the two nucleons, the neutron is very slightly the more massive: hence its decay into a proton and other products does not violate the principle of conservation of mass-energy. The neutron's intrinsic angular momentum, or *spin*, is (in units of $h/2\pi$, where h is Planck's constant, a fundamental physical constant in quantum theory that has the dimensions of energy times time) $s = \frac{1}{2}$; this "quantum number" is numerically the same for the other end-product particles of β-decay, which accordingly is consistent with the conservation of angular momentum, schematically expressed as $\frac{1}{2} \rightarrow \frac{1}{2} + \frac{1}{2} - \frac{1}{2}$. The most recent concensus for the mass of a neutron is

$$m_n = (1.6749542 \pm 0.0000048) \times 10^{-27} \text{ kg}$$

$$= 1.0086650 \pm 0.0000028 \text{ u},$$

where the atomic mass unit (u) is defined as $1/12$ of the mass of the neutral ^{12}C atom. When expressed in energy units (million electron-volts, or MeV), this is equivalent to a rest-mass-energy of 939.5731 ± 0.0027 MeV, as against the proton's rest-mass-energy of 938.2796 ± 0.0027 MeV, which corresponds to

$$m_p = (1.6726483 \pm 0.0000048) \times 10^{-27} \text{ kg}$$

$$= 1.0072764 \pm 0.0000029 \text{ u}.$$

The neutron's mass m_n is about 1839 times that of the electron, and the proton's mass is about 1836 times that of the electron:

$$m_e = (9.1095339 \pm 0.0000250) \times 10^{-31} \text{ kg}$$

$$= (5.4858027 \pm 0.0000015) \times 10^{-4} \text{ u},$$

or, in energy units, 0.5110034 ± 0.0000014 MeV. The rest mass of the antineutrino is zero (or, at all events, vanishingly small relative to the above, bearing in mind that the most recent theories suggest that it may not necessarily be precisely zero). Thus, the mass excess of the neutron over the proton, as expressed in energy units, is

$$m_n - m_p = 1.29343 \pm 0.00004 \text{ MeV}.$$

The decay, after a mean half-life of roughly 10.7 min, of neutrons into protons (which are completely stable or, at any rate, have extremely long mean lifetimes, on the order of 10^{31} years if the latest unified theories of particles and their interactions prove to be valid) would seem to preclude the existence of neutrons within stable or long-lived nuclei. On the other hand, their presence in conjunction with protons as the ingredients that make up a nucleus successfully explains all the known properties of nuclei and, in particular, overcomes the numerous, otherwise insurmountable, difficulties posed by earlier, pre-1932, models of nuclear structure, such as the proton-electron model, in which the nucleus was visualized to be composed of protons (making up the mass) and electrons (serving to neutralise the excess

electric charge). Not only can the presence of electrons within the confines of a nucleus be excluded by quantum-theoretical arguments, but simple reasoning based upon spin-conservation can rule out such a proton-electron configuration in the case of such nuclei as nitrogen-14, for which measurements indicate the positive nuclear charge to be 7 units ($^{14}_{7}N$, a designation in which the superscript on the left denotes the mass number A and the subscript on the left denotes the atomic number Z) and the ground state to have an *integer* spin (of value 1 in the conventional $h/2\pi$ units): no aggregate of 14 protons (each having *half*-integer spin, $s = \frac{1}{2}$) and 7 electrons (each having a single negative electric charge, and half-integer spin, $s = \frac{1}{2}$), making up a composite of an *odd* total number (21) of half-integer-spin particles, can yield a system having *integer* net spin, whereas of course a cluster of 7 protons with 7 neutrons constitutes an *even*-number aggregate that can give rise to an *integer* net spin.

In adopting the neutron-proton model of nuclear structure, with meson exchange to promote the binding, it is evident that the continual acquisition and loss of mesons by neutrons must be the factor responsible for inhibiting the neutrons' natural tendency to decay. By way of a somewhat oversimplified picture, one might conceive of the momentary combination of an uncharged neutron with a positively charged companion meson as akin to an overweight, but stable, proton: hence, whenever the bound neutron gains its accompanying π^+ (or ρ^+) exchange meson it assumes the character of a long-lived (pseudo-proton) particle and so enhances the nuclear lifetime beyond that of the normal free neutron. The mesic "clouds" in a nucleus thus participate in an ever-ongoing rapid exchange process, binding the nucleons and augmenting the stability of the nucleus.

As the number of neutrons (N, the *neutron number*) in a nucleus increases, up to and beyond the atomic number Z (equal to the number of protons in the nucleus, and hence to the number of counterbalancing electrons in the electrically neutral atom), the nucleus becomes more stable—up to a reasonable limit, beyond which, with a growing neutron excess ($N - Z$), the instability again progressively grows worse. For stable nuclei of odd mass number $A = N + Z$ it is found that to a good approximation the neutron number is

$$N_{(\text{most stable})} = \frac{490A + 7.75A^{5/3}}{990 + 7.75A^{2/3}}.$$

Whereas the number of nuclear protons (and thus of atomic electrons) determines the particular chemical element (e.g., $Z = 1$ for hydrogen, $Z = 2$ for helium, $Z = 27$ for cobalt, $Z = 92$ for uranium, etc.), it is the number of *neutrons* that determines the particular *isotope* (Greek, "same place") of that element. For example, the stable nuclide ^{59}Co has $N = A - Z = 59 -$

27 = 32, while the unstable isotopes on either side, ^{57}Co and ^{61}Co, respectively have N = 30 and 34; according to the above stability equation, one would expect ^{59}Co to be the most stable of these, since substitution of A = 59 yields the desired $N_{(most\ stable)}$ = 32, whereas with A = 57 one would obtain $N_{(most\ stable)}$ = 31, and with A = 61 one would find that $N_{(most\ stable)}$ = 33.

The transformation of nuclei by neutron capture to form new isotopes was first accomplished in Rome by Enrico Fermi in 1934 (the year in which he also put forward his detailed theory of β-decay). This opened up an altogether new branch of nuclear physics (proton-induced transmutation had been discovered by John Cockcroft and Ernest Walton in 1932, for which they were jointly awarded the Nobel Prize in physics in 1951; Chadwick had received the Nobel Prize in physics in 1935, and Fermi in 1938). The importance and breadth of the subsequent developments hardly need to be stressed; on the cosmic scale, neutron capture processes in stars have profound astrophysical significance, the discovery of fission in Berlin by Otto Hahn (recipient of the Nobel Prize in chemistry in 1944) and Fritz Strassmann in 1939 through the capture of slow neutrons by uranium literally had earth-shaking consequences, and the seemingly boundless range of neutron-induced nuclear transmutation reactions under widely differing conditions offered new vistas in the rapidly expanding field encompassed by neutron science.

The discovery of its conjugate antiparticle, the antineutron ($\bar{n}$, uncharged), by B. Cork, G. R. Lambertson, O. Piccioni, and W. A. Wenzel[15] in 1956 (one year after the 1955 identification of the antiproton by O. Chamberlain, E. Segrè, C. Wiegand, and T. Ypsilantis), the recognition of the neutron's having a complicated internal structure (since, as discussed below, it has a magnetic moment, and therefore acts like a magnet, there must be internal electric charges circulating within it, even though its net electric charge is found from external measurements to be zero), the interpretation of the make-up of the neutron as a baryon in terms of a triad of quarks (e.g., the pnn or, in more recent parlance, udd quark combination, comprising an "up" quark and two "down" quarks, the former of which carries a positive charge, $+\frac{2}{3}e$, and the latter of which each carry a negative charge, $-\frac{1}{3}e$, where e is the electronic charge), and other related particle aspects, have all collectively made this a particularly interesting, if difficult, particle to study in its own right. Lately,[1] increasing consideration is being given to the possible existence of "neutron oscillations," namely, the likelihood that free neutrons could be represented as a superposition of n and $\bar{n}$ states, with an interconversion (oscillation) period $\tau_{n\bar{n}}$ in excess of 10^5 sec. The understanding of neutron-neutron forces is still far from complete in its quantitative details; studies with polarized neutrons, the acquisition of explicit information on neutron matter and neutron stars, and application of neutrons for technological, biological, and medical purposes as well as for purely scientific ends, the development of more intense neutron sources and more efficient means of detection, the potentialities offered by, e.g., ultracold nuetrons, or at the other extreme, by ultrahigh-energy neutrons: all these have the ability to provide fruitful, valuable insights into the physical world that would otherwise remain unexplored and unexploited.

Discovery and Classification No finer "coming-of-age present" for nuclear physics could be imagined than the 1932 discovery of neutrons by Chadwick,[2] precisely twenty-one years after the birth of nuclear physics following the recognition by (Sir) Ernest Rutherford in 1911 that at the central core of atoms there lay a massive, dense "nucleus." To set this neutron discovery in perspective, it should be borne in mind that with the discovery of natural radioactivity by Henri Becquerel in 1896 and of electrons as the "β-ray" component of such radiation) by (Sir) Joseph Thomson in 1897, the groundwork was laid for the elucidation of the make-up of atoms. This received its major impetus in 1911 from Rutherford's interpretative studies and in 1913 from Niels Bohr's quantum-theoretical explanation, which rendered an admirable account of many observations of atomic phenomena. With the definite identification of the proton in 1919, it appeared reasonable to contend that nuclei consisted of protons and electrons in intimate combination. Upon further scrutiny, however, this model became increasingly difficult to reconcile with the findings, and by 1930 it was evident that it was untenable. Speculation was rife that the nucleus might possibly harbor a neutral ingredient, a suggestion that Rutherford espoused. With the finding of the neutron, having all the desired properties, the difficulties were solved in 1932.

Stemming from the 1930 observation by Walther Bothe and H. Becker in Berlin that a penetrating radiation resulted from the bombardment of various target elements by naturally produced α-particles (alpha-particles are helium-4 nuclei, emitted spontaneously by a radioactive polonium source and by some other radioactive sources), and that the reaction with beryllium was particularly intense, investigations were undertaken at various centers into the nature of this radiation, unusually rich in energy and of novel penetrative capability. Among these was the Cavendish Laboratory in Cambridge (UK) where, first, H. C. Webster and, subsequently, Chadwick pursued these studies to higher precision than had been attained previously, which ruled out the possibility of the radiation's being high-energy γ-rays (gamma-rays are the third component of the α, β, γ triad of radiations observed by Becquerel and separated by the Curies; they are identified as uncharged photons, i.e., electromagnetic radiation of short wavelength). Rather, Chadwick reached the conclusion that the uncharged radiation was composed of neutral massive particles, whose mass he showed to be comparable with, and slightly larger than, that of the proton. Following Rutherford's suggestion, he designated these particles as "neutrons" and concluded his exposition of their discovery with the following words: "The neutron hypothesis gives an immediate and simple explanation of the experimental facts; it is consistent in itself and it throws new light on the problem of nuclear structure."

Thus was the foundation laid for the neutron-proton model of the nucleus in 1932. That same year, Werner Heisenberg, drawing on an analogy with chemical bonding, put forward his theory of the exchange nature of binding forces; although his first (incorrect) surmise was that *neutrino* exchange was responsible for the strong internucleon forces, this was subsequently emended to *meson exchange*, putting the n-p-(π) model on a sound footing. Another contribution by Heisenberg, also in 1932, lay in the assignment of *isospin* $I = \frac{1}{2}$ to the n-p nucleon doublet, treating the neutron and the proton as merely different isospin states, with respective "projected third components of isospin" $I_3 = -\frac{1}{2}$ and $+\frac{1}{2}$, of the basic "nucleon." Thus were sown the seeds of particle classification schemes, brought to their full flower in more recent years.

Properties As well as having the mass stated above, the neutron has a spin $s = \frac{1}{2}$ (hence is subject to the Pauli exclusion principle in its quantum states, and is governed by Fermi-Dirac statistics, therefore being termed a *fermion*, as opposed to a particle that has *integer* spin and so obeys Bose-Einstein statistics, termed a *boson*). As a member of the nucleon family it has isospin $I = \frac{1}{2}$ (and, by convention, $I_3 = -\frac{1}{2}$), positive intrinsic parity ($\pi = +$), and baryon number $B = +1$ (the antineutron has $B = -1$). Its strangeness quantum number is zero ($S = 0$: it is not a so-called "strange" particle), as is also its lepton number ($L = 0$). Its hypercharge quantum number is $Y = +1$. Its net electric charge is zero ($Q = 0$, or at any rate, less than 10^{-21} of the magnitude of the electron's unit charge). As a baryon, it is composed of three quarks, arranged in the *udd* ("up-down-down") combination. Multineutron states or particles, such as ^{2}n, ^{3}n, or ^{4}n, do not exist. There is only one known decay mode of the neutron, namely the ($pe\bar{\nu}$) β-decay mode, with a mean half-life of $T_{1/2} = 10.7 \pm 0.1$ min and a maximum energy release per decay event of $E_{\beta\,\text{max}} = 0.78245 \pm 0.00007$ MeV. The probability (branching fraction) for the charge-non-conserving, but baryon-conserving and lepton-conserving, decay mode ($p\nu\bar{\nu}$) has been found to be less than 9×10^{-24}.

A property of the neutron that has particular significance is its anomalous magnetic moment, first accurately measured by Luis Alvarez and Felix Bloch.[16] The presently accepted value of this quantity,[17] expressed in units of the nuclear magneton ($\mu_N \equiv eh/4\pi m_p c = 3.1524515 \times 10^{-18}$ MeV/gauss, where c is the vacuum velocity of light) is

$$\mu_n = -1.91304184 \pm 0.00000088.$$

The negative sign implies that the magnetic-moment vector points in a direction opposite to that of the spin (i.e., indicates the magnetic polarity), while the nonzero (anomalous) value implies the existence of inner circulating positive and negative, self-cancelling electric charges in the interior of the neutron. Also of significance is the electric dipole moment (also termed the El moment, or the EDM) of the neutron, which has to be zero if the time-reversal invariance is to hold. The experimentally determined value is consistent with zero; in the conventional units of $10^{-23} e \cdot$ cm, it is[18]

$$D_n = 0.04 \pm 0.15,$$

which implies that, within 90% confidence limits, $D_n/e < 0.3 \times 10^{-23}$ cm. An estimate[19] of this value in Steven Weinberg's model of "CP violation" (akin to time-reversal violation) invoking contributions from strange quarks yielded the result $D_n/e \simeq 0.07 \times 10^{-23}$ cm, which is very close to the current experimental limit (see under "Ultracold Neutrons" below).

An illustrative instance of its usefulness as a probe of interaction characteristics, apart from its strong binding interaction and weak β-decay interaction and, indeed, from its participation in the intermediate-strength electromagnetic interaction during neutron-induced radiative capture (n, γ) reactions, is exemplified by its having been chosen[20,21] as the first massive non-self-conjugate elementary particle to test gravitation at the fundamental microscopic level: a well-collimated beam of low-energy neutrons sent along a lengthy horizontal evacuated pipe was observed to evince the expected downward deflection as a function of horizontal distance. Here, neutrons were used in order to avoid electromagnetic complications and because their path could be well determined by Bragg reflection to high precision. The successful outcome prompted the proposal to undertake an analogous long-flight-path experiment with antineutrons[22,23] as a means of testing the effect of gravity on antimatter at very high sensitivity (a 1-MeV beam of antineutrons traversing a horizonal distance of 1 km would be expected to fall a vertical distance of only 2.5×10^{-8} m). More recently, however, an alternative arrangement using antiprotons (and, for comparison, protons) in a very long drift tube, avoiding some of the inherent difficulties of the antineutron experiment, has been suggested.[24]

Details of other neutron properties and numerical data have been given in handbooks,[25] reviews of particle properties,[26] and compilations of measured data.[27] For neutron-induced reaction data, especial attention is drawn to the Evaluated Nuclear Data Files, the ENDF bank, established and maintained by the National Neutron Cross Section Center at the Brookhaven National Laboratory, Upton, NY 11973.

Production   Neither neutrons nor protons are emitted spontaneously in the course of nuclear radioactive decay, which is confined solely to α, β, or γ-emission. Free neutrons must therefore be produced artificially, through nuclear reaction processes which may be induced by charged particles (α, d, p principally), γ-rays ("photoneutron production"), or other neutrons (as in fission chain reactions). A useful more recent alternative is the spontaneous-fission source, containing a nuclide such as plutonium-240 (^{240}Pu) or californium-252 (^{252}Cf) which has a reasonably long half-life ($T_{1/2}$ = 6760 yr or 2.55 yr, respectively) and, in undergoing spontaneous fission produces fairly copious amounts of neutrons among the disintegration products (the more intense source, ^{252}Cf, produces 3 × 10^{12} neutrons/sec per gram). Other self-contained neutron sources employ natural α-emitting nuclei such as ^{226}Ra or ^{230}Pu mixed with ^{9}Be to bring about the neutron-producing reaction

$$^{9}\text{Be} + \alpha \rightarrow n + {}^{12}\text{C} + 5.7 \text{ MeV}$$

within a sealed capsule. In polonium-beryllium (Po-Be) sources, the α particles emanate from ^{210}Po (which decays to stable lead, ^{206}Pb, with a half-life $T_{1/2}$ = 0.3789 yr, and hence has less durability). Photoneutron sources employ fairly energetic spontaneous γ-ray emitters such as sodium-24 (^{24}Na) to obtain neutrons from the

$$^{9}\text{Be} + \gamma \rightarrow n + 2\alpha \quad \text{or} \quad {}^{2}\text{H} + \gamma \rightarrow n + {}^{1}\text{H}$$

reactions. In the former, the produced α-particles may initiate additional neutron production from the ^{9}Be content; in the latter, a deuterium target is disintegrated into neutrons and protons.

Nowadays, one of the most common means of producing neutrons is to direct accelerated charged particles, such as deuterons (d) or protons (p), from an accelerator onto a gaseous or solid target (of deuterium, tritium, helium-3, lithium-7, or beryllium-9) to bring about a neutron-producing reaction such as

$$^{2}\text{H} + d \rightarrow n + {}^{3}\text{He} + 3.269 \text{ MeV},$$

"D-D reaction"

$$^{3}\text{H} + d \rightarrow n + {}^{4}\text{He} + 17.590 \text{ MeV},$$

"D-T reaction"

$$^{3}\text{H} + p \rightarrow n + {}^{3}\text{He} - 0.764 \text{ MeV},$$

"P-T reaction"

$$^{7}\text{Li} + p \rightarrow n + {}^{7}\text{Be} - 1.644 \text{ MeV},$$

or

$$^{9}\text{Be} + d \rightarrow n + {}^{10}\text{B} + 4.362 \text{ MeV}.$$

Neutrons from the D-D reaction typically have energies around 2–4 MeV; those from the D-T reaction have energies around 14 MeV (the residual nuclei take up some recoil energy), while neutrons having energies below 1 MeV can be produced from the endothermic reactions having a *negative* energy term on the right-hand side. The D-D and D-T reactions are instances of thermonuclear fusion processes, such as account for energy generation in stars, in the hydrogen bomb, and in controlled thermonuclear reaction (CTR) devices for power production. An overview of some accelerator-based neutron installations and facilities has been given by S. W. Cierjacks,[3] and a survey of the high-flux production of neutrons has been presented by Lawrence Cranberg.[3] Methods for producing intense sources of fast neutrons have been reviewed by Henry Barschall;[28] in particular, the generation of high-intensity ($\gtrsim$ 10^{13} neutron/sec) beams of 14-MeV neutrons for cancer therapy[29] has been the subject of increasing attention over the past decade.[3] Of course, intense neutron fluxes can also be produced in nuclear explosions underground. Neutrons are being constantly produced in the atmosphere as one of the secondary effects of cosmic rays; typically approximately 2 × 10^{-6} neutrons per m^2 are so produced every second in the 1–2 MeV energy range. From pulsed accelerators, such as the Van de Graaff generator, pulsed neutron beams (which can be bunched, using the Mobley principle) can be produced; a text on pulsed-neutron scattering technology has recently been issued.[30] Reactors are, likewise, prolific (continuous) sources of neutrons. However, neutrons of different energies are produced at different locations within the assembly, from which they may be extracted via "canals" or beam tubes; the energy spread of the fission neutrons is considerable and although their actual energy can be reduced by the insertion of "moderator" materials such as graphite (which also acts as a blanketing shield), it is impossible without special ancillary methods to acquire monoenergetic neutron beams from a reactor.

To get well-defined beams of definite energy one may use velocity-selector devices,[8] such as filters (an Sc-Ti combination transmits only 0.002-MeV neutrons, while a ^{56}Fe-Ti combination passes 0.024-MeV beams, and a ^{28}Si filter passes only 0.144-MeV beams) or "choppers" (cadmium-steel cylinders, absorptive cylinders containing one or more nonabsorbing, transparent channels, set into rotation at such a speed as to allow only a preselected velocity to pass through). Alternatively, one may utilize the γ-ray flux of a reactor to produce mono-energetic photoneutrons resonantly by directing it onto a suitable nuclear target material. High-energy, relativistic, neutron beams would, of course, be produced from the bombardment of appropriate targets by high-energy charged-particle beams from a high-energy accelerator.

Detection The methods used for the detection of neutrons depend upon their energy, and in all instances require indirect means of detection, since the absence of an electric charge prevents the neutron from being detectable directly. Such secondary processes include the observation of charged particles or, e.g., radiation, emanating from neutron-induced reactions, the detection of recoil protons or recoil nuclei from collisions, the registration of fission fragments, or measurement of the heat produced in a given neutron-initiated reaction. The requisite instrumentation[31] that responds to these secondary charged particles includes ionization or fission chambers, Geiger or proportional counters, scintillators or solid-state surface-barrier detectors, photographic emulsions, cloud, bubble, or spark chambers, with appropriate shielding. The latter usually takes the form of copper shadow bars, borated paraffin wax (sometimes impregnated with lithium carbonate), lead, etc. Air scattering of neutrons is considerable; electronic means of discrimination and background correction have to be utilized. As an instance of energy-matching in the selection of a detector, fission chambers clad with ^{235}U are sensitive to slow neutrons; those clad with ^{238}U respond to fast neutrons. By modifying the design of a counter, its energy region of best response can be changed: for the detection of slow neutrons, boron counters are frequently employed; by surrounding these with a thick sheath of paraffin wax or other hydrogeneous material to slow down any fast primary neutrons they may, as so-called "long counters," be adapted to respond to fast neutrons. Their response is triggered by the α-particles that ensue from the $^{10}B + n \rightarrow \alpha + {}^7Li + 2.78$ MeV reaction which takes place within the boron-filled counter (usually containing gaseous BF_3 enriched in the ^{10}B isotope). The neutron-activation technique also finds many applications. Since the total neutron energy cannot be directly registered it is customary to adopt time-of-flight spectrometry methods[32,33] in which subnanosecond time resolutions over protracted measuring runs are attainable. High-energy neutron detection methods, constantly being improved, have been discussed by R. T. Siegel.[34]

Categorization Just as it has become customary to distinguish between "hard" and "soft" X-rays on the basis of their energy (i.e., wavelength), so has a nomenclature for different neutron classes come into usage. The principal categories are:

(a) Ultracold neutrons, having energies around 10^{-7} eV, whose velocities are therefore but a few meters/sec, the corresponding temperature being only 10^{-3} K, i.e., one-thousandth of a degree above absolute zero (in Kelvin degrees);

(b) Cold neutrons, having kinetic energies $\lesssim 0.002$ eV;

(c) Thermal neutrons, an important class whose kinetic energy of about 0.025 eV corresponds to velocities that are compatible with the velocity of molecules in thermal motion, i.e., at "room temperature" ($T \simeq 20°C \simeq 293$ K) $v \simeq 2.2$ km/sec;

(d) Epithermal neutrons, having somewhat higher energies, in the region $\gtrsim 0.5$ eV;

(e) Resonance neutrons, another especially important class, having energies in the range of 1–100 eV that correspond to the region of strong (resonant) absorption by nuclei.[35]

Neutrons within the categories (a)–(e) are all considered "slow neutrons." Others are

(f) Intermediate neutrons, having energies from 0.001 to 0.5 MeV;

(g) Fast neutrons, whose energies lie within the 0.5 to 15 MeV range;

(h) Very fast neutrons, from 15 to 50 MeV;

(i) Ultrafast neutrons, from 50 to 10,000 MeV;

(j) The rest, with energies above 10,000 MeV, constitute highly relativistic neutrons.

Ultracold Neutrons Interesting recent developments have opened up a new branch of knowledge and application in neutron physics.[1,3,8,12,36] Ultracold neutrons, whose motion is comparable with human running speed, have the remarkable property of being totally reflected from solid surfaces, which enables them to be isolated within suitable metal (usually, aluminum or copper) or glass containers, or trapped within magnetic storage rings (now constructed with superconducting windings), after traveling along a selective guide tube. Because of their low energy, they can rise only about a meter against the earth's gravity before falling back. A proposal to create and trap them from cold neutrons in refrigerated superfluid helium is currently being tried out.

The wavelength (in meters) of a neutron is related to its energy (in eV) as

$$\lambda = \frac{2.86 \times 10^{-11}}{\sqrt{E}}$$

and hence if E is equal to 10^{-7} eV, the wavelength is $\lambda = 9 \times 10^{-8}$ m (i.e., 900 Ångstrøm units), about two orders of magnitude larger than the spacing between the atoms in solids, which are thus impervious to a wave of ultracold neutrons. In effect, the wave is repelled by the solid, as predicted and demonstrated (for higher-energy neutrons) by Fermi in 1945. The reflection is total, provided the velocity is below a critical value which varies from one material to another, being roughly 6 m/sec for nickel, iron, beryllium, or copper. Ultracold neutrons are defined as those having a total velocity below this critical value for a given material. Thus, v = 6 m/sec corresponds to $E = 2 \times 10^{-7}$ eV, which in turn represents a temperature $T = 2 \times 10^{-3}$ K (two-thousandths of a degree above absolute zero). Though only sparsely available in the neutron flux from reactors, they have been successfully extracted and investigated. An obvious application is in the determination of the mean lifetime of free neutrons to a higher accuracy than has hitherto been feasible. However, unforeseen and as yet still not fully understood problems in maintaining the necessary long containment times have hampered these measurements, although refinement of the isolation and storage arrangements is leading[38] to the desired conditions for the experiment to be completed satisfactorily. Similarly, ultracold neutrons offer an ideal means of establishing finer limits to the value of the electric dipole moment (EDM) of the neutron. The accepted value[18] for D_n was determined at the ultracold-neutron facility situated in the Institut Laue-Langevin in Grenoble (France), while a result with even smaller error limits, namely $D_n/e = (4 \pm 7.5) \times 10^{-25}$ cm, has been obtained[39] at the B. P. Konstantinov Institute (USSR). This indicates that, at the 90% confidence level, $|D_n/e| < 0.16 \times 10^{-23}$ cm.

Because of their comparatively long wavelength (longer than that of electron beams, albeit shorter than that of visible light), ultracold neutrons lend themselves to use in a neutron microscope, complementing the familiar electron microscope, or to examination of interatomic forces and the motions of atoms in solids, through the use of a neutron spectrometer arrangement. Among other potentialities that they present is the possibility for the first time of obtaining a pure neutron target for atomic and nuclear experiments, without the need to employ subtraction procedures to correct for the presence of impurities or other nuclei.

Slow Neutrons In this category, the most important classes are the thermal neutrons (which initiate fission reactions in, e.g., ^{235}U or ^{239}Pu) and resonance neutrons (whose likelihood of being captured by a nucleus in, e.g., a radiative capture process, peaks sharply at certain resonance energies, rising dramatically to a very high, narrow, maximum). Slow-neutron capture in particular is a widely used process for the production of radioactive isotopes from stable nuclides. Whereas the probability for elastic scattering of low-energy neutrons does not change greatly with energy or emergence angle, that for neutron capture, in which radiation or particles are emitted, follows a characteristic diminishing trend that is inversely proportional to the incident neutron velocity: the "$1/v$ law" holds for such exothermic capture

reactions, except at resonance energies, when the probability may increase a hundredfold or more over a narrow incident energy range. For example, radiative capture of neutrons on indium-115(^{115}In) occurs reasonantly at E_n = 1.457 eV; the capture probability at just this energy is about 180 times larger than when measured at about $\frac{1}{2}$ eV on either side of this resonance value. For thermal neutrons (E_n = 0.025 eV), cadmium has a high capture probability: natural cadmium is, to 12.3%, comprised of the isotope ^{113}Cd, which displays a broad, powerful resonance for the absorption of neutrons about the energy E_n = 0.176 eV, overlapping with the thermal region. For this reason, cadmium is used in the fabrication of choppers and in the safety control rods of nuclear reactors.

Neutrons in Fission and Parity Violation The general subject of fission is too broad for coverage here;[1,3,40–43] this Encyclopedia discusses it elsewhere (see FISSION). Most reactors operate through fission induced in a core containing an overcritical amount of ^{235}U or ^{239}Pu, whose break-up into fission fragments (medium-heavy nuclei, more neutrons, and radiation) when bombarded by thermal neutrons (i.e., neutrons slowed down by a graphite or deuterium "moderator" to "thermal" speeds) liberates about 200 MeV of energy per fission event. Because some 2.5 neutrons are released from each fission event initiated by a single thermal neutron, capable of bringing about further fission, a "chain reaction" can occur readily, triggered by this multiplication of neutrons. Through the use of absorptive cadmium control rods, the controlled operation of a fission reactor can be maintained. In so-called "breeder reactors," based upon a plutonium or thorium cycle, as much or more fissionable material is produced in the course of operation as is used up from the primary fuel; hence a self-sustaining power-producing plant, such as that in the Superphénix installation in France can be devised for the more conserving use of nuclear fuel resources.

Parity violation, i.e., the inequivalence of coordinate-system "handedness" in the description of nuclear processes, evoked much attention when first identified in weak interactions in 1956. Although the weak interactions were found to manifest maximal parity violation, it was not considered to play any role in strong interactions. Recently, however, some indications of its presence even in strong-interaction phenomena has come to light. A mechanism for parity violation in thermal-neutron-induced fission[44] has been worked out by O. P. Sushkov and V. V. Flambaum.[45] Similarly, an explanation for parity violation in the radiative capture of polarized neutrons has been provided by D. F. Zaretskii and V. K. Sirotkin,[46] while that for neutron scattering at threshold (e.g., on ^{124}Sn) has been discussed by G. Karl and D.

Tadic.[47] The first clear evidence for parity-nonconserving spin-rotation of neutrons passing through matter has been observed in tin isotopes (^{124}Sn, ^{117}Sn, and natural Sn) by M. Forte et al.[48]

Fast Neutrons Above the threshold energy, a wide variety of interaction processes is available for neutrons to undergo. These include elastic and inelastic (i.e., excitation) scattering from nuclei, from which the neutrons emerge with effectively unchanged, or diminished, energy, multineutron emission, in which more than one neutron is ejected by each incident projectile neutron, particle or multiparticle release from nuclei following neutron capture, radiative capture (whose likelihood diminishes with energy), even fast-neutron fission, induced, e.g., in ^{232}Th or ^{238}U. A major branch of nuclear-reaction studies deals with neutron-induced processes which play a vital role in nuclear physics and astrophysics.

Ultrafast Neutrons Already at E_n = 6.32 MeV the (classical) kinetic energy of a neutron differs by 1% from that calculated relativistically; above about 50 MeV the neutrons are distinctly relativistic. At these high energies, multiple particle emission is, with increasing frequency, found to occur when neutrons in the 50–500 MeV range are directed onto nuclear targets. The mechanism is in general complicated: in an initial step, knock-on nucleons are likely to be ejected directly, followed by the redistribution of the surplus excitation energy in the residual nucleus, which deexcites by the emission of γ-rays and one or more additional nuclear particles. Spallation is often caused by ultrafast neutrons.

Neutron Polarization As spin-$\frac{1}{2}$ particles with a nonzero spin magnetic moment, neutrons can be polarized,[16] i.e., oriented in the direction of an applied external magnetic field. This magnetic moment may, at least in part, account for the small attractive force that acts between a neutron and an electron, the first evidence of which Fermi discovered in 1947. Thus, polarized neutron beams can be produced, and used in the derivation of basic information on nuclear forces or other characteristics.[1,3–5,12,49–51] Several methods exist for the production of polarized neutron beams: partial polarization can be induced by the passage of a neutron beam through a slab of magnetized iron (the "Bloch effect," first established in 1936/37 by Felix Bloch). Neutrons can be partially polarized by scattering: the use of double-scattering experiments (in which the first scattering causes the polarization, and the second scattering analyzes it) is particularly noteworthy. The neutrons produced in reactions can also be partly or almost wholly polarized.[50] The β-decay of polarized neutrons has been used to test the time-reversal invariance of the weak interaction; polarized neutrons have also been employed in other investigations of time-reversal invariance in nuclear reactions.

The values of the neutron's electric and magnetic polarizabilities have not yet been established definitively;[1] theoretical estimates of the electric polarizability indicate typically α_n = 8.5×10^{-49} m^3 but experimental values are inconclusive, whereas up to the present the magnetic polarizability has not been measured at all.

Condensed Matter (Neutron Stars) The extreme densities associated with nuclear matter, averaging about 10^{17} kg/m^3, are encountered in practice only in the highly compressed matter of neutron stars, in which the immense inward gravitational attractive force effectively "squeezes" atomic electrons onto nuclear protons to produce a highly condensed "neutron fluid" as the stellar material. Such neutron stars[52,53] constitute the end product of a stellar evolution sequence from protostar to main-sequence object, red giant and variable phase, thence to the white-dwarf stage and, via a supernova explosion if the mass transcends the Chandrasekhar limit of 1.44 solar masses, to the neutron-star condition, having a radius of but a few kilometers and rotating rapidly, as a "pulsar" emitting radio signals.

Neutron Diffraction and Interferometry Over the energy interval E_n = 0.001–10 eV, the wavelength of neutrons, according to the formula above, ranges from about λ = 0.1 to 10 Å. Such a wavelength is well suited to the examination of the structure of materials, akin to investigations pursued with X-ray (Bragg) diffraction. Neutron diffraction on the atomic lattice planes of a crystal can occur by way of reflection or transmission; many designs of neutron spectrometers have been employed in a wide variety of solid-state and material studies, making use of interferometry.[1,3,14,54,55]

Other Applications Neutron applications are too numerous to list exhaustively.[1,3,56] By way of example, one might make mention of neutron radiography in studies of paintings, in metal science and in medical investigations. Neutrons have been used extensively for other biomedical applications, including biotherapy, in the production of radioisotopes and infrared detectors or the enhancement of computer memory devices or food preservation, in many analytical applications, including activation analysis, and for such diverse purposes as assaying, moisture gauging and well-logging, forensic and archaeological investigations, solid-state research and chemical molecular studies. Their use to such ends in benefitting mankind "has been both beneficial and benevolent"[56] and has far outweighed the potential misuse in explosive fission devices.

ERIC SHELDON

References

1. Schofield, P. (Ed.), "The Neutron and its Applications 1982," Proceedings of an International Conference to commemorate the 50th anniver-

sary of the discovery of the neutron, Cambridge (UK), Sept. 13–17, 1982; Conference Series Number 64, Bristol and London, Institute of Physics, 1983.

2. Chadwick, J., *Proc. Roy. Soc. London Ser. A* 136, (1932); reproduced in "Foundations of Modern Physics," (R. T. Beyer, Ed.), p. 5, New York, Dover, 1949.

3. Sheldon, E., (Ed.), "Proceedings of the International Conference on the Interactions of Neutrons with Nuclei," Lowell, July 6–9, 1976, Vols. I & II, ERDA Conference Report CONF-760715-P1 & P2, Oak Ridge, TIC-ERDA, 1976.

4. Nève de Mévergnies, M., Van Assche, P., and Vervier, J. (Eds.), "Nuclear Structure Study with Neutrons, Proceedings of an International Conference, Antwerp, 19–23 July, 1965," Amsterdam, North-Holland Publ. Co., 1966.

5. Erö, J., and Szücs, J. (Eds.), "Nuclear Structure Study with Neutrons," Proceedings of the International Conference, Budapest, 31 July–5 Aug., 1972, London and New York, Plenum Press, 1974.

6. Feld, B. T., in "Experimental Nuclear Physics," (E. Segrè, Ed.), Vol. 2, p. 209, New York, John Wiley & Sons, 1953.

7. Amaldi, E., in "Encyclopedia of Physics/Handbuch der Physik," (S. Flügge, Ed.), Vol. 38/2, p. 1, Berlin, Springer-Verlag, 1959.

8. Feshbach, H., and Sheldon, E., *Phys. Today* 30(2), 40 (1977).

9. Curtiss, L. F., "Introduction to Neutron Physics," New York, Van Nostrand Reinhold, 1958.

10. Marion, J. B., and Fowler, J. L., (Eds.), "Fast Neutron Physics," Part I: Techniques, 1960; Part II: Experiments and Theory, 1963; New York, Wiley-Interscience.

11. Beckurts, K. H., and Wirtz, K., "Neutron Physics," Berlin, Springer-Verlag, 1964 (completely revised English edition of "Elementare Neutronenphysik").

12. Gurevich, I. I., and Tarasov, L. V., "Low-Energy Neutron Physics," (R. I. Sharp and S. Chomet, Eds.), Amsterdam, North-Holland Publ. Co., 1968.

13. Bacon, G. E., "Neutron Physics," London, Wykeham Publications, 1969.

14. Dachs, H. (Ed.), "Neutron Diffraction," Berlin and New York, Springer-Verlag, 1978.

15. Cork, B., Lambertson, G. R., Piccioni, O., and Wenzel, W. A., *Phys. Rev.* 104, 1193(L) (1956).

16. Alvarez, L. W., and Bloch, F. *Phys. Rev.* 57, 111 (1940).

17. Greene, G. L., Ramsey, N. F., Mampe, W., Pendlebury, J. M., Smith, K., Dress, W. D., Miller, P. D., and Perrin, P., *Phys. Rev.* D20, 2139 (1979).

18. Dress, W. B., Miller, P. D., Pendlebury, J. M., Perrin, P., and Ramsey, N. F., *Phys. Rev.* D15, 9 (1977).

19. Zhitnitskii, A. R., and Khriplovich, I. B., *Sov. J. Nucl. Phys.* 34(1), 95 (1981).

20. McReynolds, A. W., *Phys. Rev.* 83, 172 & 233 (1951).

21. Dabbs, J. W. T., Harvey, J. A., Paya, D., and Horstmann, H., *Phys. Rev.* 139, B756 (1965).

22. Kalogeropoulos, T., in "Proceedings of the Workshop on Nuclear and Particle Physics at Energies up to 31 GeV," Los Alamos National Laboratory report LA-8775-C, p. 499 (1981).

23. Brando, T., Fainberg, A., Kalogeropoulos, T., Michael, D., and Tzanakos, G., *Nucl. Instrum. Methods* 180, 461 (1981).

24. Goldman, T., and Nieto, M. M., *Phys. Lett.* 112B(6), 437 (1982).

25. Goldberg, M. D., and Harvey, J. A., "Neutrons," Section 8f, pp. 8-218 to 8-253, in the "American Institute of Physics Handbook," 3rd Ed. (D. E. Gray, Ed.), New York, McGraw-Hill, 1972.

26. Particle Data Group (M. Roos, F. C. Porter, M. Aguilar-Benitez, L. Montanet, Ch. Walck, R. L. Crawford, R. L. Kelly, A. Rittenberg, T. G. Trippe, C. G. Wohl, G. P. Yost, T. Shimada, M. J. Losty, G. P. Gopal, R. E. Hendrick, R. E. Shrock, R. Frosch, L. D. Roper, and B. Armstrong (Technical Associate)), *Phys. Lett.* 111B, 1 (1982).

27. References in Ref. 25, p. 8-253.

28. Barschall, H. H., *Ann. Rev. Nucl. Sci.* 28, 207 (1978).

29. Walko, R. J., Bacon, F. M., Bickes, R. W. Jr., Cowgill, D. F. Jr., Boers, J. E. Jr., Riedel, A. A. Jr., and O'Hagan, J. B. Jr., *J. Vac. Sci. Technol.* 18(3), 975 (1981).

30. Windsor, C. G., "Pulsed Neutron Scattering," London, Taylor & Francis, 1981.

31. Barschall, H. H., in "Encyclopedia of Physics/Handbuch der Physik," (S. Flügge, Ed.), Vol. 45, p. 437, Berlin, Springer-Verlag, 1958.

32. Cranberg, L., and Rosen, L., in "Nuclear Spectroscopy," (F. Ajzenberg-Selove, Ed.), Part A, p. 358, New York, Academic Press, 1960; Cranberg, L., *Scientific American* 210(3), 79 (1964).

33. Finckh, E., in "Nuclear Spectroscopy and Reactions," (J. Cerny, Ed.), Part B, p. 573, New York, Academic Press, 1974.

34. Siegel, R. T., in "Encyclopedia of Physics/Handbuch der Physik," (S. Flügge, Ed.), Vol. 45, p. 487 (1958).

35. Rainwater, J., in "Encyclopedia of Physics/Handbuch der Physik," (S. Flügge, Ed.), Vol. 40, p. 373 (1957).

36. Luschikov, V. I., *Phys. Today* 30(6), 42 (1977).

37. Golub, R., Mampe, W., Pendlebury, J. M., and Ageron, P., *Scientific American* 240(6), 134 (1979).

38. Kosvintsev, Yu. Yu., Kushnir, Yu. A., Morozov, V. I., and Terekhov, G. I., *JEPT Lett.* 31(4), 236 (1980).

39. Altarev, I. S., Borisov, Yu. V., Brandin, A. B., Egorov, A. I., Ezhov, V. F., Ivanov, S. N., Lobashev, V. M., Nazarenko, V. A., Porsev, G. D., Ryabov, V. L. Serebov, A. P., and Tal'daev, R. R., *JEPT Lett.* 29(12), 730 (1979); see also Altarev, I. S., Borisov, Y. V., Borovikova, N. V., Brandin, A. B., Egorov, A. I., Ezhof, V. F., Ivanov, S. N., Lobashev, V. M., Nazarenko, V. A., Porsev, G. D., Ryabov, V. L., Serebov, A. P., and Taldaev, R. R., *Nucl. Phys.* A341, 269 (1980) and *Phys. Lett.* 102B, 13 (1981), and N. F. Ramsey, *Rep. Prog. Phys.* 45, 96 (1982).

40. Loveland, W. D., "Nuclear Fission," Section 8g, p. 8-253ff, in the "American Institute of Physics

Handbook," 3rd Ed. (D. E. Gray, Ed.), New York, McGraw-Hill, 1972.

41. Vanderbosch, R. and Huizenga, J. R., "Nuclear Fission," New York and London, Academic Press, 1973.

42. Michaudon, A., "Nuclear Fission," in *Adv. in Nucl. Phys.* **6**, 1 (1973).

43. Michaudon, A., *Phys. Today* **31**(1), 23 (1978).

44. Danilyan, G. V., Vodennikov, B. D., Dronyaev, V. P., Novitskii, V. V., Pavlov, V. S., and Borovlev, S. P., *Sov. J. Nucl. Phys.* **27**(1), 21 (1978).

45. Sushkov, O. P., and Flambaum, V. V., *Sov. J. Nucl. Phys.* **33**(1), 31 (1981).

46. Zaretskii, D. F. and Sirotkin, V. K., *Sov. J. Nucl. Phys.* **32**(1), 54 (1980).

47. Karl, G., and Tadic, D., *Phys. Rev.* **C20**(5), 1959 (1979).

48. Forte, M., Heckel, B. R., Ramsey, N. F., Green, K., Green, G. L., Byrne, J., and Pendlebury, J. M., *Phys. Rev. Lett.* **45**(26), 2088 (1980).

49. Breit, G. and McIntosh, J. S., in "Encyclopedia of Physics/Handbuch der Physik," (S. Flügge, Ed.), Vol. 41/1, p. 466, Berlin, Springer-Verlag, 1958.

50. Walter, R. L., "Polarization Phenomena in Nuclear Reactions Observed in Neutron Studies," in "Nuclear Spectroscopy and Reactions," Part B, (J. Cerny, Ed.), Section VI.C, p. 636, New York and London, Academic Press, 1974.

51. Alfimenkov, V. P. Pikel'ner, L. B., and Sharapov, E. I., *Sov. J. Part. Nucl.* **11**(2), 154 (1980).

52. Irvine, J. M., "Neutron Stars." Oxford, Clarendon Press, 1978.

53. Sheldon, E., *Nukleonika* **23**(12), 1091 (1978).

54. Bacon, G. E., "Neutron Diffraction," 2nd Ed., Oxford, Clarendon Press, 1962.

55. Werner, S. A., *Phys. Today* **33**(12), 24 (1980).

56. Bromley, D. A., *Phys. Today* **36**(12), 30 (1983).

Cross-references: COLLISIONS OF PARTICLES, CROSS SECTIONS AND STOPPING POWER, ELECTRON, FERMI-DIRAC STATISTICS AND FERMIONS, FISSION, NEUTRON ACTIVATION ANALYSIS, NEUTRON DIFFRACTION, NUCLEAR REACTIONS, NUCLEAR REACTORS, PARITY, PROTON, QUARKS, RADIOACTIVITY, RESONANCE.

NEUTRON ACTIVATION ANALYSIS

Neutron activation analysis (NAA) is a powerful and sensitive method of elemental analysis based upon the quantitative detection of radionuclides produced in samples via nuclear reactions resulting from the high-flux thermal-neutron bombardment of the samples. There are also other types of nuclear activation analysis in which the bombarding particles are energetic charged particles (e.g., protons from a cyclotron, energetic photons (e.g., bremsstrahlung photons from an electron linear accelerator), or fast neutrons (e.g, 14 MeV neutrons from a deuteron accelerator). However, activation with the high fluxes of thermal neutrons produced by a research-type nuclear reactor provides the best sensitivities of detection for most elements, and is the most fully developed and most widely used form of the nuclear activation analysis method—hence this discussion will be limited to NAA, particularly with reactor thermal neutrons.

Neutron Sources Lower fluxes of thermal neutrons (but still useful for teaching purposes, and for the determination of major, minor, and a few trace elements) can be produced with commercially available isotopic sources—(α, n) sources or ^{252}Cf spontaneous-fission neutron sources, or with modest-sized Cockcroft-Walton or Van de Graaff accelerators. Since all sources of neutrons produce neutrons of appreciable kinetic energies, they must be slowed down to thermal velocities (a mean kinetic energy of about 0.025 eV, at 20°C) by use of a suitable moderator—usually water or some other hydrogenous material. Where activation by fast neutrons is desired, no moderator is used.

Far higher fluxes of thermal neutrons are produced in typical research-type nuclear reactors. For reactors operating at steady power levels of 0.1, 1, or 10 megawatts, thermal-neutron fluxes of, respectively, about 10^{12}, 10^{13}, or 10^{14} n cm^{-2} s^{-1} are available for the activation of samples. Most such reactors are of the pool type, in which the reactor core is located at the bottom of a deep pool of circulating high-purity water. The water serves as a transparent coolant, neutron moderator, and biological shield. The neutrons produced by the thermal-neutron fission of ^{235}U have high initial energies (ranging from about 0.5 MeV up to many MeV), but collisions with water protons rapidly slow them down to thermal energies. Where irradiation with only epithermal plus fast neutrons is desired, samples are enclosed in a thin cadmium container—the Cd serving to remove almost quantitatively all neutrons with energies less than about 0.4 eV. Using a fast-transfer pneumatic tube (for production and detection of induced radionuclides with half lives in the range of seconds to minutes), samples and standards are activated and rapidly counted one at a time. Where the interest is in the production and detection of longer-lived induced activities (half lives of an hour or longer), longer irradiations of many samples and standards are carried out, simultaneously, followed by sequential counting of the activated specimens at selected longer decay times.

Types of Neutron Reactions With only a few exceptions, thermal neutrons interact with atomic nuclei by only the (n, γ) reaction, producing a nucleus of the same atomic number (Z) as the target nucleus, but now one unit larger in mass number (A). In most cases (those of use in the conventional radioactive-decay NAA method), such (n, γ) products are radionuclides. In most instances, being neutron-rich radionuclides, these decay by β^- emission, and in all but a few cases the β^- emission is accompanied by the emission of gamma-ray photons of one or more characteristic, sharply defined energies. In (n, γ) reactions, the γ refers to the prompt gamma rays emitted immediately after the neutron capture, as the highly excited product nucleus cascades down to its ground state. These prompt gammas are quite different in their energies from the subsequently emitted radioactive-decay gamma rays. Prompt gammas are also used, somewhat, in "prompt-gamma NAA," with a sample being concurrently bombarded with a much lower beam-tube flux of thermal neutrons and counted on a γ-ray spectrometer. For almost all elements prompt-gamma NAA is much less sensitive than radioactive-decay NAA, and hence this discussion will be limited to the latter.

With sample bombardment by fast neutrons (e.g., 1–15 MeV neutrons), the cross sections for (n, γ) reactions are very small, whereas the cross sections for many (n, n'), (n, p), (n, α), and $(n, 2n)$ reactions become appreciable—ranging from millibarns to about 1 barn (1 barn is a reaction cross section of 10^{-24} cm² per nucleus). For most target nuclei, fast-neutron reaction cross sections are 1, 2, or more orders of magnitude smaller than thermal-neutron (n, γ) cross sections.

Of these five main kinds of neutron reactions, the (n, γ) reaction is always exoergic (usually by 5–15 MeV), has no Coulomb barrier or threshold energy, and produces a neutron-rich product of the same element, one unit larger in mass number. The (n, n') neutron inelastic-scattering reaction is always somewhat endoergic and hence has a threshold energy, has no Coulomb barrier, and produces a metastable isomer of the target nucleus (same Z and A). Metastable isomers usually decay by isomeric transition, emitting characteristic gamma radiation. The $(n, 2n)$ reaction is always considerably endoergic, has no Coulomb barrier, and produces a neutron-deficient product of the same element (same Z) but one unit lower in mass number. Neutron-deficient radionuclides usually decay by orbital electron capture (EC), by β^+ emission (if energetically possible), or both. (n, p) and (n, α) reactions may be either endoergic or exoergic, but both exhibit effective threshold energies because of their Coulomb barriers. The (n, p) reaction forms a neutron-rich product nucleus of the element one unit lower in Z than the target nucleus, but of unchanged mass number. The (n, α) reaction generally produces a neutron-rich product nucleus—that of the element 2 units lower in Z than the target nucleus, and 3 units lower in mass number.

Theory of the NAA Method When a sample containing a particular kind of target nucleus (of a particular Z and A) is bombarded with a steady flux ϕ of thermal neutrons, the steady rate of formation of (n, γ) product nuclei from it is simply:

$$dN^*/dt = -dN/dt = N\phi\sigma, \qquad (1)$$

in which N is the number of such target nuclei in the sample, N^* is the number of (n, γ) product radionuclide nuclei it produces, and σ is the thermal-neutron (n, γ) cross section of the target nucleus. If the product nuclei are radioactive, they will be decaying, even during the irradiation, by the usual first-order radioactive decay process:

$$\frac{-dN^*}{dt} = \lambda N^* = \frac{\ln 2}{T} N^* = \frac{0.69315}{T} N^*,$$

$$(2)$$

in which λ is the radioactive-decay rate constant of the product radionuclide, and T is the half-life of the radionuclide. The net rate of formation of the radionuclide during the irradiation is thus its steady rate of formation ($N\phi\sigma$) minus its ever-increasing rate of decay (λN^*):

$$\text{net } dN^*/dt = N\phi\sigma - \lambda N^*. \qquad (3)$$

This equation, when integrated by means of an integrating factor, yields the basic equation of NAA:

$$A_0 = (-dN^*/dt)_0 = N\phi\sigma(1 - e^{-0.69315 t_i/T}),$$

$$(4)$$

in which A_0 represents the radionuclide disintegration rate (in disintegrations per second, dps) at the end of the irradiation (i.e., at zero decay time), and t_i is the duration of the irradiation period, expressed in the same units of time as the half-life, T.

The N term may be expressed in terms of the weight of that element present in the sample, w (the sought-for unknown in the analysis of a sample for that element):

$$N = \frac{waN_A}{AW}, \qquad (5)$$

in which w is the weight of the element present (in grams), a is the fractional isotopic abundance of the target stable nuclide (stable isotope) among the various stable isotopes of that element (i.e., $a =$ exactly 1 for the 20 elements that are monoisotopic in nature), N_A is Avogadro's number (6.022×10^{23} atoms per gram atom), and AW is the regular chemical atomic weight of the element, in atomic mass units (i.e., the abundance-weighted mean of the atomic masses of its various stable isotopes).

The parenthetical term in Eq. (4),

$$1 - e^{-0.69315 t_i/T},$$

is called the saturation term S, the only time-dependent term in Eq. (4), other than A_0. It is a dimensionless term, ranging only from 0 (for $t_i/T = 0$) to 1 (for $t_i/T = \infty$). For t_i/T values of 0, 1, 2, 3, 4, 5, ..., S acquires values of, respectively, 0, $\frac{1}{2}$, $\frac{3}{4}$, $\frac{7}{8}$, $\frac{15}{16}$, $\frac{31}{32}$, ..., i.e., it asymptotically approaches the limiting value of 1 with increasing t_i/T (at $t_i/T = 10$, $S = 1023/1024$, or 0.999). The saturation term is very important in NAA work, since it enables one to accentuate the production and detection of short-lived species by employing very short irradiation periods, followed rapidly by a short counting period. To accentuate the production and detection of medium-lived species, one employs a longer irradiation period, followed by a decay period long enough for the shorter-lived species to decay out, prior to a somewhat longer counting period—and similarly even longer irradiation, decay, and counting periods to accentuate the production and detection of even longer-lived species.

All of the above equations apply to each radionuclide being produced in the thermal-neutron irradiation of a multi-element sample, independently of one another. Once the irradiation is ended, each radionuclide present decays away according to its particular half-life (Eq. (2)). When Eq. (2) is integrated, one obtains the equation for the number of radionuclei of a given type (N^*) present after a decay period t, compared with their number (N_0^*) present at the end of the irradiation:

$$N^* = N_0^* \, e^{-0.69315t/T}. \qquad (6)$$

Since $-dN^*/dt = \lambda N^*$, and $(-dN^*/dt)_0 = \lambda N_0^*$, this equation can also be expressed in terms of disintegration rates (A) instead of in terms of number of nuclei present:

$$A = A_0 e^{-0.69315t/T}. \qquad (7)$$

In practice, one does not measure absolute disintegration rates, but instead (usually) one measures the photopeak counting rate (PPcps) of the principal decay gamma-ray photon emitted by the radionuclide. For counting with a particular γ-ray detector, at a particular counting geometry, PPcps $= \epsilon A$, where ϵ is the overall photopeak detection efficiency of the detector, at that geometry, for gamma rays of that energy (also taking into account the fraction of the decays of that radionuclide that produce a gamma ray of that energy). Thus, Eq. (7) can also be expressed in terms of PPcps:

$$\text{PPcps} = (\text{PPcps})_0 \, e^{-0.69315t/T}, \qquad (8)$$

or, integrating over the counting period t_c, since the number of net photopeak counts NPPC accumulated during the counting period is NPPC $= [(\text{PPcps at SOC})T/0.69315] \, (1 - e^{-0.69315t_c/T})$:

$$\text{NPPC} = (\text{NPPC})_0 \, e^{-0.69315t/T}, \qquad (9)$$

in which SOC refers to the PPcps at the start of the counting period.

In practice, one seldom uses an "absolute" method in NAA work, but instead employs a "comparator" method, in which samples and standards of the elements of interest are activated and counted identically. Then, if the basic equations above are written for a sample and for a standard of the element of interest, and one equation is divided by the other, all of the parameters that are common to both (i.e., ϕ, σ, t_i, T, S, a, N_A, AW, ϵ, and t_c) cancel out, leaving only the very simple comparator equation (for the same decay time and same counting period):

$$\frac{\text{NPPC of sample peak}}{\text{NPPC of standard peak}} = \frac{w \text{ in sample}}{w \text{ in standard}}. \qquad (10)$$

Since sample and standard are counted at somewhat different decay times (on the same detector), one must be corrected to the decay time of the other, via a slight modification of Eq. (7). The main advantages of the comparator method over the "absolute" method are (1) it is much simpler, and (2) it is more accurate, since it does not depend upon accurate knowledge of the parameters common to both sample and standard, many of which are not accurately known (e.g., σ values) or are difficult to measure accurately (e.g., ϕ and ϵ values).

Detection Sensitivities for Various Elements Because of the wide range of each of the variables involved in NAA, among the different elements and their various (n, γ) products, the lower limits of detection (LOD's) of the different elements cover a wide range, for any given set of conditions. The LOD's for 68 elements (in the absence of significant levels of other induced activities) are summarized in Table 1. These are LOD's calculated for a fairly typical set of conditions: $\phi = 10^{13}$ n cm^{-2} s^{-1} (thermal neutrons), $t_i \leqslant 5$ hours, $t_{\text{decay}} = 0$, $t_c \leqslant 100$ minutes with a 40 cm^3 Ge(Li) detector at a distance of 2 cm. In each case, the LOD is based upon the largest photopeak of the radionuclide formed to the largest extent by the element, and is based upon a defined minimum detectable (to a reasonable relative precision, i.e., about $\pm 20\%$ of the value) number of NPPC, 30.

Table 1 includes all the nonradioactive or nearly nonradioactive (i.e., K, Th, and U) elements for which the NAA method is highly or fairly sensitive, except the inert-gas elements (for which NAA is also very sensitive, except for He)—since the noble gases are not significantly present in most solid or liquid samples. Noticeably absent from the table are the lowest atomic number elements (H, Li, Be, B, C, N, and O), which are scarcely activated by thermal neutrons, and the elements P, Tl, Bi, which

TABLE 1.

LOD (μg)	Elements
10^{-7}–10^{-6}	In, Eu, Dy, Ho
10^{-6}–10^{-5}	Mn, Sm, Au, Rh, Lu, Re, Ir
10^{-5}–10^{-4}	Co, Cu, Ga, As, I, Cs, La, Er, W, Hg, U, Na, V, Br, Ru, Pd, Sb, Yb, Th
10^{-4}–10^{-3}	Sc, Ge, Sr, Te, Ba, Nd, Ta, Cl, Se, Cd, Gd, Tb, Tm, Hf, Pt
10^{-3}–10^{-2}	Al, Zn, Mo, Ag, Sn, Ce, Os, K, Ti, Cr, Ni, Rb, Y, Pr
10^{-2}–10^{-1}	Mg, Zr
10^{-1}–1	F, Ca, Nb
1–10	Fe, Si
10–100	S, Pb

form only pure β^- emitters, with no accompanying gamma radiation.

Forms of the Method There are two forms of the NAA method: (1) the purely instrumental, nondestructive form, based upon gamma-ray spectrometry only, and (2) the radiochemical separation, destructive, form which utilizes post-irradiation radiochemical separations with carriers, of the induced activities of interest. Nowadays, instrumental NAA usually involves the counting of activated samples and standards, at selected decay times, with a high-resolution Ge(Li) or intrinsic Ge γ-ray detector, coupled to a multichannel pulse-height analyzer (typically of 4096 analysis and storage channels). The instrumental form of the method has the advantages of detecting and measuring many elements simultaneously, of not destroying the sample, and of requiring a minimum of the analyst's time. However, in some kinds of samples high levels of other induced activities can raise the LOD for an element of interest by one or two orders of magnitude above its interference-free LOD (by producing a very high Compton continuum level in the pulse-height spectrum, superimposed upon which is the photopeak of interest). In such cases, one may have to resort to the more time-consuming, sample-destructive radiochemical-separation technique, prior to counting.

Sample Sizes, Precision, and Sources of Error In NAA work, samples (any kind of solid or liquid material) can be as small as micrograms to as large as grams, depending upon how much sample is available or the maximum amount that can be used without resulting in counting rates so high that the pulse-height spectrum is markedly distorted by pulse pileup. Since one part per million (ppm) by weight corresponds to 1 μg of element per gram of sample, it is evident that the μg LOD's tabulated above are numerically the same as the various ppm LOD's in a one-gram sample.

The precision of NAA measurement of an element in a sample can be estimated, if other sources of variance are carefully minimized, from the counting statistics. For N observed counts, the standard deviation (resulting from the random nature of radioactive decay) is simply $\pm N^{1/2}$. Thus, if a certain number of gross photopeak counts (GPPC) are measured, for a particular γ-ray peak, its standard deviation is $\pm(\text{GPPC})^{1/2}$. If the underlying Compton continuum baseline counts (BLC) are estimated by summing the counts in channels on either side of the photopeak (using as many baseline channels as the number of channels included in the peak), its subtraction from the GPPC gives the desired NPPC and its standard deviation: $\pm[\text{GPPC} + \text{BLC}]^{1/2}$.

As with any analytical method, errors are possible. Discussion of these is beyond the scope of this short summary, but a few such sources of error deserve mention. They include particularly thermal-neutron self shielding and sample γ-ray self-attenuation. The experienced activation analyst, however, recognizes situations in which such sources of error can be appreciable, and knows how to avoid them, minimize them, or correct for them.

Applications A summary of the extensive applications of the NAA method that have been made, and are being made, is also beyond the scope of this short summary. However, it should be noted that NAA has found extensive application in essentially every field of physical science, biological science, and industry—as well as in such diverse fields as art, archaeology, and crime investigation.

VINCENT P. GUINN

References

1. DeSoete, D., Gijbels, R., and Hoste, J., "Neutron Activation Analysis," New York, Wiley-Interscience, 1972, 836 pp.
2. Guinn, V. P., "Activation Analysis," in "Treatise on Analytical Chemistry," Part 1, Vol. 9, (I. M. Kolthoff and P. J. Elving, Eds.), New York, Wiley, 1971, pp. 5583–5641.
3. Guinn, V. P., "Neutron Activation Analysis," in "Physical Methods of Chemistry," Vol. 1, Part III D (A. Weissberger and B. W. Rossiter, Eds.), New York, Wiley, 1972, pp. 447–500.
4. Guinn, V. P., and Hoste, J., "Neutron Activation Analysis," in "Elemental Analysis of Biological Materials," (R. M. Parr, Ed.), Vienna, Intl. Atomic Energy Agency, 1980, pp. 105–140.
5. Kruger, P., "Principles of Activation Analysis," New York, Wiley-Interscience, 1971, 522 pp.
6. Lyon, W. S. (Ed.), "Guide to Activation Analysis," New York, Van Nostrand-Reinhold, 1964, 186 pp.

Cross-references: ACCELERATOR, PARTICLE; ISOTOPES; NEUTRON; NUCLEAR REACTIONS; NUCLEAR REACTORS; RADIOACTIVITY.

NEUTRON DIFFRACTION

An experiment by Laue, Friedrich, and Knipping in 1912 demonstrated that x-rays were a form of electromagnetic radiation, with a wavelength of the same order of magnitude as the distance apart (10^{-8} cm) of atoms in crystals. This meant that beams of x-rays could be diffracted by crystals in a rather similar way to that in which an optical diffraction grating, in which the elements are separated by about 10^{-5} cm, will produce a spectrum for visible light. As a result of Laue's discovery, a technique for studying the underlying structure of solids by "x-ray diffraction" has grown up. For any given solid, the end product of this technique is a specification of the shape and content of the building block, or "unit cell," out of which the solid is built. The content is specified in terms of "electron density," and it follows that the various atoms or ions which make up the molecule of the substance can be identified.

A rather similar, but in some respects a much more powerful, technique has grown up using beams of neutrons instead of x-rays. A neutron is often thought of simply as a particle, with a mass approximately equal to that of a hydrogen atom, but in terms of wave mechanics a beam of neutrons can be regarded as a wave motion. If the neutrons are moving with velocity v, then they can be considered to have a wavelength equal to h/mv, where m is the mass of the neutron and h is Planck's constant. If such a neutron beam is scattered by a solid, it will be distributed in space as if it were radiation of this wavelength. It so happens that for neutrons having energies equivalent to a temperature of a few hundred degrees centigrade, which are readily obtainable from nuclear reactors, the wavelength is about 10^{-8} cm, i.e., 1Å, which, as we have seen above, is about equal to the interatomic distance in solids. It was shown in 1936 that neutrons, then obtainable only from a radium-beryllium source, could indeed by diffracted by solids. However, it is only since nuclear reactors have produced *intense* beams of suitable neutrons that the application of diffraction techniques to the study of solids has proved worthwhile.

Since high-intensity neutron beams are only available at a limited number of research institutions throughout the world, we shall be concerned only with their application to problems which cannot be solved by any other method. In particular, we shall enquire what can be achieved with neutrons which cannot be found out by using a beam of x-rays, and we shall see the answers to this question by making a comparison of the ways in which atoms and solids scatter x-rays and neutrons. *X-rays* are scattered by the outer, extranuclear, electrons in an atom, and it is for this reason that x-ray diffraction studies produce a picture of electron density. It follows that heavy atoms, such as lead and uranium which contain many electrons, will pre-

dominate in these pictures and that a one-electron atom, i.e., hydrogen, can be located and detailed with much less accuracy. On the other hand, *neutrons* are scattered not by electrons but by the neucleus of an atom, and the way in which the scattering power increases with the mass of the atom is very far from being a steadily increasing function. The scattering power or, more precisely, what we call the "scattering length" arises from the summation of two quite separate effects. The first of these depends on the size of the nucleus, which has a radius proportional to the cube-root of the atomic weignt, so that this effect does indeed increase with atomic weight, but nevertheless fairly slowly. Superimposed on this scattering, however, is resonance scattering, which depends in a complicated way on the actual structure of the nucleus and on its energy levels. This additional scattering often varies quite considerably from atom to atom, and sometimes from isotope to isotope, as we advance up the periodic table. When we combine together the two effects, and thus assess the resultant scattering by a nucleus, we find that it varies quite irregularly from atom to atom and this is illustrated for elements at the lower end of the periodic system in Fig. 1. It will, however, be noted that there is a relatively small spread of values among these scattering lengths. The mean value for all the nuclei which have so far been measured is 0.62×10^{-12} cm, and practically all elements have values which lie between a half and twice this average. As a result of this we find that most elements are roughly equally "visible" to neutrons, though there are a few very interesting exceptions. The practical outcome of this is that hydrogen atoms can be located quite accurately in whatever environment they are found, and this has meant important advances in our knowledge of the role of hydrogen bonds and molecules of water of crystallization in building up the structures of both inorganic and organic crystals. At the same time, we have often been able to get much improved information on the thermal motion of molecules, particularly in those common cases where hydrogen atoms are found on the outside of molecules and which, therefore, provide a very good index of the molecular movement. The technique of detection becomes much more powerful if we can use *deuterated* material, instead of ordinary hydrogen. Deuterium has a neutron scattering length of 0.65×10^{-12} cm and is, therefore, a "good average" atom, whereas ordinary hydrogen, at 0.38×10^{-12} cm, is somewhat below average. This comparison provides a very good example of a difference between the scattering behavior of two different isotopes of an element, arising from differences in the nuclear structures.

Another important field of chemistry to which neutron diffraction has contributed some useful results is in the study of the compounds of ura-

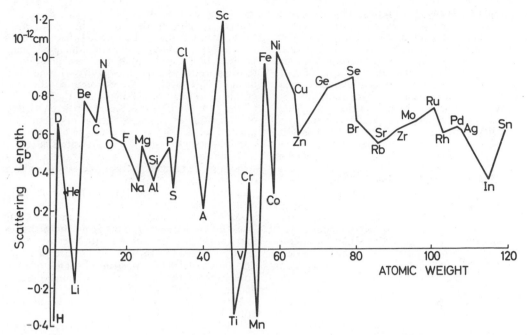

FIG. 1. The variation of the nuclear scattering amplitude of elements for neutrons, shown as a function of atomic weight, in units of 10^{-12} cm.

nium, and post-uranic elements, with nitrogen and oxygen. In the case of x-ray studies, the 92 electrons of a uranium atom completely overshadow the seven and eight of nitrogen and oxygen respectively. For neutrons, however, the value of the scattering length for uranium (0.85×10^{-12} cm) is actually less than the value for nitrogen (0.94×10^{-12} cm) and is only fractionally greater than that of oxygen (0.58×10^{-12} cm).

Of intrinsic interest, and having particular significance in the growing field of study of chemicals of biological importance is the so-called "anomalous scattering" of neutrons by a few elements, such as cadmium and samarium, for which the nuclear scattering amplitude is a complex quantity, with real and imaginary components. This can provide crucial information for determining the structure of large molecules such as insulin if one of these anomalous scatterers can be incorporated in the molecule.

So far we have been considering the process whereby the neutron is scattered by the atomic *nucleus*, and this is a process which occurs for all atoms. There is, however, an additional scattering which takes place for *magnetic* atoms, i.e., for atoms which have a resultant magnetic moment on account of the fact that the atoms contain unpaired electrons. Examples of this are an atom of iron in metallic iron, which appears to contain 2.2 unpaired electrons, and the doubly-charged manganese ion Mn^{++}, which con-

tains 5 unpaired electrons, in manganese salts. Such atoms or ions scatter additional neutrons, making an additional contribution to the scattering length by an amount which is proportional to the magnetic moment. If the magnetic moments in such a material are not arranged in any regular single direction, but point haphazardly as in a *paramagnetic* material, then there will not be any well-defined diffracted beams but there will be a broadly distributed contribution to the scattered background. This contribution may be a little difficult to identify but, nevertheless, the identification can be achieved and the phenomenon can be confirmed. In other magnetic materials, however, all the magnetic moments in a single domain lie parallel to a single direction, and in the particular case of a *ferromagnetic* material they all point *algebraically* in the same sense. In this circumstance the magnetically scattered neutrons contribute specifically to the diffracted beams and the intensity of these is observed to vary with increase of temperature, falling to a minimum at the approach of the Curie temperature, above which no ferromagnetic alignment takes place. In the case of antiferromagnetic materials, in which the moments lie parallel to a single direction but alternately up and down with opposite algebraic sense, the neutron data are extremely informative. In such a material it will be appreciated that, from a *magnetic* point of view, the repeat distance (considering the alternate + and

— moments) is twice the repeat distance which is apparent when only the *chemical* nature of the atoms is considered. This means that extra diffraction spectra will be produced at smaller angles of scattering, corresponding to what would happen if the inter-line spacing of an optical diffraction grating were doubled. The existence of antiferromagnetism can, therefore, be detected very directly by noting the appearance of these extra spectra, particularly if the neutron diffraction pattern is compared either with an x-ray pattern or with a neutron pattern taken at a higher temperature at which the regular magnetic arrangement has broken down. Such a comparison of results obtained at two different temperatures is illustrated in Fig. 2. Results such as these have established the antiferromagnetic structures of a variety of materials and have demonstrated the true nature of ferrimagnetism, as for example in the ferrites in which moments are directed in both positive and negative direc-

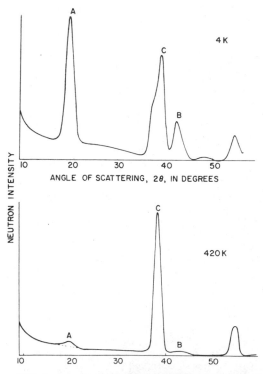

FIG. 2. A comparison of neutron diffraction patterns taken at 4 and 420 K for an antiferromagnetic alloy, $Au_2Mn_{1.7}Al_{0.3}$. At the lower temperature, intense magnetic lines A and B appear, but at the higher temperature, where very little magnetic order remains, these extra lines have practically disappeared. The patterns also show the composite nature of the nuclear scattering line C at low temperature, which occurs because the crystal symmetry changes from cubic to tetragonal when the magnetic order becomes established.

tions but with a net balance in one direction. Moreover, with further research, it has been demonstrated that these structures are only the simplest examples of a wide range of "magnetic architecture" which it is now possible to draw in detail as a result of study with neutron beams. This later work, devoted to the iron group of transition elements and the elements of the rare earth group, has identified a variety of noncollinear arrangements of magnetic moments such as the spiral spins in $MnAu_2$, the umbrella structure in CrSe, composite structures in holmium and erbium and the complicated structures, not yet fully understood, which occur in metallic chromium.

In our discussion so far, we have implicitly assumed that when a neutron is scattered by an atom it is scattered *elastically* and does not lose any of its energy. This is no more than a first approximation to the truth, because atoms are by no means rigidly fixed but are in vibration about their mean positions because of their possession of thermal energy. A neutron which makes collision with an atom may, therefore, lose or gain a quantity of energy. If we keep in mind the fact that atoms in a solid are not isolated, and that the movement of one atom will invariably affect to some degree the motion of its neighbors, it becomes fruitful to regard the interchange of energy as occurring between the neutron and the lattice vibrations of the solid. Indeed we speak of "phonons" which are the embodiment in *particle* form, from the point of view of wave mechanics, of the quanta of energy among the crystal vibrations. If we could measure accurately the interchanges of energy, then we could learn about, and indeed study in detail, the phonon spectra and the dispersion law for the solid. In fact, for *neutron* scattering, but not for x-rays, such a measurement can be made and this gives a quite unique value to the use of inelastic neutron scattering for studying solids. The particular supremacy of neutrons becomes clear if we consider the actual energies of a neutron and x-ray quantum which possess the same wavelength. We find in fact that the latter is roughly 10^5 times larger. Thus, whereas the energy of a neutron of wavelength 1Å is about equal to that of a quantum of crystal energy, yet the energy of 1Å x-ray is 10^5 times greater. It follows, therefore, that if a neutron gains or loses such a quantum, then its own energy will be greatly changed; for example, it could easily be roughly doubled or halved. On the other hand such an interchange for an x-ray would be quite insignificant and the resulting change of wavelength could not be detected. It is in fact possible, therefore, to measure both momentum and energy changes in neutron-phonon interchanges, and this information leads directly to the details of the dispersion law in the solid. The full power of these methods can be achieved only by detailed and extensive observations with single crystals. Nevertheless a relatively simple technique of neutron spectroscopy is available for powdered and polycrystalline samples, which has rather similar aims to conventional infrared and Raman spectroscopy, but with some distinctive advantages. In particular, any vibrations in which hydrogen atoms are involved are greatly enhanced, and moreover, the selection rules which limit the observation of transitions in optical spectra do not generally apply to the neutron spectra of inelastic scattering.

As the intensity of the beams of neutrons available from research reactors has steadily increased, with roughly a tenfold increase each ten years, these techniques have become progressively more powerful and determinative, leading to a steadily widening view of solids and liquids in the several unique respects which we have discussed. A neutron flux of 1.2×10^{15} cm^{-2} sec^{-1} is now available but it may be difficult to increase the flux beyond this. Further progress is being made using linear accelerators from which pulsed beams of protons generate neutrons by reactions with heavy-element targets. However, it is likely that the limitations in the use of neutron-diffraction techniques will continue to be set by the limited availability of adequate neutron sources. It may fairly be said that many promising applications have never been tested.

G. E. BACON

References

Bacon, G. E., "Neutron Diffraction," Third Edi' London, Oxford Univ. Press, 1975.
Dachs, H., "Neutron Diffraction," Berlin, Springer-Verlag, 1978.
Izyumov, Yu. A., and Ozerov, R. P., "Magnetic Neutron Diffraction," New York, Plenum Press, 1970.
Marshall, W., and Lovesey, S. W., "Theory of Thermal Neutron Scattering," London, Oxford Univ. Press, 1971.
Squires, G. L., "Introduction to the Theory of Thermal Neutron Scattering," Cambridge, Cambridge Univ. Press, 1978.

Cross-references: DIFFRACTION BY MATTER AND DIFFRACTION GRATINGS, MAGNETISM, NEUTRON, PARAMAGNETISM, X-RAYS.

NOISE, ACOUSTICAL

Strictly defined, noise is any unwanted sound, whether pleasant or unpleasant. More commonly, however, sounds that are unpleasant and disturbing, or that mask desired sound, are termed noise. Thus, noise, in a general sense, may be thought of as any sonic disturbance. Depending upon the degree of pitch distribution, intensity, and persistance, noise can range from being merely annoying, to hazardous or injurious. In our highly industrialized society, with its rapid growth of energy-producing and converting systems, noise has become a major problem. Some of its harmful effects are interference with mental and skilled work, impairment of sleep, creation of emotional disturbances, damage to hearing and a deterioration of health and well-being. Consequently, the control and reduction of noise has become an important science.

Adequate measuring means are a prime requirement in the scientific control and reduction of noise. Even then, the problem of establishing a true relationship between the subjective and objective properties of noise is difficult because of the many different aspects of human reaction to noise. The first relationship between the subjective and objective measurement of sound is the simplified rule relating *loudness in sones* to the *loudness level in phons.* The loudness scale in sones is proportional to the average person's estimate of the loudness. Also, the loudness level, P, of a given sound, in phons, is numerically equal to the median sound pressure level, p_L, of a free progressive wave at a frequency of 1000 Hz presented to listeners facing the source, which is judged by the listeners to be equally loud. The sound pressure level, p_L, is defined as

$$p_L = 20 \log_{10} \frac{p}{p_0} \text{ (decibels)} \qquad (1)$$

where p is measured sound pressure, in microbars, and $p_0 = 2 \times 10^{-4}$ microbars.

The relation between sones, S, and phons, P, is given by

$$S = 2^{(P-40)/10} \qquad (2)$$

Referring to Eq. (2), a loudness level of 40 phons produces a loudness of one sone, and a loudness level of 80 phons produces a loudness level of sixteen sones, etc.

The simplest means for the measurement of noise is the sound level meter, an instrument comprising a microphone, an amplifier, frequency weighting networks, and an output meter. The characteristics of the frequency weighting networks in the meter are based upon the equal loudness contours of hearing for different levels.

More sophisticated means for measuring noise include octave band and one-third octave band sound analyzers that supply information on the sound level in various frequency ranges. These analyzers are used for research on the reduction of machine noise, transmission and other areas where information on the sound levels in specific frequency bands is required. Narrow-band analyzers may be used to obtain the spectrum of a noise. The sound-pressure spectrum level is that level within a frequency band of 1 Hz. This level is plotted against frequency to obtain the spectrum frequency characteristic of the noise. If the spectrum level of a noise is known, sophisticated means may be used to relate the objective to the subjective qualities of the noise.

As noted earlier, noise abatement has become an important science. For instance intensive research on the quieting of automobiles has been in progress for three decades, with outstanding results; some of the major problems remaining to be solved involve wind- and road-induced noise. Similarly, research has been carried out on the reduction of noise of all types of household appliances employing motors, fans, compressors, pumps, gears, and other moving parts. Another phase of acoustical engineering involves methods for reducing the transmission of sound through the walls, floors, ceiling and partitions in all manner of buildings or enclosure by the use of construction and materials based on fundamental acoustical principles.

The noise in typical environments and noise produced by various sources are given in Table 1.

The masking effect produced by noise reduces the intelligibility of speech. For example, if the speaker and listener are separated by 5 feet, the levels of noise that will barely permit reliable word intelligibility are 51 dB for normal conversation, 57 dB for raised speech, 63 dB for very loud speech, and 69 dB for shouting.

A person subjected to high noise levels for long periods of time may suffer considerable impairment of hearing. The use of ear protectors

TABLE I. NOISE LEVELS FOR VARIOUS
SOURCES AND LOCATIONS[a]

Source or Description of Noise		Noise Level (dB)
Threshold of pain		130
Hammer blows on steel plate	2 ft	114
Riveter	35 ft	97
Factory		78
Busy street traffic		68
Large office		65
Ordinary conversation	3 ft	65
Large store		63
Factory office		63
Medium store		62
Restaurant		60
Residential street		58
Medium office		58
Garage		55
Small store		52
Theatre		42
Hotel		42
Apartment		42
House, large city		40
House, country		30
Average whisper	4 ft	20
Quiet whisper	5 ft	10
Rustle of leaves in gentle breeze		10
Threshold of hearing		0

[a]Olson, Harry F., "Acoustical Engineering," p. 256, New York, Van Nostrand Reinhold, 1957.

may provide sufficient insulation under such conditions.

There are now federal regulations on permissible noise in industry as given in the reference below. Examples of the permissible noise exposures are as follows: for 8 hours per day the permissible sound level is 90 dBA, for 4 hours per day 95 dBA, for 1 hour per day 105 dBA, etc.

HARRY F. OLSON

References

Federal Register, Saturday, May 29, 1971, Vol. **36**, No. 105, Part II, Department of Labor, "Occupational Safety and Health Administration." Article 1910.95, "Occupational Noise Exposure."

Goodfriend, Lewis S., "Noise Pollution," CRC Scientific Publications, Cleveland, Ohio, 1972.

Harris, G. M., "Handbook of Noise Control," New York McGraw-Hill Book Company, 1957.

Rettinger, M., "Vol. II Noise Control," New York, Chemical Publishing Co., 1977.

Cross-references: ACOUSTICS; ARCHITECTURAL ACOUSTICS; HEARING; MEASUREMENTS, PRINCIPLES OF; MUSICAL SOUND.

NUCLEAR INSTRUMENTS

Nuclear instruments are detectors of ionizing radiation and associated apparatus for amplification and analysis of the signals. Nuclear instruments are used in measuring the activity of a radioactive sample and identifying radioisotopes by their characteristic radiations, in nuclear physics experiments with particle accelerators, control of nuclear reactors, dosimetry for radiation safety and cancer radiotherapy, and for medical diagnosis in radiology and nuclear medicine.

Ionizing radiation includes energetic charged particles such as electrons, protons, alpha particles, and fission fragments; uncharged particles (neutrons); and short-wavelength electromagnetic radiation (x-rays and gamma rays). The charged particles ionize directly, while neutrons must first undergo a nuclear reaction that produces charged particles. X-rays and gamma rays interact primarily by the photoelectric effect, Compton scattering, or pair production, which generate secondary electrons. It is interesting that a single nuclear particle or quantum of electromagnetic radiation can be detected, and in many cases its energy and time-of-arrival measured.

Detection depends on interaction of the radiation with the detecting medium, and transfer of energy by ionization and excitation. The signal-generation process may involve generation of many electrons and ions (or holes in semiconductors) and the movement and collection of these charges; emission of light (luminescence); induction of chemical changes, or heating.

Gas ionization detectors include the ionization chamber, proportional counter, and Geiger-Müller (GM) counter. Fig. 1 illustrates some general features of these detectors. The detecting medium is a gas such as argon (or a special mixture for the GM counter), often at or below atmospheric pressure but sometimes at a few atmospheres pressure for increased interaction efficiency. The gas is contained in a vessel fitted with high-quality insulated electrodes in either parallel plate or coaxial cylinder geometry as shown. A grounded guard electrode may be added to intercept leakage current across the insulators. The applied dc potential V_0 may be a few hundred to a couple thousand volts. Normally the gas is an insulator, no current flows, and a steady electric field is established between anode and cathode. When ionizing radiation interacts in the gas or electrode, many low-energy ions and electrons are released in the gas. The energy required to generate an ion-electron pair, W, depends on the gas but is typically around 30 electron volts. Thus a 3-MeV alpha particle would release 100,000 ions and electrons. These charges drift to the oppositely charged electrode under the

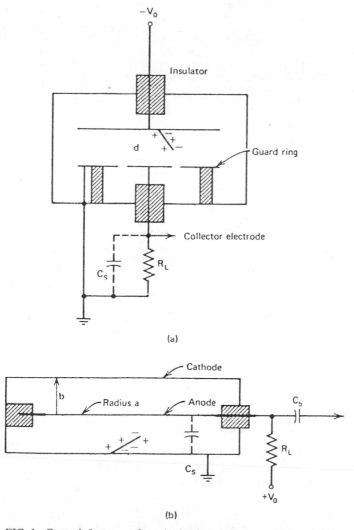

FIG. 1. General features of gas-ionization detectors: (a) parallel-plate ionization chamber with guard ring to define field and intercept leakage current; (b) cylindrical proportional or GM counter. (From Profio, A. Edward, "Experimental Reactor Physics," New York, Wiley, 1976, p. 244; reproduced in Profio, A. E., "Radiation Shielding and Dosimetry," New York, Wiley, 1979, p. 266.)

influence of the electric field. This motion constitutes a current which can be amplified, or the voltage drop across the load resistor R_L can be amplified and measured. Eventually (microseconds for electrons, milliseconds for ions) the charges are collected on the stray capacitance C_s, until drained off through R_L.

If the time constant $R_L C_s$ is small compared to the drift time of the electron or ion, that component of the signal voltage will not reach full amplitude. Furthermore, a detailed analysis shows that the contribution of each component depends on the distance from ionization point to electrode. Only if both electrons and ions are collected does the charge and corresponding voltage reflect the energy deposited by ionization in the gas. For simple counting without energy analysis, only the fast (electron) component is needed. If the time constant is long compared to the mean interval between ionizing nuclear particles, the detector will not resolve individual pulses of charge, but the current will indicate the mean rate of detection of the nuclear radiation. If the time constant is

very long, the charge is integrated and read out at the end of the irradiation. The pulse ionization chamber has been replaced by the semiconductor detector except where radiation damage is a consideration. The mean-level and integrating ionization chambers are commonly used for radiation monitoring and for reactor control. The detector is made sensitive to neutrons by coating the wall with boron or boron-10, or a fissionable material such as uranium-235, or by filling the chamber with boron trifluoride gas.

The proportional counter and GM counter differ from the ionization chamber in that the charge is increased by gas multiplication. The amount of gas multiplication is a function of $\mathscr{E}/p$, the ratio of the electric field strength to the gas pressure. The ionization chamber operates at low $\mathscr{E}/p$, where $\mathscr{E}$ is large enough to separate the ions and electrons before they can recombine, but there is no gas multiplication. Thus the ionization chamber operates in region 2 in Fig. 2, which plots the number of charges collected as a function of applied voltage, for

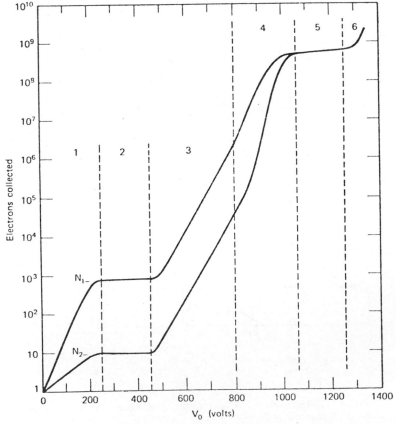

FIG. 2. Number of electrons collected as function of applied voltage for energy depositions $E_1 = W N_1$ and $E_2 = W N_2$. Region 2 corresponds to ionization chamber, region 3 to proportional counter, region 5 to GM counter. (From Profio, A. E., "Radiation Shielding and Dosimetry," New York, Wiley, 1979, p. 267.)

two different amounts of primary ionization, $N_1 = 10^3$ and $N_2 = 10$. At medium $\mathcal{E}/p$ and voltage, the electrons are accelerated sufficiently between collisions with neutral gas atoms to release more electrons on impact. These electrons can in turn release more electrons, for an overall typical gain or multiplication factor M of 10–1000. If M is not too large, gain is independent of the initial ionization, hence the output signal charge is proportional to the initial ionization and the detector can be used for energy analysis. This corresponds to region 3 in Fig. 2, the region of the proportional counter. (Actually a proportional counter may also have a different gas filling, to stabilize the gas multiplication.) In order to achieve the required $\mathcal{E}/p$ at reasonable applied voltage, it is standard to make the anode a small-diameter wire, because $\mathcal{E}$ is inversely proportional to radius in cylindrical geometry. Most of the multiplication occurs very close to the anode and collecting the electron component alone is sufficient. The low-noise gain afforded by gas multiplication makes the proportional counter suitable for detection of low energy x-rays and electrons. Gain is usually not needed for alpha particles or fission fragments, and detection efficiency for gamma rays is low although efficiencies of a few percent can be obtained with walls of a dense and high atomic number material such as bismuth. Neutrons can be detected by filling the tube with BF_3 or helium-3.

GM counters operate at high $\mathcal{E}/p$ and the gas multiplication may be 10^6 or more, corresponding to region 5 of Fig. 2. Multiple gas discharge avalanches are created and propagated by ultraviolet photons as well as electrons. The Geiger discharge can be initiated by a single primary ionization, and the amplitude of the resulting pulse is no longer proportional to the initial ionization by a nuclear particle: all pulses have the same amplitude, at a given applied voltage. In order to terminate or "quench" the discharge, the GM tube is filled with a mixture of gases, including either an organic quencher such as ethyl alcohol or ethyl formate, or a halogen (chlorine or bromine). Quenchers suppress multiple discharges otherwise started by positive ions hitting the cathode, by absorbing energy in dissociation of the molecule. An organic quencher is consumed in some 10^9 counts, but halogens are reconstituted and the tube lifetime is much longer. Because of the hundreds of microseconds for the ions to be collected and the discharge terminated, the GM counter has a correspondingly long deadtime and is limited to rather low count rates (hundreds of counts per second or less). With its large internal gain, electronics can be simple and cheap, and the GM counter is still used for simple counting of beta particles, and for monitoring gamma rays fields at low detection efficiency.

A semiconductor detector is similar to a pulse ionization chamber with the gas replaced by a semiconductor, usually silicon or germanium. (Newer materials such as cadmium telluride and mercuric iodide are under development.) A true solid insulator cannot be used, because the charges (electrons and holes) produced by ionization would not move in the applied electric field and generate a signal. On the other hand, the device must have high resistivity or the noise would be too great from the current of thermally released electrons. These requirements have been met by: (a) doped silicon operated as a reverse-biased p-n junction diode at room temperature; (b) p-type silicon or germanium compensated by drifting an electron donor, lithium, and operated at low temperature as a reverse-biased p-i-n junction diode; (c) ultra-high-purity germanium operated at low temperature as a reverse-biased p-i-n diode; and (d) mercuric iodide and other room temperature diodes. Semiconductor detectors generally have faster response times than gas detectors (order of tens of nanoseconds), and higher efficiency because of their density (2.33 g/cm^3 for Si, 5.33 g/cm^3 for Ge). Their biggest advantage over other types of detector is the low energy required per hole-electron pair (3.6 eV for Si, 2.96 eV for Ge), resulting in a larger signal amplitude and better energy resolution because of smaller statistical fluctuations in the charge collected. Their main disadvantages are sensitivity to radiation damage from heavy charged particles or fast neutrons, the relatively small sensitive volumes available, and cost in the larger sizes.

A common detector for alpha particles and other charged particles is the silicon surface barrier detector shown in Fig. 3(c). This is a type of p-n junction diode, where the junction occurs between the body of n-type (electron-donor-doped) silicon and a very thin layer of p-type (electron-acceptor-doped) material produced by controlled oxidation of the surface. The surface is protected by a thin evaporated layer of gold, but even so the surface barrier detector is sensitive to light and must be operated in the dark. When the diode is reverse biased (positive polarity to the n-type silicon contact), by a hundred volts or so, the thermally released charge carriers are swept out and a "depletion" region is left which can sustain a large electric field. The depletion region corresponds to the sensitive volume of the detector; it may be on the order of one cm^2 in area and a few hundred micrometers thick. This thickness is sufficient to stop alpha particles of a few MeV, so their energy can be measured. A small correction is needed for the loss of energy in the "window" or dead layer of gold and p-type material. Similar detectors are made with ion-implanted or diffused junctions. The depletion region can be made to extend entirely through a detector 1–2 mm thick, but this is still too small for stopping energetic electrons and efficiency for gamma rays is poor.

The maximum thickness of p-n junction depletion regions is limited by voltage breakdown. The technique of lithium drifting has been developed to make a compensated or quasi-intrinsic resistivity region up to some 10 mm thick. One starts by thermally diffusing some lithium into the surface of a crystal of rather pure but *p*-type silicon or germanium. The *p-n* junction thus formed is reverse biased and the small lithium ions caused to drift into the bulk of the crystal, at elevated temperature, by the applied electric field. After some days or weeks, nearly exact compensation is achieved automatically because the space charge remaining with imperfect compensation is sufficient to drive the lithium ions until the space charge is canceled. Figure 3(a) shows the planar geometry typical of a lithium-drifted silicon diode, designated Si(Li), where the compensated region may be a few mm thick. Such detectors

are suitable for detection and spectroscopy of electrons, where the low atomic number of silicon minimizes backscattering. Si(Li) detectors are also used for low energy x-ray and gamma ray spectroscopy (below 30 keV). Figure 3(b) illustrates the coaxial design typical of lithium-drifted germanium, Ge(Li), detectors where the compensated region may have a volume of 30–150 cm^3. The volume, density, moderately high atomic number of germanium ($Z = 32$), and most of all the excellent energy resolution (about 0.1%) of germanium detectors have made them the instrument of choice for gamma ray spectroscopy, although efficiency is less than can be obtained with the sodium iodide scintillation detector. The Si(Li) and Ge(Li) detectors are always operated at low temperature, usually at liquid nitrogen temperature (77 K) to reduce thermal noise. A Si(Li) detector is usu-

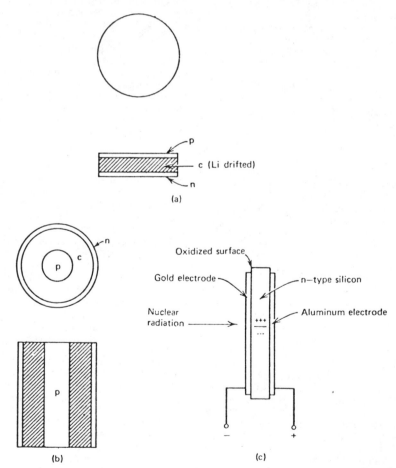

FIG. 3. Types of semiconductor detectors; (a) planar lithium-drifted; (b) coaxial cylindrical lithium-drifted; (c) surface barrier. (From Profio, A. Edward, "Experimental Reactor Physics," New York, Wiley, 1976, p. 220; reproduced in Profio, A. E., "Radiation Shielding and Dosimetry," New York, Wiley, 1979, p. 261.)

ally stored at liquid nitrogen temperature, too, but can survive an occasional warmup. The Ge(Li) detector must always be stored and used at liquid nitrogen temperature to avoid redistribution of the lithium. One warmup is sufficient to ruin a Ge(Li) detector, and while it can be redrifted, energy resolution is likely to be worse.

Figure 4 illustrates the electrical properties of a lithium-drifted *p-i-n* junction detector, as influenced by the Li concentration and starting material. As shown, after compensation there is no net space charge density within the so-called intrinsic region, but there is a potential gradient and a uniform electric field (in planar geometry) which moves and collects the electrons and holes released in this region by ionizing radiation.

The ultra-high-purity germanium detector is very expensive, but has the advantage that there is no lithium to redistribute and it can be al-

lowed to warm up occasionally. It is still usually stored at liquid nitrogen temperature to minimize contamination of the surface by vapors, and is operated at low temperature to minimize noise. The crystal is housed in a vacuum cryostat, on a good heat conductor or "cold finger" dipping into a reservoir of liquid nitrogen in a dewar. Weekly replenishment is usually required even with a 30-liter dewar. The trouble and expense of liquid nitrogen and the bulky dewar and cryostat has led to interest in materials such as cadmium telluride and mercuric iodide, which have large enough band gaps to allow operation at room temperature. Density and their high-atomic-number constituents make them candidates for x-ray and gamma ray spectroscopy. However, trapping centers and crystal growing difficulties have so far limited the detectors to about 1 mm thickness if good energy resolution is desired. With careful atten-

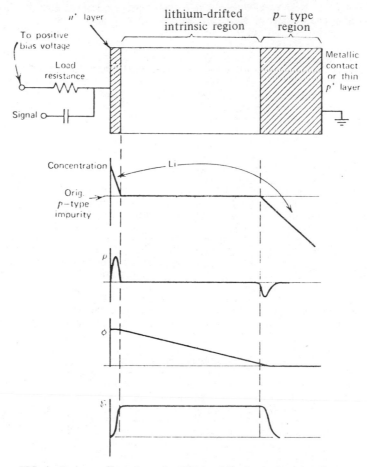

FIG. 4. Basic configuration of a lithium-drifted *p-i-n* junction detector. Also shown are the corresponding profiles for impurity concentration, charge density ρ, electric potential φ, and electric field $\mathscr{E}$. (From Knoll, G. F., "Radiation Detection and Measurement," New York, Wiley, 1979, p. 417.)

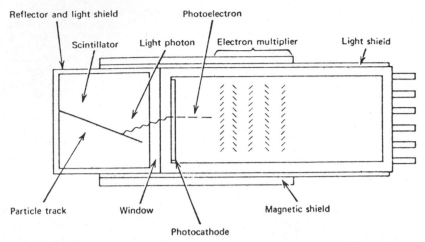

FIG. 5. Elements of a scintillation detector. (From Profio, A. Edward, "Experimental Reactor Physics," New York, Wiley, 1976, p. 274; reproduced in Profio, A. E., "Radiation Shielding and Dosimetry," New York, Wiley, 1979, p. 270.)

tion to minimizing noise sources, resolution may approach that of the Si and Ge detectors.

A scintillation detector consists of a transparent substance that fluoresces under ionization and excitation by a charged particle, optically coupled to a photomultiplier tube that detects and amplifies the weak flash of light, as shown in Fig. 5. The photomultiplier is shielded from outside light and from stray magnetic fields that could deflect the electrons in the multiplier. The scintillator is usually coated with a white reflective material and contained in a light-tight can except for the window in contact with the photomultiplier.

Table 1 summarizes properties of commonly used scintillators. Organic crystals such as anthracene may be used for detection of electrons, but are fragile and available only in thin sections. Organic liquid scintillators are used for detection and spectroscopy of fast neutrons by means of the recoil protons generated by elastic scattering on hydrogen, and for counting low-energy beta particles (e.g., from tritium and

carbon-14) when the radioactive material is dissolved in the scintillator. NE213 is a proprietary formulation which is useful for separation of gamma ray and neutron induced counts by pulse shape discrimination (PSD). Plastic scintillators are made of scintillating chemicals in a solid solution with polyvinyltoluene (PVT) or other polymer. Because of their low atomic number and density, organic scintillators are not very good for gamma rays.

The inorganic crystal scintillators, especially sodium iodide doped with thallium, NaI(Tl), are well suited for detection and spectroscopy of gamma rays because of their high effective atomic number and density. A drawback to NaI(Tl) is that it is brittle and hygroscopic, and must be protected from shock and moisture. Fluorescence decay time constant is longer than for the organic scintillators (significant in fast timing, coincidence, or high count rate applications) but light output is better. Until the advent of the germanium semiconductor detector, NaI(Tl) was widely used for gamma ray spec-

TABLE 1. PROPERTIES OF SCINTILLATORS.

Type	Scintillator, Composition	Density (g/cm^3)	Relative Output (electron)	Fluor. Decay (ns)	Remarks
Organic crystal	anthracene, $C_{14}H_{10}$	1.25	1.00	30	68 eV/photon
Organic liquid	NE213, $CH_{1.21}$	0.88	0.78	4, 100	PSD
Plastic	NE102, $CH_{1.10}$	1.03	0.65	4	PVT base
Inorganic	NaI + 0.1% Tl	3.67	2.10	230	hygroscopic
Crystal	$Bi_4Ge_3O_{12}$	7.13	0.17	300	High Z
	ZnS + 0.01% Ag	4.09	3.00	200	polycrystal
Glass	NE908, glass + 8% Li	2.67	0.20	5, 75	0.1% Ce_2O_3
Gas	xenon (1 atm)	0.006	1.5	2	UV emission

troscopy, even though its energy resolution is not very good (typically 8% at 662 keV). Some 32 eV is required to generate a light photon, but only about 10% of the photons generate a usable electron in the photomultiplier, and with other inefficiencies, the energy per electron may approach 1000 eV. Bismuth germanate gives much less light than NaI(Tl), but because of its high density and atomic number, is even more efficient in detecting gamma rays in a small volume.

Silver-activated zinc sulfide, ZnS(Ag), is useful for counting alpha particles but is available only in polycrystalline form and light scattering limits its use to thin layers. The cerium-activated, lithium-loaded silicate glass scintillators such as the proprietary NE908 are used for slow neutron detection, but have the drawback of also being sensitive to gamma rays. The xenon gas scintillator has a fast fluorescence decay time, but emits in the ultraviolet requiring a special photomultiplier tube, and being a gas is not very efficient in stopping radiation.

The photomultiplier is a vacuum tube, usually with a thin, semitransparent layer deposited on the inside of a flat glass faceplate or window at one end. The photocathode itself may be oxidized cesium antimonide, or a multialkali (K-Cs-Sb). The low-energy photoelectrons released from the cathode are collected, accelerated, and focused on the first electrode (dynode) of the electron multiplier by an applied electric field. The dynode is treated so that two or more electrons are generated per incident electron, by secondary emission. The electrons from the first dynode are accelerated and focused on the second dynode, and so forth for perhaps 9 to 14 dynode stages. Current amplification can easily exceed 10^5. The required 100 volts or so per stage is usually supplied from a resistive voltage divider connected to a high voltage supply. The supply must be very well regulated because gain is very sensitive to voltage. The photomultiplier is selected for size of the scintillator, spectral response (e.g., to match the blue emission of NaI(Tl)), gain required hence number of stages, gain stability for spectroscopy, and pulse risetime for fast timing applications.

Another type of detector emitting light is the thermoluminescent dosimeter (TLD). Certain materials such as LiF or CaF_2 doped with Mn, trap electrons and holes generated by ionization. Upon subsequent warming the electrons and holes recombine with emission of light, which is detected by a photomultiplier tube. TLD's are widely used for personnel dose monitoring, although the materials do not respond exactly as tissue does, especially at low energies.

X-ray film and other silver halide emulsions are used for imaging and for personnel dose monitoring. The ionizing radiation activates centers in the grains, much as a latent image is formed by light. Upon later chemical develop-

ment, the halide grains are reduced to metallic silver, darkening the film. The process may be considered as chemical amplification. The darkening must be calibrated against known doses measured by another instrument, and the response at low energies may be much greater than in tissue. Other chemical dosimeters exist, but are limited to high dose applications. Radiation also causes heating, which can be measured by temperature rise in a calorimeter, but heating is seldom used for radiation dosimetry because more sensitive instruments are available.

Absorbed dose, defined as energy absorbed by ionization per unit mass of material, is normally measured by an integrating or mean-level type of cylindrical ionization chamber. The walls are made of an electrically conducting plastic whose elemental composition is close to that of average soft tissue ($C_5 H_{40} O_{18} N$), and the wall is made slightly thicker than the range of the most energetic electrons involved. The gas filling also has soft tissue elemental composition, and the gas pressure and electrode spacing are chosen so that the dimensions are small compared to the range of the electrons. Under these conditions, the energy absorbed per kg of wall material (and tissue) is equal to the energy absorbed per kg of gas. The latter is obtained from the measured charge, the known energy per ionization, and the mass of the gas. The special units of absorbed dose are the gray (1 Gy = 1.0 J/kg) and the rad (1 rad = 0.01 J/kg). Depending on dose rate and mass of gas, the current may be 10^{-12} A or less, and a low-noise, high-gain, high-input-impedance dc amplifier (electrometer) is required to drive a meter.

Neutrons, protons and alpha particles are more damaging per unit of absorbed dose than electrons, x-rays and gamma rays. This is taken into account by defining the dose equivalent (measured in sievert or rem) as the absorbed dose times a dimensionless quality factor that varies from 1 (for gamma rays) to 10 (for fast neutrons). The dose equivalent in a mixed radiation field can be derived from a special ionization chamber that gives information on the microscopic distribution of ionizing events along the particle path, the so called LET (linear energy transfer). Particles with higher LET have larger quality factors associated with them.

Reactor instrumentation includes several neutron-sensitive ionization chambers and associated dc amplifiers, which can be calibrated in terms of reactor power. In addition to linear, range-switched amplifiers, it is convenient to have logarithmic amplifiers and meters displaying the power over many decades during reactor startup. Also, the output of the logarithmic amplifier can be differentiated to give a signal inversely proportional to the time constant (called reactor period) of the exponential increase in power. Electronic trips are provided to shutdown the reactor, by inserting control rods, if the reactor power gets too high or the

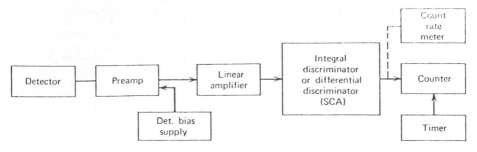

FIG. 6. Elements of a typical signal chain for pulse counting. (From Knoll, G. F. "Radiation Detection and Measurement," New York, Wiley, 1979, p. 660.)

reactor period becomes too short (rapid increase in power).

Semiconductor and scintillation detectors and the proportional counter are operated as pulse instruments. Figure 6 is a block diagram of a typical pulse counting system. The small signal pulses from the detector are amplified in the preamplifier, transmitted by coaxial cable to the main linear amplifier for further amplification and pulse shaping, and then to the integral discriminator or single channel pulse amplitude analyzer (SCA). The integral discriminator generates an output pulse only if the input pulse exceeds a set amplitude, thus rejecting noise and low-amplitude background. An SCA also has an upper-level discriminator and anticoincidence circuit, such that pulse amplitude must lie between the discriminator settings for an output pulse to be generated. The number of pulses in a given timed interval is accumulated in an electronic counter (also called a scaler), or the average count rate derived and displayed by a count rate meter.

Although pulse amplitude hence energy analysis can be performed by sequentially stepping the lower discriminator of an SCA, it is preferable to use a multichannel analyzer (MCA) instead. An MCA consists of an analog-to-digital converter for randomly arriving pulses, control unit, memory, oscilloscope display of the spectrum, and usually a printer. The ADC digitizes the pulse amplitude and the control unit adds a count to the appropriate address or channel in memory, usually for 1024 channels or 512 channels.

For further information consult the references.

A. Edward Profio

References

Knoll, G. F., "Radiation Detection and Measurement," New York, John Wiley & Sons, 1979.

Profio, A. E., "Radiation Shielding and Dosimetry," New York, Wiley-Interscience, John Wiley & Sons, 1979.

Cross-references: LUMINESCENCE; MEASUREMENTS, PRINCIPLES OF; NUCLEAR RADIATION; RADIATION, IONIZING, BASIC INTERACTIONS; RADIOACTIVITY; SPARK AND BUBBLE CHAMBERS.

NUCLEAR RADIATION

Nuclear radiation results from the transitions of atomic nuclei. The two chief types of transition in natural radioactivity are those in which the number of constituent particles of a given nucleus (nuclide) is changed by the emission of one or more particles such as an alpha or beta particle and those in which there is a rearrangement of the particles of a given nuclide such that the nuclide passes to a lower energy state with the emission of high frequency electomagnetic radiation called gamma rays.

The study of nuclear radiation together with the results of high energy bombardment of nuclei has become our most important source of information about the structure of atomic nuclei and the fundamental nature of matter. And such studies now extend to the internal processes of our sun and the stars as well as our more immediate problems concerned with emission from nuclear reactors and nuclear weaponry.

The chief sources of nuclear radiation are; (1) naturally radioactive substances; (2) those substances in which radioactivity has been induced; (3) emissions resulting from collisions with cosmic rays or by particles from high energy accelerators; (4) nuclear reactors and weaponry; and (5) nuclear processes in the sun and stars. There is even a small amount of radiation from spontaneous fission of uranium in the earth.

In 1896 Henri Becquerel found that a new and penetrating type of radiation from uranium had darkened a photographic plate and he called the new phenomenon radioactivity. His associates Marie Curie and husband Pierre after painstaking search found other far more intensely radioactive substances including radium and polonium. Soon various radioactive substances were arranged in series showing how one type would decay into a different but related type.

Early studies of rays from different substances revealed that an alpha (α) particle was the doubly charged nucleus of a helium atom; beta (β) particles were identified as the then newly discovered negative electrons; and gamma (γ) rays were found to be similar to x-rays but could be even more penetrating. Alpha rays were emitted with speeds up to $\frac{1}{15}$ the speed, c, of light and had the unusual property of producing scintillations on a fluorescent screen by which they could be counted. Beta rays could have a range of speeds up to a maximum speed of $0.96c$, a hitherto almost unbelievable speed for any particle. Beta and gamma rays could be readily detected by the ionization they produce in a gas. The rest-masses in kilograms of the two particles and their energy equivalents in million of electron volts (MeV) are

$$M_0(\text{alpha particle}) = 6.645 \times 10^{-27} \text{kg}$$
$$= 3727.2 \text{ MeV}$$
$$m_0(\text{beta particle}) = 9.109 \times 10^{-31} \text{kg}$$
$$= 0.511 \text{ MeV}$$

The three types of rays could be readily separated by means of a magnetic field (Fig. 1). In crossing the field the alpha rays are deflected in a direction that indicates they carry a positive charge. The beta rays are deflected in the opposite direction and the gamma rays are not deflected at all as they have no charge and are of the nature of light waves.

The penetrating power of any nuclear radiation depends on the nature of the ray, the energy of its emission and the nature of the absorption process. In general it may be said that alpha rays usually are stopped by a few sheets of paper and beta rays by a few millimeters of aluminum.

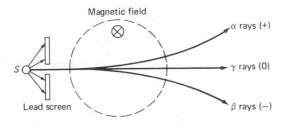

FIG. 1. Separation of alpha (α), beta (β) and gamma (γ) rays from a radioactive source (S) by a magnetic field directed into the paper. This indicates alpha rays possess a positive electric charge, beta rays a negative charge, and that gamma rays are uncharged.

Alpha and beta rays, because of their particle character, have a more sharply defined range than gamma rays, and a significant fraction of the gamma radiation may pass through a number of centimeters of metal shield. This is because particles lose speed and kinetic energy in both ionizing and non-ionizing collisions with other particles, whereas a gamma-ray-photon always travels at the same speed, and the beam of photons is usually weakened by some process such as scattering or absorption of photons. And the method of magnetic separation and the use of various absorbing materials are still basic in our far more complicated experimental procedures today.

An alpha ray can produce between 20,000 and 70,000 ion pairs per centimeter of path in a gas depending on the initial energy of the ray, maximum ionization being reached near the end of the path. A beta ray on the other hand may only produce 200 ion pairs or less per centimeter. Thus such rays lose energy rapidly in passing through a gas, and rays of a particular energy have a rather sharply defined range (see IONIZATION).

The range of alpha particles in air at 76 cm mercury pressure ($15°C$) varies from 2.7 cm for alpha particles from uranium to 8.62 cm for alpha particles from thorium C' (Po^{212}). The former are emitted with energies of 4.2 MeV while the latter, the most energetic of any from a naturally radioactive substance, are emitted with energies of 8.6 MeV.

How these radiations could be emitted from an atom was unknown until 1911, when Rutherford, in his famous alpha-ray scattering experiment, deduced that nearly all of the mass of an atom resides in a tiny, positively charged central region called the nucleus having a diameter in meters (m) of the order of 10^{-14} m whereas the atom itself with its encircling electrons was of the order of 10^{-10} m. Newton's idea that atoms are solid little bits of mass that last for ever was overthrown and the study of the nucleus as a complex, highly organized system began.

In 1929, Rosenblum discovered the fine structure of alpha rays. That is, alpha rays from a single type of nuclide may not all have exactly the same energy and range but often consist of two or more groups with slightly different but sharply defined ranges and consequently different initial energies. This led to the ultimate recognition of the existence of different energy levels in the nucleus.

Now the origin of gamma radiation could be explained. When an alpha particle is emitted as for instance from a radium nucleus (Ra 88) it loses four units of mass and two units of charge and forms a new nucleus in this instance emanation (Em 86). If the alpha particle does not possess enough energy to leave the product nucleus in its lowest energy state (ground state) then it can return to the ground state by emitting a gamma ray (Fig. 2) of the required energy.

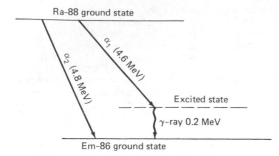

FIG. 2. Energy levels showing origin of gamma (γ) rays from radium (Ra-88) when emission of alpha ray (α₁) leaves product nucleus in an excited energy state above its ground state. It then returns to the ground state by emission of a gamma ray. By comparison, emission of α₂ carries off enough energy for the nucleus to go directly to the ground state.

The discovery of the neutron by Chadwick in 1932 made it possible to describe all atomic nuclides as simple groupings of protons and neutrons. The number of protons would be the same as the atomic number in order to give the nuclide its required charge, and the number of neutrons would be those required to give the nuclide its characteristic mass. Heisenberg then suggested that the proton and neutron be called *nucleons* and that they be considered different energy states of the same particle such that in certain nuclear changes each·could decay into the other.

Since there are not beta particles (electrons) as such in the nucleus and their existence in so small a volume would not be in accord with quantum mechanics, it was assumed that the beta particle was produced at the moment of emission by conversion of a neutron to a proton. Useful as the proton-neutron picture of the nucleus was, it became increasingly evident that the nucleus was not a single system and that the nuclear forces and modes of excitation must be investigated. Neither protons nor neutrons are ever emitted by a radioactive nuclide, but certain nuclides emit alpha particles and each alpha particle consists of two protons and two neutrons closely bound together. It must be assumed that only such a grouping can gain enough energy to escape from an unstable type of nuclide.

With the discovery of what has been called "artificial" or induced radioactivity by Irene Curie-Joliot and her husband Frederick Joliot in 1933, it was found that positive electron (positron) emission may occur in nuclides whose instability results from the nucleus possessing "too much charge for its mass." This is, of course, only another form of beta emission represented as β^+, whereas ordinary electron emission is represented as β^-. Since the positron is the antiparticle of the electron it was evident

that other antiparticles could also be involved in nuclear processes.

When induced radioactivity was discovered, only some forty naturally radioactive substances were known and the only known types of nuclear radiation were the alpha (α), beta (β^-), and gamma (γ) rays. To these now had to be added the positron (β^+).

In 1933 Pauli suggested that still another new particle of zero charge might be required to explain how a neutron might decay to a proton in the nucleus. The following year Fermi incorporated this idea into his theory of beta emission and assumed that a *little neutral particle* which he called the *neutrino* (ν) is also emitted. Since the beta particle is an electron and the electron had been found to spin on its axis like a top the new particle seemed required to possess spin in order not to violate conservation of angular momentum when a beta particle is emitted.

The new particle also seemed necessary to prevent violation of the law of conservation of energy. Unlike the more sharply defined energies of alpha particles the beta rays from a single type of nuclide have a wide distribution of ranges and energies, the distribution being approximately continuous up to a certain maximum or "end-point" characteristic of a particular type of nuclide. It was assumed that whenever a beta particle carried off less than its characteristic end-point energy the missing energy must also be carried off by the neutrino. This gave the neutrino the properties of spin and energy but since at that time there appeared to be no measurable loss of mass it was assumed that the new particle had zero mass along with zero charge.

The great difficulty in detecting a particle so elusive and frustrating that most would pass through the earth without hitting anything delayed experimental confirmation until 1956 when Reines, Cowan, and co-workers at Los Alamos with an intense beam of neutrinos (later found to be antineutrinos), ($\bar{\nu}$) finally achieved success. The only difference between the neutrino and antineutrino is the direction of spin.

In the meantime the discovery of two new nuclear particles revealed further complexities in nuclear processes. They were called *mesons* because their masses were intermediate between the masses of the electron and the proton. The first one, the μ meson or *muon* (μ) was discovered in cloud-chamber tracks by Anderson and Neddermeyer in 1937 while studying effects of cosmic rays. The new particle was found to have a mass 137 times that of an electron and it could have either a positive (μ^+) or negative (μ^-) electronic charge. After the war a closely related but still more massive particle the pi meson or *pion* (π) was detected by the tracks it left in nuclear emulsion plates sent aloft in balloons. It had a mass 273 times that of the electron and could have either a positive or negative charge, and could even be neutral, as found later. Both of these new particles are unstable and decay with a mean life of 2.2×10^{-6} sec for the muon and 2.5×10^{-8} sec for the charged pion, the neutral pion being somewhat shorter lived. Since these particles are found in the atmosphere as the result of collision by incoming cosmic rays they may be classed as secondary nuclear radiation. They can now be readily produced by high energy bombardment in the laboratory along with many other new particles.

The close relationship between the electron, the muon, and the pion is indicated by the fact that pions decay in several ways to form muons as follows. The positively charged pion π^+ can decay to a positive muon plus a neutrino:

$$\pi^+ \to \mu^+ + \nu$$

and the negative pion can decay to a negative muon plus an antineutrino:

$$\pi^- \to \mu^- + \bar{\nu}.$$

The muon of either charge in turn decays to an electron of the same charge plus a neutrino (ν) and an antineutrino $(\bar{\nu})$:

$$\mu^\pm \to e^\pm + \nu + \bar{\nu}.$$

In 1962, although the two neutrinos seemed to be the same they were found to not be quite identical, and this presented a problem. They seem to "remember" their parenthood and react only with their respective closely related particles. In order to distinguish them we call one the *electron-neutrino*, represented by ν_e, and the other we call the *muon-neutrino*, represented by ν_μ. A third neutrino, the tau-neutrino (ν_τ) is now recognized. Thus, with their antiparticles, we add six new particles to the list of basic nuclear radiation particles. These particles can now readily be detected in collision processes or in beams from nuclear reactors.

With the increasing complexity of nuclear structure had come recognition that two new forces exist that operate only in the confines of the nucleus. They are called simply the *weak force* and the *strong force* and are required to explain particle emission. These are now added to the well-known electromagnetic force and the gravitational force of classical physics. The strong force acts to bind nucleons together; it must be charge-independent because it binds neutrons to neutrons as well as to protons. The emission of light-weight particles such as the beta particle and the muon (now classed along with neutrinos as *leptons*), is now recognized as connected with the weak force. Those heavy particles such as neutrons and protons that are bound by the strong force are now classed as *hadrons*. And the emission of such particles either spontaneously or by bombardment is therefore concerned with overcoming these basic nuclear forces.

Since hadrons are believed to be composed of quarks, and since atoms in general can be described as combinations of leptons and hadrons, then according to one theory, leptons and quarks may be the ultimate particles from which all matter is made. Quarks are not radiated, being bound too strongly to escape (see QUARKS).

When the muon was discovered it was thought to be the particle predicted by Yukawa that would hold nucleons together by means of an *exchange force* derived from the newly developed theory of quantum electrodynamics. But muons failed to interact sufficiently with nuclei. Than when the pion was discovered it appeared to be the desired particle because it did interact. However, matters soon became more complicated with the discovery of K mesons, found by Rochester and Butler in 1947 as secondary nuclear radiation in cosmic ray experiments. The masses of K particles were found to be a little more than half that of a proton or neutron, and they could be either positively or negatively charged or neutral. Along with pions they are our most important particles now classed as mesons. The charged K mesons have a mean life of 1.2×10^{-8} sec before usually decaying in various ways into one or more π mesons plus a few other particles. The mean life of the neutral K particle is somewhat less.

Following the discovery of K mesons came a great profusion of new particles. Although the mean lives of K mesons seem short on our time scale they actually live an enormously long time compared to the many far more evanescent particles heavier than a proton that have mean lives as little as 10^{-24} or 10^{-26} sec. Such particles obtained by very high energy bombardment may be created with their antiparticles, and some may be identified as higher excited states of other particles.

The determination of which particles are *fundamental* and which ones are not became more complicated and seemed to thwart all attempts at a simplified theory of the nature of matter, until a new development by the theorist Gell-Mann and others suggested all matter may be composed of still smaller basic particles called *quarks*. Although quarks may never be found outside the nucleus they have led to much success in theory by showing how it is possible that many observable particles can have a simple and basic underlying relationship (see QUARKS).

Recent attempts to measure the mass of the neutrino have given more positive results than earlier ones. The newer experiments now indicate the neutrino may possibly have a small but measurable mass of a few electron volts in terms of its energy equivalent. This is to be compared with the electron, itself a tiny particle, which in the same units has an energy equivalent mass of 511,000 eV. The neutrino mass must be a very small quantity, but if the universe is full of neutrinos it could add up to a very large amount.

We are continually bathed in the streams of neutrinos from the sun which result from the nuclear processes that sustain the sun's heat, and the most noteworthy attempt to measure some of the billions of billions of neutrinos that must be radiating from the sun is that of Davis, using a 100,000 gallon tank of detecting fluid a mile deep in a South Dakota mine. The smaller than expected result which has so far been obtained may result from difficulties of detection or other reasons not yet clear.

Although 90% of the neutrino radiation which we receive most likely comes from our sun the number of stars radiating into space could fill all of space with neutrino radiation. Such a profusion of neutrinos, if they possess appreciable mass, might even affect our ideas about the expanding universe: it presents the astronomers with another source of mass to consider in figuring if there is enough mass to cause the expansion to ever stop and reverse itself.

Nuclear radiation presents both hazards and benefits. Radiation from experimental nuclear reactors at various centers now furnishes rich opportunities for further study of the nature of matter and the properties of different radiations. At the same time the hazards of radiation from nuclear power plants and nuclear warfare are becoming more widely recognized and have led to an intense study of such radiation including the long-range as well as the short-range physiological effects. Closely related are the dangers resulting from aging and deterioration of the materials in a reactor plant itself that are continuously exposed to radiation.

The rapid development of nuclear weaponry has led not only to the many problems of the physiological effects but to military problems such as the development of a "clean" bomb and to the possible enhancement of neutron radiation at the expense of explosive power, or the reverse. With all of this the subject of nuclear radiation has expanded out of the laboratory and has become more and more deeply involved in our lives.

ROGERS D. RUSK

References

Feinberg, G., "What Is the World Made Of?" Garden City, N.Y., Anchor-Doubleday, 1977.

Lapp, R. E., and Andrews, H. L., "Nuclear Radiation Physics," Fourth edition, Englewood Cliffs, N.J., Prentice-Hall, 1972.

Segrè, E., "Nuclei and Particles: An Introduction to Nuclear and Subnuclear Physics," Second edition, Reading, Mass., W. A. Benjamin, 1977.

Sorensen, J. A., and Phelps, M. E., "Physics in Nuclear Medicine," New York, Grune and Stratton, 1980.

Early Work

Rutherford, E., Chadwick, J., and Ellis, C. D., "Radiations from Radioactive Substances," New York, The Macmillan Co., 1930.

Cross-references: ELECTRON; ELEMENTARY PARTICLES; FISSION; NEUTRON; NUCLEAR REACTIONS; NUCLEAR RADIATION SHIELDING; NUCLEAR STRUCTURE; POSITRON; PROTON; QUARKS; RADIATION, IONIZING, BASIC INTERACTIONS; RADIOACTIVITY.

NUCLEAR RADIATION SHIELDING

Nuclear reactors, based upon the fission of heavy nuclei, have become an important power source for the generation of electricity. The great intensity of nuclear radiations produced in the reactors requires radiation shielding. Fast neutrons and gamma rays are by far the most penetrating of the radiations produced; therefore, most shield design is concerned with reducing their levels by factors of as much as 10^{15}.

A variety of other systems gives rise to shielding problems which are related more or less closely to those of the fission reactor and these are treated with many of the same methods which are used in reactor shielding. After discussing these various areas of application, the problem of calculating and confirming the design or performance of a radiation shield is outlined.

Areas of Application *Fission Reactors.* In the United States and much of the rest of the world, the fission power plants now being installed for electricity generation (or ship propulsion) are based upon the use of water as the reactor moderator and coolant. As a shield, water has no cracks and it offers simplifications following from the fact that it tends to produce an equlibrium neutron energy distribution due to the inverse dependence upon energy of the hydrogen neutron cross section. This equilibrium fast-neutron spectrum has allowed the use of simple attenuation methods with empirically derived constants to solve many design problems which do not require much detail in the answer. The outstanding problem for light-water reactors has been the streaming of neutrons in the cavity outside the reactor pressure vessel, where rather detailed calculations are required. A summary of approaches to this problem has recently been made available.[1]

Current interest in the development of advanced reactor types centers upon the breeders, especially the liquid-metal-cooled fast breeder reactor (LMFBR) using the ^{238}U to ^{239}Pu breeding cycle. Breeders are essential if nuclear fission is to make a long-term contribution to our energy needs. The LMFBR operates at high temperatures (~500 to 600°C) and with neutrons of high energy (~100 keV average energy for neutrons producing fission). Both conditions tend to exclude hydrogenous materials from the vicinity of the reactor, and sodium, the liquid-metal coolant usually considered, is the dominant medium near the core. The fast-neutron spectrum does not attain an equilibrium after attenuation in any thickness of sodium (to 15 ft) so far as is presently known, and much more detail in the neutron spectra must be predicted for LMFBR shield design.

Fast-neutron streaming through clearance gaps and sodium-filled pipes, which necessarily surround or penetrate the LMFBR shields, must be calculated for geometries which become quite complex in actual practice. For the coolant pipes, both the sodium and the insulation required by the high temperatures are of low density and constitute significant gaps in the shield. Specific shielding problems of liquid metal fast breeder reactors have been reviewed by Farinelli and Nicks.[2]

Finally, for fission reactors used to propel or to provide auxiliary power for space vehicles, the weight of the power plant becomes crucial to the feasibility of the mission. Since the radiation shield may account for as much as two-thirds of the total power-plant weight, the radiation shield design must be made with the highest attainable accuracy. Efforts are centered upon methods of optimizing the shield design with respect to weight. This can now be done directly and accurately insofar as the power plant can be considered one-dimensional. Two-dimensional optimization methods are less developed.

A thorough review of radiation shielding for fission reactors was published by the U.S. AEC in 1973.[3]

Fusion Reactors. Potential advantages of safety and unlimited fuel supply justify current efforts to develop controlled thermonuclear reactors (CTR) based upon the fusion of light nuclei. The most commonly considered reaction of deuterium and tritium produces 14-MeV neutrons which must be absorbed to prevent radiation damage, to breed more tritium as a fuel, and to gain the energy (heat) which the fast neutrons carry from the reaction. A major requirement of the radiation-transport calculations for CTR systems is to insure radiation levels that will not destroy components of the system; e.g., the cyrogenic magnet used to contain the fusion reaction. Radiation-shielding calculations for fusion systems are closely related to the neutronic design of the lithium-containing blanket which is used to breed tritium. An example of a calculation of this type is given in Ref. 4 for a Tokomak fusion reactor, one of the most highly developed types. Neutron streaming through necessary penetrations of the reactor shield for any type of fusion reactor now visualized will limit the attenuation

of the shield and require careful study.[5] Validation of the analytical methods used will be based upon specifically designed shielding experiments.

Nuclear Weapons Radiations. Nuclear explosives may produce radiations by both the fission and fusion processes and there are substantial overlaps in the radiation-shielding problems with the reactors considered above. For the protection of military or civilian systems against radiations from nuclear weapons, initial attention must be given to the radiation transport through air from the source to the system. As a result of many years of study of this basic problem, air-transport methods and results are now available in detail with accuracies (except for problems with the cross sections for the production of secondary gamma rays by fast neutrons) of the order of 10 to 20 percent.[6]

An overview of the problem of shielding military systems against weapons radiations is given in the multivolume DNA Handbook.[7] For the specific consideration of the shielding protection offered by structures against weapons fallout radiation, a comprehensive review has recently been published.[8]

Accelerators. Radiation of virtually all types (including neutrons and gamma rays) and energies can be produced by the many types of charged-particle accelerators. Over the energy range to ~20 MeV there may be quite direct overlap in the radiation-shielding problems with reactors or weapons radiations. For higher-energy accelerators additional reactions and new particle types such as mesons may be involved. High-energy particles which are slowed down (reduced in energy) in the shield eventually pass through or give rise to additional particles in the lower energy range applicable to fission and fusion reactors. Thus methods developed at low energy may be incorporated as part of the solution of accelerator shielding problems. In any case, the methods of solution tend to be closely related.

Space Vehicles. The space environment includes charged particles ranging from the electrons (trapped by the magnetic field of the earth) through protons and alpha particles (helium nuclei) to heavier nuclei. The protons tend to be the most numerous and their interactions the best understood. Nuclear interactions of heavy nuclei are little understood and may constitute a major problem for future manned interplanetary space travel.

All of the charged particles lose energy dominantly by ionization which creates essentially no transport or shielding problems. Nuclear reactions do occur, however, at energies above a few MeV and ignoring these may lead to dosimetry estimates in error by 25 per cent or much more depending upon the incident-particle energy spectrum and system geometry.

The methods employed for space radiation shielding are largely those also used for accelerator shield design. However, in general, the requirements in space radiation shielding are for highly accurate calculations for thin shields. For accelerators, less accuracy is usually required but much thicker shields (approaching or surpassing the attenuations of reactor shields) may be required. Space radiation shielding is treated in a comprehensive handbook.[9]

Nature of the Problem Stated generally, the problem of radiation shielding is to solve the Boltzmann transport equation which describes the radiation transport. It is possible in principle, of course, to make direct measurements of the necessary radiation attenuations, but in practice this is seldom practical. Thus, given the necessary nuclear data, radiation-shielding analysis reduces largely to (1) developing necessarily approximate methods of solving the Boltzmann equation, (2) testing their validity, usually against measurements in simplified geometries, and (3) applying these methods to design problems.

The Boltzmann equation may be considered simply as a balance equation tabulating the sources and disappearances of particles. In this sense, and ignoring time dependence, it may be written as:

$$\nabla \cdot \overline{\Omega} \phi(\bar{r}, E, \overline{\Omega}) + \Sigma_t(\bar{r}, E) \phi(\bar{r}, E, \overline{\Omega})$$

$$= S(\bar{r}, E, \overline{\Omega}) + \iint \Sigma_s(\bar{r}, \overline{\Omega}' \to \overline{\Omega}, E' \to E) \times$$

$$\phi(\bar{r}, E', \overline{\Omega}') \, dE' \, d\overline{\Omega}'$$

where ϕ is the number flux of particles of energy E, at position $\bar{r}$, with angular direction $\overline{\Omega}$. The left side of the equation represents all losses including leakage out of the region of interest and disappearance by reactions of all types with probability $\Sigma_t \phi$. On the right side, S represents all neutron sources such as fission and the integral includes in-scattering into the spatial and energy region of interest of particles scattered from elsewhere. This "scattering" process may in the most general sense include the production of gamma rays upon the interaction of neutrons.

Solving this integro-differential equation requires input for the radiation sources S and interaction probabilities Σ including details about secondary particles which may be produced. The allowable radiation levels corresponding to $\phi(E)$ are generally provided as constraints determined by the radiosensitivity of man, materials, or systems which may be exposed to the radiations. Then one solves for the minimum shield thickness (or weight or cost) consistent with the constraints.

Nuclear Data *Sources of Radiation.* Fission produces neutron and gamma-ray sources important for shield design. The characteristic

and much-studied distribution in energy of neutrons ($\sim$5 MeV/fission total energy release with $\overline{E} \sim$2 MeV) varies with the isotope undergoing fission and slowly with the energy of the neutron producing fission. Delayed neutrons, which follow β-decay of fission fragments, have an intensity relative to the "prompt" neutrons of the order of 1 to 2 per cent and become important largely for fluid fuel reactors where they may be emitted beyond part of the primary shield and thus enhanced in importance. Gamma rays are emitted at essentially all times after fission but a major share ($\sim$8 MeV/fission) appear within a few nanoseconds. Gamma rays following isomeric transitions in fission-fragment nuclei (important for times after fission $\lesssim$ 1 sec) and following β-decay (for $t > 1$ sec) together amount to another $\sim$7 MeV/fission.

The 14-MeV neutrons produced in fusion of light nuclei constitute the dominant radiation for a fusion source. For high-energy accelerators or space vehicles, the sources of penetrating radiations are dominantly due to nuclear interactions of the primary charged particles. μ-mesons, formed from the decay of other particles, are extremely penetrating due to their small interaction cross sections, and they tend to dominate parts of the shield for accelerators with energies > 10 GeV.

A broad review of radiation sources is given in the "*Engineering Compendium of Radiation Shielding.*"[10]

Cross Sections. The probability of interaction or cross section must be treated in great detail in radiation-shielding problems, especially for neutrons. The cross sections are strongly energy-dependent, in part with rapid variations or fluctuations with neutron energy. Neutrons may (1) disappear, giving rise only to charged particles of no consequence for the radiation transport; (2) produce fission or fusion with the sources these reactions imply; (3) scatter, changing energy and direction, thus introducing "new" particles into the transport equation; (4) scatter as in (3) with the additional production of deexcitation gamma rays; or (5) be "captured" with the emission of capture gamma rays. For shielding, one requires the sum of gamma rays produced in fission, capture, and inelastic scattering.

The vast amount of cross-section information required for accurate shielding calculations constitutes a central problem, and perhaps more effort is applied altogether in cross-section preparation than in transport calculations. Neutron cross sections are measured at many "points" in energy. These detailed point sets are then commonly reduced to a limited number of energy groups ($\sim$30 to $\sim$300 groups) with the energy variation in cross section within each group represented by a weighted average. Weighting schemes approximate the proper but unavailable weighting by $\phi(E)$ for the problem being considered. Since $\phi(E)$ is a function of position, no single set of weights can be exact. The group structure must be especially selected for difficult problems. The validity of sets of group cross sections must be established by comparing the results of calculations using them with values obtained from calculations with point cross-section sets or from benchmark experiments.

The use of group cross sections allows treating scattering by introducing group-to-group transfer matrices with a large reduction in necessary information storage. Similarly, group cross sections can be readily generated to handle secondary gamma-ray production.

As for neutrons, gamma-ray cross sections can be considered as point or group-averaged functions of energy. Gamma-ray cross sections tend to be known more accurately than those for neutrons, and the production of other particle types (e.g., neutrons) by gamma rays can usually be ignored for low-energy transport problems.

Nuclear Data Stores. The wealth of cross-section detail demands computer-based handling and efficient storage. A National Nuclear Cross Section Center (NNCSC) has been established at the Brookhaven National Laboratory for storing neutron and gamma-ray cross sections, and this is the appropriate primary source of information in the United States.[11] Very important is the fact that a cooperative group (The Cross Section Evaluation Working Group or CSEWG) has been set up to identify the "best" set of data for each reaction type and nuclei. This choice may require choosing among sets of experimental measurements with different values for the same quantities, or may require filling gaps covered by no measurements, on the basis of nuclear-model calculations or systematics. This Evaluated Nuclear Data File (ENDF), which changes with time as new data become available, is also available from NNCSC.[11] Similar, but more preliminary, data and sets of groups of cross sections derived from them are available from the Radiation Shielding Information Center (RSIC) at the Oak Ridge National Laboratory.[12]

Methods of Solution Exact analytic solutions of the transport equation are not possible for practical shielding problems, but numerous approximate, basically numerical methods are available. Two of these have assumed the greatest importance today, the discrete ordinates and Monte Carlo methods. Applied in their most highly developed forms, the residual approximations become unimportant. The effort and cost required for such applications may be substantial, however. Other methods which are easier to use offer approximations to the desired solution which have more or less vailidity depending upon the complexity of the problem and its similarity to those from which the approximate method has been derived. A review of transport

methods is available in the revised Chapter 3 of the "Weapons Radiation Shielding Handbook."[7]

Virtually all solutions of problems of practical interest depend upon use of a digital computer, and the type of computer available and its cost of operation may constrain the choice of methods. The availability of computer codes developed by others is very important. Codes of shielding interest are collected and made available by the Radiation Shielding Information Center,[12] which also assists new users in making them operable.

Discrete Ordinates. The nature of the Boltzmann transport equation as a balance of particle flow is closely matched by the discrete ordinates formulation as a finite difference equation, and the equivalence has been demonstrated for one-dimensional geometries. Thus, few approximations are necessarily required in the application of this method, which obtains the desired flux $\phi(\overline{r}, E, \overline{\Omega})$ by successive iterations over the difference cells. A derivation and a method of solution of the finite-difference equations which are required for a one-space dimension time-independent problem are given in Ref. 7.

The adaptation of the discrete ordinates method, which was originally developed for other purposes, to radiation shielding required special attention to several features of the method. In particular, the steep flux gradients characteristic of good shields have required modified schemes for iteration within the difference cells. For deep penetration in a shield, the angular distributions of scattered particles become very important since the forward-scattered radiation can penetrate most readily. To handle the anisotropic scattering in a practical manner, the angular distributions after scattering from energy group to energy group are approximated by a Legendre series. A low order of expansion, P_3, usually suffices.

The discrete ordinates methods have been expanded to two dimensions and widely applied to practical design problems. Demands upon computer time have provided incentives for developing improved forced-convergence techniques including those which concentrate upon the spatial or energy region of interest. Flux aberrations have been frequently observed in two-dimensional problems for small radiation sources. These "ray effects" preserve indications of the finite angular directions in the flux solution, and various methods have been devised to eliminate or control them.

Practical computer codes are available[12] which incorporate the discrete ordinates method. One-dimensional calculations are very fast and have been used as the basis of iterative shield optimization procedures which perform a series of flux solutions in one pass on the computer. Two-dimensional calculations require fast computers with large memory stores. It should be noted that the division of neutron energies into groups allows a convenient method of handling the production of secondary gamma rays by neutron capture and inelastic scattering. Proper treatment of these secondary gamma rays is essential in almost all radiation-shielding problems which include fission or fusion sources.

Overall, discrete ordinates is the method of choice for most one- or two-dimensional radiation-shielding problems. The major complication which is encountered in its use is that there is no clear a priori approach to defining the best (or even certainly adequate) sets of directions, space meshes, and energy groups for a given problem. Experience is the best guide to these choices and it must be based upon detailed comparison of test calculations with benchmark experiments which are designed for for the purpose.

Monte Carlo. Under the pressure of necessity, many components of radiation shields may be approximated in one or two dimensions. There remain, however, complicated problems, such as the streaming of neutrons through holes, which require a three-dimensional treatment. At present, the Monte Carlo method is the only approach to such problems which is free of serious approximations.

The Monte Carlo method depends upon stochastic estimation or random sampling from the probabilities describing the stochastic processes that determine the solution of the transport equation. This mathematical analog for a problem of interest may be made as exact as desired, limited only by the investment of effort and computing time. The size of the sample of particles of interest (i.e., those penetrating the shield) determines the statistical accuracy achieved.

The major determinant of the success of the Monte Carlo method for problems whose important parameters depend on very unlikely events, such as penetration through thick shields, is the method of sampling. Through "importance sampling," larger numbers of particles can be chosen which contribute to the unlikely events of interest.[7] Thus the sampling is biased without, hopefully, biasing the result, since correction factors are applied. The most powerful biasing technique uses an importance or value function based upon an adjoint solution of the transport equation, using the best available method, which will usually be one- or two-dimensional discrete ordinates. Such biasing in energy and space is routine but angular biasing has usually been limited to one-dimensional problems. Estimates of the variance calculated upon the basis of a normal distribution may be grossly in error, since Monte Carlo distributions are frequently nonnormal, and unfortunately, error estimates are usually too low. This difficulty must be kept in mind in using the Monte Carlo method for shielding problems.

Several powerful and general Monte Carlo codes are available[12] for a variety of computer types. The inherent flexibility of the method, however, requires considerable effort on the part of the user for solving a given problem. Recent developments in Monte Carlo methods have tended toward systems, such as those using standard neutron energy groups, which ease the user's burdens. Using the same energy groups as those of the discrete ordinates method greatly facilitates comparisons of results. The ultimate verification of results from Monte Carlo calculations as for all other methods must rest upon comparisons with benchmark experiments. A current review of Monte Carlo methods gives most emphasis to shielding applications.[13]

For reactor-design problems the attenuations may exceed those which may be practically treated with Monte Carlo techniques, but parts of the geometry may require three-dimensional analysis. Therefore, coupling codes have been developed[12] which enable solving part of the problem using two-dimensional discrete-ordinates and part in three-dimensional Monte Carlo.

Other Transport Methods. Many other methods have been considered and used to a limited extent for solving the transport equation. In one approach, the angle-dependent terms are represented by expansions in spherical harmonic polynomials, P_n. Development of the method has been restricted by the requirement to consider many terms in the scattering representation and the difficulty of treating multidimensional shielding problems. Limited use of the P_3 approximation has been made for one-dimensional problems, but no two-dimensional calculations are known. Diffusion theory, which corresponds to the P_1 approximation, is widely used for reactor criticality calculations because of its relative simplicity, and multidimensional versions are available. Its validity is limited to nearly isotropic fluxes, however, as opposed to the highly forward-directed fluxes characteristic of shields. Diffusion theory can sometimes be used for the inner portions of a reactor shield or to provide a source for transport shielding calculations, but the current trend is to use the transport calculation throughout the system.

Another approach to solving the transport equation consists of transforming the problem by constructing spatial moments of the flux and calculating in the transform space. This moments method has a number of computational advantages and it was used widely for gamma-ray transport. It is, however, greatly restricted in the geometries that can be considered, and has been applied largely to infinite homogenous media. Furthermore, for neutron transport, difficulties persist in reconstructing the flux from the moments.

The so-called invariant imbedding, or matrix, method does not solve the transport equation directly, but rather solves for reflection and transmission functions throughout the medium. A particular (shield) configuration is imbedded in a larger class of configurations, and solutions are obtained for all in one calculation, an obvious efficiency. Definition of reflection and transmission functions and their solution by numerical methods do not differ greatly from other approaches to the transport equation. Boundaries are precisely included in the formulation, it is well suited for heterogeneous shields, it is efficient for thick shields and inefficient for thin. The outstanding limitation is that applications have been made only for slab geometries.

Approximate Methods. Before large computers made transport solutions practical, the kernel technique for radiation transport based upon the use of a Green's function was widely used. The point kernel, $K(|\bar{r} - \bar{r}'|)$, relates the desired response of a detector at point $\bar{r}$ to a radiation source at point $\bar{r}'$. Integration over a source volume yields the solution to problems of arbitrary geometry. However, values for the kernels have been obtained from solutions by methods valid only for infinite homogeneous media, and these kernels give results which are in error when applied to finite media. Scattering may be approximated by adding one of a variety of build-up factors as a function of energy E and separation $|\bar{r} - \bar{r}'|$. For gamma rays, useful build-up factors were derived from moments-method calculations.

For neutron attenuation, many shields contain large regions of water following other materials. This water has the effect of filtering out scattered neutrons, making the material appear as an absorber. The absorption or removal cross section, as derived empirically from experiment, was used in defining a kernel applicable for shields with outer hydrogenous regions. Other kernel approaches have been used in order to parameterize the results of transport solutions.

A logical extension of the removal concept was to follow the deep penetration of neutrons as described by the removal process by moderation in energy using diffusion theory. This removal-diffusion method has been highly developed and widely applied, especially in Europe, in the design of shields for power reactors using fission induced by thermal neutrons. The basic assumptions upon which the method is based are not fulfilled for fast reactors. The removal process is considered for many energy groups (to 18) from which neutrons can be transferred to other energy groups used to describe the diffusion process. The key step then becomes derivation of the group-transfer matrices from experiments which determine the penetration of monoenergetic neutrons in various materials. A modern reference to approximate shield-design methods is available in a survey of nuclear reactor engineering.[14]

Validation of Calculations Mock-up experiments, which are universally used to validate reactor-core criticality calculations, are much less common for shielding calculations. The penalty for error or overdesign tends to be less and the costs for meaningful mockups excessive. An obvious exception would be the space reactors for which the shield design is crucial. Measurements on reactors as built can in principle be used to validate calculation methods. In practice, however, access to interesting portions of reactor or accelerator shields is limited and down time required for performing meaningful measurements is very expensive. The test is also rather late for judging methods for anything other than serial production of power plants.

The most useful tests of transport methods follow from comparisons with experiments specifically designed for the purpose. Such experiments attempt to reproduce in the simplest arrangement attainable the essential features of the system of interest (e.g., a fast-neutron spectrum for applications to fast reactors or large attenuations in the pertinent materials for thick shields). The experimental results must be accurate, reliable, and unambiguously interpretable above all else. Measurements include the neutron energy distributions penetrating the shield, using one or more of the various neutron spectrometers which have been developed for the purpose. An example of the test of a two-dimensional discrete-ordinates code against measurements of neutron spectra serves to illustrate the process.

Sensitivity Analysis and Channel Theory The complexity of radiation shielding problems has made worthwhile developing sensitivity analysis. The sensitivity of a desired response R (such as a radiation dose outside a complex shield) to changes in input parameters σ_x (such as the neutron cross sections) is just the functional derivative $\partial R/\partial \sigma_x$. Using generalized perturbation theory, it is possible to obtain for a single response sensitivities to all input parameters. This allows identification of those data which are most significant in obtaining accurate attenuation predictions. When combined with estimates of uncertainties in the input data, knowing the sensitivities allows estimating the uncertainty in the desired response. Uncertainties in the model (i.e., effects of choice of energy groups or angular quadrature for discrete-ordinates calculations) can also be taken into account.

Examination of the sensitivity profile (sensitivity as a function of energy, for example) may offer insights into the physical processes (such as "holes" in the cross-section variations with energy) which are important or crucial in determining shield penetration. Sensitivity analysis can and should be used in the design of shielding experiments to demonstrate that the important parameters of the experiment correspond to those in the shield design of interest. A session was devoted to this topic at the last international conference on reactor shielding.[15]

While sensitivity analysis has tended to consider energy regions of greatest importance in attenuation, spatial channel theory is used to locate regions in space that may be of dominant importance. Thus streaming paths for neutrons can be identified and visualized, even for the very complicated geometries of fast breeder reactors.[16] The additional insight into the radiation-transport process thus gained may be used to determine weaknesses in shield designs and/or the optimum placement of shielding materials.

For Further Information The development of radiation-shielding technology is almost entirely dependent upon high-speed digital computers. This dependence of reliable solutions to transport problems upon computer codes with masses of nuclear data required as input precludes the handbook approach to practical design problems. Therefore, a Radiation Shielding Information Center has been established to assist in providing the necessary tools for solving shielding problems and in providing means of retrieving the pertinent literature. The Center does not attempt to solve problems directly. It does provide coverage of the literature for all of the indicated areas of application. It also collects, packages, evaluates, and makes available to users the computer-based methods and data which have been developed. Feedback from users regarding their problems and successes in applying the methods is important in leading to improvements. The Center may be contacted at the address given.[12] The Center was responsible for publication of the "Proceedings of the Fifth International Conference on Reactor Shielding," held in Knoxville in 1977.[15] Included here are a wide range of relevant papers.

A general text on radiation shielding and dosimetry was written by A. E. Profio and published in 1979.[17]

FRED C. MAIENSCHEIN

References

1. Lahti, G. P., et al. (Eds.), "Radiation Streaming in Power Reactors," RSIC-53, February 1979.
2. Farinelli, U., and Nicks, R., "Physics Problems of Fast Reactor Shielding," *Atomic Energy Review* 17, 1 (1979).
3. Schaeffer, N. M. (Ed.), "Reactor Shielding for Nuclear Engineers," U.S.A.E.C. TID-25951, 1973. Available from National Information Service, Springfield, Virginia 22151.
4. Santoro, R. T., et al., "Neutronics and Photonics Calculations for the Tokamak Experimental Power Reactor," *Nucl. Tech.* 37, 274 (1978).
5. Santoro, R. T., et al., "Monte Carlo Analysis of the Effects of a Blanket-Shield Penetration on the Performance of a Tokamak Fusion Reactor," *Nucl. Tech.* 37, 65 (1978).
6. Straker, E. A., "The Effect of the Ground on the Steady-State and Time-Dependent Transport of Neutrons and Secondary Gamma Rays in the Atmosphere," *Nucl. Sci. Eng.* 46, 334 (1971).
7. Abbott, L. S., Claiborne, H. C., and Clifford, C. E. (Eds.), "Weapons Radiation Shielding Handbook," DNA-1892 (formerly DASA 1892). Chapter 3 has been issued in revised form as DNA-1892-3, Rev. 1 (March 1972). Available from National Technical Information Service, Springfield, Virginia 22151.
8. Spencer, L. V., et al., "Structure Shielding Against Fallout Gamma Rays from Nuclear Detonations," NBS Special Publication 570, U.S. Government Printing Office, 1980.
9. Alsmiller, R. G., Jr., et al., "Shielding of Manned Space Vehicles Against Protons and Alpha Particles," ORNL-RSIC-35, Nov. 1972.
10. Jaeger, R. G., et al. (Eds.), "Engineering Compendium on Radiation Shielding," Vol. I, "Shielding Fundamentals and Methods," New York and Berlin, Springer-Verlag, 1970.
11. National Nuclear Data Center, Brookhaven National Laboratory, Upton, L. I., N.Y. 11973 (516/345-2091).
12. Radiation Shielding Information Center, Oak Ridge National Laboratory, Post Office Box X, Oak Ridge, Tennessee 37830 (615/574-6176).
13. "Proceedings of a Seminar-Workshop, A Review of the Theory and Application of Monte Carlo Methods, Oak Ridge, TN, April 21–23, 1980," ORNL/RSIC-44, August 1980.
14. Glasstone, S., and Sesonske, A., "Nuclear Reactor Engineering," New York, Van Nostrand Reinhold, 1981.
15. Roussin, R. W., et al. (Eds.), "Proceedings of the Fifth International Conference on Reactor Shielding, Knoxville, Tennessee, April 18–22, 1977," Science Press, 1977.
16. Williams, M. L., and Engle, W. W., Jr., "Channel Theory in Shielding Analysis," *Nucl. Sci. Eng.* 62, 92 (1977).
17. Profio, A. E., "Radiation Shielding and Dosimetry," New York, Wiley, 1979.

Cross-references: FISSION, FUSION, NUCLEAR INSTRUMENTS, NUCLEAR RADIATION, NUCLEAR REACTIONS, NUCLEAR REACTORS, NUCLEAR STRUCTURE, TRANSPORT THEORY.

NUCLEAR REACTIONS

Several years after the discovery of the nucleus by scattering α-particles from gold, Rutherford and his collaborators noticed that if air were exposed to the flux of α-particles, occasionally a very penetrating particle was observed. After some nuclear detective work, this phenomenon was explained in the following way: the nitrogen nucleus and the α-particle react to produce an isotope of oxygen and an energetic proton. In chemical notation, such a reaction may be written as $^{14}N + \alpha \rightarrow {}^{17}O + p$, where the superscripts are the atomic mass numbers of the elements in question.

Nuclear reactions may take place only when a target nucleus and a projectile come close enough together for the nuclear forces to take effect. The range of nuclear forces is very short, about 1.5 fermis (1 fermi = 10^{-13} cm). Since nuclei and all the massive projectiles except neutrons are charged positively, they repel each other; and if they are to be brought into sufficiently intimate contact to interact, the energy of the projectile must equal or surpass the repulsive electrostatic force. Typically, for protons on light nuclei, the energy required for a reaction to take place is 1 or 2 MeV rising to 15 MeV for protons on very heavy, highly charged nuclei. Thus to produce nuclear reactions by the collision of charged particles, we must first accelerate one of them to an energy sufficient to overcome the electrostatic repulsive force. In all nuclear reactions, energies are usually given in million electron volts (MeV) and masses in atomic mass units (amu). One MeV is the energy acquired by a particle of one electronic charge as it is accelerated by an electrical potential of one million volts. An atomic mass unit is defined as one-twelfth the mass of a neutral carbon-12 atom.

A nuclear reaction, like its chemical counterpart, may be exoergic (kinetic energy is liberated) or endoergic (kinetic energy is absorbed). For example, Rutherford's original reaction is endoergic, consuming 1.19 MeV which is converted into mass of the product particles according to Einstein's $E = mc^2$. In the case of endoergic reactions, energy must be supplied in the form of projectile kinetic energy to make the reaction possible.

In the course of a nuclear reaction, the projectile and the target may fuse completely or parts of nuclear matter, neutrons, protons, or clusters of these may be transferred from the target to the projectile or vice versa.

The projectiles used to produce nuclear reactions are usually energetic charged particles. In the early studies of nuclear reactions, the most common projectiles were the ISOTOPES of hydrogen, protons or deuterons, as well as α-particles. Because these projectiles have small electric charges, they feel the lowest electrostatic repulsion from the target and need the least amount of energy to accomplish their purpose. Recently, however, increasing emphasis is being placed on heavy ion projectiles, from lithium to uranium. Because the heavy nuclei carry a large charge, they require much more energy to approach the target closely enough for the nuclear matter to interact. Usually the energy is expressed in terms of MeV per nucleon (MeV/A); and for heavy ions the desirable energy for nuclear reactions is around 10 MeV/A; thus, for example, the total energy for uranium nucleus would be 2,380 MeV. Another field of nuclear reactions that is increasing in popularity uses very high energy electrons (500 MeV to 5 GeV) usually produced by linear accelerators.

Neutrons form a special class of projectiles, since they are not charged and need not possess any large amount of energy to overcome a repulsive electrostatic barrier. In fact, the slower the neutron, the more likely it is to interact with a target simply because it spends more time in the vicinity of the target nucleus. Nuclear reactions may also be initiated by PHOTONS (electromagnetic radiation quanta) of very high energy, greater than 5–10 MeV. The photons are produced as x-rays when high-energy electrons impinge on a target. Since photons are not charged electrically, there is no barrier; but all photonuclear reactions are endoergic, that is, they require several-million-electron-volt x-rays to take place. High energy accelerators produce copious amounts of unusual projectiles: π- and K-mesons and other exotic particles with a sufficiently long half-life. While research with these projectiles is not terribly common at the present time, there are indications that their use will increase.

Rutherford's original reaction would now be written thus: $^{14}N(\alpha, p)^{17}O$. The target nucleus comes first, the projectile and the emitted particle or particles appear in that order inside the parentheses, and the residual nucleus is last. The superscript gives the atomic mass of the isotope. Some common abbreviations are p for proton, n for neutron, d for deuteron (^2H), t for triton (^3H), γ for γ-ray or photon.

The projectile energy range over which nuclear reactions are studied varies from a fraction of an electron volt for neutrons to several hundred million electron volts per atomic mass unit for charged particles. At very high energies, lighter projectiles appear to interact with the individual neutrons and protons of the target rather than with the nucleus as a whole, and we leave the domain of nuclear reactions to enter the field of ELEMENTARY PARTICLE interactions.

Nuclear reactions are literally the foundation on which our world is built. The energy of the sun is nuclear in origin deriving principally from the fusion of four protons into a helium nucleus in the course of which 22.7 MeV are released in each fusion. The fusion is not a direct four-body reaction but rather proceeds by stages through many two-body reactions (see SOLAR ENERGY SOURCES). Not only is our chief source of energy of nuclear origin, but the constituents of the earth are also the result of long-gone nuclear reactions, principally a series of (n, γ) processes which served to build up the elements in the earth as we now find them. One particularly pleasing success of nuclear reaction theory and experiment is the fact that the abundance of the elements and their isotopic ratios as they occur in nature can be calculated simply on the basis of the probability with which neutrons are captured by various nuclei, these probabilities having been measured in nuclear research centers.

Not only nature's energy source is of nuclear

origin. The two mightiest sources of man-made energy, nuclear FISSION and nuclear FUSION, derive from nuclear reactions.

In fission, a neutron is captured by a uranium nucleus which splits (fissions) into, say, a Ba and Zr nucleus, in the process releasing about 200 MeV of energy and some neutrons which in turn split other U nuclei in the vicinity. This leads to the familiar chain reaction which, if controlled, is used to produce power by converting the fission energy (heat) into electricity. If the chain reaction is allowed to proceed without control, a violent explosion results, i.e., an atomic bomb. (We note parenthetically that what is commonly known as atomic energy should really be called nuclear energy since its source is not the entire atom, but only its nucleus. Burning of coal, on the other hand, is atomic energy since heat is derived from the combination of a carbon atom with two oxygen atoms to form CO_2.)

Fusion as a source of energy derives from a reaction such as $^2H(d, n)^3He$ in which 3.3 MeV are released. This process is similar to the source of solar energy in that hydrogen nuclei fuse to produce helium. Controlled fusion reactions that proceed slowly and are contained spatially are being studied at many laboratories, and much progress has been made to the ultimate goal: a controlled fusion reactor, which however is still several decades in the future. An uncontrolled fusion reaction has been achieved—it constitutes the energy source of the hydrogen bomb.

It is clear that reserves of coal and oil must someday be exhausted. When this happens, mankind will require new sources of energy. Two candidates for this are derived from nuclear reactions, fission or fusion. With due care, either one could become the source of energy for the next several millennia; and in fact, fission energy is likely to be the major source of electrical power in many countries by the end of the century.

The terrors of war and the blessings of abundant power both come from nuclear reactions. But that is not all. Perhaps the most significant contribution to science has been the use of radioactive elements produced in nuclear reactions. Let us take a typical example. Consider the reaction $^{13}C(n, \gamma)^{14}C$. This reaction produces radioactive ^{14}C which has a half-life of 5700 years. The radioactive carbon decays with the emission of an electron, which can be counted with a suitable detector. Thus we are able to locate individual atoms of carbon and separate them from all others which are not radioactive. Such radioactive atoms are called tracers; and by using a variety of them, ^{13}C, ^{18}F, ^{32}P, ^{35}S, ^{131}I, ^{198}Au, all produced by some sort of nuclear reaction, unprecedented advances have been made in biology, geology, chemistry, metallurgy, physiology, and medicine. Furthermore, radioactive isotopes in large amounts can be used instead of x-rays for treatment of cancer, for metallography, and for food preservation.

By means of nuclear reactions such as ^{241}Am $(^{11}B, 4n)^{248}Fm$, scientists have been able to make new elements not found in nature. Some of these, for example plutonium and californium, have found important uses as reactor fuels or portable power sources. The preparation of new elements has played a decisive role in our understanding of the chemistry of heavy elements.

We turn now to a description of nuclear reactions. These may be regarded as proceeding in two principal ways. First, the colliding particles may fuse, their components get thoroughly mixed in a very "hot" compound nucleus. This compound nucleus may exist in a heated state for a period varying between 10^{-16} and 10^{-20} second, a time we normally consider imponderably short but which is nevertheless long on the nuclear time scale when compared with the transit time of a nucleon across the nucleus (10^{-22} second). The compound nucleus boils off fragments, mainly neutrons, protons, α-particles, and γ-rays, and in this way cools down to a normal energy content. In its final state, it is called a *residual nucleus;* and it may be radioactive or stable depending on the details of the reaction. If the target, the projectile, and the energy of the projectile are known, it is possible to predict what the residual nucleus will be; and thus, if certain isotopes are wanted, one can tailor the reaction accordingly. We note here, however, that the exact mechanism of these reactions is very complicated and is not really understood in a fundamental way.

In the other kind of reaction, the nuclei do not fuse but only part of the nuclear matter is transferred from one nucleus to the other. This is often called a *direct reaction* and includes stripping reactions, pickup reactions, and transfer reactions. A typical example is $^{27}Al(d, p)^{28}Al$, where a neutron is stripped from the deuteron and caught by the Al nucleus while the remaining proton stays relatively undisturbed on its original course. The nuclei in such reactions do not come into intimate contact, and the reaction is fast on the time scale of 10^{-22} second.

In general, light projectiles are used to study individual states of nuclei. For example, if a proton is scattered inelastically, it will deposit a certain amount of energy in the target nucleus before it emerges from the collision. By studying the amount of energy deposited, the angle at which the proton leaves its target, and sometimes by measuring the spin of the proton before and after collision, physicists are able to obtain detailed information about the energy levels in the nucleus such as the spin and parity of a particular state. The most common reactions used for this kind of study are inelastic scattering, stripping and pickup reactions such as (d, p) or (d, t) reactions. In order to interpret reactions of this type, physicists employ an *optical model*, in which the nucleus in all its complication is replaced by a complex potential containing a real part, an imaginary part responsible for nuclear absorption, and a spin-orbit part that accounts for the fact that many of the projectiles have spin which influences the reaction process. The optical model and the distorted wave Born approximation calculations for direct reactions have been very successful in analyzing the details of the nuclear reactions and enable physicists to interpret reaction data and thus learn more about the structure and properties of nuclei, as well as about the reaction process itself.

Heavy ion reactions are used to study an entirely different set of nuclear properties. Instead of learning about details of the nucleus, with heavy ions physicists can study the nucleus as a whole, learn about its viscosity, about oscillations and deformability of nuclei, about the compressibility of nuclear matter and even about the possibility of observing nuclear shockwaves. The nuclear model that is most useful here is the *liquid drop model* invented in the early 1940s to describe fission. In this model, the nucleus is represented by a liquid drop that rotates, changes shape, has a surface tension, and in other ways mocks up a nucleus. A particularly interesting heavy ion reaction is *deep inelastic scattering*, a process in which two nuclei just touch then rotate together for a fraction of a full rotation while they exchange between them significant amounts of nuclear matter and energy. These reactions are intermediate between compound nucleus and direct reactions. Another useful property of heavy ion reactions is the capability to produce isotopes far from stability simply by the appropriate selection of the target and projectile. For example, nuclides such as tin-100 are now being looked for in heavy ion experiments. (Recall that the most stable tin isotope is ^{120}Sn, and the lightest stable tin nuclide is ^{112}Sn.)

Electrons are used almost exclusively for the detailed study of nuclear structure. Electrons have the advantage that their interaction with nuclei is purely electromagnetic and thus understood theoretically. In principle, then, when the interaction is understood the information about nuclei is much cleaner and more readily interpreted. There is also hope that with high-energy electrons one can probe the subnuclear structure due to quarks in nuclear matter.

Exotic projectiles such as π- or K-mesons at the present time provide interesting information about the nuclear surface, or what one might even call the nuclear stratosphere. It is well known that nuclear densities fall off rapidly with radius, and it is a matter of considerable controversy just what the structure of the surface is, whether there are mostly neutrons or mostly protons, and how the density decreases for various nuclei. π- and K-mesons form mesic atoms in quasi-electronic states. If the state is known, its radius is known, and its interaction with the nucleus then provides information about nuclear density at that radius.

In spite of many years of research, a deep understanding of nuclei and a theoretical ability to predict their properties still elude us. Nuclear reactions are a prime tool for the investigation of nuclei. By careful study of nuclear reactions in all their multiplicity of detail, including the precise knowledge of the projectile, its energy and its spin direction, the nature, as well as energy, angle of emission, and the spin of the reaction product, a good deal of information has been gained about this realm. A deep understanding of nuclei yet eludes us. We have no accurate predictive theory, even though we have searched for it for many decades. The prize, however, is worth reaching for: its promise is to reveal the nature of nuclei, the smallest stable constituents of matter, and the laws which govern their interactions to give man immense power for war or for peace.

ALEXANDER ZUCKER

References

Two advanced books which deal with the nuclear reactions are:

Cohen, B. L., "Concepts of Nuclear Physics," New York, McGraw-Hill, 1971.
Satchler, G. R., "Introduction to Nuclear Reactions," New York, Wiley, 1980.

Cross-references: ACCELERATOR, PARTICLE; ELEMENTARY PARTICLES; FISSION; FUSION; ISOTOPES; MESONS; NEUTRONS; NUCLEAR STRUCTURE; PHOTON; PROTON; RADIOACTIVITY; SOLAR ENERGY SOURCES; TRANSURANIUM ELEMENTS.

NUCLEAR REACTORS

Definition A nuclear reactor is a device in which the fission of a nuclear fuel releases large amounts of energy in a controlled chain reaction involving neutrons. Reactors serve as important heat sources for the commercial generation of electricity and can as well supply neutrons and useful radioactive isotopes.

Fission and the Chain Reaction When a uranium-235 nucleus absorbs a neutron, an excited compound nucleus of uranium-236 is formed. In about 15% of the cases this results in gamma ray emission, while in 85% there is a splitting of the nucleus into two fragments. These fly apart with high speed, with total kinetic energy around 200 million electron volts (MeV). In this fission process beta particles, gamma rays, and an average of about 2.5 high-energy neutrons are emitted. The neutrons induce additional fissions, and under proper conditions a neutron chain reaction results.

Since only one neutron is needed to continue the chain, extra neutrons are available. Depending on the size and materials of the reactor, these neutrons will be captured in nonfissionable materials or will escape from the reactor, a process called *leakage*. If uranium-238 is present, neutron capture can lead to the production of plutonium-239.

Fission fragments consist of a great variety of isotopes, with the most likely split in the mass ratio 3:2. They bear most of the kinetic energy that can be transformed into thermal energy and then into electrical energy. The heavy particles resulting from fission constitute the collection of material called *fission products*. Many of the isotopes are radioactive, with half-lives ranging from a fraction of a second to millions of years. Although there are some useful radioisotopes among them, the bulk must be regarded as a waste, to be disposed of in some safe manner.

When on the average each neutron reproduces another neutron, the assembly is said to be *critical*. Similarly, if fewer or more neutrons are produced per cycle, the system is *subcritical* or *supercritical*. These three cases correspond to a neutron multiplication factor of exactly 1, less than 1, and greater than 1, respectively.

The chance of capture or fission of nuclei tends to vary inversely with neutron energy. Low energy neutrons (~ 0.025 eV) are thus said to have a higher cross section than high energy neutrons (~ 1 MeV). A target isotope is classed as *fissile* if slow neutrons can induce fission.

History The discovery of fission in Germany in 1939 led to the initiation of a major effort in the United States in World War II to exploit the new energy source for military purposes. Two paths of research were followed—(a) to separate the desired uranium-235 (0.7% of uranium as found in nature) from uranium-238 and (b) to produce plutonium by means of neutron capture in uranium-238. Each of the programs was successful and eventually led to peacetime applications.

The first "pile" (nuclear reactor) was built at the University of Chicago by Enrico Fermi and his colleagues. The reactor consisted of lumps of natural uranium metal and oxide imbedded in a large block of graphite. On December 2, 1942 it "went critical" and demonstrated that a self-sustained chain reaction was possible. The reactor served as a model for plutonium production reactors constructed at Hanford, Washington. Uranium isotope separation was achieved at Oak Ridge, Tennessee by two different processes—electromagnetic and gaseous diffusion—leading to uranium enriched to about 90% in uranium-235. Atomic bombs were developed at Los Alamos, New Mexico, in the period 1942–1945. One was composed of U-235, the other of Pu-239. The war with Japan was terminated through the use of these weapons, and a new era of peacetime applications of nuclear energy began.

During the period 1945–1960 many reactor concepts were studied by industry in cooperation with the U.S. Atomic Energy Commission. Out of these, three favored types emerged: the pressurized water reactor (PWR), developed and marketed by companies such as Westinghouse Electric, Babcock and Wilcox, and Combustion Engineering; the boiling water reactor (BWR), mainly a product of General Electric; and the high-temperature gas-cooled reactor (HTGR), built by General Atomic. In Canada, the heavy-water reactor CANDU was developed, and in Great Britain the main power source was the gas-cooled natural uranium graphite reactor.

The number of power reactors in operation throughout the world, as of June 30, 1982, is given in Table 1. The United States has 77 in operation and a comparable number on order or under construction. Worldwide, 23 countries have 265 reactors in use.

Reactor Components The only ingredients required for a chain reaction are a fissile isotope and neutrons. The main fissile materials are uranium-235, plutonium-239 and -241, and uranium-233, produced by neutron capture in thorium-232. Neutrons of high energy (~ 0.9 MeV) are required to cause fission in uranium-238. Practical operation demands as well a variety of components for heat removal, radiation protection, and control.

Fuel. Various combinations of heavy isotopes comprise the "fuel" of a reactor. This fuel is "burned" in the sense that it is consumed and gives up energy. Examples of fuel are natural uranium, uranium slightly enriched ($\sim 3\%$) in uranium-235, and highly enriched uranium ($\sim 90\%$).

Control Absorption. A reactor that is to be operated steadily for long periods of time requires fuel in excess of the amount required to be critical. Thus, at the start of operation, an excess of neutron-absorbing material such as boron must be present. Typically this control absorber is in solution or in the form of movable rods.

TABLE 1. WORLD REACTORS
(As of 30 June 1982).

	Number of Operable Reactors	Electrical Capacity, Megawatts
United States	77	57,977
France	29	19,988
Japan	25	16,652
USSR	31	16,515
Germany FR	12	9,801
United Kingdom	32	8,048
Sweden	9	6,400
Canada	10	5,476
All Others	40	17,492
TOTAL	265	158,349

Moderator. The neutron fission cross section for uranium-235 for neutrons of energy 0.025 eV is around 580 barns (1 barn = 10^{-24} cm^2). At 2 MeV, the initial energy of fission neutrons, the cross section is only 1.3 barns. Thus to maintain a chain reaction with a small amount of fuel, it is desirable to reduce ("moderate") the neutron energy through collisions with nuclei of a light element such as hydrogen or carbon. Moderators such as light water (H_2O), heavy water (D_2O), graphite (C), and beryllium oxide (BeO) slow neutrons to the energy corresponding to thermal agitation in reactors called "thermal." In contrast, a reactor without moderator is called "fast" because the neutrons remain at high speed.

Coolant. In order to prevent melting and to extract useful heat, a cooling agent is needed. Most reactors use ordinary water as coolant; a few use heavy water, liquid sodium, or helium. Reactor types are sometimes distinguished by the coolant used, e.g., light water reactors (LWR's). These are of two types. In pressurized water reactors (PWR's) the coolant is at pressure over 2000 pounds per square inch and at temperature around 600 degrees Fahrenheit. Heat is transferred from a primary circulating coolant loop by use of a heat exchanger (steam generator). In boiling water reactors (BWR's) at somewhat lower pressure and temperature, steam is generated within the reactor.

Structure. The fuel in typical reactors is formed into small cylindrical pellets of UO_2 of volume about 1 cm^3. These are placed in long (3 m) metal tubes to form what are called fuel rods. The tubes provide structural support and also prevent fission products from escaping into the coolant. Fuel rods are spaced to allow coolant flow between them to remove fission heat generated in the UO_2. Around 200 rods form a bundle called a *fuel assembly*; around 180 assemblies form the reactor *core*, the region of the system in which the chain reaction takes place. The core is supported within a thick-walled steel container called a *pressure vessel*.

Shielding and Containment. To protect workers against radiation released by the reactor, concrete shielding is erected. The reactor vessel and coolant system are located in a large cylindrical building with steel-reinforced and steel-lined concrete walls. This containment is designed to withstand pressure in the event of an accident to the coolant system.

Reactor Physics The specialized branch of physics that deals with experiments and theory of reactors is called *reactor physics*. This consists of measurements and calculations of neutron cross sections, the number and energy distribution of fission neutrons, multiplication factors, critical sizes and masses, and power reactor performance. The latter involves the amount of control absorption required, neutron and power spatial distributions, fuel consumption and production, and isotope generation.

Studies are made of both static (steady state) and dynamic (transient) behavior, using neutron diffusion theory based on Fick's law for approximate calculations, and transport theory based on the Boltzmann equation for more precise analysis. The quantity of interest is called neutron flux, which is the product of neutron number density and neutron velocity. In combination with material cross sections, it determines reaction rates.

The simplest mathematical description of a reactor was devised in the early 1940s. Neglecting neutron losses by leakage, the number of neutrons produced per initial neutron is given by the (infinite) multiplication factor k, which is a product of four components: ϵ, the fast fission factor, representing the effect of high energy neutrons; p, the resonance escape probability, the chance of not being absorbed during slowing; f, the thermal utilization, as the fraction of neutrons absorbed in fuel; and η, the reproduction factor, as the number of neutrons per absorption in fuel. Thus if $k = \epsilon p f \eta$ is greater than 1 a chain reaction is possible. If not, as is the case of natural unmoderated uranium, a chain reaction is impossible. The earliest task of researchers during World War II was to find the combinations and purity of materials that would yield a k value greater than 1. The efforts of those who design, operate, and analyze modern reactors are still directed toward the optimization of k during the operating cycle.

Nuclear Fuel Cycle The principal raw material for nuclear fuel is uranium. It is mined, chemically purified, enriched in the isotope uranium-235, and fabricated into fuel assemblies. These are shipped in special casks by truck or rail to the power plant for storage until needed.

A typical reactor is operated on a one-year cycle, in which it produces steady power except for occasional interruptions for repairs. At the end of the period the reactor is shut down, and about one-third of the assemblies are removed from the core as spent fuel. The remaining two-thirds of the assemblies are relocated in the core and fresh fuel is added. The loading pattern is designed to give a power distribution that is as uniform as possible and to obtain maximum energy from the fissile material. The reactor is brought again to operating pressure and temperature.

The spent fuel, which is highly radioactive, is stored in metal racks in water-filled pools. In some countries such as Great Britain, France, and Japan, the spent fuel is reclaimed by *reprocessing*, which consists of cutting fuel rods into small pieces, dissolving out the uranium oxide, and separating uranium and plutonium from the fission products by chemical treatment. In the United States, spent fuel continues to be stored.

Reactor Safety The physical processes that govern the operation of a reactor in themselves provide inherent safety against accident. One effect is the delayed emission of a small fraction of neutrons, as a consequence of beta-decay governed by a half-life. The time constant of power rise in a slightly supercritical system is very much longer (of the order of seconds) than it would be without delayed neutrons (of the order of milliseconds or microseconds). Another is the negative feedback effect of temperature increase of fuel, as in a power rise. Increased thermal agitation of fuel molecules causes a broadening of neutron resonance cross sections in what is labeled the *Doppler effect*. This has the effect of reducing neutron multiplication, which tends to counteract the initiating power increase.

Other safety factors are provided by equipment design. If it becomes necessary to reduce the power of the reactor quickly, "safety rods" are rapidly driven into the reactor core. These neutron absorbing components are composed of boron carbide or an alloy of cadmium, silver, and indium.

There is residual heat being produced, however, by the radioactive decay of the fission products. Although this heat power amounts to only a small percentage of the fission power, it is high enough to melt fuel if adequate cooling is not provided. In the event of a loss-of-coolant accident (LOCA) in which a pipe in the primary reactor cooling loop breaks, equipment called the *engineered safeguards* (ES) *system* comes into play. Its central feature is the emergency core cooling system (ECCS) which injects cooling water from various sources using special pumps.

Much attention has been given to an accident that occurred in March 1979 at the Three Mile Island reactor in Pennsylvania. The cause was a combination of inadequate design, equipment failure, and reactor operator misinterpretation of events and consequent error in handling the situation. As a result, some radioactive material was released, the reactor was damaged, and the building was seriously contaminated. The incident prompted corrective action in all nuclear power plants by industry and the U.S. Nuclear Regulatory Commission.

Radioactive Wastes from Reactors No use has yet been found for most of the radioactive materials produced in a reactor. Among the fission products are isotopes of short half-life such as iodine-131 (8 days) which constitute a hazard only if radioactive gases are released from the reactor. The longer-lived krypton-85 (about 11 years) does not pose a major problem since it is a noble gas, and being essentially inert it disperses throughout the earth's atmosphere. Two isotopes provide the main radioactivity for the first hundred years after nuclear fuel is removed from a reactor. These are strontium-90 ($\sim$29 years) and cesium-137 ($\sim$30 years), which are beta and gamma emitters, respectively. For the long term, isotopes such as plutonium-239 ($\sim$24,000 years), technetium-99

(214,000 years), and iodine-129 (15.7 million years) are of principal concern.

Several techniques for disposal of radioactive wastes obtained by the chemical reprocessing of spent fuel have been studied. The most likely method will involve immobilizing the fission products by solidifying them in glass and burying them in corrosion-resistant waste containers deep in the earth. Favored geologic media are rock salt, basalt, or granite.

Purpose and Uses of Reactors The principal use of nuclear reactors is in commercial electrical power generation, serving as an alternative to heat sources using coal, oil, or natural gas. These reactors typically generate around 1000 megawatts of electrical power. Figure 1 is a schematic diagram of a pressurized water reactor, showing the reactor vessel, the primary loop, the heat exchanger, turbine, generator, and condenser. Figure 2 shows a group of three reactors at a site in South Carolina.

Reactors are used for propulsion of naval vessels, including submarines and aircraft carriers. Reactors are also used to generate plutonium for military purposes.

Research/training reactors are generally of low power, up to a few megawatts. They are used in universities and government laboratories throughout the world for education and nuclear research, including studies of the effects of neutrons and gamma rays on materials.

Reactors serve as a major source of radioisotopes for a variety of peaceful and beneficial purposes. Neutrons produced by fission are the primary agents by which isotopes are produced, either through capture or through fission.

Uses of Radioisotopes Some examples of radioisotopes and their uses are listed. Cobalt-60, of half-life 5.3 years, emits two gamma rays of average energy 1.25 MeV. These rays are alternative to x-rays for medical diagnosis and treatment of cancer and for industrial radiography. The isotope is also a very promising source for irradiation of food to eliminate pests or to sterilize the food for long-term preservation. Plutonium-238, half-life 86 years, is an alpha emitter produced by neutron irradiation of a neptunium isotope. This isotope serves as an excellent source of heat from which electric power is produced for space vehicles. Tritium, the heaviest isotope of hydrogen (H-3) has a half-life of 12.3 years. Its low-energy beta particles make it a good "tracer" to follow biological processes by measurement of the radiation. Reactors are widely used to irradiate samples to induce radioactivity and thence to determine chemical composition by measurement of characteristic radiation. This process, called *neutron activation analysis*, is valuable for studies of environmental pollution because it can detect minute amounts of impurities.

Breeder Reactor Most reactors in current operation are based on thermal neutrons and consume slightly enriched uranium, obtained by an

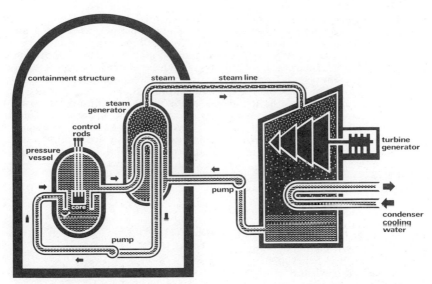

FIG. 1. Fluid flow diagram of pressurized water reactor system. (Courtesy of Atomic Industrial Forum, Inc.)

isotope separation process. Since the main fissile material is the U-235 in natural uranium, large amounts of uranium ore are required. These reactors are classed as converters, with a rather small amount of plutonium production.

In contrast, reactors that do not have a moderator can operate with fast neutrons, where the number of neutrons per absorption in fuel is high ($\sim$3). For each neutron used to perpetuate the neutron multiplication cycle, slightly more than one neutron on the average is available to produce Pu-239 from capture in U-238. The plutonium then can serve as new fuel. The breeding ratio (fissile atoms produced)/(fissile atoms consumed) is greater than 1 and the reactor is classed as a breeder. Such a reactor uses

FIG. 2. The three-reactor Oconee nuclear power plant near Seneca, S.C. (Courtesy of Duke Power Company.)

all the scarce U-235 and indirectly the abundant U-238 as fuel. Employment of breeders instead of converters would reduce the amount of uranium ore required by a factor of about 50. The length of time before resources are seriously depleted could be increased from less than a hundred years to thousands of years. Prototype breeder reactors have been built and development is proceeding, notably in France.

Safeguards and Proliferation Several nations have nuclear weapons capability while many others do not. On the premise that the more countries that have weapons the greater the chance they will be used, there is general apprehension about nuclear proliferation. Reactors can be employed to produce plutonium of weapons grade; plutonium separated from commercial reactor fuel can be made into a crude weapon. Other avenues to nuclear arms capability are available, however, including uranium-235 isotope enrichment. It is believed that any country can go nuclear if it wishes to dedicate enough of its resources, but it is generally felt that every effort should be made to discourage countries from doing so.

Public Attitudes toward Nuclear Reactors The rate of development of nuclear power has been influenced by diverse public opinion over the years. The use of nuclear weapons to end World War II was acclaimed at the time as serving to save lives, but in recent years the continued expansion of nuclear arsenals has become a source of great concern. This fear has been carried over to the peaceful use of fission, with many people believing incorrectly that a reactor can explode like a bomb.

The public generally concedes a need for alternative sources of energy in view of the high cost of oil, but many believe that reactors are unacceptable because of the chance of accident that could release radioactive material.

Concern is also expressed about the accumulation of radioactive wastes and the apparent inability of government and industry to implement a plan for disposing of the wastes.

When nuclear energy first became available, it was believed that electrical power would be extremely inexpensive. Over the years, however, total costs increased because of the introduction of additional safety equipment, escalated construction expense, delays in licensing, and changes in regulatory policy. Nuclear power is cheaper than power from oil-fired plants but about the same as that from coal-fired plants. The high capital cost of modern nuclear power plants—well over a billion dollars for a 1000 MWe facility—is regarded by many consumers of electricity as too large a burden for them to bear.

Because of uncertainties in the regulatory situation, public attitude, and economic factors, new nuclear plants have not been ordered by utilities in the U.S. for several years and construction has been delayed or suspended on some reactors already ordered. In contrast, nuclear plant construction is going ahead in several foreign countries where energy needed cannot be met by the use of coal and where the regulation and construction situation permits more rapid completion of facilities than is possible in the U.S.

It is likely that nuclear reactors will provide a significant amount of the world's electric power for several centuries. For the longer term, fusion reactors using deuterium as fuel might be found to be feasible and practical as an alternative nuclear energy source.

RAYMOND L. MURRAY

References

Henry, Allan F., "Nuclear-Reactor Analysis," Cambridge, MA, MIT Press, 1975.

Hewlett, Richard G., and Anderson, Oscar E., Jr., "The New World," Vol. 1, 1939/1946, U.S. Atomic Energy Commission, 1972.

Hewlett, Richard G., and Duncan, Francis, "Atomic Shield," Vol. 2, 1947/1952, U.S. Atomic Energy Commission, 1972.

Murray, Raymond L., "Nuclear Energy," Elmsford, NY, Pergamon Press, 1980.

Murray, Raymond L., "Understanding Radioactive Waste," Columbus, OH, Battelle Press, 1983.

Cross-references: ATOMIC ENERGY, FISSION, FUSION, ISOTOPES, NEUTRON, NUCLEAR REACTIONS, RADIOACTIVITY, NUCLEAR RADIATION SHIELDING.

NUCLEAR STRUCTURE

The basic problem of nuclear structure was first sharply defined in 1932 when Chadwick discovered the neutron. The nucleus was then recognized to be a system of A particles (nucleons) of nearly equal mass (Z protons and $N = A - Z$ neutrons). The forces binding the "nucleons" together must be quite distinct from the well-known electromagnetic interactions which will evidently operate to push the nucleons apart, as well as gravitational forces which appear to be too weak to have any influence in the nucleus at all.

The nuclear forces are found to be of short range ($\sim 10^{-13}$ cm). Within their short radius of interaction, they become strongly attractive (depths near 100 MeV), but at very small distances the interaction becomes so powerfully repulsive that nucleons seldom come closer than about 0.4×10^{-13} cm from one another. This character of the nuclear force produces "saturation." The binding energy per nucleon is nearly constant ($\sim 8.0 \pm 1$ MeV) throughout the periodic table, while each nucleon occupies a volume which may be approximated by a sphere of radius 10^{-13} cm.

The forces described above are assumed to operate between each pair of nucleons. Their exact nature is rather complicated and depends upon the motion, and even the orientation of the nucleons. The two-body portion of the nuclear force is now reasonably well determined through studies of nucleon-nucleon scattering, meson-nucleon scattering, and basic quantum field theory. Considerable attention is now being focused on the investigation of multi-nucleon interaction operators. In particular, since two-pion exchange is now known to make a very important contribution to the two-body part of the interaction, the three nucleon force arising from two-pion exchange must be carefully investigated.

Even if the nuclear force were precisely known, the structure of the nucleus would not be immediately understood in its entirety. Methods are not at present available for solving the three-body problem in closed form for simple interactions, and for more than three bodies, drastic approximations must be used to make the problems tractable. For this reason, considerable attention in physics is devoted to constructing simplified models of the nucleus. The purpose of these models is to isolate the salient characteristics of the nucleus, and thereby obtain some understanding of its behavior.

Two very general classes of nuclear models have received considerable attention. The simplest may be described as "powder models." Here it is assumed that the motion of nucleons within the nucleus is so complicated that statistical mechanics may be employed. In one version of this model, the nucleus is treated in analogy to a drop of liquid (the nuclear forces described above are qualitatively quite similar to forces between the molecules of a liquid). Such a chaotic state seems to ensue if the nucleus is highly excited, so that the nucleons have a great deal of kinetic energy. For this reason, powder models have found their greatest usefulness in describing the nucleus during a nuclear reaction, and especially in describing nuclear fission.

In the shell model (sometimes called the independent particle model), the motion of the individual nucleons is assumed to be much more simple. Each nucleon is, in fact, assigned a definite orbit within the nucleus. The resultant motion is then reminiscent of our usual picture of electrons within the atom, all revolving nearly independently of each other about the nucleus. The motivation for introducing this model into nuclear physics came from the early (1934) observation that nuclei that contained certain numbers of protons or neutrons were unusually stable and abundant. These numbers are generally referred to as magic numbers. The set we recognize today as magic are

N or Z = 2, 6, 8, 14, 20, 28, 50, 82, and 126

The major details which focus attention on these numbers may be summarized as follows:

(1) *Stability and abundance.* Nuclei with N or Z magic are unusually tightly bound and correspondingly very abundant. If both N and Z are magic ($_2He_2$, $_6C_6$, $_8O_8$, $_{14}Si_{14}$, $_{20}Ca_{20}$, $_{20}Ca_{28}$, and $_{82}Pb_{126}$), then the nucleus is even more tightly bound (here, the left subscript is the number of protons; the right is the number of neutrons). The last nucleon to complete a magic number has sometimes nearly twice the binding energy of an average nucleon. Furthermore, nuclei with Z magic possess an unusually large number of stable isotopes, and nuclei with N magic have an unusually large number of stable isotones.

(2) *Neutron capture cross sections.* Nuclei with neutron number one short of a magic number have a large neutron capture cross section, while nuclei with a magic number of neutrons exhibit a small neutron capture cross section.

(3) *Islands of isomerism.* Long lived γ-active nuclear states (half-life~1 second) appear just prior to the completion of a magic number.

(4) *Electric quadrupole moments.* Nuclear quadrupole moments tend to be small near magic numbers and large far from magic numbers. A nucleus with N or Z magic appears, therefore, to prefer a spherical shape.

(5) *Delayed neutron emission.* Delayed neutron emitters (e.g., $_{36}Kr_{51}$, $_{54}Xe_{83}$, and $_8O_9$) appear when N is one greater than a magic number.

The data clearly point to an interpretation of the magic numbers in terms of shell closures, analogous to the atomic structure of noble gases. Thus it appears that the nucleon-nucleon interaction somehow averages out within the nucleus so that each nucleon, to a reasonable first approximation, sees a fairly smooth potential. Such a potential should be flat near the origin (like an harmonic oscillator), and should go rapidly to zero outside the nucleus (like a square well). A suitable form which has undergone much study is the Fermi function.

$$V(r) = V_0 [1 + \exp \alpha(r - a)]^{-1} \qquad (1)$$

In order to reproduce the magic numbers one must add a spin-orbit term:

$$f(r)\mathbf{l}\cdot\mathbf{s} \qquad (2)$$

to this central potential. Here l is the orbital angular momentum of a nucleon, and s its spin. The effect of this l·s interaction is to couple the orbital and spin angular momenta so that the states with total angular momentum $j = l + \frac{1}{2}$ will be lower in energy than states with $j = l + \frac{1}{2}$. Shell closures obtained with such a potential are found to reproduce the observed magic numbers in a simple and striking manner.

The fact that the shell model works at all in nuclei is somewhat surprising since the theoretical motivation that enhances its success in atomic structure is lacking. One has no strong central field originally in nuclei, and furthermore the nucleons are much more closely packed. Consequently, collisions, which should tend to scatter a nucleon out of its orbit, should be more frequent. It is found that the Pauli exclusion principle plays a vital role here. When such a collision occurs, the orbit into which a nucleon would tend to be scattered is actually occupied by another nucleon. This fact makes the shell model a far better first approximation than one would originally suspect.

The main effect of the nucleon-nucleon interaction is to produce the average shell model (single-particle) potential. This, of course, is not the sole effect. Many refinements are required before an adequate description of nuclei may be attained.

Consider, for example, a nucleus with just two nucleons beyond a double closed shell. One of the nucleons will go into an orbit j, and the other into an orbit j'. These two orbits may orient themselves such that the nucleus may have total angular momentum J anywhere within the range

$$|j - j'| \leqslant J \leqslant j + j' \qquad (3)$$

so long as one is consistent with the Pauli exclusion principle. If one only had single-particle potentials all states of this configuration would have the same energy independent of J. To determine the level order actually observed, one must recognize that not all of the original nucleon-nucleon interaction is used up in constructing the single-particle potential. Some "residual interaction" is left over. This residual interaction will still be of the two-body type, and it will remove the degeneracy between states of the same configuration.

The most important feature of the residual interaction is that it produces a pairing force for like orbits (i.e., $j = j'$). If $j = j'$, and we have like nucleons, the interaction is by far the strongest if $J = 0$. This has the important consequence that for nuclei with both an even number of protons and an even number of neutrons, one must have net angular momentum equal to zero. This rule is never violated. If one adds one nucleon to this even-even nucleus, the result will be a J equal to the j of the odd nucleon. This rule is violated in only a very few cases.

There are a few nuclear phenomena which the shell model cannot describe at all. The sign of the static, nuclear, electric quadrupole moments are generally given correctly by a single-particle model, but the magnitude is frequently found to be larger than that predicted by more than a factor of ten. Similarly, electric quadrupole transition rates are frequently much faster than those given by the shell model. It seems that even in low-lying nuclear levels it is not possi-

ble to attribute all of the properties of the nucleus to the nucleons in unfilled orbits. One must take into account possible distortions of the orbits of nucleons in the closed shells.

This line of thought gives rise to the "collective model," in which the nucleons in the core (closed shells) are treated as an incompressible, irrotational fluid capable of surface oscillations. Nucleons in unfilled shells will exert a centrifugal force on this fluid and tend to deform it into a nonspherical shape. In regions of the periodic table where several shell model orbitals are nearly degenerate in energy ($90 \leqslant N \leqslant 114$ and $Z > 88$), the nuclear core takes on a rather large spheroidal deformation. The formalism of the theory is reminiscent of that for diatomic molecules. Low-lying excited states are generated through a rotation of the entire system about an axis perpendicular to the axis of symmetry. This gives rise to a band of energy levels:

$$E_J = E_0 + (\hbar^2/2I)J(J+1) \qquad (4)$$

where I is a moment of inertia. The identification of spectra in nuclei which could be empirically fitted to Eq. (4) was a great truimph of the collective model.

It should be noted that I is not the moment of inertia of the entire nucleus. It is only the moment of inertia of the part of the nucleus which participates in the rotation, and this is, of course, always less than the "rigid body" value.

We note that the shell model and collective model appear to be quite different in their basic assumptions. The collective model presumes that the nucleon motions are so closely correlated that we can treat a rotation of the nucleus on the whole, while the shell model begins with independent orbitals. It is very important to remember, in this regard, that the shell-model orbitals must be properly coupled together in the nucleus. The Pauli exclusion principle must be satisfied, angular momentum must be a good quantum number, and so on. These orbitals are, therefore not so independent as it might seem. They may very well give rise to collective effects depending upon the exact way in which they are coupled together to obtain the final nuclear wave function.

Various coupling schemes have recently been examined to investigate this, but these are beyond the scope of this article. We present here a simple illustrative example. Consider a set of identical particles attracted by means of identical springs to a common origin. At time $t = 0$ each particle is at the origin and is given some arbitrary velocity. Even though each particle has a different initial velocity, and moves independently of every other particle, the system will undergo a periodic dilation and contraction due to the fact that the frequency of vibration is the same for each particle. Thus we have a collective motion exhibited by a set of independently moving particles.

The next step in the investigation must be to attain a quantitative interpretation of the individual particle motion itself. This must come directly from a full treatment of the many-body nuclear problem using realistic nucleon-nucleon interactions. This problem is complicated by the fact that nuclear forces become sharply repulsive at very short nucleon-nucleon separations. Thus the nuclear wave function exhibits strong correlations between nucleons which must be accounted for in any final theory. The two-body correlations seem well accounted for by a formalism devised by Bruechner, Bethe, and Goldstone, but the effects attained when three or more nucleons all come into close proximity are yet to be completely evaluated in finite nuclei.

PAUL GOLDHAMMER

References

Baranger, M., "Recent Progress in the Understanding of Finite Nuclei from the Two-Nucleon Interaction," *Proceedings of the International School of Physics,* Course 40, p. 511 (Academic Press, 1969).

Eisenbud, L., and Wigner, E. P., "Nuclear Structure," Princeton, N.J., Princeton University Press, 1958.

Elliott, J. P., and Lane, A. M., "Handbuch der Physik, Vol. 39, p. 241, Berlin, Springer-Verlag, 1957.

Feenberg, E., "Shell Theory of the Nucleus," Princeton, N.J., Princeton University Press, 1955.

Goldhammer, P., *Rev. Mod. Phys.,* **35**, 40 (1963).

Inglis, D. R., *Rev. Mod. Phys.,* **25**, 390 (1953).

Mayer, M. G., and Jensen, J. H. D., "Elementary Theory of Nuclear Shell Structure," New York, John Wiley & Sons, 1955.

Rajaraman, R., and Bethe, H. A., *Revs. Modern Phys.,* **39**, 745 (1967).

Cross-references: ATOMIC PHYSICS; NUCLEAR REACTIONS; NEUTRON; PROTON; STRONG INTERACTIONS; WEAK INTERACTIONS.

O

OCEAN ACOUSTICS

Introduction Ocean Acoustics is the study of the correlation between oceanic variability and its effect on underwater acoustic signals. A vast range of scales is involved: time scales from seconds to months, acoustic frequencies from a few hertz to several hundred kilohertz, both deterministic and random ocean inhomogeneities ranging from centimeter size to circulations and currents extending hundreds of kilometers. It encompasses not only physics but also aspects of biology and chemistry. Signal propagation ranges can vary from meters to thousands of kilometers, through a strongly inhomogeneous medium whose effects alter the sound speed of the acoustic signal in many ways. We separate the description, not inconveniently, into the effects of the ocean surface, its bottom, and the interior domain on acoustic propagation and scattering.

The Ocean Surface The ocean surface is an example of a complex wave motion formed through the action of the wind. Its properties can be understood in terms of the statistical properties of wave heights. Both the wave height probability distribution and its correlation function, or the Fourier transform of the latter, the surface (spatial) spectral function, play an important role. The major height parameter is the Rayleigh roughness parameter, $\Sigma = k\sigma \cos\theta$, where k is the wavenumber, σ the rms surface height, and θ the angle of incidence of an acoustic wave (measured from the normal). Values of $\Sigma > 1$ describe a very (strong) rough surface.

For a fully developed sea (one where the wind has blown long enough so that the surface is in equilibrium) the surface height and slope are nearly Gaussian distributed, although the distribution is skewed due to the creation of broad troughs and narrow swells on the surface. Additional discrete spectral contributions can result from periodic surface components. In regions where the surface is composed of ice, other models must be considered.

For shallow surfaces the surface spectral function can be directly related to the scattering coefficient. Little direct data is available on the spatial spectrum itself, but it can be inferred from frequency spectra and the ocean gravity wave dispersion relation. In addition, the scattering coefficient depends on wind speed and direction, the direction of propagation of a wave on the surface, and the angular distribution of ocean wave energy.

Acoustic signals propagating in the ocean scatter from the surface both coherently and incoherently. Models of this scattering include perturbation methods for small wave height surfaces, tangent plane or Kirchhoff scattering for surfaces with large radii of curvature, and multiple scattering techniques for strongly rough surfaces. Multiscale surface models containing a long-scale periodic component on which is superimposed a small-scale random surface have met with success. The coherent wave propagates in the specular direction and is formed by averaging the wave field over the statistical ensemble of random surfaces. For shallow surfaces it is a Gaussian function of the Rayleigh roughness parameter. For rougher surfaces multiple scattering must be incorporated. Backscattered sound can be modeled as arising from the glitter due to a facet model of the surface. At very high frequencies trapped air bubbles near the surface are the main mechanism of surface sound scattering. The frequency spectrum of the scattered field is thus composed of a single spectral line (due to specular reflection) and a broad background due to the incoherent contributions.

The Ocean Interior The ocean is a strongly inhomogeneous medium. The inhomogeneities are of two types, regular and random. The fundamental cause of a regular inhomogeneity is the dependence of the ocean sound speed on position. In fact, because of this inhomogeneous structure, the sound speed (or index of refraction) in the ocean acts like a converging lens to the propagating acoustic field, and results in the creation of an effective ocean duct or waveguide. This is called the SOFAR channel (for sound fixing and ranging) or the underwater sound channel (USC). Its net effect is to refract and focus the acoustic signal. Sound trapped in this duct can propagate for thousands of kilometers.

The mean speed of sound in the ocean is about 1500 m/s and can vary approximately ±50 m/s. Even though changes from the mean are small, they can significantly alter sound propagation over long ranges because their effects are cumulative. The sound speed c can be parameterized in two ways. The first way is as a function of temperature T, salinity S, and pressure P, that is, $c(T, S, P)$. Near the ocean surface, c varies primarily due to the seasonal or diurnal temperature changes. Near the ocean bottom (below about 1.5 km in the Atlantic for example) the temperature and salinity are nearly constant and the major functional dependence is on pressure. In the deep ocean this latter yields a stable monotonic increase in c as a function of depth. In general then the sound speed is largest near the surface and bottom with a minimum at what is referred to as the SOFAR axis. Examples are illustrated in Fig. 1.

The second parametrization of c is a functional dependence on spatial coordinates (depth z, range r, angle θ) and possibily on time t, that is, $c(z, r, \theta, t)$. The temporal dependence is often ignored since the ocean is usually considered to be temporally frozen on the time scale of most ocean acoustic experiments. Although surface ducting can occur due to temperature variability and, if an experiment occurs over long periods of time, longer-scale effects on the system (such as tides) must be considered, the main effect arises from the depth dependence. There is a slower (adiabatic) range dependence, and a slower still angular variability.

Other examples of regular inhomogeneities in the ocean are bottom protrusions such as seamounts, which can occur on a horizontal scale of hundreds of kilometers and a vertical scale on the order of kilometers, and, in general, sea depth variability even when more gradual such as near coasts and shores. Large-scale highly energetic variability also occurs in the upper ocean regions in the form of circulating spin offs from large current flows such as the Gulf Stream and Kuroshio current near Japan. Due to their size they are generically called *mesoscale phenomena*. Examples include the synoptic mesoscale eddies and rings which can move as well-defined rotating water masses for long times at speeds of about 10 cm/s with the actual circulation speeds about an order of magnitude larger. Their scale sizes are roughly 100 km in the horizontal and about 1 km in the vertical. Thus they are highly anisotropic, as is much of the oceanic variability. Their net effect on sound propagation is again due to the fact that they modify the sound speed structure. For example, counterclockwise rotating eddies pull colder water up towards the surface thus lowering the effective sound speed and producing a deeper refraction.

Additional ocean inhomogeneities can be

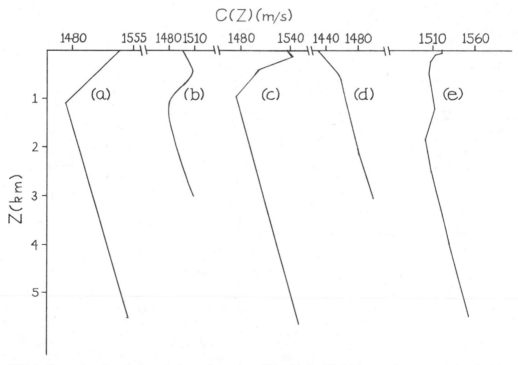

FIG. 1. Examples of typical vertical sound speed profiles $c(z)$. Profile (a) is a two-layer sound speed with each layer approximated by a linear segment in depth; (b) illustrates a sound speed which can form a surface sound channel; (c) is a typical profile from the Pacific; (d) is an Arctic profile illustrating a steep gradient and the effect of colder water; and (e) is a representative northeastern Atlantic profile.

classed as randomly varying on the acoustic time scale. The ocean can be considered as a density-stratified fluid. Displacement of a fluid particle from equilibrium and the action of a restoring (gravitational) force on the particle produces a wavelike motion in the ocean called an internal wave. Scale sizes of these waves are as large as 10 km in the horizontal and 100 m in the vertical. They have periods of anywhere from a few minutes to about a day. For fixed source and receiver they are a major source of fluctuations of the sound speed and hence of the acoustic intensity within a frequency interval of cycles per day and cycles per hour. Considerable effort has been expended in attempting to understand the fluctuation effects of internal waves on acoustic propagation and extensive results are available on the resulting acoustic amplitude and phase-rate spectra. Internal waves contain much less energy than the mesoscale phenomena.

For many purposes it is useful to treat the ocean as a layered medium. It is sometimes nearly exactly so. For example, the vertical profiles of temperature, density or salinity can have an almost steplike structure, with layer thicknesses from tens of centimeters to meters and horizontal extent on the order of kilometers. This ocean *fine structure*, as it is termed, affects acoustic propagation, since the vertical layer gradients can alter the sound speed field. The trajectories of acoustic rays can be changed due to the gradients, or wavelike reflection properties from the layers can be significantly altered. Currents in the ocean can also contribute to these effects. Periods for the fine structure are on the order of minutes, and fine structure is less energetic than internal waves.

Note that both the internal waves and fine structure as well as the deterministic mesoscale phenomena are highly anisotropic. An example of random isotropic ocean variability is the homogeneous isotropic turbulence due to ocean microstructure. Here horizontal and vertical scales are about 10 cm with periods of about a minute. This turbulence is generally confined to the upper mixed layer of the ocean and is due to atmospheric effects. Examples include wave breaking and bubble cavitation. It is the least energetic of the inhomogeneities we have described.

There is an added problem in treating many of the above phenomena, since they often do not occur with any particular regularity in the ocean. Not only is there the problem of understanding the effects of mesoscale, internal waves, etc., there is the additional problem of identifying the existence of the particular phenomenon or combinations of phenomena which may be contributing to an ocean experiment or interrupting or modifying the performance of some system. As we have mentioned, all the above inhomogeneities affect the sound speed.

The sound speed in the ocean can thus be modeled as a large deterministic refraction (due to the depth variation of temperature, salinity, and pressure), additional multidimensional deterministic variability due to mesoscale phenomena, and random modifications due to smaller scale phenomena such as internal waves, fine structure and turbulence. Functionally it can be written as

$$c(\mathbf{x}, t) = c_0(z) + c_1(\mathbf{x}) + c_2(\mathbf{x}, t).$$

Here c_0 is the purely depth-refractive contribution; c_1 the deterministic spatial mesoscale addition depending on the spatial vector $\mathbf{x}$ and smaller than c_0 by about two orders of magnitude; and c_2 the spatial and possibly temporal variation due to random effects, which is smaller than c_0 by about four orders of magnitude.

The ocean is also a very noisy environment. At low frequencies (below 10 Hz) the major noise source is geophysical (earthquakes, volcanic activity, etc.), with additional contributions from turbulence and nonlinear wave interactions. At higher frequencies (40–400 Hz) ship traffic produces a steady background noise because of low sound absorption. In the lower kilohertz regime the wind and ocean surface interactions (such as breaking waves and trapped air bubble collapse) produce the majority of the noise. At still higher frequencies molecular effects can produce some thermal noise. In addition there is biological noise produced at various ocean locations due to the presence of marine organisms.

There are several different mathematical models for acoustic propagation in the ocean. They depend of the dimensionality of the problem and on whether the oceanic variability is deterministic of random. The acoustic field in the ocean is the solution of a Helmholtz equation which can be derived from the perturbation solution of the fundamental fluid equations. If this solution is represented as an amplitude and phase term and diffraction is neglected, then the most widely used propagation method, ray theory, can be derived. It is valid at high frequencies, and the wavelength should be much less than any significant length scale. It is invalid when the phase curvature does not change as rapidly as the projection of the amplitude gradient on the phase gradient (e.g., shadow regions, edges of obstacles, caustics, or focal points) and when the fractional change in the refraction index is not small with respect to a wavelength (as at the ocean surface and bottom). Its advantages are that the rays can be found independently of the amplitude, and the amplitude derived from the intensity along a ray where the ray coupling (diffraction) is neglected. It is computationally simple to implement and yields at least a good qualitative picture of the acoustic field. It also can be used multidimensionally.

If the sound speed profile is one-dimensional

(depth-dependent) and the propagation regime can be modeled as a waveguide, then the z-dependent part of the acoustic field constitutes an eigenvalue problem or a normal mode problem. The boundary conditions on the acoustic velocity potential at the air-water interface are nearly perfectly reflecting (soft or Dirichlet conditions) and at the bottom either perfectly reflecting (hard or Neumann conditions) or are varied to include bottom loss. The discrete modes can be found numerically or, in certain cases for exactly solvable profiles, analytically. Statistically rough boundaries can be incorporated using perturbation methods, and, depending on the model chosen for the bottom, a continuous spectral contribution can be included. The latter is important at short ranges, while the discrete spectrum is the dominant contribution at long ranges. If the profile has an additional slow variability in range, an adiabatic mode theory has been developed which treats the range variation as a succession of static depth-dependent problems whose solutions are matched at successive range interfaces.

General range dependence of the profile can be modeled using the parabolic approximation to the Helmholtz equation coupled with a fast Fourier transform algorithm so that the solution can be computationally marched in range. The approximation consists in neglecting the term involving the ratio of the second to first range derivative of the field divided by the wavenumber. It has been shown that this is a narrow spectral approximation centered about the wavenumber term (which is arbitrary). The latter can be used to define an equivalent ray angle about which the parabolic approximation is also a narrow angle approximation in a forward direction. In most presently available codes the ray angle is along the horizontal. It is also possible to relate the parabolic approximate solution rigorously to the Helmholtz solution and from this to correct the parabolic solution particularly in the phase term where the original approximation fares badly. Other corrections can also be included, such as the inclusion of the earth's curvature for long-range propagation. The small dependence on angular variability can also be incorporated into a three-dimensional parabolic code.

When the ocean is random, propagation codes become more complicated. Unlike the atmosphere, the major propagation effect in the ocean is refraction which is deterministic. Thus perturbation methods for random oceanic variability must be based on ray- or modal-type solutions which include the deterministic depth dependence. Random propagation methods have also been applied to investigate the effect of internal waves on acoustic propagation, and to explain acoustic signal fading.

Experimental ocean frequencies vary over six orders of magnitude, from a few hertz up to the megahertz regime. The lower the frequency the longer the propagation range and the lower the absorption. The absorption is mainly due to liquid shear viscosity and oceanic chemical relaxation processes. Inhomogeneities (air bubbles, marine organisms) also scatter the sound and it is usually the combined effect of absorption and scattering (called *attenuation*) which is measured. The attenuation coefficient is roughly proportional to frequency squared so that low frequency sound can, as we remarked, propagate thousands of kilometers. As a comparison, laser energy in the ocean would be totally absorbed in less than a kilometer.

One final ocean topic should be included. We have mentioned direct propagation problems where the sound speed is given and the resulting acoustic field is computed. Potentially far more important are inverse methods, where from a knowledge of the source and received signal (or its travel time in ray theory) one infers the intermediate oceanic properties such as the sound speed. This latter research in ocean tomography is presently in progress and is a precursor of much of the future oceanic monitoring effort.

The Ocean Bottom The acoustic field interaction at the ocean surface can be generally decoupled from the effects in the air. That is, the major air-water interface effect is a geometric effect. This is not so with the ocean bottom interaction. Bottom topography, sub-bottom layering, and the elastic properties of the bottom can significantly alter low frequency propagation in the ocean. Thus, not only is there the bottom geometry to consider, but also its material properties such as density and compressibility. Shear as well as compressional waves are present.

Bottom geometry varies on the horizontal scale from the size of small ripples due to currents on sand to hundreds of kilometers (e.g., large ocean ridges). Vertical protrusions occur on the scale of kilometers (e.g., seamounts). The sound speed in the shallow bottom is near that in the water bottom (about 1550 m/s). It can be less (for clay, 1530 m/s) or greater (sandy clays, about 1700 m/s) or much greater (about 1800 m/s for low-porosity sediment). In the sub-bottom substrate the compressional speed is 3-4 times greater than in the water, with shear speeds roughly twice as large (3000 m/s). Estimates of the bottom sound speed can be obtained using seismic reflection techniques. Sound absorption in the bottom is roughly proportional to frequency. At low frequencies the material bottom properties dominate the sound reflection whereas above a few kilohertz the bottom topography plays an important role.

Forward scattering from the bottom can significantly contribute at low frequencies since the sound can first penetrate the bottom and, at a distance couple back into the water. It is generally insensitive to correlation lengths of the topography. Backscattering at normal incidence is quite large, and for larger incidence angles roughly follows the cosine squared dependence of Lambert's law.

JOHN A. DeSANTO

References

1. Brekhovskikh, L. M., and Lysanov, Yu. P., "Theoretical Fundamentals of Ocean Acoustics," Berlin, Heidelberg, New York, Springer-Verlag, 1982.
2. Clay, C. S., and Medwin, H., "Acoustical Oceanography," New York, Wiley, 1977.
3. DeSanto, J. A. (Ed.), "Ocean Acoustics," Topics in Current Physics, Vol. 8, Berlin, Heidelberg, New York, Springer-Verlag, 1979.
4. Flatté, S. M. (Ed.), "Sound Transmission Through a Fluctuating Ocean," Cambridge, U.K., Cambridge Univ. Press, 1979.
5. Keller, J. B., and Papadakis, J. S. (Eds.), "Wave Propagation and Underwater Acoustics," Lecture Notes in Physics, Vol. 70, Berlin, Heidelberg, New York, Springer-Verlag, 1977.
6. Kinsman, B., "Wind Waves," Englewood Cliffs, New Jersey, Prentice-Hall, 1965.

Cross-references: ACOUSTICS; CAVITATION, ELECTROACOUSTICS; FOURIER ANALYSIS; NOISE, ACOUSTICAL; PHYSICAL ACOUSTICS; ULTRASONICS; VIBRATION; WAVE MOTION.

OPTICAL INSTRUMENTS

Optical instruments create images of objects by altering the direction of travel of light rays. They are of two types. One type creates a real image that is projected onto a surface such as a screen or a piece of film. The other type creates a virtual image that can be perceived only by an observer. The observer's eye further redirects the light rays so that a final image of the object falls across the retina. The major advantage of optical instruments is that they afford an image of an object that could not be easily perceived with the unaided eye. With microscopes one can see or photograph details of an object totally inaccessible with unaided viewing. With telescopes one can see details of truly distant objects whose image to the observer would otherwise be imperceptible. Here simple microscopes and refracting telescopes will be examined.

Magnification in Microscopes The fineness of the detail that one can see in an object with unaided sight depends in part on how close the object can be brought to the eye. However, for a normal eye an object can be no closer than about 25 cm if a clear image is to form on the retina. If an optical lens is positioned appropriately between the object and the eye, the lens produces a virtual image at a distance of 25 cm. This image subtends a larger angle in the observer's field of view than the object

would have at 25 cm. The ratio of the angle subtended by the virtual image to the angle subtended by the object in the first case is called the *magnifying power* of the lens. It depends on the focal length of the lens. For example, if the lens has a focal length of 2.5 cm and if the virtual image is created 25 cm from the observer, then the lens has a magnifying power of $25/f + 1$, or 11. This value is approximately the upper limit for a single lens providing convenient viewing without major aberrations to the image.

Compound microscopes employ two lenses (the objective and the eyepiece) in order to further magnify the object. The objective is closer to the object and provides a real and inverted image that falls within the focal distance f_2 of the eyepiece. In Fig. 1, u is the distance between the object and the objective and v is the resulting image's distance. The ratio of the size of the image to the size of the object is called the *magnification by the objective* and is equal to v/u. The image serves as an object to the eyepiece which functions as a magnifying glass since it produces a virtual image (the final image of the microscope). It is this image that an observer sees. The combined magnifying power of the instrument is the multiplication of the magnification of each component: $(v/u)(25/f_2 + 1)$ where f_2 is the focal distance of the eyepiece. The magnification is increased if u is decreased and v increased. However, a practical limit of 16 cm for the length of the microscope tube restricts improvement. Also, resolution is eventually lost. The magnifying power of most microscope designs is limited to about 2000.

Magnification in Refracting Telescopes Refracting telescopes are normally employed to see objects that cannot be brought close to the objective. The magnifying power is defined as the ratio of the angle subtended by the image provided by the telescope to the angle that would be subtended in the observer's field of view with normal sight. A refractive telescope employs two lenses as in a microscope: an objective and an eyepiece. In the usual adjustment of a telescope the objective produces a real image at the focal point of the eyepiece, which then creates a virtual image that is essentially infinitely far away. The magnifying power is equal to the ratio of the objective's focal length to the eyepiece's focal length.

Astronomical telescopes are often classed according to a magnifying power that depends on the diameters of the apertures through which the light must pass. The entrance pupil is simply the objective lens. The exit pupil is defined, in principle, with an experiment in which a diffuse light source (such as an illuminated ground glass screen) is placed in front of the objective. The light rays passing through the eyepiece are focused on a surface to form an image called the exit pupil (Fig. 2). (Often a metal cap with

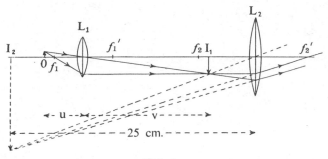

FIG. 1

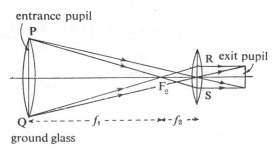

entrance pupil

exit pupil

ground glass

FIG. 2

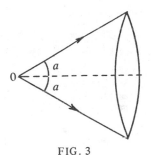

FIG. 3

a circular hole lies at the site of the exit pupil.) The magnifying power is defined as the ratio of the diameter of the entrance pupil to the diameter of the exit pupil. In a simple refracting telescope the light must pass through the pupil of the eye. In a good design the exit pupil of the telescope is no larger than the diameter of pupil in the eye.

For example, at night the diameter of the pupil in the eye is about $\frac{1}{3}$ in. For a 6-in telescope (telescopes are often classed according to the diameter of their objective) the magnifying power should be

$$\text{Magnifying power} = 6/\tfrac{1}{3} = 18.$$

If the exit pupil of the telescope is larger than the diameter of the pupil, the full light gathering capability of the objective is wasted.

Resolving Power and Limit of Resolution of Microscopes The limit of resolution of the objective of a microscope is the least distance x between two points that can just be resolved. Abbe calculated this limit as

$$x = \frac{\lambda}{2\mu \sin a}$$

where λ is the wavelength of the light, μ is the refractive index between the points and the objective, and $2a$ is the angle subtended by the objective in the field of view from the points (Fig. 3). The term $\mu \sin a$ is often called the numerical aperture of the objective. The maxi-

mum value of $\sin a$ is 1. For air μ is 1. Thus the limit of resolution in air is $\lambda/2$, or half the wavelength of the light employed. Closer spacing of the point objects would eliminate one's resolving them separately.

In practice the resolution is partially limited by how close the object can be brought to the objective so that $\sin a$ is a large as possible. Therefore objectives with short focal lengths ($\frac{2}{3}$, $\frac{1}{6}$, or $\frac{1}{12}$ inch) are employed in microscopes.

The limit of resolution can also be improved by increasing the index of refraction of the medium between the object and the objective. Commonly oil such as cedar oil is placed between a cover slip lying over the object and the objective (Fig. 4).

Another technique for improving the limit is to employ ultraviolet light whose wavelength

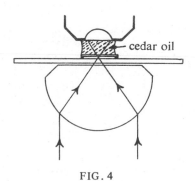

cedar oil

FIG. 4

is about half that of visible light. With UV the resolving power is doubled but a camera is needed to capture the image since the eye does not respond to UV.

Resolving Power and Limit of Resolution of Telescopes The limit of resolution of a telescope is given as an angle ϕ subtended at the objective by two points on an object. The limit is the smallest angle such that the two points can just barely be resolved and is given by

$$\phi = 1.22\lambda/a$$

where λ is the wavelength of the light and a is the diameter of the objective. Thus telescopes with larger objective diameters yield better resolving power. However, to insure full advantage of the objective in a simple refracting telescope, one must be certain that the exit pupil is not larger than the diameter of the pupil of the eye.

For example the Yerkes telescope has an objective of 40 inches in diameter. With 6×10^{-7} m as a typical wavelength for visible light its resolution limit is given by

$$\phi = \frac{1.22 \times 6 \times 10^{-7}}{40} = 0.15 \text{ second of arc.}$$

The resolution limit of the unaided eye is about 1 minute of arc.

Depth of Focus and Depth of Field The *depth of focus* for a camera is the distance the plate (or film) in the camera can be moved from the object without spoiling the resolution of the image. The *depth of field* is the corresponding distance the object can be moved. In practice these two limits depend on how the final photograph is to be viewed. Suppose that a point object lies at a distance from the camera such that it creates a well focused point on the plate. Another point object at a different distance creates a blurred circle on the plate. Whether or not these adjacent images can be resolved depends on the resolving power of the eye, whose limit is about 1 minute of arc. Assuming that the eye is normal and that the closest separation between the print and the eye is 25 cm, the two images must be no closer than

$$\frac{x}{25} = 1 \times \frac{\pi}{180 \times 60} \text{ radians}$$

$$x = 0.007 \text{ cm}$$

if they are to be at the limit of resolution. Usually a larger value of 0.025 cm is employed in cameras as being a useful limit. However, if the print is enlarged, then the compromise on the resolution becomes apparent.

The depth of focus of a lens can be increased by reducing its aperture with a stop. In Fig. 5 a point object creates a circular image on a

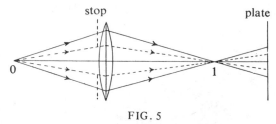

FIG. 5

plate after passing through a lens. The stop can be used to reduce the size of the image to an acceptable diameter of 0.025 cm.

The depth of focus of a microscope is approximately inversely proportional to the square of the numerical aperature. At a magnification of $1000\times$ an objective with a numerical aperture of 1.40 has a depth of focus of only 0.0005 mm.

Field of View The field of view of an instrument is the angle subtended at the observer's eye by an object when the whole of the object can just barely be seen. In the case of a simple refracting telescope having two lenses this angle is given by $a_2/(f_1 + f_2)$, where a_2 is the aperture of the eye lens and f_1 and f_2 are the focal lengths of the objective and eyepiece respectively. (See Fig. 6.) If f_1 is doubled so as to double the magnifying power of the telescope, then the field of view is nearly halved.

Vignetting is the fall off in brightness at the edge of the field of view provided by a telescope or microscope. It is due to the loss of the rays that pass through the objective but which do not pass through the eyepiece. Vignetting is reduced by inserting a circular opening (called a field stop) at the point where the image from the objective is formed.

Illumination in Microscopes The final image in a microscope magnifying $1000\times$ has an area one million times that of the object. The image is therefore only one-millionth as bright, assuming no loss of light in the instrument. To improve the illumination on a transparent object the light is concentrated on it by means of a *substage condenser*.

Dark ground illumination is used for viewing tiny particles or very fine lines. The illumination is made too oblique for light to pass directly from the condenser into the objective. The objects are seen by scattered and diffracted light against a dark background.

Phase contrast illumination is a method of making the structure of transparent objects visible. The arrangement consists of an annular stop at the condenser. Beyond the objective lies a phase plate consisting of a glass plate on which there is an annular layer of material to increase the optical path of the light by a quarter of a wavelength. Interference between direct and diffracted light augments the diffraction pattern of the image.

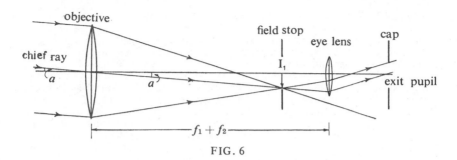

FIG. 6

Interference microscopy is another method of making fine structure visible. The transparent object is placed between two semi-silvered surfaces. The interference between light passing through the object and light passing by it is then observed.

Polarized light, obtained by a suitable polarizer below the condenser, is used in microscopes designed for geological testing of constituents in rock specimens.

Other Instruments The range of modern optical instruments and technology is vast. Here only a few items can be mentioned. Photography itself is a large area of ongoing research with continuous improvement in picture quality and reduction in size of the equipment. (See PHOTOGRAPHY.) Computer-designed lenses have revolutionized popular cameras. Highly sophisticated photography is currently employed in weather monitoring, earth resource research, and intelligence gathering. High-speed photography has captured images of events far too fleeting to be seen otherwise. Even a short pulse of light has been photographed in midflight.

The medical profession employs optical instrumentation to view inside the human body without surgery. Bundles of optical fibers illuminate the interior of a patient. They also carry an image of the interior out to a monitor to allow the doctor to see the interior. A number of ophthalmic instruments now enable a doctor to see into and perform various measurements on the human eye. The military employs telescopic gun sights, binoculars, periscopes and range finders. Many of these designs use infrared radiation and lasers in both the visible and infrared ranges. Civil engineers use theodolites and lasers for surveying. There is a variety of optical instruments used in the machine tool industry for angular measurements, aligning, and inspection for faults in the products.

Especially important for science and technology have been optical instruments designed to disperse light into its constituent wavelengths. These spectrometers have been crucial in the study of the fundamental nature of atoms and molecules and continue to play a role in many areas of research in chemistry, physics and astronomy. Spectrophotometers split the light from a source into its constituent wavelengths while measuring the intensity at each wavelength. Interferometers are used for fine distance measurement and instrument testing. Refractometers measure the index of refraction of a substance. Radiometers are employed for measurements of high temperatures.

One area of optical instrumentation that has exploded into a tremendous number of applications is that using coherent radiation in the infrared, visible, and ultraviolet. Lasers have been used in diverse applications ranging from cutting fabric to reattaching loose retinas. Laser beams allow one to measure precisely the distance to the moon as a check on the current theories of gravitation. They also allow one to monitor shifts in the tectonic plates as a prewarning of earthquakes.

Holography is an imaging process using coherent radiation. Once a novelty for its ability to present three-dimensional pictures, it has since proved valuable in industrial testing such as in searching for defects in rapidly rotating tires. Although not yet fully realized, three-dimensional movies may incorporate it.

A. E. E. McKenzie
N. C. McKenzie
J. D. Walker

References

Cooper, H. J. (Ed.), "Scientific Instruments," Vols. I and II, London, 1946 and 1948.

Martin, L. C., "Technical Optics," Second edition, Vols. I and II, London, 1961.

Kingslake, R. (Ed.), "Applied Optics and Optical Engineering," Vols. IV and V, London, Academic Press, 1967 and 1969.

Jenkins, F. A., and White, H. E., "Fundamentals of Optics," New York, McGraw-Hill, 1976.

Walker, J. (Ed.), "Light and Its Uses," San Francisco, W. H. Freeman, 1980.

Cross-references: ABERRATIONS; HOLOGRAPHY; INTERFERENCE AND INTERFEROMETRY; LASERS; LENS; LIGHT; OPTICS, GEOMETRICAL; PHOTOGRAPHY; REFLECTION; REFRACTION; SPECTROSCOPY.

OPTICAL PUMPING

Optical pumping describes the transfer of order from a beam of light or other electromagnetic radiation to matter. Some examples are the use of optical pumping to produce spin polarization of ground-state or excited-state atoms, molecules or ions; the optical production of metastable populations in matter, the optical excitation of spin-polarized conduction electrons in semiconductors, and the optical pumping of masers and lasers.

The antecedents of optical pumping go back to the original studies of atomic resonance absorption by Bunsen and Kirchhoff in the 19th Century. Zeeman, Wood, Hanle, Heydenburg, and others contributed to the early development of optical pumping. Kastler in 1949 coined the term *pompage optique* and suggested the combination of optical pumping with radiofrequency or microwave fields which resulted in the fruitful spectroscopic technique of optical double resonance, pioneered by Brossel and Bitter. The use of optical pumping to produce large numbers of spin-polarized ground-state atoms was also suggested by Kastler, who received the Nobel Prize in 1966 for his work.

To be effective as an optical pumping source, the light must differ from the normal blackbody radiation at the temperature of the mat-ter. For example, the optical pumping source may produce a beam rather than isotropic radiation, the light may be polarized, or the spectral profile of the light may consist of a single narrow frequency band rather than the broad spectral distribution of thermal radiation.

Light which is effective for producing order in matter is also useful for detecting the existence of order in matter. For example, spin-polarized atoms attenuate and scatter resonance radiation differently than unpolarized atoms. Since light can be detected as single photons, the sensitivity of optical pumping experiments often is limited by photon counting noise (shot noise) rather than thermal noise as is the case with conventional microwave or nuclear magnetic resonance spectroscopy. Thus, optical pumping methods can be sensitive to only a few atoms in the pumping beam when optical detection is used.

A simple optical pumping experiment to produce spin-polarized excited atoms is illustrated in Fig. 1(a). There atoms are contained in a transparent cell C which is located in a static magnetic field H which defines the z axis of a coordinate system. The atom has angular momentum $J_g = 0$ in the ground state g and $J_e = 1$ in the excited state e, and the energy levels are split by the magnetic field as shown in Fig. 1(b). Resonance radiation, for example the 2537-Å line of mercury, is

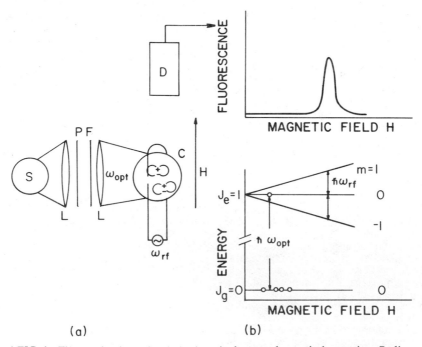

(a)   (b)

FIG. 1. The production of polarized excited atoms by optical pumping. Radio-frequency spectroscopy is used for the Doppler-free measurement of excited-state energy level splitting with fluorescence monitoring of the resonance.

isolated from the sources with the filter F, is linearly polarized with the polarizer P along the magnetic field and is used to pump the atom from the ground state to the sublevel of the excited state with azimuthal quantum number $m = 0$. Angular momentum selection rules prevent the light from exciting atoms into sublevels with $m = \pm 1$.

The fluorescent radiation has the characteristic angular distribution and polarization of an electric dipole antenna, so little radiation is received by a detector D located in the null of the antenna pattern. The pattern can be perturbed by driving transitions from the excited-state sublevel with $m = 0$ to the sublevels with $m = \pm 1$ with radiofrequency (rf) radiation of sufficient intensity and appropriate frequency. The atoms in the sublevels $m = \pm 1$ radiate strongly along the direction of the magnetic field and the rf transition is therefore detected as an increase in the intensity of light received by the detector. By observing the detected intensity as a function of the radiofrequency, a resonance curve can be plotted out which determines the splitting of the energy levels. The arrangement in Fig. 1 is typical of optical

double resonance experiments which have been widely used to determine small energy splittings of excited atoms or molecules. It is important to note that the rf resonance curve is not affected by the doppler broadening of the optical lines used to detect the energy splittings.

Closely related to the optical double-resonance experiment of Fig. 1 are the methods of level-crossing spectroscopy, light-modulation spectroscopy, decoupling spectrosopy, the Hanle effect, and various other spectroscopic techniques which may be described by the term *perturbed fluorescence spectroscopy*, by analogy with the related nuclear field of perturbed angular correlations.

The production of ground-state or meta-stable-state polarized atoms is illustrated in Fig. 2. The atom has angular momenta $J = \frac{1}{2}$ in both the ground state and in the excited state. The atom is situated in a magnetic field H and is pumped by circularly polarized light which carries one unit of spin angular momentum along the magnetic field direction. Atoms in the ground state sublevel with azimuthal quantum number $m = -\frac{1}{2}$ can absorb light, but atoms in the sublevel with $m = +\frac{1}{2}$ cannot be-

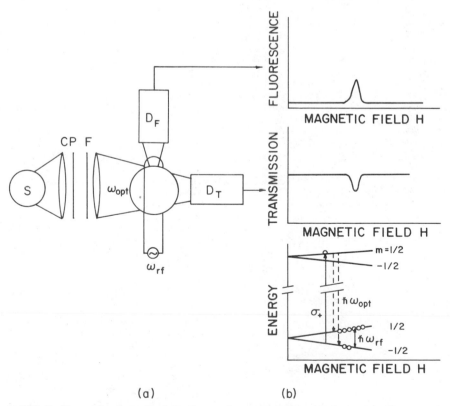

(a)   (b)

FIG. 2. The production of polarized ground-state atoms by optical pumping. The ground-state energy level splittings are measured by radio-frequency spectroscopy with fluorescence or transmission monitoring of the resonance.

cause of angular momentum selection rules. When the excited atoms decay they may fall back to the sublevel with $m = +\frac{1}{2}$ where they no longer absorb light. The result of repeated absorption and reemission of light is to pump the atoms into the ground-state sublevel with $m = +\frac{1}{2}$ where they no longer attenuate or scatter light. The pumping can be detected by a decrease in the fluorescence recorded by detector D_f. In contrast to the situation shown in Fig. 1, where only a small fraction of the atoms are maintained as polarized excited atoms, under favorable conditions substantially all of the atoms in the sample can be polarized for situations like Fig. 2 where ground-state atoms are pumped. However, the pumpup times in Fig. 2 are limited by the light intensity and can be as long as seconds, while the pumpup times in Fig. 1 are equal to the spontaneous decay times of the excited atoms and are typically some tens of nanoseconds.

Optically pumped atoms lose their polarization by various relaxation mechanisms. The relaxation is slowest for spin-polarized nuclei of atoms with no electronic angular momentum in their ground states (e.g., He^3, Xe^{129}, Hg^{199}). Relaxation times ranging from many seconds to hours have been observed for such species, and much of the relaxation is due to hyperfine interactions of atoms adsorbed on the container walls. For atoms with electronic spin angular momentum but no orbital angular momentum (e.g., H, Na, N), relaxation times as long as several seconds are observed. The relaxation is due to weak, collisionally induced spin-orbit couplings in gas-phase collisions, in weakly bound van der Waals molecules, or while the atoms are adsorbed on the container walls. For atoms with nonzero orbital angular momentum nearly every collision causes substantial changes in the polarization, and relaxation rates are comparable to gas kinetic collision rates.

Not all slowly relaxing atoms can be polarized easily by optical pumping. An important example is atomic hydrogen. Since the resonance line of atomic hydrogen is in the vacuum ultraviolet region of the spectrum, and since the fine structure interaction which transfers the polarization from the photon to the spin is small, it is very difficult to pump hydrogen directly. However, hydrogen atoms can be readily polarized by spin exchange with optically pumped rubidium according to the reaction

$$Rb(\uparrow) + H(\downarrow) \rightarrow Rb(\downarrow) + H(\uparrow)$$

Thus, spin exchange can be used to extend the usefulness of optical pumping to many species (e.g., H, N, e^-) for which direct optical pumping is impractical.

Optical pumping can also produce coherence between the energy sublevels of matter. The coherence ρ_{ij} between a pair of energy levels i and j is defined to be the average value of the product of the quantum mechanical amplitude a_i for one sublevel with the complex conjugate amplitude a_j^* of the other sublevel, i.e., $\rho_{ij} = \langle a_i a_j^* \rangle$ where the average is taken over all components (e.g., atoms) in the system. An example of a physical quantity described by coherence is the transverse component of the spin of an atom. Coherence will ordinarily oscillate at the difference frequency $\omega_{ij} = (E_i - E_j)/\hbar$ of the energy levels, and light modulated at the difference frequency, or unmodulated light in conjunction with radiofrequency or other radiation tuned to the difference frequency, is often used to generate coherence. Transient coherence can also be produced whenever there is a sudden change in the optical pumping conditions. Coherence is conveniently detected by observing the intensity or polarization modulation of the transmitted or scattered light in an optically pumped system.

Optical pumping of gallium arsenide is used to produce intense sources of spin-polarized electrons. The conduction band of gallium arsenide is S-like in character, while the top of the valence band has $P_{3/2}$ character. Thus conduction electrons generated by the absorption of circularly polarized light near the band edge of GaAs are highly polarized in analogy to the polarization of the photon-excited atoms of Fig. 2. By appropriate treatment of the semiconductor surface, the electron affinity of the GaAs conduction electrons can be made negative and the spin-polarized, photogenerated electrons can be efficiently extracted from the semiconductor into a vacuum.

Atoms polarized by optical pumping (Rb, Cs) are used as precise frequency standards and magnetometers. Optically pumped mercury and xenon have been used in experimental nuclear magnetic resonance gyroscopes. Optical pumping is used to provide population inversions for many masers and lasers, and optical pumping by starlight may drive some celestial masers. Finally, optical pumping and closely related techniques have been an extremely effective spectroscopic tool for investigating the fine structures, hyperfine structures, lifetimes, and other properties of molecules, atoms, ions, and optical centers in solids.

W. HAPPER

References

Bernheim, R. A., "Optical Pumping; an Introduction," New York, W. A. Benjamin, 1965.

Bloom, A. L., *Sci. Am.* **203**, 72 (1960).

Carver, T. R., *Science* **141**, 599 (1963).

De Zafra, R. *Am. J. Phys.* **28**, 646 (1960).

Happer, W., "Progress in Quantum Electronics," Vol. 1, p. 51, New York, Pergamon Press, 1970.

Happer, W., *Rev. Mod. Phys.* **44**, 169 (1972).

Happer, W., and Gupta, R., "Progress in Atomic Spectroscopy," Part A, (W. Hanle and H. Klempappen, Eds.), p. 391, New York, Plenum, 1978.

Kastler, A., and Cohen-Tannoudji, C., "Progress in Optics," Vol. 5, p. 3, Amsterdam, North-Holland, 1966.

Kastler, A., *Opt. Soc. Am.* **47**, 460 (1957).

Pierce, D. T., Meier, F., and Zurcher, P., *Phys. Lett.* **51A**, 465 (1975).

Series, G. W., *Rept. Prog. Phys.* **23**, 280 (1959).

Skrotskii, G. V., and Izyumova, T. G., *Soviet Phys.–Usp. (English transl.)*, **4**, 177 (1961).

Cross-references: ATOMIC AND MOLECULAR BEAMS, ATOMIC CLOCKS, ELECTRON SPIN, LASER, MAGNETIC RESONANCE, MAGNETOMETRY, MASER, MODULATION, POLARIZED LIGHT, SPECTROSCOPY, ZEEMAN AND STARK EFFECTS.

OPTICS, GEOMETRICAL

The radiant energy emitted from a point of a luminous source situated in a homogeneous medium free from obstacles travels through the medium as a spherical wave front whose velocity of advance is a characteristic of the medium. A radius of this wave front is a light ray. If a ray experiences a change of medium or encounters an obstacle, it will, in general, deviate from the rectilinear path defined by the radius. The science of geometrical optics is concerned with controlled deviations.

The Law of Reflection (a) The directions of incidence and reflection and of the normal to the reflecting surface are coplanar.

(b) The angle of reflection is equal to the angle of incidence, both angles being taken with respect to the normal (see Fig. 1(a)).

The Law of Refraction (a) The directions of incidence and refraction and of the normal to the refracting surface are coplanar.

(b) The ratio of the sines of the angles of incidence and refraction, both angles being taken with respect to the normal, is a constant whose magnitude is a function of the properties of the two media on either side of the refracting surface (see Fig. 1(b)). The function is

$$\sin\theta_1/\sin\theta_2 = \mu_{12}$$

The constant μ_{12} is the index of refraction for the two media when light travels from medium 1 to medium 2. If the direction of the ray is reversed, as in Fig. 1(c), it traverses exactly the same path so that

$$\sin\theta_2/\sin\theta_1 = \mu_{21}, \quad \therefore \mu_{21} = 1/\mu_{12}$$

If the first medium is air, it is usual to drop the suffixes and to replace μ_{12} by n. If the first medium is a vacuum then μ_{12} is the absolute refractive index of the second. For air at NTP the value is 1.00028 so that, for most practical purposes the distinction between μ_{12} and n, with air as the first medium, is negligible for substances, such as glass, commonly used for light control. The value of n depends on the wavelength of the light. For sodium yellow of 5890.6Å, the values of n for a number of media are:

Crown glass	1.517
Flint glass	1.650
Quartz	1.544
Fused silica	1.459
Water at 20°C	1.333
Diamond	2.42

For a mathematically plane surface, the coefficient of reflection for normal incidence is dependent on the value of μ_{12}. Fresnel's law gives for this coefficient

$$\rho = [(\mu_{12} - 1)/(\mu_{12} + 1)]^2$$

Clean polished surfaces give results largely in accord with Fresnel's law. Films of grease or slight roughness appreciably reduce the value of ρ.

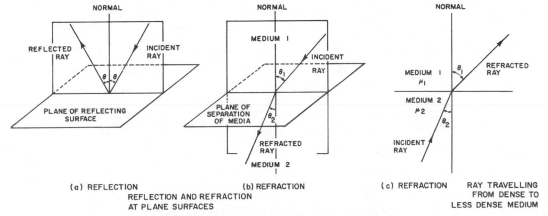

(a) REFLECTION   (b) REFRACTION   (c) REFRACTION   RAY TRAVELLING
REFLECTION AND REFRACTION   FROM DENSE TO
AT PLANE SURFACES   LESS DENSE MEDIUM

FIG. 1. Reflection and refraction at plane surfaces.

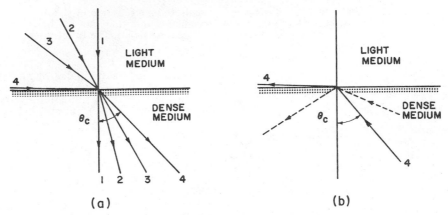

FIG. 2. Illustrating total reflections.

Total Reflection Figure 2(a) shows a series of rays passing from air into a denser medium. Ray No. 1 experiences no deviation. The others experience progressively increasing deviations until, with the tangential ray, No. 4, the angle of refraction θ_c is the greatest possible for that particular medium. Since the angle of incidence is 90°

$$1/\sin \theta_c = n, \quad \therefore \sin \theta_c = 1/n$$

By reversing the directions of the rays, we see that an angle of incidence θ_c is the maximum possible angle for emergence into the air. For angles $> \theta_c$ there is internal reflection, as in Fig. 2(b), governed by the law of reflection. The angle θ_c is called the critical angle for the particular combination of media. Figure 3 shows a number of practical applications of total reflection.

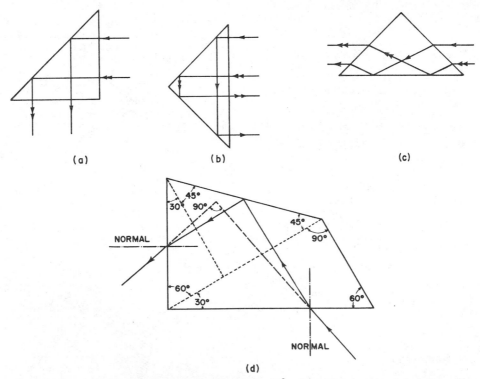

FIG. 3. (a), (b) and (c) show three methods of using the 45° prism. (d) is a constant deviation prism as used in spectrometers.

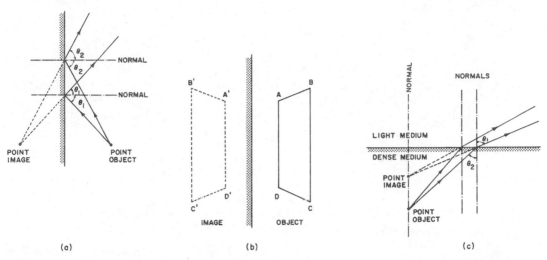

FIG. 4. Image formation by plane surfaces: (a) point object, (b) extended object, (c) image by diffraction of a point object.

Image Formation The geometry of Fig. 4(a) shows that the image in a plane mirror of a point in front of it is another point which lies behind the mirror along the normal from the object point, and is as far behind the mirror as the object is in front. The image of an extended object is identical in size and shape but experiences lateral inversion (see Fig. 4(b)). The image is virtual in the sense that it cannot be received on a screen.

The geometry of Fig. 4(c) shows that the image of a point in a dense medium is another point situated on the same normal to the surface and at a distance from the surface of $1/n$ of the distance of the object point. For large viewing angles, this relationship does not apply, the rays forming the envelope of a caustic as in Fig. 5(a). If the object is extended, a straight line for example, the image is curved and concave towards the surface as in Fig. 5(b).

Refraction by Prism Figure (6a) shows the general case, the incident angle θ_1 being different from the emergent angle θ_1'. The deviation D is given by

$$D = \theta_1 + \theta_1' - A$$

Figure 6(b) shows the conditions for minimum deviation. $\theta_1 = \theta_2 = \theta$, say. The deviation is now given by

$$n = \sin \frac{A + D_{min}}{2} \bigg/ \sin \frac{A}{2}$$

Thus for a 60° prism for which $n = 1.517$

$$\sin \frac{A + D_{min}}{2} = n \sin A/2 = 1.517 \times 0.5 = 0.7585$$

$$\therefore A + D_{min} = 2 \times 49°20'; \quad D_{min} = 38°40'$$

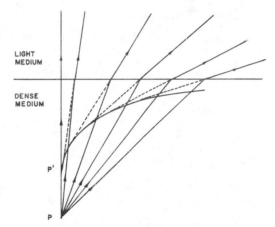

FIG. 5(a). Formation of caustic by refraction. P, point-object. P', position of point-image for small angles of incidence.

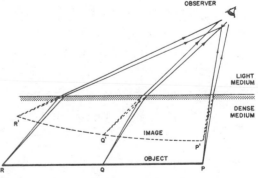

FIG. 5(b) Image by refraction of an extended object.

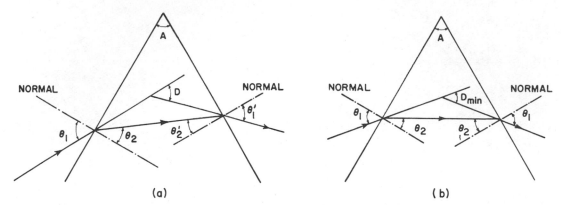

FIG. 6. Deviation by prism: (a) General case, (b) condition for minimum deviation.

A "thin" prism is one for which $A \not> 15°$, so that θ(radians) $\simeq \sin \theta$. This gives

$$D_{\min} \simeq (n - 1)A$$

Only if the incident ray is monochromatic will the emergent ray be monochromatic. If the incident ray is heterochromatic, then the various components in the emergent beam will have different deviations according to their wavelengths. Deviation without dispersion can be achieved by using a compound prism. The main prism, of apex angle A, is of crown glass of refractive index n, and the correcting prism is of flint-glass, of apex angle A' and refractive index n'. Denote the light of minimum wavelength by the suffix b and that of maximum wavelength by r, then for the main prism

$$D_b - D_r = (n_b - n_r)A,$$

and for the correcting prism

$$D_b' - D_r' = (n_b' - n_r')A'.$$

For zero dispersion we must have

$$(D_b - D_r) = (D_b' - D_r'),$$

$$(n_b - n_r)A = (n_b' - n_r')A',$$

therefore

$$\frac{A}{A'} = \frac{n_b' - n_r'}{n_b - n_r}.$$

If the deviation is to be zero, $D = D'$. Now

$$D - D' = (n - 1)A - (n' - 1)A',$$

$$(n - 1)A = (n' - 1)A'$$

since $D - D' = 0$; therefore

$$\frac{A}{A'} = \frac{n' - 1}{n - 1}.$$

Reflection at Curved Surfaces The element of area round a point on the surface is regarded as part of the tangent plane at that point. First consider a spherical surface, Figs. (7a) and (b), and let the aperture be small compared with the radius of curvature. In each Figure, O is the center of curvature, OP is a radius, and O' is the "pole" of the mirror. A ray parallel to the axis is reflected to a point F, the focus. In Fig. 7(a), F is in front of the mirror; in Fig. 7(b) it is

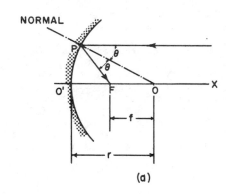

(a)

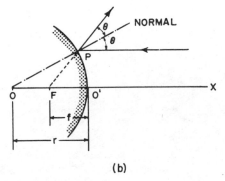

(b)

FIG. 7. Reflection at spherical mirrors. (a) Concave, (b) convex.

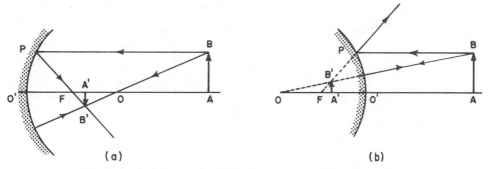

FIG. 8. Image formation in spherical mirrors. (a) Concave, (b) convex.

Concave Mirror	Convex Mirror
u and r negative	u negative and r positive
$u = -8, r = -6, f = -3$	$u = -8, r = +6, f = +3$
$\therefore \frac{1}{v} = -\frac{1}{3} - \left(-\frac{1}{8}\right) = -\frac{1}{4.8}$	$\therefore \frac{1}{v} = \frac{1}{3} - \left(-\frac{1}{8}\right) = \frac{11}{24}$
$v = -4.8$	$v = +2.18m$

behind, so that the reflected ray *appears* to come from F. For small apertures $O'F \approx PF$ and the position of F is independent of O within this limitation. This gives for $O'F$, the focal length $f = r/2$.

Image Formation For an object of small dimensions, the image can be obtained by drawing two rays only: (a) a ray through O—this strikes the mirror normally and is returned along the same path; (b) a ray parallel to the axis—this is reflected through the focal point F.

In Fig. 8, AB is the object and $A'B'$ the image. In Fig. 8(a) the image is real, in Fig. 8(b) it is

virtual. An image is real when the reflected rays actually pass through points in it; it is virtual if they have to be traced back, so that they only appear to come from the image.

Conjugate Foci In Fig. 9(a) a point object A produces a point image A'. If A and A' are interchanged, the rays take identical paths but the arrows are reversed. A and A' are called conjugate foci. In Fig. 9(b) the image A' in the convex mirror is virtual. The relationship between the angles is as follows:—

$$\gamma = \theta + \beta, \ \beta = \theta + \alpha; \ \therefore \alpha + \gamma = 2\beta$$

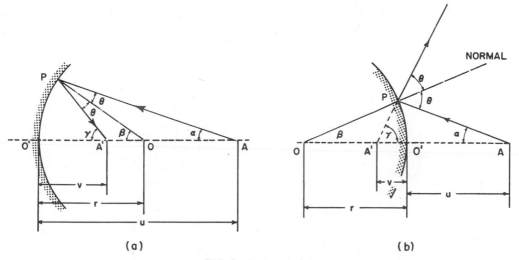

FIG. 9. Conjugate foci.

For small apertures

$$\alpha = PO'/O'A, \quad \beta = PO'/O'O, \quad \gamma = PO'/O'A'$$

$$\therefore \frac{1}{v} + \frac{1}{u} = \frac{2}{r} = \frac{1}{f}$$

This is purely quantitative relationship. For general application it is necessary to adopt a convention regarding signs. Distances are always measured from the mirror, and the direction of the incident light is reckoned positive. As an example let $u = 8$ and $r = 6$. (See calculations under Fig. 8.)

The magnification

$$m = \frac{A'B'}{AB} = \frac{v}{u}$$

This is positive if v and u have the same sign as with the real inverted image of a real object, Fig. 8(a). It is negative if u and v are of opposite sign as with the virtual erect image of a real object, Fig. 8(b).

The axial magnification of an object of finite axial dimension is given by differentiating the above equation.

$$-\frac{dv}{v^2} - \frac{du}{u^2} = 0, \quad \therefore \frac{dv}{du} = -\frac{v^2}{u^2}$$

This shows that if an object is moved towards the mirror a distance du, the image will move $dv = -v^2 du/u^2$. Thus the image will move more rapidly when u is small.

Parabolic Mirror For large apertures for which the angle β is not small, the rays of a parallel axial beam are not all brought to a focus at F, but form the envelope of a cusp, as in Fig. 10(a). A parabolic mirror brings all the rays of such a beam to the same focus. [see Fig. 10(b)].

Refraction at Spherical Surfaces The geometry of Fig. 11(a) gives for a surface of small aperture

$$\frac{n-1}{r} = \frac{n}{v} - \frac{1}{u}$$

If we put $v = \infty$, so that the rays are parallel after refraction and put f_1 for the corresponding value of u

$$\frac{n-1}{r} = -\frac{1}{f_1}$$

If we put $u = \infty$, so that the incident rays are parallel, the corresponding value of v (call it f_2) is given by

$$\frac{n-1}{r} = +\frac{n}{f_2}, \quad \therefore \frac{n}{f_2} + \frac{1}{f_1} = 0$$

For an object AB, Fig. 11(b), we obtain the position of the image by drawing two rays from B; the first parallel to the axis, thereby passing through F_2 after refraction; the second passing through F_1, thereby becoming parallel to the axis after refraction.

$$m = \frac{f_1}{u - f_1} \quad \text{or} \quad \frac{v - f_2}{f_2}$$

Lenses Figure 12 shows three media separated by spherical surfaces. Considering each surface in turn

$$\frac{n_2}{n_1} \cdot \frac{1}{w} - \frac{1}{u} = \left(\frac{n_2}{n_1} - 1\right)/r_1$$

$$\therefore \frac{n_2}{w} - \frac{n_1}{u} = \frac{n_2 - n_1}{r_1}$$

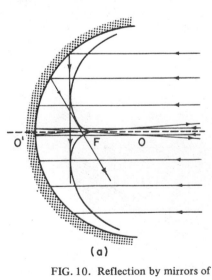

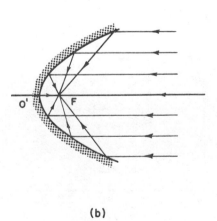

(a)   (b)

FIG. 10. Reflection by mirrors of wide aperture. (a) Spherical, (b) parabolic.

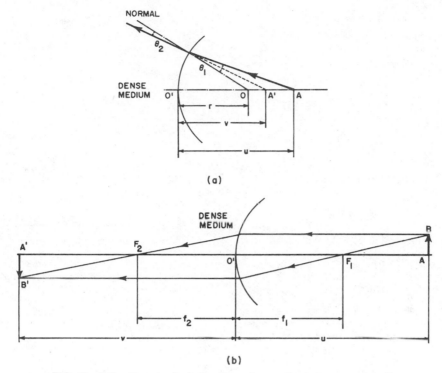

(a)

(b)

FIG. 11. Refraction at spherical surface of separation between two media.

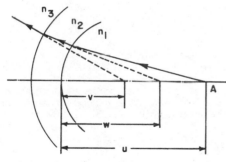

FIG. 12. Lens.

Similarly

$$\frac{n_3}{v} - \frac{n_2}{w} = \frac{n_3 - n_2}{r_2}$$

adding,

$$\frac{n_3}{v} - \frac{n_1}{u} = \frac{n_3 - n_2}{r_2} + \frac{n_2 - n_1}{r_1}$$

In the case of a lens, $n_3 = n_1$, giving

$$\frac{1}{v} - \frac{1}{u} = \left(\frac{n_2}{n_1} - 1\right)\left(\frac{1}{r_1} - \frac{1}{r_2}\right)$$

with air on either side of the central section, $n_1 = n_3 = 1$, and we can put $n_2 = n$.

$$\therefore \frac{1}{v} - \frac{1}{u} = (n - 1)\left(\frac{1}{r_1} - \frac{1}{r_2}\right)$$

If we put $u = \infty$ then $v = f$, the focal length,

$$\therefore \frac{1}{f} = (u - 1)\left(\frac{1}{r_1} - \frac{1}{r_2}\right)^* ; \quad f = \frac{r_1 r_2}{(n - 1)(r_2 - r_1)}$$

Lateral magnification $m = \dfrac{v}{u}$ or $\dfrac{f}{u - f}$

Axial magnification $\dfrac{dv}{du} = -\dfrac{v^2}{u^2}$

*Some authors give this equation with a + sign in the second bracket. In such a case it is necessary to use different signs for convex and concave surfaces, a confusion which is avoided by the simple convention adopted in this article.

Example $r_1 = 10$, $r_2 = 15$, $\mu = 1.6$

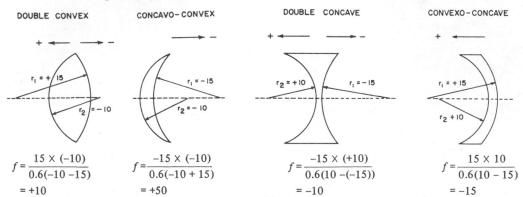

DOUBLE CONVEX

$r_1 = +15$

$r_2 = -10$

$f = \dfrac{15 \times (-10)}{0.6(-10 - 15)}$

$= +10$

CONCAVO-CONVEX

$r_1 = -15$

$r_2 = -10$

$f = \dfrac{-15 \times (-10)}{0.6(-10 + 15)}$

$= +50$

DOUBLE CONCAVE

$r_2 = +10$

$r_1 = -15$

$f = \dfrac{-15 \times (+10)}{0.6(10 - (-15))}$

$= -10$

CONVEXO-CONCAVE

$r_1 = +15$

$r_2 + 10$

$f = \dfrac{15 \times 10}{0.6(10 - 15)}$

$= -15$

With the convex lenses the focus is on the far side of the lens from the source; with the concave lenses it is on the same side as the source. With plano-convex and plano-concave lenses, $1/r$ for the plane surface is zero.

Defects in Mirrors and Lenses An aperture which is large compared with the focal length introduces errors in both mirrors and lenses. One of these is the axial spread of the focus as shown in Fig. 10(a). The amount of spread for a mirror of aperture 2θ is

$$\frac{r}{2}\left(\frac{1}{\cos\theta} - 1\right)$$

Astigmatism results when the object is situated off the axis. The reflected pencil is brought to a focus at a point I, also off the axis, and is spread into an axial line Q_1Q_2, as in Fig. 13(a). If the mirror is turned through a small angle, the point I describes a line called the first focal line. Q_1Q_2 is the second focal line and is perpendicular to the other. In between there is a region where the reflected pencil is the nearest approach to a circle; this is called the circle of least confusion.

Figure 13(b) shows longitudinal spherical aberration in a lens; rays very close to the axis pass through the focal point as defined for a thin lens, while increasing divergence brings the focus nearer to the lens. Figure 13(c) shows chromatic aberration—the focal length, even for a thin lens, being a function of the wavelength. This is because n varies with the wavelength and increases as the wavelength decreases. Thus the focal length for violet is less than for red.

Curvature of the Field The image of a plane object lies, in general, on a slightly curved field (Fig. 14), and this combines with the effects of astigmatism. Pincushion distortion results when a lens is used as a magnifying glass; barrel distortion results when an object is viewed through a lens at some distance from the eye.

Correction of Spherical Aberration (1) *Spherical mirror.* The distortion due to a spherical concave mirror can be corrected by imposing a similar distortion, but in the opposite direction, on the incident light. An elegant solution, used in the Schmidt telescope, is to place at the center of curvature of a mirror—and therefore between the incident light and the mirror—a transparent plate having one surface aspherical. The contours of this plate are such that the beam transmitted by the plate is distorted from true parallelism by the required amount. The paths of the extreme rays are indicated in Fig. 15. The cone of rays reflected by the mirror is, by this device without distortion; thereby form-

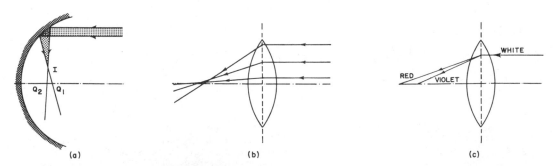

FIG. 13. Image defects. (a) Astigmatism, (b) longitudinal spherical aberration, (c) chromatic aberration.

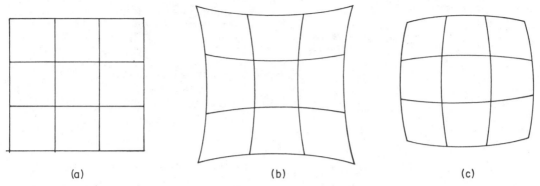

FIG. 14. Image distortion. (a) Object, (b) pincushion distortion, (c) barrel distortion.

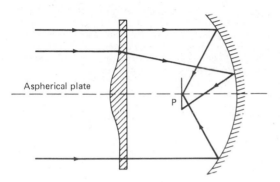

FIG. 15. Ray diagram for Schmidt Telescope.

ing a very sharp image on the photographic plate P.

In conventional reflecting telescopes of the Newtonian kind, uncorrected aberration limits the permissible inclination of rays to axis to a few degrees. In the Schmidt telescope the absence of aberration means that the angular diameter of the effective field is proportional to the ratio of the aperture to the focal length of the mirror.

(2) *Thick Lens.* Close to the axis the two opposite faces of the lens have only a small inclination to one another, whereas away from the axis this inclination is greater. This increase with distance from the axis brings the transmitted rays to a focus progressively nearer to the lens, as shown in Fig. 13(b). With large diameter lenses the aberration is corrected by means of a stepped construction in which the lens consists of a series of concentric zones. Figure 16(a) shows a single stepped lens and also a double lens in which the second component is inverted. The double lens allows a much greater angle of collection. Cheap lenses can be of moulded glass, but where precision is essential the surfaces must be polished and the cost therefore high. Figure 16(b) shows what is probably the supreme example of light control by refraction and internal reflection; it is a schematic drawing of a lighthouse lens system. In this case every prism has to be optically perfect, and also the refracting surface of each dioptic (stepped) lens. For this reason the various components have to be separate, as indicated in the figure.

Systems of Lenses The inverse of the focal length of a lens is called the power of the lens. If the unit of length is the centimeter, the unit

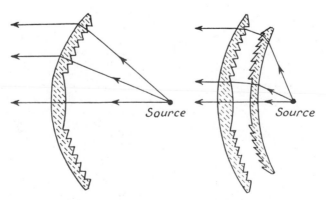

FIG. 16(a). Stepped or dioptic lens.

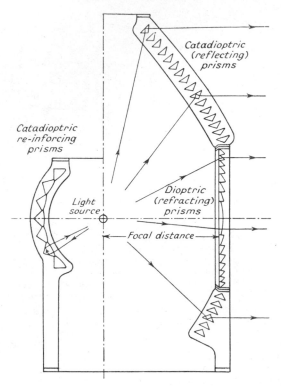

FIG. 16(b). Schematic drawing of a lighthouse lens.

of power is the diopter:

$$D = \frac{1}{f}$$

If a number of thin lenses are in contact, the power of the combination is the *algebraic sum* of the individual powers. As an example, consider two thin lenses for which $f_1 = 10$ and $f_2 = -15$ cm.

$$D_1 = \frac{1}{10} = 0.1 \text{ diopter;}$$

$$D_2 = -\frac{1}{15} = -0.0667 \text{ diopter}$$

$$D = 0.1 - 0.0667 = +0.0333$$

$$\therefore f = +30 \text{ cm}$$

If two coaxial lenses are separated a distance d, then the focal length of the combination is given by

$$\frac{1}{f} = \frac{1}{f_1} + \frac{1}{f_2} - \frac{d}{f_1 f_2}$$

For a minimum longitudinal spherical aberration, the distance d should be

$$d = f_1 - f_2$$

For minimum spherical aberration,

$$f_1 - f_2 = d.$$

For minimum achromatism,

$$f_1' + f_2' = 2d.$$

Combining these two gives

$$f_1 = \tfrac{3}{2}d \quad \text{and} \quad f_2 = \tfrac{1}{2}d.$$

These are the conditions for the Huygens eyepiece. In this eyepiece the combination is equivalent to a single positive lens. (A single negative lens can also be used as an eyepiece.) Consider the case of the compound microscope. Figure 17(a) gives the ray diagram for a positive eyepiece. The objective forms an image of the object behind the eyepiece, that is, within the instrument tube, and the rays are brought to a focus at what is called the Ramsden circle outside the eyepiece. The magnified image, as seen by the observer is inverted. Figure 17(b) shows that the Ramsden circle is between the eyepiece and the objective and that the image is erect. For the same magnifying power the working distance, that is, the distance between object and objective, is larger with the negative eyepiece. Its disadvantage is that the field of view is smaller so that a smaller portion of the object is visible. In order to ensure the largest possible field of view for given powers of the objective-eyepiece combination the Ramsden circle must exist between the eyepiece and the eye. The Ramsden eyepiece as actually manufactured is a combination of two plano-convex lenses of equal focal lengths. Their distance apart is decided by the requirements and can vary between the focal length of a single lens and two-thirds of this distance. According to the previously determined relationships for two-lens combinations a separation equal to the focal length of a single lens would require that one of the lenses would be at the first principal focus. As this is where the image is formed the separation has to be reduced in order that the focal plane may be removed from the lens surface. The ray diagram for this eyepiece is given in Figure 17(c). If the eyepiece is used as a micrometer, the cross-wires are placed between the two lenses and indicated in the diagram by AB. The image A′B′ is between the lenses and is upright as with a single double-concave lens. The shortening of the lens distance introduces a small decrease in achromatism, but this can be remedied by the use of a combination lens for the eye lens.

Thick Lens Figure 18(a) shows a thick converging lens receiving a parallel beam of light.

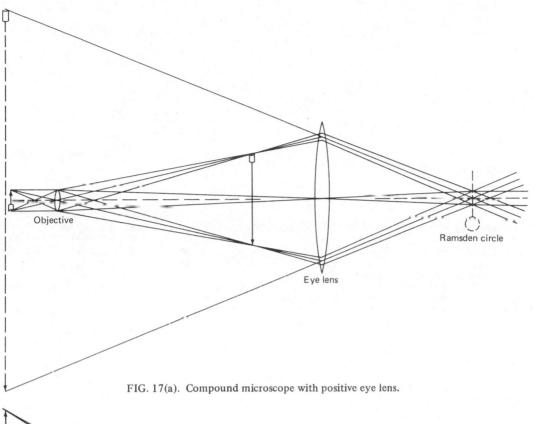

FIG. 17(a). Compound microscope with positive eye lens.

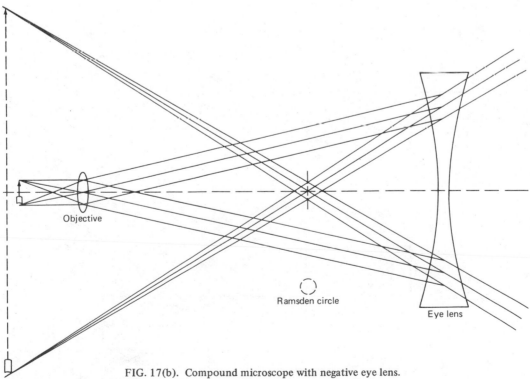

FIG. 17(b). Compound microscope with negative eye lens.

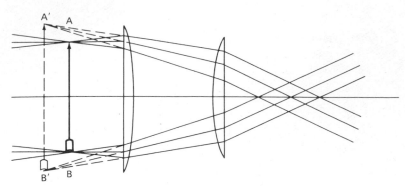

FIG. 17(c). Ramsden eyepeice.

The axial ray 1 passes through the lens without deviation. Any other ray, such as 2, on emerging intersects the axis at some point F'. If we extend the incident ray forward and the emergent ray backwards, as shown by dotted lines, they will intersect at some point P'. The point F' is called the *second focal point* and the plane through P' to which the axis is perpendicular is called the *second principal plane*.

Figure 18(b) shows a system of rays starting from a point F such that the emergent beam is parallel. Then point F is the *first focal point* and the plane through P, determined as before, the first principal plane.

The solution of the thick lens is therefore dependent on the following: The lens behaves

as though the deviation of the rays leaving point F all takes place at the plane P; also the deviation of the emergent rays appears to take place at the plane P^1. Image formation is therefore in accordance with Fig. 19.

From the geometry of the figure we see that

$$ff' = xx' \quad \therefore m = -\frac{f}{x} = -\frac{x'}{f'}$$

also

$$\frac{f'}{\ell'} + \frac{f}{\ell} = 1, \quad \therefore m - \frac{f\ell'}{f'\ell}$$

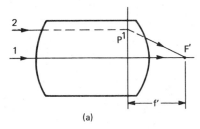
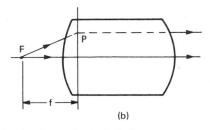

(a)   (b)

FIG. 18. Thick lens: principal focal points and principal planes.

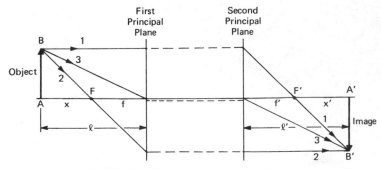

FIG. 19. Thick lens: image formation.

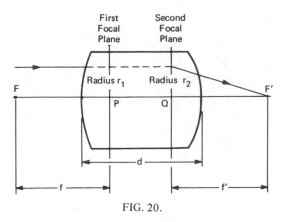

FIG. 20.

The focal length of a thick lens is reckoned from the second focal plane, namely, the distance OF' in Fig. 20. In terms of the radii it is given by

$$f' = (\mu - 1)\left(\frac{1}{r_1} - \frac{1}{r_2} + \frac{\mu - 1}{\mu}\frac{d}{r_1 r_2}\right).$$

It differs from the expression for the focal length of a thin lens by the term

$$\frac{\mu - 1}{\mu}\frac{d}{r_1 r_2},$$

and this, of course, is zero when $d = 0$.

For light traveling from right to left and for which a parallel beam is brought to a focus at F

$$f = (\mu - 1)\left(\frac{1}{r_2} - \frac{1}{r_1} + \frac{\mu - 1}{\mu}\frac{d}{r_1 r_2}\right)$$

$$f = -f'.$$

Optical Distance The optical distance between two points is the minimum possible time required for light to travel from one point to the other. Since a ray of light is the normal to an element of wave front, this concept involves velocity. In a homogeneous medium the optical distance between two points separated by a distance ℓ is $\ell n/c$ where c is the velocity of light in vacuo. Since, in all problems we are concerned with the paths between two given points we can give to c the arbitrary value of unity thus giving ℓn for the optical path. As an example consider again the equation for a thin lens (Fig. 21). For the axial ray we have

$$PS + SQ = PC + CQ + (n - 1)MN$$

$$PS = (PC^2 + CS^2) = PC + CS^2/2PC$$

the approximation being relevant to a thin lens. This gives

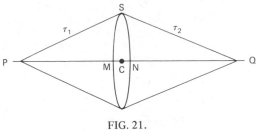

FIG. 21.

$$CS^2/2PC + CS^2/2QC = (n - 1)MN$$

$$\frac{1}{PC} + \frac{1}{QC} = \frac{1}{u} + \frac{1}{v} = 2(n - 1)MN.$$

Considering the points P and Q as the centers of curvature r_1 and r_2 of the two lens faces,

$$\frac{1}{r_1} + \frac{1}{r_2} = 2MN/CS^2,$$

giving

$$\frac{1}{u} + \frac{1}{v} = (n - 1)\left(\frac{1}{r_1} + \frac{1}{r_2}\right).$$

H. COTTON

References

Curry, G., "Geometrical Optics," London, Arnold, 1953.

Jenkins, F. A., and White, H. E., "Fundamentals of Optics," New York, McGraw-Hill Book Co., 1957.

Kingslake, R., Ed., "Applied Optics and Optical Engineering," Vol. 1, New York, Academic Press, 1965–1969.

Kline, M., and Kay, E. W., "Electromagnetic Theory and Geometrical Optics," New York, Interscience Publishers, 1965.

Longhurst, R. S., "Geometrical and Physical Optics," London, Longmans, 1967.

Martin, L. C., "Geometrical Optics," London, Pitman, 1955.

Smith, F. G., and Thomson, J. H., "Optics," London, John Wiley & Sons, 1971.

Welford, W. T., "Geometrical Optics: Optical Instrumentation," Amsterdam, North-Holland, 1962.

Cross-references: ABERRATIONS; LENS; LIGHT; OPTICAL INSTRUMENTS; OPTICS, GEOMETRICAL, ADVANCED; OPTICS, PHYSICAL; REFLECTION; REFRACTION.

OPTICS, GEOMETRICAL, ADVANCED

Geometrical optics deals with light regarded as rays, an approximation which is valid when the wavelength of the light is much smaller than the dimensions of the optical system. (Otherwise, diffraction effects are significant; see FOURIER OPTICS and DIFFRACTION BY MATTER AND DIFFRACTION GRATINGS.) The other physical principle needed to describe optical systems is *Snell's law*, in the form

$$n_1 \sin \theta_1 = n_2 \sin \theta_2 \qquad (1)$$

where n_1, n_2 are the indices of refraction of two media separated by a plane interface, and θ_1, θ_2 are the incident and refracted angles, respectively. (See OPTICS, GEOMETRICAL for a more extended treatment.)

Paraxial Matrix Optics An extremely ingenious way of simplifying the mathematics associated with ray tracing is the matrix method proposed by Sampson (1913) and Smith (1945), and extended by Brouwer (1964). Figure 1 shows a ray leaving an arbitrary object point P and striking the first surface of a typical lens at P_1. If the angles α_1, α_1', θ_1, and θ_1' shown are all small (0.1 radian or $5°$ or less), then Snell's law (1) in *paraxial* form (this term means "close to the axis") becomes

$$n_1 \theta_1 = n_2 \theta_2. \qquad (2)$$

The figure shows that this is

$$n_1 (\alpha_1 + \phi) = n_1' (\alpha_1' + \phi). \qquad (3)$$

But the paraxial approximation also leads to

$$\sin \phi = \phi = x_1 / r_1. \qquad (4)$$

Define the *refracting power* of the surface with vertex V_1 as

$$k_1 = (n_1' - n_1)/r_1. \qquad (5)$$

Combining (3), (4), and (5) with the identity

$$x_1' = x_1 \qquad (6)$$

gives

$$\begin{pmatrix} n_1'\alpha_1' \\ x_1 \end{pmatrix} = \begin{pmatrix} 1 & -k_1 \\ 0 & 1 \end{pmatrix} \begin{pmatrix} n_1\alpha_1 \\ x_1 \end{pmatrix} \qquad (7)$$

where the 2 × 2 matrix

$$R_1 = \begin{pmatrix} 1 & -k_1 \\ 0 & 1 \end{pmatrix} \qquad (8)$$

is the *refraction matrix* for surface 1.

Next, consider the ray $P_1 P_2$. Its position at P_2 is

$$x_2 = x_1' + t_1 \tan \alpha_1' \sim x_1' + t_1'\alpha_1' \qquad (9)$$

and combining this with the identity $\alpha_2 = \alpha_1'$ gives

$$\begin{pmatrix} n_2\alpha_2 \\ x_2 \end{pmatrix} = \begin{pmatrix} 1 & 0 \\ t_1'/n_1' & 1 \end{pmatrix} \begin{pmatrix} n_1'\alpha_1' \\ x_1' \end{pmatrix} \qquad (10)$$

where

$$T_{21} = \begin{pmatrix} 1 & 0 \\ t_1'/n_1' & 1 \end{pmatrix} \qquad (11)$$

is the *translation matrix* for the passage from surface 1 to surface 2.

The matrices of (8) and (11) are all that are needed, for the refraction at surface 2 can be expressed by a matrix R_2, the translation to an image plane or a second lens by T_{32}, etc. In addition, the left-hand side of (7) can be substituted into the right-hand side of (10) and the combined translation-refraction process is specified as a matrix product. Thus, we have a simple, virtually automatic procedure for tracing paraxial rays through an optical system of arbitrary complexity. The constants of the

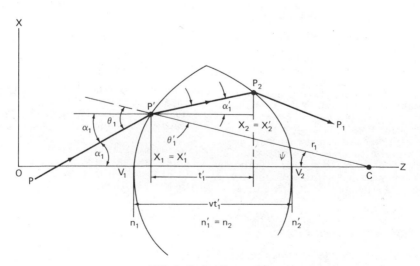

FIG. 1. Refraction matrix.

system which must be specified are the indices n_i, n_i' and the radius r_i of each refracting surface $(i = 1, 2, \ldots)$ and the axial distance t_i' between each pair of surfaces. In addition, it is a straightforward matter to derive by matrix multiplication the formulas quoted in GEOMETRICAL OPTICS, and many others which are rather difficult when using ordinary algebra; examples and a more extended discussion of sign conventions will be found in Nussbaum and Phillips (1976).

Meridional Matrices Figure 1 incorporates the usual convention that the symmetry axis of the optical system is taken as OZ. A plane containing this axis is said to be *meridional*, and we consider the problem of ray tracing in such a plane with the paraxial restriction removed. It is not difficult to show [again see Nussbaum and Phillips (1976)] that the matrix approach outlined above is preserved, provided some moderately obvious changes are made. These involve (1) defining a *skew power K* for surface 1 as

$$K_1 = (n_1' \cos \theta_1' - n_1 \cos \theta_1)/r_1 \qquad (12)$$

where θ_1 and θ_1' are the Snell's law angles before and after refraction, respectively; (2) writing (7) in the form

$$\begin{pmatrix} n_1' \sin \alpha_1' \\ x_1 \end{pmatrix} = \begin{pmatrix} 1 & -K_1 \\ 0 & 1 \end{pmatrix} \begin{pmatrix} n_1 \sin \alpha_1 \\ x_1 \end{pmatrix};$$
$$(13)$$

and (3) writing (10) as

$$\begin{pmatrix} n_2 \sin \alpha_2 \\ x_2 \end{pmatrix} = \begin{pmatrix} 1 & 0 \\ T_1'/n_1' & 1 \end{pmatrix} \begin{pmatrix} n_1' \sin \alpha_1' \\ x_1' \end{pmatrix},$$
$$(14)$$

where T_1' is the actual length of the ray P_1P_2. We note that when the paraxial approximation is valid—i.e., $\sin x = x$—Eqs. (12), (13), and (14) simplify to (5), (7), and (10), respectively, as they should do.

One complication associated with these meridional matrices is that T_1', unlike t_1', is a variable; it may readily be determined from the geometry in the form of an auxiliary formula. We shall refer to a more general form of this formula in a later section. In addition, the angle θ_1 requires an auxiliary formula and θ_1' then follows from Snell's law. Another complication is the fact that the matrices are no longer constants. All these difficulties are surmounted by using numerical, rather than algebraic calculations, and the procedure outlined here is easily programmed on hand-held calculators and computers.

The consequences of the removal of the paraxial approximation for meridional rays is shown in Fig. 2. The rays close to the axis meet at the *paraxial focus F'*, those farther out come together at an *intermediate focus F_I*, and the rays passing through the edge of the lens (the *margin*) meet at the *marginal focus F_M*. The situation of Fig. 2 is what the matrix equations would reveal for a double convex lens; the existence of a set of focal points and the resulting blurry image is called a *third-order aberration*, where the designation refers to the fact that the paraxial approximation $\sin x = x$ implies that we had cut off the Taylor series for the sine function at the first-order term and now we are including third-order terms. [There are also fifth-order and higher aberrations which are primarily of interest to lens designers; see Cox (1964).] The significant feature of Fig. 2 is that it indicates that the blurred image is a consequence of the failure of marginal meridional rays to conform to the paraxial approximation; this failure is called

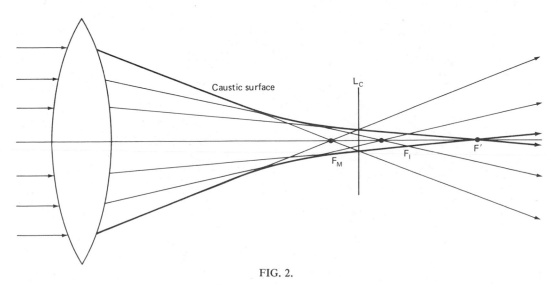

Caustic surface

L_C

F_M F_I F'

FIG. 2.

spherical aberration. If the object at infinity
producing the parallel rays of light is replaced
by a point-object on the axis, we still find a
similar aberration situation on the image side;
hence, spherical aberration is one of three third-
order point aberrations; the other two will be
introduced immediately below.

Skew Matrices We now remove the meridional restriction and introduce the possibility of
skew rays—those not confined to the plane of
the paper in Fig. 2—leaving an object. We specify
such rays in terms of their point of origin
(x, y, z) and their direction cosines γ, δ, ϵ with
respect to the axes $OXYZ$. This implies that the
single equations (13) and (14) must each be
generalized to a three-dimensional set of the
form

$$\begin{pmatrix} n_1'\gamma_1' \\ x_1' \end{pmatrix} = \begin{pmatrix} 1 & -K_1 \\ 0 & 1 \end{pmatrix} \begin{pmatrix} n_1\gamma_1 \\ x_1 \end{pmatrix} \quad (15)$$

with equations in δ, y and ϵ, z, and

$$\begin{pmatrix} n_2\gamma_2 \\ x_2 \end{pmatrix} = \begin{pmatrix} 1 & 0 \\ T_1'/n_1' & 1 \end{pmatrix} \begin{pmatrix} n_1'\gamma_1' \\ x_1' \end{pmatrix} \quad (16)$$

again with two similar equations.

Using the auxiliary formulas and numerical
approach mentioned above, we are now in a
position to trace skew rays. Consider the two
rays of Fig. 3, generating an intermediate focus.
Then incline the rays (Fig. 3b), shifting the
focus as shown and rotate the rays through an
angle of $180°$ about a parallel axis halfway between them. All the new rays produced in this
fashion are skew, and since the two sets of three
matrix equations (15) and (16) will not in gen-

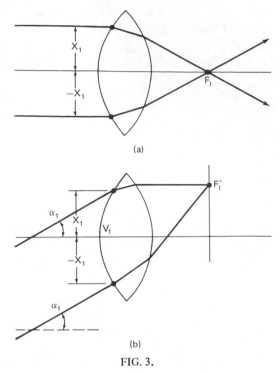

(a)

(b)

FIG. 3.

eral produce the same results as the meridional
equations (13) and (14), the location of the
focal point is no longer F_1'. What happens, in
fact, is that it must move in a closed path on
the image, returning to its original position at
the end of the $180°$ rotation, since the rays are

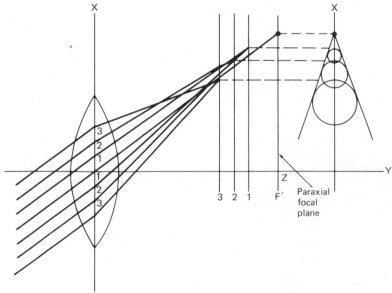

FIG. 4.

once again meridional. This path would be
expected to be a circle in the image plane, since
the rays generate such a path on the front of
the lens, and this is indicated as circle 3 in Fig.
4. A second 180° rotation gives a second circle
coincident with the first, and closer pairs of
rays give the smaller circles 2 and 1. This blurring of the image due to failure of the skew rays
to match the behavior of the meridional rays is
called *coma*, the name coming from the comet-
like shape of the image in Fig. 4. Such a shape
is seldom observed, because we have assumed
that circles are imaged perfectly. For a lens
which is not well-corrected, the double circles
in the image plane will be asymmetrically arranged, as shown in Fig. 5, and the resulting
pattern will be a cardiod [Fig. 5(b)].

Coma can be associated only with off-axis
object points, since all rays from an axial point
are automatically meridional. Let us assume
that a lens system has been corrected both for
spherical aberration and for coma. Since these
are due to essentially different causes, we have
no guarantee that the image point of the meridional and of the skew rays is identical and this
failure to agree is then the third point aberration, *astigmatism*. Figure 6 shows a meridional
or *tangential* fan of rays leaving an off-axis
object point P and forming a sharp image at
P_T'. Another set, the *sagittal* fan, is at right
angles, and all of these rays (except for one
in the vertical plane) have to be skew. The
sharp image is at P_S', and the tangential fan at
this point images as a line. In between the two
lines shown, we have the ellipses and circle of

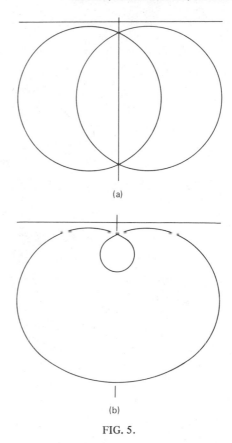

(a)

(b)

FIG. 5.

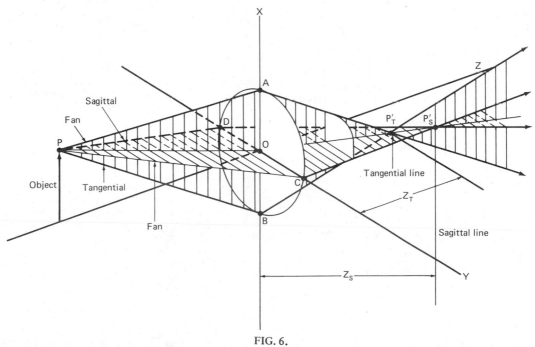

FIG. 6.

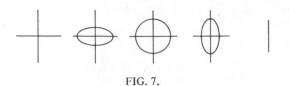

FIG. 7.

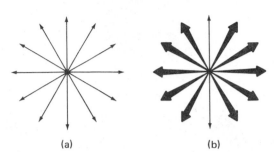

(a)   (b)

FIG. 8. Astigmatism: (a) object; (b) image.

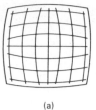
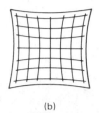

(a)   (b)

FIG. 10. Barrel (a) and pincushion (b) aberrations.

Fig. 7, so that again we have blurring. Note that a bundle of arrows [Fig. 8(a)] would have images going from sharp to blurry, as shown in Fig. 8(b); this also explains the use of the term *sagittal*, from the Latin *sagitta*, arrow. This 90° transition explains why cylindrical corrections are used for astigmatism of the eye; after the original prescription is determined, this defect requires further correction in one plane only.

The last two aberrations are associated with objects of finite extent. If we correct for astigmatism by having the tangential and sagittal focal surfaces merge to form a common *Petzval surface* (Fig. 9), this image surface will in general be curved, since the amount of astigmatism can depend on the distance of the object point from the axis. The resulting aberration, associated with the *OZ* axis, is called *Petzval curvature* or *curvature of field*. A similar effect can occur for distances measured in the *XOY* plane, producing the *barrel* or *pincushion* distortion of Fig. 10. In addition to, and independent of these five geometric aberrations, the fact that the index *n* is a function of wavelength leads to *chromatic aberrations*, which are discussed in the article OPTICS, GEOMETRICAL.

Finally, we point out that the matrix equations as outlined above are applicable only to spherical lenses (or mirrors), with plane surfaces as a special case. This treatment, however, can be readily extended to aspheric surfaces (hyperboloids, paraboloids, and ellipsoids, with spheres as a special case) by a straightforward generalization of the auxiliary relations [Nussbaum (1979)]. A brief computer program will handle the numerical aspect of ray tracing in such systems.

ALLEN NUSSBAUM

References

Sampson, R. A., *Phil. Trans. Roy. Soc. Lond.* **A212**, 149 (1913).

Smith, T., *Proc. Phys. Soc. Lond.* **57**, 286 (1945).

Brouwer, W., "Matrix Methods in Optical Instrument Design," Menlo Park, CA, W. A. Benjamin, 1964.

Nussbaum, A., and Phillips, R. A., "Contemporary Optics for Scientists and Engineers," Englewood Cliffs, NJ, Prentice-Hall, 1976.

Cox, A., "A System of Optical Design," Focal Press, 1964.

Nussbaum, A., *Am. J. Phys.* **47**, 351 (1979).

Cross-references: DIFFRACTION BY MATTER AND DIFFRACTION GRATINGS; FOURIER OPTICS; LENS; MATRICES; OPTICAL INSTRUMENTS; OPTICS, GEOMETRICAL; OPTICS, PHYSICAL.

OPTICS, PHYSICAL

Physical optics deals with those phenomena that are described using the *wave nature of light*. These are *interference, diffraction and polarization*. Of these, interference is the most important; interference phenomena require the wave concept for their explanation. Diffraction and polarization can, at least in principle, be discussed using the particle nature of light.

The unique property of a wave is that it is in two or more places at the same time (a particle, on the other hand, is well localized). A light wave has a phase and can either have a positive or negative value. A wave can be divided into two or more parts.

The *principle of superposition* states that several individual waves at any location can be

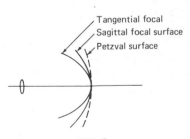

Tangential focal
Sagittal focal surface
Petzval surface

FIG. 9.

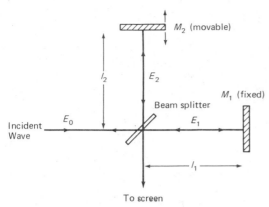

FIG. 1. Michelson interferometer.*

superimposed to determine the resultant. Because the phase of the wave is important, leading to negative as well as positive amplitudes, the summation of several waves can be zero.

Interference There are two broad categories of interference: *division of amplitude* and *division of wavefront*. The Michelson interferometer, Fig. 1, illustrates division of amplitude and two-beam interference. The incident wave with amplitude E_0 strikes a glass plate at $45°$. The back side of the plate is lightly coated with silver so that it serves as a beam splitter; part of the wave is transmitted to mirror M_1 and part is reflected to mirror M_2. The coating on the beam splitter is made with a density that will give the two parts of the beam equal amplitudes.

After reflection by mirrors M_1 and M_2, the components are recombined at the beam splitter where, on average, half the energy is reflected onto a screen and the other half is transmitted back toward the source. The path difference Δ for the two waves, which may be varied by changing the position of M_2, is given by $\Delta = 2(l_1 - l_2)$, where l_1 and l_2 are the lengths of each arm of the interferometer. The phase difference δ is $2\pi\Delta/\lambda$, where λ is the wavelength. For a monochromatic wave the total field at the output is:

$$E_T = E_1 + E_2$$
$$= 0.5E_0[1 + \cos\delta]$$
$$= 0.5E_0[1 + \cos 2\pi\Delta/\lambda].$$

Here we see an important property of interference; when $\delta = n\pi$ (where $n = 1, 3, 5, \ldots$) the two axial components will combine in such a manner that there is zero amplitude on the

*Figures 1–5 and 7–10 adapted from *Contemporary Optics for Scientists and Engineers*, by A. Nursbaum and R. Phillips, © Prentice-Hall, 1976. Reprinted with permission of the publisher.

screen. Yet if we place a card in either arm of the interferometer, blocking one beam, we will see one half of the initial amplitude. With no card (i.e., using both beams), we get zero because one beam is $180°$ out of phase with the other. This is the essence of interference. To explain the phenomenon, we have to invoke the wave nature of light.

In the preceding analysis we assumed that the source was perfectly monochromatic. When Michelson used his interferometer to examine real sources, he found a very interesting phenomenon. As the path difference between the two arms of the interferometer was increased, the visibility of the fringes decreased and finally disappeared. A simple interpretation of this result is that light from an ordinary spectral lamp consists of wave trains of finite length. When the path difference is less than the length of the wave train, part of the train coming from the other arm, interference occurs [Fig. 2(a)]. When the path difference is greater than the length of the wave train [Fig. 2(b)], the light from one arm is added to the light from the other arm which consists of a different wave train. Since the phase relation between two trains is a random variable (the trains come from independent atoms) and since many such superpositions take place in an observation time, no interference will be observed, and the intensity will be a constant. The length of the wave train is related to the temporal coherence length of the light. Ordinary spectral-line sources have coherence lengths on the order of millimeters, whereas lasers have coherence lengths on the order of kilometers.

An alternative but equally correct explanation of the disappearance of fringes can be made if we consider the spectral distribution of the source. If more than one wavelength is emitted, each will form its own interference pattern. At zero path difference, all wavelengths will form a maximum and the fringes will be very sharp, but at large path differences the fringe maxima for the different wavelengths will not coincide and the fringes will vanish. Since one may Fourier analyze a wave train and obtain its spectral distribution, it turns out that these two explanations are equivalent.

After observing the loss in visibility of fringes with increasing path difference, Michelson went on to explain this quantitatively, and he ob-

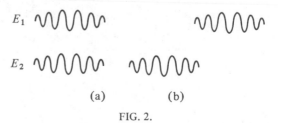

FIG. 2.

tained an extremely important result which we shall consider next. Let the incoming wave contain a number of frequencies, rather than being monochromatic.

Most optical detectors including the eye, photocells, and photoconductors detect energy, which is porportional to the square of the amplitude, i.e., E^2. In optics, this is also designated intensity, I. The total intensity for a polychromatic wave or group of waves, then, is given by a summation over frequency. When we compute intensity by squaring an amplitude, which itself involves a summation, the cross-product terms are converted into functions such as $\cos\left[(\omega_1 - \omega_2)t\right]$ and these average to zero over one or several cycles. This will be the case here for all the products of the form $E_1(\omega_1)E_2(\omega_2)$, since any physically realistic observation time will extend over many cycles. The total intensity I_T will then be

$$I_T = \sum_m I(\omega_m).$$

We assume that the distribution of frequencies is continuous rather than discrete, therefore this sum should be replaced by an integral. References 1 or 2 show that when this expression is evaluated the result is

$$I_T = \int_{-\infty}^{\infty} I(\sigma)\left[1 + \cos\left\{2\pi\Delta(\sigma_0 + \sigma)\right\}\right]d\sigma$$

where

$$\frac{1}{\lambda} = \sigma_0 + \sigma.$$

This extremely important formula states that, in a Michelson interferometer, the average intensity $I_T(\Delta)$, a function of spacing, is the Fourier transform of the intensity $I(\sigma)$, a function of wavelength (spectrum). We may thus regard this instrument as a form of computer which performs a Fourier analysis of the spectrum of the source. (We shall see that a lens behaves in a similar way. In fact, it can instantly produce a two-dimensional Fourier transform of the spatial distribution of a source, and this is something which is extremely complicated when done numerically or by an electrical network.)

An interferometer can be used as a spectrometer, by measuring $I_T(\Delta)$ and taking its Fourier transform to yield $I(\sigma)$. One may then ask about the advantages of the interferometer as compared to a conventional grating spectrometer. In the latter, light from the source is dispersed over a wide area, and a narrow spectral band is detected. It is difficult to detect a signal from weak sources and simultaneously attain high resolution because most of the light available at a given moment is not used. On the other

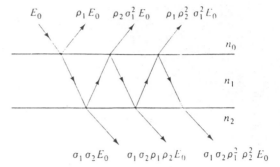

FIG. 3. Interference in thin films.

hand, in Fourier transform spectroscopy, only one-half of the light, on average, is discarded. This gives a better signal-to-noise ratio for a given detector. This technique is widely used for high-resolution analysis of weak sources, particularly in the infrared region.

Thin films illustrate multiple beam interference. (See Fig. 3.) Consider reflection from two parallel surfaces separated by a distance t. The index of refraction of the film is n, and that of the surroundings is n_0 and n_2. The amplitude of the incident wave is taken as E_0. At the first interface, the coefficients of amplitude transmission and reflectance are σ_1 and ρ_1, respectively, and at the second interface, they are σ_2 and ρ_2, respectively, where

$$-\rho_1 = \frac{n_0 - n_1}{n_0 + n_1}, \qquad \rho_2 = \frac{n_1 - n_2}{n_1 + n_2}.$$

The negative sign in $-\rho_1$ means that it is measured *inside* the layer of index n_1. (As explained below, we want to calculate the *transmitted* wave.) The phase difference introduced by two consecutive reflections inside the middle layer for a wave incident perpendicular to the surface is

$$\delta_1 = \left(\frac{2\pi n_1}{\lambda}\right)2t.$$

We will calculate the expression for the transmission coefficient and then apply conservation of energy to obtain the expression for the reflection coefficient. The amplitudes of the transmitted components are:

$$E_1 = \sigma_1\sigma_2 E_0$$
$$E_2 = \sigma_1\sigma_2\rho_1\rho_2 E_0$$
$$E_3 = \sigma_1\sigma_2{\rho_1}^2{\rho_2}^2 E_0,$$
$$\vdots$$

and the total transmitted amplitude E_t is

$$E_t = E_1 + E_2 + E_3 + \cdots + E_n.$$

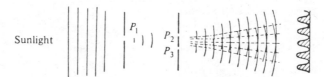

FIG. 4. Young's divided wavefront experiment.

Combining the amplitude and phase expressions and using $I_0 = \frac{1}{2}E_0^2$ yields

$$I_T = \frac{I_0 \sigma_1^2 \sigma_2^2}{1 + \rho_1^2 \rho_2^2 - 2\rho_1 \rho_2 \cos \delta}.$$

In the absence of absorption, the reflection coefficient R and the transmission coefficient T add to unity, or

$$R = 1 - T$$

where

$$T = I_T/I_0.$$

Combining the above three equations with $1 = \sigma^2 + \rho^2$ gives

$$R = \frac{\rho_1^2 + \rho_2^2 - 2\rho_1 \rho_2 \cos \delta}{1 + \rho_1^2 \rho_2^2 - 2\rho_1 \rho_2 \cos \delta}.$$

It can be seen by inspection of this equation that, with proper selection of values for n_1, n_2, t, and λ, R can be zero, i.e., reflection can be eliminated from the surfaces.

Today multiple-layer thin films are used to achieve many interesting and useful results. For example, multiple layer interference filters can achieve very low reflection over a broad band, or increased reflection, or high reflection at one waveband accompanied by low reflection at another waveband.

The two-slit experiment illustrates the other type of interference—*division of wavefront*. It was first performed in 1802 by Thomas Young, who used pinholes. In repeating the experiment, slits are usually used because they transmit more light and the resulting pattern is easier to observe. In the arrangement used by Young (Fig. 4) sunlight pass through a pinhole P_1, which acts as a point source. This assures that the wave on the right-hand side of the pinhole is spatially coherent. A screen with two adjacent pinholes, P_2 and P_3, divides the wave front into two portions, and the openings act as mutually coherent point sources of spherical waves. At a spot on the screen where the path difference for the two waves is zero, they are in phase, and a maximum in the intensity exists. If the path difference is $\lambda/2$, the two waves are $180°$ out of phase, and a minimum exists. The alternating maxima and minima constitute a fringe system, each of the maxima corresponding to a different order of interference.

We shall calculate the interference pattern produced by a single point source emitting monochromatic waves and later generalize this to polychromatic and extended sources. In Fig. 5 a monochromatic point source S is located on the line perpendicular to the midpoint between the two slits, S_1 and S_2. The separation of the slits is a. The intensity distributions of the pattern which would be observed on a screen at a distance d from the slits is given by

$$I_T = 4I_0 \cos^2 \frac{\delta}{2}$$

where δ is the phase difference:

$$\delta = \frac{2\pi}{\lambda} \frac{ay}{d}.$$

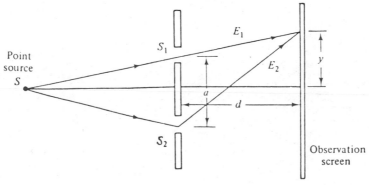

FIG. 5.

For an extended source in place of the pin-hole source, the contribution of each point on the source must be summed. A point j is located a distance u from the center of the source which is at a distance D from the slits. The phase difference between the two beams from passing through the slits is:

$$\delta = \frac{2\pi}{\lambda} \left(\frac{ay}{d} + \frac{au}{d} \right) .$$

Then the contribution to the intensity on the screen from the jth point is given by

$$dI_j(y) = 2I(u)\,(1 + \cos\delta)\,du.$$

For incoherent sources, the total intensity at point y is the sum of the intensities from each point j on the sources; therefore

$$I_T(y) = 2 \int_{-\infty}^{\infty} I(u)\,(1 + \cos\delta)\,du.$$

Again we see a Fourier transform relationship between the intensity distribution of the pattern and the intensity distribution of the source. As the source becomes larger the pattern becomes less distinct. The visibility of the fringes is related to the spatial coherence of the source just as the visibility of fringes in the Michelson interferometer is related to the temporal coherence of the source.

Division of wavefront has been used to measure the angular diameter of stars with an accuracy exceeding that attainable from telescopes.

Diffraction Diffraction is the bending of light as it passes an object. Under certain conditions examination of a shadow of an object on a screen behind it reveals that some light did not travel in a straight line from source but was bent. A close examination of the edge of certain shadows reveals that they are composed of alternating light and dark bands. The patterns can be explained and predicted using wave theory.

Division of wave front leads naturally into diffraction, for the two phenomena are closely related. When parts of a wave front come from different openings and combine, we have interference; but when parts of a wave front come from a single opening and combine, we have diffraction.

There are two categories of diffraction: Fraunhofer and Fresnel. Fraunhofer diffraction occurs when both the source of light and the observation screen are located at distances from the aperture which are very large compared to the size of the opening. It is the simpler of the two to evaluate and is the condition normally encountered. Fresnel diffraction occurs when the source, observation screen, or both are at distances such that the ratio of aperture size to distance cannot be neglected.

FIG. 6. Fraunhofer diffraction.

A classical example of Fraunhofer diffraction is produced by a circular aperture of radius R. The pattern (Fig. 6) was first calculated by G. B. Airy in 1835 and is often referred to as the *Airy disk*. The derivation can be found in many references.[2] The intensity is given by

$$I = I_0 \left[\frac{2J_1(x)}{x} \right]^2$$

where J_1 is the first order Bessel function and $x = (2\pi/\lambda)R \sin\theta$. The angle θ is the measurement between the normal at center of the aperture and the point of observation. (See Fig. 7.)

Diffraction is very important in optical imaging. The limit on the performance of many optical systems is set by their diffraction properties.

Polarization Polarization of light was discovered in 1808 by E. Malus, a French engineer, while looking at sunlight reflected from the windows of the Palace of Luxembourg through

$$I/I_0 = [2J_1(x)/x]^2$$

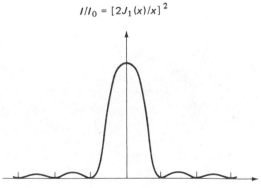

FIG. 7. Plot of Airy disk function.

a calcite crystal. He saw two images which varied in intensity as he rotated the crystal.

This and other polarization phenomena establish that light waves are transverse rather than longitudinal. Only a transverse wave i.e., one in which the amplitude is perpendicular to its direction of travel, can be two sided.

Waves of perpendicular polarization do not interfere. If, for example, polarizers are placed in each arm of a Michelson interferometer and oriented at right angles to each other, the interference pattern disappears.

Light waves are electromagnetic and usually interact most strongly with materials via the electric field. The magnetic interaction is proportional to v/c, where v is the velocity of the electrons in the material and c is the velocity of light. Hence, it is small.

An early demonstration of the polarization of electromagnetic waves was performed by Hertz. His demonstration can be repeated using a set of parallel wires. When the wires are oriented along the direction of the electric field of a radio wave, as determined by the orientation of the transmitting antenna, currents are set up along their length. By conservation of energy, the wave must be absorbed in the process. When the wires are rotated 90° so that they are perpendicular to the direction of the electric field, the wave is not absorbed.

This experiment was repeated in 1963 by Bird and Parrish using light waves. They deposited gold strips onto a plastic diffraction grating have 50,000 lines/inch. The gold film was evaporated onto the grating from the side. Viewed from the top, the gold looked like wires, each less than a wavelength wide. When incident light was polarized parallel to the gold "wires," it was absorbed, and when polarized in the perpendicular direction, it was transmitted.

It should be pointed out that the usual analogy of polarized light with waves on a string and a picket fence is off by 90° because of the orientation of the electric vector E. When the amplitude of string waves is polarized parallel to the openings of a picket fence, the waves are transmitted through the fence. But electromagnetic waves polarized parallel to wires are absorbed.

The amplitude of a light wave is a vector—it has both magnitude and direction—and the simplest type of polarization is linear polarization. In this case, the amplitude vector points in a fixed direction normal to the direction of propagation k (also a vector). This is also known as *plane-polarized light*, because the amplitude and the wave vector define a plane.

Light can also be *circularly polarized*. In this case, the direction of the amplitude rotates in time. The most general state is *elliptical polarization*, where the amplitude pattern at any one location is an ellipse. Both linear and circular polarization are special cases of elliptical polarization. There are always two orthogonal states of polarization, as shown in Fig. 8.

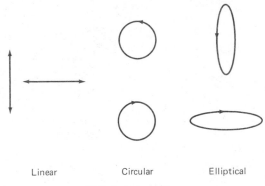

| Linear | Circular | Elliptical |

FIG. 8. Polarization.

Polarization manifests itself through the interaction of light with matter. To exhibit the polarized nature of light effects, the matter must be anisotropic. Crystals such as quartz, calcite, and cordierite can be used to display polarizing effects.

Double Refraction This phenomenon is exhibited by certain materials. A beam of light can be refracted or deviated into two portions, each of which travels in a different direction. Double refraction was discovered in 1669 by Bartholinus, who observed two images when he looked at an object through a crystal of Iceland spar (calcite, $CaCO_3$) (see Fig. 9). The effect can be explained by assuming that the crystal had two different indices for this double refraction. Therefore, light is refracted in two different directions as it enters the crystal (hence the name, double refraction).

The indices for this double refraction are referred to as *ordinary* (n_0) and *extraordinary* (n_e) (Fig. 10). We then apply Snell's law twice, namely

$$\frac{\sin \theta}{\sin \phi_0} = n_0$$

and

$$\frac{\sin \theta}{\sin \phi_e} = n_e(\phi_e).$$

There are many interesting phenomena which depend on the interaction of polarized light and anisotropic materials. A convenient way to

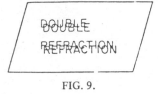

FIG. 9.

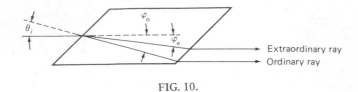

FIG. 10.

categorize them is by the general expansion of the material response P in a power series in the electric and magnetic fields and the spatial derivatives.

$$P_i^{\omega} = P_i^0 + X_{ij}E_j^{\omega} + X_{ijl}\nabla_l E_j^{\omega}$$
$$+ X_{ijl}E_j^{\omega_1}E_l^{\omega_2} + X_{ijlm}E_j^{\omega_1}E_l^{\omega_2}E_m^{\omega_3}$$
$$+ X_{ijl}E_j^{\omega_1}B_l^{\omega_2} + X_{ijlm}E_j^{\omega_1}B_l^{\omega_2}B_m^{\omega_3},$$

where the superscripts denote the frequency of the field and the subscripts denote Cartesian components. Each term in this series describes one or more effects; for example, the first term describes pyroelectricity and ferroelectricity; the second double refraction; the third, optical activity; and the fourth, the linear electrooptic effect and second harmonic generation. This last is the generation of a light wave at twice the frequency of an incident wave; it is analogous to the generation of harmonics in electronic circuits due to the nonlinear response of circuit elements.

RICHARD A. PHILLIPS

References

1. Michelson, A., "Light Waves," Chicago, Univ. Chicago Press, 1927.
2. Nussbaum, A., and Phillips, R., "Contemporary Optics for Scientists and Engineers," Englewood Cliffs, NJ, Prentice Hall, 1976.
3. Shurcliff, W., and Ballard, S., "Polarized Light," New York, Van Nostrand, 1964.

Cross-references: DIFFRACTION BY MATTER AND DIFFRACTION GRATINGS; FOURIER ANALYSIS; INTERFERENCE AND INTERFEROMETER; OPTICS, GEOMETRICAL; POLARIZED LIGHT; REFLECTION; REFRACTION; THIN FILMS.

OSCILLOSCOPES

The oscilloscope is a measurement instrument that presents electrical phenomena (signals) in the form of an X-Y graph on a cathode ray tube (CRT) for analysis. This CRT, much similar in appearance to the one in a small television, typically presents signals in a *voltage* (Y-axis) versus *time* (X-axis) relationship. (See Fig. 2(B).) As *voltage* increases, the corresponding location on the CRT moves upward. *Time*, with relation to the displayed signal, starts at the left side of the CRT and increases as the display travels to the right. Signals measured by oscilloscopes are usually generated electrically. Other physical signals must be developed mechanically and converted to electrical energy for display on the oscilloscope through the use of transducers.

Today, most oscilloscopes require manual operation. Each has a CRT to display data, a vertical input to condition the signal, a trigger section to select the signal parameter to begin the display and a horizontal section to provide an accurate time comparison to the signal being displayed. A description of how these fundamental sections of the oscilloscope function together follows. (See Fig. 1.)

Display The oscilloscope develops a graphic representation of the signal by moving the CRT's electron beam across a phosphor coating on the inside of the viewing screen. This results in a bright, visible display that traces the path of the electron beam. (See Fig. 2.) The display brightness is controlled by the oscilloscope's *intensity* control [Fig. 1(A).] Proper *focus* provides a crisp display. The display is turned on at the time the trigger section has sensed the correct input characteristic that a user has selected as the starting point (usually a voltage level).

Vertical Section The primary functions of the vertical section are to supply the display section with a Y-axis and to condition the signal being captured. Conditioning includes display size, input coupling, display position, signal delay to allow viewing of leading edge transient responses and trigger signals. (See Fig. 3.) The vertical section is further broken down into key blocks: an attenuator to set display size, input coupling to select fundamental signal characteristic, trigger pickoff to trigger display start, and a delay to allow user to see leading edge of signal.

The vertical sections's attenuator reduces large signals to a size necessary for display on the CRT. Attenuator sections also provide three input coupling sections: AC, DC, and GND. AC is selected for changing or AC (alternating current) signals. DC is selected for measuring DC levels like those in power supplies. And GND is selected to identify where ground or zero voltage is located on the display.

The vertical amplifier increases the display size for extremely small signals. In addition the amplifier generates a trigger pickoff signal to drive the trigger circuitry.

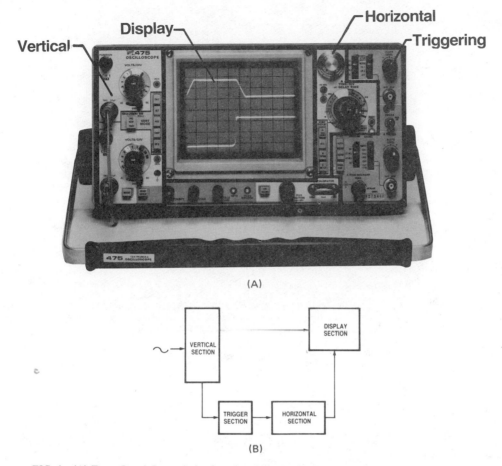

FIG. 1. (A) Front Panel Controls for functional blocks. (B) Functional relationship of blocks.

The delay compensates for normal propagation delays in any electronic circuit. With the delay line the CRT is able to display leading edges of the signal that are lost otherwise. Figure 4 illustrates how a delay line enables the CRT to display the leading edge of a positive step response.

Horizontal Section To develop a graphic representation of the signal being captured, an oscilloscope needs an accurate time relationship on the X-axis of the CRT. This is accomplished by using a very linear sweep generator which produces a *sawtooth* (ramp) waveform. Many different ramp (time/div sweep) rates are available for selection so the user can choose the one that will develop the required graph of the signal. (See Fig. 5.) The horizontal section incorporates an extremely linear and accurate timing ramp that turns the display on at the trigger point. As the ramp voltage increases the X-axis of the signal, it travels left to right on the display. At the end of the selected sweep (ramp) rate the horizontal section generates

a retrace signal to move the electron beam back to the left side of the CRT. An internal holdoff time is then generated to reset the trigger circuit to accept another trigger signal, thus completing the horizontal signal loop. (See Fig. 6.) The horizontal section establishes an accurate timing reference at several different possible time selections, trigger holdoff, X-axis signals, and positioning.

Triggering Section Triggering determines when the oscilloscope draws the graph. The trigger circuit tells other oscilloscope circuits when to start drawing the graphic representation of the input signal (Fig. 7). Controls in the trigger section that set up the required trigger condition are coupling, slope, level, and source. The control circuit is much like the vertical input coupling with possible AC and DC selections. Positive and negative trigger slopes are available on almost any oscilloscope. [The trigger slope represented in Fig. 7 is positive (+).] The level control determines the trigger signal amplitude required to activate the trigger circuit. Most

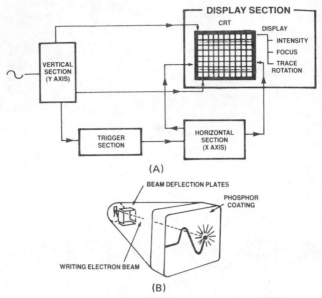

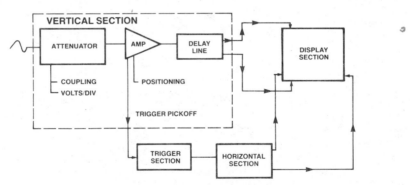

FIG. 2. (A) The basic sections of the oscilloscope work together to create a graph. (B) Changes in voltage on the CRT's deflection plates move the electron beam across the display, causing the phosphor coating to glow.

FIG. 3. Blocks of the vertical section.

oscilloscopes provide internal source (from vertical input signal) or external to trigger on a different signal than the one displayed.

Those who use an oscilloscope span many different industries and several job functions.

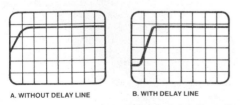

FIG. 4. The horizontal sections's delay line allows viewing of a signal's leading edge.

Applications for oscilloscopes are limited only by one's imagination. Researchers in the computer field are using oscilloscopes in developing smaller, high-speed data processing units. Also using oscilloscopes, the development of electric control systems necessary to continue the exploration of deep space is underway. Faster, more efficient digital communications systems are being designed with the help of scopes. Using scopes, designers in the automobile industry are developing safer cars. Advancement in LSI circuitry, new digital high-speed memories, and A-D converters are aided by the oscilloscope. Using scopes, medical researchers continue to develop methods of restoring eyesight. Work on robotics is advancing with the help of scopes, and eventually will

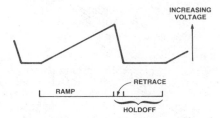

FIG. 5. Horizontal sections sweep ramp.

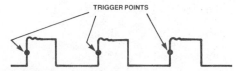

FIG. 7. Trigger points tell the oscilloscope when to begin drawing a graph. Also, for repetitive signals the trigger signal is sent at the same time to produce a stable "repetitive" display.

improve manufacturing productivity in all industries.

The Development of the Oscilloscope The use of CRT's began around the turn of the 20th Century. Their use in advanced oscilloscopes began about 40 years later, with the development of the first triggered oscilloscope. Since the late 1940's demands have increased for higher bandwidth, more accurate voltage and time measurements, better triggering, and the

ability to capture and display signal shot or low repetitive rate signals.

Until the 1970s, oscilloscope designs required several dozen power-hungry vacuum tubes. As many as 20 different oscilloscope types were available. As bandwidth pressure continued, the vacuum tube became economically unfeasible. The 1970s saw increased usage of semiconductors in the oscilloscope, and eventually vacuum tubes were rendered obsolete. Semiconductor technology provided the in-

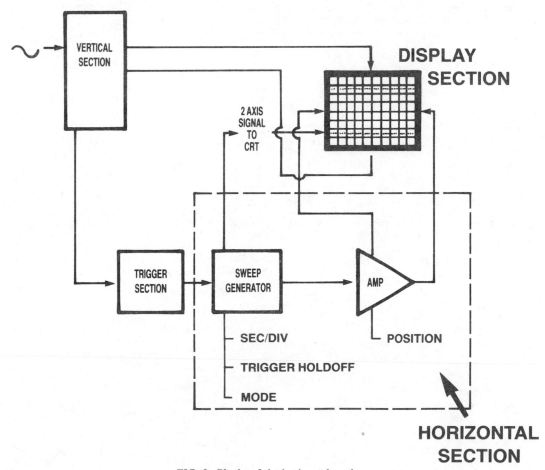

FIG. 6. Blocks of the horizontal section.

crease in accuracy and stability and the higher bandwidth demanded by users. The semiconductor also provided small size and power consumption, which made possible more than 100 different sophisticated types of scope.

Storage Scopes In addition to advancements in traditional oscilloscopes that have historically addressed repetitive signal measurements, new "storage" oscilloscopes were developed that could capture, store, and display single events. This storage technique was made possible by the use of special phosphors and processes that stored the signal for long periods of time on the inside of the CRT viewing screen in a bistable storage mode. (*Bistable* is the name given to a storage process that provides a long viewing time.) Soon, faster storage oscilloscopes became available as variable persistence and fast transfer storage CRT's were invented. People requiring storage products used the storage speed as a major criterion for selection, second only to vertical bandwidth.

With the development of storage CRT's, the oscilloscope market took two basic directions in the 1970s: (1) improved bandwidth and accuracy for making repetitive signal measurements, and (2) improved bandwidth and storage speeds for those requiring nonrepetitive signal measurements. Nonstorage oscilloscopes passed the 1 GHz mark. During the same decade, conventional CRT storage speeds became sufficient to capture and display 400 MHz signals.

Advancements in digital memory speeds, and faster and lower priced A-D converters, resulted in a new type of storage oscilloscope in the late 1970s that did not require special CRT storage. In addition, digital storage provides an effective method of securing signal data and transferring it to a computer for processing of long-term documentation.

Special Purpose Scopes Over the years, special purpose oscilloscopes have made possible totally new measurement methods. The first, the *spectrum analyzer*, was developed for specialized measurements made predominantly in the communications industry. Instead of presenting graphic data in the traditional voltage (Y-axis) versus time (X-axis) format, the spectrum analyzer displays dB (Y-axis) versus frequency spectrum (X-axis).

A second type, the *sampling oscilloscope*, maintains the traditional oscilloscope graphic display in voltage versus time, but higher bandwidths (to 14 GHz) are made available using a random sampling method. However, sampling oscilloscopes are only good for making repetitive signal measurements.

The latest development is the *logic analyzer*. Like the traditional oscilloscope, some logic analyzers show time on the X-axis. Instead of voltage on the Y-axis, however, logic analyzers display only digital wave shapes without reference to voltage. This new type of instrument has been used primarily to make high-resolution timing measurement of digital logic circuits. Other types of display (e.g., state tables, mapping, etc.) are also available in many logic analyzers.

The Future Prior to the 1980s oscilloscopes were developed primarily for manual operation. Future developments will, with continued advancements in LSI, A-D memory speeds, and programmable instrumentation, focus on the oscilloscope's interface with a computer. New oscilloscopes of the 1980s will provide improved productivity through the use of programmability, digital storage, and increased measurement specialization.

Oscilloscopes not tied to the external controllers/computers will offer the user more than just graphic CRT displays. Instead, answers (e.g., risetimes, falltimes, pulse width, frequency, period, etc.) will be available automatically with a higher degree of accuracy and repeatability. Vast improvements in the traditional oscilloscope parameters (e.g., bandwidth, stored writing speed, accuracy, and timing resolution) will occur. Oscilloscope numbers will expand beyond the 100 types available in 1980.

DAVID E. McCULLOUGH

References

Tektronix, "The XYZ's of Using a Scope," 1981.
Tektronix, "The Digital Storage Oscilloscope," 1980.

Cross-references: ELECTRON, ELECTRON OPTICS, CIRCUITRY, PULSE GENERATION.

OSMOSIS

If, into the bottom of a jar containing water, a solution of cane sugar is introduced with care so as to avoid mixing, not only will the molecules of cane sugar diffuse into the water but the molecules of water will diffuse into the sugar solution. These processes will go on until the concentration of sugar, and of water, is the same throughout.

If the solution is placed in a container, whose walls are relatively impermeable to the sugar while being permeable to the water, and the container is placed in water, the water will pass from the outside into the container. The term osmosis is usually restricted to the passage of water. If the influx of the water results in an overflow of solution to somewhere other than the surrounding water this overflow will continue until all the sugar is removed from the container. If the container is closed, water will continue to enter until there is sufficient stress in the stretched walls to cause a pressure on the solution inside; this will eventually stop the influx. Of course, if the walls of the container are not completely impermeable to sugar, then the sugar will be escaping into the water outside the container and this will go on until the concentration of sugar is the same outside and inside. If the walls of the container were impermeable to water but permeable to solute, the latter would escape. The cause of this osmosis, this "pushing," of water into the solution is that the tendency of the water molecules to escape from the pure water is greater than that of the water molecules in the solution. Consider water in contact with a limited volume of air. Of those molecules of water striking the surface some will have sufficient energy to escape into the air and this escape will result in net loss to the air which will continue until the concentration of water vapor molecules there is such that the rate of escape from the air (into the water) equals the rate of escape from the water (into the air). If the volume of the air space is fixed, the pressure will rise. Just as the temperature of all bodies is the same when they are in thermal equilibrium, although their heat content per unit volume varies with their specific heat, so the escaping tendency of the water is the same in all systems when they are in aqueous equilibrium, whether the system is pure water, solution, gas phase, wettable solid, etc. The same concept can be applied to any substance, say mercury in pure mercury, in air containing mercury vapor, and in an amalgam with another metal such as zinc. The term osmosis is usually restricted to the passage of water from a solution where the escaping tendency is higher to a solution where it is lower. Moreover it is usually restricted to the passage through a solid or liquid barrier which prevents the solutions from rapidly mixing. It is not used for the passage of water in the form of vapor through the air from a dilute solution to a stronger solution in the same confined space, although the process is fundamentally the same. It is sometimes restricted to the case where the barrier is semipermeable, that is, lets through water but not solute.

The escaping tendency of water is lowered by the addition of a solute. If the molecules of the solute have no other effect than to reduce the number of molecules of water in unit volume, then the escaping tendency of the water will be reduced proportionately to the reduction in the mole fraction of water, N_1, the ratio of the moles of water to the sum of the moles of water and solute. Such is a "perfect" solution. If, however, there is some attraction between the solute and water molecules, a smaller fraction of the latter will have energy sufficient to escape—a "nonperfect" solution. The escaping tendency is increased by pressure. Hence a solution in which the water has lower escaping tendency than it has in pure water at the same pressure, P^0, can be brought to water equilibrium by a sufficient increase in the

pressure on the solution to a value P. This sufficient increase, $P - P^0$, is the osmotic pressure of the solution. In general we cannot state $P - P^0$, the osmotic pressure, knowing only N_2, the ratio of moles of solute to the sum of the moles of water plus solute, the mole fraction of solute. ($N_2 = 1 - N_1$).

What we can say is, that if in a solution with a mole fraction N_2 of solute under a pressure P the water has the same escaping tendency as it has in pure water at the same temperature and at a pressure P^0, then $dP/dN_2 = A/B$ where dP/dN_2 is the increase of P relative to increase of N_2 to keep the escaping tendency unchanged; A is the decrease of escaping tendency relative to increase of N_2 when P is unchanged; and B is the increase in escaping tendency relative to increase in P when N_2 is unchanged. For dilute solutions A/B approximates to RT/V_1 and so $P - P^0$ approximates to N_2RT/V_1 where V_1 is the volume of one mole of water, R is constant 82.07 cm^3 atm/deg, and T is the absolute temperature. For very dilute solutions N_2/V_1 approaches n_2/V, the number of moles of solute in a volume V of solution and $P - P^0 = n_2RT/V$ (van't Hoff's equation). This gives an osmotic pressure of 1 atm for one mole of solute in 22.4 liters at 0°C. There is a departure from both these equations for stronger solutions. The fact that one mole of a perfect gas in 22.4 liters at 0°C exerts a pressure of 1 atmos, coupled with the above, has led some to say that the osmotic pressure is the bombardment pressure of the solute molecules. It is correct to say that for very dilute solutions the osmotic pressure of a solution is equal in magnitude to the pressure the solute molecules would exert if they were alone in the same volume and behaved as a perfect gas, but that is another matter.

To measure the osmotic pressure, a semipermeable membrane must be prepared which itself can stand sufficient pressure, or it must be deposited in the walls of a porous pot so that the pressure can be sustained. With the solution being inside and water out, pressure is applied to the former until there is no net movement of water.

Observations by Berkeley and Hartley showed that for 3.393 gms of cane sugar per 100 gms H_2O, the osmotic pressure at 0°C is 2.23 atmos while the van't Hoff equation gives 2.17 atmos since $n_2/V = 9.27 \times 10^{-15}$. If N_2/V_1 is used instead of n_2/V, the value of 2.22 is obtained. With stronger solutions the measured osmotic pressure exceeds that calculated: with 33.945 gms of sugar, 24.55 atmos is the value measured, while van't Hoff's equation gives 18.41 and the other 21.8 atmos. The observed value is given if, in calculating N_2/N_1, it is assumed that each sugar molecule immobilizes five molecules of water.

The solutes in the vacuole of a plant cell are exposed to the inward pressure of the distended cell wall and that of the turgid surrounding cells. Water will pass into the cell vacuole from water outside as long as the total inward pressure on the vacuole falls short of the osmotic pressure of the solution in the vacuole. Passage of water into the vacuole dilutes the contents and lowers the osmotic pressure and increases the inward pressure by distension. The amount by which the inward pressure falls short of the osmotic pressure is called by some the suction pressure.

A substance such as cellulose or gelatin tends to take up water, the tendency decreasing with increase in water content until the stress in the substance causes a sufficient rise in the escaping tendency of the water in the substance. This process, which like osmosis is a movement from higher to lower escaping tendency, is called imbibition, and the pressure on the substance sufficient to stop the uptake is the imbibitional pressure. Hence, if a plant cell with a cellulose wall, after coming to equilibrium with a solution, is transferred to water, the wall takes up water by imbibition and the vacuole by osmosis. The latter considers only the over-all movement from outside to vacuole and does not consider the movement from cellulose to vacuole, a process which is the reverse of imbibition. A plant cell in equilibrium with a solution having an osmotic pressure of 25 atmos would also be in equilibrium with air about 98 per cent saturated with water vapor. If the cell were transferred to a saturated atmosphere, it would take up water. We lack precise terms for the passage of water from air into the cellulose and into the vacuole. Condensation, which might be used, ranges more widely.

The escaping tendency of water is affected by factors other than concentration of solute and pressure. Increase of temperature increases escaping tendency. This is a complex problem involving not only transfer of water but also of heat. To a minor extent, the passage of water from pure water to a solution involves a heat transfer.

For many naturally occurring membranes which are not completely semipermeable, i.e., they let solute molecules through slowly, electro-osmosis is important. If the membrane tends to lose negative charges to, or take negative charges from, water or solutions, then the water molecules, in the pores of the membrane, will tend to take on an opposite charge to the membrane. If there is a gradient of electric potential across the membrane, the charged water will move in the appropriate direction. If the potential difference is established by the use of electrodes, this is electro-osmosis.

If, for any reason such as the greater solubility of the solute relative to that of water in the membrane, the volume flow of the solute from the side of higher concentration is greater than that of water in the opposite direction, then there is negative osmosis. This occurs with some plant cells with lipid-soluble substances of small molecular weight.

The rate of osmosis depends, not only on the excess of the escaping tendency of the water in the phase from which it moves, over that in the phase to which it moves, but also upon the area of surface of interchange and the over-all resistance experienced by the water. The rate of shrinkage of the vacuole of a plant cell when it is placed in a strong solution at first seems surprisingly high. When allowance is made for the fact that the ratio of surface to volume increases as the linear dimension is reduced then it is realized that when the vacuole of a spherical cell of radius 30 μ shrinks to half its volume in say 5 minutes the passage of water is only 1 ml per 10 000 cm^2 per minute although the thickness of the layer between vacuole and external solution is of the order of 1 μ in thickness. Under other circumstances, this layer might be said to be relatively impermeable to water. It seems probable that much of the resistance resides not in the cellulose wall or cytoplasm but in the tonoplast which separates the latter from the vacuole.

G. E. BRIGGS

P

PARACRYSTALS*

There are three states of aggregation: solid, liquid and gaseous. This categoric division, however, does not explain why there are, within the solids category, such different substances as metals, minerals, and glasses, or why liquids, when solidifying, convert partly into a crystalline and partly into a glassy state. At the same time substances exist which do not belong in any one of these groups. The word "paracrystal" was used for the first time by Stanley (1) in connection with biological substances and by Rinne (2) in connection with liquid crystals. It is the task of this contribution to find an atomistic definition of the paracrystal (3) and to demonstrate the fundamental importance of this idea relative to concept and application. In doing this one cannot neglect discussing at the same time the x-ray interferences of all these substances, because without the discovery by von Laue (4) that one can gain direct information on shape and ordering of atoms and molecules, everything would remain pure philosophy. One can understand the paracrystalline state only if one simultaneously has knowledge of the other existing modes of solids.

Gases The gaseous state is characterized by a maximum of disorder. The molecules are distributed in space completely irregularly according to purely statistical laws. Their distribution of velocity and proximity obeys, in the ideal case, the Boltzmann statistics. The more compressed the gas, the narrower become these distribution functions.

Liquids The mobility of molecules in liquids is considerably less than that in gases; opinions on their spatial order, however, are at present still divided. The school of Bernal (6) advocates the extreme of highest disorder; whereas the other extreme of quasicrystalline order is supported by Kaplow, Strong and Averbach (7).

Liquid Crystals Whereas liquids generally exhibit equal properties in all three dimensions, i.e., they are isotropic; so-called "crystalline liquids"† are also known, which in the vicinity

*The editor is indebted to Mr. Walter Trapp of the Air Force Materials Laboratory for his careful translation of this article.

†See the essay by B. Böttcher and D. Gross in *Umschau in Wissenschaft und Technik,* 69, 574 (1969).

of the interface with their container exhibit anisotropic optical and electrical properties. The explanation for this phenomenon is derived from the shape and the dipole moment of the molecules of the (mostly organic) liquids. They are oblong or platelet-shaped as in soap solutions.

Crystals Today we have much more knowledge of the crystalline state. The regular and plane-facets-limited shape of a polyether, which exhibits many single crystals, has inspired Haüy (8) to speak of "elementary building blocks" of equal size in the crystal, in the shape of cubes or parallelepipeds, which accumulate densely like a three-dimensional chessboard. In present terminology these are the "lattice-cells," and Seeber (9) recognized very early that they represent the space in which the basic unit of atoms, ions, or molecules of the crystal is housed. Since the distance between centers of gravity (CG) of the basic units is absolutely constant, i.e., 4 Å in certain lattice directions,* we can safely say that in a range of 1 μ† in these directions, exactly 2500 building blocks are situated, or that the distance from one CG to the 2501st neighbor is 10004 μ. Thus, because the building blocks are lined up in a countable fashion, one uses the expression "ideal remote order."

Amorphous Solids Almost all glasses, ceramics, and plastics have properties, which can be explained by the presence of "amorphous" components, in which the atoms are particularly disordered. How one visualizes the order of the atoms is open to discussion. It is generally agreed upon, however, that no lattice regions and no remote order exist here.

The Idea of Paracrystals The status today is that we have lattice structures with remote order on one hand and amorphous materials, liquids, and melts as structures without lattice and without remote order on the other hand opposing each other. On the basis of some fundamental work by J. J. Hermans (10) it was shown (3) that the idea of a statistical close-order for all building blocks, as it is advocated by Zernike (11) and Debye (12), definitely allows three-dimensional lattices of

*1 Å(named after the Swedish physicist Ångström) is equal to one ten-thousandth of a micron.

†1 μ(micron) is equal to one ten-thousandth of a centimeter.

the CG of the building blocks. These lattices consist of rows of countable building blocks (running number p), spaces (running number q), and columns (running number r); however, these are not straight lattice lines, but somehow curved in such a way that, for instance, along the rows of building blocks the distances between adjacent CGs (p, q, r) and $(p + 1, q, r)$ are not constant but adopt different values y with the frequency $H(y)$. This means also that this frequency distribution guarantees the same close-order ratios for all building blocks (p, q, r).

From the building block $(p + 1, q, r)$ to the adjacent CG $(p + 2, q, r)$, one finds a completely different distance vector z, which once more obeys the same probability level $H(z)$. According to the rules of probability the combination of the quantities y and z occurs with the frequency $H(y) \cdot H(z)$.

If one is interested in the statistics of the distance between nearest neighbors (p, q, r) and $(p + 2, q, r)$, one has to calculate the frequency $H_2(x)$ for the vector $y + z = x$, taking into consideration at the same time all possible

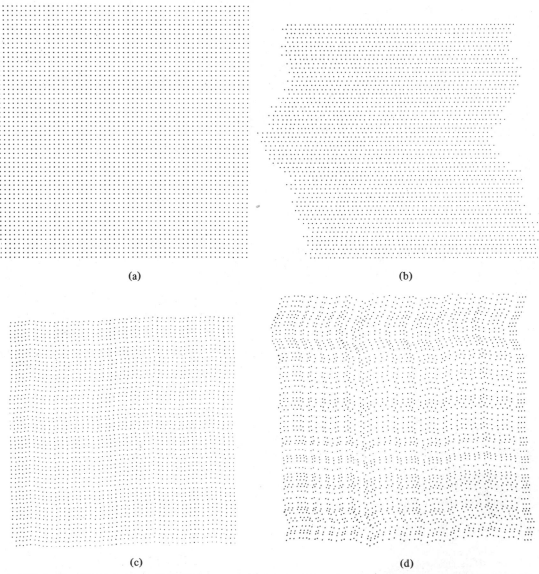

(a)

(b)

(c)

(d)

FIG. 1. Two-dimensional point-lattices; computer-drawn, with disturbances increasing from left to right. (a) Crystal. (b) Fibrous material (fiber axis vertical) in which, especially, the horizontal lines are not distorted and proceed equidistantly. (c) Metal melt with relative distance variations of 10 percent and without any partial crystalline residue. (d) Compressed gas, idealized by building blocks sorted in lines and columns.

combinations of y with z as long as they lead
to the same x. This means that one has to
ensure that z = x - y and only y varies. This
leads to the following "convolution integral"*

$$H_2(\mathbf{x}) = \int H(\mathbf{y}) H(\mathbf{x} - \mathbf{y}) \, d\mathbf{y}.$$

In order to derive the distance statistics between third neighbors one has to apply this
folding process twice, etc. In this manner the
statistics of the whole distorted lattice is derived
from knowledge of the neighbor relationship.
Since the scatter of distance increases with
the distance of the building blocks from each
other, the remote order is lost in the paracrystalline lattices. In Fig. 1 a few two-dimensional
examples are given, which demonstrate that the
paracrystal contains the gaseous as well as the
crytalline state as degenerated exceptions.

Macroscopic Examples Nature, as it is directly visible to us, offers many more examples
of paracrystalline structures than one would
normally believe. Figure 2, for instance, shows
a paracrystalline laminar lattice created by water
running off of a sand beach. Repeatedly new
waves are introduced, which correspond to
step dislocations in solid-state physics. The
statistical deviations in the direction and in the
spacing of the waves obey the laws of paracrystals. The same holds for the two-dimensional lattices, which are exhibited by the grains

*In pure mathematics "convolution integrals" have
been well known for a long time, particularly in the
theory of the Laplace transformation. (compare, i.e.,
(15)). Their importance for mathematical physics has
become obvious only in the last few years, specifically
since more and more numerical problems have become
solvable through the use of modern computers.

FIG. 2. Beach at the Tagliamento as an example of a
paracrystalline laminate lattice, as it appears also in
biological and synthetic fiber structures transverse to
the fiber axis.

of the corncobs in Fig. 3. Since they are
statistically of various sizes, only a paracrystalline lattice can originate. This is clearly
evidenced in Fig. 4. by the experiment with
steel balls of two different sizes located in the

FIG. 3. Two corn cobs, wrapped in cellophane, the grains of which build a multitude of two-dimensional paracrystals with disturbances.

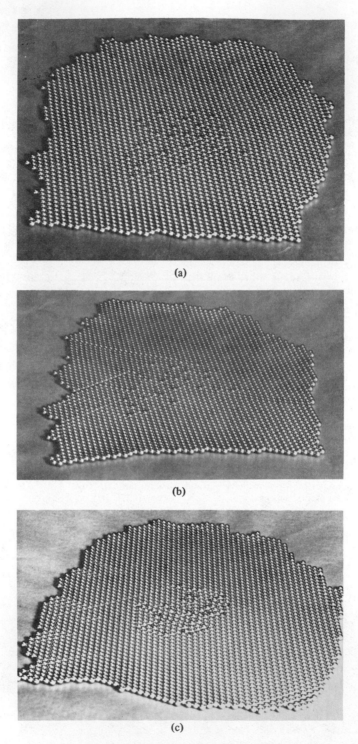

(a)

(b)

(c)

FIG. 4. Identical steel balls on a slightly sagging support. (a) In the middle section, a few balls are carefully replaced with a pair of tweezers by slightly larger ones, without disturbing the surrounding area. (b) The same after shaking lightly. The large balls make space for themselves, which is done by way of step dislocations in the surrounding host-lattice. (Two can be seen in upper left corner.) (c) The large balls have reached the support and thus created a paracrystalline lattice in the center.

center of the structure. Outside all balls have the same diameter and they form the crytalline lattice in whose center the paracrystal is embedded.

Electron-microscopic Examples   Synthetic high polymers in colloidal dimensions (below 1 μ to $\overline{1}/100$ μ ~100Å) supply us with excellent examples of paracrystalline laminar structures analogous to Fig. 2. As soon as one transforms them into fibers, for instance in a spin process, one recognizes in the electron microscope that they consist of subfibers ("ultrafibers") of 100

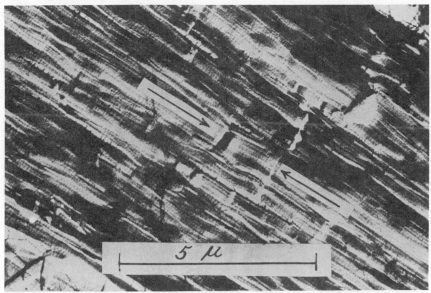

FIG. 5. Electron micrograph of a paracrystalline laminar structure, transverse to fiber, in Teflon (14). Chain molecules are located in direction of arrow. Fibrils are cross-striped. Interval of laminates is approximately 500 Å.

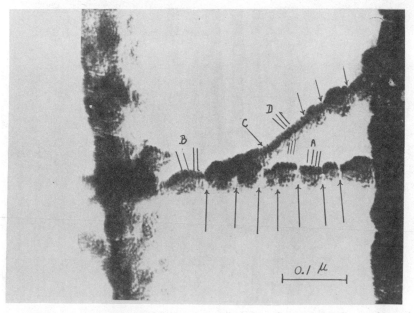

FIG. 6. Electron micrograph (light-field) of a small section of a scraped Teflon surface, where a few ultrafibrils have splintered off (14). The arrows, in an interval of approximately 500 Å, designate the boundaries of the individual blocks. These blocks each consist of about 8 disks that either are oriented vertically to the fiber (A) or are partly inclined due to paracrystalline disturbances (B) or bent (C) or buckled.

to 500Å. In these, crystalline and amorphous regions, with a paracrystalline period between 50 and 1000Å (depending on treatment), follow each other as in a string of pearls. By the assembling of ultrafibers into a microfiber the laminar structure originates (Fig. 5). But even in isotropic material (cast in the form of plates) one finds these ultrafibers (14) as is shown in Fig. 6. Here, the crystalline ranges, being approximately 500 Å thick and about equally high, consist of eight lamellae. These cross striations can be found in almost all biological fibers. However, here one reaches the limit of resolution of the electron microscope.

Interferences The only route to take from here is to investigate the interferences generated

by the structures. As an example, Fig. 7 shows the Fraunhofer diffraction patterns of the greatly reduced, transparent negatives of the models in Fig. 1, which have been produced with monochromatic visible light. Through their study the static parameters of the paracrystalline lattices can be calculated.

X-ray Interferences In order to generate interference patterns of atomic structures, the wavelength used must be smaller than the length of the edges of the lattice cells. This, for instance, is the case for x-rays which emanate from a copper anode when it is irradiated with electrons. The larger the lattice cells the closer the interference points move together and they finally yield, as for instance

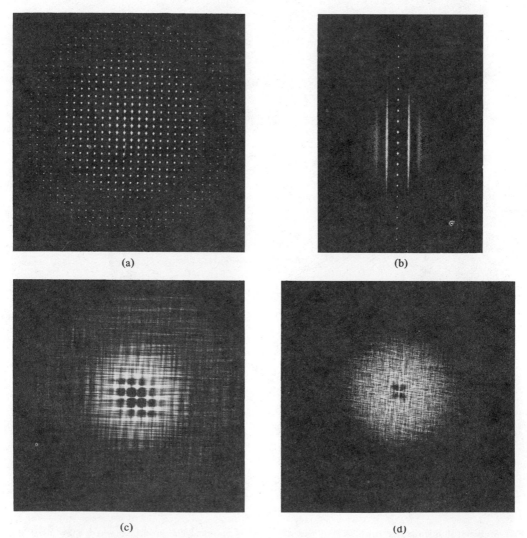

(a)

(b)

(c)

(d)

FIG. 7. Fraunhofer diffraction pattern made from the reduced transparent negatives of Fig. 1 with a He, Ne laser (13).

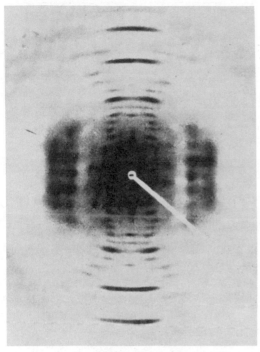

FIG. 8. X-ray small angle diagram of β-keratin of the quill of a seagull (23). Observe the similarity with Fig. 7(b). From the angle position of the interferences one can calculate that the bent lines parallel to the fiber have a distance from each other of about 12 Å, whereas the points along these lines, i.e., the centers of gravity of the basic building blocks of the protofibrils, have a distance of 183Å. This means that the basic building block is very long and thin.

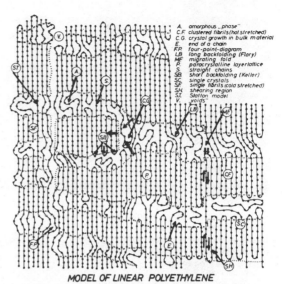

MODEL OF LINEAR POLYETHYLENE

FIG. 9. Principal figure of the structure of a synthetic material consisting of long molecules. The paracrystals consist of 100 to 500 Å thick bundles of parallel-oriented chain molecules. On the front planes, part of the molecules is folded back in larger or smaller loops; some of these run, stretched out, into the next para-crystallite and some are more or less bent. On cold drawing, chains of paracrystallites split into ultra-fibrils, separating from each other (left); whereas, with heat-treatment they join again close to each other, whereby the paracrystallites form a para-crystalline, laminate-super-lattice (center). When grown from solution, the portion of the chains which does not fold back decreases with the degree of dilution (right).

in Fig. 8, a so called x-ray angle diagram. The one shown here was the first with which it could be proven that in biological materials paracrystals exist in submicroscopic dimension (17). Their building blocks, being long, thin structures, have the characteristics of fibers, since they can deviate along the fiber only in direction and not in distance from each other.

Superstructures in Polyethylene Similarly, large lattice cells are exhibited also by the so-called "superstructures" which frequently occur in synthetic polymers. As an example, a few results from polyethylenes are given which are representative of many plastic products with long molecules and which have been obtained from the analysis of large and small angle x-ray interferences (18). The fundamental Fig. 9 shows that a lattice cell of the superstructure consists (in the fiber direction) of many basic chemical building blocks C_2H_4 (their CG marked by a dot) and is limited in the fiber direction by amorphous regions in which the chain-back-folding takes place. The lateral extension in the extreme left is particularly evident during cold

drawing, where the fibrils separate spatialy. In these lateral grain boundaries, as in the grain boundaries of a crystal, an increased disorder exists. This increased disorder is thought to be caused mainly by a high concentration of "kinks" as displayed in Fig. 10 [according to Pechhold (19)].

Influence of Mechanical Properties In the lateral grain boundaries an increased mobility of the chains relative to each other exists, because kinks can move along the chain by means of the simple flipping-over of some of the CH_2 groups, and the chain therefore creeps on like a caterpillar according to Renecker (20). If one clamps, for instance, both ends of a synthetic fiber and moves them parallel to each other as shown in Fig. 11, one can see in the small inserted diagram that the fibrils glide relative to each other. This explains why metal fibers are much less flexible, and glass fibers can only be bent elastically. In Fig. 12 the influence of heat treatment on the superstructure of stretched polyethylene is shown, by which the mechanical properties are changed. With increasing temperature the crystalline re-

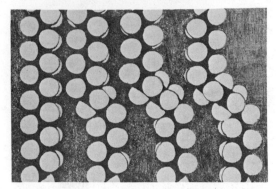

FIG. 10. Model for various kinds of "kinks" in linear polyethylene after Pechhold (19). The white half-spheres represent the hydrogen atoms. They mask the black-symbolized Carbon atoms to a great extent. In the crystalline state (extreme left) the CH2-groups are arranged in zig-zag fashion.

FIG. 11. Bending of a hot-worked piece of poly-ethylene (schematic). From the x-ray small-angle diagram (at the top) one calculates the superstructures. On the left, before, and on the right, after, bending of both ends of the piece. The crystalline ranges are marked black.

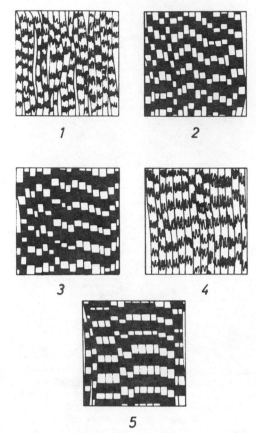

FIG. 12. Influence of heat treatment on the super-structures of polyethylene (18). 1, Cold-stretched. Splintering into ultrafibrils. 2, 500 hours tempered at 110°C. The empty spaces or vacancies have disap-peared; the front planes are more distinctly defined. No other changes have occurred. 3, 5 minutes at 120°C produces additionally an elongation in cross-section of the paracrystals. 4, 10 minutes at 120°C causes growth of the average value of the lengths at constant density, but gives a more diffuse front plane. 5, 500 hours at 120°C recrystallizes the adjacent ultrafibrils at the same time generating thicker paracrystallites with smaller disturbances. The density and the average laminate interval grow, and the front planes are again well defined.

regions grow and the fiber finally loses its desirable properties.

Paracrystalline Atomic Lattices As Fig. 7(c) shows, the widths of the reflexes increase with growing scatter angle the larger the paracrystal-line disturbances are.* These disturbances can qualitatively be determined by accurate mea-surements of line profiles of x-ray angle inter-

*Unlike the widening of reflexes through internal stresses, these vary not in a linear fashion, but with the square of the sine of the Bragg angle.

ferences. According to Fig. 4 they grow with the mixing ratio and the differences in size of the building blocks. Thus, one gets information on their shape when one knows their concen-tration. In this manner, the ammonia catalyst (enriched with 3 weight per cent aluminum) was investigated. It distinguished itself by the fact that its inner surfaces do not decline even with long service times, thus keeping the con-nected regions equally small. The result is shown in Fig. 13: A considerable part of the aluminum is built into the cubic-centered α-iron lattice in form of $FeAl_2O_4$ "motives" or units. One unit displaces seven iron atoms. It has the

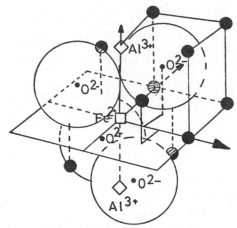

FIG. 13. Atom model of the ammonia catalyst, promoted with 3 weight per cent aluminum. $FeAl_2O_4$ units are built into the metallic iron atom lattice (•) under the substitution of seven atoms each, which fit into the lattice only in respect to their volume but not their shape. The unit consists therefore of one iron ion (□), two aluminum ions (◊), and four oxygen ions (O).

same volume, but not the same shape as these, a fact which explains the measured paracrystalline disturbances quantitatively (21). Similar measurements on polyethylenes demonstrated that the disturbances in the transverse direction of the paracrystalline microfibrils become smaller with their extension. In all the cases mentioned here, the distance variations are at the largest, 3 per cent, which is hardly recognizable in models. Figure 1(c), in comparison, with 10 per cent relative paracrystalline disturbance variations, corresponds to those values which are found for the very small lattice ranges in metallic melts (22).

Outlook The sketched examples may demonstrate that the theory of paracrystals can be applied to a multitude of groups of substances, and that it furnishes quantitative criteria for the definition of these substances through a series of novel statistical structure parameters. One result, already of interest in colloid-chemistry, is that the lattice ranges are, and remain, smaller the larger their paracrystalline disturbances are. The large and stable inner surface of the ammonia catalyst and the remarkable similarity between many biological and synthetic super-structures, as well as the similarity of a metal structure above and below the melting point, indicate that new fundamentals have been found upon which one could build.

ROLF HOSEMANN

References

1. Stanley, W. M., *Science*, 81, 644 (1935).
2. Rinne, F., *Trans. Farad. Soc.*, 29, 1016 (1933), "Investigations and Considerations concerning Paracrystallinity."
3. Hosemann, R., *Zs. f. Phsy.*, 128, 1 (1950).
4. Laue, M., F. Friedrich und P. Knipping (1912) *Sitzungs Ber. Bayer. Akad. Wiss. Math. Phys.* K1, 303, 363.
5. Boltzmann, L. (1895) "Vorlesungen über Gastheorie," Johann Ambrosius, Leipzig.
6. Bernal, J. D., Nature, 185, 68 (1960).
7. Kaplow, F., Strong, S. L., and Averbach, B. L., *Phys. Rev.*, 138 A, 1336 (1965).
8. Haüy, R. J., *Journ. de Phys. Paris* (1782).
9. Seeber, L. A., *Gilberts Annalen*, 76, 349 (1824).
10. Hermans, J. J., *Rec. Trav. Chim. Pays.−Bas.*, 63, 5 (1944).
11. Zernike, F., and Prins, J. A., *Zs. Phys.*, 41, 184 (1927).
12. Debye, P. P., *Phys. Zs.*, 31, 348 (1930).
13. Hosemann, R., and Müller, B., *Mol. Cryst. and Liqu. Cryst.*, 10, 273 (1970).
14. O'Leary, K. J., "Dissertation," Case Institite, Cleveland, Ohio (1965).
15. Doetsch, G., "Theorie und Anwendung der Laplace-Transformation," Springer Verlag, Leipzig, 1937.
16. Hosemann, R., and Bagchi, S. N., "Direct Analysis of Diffraction by Matter," North. Holl. Publ. Comp., Amsterdam, 1962.
17. Hosemann, R., "Die Erforschung der Struktur hochmolekularer und Kolloider Stoffe mittels Kleinwinkelstreuung," *Erg, d, Ex. Nat. Wiss.*, 24, 142–221, Springer Verlag, Berlin, 1951.
18. Loboda, J., Hosemann, R., and Wilke, W., *Koll. Zs. u. Zs. Polym.*, 235, 1162 (1969).
19. Pechhold, W., and Blasenbrey, S., *Koll. Zs. u. Zs. Polym.*, 235, 216 (1967).
20. Renecker, D. H., *J. Polym. Sci.*, 59, 39 (1962).
21. Preisinger, A., Hosemann, R., and Vogel, W., *Ber. Buns. Ges. Phys. Chem.*, 70, 796 (1966).
22. Lemm, K., *Mol. Cryst. and Liqu. Cryst.*, 10, 259 (1970).

Cross-references: COLLOIDS, THERMODYNAMICS OF; CRYSTALLIZATION; CRYSTALLOGRAPHY; DIFFRACTION BY MATTER AND DIFFRACTION GRATINGS; ELECTRON MICROSCOPE; INTERFERENCE AND INTERFEROMETRY; LIQUID STATE; MICROPARACRYSTALS; MICROPARACRYSTALS, EQUILIBRIUM STATE OF.

PARAMAGNETISM

Classically, paramagnetism is defined as the acquisition of a magnetization **M** that lies along the direction of an applied magnetic field **H**. This definition implies that when no field is applied, the magnetization is zero, which in essence eliminates ferromagnetism and ferrimagnetism. In the light of today's knowledge, at least one caveat must be added; specifically, antiferromagnetism must be excluded. However, ferro-, ferri-, and antiferromagnets above their ordering temperature behave as paramagnets. Temperature-dependent and temperature-

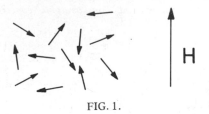

FIG. 1.

independent paramagnetism have different origins, and require separate discussion.

Temperature-dependent paramagnetism can be conveniently discussed by considering a collection (ensemble) of magnetic dipoles, as indicated in Fig. 1. In general the dipoles may have different magnetic moments, but for simplicity they will be assumed to have the same moment μ. Then the vector sum over a unit volume V yields the magnetization, that is, $\Sigma_V \, \mu = M$. The dipoles may be located at random positions in configurational space, as in a gas or vapor, or at sites with some short-range order, as in a liquid or an amorphous solid, or at sites with translational symmetry, as in a crystalline solid. Thermal energy, kT, where k is Boltzmann's constant and T is the absolute temperature, will then ensure that the moments are distributed over the possible orientations. Then if $H = 0$, it follows that $M = 0$. However, if a magnetic field is applied, the dipoles will tend to align along the field direction, in order to minimize the interaction (Zeeman) energy, $-\mu \cdot H$. Thermal agitation will normally ensure that complete alignment of the dipoles will not occur.

Classically all dipole orientations are allowed, whereas quantum mechanics requires spatial quantization. If no interactions occur between

the dipoles, then either a classical or quantum mechanical calculation shows that M is linearly related to H provided $\mu \cdot H \ll kT$. The susceptibility, defined as $\chi = M/H$, is found to be given by $\chi = C/T$, where C is a constant. This result is known as *Curie's law*, and indeed several systems are found to follow this relationship. If there are interactions between the dipoles, an elementary and oversimplified approach yields the *Curie-Weiss law*, $\chi = C/(T - \theta)$, where θ is another constant which is a measure of the interaction strength; some materials have susceptibilities that can be fitted to this equation. If H is very large, or T is very low, then the dipoles will tend toward complete alignment. The magnetization at the temperature T, $M(T)$, compared to its value at absolute zero, $M(0)$, is then given by the *Brillouin function*, $B_J(x)$, viz.,

$$M(T)/M(0) = B_J(x)$$

$$= \frac{2J+1}{2J} \coth \frac{2J+1}{2J} x$$

$$- \frac{1}{2J} \coth \frac{x}{2J}$$

where J is the total angular quantum number, and $x = \mu \cdot H/kT$. In the classical limit ($J \to \infty$), $B_J = L(x) = \coth x - 1/x$, where $L(x)$ is called the *Langevin function*. For $J = S = \frac{1}{2}$, where S is the spin quantum number, $B_{1/2}(x) = \tanh x$. Plots of $B_{1/2}(x)$, $B_{5/2}(x)$, and $L(x)$ are shown in Fig. 2. For small x the curves approach straight lines; this is the Curie Law region.

The dipoles of the ensemble can be realized in a variety of ways. Atoms, ions, or molecules with unpaired electrons, that is, with nonzero angular momentum G, (orbital and/or spin),

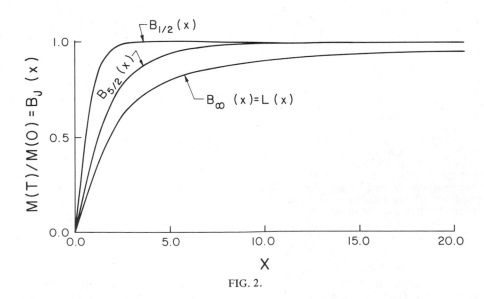

FIG. 2.

have magnetic dipole moments $\boldsymbol{\mu}$. These moving electrons are essentially equivalent to an electric current, and give rise to a permanent magnetic moment given by

$$\boldsymbol{\mu} = \frac{e}{2mc} \, \mathbf{G},$$

where e and m are the charge and mass of the electron, respectively, and c is the velocity of light. Since $\mathbf{G}$ is quantized in integral or half-integral units of Planck's constant, $\hbar$, it is convenient to introduce a unit of magnetic moment, called the Bohr magneton, μ_B, and given by $\mu_B = e\hbar/2mc$, with a value of 9.2741×10^{-21} erg G^{-1} (9.2741×10^{-24} J T^{-1} in SI). For a free atom or ion the dipole moment is given by $\boldsymbol{\mu} = g\mu_B \mathbf{J}$, where g, called the *g-factor* or *spectroscopic splitting factor*, is the Landé formula of atomic spectroscopy. For ions in a solid, the same form of equation may be used, except that g is essentially a phenomenological constant that reflects the splitting of the orbital and spin states by the internal interactions. Atomic and molecular electronic dipole moments are usually of the order of one Bohr magneton.

Examples of paramagnetic gases are sodium vapor and nitric oxide. Even though molecular oxygen, O_2, has an even number of electrons, its ground state has $S = 1$, and consequently it is a paramagnet. Other molecular paramagnets are organic free radicals and biradicals.

By far the largest number of substances with permanent dipole moments are liquids or solids containing ions of transition group elements, that is, those with partially filled $3d$, $4d$, $5d$, $4f$, or $5f$ electron shells. The lanthanides, comprising the elements from La to Lu in the periodic table, usually in the trivalent state, have 0–14 $4f$ electrons. The $4f$ electrons, which are surrounded by outer $5s$ and $5p$ electrons, are situated well to the interior of the ions. Hence, the dipoles almost behave as if the ions were free, and often obey Curie's Law fairly well, at least provided the temperature is not too low. However, the electric crystalline fields of the diamagnetic ions (ligands) in a solid do interact with the $4f$ electrons, and lead to susceptibility changes that are important at low temperatures. In recent years, interest in the actinides or $5f$ group has increased. The trivalent ions, Th^{3+}, Pa^{3+}, U^{3+}, Np^{3+}, Pu^{3+}, Am^{3+}, and Cm^{3+} have 1–7 $5f$ electrons, respectively. Complex ion groups, such as $(UO_2)^{2+}$, $(NpO_2)^{2+}$, and $(PuO_2)^{2+}$ are also formed with 0, 1, and 2 unpaired $5f$ electrons, respectively. Both spin-orbit coupling and crystal-field interactions tend to be more important in the actinides than the lanthanides.

The $3d$ group are usually divalent, but may be in trivalent or other states; this group includes Cr^{2+}, Cr^{3+}, Mn^{2+}, Fe^{2+}, Fe^{3+}, Co^{2+}, Ni^{2+}, and Cu^{2+}. The relatively exposed $3d$ electrons inter-

act with the ligand fields much more strongly than the $4f$ electrons of the rare earths. As a result, the orbital energy levels are split. Then the dipole moments are largely determined by the spin only, and often a Curie or Curie-Weiss Law is observed, again if the temperature is not too low. Indeed, the susceptibility is a measure of the paramagnetic ion's energy levels and the number of ions in each level, as determined by the principles of statistical mechanics. In principle, a susceptibility measurement provides information on the energy levels of the paramagnetic ion. In practice, details of these energy levels can be deduced from χ only for a few special cases. Instead, the procedure today is to determine these energy levels by other methods, usually electron paramagnetic (spin) resonance, and then to calculate the susceptibility. Other techniques used to deduce the magnetic energy levels include specific heats, magnetooptics, and Mössbauer spectroscopy.

In some compounds, the paramagnetic ions may be covalently bound, that is, electrons are shared with the diamagnetic ions. The dipole moments are then somewhat spread out spatially, and the localized permanent dipole moments of Fig. 1 become somewhat fuzzy. Covalent interactions are usually important for ions of the palladium ($5d$) and platinum ($6d$) groups.

Other examples of electronic permanent dipole systems include electrons trapped at vacant lattice sites (*F*-centers), radiation damaged bonds, and localized donors and acceptors in semiconductors. Next, the nucleus may also have a nonzero angular momentum, and hence a permanent magnetic moment. It is much smaller ($\sim 10^{-3}$) than the electronic magnetic moment, and consequently the nuclear contribution to the susceptibility is usually negligible. However, the nuclear susceptibility of solid (atomic) hydrogen, water, and $C_6(CH_3)_6$ has been detected. Nuclear magnetic energy levels are usually investigated by nuclear magnetic resonance (NMR).

Small particles, about 10 nm in diameter and containing transition atoms, may be ferro- or ferrimagnetic; examples are iron or iron oxide (Fe_3O_4 or γ-Fe_2O_3). Thus, in the context of Fig. 1, the permanent magnetic moment of each particle is about $\mu = 10^6 \, \mu_B$. Thermal agitation may still be sufficient to overcome any anisotropy in each particle, and hence cause oscillations of the magnetic moment between various energy minima. Since the time average of the moment of a particle is zero, the particle is said to be a *superparamagnet*. Application of a field to the system of particles will then produce a magnetization given by the Langevin function.

Electrons in an unfilled energy band, as in a metal or alloy, give rise to a weak paramagnetism that is essentially temperature independent. These electrons have a magnetic moment

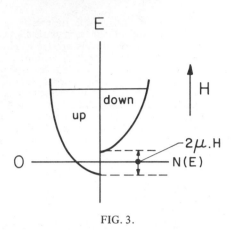

FIG. 3.

$\mu \simeq \mu_B$ since $S = \frac{1}{2}$ and $g \simeq 2$. Suppose the electron gas is considered as two subsystems, with half the moments lying along a given direction, say up, and half in the opposite direction, down. The net moment is of course zero until a magnetic field is applied. Then the energy of the electrons with magnetic moments parallel to $\mathbf{H}$ is decreased by $\mu_B H$ and the energy of those antiparallel is increased by $\mu_B H$. Thermodynamic equilibrium then requires that electrons with antiparallel moments enter the parallel-moment subsystem until the two subsystems are filled to the same energy levels, that is, the Fermi levels of the two systems become equal, as shown in Fig. 3. The susceptibility, first calculated by Pauli, is

$$\chi = \frac{3N\mu_B{}^2}{2E_F(0)}$$

at absolute zero, where N is the number of conduction electrons per unit volume and $E_F(0)$ is the Fermi energy at $T = 0$ K. Since the Fermi energy changes slightly with temperature, then so also does the susceptibility.

Second-order quantum mechanical perturbation theory applied to atomic electron wave functions also yields a small temperature-independent susceptibility. This second-order Zeeman effect was first derived by Van Vleck. The order of magnitude of the Van Vleck susceptibility is given by

$$\chi \propto \frac{N\mu_B{}^2}{E_n - E_0}$$

where E_n is the energy of the nth excited state and E_0 is the ground state.

In addition to the paramagnetic susceptibility, there is always a diamagnetic one, and it is the sum of these two that is measured in an experiment. For permanent moments of about one Bohr magneton, the diamagnetic contribution is small, and may often be neglected. The dia-

magnetic susceptibility must be considered when the temperature is high, when the relative numbers of atoms or ions with permanent moments are small, or when Pauli or Van Vleck paramagnetism dominates. If $\mu = \mu_B$, then the paramagnetic susceptibility for a mole of dipoles is about 10^{-4} at room temperature, and 10^{-2} at 1 K (cgs units). Pauli susceptibilities are of the order of 10^{-6} to 10^{-5} emu cm^{-3}.

A. H. MORRISH

References

Boudreaux, E. A., and Mulay, L. N., "Theory and Applications of Molecular Paramagnetism," New York, John Wiley & Sons, 1976.

Morrish, A. H., "Physical Principles of Magnetism," Huntington, N.Y., Krieger, 1980 (reprinted).

Vonsovskii, S. V., "Magnetism," Vol. 1, New York, Halsted Press, 1974. (Translated from the Russian, "*Magnetizm*," Moscow, Nauka, 1971.)

Cross-references: ANTIFERROMAGNETISM, CALORIMETRY, FERRIMAGNETISM, FERROMAGNETISM, MAGNETIC RESONANCE, MAGNETISM, MÖSSBAUER EFFECT.

PARITY

The theoretical description of any physical process in the relativistic domain must remain invariant under a Lorentz transformation. In a classical theory, where the interactions depend on the relative spatial separation of the position coordinates and the relative temporal separation of two instants of time implying a finite velocity of propagation of the interaction, the description of the interaction must be independent of the coordinate frame in which the process is described. A proper Lorentz transformation from one frame S to another S' is given by

$$x_\mu{}' = a_{\mu\nu} x_\nu$$

where

$$x_\mu \equiv (x_1, x_2, x_3, x_4) = (x, y, z, ict),$$

and a summation is to be made over repeated indices. (This summation convention applies throughout the present article.) Similarly,

$$p_\mu{}' = a_{\mu\nu} p_\nu$$

where

$$p_\mu \equiv (p_1, p_2, p_3, p_4) = (p_x, p_y, p_z, iE/c).$$

The four coordinates x_μ are collectively called the *space-time coordinates* and the four quantities p_μ are collectively called the *four momenta*, the first three being the spatial components of the momentum and the fourth being proportional to the energy within a factor i/c. The $a_{\mu\nu}$ are constants independent of the space-time coordinates such that the relativistic interval is to remain invariant:

$$x_\mu x_\mu = x_\mu' x_\mu' = a_{\mu\nu} x_\nu a_{\mu\tau} x_\tau$$

$$p_\mu p_\mu = p_\mu' p_\mu' = a_{\mu\nu} p_\nu a_{\mu\tau} p_\tau = -m^2 c^2.$$

This invariance requires that

$$a_{\mu\nu} a_{\mu\tau} = \delta_{\nu\tau}$$

and, in order to preserve the reality conditions of the space-time coordinates, $a_{\mu\nu}$ with $\mu, \nu = 1$ 2, 3, and a_{44} are real, while $a_{\mu4}$ and $a_{4\mu}$ are imaginary.

These transformations are continuous, being obtained from a series of infinitesimal transformations from unity, and are characterized by $\det(a_{\mu\nu}) = 1$. They do not change the direction of the time axis.

Such a formalism can be used to describe proper rotations of the physical system and boosts to a different velocity of the physical system with respect to the space-time coordinate system. The requirement of equivalence between different Lorentz frames leads directly to the laws of conservation of four-momentum and angular momentum.

In addition, there are the improper Lorentz transformations of space inversion and time inversion, which are discontinuous, unlike the proper Lorentz transformations. These improper transformations each have the property that $\det(a_{\mu\nu}) = -1$. The transformed system under space inversion is

$$x_i' = -x_i, \quad \text{where} \quad i = 1, 2, 3,$$

and that under time inversion is

$$x_4' = -x_4.$$

Clearly, $x_\mu' x_\mu' = x_\mu x_\mu$ for either case or for both taken together. It should be noted that the reversal of the space coordinates changes a right-handed coordinate system to a left-handed system, or vice versa. This implies that if there is a physical system in nature that has a definite handedness it is also possible to have a system of the opposite handedness obeying the same physical laws. Classically, this invariance under reflections leads to no conservation law in the way that invariance under rotations leads to the conservation of angular momentum. This is not so in quantum mechanics.

In a relativistic theory time inversion is not equivalent to an operation where every velocity is replaced by the opposite velocity so that the position of the particle at $+t$ becomes the same as it was, without time inversion, at $-t$. This latter operation may properly be called velocity reversal or motion reversal, but in quantum mechanics or quantum field theory it is usually, perhaps illogically, called *time reversal* or *Wig-*

ner time reversal. The laws of classical physics, apart from statistical effects such as frictional forces and electrical resistance, are invariant under time reversal where one has the normal motion in reverse.

Time inversion is actually the product of time reversal and charge conjugation, an operation where all particles are changed into their antiparticles. Charge conjugation is a relativistic quantum-mechanical concept only, and like space and time inversion, is a discontinuous operation. In the limit where a nonrelativistic approximation is valid the operation of charge conjugation is not meaningful, so only space inversion and time reversal have physical meaning there.

In quantum mechanics, the interacting objects such as elementary particles, atoms, and nuclei are described by a wave function $\psi(\mathbf{r}, t)$ which is itself a function of the space-time coordinates. Here a symmetry property called *parity* is introduced to express the equivalence between left and right. If we denote by $\psi(-\mathbf{r}, t)$ the wave function in the space-reflected system, then

$$\psi^P(\mathbf{r}, t) = \eta_P \psi(-\mathbf{r}, t).$$

The space inversion quantum number η_P is called *parity* and has two eigenvalues, ± 1. When $\eta_P = 1$ the wave function is said to have even parity and when $\eta_P = -1$ the wave function is said to have odd parity.

The supposition that any nondegenerate stationary quantum-mechanical state of an object, such as an atom or nucleus, will have a definite parity has led to very powerful conservation laws in the description of physical processes involving interactions of atoms or nuclei. The principle that the total parity of the system was conserved was regarded until recently as a fundamental principle of physics.

Stated formally, if P is a unitary operator that inverts the space coordinates through the origin such that

$$P\psi(\mathbf{r}, t) = \psi'(\mathbf{r}', t) = \eta_P \psi(-\mathbf{r}, t)$$

then the wave equation

$$H\psi(\mathbf{r}, t) = E\psi(\mathbf{r}, t)$$

can be transformed into

$$PHP^{-1} \psi'(r', t) = E\psi'(r', t).$$

The wave equation in the transformed coordinate system is unchanged provided that $PHP^{-1} = H$ or $PH - HP = 0$. The Heisenberg relation for a time rate of change of an operator P is

$$\frac{dP}{dt} = \frac{\hbar}{i} (HP - PH).$$

Therefore P is a constant of motion provided it commutes with the Hamiltonian operator.

Operating twice with P on any wave function gives the same wave function, i.e., $P^2 = 1$, so the only possible eigenvalues of P are +1 and -1.

It is now well established that the strong interactions between elementary particles and the electromagnetic interactions are invariant under space inversion, so any nondegenerate energy eigenstate of an atom or nucleus will be also an eigenstate of the parity operator P. Similar statements can be made for the time reversal operator T and the charge conjugation operator C.

A quantum-mechanical system such as an atom or nucleus is usually described by wave functions in the angular momentum representation. The one-particle wave functions of the individual constituents are a product of an intrinsic part that may be also an eigenfunction of the spin angular momentum operator and an orbital part that is an eigenfunction of the orbital angular momentum operator ℓ_z. The latter eigenfunction is a spherical harmonic $Y_\ell^m(\theta, \phi)$ having an eigenvalue $(-1)^\ell$ of the space inversion operator P. The total parity of the system is then the product of the parities of the constituents. The intrinsic parity of any elementary particle cannot be determined theoretically, nor can the eigenvalues of the time-reversal operator T and the charge-conjugation operator C, often called *time parity* and *charge parity*.

For a meaningful and satisfactory formulation of charge conjugation and time reversal it is essential to consider the field operators in a quantized theory, taking into account the ordering of operators and their commutation relations appropriate to Bose statistics for integer spins and Fermi statistics for half-integer spins. The wave function for a particle is the one-particle expectation value of the field operator. In Table 1 the transformation properties of various kinds of field operators are listed for the linear transformations P and C and the antilinear transformation T. These transformation properties are obtained by considering the invariance of the free-field equations of motion.

TABLE 2. TRANSFORMATION OF OBSERVABLES UNDER SPACE INVERSION, TIME REVERSAL, CHARGE CONJUGATION, AND TIME INVERSION

Observable	P	T	C	P_4
Position	x	-x	x	x	x
Time	t	t	$-t$	t	$-t$
Velocity	v	-v	-v	v	-v
Momentum	p	-p	-p	p	-p
Energy	E	E	E	E	E
Mass	m	m	m	m	m
Orbital angular momentum	ℓ	ℓ	$-\ell$	ℓ	$-\ell$
Spin angular momentum	s	s	-s	s	-s
Charge	e	e	e	$-e$	$-e$
Force	f	-f	f	f	f
Acceleration	a	-a	a	a	a

In Table 2 the transformation properties of various physical observables are listed. The transformed observables are independent of the values of η_P, η_T, and η_C, of the fields.

The phases η_P, η_T, η_C, etc. for those particles that can be created or destroyed can be found experimentally from the properties of their mutual interactions, assuming conservation of P, T, or C in these interactions. For example, the π^0 neutral meson decays into two photons; an analysis shows that its intrinsic parity $\eta_{P'} = -1$ independently of the $\eta_{P''}$ for the electromagnetic field. The π^0, having odd parity and zero spin, is therefore a *pseudoscalar particle*, as opposed to a scalar particle having even parity and zero spin. The charged mesons $\pi^\pm$ also have $\eta_P = -1$, but charge conservation prevents their decay into two photons so that an interaction where the $\pi^\pm$ are created or absorbed through interactions with nucleons must be considered. The number of nucleons is conserved, so only the product $|\eta_P|^2 = 1$ is determined for the nucleons. Similarly, the

TABLE I. TRANSFORMATION OF INTRINSIC FIELDS UNDER SPACE INVERSION, TIME REVERSAL, AND CHARGE CONJUGATION

Intrinsic Field		P	T	C
Scalar	$\varphi(\mathbf{r}, t)$	$\eta_{P'}\varphi(-\mathbf{r}, t)$	$\eta_{T'}\varphi(\mathbf{r}, t)$	$\eta_{C'}\varphi^+(\mathbf{r}, t)$
Dirac spinor	$\psi(\mathbf{r}, t)$	$\eta_P \gamma_4 \psi(-\mathbf{r}, t)$	$\eta_T \gamma_3 \gamma_1 \psi^*(\mathbf{r}, -t)$	$\eta_C \gamma_2 \psi^*(\mathbf{r}, t)$
Four vector	$A_\mu(\mathbf{r}, t)$	$\eta_{P''}(-1)^{\delta\mu 4} A_\mu(-\mathbf{r}, t)$	$\eta_{T''}(-1)^{\delta\mu 4} A^*(\mathbf{r}, -t)$	$\eta_{C''}A_\mu^*(\mathbf{r}, t)$
Pseudoscalar	$\pi(\mathbf{r}, t)$	$\eta_{P'''}\pi(-\mathbf{r}, t)$	$\eta_{T'''}\pi^*(\mathbf{r}, -t)$	$\eta_{C'''}\pi^*(\mathbf{r}, t)$
Electric field	$\mathcal{E}(\mathbf{r}, t)$	$\eta_{P''}\mathcal{E}(-\mathbf{r}, t)$	$-\eta_{T''}\mathcal{E}(\mathbf{r}, -t)$	$\eta_{C''}\mathcal{E}(\mathbf{r}, t)$
Magnetic field	$\mathcal{B}(\mathbf{r}, t)$	$-\eta_{P''}\mathcal{B}(-\mathbf{r}, t)$	$\eta_{T''}\mathcal{B}(\mathbf{r}, -t)$	$\eta_{C''}\mathcal{B}(\mathbf{r}, t)$
Two-component neutrino	$\psi_{(\nu)}(\mathbf{r}, t)$	does not exist	$\eta_T \gamma_3 \gamma_1 \psi_{(\nu)}(\mathbf{r}, -t)$	does not exist

electromagnetic field interacting with an electron current determines $\eta_P'' = -1$ for the electromagnetic field and $|\eta_P|^2 = 1$ for the electrons. In analogous ways all the phases for those particles with integer spin (bosons) can be found by experiment, while those with half-integer spin (fermions) cannot and, indeed, are indeterminate. Although the intrinsic parity is indeterminate for fermions, the intrinsic parity for an antiparticle is opposite to that for a particle, in contrast to the boson case; e.g., the intrinsic parity is odd for e^+e^- in an S state and even for $\pi^+\pi^-$ in an S state. Considering the conservation of C, P, and angular momentum J, the only allowed modes of annihilation of an electron-positron system are a singlet-S system decaying into two photons and a triplet-S system decaying into three photons.

Conventional usage assigns $\eta_P = +1$, i.e., positive parity, for fermions. Any fermion state of total angular momentum j can be specified by $\psi_{j\ell s}$ where $j = \ell + s$ and has total parity $(-1)^\ell$ by convention. Similar arguments define a "time parity" and a "charge parity" for each quantum-mechanical state. Likewise, the electromagnetic field can be expanded into multipoles in an angular momentum representation; however, here the intrinsic parity is specified. A photon of angular momentum J can have two possible parities, characterized by electric multipole radiation EJ, with parity $(-1)^J$, and magnetic multipole radiation MJ, with parity $(-1)^{J\pm1}$. These different types result from possible vector products of the intrinsic spin and the orbital angular momentum to give the same J.

Several examples of the use of parity conservation follow. Consider the interaction of the electromagnetic field with a quantum-mechanical system, a transition from an initial state of definite angular momentum, I_i, and parity, π_i, to a final state of definite I_f and π_f, with the emission or absorption of electromagnetic radiation of angular momentum J and parity π. The conservation of angular momentum and parity requires that

$$|I_i - I_f| \leqslant J \leqslant I_i + I_f$$

$$\pi_i \pi_f \pi = 1.$$

The possible transitions are shown in Table 3.

In atoms, the predominant electromagnetic transitions are of electric dipole type. The characteristic lifetimes of the other multipoles are long compared with the typical atomic collision times, an alternative method of deexcitation. The first concept of the use of parity in atomic physics resulted from the experiments of Laporte showing that the initial and final states were always of opposite parity. This result is predicted in Table 3.

In nuclei, collision plays little role, so all electromagnetic multipoles can contribute, but usually the lowest-order multipole comparable with the conservation of angular momentum and parity predominates. These restrictions are known as angular momentum and parity selection rules.

The parity of the probability density $\psi^\dagger\psi$ is always even. The expectation value of an operator θ is $\psi^\dagger\theta\psi$ and vanishes if θ is an odd function under the parity operation. From the properties of the multipole operators, all odd electric multipole moments and all even magnetic multipole moments vanish. The restriction to nondegeneracy is essential and is satisfied for all atomic and nuclear systems. In molecular systems there are many observed static electric dipole moments. These usually result from the accidental degeneracy of coulomb wave functions with different orbital angular momenta.

One of the best evidences for the conservation of parity and invariance under time reversal in electromagnetic interactions is the lack of an electric dipole moment in the neutron.

A plane wave of definite linear momentum can be decomposed into a sum of degenerate, even and odd, angular momentum states and is not an eigenstate of parity. Localized wave packets can be obtained by superimposing many plane waves. Any macroscopic classical body, so described, will be degenerate in states of opposite parities and thus can have such properties as electric dipole moments.

In 1956 T. D. Lee and C. N. Yang, after a careful examination of experimental evidence, concluded that there is abundant evidence for parity conservation in strong and electromagnetic interactions, but could not find any evidence for conservation in weak interactions, including nuclear beta decay. They suggested several experiments to test parity conservation in weak interactions.

An experiment performed by C. S. Wu, E. Ambler, R. W. Hayward, D. D. Hoppes, and R. P. Hudson in 1956 at the National Bureau of Standards and published in 1957 demonstrated that parity conservation was violated in nuclear beta decay. The experiment measured the angular distribution of beta and gamma radiation from radioactive ^{60}Co nuclei that had been cryogenically oriented. The angular distribution for the beta radiation obeyed a relation

$$W(\vartheta) = 1 + a\,\frac{\langle j \rangle}{|j|} \cdot \frac{p}{|p|}$$

TABLE 3. ELECTROMAGNETIC MULTIPOLE
SELECTION RULES

$\Pi_i\Pi_f$ \ ΔI	0 $I \neq 0$	1	2	3	4	5
+	M1	M1	E2	M3	E4	M5
−	E1	E1	M2	E3	M4	E5

where **j** is the nuclear spin and **p** the momentum of the beta particle. The angular distribution of the gamma radiation is a known function that depends only on even powers of **j** · **k**, where **k** is the momentum of the photon, and on numerical factors depending on the spin of the nuclear states involved in the gamma transition.

A measurement of the gamma ray anisotropy, in turn, determines the amount of nuclear polarization $\langle \mathbf{j} \rangle / |\mathbf{j}|$. Inspection of Table 2 indicates that momenta change sign under the parity operation while the spins do not. Neither the momenta nor spins change sign under the charge conjugation operation. The beta ray angular distribution will change under the operations P or CP but not C, while the gamma ray angular distribution is unaltered under C, P, or CP. The measured beta ray angular distribution from ^{60}Co gave a value for the coefficient $a = -v/c$, apart from some small coulomb corrections. A subsequent experiment measuring the angular distribution of positrons from polarized ^{58}Co gave similar results but with the opposite sign for the coefficient a. The nonzero value for the coefficient a indicates that space inversion invariance is violated. The opposite signs for the coefficient a, depending on whether electrons or positrons are emitted, indicate that charge conjugation invariance is violated. An analysis shows that both the P and C violations are maximal but that the product CP is conserved.

In the beta-decay process within the nucleus we have a neutron making a transition to a proton with the emission of an electron and an antineutrino. The inverse process has a proton making a transition to a neutron, emitting a positron and a neutrino. One of the most satisfactory explanations of the violation of C and P in the beta-decay process can be ascribed to the properties of the neutrino. Rather than being a particle which can be described by the Dirac equation with a four-component wave function, its zero mass permits a modification of the Dirac equation to one that has a two-component wave function. This two-component description, although invariant under proper Lorentz transformations, is not invariant under C or P.

Experiment shows that the antineutrino emitted in beta decay is right-handed, i.e., its intrinsic spin is parallel to its momentum. Noninvariance under P implies that there is no left-handed antineutrino, while noninvariance under C implies that there is no right-handed neutrino. Invariance under CP implies the existence of a left-handed neutrino as well as the right-handed antineutrino. The symmetry of right- and left-handedness in matter is broken and replaced by a symmetry where matter has a characteristic handedness and antimatter the opposite handedness.

A left-handed, massless neutrino appears as a left-handed neutrino in all Lorentz frames which can be continuously transformed into one another. The postulate of space inversion symmetry allows us to infer the existence of a right-handed massless neutrino; however, there is no reflection symmetry, so the existence of both right- and left-handed neutrinos in nature would be coincidental. Only if the neutrino had mass could the existence of both helicity states follow from the existence of a single state and from the properties of proper Lorentz transformations.

At present, three charged leptons are known with nonzero mass: the electron, the μ lepton, and the τ lepton, each having associated with it a left-handed neutrino, ν_e, ν_μ, and ν_τ, respectively. The ν_τ has not been observed directly but may be conjectured from the dynamical properties of the weak interaction.

In all the weak-interaction phenomena, which includes all decays of the elementary particles, except the electromagnetic decays $\pi^0 \to 2\gamma$ and $\Sigma^0 \to \Lambda + \gamma$ that were experimentally investigated prior to 1964, the violation of P and C invariance and conservation of CP and T invariance was fully established. In 1964, experiments at the Brookhaven National Laboratory showed that the K^0 meson, which should decay only into three π-mesons according to CP invariance, decayed a small fraction of the time into two π-mesons, a process violating CP invariance. Vigorous experimental research since that time has fully confirmed this CP violation result, but no other process has been observed where CP is violated. There have been many theoretical attempts to account for this small CP violation, but no fully satisfactory explanation has emerged. However the small amount of CP violating interaction in the decay of the K mesons may account for the present preponderance of particles over antiparticles, assuming that both were created in equal amounts initially after the big bang occurring at the origin of the universe.

Whether or not there is invariance under the operations C, P, T, or CP individually, any local field theory must be invariant under the product CPT. This invariance is known as the CPT theorem. The combined operation is equivalent to the product of space inversion and time inversion and is often referred to as *strong reflection*. This transformation in which $x_\mu' = -x_\mu$, although discontinuous, is characterized by $\det(a_{\mu\nu}) = 1$, and an equivalent transformation can be achieved by a proper Lorentz transformation, i.e., continuous rotations in space-time. At present only the CPT invariance remains unbroken experimentally.

RAYMOND W. HAYWARD

References

Sakurai, J. J., "Invariance Principles and Elementary Particles," Princeton, Princeton Univ. Press, 1964.

DeBenedetti, Sergio, "Nuclear Interactions," New York, John Wiley & Sons, 1964.

Lee, T. D., and Yang, C. N., "Question of Parity Conservation in Weak Interactions," *Phys. Rev.* **104**, 254 (1956).

Wu, C. S., Ambler, E., Hayward, R. W., Hoppes, D. D., and Hudson, R. P., "Experimental Test of Parity Conservation in Beta Decay," **Phys. Rev. 105**, 1413 (1957).

Christenson, J. H., Cronin, J. W., Fitch, V. L., and Turlay, R., "Evidence for the 2π Decay of the K_W^0 Meson," *Phys. Rev. Lett.* **13**, 138 (1964).

Gatto, R., "A Basic Course in Modern Weak Interaction Theory," in "Weak Interactions" (M. Baldo Ceolin, Ed.), Amsterdam, North-Holland Publishing Co., 1979.

Cross-references: ANTIPARTICLES, CONSERVATION LAWS AND SYMMETRY, ELECTROMAGNETIC THEORY, LORENTZ TRANSFORMATIONS, NEUTRINO, QUANTUM THEORY, RELATIVITY, STRONG INTERACTIONS, WEAK INTERACTIONS.

PERIODIC LAW AND PERIODIC TABLE

When the chemical elements are compared in order of increasing atomic number, many of their physical and chemical properties are observed to vary periodically rather than randomly or steadily. This relationship, recognized empirically a century ago by de Chancourtois in France, Newlands in England, Lothar Meyer in Germany, and Mendeleev in Russia, is now known to be the logical and inevitable consequence of the fundamental periodicity of atomic structure. The familiar statment of the periodic law is this: "The properties of the chemical elements vary periodically with their atomic number." A more informative statement of this same law is: *The atomic structures of the chemical elements vary periodically with their atomic number; all physical and chemical properties that depend on atomic structure therefore tend also to vary periodically with atomic number.*

The periodicity of atomic structure (see ATOMIC PHYSICS) is described by quantum theory as developed through modern wave mechanics. Each successive electron, beginning with the first, that comes within the field of an atomic nucleus, occupies the most stable position available to it. The number of possible positions is limited by quantum restrictions which describe each position in terms of four quantum numbers, and by the Pauli exclusion principle that no two electrons within the same atom may have the same four quantum numbers. These electron positions, or energy levels, are grouped with respect to their average distance from the nucleus as "principal quantum levels or shells," designated by the "principal quantum number" $n = 1, 2, 3, 4, \cdots$, successive

integral values increasing in order of increasing average distance from the nucleus. The total capacity of each shell can be expressed as $2n^2$, being 2 for $n = 1$, 8 for $n = 2$, 18 for $n = 3$, and 32 for $n = 4$; no higher level actually contains more than 32 electrons.

These total capacities can easily be accounted for by the several quantum number restrictions and the Pauli principle. Within each principal energy level are differently shaped regions called "orbitals," that can be occupied by electrons. The shape of each orbital is designated by the "orbital quantum number" l, which may only have integral values from 0 up to $n - 1$. The number of orbitals of each shape that can exist within a principal quantum level depends on the fact that an electron in an orbital is a charge in motion and therefore has magnetic properties which influence the orientation of the orbital in an external magnetic field. The possible orientations are designated by the "orbital magnetic quantum number" m_l, which may have values from 0 to plus or minus the orbital quantum number l. Thus when $n = 1$, l can only have the value 0, which means that only one orbital is possible, having orbital magnetic quantum number 0. When $n = 2$, l can have the values 0 and 1. For the value 0, one orbital is possible, but when $l = 1$, m_l can have values 0, +1, and -1, corresponding to three orbitals. Four orbitals are therefore possible in the principal quantum level $n = 2$. When $n = 3$, the same kinds of four orbitals are possible, and in addition, l can equal 2. This gives five possible values, for m_l: 0, +1, +2, -1, and -2, corresponding to five more orbitals for a total of 9. When $n = 4$, the same kinds of 9 orbitals are possible, and in addition l can equal 3. This gives seven possible values for m_l: 0, +1, +2, +3, -1, -2, and -3, corresponding to 7 more orbitals for a total of 16. No principal quantum level uses more than 16 orbitals even though more are theoretically possible when $n = 5$ or more.

One additional property of an electron in an atom needs to be considered. This is its property as a magnet, irrespective of its orbital motion. This is designated by the "spin magnetic quantum number," which can have only the values $+\frac{1}{2}$ and $-\frac{1}{2}$. Since each orbital is uniquely specified by the first three quantum numbers, n, l, and m_l, the capacity of each orbital is thus limited to 2 electrons, and these only if, according to the Pauli principle, they are of opposed spins (differ in the fourth quantum number). The total capacity of each principal quantum level is therefore twice the number of orbitals within it, because this represents the total number of permissible combinations of four quantum numbers within that level. For example, the capacity of the $n = 4$ level is limited to 32 electrons by the fact that only 32 different combinations of the four quantum numbers are possible within that level; these electrons will occupy 16 orbitals.

The differently shaped orbitals having orbital quantum numbers $l = 0$, 1, 2, and 3 are commonly called s, p, d, and f orbitals. From the above discussion, it should be clear that within any principal quantum shell, there can be only one s orbital, three p orbitals, five d orbitals, and seven f orbitals. The p orbitals do not appear until $n = 2$, the d until $n = 3$, and the f until $n = 4$. Within any given principal quantum shell, the order of decreasing stability, and therefore the order of filling with electrons, is always s-p-d-f.

The periodicity of atomic structure arises from the recurrent filling of new outermost principal quantum levels, but it is complicated by the fact that, although the principal quantum levels represent very roughly the general order of magnitude of energy, there is considerable overlapping. This overlapping is such that the outermost shell of an isolated atom can never contain more than 8 electrons. In the building up of successively higher atomic numbers, electrons always find more stable positions, once a set of p orbitals in a given principal quantum level is filled, in the s orbital of the *next higher* principal quantum level rather than the d orbitals of the same principal quantum level. When this s orbital is filled, electrons then go into the underlying d orbitals until these are filled, before continuing to fill the outermost shell by entering its p orbitals. The building-up of the atoms of successive atomic numbers may be represented by the following sequence: $1s$, $2s$, $2p$, $3s$, $3p$, $4s$, $3d$, $4p$, $5s$, $4d$, $5p$, $6s$, $5d$, $4f$, $6p$, $7s$, $6d$, $5f$. *The periodicity of atomic structure thus consists of the recurrent filling of the outermost shell with from one to 8 electrons that corresponds to the steady increase in nuclear charge.*

A *period* is considered to begin with the first electron in a new principal quantum shell and to end with the completion of the octet in this outermost shell, except, of course, for the very first period, in which the outermost shell is filled to capacity with only two electrons. From the order of orbital filling given above, it should be apparent that periods so defined cannot be alike in length. The first period, consisting of hydrogen and helium, has only two elements. The second period, beginning with lithium (3) and ending with neon (10), contains 8 elements, as does the third period, which begins with sodium (11) and ends with argon (18). The fourth period begins with potassium (19), but following calcium (20), the filling of the outermost (fourth) shell octet is interrupted by the filling of the d orbitals in the *third* shell. Thus 10 more elements enter this period before filling of the outermost shell is resumed, making the total number of elements in this period 18. In the fifth period, the first two outermost electrons are added in rubidium (37) and strontium (38), but then this outer shell filling is interrupted by the filling of penultimate shell d orbitals, which

again adds 10 elements before filling of the outermost shell is resumed; this period also contains 18 elements. The sixth period begins as before with two electrons in the outermost shell, but then there is an interruption at lanthanum (57) to begin filling the $5d$ orbitals. Here, however, occurs an additional interruption, in which 14 elements are formed through filling of the $4f$ orbitals, before the remaining $5d$ orbitals can be filled, and in turn before the outermost shell receives any more electrons. Consequently here it takes 32 elements to bring the outermost shell to 8 electrons and thus end the period. The seventh period is similar but incomplete. In principle it would end with element 118, but artificial element 105 is the highest in atomic number known at the time of writing.

These elements which represent interruptions in the filling of the outermost s and p orbital octet are called "transitional elements" (where d orbitals are being filled), and "inner transitional elements" (where f orbitals are being filled). This is to distinguish them from the other, "major group," elements in which underlying d or f orbitals are either completely empty or completely filled.

Physical properties of the elements that depend only on the electronic structure of the individual atom, such as the ionization potential and atomic, radius, vary periodically with atomic number simply because of the recurrent filling of the outermost shell. Increasing the atomic number by increasing the nuclear charge while adding to the outermost shell of electrons maintains the electroneutrality of the atom but has a highly significant effect on that part of the nuclear charge which can be sensed by outermost electrons. Underlying electrons are quite effective at shielding the nucleus, but in the outermost shell, each additional electron is evidently too involved in avoiding the others for them to be very effective in shielding one another from the additional nuclear charge. In fact each additional electron is successful in blocking off only about one-third of the positive charge that is simultaneously added. Therefore, with each unit increase in atomic number, when electrons are being added to the outermost shell to form the major group elements, the *effective nuclear charge*—that part of the total nuclear charge that is sensed by an electron in the outermost shell—increases by about two-thirds. This causes the electronic cloud to be drawn closer to the nucleus, diminishing the atomic radius and holding the electrons more tightly, as generally indicated by increasing ionization energy. The combined effect of increased effective nuclear charge and shorter radius increases the attraction between nucleus and outermost electron, defined as the *electronegativity*. Thus from left to right across the major groups, within each period, the radius decreases and the electronegativity increases.

Similar but smaller effects are observable for the addition of *d* or *f* electrons to underlying shells, in the transitional and inner transitional series.

The bonding properties of elements also depend on the electronic structure of the individual atom and therefore likewise vary periodically. For example, each period (except the first) begins with an alkali metal, lithium (3), sodium (11), potassium (19), rubidium (37), cesium (55), and francium (87), all of which show similar metallic bonding, crystallizing in a body-centered cubic lattice. Each has but one outermost electron per atom and can therefore form but one covalent bond. Each is very low in electronegativity and thus tends to become highly positive when bonded to another element. Crossing each period the elements become less metallic, higher in electronegativity, and able to form a greater number of bonds until limited by the number of outer shell vacancies rather than the number of electrons. The halogens, fluorine (9), chlorine (17), bromine (35), iodine (53), and astatine (85), each of which is next to the end of its period, are all nonmetals, highest of their respective periods

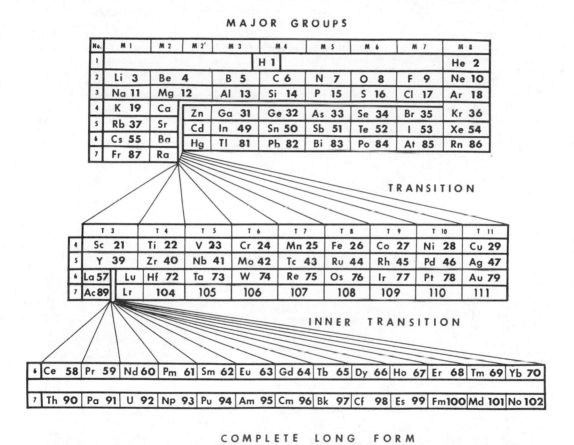

FIG. 1. Periodic chart of the chemical elements (reprinted from *J. Chem. Educ.*, 188 (1964); copyright 1964 by Division of Chemical Education, American Chemical Society, and reprinted by permission of the copyright owner, with modifications by the author).

in electronegativity and thus tending to become highly negative when bonded to other elements. Having seven outermost electrons, each has but one vacancy, permitting but one covalent bond.

Physical properties of the elements that are of greatest interest are usually properties that depend indirectly on the atomic structure but directly on the nature of the aggregate of atoms which results from the atomic structure. Such properties are melting point, density, and volatility. They may tend to vary periodically, but the periodicity is not necessarily consistent or even evident, because of abrupt differences in the type of poly-atomic aggregate. For example, the identical atoms of carbon may form graphite or diamond, depending on the kind of bonding, with entirely different properties resulting. Nitrogen follows carbon in atomic number, but because it forms N_2 molecules instead of giant three-dimensional structures like diamond or graphite, it is a gas with physical properties strikingly different from those of either form of carbon.

Among the most useful applications of the periodic law is to an understanding of the differences among compounds, whose properties also vary in a periodic manner. For example, oxides of elements at the beginnings of periods tend to be very stable, high-melting, nonvolatile solids of strongly basic character and practically no oxidizing power. Oxide properties change progressively across each period until toward the end the oxides tend to be unstable, low-melting, volatile compounds, acidic in nature and of high oxidizing power. Such periodicity is recognizable throughout a very large part of chemistry.

In order to erect a framework upon which the myriad facts that accord with the Periodic Law can be organized, the chemical elements can be arranged in an orderly manner called a "periodic table." Any such arrangement[1] can be satisfactory if it organizes the elements in some order of increasing atomic number, showing the separate periods and at the same time grouping elements of greatest similarity together.

In Fig. 1 is shown a modern version of the periodic table[2], together with the "long form." Figure 2 shows the "long form" as currently most widely used. From left to right across the table are the "periods," representing the elements in order of successively increasing atomic number and therefore progressive change in atomic strucutre. Across the period, the properties of the elements and their compounds tend to change from one extreme to the opposite. From top to bottom the periods are placed so that the most similar elements are grouped together. A distinction (shown by a physical separation) is made between "major groups" and

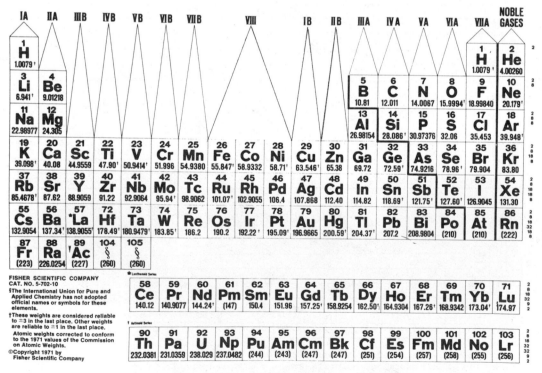

FIG. 2. Periodic chart of the elements (reproduced with permission of copyright owner, Fisher Scientific Company).

"transitional groups" because of the bonding dissimilarities originating in the availability in the transitional elements of underlying d orbitals that are not available in the major group elements. In older tables, major groups are usually, although inconsistently, designated as "A" and transitional groups as "B." At this time of writing (early 1982) a number of the world's chemical societies, in long overdue recognition of these inconsistencies, are seeking a generally acceptable alternative. In Fig. 1, the designations are consistently "M" (major) and "T" (transitional). But whether in the major group or the transitional group, the elements are arranged vertically on the basis of similarity in electronic configuration, which is then reflected in similarities in the physical and chemical properties of the elements and their compounds.

No amount of organization or correlation will ever alter the fact that each chemical element is an individual and unique, nor will the properties of any element be changed one iota by placing that element in any special position in a periodic table. Nevertheless, there is enough consistency to the structure and behavior of atoms to make any reasonable form of the periodic table an extremely useful framework upon which to organize and correlate an enormous quantity of chemical information[3]. The periodic law is truly one of the great generalizations of science.

R. T. SANDERSON

References

1. Mazurs, E., "Types of Graphic Representation of the Periodic System of Chemical Elements," published by the author, 6 S. Madison Ave., La Grange, Ill. 1957.
2. Sanderson, R. T., *J. Chem. Educ.*, **41**, 187 (1964).
3. Sanderson, R. T., "Inorganic Chemistry," New York, Van Nostrand Reinhold, 1967.

Cross-references: ATOMIC PHYSICS; ELECTRON; ELEMENTS, CHEMICAL; ISOTOPES; QUANTUM THEORY; TRANSURANIUM ELEMENTS.

PHASE RULE

The phase rule is a general equation $F = n - r - 2$, stating the conditions of thermodynamic equilibrium in a system of chemical reactants. The number of degrees of freedom or variance (F) allowed in a given heterogeneous system may be examined by analysis or observation and plotted on a graph by proper choice of the components (n), the phases (r), and the independently variable factors of temperature and pressure.

Josiah Willard Gibbs propounded the rule about 1877, and H. W. B. Roozeboom about 1890 began pioneering in specific cases. This rule has been an important instrument in the study of stability and metastability in geology, physics, metallurgy, ceramics, mineralogy, and engineering (e.g., in the exploitation of the salt deposits at Stassfurt, Germany, and Searles Lake, California, in the development of alloys, solid-state devices, and large crystals). Its use in the sophistication of techniques continues to expand; as an example, the calculation of reactions from basic structural data by computer techniques.

A phase is a homogeneous, physically distinct and mechanically separable portion of a system. H_2O has the three phases: water, vapor and ice. Each crystal form present is a phase. The relative amounts of each phase do not affect the EQUILIBRIUM.

The one-component system, water-vapor-solid, is unary. The components of a system are the smallest number of independently variable constituents by means of which the composition of each phase taking part in the equilibrium can be expressed in the form of a chemical equation.

With the components fixed in a system, variance—the degrees of freedom (F)—depends on the number of phases present. If water vapor alone is present, the system is bivariant since both temperature and pressure can vary within limits without affecting the number of phases; but if a second phase, liquid water, is present, the system is univariant and if either the temperature or pressure of the system in equilibrium is set, the other is automatically fixed as long as a second phase is present. A third phase, ice, makes the system invariant (the triple point), and any change in temperature or pressure, if maintained, results in the disappearance of one phase. Addition of another component forms a binary system, one degree of freedom is added, and the system is univariant until a fourth phase appears and the system becomes invariant (the quadruple point).

Schematic phase equilibrium diagrams outline experimental observations of physical and chemical changes in the system as the conditions of temperature, pressure and composition are varied. For unary systems, the diagram has two dimensions, for binary systems, three, and for ternary systems, four, etc. Binary systems are easily plotted with pressure or temperature constant, while ternary systems may be treated similarly as condensed systems with both constant if the vapor pressure is less than atmospheric. This added restriction reduces the variance by one, and at constant temperature, composition relationships may be plotted on a triangular diagram. Quarternary or quinary systems and ternary systems above atmospheric pressure may be treated by projections of surfaces of thermodynamic stability, but more complex systems require a mathematical approach.

The simple one-component system, water, plotted with rectangular coordinates, i.e., pressure and temperature, shows a variety of concepts which may be extended to more complex systems (see Fig. 1 in STATES OF MATTER). Each area in the diagram is a bivariant, one-phase state. Each curve separating the areas is a univariant, two-phase state showing the conditions under which a transition of the phase occurs. These are also called "indifferent phase reactions," where the amount of the phases may vary and are contrasted with "invariant equilibrium," where the degree of freedom is zero. The fusion curve for the equilibria between the solid phase and the liquid phase, the sublimation curve for solid and vapor, and the vaporization curve for liquid and vapor meet at the triple point. The three distinct phases differ in all properties except chemical potential. The end of the vaporization curve is a singular point, the critical point where liquid and vapor become identical, a restriction which reduces the variance by one so that F becomes zero.

In a binary (or higher) system, a liquidus is a curve representing the composition of the equilibrium liquid phase and a solidus represents the composition of the solid phase. The conjugate vapor phase is represented by a vaporous with tie-lines or conodes to the liquidus or solidus points in equilibrium. A minimum point for the existence of a liquid is the eutectic, sometimes (in an aqueous system) called the cryohydric point. In a ternary system the eutectics of three binary systems initiate curves leading to a ternary eutectic. If two liquids form a miscibility gap in the system, multiple quadruple points are possible. At a peritectic, a phase transition occurs at other than a minimum, i.e., one solid melts to another solid and a liquid. The solid may be a compound or a solid solution (mixed crystals) in which the composition of the solid varies with relative proportion of components and is shown by the solidus. A congruent melting point is a maximum in its curve, i.e., the solid melts to a liquid of the same composition. Where two phases become identical in composition, an indifferent or critical point exists. Where two conjugate phases become identical, the point may also be called a consolute point.

JOHN H. WILLS

References

Ricci, John E., "The Phase Rule and Heterogeneous Equilibrium," New York, Van Nostrand Reinhold, 1951.

Levin, E. M., McMurdie, H. F., and Hall, F. P., "Phase Diagrams for Ceramists," Columbus, Ohio, The American Ceramic Society, 1956, 1959.

Masing, G., "Ternary Systems," New York, Van Nostrand Reinhold, 1944.

Wetmore, F. E. W., and LeRoy, D. J., "Principles of Phase Equilibria," New York, McGraw-Hill Book Co., 1951.

Haase, R., and Schönert, H. (translated by E. S. Halberstadt), "Solid-Liquid Equilibrium," Topic 13,

Vol. 1 of "International Encyclopedia of Physical Chemistry and Chemical Physics," New York, Pergamon Press, 1969.

Refractory Materials Series, J. L. Margrave, Editor, Volume 4, Kaufman, L. K., and Bernstein, H., "Computer Calculation of Phase Diagrams," New York, Academic Press, 1970. Volume 6, Alper, A. M., "Phase Diagrams: Materials, Science and Technology" (6 volumes), New York, Academic Press, 1970-1976.

Hoffman, E. J., "Phase and Flow Behaviour in Petroleum Products," Energon Co., 1981.

Ginell, R., "Association Theory. The Phases of Matter and their Transformation," New York, Elsevier, 1979.

Cross-references: EQUILIBRIUM, STATES OF MATTER, THERMODYNAMICS.

PHONONS

Phonons are quanta of vibrational energy, and are in many ways analogous to photons, which are the quanta of light energy. The vibrations of a solid can be considered as the sum of oscillations of many traveling waves of different frequencies. Quantum theory restricts the energy of a wave of frequency ω to the possible values

$$E_n = n\hbar\omega \qquad (1)$$

where $\hbar$ is Planck's constant (6.65×10^{-27} erg · sec) divided by 2π, and n is a positive integer or zero. The total energy E_n of the wave may be considered to be the sum of the energies of n phonons, each of energy $\hbar\omega$.

A solid contains phonons having a wide distribution of frequencies. For each phonon the frequency depends on the wavevector q. In a crystalline solid with spacing a between the molecules, the shortest wavelength that phonons may have is $2a$, and hence the maximum wavenumber is π/a ($\sim 10^8$ cm^{-1}). In a typical case the relation between frequency and wavenumber (phonon dispersion relation) is as shown in Fig. 1. For each value of q there are $3r$ phonons, where r is the number of atoms per unit cell. These different phonons may be labelled by a *polarization index* j which runs from 1 to $3r$, and it is usual to indicate this by writing the phonon frequency as ω_{qj}. Figure 1 therefore represents the case $r = 2$.

For three of the polarizations (i.e. for three values of j) the frequency varies approximately linearly with wavenumber for small enough q. These are called *acoustic phonons* (A) because the motion of the atoms which occurs when these phonons are present is the same as is caused by an acoustic wave. Thus, neighboring atoms oscillate approximately in phase with each other. The remaining phonons are called *optical phonons* (O). For these phonons the atoms within each unit cell vibrate in different

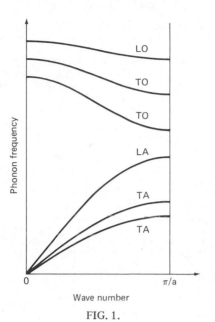

FIG. 1.

directions, and the center of mass of each cell remains approximately stationary. In ionically bonded crystals each atom has an electrical charge (positive or negative). The optical phonons in these crystals cause the atoms to move relative to each other, thereby producing strong oscillating electric fields inside the crystals. For this reason there is a large coupling of these phonons to light waves, and this is the origin of the name *optical phonon*. Note that in a crystal with only one atom per unit cell there are only acoustic phonons.

The different types of phonons may be further classified into longitudinal (L) and transverse (T). For longitudinal phonons the atoms vibrate in the same direction that the phonon is travelling, whereas for the transverse phonons the vibration is perpendicular to the direction of propagation.

For any crystal there is maximum frequency ω_m that the phonons may have, and this is usually between 10^{13} and 10^{14} sec^{-1}. Information about phonon frequencies has been obtained by many different experimental techniques. The most detailed measurements have been made using the technique of inelastic neutron scattering. A neutron of known energy E_0 and momentum p_0 enters the crystal. This neutron may create a phonon in the crystal and emerge from the crystal with a different energy E_1 and momentum p_1. From conservation of energy it is known that

$$E_0 = \hbar\omega_{qj} + E_1 \tag{2}$$

where ω_{qj} is the frequency of the phonon produced. Thus, if E_1 and E_0 are both known the phonon frequency can be determined. The

wavevector of the phonon can be found from the equation

$$p_0 = \hbar q + p_1 \tag{3}$$

which represents the condition of conservation of momentum, $\hbar q$ being the momentum of the phonon. By this means the phonon spectra of most of the common crystals have been determined.

Other techniques can be used to study the phonon dispersion relation, but these usually give less complete information. Measurements of the velocity of sound give results for acoustic phonons of small q. The frequency of optical phonons with $q = 0$, and some phonon frequencies at the zone boundary ($q = \pi/a$) can be found by light scattering or absorption experiments. Measurement of the temperature dependence of the specific heat can be used to estimate the maximum phonon frequency.

Phonons and Interatomic Forces If the forces which act between the atoms in a crystal are known, it is possible to calculate theoretically the phonon spectrum. It is therefore also possible to use measurements of the phonon spectrum to study the strength and "character" of the interatomic forces. The strength of the forces is related, roughly speaking, to the average phonon frequency. By "character" one means the range of distance over which the interatomic forces act, whether or not they are simple central forces dependent only on the atomic separation, etc. To study these features requires a more detailed analysis of the experimental results for the phonon spectra. This analysis has been carried out for many crystals, and much interesting and useful information has been obtained.

Specific Heat At a temperature T the average number $\langle n_{qj} \rangle$ of phonons of a particular type with wave vector q and polarization index j is given by the Bose-Einstein law

$$\langle n_{qj} \rangle = [\exp(\hbar\omega_{qj}/kT) - 1]^{-1} \tag{4}$$

where k is Boltzmann's constant (1.38×10^{-16} erg K^{-1}). Thus, the number of phonons increases with increasing temperature. Note also that because of the exponential factor in (4) most of the phonons which exist at temperature T have frequency less than or comparable to kT/h. The total energy of the phonons is

$$E_{ph} = \sum_{qj} \langle n_{qj} \rangle \hbar\omega_{qj} \tag{5}$$

where the sum is over all possible values of q and j. The phonon contribution to the specific heat is thus

$$C_{ph} = \frac{dE_{ph}}{dT}. \tag{6}$$

An approximate calculation of C_{ph} was made by Debye with result

$$C_{ph} = 9Nk(T/\theta_D)^3 \int_0^{\theta_D/T} \frac{x^4 e^x \, dx}{(e^x - 1)^2} \quad (7)$$

where N is the number of atoms per unit volume, and θ_D is a certain characteristic temperature called the *Debye temperature*. This temperature varies from crystal to crystal, and the value of θ_D is very roughly equal to $\hbar\omega_{max}/k$, where ω_{max} is the maximum phonon frequency. For $T \ll \theta_D$, C_{ph} is proportional to T^3, and for $T > \theta_D$, C_{ph} has a constant value.

Thermal Conduction When one part of a solid is hotter than another it contains more phonons. These excess phonons gradually spread out from the hot region and eventually the distribution of phonons throughout the solid become uniform. Macroscopically, this is described by saying that there is a flow of heat from the hot region to the cold. One can show that the thermal conductivity arising from the flow of the phonons is given approximately by

$$k = \tfrac{1}{3} C_{ph} v_{ph} l_{ph}, \quad (8)$$

where v_{ph} is the average velocity of a phonon, which is of the same order of magnitude as the sound velocity; and l_{ph} is the distance a phonon travels in a straight line before it is scattered. Phonons may be scattered by other phonons, by electrons, by defects of impurity atoms, and by the walls of the solid. The measurement of the thermal conductivity provides rough information about the strength of these scattering processes and how they depend on temperature.

Phonon Interactions The scattering processes that phonons undergo have been studied in great detail in the last few years. It is now possible to produce artificially phonons of a definite frequency and to study their interactions with other phonons, with electrons, defects, etc. One method for the generation of phonons which has proven particularly important uses superconducting tunnel junctions. These devices can also be employed as tunable detectors of phonons. However, the development of other devices for the generation and detection of phonons remains a very active field of research.

HUMPHREY J. MARIS

References

Kittle, C., "Introduction to Solid State Physics," 5th Edition, New York, John Wiley, 1976.

Klemens, P. G., "Thermal Conductivity and Lattice Vibrational Modes," in Seitz, F., and Turnbull, D. (Eds.), "Solid State Physics," Vol. 7, p. 1, New York, Academic Press, 1958.

Maris, H. J., "Phonon Scattering in Condensed Matter," New York, Plenum, 1980.

Bilz, H., and Kress, W., "Phonon Dispersion Relations in Insulators," Berlin, Springer-Verlag, 1979.

Cross-references: CRYSTALLOGRAPHY, ELECTRON, HEAT CAPACITY, HEAT TRANSFER, NEUTRON, PHOTON, QUANTUM THEORY, ULTRASONICS.

PHOTOCHEMISTRY

Photochemistry is that branch of chemistry which deals with the permanent, chemical effects of the interaction of electromagnetic radiation and matter, as distinct from the usually temporary physical effects. Upon absorption of a quantum of radiant energy a molecule is raised to an excited state, stable or unstable, in which, depending upon the particular wavelength absorbed, the energy may be retained in rotational, vibrational, and electronic degrees of freedom. Photoionization following the absorption of x- or γ-rays is usually referred to as *radiation chemistry*. Excitation by longer wavelengths, the infrared, at the intensities available from standard sources leads only to an increase in the population of the higher vibration-rotation levels of the ground electronic state; since no chemical effects are produced, such excitation is of little interest to the photochemist. Thus until recently there has existed, near 700 nm, a theoretical upper limit to the wavelength region of interest. The lower limit of this region, which theoretically extends to the wavelengths of soft x-rays, is defined in practice by the availability of materials transparent to the far ultraviolet—~160 nm for synthetic fused silica and ~120 nm for lithium fluoride.

A molecule in an upper electronic state may undergo one of several processes. (1) Almost immediately, or after losing or gaining a number of vibrational quanta by collision, it may return to the ground electronic state with emission of light (fluorescence). (2) A major portion of the energy may be transferred to rotational, vibrational and translational degrees of freedom (rarely to electronic levels in molecular energy transfer) of the neighboring molecules or to the walls in collisions of the second kind (thermal degradation or collisional deactivation). (3) It may, within a period of one-half a vibration ($\sim 10^{-13}$ sec) dissociate into mono- or polyatomic fragments, each of which may retain part of the excitation energy in any of its degrees of freedom (dissociation). (4) After a period of several vibrations in a vibrational level below the dissociation limit of the upper electronic state, but before a single rotational period ($\sim 10^{-11}$ to 10^{-12} sec.), it may, with no change in total energy, undergo a radiationless transition to a vibrational level in a second upper electronic stage; if the second vibrational level is above the dissociation limit of the second state, dissociation will occur after one-half a vibration. The transition probability may be sometimes enhanced by collisions (predissociation and induced predissociation). If the transfer occurs to bound vibrational levels of the second excited electronic state, the molecule in this excited state may undergo all the processes cited for the excited state formed directly by absorption. Often this second electronically excited state is of a different multiplicity than either the ground state or the initially formed state, and the process is termed *intersystem crossing*. The emission from this second state (if it occurs) is termed *phosphorescence* and usually exhibits a much longer lifetime than the fluorescence described in (1). (5) It may be transferred to an upper state of an isomeric molecule, which is then usually stablized by (2). This is called *rearrangement.* (6) It may react directly with other normal or excited molecules utilizing the excitation energy to overcome all or part of the activation energy barrier.

One or more of these processes must occur in any assembly of molecules absorbing light energy in the gaseous or liquid state, but not all need occur concurrently. The interpretation of the absorption spectrum of diatomic molecules and, to an increasingly greater extent, of polyatomic molecules may, in some cases, give definite information regarding (3) and (4). Advances in spectroscopy have been reflected in photochemistry; indeed, until the advent of quantum theory no satisfactory explanation of photochemical phenomena existed.

The advent of lasers, most commonly pulsed, which provide highly monochromatic beams of great intesity in many regions of the spectrum, has stimulated interest in photochemical isotope enrichment. It often is possible to stimulate reaction almost exclusively of those molecules which contain a rare isotope, thus leading to products highly enriched in that isotope, or conversely to cause almost exclusive reaction of those molecules containing an abundant isotope so that the rare isotope is enriched in the unreacted starting material. The very high radiation field provided by lasers, especially the CO_2 laser in the infrared, may cause a single molecule to absorb many photons (10–30 is not uncommon) in a single pulse. Such excitation leads to the phenomenon of *multiphoton decomposition* from the ground electronic state and whole new areas of non-steady-state kinetics have become available for study.

One of the major interests in photochemistry has been the production of free radicals by processes (3) and (4) and the study of their reactions. Studies of the photolyses of various aldehydes, ketones, alkyl halides, azocompounds, and organometallic compounds have been most fruitful in this respect. Intermittent illumination may be used to determine the lifetimes of the radicals in reactions in which the radical concentration is proportional to the absorbed intensity raised to a power

n $(0.5 \leqslant n < 1.0)$; this leads to the evaluation of rate constants for radical association reactions. A special case of (2) has also led to useful results. Metal vapors (Hg, Cd, or Zn) are raised to the excited $^3P_1$ and $^1P_1$ states upon absorption of resonance radiation from lamps containing these metals. Resonance fluorescence is efficiently quenched by the vapors of a large variety of compounds usually resulting in the production of atoms, radicals and metal atoms in the ground state. Such photosensitized methods are not confined solely to atomic vapors, but photosensitization by polyatomic molecules through collisions of the second kind has not been well developed.

The initial act of absorption and the various reactions of the excited state are usually referred to together as the primary process. The *primary quantum yield* ϕ, i.e., the number of molecules changed per quantum absorbed, is given by the rate of processes (3) to (6) divided by the rate of absorption of light; in the absence of fluorescence and deactivation, $\phi = 1$. This is merely a statement of the *Stark-Einstein law of photochemical equivalence*, known as the first law of photochemistry. The concept of the quantum yield may be extended to include secondary processes [e.g., the reverse of (3)]. Overall quantum yields are measurable directly; ϕ must be inferred from kinetic studies of the secondary processes. Thus, any photochemical investigation is immediately separable into a chemical and an optical problem.

That fraction of the incident light which is absorbed by the reactant is readily measured by standard photometric procedures. The measurement of the absolute intensity is usually more difficult; all such determinations are based ultimately upon thermopile measurements of radiant energy, although a photocell may be calibrated as a more convenient secondary standard. To obviate the use of a thermopile or photocell, chemical actinometers have been used; these are photochemical reactions whose quantum yields have been established accurately as a function of wavelength, etc. Hence the measurement of absolute intensity reduces to the determination of the amount of chemical change. Since the response of all actinometric reactions and photocells is sensitive to wavelength, monochromatic radiation is desirable for this reason as well as for the theoretical interest of the effect of various wavelengths on the photochemical reaction. This is obtained by removing the unwanted lines from sources of discontinuous spectra by means of monochromators or gaseous, liquid, and glass filters; unfortunately, the intensity of the wavelength of interest is usually reduced by this procedure.

The last factor has a profound influence on the problem of determining the amount of chemical change. Under the most advantageous conditions with unfiltered light, the rate of production of product is rarely as large as 10^{-4}

mole/hr if the quantum yield is approximately unity; with filtered light and small fractional absorption, a few micromoles must be estimated. Inorganic compounds in solution can usually be handled in such quantities, and techniques of gas phase chromatography, infrared spectroscopy, and gas chromatography coupled with mass spectrometric detection now permit the qualitative and quantitative analysis of all but the most intractable of product mixtures. In certain instances (photopolymerizations, the photolysis of HCl, etc.) long reaction chains are set up with the quantum yield reaching 10^6 in the second example quoted. Measurements of physical properties (pressure of gases, density of solutions, etc.) then become practical and convenient methods of following the reaction.

K. O. KUTSCHKE

Cross-references: CHEMISTRY; ISOTOPES; PHOTO-SYNTHESIS; RADIATION CHEMISTRY; RADIATION, IONIZING, BASIC INTERACTIONS.

PHOTOCONDUCTIVITY

Introduction Photoconductivity is the increase in electrical conductivity of nonmetallic solids produced by the motion of additional free electronic carriers created by absorbed electromagnetic radiation. Solids are characterized by the existence of bands of energy levels which electronic carriers may occupy and in which conduction takes place. The highest occupied band is called the *valence band* and the next empty (at absolute zero of temperature) band is called the *conduction band*. These conduction and valence bands are separated by an energy E_g in which, in the absence of any atomic defects or impurities, no available states exist for occupancy by photoexcited carriers from the filled valence band. As a result the spectral dependence of photoconductivity is characterized by the existence of an energy threshold E_g. Only for photon energies $h\nu > E_g$ is photoconductivity, associated with the simultaneous excitation of an electron into the conduction band and the creation of a hole in the valence band detectable. This intrinsic photoconductivity characterized by a sharp spectral threshold is one of the most direct experimental confirmations of the band theory of solids. In actual solids, of course, both defects and impurities exist and these create localized states or traps which lie within the energy gap. In particular in amorphous or noncrystalline solids the localized states exist in very high densities and are related to the disordered nature of these materials. In both crystalline and noncrystalline solids these *gap states* play a critical and usually dominant role in determining the recombination processes and rates and can also result in photoconductivity for photon energies $< E_g$,

called *extrinsic photoconductivity*. Under constant illumination a new thermal equilbrium is reached such that a steady state photocurrent can be achieved. Under these conditions the rate of photogeneration is balanced by the various recombination processes by which the carriers tend to relax to their normal dark equilibrium distribution. Upon termination of the illumination the excess photoconductivity decays until the thermal equilibrium dark conductivity is recovered. The phenomenon of photoconductivity therefore involves absorption, photogeneration, recombination and electronic transport processes. For this reason photoconductivity has played and continues to play an important role in the development and understanding of the science of the solid state. Usually two general types of photocurrents can be observed depending on the role that the contact electrodes play and these will be discussed in turn.

Primary Photocurrents Figure 1 shows a schematic representation of a primary photocurrent produced by uniform volume photoexcitation. The dashed areas represent the electrodes by which electrical contact to the solid under study is made. The contacts are blocking, i.e., carriers may exit the sample but not enter. The primary photocurrent Δi is given by

$$\Delta i = \frac{\eta Fe(X_n + X_p)}{L} \quad (1)$$

where η is the photogeneration efficiency, F is the number of absorbed photons per second, and X_n, X_p are the distances that electrons and holes, respectively, move before becoming immobilized by being captured in a localized state or trap. These distances increase as the applied field increases, but since the maximum value of $X_n + X_p = L$ the photocurrent will eventually saturate. The photoconductive gain G is $\Delta i/Fe$, which is the photogeneration efficiency and which has a maximum value of unity. In practice this simplified view is complicated by the establishment of polarization fields due to the trapped charge. This polarization field opposes the externally applied field and characteristically the photocurrents decrease with

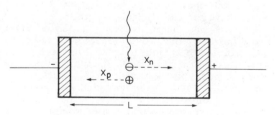

FIG. 1. Schematic representation of primary photocurrent. The electrodes are blocking so no carriers can enter the photoconductor.

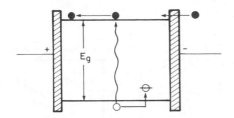

FIG. 2. Schematic representation of secondary photocurrents. Only the electrons are mobile and replenished at the electrode.

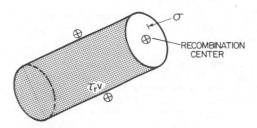

FIG. 3. Representation of recombination in a high-mobility photoconductor.

time. Thus, primary photocurrents are characterized by a photoconductive gain which has a maximum value of unity. Although this may seem hypothetical, under certain conditions the requirement of actually achieving the necessary blocking electrodes can be realized. A notable example is electrophotography, where corona charge is used to achieve a very good blocking electrode. In fact, because the electrophotographic process is such a good case of primary photocurrent, it is, where applicable, a direct and unambiguous way to study photogeneration efficiencies, since in the absence of trapping, Δi is directly controlled by η.

Secondary Photocurrents Figure 2 shows schematically an idealized photoconductor which nevertheless is a good approximation for many materials. One of the carriers, the electron, is mobile and is replenished at the electrodes. The hole created by light is considered to be immediately captured into a localized state and is effectively immobile. The photocurrent thus observed is exclusively due to the flow of electrons (the converse could of course be true). For this n-type photoconductivity the gain is given by

$$G = \frac{\tau_r \mu_n E}{L} \qquad (2)$$

μ_n is the electron mobility and τ_r is the recombination lifetime of the electrons with the previously trapped holes. Since $L/\mu_n E$ is the time it takes an electron to transit the sample thickness L, the gain is clearly the number of transits, accompanied by replenishment at the contact, that can occur within the recombination lifetime. In some materials it has proved possible to increase τ_n by doping so as to produce gains $\sim 10^6$ and hence highly sensitive photodetectors.

Recombination A detailed discussion of recombination processes is impossible within a short space and in any event rapidly becomes very model specific. It is however possible to delineate the general concepts which are common to the most complex case. Figure 3 represents a simple physical picture which describes the general process of recombination. For conceptual ease it is assumed that a recombination center is already occupied by a hole and has a capture cross section σ for the capture of an electron. The electron is moving with an average thermal velocity $v \sim 10^7$ cm/sec (typically much greater than any drift velocity $\mu_n E$). In a time τ_r the electron will have moved a distance $\tau_r v$ will have swept out a volume $\tau_r v \sigma$. If a recombination center lies within this volume (this is clearly a function of the density of recombination centers N_r) then recombination is certain to occur. Therefore the condition for recombination is

$$1/N_r = \tau_r v \sigma \qquad (3)$$

so that

$$\tau_r = 1/N_r v \sigma. \qquad (4)$$

The dependence of photoconductivity on light intensity and temperature that can be observed is virtually unlimited. Thus, sublinear, and supralinear light intensities have been reported. With the right combination of intensity and temperature it is possible to go through these regimes with the same sample. Again a detailed attempt to explain these observations is very model dependent and a particular model may not be unique. Some general features can be understood however with reference to Eq. (1)-(4). Generally speaking, the photogeneration efficiency and carrier mobility are not intensity dependent so that any observed intensity dependence of the photoconductive gain is attributable to τ_r. In a material containing traps, if the light intensities are sufficiently low such that the occupancy of traps during illumination is substantially unchanged from that in the dark, τ_r will be independent of light intensity and the photocurrent Δi will be linear in intensity. If the intensity is increased such that the occupancy of traps changes substantially then previous trapping centers can be converted into new recombination centers. Therefore, in reference to equation (4) N_r effectively becomes intensity dependent. If the capture cross-sections of the new states are radically different, then highly non-linear photocurrents can be accounted for. Ultimately at a sufficiently high light intensity that the density of photoexcited free carriers exceeds the density of trapped carriers then bimolecular recombination will predominate and the photocurrent will vary as $F^{0.5}$.

From the preceding discussion it is clear that the phenomenon of photoconductivity is a particularly powerful tool for studying electronic transport properties and the nature and distribution of localized states in solids. This power stems from the ability to transiently produce a nonequilibrium excess of photogenerated carriers and observe their evolution and disappearance with time as they move, trap and then recombine. Indeed, in highly insulating materials such as amorphous materials, transient photoconductivity has emerged as the canonical technique, since more conventional methods such as electrical dark conductivity or Hall effect studies are either ambiguous in interpretation or impossible.

Field-Controlled Photogeneration It has so far been assumed that each photon absorbed for which $h\nu > E_g$ creates an electron-hole pair which instantaneously dissociates into a free electron and hole. For crystalline photoconductors this is a valid assumption. It has recently become clear that this is not a good assumption in low-mobility photoconductors, whether they be crystals such as anthracene or amorphous materials. Low mobility implies small mean free paths due to scattering of mobile carriers by potential fluctuations comparable to carrier energies. Thus excited electron-hole pairs may thermalize within a distance that is comparable to or less than the radius of their mutual coulomb well. In this case the carriers experience an attractive potential which results in the recombination of some of the pairs before they ever become free carriers. This *geminate recombination* (from Latin *gemini*, meaning twins), therefore, limits the photogeneration efficiency to less than unity. By application of an external electric field the amount of geminate recombination can be decreased since the field can aid the carriers in overcoming their mutual coulomb attraction. This leads to a field-dependent photogeneration process which is not seen in inorganic crystalline semiconductors such as silicon, cadmium sulfide, etc. Geminate field-controlled photogeneration is seen in amorphous selenium and in molecular crystalline solids such as anthracene. In the latter case, even with the periodicity of the aromatic molecular building blocks the very weak van der Waals binding forces lead to very narrow energy bands (a few kT) and therefore low electronic carrier mobilities.

In addition to straightforward photoconductivity a wide range of related photo-effects are potentially observable. In the absence of an applied field and for widely different diffusion constants for electrons and holes a voltage (Dember potential) can be developed [see Fig. 4(a)]. The sign of the voltage indicates which sign of carrier has the larger diffusion constant.

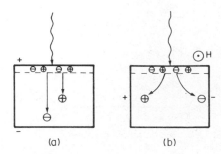

FIG. 4. Schematic representation of photovoltages developed due to diffusion of nonuniformly photogenerated carriers: (a) Dember effect, (b) in the presence of an applied magnetic field, the photomagnetoelectric (PME) effect.

Photomagnetoelectric effects are also potentially observable, Fig. 4(b). In materials with built-in internal fields due to Schottky barriers of *p-n* junctions, the photovoltaic effect may be produced. This is currently under intense investigation for the direct conversion of solar energy into electricity.

Photoconductivity has been observed to some degree or another in the majority of solids ranging from crystals to polymers and amorphous materials and from silicon crystals to polystyrene. Practical applications are both widespread and of enormous commercial value. These range from photodetectors and exposure meters to electrophotography and in total constitute a multibillion dollar industry.

J. MORT

References

Bube, R. H., "Photoconductivity of Solids," New York, John Wiley & Sons, Inc., 1960.

Rose, A., "Concepts in Photoconductivity and Allied Problems," New York, Interscience Publishers, 1963.

Mort, J. and Pai, D. (Eds.), "Photoconductivity and Related Phenomena," Amsterdam, Elsevier Scientific Publishing Co., 1976.

Seeger, K., "Semiconductor Physics," Vienna, Springer-Verlag, 1973.

Mort, J., and Pfister, G. (Eds.), "Electronic Properties of Polymers," New York, John Wiley & Sons, Inc., 1982.

Cross-references: CONDUCTIVITY, ELECTRICAL, ENERGY LEVELS, PHOTOELECTRICITY, PHOTOVOLTAIC EFFECT, SEMICONDUCTORS, SOLID STATE PHYSICS, SOLID STATE THEORY.

PHOTOELASTICITY

When a transparent material is subject to a mechanical stress, optical properties are in general functions of the stress. In particular, the strain caused by the application of a stress may change the refractive index of the material. This property of transparent materials is called *photoelasticity*, and was first found by Brewster in 1816. Photoelasticity is now widely used as a useful, practical method to determine distributions of stresses or strains in materials.

The photoelastic effect can be explained in the following manner. When a stress T is applied to an elastic medium, the produced strain S is linearly related to T by Hooke's law: $S = sT$, where s is a proportionality constant called the *elastic compliance*. The change in the refractive index induced by the strain S is then expressed to good approximation by

$$n = n_0 - \frac{1}{2} p n_0{}^3 S, \qquad (1)$$

where n is the refractive index under the influence of the strain S; n_0 is the refractive index in the absence of the strain, i.e., when $S = 0$; and p is a material constant called the *photoelastic constant*, which describes the strength of the photoelasticity of a given material. If a transparent material is optically isotropic in the absence of stress, the material in general becomes optically anisotropic by the application of a uniaxial stress. This phenomenon is called *photoelastically induced birefringence* (or double refraction). The transmission of light in such anisotropic media has some unique features not present in isotropic materials.

Consider a transparent sheet of optically isotropic material of thickness t. Suppose that a uniaxial stress is uniformly applied so that the direction of the stress lies in the plane of the sheet. A plane-polarized light ray entering the sheet perpendicular to the sheet face is in general broken up into two components, called the *ordinary wave* and the *extraordinary wave*. These waves have polarization vectors at right angles to each other, corresponding to the principal stresses created by the uniaxial stress, and in general they have different velocities which are determined by Eq. (1). The ordinary wave has the polarization vector perpendicular to the direction of the uniaxial stress. The polarization vector of the extraordinary wave is parallel to the direction of the stress.

Since the two components propagate in the medium with different velocities, the phase difference δ will be produced between the two components upon emerging from the sheet. In principle, δ is proportional to the magnitude of the shear stress, which is defined as one-half the difference between the two principal stresses. The magnitude of δ can be effectively observed by passing the emergent light through a second polarizing unit, called the *analyzer*. The principal plane of the analyzer is usually put at right angles to the polarizer, which is the first polarizing unit used to create the incident plane-polarized light. The two wave components, upon emerging from the analyzer, are reduced into one plane and give rise to inter-ference. When $\delta = (2m + 1)\pi$, where m is an arbitrary integer, constructive interference takes place and one may see the brightest field of view. When $\delta = 2m\pi$, the interference becomes destructive and the field becomes dark. In general, δ may take any value depending on the actual shear stress in the medium. If there is a distribution of stresses (or strains) in a medium, bright and dark fringes will be seen, and these fringes are contours of equal shear stress. The magnitude of the shear stress can be calculated from the interference order of the fringes. When the polarization vector of the incident lgith happens to be parallel to one of the principal stresses, the interference fringes disappear and the field of view becomes dark. Thus, the orientation of the principal stresses can be found by rotating the crossed polarizer and analyzer. In this way, one can determine the distribution of the shearing stress in a transparent material by making use of photoelasticity.

To detect stress patterns in opaque solid objects such as metallic machine parts, models are constructed from appropriate transparent materials and subjected to stresses in the same fashion as in the original object. It has been proved that the stress patterns observed in such models replicate those in the original objects, provided that the models and original objects are both elastic. The model materials most widely used today are glass, celluloids, and epoxy resins.

Since the advent of lasers for coherent light sources in the 1960s, the photoelastic effect has found another important application in electrooptic engineering. When an ultrasonic wave propagates in a transparent medium, it produces a sinusoidally periodic distribution of mechanical strain as given by $S(t, x) = S_0 \sin (2\pi f t - kx)$, where S_0 is the amplitude of the strain, f is the frequency of the ultrasonic wave, t is the propagation time, k is the wavenumber, and x is the distance in the propagation direction. This strain wave in turn produces a periodic modulation of the refractive index via the photoelastic effect described by Eq. (1). This provides a moving phase grating (or diffraction grating) which may diffract portions of an incident light into one or more directions. This phenomenon, known today as the *acoustooptic effect*, was originally predicted by Brillouin as early as in 1922, and since his discovery, extensive studies were done by a number of researchers toward the understanding and application of the phenomenon. Today, applications of the acoustooptic phenomenon to light modulation and deflection for optical communication and signal processing have become particularly important.

Figure 1 is a conceptual illustration of an acoustooptic light deflector. A piezoelectric transducer bonded on one of the ends of a transparent medium launches an ultrasonic

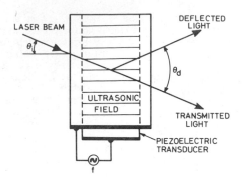

FIG. 1. Light deflector.

wave into the medium, and creates a phase grating with a spatial period equal to the wavelength of the ultrasonic wave. When the incident angle of the laser beam is adjusted so as to coincide with $\theta_i = \lambda f / 2v$, where v is the sound velocity, an appreciable portion of the emergent light is deflected into the new direction. The frequencies usually used range from 10 to 1,000 MHz. The deflection angle, given by $\theta_d = \lambda f / v$, can be varied by changing the frequency f. The intensity of the deflected light can be controlled by the electrical power applied to the transducer, and thus the same device can also be used as a light modulator. The photoelastic materials most often used for this application are fused quartz, paratellurite (TeO_2), lithium niobate ($LiNbO_3$), and lead molybdate ($PbMoO_4$).

SUSUMU FUKUDA

References

Coker, E. G., and Filon, L. N. G., "A Treatise on Photo-Elasticity, Cambridge, U.K., Cambridge Univ. Press, 1931.

Born, M., and Wolf, E., "Principles of Optics," Oxford, Pergamon Press, 1975.

Aben, H., "Integrated Photoelasticity," New York, McGraw-Hill, Inc., 1979.

Nelson, D. F., "Electric, Optic, and Acoustic Interactions in Dielectrics," New York, John Wiley & Sons Inc., 1979.

Uchida, N., and Niizeki, N., "Acousto-Optic Deflection Materials and Techniques," Proc. IEEE, 61 (1973).

Fukuda, S., Karasaki, T., Shiosaki, T., and Kawabata, A., "Photoelasticity and Acousto-Optic Diffraction in Piezoelectric Semiconductors," Phys. Rev. B 20, No. 10 (1979).

Cross-references: ELASTICITY, INTERFERENCE AND INTERFEROMETRY, MECHANICAL PROPERTIES OF SOLIDS, POLARIZED LIGHT, POLYMER PHYSICS.

PHOTOELECTRICITY

In its broadest sense, the photoelectric effect is defined as the emission of electric charges (electrons, ions) resulting from the absorption of electromagnetic radiation (infrared, visible, ultraviolet, x-ray or gamma radiation) in a material (solid, liquid, or gas). Usually photoelectricity, or photoemission, is associated with the ejection of electrons from a solid surface.

Historically, the photoelectric effect played an important role in developing quantum mechanics in the beginning of the 20th Century and explaining the duality of electromagnetic radiation, namely, that light can be described as wavelike and still have the properties of discrete particles. The discovery of the photoelectric effect is ascribed to the German physicist H. R. Hertz, who in 1887 observed that ultraviolet radiation lowers the voltage at which sparking takes place between two electrodes. About a decade later, it was firmly established that cathode rays, produced by electric discharges, were nothing but the newly discovered electrons (particles with the elementary negative charge). In 1900 E. A. Lenard devised an experiment which directly showed that electrons were emitted from a metal surface under illumination of ultraviolet light. The name of the phenomenon, photoelectricity or photoelectric effect, was coined at this time. The most important findings during the first couple of years after Lenard's discovery can be summarized as follows: the number of ejected photoelectrons at a given excitation frequency is proportional to the intensity of the incident radiation; the photoelectrons have a kinetic energy distribution from zero up to cutoff energy, which is proportional to the excitation frequency but independent of the intensity; within the detection limits the photoelectrons seemed to be emitted instantaneously.

The characteristics described above could not be described by the classical theory of electromagnetic radiation by J. C. Maxwell and led A. Einstein to propose a fundamentally new theory of light in 1905. Einstein postulated that light (electromagnetic radiation) had properties like those of particles when it interacted with electrons. The discrete light particles are called photons, each with an energy $h\nu$, where h is Planck's constant and ν is the frequency of the photon. The electron absorbs the entire energy of the photon upon interaction, i.e., the radiation is quantized. The minimum energy required to eject an electron, Φ, is called the *photoelectric threshold energy* or *work function*. Einstein's theory of the photoelectric effect now postulates that the maximum energy of the photoemitted electrons E_k is given by

$$E_k = h\nu - \Phi.$$

Einstein's photoelectric law can account for the observations described above. A direct experimental proof for Einstein's law was given in 1916 by Robert A. Millikan, who was able to determine h with high accuracy and showed that it agreed with the constant M. Planck used in his theory of blackbody radiation.

The description of the photoelectric effect in this article will be limited to fairly low photon energies. It should be pointed out that at higher photon energies (say above 10,000 eV) it is necessary to consider conservation of both energy and momentum in the electron-photon interaction. A photon interacting with an electron will thus result in a wavelength shift between the incident and scattered photon. This phenomenon (established in 1923) is called the *Compton effect* after its discoverer, A. H. Compton. The wavelength shift, $\Delta\lambda$, can be written

$$\Delta\lambda = \frac{h}{m_0 c} (1 - \cos\theta)$$

where m_0 is the rest mass of the electron, c is the velocity of light, and θ is the scattering angle of the photon. At still higher photon energies ($h\nu > 10$ MeV), with the wavelength comparable to the nuclear dimensions, so-called *photonuclear effects* occur, which in the simplest case can be described as photoemission of a neutron or proton from the nucleus. Finally, a short remark should be made on the concept of compound photoelectric effect or *Auger effect* (named after its discoverer in 1925, P. V. Auger). The Auger effect arises from the photoelectric excitation of an electron from inner core levels in atoms. The produced vacancy level can be filled by the deexcitation of a less tightly bound electron, and the released excess energy may in its turn result in the emission of electrons from other states. The latter electrons are called *Auger electrons*. Since the original vacancy can be filled in a number of different ways the Auger spectra can be quite complex and rich in structure.

As pointed out earlier, Einstein's law for the photoelectric effect, $E_k = h\nu - \Phi$, predicts that the electron emission is directly proportional to the photon flux, i.e., the number of photons absorbed in the material. The photoemission process can in a simplified fashion be described as a three-step event:

Step 1 (Excitation). The absorption of a photon by an atom and the emission of an electron; this process may occur anywhere within the penetration depth of the incident photons.

Step 2 (Transport). The photoexcited electron is moving through the material and may undergo various scattering processes (electron-electron scattering, electron-phonon scattering, etc.); thus only some of the photoexcited electrons will reach the surface.

Step 3 (Escape). If the electron reaching the surface has sufficient energy to overcome the surface barrier, or the work function Φ, it will escape from the surface. This is the photoemitted electron.

The penetration depth of the photons into the material is determined by the absorption coefficient, which depends on the wavelength and the absorbing media. In the visible and ultraviolet regions it is typically a few hundred angstroms, Å (1 Å = 10^{-10} m). However, the distance into the surface from which the photoemitted electrons are coming is almost exclusively determined by the scattering length of the electrons during the transport process (step 2). The scattering length is a strong function of the kinetic energy of the electron. It starts out being several hundred angstroms close to threshold, decreases to about 30–50 Å at about 5 eV kinetic energy, and continues through a minimum of about 5 Å at 20–200 eV and then increases to about 30 Å at 100 eV. The scattering length is thus U-shaped as a function of electron kinetic energy. It is further noteworthy that the scattering length is extremely small, only 1–2 atomic layers at its minimum, making the photoelectric effect in this spectral region very surface sensitive. The surface barrier the electron has to overcome (step 3) is typically a few electron volts, 2–5 eV, though certain materials can have a work function as low as 1–1.5 eV. Photoemission is a very fast process. The time between the absorption of the photon (step 1) and the emission of the electron (step 3) is less than 10^{-10} seconds.

It is now appropriate to comment on the temperature dependence of the photoemission process. As is apparent from the discussion above, Einstein's law for the photoelectric effect does not include the temperature as a parameter. However, it was recognized early that the energy of the emitted electrons is influenced by the temperature. The situation was resolved in 1931 by R. H. Fowler, who derived a fairly complex relationship between the photoelectric current and the temperature. According to this relationship the photoelectric current I is proportional to the square of the temperature, T^2, multiplied by a photon-energy-dependent function. Using standard notation this equation can be written

$$I = \alpha A T^2 f\left(\frac{h\nu - \Phi}{kT}\right)$$

where α and A are constants and k is Boltzmann's constant. Note that $h\nu - \Phi$ is the kinetic energy of the emitted electron. The numerical values of $f[(h\nu - \Phi)/kT]$ have been tabulated for a large number of materials. The relationship derived by Fowler can be used in a convenient way to determine the threshold frequency (work function, Φ). This is quite simply done by plotting the experimental results of photocurrent, temperature, and incident photon frequency [written as $(h\nu - \Phi)/kT$]. This is known as the *Fowler plot*. For metals, it has been found that the photoelectric current is to a good approximation proportional to $(h\nu - \Phi)^2$ for photon energies $h\nu$ within 1 eV of the threshold. As was mentioned earlier, the threshold energies range from about 2 to 5 eV (Cs: 2 eV; Pt: 5 eV). The threshold energy varies for different crystal surfaces of the same material and is also very sensitive to adsorbed gases or other contaminants on the metal surface. A typical value of the photoelectric yield is 10^{-3} electrons per incident photon at 1 eV above threshold, i.e., when $h\nu - \Phi = 1$ eV.

In the discussions up to this point, it has been assumed that the photoemission occurs from a metal surface, in which case the work function and the threshold energy are two identical concepts. It has also been pointed out that the work function depends on many factors. The purity and surface conditions of the metal are important. But, even for a perfectly pure surface, the work function depends on the crystallographic orientation. The surfaces with the orientation of the most closely packed atomic planes have the highest work functions. A polycrystalline surface thus has a work function which is an average of the most common crystallographic orientations.

The photoelectric behavior of a semiconductor (and an insulator) has additional complications. In a semiconductor, there is a so-called *forbidden energy gap* between the occupied valence band and the empty conduction band. As a result, the photoelectric threshold energy is larger than the work function. For a metal and a semiconductor with the same work function, the semiconductor thus has a larger photoelectric threshold.

The photoelectric effect has found a number of very important technological applications. Many photoemissive devices are based on the fundamental properties of the photoelectric effect, namely, that the response (i.e., emitted electrons) is directly proportional to the light intensity and that the response time (i.e., the time lag between the absorption of a photon and emission of the electron) is very short. One such photoemissive device (photodiode) can be used to detect single photons. Another class of very common devices, the so-called photomultipliers, are equipped with secondary emitters.

Much effort has gone into the development of photoemissive materials for various applications. Still, most photoemissive materials are limited to a narrow spectral region, and even at peak performance the quantum efficiency is typically only about 10%.

A common goal for all photoemissive materials is to combine low work functions with high photoelectric yields. The development of low-work-function surfaces has been centered around compound materials. Some of the most successful photoemissive materials are: silver-oxygen-cesium, cesium-antimony, bismuth-silver-oxygen-cesium, and multialkali antimonides, illustrating the complexity of these systems. A few more details about one particular system, cesium antimonide (Cs_3Sb) can emphasize this further. Cs_3Sb is a semiconductor with a band gap of about 1.5 eV. The photoelectric threshold is a little higher than this (see above). Electrons can be excited from the occupied valence band to the empty conduction only as long as the photon energy is larger than the band gap, i.e., 1.5 eV, which is slightly lower than the photoelectric threshold. Thus, even the least energetic electrons have energies only slightly less than what is required for escape. This results in a higher probability for photoemission than is the case for metals or for semiconductors with a photoelectric threshold more than twice the band gap. Cs_3Sb can give photoyields of more than 0.2 electrons per incident photon and has good response over most of the visible region. Other compounds can be tailored to have the optimal response in the red or ultraviolet regions of the spectral distribution.

The photoelectric effect has also found very widespread application in a technique for surface analysis and diagnostics, called *electron spectroscopy*. In a broad sense, electron spectroscopy has during the last decade provided important insights into the electronic structure of surfaces, interfaces, and bulk materials. The energy distributions of the photoemitted electrons contain information about the electron population of occupied states, the so-called *density-of-states*, i.e. the electronic structure can be determined. Since the scattering length of the photoexcited electrons is very short, electron spectroscopy is particularly suited for various surface studies: chemisorption, oxidation, corrosion, catalysis, etc. The possibility to vary and control different parameters involved in the photoemission process has recently found extended use. In addition to the photon energy, the polarization and direction of the incident light can be varied, and the energy, direction, and spin-polarization of the emitted electron can be monitored. The binding energy for deeper lying core levels can be determined with high accuracy and are a sensitive probe of the local chemical environment. The field of electron spectrosopy, using visible, ultraviolet, and X-ray radiation as the excitation source, has seen the development of a number of commerical instruments which are now used routinely in a large number of laboratories. At the same time, there are still vital research efforts in understanding the basic mechanisms behind the photoionization processes in solids, liquids, and gases.

INGOLF LINDAU

References

1. Hughes, A. L., and Dubridge, L. A., "Photoelectric Phenomena," New York, McGraw-Hill, 1932.
2. Weissler, G., "Photoionization in Gases and Photoelectric Emission from Solids," in "Handbuch der Physik," (S. Flügge, Ed.), Vol. XXI, pp. 304-382 Berlin, Springer-Verlag, 1956.
3. Brundle, C. R., and Baker, A. D., (Eds.), "Electron Spectroscopy Theory: Techniques and Applications," Vols. 1-4, New York, Academic Press, 1977-81.
4. Cardona, M., and Ley, L. (Eds.), "Photoemission in Solids," Vols. I-II, New York, Springer-Verlag, 1978-79.
5. Winick, H., and Doniach, S., (Eds.), "Synchrotron Radiation Research," New York, Plenum, 1980.
6. Sommers, A. H., "Photoemissive Materials: Preparation, Properties, and Uses," New York, John Wiley & Sons, 1980.

Cross-references: AUGER EFFECT, COMPTON EFFECT, ELECTRON, METALLURGY, PHOTOCONDUCTIVITY, PHOTON, PHOTOVOLTAIC EFFECT, SEMICONDUCTOR.

PHOTOGRAPHY

The history of photography spans more than a century, and embraces many techniques for producing photographic images. During the last half century, technological innovation has proceeded at an ever increasing pace. The most important advance of the past decade has been the computerized authomation of the basic photographic processes.

Every photographic technique includes an optical system, a photosensitive material, and a development process. Light from the subject passes through the optical system or camera and impinges on the photosensitive material to form a latent or invisible image. The latent image is transformed by chemical or physical means into a visible negative or positive image.

Most photographic processes use mixtures of the silver halides, i.e., silver bromide, iodide, and chloride, as the photosensitive materials. (The "speed" of a material depends on their relative proportions, with "fast" materials having a high percentage of silver bromide and little or no silver chloride.) These processes include conventional black-and-white and color photography, "instant" black-and-white and color photography, and lensless photography. A number of other processes are based on photochemical reactions of organic dyes, and of inorganic metal salts other than silver halides.

Still other processes employ light-induced changes in the physical properties of materials, such as photoconductors, magnetic discs, or thermoplastics.

In conventional silver halide photography, the camera consists of a light-proof box with a lens for admitting and focusing light, a diaphragm for controlling the size of the effective aperture, a shutter for regulating length of exposure, and the photosensitive film. There must also be film-holding and film-transporting devices and a viewfinder. A simple box camera (Fig. 1) has only one lens aperture and one shutter speed. The distance from lens to film is fixed, and the lens has a short focal length so that all objects within the range six feet to infinity are in focus on the film. Other types of cameras have systems for varying the lens-to-film distance, the effective aperture, and the shutter speed.

The two most common types of shutter (Fig. 2) are the leaf shutter consisting of three or five interleaved blades, each pivoted on the outer end, and the focal plane shutter consisting of a slit which traverses the film plane during exposure.

Holography, or lensless photography, uses an unconventional image-forming system (Fig. 3a). Monochromatic light from a laser is split into two beams, one of which enters the lensless camera after reflection from a plane mirror, while the other enters after reflection from a three-dimensional object. The two beams combine to form a Fresnel diffraction pattern of the image, which is recorded on film. This recorded pattern is called a *hologram*. To reconstruct the image, a beam of laser light is transmitted through the hologram to generate a second diffraction pattern, a component of which is an exact duplicate of the original wave reflected from the object. This wave appears to come from the object, i.e., it yields a three-dimensional virtual image which appears suspended in space behind the hologram (Fig. 3b). Another component of the diffraction pattern forms a three-dimensional real image which appears suspended in front of the hologram. Holography has become an important scientific tool, with many unique applications (see HOLOGRAPHY).

Another process makes it possible to mass-produce a three-dimensional image known as a *parallax panoramogram* on a two-dimensional card.[1] The picture is coated with a plastic layer composed of vertical rows of cylindrical lenses which give the illusion of depth.

Ordinary photographic film is a suspension of microscopic silver halide crystals in gelatin coated on a supporting layer. Gelatin is a stable porous medium which permits processing solutions to reach the dispersed crystals. Gelatin contains organic sulfide impurities which react with silver halides to form sensitivity centers. These are tiny silver sulfide centers of at least

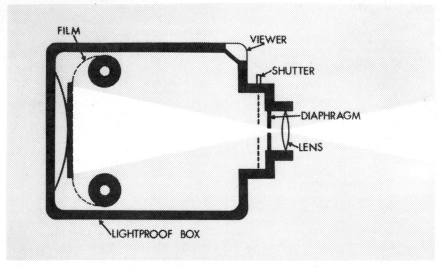

FIG. 1. Box camera. (Figures 1, 2 and 3 are redrawn from the "Encyclopedia Americana," International Edition, 1971.)

ten molecules each, which form on the surface of the crystals and catalyze the formation of the latent image by light.[2] When the film is exposed to light, the crystals absorb photons which generate mobile electrons. These electrons migrate to the sensitivity centers where they are trapped. They attract and neutralize silver ions, forming specks of metallic silver

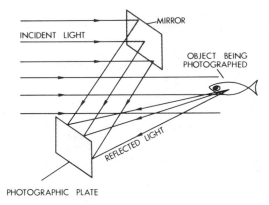

FIG. 3(a). Making a Hologram.

LEAF SHUTTER

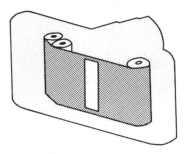

FOCAL PLANE SHUTTER

FIG. 2. Shutters.

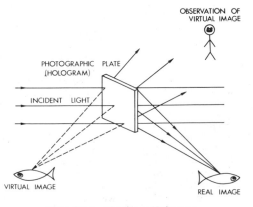

FIG. 3(b). Viewing a Hologram.

on the silver halide crystals at the sites of the sensitivity centers. These specks constitute the latent image. The smallest latent image speck that can make an exposed crystal develop during processing has from four to ten atoms of silver. Not all absorbed photons contribute to the latent image. Assuming that a typical speck of ten atoms formed by absorbing one hundred photons can cause development of a crystal with 10^{10} ion pairs, the amplification factor is about 10^8, which is higher than that of any other photographic material.

The developer reduces all the silver ions in an exposed crystal to silver atoms which deposit on the specks. The unexposed crystals are removed from the gelatin emulsion by a "fixing" solution to form the negative. A positive print is obtained by exposing an emulsion-coated paper to light through the negative, and developing it.

During development, the silver specks may grow until they merge to form irregular clusters. The image then is grainy; its contours are blurred, and its fine detail lost. This problem is amplified if enlarged prints are subsequently made. To solve the problem, the silver halide crystals are coated with developer inhibitor releasing (DIR) couplers by means of a process due to Kodak. As the metallic specks grow, the developer is oxidized in proportion to the silver reduced. The oxided developer reacts or "couples" with the DIR to release agents which limit the size of the specks, and thus prevent full development of the crystals. To compensate for this, a larger number of finer, more tightly packed crystals is now used in the emulsion. The image is well developed, but less grainy. Cubic crystals are preferred to hexagonal crystals, being smaller and more uniform in size and shape. Image sharpness is further enhanced by using very thin emulsions which minimize light scattering in the film.[3]

The Polaroid camera developed by E. H. Land in 1950 produces a positive black-and-white print ten seconds after exposure. In 1976, Kodak also entered the instant camera field. Instant photography requires rapid extraction of unexposed silver halide crystals from the exposed film, and rapid development of both the negative and positive images. Thus it was necessary to develop high speed film and very reactive agents. The negative emulsion layer is pressed against a nonsensitive layer of paper between metal rollers in the camera. Pods containing a viscous developer are ruptured as the two layers are pressed together, squeezing the developer uniformly between the layers. After forming the negative image, the developer then dissolves the unexposed crystals and transfers them to the paper layer which contains dry developer to produce the positive print.[4]

Silver halide emulsions are sensitive to ultraviolet and visible light, with maximum response in the blue at about 4200 Å. Their range may be extended to any wavelength up to 13,000 Å in the near infrared by sensitizing dyes which absorb light in the desired spectral regions and transfer the energy to the crystals. A mixture of two dyes often yields an increase in spectral sensitivity greater than the sum of increases due to each dye separately.[5] This is known as *supersensitization*. A common emulsion is that used in panchromatic film, which is sensitized to the visible spectrum up to 6200 Å.

In modern color photography, integral tripacks are used. An integral tripack is a film of three sensitive layers separated by gelatin. Each layer contains crystals sensitized to one of the primary colors for light mixing, i.e., red, green, or blue. After exposure, the three silver negatives in the nonseparable multilayered film are developed.

The most common method of forming the dye image is the dye coupling process. During development, the oxidized developer couples with a chemical to form a dye which produces a negative color image. A cyan dye in the red-sensitized layer absorbs red light, and reflects its complement, i.e., cyan, while the blue-sensitized emulsion acquires a yellow dye, and the green sensitized emulsion a magenta dye. The superposition of the three color negatives then transmits the colors complementary to those of the original image. To make color prints, the colors are reversed by exposing a three-layer emulsion coated on paper to white light through a color negative. To make positive transparencies, the color reversal is performed directly on the integral tripack film by changing the development procedure.

Infrared false color photography[6] is a process in which the film has three layers sensitized to green, red, and infrared (to 13,000 Å), instead of the normal blue, green, and red. A positive print or transparency then depicts the green, red, and infrared in the original scene as blue, green, and red, respectively. This process has important applications to mapping of earth resources from space, aerial reconnaissance, medicine, and many other disciplines.

Cameras come in many models employing various film sizes.[7] There are rangefinder cameras, single lens reflex (SLR), and double lens reflex cameras, each with different optics for viewing the subject scene. There are small cameras for hobbyists, large cameras for professionals, and special cameras for special applications.

Since 1970, advances in film chemistry, optics, electronics, and computer science have revolutionized camera design. Miniature cameras, simplified film loading, and very high-speed film are now available. In taking a picture, the effective aperture, shutter speed, and even the focal length may be adjusted in a fraction of a second by computer automation. Some cameras flash automatically under low ambient light conditions, while others flash during each exposure, thus softening harsh shadows in brightly lit scenes. Flash duration and intensity may be automatically controlled. The film may advance after each exposure, and rewind at the end. With a fully automatic camera, the user has only to compose the picture in the viewer, and press the shutter release button. Most such cameras are inexpensive, being priced between $30 and $150. A recent addition to the automatic camera field is the Canon line in the popular 35-mm film format, which features a full range of options in automation, and interchangeable accessories. These cameras were introduced during 1979–1982.

Techniques of automation vary. For example during auto-focus, the Kodak 980L instant camera and the Canon Sure Shot measure the object distance by infrared ranging, i.e., by sending an infrared pulse toward the object and timing its return. Fuji uses a pulse of white light for ranging, while Polaroid instant cameras now employ ultrasound. The Canon Super Sure Shot measures distance by triangulation, using ambient light reflected from the object.

Miniaturization has dictated the need for a short lens-to-film distance, i.e., a short focal length. This has the added advantage of increased depth of field, as illustrated in Fig. 4 for two focal lengths. Both lenses shown have the same *f*-number, which is defined as the ratio of focal length to aperture diameter. Point objects which are in focus on the film form point images; all other point objects form "circles of confusion," so that their images are blurred.

A human observer can tolerate some degree of blurring before perceiving the image as "out of focus." The distance range for which objects appear in focus is called the *depth of field*.

In Fig. 4, the diameter of the circle of confusion formed on the film by the 12.5-mm lens is about one-tenth that of the 38-mm lens. The 12.5-mm lens also gives a smaller negative image. In making positive prints, this image must be magnified three times as much as the other to yield a print of equal size; however, the diameter of its circle of confusion is still only one-third of that on the other print. Thus the lens of shorter focal length provides greater depth of field. Depth of field may also be increased by using smaller apertures, i.e., larger *f*-numbers, which yield smaller circles of confusion. Faster film must then be used to compensate for the reduced exposure to light.

In 1972, Dr. Land announced a system of instant color photography which features the Polaroid SX-70, a miniature SLR camera. All operations except composing and focusing the picture are controlled electronically. There are few moving parts, so that essentially "only the electrons move." Within a minute of exposure, a hard, dry, scratch-resistant print emerges from the camera. It has a three-dimensional quality, "true" colors, and a luminous appearance as if internally lighted. Timing and waste disposal are eliminated. Pictures may be taken at the rate of about one per second. Provision is made for time exposures, and for flash bulb lighting. Fresh batteries are included

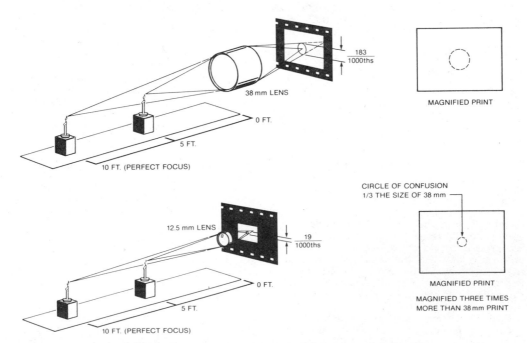

FIG. 4. Circle of confusion benefit with short focal length.

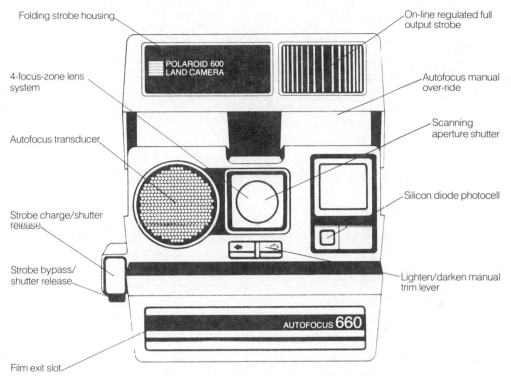

Folding strobe housing

On-line regulated full output strobe

4-focus-zone lens system

POLAROID 600
LAND CAMERA

Autofocus manual over-ride

Scanning aperture shutter

Autofocus transducer

Silicon diode photocell

Strobe charge/shutter release

Strobe bypass/ shutter release

Lighten/darken manual trim lever

AUTOFOCUS 660

Film exit slot

FIG. 5. Polaroid 660 camera features.

in each film pack. The system was developed by a large team of researchers over a 20-year period. In 1978, ultrasonic echo ranging and automatic focusing were added to the SX-70 system.

In 1981, Polaroid introduced the 600 line of folding instant cameras, featuring full automation, and one-button, no-decision operation. They flash during each exposure. Flash duration is adjusted to provide 75% ambient light and 25% strobe light for bright outdoor scenes, and 100% strobe light for dark scenes. All lenses are aspheric to minimize aberrations. Model 660 (Fig. 5) has ultrasonic echo ranging during autofocus, and automatic focal length selection, with one fixed lens and four supplementary lenses on a rotating disk. Exposure adjustments are based on detected ambient light levels. Model 640 has a fixed focal length. At low ambient light levels, its exposure adjustments are based on detected ambient infrared light levels, which give more accurate exposure information than low-level visible light.

A new line of instant cameras was marketed by Kodak in 1982. Three of the four models are folding cameras which flash during each exposure. Flash duration is adjusted so that strobe light mixes with ambient light. Model 980L features autofocus with infrared echo ranging, while Model 970L features a built-in auxiliary lens for object distances of 2–4 ft.

In 1982, Kodak introduced three miniature automatic "disc" cameras with a revolutionary film format, i.e., a rotating disc of film (Fig. 6). The disc has film for 15 exposures; each negative measures 8 × 10 mm, or one-tenth the area of a 35-mm negative. The disc has a rigid base which permits accurate film positioning during exposure and processing (Fig. 7). Thus images

FIG. 6. Kodak rotating film disc.

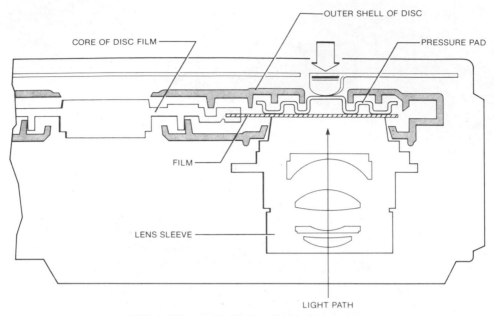

FIG. 7. Film positioning in a Kodak disc camera.

are sharper, and enlargements are of better quality. The disc is enclosed in a shell which is loaded into the camera in a fixed orientation (Fig. 8). The film is advanced automatically by a small motor, and is locked in place after a complete revolution. There are no rollers to damage the film. It is easily stored, and is processed outside the camera in the normal way. The disc has an iron oxide magnetic core which is used to store reorder information.

Each camera flashes automatically at low light levels. Each has a fixed focus, four-element lens (Fig. 7), one element of which is aspheric. The short (12.5 mm) focal length increases depth of

field. Lens surfaces are computer-designed, and checked for microroughness by laser interferometry. The quality of the lens approaches diffraction limits, and the large aperture ($f/2.8$) reduces diffraction effects.

The disc film format has made possible a pocket-sized automatic camera. Model 4000 (Fig. 9) has dimensions of $7.5 \times 12.5 \times 2$ cm (i.e., $3 \times 5 \times 0.8$ in.). The somewhat larger Models 6000 and 8000 can accommodate object distances down to 45 cm (i.e., 18 in.). Model 8000 also allows the user to hold the shutter button down and thus to expose about three frames per second in daylight; at low light

FIG. 8. Loading a Kodak disc camera.

FIG. 9. Kodak pocket-sized camera.

levels, the exposure rate is limited to one per second by the recharging time of the automatic flash.

Many practical innovations have been made for darkroom hobbyists. Development trays have now been replaced by drums or tanks which permit development in full daylight. Kits are available for developing color slides, prints, or negatives under noncritical environmental conditions. Liquid concentrates with almost unlimited shelf life may be used to prepare small quantities of development solutions by simple methods. Instant photography techniques have been transferred to the darkroom, so that one solution may be used in a compact table top processor to develop negatives, slides or prints in minutes. Commercial developing systems are adapted for different volumes of picture processing, up to 10,000 prints per hour. They are controlled by microcomputers.

Photographic processes using photochemical systems other than silver halides have lower sensitivity to light, but may be used to achieve special effects.[8]

A class of processes known as *electrophotography* uses materials which show an increase in electrical conductivity during light exposure. Electrophotography is widely used for office copying machines. In one type of electrophotography, charges are sprayed on a selenium or zinc oxide layer by a corona wire charged to five or ten thousand volts.[2] During exposure, the charge leaks off image-wise through the photoconductive layer. The remaining charge pattern is the latent image, which is developed by applying an oppositely charged black resin

powder which adheres to the charged areas. The image is fixed by slight heating to fuse the resin. This basic process has been modified by many recent innovations.

The Sony Corporation's expertise in magnetic tape technology has been applied in developing the Mavica camera, which records the latent image on a magnetic disc. In size and shape, the Mavica resembles a conventional 35-mm SLR camera. However, this new photographic process requires no film, and no chemical developers. Instead, the image is focused on a solid state imager called a *charge coupled device* or CCD, and converted into electrical signals which are recorded on the magnetic disc, called a Mavipak. The pictures may be viewed immediately on a home TV screen by means of a Mavipak viewer. They may also be transmitted over a telephone line at the rate of ten per second. Color prints may be made using the Mavigraph color video printer. The pictures may also be stored on video tape.

The new system, whose specifications are subject to change, will be marketed late in 1983. It is illustrated schematically in Fig. 10. It affords an image exposure rate of up to ten pictures per second, with a possible future rate of sixty per second. Through electronic means, the user can control the color tones, or produce composite pictures.

The magnetic disc is enclosed in a case measuring $6 \times 5.4 \times 0.3$ cm (i.e., $2\frac{3}{8} \times 2\frac{1}{8} \times \frac{1}{8}$ in.). It may be loaded, removed from the camera, and reinserted in daylight. It may be easily stored or mailed. The disc can store up to fifty color pictures; it has a frame counter, and a

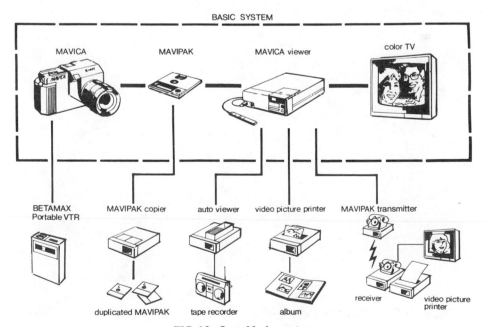

FIG. 10. Sony Mavica system.

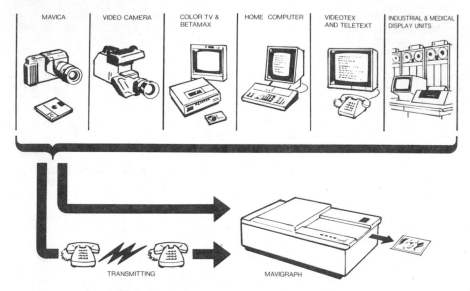

FIG. 11. Applications of Sony Mavigraph printer.

memory system which affords the user easy access to each picture. Although it is protected from accidental erasure, it may be erased and used again.

The color printer requires no chemical processing; instead, the dye is transferred to the paper image-wise by thermal evaporation. Color prints may be made from latent images stored on the magnetic disc or on videotape, and also from images transmitted by telephone. With the addition of a video camera, the printer can convert images on a TV, X-ray, or CAT scan-

ning screen into hard copies. It may also be used as a printer for an office computer terminal. These applications are shown schematically in Fig. 11.

An image forming system for infrared wavelengths beyond the limit for sensitized photographic emulsions is now used in remote sensing from aircraft and spacecraft. This system requires no image-forming optics. Instead, a detector sensitive to infrared or microwave radiation scans the field of view, and the received intensity from each point in the scene is elec-

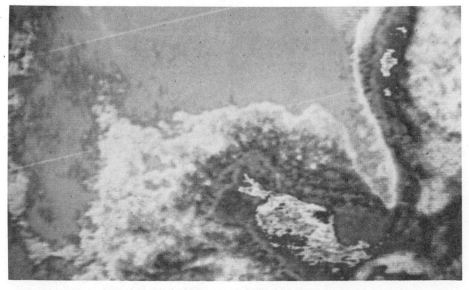

FIG. 12. Loop Current and Gulf Stream; infrared image.

tronically converted to a visible intensity at a corresponding point on a cathode ray screen. The visible image of the infrared or microwave scene is then recorded by conventional or instant photography. This system has many applications to medicine, geology, agriculture, forestry and oceanography.

Fig. 12 represents such an infrared photograph showing the warm Loop Current in the Gulf of Mexico, and the Gulf Stream. It has been processed so that bright areas in the center of these currents represent *very* warm water and dark areas represent *quite* warm water. Bright regions along the shore and outside of these currents represent very cold water. Thus bright areas depict both hot and cold water, while dark areas show the location of water at an intermediate temperature.[9]

The past decade has brought about these many innovations in photography. One can only speculate on the greater wonders which the future promises.

MIRIAM SIDRAN

References

1. Lipton, L., "Color 3-D Printing Process Permits Mass Press Run, Glassless Viewing," *Popular Photography*, 92 (May 1964).
2. Neblette, C. B., "Photography, Its Materials and Processes," Sixth Edition, New York, Van Nostrand Reinhold (1962).
3. Adams, A., "The Negative," Boston, New York Graphic Society, 1981.
4. Adams, A., "Polaroid Land Photography," Boston, New York Graphic Society, (1978).
5. Gilman, P. B., Jr., "A Review of the Electrochemical Boundaries for the Photochemistry of Spectrally Sensitized Silver Halide Emulsions," *Pure and Appl. Chem.*, 49, 357–377 (1977).
6. "Applied Infrared Photography," Eastman Kodak Publication M-28 (1972).
7. Blaker, A. A., "Handbook for Scientific Photography," San Francisco, W. H. Freeman and Company, 1977.
8. Hepher, M., "Not by Silver Alone," *Perspective*, 1, 28 (1959).
9. Sidran, M., and Hebard, F., "Charting the Loop Current by Satellite," Proceedings of the Eighth International Symposium on Remote Sensing of Environment, Willow Run Laboratories, Ann Arbor, Michigan, October 2–6, pp. 1121–1126 (1972).

Cross-references: COLOR, HOLOGRAPHY, INFRARED RADIATION, LENS, LIGHT, OPTICAL INSTRUMENTS, PHOTOCONDUCTIVITY, ULTRAVIOLET RADIATION.

PHOTOMETRY

Photometry is concerned with the measurement of light, or much more frequently, of the time rate of flow of light. Light is the aspect of radiant energy to which the human eye responds, or more precisely it is Q_v in the equation:

$$Q_v = K_m \int V(\lambda) Q_{e\lambda} d\lambda$$

in which K_m is the maximum luminous efficacy of radiant energy (683 lumens per watt), $V(\lambda)$ is the CIE (International Commission on Illumination) spectral luminous efficiency function[1] which is curve 4 of the figure shown in the article on COLOR in this Encyclopedia, and $Q_{e\lambda}$ is the spectral distribution of radiant energy. Reference 1 is highly recommended as a short discussion of the basic principles of photometry.

The quantities with which photometry is concerned are luminous intensity and flux, illuminance and LUMINANCE, and luminous reflectance and transmittance.

The devices that are used to measure these quantities are called photometers. They fall into two general categories, namely, visual and physical. The making of brightness matches is a requisite in the use of visual photometers, while in physical photometers, which are usually of the photoelectric or thermoelectric type, luminous energy incident on a receiver is transformed into electric energy and the latter is measured by means of sensitive electric meters or devices. In visual photometry, observers with so-called normal color vision are used, it being tacitly assumed that the average spectral response of the observers used to make the observations approximates the standard CIE luminous efficiency curve. In physical photometry, the receiver should be equipped with a filter, the spectral transmittance of which when multiplied by the relative spectral response of the receiver, wavelength by wavelength, closely approximates the CIE luminous efficiency curve. In all photometric measurements that involve differences between the spectral distribution of the standard source used to calibrate the photometer and that of the test source, the accuracy of the results obtained is dependent on the degree of approximation of the spectral response curve of the photometer to the CIE spectral luminous efficiency curve.

The unit of luminous intensity (often called candlepower) is the candela. From 1948 until 1979 the candela was defined as being the luminous intensity of one-sixtieth of one square centimeter of projected area of a blackbody at the freezing point of platinum.[2] By action of the 16th General Conference on Weights and Measures[3] in 1979 "the candela is the luminous intensity, in a given direction, of a source that emits monochromatic radiation of frequency 540×10^{12} hertz and that has a radiant intensity in that direction of 1/683 watt per steradian." Radiation at other frequencies is also measured in candelas in accordance with the standard luminous efficiency, $V(\lambda)$, curve that peaks at 540×10^{12} Hz (yellow green). National standardizing laboratories, like the National Bureau of Standards in the United States, calibrate incandescent-lamp reference standards of luminous intensity in conformity with the international standard. The calibration for luminous intensity of other incandescent lamps relative to these reference standards or to other standards of luminous intensity is usually done on a bar photometer, which consists of a horizontal bar equipped with movable carriages, on one of which the standards and lamps to be calibrated are mounted in turn, while the photometric measuring device is mounted on the other carriage. The calibration involves the use of the inverse-square law which states that the illuminance at a point on a surface varies directly with the luminous intensity I of the source and inversely as the square of the distance d between the source and the point on the surface. If the surface at the point is perpendicular to the direction of the incident light, the law may be expressed as follows: $E = I/d^2$. In the measurement of the intensity I_2 of a source relative to the intensity I_1 of a standard, the illumination at the photometer from the two sources is made equal by varying the distance d, so that $I_2 = I_1 d_2{}^2/d_1{}^2$ in which d_2 and d_1 are the distances for the unknown and standard, respectively. The important precautions that must be taken in calibrating and using standards of luminous intensity are (1) that the distance between the standard and photometric receiver is sufficiently large relative to the size of the source and of the receiver so that the source and receiver act approximately as points,[4] (2) that the standard is accurately oriented with respect to the receiver,[5] and (3) that by the use of baffles no flux is incident on the receiver other than that which comes directly from the standard.

Illuminance is incident luminous flux per unit area. In the measurement of illuminance, a photometer, in this application often called an illuminometer, is used. The illuminometer is calibrated at n points by the use of a standard of luminous intensity I placed at n distances d from the test plate of a visual photometer or the surface of the receiver, thus yielding n illuminances E computed from the inverse-square law $E = I/d^2$. It is important that the geometric characteristics of the receiver of the illuminometer be such as to enable it to evaluate, without bias, luminous flux that is incident on the receiver from all directions; special precautions are usually necessary if flux incident at large angles to the perpendiuclar to the receiver is to be properly evaluated.

The unit of luminous flux is the lumen which is defined as the luminous flux in a unit solid angle (steradian) from a point source having a directionally uniform intensity of one candela.

Standards of luminous flux are calibrated in terms of standards of luminous intensity by the use of a two-step procedure. In one step, the luminous intensity in a specified direction of the lamp to be calibrated is determined, and in the other step, by the use of a distribution photometer, the relative luminous intensity of this lamp in a multiplicity of directions is determined. To enable one to see how each of these intensities is effective in contributing luminous flux, one notes that each direction intersects a hypothetical sphere circumscribed about the lamp and that around the point of intersection on the sphere surface an area can logically be assigned which defines the solid angle in which the luminous intensity in that direction is effective in supplying luminous flux. These solid angles in steradians, by which the luminous intensity values must be multiplied to obtain the lumens incident on the respective areas, are called zonal constants. Such constants for various patterns of distribution measurements have been published and are thus readily available.[6,7] Reference standards of luminous flux are calibrated by this procedure, but in national standardizing laboratories reference standards of luminous flux are now generally calibrated by a similar procedure in which spectral radiant intensity is measured in step 1 and spectral irradiances are measured in step 2. The computed spectral radiant flux is then converted to luminous flux by the use of the $V(\lambda)$ function. Other standards are calibrated in terms of the reference standards or of other standards of luminous flux by the use of an integrating sphere, sometimes called an Ulbricht sphere. It is a spherical hollow enclosure with a uniform, diffusely reflecting inner surface. In such an enclosure, the illuminance due to reflected flux is the same at every point on the surface and is directly proportional to the flux emitted by a source in the sphere, independent of its angular distribution. The ratio of the fluxes emitted by two sources is thus the ratio of the illuminances at any point on the sphere surface that is shielded from receiving flux directly from the sources. The principal precautions in the use of spheres are (1) that the inner surface be a good diffuser, uniform in reflectance from point to point, and (2) that the spectral selectivity of the sphere throughout the visible region of the spectrum (0.4 to 0.7 μm) be compensated. This compensation is usually accomplished by the use of filters. It must be remembered that the effect of the spectral selectivity is greatly amplified because of the interreflections within the sphere so that the illuminance on the sphere wall by reflection only is $\rho/(1 - \rho)$ times the average illuminance by directly incident flux, where ρ is the reflectance of the sphere wall; thus if the reflectances at two wavelengths are, for example, 0.80 and

0.72, the spectral irradiance by reflection only at the wavelength of lower reflectance will be only 64 per cent of that at the wavelength of higher reflectance.

There are two related systems of units of luminance that have been and are still being used. In one of these systems the luminance is expressed in terms of luminous intensity per unit projected area or in terms of luminous flux per unit solid angle and unit projected area. The other system is that in which luminance is expressed in terms of the flux per unit area that would leave a perfect diffuser having the same luminance; the lambert and footlambert are units of this system and their magnitudes are such that one lumen per square centimeter and per square foot, respectively, would leave a perfect diffuser of unit luminance. Standards of luminance are usually combinations of luminous intensity standards and surfaces of known transmittance factor or reflectance factor (see below). The product of the illuminance of a surface by the appropriate one of these factors gives its luminance; for example, for each lumen per square centimeter incident perpendicularly on a freshly prepared surface of MgO, the luminance at an angle of $45°$ from the perpendicular will be approximately one lambert because the reflectance factor of MgO for this geometry is approximately 1.00. The reflectance factor of a more easily prepared $BaSO_4$ surface is approximately 1.01. In lieu of the MgO or $BaSO_4$ surface, use is more generally made of a highly diffusing glass or plastic plate of known transmittance factor for perpendicular incidence on one side of the plate and perpendicular viewing on the other side. The measurement of luminance relative to a luminance standard by visual photometry introduces no serious problems other than those inherent in photometry generally, because the receptor, in this case the eye, intercepts flux in a very small solid angle which can, for all practical purposes, be considered to be infinitesimally small. In physical photometry, however, because of the need to collect flux in an amount that will result in adequate sensitivity, the solid angle of reception may be relatively large and cannot in general be considered to be infinitesimally small, so that there is introduced into the measurements an inaccuracy whose magnitude is related to the magnitude of the variation of the flux per infinitesimal solid angle within the solid angle of reception.

In addition to the measurement of the dimensional quantities discussed above, photometry is concerned with the measurement of the dimensionless quantities transmittance, transmittance factor, reflectance, and reflectance factor. Transmittance and reflectance are the ratio of transmitted and reflected flux, respectively, to incident flux. For a nondiffusing specimen,

the measurement of transmittance or reflectance usually poses no great problem. For a diffusing specimen, the measurement of totally diffuse transmittance or reflectance (reception over a solid angle of 2π steradians) for specified modes of illumination of the specimen is usually made with instruments that incorporate integrating spheres;[8] for solid angles of reception smaller than 2π steradians, the measurement of the ratio of transmitted or reflected flux to incident flux yields what is designated as fractional transmittance or fractional reflectance, respectively, quantities which generally are not of as much interest as are the transmittance factor and the reflectance factor (formerly widely but inappropriately called directional transmittance and directional reflectance). These factors are defined[9] as the ratio of the flux transmitted (or reflected) in the solid angle of interest to that transmitted (or reflected) in the solid angle by the ideal perfect diffuse transmitter (or reflector) identically illuminated; the ideal perfect diffuse transmitter (or reflector) is one that transmits (or reflects) all of the luminous flux incident on it in accord with the Lambert cosine law, i.e., so that the flux per unit solid angle in any direction from it varies as the cosine of the angle between that direction and the perpendicular to the transmitter (or reflector). For a solid angle of reception of 2π steradians, the term transmittance (or reflectance) factor is synonymous with transmittance (or reflectance). For infinitesimal solid angles of reception and nonfluorescing specimens, the term transmittance (or reflectance) factor is synonymous with luminance factor which for any specimen is defined as the ratio of the luminance of the specimen to the luminance of a perfect diffuser identically illuminated.

The most commonly used photometric and related radiometric quantities, their defining equations and units, and symbols for them which are consistent with those agreed upon to date by the International Commission on Illumination, the International Electrotechnical Commission, the International Organization for Standardization, and the Commission for Symbols, Units, and Nomenclature of the International Union of Pure and Applied Physics are listed in Table 1.

L. E. BARBROW

References

1. "Principles of Light Measurements," Bureau Central de la CIE, 52, Boulevard Malesherbe, Paris, France (1970).
2. Wensel, H. T., Roeser, W. F., Barbrow, L. E., and Caldwell, F. R., "The Waidner-Burgess Standard of Light," J. Res. *Natl. Bur. Std.*, 6, 1103 (June 1931).

TABLE 1. PHOTOMETRIC AND RELATED RADIOMETRIC QUANTITIES, DEFINING EQUATIONS, UNITS AND SYMBOLS

Quantity[a]	Symbol[a]	Defining Equation	Unit	Symbol
Radiant energy	Q		Erg	
			Joule[e]	J
			Calorie	cal
			Kilowatt-hour	kWh
Radiant density	w	$w = dQ/dV$	Joule per cubic meter[e]	J/m^3
			Erg per cubic centimeter	erg/cm^3
Radiant flux	Φ	$\Phi = dQ/dt$	Erg per second	erg/s
			Watt[e]	W
Radiant flux density at a surface				
Radiant emittance[b] (Radiant exitance)	M	$M = d\Phi/dA$	Watt per square centimeter	W/cm^2
Irradiance	E	$E = d\Phi/dA$	Watt per square meter[e], etc.	W/m^2
Radiant intensity	I	$I = d\Phi/d\omega$ (ω = solid angle through which flux from point source is radiated)	Watt per steradian[e]	W/sr
Radiance	L	$L = d^2\Phi/d\omega(dA\cos\theta)$ $= dI/(dA\cos\theta)$ (θ = angle between line of sight and normal to surface considered)	Watt per steradian and square centimeter Watt per steradian and square meter[e]	$W \cdot sr^{-1}\,cm^{-2}$ $W \cdot sr^{-1}\,m^{-2}$
Absorptance	α	$\alpha = \Phi_a/\Phi_i$[d]	None (dimensionless)	–
Reflectance	ρ	$\rho = \Phi_r/\Phi_i$[d]	None (dimensionless)	–
Transmittance	τ	$\tau = \Phi_t/\Phi_i$[d]	None (dimensionless)	–

NOTE: The symbols for photometric quantities (see below) are the same as those for the corresponding radiometric quantities (see above). When it is necessary to differentiate them the subscripts v and e respectively should be used, e.g., Q_v and Q_e.

Quantity	Symbol	Defining Equation	Unit	Symbol
(quantity of light)			Lumen-second (talbot)[e]	$lm \cdot s$
Luminous density	w	$w = dQ/dV$	Lumen-second per cubic meter[e]	$lm \cdot s \cdot m^{-3}$
Luminous flux	Φ	$\Phi = dQ/dt$	Lumen[e]	lm
Luminous flux density at a surface				
Luminous emittance[c] (Luminous exitance)	M	$M = d\Phi/dA$	Lumen per square foot	lm/ft^2
Illumination (Illuminance)	E	$E = d\Phi/dA$	Footcandle (lumen per square foot)	fc
			Lux (lm/m^2)[e]	lx
			Phot (lm/cm^2)	ph
Luminous intensity (candlepower)	I	$I = d\Phi/d\omega$ (ω = solid angle through which flux from point source is radiated)	Candela (lumen per steradian)[e]	cd
Luminance (photometric brightness)	L	$L = d^2\Phi/d\omega(dA\cos\theta)$ $= dI/(dA\cos\theta)$ (θ = angle between line of sight and normal to surface considered)	Candela per unit area Stilb (cd/cm^2) Nit (cd^e/m^2) Footlambert $(cd/\pi ft^2)$ Lambert $(cd/\pi cm^2)$ Apostilb $(cd/\pi m^2)$	cd/in^2, etc. sb nt fL L asb
Luminous efficiency	V	$V = K/K_{max}$	None (dimensionless)	–
Luminous efficacy	K	$K = \Phi_v/\Phi_e$	Lumen per watt[e]	lm/W

[a]Quantities may be restricted to a narrow wavelength band by adding the word spectral and indicating the wavelength. The corresponding symbols are changed by adding a subscript λ, e.g., Q_λ, for a spectral concentration or a λ in parentheses, e.g., $K(\lambda)$, for a function of wavelength.
[b]Should be deprecated in favor of emitted radiant exitance.
[c]Should be deprecated in favor of emitted luminous exitance.
[d]Φ_i = incident flux; Φ_a = absorbed flux; Φ_r = reflected flux; Φ_t = transmitted flux.
[e]The International System (SI) unit.

3. The International System of Units (SI), Natl. Bur. Std. Spec. Publ. 330 (December 1981).
4. Walsh, J. W. T., "Photometry," 3rd Ed., New York, Dover Publications, 1958.
5. Barbrow, L. E., Wilson, S. W., "Vertical Distribution of Light from Gas-Filled Candlepower Standards," *Illum. Eng.*, 53, 645 (December 1958).
6. "IES Lighting Handbook," 1981 Reference Volume, New York, Illuminating Engineering Society, 1981.
7. Cotton, H., "Principles of Illumination," New York, John Wiley & Sons, 1961.
8. Goebel, D. G., Caldwell, B. P., and Hammond, H. K., III, "Use of an Auxiliary Sphere with a Spectroreflectometer to Obtain Absolute Reflectance," *J. Opt. Soc. Amer.* 56, 783 (1966).
9. "International Lighting Vocabulary," Bureau Central de la CIE, 52, Boulevard Malesherbes, Paris, France (1970).

Cross-references: COLOR; MEASUREMENTS, PRINCIPLES OF; OPTICS, GEOMETRICAL; REFLECTION.

PHOTON

The existence of the photon was first suggested by Planck's famous research, about 1900, into the distribution in frequency of blackbody radiation. He arrived at agreement with the experimental distribution only by making the drastic (for those times) assumption that the radiation exists in discrete amounts with energy $E = hf$. Here f is the frequency of the radiation and h is Planck's constant, $6.626(10)^{-27}$ erg sec. Confirmation of the existence of these quanta of electromagnetic energy was provided by Einstein's interpretation of the photoelectric effect (1905); he made it clear that electrons in a solid absorb light energy in the discrete amounts hf. The full realization that the photon is a particle with energy and momentum was provided by the Compton effect (1922), an aspect of the scattering of light by free electrons. Compton showed that features of the scattering are understood by balancing energy and momentum in the collision in the usual way, the light considered as a beam of photons each with energy hf and momentum hf/c.

The modern point of view is that, for every particle that exists, there is a corresponding field with wave properties. In the development of this point of view the particle aspects of electrons and nuclei were evident at the beginning and the field or wave aspects were found later (this was the development of quantum mechanics). In contrast, the wave aspects of the photon were understood first (this was the classical electromagnetic theory of Maxwell) and its particle aspects only discovered later.

From this modern point of view the photon is the particle corresponding to the electromagnetic field. It is a particle with zero rest mass and spin one. For a photon moving in a specific direction, the energy E and the momentum q of the particle are related to the frequency f and wavelength λ of the field by the Planck equation $E = hf$ and the de Broglie equation $q = h/\lambda$. As for all massless particles, the energy and momentum are related by $E = cq$ and the photon can only exist moving at light speed c. Another property of all massless particles is this: given the momentum, the particle can exist in just two states of spin orientation. The spin can be parallel or antiparallel to the momentum but no other directions are possible. The photon state with the spin and momentum parallel (antiparallel) is said to be right- (left-) handed and is a right- (left-) hand circularly polarized wave. In analogy with the neutrino, one can say that the state has positive (negative) helicity and can call the right-handed particle the antiphoton, the left-handed particle the photon. There is an operation, CP conjugation, that converts a photon state into an antiphoton state and vice versa. It is possible to superpose photon and antiphoton states in such a way that the superposition is unchanged by CP conjugation and so gives a type of photon that is its own antiparticle. The photons produced by transitions between states of definite parities in atoms or nuclei are their own antiparticles in this sense. As for all particles with integer spin, the photon follows Bose-Einstein statistics. This means that a large number of photons may be accumulated into a single state. Macroscopically observable electromagnetic waves, such as those resonating in a microwave cavity for example, are understood to be large numbers of photons all in the same state. The photon, among all the particles, is unique in having its states be macroscopically observable this way.

The electric and magnetic fields **E** and **B** describe the state of the photon and make up the wave function of the particle. Maxwell's equations give the time development of the fields and take the place, for the photon, that Schrödinger's equation takes for a material particle. Many of the remarks above follow as direct consequences of Maxwell's equations. In Gaussian units, where both **E** and **B** are measured in gauss or dynes per electrostatic unit of charge, the equations for the free fields are

$$\epsilon_{jkl}\partial E_l/\partial x_k + c^{-1}\partial B_j/\partial t = 0 \qquad (1)$$

$$\epsilon_{jkl}\partial B_l/\partial x_k - c^{-1}\partial E_j/\partial t = 0 \qquad (2)$$

$$\partial E_j/\partial x_j = \partial B_j/\partial x_j = 0 \qquad (3)$$

The particle aspect of the equations becomes evident when the equations are written in terms of the complex three-vector

$$\psi_j = E_j - iB_j \qquad (4)$$

in which case they become

$$\epsilon_{jkl}\partial\psi_l/\partial x_k + ic^{-1}\partial\psi_j/\partial t = 0 \qquad (5)$$

$$\partial\psi_j/\partial x_j = 0 \qquad (6)$$

Equation (6) is to be considered as an initial condition rather than as a equation of motion since it follows from Eq. (5) that

$$\partial(\partial\psi_j/\partial x_j)/\partial t = ic\epsilon_{jkl}\partial^2\psi_l/\partial x_j\partial x_k = 0$$

so if $\partial\psi_j/\partial x_j$ is zero at the start, it is zero forever. Equation (5) can be cast into Hamiltonian form. One writes the three components ψ_j as a column matrix ψ and introduces three, three-by-three, matrices by

$$(s_k)_{jl} = i\epsilon_{jkl} \qquad (7)$$

With this notation, Eq. (5) becomes

$$ic(s_k)_{jl}\partial\psi_l/\partial x_k = i\partial\psi_j/\partial t$$

or

$$H\psi = i\hbar\partial\psi/\partial t \qquad (8)$$

where

$$H = -cs \cdot p \qquad (9)$$

and p is $-i\hbar\nabla$. The Hamiltonian for the photon is thus $-cs \cdot p$. In detail, the matrices that occur here are

$$s_1 = \begin{pmatrix} 0 & 0 & 0 \\ 0 & 0 & -i \\ 0 & i & 0 \end{pmatrix}, \qquad s_2 = \begin{pmatrix} 0 & 0 & i \\ 0 & 0 & 0 \\ -i & 0 & 0 \end{pmatrix},$$

$$s_3 = \begin{pmatrix} 0 & -i & 0 \\ i & 0 & 0 \\ 0 & 0 & 0 \end{pmatrix} \qquad (10)$$

They are Hermitian and, as is easily verified, they fulfill the commutation rules

$$[s_i, s_j] = i\epsilon_{ijk}s_k$$

and so are a set of angular momentum matrices. Evidently each has eigenvalues $0, \pm 1$ so they are a representation of spin one.

Next consider the plane wave solutions. Let them be propagating in the 3-direction; so substitute

$$\psi = u \exp[i\hbar^{-1}(p_3 z - Wt)]$$

into Eq. (8). Here the same symbol p_3 is used for the eigenvalue as for the operator. The

system reduces to the matrix eigenvalue problem

$$-c\begin{pmatrix} 0 & -ip_3 & 0 \\ ip_3 & 0 & 0 \\ 0 & 0 & 0 \end{pmatrix}u = Wu$$

The eigenvalues are found to be $W = 0, \pm cp$, where p is $|p_3|$, and the corresponding eigenvectors are

$$u_0 = \begin{pmatrix} 0 \\ 0 \\ 1 \end{pmatrix}, u_\pm = \frac{1}{\sqrt{2}}\begin{pmatrix} \pm p_3/p \\ -i \\ 0 \end{pmatrix}$$

The $W = 0$ possibility does not satisfy the initial condition, Eq. (6), and so must be discarded. The solutions $u_\pm$ are valid for either sign of p_3; choose $p_3 = \pm p$ so both waves are propagating in the positive z direction. The two solutions of the problem are then

$$\psi_\pm = \frac{1}{\sqrt{2}}\begin{pmatrix} 1 \\ -i \\ 0 \end{pmatrix}\exp[\pm ip\hbar^{-1}(z-ct)] \qquad (11)$$

The subscript $+(-)$ denotes a particle (antiparticle) solution with positive (negative) frequency of $W/h = + cp/h$ $(- cp/h)$. Also $\psi_\pm$ are evidently eigenstates of the helicity operator $s \cdot p/p$ with eigenvalues ∓ 1. The electric and magnetic fields are the real and imaginary parts:

$$E_{\pm,x} = 2^{-1/2}\cos[p\hbar^{-1}(z-ct)] \qquad (12a)$$

$$E_{\pm,y} = \pm 2^{-1/2}\sin[p\hbar^{-1}(z-ct)] \qquad (12b)$$

$$B_{\pm,x} = \mp 2^{-1/2}\sin[p\hbar^{-1}(z-ct)] \qquad (12c)$$

$$B_{\pm,y} = 2^{-1/2}\cos[p\hbar^{-1}(z-ct)] \qquad (12d)$$

$$E_{\pm z} = B_{\pm z} = 0 \qquad (12e)$$

Here it is seen that the $-1(+1)$ helicity solution is left- (right-) hand circularly polarized with respect to the propagation direction.

The allowed states of the photon are eigenstates of the Hamiltonian H with eigenvalues $\pm cp$. Let $|H|$ be the operator which, applied to the same states, gives eigenvalue cp. The operators for the physical energy, momentum, and angular momentum of the photon are $|H|$, $(H/|H|)\mathbf{p}$, and $(H/|H|)(\mathbf{x} \times \mathbf{p} + \hbar\mathbf{s})$. One can understand these assignments for the energy and momentum by considering the plane wave states of Eq. (11). The states are eigenstates of the operators with energy eigenvalue cp and with momentum eigenvalue p in the positive z direction. As further justification for these operator assignments, the expectation values of the operators are directly related to the classical formulas for energy, momentum, and angular momemtum in the electromagnetic field:

$$(\psi, |H| \psi) = (8\pi)^{-1} \int d^3 x (E^2 + B^2)$$

$$(13a)$$

$$(\psi, \frac{H}{|H|} \mathbf{p}\, \psi) = (4\pi c)^{-1} \int d^3 x (\mathbf{E} \times \mathbf{B})$$

$$(13b)$$

$$(\psi, \frac{H}{|H|} (\mathbf{x} \times \mathbf{p} + \hbar \mathbf{s})\, \psi) =$$

$$(4\pi c)^{-1} \int d^3 x\, \mathbf{x} \times (\mathbf{E} \times \mathbf{B}) \quad (13c)$$

where the rule for taking the inner product is

$$(\psi_1, \psi_2) = \frac{1}{8\pi c} \int d^3 x\, \psi_1^\dagger \frac{1}{p} \psi_2 \quad (14)$$

These equalities apply for any solution ψ of Eqs. (6) and (8). The dagger denotes the Hermitian conjugate. The operation $(1/p)\psi$ in Eq. (14) is to be carried out by expanding ψ in the plane wave components like $\psi_\pm$ and replacing the operator $(1/p)$ by the number $(1/p)$ in each component. Proofs of Eqs. (13) will not be given here; they can be made by expressing each side of the equations in terms of the plane wave expansion coefficients. Accepting these operator assignments, one sees that the helicity operator $\mathbf{s} \cdot \mathbf{p}/p$ is the component of the spin of the photon $(H/|H|)\mathbf{s}$ in the direction of its momentum $(H/|H|)\mathbf{p}$.

The CP conjugation operation is related to the space reflection covariance of Maxwell's equations. Consider a primed and an unprimed coordinate system such that the coordinates of any point in space referred to the two axes are related by $\mathbf{x}' = -\mathbf{x}$. Suppose the electric field is axial and the magnetic field is polar so that the functions describing the fields are related by $\mathbf{E}'(\mathbf{x}', t) = \mathbf{E}(\mathbf{x}, t)$ and $\mathbf{B}'(\mathbf{x}', t) = -\mathbf{B}(\mathbf{x}, t)$. It is evident that Maxwell's equations have the same form in both coordinate systems and that the transformation rule for ψ is $\psi'(\mathbf{x}', t) = \psi^*(\mathbf{x}, t)$ where the asterisk denotes the complex conjugate. The fact that the equations have the same form in both systems implies further that if $\psi(\mathbf{x}, t)$ is any solution then $\psi'(\mathbf{x}, t)$, or equivalently $\psi^*(-\mathbf{x}, t)$, is also a solution. The operation that carries $\psi(\mathbf{x}, t)$ into $\psi'(\mathbf{x}, t)$ is called CP conjugation and one writes

$$\psi^{CP} = KP\psi \quad (15)$$

where K is the operation "take complex conjugate" and the operator P changes $\mathbf{x}$ into $-\mathbf{x}$. If ψ is a solution of Maxwell's equations so also

is ψ^{CP}. However KP anticommutes with $\mathbf{s} \cdot \mathbf{p}$ so if the solution ψ has $\mp$ helicity, then ψ^{CP} has $\pm$ helicity. The CP conjugation thus converts the particle into the antiparticle. The KP operator also anticommutes with the physical momentum operator $(H/|H|)\mathbf{p}$ so for a state $\psi\pm(\mathbf{q})$ with definite helicity $\mp$ and physical momentum $\mathbf{q}$ one has

$$KP\psi_\pm(\mathbf{q}) = \psi_\mp(-\mathbf{q}) \quad (16)$$

Instead of the two states ψ_+ and ψ_-, one may consider the superpositions

$$\psi_1(\mathbf{q}) = 2^{-1/2} [\psi_+(\mathbf{q}) + \psi_-(\mathbf{q})] \quad (17a)$$

$$\psi_2(\mathbf{q}) = 2^{-1/2} [\psi_+(\mathbf{q}) - \psi_-(\mathbf{q})] \quad (17b)$$

The reason for introducing them is the property

$$KP\psi_1(\mathbf{q}) = \psi_1(-\mathbf{q}) \quad (18a)$$

$$KP\psi_2(\mathbf{q}) = -\psi_2(-\mathbf{q}) \quad (18b)$$

Thus the KP operation applied to ψ_1 or ψ_2 reproduces the state, only traveling in the opposite direction and with a change of phase for ψ_2. The states ψ_1 and ψ_2 in this way are their own antiparticles. These self-antiparticle states are plane polarized in perpendicular directions. For the states with momenta in the positive z direction, as given by Eqs. (11) and (12), the fields are seen to be

$$E_{1x} = \cos [p\hbar^{-1}(z - ct)] \quad (19a)$$

$$B_{1y} = \cos [p\hbar^{-1}(z - ct)] \quad (19b)$$

$$E_{2y} = \sin [p\hbar^{-1}(z - ct)] \quad (19c)$$

$$B_{2x} = -\sin [p\hbar^{-1}(z - ct)] \quad (19d)$$

with all other components zero.

The final point to be demonstrated here is that only a self-antiparticle type of photon is emitted or absorbed when a system makes a transition between states of definite parity. Consider for simplicity a spinless charged particle described by a Schrödinger wave function $\psi_m(\mathbf{x}, t)$. [The subscripts m and γ are used for the material particle and the photon.] Suppose the particle is bound in some system and makes a transition from an initial state i to a final state f, both eigenstates of parity P, with emission or absorption of a photon. As is well known, the transition probability is determined by the interaction integral

$$I = -(e/Mc) \int d^3 x [\psi_{mf}^*(\mathbf{x}, t) \mathbf{p} \psi_{mi}(\mathbf{x}, t)]$$

$$\cdot \mathbf{A}(\mathbf{x}, t) \quad (20)$$

where e and M are the charge and mass of the particle and $\mathbf{A}$ is the vector potential of the photon in the Coulomb or radiation gauge,

$$\nabla \cdot \mathbf{A} = 0 \qquad (21)$$

Here and below, the integrals extend over all space. The fields are found from the potential by the relations

$$\mathbf{E} = -c^{-1}\partial\mathbf{A}/\partial t \qquad (22)$$

$$\mathbf{B} = \nabla \times \mathbf{A} \qquad (23)$$

To make the argument, one first expresses the interaction explicitly in terms of the fields. The potential is found from the fields by integrating this way:

$$\mathbf{A}(\mathbf{x}, t) = \frac{1}{4\pi} \nabla \times \int d^3y \, \frac{\mathbf{B}(\mathbf{y}, t)}{|\mathbf{x} - \mathbf{y}|} \qquad (24)$$

It is easily verified that this expression for $\mathbf{A}$ satisfies Eqs. (21), (22), and (23) by using Eqs. (1), (3), and the fact that $\nabla^2|\mathbf{x} - \mathbf{y}|^{-1} = -4\pi\delta(\mathbf{x} - \mathbf{y})$. In the verification it is assumed that fields of interest will be zero outside a finite region of space so that in making partial integrations there are no contributions from infinity. Then by using Eq. (24) and replacing $\mathbf{B}$ by $i(\psi_\gamma - \psi_\gamma{}^*)/2$, one can rewrite the interaction integral as

$$I = \frac{-ie}{8\pi Mc} \int d^3x \, [\psi_{mf}{}^*(\mathbf{x}, t)\mathbf{p}\psi_{mi}(\mathbf{x}, t)]$$

$$\cdot \nabla \times \int \frac{d^3y}{|\mathbf{x} - \mathbf{y}|} [\psi_\gamma(\mathbf{y}, t) - \psi_\gamma{}^*(\mathbf{y}, t)]$$

However, if i and f are eigenstates of parity, then, by changing integration variables from $\mathbf{x}$ and $\mathbf{y}$ to $-\mathbf{x}$ and $-\mathbf{y}$ in the $\psi_\gamma{}^*$ term, one sees that

$$I = \frac{-ie}{8\pi Mc} \int d^3x \, [\psi_{mf}{}^*(\mathbf{x}, t)\mathbf{p}\,\psi_{mi}(\mathbf{x}, t)$$

$$\cdot \nabla \times \int \frac{d^3y}{|\mathbf{x} - \mathbf{y}|} (1 \mp KP)\psi_\gamma(\mathbf{y}, t)$$

where the factor is $(1 - KP)$ if i and f have the same parity, $(1 + KP)$ if i and f have opposite parity. Since

$$KP(1 \mp KP)\psi_\gamma = \mp (1 \mp KP)\psi_\gamma$$

only a type of photon that is its own antiparticle can be involved in the transition in either case. As examples, the electric dipole radiation field has $KP = +1$ and the magnetic dipole field has $KP = -1$.

For many calculations, especially in atomic physics, the electromagnetic field is quantized in the Coulomb gauge. Still in Gaussian units, the vector potential operator for the free field is written as

$$\mathbf{A}(\mathbf{x}, t) = \frac{1}{\sqrt{V}} \sum_{\mathbf{k},\alpha} c \sqrt{\frac{2\pi\hbar}{\omega}} \, [a_{\mathbf{k},\alpha}\,\boldsymbol{\epsilon}^{(\alpha)}$$
$$\cdot e^{i(\mathbf{k}\cdot\mathbf{x}-\omega t)} + a_{\mathbf{k},\alpha}{}^\dagger \, \boldsymbol{\epsilon}^{(\alpha)} e^{-i(\mathbf{k}\cdot\mathbf{x}-\omega t)}].$$

$$(25)$$

Here $V = L^3$ is the normalization volume, $\omega = ck$, and the wave vector $\mathbf{k}$ is quantized such that

$$(k_x, k_y, k_z) = (2\pi/L)(n_x, n_y, n_z),$$

where the n_i are positive or negative integers. The $\boldsymbol{\epsilon}^{(\alpha)}$ are two real polarization vectors such that $\boldsymbol{\epsilon}^{(1)}$, $\boldsymbol{\epsilon}^{(2)}$, $\mathbf{k}/|\mathbf{k}|$ form a right-hand orthogonal triplet of unit vectors. The $a_{\mathbf{k},\alpha}$ and $a_{\mathbf{k},\alpha}{}^\dagger$ are boson destruction and creation operators with commutation rules

$$[a_{\mathbf{k},\alpha}, a_{\mathbf{k'},\alpha'}{}^\dagger] = \delta_{\mathbf{k}\mathbf{k'}}\,\delta_{\alpha\alpha'}$$

$$[a_{\mathbf{k},\alpha}, a_{\mathbf{k'},\alpha'}] = 0. \qquad (26)$$

To understand the connection between this quantization and the previous discussion of photons, one starts with Eq. (25) and finds an expression for the field operator $\psi = E - iB$ in the standard form $\sum (a_+ \psi_+ + a_-{}^\dagger \psi_-)$ with normalization $(\psi_\pm, \psi_\pm) = 1$. Equations (22) and (23) provide the electric and magnetic field operators. For simplicity consider just one of the terms in the sum on $\mathbf{k}$ and choose $\boldsymbol{\epsilon}^{(1)}$, $\boldsymbol{\epsilon}^{(2)}$, $\mathbf{k}$ in the x, y, z directions respectively. The result is, for the sum on the two values of α,

$$\psi = i\sqrt{\frac{8\pi\hbar\omega}{V}} \left[\frac{(a_{\mathbf{k},1} + ia_{\mathbf{k},2})}{\sqrt{2}} \frac{1}{\sqrt{2}} \begin{pmatrix} 1 \\ -i \\ 0 \end{pmatrix} \right.$$

$$\cdot e^{i(\mathbf{k}\cdot\mathbf{x}-\omega t)} + \frac{(-a_{\mathbf{k},1} + ia_{\mathbf{k},2})^\dagger}{\sqrt{2}} \frac{1}{\sqrt{2}}$$

$$\left. \begin{pmatrix} 1 \\ -i \\ 0 \end{pmatrix} e^{-i(\mathbf{k}\cdot\mathbf{x}-\omega t)} \right].$$

This is of the required form, as is seen by comparison with Eq. (11). The operators for destruction of photon or antiphoton with energy $\hbar\omega$, momentum $\hbar\mathbf{k}$, are identified as

$$a_{\mathbf{k},+} = (a_{\mathbf{k},1} + ia_{\mathbf{k},2})/\sqrt{2},$$

$$a_{\mathbf{k},-} = (-a_{\mathbf{k},1} + ia_{\mathbf{k},2})/\sqrt{2}. \qquad (27)$$

On the other hand, $a_{\mathbf{k},1}$ and $a_{\mathbf{k},2}$ are operators for destruction of self-antiparticle photons plane-polarized in the $\boldsymbol{\epsilon}^{(1)}$ and $\boldsymbol{\epsilon}^{(2)}$ directions. Many aspects of the interaction of photons with atoms may be understood on the basis of this quantization of the electromagnetic field, using the interaction as given in Eq. (20). For example, emission and absorption of light, Rayleigh and Thomson scattering of light by atomic electrons, the photoelectric effect, and the Raman effect may be treated this way. Limitations of this approach are that the electrons are considered nonrelativistic and that their spin is disregarded.

Relativistic calculations involving photons are almost always done with Feynman diagram techniques. For internal photon lines in the diagrams the propagator is

$$D_F(q^2)_{\mu\nu} = \frac{-g_{\mu\nu}}{q^2 + i\epsilon} . \qquad (28)$$

A simple example is the one-photon-exchange contribution to electron-proton scattering. The diagram is

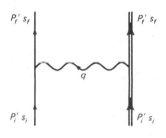

and the corresponding invariant matrix element is

$$M_{fi} = [\bar{u}(p_f s_f) \gamma_\mu u(p_i s_i)] \frac{e^2}{q^2 + i\epsilon}$$
$$\cdot [\bar{u}(P_f, S_f) \gamma^\mu u(P_i, S_i)] . \qquad (29)$$

For external photon lines the polarization four-vector ϵ_μ, defined to be $(0, \boldsymbol{\epsilon})$, is used. As an example, in the Compton effect the two diagrams

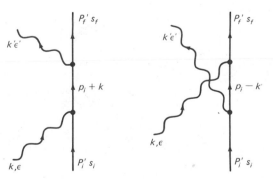

contribute and the invariant amplitude is

$$M_{fi} = \epsilon_{\mu'} \bar{u}(p_f, s_f) \left[(-ie\gamma^\mu) \frac{i}{\not{p}_i + \not{k} - m + i\epsilon} \right.$$
$$\cdot (-ie\gamma^\nu) + (-ie\gamma^\nu) \frac{i}{\not{p}_i - k' - m + i\epsilon}$$
$$\left. \cdot (-ie\gamma^\mu) \right] u(p_i, s_i) \, \epsilon_\nu. \qquad (30)$$

These two rules are the basis of a complete understanding of all photon interaction processes, including forces between charged particles, bremsstrahlung, pair production, the anomalous magnetic moment of the electron, and the Lamb shift.

R. H. GOOD, JR.

References

Heitler, W., "The Quantum Theory of Radiation," London, Oxford University Press, 1954. Heitler discusses the properties of photons from different points of view than used here and especially shows various techniques for the quantization of the electromagnetic field.

Good, R. H., Jr., *Am. J. Phys.*, **28**, 659 (1960). A nonmathematical pictorial discussion of the different types of photons is given.

Good, R. H., Jr., and Nelson, T. J., "Classical Theory of Electric and Magnetic Fields," New York, Academic Press, 1971, Chap. XI. A more complete treatment of the unquantized free field is given here.

Sakurai, J. J., "Advanced Quantum Mechanics," Reading, Mass., Addison-Wesley, 1967. The quantization of the electromagnetic field and the application to many processes is reviewed.

Bjorken, J. D., and Drell, S. D., "Relativistic Quantum Mechanics," New York, McGraw-Hill, 1964. This gives a straightforward development of Feynman diagram techniques with many applications.

Cross-references: BOSE-EINSTEIN STATISTICS AND BOSONS, ELECTROMAGNETIC THEORY, FEYNMAN DIAGRAMS, LIGHT, MATRICES.

PHOTOSYNTHESIS*

The light-driven synthetic reactions of plant chloroplasts, blue-green algae, and photosynthetic bacteria are briefly summarized and somewhat oversimplified by the van Niel equation:

$$2H_2A + CO_2 \xrightarrow{\text{light}} (CH_2O) + H_2O + 2A$$

where H_2A represents a hydrogen donor and (CH_2O) represents carbohydrate. For chloro-

*Support by the Solar Energy Research Institute (Golden, Colorado) is acknowledged.

plasts and blue-green algae H_2A is water and A is oxygen; for photosynthetic bacteria H_2A is often H_2S or other sulfur compounds, but is *never* water.

The primary action of light in photosynthetic systems is the excitation of pigment molecules. Singlet-state excitation is transferred from the light-harvesting pigment molecules to those few specialized molecules which participate in primary photochemistry at reaction centers. When excitation becomes localized in a reaction center, there is a charge separation in which an electron moves from an excited chlorophyll molecule to a primary acceptor molecule, thus leaving a chlorophyll cation radical which is easily detected by electron spin resonance techniques. All the known primary acceptors are either chlorophyll or pheophytin molecules. The array of light-harvesting pigments which deliver excitation energy to a single reaction center is called a photosynthetic unit. The size of photosynthetic units varies from roughly 50 to 2000 pigment molecules per reaction center in various systems; for plant chloroplasts 200 is a typical figure. The physical description of excitation transfer depends upon the structure of the photosynthetic unit in question. In most photosynthetic systems, there appears to be a heterogeneous arrangement of pigments with weak exciton coupling between closely spaced molecules (10 to 15 Å) and Förster resonance transfer between more widely separated molecules (e.g., 50 Å). In all cases energy transfer is a nonradiative process.

The pigments universally associated with photosynthesis are the chlorophylls. Chlorophyll *a* is essential for oxygen-evolving systems, and its close relatives, bacteriochlorophylls *a* and *b*, are characteristic of bacterial systems. Other types of chlorophyll as well as phycobilins and carotenoids function as accessory pigments which harvest light and transfer excitation to either chlorophyll *a* or bacteriochlorophylls *a* or *b*.

Reaction centers have been characterized most thoroughly in purple photosynthetic bacteria. The primary photochemical electron transfer shown in Fig. 1 can be driven by a single 880-nm photon (1.4 eV) with the storage of about 1.0 eV as electrochemical free energy in the primary products. The return of the electron to bacteriochlorophyll *a* via the intermediates (secondary acceptors Q_1 and Q_2, UQ·cytochrome *b*, and cytochrome *c*) in the cyclic transport chain is coupled to the formation of adenosine triphosphate (ATP) from adenosine diphosphate (ADP) and inorganic phosphate in which 0.5 eV per ATP is stored. Many bacteria also generate reducing power in the form of reduced nicotinamide adenine dinucleotide (NADH) from external hydrogen donors such as H_2S.

The formation of reduced organic compounds from CO_2 is driven by reducing power and ATP

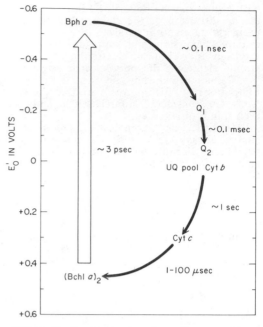

FIG. 1. Cyclic electron transfer pathway associated with the reaction center of a purple photosynthetic bacterium. Bchl = bacteriochlorophyll, Bph = bacterio-pheophytin, Q = quinone, UQ = ubiquinone, Cyt = cytochrome.

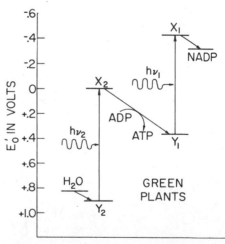

FIG. 2. Linear electron transfer pathway associated with the two reaction centers of a chloroplast or blue-green alga. X_1 = ferrodoxin, X_2 = plastoquinone, Y_1 = chlorophyll *a* (P700), Y_2 = chlorophyll *a* (P680). The primary electron acceptors (chlorophyll *a* and pheophytin *a* for systems 1 and 2 respectively) are not shown.

in a series of enzymic reactions. No light is required for these reactions.

In blue-green algae and plant chloroplasts two distinct photochemical reactions in physically separate reaction centers are required to generate reducing power (NADPH) from water. Photochemical system 2 generates an oxidant powerful enough to extract electrons from water and evolve oxygen. Photochemical system 1 generates the powerful reductant for CO_2 reduction. As shown in Fig. 2, the flow of electrons from the top of system 2 to the bottom of system 1 is coupled to the formation of ATP. In addition to the noncyclic electron transport pathway shown in Fig. 2, most oxygen-evolving systems also operate a cyclic pathway similar to that shown for bacteria in Fig. 1.

JOHN M. OLSON

References

1. Clayton, R. K., "Photosynthesis: Physical Mechanisms and Chemical Patterns," Cambridge, U.K., Cambridge Univ. Press, 1980 (281 pp.).
2. Wolstenholme, G., and Fitzsimons, D. W. (Eds.), "Chlorophyll Organization and Energy Transfer in Photosynthesis," Amsterdam, Excerpta Medica, 1979 (374 pp.).
3. Olson, J. M., and Hind, G. (Eds.), "Chlorophyll-Proteins, Reaction Centers, and Photosynthetic Membranes," Upton, New York, Brookhaven National Laboratory, 1977 (385 pp.).
4. Govindjee (Ed.), "Bioenergetics of Photosynthesis," New York, Academic Press, 1975 (698 pp.).

Cross-references: LIGHT,   PHOTOCHEMISTRY, PHOTOVOLTAIC   EFFECT,   SOLAR   ENERGY UTILIZATION.

PHOTOVOLTAIC EFFECT

The photovoltaic effect (PVE) is the generation of an emf as a result of the absorption of light. Three phenomena are involved in the effect. The first of these is photoionization, i.e., the generation of equal numbers of positive and negative charges by light absorption. The second is the migration of one or both of the photo-liberated charges to a region where separation of the positive and negative charges can occur. The third is the presence of a charge-separation mechanism. The photovoltaic effect can occur in gases, liquids and solids, but it has been studied most intensively in solids and, therefore, this discussion will be limited to solids, especially semiconductors.

Photoionization Semiconductors and insulators are characterized by a threshold energy for photoionization equal to the energy difference between the bottom of the conduction band and the top of the valence band, i.e., the forbidden energy gap E_g. Only those photons whose

energy exceeds E_g can cause photoionization. Values of E_g range from several electron volts, corresponding to an ultraviolet threshold, to small fractions of an electron volt corresponding to an infrared threshold.

The absorption of light follows Lambert's law, which states that

$$N(x) = N(0)\,e^{-\alpha x}$$

where $N(0)$ and $N(x)$ are the numbers of photons crossing unit area per second at a reference point (0) and at a distance x from the reference point along the direction of propagation of the light beam, and α is the absorption constant. Figure 1 shows how α changes with photon energy in a number of semiconductors used in photovoltaic cells. This dependence of α on photon energy $h\nu$ is intimately related to the dependence of the photovoltaic effect parameters on $h\nu$. It also has an important effect on the choice of semiconductors for solar cells. As is evident from Figure 1, the α vs $h\nu$ curves of semiconductors like silicon rise less rapidly toward values in the 10^4–10^5 cm^{-1} range than do the α vs $h\nu$ curves of semiconductors like GaAs and CdTe. Since the thickness of material x needed to absorb a photon is inversely proportional to the value of α, this means that a smaller thickness of a direct gap semiconductor is needed to absorb most of the photons with $h\nu > E_g$. Specifically, several microns of GaAs are sufficient, whereas a hundred microns of Si are needed for absorption of most of the solar photons absorbable in each of these two semiconductors. Consequently, direct gap semiconductors are preferred as the photovoltaically active semiconductors (PVAS) of thin film solar cells.

Migration of the Photo-Liberated Charges Because of Lambert's law, the photoliberated

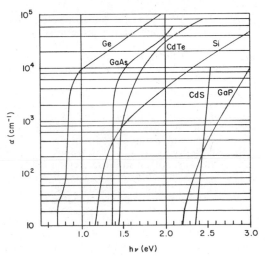

FIG. 1.

charges are distributed along the path of the light beam. Normally, they would move about at random until they recombine with carriers of opposite sign. This recombination process is characterized by a mean free lifetime τ of a pair which is of the order of microseconds in indirect gap semiconductors and of nanoseconds in direct gap semiconductors. As noted above, the PVE requires the presence of a charge separation region in the semiconductor. The magnitude of τ must be large enough to permit the carriers to move to this charge separation site before recombination occurs. Unproductive recombination can occur either in the bulk of the semiconductor or at its surface.

Charge-Separation Mechanism Charge separation requires a change in electrostatic potential between two regions of the solid, so that when a pair of opposite charges migrates to the region of the potential change, one of them can lower its potential by moving across this region. A large photovoltaic effect requires that the change in potential should be large and that it should occur over a distance which is short compared to the mean distance a free carrier can travel before recombination. These requirements imply the presence of a dipole layer which, in turn, implies an abrupt change in some property of the material. Such a dipole layer can occur either at the surface or in the interior of the material. A surface barrier usually involves a metal-simiconductor contact, a contact between the semiconductor and its oxide or, more generally, a contact with some other semiconductor. A barrier inside a material implies an abrupt change in the conductivity, which in the extreme case involves a change of conductivity type, as in a *p-n* junction. Figure 2 shows the diagrams of electron energy vs distance through the barrier for three such systems. The strongest photovoltaic effects are those which arise at *p-n* junctions, and therefore, the remainder of this article will be devoted to the *p-n* junction photovoltaic effect.

Photovoltaic Effect at p-n Junctions A *p-n* junction is formed by arranging the chemical impurity distribution in a single crystal of a semiconductor so that electric current is carried primarily by electrons on one side of the junction (the *n*-side) and primarily by holes on the other side (the *p*-side). The resulting electrostatic potential profile (Fig. 2) is such that excess holes can lower their energy by moving from the *n*- to the *p*-side while excess electrons lower their energy by moving in the opposite direction. Light absorption in either region leads to an increase in the concentrations of both holes and electrons, which are separated at the *p-n* junction. If a resistive load is connected between ohmic contacts to the *p*- and *n*-regions of a *p-n* junction illuminated by photons with energy $h\nu > E_g$, the current through the load I_L is related to the voltage across the load by the relation

$$I_L = I_s - I_0(e^{qV/kT} - 1)$$

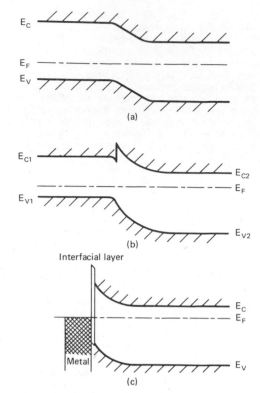

FIG. 2. Energy band diagrams of three charge collection barriers: (a) *p-n* homojunction; (b) *p-n* heterojunction; (c) MIS structure. The symbol E_c is the bottom of the conduction band; E_V the top of the valence band; and E_F is the Fermi energy.

where I_0 is the reverse saturation current of the *p-n* junction; q is the charge on the electron, k is Boltzmann's constant, T is the absolute temperature, and I_s is the photogenerated current which would flow if the load were a short circuit. The current I_s is a function of the absorption constant α, the spectral composition of the incident light, the pair lifetime and of the junction geometry. The $I_L - V$ characteristic of an illuminated PV cell is shown in Figure 3.

Applications The photovoltaic effect can be used to convert sunlight to electricity, and indeed *p-n* junction silicon solar cells have been the principal power sources on artificial earth satellites. Single crystal silicon solar cells having solar energy conversion efficiencies of about 12% are available commercially for prices in the $5–10 per peak watt range. A peak watt is an amount of solar cells which, illuminated by unconcentrated, clear day sunlight having an intensity of 1 kW/m^2, produces 1 watt of electrical power. In the case of 12% efficient solar cells, the area covered by 1 peak watt is 83 cm^2. In 1983, the world production of such silicon cells was about 8 peak megawatts. While this is a substantial increase over the 50 peak

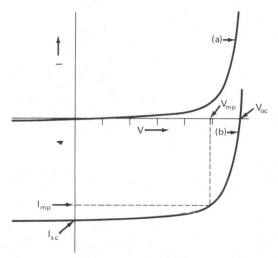

FIG. 3. Current-voltage characteristic of a solar cell in the dark (a) and under illumination (b). The symbol V_{oc} is open circuit voltage; I_{sc} is the short circuit current; V_{mp} is the voltage at maximum power and I_{mp} is the current at maximum power.

kilowatts produced in 1972, this amount of cells can produce an amount of electricity which is insignificant when compared to the world's production of electricity by fossil fuel and nuclear power plants.

Techno-economic studies of the feasibility of large-scale power generation by PV solar cells indicate that the price of solar cells must drop into the $1–2 per peak watt range before electricity produced by such cells can compete with electricity produced by conventional sources. Research programs aimed at achieving these price goals are underway, principally in the U.S., Japan and Western Europe. The research proceeds along two main paths. The first of these, which has had the greatest effort expended on it, aims at reducing the cost of single crystal silicon cells by introducing new methods for producing pure silicon and for producing sheets of silicon crystals suitable for solar cells. The second main path is aimed at alternative semiconductors for PVAS in solar cells. Among the most promising are gallium arsenide (GaAs), copper indium selenide (CuInSe$_2$), copper sulfide (Cu$_2$S) and amorphous silicon hydride. Cells having conversion efficiencies up to 22% have been fabricated from thin single-crystal GaAs, up to 10% from thin polycrystalline films of CuInSe$_2$ and Cu$_2$S, and up to 10% from thin films of amorphous silicon hydride. Research is also underway on tandem cell systems which require efficient cells made from PVAS having energy gaps between 1.0 and 2.0 eV. Calculations show that solar energy conversion efficiencies in excess of 50% are possible for such systems, whereas the upper limit efficiencies for solar cells utilizing a single PVAS are about 25%.

While first generation terrestrial solar cells will be based on single PVAS, future cells will probably be of the tandem cell type because there is a high premium for high efficiency in photovoltaic solar energy conversion systems.

It should also be pointed out that PV cells have applications other than as solar energy conversion devices. The photovoltaic effect can also be used to detect small intensities of light. Since the ionizing radiation need not be light, the effect can be used to detect x-rays, beta rays, protons, etc. Studies of the spectral response of the photovoltaic effect can yield information about basic material parameters such as α, τ, and E_g. It can be used to study surface effects, radiation damage, and other phenomena in semiconductors.

JOSEPH J. LOFERSKI

References

Wolf, M., "A New Look at Silicon Solar Cell Performance," *Energy Conversion*, **11**, 63 (1971).
van Aerschodt, A. E., et al., "The Photovoltaic Effect in the Cu-Cd-S System," *IEEE Trans. Electron Devices*, **ED-18**, 471 (1971).
Angrist, S. W., "Direct Energy Conversion," Boston, Allyn and Bacon, 1971.
Articles by Loferski, J. J.; Backus, C.; Wolf, M. in IEEE Spectrum, **17**, 1980. Conference Record of the Fourteenth (1980), Fifteenth (1981) and Sixteenth (1982) IEEE Photovoltaic Specialists Conferences, IEEE, New York. See chapter by J. J. Loferski, "Photovoltaic and Photoelectrochemical Solar Energy Conversion," Edited by F. Cardon et al, Plenum Press, 1981.

Cross-references: ENERGY LEVELS, PHOTOCONDUCTIVITY, PHOTOELECTRICITY, SEMICONDUCTORS, SOLAR ENERGY SOURCES, SOLAR ENERGY UTILIZATION.

PHYSICAL ACOUSTICS

Sound waves in the ideal sense are mechanical disturbances propagating in a continuous medium. The basic fact of interest from a physical standpoint is that these waves carry momentum and energy without a net transport of mass. In addition there are two other facts worth mentioning which are that sound waves propagate in almost all substances and that sound waves can be used to study the physical properties of, or to cause physical changes in the substances through which they propagate.

In practice, the approximation that natural materials are continuous fails when the wavelength is short enough to be comparable to interatomic or intermolecular distances—in a gas, the mean-free-path of a molecule; in a solid, the distance between atoms. The region of validity of the approximation that the waves are of small amplitude is more difficult to circumscribe since it depends on more than one property of the medium. In most cases, the approximation is valid when the disturbance in the velocity of the medium as the wave passes is small compared to the speed of sound in the medium. In air under normal conditions, this will occur when the pressure disturbance becomes an appreciable fraction of atmospheric pressure. In liquids and solids, however, pressure disturbances of hundreds or thousands of atmospheres propagate as small-amplitude waves. The low compressibility, or high "stiffness," of these media increase the speed of sound and at the same time require a larger pressure fluctuation to produce a given velocity fluctuation. If the medium is in a critical state, however, the definition of what constitutes a small disturbance can depend on other factors. For example, cavitation in a liquid occurs at particular sites that are called cavitation nuclei. While the reasons for the existence of these sites in a fluid are not completely understood, it is clear that they are regions that are subject to fracture at pressures far below those that would fracture the pure liquid.

The energy transported by the wave per unit area normal to the direction of propagation is called the intensity and can be measured in watts per square meter (the mixed unit, watt per square centimeter, is also in common use). A sound wave in air of intensity comparable to that of sunlight would be just beyond the pain threshold of the ear, and acoustic waves within the range of human tolerance do not produce a noticeable warming, as does sunlight. The heat generated when sound is absorbed can be measured, however, and in the design of absorbers for very high intensity sound the effect of the resulting heat production on the absorber may have to be taken into account. The principle of conservation of energy applied to energy transport by wave motion is a very useful tool, just as it is in mechanics.

The momentum transport per second per unit area is the intensity divided by the wave speed. Since this is also true for electromagnetic waves, it follows that a sound wave in air (speed 330 m/sec) of a given intensity transports about a million times more momentum than an electromagnetic wave of the same intensity. When the wave is absorbed or reflected, there is a radiation pressure on the absorber or reflector equal to the rate of change of momentum per unit area that is I/c on a perfect absorber and $2I/c$ on a perfect reflector, where I is the intensity and c the speed of sound. The measurement of this pressure is a fundamental method for measuring the amplitude of the wave.

The basic equations for a continuous medium are the continuity equation, which is an expression of the law of conservation of mass, and the Navier-Stokes equation, which relates the time rate of change of momentum in each volume element to the forces on the element. In differential form these equations are

$$\frac{\partial \rho}{\partial t} + \rho_0 \frac{\partial u}{\partial x} = 0 \quad \text{(continuity)}$$

$$\rho_0 \frac{\partial u}{\partial t} + \frac{\partial p}{\partial x} = 0 \quad \text{(conservation of momentum)}$$

The equations are presented here in one-dimensional form, and viscosity has been neglected. The inclusion of the two additional spatial dimensions and viscous effects is straightforward, and the complete equations can be found in most texts on acoustics. The dependent variables are p and ρ, the disturbances in pressure and density associated with the wave, and u, the disturbance in velocity. The parameter ρ_0 is the ambient density of the medium. In metric units (cgs) pressure is measured in dynes per square centimeter, density in grams per cubic centimeter, and velocity in centimeters per second. An additional equation is required that relates the scalar pressure to the other variables. While this third equation must be derived from the laws of thermodynamics, a relationship that is widely applicable is

$$\rho = \kappa \rho_0 p$$

where κ is the compressibility of the medium in units of inverse pressure. This relationship is useful as long as the wave does not change the value of κ appreciably. If there is no dissipation in the medium or if the dissipation is viscous, p and ρ will be in phase. In cases where dissipation exists due to irreversible heat flow, or the stress-strain relationship contains time derivatives, the equation can be retained by allowing κ to become complex. The three equations can be combined to give the wave equation

$$\left(\frac{\partial^2}{\partial x^2} - \frac{1}{c^2} \frac{\partial^2}{\partial t^2} \right) f = 0$$

where $c = 1\sqrt{\kappa \rho_0}$ is the speed of propagation of the wave, and f stands for any of the disturbances, p, ρ, or u.

The wave equation is a second-order partial differential equation and as such has a unique solution only if certain boundary conditions are applied. In the case of an unbounded medium, the boundary condition is that all sources of sound must radiate energy outward, and this is sufficient to produce a unique solution. In a bounded region, reflections must be taken into account. A sufficient boundary condition is that the impedance of the boundary, defined as the ratio of the pressure disturbance to the normal component of the velocity disturbance, be specified at each point on the boundary. In simple cases, the boundary conditions can be satisfied by a superposition of elementary solutions. The impedance tube is an example of a case where plane waves are used to measure the reflective and absorptive properties of a sample placed at one end of the tube in terms of the incident and reflected waves. The details of the distribution of sound in an auditorium, however, can be a complex problem. Another approach is to reformulate the equations as an integral equation containing the boundary conditions explicitly. This integral equation, a mathematical expression of Huygens' principle, is known in acoustics as Kirchhoff's formula. A third approach is to include the boundary conditions in the differential equations. An example of this procedure is the propagation of sound in a duct of varying cross section. If the change in width of the duct per wavelength is small compared to the wavelength of sound in the duct, the shape of the duct can be introduced into the continuity equation. This procedure leads to the "horn equation" which describes sound propagation in flared ducts. The subject of boundary conditions is further complicated by the fact that waves may propagate in the boundary itself. A boundary characterized by the fact that each point behaves independently is called a locally reacting boundary. If neighboring parts of a boundary interact to support wave motion, the boundary is non-locally interacting, and if the wave speed in such a boundary is higher than in the medium, energy can propagate along the boundary and leak back into the medium. In such a case, the wave may find a shorter time travel path through the boundary than through the medium itself.

Sound waves are generated in regions where time-varying forces act on the medium. While in some cases it may be convenient to think in terms of boundary conditions, this is not always convenient or even possible. A moving piston, such as the cone of a loudspeaker, can be thought of as a boundary condition in that the normal velocity, or pressure gradient, is specified. Sound generated by a region of turbulence, however, cannot be treated in this fashion because the source strength is distributed over the entire turbulent volume and not restricted to a surface. In fact, in the region of turbulence it is difficult to separate the motion of the medium into hydrodynamic and acoustical components. If the mechanism of sound generation is confined to a definite region, it is natural to describe the sound field at a distance in terms of spherical harmonics, and the amplitudes of these harmonics can be related to integrals of moments of the motion of the medium taken over the source region. The most simple natural source, the monopole, can be described as a periodic injection and removal of mass at some point in the medium. In a stationary medium, the wave fronts are concentric spheres and the pressure disturbance is proportional to the time rate of change of mass flow. The next higher spherical harmonic, the dipole, is related to the net fluctuating force on the medium. For example, if an airfoil is subject to fluctuating lift and drag forces, there are equal and opposite forces exerted on the medium. These forces give rise to dipole sound radiation, and the sound pressure is proportional to the time rate of change of the forces. The sound fields generated by turbulence and by earthquakes are related to the integral of shear motions over the source volume and have a quadrupole distribution.

If the ambient properties of the medium are everywhere the same, the medium is said to be homogeneous. A uniform translational motion of the medium, however, deserves special attention. While a suitable coordinate transformation can remove a translational motion without changing the basic equations, such a transformation does not remove relative motion between the medium and any sources or receivers, and such motion causes the medium to be anisotropic. Although a receiver is never completely passive, the usual approximation is to assume that it is. In this case, the receiver has no physical effect on the medium and simply registers what it sees. A source, on the other hand, is contributing energy and momentum to the medium, and the distribution of these quantities is affected by the motion. Surfaces of constant phase are carried along by the medium in the same way as ripples from a pebble dropped in moving water, and for a monopole source at rest in a moving medium, they form spheres centered a distance RU/c downstream from the source, where R is the radius of the sphere, U the speed of the medium, and c the speed of sound in the medium at rest. The surfaces of constant pressure for such a source, on the other hand, are ellipsoids of revolution centered at the source with the minor axis in the flow direction. It follows that the pressure is greater to one side than it is at an equal distance either upstream or downstream. In general, the problem is complicated by the fact that the energy flow vector is not perpendicular to a surface of constant phase.

The dispersion relationship for acoustic plane waves relates the phase velocity and attenuation (or growth) of these waves to physical properties of the medium. Classically, the important parameters for a fluid are viscosity and heat conduction. While a general solution including these parameters leads to a sixth-order polynomial, in cases where they are small the problem can be solved to a good approximation by three pairs of roots representing two plane waves, two viscous shear waves, and two thermal waves. The plane waves, traveling in opposite directions, are attenuated slightly by viscosity and heat conduction. The shear waves and thermal waves are not excited in the bulk of the medium. At boundaries, the shear and thermal waves will exist, however, to satisfy tangential velocity and thermal boundary conditions. While the viscous and thermal waves exist only in thin boundary layers (the real and imaginary parts of the propagation constant are equal), they may be responsible for a major part of the absorption. For example, in the case of plane waves propagating in the air in a pipe the ratio of boundary absorption co-efficient to bulk absorption coefficient is about $10^8/Lf^{3/2}$ where L is the ratio of the area of the pipe to its perimeter and f is the frequency. For very narrow pipes, the ratio is larger.

In general, the measured value of attenuation of plane waves exceeds the classical predictions based on heat conductivity and viscosity. The discrepancy varies not only from one substance to another but also may depend on the past history of the substance. The additional attenuation is caused by (and reflects) interactions between particles on a small scale compared to a wavelength—in other words, between individual atoms, molecules, or groups of molecules. The problems are as various as the chemical properties of matter.

In solids, it has been shown that dissipation usually is related to departures from ideal crystalline structure on a relatively large scale. Annealing greatly reduces the attenuation of sound. In fluids, it has been shown, however, that attenuation can be related to interatomic binding forces as well as the forces binding clusters of molecules. The vibrational relaxation of O_2 in air and of magnesium salts in water, for example, can account for the excess attenuation of sound waves in the atmosphere and in seawater.

L. WALLACE DEAN, III

References

Morse, P. M., "Vibration and Sound," Second edition, New York, McGraw-Hill Book Co., 1948.

Rayleigh, "Theory of Sound," New York, Dover, 1945.

Blokhintzev, D., "Acoustics of an Inhomogeneous, Moving Medium," NACA TM1399 (1956).

Lighthill, M. J., "On Sound Generated Aerodynamically," *Proc. Roy. Soc., A*, **211**, 564 (1952).

Mason, W. P., Ed., "Physical Acoustics," 7 Volumes, New York, Academic Press, 1964–1971.

Morse, P. M., and Ingard, K. U., "Theoretical Acoustics," New York, McGraw-Hill Book Co., 1968.

Cross-references: ACOUSTICS; ARCHITECTURAL ACOUSTICS; CAVITATION; ELECTROACOUSTICS; MUSICAL SOUND; RESONANCE; ULTRASONICS.

PHYSICAL CHEMISTRY

Physical chemistry is that branch of chemistry concerned with the rationalization of chemical properties in terms of the underlying physical processes. A distinction is sometimes made between physical chemistry and chemical physics, the former being the behavior of matter in bulk, and the latter the study of individual molecules. The physical chemist is closely involved in devising techniques for and then interpreting the results from spectroscopy, measuring and accounting for the rates of chemical reactions and their response to the conditions, and determining (both experimentally and theoretically) the structures and properties of molecules. A convenient division of the subject is into studies of (a) equilibria, (b) structure, and (c) change, and this article is arranged accordingly.

The subject may be traced back to the earliest quantitative measurements on bulk matter (by Boyle). It developed particularly during the nineteenth century, largely through studies of solutions and the rates of reactions, and the first Nobel Prize in chemistry was awarded to a physical chemist (the Latvian Ostwald). The subject took its modern form only when quantum mechanics became established, and modern chemistry draws heavily on the techniques of quantum mechanics, statistical mechanics, and techniques of investigation often drawn from physics, but normally heavily adapted in the process.

Equilibrium By "equilibrium" is meant the application of classical thermodynamics to chemical systems. Initially this took the form of assembling a large body of calorimetric information on enthalpies and entropies of compounds. The justification and motivation of the former lies in the First Law of thermodynamics (that energy is conserved), and its implication that the energy change associated with a chemical reaction is independent of the reaction pathway. This means that from tabulated values of the internal energies (more usually, the enthalpies) of the compounds on each side of a chemical reaction (e.g., of the enthalpies of methane, oxygen, water and carbon dioxide in the reaction $CH_4 + 2O_2 = CO_2 + 2H_2O$) it is possible to predict the quantity of energy required for or available from the reaction, even though the reaction itself has not been studied. This is of importance for the analysis of biological and industrial processes. Modern techniques of calorimetry now permit the measurement of the enthalpies of microgram quantities of substances, a sensitivity important for the investigation of biochemically important compounds. Calorimetry is also an important technique for the measurement of the entropies of substances, a procedure motivated by the Second Law of thermodynamics (that entropy increases during a spontaneous change). This time the heat capacity of the substance is measured down to as low a temperature as possible, and the data manipulated to give a value of the entropy of the substance relative to its value at the absolute zero of temperature. The Third Law of thermodynamics (that the entropies of perfect crystals are identical at $T = 0$) enters chemistry at this point, for not only does it enable the entropies of perfect substances at absolute zero to be chosen as zero, it also permits the establishment of tables of the Gibbs function (G, which is also called the "free energy").

The Gibbs function is the most important thermodynamic function in chemistry. Its formal definition is $G = H - TS$, where H and S are the enthalpy and the entropy of the system and T its thermodynamic temperature (i.e., its temperature in the Kelvin scale). Since H and S may be measured calorimetrically, G may be tabulated. (The fact that we do not know H on an absolute scale is circumvented by dealing with the enthalpies of compounds relative to their elements: this entails the technicality of dealing with enthalpies of formation, and hence Gibbs functions of formation; but this is a detail that need not affect the present discussion.) The reason why G is so important is that it carries information about the direction of natural change of a reaction (e.g., as to whether the reaction $CH_4 + 2O_2 \rightarrow CO_2 + 2H_2O$ or $CO_2 + 2H_2O \rightarrow CH_4 + 2O_2$ is the spontaneous direction) under conditions of constant temperature and pressure. The basis for its ability to act as a signpost for spontaneity is the Second Law. This law, however, expresses the criterion for spontaneous change in terms of the entropy of the universe (in practice, the reaction system and its surroundings). The Gibbs function transforms this criterion into a statement involving the properties of the system itself, and the change of entropy of the surroundings does not have to be taken into account explicitly. This means that the table of values of Gibbs functions can be used, almost at a glance, to decide the natural direction of a reaction. The usefulness of this is partly that the ability can be assessed of one reaction to drive another which is spontaneous in the "wrong" direction, and this is particularly fruitful in the analysis of processes within living cells.

Another role for G is that it can be used to predict the composition of systems that have come to equilibrium (and show no further tendency to react). The explicit criterion is so important that it deserves to be stated here. First, the composition of a reaction $aA + bB = cC + dD$ that has reached equilibrium is expressed in terms of the quotient $K = [C]^c[D]^d/[A]^a[B]^b$, where $[X]$ is the concentration of species X. K is called the *equilibrium constant* for the reaction. Next, one constructs from tables of Gibbs functions the quantity $\Delta G = (cG_C + dG_D) - (aG_A + bG_B)$, but one uses the values appropriate to a particular set of standard conditions, and hence obtains a standard value of ΔG, denoted $\Delta G^\ominus$. The relation between K and $\Delta G^\ominus$ at the temperature of interest is then given by the *reaction isotherm:*

$$K = \exp(-\Delta G^\ominus/RT), \quad \text{or} \quad \Delta G^\ominus = -RT \ln K,$$
$$(1)$$

where R is the gas constant. Therefore, not only can K be predicted from tables of data, since standard thermodynamic relations give the dependence of ΔG on the conditions, the dependence of K on the conditions of the reactions (e.g., its temperature) can be predicted. For example, the temperature dependence of K is given by the *van't Hoff equation:*

$$d \ln K/dT = \Delta H^\ominus/RT^2, \quad (2)$$

where $\Delta H^\ominus$ is the enthalpy change under the appropriate standard conditions. Hence the effect of temperature (and other factors) can be predicted.

One complication with the direct application of these expressions is that the statement of the equilibrium constant in terms of the concentrations is only an approximation. The assumption has been made that the Gibbs function of a substance in a mixture depends on its concentration as $G = G^\ominus + RT \ln[X]$. (The equation gives only the general form of the relation: we are avoiding precise details.) This relation is valid when the species do not interfere with each other; the more general relation is $G = G^\ominus + RT \ln a_X$ where a_X is the *activity* of species X. When K is expressed in terms of activities Eq. (1) is thermodynamically exact; but it becomes useful only when a_X and $[X]$ can be related. Much modern work has explained the connection between activities and concentrations, and has sought to find relations in terms of the intermolecular forces. The problem is particularly vexatious for reactions (and thermodynamic properties in general) of ions in solution, because the coulombic interactions between ions has a long range, and so deviations from ideality are large. An early attempt to relate a_X and $[X]$ was made by Debye and Hückel, and in the case of a salt of the form M^+A^- they deduced that, in water at room temperature,

$$a(M^+)a(A^-) = [M^+][A^-]e^{-1.02\sqrt{c}}$$
$$c = [M^+] = [A^-] \quad (3)$$

in the limit of low concentration. (In fact they deduced a more general limiting law; this is a special case.) Little further real progress of practical importance has been made in this direction, although techniques based on those developed for plasma physics have been explored.

The Gibbs function also plays a central role in the branch of equilibrium studies known as *electrochemistry*. A chemical reaction involving ions can be thought of as proceeding by the transfer of electrons from ions of one species to ions of another. When this takes place in bulk solution the direction of spatial motion is random. In an electrochemical cell the reacting species are separated, and the reaction proceeds when the donor ions deposit their electrons in an electrode, travel through an external circuit, and attach to the acceptor ions in its vicinity. The reaction proceeds as before, but the electronic motion has been channeled through a wire, and appears as an electric current. The Gibbs function determines the natural direction of change. In this application it is normally expressed in terms of the electromotive force (e.m.f.) of the cell, and that in turn is normally expressed in terms of the electronic potentials of the electrodes (and electrolyte) it is built from. Standard electrode potentials are tabulated, and so the standard e.m.f. of any cell can be predicted. Its e.m.f. under different conditions of concentrations of species in the electrolyte is expressed in terms of the *Nernst equation*, which for an electrode at which the reaction is $X^+ + e^- \rightleftharpoons X$ is

$$E = E^{\ominus} + (RT/F) \ln (a_{X^+}/a_X), \qquad (4)$$

F being Faraday's constant ($F = eL$), e the magnitude of the electron's charge, and L Avogadro's constant). The thermodynamic consequences of this equation are exact; the practical applications (the prediction and interpretation of cell e.m.f.'s) depend on the ability to relate a_X and [X].

Equilibrium e.m.f. studies are of wide importance. Note only do they permit the elucidation and prediction of the potential difference between cell electrodes, but they also allow the assessment of the relative reacting power of ions (which is important, among other things, for studies of corrosion), and the measurement of the thermodynamic properties of ions in solution. Many biological processes depend on the electrical potential difference across membranes (even thinking about electrochemistry involves electrochemistry). The development of fuel cells, an important feature of modern technology, depends at least in part on an assessment of their electrical potential in terms of equilibrium-processes; but, as in all working devices, their effectiveness can be improved only on the basis of studies of their nonequilibrium behavior, and to that we shall return.

Structure The quantitative discussion of molecular structure (and the properties of individual molecules) required two stepping stones for its emergence. The first was the development of the concept of the atom, and the realization that it was not indivisible but had structure. The second was the establishment of quantum mechanics, and its application to calculations of atomic and molecular structures and properties.

Atomic structure is a part of physical chemistry in so far as a knowledge of the distribution of electrons and their binding energies is important for assessing their role in chemical bonding. The theory of the chemical bond— its strength, why atoms form a characteristic number of bonds, and the shapes of the resulting molecules—is called *valence theory*. It emerged with the emphasis put on the role of

electron pairs by G. N. Lewis, and then took quantitative form with the theories of molecular structure known as *valence bond theory* and *molecular orbital theory*. Both theories are models of the structures of molecules; the latter has been much more richly developed, and we shall confine attention to it.

Molecular orbital theory adopts the view that the distributions of electrons in molecules can be regarded as described by wavefunctions (orbitals) spreading throughout the nuclear framework (which, as a first approximation, is assumed to be frozen into a static array). These orbitals are then obtained by solving the Schrödinger equation; but since there are numerous electrons present, approximation procedures are essential. A common procedure is to model the molecular orbital as a linear combination (sum) of atomic orbitals, the coefficients being determined by a self-consistency procedure. This gives rise to the Hartree-Fock (HF) self-consistent field (SCF) linear combination of atomic orbitals (LCAO) molecular orbitals (MO); or, in a word, HF-SCF-LCAO-MO's. While excellent progress has been made with absolute calculations of structures (the so-called *ab initio* approach, drawing on no empirical data), *semi-empirical techniques* (which use empirical data to estimate the magnitudes of some quantities, and neglect others) are widely and successfully used to account for molecular shapes, electron distributions, properties, and, increasingly, reactivities. This field is the branch of physical chemistry generally called *quantum chemistry*. Sometimes quantum chemistry itself is regarded as divided into computational quantum chemistry, which deals with the computational aspects of molecular structure, and noncomputational quantum chemistry, which is concerned with the mathematical as distinct from the numerical analysis of molecules.

Quantum mechanics is also inextricably interwoven with one of the most important of the modern branches of physical chemistry, the identification of molecules and the determination of their structures by spectroscopy. The physical chemist contributes to about half a dozen different classes of spectroscopic technique. *Microwave spectroscopy* is used to study the rotational transitions of molecules in the gas phase, and is used to obtain precise information about bond lengths and bond angles. *Infrared spectroscopy* is the study of vibrational transitions: its qualitative application is to the identification of species by interpreting their infrared fingerprint (their characteristic vibrational absorption spectrum), and its quantitative application is to the determination of the rigidities of bonds. *Optical and ultraviolet spectroscopy* studies electronic transitions, and is used to investigate excited electronic states of species. *Mass spectrometry* examines ions emerging from the fragmentation of molecules and, by monitoring their deflection in electric and magnetic fields, is used to determine their masses, and hence to infer the identity of the parent species. An offshoot of mass spectrometry is *photoionization spectroscopy*, in which the energies of electrons are measured after they have been ejected from a molecule by light of precisely controlled energy. Photoelectron spectra give information about the binding energies of electrons. The final principal class of spectroscopy consists of the *resonance techniques*. In a resonance spectroscopy experiment, a system of energy levels is tuned to the frequency of a transmitter and the power absorption increases dramatically when they match exactly (as in tuning a radio). Two cases involve using a magnetic field to tune the system. When the field acts on the magnetic moments of nuclei the technique is called *nuclear magnetic resonance* (NMR). When it is applied to electrons it is called *electron spin resonance* (ESR). The former has been richly developed to give a technique for identifying organic molecules (through the absorption patterns their protons give rise to), and its present development is to the determination of the structures of biological macromolecules, making use of the resonance of an isotope of carbon (^{13}C). Another resonance technique is *Mössbauer spectroscopy*, where the Doppler effect is used to tune nuclear energy levels to γ-ray emissions.

Apart from spectroscopy, the major techniques for studying molecular structures are those based on diffraction. Of these the most important is *X-ray diffraction*, which although it began with simple applications to simple inorganic crystals, is now (thanks to increases in computing power and improvements in techniques) capable of establishing the structures of biologically important molecules. Present day molecular biology could not have emerged without the existence of this technique. In one sense, diffraction techniques (which include electron diffraction and neutron diffraction) are complementary to spectroscopy, because while the later establishes energy levels (technically, eigenvalues) the former establishes electron distributions (technically eigenfunctions), and eigenvalues and eigenfunctions are the two aspects of the solution of the Schrödinger equation of a molecule.

Physical chemistry also provides techniques for the study of macromolecules, either synthetic (as in polymers) or natural (as in proteins). Size and shape are the principal targets of investigation, and both can be explored, at least in broad outline, by methods based on sedimentation rates and equilibrium, by light scattering, and by their effects on the viscosity of the solution. While these have been classical techniques of great usefulness, they are being displaced by techniques based on diffraction and magnetic resonance, for they can yield information of much better quality.

At this point in the survey we have two streams of information. One concerns the thermodynamic properties of bulk systems; the other concerns the characteristics of individual molecules. The fusion of the two streams is the domain of *statistical thermodynamics* (or *statistical mechanics*). Statistical mechanics deals with the energy levels of systems, and uses them to arrive at the thermodynamic properties. In effect, this means that it is possible to predict equilibrium constants of reactions (a typical "physical chemistry" property) on the basis of the spectroscopic data on the molecules involved (typical "chemical physics" properties). The key concept linking the two is the *partition function Q*, which can be expressed in terms of energy levels. For instance, if the system is a gas of N molecules with energy levels $E_1, E_2, \ldots$, the partition function at a temperature T is

$$Q = \left(\sum_n e^{-\beta E_n} \right)^N \Big/ N!, \qquad \beta = 1/kT \qquad (5)$$

where k is Boltzmann's constant. With Q known it follows that the internal energy and entropy are given by

Internal energy:

$$U(T) = U(0) - (\partial \ln Q / \partial \beta)_V \qquad (6)$$

Entropy:

$$S(T) = [U(T) - U(0)]/T + k \ln Q, \qquad (7)$$

and hence all the thermodynamic functions (e.g., G, and then the equilibrium constant) may be calculated. A particularly important field of statistical thermodynamics involves the evaluation of thermodynamic properties for systems consisting of interacting molecules. In particular, statistical techniques are used to study gas imperfections and, at higher densities, liquids. The interface between these two phases is of increasing interest in modern physical chemistry, where by "interface" we mean both the physical interface, the surface, between them and the thermodynamic interface, the characteristics of the transition between them (the *phase transition*).

Change Change takes a variety of forms. There is physical change, when a property is transported from one region to another. This is the domain of *transport properties*, the medium (for physical chemistry) being a gas (and giving rise, for instance, to viscosity and its dependence on intermolecular forces), an electrolyte solution (giving rise to ionic conductivity and being interpreted in terms of ionic mobility), and the transport of one phase past another (giving rise to the processes involved in the important preparative and analytical technique of *chromatography*). There is also the domain of chemical change, in particular, the measurement, and interpretation, and predictions of the rates of chemical reactions.

The rate of a chemical reaction is defined as the rate of change of the concentration of species (there are more sophisticated definitions). In general it varies during the course of the reaction, but extensive studies have shown that the rate is often proportional to the concentrations raised to various powers; for example,

$$d[A]/dt = -k_2[A][B]. \qquad (8)$$

The quantity k_2 is called the *rate coefficient*. The role of empirical chemical kinetics is to establish the *rate law* for the reaction, to measure the rate coefficient, and to account for the rate law in terms of the *mechanism* of the reaction, the sequence of individual steps through which the reaction proceeds. The temperature dependence of the rate is also an important quantity, and it normally found that it can be expressed as the *Arrhenius law*:

$$k_2 = Ae^{-E_a/kT}, \qquad (9)$$

where E_a is the *energy of activation*. Modern physical chemistry is increasingly concerned with observations on very fast reactions, and its employment of laser techniques has enabled the time scale of studies to be reduced to the order of 10^{-12} s.

The parameters k_2, E_a, and A are investigated theoretically in a variety of ways. One way is to use statistical thermodynamics. This leads to the *Eyring equation* for the rate coefficient:

$$k_2 = (kT/h)K' \qquad (10)$$

where K' is related to the partition functions of the reactants and a weakly bound complex of them, the *activated complex*, they are presumed to form on the route to products (K' is closely related to an equilibrium constant between the activated complex and the reactants). There are numerous difficulties associated with the application of this expression (as well as doubts about its basis), and attention now centers on treating a chemical reaction in the gas phase as an aspect of a scattering event, when groups and atoms are exchanged between molecules during an energetic collision. The theoretical studies of such events are called *molecular reaction dynamics*, and the experimental studies center on *molecular beams*, which are low-density streams of molecules shot either at a target or at another beam.

More specialized rate studies include the role of surfaces (in *catalysis*), and it is now possible to discern the structure of species attached (*adsorbed*) to surfaces, often in a state (e.g., with bonds broken) where they are ripe for reaction. Studies of the effect of using surfaces as targets for molecular beams are also being used to discover details about the relation of surface structure and surface activity. A special class of reactions at solid surfaces are those involved in electrochemistry: an electrode in contact with an electrolyte is the site of the processes involved in power generation, and in recent years attention has turned to this aspect of electrochemistry, the details of the relation between the structure of the electrode/electrolyte interface and its role in the chemical (and photochemical) generation of electricity under conditions far from equilibrium.

Conclusion Physical chemistry stands squarely between physics and chemistry. It imports techniques and fundamentals from physics, and applies them to problems of chemistry. Increasingly it is becoming involved in biology, for its special ability lies in the quantitative procedures and understanding it can bring to the discussion of chemical, industrial, and now biological processes.

P. W. ATKINS

References

Atkins, P. W., "Physical Chemistry," 2nd Ed., Oxford and San Francisco, Oxford Univ. Press and W. H. Freeman and Co., 1982.
Atkins, P. W., and Clugston, M. J., "Principles of Physical Chemistry," London, Pitman, 1982.
Berry, R. S., Rice, S. A., and Ross, J., "Physical Chemistry," New York, Wiley, 1980.
Eyring, H., Henderson, D., and Jost, W., "A Treatise in Advanced Physical Chemistry," New York, Academic Press, 1967 *et seq.*
Annual Reports, Royal Society of Chemistry, London, anually.
Ann. Review of Physical Chemistry, Annual Reviews Inc., Palo Alto, annually.

Cross-references: CHEMICAL KINETICS, CHEMICAL PHYSICS, CHEMISTRY, ELECTROCHEMISTRY, MAGNETIC RESONANCE, MICROWAVE SPECTROSCOPY, OSMOSIS, QUANTUM THEORY, STATISTICAL MECHANICS, THERMODYNAMICS.

PHYSICS

Physics can be defined as the branch of natural science that treats those phenomena of material objects included in the subjects of mechanics, properties of matter, heat, sound, light, electricity and magnetism, and molecular and atomic processes. It describes and correlates energy and radiation, and has been defined as the study of matter and energy. It usually omits animate matter and processes involving changes in chemical composition. While such definitions attempt to distinguish between physics and other sciences, it should be stressed that physics is not set off by itself, but is very closely related to other sciences, and there is a good deal of overlap with their spheres of interest. In some cases the differences are partially matters of emphasis. Chemistry stresses the regrouping of atoms to form new compounds, while physics selects a particular substance and studies its behavior.

Biology restricts its interest to living organisms, while physics largely omits them from its consideration. Geology confines its investigations to the earth itself and the inanimate material of which it is composed. Astronomy, on the other hand, finds its major interest in the vast expanses of space that run on and on away from the earth.

The extent to which these sciences overlap is shown by the special branches of science dealing with borderline areas. Astrophysics treats physical phenomena as they occur in regions beyond the immediate vicinity of the earth. It includes studies of the physical processes in stars and of the nature of the radiations they emit. Physical chemistry, as its name implies, covers phenomena common to both physics and chemistry. Energy relations in chemical reactions and the effects of changes in temperature or pressure on the reactions are samples of the topics included. Geophysics relates physics and geology, and involves studies of the earth's magnetism and the effects of high pressure in the earth's interior. Biophysics deals with the physical processes that are of specific interest to the study of living organisms. Each of these fields has grown to include a vast array of scientific knowledge, and each is discussed in a separate article in this book.

The field of engineering is closely related to physics and other sciences. In general, it can be said that as knowledge about some specific topic grows to the point where use can be made of it in every-day life or in industrial processes, the topic passes over to engineering, and engineers develop and improve things using the basic information. Scientists concern themselves with the basic phenomena themselves, while engineers direct their efforts toward the solution of problems dictated by specific applications.

Historical Background The word "physics" can be traced to the Greek word *physos,* meaning nature. All studies about nature were grouped together by the Greek philosophers, but in many cases theories suffered from a lack of experimental research, so that the name, "natural philosophy" was more descriptive of their efforts than a name such as "natural science" would have been.

A summary of the important features in the development of physics from its early beginnings to the present advanced state can be found in the article on HISTORY OF PHYSICS.

Major Divisions of Physics For purposes of study, physics can be reasonably well divided into major areas, which will be discussed briefly here and in greater detail in other articles in this book.

Mechanics is the science of the motions of material bodies, the forces which produce or change these motions, and the energy relations involved. Newton's three laws of motion and his law of gravitation form the foundation of this study. Work, momentum, vibration, wave motion, pressure, elasticity, and viscosity are other topics considered to be parts of mechanics.

Heat is the energy which an object possesses by virtue of the motions of the molecules of which it is composed plus the potential energy resulting from interatomic forces. The field is of great importance to other areas of learning, as well as to physics, and appears as a topic in chemistry and engineering. The term heat is also used in a different but related sense to indicate energy in the process of transfer between an object and its surroundings because a difference exists between their temperatures. Thermodynamics is the name given to the study of the relationships between heat and mechanics.

Acoustics is the science of sound. Sometimes wave motion is studied under acoustics rather than under mechanics because sound provides good illustrations of wave motion. Objects which are in vibrating motion in a medium can set that medium in motion, and the disturbance can travel through the medium in the form of a wave. Energy is transferred without the transfer of particles of the medium. Sound is the type of wave motion which has such a frequency that, if it reaches a human ear, it can cause a stimulus to reach the brain and hearing results. Thus, to the physicist, sound is the wave motion itself. Sound can also be defined as the sensation produced when sound waves reach the ear. Because both definitions are in common use, care must be taken to avoid confusion.

Light is the particular part of the gamut of electromagnetic waves to which the eye responds. Here, varying electric and magnetic fields travel through a vacuum or a transparent medium. The energy travels in even multiples of specific small amounts called "quanta." Related to light are infrared radiations, those frequencies which are just too low to cause the sensation of light in the normal human eye, and ultraviolet rays, similar radiations whose frequencies are just too high for the human eye. Optics, which is the study of light, often includes infrared and ultraviolet radiation.

Electricity and magnetism are so closely related that neither topic can be studied in any depth without involving the other. Electricity deals with the forces which charged particles exert upon each other, the effects of such forces, and the phenomena caused by the motions of charged particles. Magnetism was first known in the form of the peculiar attraction which a mineral called lodestone exerted because of the particular electron orientations in the material. It was found that magnetic effects could be caused or altered by electric currents. Further, it is possible to use magnetic materials to advantage in producing electric currents, which are streams of charged particles. An important branch of electricity is electronics, and this area has attracted the attention of more physicists in this country than any other part of physics. This subject deals with the flow of electric currents in vacuum tubes, semiconductors, and associated circuits, and with the use of these circuits to form devices of many kinds.

Solid-state physics is a well established area cutting across some of the other areas mentioned above. It includes those phenomena which are exhibited by materials in the solid state, with emphasis on electronic properties and their relations to the composition of crystalline substances and to energy levels in these materials. The designation *condensed matter physics* has found favor recently to indicate solid state physics plus liquid state physics.

Modern physics is a title under which physicists sometimes group many topics of comparatively recent development involving or relating to atoms and other submolecular particles or radiation resulting from atomic processes. Subjects which are considered here include radioactivity (both natural and artificially induced), x-rays, atomic and molecular structure, the quantum theory, wave mechanics and matrix mechanics, and nuclear fission and fusion.

Experimental physics and theoretical physics are two general categories into which the entire field of physics can be divided. The experimentalist attempts to discover new phenomena through the manipulation of apparatus and to make measurements of old or newly discovered properties. The theoretical physicist attempts to correlate measurements and to simplify theories. He tries to predict new phenomena and to relate one effect to another. Mathematical physics uses mathematics to describe physical phenomena, and extends or adapts mathematics to make it applicable to specific theories.

General discussions of major areas of physics can be found in the articles listed as cross-references at the end of this article.

Future Trends It is of course impossible to predict major discoveries in science or to foretell which areas of it will receive increased emphasis. Theoretical physicists can suggest what the experimentalist should look for, and the results of experiments can provide a confirmation of a theory or suggest where theories need revision, but unexpected discoveries continue. Present trends lead one to expect that certain areas will receive emphasis, but a new discovery or even gradual progress in an area can open up vast new vistas. Computers and computer techniques are speeding calculations of nuclear forces and some of the riddles of their existence are yielding to intensive research. Fundamental-particle physics is an area in which recent progress has been highly significant. We are just now approaching an adequate theory for these particles. (See GRAND UNIFIED THEORY). Which, if any, of the particles we know now are truly "fundamental"? And what does "charged" imply? What will even more powerful accelerators reveal? The emission of coherent radiation

and its behavior combine to form an area in which effort is being accelerated. The Fermi surfaces of metals, semiconductivity and superconductivity, and other cryogenic phenomena are fruitful, lively and interesting areas. Astrophysics is a very active area and many new experiments, discoveries, and theories may be expected from this field. Pulsars, quasars, neutron stars, and black holes are new concepts, and theories about them are still fairly nebulous, and mainly unproven. Undoubtedly more and more surprises are in store from exciting new research programs.

Physics Organizations Throughout the world there are many societies and other organizations that have the objectives of advancing physics and distributing knowledge of that science. In addition, other organizations deal with broader or related fields and include physics as part of their interests. Nine societies are members of the American Institute of Physics: The American Physical Society, Optical Society of Ameria, Acoustical Society of America, Society of Rheology, American Association of Physics Teachers, American Crystallographic Association, American Astronomical Society, American Association of Physicists in Medicine, and American Vacuum Society. Other groups are affiliated. The Institute was set up to assist the societies in their activities. Many other organizations are associated or affiliated with the Institute. It maintains an Information Center on International Physics Activities and publishes an "Information Booklet on Physics Organizations Abroad." Inquiries should be sent to the director of the Information Center, American Institute of Physics, 335 East 45th Street, New York, N.Y. 10017, U.S.A. In addition, scientific personnel attached to embassies in Washington, D.C. can provide information about activities in their respective countries.

ROBERT M. BESANÇON

Cross-references: ACOUSTICS, ASTROPHYSICS; ATOMIC PHYSICS; BIOPHYSICS; CHEMISTRY; ELECTRICITY; ELECTRONICS; ELEMENTARY PARTICLES; GEOPHYSICS; GRAND UNIFIED THEORIES; GRAVITATION; HEAT; LIGHT; MAGNETISM; MATHEMATICAL PHYSICS; MEASUREMENTS, PRINCIPLES OF; MECHANICS; MOLECULES AND MOLECULAR STRUCTURE; PHYSICAL CHEMISTRY; RADIOACTIVITY; RELATIVITY; THEORETICAL PHYSICS; WORK, POWER, AND ENERGY.

PLANETARY ATMOSPHERES

Introduction All of the planets and a few satellites have atmospheres of some kind. Their diversity reflects the wide range of planetary radii, densities and rotation rates as well as the range of planet-Sun distances. As a rule, planets farther from the Sun are cooler. Temperature determines atmospheric bulk, composition, state and structure. Whether a volatile substance exists as a frost on the planet's surface, an ocean, a cloud, or a gas in the atmosphere depends on the temperature. Hot, small planets with low surface gravities are least likely to possess atmospheres and any long-term atmosphere on such bodies will not include gases of low molecular weight. This explains why Mercury and the Moon have no more than transitory atmospheres and why most satellites and all asteroids do not have detectable atmospheres. Mercury is too hot owing to its proximity to the sun, the Moon is too warm and too small, and the asteroids and most satellites are too small to possess appreciable atmospheres. High temperatures in the inner solar system explain why the terrestrial planets lost their primeval atmospheres during their formation and cooling. Their present atmospheres have been outgassed from their interiors after these planets had cooled substantially. By contrast, the Jovian planets formed in the cool outer solar system and so retained essentially all of their primordial gases and grew to large sizes.

Mercury Mercury, the closest planet to the Sun, rotates once relative to the Sun for every two revolutions about its elliptical orbit. Its solar day is therefore twice the length of its year. Because of its long day and proximity to the Sun (46 to 69 million km), Mercury is too warm to retain more than a very tenuous atmosphere. While the surface temperature drops to 90–100 K during Mercury's long night (88 Earth days), the dayside temperature gets as high as 700 K. At such high temperatures, most atmospheric gases which were present when Mercury was young should have since escaped.

Mariner 10 has detected an extended, tenuous, and transient atmosphere for Mercury consisting of at least H and He.[1] It is supplied by the solar wind and escapes rapidly. The residence time for the He atoms trapped in the atmosphere is only 200 days. Surface densities of He derived from the Mariner 10 measurements are 1.3×10^3 atoms cm^{-3} on the dayside and 2.6×10^5 atoms cm^{-3} on the nightside. The corresponding values for the Moon are 2×10^3 and 4×10^4 atoms cm^{-3}, respectively, so Mercury and the Moon have exospheres of similar density. Hydrogen is present only in trace amounts (~ 8 cm^{-3}). Mariner 10 set stringent upper limits on Ar, Ne, Xe, C, and O. Upper limits on CO_2 and H_2O emissions imply that Mercury's outgassing rates for these gases are four orders of magnitude less than the Earth's rates. Mercury's total surface pressure is less than 2×10^{-10} mbar and atmospheric column density is less than 2×10^{13} atoms cm^{-2}.

Venus Of all the planets and satellites in the solar system with solid surfaces. Venus has the thickest atmosphere (see Fig. 1). The Venera and Pioneer Venus probes[2,3] agree that the surface pressure is in the neighborhood of 90 atm, depending on the topography. The turbopause, where the lighter gases begin to separate from the heavier ones is 140 km above the surface. The exobase, where atoms escape to space, is at an altitude of 160 km. The dominant gas of Venus' atmosphere is CO_2; its concentration is 96.4% in the lower atmosphere and 95.4% in the upper atmosphere. The next most abundant gas is N_2 with concentrations of 3.4 and 4.6%, respectively. Water vapor is the third most abundant gas but its concentration varies with altitude. Results of the Pioneer Venus sounder probe suggests a maximum concentration near 42 km altitude of 0.52%. The Venus atmosphere is therefore very dry. Small amounts of O_2 have been detected as well as trace amounts of NH_3, SO_2, CO, Ar, Ne, H, HCl, HF and H_2SO_4. The atmospheric gases appear to be fairly well mixed. As for Mercury, the H in Venus's upper atmosphere is deposited by the solar wind rather than by photodissociation of H_2O. Unlike the Earth, Venus has no ozone layer; the solar UV radiation penetrates deep into the atmosphere and is absorbed mostly in and above the cloud regions. Nonradiogenic Ar in Venus' atmosphere is 200–300 times more abundant than in the Earth's atmosphere. Its abundance on Earth, in turn, is greater than its abundance on Mars. This trend suggests that the primordial nebula out of which these planets formed was not as hot near its center as supposed. Sulfur dioxide (SO_2) is the dominant sulfur-bearing gas. Solutions of sulfuric acid in water form a prime component of the clouds.

The atmosphere of Venus is one of the most chemically complex and dynamically active atmospheres in the solar system. Photochemical processes dominate in the upper atmosphere (at and above the clouds) owing to the relatively deep penetration of solar UV radiation as compared with the Earth's atmosphere with its protective ozone layer. Because of the thickness of the clouds, however, only 2% of the incident sunlight reaches the surface. Photochemical processes are therefore not important below the clouds. On the other hand, a very effective greenhouse mechanism operates in Venus' atmosphere as a result of the thermal opacity of the CO_2, SO_2, and H_2O present. They drive the surface temperature to 750 K, hotter than the sunlit side of Mercury! The high temperatures of the lower atmosphere cause thermochemical processes to dominate there. End products of these reactions may be a very corrosive acid rain capable of chemical weathering of the surface topographic features.

The clouds of Venus are fundamentally different from the clouds of Earth. They are produced by chemical processes rather than by convective uplifting coupled with phase changes. They are related more to smog than to the

FIG. 1. Pioneer Venus image of Venus obtained on February 11, 1979 when the Pioneer spacecraft was at a point in its orbit directly above latutude 30°S. The cloud and haze patterns show the interaction between the high zonal and lower meridional winds.

stratus or cumulus clouds found on Earth. Much of the dynamics, weather, and climate of the Earth's atmosphere is driven by pressure gradients resulting from temperature variations associated with variations in the insolation. By contrast, the surface temperature of Venus varies diurnally by less than 1 K and latitudinally by only several percent. Also, the coriolis forces on Venus are very low owing to the long rotation period (243 days, retrograde). The solar day is 117 days long so changes in insolation are very gradual. Therefore, temperature effects play a relatively minor role compared to chemical processes in the formation of the Venus clouds or weather.[4]

Most of the atmosphere is in radiative equilibrium and therefore stable against convection. Convection is found in two altitude regimes but only one of these is within the clouds. There appear to be three cloud layers: the upper cloud region, at 58–68 km altitude; the middle cloud region, at 52–58 km; and the lower cloud region, at 48–52 km. Below these, from 31 to 48 km altitude, a thin haze region exists. The regions of convective instability are 20–29 km and 52–56 km above the surface. The first is below the haze layer and produces no clouds. The latter is in the middle cloud layer and may be only indirectly related to the cloud formation by recycling evaporated aerosols to higher altitudes where chemical processes and coalescence produce large cloud droplets which precipitate to lower, hotter altitudes where they evaporate.

Cloud particle sizes appear to be trimodal in the middle and lower clouds and bimodal in the upper cloud. The biggest particles or droplets are found in the lower layers and have diameters of 7–8 μ. All layers have aerosols of 1.5–5 μ and up to 1.5 μ in separate modal distributions. The lower haze region has the smallest aerosols. The clouds consist of sulfur, various aerosols, H_2SO_4 solutions, and impurities.

Other features of the Venus clouds are their

persistence and global uniformity. These reflect their chemical origin and the global near-uniformity of temperature at a given pressure level deep in the atmosphere. Venus' surface is never visible. There is probably little or no surface precipitation; the chemical products of the clouds apparently are not transported to the surface.

Lightning and thunder have been observed in Venus' atmosphere but these probably originate from processes other than thunderstorms since atmospheric processes which produce strong updrafts are unlikely on Venus. Again, chemical processes are probably the cause.

Above 90 km, Venus' winds are predominantly easterly and of high speed, 200 m s^{-1}. They flow around the planet in about four days. Within the clouds, where most of the solar energy is deposited, the circulation is meridional, as in Hadley cells, with speeds of about 25 m s^{-1}. Near the surface, the winds are calm. Near the top of the clouds, the zonal and meridional components of the wind cause these clouds to spiral around the planet towards the poles. The high speed zonal winds transport products of photodissociation of nitrogen and oxygen compounds to the night side of the planet where they recombine and cause the weak UV night-glow which is observed there.[4]

Mars The best information on the atmosphere of Mars has been obtained by Viking 1 and 2 orbiters and landers.[4,5] They measured an atmospheric composition of 95.3% CO_2, 2.7% N_2, and 1.6% ^{40}Ar. In addition, small quantities of O_2, CO, and H_2O exist in the Martian atmosphere. Trace gases include Ne, Kr, Xe, and O_3. The proportions of Ne, Kr, and Xe are the same as in the Earth's atmosphere. The heavier isotopes of Xe are probably absorbed onto surface material. Greater amounts of gas than are presently observed probably outgassed from the planet. Present escape rates of H to space from photodissociation of H_2O imply that as much as 10^4 g cm^{-2} of H_2O may have outgassed from the planet. Like the Earth, Mars has an upper layer of O_3 in its atmosphere. This results from the photodissociation of CO_2 and reaction of O atoms.

In contrast to the Earth's atmosphere, the atmosphere of Mars has very little solar heating of the stratosphere or thermosphere. Near the surface, the temperature ranges from 133 to 250 K over the globe. At the lander sites, the vertical temperature profiles are subadiabatic (fall off more slowly with altitude than the adiabatic lapse rate of 4.5 K/km) and are warmer than the CO_2 condensation temperature. Absorption of solar radiation by dust at low altitudes may help to stabilize the subadiabatic lapse rate. Over the polar caps, the surface temperature is essentially the temperature of CO_2 ice in the polar caps, except in northern summer when the residual north polar cap is mostly H_2O ice. This implies that the bulk CO_2 content of

Mars' atmosphere is controlled by the temperature of the polar caps. This is supported by the seasonal variation in the daily average surface pressure at the Viking lander sites. Lander 1, located at latitude 22.5° N, observed the pressure to vary from a minimum of 6.7 mbar just before Mars' autumnal equinox to a maximum of 9.0 mbar at the winter solstice, which is close to perihelion. Lander 2, located at latitude 48° N, observed a parallel pressure variation of 7.4 mbar and 10.0 mbar, respectively. The terrain height difference explains the pressure differences. The minimum corresponds to the time of maximum CO_2 accumulation in the southern polar cap at the end of its winter. The seasonal variation in the pressure is asymmetrical because of Mars' large orbital eccentricity (0.093). It causes Mars to receive 40% more sunlight at perihelion than at aphelion. A long, cold southern winter and a short, warm northern winter result. Consequently, the southern polar cap grows to a greater size than the northern one does and the atmosphere actually has its greatest bulk during the northern winter! The diurnal pressure variation at the lander sites was typically 1–2% but doubled during the planetwide dust storms which reached a vertical optical depth of 3 and greater. These reflected changes in the amplitude of the thermal tides are due to heating by the dust and due to the winds generated.

The surface winds are diurnally variable and typically have small daily mean values. They vary from about 2 m s^{-1} at night to typically 7 m s^{-1} during the day. Dust storms imply wind speeds in excess of 50 m s^{-1} in order to raise the dust off of the surface. Wind speeds in excess of 75 m s^{-1} are required to form the sand dunes observed at high latitudes. Summer cloud motions at 15–25 km altitude reveal westward winds of 30–55 m s^{-1} in the tropics. An intense eastward jet at 20 km altitude and 48°S reaching 120 m s^{-1} was inferred from Mariner 9 orbiter infrared measurements. Meridional winds also exist. Indirect evidence supports an inferred circulation analogous to that in the Earth's tropics. During the solstice, a low flow towards the summer pole may occur, rising in the subtropics and returning to descend in winter midlatitudes.

The moisture content varies seasonally and latitudinally; it is maximum at high northern latitudes during its early summer. There appears to be a supply of H_2O somewhere in the northern hemisphere, perhaps in a permafrost reservoir. Although the south polar cap is essentially all CO_2 ice, the residual north polar cap is largely H_2O ice. Water clouds, fogs, and frost have been observed along with CO_2 frosts. Clouds of H_2O and CO_2 form hoods over the polar caps in winter. Clouds sometimes form in the lee of mountains or volcanoes. Fogs and frost are transitory; they disappear soon after sunrise. Fogs form occasionally after sunrise as a result of H_2O being heated out of the soil and then condensing when it comes in contact with the cool air above.

The dust storms originate in the southern hemisphere during its summer where the heating rates are especially high. They may coalesce to become global storms, spreading to the northern hemisphere. They aid in the precipitation of H_2O over the polar caps by providing a nucleus for condensation. Dust storms are responsible for the dirty layers in these caps.

Large cyclonic storms have been observed near the edge of the north polar cap during early summer. These clouds were composed of H_2O ice and extended up to 6–7 km.

The presence of channels in the surface topography suggest the flow of H_2O at some time in the past. If this is the case, then the atmosphere of Mars must have been considerably denser at some past time. Long-term climatic changes may occur. As a result of gravitational perturbations of the Sun and other planets, the shifting longitude of perihelion and the precession of the equinoxes combine to give a climate cycle with period 51,000 years, which strongly affects the climate extremes between summer and winter.

Compounding these effects is a cyclic variation of the orbital eccentricity which varies from a nearly circular orbit to one which is highly elliptical where one hemisphere may receive 1.5 times more solar energy at perihelion than at aphelion. These effects combine to give climates much different than is observed today. In the past, H_2O could have been abundant.[4]

The Major Planets It is no coincidence that the major planets (Jupiter, Saturn, Uranus, and Neptune) all lie in the outer solar system (see Table 1). Their large mass is a consequence of the lower temperatures which have always existed in the outer solar system as a result of the great distance from the Sun. The outer protosolar nebula was cooler than the inner nebula so that the planets which condensed

in the outer regions accreted even the lightest and most cosmically abundant elements, hydrogen and helium. The higher temperatures of the inner solar system caused the terrestrial planets to accrete the heavier elements preferentially. These are much less cosmically abundant than hydrogen or helium, so that the terrestrial planets are much smaller and denser than the major planets. The higher temperatures also caused the terrestrial planets to lose their primordial atmospheres and fluid envelopes, exposing their rocky crusts. As secondary atmospheres outgassed from the rocks, the higher temperatures caused gases with the lowest molecular weight to escape into space. Therefore, a distinguishing characteristic of the terrestrial planets is the presence of a solid surface underlying a relatively thin atmosphere in which gases of low molecular weight are absent. By contrast, the major planets have no solid surface. They are fluid except for possibly small, rocky cores. Molecular hydrogen and helium in approximately the solar abundance ratio make up most of their atmospheres and fluid interiors. Evidence for the lack of any solid surface is the differential rotation of these planets. The rotation periods are a function of the latitude with the shortest periods at the equator. The period of Saturn's atmosphere, for example, varies from 10 hours 2 minutes at the equator to 11 hours at 57° latitude.

The rotational periods of Jupiter and Saturn lie between 10 and 11, hours while those of Uranus and Neptune lie between 15 and 24 hours. The rapid rotation rates combined with the large size of the major planets causes coriolis forces to be strong at all but the lowest latitudes in these planets. Consequently, their atmospheres exhibit a zonal banded structure with the band widths typically decreasing with increasing latitude. At high latitudes, where the coriolis forces are strongest, the banded structure gives way to highly varied and less orderly structure. The zonal structure is characterized by dark belts and light zones with various

TABLE 1. PLANETARY DATA

Planet or Satellite	Distance from Sun (Earth = 1)	Orbital Eccentricity	Mass (Earth = 1)	Radius (Earth = 1)	Density (g cm^{-3})	Period of Rev.	Period of Rot.	Escape Velocity (km/sec)
Mercury	0.387	0.206	0.0554	0.381	5.53	87.97 d	58.6 d	4.3
Venus	0.723	0.007	0.815	0.950	5.25	224.70 d	−244 d	10.5
Earth	1	0.017	1	1	5.52	365.26 d	23 hr 56 m	11.3
Mars	1.52	0.093	0.1075	0.533	3.92	686.98 d	24 hr 37 m	5.1
Jupiter	5.20	0.048	317.83	11.20	1.34	11.86 y	9 hr 50 m	60.2
Saturn	9.54	0.056	95.15	9.47	0.71	29.46 y	10 hr 14 m	35.8
Uranus	19.18	0.047	14.54	4.04	1.20	84.01 y	15–24 hr	21.4
Neptune	30.06	0.009	17.23	3.92	1.60	164.79 y	19 hr	23.7
Pluto	39.44	0.250	0.0023	0.24	0.9	247.7 y	6 d 9 hr	1.1
Titan	9.54	0.029	0.0237	0.40	2.0	15.95 d	15.95 d	2.8
Io	5.2	0	0.0126	0.29	2.85	1.77 d	1.77 d	2.4

colorations, which are particularly evident for Jupiter. The belts are warmer regions, thought to be subsiding and representing relatively clear regions of the atmosphere. The zones are thought to be hazier, cooler regions produced by convective upwelling in the atmosphere. In addition to the zonal structure, there are vortex features such as white ovals, red spots and brown spots in these atmospheres which usually represent cyclonic or anticyclonic storms. Jupiter's Great Red Spot (see Fig. 2), located at $23°$ S latitude, is an anticyclonic feature currently 25,000 km by 14,000 km which has been in existence for over 300 years. Like the zones, it is cooler, higher, and hazier than its surroundings. The process causing the colors in these atmospheres is not understood but may involve red phosphorous formed photochemically from phosphine, PH_3. Alternatively, it may involve polmerized carbon suboxides, $(C_3O_2)_x$. Temperature differences may also explain the color differences.

Because the major planets condensed out of the protosolar nebula with relatively little fractional loss of the lighter elements, their composition reflects the solar abundance ratios of the elements. However, temperature and chemistry play an important role in governing the composition of the outer layers. The low temperature causes some of the elements to exist in molecular form and causes many common gases to freeze out. Consequently, their atmospheres consist largely of gases which have low condensation and freezing temperatures and which are made up of the more cosmically abundant elements. This explains the predominance of H_2, He, CH_4, and NH_3. Of these, NH_3 has the highest freezing point; it is frozen out of each upper atmosphere of these planets to a depth which increases with the planet's distance from the Sun. This explains its progressively lower visibility with increasing planetary distance from the Sun. In the atmospheres of Uranus and Neptune, the NH_3

FIG. 2. Voyager 1 image of Jupiter obtained on January 26, 1979 showing the belts, zones, white ovals, dark spots, and Great Red Spot. The latter is near the central meridian at $23°$ S latitude.

abundance may be further reduced by conversion to NH_4HS clouds. Frozen NH_3 particles probably make up the upper cloud layers of Jupiter and Saturn.

In addition to the gases above, Jupiter's atmosphere contains HD, PH_3, H_2O, CO, C_2H_6, GeH_4, and probably H_2S and HCN. In Saturn's atmosphere, C_2H_4 has also been detected but H_2O, CO, GeH_4 and HCN have not yet been detected. In Uranus' atmosphere, the detected gases are H_2, HD, CH_4, and C_2H_6. The presence of He is inferred from observations. Many other gases, not yet detected, are probably also present in both atmospheres. Photochemical products and products of ion chemistry in the upper atmospheres of Jupiter, Saturn, and Titan are thought to be responsible for the dark UV albedos of these planets.[4]

All the major planets except Uranus have strong temperature inversions above the 10 mbar level. The temperature rises to over 100 K in some cases. This increase in temperature with altitude is caused by the heating of the upper atmosphere resulting from the absorption of sunlight by CH_4, C_2H_6, C_2H_2, and PH_3. The gases in this region of the atmosphere are poor thermal absorbers or radiators so the atmosphere heats up. Uranus' atmosphere may contain a thermal radiator which the other planets do not have.

Furthermore, all of the major planets except Uranus have strong internal heat sources which cause these planets to radiate about 1.7 times the absorbed incident sunlight (see Table 2). Uranus may have no internal heat source at all. These cause the temperature to increase with depth below the levels where sunlight penetrates, all the way to the source of the heat. In the tropospheric layers where sunlight penetrates, the thermal opacity from H_2-H_2 and H_2-He collisions causes the temperature to increase with depth at an ever increasing rate (the greenhouse effect) until the atmosphere becomes convectively unstable. In the convecting regions, the temperature lapse rate is about 2 K/km. Strong convection is especially evident in the cloud patterns of Jupiter. Lightning has been observed on Jupiter's night side by Voyager. Significant changes in the cloud features were observed in the four-month interval between the Voyager 1 and 2 flybys.[6] This occurred in spite of the much longer radiative time constant for the atmosphere because the convective time constants are quite short. Convection probably continues from the deepest layers penetrated by sunlight all the way to the source of the internal heating. For Jupiter, gravitational contraction suffices to explain the internal heating, but for Saturn, separation of He from H_2 is probably an additional important source of heat.

The internal heat sources moderate latitudinal and seasonal temperature variations. Jupiter has a small obliquity so its seasonal variation is small anyway. Saturn and Neptune have moderate obliquities but their strong internal heating limits seasonal effects to the uppermost layers of their atmospheres. Uranus has a large obliquity; its spin axis is nearly parallel to its orbital plane. Because of this and the lack of an appreciable internal heat source, Uranus' atmosphere should exhibit strong seasonal variations over its long orbital cycle.

Pluto In spite of its small size, Pluto has an atmosphere consisting of 27 m-Amagat CH_4 (1 cm-Amagat = 2.69×10^{19} molecules cm^{-2}). This gas is slowly escaping but is being replenished by the sublimation of the surface CH_4 ice. Adiabatic cooling of the escaping gas greatly slows the escape rate. Other gases have not yet been detected in Pluto's atmosphere; however, the presence of a heavier gas would help to stabilize the CH_4 atmosphere further. Since CH_4 gas on Pluto exists below its freezing point, its surface pressure is governed by the surface temperature. Pluto's atmosphere is apparently not cloudy because the light curve, which has a very large amplitude, has a very regular period.[7] The observed abundance therefore implies a surface temperature of 58 K, a value considerably higher than the solar equilibrium temperature. This suggests that a greenhouse mechanism is operating. However, CH_4 has no thermal opacity so the presence of a greenhouse would imply the presence of another bulk gas on Pluto.

TABLE 2. INTERNAL HEAT SOURCES FOR THE MAJOR PLANETS

Planet	Ratio of Internal Heat Source to Absorbed Solar Radiation
Jupiter	1.67 ± 0.09
Saturn	1.79 ± 0.10
Uranus	1.6 ± 1.0
Neptune	$\leqslant 0.03$

TABLE 3. TITAN'S ATMOSPHERE

Gas	Mixing Ratio (%)
N_2	90
Ar	$\leqslant 10$
CH_4	1
Ne	1
H_2	0.12
CO_2	7×10^{-10}
C_2H_2	2×10^{-4}
C_2H_4	4×10^{-5}
C_2H_6	2×10^{-3}
HCN	2×10^{-5}
C_3H_8	2×10^{-3}
C_3H_4	3×10^{-6}
C_4H_2	$10^{-6} - 10^{-5}$
HC_3N	$10^{-6} - 10^{-5}$
C_2N_2	$10^{-6} - 10^{-5}$

Unlike the terrestrial planets, Pluto's interior is largely icy rather than rocky; its density is low. It probably contains ices of volatile materials which supply volatile gases to the atmosphere.

The large ellipticity of Pluto's orbit (0.25) and Pluto's large obliquity cause a strong seasonal variation of Pluto's atmospheric bulk because of the sensitivity of the saturation vapor pressure to temperature for the volatile gases likely to constitute its atmosphere. Pluto's distance from the Sun varies from a current value of 30 a.u. near perihelion to 50 a.u. at aphelion 124 years from now. During this time, essentially all of the observed CH_4 and other volatile gases should freeze out of the atmosphere.

Titan The only satellite in the solar system known to possess a dense atmosphere is Titan, Saturn's largest satellite. Although a satellite, Titan is larger than the planet Mercury. Titan's surface temperature is 95 K, much cooler than Mercury's, so it is not surprising that Titan can retain an atmosphere. The Voyager flybys[8] obtained a surface pressure of 1.5 atm so that the column abundance of the whole atmosphere is about 100 km-Amagats. The major constituent is N_2, which accounts for 90% of the atmospheric molecules. Table 3 lists the other gases detected. The abundance of Ar could be as large as 10%. This implies that Ar is locked up in clathrates in Titan's interior which must have accreted from the protosolar nebula. The higher hydrocarbons arc probably the products

FIG. 3. Voyager 1 image of Io obtained on March 4, 1979 showing surface deposites of SO_2 frost and other material spewed from Io's volcanoes. The light circular region in the center of the satellite is the deposition site of an erupting volcano.

of photochemistry. Titan has no magnetosphere to protect it from the solar wind so charged particles and cosmic rays support ionospheric chemistry in Titan's spherically extended upper atmosphere. Owing to the abundance of H, C, and N in Titan's atmosphere, this leads to complex chemistry and possibly to organic molecules being formed. Titan's atmosphere may provide insights into the evolution of life or its precursors.

The atmosphere appears to be in radiative equilibrium except possibly for the bottom 2 km. The radiative time constant is long enough that diurnal effects should be negligible (Titan's day is 16 Earth days). There are, however, seasonal effects which cause a hemispherical asymmetry and seasonal cyclostrophic winds in the stratosphere. The dominant circulation pattern is Hadly cells, which transport heat polewards.

A detached haze layer appears high in the atmosphere over a deeper opaque cloud layer. It lies about 315 km above the surface, is nearly transparent, and is composed of aerosol particles with mean size 0.3 μ. These are probably the volatile products of photochemical reactions with CH_4 and N_2. At the time of the Voyager 1 flyby, the opaque cloud layer was 185 km above the surface in the north and 50 km higher in the south. It probably consists mainly of CH_4 crystals. The haze layer transitions into a north polar hood at high latitudes. Nowhere is Titan's surface visible.

Io Pioneer 10 detected an ionosphere on Io with a scale height of 100 km. This was the first compelling evidence that Io has an atmosphere. The innermost of the Galilean satellites, Io is immersed deep within Jupiter's magnetosphere and trapped radiation belts. Charged particles bombard its thin atmosphere and surface, causing atoms to sputter from Io into orbit around Jupiter, where they form thin toroidal clouds. Some of the atoms become ionized so that they are swept up by Jupiter's corotating magnetic field and join a plasma torus. These tori imply that S and O are the main constituents of Io's atmosphere and possibly its surface.

Io's atmosphere is very thin; the surface pressure is on the order of 0.1 μbar and is greatest on the sunlit side near the volcanoes discovered by Voyager.[6] These arise from the tidal heating of Io by interaction between Io's orbit and the other Galilean satellites. Voyager 1 observed eight large active volcanoes on Io. (See Fig. 3.) Four months later, Voyager 2 observed seven of these volcanoes and found six of them to be active. These volcanoes are believed to be the source of Io's atmosphere. They discharge primarily SO_2 gas into the atmosphere. A surface frost of SO_2 forms from the gas. Because of the rapid deposition rates, Io's surface is the youngest in the solar system.[6]

The column abundance of SO_2 observed near one of the hot vents was 0.2 cm-Amagat, a value consistent with the 130 K temperature of the sunlit surface for SO_2 gas in equilibrium with the surface frost. However, the IUE satellite observed only 0.008 cm-Amagats of SO_2, averaged over the sunlit hemisphere. Therefore, the global SO_2 atmosphere may be less abundant than if it were in equilibrium with surface frosts. The frosts, however, must have some "buffering" effect on the atmosphere bulk so that it cannot depart greatly from the saturation value.

L. Trafton

References

1. Kumar, S., "Mercury's Atmosphere: A Perspective from Mariner 10," *Icarus* 28, 579 (1976).
2. Marov, M. Y., "Results of Venus Missions," *Ann. Rev. Astron. Astrophy.* 16, 141 (1978).
3. "Pioneer Venus Results," *Science,* 205 41–119, (6 July 1979); articles by several authors.
4. Barbato, J. P., and Ayer, E. A., "Atmospheres," New York, Pergamon Press, 1981.
5. Leovy, C. B., "Martian Meteorology," *Ann. Rev. Astron. Astrophys.* 17, 387 (1979).
6. "Voyager 1 Encounter with the Jovian System," *Science,* 204, 945–1008, (1 June 1979); "Voyager 2 Encounter with the Jovian System," *Science* 206, 925–996, (23 November 1979); articles by several authors.
7. Trafton, L. M., "The Atmospheres of the Outer Planets and Satellites," *Rev. Geophys. Space Phys.* 19, 43 (1981).
8. "Voyager 1 Encounter with the Saturn System," *Science* 212, 159–239, (10 April 1981); "Voyager 2 Encounter with the Saturn System," *Science* 215, 449–587, (29 January 1982); articles by several authors.

Cross-references: ASTRONOMY, ASTROPHYSICS, IONOSPHERE, METEOROLOGY, RADIATION BELTS.

PLASMAS

Solids, liquids and gases are the three familiar states of matter. When heated, solids, in general, turn to liquids and the liquids eventually become gases. When a gas is heated to sufficiently high temperatures such that the thermal energy of the atoms (molecules) is comparable to the binding energy of the electrons, violent interatomic (intermolecular) collisions tend to ionize the gas, i.e., the atoms are split into free electrons and ions. The dynamical properties of this gas of free electrons and ions are sufficiently different from the normal unionized gas that it can be considered a fourth state of matter, and is given a new name, *plasma*. It must be mentioned that the normal state of matter in this universe is plasma, although on the planet earth it is produced only under somewhat special and often extreme laboratory conditions.

The interiors of the stars, the Solar Wind, the Van Allen belts, and the ionosphere are a few examples of the plasma state.

The study of plasma state is needed to understand the properties of a large part of the universe. Thus the astrophysicists were the first to start investigating the phenomena peculiar to plasmas. In addition to this general scientific interest, manifold technical possibilities have given a tremendous boost to research in the physics of plasmas. A list of industrial applications must include welding and cutting through heliarcs, TIG welding and plasma torches, fluorescent lamps, dry plasma etching, plasma sputtering to deposit coatings, opening and closing power switches, generation of electromagnetic waves, and generation of useful energy.

The most important area of research, however, is the physics of confining and heating plasmas with a view to generating useful energy via controlled thermonuclear fusion of light elements (hydrogen isotopes, typically). This research, which has spanned theory as well as experiment, has played a large role in the considerable advances made in the past three decades in the description and understanding of plasmas, especially magnetically confined plasmas.

In general, the plasmas are overall charge neutral, that is, the total negative electronic charge equals the total positive ionic charge. However, the definition of a plasma has been extended to include even those systems where charge neutrality does not pertain. Thus, inherently unneutralized systems like electrons and ion beams also fall within the domain of plasma physics.

The primary characteristic of the plasma state is that the long-range Coulomb forces dominate the dynamics of the electrons and ions. Thus the constituent particles constantly interact with one another through the electromagnetic fields, which in turn are self-consistently generated by the motions of the plasma particles. Because of this constant interaction, the plasma particles tend to respond collectively to any perturbation. Collective behavior is, in fact, what distinguishes a plasma from a dilute neutral gas, whose particles move about essentially free, save for an occasional catastrophic collision with another particle. The immense extent and richness of plasma dynamics owes much to these very collective modes.

A theoretical formulation of the physics of plasma requires the simultaneous specification of the dynamics of the interacting electrons, ions and the electromagnetic fields. The electromagnetic fields are described by Maxwell's equations

$$\nabla \cdot \mathbf{B} = 0, \qquad \nabla \cdot \mathbf{E} = 4\pi\rho$$

$$\nabla \times \mathbf{E} = -\frac{1}{c}\frac{\partial \mathbf{B}}{\partial t}, \qquad \nabla \times \mathbf{B} = \frac{4\pi}{c}\mathbf{J} + \frac{1}{c}\frac{\partial \mathbf{E}}{\partial t} \qquad (1)$$

where $\mathbf{E}$ and $\mathbf{B}$ are the electric and the magnetic fields respectively, c is the velocity of light in vacuum, ρ is the charge density, and $\mathbf{J}$ is the current density. Notice that ρ and $\mathbf{J}$ are due to the motion of plasma particles in response to the electric and magnetic fields $\mathbf{E}$ and $\mathbf{B}$ (externally applied as well as self-consistently generated). The main task of the study of the dynamics of plasma particles then is to provide the so-called constituitive laws: expressions which determine ρ and $\mathbf{J}$ as functions of $\mathbf{E}$ and $\mathbf{B}$, i.e.,

$$\rho = \rho(\mathbf{E}, \mathbf{B})$$

$$\mathbf{J} = \mathbf{J}(\mathbf{E}, \mathbf{B}).$$

Notice that ρ, $\mathbf{J}$, $\mathbf{E}$, and $\mathbf{B}$ are all functions of space $(\mathbf{r})$ and time (t).

There are several levels of sophistication for the description of the dynamics of plasma particles. A microscopic theory would require a proper treatment as a problem in many-body physics. A fair number of phenomena, however, depend only on the macroscopic behavior of the plasmas, and can be studied within the framework of much simpler models. We describe two such frequently used models.

One-Fluid Model The plasma is treated as a single charged highly conducting fluid, and is described by a set of hydrodynamic equations. This is generally possible when the plasma is embedded in a strong external magnetic field. This of course is the case of greatest practical importance because the strong magnetic fields are the principal confining agents used for fusion plasmas. The set of hydrodynamic and Maxwell's equations together define the discipline of magnetohydrodynamics, hydromagnetics, or MAGNETO-FLUID-MECHANICS, and is characterized by a vastly rich variety of waves and flows that are not encountered in ordinary fluid mechanics. Historically, this is the oldest and the most well studied branch of plasma physics. The well known Alfvén wave, which is a transverse electromagnetic wave propagating along the direction of the ambient magnetic field, was discovered within this model. In toroidal confinement systems, one of the most notable discoveries called the *Kruskal-Shafranov criterion of stability* was also derived in magnetohydrodynamics. It states that if a certain quantity q, called the *safety factor*, fell below unity in the plasma, an instability will develop resulting in loss of confinement. This was subsequently verified experimentally in tokamaks. (The tokamaks is a doughnut-shaped machine which is, at present, the most advanced fusion system.)

Two-Fluid Description A slightly more detailed description of the plasma would be to treat the ions and electrons each as separate conducting fluids coupled through momentum transfer Coulomb collisions and, of course, Maxwell's equations. All results of the one-fluid theory are automatically contained in this model, and several new phenomena emerge. As an example, in this model one can show the existence of high-frequency longitudinal oscillations which are characteristic of a field-free, infinite, homogeneous plasma. These oscillations have a frequency $\omega = \omega_{pe} = (4\pi n e^2 /m)^{1/2}$, where n is the density, m is the mass, and e is the charge of the electron. These oscillations are called *plasma waves* or *Langmuir waves*. (It was to describe these waves that Langmuir first coined the word *plasma*.)

Microscopic Description Fluid theories are unable to account for many microscopic phenomena, for which a detailed kinetic description is essential. Because of the long-range character of the Coulomb force, the plasma kinetic theory has to be quite different from that of a dilute gas in which the interaction is short range. However, it has been possible to systematically develop the statistical many-body description of a plasma as orders in an expansion in the small parameter $\epsilon = (1/n\lambda_D^3)$, where n is the average number density. The Debye shielding distance λ_D is given by

$$\lambda_D = \left(\frac{2kT}{4\pi ne^2}\right)^{1/2}$$

where k is the Boltzmann constant and T is the absolute temperature. The Debye shielding distance is a measure of the effective range of the Coulomb force acting on a test particle in the plasma. The potential seen by a test particle is given by

$$\phi(r) = \frac{q}{r} e^{-r/\lambda_D}$$

where r is the distance of the test particle from the charge q. The expansion parameter ϵ is the inverse of the number of particles in a Debye sphere, and for most plasmas of interest ranges from 10^{-2} to 10^{-9}. The smallness of this parameter really typifies the plasma state. It can be easily seen that the expansion parameter ϵ signifies the ratio of potential energy of interaction of one particle with its neighbor (separated typically by a distance $n^{-1/3}$) and the mean kinetic energy per particle,

$$PE/KE \simeq \frac{e^2 n^{1/3}}{kT} \sim (n\lambda_D^3)^{-2/3} = \epsilon^{2/3} \ll 1.$$

Thus the smallness of ϵ implies the smallness of the binary interaction as compared to the average collective many-body effect of all its neighbors within a Debye sphere.

How does the infinite-range Coulomb force become finite ranged? The answer can be given as follows. Let us suppose that a negatively charged test particle was introduced in the plasma. This test charge will immediately be surrounded by positively charged ions because they are attracted toward it. This cloud of surrounding positive charges tends to shield the test particle from other charges, so that it hardly interacts with charges separated from it by a distance larger than λ_D.

The plasma kinetic theory can be developed to the required order in ϵ. To order zero in ϵ, the plasma particles only interact collectively and no binary encounters are included. The binary encounters will come in order ϵ, and tertiary encounters in order ϵ^2, and so on. The leading order plasma dynamics, therefore, is an expression only of the collective behavior and is determined by the well known Vlasov or collisionless Boltzmann equation:

$$\frac{\partial f_s}{\partial t} + \mathbf{u} \cdot \frac{\partial f_s}{\partial \mathbf{r}} + \frac{e_s}{m_s}\left[\mathbf{E} + \frac{\mathbf{u} \times \mathbf{B}}{c}\right]$$

$$\cdot \frac{\partial f_s}{\partial \mathbf{u}} + \frac{\mathbf{F}_{ext}}{m_s} \cdot \frac{\partial f_s}{\partial \mathbf{u}} = 0 \qquad (2)$$

where $f_s = f_s(\mathbf{r}, \mathbf{u}, t)$ is the single-particle distribution function of s-species; e_s and m_s are the charge and mass, respectively; $\mathbf{E}$ and $\mathbf{B}$ are the self-consistent fields; and $\mathbf{F}_{ext}$ is an externally applied force, including, for example, an applied magnetic field. Notice that $f_s(\mathbf{r}, \mathbf{u}, t)$ is the particle density in six-dimensional phase space, and the Vlasov equation is an expression that f_s is conserved along the trajectory of the particle in phase space. From f_s one can construct the current and charge densities (needed in Maxwell's equations) by integrating over velocity space:

$$\rho(\mathbf{r}, t) = \sum_s e_s \int d^3u\, f_s$$

$$\qquad (3)$$

$$\mathbf{J}(\mathbf{r}, t) = \sum_s e_s \int d^3u\, \mathbf{u} f_s.$$

Observe now that substituting Eq. (3) into Maxwell's equations (1) will allow us to determine $\mathbf{E}$ and $\mathbf{B}$ in terms of f_s. Thus Eq. (2) (where we have products of $\mathbf{E}$ and $\mathbf{B}$ with f_s) is essentially nonlinear, implying that even in the lowest order, the plasma dynamics is described by partial nonlinear integro-differential equations. It must be stressed that even in the simplified fluid models described earlier, plasma behavior is nonlinear.

The exact solutions of the Vlasov equation are exceedingly difficult to find. However, it is possible to determine the behavior of a small departure f_{s1} from some plasma equilibrium state f_{s0} by solving the linearized Vlasov equation. Much of the modern plasma theory is based on the investigation of the linearized Vlasov-Maxwell equations.

The dominant result of the linear theory (which is also called the *stability theory*) is the existence of a plethora of wave motions the plasma seems to be able to sustain. For example, a field-free plasma allows transverse electromagnetic waves in addition to the Langmuir waves and ion sound waves; the last requiring electron temperature for its existence. In the presence of electron temperature, the Langmuir wave becomes a propagating wave with a dispersion relation

$$\omega^2 = \omega_{pe}^2 + 3\left(\frac{kT}{m}\right)\kappa^2$$

where κ is the wave vector. The presence of an external confining magnetic field vastly increases the number of possible excitations. In addition to the Alfvén waves already mentioned, one can have ion and electron cyclotron waves, lower hybrid and upper hybrid waves, whistler waves, and so on. In confined fusion plasmas, because of the ever present temperature and density gradients, new wave motions which are generically called *drift waves* also become possible. A very important result of the linear theory is that in a plasma in thermal equilibrium (say a Maxwellian) the waves are damped even in this collisionless approximation. This phenomenon, called *Landau damping*, is one of the most important discoveries of plasma physics and is due to the interaction of the wave with particles traveling almost with the phase velocity of the wave. Landau damping has been experimentally confirmed.

If the equilibrium distribution f_{s0} has nonthermal features (an equilibrium current, a bump in its tail, anistropy in velocity) then the wave-particle interaction can lead to a growth of the waves. In inhomogeneous plasmas, there is always additional free energy available in the gradients to drive an instability. If the waves do grow in amplitude, the linear theory becomes invalid, and a nonlinear theory has to be invoked to determine the fate of the wave as well as the distribution function. The mechanism of instability can be understood as follows. The waves are created by perturbed motion of the plasma. If the electromagnetic fields thus generated happen to accentuate the perturbation, then the increased perturbation will create larger fields and an instability can build up.

The existence of the instabilities is one of the major difficulties in the controlled nuclear fusion program. It is believed that the very large electron energy losses observed in tokamaks are caused by the instabilities.

Hot plasmas are also capable of radiating in various modes. The magnetically confined hot thermonuclear plasmas are expected to copiously radiate at the cyclotron frequency $\omega_c = eB/mc$ and its harmonics and can be a serious source of energy loss.

Approach to Equilibrium In the linear theory, it was supposed that an equilibrium state f_{s0} existed. How does a plasma approach equilibrium? An ordinary gas does it through binary collisions, which for a plasma, can be investigated by including terms of order ϵ in the dynamics. These terms reflect the effects of particle discreteness, i.e., collisions between plasma particles. Since a plasma is characterized by numerous small angle collisions, a modified Fokker-Planck collision term is needed. In plasmas, there also exists another mechanism: the instability (an expression of the collective effects) which can rapidly take a nonequilibrium system towards equilibrium. It is instructive to mention here the "Langmuir paradox." In his experiments using electrostatic probes, Langmuir discovered that the electron distribution function was far more Maxwellian than warranted by the collisions. It is surmised that Langmuir oscillations described earlier may be behind this phenomenon.

S. M. MAHAJAN

References

Krall, N. A., and Trivelpiece, A. W., "Principles of Plasma Physics," New York, McGraw-Hill Book Company, 1973.
Miyamoto, K., "Plasma Physics for Nuclear Fusion," Cambridge, Mass., MIT Press, 1980.

Cross-references: ASTROPHYSICS, COLLISION OF PARTICLES, FUSION POWER, IONIZATION, IONOSPHERE, KINETIC THEORY, LASER FUSION, MAGNETO-FLUID-MECHANICS, PLANETARY ATMOSPHERES, RADIATION BELTS.

POLAR MOLECULES

The term "polar" is applied to molecules in which there exists a permanent spatial separation of the centroids of positive and negative charge, or dipole moment. Such a moment was first postulated by P. Debye for molecules having structural asymmetry in order to explain certain of the observed electrical properties. He chose for his model an electrical dipole contained in a spherical molecule, free to rotate into alignment with an applied electric field, but subject to disorientation by collisions with other molecules due to thermal motion. At ordinary temperatures and field strengths, the electrical energy involved in the orientation is much smaller than the thermal energy kT so that only a small fraction of the dipoles are aligned with the field (k is Boltzmann's constant and T is the temperature in degrees Kelvin). The net dipole moment per mole (*molar polarization*) is then calculated statistically to be

$$P = \frac{4\pi}{3} N\left(\alpha + \frac{\mu^2}{3kT}\right) \tag{1}$$

where μ is the permanent dipole moment per molecule, i.e., the charge multiplied by the distance of separation, and N is Avogadro's number. The polarizability α represents the induced moment per molecule resulting from the temporary distortion of the electron orbits by the applied field. For nonpolar molecules, it is the only contribution; it corresponds to the optical polarizability as measured by the refractive index.

The molar polarization may be related to the dielectric constant ϵ by the approximate equation of Clausius and Mosotti

$$\frac{\epsilon - 1}{\epsilon + 2}\frac{M}{d} = P \tag{2}$$

where M is the molecular weight and d is the density. The permanent dipole moment of a molecule may therefore be determined by measuring the temperature coefficient of the dielectric constant as seen by combining Eqs. (1) and (2). Alternatively, a companion measurement of the optical refractive index may be made to determine α, and the dipole moment is then obtained as the difference between the total polarization and the optical contribution. The measurements are made in dilute vapor or solution phase, so that the individual dipoles are sufficiently far apart that they do not influence one another (see REFRACTION).

Equation (1) is valid in the low-frequency region where the dipoles are able to rotate in phase with the applied electric field. This rotation is subject to various restraints; in the simple case of a spherical molecule of radius a rotating in a fluid of viscosity η, this leads to a relaxation time

$$\tau = \frac{4\pi a^3 \eta}{kT} \tag{3}$$

corresponding to a frequency range of *anomalous dispersion* in which the molecules become unable to follow the oscillations of the applied field. This gives rise to an out-of-phase component of the dielectric constant representing a conductivity or *dielectric loss*, ϵ'', i.e., a dissipation of energy in the form of heat. Mathematically this is expressed as a complex dielectric constant.

$$\epsilon = \epsilon' - j\epsilon'' \tag{4}$$

$$\epsilon' - \epsilon_\infty' = \frac{\epsilon_0' - \epsilon_\infty'}{1 + \omega^2 \tau^2} \tag{5}$$

$$\epsilon'' = \frac{(\epsilon_0' - \epsilon_\infty')\omega\tau}{1 + \omega^2 \tau^2} \tag{6}$$

ω is $2\pi \times$ the frequency, and the subscripts 0 and ∞ refer to dielectric constant measured at very low and very high frequency, respectively. Cole and Cole showed for polar molecules having a unique relaxation time that a plot ϵ'' vs ϵ' is a semicircle centered on the ϵ'-axis. For more complicated molecules or high polymers, the

center becomes depressed below the ϵ'-axis and the curve may be further distorted. This behavior is generally characterized by a distribution of relaxation times as a result of the orientation of molecular segments of various shapes and sizes. Information regarding the freedom of orientation within the molecules may thus be gained.

These simple relationships between dipole moment and dielectric constant fail for concentrated solutions or pure polar liquids because they do not take into account the interaction of the dipoles, both permanent and induced, with one another. The calculations have been extended by Onsager, Kirkwood, and others, to include these effects. It becomes necessary to include in the theory a correlation factor which is a measure of the extent of nonrandom orientation of the dipoles, i.e., the tendency of the dipoles to aggregate parallel or antiparallel to one another. Experimental determination of this quantity yields further insight into the structure of polar liquids.

In the solid state, most crystalline polar compounds exhibit low dielectric constants because the rotational freedom necessary for dipole orientation has been frozen out. A few compounds, however, do show a persistence of high dielectric constant to temperatures below the melting point, indicating rotational freedom in the solid. At some lower temperature the rotation ceases and the dielectric constant drops. This change in dielectric constant has frequently been used as a method of detecting second-order transitions in polar compounds.

Some success in the correlation of dipole moment with molecular structure has been achieved. A series of moments assigned to individual bonds was developed empirically by Smyth, Pauling, and others from measurements on simple molecules. They are intended for approximate calculation of dipole moments of complex molecules by vectorial addition along the bond directions as determined by other means. It is assumed that there is no interaction between bonds; this generally results in the calculated values being higher than the measured moment because of inductive effects, i.e., electrons being shared in a bond between two atoms are not wholly available to a neighboring bond, so that neither bond attains its full moment. An approximate correction for this effect has been made by Eyring and co-workers. Despite these inadequacies, the calculated moments are often helpful in determining molecular structures and have often applied in the case of various substituted benzene-ring compounds. Of special importance is the ability to decide between a polar or nonpolar, i.e., symmetrical, structure.

More detailed quantum mechanical calculations of dipole moments have been less successful. Rather the experimentally determined moments have been used to assign varying degrees of ionicity and covalency to the bonds in establishing the electronic hybridization structure. Quantum mechanics has also shown how it is possible to obtain extremely accurate measurements of the dipole moments from spectroscopic Stark splittings.

D. EDELSON

Cross-references: DIELECTRIC THEORY, DIPOLE MOMENTS, MOLECULES AND MOLECULAR STRUCTURE, POLYMER PHYSICS, QUANTUM THEORY, REFRACTION, ZEEMAN AND STARK EFFECTS.

POLARIZED LIGHT

Polarized light is an especially simple form of light and can be defined easily in terms of either of the two prevalent theories of light: the wave theory and the photon theory (see LIGHT). According to the former, light consists of trains of electromagnetic waves whose wavelengths lie in the range from about 4×10^{-7} to 7×10^{-7} meters. A noteworthy feature of the waves is that two kinds of displacements are involved: electric and magnetic. Another significant feature is that, when the waves are traveling in empty space (or in glass, water, or other isotropic medium), the electric and magnetic displacements are perpendicular to the direction of propagation of energy; i.e., they are transverse, not longitudinal. Since the electric and magnetic displacements are always perpendicular to one another and have equivalent magnitudes, it is sufficient to specify just one of these quantities; most authors choose to deal with the electric displacement.

Consider, now, a slender beam of light that is traveling east, i.e., from left to right in Fig. 1. If it happens that the direction of electric displacement is everywhere up or down, but not north or south, the beam is said to be linearly polarized in the vertical plane. If, alternatively,

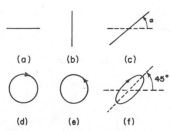

FIG. 2. Variety of sectional patterns of polarized light: (a) horizontally polarized; (b) vertically polarized; (c) linearly polarized at azimuth α; (d) right circularly polarized; (e) left circularly polarized; (f) right elliptically polarized at 45° azimuth and with ellipticity, or ratio of semi-axes, of approximately 3.

the displacements were north and south, the beam would be called *horizontally* linearly polarized. The displacement might, of course, lie in some tilted plane specified by an angle α; in such case, any given displacement may be regarded as the resultant of a vertical displacement and a horizontal displacement. The *sectional pattern* of any such beam is conventionally indicated by a straight line segment at the appropriate azimuth α, as indicated in Fig. 2(c). Linearly polarized beams that have azimuths differing by 90° are said to have *orthogonal* polarization forms.

Circular and elliptical polarization forms exist also. Circularly polarized light may be regarded as the result of combining two linearly polarized beams that have the same wavelength and same intensity, are polarized in orthogonal directions (e.g., horizontal and vertical) and differ in phase by 1/4 cycle, or 90°. If the horizontally polarized component *lags* in phase by 90°, the sectional pattern of the combined beam is drawn as a circle executed clockwise, as judged by an observer facing towards the light source. If the horizontally polarized component *leads* by 90°, the circle has a counterclockwise sense and the light is called left circularly polarized. In the general case, the components may differ in magnitude and the phase difference may have any value; the general sectional pattern is an ellipse.

A different description of polarized light is required when the light is regarded as a stream of photons. This is the case when the photons are so infrequent and so energetic that they can be detected individually. In such case, it is the spin, i.e., the angular momentum, of the photon that constitutes the polarization. A right circularly polarized photon has a spin of +1 unit, and a left circularly polarized photon has a spin of −1 unit. Theory and experiment are in agreement that the magnitude of the unit in question is $h/2\pi$, where h is Planck's constant (see QUANTUM THEORY). Linearly and elliptically polarized photons may be regarded as combinations, in suitable proportions, of positive and negative spins.

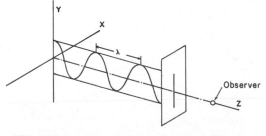

FIG. 1. Monochromatic beam of wavelength λ traveling horizontally to the right. Since, in this example, the electric displacement is vertical, the sectional pattern (indicated at right) consists of a vertical line and the beam is said to be vertically linearly polarized. The observer is situated far to the right, facing the light source.

Unpolarized light is more complex than polarized light, hence it is harder to describe. It consists of light in which the azimuth, ellipticity, and handedness of polarization vary rapidly and at random, so that no one type of polarization predominates. No simple diagram can depict the chaos and impartiality of the sectional pattern. If, in a given beam, one particular sectional pattern slightly outweighs all other patterns, the beam is said to be partially polarized; the degree of polarization may lie anywhere in the range from 0 to 100 per cent.

Polarization is not confined to visible light, but applies also to longer-wavelength radiations including the infrared and radio ranges and to shorter-wavelength radiations including ultraviolet light and x-rays. Nevertheless, the visual range deserves special attention in view of the variety of phenomena observed and the nicety of observation and wealth of applications.

Polarization was discovered by the Dutch scientist Christian Huygens in 1690, but it was not well understood until the transverse nature of the vibration, suggested by the English Physicist Robert Hooke in 1757, was confirmed by Thomas Young in 1817. Further clarification came in 1873 when Maxwell showed that light waves belong to the family of electromagnetic waves.

Today many simple methods of producing polarized light are known. Usually, an investigator starts with a beam of unpolarized light and then polarizes it by inserting a suitable optical device–a polarizer. However, the invention of the LASER makes it feasible to generate light that is polarized from the outset; no polarizer is needed. Various natural sources of polarized light exist, e.g. rays of light from a portion of the blue sky that is viewed in direction at 90° to the direction of the sun, also light from certain distant galaxies, such as the Crab nebula.

Conceptually, the simplest polarizer is the *micro-wire grid*, which consists of an array of parallel metallic wires each of which is less than a wavelength in diameter and is separated from its neighbors by comparably small distances. When a beam of unpolarized light strikes the grid, the component of the electric vibration that is perpendicular to the wires passes through readily, while the component that is parallel to the wires induces electric currents along them and is reflected or absorbed. The transmission axis of the device, defined in terms of the electric vibration of the transmitted component, is perpendicular to the wires. A far more economical type of polarizer is one that employs long, thin absorbing molecules, rather than wires. The most popular of the commercially produced polarizers, called H-sheet, contains large numbers of long, thin polymeric molecules consisting mainly of iodine atoms. These molecules are embedded in a plastic film that has previously been stretched unidirectionally so as to have a pronounced "grain";

the long slender molecules of iodine conform to this grain. Polarizers containing small absorbing units (whether wires or molecules) that show markedly different extents of absorption for different directions of electric vibration in the incident beam are called *dichroic* polarizers.

The first highly efficient polarizer was of birefringent type: it was made of the crystal *calcite* ($CaO \cdot CO_2$) which has two refractive indices and thus divides any incident beam into two beams. Each of these is linearly polarized, and the sectional patterns are orthogonal. Usually the crystal is artificially shaped so that one of the polarized beams is transmitted straight ahead and the other is deviated and disposed of by total internal reflection. The calcite prism designed by the Scottish physicist Nicol in 1828 was a basic piece of equipment in optics laboratories for over a hundred years. Types devised subsequently by Wollaston, Ahrens, and Foucault have proven to be superior in several respects.

Polarizers of reflection type are also well known. A typical reflection polarizer consists of a plate of glass that is mounted obliquely in the given beam of unpolarized light. The component that is transmitted is found to be partially polarized, and the reflected component is even more highly polarized—with the orthogonal sectional pattern. If the obliquity of the incident beam (measured from the normal to the plate) corresponds to *Brewster's angle*, defined as the angle that has a tangent of n, where n is the refractive index of the plate, the reflected beam is 100 per cent polarized. For ordinary glass, n has the value 1.5 and Brewster's angle is about 56°. Polarization by reflection is of common occurrence; yet few persons are aware of it; they are unaware, for example, that light reflected obliquely from the surface of a pond, a wet road, or a glossy sheet of paper is partially linearly polarized.

Asymmetric scattering is another process that polarizes light. The polarization observed in light from the blue sky is a consequence of scattering of the sun's rays, especially the short-wavelength or blue component thereof, by the molecules of the air.

Perhaps the simplest application of polarizers is in controlling the intensity of a light beam. For this purpose two polarizers are used, in series. If they are oriented so that their transmission axes are parallel to one another, the over-all transmittance is large. If they are oriented so that the two axes are crossed, i.e., at an angle $\theta = 90°$ to one another, the transmittance is zero; the beam is said to be extinguished. For intermediate angles-of-crossing the transmission is easily calculated from Malus' law, which affirms that the transmittance is proportional to $(\cos \theta)^2$. If the incident beam is already polarized, a single polarizer suffices to reduce the intensity to any desired extent. Polarizing sunglasses are effective in this manner thanks to the fact that light reflected obliquely from roads and most other nearly horizontal surfaces is partially *horizontally* polarized; since the polarizer lenses of the sunglasses are oriented with their transmission axes *vertical*, much of this reflected "glare" light is blocked. A polarizer that is employed to block an already polarized beam is called an analyzer.

Much of the interest in polarized light stems from the surprising convertibility of polarization form. By interposing an appropriate retardation plate, or *retarder*, in a given polarized beam, an experimenter can alter the polarization form at will, and with almost 100 per cent efficiency. A typical retarder consists of a thin flat crystal that exhibits birefringence, i.e., has two different refractive indices. Mica, being birefringent and being easily cleaved into thin plates, is often used. When a beam of polarized light enters a plate of mica, the beam is divided into two components and the phases of these are affected ("retarded") to different extents; thus when the components emerge from the plate and unite to form a single beam again, this latter is found to have a drastically altered sectional pattern. Especially versatile and accurate control of polarization form can be achieved with retarders of calcite or quartz. The effect of any given retarder on any given beam can be predicted accurately by various conventional means and, more recently, by a matrix algebra perfected by the American scientist, Hans Mueller; the procedure is to multiply the four-element vector representing the beam by the sixteen-element matrix representing the retarder. Since the vectors and matrices are tabulated in various books, the procedure entails little effort; indeed it is readily extended to cases in which there are several retarders arranged in series. In cases where high accuracy is not needed, predictions can be made especially rapidly with the aid of a kind of map, or spherical slide-rule, called the Poincaré sphere after its inventor Henri Poincaré.

When an object of glass or transparent plastic is subjected to a unidirectional stretching or compressing force, it becomes birefringent and thus acts like a retarder. Accordingly an engineer who wishes to evaluate the unidirectional strain within such an object can do so by directing a beam of polarized light through it and, with the aid of a calibrated retarder and an analyzer, measuring the change in the sectional pattern of the beam. Using conversion factors published in books on photoelastic analysis, he can interpret the change in terms of the direction and magnitude of the strain (see PHOTOELASTICITY).

Many microscopic biological objects, such as components of living cells, appear transparent and virtually invisible under a microscope. Yet such components often contain groups of aligned birefringent molecules and thus are capable of acting like miniature retarders. If the biologist illuminates a living cell with polarized light and examines it under a microscope that is equipped with an analyzer, he finds the birefringent components to be highly visible. Thus the use of polarizers renders visible a microscopic world that is normally invisible. Similarly, mineralogists find that the polarizing microscope greatly increases the visibility of small birefringent crystals.

There are many other applications of polarizers. Photographers use them to increase the contrast between white clouds and the (polarized) blue sky. Chemists use them to measure the extents to which various liquid solutions rotate the sectional pattern of a linearly polarized beam; the extent of rotation is a measure of the concentration of the solution. Electronics engineers use circular polarizers to trap and thus eliminate reflected flare from radar screens. Illumination engineers have devised systems of polarizing filters for automobile headlights and windshields that eliminate glare in night-time driving. Biologists have found that the direction of growth of certain algae can be controlled by illuminating the algae with polarized light of controlled sectional pattern. Bees and ants can detect linear polarization directly by eye, and they employ the polarization of blue sky light as a navigational aid. Even man can learn to detect the polarization of white light with the naked eye, and he can also distinguish right from left circularly polarized light.

Physicists use polarizers to study the emission of polarized light by atoms situated in regions of strong electric or magnetic field, to determine the strength of the sun's magnetic field by measuring the polarization of certain solar spectral lines, and to determine the pattern of magnetic fields within the Crab Nebula. They use polarizers to analyze the behavior of the remarkable light source *the laser* and to verify theoretical predictions as to the polarization inherent in the synchrotron radiation emitted by certain high-energy accelerators.

Because of light's puzzling dual character (waves and photons) and its central position in the growing field of physics, and because *polarized* light is a most elemental form of light, physicists are confident that polarized light will continue to be an outstandingly challenging enigma as well as a most versatile tool for many generations to come.

WILLIAM A. SHURCLIFF

References

Ditchburn, R. W., "Light," Second edition, New York, Interscience Publishers, 1963, 833 pp.

Jenkins, F. A., and White H. E., "Fundamentals of Optics," Third edition, New York, McGraw-Hill Book Co., 1957, 639 pp.

Land, E. H., "Some Aspects of the Development of Sheet Polarizers," *J. Opt. Soc. Am.*, **41**, 957 (1951).

Shurcliff, W. A., "Polarized Light: Production and

Use," Cambridge, Mass., Harvard University Press, 1962, 207 pp.

Shurcliff, W. A., and Ballard, S. S., "Polarized Light," New York, Van Nostrand Reinhold, 1964.

Cross-references: ELECTROMAGNETIC THEORY, LASER, LIGHT, PHOTOELASTICITY, PHOTOGRAPHY, PHOTON, QUANTUM THEORY.

POLARON

An electron in the conduction band of an insulator (or a hole in the valence band) polarizes the medium in its neighborhood. This effect is particularly important in ionic crystals, on account of their high polarizability. The electron together with its associated cloud of lattice polarization is called a *polaron*. The subject has been of interest ever since Landau[1] suggested that an electron could become self-trapped by the lattice polarization it induces; this would mean that its effective mass would be very large compared with its "bare" mass (as given by band theory). The name *polaron* was proposed by Pekar.[2] Theoretical treatments of the polaron fall into two classes: the "dielectric" polaron of Pekar[2] and of Fröhlich, Pelzer and Zienau[3] (FPZ) and the "molecular " polaron, first described by Fröhlich and Sewell[4] and by Holstein.[5] (This terminology is due to Mott;[6] the names "large" and "small" polaron respectively are common in the literature, but are somewhat misleading.)

In FPZ, the dielectric polaron is described by a simple model Hamiltonian. The ionic medium is represented by a set of oscillators; each oscillator is characterized by a wave vector $\mathbf{q}$. However, the oscillator frequency ω is simply related to the Reststrahl frequency ω_t, and does not depend on $\mathbf{q}$. If $\mathbf{p} = -i\nabla$ is the electron momentum, and $\mathbf{r}$ its position, m the band mass, Ω the normalization volume, and if $a_{\mathbf{q}}^+$, $a_{\mathbf{q}}$ are canonical creation and annihilation operators for oscillator quanta ("optical phonons") then the FPZ Hamiltonian is

$$H = -\frac{1}{2}\nabla^2 + \sum_{\mathbf{q}} a_{\mathbf{q}}^+ a_{\mathbf{q}} + i\left(\frac{2\sqrt{2}\pi\alpha}{\Omega}\right)^{1/2}$$
$$\cdot \sum_{\mathbf{q}} (a_{\mathbf{q}}^+ e^{-i\mathbf{q}\cdot\mathbf{r}} - a_{\mathbf{q}} e^{i\mathbf{q}\cdot\mathbf{r}}), \qquad (1)$$

in units such that the electron band mass $m = 1$, $\hbar = 1$; the frequency ω of the "optical phonons" is assumed to be independent of the wave vector $\mathbf{q}$, and is taken as the frequency unit. The parameter α is defined as

$$\alpha = \frac{e^2}{\hbar}\left(\frac{1}{\epsilon_\infty} - \frac{1}{\epsilon_0}\right)\sqrt{\frac{m}{2\hbar\omega}}, \qquad (2)$$

where ϵ_∞ is the high frequency dielectric constant. The FPZ model is one of a particle

interacting in a very simple way with a (non-relativistic) field—the only parameter is the dimensionless coupling constant α. The problem is therefore of fundamental field-theoretical interest, as well as being interesting for its applications to real ionic crystals.

For $\alpha \lesssim 1$, a perturbation-theoretical and a Tamm-Dancoff type of variational calculation agree that Landau self-trapping does *not* occur. The binding energy (in units of $\hbar\omega$) is

$$E_0 = -\alpha + O(\alpha^2), \qquad (3)$$

and the effective mass is

$$m^* = 1 + \tfrac{1}{6}\alpha. \qquad (4)$$

The "intermediate coupling" theory of Lee, Low, and Pines[7] extends the range of validity to $\alpha \lesssim 5$, without changing the essential character of the results. It is based on a canonical transformation which partially decouples the electrons from the phonons.

For very large α ("strong coupling," $\alpha \gtrsim 10$), a variational treatment by Pekar[2] *does* lead to very large effective masses:

$$E_0 = -\gamma\alpha^2, \qquad (5)$$
$$m^* = 16\alpha^4/81\pi^2, \qquad (6)$$

where[8]

$$\gamma = 0.108513. \qquad (7)$$

However, for $5 \lesssim \alpha \lesssim 10$, neither the LLP nor the Pekar method is good. A method due to Feynman[9] provides a bridge between the strong- and intermediate-coupling regimes. The electron propagator (for states with no free phonons) is expressed as an integral over all possible paths of the exponential of an action functional. The oscillator coordinates do not appear explicitly, but the action contains a term which represents the Coulomb interaction of the electron *with itself at earlier times*:

$$\langle \mathbf{r}_1, t'|t_2, t''\rangle = \int \mathfrak{D}(\mathbf{r}(t))\exp S; \qquad (8)$$

$$S = -\int_{t'}^{t''}\tfrac{1}{2}\dot{\mathbf{r}}^2\,dt + 2^{-3/2}\alpha\iint_{t'}^{t''}dt_1\,dt_2$$
$$\cdot |\mathbf{r}(t_1) - \mathbf{r}(t_2)|^{-1}\exp - (t_1 - t_2). \qquad (9)$$

(For convenience, Feynman uses an imaginary time variable.) Feynman makes the variational *Ansatz* for the action

$$S_0 = -\tfrac{1}{2}\int_{t'}^{t''}\dot{\mathbf{r}}^2\,dt - C\iint_{t'}^{t''}dt_1\,dt_2$$
$$\cdot (\mathbf{r}(t_1) - \mathbf{r}(t_2))^2\exp - w(t_1 - t_2), \qquad (10)$$

where C and w are parameters to be determined. This trial action is the exact action for a model system in which the electron is coupled to a fictitious particle of mass $M = 4C/w^3$ through a spring constant $K = 4C/w$, after the coordinates of the fictitious particle have been eliminated. An upper bound for the energy is found and minimized with respect to w and C. For general α, this bound has had to be evaluated numerically. However, analytic expressions exist in the limit of small and large α. When $\alpha \lesssim 3$, Feynman's expression reduces to the LLP solution (3). When $\alpha \gtrsim 10$ it reduces to Pekar's[2] solution (5), except that the constant γ takes the value $1/3\pi$, about 2% too small. Gerlach and Leschke[8] have shown how the Feynman theory can be refined in the strong-coupling limit, to bring it into complete agreement with Pekar.

The effective mass and the mobility of the Feynman polaron have been calculated;[10] they vary smoothly with α. Higher-order corrections[11] are $\lesssim 2\%$ for all α.

When α is large, the radius of the dielectric polaron is small. A measure of the radius R is the amplitude of the zero-point oscillations of Feynman's model system:

$$R = \sqrt{3}\,(4Cw + w^4)^{1/4}/2\sqrt{C}, \qquad (11)$$

where the unit of length is $\sqrt{(\hbar/m\omega)} \simeq 10$ Å. For large α, R is found to be $\sim\alpha^{-1}$, which can become less than the lattice spacing. When this occurs the dielectric model is no longer appropriate and we have the "molecular" (or "small") polaron. Frohlich and Sewell[4] have proposed the use of a Bloch tight-binding model; polarons localized on adjacent lattice sites are assumed to have small overlap integrals. The polaron mass can then become very large. Conduction occurs by a mixture of quantum-mechanical tunneling and thermally activated electron hopping. Holstein[5] has developed a perturbation theory in which the nearest-neighbor overlap integral J is the expansion parameter. His expression for the rate of hopping from site i to site $i + 1$ is

$$\nu_{i \to i+1} = \frac{J^2}{2\hbar} \sqrt{\left\{ \frac{2\pi}{E_a \hbar \omega_0 \ \mathrm{csch}\,(\hbar\omega_0/2kT)} \right\}}$$

$$\cdot \exp\left\{ -\frac{4E_a}{\hbar\omega_0} \tanh \frac{\hbar\omega_0}{4kT} \right\}, \qquad (12)$$

where E_a is the barrier height and ω_0 is the characteristic frequency of the bound electron (the binding being modeled by a harmonic-oscillator potential). At high temperature ($kT \gg \hbar\omega_0$), thermally activated hopping is dominant, and the exponential function in (12) reduces to the usual thermal-activation expression $\exp(-E_a/kT)$. In this high-temperature re-

gime, the molecular polaron is characterized by a mobility $\mu \propto T^{-1}$.

Experimental evidence for dielectric polarons is extensive—especially for electrons in alkali halides[12]—but it has been suggested that holes in alkali and alkaline earth halides and in some oxides[13] show polaron localization. Mott[6] suggests that in amorphous SiO_2, dielectric polarons are formed rapidly (i.e., a time scale of $\sim 10^{-12}$ s), but that formation of molecular polarons is delayed ($\sim 10^{-5}$ s).

C. G. KUPER

References

General references: Kuper, C. G., and Whitfield, G. D. (Eds.), "Polarons and Excitons," Edinburgh, Oliver & Boyd Ltd., 1963. J. T. DeVreese (Ed.), "Polarons in Ionic Crystals and Polar Semiconductors," Amsterdam, North Holland Publ. Co., 1972.

1. Landau, L. D., *Phys. Z. Sowjetunion* **3**, 644 (1933).
2. Pekar, S. I., *Zh. Eksp. i Teor. Fiz.* **16**, 335, 341 (1946); Allcock, G. R., "Polarons and Excitons," p. 45, 1963.
3. Fröhlich, H., Pelzer, H., and Zienau, S., *Phil. Mag.* **41**, 221 (1950).
4. Fröhlich, H., *Arch. Sci. Genève* **10**, 5 (1957); Fröhlich, H., and Sewell, G. L., *Proc. Phys. Soc.* **74**, 643 (1959); Sewell, G. L., *Phil. Mag.* **3**, 1361 (1958).
5. Holstein, T., *Ann. Phys.* **8**, 343 (1959); *Phil. Mag.*, **37B**, 49, 499 (1978).
6. Mott, N. F., *Advanc. Phys.* **26**, 363 (1977).
7. Lee, T.-D., Low, F., and Pines, D., *Phys. Rev.* **90**, 297 (1953), hereafter quoted as LLP; Gurari, M., *Phil. Mag.* **44**, 329 (1953); Tiablikov, S. V., *Zh. Eksp. i Teor. Fiz.* **25**, 688 (1953).
8. Gerlach, B., and Leschke, H., "Recent Developments in Condensed Matter Physics," Vol. 3, p. 365 (Devreese, J. T., Lemmens, L. F., van Doren, V. E., and van Royen, J., Eds.) New York, Plenum Press, 1981; Miyake, S. J., *J. Phys. Soc. Japan* **38**, 181 (1975).
9. Feynman, R. P., *Phys. Rev.* **97**, 660 (1955); Schultz, T. D., *Phys. Rev.* **116**, 526 (1959).
10. Feynman, R. P., Hellwarth, R. W., Iddings, C. K., and Platzman, P. M., *Phys. Rev.* **127**, 1004 (1962).
11. Marshall, J. T. and Mills, L. R., *Phys. Rev.* **B2**, 3143 (1970); Marshall, J. T. and Chawla, M. S., *Phys. Rev.* **B2**, 4209 (1970).
12. Brown, F. C., "Polarons and Excitons," p. 323, 1963; Ascarelli, G., "Polarons and Excitons," p. 357, 1963.
13. Castner, T. G., and Känzig, W., *J. Phys. & Chem. Solids* **3**, 178 (1957); Cox, R. T., "Recent Developments in Condensed Matter Physics," Vol. 3, p. 355, (Devreese, J. T., et al., Eds.) New York, Plenum Press, 1981.

Cross-references: ENERGY LEVELS, EXCITONS, POLAR MOLECULES.

POLYMER PHYSICS

Polymers are long-chain molecules with molecular weights of from thousands to many millions (generally between 20,000 and 10^7 for materials of practical interest). The molecules may be linear, branched, or crosslinked to give a gel structure.

The size and shape of polymer molecules are generally determined from measurements in dilute solution. Molecular weights are obtained from osmotic pressure (number average molecular weight), light scattering (weight average molecular weight), and intrinsic viscosity measurements. Dissymmetry of light scattered by dilute solutions gives the size of molecules. Extent of chain branching can be estimated from light scattering or solution viscosity measurements by comparing the branched polymer with a linear polymer of the same molecular weight. The degree of cross-linking can be determined from the extent of swelling in a solvent or from the elastic modulus by using the kinetic theory of rubber. Swelling decreases and modulus increases as cross-linking increases. Solution properties are sensitive not only to molecular weight but also to the interaction between the polymer and solvent molecules. Most polymers have a distribution of molecular weights. In addition to the results from fractionations, the width of the distribution can be estimated from the ratio of weight average to number average molecular weights. This ratio is 1.0 if all the molecules are the same; it is around 2 for most polymers, but may be much greater for some highly branched polymers.

More recently it has been possible to determine the size and shape of individual polymer molecules in the bulk using the technique of small angle neutron scattering. The successful application of this technique depends on the fact that deuterium and hydrogen have very different coherent neutron scattering cross sections. Thus a small amount of deuterated polymer is dissolved in a protonated matrix and this provides contrast. One of the most important results from these studies is that polymer chains in the bulk assume their "unperturbed" dimensions as predicted by theory.

The molecular structure of polymers is unusually complex since the molecules can assume many conformations; more than one type of monomeric unit can make up the chains to give an infinite variety of distribution of sequence lengths, or the monomeric units can be arranged in different types of stereoregularity—isotactic, syndiotactic, or atactic forms. Nuclear magnetic resonance and infrared spectroscopy are especially powerful techniques for studying the structure of polymers. For polymers capable of crystallizing, x-ray diffraction is another useful tool.

The most important quantity determining the mechanical and many other physical properties of polymers is the *glass transition temperature*. If the glass transition temperature T_g is below ambient temperature, the molecules have extensive freedom of movement, so the material is either a viscous liquid or a rubbery material with a low elastic modulus. If T_g is above ambient temperature, the movement of the molecules is frozen in, so that the polymer is a rigid solid with a high elastic modulus of the order of 10^9 Pa. The glass transition temperature is not sharply defined but depends to some extent on the time scale of the experiment—the faster the experiment, the higher the apparent T_g. Glass transitions may be measured by many techniques, such as where breaks occur in the slope of volume or refractive index vs temperature curves or by the rapid change in elastic modulus with temperature in the transition region. The position of T_g on the temperature scale is largely due to the stiffness of the polymer chains. Flexible molecules such as polybutadiene and silicone rubbers have low T_g, while stiff molecules such as polystyrene and polymethyl methacrylate have high transition temperatures. Cohesive energy density or polarity is another important factor in determining T_g. Symmetry plays a secondary role. Glass transitions can be regulated by copolymerization or by addition of a plasticizer which lowers T_g.

Many polymers including polyethylene and isotactic polypropylene are semicrystalline. In their bulk behavior such polymers behave as though they are a mixture of amorphous and crystalline materials, but the exact nature of the crystalline state is not yet clearly defined for such materials. In the crystal lattice, some types of polymer chains assume a zig-zag conformation while others crystallize in the form of helices. Single crystals of some polymers have been grown from dilute solutions. In these single crystals the chains are perpendicular to the faces making up the thin lamellar crystals, so that each polymer chain must fold back on itself several times. There is some morphological evidence, based on electron microscopy studies, that even in the bulk polymer cooled down from the melt there is extensive chain folding in the crystalline phase. In terms of a two-phase model, the degree of crystallinity may be determined by x-ray, density or heat capacity measurements. Different techniques generally give similar but not identical values for the degree of crystallinity; typical values vary from 40% crystallinity for low-density polyethylene to 85% for high-density polyethylene. Crystallinity is greatly affected by chain perfection. Copolymerization and branching greatly reduce crystallinity. High stereoregular polymers such as isotactic polystyrene tend to be crystalline, while the random atactic polymers are noncrystalline. The melting point also depends upon chain perfection—the greater the degree of imperfection, the lower the melting point.

Many polymers of commercial importance are not linear polymers consisting of a single type of monomeric unit. Copolymers contain two or more kinds of monomers. If one type of monomer makes up the backbone and another type side chains, the polymers are called *graft polymers*. If two polymers are mechanically mixed together, the mixture is called a *polyblend*. Most, but not all, polyblends are two-phase systems. Block polymers, consisting of long sequences of one kind of polymer chain attached to the end of another kind of polymer chain, are also generally two-phase systems in which parts of a single molecule can be in two phases simultaneously. Two-phase systems are important commercially because of their great toughness.

Even at high temperatures where linear polymers are liquid, they tend to be very viscous. The melt viscosity is especially high if the molecular weight is above a critical value where chain entanglements can occur. At molecular weights above which entanglement occurs, the melt viscosity at low rates of shear depends approximately on molecular weight. These viscous polymer melts are also more or less elastic in nature and behave somewhat like rubber. If the molecules are crosslinked to one another, the elastic behavior becomes dominant and true vulcanized rubbers result. Polymer melts are generally non-Newtonian, and the properties are very dependent upon the rate of shear. Molecular theories have been developed by Rouse, Zimm, and Bueche which explain quite well many of the rheological properties of melts and solutions.

The Rouse, Bueche, and Zimm theories do not explain the long-range motions of entangled polymer chains such as those that occur in the melt under processing conditions. De Gennes, Edwards, and Doi have developed a theory based on scaling laws taken from the renormalization group approach which has proved so successful in elementary particle physics. In this theory the motion of an entangled polymer chain is confined along its contour length so that it moves as if it were inside a tube. This process is called *reptation*. The reptation concept has been successful in explaining the molecular weight dependence of the flow of entangled polymer molecules.

The usefulness of polymers depends primarily upon their mechanical properties. The elastic modulus of rubbers is explained quite satisfactorily by the kinetic theory of rubber elasticity. No satisfactory theory is yet available for rigid polymers. Dynamic mechanical measurements using oscillating stresses or strains have been especially useful in relating mechanical properties to molecular structure; such tests are generally made to measure the elastic modulus and mechanical damping over a wide range of frequencies and temperatures. Generally the effects of temperature and frequency (or time) can be made equivalent by a superposition treatment such as the one developed by Williams, Landel, and Ferry. The phenomenological theory of viscoelasticity has developed to the stage where it is possible to interconvert data from one type of test (say dynamic mechanical) to other types of tests such as creep, stress relaxation and, to a lesser extent, stress-strain data. On heating a rigid organic polymer, the modulus drops from about 10^9 Pa to a low value of about 10^6 Pa in a small temperature interval near the glass transition temperature, unless the polymer is highly crosslinked or crystalline. Both crystallinity and crosslinking can greatly increase the modulus above T_g, but they have little effect on the modulus below T_g. Polymers are unique in that some of them can be elongated over 1000% before they break.

The dielectric constant and power factor or electrical loss depend upon the number and type of dipoles. At low temperatures (or high frequencies) where the dipoles are frozen in, the dielectric constant and electrical loss are both low. At high temperatures (or low frequencies) where the molecules have high mobility, the dielectric constant is high while the electrical loss is often low. At intermediate temperatures or frequencies, where the main relaxation times for dipolar motion are approximately the same as the applied electrical frequency, the electrical loss goes through a pronounced maximum, and the dielectric constant changes rapidly with either frequency or temperature. Since the mobility of the chain backbone is related to the ease with which dipoles can move, there is often a good correlation between electrical and mechanical properties. Impurities in polymers can be very detrimental to good electrical properties, especially at high temperatures where conductivity can be relatively high.

Most pure amorphous polymers are transparent. Crystallinity often makes a material milky or white in appearance. Long chain molecules can be oriented by stretching in the molten state or by cold-drawing in some cases. Such oriented materials are generally highly birefringent, since the polarizability along the chain is usually quite different from the polarizability perpendicular to the chain. The mechanical properties of oriented polymers such as fibers are also highly anisotropic. For instance, the modulus and tensile strength are generally much greater parallel to the chain axis than perpendicular to it.

An interesting development has been to ultraorient polymers by various techniques to produce high modulus, high strength fibers. In some cases this has led to the production of materials with a strength greater than that of steel. In order to accomplish this, liquid-crystal-forming polymers may be spun into fibers from anisotropic solutions or various high drawing processes may be used, including solid state extrusion.

Polymers have very high coefficients of thermal expansion compared to most rigid materials. The coefficients of expansion shows a distinct break at the glass transition temperature—the coefficient being greater above Tg. Most polymers would be classed as thermal insulators rather than as thermal conductors.

The discovery of high electrical conductivity in organic polymers such as polyacetylene when doped by arsenic pentafluoride has led to a great deal of experimental and theoretical activity to explain the phenomenon. Conduction is postulated to proceed by means of solitons but many questions remain obscure. This is at present a very active field and at least one superconducting polymer, sulfur nitride, $(SN)_x$, has been identified. Although major technological developments involving conducting polymers remain to be realized, a storage battery using polyacetylene has been described.

L. Nielsen
William J. MacKnight

References

Bueche, F., "Physical Properties of Polymers," New York, Interscience Publishers, 1962.

de Gennes, Pierre-Gilles, "Scaling Concepts in Polymer Physics," Ithaca, New York, Cornell Univ. Press, 1979.

Ferry, J. D., "Viscoelastic Properties of Polymers," 3rd. Ed., New York, Wiley, 1980.

Jenkins, A. D. (Ed.), "Polymer Science," Volumes I & II, Amsterdam, North Holland, 1972.

Nielsen, L., "Mechanical Properties of Polymers," New York, Van Nostrand Reinhold, 1962.

Tobolsky, A. V., and Mark, H., "Polymer Science and Materials," New York, Wiley-Interscience, 1971.

Cross-references: DIELECTRIC THEORY, LIGHT SCATTERING, MECHANICAL PROPERTIES OF SOLIDS, MOLECULAR WEIGHT, MOLECULES AND MOLECULAR STRUCTURE, NEUTRON DIFFRACTION, OSMOSIS, VISCOELASTICITY, VISCOSITY.

POSITRON

The positron is one of many fundamental bits of matter. Its rest mass (9.109×10^{-31} kg) is the same as the mass of the electron, and its charge ($+1.602 \times 10^{-19}$ coulomb) is the same magnitude but opposite in sign to that of the electron. The positron and electron are antiparticles for each other. The positron has spin 1/2 and is described by Fermi-Dirac statistics as is the electron (see ELECTRON).

The positron was discovered in 1932 by C. D. Anderson at the California Institute of Technology while doing cloud chamber experiments on cosmic rays. The cloud chamber tracks of some particles were observed to curve in such a direction in a magnetic field that the charge had to be positive. In all other respects, the tracks resembled those of high-energy electrons. The discovery of the positron was in accord with the theoretical work of Dirac on the negative energy states of electrons. These negative energy states were interpreted as predicting the existence of a positively charged particle.

Positrons can be produced by either nuclear decay or the transformation of the energy of a gamma ray into an electron-positron pair. In nuclei which are proton-rich, a mode of decay which permits a reduction in the number of protons with a small expenditure of energy is positron emission. The reaction taking place during decay is

$$p^+ \rightarrow n^0 + e^+ + \nu$$

where p^+ represents the PROTON, n^0 the NEUTRON, e^+ the POSITRON, and ν a massless, chargeless entity called a NEUTRINO. The positron and neutrino are emitted from the nucleus while the neutron remains bound within the nucleus. Although none of the naturally occuring radioactive nuclides are positron emitters, many artificial radioisotopes which decay by positron emission have been produced. In fact the first observed case of positron decay of nuclei was also the first observed case of artificial radioactivity. An example of such a nuclear decay is

$$_{11}Na^{22} \rightarrow {}_{10}Ne^{22} + e^+ + \nu \text{ (half-life} \approx 2.6 \text{ years)}$$

This particular decay provides a practical, usable source of positrons for experimental purposes.

The process of pair production occurs when a high-energy gamma ray interacts in the electromagnetic field of a nucleus to create a pair of particles—a positron and an electron. Pair production is an excellent example of the fact that the rest mass of a particle represents a fixed amount of energy. Since the rest energy ($E_{rest} = m_{rest}c^2$) of the positron plus electron is 1.022 MeV, this energy is the gamma energy threshold and no pair production can take place for lower-energy gammas. In general, the cross section for pair production increases with increasing gamma energy and also with increasing Z number of the nucleus in whose electromagnetic field the interaction takes place.

The positron is a stable particle (i.e., it does not decay itself), but when it is combined with its antiparticle, the electron, the two annihilate each other and the total energy of the particles appears in the form of gamma rays. Before annihilation with an electron, most positrons come to thermal equilibrium with their surroundings. In the process of losing energy and becoming thermalized, a high-energy positron interacts with its surroundings in almost the same way as does the electron. Thus for positrons, curves of distance traversed in a medium as a function of initial particle energy are almost identical with those of electrons.

It is energetically possible for a positron and an electron to form a bound system similar to the hydrogen atom, with the positron taking the place of the proton. This bound system has been given the name "positronium" and the chemical symbol Ps. Although the possibility of positronium formation was predicted as early as 1934, the first experimental demonstration of its existence came in 1951 during an investigation of positron annihilation rates in gases as a function of pressure. The energy levels of positronium are about one-half those of the hydrogen atom since the reduced mass of positronium is about one-half that of the hydrogen atom. This also causes the radius of the positronium system to be about twice that of the hydrogen atom. Thus positronium is a bound system with a radius of 1.06Å and a ground state binding energy of 6.8eV. As mentioned previously, the positron has an intrinsic magnetic moment and an intrinsic angular momentum, or spin. In positronium, the spins of the electron and positron can be oriented so they are either parallel or antiparallel. These two states, called ortho-positronium and para-positronium respectively, have very different annihilation characteristics. Most positrons entering a medium do not form positronium, but the general annihilation characteristics show the same dependence on orientation of the spins, regardless of whether the annihilation occurs in a collision or from the bound state.

For many years all attempts to observe positronium through the emission of its characteristic spectral lines were unsuccessful. These lines are similar to hydrogen's except that the wavelengths of all corresponding lines are doubled. Finally in 1975 the Lyman-α line (the radiation emitted when the first excited state of positronium decays to the ground state) was observed in the laboratory. In another recent development the long-predicted positronium negative ion, consisting of two electrons and one positron bound together by electrical attraction, was detected experimentally in 1981. Both of these experiments were made possible by the discovery that positronium can be formed efficiently in a vacuum when a solid target is bombarded by a slow positron beam.

Positronium is also the ideal system in which the calculations of quantum electrodynamics can be compared with experimental results. In fact measurements of the fine-structure splitting of both the ground state and the first excited state have served as important confirmations of the theory of quantum electrodynamics.

It is possible for a positron-electron system to annihilate with the emission of one, two, three, or more, gamma rays. However, not all processes are equally probable. One-gamma annihilation requires another particle to partici-

pate to conserve momentum. This process is a very infrequent type decay. The most probable decay is by the emission of two gamma rays, directed in opposite directions, with each possessing about one-half the energy of the system. The presence of these 0.511-MeV ($=m_e c^2$) gamma rays is always found when positrons are present. Whether annihilation is to be by the emission of one, two or three gammas depends on the orientation of the spins of the positron and electron. Conservation of charge parity and angular momentum requires that the number of gamma rays emitted be an even number (most likely 2) if the spins are antiparallel, and be an odd number (most likely 3, but possibly 1) if the spins are parallel.

If formed in free space, positronium exhibits two characteristic lifetimes against self-annihilation. These are $\tau_1 = 1.25 \times 10^{-10}$ second for the anti-parallel spin case (also called the singlet state) and $\tau_3 = 1.39 \times 10^{-7}$ second for the parallel spin case (called the triplet state). Another lifetime, characteristic of the physical surroundings of the positron, is found when positronium is formed in certain condensed materials. This lifetime (known as τ_2) is longer than the singlet free space lifetime τ_1, but is much shorter than the triplet free space lifetime τ_3. In general, this τ_2 lifetime is a measure of the rate of "pickoff" of atomic electrons with antiparallel spins by the positrons in triplet positronium. In this process, the positron enters the material and forms triplet positronium with an electron, but then annihilates with an electron belonging to one of the surrounding atoms, whose spin is oriented opposite to that of the positron. That the probability of "pickoff" and the subsequent two-gamma annihilation depend on the properties of the surroundings is to be expected. Indeed, the τ_2 lifetime is a function of the material, the temperature, the density, the degree of crystallinity, the phase, etc. It has been found in some cases that even in the same material not all positrons in triplet positronium have the same "pickoff" probability, a fact revealed by the presence of more than one τ_2 component. Positron lifetimes have been measured in a great variety of substances in an effort to correlate trends in the τ_2 lifetime values with chemical or physical properties.

The angular correlation of the annihilation gammas has been measured for both the two-gamma and three-gamma cases. In three-gamma annihilation, the gammas are coplanar as predicted, and azimuthally correlated such that their energies and directions are consistent with the conservation laws of energy and momentum. Two-gamma annihilation studies have been extensive, the results showing that the two gammas are emitted within a few milliradians of 180° from each other. The width of the two-gamma angular distributions is a measure of the linear momentum of the positron-electron system when annihilation occurs. These measurements can be used to gain information on the momentum distribution of the electrons with which the positrons annihilate. The development of high resolution semiconductor gamma-ray detectors has provided a faster but less accurate method of investigating the electron momentum distribution in materials by positron annihilation. This method is based on measurement of the Doppler shift in the energy of the annihilation gamma rays caused by the motion of the positron-electron pair at annihilation.

In recent years several applications of positrons have been developed. For example, a positron annihilation has been used to determine Fermi surfaces and to study defects in metals and metallic alloys. It has also become possible to produce slow monochromatic positron beams which show promise of being useful in the study of the solid surfaces to complement the well-established technique of low-energy electron diffraction (LEED). A recent medical application of positrons is positron-emission tomography (PET). This is a process by which a patient's metabolic activity, often metabolism in the brain, is observed. This technique gives a fundamentally different result from that obtained by x-rays or CAT scans, which display density differences rather than the chemical activity levels as does PET.

B. CLARK GROSECLOSE
WILLIAM W. WALKER

References

1. Goldanskii, V. I., *Atomic Energy Review* 6, 1 (1968).
2. West, R. N., "Positron Studies of Condensed Matter," London, Taylor and Francis, 1974.
3. Ache, H. J. (Ed.), "Positronium and Muonium Chemistry," Advances in Chemistry Series, Vol. 175, Washington, D.C., American Chemical Society, 1979.
4. Hautojärvi, P. (Ed.), "Positrons in Solids," Berlin, Springer-Verlag, 1979.

Cross-references: ANTIPARTICLES, ATOMIC PHYSICS, BIOMEDICAL INSTRUMENTATION, ELECTRON, ELEMENTARY PARTICLES, NEUTRINO, NEUTRON, NUCLEAR STRUCTURE, PROTON.

POTENTIAL

The concept of potential has developed in several fields in physical phenomena from either one or another of two viewpoints. One idea is that there can be the storage of some entity which results in energy storage in the system. This stored energy may be released for dynamic use. The energy so stored can thus be considered as potential energy.

The other idea from which the concept of potential has been developed is that it is some scalar function whose space rate of change yields a vector force which may be useful for mathematical analyses or for the development of other physical quantities. As will be shown, these two ideas are compatible.

From one, the other, or both of these general viewpoints has developed the concept of potential as an entity in the subjects of gravitation, electricity, magnetism, heat conduction, fluid flow, elastic stress, and others. The early analysis of entities in many of these subjects began through experimental observations and mathematical hypotheses of various sorts, but the concept of potential eventually became apparent if it was not in the original analysis.

The concept of potential as a significant entity for the analysis of electric phenomena was developed by Simeon Poisson, George Green, and others in the period from about 1813 to 1827. It was realized that energy could be released from electrically charged bodies, and the concept of potential was developed as one characteristic of such a charged system that measured the ability of the system to release this energy. The energy stored in the system was and is called *potential energy* and is a scalar quantity.

The discussion of this article is phrased in electrical terminology, although many of the interrelationships among entities discussed here are applicable to the subjects of gravitation, heat conduction, fluid flow, and other subjects. The concepts of scalar magnetic potential and vector magnetic potential are included here so that the basic concepts in electromagnetism are complete.

Closely associated with the concept of electric potential was the concept of *charge* that had been formulated by Charles Augustin de Coulomb and others over about a 100-year period starting about 1737. The electric scalar potential of a macroscopic system is defined as the electric energy of the system divided by the electric charge. This simple definition presumes a basic two-conductor, statically charged system. The mathematical expression for the definition of the potential Φ is

$$\Phi = \frac{W}{Q}$$

where W is the energy of the system and Q is the charge (see list of units at end of article).

Modern electric systems employ the concept of potential for much more sophisticated forms through the use of summations of charge effects either in discrete or in distributed forms. The complete expression for electric potential caused by the accumulation of a number of concentrated charges Q_k, surface charge density σ over a surface S, and volume charge density ρ in a volume V, all being located in a dielectric medium of uniform permittivity ϵ, is

$$\Phi = \sum_{k=1}^{n} \frac{Q_k}{4\pi\epsilon r} + \frac{1}{4\pi\epsilon} \int_S \frac{\sigma ds}{r} + \frac{1}{4\pi\epsilon} \int_V \frac{\rho dv}{1}$$

where r measures the magnitude of the distance from the point at which the electric potential is evaluated to each charged particle or element of charge.

The difference in electric potential from one point a to another point b in a static electric field is related to a vector function called the electric field intensity $\mathcal{E}$ (a characteristic of the space related to the negative of the gradient of the electric potential) through the line integral relationship

$$\Phi_{ab} = -\int_a^b \mathcal{E} \cdot d\mathbf{l},$$

where $\mathbf{l}$ is a vector direction measured along the path in the direction from the point a to the point b, and the scalar product between $\mathcal{E}$ and $d\mathbf{l}$ is denoted by the dot or scalar product notation.

The inverse form of this integral expression is the gradient form for the static electric field intensity, namely

$$\mathcal{E} = -\nabla\Phi$$

where the ∇ symbol is a vector differential space operator which, when applied to the scalar electric potential Φ, yields the electric potential gradient.

An extension of spatial derivative operations yields another function of the scalar electric potential Φ, known as the scalar Laplacian, as $\nabla^2\Phi$. This term, when equated to the negative of the volume space charge density ρ divided by the permittivity of the space, is known as Poisson's equation. Thus,

$$\nabla^2\Phi = -\rho/\epsilon$$

If the electric volume charge density is zero, the above relationship becomes

$$\nabla^2\Phi = 0$$

and is known as Laplace's equation.

Electric charges in motion produce magnetic effects, which result in a magnetic field intensity vector $\mathbf{H}$, somewhat analogous to the corresponding vector $\mathcal{E}$ in electric field phenomena. The line integral relationship between this vector $\mathbf{H}$ and the vector direction measured along a path in space from point a to b establishes an analogous scalar magnetic potential Ψ_{ab} as

$$\Psi_{ab} = -\int_a^b \mathbf{H} \cdot d\mathbf{l}$$

This function is useful in the evaluation of magnetic field geometric relations for locations in space where the electric current density is zero.

For regions in magnetic fields where current densities exist, a designation of a magnetic

potential relation can only be made through a function called the *vector magnetic potential*. This vector is defined in a form similar to the expression for scalar electric potential caused by a volume distribution of electric charge density as given earlier. Thus, the vector magnetic potential $\mathbf{A}$ at a point in space, caused by a distribution of current density $\mathbf{J}$ over a volume V is

$$\mathbf{A} = \frac{\mu}{4\pi}\int_V \frac{\mathbf{J}\,dv}{r}$$

where μ is the permeability of space and r is the magnitude of the distance from the point at which $\mathbf{A}$ is evaluated to each element of current density of the system. The point under consideration can be a point within the region of current density.

The vector magnetic flux density $\mathbf{B}$ in such a space is related to the vector magnetic potential $\mathbf{A}$ through a spatial differential function called the curl and symbolized by the operational form as

$$\mathbf{B} = \nabla \times \mathbf{A}$$

An extension of spatial derivative operations upon the vector magnetic potential also yields another function of this potential known as the vector Laplacian, $\nabla^2\mathbf{A}$. This term is related to the current density $\mathbf{J}$ at a point in the field as

$$\nabla^2\mathbf{A} = -\mu\mathbf{J}$$

This expression is analogous to the similar scalar Laplacian of the electric potential which was related to the electric charge density at the point of evaluation.

For systems in which the charges are moving in such a manner that the vector magnetic potential is not constant with respect to time, the elementary form of the relationship between the electric field intensity $\mathcal{E}$ and the scalar electric potential Φ must be modified to include a function of the vector magnetic potential, namely

$$\mathcal{E} = -\nabla\Phi - \frac{\partial\mathbf{A}}{\partial t}$$

This is the general expression that is valid at a point in space for all conditions. As action at a distance is considered, the elementary forms for the evaluation of the scalar electric potential Φ and of the vector magnetic potential $\mathbf{A}$, however, must recognize the time delay in action with respect to the causes.

If the scalar electric potential at a point P is to be evaluated at some time t, the *retarded potential* must consider the finite velocity of propagation that occurs in the path length r between the point P and the location of the charge that causes the electric potential.

Depending upon the nature of the charge, the

retarded potential expression can be expressed in terms of the retarded time $t - r/c$, where c is the velocity of the propagated effect in free space. For the case in which the charge is distributed over a volume, the expression becomes

$$\Phi_{P,t} = \frac{1}{4\pi\epsilon} \int_V \frac{[\rho]_{t-r/c} dv}{r}$$

The symbol $[\rho]_{t-r/c}$ indicates that the charge density at the source of the field is that evaluated at an earlier time $t - r/c$. If the charges are discrete or distributed over surfaces, the corresponding forms of the potential function for these geometries would be used.

In a similar manner, the retarded vector magnetic potential can be expressed for a volume distribution of current density as

$$A_{P,t} = \frac{\mu}{4\pi} \int_V \frac{[J]_{t-r/c} dv}{r}$$

All preceding relations are expressed in a form that results if a rationalized system of units is used. The internationally accepted metric (SI) units that conform with these preceeding relationships are:

Entity	Symbol	Unit
Energy	W	joule
Potential (electric)	Φ	volt
Charge	Q	coulomb
Length	$r, 1$	meter
Area	S	meter^2
Volume	V	meter^3
Surface charge density	σ	coulomb/meter^2
Volume charge density	ρ	coulomb/meter^3
Permittivity	ϵ	farad/meter
Electric field intensity	$\mathcal{E}$	volt/meter
Magnetic field intensity	H	ampere/meter
Scalar magnetic potential	Ψ	ampere
Vector magnetic potential	A	weber/meter
Permeability	μ	weber/meter-ampere
Magnetic flux density	B	tesla
Current density	J	ampere/meter^2
Time	t	second
Velocity	c	meter/second

WARREN B. BOAST

References

Boast, W. B., "Vector Fields," New York, Harper & Row, 1964.

Bradshaw, M. D., and Byatt, W. J., "Introductory Engineering Field Theory," Englewood Cliffs, N.J., Prentice-Hall, 1967.

Durney, C. H., and Johnson, C. C., "Introduction to Modern Electromagnetics," New York, McGraw-Hill, 1969.

Green, G., "Mathematical Papers of the Late George Green," (N. M. Ferrers, Ed.), London, Macmillan and Co., 1871.

Holt, C. A., "Introduction to Electromagnetic Fields and Waves," New York, Wiley, 1963.

Javid, M., and Brown, P. M., "Field Analysis and Electromagnetics," New York, McGraw-Hill, 1963.

Paris, D. T., and Hurd, F. K., "Basic Electromagnetic Theory, " New York, McGraw-Hill, 1969.

Plonsey, R., and Colin, R. E., "Principles and Applications of Electromagnetic Fields," New York, McGraw-Hill, 1961.

Ramo, S., Whinnery, J. R., and Van Duzer, T., "Fields and Waves in Communication Electronics," New York, Wiley, 1965.

Rao, N. N., "Basic Electromagnetics with Applications," Englewood Cliffs, N.J., Prentice-Hall, 1972.

Silvester, P., "Modern Electromagnetics," Englewood Cliffs, N. J., Prentice-Hall, 1968.

Wangsness, R. K., "Electromagnetic Fields," New York, Wiley, 1976.

Wermer, J., "Potential Theory," Lecture Notes in Mathematics, No. 408, Second Edition, Berlin-Heidelberg-New York, Springer-Verlag, 1981.

Cross-references: ELECTRICITY, ELECTROMAGNETIC THEORY, FIELD THEORY, HIGH VOLTAGE RESEARCH, STATIC ELECTRICITY.

PRESSURE, VERY HIGH

In 1950 high pressure research in physics was associated primarily with the macroscopic measurements of classical physics and the geophysics of the outer crust of the earth. The higher pressures now available permit extensions of these studies. However, the most far reaching development has been the use of modern instrumentation to study high pressure effects on atomic and molecular properties. Thus high pressure, like low temperature, has become an integral tool of modern science. This transformation in high pressure research took place first in solid state physics and geophysics. In recent years it has extended to chemical physics and biophysics. (See FLUID STATICS for a discussion of the concept of "pressure" and methods for measuring it.)

For the purposes of this article, "very high pressure" will imply pressures above a few kilobars (1.0 kbar = 987 atm. = 0.1 GPa.). This lower limit permits the discussion of recent significant experiments in chemical and biophysics in the 10–12 kilobar range.

Only a brief outline of techniques for generating pressure is included. Piston and cylinder devices are used to virtually all work below ∼20 kilobars and for many applications to 50 kilobars where samples larger than a few milligrams are required. For higher pressure work the tapered anvil pistons developed by P. W. Bridgman are very widely employed, as well as the modification wherein support is applied also to the taper. The use of diamond anvils in a Bridgman configuration is very wide spread. Improved alignment techniques permit pressures of the order of 500 kilobars, though with frequent breakage of the diamonds. With appropriate liquid mixtures pressures are hydrostatic to 90 kilobars, and the use of solidified rare gases extends the hydrostatic range. An important development in calibration has been the discovery (at the National Bureau of Standards) that the emission of ruby at 692 nm shifts linearly with pressure to several hundred kilobars at least. The diamond cells are especially useful for x-ray, Raman, and Brillouin scattering, and optical absorption at wavelengths longer than 300–350 nm. Since essentially all diamonds luminesce when exposed to intense UV radiation, their usefulness for fluorescence studies is limited to systems which can be excited in or near the visible region of the spectrum.

Bundy and coworkers at General Electric have used a supported taper cell with sintered diamond pistons to make extensive electrical resistance studies to 500 kilobars pressure. Other devices claim static pressures of a megabar or so, but experimental results in this range are very limited at present.

The use of dynamic pressure devices (shockwave) permits pressures in the multimegabar region, accompanied by very high temperatures, but a discussion of shock experiments is beyond the scope of this article.

To give some flavor of the breadth of current high pressure research the discussion contains a section on electronic phenomena, with briefer outlines on structural studies, the liquid state, geophysics, and biophysics.

Electronic Phenomena Probably the most generally significant application of high pressure research in the past 30 years has been the study of electronic phenomena. The basic discovery is that with increasing compression one perturbs different electronic energy levels in different degrees. This "pressure tuning" of electronic energy levels has proved to be a very powerful tool in understanding the behavior of matter and in discovering new electronic phenomena.

High pressure studies have been essential to the testing of numerous theories of electronic phenomena. A few examples are given here:

(1) In the simple form first proposed by Bethe, ligand field theory predicts that the d orbitals of transition metal ions will be split in energy by an amount which increases as R^{-5}, where R is the ion-ligand distance. High pressure studies where R is decreased continuously have established the possibilities and limitations of this analysis. Van Vleck predicted that when this splitting gets sufficiently large the normal high spin configuration would rearrange to permit spin pairing. As discussed below, this electronic transformation has been confirmed by high pressure studies.

(2) Electron donor-acceptor complexes occur widely in solutions of chemical and biological interest. Mulliken developed a theoretical description in which the ground state is held together primarily by van der Waals forces, whereas the excited state involves a transfer of charge from the donor to the acceptor. Compression should stabilize the charge transfer state. Indeed, for a number of complexes this stabilization has been verified by the observation that the charge transfer peak shifts rapidly to lower energy with increasing pressure and increases dramatically in intensity.

(3) Many luminescent and related processes have an efficiency which depends strongly on the transfer of excitation between species in a crystal, a solution, or a macromolecule. This transfer is basic in fluorescent lighting, rare earth lasers, photosynthesis, and the photographic process. The effectiveness of the energy transfer depends strongly on the difference in energy between the donor and acceptor (or activator and sensitizer) states. The pressure tuning of energy levels is an ideal scheme for testing theories of this efficiency. The Förster-Dexter theory has been confirmed quantitatively and a number of theories involving lightning and laser materials have been tested effectively by means of high pressure luminescence measurements.

The relative shift of electronic energy levels can provide a new electronic ground state for a material with changes in electrical, magnetic, optical or chemical properties. One of the most dramatic results of "pressure tuning" is that these electronic transitions are found to be as ubiquitous as structural phase transitions.

The most widely studied transitions involve a change in electrical properties. In the early 1960s, insulator-conductor transitions were observed in molecular crystals such as I_2, Se, As, and some organic materials; in ionic compounds like the thallous halides; and in Si, Ge, GaAs, GaSb, InP, InAs, InSb, ZnS, ZnSe, ZnTe, CdS, CdSe, CdTe, etc., all having the zinc blende or related structures. A few more such transitions have been observed since, but much of the more recent effort has been towards a more detailed understanding of the insulator-conductor transition. Groups at Bell Telephone Laboratories have made thorough studies of such transitions in transition metal oxides and rare earth chalcogenides.

A recent extension of the early studies on I_2 indicates that at ~160 kilobars it becomes a diatomic metal and at 200–210 kilobars it becomes a monatomic metal. These results have been used by theorists as a model for the behavior of diatomic molecular crystals, in particular for hydrogen. Calculations now indicate that hydrogen will become a diatomic metal near 2–2.5 megabars and a monatomic metal at ~5.5 megabars. These pressures, especially the latter, are beyond the foreseeable range of static experiments.

Other types of electronic rearrangements which are detected from electrical properties involve the alkali, alkaline earth, and rare earth metals. In the heavy alkali metals an s electron is promoted to the d band, giving transition metal character to these materials. In the rare earths a $4f$ electron is delocalized into the d band, while calcium and strontium transform from metals to semiconductors at high pressure.

It is of interest that many of these metallic phases which appear at high pressure are superconductors. The study of these materials has, in fact, contributed measurably to the understanding of superconductivity.

Electronic transitions involving magnetic properties include transformations of transition metal ions from high spin to low spin configuration with increasing ligand field, transformation of iron from a ferromagnet to a paramagnetic phase at 115 kilobars, the appearance of ferromagnetism in FeO near 60–80 kilobars, and a variety of transformations in ferromagnetic insulators. High pressure neutron diffraction has proved to be a very elegant technique for these latter studies.

Electronic transitions can also have chemical consequences. As mentioned earlier the excited (charge transfer) state of electron donor-acceptor complexes is stabilized vis-à-vis the ground state with pressure. At sufficiently high pressure a significant fraction of the molecules are in this state, forming radicals which can react in the solid state geometry to form compounds which have never been made by any other approach.

Structural Research The measurement of the effect of high pressure on lattice parameters for simple structures from x-ray powder patterns is an art which is at least 20 years old. The advent of rotating anode electrodes and especially synchrotron radiation, plus improvements in the hydrostaticity of the diamond cell, have made possible single crystal x-ray studies to 100 kilobars or higher, as well as resolution of degrees of ordering in compounds at high pressure, etc. Most of the studies possible on inorganic crystals at one atmosphere can now be performed at high pressure. Organic crystallography under pressure is less advanced.

Recently Brillouin scattering has proven to be a very useful tool for establishing elastic constants of materials under extreme conditions. It has been applied primarily to materials of geological interest.

P. W. Bridgman established many years ago that first-order structural phase transitions were common phenomena at high pressure. With x-ray crystallography it has been possible to establish the structures of the new phase in situ. The most exciting advance in the past decade has been the use of laser Raman scattering to follow the lattice vibrational modes as a phase transition as approached. With increasing pressure most vibrational modes stiffen (shift to higher energy). However, as one nears the pressure at which a structural phase transition occurs, one mode softens and the lattice "comes apart" and rearranges by this mechanism. There has been a great deal of study of various kinds of lattices and characterizations of the relevant

soft modes so that the mechanics of structural phase transitions is, for the first time, understood. This research is an excellent example of how a new tool extracts new and exciting science from an old field.

Liquid State Studies High pressure research has impinged most strongly on modern chemical physics in the areas of properties of and processes in the liquid state. Studies include molecular motion and relaxation processes, luminescence, and chemical reaction rates and synthesis. Much of this research has become possible in the past decade due to the availability of high field NMR equipment, lasers, photon counters, and high speed electronics.

With respect to diffusion and relaxation processes the major conclusion is that in order to test theories adequately it is necessary to study the temperature coefficient of these processes *at constant density*. This coefficient is, in general, very different from the coefficient at constant pressure which is obtained from experiments at one atmosphere.

One of the important aspects in understanding fluids involves the relationship between macroscopic properties such as viscosity or dielectric constant and molecular properties. Pressure induces very large changes in the bulk properties without changing the temperature (thermal energy) or the chemical composition of the system. It has been demonstrated from pressure experiments that the Stokes-Einstein relation between viscosity and molecular diffusion holds remarkably well over ranges of 10^4 in viscosity. Unequivocal relations between dielectric constant and rate of emission of luminescence, and between viscosity and thermal dissipation of molecular excitation have also been established.

The study of pressure effects on chemical reaction rates has long been used to establish the mechanism of chemical reactions. The analyses have been based on the classical Eyring picture of the central role of the transition state. A stochastic theory whose modern development is based on early work of Kramers predicts a different rate controlling step in fluids of moderate viscosity. Recently, high pressure experimentation has provided a clear proof of the applicability of the stochastic theory.

Geophysics The field of high pressure geophysics is one of the oldest aspects of high pressure research and one which is continuing and increasing in intensity. There are numerous active laboratories in the United States as well as in Japan, Australia, and Europe. It is beyond the scope of this article as well as the expertise of the author to review this activity thoroughly so only the barest outline is given.

In the period up to about 1965–70 most high pressure geophysics was concerned with pressure-volume-temperature measurements and changes in crystal structure in the crust and top of the mantle. The usual method was by quenching in the high pressure–high temperature phase. Today one can investigate these phenomena in situ via x-ray diffraction to pressures which correspond to those in the lower mantle. In addition Brillouin scattering permits the determination of elastic constants at high pressure—information of importance in seismology. Magnetic properties of high pressure forms can be established via Mössbauer resonance. High pressure high temperature studies involving laser heating have shown that complex silicates tend to decompose into SiO_2, FeO, MgO, etc. under conditions near the upper mantle–lower mantle boundary. High pressure studies have also contributed to our understanding of mantle dynamics. Finally, recent studies have been made on the equation of state of rare gas solids to ~500 kilobars. This information contributes to our picture of conditions in the outer layers of the Jovian planets.

Biophysics Optical, and, in particular, fluorescent properties of proteins have proved over the past 25 years to provide a powerful probe of protein conformation, of unfolding during denaturation and of rapid localized motion of protein segments. In the past decade high pressure fluorescence studies of proteins in solution have been undertaken and have revealed both reversible and irreversible changes in conformation. These experiments have contributed considerably to our understanding of segmental motion and the denaturation process. High pressure studies of rates of chemical reactions at active sites have also been significant. Biophysics (and biophysical chemistry) constitutes one of the newest but most rapidly growing areas of modern high pressure research.

This brief outline of a variety of high pressure studies is intended to convey some feeling for the power and versatility of high pressure as a tool for understanding the physics of condensed systems.

H. G. DRICKAMER

References

1. Bridgman, P. W., "Physics of High Pressure," Second Edition, New York, Bell and Co., 1949.
2. Drickamer, H. G., and Frank, C. W., "Electron Transitions and the High Pressure Chemistry and Physics of Solids," London, Chapman and Hall, New York, Halsted Press, 1973.
3. Timmerhaus, K. D., and Barker, M. S. (Eds.), "High Pressure Science and Technology," New York, Plenum Press, 1979.
4. Osugi, J. (Ed.), "Modern Aspects of Physical Chemistry at High Pressure," Kyoto, PhysicoChemical Soc. of Japan, 1980.
5. Skelton, K. F. (Ed.), "High Pressure as a Reagent and as an Environment," Washington, D.C., Amer. Chem. Soc., 1981.
6. Schilling, J. S., and Shelton, R. N. (Eds.), "Physics of Solids under High Pressure," Amsterdam, North Holland Pub. Co., 1981.
7. Samara, G. A., and Peercy, R. S., "The Study of Soft Mode Transactions at High Pressure," in "Solid State Physics," Vol. 36, (Echrenreich, H., Seitz, F., and Turnbull, D., Eds.) New York, Academic Press, 1981.

Cross-references: FLUID STATICS; GAS LAWS; GASES, THERMODYNAMIC PROPERTIES; LIQUID STATE; SEMICONDUCTORS; SOLID STATE PHYSICS; SOLID STATE THEORY.

PROPAGATION OF ELECTROMAGNETIC WAVES

The equations governing electromagnetic theory—Maxwell's equations—have special solutions that describe electromagnetic wave propagation. Such solutions provide for the transport of energy by wave propagation from regions of space containing moving charged particles or currents through regions completely devoid of matter. Wave propagation in free space (vacuum) occurs with a wave velocity $c = 2.9979 \times 10^8$ m/s; this velocity is independent of frequency and thus applies to the complete electromagnetic spectrum: radio waves, microwaves, infrared radiation, visible light, x-rays, etc.

In the presence of matter, wave propagation can be substantially altered, in some cases propagating with a velocity much smaller than c. The type of wave propagation is characterized by its dispersion relation (relation between wave number and frequency); the various types of wave behavior are referred to by names: e.g., *helicon waves*, *Alfvén waves*, *surface plasma waves*, etc. Propagation through inhomogeneous materials or through regions containing different materials gives rise to the phenomena of *reflection*, *refraction*, and *diffraction*. Wave propagation in anisotropic solids is the source of phenomena such as *double refraction* and *birefringence*.

Maxwell's equations of electromagnetic theory provide a relationship between the electric and magnetic field vectors **E**, **B** and the charge and current densities ρ, **J**. These equations, together with appropriate boundary conditions are sufficient to describe a full range of electromagnetic phenomena, including wave propagation. In the presence of a material medium some of the charge and current density arises from atoms, molecules, ions and/or electrons making up the medium; these latter densities both influence and are influenced by the electromagnetic field quantities (**E** and **B**). The situation is most conveniently handled by introducing two auxiliary field quantities, the electric displacement **D** and magnetic intensity **H**, and constitutive parameters ϵ, μ, and σ which characterize the medium. The *permittivity* ϵ, the *permeability* μ and the *conductivity* σ will be defined and discussed in more detail below.

Maxwell's equations are conventionally written:

$$\nabla \times \mathbf{E} = -\dot{\mathbf{B}} \qquad (1)$$

$$\nabla \times \mathbf{H} = \mathbf{J} + \dot{\mathbf{D}} \qquad (2)$$

$$\nabla \cdot \mathbf{D} = \rho \qquad (3)$$

$$\nabla \cdot \mathbf{B} = 0. \qquad (4)$$

These differential equations, supplemented by the constitutive relations for the medium in which the fields are to be calculated,

$$\mathbf{D} = \epsilon \mathbf{E} \qquad (5)$$

$$\mathbf{B} = \mu \mathbf{H} \qquad (6)$$

$$\mathbf{J} = \sigma \mathbf{E} \qquad (7)$$

provide a macroscopic phenomenological description of electromagnetic phenomena which is the most appropriate formalism for treating wave propagation. The effect on the electric field by free and bound charges of the medium is accounted for by the permittivity ϵ. The transport currents produced by the motion of free charges in the medium are taken into account through the conductivity σ, whereas the magnetic field produced by these currents as well as by atomic and molecular currents derives from the permeability μ.

Plane Wave Fields; Dispersion Relations A simple medium is defined as one containing no charge or current density (ρ, $\mathbf{J} = 0$) and being characterized by constant values of ϵ and μ (vacuum is a particular simple medium with $\epsilon = \epsilon_0$ and $\mu = \mu_0$). Taking the curl of Eq. (1), combining with the time derivative of (2), and eliminating $\mathbf{D}$ and $\mathbf{B}$ by means of (5) and (6) yields

$$\nabla \times \nabla \times \mathbf{E} = -\mu\epsilon\ddot{\mathbf{E}}. \qquad (8)$$

By substituting a vector identity and noting that $\nabla \cdot \mathbf{E} = 0$ in a simple medium, we find that (8) reduces to the *wave equation*. It is convenient to introduce a specific time dependence for $\mathbf{E}$, namely, $\exp(-j\omega t)$, where ω is the impressed frequency of the electromagnetic disturbance:

$$\nabla^2 \mathbf{E} + \omega^2 \mu\epsilon \mathbf{E} = 0. \qquad (9)$$

The wave or phase velocity is given by $v_p = 1/(\epsilon\mu)^{1/2}$; in the case of free space or vacuum the wave velocity is $c = 1/(\epsilon_0\mu_0)^{1/2}$.

Equation (9) has a solution in the form of a plane wave, i.e., one in which the field quantity $\mathbf{E}$ (or $\mathbf{H}$) has a space and time dependence given by $\exp(j\mathbf{k} \cdot \mathbf{r} - j\omega t)$, with $\mathbf{k}$ equal to the wave vector, $\mathbf{k} = k\mathbf{n}$, where $\mathbf{n}$ is a unit vector normal to the wavefront (i.e., the direction in which the wave is propagating) and $k \equiv |\mathbf{k}| = 2\pi/\lambda$ (λ is the wavelength). When we substitute the exponential solution into (9), we get the dispersion relation $\omega^2 = v_p^2 k^2$ (an algebraic relationship between ω and k). Examination of the Maxwell equations shows that $\mathbf{H}$ also satisfies (9), and that $\mathbf{E}$, $\mathbf{H}$, and $\mathbf{n}$ are mutually perpendicular. The energy transported per unit time per unit area is most conveniently expressed in terms of the Poynting vector $\mathbf{S} = \mathbf{E} \times \mathbf{H}$; for a plane electromagnetic wave $\mathbf{S}$ has the direction of the unit normal $\mathbf{n}$.

The case of a conducting medium with constant conductivity σ is not much more difficult to handle. Eq. (9) gets an extra term

$$\nabla^2 \mathbf{E} + \omega^2 \mu\epsilon \mathbf{E} + j\omega\mu\sigma \mathbf{E} = 0 \qquad (10)$$

or

$$\nabla^2 \mathbf{E} + \omega^2 \mu\epsilon^* \mathbf{E} = 0 \qquad (11)$$

with

$$\epsilon^* \equiv \epsilon + j\sigma/\omega. \qquad (12)$$

Thus through the introduction of the complex permittivity ϵ^*, Eq. (11) takes on the same form as (9). Using the plane wave expression $\exp(j\mathbf{k} \cdot \mathbf{r})$ in Eq. (11) gives the dispersion relation

$$k^2 = \omega^2 \mu\epsilon(1 + j\sigma/\omega\epsilon). \qquad (13)$$

Wave Propagation in Plasmas; the Ionosphere Gases which are highly ionized are good conductors of electricity. An ionized gas in which the positive and negative space charges are nearly balanced is called a *plasma*. Plasmas play an important role in many areas of physics: e.g., physics of the upper atmosphere, theoretical astrophysics, the problem of ion containment in thermonuclear (fusion) reactors, and solid state plasmas.

Wave propagation in a plasma is governed by Eqs. (11) and (12), or equivalently by (13). The conductivity of the plasma is due primarily to free electrons, since the more massive ions contribute much less to electrical transport; for most non-solid-state plasmas of interest, $\epsilon \approx \epsilon_0$ and $\mu \approx \mu_0$. The conductivity contains a contribution from the damping force of electron-ion collisions as well as an out-of-phase inertial contribution. In certain cases, such as the ionosphere or a tenuous plasma, the damping term is very small; then the "conductivity"

$$\sigma_{\text{plasma}} \approx jN_0 e^2/m\omega, \qquad (14)$$

where N_0 is the free electron density, m and e are mass and charge of the electron, respectively. The "conductivity" is purely imaginary in this case (with the electric field and current out of phase). Substituting into (12) and (13), we obtain

$$\epsilon^* = \epsilon_0(1 - \omega_p^2/\omega^2), \qquad (15)$$

and

$$k^2 = (\omega^2/c^2)(1 - \omega_p^2/\omega^2), \qquad (16)$$

where the plasma frequency

$$\omega_p \equiv (N_0 e^2 / m \epsilon_0)^{1/2}.$$

This simple picture of a plasma is sufficient to explain the role of the ionosphere in the propagation of radio waves. The ionosphere is a plasma with typical maximum electron density of $10^{12}/m^3$ (corresponding to a plasma frequency of the order of $\omega_p \approx 6 \times 10^7 \ sec^{-1}$) at a height of some 100 km above the earth. According to Eq. (15), however, the plasma turns into an ordinary dielectric at frequencies above the plasma frequency; below ω_p the ionosphere has a negative permittivity (equivalently, a negative dielectric constant), thus acts like any electrical conductor and is capable of reflecting electromagnetic waves. For vertical incidence, the cutoff is about 10 MHz, but with oblique incidence this is extended out to about 40 MHz. Reflection from the ionosphere was responsible for all long-distance radio communication in the world up to the time when a sufficient number of communication satellites were placed in orbit to provide direct line-of-sight radio transmission links. For higher electromagnetic frequencies—microwaves, visible light—the ionosphere is essentially transparent.

Propagation in Bounded Regions At the boundary or interface between two media, Maxwell's equations require a definite relationship between the field components in the two media. Let us call the two media 1 and 2. The tangential components of **E** and **H** are continuous across the boundary, the normal component of **B** is continuous, and

$$\epsilon_1 {}^* E_{1n} = \epsilon_2 {}^* E_{2n} \qquad (17)$$

at the boundary (E_n stands for normal component). Imposition of these boundary conditions gives rise to the well-known phenomena of wave *reflection* and *refraction*.

In general when an electromagnetic wave is incident on a boundary separating two media, part of the wave is reflected and part is transmitted. If the wave is incident on the boundary obliquely, then the transmitted wave is refracted, i.e., its propagation direction is deflected (toward the normal to the boundary if the second medium is optically more dense, i.e., if ϵ^* is larger in this medium). If the second medium is highly conducting then very little energy is transmitted and almost all of the energy is reflected back into the first medium.

Electromagnetic waves can be propagated in a bounded region by surrounding the propagation medium with walls made from a good electrical conductor. Such a bounded region is called a *waveguide*. Waveguides with rectangular or circular cross sections are used extensively for the transmission of microwave power. Propagation of the electromagnetic wave in the guide can be visualized as a wave which is being continuously reflected by the walls back into

the bounded region. Consider, for example, a rectangular waveguide of cross section $a \times b$ where $b \gg a$; further, consider wave propagation in a plane perpendicular to the dimension b and which is transversely polarized (i.e., **E** parallel to the b dimension). Detailed analysis shows that all frequencies with wavelength $\lambda < 2a$ are propagated; but if $\lambda > 2a$ the wave is exponentially damped. $\lambda_c = 2a$ is called the *cutoff wavelength*.

Electromagnetic waves can also be confined by a boundary between two media with substantially different indices of refraction. (The refractive index is defined as the square root of the normalized permittivity, ϵ/ϵ_0.) A wave propagating in a high index material, surrounded by a low index medium, will be totally internally reflected for a range of obliquely incident rays. Thus, light can be confined and transmitted with very little attenuation by thin optical fibers of glass or by channels of high refractive index; this forms the basis of the modern technologies of "fiber optics" and "dielectric waveguides."

Surface Plasma Waves An interesting case of wave propagation occurs at the boundary between a dielectric and a conductor. Consider, for example, a solid-state plasma (a metal or a semiconductor) in contact with vacuum ($\epsilon = \epsilon_0$). The permittivity of the plasma is given by (15) and we consider frequencies below ω_p where ϵ^* is negative. We assume that a wave—exp ($jkx - j\omega t$)—is propagating in each medium parallel to the boundary; the E-field has one component in the direction of propagation and one normal to the boundary, the H-field has only one component—parallel to the boundary. Ensuring that the fields satisfy Maxwell's equations requires, among other things, that the fields on both sides of the interface be damped exponentially as one moves away from the boundary. Using the boundary conditions of the preceding section yields the dispersion relation:

$$(k^2 - \omega^2/c^2)^{1/2} \epsilon^* = -\epsilon_0 (k^2 - \epsilon^* \omega^2/c^2 \epsilon_0)^{1/2}.$$
$$(18)$$

At small values of k, $\omega \approx ck$; but at large values of k, $\epsilon^* \approx -\epsilon_0$, or

$$\omega^2 \rightarrow \omega_p{}^2/2. \qquad (19)$$

This type of wave propagation is called a *surface plasma wave*.

Propagation in the Presence of a Static Magnetic Field In the presence of a static magnetic field B_0, electromagnetic wave propagation through materials takes on a new dimension. The reason for this is that the conductivity (and hence the complex permittivity) of the medium becomes anisotropic due to the fact that all current densities **J** experience magnetic forces at right angles to both current direction and B_0: $F_{mag} = J \times B_0$. For an anisotropic material, the conductivity σ can no longer be treated as a scalar quantity, but is in fact a tensor; assuming the z-direction to be that of the magnetic field, we find:

$$\sigma = \begin{pmatrix} \sigma_1 & \sigma_{12} & 0 \\ -\sigma_{12} & \sigma_1 & 0 \\ 0 & 0 & \sigma_z \end{pmatrix}. \qquad (20)$$

The conductivity in the direction of the magnetic field, σ_z, is unaltered; the off-diagonal element σ_{12} is often referred to as the *Hall conductivity*.

Two types of electromagnetic waves are commonly encountered in plasmas in a magnetic field; these are called *helicon waves* and *Alfvén waves*, and they are seen both in gaseous (atmospheric or galactic) plasmas and solid-state plasmas. Helicon waves are dominated by the off-diagonal Hall conductivity; the waves are circularly polarized and they propagate in the direction of the static magnetic field $\mathbf{B}_0$. Atmospheric helicons have been called "whistlers;" these propagate in the ionosphere along the earth's magnetic field. Alfvén waves also propagate along the direction of $\mathbf{B}_0$ but are governed by a different dispersion relation; they occur in compensated plasmas (e.g., electron-hole systems with equal numbers of electrons and holes) and in systems so dominated by collisions that the Hall term is very small. They are named for Nobel laureate Hannes Alfvén who first predicted their behavior.

We consider solutions to Eq. (11) for waves propagating along the z-direction (the direction of $\mathbf{B}_0$), but note that ϵ^* is a tensor quantity in this case. Since displacement currents play no role in helicon and Alfvén wave propagation (these are low frequency waves), we can neglect $\omega^2\epsilon$; Eq. (12) thus becomes $\epsilon^* \approx j\sigma/\omega$. μ will be taken as μ_0. Using Eq. (20) for the conductivity, we obtain the general dispersion relation:

$$k^2 = \omega\mu_0(j\sigma_1 \pm \sigma_{12}). \qquad (21)$$

Let us consider an electron-ion plasma in a magnetic field. Equation (14) is no longer adequate for σ_1; we must keep the collisional term, as well as new terms depending specifically on the magnetic field $\mathbf{B}_0$:

$$\sigma_1 \approx \sigma_0(1 + j\omega\tau)/[(1 + j\omega\tau)^2 + \omega_c^2\tau^2], \qquad (22)$$

$$\sigma_{12} \approx \sigma_0\omega_c\tau/[(1 + j\omega\tau)^2 + \omega_c^2\tau^2]. \qquad (23)$$

Here $\sigma_0 \equiv N_0 e^2\tau/m$, with N_0, e, and m as defined in (14), τ the electron-ion collision time, and $\omega_c \equiv eB_0/m$ the Larmor frequency in the magnetic field B_0. $\omega_c\tau$ may also be written as σ_0RB_0, where R is the Hall coefficient.

Helicon wave propagation is dominated by the σ_{12} term; if we neglect the $j\omega\tau$ term in both numerator and denominator of (22) and (23), and then substitute these conductivity terms into (21), we obtain the helicon dispersion relation:

$$k^2 = j\omega\mu_0\sigma_0/(1 \pm j\omega_c\tau), \qquad (24)$$

or

$$\omega = (k^2/\mu_0)(RB_0 - j/\sigma_0). \qquad (25)$$

The waves are circularly polarized: $E_x = \pm jE_y$.

For Alfvén waves the Hall term either vanishes or is negligible. Since we are interested in large magnetic fields, we neglect everything in the denominator of (22) except $\omega_c^2\tau^2$. Substituting in (21), we obtain the Alfvén dispersion relation:

$$\omega^2 = k^2B_0/\mu_0N_0m(1 - j/\omega\tau). \qquad (26)$$

The dominant term in (26) gives a wave velocity $v_p = (B_0/\mu_0N_0m)^{1/2}$. Since there are usually more than one group of charge carriers participating in Alfvén wave propagation, it is more appropriate to use the total plasma mass density $\sum N_i m_i$ for N_0m.

Propagation in Anisotropic Media In the preceding section we encountered one example of an anisotropic medium, namely, a conducting medium in the presence of a static magnetic field. However, many naturally occurring crystalline materials are anisotropic as a consequence of their crystal structure. Although both cubic crystals and amorphous materials are isotropic for electromagnetic wave propagation, crystalline materials with lower symmetry (tetragonal, orthorhombic, monoclinic, etc) show anisotropic behavior in their constitutive parameters ϵ, μ, and σ.

With the exception of magnetic-field-induced anisotropy, the most important examples of anisotropic behavior are those associated with the permittivity (or dielectric behavior) of nonconducting, nonmagnetic, crystalline solids. The electromagnetic (or optical) properties of such crystals depend primarily upon the symmetry of the permittivity tensor ϵ_{ij}. In this regard, the various crystals fall into three types: cubic, uniaxial and biaxial. In uniaxial crystals, one of the principal axes of ϵ_{ij} coincides with a symmetry axis of the crystal; this is called the *optic axis* of the crystal. A uniaxial crystal has two independent permittivity values, one for propagation along the optic axis and one for propagation at right angles to it; the difference between these two values is a measure of the *birefringence* of the crystal. Two types of electromagnetic waves can be propagated in a uniaxial crystal; with one, called the *ordinary wave*, the crystal behaves like an isotropic body. With waves of the second type, called *extraordinary waves*, the magnitude of the wave vector k depends on the direction it makes with the optic axis.

Two different refracted waves are formed when a wave impinges on a uniaxial crystal, a phenomenon known as *double refraction*. The two waves have different polarizations, so that by excluding the transmission of one by geometric means, one can produce a polarized wave. This is the principle of the *Nicol prism*.

Biaxial crystals have lower symmetry than uniaxial crystals; they are characterized by three independent permittivity values.

Propagation in Inhomogeneous and Composite Media; Diffraction If the medium is inhomogeneous, propagation of electromagnetic waves through it is more complicated, since the parameters ϵ, μ, and σ are no longer independent of position. Consider, for example, a medium in which $\sigma = 0$, $\mu = \mu_0$, and ϵ is a function of position. Assuming an exp $(-j\omega t)$ time dependence, and eliminating $\mathbf{H}$ between Eqs. (1) and (2), we obtain

$$\nabla^2 \mathbf{E} - \nabla(\nabla \cdot \mathbf{E}) + \mu_0 \omega^2 \epsilon \mathbf{E} = 0. \quad (27)$$

Elimination of $\mathbf{E}$ gives

$$\nabla^2 \mathbf{H} + \mu_0 \epsilon \omega^2 \mathbf{H} + \epsilon^{-1} \nabla \epsilon \times (\nabla \times \mathbf{H}) = 0. \quad (28)$$

Equations (27) and (28) can be solved in a general form for the case where propagation conditions approximate those corresponding to geometrical optics. For this approximation to be valid the effective "wavelength" λ must be much smaller than some characteristic dimension of the medium and should not change rapidly throughout space (i.e., $|\nabla \lambda| \ll 1$). This type of approximation is called *ray theory*; a one-dimensional version is equivalent to the W.K.B. approximation used extensively in the semiclassical solution of *wave mechanical* problems. If, e.g., propagation occurs in the *x-z* plane, and ϵ varies in one dimension only (so that $\epsilon = \epsilon(z)$), then the k dependence for the *x*-direction is a constant, k_1. Defining

$$f(z) = \epsilon(z) \mu_0 \omega^2 - k_1^2, \quad (29)$$

we find two independent solutions to Eq. (27) of the form:

$$(\text{constant}/f^{1/4}) \exp\left(\pm j \int f^{1/2} \, dz \right). \quad (30)$$

A composite medium can be viewed either as a structured, inhomogeneous medium, or as a combination of individual (single-medium) elements. The most interesting composites are those showing a regular or uniform modulation, such as, e.g., a medium made up of layers with differing physical properties and which alternate in a regular way, or an array of equally spaced spherical or cylindrical elements imbedded in a dielectric matrix. Wave propagation through such composite structures gives rise to the phenomenon of *diffraction*; for certain directions in the transmitted wave all of the individual rays making up the wave are in phase and a diffracted beam develops, for other directions the rays are out of phase and tend to interfere destructively with each other.

A plane diffraction grating can be made from a thin film of metal (on a transparent substrate) by cutting out parallel stripes of the metal at a uniform spacing d. If the grating is illuminated from behind by a plane electromagnetic wave, each stripe or transparent element of the grating acts as a source for the transmitted wave. The diffracted beam has a maximum when the various transmitted wavelets are in phase, i.e., when

$$n\lambda = d \sin \theta, \quad (31)$$

where n (an integer) is the diffraction order, λ is the wavelength, and θ is the angle which the diffracted beam makes with a plane normal to the grating and containing one of the stripes. Diffraction effects are most pronounced when the stripe spacing d is of the same order as the wavelength λ. The angle θ of the diffracted beam depends upon λ; if the incident electromagnetic wave contains components with different wavelengths, these will be separated into component wavelengths (or colors) by the grating.

A crystal acts like a three-dimensional diffraction grating for short-wavelength electromagnetic waves; the regularly spaced atoms making up the crystal can be viewed as sources for the wavelets contributing to the diffracted wave. Crystals are most effective in diffracting waves with wavelengths comparable to the interatomic spacing; for this reason they are used to diffract x-rays. The early experiments of Friedrich, Knipping, and von Laue (1912), in which they passed an x-ray beam through a crystal, established that both (1) x-rays are electromagnetic waves with relatively short wavelengths, and (2) the atoms of a crystal are arranged in a regular three-dimensional order:

J. R. Reitz

References

1. Reitz, J. R., Milford, F. J., and Christy, R. W., "Foundations of Electromagnetic Theory," Third Edition, Reading, MA, Addison Wesley Publishing Co., Inc., 1979.
2. Roller, D., and Blum, R., "Physics, Vol. Two: Electricity, Magnetism and Light," San Francisco, Holden-Day, Inc., 1981.
3. Read, F. H., "Electromagnetic Radiation," New York, John Wiley & Sons, Inc., 1980.
4. Brekhovskikh, L. M., "Waves in Layered Media," New York, Academic Press, 1980.
5. Ostrowsky, D. B., "Fiber and Integrated Optics," New York, Plenum Press, 1979.

Cross-references: DIFFRACTION BY MATTER AND DIFFRACTION GRATINGS, ELECTROMAGNETIC THEORY, FARADAY EFFECT, IONOSPHERE, MAGNETIC-FLUID-MECHANICS, PLASMAS, REFRACTION, WAVE MOTION.

PROTON*

Protons in Atoms The proton is the atomic nucleus of the element hydrogen, the second most abundant element on earth. Positively charged hydrogen atoms or "protons" were identified by J. J. Thomson in a series of experiments initiated in 1906. Although the structure of the hydrogen atom was not correctly understood at that time, several properties of the proton were determined. The electric charge on the proton was found to be equal but opposite in sign to that of an electron, and the measured value for the mass was much greater than that of the electron. The currently accepted proton mass is 1836 times the electron rest mass, or 1.672×10^{-24} grams.

A correct estimate of the size of the proton and an understanding of the structure of the hydrogen atom resulted from two major developments in atomic physics: the Rutherford scattering experiment (1911) and the Bohr model of the atom (1913). Rutherford showed that the nucleus is vanishingly small compared to the size of an atom. The radius of a proton is on the order of 10^{-13} cm as compared with atomic radii of 10^{-8} cm. Thus, the size of a hydrogen atom is determined by the radius of the electron orbits, but the mass is essentially that of the proton.

In the Bohr model of the hydrogen atom, the proton is a massive positive point charge about which the electron moves. By placing quantum mechanical conditions upon an otherwise classical planetary motion of the electron, Bohr explained the lines observed in optical spectra as transitions between discrete quantum mechanical energy states. Except for hyperfine splitting, which is a minute decomposition of spectrum lines into a group of closely spaced lines, the proton plays a passive role in the mechanics of the hydrogen atom. It simply provides the attractive central force field for the electron.

The proton is the lightest nucleus with atomic number one. Other singly charged nuclei are the deuteron and the triton which are nearly two and three times heavier than the proton, respectively, and are the nuclei of the hydrogen isotopes deuterium (stable) and tritium (radioactive). The difference in the nuclear masses of the isotopes accounts for a part of the hyperfine structure called the "isotope shift."

*Supported in part by The U.S. Air Force Office of Scientific Research.

In 1924, difficulties in explaining certain hyperfine structures prompted Pauli to suggest that a nucleus possesses an intrinsic angular momentum or "spin" and an associated magnetic moment. The proton spin quantum number (I) is 1/2, and the angular momentum is given by $[I(I+1)h^2/(2\pi)^2]^{1/2}$ where h is Planck's constant. The intrinsic magnetic moment is 2.793 in units of nuclear magnetons (0.50504×10^{-23} erg/gauss) which is about a factor of 660 less than the magnetic moment of the electron.

Two types of hydrogen molecule result from the two possible couplings of the proton spins. At room temperature, hydrogen gas is made up of 75 per cent orthohydrogen (proton spins parallel) and 25 per cent parahydrogen (proton spins antiparallel). Several gross properties, such as specific heat, strongly depend on the ortho or para character of the gas.

Protons in Nuclei Protons and neutrons are regarded as "nucleons" or fundamental constituents of nuclei in most theories of nuclear structure and reactions. The nuclear forces operating between them are much stronger than the electrostatic forces which govern atomic and molecular systems but operate over very short ranges, the order of several times 10^{-13} cm. Of particular significance in the structure of nuclei is the apparent charge independence of the forces. That is, the nuclear force between two nucleons may be considered separately from the electrostatic forces due to electric charges the nucleons may carry. In addition to mass and charge, other properties such as spin and parity play important roles in determining the mechanics of nuclei.

The mass of a nucleus is the sum of the masses of the nucleons contained, plus a correction due to the total binding energy of the nucleons. This correction is an application of the Einstein mass-energy equivalence ($E = mc^2$). The atomic number or positive charge of a nucleus is given by the number of protons.

A detailed description of the motion of a proton in a nucleus is complicated by the many-body, quantum mechanical nature of the problem, but several simplified theories or models have been very successful in predicting many of the properties of nuclei. One of the best known nuclear structure models is the shell model in which a nucleon is assumed to move in a central force field. This field represents the average interaction of the proton (or neutron) with all other nucleons. An essential additional assumption is a coupling of the orbital angular momentum of the independent nucleon with its spin.

The success of a theory in which a nucleon moves among its close-packed neighbors as if they were not present is due in part to quantum mechanical restrictions, in particular to the Pauli exclusion principle.

The optical model for the scattering of protons by nuclei also rests on the assumption that the interaction with many nucleons may be represented by an average potential well. An imaginary potential term is included which accounts for reactions other than elastic scattering.

Much of nuclear physics may be understood with a picture in which the proton exists as an independent particle in the nucleus. Refinements of an independent particle model to include collective effects and deformation involve a consideration of the residual interaction between nucleons and the details of the individual nucleon interaction with the average potential well.

Certain aspects of the very strong nuclear forces are understood. These forces involve π mesons in somewhat the same way that electrostatic forces responsible for atomic structure involve photons. Yukawa introduced the π meson as the field quantum for nuclear forces in 1935, and the interaction potential derived from this early theory is commonly used in nuclear physics.

Structure of the Proton A major objective of physics is to identify elementary particles and determine their properties. The proton is important in these investigations in several connections. It is itself an elementary particle. It is used as a projectile in the production and study of other elementary particles. It has been used as a target in the study of nucleon structure.

One view is to consider all fundamental particles as states of excitation of a limited number of particles. The resulting simplification correlates a large body of experimental data.

The internal structure of the proton is still being studied. In one of the experiments, the proton is a target which is probed by very short-wavelength electrons. Electrons produced by a high-energy linear accelerator are scattered from protons to study the electric charge distribution in a proton.

Questions concerning the structure of the proton or the neutron are difficult due to the ambiguity of experimental results. One cause of uncertainty is that the probe is not defined in sufficient detail. It appears desirable to use a strong interaction probe, that is, another nucleon or meson, but the structure of the probe may be no better understood than the structure of the target.

A mantel of mesons has been found about the proton. A current question is whether or not there is an internal or intrinsic core of a proton.

R. H. DAVIS

References

Shortley, George, and Williams, Dudley, "Elements of Physics," Englewood Cliffs, N.J., Prentice-Hall, 1961.
Born, Max, "Atomic Physics," Glasgow, Blackie & Son, Ltd., 1946.
Weidner, Richard T., and Sells, Robert L., "Elementary Modern Physics," Boston, Allyn and Bacon, 1960.
Richtmyer, F. K., Kennard, E. H., and Lauritsen, T.,
"Introduction to Modern Physics," New York, McGraw-Hill, 1955.

Cross-references: ACCELERATORS, PARTICLE; ATOMIC AND MOLECULAR BEAMS; ATOMIC PHYSICS; ELECTRON; ELEMENTARY PARTICLES; ISOTOPES; NEUTRON; NUCLEAR STRUCTURE; PARITY; QUANTUM THEORY; RADIOACTIVITY; RELATIVITY.

PULSARS

In November 1967 a group of radioastronomers at Cambridge, England, discovered a new type of star in our galaxy. The star emitted radio waves at frequencies ranging from 100 to 10,000 MHz and the radiation was confined to short pulses a few hundredths of a second in duration which repeated at very regular intervals of about one second. Systematic searches soon led to the discovery of further pulsars, and 300 had been located at the end of 1980. Attempts to detect visible light from pulsars have been unsuccessful except in two cases. Both the well known Crab Nebula, and a nebula in the southern sky called Vela, contain pulsars emitting synchronized flashes of light and γ-rays as well as radio pulses. These are rapid pulsars, having periodicities of 33 msec and 89 msec respectively.

Pulsar distances can be estimated from the different arrival times of a given pulse when recorded at different radio frequencies. This pulse dispersion is caused by ionized gas in interstellar space, and the gas density must be known to calculate the distance. Estimates, accurate within a factor of 2 or 3, place the pulsars at distances typically between 0.1 and 5 kiloparsec; since the total extent of our galaxy is about 30 kiloparsec it follows that only a small fraction of the pulsars which it contains have so far been detected.

No satisfactory explanation of pulsar radiation yet exists but it is generally believed that they are rapidly spinning neutron stars that emit a beam of radiation in some preferred direction so that a "lighthouse" effect accounts for the pulses. Only the intense gravitational force in neutron stars, which are predicted to have radii of about 10 km, would be sufficient to prevent stars' disrupting when they spin rapidly enough to explain the pulsars of shortest period. Pulsars therefore provide evidence for the existence of neutron stars. About five years after the discovery of pulsars, X-ray telescopes carried on spacecraft recorded periodic flashes of radiation from neutron stars in orbit about companion stars. These X-ray binaries emit thermal X-rays caused by gas dragged gravitationally from the companion star to the surface of the neutron star.

Neutron stars represent the end-point of stellar evolution and contain matter in which thermonuclear processes have been completed. Sufficiently low-mass stars are stable, in this condition, as white dwarfs, but electron pressure in degenerate matter cannot balance gravity in stars heavier than 1.4 solar masses. Heavy stars must therefore collapse, when nuclear energy generation within them ceases, and such a catastrophic collapse has been suggested as the origin of stellar explosions visible as supernovae.

The fact that the Crab Nebula (perhaps the most famous supernova of all; the explosion was documented by Chinese astronomers in the year 1054) contains a pulsar provides strong circumstantial evidence of the correctness of the neutron star theory. Certainly the estimated occurrence rate of supernovae in the galaxy accounts reasonably well for the total number of pulsars which are found. Moreover the observation that pulsars lie close to the galactic plane is in agreement with the distribution of massive OB stars that are likely to develop into supernovae.

Extended timing measurements on many pulsars have revealed, without exception, a systematic increase of period entirely in accordance with the rotating-star theory. Typical pulsars, which have periods in the range 0.5 to 1.5 sec, are slowing down at such a rate that 10^6 to 10^7 years must elapse before the pulse interval is doubled. For the Crab pulsar, however, this time is only 2200 years. If a neutron star model is assumed, it is simple to compute, from the observed lengthening of the period, the rate at which the spinning star is losing energy. Rotational energy, and a small amount of residual heat, comprise the principal energy reserves of a neutron star. Hence the spin-down energy loss should be directly related to the energy radiated. Applying this argument to the Crab pulsar shows that the rotational energy loss is sufficient to explain not only the pulsar radiation, but also all the visible light from the remainder of the nebula. It has been known for many years that some source that provided a large output of high-speed electrons was necessary to maintain the luminosity of the nebula, but its origin was a mystery. The fact that a central neutron star can provide this source gives further evidence favoring the neutron star theory.

The mechanism by which a neutron star converts rotational energy into pulsed electromagnetic radiation is a topic of speculation at present. Little is known about physical conditions at the surface, and in the environment, of a neutron star. It is probable that a neutron star is encased in a rigid shell of crystalline ^{56}Fe nuclei permeated by a Fermi sea of degenerate electrons. Somewhere beneath the surface, where the density rises above 10^{13} g/cm^3, the nuclei become unstable and matter is largely in the form of a neutron fluid. It is likely that magnetic flux is conserved during gravitational collapse, which leads to immense magnetic

field strengths of 10^{12} to 10^{15}G, and that conservation of angular momentum gives initial rotation speeds up to 10^3 rotations per second. Solid state physicists have been very interested in the fact that careful timing observations reveal some irregularities in the spin rate. These effects can be understood by the exchange of angular momentum between the superfluid interior of a neutron star, and its rigid outer crust, and they provide a means of checking some of the physical properties of neutron matter.

The rapid motion of the collapsed star, in the presence of its large magnetic field, generates an intense electric field which causes electrons and protons to be torn off the surface and distributed in space about the star. The resulting atmosphere must rotate with the star, and at a critical distance—the velocity of light cylinder—its tangential speed will approach the velocity of light. Beyond this region the atmosphere must stream outwards as a stellar wind, and it is likely that the flow-lines connect with the star's surface near the magnetic poles. In addition to these processes other phenomena arise if the magnetic axis of the star is not aligned with the rotation axis. A misaligned magnetic axis causes the star to radiate electromagnetic waves at the rotation frequency since it then acts as a giant oscillating magnetic dipole. Such outgoing fields can accelerate charged particles to relativistic energies in a manner analogous to linear accelerators in the laboratory and this may, for example, account for the injection of energetic electrons into the Crab Nebula which subsequently emit visible light by synchrotron radiation.

An important feature of the radiation mechanism is that it must produce a directed beam to yield the lighthouse effect. The duration of the observed pulses is almost independent of the radio wavelength and one idea common to several theories is that the beaming is a phenomenon of relativity. One well-known consequence of Special Relativity concerns a moving source of radiation; if the source radiates equally in all directions when it is stationary, the emitted waves must be cast into a narrow beam along the line of motion when the velocity is near that of light. Applying this to pulsars, it has been suggested that radiation emitted by charged particles moving outwards from the (misaligned) magnetic poles might be relativistically beamed along the magnetic axis, or alternatively, that the large tangential speed of the stellar atmosphere near the velocity of light cylinder could produce beaming in directions perpendicular to the rotation axis. Another feature of the radiation is that it must be coherent, involving the collective motion of groups of similarly charged particles, since otherwise, ridiculously high temperatures would be involved. Various possibilities such as wave amplification induced by plasma instabilities, or "antenna" mechanisms in which radiation is generated by the acceleration of bunched charges, have been suggested. One intriguing theory is that electron-positron pairs are generated by γ-radiation in a vacuum-gap zone above the magnetic poles. In the presence of electric fields caused by unipolar induction an initial "seed" pair can induce a cascade of further pairs, giving rise to an electron-positron "spark" discharge. Such discharges could be the seat of particle streams which ultimately bunch, due to plasma instabilities, and thereby generate the required conditions for coherent radio emission.

Unfortunately, no theory has yet been worked out in sufficient detail to allow comparison with the wealth of observational evidence now available. The radiated pulses have complex shapes, which differ from one source to the next, and may change rapidly with time. Usually the radiation is found to be highly polarized and the sense of polarization may vary systematically through the pulse; this shows that the radiation is emitted in a region containing a highly organized magnetic field, although different zones may be active from one instant to the next. In effect, during each pulse a "window" is opened through which complex time-dependent phenomena in the neutron star may be scrutinized. An extremely puzzling phenomenon is that of drifting subpulses, in which distinctive features in the envelope of individual pulses move steadily across the pulse window from pulse to pulse. It appears that some kind of differential rotation may be involved. Occasionally, pulsars showing this effect cease to radiate completely for a short time, and when they start again the drifting pattern reappears in the same position as where it was last seen. Some kind of memory is needed to explain this behavior, and heating of the stellar surface at the foot of an electron-positron spark may be involved.

Pulsars pose a challenge in many fields of physics including solid-state theory, relativistic plasma physics, and electrodynamics. They also provide astronomers with a new means of probing the galaxy and determining quantities such as the interstellar magnetic field strength and gas density. One pulsar has been of outstanding physical interest because it has an unseen binary companion which is almost certainly another neutron star. The system provides a unique test of theories of General Relativity, and timing observations have revealed a slight decrease in the binary orbital period which has been explained as a loss of energy in the form of gravitational waves.

ANTONY HEWISH

References

Hewish, A., *Ann. Rev. Astron. Astrophys.*, 8 (1970).
Ostriker, J. P., and Gunn, J. E., *Astrophys. J.*, 157, 1395 (1969).
Goldreich, P., and Julian, W. H., *Astrophys. J.*, 160, 971 (1970).
Davies, R. D., and Smith, F. G., "The Crab Nebula," Reidel, Dordrecht, 1971.
Goldreich, P., Pacini, F., and Rees, M. J., *Comments Astrophys. Space Phys.*, 3, 185 (1971).

Cross-references: ASTROMETRY, ASTROPHYSICS.

PULSE GENERATION

Pulse waveforms are distinguished by abrupt, often almost discontinuous features. These features favor pulses for two broad categories of uses: (1) information processing, and (2) supplying very large momentary bursts of power to devices whose average rating is relatively low.

The fundamental element in pulse generation is the *switch*. Opening or closing a switch generates an abrupt change of circuit conditions, resulting in the formation of an approximate step-function pulse. One or more such steps, together with suitable wave-shaping, may yield the desired pulse waveform.

Information can be coded in pulse form either in the time domain (event counting, synchronization in process controllers, time-coded analogs, sequencing in computers) or in the domain of pulse amplitude (voltage or current). The power level is usually low or moderate, and the circuit elements used as switches are most commonly bipolar junction transistors (BJTs) or field-effect transistors (FETs) employed in an overdriven mode. Such elements are incorporated into popular families of "logic"-processing integrated circuits. The relevant performance aspects are the *transition times* of the switching action and the rated voltage or current levels. The best established logic families and their typical performances are presently CMOS[1] (insulated-gate FETs; ~50 ns; 5-15 V) and TTL[2] (BJTs; ~3-50 ns, depending on family; 5 V). Some building blocks suitable for pulse generation are shown in Fig. 1 (the code numbers are designations widely shared among manufacturers). The *astable* (free running) multivibrator (a) serves as primary pulse source, either directly or by further processing of the steps contained in its rectangular output. The *monostable* (b) produces a single output pulse in response to an input trigger signal, which therefore dictates the time of occurrence. These circuits define their own pulse durations (t_0, t_1, t) by reference to exponential transients in the attached RC networks. The *comparator* (c) produces an output step when a variable input signal V_{in} crosses an adjustable threshold level V_{ref}; the step thus marks the time when $V_{in} = V_{ref}$.

Steps and rectangles can be shaped into other forms by many techniques.[3,4] Three classes, linear, nonlinear, and delay-line shaping, are typified by the examples in Fig. 2. The first

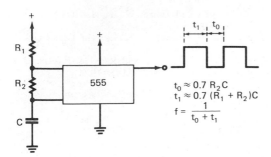

(a) Astable

$$t_0 \approx 0.7 \, R_2 C$$
$$t_1 \approx 0.7 \, (R_1 + R_2)C$$
$$f = \frac{1}{t_0 + t_1}$$

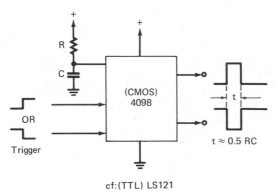

cf:(TTL) LS121

(b) Monostable

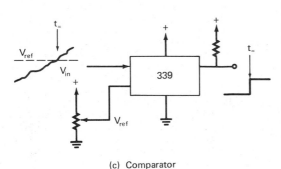

(c) Comparator

FIG. 1. Some integrated-circuit pulse sources.

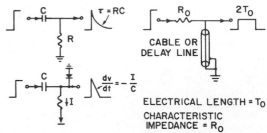

CABLE OR DELAY LINE

ELECTRICAL LENGTH = T_0
CHARACTERISTIC IMPEDANCE = R_0

FIG. 2. Examples of shaping circuits to produce short pulses from a step.

two classes produce "tail pulses" with a sharp transition only at their leading edge. The delay line produces a narrow rectangle from a step; the trailing edge is formed by the reflection of the leading edge in the unmatched line, and the pulse falltime is thus as short as its risetime (within the limitations of the delay line's response).

For applications requiring *high peak power*, switching and shaping are carried out directly at the output power level wherever possible.[5] An ideal switch dissipates no power, since voltage and current are not present simultaneously in it. The energy of the delivered pulse may come from energy *stored* in the pulse-forming network, the power supply recharging the network slowly between pulses (cf. Fig. 3). For some pulse shapes the forming process would be too inefficient to be carried out at high level; a linear output amplifier fed by the low-level shape is then required. Vacuum-tube amplifiers are common at the highest levels (up to 50 kV and several tens of A). FETs, especially such high-current types as VMOS, serve in lower-voltage ranges (up to 150 V), and BJTs may be preferred when very short transition times (below 100 ns) are needed. BJTs have restricted ratings in switching service because of "second breakdown."[6]

Some other types of switching element are listed below with their special performance aspects:

(a) *Mechanical switches.* Beyond their use in moving spark gaps in extreme-high-voltage service,[5] mechanically operated switches can excel with very short transition times, e.g., in relays with mercury-wetted contacts. Times ~100 ps are readily achieved, limited mostly by the transmission-line properties of the switch enclosure. The electromechanical drive limits the repetition rate to ~1 kHz.

(b) *Spark gaps.* Externally initiated or spontaneous flashover between electrodes in a gas near atmospheric pressure has a very short transition time (~1 ns) and can form pulses at extremely high voltages and currents. In some applications the *jitter* (irregular delay between initiating signal and flashover) can pose a problem.

(c) *Thyratrons.* These are hot-cathode gas or vapor filled tubes in which a discharge is initiated by a trigger signal applied to a grid. The discharge extinguishes only when the switch current is brought to zero by the external circuit; deionization time of the gas (~50 μs or more) limits the minimum pulse spacing. Average power limitation may also play a part; however, peak power levels can be extremely high (e.g., 50 kV and 5 kA). Risetimes as short as a few ns are common in hydrogen-filled thyratrons.

(d) *Controlled rectifiers (SCRs and TRIACs).*[7] These have regenerative switching action and

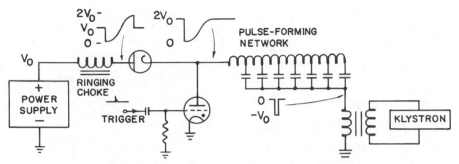

FIG. 3. High-power pulser using hydrogen thyratron.

thus require only small trigger signals for high output power. Transition times < 1 μs are typical; output levels range up to several kA and a little over 1 kV. In all but low-power devices the switch current must be brought to zero externally to restore the nonconducting state.

(e) *Avalanche transistors.*[8] Regenerative breakdown in a BPT can occur with sub-ns transition time; it may be initiated by simply overvolting the collector, or by a trigger signal applied to the base. Fast, relatively large ($\sim$30 V) pulses can be generated at high repetition rates ($\sim$5 MHz).

(f) *Unijunction transistors.* These use conductivity enhancement by injected carriers for regenerative switching. Moderate performance ($\sim$1 μs, $\sim$10 V) is balanced by attractively simple circuit configurations, especially in the generation of trigger pulses for SCRs working at power-line frequency.

(g) *Tunnel diodes.*[9] Quantum-mechanical tunneling through an extremely thin depletion layer has a short transition time because it involves no semiconductor charge storage. The tunneling occurs over a narrow forward voltage range ($\sim$0.1 V); at higher voltages, after a sharp drop of tunnel current, ordinary injection current sets in ($\sim$0.5 V). Switching from the low to the higher voltage occurs in sub-ns times. Choice of external circuitry results in astable, monostable, or bistable operation. Transitions are provoked by suitable current signals. The voltage level of the pulses is evidently very low.

(h) *Four-layer diodes (DIACs).* These are similar to SCRs, but switching occurs spontaneously at a given voltage threshold. Their chief use lies in trigger generation for SCRs.

(i) *Step-recovery (snap) diodes.*[10] Stored charge resulting from a period of forward conduction causes a diode to conduct freely when reverse bias is first applied. In specially graded junctions this reverse current ceases abruptly, with transition times $<$50 ps. The resulting opening of the switch can generate fast pulses with high repetition rates ($>$100 MHz) at relatively high level (tens of V).

R. M. LITTAUER

References

1. Lancaster, D., "CMOS Cookbook," Indianapolis, Howard W. Sams & Co., 1977.
2. Lancaster, D., "TTL Cookbook," Indianapolis, Howard W. Sams & Co., 1974.
3. Pettit, J. M., and McWhorter, M. M., "Electronic Switching, Timing, and Pulse Circuits," New York, McGraw-Hill Book Co., 1970.
4. Horowitz, P., and Hill, W., "The Art of Electronics," Cambridge, Cambridge Univ. Press, 1980.
5. Früngel, F. B. A., "High Speed Pulse Technology," New York, Academic Press, 1965 (Vols. I & II), 1976 (Vol. III).
6. Holt, C. A., "Electronic Circuits, Digital & Analog," John Wiley & Sons, New York, 1978; "Avoiding Second Breakdown," Application Note AN-415A, Motorola, Inc., Phoenix, AZ, 1968.
7. "Silicon Controlled Rectifier Manual," Fifth Edition, Syracuse, NY, General Electric Company, 1977.
8. Henebry, W. M., "Avalanche Transistor Circuits," *Rev. Sci. Instr.* **32**, 1198 (1961).
9. "RCA Tunnel Diode Manual," Somerville, NJ, Radio Corporation of America, 1963.
10. "Pulse & Waveform Generation with Step Recovery Diodes," Application Note 918, Hewlett-Packard, Inc., Palo Alto, CA, 1968.

Cross-references: DIODE (SEMICONDUCTOR), ELECTRICITY, ELECTRON TUBES, FEEDBACK, RADAR, SEMICONDUCTOR DEVICES, TRANSISTOR, TUNNELING.

PYROMETRY, OPTICAL

Optical pyrometry is that branch of thermometry in which the temperature of a solid or liquid is determined by measuring the radiation emitted in a relatively narrow spectral region. The International Practical Temperature Scale of 1968 (IPTS) above the melting point of gold (1337.58 K or 1064.43°C) is defined in terms of such radiation, and an optical pyrometer is usually used to realize the IPTS above this temperature.

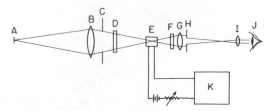

FIG. 1. Schematic diagram of a disappearing filament optical pyrometer. A. Source; B. Objective lens; C. Objective aperature; D. Absorption filter (used for temp. above 1300 C); E. Pyrometer lamp; F. Red filter; G. Microscope objective lens; H. Microscope aperture stop; I. Microscope ocular; J. Eye; K. Current measuring instrument.

Figure 1 is a schematic diagram of a common type of optical pyrometer. The objective lens (B) images that part of the source (A) for which the temperature is to be determined in the plane of the filament of the pyrometer lamp (E). The microscope (G, H, I) magnifies the source image and the small pyrometer lamp filament by a factor of about fifteen. The red filter (F) and the spectral response of the eye limit the spectral band pass of the instrument to wavelengths from about 6200 to 7100Å with the effective value about 6500Å. The absorption filter (D) is inserted at source brightness temperatures above about $1300°C$ so that the pyrometer lamp need not be operated at higher temperatures where its stability is poor.

The optical pyrometer shown in Fig. 1 is operated by adjusting the current in the pyrometer lamp filament until the brightness of the filament equals the brightness of the image of the source. From this current and a previous calibration of the pyrometer, the brightness temperature of that part of the source or object sighted upon can be determined. The brightness temperature of an object is defined as the temperature of a blackbody which emits the same spectral radiance as the object. Spectral radiance is defined as the limit of the quotient of the radiant power emitted in a particular wavelength interval and in a particular direction by the product of the wavelength interval, the solid angle and the emitting area projected perpendicular to the direction of sighting as the latter three quantities tend to zero. If the object is a blackbody, the brightness temperature of the blackbody is its temperature. If the object is not a blackbody, the temperature of the object can be obtained from its brightness temperature and the equation

$$\epsilon_\lambda = \frac{e^{c_2/\lambda T} - 1}{e^{c_2/\lambda T_B} - 1} \qquad (1)$$

where ϵ_λ is the spectral emissivity (also called spectral emittance), λ is the wavelength, in centimeters, at which the pyrometer is effec-

tively operating (usually about 0.65×10^{-4} cm), c_2 is the second radiation constant (1.4388 centimeter degrees on the IPTS), T_B is the brightness temperature and T is the temperature, both in Kelvins. The spectral emissivity of a surface is the fraction of blackbody spectral radiance emitted by the surface when it has the same temperature as the blackbody. Spectral emissivities for a wavelength of 0.65×10^{-4} cm have been determined for a large number of materials. When $C_2/\lambda T$ is greater than about 5, Eq. (1) can be replaced, with negligible error, by the simpler equation

$$\frac{1}{T} = \frac{1}{T_B} + \frac{\lambda}{C_2} \ln \epsilon_\lambda \qquad (2)$$

where ln is the symbol for the natural logarithm.

The calibration of an optical pyrometer from basic principles is called a primary calibration and is usually performed in national standard laboratories such as the National Bureau of Standards. The first step of a primary calibration of an optical pyrometer is making a brightness match while sighting on a blackbody surrounded by freezing gold. The temperature of freezing gold is defined to be 1337.58 K or $1064.43°C$ on the International Practical Temperature Scale of 1968. Higher brightness temperature points are obtained by using the defining equation of the IPTS,

$$R = \frac{e^{c_2/\lambda T_B} - 1}{e^{c_2/\lambda T_{Au}} - 1} \qquad (3)$$

Experimentally, the brightness temperature of a stable source is adjusted until the source, as seen through a rotating sectored disk with transmittance R, has the same brightness as a gold blackbody. Independent measurement of R and λ then permit, from Eq. (3), a calculation of T_B. The pyrometer lamp current required for making a brightness match while sighting on the source without the sectored disk completes the calibration at this higher temperature. This process is repeated with a sufficient number of sectored disks and temperatures so that a smooth curve can be drawn relating current in the pyrometer lamp to the brightness temperature of the source.

With a well-designed optical pyrometer of the type in Fig. 1, it is possible to realize the IPTS in a primary calibration with an estimated uncertainty of ±0.4 deg at 1337 K, ±2.0 deg at 2300 K and ±10.0 deg at 4300 K. The National Bureau of Standards also calibrates commercial pyrometers of the type in Fig. 1 by comparison to a primary calibrated pyrometer. The uncertainty of this comparison calibration is estimated to be ±3 deg, ±6 deg, and ±40 deg respectively at the temperatures given above. These figures apply to brightness or blackbody temperatures. The accuracy with which an optical pyrometer can determine the temperature of a non-blackbody depends not only on the pyrometer calibration in terms of brightness temperature, but also on the uncertainty of the spectral emissivity required. This often produces an error greater than the calibration uncertainties given above.

Many of the national standards laboratories throughout the world have developed photoelectric pyrometers to replace the visual instrument shown in Fig. 1. These modern, high-precision pyrometers use a photomultiplier tube rather than the eye as detector and an interference filter or monochromator to limit the spectral band pass. With the resulting increased sensitivity and smaller band pass, these instruments can realize (1982) the IPTS above the melting point of gold with about one tenth the uncertainty possible with the best visual instruments. They have been used in the national standards laboratories primarily in their calibration services and in determining thermodynamic temperature differences between various thermometric fixed points (melting point of gold, silver, and copper). The latter has been possible with an uncertainty of only a few one hundredths of a degree. Portable photoelectric pyrometers are available commercially, but their calibration uncertainty is much greater than that of the instruments developed by the national standards laboratories. This is due largely to the instability of the pyrometer lamp used as a reference standard in the portable instruments. The national laboratory instruments avoid this problem either by conducting frequent primary calibrations or by using a gold-point blackbody as the reference standard.

In addition to optical pyrometers, total radiation pyrometers and two color pyrometers are sometimes used for measuring temperatures. Usually, however, these instruments are less accurate than optical pyrometers and are primarily intended for controlling temperature rather than for measuring it.

H. J. KOSTKOWSKI

References

Forsythe, W. E., "Optical Pyrometry," in "Temperature: Its Measurement and Control in Science and Industry," p. 115, New York, Van Nostrand Reinhold, 1941.

Kostkowski, H. J., and Lee, R. D., "Theory and Methods of Optical Pyrometry," Natl. Bur. Std. Monograph 41 (1962).

"The International Practical Temperature Scale of 1968. Amended Edition of 1975," *Metrologia* 12, 7 (1976).

"Temperature: Its Measurement and Control in Science and Industry," Vol. 5, (J. F. Schooley, Ed.), New York, American Institute of Physics, 1982.

"Temperature Measurement, 1975," Conference Series No. 26, (B. F. Billing and P. J. Quinn, Eds.), London, Institute of Physics, 1975.

"Temperature: Its Measurement and Control in Science and Industry," Vol. 4, Part 1, (Harman H. Plumb, Ed.), Pittsburg, Instrument Society of America, 1972.

Cross-references: RADIATION, THERMAL; TEMPERATURE AND THERMOMETRY.

Q

QUANTUM CHROMODYNAMICS

Quantum chromodynamics is a theory of the strong interactions of elementary particles, including most notably the interaction that binds protons and neutrons to form nuclei. It is widely believed to be the correct theory of the strong interactions, and has survived some experimental tests at high energy accelerator laboratories.

The theory can be developed by examining the spectrum of elementary particle states discovered experimentally in the last thirty years. One finds that this spectrum can be largely explained by supposing that the strongly interacting particles are made of smaller, spin $\frac{1}{2}$ particles called *quarks*. (See QUARKS for a more detailed discussion.) For instance, a proton is made up of two quarks of a type, or *"flavor,"* called *up* and one quark of a flavor called *down*. The neutron consists of two *down* quarks and one *up* quark. Since protons and neutrons are fermions and consist of an odd number of quarks, the quarks must be fermions.

A most important feature of the quark model can be understood by considering a charge +2, spin $\frac{3}{2}$ cousin of the proton called the Δ^{++}, which consists of the three *up* quarks in the model. The most natural ground state wave function of the three quarks in the Δ^{++} is symmetric under interchange of any two quark position and spin variables. Since the quarks are fermions, a contradiction appears unless the three *up* quarks in a Δ^{++} are *not* identical. One supposes instead that the quarks are distinguished by a property called "color:" one quark is *red*, one is *yellow*, one is *blue*.

A useful mathematical description of this model is obtained by treating quarks as identical particles that carry a color quantum number I which may take the three values R, Y, and B. Then the color wave function of the Δ^{++} is the completely antisymmetric combination

$$|RYB\rangle + |BRY\rangle + |YBR\rangle - |BYR\rangle$$
$$- |RBY\rangle - |YRB\rangle. \qquad (1)$$

It is attractive to suppose that the strong force is blind to color. To be more precise, one considers an arbitrary rotation of all quarks in color space. A *red* quark is rotated into a new *red'* state, etc., with

$$|I\rangle' = \sum_J U_{IJ}|J\rangle, \qquad I = R, Y, B. \qquad (2)$$

Here U_{IJ} is an arbitrary 3×3 unitary matrix with determinant equal to 1. (Notice that the completely antisymmetric color state of these quarks given in Eq. (1) is unchanged under such a rotation. Such states may be called *colorless*.) The attractive hypothesis mentioned above is that the strong interaction Hamiltonian is invariant under rotations U_{IJ} in color space.

In quantum chromodynamics, the idea of symmetry under color rotations U_{IJ} is carried much further. One supposes that the Hamiltonian is invariant under color rotations $U_{IJ}(x)$ that depend on the space-time coordinate $x = (t, x, y, z)$ in an arbitrary way. This invariance is called $SU(3)$ *gauge invariance*. The requirement of gauge invariance is so stringent that there is essentially only one relativistic quantum field theory that has this invariance. The theoretical development of the theory is described in the article on GAUGE THEORY. In the context of quarks and color gauge symmetry, the $SU(3)$ gauge theory is called *quantum chromodynamics*. The electroweak interactions of elementary particles are thought to be described by another gauge theory, which is described in the article on ELECTROWEAK THEORY.

The content of quantum chromodynamics can be described on an elementary level quite simply. The color gauge symmetry implies the existence of a spin 1 particle, which is called a *gluon* because of its role in binding quarks together. The gluon comes in eight color states: $|R\bar{Y}\rangle$, $|R\bar{B}\rangle$, $|Y\bar{B}\rangle$, $|Y\bar{R}\rangle$, $|B\bar{R}\rangle$, $|B\bar{Y}\rangle$, $|R\bar{R}\rangle - |Y\bar{Y}\rangle$, $|R\bar{R}\rangle + |Y\bar{Y}\rangle - 2|B\bar{B}\rangle$. Here $\bar{R}$ denotes the color *anti-red*, which is the color carried by the antiparticles to a red quark. An R quark can change into a Y quark by emitting an $R\bar{Y}$ gluon, as depicted in Fig. 1(a), or by absorbing a $Y\bar{R}$ gluon. This process is mathematically very similar to the emission of a photon by an electron in quantum electrodynamics. The similarity arises from the fact that quantum electrodynamics is also a gauge theory. In quantum chromodynamics, the gluons can also interact with each other. For instance, a $R\bar{Y}$ gluon can change into an $R\bar{B}$ gluon by emitting a $B\bar{Y}$ gluon,

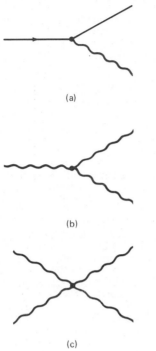

(a)

(b)

(c)

FIG. 1. Elementary processes in quantum chromodynamics. Straight lines are quarks, wavy lines are gluons.

as depicted in Fig. 1(b). There is also a four-gluon interaction, which is depicted in Fig. 1(c).

The theory gives a simple formula for calculating the probability amplitude for each of the elementary interactions shown in Fig. 1 to occur. As in all quantum field theories, physical probability amplitudes are built up from the amplitudes for the elementary interactions. For example, the scattering of two quarks can proceed by the exchange of two gluons, as shown in Fig. (2). This interaction involves four elementary interactions.

The probablity amplitude for an elementary interaction is proportional to a coupling $g(\lambda)$, in the case of Fig. 1(a) and (b), or to $g(\lambda)^2$ in the case of Fig. 1(c). The numerical value of g

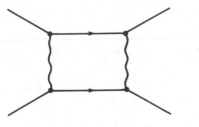

FIG. 2. Quark-quark scattering via two gluon exchange.

depends on a length λ that is taken to be a typical wavelength of the interacting particles. (There is a certain ambiguity in the choice of the value of λ to use in a calculation, but the effects of this ambiguity vanish when one sums over all possible combinations of elementary amplitudes that can contribute to the desired physical amplitude.) The coupling g is analogous to the elementary charge e in quantum electrodynamics. In quantum electrodynamics, e is small and approximately independent of λ: $e(\lambda)^2/4\pi\hbar c \cong \alpha \cong 1/137$.

The fact that e is small in quantum electrodynamics implies that only a few elementary interactions are likely to occur in a scattering process, so that calculation of the scattering probability is relatively simple. In quantum chromodynamics, $g(\lambda)$ is *not* small when λ is roughly equal to the distance between nucleons in a nucleus. This is the reason that the nuclear force is so much stronger than the electromagnetic force. The fact that $g(\lambda)$ is not small when $\lambda \gtrsim 1$ fm makes calculation difficult in many important cases. For instance, in quantum chromodynamics, the force that binds the quarks in a proton arises from the exchange of gluons among the quarks, in much the same way that the exchange of photons between an electron and a proton holds a hydrogen atom together. However, it has not so far been possible to reliably calculate this force from the theory.

The color force is strong. It is widely believed, although not definitely proven, that in quantum chromodynamics the force between quarks becomes so strong at large distances that when, say, two protons collide, the quarks involved in the collision can never escape as isolated particles. Instead, the only particles that can escape from a collision are colorless combinations of quarks and gluons. This phenomenon is called *quark confinement*. It explains why isolated quarks and gluons have not been seen in accelerator experiments.

Quantum chromodynamics has an important property that makes the theory simple under special circumstances: the coupling $g(\lambda)$ decreases as the length scale decreases. The decrease is slow, with $g^2(\lambda) \propto [\ln(\lambda_0/\lambda)]^{-1}$ when $\lambda \ll \lambda_0 \approx 1$ fm. Nevertheless, by the time λ equals the wavelength of a particle with 30 GeV of momentum, $g^2(\lambda)/4\pi\hbar c$ has decreased to a value of about 0.2. Thus when quarks and gluons have high momenta, the probability that they will interact is low: they behave almost as free particles. This property is called *asymptotic freedom*.

Asymptotic freedom allows one to make predictions from quantum chromodynamics that are testable in high energy experiments. We will mention only one such experiment. The theory leads one to the conclusion that the three quarks in a proton are really assemblages of elementary quarks and gluons; when probed on a sufficiently small distance scale λ, these elementary quarks and gluons should appear to be nearly free and noninteracting. Such a short distance probe can be implemented by scattering high energy electrons from protons, breaking the protons apart in the process. One counts the electrons that scatter through a wide angle from their initial direction. Each of these electrons can scatter from an elementary quark by exchanging a photon with the quark. When the electron energy is high, the photon wavelength λ is small. This experiment was performed at the Stanford Linear Accelerator Center in 1979, before the development of quantum chromodynamics. Surprisingly, the measured cross section agreed with what would be expected if the proton consisted of noninteracting pointlike particles. The fact that the quarks appeared to be nearly non-interacting when λ was small led to the idea of asymptotic freedom and the theory of quantum chromodynamics.

DAVISON E. SOPER

References

1. Bloom, E. D., et al., *Phys. Rev. Letters* **23**, 930 (1969).
2. Gross, D. J., and Wilczek, F., *Phys. Rev. Letters* **30**, 1343 (1973); Politzer, H. D., *Phys. Rev. Letters* **30**, 1346 (1973).
3. Balian, R., and Zin-Justin, J. (Eds), "Methods in Field Theory," Amsterdam, North Holland, 1976; Itzykson, C., and Zuber, J. B., "Quantum Field Theory," New York, McGraw Hill, 1980.

Cross-references: CONSERVATION LAWS AND SYMMETRY, ELECTROWEAK THEORY, ELEMENTARY PARTICLES, FERMI-DIRAC STATISTICS AND FERMIONS, FEYNMAN DIAGRAMS, GAUGE THEORIES, GRAND UNIFIED THEORY, NUCLEAR STRUCTURE, QUANTUM ELECTRODYNAMICS, QUARKS, STRONG INTERACTIONS.

QUANTUM ELECTRODYNAMICS*

Introduction Quantum electrodynamics is the fundamental theory of electromagnetic interactions. As far as we know, its equations provide a mathematically exact description of the interactions of electrons, muons, and photons (the quanta of the electromagnetic field). By extension, quantum electrodynamics (QED) is the underlying theory of all electromagnetic phenomena including atomic and chemical forces.

QED is without question the most successful dynamical theory in all of physics. Because of the pioneering theoretical efforts of Dirac, Feynman, Schwinger, and Tomonaga, and heroic

*Supported in part by the U.S. Department of Energy and the U.S. National Science Foundation.

labors by many others, QED possesses a systematic calculational scheme which has met with brilliant quantitative successes when faced with all experimental challenges. Its range of application has over the years been extended from atomic ($\sim 10^{-8}$ cm) to electron ($\sim 10^{-11}$ cm) to nuclear ($\sim 10^{-14}$ to 10^{-16} cm) dimensions. Adding in the classical aspects of electrodynamics, the total range of verification actually covers more than 25 decades out to 10^9 cm, where satellite measurements have verified the predicted cubic power law fall-off of the earth's magnetic field. In several cases the predictions of the theory have been confirmed at the level of parts per million or better.

In spite of this phenomenal success, there is growing evidence that electromagnetic and weak interactions are intricately related to each other and are just different manifestations of a unified electroweak interaction. It is plausible that even the strong interaction is involved in such a unification (grand unification). Experiments are in progress to verify these possibilities.

All this means is that QED, which deals with the electromagnetic force only, must be regarded as part of a more comprehensive theory unifying diverse forces. In practice, however, the modifications of the QED predictions due to weak and strong forces are generally very small until one explores the region smaller than $\sim 10^{-16}$ cm. To the extent that these small effects can be understood, either by calculations based on a (unified) theory or determined by appropriate measurements, it is possible to push the test of the validity of QED to even higher precision

Basic Features and Early History Quantum electrodynamics, the quantum theory of photons and electrons, has a simple conceptual basis. Shortly after the birth of quantum mechanics in 1925–26, P. A. M. Dirac, W. Heisenberg, and W. Pauli noted that the energy in the electromagnetic field amplitudes $\vec{E}(x, t)$ and $\vec{B}(x, t)$ described by Maxwell's equation must be quantized—as required by quantum mechanics for the energy of any physical system. The quanta of the radiation field could be identified exactly with Planck's photon, the fundamental carrier of light and electromagnetic radiation. The quantized form of Maxwell's equations require that the photon be a particle of zero mass, with velocity c and one unit of angular momentum $\hbar$ (Planck's constant divided by 2π)—directed either along or against its direction of motion. Upon quantization, the classical field amplitudes $\vec{E}$ and $\vec{B}$ (and the four-vector potential $A^\mu(x)$) become quantum-mechanical operators which have nonzero matrix elements only between states which differ by exactly one photon. This was the foundation stone of quantum electrodynamics, which as developed by Dirac, successfully described the emission and absorption of radiation in atomic systems.

During this exciting period in which developments in quantum theory came at an incredible rate, Dirac also presented his famous equation

$$[i\gamma_\mu(\partial^\mu - eA^\mu) - m]\,\psi(x) = 0,$$

which described the motion and spin (internal angular momentum) of an electron or positron in an electromagnetic potential. The application of this equation to an electron bound in a Coulomb field led to a successful prediction of the fine structure of the energy-level separations in the hydrogen atom. Soon afterwards P. Jordan and E. Wigner applied the quantization procedure to the energy contained in the ψ amplitude just as had been done for Maxwell's equation. The quanta of the ψ field can, in fact, be exactly identified with the physical electrons and positrons—particles (and antiparticles) which have charge $\pm e$, mass m, and angular momentum $\tfrac{1}{2}h$ directed along or against their motion, which must obey the Pauli principle. The Dirac ψ field thus became a quantum-mechanical operator which connects states that differ by one electron or one positron.

The Dirac theory also determines the basic interaction of electrons, positrons and photons and thus provides the fundamental dynamical assumptions of quantum electrodynamics. The form of the interaction density is

$$e\overline{\psi}(x)\gamma_\mu\,\psi(x)A^\mu(x)$$

which, in fact, is the simplest form consistent with the principles of quantum mechanics and relativity. Because of quantization of ψ and A, the interaction only has nonzero matrix elements for the basic electron-positron-photon vertex shown in Fig. 1. The vertex conserves charge and is "local," i.e., it describes the creation or absorption of a photon at the same point $x = (\vec{x}, t)$ in space and time where an electron (or positron) scatters or where an electron-positron

pair is created or destroyed. The amplitude for more complicated physical processes in QED can then be obtained from the iteration of the basic three-point interaction (i.e., a perturbation expansion in powers of e). The absolute square of the sum of all the possible amplitudes for the process gives the quantum mechanical probability, or in the case of a collision process, the cross section.

Rules for calculating the contributing amplitudes have been given in a very elegant and simple form by R. P. Feynman. The rules are explicitly covariant (i.e., independent of the choice of reference frame). For example, the contributing amplitudes (through order e^4) for electron-positron collisions are computed from the "Feynman diagrams" shown in Fig. 2. The lowest order contributing diagrams are of order e^2 since only the iteration of two basic three-point interactions is required. The amplitude for the first diagram in Fig. 2(a) is of the form

$$e^2\,J_\mu(e^-)J^\mu(e^+)\,\frac{1}{q^2}$$

which contains factors from Dirac current of the electron $J_\mu(e^-) = \overline{u}(p')\gamma^\mu u(p)$, the Dirac current of the positron, and a factor $(q^2)^{-1} = [(E - E')^2 - (\vec{p} - \vec{p}')^2]^{-1}$ for the propagation of the photon carrying momentum $\vec{q} = \vec{p} - \vec{p}'$ and energy $E - E'$ from the electron to the positron. The last factor for the photon (the "Feynman propagator") automatically accounts for the process in which the photon is emitted by the positron and then absorbed by the electron as well. A similar propagator $(\gamma_\mu q^\mu - m)^{-1}$ occurs in diagrams such as Fig. 2(b) for the propagation of the internal electron or positron. The diagrams in Fig. 2(b) are of order e^4, and give contributions to the collision cross section which are smaller by a factor of

$$\alpha = e^2/4\pi\hbar c = 1/137.035963\ (15),$$

which is the fundamental dimensionless constant of QED. (The quantity enclosed in parentheses represents the uncertainty in the final digits of numerical value.) Because of the smallness of the fine-structure constant α (so named since its square determines the ratio of the fine structure separation to the Rydberg in the Dirac theory for the hydrogen atom energy spectrum), it is seldom necessary to retain terms in perturbation theory beyond the first few orders in order to compare with high-energy scattering experiments.

Recently, very high energy (short-distance) tests of the predictions of quantum electrodynamics for electron-positron scattering have been performed at colliding electron-positron beam facilities at Stanford, Orsay, Novosibirsk, Frascati, and Hamburg. The cross section is predicted to fall inversely with the square of the center of mass energy. The measurements have directly confirmed this prediction at cen-

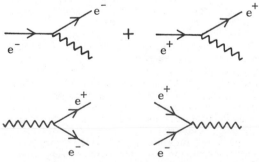

FIG. 1. The basic interaction vertex of quantum electrodynamics. The photon (denoted by a wavy line) is emitted or absorbed by a scattered electron or positron, or by the creation or annihilation of an electron-positron pair. The interaction conserves energy, momentum, and charge. The coupling strength is proportional to the electron charge e.

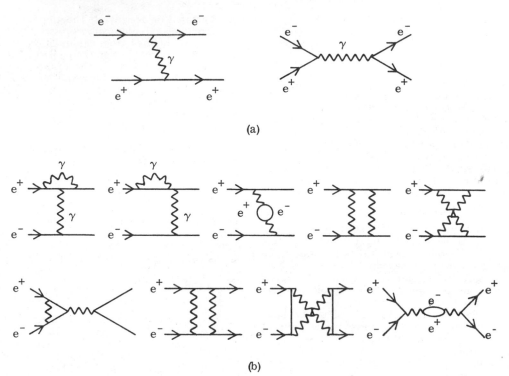

(a)

(b)

FIG. 2. Feynman diagrams for electron-positron scattering. Figure 2(a) represents the amplitudes of order e^2 obtained from the application of the basic interaction vertex of Fig. 1 to second order in perturbation theory. Figure 2(b) shows representative amplitude in order e^4 for electron-positron scattering.

ter of mass energies as high as 37 billion electron volts. The measurements rule out any modification or breakdown of QED unless it occurs at distances smaller than 10^{-16} cm where the weak boson Z^0 predicted by the theory of electroweak interaction due to Glashow, Weinberg, and Salam will modify the prediction of QED. Colliding-beam experiments

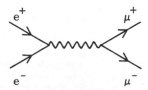

FIG. 3. The lowest-order Feynman diagram for the annihilation of an electron-positron pair into a muon pair. This process can now be studied experimentally at electron-positron colliding-beam facilities with beam energies up to nineteen billion electron volts. In these experiments, the electron and positron collide with equal and opposite momenta, producing an intermediate photon at rest in the laboratory, which then decays to the final muon pair. At higher energies this diagram interferes with one in which the photon is replaced by a weak boson Z^0.

in which a muon pair is produced from the annihilation of an electron pair (see Fig. 3) have provided dramatic verification of the correctness of the application of QED to the electromagnetic interactions of the muon as well.

Other high-energy collision measurements including electron or muon pair production and other inelastic processes have shown complete agreement with theory and no sign of any fundamental length at small distances, or additional heavy leptons or photons.

Complementary to these high-energy tests are the precision atomic physics and magnetic moment measurements, which have confirmed the validity of quantum electrodynamics at the level of billionths of an electron volt. For these tests, which are performed at the ppm level, an expansion through high orders in e is required, and the precise value of the fine-structure constant α is essential. The numerical value of α can be obtained independently of QED from a combination of measurements from very diverse fields, expressed via the relation

$$\alpha^{-2} = \frac{1}{4\mathrm{Ry}_\infty} \frac{1}{\gamma_{p'}} \frac{\mu_{p'}}{\mu_B} \frac{2e}{h} \frac{c\Omega_{\mathrm{abs}}}{\Omega_{\mathrm{NBS}}}$$

The Rydberg Ry_∞, the proton gyromagnetic ratio in water $\gamma_{p'}$, the magnetic moment

$\mu_{p'}/\mu_B$ of the proton (in a water sample) in units of the electron Bohr magnetron, and $c\Omega_{abs}/\Omega_{NBS}$, the ratio of absolute to NBS ohm (required for standard voltage measurements), are known to 0.2 parts per million, or better, and the ratio $2e/h$ has been determined recently to better than 0.03 ppm via the alternating current Josephson effect in superconductors by E. R. Williams and P. T. Olsen. Very recently a completely different way of determining α has been discovered by K. von Klitzing et al. This is based on the quantized Hall effect in a semiconductor which measures e^2/h directly. It has already provided an α value almost as accurate as that described above, and further improvement is expected shortly.

The Self-Energy Problem and the Lamb Shift Despite the apparent correctness of the basic structure of the interaction between the photon and the electron, and the elegance of its equations, there is a central problem in quantum electrodynamics, which reflects on its fundamental validity. This problem involves the "self-energy" corrections of the theory. For example, the calculation of the energy shift of a free electron due to its emission and subsequent reabsorption of a single photon, gives an infinite result! This divergence is a consequence of the point-like (local) nature of the basic three-point vertex of QED.

Progress on this problem came only because of an extraordinary experiment performed by W. Lamb and his co-workers in 1947. The Dirac equation for an electron bound in a Coulomb field predicted that the energy levels with the same quantum number n and total angular momentum j should be degenerate; in particular, negligible energy separation between the $2P_{1/2}$ and $2S_{1/2}$ excited levels in atomic hydrogen was predicted. The Lamb experiment, however, conclusively demonstrated that the levels were in fact separated by ~ 1000 MHz, a full tenth of the fine-structure separation of the $n = 2$ levels (the Lamb shift). Then, in a historic paper, H. Bethe was able to show that not only could quantum electrodynamics explain the Lamb shift, but also that the divergence problem in the self-energy calculation could be circumvented.

In his 1947 paper, Bethe pointed out that when one calculates the self-energy correction to a bound electron from the emission and absorption of a photon, part of the energy shift—in fact, the divergent part—should be identified with the energy or "self-mass" of the electron itself. When the self-mass term is identified and canceled out, or alternately if the difference of energy shifts for two bound state levels is calculated, the result is a finite energy shift. Bethe's calculations showed that the net energy shift due to the electron in the hydrogen atom emitting and absorbing one photon causes the $2S_{1/2}$ level to be displaced above the $2P_{1/2}$ level by about 1080 MHz. This can be understood physically by the fact that the emission and subsequent reabsorption of the photon

causes the position of the electron charge to fluctuate over a distance of the Compton wavelength of the electron ($\hbar/m_ec$) a percentage α of the time. The degeneracy predicted by the Dirac equation is a strict consequence of the inverse r dependence of the electron-nucleus Coulomb potential, and this is destroyed by the orbital dependence of the effective spreading of the electron's position. Another important effect is "vacuum polarization," which increases the binding of the electron and reduces the $2S_{1/2} - 2P_{1/2}$ separation by 27 MHz. This quantum-electrodynamical effect, which was discussed by R. Serber and E. A. Uehling in 1935, is caused by the modification of the Coulomb potential due to the virtual electron-positron pairs. In fact, because two charged particles always see each other through a cloud of virtual particle-antiparticle pairs, the charge e_0 that appears as a parameter in the equations of quantum electrodynamics is not precisely the physical charge e of the electron (as defined classically for an electron scattering at large distances from a test electric field).

Over the years both the theory and measurements of the Lamb shift have been developed to extraordinary precision. The theory accounts for higher-order photon processes and the relativistic effects of a two-particle bound state system (relativistic recoil corrections) using a covariant equation developed by H. Bethe and E. Salpeter. The latest theoretical prediction (by J. Sapirstein) is

$$\Delta E(2S_{1/2} - 2P_{1/2}) = 1057.860(9) \text{ MHz}.$$

The most recent measurement (by S. R. Lundeen and F. M. Pipkin) is

$$\Delta E(2S_{1/2} - 2P_{1/2}) = 1057.845(9) \text{ MHz}.$$

Other precision atomic physics tests of QED are described below.

Renormalization Theory Within a few years after Bethe's work, R. Feynman, J. Schwinger, S. Tomonaga, and F. Dyson demonstrated that all the divergences in quantum-electrodynamic calculations could be circumvented order by order in perturbation theory. This procedure, which is called the renormalization program, was shown to be covariant (independent of Lorentz frame), and despite the presence of infinite quantities in intermediate expressions, the calculations were shown always to lead to unambiguous finite results.

Briefly, one notes that the "bare charge" e_0 and "bare mass" m_0 which appear as parameters in the equations of quantum electrodynamics are not the physical charge and mass of the electron, but must be chosen correctly. Thus (to any given order in perturbation theory) one must adjust the values of m_0 and e_0 so that after calculation of all the contributing diagrams, electron-electron scattering at large distances will agree with classical Coulomb scattering for two particles of charge e, and that Compton scattering (elastic electron-photon scattering) at zero frequency will agree with the Thomson limit which defines the ratio e^2/m. After this adjustment is made, calculations of all *physical* quantities are finite, although the values in perturbation theory for the quantities e_0 and m_0 are mathematically divergent. In recent years, much theoretical work has been done in order to remove this last mathematical problem —either by attempts to demonstrate that e_0 and m_0 are actually finite and calculable when all orders in perturbation theory are taken into account, or by modifications of quantum electrodynamics at very small distances where unification with other interactions may occur.

Other Precision Tests of QED The classic and most basic test of quantum electrodynamics is the measurement of the anomalous magnetic moment of the electron. Thus far it has been one of the most stunning triumphs of experimental and theoretical analysis. The measurements are based on the following facts: An electron in a uniform magnetic field moves in a helical orbit with a rotational frequency $\omega_L = eB/mc$. If the electron's magnetic interactions conformed to the simple Dirac equation, the direction of its internal spin would precess with exactly the same frequency. The first experimental determination of a deviation of the electron moment from the Dirac value was made by H. M. Foley and P. Kusch in 1947; their results for the electron's "anomalous moment" were confirmed by J. Schwinger's calculation of the quantum-electrodynamic effect on the electron's spin precession frequency due to the emission and absorption of a single photon. Small corrections to the normal Dirac value can in fact be measured to tremendous precision since the net angle between the momentum and spin vectors of the electron can be determined to a few parts per million, as was demonstrated by D. T. Wilkinson and H. R. Crane in 1963. The most recent result by A. Rich and J. C. Wesley (1971) is

$$a_e{}^{\exp} = 0.001\ 159\ 656\ 7\ (35).$$

Recently, based on an entirely different technique which utilizes *rf* resonance transitions between quantized levels of electrons in a magnetic field, H. G. Dehmelt and his collaborators have improved the value of a_e by two orders of magnitude. Their latest results for the electron and positron are

$$a_{e^-}{}^{\exp} = 0.001\ 159\ 652\ 200(40)$$

and

$$a_{e^+}{}^{\exp} = 0.001\ 159\ 652\ 222(50),$$

respectively, which are in good agreement with the theoretical value of Kinoshita and Lindquist:

$$a_e{}^{\text{th}} = \frac{\alpha}{2\pi} - 0.328\ 478\ 966\left(\frac{\alpha}{\pi}\right)^2$$
$$+ 1.176\ 5(13)\left(\frac{\alpha}{\pi}\right)^2 - 0.8\ (2.5)\left(\frac{\alpha}{\pi}\right)^4$$
$$+ 4.4 \times 10^{-12}$$
$$= 0.001\ 159\ 652\ 460\ (150),$$

where the contribution 4.4×10^{-12} is due to the muon loop, hadronic, and weak interaction effects.

The gyromagnetic ratio of the electron $g = 2(1 + a)$, the ratio of the spin and rotation frequencies, which is a fundamental constant of nature, is thus correctly predicted by QED to 12 significant figures!

The theoretical result for the anomalous moment includes contributions due to emission and absorption of up to four photons by the electron. The calculations are carried out eight orders in perturbation theory beyond the lowest order and are the most complex of any in theoretical physics. The application of computerized algebraic and numerical computation techniques has been required.

Measurements of the muon anomalous magnetic moment by F. J. M. Farley and his co-workers are also in agreement with quantum electrodynamics if corrections due to the photon-hadron interaction are included. In this and in other precision measurements, the muon has been found to have symmetrical properties to the electron but for its heavier mass:

$$m_\mu/m_e = 206.768\ 60\ (47).$$

Other precision tests of quantum electrodynamics which have confirmed the correctness of the theory, including the renormalization procedure, involve the hyperfine separation of the ground state of fundamental atoms such as hydrogen, muonium, and positronium. The hyperfine separation results because of the difference in energy which occurs when the electron and nuclear magnetic moments are parallel and antiparallel. Measurements (based on the properties of the hydrogen maser) by H. Helvig et al. and L. Essen et al. have determined this separation to 13 significant figures in atomic hydrogen. The results agree with the theoretical predictions to six significant figures. Measurements of hyperfine splittings in positronium (the positron-electron atom) by V. Hughes et al. are in good agreement with the quantum electrodynamics predictions at nearly the same precision. Very accurate measurements of hyperfine splitting in muonium (the positive muon-electron atom) by V. W. Hughes et al. and V. Telegdi, et al., and recent theoretical efforts by D. R. Yennie, G. P. Lepage, et al. are pushing the test of QED to a precision approaching 0.1 ppm.

Other applications of quantum electrodynamics include high Z muonic atoms, coherent and incoherent electromagnetic processes, and general electromagnetic phenomena. Quantum electrodynamics has also served as a prototype for the development of various non-Abelian local gauge theories. There is now considerable evidence that these renormalizable gauge theories can account for the properties of the strong and weak interactions in elementary particle physics.

Conclusions As far as we know, quantum electrodynamics provides a mathematically exact description of the electromagnetic properties of the electron and muon. To the extent that the electromagnetic properties of the nucleus are known, the theory provides the fundamental dynamical theory of the relativistic atom, including its external electromagnetic interactions. At present there is no outstanding discrepancy with any of its predictions provided that the experimental data are corrected for effects due to weak and hadronic interactions, and despite our pursuing the limits of the theory to very high accuracy and very small distances.

Breakdown of the theory could still occur at short distances or at high precision if any of the following existed: (a) intrinsic lepton size or nonlocal currents, (b) heavy leptons, (c) heavy photons, (d) magnetic charge, (e) breakdown of perturbation theory or anomalous subtraction constants not given by the renormalization procedure. All of these modifications are ruled out to some extent by the various tests discussed in this article. In addition, the basic space-time symmetries of QED, conservation laws, the constancy in time of α, c, etc. have all been checked to various degrees.

All of the suggested modifications would mar the essential simplicity of Maxwell's equations and the Dirac form of the vertex interaction. Despite this simplicity, however, and despite its phenomenal success, the fundamental problems of renormalization in local field theory and the nature of the exact solutions of quantum electrodynamics are still to be resolved.

STANLEY J. BRODSKY
TOICHIRO KINOSHITA

References

1. Schwinger, J., Ed., "Quantum Electrodynamics," New York, Dover Publications, 1958. (The historical development of QED can be traced through this literature collection.)
2. Brodsky, S. J., and Drell, S. D., "The Present Status of Quantum Electrodynamics," *Ann. Rev. Nucl. Sci.*, **20**, 177 (1970).
3. Bjorken, J. D., and Drell, S. D., "Relativistic Quantum Mechanics" and "Relativistic Quantum Fields," New York, McGraw-Hill, 1965.
4. Taylor, B. N., Parker, W. H., and Langenberg, D. N., "The Fundamental Constants and Quantum Electrodynamics," *Rev. Mod. Phys.*, **41** (1969).
5. Klitzing, K. von, Dorda, G., and Pepper, M., *Phys. Rev. Lett.* **45**, 494 (1980).
6. Schwinberg, P. B., Van Dyck, R. S., Jr., and Dehmelt, H. G., *Phys. Rev. Lett.* **47**, 1679 (1981).
7. Kinoshita, T., and Lindquist, W. B., *Phys. Rev. Lett.* **47**, 1573 (1981).

Cross-references: ANTIPARTICLES, ELECTROMAGNETIC THEORY, ELEMENTARY PARTICLES, FEYNMAN DIAGRAMS, GAUGE THEORY, MANYBODY PROBLEM, MATRIX MECHANICS, POSITRON, QUANTUM CHROMODYNAMICS, QUANTUM THEORY, WAVE MECHANICS.

QUANTUM THEORY

1. Early History The theory of quantum mechanics has progressed by a series of discontinuous jumps from rather elementary beginnings through increasingly complex levels of abstraction. In this article we review this evolution and discuss the conceptual difficulties that have arisen, been resolved, or which have been ignored and remain to flaw the theory.

The break with classical mechanics came rapidly with two major observational discoveries. First, the spectral distribution in blackbody radiation was found to be inexplicable in terms of classical models; this led Planck (1900) to postulate the particulate nature of energy propagation in electromagnetic waves. Second, Davisson and Germer (1927) and G. P. Thomson (1928) discovered that electron beams passing through crystals were diffracted in the same way as X-ray electromagnetic (e.m.) waves. Both these observations were "explained" by the simple formulae

$$E = \hbar w, \qquad p = \hbar k$$
$$(w = 2\pi\nu, \qquad k = 2\pi/\lambda, \qquad \hbar = h/2\pi) \qquad (1.1)$$

together with the statement that electromagnetic waves of frequency w can have only energies $n\hbar w$ where n is any positive integer, and electrons having momentum p interact with crystalline matter like waves having wavelength λ equal to h/p. The value of Planck's constant $\hbar$ is the same for both phenomena.

The wave-particle paradox was the major conceptual problem for quantum mechanics for many years—and still is treated as such in many semipopular discussions of the theory.

It is worth going through the calculations leading to the Planck black-body radiation formula, using the classical Gibbs expression for the mean energy of a gas (of photons) in thermodynamic equilibrium.[1] Because the photons are indistinguishable particles, the spectrum of energies of distinguishable states of the radiation having frequency w is $n\hbar w$, $n = 1, 2, \ldots$ so the mean energy of the radiation is given by the classical formula

$$\langle E(w) \rangle = \sum_n n\hbar w e^{-n\hbar w/kT} \Big/ \sum_n e^{-n\hbar w/kT}$$

$$(1.2)$$

$$= \hbar w/(e^{\hbar w/kT} - 1). \qquad (1.3)$$

If instead of taking the discrete spectrum of energy states, we allow n to be a continuous variable, the sums of Eq. (1.2) are replaced by integrals and we find instead of Eq. (1.3) the result

$$\langle E(w) \rangle = kT \quad \text{independent of } \hbar. \qquad (1.4)$$

If the radiation field is expressed as a superposition of standing modes in a cubical box of side L, the number of modes having frequency between w and $w + dw$ is

$$N(w)\, dw = 2(L/c)^3 w^2\, dw, \qquad (1.5)$$

so the total energy in the field within this frequency range is on Planck's model, the product of (1.3) and (1.5):

$$2(L/c)^3 \hbar w^3\, dw/(e^{\hbar w/kT} - 1), \qquad (1.6)$$

and on the continuous model, the product of (1.4) and (1.5):

$$2(L/c)^3 w^2 kT\, dw. \qquad (1.7)$$

Planck's model agrees with the observed distribution, while the continuous model suffers from the "ultraviolet catastrophe"—the energy density becomes infinite at very high frequencies.

In Planck's model we picture the photons as relativistic particles with zero rest mass and energy $E = pc$, so that $E = \hbar w$ is equivalent to $p = \hbar k$. Just how the photons are moving under the guidance of the standing waves is an essentially unanswerable question at this stage of the theory. It must be borne in mind that Eq. (1.2) derives from the statistical theory (J. W. Gibbs, 1893-1903) of the distribution having maximum probability in an ensemble of systems (gases), and identified with thermodynamic equilibrium in any one of the members of the ensemble. The mechanism for establishing equilibrium—the "approach to equilibrium"—is left very much an open question here. In the present case the normal modes of electromagnetic radiation have no direct interaction with each other, and the process of absorption and of emission of energy between the walls of the enclosure and the radiation must be invoked to cause adjustments in the energy distribution of the radiation field. The "blackness" of the walls ensures a continuous frequency spectrum for these processes, and the structure (normal modes) of the radiation field gives rise to the spectral distribution, and only if the exchanges of energy between the walls and the radiation are effected in discrete photons does one obtain the correct Planck distribution. The photons

essentially only need to exist at the energy exchanges between matter and radiation field, and one has no need to try to visualize the photons as a gas of particles in space guided by the electromagnetic waves. Clearly there are serious conceptual difficulties here which are still subject to philosophical controversy.

In sharp contrast to this picture, the fact that electron beams can be diffracted supported de Broglie's idea (1926) that all material particles may be guided by wave fields. Matter waves were to be regarded as just as "real" as electromagnetic waves. Schrödinger (1887–1961) discovered a successful formula to express this concept. Starting with the classical Hamiltonian function $H(p, q)$ in phase space, and adopting the transcription formulae

$$\mathbf{p} \rightarrow -i\hbar\nabla, \qquad E \rightarrow i\hbar\partial/\partial t \qquad (1.8)$$

from the classical variables p and E to differential operators, and writing $\Psi(x, y, z, t)$ for a wave amplitude, the classical relation

$$H = p^2/2m + V(x, y, z) = E \qquad (1.9)$$

becomes an operator which operating on the function Ψ yields the Schrödinger wave equation

$$i\hbar\partial\Psi/\partial t = -(\hbar^2/2m)\nabla^2\Psi + V\Psi. \qquad (1.10)$$

This differential equation seems to describe the matter waves of the de Broglie picture. Indeed when solved for the case $V = 0$ everywhere—the free particle in empty space—one finds plane wave solutions:

$$\Psi_k = C \exp(-iE_k t/\hbar + i\mathbf{k} \cdot \mathbf{r}) \qquad (1.11)$$

where $\mathbf{r}$ means the vector (x, y, z), $\mathbf{k}$ is an arbitrary wave number vector, $E_k = \hbar^2 k^2/2m$ has the dimensions of energy, and C is an arbitrary amplitude. Ψ_k is what is called the *eigenfunction* of the operators p and E with eigenvalues $\hbar k$ and E_k because using (1.11) we find

$$E\Psi_k \equiv i\hbar\partial\Psi_k/\partial t = E_k\Psi_k \qquad (1.12)$$

and

$$P\Psi_k \equiv -i\hbar\nabla\Psi_k = \hbar\mathbf{k}\Psi_k. \qquad (1.13)$$

Note that the wavelength in Eq. (1.11) is $\lambda = 2\pi/k$ and the particle momentum is $\hbar k$, so that the de Broglie relation $p = h/\lambda$ is satisfied. However the speed of propagation of the Ψ-wave is $E_k/\hbar k = \hbar k/2m = p/2m$, which is only half the particle speed. Instead, the "group" velocity of a wave packet with a Gaussian distribution of frequencies does agree with the particle speed, so a rather clear picture emerged for the association between wave and particle aspects of the phenomenon, which we may call the *tentative probability interpretation* of the Ψ wave, discussed in Section 2.

Using the appropriate Hamiltonian for the motion of an electron in the Bohr-Sommerfeld model of the hydrogen atom, and noting that the Schrödinger transcription yields an operator also for orbital angular momentum, one can derive the familiar H-atom eigenfunctions and the fact that the angular momentum of an orbit is quantized in integral numbers of $\hbar$. But to obtain the correct Zeeman effect, Pauli, Uhlenbeck, and Goudsmit discovered (1925-1927) that it was necessary to postulate angular momentum to the electron itself, equal in magnitude to one half the quantum $\hbar$, and Pauli invented the *spin operator* with the eigenvalues plus and minus $\frac{1}{2}\hbar$, an operator with no classical analog—not derived from any classical variable by any transcription formula like Eq. (1.8). The result was to split every atomic state into two states of opposite spin; the wave functions became "spin degenerate" in the absence of a magnetic field.[a] The spin operator was represented by three two-by-two matrices, and seemed completely alien in nature to the differential operators for energy and linear momentum.

Generally ignoring the conceptual difficulties of reconciling the Schrödinger transcription formulae with common sense, theoretical physicists proceeded to apply the rules with widespread spectacular success. The entire chaotic field of spectroscopy was brought into a beautiful orderly pattern, the periodicity of the table of the elements was rationalized, the basic elastic properties of the solid state were explained.[3] Such a vast array of agreements between new and old experimental data on the one hand and the predictions of quantum mechanics on the other was established that the essential truth of the theory could not be denied.

2. Tentative Statistical Interpretation Conceptual difficulties meanwhile continued to accumulate.[4] Success of the theory was possible only after abandoning the naive interpretation of Schrödinger's wave function as the "real" de Broglie matter wave. The classical Hamiltonian of a many-particle system, like a crystalline solid or a complex molecule, involves an abstract phase space with at least $6N$ dimensions, where N is the number of particles. To carry through the Schrödinger transcription in such a phase space is possible only if one takes the overall wave function to be a product of functions, each factor a function of the phase coordinates of one of the particles. This is quite incompatible with the de Broglie picture of each particle being accompanied by a wave packet, all packets being defined on the same physical space. Even a system as elementary as a single pair of indistinguishable particles must be represented by a wave function of 12 variables, a function that cannot be visualized as two wave packets in the same three-dimensional space. Abstract interpretation of Ψ as a complex probability amplitude was first adopted, tentatively, leading to the following basic rule: When a system is in a state represented by the wave function Ψ, then measurements of any observable attribute of the system, represented by the operator A, produce on average the result

$$\langle A \rangle = \int \Psi^* A \Psi \, d\tau \qquad (2.1)$$

asserting that the function Ψ is a probability density amplitude. The arbitrary amplitude C in Eq. (1.11) has to be chosen so that the probability is normalized:

$$\int \Psi^* \Psi \, d\tau = 1. \qquad (2.2)$$

Note the essential distinction between the "superposition" of different functions Ψ involved in the wave packets mentioned in Section 1 above and the product of different Ψ functions in different spaces required to describe different particles. The wave packets result in a dispersion of measurement probabilities on a single-particle system; the wave products result in a single function describing a state of many particles, and may well be an eigenstate with well defined measurement probabilities.

One labels this interpretation as "tentative" advisedly, because many of the more creative thinkers in the field—Einstein,[5] de Broglie, and E. Wigner, to name a few—felt with good reason that a true quantum mechanics should have a deeper than purely statistical meaning. Louis de Broglie in fact established an active school of thought that sought escape from statistics by postulating the existence of "hidden variables"[6] and the use of nonlinear equations of motion. Einstein was never convinced that statistical quantum mechanics could be a complete theory, but failed to discover any alternative. Wigner, contemplating the meaning of Heisenberg's uncertainty principle has been interested in the essential part played by the consciousness of the physical observer during the process of measurement, in part because the concept of "probability" is itself difficult to divorce from purely subjective considerations.

The *Heisenberg uncertainty principle* arises directly from the wave-packet picture of the single particle: The more precisely one locates the particle by sharpening the spatial dimensions of the packet, the wider the spread of the wave numbers of the component waves, and hence the less precisely can one determine the momentum. Crudely written the principle states:

$$\Delta x \, \Delta p \geqslant \hbar/2 \qquad (2.3)$$

and identical statements pertaining to many other "conjugate pairs" of classical variables. Innumerable "Gedankenexperiments" were described to test the principle, and much discussion was concerned with whether the particle really "possessed" the quantities that could not be measured precisely in one and the same procedure ("simultaneously"). Perhaps physical observables should be thought of as attributes belonging rather to the measurement procedures than to the object measured. The mathematical content of the principle was quite definite,[7] whatever the philosophical implications. Let A and B be any two operators. Schwartz's inequality can be applied to two functions u and v defined in terms of some wave function Ψ and operators A and B as follows. Let

$$u \equiv (A - \langle A \rangle), \qquad v \equiv i(B - \langle B \rangle), \quad (2.4)$$

where $\langle A \rangle$ and $\langle B \rangle$ are defined as in Eq. (2.1). The result is the following form of Schwartz's inequality:

$$\langle (A - \langle A \rangle^2) \rangle - \langle (B - \langle B \rangle^2) \rangle \geqslant - \tfrac{1}{4} \langle (AB - BA)^2 \rangle. \quad (2.5)$$

Then if $A = P$ and $B = Q$, the momentum and position operators, respectively, this inequality reads

$$\Delta p^2 \Delta q^2 \geqslant - \tfrac{1}{4} \langle (PQ - QP)^2 \rangle. \quad (2.6)$$

The commutator $PQ - QP$ is to be evaluated directly from the definitions of the two operators, $P\Psi = -i\hbar \partial \Psi / \partial q$, $Q\Psi = q\Psi$, when $PQ\Psi - QP\Psi = -i\hbar\Psi$ and therefore, as operators

$$PQ - QP = -i\hbar \quad (2.7)$$

and Eq. (2.6) becomes

$$\Delta p^2 \Delta q^2 \geqslant \tfrac{1}{4} \hbar^2 \quad (2.8)$$

equivalent to Eq. (2.3). The meaning is taken to be that the dispersion in measurements of momentum and the dispersion in measurements of position of a particle in any "state" are necessarily related in such a way that the product of two dispersions cannot be less than $\tfrac{1}{4}\hbar^2$. This result cannot be understood in terms of classical mechanics.

3. Hilbert Space Concepts Without going into all the technical mathematical details, we may state the following abbreviated definition of a Hilbert space:[8] A set of elements, for example a set of complex functions χ defined on the Cartesian space of real numbers (x, y, z), is said to form a Hilbert space $\mathcal{H}$ if:

(a) A finite scalar product of every pair of members of the set can be defined, writing for the scalar product of the nth and mth member

$$(\chi_n, \chi_m) \quad \left(\text{for example} \right.$$
$$\left. (\chi_n, \chi_m) \equiv \left(\int \chi_n^* \chi_m \, d\tau \right) \right); \quad (3.1)$$

(b) A distance between any two members, written $d(n, m)$, can be defined with the metric

properties

$$d(n, m) \geqslant 0, \qquad d(n, m) = 0$$
$$\text{if and only if} \quad n = m$$
$$d(n, m) \leqslant d(n, k) + d(k, m) \quad (3.2)$$

(for example

$$d(n, m) \equiv \sup_{\text{all } x} \{ |\chi_n(x) - \chi_m(x)| \}).$$

There are other prerequisites which need not concern us here.

To be of physical interest, function spaces must be "separable," meaning that they can be "spanned" by a complete set of mutually orthogonal functions chosen from among the functions forming the space. Separable Hilbert spaces embrace almost all functions of physical interest. If we write $\{|n\rangle\}$ for the set of functions in $\mathcal{H}$ that span the space, we have $(|n\rangle, |m\rangle)$ for the scalar product, and writing $\langle m |$ for the complex conjugate of $|m\rangle$ we can match the definition (3.1) by writing the scalar product in the form $(\langle n | m \rangle)$ and require

$$(\langle n | m \rangle) = \delta_{mn}, \quad 0 \text{ or } 1. \quad (3.3)$$

The basis vectors $|m\rangle$ and $|n\rangle$ are mutually orthogonal and normalized to unity.

Simply stated, the spanning concept means that any member of $\mathcal{H}$, for example χ, can be expressed as a linear combination (or superposition) of the members of the basis set spanning the space:

$$\chi = \sum_n a_n |n\rangle. \quad (3.4)$$

It also means that any operator, say A, which maps any member of the space into some other member of the same space, can be represented by a matrix: thus let A map χ into ϕ:

$$A\chi = \phi \quad (3.5)$$

and expand both χ and ϕ in terms of the basis $\{|n\rangle\}$

$$\chi = \sum_n a_n |n\rangle \quad \text{and} \quad \phi = \sum_n b_n |n\rangle. \quad (3.6)$$

Then Eq. (3.5) becomes

$$\sum_n a_n A |n\rangle = \sum_m b_m |m\rangle. \quad (3.7)$$

Taking the scalar product from the left with the basis element $|k\rangle$ Eq. (3.7) yields

$$\sum_n a_n \langle k | A | n \rangle = \sum_m b_m \langle k | m \rangle = b_k \quad (3.8)$$

because of Eq. (3.3). This permits one to determine the function ϕ in Eq. (3.6) when we know the operator A and operand χ. The quantities $\langle k | A | n \rangle$ in Eq. (3.8) are collectively called the matrix of A on the "$\{|n\rangle\}$ representation." An abbreviated notation is often used:

$$\langle k | A | n \rangle = A_{kn}, \qquad (3.9)$$

and explicitly we could write out the matrix in the form

$$\{A_{kn}\} = \begin{pmatrix} \langle 1|A|1\rangle & \langle 1|A|2\rangle & \langle 1|A|3\rangle & \cdots \\ \langle 2|A|1\rangle & \langle 2|A|2\rangle & \langle 2|A|3\rangle & \cdots \\ \langle 3|A|1\rangle & \langle 3|A|2\rangle & \langle 3|A|3\rangle & \cdots \\ \cdots\cdots\cdots\cdots\cdots\cdots\cdots\cdots \end{pmatrix}$$

$$(3.10)$$

with as many rows, or columns, as there are members in the basis set, which in some cases may be infinite. Using the scalar product as defined in Eq. (3.1), the matrix elements displayed in Eq. (3.10) are

$$\langle k | A | n \rangle = \int f_k^* A f_n \, d\tau \qquad (3.11)$$

where f_n is written in place of $|n\rangle$ to emphasize the familiar form of integrals.†

There is one other general theorem we shall quote here: The complete set of eigenvectors of any operator with a discrete point spectrum of eigenvalues, forms an acceptable basis of a separable Hilbert space.

We saw in Eqs. (1.11) and (1.13) that Ψ_k of Eq. (1.11) was an eigenfunction of the momentum operator with eigenvalue $\hbar k$. In free space this is a continuous variable and therefore the functions Ψ_k cannot form the discrete basis of a Hilbert space. But if the problem is restricted to a finite volume the values of k do become discrete, and the set of functions Ψ_k reduce to the discrete set of trigonometric functions which vanish at the boundaries of the volume of interest. If this is a line of length $2L$, the functions are $\sin \pi n x/L$ when n is even and $\cos \pi n k/L$ when n is odd, and the expansion of Eq. (3.4) is the familiar Fourier expansion of any function of x which vanishes outside the domain $-L \leqslant x \leqslant L$.

A similar difficulty occurs with the eigenvectors of the position operator Q which presents the eigenvector equation

$$Q f(x) = x f(x). \qquad (3.12)$$

This operator maps any function $f(x)$ on to the function $x f(x)$, and x is clearly the eigenvalue; any function whatever is an eigenfunction, and the eigenvalue x can have any value in the space, so is continuous, and there is no acceptable basis for the Hilbert space on these terms.[8]

4. Simple Harmonic Oscillator Representation[9] The classical Hamiltonian for a one-di-

†An excellent elementary treatment of matrix algebra, as needed for quantum mechanics, can be found in Ref. 7.

mensional simple harmonic oscillator (S.H.O.) is

$$H = (1/2m)(p^2 + m^2 w^2 x^2). \qquad (4.1)$$

Schrödinger's wave equation derived from this by the transcription of Eq. (1.8) reads

$$i\hbar \partial \Psi/\partial t = (1/2m)(m^2 w^2 x^2 \Psi - \hbar^2 \partial^2 \Psi/\partial x^2). \qquad (4.2)$$

If we write

$$\Psi = u(x) e^{-iEt/\hbar} \qquad (4.3)$$

Eq. (4.2) reads

$$(1/2m)(m^2 w^2 x^2 - \hbar^2 \partial^2/\partial x^2) u(x) = E u(x) \qquad (4.4)$$

which one recognizes as an eigenvalue relation for the Hamiltonian operator

$$H = (1/2m)(m^2 w^2 x^2 - \hbar^2 \partial^2/\partial x^2). \qquad (4.5)$$

Eq. (4.4) is a differential equation with well-known solutions if and only if E is chosen as one of the spectrum of values

$$E = E_n = (n + \tfrac{1}{2})\hbar w, \qquad n = 1, 2, \ldots \qquad (4.6)$$

The solutions are Hermite functions

$$u_n(x) = (2^n n! \sqrt{\pi})^{-1/2} e^{x^2/2} H_n(x), \qquad (4.7)$$

$H_n(x)$ being the nth Hermite polynomial defined by

$$H_n(x) = (-1)^n e^{x^2} \partial^n (e^{-x^2})/\partial x^n. \qquad (4.8)$$

From the point of view of Hilbert space concepts, one can say that the functions $u_n(x)$ form a closed orthonormal basis set for the whole space of Lebesgue square integrable functions of x. Any such function can therefore be written in the form

$$f(x) = \sum_n f_n u_n(x) \qquad (4.9)$$

where f_n are complex constants. For convenience we can write

$$u_n(x) = |n, x\rangle, \qquad (4.10)$$

and any operator or mapping in the space onto itself can be written as a matrix as in Eqs. (3.9) and (3.10):

$$A \to A_{mn} = \int \langle n, x | A | m, x \rangle \, dx. \qquad (4.11)$$

In particular the matrix for the Hamiltonian operator is diagonal:

$$\langle m, x | H | n, x \rangle = (n + \tfrac{1}{2})\hbar w \langle m, x | n, x \rangle$$
$$= (n + \tfrac{1}{2})\hbar w \delta(m, n) \qquad (4.12)$$

which has the diagonal (eigen)values $(n + \frac{1}{2})\hbar w$ and zeroes elsewhere. In general those operators which have real eigenvalues—like the Hamiltonian—are called *Hermitian*. All physical observables are presumed to be represented by Hermitian operators. A Hermitian operator need not be diagonal on any given representation, but will have real mean values—the diagonal elements are real. The momentum and position operators discussed above have interesting matrices on the S.H.O. basis. They are zero everywhere except on two parallel rows once removed from the diagonal.

(1) Momentum operator P: the only non zero elements are, omitting the x-indicator as superfluous,

$$\langle n|P|n-1\rangle = -\langle n-1|P|n\rangle = i\sqrt{\tfrac{1}{2}n\hbar mw},$$

$$n = 1, 2, \ldots . \qquad (4.13)$$

(2) Position Q: the only non zero elements are

$$\langle n|Q|n-1\rangle = \langle n-1|Q|n\rangle = \sqrt{\tfrac{1}{2}n\hbar mw},$$

$$n = 1, 2, \ldots . \qquad (4.14)$$

The diagonal elements—mean values—are all zero. Using elementary matrix algebra it is not difficult to derive the theorem that the product of the matrices for P and Q satisfy the relation

$$PQ - QP = -i\hbar\mathbf{I} \qquad (4.15)$$

where $\mathbf{I}$ is the identity matrix (unity on every diagonal element and zero elsewhere), matching Eq. (2.7) derived from the differential form of the operator P.

So far we have discussed motion only in one dimension. If we wish to discuss functions of position in three-dimensional space we may write out the classical Hamiltonian for the isotropic three-dimensional oscillator and transcribe it into the Schrödinger equation:

$$-(\hbar^2/2m)\nabla^2 u + \tfrac{1}{2}mw^2(x^2 + y^2 + z^2)u$$

$$= Eu(x, y, z) \qquad (4.16)$$

and note that the variables separate, giving three equations each identical in form with Eq. (4.4), independent of one another except for the fact that the sum of the three energies has to equal E. Since the individual energies must each have one of the values $(n + \frac{1}{2})\hbar w$, the total energy must be

$$E = (n_x + n_y + n_z + \tfrac{3}{2})\hbar w \qquad (4.17)$$

where the set of integers n_x, n_y, n_z is an arbitrary vector $\mathbf{n}$. Writing the eigenstates

$$u_{n_x}(x) = |n, x\rangle, \quad \text{etc.} \qquad (4.18)$$

the general solution of Eq. (4.16) can be written

$$\Psi = \sum_{\mathbf{n}} C_{\mathbf{n}} \prod_j |n, j\rangle, \quad j = x, y, z. \ (4.19)$$

Operating on this with the Hamiltonian

$$H = -(\hbar^2/2m)(\partial^2/\partial x^2 + \partial^2/\partial y^2 + \partial^2/\partial z^2)$$

$$+ \tfrac{1}{2}mw^2(x^2 + y^2 + z^2)$$

$$= \sum_k H_k, \quad k = x, y, z, \qquad (4.20)$$

we have

$$\sum_k H_k \Psi = \sum_{\mathbf{n}} C_{\mathbf{n}} \sum_k H_k \prod_j |n, j\rangle. \ (4.21)$$

Where H_k operates on a product like $\prod_j|n, j\rangle$, it is to be understood that it "sees" only the kth factor:

$$H_k \prod_j |n, j\rangle = E_k \prod_j |n, j\rangle, \quad k = x, y, \text{ or } z.$$

$$(4.22)$$

Thus we have

$$H\Psi = \sum_{\mathbf{n}} C_{\mathbf{n}} \sum_k E_{n_k} \prod_j |n, j\rangle. \qquad (4.24)$$

The mean energy in the state—see Eq. (2.1)—is

$$\langle H \rangle = \langle \Psi^* | H | \Psi \rangle$$

$$= \sum_{\mathbf{m}} C_{\mathbf{m}}^* \sum_{\mathbf{n}} C_{\mathbf{n}} \sum_k E_{n_k} \prod_j \langle m, j | \prod_i |n, j\rangle;$$

$$(4.25)$$

and because of the orthonormality of eigenvectors this reduces to

$$\langle H \rangle = \sum_{\mathbf{n}} C_{\mathbf{n}}^* C_{\mathbf{n}} \sum_k E_{n_k}. \qquad (4.26)$$

The square modulus $C_{\mathbf{n}}^* C_{\mathbf{n}}$ is then interpreted as the probability that, in the state Ψ, the energy "has" the set of energy eigenvalues determined by Eq. (4.17) and the set n. In the back of one's mind in making this interpretation is the idea—sometimes enunciated as a principle—that any measurement of the energy must result in some eigenvalue of the energy. We must return to the quantum theory of measurement later (Section 9).

5. Spherical Harmonic Representation A very different and frequently employed representation of the same Hilbert space of Lebesgue square integrable functions on three-dimensional space—the variable being r, θ, and ϕ instead of x, y, and z—is that based on the eigenstates of the classical Hamiltonian of the Bohr-Sommerfeld model of the hydrogen atom. Thus, the Hamiltonian of the relative motion of the electron and proton in the H atom leads to the Schrödinger equation[9]

$$[-(\hbar^2/2\mu)\nabla^2 - Ze^2/r]\, u(r, \theta, \phi)$$

$$= Eu(r, \theta, \phi). \qquad (5.1)$$

The variables separate, and solutions exist of the form

$$u(r, \theta, \phi) = R(r) Y(\theta, \phi). \qquad (5.2)$$

The factor $Y(\theta, \phi)$ may be any one of the set of spherical harmonic functions, normalized to unity under integration over a unit sphere:

$$Y_{lm}(\theta, \phi) = \left[\frac{2l+1}{4\pi} \frac{(l-m)!}{(l+m)!}\right]^{1/2}$$
$$\cdot P_l^m(\cos\theta) e^{im\phi}. \qquad (5.3)$$

Here $P_l^m(\cos\theta)$ are the associated Legendre functions described in many texts on differential equations, l is any integer, and m is an integer between $+l$ and $-l$.

The factor $R(r)$ is some member of a set of functions defined in terms of the associated Laguerre polynomials; which member it is depends in part on the choice of which spherical harmonic is chosen—explicitly on the integer l. It also depends on another quantum number, n, indexing which energy eigenvalue is chosen. The solution of Eq. (5.1) exists only when the energy E has one of the eigenvalues

$$E_n = -Z^2 e^4/2\hbar^2 n^2, \quad n = 1, 2, \ldots. \qquad (5.4)$$

The general solution of Eq. (5.1) is then

$$u(r, \theta, \phi) = \sum_{n,l,m} C_{nlm} R_{nl}(r) Y_{lm}(\theta, \phi)$$

$$n, l \text{ arbitrary integers}, \quad -l \leqslant m < l. \qquad (5.5)$$

The constants C_{nlm} are chosen to normalize $\int u^* u \, d\tau$ to unity over an infinite sphere.

The set of eigenvectors $R_{nl} Y_{lm} \equiv |nlm\rangle$ determine the spherical representation matrix of the Hermitian operator corresponding to any physical observable, being particularly useful when axial symmetry exists in the situation. For example, this is the case for the Schrödinger operators corresponding to orbital angular momentum, derived from the classical expressions

$$M_x = yp_z - zp_y \rightarrow -i\hbar(y\partial/\partial z - z\partial/\partial y),$$
$$= i\hbar [\sin\phi \, \partial/\partial\theta + \cot\theta \cos\phi \, \partial/\partial\phi],$$
$$M_y = -i\hbar [\cos\phi \, \partial/\partial\theta - \cot\theta \sin\phi \, \partial/\partial\phi],$$
$$M_z = -i\hbar \, \partial/\partial\phi, \qquad (5.6)$$

and

$$M^2 = M_x^2 + M_y^2 + M_z^2 \rightarrow$$
$$-\hbar^2 [(1/\sin\theta)(\partial/\partial\theta)(\sin\theta \, \partial/\partial\theta)$$
$$+ (1/\sin^2\theta)\partial^2/\partial\phi^2].$$

The matrices as defined in Eq. (4.11) are given by the integrals

$$(M_j)_{nlm, n'm'l'} = \iiint \langle nlm | M_j | n'l'm' \rangle$$
$$\cdot \sin\theta \, d\theta \, d\phi \, r \, dr. \qquad (5.7)$$

The radial factor integrates to $\delta(n, n')$, integration over θ yields $\delta(l, l')$ and the final integration over ϕ results in a matrix the number of whose elements depends on l. For any given l value, the number of eigenvectors, i.e., the number of rows or columns in the matrix is the number of values of m, namely $2l+1$. This for $l=1$ we find the matrices:

$$M_x = (\hbar/\sqrt{2}) \begin{pmatrix} 0 & 1 & 0 \\ 1 & 0 & 1 \\ 0 & 1 & 0 \end{pmatrix}$$

$$M_y = (i\hbar/\sqrt{2}) \begin{pmatrix} 0 & -1 & 0 \\ 1 & 0 & -1 \\ 0 & 1 & 0 \end{pmatrix}$$

$$M_z = \hbar \begin{pmatrix} 1 & 0 & 0 \\ 0 & 0 & 0 \\ 0 & 0 & -1 \end{pmatrix}$$

$$M^2 = 2\hbar^2 \begin{pmatrix} 1 & 0 & 0 \\ 0 & 1 & 0 \\ 0 & 0 & 1 \end{pmatrix}. \qquad (5.8)$$

For $l=2$ there are five rows or columns, etc. In general, for all values of l the matrices for M_z and M^2 are diagonal, meaning that the basis vectors are eigenvectors of the operators, the eigenvalues being displayed on the diagonal terms. The same basis vectors are not eigenvectors of either M_x or M_y, and the three operators M_x, M_y, and M_z do not commute with one another under the rules of matrix multiplication. Indeed, it is easy to verify the following:

$$M_x M_y - M_y M_x = [M_x, M_y] = i\hbar M_z,$$
$$[M_y, M_z] = i\hbar M_x, \quad \text{and} \quad [M_z, M_x] = i\hbar M_y. \qquad (5.9)$$

These commutation relations can also be derived independently of any matrix representation directly from the definitions of M_x, etc., namely $M_x = yp_z - zp_y$, etc., and the known commutation relations between P and Q, as in Eq. (2.7):

$$[x, p_x] = [y, p_y] = [z, p_z] = i\hbar. \quad (5.10)$$

The existence of the Pauli spin, or intrinsic angular momentum can be incorporated into this formalism when it is noticed that the set of matrices typified by Eq. (5.8) for all integral values of l is not the only set that satisfies the commutation relations Eq. (5.9). A set of matrices having even numbers of rows or columns also satisfy Eq. (5.9), in particular, corresponding to $l = \frac{1}{2}$ we have

$$M_x = \frac{1}{2}\hbar \begin{pmatrix} 0 & 1 \\ 1 & 0 \end{pmatrix}, \qquad M_y = i\frac{1}{2}\hbar \begin{pmatrix} 0 & -1 \\ 1 & 0 \end{pmatrix},$$

$$M_z = \frac{1}{2}\hbar \begin{pmatrix} 1 & 0 \\ 0 & -1 \end{pmatrix}, \qquad M^2 = \frac{3}{4}\hbar^2 \begin{pmatrix} 1 & 0 \\ 0 & 1 \end{pmatrix}$$

$$(5.11)$$

which are recognized at once as the Pauli spin matrices. When a system such as the electron in an H-atom possesses both orbital and intrinsic angular momentum a technical problem presents itself: how to add operators which receive matrix representations having different numbers of rows or columns. This leads to a fascinating and highly abstract field of mathematics usually known as the *theory of irreducible tensors*, which is far beyond the scope of this article, but which is an essential chapter in the application of quantum mechanics to all atomic problems. Some valuable references to the theory of angular momentum generally are given at the the end of this article.[10]

6. Dirac's Relativistic Wave Equation The classical relativistic Hamiltonian for a free particle is

$$H = \sqrt{c^2 p^2 + m^2 c^4} \qquad (6.1)$$

where c is the velocity of light and m the mass of the particle, but when $\mathbf{p}$ is transcribed into the operator $-i\hbar\nabla$, there is trouble assigning a meaning to the square root. To avoid this one may take the square of Eq. (6.1) and then use the transcription of Eq. (1.8) to derive a relativistic invariant second-order differential equation known as the Klein-Gordon equation (1926):

$$\left(\nabla^2 - \frac{1}{c^2}\frac{\partial^2}{\partial t} - \frac{m^2 c^2}{\hbar^2} \right)\Psi = 0. \qquad (6.2)$$

However it was found that this equation is incapable of describing the behavior of a particle with spin, and so will be of no use in electron theory. Dirac (1928) discovered that the only way to achieve a linear differential equation—equivalent to the "square root" of Eq. (6.2)–in the form

$$i\hbar\,\partial\Psi/\partial t = H\Psi \qquad (6.3)$$

is to write H as a 4×4 matrix having the form

$$H_{\mu\nu} = c\sum_k (\alpha_k)_{\mu\nu} p_k + \beta_{\mu\nu} mc^2, \qquad (6.3)$$

where $k = 1, 2, 3, \mu, \nu = 1, 2, 3, 4$, and β, α_k are four 4×4 dimensionless constant matrices which have to satisfy several algebraic conditions in order to agree with Eq. (6.1), in the sense that matrix multiplication yields

$$H^2 = c^2 p^2 + m^2 c^4. \qquad (6.4)$$

The conditions on the four matrices are stated as matrix products:

$$\alpha_j\alpha_k + \alpha_k\alpha_j = 2\delta_{jk}$$
$$\alpha_j\beta + \beta\alpha_j = 0 \qquad (6.5)$$
$$\beta^2 = \mathbf{I}.$$

A set of four matrices in which β is diagonal was found to be

$$\alpha_j = \begin{pmatrix} 0 & \sigma_j \\ \sigma_j & 0 \end{pmatrix}, \qquad \beta = \begin{pmatrix} \mathbf{I}_2 & 0 \\ 0 & \mathbf{I}_2 \end{pmatrix} \qquad (6.6)$$

where each element in the brackets is itself a 2×2 matrix:

$$\mathbf{0} = \begin{pmatrix} 0 & 0 \\ 0 & 0 \end{pmatrix}, \qquad \mathbf{I}_2 = \begin{pmatrix} 1 & 0 \\ 0 & 1 \end{pmatrix},$$

and $\frac{1}{2}i\hbar\sigma_j$ is the jth Pauli spin matrix of Eq. (5.11). To give Eq. (6.3) a meaning with this kind of Hamiltonian, it is clearly necessary to write Ψ as a column of four separate functions, and the equation becomes in effect four equations for four different wave functions. Taking the time factor $e^{-iEt/\hbar}$ to be the same on each function, and seeking four-particle solutions analogous to Eq. (1.11), one finds

$$\Psi_\mu(r, t) = C_\mu(E, k)\, e^{i\mathbf{k}\cdot\mathbf{r}}\, e^{-iEt/\hbar},$$
$$\mu = 1, 2, 3, 4, \qquad (6.7)$$

where $\hbar k$ is the eigenvalue of $\mathbf{p}$, and C_μ are constants. Because of the "square root" nature of the equation, for each value of E there is another set of solutions with $-E$ for the energy—the negative energy states. This was eventually understood in terms of the existence of a positive electron and its interaction with electric fields.

There is another important "degeneracy" in the solutions. For each value of E, and correspondingly also for each negative energy state, there are two sets of four wave functions. This was quickly identified with the "spin degeneracy" postulated in the Pauli theory of the electron. When one applies the time-evolution law with the relativistic Hamiltonian (see Eq. (7.8) in the next section) the orbital angular momentum has to be augmented with spin angular momentum before one obtains a "constant of the motion." Thus Dirac's equation was the first theory of the electron that actually required the spin logically, rather than simply including it as an ad hoc hypothesis. The fundamental significance of Dirac's discoveries for the subsequent development of quantum mechanics can hardly be exaggerated, but to go further in that direction would take us beyond the scope of this article. The serious student is referred to Ref. 11.

7. **Time-Dependent Wave Functions** Up to this point we have been ignoring the time dependence of the Schrödinger wave function implied by Eq. (1.10), and included explicitly in Eq. (1.11) and Eq. (4.3). We shall now rewrite Eq. (4.3), in view of Eq. (4.6) and (4.10), in the form

$$|n, x, t\rangle = e^{-iEnt/\hbar}|n, x\rangle, \qquad E_n = (n + \tfrac{1}{2})\hbar.$$
$$(7.1)$$

The general time-dependent wave function is

$$\Psi(x, t) = \sum_n C_n(t)|n, x\rangle$$

where

$$C_n(t) = C_n(0)\, e^{-iEnt/\hbar}. \qquad (7.2)$$

On the other hand, because $H|n, x\rangle = E_n|n, x\rangle$, we have

$$|n, x, t\rangle = e^{-iHt/\hbar}|n, x\rangle, \qquad (7.3)$$

which is the fundamental dynamical law on what is traditionally known as "the Schrödinger picture." On this same picture, any observable A, defined as a time-independent mapping on $\mathcal{H}$, has matrix elements formed on the time-dependent Schrödinger eigenstates, using Eq. (7.3):

$$A_{mn}(t) = \langle m|e^{iHt/\hbar} A\, e^{-iHt/\hbar}|n\rangle. \ (7.4)$$

We may define a time-dependent operator

$$A(t) \equiv e^{iHt/\hbar} A\, e^{-iHt/\hbar}, \qquad (7.5)$$

called the Heisenberg operator, corresponding to the Schrödinger operator A. The matrix elements of $A(t)$ formed on the time-independent eigenstates $|n, x\rangle$ are the same as Eq. (7.4). The time derivative of $A(t)$ is clearly

$$\partial A(t)/\partial t = (i/\hbar)\, [H, A(t)], \qquad (7.6)$$

and this equation is the fundamental dynamical law on the Heisenberg picture.

In both pictures, time evolution is seen to correspond to a unitary transformation in $\mathcal{H}$ with the Hamiltonian H acting as the generator. In the Schrödinger picture it is the wave function or the state of the system that is evolving in time, while in the Heisenberg picture it is the observables that are evolving in time—two alternative pictures of the same essential situation.

Note that if A commutes with H, the product,

$$e^{iHt/\hbar} A\, e^{-iHt/\hbar} = A, \qquad (7.7)$$

so the Schrödinger matrix elements of Eq. (7.4) are time independent even when formed on the time dependent wave functions, while the time-dependent Heisenberg operator of Eq. (7.5) is identical with the Schrödinger operator.

There is one other important implication of these considerations. If we take the average (expectation) value of any observable A when the system is in a state represented by some wave function Ψ, then, using Eq. (2.1) we have from Eq. (7.5)

$$\langle \partial A/\partial t\rangle = (i/\hbar)\, [\langle\Psi|HA|\Psi\rangle - \langle\Psi|AH|\Psi\rangle].$$
$$(7.8)$$

If Ψ happens to be an eigenfunction of H,

$$\Psi = \Psi_E, \qquad H\Psi_E = E\Psi_E, \qquad (7.9)$$

then

$$\langle \partial A/\partial t\rangle = (i/\hbar)\, [E\langle\Psi_E|A|\Psi_E\rangle$$
$$- \langle\Psi_E|A|\Psi_E\rangle E].$$
$$= 0. \qquad (7.10)$$

We conclude that any eigenstate of the Hamiltonian is an equilibrium state of the system, such that the average time derivative of any observable A, whether it commutes with H or not, averages zero when measured in that state.

For lack of space, we refrain from discussing either perturbation theory or collision theory,[8,12] both of which have very wide uses in practical applications, and proceed instead with subjects that have more impact on the theoretical concepts of quantum theory.

8. **Many-Body Problems** In dealing with a single particle system we have seen in Section 3, Eqs. (3.4) and (3.6), that the wave function can be expressed as a linear "superposition" of eigenstates of any Hermitian operator—in particular the energy eigenstates of a harmonic oscillator, Eqs. (4.19) and (4.26)—when the square moduli of the superposition coefficients are interpreted in some sense as the "probabilities" that measurement of the observable results in the corresponding eigenvalue. In dealing with a more complex system—two or more particles— we have already stated (Section 2) that the sum of the individual wave functions does not result in a suitable wave function of the pair, but that a product is required, and that this fact forces a further increase in the level of abstraction needed to interpret the symbolism physically.

Consider two "classical" particles, i.e., particles that are the "same" in the normal sense, like two billiard balls, but may be distinguished from one another by otherwise neutral indicators, like "color" or some simple numbered label. The wave function for such a pair is a function of at least the six center-of-mass coordinates of the two particles, and the simplest such function would be the product

$$\Psi(x_A, x_B) = f(x_A)g(x_B) \qquad (8.1)$$

where $f(x_A)$ is the wave function for particle A alone, and $g(x_B)$ that for particle B alone, and x_A, x_B are the center-of-mass positions of the two particles. Plotted on a hyperspace spanned by x_A and x_B, this function $\Psi(x_A, x_B)$ is different from the function

$$\tilde{\Psi}(x_A, x_B) = g(x_A)f(x_B) \qquad (8.2)$$

formed from Ψ by interchanging the two particles.

In Section 1, in developing Eq. (1.2) for blackbody radiation, we noted that photons were treated as indistinguishable particles. It is not possible to attach distinguishing labels to them as individuals, and it is reasonable to make the same assumption regarding all the fundamental particles—in particular electrons, protons, etc. For the compelling observational evidence for this assumption see BOSE-EINSTEIN STATISTICS AND BOSONS and FERMI-DIRAC STATISTICS AND FERMIONS in this Encyclopedia. To embody this assumption in the appropriate wave function one must form a linear superposition of products such that no exchange among the particles can make any difference that could be detected experimentally. For a pair, the only possible such wave functions are

$$\Psi_\pm(x_A, x_B) = (1/\sqrt{2})\,[\,f(x_A)g(x_B)$$
$$\pm g(x_A)f(x_B)]. \qquad (8.3)$$

With the $+$ sign the function is symmetrical; with the $-$ it is antisymmetrical, but its square modulus—its only experimentally significant feature at this stage of the theory—is unchanged by the exchange of particles.

To generalize this to N particles, we may replace f and g by the set of eigenstates written $|m, x_j\rangle$ to show the mth energy eigenstate and the jth particle, $j = 1, 2, \ldots, N$. Then the normalized symmetrical product is a function of the set of coordinates $\{x\}$ and the set of eigenvalues $\{m\}$:

$$\Psi_S(\{m\}, \{x\}) = (1/\sqrt{N!}) \sum_P \prod_{j=1}^{N} |m_j, x_{Pj}\rangle$$
$$(8.4)$$

where the state $|m_j, x_{Pj}\rangle$ assigns the quantum number m_j to the particle $j' = Pj$ obtained by a permutation P among the N particles. The product is over all N particles, the sum is over all $N!$ permutations among them. Clearly this wave function is unchanged by any further permutation among the particles; all the permutations are already included with equal weight.

Similarly for the normalized antisymmetrical product we write

$$\Psi_A(\{m\}, \{x\}) = (1/\sqrt{N!}) \sum_P (-1)^P$$

$$\cdot \prod_{j=1}^{N} |m_j, x_{Pj}\rangle, \qquad (8.5)$$

where the sum alternates in sign, positive for even permutations, negative for odd permutations—it is in fact an $N \times N$ determinant. The interchange of any two particles means interchanging two rows in the determinant, which simply changes its sign without changing the modulus of the wave function. The most important feature of this is that if any two particles are assigned the same eigenstate, their two rows are identical and the determinant vanishes. In this case no more than one particle can be assigned to any one-particle state. This is the feature of the antisymmetrical wave function that leads to the Fermi-Dirac statistics. The symmetrical wave function leads to the Bose-Einstein statistics where there is no upper limit to the population of any one-particle state—except of course N.

If the number of m-values available in the one-particle energy spectrum is greater than N, some of the m-values cannot be assigned any occupants, but every particle must be assigned to some m-value. In the antisymmetrical case these occupation numbers are either 1 or 0. In the symmetrical case the only restriction is that their sum be N. Each assignment of all N particles to the spectrum characterized by the set of occupation numbers determines the overall symmetrical or antisymmetrical wave function. Writing n_m for the number of particles assigned to the mth eigenvalue, the occupation numbers n_m determine the state, and in place of Eq. (8.4) or (8.5) we could equally well write

$$\Psi(S, A) = |n_1, n_2, \ldots, n_m, \ldots\rangle_{(S,A)}$$
$$(8.7)$$

with the understanding that the one-particle Hamiltonian is used as the initial basis. These "vectors," identified by the set of n-values, are clearly energy eigenvectors for the whole system, and can be used as the basis spanning the Hilbert space having the "dimensionality" equal to the total number of such sets of occupational numbers. If the one-particle spectrum has a countable infinity of states, aleph-zero in number, the number of dimensions in this Hilbert space is aleph-zero to the power N. This huge multitude does not alter the theorem that any member of the space can be expressed as a linear superposition of the basis vectors written out in Eq. (8.7). This means that the general state of an N-particle system can be expressed in the form

$$\Psi(\{x\}, t) = \sum_{\{n\}} C(\{n\}, t)|n_1, n_2, \ldots, n_m, \ldots\rangle$$
$$(8.8)$$

where $C(\{n\}, t)$ is the time-dependent amplitude of the probability that the system be in the state specified by the set $\{n\}$ and have therefore, the energy

$$E = E\{n\} = \sum_m n_m (m + \tfrac{1}{2})\hbar w, \qquad (8.9)$$

assuming here that the Hamiltonian used for the initial one-particle basis was that of the simple harmonic oscillator as in Section 4. In the antisymmetrical case $n_m = 0$ or 1. In the symmetrical case the only restriction on n_m is that the sum

$$\sum_m n_m = N. \qquad (8.10)$$

The scheme here outlined is usually known as the "N-representation," and was historically derived in the process known as "second quantization," which gets its name from the following: First quantization yields Schrödinger's wave equation; then, regarding Schrödinger's function Ψ as a field, the classical Hamiltonian theory of fields is quantized and applied to Ψ, which then has to be taken as an operator whose domain is the Hilbert space spanned by the basis set we have written in Eq. (8.7). Having already produced this Hilbert space, which we may call $\mathcal{H}_N$, from the obvious N-particle wave function based on the "first" quantization of the single particle, we do not need to go through that painfully obscure reasoning. The second quantization operators can now be defined directly from the form of the basis vectors, Eq. (8.7).

The "destruction," or "lowering" operators a_j are defined by*

$$a_j|n_1, n_2, \ldots, n_j, \ldots, n_m, \ldots\rangle$$
$$= \sqrt{n_j}\,|n_1, n_2, \ldots, n_j - 1, \ldots, n_m, \ldots\rangle, \qquad (8.11)$$

and the "creation" or "raising" operator $a_j{}^\dagger$ are defined by*

$$a_j{}^\dagger|n_1, n_2, \ldots, n_j, \ldots, n_m, \ldots\rangle$$
$$= \sqrt{n_j + 1}\,|n_1, n_2, \ldots, n_j + 1, \ldots, n_m, \ldots\rangle. \qquad (8.12)$$

In each case $j = 1, 2, \ldots$ over the whole spectrum. There is a creation and a destruction operator associated with each one-particle state. The ordered product of operations, first a_j then $a_j{}^\dagger$, results in

$$a_j{}^\dagger a_j|n_1, n_2, \ldots, n_j, \ldots\rangle$$
$$= n_j|n_1, n_2, \ldots, n_j, \ldots\rangle. \qquad (8.13)$$

The basis states Eq. (8.7) are eigenstates of this operator with eigenvalue equal to the occupation number. We may write

$$N_j = a_j{}^\dagger a_j \qquad (8.14)$$

for the number operator for the jth state.

This number operator is Hermitian, and so is a good observable. The creation and destruction operators are not Hermitian, but the following

_____
*These definitions are appropriate only for the symmetrical, Bose-Einstein case.

linear combinations of them are:

$$\kappa_j \equiv a_j + a_j{}^\dagger, \qquad \pi_j \equiv i(a_j{}^\dagger - a_j), \qquad (8.15)$$

and could be identified with observables. To check for Hermiticity one can write down the matrix elements by the rule of Eq. (3.9), omitting the unaffected occupation numbers, and dropping the then redundant subscript, thus:

$$\langle n'|N|n\rangle = n\delta(n', n), \quad \text{purely diagonal,}$$
$$\langle n'|(a + a^\dagger)|n\rangle = \langle n|(a + a^\dagger)|n'\rangle$$
$$= \sqrt{n}\,\delta(n', n - 1)$$
$$\quad + \sqrt{n + 1}\,\delta(n', n + 1)$$
$$i\,|\langle n'|(a^\dagger - a)|n\rangle = -i\langle n|(a^\dagger - a)|n'\rangle$$
$$= -i[\sqrt{n}\,\delta(n', n - 1)$$
$$\quad - \sqrt{n + 1}\,\delta(n', n + 1)].$$
$$(8.16)$$

We must emphasize that this formalism is the moment purely mathematical. One is not obliged to think of the particles as harmonic oscillators. Any wave function in the space $\mathcal{H}_N$ can be expressed in the form Eq. (8.8), and we could have chosen spherical representation for the basic one particle states just as effectively. Even box normalized plane wave states are also available, provided we are satisfied with functions existing only inside the "box."

Quantum field theory, or for that matter, many-body problems in solid state physics, depends on a systematic application of this formalism to a wide variety of different Hamiltonians. The serious student should refer to the texts listed in Ref. 13.

9. The Quantum State In classical mechanics the state of a system is defined by the generalized coordinates of the system in its phase space—e.g., the position and momentum of every particle in the system. If this is known, the state is completely specified and both the future and the past of the path of the system in its phase space is determined. Ever since the earliest form of quantum theory it has been recognized that the quantum state must be less specific, and it is still traditional to identify the wave function with the quantum state, its time evolution being determined by the Hamiltonian operator, Eq. (1.10), and by Eq. (6.3). The statistical interpretation of the wave function signifies that the present quantum state can determine at most the probabilities of future measurement results, no matter how precisely one determines the present wave function.

This tradition has one serious flaw that is generally overlooked—there has so far in the theory been no clear prescription for observationally determining the wave function representing the quantum state. Almost all problems posed traditionally have been of the form: given Ψ, find the behavior of the system. In this section we address the reverse problem: given some system, what, if any, measurements can we make to determine the quantum state it happens to find itself in? The question is not nearly as trivial as it may sound.

Some tentative work in this direction was done during the very early development of the theory, but the only attempt to include the idea in the structure of the theory was Von Neumann's[14] "measurement of the first kind."* This was a proposition that a measurement of any observable A that results in an eigenvalue a of A leaves the system in the eigenstate of A following the measurement. This was, from a mathematical point of view, a very neat axiom, and undoubtedly because Von Neumann's other contributions to the theory were tremendously important, it still carries weight in the literature. But from a physicist's point of view it is utter nonsense. The typical measurement performed on any fundamental particle, e.g., the energy of a photon, or the momentum of an electron, results either in the complete destruction of the particle or at least a violent disturbance in its state: the best any real measurement can do is to ascertain the properties of the particle just before the measurement; what happens afterwards is practically impossible to predict. This is known as a "measurement of the second kind."

To make sense of the "first kind," the axiom must be rephrased as follows: It is postulated that there exist procedures that can leave a system in an eigenstate of any observable of interest—a strategically designed preparation procedure which will produce the desired quantum state. This is not a measurement. To give the postulate a meaning, one must still make measurements on the emerging system, and repeat the process a large enough number of times with the identical preparation procedure to convince ourselves that the scheme "works." In simpler words we must set up a mechanism for determining the quantum state; otherwise any proposition concerning such states is empty.

As an example, consider a nonideal gas of N particles weakly interacting, given some total energy at time $t = 0$, and then allowed to remain undisturbed and isolated for a sufficient time to reach what physicists call "thermodynamic equilibrium." It is a familiar experience in the laboratory, and the thermodynamic state is well known. How shall we describe this state quantum mechanically? Clearly it must be an equilibrium quantum state, and therefore an eigenstate of the system's Hamiltonian. We can express these eigenstates in the form of Eq. (8.8). In a large system there is a great deal of "accidental degeneracy," numerous sets $\{n\}$ which correspond to the same total energy, Eq. (8.9). The "vectors," Eq. (8.8), all corresponding to the given total energy may be written $\Psi_s(C_s\{n\}, E)$,

*The names "measurement of the first kind" and "measurement of the second kind" were coined by W. Pauli.

and for each value of s, the degeneracy index, there exists a set of coefficients $C_s\{n\}$, a different set for each index. Among these different wave functions there seems to be no way, on the basis of thermodynamic information, to decide which is the "true" state of the gas. Traditionally, this problem has been resolved by assigning a "reasonable" (on the basis of information theory[15]) probability to every member of the set of possible states, and combining them to form what has been termed a "mixed state."

The distinction between the mixed state and the superposition state where the coefficients are probability *amplitudes* was emphasized by writing it as a diagonal matrix

$$\Psi_{\text{mixed}} = \sum_s \Psi_s(C_s\{n\}, E)\rangle W_s \langle \Psi_s{}^*(C_s\{n\}, E)$$

(9.1)

which has rows and columns equal in number to the "degeneracy" number S:

$$s = 1, 2, 3, \ldots, S; \qquad W_s = \text{real probability}.$$

To accommodate this generalization, some of the basic rules of quantum theory have to be modified. In particular Eq. (2.1) must be replaced as follows: The expectation value of an operator A on the state Ψ (mixed) is

$$\langle A \rangle = \text{Trace}(\Psi_{\text{mixed}} A). \qquad (9.2)$$

The Trace operation involves the following sum:

$$\langle A \rangle = \sum_{s'} \sum_s \langle \Psi_{s'} | \Psi_s \rangle W_s \langle \Psi_s | A | \Psi_{s'} \rangle \quad (9.3)$$

where we have omitted the detailed arguments of the functions $\Psi_{s'}$ and Ψ_s. Then because the functions Ψ_s can in principle be made an orthonormal set this reduces to

$$\langle A \rangle = \sum_s W_s \langle \Psi_s | A | \Psi_s \rangle \qquad (9.4)$$

which is the weighted average of the expectation values of A on the individual members of the degenerate set of eigenvectors of H, as given by the traditional formula (2.1) on the "pure states" Ψ_s.

The procedure just outlined has been widely adopted and has been remarkably useful in applying quantum mechanics to practical problems. For advocates of the application of information theory to explain the laws of thermodynamics the situation is completely satisfactory. However, other physicists do not accept the implication that the quantum state of any system—in particular the "mixed" state—depends essentially on a "gentleman's agreement" to avoid seeking technological advances which could provide more precise information than is currently available regarding the "likelihood" of the various members of the degenerate set of eigenstates in the state of the system.

The more "realistic" school of thought has adopted a straightforward postulate which denies that the wave function by itself is an adequate quantum state except under the most unusual conditions. The new postulate reads:[16]

I. To every reproducible empirical scheme for the preparation of the physical system, there corresponds an Hermitian operator ρ with unit Trace and only nonnegative eigenvalues, such that, given any observable A on the domain $\mathcal{H}$ of the operator, the arithmetic mean value of a statistical collective of A-measurements obtained from an ensemble of systems all prepared identically in the manner characterized by ρ is equal to

$$\langle A \rangle = \mathrm{Trace}(\rho A). \qquad (9.5)$$

The "mixed state" of Eq. (9.1) is an example of ρ, which is already in diagonal form and displays its nonzero positive eigenvalues W_s, with unit Trace equal to

$$\sum_s W_s = 1. \qquad (9.6)$$

Any "pure" state Ψ_m of the Schrödinger theory is also equivalent to the density matrix $|\Psi_m\rangle\langle\Psi_m|$—it is a matrix with only one eigenvalue (equal to unity). The rule (9.5) in this case states

$$\langle A \rangle = \mathrm{Trace}\,|\Psi_m\rangle\langle\Psi_m|A$$
$$= \sum_n \langle\Psi_n|\Psi_m\rangle\langle\Psi_m|A|\Psi_n\rangle \quad (9.7)$$

where the set $\{\Psi_n\}$ is an orthonormal set including Ψ_m. Thus

$$\langle A \rangle = \langle\Psi_m|A|\Psi_m\rangle \qquad (9.8)$$

which is identical with Eq. (2.1).

It must be emphasized that the quantum state operator ρ is a feature not of any single system, but of the procedures employed to produce the ensemble of systems. It does not depend on the amount of information available to the observer, but can in principle be determined by a sufficient number of measurements of observables in the ensemble. We discuss this in detail in the next section.

The more "realistic" statistical interpretation of the quantum state based on postulate I has important implications for scientific epistemology. The proposition itself is simply an extension of the purely classical statement that the accepted result of a series of measurements has to be obtained as an average over an ensemble of results obtained from a system prepared in the same way before each measurement. There are two differences between the quantum and classical theories: (a) The quantum system is usually so small that measurement itself disturbs it, and the preparation must be fully repeated

before each measurement; whereas the classical system is usually so bulky that the repreparation can be quite perfunctory. (b) In classical physics one accepts the hypothesis that every system "possesses" observables with unique and precise numerical values that measurements are designed to discover; the classical laws of motion refer only to these ideal values; whereas in quantum theory no such hypothesis is entertained, and the quantum laws refer only to the statistics of the measurement results.

It was essentially this difference that caused Einstein[5] to proclaim that quantum theory was an incomplete theory; like most classical physicists Einstein hoped for a deeper meaning behind observed data which quantum theory explicitly denies being able to produce.

10. The Quorum Concept To postulate I of Section 9 we now append another postulate:[11]

II. There exists a "quorum" of observables, Q_m, $m = 1, 2, 3, \ldots$, such that, from the collective of mean values of the Q_m-measurements

$$\langle Q_m \rangle = \mathrm{Trace}\,\rho Q_m \qquad (10.1)$$

it is possible to calculate a matrix representation of ρ which is thereby uniquely determined.

The simplest example of this is to be found in the "spin $\frac{1}{2}$" system. This is a traditionally standard example of a nonclassical system, and is essentially a "disembodied electron," i.e., one ignores all other physical parameters such as position, orbital momentum, and electric charge, and treats the Hamiltonian as a two-eigenvalue operator describing the interaction of the spin with a uniform magnetic field in the z-direction. The Hilbert space for this system is a two-dimensional continuum, and the "wave functions" are two-element "spinors"—compare the remarks in Section 6 about the Dirac electron. The familiar observables are the Pauli spin matrices, Eq. (5.11). The Hamiltonian is the diagonal

$$H = E \begin{pmatrix} 1 & 0 \\ 0 & -1 \end{pmatrix}$$

where E is the product of spin and magnetic field.

The matrix representation of the operator ρ has to be of the form

$$\rho = \begin{pmatrix} \rho_{11} & \rho_{11} \\ \rho_{12} & 1 - \rho_{11} \end{pmatrix}, \qquad \rho_{12} = \rho_{21}{}^*, \quad \rho_{11} \leqslant 1.$$
$$(10.3)$$

Clearly only three real numbers are needed to determine ρ. After some elementary algebra we find,[18]

$$\rho_{11} = \tfrac{1}{2} + \langle\sigma_z\rangle, \qquad \rho_{22} = \tfrac{1}{2} - \langle\sigma_z\rangle$$
$$\rho_{12} = \langle\sigma_x\rangle - i\langle\sigma_y\rangle, \qquad \rho_{21} = \langle\sigma_x\rangle + i\langle\sigma_y\rangle$$
$$(10.4)$$

where $\langle\sigma_x\rangle$, $\langle\sigma_y\rangle$ and $\langle\sigma_z\rangle$ are the mean values of the spin component measurements. There is no question here about "simultaneous" measurements of noncommuting observables. Each individual measurement is preceded by a new (identical) preparation, the process being repeated on each observable often enough to ensure statistical validity of the results. The three spin-component observables therefore constitute a quorum. From the measurements one can decide exactly what the density matrix is corresponding to any prescribed preparation procedure. The algebra shows, in fact, that

$$\langle\sigma_x\rangle^2 + \langle\sigma_y\rangle^2 + \langle\sigma_z\rangle^2 \leqslant \tfrac{1}{4}. \qquad (10.5)$$

Only if the equality sign agrees with the data is the density matrix "pure;" otherwise it is "mixed." There is nothing in this "mixed" case that asks for a confession of ignorance on the part of the observer.

Another example that has been worked out in detail—this one a "classical problem"—is the motion of a free particle in one dimension, where the wave functions are box-normalized plane waves. The density matrix must have a countably infinite number of rows or columns, and a countably infinite sequence of time derivatives $d^n Q^m/dt^n$ to all orders n of powers m of the position operator form a quorum that could in principle determine the density matrix.[19]

Other examples have been worked out in detail, but no general proof of postulate II has been found. However, while it is obvious that the practical determination of the quantum state ρ of a complicated system may be even more difficult than in classical mechanics, the theoretical existence of an unambiguous result is at least plausible.

11. Quantum Dynamics Having taken the quantum state to be a Hermitian statistical operator ρ in $\mathcal{H}$, we are obliged to regard it as in some sense an observable, although it is not to be assumed that its time evolution be subject to the same laws as are conventional physical observables. To settle this question we recall first that from Eq. (7.3) we have the time evolution of the "state vector"

$$\Psi(t) = e^{-iHt/\hbar}\,\psi(0), \qquad (11.1)$$

and that the diagonal form of ρ is essentially the same as Eq. (9.1):

$$\rho = \sum_s \Psi_s W_s \Psi_s{}^* \qquad (11.2)$$

which now implies

$$\rho(t) = e^{-iHt/\hbar}\,\rho(0)\,e^{iHt/\hbar}, \qquad (11.3)$$

and from this we derive

$$\partial\rho/\partial t = -(i/\hbar)\,[H,\rho(t)]. \qquad (11.4)$$

This is adopted as the dynamical law of evolution on the Schrödinger picture. It has the opposite sign compared with Eq. (7.6), which is the dynamical law on the Heisenberg picture.

The time dependence of an observable A on the Heisenberg picture can actually be derived from Eq. (11.4). Applying Eq. (11.4) to Eq. (9.5) we have

$$\partial\langle A\rangle/\partial t = \text{Trace}\,(A\,\partial\rho/\partial t)$$

$$= -(i/\hbar)\,\text{Trace}\,(A\,[H,\rho(t)]),$$

$$(11.5)$$

and because the Trace of a product is unchanged by cyclic permutation among the factors this becomes

$$\partial\langle A\rangle/\partial t = (i/\hbar)\,\text{Trace}\,(\rho(t)\,[H,A]),$$

$$(11.6)$$

which is not inconsistent with Eq. (7.6) of the Heisenberg picture if we accept the equivalence of $\partial\langle A\rangle/\partial t$ with $\langle\partial A/\partial t\rangle$.

From Eq. (11.3) it is clear that the dynamical evolution of ρ is equivalent to a unitary transformation generated by H and parametrized by the time t. The changes created by a unitary transformation are very limited. First, consider equilibrium states under which no observable measurement results can change with time. Evidently all this would require is that $\partial\rho/\partial t$ be zero, which in turn requires from Eq. (11.4) that ρ commutes with the Hamiltonian at any one time—when it would continue to commute for all time. Therefore the equilibrium state ρ may be taken as some function of the Hamiltonian, and for compelling reasons—chiefly by comparison with well established classical thermodynamics—the equilibrium ρ is assumed to be

$$\rho_{\text{eq}} = e^{-\beta H}/\text{Trace}\,e^{-\beta H}, \qquad \beta = 1/kT.$$

$$(11.7)$$

Using this equilibrium ρ in Eq. (9.5) we have the equilibrium "value" of the observable A:

$$\langle A\rangle_{\text{eq}} = \text{Trace}\,Ae^{-\beta H}/Z \qquad (11.8)$$

where $Z = \text{Trace}\,e^{-\beta H}$ is the "partition function." If Eq. (11.7) is used for ρ in Eq. (11.5) we see that

$$\partial\langle A\rangle_{\text{eq}}/\partial t = 0 \quad \text{for all } A. \qquad (11.9)$$

If on the other hand ρ does not commute with H we shall have a nonequilibrium situation; we would then expect physically that the dynamical laws would propel the system toward equilibrium, but this result does not emerge from the theory. According to traditionally accepted axioms, the time evolution of ρ would still be governed by Eq. (11.3) whether ρ is an equilibrium state or not, and we can prove that this unitary transformation will not produce the desired approach towards equilibrium. The proof is outlined below.

The equlibrium state Eq. (11.7) is derived by maximizing the quantity

$$S = -k \text{ Trace } (\rho \ln \rho). \qquad (11.10)$$

For any state differing from ρ_{eq} the quantity S is less than the maximum. (For this reason S is identified with the thermodynamic entropy.) But Trace ($\rho \ln \rho$) is invariant under any unitary transformation of ρ, so the dynamical law cannot change the entropy of the system, whatever its initial state. Quantum mechanics is no more successful than classical mechanics in "explaining the second law of thermodynamics."

Over the past twenty years a number of mathematical physicists have made sporadic attempts to generalize quantum dynamics in some way that would escape this dilemma.[20] Ths most recent attempts have made use of the theorem that the set of all Hermitian operators on a Hilbert space $\mathcal{H}$ themselves form a super Hilbert space, $\mathcal{H}_S$. The statistical operators ρ form a special subset of the elements of $\mathcal{H}_S$, and a "super operator" is sought, acting on the elements of $\mathcal{H}_S$, parametrized by t, which would carry $\rho(0)$ to $\rho(t)$ in a more general way than that given by the standard unitary transformation. There are so many possibilities here that it becomes a matter of taste what particular hypothesis to espouse.

The majority of specialists[21] in this field have been convinced that the unitary transformation is correct, and they have sought alternative ways of escape from the dilemma. The most recent consensus among these workers seems to be to blame the second law of thermodynamics on the long-range interactions of any apparently isolated system with the rest of the universe: gravitational fields, the pervasive background of electromagnetic radiation, beams of neutrinos, and so on. The mathematical decoration of these ideas runs something like this: Let $\mathcal{H}_{SL}$ be the super Hilbert space accommodating all the observables in the lab system, supposed isolated. The unitary evolution law then causes the state "point" ρ in $\mathcal{H}_{SL}$ to wander all over the place but never to cross its own path. If two paths start from two different points they can never meet, so the approach to equilibrium is forbidden. But if we admit the existence of long-range interactions, no matter how weak, we must enlarge the super Hilbert space to accommodate more observables. Call the enlarged space $\mathcal{H}_{SO} \supset \mathcal{H}_{SL}$. Then the true unitary evolution in $\mathcal{H}_{SO}$ will again be a path which cannot intersect itself in $\mathcal{H}_{SO}$, but the projection of the path onto the lab space $\mathcal{H}_{SL}$ could intersect itself, and so a state of equilibrium be approached from different starting points. Also the existence of the difference space $\mathcal{H}_{SO} - \mathcal{H}_{SL}$ provides an escape for the entropy balance—what is gained in the lab is taken from the rest of the universe. Probably the most fanciful of these speculations is the hypothesis that there is an entropy path connecting the universe with a mirror image antimatter universe, and when entropy increases in the universe, negative entropy increases in the anti-universe, so that the overall unitary evolution causes the second law to hold both in the universe and in the anti-universe.

WILLIAM BAND

References

1. Band, William, "Introduction to Quantum Statistics," Princeton, NJ, D. Van Nostrand, 1955.
2. "The Feynman Lectures on Physics," Vol. III, "Quantum Mechanics," Reading, MA, Addison-Wesley, 1965.
3. Kiltel, Charles, "Quantum Theory of Solids," New York, John Wiley & Sons, 1967.
4. Jammer, Max, "The Philosophy of Quantum Mechanics," New York, John Wiley & Sons, 1967.
5. Einstein, Podolsky, and Rosen, *Phys. Rev.* 47, 777 (1935).
6. Belinfante, F. J., "A Survey of Hidden Variables Theories," London, Pergamon Press, 1973; Vigier, J. P., *Nuovo Cimento Letters* 29, 467 (1980).
7. Margenau, Henry, and Murphy, George M., "The Mathematics of Physics and Chemistry," Vol. I, Princeton, NJ, D. Van Nostrand, 1956.
8. Roman, Paul, "Some Modern Mathematics for Physicists," London, Pergamon Press, 1975; Berberian, S. K., "Introduction to Hilbert Space," Oxford, Oxford Univ. Press, 1961.
9. Schiff, L. I., "Quantum Mechanics," New York, McGraw-Hill, 1955.
10. Rose, M. E., "Elementary Theory of Angular Momentum," New York, John Wiley & Sons, 1951; Fano, U., and Racah, G., "Irreducible Tensorial Sets," New York, Academic Press, 1959.
11. Rose, M. E., "Relativistic Electron Theory," New York, John Wiley & Sons, 1961.
12. Goldberger, M. L., and Watson, K. M., "Collision Theory," New York, John Wiley & Sons, 1967; Band, W., and Park, J. L., *Foundations of Physics*, 8, 677–694, (1977).
13. Schweber, S. S., "Relativistic Field Theory," Row Peterson, 1961; Abrikov, Gorkov, and Dzyaloshinski, "Methods of Quantum Field Theory in Statistical Physics" (English transl.), Englewood Cliffs, NJ, Prentice Hall, 1963.
14. von Neumann, J., "Mathematical Foundations of Quantum Mechanics," Princeton, NJ, Princeton Univ. Press, 1955.
15. Katz, Amnon, "Principles of Statistical Mechanics: The Information Theory Approach," San Francisco, W. H. Freeman, 1967; Jaynes, E. T., *Phys. Rev.* 108, 171 (1951); Band, W., and Park, J. L., *Foundations of Physics*, 6, 259–262 (1976), 7, 232–244, 705–721 (1977).
16. Roman, Paul, "Advanced Quantum Theory," Reading, MA, Addison-Wesley, 1965.

17. Park, J. L., and Band, W., *Foundations of Physics*, 1, 211–266, 339–357 (1971).
18. Park, J. L., and Band, W., *Foundations of Physics*, 1, 133–144 (1970).
19. Band, W., and Park, J. L., *Amer. J. Phys.* 47, 188–191 (1979).
20. Park, J. L., and Band, W., *Foundations of Physics* 7, 813–825 (1977), 8, 45–58, 239–254 (1978); Ingarden, R. S., and Kossakowski, A., *Ann. Phys. (N.Y.)*, 451 (1975).
21. Prigogine, I., George, C., Henin, F., and Rosenfeld, L., *Chem. Scripta*, 4 (1973); Sudarshan, E. C. G., Mathews, P. M., and Rau, J., *Phys. Rev.* 121, 920 (1961).

Cross-references: HEISENBERG UNCERTAINTY PRINCIPLE; MATHEMATICAL PRINCIPLES OF QUANTUM MECHANICS; PHOTON; QUANTUM ELECTRODYNAMICS; RADIATION, THERMAL, SCHRÖDINGER EQUATION; WAVE MECHANICS.

QUARKS

Introduction It seems that protons and neutrons (in fact, the various hadrons) are not the fundamental constituents of nuclear matter. These particles, themselves, have a detailed, and possibly complicated, structure. Several theoretical physicists have suggested that they may be composed of various combinations of a few much more "fundamental" objects. These objects have been called "quarks" by Gell-Mann (1964); "aces" by Zweig (1964). In both these schemes the new "fundamental" particles have electric charges less than the charge on the electron. In other schemes, developed later, the charges are normal.

Chemical experiments in the late eighteenth and early nineteenth centuries led Dalton to revive the atomic hypothesis. Atoms, at first, were thought of as small impenetrable spheres. By the end of the nineteenth century it had become obvious that the atoms themselves had a detailed structure. Research, particularly on radioactivity, showed that they consist of a central, small ($\sim 10^{-12}$ cm diameter), massive, positively charged nucleus surrounded by a light, negative, and comparatively large ($\sim 10^{-8}$ cm diameter) electron cloud. The structure of the nucleus itself then came under investigation. By 1932 it was known that nuclei are made of comparatively small numbers of neutrons and protons. A new force was discovered (in addition to the electromagnetic and gravitational forces) that held the positive protons and electrically uncharged neutrons together in the nucleus. This nuclear force was very strong but of limited range. Its "quantum," the particle analogous to the photon in the electromagnetic field, was of nonzero rest mass. This particle, now called the π-meson, or pion, was predicted by Yukawa in 1936 and discovered in *Cosmic Radiation* by Lattes, Occhialini, and Powell in 1947. For a short time in that year it seemed that physicists had achieved a clear, simple, and correct theory of the fundamental constitution of matter. However, later in the same year two new and unpredicted particles were reported, again in *Cosmic Radiation*, by Rochester and Butler. The first of these was another meson, somewhat like the pion but more massive. The second was probably a hyperon, i.e., a strongly interacting particle heavier than the neutron. Since then many more examples of both classes have been found.

To describe these particles one gives the values of various parameters, or quantum numbers. Some of these are familiar. For instance, the particles have different masses, and some relationships between these masses are now quite well understood. They have various electric charges which are integral multiples of the electronic charge. They possess intrinsic angular momentum or spin (sometimes this is zero). They have an analogous property called isotopic spin (or isospin). The mesons are distinguished from particles like the proton, neutron, and various hyperons (which are collectively called baryons) by a quantity, baryon number, which is an extension of the old atomic weight. The parity of the wave function describing the particle is an important property. There is also a quantum number called "strangeness." This has zero value for the "familiar" particles such as the proton, neutron, and pion, ± 1 for the early strange particles discovered by Rochester and Butler, and as high as ± 3 for some fairly recently discovered particles.

There are about one hundred of these "fundamental" particles now known. They interact with each other via various fields. Two of these have been known for some time, the gravitational and electromagnetic fields. Two others, the weak nuclear field and the strong nuclear field, were discovered in the 1930s. It seems likely from the results of experiments in high-energy cosmic radiation that there is at least one more, the superstrong field.

The large number of these particles has made physicists suspect that they are not "fundamental" but must themselves have structure, just as in the nineteenth century the large number of different types of atoms discovered suggested that atoms had structure. Also, many of the properties of these particles point to an internal structure. For instance, the neutron has a total electric charge which is indistinguishable from zero down to very fine limits, yet the neutron has a sizeable magnetic moment.

The Quark Hypothesis In 1964 M. Gell-Mann and G. Zweig independently pointed out that all the known hadrons (i.e., particles that interact via the strong nuclear force) could be constructed out of simple combinations of three particles (and their three antiparticles). These hypothetical particles had to have slightly peculiar properties (the most peculiar being a fractional electric charge). Gell-Mann called them quarks (referring to a sentence in James

TABLE 1. SOME OF THE SUGGESTED CHARACTERISTICS OF THE
THREE QUARKS. M_p IS NOT DEFINED BY THE THEORY OTHER
THAN THAT IT SHOULD BE CONSIDERABLE; THE UNITS
OF MASS ARE MeV/c^2.

Particle	Charge	Spin	Baryon No.	Strangeness	Mass	Isospin	I_Z
u	$\frac{2}{3}e$	$\frac{1}{2}$	$\frac{1}{3}$	0	M_u	$\frac{1}{2}$	$-\frac{1}{2}$
d	$-\frac{1}{3}e$	$\frac{1}{2}$	$\frac{1}{3}$	0	$\sim M_u$	$\frac{1}{2}$	$\frac{1}{2}$
s	$-\frac{1}{3}e$	$\frac{1}{2}$	$\frac{1}{3}$	-1	$M_u + 146$	0	0

Joyce's work *Finnegan's Wake*, "Three quarks for Muster Mark"). The three quarks are designated u, d, and s because they somewhat resemble the proton, neutron, and Λ° hyperon. Some of their quantum numbers are given in Table 1.

The theory supposes that three quarks bind together to form a baryon, while a quark and an antiquark bind together to form a meson. If one supposes that the binding is such that the internal motion of the quarks is nonrelativistic (which requires that the quarks be massive and sit in a broad potential well), then many quite detailed properties of the hadrons can be explained.

Predictions and Verification of the Nonrelativistic Quark Theory (1) *The Predicted and Observed Particles.* The particles which can interact via the strong nuclear force are called hadrons. Hadrons can be divided into two main classes, the mesons (with baryon number zero) and the baryons (with nonzero baryon number). Within each of the classes there are small subclasses. The subclass of baryons which has been known longest consists of those particles with spin $\frac{1}{2}$ and even parity. The members of this class are the proton, the neutron, the Λ° hyperon, the three Σ hyperons and the two Ξ hyperons. There are no other baryons with spin $\frac{1}{2}$ and even parity (or, to use the usual notation, $J^P = \frac{1}{2}^+$). The next "family" of baryons has ten members, each with $J^P = \frac{3}{2}^+$. The mesons can be grouped into similar families. One of the first successes of the quark model was to explain just why there should be eight baryons with

$J^P = \frac{1}{2}^+$, 10 with $\frac{3}{2}^+$, and so on and why the various members of these families have the particular quantum numbers observed. The explanation is most easily understood if we start with the $\frac{3}{2}^+$ baryon family.

On the left-hand side of Fig. 1 the members of the family are arranged on a plot on which the ordinate is the value of their strangeness and the abscissa that of their electric charge. Most of the particles have rather short lifetimes but otherwise are not very different from "ordinary" baryons like the neutron. On the right-hand side of the diagram we see the way in which quark theory suggests that the particles arise. For instance, the Ω^- hyperon is supposed to consist of three s quarks. The spin of all these baryons is $\frac{3}{2}$, so all three quark spins must be aligned. One then has ten, and only ten, possible combinations of three quarks and all these combinations exist. Moreover the quark model gives just the right quantum numbers for all particles. If we assume, in addition, that s is 146 MeV/c^2 more massive than u and d (which we take to be about equally massive) then we have at once an explanation of the mass differences. The "apex" particle, the Ω^-, had its existence and detailed properties predicted before its discovery.

The $J^P = \frac{1}{2}^+$ baryon family is similarly treated in Fig. 2. In this case, to get spin $\frac{1}{2}$ we must always have one quark with its spin antiparallel. This explains the absence of "corner" particles like sss and also the occurrence of two different particles in the center position. For, if we consider both groups, we must obviously have

	Charge				Mass in	
Strangeness	+2	+1	0	−1	MeV/c²	Quark Scheme
−3				Ω^-	1672	sss
−2			$\Xi^{\circ *}$	Ξ^{-*}	1526	uss, dss
−1		Σ^{+*}	$\Sigma^{\circ *}$	Σ^{-*}	1380	uus, uds, dds
0	Δ^{++}	Δ^+	Δ°	Δ^-	1236	uuu, uud, ddu, ddd

FIG. 1. The family of baryons with $J^P = \frac{3}{2}^+$ and their quark groupings that explain them.

Charge Strangeness	+1	0	−1	Quark Scheme		
−2		Ξ°	Ξ^-		uss	dss
−1	Σ^+	Σ°	Σ^-	uus	uds	
		Λ°			uds	dds
0	P	N		uud	udd	

FIG. 2. The baryon family with $J^P = \frac{1}{2}^+$.

Charge Strangeness	+1	0	−1	Quark Scheme		
−1		$\overline{K}^\circ$	K^-		$s\bar{d}$	$s\bar{u}$
0	π^+	$\pi^\circ, \eta^\circ, \chi^\circ$	π^-	$u\bar{d}$ $(u\bar{u}, d\bar{d}, s\bar{s})$		$d\bar{u}$
+1	K^+	K°		$u\bar{s}$	$d\bar{s}$	

FIG. 3. The meson family with $J^P = 0^-$.

a particle with spin $\frac{3}{2}$ and spin projection on the z asix (s_z) equal to $\frac{3}{2}$. For the u quark, for instance, this is the particle $\Delta^{++} = u \uparrow u \uparrow u \uparrow$. This particle with $s = \frac{3}{2}$ must equally obviously have a state with spin projection $\frac{1}{2}$ (i.e., $s = \frac{3}{2}$, $s_z = \frac{1}{2}$). This is the case $u \uparrow u \downarrow u \uparrow$. But that is the only possibility, because $u \uparrow u \downarrow u \uparrow = u \downarrow u \uparrow u \uparrow = u \uparrow u \uparrow u \downarrow$, since the u's are indistinguishable. So no $s = \frac{1}{2}$ state is possible from three u's (or 3 d's or 3 s's). However, for the combination of u, d, and s we can have either an $s = \frac{3}{2}$ state or an $s = \frac{1}{2}$ state with either the d and s spins parallel, or u and s parallel. So two different $s = \frac{1}{2}$, $s_z = \frac{1}{2}$ combinations of these three can occur, and indeed, are found (the Σ° and the singlet Λ° hyperons.)

The various mesons are constructed from pairs of quarks and antiquarks. The scheme for the pseudoscalar, $J^P = 0^-$, mesons is shown in Fig. 3. These include the pions and kaons.

The theory predicts all the known particles. Extensive searches have been made for particles *not* predicted by the theory (the so called "exotic" particles), so far without success.

(2) *Particle Masses*. The masses of the various particles can be calculated using the nonrelativistic quark model. Some of these are given

in Table 2. The model predicts the masses both of mesons and baryons with very fair accuracy. Note that this table gives the mean masses of various subfamilies. Thus proton and neutron are taken together and designated by N (for nucleon) and have a mean mass of 939 MeV/$_{c^2}$.

The quark model also predicts the mass differences within these subgroups rather accurately. A few of the predicted and observed differences are given in Table 3.

(3) *The Electromagnetic Properties of Hadrons*. If the uncharged neutron were a "simple" elementary particle with no substructure, then it should not have a magnetic moment. On the other hand the quark model supposes that the neutron is composed of three charged particles (two with charge $-\frac{1}{3}e$ and the third with charge $+\frac{2}{3}e$). Thus, on this theory, we expect it to have a magnetic moment. The magnetic moments of various particles calculated using the quark model are given in Table 4 and compared with observation. The two are obviously in good agreement.

Many other aspects of electromagnetic behavior also can be calculated with good accuracy using the quark model.

(4) *Scattering of High-energy Hadrons*. The

TABLE 2. THE MASSES IN MeV/c^2 OF VARIOUS PARTICLES PREDICTED BY THE QUARKS MODEL AND THEIR EXPERIMENTALLY DETERMINED VALUES.

Particle	N	Λ	Σ	Δ	Ξ	Ω^-	π	K	ρ	ω	ϕ
Quark theory prediction	928	1108	1195	1238	1340	1675	133	490	753	753	1010
Experimental	939	1116	1193	1236	1318	1672	137	445	765	783	1020

TABLE 3. MASS DIFFERENCES OF VARIOUS PARTICLES

| Particles | Mass Difference in MeV/c^2 | |
	Prediction of Quark Model	Experiment
Neutron-proton	1.3	1.3
$\Sigma^- - \Sigma^0$	4.8	4.8
$\Sigma^- - \Sigma^+$	7.9	7.9
$\Xi^- - \Xi^0$	6.6	6.5
$\Delta^- - \Delta^{++}$	3.9	7.9 ± 6.8
$\Delta^0 - \Delta^{++}$	0.89	5.8 ± 3.9

TABLE 4. THE CALCULATED AND OBSERVED MAGNETIC
MOMENTS (IN NUCLEAR MAGNETONS)
OF VARIOUS PARTICLES.[a]

	Particle	Proton	Neutron	Λ^0	Σ^+
Magnetic Moment	Calculated	+2.79	−1.86	−0.93	+2.79
	Observed	+2.79	−1.91	−0.73 ± 0.16	+2.5 ± 0.5

[a]Many other aspects of electromagnetic behavior also can be calculated
with good accuracy using the quark model.

total cross sections for proton-proton and pion-proton scattering at high energies (> 20 GeV) are predicted by the quark model to be in the ratio 3:2 (crudely, because a proton is supposed to contain three quarks and a pion two quarks). The observed value is (1.58 ± 0.05):1. This result has not been predicted by any other model.

The nonrelativistic quark model has also had considerable success in explaining the scattering of positrons by negative electrons at high energies and the deep inelastic scattering of electrons by protons.

Defects and Extensions of the Simple Quark Model Quarks are fermions (i.e., they have spin $\frac{1}{2}$) so one expects them to obey the Pauli exclusion principle. But the Ω^- particle (whose prediction, as we have seen, was one of the great early triumphs of quark theory) requires three s quarks with their spins aligned. If the three quarks were identical this would be forbidden by the Pauli exclusion principle. To get out of this and other similar difficulties Greenberg suggested in 1964 that quarks have a previously unknown property which he called "color." Each "flavor" of quark (i.e., u, d, or s) comes in three different colors (and the anti-quarks in three different anti-colors). These "colors" have become known as *red*, *blue*, and *green* (and the anti-colors as *magenta* etc.). All hadrons are colorless, i.e., they are made of one *red*, one *blue*, and one *green* quark, or one quark and its oppositely colored anti-quark.

A further extension of the quark scheme was suggested (also in 1964) by Glashow, Bjorken, and others. This was the addition of at least one extra flavor which has since been given the symbol c (for *charm*). The clinching evidence for this extra quark (and anti-quark, also in three colors) came with the discovery of the J-ψ particle in 1970 by Ting and his co-workers at Brookhaven, and Richter and his team at Stanford. Since then, evidence for at least two other flavors, b (*bottom* or *beauty*) and t (*top* or *truth*) has accumulated.

So now the scheme requires at least 18 quarks (6 flavors each in 3 colors) and 18 anti-quarks plus a number of particles called *gluons* which serve to carry the force binding the quarks together.

Another surprise in quark physics has been the failure of many experiments to find free quarks. If these exist, their remarkable properties (including electric charges of $\frac{2}{3}e$ and $\frac{1}{3}e$ should make them easy to detect. They have been looked for at all large accelerators, in low energy and high energy cosmic radiation, and in condensed matter (water, iron, niobium, oyster shells, deep sea cores, moon dust, meteorites, etc.). No certain free quark has been seen in many experiments using accelerators or low energy cosmic radiation. Only three of the many searches in condensed matter have found apparently fractionally charged particles (though in one case, at least, the evidence is very solid). Two experiments have found lightly ionizing tracks in the cores of the air showers generated by high energy cosmic radiation (including one by the present author) and other experiments in this field give supporting evidence for the existence of free quarks. The failure of many experiments to find free quarks has led to the promulgation of "confinement" theories which assert that quarks are permanently confined in hadrons and can never get free. However, it has not been possible, so far, to prove that this is necessarily so.

The quark families seem to have some relation to the leptons. (See ELEMENTARY PARTICLES.) The quark family of lowest mass is the u-d family. Corresponding to it is the electron group (electron, positron, electron neutrino, and electron anti-neutrino). Next is the s-c quark family and the corresponding leptons (positive and negative muons and the muon neutrino and the anti-neutrino). Finally (at present) the b-t quark family and the τ mesons and their neutrinos. Thus, if these are the fundamental particles out of which the Universe is constructed there are a lot of them— 36 quarks and anti-quarks, 12 leptons and anti-leptons, the gluons, the photon, the graviton, and some other assorted particles. The picture is reminiscent of the many different types of atom that Daltonian chemistry required, which, of course, turned out to be not fundamental, but compound particles. One is driven to consider Heisenberg's rejection of the atomic hypothesis and believe, instead, that the Universe is one, whole, and indivisible.

C. B. A. MCCUSKER

References

Chew, G., Gell-Mann, M., and Rosenfeld, A., "Strongly Interacting Particles," *Sci. Am.*, Feb. 1964.

Davies, P. C. W., "The Forces of Nature," Ithaca, NY, Cornell Univ. Press, 1979.

Heisenberg, W., "Physics and Philosophy," London, George Allen & Unwin, 1959.

Cross-references: ELEMENTARY PARTICLES, ISOSPIN, NUCLEAR STRUCTURE, QUANTUM CHROMODYNAMICS, STRONG INTERACTIONS.

R

RADAR

Radar is the name given to the use of electromagnetic energy for the detection and location of reflecting objects. It operates by transmitting an electromagnetic signal and comparing the echo reflected from the target with the transmitted signal.

The first demonstration of the basic radar effects was by Heinrich Hertz in his famous experiments in the late 1880's in which he verified Maxwell's electromagnetic theory. Hertz showed that short-wave radiation could be reflected from metallic and dielectric bodies. Although the basic principle of radar was embodied in Hertz's experiments, the practical development of radar had to wait more than 50 years for radio technology to advance sufficiently. It wasn't until the late 1930's that practical models of radars appeared. The rapid advance in radar technology during World War II was aided by the many significant contributions of physicists and other scientists pressed into the practical pursuit of a new technology important to the military. In addition to its military application, radar has been applied to the peace-time needs of air and ship navigation, air traffic control, rainfall observation, tornado detection, hurricane tracking, surveying, radar astronomy, remote sensing of the environment, and the familiar speed measuring meter of the highway police.

The measurement of distance, or range, is probably the most distinctive feature of radar. Range is determined from the time taken by the transmitted signal to travel out to the target and back. The distances involved might be as short as a few feet or as long as interplanetary distances.

If the target is in motion relative to the radar, the echo signal will be shifted in frequency by the DOPPLER EFFECT and may be used as a direct measurement of the relative target velocity. A more important application of the Doppler shift is to separate moving targets from stationary targets (clutter) by means of frequency filtering. This is the basis of MTI (moving target indication) radar and pulse Doppler radar.

Radar antennas are large compared to the wavelength so as to produce narrow, directive beams. The direction of the target may be inferred from the angle of arrival of the echo. Radar antenna technology has profited greatly from the theory and practice of optics. Both the lens and the parabolic mirror have their counterpart in radar, and the analysis of antenna radiation patterns follows from diffraction theory developed for optics.

The external appearance of a radar is dominated by its ANTENNA. Most radars use some form of parabolic reflector. The radar antenna can also be a fixed array of many small radiating elements (perhaps several tens of thousands or more) operating in unison to produce the desired radiation characteristics. Array antennas have the advantage of greater flexibility and more rapid beam steering than mechanically steered reflector antennas because the beam movement can be accomplished by electrically changing the relative phase at each antenna element. High power can be radiated since a separate transmitter can be applied at each element. The flexibility and beam switching speed of an array antenna make it necessary in many instances to control its functions and analyze its output by automatic data processing equipment rather than with an operator using a grease pencil to mark the face of a cathode ray tube.

Radars are generally found within the microwave portion of the electromagnetic spectrum, typically from about 200 MHz (1.5 meters wavelength) to about 35,000 MHz (8.5 mm wavelength). These are not firm bounds. Many radars operate outside these limits. The famous British CH radar system of World War II which provided warning of air attack, operated in the high-frequency region of the spectrum in the vicinity of 25 MHz. Radars have also been demonstrated in the millimeter wavelength region where small physical apertures are capable of narrow beam widths and good angular resolution. The radar principle has also been applied at optical frequencies with LASERS for the measurement of range and detection of small motions (using the Doppler effect).

The detection performance of a radar system is specified by the *radar equation* which states

$$P_{\text{rec}} = \frac{P_t G}{4\pi R^2} \times \quad \sigma \quad \times \frac{1}{4\pi R^2} \times A$$

Received = Power × Target × Space × Antenna
power   density back-   atten-   collect-
           at a   scatter uation   ing
           distance cross   on   area
           R   section return
                             path

where P_t is the transmitted power, G is the transmitting antenna gain, R is the range, σ is the target backscatter cross section, and A is the effective receiving aperture of the antenna. The wavelength λ of the radar signal does not appear explicitly in this expression, but it can be introduced by the relationship between the gain and effective receiving area of an antenna which states $G = 4\pi A/\lambda^2$.

The detection capability and the measurement accuracy of a radar are ultimately limited by noise. The noise may be generated within the radar receiver itself, or it may be external and enter the receiver via the antenna along with the desired signal. External noise is generally small at microwave frequencies, but it can be a significant part of the overall noise if low-noise receiving devices are used (see MEASUREMENTS, PRINCIPLES OF).

Since the effects of noise must be considered in statistical terms, the analysis and understanding of the basic properties of radar have benefited from the application of the mathematical theory of statistics. The statistical theory of hypothesis testing has been applied to the radar detection problem where it is necessary to determine which of two hypotheses is correct: the output of a radar receiver is due to (1) noise alone or (2) signal plus noise. One of the results is the quantitative specification of the signal-to-noise ratio required at the receiver for reliable detection. The statistical theory of parameter estimation has also been applied with success to analyze the accuracy and theoretical limits of radar measurements.

Radar receivers are usually designed as *matched filters* which maximize the ratio of peak-signal power to mean-noise power at the output. The magnitude of the frequency response function of a matched filter is equal to the magnitude of the spectrum of the radar signal, and the phase is the negative of the phase of the radar signal.

Reliable detection of targets requires signal-to-noise power ratios of the order of 10 to 100 at the receiver, depending on the degree of error that can be tolerated in making the decision as to the presence or absence of a target. Even larger values are generally needed for the accurate measurement of target parameters. (These values may seem surprisingly high but, for comparison, the minimum signal-to-noise ratio of quality television signals is usually of the order of 10 000.)

The rms error δT in measuring the time delay to the target and back (the range measurement) can be expressed as

$$\delta T = \frac{1}{\beta\sqrt{2E/N_0}}$$

where β is defined as the effective signal bandwidth, E is the total energy of the received signal, and N_0 is the noise power per unit cycle of bandwidth assuming the noise has a uniform spectrum over the bandwidth of the receiver. The square of β is equal to $(2\pi)^2$ times the second central moment of the power spectrum normalized with respect to the signal energy. For a simple rectangular pulse, β is approximately equal to $\sqrt{2B/\tau}$, where B = receiver bandwidth and τ = pulse width. Also, E/N_0 is approximately equal to the signal-to-noise (power) ratio. To obtain an accurate range measurement, E/N_0 and the signal bandwidth must be large. A similar expression applies to the accuracy of the measurement of Doppler frequency if the rms time delay error is replaced by the rms frequency error and the effective bandwidth is replaced by the effective time duration of the signal. Thus, the longer the signal duration and the greater the ratio E/N_0, the more accurate is the Doppler frequency measurement. Likewise the angular measurement accuracy also depends on the ratio E/N_0 and the effective aperture size.

In addition to noise, radar can be limited by the presence of unwanted interfering echoes from large nearby objects such as the surface of the ground, trees, vegetation, sea waves, birds, insects, and weather. Although these "clutter" echoes may be troublesome in some applications, they are sometimes the echoes of interest as, for example, in ground mapping and meteorological applications.

Radar pulses can be as short in duration as a nanosecond or less, giving a range resolution of several centimeters. The shorter the pulse width the greater must be the peak power in order to maintain sufficient energy in the pulse for adequate detection sensitivity and accurate measurement in the presence of noise. Unfortunately, peak power is limited by voltage breakdown. It is possible, however, to achieve the resolution of a short pulse and the energy of a long pulse by means of *pulse compression* in which a pulse with peak power and pulse width T sufficient to give the desired energy is internally modulated in either phase or frequency to increase its bandwidth B. The product BT is called the *pulse compression ratio*. After passing through a matched filter on reception, the pulse of width T is compressed to a width $1/B$, and the peak power of the received signal is increased by a factor BT. A common form of pulse compression modulation is linear FM, also called *chirp*. Pulse compression ratios of practical radars are typically from 100 to 200, but can be as great as several hundreds of thousands.

Recent advances in digital processing technology have allowed the practical employment of automatic detection and tracking (ADT) in radars used for the surveillance of aircraft. ADT systems relieve the burden on the radar operator by automatically making the detection decision as to the presence or absence of targets and establishing target tracks (course or speed). ADT systems generally employ some form of CFAR (constant false alarm rate) receiver so as to prevent false targets due to noise or clutter echoes from overloading the tracking computer.

Radar has been used for the remote sensing of the environment. One of the oldest and most important applications of radar as a remote sensor has been for weather observation. In addition to providing rainfall rate, weather radar is able to determine wind conditions by extracting the relative velocity (measurement of the Doppler frequency shift). Remote sensing radar on aircraft or spacecraft have been employed for geological prospecting, mapping of remote regions under cloud cover, sea state measurement, measurement of the geoid, and the mapping of ice conditions. Other potential applications include crop census and measurement of soil moisture. The synthetic aperture radar (SAR) has been employed for remote sensing because of its ability to produce maplike images with high resolution at long ranges. Resolution can be as small as several meters and, unlike optical imaging, the resolution is independent of the range. In the range dimension good resolution is obtained by pulse compression, and in the cross-range dimension it is obtained by synthesizing the effect of a large antenna by storing the received signals over the time it takes for the radar to travel the length of the synthesized antenna.

The HF over-the-horizon (OTH) radar operates at frequencies (typically from about 5 MHz to perhaps 30 MHz) which are refracted by the ionosphere so as to provide illumination of regions from about 500 to 2000 nautical miles from the radar. Such radars can provide detection and tracking of aircraft and other targets at distances far greater than can be obtained with line-of-sight microwave radar. One interesting application of HF OTH radar is the determination of sea conditions and the winds generating the sea, by proper interpretation of the radar echo's Doppler spectrum induced by the motion of the sea waves.

Radar comes in many sizes and shapes. The smallest can be held in the hand and might radiate as little as a few milliwatts and be used to detect the movement of people at short range. The largest might radiate megawatts of average power and operate with antennas the size of a football field and would be used for the detection of space objects at ranges of many thousands of miles or more.

MERRILL I. SKOLNIK

References

Skolnik, M. I., "Introduction to Radar Systems," 2nd Edition, New York, McGraw-Hill Book Co., 1980.

Skolnik, M. I., "Radar Handbook," New York, McGraw-Hill Book Co., 1970.

Barton, D. K., "Radar Systems Analysis," Dedham, Mass., Artech House, 1976.

Barton, D. K., and Ward, H. R., "Handbook of Radar Measurement," Englewood Cliffs, N.J., Prentice-Hall, Inc., 1969.

Nathanson, F., "Radar Signal Processing and the Environment," New York, McGraw-Hill Book Co., 1969.

Blake, L. V., "Radar Range-Performance Analysis," Lexington, Mass., Lexington Books, 1980.

Brookner, F., "Radar Technology," Dedham, Mass., Artech House, 1977.

Ulaby, F. T., Moore, R. K., and Fung, A. K., "Microwave Remote Sensing, Active and Passive," Reading, Mass., Addison-Wesley Publishing Co., 1981.

Cross-references: ANTENNAS; DOPPLER EFFECT; LASER; MEASUREMENT, PRINCIPLES OF; MICROWAVE TRANSMISSION; PROPAGATION OF ELECTROMAGNETIC WAVES; PULSE GENERATION.

RADIATION CHEMISTRY

Radiation chemistry is the study of the chemical effects produced by the absorption of ionizing radiation. It includes chemical effects produced by the absorption of radiation from radioactive nuclei (α, β, and γ rays), of high-energy particles (electrons, protons, neutrons, recoil nuclei, etc.), and of electromagnetic radiation of short wavelength (x-rays with a wavelength less than about 100Å and an energy greater than about 100 eV, for example). Electromagnetic radiation of rather longer wavelength, in the ultraviolet and visible regions of the spectrum, may also initiate chemical reactions, though normally without producing ions as reactive intermediates; such reactions are the province of photochemistry. Reactions chemically similar to those caused by the absorption of ionizing radiation can be initiated by electric discharges; others, initiated in various ways, occur in the upper atmosphere.

Radiation chemistry originated with the observations by Röentgen (1895) and by Becquerel (1896) that led to the discovery of x-rays and of radioactivity, namely that photographic plates become fogged when placed near discharge tubes and uranium salts respectively. The subject was studied to only a limited extent until about 1942, when the advent of nuclear reactors and the increased interest in high-energy physics provided both the incentive and the means (in the form of relatively cheap artificial radioactive isotopes and of large particle accelerators) for a more intensive study. In the following two decades, earlier work was consolidated and the basic mechanisms for radiation-induced action were established in outline, making possible an extremely rapid development of the subject since about 1960.

Absorption of any form of ionizing radiation by matter (A) produces positive ions (A+), electrons (e^-), and electronically excited atoms or molecules (A*) distributed along the tracks of charged particles. The charged particles may be those which comprise the radiation (e.g., electrons or helium nuclei with fast electron or α-irradiation) or secondary particles produced by interactions of the primary radiation (e.g., fast secondary electrons formed by the absorption

photochemistry, where excited molecules are produced with an essentially uniform distribution in any plane at right angles to the direction of the beam of light. The excited molecules involved in radiation chemistry include excited states similar to those formed by the absorption of ultraviolet or visible light and also other states, formed by optically forbidden transitions or with more intrinsic energy, that are not produced photochemically.

Chemical changes in the irradiated material are brought about by breakdown or reaction of the ions and excited molecules and via free radicals formed by these primary species. The sequence of events in many radiation-induced (radiolysis) reactions may be represented as:

$$
A \xrightarrow{\text{irradiation}} \left\{ \begin{array}{c} A^* \\ A^+ \\ e^- \end{array} \right\} \longrightarrow \text{free radicals} \longrightarrow \text{chemical products}
$$

A (original molecules)

free radicals (including e^-_{solv} in polar media)

Approximate interval after passage of ionizing particle 10^{-15} 10^{-11} 10^{-3} sec

of x- or γ-rays, or fast protons produced by interaction of neutrons with hydrogenous materials). In addition, some of the electrons produced by ionization in the medium will have sufficient energy to produce further ionization and excitation, and will do so in slowing down to thermal energy (such electrons are known as δ-rays). Except in gases, where the ions and excited molecules can diffuse apart quite readily, these primary species are initially concentrated within about 10Å of the track of the ionizing particle. Heavy charged particles (protons and, particularly, helium nuclei and heavier particles) lose energy very rapidly in liquids and solids and leave a track densely populated with the primary species; radiations of this type are said to have a "high LET," where LET stands for linear energy transfer. In contrast, fast electrons and the secondary electrons produced by the absorption of x- and γ-rays lose energy relatively slowly and form the primary species (which are the same as those produced by the heavier particles) in small groups containing, on an average, two or three ion pairs and about the same number of excited molecules. These groups (called *spurs*), with an initial diameter of $\sim$ 20Å, are separated by a relatively great distance ($\sim 10^3$ Å) from neighboring spurs along the same track, and the radiation responsible is said to have a "low LET." However, the spurs along an electron track become closer together as the electron loses energy, and eventually overlap each other. The initial localization of the ions and excited molecules in the track of a high-LET particle or in spurs produced by a low-LET particle causes spatial, or track, effects in radiation-induced reactions which are absent in

Electronically excited molecules and ions are produced by passage of an ionizing particle. The excited molecules may return to the ground state without producing chemical change, dissociate to free radicals, or dissociate directly to stable chemical products. Positive ions also give rise to free radicals or stable products by dissociation, reaction with the substrate (in ion-molecule reactions), or upon neutralization by an electron. In polar media the neutralization may be delayed by solvation (association of the ion with a group of solvent molecules held by electrostatic forces), and the solvated ions, particularly the solvated electron, may react in a similar manner to free radicals. The events outlined above are often considered to occur in three stages: a *physical stage* lasting about 10^{-14} sec corresponding with the initial dissipation of energy in the system; a *physiochemical stage* lasting from $\sim 10^{-14}$ to $\sim 10^{-10}$ sec during which thermal equilibrium is established, dissociation of excited molecules and positive ions occurs, and (in polar media) ions become solvated; and a *chemical stage* lasting about 10^{-3} sec during which radical reactions occur and give rise to chemically stable molecules. Processes occurring during the late physiochemical and the chemical stages can sometimes be observed directly by absorption spectroscopy following a brief radiation pulse. This technique, known as *pulse radiolysis*, is routinely used to investigate reactions occurring microseconds after the radiation pulse and can be extended into the nanosecond and picosecond regions.

Free radicals, which are formed when most materials except metals are irradiated, are atoms or molecules which have unpaired valence electrons; the solvated electron behaves like, and is often described as, a free radical. Free radicals are generally chemically unstable species and react rapidly in such a manner that the unpaired electrons become paired, for example by combination of two free radicals (shown below with a · to represent the unpaired electron)

$$2 \cdot OH \longrightarrow HO\text{—}OH \ (or \ H_2O_2)$$

or transfer of an atom (disproportionation)

$$CH_3CH_2 \cdot + \cdot CH_2CH_3 \longrightarrow$$
$$CH_3\text{—}CH_3 + CH_2\text{=}CH_2$$

or an electron

$$\cdot OH + Fe^{2+} \longrightarrow OH^- + Fe^{3+}.$$

Other common radical reactions are addition to oxygen and doubly bonded (unsaturated) compounds and abstraction of hydrogen or halogen atoms from compounds, e.g.,

$$e_{aq}^- + O_2 \longrightarrow O_2{}^-$$
$$\cdot CH_3 + CH_2\text{=}CH_2 \longrightarrow CH_3\text{—}CH_2\text{—}CH_2 \cdot$$
$$H \cdot + CCl_4 \longrightarrow HCl + \cdot CCl_3.$$

The latter reactions do not result in electron pairing but instead give rise to less reactive (more stable) radicals that may react further before they are stabilized by electron pairing. Many free radicals have been identified in irradiated systems by pulse radiolysis, and in solid or frozen systems, by electron spin resonance.

Quantitative studies in radiation chemistry are based either on the number of ions (N) formed in the irradiated medium or on the energy absorbed from the radiation. Ionization measurements in irradiated gases allow N to be estimated, and radiation yields in gases are best expressed as the *ionic* (or *ion-pair*) *yield*, M/N, where M is the number of molecules of product formed by radiation which produces N ion pairs. Reliable estimates of N are, at present, only available for gaseous systems, and for condensed materials it is usual to express the radiolysis yield as a G value (the number of molecules of product formed, or of starting material changed, per 100 eV of energy absorbed). Ionic yields and G values are related by $G = M/N \times 100/W$, where W (electron volts) is the mean energy required to form an ion pair in the material being irradiated. The determination of N, or of the energy absorbed, is termed "dosimetry" and may involve ionization measurements, calorimetry, measurement of the charge carried by a beam of charged particles of known energy, or measurement of a chemical change produced by irradiation [the oxidation of ferrous iron to ferric iron in 0.4 M sulfuric acid solution is often used (Fricke dosimeter)]. However, chemical dosimeters must first be calibrated against some absolute physical measurement. The energy absorbed by an irradiated system is generally expressed in units of *rads* (1 rad is an energy absorption of 100 ergs g^{-1}) or of eV g^{-1} or eV ml^{-1}, though these units are being displaced by a new SI unit, the gray (Gy). 1 Gy = 1 J · kg^{-1} = 100 rad.

Radiation-induced reactions have been studied in the gas, liquid and solid phases and in inorganic, organic, and biological systems. They have also been studied over a wide range of temperatures.

Most thoroughly studied of all radiation-induced reactions is the radiolysis of water and aqueous solutions, where the following have been identified as the major products formed in the tracks or spurs: hydrogen (H_2), hydrogen peroxide (H_2O_2), hydrogen atoms (H·), hydroxyl radicals ($\cdot OH$), and solvated, or hydrated, electrons (e_{eq}^-). These products are formed by the primary species, H_2O^+, e^-, and H_2O^*, and, in the case of hydrogen and hydrogen peroxide, by H·, $\cdot OH$, and e_{aq}^-. Relatively more hydrogen and hydrogen peroxide are formed in the track of a high-LET α-particle [$G(H_2) \sim 1.57$, $G(H_2O_2) \sim 1.45$] than in the spurs associated with a high-energy, low-LET, electron [$G(H_2) \sim 0.40$, $G(H_2O_2) \sim 0.80$] since the concentration of the precursors is greater in the track than in the spurs. However, relatively more of the other products (H·, $\cdot OH$, and e_{aq}^-) escape from the spurs [G(radical) ~ 6.5] than from the α-particle track [G(radical) ~ 1.2]. These three species are free radicals and react very readily with substances present in solution, the hydroxyl radical generally producing oxidation, and the hydrogen atom and the solvated electron producing reduction, of the solute if the solution is free of air. Both the hydrogen atom and the solvated electron react rapidly with oxygen to give oxidizing species, so that oxidation reactions predominate when aerated aqueous solutions are irradiated. In acid solutions, solvated electrons are rapidly converted to hydrogen atoms (H·) by reaction with hydrogen ions (H^+).

Many organic materials have been irradiated and, making a very rough generalization, the products are those expected if the action of the radiation is to break the organic molecules randomly into two fragments (free radicals). The fragments then react together in pairs, again in a random fashion, either combining to form larger molecules or transferring an atom from one fragment to the other to give two stable molecules. The products from a hydrocarbon, for example, include hydrogen and hydrocarbons ranging in size from methane (CH_4) to compounds containing twice as many carbon atoms as the original compound; unsaturated materials, containing relatively less hydrogen than the original compound, will also be formed. Hexane, $CH_3\text{—}(CH_2)_4\text{—}CH_3$, forms at least 16 products upon irradiation, of which the two most abundant are hydrogen ($G = 5.0$) and dimeric C_{12} hydrocarbons ($G = 2.0$); the remaining products are formed with G values of 0.5 or less. More complex compounds may break preferentially (though generally not exclusively) at a particular point in the molecule; thus the carbon-iodine bond breaks most frequently when methyl iodide CH_3I is irradiated. Some classes of compound can enter into chain reactions in which the reaction, once started, continues on its own. Typical of such substances are the unsaturated "monomers" which polymerize to produce polymers such as polymethyl methacrylate ("Lucite") and polyvinyl chloride (PVC). The chain reactions here are initiated by free radicals, and other means of producing free radicals besides irradiation bring about the same effect. G values for chain reactions may run from a few hundred to many thousands.

Aromatic compounds such as benzene are more resistant to radiation than most compounds lacking the "benzene" ring. Thus $G(H_2)$ from liquid benzene is only 0.036, and the total yield of radiolysis products does not exceed $G = 1$. In favorable instances, aromatic compounds can reduce the radiation damage in non-aromatic compounds mixed with them, energy absorbed by the second component being transferred in part to the "protecting" aromatic compound.

Apart from its intrinsic interest, the importance of radiation chemistry has rested largely upon its application to problems of reactor technology and its close relationship to radiation biology and radiation medicine. More recently several commercial applications have been realized, such as irradiation of polyethylene to produce a higher-melting polymer and the synthesis of ethyl bromide via a radiation-induced chain reaction between ethylene and hydrogen bromide. However radiation energy is relatively expensive, and the commerically viable processes have generally involved either reactions having a high G value (i.e., chain reactions) or reactions producing a high molecular weight product such as a polymer. Both G value and molecular weight are related to the mass of product formed per unit radiation energy absorbed, for example the yield of product in pounds produced per kilowatt hour radiation energy absorbed is given by $G \times M \times 8.23 \times 10^{-4}$. The ethyl bromide synthesis is an example of a reaction with a high G value ($\sim 10^4$–10^5), other examples being the polymerization of unsaturated monomers absorbed in wood or concrete, or present in printing inks and coatings, and the production of graft polymers. Crosslinking of bulk polymer, such as polyethylene wire and cable insulation, to increase heat resistance and toughness is an example where M is large and a relatively small degree of chemical change brings about a large change in physical properties. Other well established applications of radiation technology are the sterilization of bulk medical supplies such as sutures and disposable plastic items, and radia-tion treatment of foodstuffs to extend storage life (the key chemical changes in these instances presumably involve large protein or nucleic acid molecules).

J. W. T. SPINKS
R. J. WOODS

References

Attix, F. H., Roesch, W. C., and Tochilin, E. (Eds.), "Radiation Dosimetry," Second Ed., New York, Academic Press, 1966.

Ausloos, P. (Ed.), "Fundamental Processes in Radiation Chemistry," New York, Interscience, 1968.

Burton, M., and Magee, J. L. (Eds.), "Advances in Radiation Chemistry," Vol. 1, New York, Wiley-Interscience, 1969, and succeeding volumes.

Chapiro, A., "Radiation Chemistry of Polymeric Systems," New York, Interscience Publishers, 1962.

Draganić, I. G., and Draganić, Z. D., "The Radiation Chemistry of Water," New York, Academic Press, 1971.

Haissinsky, M. (Ed.), "Actions Chimiques et Biollogiques des Radiations," Vol. 1, Paris, Masson et Cie., 1955, and succeeding volumes.

Hedvig, P., and Schiller, R., "Proceedings of the Fourth Tihany Symposium on Radiation Chemistry," Budapest, Akadémiai Kiadó, 1977.

Matheson, M. S., and Dorfman, L. M., "Pulse Radiolysis," Cambridge, Mass., M.I.T. Press, 1969.

O'Donnell, J. H., and Sangster, D. F., "Principles of Radiation Chemistry," London, Edward Arnold Ltd., 1970.

Silverman, J. (Ed.), "Advances in Radiation Processing," New York, Pergamon Press, 1979. [*Radiat. Phys. Chem.* **14**, 1–961 (1979).]

Silverman, J. (Ed.), "Trends in Radiation Processing," New York, Pergamon Press, 1981. [*Radiat. Phys. Chem.* **18**, 1 (1981).]

Silverman, J., and Van Dyken, A. R. (Eds.), "Radiation Processing," New York, Pergamon Press, 1977. [*Radiat. Phys. Chem.* **9**, 1–885 (1977).]

Spinks, J. W. T., and Woods, R. J., "An Introduction to Radiation Chemistry," Second Ed., New York, Wiley-Interscience, 1976.

Swallow, A. J., "Radiation Chemistry: An Introduction," London, Longmans, 1973.

Cross-references: IONIZATION; MAGNETIC RESONANCE; PHOTOCHEMISTRY; RADIATION, IONIZING, BASIC INTERACTIONS; SECONDARY EMISSION.

RADIATION, IONIZING, BASIC INTERACTIONS

Radiations of a large class, called *ionizing*, which interact with matter in its many forms and lead to a wide variety of "observed" or expressed effects have similar basic interaction pathways. Sources of ionizing radiations are varied and include radioactive isotopes, fission and fusion reactions, particle accelerators, and cosmic rays. Regardless of the source of any given radiation (applied to a given target), the basic interaction depends only upon the fundamental properties of the radiation itself.

Ionizing Radiations The principal ionizing radiations are summarized in Table 1. Although only the gamma or x-rays are electromagnetic in character and thus "radiations" in the classical sense, the distinction between "radiations" and ionizing "particles" is often not made (x-rays are distinguished from gamma rays only with respect to their origins; gamma rays result from nuclear interactions or decays; x-rays result from transitions of atomic or free electrons, produced artificially by bombarding metallic targets with energetic electrons). It is sometimes difficult to make a clear distinction between ionizing and nonionizing electromagnetic radiations, particularly in condensed phases. The ionization potential of *gaseous* elements, i.e., the energy required for removal of the first electron, varies from 3.9 eV (Cs) to 24.6 eV (He). Comparable values are not well known for most complicated molecular systems or liquid- and solid-state systems, but a few selected additional values are included in Table 2, and others are probably in or near this general range. Although ultraviolet and even visible light can in special cases cause ionizations, the general assumption is that more energetic x- or gamma radiation is required to insure ionization, especially in substances of practical importance.

Hence the name "ionizing radiation" is reserved for electromagnetic radiation at least as energetic as x-rays and for charged particles of similar energies. Neutrons also lead to ionization, but for other reasons, described below. *Ionization* is, of course, not the sole interaction of high-energy particles and radiations with matter. The *excitation* of atomic electrons into higher-energy states always accompanies ionization.

Basic Action The fundamental processes leading to ionization differ for charged particles, electromagnetic radiations, and neutrons.

Charged Particles. Fast charged particles are produced in radioactive decay processes, particle accelerators, nuclear reactions, and extraterrestrial sources. They undergo *coulombic* and *nuclear* interactions. The latter are far less probable and are discussed in other parts of this encyclopedia (see NUCLEAR REACTIONS). There are two principal means whereby charged particles can lose energy by coulomb interaction: radiative loss and direct ionization. The probability of radiative energy loss (BREMSSTRAHLUNG) is roughly proportional to

TABLE 2. IONIZATION POTENTIALS OF SOME SELECTED ELEMENTS (WEAST, 1975).

Element	Ionization Potential (Volts)
Argon	15.6
Carbon	11.3
Cesium	3.9
Hydrogen	13.6
Helium	24.5
Nitrogen	14.5
Neon	21.6
Oxygen	13.6
Phosphorus	10.5
Silicon	8.1
Sodium	5.1

TABLE 1. PROPERTIES OF THE MOST COMMON IONIZING RADIATIONS.

Name	Symbol	Location in Atom	Relative Rest Mass	Charge
Proton $(H^1)^+$	p	Nucleus	1	+1
Neutron	n	Nucleus	1	0
Electron	e	Outer shells	0.00055	−1
Beta− (electron)	β	Emmitted during decay	0.00055	−1
Beta+ (positron)	β^+	processes	0.00055	+1
Alpha $(He^4)^{++}$	α		4	+2
Gamma[a] (photon)	γ	Emitted during decay processes	0.0	0

[a]X-rays of equal energy are identical, but of extranuclear origin.

$$\frac{z^2 Z^2 T}{M_0{}^2}$$

where z is the particle charge in units of the electron charge e, Z is the atomic number of the target material, T is the particle kinetic energy, and M_0 is the rest mass of the particle. The ratio of energy lost by bremsstrahlung to that by ionization can be approximated by

$$\left(\frac{m_0}{M_0}\right)^2 \frac{ZT}{1600\, m_0 c^2}$$

in which m_0 is the rest mass of the electron and c^2 is the speed of light squared, or 931 MeV per atomic mass unit. Electrons in the 10-MeV region lose about half of their energy by bremsstrahlung (in high-Z material), whereas heavier charged particles lose nearly all of their energy by ionization.

Loss of energy by ionization results when the particle undergoes coulomb collision with the *electrons* of the target. From the ratio of total energy lost to the number of ion pairs produced in cloud chambers (see Fig. 1), it has been estimated that approximately 100 eV are dissipated in each *primary ionization event* and that each event results in the production of, on the average, three ion pairs, each consisting of a free electron and a positive ion. If a resulting electron carries more than a certain amount (usually considered to be about 100 eV) of kinetic energy, it is called a *delta ray*, because it is capable of further ionizations. "Delta ray" and "track core" ionizations are depicted in Fig. 1. The experimentally determined energy required to produce each *observed* ion pair lies in the range of 20 to 40 eV per ion pair for gases. A figure in the neighborhood of 30 eV per ion pair is commonly assumed for condensed phases of matter, with the exception of semiconducting solids. This value is called W and is generally slightly different for electrons and low-velocity heavy particles in the same material.

For example, although $W = 26.2 \pm 0.2$ eV per ion pair for electrons in argon, and $W = 26.2 \pm 1.8$ for fast (relativistic) nitrogen ions in argon, recoiling nuclei of polonium and thorium (with very low energies) ionize argon gas with a W value exceeding 85 eV (Myers, 1968; Schimmerling et al., 1976). Theoretically, for heavy ions of low velocity W' takes the place of W and is found from

$$W' = \frac{W}{1 + (\frac{1}{2} A / W T^{1/2})}$$

where A is an empirical constant, 9500 for nitrogen ions in argon gas, for example. As T approaches the Bohr-orbit energy W' becomes large. The value of W in liquid argon is 26.0 eV per ion pair, not significantly different from that of gaseous argon, 26.3 eV per ion pair. Ionization in a few solids has been studied, and $W = 3.57 \pm 0.05$ eV per electron-hole pair in silicon, for example, and a range of about 1–10 eV per electron-hole pair has been found in about a dozen semi-conducting solids (Myers, 1968). Additional useful W values are included in Table 3 for materials often used in radiation dosimetry.

On the basis of coulomb scattering theory, it is possible to calculate the energy loss per unit path length ($-dT/dx$) for charged particles. For electrons and positrons the theory is complicated by the quantum mechanical effects of spin, identity, and relativity. In general, the *total* expectation energy loss per unit path length (also termed *mass stopping power* when path length is expressed in grams per square centimeter) for electrons and positrons can be expressed by

$$-\frac{dT}{dx} = \frac{2\pi e^4}{m_0 V^2} NZ \left\{ \ln\left[\frac{m_0 V^2 T}{2I^2(1 - \beta^2)} \right] - \beta^2 \right\}$$

in which V is the particle velocity, beta is V/c, N is the number of atoms per unit mass, and I is the geometric mean ionization and excitation energy of the atoms of the target material. Classical theory is more adequate in describing the

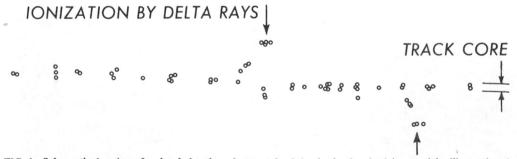

FIG. 1. Schematic drawing of a cloud-chamber photograph of the track of an ionizing particle, illustrating that ion pairs (illustrated by small circles) occur in clusters. Spurs indicated by arrows are ionizations due to *delta rays*. The arbitrary identification of the *track core* is also indicated.

TABLE 3. VALUES OF W, THE AVERAGE
ENERGY LOST IN PRODUCING AN ION
PAIR, FOR MATERIALS OFTEN USED
IN RADIATION DOSIMETRY.

Substance (Gas)	W (eV/ion pair)
Air	33.7
Ammonia	34.8
Argon	26.2
Boron Trifluoride	35.6
Carbon dioxide	32.9
Cesium	16.0
Ethane	24.6
Ethanol	32.6
Helium	41.5
Hydrogen	36.6
Krypton	28.8
Methane	27.3
Neon	36.2
Nitrogen	34.6
Oxygen	31.8
Tissue-equivalent gas	30.5
Water	30.1
Xenon	21.9

loss of energy by heavy charged particles, and the mass stopping power is determined from

$$-\frac{dT}{dx} = \frac{4\pi e^4 z^2}{m_0 V^2}\, NB$$

where B is called the *stopping number* and assumes various forms.

The complete stopping number for a positively charged particle with effective charge Z, velocity V, and mass m traversing a substance having N atoms/cm^3, atomic number Z, and mean excitation energy I is

$$B = Z \left[\ln \frac{2mV^2}{I(1-\beta^2)} - \beta^2 - \frac{C(V,Z)}{Z} \right.$$
$$\left. - \frac{1}{2}\delta(V,Z) + z\phi(V,Z) + z^2\psi(V,Z) \right].$$

The terms within the brackets are as follows:

$C(V,Z)/Z$ is the *shell correction term* and adjusts the field of the target atom on the basis of the electric fields of outer shell electrons; there is one such term for each significant shell.

$\frac{1}{2}\delta(V,Z)$ is the *Fermi density effect* and adjusts the field of the target atom on the basis of atoms that actually lie between the projectile and the target atom when interaction distances are very large.

$Z\phi(V,Z)$ is the Z^3 term.

$Z^2\psi(V,Z)$ is the Z^4 term, and these account for higher-order electric force fields related to the shapes of the target atom and projectile.

The most important parameter in the stopping equation is the mean excitation energy I. These have been determined by theory and/or by experiment for the elements and numerous compounds. In the case of compounds the theoretical determinations may include the effects of chemical binding and the physical state of aggregation of the medium. Mean excitation energies for 50 common substances are compiled in Table 4, using the data compiled by S. M. Seltzer and M. J. Berger which lists 278 materials.

A complete account of the energy lost by fast charge particles includes energy loss events over the complete range of amount of energy transferred. Some events only result in excitation of atoms and molecules and can be characterized through optical studies of the absorbing material. The probabilities (or *cross sections*) of such events depend in a very irregular fashion on the amount of energy transferred per event, since quantum level transitions of orbital electrons absorb the energy. Their transition probabilities are called *oscillator strengths;* the interaction cross sections depend strongly on the molecular structure of the target. Bethe-Born theory provides a reliable extrapolation of emission spectra of low energy electrons from optical spectroscopic data. Thus a means of determining I from first principles is to consider *oscillator strength df/dT* and dielectric response ϵ determined from optical properties of the material:

$$\ln I = \frac{\displaystyle\int \ln T\,(df/dT)\,dT}{\displaystyle\int (df/dT)\,dT}$$

$$= \frac{\displaystyle\int T \ln T \, \mathrm{Im}\,[-1/\epsilon(T/\hbar)]\,dT}{\displaystyle\int T\,\mathrm{Im}\,[-1/\epsilon(T/\hbar)]\,dT}$$

where Im refers to the imaginary part of the dielectric response, which is responsible for energy absorption.

Instead of stopping power, the term *linear energy transfer* (LET) is frequently used. The two quantities are not strictly interconvertible, according to a recent decision by the international committee on radiation units (see *National Bureau of Standards (U.S.) Handbook* 84). Mass stopping power should be used only in the sense described above, whereas LET should be used only for energy lost by the particle within a specified distance of the track core. (Stopping

TABLE 4. MEAN EXCITATION ENERGIES FOR 50 COMMON SUBSTANCES.

Element	State	I + S.D. (eV)	Material	I (eV)
1 hydrogen	H_2 gas	19.2 ± 0.4	air, dry (STP)	85.7
2 helium	He gas	41.8 ± 0.8	bone-equivalent plastic	85.8
6 carbon	graphite	78.0 ± 7.0	compact bone	92.1
7 nitrogen	N_2 gas	82 ± 2	cortical bone	106.7
8 oxygen	O_2 gas	95 ± 2	ferrous sulfate dosimeter	76.4
10 neon	Ne gas	137 ± 4	methane	41.7
13 aluminum	solid	166 ± 2	muscle-equivalent liquid	74.1
14 silicon	solid	173 ± 3	mylar	78.7
15 phosphorus	solid	173	paraffin wax	55.9
16 sulfur	solid	180	photoemulsion	331.0
18 argon	Ar gas	188 ± 10	polycarbonate	73.1
20 calcium	solid	191 ± 8	polyethylene	57.4
26 iron	solid	286 ± 9	polymethylmethacrylate	74.0
29 copper	solid	322 ± 10	polystyrene	68.7
32 germanium	solid	350 ± 11	propane	47.1
36 krypton	Kr gas	352 ± 25	silicon dioxide	139.2
47 silver	solid	470 ± 10	skeletal muscle	75.0
49 indium	solid	488 ± 20	sodium iodide	468.6
50 tin	solid	488 ± 15	teflon	99.1
54 xenon	Xe gas	482 ± 30	tissue equivalent gas	61.5
74 tungsten	solid	727 ± 30	tissue-equivalent plastic	65.1
78 platinum	solid	790 ± 30	toluene	62.5
79 gold	solid	790 ± 30	vinyltoluene scintillator	64.7
82 lead	solid	823 ± 30	water liquid	75.0
92 uranium	solid	890 ± 33	water vapor	71.6

power is usually expressed in units of million electron volts per gram per square centimeter, and LET in units of kiloelectron volts per micron.) These are both measures of linear *density of ionization* in the target material.

From the above equations, it is possible to derive range-energy relationships for charged particles by simple integration. Theoretical and experimental range-energy curves are now available in the literature for nearly all charged particles.

Electromagnetic Radiation (X, gamma). For fast, charged particles interacting with matter we had, above, relations of the form

$$- \frac{dT}{dx} \propto \frac{(\text{charge})^2}{(\text{velocity})^2}$$

If it were permissible to extend this idea to electromagnetic radiation, one might expect photons with no charge (and with the velocity of light) not to ionize at all. They indeed do ionize sparsely though for different reasons. The three principal mechanisms of interaction can be summarized as follows:

(1) Photoelectric effect. Low-energy photons can give up all their energy to a bound electron, forming an ion pair and disappearing in the process. (Generally unimportant above 1 MeV.)

(2) Compton scattering. For medium-energy photons (0.5 to 5 MeV), this elastic collision process predominates, leading to ejection of a recoil electron plus the partially degraded (longer-wavelength) photon.

(3) Pair production. Photons of highest energy most often interact by forming an electron-positron pair in the field of a nucleus and disappearing in the process. The absolute energy threshold for this process is the rest mass energy of the pair: 1.02 MeV.

The net result of all three processes is the formation of (charged) ion pairs and in particular of electrons having energies ranging up to the photon energy but on the average only a fraction of this maximum. The discussion on charged particles thus applies to x- or gamma-ray action also. Each of these processes will now be considered in more detail.

In the *photoelectric process*, the photon collides directly with an atomic electron, imparting kinetic energy

$$T = h\nu - B_e$$

where B_e is the atomic binding energy of the electron and $h\nu$ is the photon energy. For a considerable range of photon energies and target material Z's, the probability of this process is approximately dependent upon the fourth power of Z and the inverse third power of the photon energy. There are, however, very specific energies at which photoelectric absorption is very probable ("absorption edges"), due to nearness of the photon energy to the binding energy of a specific electron (K, L, M, etc.). The probability per unit thickness of absorber of a photon undergoing photoelectric absorption is denoted by τ. An initial photon intensity I_0 is reduced by this process to an intensity I_x after traversing a thickness of material, x, according to the relation

$$I_x/I_0 = e^{-\tau x}$$

The *Compton process* differs from the photoelectric process in two important ways: the photon usually loses less than all of its energy to the electron with which it collides, giving rise to a lower-energy scattered photon, and the process may occur with a free or loosely bound electron. Momentum and energy are conserved according to the laws of classical mechanics, and the energetics can be described by

$$h\nu = h\nu' + (m - m_0)c^2$$

where $h\nu'$ is the energy of the secondary photon, and $(m - m_0)c^2$ is the kinetic energy of the recoil electron (expressed in terms of the relativistic mass increase). At high photon energies, the collision probabilities are governed by the laws of quantum mechanics. Photons are lost by the Compton process according to the relation

$$I_x/I_0 = e^{-\sigma x}$$

where σ is the absorption probability per unit thickness.

If the energy of the photon is greater than two electron masses (1.02 MeV), there is a finite probability that it will interact with the nuclear field giving rise to electron-positron *pair production*. In pair production, energy is conserved according to

$$h\nu - 2m_0 c^2 = T_+ + T_-$$

where T_+ and T_- are the positron and electron kinetic energies, respectively. The positive electron is ultimately annihilated by a negative electron, and the masses of both are converted into photons with energies distributed about 0.51 MeV (one electron mass each). Very high-energy electron-positron pairs may, in turn, produce further photons by bremsstrahlung, initiating a sequence of photon-electron-photon, etc., interactions, known as a *cascade*. (If the photon energy is greater than four electron masses, pair production may occur in the field of an electron, under which condition the original electron is set in motion, and the process is called "triplet production.") The probability of pair production increases very rapidly with increasing photon energy above 1.02 MeV and increases as Z^2, the square of the atomic number of the target material. The absorption probability per unit thickness is denoted by κ, so that high-energy photons are absorbed according to

$$I_x/I_0 = e^{-\kappa x}$$

It is usually desirable to know the *total* photon absorption per unit thickness of a given material,

so one simply states

$$I_x/I_0 = e^{-\mu x}$$

where

$$\mu = \tau + \sigma + \kappa.$$

μ is called the "total linear absorption coefficient" and is expressed in units reciprocal to those in which x is measured.

Neutrons. Because of their lack of charge, neutrons do not interact electrostatically either with orbital electrons or with nuclei. They do interact with nuclei, however, in various other ways.

Fast neutrons (up to a few MeV) lose energy primarily by elastic collisions with other nuclei. From considerations of momentum transfer, this process is most efficient for target nuclei of about the same mass (i.e., protons in hydrogenous materials), though other light nuclei are also effective. On the *average*, about half the initial neutron kinetic energy is transferred to the protons, so that fast-neutron bombardment looks (to a hydrogenous material) like bombardment with fast protons of half the neutron energy. In addition to such simple collision processes, fast neutrons also induce nuclear reactions in certain elements, leading to emission of particles or photons with their previously described interactions.

After about 20 collisions, neutrons are no longer sufficiently energetic to eject recoil protons but have become "thermalized," i.e., they act (for a short time) like a gas in thermal equilibrium with its surroundings (energies of about 1/40 eV).

When a neutron has become thermalized and wanders into a nucleus, it is quite often captured, momentarily yielding an excited isotope of the original nucleus. Nuclei usually lose their excitation by emission of particles or characteristic gamma rays. Thus, even an uncharged slow neutron gives rise to the release of ionizing radiation inside a material being irradiated. In living tissue, slow neutrons commonly are captured by H^1 nuclei with emission of an energetic gamma ray, and by N^{14} nuclei with emission of an energetic proton.

Associated or Post-ionization Events Subsequent to the primary and secondary molecular ionizations (and excitations), a number of events can occur that depend rather strongly on the form of the target material, including

(1) dissociation of molecules and formation of free radicals (species with unpaired electrons, hence great chemical reactivity, e.g., in water, H, OH, HO_2);

(2) recombination of ions and radicals, leading to no net change;

(3) dispersion of energetic ions and radicals, and reaction with other species present or with each other;

(4) nondiffusion migration of electronic excitation to energy "sinks," e.g., in macro-molecules or crystals; and

(5) eventual degradation of the excess absorbed energy to heat (insignificant from the standpoint of effects).

For many systems the basic interactions must be considered only the initiators of a complex sequence of later events.

Comparison of the Radiations Diverse electromagnetic and particulate radiations thus have in common as their basic interaction (or closely following upon it) the production of molecular ionizations (and excitations, dissociations, free radicals) inside target matter. They differ in the *geometry* of these events (especially in LET), a difference that leads to wide variations in range or penetrating ability and, thus, in the subsequent reactions leading to the final expression of the radiation effect.

The slower, more highly charged, heavy particles (such as alphas) travel in straight-line tracks ionizing densely along the track and exhibiting discrete ranges characteristic of the particle energy. Lighter, less highly charged particles such as electrons have their tracks more easily deflected and therefore have less precisely specified ranges in matter (although they have a maximum range). With a lower ionization density, however, they travel much farther than heavy charged particles of the same energy. Gamma or x-rays interact causing release of electrons in matter but at widely spaced intervals and in a random fashion; they have a still lower ionization density (LET) and much longer "range." (Since the resultant of all their absorption processes is a roughly exponential attenuation in matter, their "range" must be described in terms of a parameter such as "half thickness," $x_{1/2}$, the thickness of material that reduces incident intensity to 50 per cent. The relation $x_{1/2} = 0.693/\mu$ is seen to follow from the above equation for total photon absorption.) X-ray interaction has been likened to a "shotgun" effect in contrast to the "rifle" effect from incident heavy charged particles. Neutrons, which only interact with nuclei, may have either high or low LET. Their penetration in matter is great though difficult to specify well except in terms of specific materials. For much of their path, *fast* neutrons also are attenuated roughly exponentially.

The approximate range r or half thickness $x_{1/2}$ is given in Table 5 for several 1-MeV radiations in water.

Finally, a variety of conventional particle and wave interactions (such as reflection, transmission, and refraction) are also experienced by the above radiations. Moreover, many other more-or-less-ionizing radiations have been omitted from this discussion, including mesons, hyperons, heavy cosmic particles, large fission and spallation products, and anti-particles. Their relative interactions can be quite well predicted from their composition, charge, and velocity, and the above considerations.

Recommended general references include *Radiation Dosimetry*, edited by Attix et al., and the detailed review by Birkhoff.

PAUL TODD
HOWARD C. MEL

References

Attix, F. H., Roesch, W. C., and Tochilin, E. (Eds.), "Radiation Dosimetry," 2nd Ed., Vols. 1, II, III, New York, Academic Press, 1968.

Birkhoff, R. D., "The Passage of Fast Electrons Through Matter," in "Handbuch der Physik," (S. Flügge, Ed.), Vol. 34, pp. 53–138, Berlin, Springer-Verlag, 1958.

Dennis, J. A., "Further Thoughts on the *W* Values for Heavy Ions," in "Proc. 3rd Symp. Microdosimetry" (H. G. Ebert, Ed.), pp. 403–415. Luxembourg, Euratom, 1972.

Myers, I. T., "Ionization," in "Radiation Dosimetry," 2nd Ed., Vol. 1 (F. H. Attix and W. C. Roesch, Eds.), pp. 317–328, New York, Academic Press, 1968.

Schimmerling, W., Vosburgh, K. G., Todd, P., and Appleby, A., "Apparatus and Dosimetry for High-Energy Heavy-Ion Beam Irradiations," *Radiat. Res.* 65, 389–413 (1976).

Seltzer, S. M., and Berger, M. J., "Mean Excitation Energies for Elemental Substances, Compounds, and Mixtures," *Intern. J. Applied Radiation Isotropes*, 33, 1189–1218 (1982).

Weast, R. C. (Ed.), "Handbook of Physics and Chemistry," Cleveland, Chemical Rubber Co. Press, 1975.

Cross-references: ATOMIC PHYSICS, BREMSSTRAHLUNG, COMPTON EFFECT, ELECTRON, IONIZATION, NEUTRON, NUCLEAR REACTIONS, NUCLEAR STRUCTURE, PHOTOELECTRICITY, PHOTON, POSITRON, PROTON, RADIATION CHEMISTRY, RADIOACTIVITY, X-RAYS.

RADIATION, THERMAL

The term *thermal radiation* refers to the electromagnetic energy that all substances radiate, by transformation of their thermal energy. More exactly the term implies that the radiative degrees of freedom of a system are in equilibrium with the material ones, so that a common temperature has been achieved. For a closed system the radiation spectrum is a continuum, and the intensity at any wavelength is determined solely by the temperature. Viewing the universe as a closed system, its temperature is 3 K, based on observations of the isotropic background radiation at microwave and far infrared wavelength.

TABLE 5.

$\alpha(r)$ (cm)	β^-(max r) (cm)	γ ($x_{1/2}$) (cm)	Neutrons ($x_{1/2}$) (cm)
0.0007	0.4	10	5–10

Radiative cooling occurs when the energy radiated by a body is not replenished; thus it might cool to a limiting temperature of 3 K in space.

When a body's temperature is raised to approximately 500 C it appears "red-hot," due to the visible part of its thermal radiation. At lower temperatures almost all the radiation is in the infrared.

The discovery that radiation existed outside the visible spectral region was made by Sir William Herschel in 1800. He used a prism to disperse the solar spectrum onto a table in a dark room, and scanned it with a sensitive thermometer. The latter revealed the presence of infrared solar radiation beyond the red end.

Sources other than the sun were also studied. One that is of fundamental importance is an isothermal enclosure, or hohlraum, with a small hole for viewing the radiation escaping from the interior. Kirchhoff had proved, from the second law of thermodynamics, that the flux and the spectral distribution of the radiation are the same in all such enclosures at a given temperature, irrespective of the materials composing them. A related fact, known as Kirchhoff's law, is that the ratio of radiant emittance to absorptance is the same for all surfaces at the same temperature. (The radiant emittance of a surface is the integrated power radiated in all directions per unit area of surface. Definitions of other radiometric quantities, symbols and units will be found in the references.) In accordance with this law, metallic surfaces, having high reflectance and hence low absorptance, also have a low radiant emittance. Hence, the inner surfaces of the double walls of a Dewar flask are silvered, to minimize radiant heat loss.

At a given temperature, the surface having maximum absorptance will also have maximum emittance of radiant energy. Since such a surface absorbs all incident energy, it appears black, and the radiation it emits is known as "blackbody radiation." Like the pupil of the eye, an opening in the surface of a hollow body appears to be perfectly black (as long as there is no reflection back out) and the opening acts as a blackbody radiator as well as absorber. No surface can have a larger radiant emittance than a blackbody, at a given temperature.

Toward the end of the nineteenth century, quantitative studies of the magnitude and spectral distribution of blackbody radiation were made (see INFRARED RADIATION for typical curves). Stefan had found experimentally in 1879 that the radiant emittance of a blackbody, integrated over all wavelengths, is proportional to the fourth power of the absolute temperature. And, in 1884, Boltzmann gave a theoretical derivation for what is now known as the Stefan-Boltzmann law: $W = \sigma T^4$. The value of the constant σ is 5.67×10^{-12} watts cm^{-2} deg^{-4}. In 1893, Wien derived a "displacement law," one of whose implications is that $\lambda_{max} T = 2898$, where λ_{max} is the wavelength (in microns) at which maximum radiance occurs for a blackbody at absolute temperature T.

In 1896, Wien derived the following distribution law for the spectral radiant emittance W_λ of a blackbody: $W_\lambda = c_1 \lambda^{-5} \exp(-c_2/\lambda T)$. This fits the experimental observations within 1 per cent, provided λT is less than 3100, i.e., $\lambda \leqq \lambda_{max}$; but at larger values, the predicted values rapidly become too low (see Fig. 1). On the other hand, by applying the classical equipartition theorem of statistical mechanics to the radiation, Rayleigh and Jeans derived the following formula: $W_\lambda = (c_1/c_2)\lambda^{-4} T$. For $\lambda \geqq$

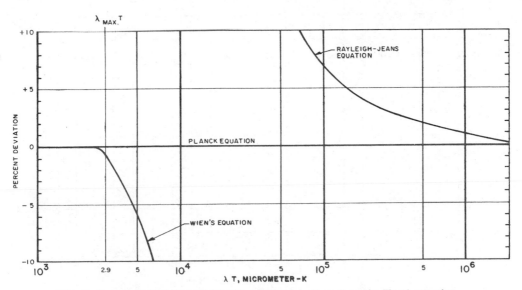

FIG. 1. Fractional deviation of classical radiation equations from the Planck equation.

$250\lambda_{max}$, the Rayleigh-Jeans equation matches the experimental data to 1 per cent, but it diverges, leading to the "ultraviolet catastrophe," as $\lambda \to 0$. It remained for Max Planck in 1900 to find an expression that is valid at all wavelengths and temperatures, namely: $W_\lambda = c_1 \lambda^{-5}/[\exp(c_2/\lambda T) - 1]$. c_1 and c_2 are known as the first and second radiation constants, respectively, and have the values $c_1 = 3.74 \times 10^{-12}$ watt cm^2; $c_2 = 1.44$ cm degree. Planck's theory leads to the following expressions for the radiation constants in terms of fundamental physical quantities: $c_1 = 2\pi hc^2$; $c_2 = hc/k$, c being the velocity of light, and k Boltzmann's constant. The validity of these expressions has been experimentally justified.

Planck first developed this law empirically, and tried unsuccessfully to justify it on the basis of classical physics. He was forced to postulate that the elementary oscillators, of which a radiating body consists, did not have a continuous distribution of energy, but only quantized values. According to the "quantum hypothesis," the energy E that an oscillator may assume is given by $E = nh\nu$, where n is an integer, and the proportionality constant h (now known as Planck's constant) $= 6.63 \times 10^{-34}$ watt sec^2. Though forced to assume that the energies of the elementary oscillators, of which the radiator is composed, could take on only discrete values, Planck considered the emitted radiation to be propagated according to classical electromagnetic theory. The quantization of the radiant energy into photons was conceived by Einstein and used by him to explain the phenomena of photoelectricity. These quantum concepts were later extended by Bohr and Sommerfeld to explain atomic spectra and by Schrödinger, Heisenberg, Dirac, and others to develop quantum mechanics. Thus, they stand as one of the most important milestones in the history of theoretical physics.

Integration of Planck's equation leads to the Stefan-Boltzmann equation. Wien's laws and the Rayleigh-Jeans law may also be derived from Planck's law under appropriate conditions. For example, when dealing with photons whose energy $h\nu$ is sufficiently small compared to the thermal energy kT, the product λT is large enough so that the exponential in the denominator of Planck's law may be replaced by the first two terms of a series expansion. This leads to the Rayleigh-Jeans law.

High-speed computers have been used to compile tables of spectral radiance and related functions for a blackbody over a broad range of wavelengths and temperatures. There are also several radiation calculators, or slide rules, that are convenient for engineering use. Some of these make use of the fact that a log-log plot of blackbody radiation vs wavelength has the same shape at all temperatures. The radiance curve for a given temperature blackbody may then be obtained by sliding the universal curve in such a way that its peak falls on the line corresponding to the Wien displacement law (see Fig. 1 in INFRARED RADIATION). It is convenient to remember that the peak radiance varies as T^5; that a quarter of the total power radiated lies between $\lambda = 0$ and $\lambda = \lambda_{max}$ (see Fig. 2); and that this power, on the short-wavelength side of the peak of the radiation curve varies as $T^{6.4}$.

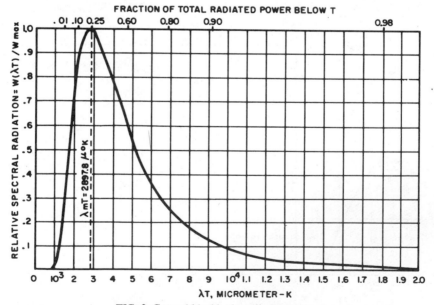

FIG. 2. General blackbody radiation curve.

At wavelengths less than λ_{max}, the monochromatic radiance changes as T^n, where $n \cong 15000/\lambda T$. For an object at 1500 K, for example, observed through an optical pyrometer with a filter transmitting at $0.6\mu m$, $n \cong 17$. This relatively high value of n is advantageous in reducing the temperature error resulting from uncertainty in the observed radiance or the emittance of a surface.

The ratio by which the surface radiation falls short of that of an ideal blackbody, is denoted as the surface's emissivity or emittance. Some authorities prefer to use "emissivity" as a material property characterizing an ideally pure and polished surface of the material, as distinguished from the properties of engineering samples. The emittance of some substances, though less than one, is essentially the same at all wavelengths. Such radiators are referred to as gray bodies. In most real substances, however, emittance varies as a function of wavelength as well as temperature.

If one can make a reasonable estimate of the emittance of a surface, a measurement of its thermal emission will give its temperature. This is the basis of contactless, radiometric temperature measurement. Should the emittance be unknown, but assumed to be relatively constant, a temperature may be inferred from the shape of the spectral distribution curve. Other schemes have also been used to determine the temperature of surfaces of undetermined characteristics.

The *color temperature* of a body is that temperature of a blackbody which has the same ratio of radiances as the selective radiator in two spectral intervals. In general, the value of the color temperature depends on the choice of the two spectral intervals. However, if these wavelengths are in the visible region, the color temperature is relatively insensitive to the specific values chosen. The color temperature of a gray body equals its true temperature. When dealing with semitransparent materials such as glass in the near infrared, one speaks of volume emissivity, which in turn is related to the optical constants of the material.

In some solids, and especially in gases and flames, the emittance in some spectral regions is much larger than in others. Kirchhoff's law tells us that gases radiate well in the same spectral regions where they have strong absorption bands. One must bear in mind, however, that the shape of an absorption band changes and new "hot" bands may appear as the gas temperature rises, so emittance and absorptance must be equated at the same temperature. The thermal radiation from a cool, low-pressure gas, is resolvable into discrete emission lines, as explained by the quantum theory. Increased temperature and pressure cause the lines to broaden as a result of molecular interactions and perturbations of the energy levels, and new lines also appear. As the optical path length increases, the emissivity of the gas approaches

unity, at first at the line centers, where the absorptance is strongest, and then extending into the line wings. Eventually, when the gas density and thickness are great enough, as on the sun, the original line spectrum assumes the appearance of a blackbody distribution. A good example of this is seen when one compares two infrared emission spectra of earth's atmosphere—first looking overhead from a mountain top, and then along the horizon at a humid seacoast. The latter spectrum's blackbody-like shape (from which, incidentally, the atmosphere's temperature may be inferred) contrasts strongly with the peaks and valleys of the former.

In strongly absorbing spectral regions, only a relatively thin layer of gas, nearest the viewer, contributes to the observed radiance, the radiation by the more distant molecules having been almost completely absorbed by the nearer molecules. For this reason, the Fraunhofer lines in the solar spectrum, occurring at strongly absorbing atomic wavelengths, originate in the cooler outer part of the sun and appear relatively dark.

In exceptional situations, it is possible to circumvent the consequences of Kirchhoff's law. The Doppler shift, due to the relative motion of the earth and Mars, prevents the narrow line radiation by planetary H_2O from being reabsorbed by terrestrial water vapor. As a result, it has been possible to estimate the water vapor concentration on Mars (see DOPPLER EFFECT and PLANETARY ATMOSPHERES).

The spectral emittance of gases is an important area of investigation, with respect to such topics as the heat budget of the earth; the composition of planetary atmospheres; and radiation by flames and rocket engines. The analysis of the gas radiation transfer in many of these applications is complicated by the fact that conditions are nonisothermal.

The Welsbach mantle (or the Coleman lantern) is an example of selective radiation in a solid; in this case, its efficiency as a source of visible light is enhanced by the high emittance in the visible, and low emittance in the near infrared, of the mixture of thoria and cerium oxide of which the source is composed. For some metals, the emittance varies as $\sqrt{T/\lambda}$, while in others it may vary in a more complicated way. In either case, the form of the spectral distribution curve differs from that of a blackbody, and the total emission will usually vary more nearly as T^5 than as T^4. The emittance of a tungsten filament is approximately 0.45 in the visible spectrum, decreasing to less than 0.2 in the infrared. Its visible efficiency increases as its operating temperature increases.

By considering the equilibrium between incoming and outgoing radiation, one can show that the sum of the radiant absorptance a, reflectance r, and transmittance t, is unity. These quantities refer to monochromatic, hemispherical radiation from the surface and do not take into account the way it is distributed geometrically. For an opaque surface, $a + r = 1$, and since by Kirchhoff's law $a = e$, the emittance, we have $e = 1 - r$. Errors may result if due account is not taken of angular and spectral factors in the application of this equation. Early studies of the angular distribution of radiation from surfaces led to the formulation of Lambert's law, i.e., $J_\theta = J_n \cos \theta$, where J_θ is the source intensity or radiant power per unit solid angle in a direction making an angle θ with the normal to the surface, and J_n is the intensity along the normal. A surface that obeys Lambert's law is said to be perfectly diffuse, like a sheet of blotting paper in the visible. Its radiance, N (intensity per unit of projected area of source), is constant and independent of θ. Integration over the hemisphere leads to the relationship $W = \pi N$. A truly black surface obeys Lambert's law exactly. Although the law is a useful approximation for many radiators and reflectors, there are numerous exceptions. The emissivity of clean, smooth surfaces of some materials has been studied from a basic theoretical viewpoint and related to their optical constants. The analysis, whose results agree generally with experimental determinations, indicates that for electrical conductors, the normal emissivity is quite low and increases with θ; whereas insulators have relatively high normal emissivity, decreasing at large values of θ. Insulators, as well as metals covered with thick oxide films, behave approximately as diffuse radiators. In the case of many practical materials, however, the surfaces are either rough or chemically complex, and it is necessary to rely upon empirically determined emittances. Many substances, such as concrete, porcelain, and paper, have a higher absorptance for long-than for short-wave radiation. Many more data are needed on the spectral emittances of various materials under different conditions. In addition to their engineering uses in radiant heat transfer calculations, emittance and reflectance data have been used to deduce the chemical composition of the moon and planets.

Space vehicles absorb energy from the sun and lose it by radiation to space. By the application of coatings with suitable radiative characteristics, the internal temperature of a vehicle may be controlled within desired limits. The quantity a/e is often used to characterize such coatings. It refers to the ratio of the absorptance of solar radiation (approximated by a 6000 K blackbody) to the emittance at the temperature of the vehicle's surface. High-temperature emittances of many materials are being studied in connection with the design of ablative nose cones, and reentry vehicles.

Although all objects are continuously emitting and absorbing thermal radiation, specially designed sources are available for particular applications. For industrial heating and drying, for example, there are numerous varieties of heaters and infrared lamps. The latter are similar to incandescent lamps used for lighting purposes, but designed for operation at lower temperatures, with reduced visible output. In recent years, high-temperature lamps with quartz envelopes have been used increasingly for both heating and illumination. The clarity of the envelope is maintained, and the filament life prolonged, by incorporating a small amount of iodine within the tube. Tungsten that has evaporated onto the tube walls combines with the iodine vapor to form tungsten iodide. This is a gas which decomposes thermally when it comes in contact with the hot filament, thus replenishing the latter and liberating the iodine for use again.

For scientific purposes, such as calibrating a radiometer or spectrometer, a reproducible source of known characteristics is essential. Sources with characteristics approaching those of an ideal blackbody, have been built for operation from cryogenic temperatures to about 3000 K. Most of the commercially available blackbodies, which come with a variety of aperture sizes and in many configurations, use a conical cavity (Fig. 3), the interior walls of which are oxidized or blackened to decrease their reflectance. Electrical heating is most often used, in conjunction with a thermostatic controller. A number of authors have discussed the considerations involved in designing a blackbody, and it is possible to calculate how closely a given design approximates an ideal blackbody. It is sometimes necessary to have a black radiating surface whose area is impractical to obtain with a cavity-type body. Flat surfaces have been coated with "blacks" whose emittances are close to unity over a broad spectral

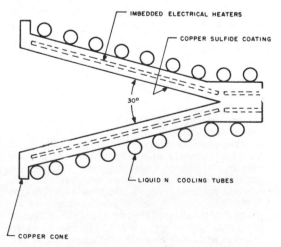

FIG. 3. Schematic representation of a conical blackbody that can be used at high or low temperature.

range. These are generally not suitable for high-temperature use, however. Other methods that have been used to obtain high-emittance surfaces include: wedges or closely stacked razor blades viewed edge on, a telescoped cone pattern similar to that of a Fresnel lens impressed on a flat surface, a vortex in a liquid or molten metal, etc. The National Bureau of Standards has calibrated some special tungsten filament lamps for use as spectral irradiance standards from about 0.3 to $2.6\mu m$. Beyond $2.6\mu m$, strong, but variable, atmospheric absorption becomes troublesome. When using infrared sources for quantitative thermal radiation measurements, one must make due allowance for the absorption at different infrared wavelengths by the CO_2 and water vapor in the atmosphere between the source and the instrument being calibrated.

For reasons of compactness and convenience, it is sometimes desirable to use non-blackbody sources of thermal radiation. Among the most common of these are the Nernst glower, which is a hollow rod approximately 25 mm long and 2 mm in diameter, made of a mixture of oxides of zirconium, yttrium, and thorium and the globar, a rod of bonded silicon carbide. These are most useful at the shorter infrared wavelengths. In the far infrared, mercury discharge tubes and other sources have been used. At long wavelengths, the power output of a thermal radiator increases almost linearly with temperature, as indicated by the Rayleigh-Jeans law, and inordinately high temperatures would be required for a significant increase in power. Nonthermal sources, in particular lasers, or other coherent radiators can generate much greater power. Lasers that can be tuned over a considerable spectral range are coming into increasing use.

In order to study the properties of materials at very high temperatures without contamination, it is convenient to heat them by thermal radiation rather than convection. Solar furnaces have been used for this purpose, as well as arc-imaging furnaces, in which a specially designed high-intensity carbon arc replaces the sun as the source.

Thermal radiation can be detected by eye when the source temperature is sufficiently high. Figure 4 shows the luminous efficiency of a blackbody as a function of temperature. And although its sensitivity is greatly reduced at longer wavelengths, the human eye has some sensitivity to radiation at wavelengths as long as $1.2\mu m$. Some insects, such as moths, respond to thermal radiation at even longer wavelengths, and some snakes can detect the heat of a warm-blooded animal at an appreciable distance.

Many types of detectors have been developed for thermal radiation. In some of these, known as quantum detectors, the energy of the incident photon must exceed some lower limit, but its effects are very rapid. An example is the lead

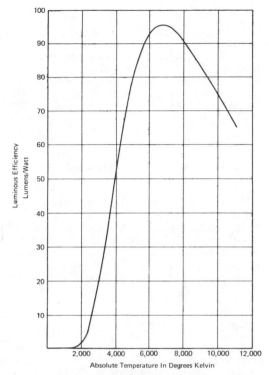

FIG. 4. Luminous efficiency of blackbody radiation (based on $K_{max} = 683$ lumens/watt). (This K_{max} corresponds to a choice of 2041 K for the freezing point of platinum.)

sulfide detector, which responds to wavelengths less than $3\mu m$, or HgCdTe, which is most sensitive to radiation around $10\ \mu m$, where objects at ambient temperatures emit most copiously. In others, thermal radiation of any wavelength can be detected by virtue of its effect on some physical property of the detector. The response of the latter kind of detector is slower than that of the former. Among the thermal detectors are: metal and thermistor bolometers, in which the absorbed radiation causes a slight temperature rise and consequent change in the resistance of the detector; thermocouples and thermopiles in which the differential heating of two junctions of dissimilar materials generates an emf; pyroelectric crystals whose polarization changes with temperature; the Golay cell or pneumatic detector, where the absorbed radiation heats a gas and distorts a reflecting optical element; the evaporograph in which the incident radiation causes differences in the rate of evaporation of a thin oil film and the resulting differences in thickness give rise to interferometric patterns. Numerous other physical mechanisms have been exploited for radiation detection. (The more usual infrared detectors are discussed under INFRARED RADIATION.)

Recent developments in detector technology include: (a) charge-coupled detectors (CCD). These are devices that arrange a relatively large number of discrete detectors in a self-scanned array, requiring only one output channel. (b) Negative-electron-affinity photoemitters, whose uses may extend the use of photomultipliers to longer wavelengths in the near IR. (c) Nonlinear heterodyne detectors for both optical and IR.

For coverage of an extended field of view, mosaics of conventional detectors have been used, as well as electronic imaging tubes such as infrared vidicons and orthicons. Whether by such detectors or by raster-like scanning of a small elemental field of view, it is possible to study the temperature pattern of an extended source of thermal radiation. Thermography is such a process, in which the temperature gradients are displayed in visible form, e.g., on photographic film, or in real time, on a cathode ray tube. When properly interpreted, the thermograph can be a valuable diagnostic aid to the physician. Thermography is also used industrially, e.g., for aerial surveys and energy audits. In Sweden, thermal scanning is required before final acceptance of a new building. Microscopic thermography is used in the design and nondestructive testing of integrated circuits. In interpreting the observed patterns, one must bear in mind that radiance variations may result from differences in either or both surface emittance and surface temperature from point to point. This fact is less troublesome in medical thermography because the emittance of the body is close to unity. In a similar way, the radiometry of lakes, rivers, and oceans by an instrument looking vertically down from an aircraft, takes advantage of the near-unity emissivity of water within the range of wavelengths to which the radiometer is sensitive. Thermal pollution can thus be detected remotely. Oil films, due to pollution, or occurring naturally in association with schools of certain fish, can also be detected; but the significant factor here is the effect of the oil on the water's emissivity. In the near infrared and visible, as well as in the microwave region, water is not quite so opaque as it is around 10 μm. Microwave radiometry, therefore, can give information about the water temperature slightly (1 or 2 mm) below the surface, rather than at the surface itself. Microwaves are less subject than infrared to atmospheric absorption and are better able to penetrate overcast and clouds.

Satellite observations of the thermal radiation of the oceans and continents are of meteorological value. For example, satellite radiometers record the radiation by atmospheric CO_2 in a series of narrow wavelength intervals centered around the strong absorption band at 15 μm. Knowledge of the varying emittances of the atmosphere at these wavelengths, together with the use of mathematical inversion techniques, permit one to compute profiles of temperature vs height on a global scale, for a variety of atmospheric conditions; these are generally in good agreement with data from radiosondes. Data from another infrared interferometric spectrometer have yielded distribution profiles for atmospheric water vapor and ozone.

Earth resources are being surveyed by radiometers in LANDSAT and orbiting SKYLAB missions, with applications to agriculture, forestry, fishing, geology, mining, and tracking sea-ice.

LEONARD EISNER

References

Klein, M. J., "Max Planck and the Beginnings of the Quantum Theory," in *Archive for History of Exact Sciences*, 1 (5), 459–479 (1962).

Pivovonsky, M., and Nagel, M. R., "Tables of Blackbody Radiation Functions," New York, The Macmillan Co., 1961.

Rutgers, G. A. W., "Temperature Radiation of Solids," Flugge, S., Ed., "Handbuch der Physik," Vol. 26, pp. 129–170, Berlin, Springer, 1958.

Gubareff, G. G., Janssen, J. E., and Torborg, R. H., "Thermal Radiation Properties Survey," Second edition, Minneapolis-Honeywell Regulator Co., Honeywell Research Center, Minneapolis, Minn., 1960.

Penner, S. S., "Quantitative Molecular Spectroscopy and Gas Emissivities," Reading, Mass., Addison-Wesley Publishing Co., 1959.

Glaser, P. E., and Walker, R. F., Eds., "Thermal Imaging Techniques," New York, Plenum Press, 1964.

Baker, H. D., Ryder, E. A., and Baker, N. H., "Temperature Measurement in Engineering," 2 vols., New York, John Wiley & Sons, 1953 and 1961.

American Institute of Physics, "Temperature, Its Measurement and Control in Science and Industry," Vols. 1–3, New York, Van Nostrand Reinhold, 1941–1963.

Touloukian, Y. S., and Ho, C. Y., "Thermophysical Properties of Matter," New York, Plenum Press, Vols. 7 and 8, 1972; Vol. 9, 1973.

Svet, D. Ya, "Thermal Radiation," New York Consultants Bureau, 1965.

Goody, R. M., "Atmospheric Radiation. I. Theoretical Basis," Oxford, Oxford University Press, 1964.

Summer, W., "Ultraviolet and Infrared Engineering," New York, Interscience Publishers, 1962.

Sparrow, E. M., and Cess, R. D., "Radiation Heat Transfer," New York, Brooks-Cole, 1966.

Wiebelt, J. A., "Engineering Heat Transfer," New York, Holt, Rinehart, and Winston, 1966.

Harrison, T. R., "Radiation Pyrometry and Its Underlying Principles of Radiant Heat Transfer," New York, John Wiley & Sons, 1960.

Keyes, R. J. (Ed.), "Optical and Infrared Detectors," Vol. 1-9, New York, Springer-Verlag, 1977.

Kingston, R. H., "Detection of Optical and Infrared Radiation," Vol. 10 of Springer Series in Optical Science, New York, Springer-Verlag, 1978.

Hudson, R. D., and Hudson, J. W. (Eds.), "Infrared Detectors," New York, Wiley, 1975.

Grum, F., and Becuerer, R. J., "Radiometry," Vol. 1 of "Optical Radiation Measurements," New York, Academic Press, 1979.

Lintz, Joseph, Jr., and Simonett, David S. (Eds.), "Remote Sensing of Environment," Reading, Mass., Addison-Wesley, 1976.

Warren, Cliff (Ed.), "Proceedings of Second Biennial Infrared Information Exchange," Secaucus, N.J., AGA Corporation, 1975.

Kangro, Hans, "Early History of Planck's Radiation Law," Crane-Russak Co., 1976.

Cross-references: DOPPLER EFFECT; HEAT TRANSFER; INFRARED RADIATION; KINETIC THEORY; LASER; LIGHT; OPTICS, GEOMETRICAL; OPTICS, PHYSICAL; PHOTON; PLANETARY ATMOSPHERES; QUANTUM THEORY; RADIO ASTRONOMY; REFLECTION; SOLAR PHYSICS; SPECTROSCOPY; STATISTICAL MECHANICS; TEMPERATURE AND THERMOMETRY.

RADIO ASTRONOMY

Radio astronomy, the study of radio-frequency radiation from celestial objects, began in 1932 with the discovery of radio emission from the galaxy by K. G. Jansky at the Bell Telephone Laboratories. Fourteen years later, the U. S. Army Signal Corps and the Hungarian scientist Z. Bay independently succeeded in obtaining radar echoes from the moon—thus launching the field of *radar astronomy*. Since the 1960s, *space radio astronomy* has investigated the low frequency radio waves that cannot penetrate the earth's ionosphere; a major achievement was the 1968 launch of the Radio Astronomy Explorer satellite.

Radio Telescopes A radio telescope is composed of one or more aerials and associated radiometers. The design depends on the intended operating frequency range. High spatial resolution, which implies large antennae or at least large separations between antennae, has proven to be perhaps the most important design goal. At first, high resolution was required because observations at meter wavelengths were limited by confusion of adjacent radio sources. More recently, high resolution at wavelengths of a few cm has been crucial in investigating the nature of the core regions of radio galaxies where jet phenomena originate.

L. L. McCready, J. L. Pawsey, and R. Payne-Scott at Sydney introduced the technique of interferometry to radio astronomy in 1946. They used the Lloyd's mirror principle, with the ocean surface as reflector, to measure the angular diameter and position of a localized source of solar radio emission.

Two-element interferometers were employed for radio sky surveys at the Mullard Radio Astronomy Observatory; others, notably those at Jodrell Bank, Owens Valley, and Nançay, yielded basic data on the brightness distributions of individual sources by means of variable-baseline interferometry. This technique later was generalized to the case where there are two or more antennae, located any distance apart, and not connected to each other in any sense. In this very *long baseline interferometry* (VLBI) a given radio source is observed simultaneously from each antenna location, with the radiometer output being recorded on magnetic tape, along with an atomic clock signal. Then, the taped data from all antennae are combined in a computer, and a resolving power corresponding to the separation of the antennae is obtained. Thus, the earth's diameter is the limiting baseline. In some cases, the signals are combined in real time with the aid of a communications satellite that provides a link to the central computer; then the complications of synchronizing clocks and combining tape recorder playback data are avoided. For higher spatial resolution at the same operating frequency, the VLBI technique can be surpassed only if it is extended into space, where larger baselines are possible. One or more of the aerials would be mounted on orbiting satellites or even on another celestial body. Such experiments have been proposed to NASA.

Aperture synthesis, a technique introduced by M. Ryle and A. Hewish at Cambridge, involves the use of multiple aerials, at least some of them movable, to simulate a larger array. Data taken with the aerials in each of several different relative orientations are combined mathematically. This technique can sample a large range of spatial frequencies and thus provides more complete two-dimensional mapping than is usually possible by means of VLBI, although the resolution is not as great. Aperture synthesis is used in the Very Large Array (VLA), currently the most powerful existing radio telescope, near Socorro, New Mexico. The VLA is an array of 27 parabolic dish reflectors, each 25 meters in diameter. At its highest operating frequency, the VLA produces radio images about as sharp as photographs made with visible-light telescopes.

The 250-foot dish at Jodrell Bank, near Manchester, was the largest fully steerable parabolic antenna until the construction of the 100-meter antenna near Bonn in 1972. In the Southern Hemisphere, the major dish antenna is the 210-foot telescope at Parkes, New South Wales. The 1000-foot antenna at Arecibo, Puerto Rico is built in a natural limestone sinkhole; it is perhaps the most important of the existing fixed reflectors and is used for radio astronomy, radar astronomy, and for remote sensing of the ionosphere.

Gain stability, phase stability, and low receiver noise temperature are the prime desiderata in a radiometer. Compensation for gain variation by switching between the object signal and a reference standard was achieved by R. H. Dicke in 1946. Many refinements of this technique have been developed. Traveling-wave tubes, masers, parametric amplifiers, and tunnel diodes have all been used for low-noise amplification, particularly in the microwave frequency range, where the average intensity of cosmic sources is low. Newer detection methods that show promise for high-frequency radio astronomy include the use of Josephson junction and other solid-state, cryogenically cooled devices. Multichannel and autocorrelation receivers are used for spectral line studies.

Cosmic Radio Sources In the solar system, radio emission has been detected from the sun, from each of the planets except Pluto, from the Moon and some other planetary satellites including Saturn's rings, from a few large asteroids, and from several comets.

Solar radio waves are observed from the quiet solar atmosphere and from transient phenomena such as active regions and solar flares. Bursts of low-frequency radio emission associated with plasma ejected by flares and erupting solar prominences have been tracked with radio telescopes on orbiting satellites as the plasma moved out from the solar corona and through the inner solar system.

The radio emission from Mercury and Mars is thermal and originates at and near the planetary surfaces. Venus is a much stronger thermal radio source than expected for a blackbody in equilibrium with solar radiation; its radio emission arises in the atmosphere, which is heated by a greenhouse-type process. At microwave frequencies, Jupiter, Saturn, Uranus, and Neptune also are stronger thermal sources than expected, which indicates that their radio emission is partially self-generated in addition to its component of reradiation of energy from sunlight. At low frequencies, Jupiter is a variable, nonthermal source, as revealed by the discovery of Jovian radio bursts in 1955 by B. F. Burke and K. L. Franklin of the Carnegie Institution. The occurrence rate of the bursts is controlled by the position of the Jovian satellite Io. In the decimeter range, Jupiter is also the source of nonthermal continuum emission associated with the planetary magnetosphere.

Lunar radio emission is thermal in character and at some frequencies it varies in strength with the cycle of lunar night and day. Analysis of the signals provided basic data on the character of the near-surface soil and rock prior to the first lunar spacecraft landings. Centimeter-wave observations of the large Jovian moons Callisto and Ganymede reveal thermal emission as expected for a bare satellite surface in equilibrium with solar radiation, but similar measurements of Saturn's large moon Titan, which has an atmosphere, indicate a higher temperature. This confirms that the surface of Titan, at positions far from the subsolar point, is warmed due to the presence of the atmosphere. The rings of Saturn are weak radio emitters and indeed absorb some of the radio emission from the planet itself, but they are strong reflectors of radar at wavelengths around 12 cm. These properties of the rings are consistent with the hypothesis that they consist of numerous lumps of ice, typically of snowball size, although larger lumps may also be present.

Radio emission is detected from some of the larger asteroids at cm wavelengths. Observations of Ceres and Bamberga suggest that the surfaces of the rocky asteroids are covered with dust layers in at least some cases. Radar measurements of asteroids at S band (wavelength 12.6 cm) and X band (3.54 cm) reveal the surface properties of radar albedo, scattering phase function, and polarization as they vary while an asteroid rotates, thus presenting different areas of its surface for observation. Radar measurements have been obtained from the largest asteroids in the asteriod belt between Mars and Jupiter and for several of the small astroids whose orbits take them within the orbit of Mars and thus fairly close to earth.

Continuum radio emission from a comet was first observed in Comet Kohoutek (1973 XII) and Comet West (1976 VI) by R. W. Hobbs, J. C. Brandt, and S. P. Maran. In each case, the emission was transient and has been ascribed to thermal radiation from the icy-grain halo, a cloud of water-ice grains around the cometary nucleus. Much effort has been invested in searches for molecular line radio radiation. The most definitive work has been done on the OH radical, whose radio transitions have been observed both in emission and in absorption (the latter against the galactic background). OH is thought to be a daughter product from the dissociation of cometary water. Studies of OH radio emission in several comets confirm the high production rates of water vapor (itself not conclusively detected) by sublimation from the ice state that are predicted by the "dirty snowball" theory of the cometary nucleus.

Among stars in the galaxy, novae, flare stars, interacting binary systems, red giants, pre-main sequence stars, and pulsars have been studied by radio astronomers. In novae, radio emission originates in the gaseous envelope ejected by the stellar explosion. Flare stars are small, cool ($T = 2900$ K, typically) objects. Their radio outbursts are of much greater absolute strength than typical solar flare events. Binary stars observed at radio frequencies, notably by R. Hjellming and C. M. Wade of the National Radio Astronomy Observatory, include the famous eclipsing systems Algol and beta Lyrae and several binaries associated with x-ray sources, including Scorpius X-1 and SS 433. SS 433, noted for its twin rotating beams that produce optical radiation with enormous and rapidly changing Doppler shifts, is an especially striking radio source. VLA observations show a characteristic corkscrew pattern in the spatial distribution of radio brightness. This is interpreted as revealing the presence of discrete radio-emitting plasmoids that follow ballistic trajectories once they are ejected from the rotating beams. Continuum radio emission from red giant stars such as Betelgeuse is attributed to their extended outer atmospheres. Pulsating red giants, such as long period and semi-regular variable stars, are often associated with molecular line emission from a circumstellar distribution of small sources of maser radiation. Molecules involved in masering include OH, water, and SiO. Weak continuum radio emission has been detected from several T Tauri stars, which are thought to be objects resembling the sun as it was roughly five billion years ago, still contracting as it approached the main sequence, and possessed of a powerful stellar wind. Pulsars, discovered in 1967 by A. Hewish and S. J. Bell at Cambridge, are sources of discrete, periodic pulsed signals, with typical duty cycles of about 5%. The pulse periods of the over 300 known pulsars range from 0.00156 sec to 3.75 sec. In general, the periods are gradually lengthening. The objects responsible for the emission do not actually emit pulses; rather, the pulsar phenomenon is due to the beaming of non-thermal radiation from rotating, highly magnetized neutron stars. As a pulsar beam sweeps along the surface of the earth (at a rate, in the case of the Crab Nebula pulsar NP 0532, of 10^{24} cm/sec!) it is detected momentarily at a given observatory, and the effect is that of a "pulse." Only about 1% of the known radio pulsars seem to be members of binary star systems (see PULSARS).

Besides the types of stars mentioned above, radio observations of our Milky Way galaxy reveal emission from the interstellar medium and clouds, the galactic cosmic ray gas, H II regions such as the Orion Nebula, planetary nebulae such as the Ring Nebula, supernova remnants such as the Crab Nebula, stellar-wind-blown interstellar bubbles such as the Rosette Nebula, and several large gaseous structures of uncertain origin, the galactic spurs or "giant loops."

At a 1944 Leiden Observatory colloquium, H. C. van de Hulst predicted that a spectral line could be observed by radio astronomers, who previously had worked only in the continuum. He referred to the 1420-MHz hyperfine transition in the ground state of neutral hydrogen, the now-famous 21-cm line. Seven years later, H. I. Ewen and E. M. Purcell at Harvard made the first successful detection of this radiation. The importance of the 21-cm line is twofold: (1) it originates in an extremely abundant constituent of interstellar matter, atomic hydrogen; (2) Doppler shifts are readily measurable

in line radiation but not in the continuum. As a result of many 21-cm surveys, notably by the Harvard, Leiden, and Australian groups, the distribution of interstellar atomic hydrogen in the galaxy was mapped and it was possible to investigate the dynamics of the interstellar gas. In 1963, A. H. Barrett, M. L. Meeks, and S. Weinreb at M.I.T. detected the 18-cm lines of OH. Since then, the line radiation of several dozen molecules, including a few ions and some interesting organics such as formaldehyde and methyl alcohol, has been discovered in radio studies of the galaxy. During the mid- and late 1970s, mm-wave surveys of CO line emission were made. CO is second in abundance among interstellar molecules to molecular hydrogen. The latter is not directly observable with radio telescopes, but its abundance is inferred from CO mm-wave line radiation. The CO sky surveys revealed that in the inner portion of the Milky Way, molecular hydrogen is the most abundant constituent of the interstellar gas and comprises more than 90% of the hydrogen present, in contrast to other portions of the galaxy, where hydrogen is mostly atomic or ionized. Much of the molecular hydrogen and CO is localized in so-called giant molecular clouds, regarded as the most massive individual structures in the galaxy. In addition, a number of recombination lines have been detected in the radio spectra of H II regions. These arise from upper levels with very high principal quantum numbers, e.g., $n = 109$ in hydrogen.

Beyond the Milky Way, the known radio sources are galaxies, quasars, and the microwave background radiation, as well as individual stellar and nebular sources in some of the nearer galaxies. Radio astronomers originally considered two categories of galaxies. A weak emitter, or ordinary galaxy (such as our own system, or M31 in Andromeda) typically radiates 10^{38} erg/sec in radio waves, compared with 10^{44} erg/sec in the optical region. The strong emitters, or radio galaxies, each produce up to 10^{45} erg/sec in the radio range alone. In both types of galaxy, the bulk of the radio continuum is accounted for by synchrotron radiation from relativistic electrons traveling through an associated weak, large-scale magnetic field. D. S. Heeschen and C. M. Wade at Green Bank made a systematic study of bright galaxies; they concluded that all normal spiral and irregular galaxies are probably weak radio sources. On the other hand, the strong emitters tend to be giant elliptical galaxies, often distinguished by peculiar optical phenomena. Another, rare type of radio galaxy is the Seyfert galaxy. These objects are spiral galaxies characterized by strong optical emission line radiation from their central regions. In the early 1960s, the interferometric studies of A. T. Moffet and P. Maltby at the California Institute of Technology and of J. Lequeux at Nançay led them to divide resolved extragalactic sources (galaxies and quasars) into three groups on the basis of brightness distribution: simple, double, and core-halo sources. C. Hazard and M. B. Mackey at Sydney used the method of lunar occultations to resolve detail as small as 0.5 second of arc and detected triple and even more complex structure in some sources. Moffet found that the emission in a given lobe of a double source tends to be concentrated at the end furthest from the other lobe, in accord with an expanding model. With the exception of this last result, nearly all the findings of the early investigations of brightness distributions in extragalactic sources have been eclipsed by modern work at much higher spatial resolution, made possible by the advent of larger interferometers, aperture synthesis, and the VLBI technique. Core-halo sources, for example, have disappeared from the astronomers' lexicon. The modern work reveals that radio galaxies possess a *central source*, generally much smaller and less luminous than the associated radio sources located to either side. Jet structures or *beams* connect the central source with the lobe of emission to either side; sometimes two or more co-aligned pairs of lobes display striking symmetry about the central source. The central source is identified as the site of the "machine," the unknown generator of high-energy particles that are ultimately responsible for the radio emission from every aspect of the radio galaxy. The "machine" is widely believed to consist of a massive black hole, powered by accretion of matter from the surrounding central region of the radio galaxy. The jet structures trace the path by which energy in the form of particles and/or waves is transmitted to the lobes. The presence of an accretion disk may account for the characteristic presence of two oppositely directed jets. Deviations of some jet pairs from simple geometry are taken to suggest the encounter of material in the jets with intergalactic plasma associated with the cluster of galaxies in which the source is located, or with motion (e.g., in a binary system orbit) of the central "machine."

VLBI observations reveal superluminal motion of individual plasmoids in some radio galaxies whose distances are known from redshifts. Apparent velocities of several to ten times the speed of light are found. These are ascribed to a sort of optical illusion due to the ejection of the emission sources nearly in the direction of the observer and at a significant fraction of light speed. The superluminal effect arises from the circumstance that the light travel time from source to observer associated with a plasmoid observed at one epoch is not the same as that of the same plasmoid observed (say) a year later.

The identification in 1963 by M. Schmidt at the California Institute of Technology, C. Hazard in Australia, and others of a new class of radio source, the quasistellar objects, now called *quasars*, had a profound impact on astrophysics. Originally thought to be peculiar stars within our galaxy, quasars are now known to be, as a class, the most distant and luminous objects in the universe. Although their radio emission typically is similar to that of a strong radio galaxy and the optical emission rates are up to 100 times greater than those of any known galaxy, they occupy volumes that are extremely small compared to typical galactic dimensions. Increasing evidence from optical studies suggests that quasars are eruptions at the centers of unseen host galaxies that are lost in the glare of the quasars in telescopic photographs. Quasars exhibit many of the same jet and superluminal phenomena described above for radio galaxies. Radio observations also show that quasars are variable in luminosity, radio spectrum, and polarization. The source of quasar energy has been regarded as a major problem. Among the possibilities considered have been gravitational collapse of large masses, a high rate of supernova explosions due to collisions in dense star clusters, the mutual annihilation of matter and antimatter, and energy derived from the slowdown of rotating, magnetized stars. Currently, it is thought that the mechanism may be accretion into a giant black hole, as supposedly occurs in a strong radio galaxy. At least for nearby and thus present-epoch quasars, collisions between the host galaxies and other galaxies may provide the accreting mass. If both strong radio galaxies and quasars are energized by a similar process in the centers of galaxies, what accounts for the distinction between quasars and radio galazies? According to one reasonable theory, quasars represent the form that the strong radio galaxy phenomenon assumes when it occurs in a spiral galaxy rather than in an elliptical galaxy. Seyfert galaxies may be weak forms of the quasar phenomenon, in which the central eruption is not bright enough to drown out the surrounding galaxy in photographs.

Radio Astronomy and Cosmology Radio astronomy has provided the single most important observation available to cosmologists—the discovery of the *microwave background radiation*. Reported in 1965 by A. A. Penzias and R. W. Wilson of the Bell Telephone Laboratories, this radiation appears to arrive in nearly equal amounts from all directions in space. It has a spectrum corresponding with high precision to that of a blackbody at 2.7 K and is almost certainly the redshifted emission from the "Big Bang" that occurred at the origin of the universe. More specifically, it is interpreted as radiation from the last moment after the Big Bang when the universe was still opaque. After that moment, the ionized hydrogen that filled the universe recombined. Since the microwave background consists of emission from the last moment of the opaque early universe, it represents radiation from the most ancient epoch that can ever be observed, and thus from the greatest observable distance in space. Precisely such radiation was predicted by G. Gamow. A slight departure from isotropy in the observed background results from the motion of our galaxy with respect to distant galaxies. Surveys of extragalactic radio sources have been analyzed for comparison with the predictions of cosmological theories as to the number of sources that can be observed in each range of flux density, the relation between the angular diameters and flux densities of sources, and similar tests. All such well established studies are consistent with the universe as an expanding and evolving system.

STEPHEN P. MARAN

References

Christiansen, W. N., and Högbom, J. A., "Radiotelescopes," Cambridge, U.K., Cambridge Univ. Press, 1969.
Verschuur, G. L., and Kellermann, K. I. (Ed.), "Galactic and Extragalactic Radio Astronomy," New York, Springer-Verlag, 1974.

Cross-references: ANTENNAS, ASTROPHYSICS, COSMOLOGY, INTERFERENCE AND INTERFEROMETRY, PLANETARY ATMOSPHERES, PULSARS, SOLAR PHYSICS.

RADIOACTIVE TRACERS

The use of radioisotopes as tracers rests on the nearly indistinguishable physical and chemical properties of all the isotopes of a given element. Proper incorporation into a material of a radioisotope that can be measured with appropriate radiation detectors provides a means of studying the behavior of the material or a component thereof. Thus, a radioisotope may be used to study the chemistry or physics of an element, a chemical compound, or a mixture of substances. For example, ^{131}I, a radioisotope with an eight-day half-life, has been used to study the distribution of iodine in multiple phase systems, to study the biochemistry of ^{131}I-tagged diiodotyrosine, and to measure the flow rates of underground streams.

Shortly after World War II, the U.S. Atomic Energy Commission made a variety of radioisotopes available from the Oak Ridge National Laboratory. Subsequently, additional suppliers have been established in the United States and in other countries. One may now purchase radioisotopes of a majority of the elements. Useful radiotracers are absent only for the elements He, Li, and B. Most radioisotopes are produced in nuclear reactors, but accelerator-produced species also comprise a useful selection. Suppliers are listed in several guides, such as the one published by the American Nuclear Society.[1]

Facilities for the formation of radioactive tracers directly within a sample or test material have also become available. For example, it is possible to form radioactive iron 59 within a sample of steel by irradiating the sample in a nuclear reactor.

The wide choice of radioisotopic tracers and the availability of sensitive detection systems to fit most circumstances have made possible the use of radiotracer techniques in many branches of science, medicine, and industry. Radiotracers have a number of features that make their use generally attractive. Unlike other types of tracers, they provide unequivocal evidence of their presence by virtue of their own radiation. Most species of radioisotopes are inexpensive (although incorporation into a specific compound can be somewhat costly), and detection equipment can be obtained at moderate expense. Also, due to the excellent detection efficiency of available equipment, it is possible to carry out most experiments without undue health hazards. Investigators trained in the safe handling of radioisotopes can follow gas, liquid, solid, or mixed-state systems at the laboratory, pilot-plant or even full plant scale with complete safety.

The following examples of radiotracer applications comprise but a partial list of uses.

Absorption of gaseous or liquid-phase constituents can be readily studied with the aid of radiotracers. Either the deposition of tracer onto the substrate or the disappearance of tracer from its initial phase may be measured.

Analysis of chemical composition may be accomplished either by utilizing the reaction of tagged reagents or by "isotope dilution." An example of the former method is the use of ^{110}Ag-tagged silver nitrate reagent in chloride determinations by the precipitation of AgCl from solution; a sharp rise in liquid-phase radioactivity indicates the point of essentially complete chloride precipitation. Isotope dilution utilizes an isotope of an element to measure the amount of the same element in a sample. It is based on the fact that the amount of radioactivity per unit weight of the element or compound (specific activity) in the tracer reagent will be decreased when the reagent is added to a solution containing the naturally-occurring element or compound. The change in specific activity is an analytical measure of the amount of the element or compound originally present in the sample.

Radiotracers may be used as analytical adjuncts, also. They provide a convenient means of checking the degree of completion of analytical steps, such as precipitation and extraction, and can provide a measure of chemical losses in analytical procedures. They may be used to mark compound locations in chromatographic separation procedures.

Many other aspects of chemistry have been elucidated with radiotracer techniques. Chemical reaction rates, equilibria, and mechanisms have been studied. Diffusion rates, exchange rates, solubility products, partition coefficients, dissociation constants, and vapor pressures have been measured. Processes due to the effects of high-energy radiation, photolysis, and catalysis have been unraveled. Many of the recent advances in biochemistry would not have been possible without the use of radioactive tracers, especially in the study of biological catalysis (enzyme reactions).

Many of the above subjects have been studied in connection with fields other than chemistry. Thus, while diffusion rates are of interest in elucidating rate-limiting chemical processes, they are also of considerable interest in electronics, metallurgy, and process industries. The self-diffusion of alloy components as a function of alloy composition and grain substructure has been investigated. Diffusion in other solid-state materials, such as semiconductors, has been measured. The specific sulfide surface area of metal sulfides supported on alumina has been determined by exchange of normal surface sulfur with ^{35}S-tagged H_2S. The rate of such exchange is of interest, and the final equilibrium state gives a measure of the surface sulfide area. These are examples of problems that are not amenable to solution with techniques other than the radiotracer method.

Industry has obtained marked economic benefits from radiotracer applications. Corrosion and wear studies can be carried out with rapidity and insight otherwise impossible. The corrosion of a steel pipe containing ^{59}Fe (produced by irradiating a section of pipe) can be followed in situ by measuring the appearance of radioactivity in the corrosive medium or by following the disappearance of radioactivity of the part. Similarly, wear of an irradiated part such as a piston ring, cylinder sleeve, or gear, can be measured in situ. Prior to the availability of radiotracers, such wear studies required frequent dismantling of machinery for weight measurements. Furthermore, it is now often possible to obtain detailed wear or corrosion patterns by autoradiography. The techniques used in wear studies have also been used to study the effectiveness of lubricants and the mechanism of wear prevention.

Radioactive tracers have been used to examine fluid processes. They are highly useful in detecting leaks and are used routinely to mark the interface between two different products moving consecutively through a pipe-line. The gamma rays from an isotope such as ^{140}Ba in soluble form at the interface can be discerned easily through the walls of the pipe. In a similar fashion, flow rates may be measured by quickly injecting a tracer into a stream and noting the time required for the radioactive pulse to travel the distance between two detectors or between the injection point and a single detector. Rapid injection and accurate timing may be avoided where the amount of radioactivity injected and the detection efficiency of a downstream detector are accurately calibrated. The total signal from the tracer is inversely proportional to the

velocity with which it goes past the detector. The techniques have been used in pilot plants, refineries, chemical plants, and even to study the flow rates and patterns of rivers and ocean currents.

The disposition of materials in various process units has been evaluated. Stream splitting, recycling, residence times, entrainment in distillations, mixing, and unit inventories have been measured.

The techniques of use in industrial fluid processes also apply in other fields. Stream splitting in capillary gas chromatography sampling units has been studied. Blood flow rate, total blood volume, and heart function are examples of medical applications.

Many of the medical uses of radiotracers have been put to routine clinical practice. The radioimmunoassay (RIA) technique is a case in point.[2] In a typical RIA, antibody to a particular compound is fixed to an insoluble matrix. A known quantity of the insoluble preparation is first incubated with a given amount of unknown solution and then incubated with a known quantity of the compound that has been labeled with radiotracer. The amount of radioactivity on the insoluble phase after rinsing will be inversely proportional to the concentration of the compound in question in the unknown. This phenomenon, which is due to competitive binding, may be calibrated to give accurate determinations.

RIA is routinely used for assaying a wide variety of peptides and proteins. These include numerous hormones, enzymes, and antigens. A variation, known as the radioallergosorbent test (RAST), wherein an antigen is linked to the insoluble phase and an unknown is measured for the presence of antibody by competitive binding with labeled antibody, is also clinically useful.

A relatively new and promising application of radiotracers is positron emission transaxial tomography (PETT).[3] PETT enables measurement of fluid flow, metabolite reaction rate, surface reactions, membrane permeability, and other phenomena. The tool has been particularly useful for investigating tumors and disorders of the heart and brain. New understanding of several mental disorders of organic origin, together with promising procedures for their diagnosis, have been provided by PETT.

H. R. LUKENS

References

Cited

1. "Buyer's Guide Eighty-One," Nuclear News, Vol. 24, No. 4, 1981.
2. Grier, O. G., daSilva, W. D., Gotze, D., and Mota, I., "Fundamentals of Immunology," Springer-Verlag, New York, 1981.

3. Dagani, R., "Radiochemicals Key to New Diagnostic Tool," *Chemical & Engineering News*, p. 30, Nov. 9, 1981.

General

Broda, E., and Schonfeld, T., "The Technical Applications of Radioactivity," Vol. 11, Oxford, Pergamon Press, 1966. Describes analytical, agricultural, hydrological, and industrial applications of tracers. Contains 2500 references.

Gardner, R. P., and Ely, R. L., Jr., "Radioisotope Measurement Applications in Engineering," New York, Reinhold, 1967. Describes radiotracer uses in engineering processes. Also, includes several chapters on radiogauging.

Cross-references: ACCELERATORS, PARTICLE; ISOTOPES; NUCLEAR INSTRUMENTS; NUCLEAR RADIATION; NUCLEAR REACTORS; RADIATION CHEMISTRY; RADIATION, IONIZING, BASIC INTERACTIONS; RADIOACTIVITY.

RADIOACTIVITY*

Radioactivity is the term applied to the spontaneous disintegration of atomic nuclei. It was one of the first and most important phenomena which led to our present understanding of nuclear structure. Credit for the discovery of radioactivity is usually given to Henri Becquerel, who made the observation in 1896 that penetrating radiation was given off by certain compounds of heavy elements in the absence of any external stimulus. Many other scientists were working in the field of radiation, however, and the announcement by Becquerel led to a flood of discoveries about the nature of the radiations which were emitted. A number of workers determined that certain of the radiations from these radioactive substances could be deflected in a magnetic field, and by 1900 three separate types of rays were identified. They were given the names alpha, beta, and gamma rays.

Distinction is frequently made between natural radioactivity, which was observed by the early workers, and artificial radioactivity which was first produced by F. Joliot and I. Curie-Joliot in 1934. These workers bombarded ^{27}Al with alpha particles to produce ^{30}P. For purposes of our discussion, we shall not distinguish between the sources of the radioactive material.

A discussion of radioactivity requires that mention be made of the stability of nuclei. Stable nuclear species or nuclides exist for all elements having proton numbers in the range from 1 to 83 except for elements 43 and 61 (technetium and promethium). In general, elements having even atomic numbers have two or more stable isotopes, whereas odd-numbered nuclei never have more than two.

The assumption is usually made that all possible nuclides were formed in the original atomic production processes, and that those which remain at the present time do so because of some inherent stability. In general, this stability involves the neutron-proton ratio, and a number of theoretical studies have been undertaken to determine the conditions for the maximum stability for nuclei.

Stability may be considered from three different standpoints: relative to the size and the number of particles in the nucleus, the ratio of the neutrons and protons in the nucleus, and the ratio of the total mass-energy of the nucleus. A nucleus which is unstable with respect to its size will emit alpha particles whereas a nucleus unstable with respect to its neutron-proton ratio may emit a negative or positive electron or may capture an electron. If a nucleus is unstable with respect to its total energy, the excess energy may be given off as gamma radiation which is electromagnetic in nature. Let us consider these emissions in more detail.

Alpha Emission It is evident that there are two types of forces existing in the nucleus. The first is a disruptive force arising from the repulsion of similarly charged particles. In addition, however, there are very strong attractive forces arising from the interactions of the nucleons. These attractive forces are very strong within the nucleus, but drop off quite sharply beyond about 10^{-12} cm from the center of the nucleus. Alpha particles, consisting of two protons and two neutrons, are, with certain exceptions, observed to come only from the larger nuclei. The mechanism of the emission process, however, is not a simple force phenomenon.

If we make calculations involving only the energies of the nucleus and the ejected particle, we should find, for example, that an alpha particle should come from a ^{226}Ra nucleus with approximately 27 MeV of energy. Instead, the emerging particle is observed to have about 5.3 MeV. This difference in energy cannot be explained using classical energy computations. The explanation of the alpha emission is usually given in terms of a "tunneling" effect, which is a quantum mechanical description first developed by Gamow and by Gurney and Condon. There are two equivalent ways of looking at this effect. One is to consider the alpha particle as being in motion inside the nucleus. In this picture, we visualize the particles as striking the potential "wall" of the nucleus. According to the quantum mechanical treatment for a particle striking such a barrier, there is a finite probability of passing through the barrier and appearing on the outside. Calculations making use of this probability give correct values for the observed half life and energies of the alpha particles from radioactive nuclides.

The other description of this process considers the alpha particle as being a wave packet with a very high probability of being found within the nuclear radius, but also having a finite probability of being outside the nucleus. According to

*Editor was unable to locate author. Article is reprinted from first edition.

this notion, the probability of finding the alpha wave packet at a distance greater than the nuclear radius, likewise gives the proper lifetime and energy values. It can be shown that this type of radioactivity is more probable in elements of high atomic number.

Beta Radiation There are over 800 nuclear species which have been artificially produced in the laboratory. Nearly all of these have been produced by the reactions which give rise to a net gain or loss of neutrons from the stable nuclei. In such a case the residual nucleus is characterized by having a neutron-proton ratio higher or lower than stable nuclei of that element. Similarly, if one considers alpha-emitting nuclei as being the "stable" nuclides for elements above lead, certain nuclides may be formed after the alpha emission with neutron-proton ratios higher or lower than the original element. The general decay process for nuclides with differing n/p ratios involves the reorganization of the nucleus in a manner which will leave a nucleus having a neutron-proton ratio corresponding to that requisite for stability.

Let us first consider the case in which a nucleus has a neutron-proton ratio higher than a stable isotope of that element. For example, the nuclei of all stable phosphorous atoms contain 15 protons and 16 neutrons. If a nuclear reaction takes place which leaves a nucleus with 15 protons and 17 neutrons we have a nucleus of ^{32}P. Since only ^{31}P occurs in nature, we know that some adjustment will take place to bring about stability. In this case, one of the neutrons is transformed into a proton, a negative electron, and a neutrino (specifically an anti-neutrino) which is shown in the following reaction:

$$n \rightarrow p + e^- + \bar{\nu}$$

The negative electron cannot exist as part of the nucleus and is ejected from the nucleus along with the neutrino. When the electron is investigated, it is found to be identical to other negative electrons, but when it is formed in this process, it is given the name negatron or negative beta particle. The residual nucleus then contains 16 protons and 16 neutrons. The element possessing 16 protons is sulfur, so the nucleus resulting from this negative beta emission is ^{32}S. In general, then, nuclei with neutron-proton ratios higher than stable nuclei eject a negative electron from each nucleus and are thus transmuted into nuclei of the next higher atomic number.

There are also types of nuclear reactions which may leave a nucleus with a neutron-proton ratio lower than the corresponding stable nucleus. In this type of instability, there are two processes which may compete with one another for the production of a stable nucleus. In the first of these, an extranuclear electron may be captured by the nucleus and combined with one of the protons. The most likely electron taking part in this process is one from the K level. L or M level electrons may be "captured," however. This electron capture reaction may be shown as follows:

$$p + e^- \rightarrow n + \nu$$

A neutrino is also ejected in this process. When this reaction takes place, the nucleus then contains one less proton than before, and is thus a nucleus of one lower atomic number. For example, if ^{55}Fe "decays" by electron capture, the resulting nucleus of ^{55}Mn is stable. This process can take place whenever the neutron-proton ratio is too low for stability.

The reaction which competes with electron capture may occur if the neutron-proton ratio is low, and if there exists a certain minimum mass-energy difference between the unstable nucleus and a possible stable nucleus having one less proton. If there is at least 1.02 MeV mass-energy difference a proton may transform into a neutron, a positive electron (or positron), and a neutrino according to the following equation:

$$1.02 \text{ MeV} + p \rightarrow n + e^+ + \nu$$

The positron is then ejected from the nucleus. The characteristics of the positron are identical to those of the negatron except for its positive charge. A number of cases are known in which both electron capture and positron emission processes take place in the same nuclear species. A radioactive nuclide is characterized as having a certain "branching ratio" when more than one type of decay is possible.

Gamma Radiation The usual modes of decay which involve the reorganization of the nucleus are those described above. In many cases, however, another step is involved in attaining final stability. After one of the nuclear transformations described above takes place, the nucleus may still possess excess energy. In this case, the extra energy is given off directly as gamma radiation. These are electromagnetic radiations which have energies corresponding to the difference in energy levels in the nucleus from which they come. A particular nuclide thus exhibits a certain pattern or disintegration scheme by which it decays. For example, ^{32}P decays by negative beta emissions, which are not followed by gammas. On the other hand, ^{60}Co emits negative beta radiation which is followed in each case by two gamma rays in cascade. These gamma radiations have energy of 1.17 and 1.33 MeV respectively. In general, the gamma radiations are emitted in time periods less than 10^{-12} seconds following the first transmutation step. In some cases, however, excited energy states may exist for significantly longer periods. Experimental determinations of the lifetimes of these slower gamma ray transitions range from 10^{-8} second to several months. If such an excited state exists in a nuclide for a period long enough to be measured experimentally, the nuclide is called a nuclear isomer, and the transition process involving such gamma radiation is called an isomeric transition (I.T.).

It should be mentioned also that gamma radiation is given off following nuclear reactions. Such gamma rays called "capture" or "prompt" are discussed in conjunction with nuclear reactions, although isomeric states are often formed in this manner, and isomeric transitions may leave the nuclei in radioactive rather than stable states.

Decay Schemes and Units Information as to the radioactive transitions which take place in a given case is frequently presented in what is called a decay scheme. A group of simple decay schemes is shown in Fig. 1. These indicate the type of radioactive process which the nucleus undergoes, the energies of the radiations given off, and the branching which may take place. Most nuclei have decay schemes which are much more complex than those represented in the Figure, but they involve only multiple occurrence of the processes which have been described.

Decay Rate It can be seen from the foregoing that a nucleus may change from an unstable to a stable form by one of several decay processes. We can thus speak of the decay rate of a sample of radioactive material in disintegrations per unit time—usually disintegrations per second. This refers specifically to the transformation of the nucleus, and does not give any indication of the kind or energy of the radiations emitted. The unit of activity is the "curie" and is defined to be 3.7000×10^{10} disintegrations per second. Submultiples or multiples of this unit in common use are micro-, milli-, kilo-, and megacurie. A sample containing one millicurie of radioactive material is a sample which decays at the rate of 3.7000×10^7 disintegrations per second.

We have no way to determine when any given nucleus will decay, but some of the most important work in the early study of radioactivity involved the study of decay rates. It was shown very early that the rate of radioactive decay is proportional to the amount of the radioactive material present. This can be expressed in the following equation:

$$A = - \lambda N$$

in which A is the disintegration rate, N is the number of radioactive atoms present, and λ is a proportionality constant characteristic of each radioactive species. This relationship is sometimes important, but often a more useful relationship is the following:

$$N = N_0 e^{-\lambda t}$$

in which N_0 is the number of atoms present at some reference time, and N is the number of

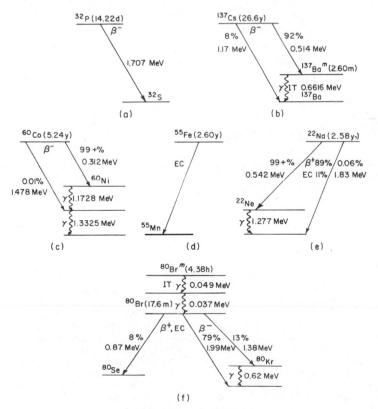

FIG. 1. Decay schemes.

atoms present at some time t later, and e is the base of the natural logarithms. This exponential decay relationship is of fundamental concern in working with radioactive materials, even though it is valid only for a statistically large number of atoms (i.e., all unstable nuclei will eventually decay).

The above equations are valid for any single radioactive species, and the exponential expression states that a given fraction of nuclei will decay in a given time period. In practice, we frequently refer to the time required for one-half of the atoms to decay. This is called the half-life of the radioactive material. It is related to the decay constant in the following way:

$$T_{1/2} = \frac{0.693}{\lambda}$$

It can also be shown that the average life, $\tau = 1.44$ times the half-life.

There are many cases in which a series of radioactive steps takes place. For example, ^{238}U undergoes 14 successive decays before arriving at the stable end product, ^{206}Pb. These steps involve the emission of successive alpha and beta rays along with gamma radiation in some cases. Although the calculations are somewhat more complex, it is possible to determine the disintegration rate of each of the daughter radioactive species formed in these processes. Details of this type of calculations are given in the references below.

RALPH T. OVERMAN

References

Evans, R., "The Atomic Nucleus," New York, McGraw-Hill Book Co., 1955.

Glasstone, S., "Sourcebook on Atomic Energy," Second edition, New York, Van Nostrand Reinhold, 1958.

Lapp and Andrews, "Nuclear Radiation Physics," Third edition, Englewood Cliffs, N.J., Prentice-Hall, 1963.

Overman, R. T. and Clark, "Radioisotope Techniques," New York, McGraw-Hill Book Co., 1960.

Cross-references: ELECTRON, NEUTRINO, NEUTRON, NUCLEAR RADIATION, NUCLEAR STRUCTURE, NUCLEONICS, POSITRON, PROTON, TUNNELING.

RADIOCARBON DATING

Radiocarbon dating is a method of deriving the age of Holocene and late Pleistocene organic materials through a measurement of residual ^{14}C activity. Radiocarbon (carbon-14, C-14, ^{14}C) measurements can be obtained on a wide range of carbon-containing samples including wood, charcoal, marine and freshwater shell, bone and antler, peat and organic-bearing sediments, carbonate deposits such as caliche, marl, and tufa, as well as dissolved CO_2 and carbonates in ocean, lake, and groundwater sources. While the impact of the ^{14}C dating method has been most significant in archaeological research (particularly prehistoric studies), extremely important contributions have also been made in hydrology, oceanography, and geophysical studies. In addition, during the decade of the 1950s, the testing of thermonuclear weapons added large amounts of artificial ^{14}C ("bomb ^{14}C") to the atmosphere, permitting ^{14}C to be used as a geophysical tracer. The dating method was developed immediately following World War II at the University of Chicago by Willard F. Libby and his co-workers James R. Arnold and Ernest C. Anderson. Libby received the 1960 Nobel Prize in Chemistry for the development of the technique.

Basis of the Method Carbon has three naturally occurring isotopes. Two (^{12}C and ^{13}C) are stable. However, the third, ^{14}C, decays by very weak beta decay (electron emission) to ^{14}N with a half-life of approximately 5700 years. Naturally occurring ^{14}C is produced as a secondary effect of cosmic ray bombardment of the upper atmosphere (Fig. 1). As $^{14}CO_2$, it is distributed on a world wide basis into various atmospheric, biospheric, and hydrospheric reservoirs on a time scale much less than its half-life. Metabolic processes in living organisms and relatively rapid turn over of carbonates in surface ocean waters maintain ^{14}C levels at approximately constant levels in most of the biosphere. To the degree that ^{14}C production has proceeded long enough without significant variation to produce an equilibrium or steady-state condition, ^{14}C levels observed in contemporary materials may be used to characterize the original ^{14}C activity in the corresponding carbon reservoirs. Once a sample has been removed from exchange with its reservoir (as at the death of an organism) the amount of ^{14}C begins to decrease as a function of its half-life.

A ^{14}C age determination is based on a measurement of the residual ^{14}C activity in a sample compared to the activity of a sample of assumed zero age (a contemporary standard) from the same reservoir. The relationship between the ^{14}C age and the ^{14}C activity of a sample is given by

$$t = 1/\lambda \ln (A_0/A_s) \qquad (1)$$

where t is radiocarbon years B.P. (B.P. = before the present, where A.D. 1950 = 0 B.P.), λ is the decay constant of ^{14}C (related to the half-life $t_{1/2}$ by the expression $t_{1/2} = 0.693/\lambda$), A_0 is the activity of the contemporary standard, and A_s the activity of the unknown age sample. "Conventional radiocarbon dates" are calculated using this formula using an internationally agreed half-life value (the "Libby half-life" of 5568 ± 30 years) and a specific contemporary or modern standard. Most laboratories define the contemporary standard value by using an

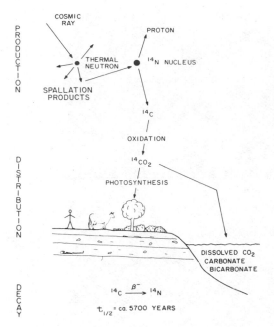

FIG. 1. Production, distribution, and decay of radiocarbon. Although the most probable half-life of radiocarbon is approximately 5730 years, by international convention, radiocarbon age determinations are calculated by laboratories using the "Libby half-life" of approximately 5570 years.

oxalic acid sample prepared by the United States National Bureau of Standards (NBS) or a secondary standard with a known relationship to the NBS standard. The ^{14}C activity of the contemporary biosphere is taken as 0.95 of the counting activity of NBS oxalic acid.

Measurement of Radiocarbon The naturally occurring isotopes of carbon occur in the proportion of approximately 98.9% ^{12}C, 1.1% ^{13}C, and 10^{-10}% ^{14}C. The extremely small amount of radiocarbon in natural materials is one reason why ^{14}C was one of the isotopes which was produced artificially in the laboratory before being detected in natural materials. The routine development of the radiocarbon method was made possible by the development by Libby of a practical method of low-level counting. In all of Libby's early work, the sample was converted to elementary carbon and deposited on a sleeve which fitted inside a screen wall type of Geiger counter. With solid carbon counting the maximum age that could be measured was about 23,000 years and required the use of from 5 to 6 grams of carbon from sample materials. Because of self-absorption of the weak betas in the sample, the efficiency of the detector was only about 5%. Because of this low efficiency and the susceptibility of the elementary carbon to contamination from airborne radioactive fallout, this technique was

replaced by either gas counting or liquid scintillation systems.

Currently gas counters are employed for ^{14}C measurements typically using carbon dioxide, acetylene or methane. As with the case with solid carbon systems, the center counter containing the sample gas is surrounded by individual Geiger tubes or an annular or continuous ring guard, all housed within an iron or lead shield assembly (Fig. 2). Because of the 90–98% efficiency in most gas counting systems, the typical maximum age limits can be extended to between 40,000 and 60,000 years depending on the experimental configuration. Counting times typically range from 1 to 4 days. Isotopic enrichment of sample gases permits the maximum age attainable to be extended several additional half-lives. Sample size requirements with gas detectors were generally reduced from that required with the solid carbon method—special detectors recently being designed to permit the measurement of as little as 10 milligrams of carbon. Liquid scintillation systems are also used to measure natural ^{14}C concentrations. This currently involves the conversion of sample material to benzene through a series of chemical steps. The addition of a scintillator chemical allows the beta decay events to be monitored by photomultiplier tubes. Typically, the liquid scintillation method has required somewhat larger amounts of sample materials. However, maximum ages can usually be extended several half-lives beyond that of the typical gas system.

Both of the current conventional decay counting methods—gas or liquid scintillation—share a common problem in that they employ an inherently inefficient means of measuring ^{14}C concentrations. In one gram of contemporary carbon, for example, there are approximately 6×10^{10} atoms of ^{14}C. However, on the average, in one minute only about 15 of them will decay. It has always been recognized that the higher efficiencies of atom-by-atom detection such as that employed in mass spectrometers would allow radiocarbon laboratories to use much smaller samples and, at the same time, extend the maximum ages resolvable beyond that typically possible with decay-counting techniques. Previous attempts to make direct counting measurements using a conventional mass spectrometer failed in practical applications because the radioactive species was obscured by background isotopes in the mass spectrum. The central idea that made direct detection possible was to accelerate sample atoms to high energy. When this occurs, the separation and identification of an isotope of interest is readily made using standard techniques developed in nuclear physics. The ionization loss of a high energy ion in matter provides a measure of the nuclear charge. When combined with magnetic or electrostatic analysis, the required separation and identification can be accomplished.

FIG. 2. Sample counter and annular anticoincidence (guard) counter used for low-level radiocarbon measurements. Gas sample is introduced into central counter through glass connection. Surrounding the sample counter is concentric lead shroud. Surrounding both the sample counter and shroud is an annular or concentric guard counter. Sheets of paraffin in which has been mixed borax acid are used to absorb cosmic ray mesons. These have been placed above and below the counter assembly. All have been encased in a lead and steel plate shield. (Photograph courtesy of T. Linick and L. E. Ford, Scripps Institution of Oceanography, University of California, San Diego.)

Most of the facilities involved in direct ^{14}C measurements currently employ tandem accelerator mass spectrometer (TAMS) systems. The elements of this type of instrument are represented in Figure 3. In such a system a significant separation of the ^{14}C from other isotopes takes place in the negative ion source before acceleration and takes advantage of the fact that ^{14}N apparently does not form negative ions that live long enough to pass through the accelerator. Samples are converted to elementary carbon to maximize the produc-

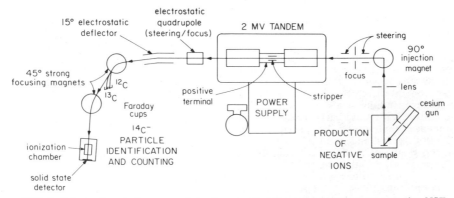

FIG. 3. Schematic of elements of tandem accelerator mass spectrometer at the NSF Accelerator Facility for Radioisotope Analysis, University of Arizona. Prepared with the assistance of D. Donahue and A. J. T. Jull, Department of Phusics, University of Arizona.

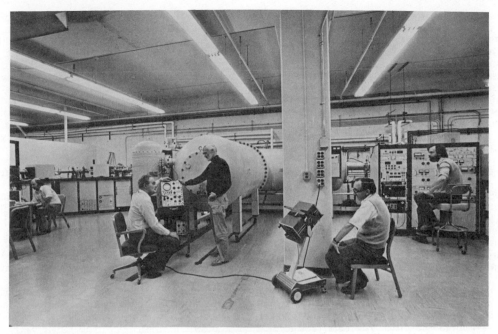

FIG. 4. Tandem accelerator mass spectrometer (TAMS) used for direct or ion counting of ^{14}C at the National Science Foundation Accelerator Facility for Radioisotope Analysis, University of Arizona, Tuscon, Arizona. Photograph oriented as in Fig. 3, with the ion beam proceeding from right to left. The 2 MV Tandem with its power supply appears in the center of the photograph. Principal personnel on the TAMS project include, from left to right, Dr. A. J. T. Jull, Dr. K. H. Purser (General Ionex Corporation), Prof. P. E. Damon, Prof. D. J. Donahue, and Dr. T. H. Zabel. (Photograph courtesy of G. Kew, University of Arizona.)

tion of ions in the accelerator. Figure 4 is the TAMS instrument operating at the University of Arizona National Science Foundation Accelerator Facility for Radioisotope Analysis.

Because of the greatly increased efficiency of ion counting systems, sample sizes can be reduced by several orders of magnitude from that typically required by decay counting instrumentation. Sample sizes down to less than 1 milligram of carbon are possible. The time needed for measurement is variable, in part depending on the amount of sample available, since, unlike decay counting systems, the sample is consumed during measurement. Maximum sample age values that can be measured using particle accelerators are limited by the background present in the instrument being used. Achieving background reduction and instrumental stability is one of the major problems currently being addressed by those operating accelerator systems.

Accuracy of Radiocarbon Determinations

The measurement of the residual ^{14}C activity in an organic sample will provide an accurate determination of its true age if it can be assumed that (1) the production of ^{14}C by cosmic rays has remained constant long enough to establish a steady-state or equilibrium in the $^{14}C/^{12}C$ ratio in the atmosphere; (2) there has been complete and rapid mixing of ^{14}C throughout the various atmospheric, hydrospheric, and biospheric carbon reservoirs; (3) the carbon isotope ratio within sample materials has not been altered except by ^{14}C decay since they ceased to be part of one of the carbon reservoirs and; (4) the total amount of carbon in all reservoirs has remained essentially constant. In addition, the half-life (or decay constant) of ^{14}C must be known with sufficient accuracy and it must be possible to measure natural levels of ^{14}C to appropriate levels of accuracy and precision. Finally, there must be a direct and specific association between a sample to be analyzed and the event or phenomena for which temporal assignment is desired.

Studies by specialists in many different disciplines over the last three decades have shown that the primary assumptions on which the method rests have been violated both systematically and to different degrees for particular sample types and environmental contexts. Various strategies have been developed to provide calibration and correction procedures for ^{14}C determinations.

Variability in Radiocarbon Production Rates. Hints of systematic discrepancies between ^{14}C

values on known-age samples were reported very early in the ^{14}C literature. Almost all ^{14}C determinations obtained on early Egyptian archaeological materials yielded age values consistently too young by about 300–600 years. In the late 1950s, ^{14}C determinations were obtained on a series of tree ring samples which showed consistent variation in ^{14}C activity in the atmosphere on the order of several percent over the last 1300 years. Primarily because of the geophysical implications of these deviations, a number of laboratories and researchers have directed their attention to the magnitude and extent of what has come to be called the *secular variation* (long term) and *De Vries* (short term) variations in ^{14}C values over time.

The data base which has contributed most directly to the intensive study of secular variation and De Vries effects has been provided by dendrochronologically dated wood from the giant California sequoia (*Sequoia gigantea*), the European oak (*Quercus sp.*) and, most importantly, the bristlecone pine (*Pinus longaeva*). Radiocarbon determinations on dendrochronologically dated wood, have been undertaken by several laboratories. The data produced as part of these studies have provided radiocarbon specialists with the ability to *calibrate* individual ^{14}C determinations. Figure 5 provides an example of a plot of such data assembled by Klein et al. (1980).

The most obvious feature of these data is the long-term *secular variation effect*. It has been suggested that this phenomenon exhibits a sine wave function with an apparent period of about 8500 to 9000 years, and a maximum deviation of ca. 8–10% (650–800 years) at about 4500 to 5000 B.C. In examining Fig. 5 one can appreciate the observation that "radiocarbon years" and "calendar years" are not always equivalent. If such had been the case, all of the data points would lie along the thin horizontal line in Figure 5. However, some of the points lie below the line, indicating that the ^{14}C values for that period are "too old" while those lying above the line are "too young."

Most researchers agree that the most probable cause for most of the secular variation effects are changes in the intensity of the earth's geomagnetic dipole field. A decrease in the intensity of the field strength, for example, would allow an increase of the cosmic ray flux in the vicinity of the earth and therefore an increase in the production rate of [14]C.

For the period where bristlecone pine calibration is currently available, several investigators have published tables or charts which are intended to assist those interested in calibrating their ^{14}C determinations in light of primarily the bristlecone pine values. Such tabulations are useful if appropriate caution is exercised. First of all, the calibration process assumes that all other physical effects responsible for affecting the accuracy of the ^{14}C values have been identified and appropriate correction values applied. For example, inasmuch as calibration data is based on samples from the terrestrial biosphere (tree-rings), it is imperative that if one wishes to calibrate samples from other car-

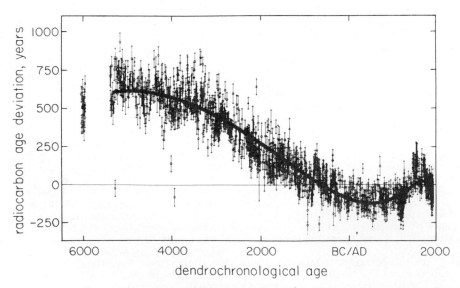

FIG. 5. Secular variation effect: relationship between ^{14}C and dendrochronological data. The ^{14}C data are plotted in terms of ^{14}C age deviation in years from the age of the dendrochronologically dated wood samples. After Klein et al. (1980), with a change in the vertical units. (P. Damon, personal communication.)

bon reservoirs that these values be first corrected in light of documented variations in initial ^{14}C activities in these reservoirs as well as any ^{13}C/^{12}C variations. Even more important is the need to take cognizance of intervals where shorter-term or *De Vries effects* made calibration efforts less precise.

If we examine closely the plots in Fig. 5 we note the short-term or high-frequency variations. It was initially suggested that these anomalies were artifacts of statistical fluctuations in the ^{14}C data. Although there is now a consensus that De Vries effects are real geophysical phenomena, there is still some uncertainty as to the cause(s) as well as the magnitude and duration of these episodes in the period before 1200 years ago. Figure 6 presents a plot of the De Vries anomalies over the last 7,000 years based on the composite bristlecone pine ^{14}C data prepared by Klein et al. (1980). This plot results from the subtraction of the main trend of the sine wave function (describing the secular variation effect) from each data point. The residual values can then be used to characterize the short term anomalies. Several researchers have obtained ^{14}C determinations on tree ring samples with shorter time segments from several other localities. These can be used to compare the magnitude and duration of these anomalies. While these show a general correspondence, there are several intervals during which some discordance seems to be indicated. Thus, some small latitude dependent variation may also be present.

A number of factors have been offered to account for much of the De Vries variations. Probably most of the effects are caused by solar or heliomagnetic modulations in ^{14}C production. It has been suggested that the process involves changes in the solar corona which affects the solar wind, which in turn deflects cosmic rays in the vicinity of the earth. The most obvious physical feature associated with variations in solar activity are changes in the number of sunspots over time. Observations made over the last few centuries have recorded sunspot numbers and these records have been compared to variations observed in the bristlecone pine tree ring record. Although good correlations can be obtained, a much better fit of the data has been achieved by correlating variations in ^{14}C activity with an index reflecting short-term low-frequency geomagnetic variations on the sun.

Variability in Radiocarbon Distribution. One of the major advantages of the ^{14}C method is its potential ability to provide directly comparable age determinations on a worldwide basis for all organic materials. For this potential to be realized, however, ^{14}C must be rapidly and uniformly mixed throughout all carbon-containing reservoirs. If such conditions prevail, the contemporary ^{14}C content of all organic samples will be essentially identical. However, in a number of situations, it has been determined that such an assumption cannot be made. The classic illustration of this problem is the determination that living organisms from a fresh water lake with a limestone bed exhibited apparent ^{14}C ages of up to 2,000 years. In this case, the ^{14}C activity of these organisms have been diluted by carbonates derived from the limestone. From a ^{14}C standpoint, this lime-

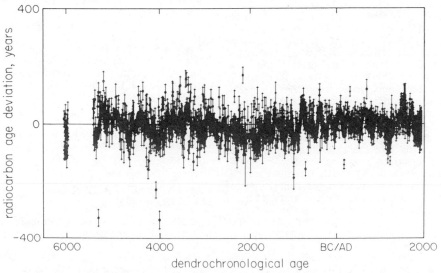

FIG. 6. The De Vries effect. High frequency or short term variation in ^{14}C activity. Plot results from the subtraction of secular variation function (Fig. 5). After Klein et al. (1980), with a change in the vertical units. (P. Damon, personal communication.)

stone was "dead" since, being of great geologic age, it did not contain any measurable ^{14}C.

One effect of the recognition that a sample's general and specific geochemical environment can affect initial ^{14}C concentrations has been to focus attention on the reliability of certain types of samples. For example, the use of shell in ^{14}C dating has been the subject of continuing discussion. Terrestrial shells or gastropods from fresh water sources certainly merit concern since they can take up carbonates which are not in equilibrium with atmospheric ^{14}C. On the other hand, well-preserved marine shells from open ocean environments have increasingly been employed in ^{14}C studies in areas where upwelling effects along continental margins have been examined.

Variability in Carbon Isotope Ratios. For the ^{14}C dating method, the basic physical measurement used to index time is the ^{14}C/^{12}C ratio. An accurate estimate of the age of sample using the ^{14}C method assumes that no change has occurred in the natural carbon isotope ratios except by the decay of ^{14}C. Several physical effects other than decay, however, have been shown to alter the carbon isotope ratio and thus have introduced the need to correct the ^{14}C values affected. One problem has to do with the fractionation of carbon isotopes under natural conditions. A second problem involves situations where carbon-containing compounds not indigenous to a sample are physically or chemically added, resulting in the sample's contamination.

While all of the isotopes of carbon follow the same chemical or physical pathways, the rate at which this occurs varies as a function of their difference in mass. The studies of Harmon Craig in the early 1950s pointed to the need to consider variations in the stable carbon isotopic ratios (^{13}C/^{12}C) if one wished to obtain precise and comparable ^{14}C/^{12}C measurements. Fortunately, most of the stable carbon isotope values on materials such as charcoal and wood exhibit little variation. Thus little, if any, correction in the ^{14}C ages values is normally required. Problems arise, however, when ^{14}C values from a variety of sample types such as grasses, grains, seeds, succulents, marine carbonates, and terrestrial organics must be compared. In such cases, it is necessary to measure the ^{13}C/^{12}C ratios and use this data to normalize the ^{14}C values onto a common scale.

Contamination of materials intended for ^{14}C analysis presents a problem for some types of samples. Obviously, the sources and effects of the introduction of foreign organics are complex depending on the nature and condition of the sample material, the characteristic(s) of the geochemical environment(s) within which the sample has been embedded, and the time frame over which such action(s) occurred. A range of sample pretreatment techniques have been developed to deal with different types of potential contamination. The high rating given to charcoal and wood is based on the fact that such samples can be subjected to pretreatment by strong acids and bases, and thus the removal of absorbed carbonates and soluble soil organic compounds can be greatly facilitated.

Development of Other Quaternary Dating Methods. ^{14}C values have been extensively employed to provide known-age reference points to facilitate the development of other dating methods applicable to Quaternary materials. Its use, for example in the development of the amino acid racemization method for late Quaternary bone and shell samples has been crucial. Concordance of ^{14}C and amino acid racemization-deduced values on bone samples was instrumental in the initial acceptance, by some scientists, of the validity of somewhat controversial racemization-deduced age values on certain samples. More recent studies, including an increasing number of clearly anomalous racemization determinations, has resulted in an increasing recognition that a number of problems with the method need to be resolved. This same process has been involved in the research on the obsidian hydration method.

R. E. TAYLOR

References

Berger, R., and Suess, H. E. (Eds.), "Radiocarbon Dating," Proceedings of the Ninth International Radiocarbon Conference, Berkeley, Univ. California Press, 1979.

Hedges, R. E. M., "Radiocarbon Dating with an Accelerator: Review and Preview," *Archaeometry* 23, 3–18 (1981).

Klein, J., Lerman, J. C., Damon, P. E., and Linick, T., "Radiocarbon Concentration in the Atmosphere: 8000-Year Record of Variations in Tree Rings," *Radiocarbon* 22, 950–961 (1980).

Stuiver, M., and Kra, R. S. (Eds.), "Proceedings of the Tenth International Radiocarbon Conference," *Radiocarbon* 22, 1–561 (1980).

Taylor, R. E., "Radiocarbon Dating: An Archaeological Perspective," in "Archaeological Chemistry II," pp. 33–69, Washington, D.C., American Chemical Society, 1978.

Cross-references: COSMIC RAYS, ISOTOPES, MASS SPECTROMETRY; NUCLEAR INSTRUMENTS, RADIOACTIVE TRACERS, RADIOACTIVITY.

RAMAN EFFECT AND RAMAN SPECTROSCOPY

The Raman effect is the phenomenon of light scattering from a material medium, whereby the light undergoes a wavelength change in the scattering process. For a given medium, the Raman scattering per unit volume is of the order of one-thousandth of the intensity of the ordinary or Rayleigh scattering, in which there is no change of wavelength (see LIGHT SCAT-

TERING). The Raman-scattered light bears no phase relationship with the incident light, whereas the Rayleigh light is a residual effect resulting from the departure of the incident and the scattered light from complete mutual coherence. The Raman intensity per molecule is thus independent of the state of the medium, apart from a certain small refractive index effect. The Rayleigh intensity per molecule, on the other hand, depends strongly on the degree of randomness of the spatial positions and orientations of the molecules of the medium; it is small for a crystal at absolute zero and greatest for a gas at low density.

Scattering of light with change of wavelength was predicted in 1923 by Smekal, inspired by the discovery of the Compton effect. The Raman effect was discovered experimentally in 1928 in India by Raman and Krishnan, who showed that the spectrum of the scattered light of liquids and solids, strongly illuminated with monochromatic light, contains frequencies which are not present in the exciting light and which are characteristic of the scattering medium. Independently, and almost simultaneously, Landsberg and Mandelstam in Russia discovered the effect in crystals.

Until the advent of lasers in the early 1960s methods of observing Raman spectra were usually modifications of the arrangement introduced originally by R. W. Wood. The specimen, for example, a liquid contained in a glass or quartz tube 1 cm in diameter and 20 cm long, was strongly illuminated along its length by mercury arcs, with filters interposed between the arcs and the tube to isolate monochromatic exciting radiation (435.8, 404.7, or 253.7 mm) if necessary. The scattered radiation was observed along the axis of the tube through a plane window at one end. The other end of the tube was usually drawn out into a cone,

which, when blackened, formed a dark background against which the weak scattered light could be observed with a minimum of stray radiation. The scattered light was analyzed by a spectrograph or spectrometer of high light power. Modifications of the Wood arrangement involving high-intensity water-cooled mercury arcs and Raman tubes with a very long pathlength achieved by a mirror arrangement were introduced around 1950, and used for extensive investigations of the rotational and vibrational Raman spectra of gases. These developments have been summarized by Stoicheff.[1]

The experimentally confirmed laws of Raman scattering are as follows:

(a) The pattern of Raman lines, expressed as *frequency* shifts from the exciting line, $\Delta\nu_i$ ($i = 1, 2, \cdots$), is independent of the exciting frequency.

(b) The pattern of Raman frequency shifts, $\Delta\nu_i$, is symmetrical about the exciting line. However, the lines on the low-frequency side of the exciting line (Stokes lines) are always more intense than the corresponding lines on the high-frequency side (anti-Stokes lines). The ratio of the intensities of corresponding anti-Stokes and Stokes lines is $I_a/I_s = \exp(-\Delta\nu_i hc/kT)$, where the Raman shift $\Delta\nu_i$ is expressed as usual in cm^{-1}. Thus anti-Stokes lines for $\Delta\nu_i > \sim 1000$ cm^{-1} are too weak to be observed at room temperature.

(c) A given Raman line shows a degree of polarization which depends on the origin of the line and on the experimental arrangement. For strictly transverse observation, i.e., observation at right angles to the incident light, the depolarization factor, ρ_n, has a value in the range 0 to 6/7 for unpolarized incident light.

Figure 1 shows a schematic diagram of the Raman spectrum of carbon tetrachloride. The four Raman shifts, $\Delta\nu_i$ ($i = 1, 2, 3, 4$), are 218,

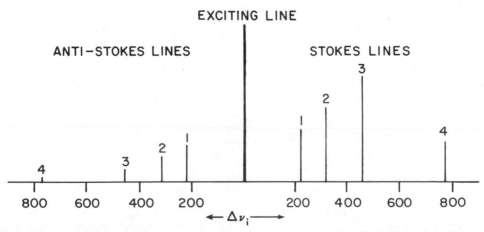

EXCITING LINE

ANTI−STOKES LINES   STOKES LINES

FIG. 1. Schematic diagram of the Raman spectrum of carbon tetrachloride. The heights of the Raman lines represent their relative intensities.

314, 459 and 775 cm^{-1}; the corresponding values of ρ_n are 6/7, 6/7, 0, and 6/7.

The Raman shifts, $\Delta\nu_i$, correspond to energy differences (in cm^{-1}) between discrete stationary states of the scattering system. Thus, in the quantum picture, the incident photons collide *elastically* with the molecules to give Rayleigh scattering, or *inelastically* to give Raman scattering, the latter process being much less probable than the former. For a Stokes Raman line, the photon furnishes energy to raise the molecule from a lower to a higher state; for an anti-Stokes line, the molecule must furnish energy to the scattered photon and move to a lower energy state. The anti-Stokes line thus originates in a less highly populated state and is weaker than the corresponding Stokes line. The Raman process can be described classically, but not as accurately, as the modulation of the scattered light wave by the internal motions of the scattering molecule, the Raman lines constituting "sidebands" of the Rayleigh "carrier" frequency.

A rigorous theory of the Raman effect, based on the quantum theory of dispersion, has been given by Placzek.[2] The process of light scattering can be visualized as the absorption of an incident photon of frequency, ν_e, by a molecule in a given initial state, thus raising the molecule to a "virtual" state from which it immediately returns to a final state emitting the scattered photon. Figure 2 illustrates the production of the Rayleigh line and Stokes and anti-Stokes Raman lines for a two-level system. If the exciting frequency ν_e is far away from any

allowed electronic absorption of the molecule, the intensity of a Stokes line $\Delta\nu_i$ varies with ν_e according to the law, $I_i \propto (\nu_e - \Delta\nu_i)^4$. As ν_e approaches an electronic absorption the intensity increases more rapidly; this is the region of so-called *resonance* Raman scattering.[3] Finally, if ν_e is such that the virtual state coincides with a real energy state of the system, the Raman effect goes over into molecular fluorescence provided the states involved can combine with one another according to the selection rules.

In principle, the Raman shifts, $\Delta\nu_i$, can represent energy differences between electronic, vibrational or rotational energy states. In practice, Raman spectroscopy has been concerned chiefly with the determination of vibrational frequencies of polyatomic molecules from the scattered light spectrum and the correlation of the observed frequencies with possible modes of vibration of the molecules. In this role, Raman spectroscopy forms an important complement to near-infrared spectroscopy which also furnishes precise information on molecular vibration frequencies.[4]

The so-called polarizability theory of Placzek[2] is an approximate theory which is particularly useful in relating observed vibrational Raman lines to the normal vibrations of the scattering molecule. In the equation, $\mathbf{m} = \alpha\mathbf{E}$, $\mathbf{m}$ is the electric dipole induced in the molecule by the oscillating light field $\mathbf{E}$. The constant of proportionality, α, is the molecular polarizability, which can be developed in a Taylor series in the vibrational coordinate q,

$$\alpha = \alpha_0 + (\partial\alpha/\partial q)_0 q + (1/2)(\partial^2\alpha/\partial q^2)_0 q^2 + \cdots$$

The intensity of the fundamental Raman band, corresponding to the $v = 0$ to $v = 1$ transition where v is the quantum number of the vibration, depends essentially on the *rate of change of polarizability*, $\alpha' \equiv (\partial\alpha/\partial q)_0$, with respect to the coordinate q. Like α_0, the polarizability of the molecule in the equilibrium position ($q = 0$), α' is a tensor quantity. The vibration is active in Raman scattering if, because of the symmetry properties of the vibration, α' is not identically zero. Also the depolarization, ρ_n, of a Raman band can be related by the polarizability theory to the symmetry of the vibration. Thus, the single polarized ($\rho_n = 0$) line of carbon tetrachloride at 459 cm^{-1} (Fig. 1) can be identified immediately with the "breathing" vibration of the tetrahedral CCl_4 molecule, and the three depolarized ($\rho_n = 6/7$) lines with the three degenerate vibration forms of the molecule.

Since the infrared activity of a molecule depends on the molecular dipole, whereas the Raman activity depends on the molecular polarizability, the selection rules for two types of spectra can be very different. For example, all four fundamental vibrations of CCl_4 are active in Raman scattering (Fig. 1), whereas only two of these ($i = 2$, 4) are active in in-

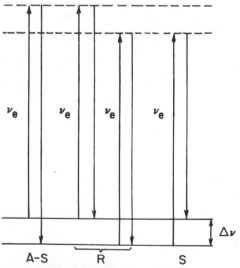

FIG. 2. Illustrating the production of Stokes (S) and anti-Stokes (A-S) Raman lines and the Rayleigh line (R) by an exciting frequency, ν_e, for a two-level system. The dashed lines represent virtual states of the system.

frared absorption. An extreme case occurs for molecules with a center of symmetry: vibrations which are active in infrared absorption are not active in Raman effect and vice versa. The selection rules for rotational transitions in infrared absorption and in light scattering also differ: they are $\Delta J = 0, \pm 1$ for the former and $\Delta J = 0, \pm 1, \pm 2$ for the latter, where ΔJ is the change in the rotational quantum number J.

Mercury lamps as sources of ordinary or spontaneous Raman scattering have now been superseded by gas lasers, particularly the He-Ne laser operating at 632.8 nm and the argon ion laser operating at 488.0 or 514.5 nm. The higher flux density, monochromaticity, directionality, and polarization of the laser beam combine to make it an almost ideal means of exciting Raman spectra. As a result there has been an extraordinary expansion of Raman studies in the last twenty years (see, for example, Refs. 5 and 6). The higher monochromaticity of laser sources has given a new impetus to molecular structure studies by Raman spectra of gases. With mercury lamp sources a practical resolution of 0.3 cm^{-1} at best could be achieved, whereas resolutions of 0.1 cm^{-1} and 0.003 cm^{-1} are obtainable with a multimode and a single mode laser, respectively. These higher resolutions have of course required the development of suitable recording equipment.[7]

The directionality and polarization properties of the laser beam are particularly advantageous in Raman studies of crystals.[8] The direction of propagation of both the incident beam and the observed scattered beam, as well as the polarization vectors of these beams, can be set in various well-defined relationships with respect to the crystal axes of the specimen; the result is that much precise information can be obtained concerning the structure and the elementary excitations of the crystal. The usual components of the Raman spectrum of a crystal can be assigned to (a) *internal* vibrations (vibrons), i.e., the vibrations of the molecules or ions of the crystal coupled by the intermolecular forces; and (b) *external* vibrations (phonons), i.e., optically active vibrations of the crystal lattice. Other elementary excitations in particular crystals have also been detected recently by laser Raman spectra: spin waves (magnons) in ferromagnetic and antiferromagnetic crystals, coupled photon-electric dipole exciton modes (polaritons) in dipole crystals, and spin waves of angular momentum (librons) due to the almost free rotation of molecules in the ordered phase of ortho-enriched solid hydrogen. Phonon excitations in metals and electronic Raman transitions of ions in solids have also been studied. The discovery of *spin-flip* Raman transitions in wide band-gap semiconductors in a magnetic field has led to the introduction of tunable spin-flip Raman lasers. A new effect is *surface enhanced Raman scattering* (SERS) in which the scattering from a monomolecular

layer of a substance absorbed on a metal surface is enhanced by a factor of as much as 10^6 over the scattering from the free molecule, the degree of enhancement increasing with the roughness of the metal surface; for recent articles on this subject see Part III of Ref. 6.

In addition to its importance in the study of ordinary or *spontaneous* Raman scattering, the laser has led to the discovery of four phenomena, which are of considerable significance in present-day Raman research; these are *stimulated* Raman scattering, the *inverse* Raman effect, the *hyper* Raman effect, and *coherent anti-Stokes* Raman scattering. Each of these effects depends on the intense electric field produced in a material medium when it is illustrated by the focussed beam of a giant pulse (Q-switched) laser such as the ruby system or the neodymium gas system. Since the electric field intensity at the focus of the beam can be as high as 10^9 V m^{-1}, nonlinear contributions to the electric moment induced in the molecule can arise and new effects are observed.

In stimulated Raman scattering the giant laser pulse causes a coherent stimulated emission of a Stokes Raman frequency $\nu_e - \Delta\nu_i$, when the incident intensity exceeds a certain threshold value.[9] As a rule, only one Raman frequency is emitted and this corresponds to the line with the greatest peak intensity in the spontaneous Raman spectrum. At higher incident intensities, harmonics (not overtones) of the first Stokes emission appear at $\nu_e - 2\Delta\nu_i$, $\nu_e - 3\Delta\nu_i$, $\cdots$ along with corresponding anti-Stokes frequencies.

When a substance is irradiated by the radiation ν_e of a giant-pulse laser and simultaneously by a continuum lying at frequencies higher than ν_e, the Raman shifts $\Delta\nu_i$ appear as absorption lines at frequencies $\nu_1 + \Delta\nu_i$; this is the so-called inverse Raman effect[10,11]. In this case all the Raman-active molecular transitions can appear in the spectrum. Since the spectrum is generated in liquids and solids in a time of ~ 40 nsec, and since there are possibilities of reducing the time much further, the effect can perhaps furnish a high-speed spectroscopic method for studying short-lived molecular species and other transient phenomena. The use of laser excitation combined with recording by image-converter tubes also has attractive possibilities for high-speed Raman spectroscopy.[12]

The hyper Raman effect[13] occurs when the incident electric field E is so large that nonlinear susceptibility terms are no longer negligible. Thus, the equation $\mathbf{m} = \alpha\mathbf{E}$ must be written

$$m_k = \alpha_{k\ell}E_\ell + \tfrac{1}{2}\beta_{k\ell m}E_\ell E_m + \tfrac{1}{6}\gamma_{k\ell mn}E_\ell E_m E_n$$

$$+ \cdots (k, \ell, m, n = x, y, z),$$

where the *hyperpolarizabilities* β, γ, $\cdots$ are tensors of the 3rd, 4th, $\cdots$ orders. The term in β gives rise to scattering at $2\nu_e$ and $2\nu_e \pm \Delta\nu_i$, and the further terms to higher harmonics. The selection rules for hyper Raman scattering are of course different from those for the ordinary case,[14] and when further developments in the recording of extremely low intensities have been made, the effect will constitute an important extension of ordinary Raman scattering.

Coherent anti-Stokes Raman scattering (CARS) was discovered some time ago;[15] however, its general usefulness has become apparent only in recent years when powerful tunable pulsed lasers became available. (Recent papers are given in Refs. 6 and 16.) The CARS experiment is a parametric process in which two waves, one at the constant frequency ν_1 and one at a continuously variable frequency ν_2, are in spatial and temporal coincidence in the scattering medium; a coherent radiation of frequency $\nu_3 = \nu_1 + (\nu_1 - \nu_2)$, among others, is produced. If ν_2 is varied so that $\nu_1 - \nu_2 = \Delta\nu$, where $\Delta\nu$ is the separation of two states of the medium, then $\nu_3 = \nu_1 + \Delta\nu$ corresponds to an anti-Stokes Raman line. The possibilities of producing coherent, directed and intense Raman lines in this way are most attractive.

H. L. WELSH

References

1. Stoicheff, B. P., in "Advances in Spectroscopy," Thompson, H. W., Ed., Vol. 1, New York, Interscience, 1959.
2. Placzek, G., "Rayleigh Scattering and the Raman Effect," in "Handbuch der Radiologie," Marx, E., Ed., Vol. 6, pt. 2, Leipzig, Akademische Verlagsgesellschaft, 1934; English translation by A. Werbin, Livermore, Lawrence Radiation Laboratory, University of California, 1959.
3. Szymanski, H. A., Ed., "Raman Scattering—Theory and Practice," New York, Plenum Press, Vol. 1, 1967, Vol. 2, 1970.
4. Herzberg, G., "Infrared and Raman Spectra of Polyatomic Molecules," New York, Van Nostrand Reinhold, 1945.
5. Long, D. A., "Raman Spectroscopy," New York, McGraw-Hill, 1977.
6. Murphy, W. F., Ed., "VIIth International Conference on Raman Spectroscopy," New York, North-Holland, 1980.
7. Weber, A., Ed., "Raman Spectroscopy of Gases and Liquids," New York, Springer-Verlag, 1979.
8. Cardona, M., Ed., "Light Scattering in Solids," New York, Springer-Verlag, 1975.
9. Hellworth, R. W., *Phys. Rev.*, **130**, 1850 (1963).
10. Jones, W. J., and Stoicheff, B. P., *Phys. Rev. Letters*, **13**, 657 (1964).
11. McLaren, R. A., and Stoicheff, B. P., *App. Phys. Letters*, **16**, 140 (1970).
12. Delhaye, M., and Migeon, M., *Compt. rendu.*, **261**, 2613 (1965).
13. Terhune, R. W., Maker, P. D., and Savage, C. M., *Phys. Rev. Letters*, **14**, 681 (1965).
14. Long, D. A., in "Essays in Structural Chemistry," Downs, A. J., Long, D. A., and Stavely, L. A. K., Eds., Chap. 2, London, Macmillan, 1971.
15. Maker, P. D., and Terhune, R. W., *Phys. Rev.* **137A**, 801 (1965).
16. McKellar, A. R. W., Oka, T., and Stoicheff, B. P., Eds., "VIth International Conference on Laser Spectroscopy," New York, Springer-Verlag, 1981.

Cross-references: ATOMIC SPECTRA, LASER, LIGHT SCATTERING, MOLECULAR SPECTROSCOPY, POLARIZED LIGHT.

RARE EARTHS

The seventeen rare-earth elements are those of group IIIb of the periodic table, including scandium (Sc), yttrium (Y), lanthanum (La), and the lanthanides: cerium (Ce), praseodymium (Pr), neodymium (Nd), promethium (Pm), samarium (Sm), europium (Eu), gadolinium (Gd), terbium (Tb), dysprosium (Dy), holmium (Ho), erbium (Er), thulium (Tm), ytterbium (Yb), and lutetium (Lu). These elements are of great interest to theoretical and experimental researchers because of the systematic manner in which many of their physical and chemical properties are expected to vary across the lanthanide series.

The rare earth yttria (actually yttrium oxide plus several of the heavy rare-earth oxides) was discovered near Ytterby, Sweden, in 1794. Ceria (containing several light rare-earth oxides) was discovered some years later. In the 19th Century, because of the difficulty in separating the rare-earth elements and in removing impurities, it was thought that many different rare earths existed. During this period, over 70 different rare earths were reported in the literature. With the advent of spectroscopy (especially the work of J. Becquerel at the turn of the century), Mendeleev's periodic table (from which he predicted the existence and properties of the as yet undiscovered scandium), and Moseley's work on X-rays, the actual number and properties of the lanthanides were better understood. The discovery of lutetium in 1907 completed the lanthanide series (except for promethium, which has no stable isotopes and was not isolated until 1947).

Rare earths are obtained commercially from the minerals monazite (for light rare earths) and xenotime (for heavy rare earths). Other minerals containing substantial quantities of rare earths are bastnaesite, samarskite, euxenite, and gadolinite, although these are, for the most part, not used commercially. Rare-earth minerals are found in Africa, Asia, and the Americas. In the United States they are found in the Blue Ridge, Piedmont, and Coastal Plain of the southeastern states, eastern Idaho through southwestern Montana, Colorado, southern California, and Alaska.

Rare earths are also obtained as by-products of fission reactions. The difficulty in separating the rare earths was once an impediment in performing research on these elements. The early fractionation methods were quite slow and inefficient, and involved large quantities of solvent and material. However, it was discovered that certain rare earths, such as cerium, europium, and ytterbium could exist in valence states other than the normal trivalent state, and that oxidation and reduction reactions could be used to separate these rare earths from the others. With the advent of the ion-exchange techniques pioneered by groups at Iowa State University and Oak Ridge National Laboratory, it is now possible to obtain high-purity samples of any of the rare earths.

Atomic Properties The rare earths are characterized by the progressive addition of electrons in the $4f^n$ subshell that is largely shielded from the environment by the outer orbitals $5s^2$, $5p^6$, $5d^{0,1}$, or $6s^2$. These features produce the interesting physical and chemical properties. The ground electronic configurations of neutral Sc, Y, and La are $[\text{Ar}]3d^14s^2$, $[\text{Kr}]4d^15s^2$, and $[\text{Xe}]5d^16s^2$, respectively; the ground configurations of the lanthanides are given in Table 1. The common oxidation state of all the rare earths is the trivalent state; however, compounds are known in which Ce, Pr, and Tb are tetravalent, others in which Sm, Eu, and Yb are divalent.

The trivalent lanthanides have a $[\text{Xe}]4f^n$ ground-state configuration (see Table 1). Their spectroscopic and magnetic properties derive from the partially filled $4f$ shell. The angular momenta of these $4f$ electrons can combine in many ways to produce a variety of electronic states within the $4f^n$ configuration. The ground electronic states of the trivalent lanthanides follow *Hund's rule*, which can be summarized as follows: if n is the number of $4f$ electrons,

then

(a) S, the spin angular momentum, is $n/2$ if $n \leqslant 7$ and $(7 - n)/2$ if $n > 7$;

(b) L, the orbital angular momentum, is the maximum value consistent with S and the Pauli exclusion principle;

(c) J, the total angular momentum, is $|L - S|$ if $n \leqslant 7$ and $L + S$ if $n > 7$.

The *Hund ground states* obtained from this rule (written in the usual spectroscopic notation $^{2S+1}L_J$) are shown in Table 1. Excited electronic states within the $4f^n$ configuration are complex and not uniquely specified, in general, by S, L, and J.

Russell-Saunders coupling (L-S coupling) is a good approximation for the ground states of the lanthanides; thus, L and S are nearly (though not exactly) good quantum numbers. The magnetic properties of the trivalent ions are summarized by the Lande g-value, given (in L-S coupling) by

$$g_J = 1 + \frac{S(S + 1) + J(J + 1) - L(L + 1)}{2J(J + 1)}.$$

Actual values of g_J obtained in intermediate coupling calculations differ slightly from the L-S values; the latter are given in Table 1.

The free-ion spectrum of Pr^{3+} has been well studied, and all states of $\text{Pr}^{3+}(4f^2)$ have been identified except for $^1S_0$. (This state has, however, been observed in some solids.) Free-ion spectra of other trivalent rare earths have been observed as well, although level identifications are far from complete. Most of the information that we currently possess concerning the spectroscopic properties of trivalent rare earths has been obtained from solution spectra and from

TABLE 1. ATOMIC PROPERTIES OF THE LANTHANIDES.

Element	Atomic Number	Atomic Weight	Neutral Atom Configuration	Trivalent Ion		
				Ground Configuration	Hund Ground State	Lande g-Value
Ce	58	140.12	$[\text{Xe}]4f^15d^16s^2$	$[\text{Xe}]4f^1$	$^2F_{5/2}$	6/7
Pr	59	140.907	$[\text{Xe}]4f^36s^2$	$[\text{Xe}]4f^2$	$^3H_4$	4/5
Nd	60	144.24	$[\text{Xe}]4f^46s^2$	$[\text{Xe}]4f^3$	$^4I_{9/2}$	8/11
Pm	61	(145)	$[\text{Xe}]4f^56s^2$	$[\text{Xe}]4f^4$	$^5I_4$	3/5
Sm	62	150.35	$[\text{Xe}]4f^66s^2$	$[\text{Xe}]4f^5$	$^6H_{5/2}$	2/7
Eu	63	151.96	$[\text{Xe}]4f^76s^2$	$[\text{Xe}]4f^6$	$^7F_0$	—
Gd	64	157.25	$[\text{Xe}]4f^75d^16s^2$	$[\text{Xe}]4f^7$	$^8S_{7/2}$	2
Tb	65	158.924	$[\text{Xe}]4f^96s^2$	$[\text{Xe}]4f^8$	$^7F_6$	3/2
Dy	66	162.50	$[\text{Xe}]4f^{10}6s^2$	$[\text{Xe}]4f^9$	$^6H_{15/2}$	4/3
Ho	67	164.930	$[\text{Xe}]4f^{11}6s^2$	$[\text{Xe}]4f^{10}$	$^5I_8$	5/4
Er	68	167.26	$[\text{Xe}]4f^{12}6s^2$	$[\text{Xe}]4f^{11}$	$^4I_{15/2}$	6/5
Tm	69	168.934	$[\text{Xe}]4f^{13}6s^2$	$[\text{Xe}]4f^{12}$	$^3H_6$	7/6
Yb	70	173.04	$[\text{Xe}]4f^{14}6s^2$	$[\text{Xe}]4f^{13}$	$^2F_{7/2}$	8/7
Lu	71	174.97	$[\text{Xe}]4f^{14}5d^16s^2$	$[\text{Xe}]4f^{14}$	$^1S_0$	—

spectra of rare-earth doped solids. Several comprehensive reviews are included among the references.

Nuclear Properties The rare-earth elements present a rich display of isotopes for study by the nuclear physicist, with between 10 and 20 isotopes known for each element. Only one or two stable isotopes exist for the odd-Z elements (none exist for promethium); in contrast, several stable isotopes are known for each of the even-Z elements.

Table 2 summarizes some of the nuclear properties of isotopes with greater than 5% abundance. About half of these have spinless nuclei; the other half have nuclear spins ranging from $\frac{1}{2}$ to $\frac{7}{2}$ (in units of $\hbar$) with accompanying electric quadrupole moments and magnetic dipole moments, which are given in the table. The last column of Table 2, giving the thermal neutron cross sections, is of practical interest. Elements having high cross sections, such as Sm, Eu, Gd, and Dy, are incorporated into control rods for regulating nuclear reactors and in other applications (such as nuclear shielding) in which high nuclear cross sections are desirable.

Rare-Earth Metals The physical properties of the rare-earth metals are summarized in Table 3. These metals exhibit many of the features common to rare-earth compounds in general. Within structures of the same type, the lattice constants decrease systematically with increasing atomic number. This decrease in the unit cell size is a manifestation of the *lanthanide contraction*, which is a decrease in the size of the trivalent ion as more electrons are progressively added to the incomplete $4f$ shell. The Eu and Yb ions in the corresponding metals are in the divalent state, with half-filled ($4f^7$) and filled ($4f^{14}$) shells, respectively. With these two exceptions, the melting points, boiling points, and densities vary systematically across the lanthanide series, providing further manifestations of the lanthanide contraction.

Rare-earth metals are relatively good conductors, with conductivities about 50 times smaller than that of silver. In addition, La becomes superconducting at approximately 5 K. Of particular interest are the magnetic properties of the rare-earth metals. Nearly all the rare earths with unpaired electrons become antiferromagnetic, with Néel temperatures T_N ranging from 19.2 to 230 K. Ferromagnetism has been observed in all the heavy rare earths with unpaired electrons, with the Curie temperature T_c of Gd approaching room temperature. The magnetic moments of the trivalent ions in the rare-earth metals are large compared with those of other magnetic materials, due to the large values of J for these ions ($\mu \sim Jg_J\mu_B$, where $\mu_B = e\hbar/2m$ is the Bohr magneton and m is the electron mass; see Table 1).

Cerium is an interesting exception to the general trend in the rare-earth metals. Three different allotropes of Ce metal are known to exist at

low temperatures and pressures, known as α, β, and γ (other forms exist at elevated temperatures and pressures). All three allotropes can exist at room temperature, although γ-Ce (whose properties are given in Table 3) is reported to be the most stable. All three forms have noninteger valence; α-Ce has valence ≈ 3.67, and the β and γ allotropes have valence slightly greater than 3. Tendency toward tetravalence is also exhibited by a number of Ce compounds and is a consequence of the stability of the [Xe] configuration of Ce^{4+}. Because of this tendency toward the magnetically inert tetravalent state, Ce metal does not exhibit a magnetic phase transition.

Rare-Earth Compounds The rare earths form complexes in aqueous solution in which the rare-earth ion is surrounded by a layer of H_2O molecules; under these conditions, the chemical properties of the rare earths are similar. This led early workers to believe (falsely) that chemical similarity was an intrinsic property of the rare earths. The hydrated ions are strongly electrolytic. Many compounds can be formed by precipitation from solution; most compounds of interest, however, are formed by other methods.

Binary halides, chalcogenides, and pnictides of the rare earths are well studied and possess a wide variety of physical and chemical properties. Several chalcogenides of Sm, Eu, and Yb undergo pressure-sensitive insulator-metal or semiconductor-metal phase transitions corresponding to a change in valence (electron delocalization). The oxides in particular are an interesting series of compounds. Examples are the sesquioxides R_2O_3 for all R, where R represents a rare earth, and R_nO_{2n-2}, where n is an integer, particularly R_7O_{12} and RO_2 ($n \to \infty$) for R = Ce, Pr, and Tb. Similarly, sulfides of the rare earths including R_2S_3, RS, and RS_2 also exhibit a wealth of varying crystalline structures and other physical properties that have received growing attention over the past several years. Other, more complex anhydrous compounds of the rare earths have applications in various areas of science and technology; some of these are $RAlO_3$, $R_3Al_5O_{12}$, RVO_4, and R_2O_2S.

The spectroscopic properties of rare-earth doped insulating solids are of interest because of their applications to numerous devices, especially lasers. A rare-earth ion doped into a solid experiences a *crystal field* due to the surrounding ions in the crystal. This crystal field (which is shielded by the outer $5s^2 5p^6$ shell of the rare-earth ion) has two effects: (1) it splits each electronic state of the free ion into a number of Stark components; (2) it breaks the inversion symmetry of the free ion, thereby permitting the parity-forbidden electric-dipole transitions within the $4f^n$ shell to take place (provided the rare-earth ion is at a site that lacks inversion symmetry). Laser action based

TABLE 2. NUCLEAR PROPERTIES OF STABLE RARE-EARTH ISOTOPES

Isotope	Abund.[a]	$I\,\pi$[b]	μ[c]	Q[d]	σ_n[e]
$_{21}Sc^{45}$	100	$\frac{7}{2}-$	+4.7564	−0.22	25
$_{39}Y^{89}$	100	$\frac{1}{2}-$	−0.1373	−	1.3
$_{57}La^{139}$	99.911	$\frac{7}{2}+$	+2.778	+0.21	8.8
$_{58}Ce^{140}$	88.48	$0+$	−	−	0.58
$_{58}Ce^{142}$	11.07	$0+$	−	−	0.95
$_{59}Pr^{141}$	100	$\frac{5}{2}+$	+4.3	−0.059	3.9
$_{60}Nd^{142}$	27.13	$0+$	−	−	18.8
$_{60}Nd^{143}$	12.20	$\frac{7}{2}-$	−1.08	−0.48	330
$_{60}Nd^{144}$*	23.87	$0+$	−	−	4.0
$_{60}Nd^{145}$	8.29	$\frac{7}{2}-$	−0.66	−0.25	50
$_{60}Nd^{146}$	17.18	$0+$	−	−	1.4
$_{60}Nd^{148}$	5.72	$0+$	−	−	2.5
$_{60}Nd^{150}$	5.60	$0+$	−	−	1.3
$_{62}Sm^{147}$*	15.07	$\frac{7}{2}-$	−0.813	−0.208	75
$_{62}Sm^{148}$*	11.27	$0+$	−	−	2.4
$_{62}Sm^{149}$*	13.82	$\frac{7}{2}-$	−0.66	+0.06	41,000
$_{62}Sm^{150}$	7.47	$0+$	−	−	102
$_{62}Sm^{152}$	26.63	$0+$	−	−	210
$_{62}Sm^{154}$	22.53	$0+$	−	−	5.5
$_{63}Eu^{151}$	47.77	$\frac{5}{2}+$	+3.464	+1.16	8,000
$_{63}Eu^{153}$	52.23	$\frac{5}{2}+$	+1.530	+2.9	450
$_{64}Gd^{155}$	14.7	$\frac{3}{2}-$	−0.27	+1.6	61,000
$_{64}Gd^{156}$	20.47	$0+$	−	−	1.8
$_{64}Gd^{157}$	15.68	$\frac{3}{2}-$	−0.36	+2	254,000
$_{64}Gd^{158}$	24.9	$0+$	−	−	3.5
$_{64}Gd^{160}$	21.9	$0+$	−	−	0.77
$_{65}Tb^{159}$	100	$\frac{3}{2}+$	+1.95	+1.3	30
$_{66}Dy^{161}$	18.88	$\frac{5}{2}+$	−0.46	+1.4	585
$_{66}Dy^{162}$	25.53	$0+$	−	−	200
$_{66}Dy^{163}$	24.97	$\frac{5}{2}-$	+0.64	+1.6	140
$_{66}Dy^{164}$	28.18	$0+$	−	−	2,600
$_{67}Ho^{165}$	100	$\frac{7}{2}-$	+4.08	+2.82	65
$_{68}Er^{166}$	33.41	$0+$	−	−	30
$_{68}Er^{167}$	22.94	$\frac{7}{2}+$	−0.564	+2.83	700
$_{68}Er^{168}$	27.07	$0+$	−	−	1.9
$_{68}Er^{170}$	14.88	$0+$	−	−	6
$_{69}Tm^{169}$	100	$\frac{1}{2}-$	−0.230	−	115
$_{70}Yb^{171}$	14.31	$\frac{1}{2}-$	+0.4919	−	50
$_{70}Yb^{172}$	21.82	$0+$	−	−	0.4
$_{70}Yb^{173}$	16.13	$\frac{5}{2}-$	−0.6776	+2.8	19
$_{70}Yb^{174}$	31.84	$0+$	−	−	65
$_{70}Yb^{176}$	12.73	$0+$	−	−	5.5
$_{71}Lu^{175}$	97.40	$\frac{7}{2}+$	+2.23	+5.68	21

[a]Natural abundance in percent. Only those isotopes having greater than 5% abundance are included.
[b]Nuclear spin (in units of $\hbar$) and parity.
[c]Nuclear magnetic moment in nuclear magnetons (1 nuclear magneton = $e\hbar/2M_p$, where M_p is the proton mass).
[d]Nuclear quadrupole moment in barns (1 barn = 10^{-28} m^2).
[e]Cross section for thermal neutrons in barns.
*Long-lived radioactive isotopes.

TABLE 3. PROPERTIES OF THE RARE-EARTH METALS.

Metal	Structure[a] (24°C)	Lattice Constants a_0(Å, 24°C)	Lattice Constants c_0(Å, 24°C)	Density (g/cm^3, 24°C)	Melting Point (°C)	Boiling Point (°C)	T_N (K)	T_C (K)
Sc	hcp	3.3088	5.2680	2.989	1541	2836	–	–
Y	hcp	3.6482	5.7318	4.469	1522	3338	–	–
La	dhcp	3.7740	12.171	6.146	918	3464	–	–
Ce[b]	fcc	5.1610	–	6.770	798	3433	–	–
Pr	dhcp	3.6721	11.8326	6.773	931	3520	19.2	–
Nd	dhcp	3.6582	11.7966	7.008	1021	3074	106	–
Pm	dhcp	3.65	11.65	7.264	1042	3000[d]	?	?
Sm	rhomb	3.6290	26.207	7.520	1074	1794	14	–
Eu[c]	bcc	4.5827	–	5.244	822	1529	90	–
Gd	hcp	3.6336	5.7810	7.901	1313	3273	–	293
Tb	hcp	3.6055	5.6966	8.230	1365	3230	230	220
Dy	hcp	3.5915	5.6501	8.551	1412	2567	178	86
Ho	hcp	3.5778	5.6178	8.795	1474	2700	133	19
Er	hcp	3.5592	5.5850	9.066	1529	2868	84	18
Tm	hcp	3.5375	5.5540	9.321	1545	1950	56	32
Yb[c]	fcc	5.4848	–	6.966	819	1196	–	–
Lu	hcp	3.5052	5.5494	9.841	1663	3402	–	–

[a] hcp = hexagonal close packed; dhcp = double-hexagonal close packed; fcc = face-centered cubic; rhomb = rhombohedral (lattice constants for equivalent hexagonal cell); bcc = body-centered cubic.
[b] γ-allotrope; see text.
[c] Divalent.
[d] Estimated.

on transitions within the $4f^n$ shell has been observed in many rare earths; a notable example is $Y_3Al_5O_{12}$ (YAG) doped with Nd^{3+}, which is the most powerful laser known at present.

Uses Aside from the uses of rare earths discussed above, mixed rare-earth compounds and metals serve a variety of industrial purposes. *Mischmetal*, composed of cerium and other light rare earths, is used in lighter flints and in alloys. Mixed rare earths are also used as catalysts in the cracking of petroleum to produce gasoline. Carbon arcs containing rare-earth mixtures are used in electrodes for extremely bright lights in making motion pictures and in searchlights.

The light rare earths find frequent applications in optics. For example, lanthanum compounds are frequently used in the manufacture of low-dispersion lenses. CeO_2 and other rare-earth oxides are used for polishing glass used in optical systems. Didymium mixtures (oxides of Pr + Nd) are used as absorbers in safety goggles worn to protect the eye from intense, harmful radiation (e.g., welder's goggles). Erbium oxide often performs the same function in the infrared. Rare-earth hydrides can store large amounts of hydrogen and thus serve as fuel-storage materials for "clean" energy. Rare-earth sulfides are efficient as the conversion materials in solar photovoltaic cells. Rare-earth chelates are useful in many biochemical reactions and participate in human physiology. Rare-earth radioactive isotopes probe the body for diseases such as cancer.

Rare-earth compounds play an important role in electronics. Rare-earth garnets are important materials for ferrites. The compounds Y_2O_2S, Y_2O_3 and YVO_4, doped with Eu^{3+}, are used as the red phosphor component in color television tubes. The unique magnetic properties of the rare earths make them attractive candidates for permanent magnets; $SmCo_5$ is but one example. Also, paramagnetic compounds of the rare earths, with their large saturation magnetic moments, are used in magnetic cooling. Onnes first produced temperatures below 1 K by this method with gadolinium sulfate.

RICHARD P. LEAVITT
JOHN B. GRUBER

References

Dieke, G. H. (ed. by Crosswhite, H. M. and Crosswhite, H.), "Spectra and Energy Levels of Rare Earth Ions in Crystals," New York, Interscience, 1968.

Gruber, J. B., "Optical Transitions in Rare Earth Crystals," in Eyring, L. (Ed.), "Progress in the Science and Technology of the Rare Earths," Vol. 3, pp. 38–61, London, Pergamon Press, 1968.

Gschneidner, K. A., and Eyring, L. (Eds.), "Handbook on the Physics and Chemistry of Rare Earths," Vols. 1–4, Amsterdam, North-Holland, 1978.

Gschneidner, K. A. (Ed.), *Rare Earth Information Center News*, Ames, Iowa, Institute of Atomic Research, Iowa State University, published quarterly.

McCarthy, G. J., Rhyne, J. J., and Silber, H. (Eds.), "The Rare Earths in Modern Science and Technology," Vols. 1, 2, New York, Plenum, 1978, 1980.

Morrison, C. A., and Leavitt, R. P., "Spectroscopic Properties of Triply Ionized Lanthanides in Transparent Host Crystals," in Gschneidner, K. A., and Eyring, L. (Eds.), "Handbook on the Physics and Chemistry of Rare Earths," Vol. 5, Amsterdam, North-Holland, 1982.

Reisfeld, R., and Jorgensen, C. K., "Lasers and Excited States of Rare Earths," Berlin, Springer-Verlag, 1977.

Spedding, F. H., and Daane, A. H., "The Rare Earths," New York, John Wiley & Sons, Inc., 1961.

Wybourne, B. G., "Spectroscopic Properties of Rare Earths," New York, Interscience, 1965.

Cross-references: ATOMIC SPECTRA, ENERGY LEVELS, LASER, MAGNETISM, PERIODIC LAW AND PERIODIC TABLE, TRANSURANIUM ELEMENTS.

RECTIFIERS

Rectification Rectification is a process which imparts unidirectional symmetry to a system. It removes the inversion symmetry, and dictates the forward and reverse directions for the system. Electrical rectification is used to convert an alternating current (ac) to a direct current (dc). Rectifiers are the devices which can be used to accomplish the electrical rectification.

Ideal Rectifier An ideal rectifier is a perfect conductor for forward current and a perfect insulator for reverse current. If R_f is the forward resistance and R_b is the reverse resistance, an ideal rectifier has the ratio of R_b/R_f approaching ∞. The ratio R_b/R_f is used as a figure of merit for rectifiers.

Vacuum Diode Vacuum diodes are used for all purposes from low-voltage signal rectification to fairly high power rectification. The current-carrying capacity of the diode is usually limited by the ability of the anode to dissipate heat. The plate dissipation in watts is obtained by multiplying the plate current by the plate-to-cathode voltage. There is a maximum peak inverse voltage above which the tube breaks down and electrons flow from anode to cathode. There is also a maximum heater-to-cathode voltage above which there is breakdown between heater and cathode. The cathode of the diode is heated, directly or indirectly, to the temperature of thermionic emission of electrons. When the anode is positive with respect to the cathode, a current is passed through the diode. When the anode is negative with respect to the cathode, no charge will flow through the tube.

Hot-Cathode Gaseous Rectifiers In gaseous rectifiers, an inert gas is contained in a closed vessel at very low pressure. The hot cathode emits electrons toward the anode when the anode is positive with respect to the cathode. The electrons collide with molecules of the gas and produce positive ions and free electrons. The electrons travel to the anode and the positive ions travel to the cathode. A current is established when the potential reverses, and the cathode is positive with respect to the anode. The electrons are drawn to the cathode. Ionization ceases and current no longer flows. Tungar and Rectigon rectifiers are examples of hot-cathode gaseous rectifiers. The anode is made of graphite, and the cathode is a coiled tungsten filament heated by an electric current. An inert gas, such as argon, is contained in a bulb at reduced pressure. In the Tungar rectifier, the gas pressure is approximately 5 cm of Hg. Usually, the gas pressure varies from 0.01 mm to several cm of Hg.

Mercury-Arc Rectifiers Mercury, in this case, is both the cathode and source of mercury vapor which can be ionized. The ionization potential of mercury is 10.39 V. If the cathode-to-anode voltage exceeds this value, the emitted electrons will have enough energy to ionize the mercury atoms. The positive mercury ions move toward the cathode. When mercury vapor condenses, it returns to the cathode pool. There is no deterioration of the cathode with use. When the rectifier is in operation, a mercury arc concentrates on the surface of the mercury at the cathode and produces the cathode spot. It is a region of high temperature at which ionization can occur readily. The cathode spot may reach a temperature of 2087°C, and the mean temperature of the cathode pool is of the order of 100°C. The anode may be of either iron or graphite. The ordinary arc drop is of the order of 12 to 18 volts in single-tank rectifiers and may go as high as 30 volts in multi-anode tank rectifiers.

Grid-Controlled Gaseous Rectifiers By means of the potential applied to the grid of a gaseous rectifier tube, its plate current can be controlled. The hot-cathode thyratron has a cathode made of a small nickel cylinder. The cathode is heated by a single tungsten filament. The cathode assembly is surrounded by three concentric nickel cylinders, with spaces between, to heat-shield the vanes and thus minimize the watts per ampere emission. The grid consists of perforated metal completely surrounding both cathode and anode. The electrons and ions go through the perforations of a disk in passing between cathode and anode. The grid controls the arc only to the extent of either initiating it or preventing it from starting. The grid controls the firing of the tube by varying the phase of the voltage applied to the grid. Thyratrons are capable of rectifying relatively large values of power with ratings such as 65 amp at 2,500 volts.

Power Rectifiers with Grid Control Large power mercury-arc rectifiers can be equipped with grid control. The energized grid controls and regulates the output voltage of the power rectifier. After a backfire has occurred, the grid restores the rectifier to normal operating conditions by the quick application of a negative potential to it. Grid control is made by variation of the magnitude of the grid voltages. Large power rectifiers can have 6 to 12 anodes with ratings up to 1,000 amp and 20,000 volts. The multianode single-tank rectifier requires a continuous arc and is susceptible to backfiring due to diffuse ionization. It also has a high arc drop, usually between 25 and 30 volts.

Ignitron The ignitron is a single half-wave mercury-arc rectifier in which the ignition of the arc and its control are both accomplished by the same element. The main anode is made of graphite, and the cathode is a mercury pool. The starting ignitor is a pointed rod of high-resistance refractory material, the pointed end of which dips into the mercury pool. When a positive current impulse, usually of 20 to 40 amp, is delivered to the ignitor, a spark occurs at the junction of the ignitor and the mercury pool and this spark instantly develops into a small cathode spot, which ionizes the mercury. If the anode is sufficiently positive, an arc strikes from anode to cathode. The entire process requires only a few microseconds. The arc is extinguished during the following half-cycle, when the anode-cathode potential reverses, and the current for the half-cycle goes to zero. The power output can be controlled by timing the firing by shifting the phase of the pulse to the ignitor. Efficiency increases with voltage rating. The efficiency at an output voltage of 100 volts of an ignitron with 20 volts arc drop cannot exceed 83.3%; at 200 volts the efficiency would be 91% and at 600 volts nearly 97%. At 2,000 volts the efficiency approaches 99%.

Solid State Rectifiers Rectification can result from the presence of potential barriers at the surface or within the bulk of the crystalline semiconductor material. These barriers impede the flow of current in one direction, which is the reverse direction. At the same time they enhance the flow of current in the opposite direction, which is the forward direction. Solid state rectifiers can be classified as *contact* rectifiers and *junction* rectifiers.

Contact Rectifiers Contact rectifiers are based on the Schottky effect. They consist of metal-to-semiconductor contacts. The semiconductor has a different work function than the metal. The metal-to-semiconductor contact has a potential barrier, i.e., Schottky barrier, which results in rectification. When the semiconductor is biased positively with respect to the metal (reverse bias), the electrons in the metal move toward the barrier but have insufficient energy to overcome it. Electrons in the semiconductor move away from the barrier, and this depletion widens the space-charge region of ionized donors. In the case of forward bias, the potential barrier is reduced, and electrons flow more easily from semiconductor to the metal. The electrons entering the metal have energies much greater than the Fermi energy in the metal. They are called *hot electrons*. The conduction in this case is done by

the majority carriers. Contact rectifiers can be either in the form of plate rectifiers or point contact rectifiers.

Plate Rectifiers Plate rectifiers take the form of thin semiconducting layers with metallic electrodes on either side. The thin layer permits the ohmic series resistance of the bulk material to be as small as possible. Plate rectifiers can be made of selenium, cuprous oxide, titanium dioxide, or cupric sulfide, among others.

Point-Contact Rectifiers The semiconducting material, in the form of a chip, either single crystal or polycrystalline, is soldered to a metal support. The point contact is a thin wire which is welded or soldered to a second external electrode. The point contact is adjustable during assembly, but remains fixed after testing. The semiconductor material may be silicon or germanium. The crystal and the whisker are mounted in a cartridge. A tapping process, which involves a number of slight mechanical taps to the cartridge assembly, improves the rectifying properties and the stability of the point contact. An electrical forming treatment can stabilize the voltage-current characteristics. For example, the passage of a large forward current, about 100 mA or more for a period of between 1 and 50 milliseconds, lowers the forward resistance permanently and increases the reverse resistance of an n-type germanium diode. Rectifiers can also be made of lead sulfide (PbS), lead selinide (PbSe), lead telluride (PbTe), aluminum antimonide (AlSb), gallium antimonide (GaSb), indium phosphide (InP), or gallium arsenide (GaAs).

Junction Rectifiers Rectification can result when there is a junction region between two semiconductor regions which have different types of majority carriers, such as a junction between a n-type layer and a p-type layer. In the p-type layer, holes are the majority carriers. They move toward the n-type layer where there are few holes, leaving behind negatively charged acceptor ions. The electrons, which are the majority carriers in the n-type layer, move toward the p-type layer where there are few electrons, leaving behind positively charged donor ions. Because the Fermi levels are different for the p-type layer and the n-type layer, the p-n junction causes a potential barrier to be formed as these two Fermi levels are brought together to be equal to one Fermi level in the junction. This potential barrier is the source of rectification in the junction rectifier. When an external reverse voltage is applied to the p-n diode, the minority carriers flow through the junction and they are the source of current flow. Silicon rectifiers can be operated at ambient temperatures up to 200°C and at current levels of hundreds of amperes and voltages as high as 1000 volts.

Silicon-Controlled Rectifiers A silicon-controlled rectifier (SCR) is basically a four-layer p-n-p-n device that has three electrodes. They are a *cathode*, and *anode*, and a control electrode called the *gate*. Under forward bias condition where the anode is positive with respect to the cathode, the SCR has two states. At low values of forward bias, the SCR exhibits a very high impedance. This is the "off" state where a small forward current flows. As the forward bias is increased to a certain voltage value, the current increases rapidly, and the SCR device switches to the "on" state. This voltage value is called the *breakover voltage*. When the SCR is in the "on" state, the forward current is limited by the impedance of the external circuit. Under reverse bias where the anode is negative with respect to the cathode, the SCR exhibits a very high internal impedance, and only a small amount of current (the *reverse blocking current*) flows. The device remains in this state with a small current until the reverse voltage exceeds the *reverse breakdown voltage limit*. Then, the reverse current increases rapidly, and the SCR suffers irreversible damage.

Thyristors Thyristors are semiconductor devices that are similar to thyratron tubes. They include bistable semiconductor devices that have three or more junctions and devices that can be switched between conducting states. SCR is one type of thyristor.

FRANKLIN F. Y. WANG

References

Henisch, H. K., "Rectifying Semi-Conductor Contacts," London, Oxford Univ. Press, 1957.

Hunter, L. P. (Ed.), "Handbook of Semiconductor Electronics," Third Edition, New York, McGraw-Hill Book Co., 1970.

Navon, D. H., "Electronic Materials and Devices," Boston, Houghton Mifflin Co., 1975.

Wang, F. F. Y., "Introduction to Solid State Electronics," Amsterdam, North-Holland, 1980.

Cross-references: CONDUCTIVITY, ELECTRICAL; DIODE (SEMICONDUCTOR); ELECTRIC POWER GENERATION; ELECTRON TUBES; ELECTRONICS; MODULATION; SEMICONDUCTORS; THERMIONICS; TRANSISTOR.

REFLECTION

If a perfectly smooth and flat interface exists between two homogeneous media, the ratio of the reflected to the incident intensity of light striking the interface is called the *reflectivity*. A real surface, however, is not ideally smooth and undistorted so that the situation is complicated by surface conditions such as films, roughness, and disorder in the crystal lattice at the surface introduced by the polishing process. The measured ratio is thus usually termed the *reflectance*

to distinguish it from the former more- or less-idealized situation.

The reflectivity of a material is intimately connected with its band structure.[1] In a single atom, the electrons surrounding the nucleus can have only certain discrete energies. In a solid or liquid, neighboring atoms interact with each other changing these discrete energy levels into energy bands. The energy band structure depends not only on the type of atoms involved but also on the interatomic spacing, and hence in a crystalline material on the lattice structure. In a noncrystalline solid or in a liquid, the correlation function takes the place of the crystalline lattice. The highest occupied energy band may be completely filled with electrons, as in the case of a dielectric or semiconductor, or only partially filled, as for a metal. If the band is not completely full, electrons may absorb energy from the incident light beam and be raised to higher energies in the band, thus affecting the reflectivity of the material. These *intraband* transitions of the conduction electrons, or, in case of a nearly filled band, holes (the absence of electrons), largely determine the reflectivity of metals and some semiconductors in the infrared region of the spectrum. At shorter wavelengths, the light has sufficient energy to raise the electrons from one energy band to another, and the reflectivity is then also affected by these *interband* transitions.

In dielectric materials, where the occupied band having the highest energy is completely filled and the gap between allowed energy bands is large, the energy in the incident light beam cannot be absorbed either in intraband transitions or, except at very short wavelengths, in interband transitions. Such materials are therefore often transparent over an extended wavelength region in the ultraviolet, visible and near infrared, and have a nonzero reflectivity in this wavelength region only because of the difference in the speed of light in the material and in the surrounding medium. However various types of imperfections, termed F centers, M centers, etc., may exist in the lattices of crystalline materials, and electrons trapped in such centers can strongly absorb certain energies, thus affecting the reflectivity. In the intermediate and far infrared, dielectrics and many semiconductors absorb because the incident light excites vibrations in their crystal lattices. Since the frequencies of these lattice vibrations are quantized, this so-called phonon absorption results in maxima in the reflectivity spectra, termed restrahlen bands.

Other mechanisms also contribute to the reflectivity. Some which should be mentioned are transitions between a localized impurity level and an energy band, absorption by a bound hole-electron pair or exciton, and indirect or phonon-assisted interband transitions.

Electromagnetic Theory The reflectivity of any material can be calculated from electromagnetic theory[2,3] if the optical constants of the material are known. This theory, which is based entirely on Maxwell's equations, is phenomenological in that it does not attempt to explain why materials behave as they do, but rather sets forth relationships which exist between various properties of the material. In order to calculate the reflectance of a material from this theory, the two optical constants, n and k, must be known. The index of refraction n is equal to the ratio of the phase velocity of light in vacuum to the phase velocity in the material. The extinction coefficient k is equal to $\lambda/4\pi$ times the absorption coefficient α of the material; α is thus a measure of the fraction of light absorbed by a unit thickness of the material. These parameters, which are frequently combined into the complex refractive index $\bar{n} = n - jk$, arise in the solution to the wave equation and completely describe all the optical properties of the material.

The equation for the propagation of an electromagnetic wave can be obtained directly from Maxwell's equations, and in Gaussian units can be written

$$\nabla^2 \mathbf{E} = \frac{\epsilon}{c^2} \frac{\partial^2 \mathbf{E}}{\partial t^2} + \frac{4\pi\sigma}{c^2} \frac{\partial \mathbf{E}}{\partial t} \qquad (1)$$

where ∇^2 is the Laplacian operator $\partial^2/\partial x^2 + \partial^2/\partial y^2 + \partial^2/\partial z^2$, $\mathbf{E}$ the electric field strength of the traveling wave, t the time, c the velocity of light and ϵ and σ the dielectric constant and conductivity of the material, respectively, at the frequency of the wave. The solution representing a plane wave traveling in the z direction is

$$E = E_0 e^{j\omega(t-\bar{n}z/c)} \qquad (2)$$

where $\omega = 2\pi\nu$ is the angular frequency of the wave, E_0 the amplitude, and $\bar{n}$ the complex refractive index. By matching E_0 and its first derivative on either side of a smooth, plane interface between a transparent material of index n_0, the medium of incidence, and a second material of index $\bar{n}_1$, one can obtain r, the ratio of the reflected to the incident amplitude, called the *amplitude reflection coefficient*:

$$r = \frac{\eta_0 - \bar{\eta}_1}{\eta_0 + \bar{\eta}_1} \qquad (3)$$

where η_0 and $\bar{\eta}_1$ are the effective indices. At normal incidence, $\eta_0 = n_0$ and $\bar{\eta}_1 = \bar{n}_1 = n_1 - jk_1$.

At non-normal incidence, Snell's law may be used to determine the angle of refraction ϕ_r corresponding to a given angle of incidence ϕ_i. For the case when both materials are nonabsorbing, Snell's law states

$$\frac{\sin \phi_i}{\sin \phi_r} = \frac{n_1}{n_0} \qquad (4)$$

It is also necessary to specify the state of polarization of the incident light. Since E is a vector quantity, it is always possible to resolve it into two components, the so-called p and s components, polarized parallel to and perpendicular to the plane of incidence (the plane containing both the incident beam and the normal to the surface). There are then two effective indices, η_p and η_s, for each medium at a given angle of incidence. For nonabsorbing materials they are defined as

$$\eta_s = n \cos \phi \qquad (5)$$

$$\eta_p = \frac{n}{\cos \phi} \qquad (6)$$

where ϕ is the angle of incidence or refraction in the medium of index n. In the medium of incidence, ϕ becomes ϕ_i while in the second medium, ϕ becomes ϕ_r. If the second medium is absorbing ($k \neq 0$), $\cos \phi$ becomes complex, making η_s and η_p also complex. They may then be most easily calculated from the following expressions:[4]

$$\bar{\eta}_s = A - jB \qquad (7)$$

$$\bar{\eta}_p = C - jD \qquad (8)$$

where

$$A^2 - B^2 = n_1^2 - k_1^2 - n_0^2 \sin^2 \phi_i \qquad (9)$$

$$AB = n_1 k_1 \qquad (10)$$

$$C = A \left[1 + \frac{n_0^2 \sin^2 \phi_i}{A^2 + B^2} \right] \qquad (11)$$

$$D = B \left[1 - \frac{n_0^2 \sin^2 \phi_i}{A^2 + B^2} \right] \qquad (12)$$

The *intensity reflection coefficient* or *reflectivity* R, defined as the ratio of the intensities of the reflected and incident light, is obtained by multiplying the amplitude reflection coefficient of Eq. (3) by its complex conjugate. At normal incidence for an absorbing material

$$R = \frac{(n_1 - n_0)^2 + k_1^2}{(n_1 + n_0)^2 + k_1^2} \qquad (13)$$

At non-normal incidence, the expressions for R_s and R_p for absorbing materials become quite complicated. However, for nonabsorbing materials

$$R_s = \frac{\sin^2(\phi_i - \phi_r)}{\sin^2(\phi_i + \phi_r)} \qquad (14)$$

and

$$R_p = \frac{\tan^2(\phi_i - \phi_r)}{\tan^2(\phi_i + \phi_r)} \qquad (15)$$

Equations (14) and (15) are the intensity expressions for Fresnel's equations. From Eq. (15) it is seen that when $\phi_i + \phi_r = 90°$, $R_p = 0$ so that all of the reflected light is polarized in the s direction. The angle of incidence for this case is called the polarizing angle and is given by

$$\tan \phi_i = \frac{n_1}{n_0} \qquad (16)$$

A plot of the reflectivity R_p and R_s as a function of angle of incidence is shown in Fig. 1(a) for a transparent material in air and in Fig. 1(b) for an absorbing material. Note that the R_p curve for the transparent material goes to zero while the R_p curve for the absorbing material does not.

The phenomenon of total internal reflection is important for transparent materials. If light passes from a more optically dense (higher index) medium of index n_0 to a less optically dense (lower index) medium of index n_1, when $\sin \phi_i = n_1/n_0$, the angle of refraction in the less dense medium is 90°, as can be seen from Eq. (4). At larger angles of incidence, the light is totally internally reflected in the more dense medium. This method of obtaining a surface whose reflectivity is 100 per cent is widely used in optical instruments such as binoculars and periscopes, which contain internally reflecting prisms.

Whenever light is reflected from a surface, a change in phase occurs in the electric vector. For the case of a normal incidence reflection in air from a dielectric, the phase change β is 180°, while for a reflection in the more dense medium,

β is 0°. When the reflection is from an absorbing material, β depends on the optical constants of the material,[5] and tan β is given by the ratio of the imaginary to the real part of Eq. (3). If the medium of incidence is air,

$$\tan \beta = -\frac{2k}{n^2 + k^2 - 1} \qquad (17)$$

for a normal incidence reflection from an absorbing material with $\bar{n} = n - jk$. At non-normal incidence, the phase change on reflection is different for the p and s components. By a technique called ellipsometry,[6] this difference in phase can be measured and used, along with other data, to determine the optical constants of a material. If a surface film is present, the phase difference between the p and s components can also be used to measure its growth. Films as thin as a monolayer or less can be detected in this way.

As was indicated at the beginning of this section, Maxwell's equations by themselves do not give sufficient information to relate the optical constants of a material to basic atomic parameters. A fundamental problem in the theory of the optical properties of solids is thus to supplement electromagnetic theory in such a way as to relate the optical constants of a material to nonoptical quantities which can be experimentally determined. If Eq. (2) is substituted into Eq. (1), one obtains the basic equations relating the optical constants to the dielectric constant and conductivity measured at optical frequencies:

$$n^2 - k^2 = \epsilon \qquad (18)$$

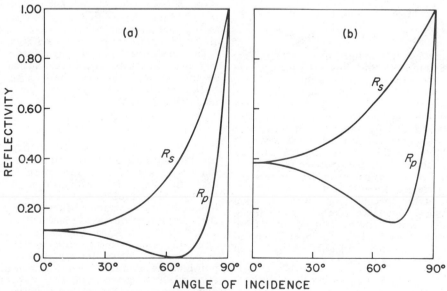

FIG. 1. Graph of the reflectivity R_p and R_s as a function of angle of incidence for (a) a transparent material ($n = 2$) in air, and (b) an absorbing material ($n = 2$ and $k = 2$) in air.

$$nk = \sigma/\nu \qquad (19)$$

The problem is how to relate ϵ and σ to the dielectric constant and conductivity at zero frequency, ϵ_0 and σ_0, or to atomic parameters such as the effective mass m^*, number of free carriers per cubic centimeter N, oscillator strength f, etc. A classical simple harmonic oscillator theory first proposed by H. A. Lorentz in 1880 has been successfully used to solve this problem for materials in which the absorption is caused by forced vibrations, while a companion classical theory proposed by P. Drude in 1901 has had similar success for materials where the absorption is due to free electrons and holes. Quantum mechanical treatments give rise to equations having the same form as the classical ones, but containing parameters whose values can be calculated rather than having to be empirically determined. Although these theories will not be discussed here, it can be pointed out that, to a very good approximation, the reflectivity of good conductors in the infrared region is given by[7]

$$R = 1 - \left[\frac{2\omega}{\pi\sigma_0}\right]^{1/2} [(1 + \omega^2\tau^2)^{1/2} - \omega\tau]^{1/2} \qquad (20)$$

where the relaxation time τ is given by

$$\tau = \frac{m^*\sigma_0}{Ne^2} \qquad (21)$$

and e is the electronic charge. At sufficiently long wavelengths, $\omega\tau \ll 1$ and the reflectivity is then given by the Hagen-Rubens relation

$$R = 1 - \left[\frac{2\omega}{\pi\sigma_0}\right]^{1/2} \qquad (22)$$

which may also be obtained directly from Eqs. (13), (18) and (19). Both Eqs. (20) and (22) are in good agreement with experiment.

Reflectance of Real Surfaces In the preceding discussion, it has been assumed that there are two homogeneous media separated by a perfectly smooth and flat interface. In actual fact this situation does not occur, and it is necessary to consider how the deviation of real surfaces from this model affects the observed reflectance. Consider first the effect of surface roughness. For a perfectly smooth surface, all light is reflected at the specular angle, which is equal to the angle of incidence. At the other extreme, the angular dependence of light reflected from a perfectly diffuse reflector is independent of the angle of incidence, and Lambert's law then holds. This law may be stated

$$I_d(\theta) = I_0 \cos\theta \qquad (23)$$

where I_0 is the total amount of light reflected per unit area normal to a perfectly diffuse plane reflector, and $I_d(\theta)$ that reflected per unit area at an angle θ to the normal. The reflectance of actual rough surfaces can be separated into "specular" and "diffuse" components, although the diffuse component usually does not obey Lambert's law exactly. If the heights of the surface irregularities are of the order of a wavelength or more, the reflectance is mostly diffuse and depends strongly on the shape of the irregularities. However, if they are very small relative to the wavelength, most of the light is reflected specularly, the amount depending only on the height of the irregularities. For a random height distribution of surface irregularities, the specular reflectance at normal incidence R_s is given by[8]

$$R_s = R_0 e^{-(4\pi h/\lambda)^2} \qquad (24)$$

where R_0 is the total reflectance of the surface, h the rms height of the surface irregularities, and λ the wavelength. R_s/R_0 is very sensitive to small values of h/λ. For example, if h is as small as $0.025\,\lambda$ (in the visible h would then be only 12 nm, or one-half a microinch), the specular reflectance will be decreased by 10 per cent. The surfaces of mirrors used in optical instruments must thus be extremely smooth. Typically the rms roughness of such mirrors is about 2.5 nm (0.1 microinch) rms and may be as small as 0.8 nm (0.03 microinch) rms. Roughness of 2.5 nm are large enough to reduce the reflectance of metals slightly in the infrared via the anomalous skin effect.[9] In addition, at wavelengths near the surface plasmon frequency for a metal the reflectance may be reduced significantly for 2.5 nm rms surfaces by optical excitation of surface plasmons.[10] This effect is of primary importance in the ultraviolet, but for metals such as silver it extends into the visible region. Since the excited plasmons may decay by incoherent reemission, scattered light may also increase where plasmon excitation occurs.

The reflectance may also be affected by distortion of the crystal lattice at the surface caused by the polishing process. The amplitude penetration depth δ of light into an absorbing medium, given by

$$\delta = \lambda/2\pi k \qquad (25)$$

is usually only a few tens of nanometers or roughly one microinch for metals or semiconductors in the intrinsic absorption region. On the other hand, lattice distortion introduced by optical polishing may extend for many hundreds of nanometers below the surface, so that the reflection takes place entirely in the disturbed surface layer. Fortunately, the lattice distortion on the surface can be nearly eliminated in many cases by using proper electropolishing techniques.[11] Optical polishing can also produce changes in the reflectance of dielectrics and noncrystalline materials such as optical glass.

Finally, surface films can have a large effect

on the reflectance. Although naturally occurring oxide films are important mainly in the ultraviolet, the reflectance may be substantially modified at any wavelength by overcoating the material with one or more evaporated thin films. Over a limited wavelength region, nearly any desired reflectance characteristic may be obtained in this way, and this technique is widely used in the optical industriy.[12] Perhaps the most familiar example of the application of a thin film is the antireflection coating on lenses. Others are the "cold" mirrors used in projection systems, the multilayer coatings which control the temperature of space vehicles by adjusting the reflectance of their outer surfaces, and finally the highly reflecting, low-absorbance dielectric multilayer films used with lasers. Some of these latter coatings have reflectances well over 99% at the laser wavelength.

Often the multilayer coatings consist of alternating high and low refractive index films, each a quarter wavelength in optical thickness. (The optical thickness of a film is the product of its physical thickness and its refractive index.) The limiting reflectance R for very high reflectance coatings is determined by the absorption in the film materials and at alternate film interfaces. It is given by[13]

$$R = 1 - \left[\frac{\lambda}{2}(\alpha_L + \alpha_H) + 4A_{HL}\right]$$

$$\cdot \left[1 - \left(\frac{n_L}{n_H}\right)^{2N+1}\right]\frac{1}{n_H{}^2 - n_L{}^2}$$

$$- 4\frac{n_S}{n_H{}^2}\left(\frac{n_L}{n_H}\right)^{2N}(1 + A_{HS}) - \left(\frac{4\pi h}{\lambda}\right)^2$$

$$\tag{26}$$

where n_S is the (nonabsorbing) substrate refractive index and n_H and n_L are the real parts of the refractive indices for the high and low index films making up the multilayer coating. The absorption coefficients of the high and low index films are α_H and α_L, respectively. As mentioned previously, the absorption coefficient α of a material is related to its extinction coefficient k by the relation

$$\alpha = 4\pi k/\lambda. \tag{27}$$

As a result, α gives the power loss per unit distance in the material traversed by a light ray, and is often expressed in units of cm^{-1}. In deriving Eq. (26) it is assumed that the optical thickness of each of the layers is $\lambda/4$ or an odd multiple thereof, that the layers are alternately of high and low refractive index, N being the number of high/low layer pairs, and that a high index layer is next to the substrate as well as being the outermost layer of the coating. The interface absorption A_{HL} is found between each high index layer and the next inner low index layer. Similarly A_{HS} is the interface absorption between the innermost high index layer and the substrate. Since a standing wave electric field is set up in the coating and goes nearly to zero at the surface and at each low index-high index interface, absorption does not occur at the outer surface of the coating and at each alternate interface. If the low refractive index layer is next to the substrate, the next to the last term in Eq. (2) involving the substrate refractive index n_S is replaced by[13]

$$\frac{-4}{n_S}\left(\frac{n_L}{n_H}\right)^{2N}.$$

The last term in Eq. (26) represents light scattered by the substrate-film combination. If the films contour the surface exactly, as is usually the case to a good approximation, then the single interface expression for scattered light given by Eq. (24) can be used to calculate scattering from a multilayer coating by simply substituting the multilayer reflectance for the single surface reflectance R_0.

Equation (26) represents only one of a great number of possible interference coatings which could be used to modify the reflectance of a surface. Another common design is the antireflection quarter-wave coating, which is commonly used on camera lenses and other transmissive optics to minimize reflectance. Both single and multilayer coatings are used for this purpose. If a single layer coating of a quarter-waver optical thickness is used and the absorption is assumed to be negligible, the reflectance R of the interface is

$$R = \left(\frac{n_S - n_L{}^2}{n_S + n_L{}^2}\right)^2. \tag{28}$$

This expression approaches zero if $n_L \simeq \sqrt{n_S}$. In this case the normal incidence reflectance of the surface completely disappears.

H. E. BENNETT

References

1. Callaway, J., "Energy Band Theory," New York, Academic Press, 1964; Bube, R. H., "Photoconductivity of Solids," New York, John Wiley & Sons, Inc., 1960.
2. Born, M., and Wolf, E., "Principles of Optics," New York, Pergamon Press, 1959; Ditchburn, R. W., "Light," Second edition, London, Blackie and Son Ltd., 1963.
3. Stratton, J. A., "Electromagnetic Theory," New York, McGraw-Hill Book Co., 1941; Slater, J. C., and Frank, N. H., "Electromagnetism," New York, McGraw-Hill Book Co., 1947.
4. Abelès, F., "Progress in Optics," Vol. 2, Amsterdam, North-Holland Publishing Co., 1963.
5. Bennett, J. M., J. Opt. Soc. Am., 54, 612 (1964).

6. McCrackin, F. L., Passaglia, E., Stromberg, R. R., and Steinberg, H. L., *J. Res. Natl. Bur. St.*, 67A, 363 (1963).
7. Seitz, F., "The Modern Theory of Solids," New York, McGraw-Hill Book Co., 1940; Bennett, H. E., Silver, M., and Ashley, E. J., *J. Opt. Soc. Am.*, 53, 1089 (1963).
8. Bennett, H. E., and Porteus, J. O., *J. Opt. Soc. Am.*, 51, 123 (1961); Porteus, J. O., *J. Opt. Soc. Am.*, 53, 1394 (1963); Beckmann, P., and Spizzichino, A., "The Scattering of Electromagnetic Waves from Rough Surfaces," New York, The Macmillan Co., 1963.
9. Bennett, H. E., Bennett, J. M., Ashley, E. J., and Motyka, R. J., *Phys. Rev.*, 165, 755 (1968).
10. Jasperson, S. N., and Schnatterly, S. E., *Phys. Rev.*, 188, 759 (1969); Endriz, J. G., and Spicer, W. E., *Phys. Rev.*, 4B, 4144 (1971); Elson, J. M., and Ritchie, R. H., *Phys. Rev.*, 4B, 4129 (1971).
11. Donovan, T. M., Ashley, E. J., and Bennett, H. E., *J. Opt. Soc. Am.*, 53, 1403 (1963); Holland, L., "The Properties of Glass Surfaces," London, Chapman and Hall, 1964.
12. Heavens, O. S., *Rep. Progr. Phys.*, 23, 1 (1960); Heavens, O. S., "Optical Properties of Thin Solid Films," London, Butterworths Scientific Publications, 1955.
13. Bennett, H. E., and Burge, D. K., *J. Opt. Soc. Am.* 70, 268 (1980).

Cross-references: ENERGY LEVELS; OPTICS, GEOMETRICAL; OPTICS, PHYSICAL; POLARIZED LIGHT; REFRACTION; SEMICONDUCTORS.

REFRACTION

Refraction is the name given to the bending of a ray of light as it crosses the boundary separating two transparent media having differing propagation velocities.

In an attempt to discover a law connecting the directions of the light rays in the two media, W. Snell (1621) observed that there is a constant ratio between the lengths PB, PC of the two rays (Fig. 1) measured from the point of incidence P to any line such as DD' drawn parallel to the normal PN at the point of incidence. Later Descartes recognized that Snell's construction is equivalent to the mathematical expression

$$n \sin I = n' \sin I'$$

when I, I' are the angles of incidence between the two rays and the normal, and n, n' are the *refractive indices* of the two media respectively. The ratio of n' to n is called the "relative" refractive index of the two media.

Huygens' Wavelets C. Huygens (1690) attempted to explain how a wave front progresses through a transparent medium. He supposed that each point on the wave front acts as an independent source of wavelets which expand at the velocity of light, the new wave front being the common envelope of all the little wavelets. After an instant of time, each point in the new wave front becomes a source of new wavelets, and so on. In a homogeneous isotropic medium, the wavelets will be spheres, and the new wave front is a parallel curve to the original wave front. The light energy travels along *rays* which are everywhere perpendicular to the wave fronts. Malus (1808) showed that the rays and wave fronts remain always orthogonal as the beam of light from an object point traverses an optical system.

The absolute refractive index of a medium is defined as the ratio of the velocity of light in vacuum to its velocity in the medium. Because the presence of matter has the effect of making light travel more slowly, all absolute refractive indices are necessarily positive and greater than unity. Since the frequency of the light waves must remain constant, the wavelength in a dense medium will be less than the wavelength in vacuum in proportion to the refractive index of the medium. Refractive indices of liquids and crystals generally drop with increasing temperature; glasses are, however, often exceptional in this regard.[1]

Refractive indices range from as low as 1.3 for some liquids such as water, through 1.5 to 1.9 for various types of glass, 2.5 for diamond, up to as high as 4.0 for some materials (germanium) in the infrared. These exceptionally high indices are generally associated with complete opacity in the visible part of the spectrum.

Dispersion It is found that the refractive index of all common materials rises with increasing frequency of the light (shorter wavelengths) leading to the phenomenon of *dispersion*. Because of this, a glass prism bends blue light more than red, thus spreading a ray of white light into its component colors (Fig. 2). For a prism of very small angle A, the deviation angle D is given by

$$D = A(n - 1)$$

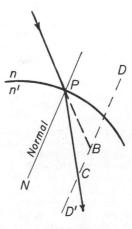

FIG. 1. Illustrates Snell's construction.

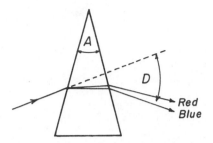

FIG. 2. The dispersion caused by a prism.

and the angular dispersion Δ between wavelengths a and b is given by

$$\Delta = A(n_b - n_a).$$

It has been customary to define the dispersive power w of a material as the ratio of the dispersion between the F and C Fraunhofer lines to the mean deviation, i.e., the deviation for the D Fraunhofer line. Thus

$$w = (n_F - n_C)/(n_D - 1).$$

The vacuum wavelengths of these lines are $C = 0.6563\mu$, $D = 0.5893\mu$, and $F = 0.4861\mu$. In the optical industry, the reciprocal of the dispersive power is, however, more generally used; i.e.,

$$V = 1/w = (n_D - 1)/(n_F - n_C).$$

This so-called "V-value" or "Abbe number" of optical glass ranges from about 25 for the densest flints up to about 70 for the lightest fluor crowns. Liquids and crystals are known in which the V-numbers range from about 16 (methylene iodide) to 95 (calcium fluoride).

Many attempts have been made to develop a formula connecting the refractive index of a material with the wavelength of the light. The best known and most comprehensive is that proposed by Sellmeier,[2] namely,

$$n^2 = 1.0 + \sum \frac{b\lambda^2}{c^2 - \lambda^2}.$$

Here n is the refractive index corresponding to wavelength λ, while b and c are constants. The constant c represents the center of an absorption band for which the refractive index becomes infinite; there are thus as many terms under the summation sign as there are absorption bands to be considered. For most transparent materials, it is sufficient to include one absorption band in the ultraviolet and one in the infrared.

This general formula may be reduced to a simpler form for a limited spectral range, some well-known simplifications being

$$n = a + b/\lambda^2 + c/\lambda^4 + \cdots \quad \text{(Cauchy)}$$

$$n = 1 + b/(c - \lambda) \quad \text{(Hartmann)}$$

$$n = a + b/\lambda + c/\lambda^{7/2} \quad \text{(Conrady)}$$

Herzberger[3] has shown that the refractive index of a transparent substance such as glass can be accurately represented by a four-constant formula of this type:

$$n = a + b\lambda^2 + cL + dL^2$$

where $L = 1/(\lambda^2 - 0.028)$.

Birefringence, or Double Refraction Glasses, liquids, and some (e.g., cubic) crystals are *isotropic*, meaning that the velocity of light in the medium is independent of direction and the Huygenian wavelets are spherical. However, there are other crystals which are *anisotropic*. These have the property that a beam of light on entering the crystal is split into two perpendicularly polarized beams, one of which (the "ordinary") behaves normally, while the other (the "extraordinary") behaves abnormally in that the wave front is not perpendicular to the direction of propagation.

These two wave fronts can be explained by supposing that two Huygenian wavelets are formed at any point of incidence, the ordinary wavelet being spherical while the extraordinary wavelet is an ellipsoid, (Fig. 3).[4] The advancing wave front is the common tangent to a row of wavelets, and the ray, or direction of travel, is found by joining the point of origin of the wavelet to the point of contact between the wavelet and the advancing wave front. The spherical wavelets thus yield a ray which is perpendicular to the wave front, but the ellipsoidal wavelets do not.

All anisotropic crystals possess either one or two directions ("optic axes") along which a ray of light is not divided into two. These are represented by the axes of symmetry of the ellipsoidal wavelets. Evidently if parallel light is travelling along an optic axis, the two advancing wave fronts will be parallel to one another and the two rays will coincide (see POLARIZED LIGHT).

Refraction by an Inhomogeneous Medium Suppose we have a parallel plate of material of which the refractive index varies laterally across the plate. A ray entering such a plate perpendicularly to its surface will suffer no refraction itself since the incidence angle is zero, but because there is a gradient of refractive index at right angles to the ray, the ray inside the plate will be bent towards the high-index region and will follow a curved path. This is readily seen if we remember that the ray will be perpendicular to the wave front and that each point on the wave front travels at a rate which is inversely proportional to the refractive index at that point. To determine the curvature of the ray inside the medium, we may refer to Fig. 4 in which CC' represents the surface of a nonhomogeneous medium, and A'A, B'B represent two very close rays entering perpendicular to the surface. We shall suppose the velocity of light in the material to be v at A and $(v + dv)$ at B, the refractive indices being $(n + dn)$ and n respectively. Typical wavelets have been added at A and B, having radii AD and BE which are proportional to the

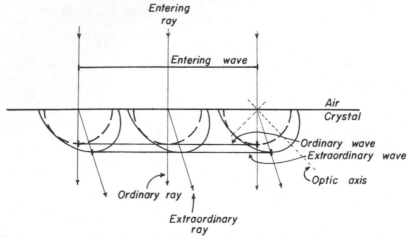

FIG. 3. Explanation of double refraction.

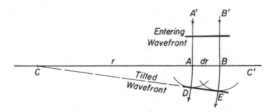

FIG. 4. Bending of a light ray in an inhomogeneous medium.

velocities v and $(v + dv)$. Hence the refracted wave front is ED; this intersects the surface at C, which is therefore the instantaneous center of curvature of the rays within the medium.

If r is the radius of ray AD,

$$\frac{CA}{CB} = \frac{r}{r+dr} = \frac{AD}{BE} = \frac{v}{v+dv} = \frac{n+dn}{n}.$$

Neglecting second-order infinitesimals this gives

$$\frac{1}{r} = -\frac{1}{n}\left(\frac{dn}{dr}\right).$$

The curvature of the ray is, therefore, proportional to the rate of change of refractive index in a direction perpendicular to the ray itself.

This property of inhomogeneous media provides the explanation of mirages and also the well-known atmospheric refraction which makes celestial objects appear to be raised up by as much as half a degree for objects situated near the horizon.

An extensive literature[5] has recently appeared dealing with the passage of light through gradient-index materials, as these may eventually replace aspheric surfaces in lenses. Short glass rods having a radial index gradient (so-called GRIN rods) have the property of forming images of external objects. The increasing use of fiber-optic communication channels over long distances requires fibers having a gradient of refractive index across the fiber to equalize the time of transit of light along all possible paths within the fiber.

R. KINGSLAKE

References

1. Molby, F. A., *J. Opt. Soc. Am.*, 39, 600–611 (1949)
2. Wood, R. W., "Physical Optics," page 470, New York, Dover, 1967.
3. Herzberger, M. J., *Optica Acta*, 6, 197–215 (1959).
4. Wood, R. W., "Physical Optics," Third Edition, pp. 365–387, New York, Dover, 1967.
5. E. W. Marchand, "Gradient Index Optics," New New York, Academic Press, 1978.

Cross-references: FIBER OPTICS; LENS; LIGHT; OPTICS, GEOMETRICAL; OPTICS, GEOMETRICAL, ADVANCED; OPTICS, PHYSICAL; POLARIZED LIGHT; REFLECTION; WAVE MOTION.

REFRIGERATION

Refrigeration is the process of cooling or freezing a substance to a temperature well below that of the surroundings and maintaining it in the cold state. The simplest device for this purpose consists of a box or a cryostat immersed in a cooling agent such as cold running water, circulating air cooled by melting ice, liquid air, liquid helium, etc. Generally, the coolant must be replenished periodically, an inconvenient and often inefficient procedure. Since the invention of vapor-compression and vapor-absorption refrigeration methods, recirculating mechanical refrigerators have become commonplace. Other refrigeration methods exploit the thermoelectric properties of semiconductors, the magneto-thermoelectric effects in semi-metals, or the diffusion of ^{3}He atoms across the interface between distinct phases of liquid helium having high and low concentrations of ^{3}He in ^{4}He.

A refrigerator operating in a cyclic process may be considered a heat pump, for it continually extracts heat from a low-temperature region and delivers it to a high-temperature region. It is rated by its *coefficient of performance*, defined as the ratio of the heat removed from the cold region (room, box, cryostat) per unit time to the net input power for operating the device, in symbols $K = Q_t/P$. Vapor-absorption and thermoelectric refrigerators have lower coefficients of performance than vapor-compression refrigerators, but they have other characteristics that are superior, such as quietness of operation and compactness.

Vapor-compression Refrigerator This machine consists of a compressor, a condenser, a storage tank, a throttling valve, and an evaporator connected by suitable tubes with intake and outlet valves, as shown schematically in Fig. 1. The refrigerant is a liquid which partially vaporizes and cools as it passes through the throttling valve. Among the common refrigerants are ammonia, sulfur dioxide, and certain halide compounds of methane and ethane. Perhaps the most widely used of these in industry is ammonia and in the household dichlorodifluoromethane ("Freon-12"). Nearly constant pressures are maintained on either side of the throttling valve by means of the compressor. The mixed liquid and vapor entering

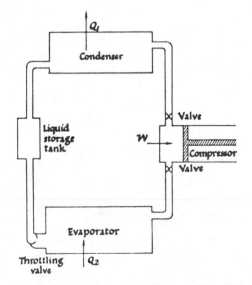

FIG. 1. Vapor-compression refrigerator (from "Thermophysics," by Allen L. King, San Francisco, W. H. Freeman and Co., 1962).

the evaporator is colder than the near-surround; it absorbs heat from the interior of the refrigerator box or cold room and completely vaporizes. The vapor is then forced into the compressor where its temperature and pressure increase as it is compressed. The compressed vapor finally pours into the condenser where it cools down and liquefies as the heat is transferred to cold air, water, or other fluid flowing by the cooling coils.

Comparative tests have shown that the coefficient of performance of vapor-compression refrigerators depends very little on the nature of the refrigerant. Because of mechanical inefficiencies, its actual value may be well below the ideal value—ordinarily, between 2 and 3.

Vapor-absorption Refrigerator In this system there are no moving parts; the added energy comes from a gas or liquid fuel burner or from an electrical heater, as heat, rather than from a compressor, as work. A simplified diagram of it is shown in Fig. 2. The refrigerant is ammonia gas, which is liberated from a water solution and transported from one region to another by the aid of hydrogen. The total pressure throughout the system is constant and therefore no valves are needed.

Heat from the external source is supplied to the generator where a mixture of ammonia and water vapor with drops of ammoniated water is raised to the separator in the same manner as water is raised to the coffee in a percolator. Ammonia vapor escapes from the liquid in the separator and rises to the condenser, where it cools and liquefies. Before the liquefied ammonia enters the evaporator, hydrogen, rising from the absorber, mixes with it and aids in the evaporation process. Finally, the mixture of hydrogen and ammonia vapor enters the absorber, where water from the separator dissolves the ammonia. The ammonia water returns to the generator to complete the cycle. In this cycle heat enters the system not only at the generator but also at the evaporator, and heat leaves the system at both the condenser and the absorber to enter the atmosphere by means of radiating fins.

No external work is done, and the change in internal energy of the refrigerant during a complete cycle is zero. The total heat $Q_a + Q_c$ released to the atmosphere per unit time by the absorber and the condenser equals the total heat $Q_g + Q_c$ absorbed per unit time from the heater at the generator and from the cold box at the evaporator; so $Q_e = Q_a + Q_c - Q_g$, and therefore the coefficient of performance is $K = Q_e/Q_g = \{(Q_a + Q_c)/Q_g\} - 1$.

The vapor-absorption refrigerator is free from intermittent noises; but it requires a continuous supply of heat, as from bottled gas or electrical generators. Refrigerators of this type are found in camps and farm houses not supplied with commercial electric power and in apartment houses where unnecessary noise is prohibited.

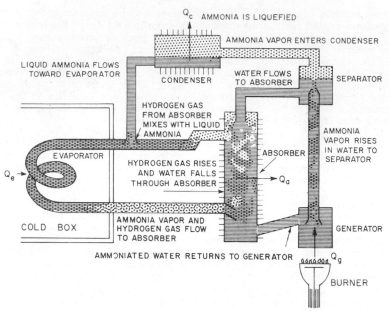

Q_c AMMONIA IS LIQUEFIED

AMMONIA VAPOR ENTERS CONDENSER

LIQUID AMMONIA FLOWS
TOWARD EVAPORATOR

CONDENSER

WATER FLOWS
TO ABSORBER

SEPARATOR

HYDROGEN GAS
FROM ABSORBER
MIXES WITH LIQUID
AMMONIA

ABSORBER

AMMONIA
VAPOR RISES
IN WATER TO
SEPARATOR

EVAPORATOR

Q_e

HYDROGEN GAS RISES
AND WATER FALLS
THROUGH ABSORBER

Q_a

COLD BOX

AMMONIA VAPOR AND
HYDROGEN GAS FLOW
TO ABSORBER

GENERATOR

AMMONIATED WATER RETURNS TO GENERATOR Q_g

BURNER

FIG. 2. Vapor-absorption refrigerator (from "Thermophysics," by Allen L. King, San Francisco, W. H. Freeman and Co., 1962).

Dilution Refrigerator Below a temperature of 0.87 K liquid mixtures of ^{3}He and ^{4}He at certain concentrations separate into two distinct phases, a concentrated (^{3}He-rich) phase floating on a denser (^{4}He-rich) phase with a visible surface between them. The concentrations of ^{3}He in the two phases are functions of temperature, approaching 100 per cent in the concentrated phase and about 6 per cent in the dilute phase at 0 K. The transfer of ^{3}He atoms from the concentrated to the dilute phase, like an evaporation process, entails a latent heat, an increase in entropy, and a lowering of temperature. This effect is utilized in the dilution refrigerator for obtaining temperatures of extremely low values.

The diagram in Fig. 3 shows the main components of a recirculating dilution refrigerator. The pump forces helium vapor (primarily ^{3}He) from the still into the condenser, where it is liquefied at a temperature near 1 K in a bath of rapidly evaporating ^{4}He, through a flow controller consisting of a narrow tube of suitable diameter to obtain an optimum rate of flow, and then through the still where its temperature is further reduced to about 0.6 K. The liquefied ^{3}He next passes through a heat exchanger (sintered copper or sintered silver) so as to reduce its temperature to nearly that of the dilution chamber, by giving up thermal energy to the counterflowing dilute phase, before entering the concentrated phase therein. The diffusion of ^{3}He atoms from the concentrated into the dilute phase within this chamber can produce steady temperatures of very low values

(2.5 mK or less). Liquid ^{3}He from the dilute phase then passes through the heat exchanger to the still where it is warmed to transform the liquid to the vapor phase that goes to the pump thus completing the cycle. Modified versions of this system have been constructed, sometimes with an added single-cycle process for produc-

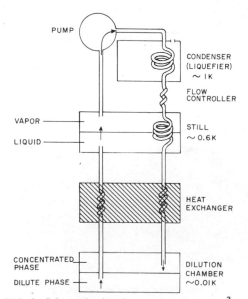

PUMP

CONDENSER
(LIQUEFIER)
~ 1 K

FLOW
CONTROLLER

VAPOR

LIQUID

STILL
~ 0.6 K

HEAT
EXCHANGER

CONCENTRATED
PHASE

DILUTE PHASE

DILUTION
CHAMBER
~0.01 K

FIG. 3. Schematic showing components of a ^{3}He–^{4}He recirculating dilution refrigerator.

ing temporarily even lower temperatures. The low-temperature limit in any of these systems is determined largely by two important sources of inefficiency that cannot be completely eliminated; heat leakage, especially severe because of the extreme range of temperatures, and recirculation of some ^{4}He with ^{3}He.

Thermoelectric Refrigerator This device utilizes the thermoelectric effect, discovered by Peltier in 1834 (see THERMOELECTRICITY), wherein heat is either absorbed or generated at the junction of two different conductors, or semiconductors, depending on the direction of an electrical current through it. For satisfactory operation in refrigerating devices the thermojunctions should be made of materials having not only high thermoelectric coefficients but also high electrical conductivities, so as to reduce Joule heating, and low thermal conductivities, so as to minimize heat losses by conduction. These requirements are best met by semiconductors such as lead telluride and bismuth telluride.

A thermoelectric refrigerator unit consists of one or more stages of series-connected n-type and p-type semiconductors as illustrated in Fig. 4(a). The charge carriers are electrons in the n-type semiconductor and holes in the p-type semiconductor. When a difference of potential is maintained across AB with B at the higher potential, the negative electrons carry both kinetic and potential energy away from C as they move toward B in the n-type semiconductor; the positive holes do the same as they move away from C toward A in the p-type semiconductor. Since energy is carried away from C to AB in both arms of the thermocouple, junction C becomes cold and junctions AB become warm. Temperature differences as large as 75 C have been obtained in a single-stage unit. Larger temperature differences may be produced by arranging several stages in cascade as illustrated by the two-stage system in Fig. 4(b). In this way a temperature of $-118°C$ was reached in a small seven-stage thermoelectric refrigerator only 38 mm high, employing bismuth telluride alloys in the thermojunctions and water at 27°C as the heat sink.

The coefficient of performance of a multistage thermoelectric refrigerator is no greater than that of one of its units. Let the temperature

difference between AB and C in Fig. 4(a) be ΔT. As a result of the Peltier effect, the cold junction cools down at the rate $\bar{\alpha}(T_m - \frac{1}{2}\Delta T)I$, where $\bar{\alpha}$ is the mean value of the Seebeck coefficients for the n-type and p-type semiconductors, T_m is the mean temperature of the thermocouple, and I is the current through it. But due to Joule heating and thermal conduction from the warm to the cold junction, the rate of cooling at C is only $Q_t = \bar{\alpha}(T_m - \frac{1}{2}\Delta T)I - \frac{1}{2}I^2R - \lambda\Delta T$, where the total resistance of the couple $R = (l_p/A_p\sigma_p) + (l_n/A_n\sigma_n)$ and its thermal conductance $\lambda = (A_pk_p/l_p) + (A_nk_n/l_n)$ in which A_p and A_n are the cross-sectional areas of the p-type and n-type semiconductors, l_p and l_n are their lengths, σ_p and σ_n are their electrical conductivities, and k_p and k_n are their thermal conductivities. The power supplied externally must just equal the total Seebeck and Joule terms, namely, $P = \bar{\alpha}I\,\Delta T + I^2R$. The coefficient of performance, therefore, is

$$K = Q_t/P = \frac{\bar{\alpha}(T_m - \frac{1}{2}\Delta T)I - \frac{1}{2}I^2R - \lambda\Delta T}{\bar{\alpha}I\,\Delta T + I^2R}$$

It reaches a maximum value when the products of the thermal and electrical conductances for the two semiconductors have the same value and when the electrical resistance at the junctions is much smaller than R. The optimum current then is given by the equation $(IR)_{opt} = \bar{\alpha}\,\Delta T[\sqrt{1 + ZT_m} - 1]^{-1}$ where Z is the *figure of merit* of the thermocouple,

$$Z = \bar{\alpha}^2/[(k_p/\sigma_p)^{1/2} + (k_n/\sigma_n)^{1/2}]^2$$

The maximum value of K can now be written in the form

$$K_{max} = \frac{T_m(\sqrt{1 + ZT_m} - 1)}{\Delta T(\sqrt{1 + ZT_m} + 1)} - \frac{1}{2}$$

Thus K_{max} increases with an increase in the figure of merit, reaching the value $(T_m - \frac{1}{2}\Delta T)/\Delta T$ for very large Z. This is the coefficient of performance for an ideal thermodynamic machine.

The thermoelectric refrigerator is useful where space is at a premium or where, for other reasons, mechanical refrigerators are inconvenient.

Thermomagnetic and Magnetothermoelectric Refrigerators In 1958 O'Brien and Wallace suggested that by means of the Ettingshausen effect, one should be able to achieve cooling for refrigeration purposes (see HALL EFFECT AND RELATED PHENOMENA). This suggestion has been followed up with some success at low temperatures and even with the heat sink near room temperature by a special "cascading" device. In 1962, Smith and Wolfe discovered that the thermoelectric figure of merit for bismuth-antimony alloys can be increased by means of a magnetic field and that this enhancement is especially pronounced at low temperatures. These

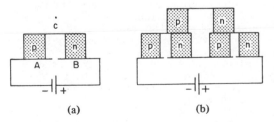

FIG. 4. Thermoelectric refrigerator: (a) single stage, (b) two stage cascade.

effects may be applied advantageously to refrigeration at low ambient temperatures.

ALLEN L. KING

References

Betts, D. S., "Refrigeration and Thermometry below One Kelvin," London, Sussex Univ. Press, 1976.

Goldsmid, H. J., "Applications of Thermoelectricity," London, Methuen and Co., Ltd., 1960.

King, A. L., "Thermophysics," San Francisco, W. H. Freeman and Co., 1962.

Wolfe, R., "Magnetothermoelectricity," *Sci. Am.*, 210, 70 (1964).

Worthing, A. G., and Halliday, D., "Heat," New York, John Wiley & Sons, 1948.

Zemansky, M. W., "Heat and Thermodynamics," New York, McGraw-Hill Book Co., 1968.

Zimmerman, James E., and Flynn, Thomas M. (Eds.), "Applications of Closed-Cycle Cryocoolers to Small Superconducting Devices," Special Publication No. 508 of the National Bureau of Standards, Washington, U.S. Government Printing Office, 1978.

Cross-references: CRYOGENICS, GAS LAWS, HALL EFFECT AND RELATED PHENOMENA, HEAT, HEAT TRANSFER, LIQUEFACTION OF GASES, SEMICONDUCTORS,   THERMOELECTRICITY, THERMODYNAMICS.

RELATIVITY

The basic ideas of modern relativity theory are largely due to one man: Albert Einstein (1879–1955). Both main branches of pre-Einstein physics (mechanics and electromagnetism) had relied on an absolute space. To Newton this had served as the agent responsible for a particle's resistance to acceleration; to Maxwell—in the guise of an "aether"—it was the carrier of electromagnetic stresses and waves and thus, in particular, of light. Relativity may be defined briefly as the abolition of absolute space. Special relativity (1905) abolished it in its Maxwellian sense, and general relativity (1915) in its Newtonian sense as well.

Before looking at the theoretical background of relativity, we shall mention some of its more striking practical implications. According to special relativity, for example, a rod moving longitudinally at speed v through an inertial frame is shortened, relative to that frame, by a factor $\gamma = (1 - v^2/c^2)^{-1/2}$, where c is the speed of light. This factor increases with v; when v is as large as $\frac{1}{7}c$, γ is only 1.01, but at higher speeds it grows rapidly and becomes infinite when $v = c$. The rate of a clock moving at speed v is decreased by the same factor γ; this is one aspect of the revolutionary prediction that time, also, is not absolute and that, for example, after journeying at high speed through space, one could, upon return, find the world aged very much more than oneself. In fact, time and space

become merged in a four-dimensional continuum in which neither possesses more absoluteness than, e.g., the x-separation between points in a Cartesian plane, which depends on the choice of axes. According to special relativity, time- and space-separations between events are similarly non-absolute and depend, in fact, on the motion of the reference frame. The mass of a body moving at speed v is also increased by the factor γ and thus becomes infinite at the speed of light. This illustrates another prediction of the theory—that no body or physical effect can travel faster than light. But the single most important result of special relativity, in Einstein's opinion, was the equivalence of mass m and energy E according to the formula $E = mc^2$.

Although the original impact of special relativity was mainly theoretical and philosophical—its main effects show up only under certain extreme conditions—technology since 1905 has made such vast strides (atomic power, particle accelerators, Mössbauer effect, etc.) that today special relativity is one of the most practical and, at the same time, best verified branches of all physics.

The same cannot yet be said for general relativity, whose results are still largely theoretical and speculative. But its ideas are hardly less striking. General relativity is the modern theory of gravitation. Like special relativity, it pictures the world as a four-dimensional space-time continuum, but according to general relativity this is curved by the matter present in it. Particles and light rays are postulated to move along geodesics ("straightest possible" curves) in this four-space, and all reference to a "force" of gravity or to an absolute three-space as the standard of nonacceleration has disappeared. It is one of the marvels of this theory that, starting from such utterly different premises, it nevertheless reproduces within experimental accuracy almost all the well-established results of Newton's (inverse square) gravitational theory. In the few cases where its predictions differ to a presently measurable extent from Newton's theory (as for the advance of the perihelia of the planets), general relativity has been borne out by observation. Furthermore, general relativity first led to the construction of many interesting cosmological models, such as closed and finite universes; it implies the existence of gravitational waves and thus the need to quantize the gravitational field; and it contains such exotic possibilities as "black holes" (into which space itself falls at the speed of light, so that *nothing*, not even light, can come out).

That gravity can be "explained" by theories as diverse as Newton's and Einstein's well illustrates the character of all physical theories: they are no more than mathematical *models* of certain parts of nature; they cannot aspire to be ultimate truths.

The theoretical basis of special relativity is Einstein's *principle of relativity* which asserts that all the laws of physics are equally valid in all inertial frames of reference. This is an extension to the whole of physics of a relativity principle which the laws of mechanics have long been known to obey. Newton, as Galileo before him, illustrated this with the familiar example of a ship, "where all motions happen after the same manner whether the ship is at rest or is carried forward in a right line." The reason why Einstein's principle was revolutionary is that the known properties of light seem to contradict it at once. In our quasi-inertial terrestrial reference frame (which was assumed to coincide more or less with Maxwell's aether), light is propagated rectilinearly in all directions at constant speed, in accordance with Maxwell's theory. This fact is often called the *law of light propagation*. But the validity of this law in *all* inertial frames would imply, for example, that a light signal emitted at the instantaneous coincidence of two observers O and O' who are moving uniformly relative to each other, each fixed in an inertial frame, spreads spherically with *both* observers considering themselves to remain permanently at the center of the expanding signal-sphere. Hence a light signal would always recede from an observer at the same speed, no matter how fast he chases it. The adoption of the principle of relativity together with the law of light propagation thus seems to lead to absurdities. But, in fact, that is not so: it merely leads to the downfall of the classical ideas of space and time. It was part of Einstein's genius to recognize that these ideas were dispensable and could be replaced by others.

Two types of argument can be made in support of Einstein's principle. The first is experimental: all experiments devised to discover the frame of Maxwell's aether (in which *alone* the law of light propagation had been expected to hold), such as the well-known MICHELSON-MORLEY EXPERIMENT (1887), failed to give positive results, though such results would have been well within range of observability. The second argument is theoretical, and rests on the unity of physics. For example, mechanics involves matter, which is electromagnetically constituted; electromagnetic apparatus involves mechanical parts; and so forth. If then, physics cannot be separated into strictly exclusive branches, it would seem unlikely that the laws of different branches should have different transformation properties, i.e., relativity for mechanics but not for electromagnetism.

Consider now two observers O and O' like the ones mentioned earlier, and a light signal emitted at their coincidence. If each observer remains at the origin of a Cartesian reference system and sets his clock to read zero when the signal is emitted, the events on the light front must satisfy both the equations

$$x^2 + y^2 + z^2 - c^2t^2 = 0$$

$$x'^2 + y'^2 + z'^2 - c^2t'^2 = 0 \qquad (1)$$

where primes distinguish the space and time coordinates used by O' from those used by O. Now suppose the two observers arrange their corresponding y and z axes to remain parallel, and their x axes to coincide. In classical kinematics, with this configuration of reference systems, the so-called *Galilean transformation equations*

$$x' = x - vt, \quad y' = y, \quad z' = z, \quad t' = t \qquad (2)$$

relate the corresponding coordinates of any event, where v is the relative velocity between O and O'. But under this transformation, the two equations (1) are not equivalent. Einstein showed that for these equations to be equivalent, the transformation equations must necessarily be

$$x' = \gamma(x - vt), \quad y' = y, \quad z' = z,$$
$$t' = \gamma(t - vx/c^2) \qquad (3)$$

where $\gamma = (1 - v^2/c^2)^{-1/2}$. These are the well-known *Lorentz equations* which constitute the mathematical core of the special theory of relativity. They replace equations (2), to which they nevertheless approximate when v is small. The most striking of equations (3) is the last. It implies that events with the same value of t do not necessarily correspond to events with the same value of t', which means that *simultaneity is relative*. Setting $x = 0$ in that equation also shows that the clock at the origin of O goes slow by a factor γ in the frame of O'. But, setting $x = vt$, we see that the clock at the origin of O' similarly goes slow in the frame of O. Setting $t = 0$ in the first of equations (3), we see that a rod, fixed in the frame of O' along the x' axis, appears shortened by a factor γ in the frame of O; this phenomenon too can be shown to be symmetric between the frames.

An important property (and, in fact, the *defining* property) of equations (3) is that they leave invariant the differential quadratic

$$ds^2 = dx^2 + dy^2 + dz^2 - c^2dt^2 \qquad (4)$$

which is, of course, related to the equivalence of equations (1). This leads to the possibility of mapping events in a four-dimensional pseudo-Euclidean *space-time* in which an absolute *interval ds* exists, and in which the language and results of four-dimensional geometry can thus be applied. For example, a uniformly moving particle is described simply by a straight line in this space-time, while a nonuniformly moving particle corresponds to a curve in space-time, whose curvature represents its acceleration.

Equation (4) determines the new space-time background to all physics, replacing the older concepts of absolute 3-space and absolute time. All physics must be made consistent with this new background. Special relativity is therefore often called a "metatheory," since it imposes certain requirements on *all* physical theories.

Thus it was the first task of special relativity to review the existing laws of physics and to subject them to the test of the new relativity principle by seeing whether they were invariant under Lorentz transformations. Any law found lacking had to be modified accordingly.

Since Newton's laws of mechanics are invariant under the transformation (2) and *not* (3), it was necessary to amend these laws so as to make them "Lorentz invariant." It was found possible to do this by retaining the classical laws of conservation of mass and momentum *as axioms*, but postulating that the mass of moving bodies increases by the factor γ, a fact amply borne out by modern particle accelerators. This led to the theoretical discovery of the equivalence of mass and energy—most spectacularly exemplified by the atomic bomb, which *transmutes* mass into energy.

In contrast to Newton's theory, Maxwell's vacuum electrodynamics was compatible with Einstein's theory. Lorentz, independently of Einstein, but without realizing the full significance of his result, had already discovered equations (3) as precisely those which leave Maxwell's equations invariant. In other words, Maxwell's equations already were "Lorentz invariant" and needed no modification. (Hence the validity of the law of light propagation in all inertial frames.) Nevertheless relativity has considerably deepened our understanding of Maxwell's theory. Other branches of physics, like optics, hydrodynamics, thermodynamics, nonvacuum electrodynamics, etc., all underwent modifications to make them Lorentz invariant. Only Newton's inverse square gravitational theory proved refractory; several Lorentz invariant modifications of it were proposed but none was entirely satisfactory.

Einstein eventually solved the gravitational problem in an unexpected way. He rejected Newton's absolute space as the cause of inertia on the grounds that "it is contrary to the spirit of science to conceive of a thing which acts but cannot be acted upon." His general theory of relativity ascribes to the space-time continuum discovered by special relativity the role of an inertial guiding field (free particles and light follow geodesics) but allows this field to be affected (curved) by the matter in it.

This extension was made possible by the so-called *principle of equivalence*. To Newton, an inertial frame was, primarily, the (infinite) frame of "absolute space" in which the stars were assumed to be fixed, and, secondarily, any frame moving uniformly relative to absolute space. Thus a Newtonian inertial frame exhibits its defining property, viz., that in it free particles move uniformly and rectilinearly (Newton's first law), only in the regions far from attracting masses, since elsewhere there *are* no free particles. In 1907 Einstein replaced this global definition with a local one: a *local* inertial frame is a freely falling non-rotating reference system. (The meaning of "local" is here determined by

the extent to which the nonuniformity of the gravitational field is negligible.) We have all had vicarious experience of local inertial frames by watching astronauts in their freely falling space vehicles. In Newton's theory such frames, though accelerating relative to absolute space, are nevertheless equivalent to inertial frames from which gravity has been eliminated. For example, Newton's first law is here in full evidence. The reason for this is the fact that in a gravitational field *all* particles accelerate equally. And this, in turn, is due to the equality of "gravitational mass" (the analog of electric charge) and "inertial mass" (the measure of a particle's resistance to acceleration)—a puzzling coincidence in Newton's theory. Einstein once again made the generalization from mechanics to the whole of physics. His principle of equivalence asserts that all the laws of physics are the same in each local inertial frame. It is these frames, therefore, which are the real province of the principle of relativity. Special relativity thereby becomes a local theory.

An elementary consequence of the principle of equivalence is the bending of light in a gravitational field. For, if light travels rectilinearly in the local inertial frame, and *that* accelerates freely in the gravitational field, the light path is evidently curved in the field. No property of light other than its uniform motion in a local inertial frame has been used in this argument. This, in turn, suggests that one might ascribe the bending of light to an inherent space curvature, rather than to the nature of light. In much the same way, the characteristic motion of free particles in a gravitional field suggests that they follow "natural" paths (geodesics) in a curved space-time. Their motion is independent of everything except their initial position and velocity. It should be noted, however, that for a geodesic law of motion to be possible, space and time *must* be welded into four-dimensional space-time. Free motions could not be represented by geodesics in a *three-*dimensional curved space. For a geodesic is uniquely determined by a point and a direction at that point. But an initial point and an initial direction do *not* uniquely determine a free path in a gravitational field. *That* depends also on the initial speed. In space-time, on the other hand, a direction is equivalent to a (vector-) velocity. And it *is* the case in Newton's theory that an initial point and velocity uniquely determine a free path in a gravitational field. Note that all this depends on the equality of inertial and gravitational mass, which is therefore a *sine qua non* of Einstein's general relativity. Consequently this equality has been subjected to the most stringent tests (e.g., by R. H. Dicke et al. in the 1960s) and is now considered verified to at least one part in 10^{11}.

As we have seen, special relativity forces a four-dimensional metric structure [Eq. (4)] on the events within an inertial frame. By patching together the structures of all the local inertial frames, we obtain the structure of the world of general relativity. Locally, i.e., in special relativity, it can be regarded as flat. But it is evident that, if the very pleasing geodesic law of motion is to hold, the presence of matter must impress a curvature on this space-time. For example, the planets move in patently curved paths around the sun (in four-dimensions these are helicoidal rather than elliptical); for these paths to be geodesics, the space-time around the sun must be curved. Just how matter curves the surrounding space-time is expressed by Einstein's field equations

$$G_{ij} = -\frac{8\pi G}{c^4} T_{ij} \qquad (5)$$

which look deceptively simple. Technically, they represent ten second order partial differential equations for the metric of space-time. This metric enters the 16 components of the "Einstein (curvature) tensor" G_{ij}, of which only 10 are independent, for $G_{ij} = G_{ji}$. G is the constant of gravitation; T_{ij} is the so-called energy tensor of the matter, and its components represent a generalization of the classical concept of mass density.

The exact solution of Eq. (5) has been possible only in a limited number of physical situations. For example, in 1916 Schwarzschild gave the exact solution for the space-time around a spherical mass m (e.g., the sun):

$$ds^2 = (1 - a/r)^{-1} dr^2 + r^2 (d\theta^2 + \sin^2\theta \, d\phi^2)$$
$$- (1 - a/r)c^2 \, dt^2 \qquad (6)$$

where $a = 2Gm/c^2$, r is a measure of distance from the central mass, t is a measure of time, and θ and ϕ are the usual angular coordinates. Note that when $m = 0$, Eq. (6) reduces simply to the flat space-time of Eq. (4), written in polar coordinates, and its geodesics would be straight lines (in space and time). But for Eq. (6), the geodesics in the symmetry "plane" $\theta = \pi/2$ are found to satisfy the equation

$$\frac{d^2u}{d\phi^2} + u = \frac{Gm}{h^2} + \frac{3Gmu^2}{c^2} \qquad (7)$$

where $u = 1/r$ and h is a constant (essentially the orbital angular momentum). This differs formally from the classical orbit equations only by the presence of the last term, which is very small. But as a consequence of that term, the solution of Eq. (7) is

$$u = Gmh^{-2}(1 + e \cos p\phi),$$
$$p = 1 + 3G^2 m^2 h^{-2} c^{-2} \qquad (8)$$

instead of the classical solution which has $p = 1$. Now r is a function in ϕ of period $2\pi/p$ instead of 2π, and therefore the orbital ellipse precesses. For the planet Mercury, for example, the secular precession predicted is $42''$ (seconds of arc),

and this agrees well with observation. In the space-time defined by Equation (6) one also finds that light-signals which pass close to the central mass are bent by an angle twice as big as that predicted on a simple Newtonian corpuscular theory of light; and again observations bear out the relativistic prediction. A third "crucial" prediction is the reddening of light as it "climbs" up a gravitational potential gradient. (The redshift in traversing as little as 70 feet of the earth's field has been measured and bears out the theory.)

A new test of general relativity was suggested in 1964 by I. I. Shapiro. This consists in bouncing a radar signal off a planet or spacecraft passing behind the sun. Theory predicts a time delay (of about 250 microseconds) when the signal grazes the sun, which adds a "bump" to the apparent orbit. By now this test has confirmed the predictions to an accuracy of 0.3%, bettered only by the redshift test. (Nevertheless the latter is considered weaker, since it only tests the *last* metric coefficient in (6).) Also begun in 1964 (but still not completed) were preparations for an experiment in which a gyroscope will be carried in an artificial satellite around the earth. According to general relativity the axis of the gyroscope should precess by a certain amount which will be measured. An advantage of this experiment is the relative absence of other effects overlaying the relativistic effect, but an evident difficulty is the manufacture of the instrument to the required tolerances.

The detection of gravitational waves would also greatly support the theory. Direct detection has been reported by the pioneer in this field, J. Weber, since the 1960s, but no other group has been able to duplicate his results. If, indeed, the pulses observed by Weber are gravitational, their generation would require huge energies—amounting to the annihilation from our galaxy of a hundred suns a year. However, an indirect indication of the existence of gravitational waves has recently been found by J. H. Taylor et al. in a binary star system that apparently loses rotational energy at just the rate at which it should radiate gravitationally.

Meanwhile the search for stellar black holes (relativistic remnants of collapsed stars) continues. In recent years the evidence has been mounting that a certain x-ray source in the constellation of the Swan, Cygnus X-1, has the required black-hole characteristics, and this may lead to the first positive identification. Astronomers believe on theoretical grounds that our galaxy should contain millions of stellar black holes.

Our account of ongoing "tests" for general relativity should not convey the impression that its validity is in doubt. *All* good physical theories are rich in predictions and it is in the nature of physicists to test these. As for its practicability, general relativity is on the way to becoming a service discipline for astrophysics and cosmology, where the fields are sufficiently extreme to preclude Newtonian gravity. On the theoretical side, enormous progress has been made in recent years in the understanding of many of the outstanding problems of classical (i.e., non-quantum) general relativity. Currently there is much interest in the problems of quantizing gravity, and of combining relativistic gravity and quantum mechanics in a consistent way. The days when general relativity was a "dormant" theory are definitely past.

W. RINDLER

References

Dixon, W. G., "Special Relativity," Cambridge, U.K., Cambridge Univ. Press, 1978.

Einstein, A., et al., "The Principle of Relativity," New York, Dover, 1923.

Hawking, S. W., and Israel, W., "General Relativity: An Einstein Centenary Survey," Cambridge, U.K., Cambridge Univ. Press, 1979.

Hoffman, B., with Dukas, H., "Albert Einstein: Creator and Rebel," New York, Viking Press, 1972.

Misner, C. W., Thorne, K. S., and Wheeler, J. A., "Gravitation," San Francisco, W. H. Freeman and Co., 1973.

Rindler, W., "Essential Relativity," New York, Springer-Verlag, 1979; "Introduction to Special Relativity," Oxford, Clarendon Press, 1982.

Tauber, G. (Ed.), "Albert Einstein's Theory of General Relativity," New York, Crown, 1979.

Zeldovich, Ya.B., and Novikov, I. D., "Relativistic Astrophysics," Chicago, Univ. Chicago Press, 1971.

Cross-references: ACCELERATORS, PARTICLE; ASTROPHYSICS; ATOMIC ENERGY; DYNAMICS; GRAVITATION; MICHELSON-MORLEY EXPERIMENT; MÖSSBAUER EFFECT; TIME; VELOCITY OF LIGHT.

RELAXATION

By relaxation is understood the phenomenon that an observable time elapses between the moment when a system in equilibrium is subjected to a momentary change in condition and the moment when the system is again in equilibrium.

A good example of a simple system showing relaxation is an unchanged capacitor with a capacitance C connected in series with a resistance R, to which circuit a constant voltage U is applied at a time $t = 0$. The charge Q of the capacitor at time t is given by the differential equation

$$U = \frac{Q}{C} + R\frac{dQ}{dt} \qquad (1)$$

the solution of which is:

$$Q = CU(1 - e^{-\frac{t}{RC}}) \qquad (2)$$

Hence the charge Q does not follow the sudden change of U but shows an exponential increase towards the final value CU (see Fig. 1). The time $\tau = RC$ is referred to as the relaxation time of the system. Obviously τ governs both the rate of charging and the rate of discharge: if the capacitor with a charge Q_0 is shortcircuited via a resistance at the time $t = 0$, then the charge at time t is given by the single exponential equation

$$Q = Q_0 e^{-\frac{t}{\tau}}.$$

Multiexponential behavior of the type

$$y = a_0 + \sum_{n=1}^{N} a_n e^{-\frac{t}{\tau n}}$$

is found in case the time dependence of a system is not described by Eq. (1) but by a set of N coupled linear differential equations.

Relaxation is met with in various fields of physics, e.g.:

(a) If a spring connected in parallel with a dash-pot is suddenly loaded with a constant force, it will take some time until this system is again in equilibrium.

(b) If an electrically or magnetically polarizable substance is suddenly placed in a constant electric or magnetic field, it will take some time until the electric or magnetic dipoles have orientated themselves in the field.

(c) If the irradiation of a phosphorescent substance is suddenly stopped, the phosphorescence does not cease immediately but continues for some time during which the intensity decreases exponentially.

An electrical analogy can be devised for all these cases. For instance, if in case (a) we substitute a voltage for the force, a charge for the displacement, the reciprocal of a capacitance for the spring constant f, and a resistance for the coefficient of friction η, we obtain exactly the differential equation [Eq. (1)]. The relaxation time τ will then be $\tau = \eta/f$, which quantity depends on temperature since as a rule η greatly

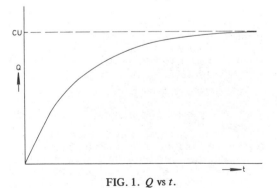

FIG. 1. Q vs t.

varies with temperature. Generally, the relaxation times in electrically and magnetically polarizable media also vary with temperature.

If we subject a system having relaxation properties to a periodically varying change in condition instead of one sudden change in condition, some characteristics are revealed which are likewise typical of relaxation. Let us again take the RC circuit as an example, this time applying to it a voltage $U = U_0 \sin \omega t$. Solution of the differential equation [Eq. (1)] yields the following expression for the charge Q of the capacitor:

$$Q = \frac{CU_0}{\sqrt{1 + \omega^2 \tau^2}} \sin (\omega t - tg^{-1} \omega \tau)$$

The charge has a component Q' which is in phase with the voltage applied and a component Q'' which is shifted $90°$ in relation to the voltage applied. Q' and Q'', usually referred to as the real and the imaginary component of Q, respectively, are given by:

$$Q' = CU_0 \frac{1}{1 + \omega^2 \tau^2}$$

$$Q'' = CU_0 \frac{\omega \tau}{1 + \omega^2 \tau^2}$$

The variation of Q' and Q'' with $\omega \tau$ is shown in Fig. 2 where $\omega \tau$ has been plotted logarithmically in order to obtain symmetrical curves.

Q'' is a measure of the energy dissipated in the circuit per cycle; it has its maximum value at $\omega = 1/\tau$ and a negligibly small value at much higher and much lower frequencies.

Furthermore, Q'' reaches half its maximum value at $\omega = 3.73/\tau$ and $\omega = 0.27/\tau$ so that the half-width value $\Delta\omega$ of the Q'' curve expressed in angular frequency units is $3.46/\tau$. At $\omega = 1/\tau$, Q'' just has half its maximum value.

Hence, by measuring Q' and Q'' vs frequency, we have various possibilities for determining the relaxation time. The quantity $Q''/Q' = \omega \tau$ is called the *loss tangent* and plays an important role in characterizing a material.

If a dielectric solid is subjected to a periodically changing electric stress and the real (in-phase) and imaginary (out-of-phase) components of the dielectric constant of the substance are determined, curves will be found having a shape as shown in Figs. 2(a) and 2(b), from which the relaxation time can be determined. The mechanical relaxation time of an elastic solid can be determined in a similar manner.

In some cases, not one but several absorption peaks with the corresponding changes of the real component are found in this kind of experiment, from which it can be concluded that several relaxation mechanisms are involved. This phenomenon is frequently observed in polymers where the various components, such as short chains, long chains, cross-linked chains, etc.,

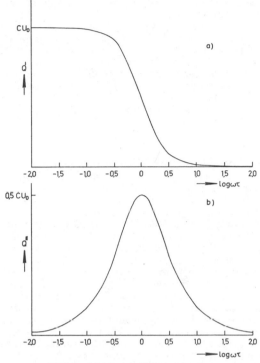

FIG. 2. Q' (a) and Q'' (b) vs log $\omega \tau$.

have different possibilities of moving and therefore have different relaxation times. There are also cases in which there is a more or less continuous distribution of relaxation times, and consequently, an absorption which is practically independent of frequency.

Since the above-mentioned electric circuit has no inductance (which corresponds to the masses of the particles in magnetic, electric and mechanical systems), there can be no resonance either.

As an example of a system having both resonance and relaxation properties, we will consider a liquid substance containing magnetic nuclei, e.g., hydrogen nuclei, which at the time $t = 0$ is placed in a magnetic field H. In this substance a magnetic moment M is built up in the direction of H by the fact that the original random orientation of the magnetic dipoles of the nuclei changes under the influence of the magnetic field into a distribution of such a nature that a resulting moment M arises in the direction of the magnetic field. M is determined by an equation analogous to Eq. (2), namely:

$$M = M_0 \left(1 - e^{-\frac{t}{T_1}}\right)$$

where M_0 is the final value of M and T_1 is a relaxation time known as longitudinal or spin-lattice relaxation time, which is associated with the interaction between the magnetic dipoles and their surroundings.

Now, if the direction of the magnetic field is suddenly changed, M will make a precessional motion about the new H due to the fact that the resulting angular momentum J is coupled to the magnetic moment M of the nuclei via the relation $\gamma = M/J$ where γ is the magnetogyric ratio of the nuclei.

The angular frequency of the precessional motion is given by:

$$\omega = \gamma H$$

On behalf of the precessional motion, the magnetic moment M has not only a longitudinal component M_l in the direction of the new magnetic field but also a transverse component M_t at right angles to the new magnetic field. Whereas it would seem logical to expect that M_l would increase and M_t decrease exponentially with the relaxation time T_1 due to the relaxation mechanism, in reality this is not so.

Experiments have shown that M_t decreases faster than the increase of M_l so that apart from the longitudinal relaxation time T_1 there must also be a separate transverse relaxation time T_2 which invariably is smaller than or equal to $2T_1$. The existence of a T_2 besides T_1 can be explained by the fact that apart from the return of individual magnetic moments of the nuclei to the direction of H (which is characterized by T_1), it is also possible for these individual magnetic moments—which together form M— to continue their precessional motion along the same conical shell though getting out of phase by mutual interaction. In the latter case, M_t decreases whereas M_l remains unchanged.

Just as with nonresonating systems, periodically changing quantities can be successfully introduced in resonating systems. In the case of a substance containing magnetic nuclei, this is done by applying, beside a stationary magnetic field H_z in the z-direction, also a periodically changing magnetic field $2H_1 \sin \omega t$ in the x-direction. This is known as a nuclear magnetic resonance experiment. In this case, a periodically changing magnetization intensity is created in the x-direction whose real (in-phase, dispersion) component M' and imaginary (out-of-phase, absorption) component M'' are given by:

$$M' = \gamma M_0 H_1 \frac{T_2^2 (\gamma H_z - \omega)}{1 + T_2^2 (\gamma H_z - \omega)^2 + \gamma^2 H_1^2 T_1 T_2}$$

$$M'' = \gamma M_0 H_1 \frac{T_2}{1 + T_2^2 (\gamma H_z - \omega)^2 + \gamma^2 H_1^2 T_1 T_2}$$

where M_0 is the magnetic moment when no periodically changing magnetic field is applied. These quantities can both be determined experimentally.

The variation of $M'/\gamma M_0 H_1 T_2$ and $M''/\gamma M_0 H_1 T_2$ has been plotted against $T_2 (\gamma H_z - \omega)$ in Fig. 3, it having been assumed that H_1 is

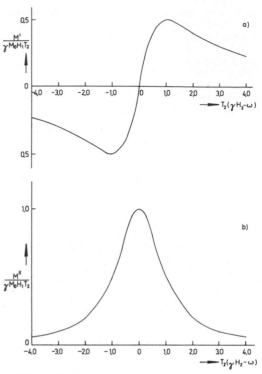

FIG. 3. Shape of the dispersion (a) and absorption (b) curves.

so small that $\gamma^2 H_1^2 T_1 T_2$ is much smaller than 1. It will be seen that the curves differ considerably from those in Fig. 2. The absorption curve has its maximum at the value $\gamma H_z = \omega$ which value has nothing to do with the relaxation times but only with the concurrence of the precessional frequency γH_z of the magnetic moment in the magnetic field H_z with the frequency of the periodically changing magnetic field. Hence it is clearly a matter of resonance absorption. The half-width value $\Delta \omega$ of the absorption curve, expressed in terms of a frequency, is $2/T_2$. The $M'/\gamma M_0 H_1 T_2$ curve is the curve of anomalous dispersion, which at $\gamma H_z = \omega$ has just the value 0, which is invariably found with resonating systems. With increasing values of H_1 both curves get wider and wider and the absorption curve eventually becomes 0 throughout. This phenomenon is called saturation. T_1 and T_2 can be calculated in principle from curves plotted for different known values of H_1; in practice, pulse techniques have to be used, often combined with Fourier transform techniques. Both relaxation times are very much dependent on temperature and frequency, and on the phase of the substance under test.

The same phenomena are observed in the case of resonance of unpaired electrons, known as electron spin resonance.

Generally speaking, the investigation of relaxation times provides information on the surroundings of relaxing particles and thus can contribute to knowledge about the structure and the internal mobility of substances, which is very important for physicists, chemists, and biologists.

Furthermore, profound knowledge of relaxation times is essential in many instances to the control of physical processes (e.g., the creation of low temperatures by magnetic means) and physical techniques (e.g., the maser technique).

J. SMIDT

References

Farrar, Th. C., and Becker, E. D., "Pulse and Fourier Transform NMR," New York, Academic Press, 1971.

Van Krevelen, D. W., "Properties of Polymers," Amsterdam, Elsevier Publishing Company, 1972. (Part III and references mentioned there.)

Slichter, C. P., "Principles of Magnetic Resonance," 2nd Ed., Berlin, Heidelberg, New York, Springer-Verlag, 1978.

Cohen, J. S., "Magnetic Resonance in Biology," Vol. 1, New York, John Wiley & Sons, 1980.

Becker, E. D., "High Resolution NMR," 2nd Ed., New York, Academic Press, 1980.

Ferry, J. D., "Viscoelastic Properties of Polymers," 3rd Ed., New York, John Wiley & Sons, 1980.

Cross-references: ABSORPTION SPECTRA; CAPACITANCE; CONDUCTIVITY, ELECTRICAL; ELECTRON SPIN; FOURIER ANALYSIS; LUMINESCENCE; MAGNETIC RESONANCE; MAGNETISM; RESONANCE; VISCOELASTICITY.

REPRODUCTION OF SOUND

History In 1807 the British physicist Thomas Young designed the first device capable of making a graphic record of sound waves. His description of the principle of sound recording is as clear and valid today as it was then, and it may serve here:

"The situation of a particle at any time may be represented by supposing it to mark its path, on a surface sliding uniformly along in a transverse direction. Thus, if we fix a small pencil in a vibrating rod, and draw a sheet of paper along, against the point of the pencil, an undulated line will be marked on the paper, and will correctly represent the progress of the vibration."

The recording stylus of Young's device had to be touched directly with the sound source. The "phonautograph" developed by Léon Scott de Martinville in 1856 was able to record sound from the air, via a horn, parchment diaphragm, and hog-bristle stylus. Both the Young and Scott devices made a helical trace on a rotating cylinder. Neither recording could be played back

because the recorded trace was not deep or stiff enough to guide the vibrations of a reproducing stylus.

In April 1877, Charles Cros deposited with the French Academy of Sciences a sealed package containing the description of a complete record-reproduce system. Cros planned to use metal records photoengraved from an original tracing in lampblack, but never carried out his plans. In the fall of that year, Edison constructed a "phonograph" whose recording stylus made indentations on tinfoil wrapped on a pre-grooved cylinder. Although these indentations were partly deformed by the playback stylus, a weak, distorted, but intelligible version of the human voice could be reproduced.

Modern disc recording is closer to the Cros system than to Edison's; commercial recordings are stampings from a hardened mold. Virtually all original recordings for home use are first made on magnetic tape. They are then cut into a master disc made of a relatively soft material, lacquer, and end up as a metal stamper from which mass impressions can be made in a vinyl composition.

Modern Recording Systems Current recording media include magnetic tape, transparent film, and grooved discs. The most widely used is the disc.

If the disc cutter head is fed with constant electrical energy over the frequency spectrum, it will produce constant average velocity in the recording stylus. Since the wave length of the recorded signal is doubled with each lower octave, constant stylus velocity would produce impractically large groove excursions in the bass range. Progressive bass attenuation is therefore introduced below a frequency which has now been standardized at 500 hertz, called the *turnover* frequency, in such an amount that the amplitude of groove modulation for signals below this frequency remains the same at a given power. A compensatory boosting of the bass frequencies must be employed in the playback amplifier.

A second problem in disc recording has to do with surface noise introduced by irregularities in the record material and picked up by the needle. The noise is distributed fairly evenly over the frequency spectrum on the basis of energy per cycle. Since each higher octave covers twice the number of cycles, this noise may be considered primarily a treble phenomenon (see NOISE, ACOUSTICAL).

The signal to the recording cutter is again altered, this time by progressive treble boost. In playback, a compensatory treble attenuation is introduced which brings the recorded signal back to normal and at the same time significantly reduces the amount of surface noise. This system of recording *preemphasis* does not change the treble content of the final reproduced sound, but it increases the amplitude of the high-frequency groove modulations relative to the random surface irregularities in the recorded material.

These changes of frequency balance in the recorded signal are called *equalization*; the particular equalization curve is called the *recording characteristic*.

In a more recent technique for reducing recording noise, the signal is subjected to amplitude compression before recording, that is, the dynamic range of the signal is reduced by an amplifier whose gain increases automatically as the amplitude of the signal decreases. The overamplified weak signal elements can then be recorded at a higher level relative to intrinsic recording noise such as tape hiss and surface noise, without increasing the recording level of strong signals and creating tape overload or groove overcutting. The signal may be divided into several frequency channels with separate compressors for each channel, so that signals in one frequency region do not control compressor action in another. In reproduction a matching expansion system (whose gain increases with signal amplitude) restores the original dynamic range to the signal, which now has reduced recording noise.

This technique substantially increases the dynamic-range capability of the recording system. It has an effect very similar to that of recording equalization, but it is controlled by signal amplitude rather than signal frequency.

Digital Recording When sound is recorded digitally the acoustic waveform is no longer represented in the recording medium by an analogous, continuously varying waveform. The original waveform is instead "sampled": Successive readings of its instantaneous amplitude are taken at short intervals, and these readings are recorded as binary-code pulses that define a series of discrete amplitude values. The current sampling rate is 44.1 thousand times a second, which means a reading every 22.7 μsec. The frequency bandwidth required for digital stereo recording is the sampling rate multiplied by a factor representing the accuracy with which each sample can be recorded, and is typically of the order of 2 MHz.

Digital playback apparatus reconstructs the sound, producing instantaneous voltages at clock-controlled intervals to match the recorded amplitude values. Digital tape recording has been used for some commercial master tapes since the late 1970s, and digital discs become available in the early 1980s. The bandwidth required is much too great for mechanical record/pickup systems, so that these have been abandoned in favor of laser disc systems.

Since the reproducing amplifier is operating at a clock-controlled rate that matches the original sampling rate, each digital sample must be delivered at exactly the right time. Inevitable small variations in tape-mechanism or turntable speed are compensated for by an electronic memory in the reproducer, which stores signals a little bit ahead, and the average speed of the playback device is controlled by a servo or clock mechanism.

The aim of digital recording is to eliminate flutter and rumble, to reduce distortion, to improve the signal-to-noise ratio, and to increase the dynamic range that the system can accommodate.

The Modern Disc Reproducing System A pickup, also called a cartridge, traces the groove modulations through a needle or stylus, whose vibrations are converted to an electrical signal. The most common types of electrical generator employed in cartridges are the ceramic (piezoelectric), moving magnet, moving coil, and variable reluctance. It is the task of the cartridge to translate faithfully the wave forms of the groove into an electrical signal. Some of the problems in cartridge design have to do with the stiffness of the moving needle system, which tends to wear the record, and with the mass of the needle tip, which tends to resonate with the semi-elastic record material at high frequencies.

The turntable must revolve the record at a constant speed. Periodic variations in this speed are called *flutter* (the onomatopoetic term for very slow flutter is *wow*). Any noise introduced into the signal by the moving parts (via the pickup) is called *rumble*. A pivoted arm holds the cartridge in place over the record groove, with the vibration axis of the cartridge approximately tangent to the groove over the entire radius of the record.

The amplifier, which may be on one or more chassis, has two sections: the *preamplifier* and the *power amplifier*. The former is a voltage amplifier which performs the functions of input program selection, tone control, volume control, compensation for the frequency equalization introduced in recording, and voltage amplification of the input signal to the point where it can drive the power amplifier. The latter builds up the electrical signal power so that the signal is able to drive the loudspeaker or speakers.

An amplifier is a device whose output energy is greater than, but in the same form as, the input signal energy. The amplifier must therefore borrow energy from an outside source, normally the electrical power line. There are many types of basic amplifying devices. The two that are used almost universally in sound reproduction are the vacuum tube and the transistor.

The loudspeaker converts the electrical output of the amplifier into acoustical energy, usually through a vibrating diaphragm. The loudspeaker and the cartridge, because they are mechanical devices with their own resonances and characteristic behavior, are more intransigent to precise control by the input signal than electronic circuits.

A reproducing system is designed for minimum noise, distortion, frequency discrimination over the audible spectrum, and transient ringing;

speakers must also have adequate treble dispersion.

Stereophony Stereophonic sound is recorded on two separate channels from separate microphone inputs. Just as each lens of a stereoscopic camera takes a complete picture, each microphone channel picks up all of the sound, one from a right-oriented perspective and the other from a left-oriented perspective. When the two channels are played back through separate right and left speakers, more of the sense of the acoustical atmosphere of the concert hall is recreated; a corollary of this is an increased clarity of inner melodic voices. The ability of the listener to determine the apparent position of different musical instruments is a less important part of the stereo effect.

In stereo disc recording, one channel is, in effect, recorded on the left groove wall, and the other channel on the right groove wall. The reproducing stylus must execute a complex motion containing both vertical and horizontal components in order to follow the modulations of both groove walls simultaneously. These complex movements are analyzed into two vectors by two separate generating elements, each at 45° to the vertical and on opposite sides. Each generating element produces the electrical signal for one channel.

Recent developments have been aimed at recapturing the acoustical atmosphere of the concert hall more accurately. In one system, additional speakers at the back and sides of the listening area are used to reproduce delayed signals that simulate reflections from more distant walls. Another system uses four separate channels, two to reproduce rear-hall sound at the rear of the listening room. As in two-channel stereo, there is sometimes a misplaced emphasis on directional effects at the expense of concert-hall ambience.

The ultimate design goal of sound reproducing equipment is not to create "better" sound, but to efface the imprint of the equipment, so that the original musical quality is recreated.

EDGAR VILLCHUR

References

Beranek, Leo L., "Acoustics," New York, McGraw-Hill Book Company, Inc., 1954.

Hunt, Frederick, V., "Electroacoustics," Cambridge, Harvard University Press, 1954.

Olson, Harry F., "Acoustical Engineering," New York, Van Nostrand Reinhold, 1957.

Villchur, Edgar, "Reproduction of Sound," New York, Dover Publications, Inc., 1965.

Villchur, Edgar, "Reproduction of Sound," *Phys. Today*, 5, No. 9 (September, 1952).

Oppenheim, Alan V. (Ed), "Applications of Digital Signal Processing," Englewood Cliffs, N.J., Prentice-Hall, 1978.

Cross-references: ACOUSTICS; ARCHITECTURAL ACOUSTICS; MUSICAL SOUND; NOISE, ACOUSTICAL; PHYSICAL ACOUSTICS.

RESONANCE

The phenomenon which scientists call resonance can be identified in many different physical systems of widely varying sizes. For instance, a father who pushes his small child on a swing finds that with each successive push the swing goes higher. An astronomer who examines the spectrum of the sun notes the appearance of a series of dark lines superposed on the continuous red to violet band of colors. The solid state physicist who observes the amount of electromagnetic radiation transmitted through a waveguide at a particular microwave frequency finds it can be sharply reduced if certain materials are placed in the guide and a magnetic field is applied. All of these effects are examples of resonance, yet the actual physical mechanisms are different and the explanations require different analytical procedures depending on whether the system is governed by classical or quantum theory.

In the case of the father pushing the swing, the resonance can be explained as a direct consequence of Newton's laws of classical mechanics. In simple terms, it occurs when an outside agent (the father) pushes the system (swing with child) with a periodic force having the same frequency as the natural frequency of the system itself. The natural frequency is the frequency with which the system oscillates if it is displaced from its normal position of equilibrium and then released to swing freely back and forth. If the external push is timed so as to be exactly in step with the natural frequency, one can think of the swing as being steadily accelerated in its natural direction of motion. When the natural direction of motion changes at the end points of the path so does the push. Elementary kinematics then predicts that such a steady acceleration will cause both the displacement and the maximum velocity to increase continuously and eventually become very large. It is this extreme magnitude of displacement and velocity which is the most noticeable aspect of resonance in a mechanical system. However, less apparent visually, yet equally important, is the rate of energy transfer between the source and the system at resonance. The external agent transfers energy in the form of mechanical work to the system. At resonance, the average rate of this energy transfer per cycle becomes a maximum. This property of maximum rate of energy transfer is of great value when analyzing the effect of radiation on microscopic systems such as atoms and nuclei whose physical behavior must be explained by the laws of quantum physics rather than classical physics. In these systems as in mechanical systems a necessary condition for resonance is that the frequency of the radiation must match a frequency which is in some way associated intrinsically with the system. But here the coordinates of position and velocity are no longer suitable ones to use in describing the response at resonance. Nevertheless, the resonant system is still characterized by the general property that the rate of energy transfer into it is a maximum. In order to bring out the essential features of resonance involving both the behavior of position and velocity and the rate of energy transfer, it is simplest to examine a mechanical system consisting of a mass on a spring with frictional damping.

Resonance in a Simple Mechanical Oscillator
We begin with a mass M attached to a fixed point by a spring with a force constant K and a damping resistance R. In addition the mass is acted upon by a periodic external force of frequency $v, F = F_0 \cos 2\pi v t$. The displacement of the mass from its normal equilibrium position is represented by x. The acceleration d^2x/dt^2 of the mass M is determined by the resultant of these three forces, all along the x-direction: $-Kx$, the elastic restoring force; $-R(dx/dt)$, the damping force proportional to the instantaneous velocity dx/dt; and the external force $F_0 \cos 2\pi v t$. Newton's second law of motion, which states that resultant force = mass × acceleration, can be expressed by a second-order inhomogeneous linear differential equation:

$$M\frac{d^2 x}{dt^2} + R\frac{dx}{dt} + Kx = F_0 \cos 2\pi v t. \quad (1)$$

However, in order to discuss resonance it is first necessary to describe the behavior of the spring system when no external driving force is being applied. If the mass is displaced from its normal equilibrium position and then released, it will move in simple harmonic motion with the natural frequency v_1 given by

$$v_1 = \frac{1}{2\pi}\sqrt{\frac{K}{M} - \frac{R^2}{4M^2}}, \quad \left(\frac{K}{M} > \frac{R^2}{4M^2}\right). \quad (2)$$

This result can be derived mathematically by setting the term on the right-hand side of Eq. (1) equal to zero thus making the equation homogeneous. It can be solved for $x(t)$ by standard methods (see references 1 and 2). The solution is:

$$x = A e^{-(R/2M)t} \sin(2\pi v_1 t + \phi) \quad (3)$$

where A and ϕ are constants depending on the initial position and velocity. For example, if the mass starts from a position $x = x_0$, with zero velocity, then $A = x_0$ and $\phi = 90°$, and Eq. (3) becomes

$$x = x_0 e^{-(R/2M)t} \cos 2\pi v_1 t. \quad (4)$$

Since the solution is the product of a negative exponential function and a sinusoidal function, the amplitude of the displacement will diminish a little bit more with each succeeding cycle of oscillation and eventually the displacement becomes zero again. This is called a damped harmonic oscillation and is physically what we see any time a real spring or pendulum is disturbed and then allowed to oscillate freely. The damping of the oscillations can also be looked upon as representing a conversion of mechanical energy into heat which is proceeding at the instantaneous rate $R(dx/dt)^2$.

We now consider the solution of the inhomogeneous equation [Eq. (1)]. From the theory of linear differential equations, the solution of an inhomogeneous equation can be expressed as the sum of solutions of the homogeneous and inhomogeneous equation. Thus a general solution can be written down in which one term is the solution [Eq. (3)] while the second is the particular integral satisfying Eq. (1).

$$x = A e^{-(R/2M)t} \sin(2\pi v_1 t + \phi)$$
$$+ \frac{F_0}{2\pi M} \frac{\cos(2\pi v t - \alpha)}{\sqrt{4\pi^2(v_0{}^2 - v^2)^2 + \frac{v^2 R^2}{M^2}}} \quad (5)$$

where $v_0 = (1/2\pi)\sqrt{K/M}$ is the natural frequency for the undamped oscillator and α, the phase angle between the applied force and the displacement, is defined by

$$\tan\alpha = \frac{vR}{2\pi M(v_0{}^2 - v^2)}. \quad (6)$$

Since the first term in Eq. (5) becomes vanishingly small and can be neglected after a period of time has elapsed, it is called the transient part of the solution. The particular integral maintains a constant amplitude as long as the driving frequency does not change, and it is called the steady-state part of the solution.

The steady-state solution for the instantaneous velocity is also of great interest

$$v = \frac{dx}{dt} = \frac{F_0 v}{M} \frac{\cos(2\pi v t - \beta)}{\sqrt{4\pi^2(v^2 - v_0{}^2)^2 + \frac{R^2 v^2}{M^2}}} \quad (7)$$

where β, the phase angle between applied force and velocity, is defined by

$$\tan\beta = \frac{2\pi M(v^2 - v_0{}^2)}{vR}. \quad (8)$$

It is seen that both displacement and velocity have amplitudes which depend on the frequency of the applied force. The form of Eq. (5) and (7) is such that there must exist frequencies for which each achieves a maximum value. The frequencies which produce these maximum values can be found by differentiating the displacement or velocity function with respect to frequency, setting the derivative equal to zero, and solving for v. In this way, it is deduced that the amplitude of x becomes a maximum when

$$v = \sqrt{v_0{}^2 - \frac{R^2}{8\pi^2 M^2}} \qquad (9)$$

while the amplitude of v becomes a maximum when

$$v = v_0 \qquad (10)$$

This latter frequency v_0 which is the natural frequency of the undamped oscillator is customarily referred to as the resonance frequency of the oscillator. The phase difference between the force and velocity at resonance is zero. Thus the velocity will achieve its largest magnitude in the part of the cycle when the push is greatest, confirming what we sense intuitively in pushing a swing. The displacement resonance occurs at a frequency slightly different from the resonance frequency, but the difference becomes smaller as R decreases. The velocity resonance is of significance in discussing energy because the rates of energy transfer into and out of the oscillator depend on velocity. The external periodic force is doing work on the oscillator at the instantaneous rate of Fv, while the oscillator is working against the damping force at the instantaneous rate of $-Rv^2$. Thus energy is being simultaneously absorbed and dissipated. Since the instantaneous rates vary periodically with time, it is more meaningful to calculate average rates per period and then compare. Such a calculation (see reference 1) shows that the average rate of energy absorption is exactly equal to the average rate of energy dissipation and has the magnitude $(F_0{}^2/2R)\cos^2\beta$. At velocity resonance $\cos^2\beta = 1$, and these rates of absorption and dissipation will have maximum magnitudes of $F_0{}^2/2R$. The average rate of energy absorption can be expressed as

$$\bar{W} = \frac{\pi v_0 F_0{}^2}{KQ\left[\dfrac{1}{Q^2} + \left(\dfrac{v}{v_0} - \dfrac{v_0}{v}\right)^2\right]} \qquad (11)$$

where Q (for quality) is defined as $2\pi v_0 M/R$. $\bar{W}$ is plotted in Fig. 1. The treatment of resonance in terms of rate of energy absorption is useful in many kinds of systems.

There is an exact analogue to this analysis in the electrical circuit consisting of an inductance L, a capacitance C, and a resistance R in series and driven by an alternating emf, $E_0 \cos 2\pi vt$. The differential equation for the variation of charge q with time is

$$L\frac{d^2 q}{dt^2} + R\frac{dq}{dt} + \frac{q}{c} = E_0 \cos 2\pi vt \qquad (12)$$

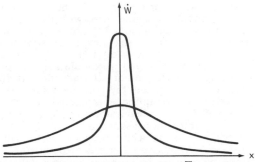

FIG. 1. Rate of energy absorption $\bar{W}$ as a function of $x = (v/v_0) - (v_0/v)$. For large Q the resonance peak is high and narrow; for small Q it is low and wide. From "Physical Mechanics," R. Bruce Lindsay, Fig. 10.6, p. 290. D. Van Nostrand Co. 1961.

and resonance will occur under the same corresponding conditions as for the mechanical oscillator (see ALTERNATING CURRENTS).

Resonance in Coupled Systems A system of particles with n degrees of freedom in which the particles are coupled together by elastic restoring forces will possess a set of n natural or normal frequencies. If the motion is started with an arbitrary distribution of initial displacements and velocities, the subsequent time-dependent motion of each particle will appear as a superposition of all the normal frequencies. However, through a proper choice of initial conditions it is possible to excite any one of n so-called normal modes. A normal mode is characterized by all of the particles oscillating with one particular normal frequency. For each normal frequency there will be a corresponding normal mode. As an example, a coupled system of two identical particles has two normal frequencies, hence two normal modes. For the lower frequency the normal mode is one in which the particle displacements are exactly in phase with each other. For the higher frequency they are exactly out of phase with each other. If a periodic external force is applied with a frequency matching one of the normal frequencies, the entire system will be excited into resonance in the normal mode appropriate to that particular frequency. Just as with the simple mechanical oscillator, the magnitudes of the maximized amplitudes are limited by whatever damping forces are acting and the rate of energy absorption equals the rate of energy dissipation. This mechanism is the basis for the excitation by resonance of standing waves in continuous elastic media such as solid structures and fluid columns. A standing wave is the configuration of the displacements of all the individual points when vibrating in a normal mode. A string held under tension between two fixed points will vibrate in a normal mode if a transverse frequency is applied which equals an integral multiple of the velocity of transverse waves in the string divided by twice the length. The instantaneous visual appearance of the standing wave corresponding to the lowest normal frequency is a sine curve with its point of maximum vibration, called an antinode, located midway between the fixed ends. This sine curve vibrates transversely back and forth between maximum positive and maximum negative displacements with the normal frequency. If the frequency is doubled, the corresponding standing wave has two antinodes separated by a point of no vibration, called a node, at the midpoint. Standing longitudinal sound waves can be set up in columns of air which are enclosed by tubes that are terminated by closed or open ends. A node will always be found at a closed end. An antinode will always be found just beyond the mouth of an open tube. The excitation of such standing waves by resonance with the appropriate external driving frequency results in an enhancement of the intensity of the sound. One of the most spectacular illustrations of the power of resonance was the destruction of the Tacoma Narrows Bridge on November 7, 1940 by large-amplitude torsional oscillations set up in its main span by strong cross winds.

Resonance Phenomena in Atomic and Nuclear-Sized Systems The exact theory explaining the processes whereby systems composed of atoms and nuclei absorb and emit energy is based on the principles of quantum theory. Nevertheless, such processes still show some physical analogy to resonance in a mechanical oscillator. A simple example is the single atom consisting of one or more electrons bound to a nucleus by predominantly electrostatic forces. Quantum theory predicts that the atom can exist only in a set of discrete energies in contrast to the continuous range of energies which in principle is available to a large-size satellite system such as the earth and moon. If the atom changes from a higher-energy state E_j to a lower-energy state E_i, energy in the form of electromagnetic radiation is given off and the frequency of the radiation ν_{ij} is related to the energy difference between the two states by the equation:

$$\nu_{ji} = \frac{E_j - E_i}{h}. \qquad (13)$$

h is the well-known Planck constant of action and has a value of 6.625×10^{-27} erg sec. This frequency and others which connect discrete energy states consistent with the so-called selection rules for the atom can be pictured as a set of frequencies characteristic of the particular atom. By analogy with resonance in a mechanical oscillator absorption of energy at a maximum rate might be expected to occur when there is radiation incident on the atom with a frequency which matches one of these frequencies. In quantum mechanics, the meaningful index for such a process taking place is the probability that an energy transition will occur a time t after being exposed to the radiation. When calculated, this probability is found to be proportional to

$$\frac{1 - \cos 2\pi(\nu_{ij} - \nu)t}{(\nu_{ij} - \nu)^2} \qquad (14)$$

(see references 2 and 8).

Even though this probability function is not of the same mathematical form as the expression for rate of energy absorption derived in the section on "Resonance in a Simple Mechanical Oscillator," because of its dependence on $(\nu_{ij} - \nu)^2$ instead of $(\nu_0^2 - \nu^2)^2$, it can be looked upon as establishing a condition for a resonance type process. It predicts that the likelihood of resonance absorption becomes very large when the external frequency approaches something resembling a natural frequency of the atom. A distinctive example of this kind of resonance absorption is the phenomenon in the solar spectrum described in the introduction. The radiation from the hot core of solar gases is characterized by a continuous spectrum. In passing through the cooler gases of the sun's outer atmosphere, radiation will be absorbed at those frequencies which match frequencies of atomic transitions in the cool gas. These blocked out lines in the continuous spectrum show up as dark lines against the background. They are called Fraunhofer lines. The rare gas helium was first discovered as a consequence of these observations.

Another important class of atomic-scale resonance phenomena concerns the behavior of atoms and nuclei in static magnetic fields when subjected to radio frequency (rf) magnetic fields. If the frequency of the rf field matches the Larmor precession frequency of the particle's magnetic moment vector about the direction of the static field, a resonance absorption of energy from the rf field can occur. The phenomenon is called electron spin resonance (ESR) if the magnetic moment is associated with unpaired electron spins of the atom and nuclear magnetic resonance (NMR) if the moment is associated with the spin of the nucleus. The ESR frequency of an atom with a single unpaired electron spin of $1/2$ in a static magnetic field H is given by ν (in megahertz) = $2.80\,H$ (in gauss). The NMR frequency of a proton is given by ν (in kilohertz) = $4.258\,H$ (in gauss). During recent years magnetic resonance has proved to be a powerful experimental technique for obtaining information about the magnetic moments of atoms and nuclei and the strengths of internal magnetic fields in solids. (See MAGNETIC RESONANCE and references 4, 5, 6 and 8 for more details.)

Resonances are also found in nuclear reactions and have proved useful in clarifying some features of nuclear structure, which in general is more complicated than atomic structure. When a particle such as a neutron, proton, deuteron, or alpha particle strikes a target nucleus a compound nucleus consisting of the target nucleus and the bombarding particle is temporarily created. This compound nucleus, the model for which was first proposed by Bohr, can exist in excited states. If the energy of the incident particle plus the energy binding it to the compound nucleus equals the energy of an excited state a resonance effect occurs. The resonance is characterized phenomenologically by a sharp increase in the rate of emission of particles or γ-rays from the target as the compound nuclei decay. From the form of the curve showing the rate of emission or the cross section for the reaction as a function of the bombarding energy the energy of the excited state and its level width can be obtained. From systematic investigations carried out over a range of bombarding energies and with different incident particles it is possible to obtain information about the lifetimes, decay modes, and spacing of the excited states. The lifetime of the state refers to the time for the compound nucleus to decay to its final nuclear state accompanied by the emission of a particle or γ-ray. It is inversely proportional to the level width. The Breit-Wigner formula, derived from quantum mechanical considerations, gives a theoretical expression for the dependence of the cross section σ on the bombarding particle energy E and the level width Γ in the vicinity of the peak resonance energy E_0. In its simplest form, σ is proportional to

$$\frac{1}{(E - E_0)^2 + \left(\frac{\Gamma}{2}\right)^2}. \tag{15}$$

This formula has been applied with considerable success to the interpretation of data from many nuclear reactions, particularly those involving neutron bombardment at energies such that the excited states do not overlap. A narrow level width implies that the lifetime of the excited state is long, which means the rate of energy loss due to emission is smaller than for an excited state with a wider level width. The plot of σ vs. E for different Γ value is analogous to the plot of the rate of absorbed energy vs. frequency for differing Q values for the simple mechanical oscillator in the neighborhood of resonance, as shown in Fig. 1. (See references 3, 7, and 9.)

Although resonance is a phenomenon which appears in many different systems with widely ranging dimensions and configurations, its fundamental characteristic of a maximum response when the frequency or energy of a perturbing influence matches a natural frequency or energy difference is universal.

ROBERT LINDSAY

References

1. Lindsay, R. B., "Physical Mechanics," Third Ed., New York, Van Nostrand Reinhold, 1962.
2. Lindsay, R. B., and Margenau, H., "Foundations of Physics," Woodbridge, Ct., Ox Bow Press, 1981.
3. French, A. P., "Vibrations and Waves," New York, Norton, 1971.
4. Kittel, C., "Introduction to Solid State Physics," Fifth Ed., New York, John Wiley, 1976.
5. Rushworth, F. A., and Tunstall, D. P., "Nuclear Magnetic Resonance," Gordon, 1973.
6. Slichter, C., "Principles of Magnetic Resonance," New York, Harper & Row, 1963.
7. Frauenfelder, H., and Henley, E. M., "Subatomic Physics," Englewood Cliffs, N.J., Prentice-Hall, 1974.
8. Morrison, M. A., Estle, T. L., and Lane, N. E., "Quantum States of Atoms, Molecules and Solids," Englewood Cliffs, N. J., Prentice-Hall, 1976.
9. Segré, E. (Ed.), "Experimental Nuclear Physics," New York, John Wiley, 1953, 1959 (3 Vols.).

Cross-references: ALTERNATING CURRENTS, DYNAMICS, ELECTRON SPIN, MAGNETIC RESONANCE, SOLAR PHYSICS.

RHEOLOGY

Rheology is the study of the response of materials to an applied force. It deals with the deformation and flow of matter.

Heraclitus, a pre-Socratic metaphysician, recognized in the fifth century B.C. that $\pi\alpha\nu\tau\alpha$ $\rho\epsilon\iota$, or "everything flows." Long before Heraclitus, the prophetess Deborah, fourth judge of the Israelites, had sung that "the mountains flowed before the Lord" in celebrating the victory of Barak over the Canaanites (1). Reiner (2) claims that the translation in the authorized version of the Bible that the mountains "melted" before the Lord misses the essential point of her wisdom, for Deborah recognized this early instance of a relaxation in the time scale provided by eternity; in recognition of her basic contribution to the primary literature of rheology, Reiner proposed the dimensionless quantity D (for Deborah) where

$$D = \frac{\text{time of relaxation}}{\text{time of observation}} = \frac{\tau}{t} \tag{1}$$

The difference between solids and liquids is found in the magnitude of D. Liquids, which relax in small fractions of a second, have small D; solids, large D. A long enough time span can reduce the Deborah number of a solid to unity, and impact loading can increase D of a liquid. Viscoelastic materials are best characterized under conditions in which D lies within a few decades of unity.

Force Balance Equation When a force f is applied to a body, four things may happen. The body may be accelerated, strained, or made to flow. If these three responses are added to each other, one can write for motion in one direction

$$f = m\ddot{x} + r\dot{x} + sx \tag{2}$$

where m is the mass, r is a damping diameter related to viscosity, and s to elasticity. Evaluation of the coefficients m, r, and s involves the measurement of displacements, x, and their time derivatives in a manner which links these kinematic variables via an equation of state such as Eq. (2) to stress, σ (force per unit area), and its time derivatives.

Scope of Rheology In contrast to the discipline of mechanics wherein the responses of bodies to unbalanced forces are of concern, rheology concerns balanced forces which do not change the center of gravity of the body. Since rheology involves deformation and flow, it is concerned primarily with the evaluation of the coefficients r and s of Eq. (2). The coefficients account for most of the energy dissipated and stored, respectively, during the process of distorting a body.

Most rheological systems lie between the two extremes of ideality: the Hookean solid and the Newtonian liquid.

Measurements of Viscosity and Elasticity in Shear *Simple Shear.* Shear viscosity η and shear elasticity G are determined by evaluating the coefficients of the variables $\dot{x}$ and x respectively, which result when the geometry of the system has been taken into account. The resulting equation of state balances stress against shear rate $\dot{\gamma}$ (reciprocal seconds) and shear strain γ (dimensionless) as the kinematic variables. For a purely elastic, or Hookean, response,

$$\sigma = G\gamma \tag{3}$$

and for a purely viscous, or Newtonian, response

$$\sigma = \eta\dot{\gamma}. \tag{4}$$

As a consequence, G can be measured from stress-strain measurements, and η from stress-shear rate measurements.

Elasticoviscous behavior is described in terms of the additivity of shear rates:

$$\dot{\gamma} = \frac{\sigma}{\eta} + \frac{\dot{\sigma}}{G} \tag{5}$$

whereas viscoelastic behavior is characterized by the additivity of stress according to Eq. (2):

$$\sigma = G\gamma + \eta\dot{\gamma}. \tag{6}$$

See reference 3 for further information on rheological bodies.

Relaxation. Numerous attempts have been made to fit simplified mechanical models to the two behavior patterns described by Eqs. (5) and (6). One can picture the elastic element as a spring arrayed in a network parallel with the viscous element to give essentially a (Kelvin) solid with retarded elastic behavior, wherein

$$\frac{\eta_k}{G_k} = \lambda \text{ sec (retardation time)} \tag{7}$$

or as a series (Maxwell) network which flows when stressed or relaxes under constant strain:

$$\frac{\eta_m}{G_m} = \tau \text{ sec (relaxation time)} \tag{8}$$

and transient experiments may be designed to measure these parameters singly. In real systems, a single relaxation (or retardation) time fails to account for experimental results. A distribution of relaxation times exists (see RELAXATION).

Dynamic Studies. When Eq. (2) is written in the form

$$\ddot{x} + 2k\dot{x} + \omega_1{}^2 x = 0 \tag{2b}$$

the equation suggests that the variation in stress be cyclic. Rheometers are designed so that the system may oscillate in free vibration of natural resonant frequency ω_1, or else so that a cyclic shearing stress of the form $f_0 \cos \omega t$ is impressed on the sample over a frequency range which spans ω_1. In neither case is the material strained beyond its range of linearity. Equation (2b) represents a damped harmonic oscillator, providing that the coefficients are constant (i.e., providing that they do not depend on the strain magnitude). Not all systems meet this requirement in the strict sense, with the result that one of the first consistency checks which the experimenter makes is for linearity. Doubling the amplitude of oscillation should double the stress and should not change the phase relationships between the cyclic stress and the deformation.

See RESONANCE and see reference 4 for more information on dynamic studies.

Time-temperature Equivalence *Steady-state Phenomena.* The creep of a viscoelastic body or the stress relaxation of an elasticoviscous one is employed in the evaluation of η and G. In such studies, the long-time behavior of a material at low temperatures resembles the short-time response at high temperatures. A means of superposing data over a wide range of temperatures has resulted which permits the mechanical behavior of viscoelastic materials to be expressed as a master curve over a reduced time scale as large as twenty decades, or powers of ten (see references 5 through 7).

Polymeric materials generally display large G values 10^{10} dynes/cm^2 or greater) at low temperatures or at short times of measurement. As either of these variables is increased, the modulus drops, slowly at first, then attaining a steady rate of roughly one decade drop per decade increase in time. If the material possesses a yield value, this steady drop is arrested at a level of G which ranges from 10^7 downward.

Dynamic Behavior. The application of sinusoidal stress to a body leads inevitably to the complex modulus G^*, where

$$G^* = G' + iG'' = G' + i\omega\eta' \tag{9}$$

where G' is the in-phase modulus (σ/γ) which represents the stored energy, and G'' is the out-of-phase modulus ($\sigma/\dot{\gamma}$) representing dissipated energy (as its relation to η' suggests); the variable against which G' and η' are determined is the circular frequency ω. Superposition of variable temperature data or variable frequency data provides a master curve of the type described for the steady-state parameters.

Problems in Three Dimensions *State of Stress.* The forces and stresses applied to a body may be resolved into three vectors, one normal to an arbitrarily selected element of area and two tangential. For the yz plane the stress vectors are σ_{xx}, and σ_{xy}, σ_{xz}, respectively. Six analogous stresses exist for the other orthogonal orientations, giving a total of nine quantities, of which three exist as commutative pairs ($\sigma_{rs} = \sigma_{sr}$). The state of stress, therefore, is defined by three tensile or normal components (σ_{xx}, σ_{yy}, σ_{zz}) and three shear or tangential components (σ_{xy}, σ_{xz}, σ_{yz}). The shear components are most readily applicable to the determination of η and G.

Strain Components. For each stress component σ there exists a corresponding strain component γ. Even for an ideally elastic body, however, a pure tension does not produce a pure γ_{xx} strain; γ components exist which constrict the body in the y and z directions.

The complete stress-strain relation requires the six σ's to be written in terms of the six γ components. The result is a 6×6 matrix with 36 coefficients, k_{rs}, in place of the single constant. Twenty-one of these coefficients (the diagonal elements and half of the cross elements) are needed to express the deformation of a completely anisotropic material. Only three are necessary for a cubic crystal, and two for an amorphous isotropic body. These parameters are discussed in reference 8.

Similar considerations prevail for viscous flow, in which the kinematic variable is $\dot{\gamma}$.

RAYMOND R. MYERS

References

1. Judges 5:5.
2. Reiner, M., *Phys. Today*, 17, 62 (1964).
3. Reiner, Marcus, "Deformation, Strain and Flow," Second edition, London, H. K. Lewis & Co., 1960; "Lectures in Theoretical Rheology," Third edition, New York, Interscience Publishers, 1960.
4. Eirich, F. R., Ed., "Rheology, Theory and Applications," Five volumes, New York, Academic Press, 1956–1969.
5. Myers, R. R., (Ed.), *Journal of Rheology*, published bimonthly since 1977. Formerly *Transactions of the Society of Rheology*, published annually 1957–1964; semiannually 1964–1966; and quarterly 1967–1977 by John Wiley & Sons, New York.
6. Ferry, J. D., "Viscoelastic Properties of Polymers," Third Edition, New York, John Wiley & Sons, 1981.
7. Tobolsky, A. V., "Properties and Structures of Polymers," New York, John Wiley & Sons, 1960.
8. Alfrey, Turner, Jr., "Mechanical Behavior of High Polymers," New York, Interscience Publishers, 1948.

Cross-references: ELASTICITY, FLUID DYNAMICS, POLYMER PHYSICS, VISCOSITY.

ROTATION—CURVILINEAR MOTION

1. Curvilinear Motion, Rotation Rotary motion such as a wheel revolving on an axel is common in physics and everyday life. The more general term *curvilinear motion* is exemplified by a satellite such as our Earth traversing its orbit about the Sun as an attracting center. The curved line followed as the satellite's path may be a circle, which is a special case, or in general a curved line of any shape. Figure 1 represents simple rotation in a circle in the plane of the page. If angle θ changes between times t_1 and t_2 from θ_1 to θ_2, the average angular speed ω_{av} of P (or r), measured usually in radians·sec^{-1} is

$$\omega_{av} = \frac{\theta_2 - \theta_1}{t_2 - t_1} = \frac{\Delta\theta}{\Delta t}; \qquad (1)$$

and the instantaneous angular speed is

$$\omega = \frac{d\theta}{dt}, \qquad (2)$$

or in vector notation

$$\boldsymbol{\omega} = \boldsymbol{\omega}_1 \frac{d\theta}{dt}, \qquad (2a)$$

where $\boldsymbol{\omega}_1$ is a unit vector perpendicular to and out from the page at O.

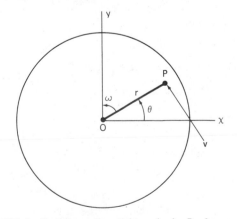

FIG. 1. Rotation of particle or body P of mass m about a fixed axis through O perpendicular to page. r, rotation radius OP; $\mathbf{v}$, tangential velocity vector.

When ω is not constant but changes from ω_1 to ω_2 as t increases from t_1 to t_2, there is angular acceleration α measured usually in radians $\cdot$ sec^{-2} and the corresponding equations are

$$\alpha_{av} = \frac{\omega_2 - \omega_1}{t_2 - t_1} = \frac{\Delta\omega}{\Delta t}, \qquad (3)$$

$$\alpha = \frac{d\omega}{dt} = \frac{d^2\theta}{dt^2}, \qquad (4)$$

$$\boldsymbol{\alpha} = \frac{d\boldsymbol{\omega}}{dt} = \boldsymbol{\omega}_1 \frac{d^2\theta}{dt^2}, \qquad (4a)$$

for the average, instantaneous, and vector expressions.

In *uniform* circular motion (both angular and linear tangential speeds constant) there is an acceleration different from that of Eqs. (3), (4), and (4a) which arises because the *velocity* vector, while not changing in magnitude, is continuously varying in *direction*, remaining always tangent to the circle of motion. This acceleration a_c is termed *centripetal* acceleration, since its cause is a central attractive or centripetal force directed always from the rotating point P toward the center of motion. a_c does not, however, involve a change in the length of the radius r, so it is not expressed here in terms of d^2r/dt^2.

The standard physics texts[1] show in simple derivation that

$$a_c = \omega v = \omega^2 r = v^2/r; \qquad (5)$$

or vectorially, employing unit vector $\mathbf{i}_r$ along r and outward from center O,

$$\mathbf{a}_c = \mathbf{i}_r\, a_c. \qquad (5a)$$

The corresponding centripetal force F_c measured in newtons is then

$$F_c = ma_c = m\omega^2 r = mv^2/r, \qquad (6)$$

or vectorially

$$\mathbf{F}_c = \mathbf{i}_r F_c = \mathbf{i}_r ma_c = \mathbf{i}_r m\omega^2 r, \qquad (6a)$$

where m is measured in kilograms.

Returning to Fig. 1, the linear, tangential speed v of P measured in m $\cdot$ sec^{-1} is

$$v = r\omega = r\frac{d\theta}{dt}, \qquad (7)$$

and the vector velocity is

$$\mathbf{v} = \mathbf{i}_\theta\, v = \mathbf{i}_\theta\, r\frac{d\theta}{dt}, \qquad (7a)$$

where $\mathbf{i}_\theta$ is a unit vector in the direction of $\mathbf{v}$, perpendicular to r and in the direction of increasing θ.

More generally, if the curvilinear motion of P

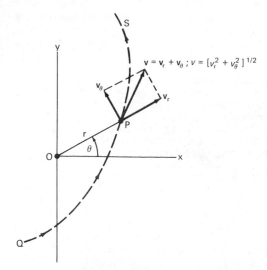

FIG. 2. Curvilinear motion in a plane of particle or body P of mass m along dashed path QPS, showing noncircular rotation about fixed axis through O perpendicular to page. $r\,(= OP)$, θ are polar coordinates of P; $\mathbf{v}$, total velocity vector tangent to curve QPS at P; v_θ, velocity component perpendicular to r; v_r, velocity component along r.

is not circular but still in the plane of the page along the dotted path QPS (Fig. 2), there is still rotation about the fixed center O but with the important difference that the linear velocity $\mathbf{v}$ has components both tangential, $\mathbf{v}_\theta$, and radial $\mathbf{v}_r$. $\mathbf{v}_\theta$ is defined by Eq. (7a) and $\mathbf{v}_r$ by

$$\mathbf{v}_r = \mathbf{i}_r\, v_r = \mathbf{i}_r\frac{dr}{dt}; \qquad (8)$$

where $\mathbf{i}_r$ is a unit vector in the direction of $\mathbf{r}$. The corresponding linear accelerations are

$$\mathbf{a}_\theta = \mathbf{i}_\theta\, a_\theta$$
$$= \mathbf{i}_\theta\, r\frac{d^2\theta}{dt^2} = \mathbf{i}_\theta\, r\alpha, \qquad (9)$$

where α is as defined by Eq. (4), and

$$\mathbf{a}_r = \mathbf{i}_r\, a_r = \mathbf{i}_r\frac{d^2r}{dt^2}. \qquad (9a)$$

Truly general 3-dimensional curvilinear motion with rotation about the center O (Fig. 2) would include, in addition to $\mathbf{v}_\theta$ and $\mathbf{v}_r$, a third component $\mathbf{v}_\phi$ of the linear velocity $\mathbf{v}$, perpendicular to the page and given by

$$\mathbf{v}_\phi = \mathbf{i}_\phi\, v_\phi = \mathbf{i}_\phi\, r\cos\theta\,\frac{d\phi}{dt}, \qquad (9b)$$

where ϕ is the angle between the plane of the page and a fixed plane (not shown by Fig. 2) passing through Oy, which should be visualized as tilted toward the reader out from the plane of the page, and i_ϕ is a unit vector perpendicular to r and both perpendicular to and out of the page at P. i_r, i_θ, i_ϕ are then the three mutually orthogonal unit vectors at P, and r, θ, ϕ are the spherical coordinates of the moving point P. The total 3-dimensional translated velocity v is

$$v = v_r + v_\theta + v_\phi = i_r\, v_r + i_\theta\, v_\phi + i_\phi\, v_\phi,$$

(10)

and the total linear speed v is

$$v = [v_r{}^2 + v_\theta{}^2 + v_\phi{}^2]^{1/2}.$$

(10a)

The corresponding 3-dimensional linear accelerations are a_r, a_θ, a_ϕ, or vectorially $i_r\, a_r, i_\theta\, a_\theta, i_\phi\, a_\phi$, which include all linear accelerations, including centripetal and, from another viewpoint (see Section 6), the *centrifugal* and *Coriolis* accelerations corresponding to the same-named forces.

2. Momentum The basic physical quantity *momentum* is of great importance in curvilinear motion. Both *linear* and *angular* momentum, of parallel importance, are possessed simultaneously by a particle or body P of mass m moving in a curve, as in Figs. 1 and 2. The defining equations are, for linear momentum M_ϱ measured in kg · m · sec^{-1},

$$M_\varrho = mv,$$

(11)

and for angular momentum M_ω measured in kg · m^2 · sec^{-1},

$$M_\omega = m\mathbf{r} \times \mathbf{v} = I\omega,$$

(12)

where I is the *moment of inertia* (defined in all physics texts) of m about its axis of rotation. Equation (12) necessarily employs vector cross-product notation, since r and v are not in general orthogonal; vector M_ω is perpendicular to the plane of r and v and in Figs. 1 and 2 points vertically upward from the page. In nonvector (scalar) form Eq. (12) becomes

$$M_\omega = mrv \sin \theta,$$

(12a)

where θ is the angle between r and v.

Both linear and angular momentum are conserved in closed (isolated) systems; this constitutes the most basic of all conversion principles. In quantum theory the "quantization" of mementa is fundamental; this means that these quantities can change by fixed amounts only.

3. Kinetic Energy As for momentum, the kinetic energy of a mass m in curvilinear motion simultaneously possesses both linear and angular kinetic energies, E_ϱ and E_ω, respectively, measured in joules, defined by

$$E_\varrho = \tfrac{1}{2} mv^2,$$

(13)

and

$$E_\omega = \tfrac{1}{2} I\omega^2.$$

(14)

Both E_ϱ and E_ω are scalar quantities, so that for an aggregate of mass particles both the total linear and angular kinetic energies of the aggregation (or body) are simply the algebraic sums of the individual kinetic energies of the separate mass particles comprising the aggregate or body.

4. Planetary Motion (Central Force, Kepler's Laws, Newtonian Motion) Some real-life examples of curvilinear motion will emphasize the importance of the subject. Consider first the rotation of the planets (*natural* satellites) of our solar system about the Sun.

(1) Johannes Kepler (German astronomer, 1571–1630) had limited mathematics available to him but was a man of rare physical insight, dedicated persistence, and keen intuitive reasoning. He had no knowledge of the calculus of Gottfried von Leibnitz (1646–1716) and Isaac Newton (1642–1727), which was devised about a half-century after Kepler's death, nor of the algebra and calculus of vectors. Employing only, along with his remarkable talents, the analytic idea of the attractive *central force* (directed from the planet toward the Sun), but not knowing the central force followed an inverse-square law (Newton's discovery), he inferred the principal analytics governing planetary motion in our solar system, known as Kepler's Laws:

First Law, the Law of Orbits (1609). All planets and other bodies of the solar system must move in plane orbits which are conic sections—ellipses for the planet, with the Sun at one focus as the center of attraction.

Second Law, the Law of Areas (1609). Equal areas are swept out by the radius vector from the Sun to the planet in equal times. This means that as the planet is approaching the orbit perihelion (the point nearest the Sun) it speeds up, whereas approaching the orbit aphelion (the point farthest from the Sun, at the opposite end of the major axis of the ellipse from perihelion) it must slow down.

Third Law, the Law of Periods (1619). The squares of the periods of the planets are proportional to the cubes of the major axes of their elliptic orbits.

Analytic details of Kepler's Three laws may be found in Joos.[2]

(2) Analytically, Kepler's central force $F(r)$ is written

$$\mathbf{F}(r) = f(r)\, \mathbf{i}_r,$$

(15)

where $f(r)$ is a function of the polar coordinate r of the planet and i_r is a unit vector along r, as used previously.

In 1671, some 40 years after Kepler's death, Isaac Newton discovered the *Law of Universal Gravitation*, which described Kepler's central force function $f(r)$ as an inverse-square dependence upon the radius vector $\mathbf{r}$; analytically, then, Eq. (15) becomes

$$F(r) = f(r)\,\mathbf{i}_r = \frac{N}{r^2}\,\mathbf{i}_r, \qquad (16)$$

where N is a constant negative number, the minus sign indicating that the force is attractive and hence directed in the negative $\mathbf{r}$ direction.

Using Eq. (16) Newton deduced analytically that the orbital paths of the planets are conic sections, specifically ellipses, with the Sun at one focus, thus confirming mathematically Kepler's First Law.

Comets, meteorites, and other bodies entering the solar system and therefore subject to the gravitational pull of the Sun may execute plane parabolic or hyperbolic orbits which are not closed curves. Therefore such objects may pass through and beyond the solar system never to return.

The possibility of parabolic or hyperbolic orbits which permit an object to leave the gravitational field of the Sun, or better the gravitational field of the solar system, and so "escape" from our planetary system elicits the interesting question of what is the Earth's *escape velocity*. More precisely, what is the minimum velocity, directed vertically opposite to the gravity pull at the Earth's surface, which would result in an object (e.g., a rocket-propelled spacecraft) leaving the Earth's gravitational field.

To determine this interesting number it is necessary simply to calculate the work done against the pull of gravity in moving a mass m from the Earth's surface to that distant point where its gravitational force is zero; that is, to an infinite distance away. Then, this amount of work or potential energy is equated to the kinetic energy that the same mass acquires if it were to fall from rest at that infinitely distant point and impact upon Earth (Fig. 3):

$$\int_{R_E}^{\infty} - F(r)\,dr = \tfrac{1}{2}\,m v_{es}^2, \qquad (17)$$

where $F(r)$ is Newton's gravitational force, defined by Eq. (16), with the minus sign now

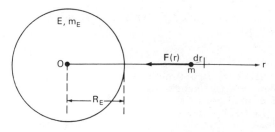

FIG. 3. Earth E (m_E, mass; R_E, radius) escape velocity diagram. m, escaping mass; r, radial distance from Earth's center O; $\mathbf{F}(r)$, gravitational attractive force.

used because it is directed toward Earth in the negative radial direction, and v_{es} is the speed with which the freely falling mass m strikes Earth, which in turn is numerically equal to the "lift-off" escape speed but is oppositely directed. $F(r)$ is usually expressed in terms of the two attracting masses m and m_E, so that

$$F(r) = -\frac{N}{r^2} = -\frac{Gmm_E}{r^2}, \qquad (18)$$

where G is the gravitational constant (6.6720×10^{-11} newton $\cdot\, m^2 \cdot kg^{-2}$). Using Eq. (18) in Eq. (17) yields

$$\int_{R_E}^{\infty} - \frac{Gmm_E}{r^2}\,dr = \tfrac{1}{2}\,m v_{es}^2. \qquad (19)$$

Dividing by m and integrating the left-hand side of Eq. (19) yields

$$\frac{Gm_E}{R_E} = \frac{v_{es}^2}{2}. \qquad (20)$$

Numerical values for G, m_E, and R_E could be substituted in Eq. (20) to calculate the numerical value of v_{es} but it is simpler to write Newton's law for the Earth's surface. Thus

$$\frac{Gmm_E}{R_E^2} = mg, \qquad (21)$$

where g is the acceleration due to gravity at Earth's surface. Again dividing by m, Eq. (21), another expression for the left-hand side of Eq. (20) is obtained:

$$\frac{Gm_E}{R_E} = gR_E. \qquad (21a)$$

Finally, equating the right-hand sides of Eqs. (20) and (21a) yields

$$v_{es} = (2gR_E)^{1/2}. \qquad (22)$$

Note that employing Eq. (21) avoids use of constants G and m_E, replacing the two constants by a familiar single constant g.

Inserting approximate but quite accurate values for g and R_E in Eq. (22) yields

$$v_{es} = \left[2 \times \frac{32\ \text{ft} \cdot \text{sec}^{-2}}{5280\ \text{ft} \cdot \text{m}^{-1}} \times 4000\ \text{mi} \right]^{1/2}$$

$$= 6.96\ \text{mi} \cdot \text{sec}^{-1}, \qquad (23)$$

which establishes the magnitude of Earth's escape velocity as $7\ \text{mi} \cdot \text{sec}^{-1}$, very nearly.

5. Elementary Dynamics of Space Flight (Artificial Earth Satellites) A second physical example of curvilinear motion is the rotation of man-made satellites placed in stable orbits about Earth via rocket launches from Earth's surface. Such *artificial* satellites are governed, of course, by the same dynamics as the planets, which are *natural* satellites of the Sun.

(1) The tangential orbital speed v_s which must be attained by the artificial satellite to achieve a stable, fixed orbit about Earth has to be adjusted critically so that the satellite neither spirals into the Earth nor out of the solar system.

To calculate v_s, consider for simplicity a circular orbit with Earth as center. The Earth's attracting force upon the satellite or spacecraft in such a stable orbit can be expressed in two ways: by Newton's law of gravitation, Eq. (18), and by a centripetal force expression, Eq. (6). Equating the two expressions yields

$$\frac{Gm_s m_E}{r_s^2} = \frac{m_s v_s^2}{r_s}, \qquad (24)$$

where m_s is the satellite mass, r_s is its radial distance, and the other symbols have the same meanings as before. Dividing both sides of Eq. (24) by m_s and solving for the critical speed v_s yields

$$v_s = \left(\frac{Gm_E}{r_2}\right)^{1/2}. \qquad (25)$$

Known values could be substituted for the three quantities of the right-hand side of Eq. (24) to yield a numerical value for v_s. Again, as was done in the escape velocity calculation in Section 4, it is easier, and instructive, to express the right-hand side of Eq. (24) in terms of weight, $m_s g_s$. Thus

$$\frac{m_s v_s^2}{r_s} = m_s g_s, \qquad (26)$$

or

$$v_s = (g_s r_s)^{1/2} \simeq (gR_E)^{1/2}, \qquad (27)$$

where g_s is the acceleration due to gravity at the satellite altitude, which for artificial satellites at altitudes of 100 mi (or several hundreds of miles) can be replaced by g (gravity at Earth's surface), and r_s by R_E (Earth's radius) to good approximations. Substitution of the numerical values in Eq. (27) yields

$$v_s = \left(\frac{32 \text{ ft} \cdot \sec^{-2} \times 4000 \text{ mi}}{5280 \text{ ft} \cdot \text{mi}^{-1}}\right)^{1/2}$$

$$= 4.92 \text{ mi} \cdot \sec^{-1} \simeq 5 \text{ mi} \cdot \sec^{-1}. \quad (28)$$

The period for such a satellite is calculated approximately, but with good accuracy, by dividing the circumference of its orbit (Earth circumference, approximately) by the round-number satellite speed from Eq. (28). Thus

$$\text{Period} = \frac{25,000 \text{ mi}}{5 \text{ mi} \cdot \sec^{-1} \times 60 \sec \cdot \min^{-1}}$$

$$= 83 \text{ min} \simeq 1\tfrac{1}{3} \text{ hr}, \qquad (29)$$

which is the time to complete one revolution (one orbit) about Earth.

(2) It should be noted that an artificial satellite moving in a fixed stable circular orbit does not "fall" toward Earth but maintains a constant radial distance from the center of the Earth. This recalls the elementary physics lecture-hall demonstration (Fig. 4) which is designed to show that a ball dropped *vertically* from rest (by deenergizing magnet M) and a projectile fired *horizontally* from a gun at the same vertical height as the ball and at the same instant the magnet is deenergized, fall vertically together so that the projectile strikes the ball at impact point A after elapsed time t.

In view of the artificial satellite situation, where the satellite moving with horizontal speed relative to Earth's surface of about 5 mi/sec *maintains* its vertical distance above Earth, it follows that the lecture-hall demonstration appears to work only because of the short distances and times and the low speed involved. The conclusion usually stated, that an initially horizontally moving object falls to Earth with the same gravity acceleration as an object dropped from rest with zero horizontal speed from the same vertical height, is, strictly speaking, incorrect. For the horizontally moving satellite speeding at about 5 mi/sec does not "fall" at all. The correct conclusion is that the horizontal speed of the gun projectile, negligible compared to the satellite speed, results in negligible *centrifugal* force upon the projectile so that it *appears* to fall with the same gravity acceleration as an object falling from rest with zero horizontal speed. (The concept of centrifugal force is treated in detail in Section 6.)

(3) It is interesting to note that Kepler's Third Law, derived in Joos[2] (pp. 92–93) for elliptic orbits, can be derived quite simply for circular orbits employing Newton's law of gravitation and algebraic manipulations only.

Thus, considering an artificial Earth satellite as appropriate in this section, the central attractive force upon the orbiting satellite can be expressed in two ways (as was done in obtaining Eq. (24) above), by Newton's law, Eq. (18),

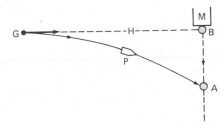

FIG. 4. Sketch of apparatus used to demonstrate "falling" of a horizontally-fired projectile and a vertically dropped ball. G, gun; H, horizontal line of fire (dashed); P, projectile; M, electromagnet; B, ball; A, impact point.

and by a centripetal force expression, Eq. (6). Equating these two expressions yields

$$\frac{Gm_s m_E}{r_s^2} = m_s \omega_s^2 r_s, \qquad (30)$$

where ω_s is the angular speed of the satellite:

$$\omega_s = 2\pi/T_s, \qquad (31)$$

with T_s the rotation period of the satellite about Earth. Using Eq. (31) in Eq. (30) and solving for T_s^2 yields

$$T_s^2 = \left(\frac{4\pi^2}{Gm_E}\right) r_s^3. \qquad (32)$$

Since the term in parentheses in Eq. (32) is constant, this equation states Kepler's Third Law for circular orbits: The squares of the periods of satellites are proportional to the cubes of the diameters of their circular orbits (remembering that the diameter of a circle, a special case of the ellipse, is its major axis and is twice its radius).

(4) "Stationary" Artificial Satellites. Artificial satellites which revolve about Earth with the same period as, and in phase with, the Earth are extremely important in communications, in direct television-antenna transmission over distances far greater than horizon-limited line-of-sight TV-tower transmissions, in observations of selected areas of Earth's surface, and so on. Such 24-hr period satellites are termed stationary because, synchronized with Earth's turning, they remain "fixed" above the same spot upon Earth's surface.

The critical altitude (above Earth's surface) of a stationary satellite in circular orbit is calculated directly by Eq. (32), in which r_s and T_s may be changed to R_{st} and T_{st}, the satellite radius (measured from Earth center) and period, respectively, for the "stationary" orbit; or

$$R_{st}^3 = \left(\frac{Gm_E}{4\pi^2}\right) T_{st}^2. \qquad (33)$$

As before, instead of using numerical values for G and m_E in Eq. (33), it is simpler to evaluate the product Gm_E using Eq. (21a). Thus

$$Gm_E = gR_E^2, \qquad (33a)$$

which substituted in Eq. (33) yields

$$R_{st}^3 = \left(\frac{gR_E^2}{4\pi^2}\right) T_{st}^2. \qquad (34)$$

Substituting numerical values for the constants in parentheses in the right-hand side of Eq. (34) yields

from which, substracting the radius of Earth (4000 mi, approximately), 22,400 mi, approximately is obtained as the critical altitude of a stationary satellite; this is very roughly equivalent to Earth's circumference, 25.000 mi.

6. Forces* (Centripetal Force; Centrifugal and Coriolis Forces; Inertial and Noninertial Reference Frames) In Section 1 the radially directed attractive force upon a point or body of mass m was discussed briefly and termed centripetal force [Eqs. (6) and (6a)] since it is directed toward the center of motion. It should be noted here that the concept of centripetal force is much preferred by physicists to that of centrifugal force. In fact, the latter term is denied and labeled as fictitious in some treatments of the subject. However, this negative attitude may be justified only when the observer of the motion is in an inertial frame of reference. Centrifugal forces have a proper place in noninertial reference frames. This matter is discussed in detail in this section.

Centrifugal and Coriolis forces are considered in somewhat more sophisticated discussions of curvilinear motion. It is important to realize that the concept of centrifugal force, and also that of Coriolis force, can be used instead of the simpler centripetal force concept, provided the reference frame for the motion is changed from an inertial to a noninertial frame.

Thus far an inertial frame of reference has been implied; that is, the observer is fixed in space as a bystander watching the motion. (A simple example: a bystander watching a wheel rotating on an axle.) An alternative and equally valid way of observing motions such as the rotary motion of the mass point P in Fig. 1 is for the observer to take his position upon the radius vector OP, as though he were on a turntable, and observe the same point P. Such an observer is in a noninertial reference frame.

Both the inertial and the noninertial observations are equally correct, as is emphasized in general relativity theory, but they are separate and distinct viewpoints; either one, but not both simultaneously, can be used to describe and analyze the motion.

To clarify these two viewpoints somewhat further consider a ball tied to a string whirled in a horizontal circle and then released while whirling. A bystander in this case is essentially

*This section is a revised summary of the corresponding discussion of forces in rotary motion by Arthur G. Rouse (Saint Louis University, deceased) which appeared in the first two editions of this Encyclopedia. The present author expresses his indebtedness to that original review of the subject.

$$R_{st} = \left[\frac{\dfrac{32}{5280} \text{ mi} \cdot \text{sec}^{-2} \, (4000 \text{ mi})^2 \, (24 \times 3600 \text{ sec})^2}{4\pi^2}\right]^{1/3}$$

$$= 2.64 \times 10^4 \text{ mi}, \qquad (34a)$$

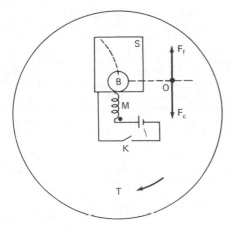

FIG. 5. Sketch of apparatus to demonstrate an "observer" in a *noninertial* frame. T, horizontal turntable; S, "magic slate," M, electromagnet energized by closing switch K to hold steel ball B; F_f and F_c, centrifugal and centripetal forces, respectively, translated for clarity from center of B to O. (Adapted from Ref. 3.)

an inertial observer and sees the ball fly off in a straight line in a tangential direction. Now consider the apparatus shown in Fig. 5. While the turntable is rotating clockwise at constant speed the steel ball can be released by deenergizing the electromagnet, causing the ball to roll on the "magic slate" and leave a trace as shown. The slate plays the role of a rotating, noninertial, observer and the trace on it is the record of the ball's path as seen by such observer. Note that the ball begins to roll in a radial direction on the slate and then curves. The inertial observer standing nearby sees a tangential motion of the ball just as it is released, whereas the noninertial observer on the turntable sees an initial radial motion.

In the example of the ball being whirled horizontally in a circle the bystander (in the inertial frame) says there is but one force acting upon the ball, the tension in the string directed toward the center of the uniform circular motion, which he terms the centripetal force. When the ball is released the centripetal force disappears, there is no horizontal force upon the ball, and so it goes off on a straight-line tangent at constant linear speed.

In contrast, the rotating observer (in the noninertial frame) on the rotating turntable says that while the ball is held by the magnet it is at rest, it has no motion and no acceleration relative to the "magic slate" and is in equilibrium (there is no net force acting on it). Hence there are two equal and opposing forces acting on the ball, the centripetal pull F_c of the magnet and the centrifugal force F_f acting radially outward. Upon release of the ball by the deenergized magnet the centripetal force disappears, the centrifugal force initially accelerates the ball radially outward, and an additional *Coriolis*

force (plus the effect of conservation of angular momentum, as will be explained in Section 8) causes the ball to follow the curved path on the slate seen by the noninertial rotating observer. The Coriolis force appears when the ball acquires velocity **v** with respect to the rotating observer and it turns out to be always perpendicular to **v**, while **v** in this experiment is always perpendicular to the angular velocity **ω** of the rotating reference frame. (Remembering, as noted in Section 1, that **ω** is perpendicular to the plane of the rotary motion.) Note that it is the rotating observer only, in his noninertial reference frame, who employs the concepts of centrifugal and Coriolis force. The discussion of Coriolis force is continued and expanded in the next section.

(NOTE: An observer at rest upon Earth is strictly not an inertial observer—one at rest with respect to the fixed stars—since Earth is rotating on its axis and traversing its orbit about the Sun at the same time. However, the motion of Earth in space may be neglected in the examples considered in this section, so that a bystander at rest upon Earth is essentially an inertial observer.)

7. Rotation of Earth; Some Effects of Centrifugal, Gravitational, and Coriolis Forces Simple yet adequate and correct discussions of the Coriolis force [named for G. G. Coriolis (1792–1843) a French scientist who first published an analysis in 1835] are not found in elementary physics texts and also are difficult to locate in the literature. An adequate discussion should include the concurrent centrifugal force and other related forces. Hence it is attempted here to supply such a treatment.

Our Earth in continuous rotation about its axis furnishes striking and convincing evidence of the reality and importance of the Coriolis force in combination with the other related forces, centrifugal and gravitational. An observer upon eastward rotating, Earth is a noninertial observer on a rotating plane disk (the circle of radius CP in Fig. 6) parallel to the equatorial plane and perpendicular to Earth's North-South axis, who plays the role of the "magic-slate observer" on the revolving turntable of Fig. 5.

Figure 6 illustrates the Earth rotating eastward with constant angular velocity **ω**$_E$ (counterclockwise, looking down the axis) about its axis, and shows the three forces F_G, F_f, and F_{cor} acting on a body freely falling from rest at P near Earth's surface in the Northern hemisphere. It should be noted that the *gravitational* force F_G which is not affected by Earth's rotation is directed toward Earth's center C, whereas the force of gravity F_g (the vector resultant of F_G and F_f) points toward O' on Earth's axis but "below" O in the Northern hemisphere.

The standard theoretical physics texts[4] derive the following expression for $\mathbf{F}_{cor}$:

$$\mathbf{F}_{cor} = -2m(\boldsymbol{\omega} \times \mathbf{v}) = -2m(\boldsymbol{\omega}_E \times \mathbf{v}), \quad (35)$$

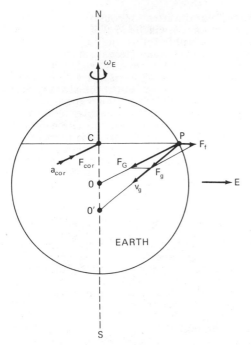

FIG. 6. Forces seen by *noninertial* observer upon a freely falling body at P near Earth rotating about geographic axis N-S (cross-sectional view through axis) with angular velocity ω_E. F_f, centrifugal force; F_G, gravitational force; F_g, gravity force (vector resultant of F_g and F_f); F_{cor}, a_{cor} Coriolis force and acceleration, initial directions (translated for clarity from P to C). C, center of plane circle of latitude through P parallel to equatorial plane.

where $\mathbf{v}$ is the instantaneous velocity, initially $\mathbf{v_g}$, of mass m at P. The corresponding Coriolis acceleration $\mathbf{a_{cor}}$ then is given by

$$\mathbf{a_{cor}} = -2(\boldsymbol{\omega} \times \mathbf{v}) = -2(\boldsymbol{\omega_E} \times \mathbf{v}). \qquad (35a)$$

The vector cross-products in Eqs. (35) and (35a) establish that both $\mathbf{F_{cor}}$ and $\mathbf{a_{cor}}$ are perpendicular to v (and to ω also) as stated in Section 6 and shown by Fig. 6.

8. Examples of Coriolis Force Although as indicated the Coriolis force is not known or understood widely, vivid examples exist for us who dwell upon a rotating platform, the planet Earth, and these follow.

(1) Figure 6 illustrates the effect of centrifugal force $\mathbf{F_f}$, in vector combination with mass attraction $\mathbf{F_G}$, in causing a plumbline's deviation from the direction PO toward Earth's center to the "vertical" $\mathbf{F_g}$ direction PO' away from Earth's center.

(2) If a downward motion is introduced, say in allowing lead balls to fall freely in the $\mathbf{F_g}$ plumbline direction by dropping them in a deep mine shaft drilled in the same vertical direction,

the balls would strike the *East* side of the shaft before reaching the bottom. Such an experiment was in fact performed in northern Michigan some years ago, the balls being dropped from the exact center of an 8-ft-square, mile-deep shaft. None of the balls ever reached the bottom; instead they were found imbedded in the East side of the shaft, i.e., the balls in falling drifted some 4 ft *eastward* with respect to the bottom of the shaft before striking the side. This eastward drift of freely falling bodies (neglecting air resistance) in the Northern hemisphere is the effect of the Coriolis force arising from the combination of the Earth's rotational velocity ω_E and the linear velocity $\mathbf{v}$, as shown by Fig. 6 and explained above.

Also involved in the mine-shaft experiment is the principle of conservation of angular momentum (see Section 2), which must be included with the Coriolis force to account for the observed eastward drift of the falling balls. Because the angular momentum of the closed system (Earth and the falling ball) must remain unchanged, the original eastward horizontal velocity of the ball, which is the same as the Earth's eastward surface velocity at the drop point, remains essentially constant while the ball is dropping. Hence the ball, dropped from "rest" relative to Earth, catches up to the East side of the shaft as it falls. It is essential to realize that the bottom of the shaft has a smaller eastward horizontal velocity than the top due to its shorter radial distance from Earth's axis through COO' (Fig. 6). Here Eq. (7), $v = r\omega$, applies; since ω_E is constant, v for the mine shaft bottom is lower because r has decreased, whereas the falling ball, conserving angular momentum, continues eastward with the same horizontal velocity as at the top of the shaft.

(3) Another everyday example of the action of the Coriolis force is available to everyone in the draining of water from his bathroom washbowl (a hemispherical, symmetric bowl best shows the effect). If the drain plug of the water-filled bowl is opened carefully when the water has quieted to a still surface, the discharging water will usually be seen to swirl into a rotating (counterclockwise in the Northern, clockwise in the Southern hemisphere) hollow whirlpool as it goes down the drain. This effect occurs because the central core of the draining water drops faster down toward and into the drain than the water near the side of the bowl, so that a central hollow is formed, which is turn causes the water immediately around the hollow to slide down the hill of the vortex cavity, thus resulting in a downward-slanted velocity $\mathbf{v_g}$ due to gravity, much as in Fig. 6. The vector combination of $\mathbf{v_g}$ with ω_E produces the Coriolis force [Eq. (35)] upon the downward-draining water, causing it to swirl as observed. The Coriolis force, and its consequent acceleration and velocity, turn out to be

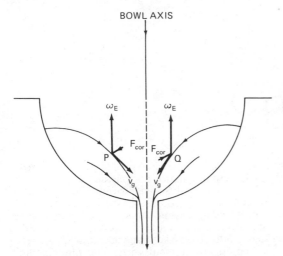

FIG. 7. Diagram to explain effect of Coriolis force inducing counterclockwise vortex motion of water draining from washbowl (vertical cross-sectional view. Northern hemisphere, ω_E, angular velocity of Earth due to eastward rotation about N-S axis; v_g, velocity due to gravity of small volumes of water at typical locations P and Q; F_{cor}, perpendicular to page (out from page at P, into page at Q). ω_E and v_g, in plane of page (see text for note on more precise directions).

directed all around the hollow vortex of the draining water in a counterclockwise direction (looking down upon the water) in the Northern hemisphere, as should be clear from Fig. 7. (Actually, in Fig. 7 ω_E is pointed parallel to Earth's North-South axis, which is tilted relative to the "bowl axis" shown in the plane

of the page so that the ω_E and F_{cor} vectors shown are really the components of those two vectors in and perpendicular to the plane of the page, respectively.)

This analysis is somewhat simplified. For example, it is evident that any small volume of water in the draining washbowl considered as the "body" in motion is not isolated, as is the lead ball falling in the mine shaft, but interacts continually with other surrounding small volumes or bodies of water. So it appears that the centrifugal force on any such small volume would be counteracted in part at least by the resisting forces of the next radially outward small volume because of liquid incompressibility and other liquid (viscous) forces. Further, once the draining washbowl water is in circular motion due to the Coriolis force effect, conservation of angular momentum acts to increase the swirling counterclockwise motion of the water as it moves *inward* from the outer edges of the whirlpool toward and down into the central drain, because, by Eq. (12), as the radius **r** of swirling decreases, the speed **v** must increase to keep the momentum **M** constant.

A. H. WEBER

References

1. Sears and Zemansky, "University Physics," pp. 146–149, Cambridge, Mass., Addison-Wesley, 1952.
2. Joos, G., "Theoretical Physics," 3rd Ed., pp. 88–93, New York, Hafner.
3. Rouse, A. G., *Am. Jour. Phys.* 27 (6), 429 (1959).
4. Ref. 2, pp. 232–233.

Cross-references: CENTRIFUGE, DYNAMICS, IMPULSE AND MOMENTUM.

S

SCHRÖDINGER EQUATION

The Schrödinger equation, first obtained in 1926[1], was an extension of de Broglie's hypothesis, proposed two years earlier,[2] that each material particle has associated with it a wavelength λ related to the linear momentum p of the particle by the equation

$$\lambda = \frac{h}{p} \qquad (1)$$

where h is Planck's constant. Since any sinusoidally varying wave motion of amplitude ψ and wavelength λ satisfies the differential equation*

$$\nabla^2 \psi + \frac{4\pi^2}{\lambda^2} \psi = 0 \qquad (2)$$

matter waves would obey the equation

$$\nabla^2 \psi + \frac{4\pi^2}{h^2} p^2 \psi = 0 \qquad (3)$$

In particular, a particle of mass m with no forces acting on it has energy $E = \frac{1}{2} mv^2 = p^2/2m$, or $p^2 = 2mE$. Equation (3) may thus be written

$$\nabla^2 \psi + \frac{8\pi^2 m}{h^2} E\psi = 0 \qquad (4)$$

This is the Schrödinger equation for a free particle. Of greater importance is the equation for a bound particle for which the binding force can be related to a potential energy V. In this case, the total energy of the particle is equal to the sum of the kinetic and potential energies, i.e., $E = p^2/2m + V$, or $p^2 = 2m(E - V)$. Equation (3) thus becomes

$$\nabla^2 \psi + \frac{8\pi^2 m}{h^2} (E - V)\psi = 0 \qquad (5)$$

This is known as the time-independent Schrödinger equation. Its great utility lies in the fact that it enables one to calculate energy levels, or eigenvalues, of the energy E (see QUANTUM THEORY). The remarkable agreement of the results of the equation with experimental fact

$$*\nabla^2 \equiv \frac{\partial^2}{\partial x^2} + \frac{\partial^2}{\partial y^2} + \frac{\partial^2}{\partial z^2}.$$

has led to its being regarded as one of the fundamental equations of physics.

The more advanced formulation of the Schrödinger equation is based on the operator concept of quantum theory together with the Hamiltonian methods of classical mechanics. Specifically, to each dynamical variable q_i, there is a conjugate momentum p_i. The fundamental postulate of quantum mechanics states that for each pair (q_i, p_i), the equation

$$p_i \psi = -i\hbar \frac{\partial \psi}{\partial q_i} \qquad (6)$$

holds, where $\hbar = h/2\pi$. That is, $-i\hbar \partial/\partial q_i$ is an operator, which, when applied to the wave function ψ, is equivalent to multiplying by p_i. In the Hamiltonian function $H(p_i, q_i)$, each p_i is replaced by its corresponding operator. The result is the Hamiltonian operator $H(-i\hbar \partial/\partial q_i, q_i)$, which, when it operates on the wave function, is equivalent to multiplying by the energy E, viz.,

$$H(-i\hbar \partial/\partial q_i, q_i)\psi = E\psi \qquad (7)$$

This is the generalized form of the time-independent Schrödinger equation. It reduces to Eq. (5) when applied to a single particle, but it has the advantage that the application to systems of many particles can be carried out in a straightforward manner.

The time-dependent form of the Schrödinger equation is obtained by replacing E by its operator $E \to i\hbar \partial/\partial t$, so that

$$H\left(-i\hbar \frac{\partial}{\partial q_i}, q_i\right)\psi = i\hbar \frac{\partial}{\partial t} \psi \qquad (8)$$

For applications of the various forms of the Schrödinger equation, the reader should consult the article on WAVE MECHANICS and references therein.

GRANT R. FOWLES

References

1. *Ann. Physik*, 79, 361, 489 (1926).
2. *Phil. Mag.*, 47, 446 (1924).

Cross-references: QUANTUM THEORY, WAVE MECHANICS, WAVE MOTION.

SECONDARY EMISSION

Secondary electron emission is the emission of electrons from a surface as the result of bombardment of the surface by incident "primary" particles. The primary particles can be electrons, ions, or neutral atoms. Unless otherwise specified, secondary electron emission is generally taken to refer to electron emission as the result of impinging primary electrons. The phenomenon of secondary emission is the basis for a wide range of practical applications. In the last decade, secondary emission has been extensively used as a method of surface atom identification (Auger electron spectroscopy). Secondary emission is important in most vacuum systems, because insulating or electrically floating surfaces in the presence of charged particles assume a potential that is determined in large part by the secondary emission properties of the material.

Secondary electrons have energies ranging from zero up to the primary electron energy (Fig. 1). The energy distribution curve peaks between 1 and 10 eV, drops down to a low value above 30 eV, and may rise again near the primary electron energy. Superimposed on this curve are discrete energy peaks, described below.

The secondary electron emission ratio or yield δ is the number of secondary electrons emitted on average per incident primary electron. This ratio is a function of the energy of the incident primary electron and of its angle of incidence (Fig. 2). For a given material, surface conditions are important, and small amounts of contamination or oxidation can alter the secondary emission ratio. Different crystalline faces of the same crystal may have different work functions and secondary emission yields.

The emission of low energy secondary electrons is a complicated multiple process. The primary electron loses its energy in one or more inelastic collisions. Some of this energy may be imparted to other electrons in a cascading process. Electrons arriving at the surface from within the solid with energy and direction enabling them to traverse the surface barrier emerge as secondary electrons. Factors tending to increase the secondary emission yield are: a long mean free path inside the solid for the cascading low energy secondary electrons, and a low surface barrier or work function. If the primary electron makes a small angle of incidence (nonperpendicular) with the surface, the yield is enhanced because relatively more secondary electrons originate close to the surface. As the primary electron energy is increases, many of the secondaries are formed too far from the surface to be able to reach it with sufficient energy to escape, and the secondary emission yield drops with primary electron energy.

Metals generally have a maximum yield between 1 and 2. This low yield is mostly a consequence of the large number of free electrons in the metal that interact with the secondaries, causing short mean free paths and short escape depths.[1] In insulators, low energy secondary electrons may have relatively long mean free paths because of the absence of nearly free electrons with which to interact. The secondaries lose energy to phonons, as can be shown by a decrease in yield with increasing temperature. The yields of insulators are generally higher than yields of metals. See Table 1.

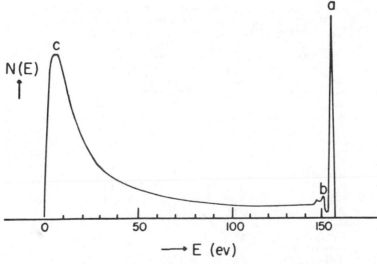

FIG. 1. Energy distribution of secondary electrons, showing a large peak of low energy electrons, a sharp peak of elastically reflected electrons at the primary electron energy, and some energy loss peaks. [E. Rudberg, Phs. Rev. 50, 138 (1937)]

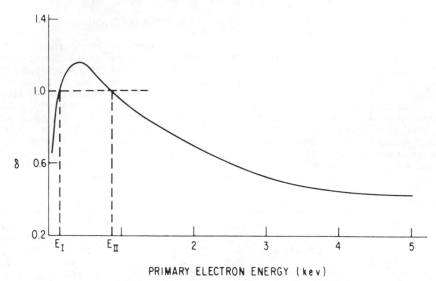

PRIMARY ELECTRON ENERGY (kev)

FIG. 2. The secondary emission yield of germanium as a function of primary electron energy. The first and second crossovers where the yield is unity are given by E_I and E_{II}. [J. B. Johnson and K. G. McKay, Phys. Rev. 93, 668 (1954)]

Backscattered Electrons Whereas most secondary electrons have lost all but a few eV of energy in emerging from the solid, some energetic secondaries have energies all the way up to the primary electron energy. By convention, backscattered electrons are denoted as those with energy of 50 eV or more. The backscattered fraction or ratio is the average number of backscattered electrons per primary electron.

Figure 3 presents the backscattered fraction for a number of elements. For low atomic number materials ($Z < 30$), the backscattered fraction is relatively constant with primary electron energy, and increases with increasing atomic number. At higher atomic numbers, the backscattered fraction increases with primary electron energy.

Elastically Reflected Electrons A number of

TABLE 1. SECONDARY ELECTRON EMISSION YIELDS
OF SOME REPRESENTATIVE MATERIALS.
(δ_{max} is the maximum secondary emission yield at
energy $E_{p\,max}$. E_I and E_{II} are the first and secondary
crossovers. For a more complete list, see Ref. 5.)

Material	δ_{max}	$E_{p\,max}$ (eV)	E_I (eV)	E_{II} (eV)
Ag	1.5	800	200	>2000
Al	1.0	300	300	300
Au	1.4	800	150	>2000
Be	0.5	200	none	none
C (graphite)	1.0	300	300	300
C (soot)	0.45	500	none	none
Cu	1.3	600	200	1500
Fe	1.3	400	120	1400
Mg	0.95	300	none	none
Pt	1.8	700	350	3000
Si	1.1	250	125	500
Si (cesiated)	900	20,000	125	–
W	1.4	650	250	>1500
NaCl (crystal)	14	1200		
NaI (crystal)	19	1300		
MgO (crystal)	25	1500		
MgO (layer)	3–15	400–1500		
BeO	3.4	2000		

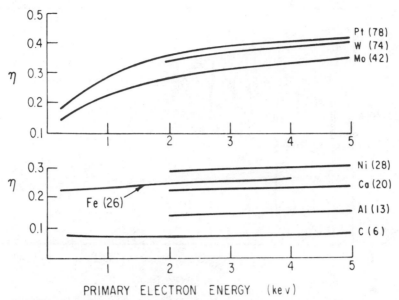

FIG. 3. The backscattered fraction as a function of primary electron energy. [E. J. Sternglass, Phys. Rev. 95, 345 (1954), P. Palluel, Compt. Rend. 224, 1492 (1947)

processes give rise to electrons with discrete amounts of energy. To retain this precise energy, these processes must occur close to the surface to reduce the chance of further inelastic losses. Electrons that are elastically scattered back out of the surface give rise to a peak with the full primary electron energy.

Energy Loss Electrons Reflected electrons may excite surface or volume plasma excitations of the electrons in the solid before leaving the surface. Vibrational modes of individual atoms may also be excited.[2] The emerging secondary electrons have the full primary energy minus the excitation energy. These energy losses cause a series of peaks in the yield curve just below the primary electron energy. Measurement of these losses is termed *electron energy loss spectroscopy,* and is used to study the vibrational properties of atoms and ordered overlayers on surfaces.

Auger Electrons Auger electrons have energies related to transition levels within the atom in the solid where they originate. This energy is independent of the primary electron energy, in contrast to plasma loss electrons that have the primary energy minus the loss energy. Auger electrons originate in processes where an inner shell electron is ejected from an atom by a bombarding particle. A vacancy is created that is filled by a higher energy level electron. The difference in energy levels may be either given to an outer shell electron or emitted as an x-ray. If such a process occurs near the surface with an electron receiving the energy, and if the electron suffers no further energy loss, it will have the Auger electron energy for that atom.

Atoms with many transition levels have a series of Auger electron energies. (See Fig. 4.)

High Yield Materials Considered as a group, insulators have higher secondary emission yield maxima than conductors, although there are exceptions. Some high yield materials are MgO, with a yield as high as 25 for a good single crystal, and NaBr, with 24. In insulators, an excited electron in the conduction band may have insufficient energy to interact with electrons in the valence band. Its main energy loss mechanism may be through interactions with lattice vibrations, with a small loss of energy per interaction. Hence the diffusion path lengths can be long, and excited electrons formed deep within the insulator may be able to reach the surface.

Interesting high yield materials can be made with negative electron affinity semiconductors. Examples of these are cesiated GaP and Cs-activated silicon.[4] Electrons at the bottom of the conduction band at the surface are able to emerge as secondaries because of the absence of a surface barrier. The electron may originate at a long distance from the surface. In cesiated silicon, a yield of 900 has been reported at a primary energy near 20 keV.

In practical high yield devices, such as electron multipliers, the emitting material is usually in thin film form on a suitable conductor, so that electrical leakage or breakdown through the film prevents the surface potential from differing appreciably from that of the substrate. An alloy of magnesium and silver is frequently used, containing a few percent magnesium. When heated in an oxidizing at-

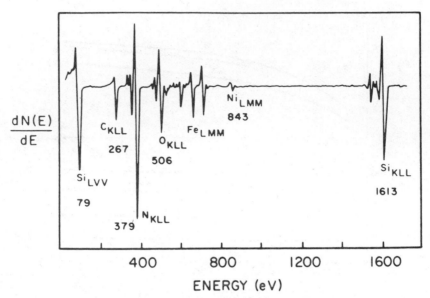

$$\frac{dN(E)}{dE}$$

FIG. 4. Auger Spectrum from the surface of Si_3N_4. Note that the derivative of the yield is measured, enhancing the structure in the curve. The various Auger electron peaks and the corresponding transition are labeled. (Ref. 3)

mosphere, a high yield magnesium oxide film forms on the surface. Cesium or cesium oxide are sometimes introduced to lower the electron affinity of the film. Yields of thin film MgO range from 3 to 15, depending on how the film was handled. An alternative material is an oxidized Be-Cu alloy containing a few percent Be, with beryllium oxide at the surface.

Thick porous oxide films occasionally exhibit field-enhanced emission. Positive charges near the surface create strong local fields, enhancing the emission of electrons. This type of emission is known as the Malter effect. It is usually unstable and nonreproducible.

Low Yield Surfaces It is frequently necessary to minimize the secondary emission yield of a surface. A roughened or porous surface often helps to do this, because many secondaries are intercepted by other surfaces before exiting the solid. Carbon in the form of soot is commonly used for this purpose.

First and Second Crossovers The potential which an insulating or floating surface attains in the presence of charged particles may depend on its secondary emission properties. The first and second crossovers of a material are the primary energies at which its secondary emission is unity (Fig. 2). If the primary energy is less than the first crossover, the surface will charge negatively toward the electron source potential. If the primary energy is between the first and second crossovers, the yield is greater than unity. The surface will then charge positively toward the second crossover provided there is a positive surface that attracts or collects the secondaries. If the primary energy is greater than the second crossover, the yield is less than unity, and the surface will charge negatively toward the second crossover.

Applications Secondary emission is always present when charged particles strike surfaces. Many devices make use of this property. Secondary emission multipliers use a series of secondary emission surfaces to multiply the number of electrons—gains of a million or more are readily achieved. Scanning electron microscopes can monitor the secondary emission yield of the surface being scanned by the primary beam—with changes in the secondary emission current showing the boundaries of each material. Auger electron spectroscopy (and scanning Auger microscopes) use the discrete energies of Auger electrons to identify surface atoms with great sensitivity. Channel plate multipliers consist of fine capillary tubes through a glass plate. The conductivity and secondary emission are such that an incident focused optical image produces photoelectrons which are multiplied in the channel plate capillaries to give proportionate electron currents at the exit side.

Historical Secondary emission was discovered by Austin and Starke in 1902. In their experiment, a primary electron beam was incident on metal targets at different angles of incident. As the incidence angle went from perpendicular to nearly grazing, they observed that the total current to the target decreased to zero and went negative. They concluded that not only are some primaries reflected, but secondaries formed in the metal also leave the surface, exceeding the primary electron current. The phenomenon was observed to be strictly proportional to the primary electron current.

N. REY WHETTEN

References

1. Ganachaud, J. P., and Cailler, M., "A Monte Carlo Calculation of the Secondary Electron Emission of Normal Metals II. Results for Aluminum," *Surface Science* 83, 519 (1979).
2. Ibach, H., *J. Vac. Sci. Techn.* 20, 574 (1982).
3. Kapoor, V. J., Bailey, R. S., and Smith, S. R., *J. Vac. Sci. Techn.* 18, 305 (1981).
4. Williams, B. F., Martinelli, R. U., and Kohn, E. S., *Adv. Electronics and Electron Phys.* 33A, 447 (1972).
5. N. R. Whetten, in "Handbook of Chemistry and Physics," 62nd Ed. (R. C. Weast and M. J. Astle, Eds.), p. E-373, Boca Raton, FL, CRC Press, 1981–1982.

Cross-references: AUGER EFFECT, COLLISIONS OF PARTICLES, ELECTRON, PHOTOELECTRICITY, THERMIONICS.

SEISMOLOGY

Most of our detailed knowledge of the interior of the Earth and Moon and the most precise geophysical explorations of the Earth's crust come from seismology. In a narrow sense, seismology is the study of earthquakes and their effects, but knowledge of seismic waves has been applied in many fields. Not only does seismology provide information on Earth structure, but it also relates to fundamental problems of economic and social importance. Earthquakes are destructive over wide areas of the world and, with the growth of industrialization and the spread of population, the consequences of strong ground shaking can be catastropic. Thus, seismologists provide engineers with risk maps which are the basis for national building codes, as well as estimates of strong ground motions for the earthquake-resistant design of large critical facilities, such as nuclear reactors, hospitals, dams, and bridges.

The first advance in quantitative seismology took place at the turn of the century, when instruments called *seismographs* were installed in various parts of the world to measure earthquake ground motion. The availability of seismograms that show the actual ground motion in earthquakes provided the basis for physical interpretations. As the century advanced, seismological studies led to the understanding of interior Earth structure and the tectonics of the crust and played a crucial role in exploration for minerals, oil, and other energy resources. In addition, in the last two decades, seismograph networks and analysis of seismic waves have provided vital information in the attempt to attain a comprehensive treaty to ban underground nuclear tests.

Most earthquakes are caused by the sudden release of elastic energy that occurs during rupture of a few km/sec along geological faults in the Earth's upper layers. Some earthquakes are caused by motion of magma under volcanoes, massive landslides, rock bursts in mines, and similar phenomena. Most significant earthquakes, called *tectonic earthquakes*, are produced, however, by fault slip. Their mechanism became clear after studies of faulting and strain of the crust associated with the great 1906 San Francisco earthquake along the San Andreas fault. A definitive study by H. F. Reid modeled the earthquake source as rapid elastic rebound of the rock on both sides of the fault after slow accumulation of elastic strain in the region. In detail, the physical processes involved are complicated, involving opening and closing of microfractures in the rock, percolation of ground water, and the distribution of strain between compression and shear. Dilatancy of the rock seems to occur before the major fault rupture; this involves an increase in the volume of the strained rocks due to various elastic and nonelastic changes. In a strained area, there may be minor percursory slip on faults producing *foreshocks*. After the main or principal earthquake the readjustment of strain around the ruptured fault often produces smaller slips that are observed as an *aftershock* sequence.

The place where slip begins on a fault is called the *hypocenter* or *focus* and the point on the Earth's surface above the focus is the *epicenter*. The main activity at the several hundred seismographic observatories around the world is to determine the location of such foci, using seismic waves that radiate outward at known speeds from the fault rupture through the Earth's interior. The amplitude of the seismic waves is also measured to give the magnitude or the seismic moment of the earthquake. A common measurement of earthquake size is the *Richter magnitude*, although there are other magnitude scales in use which depend on measurements of various parts of the seismic wave train. The magnitude scale is a logarithmic one that can range from negative values to values above 8. The largest shocks have been found to have magnitudes in excess of $8\frac{1}{2}$. The *seismic moment* is equal to the rigidity of the strained rock times the area of faulting times the amount of fault-slip. The largest earthquakes have moments in excess of 10^{26} dyne cm. Neither magnitude nor moment is a direct measure of the energy released in an earthquake. Such measurements are difficult because of energy loss by friction and other causes along the slipping fault and the damping of the waves as they travel through the Earth. An empirical relation between surface wave magnitude M and the energy E (in ergs) is $\log E = 11.8 + 1.5M$. Earthquakes of largest size probably release as much as 10^{25} ergs in the form of seismic wave energy, and the average annual seismic energy released from all earthquakes is also about 10^{25} ergs.

Earthquake waves consist predominantly of three types. Two types, called *P waves* and *S waves*, travel through the body of Earth while the third type, called *surface waves*, travel around the Earth's surface. The fastest is the P wave, involving compression and dilation of the rock. The slower body wave is the S wave, which shears the rock as it propagates. It follows that there are no S waves in liquid media. The speed of seismic waves depends on the elastic properties of the rocks through which they pass. The P wave velocity is given by $V_p = \sqrt{k + \frac{4}{3}\mu}$, and the S wave velocity is $V_s = \sqrt{\mu/\rho}$, where k and μ are elastic moduli and ρ is the density. The velocities of P waves range from about 5 km/sec in crustal rocks to a maximum of 13 km/sec (see Fig. 1). The shear wave velocities range from about 3 to 8 km/sec. Surface waves are of two main kinds, namely, *Rayleigh waves*, in which particle motions are in a vertical plane aligned with the direction of wave propagation, and *Love waves*, in which motions are horizontal at right angles to the propagation direction. Various physical complexities in the Earth, such as heterogeneity and anisotropy, couple these simple wave types to a minor extent. Seismic waves contain a wide range of frequencies (see Fig. 2). The highest frequencies recorded from near earthquakes are of the order of 10 Hz (0.1 sec period) and periods of 1–20 seconds predominate in the recordings of most distant earthquakes. The longest periods in seismic motion range up to 53 minutes which corresponds to the greatest free oscillation of the whole Earth.

A major observational improvement was the establishment of the World-Wide Standardized Seismograph Network (WWSSN) in the 1960s. By 1969, about 120 of these special stations were distributed in 60 countries. The immediate consequence was the more complete and pre-

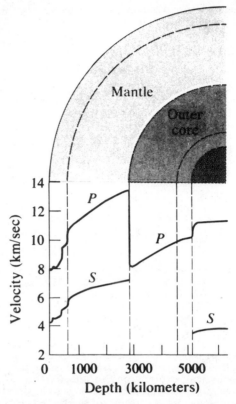

FIG. 1. A cross section of the Earth showing the main structural shells and the variation of P and S seismic velocities with depth.

cise coverage of the location of earthquakes around the globe. The network is being upgraded in the 1980s with more modern seismographs that record digitally over a wider range of frequencies. Global seismicity recorded by the WWSSN concentrates along the mid-oceanic ridges, island arcs, and deep oceanic trenches. Most foci are situated at shallow depths (less

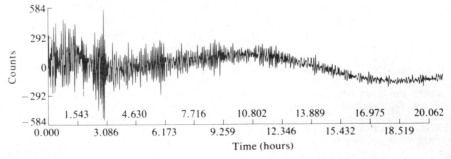

FIG. 2. A seismogram showing the vertical ground motion at Whiskeytown Station of the University of California, Berkeley, in the great Indonesia earthquake of August 19, 1977 (magnitude 7.9). The Earth tide with period of about 12 hrs can also be seen.

than 30 km), but in regions associated with deep oceanic trenches (such as western Central and South America, Alaska, Japan, the Caribbean Antilles, the Aegean Sea, Tonga, and Indonesia), earthquake foci dip to depths of up to 700 km. Typically, the zones of deep earthquakes, named after H. Benioff, dip at 30–60° toward the land. Other deep earthquakes are located in regions without trenches, such as the Hindu Kush and Romania.

A recent explanation of the strong patterns of seismically active belts surrounding largely aseismic areas is the *plate tectonic theory*. The theory describes the lithosphere of the Earth as consisting of about 10 large, and some smaller, stable regions or *plates*. Each plate extends to about 80 km depth and moves horizontally relative to neighboring plates. At plate edges, large strains form, producing earthquakes. Along the mid-oceanic ridges, the plates spread away from each other, and along the Benioff zones the lithosphere subducts downward into the Earth's interior. The mechanisms of earthquake sources are found to agree generally with this description, with normal faulting predominating along ocean ridges and thrust faulting dominating along subduction zones.

Tens of thousands of recorded seismic waves from natural earthquakes and underground nuclear explosions have given a detailed picture of the structure of the Earth's interior (see Fig. 1). It consists of a crustal shell about 5 km thick under oceans and 30–40 km thick under continents; this is the upper part of the 80-km-thick lithosphere. There is a solid but layered mantle extending to about 2900 km depth. Below the mantle is a 2259-km-thick outer core with properties resembling a liquid (no S waves) and at the center of the planet is a separate solid inner core of radius 1216 km. Present research is aimed at elucidating the interior fine structure, including the properties of transition shells and the mapping of regions of large-scale heterogeneity.

Seismograms have been obtained from both the Moon and Mars. The lunar Apollo seismograms gave preliminary information on Moon structure, detecting between 600 and 3000 moonquakes during every year of operation, although most have magnitudes less than 2.0. The lunar interior has a crust 60–100 km thick overlying a two-layered mantle with a transition zone near 450–480 km. A seismograph on Viking II operated on Mars for 546 Martian days. Although background seismic noise (microseisms) was generally high, the indication was that seismicity is very low.

Exploration of the Earth's interior represents a clear example of mathematical inverse problems. Measurements on the surface must be used to infer deep structure. Seismological applications have stimulated the development of inverse theory, particularly in the use of the free oscillations (eigenvibrations) of the Earth (see Fig. 2). The largest earthquakes vibrate the whole planet, and interference between traveling seismic waves produces two classes of eigenvibration, *spheroidal*, $_nS_l$, and *torsional*, $_nT_l$ (see Fig. 3). The subscripts n and l refer to interior and surface nodal surfaces, respectively. The fundamental spheroidal oscillations $_0S_2$ has the longest period, approximately 3,233 seconds, while the fundamental torsional $_0T_2$ has a period of 2,638 seconds. The ellipticity, rotation, and heterogeneity of the Earth split the eigenspectra into multiplets. These hyperfine spectra are now being intensively studied with a view to their inversion to finer Earth structure. Observations of the damping of both traveling and standing seismic waves give information on the rheology of the Earth's interior. The average value of the damping parameter Q for $_0S_2$ is 620 and that for $_0T_2$ is 350.

Emphasis has recently been placed on earthquake prediction. Forecasting the time and place of great earthquakes would provide many social benefits in countries where construction is generally not earthquake-resistant, such as China. Although the theoretical basis for prediction is still weak, rock mechanics experiments suggest that variations in rock properties occur before a large fault slip. These effects produce ground uplift and tilting, increased concentrations of radon emission from ground water, changes in electrical resistivity, and changes in microseismicity. Field programs with special instruments searching for such precursory effects in various countries have given conflicting results to date. Well documented claims for specific short-term earthquake predictions are few, the most important being a 24-hour prediction of a damaging earthquake on February 4, 1975, in Liaoning province, China. That evening, an earthquake, magnitude 7.3, occurred in the Haicheng-Yingkow region. Previous evacuation of the populace ensured few casualties from the damaged structures.

The second type of prediction is the specification of strong ground motion likely to occur at a particular site should a large earthquake occur. Seismic risk maps are produced that give the probabilities of exceeding specified amplitudes of seismic waves in a specified number of years. Mathematical models of fault rupture have led to plausible synthetic records that predict the pattern of seismic waves at points near and far from the seismic source. Such calculations must be scaled to the actual levels of ground acceleration, velocity, and displacement by using records from strong-motion accelerometers of moderate and large earthquakes. Accelerations up to $1.2g$ have been recorded horizontally and vertically. Because few strong-motion accelerograms have as yet been obtained, most estimates of the distribution of ground motion rely heavily on reported seismic intensity. The intensity scale is not instrumental, but describes human responses, earthquake damage to structures, etc.

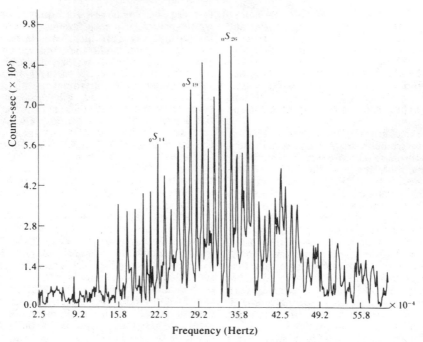

FIG. 3. Spectral peaks of free oscillations of the Earth after the 1977 Indonesia earthquake from an analysis of the motion in Fig. 2. Three peaks have been identified as spheroidal type $_nS_l$.

The most common Western scale is the *Modified Mercalli intensity*, with divisions from I to XII. Correlations have now been made between intensities and seismic magnitudes and moments and recorded ground accelerations for various soil and rock conditions and types of earthquake mechanism (strike-slip, dip-slip, etc.).

Specially designed groups of accelerometers, called *strong-motion arrays*, are now being installed in highly seismic parts of the world. The first large-scale array with digital accelerometers began operation in Taiwan in 1980. Building codes have been adopted in some earthquake areas to minimize the damage from shaking. Studies of strong-motion records and building collapse and failure mechanisms in large earthquakes have lead to code revisions.

BRUCE A. BOLT

References

Aki, K., and Richards, P. G., "Quantitative Seismology," San Franciso, W. H. Freeman, 1980.

Bolt, B. A., "Nuclear Explosions and Earthquakes, the Parted Veil," San Francisco, W. H. Freeman, 1976.

Bolt, B. A., "Earthquakes: A Primer," San Francisco, W. H. Freeman, 1978.

Bolt, B. A., "Inside the Earth," San Francisco, W. H. Freeman, 1982.

Bullen, K. E., "Introduction to the Theory of Seismology," 3d Ed., Cambridge, U.K., Cambridge Univ. Press, 1963.

Gutenberg, B., and Richter, C. F., "Seismicity of the Earth," Princeton, N.J., Princeton Univ. Press, 1949.

Hudson, J. A., "The Excitation and Propagation of Elastic Waves," Cambridge, U.K., Cambridge Univ. Press, 1980.

Jeffreys, H., "The Earth," 5th Ed., Cambridge, U.K., Cambridge Univ. Press, 1970.

Kasahara, K., "Earthquake Mechanics," Cambridge, U.K., Cambridge Univ. Press, 1981.

Pilant, W. L., "Elastic Waves in the Earth," Amsterdam, Elsevier, 1979.

Richter, C. F., "Elementary Seismology," San Francisco, W. H. Freeman, 1958.

Rothé, J. P., "The Seismicity of the Earth, 1953–1965," Paris, UNESCO, 1969.

Cross-references: GEOPHYSICS, VIBRATION, VOLCANOLOGY, WAVE MOTION.

SEMICONDUCTOR

Semiconductors are distinguished from other classes of materials by their characteristic electrical conductivity σ. The electrical conductivities of materials vary by many orders of magnitude, and consequently can be classified into the following categories: (1) the perfectly conducting superconductors, (2) the highly conducting metals ($\sigma \sim 10^6$ mho/cm), (3) the somewhat less conducting semimetals ($\sigma \sim 10^4$ mho/cm), (4) the semiconductors covering a wide range of conductivities ($10^3 \gtrsim \sigma \gtrsim 10^{-7}$ mho/cm), (5) the insulators also covering a wide range ($10^{-10} \gtrsim \sigma \gtrsim 10^{-20}$ mho/cm).

The low-conductivity semiconductors are characterized by the great sensitivity of their electrical conductivities to sample purity, crystal perfection, and external parameters such as temperature, pressure, and frequency of the applied electric field. For example, the addition of less than 0.01 per cent of a particular type of impurity can increase the electrical conductivity of a typical semiconductor like silicon and germanium by six or seven orders of magnitude. In contrast, the addition of impurities to typical metals and semimetals tends to decrease the electrical conductivity, but this decrease is usually small. Furthermore, the conductivity of semiconductors and insulators characteristically *decreases* by many orders of magnitude as the temperature is lowered from room temperature to 1 K. On the other hand, the conductivity of metals and semimetals characteristically *increases* in going to low temperatures and the relative magnitude of this increase is much smaller than are the characteristic changes for semiconductors. The principal conduction mechanism in metals, semimetals, and semiconductors is electronic, whereas both electrons and the heavier charged ions may participate in the conduction process of insulators. (See CONDUCTIVITY, ELECTRICAL.)

It is customary to classify a semiconductor according to the sign of the majority of its charged carriers, so that a semiconductor with an excess of negatively charged carriers is termed *n*-type, and an excess of positively charged carriers is called *p*-type, while a material with no excess of charged carriers is considered to be perfectly compensated. Many of the important semiconductor devices depend on fabricating a sharp discontinuity between the *n*- and *p*-type materials, the discontinuity being called a *p-n* junction. (See SEMICONDUCTOR DEVICES, TRANSISTORS.)

Most semiconductors exhibit a metallic luster upon visual inspection; nevertheless, the visual appearance of materials does not provide an adequate criterion for the classification of materials, since the electrical conductivity of all materials is frequency-dependent. Visual inspection tends to be sensitive to the conductivity properties at visible frequencies ($\sim 10^{15}$ Hz). Although materials with a high optical reflectivity tend also to exhibit high dc conductivity, these two properties are not necessarily correlated in all semiconductors and metals. An example of a metal without metallic luster is ReO_3, a semitransparent reddish solid. On the other hand, most of the common semiconductors do exhibit metallic luster primarily because electronic excitation across their fundamental energy gaps can be achieved at infrared frequencies. At low frequencies, the principal conduction mechanism is free carrier conduction, which is important in metals and is present

to some extent in semiconductors which contain impurities or are found at elevated temperatures. On the other hand, interband transitions dominate the conduction process at very high frequencies. Interband transitions contribute to the conductivity by about the same order of magnitude in semiconductors, metals, and insulators.

Since the dc conductivity due to free carriers is characteristically low in semiconductors and insulators, the generation of free carriers by exposure to light at infrared, visible, and ultraviolet frequencies can lead to a large increase in the dc conductivity. This photoconductive effect, which is not observed in metals or semimetals, can be enormous in low-conductivity semiconductors (an increase in the dc conductivity of CdS by eight orders of magnitude is observed). (See PHOTOCONDUCTIVITY.)

Because of the extreme sensitivity of semiconductors to impurities, temperature, pressure, light exposure, etc., these materials can be exploited in the fabrication of useful devices, such as the crystal diode, the transistor, integrated circuits, photodetectors, and light switches. Semiconductor devices do, in fact, date back to the infancy of the electronics industry when crystal sets were used for radio reception. With the development of reliable and efficient vacuum tubes, the interest in semiconductor devices waned for several decades. Renewed interest in crystal rectifiers was stimulated by the needs of the radar technology which developed during the World War II period at the M.I.T. Radiation Laboratory and at other laboratories. During this period and the immediate post-war years, intensive activity developed in the fabrication of very pure semiconducting materials as well as in the basic understanding of the energy level schemes and of the charged carrier transport in silicon and germanium. This intense activity culminated in the discovery of the transistor by Shockley, Bardeen, and Brattain in 1947, for which they were awarded the Nobel Prize in Physics (1956). With the invention of the transistor, the electronics industry rapidly incorporated semiconductor devices, and since the mid-1960s, solid-state components have largely replaced vacuum tubes in many technological applications, from the transistor radio to components for high-speed electronic computers and microprocessors.

The most common semiconductors are the elemental semiconductors silicon and germanium. Other common elemental semiconductors are diamond (carbon), gray tin, tellurium, selenium, and boron. Closely related to the group IV semiconductors diamond, silicon, germanium, and gray tin are the III-V compounds formed from elements in the third and fifth columns of the periodic table, such as InSb, InAs, GaAs, GaP, and GaSb. Another important class of inorganic semiconductors is the II-VI compounds, formed from elements in the second and sixth columns of the periodic table such as CdS, CdSe, CdTe, ZnS, ZnSe, ZnTe, HgS, HgSe, and HgTe. Many other varieties of compound semiconductors are found, some of the more common varieties being the IV-IV compound SiC, the IV-VI compounds PbS, PbSe, PbTe, SnSe, SnTe, $Pb_xSn_{1-x}Te$, $Pb_xSn_{1-x}Se$, GeTe and the oxides MnO, NiO, SiO_2, SnO_2, GeO_2. Another class of semiconductors is the organic semiconductors, common examples being anthracene, tetracene, free radicals such as α α-diphenyl-β-picryl hydrazyl, biologically interesting materials such as the phthalocyanines, and various organic dyes.

Certain classes of semiconductors also possess other interesting properties. For example, the europium chalcogenides form a family of magnetic semiconductors, with EuO, EuS, and EuSe undergoing a ferromagnetic phase transition, while EuTe becomes antiferromagnetic below a Néel temperature of 9.8 K. Magnetic semiconductors are of particular interest because of the close coupling between the electrical and magnetic properties, such as electrical conductivity and the magnetic susceptibility. Some semiconductors have also been found to undergo a superconducting phase transition, as for example GeTe and SnTe for carrier concentrations of $\sim 10^{21}/cm^3$ and transition temperatures below 0.3K. More recently, semimagnetic semiconductors have been prepared by doping II-IV compounts with substitutional magnetic cations, e.g., Mn. These materials are of interest because of their intriguing magnetic and magneto-optical properties.

Although most of the common semiconductor devices utilize crystalline materials, semiconductors are also found in the liquid and amorphous states. Of special interest is the fact that the electrical conductivity of an amorphous semiconductor tends to be much lower than that of its crystalline counterpart; the opposite situation prevails for amorphous and crystalline metals. Common semiconductors like silicon and germanium have been prepared in the amorphous state, although the major emphasis in recent years has been given to the amorphous chalcogenide glasses containing tellurium and selenium along with a host of other elements; these glasses have been utilized for switching and memory devices.

Semiconductors tend to be hard and brittle and become ductile only at high temperatures. The hardness of semiconductors like diamond and SiC is utilized in the manufacture of industrial abrasives. Because of this hardness, high-quality optical surfaces on semiconductors can be achieved using lapping and etching techniques.

Because of the industrial demands for the fabrication of high-quality, high-purity semiconductors, a great deal of attention has been given to the development of a sophisticated semiconductor technology. To illustrate the present state of the art, it is now possible to grow single crystals of germanium with uncompensated impurity concentrations of $2 \times 10^{10}/cm^3$, which corresponds to less than one charged impurity in 10^{12} germanium atoms; this is probably the purest and most perfect crystalline material of any type that is now available.

A number of techniques have been employed in growing single crystals of semiconductors. To produce single crystals of large size, it is common to pull the crystal from the melt by the Czochralski technique. These large boules are then cut up into wafers and appropriate impurities are diffused into the material to produce the desired device structures.

For some specific applications, it is necessary to achieve greater control in the carrier concentration than is possible with diffusion techniques, as, for example, in the production of heterojunction devices. In such cases, liquid-phase epitaxial growth is employed, using as a substrate either the same semiconductor or another material of similar lattice constant and thermal expansion coefficient. For example, in the fabrication of sharp p-n junctions in GaP for light-emitting-diode applications, a wafer of GaP grown by the Czochralski technique is used as a substrate and epitaxial layers of n- and p-type GaP are then sequentially deposited by liquid-phase epitaxy.

To produce crystals of the highest purity and crystalline perfection, the method of chemical vapor deposition is often employed. For certain semiconductors, such as SiC and SnO_2, single crystals can be prepared only by chemical vapor deposition. Although the growth rate by this technique is slow, chemical vapor deposition has been utilized in the fabrication of specific microcircuits in order to exploit the flexibility that this technique provides for varying the type and concentration of dopants which are introduced. Furthermore, chemical vapor deposition can be utilized in the growth of certain mixed compound semiconductors. For example, $GaAs_{.6}P_{.4}$, which is a desirable light-emitting-diode material, is difficult to prepare by conventional means but can be prepared using as a substrate a wafer of GaP, which can be grown easily by the Czochralski technique: on this substrate a graded growth of $GaP_{1-x}As_x$ proceeds by chemical vapor deposition until the desired composition is achieved. (See CRYSTALLIZATION.)

To develop an understanding of the electrical conduction in semiconductors, it is necessary to examine the conduction mechanisms appropriate to these materials. The flow of electric current depends on the acceleration of charges by the externally applied electric field. Only those charges that resist collisions or scattering events are effective in the conduction process. Because of collisions, charged particles in a solid are not accelerated indefinitely by the applied field, but rather, after every scattering event, the velocity of a charged particle tends to be randomized. Thus, the acceleration process must start anew after each scattering event and charged particles achieve only a finite velocity along the electric field E, the average value of this velocity being denoted by v_D, the drift velocity. The effectiveness of the charge transport by a particular charged particle is expressed by the mobility μ which is defined as $\mu = v_D/E$. The mobility of a particle with charge e and mass m can be related directly to the mean time between scattering events τ (also called the relaxation time) by the expression $\mu = e\tau/m$. The electrical conductivity σ depends on the mobility of the charged carriers as well as on their concentration n, and is simply written as $\sigma = ne\mu$ where e is the charge of the carriers. The advantage of expressing the conductivity in this form is the explicit separation into a factor n which is highly sensitive to external parameters such as temperature, pressure, optical excitation, irradiation, and into another factor μ which depends characteristically on scattering mechanisms and on the electronic structure of the semiconductor.

The classical theory for electronic conduction in solids was developed by Drude in 1900. This theory has since been reinterpreted to explain why all contributions to the conductivity are made by electrons which can be excited into unoccupied states (Pauli principle) and why electrons moving through a perfectly periodic lattice are not scattered (wave-particle duality in quantum mechanics). Because of the wavelike character of an electron in quantum mechanics, the electron is subject to diffraction by the periodic array, yielding diffraction maxima in certain crystalline directions and diffraction minima in other directions (See DIFFRACTION BY MATTER.) Although the periodic lattice does not scatter the electrons, it nevertheless modifies the mobility of the electrons through introduction of an effective mass m^* so that $\mu = e\tau/m^*$. A sensitive technique for the measurement of the effective mass and its anisotropy is the cyclotron resonance technique. (See CYCLOTRON RESONANCE.) In terms of the effective mass approximation, which provides a good description for the transport properties of many common semiconductors, the effect of a periodic potential on a conduction electron can be approximately taken into account by considering the electron to move with an effective mass m^* and in a medium with dielectric constant ϵ. For many of the common semiconductors, the effective mass of the carriers is characteristically lighter than the free electron mass m_0, though low-mobility semiconductors containing transition metal atoms (d-bands) tend to have carriers with large effective masses, often larger than m_0. (See SOLID-STATE PHYSICS.)

Although the perfectly periodic lattice does not scatter electrons, an electron in a solid does, in fact, experience scattering events through a variety of mechanisms. At room temperature and above, the principal scattering mechanism is due to lattice vibrations, which arise from the thermal motion of the lattice. Lattice vibrations cause a displacement of the atoms from their equilibrium positions, thereby destroying the perfect periodicity of the lattice.

The effectiveness of lattice vibrations (or phonons) in scattering electrons depends to some extent on the *type* of phonon which is involved in the scattering event. Phonons are classified as either *acoustic* (atoms on adjacent sites vibrate *in* phase) or *optical* (atoms on adjacent sites vibrate *out of* phase). Materials with only one atom/unit cell have only acoustic branches, three in number, while materials with multiple atoms/unit cell have, in addition to the three acoustic branches, three optical branches for every additional atom/unit cell. Every acoustic and optical phonon branch is further classified according to whether the lattice vibrations are along the direction of propagation of the lattice mode (longitudinal) or perpendicular to this direction (transverse). Phonons exhibit dispersion relations whereby the phonon frequency ω_q depends on the wave vector q for the lattice mode. Of particular interest are those ranges of q where ω_q varies slowly with q, thereby producing a high density of phonon states. Those phonons which have a high density of states (such as acoustic phonons near the Brillouin zone boundary, and the optical phonons) are most effective in scattering electrons. Since less thermal energy is required to excite an acoustic phonon than an optical phonon, it is the acoustic phonons which play a leading role in the scattering process at intermediate temperatures. At very low temperatures, the thermal energy is too small to excite significant numbers of phonons of any sort, so electron scattering is dominated by mechanisms involving crystalline imperfections such as lattice vacancies, interstitial atoms, impurity atoms, and crystal boundaries; these crystalline imperfection mechanisms do not contribute significantly to electron scattering at room temperature in solids which have a high degree of crystalline perfection and purity. On the other hand, for amorphous semiconductors, these imperfection scattering mechanisms are relatively important, even at temperatures where there are a significant number of thermally excited phonons. (See PHONONS.)

In considering the mobility of carriers in the vicinity of room temperature, it is clear that the increased probability of exciting phonons with increasing temperature T results in more frequent electron scattering events with increasing T, and consequently a decline in the electron mobility with increasing T. However, this decrease in mobility varies relatively slowly with T, having a power law dependence on T.

On the other hand, the temperature dependence of the carrier density in semiconductors is a strongly increasing function of temperature, which is closely related to the existence of a thermal activation energy E_t, or energy barrier which restricts potential carriers to bound, nonconducting states. This energy barrier can be overcome by the thermal or optical excitation of carriers, and the carrier density, n, excited at a given temperature is given by a Boltzmann factor $n = n_0 \exp(-E_t/2kT)$, where n_0 is the density of available bound carriers and k is Boltzmann's constant. With $E_t \sim 1$ eV for typical semiconductors, it is evident that $E_t \gg 2kT$ at room temperature, so the carrier density is a rapidly increasing function of temperature. On the other hand, the carrier density of metals and semimetals is essentially independent of T, so the entire temperature dependence of the electrical conductivity is related to the mobility through the temperature-dependent scattering time τ.

The origin of the energy barrier for carrier generation is directly connected with the energy levels for electrons in a solid. Considering electrons in a solid from a tight-binding point of view, the discrete energy levels of the free atom broaden in the solid to form energy bands. For materials which are well described by the tight-binding approximation, the width of the energy bands is sufficiently small so that an energy gap between the energy bands is formed; in the forbidden energy gap there are no bound states. Of particular importance to the conduction properties of a solid is the fact that *all* of the available states in each band would be filled if each atom were to contribute exactly two electrons, thereby causing every solid with an odd number of electrons per atom to be metallic, while solids with an even number of electrons per atom would be insulating or semiconducting. The occurrence of energy bandgaps is also a consequence of the weak binding approximation, whereby the periodic potential itself is responsible for creating bandgaps through the mixing of states separated by a reciprocal lattice vector. (See ENERGY LEVELS.)

The solution of Schrödinger's equation to obtain the electronic energy levels of a semiconductor can be carried out in an approximate way by exploiting the translational symmetry of the perfectly periodic lattice. Through the translational symmetry, phase factors of the form $e^{i\mathbf{k}\cdot\mathbf{R}_n}$ are introduced as eigenvalues of the operator for translation by a lattice vector $\mathbf{R}_n$, where the wave vector $\mathbf{k}$ assumes the role of the quantum number labeling the various translational states. Since the Hamiltonian for an electron in a periodic solid is invariant under translations by a lattice vector, the energy eigenvalues of Schrödinger's equation are also labeled by the quantum number $\mathbf{k}$. In fact, a knowledge of the dispersion relation for electrons in a solid $E(\mathbf{k})$ uniquely determines the conduction properties of the solid; for this reason, a great deal of attention has been given to the detailed study of the energy band structure of semiconductors.

Semiconductors are classified according to whether they are direct- or indirect-gap semiconductors. For a direct-gap semiconductor, the maximum energy of the occupied valence band occurs at the same value of k as the minimum energy of the unoccupied conduction band. Typical direct-gap semiconductors are InSb and GaAs; these materials have their energy extrema at $k = 0$. On the other hand, for indirect-gap semiconductors, the conduction band minima and the valence band maxima occur at different wave vectors. Typical indirect gap semiconductors are silicon and germanium. Both of these semiconductors have valence band maxima at $k = 0$, but the conduction band minima for germanium are located at the Brillouin zone boundary along a (111) direction and for silicon are located along a (100) direction about 85 per cent of the distance out to the zone boundary.

Whereas the minimum energy gap is the threshold for thermal excitation in the case of both direct- and indirect-gap semiconductors, the response to optical excitation is different for the two types of semiconductors. Optical transitions involve only a very small change in k because the wavelength of light is very large compared with atomic spacings in a solid. Since optical transitions are direct (essentially k-conserving), it is necessary to either absorb or emit a phonon with wave vector $(\mathbf{k}_c - \mathbf{k}_v)$ to excite carriers optically in an indirect-gap semiconductor, where $\mathbf{k}_c$ and $\mathbf{k}_v$ correspond to the wave vectors for the conduction and valence band extrema, respectively. Measurement of the threshold for phonon-assisted absorption provided one of the earliest determinations of the phonon frequencies in a semiconductor. More accurate measurements of phonon frequencies in semiconductors are now made using the techniques of phonon-assisted tunneling, Raman scattering, and inelastic neutron scattering. (See TUNNELING, RAMAN EFFECT.)

The different response of the optical carrier excitation process can be utilized in the identification of an energy gap as direct or indirect. For a direct-gap semiconductor, the onset of optically induced interband transitions across an energy gap E_g is marked by a photon energy dependence of the optical absorption coefficient α, which is $\sim (\hbar\omega - E_g)^{1/2}$. On the other hand, the threshold for interband transitions in an indirect-gap semiconductor has a characteristic photon energy dependence $\alpha \sim (\hbar\omega - E_t \pm \hbar\omega_q)^2$ in which E_t represents the indirect energy gap; for indirect transitions, the threshold occurs at a lower photon energy when a phonon of energy $\hbar\omega_q$ is absorbed than when it is emitted. Furthermore, the magnitude of the absorption coefficient for a direct-gap semiconductor is as much as two or three orders of magnitude larger than for an indirect-gap semiconductor at the threshold for the optical excitation of carriers. In the optical excitation process, electrons are introduced into the conduction band and empty states (holes) are left behind in the valence band. Both electrons and holes contribute to the electrical conductivity. Furthermore, the electron is attracted to the hole it left behind often forming a bound exciton state.

The actual calculation for the energy bands of solids is complicated by the band degeneracies which occur at high symmetry points in the Brillouin zone because of the rotational point symmetry of the crystal structure. These band degeneracies are evident in the $E(\mathbf{k})$ diagram for germanium shown in Fig. 1. Of particular interest is the band degeneracy of the valence band at $\mathbf{k} = 0$, which is typical of semiconductors which crystallize in the diamond and zincblende crystal structures. Further complications arise because these semiconductors contain two atoms per unit cell; as a result, the covalently bonded electrons form filled bonding states in the valence band, leaving the higher-lying antibonding states in the conduction band unoccupied. The basic difference between the energy band structure for a semiconductor as contrasted with a semimetal or metal is the existence of a forbidden energy gap for all k values. In a semimetal or metal, the maximum energy for a

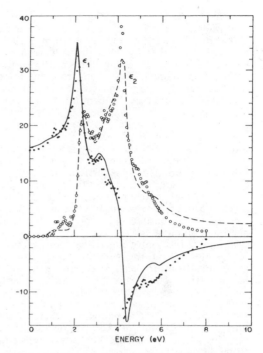

FIG. 2. The frequency-dependent complex dielectric constant ($\epsilon_1 + i\epsilon_2$). The curves are experimental results and the points represent a numerical calculation based on the energy bands shown in Fig. 1.

valence band state is larger than the minimum energy for a conduction band state, so band overlaps can occur. In a semimetal, the number of holes is equal to the number of electrons, but this need not be the case in a metal.

The figure of merit for a given energy band calculation can be determined by calculation of the frequency-dependent dielectric constant which these bands imply. The good agreement shown in Fig. 2 between the calculated (points) and experimental (curves) dielectric constants indicates confidence in the major features of the energy band structure given in Fig. 1.

It is of interest to observe that the crystal structure affects the electrical properties of a material, insofar as solids formed from a particular atomic species may be conducting or semiconducting, depending on the crystal structure. Trigonally bonded carbon forms graphite, which is highly conducting in the layer planes, while tetrahedrally bonded carbon forms the semiconductor, diamond.

The electronic energy levels of a semiconductor may be examined with profit from the chemical-bonding point of view. The elemental semiconductors, silicon, germanium, gray tin, and diamond, all crystallize in the diamond structure, in which identical atoms are located at the center and corners of a regular tetrahedron. Each group IV atom contains four valence

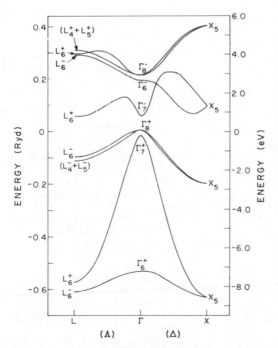

FIG. 1. Electronic band structure for Ge. The zero of energy is taken at the highest point in the valence band (Γ_8^+). The lowest level in the conduction band occurs at the Brillouin zone boundary (L_6^+).

electrons, thereby occupying exactly eight of the valence states having about equal energy. Through covalent bonding, each atom achieves a filled valence shell by sharing one electron with each of its four neighboring atoms at the corners of the tetrahedron. This strong covalent bonding is responsible for the mechanical hardness of these semiconductors. In diamond, the valence electrons are in the $n = 2$ shell, in silicon in the $n = 3$ shell, in germanium in the $n = 4$ shell, in gray tin in the $n = 5$ shell. Less energy is required to break a bond as we move down the periodic table from diamond to gray tin. The breaking of bonds freeing carriers for electronic conduction can be conveniently accomplished by thermal or optical excitation. The semiconducting properties of some amorphous materials can be most easily understood in terms of chemical bonding.

The introduction of impurities from group V (or group III) of the periodic table into the diamond structure leads to carrier generation for electrical conduction. For example, antimony has 5 valence electrons, which is one more than is necessary to satisfy the covalent bonding requirements of the diamond structure. Thus, only a small amount of energy is necessary for the liberation of this weakly bound electron from the positively charged Sb^+ ion. Impurities which contribute electrons to the conduction process are called donors and usually result in an excess of negatively charged carriers, thereby yielding an n-type semiconductor. The presence of impurities destroys the perfect periodicity of the lattice and produces energy states within the energy gap between the normally occupied valence band and the normally unoccupied conduction band. The weakly bound donor electrons occupy states close to the conduction band minima and can be treated with considerable success within the effective mass approximation by a hydrogenic model for the Coulombic attraction of the donor electron to the positively charged donor ion. In this approximation, the energy levels of the donor states are given by the bound hydrogenic states $E_n = - m^* e^4 / 2\hbar^2 n^2 \epsilon^2$, where $n = 1, 2, \cdots$, ϵ is the static dielectric constant of the semiconductor, and the donor levels are measured relative to the lowest-lying conduction band states. Ionization of this bound donor to the conduction band by thermal or optical excitation results in electronic conduction. Since the effective masses for such donors tends to be small (group V donors in germanium have $m^* \simeq 0.1 m_0$) and the dielectric constant large (in germanium $\epsilon \simeq 16$), the hydrogenic donor levels are shallow (within $\sim 6 \times 10^{-3}$ eV from the conduction band minima). Furthermore, the small effective masses and large dielectric constant result in a large effective Bohr radius $r_B = \hbar^2 \epsilon / m^* e^2$, so the donor electrons are not localized within the unit cell of the donor ion, but rather extend over many unit cells.

Introduction of group III impurities, such as indium into silicon, results in a p-type semiconductor since the group III impurities have a deficiency of one electron for the fulfillment of the bonding requirement. Since the covalent bonding in these semiconductors is strong, an electron deficiency at an impurity site results in the acquisition of an electron from a covalent bond elsewhere in the material. The covalent bond with an electron deficiency represents an effective positive charge (or acceptor state) which can hop from one covalent bond to another, thereby representing the flow of positive charge, which is characteristic of p-type semiconductors. Acceptor states, like donor states, tend to lie within the forbidden semiconducting energy gap; and weakly bound acceptors characteristically have energy levels which lie just slightly above the valence band maximum energy.

Since impurity atoms in typical semiconducting crystals tend to be relatively isolated from each other, the degree of ionization of the impurity atoms and hence the concentration of n- and p-type carriers can be determined from the law of mass action:

$$np = 4 \left(\frac{kT}{2\pi\hbar^2} \right)^3 (m_e m_h)^{3/2} \, e^{-E_t/kT}$$

where n and p are the concentration of electrons and holes and m_e and m_h are the corresponding effective masses. This relation is generally valid whether the carriers are produced exclusively by thermal excitation (an *intrinsic* semiconductor) or by the introduction of impurities (an *extrinsic* semiconductor). If the semiconductor is intrinsic, then $n = p$ and the Fermi level or chemical potential is determined by

$$E_F = \tfrac{1}{2} E_t + \tfrac{3}{4} kT \ln (m_h/m_e).$$

In these relations, the appropriate energy gap is the thermal energy gap E_t, which is smaller than the direct-band gap E_g, for indirect-gap semiconductors such as silicon and germanium. For extrinsic semiconductors, the Fermi level may be pinned to an impurity band of donors or acceptors within the band gap or may even lie in the valence or conduction bands, in which case the semiconductor is called *degenerate*.

In practice most semiconducting crystals contain measurable concentrations of impurities, whether or not these impurities are introduced intentionally. At high temperatures, the thermally excited carriers tend to dominate the conduction process and the semiconductor behaves as an intrinsic semiconductor; however, at low temperatures the impurity conduction mechanism is dominant and the semiconductor behaves as an extrinsic semiconductor. The passage between intrinsic and extrinsic

TABLE 1. ROOM TEMPERATURE PROPERTIES OF SOME SEMICONDUCTORS

Material	Band Gap (eV)	Electron Mobility (cm^2 $volt^{-1}$ sec^{-1})	Hole Mobility (cm^2 $volt^{-1}$ sec^{-1})	Dielectric Constant, $\epsilon = n^2$	Lattice Constant (Å)	Density (g/cm^3)	Melting Point (°C)
Si	1.15	1900	480	11.8	5.42	2.4	1412
Ge	0.65	3800	1800	16.0	5.646	5.36	938
GaAs	1.35	8500	400	13.5	5.65	5.31	1280
GaSb	0.69	4000	650	15.2	6.095	5.62	728
InSb	0.17	70000	1000	16.8	6.48	5.775	525
SiC	3.0	60	8	10.2	4.35	3.21	2700
PbS	0.37	800	1000	17.9	7.5	7.61	1114
ZnO	3.2	190	–	8.5	(a) 3.24 (c) 5.18	5.60	1975
CdS	2.4	200	–	5.9	5.83	4.82	685
HgTe	0.2	22000	160		6.429	8.42	670
C (diamond)	5.3	1800	1600	16.5	3.56	3.51	3800
Te	0.38	1100	700	‖ 38.0 ⊥ 23.0	(a) 4.45 (c) 5.93	6.25	452
Se (amorphous)	1.8	0.005	.015	6.6	–	4.82	–
Se (crystalline)	2.6	–	1	‖ 13.3 ⊥ 8.0	(a) 4.36 (c) 4.95	4.79	217

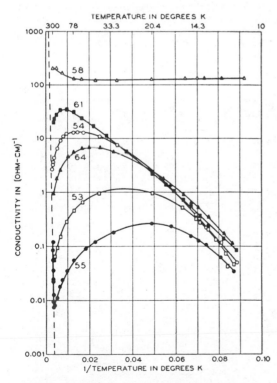

FIG. 3. Conductivity versus absolute temperature for a set of arsenic-doped germanium samples indicated in the figure by sample numbers. The data were taken by Debye at Bell Laboratories. The dashed line represents intrinsic conductivity. (After Conwell)

behavior is illustrated in the plot of ln σ vs reciprocal temperature shown in Fig. 3 for germanium. At low temperatures, conduction proceeds via impurity states, often employing a thermally activated hopping mechanism.

To separate the temperature dependence of the mobility from the temperature dependence of the carrier concentration, studies of the Hall mobility are carried out. (See HALL EFFECT.) If conduction proceeds primarily by one type of carrier, the Hall mobility μ_H is related to the conductivity by the Hall constant R_H, which is determined by Hall effect measurements, such that $\mu_H = c|R_H|\sigma$ where c is the speed of light. On the other hand, if two carriers participate in the conduction process, then the Hall constant is related to

$$R_H = \frac{1}{ec} \frac{(p\mu_h^2 - n\mu_e^2)}{(p\mu_h + n\mu_e)^2}$$

and the conductivity is given by

$$\sigma = e(n\mu_e + p\mu_h).$$

By studying the temperature dependence of the Hall constant and of the electrical conductivity, the carrier densities and mobilities of the two types of carriers can be determined. A summary of the relevant parameters for some of the more common semiconductors is given in Table 1.

MILDRED S. DRESSELHAUS

References

Kittel, C., "Introduction to Solid State Physics," 5th edition, New York, John Wiley & Sons, 1976.

Shockley, W., "Electrons and Holes in Semiconductors," Englewood Cliffs, N.J., Prentice-Hall, 1950.

Cross-references: DIODE (SEMICONDUCTOR); ENERGY LEVELS; PHONONS; SEMICONDUCTOR DEVICES; SEMICONDUCTORS, INHOMOGENEOUS; SOLID-STATE PHYSICS; SOLID-STATE THEORY; TRANSITORS.

SEMICONDUCTOR DEVICES

1. Introduction Since the invention of the bipolar transistor in 1947, the transistor and related semiconductor devices have had an unprecedented impact on our society in general and on the electronic industry in particular.[1] Semiconductor devices are compact, reliable, and inexpensive. They can be operated at ultrahigh speed with low power consumption. They lend themselves to integration into complex but readily manufacturable microelectronic circuits. The semiconductor devices are key elements, for example, in high-speed computers, in space vehicles, in communication and transportation systems, in medical equipments, and in industrial control systems.

Semiconductor devices[2] are intimately related to the energy band structures and transport properties of semiconductors, and can be affected by electric field, magnetic field, temperature, pressure, and radiation. We shall be concerned mainly with devices associated with the electric effect in which the electrical conductivity of a semiconductor device is modulated by external electrical signals. We shall divide the semiconductor devices into four groups and consider a few important devices in each group. The first group, discussed in Section 2, is bipolar devices, in which both electrons and holes are involved in the transport process. Section 3 considers the unipolar devices in which only one type of carriers predominantly participate in the conduction mechanism. Section 4 considers some special microwave devices. Most of the bipolar and unipolar devices considered in Sections 2 and 3 can be operated in the microwave region (1 to 100 GHz with corresponding wavelength from 30 to 0.3 cm). In this section we shall consider two microwave devices which are based on novel transport phenomena. Section 5 deals with photonic devices in which the basic particle of light—the photon—plays a major role.

2. Bipolar Devices *Bipolar Transistor.* A schematic diagram of an *n-p-n* bipolar transistor is shown in Fig. 1(a). The device consists of two closely coupled *p-n* junctions. The energy band diagram under normal operating condition is shown in Fig. 1(b). The emitter-base junction is forward-biased. The emitter current consists of two components—the electron component injected from the emitter into the base, and the hole component injected from the base into the emitter region. The collector-base junction is reverse-biased. The collector current also consists of the electron and hole components. The current-voltage characteristics for common-emitter configuration is shown in Fig. 1(c). For a given base current I_B the collector current I_C increases initially with the biasing voltage V_{CE}, then reaches a saturated value. For a given V_{CE} a small increase in the base current can result in a large increase of the collector current. The bipolar transistor is one of the most important devices for switching, power, and microwave applications. The device is extensively used in integrated circuits, and is capable of generating over 1 W power at 10 GHz (10^{10} Hz).

Thyristor. A schematic diagram of the thyristor is shown in Fig. 2(a). The four-layer *p-n-p-n* structure is intimately related to the bipolar transistor in which both electrons and holes are involved in the transport processes. The energy band diagrams for the forward "off" condition and the forward "on" condition are shown in Fig. 2(b) and (c), respectively. In the "off" condition most of the voltage drop is across the center junction, while in the "on condition all three junctions are forward-biased. The current-voltage characteristics of the forward and the reverse regions are shown in Fig. 2(d) for three different gate currents.[3] The

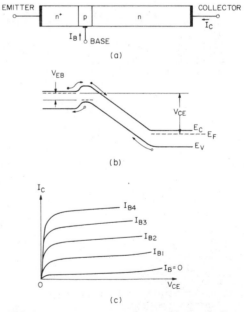

FIG. 1. (a) Bipolar transistor. (b) Band diagram under normal operating conditions. (c) Typical current-voltage characteristics for common-emitter configuration.

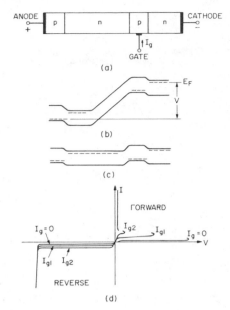

FIG. 2. (a) Thyristor. (b) Band diagram under forward "off" condition. (c) Band diagram under forward "on" condition. (d) Current-voltage characteristics.

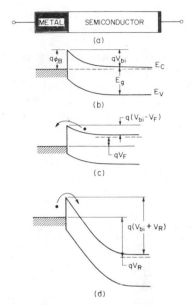

FIG. 3. (a) Metal-semiconductor (Schottky) diode. (b) Band diagram at thermal equilibrium. (c) Forward bias. (d) Reverse bias.

"forward" characteristic can be explained using the method of two-transistor analog. As the voltage increases from zero, the current will increase. This in turn will cause the current gains of both the p-n-p and n-p-n transistors to increase. Because of the regenerative nature of these processes, switching eventually occurs and the device is in its "on" state. As the gate current increases, one can reduce the threshold voltage at which switching occurs. Because of its two stable states (on and off) and the low power dissipation in these two states, the thyristor has found unique usefulness in applications ranging from speed control in home appliances to switching and power inversion in high-voltage transmission lines. Thyristors are now available with current ratings from a few milliamperes to over 5,000 A and voltage ratings extending above 10,000 V.

3. Unipolar Devices *Metal-Semiconductor (Schottky) Diode.* The metal-semiconductor diode, Fig. 3(a), is one of the earliest semiconductor devices (it was first investigated in 1874). In 1938, Schottky suggested that the potential barrier could arise from stable charges in the semiconductor. The model is known as the *Schottky barrier*, and the energy band diagram at thermal equilibrium is shown in Fig. 3(b). When a forward bias is applied (i.e., the metal electrode is positive with respect to the n-type semiconductor) electrons will be injected thermionically from the semiconductor into the metal, as shown in Fig. 3(c). Under

reverse bias, Fig. 3(d), the current is small and is limited by the barrier height $q\phi_B$. The general I-V characteristic of a Schottky diode is very similar to that of a p-n junction. The main difference is that in usual Schottky diodes the current is due to majority carriers (e.g., electrons in an n-type semiconductor) so that the storage time associated with minority carriers is negligibly small. This property makes Schottky diodes useful as microwave mixer and detector diodes as well as high-speed switching devices.

MESFET. The metal-semiconductor field-effect transistor (MESFET) is a device based on an entirely different physical principle from the bipolar transistor. While the bipolar transistor operates through the transport of injected carriers across the base, the MESFET is basically a voltage-controlled resistor, and its resistance can be varied by varying the width of the depletion layers extending into the channel. See Fig. 4(a). The device consists of a conductive channel provided with two ohmic contacts, one acting as a cathode (source) and the other as the anode (drain). The third electrode, the *gate*, forms a rectifying Schottky barrier with the channel. The gate voltage controls the width of the depletion region, and therefore varies the conductance of the channel. A typical I-V characteristic is shown in Fig. 4(b). We note that for a given V_G, the current increases initially with the drain voltage and then reaches a saturated value.

The GaAs MESFET is one of the most important power microwave devices. Power out-

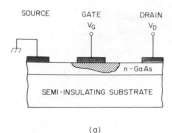

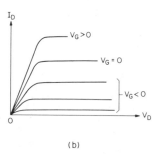

(a)

(b)

FIG. 4. (a) MESFET. (b) Typical current-voltage characteristics.

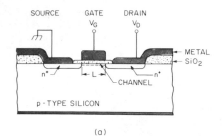

(a)

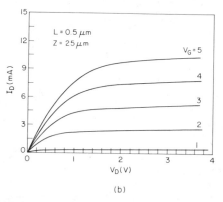

(b)

FIG. 5. (a) MOSFET. (b) Measured current-voltage characteristics for a MOSFET with 0.5-μm channel length. The channel width is 25 μm.

put up to 10 W at 10 GHz has been obtained with a noise figure less than 1 dB. The operation of a junction field-effect transistor (JFET) is identical to that of a MESFET. In the JFET, however, a *p-n* junction instead of a Schottky barrier is used for the gate electrode. Both devices offer many attractive features for applications in switching, microwave amplifications, and integrated circuits.

MOSFET. The basic structure of a metal-oxide-semiconductor field-effect transistor (MOSFET) is shown in Fig. 5(a). The device consists of a *p*-type semiconductor substrate into which two n^+ regions, the *source* and the *drain*, are formed. The metal contact on the insulator is called the *gate electrode*. When there is no voltage applied to the gate the source-to-drain electrodes correspond to two *p-n* junctions connected back to back. Therefore there is practically no current flowing from source to drain. When a sufficiently large negative voltage is applied to the gate such that a surface inversion layer is formed between the two n^+ regions, the source and the drain are connected by a conducting-surface *n* channel through which a large current can flow. The conductance of this channel can be modulated by varying the gate voltage. One may readily extend the discussion to a *p*-channel device by exchanging *p* for *n* and reversing the polarity of the voltage. A current-voltage characteristic of a MOSFET with 0.5 μm channel length is shown in Fig. 5(b). Note that as the gate voltage increases, the drain current increases accordingly; and the general behavior is quite similar to that of a MESFET.

MOSFET is the most important device for very-large-scale integrated (VLSI) circuits such as microprocessors and semiconductor memories. Figure 6(a) shows the reduction of the minimum device dimension since the beginning of the integrated circuit era in 1959. Figure 6(a) also shows that the minimum dimension will shrink continuously; the 1-μm barrier for commercial devices will be overcome by 1990. The reduction of device dimensions is driven by the requirement that integrated circuits of high complexity be fabricated.[4] The number of components per integrated-circuit chip has grown exponentially since 1959 [Fig. 6(b)]. The rate of growth is expected to slow down because of a lack of product definition and design. However, a complexity of 1 million or more devices per chip will be available before 1990 using 1-μm or submicron device geometries.

Charge-Coupled Device (CCD). In CCD, minority carriers are transferred from under one electrode to a closely adjacent electrode on the same semiconductor substrate when appropriate voltages are applied to the electrodes. Figure 7(a) shows a CCD in a situation wherein a sufficiently large negative bias has been applied to all the electrodes to produce surface depletion, and the center electrode has a slightly larger applied bias such that the center MOS structure is under greater depletion. If minor-

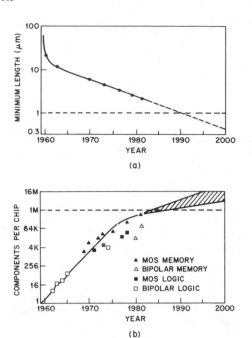

(a)

(b)

FIG. 6. (a) Minimum device dimension in an integrated circuit as a function of the year for commercial devices. (b) Complexity of integrated circuits as a function of the year.

ity carriers are introduced, they will collect at the surface in the potential minimum defined by the excess potential on the central electrode. If now the potential of the right-hand electrode is increased to exceed that of the central electrode, one obtains the potential distribution as shown in Fig. 7(b). In this case, the minority carriers will be transferred from the central to the right-hand electrode. Subsequently, the potential on the electrodes can be readjusted so that the quiescent storage site is located at

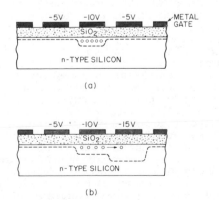

FIG. 7. (a) Charge-coupled device (CCD) with central electrode under depletion condition. (b) CCD under charge transfer operation.

the right-hand electrode. By continuing the above process, one can successively transfer the carriers along a linear array. Using this basic mechanism, CCD can perform a wide range of electronic functions including image sensing and signal processing.

4. Special Microwave Devices *IMPATT Diode*. IMPATT stands for *im*pact ionization *a*valanche *t*ransit *t*ime. IMPATT diodes employ impact-ionization and transit-time properties of semiconductor structures to produce negative resistance at microwave frequencies. A basic IMPATT diode is a p-n junction as shown in Fig. 8(a), where it is biased into reverse avalanche breakdown. There is a high-field avalanche region at the p^+n junction where electron-hole pairs are generated. The generated holes quickly enter the p^+ region; the generated electrons are injected into the drift region, Fig. 8(b). As the electric field changes periodically with time around an average value [Fig. 8(c)], the impact ionization rate per carrier follows the field change nearly instantaneously. However, the carrier density does not follow the field change in unison, because carrier generation depends also on the number already present. Even after the field has passed its maximum value, the carrier density keeps increasing because the carrier generation rate is still above the average value. The maximum carrier density is reached approximately when the field has decreased from the peak to the average value. Thus the ac variation of the carrier density lags the ionization rate by about 90°. The above situation is illustrated as the "injected" current in Fig. 8(d). The peak value of ac field (or volt-

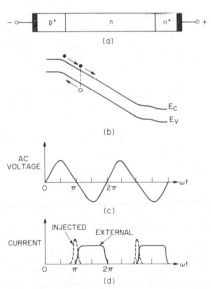

FIG. 8. (a) IMPATT diode. (b) Band diagram under avalanche condition. (c) AC voltage. (d) Injected and external current waveforms.

age) occurs at $\theta = \pi/2$, but the peak of the injected carrier density occurs at $\theta = \pi$. The injected electrons then enter the drift region, which they traverse at the saturation velocity. The induced external current is also shown in Fig. 8(d). From comparison of the ac field and the external current it is clear that the diode exhibits a negative resistance at its terminals.

IMPATT diodes are among the most powerful solid-state sources of microwave power. Up to 10 W at 10 GHz and 100 mW at 200 GHz have been obtained.

Transferred-Electron Device (TED). In GaAs, InP, and some other compound semiconductors, the energy band consists of a high-mobility low-energy valley and an adjacent low-mobility high-energy valley [inset of Fig. 9(a)]. When an electric field is applied, the electrons can gain energy and be transferred from the lower valley to the upper valley. In doing so, because of the reduction of mobility, the current is reduced as the field increases giving rise to a differential negative resistance as shown in Fig. 9(a).

The TED can be operated in various modes. One of the modes, called the *transit-time dipole-layer mode*, is shown in Fig. 9(b). When the diode is biased above the threshold E_T, periodic space-charge dipolar layers are launched at the negative terminal and propagate through

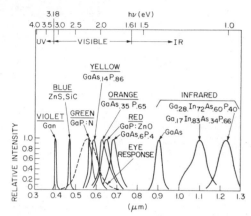

FIG. 10. Relative intensity versus wavelength of various LEDs.

the active region. A series of current pulses results, each spaced in time by the length of the active layer divided by the drift velocity. For a 10 μm sample, the frequency of oscillation is about 10 GHz. About 3 W power output at 10 GHz and over 50 mW at 100 GHz have been obtained.

In addition to the above devices, there are many other two-terminal microwave devices including tunnel diode and BARITT diode.

5. Photonic Devices *Light Emitter Diode (LED).* The LED belongs to the electroluminescent device family. Electroluminescence is the generation of light by an electrical current passed through a material under an applied electric field. The basic operation of a LED is a *p-n* junction under forward bias; holes and electrons are diffused across the depletion region where radiative recombinations take place. Figure 10 shows the relative intensity of optical radiation versus wavelength for various LEDs from ultraviolet to infrared. The visible LEDs are useful as indication lamps and for alphanumeric displays. The infrared LEDs are useful for opto-isolators and are potentially very important as sources in optical-fiber communication systems.

Laser. The *p-n* junction laser (*l*ight *a*mplification by *s*timulated *e*mission of *r*adiation) also belongs to the electroluminescent device family. The laser light has spatial and temporal coherence, and is nearly perfectly monochromatic (0.1 to 1 Å) as compared to a much broader spectrum (100 to 500 Å) in a LED. Using *double heterojunction structure*[5] as shown in Fig. 11(a) and (b), the laser can be operated continuously at or above room temperature. (A heterojunction is a junction formed between two semiconductors having different energy band gaps.) The heterostructure can confine the carriers in the active region (the central *p*-GaAs layer) by the potential barriers, and can confine the light intensity within the active region by

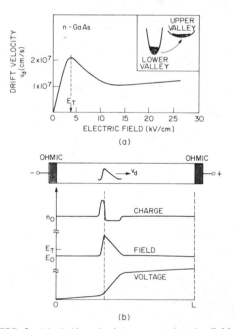

FIG. 9. (a) Drift velocity versus electric field in GaAs; insert shows electron transfer from lower valley to upper valley in the conduction band. (b) Schematic diagram of a TED, charge distributions (dipole layer), field profile, and voltage variation along the active device length.

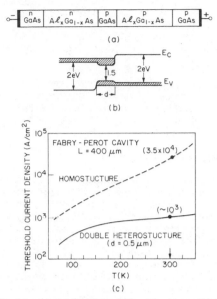

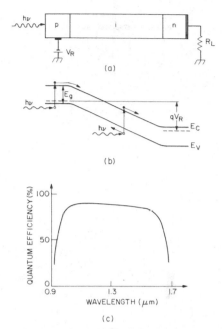

FIG. 11. (a) Double heterostructure laser. (b) Band diagrams under forward bias. (c) Comparison of threshold current density for homostructure (ordinary *p-n* junction) and double heterostructure laser.

FIG. 12. (a) *p-i-n* photodetector, (b) Band diagram under reverse-biased operating condition. (c) Quantum efficiency versus wavelength for a *p-i-n* photodetector.

the sudden reduction of the refractive index outside the active region. These confinements will enhance the population inversion and provide significant quantum amplification in the active region; therefore, the threshold current density (i.e., the minimum current density to initiate lasing) is considerably lower than that required for a homostructure, as shown in Fig. 11(c). The wavelength of coherent radiation has been extended from the near ultraviolet into the visible and then out to the far-infrared region (0.3 to 30 μm). The diode laser is considered one of the most important light sources for optical-fiber communication systems. It is also important for applications in many areas of basic research and technology, such as high-resolution gas spectroscopy and atmospheric pollution monitoring.

Photodetector. Photodetectors are semiconductor devices that can detect optical signals through electronic processes. They convert the optical variations into electrical variations that are subsequently amplified and further processed. One of the most important photodetectors for optical-fiber communication systems is the *p-i-n* diode. Figure 12(a) shows a schematic representation of a *p-i-n* diode. The corresponding energy-band diagram under reverse-bias condition is shown in Fig. 12(b). Light absorption in the semiconductor $(h\nu > E_g)$ produces electron-hole pairs. Pairs produced in the depletion region or within a diffusion length of it will be separated by the electric field, leading to current flow in the external circuit as

carriers drift across the depletion layer. A typical plot of quantum efficiency versus wavelength for a ternary compound $(Ga_x In_{1-x} As)$ *p-i-n* photodetector is shown in Fig. 12(c). The quantum efficiency is the number of electron-hole pairs generated per incident photon. The quantum efficiency for this photodetector is almost constant over the spectral range from 1.0 to 1.5 μm. In addition to the *p-i-n* diode, there are many other photodetectors including photoconductors, Schottky photodiodes, avalanche photodiodes, and phototransistors.

Solar Cell. Solar cells at present furnish the most important long-duration power supply for satellites and space vehicles. As worldwide energy demand increases, conventional energy resources, such as fossil fuels, will be exhausted in the not-too-distant future. Therefore, we must develop and use alternative energy resources, especially our only long-term natural resource, the sun. The solar cell is considered a major candidate for obtaining energy from the sun, because it can convert sunlight directly to electricity with high conversion efficiency, can provide nearly permanent power at low operating cost, and is virtually free from pollution.[6] A representative solar cell is diagrammed in Fig. 13(a). It consists of a shallow *p-n* junction formed on the surface, a front ohmic contact stripe and figures, a back ohmic contact that covers the entire back surface, and an antireflection coating on the front surface. The current-voltage characteristics under solar ra-

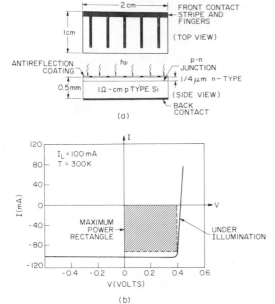

FIG. 13. (a) Schematic representation of the silicon *p-n* junction solar cell. (b) Current-voltage characteristics of a solar cell under illuminaton.

diation are shown in Fig. 13(b). The curve passes through the fourth quadrant and, therefore, power can be extracted from the device. Recently, development of low-cost thin-film solar cells, flat-panel solar cells, and concentrator systems has quickened. Typical solar-to-electrical conversion efficiencies are 5–10% for thin-film devices, 10–15% for flat-panel cells and 20–30% for concentrator systems.

6. Summary We have briefly described some of the major semiconductor devices. We expect that the semiconductor-device development in the 1980s will be vigorous, with continued breakthroughs in device concept and performance.

S. M. SZE

References

1. For a review of semiconductor devices developments in the past decade, see, e.g., Sze, S. M., "Semiconductor Device Development in the 1970s and 1980s—A Perspective," *Proc. IEEE* **69**, 1121 (1981).
2. For a general discussion of semiconductor devices, see, e.g., Sze, S. M., "Physics of Semiconductor Devices," 2nd Ed., New York, Wiley-Interscience, 1981.
4. For a discussion of integrated circuit fabrication, see, e.g., S. M. Sze, Ed., "VLSI Technology," New York, McGraw Hill, 1983.
5. For a general discussion, see, e.g., Casey, H. C., Jr.,
and Panish, M. B., "Heterostructure Laser," New York, Academic, 1978.
6. For a discussion, see e.g., Johnston, W. D., Jr., "Solar Voltaic Cells," New York, Marcel Dekker, 1980.

Cross-references: DIODE (SEMICONDUCTOR); ENERGY LEVELS, ATOMIC; SEMICONDUCTOR; SOLID STATE PHYSICS; SOLID STATE THEORY; TRANSISTOR.

SERVOMECHANISMS

Definition—A servomechanism is a feedback control system in which the difference between a reference input $r(t)$ and some function of a controlled output $c(t)$ is used to supply an actuating signal $e(t)$ to a controller and a controlled system. The actuating signal is amplified in the controller and is used to vary the output of the controlled system in such a manner that the difference between input and output is reduced to zero. A simple block diagram representation of a servomechanism is shown in Fig. 1.

The controlled system may consist of a mechanical structure, a chemical process, a heating system, an electric supply or any system in which a variable can be measured and controlled. The reference input may be a reference level, a sinusoidal or a polynomial function of time, or a discrete, sampled, or programmed set of values. The difference detector is usually matched to the form of the output and input signals. The controller contains signal amplifiers and may also contain power amplifiers which furnish power to an actuating device from an external power source. Actuating devices vary with the controlled system. Actuating devices to position or move mechanical structures may be electromechanical, hydraulic, or pneumatic.

The controller is designed to control a system with known dynamic properties to respond to a specified type of input signal with a specified steady-state and transient performance. The controller may be digital or analog and may be designed for nonlinear or linear operation or may have a linear operating region of actuating signals and a nonlinear saturation zone. Nonlinear systems can be designed with better performance characteristics than linear systems, but linear systems are easier to analyze. A simple linear system with unity feedback will be described.

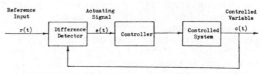

FIG. 1. Block diagram of a servomechanism with unity feedback.

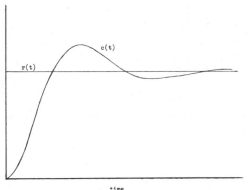

FIG. 2. Transient response of a type I servomechanism to a step input.

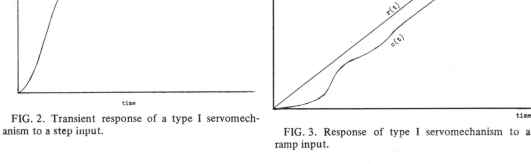

FIG. 3. Response of type I servomechanism to a ramp input.

Linear servomechanisms are analyzed in terms of their transfer function, which can be obtained by taking the Laplace transform of the differential equations of the open loop system with zero initial energy storage. The transfer function is the ratio of the Laplace transform of the output to the Laplace transform of the actuating signal.

$$G(s) = \frac{C(s)}{E(s)}.$$

In most cases, $G(s)$ is or can be approximated as a rational algebraic function and can be expressed in the following form.

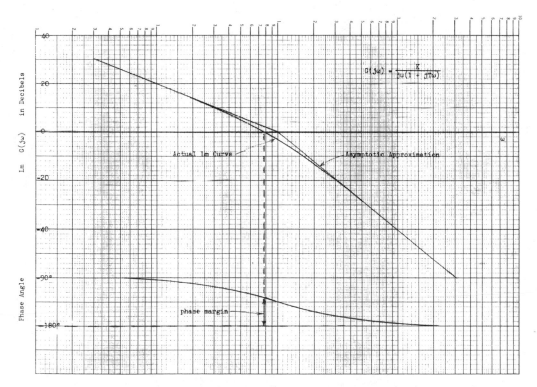

FIG. 4. Bode plots of simple type I servomechanism.

$$G(s) = \frac{K_n(T_1s+1)(T_2s+1)\cdots\left[\dfrac{s^2}{\omega_1{}^2}+\dfrac{2\zeta_1}{\omega_1}s+1\right]\left[\dfrac{s^2}{\omega_2{}^2}+\dfrac{2\zeta_2}{\omega_2}s+1\right]\cdots}{s^n(T_as+1)(T_bs+1)\cdots\left[\dfrac{s^2}{\omega_a{}^2}+\dfrac{2\zeta_a}{\omega_a}s+1\right]\left[\dfrac{s^2}{\omega_b{}^2}+\dfrac{2\zeta_b}{\omega_b}s+1\right]\cdots}$$

The value of the exponent n is used to classify the type of servomechanism. For $n = 0$ (a type 0 servomechanism), the system is often called a regulator or governor, rather than a servomechanism, the distinction being that a servomechanism must contain integration in its transfer function $(n \geqq 1)$ in order to be able to reduce its steady-state error for a position input to zero.

The control ratio is the closed-loop counterpart of the transfer function.

$$\frac{C(s)}{R(s)} = \frac{G(s)}{1 + G(s)}.$$

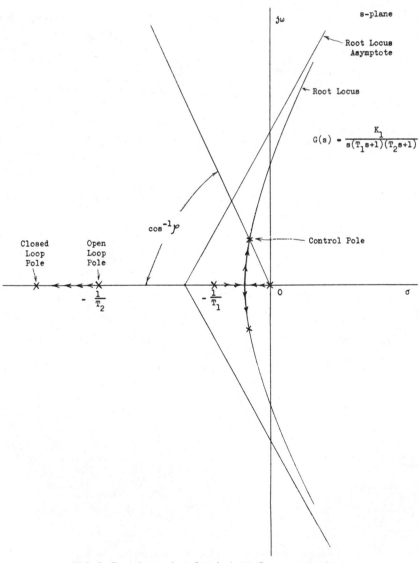

FIG. 5.　Root locus plot of typical type I servomechanism.

The ratio of error (or actuating signal) transform to the input signal transform is given by

$$\frac{E(s)}{R(s)} = \frac{1}{1 + G(s)}.$$

The characteristic equation for the closed loop system is given by

$$1 + G(s) = 0.$$

For a stable system, there must not be any roots of the characteristic equation in the right half of the complex s plane. A stable system usually has a pair of complex roots much closer to the imaginary axis than any of the other roots. These roots, known as the control poles, are largely responsible for the transient performance of the servomechanism. The controller may contain compensating networks which introduce zeros to cancel undesirable poles and substitute desirable poles.

Transient performance is often given in terms of the output response to a unit step input—the response frequency, overshoot, and settling time being specified. A value of $\zeta = 0.4$ in the control poles usually limits the transient overshoot to 25 per cent.

Steady-state performance may be specified in terms of the steady-state error during a steady ramp input.

$$r(t) = \omega_i t.$$

For a type I servomechanism ($n = 1$)

$$e_{ss} = \lim_{s \to 0} sE(s)$$
$$= \frac{\omega_i}{K_1}.$$

For a type II servomechanism ($n = 2$), the steady-state error for a steady ramp input is theoretically zero.

Servomechanisms can be complicated systems with multiple loops, feedback functions, nonlinear or time-varying components, adaptive elements, and undesired random-variable inputs. Special design techniques have been developed to handle these systems.

A very widely used design technique for linear servomechanisms is the Bode plot. The log modulus of the transfer function,

$$Lm[G(j\omega)] = 20 \log_{10} |G(j\omega)|$$

and the corresponding phase angle of $G(j\omega)$ are plotted vs frequency on semilog paper. Straight-line asymptotic approximations for the log modulus plots are usually sufficiently accurate. The phase margin, defined as the amount of phase lag less than $180°$ of $G(j\omega)$ when $Lm[G(j\omega)] = 0$ decibels, is the basic design quantity. A phase margin of $45°$ corresponds to control poles with a $\zeta = 0.4$ and a 25 per cent transient overshoot.

A useful tool in the design and analysis of servomechanisms is the root locus plot. The loci of the roots of the closed loop characteristic equation are plotted as the gain is varied. The gain is adjusted until the control poles have a value of ζ corresponding to the desired transient response.

STEPHEN J. O'NEIL

References

Eveleigh, V. W., "Introduction to Control System Design," New York, McGraw-Hill Book Co., 1972.
Melsa, J., and Schultz, D., "Linear Control Systems," New York, McGraw-Hill Book Co., 1969.
Raven, F., "Automatic Control Engineering," New York, McGraw-Hill Book Co., 1968.
Elgood, O., "Control Systems Theory," New York, McGraw-Hill Book Co., 1967.
Truxal, J. G., Ed., "Control Engineers' Handbook," New York, McGraw-Hill Book Co., 1958.
Truxal, J. G., Ed., "Automatic Feedback Control Systems Synthesis," New York, McGraw-Hill Book Co., 1955.
D'Azzo, J. J., and Houpis, C. H., "Feedback Control System Analysis and Synthesis," New York, McGraw-Hill Book Company, 1960.
Del Toro, V., and Parker, S. R., "Principles of Control Systems Engineering," New York, McGraw-Hill Book Company, 1960.
Ragazzini, J. R., and Franklin, G. F., "Sampled-Data Control Systems," New York, McGraw-Hill Book Company, Inc., 1958.
Newton, G. C., Jr., Gould, L. A., and Kaiser, J. F., "Analytical Design of Linear Feedback Controls," New York, John Wiley & Sons, Inc., 1957.

Cross-references: BIONICS, CYBERNETICS, FEEDBACK.

SHOCK WAVES*

Shock waves arise from a rapid release of chemical, nuclear, electrical, or mechanical energy in a small region. Although the end result of a shock wave experiment may appear to be caused by a violent and disorderly process, the shock process is quite orderly and amenable to mathematical analysis by the laws of physics. Shock waves do, however, subject matter to extraordinary conditions and, therefore, provide a stringent test of our understanding of the fundamental physical processes. The book by Glass (1974) provides a broad discussion of natural and man-made shock waves.

*The preparation of this article was supported by the Office of Naval Research under contract number N00014-81-K-0840.

The practical uses of shock waves date back to the invention of gun powder. Scientific work in this field started in the late nineteenth century and early contributions to the theoretical work were made by Poisson, Stokes, Riemann, Earnshaw, Hugoniot, and Rayleigh. Pioneering experiments on large amplitude stress waves were done by Vieille in gases, and by John and Bertram Hopkinson in solids. As with many other areas of physics, however, active research in shock waves started during the Second World War and has grown steadily since that time. Despite the many developments over the last four decades, there does not exist an in-depth understanding of fundamental mechanisms associated with the effects of shock propagation in condensed matter. Questions concerning atomic and molecular processes of interest to physicists and chemists are only now being addressed. There remain outstanding problems for those who enjoy challenges.

A shock wave is a large amplitude disturbance, propagating at a supersonic speed, across which the pressure, density, temperature, and related physical properties change in a near-discontinuous manner. Although mathematical discontinuities do not occur in real materials, the jump shown in Fig. 1 is used to approximate the very short rise time disturbances observed in experiments. Unlike acoustic waves, shock waves are nonlinear waves characterized by amplitude-dependent wave velocity.

The present discussion, like most shock wave studies, will be restricted to one-dimensional problems, that is, the material properties vary with respect to only one spatial coordinate and time. Two- and three-dimensional problems have received considerably less scientific attention due to the complexity in experiments and analysis. The nonlinear nature of shock problems has also stimulated vast amounts of numerical studies and developments and these will also not be addressed here.

Shock wave studies can be conveniently divided into two areas: shocks in gases and shocks in condensed matter. In gases, the interatomic (or intermolecular) forces are weak and large compressions can be achieved at modest pressures. Temperature changes are, however, large;

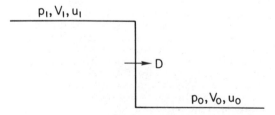

FIG. 1. Discontinuous shock front propagating at a velocity D and connecting two uniform states. The discontinuity is an idealization and actual shocks have finite widths.

temperatures of the order of 10^7 °K have been produced by strong shocks (tens of atmospheres) in gaseous plasmas. In the condensed state, the interatomic (or intermolecular) forces are strong and large pressures are required for even modest compressions (a few percent). Accordingly, temperature changes are small. At the very highest pressures (on the order of 10^7 atmosphere or 10 Mbar) reported in condensed materials, the compressions are large (approaching a factor of 2) and temperature changes are also significant (of the order of 5×10^4 °K). Thus, at very high pressures the solid starts behaving like a gas. The contributions to pressure from the thermal motion of the atoms and electrons become significant.

Recent interest in thermonuclear research has generated considerable interest in shock waves in plasmas. (See articles by Gross and Krokhin in "Physics of High Energy Density," 1971.) Another area related to shock wave studies that will not be addressed here is the study of detonations (see Fickett and Davis, 1979).

The formation of a shock wave can be qualitatively understood by considering the wave motion due to a piston moving into a compressible material. The piston motion and the resulting compression can be divided into a large number of smaller successive steps. The propagation velocity of each step can be written as $c + u$, where c represents the acoustic velocity corresponding to a given compression and u represents the velocity of the material at the end of the previous step. Because this local acoustic velocity is higher for each successive step, the wave motion due to the piston compression steepens into a shock front traveling supersonically into the undisturbed material. This approximate example illustrates shock formation due to compression in most materials. The above arguments can be used to show that wave motion due to unloading (rarefaction waves) spreads out with propagation. The remarks about shock formation assume a particular, though common, material response to be discussed later.

The conservation equations for mass, momentum, and energy for continuous one-dimensional planar flows, independent of material properties, in the absence of heat conduction are

$$\frac{\partial \rho}{\partial t} + \frac{\partial (\rho u)}{\partial x} = 0 \qquad (1)$$

$$\rho \frac{du}{dt} + \frac{\partial p}{\partial x} = 0 \qquad (2)$$

$$\frac{dE}{dt} + p \frac{dV}{dt} = 0 \qquad (3)$$

where ρ is density, u is particle or mass velocity, p is the compressive stress in the x-direction and includes viscosity and other time-dependent contributions, E is the specific internal energy; and $V = 1/\rho$ is the specific volume. The choice of spatial (or Eulerian) coordinates is preferred for fluids. For solids, the material (or Lagrangian) description is preferable. The above equations represent a restricted type of planar flow, that is, uniaxial strain deformation. Although this is the most commonly studied problem, the more general one-dimensional flow includes shear deformations.

For stress waves traveling without a change in shape, that is, discontinuous or steady shocks, the above equations can be integrated to give the Rankine-Hugoniot jump conditions:

$$\rho_0(D - u_0) = \rho_1(D - u_1) \qquad (4)$$

$$p_1 - p_0 = \rho_0(D - u_0)(u_1 - u_0) \qquad (5)$$

$$E_1 - E_0 = \frac{1}{2}(p_1 + p_0)(V_0 - V) \qquad (6)$$

where D is the shock velocity. Equations (4)–(6) relate the states on the two sides of the shock front. Hence, p in the jump conditions represents the equilibrium stress, unlike the p in the conservation equations. Within the shock front itself, deviations from thermodynamic equilibrium can be appreciable.

The locus of end states achieved through a shock transition [Eqs. (4)–(6)], shown in Fig. 2, is referred to as the *Rankine-Hugoniot (R-H) curve* or merely the *Hugoniot* or *shock adiabat*. The chord connecting the two end states is called the *Rayleigh line*. The slope of this line is a measure of the shock velocity and increases with shock amplitude. Strictly speaking, the Hugoniot curve represents only those states for which the jump conditions are applicable, that is, uniform states connected by

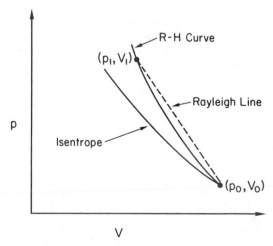

FIG. 2. R-H curve centered at p_0, V_0. The Rayleigh line and an isentrope through p_0, V_0 are also shown. The isentrope and the R-H curve have a second-order contact at p_0, V_0.

a steady shock. However, in the literature the term "Hugoniot" is often used to denote states that are achieved under rapid loading but may not be uniform or may not be connected by a steady shock.

Although the jump conditions are valid for both fluids and solids, the following thermodynamic discussion will be limited to fluids. The energy jump condition refers only to thermodynamic quantities and can be combined with an equation of state of the material to provide the Rankine-Hugoniot curve (locus of p-V states). For gases, at least in many simple systems, the E-p-V relation is known and the R-H curve can be easily calculated. For condensed matter, the pressure changes are large and the needed thermodynamic data are not available. Consequently, most of the effort in these materials has focused on the converse problem: how to obtain thermodynamic information from shock wave experiments. By determining two of the variables in Eqs. (4)–(6) from experiments, the equations can be completely solved. Generally the shock velocity D and one other variable, either p or u, are measured. A knowledge of the Hugoniot curve provides the p, V, and E values at each point of the curve. To construct the E-p-V surface requires additional information and/or assumptions. Analysis of the Hugoniot data to provide more complete thermodynamic information for condensed materials is an active area of shock wave research. (See papers by Duvall, and Royce and Keeler in "Physics of High Energy Density," 1971.) Despite the limited thermodynamic nature of the shock data, these experiments provide the principal means for obtaining information at very high pressures and temperature. Hence, these data have proved valuable both in fundamental and applied studies.

The shock transition is an irreversible process giving rise to entropy production. The entropy change can be written most easily for weak shocks by expanding the entropy in a Taylor series along the Hugoniot. The resulting expression is

$$S_1 - S_0 = \frac{1}{12 T_0}\left(\frac{\partial^2 p}{\partial V^2}\right)_{S_0, V_0}(V_0 - V_1)^3 + \cdots. \qquad (7)$$

Equation (7) shows that the entropy change is third order in compression and $(\partial^2 p/\partial V^2)_S$ has to be positive for a positive entropy change. The p-V relations for most fluids are indeed concave upward (Fig. 2) as required. For materials satisfying this restriction, shocks are restricted to compressive waves based on hydrodynamic and thermodynamic considerations. For many problems in condensed matter, the volume compression is quite small (a few percent) and the entropy change is negligible.

It is interesting to note that the magnitude of

the entropy jump across a shock is independent of the dissipative mechanisms like viscosity and thermal conductivity. The magnitude of these mechanisms determines the thickness of the shock front. Because these mechanisms can never be completely absent in a real material, the concept of a shock discontinuity (Fig. 1) is an idealization. The formation of a steady shock, with finite thickness, results from a balancing of the dissipative processes tending to diffuse the shock front and the curvature of the p-V curve which tends to steepen the shock. (See discussion about shock fronts in Chapters I and VII in the book by Zeldovich and Raizer, 1966.)

In many problems of scientific interest, deviations occur from the "normal" Hugoniot shown in Fig. 2. Thus, the question of shock stability is of considerable interest for both theoretical and experimental problems. The general theory of shock stability is complicated (see paper by Fowles in "Shock Waves in Condensed Matter," 1981). However, in nonreactive materials the stability condition can be written as

$$D - u_1 < C_1 \qquad (8)$$

where C_1 is the sound speed in the shocked state. Physically, this condition implies that the shock front is subsonic with respect to the material behind it. Hence a disturbance (e.g., rarefaction) behind the shock front catches up with the front. The above condition has been analyzed by Duvall to obtain the condition on the p-V relation cited after Eq. (7).

Another area of considerable theoretical interest is the study of wave interactions involving both shock and rarefaction waves. These interactions arise from reflections, transmission, and diffraction of shock waves from material boundaries. Wave interaction problems in one dimension are useful in analyzing laboratory experiments and are important in developing an intuitive understanding of shock wave problems. Two- and three-dimensional problems, though quite complicated, are important in applications involving gas dynamics and in aeronautics. The book of Courant and Friedrichs (1948) provides a comprehensive discussion of wave interaction problems. Problems of contemporary interest may be seen in the proceedings of the symposium on shock tube research ("Shock Tube and Shock Wave Research," 1978).

Shock waves are produced by a rapid deposition of energy, as stated at the beginning. The most common technique for producing shock waves in gases is the use of a shock tube. A long cylindrical tube is divided in two sections by means of a diaphragm. One section of the tube contains the high pressure or driver gas and the other section contains the low pressure or sample gas. Upon rupture of the diaphragm, a compression wave propagates into the low pressure side and a rarefaction wave propagates into the high pressure side. Because the gas is constrained laterally, the flow is one dimensional. Deviations from the idealized one-dimensional situation occur because of diaphragm effects and viscous forces at the wall boundary. Well designed shock tubes minimize these perturbations. The use of light driver gases, explosives, and electromagnetically driven shock tubes can produce higher compressions. In addition, the implosion techniques have been utilized to provide even larger compressions and higher temperatures. Temperatures of the order of 20,000°K are generated in these experiments.

Shock tube measurements generally consist of measuring the shock velocity and one or more of the following shock properties: density, pressure, temperature, and chemical composition. The need for minimal interference and high time resolution from the measurement technique has resulted in the wide use of optical methods. Techniques employed consist of optical interferometry, Schlieren and shadow photography, visible, ultraviolet, and infrared spectroscopy, and x-ray absorption. Spectroscopy measurements have been used for the determination of chemical composition. Developments in laser technology are rapidly expanding the realm of spectroscopy measurements. At very high temperatures, the luminosity behind the shock front is used for studying the shock front and the processes behind the shock front. The nonoptical methods consist of piezoelectric and ionization gauge measurements. Further details may be found in the references (Bradley, 1962; Glass, 1974; "Shock Tube and Shock Wave Research," 1977).

Shock wave studies in gases are of considerable interest to the fields of gas dynamics and aeronautics related to flow past high speed aircraft and reentry vehicles. Shock studies are also used in the study of chemical kinetics to probe relaxation processes involving chemical reactions, electronic vibration and rotation excitations, and even dissociation at high temperatures.

In shock studies of condensed matter, a wide variety of experimental methods have been used, including radiation deposition. In comparison to shock propagation in gases, the compressions and temperatures achieved in condensed materials are smaller, but the pressures are much higher. Deviations from idealized shock response are also more common in condensed matter. Using laser driven shocks, pressures to 35 Mbar (10 times the pressure at the center of the earth) have been reported. The two most common techniques used are in-contact explosives and colliding flyer plates. Stresses ranging from 0.1 GPa (1 kbar) to approximately 1 TPa (10 Mbar) have been achieved using these techniques. At very high stresses, temperatures approach tens of thousands of Kelvin. Time durations for the experiments range from a nanosecond to several microseconds. The development of gas guns that use projectiles driven by compressed gas have permitted very precise impact experiments. Shock production can be understood by examining the collision of flat plates shown in Fig. 3(a). The stress and particle velocity states (p, u) achieved in the experiment can be determined by imposing the condition of continuity of p and u at an interface and transforming the R-H curve to the p-u plane. In Fig. 3(b), the intersection of the p-u curves for the impactor and the target plates are shown. The slopes and positioning of the curves indicate the direction of shock propagation in each plate and the initial particle velocity of the plates, respectively. Prior to the time of arrival of rarefactions from the edges (the useful time of the experiment), the sample is exactly in a state of uniaxial strain. Hence, the one-dimensional equations indicated earlier are rigorously applicable.

Unlike the work in gases, the experiments in condensed matter have mostly consisted of measuring continuum variables like stress, particle velocity, and shock velocity. Earlier measurements concentrated on time of wave arrival (shock velocity measurements) and free surface velocity measurements (these provide particle velocity data) using optical and capacitor techniques. Recent developments using electromagnetic gauges, laser interferometry, piezoelectric, and piezoresistance gauges have given continuous time-resolved, in-material measurements of the wave profiles. These data are noteworthy in their ability to better characterize the shock front and the states behind the shock front. Time resolution to one nanosecond has been achieved. In contrast to the measurement of continuum variables, attempts to study electrical, magnetic, and optical properties of condensed materials have been minimal. These measurements will be discussed in subsequent paragraphs.

The ability to study the response of solids under one-dimensional combined compression and shear loading is an important recent development. In this experiment, the colliding plates are parallel but inclined to the direction of the impactor plate. The motion upon impact can be resolved into a compression and a shear motion which propagate at the appropriate wave velocities. Particle velocity profiles can be measured using electromagnetic velocity gauges. This experimental technique has the ability to measure the shear response of the material behind the shock and to examine the change in other material properties due to shear loading. The use of this technique to study solids and viscous fluids is just starting and the full significance of this development will not be known for some time.

The majority of the studies in condensed matter have been on the mechanical properties. The very high pressure data obtained for a number of materials are particularly noteworthy. The pressure range of these data are well beyond the range of static high pressure measurements and have made significant contributions to the fields of geophysics (quantifying material response at pressures in the interior of the earth) and solid state physics. At large compressions, the shock data permit an evaluation of the interatomic potentials.

One aspect of the dynamic high pressure work involves the study of polymorphic phase transitions in solids. Initially, it was believed that such transitions could not occur on the microsecond time scale available in the shock wave experiments. The experimental data have provided results to the contrary. The conversion of graphite to diamonds in shock wave experiments

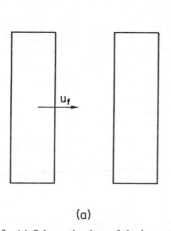

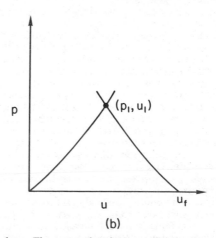

(a)   (b)

FIG. 3. (a) Schematic view of the impact of two parallel plates. The target plate is at rest and the flyer plate is moving with velocity u_f. (b) The stress and particle velocity (p_1, u_1) achieved upon impact are determined by the intersection of the R-H curves in the p-u plane. The sign of the slope indicates the direction of wave propagation.

by DeCarli and Jamieson is a noteworthy example. Despite studies of shock-induced phase transition in many materials (analysis of wave-profiles), a quantitative understanding of the mechanisms has not been achieved. An understanding of the kinetics of the transition and the role of shear strength and deformation are necessary before correlations can be confidently made with static high pressure data.

The phase change studies are a subclass of an old problem receiving considerable attention in recent years: Do shock waves merely simulate the effects of superposing the mechanical stress or pressure and temperature, or are there effects that are unique to shock waves? This question has received considerable attention in the Soviet Union in the studies of chemical reaction under shock loading. The results suggest that shock waves produce unique changes in materials. Unequivocal resolution of this question will require considerable work involving improved time-resolved measurements.

Dynamic yielding is another aspect of mechanical studies that has received considerable attention in the past two decades. Since the late 1950s it has been recognized that the use of the fluid approximation at stresses below 10–20 GPa (100–200 kbar) was incorrect. Recent work suggests that strength effects may be important even at stresses in excess of several hundred kilobars. Although considerable progress has been made in relating the observed wave structure in metals to continuum aspects of deformation, the atomic or microscopic mechanisms are not well understood. For nonmetals even the continuum response, because of a lack of knowledge of the complete stress tensor, is not well understood. Shock wave interactions have also been used to study fracture on microsecond time scales. Fracture strength shows appreciable rate dependence.

Among the thermodynamic and continuum measurements, the two big uncertainties in existing shock data are temperature and shear stresses. As indicated earlier, some progress has been made toward measurement of the latter. Temperature measurement remains a challenging problem and is perhaps the biggest limitation of shock work. Previous attempts have used optical radiation from shocked solids and thermoelectric measurements. No reliable techniques have emerged to date. The difficulty is that the techniques used to monitor temperature in other branches of physics cannot be easily extended because of the coupling between stress, loading rate, and temperature in shock experiments.

Electrical, magnetic, optical, and x-ray measurements that are commonly used in other fields of condensed matter physics and chemistry have not been extensively used in shock wave studies. The most common of these measurements have been electrical conductivity data in insulators, semiconductors, and metals.

Some studies of piezoelectric and ferroelectric materials have also been carried out. However, the mechanisms responsible for shock-induced conductivity changes, polarization, and depoling of ferroelectric materials are not well understood.

Although optical measurements of various kinds including luminescence, refractive index changes, and absorption spectroscopy have been attempted in the past, they have not lead to any comprehensive understanding. Recently, a concentrated effort on shock wave spectroscopy has started at several laboratories. These studies should lead to a better understanding of the molecular processes important under shock loading in condensed materials.

Finally, mention should be made of the x-ray diffraction work under shock loading by Johnson and Mitchell at Lawrence Livermore Laboratory. This one set of experiments confirmed the existence of crystalline order behind the shock front in two materials on time scales of the order of tens of nanoseconds. The confirmation of this important but commonly made assumption is very gratifying. Further work in this area would be beneficial to a basic understanding of shock effects.

The lack of optical and x-ray data in condensed matter studies is caused by the difficulty and costs of the experiments. In contrast to shock tube experiments, condensed matter samples cannot be probed through the sides because of edge effects. This not only makes the experimental geometry difficult, but severely limits the time available for recording. The review by Davison and Graham (1979) provides a very comprehensive list of references for shock waves in solids.

In conclusion, shock waves in condensed matter have contributed greatly to mechanical and thermodynamic information at extreme loading conditions. The use of shock waves as a tool to probe the atomic and molecular states of matter at extraordinary conditions remains a challenging problem both experimentally and theoretically. The single-event nature of the experiments and the need to make deductions from incomplete measurements in a single experiment require considerable experimental skills and theoretical innovation.

Ongoing research in shock waves in gases is published in the proceedings of the symposia on shock tube research held every two years. Contemporary problems in shock waves in condensed matter are published in the proceedings of the AIP topical conferences held every two years.

Y. M. Gupta

References

Bradley, J. N., "Shock Waves in Chemistry and Physics," London, Methuen and Co., Ltd, and New York, John Wiley and Sons, Inc., 1962.

Courant, R., and Friedrichs, K. O., "Supersonic Flow and Shock Waves," New York, Interscience Publishers, Inc., 1948.

Davison, L., and Graham, R. A., "Shock Compression of Solids," in *Physics Reports* 55, 255–379, Amsterdam, North-Holland Publishing Co., 1979.

Fickett, W., and Davis, W. C., "Detonation," Berkeley, Los Angeles, and London, Univ. California Press, 1979.

Glass, I. I., "Shock Waves and Man," Toronto, Univ. Toronto Press, 1974.

"Physics of High Energy Density," Proceedings of the International School of Physics "Enrico Fermi," Course XLVII, VARENNA, July 1969 (P. Caldirola and H. Knoepfel, Eds.), New York and London, Academic Press, 1971.

"Shock Tube and Shock Wave Research," Proceedings of the Eleventh International Symposium on Shock Tubes and Waves, Seattle, July 1977 (B. Ahlborn, A. Hertzberg, and D. Russell, Eds.), Seattle and London, Univ. Washington Press, 1978.

"Shock Waves in Condensed Matter," AIP Topical Conference Proceedings, Menlo Park, June 1981 (W. J. Nellis, L. Seaman, and R. A. Graham, Eds.), New York, American Institute of Physics, 1981.

Zeldovich, Ya. B., and Raizer, Yu. R., "Physics of Shock Waves and High-Temperature Hydrodynamic Phenomena," Vols. I and II, New York and London, Academic Press, 1966.

Cross-references: AERODYNAMICS; FLUID DYNAMICS; MECHANICAL PROPERTIES OF SOLIDS; PRESSURE, VERY HIGH; SOLID STATE PHYSICS; WAVE MOTION.

SIMPLE MACHINES

A machine is a device used to change the magnitude and or direction of a force or torque. This enables man to perform laborious work. Basically there are three types of machines: Type A, the *lever*, including simple levers, pulleys, and wheel and axle; Type B, the *inclined plane*, including the simple inclined planes, wedges, and screws; and Type C, the *hydraulic press*, which is not discussed here. All machines, however complicated, are combinations of these simple machines.

Mechanical Advantage and Efficiency Let us consider a simple machine in which an input force F_i is applied through a distance d_i so that the input work is $W_i = F_i d_i$; while the output force F_o is applied through a distance d_o, so that the output work is $W_o = F_o d_o$. In addition, the work done against friction by the machine is W_f. From the conservation of work-energy principle (see WORK, POWER AND ENERGY) $W_i = W_o + W_f$, that is,

$$F_i d_i = F_o d_o + W_f. \qquad (1)$$

The actual mechanical advantage (AMA) is defined as the ratio of the output force to the input force, that is,

$$\text{AMA} = F_o / F_i. \qquad (2)$$

Usually F_o is greater than F_i (while d_o is less than d_i) and hence AMA is greater than 1. There are situations in which it is desirable to have F_o less than F_i, but d_o greater than d_i, as in the case of scissors.

For an ideal machine, $W_f = 0$, and hence from Eq. (1), $F_i d_i = F_o d_o$. We define the *theoretical mechanical advantage* (TMA) (or ideal mechanical advantage) as

$$\text{TMA} = d_i / d_o. \qquad (3)$$

Note that when there is no friction ($W_f = 0$), TMA = AMA = $F_o / F_i = d_i / d_o$.

We define the *efficiency* η of a simple machine as the ratio of the output work to the input work, that is,

$$\eta = \frac{\text{work output}}{\text{work input}} = \frac{F_o d_o}{F_i d_i} = \frac{F_o / F_i}{d_i / d_o} = \frac{\text{AMA}}{\text{TMA}}. \qquad (4)$$

η is usually expressed as a percentage and represents the part of the input work changed into useful work by the machine. The efficiencies of simple and complex machines vary between $\sim 5\%$ and $\sim 95\%$.

The efficiency can also be calculated by applying the equilibrium principle instead of the work energy principle as discussed above.

Levers A *simple lever* consists of a nearly rigid rod pivoted about a certain point P called the *fulcrum*. A most common type is shown in Fig. 1(a) with point P between the load $W = F_o$ and the applied force F_i. By applying a small

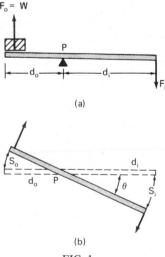

(a)

(b)

FIG. 1.

force F_i, a large load W can be lifted. Applying the condition of equilibrium that the input torque is equal to the output torque, we get $F_i d_i = F_o d_o$ (neglecting a very small friction that may be present at the pivot point). Therefore AMA = TMA, and

$$\text{TMA} = \frac{d_i}{d_o} = \frac{F_o}{F_i}. \tag{5}$$

We can obtain the same result from work considerations. As shown in Fig. 1(b), when F_i moves through a distance s_i, the force F_o moves through a distance s_o. Since the two arcs s_i and s_o subtend the same angle θ at P, we may write

$$\theta = \frac{s_i}{d_i} = \frac{s_o}{d_o}. \tag{6}$$

Therefore

$$\text{TMA} = \frac{s_i}{s_o} = \frac{d_i}{d_o}. \tag{7}$$

The above type of lever in which the fulcrum point is between the load and the applied force is called the Class I lever. In the Class II lever, the load is between the fulcrum and the applied force, as in the case of a wheelbarrow. In the Class III lever, the applied force is between the load and the fulcrum, as in the case of the human forearm held horizontally with a load in the hand.

The simple lever has a limitation in that it can operate only through a small angle. This limitation is overcome by using a continuous rotation of the lever arm as in the case of the *wheel and axle*. It consists of a wheel of radius R fixed to a shaft or axle of radius r with the same axis of rotation, as shown in Fig. 2. A force F_i applied tangentially to the edge of a wheel lifts the load $W (= F_o)$ by means of a rope wrapped around the axle. Thus when the wheel is rotated through one full rotation, the work done is $F_i d_i = F_i(2\pi R)$, the axle will rotate through one rotation doing work equal to $F_o d_o = W(2\pi r)$. We may equate these two or apply the equilibrium principle by taking torques about point

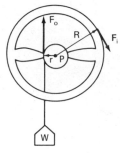

FIG. 2.

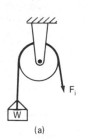
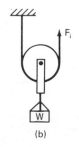

(a)   (b)

FIG. 3.

P, that is,

$$F_i r = F_o R.$$

Thus, we get

$$\text{TMA} = \frac{d_i}{d_o} = \frac{R}{r}. \tag{8}$$

This principle is used in almost every motor-driven machine. In trucks which may be equipped with power steering, $R \gg r$. Also we use the same principle every time we use a doorknob to open a door.

The Pulley. A pulley is a wheel or disk with a groove on the circumference edge. A pulley is a continuous lever in which forces are transmitted by ropes. A single fixed pulley, as shown in Fig. 3(a), is a lever with equal moment arms $(d_i = d_o)$, and hence TMA $= d_i/d_o = F_o/F_i = 1$. The only purpose this single pulley serves is to change the direction of the applied force F_i. Instead, if we have a single movable pulley as shown in Fig. 3(b), TMA = 2. In order to lift the weight W through a height $h (= d_o)$, each of the two vertical ropes must be raised through a height h, that is, F_i must be applied through a distance $2h (= d_i)$. Thus TMA $= d_i/d_o = 2h/h = 2$.

A system of pulleys, called a *block and tackle*, as shown in Fig. 4, may be considered equivalent to a lever with arms of different lengths. For example, consider the block and tackle systems shown in Fig. 4(a), where there are 4 ropes supporting the movable block, hence TMA = 4, and in Fig. 4(b) (which is the system of Fig. 4(a) inverted) where the TMA = 5, because there are five ropes supporting the movable system of block and pulleys. Hence

TMA = Number of strands of rope supporting movable block of block and tackle.

$$(9)$$

The chain hoist or *differential pulley* is based on the combination of the block and tackle and the wheel and axle. As shown in Fig. 5, it consists of two pulleys of radii R and r attached to each other and rotating about a common axle. An endless chain connects the dual pulley to another free pulley wheel, as shown. The chain

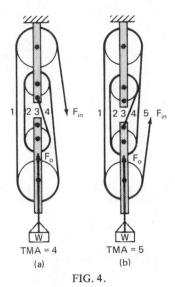

TMA = 4   TMA = 5
(a)   (b)

FIG. 4.

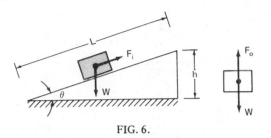

FIG. 6.

and the corresponding teeth on the dual pulleys prevents slipping between the chain and the pulleys. In one complete revolution, the input force F_i acts through a distance $d_i = 2\pi R$. The chain around the movable pulley results in shortening by $2\pi R$ and lengthening by $2\pi r$, thereby, raising the weight $W (= F_o)$ by a height $d_o = 2\pi R - 2\pi r$. Thus,

$$\text{TMA} = \frac{d_i}{d_o} = \frac{2\pi R}{2\pi R - 2\pi r} = \frac{R}{R - r}. \qquad (10)$$

Thus, the closer the values of R and r, the larger the TMA.

The Inclined Plane Ramps and staircases are two simple examples of *inclined planes*. Large loads can be moved by applying small forces. As shown in Fig. 6, when a force F_i is applied

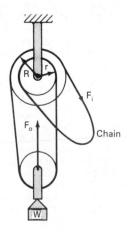

FIG. 5.

through a distance $d_i (= L)$, the load $W = F_o$ is lifted through a height h. Thus work input = work output, that is, $F_i L = Wh$, and hence, if there were no friction,

$$\text{TMA} = \frac{W}{F_i} = \frac{L}{h} = \frac{1}{\sin \theta}. \qquad (11)$$

The *wedge* is a single or double movable inclined plane, as shown in Fig. 7. For a double wedge TMA = L/t. Wedges, besides being used for splitting logs or holding doors, find their application in knives, chisels, and other cutting tools. A rotary wedge called a *cam* is used to lift valves held by heavy springs in internal combustion engines.

A *screw* is essentially an inclined plane, wrapped around a cylindrical shaft to form a continuous helix. The corrugations of a screw are called *threads*. The distance between the two adjacent threads is called the *pitch* of the screw, p. A typical example of screw is a *screw jack* shown in Fig. 8. When an applied force F_i is moved through one complete revolution $d_i = 2\pi R$, the weight $W = F_o$ is raised through a distance p. Thus

$$\text{TAM} = \frac{d_i}{d_o} = \frac{2\pi R}{p}. \qquad (12)$$

Because of friction, the AMA of a screw jack is much less than the TMA. But the presence of friction happens to be an advantage in this case, because it holds the load in place while no input force is being applied, as in the case of lifting an automobile by using a screw jack.

Power and Torque Transmission Power P is defined as $P = Fv$ or $P = \tau\omega$, where v, ω and τ are velocity, angular velocity, and torque respectively. Power and torque may be transmitted by using mechanical devices such as

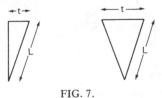

FIG. 7.

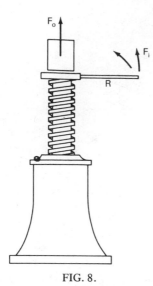

FIG. 8.

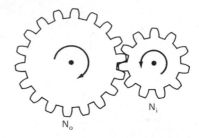

FIG. 10.

For no slipping between the belts and the pulleys ($F_i = F_o$),

$$\text{TMA} = \frac{r_o}{r_i} = \frac{D_o}{D_i}$$

where D_o and D_i are the diameters. Also input power = output power, that is,

$$\tau_i \omega_i = \tau_o \omega_o.$$

From the above two equations, if there is no friction,

$$\text{TMA} = \frac{\tau_o}{\tau_i} = \frac{\omega_i}{\omega_o} = \frac{D_o}{D_i}, \qquad (13)$$

while the *speed ratio* SR $= \omega_o/\omega_i = 1/\text{TMA}$. Thus by adjusting the diameters of the input and the output pulleys, the speed ratio and τ_o/τ_i may be adjusted. If TMA = 4, the input pulley must rotate four times as fast as output pulley, which can be achieved by having the diameter of the output pulley four times the diameter of the input pulley.

A *gear* is a notched wheel. When two gears whose notches or teeth are of the same size are meshed together, it is possible to transfer torque from one gear to the other. Gear drives are capable of transmitting very high torques without causing slipping. A typical *spur gear* is shown in Fig. 10. Note that the two gears rotate in opposite directions. Let N_i and N_o be the number of teeth on the input and the output gear respectively. As in the belt drive system, for gears

$$\text{TMA} = \frac{D_o}{D_i} = \frac{\omega_i}{\omega_o} = \frac{N_o}{N_i}. \qquad (14)$$

There are five common types of gears. *Spur gears*, Fig. 10, are those which have their teeth cut parallel to the axis of rotation. *Bevel gears* are used when the axes of rotation intersect; their teeth are cut on conical surfaces with a common apex at the point of intersection of the axes. Bevel gears are used to change the direction of rotation by 90° with or without a change of speed. In *helical gears*, the teeth are curved and cut in a spiral pattern. Helical gears

belts, shafts and gears. Figure 9(a) shows pulleys joined by a belt. The belt has a V-shaped cross section so that it cannot slide sideways, and furthermore it has a large area in contact with the pulleys thereby preventing slipping. Basically it has the same principle as a wheel and axle. A pair of step pulleys, as shown in Fig. 9(b), allows one to change shaft speeds by shifting the belt. Note that the driving pulley and the output pulley rotate in the same direction.

The mechanical advantage of a system shown in Fig. 9(a) is the ratio of the torques between the output pulley and the driving pulley, that is,

$$\text{TMA} = \frac{\tau_o}{\tau_i} = \frac{F_o r_o}{F_i r_i}.$$

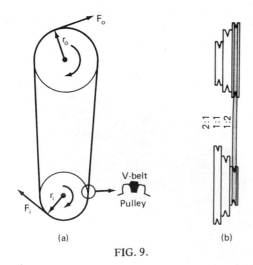

FIG. 9.

operate very smoothly. In *planetary gears*, the teeth are cut inside one wheel instead of outside. When a screw meshes with a cogged wheel, it is known as a worm and worm wheel or *worm gear*. Lastly, *gear trains* consists of more than two gear wheels.

ATAM P. ARYA

Cross-references: FRICTION; MECHANICS; STATICS; WORK, POWER, AND ENERGY.

SKIN EFFECT

For a steady unidirectional current through a homogeneous conductor, the current distribution over the cross section is uniform. However, for an alternating current, the current is not uniformly distributed over the cross section but is displaced more and more to the surface as the frequency increases. For high frequencies, practically the entire current is concentrated in a thin layer at the surface or "skin" of the conductor. This phenomenon has commonly come to be called the "skin effect."

For a physical explanation of this phenomenon, consider a cylindrical conductor that is carrying a steady unidirectional current. The current distribution over the cross section is uniform. A magnetic field is set up by the current in which the magnetic flux lines within and around the conductor are symmetrical to the conductor axis, i.e., they are in concentric circles. Consider the conductor to be composed of very small circular filaments of equal area which carry equal fractions of the total current. Consider a filament A near the axis of the conductor and another filament B near the surface. The flux linking A is greater than that linking B. If the steady unidirectional current is changed to an alternating current, the magnetic field which is set up by the current must reverse itself periodically with the result that the flux linkages change with time. Self-induced electromotive forces (emf's) are created within the conductor due to these flux changes. These emf's tend to generate currents (eddy currents) which oppose the main current. Because the emf created in filament A is greater than that in B due to a greater flux linkage, the net or resulting current in A becomes less than in B. The current in the conductor is no longer uniform but is displaced toward the surface. The current density becomes greater at the surface and decreases toward the center. This current displacement becomes more pronounced the higher the rate of change of flux.

The total resistance offered by the conductor to alternating current is greater than to a steady unidirectional current because of skin effect. When the current is displaced from the center of the conductor and crowded into the area near the surface, the effective cross section of the conductor is reduced, thereby increasing its resistance. The resistance of the conductor increases with frequency

The equations which mathematically describe skin effect can be derived from Maxwells' equations for electrodynamic problems. However, exact solutions have been obtained for only a few simple shapes of conductor. Shapes which have been amenable to analysis include the cylindrical conductor, tubular conductor, flat conductor, surface plated flat, cylindrical and tubular conductor, and coaxial conductor. Even then it is necessary to consider conductors whose material properties do not change with time or temperature. More complex shapes must be treated numerically.

The first theoretical explanation of skin effect for wires of circular cross section was given by Lord Kelvin in 1887. Early investigators were Kennelly, Laws, and Pierce,[2] who conducted comprehensive experiments on skin effect, and Dwight[3] who obtained solutions for skin effect in tubular and flat conductors. Many other investigations have been made since that time.

The solution to the skin effect equation for a flat or plane conductor carrying a sinusoidal alternating current shows that the current density is maximum at the surface and decreases exponentially in magnitude with distance from the surface into the conductor. Also, as the distance from the surface increases, the current lags in time-phase further and further behind the current at the surface. A quantity $\delta = (2/\omega\mu\sigma)^{1/2}$ is often called the "skin thickness" since it corresponds to the distance from the surface in which the current density drops to $1/e$ of its value at the surface. In this equation, σ is the conductivity, μ is the permeability of the material, ω is the angular frequency. The resistance and internal reactance of the plane conductor are equal at any frequency. Also, the ac resistance of the conductor is exactly the same as the dc resistance of a plane conductor of thickness δ.

In copper, the skin thickness δ at 60 Hz is about 8.5 mm. At 1 GHz, δ decreases to about 2×10^{-6} m.

For a cylindrical conductor carrying a sinusoidal alternating current, the solution to the skin effect equation is found in terms of the zero-order Bessel function of the first kind. If the frequency is quite high, the exact formulation in terms of Bessel functions reduces to the simple exponential solution for flat or plane conductors. A useful engineering approximation for the ratio of ac to dc resistance is $R_{ac}/R_{dc} = a/2\delta$ where a is the radius of the conductor and $\delta \leqslant 0.1a$. The exact ratio in terms of Bessel functions must be used to obtain a solution for low frequencies or small conductors where $0.1a < \delta < a$. At radio frequencies, there is practically no magnetic field inside the conductor, so the reactance is negligible.

The solution to the skin effect equation for

tubular conductors is expressed in terms of Bessel functions of the first and second kind. Since the interior portion of a cylindrical conductor carries very little current at high frequencies, thin tubular conductors are often used with a corresponding saving in material. At these frequencies, the tube acts like a solid conductor and effectively serves as an electromagnetic shield.

Conductors are often surface plated for specific applications. Since silver has high conductivity, resonant cavities and wave guides which operate at very high frequencies are silver plated to reduce $I^2 R$ loss since most of the current is concentrated at the surface. Analyses have been made of a number of surface plating combinations to determine what effect such coatings have on the over-all resistance.

The skin effect phenomenon has a very practical application in induction heating. Since the current is concentrated near the surface, a highly selective heating source is created in the surface itself. Advantage is taken of this phenomenon and electromagnetic induction to create a heating method which requires no external heat source and no physical contact with the energy source, the induction coil.

Consider an induction coil carrying alternating current which is wound uniformly around a metal cylinder. Eddy currents are induced in the metal and tend to concentrate near the surface. The problem is similar to that of the cylindrical conductor previously described except that the induced currents are in concentric circles at right angles to the axis of the cylinder instead of being in the axial direction. The solution of the skin effect equation is in terms of the first-order Bessel function instead of the previous zero-order function.

With the advent of high-speed digital computers, numerical methods have replaced analytical techniques in solving skin effect problems. Analytical methods, even approximate ones, are quite limited. Numerical methods tend to be more flexible and can be used for analysis of systems of more than one conductor and of conductors having a wide variety of cross sections. Numerical solutions such as finite difference schemes, finite elements, and integral equations have gained currency. At present, the finite-element method can be considered to be the fastest-growing technique for solving practical electrical engineering problems with complicated geometries.[11-15]

Up to this point, the concepts of classical skin effect have been discussed. However, at sufficiently high frequencies or conductivities, the properties of the individual conduction electrons must be considered. In simple electron theory, σ is proportional to the mean free path ℓ of conduction electrons. As σ or ω increases, at some point δ becomes smaller than ℓ. Incident electromagnetic radiation is now more strongly absorbed than would be calculated

from classical skin effect theory. This situation is referred to as an *anomalous skin effect*. As ω increases, other effects may become significant, namely, relaxation effects due to the inertia of the electrons. An incident wave now "sees" a layer of virtually free electrons which results in almost total reflection. This situation is referred to as *anomalous reflection*. If ω is increased still further, the *plasma frequency* of the metal is reached. Above this frequency, the metal becomes transparent to the incident radiation. Transmission takes place and skin effect no longer exists.[16]

In metals that are in a superconducting state, skin effect exists but is independent of frequency. Theory suggests that superconducting electrons are responsible for the screening (or skin) effect, even at zero frequency, while normal electrons in the skin layer thus formed are responsible for absorption that occurs only at nonzero frequency. An application of this aspect of superconductivity, in which the field penetration is limited to a skin layer, is to be found in the fabrication of high-Q resonant cavities.[17]

JOHN C. CORBIN

References

1. Thomson, W., "Mathematical and Physical Papers," Vol. 3, p. 493, Cambridge, Cambridge University Press, 1890.
2. Kennelly, A. E., Laws, F. A., and Pierce, P. H., "Skin Effect in Conductors," *Trans. AIEE*, **34**, 1953 (1915).
3. Dwight, H. B., "Skin Effect in Tubular and Flat Conductors," *Trans. AIEE*, **37**, 1379 (1918).
4. Ramo, S., and Whinnery, J. R., "Fields and Waves in Modern Radio," Ch. 6, New York, John Wiley & Sons, 1953.
5. McLachlan, N. W., "Bessel Functions for Engineers," Ch. 8, London, Oxford University Press, 1955.
6. Flugge, S., Ed., "Handbuch der Physik," Vol. XVI, "Electric Fields and Waves," p. 182, Berlin, Springer-Verlag, 1958.
7. Moon, P., and Spencer, D. E., "Foundations of Electrodynamics," Ch. 7, New York, Van Nostrand Reinhold, 1960.
8. Simpson, P. G., "Induction Heating," New York, McGraw-Hill Book Co., 1960.
9. Corbin, J. C., "The Influence of Magnetic Hysteresis on Skin Effect," Aerospace Research Laboratories Report 65-167, August 1965 (U.S. Air Force Publication).
10. Casimir, H. B. G. and Ubbink, J., "The Skin Effect; I. Introduction—The Current Distribution for Various Configurations," *Philips Technical Review* **28**, 271–283, (1967).
11. Weeks, W. T. et al., "Resistive and Inductive Skin Effect in Rectangular Conductors," *IBM J. Res. Develop.* **23**, 652–660 (1979).
12. Chari, M. V. K., and Silvester, P. P. (Eds.), "Fi-

nite Elements in Electrical and Magnetic Field Problems," New York, Wiley, 1980.

13. Konrad, A., "Integrodifferential Finite Element Formulation of Two-Dimensional Steady-State Skin Effect Problems," *IEEE Trans. Magn.*, **MAG-18**, 284–292, 1982.

14. Konrad, A., Chari, M. V. K., and Csendes, Z. J., "New Finite Element Techniques for Skin Effect Problems," *IEEE Trans. Magn.*, **MAG-18**, 450–455 (1982).

15. Brauer, J. R., "Finite Element Calculation of Eddy Currents and Skin Effects," *IEEE Trans. Magn.*, **MAG-18**, 540–509, 1982.

16. Casimir, H. B. G., and Ubbink, J., "The Skin Effect. II. The Skin Effect at High Frequencies," *Philips Tech. Rev.* **28**, 300–315 (1967).

17. Casimir, H. B. G., and Ubbink, J., "The Skin Effect. III. The Skin Effect in Superconductors," *Philips Tech. Rev.* **28**, 366–381 (1967).

Cross-references: ALTERNATING CURRENTS; CONDUCTIVITY, ELECTRICAL; INDUCTANCE; INDUCED ELECTROMOTIVE FORCE.

SOLAR CONCENTRATOR DESIGN, OPTICS OF

The concentration factor of a solar collector is the ratio of the solar flux reaching a receiver surface to the solar flux crossing the aperture of the concentrator. The geometrically simplest and least expensive concentrator would be composed of flat surfaces and as few of those as possible.

The flat-sided rectilinear trough is a simple design whose properties have been studied in detail.[1] Figure 1 illustrates how a trough reflector can be designed for one reflection before reaching the base, or two reflections by truncating the reflector at D_1 or D_2, and so on. It is desirable to keep the number of reflections small, since energy is lost at each reflection. The concentration factor is a function of α, half the apex angle formed by the extended sides of the trough. Assuming no loss at each reflection the concentration factor for normal incidence at the aperture and n reflections is given by

$$C = \frac{D}{D_0} = \frac{\sin{(2n+1)}\,\alpha}{\sin \alpha}$$

where D is aperture width and D_0 base width.

One must also consider performance when the sunlight is not normal to the aperture. It is also desirable to keep the length of the sides as small as possible in order to minimize material requirements. All these factors are studied in Ref. 1. In general it appears that commercial flat-sided trough concentrators are limited to a concentration factor of 2–3 or so. The corresponding L/D_0 ratio is about 5. L is the length of one side. It is possible to increase the con-

centration factor by forming each side of the trough from two or more flat surfaces; or by shaping the sides and extending their lengths.

Much higher concentration factors can be obtained from truncated flat sided pyramidal or hexagonal cones. A concentration factor of about 10 can be achieved with an apex angle α of $15°$ and a value of $L/D_0 = 4$ or 5 where L is the length of side and D_0 is the base width. A concentration factor of 15 or so can be achieved by decreasing α to $10°$ and increasing L/D_0 to 10. In all cases the distribution of energy over the base is remarkably uniform. For a detailed study of the effect of all parameter variations see Ref. 2.

It has been shown by Winston[3] that the maximum geometrical concentration factor possible for a given acceptance angle θ_m is

$$C \leqslant 1/\sin{(\theta_m/2)} \quad \text{(2-dimensional case)}$$

$$C \leqslant 1/\sin^2{(\theta_m/2)} \quad \text{(3-dimensional case)}.$$

$\theta_m/2$ is the angle of incidence with respect to the plane or line of symmetry at the entrance aperture. For another discussion of this subject see R. E. Jones.[4] One may not always be interested in the maximum concentration but rather in achieving a certain distribution of sunlight over the receiver surface. For example, in heating solar cells it is desirable to avoid hot spots. In the next section we describe the principles involved in designing for a given distribution of energy. Often this will still allow one to achieve close to optimum average concentration factor. The same considerations apply when designing reflectors for illumination purposes.[5] See also the book by W. B. Elmer.[6] A specified distribution can be achieved only for one direction of solar incidence, usually but not always for normal incidence at the aperture. In order to monitor deviations from the initial specified distribution as the sun's rays change direction one may employ the method for irradiance calculation described in the article IRRADIANCE (ILLUMINANCE) CALCULATION.

It is possible to shape a reflecting surface to achieve any desired distribution and concentration of solar energy, except for a theoretical upper limit which does not concern us here. We will describe the steps to be followed to design such a surface and illustrate their application to both an axially symmetric forward lighted reflector surface; and a rectilinear trough with curved sides.[7] The same procedure applies to unsymmetric configurations and also to the more general problem of changing an arbitrary distribution of incident radiation to a specified output. For example, Rhodes and Shealy[8] have designed a two-lens system that will convert a collimated laser beam of arbitrary incident irradiance profile to a uniform expanded irradiance profile. The principles involved are the same as those described

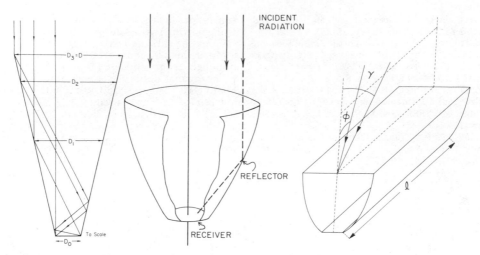

FIG. 1. Cross section of a flat-sided rectilinear trough concentrator. Axially symmetric conical concentrator. Rectilinear trough concentrator with curved sides. ϕ is an angle of incidence oblique to the length of the trough. γ is an angle of incidence with respect to the symmetry plane of the trough.

herein, except for one additional requirement, namely, that the existing beam profile be uniform not simply over a fixed receiver plane but also along the complete range of the exiting beam. They achieve this through the use of a second refracting surface, a second lens, and require that the optical path length be the same for all rays over an existing wave front that is normal to the optical axis of the system. In the latter case, the same reshaping has been achieved by means of computer generated hologram filters that replace the aspheric lenses.[9]

Axially Symmetric Geometry We first consider the rotationally symmetric geometry shown in Fig. 1. Solar energy is incident along the z-axis, left to right, upon reflector surface S_1 described by the equation $z = z(x)$. x and X are radial distances from the z-axis. The receiver surface S_2 is described by the equation $Z = Z(X)$. Let σ be the incident solar energy per unit area per unit time. The solar flux incident upon a circular ring of area about the z-axis is $2\pi\, x\, dx$. This flux is reflected to a circular ring of width dL_2 on S_2 whose area is $2\pi X\, dL_2$. The flux density at dL_2 is $\sigma x\, dx/X\, dL_2$. Since $dL_2 = [1 + (dZ/dX)^2]^{1/2}\, dX$, we have the energy balance equation

$$EX[1 + (dZ/dX)^2]^{1/2}\, dX = \sigma x\, dx. \qquad (1)$$

When the receiver surface equation $Z = Z(X)$ is specified, dZ/dX can be evaluated and both sides of Eq. (1) can be integrated.

Although E may be an arbitrary function of position over the receiver surface, it must have an adjustable parameter so that conservation of energy is satisfied. The adjustable parameter in E is obtained by integrating (1) and substituting the limits of integration.

For a flat receiver at $Z = Z_0$, we have

$$\int_{X_0}^{X} EX\, dX = \int_{x_0}^{x} \sigma x\, dx. \qquad (1a)$$

For uniform flux over a flat receiver which fills up according to Fig. 2(a), that is, integrating Eq. (1a) from $X_0 = 0$ to the upper limit of the receiver plane, then $E = $ constant $= E_0 = \sigma(x_f^2 - x_0^2)/X_f^2$. This value of E_0 is substituted into the integrated form of Eq. (1a) to yield the connection between X and x. Thus

$$X = [\sigma(x^2 - x_0^2)/E_0]^{1/2}, \qquad (2)$$

with E_0 given as above. We now write the ray trace equation for a ray connecting coordinates $x, z,$ and X, Z. If $\mathbf{a}$ is a unit vector along a ray incident upon S_1 and $\mathbf{A}$ a unit vector along the reflected ray, we have from the law of reflection

$$\mathbf{A} = \mathbf{a} - 2\mathbf{N}(\mathbf{N} \cdot \mathbf{a})$$

where $\mathbf{N}$ is a unit vector normal to the reflector surface S_1. Since $\mathbf{N} = (z'\mathbf{I} - \mathbf{K})/(1 + z'^2)^{1/2}$ where $z' = dz/dx$,

$$\mathbf{A} = [2z'\mathbf{I} + (z'^2 - 1)\mathbf{K}]/(1 + z'^2).$$

The ray trace equation is then

$$\frac{A_x}{A_z} \equiv \frac{2z'}{z'^2 - 1} = \frac{X - x}{Z_0 - z}. \qquad (3)$$

Abbreviating, $\ell \equiv (z - Z_0)/(x - X)$, Eq. (3) can be solved as a quadratic equation in the vari-

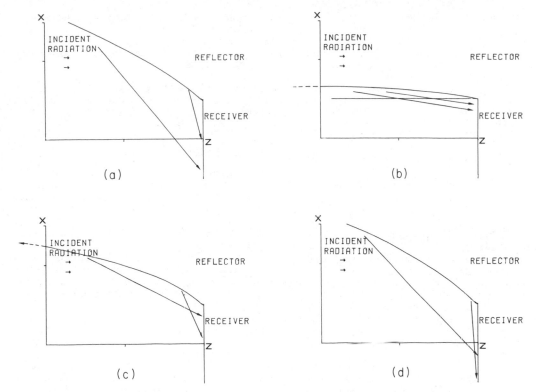

FIG. 2. Various rays in which the light can be reflected to and fill up the receiver base. The limits on the flux conservation integral determined the filling order. One would normally follow graph (a), but the options are shown to emphasize the choice involved.

able z' for the differential equation for the reflecting surface:

$$z' = \ell \pm (1 + \ell^2)^{1/2} \qquad (4)$$

where the $+(-)$ sign is used when the concave side of the reflector is oriented toward the positive (negative) z direction. In the following discussion we shall only consider the $-$ sign in Eq. (4).

Using the energy balance equation (2) to relate X to x,

$$\ell = (z - Z_0)/\{x - [\sigma(x^2 - x_0^2)/E_0]^{1/2}\}. \qquad (4a)$$

We can now solve the first-order differential equation (4), usually numerically, to any precision required. The initial conditions are $z = z_0$ when $x = x_0$; these conditions are also involved in the choice of $X_0 = 0$ for the lower limit of integration in Eq. (1a).

Figure 2 shows some of the ways in which reflected light can fill the axially symmetric collector for forward reflection. If the starting and terminating rays are as shown in Fig. 2(a),

then the limits on x are x_f to $-X_0$ while X varies from X_0 to 0. In Fig. 2(b) x goes from x_f to $-X_0$ while X goes from 0 to $-X_0$. In Fig. 2(c) x goes from x_f to $-X_0$ while X varies from $-X_0$ to 0. In Fig. 2(d) x goes from x_f to $-X_0$ while X varies from 0 to X_0. X_0 is the radius of the base. The configuration shown in Fig. 2(a) has the widest aperture. The graphs are based on exact calculations for normal incidence on a conical concentrator. The light that reaches the base without reflection must be added to the desired concentration factor.

Two-Dimensional Geometry The preceding equations are modified slightly when light impinges upon a cylindrical reflector with axis perpendicular to the direction of the incident light. The heated surface may be a pipe, for example, parallel to the axis of the reflecting cylinder. We shall write the differential equation of the reflecting surface when the receiver surface is flat. We shall require uniform heating.

The only equation which is changed is the energy balance equation, since the reflector is no longer axially symmetric about an incoming axis. The variable x is now the Cartesian x-coordinate perpendicular to the axis of the trough.

Thus for energy balance

$$\int_{X_0}^{X} E[1 + (dZ/dX)^2]^{1/2} \, dX = \int_{x_0}^{x} \sigma \, dx.$$

(5)

For a flat surface, $dZ/dX = 0$. When the surface is filled in accordance with Fig. 1(a), that is, $X_0 = 0$ when $x = x_0$, we have $X = \sigma(x - x_0)/E_0$. E_0 has the value $E_0 = (x_f - x_0)/X_f$, where x_f is the maximum width of the entrance pupil and X_f is the half-width of the collector. The ray trace equation (3) applies as before; ℓ is now given by

$$\ell = (z - Z_0)/\{x - \sigma(x - x_0)/E_0\}. \quad (5a)$$

The differential equation for the surface is Eq. (4) as before.

In the preceding discussion the sunlight is assumed to be normal to the aperture. It is possible to specify a given concentration when the sunlight is oblique to the symmetry axes. As the rays approach the symmetry axis the distribution of energy will change. As a limiting case, for the trough, one can specify a delta function concentration along one edge of the base of the trough when the sun rays make some angle, usually 10% or less, with the normal to the base. In this case the reflecting surfaces become parabolas and form the basis for the compound parabolic concentrator.[10] The edge of one side is the focal line for the parabolic surface forming the opposite side. Maximum energy reaches the base, but a hot strip is initially formed along the edge.

If the incident light is parallel to the plane of symmetry of the trough but makes an angle ϕ with respect to the normal to the base, the distribution over the base is attenuated by the obliquity factor $\cos \phi$ but is otherwise unaffected.[11]

DONALD G. BURKHARD

References

1. Burkhard, D. G., Strobel, G. L., and Burkhard, D. R., *App. Opt.* 17, 1870 (1978).
2. Burkhard, D. G., Strobel, G. L., and Shealy, D. L., *App. Opt.* 17, 2431 (1978).
3. Winston, R., *J. Opt. Soc. Am.* 60, 245 (1970).
4. Jones, R. E., Jr., *J. Opt. Soc. Am.* 67, 1594 (1977).
5. Schruben, J. S., *J. Opt. Soc. Am.* 64, 55 (1974).
6. Elmer, W. B., "The Optical Design of Reflectors," New York, John Wiley, 1980.
7. Burkhard, D. G., and Shealy, D. L., *Solar Energy* 17, 221 (1975).
8. Rhodes, P. W., and Shealy, D. L., *App. Opt.* 19, 3545 (1980).
9. Han, C. Y., Ishii, Y., and Murata, K., App. Opt. 22, 3644 (1983).
10. Welford, W. T., and Winston, R., "The Optics of Nonimaging Concentrators," New York, Academic Press, 1978.
11. Strobel, G. L., and Burkhard, D. G., *Solar Energy* 20, 25 (1978).

Cross-references: LASER; OPTICS, GEOMETRICAL; OPTICS, GEOMETRICAL, ADVANCED; REFLECTION; SOLAR ENERGY UTILIZATION.

SOLAR ENERGY SOURCES

In the past 50 years the birth, the evolution and death of stars have been actively studied. It is now believed that stars form from the interstellar medium. At times hydrogen atoms assemble into condensed clouds, or protostars, under the influence of the gravitational force. This theoretical picture has been confirmed by the observation of localized hot spots of infrared radiation inside regions containing giant molecular clouds. Some of these clouds are so large that they contain thousands of solar masses of gas and dust.

The localized gravitational contraction continues until the pressure and temperature at the center of the protostar become sufficient to ignite thermonuclear reactions in which hydrogen nuclei combine to form helium nuclei. In this fusion process a large amount of energy is released. The star reaches equilibrium when the outward radiation pressure from γ-rays given off in the nuclear reactions is equal to the gravitational force causing the contraction. For a star such as the sun the period of contraction lasts about 100 million years. During this hydrogen burning stage a star is considered to be on the main sequence portion of a Hertzsprung-Russel diagram.[1] Our sun has been a main sequence star for 4.5 billion years and should go on burning hydrogen in this manner for another 5 billion years.

Extremely high temperatures are needed to maintain this hydrogen fusion process, probably above 10 million K. This constraint on the temperature, combined with the requirement for high densities restricts the energy production to the inner 25% of the sun. Here the temperature is probably of the order of 15 million K and the density ranges from about 160 g/cc at the center to about 20 g/cc at the edge of the core region. While 72% of the total mass of the sun is hydrogen, only about 10% of the hydrogen will be used up during the sun's 10-billion-year period on the main sequence.

The pioneering work of Bethe[2] and others over 40 years ago led to our present understanding of the nuclear reactions occurring in the core of the sun. Energy is liberated through several different networks of nuclear reactions which have the end result of combining four hydrogen nuclei into one helium nucleus. The dominant energy production process is thought to be the following proton-proton chain:

PP I: $^1H + {}^1H \to {}^2D + \beta^+ + \nu,$

$^2D + {}^1H \to {}^3He + \gamma,$

$^3He + {}^3He \to {}^4He + 2{}^1H.$

The first reaction involves a positive electron and a neutrino and takes place because of the weak force in nature. The slowness of this reaction process means that this reaction has never been observed in a laboratory experiment. The first and second reactions take place twice for each one of the third reactions. The energy released in the fusion of four protons into helium is 26.7 MeV, with 0.5 taken away by the two neutrinos.

The reaction mechanism PP I is believed to account for about 91% of the sun's energy production. The third step of PP I is not the only way to form ^4He. Occasionally (about 8% of the time) ^4He is made through intermediate steps involving the formation of ^7Be and ^7Li:

PP II: $^3He + {}^4He \to {}^7Be + \gamma,$

$^7Be + \beta^- \to {}^7Li + \nu,$

$^7Li + {}^1H \to 2{}^4He.$

The total energy released in this set of reactions is also 26.7 MeV, including the energy taken away by the neutrino.

^3He can also interact with ^4He as in the first step of PP II, but then proceed through ^8Be as an intermediate state:

PP III: $^3He + {}^4He \to {}^7Be + \gamma,$

$^7Be + {}^1H \to {}^8B + \gamma,$

$^8B \to {}^8Be + \beta^+ + \nu,$

$^8Be \to 2{}^4He.$

This process occurs less than 1% of the time, so that it is of minor importance from the standpoint of energy production in the sun. However, the highly energetic (14.1 MeV) neutrinos from ^8B are of major importance to an ongoing test of the reaction processes in the sun. To date only about one-third of the predicted neutrino flux at the earth is observed.[3] Further experimental tests of this highly important method of looking at the sun's interior are being considered.

Hydrogen burning can also take place in the sun through a cycle involving carbon, nitrogen, and oxygen:

$$^{12}C + {}^1H \to {}^{13}N + \gamma,$$

$$^{13}N \to {}^{13}C + \beta^+ + \nu,$$

$$^{13}C + {}^1H \to {}^{14}N + \gamma,$$

$$^{14}N + {}^1H \to {}^{15}O + \gamma,$$

$$^{15}O \to {}^{15}N + \beta^+ + \nu,$$

$$^{15}N + {}^1H \to {}^{12}C + {}^4He.$$

This carbon cycle is a minor contributor to the energy production for the sun, but it is an important source in more massive stars. For these reactions to occur, the temperature in the star's interior must reach more than 20 million K.

What happens when all the hydrogen in the core is converted to helium? As the energy produced in these primary fusion reactions decreases, the radiation pressure from the gamma rays given off is no longer sufficient to counteract the force of gravity and the star's core will contract. As this happens, gravitational potential energy is converted to thermal energy and the temperature rises. This heats up a shell of hydrogen around the core, and the star expands. As the radius of the star increases its temperature decreases, and it takes on the characteristics of a red giant. In about 500 million years the radius of the sun will be larger than that of the earth's orbit, it will have luminosity of the order of 10,000 times its present value, and its surface temperature will be about 4000 K. The core temperature of 100 million K allows three helium nuclei to fuse and form carbon. This helium burning process rapidly uses up the available helium and the star then goes on to the final stages of its existence.

Sunspots at the surface of the sun are manifestations of the dynamic activity taking place inside. Sunspot activity varies from a maximum to a minimum in 5.5 years. Sunspots are associated with a variety of complex phenomena such as magnetic fields, solar flares, and fluctuations in the solar wind of electrified particles from the sun. The origin of sunspots is due to some combination of the sun's magnetic field with its nonuniform rotation. The British astronomer E. W. Maunder discovered that not only do sunspot numbers vary during a cycle, but their location also changes. When the sunspot locations are plotted as a function of time, the migration of spots toward the equator gives rise to a "butterfly" pattern.[5]

One suggestion that has been advanced to account for the absence of solar neutrinos is that the sun is presently undergoing a period of reduced power generation. In 1893 Maunder, after a search through old books and journals, postulated that the sun was not the regular and predictable star that had previously been believed.[6] He found that almost no sunspots were observed during the 70-year period from 1645 to 1715. This study has been verified and extended using a method based on the analysis of the ratio of ^{14}C to ^{12}C in tree rings. Over the past 5000 years it appears that there have been at least 11 other similar minima in the sunspot activity. A detailed comparison of this long-term cycle in sunspot activity with climatological information shows that the sunspot minima appear to correlate very well with lower temperatures in the earth's atmosphere.[7] The Maunder minimum time period, for example, was associated with a period of unusual cold in Europe.

In conclusion, we note that the sun will continue to provide solar energy for the remainder of our existence on earth. Our knowledge of the processes taking place in the sun's interior has grown tremendously in the past half-century. But advances in observational techniques have also raised a number of new and interesting features. Obtaining an understanding of these new dynamical phenomena will keep scientists occupied for many years to come.

DAVID K. MCDANIELS

References

1. Harwit, M., "Astrophysical Concepts," New York, John Wiley & Sons, 1973.
2. Bethe, H., "Energy Production in Stars," *Phys. Rev.* 55, 434 (1939).
3. See the article NEUTRINO by F. Reines in this Encyclopedia.
4. Zeilik, M., "Astronomy: The Evolving Universe," New York, Harper and Row, 1979.
5. Wilson, O. C., Vaughan, O. C., and Mihalas, D., "The Activity Cycles of Stars," *Scientific American*, January, 1981.
6. McDaniels, D. K., "The Sun: Our Future Energy Source," New York, John Wiley & Sons, 1979.
7. Eddy, J. A., "The Case of the Missing Sunspots," *Scientific American*, May, 1977.

Cross-references: ASTROPHYSICS, NEUTRINO, NUCLEAR REACTIONS, SOLAR PHYSICS.

SOLAR ENERGY UTILIZATION

The use of solar energy is becoming increasingly important both on earth, because of the rapidly diminishing supplies of fossil fuels, and in man's expanding adventures into space.[1] The problem of economically using solar energy for various terrestrial needs has challenged us for many years, but there are still only a few economically attractive applications. For space applications, where reliability and weight are generally much more important than cost, silicon photovoltaic power supplies already have found extensive use.

The Sun as an Energy Source Our sun is a typical main sequence dwarf of spectral class G-2. In almost every respect (size, luminosity, mass, etc.), it is an average star. The mean distance from the earth is 93 004 000 miles, at which distance the sun subtends an angle of 31 minutes 59 seconds.

The sun's total radiation output is approximately equivalent to that of a blackbody at 10 350°R (5750 K). However, its maximum intensity occurs at a wavelength which corresponds to a temperature of 11 070°R (6150 K) as given by Wien's displacement law. Figure 1(a) presents the intensity of solar radiation outside the earth's atmosphere as a function of wavelength. The total irradiance at the mean sun-earth distance is approximately 442 Btu/hr ft² (0.140 watts/cm²). Figure 1(b) shows how the energy in the sun's spectrum is distributed.

Sunlight passing through the atmosphere suffers both absorption and scattering. The solar irradiance reaching the earth depends upon the total air mass through which the sun's radiation passes and, of course, weather conditions. Figure 1(c) shows a typical plot of the solar irradiance at the earth's surface. For some applications, the radiation from the sky is also important. The day sky exhibits wide variations in brightness and in spectral distribution, depending upon the sun's position, weather conditions, and the "receiver's" orientation. Table 1 presents tyical values of direct and diffuse solar radiation for several different atmospheric conditions. A great deal of study has been made of the variation of the solar irradiance at the earth's surface with weather, atmospheric conditions, zenith angle of the sun, etc.[2]

Uses for Solar Energy Table 2 lists most types of application of solar energy of interest today. They are grouped according to the form of energy required by the application. As an example, supplying electrical power for space vehicles requires a conversion from solar to electrical energy which may be a one-step process with photovoltaic cells (solar cells) or a multistep process. A typical multistep process might involve conversion of solar energy to thermal energy by using a parabolic mirror to heat a working fluid in a boiler, then conversion to mechanical energy using a Stirling engine, and then conversion to electrical power using a generator.

Considerations in the Conversion of Solar Energy to Other Forms Table 2 indicates that most of the applications of solar energy require that the energy be delivered in one of four forms: thermal, electrical, mechanical, or chemical. Table 3(a) shows several methods of directly converting solar energy to these other forms.

Table 3(b) shows additional conversion steps that might be used in a multistep process. The devices described in Table 3(b) make use of processes which are well known and not restricted to solar energy utilization. Those in Table 3(a), on the other hand, are of prime interest to the present discussion.

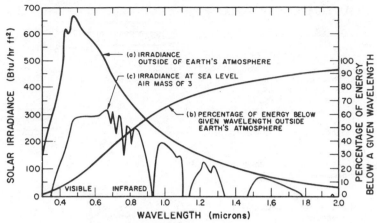

FIG. 1. Spectral distribution of solar energy.

Conversion of Solar Energy to Thermal Energy. Many of the systems which utilize solar energy first collect the energy and its heat. A solar heat collector intercepts radiation, converting this to thermal energy, and transfers this heat to a working fluid. Some collectors use mirrors or lenses to achieve high flux densities (concentrating collectors) while others do not (flat-plate collectors).

Flat-plate Collectors. The most common flat-plate collector consists of a metal plate painted black on the side facing the sun and thermally insulated on the edges and back. Above the plate, spaced on inch or so apart, are one or more glass or plastic covers to reduce upward heat losses. The absorbed energy is transferred to water or some other working fluid in tubes which are in thermal contact with the absorber plate or by circulating air past the absorber. Figure 2 displays several flat-plate collector designs. Design 2(b) is the conventional type described above; 2(a) uses air instead of water for heat transfer. In 2(c), the glass-shingle collector, heat is absorbed on the blackened bottom third of tilted glass plates and transferred to the air drawn down through these shingles.

TABLE 1. VALUES OF DIRECT AND DIFFUSE SOLAR RADIATION[a]

Solar Altitude α (degrees)	Optical air-mass path,[b] m ~ cscα	Standard, Cloudless Atmosphere			Industrial, Cloudless Atmosphere			Through Complete Overcasts, Blue Hill, Average Total Insolation on Horizon			
		Direct, Perpendicular Radiation I (B/hr ft^2)	Diffuse on Horizontal I_z, Difference (B/hr ft^2)	Total on Horizontal W_z, (B/hr ft^2)	Direct, Perpendicular Radiation I (B/hr ft^2)	Diffuse on Horizontal I_z Difference (B/hr ft^2)	Total on Horizontal W_z, (B/hr ft^2)	Cirrostratus W_z, (B/hr ft^2)	Altocumulus W_z, (B/hr ft^2)	Stratocumulus W_z, (B/hr ft^2)	Fog W_z (B/hr ft^2)
5	10.39	67	7	13	34	9	12	–	–	–	–
10	5.60	123	14	35	58	18	28	–	–	15	10
15	3.82	166	19	62	80	24	45	50	35	25	15
20	2.90	197	23	90	103	31	64	70	50	35	20
25	2.36	218	26	118	121	38	89	95	65	40	20
30	2.00	235	28	146	136	44	112	120	75	50	25
35	1.74	248	30	172	148	48	133	145	90	60	30
40	1.55	258	31	197	158	52	154	165	105	70	35
45	1.41	266	32	220	165	55	172	185	115	80	40
50	1.30	273	33	242	172	58	190	205	130	85	40
60	1.15	283	34	279	181	63	220	235	150	100	45
70	1.06	289	35	307	188	69	246	260	160	110	50
80	1.02	292	(35)	(322)	195	–	–	–	–	–	–
90	1.00	294	(36)	(328)	200[c]	–	–	–	–	–	–

[a]Data assembled by F. A. Brooks. "B" in the headings stands for British thermal units (Btu).
[b]Smithsonian Meteorological Tables, 6th rev. ed., p. 422, 1951.
[c]192 would be more consistent with the curve from 70° down.

TABLE 2. SOME APPLICATIONS OF SOLAR
ENERGY

Form of Energy Required	Application
Thermal	Concentration of brine
	Cooking
	Cooling and refrigeration
	Dehumidifying of buildings
	Distillation of water
	Drying of materials, fruits, grain, etc.
	Heating of buildings
	Heating of water
	Heating of materials to high temperatures
	Salt making
Electrical	Supplying power for special uses
	Supplying space vehicle power
Mechanical	Pumping water
Chemical	Producing food
	Producing fuel
Light	Natural lighting
	Phosphorescent markers

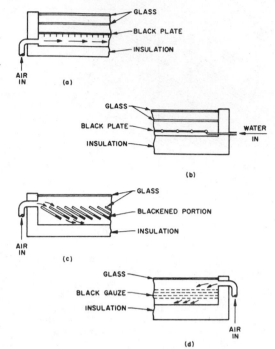

FIG. 2. Four flat-plate collector designs.

In 2(d), solar energy is absorbed on black gauze. The flat-plate collector is a simple, rugged device which, without orientation mechanisms, can efficiently collect solar energy at moderate temperature levels.

The results of some comparative calculations

TABLE 3. MATRIX OF A FEW
ENERGY CONVERSION
DEVICES

(a) Direct Conversion from Solar Energy

To \ From	Solar Energy
	Plants
Chemical	Photovoltaic cells
Electrical	Photon "sails"
Mechanical	Flat-plate collectors
Thermal	Concentrator-boilers
	Solar ponds

(b) Other Conversion Steps

To \ From	Chemical	Electrical	Mechanical	Thermal
Chemical	—	Electrolysis	(Mechanical activation of chemical processes)	(Endothermic reactions) (Thermal dissociation)
Electrical	Fuel cell battery	—	Generator	Thermopiles Thermionic diodes MHD devices
Mechanical	(Equilibrium volume and pressure)	Motor Solenoid	—	Turbine Positive displacement engine
Thermal	(Combustion)	Resistance heaters	Friction brake	—

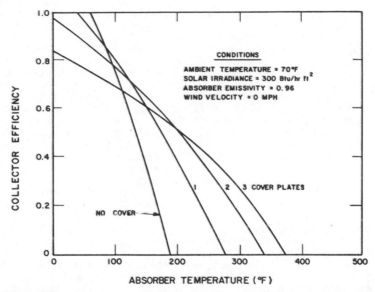

FIG. 3. Calculated performance of flat-plate collectors showing effect of multiple glass covers.

of conventional collectors are shown in Fig. 3. The emissivity and absorptance of the absorber plate was assumed to be 0.96, and the curves were plotted for zero, one, two, and three cover plates of typical single-strength window glass.[3] It is seen that for high efficiencies (with correspondingly low collector temperatures), it might be preferable to have only one or two covers. As an example, a two-cover collector is preferable to a three-cover collector out to a temperature of about 200 F (for the case analyzed); it is only at higher temperatures that the decrease in upward losses made by using one more cover is greater than the transmission loss of that cover.

Concentrating Collectors. An optical system using lenses or mirrors can be used to concentrate solar radiation into a very small area. If this energy is received into a cavity or absorbed on a metal plate, heat is generated and very high temperatures may be obtained.

The concentration ratio C is defined as the ratio of the flux density within the image of the sun formed by the optical system to the actual flux density reflected from the mirror. For a parabolic mirror, C is given by[4]

$$C = \frac{\sin^2 \theta}{\tan^2 \phi/2} \qquad (1)$$

where θ is the rim angle of the mirror as defined by the sketch in Fig. 4 and ϕ is the angle subtended by the sun from the earth. Using $\phi = 32'$,

$$C = 46.2 \times 10^3 \sin^2 \theta. \qquad (2)$$

Figure 4 plots Eq. (2) and also a parameter called concentration efficiency which is defined

as the ratio of the power received, within the sun's image to the total power reflected by the mirror. For the parabolic mirror, the concentration efficiency is given by

$$\eta_c = \left(\frac{1 + \cos \theta}{2}\right)^2. \qquad (3)$$

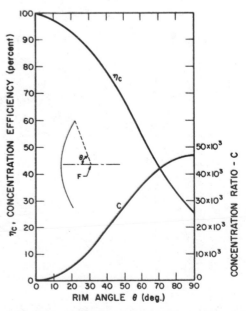

FIG. 4. Concentration ratio and concentration efficiency as a function of rim angle for a paraboloidal reflector.

The concentration ratio and the concentration efficiency are strongly influenced by the geometric perfection of the mirror and the location of the absorber with respect to the focus of the mirror.

Consider an example where the energy reflected from a parabolic mirror is collected by an ideal cavity receiver having an opening equal to the image diameter of the sun. The rate at which heat energy can be withdrawn from the cavity, P, can be calculated by substracting the radiative heat losses from the input flux. The result is:

$$P = \eta_c A \left[rH - \frac{\sigma T^4}{c} \right] \qquad (4)$$

where A is the projected area of the mirror, r is its reflectance and H is the solar irradiance. T is the temperature of the cavity. Figure 5 uses Eq. (4) to plot the ratio of P/A as a function of cavity temperature for several rim angles. The intersection of the curves with the abscissa show the maximum attainable temperature for various rim angles, corresponding to $P = 0$ in Eq. (4).

A variety of optical systems has been studied for use as solar concentrators with various absorbers.[5,6] The parabolic mirror-cavity receiver example, however, is indicative of the ultimate performance attainable.

Conversion of Solar Energy to Electrical Energy. The PHOTOVOLTAIC EFFECT, particularly in silicon, has become very important to solar energy utilization. In the silicon "solar cell," photons create hole-electron pairs by removing electrons from valence bonds. This occurs in the junction region between p-type and n-type silicon where the electric field causes a current to flow in the cell. Equilibrium is restored by charge flowing around an external circuit through a load resistor.

Each silicon nucleus shares its four valence electrons with neighboring nuclei forming a stable tetrahedral crystal. In the junction region, most of the electrons are in the lower filled band, below the forbidden energy band. The electron may be thought of as either bound to a crystal lattice region or, with the addition of enough energy, free to move about and conduct electricity. The width of the forbidden region, or energy gap, represents the threshold energy necessary to remove an electron from the bound position to the conduction band. For silicon, the energy gap is approximately 1.1 eV. It follows that only photons of 1.1 eV or over can create hole-electron pairs in silicon. Higher-energy photons can create hole-electron pairs with the excess energy being dissipated as heat (see ENERGY LEVELS and SEMICONDUCTORS).

Since, at most, one hole-electron pair can be created for each photon possessing sufficient energy to do so, the ultimate efficiency that might be achieved depends upon the threshold energy at which the hole-electron conversion takes place and upon the spectrum of the radiation. Figure 6 is the plot of the maximum power conversion efficiency as a function of threshold level. It assumes that all of the photons whose energy is greater than or equal to the threshold energy are converted into hole-electron pairs at the potential of the energy gap.

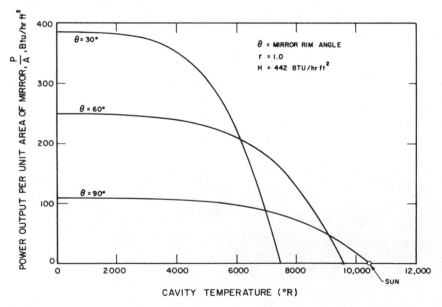

FIG. 5. Maximum power output from an ideal parabolic mirror-cavity absorber as a function of operating temperature.

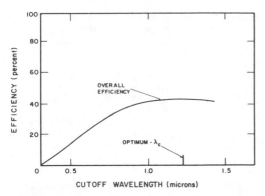

FIG. 6. Maximum solar conversion efficiency of an ideal quantum converter.

Some of the factors which limit the efficiency in a practical cell include:

(1) Some hole-electron pairs recombine before they can be separated by the field in the junction,

(2) Reflection losses from the front surface,

(3) Loss due to electrical resistance within the cell, and

(4) Loss due to leakage across the barrier (diode current).

In full sunlight, a good single-crystal silicon solar cell will operate at 12–16%; 20% cells have been demonstrated. Efficiencies well over 20% have been demonstrated for single-crystal gallium arsenide cells, although these are at present more expensive to manufacture than silicon.

For terrestrial applications, two basically different approaches are being pursued to reduce the cost per unit power output:

• Developing low-cost ways to manufacture single-crystal materials, including growth of silicon ribbons,

• Improving the efficiency of polycrystalline and thin-film cells, which are believed to be inherently cheaper to manufacture.

Thin-film cells have been demonstrated using silicon, silicon/fluorene/hydrogen compounds and mixtures of cadmium sulfide with copper sulfide.[7,8]

Figure 7 is a photograph of one of the solar panels used on the Ranger Spacecraft developed for NASA's Jet Propulsion Laboratory. The panel contains 72 strings of 68 cells each to provide 100 watts output at approximately 30 volts. The use of optical filters to cover the cells results in a lower cell temperature, hence higher efficiency, and provides improved resistance to radiation damage.[9]

Conversion of Solar to Mechanical Energy. Solar radiation incident upon a reflecting surface exerts a radiation pressure of approximately 0.8×10^{-4} dynes/cm^2 or 1.7×10^{-7} lb/ft^2. Although impractical for terrestrial applications, the use of radiation pressure for space propulsion (solar sailing) may prove to be worthwhile for some missions.

Conversion of Solar Energy to Chemical Energy. It has been estimated that only a few

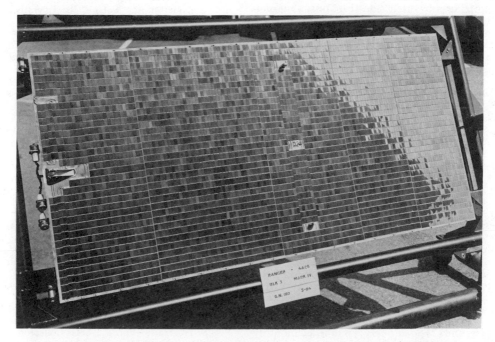

FIG. 7. Ranger spacecraft solar cell panel. (*Photograph courtesy Electro-Optical Systems, Inc.*

tenths of a per cent of a year's supply of incident solar energy is stored in an average farm crop. Research into the mass production of algae has been directed toward tenfold greater yields; research in other types of plants, toward the end of increasing photosynthetic efficiency, is also being conducted.[10]

The existence of photosynthesis raises the hope that other photochemical reactions will be found which can be effectively used to obtain chemical energy from the sun. What is needed is a reversible endothermic photochemical reaction which can utilize a large portion of the sun's spectrum. Most photochemical reactions evolve heat instead of absorbing it, or they reverse so quickly that the energy is lost even during exposure to sunlight.

Solar-hydrogen generators have long been of interest because of the potential advantages of using hydrogen as a fuel, either directly or in a fuel cell to produce electricity. Water is photochemically decomposed into hydrogen by ultraviolet light, though at very low efficiency. Two approaches that promise higher efficiency are[11,12,13]

- Chlorophyll-like metal organic compounds that create hydrogen through photosynthesis,
- Photovoltaic semiconductor cells that operate in water or in other hydrogen compounds to generate sufficient potential for hydrolysis.

The first approach uses molecules like zinc metalporphyrine as catalytic agents in the reaction. Present research efforts are directed to minimizing unwanted side reactions, thus permitting hydrogen production to continue for months or years. The second approach uses a suitable liquid electrolye and semiconductor combination. The p-n junction may be formed either by diffusion or at the liquid-semiconductor interface (e.g., cadmium selenide with a sulfaselenide solution).[14]

Unfortunately, 50% of the sun's energy is at wavelengths too long to be of much use for photochemical reactions, since the energy per photon is too low. All molecules which undergo photochemical change have a minimum threshold energy required to create a bond rupture. If a quantum is absorbed with energy above this threshold, the excess energy is dissipated (as kinetic, vibrational, etc.) and usually does not contribute to the conversion to chemical energy. The maximum theoretical efficiency of conversion of solar to chemical energy can be evaluated with the help of Fig. 6 providing the cutoff wavelength is known.

Current Progress Interested readers are directed to the references listed and current articles in two journals: (1) the *Journal of Solar Energy Engineering*, published quarterly by the American Society of Mechanical Engineers; and (2) *Solar Energy*, published monthly by Pergamon Press for the International Solar Energy Society. Progress on Federal solar energy programs is reported by the Solar Energy Research Institute in their annual reviews, published by the U.S. Government Printing Office.

In addition, solar energy handbooks have begun to appear which represent useful gatherings of information.[15,16]

A. M. ZAREM
DUANE D. ERWAY

References

1. Butte, Ken, and Perlin, John, "A Golden Thread: 2500 Years of Solar Architecture and Technology," New York, Van Nostrand Reinhold Company, 1980.
2. Becker, C. P., "Solar Radiation Availability on Surfaces in the U.S. as Affected by Season, Orientation, Latitude, Attitude and Cloudiness," New York, Arno Press, 1979.
3. Zarem, A. M., and Erway, D. D., "Introduction to the Utilization of Solar Energy," pp. 89–100, New York, McGraw-Hill Book Co., 1963.
4. Hiester, N. K., Tietz, T. E., Loh, E., and Duwez, P., "Theoretical Considerations on Performance of Solar Furnances," *Jet Propulsion* 27, 507 (1957).
5. Giutronich, J. E., "The Design of Solar Concentrations using Toroidal, Spherical, or Flat Components," *Solar Energy J.* 7(4), 162–166 (1963).
6. Masterson, K. D., "Optics Applied to Solar Energy IV," *Proc. Soc. of Photo-Optical Instrumentation Engineers* V, 161 (1979).
7. Green, Martin A., "Solar Cells: Operating Principles, Technology & Systems Applications," Englewood Cliffs, NJ, Prentice-Hall, 1982.
8. Burch, Charles G., "Solar Energy Comes out of the Shadows," *Fortune*, pp. 67–75, September 24, 1979.
9. Hamilton, Robert C., "Ranger Spacecraft Power System," in Snider, N. W. (Ed.), "Space Power Systems," p. 19, New York, Academic Press, 1961.
10. Wise, Donald L., "Probing the Feasibility of Large Scale Aquatic Biomass Energy Farms," *Solar Energy* 25(5), 455–457 (November 5, 1981).
11. Ohta, T., "Solar-Hydrogen Energy System: An Authoritative Review of Water-Splitting Systems by Solar Beam and Solar Heat; Hydrogen Production, Storage, and Utilization," New York, Pergamon Press, 1979.
12. "Summary of Papers from the 3rd International Conference on Photochemical Conversion," *Science News*, 118, 105 (August 26, 1980).
13. Bryce-Smith, David (Ed.), "Photochemistry," Vol. 10, London, The Chemical Society, 1979.
14. "Wet Solar Cells," *Scientific American*, 241, 72 (October 1979).
15. Boer, K. W., "Sharing the Sun," New York Pergamon Press, 1976. (Vol. 1, "International and U.S. Programs;" Vol. 2, "Solar Collectors;" Vol. 3, "Heating and Cooling;" Vol. 4, "Systems Simu-

lation and Design;" Vol. 5, "Solar Thermal and Ocean Thermal;" Vol. 6, "Photovoltaics;" Vol. 7, "Agriculture Bismars, Wind;" Vol. 8, "Storage, Water Heater, Data Communications;" Vol. 9, "Socioeconomics and Cultural;" Vol. 10, "Business and Commercial Implications.")

16. Kreider, Jan F., and Frank Kreith (Ed.), "Solar Energy Handbook," New York, McGraw-Hill, 1981.

Cross-references: ENERGY LEVELS; INFRARED RADIATION; INTERNATIONAL SOLAR-TERRESTRIAL PHYSICS PROGRAM; IRRADIANCE (ILLUMINANCE) CALCULATIONS; PHOTOCHEMISTRY; PHOTOVOLTAIC EFFECT; REFLECTION; SEMICONDUCTORS; SOLAR CONCENTRATOR DESIGN, OPTICS OF; SOLAR ENERGY SOURCES; SOLAR PHYSICS; SOLAR TOTAL IRRADIANCE AND ITS SPECTRAL DISTRIBUTION; WORK, POWER, AND ENERGY.

SOLAR PHYSICS

The main activity of solar physics is the interpretation of the observed flow of energy away from the sun. Most of the sun's energy output appears as a nearly constant flux of electromagnetic radiation in the photographic, visual, and infrared regions of the spectrum, but extraordinary variations are characteristic of the x-, ultraviolet, and radio radiation. A small part of the sun's energy flows outward in highly variable streams of particles (mainly protons and electrons). Since the sun appears as a rather large disk, the observations forming the bases for the solar physicist's interpretations often refer to small areas of the solar surface, as well as the integrated radiation and particle streams from the entire sun. A reasonably good solar telescope can subdivide the apparent disk of the sun into about three million smaller elements of area. Streams of particles and radiation from each of these elements should be recorded, with all possible detail and precision, nearly continuously for a complete observational record. A close approach to such observational perfection probably is unattainable, perhaps it is unnecessary, but the fragmentary and incomplete nature of all solar observations continues to be a serious barrier in the search for a satisfactory general theory of the sun. Some progress has been made, but until an adequate theory is developed, it is best to consider the physics of the sun by reviewing the observations of possible importance.

The radius, mass, and luminosity of the sun are fundamental in the physical interpretation of the sun. Luckily, they can all be deduced with reasonable directness from observations. In principle, the radius of the sun can be obtained directly from angular and linear observations made on earth, but a number of practical difficulties limit the attainable precision. Somewhat devious methods, invoking gravitational theory, give more consistent results that converge on the value,

Radius of the sun, $R = 6.9598 \pm 0.007 \times 10^{10}$ cm

The mass of the sun is also derived from the application of gravitational theory: first, to the measurement of the mass of the earth; then, by way of the moon and planets, to the sun; with the result,

Mass of the sun, $M = 1.989 \pm 0.002 \times 10^{33}$ grams

A measurement of the total radiation received at the earth's distance from the sun is the basic observation from which its luminosity can be found (once the distance to the sun, and the radius of the sun are known). This measurement is extremely difficult and is seriously distorted by the earth's atmosphere, but the value

Luminosity of the sun, $L = 3.90 \pm 0.04 \times 10^{33}$ ergs/sec

can be derived from long, independent series of observations. The difficulties caused by the earth's atmosphere can be eliminated if observations are made from a space vehicle and measurements have been made in that way. After such data have accumulated over an interval of at least decades, perhaps the observed value of L can be evaluated with greater precision.

An additional parameter, or series of numbers, is essential for the construction of an adequate theoretical model sun. These are the abundances of the chemical elements. For convenience, abundances are stated in terms of the abundance of hydrogen. In the sun, the relative abundances of the important constituents are

$$\left.\begin{array}{l} \text{Hydrogen} = 1.00 \\ \text{Helium} = 0.23 \\ Z \text{ (all other elements)} = 0.02 \end{array}\right\} \begin{array}{l} Q, \text{ the relative} \\ \text{chemical} \\ \text{abundance} \end{array}$$

The values of L, M, R, Q, deduced from observation for the sun, are somewhere near the middle of ranges of these quantities as derived for other stars. It is, therefore, sometimes said that the sun is an average star, but for the stars inside an imaginary spherical shell surrounding the sun, with a radius so small that it is reasonably certain that all stars within the enclosed volume have been observed, a different interpretation is valid. Most of the stars close to the sun are cool, red, dwarf stars, probably the most numerous of all stellar varieties. Only ten of the fifty-five nearest stars can be seen with the eye alone. Only three of these stars are brighter than the sun, most are exceedingly faint. Stars brighter than the sun are extremely rare, and the sun is outstanding in all of its properties.

Although the sun's L, M, R, and Q can be fitted into a coherent physical theory which, starting with a plausible nuclear energy source contained within a relatively small volume surrounding the sun's center, can trace the outward flow of energy from the deep interior through complicated transformations until quantities and qualities of radiation in adequate but limited agreement with observation are predicted at the solar surface, there is an outstanding exception. The flux of neutrinos measured at the earth is only about half as large as it should be in order to agree with the flux to be expected from a proton-proton reaction that is otherwise suitable as the source of solar radiation.

The synoptic general agreement of solar theory with observation, and the persistence of discordant details is well attested by almost innumerable studies concerned with the three visible subdivisions of the sun: the photosphere, the chromosphere, and the corona.

The photosphere of the sun is the surface observed directly by the eye through a protective dense black glass screen. In very transparent skies with the help of special telescopes or at times of total solar eclipses, the chromosphere and corona may also be observed visually. All three of these layers show structures in widely different sizes and rates of change. The theoretical problems connected with the photosphere-chromosphere and the chromosphere-corona interfaces are nearly intractable, and the individual layers are understood only slightly better.

Nearly three centuries of telescopic observation of the solar photosphere define this part of the sun as that imaged on a photographic plate, or seen by the eye, using the light in a spectral band at least 1000Å wide, centered near 5000Å in the blue-green part of the spectrum. The portion of the sun thus recorded is a spherical shell, 5×10^8 cm thick, whose outer radius is the edge of the apparent solar disk, 7×10^{10} cm from the sun's center. Good photospheric photographs show a granulation composed of small (10^8 cm, or less) bright, circular, or hexagonal, structures all over the sun except extremely near the edge. These are the tops of convection cells that carry much of the solar energy. They are a few hundred degrees hotter than the five-thousand-plus degrees absolute that is consistent with the sun's assigned total luminosity. Occasionally, small dark spots (pores) appear among the granulations, and still more infrequently, a pore will develop into a larger dark complex, a sunspot region. Spot regions are some thousands of degrees cooler than the granulated photosphere in which they are immersed. They mark the locations of sizable magnetic fields, and the spot regions may cover some tenths of the sun's disk. Since the spot regions are nearly fixed on the solar surface, they may also be used as markers for the measurement of the rate of rotation of the sun.

The angular rate of rotation, deduced from sunspot, or spectroscopic, observations, varies from the sun's equator to its poles. It is greatest on the equator, there corresponding to a period of rotation of twenty-five days. At the pole the period of rotation is nearly thirty-five days.

Sunspots are the main sources of difficult problems of the photosphere. How can the refrigeration of the spot regions be explained? What is the origin of the magnetic fields? How is the distribution of angular velocities established, and how is it maintained?

The chromosphere is most easily observed at times of total solar eclipse. It appears as a red-purple narrow irregular ring just after the beginning and again just before the ending of totality. The average thickness of the chromospheric layer is 10^9 cm, but it is so tenuous that it is undetectable under conditions satisfactory for observation of the photosphere. For the daily observation of the chromosphere, a filter transmitting a spectral band not wider than 1 Å must be used to produce monochromatic solar images, and the center of the spectral band transmitted by the filter must be adjusted to coincide with the center of an emission line that appears strongly in the spectrum of the chromosphere. Nearly all observations of the chromosphere are made with light from the center of just two spectrum lines: the H-alpha line of hydrogen in the red part of the spectrum; and the K line of ionized calcium in the ultraviolet. The pictures show the chromosphere as a continually seething part of the solar atmosphere, subject to spectacularly sudden changes in the neighborhood of sunspots. The changes in the chromosphere and the dominating changes that occur in the underlying sunspots are considered together as the phenomenon of *solar activity*.

Long before regular observation of the chromosphere became possible, records of the numbers of sunspots had led to the discovery of a ten-year cycle of variation. As solar observation has become more nearly continuous, nearly every aspect of the sun's activity has revealed cyclical behavior closely synchronized with the variation in the numbers of spots. This is especially true for the changes in the chromosphere. The solar flares and extensive systems of solar prominences in the chromosphere are apparently organized and controlled by the magnetic fields rooted in the sunspot regions.

However, the smallest structures in the chromosphere are not obviously connected with sunspot activity. The smallest features in undisturbed solar areas far from the dominance of spot regions undergo damped oscillations that are nearly periodic with characteristic times of about three hundred seconds. Perhaps these motions are enforced by the somewhat slower changes observed in the photospheric granulations on which the chromospheric structures are based. These and other quasi-periodic motions in the sun's outer layers can be observed continuously for long intervals and analyzed, somewhat as earthquake data are used, to yield information about subsurface solar conditions.

Like the chromosphere, the corona can only be observed with the unaided eye at the time of a total solar eclipse, but with telescopic, spectroscopic, and other instrumental aids, it can be observed every day. It is the most extensive of the divisions of the sun's atmosphere, its outer limits lying somewhere beyond the distance of the earth from the sun where the corona becomes indistinguishable from the interplanetary medium. The corona changes both gradually and suddenly in rather sensitive connection with the spot regions. It is the principal source of the sun's x-rays, extreme ultraviolet, and radio emission, and the combination of satellite observations near the short-wavelength limit of the solar spectrum and radio observations at the other end of the spectrum results in an attractive qualitative picture of solar activity.

At times when solar flares flash out in active regions, generally near sunspots, observations in the x-ray and radio sections of the solar spectrum can be interpreted as indicating motion of a disturbance starting away from the sun at the beginning of the flare and moving outward through the chromosphere and corona. Gamma radiation has been detected in some flares—certain evidence for nuclear transformations occurring near the photosphere or chromosphere. Frequently, streams of particles associated with the flare activity reach earth, and are detected as cosmic rays and by their secondary effects such as terrestrial magnetic storms, but many of the changes induced in the corona by activity in spot regions seem to be connected with variations in local solar magnetic fields and not with streams of particles.

It may be evident from this synopsis that the broad results from intensive solar observations, covering more than a century, have been accompanied by coordinating solar theories which are most satisfactory for parts of the sun inaccessible to direct observation. The attainable precision in solar observation narrowly limits the range of acceptable theoretical solutions, and especially limits attempts to combine hypotheses describing solar details into consistent synoptic theory. The neutrino anomaly indicates clearly a need for a careful reconciliation between the solar nuclear energy sources and their physical environment. Furthermore, there are two apparently fundamental phenomena for which no theoretical explanation has appeared: (1) How is the observed variation of angular rotation from the sun's equator to its poles to be explained? (2) What is the theory that includes the entire range of sunspot phenomena?

New results, flooding from new observatories and the exploratory applications of rockets and space technologies to solar studies have emphasized the persistence of unsolved problems and the remarkable opportunities for contributions to observational and theoretical solar physics.

ORREN C. MOHLER

References

Kuiper, G. P., Ed., "The Sun," Chicago, University of Chicago Press, 1953.

Thomas, R. N., and Athay, R. G., "Physics of the Solar Chromosphere," New York, Interscience Publishers, 1961.

Smith, H., and E., "Solar Flares," New York, The Macmillan Co., 1963.

Gamow, G., "A Star Called the Sun," New York, Viking Press, 1964.

Bray, R., and Loughhead, R., "Sunspots," London, Longmans, 1964.

Zirin, H., "The Solar Atmosphere," Waltham, Mass., Blaisdell Publ. Co., 1966.

Tandberg-Hanssen, E., "Solar Activity," Waltham, Mass., Blaisdell Publ. Co., 1967.

Bray, R., and Loughhead, R., "The Solar Granulation," London, Chapman and Hall, Ltd., 1967.

Gibson, E. G., "The Quiet Sun," Washington, NASA 1:21:303, U.S. Govt. Print. Off., 1973.

Bahcall, J. N., and Davis, Jr., R., "Solar Neutrinos: A Scientific Puzzle," *Science* 191, 264 (1976).

Cross-references: ASTROMETRY, ASTROPHYSICS, INTERNATIONAL SOLAR-TERRESTRIAL PHYSICS PROGRAM, NEUTRINO, SOLAR ENERGY SOURCES.

SOLAR TOTAL IRRADIANCE AND ITS SPECTRAL DISTRIBUTION

Introduction The nature of the emission of radiation by the sun is of overriding importance to every life-form on earth. Interactions of the earth's atmosphere, oceans, and land masses with the sun's radiation completely determine the earth's weather and climate. Sustained variations in the total solar energy received by the earth from the sun, the *solar total irradiance*, would have significant effects on both. Variations of solar total irradiance were probably the cause of some of the many past climate changes. Variations of the spectral distribution of solar irradiance have significance for photochemical and thermodynamic processes in the earth's atmosphere.

In spite of its importance, solar irradiance has not been systematically measured in the past. Total irradiance observations with the precision required by climate studies did not begin until 1980. The solar spectrum in the 0.4–4.0 micrometer (μm) wavelength range, containing 90% of the total solar energy, has never been measured with the accuracy or precision required by modern atmospheric science.

The following sections will discuss the present knowledge of the absolute radiation scale, the solar total and spectral irradiance, and the significance of recent discoveries of solar irradiance variability.

Pyrheliometry and Radiation Scales The science of measuring total solar irradiance is *pyrheliometry*, derived from the Greek words *pyr* (heat), *helios* (sun) and *meter* (to measure). The instruments used to "measure the heat of the sun" are referred to as *pyrheliometers* or *radiometers*. Two basic types have been commonly used in solar observations. The first is an absolute or self-calibrating detector which can accurately relate radiation measurements to the International System of units (SI) by its standalone operation. The second basic type is a relative detector that simply provides a signal in response to solar heating and must be calibrated by comparison with a self-calibrating pyrheliometer or standard radiation source to produce a quantitative result in SI units.

Several radiation "scales" have been used during the 20th Century for referencing solar irradiance measurements to SI units. Each was based on the performance of electrically self-calibrated pyrheliometers, the most accurate standards of radiation measurement at the solar total irradiance level. The Angstrom scale, defined in 1905 and based on the performance of the Swedish Angstrom pyrheliometer, was widely used in Europe to reference meteorological solar irradiance measurements. Electrically self-calibrated cavity pyrheliometers developed in the U.S. at the Smithsonian Institution were the references for the 1913 Smithsonian scale used as the standard for their long-term program to detect solar variability from mountaintop observatories. Solar observations on the two scales differed by about 5% until a downward revision of the Smithsonian scale in 1932 reconciled them to within 2.5%. The Angstrom 1905 scale was perpetuated under the name of the International Pyrheliometric Scale of 1956 (IPS56), adopted by international agreement in that year to provide a common reference for solar irradiance measurements during the International Geophysical Year (1957–58). The IPS56 was employed as the principal international reference for solar observations from 1956 until the mid-1970s.

A new generation of electrically self-calibrating cavity pyrheliometers was developed in the latter half of the 1960s at several laboratories in the U.S. and Europe. Radiation scale experiments in the U.S. at the Jet Propulsion Laboratory's Table Mountain Observatory (1968–69) discovered a −2.2% systematic error in the IPS56, placing the absolute radiation scale within 0.5% of the 1932 revision of the Smithsonian scale. The error was confirmed by the First International Comparison of Absolute Cavity Pyrheliometers, convened at the Physical Meteorological Observatory of Davos

(PMOD), in Switzerland, in 1975. A new solar irradiance reference scale was defined that incorporated this result, the World Radiometric Reference (WRR), defined by the average performance of five of the new pyrheliometers, as shown in Table 1. Measurements referenced to the WRR are uncertain by less than 0.3% in SI units. Newer versions of the WRR defining pyrheliometric instrumentation developed since 1975 have the theoretical capability of defining the radiation scale at the solar total irradiance level with less than ± 0.1% uncertainty. Experimental verification of this achievement is the present focus of pyrheliometry.

Solar Total Irradiance *Long-Term Solar Variability*. A systematic long-term solar variability monitoring program was conducted by the Smithsonian Institution during the first half of the 20th Century. Their measurements from mountaintop observatories, limited to a long-term precision of about 1% by fluctuations in atmospheric transmittance, provided no clear evidence of solar variability.

Experiments by the U.S. and U.S.S.R. on high-altitude aircraft and balloons in the 1960s were unable to produce unambiguous evidence of solar irradiance variability despite decreased uncertainties in atmospheric transmittance. The precision and accuracy of their results was not significantly better than the Smithsonian ground-based measurements, a limitation imposed mainly by the failure to employ self-calibrated flight instrumentation.

The results from solar total irradiance flight experiments that have contributed significantly to the present understanding of solar variability are described below and summarized in Table 2. The solar total irradiance outside the earth's atmosphere (S) is expressed in units of

TABLE 1. RESULTS OF THE FIRST INTERNATIONAL COMPARISON OF ABSOLUTE CAVITY RADIOMETERS (1975). The relative performance for the five cavity pyrheliometers whose average weighted results are used to define the World Radiometric Reference Scale (WRR) are shown.

Instrument	Developer	Performance Relative to WRR (%)
ACR 310	Willson	+0.04
PMO 2	Frohlich	+0.10
ACR 311	Willson	+0.11
PACRAD III	Kendall	−0.19
CROM	Crommelynk	−0.20

The affiliations of the developers of the instruments are: Kendall and Willson, Jet Propulsion Laboratory, California Institute of Technology, U.S.; Frohlich, Physical Meteorological Observatory at Davos, Switzerland; Crommelynk, Royal Meteorological Institute of Belgium.

TABLE 2.   EXPERIMENTAL DETERMINATIONS OF THE 1 A.U. TOTAL SOLAR IRRADIANCE BETWEEN 1969 AND 1982.

Year	Experiment Platform	Instrument Name	Instrument Type*	Result (W/m^2)	Variation from ESCC Avg. (%)**
1969	JPL Balloon	ACR 301	ESCC	1366.4	−0.08
1969	NASA Mariner 6, 7	TCFM	ESCC/P	1363.–	−0.33
1976	NIMBUS 6/ERB	ERB Ch. 3	Thermopile	1389.–	1.57
1976	NASA Rocket	ACR 402A	ESCC	1368.1	0.05
1976	NASA Rocket	ACR 402B	ESCC	1367.6	0.01
1976	NASA Rocket	PACRAD	ESCC/PF	1364.–	−0.25
1976	NASA Rocket	ESP	ESCC/PF	1369.0	0.11
1978	NIMBUS 7/ERB	HF***	ESCC/P	1376.0	0.62
1978	NASA Rocket	ACR 402A	ESCC	1367.6	0.01
1979	WRR Balloon	PMO6-9	ESCC	1366.0	−0.11
1980	SMM/ACRIM I	ACRIM A	ESCC	1368.2	0.05
1980	SMM/ACRIM I	ACRIM B	ESCC	1367.4	<−0.01
1980	SMM/ACRIM I	ACRIM C	ESCC	1367.8	0.02
1980	NIMBUS 7/ERB	HF***	ESCC/P	1376.0	0.62
1980	NASA Rocket	ACR 402A	ESCC	1367.6	0.01
1980	NASA Rocket	ACR 402B	ESCC	1368.7	0.09
1980	NASA Rocket	HF	ESCC/PF	1376.0	0.62
1980	NASA Rocket	ESP	ESCC/PF	1385.–	1.28
1980	WRR Balloon	PMO6-9	ESCC	1366.8	−0.04

SUMMARY

Type of Flight Experiment	Average 1 A.U. Irradiance (W/m^2)	Standard Deviation (1 sigma in %)
ESCC	1367.5	0.06
ESCC, ESCC/P, ESCC/PF	1368.4	0.26
ALL	1370.5	0.50

*Experiment approach for relating observations to the International System of Units listed in descending order of accuracy: ESCC = electrically self-calibrated cavity fully implemented in flight; ESCC/P indicates partial implementation of ESCC in flight; ESCC/PF indicates pre-flight electrical self-calibration only; the NIMBUS 6/ERB thermopile was a flat-plate sensor without self-calibration capability.

**The average of the 11 ESCC observations was chosen as the basis for the long-term variation study described in the text because they use the most accurate measurement approach and span the 11-year period covered by the table.

***The NIMBUS 7/ERB results have recently been scaled down, resulting in an average result for the 3.5 years' observations of 1372.6 W/m².

watts per square meter (W/m^2) at 1 astronomical unit (1 AU), the average annual distance between the earth and the sun. S has been referred to as the *solar constant*. In view of the solar variability discovered in recent experiments, this is no longer an appropriate characterization.

Solar monitoring with SI uncertainty smaller than 1% began in the late 1960s following development of a new generation of electrically self-calibrated cavity pyrheliometers. The first flight observations of solar total irradiance by these new instruments were made by the Jet Propulsion Laboratory in 1968–69. Willson's Active Cavity Radiometer (ACR) experiment on a high altitude balloon in 1969 yielded $S = 1366$ W/m^2, with an uncertainty less than ±0.3%. The first extended measurements by self-calibrated cavity pyrheliometers outside the

earth's atmosphere were made the same year by Plamondon's TCFM experiments on NASA's Mariner 6 and 7 spacecraft. An uncertainty of ±1% was quoted for the $S = 1363$ W/m^2 TCFM result.

The first long-term solar irradiance flight observations were made by Hickey's Earth Radiation Budget (ERB) experiment on the Nimbus 6 satellite between mid-1975 and late 1978. The $S = 1389$ W/m^2 result quoted for ERB exceeded the 1969 JPL/ACR result by 1.7%. The first of a series of NASA solar irradiance rocket experiments, conducted in 1976 to provide an independent calibration of the NIMBUS 6 ERB solar measurements, found ERB's calibration to be 1.6% higher than SI units. This reconciled the NIMBUS 6/ERB measurements with the 1969 ACR results, to within the latter's SI uncertainty, eliminating the possibility of a detectable change in solar output between the experiments.

The NIMBUS 7 ERB experiment, launched in late 1978, included a Hickey-Friedan (HF) electrically self-calibrating cavity detector capable of greater accuracy and long-term precision than the NIMBUS 6/ERB thermopile. The average result of $S = 1376$ W/m^2 quoted by Hickey for NIMBUS 7/ERB was 0.6% higher than the SI corrected NIMBUS 6 result. A 1978 NASA rocket experiment found the difference to be due to calibration error and not evidence for a change in solar irradiance (see Figs. 1 and 2).

A high-altitude balloon-borne solar total irradiance experiment was conducted by the Swiss Physical Meteorological Observatory (PMO) in mid-1979. The $S = 1366$ W/m^2 ($\pm0.2\%$) result quoted by Frohlich and Brusa for the experiment is close to the average of the JPL/ACR balloon and rocket results.

The Active Cavity Radiometer Irradiance Monitor I (ACRIM I) experiment, launched on NASA's Solar Maximum Mission (SMM) in early 1980, was the first dedicated satellite experiment for monitoring the long-term variability of solar total irradiance. The average result quoted by Willson for its first two years of operation was $S = 1367.8$ W/m^2 (see Figs. 1 and 2).

A third NASA rocket flight was conducted in May 1980 to provide the first calibration of the SMM/ACRIM I experiment and the second of the NIMBUS 7 ERB. The results for the five ACR's involved in the experiment (three on the SMM/ACRIM I and two on the rocket) were within $\pm0.05\%$ of their average value of 1367.7 W/m^2. The NIMBUS 7 ERB measurement exceeded the ACR result by 0.6%, the same difference found by the 1978 rocket experiment.

A second PMO high-altitude balloon flight experiment conducted by Frohlich and Brusa in June 1980 found $S = 1366.8$ W/m^2, less than 0.1% lower than the 1980 average result for the rocket and SMM ACR experiments.

The most accurate and precise solar irradiance observations are those made by experiments with fully implemented electrically self-calibrating cavity detectors, denoted as ESCC in Table 2. These experiments operate in the differential active cavity radiometer mode. Constant relative temperatures are sustained within the instruments at all times and self-calibration occurs during a shutter closed period between each one-minute observation period, minimizing errors due to slowly varying electrical or thermal properties.

The partially implemented self-calibrating cavity (ESCC/P) instruments calibrate periodically in flight. The Mariner/TCFM was self-

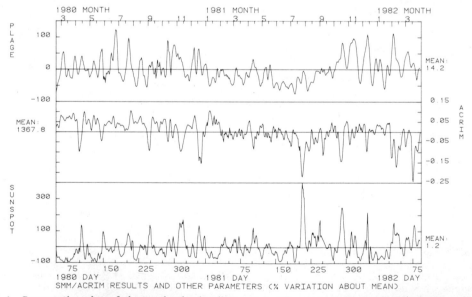

FIG. 1. Comparative plot of the total solar irradiance at 1 AU measured by the SMM/ACRIM experiment (middle panel), together with two indices of solar activity: the plage and sunspot areas (from the NOAA compilation in "Solar Geophysical Data"). The scales for the plots are in percentage variation about their means. The means are in units of W/m^2 for irradiance and thousandths of the solar disk area for plage and sunspot areas. Gaps in these data have been filled by linear interpolation over missing segments, and smoothing has been employed using a 3-day running mean to emphasize the variability on longer timescales appropriate to the evolution of solar active regions. The Calcium plage areas have been taken as an indication of the area of photospheric faculae, for which there are no regular observations.

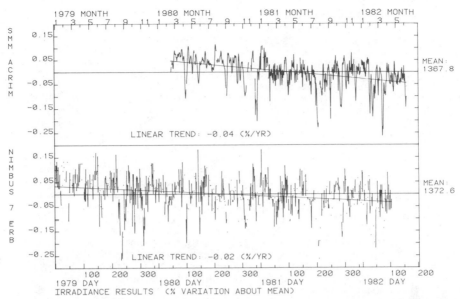

FIG. 2. Extended solar total irradiance observations from two satellite experiments. The daily mean irradiances at 1 AU measured by the Active Cavity Radiometer Irradiance Monitor experiment on the Solar Maximum Mission (SMM/ACRIM) and by the Earth Radiation Budget experiment on the NIMBUS 7 satellite (NIMBUS 7/ERB) are compared. The irradiances are plotted as the percentage variation about the means (shown in units of W/m²). The daily means have been connected by straight lines to emphasize the short-term solar variability. Isolated points indicate daily means without adjacent data on at least one day preceding and following.

calibrated once at the end of its 150-day flight; the NIMBUS 7/ERB/HF self-calibrates approximately every two weeks. These experiments cannot eliminate the effects of electrical and thermal variations in the instrument occurring between self-calibrations.

Solar irradiance flight experiments relying on preflight calibrations are the least accurate type. The rocket ESP, HF, and PACRAD instruments self-calibrate prior to but not during flight experiments. The empirical corrections applied to the results for differences between preflight and flight conditions cannot produce measurement accuracy equal to the ESCC and ESCC/P experiments. The NIMBUS 6/ERB thermopile sensor is the least accurate experiment of this type. Lacking a cavity detector and self-calibration capability, its measurements are susceptible to the full effects of detector surface degradation and depend on preflight calibration by other instruments for interpretation in SI units.

There are only a few scattered ESCC results between the 1st, in 1969, and the launch of the SMM/ACRIM I in 1980, but they can at least provide an upper limit to long-term solar total irradiance variability over this 11-year period. Although the 1969 balloon and 1976 rocket ACR instruments were never directly compared, the relativity of their measurements was defined through cross-comparisons with WRR-defining pyrheliometers. The 0.15% difference

between their results is comparable to the uncertainty of relating them, placing a probable upper limit of ±0.3% on long-term solar variability between 1969 and 1976.

The rocket ACR, SMM/ACRIM I, and PMO balloon experiments, extensively compared with each other and with WRR reference pyrheliometers, have demonstrated consistent performance throughout their various flight experiments at the ±0.1% level. The combined average of all ACR and PMO results is $S = 1367.7$ W/m² with ACR results systematically 0.1% higher. Accordingly, ±0.2% is both a likely lower limit of the SI uncertainty of any single measurement by these instruments from 1976 through 1980, and the range in which long-term solar variability over the period could have occurred without detection.

Short-Term Variability. The principal evidence from flight experiments for short-term solar irradiance variability comes from the SMM/ACRIM I experiment. Through periodic comparisons among its three independent ACR detectors, precision was sustained at the ±0.005% level through the 300-day period of normal SMM operation (Feb.–Dec., 1980). The larger range of scatter in ACRIM I results after mid-December 1980 is caused by the decreased number of daily observations in the spin-stabilized mode of SMM operation following loss of precision solar pointing capability by the spacecraft. Long-term precision throughout the 2+ years of the mission has been sustained at the ±0.02% level. The long-term decreasing trend in irradiance, averaging 0.04%/yr, has resulted in a cumulative decrease of nearly 0.1% in the total irradiance since launch.

The most striking features of the ACRIM I observations, the major irradiance decreases, demonstrate a high degree of inverse correlation with the total sunspot area (graphically shown by panels 2 and 3 in Fig. 1). This correlation, together with the theoretical prediction of the irradiance deficit by models based on the areas and lower radiative intensity of sunspots (about 50% of the normal solar photosphere), leaves little doubt about the cause of the principal irradiance variations observed by ACRIM I: The decreases are a modulation of the average irradiance by sunspots in solar active regions.

The proportionality between irradiance decreases and sunspot areas is not constant, however, and the effects of another solar activity phenomenon, the radiative excess of facular areas, must be taken into account to explain solar variability more accurately. The radiative intensity of faculae averages only 3% higher than the normal sun, but when present in sizable total area facular radiative excess can measurably moderate S. The effects of faculae can be seen in Fig. 1 (facular area is approximated by plage area in panel 1), where maxima of facular area cause irradiance peaks and decrease the amplitude of some sunspot-induced dips. Although the sunspot deficit effect dominates the days-to-weeks timescale, the irradiance appears generally to follow the longer-term (months-to-years) envelope of facular behavior.

A comparison of the results from a WRC/PMO balloon experiment in June 1980 with SMM/ACRIM I results for three hours of simultaneous solar observation has revealed a high degree of correlation between them. Solar variability on time scales of minutes to hours with amplitudes ranging from 0.005 to 0.02% were observed simultaneously by both experiments. Three filter photometers on the PMO balloon experiment detected spectral fluctuations in synchronization with the total irradiance variability. Measurements at three visible wavelengths demonstrated an inverse dependence of variation amplitude on wavelength.

A total irradiance signature for low-order p-modes of the solar 5-minute global oscillation phenomena has been detected in the SMM/ACRIM I results at the level of approximately 5 ppm of the total irradiance. No evidence of 120-minute oscillations reported from U.S.S.R. data has been found.

The multiyear record of solar measurements by the SMM/ACRIM I and NIMBUS 7/ERB experiments is shown in Fig. 2 (panels 1 and 2, respectively). The irradiance scale for the ERB results has recently been adjusted downward to yield an average value of $S = 1372.6$ W/m² (0.25% lower than the originally given value shown in Table 2). The ERB experiment measures S less frequently and with less precision than ACRIM I but shows unambiguous evidence of many of the irradiance decreases clearly seen in the ACRIM I results. The larger noise component in the NIMBUS 7/ERB results is partially due to the "quick look" quality of the data available for the period shown in Fig. 2. By 1982, final data processing had been completed for less than the first full year of the experiment, which began in 1978.

Solar Spectral Irradiance *Introduction.* No effort has been made to measure the spectral distribution of solar irradiance over the wavelength range containing most of the total energy since flight experiment opportunities above the atmosphere became available. Flight experiments during the last decade have explored relatively narrow ranges of wavelengths pertinent to specific physical processes in the earth or solar atmospheres. These have principally focused either on the ultraviolet (UV) and shorter wavelengths or infrared (IR) and radio frequencies. Extant models of the solar spectral irradiance over the 0.2–4.0 μm wavelength range (containing nearly 98% of the total solar flux) are constructed from the results of numerous experiments conducted from the earth's surface, aircraft, balloons, rockets, and spacecraft, and from models of solar emittance using graybody approximations.

The Solar Ultraviolet Spectrum (UV). The principal focus of solar spectral irradiance flight experiments in recent years has been on the UV. Most have been associated with aeronomy investigations, directed at specific upper atmospheric photochemical and thermodynamical processes. While the solar irradiance at wavelengths shorter than 0.3 μm represents only about 1% of the total flux, it is totally absorbed by radiatively active gases above 15 km, representing the primary source of energy for the upper stratosphere, mesosphere, and lower thermosphere.

The spectral variability of solar irradiance is known to be greatest at these short wavelengths, and the solar spectrum models below 0.30 μm are specified for average levels of solar activity with a range of variability on different time scales: (1) seconds to hours, resulting from solar flares and prominences; (2) days to weeks, resulting from active region evolution and solar rotation; and (3) months to decades, resulting from varying solar activity levels in the 11-year sunspot and 22-year solar magnetic cycles.

The irradiances between 0.2 and 0.3 μm of the composite solar spectrum, listed in Table 3 and shown in Fig. 3, were derived by Donnely and Pope from the results of Broadfoot, Simon, Thekaekara, Heath, and Ackerman, obtained by experiments on satellites, rockets, and balloons.

The Solar Spectrum from the Near UV to the Mid-IR. Many experiments have measured parts

TABLE 3. COMPOSITE SOLAR SPECTRAL IRRADIANCE
AT 1 A.U. OUTSIDE THE EARTH'S ATMOSPHERE FOR
THE 0.2–3.0 MICROMETER WAVELENGTH RANGES.*

Lambda	Flux	Lambda	Flux	Lambda	Flux
0.205	1.1	0.605	174	1.05	66.1
0.215	4.83	0.615	171	1.15	54.0
0.225	6.65	0.625	166	1.25	44.7
0.235	6.73	0.635	164	1.35	38.4
0.245	6.01	0.645	160	1.45	32.3
0.255	9.98	0.655	152	1.55	27.5
0.265	29.2	0.665	156	1.65	23.7
0.275	21.2	0.675	152	1.75	19.2
0.285	18.3	0.685	149	1.85	15.2
0.295	58.5	0.695	145	1.95	13.1
0.305	53.7	0.705	142	2.05	10.66
0.315	72.9	0.715	138	2.15	8.44
0.325	87.8	0.725	136	2.25	7.22
0.335	102	0.735	132	2.35	6.04
0.345	99.7	0.745	128	2.45	5.30
0.355	102	0.755	126	2.55	4.83
0.365	115	0.765	124	2.65	4.19
0.375	113	0.775	121	2.75	3.65
0.385	106	0.785	118	2.85	3.20
0.395	121	0.795	116	2.95	2.81
0.405	163	0.805	114		
0.415	170	0.815	110		
0.425	166	0.825	108		
0.435	167	0.835	105		
0.445	193	0.845	101		
0.455	201	0.855	98.6		
0.465	199	0.865	96.8		
0.475	199	0.875	94.7		
0.485	189	0.885	92.4		
0.495	196	0.895	92.0		
0.505	190	0.905	89.8		
0.515	183	0.915	87.4		
0.525	186	0.925	85.7		
0.535	192	0.935	84.1		
0.545	186	0.945	82.3		
0.555	184	0.955	80.6		
0.565	183	0.965	78.9		
0.575	183	0.975	77.3		
0.585	181	0.985	75.6		
0.595	176	0.995	73.9		

*About 98% of the total solar irradiance is included within these
wavelengths. The central wavelengths are listed for bandwidths of
0.01 μm between 0.2 and 1.0 μm, and for bandwidths of 0.1 μm
between 1.0 and 3.0 μm. The spectral irradiance for each central
band is the irradiance at 1 AU in units of milliwatts/square cm/μm.
The spectrum is derived from the results of Donnely and Pope for
wavelengths 0.2–0.3 μm; Arvesen, Griffin, and Pearson for wavelengths 0.3–0.4 μm and 1.3–2.5 μm; and Labs and Neckel for
wavelengths 0.4–1.3 μm and 2.5–3.0 μm.

of the near UV to mid-IR solar spectrum and
there have been numerous efforts to assemble
comprehensive models of it. Here discussion
will be limited to two experiments that covered
a large fraction of this wavelength range and
added significantly to the experimental data
base on solar spectral irradiance.

The NASA/AMES Experiment of Arvesen,
Griffin and Pearson: An Aircraft experiment to
measure solar spectral irradiance over the 0.3–
2.5 μm range was conducted by Arvesen,
Griffin, and Pearson of NASA's Ames Research
Center. Their experiment was flown on the
NASA CV990 research aircraft at altitudes just

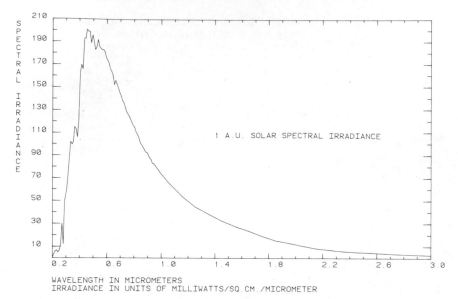

FIG. 3. Composite solar spectral irradiance between 0.2 and 3.0 μm (from Table 3).

above the troposphere. Calibration in SI units was provided by a spectral irradiance standard traceable to the US National Bureau of Standards.

The "zero air mass" or extra-atmospheric spectral irradiance was determined using an extrapolation procedure for data taken over a range of optical thickness. A solar spectrum model was developed covering the 0.2–4.0 μm wavelength range. Data from other experiments and a 5800 K graybody model were used to construct the spectrum below 0.3 μm and above 2.5 μm, respectively.

Their initial error analysis indicated uncertainties ranging from ±25% at 0.3 μm to ±3% in the visible. Following corrections for a systematic error in the standard lamp and some wavelength calibration errors, a final SI uncertainty of ±5% or less was probably achieved over the wavelength range 0.4–2.5 μm, which contains more than 90% of the total solar flux.

The Experiment of Labs and Neckel: The solar spectral irradiance experiment of Labs and Neckel was conducted at the Jungfraujoch Observatory, 3.6 km above sea level in the Swiss Alps. The disadvantages of increased atmospheric attenuation, relative to the previous aircraft experiment, were partially offset by the advantages of working in a laboratory environment.

Their well controlled experiment was comprised of observations of the spectral intensity at the center of the solar disk in the 0.33–1.25 μm wavelength range over a range of atmospheric optical thicknesses. The spectral irradiance from the full disk was then computed using empirical relationships for limb darkening.

The SI calibration reference was a standard radiance lamp calibrated to blackbody emitters located at the Happel-Laboratory and the Berlin Institute of the Physikalisch-Technisch Bundesanstalt (PTB). The solar spectral irradiance outside the earth's atmosphere was computed using an extrapolation procedure based on the Bouguer-Lambert law.

The traceability of their standard lamp to SI units was claimed to be ±2% by Labs and Neckel. This is more accurate than the irradiance standards used in the aircraft experiment described above and is one of the advantages of their laboratory experiment environment. Derivation of the full-disk spectral radiance from measurements of central intensity is a significant source of uncertainty due to photospheric variability, especially when solar active regions are present. (The angular field of view of their experiment, less than 0.01 degree, represents only 1.5% of the full solar disk.)

Labs and Neckel constructed a solar irradiance model for the 0.2–100 μm wavelength range, using the experimental results of others and modeling outside the range of their measurements. No systematic error analysis is given but an error bound of ±3.5% for wavelengths between 0.4 and 1.25 μm is likely required to accommodate the various sources of experimental and analytical SI uncertainty discussed above.

Spectral Irradiance Summary. A composite solar spectrum derived from the sources above is listed in Table 3 and shown in Fig. 3. The spectrum has a probable SI uncertainty in excess of ±10–30% for wavelengths shorter than 0.3 μm (see Table 4). The two principal con-

TABLE 4. WAVELENGTH DEPENDENT ABSOLUTE
UNCERTAINTY (SI) AND TEMPORAL VARIABILITY
OF THE SOLAR SPECTRUM OVER VARIOUS TIMESCALES.

Wavelength Range (μm)	Uncertainty (SI)	Seconds to Hours	Days to Months	Months to Decades
Lyman Alpha (0.12)	30	30	30	200
0.135–0.175	30	??	30–4	100–20
0.175–0.21	20	??	4	20–??
0.21–0.24	15	??	4–2	??
0.24–0.33	10	??	2–1	??
0.33–0.4	5	??	<1	??
0.4–1.25	5–3	0.05	??	??
1.25–2.5	5	??	??	??
2.5–3.0	5	??	??	??

Wavelengths are in units of micrometers (μm), uncertainties and variability in ±%. Entries listed as ?? indicate the absence of experimentally determined values.

tributors to UV uncertainty are errors in calibration and the effects of inherent solar variability. Variable scattering and absorption by the earth's atmosphere for solar irradiance with wavelengths between 0.3 and 0.4 μm would cause a corresponding increase in uncertainty for measurements made over longer optical paths. The AMES data were obtained through the smallest "air mass" and are probably the most accurate in the 0.3–0.4 μm region, with uncertainties ranging from (±)10–5%, respectively. The results of Labs and Neckel appear to be the most accurate in the 0.4–1.25 μm range, with SI uncertainty of about (±)3–5%. Intensities in the wavelength range from 1.25 to 2.5 μm are determined most accurately from AMES data, once again due to the smaller "air mass" associated with the experiment. The SI uncertainty in this range is probably less than ±5%. The solar spectral model at wavelengths longer than 2.5 μm is taken from Labs and Neckel, who have used a combination of modeling and an assessment of experimental results of others. The SI uncertainty between 2.5 and 3.0 μm is probably near ±5%.

RICHARD C. WILLSON

References

1. White, O. R. (Ed.), "The Solar Output and Its Variation," Boulder, CO, Univ. of Colorado Press, 1977.
2. *Solar Physics* 74, Nos. 1 and 2, (Nov. 1981)—Special Issue: Proceedings of the 14th ESLAB Symposium on the Physics of Solar Variations, Scheveningen, Holland, Sept., 1980.
3. Willson, R. C., "Active Cavity Radiometric Scale, International Pyrheliometric Scale and Solar Constant," *Journal of Geophysical Research* 76, 4325 (1971).
4. Brusa, R. W., and Frohlich, C., "Recent Solar Constant Determinations from High Altitude Balloons," Proc. of IAMAP 3rd Scientific Assembly, Hamburg, FRG, Aug., 1981, p. 35, Pub. by National Center for Atmospheric Research, Boulder, Co.
5. Hickey, J. R., et al., "Initial Solar Irradiance Determinations from Nimbus & Cavity Radiometer Measurements," *Science* 208, 281 (1980).
6. Willson, R. C., et al., "Observations of Solar Irradiance Variability," *Science* 211, 700 (1981).
7. Willson, R. C., "Solar Irradiance and Solar Variability," *Journal of Geophysical Research* 87, 4319 (1982).
8. Eddy, J. A., Hoyt, D. V., and White, O. R., "Reconstructed Values of the Solar Constant from 1874 to the Present," Proc. of IAMAP 3rd Scientific Assembly, Hamburg, FRG, Aug., 1981, p. 29. Pub. by National Center for Atmospheric Research, Boulder, Co.
9. Arvesen, J. C., Griffin, R. N., and Pearson, B. D., "Determination of Extraterrestrial Solar Spectral Irradiance from a Research Aircraft," *Journal of Applied Optics* 8, 2215 (1969).
10. Labs, D., and Neckel, H., "The Radiation of the Solar Photosphere from 2000 Angstroms to 100 Microns," *Zeitschrift für Astrophysik* 69, 1 (1968).

Cross-references: METEOROLOGY; INFRARED RADIATION, INTERNATIONAL SOLAR-TERRESTRIAL PHYSICS PROGRAM; RADIATION, THERMAL; SOLAR ENERGY SOURCES; SOLAR ENERGY UTILIZATION; SOLAR PHYSICS; SPECTROSCOPY.

SOLID-STATE PHYSICS

Solid-state physics is the study of the crystallographic and electronic properties of solids. It is concerned not only with the inherent nature of solids as collections of atoms, but also with interaction of solids with external forces: mechanical forces, heat, electromagnetic fields, bombarding particles, light, etc. It includes both theoretical and experimental aspects of solids. It is a broad field, merging with the associated sciences of chemistry, biology, and astronomy and such technologies as metallurgy, ceramics, and electrical engineering. The boundary between solid-state physics and these associated disciplines is not well defined.

The atoms comprising a solid can be considered, for many purposes, to be hard balls which rest against each other in a regular repetitive pattern called the *crystal structure*. Most elements have simple crystal structures of high symmetry, but many compounds have complex crystal structures of low symmetry. The determination of crystal structures and the location of atoms in the crystal relative to each other is an absorbing study; it has occupied the lives of many geologists, mineralogists, physicists, chemists, and other scientists for many years. (See CRYSTALLOGRAPHY.)

The rigid, hard-ball model is not adequate to explain many properties of solids. We know that solids can be deformed by finite forces, so the crystal lattice must not be completely rigid. Furthermore, atoms in a solid possess vibrational energy, so they must not be precisely fixed to mathematically defined lattice points. This deformability of solids is built into the model by the assignment of deformable bonds between nearest atom neighbors. This ball-and-spring model has had many successes; one important early use was that of Einstein to devise a reasonably successful theory of specific heat. Later incorporation by Debye of coupled motion of groups of atoms led to an even more successful theory of heat absorption by solids.

Several measures exist of the strength of these bonds. One is the size of the elastic constants—for most solids Young's modulus is about 10^{11} newtons/m^2. The other is the frequency of vibration of atoms in solids, for which values around 10^{13}–10^{14} Hz are found. From these experimental observations, the intrinsic forces of deformation between atoms in the solid may be deduced; they are about 10 newtons/meter (about 10^{-4} dynes/angstrom).

This lack of perfection demonstrated by elastic deformation of solids is but one of many evidences of lack of crystal perfection. In the way crystals are produced, either in the laboratory or in nature, defects in structure are incorporated, often in profusion. Such defects may be characterized by three principal parameters—their geometry, size, and energy of formation.

All real crystals have external surfaces. As a consequence, atoms in surface sites do not possess the correct number and arrangement of nearest neighbors. This increases their local energy, so that the surface has an intrinsic energy and is characterized by a SURFACE TENSION. In addition, internal surfaces exist, grain boundaries and twin boundaries across which atoms are incorrectly positioned. In a crystal of reasonable size, say, 1 cm^3, these two-dimensional defects, called *surface defects*, account for only about 1 atom in 10^6, a small fraction. Even so, surfaces are important attributes of solids.

Some defects have extent in only one dimension—these are called *line defects*. The most prominent of these, the *dislocation*, is a line in the crystal along which atoms have either an incorrect number of neighbors or neighbors which do not have the correct distance or angle. In 1 cm^3 of a real crystal one might find wide variation in lengths of dislocation lines—from near zero in nearly perfect semiconductor material to perhaps 10^{11} cm per cubic cm in a deformed metal.

Defects which extend only to about one atomic diameter also exist in crystals—these are called *point defects*. Vacant lattice sites may occur—these are called *vacancies*. Extra atoms (*interstitials*) may be inserted between proper crystal atoms. Atoms of the wrong chemical species (*impurities*) may also be present.

Finally, solids exist in which the local atom arrangement is not repeated regularly. Among these are the glasses, polymeric materials, and amorphous alloys. Such solids are enormously important technologically, hence considerable effort is now underway to deduce local arrangements exactly.

The importance of the various defects is intimately related to their energy of formation. A standard against which this energy can be compared is provided by the energy of sublimation—the energy necessary to separate the ions of a solid into neutral, non-interacting atoms. This energy is about 81,000 cal/mole for a typical metal, Cu, at room temperature—i.e., about 3.4 eV per atom. Energies of surfaces, both free surfaces and grain boundaries, are about 1000 ergs/cm^2, i.e., about 1 eV per surface atom. Dislocation energies are of similar size per atom length of dislocation line, i.e., about 1–5 eV, so the energy of a dislocation is about 10^{-4} ergs/cm of length. Point defects, too, possess an inherent energy of about 1 eV each. Vacancies in Cu have an energy of about 1 eV, self-interstitials 2 or 3 eV. The energies per atom of these various defects, surface, line, and point, are all much larger than the average thermal energy per atom in a solid at reasonable temperatures: kT is only about 1/40 eV at toom temperature. Thus, defects can be produced only by conditions which exist during manufacture of crystals, by external means such as plastic deformation or particle bombardment, or by large local fluctuations in thermal energy from the average.

Even so, the total amount of energy bound up in ordinary concentrations of these defects is not large compared to the total thermal energy of a solid at normal temperatures. All the vacancies at equilibrium in Cu, even at the melting point, comprise less than 10 cal of energy per mole, much less than the enthalpy at 1357 K (the melting point) of more than 7000 cal/mole. In a material with very heavy dislocation density, 10^{12} cm of line length/cm³ of material, the excess energy is only a few calories per mole. And the total energy of a free surface of a compact block of 1 mole of Cu is even less, about 10^{-3} cal. Thus the inherent energy of these defects is not large compared with the total thermal energy of the solid, because the chance of an atom being located in one of these defects is very small. Even so, defects are important in controlling many phenomena in crystals.

Crystallographic defects need not remain stationary in the crystal; they may move about. Some movements may reduce the overall free energy of the solid; others (these are chiefly movements of the point defects) may be the wandering of a random walk. Since these movements require the surmounting of a potential barrier invariably much larger than kT, the motion of defects depends upon large local fluctuations in energy. Consequently, their rate of motion depends on temperature through a Boltzmann factor exp $(-\Delta H/RT)$, where ΔH is the enthalpy change necessary to move the defect from the lowest energy site to the highest in its path of motion.

Description of the crystalline structure of solids is thus seen to consist of successive stages of approximation. First the mathematically perfect geometrical model is described, then departures from this perfect regularity are permitted. The deformability of solids is permitted by allowing the force constants between adjacent atoms to be finite, not infinite. Misplacement of atoms is permitted either by equilibrium thermodynamical justification or by the effect of external forces. Some defects have intrinsic features which directly affect properties of the crystal, others affect crystalline properties by their diffusive motion. In spite of their relatively small number, defects are of immense importance.

Some important solids do not possess long-range crystalline regularity. Glasses and polymeric solids are in this class. Although local atomic configuration may be known, the atomic arrangements are not spatially periodic. By rapid cooling or by deposition of atoms from vapor or plasma, materials which are normally crystalline can also be made amorphous. The correlation of atom positions in the usual metallic glasses or amorphous semiconductors is lost above 5–10 atom diameters. Beyond that distance the accumulated errors

of atom positioning make spatial assignments unreliable. Important features of amorphous solids are atomic arrangements, chemical composition and chemical gradients, defects, voids, inclusions, phase boundaries, and electronic character.

Pure metals can be made amorphous by evaporation techniques, but most amorphous metallic alloys are made by rapid cooling from the melt. These are commonly binary or ternary alloys such as Au-Si, Fe-Pd-P, and Ge-Sn-S. They have interesting mechanical, electrical, and magnetic properties which may make them technologically valuable. Especially interesting is the property that many amorphous metallic alloys do not undergo structural or chemical changes to temperatures far above room temperature—say 400°C. When the amorphous structure is lost at high temperatures, recrystallized grains are often formed and compositional changes occur, since amorphous metallic alloys are indeed metastable. Thus, heat treatments can be devised to produce novel properties in the semi-transformed state.

The *electronic structure* of solids is determined, in principle, solely by the electronic structure of the free atoms of which the solid is composed, modified by interactions which take place when the atoms approach each other at normal interatomic distances—a few angstroms. Since the free atom structure is known rather well, especially for atoms of lower atomic number, the electronic structure should be readily calculable, even though the calculation might be tedious. However, the variety of interactions which take place between electrons on adjacent atoms make such calculations difficult. The use of the computer to do numerical calculations has made this process more manageable. (See the article on SOLID STATE THEORY.) The close interplay between theory and experiment in deducing the electronic structure of solids has permitted many general features of the electronic structure to be deduced. As for the crystalline structure of solids, two stages are useful in understanding the electronic structure. First, the perfect electronic structure is defined, then irregularities in this structure, again termed defects, are introduced. Although both the geometry and energy of crystalline defects are defined, description of the many electronic defects has been less straightforward.

The nuclei of the atoms in a solid and the ion cores have energy levels little different from corresponding levels in free atoms. The characteristics of the valence electrons are modified greatly, however. The state functions of these outer electrons greatly overlap those of neighboring atoms, and the restrictions imposed by the Pauli Exclusion Principle and the Uncertainty Principle force modification of the state functions and development of a set of split energy levels rather than the single level characteristic of isolated atoms. This set of split levels becomes a quasi-continuous band of levels several electron volts wide for most solids. Importantly, unoccupied levels of the atoms are also split into bands. The electronic characteristics of solids are determined by the relative positions and energies of the occupied and unoccupied levels as well as by the characteristics of the electrons within a band.

A solid is called a *metal* if excitation of electrons from the highest filled levels to the lowest unoccupied levels can occur with infinitesimal expenditure of energy. The excitation can occur by means of many external influences, such as electric fields, heat, light, or radio waves. Metals are good conductors of electricity and heat, they are opaque to light, and they reflect radio waves.

Some solids have wide spacing between the occupied and unoccupied energy states—2 eV or more. Such solids are called *insulators*, since normal electric fields cannot cause motion of an extensive fraction of the outer electrons. Examples are diamonds, sodium chloride, sulfur, quartz, and mica. They are poor conductors of electricity and heat are usually transparent to visible light (when not filled with impurities or defects).

Solids with conductivity properties intermediate between those of metals and insulators are called *semiconductors*. For them, the excitation energy lies in the range 0.1 to about 1.5 eV. Thermal fluctuations are sufficient to excite a small, but significant, fraction of electrons from the occupied levels (the valence band) into the unoccupied levels (the conduction band). Both the excited electrons and the empty states in the valence band (called *holes*) may move under the influence of an electric field, providing a means for conduction of electric current. Such electron-hole pairs may be produced not only by thermal energy, but also by incident light, x-rays, and many forms of radiation. The inverse process, emission of light by annihilation of electrons and holes in suitably prepared materials, provides a highly efficient light source (light-emitting diodes).

Crystallographic defects, in general, are also electronic defects. In metals, and alloys, they provide scattering centers for electrons, increasing the resistance to charge flow. The heating element in electric heaters, in fact, consists of an alloy of an ordinary metal such as iron with additional alloying elements providing scattering centers for electrons—nickel and chromium in solid solution are commonly employed. In semiconductors and insulators, however, alloying elements and crystalline defects provide an even greater variety of phenomena, since they can change the electron-hole concentrations drastically; in addition they provide scattering centers and electron-hole recombination centers. The entire semiconductor device industry is based on alloying of silicon, germanium, and compound semiconductors with selected impurities in carefully controlled concentration and geometries. The strong interplay between solid-state scientists and design engineers has provided a myriad of devices utilizing the alloying and geometrical principles in exquisite designs. Amorphous semiconductors are under intensive study from a basic point of view. Silicon, especially, is being examined intensively—primarily because of the cheapness with which thin layers of amorphous silicon can be deposited by sputtering techniques. However, amorphous silicon has many "dangling bonds" which spoil the electrical properties by reducing the lifetime of charge carriers and by affecting surface charge; addition of hydrogen is commonly made to satisfy local bonding requirements.

Crystal defects may produce important electronic defects in ionic crystals. Vacant atom sites and impurities of valence different from that of the host may trap electrons, which can be excited by incident photons, thus producing light absorption. Since the excitation occurs in specific ranges of frequency, crystals otherwise transparent may become colored. This phenomenon, of large intellectual interest, also has immense practical importance—the color of most gemstones has its origin in this selective absorption. Likewise the photographic process is a result of these effects in AgBr and AgCl.

Other geometries of conducting media are ultrathin layers, especially of elemental and compound semiconductors grown epitaxially on stable substrate crystals. Remarkably thin layers can be achieved and high stability is claimed for many layered structures. In the technological push toward ultraminiaturization, these thin-layer structures may be extremely important. Other types of composite structures are also being examined—composite layers and fiber structures. Thin films of nearly single-crystal silicon are being produced by growth from melt for possible use in solar cells.

The chemistry and physics of surfaces are being investigated widely. A large part of this effort is aimed at understanding catalysis and at solving problems in coal conversion, in chemical processing, and in photosynthetic reactions. Both improved methods of experimentation and interaction with theory are providing advances in this investigation—important technological features being the driving force.

Solids are useful because interaction with external forces or stimuli such as magnetic fields, heat, and mechanical forces are important to us. Yet among all of these interactions none is more important than interrelations between matter and light. The earth is warmed by the sun, photosynthesis provides food, vision is the most important of our senses. Yet the physics and chemistry of absorption processes on which these phenomena depend are only partially understood. That generation of light is also of great importance may be seen from the fact that some 25% of all electrical energy consumed in the United States is used in the production of light. Small wonder that immense effort is being spent on increasing lighting efficiency by high-pressure arc discharge lamps, by more efficient phosphors for fluorescent tubes, and by improved devices using cathodoluminescence and electroluminescence. The most efficient light source, the light-emitting-diode, is highly important as used in indicator lamps, but also in its laser form for telecommunication systems. This application has been made possible by the development of optical fibers with appropriate variation of the index of refraction across the fiber diameter to permit waveguide performance; at the same time the optical loss in the fiber has been reduced to tolerable limits. Communication over distances of thousands of kilometers using fiber optics is a commercial reality.

The interaction of light with matter is so important that a chief hallmark of physics in this century is the extraordinary development of spectroscopy and its application to processes in solids. First came investigation of emission and absorption of radiation from free atoms. Later investigations have included emission and absorption by atoms in solids, MASER and LASER phenomena, Mössbauer spectroscopy, nuclear magnetic resonance, x-ray diffraction, infrared spectroscopy, fluorescence, the Raman effect, and microwave emission and absorption. These investigations have led not only to an increasingly detailed perception of the nature of solids, but also to a host of practical devices such as light sources, laser generators and amplifiers, and x-ray generators.

MAGNETISM is an inherent property of the electrons in a perfect crystal. It is important both for insights which magnetic studies provide into the intrinsic properties of materials and for its practical importance. Two weak effects, DIAMAGNETISM and PARAMAGNETISM, are purely of academic importance. Ferromagnetism, however, is of immense technological value. The basic properties of ferromagnets (saturation magnetization, Curie temperature, magnetostriction, crystalline anisotropy) are inherent properties of the perfect lattice. The technological application of ferromagnets, though, demands careful control of defects, which exert strong influence on such properties as hysteresis loss, permeability, and coercive force. (See FERROMAGNETISM.) The amorphous metals have interesting ferromagnetic characteristics—in the amorphous state, they have soft magnetic qualities, similar to those of the important transformer materials. The domain walls must move easily in these amorphous metals. Partially recrystallized amorphous materials, however, have more nearly the characteristics of permanent magnets; spatial inhomogeneities of composition and crystalline character provide this appropriate magnetic character. The most striking feature of ferromagnetism is the enormous variety of properties which can be produced in a rather few materials—alloys of iron, nickel and cobalt plus a few oxides (spinels and garnets). This variety comes from the ability of scientists and engineers to tailor the chemistry and physical arrangement of alloy phases to produce desired magnetic qualities.

The physics of computers depends increasingly on the properties of solids. Active computational devices depend on ever increasing sophistication of semiconductor devices. The pressure for memory cores of smaller physical size, faster readout time, and lower power consumption puts great demands on continued development of magnetic and electronic materials.

The search for superconductors of ever higher critical temperature continues to be of importance to physicists. High-field critical temperatures among the alloys rose rapidly from the 5–10 K range to some 20–22 K in a few years. However, the rate of increase of critical temperature with time has slowed and the long sought goal of superconductors suitable for use in liquid nitrogen seems far away.

Summarizing: Solid-state physics, in its broadest sense, is the study of the perfect and imperfect crystalline and electronic properties of solids, ranging from attempts to understand these phenomena from the most fundamental point of view to the edge of the technological applications of solids. Indeed, the field is both driven by technological demands and provides, through basic research, innovations of often unexpected nature which make rapid technological application possible.

CHARLES A. WERT

References

Ziman, J. M., "Electrons and Phonons," Oxford, Clarendon Press, 1970; "Principles of the Theory of Solids," Cambridge, The University Press, 1972.

Livingston, J. D., "Microstructure and Coercivity of Permanent-Magnet Materials," *Prog. Mat. Sci.*, 243, 1981.

Various authors: Articles in "Rapidly Quenched Metals III" (R. Cantor, Ed.), London, The Metals Society, 1978.

Various authors: Articles in "Metallic Glasses" (J. J. Gilman and J. H. Leamy, Eds.), Metals Park, Ohio, Amer. Soc. Metals, 1978.

Kao, Kwan C., and Hwang, Wei, "Electrical Transport in Solids," Oxford, Pergamon Press, 1981.

Pankove, J. I., and Carlson, D. E., "Electrical and Optical Properties of Hydrogenated Amorphous Silicon," *Ann. Rev. Mat. Sci.*, p. 43, 1980.

Cross-references: CRYSTALLOGRAPHY, DIAMAGNETISM, ENERGY LEVELS, FERROMAGNETISM, HEAT CAPACITY, MAGNETISM, PARAMAGNETISM, SEMICONDUCTORS, SOLID-STATE THEORY, SURFACE TENSION.

SOLID-STATE THEORY

The ideal, perfect crystalline solid possesses a true long-range order which is not found in other phases of matter. The periodicity of the crystalline lattice potential in a solid enables one to make detailed investigations of many properties of the solid by employing powerful mathematical techniques which exploit the long-range symmetry. In such studies the aid of high-speed digital computers is now routinely employed, and a group of computational solid state theorists is emerging. Most phenomena studied by solid state theorists are related to the *almost periodic potential*. However a substantial minority of theorists are studying systems which deviate markedly from those related to the periodic potential. Such studies include the properties of point defects and impurities (color centers), dislocations and phase boundaries between material media, solid surfaces, random alloys, phase transitions, and charge instabilities (spin-density waves or charge-density waves).

The problem of one electron in an ideal periodic potential has been solved in principle. The advent of fast digital computers has permitted many independent groups to perform calculations of the *band structure*, that is, the dependence of the energy of the one electron on its wave vector $\epsilon(\mathbf{k})$, as a routine task of numerical precision. Even if one obtains the potentials determining the bands in a self-consistent manner, the evaluation of an energy band structure is a task requiring at worst a few minutes' time on a large digital computer. These successes are tempered by the realization that the question of which "exchange" to use is as yet unresolved. There exists active investigation into the relative merits of "local," or "nonlocal exchange," and into the precise nature of self-interaction corrections. The increasing precision of the knowledge of experimental properties of solids is causing theorists to seek ever more precise theories with which to describe solids. This need is leading theorists to attempt to include the dynamical properties of electron-electron interactions into the basic band structure calculations. The band theory of solids of dimensionality less than three is of recent interest.

The effect of these electron-electron interactions, known as *many-body* or *correlation effects*, may determine both the qualitative behavior of solids and quantitative aspects as well. Currently, most researchers are using a density functional approach based upon the Hohenberg-Kohn theorem to include such effects in band structure calculations. In addition to these efforts a sizeable minority are using various many-body approaches such as *Green's functions, diagrammatic perturbation theory*, or a *coupled cluster method* to include correlation effects in one-electron band theory. These energy band calculations are being used to predict other solid state properties beyond the basic $\epsilon(\mathbf{k})$ relationship. These properties include charge density, Compton profiles, cohesive energy, and other basic properties of the ideal three-dimensional solid.

Many research groups are concentrating on properties of nonideal three-dimensional solids which are analogous to properties studied in the band theory problem. These studies include determination of the electron density of states, $N(\epsilon)$, for random alloys; properties of glasses; the two-dimensionally periodic band structures of solids with surfaces or with interfaces between two types of solid; and electronic localization in random potentials. Study of symmetry-lowering instabilities in the ground state of many solids is an active field of research. Such instabilities include the formation of charge-density waves or spin-density waves, and may be commensurate or incommensurate with the basic lattice structure.

The atomic nuclei comprising a solid are not fixed in place but vibrate. These vibrations are in fundamental quanta called *phonons*. The basic calculation of the dispersion of the frequency of these phonons in terms of their wave vector $\omega(\mathbf{q})$ is not as simple as determining the basic energy band structure of a solid. This difficulty is due to the very complex nature of the interatomic forces existing in the solid. The phonons also interact with one another in complex ways due to the fact that the interatomic forces are not truly harmonic in nature. Furthermore, phonon properties may be a function of temperature. Considerable effort remains directed at these problems.

Still more difficult problems arise when one considers that the basic electronic properties and the basic vibrational properties are not truly independent, but interact. The bulk of the theoretical work on pure systems is based on (1) a description of the electronic structure as if no phonons or even zero-point vibrational motion were present, and (2) a description of the phonon properties which ignores the detailed nature of the electronic structure of the crystal. In the main, this approximation is not as crude as it might seem; the electronic wave function is in actuality often coherent over many atomic positions, and therefore the phonons are insensitive to small changes in the electronic structure, and vice versa. Thus in a simple metal under ordinary conditions one may ignore the change in the phonon dispersion curve as the electronic state is changed, say, by absorption of light or an external electric field. Nonetheless, many interesting phenomena do occur in such situations, in which one may not ignore the coupling of electronic and vibrational properties. Such phenomena include the lattice distortions which accompany the formation of a charge-density wave in a metal. In the study of electron transport in a semiconductor such as Ge, the interaction creates a *phonon drag*. In ionic crystals the presence of electron-phonon interactions is responsible for the formation of *polarons*, mobile electrons which drag a cloud of ionic distortion around with them. Such a polaron is called a *large* polaron. If the interaction is more severe, the electron and its ionic distortion may form a localized state termed a small polaron.

Solids at finite temperature are never perfect. The common imperfections present include vacancies, interstitials, impurities, and dislocations. Such imperfections manifest themselves in a number of ways. This manifestation may be in the formation of a local mode in the phonon spectrum of the solid. This mode is not found in the perfect solid. This local mode may be associated with the imperfection and exist primarily in its vicinity. A second manifestation may be in the formation of new electronic energy levels in the system. In this case the most dramatic occurrence is when the impurity generates energy levels in forbidden energy gaps in the band structure. Such occurrences are long recognized as the source of coloration in the alkali-halide crystals, for example. A third manifestation may be by means of generating new magnetic states for the solid. Active theoretical studies on all phases of the questions raised by the presence of defects in solids are under way. The problem here is severely complicated by the often strong interaction of vibrational electronic and magnetic degrees of freedom of these imperfections.

It is generally recognized that when an imperfection occurs in a crystal, the lattice will distort in the vicinity of the imperfection. This distortion, or *relaxation*, of the lattice may not preserve the basic point symmetry of the system. In some cases the symmetry of the electronic structure requires such a relaxation (Jahn-Teller effect). Furthermore, if the imperfection changes its electronic state the lattice distortion may change in response to the change of electronic state. These relaxations and changes of symmetry often cause severe modification of optical properties, not only in producing new absorption bands but also in reordering the strengths of existing absorptions. These relaxation effects were initially recognized as significant in studies on ionic solids and are now recognized as equally important in understanding imperfections in semiconductors. Several research groups, using powerful Green's function techniques and sophisticated computer techniques, are developing accurate methods for self-consistently determining the properties of imperfections in semiconductors, including lattice relaxation. As important as such local relaxations are in determining the properties of bulk imperfections, it is becoming equally clear from the efforts of many research groups that such relaxations are at least as important in the properties of imperfect solid surfaces.

If one considers the optical properties of the perfect crystal, there exists a simple selection rule related to the initial and final wave vector of the electron being excited which determines if a given optical transition is allowed. This is often a good approximation in the actual solid. Nonetheless, in many real systems important phenomena are associated with "indirect" transitions in which both phonons and light quanta (photons) are simultaneously absorbed or emitted. The silver halide solids used in photography are excellent examples of systems in which such indirect transitions are important. Detailed theories have been developed to account for the numerous new effects in the optical properties of solids such as laser action, nonlinear optical effects, and multiphoton effects. The interest in optical theories is broadening to include photoemission theories in both the ultraviolet and the x-ray spectral regions. Of particular new importance is the use of angular resolved photoemission data to test band theories. Theories relating to the optical properties of perfect solids in which the hole may be considered localized after its creation no longer just consider the creation of excitons but also include x-ray extended fine structure and also near absorption edge structure in x-ray absorption. These latter studies now involve many active research groups.

The mechanical properties of solids continue to receive much theoretical emphasis and are now becoming understood at a fundamental level largely through the development of the theory of dislocations and also through lattice statics and molecular dynamics calculations. Many new phenomena have been catalogued, such as spiral growth "climb," dislocation networks, pinning, "whiskers," and decoration with collodial particles. The use of lattice static calculations is permitting studies of solid phase segregation at zone boundaries, of crack formation at zone boundaries or imperfect ions, and in general is providing an atomistic view of the formation of dislocations, cracks, and voids in solid systems. Such studies now may include the effect of external stresses on the solid.

Magnetic phenomena of all types continue to receive much attention and fascinate the theorist. Nuclear and electron resonance phenomena continue to be useful tools in the investigation of the magnetic properties of solids. Similarly, the Mössbauer effect continues to provide insights into magnetic properties. The theory of ferromagnetism continues to attract the greatest amount of attention in this field. However, there are substantial numbers of investigators studying antiferromagnetism as well. From time to time, reports of anomalous diamagnetism or anomalous paramagnetism in solid systems also evoke substantial theoretical response. Some recent interest has been generated by the discovery of true two-dimensional magnets. Properties of spin-glasses also occupy the attention of many theorists. In general, although the specific details of the interactions among electrons which give rise to ferromagnetism or antiferromagnetism are not completely understood quantitatively, the qualitative features responsible for these magnetic phenomena seem to be well understood.

The phenomenon of superconductivity, which has been given an elegant explanation in terms of electron-phonon interaction in the BCS theory, continues to attract the attention of many theorists. Much work in extending and generalizing the BCS theory is being performed. Specific, detailed calculations of the electron and phonon structure and their interactions in several metals have been performed and transition temperatures predicted. Theories of the destruction of superconductivity by magnetic impurities continue to be of interest. Speculation continues that high-temperature superconductors may exist. In the search for high-temperature superconductors, theorists have been postulating that interactions other than the electron-acoustic phonon interaction might produce superconductivity. Recent efforts concentrate mostly on electron-exciton interactions or on electron-optical phonon interactions.

In the area of transport theory there are substantial attempts to achieve more sophisticated theories than the conventional one-particle Boltzmann equation methods permit. Many approaches for studying electronic transport are based upon *density matrix* methods. In these approaches one may study the actual scattering probabilities without the need for drastic approximations such as spherical energy surfaces or the Born approximation to obtain the scattering amplitude. The need to include Umklapp processes in order to obtain quantitative understanding of transport processes is appreciated. In some cases, studies now include the actual band structure of the solid in question in determining the transport of electrons. Studies of the transport of atoms in solids and on their surface are also of importance. In many cases, the technique of *molecular dynamics* is used to obtain detailed results for atomic transport.

The study of quasi-particles continues to expand. The lexicon of interesting quasi-particles now includes excitons, plasmons, magnons, phonons, helicons, polarons, polaritons, and solitons, among others. These quasi-particles are the elementary excitations of a solid, and may be characterized by an energy, a wave vector, and a polarization. Much current interest is directed at understanding the dependence of energy upon wave vector and polarization and the propagation of these quasi-particles in the solid. As one better understands the individual quasi-particle it becomes interesting to study the interactions among them and their interdependences. In the area of semiconductors, the interaction of excitons with each other has attracted much attention, and theories of the fusion of two excitons into a biexciton or of many excitons into an excitonic gas or into an electron-hole droplet have been of great interest. The study of the interaction of excitons and phonons or phonons and magnons is also attracting investigators. Many theorists are studying the corresponding fusion and/or fission properties of various types of quasi-particles.

Solid state theory continues to evolve new areas of high interest on a regular basis. There is much activity in studying the critical behavior of one-, two-, or three-dimensional systems. Much of the interest in one-dimensional systems is due to the discovery of highly conducting, polymerlike, one-dimensional organic solids. In this area Peierls instabilities in the lattice structure are of chief interest. In the case of two-dimensional solids there are many current studies related to melting, roughness, wetting, formation of ordered chemisorbed layers, or surface reconstructions. In the case of three-dimensional systems virtually all types of critical phenomena are of current interest. Of these, the insulator-metal transition, especially in random systems, probably attracts the most attention. There are many attempts to apply the techniques of critical phenomena to biological systems. Although it is not a solid system, solid state theorists continue to study the properties of liquid helium. Recent techniques developed for the study of critical phenomena in general are tending to unify the studies of critical phenomena in solids and in helium.

The most recent addition to the arsenal of techniques employed by the solid state theorists is that of *renormalization group theory*, which seems to be the most far-reaching of the new techniques for studies of critical phenomena. The renormalization group technique has finally permitted solution of a long-standing problem in magnetic alloys, the Kondo effect. This technique is of use in many areas of physics other than solid state and is providing a unifying link between several subdisciplines in physics. Theorists continue to exploit recent advances in fast digital computers. The advent of the array processor (vector processor) shows promise of further substantial increase in computer speed. The techniques of group theory, to codify and exploit the symmetry properties of solids, continue to be useful. Methods of quantum field theory and Green's functions continue to be of significant importance to solid state theory. In addition there is a reawakening of interest in the techniques of quantum chemistry among solid state theorists. These techniques are of particular use for studies of surfaces and point imperfections.

Work on the theory of solids is widespread. Significant contributions to the theory of solids are not the sole province of the physicist but are being generated by ceramists, metallurgists, chemists, and electrical engineers as well. Most notable is the work in the United States, the Soviet Union, Japan, England, and Italy. However, there are individual contributions of great value to solid-state theory from all parts of the world.

A. BARRY KUNZ

References

Andrei, N., *Physics Rev. Lett.* **45**, 379 (1980); Andrei, N., and Lowenstein, J., *Phys. Rev. Lett.* **46**, 356 (1981); see also *Physics Today*, **34**(5), 21 (1981).
Duke, C. B., and Schein, L. P., *Physics Today* **33**(2), 42 (1980).
Madelung, O., "Introduction to Solid State Theory," Berlin, Springer-Verlag, 1978.
Pantelides, S. T., Mickish, D. J., and Kunz, A. B., *Phys. Rev.* **B10**, 2602 (1974).
Pines, D., *Physics Today* **34**(11), 106 (1981).
Schulter, M., and Sham, L. J., *Physics Today* **34**(2), 36 (1982).
Seitz, F., Turnbull D., and Ehrenreich, H. (Eds.), "Solid State Physics" (All Volumes), New York, Academic Press.
Smith, J. R. (Ed.), "Theory of Chemisorption," Berlin, Springer-Verlag (1980).

Cross-references: CRITICAL PHENOMENA, ENERGY LEVELS, EXCITONS, FERMI SURFACE, FERROMAGNETISM, MAGNETIC RESONANCE, MAGNETISM, PHONONS, SEMICONDUCTORS, SOLID-STATE DEVICES, SOLID-STATE PHYSICS, SUPERCONDUCTIVITY, TUNNELING.

SPACE PHYSICS

Space physics is usually taken to be the physics of phenomena naturally occurring in space, but frequently space is not clearly defined. It normally includes at least the outer portions of planetary atmospheres (in which densities are low enough to permit satellite flight), the outer portion of the sun's atmosphere, and interplanetary space. It frequently also includes galactic space and intergalactic space but these are more often considered within the context of astrophysics. It is also sometimes taken to include planetary studies and the study of the earth's atmosphere at altitudes above those attainable by balloons. With such a broad coverage, there are many special areas of study included within the scope of space physics, and several of these are described separately; for example, see ASTROPHYSICS, COSMIC RAYS, IONOSPHERE, PLANETARY ATMOSPHERES, RADIATION BELTS, RADIO ASTRONOMY, and SOLAR PHYSICS.

Earth's Outer Atmosphere The earth's atmosphere extends far out into space, something which would not necessarily be expected on the basis of atmospheric properties near the earth's surface. There are two factors involved in this great extension. First, above 100 km, the atmospheric temperature increases rapidly with altitude, causing an outward expansion of the atmosphere far beyond that which would have occurred had the temperature stayed within the bounds observed at the earth's surface. Second, above about 100 km, the atmosphere is sufficiently rarefied so that the different atmospheric constituents attain diffusive equilibrium distributions in the gravitational field; the lighter constituents then predominate at the higher altitudes and extend farther into space than would an atmosphere of more massive particles. This effect is enhanced by the dissociation of some molecular species into atoms. The pressure p at altitude h, in terms of the pressure p_0 at the earth's surface, is given by

$$p = p_0 \exp\left(-\int_0^h \frac{\overline{m}g\,dh}{kT}\right) \qquad (1)$$

where $\overline{m}$ is the average mass of the atmospheric particles, g is the acceleration of gravity, k is the Boltzmann constant, and T is the atmospheric temperature. It is clear from this expression that the high temperatures and low particle masses above 100-km altitude act to maintain pressures at still higher altitudes in excess of those that would exist if the temperature and molecular weight were constant with altitude. Where diffusion equilibrium prevails, Eq. (1) applies to each constituent separately, provided $\overline{m}$ is replaced by the particle mass for the particular constituent under consideration and p_0 is the partial pressure at some reference altitude where h is taken to be equal to zero.

The composition of the atmosphere does not change much up to 100 km; there is a region of maximum concentration of ozone (still a very minor constituent) near 20 to 30 km (see PLANETARY ATMOSPHERES), the relative concentration of water vapor falls markedly from its average sea-level value up to 10 or 15 km, and the relative abundance of atomic oxygen begins to become appreciable on approaching 100 km, due to photodissociation of oxygen by ultraviolet sunlight. Above 100 km, atomic oxygen rapidly increases in importance, due to the combined influence of photodissociation and diffusive separation in the gravitational field; above 200 km, atomic oxygen is the principal atmospheric constituent for several hundred kilometers. However, helium is even lighter than atomic oxygen, so its concentration falls less rapidly with altitude, and it finally replaces atomic oxygen as the principal atmospheric constituent above some altitude which varies with the sunspot cycle between 600 and 1500 km. At still higher altitudes, atomic hydrogen finally displaces helium as the principal constituent. The hydrogen extends many earth radii out into space and constitutes the telluric hydrogen corona, or geocorona.

The temperature of the upper atmosphere, and hence its density, varies with the intensity of solar ultraviolet radiation, and this in turn varies with the sunspot cycle and with solar activity in general. The solar radio-noise flux is a convenient index of solar activity, since it can be monitored at the earth's surface, and since it varies more or less proportionally with the extreme ultraviolet emissions most important in heating the upper atmosphere. The minimum nighttime temperature of the upper atmosphere above 300 km has been expressed in terms of the 27-day average of the solar radio-noise flux $\bar{S}$ at 8-cm wavelength, as follows,[9]

$$T = 280 + 4.6\bar{S}$$

This varies from about 600 K near the minimum of the sunspot cycle to about 1400 K near the maximum of the cycle. The maximum daytime temperature is about one-third larger than the nighttime minimum.

Another important heat source for the upper atmosphere or thermosphere exists in the polar regions due to auroral and allied processes. Under geomagnetically disturbed conditions, auroral processes are enhanced and the associated strong heating in the polar regions causes a major perturbation in the general circulation of the thermosphere.[4] The normal circulation of the mid- and upper thermosphere under geomagnetically quiet conditions transfers heat from low lattitudes to high, as well as to the night side of the earth, but under geomagnetically disturbed conditions the circulation transfers heat away from the polar regions.

Magnetosphere The magnetosphere is that region of space in which the geomagnetic field dominates the motion of charged particles. Near the surface of the earth, the geomagnetic field resembles that of a dipole; the best-fit dipole is off center about 440 km and is inclined about $11°$ to the earth's axis of rotation. Well out into space, the field is severely deformed so that it no longer resembles a dipole field. The most apparent deformation is that due to a plasma of charged particles flowing away from the sun, a flow generally referred to as the solar wind. When the kinetic energy of the directed flow of the solar wind exceeds the energy density of the geomagnetic field, the plasma displaces the field. The diamagnetic properties of the flowing plasma tend to compress and confine the geomagnetic field,[2] at least on the side facing the sun. On the side away from the sun the geomagnetic field is very extended, perhaps indefinitely. Calculations indicate that the surface of the magnetosphere facing the sun is roughly hemispherical, with dimples over the earth's magnetic poles. Observations by space probes indicate that the distance that the geomagnetic field extends towards the sun is about 10 earth radii,[7] quite variable with the intensity of the solar wind.

The magnetosphere is a region within which many geophysical phenomena occur. Most deeply embedded within it are the plasmasphere and the Van Allen belts (see RADIATION BELTS). The plasmasphere consists of an upward extension of the ionosphere and hence it is made up of relatively low-energy particles; it has a sharp outer boundary that exhibits a diurnal pattern of movement. Beyond the plasmasphere there is virtually no low-energy plasma. The Van Allen radiation belts consist of high-energy particles, both electrons and ions, trapped within the geomagnetic field. The inner belt consists of protons with energies in the 10 to 100 MeV range and it is centered about half an earth radius above the geomagnetic equator. The outer belt consists of electrons with energies from 1 keV to 5 MeV that fill most of the magnetosphere almost out to the auroral zone. The tail of the magnetosphere is divided into two parts with oppositely directed field—one half with field lines directed away from the sun and connecting with the south polar region of the earth, and one half with field lines directed toward the sun and connecting with the north polar region of the earth. The neutral sheet separates the two regions of oppositely directed field, and the plasma sheet is a broader region extending on both sides of the neutral sheet containing more energetic particles than those in the neutral sheet.

The solar plasma that compresses the geomagnetic field and limits the magnetosphere flows away from the sun at a velocity that might be described as hypersonic—the ordered velocity greatly exceeds the average random thermal velocity of the particles. Although the medium is so rarefied that collisions are rare, the particles can interact with one another through the agency of magnetic fields contained within the plasma. As a result, a collisionless shock wave develops in the flow before it reaches the surface of the magnetosphere.[16] Space probes have shown that the shock front lies about 4 earth radii beyond the surface of the magnetosphere in the direction of the sun.[7]

Aurorae are luminosities in the upper atmosphere at high latitudes caused by energetic particles that flow from the outer magnetosphere into the atmosphere. The solar wind interacts with the outer magnetosphere so as to produce a dynamo action that provides the energy source for the aurorae. Electrons driven by this source provide the most spectacular displays because they enter the atmosphere in relatively well defined structures. Protons provide more diffuse, widespread aurorae. The degree of auroral activity depends largely upon solar activity and hence tends to follow large solar flares, extending to much lower latitudes than normal displays. (See AURORA and AIRGLOW.)

Interplanetary Space Up until about 1954, interplanetary space was thought to be essentially a perfect vacuum, devoid of interesting physical phenomena except for occasional sporadic events such as the ejection of gas clouds by the sun or the passage of comets. It is now recognized that interesting physical phenomena are always present. The solar corona expands continually, giving rise to a steady outstreaming into interplanetary space of ionized gas from the sun. As was mentioned above, this outflow of ionized gas is generally referred to as the solar wind.

The steady outflow of particulate matter from the sun was first recognized by Biermann in 1954 on the basis of the deflection of comet tails away from the sun—a deflection that was too great to be explained on the basis of light pressure. Biermann at first overestimated the strength of the outflow, because of an underestimate of the strength of the interaction between the solar plasma and the cometary plasma. The solar wind was first observed directly and measured with some accuracy in the spacecraft Mariner II.[14] This showed that the concentration of solar material near the earth was of the order of 5 protons/cm^3, along with a corresponding concentration of electrons, moving with a velocity of about 500 km/sec.[14]

Parker[10] has given the most satisfactory explanation of the solar wind, describing it as a continuous hydrodynamic expansion of the solar corona, with a continuous heat input into the outflowing gas for a substantial distance near the sun. The flow can be compared to the flow of gas through a rocket nozzle, where, in the case of the sun, gravity plays the role in restricting the gas flow that the throat plays in the rocket nozzle. Parker has also shown that the outflowing, electrically conducting gas must pull the solar magnetic field out radially and that the rotation of the sun must twist the radial pattern into a spiral. This provides a magnetic connection, or a guiding path, from the western portion of the sun to the earth for any cosmic radiation produced in solar flares, something that is confirmed by observation.

A surprising property of interplanetary space as observed by space probes, is the irregularity of the magnetic field. Although the average orientation of the field agrees with the spiral pattern predicted by Parker, many irregularities are present.[7] The magnetic field energy density is mainly due to the irregular fields, and it is approximately equal to the thermal energy of the solar wind particles, which is approximately 1 per cent of the energy of ordered flow. The real significance of these observations seems not to have been recognized at this writing.

A satisfying concept for the termination of the solar wind at some finite distance from the sun has been provided by Axford et al.[1] As the solar wind moves outward from the sun, its concentration falls according to an inverse square law, and the dynamic pressure that can be generated by stopping it falls accordingly. At some point, it must become so attenuated that its continued outward hypersonic flow will be stopped by the galactic magnetic field or by charged particles in galactic space. At this point, there must be a shock front and a conversion of ordered energy of flow into disordered thermal energy. The heated gas beyond the shock front should cool mainly by charge exchange between the high-temperature protons from the solar wind and cool hydrogen atoms from galactic space. After charge exchange, the hydrogen atoms will carry the energy away. The cool proton gas left behind is less electrically conducting than was the hotter gas, and the magnetic field is gradually released from the proton gas, allowing the magnetic field lines to merge and the proton gas to drift out into galactic space.

The hydrogen atoms that are heated beyond the shock front have high enough velocities to penetrate far into the solar system, even to the vicinity of the earth's orbit, before becoming ionized by solar ultraviolet radiation. The hydrogen atoms within a few astronomical units from the sun scatter hydrogen Lyman-alpha radiation emitted by the sun, and this scattered radiation can be detected with instrumentation flown in rockets and spacecraft.[3] The rocket observations can be used to determine the concentration of the high-velocity neutral hydrogen atoms in interplanetary space, and this in turn can be interpreted in terms of the distance from the sun to the shock front beyond which the hydrogen originates. Patterson et al.[11] have shown that this interpretation indicates that the distance to the shock front is about 20 astronomical units (AU). The solar magnetic field spirals around about three times in this distance, and in the outer portion of the solar system, the magnetic field lines in the plane of the ecliptic are approximately circular with the sun at the center.

Such a pattern of magnetic field in the solar system can be expected to produce significant anisotropies in the cosmic radiation, whether of solar or galactic origin. This is most pronouncedly true when cosmic radiation is released from a point on the sun that is connected by a magnetic field to the vicinity of the earth, in which case the cosmic radiation appears to approach the earth's magnetosphere from a point forty or fifty degrees to the west of the sun. A weak anisotropy is produced in the low-energy galactic cosmic radiation, and the anisotropy becomes smaller for the higher energy radiation.

Moon The nature of the interaction of the solar wind with a planetary body depends upon a number of factors including the conductivity of the planet, the characteristics of its atmosphere, and the presence of a planetary magnetic field, with each of these able under certain conditions to cause the deflection of the solar wind around the planet. In the case of the moon, there is neither atmosphere nor magnetic field to stop the solar wind, and the conductivity of the moon, mainly near the surface, controls the interaction.[5] Spacecraft observations have shown that the solar wind impinges directly on the lunar surface,[8,15] thus indicating that the moon is a poor conductor at its surface.

Measurements on the lunar surface have shown that magnetic field variations in the solar wind are magnified near the lunar surface on the sunward side. This indicates significant conductivity somewhere in the lunar interior, which permits the induction of internal currents that resist the field change, causing an enhancement of the field charge near the lunar surface. There are residual magnetic fields in lunar rocks, indicating the presence of magnetic field in an earlier era.

Planets There is great diversity among the planets and their influences upon physical phenomena occuring in space. Mercury, the closest planet to the sun, has no atmosphere. It does have an intrinsic magnetic field sufficiently strong to deflect the solar wind most of the time;[12] however the solar wind may impinge upon the planetary surface when it is usually strong.

Venus has a massive atmosphere, while that of Mars is thin, but still sufficient to produce an ionosphere. Venus does not have sufficient magnetic field to deflect the solar wind, whereas Mars apparently has enough field to deflect the solar wind when it is weak.[12] At Venus and most of the time at Mars, the solar wind interacts with the conducting atmosphere in such a way as to produce a shock wave in the solar wind and a deflection of the flow around the planet, but the details of the interactions are not entirely clear. The interaction between the solar wind and the atmosphere is not satisfactorily understood, but it is clear from observation that the interaction does produce a shock wave in the solar wind. Both planets have warm thermospheres, though not nearly so warm as the earth's.

Jupiter has a massive magnetic field[12] and hence the solar wind is kept well away from the planet. The magnetic field also has a strong asymmetry that strongly influences the interaction with the solar wind. Jupiter emits energetic electrons into space on a cyclic basis controlled by the asymmetry. Its behavior resembles that of a pulsar.[16] The energy source is the rotation of the planet. Jupiter is a strong source of radio noise, some of it triggered by its satellite Io. Jupiter has a belt of trapped radiation much more intense than the Van Allen radiation belt around the earth.

Saturn has a weak magnetic field compared to Jupiter,[12] but still strong enough to deflect the solar wind. The magnetic field seems not to have pronounced anomalies in it, and it is closely aligned with the rotation axis of the planet. Saturn emits radio noise sporadically. The energy source for the magnetospheric dynamics is the solar wind, similar to the case on earth and unlike that for Jupiter.

Both Saturn and Jupiter have aurorae that can be observed in the ultraviolet.

FRANCIS S. JOHNSON

References

1. Axford, W. I., Dessler, A. J., and Gottlieb, B., "Termination of Solar Wind and Solar Magnetic Field," *Astrophys. J.*, 137, 1268–1278 (1963).
2. Beard, D. B., "The Solar Wind Geomagnetic Field Boundary," *Rev. Geophys.*, 2, 335–336 (1964).
3. Broadfoot, A. L., et al., "Overview of the Voyager Ultraviolet Spectroscopy Results through Jupiter Encounter," *J. Geophys. Res.*, 86, 8259–8284 (1981).
4. Dickinson, R. E., Ridley, E. C., and Roble, R. G., "A Three-Dimensional General Circulation Model of the Thermosphere," *J. Geophys. Res.*, 86, 1499–1512 (1981).
5. Johnson, F. S., and Midgley, J. E., "Notes on the lunar magnetosphere," *J. Geophys. Res.*, 73, 1523–1532 (1968).
6. Kupperian, J. E., Jr., Byram, E. T., Chubb, T. A., and Friedman, H., "Extreme Ultraviolet Radiation in the Night Sky," *Ann. Geophysique*, 14, 329–333 (1958).
7. Ness, N. F., Scearce, C. S., and Seek, J. B., "Initial Results of Imp I Magnetic Field Experiment," *J. Geophys. Res.*, 69, 3531–3569, (1964).
8. Ness, N. F., Behannon, K. W., Scearce, C. S., and Cantarano, S. C., "Early results from the magnetic field experiment on lunar Explorer 35," *J. Geophys. Res.*, 72, 5769–5778 (1967).
9. Nicolet, M., "Solar Radio Flux and Temperature of the Upper Atmosphere," *J. Geophys. Res.*, 68, 6121–6144 (1963).
10. Parker, E. N., "Interplanetary Dynamical Processes," pp. 272, New York, Interscience Publishers, 1963.
11. Patterson, T. N. L., Johnson, F. S., and Hanson, W. B., "The Distribution of Interplanetary Hydrogen," *Planetary Space Sci.*, 11, 767–778 (1963).
12. Russel, C. T., "Planetary Magnetism," *Rev. Geophys., Space Physics*, 18, 77–106 (1980).
13. Sandel, B. R., et al., "Extreme Ultraviolet Observations from Voyager 2 Encounter with Saturn," *Science*, 215, 548–553 (1982).
14. Snyder, C. W., Neugebauer, M., and Rao, U. R., "The Solar Wind Velocity and Its Correlation with Cosmic Ray Variations and with Solar and Geomagnetic Activity," *J. Geophys. Res.*, 68, 6361–6370 (1963).
15. Sonett, C. P., and Colburn, D. S., "Establishment of a lunar unipolar generator and associated shock and wake by the solar wind," *Nature*, 216, 340–343 (1967).
16. Spreiter, J. R., and Jones, W. P., "On the Effect of a Weak Interplanetary Field on the Interaction between the Solar Wind and the Geomagnetic Field," *J. Geophys. Res.*, 68, 3555–3565 (1963).
17. Vasyliunas, V. M., and Dessler, A. J., "The Magnetic Anomaly Model of the Jovian Magnetosphere: A Post-Voyager Assessment," *J. Geophys. Res.*, 86, 8435–8446 (1981).

Cross-references: AIRGLOW, ASTROPHYSICS, AU-
RORA, COSMIC RAYS, IONOSPHERE, IONIZA-
TION, PLANETARY ATMOSPHERES, RADIO AS-
TRONOMY, SOLAR PHYSICS.

SPARK AND BUBBLE CHAMBERS

When particles such as protons, neutrons, or
mesons, occurring naturally in the cosmic radiation or produced copiously by accelerators,
move through matter, they can produce nuclear
interactions which are studied in nuclear and
high-energy or particle physics. Electrically
charged particles also produce pairs of electrons
and positively charged ions in frequent *atomic*
collisions, which are used to detect and trace
the paths taken by incident and outgoing
particles in the nuclear collisions.

In spark chambers, the electrons freed in a gas
are accelerated by electric fields between electrodes, so that in further atomic collisions they
can free additional electrons, leading to electron
avalanches. Space charge effects, also due to the
positively charged ions, and energetic photons
due to excitation of atoms contribute additional
avalanches. "Streamers" result which increase in
length until they reach the electrodes at which
point spark breakdown occurs, the location of
which is recorded. The necessary high voltage
potential is applied to the electrodes only after
passage of a particle, making spark chambers
quite insensitive to background radiation, but
allowing them to be triggered by interaction
events under investigation.

Parallel-plate spark chambers consist of a number of spark gaps formed by metal plates or foils
separated by plastic rectangular frames. Chambers consisting of concentric cylindrical electrodes have also been built. The gaps of 2 to
15 mm must be maintained to accuracies of a
few per cent over the full area, up to 10 m^2, of
the plates. In thick, heavy metal plates, entering
particles can be absorbed by ionization losses so
that their ranges and thus energies can be measured, or they can produce electron showers or
nuclear interactions. More often, however, thin
aluminum foils are preferred merely to track
particles originating, for instance, from a liquid
hydrogen or deuterium target; high-energy collisions of elementary particles with individual
protons or with the one neutron in deuterons
are much easier to interpret than events in heavier nuclei where interaction products can
undergo secondary collisions.

The spark gaps are usually filled with a gas
near normal pressure. Operation at higher gas
pressures, and perhaps with liquids, is also possible if the origins of interactions are to be
viewed or if very high accuracy for spark location is required. Most frequently used is a mixture of 90 per cent neon and 10 per cent helium
which usually is continuously recirculated for
purification. If every electron produces α additional electrons per centimeter of drift space,
then an avalanche containing about 10^8 electrons can develop if a minimum drift space (or
minimum gap width) $d = 20/\alpha$ cm is available.
From this "Raether criterion" many operating
features of a spark chamber can be predicted. α
is the "first Townsend coefficient" which depends on the gas used and on the ratio E/p of
applied electric field to gas pressure. For instance, for neon at atmospheric pressure and for
a field of 10 kV/cm, $\alpha = 65$. Thus $d \geqslant 3$ mm. If
the Raether condition is not met, sparks are not
intense or are not formed at all. To obtain
spark efficiencies near 100 per cent in a gap, the
rise time, duration, and height of the voltage
pulse must also be adjusted carefully, especially
when many tracks are to be formed simultaneously.

The sensitive time in a spark chamber is defined by the maximum period between arrival
of an event to be recorded and application of
the voltage trigger pulse. It must be short
enough to avoid recording of background events
and can be decreased by application of a clearing
field of 100 to 200 V/cm between pulses or by
an admixture of, for instance, 1 per cent sulfur
dioxide or alcohol to the gas filling. For multi-
plate chambers the sensitive time can be $\geqslant 1$
μsec. The number of events that can be recorded per unit time depends on the repetition
time for the whole apparatus. If the sparks are
photographed, this time is $\geqslant 0.1$ sec. TV vidicon
cameras can be used, where the image focused
on a photoconductive surface is scanned by an
electron beam and the information stored on
magnetic tape or fed directly into a digital
computer. The Plumbicon operates similarly
but with better space resolution. Repetition
times $\geqslant 30$ msec can be expected. The space
resolution or accuracy of 0.2 to 1 mm obtainable with a spark chamber is defined by the
displacement between actual passage of a particle and resulting spark. An angle $< 45°$ between
direction of travel of the particle and electrode
surface usually results in insufficient accuracy.
Accuracy is also affected by time delay of high-
voltage pulse, gas properties, Larmor precession
in magnetic fields, etc.

Besides emitting light, sparks also emit sound
the travel time of which can be recorded, as in
acoustic chambers, and of course constitute
electric currents used in wire spark chambers,
where wires (often aluminum), $\geqslant 100 \mu$m in
diameter and 1 to 2 mm apart, are stretched and
fastened across plastic frames to form planes
or cylinders. A spark between two wires in two
planes produces a current which can be "read
out," for instance, after letting it pass through a
computer-type ferrite core at the end of every
wire, or after causing a sound wave in a magnetostrictive cobalt-iron wire laced across all
the ends of the wires in a plane, or after charging a capacitor, the latter being insensitive to

magnetic fields. So far magnetostrictive read-out is the cheapest, requiring less electronic circuitry. For unambiguous determination of the location of two or more tracks, three sets of planes with wires at different angles are needed. Wire spark chambers are relatively inexpensive, highly efficient, have repetition times ≥ 2 msec, and supply digitized information on coordinates. Besides magnetic tape records, it is possible to put at least a fraction of the data pulses into a digital computer on line to the read-out system and display output data on an oscilloscope in the form of tables or histograms so that experiments can be continuously monitored.

In wide-gap chambers the electrodes are up to 60 cm apart so that under proper conditions, including a pulse of several hundred kilovolts, a long spark is produced which follows the path of a particle if inclined $>45°$ to the electrode surface. In a magnetic field the track becomes part of a circle.

In the streamer chamber the duration of the high-voltage pulse is kept so short that a spark cannot develop, but individual streamers, a few millimeters long, are formed from electron avalanches and are photographed end-on, parallel to the direction of the electric field, through wire electrodes. Vidicon viewing usually requires image intensification. Particles must travel at *small* angles to the electrode planes. The density of the streamers along a track is a measure for the velocity of a particle, which decreases with increasing ionization density.

In multiwire proportional chambers, gas amplification proceeds only to the point of avalanche formation. Between planes of tungsten or other metal wires (≥ 10 μm diameter) a dc potential difference of a few thousand volts is maintained. The electric field is highest near the fine wires where an avalanche due to a free electron can form and produce a signal which must be amplified at the end of the wire before further transmission. For higher ionization densities, several avalanches form near a wire so that a proportionally stronger signal will be transmitted. Repetition time is about 1 μsec, several orders of magnitude faster than for other types of chambers. Since the high voltage is applied continuously and since they respond in $\geq 2 \times 10^{-2}$ μsec, multiwire proportional chambers can be used to trigger other chambers. They are practically 100 per cent efficient for any number of particles and their read-out is not sensitive to magnetic fields.

If a free electron is produced by ionization in a space with known distribution of an electric potential, the electron will "drift" through the gas toward the anode with a velocity that can be predicted accurately from measured data on drift velocities. If the time when a fast ionizing particle entered the space is known by means of a signal from a fast-acting detector, such as a scintillator, and if the time when the electron produces a signal at the anode is measured, then the location where the particle passed through the space can be determined. This has been accomplished rather recently by means of "drift chambers" which, again, consist of arrays of wires, and which can provide excellent space resolution of ≤ 100 μm at relatively low cost.

At high-energy accelerators, experimental arrangements can involve many kinds of electronic detectors simultaneously. A "spectrometer facility" might make use of a large magnet to determine particle momenta from track curvatures. The magnet gap may contain a liquid-hydrogen target, surrounded and followed by some of the mentioned electronic chambers. Additional equipment can be mounted outside the magnet, forming "arms" of the spectrometer. Events produced in the target by a beam of particles are selected by coincidences in a trigger system, combining scintillators, Čerenkov counters, or multiwire proportional or drift chambers, and very fast associated circuitry. The whole apparatus is monitored by an on-line computer and all data are stored on magnetic tape for further analysis. High-statistics experiments involving events with many secondary particles distributed over large solid angles are possible with this very powerful technique.

For comparison, the liquid in bubble chambers can act as a target so that the origins of events, as well as secondary events occurring nearby, are visible over a full 4π-solid angle. Excellent space resolution of 50 to 150 μm is provided, depending on the size of a chamber. However, sensitive times are ≥ 0.5 msec, much longer than for spark chambers, and repetition times usually >50 msec, depending on size. Furthermore, bubble chambers cannot be triggered for selected events. For these reasons high-statistics experiments are not feasible. The spark and bubble chamber techniques have complemented each other very well, the discoveries with one technique often stimulating further research with the other. For example, neutrino interactions were first studied with spark chambers, but neutrino interactions in hydrogen and deuterium have been studied with very large bubble chambers. On the other hand, many of the hundreds of strongly interacting particles were first discovered in bubble chambers and then their properties were further investigated in spark chamber experiments.

A bubble chamber consists of a pressure vessel filled with a liquid, kept at proper temperature and pressure, which becomes superheated when expanded by means of a piston or diaphragm. A low-energy electron (δ-ray), produced at this time by an atomic collision of a particle passing through the liquid, generates a number of ion pairs which upon recombination cause local heating, or a heat spike, so that a bubble can begin to grow. The heat spike would diffuse in < 1 μsec, which is insufficient to trigger the expansion system. The number of electrons, and therefore bubbles produced per unit length, again depends on the particle's velocity. After a time ≥ 0.3 msec bubbles grow large enough to be photographed. The liquid is then immediately compressed to reliquefy the bubbles. This produces some heat which must be carried off by conduction and convection through the liquid and exchanged at cooling surfaces. At least three glass or quartz windows are provided for complete stereoscopic viewing with camera lenses of high quality, to match the available space resolution. The available depth of the chamber must also be fully covered by the lenses, requiring small apertures and sensitive photographic film. While bubbles in heavy liquids can be dark-field illuminated at 90° to the viewing axis, cryogenic liquids have low indices of refraction and require dark- or bright-field illumination at small angles. Of particular interest is the wide application of a specially developed Scotchlite reflecting light within a narrow cone into the photographic lens located at the center of a ring-shaped light source. Usually, bubble chambers are mounted inside magnets providing fields as high as 40 kilogauss. Some very large magnets have employed superconducting niobium-titanium wires embedded in and stabilized by copper. Thus large amounts of electric power are saved, which will soon amortize the initial investment for required helium refrigerators operating at 4.2 K.

In heavy-liquid bubble chambers, mixtures of propane and freon have been used, which due to their short radiation lengths are especially important for detection of gamma rays by conversion to electron pairs. They are operated near room temperature at pressures up to 20 atm and must be expanded by about 3 per cent requiring very powerful actuating systems for expansions. Studies of interactions in liquid hydrogen or deuterium avoid complications due to heavy nuclei, but require cryogenic techniques at operating temperatures between 25 and 30 K. Cryogenic bubble chambers are expensive and require careful safety precautions and relatively large operating staffs. They are expanded by ≤ 1 per cent from pressures ≤ 8 atm. Liquid neon or neon-hydrogen mixtures can also be used for gamma-ray detection. "Track-sensitive targets" have been developed, consisting of transparent plastic containers filled with hydrogen or deuterium mounted in a bubble chamber filled with neon-hydrogen and expanded simultaneously with the chamber. A cryogenic bubble chamber requires multilayered aluminized plastic insulation in a vacuum chamber which may be shared with the magnet if it is superconducting. Cryogenic bubble chambers containing as much as 30 m³ of liquid are in use.

Also of interest are sonic chambers which are expanded >20 times per second by producing a standing sound wave between two plates in the liquid. Sonic chambers, as well as other rapid-cycled chambers are especially useful at rapid-pulsed linear accelerators, in contrast with synchrotrons where expanding a chamber two or three times during the less frequent accelerator pulse is usually sufficient.

Bubble chamber and spark chamber photographs store a large amount of data which must be extracted by first scanning the film for events that may satisfy the requirements of a particular experiment. For simple interactions computer-guided pattern recognition methods may be applied in conjunction with human intervention and interrogation, combining the scanning and subsequent measuring processes. More often the scanned film is transferred to measuring machines of varying complexity where positions of bubble or spark images are measured with commensurate accuracy. As an example, on a flying-spot digitizer, a light spot, 15 μm in diameter, scans a photograph in a few seconds. A bubble image causes a signal in a photomultiplier, which is compared with time pulses similarly produced by a grating scanned by a synchronized light spot, locating the image with respect to fiducial marks. Hundreds of events per hour can thus be measured with excellent precision. In the spiral reader a fine slit spirals over the photograph, again producing light signals. A flying-image digitizer has scanned many different portions of a photograph simultaneously so that thousands of events were processed per hour. The processing of spark or bubble chamber data requires large amounts of computer time, on-line to scanning and measuring equipment and off-line for spatial reconstruction of events from the three or more stereoscopic views, for kinematic fitting and selection of probable events, and finally for interpretation of results.

R. P. SHUTT

References

Shutt, R. P., Ed., "Bubble and Spark Chambers," Vols. I and II, New York and London, Academic Press, 1967.

Aleksandrov, Yu. A., Voronov, G. S., Gorbunkov, V. M., Delone, N. B., and Nechayev, Yu. I., "Bubble Chambers" (W. R. Frisken, Ed.), Bloomington, Ind., Indiana Univ. Press, 1967.

Allkofer, O. C., "Spark Chambers," Munich, Verlag Karl Thiemig KG, 1969.

Charpak, G., "Evolution of the Automatic Spark Chambers," *Ann. Rev. Nuclear Sci.*, 20 (1970).

Rice, Evans, P., "Spark, Streamer, Proportional and Drift Chambers," London, Richelieu Press, 1974.

Cross-references: ACCELERATOR, PARTICLE; IONIZATION; NUCLEAR INSTRUMENTS; NUCLEAR RADIATION; NUCLEAR REACTIONS.

SPECTROSCOPY

The term *spectroscopy* generally refers to the study of spectra of some quantum mechanical particles. The spectrum is a display of these particles arranged in order of their energy values. Traditionally, the term was used in a rather restricted sense to refer only to electromagnetic waves, i.e., the photons. In present practice, the use of the term has been substantially broadened to include various particles, as in mass spectroscopy, β-ray spectroscopy, neutron spectroscopy, etc., where the term is used jointly with the name of the particle in question. If no specific particle is mentioned, electromagnetic spectra are implied. In this article, spectroscopy is discussed as the study of the atomic and molecular properties which are revealed when they interact with an electromagnetic field. The main concern of spectroscopic study is (1) the

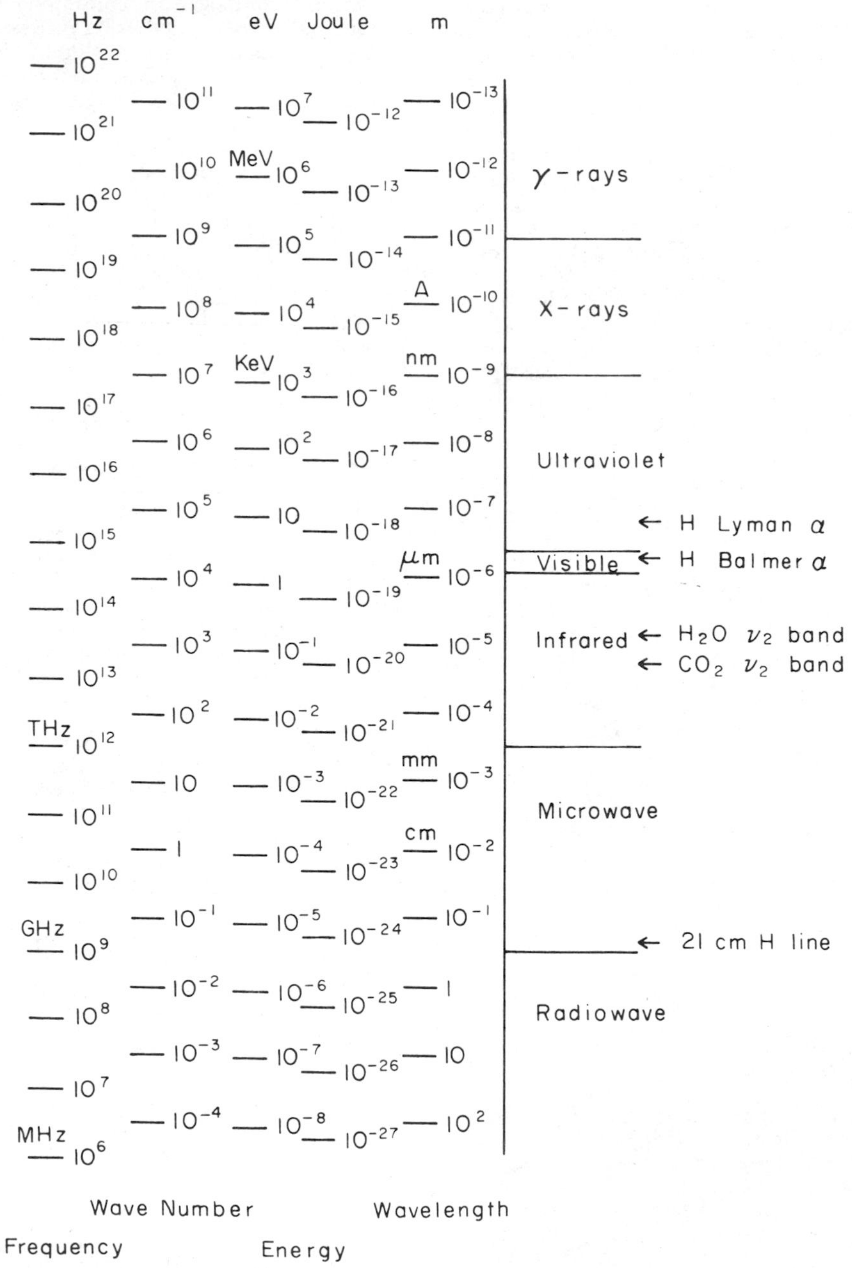

FIG. 1. Spectrum of electromagnetic waves.

atomic and molecular energy levels, and (2) the transitions which take place from a level to another as a result of their interaction with the electromagnetic field.

The electromagnetic spectrum is classified according to frequency or wavelength. The quanta of electromagnetism, photons, have a definite energy E given by the product of the frequency ν and the Planck's constant $h = 6.63 \times 10^{-34}$ J s:

$$E = h\nu.$$

Fig. 1 is a schematic diagram showing the electromagnetic spectrum, together with various units of energy, frequency, and wavelength. An atomic or molecular system interacts with an electromagnetic field through its emission or absorption process. The spectrum represents the emission or absorption of photons by the atoms or molecules. The various units which characterize the spectrum are shown in Fig. 1. Unfortunately, there is no conformity in usage of these units; spectroscopists working on a particular spectral region use their own preferred units.

General Description of Atomic and Molecular Transitions Both absorption and emission processes involve a quantum transition between two energy states of an atom or molecule. The energy difference covers a wide range, depending on the nature of the quantum transition involved (nuclear, electronic, vibrational, rotational, etc.).

Atoms and molecules have, in general, a size on the order of an angstrom (1 Å = 10^{-10} m). We can estimate the energy involved in electronic, vibrational, and rotational transitions using a simple atomic or molecular model having a size of 1 Å. The electrostatic energy which binds a nuclear charge of 1.6×10^{-19} coulomb and an electronic charge of -1.6×10^{-19} coulomb at a distance of 1 Å for a hydrogenlike atom shown (Fig. 2) is given by

$$E_e = \frac{Ke^2}{r} = \frac{9 \times 10^9 \times (1.6 \times 10^{-19})^2}{10^{-10}} \text{ J}$$

$$= 2.3 \times 10^{-18} \text{ J} = 14.4 \text{ eV}.$$

Balmer, in 1885, discovered that the four lines of the hydrogen spectrum (those marked α, β, γ, and δ in Fig. 3) formed a series. Subsequently, Rydberg found that those lines are given by

$$\bar{\nu} = R\left(\frac{1}{2^2} - \frac{1}{n^2}\right), \qquad n = 3, 4, 5, 6$$

where $\bar{\nu}$ is the *wavenumber*, which is equal to the inverse of wavelength, and R, the *Rydberg constant*, is equal to 109677 cm^{-1} = 13.6 eV, remarkably close to the value calculated above. Bohr formulated the atomic model that yields the energy level of a bounded electron in the

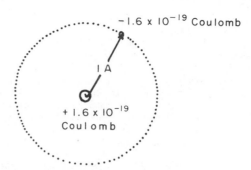

FIG. 2. The model of atom used for the calculation in the text.

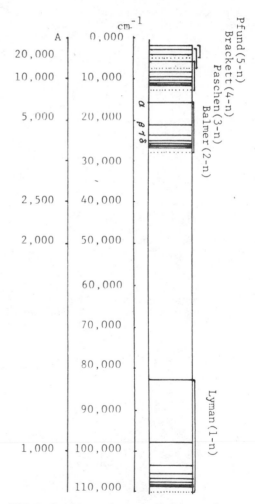

FIG. 3. Spectrum of atomic hydrogen: the wavenumber of these spectral lines is given by $\bar{\nu} = R(1/m^2 - 1/n^2)$, where $m = 1$ is for the Lyman series, $m = 2$ for the Balmer, $m = 3$ for the Paschen, $m = 4$ for the Brackett, and $m = 5$ for the Pfund.

electrostatic field of a proton by

$$E = -\frac{R}{n^2}$$

where n is the principal quantum number. The electronic energy levels in the valence shells of atoms and molecules are approximated by the Bohr model. Electronic transitions take place between two electronic energy states. They require energies ranging from several tenths of an eV to several eV. The corresponding spectral region ranges from the infrared to the extreme ultraviolet, as seen in Fig. 3.

Nuclear transitions involve the nuclear particles, the protons and neutrons. The electrostatic energy between two protons of 1.6×10^{-19} coulomb spaced at an average nuclear distance of 10^{-15} m is on the order of 10^6 eV. The nuclear binding energy is roughly equal to the electrostatic energy of this magnitude; without it, these particles would repel each other by the electrostatic force. A nuclear transition involves an energy of this magnitude, the spectrum of which falls in the γ-ray region.

The electrons in the inner shells of atoms have binding energies at least an order of magnitude higher than the Rydberg constant, since the electrons there are bound at distances that are a fractional part of the atomic size under the coulomb field of several proton charges. Inner shell electron transitions have a photon energy ranging from several hundred eV to several thousand eV, the spectrum of which falls in the X-ray region.

For a molecular vibrational transition, we can model a simple molecule of two 20-proton-mass nuclei spaced by a distance of 1 Å. The binding energy is approximated by the electrostatic energy 14.4 eV for two charges of 1.6×10^{-19} coulomb spaced at 1 Å. A quadratic function (Fig. 4) approximates the molecular potential at its minimum. We can use the quadratic potential for calculating the vibrational frequency. The vibrational force constant is then calculated by

$$\tfrac{1}{2} k x_0^2 = V_0$$

or

$$k = \frac{2V_0}{x_0^2} = \frac{2 \times (14.4 \text{ eV})}{(0.5 \text{ Å})^2} = 18.4 \times 10^2 \text{ N/m}.$$

We can calculate the vibrational frequency of the model molecule by

$$f_v = \frac{1}{2\pi} \sqrt{\frac{k}{\mu}} = \frac{1}{6.28} \sqrt{\frac{1840 \text{ N/m}}{10 \times 1.6 \times 10^{-27} \text{ kg}}}$$

$$= 5.4 \times 10^{13} \text{ Hz} = 1800 \text{ cm}^{-1}$$

where μ is the reduced mass of oscillator, 10 proton-masses of $10 \times 1.6 \times 10^{-27}$ kg. The obtained vibrational frequency 5.4×10^{13} Hz

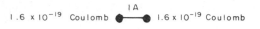

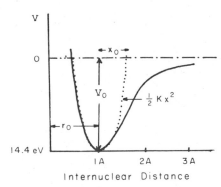

FIG. 4. The model of molecule used for the calculation in the text.

or the wavenumber of 1800 cm^{-1} falls in the mid-infrared region. A majority of simple molecules show their fundamental vibrational transition in the mid-infrared spectral range.

For molecular rotation, the model molecule has a moment of inertia

$$I = \tfrac{1}{2} m x_0^2 = \tfrac{1}{2} (20 \times 1.6 \times 10^{-27} \text{ kg})$$
$$\times (10^{-10} \text{ m})^2 = 1.6 \times 10^{-46} \text{ kg m}^2.$$

Its rotational energy is calculated by

$$E_R \approx \frac{(\text{angular momentum})^2}{(\text{moment of inertia})}$$

$$= \frac{(1.06 \times 10^{-34})^2}{1.6 \times 10^{-46}}$$

$$= 3.5 \times 10^{-23} \text{ J} = 2 \text{ cm}^{-1},$$

which falls into the far infrared or the microwave region.

The energy levels of a simple molecule can be classified into three major contributions: electronic, vibrational, and rotational. The molecular energy is expressed as a sum of these three terms:

$$E = E_e + E_v + E_R.$$

The electronic energy contribution is by far the largest. The molecular energy state is structured with an electronic energy state of fixed value E_e accompanied by many vibrational-rotational energy states. We can interpret this situation as an electronic energy state split into multiple levels characterized by different sets of E_v and E_R, the splitting being caused by the vibrational-rotational interaction. Most molecular transitions between electronic states are composed of several thousand lines, each of which is the transition characterized by a common value of the electronic energy change but a different value of the vibrational-rotational energy change. The very complex patterns exhibited by molecular transitions contrasts the simple patterns of atomic transitions, which we see for the case of the hydrogen spectrum in Fig. 3.

In general the quantum states of molecules and atoms are perturbed by various interactions. The molecular electronic energy state perturbed by the vibrational-rotational interaction is an example. The magnitude of interaction is quite different, depending on its nature. Table 1 summarizes the interactions observable in atomic and molecular systems, together with their magnitudes in Hz and cm^{-1}. Some interactions listed in Table 1 cause shifting of the atomic and molecular energy levels. In such cases, the transition frequency shifts accordingly. Most interactions, in addition to level shifting, cause level splitting. The corresponding spectral transition would exhibit a complex structure. When the resolution of the spectrometer is improved, a previously unobservable interaction may suddenly become observable, and a complicated structure may often be revealed. A challenge is always imposed to spectroscopic study (1) to discover complicated structures, and (2) to analyze the observed structures in terms of interactions hitherto unnoticed. Effort along this direction requires two elements: an improvement of spectroscopic technique and a development of theory to understand the interaction processes involved.

History of Spectroscopy Spectroscopy began long before the establishment of quantum mechanics. Isaac Newton documented the fact that the sunlight was dispersed into many colors when it was passed through a prism. Joseph Fraunhofer was instrumental in improvement of the art of spectroscopy. By the end of the

19th Century, spectroscopy had advanced to such a degree that the spectroscopic terminology now used for classifying observed atomic spectral lines was already established, even though the physics involved was not clearly understood until the advent of quantum mechanics. Indeed, the spectroscopic data collected in those days provided a driving force for development of quantum mechanics. By 1940, the quantum mechanical model of atoms and molecules as currently understood was firmly established, based on spectroscopic data taken mainly in the visible and the ultraviolet region during the 1920s and '30s.

The advent of the radiowave and microwave technology during the Second World War made it possible to extend spectroscopic measurement into these two regions. For the first time, atomic and molecular systems were investigated using a coherent monochromatic source. Spectral transitions which had previously been measured only in wavelength by applying optical techniques now became known in frequency. The development of magnetic resonance techniques was paralleled by microwave spectroscopy. The technique was applied to both electronic and nuclear spins. Transition frequencies for both spin resonances are very low, falling in the radiowave and microwave region. Spectroscopy extended to these low-frequency regions achieved a spectacular improvement in resolution, in part because the electromagnetic field applied for the spectroscopic measurement was coherent and monochromatic, and in part because the frequency itself was very low compared with the frequency resolution then available in other spectral regions. Extremely small transition frequency differences and consequently extremely weak interactions were measured with a great accuracy. W. Lamb succeeded in measuring the small energy difference between the

TABLE 1. ATOMIC AND MOLECULAR INTERACTIONS.

Magnitude of Atomic Interactions		
Electron Spin-Orbit (fine structure)	$\sim 10^{12}$–10^{13} Hz	10^2–10^3 cm^{-1}
Nuclear Spin-Orbit (hyperfine structure)	$\sim 10^{10}$–10^{11} Hz	10^{-1}–1 cm^{-1}
Nuclear Quadrupole Moment	10^{10}–10^{11} Hz	10^{-1}–1 cm^{-1}
Zeeman Effect	10^6 Hz/Gauss	10^{-4} cm^{-1}/Gauss
Stark Effect	10^7 Hz/(kV cm^{-1})^2	10^{-3} cm^{-1}/(kV cm^{-1})^2
Lamb Shift	10^{10} Hz	10^{-1} cm^{-1}

Magnitude of Molecular Interactions		
Rotation Electron Orbit	10^{10}–10^{11} Hz	10^{-1}–1 cm^{-1}
Rotation-Nuclear Spin	10^5–10^6 Hz	10^{-5}–10^{-4} cm^{-1}
Nuclear Spin-Nuclear Spin	10^5–10^6 Hz	10^{-5}–10^{-4} cm^{-1}
Zeeman Effect (Rotation-Nuclear Spin)	$\sim 10^3$ Hz/Gauss	$\sim 10^{-7}$ cm^{-1}/Gauss
First-Order Stark Effect	$\sim 10^6$ Hz/V cm^{-1}	$\sim 10^{-4}$ cm^{-1}/V cm^{-1}
Second-Order Stark Effect	10^4 Hz/(V cm^{-1})^2	10^{-6} cm^{-1}/(V cm^{-1})^2
	$\sim 10^5$ Hz/(V cm^{-1})^2	-10^{-5} cm^{-1}/(V cm^{-1})^2

hydrogen $2S_{1/2}$ and $2P_{1/2}$ levels (the *Lamb shift*), confirming the validity of quantum electrodynamics.

A question concerning coherent sources in the infrared, visible, and ultraviolet regions was finally solved during the early 1960s when laser devices were developed. The laser provided a source of extremely long coherence length, of extremely good monochromaticity, and of excellent directionality. Because of these characteristics, it provided a large energy density and power density, which could not be obtained with the use of thermal light sources. Spectroscopic measurements which were previously unthinkable became realized in the spectral region from infrared to ultraviolet.

Spectrometers The monochromator spectrometer, which disperses incident light into its spectral components, plays a primary role in spectroscopy, although the development of tunable laser devices has diminished its importance somewhat. Primitive instruments dispersed the spectral elements by means of a prism. Soon it was found that diffraction gratings used as dispersive elements were characteristically much more favorable than prisms in achieving high-quality measurement. The ultimate spectral resolution which a spectrometer can achieve is determined by the maximum optical path difference produced by the dispersive element [see Fig. 5(a)]. The resolution is in essence limited by the size of the grating or prism used. A. A. Michelson circumvented the limitation imposed by the small dispersive elements available in his days by his invention of the Michelson interferometer, shown in Fig. 5(b), and achieved remarkable results. Resolution in his interference spectroscopy technique was limited by the maximum optical distance through which the movable mirror of the interferometer traveled during the measurement of the visibility curve. Even though remarkable spectroscopic data were obtained using interference spectroscopy, the mainstream of spectroscopists in the 1920s and '30s stayed with diffraction grating monochromator instruments. The best spectral resolution achieved using grating instruments was on the order of 0.1 cm^{-1} (3 GHz). The grating spectrometer has remained predominant in the visible and the ultraviolet region, while a new

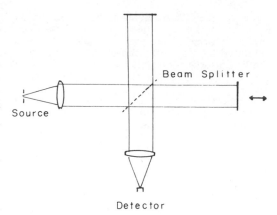

Beam Splitter

Source

Detector

FIG. 5(b). Michelson interferometer. The optical path difference is given by twice the difference between the distance to the stationary mirror and to the movable mirror measured from a common point in the incident beam.

type of spectrometer has emerged in the infrared region, where the collecting efficiency of spectral power is one of the most critical issues, because the detectors generally available for that region are rather insensitive. By 1950, two important discoveries had been made in infrared spectroscopy: (1) The Michelson interferometer turned out to be better than conventional grating instruments in collecting incoming spectral energy. (2) Two-beam interference could be used for simultaneous observation of all spectral elements in the incoming signal. The Michelson interferometer used as a spectrometer produces an output proportional to the Fourier transform of the incoming spectral signal. The spectrum is computed by applying the Fourier transformation to the output signal which is obtained as a function of the optical path difference between two interfering beams. Because of this feature, the technique is generally referred to as *Fourier transform spectroscopy*. It brought enormous success to infrared spectroscopy. The spectrum obtained for the CO_2 ν_2 vibrational-rotational band using the FTS technique is shown in Fig. 6(a). An exceptionally clean spectrum obtained with a very high resolution figure of 0.004 cm^{-1} indicates the power that this spectroscopic method has accomplished with a combination of the traditional and modern approach.

Selection Rules The atoms and molecules absorb photons to produce transitions from one energy state to another. The state reached by absorbing a photon must differ from the initial state by the energy and angular momentum of the absorbed photon. Since the photon is a Bose particle of spin quantum number 1, the angular momentum quantum number must be changed by 1 in an optically allowed transition of atoms and molecules. This condition is the

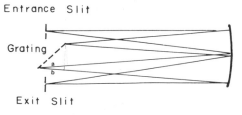

Entrance Slit

Grating

Exit Slit

FIG. 5(a). Typical grating spectrometer. This is called the Ebert-Fastie mount. The maximum optical path difference is given by $a + b$.

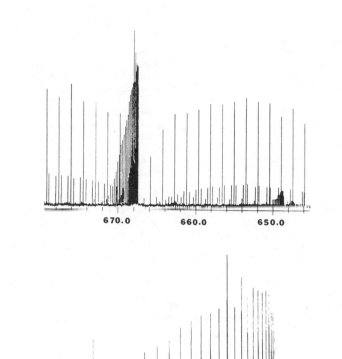

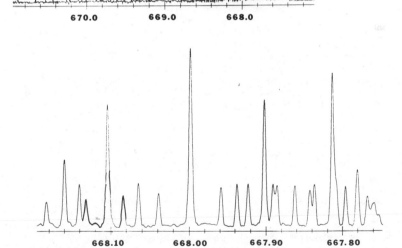

FIG. 6(a). The CO_2 spectral transitions observed in the 670 cm^{-1} region using an FTS instrument with a spectral resolution of 0.004 cm^{-1}. They are produced by its bending motion superposed by the rotational motion. In the top spectrum, the three branches (P, Q, and R) of the major band and the other minor bands are seen. A detail of the Q branch at the 667 cm^{-1} region is difficult to see. In the middle spectrum, the Q branch region is expanded to show three bands ($\nu_2 = 1 \leftrightarrow \nu_2 = 0$), ($\nu_2 = 2 \leftrightarrow \nu_2 = 1$), and ($\nu_2 = 3 \leftrightarrow \nu_2 = 2$). In the bottom spectrum, the region is further expanded. The line profile is observed to be Gaussian. (Courtesy of Henry Buijs, Bomem, Inc., Quebec, Canada.)

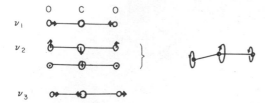

FIG. 6(b). Three vibrational modes of the CO_2 molecule. The bending ν_2 mode is a two-hold, up-and-down and in-and-out, motion shown in the figure. The combined two-hold motion produces the vibrational angular momentum parallel to the figure axis.

basis of the selection rules which formulate the possible optical transitions between two states. The strength of the transition is controlled by the electric dipole matrix elements defined between the two states. For a pure rotational transition, the dipole matrix element is proportional to the permanent dipole of the molecules. No pure rotational band is observable for a nonpolar molecule which has its electric charge distribution symmetric with respect to its center of mass. For the vibrational band, the matrix element is proportional to the first derivative of electric dipole with respect to the molecular geometry change which accompanies the vibrational motion in question.

The CO_2 molecule is linear in the ground state. No permanent electric dipole is associated with the ground state, therefore no pure rotational band is observed. The lowest vibrational frequency belongs to the bending motion shown schematically in Fig. 6(b). Since the bending motion of a linear CO_2 molecule is twofold, as indicated in Fig. 6(b), it carries an angular momentum along the molecular axis. If the optical transition may accompany a change in the vibrational angular momentum quantum number by 1, no change in the rotational angular momentum quantum number J takes place, so as to conserve the angular momentum of the whole system, the molecule and the photon. The transitions which belong to this category are classified as the Q branch ($\Delta J = 0$). If the vibrational angular momentum quantum number remains unchanged in the transition, the rotational quantum number J must change by +1 or −1 ($\Delta J = \pm 1$). The transition forms the P ($\Delta J = -1$) branch at the lower frequency side of the Q branch, and the R branch ($\Delta J = +1$) at the higher frequency side.

Practical Use of Spectroscopy The CO_2 spectrum described above demonstrates the extent of the information which can be collected by means of spectroscopy. The energy difference and the transition strength are the two quantities which are obtained in the observed spectral transition between two energy states. A collection of these observed values eventually reveals the exact nature of the entire energy states, and in turn it can be used to determine the field which specifies them. For the case of atomic spectroscopy it reveals the atomic field. For the case of molecular spectroscopy it reveals the molecular field. Spectroscopy in a particular frequency region yields a limited coverage of spectral transitions, hence the information collected is not extensive enough to reveal all aspects of the atoms or molecules in question. Nonetheless, spectroscopy with a limited frequency coverage is applied for the purpose of identifying atoms and molecules in a sample. For this purpose, characteristic transition frequencies exhibited by known atoms and molecules are sought in the sample, and identification is accomplished by finding them. Because of this feature, spectroscopy provides a vital tool in sample analysis for astronomy, chemistry, biology, and other sciences. Its practical merit lies in the fact that the information is transmitted through electromagnetic radiation, so that observation of spectra does not necessitate direct contact with the sample. The astronomer sitting on earth a million light years from an astronomical object is able to determine what atoms and molecules are in it by means of spectral observation of the electromagnetic radiation which arrives from it. Because spectroscopy can provide direct information on the nature of a substance as well as its environment without requiring direct contact, it is a very important experimental method not only for physics but for other branches of science as well.

Absorption and Emission Processes The spectral transition being observed is primarily controlled by two factors: the electric dipole matrix element and the population of atoms or molecules which occupy the initial state of the transition. A secondary parameter is the density of photons to which the atoms and the molecules interact. The atomic or molecular system establishes an equilibrium with the photon field by absorbing photons from the field and emitting photons back to the field. The absorption process is controlled by the photon density in the field, while the emission process consists of two terms: the spontaneous term, which is independent of the photon density in the field; and the induced (or stimulated) emission term, which increases its magnitude in proportion to the photon density in the field. If the upper state population is very small compared with the lower state population, the emission is predominantly spontaneous, and totally independent of neighboring atoms or molecules. Excitation to the upper states is accomplished either thermally in the case of thermal light, or by electron collisions in case of the electric discharge lamp. No laser action is present as long as the upper state population stays low. When this condition is broken, the emission is no longer dominated by the spontaneous process.

Profile of Spectral Lines The spectral lines emitted by a transition between two states are subject to the finite lifetime of the excited state. The uncertainty relation $\Delta E \Delta t \geqslant h$ demands that the spectral lines have a line profile of Lorentzian shape with half-width $\Delta E \approx h/\Delta t$ in accordance with the finite lifetime Δt of the excited state. If the emission is interrupted by collision with neighbors, the lifetime further decreases, resulting in an increase of the line width. In addition to the line broadening caused by finite lifetime, spectral lines are subject to inhomogeneous broadening processes. In a gas, each emitter moves at a different velocity and emits at a frequency slightly shifted from the value at rest because of the Doppler effect. Consequently, an ensemble of emitters produces a spectral line inhomogeneously broadened by the thermal motion of each emitter. A profile of the inhomogeneously broadened line differs from the Lorentzian profile broadened by a finite lifetime. It has a Gaussian shape because the emitters move in accordance with the Maxwell-Boltzmann velocity distribution law. If pressure is kept low, the absorption or emission spectral line exhibited by a gaseous sample is characterized by the Doppler broadened Gaussian line profile, as seen in the CO_2 spectral lines of Fig. 6.

Laser Spectroscopy The laser can play a very special role in spectroscopy because of its highly monochromatic nature and its high energy density. The tunable laser device, being highly monochromatic, provides a convenient source for absorption spectroscopy. The resolution obtained in such a measurement is not controlled by the line width of the source. It is limited normally by the Doppler width of the line being studied. Minor interactions responsible for splitting spectral lines into multicomponents may escape detection because the splitting may be completely smeared by Doppler broadening. The high energy density of laser radiation provides a solution to such a case.

The photon density in the intense field would influence the absorption process by producing a significant perturbation in the population distribution. The lower state may be pumped significantly by the laser field. If the pumping laser line is much narrower than the Doppler width of a molecular line, for example, those molecules which move at a proper speed for making a resonance with the laser frequency are pumped to their excited state. If a second probe laser is swept across the Doppler width of a spectral line, it detects the absorption reduced to a low value at the frequency of the pumping laser, and the absorption is unaffected at other frequencies. This phenomenon is referred to as the *hole-burning process*. A large perturbation in the population distribution is observable when a source of high energy density is used to excite a gaseous sample.

The case of laser beam excitation is unique because its highly monochromatic pumping produces a large population perturbation with an extremely narrow spectral range in Doppler width. If the probe beam and the pumping beam originating from a single laser source are sent into a gas cell from both sides, as shown in Fig. 7, the molecules which move at a certain speed in one direction resonate with one beam, and those which move at the same speed in the opposite direction resonate with the other beam [see Fig. 7(a)]. If the laser oscillates at the center frequency of the spectral line, the molecules which have zero velocity components along the beam direction are able to resonate with both beams. The probe beam can detect the hole burnt by the pumping beam only if the laser oscillates at the center frequency of the

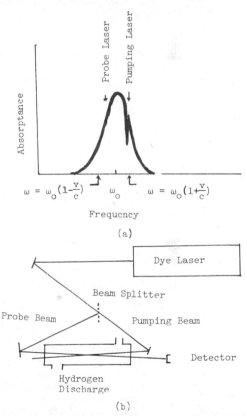

FIG. 7. The hole-burning effect (a) produced by the scheme shown in (b). The molecules moving at $+v$ make a resonance with the pump beam which is directed to the negative direction, while those moving at $-v$ make a resonance with the probe beam which is directed to the positive direction. The pump beam produces a hole at $\bar{\nu} = \bar{\nu}_0(1 + v/c)$. The molecules resonating with the pump beam see the frequency of the probe beam to be $\bar{\nu} = \bar{\nu}_0(1 - v/c)$, and they do not resonate with it.

TABLE II. ATOMIC AND MOLECULAR LASER SPECTROSCOPY.

Spectroscopy Scheme	Factor Effecting Resolving Power	Laser
(i) Absorption Spectroscopy		
Stark Spectroscopy	Doppler Effect	Molecular Gas Laser
Laser Magnetic Resonance	Doppler Effect	Molecular Gas Laser
Absorption Spectroscopy	Doppler Effect Laser Stability	Tunable Diode Laser
Photoacoustic Spectroscopy	Doppler Effect Laser Stability	Dye Laser
(ii) Fluorescence Spectroscopy		
with Absorption Cell	Doppler Effect Laser Stability	Ar^+ Laser
with Molecular Beam	Laser Stability Beam Collimation	Ar^+ Laser
Pulsed Excitation		Dye Laser
Quantum Beat Spectroscopy	Homogeneous Width	Dye Laser
(iii) Raman Spectroscopy	Spectrometer Laser	He-Ne Laser Ar^+ Laser He-Cd Laser
(iv) Hole-Burning Spectroscopy		
Saturated Absorption Spectroscopy	Homogeneous Width	He-Ne Laser
Saturated Fluorescence Spectroscopy	Homogeneous Width	CO_2 Laser
Hole-Burning Spectroscopy	Homogeneous Width Beam Angle	Dye Laser
(v) Multiphoton Spectroscopy		
Double Resonance	Doppler Effect Laser	He-Xe Laser Molecular Gas Laser
Multiphoton Absorption Spectroscopy	Doppler Effect Laser	Dye Laser
Two-Photon Absorption High-Resolution	Homogeneous Width	Dye Laser

spectral line. The absorption of the probe beam drops sharply at the line center if the laser frequency is scanned across the Doppler width of the line. Spectroscopy is then carried out free of inhomogeneous Doppler broadening. This Doppler-free technique has achieved measurements of hydrogen's Rydberg constant and Lamb shift with an accuracy of ± 10 MHz.

Excited atoms and molecules by absorbing a photon jump back to the lower state either by emitting a photon or by transferring their internal energy to kinetic energy. The first process is referred to as *fluorescence scattering* if the involved initial excitation is to a discrete bound state, and as *Raman scattering* if it is to a certain undefined level. For either case, the emitted photon may or may not have the same frequency as the incident photon. The second process is called the *photoacoustic* phenomenon. Study of these processes began long before the development of lasers. Now, use of a laser beam as the excitation source for these relaxation processes is common practice, taking advantage of high photon density concentrated within an extremely narrow spectral width.

Nonlinear absorption of photons by atoms or molecules becomes pronounced when the density of incident photons becomes very large. The use of the laser has been essential for study of the nonlinear effects. Several spectroscopic schemes have been developed based on the nonlinear effect exhibited by atomic or molecular systems. The hole-burning spectroscopy mentioned above is an example. Another category which uses the nonlinear optical effect is multiphoton spectroscopy, where more than a single photon contributes to the absorption process. The double resonance scheme uses two photons of different frequencies to create a single transition, while the two-photon or multiphoton absorption spectroscopy scheme relies on excitation by two or more photons of the same frequency.

Spectroscopy performed with incorporated laser devices is generally classified as laser spectroscopy. The technique has been and is progressing rapidly. Table 2 summarizes various schemes practiced nowadays. Included in the Table are their salient features.

HAJIME SAKAI

References

Herzberg, G., "Atomic Spectra and Atomic Structure," New York, Dover, 1944.

Herzberg, G., "Molecular Spectra and Molecular Structure," Vol. I, "Spectra of Diatomic Molecules," New York, Van Nostrand Reinhold, 1950.

Herzberg, G., "Molecular Spectra and Molecular Structure," Vol. II, "Infrared and Raman Spectra of Polyatomic Molecules," New York, Van Nostrand Reinhold, 1945.

Townes, C. H., and Schawlow, A. L., "Microwave Spectroscopy," New York, Dover, 1975.

Shinoda, K., "Topics in Applied Physics," Vol. 13, "High Resolution Laser Spectroscopy," Berlin, Springer-Verlag, 1976.

Bloembergen, N., "Nonlinear Optics and Spectroscopy," *Rev. Mod. Phys.* 54, 685 (1982).

Schawlow, A. L., "Spectroscopy in a New Light," *Rev. Mod. Phys.* 54, 697 (1982).

Vanasse, G., "Spectrometric Techniques," Vol. 1, New York, Academic Press, 1977.

Vanasse, G., "Spectrometric Techniques," Vol. 2, New York, Academic Press, 1981.

Cross-references: ABSORPTION SPECTRA; ATOMIC SPECTRA; ELECTRON SPIN; ENERGY LEVELS; LIGHT; MASS SPECTROMETRY; MICROWAVE SPECTROSCOPY; MOLECULAR SPECTROSCOPY; RAMAN EFFECT AND RAMAN SPECTROSCOPY.

SPIN WAVES (MAGNONS)

General Properties In magnetically ordered solids such as ferromagnets, antiferromagnets, and ferrimagnets, the spins of the magnetic ions are oriented along certain preferred directions. Coherent deviations of the spins from their preferred direction which propagate in space and time with wave-like characteristics are called spin waves. They were first studied by F. Block in 1930.[1] Spin waves are similar to lattice waves in solids. The departure of the spin vector from its preferred direction is analogous to the atomic displacement of a lattice vibration. The spins of a ferromagnetic insulator, for example, may be pictured classically as precessing about the direction of magnetization as illustrated in Fig. 1. The radius of the precessional circle is proportional to the amplitude of excitation of the spin wave and the phase on the precessional circle varies in a wave-like manner. Spin waves are characterized by a propagation vector $\mathbf{k}$ and a frequency vs wave-vector dispersion law $\nu(\mathbf{k})$. According to quantum theory, spin waves are elementary (Bose) excitations called magnons which possess linear momentum $\hbar\mathbf{k}$, a quantum of energy $h\nu(\mathbf{k})$, and one ($\hbar$) unit of spin angular momentum. At low temperatures the magnetic contribution to thermodynamic quantities such as internal energy, specific heat, magnetization, and thermal conductivity is determined by the number and energy of thermally excited magnons. Spin waves have been observed and studied by a variety of experimental techniques. Some of the important physical phenomena associated with spin-wave excitations are presented in Table 1.

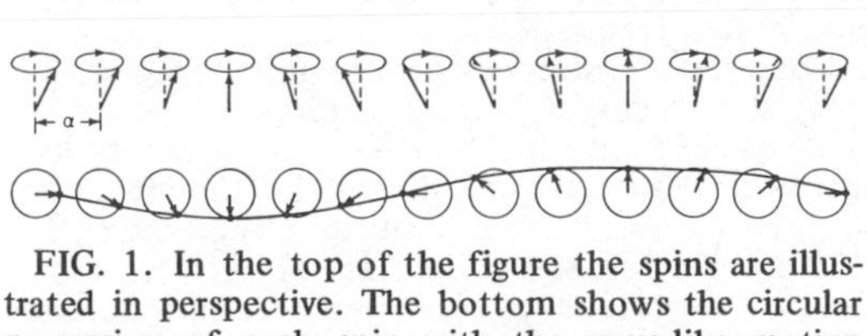

Microscopic Theory of Spin Waves Magnetic ordering in solids is a consequence of the Pauli exclusion principle, which leads to a dependence of the coulomb energy on the spin directions of the electrons comprising a quantum-mechanical system. The energy associated with this effect is called the exchange energy, and it may be represented by an exchange Hamiltonian[9]

$$\mathcal{H} = -\tfrac{1}{2} \sum_{\ell \neq m} J(\ell, m) \, S_\ell \cdot S_m, \qquad (1)$$

where S_ℓ is the spin operator of a magnetic ion located at the lattice position $\mathbf{R}_\ell$. The exchange integrals $J(\ell, m)$ depend on $\mathbf{R}_\ell - \mathbf{R}_p$ but are nonnegligible only for $|\mathbf{R}_\ell - \mathbf{R}_p|$ less than a few atomic lattice spacings. For a ferromagnet the J's are predominantly positive so that the minimum exchange energy results for a ground state $|\psi_g\rangle$ in which all spins are parallel. For the antiferromagnet or ferrimagnet the J's are predominantly negative leading to minimum energy for approximately antiparallel (or complex) spin arrangements in the ground state. The low-lying excited states $|\psi_k\rangle$ of the exchange Hamiltonian are spin waves which correspond to a uniform probability of finding one quantum of spin disorder in the system. For a ferromagnet the magnon state is

$$|\psi_{\mathbf{k}}\rangle = \frac{1}{\sqrt{N}} \sum_\ell e^{i\mathbf{k}\cdot R_\ell} \, S_\ell^- \, |\psi_g\rangle \qquad (2)$$

where S_ℓ^- is the operator which lowers the spin at R_ℓ by one unit. These states are solutions of the exchange Hamiltonian only when the number of spin waves in the system is small. Neglect of magnon-magnon interactions is referred to as the spin-wave approximation. This approximation is valid at low temperatures where the energy of the system associated with the magnons is

$$\mathcal{E} = \sum_{\mathbf{k}} \bar{n}_{\mathbf{k}} h\nu(\mathbf{k}) \qquad (3)$$

with $\bar{n}_{\mathbf{k}}$ the number of thermally excited magnons. The quantity $\bar{n}_{\mathbf{k}}$ is given by the Bose dis-

FIG. 1. In the top of the figure the spins are illustrated in perspective. The bottom shows the circular precession of each spin with the wave-like motion drawn through the ends of the spin vectors.

TABLE 1. MAGNON-PARTICLE INTERACTIONS

Particles	Effect	Source of Interaction	Information	References
Neutron	Neutron inelastic scattering	Dipole-dipole	Magnon dispersion and relaxation, critical behavior of the magnetization	2 3
Electron	Electron-magnon interaction	Exchange interaction	Thermodynamic transport and optical properties in magnetic semiconductors	4
Photon	Raman scattering Infrared absorption Magnon sidebands	Spin-orbit interaction Symmetry-breaking interactions	Magnon dispersion and relaxation	5 6 7
Phonon	Magnon-phonon interactions Magnetostatic waves	Dependence of exchange on lattice position	Magnetoelastic dispersion law, parametric processes	8
Magnon	Magnon-magnon interactions	Nonlinearity of the spin Hamiltonian	Magnon-lifetime Parametric processes	8

tribution function. The fractional decrease ΔM in the saturation magnetization M with increasing temperature is given by $\Delta M/M \alpha \sum_{k} n_{k}$ which is proportional to $T^{3/2}$ for magnons whose dispersion law is of the form $h\nu(k) \alpha k^2$ for small k. For a simple cubic ferromagnet with only nearest-neighbor exchange integrals the dispersion law is of the form

$$h\nu(\mathbf{k}) \alpha JM(T)(3 - \cos k_x a - \cos k_y a - \cos k_z a)$$

(4)

where k_α are the cartesian components of the wave vector, a is the lattice spacing, and $M(T)$ is the temperature-dependent saturation magnetization. The low-temperature specific heat is also proportional to $T^{3/2}$.

The exchange Hamiltonian described above applies to magnetic insulators in which the magnetic electrons are localized at lattice sites. An appropriate quantum-mechanical Hamiltonian for magnetic metals in which the magnetic electrons are itinerant can be defined. Long-wavelength spin waves in metallic systems have essentially the same qualitative features as those of insulating magnets with localized electrons. Short-wavelength magnons in metals, however, have a short lifetime due to decay into electron-hole pairs called Stoner excitations.[10]

Antiferromagnets In a simple two-sublattice antiferromagnet the spins on one sublattice are oriented opposite to those of the other sublattice so that the net magnetization of the solid vanishes. Two degenerate spin-wave branches occur for such systems. The spin waves of either branch involve excitations of the spins of both sublattices but in different proportions. The degeneracy of the two spin-wave branches is removed by an external magnetic field. The frequency of one branch is raised while that of the other is lowered. At a critical magnetic field called the spin-flop field, the frequency of the lower branch vanishes. For external fields in excess of the spin-flop field the spin system is unstable and the spins reorient approximately perpendicular to the external field. Long-wavelength spin waves can be excited by electromagnetic radiation with a frequency equal to the spin-wave frequency. This phenomenon is called antiferromagnetic resonance (AFMR) and ferromagnetic resonance[8] (FMR) in antiferromagnets and ferromagnets, respectively.

Special Types of Spin Waves In real crystals effects of impurities and surfaces lead to spin-wave oscillations having local properties. These localized excited states of the spin system are called localized magnons[11] (the amplitude of the spin deviation is localized near an impurity atom), and surface magnons (the excitation is localized near the surface of a magnetic solid but propagates parallel to the surface[12]).

Phenomenological Theory In a complete description of the magnetic system there exists, in addition to the exchange interaction, contributions which are relativistic in origin (dipole-dipole, spin-orbit). These interactions are weak but long range, and they are important in kinetic and relaxation processes, and microwave absorption at low frequencies.

Long-range interactions are most frequently treated by using a phenomenological theory of spin waves.[8] In this treatment spin waves are introduced as wave oscillations of the magnetic moment around the equilibrium value $\mathbf{M(r)}$. In this description the properties of the ferromagnet are described by Maxwell's equations and the equation of motion for the magnetization. Among the predictions of this model are magnetostatic modes of oscillations (magnetic modes principally characterized by the dipolar energy) and standing spin waves (dominated by the exchange contribution).[8] These long-wavelength modes have been studied by ferromagnetic resonance techniques. Ferromagnetic resonance of the $k = 0$ (homogeneous resonance) was first investigated by Griffiths[13] and is an important method for characterizing magnetic materials. The exchange interaction can be determined from measurements of the standing spin wave frequencies.[14]

R. E. DE WAMES
T. WOLFRAM

References

1. Block, F., Z. Physik, 61, 206 (1930).
2. Sinclair, R. N., and Brockhouse, B. N., Phys. Rev., 120, 1638 (1960).
3. Marshall, W., "Critical Phenomena," Proceedings of a conference held in Washington, D.C., April 1965; Edited by M. S. Green and J. V. Sengers, National Bureau of Standards, Miscellaneous Publication 273.
4. Hass, C., CRC, VI, Issue 1, p. 47 (March 1970).
5. Fleury, P. A., Porto, S. P., Chiesman, L. E., and Guggenheim, H. J., Phys. Rev. Letters, 17, 84 (1966).
6. Halley, J. Woods, and Silvera, I., Phys. Rev. Letters, 15, 654 (1965).
7. Green, R. L., Sell, D. D., Yen, W. M., Schawlow, A. L., and White, R. M., Phys. Rev. Letters, 15, 656 (1965).
8. Akheizer, A. I., Bar'Yakhtar, V. G., and Peletminskii, S. V., "Spin-Waves," Amsterdam, North-Holland Publishing Co., 1968.
9. Heisenberg, W., Z. Physik, 49, 619 (1928).
10. Izuyama, T., Kim, D. J., and Kubo, R., J. Phys. Soc. Japan, 18, 1025 (1963).
11. Wolfram, T., and Callaway, Phys. Rev., 130, 2207 (1963).
12. De Wames, R. E., and Wolfram, T., Phys. Rev., 185, 720 (1969).
13. Griffiths, J., Nature, 158, 670 (1947).
14. Seavey, M., and Tannenwald, P., Phys. Rev. Letters, 1, 168 (1958).

Cross-references: ANTIFERROMAGNETISM, FERRIMAGNETISM, FERROMAGNETISM, MAGNETISM.

STATES OF MATTER

In writing about the "states of matter" it would seem reasonable to begin by deciding what is meant by "matter" and by "state." Let, then, a short space be devoted to those questions.

As for "matter" consider it this way. Consider a brazen sphere, cube, and pyramid. What have they in common? They are brazen, or "of bronze." Now let the sphere be brazen, the cube golden, the pyramid of iron. What have

they now in common? They are all metallic, or "of metal." Now let the sphere be brazen, the cube wooden, and the pyramid of earthenware. What have they in common? They are all solid, or "of solid stuff." Finally, let the sphere be brazen, the cube water, the pyramid outlined with smoke in the air. What have they in common? They are all material, or "of matter." So matter is that which is common to all material objects, but is not involved with immaterial things—the binomial theorem, for instance, or the right-angled isosceles triangle is not, as an abstract idea, material. This argument shows us, moreover, that we never have any experience of matter in the abstract. It is always "this ingot of iron," "that block of stone," "the pile of clay over there." However, just as, while we never see "man" but always John Jones or Sam Smith or some such individual, it is convenient to talk about the abstraction "man"; even so, the abstraction "matter" can be a convenient one.

As for "states," suppose we think this way. What is it that, above all else, distinguishes material objects? Is it not that they are tangible? For, on the one hand, a rainbow is visible, but we should not call it "material"; on the other hand, if we had something invisible, but which we could feel, we should declare it to be "material." So tangibility, it would seem, is the criterion. Now as to tangibility we might divide objects into: (a) unyielding, (b) yielding but quite tangible, and (c) hardly tangible. And these three classes, of course, we call solid, liquid, and vapour or gas: the three states of matter in which we are interested. This argument shows why we distinguish matter primarily into these classes, and not, for instance, according to color. It is because it is above all else tangibility in which we are interested.

It is true, of course, that the division into three classes will leave some doubtful cases at the borders, just as does the division of living things into animal and vegetable. And we can always deal with a thing according to the aspect which predominates in the circumstances in which we are interested, so that we might think of pitch, say, as a solid at one time, a liquid at another.

Gases A gas, of course, must be kept in some container, and it tends to expand the container, unless prevented from outside. We say that it exerts pressure; for instance, the pressure of air near sea level is about 14.7 psi, or just over 1×10^6 dynes/cm^2. It is found that the pressure of a given amount (by weight) of a gas in a container depends on the volume of the container, and on the temperature. The relation may be a very complicated one, but at sufficiently low pressure and sufficiently high temperatures, it is approximately true for all gases that

$$\frac{PV}{T} = \text{Constant}$$

P represents the pressure of the gas, V the volume and T the temperature measured from absolute zero, which is at $-273°C$ or $-460°F$, approximately. This kind of equation is called an equation of state, and this particular one is the law of ideal gases, and it is said that a gas behaving thus is an ideal gas. Actually, of course, there is no such thing as an ideal gas, but any gas under suitable conditions will behave nearly ideally.

Liquids It is sometimes said that there is no simple equation for liquids corresponding to the ideal law for gases. However, it is close to the truth, under many circumstances, to write

$$V = \text{Constant}$$

i.e., the volume of the liquid is constant independently of pressure and temperature. In fact, this is not true; the liquid is compressed a little by pressure, and the volume usually increases a little with increasing temperature. Nevertheless, this simple law is probably as close to the truth as is the law of ideal gases in many of the circumstances in which it is used.

Solids Likewise a simple law can be given for solids, i.e.,

$$L = \text{Constant}$$

L being the length of the line joining any two points in the solid. That is to say, not only does the volume not change, but neither is the solid distorted by any action on it. Again, this is only an approximation to the truth, and probably about as valid.

Changes of State We often tend to think of a given kind of stuff as being in a particular state; for instance we think of water as liquid, iron as solid, and air as gas. However we know that almost anything may exist in any of the states; water, for instance may be liquid, solid, or vapor. The ways in which things change from one state into another can be very interesting. (In books on THERMODYNAMICS, it may be mentioned here, in which these things are discussed in great detail, the word "phase" is often used instead of "state" as used here.) In general, the division between the different states depends on the pressure and temperature of the sample. Increasing the temperature makes the material go to a "looser" state; e.g., if ice is heated somewhat, it melts, and if the water is heated more, it evaporates. Increasing the pressure tends to squeeze the material into a "denser" state, either liquid or solid according to circumstances.

The way in which the state of a given material, say water, varies with pressure and temperature can be shown well in a graph such as that in Fig. 1. Every point on the graph will represent a certain combination of pressure and temperature, in the way of analytic geometry. The graph is divided into regions marked "solid," "liquid" and "vapor." At temperatures and pressures such as T_1 and P_1, which give a point in the region marked "solid," the material is solid. Like-

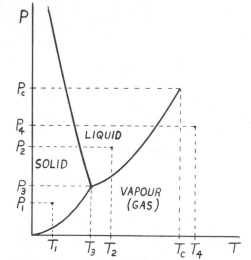

FIG. 1. Phase diagram: plot of pressure vs temperature for a typical material.

wise at temperatures and pressures such as T_2 and P_2, giving a point in the "liquid" region, the material is liquid. And at temperatures and pressures such as T_4 and P_4, it is a vapor. Thus, this graph indicates the whole behavior of the material with respect to temperature and pressure. Note that the line dividing the "liquid" from the "vapor" region just ends at a point at temperature T_c and pressure P_c. These are called the "critical point," and the "critical temperature" and "critical pressure", respectively. At temperatures higher than critical, there is no distinction between liquid and vapor. For water, the critical temperature is $374°C$; critical pressure, 219 atmospheres.

It can be seen that three parts of the curve meet as it were, at the point given by T_3 and P_3. This point is called the "triple point." For water, the triple point is at about $0.01°C$ and 4.6 mm of mercury. Note that if you warm a solid at pressures lower than the pressure at the triple point, it will change to vapor without first melting to a liquid at all. Such "dry evaporation" is called "sublimation." For carbon dioxide, for instance, the pressure at the triple point is about 5.1 times atmospheric. Thus, solid carbon dioxide, i.e., "dry ice," when allowed to warm up at atmospheric pressure, sublimes directly into vapor, as is well known. A process the reverse of sublimation happens when frost forms on cold objects without any intervening stage of liquid.

Vapor Pressure In Fig. 1, along the curve joining the triple point and critical point, the "liquid" and "vapor" regions meet; i.e., liquid and vapor can exist together. Note, however, that at a given temperature, they exist together only at a certain pressure. This pressure is called the vapor pressure of the liquid at the temperature concerned.

An analogous thing could be said about liquid and solid, or solid and vapor, along the appropriate curves.

Vapor Pressure and Relative Humidity In a certain volume, say one cubic foot, of air, there is, along with the air, ordinarily a small amount of water vapor. Suppose that this same amount of water vapor were put into a vessel of the same volume, i.e., one cubic foot, which contained nothing else at all. There would be a small pressure, perhaps about 1/50 of atmospheric pressure. This pressure is called the partial pressure of the water vapor.

The ratio of the partial pressure of water vapor in the air to the vapor pressure of water at the temperature of the air is called the relative humidity. (There is a bit of approximation here, because water vapor is not really an ideal gas.) For instance, at 70°F or 21°C, the vapor pressure of water is about 20 mm of mercury. (Standard atmospheric pressure is 760 mm of mercury.) If the actual partial pressure of water vapor were 10 mm the relative humidity would be 0.5 or 50 per cent.

Changes of Volume and Shape of Liquids and Solids The "equations of state" proposed above for liquids and solids are really too drastic; they ordinarily expand when heated. Typically, the volume of a liquid increases by 0.05 per cent for each degree Fahrenheit increase in temperature. (About 0.09 per cent for each degree C.) (Water may be anomalous in this, as well as other, respects.) They are compressed by the application of pressure; typically, the volume is decreased by 0.01 per cent for each atmosphere of pressure.

Solids are generally compressed only about 1/50 as much as liquids by the same pressure. It is more interesting to know how much force is needed to stretch a bar of solid material. For steel, it takes about 30 000 pounds force for each square inch of cross section of the bar to stretch it by 0.1 per cent. (About 2×10^9 dynes for a bar of cross section 1 cm^2.) For other solids, the force needed is mostly less, typically one-third to one-tenth of that.

The Surface of Liquids A liquid, if it does not fill the container entirely, has an exposed surface. If an attempt is made to stretch the surface, for instance by withdrawing a horizontal piece of wire from the liquid, the surface acts as if it were an elastic sheet. The force necessary to stretch a unit length of surface is called the SURFACE TENSION. For instance, the surface tension of pure water around room temperature is about 72 dynes/cm. The addition of impurities, e.g., soap to water, reduces the surface tension greatly. The rising of liquids in fine tubes or in porous materials is connected with surface tension.

Molecular Theory of the States of Matter The molecular theory considers all matter to be built up of molecules in some arrangement (or lack thereof!), each molecule, in turn, consisting of one or more atoms. The molecules attract one another with forces of electrical origin, and thus may stick together. However, they are also in motion or at least vibration, and this tends to make them break apart. The motion increases with increasing temperature, hence the structure "loosens." In a gas, around atmospheric pressure and room temperature, the molecules are about 3×10^{-7} cm apart; the forces between them are very small and they move almost freely and randomly. In so doing, they bombard the walls of the vessel; the effect of this bombardment is the pressure.

In a liquid or solid, the molecules are typically only about one-tenth as far apart. The forces then hold the molecules together; their motion is reduced to a vibration which never gets them far from one place. Occasionally, one molecule gets an extra hard push and flies away; this is evaporation or sublimation. According to the nature of the molecules, they may pile together as would a heap of spheres, or they may stick together in chains or other arrangements.

In a liquid the distance between molecules is about the same as in a solid. The different behavior may be because of a few "holes" in the liquid, i.e., places, a few per cent in all, where a molecule is missing. These spaces allow the liquid to flow, just as on a checker board when a few of the men are missing, the whole pattern can be moved by shifting men into and out of the holes. Of course this takes time, and it shows up in the viscosity or "slowness" of liquids. Molasses, for instance, is very viscous, water not very.

Polymers and Other Special Kinds of Solid Simpler solids, such as many metals, may be thought of as composed of atoms or molecules stacked together, as was mentioned above. The atoms may be considered spherical; then the whole arrangement is rather like a pile of oranges.

In other materials, the molecules are arranged in long chains. Often in such chains a simple sequence of molecules will be repeated many times, maybe hundreds. These materials are polymers. ("Polymer" means "many parts," i.e., the chains are made up of many, often identical, links.)

Some polymers, or polymerlike materials, consist of a whole jumble of these chains. Since the chains can move around with respect to one another to a considerable extent, such materials can be deformed easily. The elasticity of rubber, for instance, can be considered in terms of its construction from such chains. Ordinarily the chains are much curled and tangled. When the rubber is stretched, the chains can uncurl. It is for this reason that rubber can be stretched so readily.

In some cases, however, there are crosslinkages; the chains are connected to one another at many places, so that the result is like a net, or a three-dimensional lattice. Obviously, a material with many crosslinkages will be much more rigid, and less easily deformed. Treatments of plastics, or processes such as the vulcanization of rubber, can introduce such partial rigidity, to an extent suitable for the intended use.

Many biological materials are rather like polymers, and it may be that some useful insights will be gained by considering them from that viewpoint.

H. L. ARMSTRONG

References

General

Slater, J. C., "Introduction to Chemical Physics," New York, McGraw-Hill Book Co., 1939.

Flowers, B. H., and Mendoza, E., "Properties of Matter," New York, John Wiley & Sons, 1970.

Goldstein, D. L., "States of Matter," Englewood Cliffs, N.J., Prentice-Hall, 1975.

On the Forces Which Hold Matter Together

Moelwyn-Hughes, A. E., "States of Matter," London, Oliver and Boyd, 1961.

On Thermodynamics

Pippard, A., "Elements of Classical Thermodynamics," Cambridge, Cambridge Univ. Press, 1957.

Porter, A. W., "Thermodynamics," London, Methuen, and New York, John Wiley & Sons, 1951.

Cross-references: COMPRESSIBILITY, GAS; CONDENSATION; GAS LAWS; KINETIC THEORY; LIQUID STATE; PHASE RULE; SOLID-STATE PHYSICS; VAPOR PRESSURE AND EVAPORATION.

STATIC ELECTRICITY

Until it was accepted that galvanic currents are identical with moving electric charges, *static electricity* referred to all electric charges that were stationary or nearly so, including such charges as those on Leyden jars and pyroelectric crystals. Nowadays the meaning encompasses only the more or less immobilized electrification due to (1) charge redistributed on a single body through *induction* caused by the presence of charge on neighboring bodies; (2) charge transferred from one material (solid or liquid) to another by *contact* and subsequent separation; (3) charge accumulated on a body (solid or liquid) by *deposition* of electrons or ions from the surrounding atmosphere, as in charging by corona, flame, or radiation. One generally excludes separation and redistribution of charge in pyroelectrics, piezoelectrics, thermoelectrets, photoelectrets, capacitors, and electrolytic cells.

In scientific and technical work, it is often important to control static electricity, either to exploit its effects or to avoid them. For convenience we divide the subject into *generation* of charge and *dissipation* of charge. So far as

generation is concerned, *induced electrification* is largely understandable in terms of classical physics; the fundamentals are well understood, and the applications are straightforwardly made as part of electrical engineering and safety engineering (see ELECTRICITY). *Contact electrification*, on the other hand, is far from being understood, and immediately comes to frontier problems in the modern theory of liquids and solids. *Deposition electrification* [from a gaseous ion cloud] is in an intermediate position. The fundamental processes of electrical discharge in gases are understood in principle, but real situations are so complex that theory is of limited use. As far as dissipation is concerned, the origin of the charge is largely irrelevant; a wide variety of physical and chemical processes of all degrees of complexity enter in analyzing the decay or neutralization of static charge.

Generation *Induction Charging.* Induction phenomena, as mentioned, are adequately described as a branch of classical electrostatics. Hence we devote the main part of our discussion here to the more poorly understood topic of contact phenomena, plus a brief overview of deposition from corona discharges or other ion clouds.

Contact Charging. When two materials differing at their surfaces in chemical composition, or even temperature or state of strain, are placed in contact, charge tends to flow from one to the other until their electrochemical potentials are identical. If the materials are subsequently separated, some portion of the transferred charge is retained, the potential difference between the materials increasing as the capacitance decreases. As a rule, this net transfer produces no noteworthy effects; in fact, a sensitive electrometer is ordinarily needed to measure it. With metal-metal contacts, the charge retained is invariably very small and is not detected by the senses. With metal-insulator or insulator-insulator contacts, the charge transferred while the materials remain in contact may be large, but it decreases during separation to an unimportant amount unless the insulator has quite high resistivity. At the speeds encountered in ordinary events, say about 100 cm/sec, the charge leaks back too fast to give noticeable effects whenever the resistivity of the more poorly conducting material falls below about 10^9 ohm-cm. At higher speeds, static effects become noticeable even at low resistivity, whereas at lower speeds they appear only at very high resistivity.

The combination of moderate or high speed and high resistivity frequently produces large enough charge transfer that the surrounding medium breaks down electrically. In air at atmospheric pressure, the requisite charge densities on a uniformly charged plane conducting surface are about 8 esu (statcoulombs/cm^2) or 25 $\mu C/m^2$, to give a field of about 30 kV/cm just above the surface. This density represents about 2×10^{10} electronic charges per square centimeter, so that only one in perhaps 10^5 surface atoms is charged even at the highest electrifications ordinarily occurring.

For practical purposes, the phenomena of interest in static electrification are the forces of attraction or repulsion resulting from excess charge (e.g., in textile processing, ore separating, and electrostatic copying), the occurrence of sparks and their consequences (e.g., in transfer of flammable liquids, or in handling of photographic films), and so on. From a fundamental point of view, these phenomena are simply consequences of transfer of charge and its subsequent behavior. Hence we may take as the central question in static electrification the following: Given two materials of specified chemical composition and physical state, what is the charge—in sign and amount—transferred when they are placed in contact under specified mechanical and ambient conditions, and then separated?

At the present stage of theories of matter, answers are available only for some quite special situations. For the rest of the time, we must content ourselves with trying to get some guidance from rather crude models of the type to be described.

Solid-Solid Contact. Let us restrict our treatment to bodies that are initially uncharged, and examine the charge q_0 that is transferred between two objects while they are touching. For many materials, q_0 will increase with duration of contact towards an equilibrium value q_∞. We designate as α the degree of attainment of this value, writing $q_0 = \alpha q_\infty$. We should expect q_∞ to depend on the chemical nature of the materials, as well as on the mechanical nature of the contact, i.e., on the size and shape of the objects, and the normal force between them. Let us write q_∞ as a factor b ("b" for band structure, in the case of solids) dependent on the electrochemical properties of the materials, multiplied by a factor g ("g" for geometry) dependent on the mechanical parameters of the contact. Upon separation some of the charge returns to its origin, either through conduction processes such as tunneling, or through atmospheric breakdown; let us call f that fraction of the initial charge q_0 that remains to give the observed charge $q = q_0 f$. Upon combining all these relations, we get a four-factor formula

$$q = \alpha bgf$$

The utility of this expression can be assessed only by experience. Note that α and f are both positive numbers between zero and unity, and that g is intrinsically positive; the sign of the charge enters in b.

An important qualitative consequence of the above scheme is the existence of a *triboelectric series*, i.e., a listing of materials such that any one in it becomes positive when rubbed against another lower in the series. When the materials, the ambient conditions, and the mode of contact are reasonably well defined, such series are generally conceded to exist. An example is the following: Wool, nylon, viscose rayon (regenerated cellulose), cotton, silk, cellulose acetate, polymethylmethacrylate, polyvinylalcohol, polyethylene, polytetrafluoroethylene. In principle, one ought to be able to predict the position of a given substance from its chemical properties, but as yet the attempts have been more suggestive than successful.

So far as quantitative results are concerned, prediction of α and b is possible in principle when detailed information is available on the energy levels of the materials. As yet, only metal-metal systems have been found to be simple enough to be analyzed successfully. With insulators, surface states of high complexity may occur.

Prediction of g is made, one hopes, by estimating the area of "true contact," namely, the area wherein atomic fields interpenetrate. The size of this area depends on the roughness of the contacting objects and on the normal forces between the objects up to the point where the area of "true contact" approaches a maximum approximately equaling the area of apparent contact. In the case of rolling contacts (and some sliding contacts) between cylinders and spheres, the area of apparent contact increases with normal force, in a fashion well understood for elastic deformation and pretty well understood for plastic deformation. Thus it is not uncommon in such cases to find reports of fractional-power dependence on normal force between contacting objects, the power taking values from near zero to unity. In the case of sliding contacts where one or both objects are planar, the "true contact" can apparently be increased by two or three orders of magnitude, as deduced from the much larger charge transfers sometimes observed in "frictional electrification," a not precisely defined concept in which "rubbing"—as opposed to mere "touching"—predominates. Here effects beyond mere "contact" enter, such as local heating, wear, and material transfer; then the Helmholtz theory that sliding simply serves to multiply the contact region loses much of its utility. Even so, in many cases of rolling or sliding contact, the effective area for given normal force is proportional to the *length* of the stroke.

Prediction of f is made, in the absence of atmospheric breakdown, by analyzing the time scale of the experiment. We expect f to be a function of the ratio of T, a time characteristic of the speed of separation, to τ, a relaxation time characteristic of the material of higher resistivity. The separation time T may be taken as l_0/v, where v is a speed of separation, and l_0 is some characteristic length; the relaxation time τ should be related to the time constant for redistribution of charge in a medium of dielectric constant ϵ and volume resistivity ρ. We have then $T/\tau = (l_0/v)/(\epsilon\rho)$, and we see that velocity may be traded for resistivity, since

$f = f(T/\tau) = f(l_0/\epsilon\rho v)$. For ordinary materials at ordinary speeds, f is very small for ρ lower than 10^9 ohm-cm. For metal-metal contacts in particular, where $\rho \sim 10^{-5}$ ohm-cm, static effects disappear, their only remnant being the slight transfer due to contact potential difference at the last points of contact. In the presence of atmospheric breakdown, a saturation value of charge q is reached, the magnitude depending on the properties of the surrounding atmosphere. Estimating f here is largely a matter of bookkeeping, and without much point.

Deposition Charging. When a gas is exposed to a high electric field set up between a pair of electrodes, it breaks down to produce a continuous electrical discharge when a threshold voltage between the electrodes is reached. This critical voltage depends on the nature of the gas, its density, and the geometry of the electrodes. To maximize field strength at a given potential, one of the electrodes at least should have a very small radius of curvature, as with a set of points or a fine wire. In a typical application in electrophotography, a cylindrical wire of 0.1 mm diameter might be mounted at a distance of 1 cm from a flat grounded surface. The threshold voltage to start the discharge would be about 3 kV, and the operating voltage might be about 5kV, to yield a corona current of say 100 μA per cm electrode length. The threshold voltage is nearly proportional to air density, and hence to pressure at constant temperature. It increases with increasing diameter of the wire. Details may be found in references on electrical discharges in gases. Other sources of free charge are ionizing radiation and flames. The copious and controllable amounts of charge provided by corona, however, have made it by far the most common source.

A plane surface (or other large surface) maintained in the neighborhood of the wire at a potential different from it will eventually accumulate electrostatic charge from the corona until the field between the wire and the surface vanishes. In practice one waits only long enough to charge the surface to several hundred volts, corresponding to 10^{11} to 10^{12} electron charges/ cm^2. A particle in the charge cloud and hence in an electric field will pick up charge Q to the limiting amount

$$Q = pE_{corona} a^2 ,$$

where a is the "diameter" of the particle, E_{corona} is the electric field strength in the region of the corona, and p is a factor—of the order of unity— dependent on the dielectric properties of the particles and on the properties of the corona discharge. Since the electric force on the particle equals the charge times the electric field strength, and since the viscous drag in the Stokes' law regime is proportional to the diameter and the viscosity of the medium, the terminal velocity of a particle within the corona is proportional to the electrical field strength and the diameter, and inversely proportional to the viscosity.

Liquid-Solid (and Liquid-Liquid) Contact. The two major phenomena are *spray electrification* and *flow electrification*. The former is seldom of practical significance (except when deliberately induced by charged nozzles in electrostatically aided spray painting). The latter is often of great concern, notably in the handling of inflammable liquids. Although neither phenomenon is completely understood, there is available enough empirical knowledge to provide a basis for control.

Spray electrification occurs when a chemically homogeneous liquid is dispersed into a mist, the larger droplets on average charged with one polarity, the smaller ones oppositely. The source of differential charging is the dependence of the electrochemical potential on the droplet radius, as a consequence of the difference in molecular environment of the material near the surface. The effect would be expected to be most pronounced with polar molecules, as is indeed observed in the conspicuous example of water ("waterfall electrification"). The magnitude of the effect depends on the mechanism of formation of the droplets, and is difficult to predict.

Flow electrification appears when liquid moves relative to a solid with which it is in contact, as in pouring, wiping, pumping, or sedimentation; or relative to another liquid, as in settling of water drops in a petroleum tank. The charge-transfer mechanism appears to be the same as in the contact between solids, at least for field strengths below breakdown. (The language, however, is that of electrokinetic chemistry rather than that of solid-state physics.) The charge-separation mechanism, in contrast, proceeds differently. In charging, a double layer is set up between dissimilar materials in either case, because of the tendency for equalization of the electrochemical potential. With a liquid, part of the transferred charge is carried away within liquid layers, whereas with solids the charge is carried away in or on the separated object. In either case the transferred charge can leak back through low-conductivity paths and in some circumstances reduce the net electrification to a negligible amount.

The calculation of the streaming current, that is, the motion of charge entrained in a moving fluid, is on uncertain ground, embracing as it does a wide spectrum of complicated electrokinetic effects together with imperfectly known boundary conditions and empirical coefficients. Nevertheless the theory does predict, and experiment does confirm, that the streaming current rises as the square of the velocity at first, then varies linearly; and that it rises with conductivity from a negligible value at very low conductivity (say, 1 picomho/meter, corresponding to a resistivity of 10^{14} ohm-cm) where the double layer is very thick and hence of low volume charge density, through a maximum at intermediate conductivity (say, 100 picomho/meter), and finally decreases to perhaps a hundredth of the maximum at high conductivity (say, 10,000 picomho/meter). The higher streaming currents can carry enough charge into a container to cause sparking in it, even though grounded. To decrease the extent of electrification, then, the liquid velocity must be kept small, and the generation of charge must be kept very low (by avoiding ionizing impurities such as water in gasoline) or the conductivity must be made high (by addition of an antistatic agent). Piping size and mass-flow rate, impurity removal and antistatic-agent addition, are to be engineered for prudent and economical operation.

Dissipation Charge may be dissipated by currents within the body of the object or over its surface, or by currents within the medium in which the object is immersed. Charge in the interior of a body ultimately reaches the boundaries, decaying exponentially with a well-defined time constant $\epsilon\rho$. Charge on the surface of a body moves to attain an equilibrium distribution, decaying in a complicated fashion that is only approximately described by a time constant proportional to $\epsilon\rho$. Currents within the medium surrounding the object are described by the laws of electrical conduction in liquids or gases, as the case may be, in all their complexity.

Control of Static Electrification Effects of static charge, beneficial or harmful, can be controlled by influencing either the production of charge or its neutralization.

Production of Charge. In the case of induced electrification, the fields between objects may be controlled by altering the potentials of neighboring conductors, usually by screening and grounding. In the case of contact electrification, the four-factor formula may be used as a basis for discussion:

Control of α: The only practical control is through fixing of the charge state of the object. (We must accordingly generalize our analysis to include objects initially charged.) If one object has lost so much charge that it can lose no more upon contact with another object, this second object will remain in its initial charge state. (As an illustration, a yarn running through an insulating or insulated guide may not pick up more charge after a brief initial period in which the guide electrically "saturates.")

Control of b: In principle, the electrochemical potential for one given material may be matched against that for some other material so that charge transfer between the two is negligible; in practice the potentials cannot be controlled closely enough, particularly over a range of ambient and mechanical conditions. More significantly, objects usually must work against a variety of materials which will have a variety of electrochemical potentials. There have been proposed composite surfaces wherein a material high in the triboelectric series is interspersed with one low in the series to give a small average electrification. The efficacy of such blending is high in some applications.

Control of g: Decrease of the normal force almost always produces a decrease in charge transfer. In cases where the dependence follows a fractional power law, it may not be useful to go to great lengths to decrease the normal force to very small values. With respect to path length, action is obvious though seldom practical. If the nature of the contact can be changed somewhat, however, by modifying the extent of slip between contacting objects, the amount of charge transferred can sometimes be modified drastically.

Control of f: In practice, charge transfer is reduced by decreasing the product $\epsilon\rho v$. It is difficult to alter the dielectric constant by more than a small factor and hence to affect the rate of dissipation strongly in this way. The resistivity, on the contrary, may be changed greatly, either by changing the molecular structures of the body (say by grafting conducting segments onto polymers) or by adding conducting materials within or onto the object (say by adding moisture or various antistatic agents). With hydrophilic materials, one traditionally alters the moisture content of the surrounding air, with that of the object following. The resistivity varies as a high negative power of the moisture content, and a change in relative humidity of a few per cent may bring about a tenfold difference in charge observed at a given time, or a tenfold difference in the time required to attain a given charge state. In a few special applications, the resistivity may be lowered by increasing the temperature. The factor v can be decreased by simply slowing down the process, of course, and also by changing the mode of contact, say from sliding to rolling (with concomitant change in g), or by lowering the relative speed of the two contacting objects.

In the case of deposition electrification, the relation $Q \sim E_{corona}a^2$ shows that the limiting charge can be increased through increase of the electric field strength by increasing the voltage difference across the electrodes, or decreasing the distance between them, on the one hand; and working with larger particles, on the other hand. In practice the ultimate values are seldom approached, and the relevant design factors are those determining the time required for charging and for moving particles along a trajectory. We note in general that gravitational and inertial forces vary as the cube of the diameter, electrical forces as the square, and viscous forces as the first power. It is inappropriate to pursue these matters further here.

Neutralization of Charge. Surplus charge inexorably is neutralized, since the resistivities of even the best nongaseous insulators seldom exceed 10^{18} ohm-cm or so, and since natural radioactivity and cosmic rays produce mobile ions in surrounding fluids. When it is desired to hasten neutralization, charge must be supplied by other means. The conductivity of the body can be increased, especially at the surface, according to some of the ideas expressed earlier. More commonly, charge can be supplied through the surroundings. The medium in which the object is immersed can sometimes be rendered conductive by the electric field set up by the body itself, especially at surface regions of small radius of curvature, for example, at the points of tinsel or needles. Although the charge cannot be completely eliminated in this way, it can be made quite small if the ambient pressure can be brought near the minimum in the Paschen law for electrical breakdown in gases.

More often the surrounding atmosphere is ionized with the aid of external agents. Various commercial static eliminators have been developed to produce glow or spark discharges from high-tension wires or points. Electromagnetic radiation, in the form of x-rays or gamma rays, will ionize surrounding media, but it is often objectionable because of its hazards to personnel. Particle radiation, in the form of alpha or beta rays is very effective in producing ions, and is easier to control with respect to health hazard. Plutonium 240 with a half-life of 6600 years and americium 241 with a shorter half-life of 462 years are alpha emitters nearly free from gamma rays. They are modern substitutes for the nearly gamma-free polonium 210 with half-life of 138 days, and the gamma-active radium 226 with half-life of 1620 years. Flames produce copious ionization, but their action is usually only ancillary, as in flame-driers at the take-off of some printing presses.

R. G. CUNNINGHAM
D. J. MONTGOMERY

References

Even if dated, the following three works are useful as an introduction to electrostatic phenomena:

Loeb, Leonard B., "Static Electrification," Berlin, Goettingen, and Heidelberg, Springer-Verlag, 1958. Covers solids, liquids, and gases.

Klinkenberg, A., and Van der Minne, J. L., "Electrostatics in the Petroleum Industry," Amsterdam and New York, Elsevier Publishing Company, 1958. An excellent coverage of liquid-solid contacts.

Montgomery, D. J., "Static Electrification of Solids," Solid State Physics 9, 139 (1959). An introduction into solid-solid contacts.

Some books of more or less generality are:

Gross, B., "Charge Storage in Dielectrics," Amsterdam and New York, Elsevier Publishing Co., 1964. An excellent source of earlier references.

Dessauer, J. H., and Clark, H. E. (Eds.), "Xerography and Related Processes," London and New York, The Focal Press, 1965. An extensive treatment of fundamentals of electrostatics and the application thereof to electrostatic printing.

Schaffert, R. M., "Electrophotography," London and New York, The Focal Press, 1965; 2nd Edition, 1975. An analysis of the theory of electrophotographic processes, and a list of patents pertaining thereto.

Harper, W. R., "Contact and Frictional Electrification," London, Oxford Univ. Press, 1967. A careful treatment of the significant phenomena in contact charging between solids.

Jowett, Charles E., "Electrostatics in the Electronics Environment," New York, John Wiley & Sons, 1976. A useful practical treatment of electrostatic problems.

Textbooks and treatises have difficulty in keeping up with the rapid developments in the field. Substantial surveys are available in proceedings of conferences. A highly useful series is the following:

Lowell, J. (Ed.), "Conference Series No. 48, Proceedings of the 5th Conference on Electrostatic Phenomena, 1979," Bristol and London, The Institute of Physics, 1979.

Blythe, A. R. (Ed.), "Conference Series No. 27, Proceedings from the 4th Conference on Static Electrification, 1975," London and Bristol, The Institute of Physics, 1975. (Previous reports are Conference Series No. 11, 3rd Conference, 1971; Conference Series No. 4, 2nd Conference, 1967; British Journal of Applied Physics, Suppl. No. 2, 1st Conference, 1953.)

Other series are described in:

Ong, P. H. (Ed.), "Fourth International Conference on Electrostatics, 1981," Journal of Electrostatics 10, special issue (1981).

Asano, Kazutoshi, "Review of the 1980 Annual Meeting of the Institute of Electrostatics, Japan," Journal of Electrostatics 9, 367 (1981).

In general, the Journal of Electrostatics (published by the Elsevier Scientific Publishing Company, Amsterdam) serves as the current center for dissemination of reports on static electrification.

Persons concerned with protection against ignition arising from static electrification may wish to read publications of the American Petroleum Institute (e.g., API RP 2003, API Bull. 1003), and of the National Institute for Occupational Safety and Health (e.g., DHEW (NIOSH), Publn. No. 78-206). A recent brief but cogent paper is "Electrostatic Hazards," a report prepared by Arthur Conway and published by Oyez Publishing Ltd., London, 1980 (ISBN 0 85120 442 2).

Cross-references: CONDUCTIVITY, ELECTRICAL; DIELECTRICS; ELECTRICAL DISCHARGES IN GASES; ELECTRICITY; HIGH VOLTAGE RESEARCH; SOLID-STATE PHYSICS.

STATICS

Statics is the branch of MECHANICS which studies the conditions of equilibrium of forces acting on particles or rigid bodies, or on inextensible cords, belts and chains. Hydrostatics, the study of the equilibrium of fluids, is usually not regarded as a part of statics in the conventional sense of the term.

Statics is the oldest branch of mechanics, some of its principles having been used by the ancient Egyptians and Babylonians in their constructions of temples and pyramids. As a science it was established by Archytas of Taras (ca. 380 B.C.) and primarily by Archimedes (287–212 B.C.); it was further developed by medieval writers on the "science of weights" such as Jordanus de Nemore (thirteenth century) and Blasius of Parma (fourteenth century). In the sixteenth century, it was revived by Leonardo da Vinci, Guido Ubaldi and especially by Simon Stevin (1548–1620) who laid the foundations of modern statics (inclined plane, equilibrium of pulleys, parallelogram of forces).

Although the laws of statics can in principle be derived from those of dynamics as a limiting case for vanished velocities or accelerations, statics has been developed, since the end of the eighteenth century, independently of dynamics. Its fundamental notion, like that of dynamics, is the concept of *force*, representing the action of one body on another and characterized by its point of application, its magnitude and its direction (line of action) or briefly by a VECTOR f. Two equal and opposite forces whose lines of action are parallel and non-coinciding are said to form a *couple*. The *moment* or *torque* m_0 of a force f about a point O is a vector whose magnitude is the product of the magnitude of f and the length of the perpendicular distance of O from the line of action of f, or in VECTOR notation $m_0 = r \times f$ (vector product), where r denotes the vector from O to the point of application of f.

The following four principles may serve as the basic postulates for statics. (1) *The principle of composition* (*addition*) *of forces*: two forces, f_1 and f_2, with a common point of application A can be replaced by a third force, the resultant f, which is obtained graphically (geometrically) as the diagonal, from A, in the parallelogram determined by the two given forces, or analytically (algebraically) as the vector whose components, usually with reference to a rectangular reference system, are the sum of the corresponding components of the two given forces, $f_x = f_{1x} + f_{2x}$, etc. (vector addition). The resultant of more than two forces is independent of the order of addition. (2) *The principle of transmissibility of force*: the point of application of force acting on a rigid body can be transferred to any point on the line of action of the force provided the point is rigidly connected with the body (sliding vector). (3) *The principle of equilibrium*: the necessary and sufficient condition for the EQUILIBRIUM, that is, absence of accelerated motion, of a particle is the vanishing of the resultant of all forces acting on the

particle, or $F = \sum f_i = 0$. The condition for the equilibrium of a rigid body is the vanishing of the resultant of all forces as well as the vanishing of the resultant of their moments about an arbitrary point O, or $M_0 = \sum (r_i \times f_i) = 0$. If $F = 0$, M_0 is independent of the choice of O. (4) *The principle of action and reaction* (Newton's third law): the force exerted by one body on another is equal and opposite to that exerted by the second body on the first and both forces lie along the same line of action.

These principles imply the following results. Two parallel forces can be added if additional compensating forces are introduced. Any set of coplanar forces, with the exception of couples, can be reduced to a single resultant. The sum of the moments of any two intersecting forces about any point in their plane equals the moment of their resultant about the same point (Varignon's Theorem). Any system of forces acting on a rigid body can be reduced, in an infinite number of ways, to a single resultant F and a single couple M. In all these reductions F is uniquely determined but M depends on the position of F. There is one, and only one, line of action for F, called *Poinsot's central axis of the system*, for which M is parallel to F. Hence every system of forces is equivalent to a *wrench* as this particular force-couple combination is called.

Statical analysis of framed structures or trusses, collections of straight members pinned or jointed together at the ends, is based on the preceding theorems. Such structures rest upon supports whose reactions or pressures have usually to be determined in practical applications. The equilibrium conditions, according to (3), are two vector equations or, equivalently, six scalar equations and hence can be solved for no more than six unknowns (the reactions at supports and connections). If the reactions involve more than six unknowns, some of the reactions are statically indeterminate; if less, the body is said to be unstable. In case the unknown reactions arise from constraints, i.e., conditions restricting possible motions, a convenient method for the elimination of unknown reactions is the use of the *principle of virtual work* according to which the total virtual work (work due to a possible small displacement which need not necessarily take place) of the external forces acting on the body vanishes for any virtual displacement of the body. In general, internal forces, holding together the various parts of the structure, also have to be taken into account, e.g., for trusses which consist of straight members connected by joints.

In particular, parallel forces can always be replaced by a single resultant whose point of application is called the *center of the system of parallel forces*; it is invariant if all forces change their directions but remain parallel to each other. In this case the sum of the moments of these forces about any point equals the moment of their resultant about the same point; in particular, the sum of the moments about any point on the resultant is zero (generalization of the law of the lever). An important case of this kind is that of the earth's gravitational forces which act at a given place in practically parallel lines on every element of a not too voluminous body. The center of the system of forces, in this case, is the *center of gravity* and is identical with the center of mass or *centroid* which can easily be determined by summation for a system of discrete particles or by integration for continuous masses.

The study of friction, the resistance of a surface to the motion of a body upon it, belongs properly to applied mechanics. Since however many problems in statics involve, at least, considerations concerning *static friction* (the frictional force which just prevents motion), the study of FRICTION is often included in the science of statics.

MAX JAMMER

References

Becker, R. A., "Introduction to Theoretical Mechanics," (Chapters 2 and 5), New York, McGraw-Hill Book Co., 1954.

Eliezer, C. J., "A Modern Textbook on Statics," New York, MacMillan and Oxford, Pergamon Press, 1964.

Lamb, H., "Statics," Cambridge, U.K., Cambridge Univ. Press, 1913, 1945.

Lindsay, R. B., "Physical Mechanics," (Chapter 5), New York, Van Nostrand, 1950.

MacMillan, W. D., "Statics and Dynamics of a Particle," New York, McGraw-Hill Book Co., 1927.

Synge, J. L., and Griffith, B. A., "Principles of Mechanics," (Chapters 2, 3, and 10), New York, McGraw-Hill Book Co., 1959.

STATISTICAL MECHANICS

The object of statistical mechanics is the explanation of the macroscopical physical phenomena as consequences of the laws of motion of the atoms and molecules. Equivalently, statistical mechanics can be defined as the mechanics of systems of a very large number of degrees of freedom. Whereas in "ordinary" mechanics even the three-body problem cannot be solved exactly, statistical mechanics takes advantage of the large number N of degrees of freedom and tries to formulate exact *asymptotic* results in the limit $N \to \infty$ (in a certain well-defined way).

The fundamental laws of motion of the atoms and molecules are those of quantum mechanics; however, in many statistical mechanical problems, classical mechanics provides a sufficient approximation.

In ordinary classical mechanics, a problem is completely specified when the initial positions and momenta of all its particles are specified. Such information is impossible to obtain, and moreover is completely useless, for systems consisting of about 10^{24} particles. The only initial data which are interesting for such systems are of a macroscopic nature: density, local velocity, temperature at each point of a fluid, correlations between the density fluctuations in two points of the system, etc. There is a very large number of microscopic initial configurations of the molecules which is compatible with a given macroscopic specification. Hence, in order to describe such systems, Gibbs introduced the concept of an *ensemble*. This is defined as a set of a very large number of systems, all dynamically identical with the system under consideration (i.e., having the same Hamiltonian H), differing in the initial conditions of the molecules but compatible with the macroscopic specification of the system.

The natural mathematical framework of such a description is the *phase space*, a many-dimensional space whose coordinates are the positions $x_1, \cdots, x_N$ (shortly: x) and the momenta $p_1, \cdots, p_N$ (shortly: p) of all the particles of the system. A point in phase space therefore represents a complete dynamical system in a definite microscopic configuration. A Gibbs ensemble corresponds to a cloud of points in phase space, which can usually be considered as a continuous distribution. The basic concept is therefore the *distribution function* in phase space $\rho(x, p; t)$, giving the density of the ensemble as a function of the positions and momenta of the particles (i.e., of the coordinates of the phase space) at time t. The connection between microscopic and macroscopic physics is then given by the following assumption: The observable value of a dynamical property of the system, (e.g., density, local velocity, average energy, etc.), $\bar{A}(t)$, is the average value of the corresponding microscopic dynamical function $A(x, p)$, weighted by the distribution function:

$$\bar{A}(t) = \int dx \int dp \, A(x, p) \rho(x, p; t) \qquad (1)$$

Practically all functions $A(x, p)$ of physical interest are sums of functions involving only one or two particles. Hence it follows from Eq. (1) that the functions of real importance are the integrals of $\rho(x, p; t)$ over all but one or two particles: these are called reduced (one- or two-body) distribution functions. Their main importance comes from the fact that they remain finite in the limit $N \to \infty$.

The evolution of the system in time is described by the change in time of the distribution function. According to the laws of classical mechanics, the latter obeys a partial differential equation called the *Liouville equation*:

$$\frac{\partial \rho}{\partial t} + \sum_{i=1}^{N} \left[\frac{\partial H}{\partial p_i} \frac{\partial \rho}{\partial x_i} - \frac{\partial H}{\partial x_i} \frac{\partial \rho}{\partial p_i} \right] \equiv \frac{\partial \rho}{\partial t} + [\rho, H] = 0$$

$$(2)$$

The bracketted expression $[\rho, H]$ is called the Poisson bracket of the Hamiltonian H and the distribution function. The purpose of classical statistical mechanics is the solution of the Liouville equation in the limit $N \to \infty$.

In quantum statistical mechanics, the conceptual situation is much the same, but it is more complicated because of the proper statistical character of the quantum description of even a single system. Indeed, due to the HEISENBERG UNCERTAINTY PRINCIPLE, the momentum and the position of a particle can never be measured simultaneously with arbitrary accuracy. Hence the concept of a phase space has no meaning in quantum mechanics. The maximum information which can be obtained about a single system (in a "pure" state) is contained in its wave function $\Psi(x; t)$. The observable value of a dynamical variable in such a state is given in terms of the corresponding operator $\hat{A}$ by the expression

$$\bar{A}(t) = \int dx \, \psi^*(x; t) \hat{A} \psi(x, t) \qquad (3)$$

To this statistical aspect of quantum mechanics is added the proper indeterminism of statistical mechanics. Suppose the wave function of a single system (n) can be expanded in a series of orthonormal functions $\varphi_i(x)$ [see WAVE MECHANICS]

$$\Psi^{(n)}(x; t) = \sum_i a_i^{(n)}(t) \varphi_i(x) \qquad (4)$$

In the statistical mechanical description, the single system is replaced by an ensemble, in which each system (n) is weighted by a density p_n. The role of the classical distribution function is now played by the *density operator* ρ introduced by J. von Neumann (1932). The matrix elements of this operator in the present representation are defined by

$$\rho_{ij} = \sum_n p_n a_j^{*(n)} a_i^{(n)} \qquad (5)$$

the sum over n running over all the systems of the ensemble.

The averaging prescription which replaces Eqs. (1) and (3) is now:

$$\bar{A}(t) = \text{Trace } \rho A \qquad (6)$$

It is easily seen that this rule embodies the double averaging necessary for quantum and statistical mechanics. The evolution in time of the density operator is given by von Neumann's

equation:

$$\frac{\partial \rho}{\partial t} + \frac{1}{i\hbar} [\rho H - H\rho] \equiv \frac{\partial \rho}{\partial t} + \frac{1}{i\hbar} [\rho, H] = 0 \quad (7)$$

H again being the (quantum) Hamiltonian of the system. The classical Poisson bracket has been replaced by $(i\hbar)^{-1}$ times the commutator of the operators ρ and H.

Quantum statistical mechanics shares with the usual quantum mechanical many-body problem the following characteristic feature. It is known that in a system of several identical particles, the latter are undistinguishable. In order to satisfy this requirement, the wave function of the system must be either symmetrical or antisymmetrical with respect to a permutation of any two particles. This symmetry requirement introduces the classification of quantum statistical systems into systems of *bosons* (Bose-Einstein statistics, symmetric wave functions) and systems of *fermions* (Fermi-Dirac statistics, antisymmetric wave functions). The purpose of quantum statistical mechanics is the solution of Eq. (7) with the proper symmetry condition.

The simplest solutions of Eqs. (2) or (7) are the time-independent solutions: these are functions of the Hamiltonian H alone. They describe systems in equilibrium. A particular function of the Hamiltonian is the *canonical distribution*:

$$\rho = Ce^{-H/kT} \qquad (8)$$

where C is a normalization constant and k is called the Boltzmann constant. It can be shown that this distribution represents a system in thermodynamical equilibrium at temperature T. The main result in classical equilibrium statistical mechanics is the following. Consider the function

$$Z(V, T) = (N! \, h^{3N})^{-1} \int dp \int dx \, e^{-H(p, x)/kT}$$

$$(9)$$

the integration extending over the whole volume occupied by the system in phase space. $Z(V, T)$, as a function of the volume and of the temperature T, is called the *partition function*. It can be shown that the Helmholtz free energy $F(V, T)$ of the system [see THERMODYNAMICS] is given by

$$F(V, T) = - kT \ln Z(V, T) \qquad (10)$$

The importance of this formula lies in the fact that the knowledge of a thermodynamic potential such as $F(V, T)$ enables one to calculate all thermodynamic properties (pressure, entropy, specific heat, etc.) by simple differential operations. A completely analogous result holds in quantum statistical mechanics.

Although the basic problem of equilibrium statistical mechanics is solved in principle (i.e., it is reduced to quadratures), the explicit evaluation of the $6N$-fold integral occuring in the partition function poses a formidable mathematical problem. Only in the case of systems of noninteracting degrees of freedom can one calculate the partition function exactly. Moreover, essentially three main groups of systems of interacting particles are thoroughly understood at the present time. All of these systems are characterized by the following Hamiltonian:

$$H = H_0 + \lambda V \qquad (11)$$

where H_0 is the Hamiltonian of a set of independent particles, whereas V describes the interactions and λ characterizes the size of the interactions. The three cases mentioned are the following:

(a) *Weakly coupled systems.* In such systems the interactions between components are weak compared to some intrinsic energy. The components may be particles of a gas, or quantized oscillations of the atoms around their equilibrium positions in a crystal lattice (see PHONONS), or spins attached to atoms located at the sites of a ferromagnetic crystal (see FERROMAGNETISM), etc. In all these cases the partition function can be expanded in a power series in λ.

(b) *Dilute gases.* In real gases, the particles interact through forces which have a very short range (the molecules can usually be idealized as hard spheres). In a dilute gas the molecules move most of the time in straight lines and occasionally suffer collisions, which are more and more frequent and involve more and more particles simultaneously as the density of the gas increases. H. D. Ursell (1927) and J. E. Mayer (1937) have shown that the partition function, and hence the pressure, can be expanded as a power series in the density; the result is the famous *virial equation of state* (or *cluster expansion*). The original derivation of this equation made extensive use of a diagram technique, a procedure which establishes a one-to-one correspondence between certain mathematical expressions and certain graphs; such diagram techniques proved to be of major importance in modern perturbation theory. The coefficients of the various powers of the density (or virial coefficients) are expressed in terms of so-called cluster integrals, describing correlations of various types between a given number of particles (see KINETIC THEORY).

As the density increases, the cluster expansion breaks down and more powerful methods have to be used. The statistical theory of liquids is now well accepted, but it is based on more or less ad hoc assumptions, the justification of which from first principles is impossible. The advent of fast computers played a very important role in this field. The development of the methods of molecular dynamics allows one to perform detailed "numerical experiments" in order to determine the structure of compli-

cated many-body problems (up to 1000 particles interacting with realistic potentials).

The theory of critical phenomena is one of the "hot points" of present research. The problem here is to understand the occurrence of singularities in a number of physical phenomena as well as the relations among them. Although no final solution of this problem has been reached as yet, considerable progress has been made in recent years. (See CRITICAL PHENOMENA.) A considerable amount of work in the last fifteen years has culminated in the so-called *renormalization group* theory of critical phenomena (originally due to K. Wilson). The latter provides a very elegant and original method of studying the partition function and yields a detailed description of the properties of systems in the neighborhood of the critical point.

We must also mention that, after Onsager's masterwork on the exact solution of the two-dimensional Ising model of a ferromagnet (1944), an increasing number of sophisticated models have been solved exactly, mainly by E. Lieb and by R. Baxter. These models have provided important information about phase transitions and critical phenomena.

(c) *Plasmas.* The ionized gases, or plasmas, have a radically different behavior as compared to "usual" gases. This is due to the long range of the Coulomb forces. As a result, one cannot speak of collisions in the ordinary sense: the interactions have a markedly collective character, involving many particles simultaneously. Mathematically, the various virial coefficients turn out to be divergent. J. E. Mayer (1950) has shown that by rearranging the cluster expansion and by performing summations of certain subseries, one can obtain a convergent equation of state for a plasma, the first term of which agrees with the one calculated by Debye and Hückel in their famous theory of electrolytes (which is semiphenomenological) (see PLASMAS).

Whereas equilibrium statistical mechanics is in a state where the difficulties are mainly mathematical, nonequilibrium statistical mechanics is still in a state where the principles and the ideas are not yet completely clarified and unified. Most of our present knowledge has been achieved in the last thirty years, and this field is still in rapid growth.

The main problem here is to understand the basic paradox of irreversibility: whereas the laws of mechanics are invarient with respect to an inversion of time, the macroscopic evolution (e.g., heat conduction, dissipative phenomena, etc.) is irreversible. In older theories (Boltzmann), it was argued that due to the large number of particles and their complicated motions, one could invoke a probabilistic argument which, superposed to mechanics, would readily yield an explanation of irreversibility.

The basic idea in the modern theories is to

avoid the probability arguments. It is now widely accepted that the origin of irreversibility has to be sought in the *dynamical instability* of many-body systems. Two systems which are initially arbitrarily close together in phase-space may evolve after a finite time into widely separated positions. As a result, any finite domain in phase space becomes extraordinarily distorted in its motion, and the flow in phase space becomes highly chaotic, reminiscent of a turbulent flow (see FLUID DYNAMICS). Hence the concept of individual trajectories loses its meaning, and is to be replaced by a statistical description which eventually leads to irreversibility.

In a recent and quite illuminating version of nonequilibrium statistical mechanics, the theory is presented as follows. The distribution function $\rho(x, p; t) \equiv \rho(t)$ can be split into two terms

$$\rho(t) = \bar{\rho}(t) + \hat{\rho}(t) \tag{12}$$

If a certain number of very mild symmetry requirements are imposed [such as: $\bar{\rho}(t)$ must remain at all times in the subspace $\bar{\rho}$ as it evolves under the laws of exact mechanics], and if certain conditions on the nature of the interactions and the size of the system are met, the separation is unique. It can then be shown that the component $\bar{\rho}(t)$ has a "kinetic behavior." The reduced one-particle function determined from it, $F_1(x, p; t)$, obeys a *closed* equation, independently of the correlations. This closed equation, called the *kinetic equation*, describes the approach of the system to thermal equilibrium. For dilute gases the kinetic equation is none other than the classical Boltzmann equation (see KINETIC THEORY).

As for the complementary (and complicated) component $\hat{\rho}(t)$, it can be shown that, even if present, it does not contribute to the calculation of some important physical quantities, such as the static transport coefficients (see below). For all these problems the kinetic equation provides a complete and self-contained description.

One of the practical purposes of nonequilibrium statistical mechanics is the calculation of transport coefficients (thermal and electrical conductivity, viscosity, etc.) from molecular data. This can be done by starting from the kinetic equations. Alternatively, it has been shown recently (M. S. Green, R. Kubo) that a compact expression for most transport coefficients can be obtained from general arguments. The "Kubo formulas" express these coefficients in terms of an autocorrelation function of two microscopic currents (electrical currents, heat flow, etc.) averaged over the equilibrium distribution function. These two equivalent approaches can be combined in order to form a basis for the rigorous calculation of transport coefficients (see TRANSPORT THEORY).

Many other important results have been ob-

tained recently in statistical mechanics; they cannot be reviewed in such a short article. The interested reader is referred to the existing textbooks, such as those given below.

RADU C. BALESCU

References

Huang, K., "Statistical Mechanics," New York, J. Wiley & Sons, 1963.

de Boer, J., and Uhlenbeck, G. E., Eds., "Studies in Statistical Mechanics," Vol. 1, New York, Interscience Publishers, 1962.

Prigogine, I., "Non Equilibrium Statistical Mechanics," New York, Interscience, 1963.

Balescu, R., "Statistical Mechanics of Charged Particles," New York, Interscience Publishers 1963.

Stanley, H. E., "Introduction to Phase Transitions and Critical Phenomena," London, Oxford University Press, 1971.

Balescu, R., and Wallenborn, J., "On the structure of the time evolution process in many-body systems," *Physica*, 54, 477 (1971).

Balescu, R., "Equilibrium and Nonequilibrium Statistical Mechanics," New York, Wiley, 1975.

Cohen, E. G. D. (Ed.), "Fundamental Problems in Statistical Mechanics," Vol. V, Amsterdam, North Holland, 1980.

Feynman, R. P., "Statistical Mechanics," Reading, Mass., W. A. Benjamin, 1975.

Misra, B., Prigogine, I., and Courbage, M., "From Deterministic Dynamics to Probabilistic Descriptions," *Physica*, 98A, 1 (1979).

Cross-references: BOLTZMANN'S DISTRIBUTION LAW, BOSE-EINSTEIN STATISTICS AND BOSONS, CRITICAL PHENOMENA, FERMI-DIRAC STATISTICS AND FERMIONS, FLUID DYNAMICS, KINETIC THEORY, HEISENBERG UNCERTAINTY PRINCIPLE, LIQUID STATE, MATHEMATICAL PRINCIPLES OF QUANTUM MECHANICS, PLASMAS, STATISTICS, THERMODYNAMICS, TRANSPORT THEORY, WAVE MECHANICS.

STATISTICS

Statistics has been called the art and science of making reasonable decisions in the face of uncertainty or incomplete information. Such a wide ranging description suggests that the basic procedures of statistics must be very much involved with what is usually called simply the scientific method. This is indeed the case.

Descriptive statistics deals with the problem of how to summarize and present masses of data so that they will be useful to the scholar and understandable to his reader. In recent years, new techniques of *computer graphics*, the growth of *principal component analysis* and *factor analysis* as methods of dealing with multidimensional data in lower dimensions, the development of *cluster analysis* or *numerical taxonomy*, and some recognition of the importance of *exploratory data analysis* have all led to a rebirth of interest in descriptive methods. Such methods often make use of matrix algebra and the mathematics of euclidean vector spaces.

Building on previous work in his field and, perhaps, using some descriptive statistical methods, a scientist builds a theory (model) that he hopes will explain some aspects of nature. He deduces from this theory predictions which he then checks against experimental data, thus confirming or repudiating his theory. When the theory has factors of uncertainty built into it, by way of probabilities, statistical methods must come into play.

Suppose we let X be a *random variable*. That is, X stands for the outcome of a certain experiment. We call the collection E of all possible values for X (i.e., the collection of all possible outcomes of the given experiment) the *sample space* for the random variable. We may well be uncertain as to which of the possible values of X will actually occur. But, in such a case, we can imagine a *probability distribution* for X. If E is a discrete set, this probability distribution assigns a probability to each point in E. If E is not discrete (e.g., E might be an interval on the real line) this probability distribution assigns probabilities to certain sets by way of a *probability density function*. The probability assigned to a set A is obtained by integrating the probability density function over A, much as integrating a mass density function yields a mass. We think of the probability assigned to a point or to a set as the long-run relative frequency with which that point or some point in that set will occur in repeated trials of the experiment. If we make n independent runs of the given experiment, we obtain the outcome $(X_1, X_2, \cdots, X_n)$ which is called a *random sample* for X. The process of drawing conclusions about the probability distribution from information about a sample is called *statistical inference*.

Quantities that we may compute from a given sample to be used in this process are called *statistics*. Some of the useful statistics that may be found for a sample are the mean, median, mode, range, and standard deviation. This last is the positive square root of the variance S^2 which is given by

$$S^2 = \frac{1}{n} \sum_{i=1}^{n} (X_i - \bar{X})^2$$

where $\bar{X}$ is the sample mean. Any statistic is itself a random variable in the sense that its value is obtained by performing an experiment—the experiment of drawing a random sample for X and then computing the value of the statistic from these sample values.

On the other hand, any probability distribution for a random variable will, itself, usually have a mean, median, mode, range, and standard deviation. Such numerical properties of probability distributions are called *parameters*. Parameters are *not* statistics.

A standard example of this situation would be the case where $X_1, X_2, \cdots,$ and X_n are repeated measurements of some physical quantity and there is a random error in our measurement. (In this case, the standard deviation is often called the standard error.) The mean $\bar{X}$ of all our measurements would be a statistic and we would probably use $\bar{X}$ to estimate the underlying true value we are trying to measure. This unknown true value would be a parameter of the distribution for our measurements.

This is an instance of a broad class of problems where we wish to *estimate* one or more *parameters*. Another instance would be an opinion poll. The pollster is estimating certain parameters (the proportions of people in the population at large that would respond in certain ways to his questions) on the basis of the responses of people in his sample.

If θ is an unknown parameter for a random variable X and $(X_1, X_2, \cdots, X_n)$ is a random sample for X, we may consider a certain statistic $\hat{\theta}$, which will be a function of the sample values, as an *estimator* for θ. We would call θ an *unbiased* estimator if θ is the mean of the probability distribution for $\hat{\theta}$. The variance of the distribution for $\hat{\theta}$ is a measure of the *efficiency* of the estimator. A *best* estimator is one which is unbiased and of minimum variance (i.e., as efficient as possible) for the given sample size.

It is often desirable to give an *interval estimate* for an unknown parameter instead of giving a *point estimate* as above. Suppose it is possible to find two functions L and U which have the property that, no matter what the value of the unknown parameter θ, the probability that θ will be between $L(X_1, X_2, \cdots, X_n)$ and $U(X_1, X_2, \cdots, X_n)$ is, say, 0.95. If we then observe the sample values $X_1 = x_1, X_2 = x_2, \cdots,$ and $X_n = x_n$, we say that the interval from $L(x_1, x_2, \cdots, x_n)$ to $U(x_1, x_2, \cdots, x_n)$ is a 95 per cent *confidence interval* for θ. Similar interpretations are given for other levels of confidence or for one-sided confidence intervals.

Aside from problems of parameter estimation, there is the broad question of which probability distributions seem feasible in view of some experimental data. The classical *Neyman-Pearson theory* of hypothesis testing is a widely accepted approach to such questions. Any assertion about a probability distribution may be considered to be a *statistical hypothesis*. In many scientific endeavors, the basic questions can be put in terms of deciding between competing statistical hypotheses. For example, we might wish to decide between the hypothesis H_0 that a new medical treatment produces no higher a proportion of cures than an older treatment, and the hypothesis H_1 that the new treatment increases the proportion of cures. Or we might consider the hypothesis H_0 that a rate of radioactive decay is independent of the temperature, against the hypothesis H_1 that it is not. Or we might want to look at the hypothesis H_0 that the proportion of physicists who are unemployed is no higher this year than it was last, against the hypothesis H_1 that it has increased.

When we wish to decide between two hypotheses H_0 and H_1 about the probability distribution for a random variable X, we usually may arrange it so that we base our decision on the observation of a random sample for X. A *test* for the given hypotheses will be a specification of the conditions under which we will accept one of the hypotheses and reject the other. Since there are only two hypotheses, H_0 and H_1, being considered, such a test is given completely by specifying exactly which samples will cause us to reject H_0. The collection of such samples is called the *critical region* for the test. If H_0 happens to be true and yet when we draw our sample and use the test we find that we reject H_0 (and thus accept H_1), then we have committed an error. Such a mistake is traditionally called an *error of the first kind*. To accept H_0 when actually H_0 is false is to commit an *error of the second kind*. In some cases, it is possible to compute the probability of rejecting H_0 when H_0 is true. This probability is called the *significance level* of the test. Thus, to say that we reject H_0 at the 1 per cent significance level is to say that our observations led us to reject H_0 using a test which has the property that it will reject H_0 only 1 per cent of the time when H_0 is true. When it is not possible to compute the probability of rejecting H_0 given that H_0 is true, it is customary to take the *significance level* of the test to be the maximum possible probability of rejecting H_0 when H_0 is true. If both H_0 and H_1 are concerned with the value of a parameter θ we may judge the value of a proposed test by considering its *power function* f. $f(y)$ is defined to be the probability of rejecting H_0 when y is the true value of θ. For an effective test we would want $f(y)$ to be small for values of y such that $\theta = y$ makes H_0 true, and large for values of y such that $\theta = y$ makes H_1 true. The function whose value at y is $1 - f(y)$ is sometimes called the *characteristic function* for the test.

We note that in view of the intimate connections between the mathematical models we are using, the decisions we want to make, and the experiments we carry out, it is of the utmost importance to plan our experiment in a manner which takes these factors into consideration. There are cases on record where large-scale experiments have been carried out without such planning and were of such a nature that no decision could be made on the basis of the experiment—no matter what the data came out to be. Common sense dictates that planning should take place *before* the experimentation. The broad field of study known as *design of experiments* attempts to give guidance in such planning.

Most of the ideas about estimation and hypothesis testing are useful only when used in conjunction with facts from the theory of probability distributions. Some of the most useful distributions can be described as follows. If X is the number of "successes" in n independent trials where the probability of a "success" on each trial is p, then X is said to have the *binomial* distribution $B(n, p)$. If X has the probability density function

$$f(x) = \frac{1}{\sigma\sqrt{2\pi}} e^{-(1/2)[(x - \mu)/\sigma]^2}$$

where μ (the mean of the distribution) and σ (the standard deviation of the distribution) are real numbers with σ positive, then X is said to have the *normal distribution* $N(\mu, \sigma^2)$. If

$$Y = \sum_{i=1}^{n} X_i^2$$

where $(X_1, X_2, \cdots, X_n)$ is a random sample for X and X is $N(0, 1)$, then Y is said to have the *chi-square distribution* $\chi^2(n)$ with n degrees of freedom. If X is $N(\mu, \sigma^2)$ and $\overline{X}$ and S are the mean and standard deviation for a random sample of size n for X, then

$$T = \frac{\overline{X} - \mu}{S/\sqrt{n - 1}}$$

has the *Student's distribution* $t(n - 1)$ with $n - 1$ *degrees of freedom*. If Y is $\chi^2(n)$ and Z is $\chi^2(m)$ and Y and Z are independent, then $W = (Y/n)/(Z/m)$ has the *F-distribution* $F(n, m)$ with n and m *degrees of freedom*. Handbooks of mathematics and statistics usually contain tables that give certain probabilities for these standard distributions. Often these tables give only *critical values* that are the most useful. These are the values that cut off, on one side or the other, intervals over which, say, 5 per cent or 1 per cent of the probability lies.

Measurements that involve random errors are often observed to have distributions that are normal or approximately so. (In fact, a probability density function for a normal variable is often called a *Gaussian error function*.) Many other quantities that have approximately normal distributions seem to occur in nature. But even if this did not happen, normal distributions would be extremely useful because of the following powerful result which is called the *central limit theorem*: If X has a mean μ and a standard deviation σ, and $\overline{X}$ is the mean of a random sample of size n for X, then $(\overline{X} - \mu)/(\sigma/\sqrt{n})$ has a distribution which is approximately $N(0, 1)$ when n is large. Notice that there is no assumption about the underlying distribution for X aside from the fact that μ and σ exist. Many standard statistical techniques rely heavily on the central limit theorem. These techniques would be part of what is called *large-sample theory* as opposed to *small-sample theory* where other results must be used. Many useful small-sample techniques are concerned with so-called *nonparametric tests*, which involve hypotheses which are not stated in terms of the values of certain parameters.

Student's distributions are of special importance in small-sample theory. Chi-square distributions are useful for certain *goodness-of-fit* tests which include standard procedures for handling *contingency tables*. F distributions come into play in *analysis of variance* where the underlying variables are assumed to be normal and we are interested in detecting differences between their means. Analysis of variance is an example of a multivariate technique that may be explained in terms of the *general linear model*, where we assume that a random n-vector $\mathbf{Y}$ can be written as $\mathbf{Y} = \mathbf{AB} + \mathbf{E}$ where $\mathbf{A}$ is an $n \times k$ structure matrix of known constants, $\mathbf{B}$ is a $k \times 1$ matrix of unknown parameters, and the entries in the n-vector $\mathbf{E}$ are independent normal deviates with mean 0 and a common variance. *Regression analysis* is also an application of the general linear model. A special case of regression analysis is *least-squares curve fitting*.

Statistical software packages are becoming common. These have built-in routines to carry out the more or less standard chores of description, estimation, and hypothesis testing. Regression analysis packages may be used for polynomial curve fitting. The wide availability of computers has encouraged a resurgence of interest in discrete (rather than continuous, as with normal variables) statistical situations such as multidimensional contingency tables. Some of the standard statistical calculations can easily be carried out using statistical packages on microcomputers or even on hand-held calculators.

Since statistical methods are so closely intertwined with basic questions about scientific methods it is not surprising to find that there are controversies about these methods. The *decision theoretic* approach to statistics questions some of the basic ideas of classical hypothesis testing and suggests that, where possible, we should take into account the *cost* of committing the different kinds of errors we may make. Here the proper approach would be to design a procedure that would, in some sense, minimize our expected loss. In a laboratory situation, however, it may be difficult to assign costs to the sorts of errors. In managerial or social situations it may make a great deal of sense to consider seriously how much we will lose if we follow certain paths.

Statistics relies heavily on the theory of probability, and the question of the exact nature (or natures) of probability is an old one in the philosophy of science. A growing group of theoretical and practicing statisticians takes a somewhat different view of probability than is usual. These people call themselves *Bayesians* or *subjectivists* and believe that it is most useful to consider a probability to be a subjective, personal thing which is based on an individual's knowledge (and ignorance) and is subject to modification (via what is known as Bayes theorem) in view of empirical information. We may determine an individual's probability for an event by offering him bets and discovering at what odds he will take either side of the bet. A Bayesian point of view has considerable impact on the question of what constitutes an acceptable statistical procedure and articles about this impact are common in the current literature.

This ferment in statistics happens to coincide in time with a great increase in the use of statistical methods in the natural and social sciences. With the continual pressures caused by a reconsideration of the foundations and a demand for useful tools for applications, the field of statistics will continue to grow in both depth and usefulness.

FRANK L. WOLF

References

Beyer, W. H. (Ed.), "Chemical Rubber Company Handbook of Tables for Probability and Statistics," Cleveland, Chemical Rubber Company, 1966.

Feller, W., "An Introduction to Probability Theory and its Applications," Vol. I, 3rd ed., New York, John Wiley & Sons, 1968.

Gnanadesikan, R., "Methods for Statistical Data Analysis of Multivariate Observation," New York, John Wiley & Sons, 1977.

Hicks, C. R., "Fundamental Concepts in the Design of Experiments," New York; Holt, Rinehart, and Winston, 1965.

Hogg, R. V., and Craig, A. T., "Introduction to Mathematical Statistics," 3rd ed., New York, Macmillan, 1970.

Larsen, R. J., and Marx, M. L., "An Introduction to Mathematical Statistics and its Applications," Englewood Cliffs, New Jersey, Prentice-Hall, 1981.

Lindgren, B. W., "Statistical Theory," 2nd ed., New York, Macmillan, 1968.

Nie, N. H., et al., "SPSS: Statistical Package for the Social Sciences," New York, McGraw-Hill, 1975.

Owen, D. B., "Handbook of Statistical Tables," Reading, Mass., Addison-Wesley, 1962.

Ryan, T. A., Joiner, B. L., and Ryan, B. F., "MINITAB Student Handbook," North Scituate, Mass., Duxbury Press, 1976.

Tukey, J. W., "Exploratory Data Analysis," New York, Addison-Wesley, 1977.

Wolf, F. L., "Elements of Probability and Statistics," 2nd ed., New York, McGraw-Hill, 1974.

Cross references: CALCULUS OF PHYSICS; MEASUREMENTS, PRINCIPLES OF; MATHEMATICAL PRINCIPLES OF QUANTUM MECHANICS.

STRONG INTERACTIONS

Only two types of fundamental interaction operate in classical physics, which account for all macroscopic and chemical phenomena including atomic structure. There are the gravitatonal forces which control the motion of bulk matter, which is electrically neutral—particularly the motion of the planets in the solar system. Then there are the electromagnetic forces which govern the motion of the electrons in atoms, and also give rise to the interactions between atoms and, hence, form the basis of all chemical reactions.

An atom has a radius of about 10^{-10} m. In an atom, a cloud of electrons orbit about a central nucleus with a radius of about 10^{-14} m. These nuclei are made up of nucleons—protons and neutrons. The total charge, which is determined by the number of protons, fixes the chemical nature of the atom. The nuclei are extremely stable, remaining unchanged through the most violent chemical reactions. The gravitational forces within a nucleus are completely negligible compared with the electrical repulsion between the charges on the tightly packed protons. Since the electrical forces tend to blow the nucleus apart, it follows that a completely new, specifically nuclear, force must be operating to form these stable structures. This force is of short range, only effective when the nucleons are less than 10^{-14} m apart. Within this range, it is strong enough to overcome the powerful electrical repulsions. It is known as the *strong nuclear interaction*, to distinguish it from the other specifically nuclear interaction—the *weak* interaction—which causes the spontaneous disintegration of subnuclear particles (see WEAK INTERACTIONS). It is the strong nucleon interaction which is the source of energy in nuclear weapons and in nuclear reactors.

To investigate the workings of the strong nuclear interaction a beam of high-energy protons from a proton accelerator is directed at a target of liquid hydrogen (more protons) and scattered. At low energies, below 300 MeV in the laboratory, the protons are merely deflected by the strong nucleon-nucleon interaction. Above this energy, new particles are produced in the collision. The first to appear are the π-mesons, or pions. As the energy increases, more and more particles are found. The least massive are stable as far as the strong interaction is concerned, but they disintegrate through the weak interaction with mean lifetimes of about 10^{-10} second. The more massive particles decay via the strong interaction into other strongly interacting particles in about 10^{-23} second. These are tabulated in the article on ELEMENTARY PARTICLES. These strongly interacting subnuclear particles, which include the proton and neutron, are known collectively as *hadrons*. This distinguishes them from the *leptons* (electrons, muons and neutrinos) which are subject only to the weak and—if they are charged—the electromagnetic interactions.

Because of the strength of the strong interaction, it has not been possible to determine the details of the corresponding dynamics. However, considerable progress has been made by studying the conservation laws which govern the collision of hadrons and by analogy with the theory of electromagnetism.

The hadrons are specified according to their properties which correspond to those physical quantities which are conserved (i.e., do not change) in any collision which is dominated by the strong interaction. These are typically the mass, which contributes to the energy, and the spin which is part of the total angular momentum. If a particle has spin J (in units of $h/2\pi$) there are $2J + 1$ possible states, corresponding to the orientations of the spin axis allowed by quantum mechanics. These states are distinguished by the magnitude of the component J_z which ranges by integers from $+J$ to $-J$.

Other conserved quantities are the electric charge, the so-called hypercharge, and the baryon number. These are also tabulated in the article ELEMENTARY PARTICLES.

These hadrons appear in multiplets of nearly equal mass, but differing charge. Thus there are three pions (positive, negative and neutral) and two nucleons, one positive (the proton) and one neutral (the neutron). These are closely analogous to the $2J + 1$ spin states of a particle of spin J and may be explained by attributing to each mass multiplet a spin I (isotopic spin) in an abstract space. The different charge states are then interpreted as the different "orientations" in this abstract space and are specified by the $2I + 1$ values of the component I_3. Thus the pion has isotopic spin 1, with the three charge states specified by $I_3 = \pm 1$, 0. The nucleon has isotopic spin $I = 1/2$, with two charge states $I_3 = \pm 1/2$. Isotopic spin is conserved in strong interactions, just as total angular momentum is conserved. The conservation of angular momentum is related to the invariance of the whole colliding system with respect to its orientation in space. The conservation of isotopic spin is similarly equivalent to the statement that the strong interactions are invariant under a group of two-dimensional unitary transformations, known to group theorists as SU(2).

It has been conjectured that the strong interactions are approximately invariant under a wider group of three-dimensional unitary transformations, SU(3), which combines the notion of isotopic spin with that of hypercharge. According to this theory, the isotopic multiplets of particles of the same spin J can be combined into supermultiplets of roughly similar mass, which form hexagonal or triangular patterns when they are exhibited on a graphical plot of I_3 against Y. By 1963, three such supermultiplets had been established. They form octets of particles, with $J = 0, \frac{1}{2}$ and 1. (These are $\pi^+ \pi^- \pi^0 \eta^0 K^+ K^0 K^- \bar{K}^0$; $\Sigma^+ \Sigma^- \Sigma^0 \Lambda^0 p n \Xi^- \Xi^0$; $\rho^+ \rho^- \rho^0 \phi^0 K^{*+} K^{*0} K^{*-} \bar{K}^{*-}$.) At that time, the known particles of spin $J = 3/2$ formed a tenfold triangular multiplet, with one particle missing—the Ω^-. This missing particle was an isotopic singlet ($I = 0$) with hypercharge minus two. It had remarkable physical properties very well defined so that its mass, mode of production, mean life time, and decay could all be predicted. This particle was discovered in 1964 exactly as forecasted, thus confirming the wider unitary scheme.

This scheme provides a kind of "periodic table" for the subnuclear particles which can clearly no longer be regarded as "elementary." The simplest explanation of these regularities is that all the hadrons are themselves composed of yet more basic entities—called QUARKS. Three quarks are required, corresponding to the three dimensions of the $SU(3)$ group, leading to the three conserved quantities: electric charge, hypercharge, and baryon number. The quarks form an isotopic doublet and an isotopic singlet. If they are assumed to be of approximately equal mass, the observed patterns emerge if the hadrons with baryon number zero are the various possible combinations of quark-antiquark pairs, while those with unit baryon number are bound states of three quarks. If the quarks are assumed to have an ordinary spin of $\frac{1}{2}$, they also provide a natural explanation of the spins of all the particles appearing in Fig. 1. This three-quark theory provided a very adequate and economic explanation of the known properties of hadrons for ten years following the discovery of Ω^-, but it has one very remarkable feature. For the hadrons to have the observed integer charges (in units of the electron charge) it is necessary for the quarks to have fractional charges of $\frac{2}{3}$ and $-\frac{1}{3}$. This should make them easy to detect and, if the known structure of atoms and nuclei is taken as a guide, the theory implies that it should be possible, given sufficient energy, for hadrons to be disintegrated and for free, fractionally charged, quarks to be observed directly. In spite of intensive searches no convincing evidence for such objects has ever been found.

The next important development came in 1974 with the discovery of the ψ/J particle with a lifetime of about 10^{-20} secs, with approximately three times the mass of the proton. This was found independently in proton-proton collisions and in the hadronic matter produced in high energy collisions of electrons and positrons. This particle lies outside the scope of the earlier three-quark model. It was not unexpected, as it had already been shown that if hadrons were to be incorporated in a unified theory of electric and weak interactions (see ELECTROWEAK THEORY), there must be a fourth, heavier quark carrying a new quantum number known as *charm*. The ψ/J is interpreted as the charm-anticharm quark combination, and

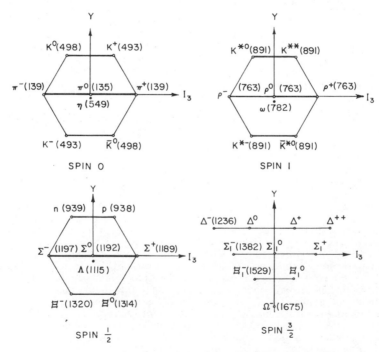

FIG. 1. The well-established SU(3) multiplets. In each of these, particles lying together on horizontal lines form sub-multiplets of SU(2) (isotopic spin). The figures in brackets indicate the masses of the particles in MeV/c².

the existence of three other hadrons expected from the combination of an anti-charm quark with the three quarks of the older theory was established within two years. Similar evidence now exists for a fifth quark and there are strong theoretical arguments for believing that there may be a sixth, corresponding to the six known leptons—e, ν_e, μ, ν_μ, τ, and ν_τ.

None of these five (or six) quarks has been observed as an independent entity. A qualitative explanation of this, which also makes the whole quark theory consistent with the Pauli exclusion principle, is contained in QUANTUM CHROMODYNAMICS (QCD). In this theory each quark appears in three forms, picturesquely denoted by *color*—red, blue, and yellow, say. Each quark carries a unit of color and the net total of each color is exactly conserved due to a new $SU(3)$ invariance with regard to the color variables. The forces that bind the quarks together to form hadrons are related to this invariance and these conservation laws by the same mechanism (local gauge invariance) which relates electromagnetic forces to the conservation of electric charge in QUANTUM ELECTRODYNAMICS (QED). In this picture, the hadrons in QCD are the subnuclear analogues of the atoms in QED and the strong interactions between hadrons are the analogues of chemical binding between atoms. It is conjectured that the QCD forces binding the quarks to form hadrons are so strong that individual quarks cannot be separated, except possibly under the most extreme conditions of temperature and pressure, which would explain why individual quarks have not been detected.

It should be emphasized that belief in QCD theory is based largely on aesthetic arguments. It is an elegant theory which seems to contain the general features required to explain the observed properties of hadronic matter. However, the calculation of its implications is so complicated that even some of its most important qualitative features, such as the containment of quarks, have yet to be established mathematically. It is also evident, with six different types of quark, each appearing in three different forms, that the substructure of hadronic matter is itself becoming rather complicated, suggesting a possible further synthesis of the strong nuclear interaction with the weak interaction and electromagnetism.

P. T. MATTHEWS

References

't Hooft, G., "Gauge Theories of the Forces between Elementary Particles," *Scientific American* **242**(6), 104 (1980).
Georgi, H., "A Unified Theory of Elementary Particles and Forces," *Scientific American* **244**(4), 48 (1981).
Weinberg, S., *Scientific American* **244**(6), 64 (1981).

Cross-references: CONSERVATION LAWS AND SYMMETRY, ELECTRON, ELECTROWEAK THEORY, ELEMENTARY PARTICLES, GAUGE THEORIES, GRAND UNIFIED THEORY, ISOSPIN, NEUTRINO, NEUTRON, PROTON, QUARKS, WEAK INTERACTIONS.

SUPERCONDUCTIVITY

This article describes the experimental facts relating to superconductivity, our present theoretical understanding, and some practical applications. No attempt is made to follow historical order or to provide an exhaustive treatment; the references at the end of the article will help those who wish to pursue the subject further.

The most spectacular property of a superconductor is the total disappearance of its electrical resistance when it is cooled below a critical temperature T_c. Very careful measurements show that the electrical resistance of a superconductor is at least a factor of 10^{17} smaller than the resistance of copper at room temperature and may therefore for all practical purposes be taken to be zero. Some 27 elements, and a vast number of alloys and compounds, have been discovered to be superconducting. In addition, a number of materials which are non-superconducting under ambient pressure in the bulk form, can be made superconducting by subjecting them to high pressures or by preparing them in the form of amorphous films. Even some organic chain compounds have been found to be superconducting. To name just a few from each category: elements In, Sn, Mo, Hg; Nb-Ti alloys; the intermetallic compound Nb_3Sn; As, Ge, Si under high pressure; amorphous films of Be, Bi, Ba; ditetramethyltetra selenafulvaleneperchlorate, $(TMTSF)_2ClO_4$ for short. Transition temperatures range from a few thousandths of a kelvin (tungsten, for example at 0.015 K) to the highest reported to date, about 23 K for Nb_3Ge. The search for superconductors with yet higher transition temperatures continues; there are occasional reports of such finds, but none has yet been convincingly confirmed.

Another important property is the destruction of superconductivity by a large enough applied magnetic field. The magnetic behavior of a superconductor falls into either one of two distinct classes, depending on the particular material. These are called *type I* and *type II* superconductors. The difference can be brought out by considering the response of a superconducting sample which is in a magnetic field swept from zero to a value large enough to make it normal, i.e., destroy the superconductivity. For a type I sample, the magnetic flux is almost totally excluded from the interior of the sample up to a *critical field* H_c. At H_c, an abrupt transition occurs, superconductivity is destroyed, and flux penetrates the sample fully. For a type II sample also, the initial behavior is the same; the flux is excluded up to a critical field H_{c1}. But now the transition is gentler. As the applied field is increased, some flux begins to creep into the sample. The entry of flux is completed at a higher critical field H_{c2}, at which point the sample becomes normal. These three characteristic fields H_c, H_{c1}, and H_{c2} vary from material to material, and depend on temperature. They all tend to zero as T tends to T_c, and go to finite limiting values as T tends to zero.

The state of a superconductor in fields up to H_c (for type I) or H_{c1} (for type II) is called the *Meissner state*, after the co-discoverer of the effect. The original paper in fact was by W. Meissner and R. Ochsenfeld, a fact which may interest social historians of science. The state of a type II superconductor between H_{c1} and H_{c2} is called the *mixed state*, because it turns out that the sample can be considered in some sense as breaking up into an intimate ordered mixture of normal and superconducting regions. Both types of magnetic behavior, in suitably prepared samples, are thermodynamically reversible and cannot therefore be deduced from the perfect conductivity.

In the Meissner state, the magnetic field decays exponentially to zero as one goes in from the surface of the sample. This characteristic decay length, denoted by λ, is called the *penetration depth* and is typically a few times 10^{-5} cm. The precise value varies with material. Thin films of a type I superconductor, of thickness d less than λ, have critical fields higher than the bulk critical field H_c, approximately in the ratio of λ to d. This result follows qualitatively from the thermodynamics of the Meissner effect: the metal in the superconducting state has a lower free energy than in the normal state, and the transition to the normal state occurs when the energy needed to keep the flux out becomes equal to this free energy difference. But in the case of a thin film with $d < \lambda$, there is partial penetration of the flux into the film, and thus one must go to a higher applied field before the free energy difference is compensated by the magnetic energy. The situation is more complicated for type II superconductors, because flux penetration begins at H_{c1} even for bulk samples.

All the elemental superconductors are type I, with the exception of Nb and possibly V. Practically all the alloys and compounds are type II, which is therefore by far the preponderant class. It is interesting that even though the phonomenon of superconductivity has been known for over 70 years, it is only within the past 25 years that type II behavior was clearly recognized, described and understood. The highest values of H_c, extrapolated to zero kelvin, are of the order of 10^3 gauss; values of H_{c2}, on the other hand, have been reported for some Chevrel phases in excess of 5×10^5 gauss. It may be noted that the typical field in an iron-core electromagnet is 10^4 gauss. It is clear that the discovery of such high critical field materials has opened up some exciting possibilities in technologies where high magnetic fields are desirable.

Since a magnetic field can destroy superconductivity, so should an electrical current flowing through a superconducting wire, through the self-generated magnetic field. One can therefore talk about a *critical current*, above which the wire becomes normal, and which should depend upon temperature and an externally applied magnetic field. But again, there are significant differences between type I and type II superconductors. A given wire of lead (type I) may carry 10^3 amp/cm^2 in a field of 100 gauss before going normal. But one can make a wire of Nb_3Sn (type II) which can carry 10^5 amp/cm^2 in 10^5 gauss, and still be superconducting. The technological implications are obvious.

The existence of two types of superconductivity does not at all mean that two separate theories are needed. There is a single comprehensive microscopic theory, as will be seen shortly, which gives a beautiful explanation of how the two types arise. Everything from here on, in fact, applies to all superconductors, unless explicitly stated otherwise.

All of the above properties distinguish superconductors from "normal" metals. There is another very important distinction, which contains a clue to the understanding of some of the properties of superconductors. In a normal metal at 0 K, the electrons, which obey Fermi statistics, occupy all available states of energy below a certain maximum energy called the *Fermi energy* ζ. Raising the temperature of the metal causes electrons to be singly excited to states just above the Fermi energy. There is for all practical purposes a continuum of such excited energy states available above the Fermi energy.

The situation is quite different in a superconductor; in a superconductor, the lowest excited state for an electron is separated by an energy gap Δ from the ground state. The existence of this gap in the excitation spectrum has been confirmed by a variety of experiments: electronic heat capacity, thermal conductivity, ultrasonic attenuation, far infrared and microwave absorption, and tunneling. It is to be noted that the excitation of electrons across the gap by photons requires a minimum energy of 2Δ, which is consistent with the description of the superconducting ground state in terms of Cooper pairs, as described further below. The energy gap decreases monotonically with increasing temperature, having a value $\Delta_0 = 1.75\,k_B T_c$ at 0 K (where k_B is the Boltzmann constant) and vanishing at T_c.

The superconducting state has a lower entropy than the normal state, and therefore one concludes that superconducting electrons are in a more ordered state. Without, for the present, inquiring more deeply into the nature of this ordering, one can state that a spatial change in this order produced say by a magnetic field will occur, not discontinuously, but over a finite distance ξ, which is called the *coherence length*. The coherence length represents the range of order in the superconducting state and in elemental superconductors is about 10^{-4} cm, though we shall see later that it can in many superconductors take much lower values, and lead to type II behavior.

Measurements of the transition temperature on different isotopes of the same superconductor showed that T_c is proportional to $M^{-1/2}$, where M is the isotopic mass. This isotope effect suggests that the mechanism underlying superconductivity must involve the properties of the lattice in addition to those of the electrons. Another indication of this is given by the behavior of allotropic modifications of the same element: white tin is superconducting, while gray tin is not, and the hexagonal and face-centered cubic phases of lanthanum have different transition temperatures. A third, and most striking, indication is that the current vs. voltage characteristic of a superconducting tunneling junction shows a structure which is intimately related to the phonon spectrum of the superconductor.

The theory of superconductivity has developed along two lines, the phenomenological and the microscopic. The phenomenological treatment was initiated by the brothers F. and H. London, who modified the Maxwell electromagnetic equations so as to allow for the Meissner effect. Their theory explained the existence and order of magnitude of the penetration depth, and gave a qualitative account of some of the electrodynamic properties. The treatment was extended by V. L. Ginzburg and L. D. Landau, and by A. B. Pippard, who in particular emphasized the concept of the coherence length ξ and showed that if ξ is greater than λ many of the properties of type I superconductors could be explained. A. A. Abrikosov used these ideas and showed that if the electronic structure of the superconductor were such that the coherence length ξ becomes smaller than the penetration depth λ, one would get magnetic behavior similar to that observed in type II superconductors, with two critical fields H_{c1} and H_{c2}. In particular, Abrikosov's work, as extended by others, showed that the state of such a superconductor in applied fields between H_{c1} and H_{c2}, the mixed state, consists of a triangular lattice of magnetic flux lines uniformly penetrating the superconductor. The core of each flux line, of approximate radius ξ, consists essentially of normal material, while the surrounding regions are superconducting. Each flux line core is surrounded by persistent supercurrents, and contains a quantum of flux $\phi_0 \equiv h/2e \simeq 2 \times 10^{-7}$ gauss-cm^2. The existence of such a lattice of flux lines received brilliant experimental confirmation through the work of Essmann and Träuble, who used a Bitter technique involving ferromagnetic powder to show up the magnetic lattice on the surface of a superconductor in the mixed state. The problem of high critical currents in unannealed (or otherwise metallurgically imperfect) alloys and compounds is more complicated because it involves the interaction between the microscopic metallurgical structure and the superconducting properties.

The microscopic theory of superconductivity was initiated by H. Fröhlich, who first recognized the importance of the interactions of electrons with lattice vibrations and in fact predicted the isotope effect before its experimental observation. The detailed microscopic theory was developed by J. Bardeen, L. N. Cooper, and J. R. Schrieffer in 1957, and represents one of the outstanding landmarks in the modern theory of solids. The *BCS theory*, as it is called, considers a system of electrons interacting with the phonons, which are the quantized vibrations of the lattice. There is a screened coulomb repulsion between pairs of electrons, but in addition there is also an attraction between them via the electron-phonon interaction. If the net effect of these two interactions is attractive, then the lowest energy state of the electron system has a strong correlation between pairs of electrons with equal and opposite momenta and opposite spin and having energies within the range $k_B\theta$ (where k_B is the Boltzmann constant and θ the Debye temperature) about the Fermi energy. This correlation causes a lowering of the energy of each of these *Cooper pairs* (named after L. N. Cooper, who first pointed out their existence on the basis of some general arguments) by an amount 2Δ relative to the Fermi energy. The energy 2Δ may be regarded as the binding energy of the pair, and is therefore the minimum energy which must be supplied in order to raise a pair of electrons to excited states. We see thus that the experimentally observed energy gap follows from the theory. The magnitude Δ_0 of the gap at 0 K is

$$\Delta_0 \simeq 2k_B\theta \exp\left(-\frac{1}{NV}\right)$$

where N is the density of electronic states at the Fermi energy and V is the net electron-electron interaction energy. The superconducting transition temperature T_c is given by

$$3.5 k_B T_c \simeq 2\Delta_0.$$

It has been shown that the BCS theory does lead to the phenomenological equations of London, Pippard, and Ginzburg and Landau, and one may therefore state that the basic phenomena of superconductivity are now understood from a microscopic point of view, i.e., in terms of the atomic and electronic structure of solids. It is true, however, that we cannot yet, easily, calculate V for a given metal and therefore predict whether it will be superconducting or not. The difficulty here is our ignorance of the exact wave functions to be used in describing the electrons and phonons in a specific metal, and their interactions. However, we believe that the problem is soluble in principle at least.

The range of coherence follows naturally from the BCS theory, and we see now why it becomes short in alloys. The electron mean free path is much shorter in an alloy than in a pure metal, and electron scattering tends to break up the correlated pairs, so that for very short mean free paths one would expect the coherence length to become comparable to the mean free path. Then the ratio $\kappa \simeq \lambda/\xi$ (called the *Ginzburg-Landau parameter*) becomes greater than unity, and the observed magnetic properties of alloy superconductors can be derived. The two kinds of superconductors, namely those with $\kappa < 1/\sqrt{2}$ and those with $\kappa > 1/\sqrt{2}$ (the inequalities follow from the detailed theory) correspond respectively to type I and type II superconductors.

As the implications of the BCS theory were better understood, and its relationship to the Ginzburg-Landau and London theories further clarified, during the past few years, a profound and very beautiful new understanding of the nature of the superconducting state has been obtained. F. London had in his earliest work described the superconducting state as a *macroscopic quantum state*, and used a many-body wave function to describe this state. But it has now become clear that, while the full statistical-mechanical description of the condensation of the electrons into the superconducting state is indeed very complicated, the resulting macroscopic quantum state can be described by a complex "wave function" $\Psi(\mathbf{r}, t)$ which is a function of position $\mathbf{r}$ and time t. The amplitude of Ψ gives the fraction of the electron fluid which is in the condensed superconducting state, and its phase contains information about its dynamics. One notes the formal similarity with the Schrödinger wave function of a free particle,

$$\psi = \psi_0 \exp \frac{1}{\hbar} (px - Et)$$

where p is the momentum and E is the energy. The wave function of the superconductor, which is referred to usually as its *order parameter*, can be written

$$\Psi = \Psi_0 \exp (i\phi)$$

where for a uniform superconductor Ψ_0 is a constant and ϕ is in general a function of space and time, and also involves the magnetic field or more precisely the vector potential. The single-valuedness of the order parameter leads to the quantization of magnetic flux in units of ϕ_0, defined earlier, in any area which is fully surrounded by a superconducting region, such as the core of an Abrikosov flux line or a superconducting ring carrying a persistent current. It should be mentioned here that this concept of a complex order parameter provides a fundamental link between superfluid helium and superconductivity, which are now seen as two manifestations of the basic phenomenon of SUPERFLUIDITY, even though the microscopic origins are quite different in the two cases. In terms of the order parameter, one can now give a unified description of similar phenomena in the two areas, such as critical velocities, quantized flux/vortex lines, and the influence of magnetic field/rotation.

B. Josephson's important contribution was to recognize the implications of the complex order parameter for the dynamics of the superconductor, and in particular when one considers a system consisting of two bulk superconductors 1 and 2 connected by a "weak link." The basic requirement for the weak link is that the amplitude of the order parameter there should be substantially smaller than in the bulk regions. Experimentally such a situation has been realized in a variety of ways: two evaporated films separated by a thin (< 20 Å) oxide layer, a light point contact between two bulk superconductors, a single hourglass-shaped evaporated film with the constriction of dimensions small compared to the coherence length, or indeed even a bare niobium wire with a pendant frozen blob of soft solder, where the weak links, indeterminate in number, were formed by solder bridges through pinholes in the surface oxide. Collectively, all such weak link junctions are referred to as *Josephson junctions*.

The electrodynamic behavior of such a Josephson junction can be described at several levels of sophistication: as the solutions of the time-dependent Schrödinger equations for ψ_1 and ψ_2 with an assumed weak coupling between 1 and 2, by perturbation-theoretic methods, or by Green's function methods. All methods fortunately give the same two simple equations, the famous Josephson equations:

$$j = j_0 \sin \left(\phi_2 - \phi_1 - \frac{2e}{\hbar} \int_1^2 A \, dx \right). \quad \text{(J1)}$$

$$\frac{\partial}{\partial t} \left(\phi_2 - \phi_1 - \frac{2e}{\hbar} \int_1^2 A \, dx \right)$$

$$= -\frac{1}{\hbar} (\mu_2 - \mu_1) = \frac{2eV}{\hbar} \quad \text{(J2)}$$

where for simplicity a one-dimensional junction has been treated. In the above equations j is the supercurrent through the junction, ϕ is the phase of the order parameter, A is the vector potential, μ is the chemical potential, V is the voltage bias across the junction, and e and $\hbar$ have their usual meanings. The subscripts 1 and 2 refer to the bulk superconductors on either side of the weak link respectively.

From Eqs. (J1) and (J2) one can see easily that, in the absence of a magnetic field, i.e., $A = 0$, a maximum supercurrent j_0 can flow (which is a function of the junction parameters), and the application of a constant bias voltage V to the junction produces an alternating current at a frequency $\nu_J = 2eV/h$. These two effects constitute the dc and ac *Josephson effects* respectively. In addition, when a Josephson junction biased at a voltage V is exposed to electromagnetic radiation at frequency f, the beating of the two frequencies f and ν_J produces a zero-frequency current component, and consequently a constant-voltage step in the dc junction current, whenever $\nu_J = nf$ where n is an integer. Thus, at values of the voltage $V_n = nhf/2e$, the current-voltage characteristic of the junction shows vertical current steps.

Space does not permit a detailed discussion here of the various effects of a magnetic field on Josephson junctions in different circuit configurations. But it is apparent from Eq. (J1) that the integral on the right can in certain configurations be made to correspond to the magnetic flux through a portion of the circuit. Then j would show an oscillatory dependence upon the field, going through one cycle each time the flux changes by one quantum ϕ_0. For example, in the case of two junctions in parallel enclosing an area A, the maximum supercurrent goes through a cycle as the applied field changes by ϕ_0/A. If A is of the order of a square centimeter it is obvious from the magnitude of $\phi_0 \sim 2 \times 10^{-7}$ cm^2 · gauss that field changes of less than a microgauss can be detected.

Both the dc and ac Josephson effects have found some exciting and novel applications. The high sensitivity to magnetic field of the dc Josephson current in certain configurations has been used to develop a family of devices called "squids" (superconducting quantum interferometric devices) which are used to measure extremely small currents, voltages, and magnetic fields. The ac effect has been used to obtain a very precise measurement of $2e/h$ and thence the fine structure constant α, in terms of the standard volt and a frequency, by D. N. Langenberg and his collaborators. Their work has helped to resolve a number of discrepancies between experiment and the predictions of quantum electrodynamics. Since 1972, the U.S. legal volt has been maintained in terms of the Josephson frequency by the National Bureau of Standards, using the relation

$$1 \, \mu V = 483.593420 \text{ MHz.}$$

The intriguing question of how high a transition temperature is possible has not yet been given a satisfactory theoretical answer. The BCS theory suggests as an upper limit the Debye temperature, which is a few hundred degrees Kelvin for most materials; but actual T_c's are a tenth of this or less. It appears that as one progressively modifies a given material in the direction of higher T_c's, lattice instabilities set in which inhibit the superconductivity. It is possible that these instabilities can be avoided by using metastable high-temperature phases of complex alloys and compounds.

There have been several attempts at technological applications of superconductors. The most spectacularly successful one is the use of certain type II superconductors like Nb-Ti alloys and $Nb_3 Sn$ in making electromagnets. In a conventional electromagnet employing normal conductors, the entire electric power applied to the magnet is consumed as Joule heating. For a magnet to produce 100 kilogauss in a reasonable volume, the power requirement can run into megawatts. In striking contrast, a superconducting magnet develops no Joule heat because its resistance is zero. Indeed, if such a magnet has a superconducting shunt placed across it after it is energized, the external power supply can be removed, and the current continues to flow indefinitely through the magnet and shunt, maintaining the field constant. Superconducting magnets have already been constructed producing fields of over 100 kilogauss in usable volumes. There is a natural upper limit to the critical field possible in such superconductors, given by the paramagnetic energy of the electrons (due to their spin moment) in the normal state becoming equal to the condensation energy of the Cooper pairs in the superconducting state. This leads to a limit of about 360 kilogauss for a superconductor with a T_c of 20 K. Coupling between the orbital motion and spin of the electrons, however, can sometimes raise this limit.

Once the technological problems of building superconducting magnets were overcome, they became natural candidates to replace conventional magnets which can produce only a tenth of the magnetic fields that the superconducting ones can. The past ten years have seen a burgeoning of activity in such applications, and the following are just a few illustrative examples. The Fermi National Accelerator Laboratory is constructing over one thousand large superconducting magnets to accelerate elementary particles to higher energies than ever before. In the U.S., a 500 MW generator is on the drawing boards, while smaller generators have been successfully built. Great Britain has a superconducting motor driving a ship, and Japan and West Germany have programs on levitated ground transportation using superconducting magnets. There is consideration of superconducting cable for power transmission, and magnets for fusion reactors.

Superconductors are beginning to find applications at the level of electronic circuits too. Reference has already been made to the use of squids for low-level electrical and magnetic measurements. The properties of thin films and Josephson junctions are being wedded by a well-known computer company in the U.S. and Switzerland to produce a prototype computer which will be smaller, lighter, and more than ten times faster than existing semiconductor-based computers.

B. S. CHANDRASEKHAR

References

Lynton, Ernest A., "Superconductivity," New York, John Wiley & Sons 1969 (3rd Edition). A concise, extremely readable book in which most of the ideas in this article are elaborated.

Shoenberg, D., "Superconductivity," New York, Cambridge Univ. Press, 1952. An excellent account of the field up to about 1952.

London, F., "Superfluids," Vol. 1, New York, John Wiley & Sons, 1950. A classic account of the early theory of superconductivity; it and its companion Volume 2 on superfluid helium should be required reading for every cryophysicist.

Rose-Innes, A. C., and Rhoderick, E. H., "Introduction to Superconductivity," New York, Pergamon, 1969. About the same level as Lynton, but some topics discussed in more detail; all equations are given in mks units, which should please some.

There are several books on the modern theory of superconductivity, some of which are Schrieffer, J. Robert, "Theory of Superconductivity," New York, W. A. Benjamin, 1964; Rickayzen, G., "Theory of Superconductivity," New York, Interscience 1965; de Gennes, P. G., "Superconductivity of Metals and Alloys," New York, W. A. Benjamin, 1966; Kuper, C. G., "An Introduction to the Theory of Superconductivity," New York, Oxford Univ. Press, 1968; Saint-James, D., Sarma, G., and Thomas, E. J., "Type II Superconductivity," New York, Pergamon, 1969; Tinkham, M., "Introduction to Superconductivity," New York, McGraw-Hill, 1975.

Parks, R. D. (Ed.), "Superconductivity," Volumes I and II, New York, Marcel Dekker, 1969. A truly encyclopedic treatment of the subject, written by about thirty international experts.

Rev. Mod. Phys., 36, 1-331 (1964). This contains the proceedings of an international conference on superconductivity held in 1963.

Chilton, F. (Ed.), "Superconductivity," Amsterdam, North-Holland, 1971. This contains the proceedings of an international conference on superconductivity held in 1969.

Suhl, H., and Maple, M. B. (Eds.), "Superconductivity in d- and f-Band Metals," Academic Press, 1980. This contains the papers from the next to the latest of a series of international conferences devoted to the basic superconductivity (i.e., no applications) of transition metals, alloys, and compounds, which constitute the overwhelming majority of known superconductors, and are also the ones mostly used in applications.

Journal of Physical and Chemical Reference Data 5, 581-821 (1976) (American Chemical Society and American Institute of Physics) contains a superb compilation, by B. W. Roberts, of basic properties of all the then known superconductors. Supplements to bring the data up to date may be expected to appear in the same journal.

Proceedings of the LT conferences, various publishers. These contain papers, many of them on superconductivity, from the biennial (recently triennial) International Conference on Low Temperature Physics.

"Advances in Cryogenic Engineering," New York, Plenum. This is an annual series containing the proceedings of the annual cryogenic Engineering Conference, dealing with many applications of superconductivity and techniques of low temperatures.

Schwartz, B. B., and Foner, S. (Eds.), "Superconducting Applications: Squids and Machines," 1977, and "Superconductor Materials Science: Metallurgy, Fabrication and Applications," 1981, both published by Plenum Press. NATO Advanced Study Institutes, with experts giving tutorial lectures.

IEEE Transactions on Magnetics, 17 (1981). Contains the proceedings of the latest of a series of conferences on applied superconductivity.

Cross-references: AMORPHOUS METALS; CONDUCTIVITY, ELECTRICAL; ELECTRON SPIN; ENERGY LEVELS; FERMI-DIRAC STATISTICS AND FERMIONS; HEAT CAPACITY; HEAT TRANSFER; MAGNETISM; PHONONS; SEMICONDUCTORS; SOLID-STATE PHYSICS; SUPERFLUIDITY.

SUPERFLUIDITY

Superfluidity is a term used to describe a property of condensed matter in which a resistance-less flow of current occurs. The mass-four isotope of helium in the liquid state, plus about one-half of the elements in the periodic table and a variety of alloys and compounds, are presently known to exhibit this phenomenon. In the case of liquid helium, these currents are hydrodynamic; for the metallic elements, they consist of electron streams. The effect occurs only at very low temperatures in the vicinity of the absolute zero ($-273.16°C$ or 0 K). In the case of helium, the maximum temperature at which the effect occurs is about 2.2 K; for metals the highest temperature is in the vicinity of 23 K.

If one of these metals (called superconductors) is cast in the form of a ring and an external magnetic field is applied perpendicular to its plane and then removed, a current will flow round the ring induced by Faraday induction. This current will produce a magnetic field, proportional to the current, and the size of the current may be observed by measuring this field. Were the ring (e.g., one made of Pb) at a temperature above

7.2 K, this current and field would decay to zero in a fraction of a second. But with the metal at a temperature below 7.2 K before the external field is removed, this current shows no signs of decay even when observations extend over a period of a year. As a result of such measurements, it has been estimated that it would require 10^{99} years for the supercurrent to decay! To the best of our knowledge, therefore, the lifetime of these "persistent" currents is infinite. The persistent or frictionless currents in superconductors are not a recent discovery; they were observed first in 1911 (see SUPERCONDUCTIVITY).

In the case of liquid helium, these currents are, as mentioned, hydrodynamic, i.e., they consist of streams of neutral (uncharged) helium atoms flowing in rings. Since, unlike electrons, the helium atoms carry no charge, there is no resulting magnetic field. This makes such currents much more difficult to create and detect. Nevertheless, as a result of research carried out here and in England starting about 1960, the existence of these supercurrents in liquid helium has definitely been proved. These currents have been observed (1964) for periods as long as 12 hours, a time of the order of 10^3 shorter than is the case for electron currents. Nevertheless, our present belief is that these hydrodynamic currents also possess an infinite lifetime.

As mentioned, the empirical discovery of infinite-lifetime electron currents is of considerable antiquity, and from the beginning, many attempts have been made to explain the effect theoretically. Until recently all such attempts have failed completely, and as a matter of fact, the theoretical picture is still not completely satisfactory. Nevertheless, immense progress in this direction has occurred in the past three decades largely as the result of work in the U.S.A., England and Russia.

Although superfluidity in liquid helium is important to our basic understanding of the phenomenon, it is the effect in superconductors which arouses the most interest. This is due, in part at least, to the possible practical applications to which the effect might lead. In ordinary conductors (e.g., copper), the flow of an electric current is always accompanied by energy dissipation. The supercurrents propagate with no such power loss; superconducting transmission lines would be an economic advance of the first order. However, they would require "room temperature" superconductors, and unfortunately, modern theories, while they do not absolutely prohibit such, render their occurrence most unlikely.

In other ways, however, superconductors have already proved of practical value. Since the currents once created are there for all time, they have fairly obvious use as memory elements in computers. Again the persistent currents form a sort of super gyroscope more perfect than any so far devised. Alloy superconductors have been found (e.g., Nb_3Sn) which can support persistent currents even in the presence of intense magnetic fields of the order of 100 000 gauss. It is, generally speaking, a property of elemental superconductors that the persistent current is quenched in fields of a few hundred gauss; the situation in some alloys like the above is very different.

This has led to the development of intense field solenoidal magnets which maintain their magnetic fields with zero energy dissipation. A conventional electromagnet of the same size would consume many hundreds of kilowatts. This discovery has important consequences in several areas of physics.

Both aspects of superfluidity find their theoretical explanation in what is called a two-fluid theoretical model. We suppose that liquid helium and superconductors are quantum systems possessing a zero energy ground state plus a series of available states of higher energy called normal or excited states. The occupancy of the ground state is zero above the superfluid transition temperature (2.2 K for liquid helium) but grows steadily as the temperature is lowered. At absolute zero, the whole system is in the ground state. At any finite temperature, below transition, the system is in a mixture of ground and excited states. It is the particles in the ground state which form the persistent current.

A simple, though not entirely accurate, *raison d'etre* for the persistent currents lies in the fact that in the ground state, all the very many particles possess the same single wave function. It follows from this that such particles are not easily scattered out of the ground state—a finite amount of energy is required. Such an assembly will not readily interact with outside particles including those in excited states. Since fluid friction or viscosity is due to particle momentum interchange between neighboring layers of liquid, it follows that particles in the ground state will possess zero viscosity. This, in turn, means frictionless flow.

An assembly of particles obeying Bose-Einstein statistics will, below a certain temperature, possess a ground state like the one postulated above. It is known that Bose-Einstein statistics apply only to particles which possess zero (or integral) spin. The neutral He^4 atom possesses zero spin. The isotope He^3 (spin $\frac{1}{2}$) does not; and, in fact, liquid He^3 is not a superfluid. But this is also true of electrons which possess half integral spin and obey a very different statistic (Fermi-Dirac). In this statistic no more than two electrons can possess the same wave function. Hence a ground state, in the above sense, can clearly not exist. Thus according to our model, superfluid flow of electrons cannot occur—but it does!

A way out of this dilemma was suggested nearly 40 years ago. Namely, combine the electrons in pairs, with opposite spins. The resulting "particle" is then a boson and may properly reside in the ground state. The reason why this idea was not accepted by theoretical physicists until recently was because no mechanism could be found by which the electrons could be induced to form pairs. It turns out that the required mechanism arises as a result of a phonon (quantized lattice wave) emitted by one electron and absorbed by another at some other place. This couples the two together.

As mentioned, none of the current theories clearly explains persistent currents. It is not at all evident that their lifetimes should be infinite. It is thought by some that a better formulation for statistical mechanics than presently exists may be necessary before this becomes possible.

A system of persistent currents distinguished by the fact that many particles possess the same wave function constitutes a quantum effect on a hitherto unknown macroscopic scale. In other words, superfluids should exhibit quantum effects of such size that they are readily amenable to experimentation. Two such effects, one in each of the superfluids, have been found. A long thin cylinder with a hole along the axis (i.e., a tube) is similar to the previously mentioned superconducting ring in that persistent currents may also be produced in it. If the length of the cylinder is large compared to the diameter of the hole, the "trapped" magnetic field due to the persistent current is substantially uniform. The flux is the product of this field by the area of the hole.

As a consequence of the fact that the particles producing the current all possess the same wave function, it may be shown that this flux is quantized in integral multiples of hc/q ($h =$ Planck's constant, $c =$ velocity of light, $q =$ charge on the particles).

This prediction has been confirmed experimentally (1961). Further, these experiments show that $q = 2e$, where e is the charge on an electron. Thus the pair hypothesis is very nicely proven.

An interesting consequence of the above effect is that it is impossible to have any field (below a certain value depending on the size of the hole) at all in the cavity. Thus the measurements show that $hc/q \cong 2 \times 10^{-7}$ so that with a hole 1μ in diameter, a field less than about 13 gauss could not exist. We should therefore have a truly field-free space.

A very similar situation, due fundamentally to the same cause, exists for the hydrodynamic currents in liquid helium. In this case, it is the hydrodynamic "circulation" which is quantized in units of h/m ($h =$ Planck's constant, $m =$ helium atom mass). By circulation we mean $\oint \mathbf{v} \cdot \mathbf{dl}$ where $\mathbf{v}$ is the velocity of the atom in the ring-current and $\mathbf{dl}$ is a line element of the periphery of the ring. This effect has also been observed in laboratory experiments (1958).

There is also a surprising consequence connected with this effect. Suppose the helium was being rotated in its containing vessel at a tem-

perature above 2.2 K. Here the helium is "classical" and would rotate, like any other familiar liquid, at the same angular velocity as the vessel. Suppose, now, that the helium was cooled down to near 0 K with the vessel still rotating. If the initial speed were less than that required to produce one quantum of circulation, the rotating helium would come to rest while the container continued to rotate!

This is analogous to the behavior of the magnetic flux on the superconductor, the angular velocity being the quantity analogous to the magnetic field. To be sure, the above experiment has, to date, not been performed, but this is entirely due to difficulties with the existing instrumentation. With advances in technique, it seems very likely that the experiment can eventually be performed, and there is little doubt that the result perdicted will be observed.

C. T. LANE

Cross-references: BOSE-EINSTEIN STATISTICS AND BOSONS; CONDUCTIVITY, ELECTRICAL; ELECTRON SPIN; FERMI-DIRAC STATISTICS AND FERMIONS; PHONONS; SUPERCONDUCTIVITY.

SURFACE PHYSICS

Surface physics is the study of the composition, structure, and electronic properties of surfaces and interfaces. Its major goal is to develop an understanding of the macroscopic mechanical, chemical, and electrical properties of surfaces and interfaces in terms of fundamental quantum mechanical interactions. Whenever two distinct phases of matter are in contact, a boundary region is formed with physical and chemical properties which are different from those of the bulk phases on either side. This boundary region is called the *surface* or *interface*. Its thickness can be as small as a few atomic layers, as is the case for the metal-vacuum interface where screening by the conduction electrons is efficient, or as large as thousands of Ångstroms in the case of dielectrics.

Many processes which occur at surfaces or are influenced by surface properties are of great practical importance.[1] Examples include heterogeneous catalysis, crystal growth, thermionic and photoelectron emission, semiconductor devices, colloidal suspension, oxidation and corrosion, brittle fracture, and adhesion and friction.

Surface processes have also played a significant role in the development of modern physics. Photoelectron emission was first observed by Hertz (1887) and its properties were shown by Einstein (1905) to demonstrate the quantum nature of light. Diffraction of low-energy electrons from a nickel crystal surface by Davisson and Germer (1927) and of He atoms from LiF by Stern and Estermann (1930) demonstrated the wave-particle duality of matter. Individual atoms have been observed using the technique of field ion microscopy invented by E. W. Muller (1936).

There are several important reasons why our understanding of surface properties did not begin to increase rapidly until the 1960s. First, the technological importance of surfaces has increased, as is particularly evident in the semiconductor electronic industry. Second, high purity single crystal samples required for well-characterized experiments have become available. Third, the theoretical and experimental understanding of bulk properties has been sufficiently successful that it is reasonable to attempt to extend the results to the more complex case of surfaces which do not have the three-dimensional symmetry that greatly simplifies the description of many bulk properties. Fourth and perhaps most important, techniques have been developed to routinely obtain pressures less than 10^{-8} Pa. Such an ultra-high vacuum is crucial because many experimental measurements require 10–100 minutes during which the surface must not be contaminated from its environment. To see this, consider the rate (ν) at which molecules from the residual gas strike a surface

$$\nu = \frac{P}{(2\pi m k T)^{1/2}}$$

where P is the pressure, m the molecular mass, T the temperature, and k the Boltzmann constant. For a pressure of only 10^{-4} Pa at 300 K, about one monolayer of N_2 molecules will interact with a surface every second. Thus, the requirement of negligable contamination for 10–100 minutes requires background gas pressures three or four orders of magnitudes below 10^{-4} Pa.

The thermodynamic properties of surfaces were originally treated by Gibbs.[2] In this approach, the surface is defined by a reference plane and any extensive thermodynamic property such as the internal energy E^s associated with a surface of area A is given by

$$E^s = E^{\text{total}} - E^1 - E^2$$

where E^{total} is the internal energy for the entire system, E^1 and E^2 are the internal energies *if* the bulk phases on either side of the surface retained their properties up to the reference plane. E^s is called the *surface excess* internal energy. All other thermodynamic quantities associated with the surface can be defined in terms of similar excess quantities. The development of surface thermodynamics proceeds along standard lines following these definitions. The reversible work required to form a unit area of new surface of given orientation $\hat{n}$ at constant volume, temperature, and chemical potentials μ_i of each of the components is called

the *surface tension* $\gamma(\hat{n})$. The equilibrium shape of a solid or liquid in contact with its vapor at constant T and V is that shape for which the total energy associated with the surface is minimized and is given by the minimum of the integral

$$\int_A \gamma(n) \, dA.$$

Surface tension is related to other surface excess quantities by the *Gibbs adsorption equation*

$$d\gamma = -s^s \, dT - \sum \Gamma_i \, d\mu_i$$

where s^s and Γ_i are the excess surface entropy and number of moles of each component per unit area, respectively.[2,3]

Surface statistical mechanics seeks to describe the macroscopic thermodynamic properties of a surface and interface in terms of the forces between the atoms composing the system. As for bulk systems this requires a calculation of the canonical partition function Z from which the Helmholtz free energy

$$F = -kT \ln Z$$

and thus all other properties of the system can be determined. Calculations of the surface composition of alloys, equilibrium defect concentrations, and the character of phase transitions and phase boundaries have been carried out for simple models of surfaces.[1,3,4]

A one-centimeter cube of matter contains about 10^{23} atoms; only about 10^{16} of these are exposed on the surface. The surface to volume ratio is thus about 10^{-7}. Clearly, experiments which are sensitive to surface properties must have a sample prepared in such a way as to increase its surface area, or an experimental probe sensitive only to the surface must be used. The second approach using techniques with enhanced surface sensitivity is more popular because these techniques can be applied to many more types and forms of material. Electrons, ions, atoms, and photons have all been used as probes.[1] The most widely used techniques employ electrons with energies in the range 10–1000 eV because these electrons penetrate only 5–50 Å into the substrate. Thermal atoms and low-energy ions are also sensitive surface probes but beams are more difficult to prepare. Techniques utilizing photons which generally penetrate far into a material achieve surface sensitivity by involving a second particle which has shallow penetration. Examples include photoelectron emission (photons incident, electrons emitted) and photon stimulated desorption (photons incident, atoms or ions emitted).

One of the most important properties of clean crystal surfaces, is that they exhibit crystalline order. In many cases, especially for metals, the surface atoms are at the positions which would be expected from a simple truncation of the bulk lattice; the only difference being a slight inward or outward relaxation of the surface layer spacing due to the absence of neighbors on one side. For a second, fairly large group of surfaces including most semiconductors and the metals platinum, iridium, and gold, the surface lattice is completely different from that of the underlying bulk. Such surfaces are said to be *reconstructed*. The most widely known examples are the (7×7) Si(111) surface and the (5×20) Pt(100) surface. (The notation (7×7) and (5×20) gives the size of the reconstructed unit surface mesh measured in units of the primative unit mesh for the unreconstructed surface.) Although extensive research has been directed toward determination of the structures of reconstructed surfaces, the (110) surface of GaAs is perhaps the only one whose structure is generally believed to be known.[5] A third, small group of clean surfaces undergoes *displacive phase transitions* in which the surface atoms are only slightly displaced from their bulk lattice positions to form a new structure at low temperatures. These transitions are reversible, but whether they are order-order or order-disorder transitions remains to be determined. One proposed mechanism for displacive transitions is that a soft surface phonon mode drives surface atoms into new positions which have slightly lower electronic energy. The only examples of displacively reconstructed surfaces are W(100) and Mo(100).[5]

When a molecule impinges on a surface, many events are possible.[1,3,4] The molecule may be elastically reflected or inelastically scattered into new electronic or vibrational states. If the molecule is to be absorbed it must discard some of its initial energy. This can be accomplished by dissociation, by excitation of internal electronic or vibrational modes, and by interactions with phonons and electronic excitations in the substrate.

A molecule is said to be physically absorbed, or *physisorbed*, when no charge is transferred between it and the substrate; bonding is via weak van der Waals forces and the heat of adsorption is on the order of a few tenths of an electron volt. Adsorption of inert gases is an example of physisorption. Chemical adsorption, or *chemisorption*, occurs when charge is transferred onto or away from the adsorbing atom or molecule due to the formation of a chemical bond with the substrate. Bond energies are typically an electron volt or larger.

Segregation occurs when atoms reach the surface by diffusing from the substrate rather than impinging from the gas or liquid phase.

Theoretical descriptions of adsorption are very complex, and although many aspects of adsorption are understood qualitatively, complete general theories have not yet been developed.[6] This is partly because adsorption is an "intermediate coupling" problem with no obvious single physical parameter dominating the process. Furthermore, any theoretical treatment needs to be self-consistent, particularly in cases involving charge transfer and/or electronic screening.

When clean surfaces adsorb gases, the adsorbed atoms or molecules take on a variety of ordered structures. In their pioneering experiments, Davisson and Germer observed ordered surface impurities, probably carbon or oxygen, on a Ni(111) surface. In some cases, such as oxygen adsorption on aluminum or silicon, the surface layer has no long-range crystalline order. However, in many important cases where adsorbate-substrate bonds are strong, the adsorbed species form ordered overlayers which are in simple registry with the substrate. In these cases, the lattice of the ordered overlayer is built up from a *unit mesh*, analogous to the unit cell in bulk crystal structures, whose dimensions are simply related to those of the underlying substrate lattice. This type of overlayer is said to be *commensurate*.

If the interactions between the adsorbed species are stronger than those between the adsorbate and substrate, the overlayer unit mesh may not be simply related to that of the substrate but instead becomes *incommensurate*. In several cases, an adsorbate which is commensurate at low coverages becomes incommensurate as the coverage is increased beyond a critical value. The properties of the commensurate-incommensurate transition have been extensively investigated in recent years but are still hotly debated.[4]

Adsorbed atoms can form several phases as their coverage on the surface increases. These phases are analogous to the solid, liquid, and gas phases of a three-dimensional system and are typically characterized using a temperature vs. adsorbate coverage phase diagram. Oxygen adsorbed onto the (111) surface of nickel[7] is the most carefully characterized chemisorption system, and inert gases adsorbed on the graphite basal plane are the most extensively studied physisorption system.[1,4] Only a few complete phase diagrams of this type have been determined. However, they are extremely valuable for testing statistical mechanical descriptions of phase transitions in two dimensions[4] and more work is needed.

Many experimental methods have been used to determine surface structures. The methods can be divided into three groups according to the type of information obtained: (1) long-range order; (2) local geometry; (3) single atom resolution.

Low-energy electron diffraction (LEED) is the predominant technique for studies of long-range order. The size and shape of the unit mesh are readily obtained and in favorable cases atomic spacings can be determined with an accuracy of about 10%. Other techniques with sensitivity to long-range order include photoelectron spectroscopy and atom diffraction.

Recently, it has been demonstrated that x-ray diffraction and Rutherford backscattering of high-energy ions can be applied to structure determinations for both surfaces and interfaces.[8] This development promises to revolutionize surface crystallography because both x-ray diffraction and Rutherford backscattering are easier to interpret than techniques previously used.

The local geometry of an adsorbed atom or molecule can be determined from measurements of its vibrational mode using infrared absorption spectroscopy or high-resolution electron energy-loss spectroscopy. The angular distributions of ions which desorb during electron or photon irradiation yield similar information. Bond lengths can be extracted from the energy dependence of electron or photon absorption near absorption edges. Several of these techniques are particularly important for studies of absorption on amorphous or polycrystalline surfaces where diffraction methods are not useful. Such surfaces are technologically important, for example, as catalysts.

Two techniques make possible the observation of individual atoms on surfaces. These are field ion microscopy and high-resolution electron microscopy.[1] Field ion microscopy has been particularly valuable for studying diffusion on surfaces.

Real surfaces are not perfect. Both point defects, such as vacancies and impurities, and extended defects such as steps, dislocations, and grain boundaries are always present. These defects can act as sites for preferential adsorption, alter surface transport properties, and influence crystal growth.[1,3] Only the properties of ordered arrays of steps have been studied in any detail because such arrays can be observed by LEED. These experiments show that steps influence the work function, change the energy distribution of electronic states, alter absorption kinetics, affect catalytic processes which proceed via bond breaking reactions, and induce asymmetries in surface diffusion.

Surface composition and electronic properties are studied using various spectroscopic methods.[1] These include Auger electron spectroscopy, ultraviolet and X-ray photoelectron emission, and electron energy-loss spectroscopies. Other methods for determining surface compositions include ion sputtering and scattering, nuclear reactions, thermal desorption, and radioactive tracers.

Surface physics is an important subdiscipline of physics with tremendous technological applications. Although, our knowledge about the basic properties of surfaces has advanced significantly in recent years, our level of understanding remains crude compared to that in condensed matter physics. Thus, the field of surface physics will have a brilliant future with many exciting advances.

W. N. UNERTL

References

1. Vanselow, R. (Ed.), "Chemistry and Physics of Solid Surfaces," Vols. I and II, Boca Raton, FL, CRC Press, 1979.
2. Gibbs, J. W., "Collected Works," Vol. I, New Haven, Yale Univ. Press, 1948.
3. Blakely, J. M. (Ed.), "Surface Physics of Materials," Vols. I and II, New York, Academic Press, 1975.
4. Dash, J. G. and Ruvalds, J. (Eds.), "Phase Transitions in Surface Films," New York, Plenum Press, 1979.
5. Inglesfield, J. E., "Electronic Surface Structure," *Rep. Prog. Phys.* **45**, 224 (1982).
6. Smith, J. R. (Ed.), "Theory of Chemisorption," Berlin, Springer-Verlag, 1980.
7. Kortan, A. R., and Park, R. L., "Phase Diagram for Oxygen Chemisorbed on Nickel (111)," *Phys. Rev. B* **23**, 6340 (1981).
8. Eisenberger, P., and Feldman, L. C., "New Approaches to Surface Structure Determinations," *Science* **214**, 300 (1981).

Cross-references: ADSORPTION AND ABSORPTION, CRYSTALLOGRAPHY, FIELD EMISSION, PHOTOELECTRICITY, SOLID STATE PHYSICS, SOLID STATE THEORY, SURFACE TENSION, THIN FILMS.

SURFACE TENSION

Surface tension results from the tendency of a liquid surface to contract. It is given by the tension σ across a unit length of a line on the surface of a liquid. The surface tension of a liquid depends on the temperature and diminishes as temperature increases and becomes 0 at the critical temperature. For water σ is 0.073 newtons/m at 20°C, and for mercury it is 0.47 newtons/m at 18°C.

Surface tension is intimately connected with capillarity, that is, rise or depression of liquid inside a tube of small bore when the tube is dipped into the liquid. Another factor which is related to this phenomenon is the angle of contact. If a liquid is in contact with a solid and with air along a line, the angle θ between the solid-liquid interface and the liquid-air interface is called the angle of contact (Fig. 1). If $\theta = 0$, the liquid is said to wet the tube thoroughly. If θ is less than 90°, the liquid rises in the capillary, and if θ is greater than 90°, the liquid does not wet the solid but is depressed in the tube. For mercury on glass, the angle of contact is 140°, so that mercury is depressed when a glass capillary is dipped into mercury. The rise h of the liquid in the capillary is given by $h = 2\sigma \cos \theta / r\rho g$, where r is the radius of the tube, ρ the density of the liquid, and g is the acceleration due to gravity.

Surface tension can be explained on the basis of molecular theory. If the surface area of liquid is expanded, some of the molecules inside the liquid rise to the surface. Because a molecule inside a mass of liquid is under the forces of the surrounding molecules while a molecule on the surface is only partly surrounded by other molecules, work is necessary to bring molecules from the inside to the surface. This indicates that force must be applied along the surface in order to increase the area of the surface. This force appears as tension on the surface and when expressed as tension per unit length of a line lying on the surface, it is called the surface tension of the liquid.

The molecular theory of surface tension has been dealt with since the time of Laplace (1749–1827). As a result of the clarification of the nature of intermolecular forces by quantum mechanics and of the recent development in the study of molecular distribution in liquids, the nature and the value of surface tension have come to be understood from a molecular point of view.

Surface tension is closely associated with a sudden but continuous change in the density from the value for bulk liquid to the value for the gaseous state in traversing the surface (Fig. 2). As a result of this inhomogeneity, the stress across a strip parallel to the boundary—p_N per unit area—is different from that across a strip perpendicular to the boundary—p_T per unit area. This is in contrast with the case of homogeneous fluid in which the stress across any elementary plane has the same value regardless of the direction of the plane.

The stress p_T is a function of the coordinate z, the z-axis being taken normal to the surface and directed from liquid to vapor. The stress p_N is constant throughout the liquid and the vapor. The figure shows the stress p_N and p_T.

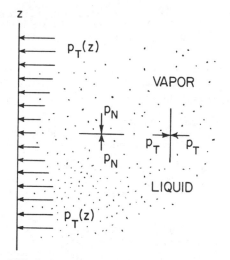

FIG. 2. Stress relationships in surface tension.

The stress $p_T(z)$ as function of z is also shown on the left side of the figure.

The surface tension is given by integrating the difference $p_N - p_T(z)$ over z:

$$\sigma = \int_{(\text{liquid})}^{(\text{vapor})} [p_N - p_T(z)]\,dz$$

A statistical mechanical treatment of the system leads to the expression of p_N and p_T in terms of intermolecular forces, density distribution, and the distribution of other molecules around a molecule which is located at a position z. The

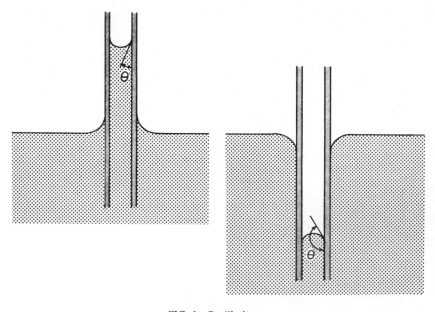

FIG. 1. Capillarity.

change of the number density and the distribution of other molecules around a central molecule at z are problems which have not yet been completely solved. Some simplifying assumptions such as to assume the transition layer to be a mathematical surface of density discontinuity have made the theory more amenable to numerical calculations. However, in spite of attempts by various authors, agreement between theoretical and observed values has not yet been attained.

AKIRA HARASIMA

References

Hirschfelder, J. O., Curtiss C. F., and Bird, R. B., "Molecular Theory of Gases and Liquids," p. 336, New York, John Wiley & Sons, 1954.

Harasima, A., "Molecular Theory of Surface Tension," in Prigogine, Ed., "Advances in Chemical Physics," Vol. I, New York, Interscience Publishers, 1958.

Ono, S., and Kondo, S., "Molecular Theory of Surface Tension in Liquids," in Flügge, S., Ed., "Encyclopedia of Physics," Vol. X, "Structure of Liquids," p. 134, Berlin, Springer, 1960.

Kirkwood, J. G., "Theory of Liquids," p. 127, New York, Gordon and Breach, 1968.

Croxton, C. A., "Statistical Mechanics of the Liquid Surface," New York, John Wiley & Sons, 1980.

Cross-references: LIQUID STATE, STATISTICAL MECHANICS.

SYMBOLS, UNITS, AND NOMENCLATURE IN PHYSICS

Introduction International communication and cooperation in science continue to grow in importance. Not the least important aspect of this cooperation is uniformity of international usage of symbols, units, and nomenclature in physics. The proliferation of research in countries throughout the world makes the problem of uniformity of usage a vital one in the dissemination of scientific literature, for it is obvious that much time and effort can be wasted in misunderstandings arising from terminology.

The recommendations given here are primarily those of the Commission on Symbols, Units and Nomenclature of the International Union of Pure and Applied Physics* as approved by the General Assemblies of IUPAP, 1960–1972. There is agreement on these recommendations among the following international organizations:

(1) International Organization for Standardization;

(2) General Conference on Weights and Measures;

(3) International Union of Pure and Applied Chemistry;

*UIP 11 (SUN 65-3), 1965.

(4) International Electrotechnical Commission, Technical Committees 24, 25;

(5) International Commission on Illumination.

Physical Quantities—General Recommendations A physical quantity, represented by a symbol, is equivalent to the product of a *numerical value* and a *unit*. For dimensionless physical quantities, the unit often has no name or symbol and is not explicitly indicated.

EXAMPLES:

$$E = 200 \text{ J} \qquad n_{\text{qu.}} = 1.55$$

$$F = 27 \text{ N} \qquad f = 3 \times 10^8 \text{ Hz}$$

Symbols for Physical Quantities—General Rules. (1) *Symbols for physical quantities* should be *single letters* of the Latin or Greek alphabet with or without modifying signs: subscripts, superscripts, dashes, etc.

REMARK: (a) An exception to this rule consists of the two-letter symbols, which are sometimes used to represent dimensionless combinations of physical quantities. If such a symbol, composed of two letters, appears as a factor in a product, it is recommended to separate this symbol from the other symbols by a dot or by brackets or by a space.

(b) Abbreviations, i.e., shortened forms of names or expressions, such as p.f. for partition function should not be used in physical equations. These abbreviations in the text should be written in ordinary Roman type.

(2) *Symbols for physical quantities* should be printed in *italic* (i.e., sloping) *type*.

REMARK: Subscripts and superscripts should be in italic type when they are symbols for physical quantities or when they are running indices: e.g., C_p where p represents pressure but C_g where g means gas; F_{ik} where i and k are running indices but E_k where k means kinetic.

(3) *Symbols for vectors and tensors*: Special type fonts are recommended for these quantities but not for their components:

(a) Vectors should be printed in boldface type.

(b) Tensors of the second rank should be printed in sans serif type.

Simple Mathematical Operations. (1) *Addition and subtraction* of two physical quantities are indicated by

$$a + b \quad \text{and} \quad a - b$$

(2) *Multiplication* of two physical quantities may be indicated in one of the following ways:

$$ab \quad a\,b \quad a \cdot b \quad a \times b$$

REMARK: The various products of vectors and tensors may be written in the following ways:

Scalar product of vectors **A** and **B**: **A** · **B**
Vector product of vectors **A** and **B**: **A** × **B**

Dyadic product of vectors **A** and **B**: **AB**
Scalar product of tensors S and T
$(\Sigma_{ik} S_{ik} T_{ki})$ S : T
Tensor product of tensors S and T
$(\Sigma_k S_{ik} T_{kl})$ S $\cdot$ T
Product of tensor S and vector **A**
$(\Sigma_k S_{ik} A_k)$ S $\cdot$ **A**

(3) *Division* of one quantity by another quantity may be indicated in one of the following ways:

$$\frac{a}{b} \quad a/b \quad a\,b^{-1} \quad a(1/b)$$

These procedures can be extended to cases where one of the quantities or both are themselves products, quotients, sums or differences of other quantities

If necessary, brackets have to be used in accordance with the rules of mathematics.

If the solidus is used to separate the numerator from the denominator and if there is any doubt where the numerator starts or where the denominator ends, brackets should be used.

EXAMPLES:

Expressions with a Horizontal bar	Same Expressions with a Solidus
$\dfrac{a}{bcd}$	a/bcd
$\dfrac{2}{9}\sin kx$	$(2/9)\sin kx$
$\dfrac{1}{2}RT$	$(1/2)RT$ or $RT/2$
$\dfrac{a}{b} - c$	$a/b - c$
$\dfrac{a}{b-c}$	$a/(b-c)$
$\dfrac{a-b}{c-d}$	$(a-b)/(c-d)$
$\dfrac{a}{c} - \dfrac{b}{d}$	$a/c - b/d$

REMARK: It is recommended that in expressions like:

$$\sin\{2\pi(x - x_0)/\lambda\}, \quad \exp\{(r - r_0)/\sigma\},$$
$$\exp\{-V(r)/kT\}, \sqrt{(\epsilon/c^2)}$$

the argument should always be placed between brackets, except when the argument is a simple product of two quantities, e.g., sin kx. When the horizontal bar above the square root is used no brackets are needed.

Units—General Recommendations *Symbols for Units—General Rules.* (1) *Symbols for units* of physical quantities should be printed in *roman* (upright) *type*.

(2) *Symbols for units* should not contain a final full stop and should remain unaltered in the plural, e.g., 7 cm and *not* 7cms.

(3) *Symbols for units* should be printed in *lower case* roman (upright) type. However, the symbol for a unit, derived from a proper name, should start with a capital roman letter, e.g., m (meter); A (ampere); Wb (weber); Hz (hertz).

Mathematical Operations. (1) *Multiplication* of two units may be indicated in one of the following ways:

newton meter: N $\cdot$ m (preferred), N m

(2) *Division* of one unit by another unit may be indicated in one of the following ways:

meter per second: $\dfrac{m}{s}$, m/s, m s^{-1}

or by any other way of writing the product of m and s^{-1}. Not more than one solidus should be used, e.g., not J/K/mol, but J/K $\cdot$ mol.

Numbers and Figures (1) *Numbers* should be printed in *upright type*.

(2) *Division* of one number by another number may be indicated in the following ways:

$$\frac{136}{273.15} \quad 136/273.15$$

or by writing it as the product of numerator and the inverse first power of the denominator. In such cases, the number under the inverse power should always be placed between brackets.

REMARK: When the solidus is used and when there is any doubt where the numerator starts or the denominator ends, brackets should be used, as in the case of physical quantities.

(3) To facilitate the reading of large numbers, the figures may be grouped in *groups of three*, but no comma should be used, since European convention uses the comma as a decimal point.

EXAMPLE:

2 573 421 736.01

Symbols for Chemical Elements, Nuclides, and Particles (1) *Symbols for chemical elements* should be written in *roman* (upright) *type*. The symbol is not followed by a full stop.

EXAMPLES:

Ca C H He

(2) A nuclide is specified by the chemical symbol and the mass number, which should appear as a left superscript, e.g., ^{14}N. The atomic number may be shown too as a left subscript, e.g., $_7^{14}$N, if needed. The right superscript position may be used to indicate a state of ioniza-

tion, e.g., Ca^{2+}, OH^-, or a state of excitation, e.g., $^{110}Ag^m$, $^4He^*$. The right subscript position is used to indicate the number of atoms of the specified nuclide or chemical element in a molecule, e.g., H_2SO_4.

(3) Symbols for particles and quanta

neutron	n
triton	t
leptons	e (electron), ν (neutrino), μ (muon)
mesons	π (pion), K, η
baryons	Λ, Σ, Ξ, Ω
proton	p
helion	h
α-particle	α
deuteron	d
photon	γ

The charge of particles may be indicated by adding the superscript +, −, or 0.
EXAMPLES:

$$\pi^+, \pi^-, \eta^0, p^+, p^-, e^+, e^-$$

If in connection with the symbols p and e no charge is indicated, these symbols should refer to the positive proton and the negative electron respectively. The bar or tilde above the symbol of a particle is ofted used to indicate the antiparticle of that particle (e.g., $\tilde{\nu}$ for antineutrino).

Quantum States (1) A symbol indicating the quantum state of a *system* such as an atom should be printed in capital roman (upright) type. The right subscript indicates the total angular momentum quantum number, and the left superscript indicates the multiplicity.
EXAMPLE:

$$^2P_{3/2}(J = \frac{3}{2}, \text{spin-multiplicity: 2})$$

(2) A symbol indicating the quantum state of a single *particle* such as an electron should be printed in lower-case roman (upright) type. The right subscript may be used to indicate the total angular momentum quantum number of the particle.
EXAMPLE:

$$p_{3/2}(\text{electron state})$$

(3) The letter symbols corresponding to the *angular momentum quantum numbers* should be:

0	S, s	4	G, g	8	L, l
1	P, p	5	H, h	9	M, m
2	D, d	6	I, i	10	N, n
3	F, f	7	K, k	11	O, o

Nomenclature (1) *Use of the words specific and molar.* The word "specific" in English names

for physical quantities should be restricted to the meaning "divided by mass."
EXAMPLES:

Specific volume	volume/mass
Specific energy	energy/mass
Specific heat capacity	heat capacity/mass

The word "molar" in the name of an extensive physical quantity should be restricted to the meaning "divided by amount of substance."
EXAMPLE

molar volume volume/amount of substance

(2) *Notation for covariant character of coupling:*

S Scalar coupling	A Axial vector coupling
V Vector coupling	P Pseudoscalar coupling
T Tensor coupling	

(3) *Abbreviated notation for a nuclear reaction.* The meaning of the symbolic expression indicating a nuclear reaction should be the following:

$$\text{Initial nuclide}\begin{pmatrix}\text{incoming particle or quantum} & \text{outgoing particle(s) or quanta}\end{pmatrix}\text{Final nuclide}$$

EXAMPLES:

$$^{14}N(\alpha, p)^{17}O \qquad ^{59}Co(n, \gamma)^{60}Co$$
$$^{23}Na(\gamma, 3n)^{20}Na \qquad ^{31}P(\gamma, pn)^{29}Si$$

(4) *Character of transitions.* Multipolarity of transition:

Electric or magnetic monopole	E0 or M0
Electric or magnetic dipole	E1 or M1
Electric or magnetic quadrupole	E2 or M2
Electric or magnetic octupole	E3 or M3
Electric or magnetic 2^n pole	En or Mn

(5) *Nuclide:* A species of *atoms*, with specified atomic number and mass number should be indicated by the word *nuclide*, not by the word isotope.
Different nuclides having the same atomic number should be described as *isotopes*.
Different nuclides having the same mass number should be described as *isobars*.

SI Units The General Conference on Weights and Measures, to which the United States adheres by treaty, has established the International System of Units, called SI units. The seven base quantities of this system and the corresponding units and their symbols are:

length	meter	m
mass	kilogram	kg

time	second	s
electric current	ampere	A
thermodynamic temperature	kelvin	K
luminous intensity	candela	cd
amount of substance	mole	mol

10^{-2}	centi	c
10^{-3}	milli	m
10^{-6}	micro	μ
10^{-9}	nano	n
10^{-12}	pico	p
10^{-15}	femto	f
10^{-18}	atto	a

The coherent SI unit system consists of the above units plus all of the units derived from them by multiplication and division without introducing numerical factors. The units radian (rad) for plane angle and steradian (sr) for solid angle have historically been designated as supplementary units. In 1980, the International Committee for Weights and Measures determined that angle and solid angle are to be regarded as dimensionless derived quantities. The units radian and steradian are equivalent to the number one (1) and may be omitted in the expressions for derived units or used as a reminder of the role of angle or solid angle in the physical quantities.

A number of derived units have been given special names and symbols.

The symbol of a prefix forms, with the unit symbol to which it is directly attached, a compound symbol which may have a positive or negative exponent, and which may be combined with other unit symbols. Thus:

$$1 \text{ cm}^3 = (10^{-2}\text{ m})^3 = 10^{-6}\text{ m}^3 \neq 10^{-2}\text{ m}^3$$

$$1 \text{ C} \cdot \mu\text{s}^{-1} = 1 \text{ C}(10^{-6}\text{ s})^{-1} = 10^6 \text{ C/s}.$$

Note 1. Because the base unit kilogram contains the prefix "kilo," the other prefixes are attached to the name "gram," e.g., milligram (mg) instead of microkilogram (μkg).

Note 2. Compound prefixes should not be

Quantity	Name of SI Derived Unit	Symbol	Expressed in Terms of SI Base or Derived Units
frequency	hertz	Hz	$1 \text{ Hz} = 1 \text{ s}^{-1}$
force	newton	N	$1 \text{ N} = 1 \text{ kg} \cdot \text{m/s}^2$
pressure and stress	pascal	Pa	$1 \text{ Pa} = 1 \text{ N/m}^2$
work, energy, quantity of heat	joule	J	$1 \text{ J} = 1 \text{ N} \cdot \text{m}$
power	watt	W	$1 \text{ W} = 1 \text{ J/s}$
quantity of electricity	coulomb	C	$1 \text{ C} = 1 \text{ A} \cdot \text{s}$
electromotive force, potential difference	volt	V	$1 \text{ V} = 1 \text{ W/A}$
electric capacitance	farad	F	$1 \text{ F} = 1 \text{ A} \cdot \text{s/V}$
electric resistance	ohm	Ω	$1 \text{ }\Omega = 1 \text{ V/A}$
electric conductance	siemens	S	$1 \text{ S} = 1 \text{ }\Omega^{-1}$
flux of magnetic induction, magnetic flux	weber	Wb	$1 \text{ Wb} = 1 \text{ V} \cdot \text{s}$
magnetic flux density, magnetic induction	tesla	T	$1 \text{ T} = 1 \text{ Wb/m}^2$
inductance	henry	H	$1 \text{ H} = 1 \text{ V} \cdot \text{s/A}$
luminous flux	lumen	lm	$1 \text{ lm} = 1 \text{ cd} \cdot \text{sr}$
illuminance	lux	lx	$1 \text{ lx} = 1 \text{ lm/m}^2$

Decimal multiples of the coherent base and derived SI units are formed by attaching to these units the following prefixes.

Factor by which the Unit is Multiplied	Prefix Name	Symbol
10^{18}	exa	E
10^{15}	peta	P
10^{12}	tera	T
10^9	giga	G
10^6	mega	M
10^3	kilo	k
10^2	hecto	h
10	deca	da
10^{-1}	deci	d

used, e.g., use nanometer (nm) and not millimicrometer (mμm).

There are additional units, not part of the coherent system, which are generally accepted for use with the SI units or accepted for use in special fields.

The following units are accepted by the International Committee of Weights and Measures for use with the SI units for a limited transitional period: angstrom (1 Å = 10^{-10} m), barn (1 b = 10^{-28} m^2), bar (1 bar = 10^5 Pa), standard atmosphere (1 atm = 101 325 Pa), curie (1 Ci = 3.7 × 10^{10} s^{-1}), roentgen (1 R = 2.58 × 10^{-4} C/kg), rad (1 rad = 10^2 J/kg).

Other Unit Systems The cgs system of units, based on the centimeter, gram, and second, as

Quantity	Name of Unit	Unit Symbol	Magnitude in SI Units
time	minute	min	60 s
	hour	h	3600 s
	day	d	86400 s
plane angle	degree	°	$\pi/180$ rad
	minute	′	$\pi/10\ 800$ rad
	second	″	$\pi/648\ 000$ rad
volume	liter	L	$1\ L = 1\ dm^3$
mass	tonne	t	$1\ t = 10^3\ kg$
energy	electronvolt	eV	approx. 1.60219×10^{-19} J
mass of an atom	atomic mass unit	u	approx. 1.66053×10^{-27} kg
length	astronomical unit	AU	$149\ 600 \times 10^6$ m
	parsec	pc	approx. $30\ 857 \times 10^{12}$ m

units in mechanics, is a metric system which continues to be used in some branches of physics. In daily life the customary units in the United States are those based on the foot, pound-force, and second, but these units are almost never used in physics except for the description of equipment, e.g., a 2-inch pipe.

Symbols and Units for Physical Quantities The table below provides an extensive but not exhaustive list of physical quantities with recommended symbols and with the symbol for the SI unit of each quantity. It is understood that deviations from the recommended quantity symbols may sometimes be necessary or desirable. On the other hand, the unit symbols are *standard* symbols for the units they represent, although other units with other symbols may, of course, be used.

Table of Symbols for Physical Quantities

Quantity	Symbol	International Symbol for Unit
SPACE AND TIME		
length	l	m
breadth	b	m
height	h	m
radius	r	m
diameter: $d = 2r$	d	m
path: $L = \int ds$	L, s	m
area	A, S	m^2
volume	$V, (v)$	m^3
plane angle	$\alpha, \beta, \gamma, \theta, \vartheta, \phi$	rad
solid angle	ω, Ω	sr
wave length	λ	m
wave number: $\sigma = 1/\lambda$	σ	m^{-1}
circular wave number: $k = 2\pi/\lambda$	k	m^{-1}
time	t	s
period	T	s
frequency: $\nu = 1/T$	ν, f	Hz
circular frequency: $\omega = 2\pi\nu$	ω	s^{-1}
velocity: $v = ds/dt$	c, u, v	m/s
angular velocity: $\omega = d\phi/dt$	ω	rad/s
acceleration: $a = dv/dt$	a	m/s^2
angular acceleration: $\alpha = d\omega/dt$	α	rad/s^2
gravitational acceleration	g	m/s^2
standard gravitational acceleration	g_n	m/s^2
speed of light in vacuum	c	m/s
relative velocity: v/c	β	
MECHANICS		
mass	m	kg
density: $\rho = m/V$	ρ	kg/m^3

Quantity	Symbol	International Symbol for Unit
MECHANICS (*Cont.*)		
reduced mass	μ	kg
momentum: $p = mv$	$\mathbf{p}, p$	kg · m/s
moment of inertia: $I = \int r^2\,dm$	I, J	kg · m^2
force	$\mathbf{F}, f$	N
weight	$\mathbf{G}, \mathbf{W}$	N
moment of force	$\mathbf{M}, M$	N · m
pressure	p	Pa
normal stress	σ	Pa
shear stress	τ	Pa
gravitational constant: $F(r) = Gm_1 m_2/r^2$	G	N · m^2/kg^2
modulus of elasticity, Young's modulus: $\sigma = E\Delta l/l$	E	N/m^2
shear modulus: $\tau = G\mathrm{tg}\gamma$	G	N/m^2
compressibility: $\kappa = -(1/V)\mathrm{d}V/\mathrm{d}p$	κ	m^2/N
bulk modulus: $K = 1/\kappa$	K	N/m^2
viscosity	η	Pa · s
kinematic viscosity: $\nu = \eta/\rho$	ν	m^2/s
friction coefficient	f	—
surface tension	γ, σ	N/m
energy	E, U	J
potential energy	V, E_p	J
kinetic energy	T, E_k	J
work	W, A	J
power	P	W
efficiency	η	—
Hamiltonian function	H	J
Lagrangian function	L	J
relative density	d	—
MOLECULAR PHYSICS		
number of molecules	N	
number density of molecules: $n = N/V$	n	m^{-3}
Avogadro's constant	$N_0, (L)$	mol^{-1}
molecular mass	m	kg
molecular velocity vector with components	$\mathbf{c}, (c_x, c_y, c_z)$	m/s
	$\mathbf{u}, (u_x, u_y, u_z)$	m/s
molecular position vector with components	$\mathbf{r}, (x, y, z)$	m
molecular momentum vector with components	$\mathbf{p}, (p_x, p_y, p_z)$	kg · m/s
average speed	$\bar{c}, \bar{u}$	m/s
most probable speed	$\hat{c}, \hat{u}$	m/s
mean free path	l	m
molecular attraction energy	ϵ	J
interaction energy between molecules i and j	ϕ_{ij}, V_{ij}	J
velocity distribution function: $n = \int f\,\mathrm{d}c_x\mathrm{d}c_y\mathrm{d}c_z$	$f(c)$	—
generalized coordinate	q	—
generalized momentum	p	—
volume in γ phase space	Ω	—
Boltzmann's constant	k	J/K
$1/kT$ in exponential functions	β	J^{-1}
gas constant per mole	R	J/mol · K
partition function	Q, Z	—
diffusion coefficient	D	m^2/s
thermal diffusion coefficient	D_T	m^2/s
thermal diffusion ratio	K_T	—
thermal diffusion factor	α_T	—
characteristic temperature	Θ	K

Quantity	Symbol	International Symbol for Unit

MOLECULAR PHYSICS (*Cont.*)

Debye temperature: $\Theta_D = h\nu_D/k$	Θ_D	K
Einstein temperature: $\Theta_E = h\nu_E/k$	Θ_E	K
rotational temperature: $\Theta_r = h^2/8\pi^2 Ik$	Θ_r	K
vibrational temperature: $\Theta_v = h\nu/k$	Θ_v	K

THERMODYNAMICS

quantity of heat	Q	J
work	W, A	J
temperature	$t, (\vartheta)$	K
thermodynamic temperature	$T, (\Theta)$	K
entropy	S	J/K
internal energy	U	J
Helmholtz function: $F = U - TS$	F, A	J
enthalpy: $H = U + pV$	H	J
Gibbs function: $G = U + pV - TS$	G	J
linear expansion coefficient	α_l	K^{-1}
cubic expansion coefficient	α, γ	K^{-1}
thermal conductivity	λ	$W/m \cdot K$
specific heat capacity	c_p, c_v	$J/kg \cdot K$
heat capacity	C_p, C_v	J/K
Joule-Thomson coefficient	μ	$K \cdot m^2/N$
ratio of specific heats	γ, κ	—

ELECTRICITY AND MAGNETISM

quantity of electricity	Q	C
charge density	ρ	C/m^3
surface charge density	σ	C/m^2
electric potential	V, Φ	V
electric field strength	$\mathbf{E}, E$	$N/C, V/m$
electric displacement	$\mathbf{D}, D$	C/m^2
capacitance	C	F
permittivity: $\epsilon = D/E$	ϵ	F/m
permittivity of vacuum	ϵ_0	F/m
relative permittivity: $\epsilon_r = \epsilon/\epsilon_0$	ϵ_r	—
dielectric polarization: $\mathbf{D} = \epsilon_0\mathbf{E} + \mathbf{P}$	$\mathbf{P}, P$	C/m^2
electric susceptibility	χ_e	—
polarizability	α, γ	$C \cdot m^2/V$
electric dipole moment	$\mathbf{p}, p$	$C \cdot m$
electric current	I	A
electric current density	j, J	A/m^2
magnetic field strength	$\mathbf{H}, H$	A/m
magnetic induction	$\mathbf{B}, B$	T
magnetic flux	Φ	Wb
permeability: $\mu = B/H$	μ	H/m
permeability of vacuum	μ_0	H/m
relative permeability: $\mu_r = \mu/\mu_0$	μ_r	—
magnetization: $\mathbf{B} = \mu_0(\mathbf{H} + \mathbf{M})$	$\mathbf{M}, M$	A/m
magnetic susceptibility	χ_m	—
electromagnetic moment	$\mu, \mu, \mathbf{m}, m$	$A \cdot m^2$
magnetic polarization: $\mathbf{B} = \mu_0\mathbf{H} + \mathbf{J}$	$\mathbf{J}$	T
magnetic dipole moment	$\mathbf{j}, j$	$Wb \cdot m$
resistance	R	Ω
reactance	X	Ω
impedance: $Z = R + iX$	Z	Ω
admittance: $Y = 1/Z = G + iB$	Y	S
conductance	G	S
susceptance	B	S
resistivity	ρ	$\Omega \cdot m$

Quantity	Symbol	International Symbol for Unit
ELECTRICITY AND MAGNETISM (*Cont.*)		
conductivity: $1/\rho$	γ, σ	S/m
self-inductance	L	H
mutual inductance	M, L_{12}	H
phase number	m	—
loss angle	δ	rad
number of turns	N	—
power	P	W
Poynting vector	$\mathbf{S}, S$	W/m²
magnetic vector potential	$\mathbf{A}$	Wb/m
LIGHT, RADIATION		
quantity of light	$Q(Q_v)$	lm · s
luminous flux	$\Phi(\Phi_v)$	lm
luminous intensity	$I(I_v)$	cd
illuminance	$E(E_v)$	lx
luminance	$L, (L_v)$	cd/m²
luminous exitance	M	lm/m²
radiant energy	$Q, W, (Q_e)$	J
radiant flux	$\Phi, (P, \Phi_e)$	W
radiant intensity	$I, (I_e)$	W/sr
irradiance	$E, (E_e)$	W/m²
radiance	$L, (L_e)$	W/sr · m²
radiant exitance	$M, (M_e)$	W/m²
absorptance	$\alpha(\lambda)$	—
reflectance	$\rho(\lambda)$	—
transmittance	$\tau(\lambda)$	—
linear attenuation coefficient	μ	m⁻¹
linear absorption coefficient	a	m⁻¹
refractive index	n	—
Stefan-Boltzmann constant: $M = \sigma T^4$	σ	W/m² · K⁴
luminous efficacy: Φ_v/Φ_e	K	lm/W
luminous efficiency: K/K_{max}	V	—
ACOUSTICS		
velocity of sound	c	m/sec
sound energy flux	P	W
reflection factor	ρ	—
acoustic absorption factor: $1 - \rho$	$\alpha, (\alpha_a)$	—
transmisssion factor	τ	—
dissipation factor: $\alpha - \tau$	δ	—
loudness level	$L_N, (\Lambda)$	—
reverberation time	T	s
specific acoustic impedance	$Z_s, (W)$	N · s/m³
acoustic impedance	$Z_a, (Z)$	N · s/m⁵
mechanical impedance	$Z_m, (\omega)$	N · s/m
ATOMIC AND NUCLEAR PHYSICS		
atomic number, proton number	Z	—
mass number	A	—
neutron number: $N = A - Z$	N	—
elementary charge	e	C
electron mass	m, m_e	kg
proton mass	m_p	kg
neutron mass	m_n	kg
meson mass	m_π, m_μ	kg
atomic mass	m_a	kg
(unified) atomic mass constant: $m_u = m_a(^{12}C)/12$	m_u	u
magnetic moment of atom or nucleus	μ	A · m²

Quantity	Symbol	International Symbol for Unit
ATOMIC AND NUCLEAR PHYSICS (*Cont.*)		
magnetic moment of proton	μ_p	$A \cdot m^2$
magnetic moment of neutron	μ_n	$A \cdot m^2$
magnetic moment of electron	μ_e	$A \cdot m^2$
Bohr magneton	μ_B	$A \cdot m^2$
Planck constant $\left(\dfrac{h}{2\pi} = \hbar\right)$	h	$J \cdot s$
principal quantum number	n, n_i	—
orbital angular momentum quantum number	L, l_i	—
spin quantum number	S, s_i	—
total angular momentum quantum number	J, j_i	—
magnetic quantum number	M, m_i	—
nuclear spin quantum number	I, J	
hyperfine quantum number	F	—
rotational quantum number	J, K	—
vibrational quantum number	v	—
quadrupole moment	Q	m^2
Rydberg constant	R_∞	m^{-1}
Bohr radius: $a_0 = 4\pi\epsilon_0\hbar^2/m_e e^2$	a_0	m
fine structure constant: $\alpha = e^2/4\pi\epsilon_0\hbar c$	α	—
mass excess: $m_a - Am_u$	Δ	kg
mass defect	B	kg
packing fraction: Δ/Am_u	f	
nuclear radius	R	m
nuclear magneton	μ_N	$A \cdot m^2$
g-factor of nucleus: $\mu = gI\mu_N$	g	—
gyromagnetic ratio: $\gamma = \mu/I\hbar$	γ	$A \cdot m^2/J \cdot s$
Larmor (circular) frequency	ω_L	s^{-1}
cyclotron (angular) frequency	ω_c	s^{-1}
level width	Γ	J
mean life	τ	s
reaction energy	Q	J
cross section	σ	m^2
macroscopic cross section	Σ	m^2
impact parameter	b	m
scattering angle	$\vartheta, \theta, \varphi$	rad
particle flux density	φ	$s^{-1} \cdot m^{-2}$
particle fluence	Φ	m^{-2}
energy flux density	ψ	W/m^2
energy fluence	F	J/m^2
internal conversion coefficient	α	—
half-life	$T_{1/2}$	s
decay constant, disintegration constant	λ	s^{-1}
activity	A	s^{-1}
Compton wavelength; $\lambda c = h/mc$	λc	m
electron radius: $r_e = e^2/mc^2$	r_e	m
linear attenuation coefficient	μ, μ_l	m^{-1}
atomic attenuation coefficient	μ_a	m^2
mass attenuation coefficient	μ_m	m^2/kg
linear stopping power	S, S_l	J/m
atomic stopping power	S_a	$J \cdot m^2$
linear range	R, R_l	m
recombination coefficient	α	m^3/s
linear ionization by a particle	N_{il}	m^{-1}
total ionization by a particle	N_i	—
number of neutrons per fission	ν	—
number of produced neutrons per absorption	η	—
absorbed dose	D	J/kg
linear energy transfer	L	J/m

Quantity	Symbol	International Symbol for Unit
CHEMICAL PHYSICS		
amount of substance	n, ν	mol
molar mass of substance B	M_B	kg/mol
molar concentration of substance B	c_B	mol/m^3
mole fraction of substance B	x_B	−
molar internal energy	$U_m, (E_m)$	J/mol
mass fraction of substance B	w_B	−
molality of solute component B	m_B	mol/kg
chemical potential of component B	μ_B	J/mol
activity of component B	λ_B	−
relative activity	a_B	−
activity coefficient of component B	f_B	−
osmotic pressure	Π	Pa
osmotic coefficient	g, ϕ	−
stoichiometric number of component B	ν_B	−
affinity	A	J/mol
equilibrium constant	K_p	Pa
charge number of ion	z	−
Faraday constant	F	C/mol
ionic strength	I	mol/kg
fugacity of component B	f_B	Pa
electrolytic conductivity	σ, κ, γ	S/m
transport number	t	−
degree of dissociation	α	−

Mathematical Symbols

GENERAL SYMBOLS

equal to	$=$
not equal to	$\neq$
identically equal to	$\equiv$
corresponds to	$\triangleq$
approximately equal to	$\approx$
proportional to	$\sim, \propto$
approaches	$\to$
greater than	$>$
less than	$<$
much greater than	$\gg$
much less than	$\ll$
greater than or equal to	$\geqslant$
less than or equal to	$\leqslant$
plus	$+$
minus	$-$
plus or minus	$\pm$
a raised to the power n	a^n
magnitude of a	$\lvert a \rvert$
square root of a	$\sqrt{a}, \sqrt{a}, a^{1/2}$
mean value of a	$\bar{a}, \langle a \rangle \langle a \rangle_{av}$
factorial p	$p!$
binomial coefficient: $n!/p!(n-p)!$	$\binom{n}{p}$
infinity	∞

LETTER SYMBOLS AND LETTER EXPRESSIONS FOR MATHEMATICAL OPERATIONS

These should be written in roman (upright) type

exponential of x	$\exp x, e^x$
base of natural logarithms	e

logarithm to the base a of x	$\log_a x$
natural logarithm of x	$\ln x$
common logarithm of x	$\lg x, \log x$
summation	Σ
product	Π
finite increase of x	Δx
variation of x	δx
total differential of x	dx
function of x	$f(x), \mathrm{f}(x)$
limit of $f(x)$	$\lim\limits_{x \to a} f(x)$

TRIGONOMETRIC FUNCTIONS

sine of x	$\sin x$
cosine of x	$\cos x$
tangent of x	$\tan x$
cotangent of x	$\cot x$
secant of x	$\sec x$
cosecant of x	$\operatorname{cosec} x$

REMARKS: (a) It is recommended to use for the *inverse trigonometric functions* the symbolic expressions for the corresponding trigonometric function preceded by the letters: arc.
EXAMPLES:

$$\arcsin x, \ \arccos x, \ \arctan x, \ \text{etc.}$$

(b) It is recommended to use for the *hyperbolic functions* the symbolic expressions for the corresponding trigonometric function, followed by the letter: h.
EXAMPLES:

$$\sinh x, \ \cosh x, \ \tanh x, \ \text{etc.}$$

(c) It is recommended to use for the *inverse hyperbolic functions* the symbolic expression for the corresponding hyperbolic function preceded by the letters: ar.

$$\text{arsinh } x, \text{ arcosh } x, \text{ etc.}$$

COMPLEX QUANTITIES

imaginary unit ($i^2 = -1$)	i, j
real part of z	Re z, z'
imaginary part of z	Im z, z''
modulus of z	$|z|$
argument of $z : z = |z| \exp i\varphi$	arg z, φ
complex conjugate of z	z^*, $\bar{z}$

VECTOR CALCULUS (SEE ALSO P. 916)

absolute value of **A**	$|\mathbf{A}|$, A
differential vector operator	$\partial/\partial\mathbf{r}$, ∇
gradient of φ	grad φ, $\nabla\varphi$
divergence of **A**	div **A**, $\nabla \cdot \mathbf{A}$
curl of **A**	curl **A**, rot **A**, $\nabla \times \mathbf{A}$
Laplacian of φ	$\Delta\varphi$, $\nabla^2\varphi$
d'Alembertian of φ	$\Box\varphi$

MATRIX CALCULUS

transpose of matrix A	$\tilde{A}_{ij} = A_{ji}$ $\tilde{A}$
complex conjugate of A	$(A^*)_{ij} = (A_{ij})^*$ A^*
Hermitian conjugate of A	$(A^\dagger_{ij}) = A_{ji}^*$ $A^\dagger$

HUGH C. WOLFE

Cross-references: CONSTANTS, FUNDAMENTAL, PHOTOMETRY.

SYNCHROTRON

The term synchrotron has come to mean "a ring shaped device for accelerating charged particles, e.g., electrons, protons, or heavier ions, to relativistic energies by the repeated passage of the particles, at essentially constant radius, through a time-varying electric field which alternates in direction at a fixed or variable frequency." Special examples are: (1) weak focusing synchrotrons, ie., constant gradient (CGS); (2) alternating-gradient synchrotrons (AGS); (3) fixed-field alternating-gradient synchrotrons (FFAG), or azimuthally varying-field synchrotrons (AVF); (4) separated-function, alternating-gradient synchrotron (SFAGS). All these machines have evolved from the early accelerators invented by E. O. Lawrence (cyclotron—1930), D. W. Kerst (betatron—1940), the synchronous acceleration principle discovered by E. M. McMillan and V. Veksler (1945), and the alternating-gradient strong focusing principle of Christofilos (1950), and Livingston, Courant and Snyder (1952); and the separated-function concept of M. G. White (1952) and T. Kitagaki (1953).

There are basically six components or functions which a synchrotron must provide: (1) A source of particles, e.g., electrons or protons

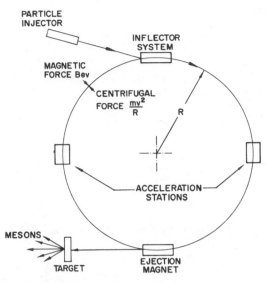

FIG. 1. Synchrotron.

with energy and direction suitable for injection; (2) a vacuum chamber; (3) a magnet to bend particles in a circle; (4) focusing of particles; (5) acceleration of particles; (6) ejection of particles.

Ion Source Electrons or ions to be accelerated are generally produced in an external device by a hot cathode or a gaseous discharge and then, after being electromagnetically focused into a narrow pencil, they are accelerated up to several tens of MeV before injection into the synchrotron magnet gap. The optimum injector energy depends upon a number of factors, e.g., minimum magnet field strength at which the magnetic field shape will produce satisfactory focusing, space charge limitations, design of the radio-frequency acceleration system, and "inflection" problems associated with bending the particles into orbit. Generally speaking, a fairly high injector energy is preferred since most of the above problems are eased as the injector energy increases. Very close tolerances must be placed on the energy and angular spread of the injected beam; otherwise the particles will strike the walls of the vacuum chamber after only a few turns. Injector accelerators generally in use are: (1) Cockroft Walton (750 kV); (2) Van de Graaff (3 to 5 MeV); (3) linear accelerators (15 to 200 MeV).

Injection into extremely high energy synchrotrons, such as the 200/500 GeV National Accelerator Laboratory machine or the similar device now under design at CERN (Geneva, Switzerland), is accomplished in three steps: A 750 KeV Cockcroft-Walton, followed by a 200 MeV linear accelerator, followed by an 8 GeV, fast-cycling booster synchrotron which pulses at 15 Hz and injects into the final, separated-function synchrotron. The use of the intermediate booster synchrotron permits a smaller mag-

TABLE 1. SOME TYPICAL SYNCHROTRONS[a]

	Princeton Heavy Ion Accelerator, Princeton, N.J.	Cornell Electron Synchrotron, Ithaca, N.Y.	Brookhaven National Laboratory, Upton, L.I.	National Accelerator Laboratory, Batavia, Ill.
1. Focusing scheme	CGS	AGS	AGS	SFAGS
2. Particle	neon	electrons	protons	protons
3. Maximum energy, GeV	9 (Ne^{7+})	12.0	33	200/500
4. Pulse rate, per min	1110	3600	30	15/6
5. Particles per pulse	10^5	3×10^{10}	10^{13}	5×10^{13}
6. Particles per sec	2×10^6	1.8×10^{12}	5×10^{12}	10^{13}
7. Orbit diameter, m	25	250	257	2000
8. Magnet wt., kG (tons)	3.8×10^5 (415)	1.1×10^5 (125)	4×10^6 (4400)	8.9×10^6 (9850)
9. Magnet power, kW (ave)	2500	1078	5000	28 000
10. Vacuum chamber, cm (aperture)	18×6	5.5×2.5	13×6	12×5
11. Acceleration system	4 drift tubes 4 ferrite cavities	6 cavities	10 4-gap cavities	16 cavities
12. Acceleration frequency, MHz	1.0–30.0	714.0	2.5–4.5	53.08–53.10
13. rf cycles per revolution	8–10–12–16	1800	12	1113
14. Energy gain per turn, keV	60	2×10^4	192	2.5×10^3
15. rf power, peak kW	320	2000	1400	1800
16. rf power, ave kW	80	400	350	800
17. Magnetic field (injection) gauss	300	50	251	396
18. Magnetic field (max) gauss	13 900	3960	13 100	9000/22 500
19. Injector, MeV	Van de Graaff 4.0	Linac 150	Linac 400	Linac 200
20. Booster Injector, GeV	none	none	none	synchrotron 8

[a]Parameters listed for a given synchrotron do not necessarily occur simultaneously but represent maximum capabilities.

netic aperture in the final magnet and also raises the injection field to a level at which variations in remanent fields play no disturbing role.

Vacuum Chamber Since particles are scattered by collision with air molecules, it is necessary to provide a good vacuum over the entire region traversed by the particles while they are undergoing acceleration. The most critical period is when the particles are at low velocity, for then the Rutherford nuclear scattering and small angle atomic scattering are most probable. A vacuum of 10^{-6} mm of Hg is generally quite sufficient to reduce gas scattering losses to negligible proportions for electrons, protons, and deuterons.

For heavy ions, e.g., neon or heavier, the vacuum requirements are much more severe since the ion, if only partially ionized, can either lose or gain electrons by collisions with the residual gas and consequently be immediately lost to the walls. The dependence of beam attenuation on ion species, velocity, gas pressure, and synchrotron parameters is quite complex. As an example of the heavy ion vacuum requirement, the Princeton Heavy Ion

Synchrotron (see Table 1) successfully accelerated neon (Ne^{7+}) at a pressure of 4×10^{-8} Torr, but the beam was attenuated by a factor of 100. Calculations show that this synchrotron, at a pressure of 10^{-9} Torr, could accelerate any ion in any charge state with very little attenuation. A vacuum of even 10^{-10} Torr is now feasible.

In order that the vacuum chamber walls should not interfere with the magnetic field, or its space and time derivatives, it is necessary to employ either insulating walls of ceramic or laminated metal structures made vacuum tight by an outer skin of epoxy fibre glass. A thin-walled stainless steel pipe is satisfactory for slowly varying magnetic fields as in the Brookhaven synchrotron or the National Accelerator Laboratory final ring.

Bending A particle of rest mass m_0, charge $Z \cdot e$, and kinetic energy T, moving in a direction perpendicular to a magnetic field of B gauss, will move in a circle of radius R given by:

$$B \cdot R = \frac{1}{2.8Z} (T^2 + 2TE_0)^{1/2}$$

(kilogauss, meters, MeV)

where $E_0 = m_0 c^2$, the rest energy of the particle. From this equation it can be seen that as the particle energy is increased from a very low initial energy to the final energy of several thousand MeV, the product $B \cdot R$ must increase manyfold. If B is held constant in time during the acceleration cycle and if the space average of B is also relatively constant, then R must increase with energy thus leading to the need for a very large radial aperture over which the field must be maintained. This is the case with cyclotrons. The radial aperture can be drastically reduced in either of two ways: (1) time modulate (e.g., pulse) the magnetic field so that it increases from a few tens or hundreds of gauss at ion injection time up to 10 to 20 kilogauss at which field the maximum particle energy is reached; (2) shape the magnetic field such that it increases sharply with increasing radius. The latter approach, while seemingly obvious, actually requires a very sophisticated angular and radial variation of the field in order to retain focusing in the direction parallel to the magnetic field. (It has been done by L. H. Thomas, 1938; Ohkawa, 1955; Kolomenskji, Petukhov, and Rabinovitch, 1955; Symon, 1955.) Accelerators based on this time-independent but spatially varying field are generally referred to as FFAG or Spiral Sector Ring Accelerators. Their major utility appears to be in the field of high ion currents, which they can achieve by virtue of their non-pulsed character, but they appear to be practicable up to only a few thousand MeV of energy since they still require considerable radial aperture. By far the most widely exploited type of synchrotron is that first mentioned, i.e., the time-varying magnetic field, which holds the particles at an essentially constant radius while gaining energy. This technique, which leads to a simpler magnet design and a minimum of weight, will be described in detail in the remainder of this article.

Focusing A magnetic field which is perfectly uniform over the entire circular path of the particle and over the radial aperture leads to stable motion only for radial displacements of the particle. Particles displaced parallel to B (i.e., out of the orbit plane) are not refocused back on to the original circle and are quickly lost to the chamber walls. This can be readily understood by noting that motion parallel to B generates no $e\mathbf{V} \times \mathbf{B}$ force, and therefore such motion persists until the particle strikes the chamber walls. However, by deliberately introducing curved lines of force, it is possible to produce $e\mathbf{V} \times \mathbf{B}$ forces whose net effect results in restoring forces for all displacements from the ideal circular orbit. Oliphant's (1945) pioneering proton synchrotron in Manchester, England achieved weak focusing by employing a small, constant, radial *decrease* of magnetic field with *increasing* radius (constant-gradient synchrotron, GGS). Subsequent to the use of this principle in several large synchrotrons there has evolved a wide variety of field shapes which

have a much stronger focusing action. All these strong-focusing fields are characterized by rapid spatial changes in the field strength, periodic reversal of sign of field gradients and even of the field itself. The most widely employed alternating gradient scheme is one in which a magnet sector that focuses *radially*, but defocuses *axially*, is followed by a sector which has the reverse property, i.e., *defocuses* radially and focuses axially. The net effect is strong focusing in *both* directions. There is a close analogy between alternating-gradient synchrotron focusing and an alternating series of convergent and divergent optical lenses whose net action is focusing. As a result of this arrangement of magnetic lenses, it is possible to use larger field gradients than in the weak-focusing case and thereby to achieve very strong focusing action. For example, an alternating-gradient 35 000-MeV accelerator, 560 feet in diameter, requires a vacuum chamber with a cross section of only 2.7 X 6 inches, whereas a weak-focusing, constant-gradient synchrotron would require at least ten times more aperture for equivalent performance.

The separated-function synchrotron (SFAGS) also employs alternating gradient focusing, but the bending function of the magnet is separate from the focusing. This is achieved by employing separate quadrupole and sextupole magnets for focusing only, while the bending magnets, having zero radial gradient, provide no focusing. The advantage of this separation lies in the possibility of achieving much higher bending fields than in a combined function magnet. For example, the National Accelerator Laboratory separated-function main ring is designed to operate at 22 kG whereas the usual AGS is limited to about 13 kG. The SFAGS principle also permits the exploitation of very high field, superconducting, iron-free bending magnets which do not lend themselves to the combined-function approach. Brookhaven National Laboratory is currently designing a superconducting, separated-function, combination intersecting storage ring and accelerator (ISA) which is to operate at 40 kG. Counter-rotating beams of 200 GeV protons will collide head-on releasing 400 GeV in the center-of-mass, or the equivalent of 80 000 GeV in a one-way machine with the target nucleus at rest.

Acceleration The increase in particle energy actually takes place when the circling charged particles pass through electric fields spaced around the circumference of the magnet. These fields must be time varying since a complete, repetitive traversal of a static field leads to zero energy gain. The usual arrangement is one in which the acceleration electrodes are excited by a sinusoidal voltage whose frequency is a harmonic (one to several thousand) of the particle rotation frequency and whose amplitude is such that particles can gain sufficient energy to match the rising magnetic field. Because of the principle of synchronous phase stability, discovered by McMillan and Veksler, particles with a wide spread in phase angle relative to the radio frequency are captured in stable "buckets" in phase space and are accelerated at just the correct rate, on the average, to stay in the middle of the vacuum chamber. Of course, the frequency of the accelerating field must be steadily and precisely increased as the particle rotation frequency increases. In some weak-focusing synchrotrons, this is one of the most difficult engineering tasks.

Electron synchrotrons pose special problems for the radio-frequency system. Because it is easy to inject electrons with a velocity very close to the velocity of light, the frequency of the required acceleration voltage is constant. This greatly simplifies the rf high voltage system and associated controls as compared with a proton synchrotron. However, a steep price is paid since relativistic electrons copiously radiate electromagnetic energy when traveling in a circle. [Radiation loss = $8.85 \times 10^4 (W^4/R)$ (eV/turn) where W is particle energy in GeV and R is magnet radius in meters.] For example, the Cornell electron synchrotron (see Table 1) radiates 20 MeV/turn at 12.0 GeV. This energy must be replaced by the rf cavities during each revolution in addition to increasing the kinetic energy of the electrons by a relatively small amount.

Ejection When the particles have been accelerated to full energy, they are caused either to run into an internal target or they are ejected by one of several schemes which usually involve exciting strong radial betatron oscillations with the last oscillation carrying the particles into an ejection magnet which deflects the beam clear of the synchrotron magnet structure.

In Table 1 are listed the important parameters of several typical existing synchrotrons. The principles discussed above for the separated-function, alternating-gradient synchrotron (SFAGS) are believed to be capable of extension to extremely high energy limited only by financial considerations. For protons and heavier particles, the electromagnetic radiation loss, which sets a practical limit for electron synchrotrons at about 60 GeV, will not be troublesome at any conceivable energy.

MILTON G. WHITE

References

Livingood, J. J., "Cyclic Particle Accelerators," New York, Van Nostrand Reinhold, 1961.

Livingston, M. S., and Blewett, J. P., "Particle Accelerators," New York, McGraw-Hill Book Co., 1962.

Wilson, R. R., and Littauer, R., "Accelerators, Machines of Nuclear Physics," Science Study Series, Garden City, N. Y., Doubleday Anchor, 1960.

Green, G. K., and Courant, E. D., in F. Flügge, Ed., "Handbuch der Physik," Vol. 44, Berlin, Springer Verlag, pp. 218–340, 1959.

Cross-references: ACCELERATORS, VAN DE GRAAFF; ACCELERATORS, LINEAR; ACCELERATORS, PARTICLE; BETATRON; CYCLOTRON; IONIZATION.

T

TELEMETRY

The terms telemetry and telemetering imply both distance and measurement, but beyond this there is not universal agreement regarding the use and meaning of the terms. The American Standards Association has defined telemetering as: "The measurement with the aid of intermediate means which permit the measurement to be interpreted at a distance from the primary detector." The distance involved may be anywhere from a few inches in the case of certain test projectiles to many millions of miles in the case of space probes to other planets. Other terms such as data transmission system or data link are frequently used. A distinction in terminology is sometimes made between the transmission of a measurement for observation and interpretation and the transmission of a measurement to directly govern a controlling action. In modern applications, the quantity transmitted may not be a measurement at all, but the result of some complex calculation based on numerous measurements over a time interval.

The dominant transmission means for telemetry has been electrical, and a few applications were already reported during the first half of the nineteenth century. During the latter half of the century, applications included regular transmission of meteorological and other measurements. During the early part of the twentieth century, there were extensive applications of telemetry in connection with the Panama Canal followed by applications in many other areas such as electrical power and pipeline distribution systems. Soon after 1930, telemetry through the transmission medium of radio links was in use for meteorological measurements from small unmanned balloons. Soon after 1940, the first radio telemetering systems for testing aircraft were being designed and used. This occurred in response to the need for more complete and reliable measurement during experimental flights of high-performance military aircraft than could be provided by onboard recorders and pilot's observations, sometimes lost. By 1945, radio telemetry was being used for measurements in the then small rockets, and before the end of the decade, telemetry carried by rocket vehicles was being used to investigate the upper reaches of the earth's atmosphere. In the 1950s, the development of advanced radio telemetering systems

was spurred by the well-known rapid development of the field of rocket-propelled vehicles including, particularly, the development of ballistic missiles for military application.

In the 1960s there were developed and launched an increasing number of Earth satellites for both military and civilian application, all of which required radio telemetry. Civilian applications included communication, geophysics, meterology, oceanography, earth resources, and others. During this decade the manned space programs Mercury, Gemini, and Apollo came into being with their demanding radio telemetry requirements. Also during this decade came the first interplanetary probes with radio telemetry over distances measured in the hundreds of millions of miles. During the 1960s biomedical applications of radio telemetry developed rapidly, including "radio pills," implantable radio telemeters, and noninvasive patient monitoring.

It is sometimes convenient to divide telemetry applications into two areas which might be called operational and test. By operational we mean the application of telemetry as a permanent part of some system, in which case the telemeter must be compatible with the rest of the system in such aspects as cost, reliability, and maintainability. By test we mean the temporary application of telemetry to obtain test information during the developmental phase of some system, in which case the telemeter need not be completely compatible with the rest of the system.

Telemetry systems for industrial purposes tend to fall more under the operational area and formed the majority of applications before 1940. Industrial telemetering systems have been frequently characterized by modest requirements for speed of response and accuracy which fall well within the physical limitations of the transmission link (frequently wire circuits) and are therefore accomplished by rather straightforward and uncomplicated signal handling methods. On the other hand, these applications have been characterized by demanding requirements for low first cost and high reliability during long periods of unattended operation. Geophysical (including oceanographic) applications often have similar requirements. Industrial telemetry is well exemplified by the many thousands of telemeters in constant use by the electrical power industry throughout the world to measure electrical quantities and plant conditions.

The majority of aerospace telemetry during the 1940's and '50's was in the test area. Because of the destructive nature of tests and/or the difficulty in repeating them and because of the marginal nature of many of the aerospace systems, aerospace telemetry has been characterized by demanding requirements for speed of response, accuracy, and number of channels which are frequently not well within the physical limitations of the radio transmission link. This .together with severe weight and space restrictions and difficult environmental conditions for equipment operation has resulted in a field of telemetry activity more or less separate from the industrial area. Much attention has been given to the statistical efficiency of signal processing in the presence of random errors of various kinds and to the specialized hardware techniques needed to satisfy space, weight, power supply, and environmental conditions. The advent of long-lived satellites and deep space probes, among many systems whose mission is wholly dependent on telemetered measurements, has greatly emphasized telemetry reliability considerations.

The deeper understanding of the nature of measurement and information that occurred shortly after World War II by application of statistical methods had its effect on the telemetering field even before the general communication field. In the beginning, the design and classification of telemetering systems tended to be in terms of the physical quantity measured, the physical quantities used for transmission and interpretation, and the transducers for conversion between these physical quantities, i.e., in terms of the hardware. Terms such as current, voltage, frequency, ratio, phase, impulse, etc., were and are used to describe some systems. Multiplexing (the transmission of many measurements over the same channel) was either frequency division (assignment of a frequency band to each measurement, non-overlapping with those of the others) or time division (assignment of a periodic time interval to each measurement, non-over-lapping with those of of the others).

The more recent statistical systems point of view in telemetry is concerned with *what* is done to the signal rather than *how* it is accomplished physically. From this point of view a telemetering system may do any or all of the following in logical sequence: (1) make a measurement; (2) abstract from the measurement in the form of a signal those characteristics which are needed for the eventual interpretation or decision; (3) store this signal; (4) code (modulate) this signal to give it greater immunity to the errors in the transmission link; (5) receive the signal; (6) store the signal; (7) decode (demodulate) the received signal

to best preserve those characteristics which are needed for interpretation and decision; (8) store the result. A measurement is said to have been made on a system if after the measurement, the uncertainty regarding the quantitative state of the system is less than the uncertainty existing before the measurement. Uncertainty in the sense of information theory has an exact quantitative measure in terms of the probability distributions involved and can be replaced by a distortion or cost index when such a criterion is available. Similarly, a measurement is said to have been telemetered when there is a reduction in uncertainty or cost on the basis of the data produced by the telemetering system.

Radio telemetering systems use many methods of modulation in various combinations. These include amplitude, phase, frequency, pulse amplitude, pulse position, pulse duration, pulse frequency, and pulse code. Multiplexing is sometimes accomplished by more general orthogonal methods than frequency- and time-division. During the last decade aerospace telemetry has become predominantly digital, with the use of error-correcting block codes and convolutional codes in various combinations. The reason for this is twofold: (1) the very rapidly decreasing cost of electronic digital hardware, including microprocessors and dedicated designs; (2) the ease of encryption of digital signals to keep radio-telemetry transmissions private.

In the 1970s, although the applications of telemetry continued to grow rapidly and to embrace the rapidly developing electronic technology, the development of the general communications field caught up with the previously more sophisticated applications of communication principles by telemetry. The result has been a loss of visibility for telemetry as a special field; instead, it has tended to take a place as one of a number of applications in the communication field. For this reason, although a number of good books on telemetry appeared in the 1950s and '60s, there were none of general significance in the 1970s. An exception for a special area of application is given by Ref. 1. The most comprehensive of the books of the 1960s is probably Ref. 2. Many papers on telemetry and related matters up to the present time are to be found in the volumes of Ref. 3. Non-application-oriented communication theory and technology, although sometimes developed for telemetry, are now usually to be found in the general communication books and journals, often without reference to the telemetering application.

LAWRENCE L. RAUCH

References

1. Valentich, Joseph "Short Range Radio Telemetry for Rotating Instrumentation," Pittsburgh, Instrument Society of America, 1977.

2. Gruenberg, Elliot L. (Ed.), "Handbook of Telemetry and Remote Control," New York, McGraw-Hill Book Company, 1967.
3. "Proceedings of the International Telemetering Conference/USA," issued annually beginning with Vol. 1 in 1965. Published by the International Foundation for Telemetering and available through the Instrument Society of America, Pittsburgh, PA.

Cross-references: FEEDBACK; MEASUREMENT, PRINCIPLES OF; MODULATION.

TEMPERATURE AND THERMOMETRY

Temperature is that attribute or state property which describes the thermodynamic state of a system and is a measure of the system's hotness, as expressed in terms of any of several arbitrary scales; it is an indicator of the direction in which energy will flow spontaneously when two bodies are brought into contact, i.e., from the hotter body (the one at a higher temperature) to the colder one (the one at a lower temperature). Temperature is not the equivalent of the total energy of a thermodynamic system; e.g., liquid water in thermal equilibrium with solid water (ice) has a greater energy per unit mass than the ice, although the two are at the same temperature. Temperature (like pressure and density) is an intensive property, i.e., it is independent of the quantity of matter, whereas properties such as mass and volume are extensive. Any device or system which has one or more physical properties (e.g., electrical resistance, electrical potential, pressure at constant volume or volume at constant pressure, length) that varies monotonically and reproducibly with temperature may be used to measure temperature; such a temperature measuring device is known as a *thermometer*. The science of temperature measurements is called *thermometry*. In the past, the science of the measurement of high temperature was known as pyrometry, but now that term usually refers to radiation thermometry at any temperature.

Temperature is one of the seven basic physical quantities of the International System (SI) of Units. All other physical quantities are defined in terms of these seven. Temperature measurement and control play a very important role in a wide variety of scientific, industrial, and domestic activities. Applications with which most people are familiar occur in medicine (patient and clinical laboratory), automobiles, heating and air conditioning, refrigeration, metallurgy, manufacturing, and electronics. Almost every household in the United States has at least one thermometer, whether it is a fever thermometer or is a thermostat on a furnace, air conditioner or kitchen range.

Temperature Scales A temperature scale is a system of assigning numerical values to temperatures. Scales are usually defined in terms of reproducible physical phenomena, such as the freezing and boiling points of water, to which temperature values are assigned, the subdivision of the interval between those defined points into units (e.g., degrees, kelvins), and by specifying the instruments for interpolating between the defining points.

Although sensitive thermometers were available by the middle of the 17th Century, no significant effort had been made to develop a universal temperature scale. Robert Hooke changed that situation in the 1660s, however, when he began divising a scale for his modified Florentine spirit thermometers. The scale he developed was based on equal increments of volume and used the freezing point of water as its starting point.

Gabriel Daniel Fahrenheit, a German instrument maker residing in Holland, was the first person to make reliable mercury-in-glass thermometers and the first to investigate the method of using two fixed points to construct a temperature scale, dividing the interval between the points into a convenient number of degrees. His scale (first proposed in 1714) was based on the lowest temperature obtainable with a freezing mixture of ice and salt as one fixed point, designated by him to be 0 degrees, and the temperature of the human body as the other, which he called 96 degrees. With this definition, he found the freezing point of water to be 32 degrees and its boiling point to be 212 degrees. Later, the freezing and boiling points of water became the defining fixed points of Fahrenheit's scale (as suggested by him in 1724).

The French physicist Guillaume Amontons proposed in 1688 that temperature could be defined as being directly proportional to the pressure of a fixed mass of gas contained in a constant volume, the lowest temperature which could exist corresponding to zero gas pressure. Only one fixed point (besides the absolute zero of temperature) would be needed then to define a scale.

Since the time of Amontons and Fahrenheit, there have been parallel developments of gas thermometry and of practical scales. The most common of these scales, which bear the name of their originators, were proposed by Fahrenheit in 1724, Rene Réaumur in 1730, Anders Celsius in 1742, William Thomson (later, Lord Kelvin; hence, the Kelvin scale) in 1848, and W. J. M. Rankine in 1850. Values on the Fahrenheit ($°F$), Réaumur ($°Re$), and Rankine ($°R$) scales are fixed by defining the freezing point of water as $32°F$, $0°Re$ and $491.67°R$ and the boiling point of water (at one standard atmosphere) as $212°F$, $80°Re$, and $671.67°R$, respectively. The Rankine scale is an absolute temperature scale with a degree of the same size as the degree on the Fahrenheit scale. The absolute zero is then $0°R$ or $-459.67°F$. The

Rankine (or absolute Fahrenheit) scale, defined by the relation $T(^\circ R) = t(^\circ F) + 459.67$, is one of the many possible thermodynamic scales. The Kelvin scale is a thermodynamic temperature scale and, as Thomson showed, is the same as the absolute scale based on the ideal-gas laws. In 1954, a thermodynamic temperature scale, with as unit the Kelvin ($^\circ K$), was adopted by assigning a value of $273.16^\circ K$ to the triple point of water. (The zero of the thermodynamic Celsius scale had already been adopted in 1948 as being $0.0100^\circ C$ below that of the triple point of water.) The magnitude of the unit of temperature was thus defined as 1/273.16 of the thermodynamic temperature of that point. The choice of $273.16^\circ K$ was made from a consideration of the results of measurements of the mean expansion coefficient of gases between $0^\circ C$ and $100^\circ C$ which led, after suitable correction for nonideal behavior, to an absolute zero of $-273.15^\circ C$. For those measurements, the interval between the ice point and the steam point of water was taken to be $100^\circ C$ exactly. Having made those measurements and defined the unit of temperature, the interval of exactly $100^\circ C$ between the ice point and the steam point is no longer required or defined as such. In fact, recent primary gas thermometry results indicate that the Celsius thermodynamic temperature, in terms of this temperature unit, for the steam point is about $99.975^\circ C$. As suggested by Celsius, values on the Centigrade ($^\circ C$) scale were based on the freezing point of water as $0^\circ C$ and the boiling point of water as $100^\circ C$, an interval of $100^\circ C$ exactly (hence, the term centigrade). The Celsius scale is defined in terms of the Kelvin scale: $t(^\circ C) = T(K) - 273.15$ K. (The degree symbol $^\circ$ was dropped from the symbol for Kelvin in 1968.) The size of the units of the Celsius and the Kelvin scales are the same.

For the scales named, the temperature interval between the freezing and the boiling points of water are divided into degrees, the number of them being $180^\circ F$, $80^\circ Re$, $180^\circ R$, $100^\circ C$ and 100 K. The Fahrenheit scale has been discontinued everywhere except in the United States where its use is being phased out as the metric system is being adopted. The Rankine scale is still used to some extent in some aspects of engineering, e.g., in the calculation of the theoretical efficiency of engines. Apart from the limited use of these two scales, the only ones in use today are the Celsius and the Kelvin scales. The latter have been widely used in scientific work for many years.

Scale Conversions To convert a value of temperature on a given scale to that on another scale involves consideration of the size of the degree of each scale and of the temperature values of the scales' zeros. Some examples of conversions follow.

To convert a temperature value on the Celsius scale to that on the Kelvin scale, add 273.15.

Conversely, to convert a temperature value from the Kelvin scale to the Celsius scale, subtract 273.15 from the value in kelvins. ($t(^\circ C) = T(K) - 273.15$).

To convert a temperature value on the Fahrenheit scale to that on the Celsius scale, subtract 32 and multiply by $\frac{5}{9}$. Conversely, to convert a temperature value from the Celsius scale to that on the Fahrenheit scale, multiply by $\frac{9}{5}$ and then add 32. ($t(^\circ F) = \frac{9}{5} t(^\circ C) + 32$).

To convert a temperature value on the Kelvin scale to that on the Rankine scale, multiply by $\frac{9}{5}$. Conversely, to convert a value from the Rankine scale to that on the Kelvin scale, multiply by $\frac{5}{9}$. ($T(K) = \frac{5}{9} T(^\circ R)$).

To convert a temperature value on the Fahrenheit scale to that on the Rankine, add 459.67. Conversely, to convert a temperature value from the Rankine scale to that on the Fahrenheit scale, subtract 459.67. ($t(^\circ F) = T(^\circ R) - 459.67$).

To convert a temperature value on the Celsius scale to that on the Réaumur scale, multiply by $\frac{4}{5}$. Conversely, to convert a value on the Réaumur scale to one on the Celsius scale, multiply by $\frac{5}{4}$. ($t(^\circ C) = \frac{5}{4} t(^\circ Re)$).

Other conversions among the scales can be performed by similar methods, or one can use the methods given above to convert a temperature value on one scale to that on a second scale, and then from that scale to a third scale, and so forth until the value on the desired scale is obtained.

The Ideal-Gas Temperature Scale Although there had been very little work on the properties of gases prior to about 1800, considerable progress was made in understanding gases in the 19th Century and the methods of gas thermometry continued to be improved during that period. It was found that the laws of the expansion of gases were much simpler than those of liquids and that very high precision was required to detect any difference in the expansion behavior of gases, excluding the easily condensable vapors. In 1847, H. Regnault showed that a gas thermometer could provide high precision in the range from $0^\circ C$ to $100^\circ C$. Using such a thermometer, he showed that, although he could detect differences in the expansion coefficients of the easily condensable gases, he was unable to detect any difference in the expansion behavior of hydrogen and nitrogen, each having an expansion of about one part in 273 per degree *centigrade* (the constant-volume hydrogen thermometer later became the basis for a practical temperature scale). Later it was found that all gases at low pressures and at temperatures well above their critical temperatures had identical thermal behavior to within their accuracy of measurement, the pressure at constant volume and the volume at constant pressure for a given quantity of gas varying linearly with temperature (e.g., $v = v_0(1 + \alpha t)$, where v is the volume at temperature $t°C$, v_0 is the volume at $0°C$, and α is the coefficient of thermal expansion). A mercury-in-glass thermometer calibrated on the Centigrade scale was generally used for the indication of temperature in those measurements. The coefficient of thermal expansion was determined to have a best value of $(1/273.15)°C^{-1}$, so that by extrapolation, the lowest possible temperature would be $-273.15°C$ (the temperature at which the gas would have no volume). Consequently, $-273.15°C$ is called the *absolute zero* of temperature. One should not take the statement of zero volume literally; it just means that at $-273.15°C$, all systems are in their ground states and no energy can be extracted from them. Some systems at or near absolute zero have what is referred to as *zero-point motion*. This is not a reservoir of energy which can be extracted, but is just a manifestation of the Heisenberg uncertainty principle.

By introducing the concept of *absolute temperature*, we may set up a new temperature scale based upon the behavior of ideal gases and avoid negative temperatures. Temperatures θ on that scale are given by $\theta = t(°C) + 1/\alpha = t(°C) + 273.15°C$. Through the use of θ, the general gas law (the law obeyed by an ideal gas) becomes

$$\frac{PV}{\theta} = \text{constant}.$$

This is the basis of the gas thermometer, and the scale on which the temperature θ is defined is the ideal-gas or absolute temperature scale. If either the pressure or the volume of a fixed quantity of gas is held constant, then the measurement of the other determines the temperature. No gas is ideal, however, so the gas thermometer is actually based on an approximately ideal gas and the assumption that the departures from ideality can be accurately measured and taken into account. By this means, the thermometer is made to be independent of the properties of real gases and becomes effectively ideal.

Thermodynamic Temperature Scale Since real gases don't behave *exactly* as ideal gases, gas thermometers using different principles of operation (constant volume or constant pressure) or different gases gave early experimenters somewhat different scales. Consequently, the need remained for a fundamental scale independent of the properties of the particular substance being used. In 1848, W. Thomson (later, Lord Kelvin) proposed such a scale based on the efficiency of an ideal reversible heat engine, which is dependent only on the limits of temperature between which it works. S. Carnot had described the ideal heat engine in 1824; its cycle consists of two isothermal and two adiabatic paths. Thomson's scale of thermodynamic temperature T was defined by the relation

$$-\frac{Q_1}{Q_2} = \frac{T_1}{T_2}$$

where Q_1 refers to the quantity (joules) of heat extracted from the hot reservoir at a temperature of T_1 and Q_2 refers to the quantity of heat returned to the cold reservoir at a temperature of T_2. T_1 and T_2 are the temperatures on the thermodynamic scale. If a perfect (ideal) gas (internal energy restriction is that $(\partial U/\partial V)_\theta = 0$, and equation of state is $PV = nR\theta$, where n is the number of moles of gas and R is the universal gas constant) is considered as the working substance in a Carnot cycle, then

$$-\frac{Q_1}{Q_2} = \frac{\theta_1}{\theta_2}.$$

Thus, the thermodynamic and the absolute, ideal gas, temperature scales are the same if the values are selected to be identical at one finite temperature. It is noteworthy that the universal gas constant R forms the link between energy and thermodynamic temperature; given Avogadro's number, the value of R has been determined solely by the arbitrary choice of the magnitude of the unit of temperature, the kelvin.

Statistical Mechanical Temperature Scale In the statistical mechanics of a many-particle system at equilibrium, there is a parameter that defines the state of the system which, when assumed to be temperature, permits the derivation of the laws of thermodynamics and, hence, confirms the validity of the assumption. The definition of the kelvin then determines the value of the Boltzmann constant, k (which with Avogadro's number, gives the gas constant R), and the statistical mechanical temperature is by definition equal to the thermodynamic temperature. Consequently, any temperature-dependent phenomenon, which can be exactly described by statistical mechanics, can be used (as a thermometer) to determine the thermodynamic temperature of the system. Four well-known examples of such phenomena are the distribution of radiant energy from a surface (obeying Planck's radiation law), the magnetic susceptibility of weak paramagnets (nuclear and electronic, obeying the Curie law) at temperatures high compared to their critical (magnetic ordering) temperatures, nuclear orientation (γ-ray anistropy being the most widely used technique) of magnetically (electronic) ordered systems at ultralow temperatures, and electrical noise in a resistor (obeying the Nyquist relation).

The Need for a Practical Temperature Scale Precision thermometry based on the thermodynamic temperature scale had its beginnings with the work of P. Chappuis and of H. L. Callendar during the period from the late 1880s to the early 1900s. Chappuis transferred the hydrogen (Centigrade thermodynamic) scale in the range from 0°C to 100°C, provided by his constant-volume hydrogen gas thermometer, to several carefully made mercury thermometers, constructed of French hard glass. These were then used to calibrate many other mercury thermometers which in turn were to be used in many countries to put temperature measurements on the same scale. The probable inaccuracy of those thermometers was stated to be 0.002°C. Callendar also developed a constant-volume gas thermometer which he used to calibrate a platinum resistance thermometer up to about 550°C. (W. Siemens had previously proposed the use of the electrical resistance of platinum as a thermometer.) The choice of platinum as a sensor was excellent as it has turned out to be the most stable and accurate thermometer available; indeed, it is now a standard instrument of the International Practical Temperature Scale (IPTS). Using the equations which he developed for expressing the relationship between the resistance of platinum and the temperature, Callendar determined the melting points of silver and gold by extrapolation.

In 1887, the Comité Internationale des Poids et Mesures (International Committee on Weights and Measures) (CIPM) adopted a resolution stating that the standard thermometric scale for the international service of weights and measures was to be the Centigrade scale of the hydrogen gas thermometer, having for fixed points the temperature of melting ice (0°C) and that of the vapor of boiling distilled water (100°C) under a pressure of one standard atmosphere.

Although gas thermometers could provide thermodynamic temperatures, they were cumbersome and unsuited for many applications. The more practical thermometers were mercury and platinum resistance thermometers, which were much more reproducible, simple to use, and generally provided a greater range of operation. Since thermodynamic temperatures are so difficult to measure accurately by gas thermometry and because of the increasing demands of science and technology, discussions, which were designed to lead to an agreement on an international practical scale covering as wide a temperature range as possible, were begun before World War I, discontinued during the war, and then held again in the 1920s. Some desired features of such a scale were that temperatures on the scale should agree as closely as possible with thermodynamic temperatures, that it should be precisely reproducible, and that it should be conveniently and accurately realizable. This would then give the users a single, internationally accepted basis for measurements which was in close agreement with the thermodynamic scale. The deliberations eventually led to the adoption of the first international temperature scale.

International Temperature Scales The needs as given above for a practical scale of temperature and the work on practical thermometry eventually led to the adoption of the International Temperature Scale of 1927 (ITS-27) by the 7th General Conference on Weights and Measures. That scale was based on the freezing point of water (0°C) and the boiling point of water (100°C), a fundamental interval of 100°C exactly, and on some fixed points outside this range. In the range of –190°C to 660°C of that scale, the platinum resistance thermometer was specified as the interpolation instrument; a Pt–10% Rh/Pt thermocouple was specified as the interpolation instrument for the range 660°C to 1063°C; for higher temperatures, an optical pyrometer was specified. As a consequence of improvements in measurements and techniques, the scale was revised and the revision adopted in 1948 by the 9th General Conference. The revised scale (called the ITS-48) was almost identical with the ITS-27 in form. The value of only one fixed point, the freezing point of silver, was changed, and that merely to make the scale more uniform. The zero of the thermodynamic Centigrade scale was adopted as being 0.0100°C below the triple point of water. Planck's radiation formula instead of Wien's formula was specified to make the scale consistent with the thermodynamic scale above the freezing point of gold. A new value of the second radiation constant, C_2, closer to that derived from atomic constants was also specified. Other changes were the specification of higher purity platinum for the standard resistance thermometers and thermocouples, and the recommendations for methods of realizing the fixed points. Additionally, the designation of the unit of temperature, the degree Centigrade, was replaced by the degree Celsius. By so doing, all temperature scales were then named after their originators.

Accepting the 1854 recommendation of Lord Kelvin that when a single fixed point was sufficiently stable, it would be preferable to define the scale using that one point, a redefinition of the Kelvin thermodynamic scale, unit of temperature being °K, was adopted in 1954 by the 10th General Conference, as already mentioned. The redefinition of the scale was accomplished by assigning a value of 273.16°K to the temperature of the triple point of water. The unit, °K, was defined as 1/273.16 of the thermodynamic temperature of that triple point. The zero of the thermodynamic Celsius scale was defined to be 0.01°C below the triple point.

In 1960, the 11th General Conference changed the name of the ITS-48 to the International Practical Temperature Scale of 1948 (IPTS-48); this was not a revision of the 1948 scale but merely a revision of its text. A list of the defining fixed points of the IPTS-48 is given in Table 1.

In order to have a scale that gave tempera-

TABLE 1. DEFINING FIXED POINTS
OF IPTS-48*

Fixed Points	Temperature, °C
Temperature of equilibrium between liquid oxygen and its vapor (oxygen point)	-182.97
Temperature of equilibrium between ice, liquid water, and water vapor (triple point of water)	+0.01
Temperature of equilibrium between liquid water and its vapor (steam point)	100
Temperature of equilibrium between liquid sulfur and its vapor (sulfur point)	444.6**
Temperature of equilibrium between solid silver and liquid silver (silver point)	960.8
Temperature of equilibrium between solid gold and liquid gold (gold point)	1063

*Exact values assigned. The pressure is 101,325 Pa (1 standard atmosphere), except for the triple point of water.
**In place of the sulfur point, it is recommended to use the temperature of equilibrium between solid zinc and liquid zinc (zinc point) with the value 419.505°C. The zinc point is more reproducible than the sulfur point and the value which is assigned to it has been so chosen that its use leads to the same values of temperature on the International Practical Temperature Scale as does the use of the sulfur point.

ture as close as possible to thermodynamic temperature, a completely revised and extended version of the IPTS was adopted in 1968. The result was the IPTS-68. That is the present version of the international temperature scale and it is defined by: (1) assigning values to the temperatures of eleven fixed points, extending from the triple point of hydrogen (13.81 K) to the freezing point of gold (1337.58 K); (2) specifying interpolation formula for the specified standard instruments, namely, the platinum resistance thermometer between 13.81 K and 630.74°C, and the platinum–10% rhodium/ platinum thermocouple for the range from 630.74°C to 1064.43°C; and (3) specifying the Planck law of radiation for temperatures higher than 1064.43°C, with 1064.43°C (1337.58 K) as the reference temperature and the value 0.014388 m · K for the second radiation constant C_2. The IPTS-68 recognizes the thermodynamic temperature as the basic temperature and defines its unit (the kelvin, K) to be 1/273.16 of the thermodynamic temperature of the triple point (equilibrium among the solid, liquid, and vapor phases) of water. The IPTS-68 is not defined below -259.34°C. The

11 fixed points of the IPTS-68 and their assigned temperatures are listed in Table 2.

The major difference between the IPTS-48 and the IPTS-68 is that on the IPTS-68, the temperature range was extended from the boiling point of oxygen, -182.97°C (90.18 K), (the lowest temperature of IPTS-48) to the triple point of hydrogen, 13.81 K. The IPTS-68 extended the scale to lower temperatures in order to unify the existing national scales in the region 10 K to 90 K and brought the values measured on the scale into agreement with thermodynamic temperatures within the limits of the accuracy of measurement at that time. The equations specified for the interpolating instruments were modified also. The IPTS-68 distinguishes between the International Practical Kelvin Temperature, symbol T_{68}, and the International Practical Celsius Temperature, t_{68}.

$$t_{68} = T_{68} - 273.15 \text{ K.}$$

The units of t_{68} and T_{68} are the same as for t and T, respectively, where T is the thermodynamic temperature (unit is the kelvin, K) and t is the Celsius thermodynamic temperature (unit is the degree Celsius, °C). The degree Celsius is by definition equal in magnitude to the kelvin.

An amended version of the IPTS-68 was adopted in 1975. Any measured temperature, T_{68}, is unchanged by that amended version. It differs from the 1968 version only in that an alternative fixed point was introduced (the argon triple point as an alternative to the oxygen boiling point), the specified natural isotopic composition of neon was changed slightly, the reference function for the standard platinum thermometer was given in an improved form, the criteria for selection of thermocouples were changed, the values of some of the secondary reference points were changed, a table of estimated uncertainties of the assigned values of the defining fixed points was deleted, and some inconsistencies and deficiencies were removed from and additional information added to the section on supplementary information.

An important but subtle point regarding practical temperature scales should be noted. By definition, the unit for thermodynamic temperatures is the same size in all temperature regions; this however, is not necessarily true for the IPTS-68 and the other practical scales. Even if the temperatures assigned to the fixed points are their true thermodynamic values, it has not been established that the specified interpolating formulas will exactly reproduce thermodynamic temperatures, except above the gold point. Consequently, the unit on the practical scale may not be constant; a unit of temperature (K or °C) in one temperature region may be of different size from that in

TABLE 2. DEFINING FIXED POINTS OF THE IPTS-68.[a]

Fixed Points	Assigned Value of International Practical Temperature	
	T_{68}(K)	t_{68}(°C)
Equilibrium between the solid, liquid, and vapor phases of equilibrium hydrogen (triple point of equilibrium hydrogen)[b]	13.81	−259.34
Equilibrium between the liquid and vapor phases of equilibrium hydrogen at a pressure of 33 330.6 Pa (25/76 standard atmosphere)[b,c]	17.042	−256.108
Equilibrium between the liquid and vapor phases of equilibrium hydrogen (boiling point of equilibrium hydrogen)[b,c]	20.28	−252.87
Equilibrium between the liquid and vapor phases of neon (boiling point of neon)[c]	27.102	−246.048
Equilibrium among the solid, liquid, and vapor phases of oxygen (triple point of oxygen)	54.361	−218.789
Equilibrium between the liquid and vapor phases of oxygen (condensation point of oxygen)[c,d]	90.188	−182.962
Equilibrium among the solid, liquid, and vapor phases of water (triple point of water)[e]	273.16	0.01
Equilibrium between the liquid and vapor phases of water (boiling point of water)[e,f]	373.15	100
Equilibrium between the solid and liquid phases of zinc (freezing point of zinc)	692.73	419.58
Equilibrium between the solid and liquid phases of silver (freezing point of silver)	1235.08	961.93
Equilibrium between the solid and liquid phases of gold (freezing point of gold)	1337.58	1064.43

[a]Except for the triple points and the equilibrium hydrogen point at 17.042 K, the assigned values of temperature are for equilibrium states at a pressure of 101,325 Pa (1 standard atmosphere). If differing isotopic abundances could significantly affect the fixed point temperatures, the abundances are specified.

[b]Equilibrium hydrogen means that the hydrogen has its equilibrium *ortho-para* composition at the relevant temperature. *Ortho* and *para* are the designations for the molecular configurations (nuclear spin arrangements) of hydrogen.

[c]Fractionation of isotopes or impurities dictates the use of boiling points (vanishingly small vapor fractions) for hydrogen and neon and condensation point (vanishingly small liquid fraction) for oxygen.

[d]The equilibrium state among the solid, liquid and vapor phases of argon (triple point of argon) at T_{68} = 83.798 K (t_{68} = −189.352°C) may be used as an alternative to the condensation point of oxygen.

[e]The water used should have the isotopic composition of ocean water.

[f]The equilibrium state between the solid and liquid phases of tin (freezing point of tin) has the assigned value of T_{68} = 505.1181 K (t_{68} = 231.9681°C) and may be used as an alternative to the boiling point of water.

another. As a result, the accurate determination of thermodynamic temperatures in all temperature regions is the principal task of thermometry today.

Future Improvements and Extensions of the IPTS In view of the elaborate experimental techniques usually required to make accurate thermodynamic temperature measurements, the need for a practical scale above about 0.5 K that is close to the Kelvin thermodynamic temperature scale remains great. There are several modifications which can be anticipated for a future IPTS. They include assigned values of fixed points which are in closer agreement with thermodynamic temperatures (as determined by recent experiments), extension of the range covered to lower temperature (0.5 K would seem to be a reasonable choice for the lower end of a practical scale), improved standard instruments for interpolation between the defining fixed points, and improved interpolation procedures. It is expected that there will be a scale revision, that it will encompass the above, and that the new scale will be adopted by about 1987.

A new (provisional) scale, the 1976 Provisional 0.5 K to 30 K Temperature Scale (EPT-76), was approved in 1976 by the CIPM. It is based on low temperature gas thermometry, on magnetic thermometry, and on fixed points to which temperature values have been assigned. The defining fixed points of the EPT-76 and their assigned temperatures are given in Table 3. This scale is expected to provide the information on which to base the downward extension of the future IPTS. Gas thermometry has recently been completed in certain regions and is in progress in others to provide refined thermodynamic temperatures throughout the entire range from about 2 K to about 1337 K. The standard thermocouple will probably be replaced with high temperature platinum resistance thermometers, which appear to be superior in sensitivity and stability. It is possible that the radiation scale will be extended downward, perhaps to the silver point (961.93°C) or even the aluminum point (660.37°C). Previously, it was thought this was unlikely, but recent developments in instrumentation now appear to make this more feasible.

Practical Thermometry Since temperature measurements are required over such a wide range and diversity of situations, a large number of different types of thermometers with varying levels of accuracy and convenience have been developed over the years. Those most frequently used are based on the expansion of gas, liquid or solid; on changes in electrical resistance; on the thermoelectric effect; on changes in the thermal radiation of a system; on changes in the thermal (Johnson) noise of electrical resistors; on changes in the resonant frequency of some materials; on spectroscopic changes; on changes in voltage of semiconductor p-n junction diodes; and on changes in the magnetic susceptibility of paramagnets.

The most familiar thermometer, the mercury-in-glass thermometer, is based on the fact that mercury expands much more rapidly than its glass container. Liquid-in-glass thermometers, other than mercury-filled ones, are used for measurements at temperatures below the freezing point of mercury. Usually a dye is put in the liquid to improve readability. Liquid-in-glass thermometers may be designed for partial-immersion, total-immersion, or complete-immersion operation. These thermometers can be made as maximum and/or minimum-indicating devices. One maximum-indicating thermometer is the clinical or fever thermometer. It contains a narrow constriction in the capillary just above the bulb which breaks the mercury column and prevents the mercury from retreating into the bulb when the thermometer is re-

TABLE 3. REFERENCE POINTS OF THE EPT-76.

Reference Point	Assigned Temperature, $T_{76}(K)$
Superconducting transition point of cadmium	0.519
Superconducting transition point of zinc	0.851
Superconducting transition point of aluminum	1.179 6
Superconducting transition point of indium	3.414 5
Boiling point of ^{4}He[a]	4.222 1
Superconducting transition point of lead	7.199 9
Triple point of equilibrium hydrogen[b]	13.804 4
Boiling point of equilibrium hydrogen at a pressure of 33,330.6 Pa (25/76 standard atmosphere)[b]	17.037 3
Boiling point of equilibrium hydrogen[a,b]	20.273 4
Triple point of neon[c]	24.559 1
Boiling point of neon[a,b,c]	27.102

[a]Boiling point under a pressure of 101 325 Pa (1 standard atmosphere).

[b]These are the four lower defining points of the IPTS-68. (NOTE: the values of temperature assigned to these points in EPT-76 are not the same as those assigned in IPTS-68.) The term equilibrium hydrogen means here that the hydrogen should have its equilibrium *ortho-para* composition at the relevant temperature.

[c]The two neon points are for neon with the natural isotopic composition of 2.7 mmol of ^{21}Ne and 92 mmol of ^{22}Ne per 0.905 mol of ^{20}Ne.

moved from the body and the temperature of the thermometer begins to decrease; this leaves the column above the constriction to indicate the maximum temperature reached by the thermometer.

There are many types of dial thermometers, including liquid, liquid-vapor, or gas filled, and their use is very widespread, particularly in industry. These thermometers usually consist of a bulb connected via a capillary tube to a Bourdon tube, mounted in a case and attached through some mechanism to a rotatable pointer on a scale. The usual liquid used in dial thermometers is mercury.

The principle of the expansion of a solid is employed in bimetallic-strip thermometers, which are comprised of strips of two different metals bonded together, side by side. If the metals have different coefficients of expansion, then the bimetallic strip will bend with a change in temperature; this can be used to move an indicator along a scale or to open and close electrical contacts, the latter operation forming a thermostat for a furnace, oven, or air conditioning system.

The most frequently and conveniently used gas thermometer is the constant-volume one which utilizes the changes in pressure to indicate changes in temperature. Another type of gas thermometer is the dielectric-constant gas thermometer. Measurements in this case depend on the change of the dielectric constant with gas density and, thus, are intensive in their nature, in contrast to the extensive quality of regular gas thermometry. A reference temperature is required for the operation of gas thermometers.

Saturation vapor-pressure thermometry is commonly used for the measurement of temperature in the liquid helium (^{4}He and ^{3}He) and liquid hydrogen regions, and to a lesser extent in the regions of other cryogenic fluids, because of the sensitivity and convenience of this type of measurement.

There are many different types of resistance thermometers, with each type being most suited for use in a particular temperature region. In all cases, however, a circuit for measuring resistance with change in temperature is required. Included among the resistance thermometers are the platinum, copper, and nickel resistance thermometers, the rhodium-iron resistance thermometers, and the semiconductor, thermometers (arsenic-doped germanium, thermistors, carbon).

Thermocouples are used for measurements of temperature from a few millikelvins to above 2800°C. The basic thermocouple consists of two wires of different materials which, when joined together at one end (the hot junction) and connected to a voltage measuring instrument at the other end (the cold junction), will produce a voltage as a smooth function of the temperature difference between the two junctions. Various pure elements and alloy combinations can be used to form couples which are best suited to particular temperature regions. Thermocouples are probably the most widely used industrial thermometers because they are inexpensive, rugged, long-lasting, and suitable for continuous recording and/or control of temperature. Since thermocouples can be made very small, they can respond very rapidly to fast changes in temperature.

Radiation thermometers were developed for measuring high temperatures (greater than 1064°C) and they have the advantage that they are noncontact thermometers. Optical pyrometers measure apparent temperatures of objects by comparing the radiation from the objects over a narrow wavelength band with that of a standard, preferably using a photoelectric detector for the comparison. Corrections for the emissivity of the source must be made to determine the temperature. Total-radiation pyrometers measure the whole spectrum of energy radiated by the source and that is used to determine the temperature. They are less accurate than optical pyrometers but can measure much lower temperatures (of the order of 100°C). This type of pyrometer also requires emissivity corrections. Also utilizing radiation from an object to measure its temperature is the technique of thermography (mapping of surface temperature distributions over extended areas). This is widely used in the medical field for the detection of tumors near the surface of the skin, and in industrial applications for detection of hot spots (defective insulators on power lines, furnace walls, areas of heat leaks in buildings, etc.).

Another noncontact technique for measuring high temperatures involves Raman spectroscopy, in particular the nonlinear process known as coherent anti-Stokes Raman spectroscopy (CARS). This technique is finding practical applications in measurements of temperatures of flames, in internal combustion engines, in jet engines, and of hot gases.

Johnson-noise thermometers, based on the measurement of the Johnson-noise power or voltage, are now commercially available. Johnson (thermal) noise voltage is the small fluctuating voltage generated in any electrical conductor by the random motion of the electrons (charged particles). The extent of the motion of the electrons is a function of temperature and, thus, the voltage is related to thermodynamic temperature. Noise thermometers are suitable for use from a few mK to over 1000°C.

Thermometers which are based on the temperature-dependent resonant frequency of a material are very attractive since the quantity measured is frequency. One such thermometer is the nuclear quadrupole resonance (NQR) thermometer. An excellent feature of the NQR thermometer is that the thermometric property involved is a fundamental property of a substance, a unique frequency-temperature relationship which need be established only once and is forever after applicable for that substance. A commercial NQR thermometer is available; it is based on the NQR of ^{35}Cl in $KClO_3$, which has been the substance most studied. NQR thermometers are suitable for use over the range from about 50 K to 450 K.

Another resonant-frequency thermometer is the quartz crystal resonator, which, if the crystal is properly cut, is quite linear from about 190 K to 525 K. Although this thermometer has excellent resolution, it does exhibit hysteresis and drift.

Semiconductor p-n junction diode thermometers are becoming widely used throughout the range from liquid helium temperatures (·1 K) to about 200°C. The diodes are currently made of germanium, silicon or gallium arsenide. These thermometers are based on the principle that for forward-biased p-n junction diodes, the current varies approximately exponentially with V/T, where V is the voltage. At constant current, then, the junction voltage decreases approximately linearly with increasing temperature. The emitter-base junction diodes of transistors have been used also as thermometers.

Liquid-crystal thermometers have come into fairly widespread use in recent years. In their operation, they rely on the change in wavelength of reflected light with a change in temperature. The liquid crystals or their mixtures are selected such that the wavelengths of the reflected light are in the visible region, giving rise to a change in color with a change in temperature. To serve as thermometers, the liquid crystals are in their cholesteric phase and their temperature range of operation is from about 0°C to about 70°C.

In the cryogenic temperature region, magnetic thermometers are very useful and widely used. They employ the temperature dependence of the magnetic susceptibility χ of weak paramagnets (nuclear and electronic) far above their critical temperatures. χ is a parameter that indicates the extent to which a substance (paramagnetic salt or nuclear spin system) is susceptible to magnetization and which for "weak" paramagnets is given by the Curie law ($\chi = C/T$).

Each type of thermometer has its particular advantages and disadvantages, and the selection of a thermometer for a given application must be based on the requirements of that application. Among the considerations which influence the choice of the thermometer are accuracy, sensitivity, reproducibility, size, temperature range, speed of response, durability, and cost.

B. W. MANGUM

References

Schooley, James F. (Ed.), "Temperature: Its Measurement and Control in Science and Industry,"
Vol. 5, New York, American Institute of Physics, 1982.

Hudson, R. P., *Rev. Sci. Instrum.* **51**, 871 (1980).

Berry, K. H., *Metrologia* **15**, 89 (1979).

"The 1976 Provisional 0.5 K to 30 K Temperature Scale," *Metrologia* **15**, 65 (1979).

Guildner, L. A., and Edsinger, R. E., *NBS J. of Research* **80A**, 703 (1976).

"The International Practical Temperature Scale of 1968," *Metrologia* **5**, 35 (1969); Preston-Thomas, H. Amended Edition of 1975, *Metrologia* **12**, 7 (1976).

Riddle, J. L., Furukawa, G. T., and Plumb, H. H., "Platinum Resistance Thermometry," NBS Monograph 126 (U.S.G.P.O., Washington, D.C. 20402, 1972).

Plumb, Harmon H. (Ed.), "Temperature: Its Measurement and Control in Sequence and Industry," Vol. 4, Pittsburgh, Instrument Society of America, 1972.

Cross-references: CARNOT CYCLES AND CARNOT ENGINES; CONDUCTIVITY, ELECTRICAL; CRYOGENICS; DIAMAGNETISM; ELECTRON SPIN; EXPANSION, THERMAL; GAS LAWS; HEAT; MAGNETIC RESONANCE; MAGNETISM; PYROMETRY; RADIATION, THERMAL; SEMICONDUCTORS; STATES OF MATTER; THERMODYNAMICS; THERMOELECTRICITY; TRANSISTORS.

TENSORS AND TENSOR ANALYSIS

Introduction Tensors and tensor analysis represent a powerful and useful part of mathematical physics. Tensors include such familiar objects of physical interest as ordinary scalars and vectors. They also include more complicated objects and are basic to a full understanding of advanced subjects such as electromagnetic theory and general relativity. Consequently, tensors often possess an aura of mystery and a special mystique that they do not always deserve. The reader is referred to the articles of VECTOR ANALYSIS, ELECTROMAGNETIC THEORY, and RELATIVITY appearing elsewhere in this encyclopedia.

The concept of tensor can be traced back to such men as Karl Gauss, George Riemann, and Elwin Christoffel in the last century. The algebra and calculus of tensors can be traced back to Gregorio Ricci and his pupil Tullio Levi-Civita early in this century. Their names appear frequently in connection with specific concepts or special tensors. The books by Levi-Civita[1] and Eisenhart[2] in the general references at the end of this article will provide a more complete history of this subject.

The word *tensor* means "to stretch" and thus is related to the concepts of stress and strain appearing in the theory of elasticity and deformable bodies. Dyadics, which were used more frequently in the last century, may be considered to be almost synonymous with the more modern concept of tensors and to be superseded by them. Tensors and tensor analysis are important in both mathematics and physics and are gaining importance in many fields of engineering and technology. See the excellent book by Sommerfeld[3] for applications to the mechanics of deformable bodies, and the book by Page[4] for more information about dyadics.

Tensors, Rank, and Dimensionality Before wondering about the physical interpretation of a tensor, which depends entirely on the problem at hand, consider its mathematical description. The number of components in a given tensor depends on both its rank R and on its tensor dimensionality D. The lowest rank R that a tensor may have is 0. The number of components possessed by a tensor of rank 0 is always 1, no matter what the dimensionality D of the tensor may be. A common name for such a tensor is "scalar." So, an n-dimensional scalar is a one-component tensor of rank zero. One example of a scalar quantity is a thermometer reading representing temperature. A second example is ordinary mass—but not the gravitational weight associated with the mass which is, in fact, a downward-pointing force vector.

The next rank that a tensor may have is 1. The number of components possessed by a tensor or rank 1 is always n, the dimensionality of the tensor. Thus, in two-dimensional space, a tensor of rank 1 has 2 components; in three-dimensional space, 3 components; and so on. Again, a common name for such a tensor is "vector." So an n-dimensional vector is an n-component tensor of rank 1. The first example of a vector that comes to mind is velocity.

The next rank that a tensor may have is 2. The number of components possessed by a tensor of rank 2 is always n^2, the square of the dimensionality of the tensor. Thus in two-dimensional space a tensor of rank 2 has 4 components; in three-dimensional space, 9 components, and so on. A tensor of rank 2 is the first that normally would be called a true tensor.

These facts may be summarized. For what are called complete tensors the general result is that the number of components in a tensor of rank R and dimensionality D is given by the formula $Z = D^R$.

One kind of incomplete tensor is called an "antisymmetric tensor." Such a tensor always has some zero components. The number of independent nonzero components in an antisymmetric tensor in n-dimensional space is given by $\frac{1}{2}n(n-1)$.

It is convenient that in three dimensions an antisymmetric tensor of rank 2 has only three nonzero components. This coincidence gives rise to this tensor's interpretation as a kind of vector or "pseudovector." A familiar example of an antisymmetric tensor that may be thought of as a vector is magnetic field strength **H** and the associated magnetic induction **B**. A second example is angular momentum **L**. It is this last

association that has given the name axial vector to this general class of pseudovector. The coincidence is related to the fact that in three dimensions the number of planes (defined by pairs of coordinates x, y; x, z; y, z) is the same as the number of coordinates x, y, z. In any other dimension this does not happen, but the number of independent planes still is equal to the number of tensor components. Thus, in two dimensions, there is only one plane $(1, 2)$ and one tensor component, while in four dimensions, there are six independent planes $(1,2; 1,3; 1,4; 2,3; 2,4; 3,4)$ and six components. Finally, Table 1 summarizes the number of components of simple tensors for spaces with dimensions up to five. Notice that the number of components of the symmetric and antisymmetric tensors of a given rank sum up to the maximum number for that rank. For more details on the elementary properties of vectors and tensors the reader is referred to the book by Lass[5] in the general references.

Algebra of Tensors As a general rule, the algebra that has been developed for matrices may be used with tensors. In fact, second-rank tensors look like square matrices; first-rank tensors (vectors) look like 1 by n matrices. In any given situation, tensors and matrices may be multiplied together to achieve a purpose such as transforming the tensor by a linear operator. In these instances the tensor components may express the physical aspects of the problem and the matrix components the more mathematical aspects of the problem. The reader may find more information on this subject in the article on MATRICES appearing elsewhere in this encyclopedia.

All vectors are tensors, but not all tensors are vectors. A vector $\mathbf{A}_i$ has the transformation law:

$$\mathbf{A}'_j = \ell_{ij}\mathbf{A}_i,$$

whereas a tensor $\mathbf{K}_{ik}$ has a transformation law:

$$\mathbf{K}'_{jl} = \ell_{ij}\ell_{kl}\mathbf{K}_{ik}.$$

Here the symbol ℓ_{ij} or ℓ_{kl} represents the transformation matrix taking $\mathbf{A}_i$ or $\mathbf{K}_{ik}$ from the

TABLE 1. NUMBER OF COMPONENTS OF A TENSOR

	Dimensions				
	1	2	3	4	5
Scalar	1	1	1	1	1
Vector	1	2	3	4	5
Axial vector	1	2	3	4	5
Tensor—second rank	1	4	9	16	25
Symmetric tensor	1	3	6	10	15
Antisymmetric tensor	0	1	3	6	10
Tensor—third rank	1	8	27	64	125

unprimed to the primed coordinate system. The n^2 elements of the matrix ℓ_{ij} are derived from the direction cosines of the unit vectors in one system of coordinates with respect to the unit vectors in the other.

A tensor $\mathbf{K}_{ik}$ or $\mathbf{K}'_{jl}$ represents a physical object. The double index ik or jl, while similar to ij or kl for the matrix ℓ, refers to the axes in one system of coordinates. Accordingly, they are different in nature even though they may have the same form.

Covariance and contravariance refer to the transformation properties of vectors and tensors. These concepts are essential to a complete understanding of vector and tensor analysis. The rules of manipulation of the vector or tensor indices, which designate whether a vector is covariant or contravariant or whether a given tensor is covariant, contravariant, or mixed in its transformation properties, are well documented along with other mathematical and formal aspects in many textbooks. One of the best of these is Sokolnikoff[6] listed in the general references. An approach that emphasizes the relationship between the concept of reciprocal lattices (long familiar to solid-state physicists and crystallographers) and the fact that the covariant and contravariant versions of the same vector reside in different coordinate systems is explained in Stratton[7] and illustrated in Eisele and Mason[8]. In many problems the value of retaining the distinction between covariance and contravariance vanishes because, for cartesian coordinates, a coordinate lattice and its reciprocal lattice are one and the same. It follows that the covariant and contravariant components of a vector or tensor will be identical in magnitude and direction.

Calculus of Tensors Numerous applications of tensor analysis can be found in texts dealing with special and general relativity. In fact, it would not be too inaccurate to say that the subject of tensor calculus grew up with general relativity. The small book by Einstein[9], and another by Møller[10] provide further background on the subject of relativity. Deserving special consideration as an extremely simple introduction to the manipulation of tensors and their calculus is the book by Lieber and Lieber.[11]

The concepts of covariance and contravariance reappear in tensor calculus in a new and more general form than previously mentioned in connection with reciprocal lattices. The archetype contravariant vector is taken as the position vector $\mathbf{r} = (x^1 \cdots x^n)$ or differential components of $d\mathbf{r}$. It obeys the transformation:

$$dx'^i = M^i{}_j dx^j = \frac{\partial x'^i dx^j}{\partial x^j},$$

or $\mathbf{A}'^i = M^i{}_j A^j$. Contravariant tensors of higher rank follow a similar pattern of index and prime placement. The archetype covariant vector is

taken as a vector generated by the gradient operator and obeys the transformation

$$\frac{\partial \phi}{\partial x'^i} = N^j{}_i \frac{\partial \phi}{\partial x^j} = \frac{\partial x^j}{\partial x'^i} \frac{\phi}{\partial x^j},$$

or $\mathbf{B}'_i = N^j{}_i B_j$. Covariant tensors of higher rank follow a similar pattern of index and prime placement. A mnemonic device "co-lo-prime-below" has been introduced by Lieber and Lieber[11] to aid in remembering the difference between the two classes of tensors.

The metric tensor g_{ij} or $g^i{}_j$, which has a special meaning in relativity, is introduced in differential geometry in connection with the definition of differential line element:

$$ds^2 = g_{ij} dx^i dx^j.$$

Its forms for three-dimensional cartesian and spherical coordinates respectively, are:

$$\begin{vmatrix} 1 & 0 & 0 \\ 0 & 1 & 0 \\ 0 & 0 & 1 \end{vmatrix} \quad \text{and} \quad \begin{vmatrix} 1 & 0 & 0 \\ 0 & r^2 & 0 \\ 0 & 0 & r^2 \sin^2\theta \end{vmatrix}$$

The Kronecker delta and the Levi-Civita symbols are defined as:

$$\delta_{ij} = 1 \text{ if } i = j$$
$$0 \text{ if } i \neq j$$

and

$$\epsilon_{ijk} = +1 \text{ if } i, j, k = 1, 2, 3 \text{ or a cyclic}$$
$$\text{permutation}$$
$$-1 \text{ if } i, j, k = 3, 2, 1 \text{ or a cyclic}$$
$$\text{permutation}$$
$$0 \text{ if any two or three of } i, j, k$$
$$\text{have the same value.}$$

The delta symbol is useful in discarding cross-product terms that vanish under conditions of orthogonality and in indicating unity in the normalization process. The epsilon symbol is useful in simplifying multiple vector products and revealing their true tensor nature. Thus for example in tensor component form:

$\mathbf{C} = \mathbf{A} \times \mathbf{B}$ becomes $C_i = \epsilon_{ijk} A_j B_k$,

$\mathbf{A} \cdot (\mathbf{B} \times \mathbf{C})$ becomes $\epsilon_{ijk} A_i B_j C_k$, and

$\mathbf{A} \times (\mathbf{B} \times \mathbf{C})$ becomes $\epsilon_{ijm} \epsilon_{klm} A_j B_k C_l$.

Tensors and Quadratic Forms In classical mechanics, tensors arise naturally in the study of rotating objects. The moment of inertia tensor associated with an object is of rank 2 and can be represented by an ellipsoid of rotation. Many important and useful properties of the system can be obtained by applying matrix transformations to this tensor. A particularly lucid development of this subject may be found in the excellent textbook by Goldstein,[12] where eigenvalues and eigenvectors are obtained through the reduction to principal axes (i.e., diagonal form).

In quantum mechanics, a similar application exists in which it is required to diagonalize simultaneously two Hermitian quadratic forms with real eigenvalues. For the general case, a principal axis transformation can be found. However, the resulting principal axes need not be an orthogonal set. This is illustrated in Eisele and Mason.[8]

For a modern general treatment of the subject of tensors see Krogdahl.[13] A more specific treatment may be found in Goodbody.[14]

JOHN A. EISELE
ROBERT M. MASON

References

1. Levi-Civita, T., "The Absolute Differential Calculus," London, Blackie and Son, Limited, 1947.
2. Eisenhart, L., "Riemannian Geometry," Princeton, N.J., Princeton University Press, 1949.
3. Sommerfeld, A., "Mechanics of Deformable Bodies," Translated by G. Knerti, New York, Academic Press, Inc., 1950.
4. Page, L., "Introduction to Theoretical Physics," 3rd Ed., New York, Van Nostrand Reinhold, 1952.
5. Lass, H., "Vector and Tensor Analysis," New York, McGraw-Hill Book Company, Inc., 1950.
6. Sokolnikoff, I., "Tensor Analysis," New York, John Wiley & Sons, Inc., 1964.
7. Stratton, J., "Electromagnetic Theory," New York, McGraw-Hill Book Company, Inc., 1941.
8. Eisele, J., and Mason, R., "Applied Matrix and Tensor Analysis," New York, John Wiley & Sons, Inc., 1970.
9. Einstein, A., "The Meaning of Relativity," 5th Ed., Princeton, N.J., Princeton University Press, 1955.
10. Møller, C., "The Theory of Relativity," Oxford, Clarendon Press, 1952.
11. Lieber, L., and Lieber, H., "The Einstein Theory of Relativity," New York, Holt, Rinehart, and Winston, 1945.
12. Goldstein, H., "Classical Mechanics," 2nd Ed., Cambridge, Mass., Addison-Wesley Press, 1980.
13. Krogdahl, W. S., "Tensor Analysis: Fundamentals and Applications," Washington, D.C., University Press of America, Inc., 1978.
14. Goodbody, A. M., "Cartesian Tensors," New York, Halsted Press, 1981.

Cross-references: ELECTROMAGNETIC THEORY, MATRICES, QUANTUM MECHANICS, RELATIVITY, VECTOR ANALYSIS.

THEORETICAL PHYSICS

As knowledge increases and becomes more diversified, a process of arborization of subject matter ensues—various branches of knowledge branch again into new specializations; for example, natural philosophy, which embraced a general curiosity about phenomena occurring in Nature, branched into the physical and natural sciences, and these in turn into specific sciences such as physics, chemistry, and biology. Physics, in turn, has branched into many subject areas, such as nuclear physics, laser physics, hydrodynamics, and statistical physics. Another kind of branching, which is related to personality and aptitude rather than to subject matter, is the division of activity in physics into experimental and theoretical physics. A similar division has developed in chemistry and is in an early period of infancy in the biological and social sciences.

The branching of activity in physics between experiment and theory is both convenient and natural. Experiment has to do with the design of a measuring apparatus, the taking of measurements, and the comparison of these measurements with expectation. Theoretical physics has largely to do with providing a conceptual and explanatory background in the form of equations and their solutions with predictive value. The language of physics is interpreted mathematics; mathematical symbols are used to represent quantitative physical concepts and the mathematical equations provide relationships between them. For example, Robert Boyle, in 1660, found that for an enclosed mass of air, the pressure, volume, and temperature (each measurable quantities) are not independent; in fact, Boyle showed that if the temperature is held fixed, the volume varies inversely as the pressure applied. This was an empirical result which found "explanation" later in the kinetic theory of gases based on Newton's laws of particle motion. The kinetic theory, in turn, predicts that Boyle's law represents only ideal behavior and that deviations from the ideal should be measurable. This going back and forth between experiment and theory is characteristic of physics and provides a powerful process for improving and extending experimentation on the one hand, and of developing and testing theoretical hypotheses on the other.

The breadth and diversity of the human interest and intellect is such that some physicists prefer the earthy experience of designing apparatus and using it in a sophisticated measuring situation, while others prefer to work abstractly with ideas and their mathematical expression. This concrete-to-abstract spectrum extends from architects and engineers at one limit, through practical experimenters, theoretical physicists, to mathematicians or philosophers at the other. Theoreticians themselves divide into fuzzy categories, some in close and immediate contact with experiment, some relating general theoretical developments to an overall philosophic view, and some (the so-called applied mathematicians) engaging largely in the mathematics of very practical problems, such as calculating heat loss in steam pipes or the optimal shape of airfoils for maximum lift in an air stream.

The success of theoretical physics in the

modern, post-industrial era (exemplified, for example, in the large proportion of theoretical physicists among the winners of Nobel prizes) arises out of the power of the human mind to conceptualize beyond the limits imposed by everyday experience. Although we live in a world of three space dimensions, the physicist and mathematician conceive of and explore imagined or abstract worlds of other dimensions, including worlds of infinite dimension and even worlds of fractional dimensionality. In the early scientific era this power of the human concept was employed to do the seemingly impossible—to measure the radius of the earth without penetrating its depths or sailing around it; to measure the height of inaccessible mountains by triangulation; to weigh the moon without touching it or leaving the earth. The concept and design of a modern nuclear power plant is based on fantastic abstractions—theoretical nuclear physicists talking about Einstein's mass-energy conversion inside countless nuclei so tiny that no one has ever seen or touched, tasted or heard them; but the plant itself is built with metal, concrete, and other common materials by skilled and unskilled tradesmen of a variety not basically different from those employed on any construction site. Yet the day comes when the switches are thrown and the houselights come on in obscure towns a thousand miles distant from the slowly disintegrating uranium fuel rods embedded in the moderator of the nuclear pile.

Just as the discipline of physics as a whole is divided into subdisciplines or subject areas, so it is with theoretical physics. One roughly defined division is between so-called h-bar physics ($\hbar$ is Planck's constant divided by 2π) and non-h-bar physics, depending on whether the phenomenon in question can be explained on the basis of the so-called classical or pre-quantum laws, such as those of Newton or Maxwell, or whether one must resort to the new mechanics devised in the 1920s to take account of the quantized nature of matter in its microscopic form. Examples of non-$\hbar$ physics would be much of geophysics, hydrodynamics, thermodynamics and biophysics. Generally, non-$\hbar$ physics has to do with the description of rather large-scale, long-term phenomena. Even these phenomena, however, generally have a microscopic basis which requires the quantum theory for more detailed understanding or explication.

Theoretical physics as it has been practiced in the post-Renaissance period has been at once both intensely reductionistic and grandly holistic. The extrapolation by Newton of a very limited experience with planetary motions to universal gravitation between all material bodies is an example of grand holism; the behavior of one piece of matter is related to the behavior of all other matter in the universe. Pauli's exclusion principle and associated antisymmetriza-

tion of the wave function is another. Nonetheless, theoretical physicists are primarily reductionistic—looking for explanations by conceptually taking things apart and assuming that a better understanding of macroscopic phenomena will follow from an understanding at a finer, more microscopic level of description. A fundamental question today is whether there is a natural limit to this procedure of explanation of a phenomenon in terms of its "parts." One looks to the behavior of liquids, solids, and gases in terms of their atomic constituents and the interactions among them; the atomic constituents are explicable in terms of electrons and nuclei; nuclei are explicable in terms of protons and neutrons, and these in terms of the modern "elementary particles" like quarks and gluons. Do the quarks also have subworlds for their explication? or is one then pushing the concept of explanation by reduction too far? These are open and fascinating questions in contemporary physical theory.

Another division of activity in theoretical physics is the division of phenomena into relativistic and nonrelativistic categories. Loosely speaking, relativity has to do with high-velocity phenomena, that is, with relative motions of matter reaching nearly to the velocity of light. Because of Einstein's general theory and the close association between inertial and gravitational mass, relativity becomes essentially inseparable from gravitation. Special relativity is the domain of high-velocity phenomena in regions of space which have an extremely small density of matter. The so-called theory of special relativity is basically an algorithm by which observers in rapid relative motion to one another can establish overall consistency between their apparently different observations on presumably the same natural phenomenon.

In the presence of strong gravitational fields Einstein showed that the laws of physics take on their simplest form in coordinate systems which have been warped by the gravitational fields. This brilliant concept that one can replace kinematics by geometry went a long way to fulfilling ancient speculations about the essentially geometric structure of the universe. The details of Einstein's theory may yet require modification, but its essential conceptual correctness seems to be beyond question.

It is fascinating that the two successful, modern paradigms about the ultimate structure and nature of our universe, to both of which Einstein contributed in an essential way, seem at some basic level in contradiction to one another. This presents the theoretical physicist with one of the great paradoxes of our time. On the one hand the quantum theory seems to deal efficiently and successfully with the "other" basic forces of the universe—electromagnetic forces, nuclear forces, the "weak" forces associated with beta decay—but not with gravitational forces. Grand unification theories have not been successful so far, despite many sophisticated assaults on the problem. On the other hand, general relativity deals in an inherent way only with the gravitational force, its other aspect being the intimate relationships between space, time and matter and how these relationships affect our observations.

A second aspect of the apparent conflict between quantum mechanics and relativity has to do with Bell's theorem, which shows that either quantum mechanics does not obey the ordinary rules of logic and reason, or that it is inherently non-local in a sense which must be carefully defined. Non-locality would imply that information is transmitted essentially instantaneously, whereas special relativity states that no signal can travel faster than light. Bell's inequality is still under test, though many scientists would take the view that it has, in essence, already been tested.

One of the most intriguing problems in quantum mechanics, which also relates to Bell's inequality, is to define the interface between quantum phenomena and classical measurement. Schrödinger's equation gives a rather strange description of microscopic phenomena (strange because the dynamical quantity, the wave function Ψ, is a probability amplitude and cannot be measured or observed directly); it seems essential to the understanding of the results of measurements. Yet the measurements themselves are made with, and the values read from, classical instruments. Even more intriguing is the interface between the measuring apparatus and the human consciousness. From a more specific point of view, one can raise the question whether large macromolecules (e.g., enzymes) of the kind involved in living systems are essentially classical objects, essentially quantum objects, or perhaps interesting interface systems between the classical and quantum limits.

The most specific divisions of activity in theoretical physics are those relating to subfield. For example, one has the following broad areas: condensed matter, nuclear physics, elementary particle physics, field theory, quantum optics, biophysics, geophysics, plasma physics, upper atmospheric physics, astrophysics, relativity, gravitation, and cosmology. In general these areas parallel counterparts in experimental physics and reflect a division of interest and activity on the basis of size of objects considered, degree of accessibility, degree of abstraction required, etc., or some combination of these factors. Often one is influenced to enter a particular specialization through the interests of a gifted teacher or colleague; sometimes a researcher develops a peculiar, absorbing, and lifetime interest in a certain specific area such as stellar evolution in much the same way that a person might develop a special yen to visit Australia or Tanzania. Whatever factors induce the theoretician to enter a certain specialized area of teaching or research, subject sophistica-

tion and mathematical and conceptual complexity generally serve as effective trapping mechanisms. Although it is easier for the theoretician to cross subject areas than for the experimenter (who is burdened with specialized equipment and techniques), it is nonetheless difficult in practice; and this despite a great unity in theoretical physics of conceptual ideas and supporting mathematics. (For example, the concept of space-time fields and the mathematics of differential equations are common to almost all specialized areas of physical theory.)

One important technical development affecting modern theoretical physics is the high-speed computer. Most theoreticians do at least some calculations, and many do a lot. The kinds of problems that are being tackled today would have been inconceivable even 15 years ago. Computations are now done overnight that would have then involved a lifetime. Whether reliance on the computer has become the tail wagging the dog of theoretical inquiry is a question widely debated and largely unanswered. Evidence can be adduced to support the claim that computers enhance concept development; but evidence can also be adduced to argue that computers get in the way of creative thinking, perhaps largely via the illusion that numbers constitute understanding. Whatever the implication, the fact is that computers play an important, almost dominant role in modern theoretical inquiry.

In some sense physical theory is model building. All but the simplest experiments are too difficult for a complete mathematical description. Idealization and abstraction based on knowledge and driven by curiosity and interest lead to sets of equations which attempt to model experimental reality. The hallmarks of the good theorist are first the ability to see the forest as well as the trees, that is, not to get lost in detail before a perspective is developed; then to be clever at manipulating equations so that justifiable and revealing approximations can be made. The ultimate test of a theory is its ability to account for present experimental results and to give guidance as to useful new experiments. A theory may refer to a schema or a paradigm of broad scope and wide application like quantum mechanics or Maxwell's equations; but the term can also be used narrowly for some interesting detail within the paradigm, such as scattering theory, or the Landau theory of second-order phase transitions.

Theoretical physics is especially rich in models and metaphors, so that people trained in this area can make imaginative contributions to other fields of human endeavor where the tradition of using mathematics is less highly developed. Dramatic examples are the contribution of physicists such as Francis Crick and Max Delbruck to the recent dramatic advances made in molecular biology.

L. E. H. TRAINOR

References

Many modern scientific magazines carry articles bearing on major advances in modern physical theory, notably Scientific American and Physics Today. Bohm's book, listed below, deals with some interesting philosophical questions relating to the interpretation of quantum theory, relativity and the nature of human consciousness. Although the work of Prigogine is not without its strong critics, nonetheless, he deals in an interesting way with some fundamental problems bearing on time, irreversibility and the nature of living systems. Kuhn's views on the existence of scientific paradigms and their revolutionary modification has become a classic work in the philosophy of science. The Einstein-Infeld book is an old classic with contemporary value. Finally, the book by Trainor and Wise develops a perspective as to how the mathematical structures used in physical theory arise naturally out of our perception of the world and the phenomena we are trying to explain.

1. Bohm, D., "Wholeness and Implicate Order," London, Routledge and Kegan Paul, 1980.
2. Einstein, A., and Infeld, L., "The Evolution of Physics," New York, Simon and Shuster, 1927.
3. Kuhn, T., "The Structure of Scientific Revolutions," 2nd Edition, Chicago, Univ. Chicago Press, 1970.
4. Prigogine, I., "From Being to Becoming," San Francisco, W. H. Freeman, 1980.
5. Trainor, L. E. H., and Wise, M. B., "From Physical Concept to Mathematical Structure," Toronto, Univ. Toronto Press, 1981.

Cross-references: FIELD THEORY; HEISENBERG UNCERTAINTY PRINCIPLE; KINETIC THEORY; MATHEMATICAL PHYSICS; MEASUREMENTS, PRINCIPLES OF; MECHANICS; MOLECULAR BIOLOGY: QUANTUM THEORY; RELATIVITY; SPECTROSCOPY; STATISTICAL MECHANICS.

THERMIONICS

Thermionics is the study of electron emission from solids occurring at elevated temperatures (1000–2500°C). The effect is of primary importance with vacuum tubes, x-ray tubes, microwave tubes, electron microscopes, and other devices which depend on a continuous supply of electrons. With such a device electrons are provided by heating a filament of an appropriate material such as BaO, SrO, thoriated tungsten, etc., as discussed below.

Among the early workers with currents from hot electrodes were Hittorf (1869–1883) and Goldstein (1885), both drawing quite large currents, and Elster and Geitel (1882–1889) who worked with very small currents, both positive and negative, in their research on the phenomenon. Edison (1883) in his work on the incandescent lamp discovered current emitted from the hot carbon filament and proposed a use for it in a patent granted to him. This emission became known as the "Edison effect." None of these men, however, knew what they had, supposing that they dealt with ions such as occur in electrolysis or gas discharges. It was not until 1897–1899 that the work of J. J. Thomson showed that the negative carriers of cathode rays were a new species of particle with mass about 1700 times smaller than that of the hydrogen ion. This was the electron. Drude (1900) suggested that electrons rather than metallic ions are the carriers of current in metals, and Thomson proposed that they are also the negative charges emitted by hot metals. O. W. Richardson made a study on this basis (1901) and derived equations relating the current density to the absolute temperature. The theory was subsequently extended by W. Schottky (1919) and S. Dushman (1923). Latter developments utilized the Fermi-Dirac statistics, introduced by Sommerfeld (1928) to analyze electronic conduction in solids.

The theory of thermionic emission from metals will now be briefly outlined. The energy relationships are shown in Fig. 1, where μ is the Fermi level and ϕ the work function (the energy required to remove an electron, at the Fermi level, from the metal). In the following we consider electrons to be emitted only in the positive x direction taken to be perpendicular to the surface. For electrons to be emitted they must have sufficient energy, directed in the positive x direction, to overcome the barrier W $(\mu + \phi)$. Assuming that all of the electron's energy is kinetic, this condition may be stated as

$$\tfrac{1}{2} m v_x^2 > \mu + \phi$$

or

$$v_x > \left[\frac{2(\mu + \phi)}{m} \right]^{1/2} \tag{1}$$

where m is the mass of the electron and v_x is the velocity component along the positive x direction.

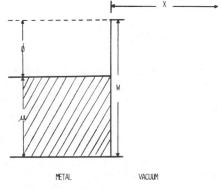

FIG. 1.

The problem now is that of calculating the number of electrons at a given temperature, which have the required velocities. In general, the number of electrons dn with a velocity in the interval v to $v + dv$ is given by the product obtained by multiplying the number of available quantum states by the probability that the states will be occupied, i.e.,

$$dn = \rho(v) f(v) \, dv_x \, dv_y \, dv_z \tag{2}$$

where $\rho(v)$ is the density of states, $f(v)$ is the Fermi-Dirac occupancy probability and v_x, v_y, v_z refer to the velocity components in the direction signified by the subscripts. For electrons we have

$$\rho(v) = \left(2\frac{m}{h}\right)^3 \tag{3}$$

where h is Planck's constant and

$$f(v) = \frac{1}{e^{(E-\mu)/kT} + 1} \tag{4}$$

where

$$E = \frac{m}{2}(v_x{}^2 + v_y{}^2 + v_z{}^2),$$

k is the Boltzmann constant and T the absolute temperature.

The thermionic current in the positive x direction is obtained by multiplying Eq. (2) by qv_x, where q is the charge per electron, integrating the result over the intervals

$$\left[\frac{2(\phi+\mu)}{m}\right]^{1/2} < v_x \leqslant \infty$$

$$-\infty \leqslant v_y \leqslant \infty$$

$$-\infty \leqslant v_z \leqslant \infty,$$

and using the assumption that the exponential term in (4) is large compared to unity.

The result is the so-called Richardson-Dushman equation for the thermionic current density I:

$$I = \left(\frac{4\pi q m k^2}{h^3}\right) T^2 e^{-\phi/kT} \tag{5}$$

or representing the term in the brackets by A

$$I = A T^2 e^{-\phi/kT}. \tag{6}$$

Evaluation of A gives

$$A = 120 \text{ amperes cm}^{-2} \text{ K}^{-2}.$$

The equation is based on the assumption that all the emitted electrons will be drawn to a collector. Actually, the current will be slightly less than that given by the above equation, since

TABLE I

Material	A (amp cm^{-2} K^{-2})	ϕ (eV)
Tungsten	60–100	4.5
Tantalum	120	4.3
Thorium on tungsten	3–5	2.7
BaO, SrO	0.01–0.1	0.45
Thorium carbide	550	3.5

From D. G. Fink, "Electronic Engineer Handbook," New York, McGraw-Hill, 1975.

there is a probability that an electron approaching the metal surface will be reflected back into the metal even though it has enough energy to escape.

Some values of A and ϕ are given in Table 1. As may be seen, the values of A are different from the theoretical value. The primary reason is the temperature dependency of the work function. This may be seen by considering a linear temperature dependency

$$\phi(T) = \phi_0 + \alpha T.$$

Substituting into (6) gives

$$I = A e^{-\alpha/T} T^2 e^{-\phi_0/kT}$$

or

$$I = A' T^2 e^{-\phi_0/kT}.$$

If $\alpha > 0$, as is often the case, then A' will be smaller than A.

The exponential term of the expression has been amply verified over a large temperature range. The work function ϕ is known for a large number of metals. It depends on the orientation of the crystal and tends to be larger for metals in which the atoms are packed closely together, smaller for open lattice metals, the range being about 1.5–6 volts. It is slightly higher on dense crystal faces than on open ones. The pure metal that is most often used as a thermionic emitter where ruggedness and high voltage are involved is tungsten, because of its strength and high melting point. For it, the values of ϕ and apparent A are about 4.5 volt and 100 amperes/cm^2 deg^2. Molybdenum, tantalum, and niobium are others used in special applications.

The thermionic properties of a metal surface are profoundly changed by thin films of foreign materials. This is the basis of the thoriated tungsten emitter used in small and medium-size vacuum tubes by Langmuir and Rodgers in 1914. The filament is made of tungsten having a small additive of thorium. In the heat treatment, some of the thorium diffuses to the surface, where it forms a quite stable deposit that is less than one atom deep. The work function of this surface is less than that of either thorium or tungsten, about 2.6 volts, and A is about 3 ampere/cm² deg². Where the tungsten filament is normally operated near 2700 K the thoriated tungsten yields a comparable emission current density at 1800–2000 K, with a very considerable saving in heating power. The surface is less rugged than that of pure tungsten.

By far the larger number of vacuum tubes use the oxide-coated filament, described by Wehnelt (1904). On nickel alloy base is deposited a relatively thick coating of the mixed oxides of barium, strontium, and calcium. Certain activation processes yield a surface with work function ϕ in the region of 1 volt. In spite of the small and variable value of the factor A, in the range of 0.01, the low work function provides a surface of high emission so that it can be used at the temperature of near 1000 K, giving still higher thermal efficiency than the thoriated tungsten. The mechanism of electron emission from this surface is considerably different from that of the pure metal. The oxide layer is normally an insulator at room temperature that becomes a semiconductor at the operating temperature. The oxides are partly dissociated so that there are metal atoms, particularly of barium, in the body of the layer and on its surface. The surface barium probably contributes to the low work function. The body barium is presumably ionized so that it contributes conduction electrons in the semiconductor. At the metal-oxide interface, there is another low-work-function surface so that electrons can pass from the metal base into the oxide, to be available for emission at the outer surface. With this modified mechanism, the emission equation still essentially holds. The reason is largely that the Boltzmann factor $\exp(\phi/kT)$ varies so rapidly with temperature that it renders other factors of little consequence experimentally.

It has been assumed that there is another electrode nearby with high enough positive voltage on it so that all of the emitted electrons are drawn to it. The current is then said to be saturated. In this condition, the current can still increase slowly with increasing voltage. The reason is that the strong electric field at the surface of the metal penetrates between the surface atoms and helps the electrons escape. The actual current then is increased above the saturation current by a factor $\exp 4.40(F^{1/2}/T)$ determined by Schottky in 1914, where F is the field strength in volts per centimeter, as verified with the refractory metals. With thoriated tungsten, the increase is more rapid than this, and with the oxide cathodes, the increase is so rapid that it is hard to say when saturation sets in.

At still higher surface fields, of the order 10 million volts/cm, another emission effect sets in, whereby the electrons are drawn out of even the cold metal. This is FIELD EMISSION, q.v.

At voltages below that required for saturation, the repulsion between the negatively charged electrons tends to limit the current. This is the space charge region, the condition in which most thermionic devices work. The current then is fairly insensitive to temperature and other conditions at the cathode so long as the anode voltage is well below the saturation value. In this condition, the current can be controlled by grids and other means. The space-charge-limited current between a plane emitting cathode and an anode at distance d from it and at voltage V, each of area 1 cm², is given fairly closely by the expression

$$i = 2.33 \times 10^{-6}\, V^{3/2}/d^2 \text{ amperes}$$

derived by Child (1911) and by Langmuir (1913). The equation is modified for a cylindrical structure, but the $V^{3/2}$ factor applies to any structure.

When the potential between emitter and plate is reversed so as to become retarding for electrons, the current is limited to the number of electrons with enough energy to overcome the retarding potential and is not limited by space charge. The current i_r is then related to the saturation current i_s by the expression

$$i_r = i_s \exp\left(-11{,}600\,\frac{V_r}{T}\right)$$

T being the temperature of the emitter and V_r the retarding potential. This may be written

$$\log i_r = 5030\,\frac{V_r}{T} + \log i_s.$$

Besides giving a means of determining the temperature of the emitter as shown by Germer (1925), the formula also is the basis of electronic devices with logarithmic response.

SUMNER LEVINE

References

1. Millman, J., and Seely, S., "Thermionics," New York, McGraw-Hill, 1941.
2. Sproull, R. W., and Phillips, W. A., "Modern Physics," 3rd Edition, New York, John Wiley, 1980. Good general reference for basic physics of thermionic emission.
3. West, R. C., and Astle, M. J., "CRC Handbook of Chemistry and Physics," Boca Raton, Florida, CRC Press, 1981. Contains an extensive listing of the work functions of the elements.
4. Giacoletto, L. J., "Electronic Designs Handbook," 2nd Edition, New York, McGraw-Hill, 1977.
5. Fomenko, V., "Handbook of Thermionic Properties," New York, Plenum Press, 1966.

Cross-references: ELECTRON, ELECTRON TUBES, FIELD EMISSION, IONIZATION, PHOTOELECTRICITY.

THERMODYNAMICS

Classical Thermodynamics is a theory which on the basis of four main laws and some ancillary assumptions deals with general limitations exhibited by the behavior of macroscopic systems. Phenomenologically it takes no cognizance of the atomic constitution of matter. All *mechanical* concepts such as kinetic energy or work are presupposed. Thermodynamics is motivated by the existence of dissipative mechanical systems. A *thermodynamic system* K may be thought of as a collection of bodies in bulk; when its condition is found to be unchanging in time (on a reasonable time scale) it is *in equilibrium*. It is then characterized by the values of a finite set of say *n* physical quantities, it being supposed that none of these is redundant. Such a set of quantities constitutes the *coordinates* of K, denoted by $x(=x_1, \cdots, x_n)$. Any set of values of these is a state $\mathfrak{S}$ of K. In virtue of these definitions, K is in a state only when it is in equilibrium. The passage of K from a state $\mathfrak{S}$ to a state $\mathfrak{S}'$ is a *transition* of K. A transition is *quasi-static* if in its course it goes through a continuous sequence of states, and if the forces which do work on the system are just those which hold it in equilibrium. A transition is *reversible* if there exists a second transition which restores the initial state, the final condition of the surroundings of K being the same as the initial condition. Reversible transitions are assumed to be quasi-static.

An enclosure which is such that the equilibrium of a system contained within it can only be disturbed by mechanical means is *adiabatic*, otherwise it is *diathermic*. For instance, stirring, or the passage of an electric current, constitute "mechanical means." A system K_0 in an adiabatic enclosure is *adiabatically isolated* but this does not preclude mechanical interactions with the surroundings. Its transitions are then called adiabatic.

For the time being, the masses of all substances present will be supposed fixed, and to achieve simplicity it will be given that (1) there are no substances present whose properties depend on their previous histories; (2) capillary forces as well as long-range interactions are absent. Further it will be supposed that of the *n* coordinates of K just $n - 1$ have geometrical character (*deformation coordinates*, e.g., volumes of enclosures), so that the work done by K in a quasi-static transition is

$$\int dW = \int \sum_{k=1}^{n-1} P_k(x)dx_k \tag{1}$$

Such a system will be called a *standard system* ($n - 1$ enclosures in diathermic contact, each containing a simple fluid, may serve as example, x_n being any one of the pressures).

The Zeroth Law Suppose two systems $K_A(x)$ and $K_B(y)$ to be in mutual diathermic contact.

Experience shows that the states $\mathfrak{S}_A$ and $\mathfrak{S}_B$ cannot be assigned arbitrarily, but that there exists a necessary relation of the form

$$f(x;y) \equiv f(x_1, \cdots, x_n; y_1, \cdots, y_m) = 0 \tag{2}$$

between them. If K_C is a third system, its diathermic equilibrium with K_B on the one hand, or with K_A on the other, is governed by conditions

$$g(y;z) = 0 \tag{3}$$

and

$$h(z;x) = 0 \tag{4}$$

respectively. That these three functions are not independent is expressed by the *Zeroth Law*: *If each of two systems is in equilibrium with a third system then they are in equilibrium with each other*. It follows that any two of Eqs. (2) through (4) imply the third, i.e., they must be equivalent to equations of the form

$$\xi(x) = \eta(y) = \zeta(z) \tag{5}$$

Thus with each system there is now associated a function, its *empirical temperature function*, such that two systems can be in equilibrium if and only if their *empirical temperatures* (i.e., the values of their empirical temperature functions) are equal. Write $t = \xi(x)$ so that one has the *equation of state* of K_A. Also, t may be introduced in place of any one of the x_k. Note that the empirical temperature is not uniquely determined since $t_A = t_B$ may be replaced by $\phi(t_A) = \phi(t_B)$ where the function ϕ is monotonic but otherwise arbitrary: one has a choice of *temperature* scales. For a system not in equilibrium, temperature is not defined.

The First Law It is obvious that one can do mechanical work upon a system (say by stirring) while its initial and final states are the same. (Nothing is being said about the surroundings!) In this sense mechanical energy is not conserved. One might however hope that it is conserved at least in a restricted class of transitions. That this is so is asserted by the *First Law: The work W_0 done by a system K_0 in an adiabatic transition depends on the terminal states alone*. Thus if $\mathfrak{S}'(x')$, $\mathfrak{S}''(x'')$ are the terminal states

$$W_0 = F(x';x'')$$

If $\mathfrak{S}'''(x''')$ is a third state, and the previous transition proceeds via $\mathfrak{S}'''$, W_0 must not depend on x''', i.e.,

$$F(x';x''') + F(x''';x'') \equiv F(x';x'')$$

It follows that there must exist a function $U(x)$, defined to within an arbitrary additive constant, such that

$$F(x'; x'') = U(x') - U(x'')(= -\Delta U \text{ say})$$

$U(x)$ is the *internal energy function* of K. (To make sure that U is in fact defined for all states, one assumes that *some* adiabatic transition always exists between any pair of given states.) The energy of a compound standard system is the sum of the energies of its constituent standard systems. Further, U must be a monotonic function of t, and it is convenient to choose the scale of t such that $\partial U/\partial t > 0$.

When the transition from $\mathfrak{S}'$ to $\mathfrak{S}''$ is adiabatic, $W_0 + \Delta U$ vanishes by definition of U. If the transition is not adiabatic and W is the work done by K, the quantity

$$\Delta U + W(=Q, \text{ say}) \tag{6}$$

will in general fail to vanish. Q is then called the *heat absorbed* by K. Every element of a quasi-static adiabatic transition is subject to $dQ = 0$, i.e., by Eqs. (1) and (6), to the differential equation

$$\sum_{k=1}^{n-1} \left(P_k(x) + \frac{\partial U(x)}{\partial x_k} \right) dx_k + \frac{\partial U}{\partial t} dt = 0 \tag{7}$$

The Second Law Experiment shows that if $\mathfrak{S}'$ and $\mathfrak{S}''$ are arbitrarily prescribed states, then it may be that no adiabatic transition from $\mathfrak{S}'$ to $\mathfrak{S}''$ exists. When this is the case one says that $\mathfrak{S}''$ is inaccessible from $\mathfrak{S}'$, but $\mathfrak{S}'$ is then accessible from $\mathfrak{S}''$, as has been already assumed. The states may of course happen to be mutually accessible. The existence of states adiabatically inaccessible from a given state is asserted precisely by the *Second Law: In every neighbourhood of any state $\mathfrak{S}$ of an adiabatically isolated system, there are states inaccessible from $\mathfrak{S}$.* (This formulation of the Second Law is known as the *Principle of Carathéodory*.) *A fortiori* this law applies to quasi-static transitions, i.e., those which satisfy Eq. (7). It asserts there are states $\mathfrak{S}''$ near $\mathfrak{S}'$ such that no functions $x_k(t)$ exist which satisfy Eq. (7) and whose values when $t = t''$ are just x_k'', ($k = 1, \cdots, n - 1$). It is merely a mathematical problem (the Theorem of Carathéodory) to prove that this is the case if and only if there exist functions $\lambda(x)$ and $s(x)$, ($x_n \equiv t$) such that the left-hand member is identically equal to λds, where ds is the total differential of s. Thus, the Second Law entails that

$$dQ = dU + dW = \lambda \, ds \tag{8}$$

(dQ is of course not a total differential). s is called the *empirical entropy function of* K. It is not uniquely determined, since it may be replaced by any monotonic function of s. If two standard systems K_A and K_B in diathermic contact make up a compound system K_C,

$dQ_C = dQ_A + dQ_B$, i.e., because of Eq. (8),

$$\lambda_A ds_A + \lambda_B ds_B = \lambda_C ds_C$$

By including s_A, s_B and the common empirical temperature t among the coordinates of K_C, one infers that

$$\lambda_A = T(t)\theta_A(s_A), \quad \lambda_B = T(t)\theta_B(s_B),$$
$$\lambda_C = T(t)\theta(s_A, s_B)$$

The common function $T(t)$ is called the *absolute temperature function*, while

$$S_A(s_A) = \int \theta_A(s_A) ds_A$$

is the *metrical entropy* of K_A. The "element of heat" dQ of any standard system thus splits up into the product of a universal function of empirical entropy and the total differential $dS(x)$ of the metrical entropy function:

$$T \, dS = dU + dW \tag{9}$$

By multiplying T by a constant and dividing S by the same constant, T can be arranged to be positive.

If one now chooses $x_n = S$ and recalls that the $x_k (k < n)$ are freely adjustable, the Second Law would be violated if S were also adjustable at will (by means of non-static adiabatic transitions.) Taking continuity requirements into account, it follows that S can either never decrease or never increase. The single example of the sudden expansion of a real gas shows that it can never decrease. One has the *Principle of Increase of Entropy: The entropy of an adiabatically isolated system can never decrease.*

So far the first three principal laws have been brought in in their traditional order. This has the disadvantage that one comes to grips with the heart of thermodynamics, namely, the existence of irreversible processes and its consequences, only after one has encountered certain laws and concomitant concepts, such as temperature, heat, and energy, which are largely irrelevant to it. It is, in fact, possible to begin with the Second Law provided one adopts an appropriate formulation, e.g., *there exists an ordering of the states of any standard system K, reflected in the existence of a function $s(x)$ such that if s', s'', are the values of $s(x)$ for arbitrarily selected states $\mathfrak{S}'$, $\mathfrak{S}''$ of K then $\mathfrak{S}''$ is adiabatically inaccessible from $\mathfrak{S}'$ if and only if $s'' < s'$.* Then $s(x)$ is the empirical entropy function already encountered. On bringing in the First Law now it follows at once that $dQ = \lambda \, ds$ for quasi-static transitions. The Principle of Increase of Entropy now follows trivially from the Second Law in the revised formulation. Further, one recognizes immediately that the First and Second Laws are independent of each other: the Second could remain valid even if the First were found to be false.

The Third Law It is known from experiment that for given values of the deformation coordinates, the energy function has a lower bound U_0. The question arises whether the entropy S has an analogous property. It is found in practice that the specific heats $\partial U/\partial T$ of all substances appear to go to zero at least linearly with T as $T \to 0$. This ensures that the function S goes to a finite limit S_0 as $T \to 0$. Experiment shows however further that as $T \to 0$, the derivatives of S with respect to the deformation coordinates also go to zero. In contrast with U_0, S_0 has therefore the remarkable property that it is independent of the deformation coordinates. One thus arrives at the *Third Law: The entropy of any given system attains the same finite least value for every state of least energy.* One immediate consequence of this is that the so-called *classical ideal gas* (the product of whose volume V and pressure P is proportional to T) cannot exist in nature. Further, no system can have its absolute temperature reduced to zero. The Third Law is therefore a statement about the properties of functions, not of systems, at $T = 0$.

The practical applications of the theory just outlined divide themselves into two broad classes: (1) those which are based on the existence and properties of the functions U and S and some others related to them—all "thermodynamic identities" being merely the integrability condition for the total differentials of these functions; and (2) those which are based on the Principle of Increase of Entropy: the entropy of the actual state of an adiabatically enclosed system being greater than that of any neighbouring "virtual" state.

The most important of the auxiliary functions just mentioned are
the *Helmholtz Function:*

$$F = U - TS \tag{10}$$

the *Gibbs Function:*

$$G = U - TS + \sum_{k=1}^{n-1} P_k x_k \tag{11}$$

the *enthalpy:*

$$H = U + \sum_{k=1}^{n-1} P_k x_k \tag{12}$$

sometimes called *thermodynamic potentials.* Then, e.g.,

$$dF = -S\,dT - dW$$

F therefore contains all available quantitative information about K, since

$$S = -\frac{\partial F}{\partial T}, \quad \text{and} \quad P_k = -\frac{\partial F}{\partial x_k} \tag{13}$$

The same is true of G for instance, since

$$S = -\frac{\partial G}{\partial T}, \quad \text{and} \quad x_k = \frac{\partial G}{\partial P_k}$$

F and G are naturally taken as functions of $x_1, \cdots, x_{n-1}, T$ and of $P_1, \cdots, P_{n-1}, T$, respectively. At times one speaks of F as the "Helmholtz free energy" and of G as the "Gibbs free energy." In an *isothermal* reversible transition, the amount W of work done by a system is equal not to the decrease of its energy U but to the decrease $-\Delta F$ of its (Helmholtz) free energy. In the presence of internal sources of irreversibility

$$W < -\Delta F$$

In considering physicochemical equilibria, that is to say, if one is interested in the internal constitution of a system in equilibrium when changes of phase and chemical reactions are admitted, one introduces the *constitutive coordinates* $n_i{}^\alpha$; this being the number of moles of the ith constituent C_i in the αth phase. The definitions of Eqs. (10) through (12) remain unaltered, for the $n_i{}^\alpha$ do not enter into the description of the interaction of the system with its surroundings. Let an amount $dn_i{}^\alpha$ of C_i be introduced quasistatically into the αth phase of the system. The work done on K shall be $\mu_i{}^\alpha dn_i$. The quantity $\mu_i{}^\alpha$ so defined is the *chemical potential* of C_i in the αth phase. It is in general a function of all the coordinates of K. Then, identically,

$$dG = \sum_{k=1}^{n-1} x_k dP_k - S\,dT + \sum_i \sum_\alpha \mu_i{}^\alpha dn_i{}^\alpha$$

Integrability conditions such as

$$\partial \mu_i{}^\alpha / \partial T = -\partial S/\partial n_i{}^\alpha$$

are applications of the first kind. On the other hand, the minimal property of G, derived from the maximal property of S, requires that

$$\sum_i \sum_\alpha \mu_i{}^\alpha dn_i{}^\alpha = 0$$

when all virtual states differ only in the values of the constitutive coordinates. If the system is chemically inert, the $dn_i{}^\alpha$ are subject only to the requirements of the conservation of matter. One then concludes that if there are c constituents and p phases, i.e., $n + pc$ coordinates in all, then the number f of these to which arbitrary values may be assigned is

$$f = c - p + n$$

This typical application of the second kind is the Gibbs PHASE RULE (for inert systems.) This rule is often stated merely for systems with only two external coordinates ($n = 2$, e.g., $x_i = P$, $x_2 = T$). There must then be no internal partitions within the system nor may it, for instance, contain magnetic substances in the presence of external magnetic fields.

The beauty and power of phenomenological thermodynamics lies just in the generality and paucity of its basic laws which hold independently of any assumptions concerning the microscopic structure of the systems which they govern. Its quantitative content is limited to conditions of equilibrium. Its conceptual framework is too narrow to permit the description of the temporal behavior of systems, except in as far as it makes it possible to decide which one of any pair of states of an adiabatically enclosed system must have been the earlier state.

Statistical thermodynamics seeks to remedy these deficiencies by making specific assumptions about the microscopic structure of the system K, and relating its macroscopic behaviour to that of its atomic constituents. K is then to be regarded as an *assembly* of a very large number of particles, which, on a non-quantal level, is a mechanical system with, say, N degrees of freedom. A *microstate* of K is a set of values of its N coordinates and its N conjugate momenta. It is out of the question to measure all these at a given time. One therefore constructs a *representative ensemble* $\mathcal{E}_K$ of K, which is an abstract collection of a very large number of identical copies of K. At any time t, the members of $\mathcal{E}_K$ will be in different microstates. Let the fractional number of members of the ensemble whose microstates lie in the range dp, dq about p, q be $\phi\, dp\, dq$. Then ϕ is the *probability-in-phase*, and with $d\Gamma = dp\, dq$

$$\int \phi\, d\Gamma = 1 \qquad (14)$$

The reason for this terminology is implicit in the *Postulate: The probability that a given assembly K will, at time t, be in a microstate lying in the range $d\Gamma$ about p,q, is equal to the probability $\phi\, d\Gamma$ that the microstate of a member of $\mathcal{E}_K$, selected at random at time t, lies in the same range.*

The mean value $\langle f \rangle$ of a dynamical quantity f is defined to be

$$\langle f \rangle = \int f \phi\, d\Gamma$$

If N is sufficiently large, fluctuations about the mean will usually be negligible.

When K is in equilibrium ϕ must be constant in time, and this will be the case if it is a function of the (time-independent) Hamiltonian H of K. Ensemble averages are now assumed to coincide with temporal averages. When, in particular, K is in diathermic equilibrium with its surroundings one can show that ϕ must have the form

$$\phi = \exp[(\Phi - H)/\theta] \qquad (15)$$

where Φ and θ are independent of p,q. Then

$$\theta \langle \ln \phi \rangle = \Phi - \langle H \rangle \qquad (16)$$

and, because of Eq. (14)

$$d \int \exp[(\Phi - H)/\theta]\, d\Gamma = 0 = \langle d[(\Phi - H)/\theta] \rangle$$

where d refers to a variation of the macroscopic coordinates of K. Using Eq. (16) and its variation, the relation

$$-\theta d \langle \ln \phi \rangle = d\langle H \rangle - \langle dH \rangle \qquad (17)$$

follows. Now $\langle H \rangle$ ($=\overline{U}$, say) is the total energy of the assembly, while $\langle dH \rangle$ is the average of the change of the potential energy, i.e., the work $-dW$ done by the external forces on K. If one writes

$$\overline{S} = -k \langle \ln \phi \rangle$$

where k is a constant, Eq. (17) becomes

$$k^{-1} \theta\, d\overline{S} = d\overline{U} + dW$$

This is identical with the phenomenological relation of Eq. (9) if one formally identifies S with $\overline{S}$, U with $\overline{U}$ and θ with kT. In this way, contact with the phenomenological theory has been established, and the quantities characteristic of the one theory have been *correlated* with that of the other. With this correlation, or interpretation, Φ becomes F. However, because of Eqs. (14) and (15)

$$F = -kT \ln \int \exp(-H/kT)\, d\Gamma$$

so that if only H is known, the integral on the right (the *partition function*), and thus F, can be calculated. The equation of state of a real gas can thus in principle be obtained from a knowledge of the forces operating within the assembly. This illustrates how the additional information put into the theory yields a correspondingly greater output. Phenomenologically such an equation of state might be written as

$$PV = \sum_{n=1}^{\infty} B_n(T)\, V^{1-n}$$

but here each of the *virial coefficients* B_1, B_2, $\cdots$ must be measured separately.

If the quantum mechanical behavior of matter is taken into account, the fact that one cannot assign precise simultaneous values to canonically conjugate quantities must produce modifications of the details of the statistical theory. However, it is not necessary to consider these here.

H. A. BUCHDAHL

Reference

Buchdahl, H. A., "The Concepts of Classical Thermodynamics," Cambridge, England, The University Press, 1966.

Buchdahl, H. A., "Twenty Lectures on Thermodynamics," Sydney, Pergamon Press, 1975.

Buchdahl, H. A., "From Phenomenological Thermodynamics to the Canonical Ensemble," *Foundations of Physics* 9, 819–829 (1979).

Sears, F. W., and Salinger, G. L., "Thermodynamics, Kinetic Theory and Statistical Thermodynamics," Third Ed., Reading, Mass., Addison-Wesley, 1978.

Cross-references: ENTROPY, HEAT CAPACITY, PHASE RULE, PHYSICAL CHEMISTRY.

THERMOELECTRICITY

Thermoelectricity is the subject dealing with the relation between electrical effects (potential differences, current flow) and thermal effects (temperature differences, heat flow) in a conducting material. The conducting material is not limited to solids in which the charge carriers are electrons. For example, thermoelectric effects can be observed in ionic conductors, where the carriers are ions, or in semiconductors where the carriers may be holes or electrons.

1. **Macroscopic Observations** (a) *Seebeck Effect*. Suppose we take two conducting wires (of different chemical composition) A and B joined together as in Fig. 1, and maintain junction d at a temperature T_1 different from the temperature T_2 of junction c. Then a difference of potential $V_B - V_A$ is observed at the terminals a and b (both at the same temperature), or, if a and b are connected through a sensitive current meter a current will be observed to flow. The effect was observed by Seebeck in 1826 and has carried his name thereafter. This combination of conductors is called a *thermocouple*. Now suppose that $T_2 = T_1 + \Delta T$, where ΔT is a small increment in T, and a voltage ΔV is observed at terminals a and b, then the thermo-power of the thermocouple is defined by

$$S_{AB} = \lim_{\Delta T \to 0} (\Delta V / \Delta T). \qquad (1)$$

Although the appearance of the Seebeck voltage requires the presence of two disimilar conductors, the thermoelectric effects are determined by intrinsic properties of the individual conductors, and indeed

$$S_{AB} = S_B - S_A \qquad (2)$$

where S_B and S_A are termed *absolute thermopowers* of the individual conductors A and B. Both S_B and S_A will generally be temperature dependent, and the total voltage observed at a and b for a finite difference in T_1 and T_2 is

$$V_b - V_a = \int_{T_1}^{T_2} (S_B - S_A)\, dT. \qquad (3)$$

If T_1 is constant then $V_b - V_a$ is a function of T_2. If this function is determined by calibration against a standard thermometer, then the voltage of the thermocouple can be used to measure temperature, or it may be used to control a switch. These are the bases of thermocouple thermometers and thermostatic controls which are the two main practical uses of the Seebeck effect.

(b) *Peltier Effect*. In the same Fig. 1 suppose that a battery is connected across terminals a and b so that an electrical current density (current divided by cross section) J flows in the circuit. Then it is observed that heat is evolved at one of the junctions and absorbed at the other. This effect is different from Joule ($I^2 R$) heating in that, when the current is reversed, the junction which previously evolved heat absorbs heat, and vice versa, whereas the Joule heat is independent of current direction. Mathematically

$$Q = \Pi_{AB} J \qquad (4)$$

where Q is the Peltier heat absorbed per unit time and unit cross-sectional area. Π_{AB} is the Peltier coefficient of the thermocouple, but again it can be written as

$$\Pi_{AB} = \Pi_B - \Pi_A \qquad (5)$$

where Π_A and Π_B are intrinsic properties of materials A and B. By using semiconductors which have high Peltier coefficients the Peltier effect can be used in a refrigerator with no moving parts.

From the above description it is clear that S_{AB} can easily be measured experimentally. Π_{AB} can also be measured but with somewhat more difficulty. It is not obvious, however, how one separates S_{AB} and Π_{AB} into parts characteristic of A and B individually. The third thermoelectric effect is important in this respect.

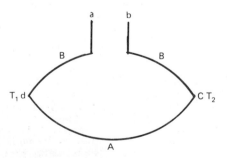

FIG. 1. A circuit consisting of two different conductors A and B. If $T_1 \neq T_2$, then a voltage $V_b - V_a$ appears between terminals a and b.

(c) *Thomson Effect*. This involves one conductor only. If we pass a current density J through a homogeneous wire then the heat production per unit volume is

$$Q = \rho J^2 - \mu J \, dT/dx \qquad (6)$$

where ρ is the resistivity of the material, dT/dx is the temperature gradient along the wire, and μ is called the *Thomson coefficient*. The first term is just the Joule $I^2 R$ term which does not reverse on heating; the second term is the Thomson heat, whose measurement depends on the fact that it changes sign when J changes directions.

Using techniques of irreversible thermodynamics it is possible to show rigorously that the three effects are connected by the Kelvin relations:

$$\mu = T \, dS/dT \qquad (7)$$

and

$$\Pi = TS. \qquad (8)$$

Hence if μ_A is measured as a function of T then S_A and Π_A can be obtained from

$$S_A = \int \frac{\mu_A(T)}{T} \, dT \qquad (9)$$

and

$$\Pi_A = TS_A. \qquad (10)$$

The measurement of $\mu(T)$ is a difficult experiment. Fortunately it only has to be performed on one material and S_A is then obtained from Eq. (9). Thereafter S_{AB} can be measured in a much simpler experiment and S_B derived from Eq. (2). Recently, first-class measurements of μ_A for Pb and Cu have been made by Roberts.[4] The derived values of S_A are now the accepted standards for use in thermocouples to derive the thermopowers of any other material. There is one other important family of standard materials, namely superconductors, which have zero thermopower. At low temperatures ($T < 22$ K) it is possible to make a thermocouple where A and B are normal and superconducting materials, respectively. Then $S_A = S_{AB}$ which can be more or less easily measured .

In Fig. 2 we show the thermopower of a simple metal—aluminum.[5,6] It is typical in the sense that at low temperature there is a large dip (or peak) which arises from a mechanism called *phonon drag*. It is superimposed on a more slowly varying background arising from a diffusion mechanism. Simple free electron theory indicates that for negative charge carriers (electrons in this case) the sign of both of these quantities should be negative. For more

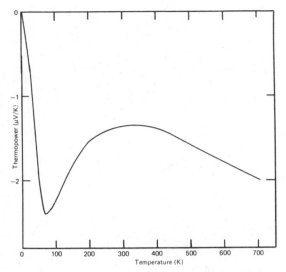

FIG. 2. Thermopower of pure aluminum.[5,6]

complicated metals (e.g., the noble metals, Fig. 3) they are frequently positive. In semiconductors the sign of the thermopower has frequently been used to decide whether the charge carriers were electrons or positive holes.

2. **Microscopic Theory** The thermopower can frequently be divided into two components, diffusion thermopower and phonon drag thermopower.

(a) *Diffusion Thermopower*. Consider a bar of metallic conductor with its ends at temperatures T and $T + \Delta T$ as in Fig. 4. Electrons at G will on the average be "hotter," i.e., have higher energies and velocities than those at F. The hot electrons at E will tend to diffuse toward F where the density of hot electrons is smaller. The cold electrons at F will tend to diffuse toward E where the density of cold electrons is smaller. If these two effects were the same they would cancel and there would be no thermopower. However, the diffusion process involves scattering of electrons from impurities, imperfections, and lattice vibrations (phonons). If the scattering is energy dependent—suppose hot electrons are scattered more than cold ones, for example—then the cold ones diffuse more rapidly than the hot ones. As a consequence the density of electrons increases at the hot end. But increased density means that an electrostatic field will be produced in the material. This field opposes further increase in electron density at the hot end. An equilibrium is reached when the net number of electrons diffusing in one direction is equalled by the number returning under the action of the electric field. Thus in a single material a thermoelectric field is set up. Its magnitude is determined by the rate of change of electron scattering cross section with energy. This can be positive or

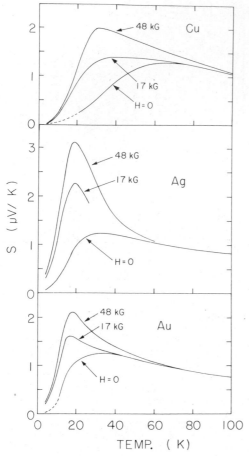

FIG. 3. Thermopower of the noble metals in a series of magnetic fields.[8]

portional to scattering cross section, thermopower depends critically on the rate of change of scattering cross sections with energy. This property makes it very sensitive to impurities, crystal imperfections, structural changes, etc. For example, we show in Fig. 5 how the thermopower changes at two phase changes, at 38 and 54 K, of the one-dimensional organic conductor tetrathiafulvalene-tetracyanoquinodimethane (TTF-TCNQ).[7]

(b) *Phonon Drag.* The simplest picture of diffusion thermopower assumes that the phonons (lattice vibrations) are always in local thermal equilibrium. In fact this will never be the case exactly and phonons will move along the temperature gradient. They will lose their momentum by interaction with crystal imperfections and with electrons. If the phonon-electron interaction predominates then the phonons will lose momentum to the electrons, pushing the electrons to one end of the conductor and thereby setting up an added contribution to the thermoelectric field. This contribution produces phonon drag thermopower. (The reason for the name phonon drag is more apparent in a parallel treatment of the Peltier effect, where the current of electrons "drag" the phonons with them.) The phonon drag contribution is most important in the temperature region where phonon-electron scattering predominates. This is for $T \sim \frac{1}{5} \theta_D$, where θ_D is the Debye temperature. At lower temperatures the number of phonons available for phonon drag decreases rapidly, and at higher temperatures the phonons tend to lose their momentum by phonon-phonon scattering rather than by phonon-electron scattering processes.

Measurements in a magnetic field (Fig. 3) indicate that the thermopower is highly field

negative, resulting in thermoelectric fields of either sign. To measure the thermoelectric field one attaches a voltage probe at each end and connects these to a sensitive voltmeter. This is just the circuit of Fig. 1 again. But if there is a temperature gradient along A there must also be one along B and therefore there is a thermoelectric field set up in B. The voltage measured is the consequence of these fields in the two conductors. That is, S_{AB} rather than S_A or S_B is measured.

The significant feature of this simple analysis is that, in contrast to resistivity, which is pro-

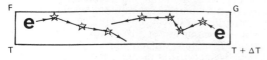

FIG. 4. Electron diffusion in a bar of metal with a temperature gradient.

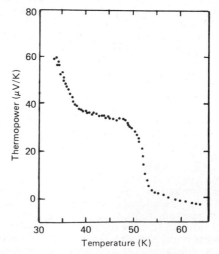

FIG. 5. Thermopower of TTF-TCNQ in the vicinity of phase transitions at 38 and 54 K.[7]

dependent in this region, but the definitive separation of the field effect into diffusion and phonon drag components has not been achieved.

PETER A. SCHROEDER

References

1. Blatt, F. J., Schroeder, P. A., Foiles, C. L., Greig, D. "Thermoelectric Power of Metals," New York, Plenum, 1976.
2. Barnard, R. D., "Thermoelecticity in Metals and Alloys," London, Taylor and Francis, and New York, Halstead Press, 1972.
3. MacDonald, D. K. C., "Thermoelectricity: An Introduction to the Principles," New York, John Wiley, 1962.
4. Roberts, R. B., *Phil. Mag.* 36, 91 (1977).
5. Roberts, R. B., and Crisp, R. S., *Phil. Mag.* 36, 81 (1977).
6. Gripshover, R. J., Van Zytveld, J. B., and Bass, J., *Phys. Rev.* 163, 598 (1967).
7. Chiang, C. K., in "Thermoelectricity in Metallic Conductors" (F. J. Blatt and P. A. Schroeder, Eds.), New York, Plenum, 1978.
8. Blatt, F. J., Caplan, A. D., Chiang, C. K., and Schroeder, P. A., *Solid State Commun.* 15, 411 (1974).

Cross-references: CONDUCTIVITY, ELECTRICAL; HALL EFFECT AND RELATED PHENOMENA; TEMPERATURE AND THERMOMETRY.

THIN FILMS

General The term "thin films" is used for a wide variety of physical structures. Self-supporting solid sheets usually are called foils when thinned from thicker material by such methods as rolling, beating, or etching; and films, when obtained by stripping a deposited layer from its substrate. Supported thin films are deposited on planar or (in special cases) curved substrates by such methods as vacuum evaporation, cathode sputtering, electroplating, electroless plating, spraying, and various chemical surface reactions in a controlled atmosphere or electrolyte. Thicknesses of such supported films range from less than an atomic monolayer to a few microns ($1\mu = 10^{-4}$ cm). A frequently used thickness measure is the angstrom ($1\text{Å} = 10^{-8}$ cm). Thin films not forming a continuous sheet are called "island films." Particularly, noble metals may condense as islands of considerable thickness (up to $\sim 10^2 \text{Å}$).

In scientific studies and technical applications, the use of well-controllable deposition methods such as vacuum evaporation and cathode sputtering are generally preferred. The film structure is markedly influenced by such deposition parameters as substrate composition and surface structure, source and substrate temperatures, deposition rate, and composition and pressure of the ambient atmosphere (where applicable). In general, the structure of films is more disordered than the corresponding bulk material. Smaller grains, higher dislocation concentrations, and deviations from stoichiometry are typical, and films approach bulk structure only as a limiting case. Under certain growth conditions, films exhibit preferential crystal orientations or even epitaxy. (Epitaxy means that the film structure is determined by the crystal structure and orientation of the underlying substrate.)

Solid thin films are common study objects in most phases of solid-state physics. They supply the samples for the study of general structural and physical properties of solid matter where special beam methods require small quantities of material or extremely thin layers, as for instance in transmission electron microscopy and diffraction, NEUTRON DIFFRACTION, UV spectroscopy, and X-RAY DIFFRACTION and SPECTROSCOPY. Thin films represent the best means for studying physical effects, where these effects are caused by the extreme thinness of the material itself. Examples are the rotational switching of ferromagnetic films, electron tunneling phenomena, electromagnetic skin effects of various kinds, and certain optical interference phenomena (see FERROMAGNETISM, SKIN EFFECT and TUNNELING). Films also are convenient vehicles for the investigation of nucleation and crystal growth, and for states of extremely disturbed thermodynamic equilibrium.

Presently, films find three major industrial uses: the decorative finishing of plastics, optical coatings of various kinds (mainly antireflection coatings, reflection increasing films, multilayer interference filters, and fluorescent coatings), and in electronic components from transistors or resistor-capacitor networks to such specialized devices as magnetic storage bits, photosensors, and cryotrons. The restricted space only permits the discussion of a few selected research and application areas.

Nucleation, Growth and Mechanical Properties of Films In vacuum evaporation, molecules or atoms of thermal energy are deposited at a uniform angle of incidence and under well-defined environmental conditions. Most nucleation and growth studies, therefore, have been made on evaporated films. A particle approaching the substrate enters close to its surface a field of attracting short-range London forces with an exchange energy proportional to $-1/r^6$. At a still shorter distance r, repulsive forces proportional to $e^{-r/\text{constant}}$ resist the penetration of the electron clouds of the surface atoms. Due to the atomic or crystalline structure of the substrate, this potential field exhibits periodicity or quasiperiodicity in the substrate plane. The freshly condensed particles migrate over the surface with a jump frequency $i_D \propto \exp(-Q_D/kT)$, or desorb with a frequency $i_{ad} \propto \exp(-Q_{ad}/kT)$, where the activation energy Q_D is often approximately one-fourth of Q_{ad}. Permanent condensation occurs in most cases at distinct nucleation centers which may consist of deep potential wells of the substrate, clusters of condensed particles, or previously deposited "seed" particles of a different material. The number of nuclei formed in the second case is strongly temperature and rate dependent.

Most metals always condense in crystalline form, but the grain size is extremely small at low temperatures (on the order of a few angstroms) and increases markedly with increasing substrate temperatures. Grain size decreases with increasing deposition rates. The condensation of amorphous or quasi-liquid phases at low temperatures has been observed for such metals as antimony and bismuth and a few dielectrics. Some of these materials, on annealing, pass through otherwise unobserved, and probably metastable, phases.

Stresses of considerable magnitude are often observed in deposited films. The main causes of these stresses are a mismatch of expansion coefficients between substrate and film, enclosed impurity atoms, a high concentration of lattice defects and, in very thin films, a variety of surface effects. Often, the stresses resulting from lattice defects can be minimized by the choice of a higher substrate temperature during deposition, or they can be reduced by a post-deposition anneal. Metal films frequently exhibit tensile strengths which are considerably larger than those of the corresponding bulk materials.

Thin-film Optics Deposited metal mirrors probably represent the oldest optical application of films. High-quality mirrors usually are produced by the vacuum evaporation of aluminium on an appropriately shaped glass substrate. Often, a glow-discharge cleaning of the substrate or a chromium undercoat is first applied to increase the adhesion of the aluminium. After deposition, the aluminium is protected by anodic oxidation or an evaporated overcoat of SiO, SiO_2, or Al_2O_3.

For SiO, maximum reflectance in the visible spectral region is achieved at a thickness of about 1400Å. Rapid SiO evaporation reduces the reflectance at shorter wavelengths.

Single or multilayer coatings find increasing use as optical interference filters. These film stacks may consist solely of transparent films of different refractive indices n_f, or a combination of absorbing and nonabsorbing layers. Common low-index materials for glass coatings in the visible region of the spectrum are MgF_2 (n_f = 1.32 to 1.37), and cryolite Na_3AlF_6 (n_f = 1.28 to 1.34); high-index materials are SiO (n_f = 1.97), ZnS ($n_f \approx 2.34$), TiO_2 (n_f = 2.66 to 2.69) and CeO_2 (n_f = 2.2 to 2.4). The indices are given for the sodium D line. Various semiconductors are used for infrared coatings.

At each air-film, film-film, or film-substrate interface, the incident light amplitude is split into a reflected and a transmitted fraction according to the Fresnel coefficients

$$f_{j-1} = (\hat{n}_{j-1} - \hat{n}_j)/(\hat{n}_{j-1} + \hat{n}_j) \text{ and}$$

$$g_{j-1} = 2\hat{n}_{j-1}/(\hat{n}_{j-1} + \hat{n}_j)$$

where j and $j - 1$ denote the number of the optical layer counted from the side of the incident beam. $\hat{n}_j = n/\cos \Theta_j$ for p polarization or $\hat{n}_j = n_j \cos \Theta_j$ for s polarization is the effective refractive index, and $n_j = n_j - ik_j$ the refractive index of the j layer.

$$\cos \Theta_j = \sqrt{(\sqrt{p_j^2 + q_j^2} + p_j)/2}$$
$$-i \sqrt{(\sqrt{p_j^2 + q_j^2} - p_j)/2}$$

$$p_j = 1 + (k_j^2 - n_j^2) [n_0 \sin \theta_0/(n_j^2 + k_j^2)]^2$$

$$q_j = -2n_j k_j [n_0 \sin \theta_0/(n_j^2 + k_j^2)]^2$$

The symbol θ_0 is the angle of incidence in the incident medium.

For nonabsorbing film stacks ($k_i = 0$; $i = 1, 2 \cdots, m + 1$), the over-all reflectance and transmittance may be obtained by summing the multiple coherent reflections between the film boundaries. A more general treatment based on electromagnetic theory yields for amplitude reflectance and transmittance the recursion formulas

$$r_{(j-1)-} = (f_{j-1} + r_{j-} \exp(-2i \hat{\Phi}_j))/$$
$$(1 + f_{j-1} r_{j-} \exp(-2i \hat{\Phi}_j))$$

and

$$t_{(j-1)-} = (g_{j-1} t_{j-} \exp(-i \hat{\Phi}_j))/$$
$$(1 + f_{j-1} r_{j-} \exp(-2i \hat{\Phi}_j))$$

$\hat{\Phi}_j = \Phi_j \cos \Theta_j$ is the effective phase thickness. $\Phi_j = (2\pi/\lambda)\hat{n}_j l_j$ where λ is the wavelength in vacuo and l_j is the geometrical film thickness. The recursion is started on the side of emergence, using the initial conditions $r_{m-} = f_m$ and $t_{m-} = g_m$. Intensities are given by $R = |r_{0-}|^2$ and $T = (\Re \hat{n}_{m+1}/n_0) |t_{0-}|^2$ where $\Re$ denotes "real part of." If A_j is the absorption in the layer j, $R + T + \Sigma_j A_j = 1$.

A single antireflection coating of $\lambda/4$ optical thickness $n_f l_f$ yields zero reflectance at $n_f = \sqrt{n_{glass}}$. A double layer coating of $\lambda/4$ films requires $n_2/n_1 = \sqrt{n_g}$. The transmission of a Fabry-Perot interference filter consisting of a dielectric spacer layer between two partially reflecting metal films is given by $I/I_0 = [(1 + A/T)^2 + (4R/T^2) \sin^2 (\delta - \Phi)]^{-1}$ where $\Phi = 2\pi nl \cos \theta/\lambda$. R, T, and A are the reflection, transmission and absorption coefficients of the reflecting layers. The refractive index and thickness of the spacer film are n and l. θ is the angle of refraction in the spacer, and δ the phase change for reflections at the spacer-metal film interfaces. $(I/I_0)_{max} = (T/(1 - R))^2$ and $(I/I_0)_{min} = (T/(1 + R))^2$. The band pass half-width is $\Delta\lambda_{1/2} \simeq \lambda(1 - R)/m\pi R^{1/2}$ for the interference order m ($m\pi = \Phi$). More complex coatings and filters, and their various applications, cannot be discussed here. It should be mentioned, however, that films play a very important role today in the accurate determination of the optical constants of many materials, but particularly of metals (see REFLECTION).

Film Electronics Deposited dielectric film materials in common use are SiO_2, Al_2O_3, Si_3N_4, and various glasses. Thin capacitive layers in the 100 to 500 Å thickness region are often produced by the anodization of tantalum and aluminum to Ta_2O_5 or Al_2O_3, respectively. The breakdown strength and dielectric constant of films approach bulk values, but might be reduced by surface roughness, structural faults, and lower density. According to the Lorentz-Lorenz formula, the dielectric constant D changes with reduced density ρ as $dD/d\rho = 3C/(1 - C\rho)^2$, where C is a constant depending on the material. On metal-dielectric-metal films, quantum mechanical tunneling through the dielectric film becomes observable below a dielectric thickness of about 100 Å. For applied voltages less than the metal-insulator work function ϕ, the tunneling current density J is proportional to the applied voltage V, demonstrating that the low-voltage tunneling resistance is ohmic. $J = (qV/h^2s)(2m^*\phi)^{1/2} \exp[-(4\pi s/h)(2m^*\phi)^{1/2}]$. At high applied voltages ($qV > \phi$), the current increases very rapidly: $J = (q^2V^2/8\pi h\phi_s{}^2) \exp[-(8\pi s/3hqV)(2m^*)^{1/2}\phi^{3/2}]$. s is the insulator thickness, m^* the electronic effective mass, and q the electron charge. Thicker dielectric films may exhibit in high fields appreciable Schottky or avalanche currents when they are greatly disordered.

Polycrystalline metal films generally show, due to their low structural order, a larger resistivity than the bulk material. According to Matthiessen's rule, the total resistivity can be expressed as $\rho = \rho(t) + \rho(i)$ where $\rho(t)$ is the temperature-dependent resistivity associated with scattering by lattice vibrations, and $\rho(i)$ is a temperature-independent resistivity caused by impurity or imperfection scattering. Very thin specimens with a thickness comparable to the electron mean free path show a $\rho(i)$ rapidly increasing with decreasing thickness. This increase is caused by an increasing contribution of non-specular electron scattering at the film surfaces. By annealing a metal film, $\rho(i)$ might be reduced permanently. A large $\rho(i)$ results in a small temperature coefficient α.

Many known superconductors can be deposited as super-conductive films (see SUPERCONDUCTIVITY). Through thin-film experiments, the energy gap in semiconductors can be measured, and material parameters, such as the penetration depth of magnetic fields, can be studied at dimensions less than the coherence range.

Studies of semiconductor films have shown many facets. The properties of epitaxial films have mainly been investigated on Ge and Si, and to a lesser degree on III-V compounds. Much work has been done on polycrystalline II-VI films, particularly with regard to the stoichiometry of the deposits, doping and post-deposition treatments, conductivity and carrier mobility, photo-conductance, fluorescence, electroluminescence, and metal-semiconductor junction properties. Among other semiconductors, selenium, tellurium, and a few transition metal oxides have found some interest.

Film resistors, capacitors, and interconnected R-C net-works on planar glass or ceramic substrates are finding widespread industrial use. Common resistor materials are carbon, nichrome and tin oxide in individual components; and nichrome, tantalum nitride, SiO-chromium cermet and cermet glazes in planar networks. Gold, copper, aluminum, or tantalum are used for termination lands, connection leads, and capacitor plates. SiO_2, Al_2O_3 and Si_3N_4 serve as film capacitor dielectrics and crossover insulation. The geometrical configuration of the desired component or circuit pattern is obtained either by deposition through mechanical masks or by removing from a continuous sheet the undesired portions after the deposition process is completed. This removal is frequently accomplished by a combination of photolithographic and etch processes.

The minimum length l and width w of a resistor are calculated from the given resistance R, the sheet resistance $\mathfrak{R}$ in ohms per square, dissipated power P, and permissible power dissipation per square inch $\mathfrak{P}$ by use of the formulas $w = \sqrt{(P \cdot \mathfrak{R})/\mathfrak{P} \cdot R}$ and $l = wR/\mathfrak{R}$. The capacitance of film capacitors is given by $C = 0.225 D(N - 1)A/t$, where C is the capacitance in picofarads, D the dielectric constant, N the number of plates, A the area in square inches, and t the dielectric thickness in inches.

Thin-film semiconductor devices have been slow to reach the production stage, mainly due to difficulties in controlling the film surface and interface properties. Various barrier layer diodes have exhibited impressive rectification ratios, but limited breakdown strength and low speed due to their large specific capacitance. Of the many film TRANSISTOR concepts proposed, the insulated gate field effect device looks the most promising and manufacturable. Its structure consists of a minute metal-dielectric-semiconductor capacitor. The semiconductor strip carries current between two terminals called source and drain. A field applied between metal "gate" and source modulates the semiconductor conductance and consequently the source-drain current. Usable semiconductor materials with a sufficiently low concentration of interface states are CdS, CdSe, and tellurium. These devices exhibit pentode-like characteristics with voltage gains ranging from 2.5 at 60 MHz to 8.5 at 2.5 MHz. The gain band width product G.B., which is equal to the transconductance divided by 2π times the gate capacitance, reaches values of about 20 MHz. It is determined by G.B. $= \mu_d V_D /$ $2\pi L^2$, where μ_d is the effective drift mobility of the electrons, V_D the source-drain potential, and L the source-drain spacing which is usually chosen between 1 and 5μ. Special film semiconductor devices in industrial use are various types of photodetectors.

Magnetic Films Magnetic thin films of nickel-iron (usually deposited at an 80:20 composition by weight) exhibit a number of unusual properties, which have led to many experimental and theoretical studies, as well as to important applications in binary storage and switching, magnetic amplifiers, and magneto-optical Kerr-effect displays.

Such "Permalloy" films have two stable states of magnetization, corresponding to positive and negative remanence. When deposited in a magnetic field or at an oblique angle, they exhibit uniaxial anisotropy. In practice, this anisotropy shows some dispersion, since it results from the alignment of local lattice disturbances. The stable states result from the minimization of the free energy $E = MH_L \cos\theta - MH_T \sin\theta + K \sin^2\theta$, where the last term represents the anisotropy energy, and θ is the angle between the magnetization M and the easy axis. From an inspection of the derivatives of this equation follows the hard-direction straight-line and the easy-direction square hysteresis loops of anistropic films. In the latter case, the magnetization is always either $+M$ or $-M$, and the change occurs at $H_L = \pm H_K$. The transitions from unstable to stable states occur at $\partial^2 E/\partial\theta^2 = 0$, resulting in a critical curve $H_L{}^{2/3} + H_T{}^{2/3} = H_K{}^{2/3}$ which has the form of an asteroid enclosing the origin (see MAGNETISM).

An important feature of magnetic films is the high speed with which the state of magnetization can be reversed. Dependent on film properties and magnetic fields, three modes of magnetization reversal occur: Domain wall motion, incoherent rotations, and the extremely fast coherent rotation of the magnetization. Wall-motion switching is expected when the driving fields are smaller than the critical values.

More recently, various magnetic garnet films have gained importance in research and industrial applications.

During the last decade, progress in thin film technology has been promoted to a large extent by the development of solid state devices, and especially integrated circuits. The requirements for defect-free epitaxial growth, extreme chemical purity, multilayering of different materials, and complex patterning of films at feature sizes at or below one micrometer have led to the development of vastly improved methods and highly automated equipment for thin film deposition, optical x-ray and electron lithography, plasma and ion beam etching, and chemical and structural analysis.

Some of the important newer deposition methods are chemical gas transport reactions, low temperature chemical vapor deposition (LTCVD), high pressure epitaxial growth, and molecular beam epitaxial growth (MBE). MBE in particular has been instrumental in the development and optimization of many novel device structures and has made possible the deposition of modulated semiconductor films consisting of stacks of thin alternating epitaxial layers ($\leqslant 100$ Å per layer) of two different semiconductor compounds. These III-V/III-V or IV/III-V structures exhibit superlattice character with very large carrier mobilities and other interesting electronic properties.

These processing methods are complemented by such analytical tools as the scanning electron microscope (SEM) for topological studies, low energy electron diffraction (LEED) for surface crystal structure, Auger analysis (AES) for elemental composition, x-ray photoelectron spectroscopy (XPS) for chemical bonds, secondary ion mass spectroscopy (SIMS), electron induced desorption (EID), photoelectron spectroscopy, and surface voltage spectroscopy.

The improved processing and analysis methods for thin films together with advances in theoretical solid state physics have led to the development of important new classes of solid state devices such as: very large scale integrated circuits (VLSI), various microwave device types and microwave integrated circuits, planar solid state opto-electronics, thin film transducers of various kinds, charge coupled devices (CCD's), magnetic bubble memories, surface acoustic wave (SAW) delay lines and filters, and many others. Further progress—albeit slower—has also been made in the technology of optical coatings, especially in the areas of transparent conductive films and metal-dielectric interference filters.

RUDOLF E. THUN

References

Dushman, S., "Scientific Foundations of Vacuum Technique," Second edition, New York, John Wiley & Sons, 1962.

Holland, L., "Vacuum Deposition of Thin Films," New York, John Wiley & Sons, 1958.

Keonjian, Edward, Ed., "Microelectronics," New York, McGraw-Hill, 1963.

Hass, G. Ed., "Physics of Thin Films," Vols. I to VI, New York, Academic Press, 1963 to 1971.

Neugebauer, C. A., et al., Ed., "Structure and Properties of Thin Films," New York, John Wiley & Sons, 1959.

Mayer, H., "Physik dünner Schichten, I and II," Stuttgart, Wissenschaftliche, 1950 and 1955.

Heavens, O. S. "Optical Properties of Thin Solid Films," London, Butterworths, 1955.

Series: "Vacuum Technology Transactions," New York, Pergamon Press, 1955–1963.

Walter, H., in Flügge, S., Ed., "Handbuch der Physik," Vol. 24, Berlin, Springer, 1956.

Maissel, L. I., and Glang, R., Eds., "Handbook of Thin Film Technology," New York, McGraw-Hill, 1970.

Hass, G., et al., "Physics of Thin Films," Vols. 7 to 10, New York, Academic Press, 1973 to 1978.

T. Tamir, Ed., "Integrated Optics," Berlin and New York, Springer-Verlag, 1975.

27th National symposium of the Am. Vac. Soc. Part II, *J. of Vac. Sc. & Techn.* 18(3) (April 1981).

Proceedings of the 8th Annual Conf. on the Physics of Compound Semiconductor Interfaces, *J. of Vac. Sc. & Techn.* 19(3) (Sept./Oct. 1981).

16th Symposium on Electron, Ion and Photon Beam Technology, *J. Vac. Sc. & Techn.* 19(4) (Nov./Dec. 1981).

A. Madhukar, "Modulated Semiconductor Structures," *J. Vac. Sc. & Techn.* 20(2), 149–161 (Feb. 1982).

Cross-references: CONDUCTIVITY, ELECTRICAL; ELECTRON MICROSCOPE; FERROMAGNETISM; MAGNETISM; NEUTRON DIFFRACTION; SEMICONDUCTORS; SKIN EFFECT; SPECTROSCOPY; SUPERCONDUCTIVITY; TRANSISTOR; TUNNELING; X-RAY DIFFRACTION.

TIME

Too many interpretations of the concept of time are based on one of the following two kinds of oversimplification. Philosophers have speculated about time on the premise that it is a primary notion and can be abstractly defined without bothering about the implementation of the definition. Conversely, the physicists, before Einstein, had a tendency to take time for granted and to use it as a parameter without further questioning its definition. Phychologists may have been the first to make a step in the right direction by trying to relate the concept of time to actual perceptions.

Nowadays, the physicist has become aware of the necessity of providing operational definitions of the concepts he uses, and it is generally acknowledged that the very concept of time depends upon the possibility of the repetition of events that may be considered identical or, at least, that have a common recognizable feature. However, the far-reaching implications of this idea are seldom realized, and it is not infrequent to read otherwise respectable discussions based on a notion so "obvious" that its vagueness remains completely unsuspected, namely, the notion of a "clock."

Assigning to time the character of a self-contained concept and assuming the existence of appropriate "clocks" showing the flow of this "time" is putting the cart before the horse. In a more refined approach, an a priori time concept is accepted and principles are formulated by virtue of which a motion taking place in some specified conditions is uniform. But even such a procedure amounts to a self-deception, as the actual definition of time is then camouflaged behind those "principles." (For example, the essential of the definition of time in classical mechanics lies in Newton's first law). In brief, the true primary operation is the *arbitrary choice* of a repeatable phenomenon that may be used in the definition of a clock. The rate of flow of "time" is then implied by this choice, that is, the choice of the fundamental clock *is* the definition of time.

For practical purposes, it is appropriate to limit the freedom inherent in the choice of a clock by specifying convenient properties to be imposed on the resulting time scale. The main such properties are the availability of a sufficiently perennial master "clock" and the possibility of devising wieldy secondary clocks, for everyday use, that give reproducible and consistent readings endowed with a property of additivity. The first master clock that suggested itself to mankind was provided by the rotation of the earth on its axis and around the sun. The unit of time thus defined and its aliquot parts gave birth to the first astronomical time scales (mean solar day, tropical year), which served as a background for the development of classical mechanics and astronomy. A large number of phenomena were discovered (e.g., the beats of a good watch) that bear a linear relationship to that time scale. Any of the "linear systems" involved could be used as a secondary clock. Then, when the measuring techniques improved, it appeared that the mutual linearity of those phenomena was only an approximation. This discovery brought about a mild crisis of the metrology of time. The crisis was readily dismissed by stating that the rotation of the earth was not really uniform (as compared to "more accurate" clocks). In fact, the situation had a deeper purport: two descriptions of the fundamental master clock, that had been hitherto considered equivalent, appeared to be inconsistent, and the question of the *choice* of the clock was brought to the foreground again. The difficulty was temporarily settled in 1955 by relating the astronomical time scale to one particular period (tropical year $1900.0 = 31\,556\,925.975$ seconds), and the following new permanent definition was adopted at the October, 1967, meeting of the 13th General Conference on Weights and Measures: "The second is the duration of 9 192 631 770 periods of the radiation corresponding to the transition between the two hyperfine levels of the fundamental state of the atom of cesium 133." This definition, which is in full agreement with the preceding one, has two advantages: It can readily be implemented by a so-called atomic clock, and it yields a standard which is reproducible to 1 part in 10^{11} or better.

Newtonian time, a basic feature of the whole body of classical mechanics, is practically the astronomical time, operationally defined as above, complemented by the following extra postulate. Let any single observer, standing still on earth, determine the (improved as above) astronomical time scale; the time scale so obtained is then to be used by every "observer," whatever his motion with respect to the first one. This postulate expresses, for each "observer," one choice of the master clock among infinitely many possible choices and as such it is legitimate. When it was stated, it was also consistent with the contemporaneous physical knowledge. Later on, the physicists grew accustomed to certain properties derived from the choice of Newtonian time and space and from other postulates of mechanics and electromagnetism, until a calamity happened which was similar to what has befallen astronomical time: the improvement of the measuring techniques showed that one empirical fact (namely, that the velocity of light in the laboratory frame of reference is independent of the motion of the emitter) was not compatible with all of those properties. Again a choice was necessary. The analysis of special relativity disclosed which of the properties at stake were incompatible. The decision as to which to drop was largely a matter of convenience. Einstein's choice (justified by strong operational reasons) was to drop the universality of time and space in order to retain more physical postulates. From then on, time and space ceased to be absolute concepts (see RELATIVITY). It is worth mentioning that the presence of such a fundamental choice at the basis of the special-relativistic theory of time is not always recognized.[1]

As the choice of the master clock may not be made any more by one "observer" on behalf of another one, the choice has to be decided upon for each "observer" separately. Special relativity's specification is that each inertial "observer" shall use the time scale defined by means of a conventional atomic "clock" at rest with respect to himself. This procedure leads to the concept of proper time, which enjoys a mathematically invariant character and plays, for inertial systems, the part formerly played by the universal time (see RELATIVITY).

When it comes to comparing the descriptions of the universe as made by "observers" whose motions relative to an inertial frame involve an acceleration, use is generally made of "general covariance," which permits a straightforward generalization of proper time. The relevant mathematical framework is that of general RELATIVITY. However, the interpretation of the general-relativistic proper time in terms of everyday experience involves instantaneous inertial frames, which are not embodied by any material system in this kind of problem. Therefore, the explicit relationship between proper time and the time actually measured by a material device called a clock is by no means clear.[2] In fact, the use of proper time in accelerated systems again implies a choice of the master clock, and in the present instance the choice has the

drawback of being fairly abstract. Notwithstanding the largely widespread opinion, general covariance may not be the most convenient tool for the study of time in noninertial frames of reference. The problem of time measurements in such frames is an open problem, on which practically everything remains to be done.

Editor's Note: For space-time see section entitled "Einstein's Theory of Gravitation" in the article GRAVITATION, and for quantization of gravitational field and space-time, see the section entitled "Quantum Theory" in that article.

JACQUES E. ROMAIN

References

1. Romain, J. E., *Nuovo Cimento*, 30, 1254 (1963).
2. Romain, J. E., *Rev. Mod. Phys.*, 35, 376 (1963); *Advances in Astronautical Sciences*, 13, 616 (1963); *Nuovo Cimento*, 31, 1060 (1964); 33, 1576 (1964); 34, 1544 (1964).
3. Finkelstein, D., *Phys. Rev.*, 184, 1261 (1969) "International Seminar on Relativity and Gravitation," Technion City, Israel, July, 1969 (Gordon and Breach, 1971), p. 159.
4. Cole, E. A. B., *Nuovo Cimento*, A66, 645 (1970).

Cross-references: ATOMIC CLOCKS, GRAVITATION, QUANTUM THEORY, RELATIVITY.

TRANSFORMER*

In elementary form, a transformer consists of two coils wound of wire and inductively coupled to each other. When alternating current at a given frequency flows in either coil, an alternating electromotive force (emf) of the same frequency is induced in the other coil. The value of this emf depends on the degree of coupling and the magnetic flux linking the two coils. The coil connected to a source of alternating emf is usually called the primary coil, and the emf across this coil is the primary emf. The emf induced in the secondary coil may be greater than or less than the primary emf, depending on the ratio of primary to secondary turns. A transformer is termed a step-up or a step-down transformer accordingly.

Most transformers have stationary iron alloy cores, around which the primary and secondary coils are placed. Because of the high permeability of iron alloys, most of the flux is confined to the core, and tight coupling between the coils is thereby obtained. So tight is the coupling

*Figures 1, 2, 4, and 5 and information contained in this article are based on the book "Electronic Transformers and Circuits" by Reuben Lee, New York, John Wiley & Sons, Inc., and are used with permission of the publisher.

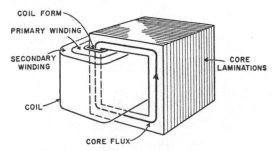

FIG. 1. Transformer coil and core.

between the coils in some transformers that the primary and secondary emf's bear almost exactly the same ratio to each other as the turns in the respective coils or windings. Thus, the turns ratio of a transformer is a common index of its function in raising or lowering potential.

A simple transformer coil and core arrangement is shown in Fig. 1. The primary and secondary coils are wound one over the other on an insulating coil tube or form. The core is laminated to reduce eddy currents. Flux flows in the core along the path indicated, so that all the core flux links both windings. In a circuit diagram, the transformer is represented by the symbol of Fig. 2.

In order for a transformer to deliver secondary emf, the primary emf must vary with respect to time. A dc potential produces no voltage in the secondary winding or power in the load. If both varying and dc potentials are impressed across the primary, only the varying part is delivered to the load. This comes about because the electromotive force e in the secondary is induced in that winding by the core flux ϕ according to the equation

$$e = -\frac{N\,d\phi}{dt}$$

This equation may be stated in words as follows: The voltage induced in a coil is proportional to the number of turns and to the time rate of change of magnetic flux linking the coil. This rate of change of flux may be large or small. For a given potential, if the rate of change of flux is small, many turns must be used. Conversely, if a small number of turns is used, a large rate of change of flux is necessary to produce a given potential.

Without transformers, modern industry could not have reached its present state of development. The highest potentials which are eco-

nomically feasible in ac generators are of the order of 20 kV. Transmission of power over long distances is most economical at high potentials which have reached levels of 750 kV and over. The higher the potential of the transmission line, the greater is the amount of power that can be transmitted over a given line conductor. The upper limit of potential is determined by insulation. Insulation research and development have resulted in completely surge-proof transformers, and have made possible power systems which are capable of withstanding lightning surges. At the utilization end of power systems, potential is successively lowered by means of step-down transformers to make power available in safe, useful form. Instrumentation and control of electrical power also require special forms of transformers.

An ideal transformer is defined as one which neither stores nor dissipates energy. Departures from the ideal transformer are caused by:

(1) Winding resistance and capacitance,
(2) Leakage inductance (due to flux which does not link both windings),
(3) Core hysteresis and eddy current losses,
(4) Magnetizing current.

Factors (1) and (2) above contribute to regulation, or the difference between secondary emf at no-load and full-load. This property is most important with variable load. Although no actual transformer is ideal, some transformers very nearly approach it. For example, in a 50-kVA rectifier transformer winding resistance amounts to 1 per cent, and leakage reactance 3 per cent, of the applied emf; core loss is 0.6 per cent of rated power, and magnetizing current is 2 per cent of rated current. Efficiency (output power divided by input power) is 98.4 per cent for this transformer.

Transformers are needed in electronic apparatus to provide the different values of potential for proper vacuum, gas or solid-state device operation, to insulate circuits from each other, to furnish high impedance to alternating current but low impedance to direct current, to change from one impedance level to another, to connect balanced lines to unbalanced loads, and to maintain or modify wave shape and frequency response at different potentials. Electronic transformers differ from power frequency transformers in the range of impedance levels, frequencies, size and weight. Categories of electronic transformers are: (1) Power, (2) frequency range, and (3) pulse.

Electronic power transformers are generally used to supply rectifiers at potentials ranging from 150 volts to 750 kV. Recent years have seen the widespread use of inverter transformers which convert dc potentials to higher dc potentials in conjunction with semiconductor or gas-filled devices. A simplified circuit is that of Fig. 3. Here the transformer output is a rectangular alternating wave which is often rectified again to produce dc output at increased potential.

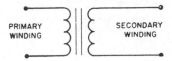

FIG. 2. Simple transformer.

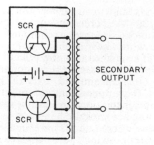

FIG. 3. Solid-state controlled rectifier and transformer for dc to ac inverter.

Frequency range transformers are used in applications where the frequency varies, including audio, video, carrier and control frequencies. Such frequencies vary from a fraction of 1 Hz to uhf (300 to 3000 MHz). Transformers may be wide-band or narrow-band in frequency response, and the core material changes accordingly. In wide-band transformers, the ratio of lowest frequency to highest frequency may be as great as 10^5. For such a wide band, the core material consists of nickel alloy laminations of high permeability which maintain uniform secondary voltage at the lowest frequency; also, this material makes possible low leakage inductance and winding capacitance, both of which are essential to good response at the highest frequency.

Narrow-band applications use mostly high frequencies and operate over a small percentage of the carrier frequency (e.g., 50 to 55 MHz). At radio frequencies, air-core transformers are used. Primary and secondary windings may be coaxial or concentric, with provision for adjusting the coupling between them. A circuit often used at radio frequencies is shown in Fig. 4. Primary and secondary circuits are tuned to resonance at the carrier frequency, and the response is shown in Fig. 5 for three conditions of coupling. Here

L_1 = self-inductance of primary coil
L_2 = self-inductance of secondary coil
L_m = mutual inductance between coils
k = coefficient of coupling = $L_m/\sqrt{L_1 L_2}$
f = carrier frequency
ω = $2\pi f$

It can be shown that maximum output occurs when

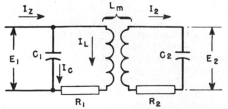

FIG. 4. Tuned air-core transformer.

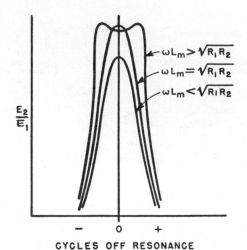

CYCLES OFF RESONANCE

FIG. 5. Response curves for circuit of Fig. 4.

$$2\pi f L_m = \sqrt{R_1 R_2}$$

and the value of the coupling coefficient that produces this output at resonance is known as the critical value. Undercoupling produces less output than critical. A certain degree of overcoupling may be advantageous, in that it causes a broad-nosed response curve such as the slight double hump in the response marked $\omega L_m > \sqrt{R_1 R_2}$. With modulated carriers, this response would offer very little attenuation to sidebands formed by the carrier and audio frequencies; yet the circuit would reject unwanted signals in adjacent carrier channels.

In tuned circuits like these, cores of powdered iron or ferrites are often used to tune the coils, minimize size, and obtain better performance than would be possible with air-core coils. The kind of core material used depends upon the frequency and Q (= $2\pi f L_1/R_1$) required. In general, the higher the frequency, the lower the permeability of the core material. Ferrites are available that range in permeability from $\mu = 10^4$ for use at low frequencies to $\mu = 10$ for use at 60 MHz.

Pulse transformers are used in radar modulators and computers. In radar applications, the pulses are usually rectangular and occur repetitively. Pulse widths range from 0.1 to 200 μsec. Peak ratings are large, from 100 to 50 000 kW. Secondary emf may range up to 300 kV for operation of high-power magnetrons and klystrons. Computer transformers are usually small, have ferrite cores and operate from current pulses to drive core matrices.

Advances in transformer technology depend largely on development of new core and insulation materials, conductor arrangements, measuring techniques and methods of application. By these means, transformers come into use at higher frequencies, with better balance, smaller size or higher power than formerly thought possible.

REUBEN LEE

References

M.I.T. Electrical Engineering Staff, "Magnetic Circuits and Transformers," New York, John Wiley & Sons, 1943.

Lee, Reuben, "Electronic Transformers and Circuits," available from University Microfilms, 300 North Zeeb Road, Ann Arbor, MI 48106, Catalog No. 41472.

Standard for Wideband Transformers No. 111, Institute of Electrical and Electronic Engineers, 345 E. 47th St., New York, NY 10017, 1981.

IEEE Standard for Computer-Type (Square loop) Pulse Transformers, No. 272, 1981.

Dowan, S. D., and Duff, D. L., "Analysis of Energy Recovery Transformer in DC Choppers and Inverters," *IEEE Trans. on Magnetics*, March 1970.

Lord, H. W., "Pulse Transformers," *IEEE Trans. on Magnetics*, March 1971.

Polydoroff, W. J., "High Frequency Magnetic Materials," New York, John Wiley & Sons, 1960.

Snelling, E. C., "Ferrites for Linear Applications," *IEEE Spectrum*, January 1972.

IEEE Standard Test Procedures for Evaluation of Insulation Systems for Electronic Power Transformers, No. 266, 1981.

IEEE Standard for Electronic Power Transformers, No. 295, 1981.

Cross-references: ALTERNATING CURRENTS, ELECTRICITY, INDUCED ELECTROMOTIVE FORCE, INDUCTANCE, MAGNETISM, PULSE GENERATION, RECTIFIERS.

TRANSISTORS

There are two generic types of transistor, the bipolar transistor and the field effect transistor. In the bipolar transistor, minority carriers are injected across a forward biased space charge region, at the emitter, diffuse across an essentially field free base region, and are collected across a high field space charge region, at the collector.* The minority carriers transit the base with little loss through recombination. Since they are emitted across a low impedance junction and collected across a high impedance junction, this results in power gain. The field effect transistor employs a channel, either built in or induced, which is modulated by a transverse gate region. This is somewhat analogous to the modulation of the plate current in a vacuum tube by the grid.

The bipolar transistor was discovered in 1947 by Bardeen and Brattain at Bell Laboratories

---

*Computer simulations have shown this to be an oversimplification, but it is helpful conceptually. This assumption is widely used in transistor design.

while doing basic research on germanium surfaces which was motivated by failure to make a field effect transistor. The invention was rapidly exploited by a group at Bell under William Shockley. The field effect transistor had been patented in 1930 by Lilienfeld, but apparently was not reduced to practice, also at Bell Laboratories, until after the bipolar transistor. Had the field effect transistor preceded the bipolar transistor, the application of bipolar transistors would probably have proceeded more slowly, since circuit engineers would have found the field effect transistor more akin to their previous experience.

The transistor has influenced almost all phases of human activity. It is prevalent in consumer goods, in industrial control, and, particularly, it has made possible the large scale computer, thus increasing the capability of the human mind. It may possibly have an even larger effect on the future than nuclear fission or space travel.

Various basic types of transistors are given in Table 1. Typical structures are shown in Figs. 1(a)–1(d).

Of the bipolar transistors, probably the original point contact transistor is the least understood. Shockley,[1]* in a classical example of applied physics, predicted almost completely the characteristics of junction transistors prior to any experimental results. To achieve the junction transistor structure it was necessary to form two *pn* junctions in close proximity. Attempts were made to do this by adding dopants during crystal growth, by alloying and regrowing junctions, and by vapor phase epitaxy. The first two resulted in commercial germanium devices. The vertical structure of the grown junction transistor was complete after crystal growth. It was then necessary to cut the transistors from this material and make emitter, base, and collector contacts, the base contact being difficult because of the thin base structure. The grown junction transistor had poor high frequency characteristics and excessive base resistance and was soon displaced. The alloy junction transistor was developed further. The original devices had base widths of the order of 50 μm and were limited to not much more than the audio frequency range. However, extremely clever junction formation techniques, including the jet etched surface barrier transistor and the microalloy diffused transistor, resulted in transistors which pushed performance up into the middle radio frequencies. These devices were widely incorporated in entertainment products, computers, and military electronics. However, they had the disadvantage that the transistors were fabricated essentially one at a time. The use of a solid state diffusion, in which a multiplicity of transistors could be

---

*Due to the huge number of publications in the semiconductor field, only this paper, general texts, and data shown will be referenced.

TABLE 1. TRANSISTOR TYPES.

GENERIC TYPE	BIPOLAR				FIELD EFFECT			UNITS
TYPE	POINT CONTACT	GROWN JUNCTION	ALLOY JUNCTION	DOUBLE DIFFUSED	JFET	MOS	MESFET	—
TYPICAL MATERIAL	Ge	Ge	Ge	Si	Si	Si	GaAs	—
MIN. VERT. DIM.	—	25	25	0.2	0.2	0.05	0.2	μm
MIN. LAT. DIM.	25	100	100	1	2	2	0.5	μm
MAXIMUM FREQUENCY	10^7	10^4	$10^4 - 10^7$	2×10^9	5×10^8	5×10^8	5×10^{10}	Hz
BATCH PROCESSING	NO	NO	NO	YES	YES	YES	YES	—
SUITABLE FOR I. C.	NO	NO	NO	YES	YES	YES	YES	—

fabricated at the same time in a wafer of germanium or silicon, produced structures of tighter geometrical control, greater uniformity, and increased performance. In silicon devices, silicon dioxide was used as a mask to localize the diffusions.

Photolithographic technique borrowed from the printing industry was used to form localized cuts in the silicon dioxide, and then the silicon dioxide was used as a mask to localize the diffusion. It became possible to decrease base widths to $\frac{1}{4}$ μm. This made possible the fabrication of transistors useful to over 100 MHz. Through the early stages of this work, although the silicon dioxide is an excellent dielectric, it was thought to be contaminated and was removed as one of the processing steps. The Fairchild group realized the usefulness of this

material to passivate junctions and developed the planar transistor. This group then made use of this extra level of insulation to extend the contacts and decrease lateral active device dimensions. This not only made possible much higher performance discrete devices, but also laid the groundwork for integrated circuits.

The choice of material has always been an emotionally debated issue in semiconductors. The original work was done on germanium partly because the surface of germanium could be treated chemically to insure low surface recombination velocity and because zone melting made possible the fabrication of high purity single crystals. Silicon offered higher performance plus the outstanding qualities of silicon dioxide as an insulator. However, the melting temperature of this material was higher

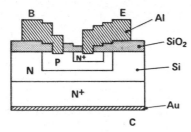

FIG. 1(a). Junction transistor.

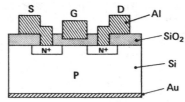

FIG. 1(c). MOS transistor.

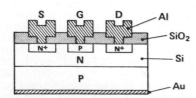

FIG. 1(b). Junction field effect transistor.

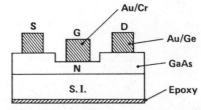

FIG. 1(d). Metal semiconductor transistor (MESFET).

and it took considerable work to produce single crystals of adequate perfection for semiconductor use. In distinction to the physical methods used with germanium, progress in silicon came through the purification of $SiCl_4$ in the gas phase prior to its reaction to elemental silicon. Once the technology of silicon had been mastered, it rapidly displaced germanium because of its better high temperature operation, the ubiquitous usefulness of silicon dioxide, and its higher saturation drift velocity. At present, gallium arsenide and silicon are vying for applications requiring the highest performance. At coarser geometries, the higher mobility of gallium arsenide is attractive. However, at very small geometries, which are now attainable through electron beam lithography, the differences tend to wash out as the ultimate scattering limited velocity of the two materials are similar. The properties of the three materials are summarized in Table 2.

Junction field effect transistors were originally fabricated in germanium. However, using the same technology developed for silicon bipolar transistors, JFETs were soon fabricated in silicon. It had been proposed that metal oxide field effect (MOS) transistors with an induced rather than a defined channel and with an insulated gate would give superior performance. However, instabilities in the oxide charge made these devices impractical due to threshold voltage shifts until it was determined that the instabilities were caused by the drift of sodium in the SiO_2. By eliminating sodium in the silicon dioxide through meticulous processing and by using mixed SiO_2-P_2O_5 glass as a sodium barrier, stable MOS field effect transistors were fabricated and this type of transistor has more and more become the predominant active device for integrated circuits. This is for three reasons. First, the source and drain contact diffusions can be extended to interconnect devices, giving in effect a second level of interconnections in addition to the metallization above the silicon dioxide. Second, the devices are self-isolating. Third, field effect transistors are ideal switches, that is, they have no offset at zero current.

Space charge regions in the junction field effect transistor modulate an existing channel. An analogous device can be made using a metal-semiconductor barrier to produce the space charge region. These have been manufactured mostly in gallium arsenide and have produced the highest performance three-terminal semiconductor devices yet fabricated.

Although discrete transistors have had a very large effect on the electronics industry, the integrated circuit has made possible levels of complexity totally unobtainable by other means. Hundreds of thousands of transistors are now incorporated into a single chip. This is done making use of transistors and diffused conductors within the bulk silicon and a multiplicity of conducting patterns separated by intervening dielectric layers above the silicon surface. Both linear and digital integrated circuits have found innumerable applications. Linear integrated circuits require a higher diversity of active devices and have led to the development of lateral as well as vertical diffused transistors and the utilization of parasitic transistors within the structure, such as substrate *pnp* devices.

The properties of semiconductors can be explained using a model in which conduction can take place by either the motion of negatively charged particles (electrons) or the absence of electrons modeled as positively charged particles (holes). Column-three impurities, such as boron, gallium, or indium, added to a column-four semiconductor such as germanium or silicon, accept electrons and provide free holes. Added column-five elements such as phosphorous, arsenic, or antimony donate electrons and provide free electrons for conduction. A semiconductor is called *n*-type if the donated electrons predominate and *p*-type if the holes predominate. An equilibrium tends to exist in which the product of the concentration of holes and electrons is a constant. A perturbation which changes this product from the constant value tends to recover to this constant value. An excess returns to equilibrium through a recombination process where holes and electrons recombine. A deviation below equilibrium is counteracted by generation processes. The time for an excess to decay to $1/e$ of its initial value is called the *minority carrier lifetime* τ, and the distance these charges can diffuse in this time is called the *diffusion length* $(D\tau)^{1/2}$.

If *p*- and *n*-type materials are adjacent, a *pn junction* is formed. At zero bias, carriers tend to diffuse across this junction but set up a built-in field which stops further diffusion by providing equal but opposite drift and diffusion currents. However, if a positive potential is applied to the *p* side relative to the *n* side, forward current flows and minority carriers are injected mostly into the more lightly doped of the two sides of the junction. If a negative potential is applied to the *p* side relative to the *n* side, little current flows. However, any minority carriers in the vicinity of the junction are swept over the potential barrier and cause

TABLE 2. TRANSISTOR MATERIAL
PROPERTIES.

MATERIAL	Ge	Si	GaAs	UNITS
BANDGAP	0.8	1.12	1.43	eV
μ_n	3900	1500	8500	cm^2/Vs
μ_p	1900	600	400	cm^2/Vs
$v_{LIM.}$	6×10^6	10^7	$2 \times 10^7/10^7$	cm/sec

reverse leakage current. Thus *pn* junctions make excellent rectifiers. If two junctions are in close proximity, a bipolar transistor is formed. The emitter is a forward biased *pn* junction. On the more lightly doped side of the junction, the base side, a nonequilibrium concentration of minority carriers is maintained, the concentration being exponentially proportional to the applied voltage. The other junction, the collector junction, is reverse biased. This maintains the carrier concentration at close to zero. Thus, within the base of the transistor there is a gradient of minority carriers, resulting in minority carrier current. If the base width is small relative to the diffusion length, little recombination occurs in the base and the minority carrier concentration profile is close to linear. Therefore, the transistor can be modeled either as a device which provides an essentially constant current with an increase in impedance or as a transconductance in which a voltage across the emitter-base junction results in a current across the collector-base junction. It is generally more convenient to have current gain as well as power gain and not have to transform impedance between stages. This can be accomplished by using the emitter rather than the base as the common element. If the common base current gain is defined as α, then the common emitter current gain is $\alpha/(1 - \alpha)$. However, this is achieved at the cost of bandwidth, the current gain bandwidth product being essentially constant. Base current provides the majority carriers which combine with the minority carriers. The components of base current are shown in Fig. 2(a). In early transistors, where the base width was of the order of fifty microns, recombination in the base was the principal cause of base current. In modern diffused silicon transistors, this term is fairly low. Recombination in emitter-base space charge regions, minority carriers injected into the emitter, and injection in the parasitic lateral base are dominant. Typical values for a modern diffused transistor[2] are given in Fig. 2(b).

In MOS transistors in saturation, the drain current is proportional to the charge induced in the channel divided by the transit time across the channel. For devices in which the gate length is relatively long, so that the fields do not approach the fields necessary to accelerate carriers to scattering limited velocity, the drain current is directly proportional to the mobility, the breadth of the channel and the square of the gate voltage above threshold divided by the oxide thickness. If the gate length is reduced to a value such that the average field over most of the gate is sufficient to accelerate the carriers to scattering limited velocity, then drain current is proportional to the breadth of the channel, the value of scattering limited velocity, and the value of the gate voltage above threshold divided by the oxide thickness. Equations for the transfer characteristics are compared in Table 3. Thus an MOS transistor

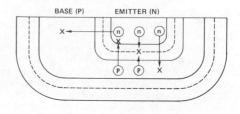

FIG. 2(a). Components of base current.

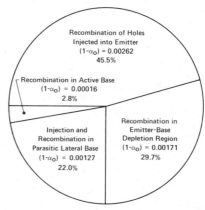

FIG. 2(b). Relative values of the components of base current for a modern diffused transistor. (See Ref. 2 for assumptions.)

with $V_D > V_{D\text{ saturation}}$ can have a transfer characteristic in which the drain current is either linearly proportional to voltage or proportional to the square of the voltage, whereas the bipolar transistor has a transfer characteristic in which the collector current is exponentially proportional to base-emitter voltage. Examples of these three characteristics are given in Fig. 3. It should be noted that although the mobility dominated case has the current proportional to the square of the voltage, because of differences in the proportionality constant the current is still less than in the saturation velocity dominated case where it is linearly proportional to voltage.

The high frequency limit of transistors is due to four factors: the charging time constant of the emitter, the transit time across the base, the drift time across the collector, and the time constant of the collector circuit. Again, modern technology has reduced the base transit time to a small fraction of the total. Figure 4(a) compares common emitter and common base frequency response. For the common emitter case, gain bandwidth product is essentially constant and the phase shift is approximately 90° over most of the frequency range. Figure 4(b) shows the reciprocal of current gain plotted as a function of the reciprocal of emitter current.

TABLE 3. COMPARISON OF TRANSFER CHARACTERISTICS,
BIPOLAR AND MOS.

	BIPOLAR (NPN)	MOS-MOBILITY DOMINATED ($V_D > V_{DSAT}$)	MOS-LIMITING VELOCITY DOMINATED ($V_D > V_{DLIM.}$)
CURRENT	$I_c = \dfrac{A_E q D_n n_i^2}{N_A w} e^{\frac{q V_{BE}}{kT}}$	$I_D = \dfrac{\epsilon \epsilon_0 Z \mu_n}{2 t_{ox} L}(V_G - V_T)^2$	$I_D = \dfrac{\epsilon \epsilon_0 Z v_{LIM.}}{t_{ox}}(V_G - V_T)$
TRANS-CONDUCTANCE	$g_m = \dfrac{A_E q^2 D_n n_i^2}{N_A w kT} e^{\frac{q V_{BE}}{kT}}$	$g_m = \dfrac{\epsilon \epsilon_0 Z \mu_n}{t_{ox} L}(V_G - V_T)$	$g_m = \dfrac{\epsilon \epsilon_0 Z v_{LIM.}}{t_{ox}}$
	$g_m = \dfrac{q I_c}{kT}$	$g_m = \left(\dfrac{2\epsilon\epsilon_0 Z \mu_n}{t_{ox} L}\right)^{1/2} I_D^{1/2}$	
	I_c = COLLECTOR CURRENT V_{BE} = BASE EMITTER VOLTAGE $g_m = \left.\dfrac{\partial I_c}{\partial V_{BE}}\right]_{V_{CE}}$ A_E = EMITTER AREA $\dfrac{q}{kT}$ = 39 VOLTS^{-1} n_i = 1.45×10^{10} cm^{-3} N_A = BASE ACCEPTOR CONC. w = BASE WIDTH	I_D = DRAIN CURRENT V_G = GATE VOLTAGE V_T = THRESHOLD VOLTAGE $g_m = \left.\dfrac{\partial I_D}{\partial V_G}\right]_{V_D}$ ϵ = 12 ϵ_0 = 8.85×10^{-14} f/cm^2 μ_n = ELECTRON MOBILITY Z = GATE BREADTH L = GATE LENGTH t_{ox} = OXIDE THICKNESS	$v_{LIM.}$ = SATURATION LIMITED VELOCITY (10^7 cm/s)

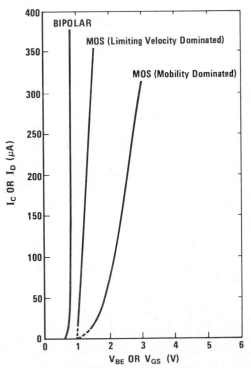

FIG. 3. Comparison of the transfer characteristics of a bipolar transistor, a MOS transistor in which carriers do not reach scattering limited velocity, and an MOS transistor in which carriers do reach scattering limited velocity.

The only one of the aforementioned time constants that is strongly current dependent is the emitter time constant, which is inversely proportional to emitter current. Therefore, over most of this range, as shown in Fig. 4(b), there is a linear relationship between the high frequency current gain and the reciprocal of the emitter current. However, at some high current value the current gain shows a maximum, and this is the practical upper current level of the transistor, typically at a current density of several thousand amperes per square centimeter. In most designs, this is due to the charge in transit modifying the collector depletion region by neutralizing the fixed charge of donors. This results in a shift of this depletion region toward the collector, increasing the base width. This can be minimized by doping the collector heavily. However, this reduces the maximum voltage which can be applied to the collector. Thus the product of gain, bandwidth, and voltage swing tends to be a constant.

The initial application of transistors was in small signal amplifiers. Commonly used equivalent circuits for bipolar transistors are given in Figs. 5(a)–5(c). It should be noted that the input and output impedances are moderate and predominantly real. This makes it easy to design wideband circuits. As shown in Fig. 5(d), the input impedance of the field effect transistor is primarily a small capacitance, where the output is predominantly resistive. This is very similar to a vacuum tube and lends itself to easy application in narrow band tuned circuits.

Both bipolar and field effect transistors make

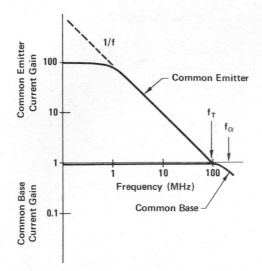

FIG. 4(a). Comparison of common emitter and common base frequency response.

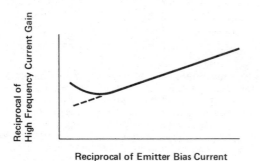

FIG. 4(b). Reciprocal of high frequency current gain as a function of the reciprocal of emitter current.

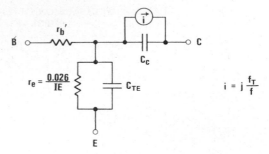

FIG. 5(a). "T" equivalent circuit.

$$i = j\frac{f_T}{f}$$

$$r_e = \frac{0.026}{IE}$$

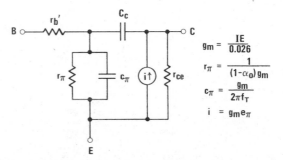

FIG. 5(b). "Pi" equivalent circuit.

$$g_m = \frac{IE}{0.026}$$

$$r_\pi = \frac{1}{(1-\alpha_0)\,g_m}$$

$$c_\pi = \frac{g_m}{2\pi f_T}$$

$$i = g_m e_\pi$$

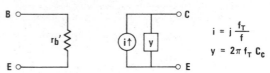

FIG. 5(c). Simplified high frequency equivalent circuit.

$$i = j\frac{f_T}{f}$$

$$y = 2\pi f_T C_c$$

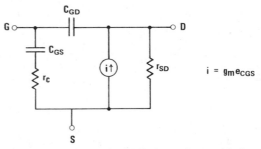

FIG. 5(d). MOS or field effect transistor equivalent circuit.

$$i = g_m e_{CGS}$$

excellent switches, the bipolar transistor having an offset voltage, the field effect transistor being more ideal. Bipolar transistor switching circuits have their lowest on-voltage drop when driven into saturation, that is, when there is sufficient charge stored in the base that both the emitter-base junction and the collector-base junctions are forward biased. This also stores minority carrier charge in the collector and tends to result in excessive turn-off times. This can be controlled by geometry control, through the use of epitaxial collectors, by adding impurities such as gold to reduce the minority carrier storage time, by clamping the collector out of saturation using metal-semiconductor diodes, or by circuit designs which avoid saturation. MOS transistors, being majority carrier devices, do not have this problem. However, simple n-channel or p-channel switching circuits have unequal turn-on and turn-off times. Complementary MOS circuits (CMOS) obviate this problem by having equal pull-up and pull-down

ability. Switching transistors have found very extensive use in digital logic circuits because the power is small and the speed high. Typical switching times are a few nanoseconds. The transistor has totally revolutionized the production of memory for computers. Sixty-four thousand bits on a chip is common. Both static and dynamic circuits are used.

As previously discussed, bipolar transistors

attain their best high frequency performance at collector current as high as possible so as to decrease the emitter time constant, limited, however, by charge in transit effects. Thus in order to obtain high frequency performance, bipolar transistors tend to require considerable power. A figure of merit which is frequently used is the speed power product, typical values are 1 pj. Gallium arsenide devices have the best yet obtained high frequency performance. This is due in part to the high mobility of GaAs and also due to an anomalously low feedthrough capacitance caused by second-order effects involving the formation of a stationary Gunn domain. Gallium arsenide field effect transistors have become the preferred three-terminal active device for very high frequency use. As geometries have tended to become finer and scattering-limited velocity effects dominant, the performance of silicon bipolar, silicon MOS, and gallium arsenide FETs have tended to converge.

An area in which transistors, particularly bipolar transistors, have become dominant is in power control. It is quite difficult to make vacuum tubes with plate current over $\frac{1}{2}$ ampere. It is very easy to make bipolar transistors with collector currents over 10 amperes, and many designs operate in the hundreds of amperes. This is due to the fact that transistor emitters can operate at thousands of amp/cm^2. It had long been thought that field effect transistors had little place in high current applications. However, as integrated circuit technology progressed, it was found that very large gate areas could be made with reasonable yields. Thus, it became possible to use this gate area to form very great gate breadth field effect transistors. At present, MOS field effect transistors are competitive with bipolar transistors at high voltages and at currents up to tens of amperes.

Although polycrystalline material is used in certain steps in transistor fabrication, particularly as a gate electrode, the heart of all transistors is single-crystal material. Single crystals, at the start of the transistor era, were laboratory curiosities. At present, 100- and 125-mm-diameter single crystals of silicon are in common use in semiconductor fabrication. As grown, the silicon crystals have extremely low defect densities. However, much work is currently underway to prevent the introduction of defects during device fabrication. The single crystals are sliced into wafers and additional layers grown by epitaxy.

Modern diffused silicon technology incorporates a series of oxidation, patterning, and diffusion steps. Silicon dioxide is grown at high temperatures in oxygen or steam. It is patterned using photo- or electron-beam-generated masks to transfer the image to an organic photosensitive resist. This is developed and the pattern is etched into silicon dioxide. This is then used as a mask for solid state diffusions. Diffusions with gaseous sources are performed at high temperatures, the depth and concentration being controlled by time and temperature. Recently, ion implantation, in which the desired species of atom is accelerated in an electrostatic accelerator, purified by mass separation and then implanted into the bulk crystal, has tended to displace diffusion. This provides better control than diffusion and produces profiles unattainable by diffusion. Contacts and interconnects are formed by evaporated or sputtered metal.

When integrated circuits were first proposed, people looked at the yields of discrete devices and predicted that integrated circuits of significant size could not be fabricated economically. This was based on two fallacies. First, the devices at that time were much larger than necessary. By refinement of lithographic technique, it proved possible to bring device sizes down by about an order of magnitude, thus reducing the probability of a defect occurring in a device. The reduction in size was possible because the interconnect capacitances being charged were on chip and thus much lower than would be encountered running wiring to other discrete devices. The second fallacy was that the defects were random. It turned out that most wafers had good areas and bad areas. Thus, while calculated random defect yield might be low, there would be areas in the wafer in which the yield would be high. Murphy's law[3] (Fig. 6) for predicting yield has been widely used for extrapolations, but is known to be inaccurate due to the assumption of random defects.

Packaging has possibly been the Achilles' heel of semiconductor technology. The purpose of the package is to protect the device from its environment, to admit power and signal leads, and to conduct heat out. As integrated circuits have gotten larger and more complex, packaging technologies have proved woefully inadequate to meet the needs of modern circuits, which require as many as 72 leads. An ingenious

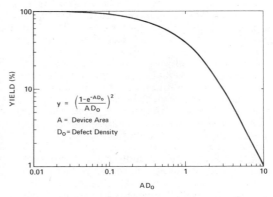

FIG. 6. Yield as predicted by Murphy's law[3] as a function of the product of the susceptible area and the defect density.

TABLE 4. TRANSISTOR FAILURE MODES.

FAILURE MODE	CURE
SURFACE INVERSION DUE TO SODIUM DRIFT IN SiO_2	1. CLEAN PROCESSING 2. PHOSPHOROUS DOPED SiO_2 BARRIER
ELECTROMIGRATION FAILURE OF METALLIZATION	1. FOR Al METALLIZATION KEEP $J_{max} < 5 \cdot 10^5$ amp/cm^2 2. ADD Si AND/OR Cu TO Al 3. USE HIGH ATOMIC NUMBER METAL SUCH AS Au, W
PENETRATION OF METAL THROUGH CONTACTS	1. USE BARRIER LAYER SUCH AS Ti, TiSi, PtSi 2. ADD Si TO Al
CORROSION FAILURE OF Al METALLIZATION AND Al BOND WIRES	1. USE Au INSTEAD OF Al 2. CONTROL COMPOSITION OF PHOSPHOROUS DOPED SiO_2 3. CONTROL PURITY OF EPOXY
FAILURE OF Au BOND WIRES DUE TO BRITTLE ALLOY FORMATION	CONTROL RATIO Au TO AVAILABLE Si
CRACKED DIE, WEAK BONDS, etc.	BETTER INSPECTION
THERMAL FATIGUE FAILURE OF DIE ATTACH	USE HARD ALLOY DIE ATTACH i.e. Si - Au
MELTING, OTHER THERMAL FAILURES	BETTER CONTROL OF DIE ATTACH, ELIMINATION OF VOIDS
BLOWN GATES IN MOS AND GaAs FETs	AVOID STATIC DISCHARGE IN DEVICE

method for discrete devices was developed at Bell Laboratories, i.e., the self-encapsulated beam lead transistor. This has attained extremely high reliability, but unfortunately, is applicable only to a limited number of leads and does not meet the needs of modern integrated circuits. Other batch interconnect processes are receiving increased usage.

The ultimate complexity possible in semiconductor systems is limited by the reliability of the components. Great strides have been made in this area, particularly through the efforts of the scientists at Bell Telephone Laboratories and through the impetus of the military in such systems as Minuteman. Under severe operating conditions, failure rates less than 0.01% per 1000 hours are commonly achieved. Fortunately, these numbers tend to refer to packages rather than individual transistors in an integrated circuit. A rule of thumb is that the failure rate increases linearly with the number of leads to the chip, not with the number of devices in the chip. Thus complex integrated circuits have shown good reliability. Common causes of failure are given in Table 4.

The semiconductor industry has become a major one, producing billions of parts per year, and is a major contributor to the economy, particularly in California and in the Northeastern United States. Present shipments in the United States amount to six billion dollars per year, with a similar volume produced for intra-company use by captive suppliers. At present, a battle is being waged between the United States and Japan for supremacy in semiconductors.

CHARLES A. BITTMANN

References

1. Shockley, W., "The Theory of *P-N* Junctions in Semiconductors and *P-N* Junction Transistors," *Bell System Tech. J.* 28, 435–489 (July 1949).
2. Bittmann, C. A., et al., "Technology for the Design of Low-Power Circuits," *IEEE J. of Solid State Circuits* SC5(1), 29–37 (February 1970).
3. Murphy, B. T., "Cost Size Optima of Monolithic Integrated Circuits," *Proceed. IEEE*, 1537–1545 (December 1964).

General References

Jonscher, A. K., "Principles of Semiconductor Device Operation," New York, John Wiley, 1960.
Lindmayer, J., and Wrigley, C. Y., "Semiconductor Devices," New York, D. Van Nostrand Co., 1965.
Grove, A. S., "Physics and Technology of Semiconductor Devices," New York, John Wiley, 1967.
Muller, R. S., and Kamins, T. I., "Device Electronics for Integrated Circuits," New York, John Wiley, 1977.

Cross-references: DIODE (SEMICONDUCTOR), ELECTRON TUBES, RECTIFIER, SEMICONDUCTOR, SEMICONDUCTOR DEVICES, SOLID STATE PHYSICS.

TRANSPORT THEORY (RADIATIVE TRANSFER)*

When light passes through an atmosphere, it may be scattered and absorbed. When neutrons move through a medium, collisions with the nuclei of the material may result in absorption of the neutrons; changes in their direction and energy; and sometimes, by the fission mechanism, production of more neutrons.

These two phenomena are examples of transport processes. The problem of specifying the radiation field in an atmosphere stems back to Rayleigh's investigations on the illumination of a sunlit sky. The astrophysicists refer to this general subject as *radiative transfer* and have studied it for well over half a century. Interest in neutron transport has perforce been of more recent origin.

If light is thought of as consisting of photons, then there is seen to be a very strong similarity between neutron transport and radiative trans-

*This work was supported in part by the United States Department of Energy and by the National Science Foundation. Reproduction in whole or in part is permitted for any purpose of the U.S. Government.

fer. The former process is both complicated and made more interesting by the possibility of fission in various materials. Other physical phenomena, such as the passage of γ-rays through a medium, possess many characteristics of the two processes that have been mentioned.

Fortunately, it is possible to develop a mathematical structure which encompasses all these phenomena. The situation is similar to that in classical diffusion theory, where the same mathematical equations may be interpreted, for example, to yield information concerning the distribution of heat in a metal or the flow of one material into another. The equations describing particle transport are, however, of a much more complicated nature than many of those of classical physics and are only now beginning to yield to the techniques of the mathematicians and theoretical physicists.

All the transport processes described above have the property that the moving particles involved may be thought of as interacting or colliding with fixed centers or nuclei of the material through which they are passing. The moving particles do not collide with each other. The interactions, in the situations which we will consider, are strictly independent and local events—a particle is affected only by a scattering center in its immediate neighborhood. Most important is the fact that a probability of such an interaction may be assigned:

Probability of interaction in moving a distance Δ

$\quad = \Sigma\Delta +$ (terms of higher order in Δ). (1)

The quantity Σ, called the *macroscopic cross section*, is dependent upon the kind and density of the medium, the type of moving particle (photon, neutron, etc.), the particle energy, etc. Determination of Σ is a complicated problem of experimental and theoretical physics.

An interaction or collision may result in a change in the direction of the moving particle (scattering), disappearance of the particle (absorption), or production of new particles of the same kind (fission). Scattering may produce no energy loss (elastic) or it may involve energy loss (inelastic). Such interactions may affect the transport medium, though for most purposes this change may be neglected. Again, the actual physical determination of the result of a collision is frequently a difficult task. Such information, together with the quantity Σ, must be considered for our purposes as already known.[14]

The central problem of transport theory, as we view the subject, may now be formulated. Given a physical medium with all parameters specified, such as the macroscopic cross section and the results of interactions, let a population of particles of given kind, direction, energy, etc., be present in the material at initial time $t = 0$. Let any internal or external sources of such particles be given. Describe as a function of position, direction, energy, etc., the expected particle population at any time $t > 0$.

The transport process clearly is probabilistic in nature. The resulting stochastic considerations can be very important for situations in which the total particle population is relatively small.[17] However, in most situations of interest these populations are sufficiently large so that stochastic variations are quite negligible. Thus it is customary to study expected value theory. Often the term "expected" is dropped in discussion, so that "expected flux" becomes simply "flux," etc. It is important to remember that this is done only as a matter of convenience in writing.

From Eq. (1) further information concerning the probability of collision may be obtained. Write

$p(x) =$ Probability that a particle moves a

$\quad\quad$ distance x without collision. (2)

Then,

$$p(x + \Delta) = (1 - \Sigma\Delta)p(x)$$
$$+ \text{(terms of higher order in } \Delta\text{)}.$$
$$(3)$$

Equation (3) gives

$$\frac{dp}{dx} = -\Sigma p(x)$$

and, if Σ is constant,

$$p(x) = e^{-\Sigma x}. \quad (4)$$

This is the well-known exponential law.

To find the average distance a particle moves without collision observe that

$F(x) =$ probability that the *first* collision occurs

$\quad\quad$ at some $X \leqslant x = 1 - e^{-\Sigma x}$. (5)

Then the average or expected value of X is

$$E(X) = \int_0^\infty x\, dF(x) = \int_0^\infty \Sigma x e^{-\Sigma x}\, dx = 1/\Sigma.$$
$$(6)$$

This average distance between collisions is referred to as the *mean free path*, usually denoted λ.

Suppose the half-space $x > 0$ is filled with a medium characterized by constant macroscopic cross section Σ. If a flux of N_0 particles impinges perpendicularly on $x = 0$, then the expected flux at $x = d$ of particles *that have made no collision* is, from Eq. (4),

$$N(d) = N_0 e^{-\Sigma d}. \quad (7)$$

When the collision process involves only absorption, $N(d)$ is the *total* expected flux at $x = d$, but if scattering and fission processes occur, $N(d)$ may be quite different from the total flux. These physical events greatly complicate transport theory.

Some idea of the complexities introduced by the fission process may be obtained by study of a very simple mathematical model in which particles are allowed to move only to the right or left in a rod (i.e., a line segment) of length a. Suppose that in an interaction the colliding particle disappears and two new ones emerge, one moving left and one moving right (binary fission). All particles have the same speed c, and Σ is constant. Finally, assume that the process is such that the average particle density is the same at one time as it is at any other, so that time dependence may be neglected. Denote by $cu(x)$ and $cv(x)$ the right and left fluxes at x.

Then[18]

$$u(x + \Delta) = (1 - \Sigma\Delta)u(x) + \Sigma\Delta u(x) + \Sigma\Delta v(x)$$

$$+ \text{(higher order terms in } \Delta), \qquad (8)$$

which leads to

$$du/dx = \Sigma v. \qquad (9)$$

Similarly,

$$-dv/dx = \Sigma u. \qquad (10)$$

If one left-moving particle per second is introduced at $x = a$, with no source at $x = 0$, then

$$u(0) = 0, \qquad cv(a) = 1, \qquad (11)$$

and the system of Eqs. (9)–(11) yields

$$cu(x) = \frac{\sin \Sigma x}{\cos \Sigma a}, \quad cv(x) = \frac{\cos \Sigma x}{\cos \Sigma a}, \quad 0 \leqslant x \leqslant a.$$

$$(12)$$

These results are obviously quite different from any that would be given by a simple attenuation law, such as Eq. (7). Indeed, because of the fission assumed, a collision results in an increase, rather than a decrease, in the particle population. The case $a = \pi/2\Sigma$ is of especial interest. For a rod of that length neither $u(x)$ nor $v(x)$ is defined. Physically, the system is just *critical*. A time-independent population cannot prevail with the source specified. Equations (9) and (10) no longer hold when $a \geqslant \pi/2\Sigma$.

This observation does not imply that supercritical systems cannot be analyzed. To so do requires explicit introduction of the time variable.[18] Equations (9) and (10) are replaced by

$$\frac{1}{c} \frac{\partial u}{\partial t} + \frac{\partial u}{\partial x} = \Sigma v(x, t) \qquad (13)$$

and

$$\frac{1}{c} \frac{\partial v}{\partial t} - \frac{\partial v}{\partial x} = \Sigma u(x, t) \qquad (14)$$

with suitable boundary and initial conditions. For large times it may be shown that the solutions to these equations can be written *approximately* as

$$u(x, t) \doteq u_0(x)e^{\alpha t}, \qquad v(x, t) \doteq v_0(x)e^{\alpha t}$$

$$(15)$$

for some α. When $a > \pi/2\Sigma$, α is positive so that the particle fluxes build up exponentially in time. This exponential increase is observed in actual experiments involving fissionable materials. When the particle population in a system is just sustained without introduction of additional particles (sources), the system is just critical (see CRITICAL MASS).

It is possible to consider other relatively simple mathematical models of transport phenomena and from them to determine much valuable information. Any attempt to solve a "realistic" physical problem, however, usually results in great difficulties. The general *transport equation* is of the form

$$\frac{DN}{Dt} (\mathbf{r}, E, \mathbf{\Omega}, t) = -\Sigma(\mathbf{r}, E, \mathbf{\Omega}, t)N(\mathbf{r}, E, \mathbf{\Omega}, t)$$

$$+ \int_{E', \mathbf{\Omega}'} K(\mathbf{r}; E', \mathbf{\Omega}' \to E, \mathbf{\Omega}; t)\Sigma(\mathbf{r}, E', \mathbf{\Omega}', t)$$

$$N(\mathbf{r}, E', \mathbf{\Omega}', t) dE' d\mathbf{\Omega}' + S(\mathbf{r}, E, \mathbf{\Omega}, t) \qquad (16)$$

subject to boundary and initial conditions dependent upon the geometry. In Eq. (16), D/Dt is the total time derivative, E is the energy of a particle, $\mathbf{\Omega}$ is a vector of unit length along the direction of motion of a particle, and K is a function that gives the density of particles emerging at energy E and in the direction $\mathbf{\Omega}$ from a collision at location $\mathbf{r}$ and time t involving a particle having energy E' and moving in the direction $\mathbf{\Omega}'$. Sources are represented by S.

It is clear that Eq. (16), while linear, has a much different structure from the classical equations of mathematical physics. This structural difference results in the necessity to use relatively unusual mathematical techniques in the analysis of Eq. (16). For example, it is clear that physically meaningful solutions of the transport equation must be nonnegative. This *positivity property* plays a key role in the analysis of the transport equation.[2] On the other hand, the transport operator does *not* have the properties of being symmetric or positive definite, both of which are widely used in analyzing the classical equations of mathematical physics.

The complexity of the general transport equation, coupled in some cases with the urgencies of designing nuclear weapons, fission and fusion reactors, etc., has resulted in a plethora of approximate methods for its treatment.[1,5,7,8,14] One class of such methods, which was developed relatively early and has been extensively applied, is based upon modifying the exponential attenuation formula Eq. (7) by multiplicative factors intended to account for contributions from collided particles. These modifying factors may be determined either computationally or experimentally. This approach has perhaps achieved its highest form in the "Standard Method" for calculation of the shielding effect of structures against fallout gamma rays.[13]

For situations in which scattering interactions occur much more often than absorptions, the transport equation can be approximated by the neutron diffusion equation.[7] The latter is identical in structure to the heat equation, allowing classical methods of solution to be used in many cases. Neutron diffusion theory often gives surprisingly good results, particularly for systems that are sufficiently large so that boundary effects are quite negligible. Variants of simple neutron diffusion theory are numerous. They include age theory and the continuous slowing-down approximation.

The advent of modern high-speed electronic computers has made it a widespread practice to generate approximations to the solution of Eq. (16) based upon replacing the continuous independent variables by discrete counterparts. Different such discretizations can be used for the four independent variables t, E, $\mathbf{\Omega}$, and $\mathbf{r}$, and these may be combined in many ways. The time variation is almost always approximated by a finite-difference method.

The multigroup method, in which all particles in several contiguous energy ranges are combined, is the most commonly used technique for discretely representing particle energies. This approach has been widely and successfully applied in practice over the past two decades, although the underlying mathematical theory has only very recently begun to be developed.[9] The multigroup method also has been successfully combined with diffusion theory. Other techniques have been developed for discretizing energy, such as the overlapping group method and the multiband approximation. The search for alternatives is largely motivated by situations involving wide variations in cross sections over small energy ranges (e.g., cross-section resonances) because direct application of the multigroup method can be prohibitively time consuming in such circumstances.

The directional variable $\mathbf{\Omega}$ in Eq. (16) can be discretely approximated in various ways. The discrete-ordinates method arises from replacing the directional integral in Eq. (16) by a numerical quadrature. In the closely related S_N method,

the directional dependence of the flux is assumed to be piecewise linear. Yet another approach is based upon using a truncated expansion in spherical harmonics to represent the directional dependence of the flux. The latter is theoretically equivalent to an appropriately defined discrete-ordinates approximation.[1] If the expansion in spherical harmonics is truncated at lowest order, then the neutron diffusion equation results.

Finite difference methods are widely used to provide a discrete representation of the spatial variable $\mathbf{r}$. In the past decade it has become increasingly popular to use finite-element methods for this purpose. The recently developed coarse-mesh nodal methods[6] seem particularly promising in terms of computational efficiency. Discretization of the spatial variable also can be accomplished by techniques such as invariant imbedding[18] or the related method of response matrices.[16] In these approaches attention is focused on determination of particles emergent from a system as a function of the incident particles. Such a viewpoint is quite natural in astrophysics and shielding, which accounts for the rather extensive use of invariant imbedding in these fields.[4,10]

Another popular attack on transport problems is through the Monte Carlo method.[12] There, the particle motion is actually simulated by machine computation. Particles suffer collisions, change direction, lose energy, etc., according to specified probabilities. In principle the particle energy and direction can be retained as continuous variables, but in practice these often are discretized. In the simplest verions of the method, the computer traces the history of a single particle from its appearance in the system until it leaves or is absorbed. This is inefficient for problems in which the primary interest is upon behavior of a relatively small fraction of the total particle population, such as transmission through a thick shield. Accordingly, various biasing techniques have been developed with the objective of reducing computational effort spent upon particles likely to be of little ultimate interest. Events in a particle's history are allowed to happen randomly, the computer deciding on the event according to a preassigned (and presumably physically correct) probability distribution. This method is hence stochastic in nature, although expected values are usually taken after a sufficiently large particle population has been examined. The Monte Carlo technique is often used when the geometry of the problem is complicated.

Another widely used approach to the transport equation is the method of singular eigenfunctions.[3] This method originally was used to obtain semianalytic solutions, but in this framework it seems applicable only in relatively simplified circumstances. In recent years this method and related techniques, such as the F_N method, have been increasingly applied numerically.[11] Singular eigenfunctions originally arose in plasma physics. In the past decade there has been increasing applications of this and other transport-theoretic methods to the transport of charged particles.

It should be pointed out that no theoretical or computational scheme can possibly give satisfactory results without good knowledge, either experimental or theoretical, of the underlying physical parameters Σ and K. Satisfactory determination of these quantities often involves considerable effort and expense. In response to this, substantial recent effort has been directed toward determination of the sensitivity of the solution of the transport equation to errors in these parameters.[15] The underlying idea is to focus efforts in computing or measuring these quantities upon the most crucial materials and energies.

Transport theory is approaching maturity as a subject. Much has been accomplished within the past two decades, but many very challenging and important problems remain, both from the viewpoint of the physicist and from that of the mathematician.

G. Milton Wing
Paul Nelson

References

1. Bell, G. I., and Glasstone, S., "Nuclear Reactor Theory," New York, Van Nostrand Reinhold Co., 1970.
2. Birkhoff, G., "Reactor Criticality in Transport Theory," *Proc. Nat. Acad. Sci.* **45**, 567–569 (1959).
3. Case, K. M., and Zweifel, P. F., "Linear Transport Theory," Reading, Mass., Addison-Wesley Publishing Co., 1967.
4. Chandrasekhar, S., "Radiative Transfer," Oxford, Clarendon Press, 1950.
5. Davison, B., "Neutron Transport Theory," Oxford, Clarendon Press, 1957.
6. Dorning, J. J., "Modern Coarse-Mesh Methods– A Development of the 70's," *Proc. Computational Methods in Nuclear Engineering*, pp. 3-1–3-32, Am. Nuclear Soc., LaGrange Park, Illinois (1979).
7. Duderstadt, J. J., and Martin, W. R., "Transport Theory," New York, John Wiley & Sons, 1979.
8. Kourganoff, V., and Busbridge, I. W., "Basic Methods in Transfer Problems," Oxford, Clarendon Press, 1952.
9. Nelson, P., and Victory, H. D., Jr., "On the Convergence of the Multigroup Approximations for Submultiplying Slab Media," *Mathematical Methods in the Applied Sciences* **4**, 206–229 (1982).
10. Shimizu, A., and Aoki, K., "Application of Invariant Embedding to Reactor Physics," New York, Academic Press, 1972.
11. Siewert, C. E., and Benoist, P., "The F_N Method in Neutron-Transport Theory, Part I: Theory and Applications," *Nuclear Science and Engineering* **69**, 156–160 (1979).
12. Spanier, J., and Gelbard, E. M., "Monte Carlo Principles and Neutron Transport Problems," Reading, Mass., Addison-Wesley, 1969.
13. Spencer, L. V., Chilton, A. B., and Eisenhauer, C. M., "Structure Shielding Against Fallout Gamma-Rays from Nuclear Detonations," National Bureau of Standards Special Pub. 570, Washington, D.C., U.S. Government Printing Office, 1980.
14. Weinberg, A. M., and Wigner, E. P., "The Physical Theory of Neutron Chain Reactors," Chicago, Univ. Chicago Press, 1958.
15. Weisbin, C. R., Peele, R. W., Marable, J. H. Collins, P., Kujawski, E., Greenspan, E., and de Saussure, G., "Sensitivity and Uncertainty Analysis of Reactor Performance Parameters," New York, Plenum Press, 1982.
16. Weiss, Z., "Some Basic Properties of the Response Matrix Equations," *Nuclear Science and Engineering* **63**, 457–492 (1977).
17. Williams, M. M. R., "Random Processes in Nuclear Reactors," Oxford, Pergamon Press, 1974.
18. Wing, G. M., "An Introduction to Transport Theory," New York, John Wiley & Sons, 1962.

Cross-references: COLLISIONS OF PARTICLES, CRITICAL MASS, CROSS SECTIONS AND STOPPING POWER, FISSION, HEAT TRANSFER, NUCLEAR RADIATION SHIELDING, NUCLEAR REACTOR.

TRANSURANIUM ELEMENTS

The transuranium elements are those elements heavier than uranium, element 92. They are all radioactive and, in general, have half-lives too short to have existed in nature since their original creation. They were all discovered and produced by nuclear synthesis. The presently known (1982) transuranium elements have the following names and symbols: 93, neptunium (Np); 94, plutonium (Pu); 95, americium (Am); 96, curium (Cm); 97, berkelium (Bk); 98, californium (Cf); 99, einsteinium (Es); 100, fermium (Fm); 101, mendelevium (Md); 102, nobelium (No); 103, lawrencium (Lr); 104, rutherfordium (Rf); 105, hahnium (Ha); 106, not named; 107, not named.

The heaviest elements of the periodic system with atomic numbers 89 (actinium) through 103 (lawrencium) are members of the actinide series, analogous to the lanthanide series or rare earths (atomic numbers 57 through 71). An inner electron shell, consisting of fourteen $5f$ electrons, is filled in progressing across the series. The electronic configurations of the actinide transuranium elements are given in Table 1.

Since the actinide series is completed with lawrencium, element 103, the succeeding "transactinide" elements with atomic numbers 104 through 108 should be chemical homologs to the known elements with atomic numbers 72 (hafnium) to 76 (osmium). This analogy should continue in the still heavier transuranium ele-

TABLE 1. ELECTRONIC CONFIGURATIONS
(BEYOND RADON) FOR GASEOUS ATOMS OF
ACTINIDE TRANSURANIUM ELEMENTS

Atomic No.	Element	Electronic Configuration[a]
93	Neptunium	$5f^4 6d 7s^2$
94	Plutonium	$5f^6 7s^2$
95	Americium	$5f^7 7s^2$
96	Curium	$5f^7 6d 7s^2$
97	Berkelium	$5f^9 7s^2$
98	Californium	$5f^{10} 7s^2$
99	Einsteinium	$5f^{11} 7s^2$
100	Fermium	$5f^{12} 7s^2$
101	Mendelevium	$(5f^{13} 7s^2)$
102	Nobelium	$(5f^{14} 7s^2)$
103	Lawrencium	$(5f^{14} 6d 7s^2$ or $5f^{14} 7s^2 7p)$

[a]Configurations in parentheses have not been determined experimentally.

ments, and element 118 should be a noble gas. Some typical predicted electronic configurations for transactinide elements are given in Table 2.

Chemically the transuranium elements which are members of the actinide series are very similar, although the observed differences are those expected and anticipated from their unique position in the periodic system as part of a second rare-earth series. All have trivalent ions, which form inorganic complex ions and organic chelates. Also in common are acid-insoluble trifluorides and oxalates, soluble sulfates, nitrates, chlorides, and perchlorates. Neptunium, plutonium, and americium have several higher oxidation states in aqueous solution (similar to uranium), but the stability of these states relative to the common trivalent ion becomes progressively less as one proceeds to the higher atomic numbers. This is a direct consequence,

indeed an identifying feature, of the actinide role as a second rare-earth type transition series.

One of the most important methods for study and elucidation of chemical behavior of the actinide elements has been ion-exchange chromatography. Adsorption on and elution from ion-exchange columns have made possible the identification and separation of trace quantities of all of the actinides. The behavior of each actinide transuranium element in this respect is very similar to its analogous rare-earth element. This has made it possible to detect as little as one or two atoms when this small a number has been made in some of the transmutation experiments.

Each of the transuranium elements (except for the last two) has a number of isotopes, some of which can be obtained in isotopically pure form. They are produced by neutron- or charged-particle-induced transmutation. The first eight have an average of about 15 isotopes (including isomers) apiece. The total number of known nuclides for the 15 elements is about 200, when the approximately 30 short-lived, so-called "fission isomers" are included. The isotopes available in weighable quantity, and hence suitable for investigation, are listed in Table 3.

The concept of atomic weight in the sense applied to naturally occurring elements is not applicable to the transuranium elements, since the isotopic composition of any given sample depends on its source. In most cases, the use of the mass number of the longest-lived isotope in combination with an evaluation of its availability has been adequate. Good choices at present are neptunium, 237; plutonium, 242; americium, 243; curium, 248; berkelium, 249; californium, 249; einsteinium, 254; and fermium, 257.

Brief descriptions of the transuranium elements follow:

Neptunium (Np, atomic number 93, after the planet Neptune.) Neptunium was the first of the synthetic transuranium elements to be

TABLE 2. PREDICTED ELECTRONIC CONFIGURATIONS
(BEYOND RADON) FOR GASEOUS ATOMS OF
SOME TRANSACTINIDE ELEMENTS
(*Undiscovered elements in parentheses*)

Atomic No.	Element	Electronic Configuration
104	Rutherfordium	$5f^{14} 6d^2 7s^2$
105	Hahnium	$5f^{14} 6d^3 7s^2$
106	Unnamed	$5f^{14} 6d^4 7s^2$
107	Unnamed	$5f^{14} 6d^5 7s^2$
(108)		$5f^{14} 6d^6 7s^2$
(112)		$5f^{14} 6d^{10} 7s^2$
(114)		$5f^{14} 6d^{10} 7s^2 7p^2$
(118)		$5f^{14} 6d^{10} 7s^2 7p^6$
(120)		$5f^{14} 6d^{10} 7s^2 7p^6 8s^2$
(126)		$5f^{14} 5g^2 6d^{10} 6f^2 7s^2 7p 7d^6 8s^2 8p$

TABLE 3. LONG-LIVED TRANSURANIUM ISOTOPES
AVAILABLE FOR INVESTIGATION.

Nuclide	$t_{1/2}$	Isotopic Composition (%)	Amounts Available
^{237}Np	2.14×10^6 y	100	kg
^{238}Pu	87.74 y		10–100 g
^{239}Pu	2.41×10^4 y	99.7	kg
^{240}Pu	6.57×10^3 y	98.3	10–50 g
^{241}Pu	14.4 y	93.4	1–10 g
^{242}Pu	3.76×10^5 y	99.9	100 g
^{244}Pu	8.1×10^7 y	88.6	10–100 mg
^{241}Am	433 y	100	kg
^{243}Am	7.37×10^3 y		10–100 g
^{242}Cm	162.8 d	100	100 g
^{243}Cm	28.5 y		10–100 mg
^{244}Cm	18.11 y	>95%	10–100 g
^{248}Cm	3.5×10^5 y	97	10–100 mg
^{249}Bk	0.88 y	100	10–50 mg
^{249}Cf	351 y	100	1–10 mg
^{250}Cf	13.1 y		10 mg
^{252}Cf	2.64 y		10–1000 mg
^{254}Cf	60.5 d		μg
^{253}Es	20.47 d		1–10 mg
^{254}Es	276 d	100	5–10 μg
^{257}Fm	100.5 d		1 pg

discovered; the isotope ^{239}Np was produced by McMillan and Abelson in 1940 at Berkeley, California, as the result of the bombardment of uranium with cyclotron-produced neutrons. The isotope ^{237}Np (half-life 2.14×10^6 years) is currently obtained in kilogram quantities as a by-product of nuclear power reactions. Trace quantities of the element are actually found in nature owing to transmutation reactions in uranium ores produced by the neutrons which are present.

Neptunium metal has a silvery appearance, is chemically reactive, melts at $637°$C, and exists in at least three structural modifications: α-neptunium, orthorhombic, density $= 20.45$ g/cm^3; β-neptunium (above $280°$C), tetragonal, density $(313°$C$) = 19.36$ g/cm^3; γ-neptunium (above $577°$C), cubic, density $(600°$C$) = 18.0$ g/cm^3.

Neptunium gives rise to five ionic oxidation states in solution: Np^{3+} (pale purple), analogous to the rare earth ion Pm^{3+}, Np^{4+} (yellow-green), NpO$_2^+$ (green-blue), NpO$_2^{2+}$ (pale pink), and NpO$_5^{3-}$ (green). These oxygenated species are in contrast to the rare earths which exhibit only simple ions of the (II), (III), and (IV) oxidation states in aqueous solution. The element forms tri- and tetrahalides such as NpF$_3$, NpF$_4$, NpCl$_3$, NpCl$_4$, NpBr$_3$, NpI$_3$, and oxides of various compositions such as are found in the uranium-oxygen system, including Np$_3$O$_8$ and NpO$_2$.

Plutonium (Pu, atomic number 94, after the planet Pluto.) Plutonium was the second transuranium element to be discovered; the isotope ^{238}Pu was produced in 1940 by Seaborg, McMillan, Kennedy, and Wahl at Berkeley, California, by deuteron bombardment of uranium in the 150-cm cyclotron. By far of greatest importance is the isotope ^{239}Pu (half-life 24 400 years), which is fissionable with thermal neutrons and produced in extensive quantities in nuclear reactors from the abundant nonfissionable uranium isotope ^{238}U:

$$^{238}\text{U}(n,\gamma)\,^{239}\text{U} \xrightarrow{\beta^-} \,^{239}\text{Np} \xrightarrow{\beta^-} \,^{239}\text{Pu}$$

Plutonium (in the form of ^{239}Pu) has assumed the position of dominant importance among the transuranium elements because of its successful use as an explosive ingredient in nuclear weapons and the place which it holds as a key material in the development of industrial utilization of nuclear energy, one pound being equivalent to about 10 000 000 kWh of heat-energy equivalent. In certain nuclear reactors called breeder reactors, it is possible to create more new plutonium from ^{238}U than plutonium consumed in sustaining the fission chain reaction. Because of this, plutonium is the key to unlocking the enormous energy reserves in the nonfissionable isotope ^{238}U.

Plutonium (in the form of ^{239}Pu) also exists

in trace quantities in naturally occurring uranium ores. It is formed in much the same manner as neptunium, by irradiation of natural uranium with the neutrons which are present. Much smaller quantities of the longer-lived isotope, ^{244}Pu, (half-life 83 000 000 years), have been found in nature; in this case it may represent the small fraction remaining from a primordial source or it may be due to cosmic rays.

Much of the early work on the determination of the chemical and physical properties of plutonium has been done employing the isotope ^{239}Pu. However, the relatively high specific alpha particle radioactivity of this isotope leads to difficulties caused by radiation damage and self-heating of the material under investigation. Hence, the longer-lived isotope ^{242}Pu (half-life 390 000 years) is better suited for such investigations; this isotope is rather readily available in high isotopic purity as the result of successive neutron capture reactions when plutonium is irradiated over sufficiently long periods of time in very high neutron flux reactors:

$$^{239}\text{Pu} (n,\gamma) \; ^{240}\text{Pu} (n, \gamma) \; ^{241}\text{Pu} (n, \gamma) \; ^{242}\text{Pu}$$

The still longer-lived ^{244}Pu is even better suited for such investigations, but the difficulty of producing it limits its availability; in this case the neutron capture reactions must continue for two additional steps:

$$^{242}\text{Pu} (n, \gamma) \; ^{243}\text{Pu} (n, \gamma) \; ^{244}\text{Pu}$$

and the yield is drastically reduced owing to the short half-life of ^{243}Pu (4.96 hours).

Plutonium metal can be prepared, in common with neptunium and uranium, by reduction of the trifluoride with alkaline-earth metals. The metal has a silvery appearance, is chemically reactive, and melts at 640°C. It exhibits six crystalline modifications: α-plutonium, primitive monoclinic, below 115°C; β-plutonium, monoclinic, 115 to 185°C; γ-plutonium, face-centered orthorhombic, 185 to 310°C; δ-plutonium, face-centered cubic, 310 to 452°C; δ'-plutonium, tetragonal, 452 to 480°C; and ε-plutonium, body-centered cubic, 480°C up to the melting point.

Plutonium also exhibits five ionic valence states in aqueous solutions: Pu^{3+} (blue-lavender), Pu^{4+} (yellow-brown), PuO_2^+ (pink), PuO_2^{2+} (pink-orange), and PuO_5^{3-} (blue-green). The ion PuO_2^+ is unstable in aqueous solutions, disproportionating into Pu^{4+} and PuO_2^{2+}; the Pu^{4+} thus formed, however, oxidizes the PuO_2^+ into PuO_2^{2+}, itself being reduced to Pu^{3+}, giving finally Pu^{3+} and PuO_2^{2+}.

Plutonium forms binary compounds with oxygen: PuO, PuO_2, and intermediate oxides of variable composition; with the halides: PuF_3, PuF_4, $PuCl_3$, $PuBr_3$, PuI_3; with carbon, nitrogen, and silicon: PuC, PuN, $PuSi_2$; in addition oxyhalides are well known: PuOCl, PuOBr, PuOI.

Because of the high rate of emission of alpha particles, and the physiological fact that the element is specifically absorbed by bone marrow, plutonium, like all of the transuranium elements, is a radiological poison and must be handled with special equipment and precautions.

Americium (Am, atomic number 95, after the Americas.) Americium was the fourth transuranium element to be discovered; the isotope ^{241}Am was identified by Seaborg, James, Morgan, and Ghiorso late in 1944 at the wartime Metallurgical Laboratory (now the Argonne National Laboratory) of the University of Chicago as the result of successive neutron capture reactions by plutonium isotopes in a nuclear reactor:

$$^{239}\text{Pu} (n, \gamma) \; ^{240}\text{Pu} (n, \gamma) \; ^{241}\text{Pu} \xrightarrow{\beta^-} \; ^{241}\text{Am}$$

Americium is produced in kilogram quantities. Since the isotope ^{241}Am can be prepared in relatively pure form by extraction as a decay product over a period of years from plutonium containing ^{241}Pu, this isotope has been used for much of the chemical investigation of this element. Better suited is the isotope ^{243}Am owing to its longer half-life (7.4×10^3 years as compared to 433 years for ^{241}Am). A mixture of the isotopes ^{241}Am, ^{242}Am, and ^{243}Am can be prepared by intense neutron irradiation of ^{241}Am according to the reactions ^{241}Am (n, γ) ^{242}Am (n, γ) ^{243}Am. Nearly isotopically pure ^{243}Am can be prepared by the reactions ^{242}Pu (n, γ) ^{243}Pu $\xrightarrow{\beta^-}$ ^{243}Am, and the ^{243}Am can be chemically separated.

Americium can be obtained as a silvery white reactive metal by reduction of americium trifluoride with barium vapor at 1300°C. It appears to be more malleable than uranium or neptunium and tarnishes slowly in dry air at room temperature. The density is 13.67 g/cm^3 with a melting point at 1176°C. It has a double hexagonal-close-packed crystalline structure at temperatures up to 1079°C (α-americium) and a face-centered cubic structure above 1079°C (β-americium).

The element exists in four oxidation states in aqueous solution: Am^{3+} (light salmon), AmO_2^+ (light tan), AmO_2^{2+} (light tan), and a fluoride complex of the 4+ state (pink). The trivalent state is highly stable and difficult to oxidize. AmO_2^+, like plutonium, is unstable with respect to disproportionation into Am^{3+} and AmO_2^{2+}. The ion Am^{4+} may be stabilized in solution only in the presence of very high concentrations of fluoride ion, and tetravalent solid compounds are well known. Divalent americium has been prepared in solid compounds; this is consistent with the presence of seven $5f$ electrons in americium (enhanced stability of half-filled

5f electron shell) and is similar to the analogous lanthanide, europium, which can be reduced to the divalent state.

Americium dioxide, AmO_2, is the important oxide; Am_2O_3 and, as with previous actinide elements, oxides of variable composition between $AmO_{1.5}$ and AmO_2 are known. The halides AmF_2 (in CaF_2), AmF_3, AmF_4, $AmCl_2$, $AmCl_3$, $AmBr_3$, AmI_2, and AmI_3 have also been prepared.

Curium (Cm, atomic number 96, after Pierre and Marie Curie.) Although curium comes after americium in the periodic system, it was actually known before americium and was the third transuranium element to be discovered. It was identified by Seaborg, James, and Ghiorso in the summer of 1944 at the wartime Metallurgical Laboratory in Chicago as a result of helium-ion bombardment of ^{239}Pu in the Berkeley, California, 150-cm cyclotron. It is of special interest because it is in this element that the first half of the transition series of actinide elements is completed.

The isotope ^{242}Cm (half-life 163 days) produced from ^{241}Am by the reactions ^{241}Am (n, γ) ^{242}Am $\xrightarrow{\beta^-}$ ^{242}Cm was used for much of the early work with macroscopic quantities, although this was difficult due to the extremely high specific alpha activity. A better, but still far from ideal, isotope for the investigation of curium is ^{244}Cm. Its somewhat longer half-life of 18 years still presents a problem of relatively high specific alpha activity, but it has been used extensively because of its availability as the result of production by the reactions ^{243}Am (n, γ) ^{244}Am $\xrightarrow{\beta^-}$ ^{244}Cm. Much better suited for such investigations are the longer-lived isotopes ^{247}Cm (half-life 16 000 000 years) and ^{248}Cm (half-life 350 000 years) which are becoming available in increasing quantities as the result of their production in high neutron-flux reactors through the successive capture of neutrons in the reactions ^{244}Cm (n, γ) ^{245}Cm (n, γ) ^{246}Cm (n, γ) ^{247}Cm (n, γ) ^{248}Cm; the difficulties of this long production chain are compounded by the necessity of separating the ^{247}Cm and ^{248}Cm from the remaining lower mass-number isotopes by mass spectrometric methods in order to obtain final products of the desired low specific alpha activity. ^{248}Cm is also produced in relatively high isotopic purity as the alpha-decay daughter of ^{252}Cf.

Curium metal resembles americium metal in crystal structure (α and β modifications) but melts at the considerably higher temperature of 1340°C. It can be prepared by heating curium trifluoride with barium vapor at 1350°C. The density is 13.51 at room temperature.

Curium exists solely as Cm^{3+} (colorless to yellow) in the uncomplexed state in aqueous solution. This behavior is related to its position as the element in the actinide series in which the 5f electron shell is half filled, i.e., it has the especially stable electronic configuration, $5f^7$, analogous to its lanthanide homolog, gadolinium. Similarly to americium (IV) a curium (IV) fluoride complex ion exists in aqueous solution. Solid compounds include Cm_2O_3, CmO_2 (and oxides of intermediate composition), CmF_3, CmF_4, $CmCl_3$, $CmBr_3$, and CmI_3.

Berkelium (Bk, atomic number 97, after Berkeley, California.) Berkelium, the eighth member of the actinide transition series, was discovered in December 1949, by Thompson, Ghiorso, and Seaborg and was the fifth transuranium element synthesized. It was produced by cyclotron bombardment of ^{241}Am with helium ions at Berkeley, California.

The only isotope available in weighable quantities for the study of the chemical and physical properties of berkelium is ^{249}Bk (half-life 314 days) produced in limited quantity by the neutron irradiation of somewhat rare ^{248}Cm by the reactions ^{248}Cm (n, γ) ^{249}Cm $\xrightarrow{\beta^-}$ ^{249}Bk. This isotope is difficult to work with because of its relatively high specific beta particle radioactivity and because of continuous self-contamination with its daughter ^{249}Cf radioactivity.

Berkelium metal can be prepared by the reduction of BkF_3 with lithium at 1025°C. It exists in a double hexagonal-close-packed α phase (presumably the low temperature phase) and a face-centered cubic β phase and melts at 986 ± 25°C. It has a density of 14.78 at room temperature.

Berkelium exhibits two ionic oxidation states in aqueous solution, Bk^{3+} (yellow-green) and somewhat unstable Bk^{4+} (yellow) as might be expected by analogy with its rare-earth homolog terbium. Solid compounds include Bk_2O_3, BkO_2 (and presumably oxides of intermediate composition), BkF_3, BkF_4, $BkCl_3$, $BkBr_3$, and BkI_3.

Californium (Cf, atomic number 98, after the state and University of California.) Californium, the sixth transuranium element to be discovered, was produced by Thompson, Street, Ghiorso, and Seaborg in January 1950, by helium-ion bombardment of microgram quantities of ^{242}Cm in the Berkeley 150-cm cyclotron.

The best isotope for the investigation of the chemical and physical properties of californium is ^{249}Cf (half-life 352 years), produced in pure form as the beta-particle decay product of ^{249}Bk, which is available in only limited quantity because its production from lighter isotopes requires multiple neutron capture over long periods of time in high neutron-flux reactors. Mixtures of californium isotopes produced by reactions such as ^{249}Bk (n, γ) ^{250}Bk $\xrightarrow{\beta^-}$ ^{250}Cf (n, γ) ^{251}Cf (n, γ) ^{252}Cf are also used but these have the disadvantage of high specific radioactivity, especially the spontaneous fission decay of ^{252}Cf.

Californium metal can be prepared by the reduction of CfF_3 with lithium. It is quite volatile and can be distilled at temperatures of the order of 1100 to 1200°C. It appears to exist in three crystalline modifications, hexagonal (α), face-centered cubic (β) and face-centered cubic (γ). It has a melting point of 900 ± 30°C. The density is 15.1 at room temperature.

Californium exists mainly as Cf^{3+} in aqueous solution (emerald green), but it is the first of the actinide elements in the second half of the series to exhibit the (II) state, which becomes progressively more stable on proceeding through the heavier members of the series. It also exhibits the (IV) oxidation state in CfF_4 and CfO_2 which can be prepared under somewhat intensive oxidizing conditions. Solid compounds also include Cf_2O_3 (and higher intermediate oxides), CfF_3, $CfCl_3$, $CfBr_2$, $CfBr_3$, CfI_2, and CfI_3.

Einsteinium (Es, atomic number 99, after Albert Einstein.) Einsteinium, the seventh transuranium element to be discovered, was identified by Ghiorso et al. in December 1952, in the debris from a thermonuclear explosion in work involving the University of California Radiation Laboratory, the Argonne National Laboratory, and the Los Alamos Scientific Laboratory. The isotope produced was the 20-day ^{253}Es, originating from beta decay of ^{253}U and daughters.

Einsteinium can be investigated with macroscopic quantities using the isotopes ^{253}Es (half-life 20.5 days), ^{254}Es (half-life 276 days), and ^{255}Es (half-life 38.3 days), whose production by the irradiation of lighter elements is severely limited because of the required long sequence of neutron capture reactions over long periods of time in high-neutron-flux reactors. Most of the investigations have used the short-lived ^{253}Es because of its greater availability, but the use of ^{254}Es will increase as it becomes more available. In any case the investigation of this element is very difficult due to the high specific radioactivity and small available quantities of the isotopes.

Einsteinium metal, which is quite volatile, can be prepared by the reduction of EsF_3 with lithium and has a face-centered cubic crystal structure. It melts at 860 ± 30°C.

Einsteinium exists in normal aqueous solution essentially as Es^{3+} (green) although Es^{2+} can be produced under strong reducing conditions. Solid compounds such as Es_2O_3, $EsCl_3$, $EsOCl$, $EsBr_3$, and EsI_3 have been made.

Fermium (Fm, atomic number 100, after Enrico Fermi.) Fermium, the eighth transuranium element to be discovered, was identified by Ghiorso et al. early in 1953 in the debris from a thermonuclear explosion in work involving the University of California Radiation Laboratory, the Argonne National Laboratory, and the Los Alamos Scientific Laboratory. The isotope produced was the 20.1-hour ^{255}Fm, originating from the beta decay of ^{255}U and daughters.

No isotope of fermium has yet been isolated in weighable amounts and thus all the investigations of this element have been done with tracer quantities. The longest-lived isotope is ^{257}Fm (half-life about 100 days) whose production in high-neutron-flux reactors is extremely limited because of the very long sequence of neutron-capture reactions that is required.

Despite its very limited availability, fermium, in the form of the 3.24-hour ^{254}Fm isotope, has been identified in the "metallic" zero-valent state in an atomic-beam magnetic resonance experiment. This established the electron structure of elemental fermium in the ground state as $5f^{12}7s^2$ (beyond the radon structure).

Fermium exists in normal aqueous solution almost exclusively as Fm^{3+} but strong reducing conditions can produce Fm^{2+} which has greater stability than Es^{2+} and less stability than Md^{2+}.

Mendelevium (Md, atomic number 101, after Dmitri Mendeleev.) Mendelevium, the ninth transuranium element to be discovered, was first identified by Ghiorso, Harvey, Choppin, Thompson, and Seaborg in early 1955 as a result of the bombardment of the isotope ^{253}Es with helium ions in the Berkeley 150-cm cyclotron. The isotope produced was ^{256}Md which decays by electron capture to ^{256}Fm, which in turn decays predominantly by spontaneous fission with a half-life of 2.6 hours. This first identification was notable in that only of the order of one to three atoms per experiment were produced. The extreme sensitivity for detection depended on the fact that its chemical properties could be accurately predicted as eka-thulium and there was a high sensitivity for detection because of the spontaneous fission decay.

All of the isotopes of mendelevium, which include the relatively long-lived ^{258}Md (half-life 56 days), are produced by the bombardment of lighter elements with charged particles provided by accelerators. The chemical properties have, of necessity, been determined solely by the use of the tracer method. Mendelevium has the ions Md^{3+} and Md^{2+} in aqueous solution with the former somewhat more stable than the latter.

Nobelium (No, atomic number 102, after Alfred Nobel.) The discovery of nobelium, the tenth transuranium element to be discovered, corresponds to a complicated history. For the first time scientists from countries other than the United States embarked on serious efforts to compete with the United States in this field. The reported discovery of element 102 in 1957 by an international group of scientists working at the Nobel Institute for Physics in Stockholm, who suggested the name nobelium, has never been confirmed and must be considered to be erroneous. Working at the Kurchatov Institute of Atomic Energy in Moscow, G. N. Flerov and co-workers in 1958 reported a radioactivity which they thought might be attributed to element 102 but a wide range of half-lives was suggested and no chemistry was performed. As the result of more definitive work performed in 1958, Ghiorso, Sikkeland, Walton, and Seaborg reported an isotope of the element produced by bombarding a mixture of curium isotopes with ^{12}C ions in the then-new Heavy Ion Linear Accelerator (HILAC) at Berkeley. They described a novel "double recoil" technique which permitted identification by chemical means of any daughter isotope of element 102 that might have been formed. The isotope ^{250}Fm was identified conclusively by this means, indicating that its parent should be the isotope of element 102 with mass number 254 produced by the reaction of ^{12}C ions with ^{246}Cm. However, another isotope of element 102, with half-life 3 seconds, also observed indirectly in 1958, and whose alpha particles were shown to have an energy of 8.3 MeV by Ghiorso and co-workers in 1959, was shown later by G. N. Flerov and co-workers (working at the Dubna Laboratory near Moscow) to be due to an isotope of element 102 with mass number 252 rather than 254; in other words two isotopes of element 102 were discovered by the Berkeley group in 1958 but the correct mass number assignments were not made until later. On the basis that they identified the atomic number correctly, the Berkeley scientists probably have the best claim to the discovery of element 102; they suggest the retention of nobelium as the name for this element.

All isotopes of nobelium are short-lived and are produced by the bombardment of lighter elements with charged particles; thus all of the chemical investigations have been done on the tracer scale. These have demonstrated the existence of No^{3+} and No^{2+} in aqueous solution with the latter much more stable than the former. The stability of No^{2+} is consistent with the expected presence of the completed shell of 14 $5f$ electrons in this ion.

Lawrencium (Lr) atomic number 103, after Ernest O. Lawrence.) Lawrencium was discovered in 1961 by Ghiorso, Sikkeland, Larsh, and Latimer using the Heavy Ion Linear Accelerator (HILAC) at the University of California at Berkeley. A few micrograms of a mixture of ^{249}Cf, ^{250}Cf, ^{251}Cf, and ^{252}Cf were bombarded with ^{10}B and ^{11}B ions to produce an isotope of element 103 with a half-life measured as eight seconds and decaying by the emission of alpha particles of 8.6 MeV energy. Ghiorso suggested at that time that this radioactivity might be assigned the mass number 257. G. N. Flerov and co-workers have disputed this discovery on the basis that their later work suggests a greatly different half-life for the isotope with the mass number 257. Subsequent work by Ghiorso and co-workers proves that the correct assignment of mass number to the isotope discovered in 1961 is 258, and this later work gives four seconds as a better value for the half-life.

All isotopes of lawrencium are short-lived, are produced by bombardment of lighter elements with charged particles, and all chemical investigations have been performed in the tracer scale. These have demonstrated that the normal oxidation state in aqueous solution is the (III) state, corresponding to the ion Lr^{3+}, as would be expected for the last member of the actinide series.

Rutherfordium (Rf, atomic number 104, after Lord Ernest Rutherford.) Rutherfordium, the first transactinide element to be discovered, was probably first identified in a definitive manner by Ghiorso, Nurmia, Harris, K. Eskola, and P. Eskola in 1969 at Berkeley. Flerov and co-workers have suggested the name kurchatovium (after Igor Kurchatov with symbol Ku) on the basis of an earlier claim to the discovery of this element; they bombarded, in 1964, ^{242}Pu with ^{22}Ne ions in their cyclotron in Moscow and reported the production of an isotope, suggested to be ^{260}Ku, which was held to decay by spontaneous fission with a half-life of 0.3 second. After finding it impossible to confirm this observation, Ghiorso and co-workers reported definitive proof of the production of alpha particle emitting ^{257}Rf and ^{259}Rf, demonstrated by the identification of the previously known ^{255}No and ^{253}No as decay products, by means of the bombardment of ^{249}Cf with ^{12}C and ^{13}C ions in the Berkeley HILAC.

All isotopes of rutherfordium are short-lived and are produced by bombardment of lighter elements with charged, heavy-ion particles. The isotope ^{261}Ru (half-life 65 seconds) has made it possible, by means of rapid chemical experiments, to demonstrate that the normal oxidation state of rutherfordium in aqueous solution is the (IV) state corresponding to the ion Rf^{4+}. This is consistent with expectations for the first transactinide element which should be a homolog of hafnium, an element that is exclusively tetrapositive in aqueous solution.

Hahnium (Ha, atomic number 105, after Otto Hahn.) Hahnium, the second transactinide element to be discovered, was probably first identified in a definitive manner in 1970 by Ghiorso, Nurmia, K. Eskola, Harris, and P. Eskola at Berkeley. They reported the production of alpha particle emitting ^{260}Ha, demonstrated through the identification of the previously known ^{256}Lr as the decay product, by bombardment of ^{249}Cf with ^{15}N ions in the Berkeley HILAC. Again the Berkeley claim to discovery is disputed by Flerov and co-workers who in 1970 reported the discovery of an isotope held to be element 105, decaying by the less definitive process of spontaneous fission, produced by the bombardment of ^{243}Am with ^{22}Ne ions in the Dubna cyclotron; later work by Flerov has confirmed the alpha particle emitting isotope of element 105 reported by Ghiorso and workers. Flerov has suggested Nielsbohrium (after Niels, Bohr, symbol Ns) as the name for element 105.

All the isotopes of element 105 are short-

TABLE 4. EXPANDED PERIODIC TABLE
(ATOMIC NUMBERS OF UNDISCOVERED ELEMENTS IN PARENTHESES).

1 H																	2 He
3 Li	4 Be											5 B	6 C	7 N	8 O	9 F	10 Ne
11 Na	12 Mg											13 Al	14 Si	15 P	16 S	17 Cl	18 Ar
19 K	20 Ca	21 Sc	22 Ti	23 V	24 Cr	25 Mn	26 Fe	27 Co	28 Ni	29 Cu	30 Zn	31 Ga	32 Ge	33 As	34 Se	35 Br	36 Kr
37 Rb	38 Sr	39 Y	40 Zr	41 Nb	42 Mo	43 Tc	44 Ru	45 Rh	46 Pd	47 Ag	48 Cd	49 In	50 Sn	51 Sb	52 Te	53 I	54 Xe
55 Cs	56 Ba	57 La	72 Hf	73 Ta	74 W	75 Re	76 Os	77 Ir	78 Pt	79 Au	80 Hg	81 Tl	82 Pb	83 Bi	84 Po	85 At	86 Rn
87 Fr	88 Ra	89 Ac	104 Rf	105 Ha	106	107	(108)	(109)	(110)	(111)	(112)	(113)	(114)	(115)	(116)	(117)	(118)
(119)	(120)	(121)	(154)	(155)	(156)	(157)	(158)	(159)	(160)	(161)	(162)	(163)	(164)	(165)	(166)	(167)	(168)

LANTHANIDES

58 Ce	59 Pr	60 Nd	61 Pm	62 Sm	63 Eu	64 Gd	65 Tb	66 Dy	67 Ho	68 Er	69 Tm	70 Yb	71 Lu

ACTINIDES

90 Th	91 Pa	92 U	93 Np	94 Pu	95 Am	96 Cm	97 Bk	98 Cf	99 Es	100 Fm	101 Md	102 No	103 Lr

SUPER-ACTINIDES

(122)	(123)	(124)	(125)	(126)			(153)

lived and are produced by bombardment of lighter elements with charged, heavy-ion particles. The isotope ^{262}Ha (half-life 40 seconds) makes it possible, with rapid chemical techniques, to study the chemical properties of hahnium. It should exhibit the (V) oxidation state like its homolog tantalum.

Element 106 Experiments leading to competing claims for the discovery of element 106 were performed essentially simultaneously at Berkeley and Dubna in 1974. Ghiorso et al. produced the new isotope ^{263}106 through the bombardment of ^{249}Cf with ^{18}O ions in the Berkeley SuperHILAC. Decaying with a half-life of 0.9 second, this isotope was identified through the observation of previously known ^{259}Rf and ^{255}No as its daughter and grand-daughter. Oganessian et al. of the Dubna Laboratory reported the observation of spontaneous fission activity with a half-life of 4–10 milliseconds, which was produced by bombarding ^{207,208}Pb with ^{54}Cr in their cyclotron and which they attributed to ^{259}106. Neither group has suggested a name for element 106. Its chemical properties should be like those of its homolog tungsten.

Element 107 In 1976, Oganessian et al. reported the observation of a 2 ms spontaneous fission activity produced in the bombardment of ^{209}Bi with ^{54}Cr ions, which they assigned to ^{261}107. In 1981, Münzenberg et al., using the linear accelerator (UNILAC) in the GSI laboratory in Germany, found an alpha-particle activity, produced in the bombardment of ^{209}Bi with ^{54}Cr ions, which could be assigned to ^{262}107 on the basis of the identification of its daughter ^{258}105 (which also undergoes alpha-particle decay). On the basis of the clear identification through the genetic relationship, as opposed to the indefinite identification through spontaneous fission, the discovery of element 107 must be credited to Münzenberg et al. Element 107 should have chemical properties like those of its homolog rhenium.

The transactinide elements with atomic numbers 104 to 118 clearly should be placed in an expanded periodic table under the row of elements beginning with hafnium, number 72, and ending with radon, number 86 (Table 4). This arrangement allows prediction of the chemical properties of these transactinide elements and it is suggested that they will have an element-by-element chemical analogy with the elements immediately above them in the periodic table. In other words, rutherfordium should chemically be like hafnium, hahnium like tantalum, element 106 like tungsten, 107 like rhenium, and so on across the periodic table to element 118, which should be a noble gas like radon. Beyond element 118, the elements 119, 120, and 121 should fit into the periodic table under the elements francium, radium, and actinium (atomic numbers 87, 88, and 89). At about this point there should start another, but a special kind of inner transition series perhaps similar in some respects to the actinide series. However, this series, which may be termed the "superactinide" series, will be different in that it will contain 32 elements, corresponding to the filling of the 18 member inner $5g$ and 14 member inner $6f$ shells. After the filling of these shells the still higher elements should again be placed in the main body of the periodic table.

It is not considered to be possible to synthesize and identify transuranium elements with atomic numbers as large as those in the superactinide series (approximately 121 to 153) or beyond. Even the earliest transactinide elements have short half-lives and are produced in small yields, and these half-lives and yields rapidly become shorter and smaller with increasing atomic number. Thus a simple extrapolation would indicate we might reach the end of the road at only a few elements beyond the heaviest now known. However, theoretical considerations suggest that elements in the region of atomic number (proton number) 114 and neutron number 184 should be stabilized against fission (the main mode of decay for the heaviest elements). This stabilization would be due to the filling of major proton and neutron shells in this region and is analogous to the stabilization of chemical elements, such as the noble gases, by the filling of their electronic shells. The elements constituting such an "island of stability" have come to be known as "superheavy elements." Many attempts have been made to synthesize and identify such elements, through bombardments with heavy ions, but success has not yet been achieved. The efforts are continuing.

The IUPAC Commission on Nomenclature of Inorganic Chemistry has recommended that temporary names be assigned to new transuranium elements until their names have been assigned in the traditional manner by their discoverers. They advocate using the "ium" ending preceded by the following roots: nil = 0, un = 1, bi = 2, tri = 3, quad = 4, pent = 5, hex = 6, sept = 7, oct = 8, and enn = 9. The corresponding chemical symbols would have three letters. Some examples from this system are: 108, Unniloctium (Uno); 112, Ununbium (Unb); 118, Ununoctium (Uuo); 140, Unquadnilium (Uqn); 200, Binilnilium (Bnn); 500, Pentnilnilium (Pnn). An alternative system, which is much easier to use, is to designate the new elements simply by their atomic numbers.

GLENN T. SEABORG

References

Keller, Cornelius, "The Chemistry of the Transuranium Elements," "Kernchemie in Einzeldarstellungen," Vol. 3, Germany, Verlag Chemie GmbH, 1971.

Seaborg, Glenn T., "Man-Made Transuranium Elements," Englewood Cliffs, N.J., Prentice-Hall, Inc., 1963.

Seaborg, Glenn T., "Transuranium Elements: Products of Modern Alchemy," Benchmark Book compilation of historic original papers, Stroudsburg, PA, Dowden, Hutchinson & Ross, Inc., 1978.

Keller, O. Lewin, Jr., and Seaborg, Glenn T., "Chemistry of the Transactinide Elements," *Ann. Rev. Nucl. Sci.* 27, 139–166, 1977.

Katz, Joseph J., Morss, Lester R., and Seaborg, Glenn T., "The Chemistry of the Actinide Elements," Second Edition, London, Chapman & Hall, Ltd., Pickwick, 1985.

Lederer, C. M., and Shirley, V. S., "Table of Isotopes," Seventh Edition, New York, John Wiley & Sons, Inc., 1978.

Cross-references: ATOMIC PHYSICS; CYCLOTRON; ELEMENTS, CHEMICAL; ISOTOPES; NUCLEAR REACTIONS; NUCLEAR REACTORS; RADIOACTIVITY; RARE EARTHS.

TUNNELING

Tunneling is a quantum mechanical process without a classical analog. An electron (or other quantum mechanical particle) incident upon a potential barrier whose height is larger than the kinetic energy of the electron will penetrate (tunnel) a certain distance into the barrier. This is most easily visualized by considering the one-dimensional Schrödinger equation for the wave function $\psi(x)$ of such an electron:

$$-\frac{\hbar^2}{2m}\frac{\partial^2 \psi}{\partial x^2} + V(x)\psi = E\psi \qquad (1)$$

If V varies relatively slowly with distance, the solution of Eq. (1) can be approximated by

$$\psi \approx \psi_0 \exp\left[\pm i \frac{\sqrt{2m}}{\hbar}\sqrt{E - V(x)}\,x\right] \qquad (2)$$

The nonclassical case is the one where the electron energy E is smaller than the potential $V(x)$. Then ψ is no longer an oscillating wave, but it is real and decays exponentially with distance into the barrier. The penetration probability through a given barrier is equal to the ratio of the probability density $|\psi(x)|^2$ at the exit from the barrier to its value at the entrance. Equation (2) shows that this probability is exponentially dependent on barrier height and thickness. The tunneling takes place at constant electron energy as there is no scattering involved.

A more accurate solution of Eq. (1), which adequately describes many real situations, may be calculated by the Wentzel, Kremers, Brillouin

(WKB) approximation. The tunneling probability through a barrier then becomes

$$P \approx \exp\left[- 2 \int_{x_1}^{x_2} \frac{\sqrt{2m}}{\hbar} \sqrt{V(x) - E} \; dx \right] \quad (3)$$

An idea of the magnitude of the tunneling probability may be obtained by calculating the value of Eq. (3) for a rectangular barrier 1 eV higher than the particle energy and 25 Å wide. The result for an electron is $P \approx 10^{-11}$; for a proton, $P \approx 10^{-400}$.

Tunneling can occur in many physical situations. The simplest case is that of FIELD EMISSION. This is the emission of electrons from a solid by the application of very high fields. The energy-band diagram for such a system is shown in Fig. 1.

The electrons in a metal surrounded by vacuum may be regarded as an ensemble of quasi-free electrons held in a potential well by the positive charges of the metal ions. The depth of the well is called the electron affinity χ. The potential well is filled with electrons up to the Fermi level (E_F) and consequently electrons at this level "see" a barrier $\phi_M = \chi - E_F$ (work function) surrounding them. They can tunnel a short distance into the walls of the potential well, but of course cannot escape from it.

If a high positive field ξ is applied to the metal (e.g., by applying a potential difference between the metal under consideration and an adjacent piece of metal), the potential energy distribution is that shown in Figure 1. There now exists a barrier through which, under appropriate conditions, the electrons may tunnel. The maximum height of the barrier is ϕ_M, and its width is approximately ϕ_M / ξ so that the current becomes

$$I \propto P \approx \exp\left[- 2 \frac{2}{3} \frac{\sqrt{2m}}{\hbar} \sqrt{\phi_M} \frac{\phi_M}{\xi} \right]$$

$$\approx \exp\left[- 10^8 \frac{\phi_M^{3/2}}{\xi} \right] \quad (4)$$

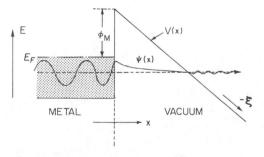

FIG. 1. Simplified energy-band diagram for field emission from a metal. $\psi(x)$ represents the wave function of an electron with energy E.

if ϕ_M is given in units of electron volts and ξ in units of volts per centimeter. Since ϕ_M generally has a value between 2 and 5 eV, an electric field of the order of 10^7 volts/cm is required before appreciable numbers of electrons will be emitted into vacuum.

Figure 1 is a somewhat simplified picture of an actual metal-vacuum interface. An electron outside the metal experiences a polarization force towards the metal, the so-called image force. While this is not a true potential, it may nevertheless be included in the potential energy diagram as a rounding-off of the well edge. When the field is applied, the barrier is lowered and so the tunneling probability increases. Particularly at high fields this causes a departure from the simple current relationship described above. If these effects are included in the calculation, good agreement can be obtained with experimental results.

Current flow at a metal-insulator interface may be treated in the same way as that at a metal-vacuum boundary. The potential barrier height is now given by the distance from the metal Fermi level to the bottom of the insulator conduction band, which is usually smaller than the work function. Also the dielectric constant of the insulator must be taken into account (e.g., in the image force). These modifications do not change the basic current-voltage relationship [Eq. (4)]. This simple picture has been extended by including electron coherency effects and detailed agreement has been obtained in semiconductor-insulator structures.

There may be localized electronic states in the barrier region, either due to impurities in the insulator layer or due to adsorbed molecules at the metal-insulator or metal-vacuum interfaces. An electron in the metal at the energy of the impurity state will then have a higher probability of tunneling to the impurity state and from there to the other side of the barrier than the probability of tunneling directly through the barrier. This means that when the applied voltage raises the Fermi level in the metal to the energy of impurity states in the barrier, there will be a sudden increase in current. The current-voltage relationship can thus be used to study the distribution of impurity states.

In case of a very thin insulator layer (< 60Å) bounded on both sides by metallic regions, tunneling occurs even in the absence of fields. The band structure may be approximated by the square potential barrier discussed initially. The top of the barrier is again formed by the conduction band of the insulator, modified appropriately by image force considerations. In equilibrium, there will be equal numbers of electrons tunneling in both directions through the insulator; no net current flows.

When a potential difference is applied, one tunneling direction is favored and net current flows. For small potential differences (much less than the barrier heights) the tunneling probability does not vary with applied field and the current flowing will be proportional to the difference in number of electrons available on the two sides of the barrier at the same energies. These numbers are in turn closely related to the density of electronic states in the metals at these energies. The tunneling current may therefore be used to investigate the density of states in certain materials, where this quantity changes rapidly with energy.

This technique for determining the density of states has been put to particularly good use in the case of superconductors. The fact that the superconductor has an energy gap in the density-of-states function at the Fermi level leads to nonlinearities and negative resistance regions in the current-voltage characteristics. This has become the most accurate method of measuring the energy gap of superconductors as a function of temperature, magnetic field and other variables. More complicated phenomena, such as simultaneous tunneling of two electrons as a pair, have also been observed. This forms the basis for the Josephson effects.

Other nonlinearities are observed at small voltages in junctions made of normal as well as superconducting metals. These are due to inelastic tunneling where the electron changes energy during tunneling by interaction with lattice excitations; e.g., it may emit or adsorb a phonon. Emission can only take place if the applied voltage is larger than the phonon energy so that a step in the current (or in its derivative) is expected at that voltage. This is observed experimentally at low temperatures. It has become one of the best tools for quantitative study of the properties of electrons and phonons in metals. In some cases, anomalies are also observed at zero voltage and they may be due to magnetic interactions.

Tunneling out of a three-dimensional well is treated similarly to tunneling through the one-dimensional barrier. The exact exponential dependence will be different from Eq. (2) and depend on the form of the barrier. One type of three-dimensional well is formed by an impurity in a semiconductor or insulator which may form a bound state in the region of the forbidden gap. An electron can only tunnel out of such a state when a high field is applied, as in field emission from metals.

A different type of three-dimensional potential well is found in the atomic nucleus. A combination of short-range nuclear attraction and Coulomb repulsion forms a potential barrier of the type shown in Fig. 2. Many heavy radioactive nuclei contain α-particles with high enough kinetic energy to tunnel through the barrier (α-decay). Because of the much heavier mass and high barrier energy, the barrier must be considerably thinner than that for an electron ($\sim 10^{-12}$ cm). Again calculations for a Coulomb barrier agree well with experimental observations of the energy dependence of the decay time (Geiger-Nuttall relation).

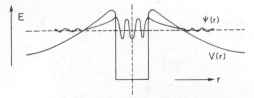

FIG. 2. Model of the nucleus, with $\psi(r)$ the wave function of an α-particle.

Tunneling of entire atoms can take place in solids, particularly when there are two possible equilibrium states (potential minima) at equal energy for the atoms which are separated by a small distance and low potential barrier. Examples of this are ferroelectric crystals of the hydrogen-bonded type (e.g., KH_2PO_4). The hydrogen atom has two possible positions in the hydrogen bond, near one or the other of the adjacent atoms. In the ferroelectric phase, all hydrogen atoms occupy the same potential well and a reversal of polarization direction is accomplished by a switch from one well to the other.

Tunneling of one or more atoms has also been used to explain the low-temperature properties of some solids. Alkali halide crystals doped with certain impurities exhibit reduced thermal conductivity at low temperatures with resonance peaks. These impurities have more than one equilibrium position in the lattice and can tunnel between these positions (either by rotation or by displacement). These low-frequency excitations produce the observed phonon scattering.

A related situation applies in amorphous solids. The thermal and acoustic properties of glasses differ significantly from those of crystalline solids at low temperatures. It has been proposed that some atoms in the disordered solid can exist in more than one potential minimum. In contrast to the crystalline case, these minima vary in energy depending on the local environment. The excitation energies related to tunneling between two such minima are randomly distributed, so that no resonances are observed in thermal spectra.

Electron tunneling is also observed in semi-

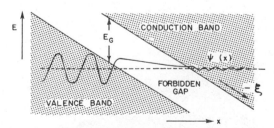

FIG. 3. Band structure of a semiconductor or insulator under an applied field.

conductors (Zener tunneling), but in this case the situation cannot be represented by the kind of barrier discussed previously. The band structure of a semiconductor under a large applied field is shown in Fig. 3. There exists no set of real energy states connecting the two regions, and an electron must always make a discontinuous step in passing from the valence band to the conduction band. Still the tunneling probability may be calculated with an approximation equivalent to that used for the simple barrier, and except for a numerical factor, the result is the same as Eq. (4) with ϕ_M replaced by E_G.

Conduction of this type is observed in very narrow p-n junctions under reverse bias (Zener breakdown) or in insulators where there are no free carriers available. In the former case, there is a large built-in field even at zero external voltage which is produced by the difference in the electrochemical potentials of the n- and p-regions. A small additional voltage (1 to 2 volts) raises the field to the value required for tunneling.

In still narrower p-n junctions ("tunnel diodes") tunneling readily takes place even where no applied voltage exists. As in the case of the very thin insulator, the electrons tunneling in opposite directions then just cancel one another out. The current-voltage curve of such a junction is drawn in Fig. 4. The band structure configurations at two values of forward bias are also shown. Net tunneling current will flow in either direction of applied voltage, as long as there are electrons on one side opposite empty states of the same energy on the other side. Under large forward current this condition is no longer fulfilled, as the bottom of the conduction band on the n-type side comes opposite the forbidden energy gap on the p-type side. This means that the current passes through a maximum value and then decreases towards zero with increasing forward bias. At still larger bias, the conventional forward diode current becomes dominating.

In actual diodes, the current never decreases completely to zero. There always exist impurity states in the forbidden gap to which an electron of appropriate energy may tunnel and from which it may drop into an empty state in the valence band giving up energy to localized lattice vibration. Alternatively, the electron need not actually occupy the impurity state but can tunnel "inelastically" directly to the valence band while simultaneously giving up the differ-

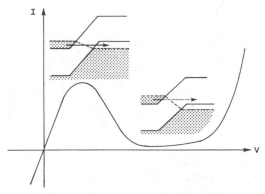

FIG. 4. Current-voltage characteristic of a tunnel diode. The two inserts depict the band structure at the maximum tunnel current and at the valley where tunneling is no longer possible.

ence energy to the impurity atom and its vibration.

DIETRICH MEYERHOFER

References

Franz, W., in Flügge, S., Ed., "Handbuch der Physik," Vol. XVII, pp. 201–219, Berlin, Springer-Verlag 1956.

Fisher, J. C., and Giaever, I., *J. Appl. Phys.*, **32**, 172 (1961).

Kane, E. O., *J. Appl. Phys.*, **32**, 83 (1961).

Burstein, E., and Lundqvist, S., Eds., "Tunneling Phenomena in Solids," New York, Plenum Press, 1969.

Duke, C. B., "Tunneling in Solids," New York, Academic Press, 1969.

McMillan, W. L., and Rowell, J. M., in Parks, R. D., Ed., "Superconductivity," pp. 561 ff, New York, Marcel Dekker, 1969.

Phillips, W. A., Ed., "Amorphous Solids: Low Temperature Properties," Berlin, Springer-Verlag, 1981.

Narayanamurti, V., and Pohl, R. O., *Revs. Mod. Phys.* **42**, 201 (1970).

Roy, D. K., "Tunnelling and Negative Resistance Phenomena in Semiconductors," Oxford, Pergamon Press, 1977.

Maserjian, J., and Zamani, N., *J. Appl. Phys.* **53**, 559 (1982).

Cross-references: ENERGY LEVELS, FIELD EMISSION, RADIOACTIVITY, SCHRÖDINGER EQUATION, SEMICONDUCTORS, SOLID-STATE PHYSICS, SOLID-STATE THEORY, SUPERCONDUCTIVITY, THIN FILMS.

U

ULTRASONICS

Ultrasonic Waves Ultrasonic waves are sound waves above the frequency normally detectable by the human ear, i.e., above about 20 kHz. The particles of matter transmitting a *longitudinal* wave move back and forth about mean positions in a direction parallel to the path of the wave. Alternate compressions and rarefactions in the transmitting material exist along the wave propagation direction. Such waves are possible in solids, liquids or gases. In solids, *shear* waves can also propagate. For these, the particles move perpendicularly to the direction of wave propagation. When material boundaries are present in the vicinity of a wave more complicated particle motions take place. Common examples are *Rayleigh* waves (at the boundary between a solid and air) and *Lamb* waves (in thin plates whose thickness is of the order of the wavelength of sound). The predominant particle displacement in these latter cases is perpendicular to the surface of the solid (see WAVE MOTION). Examples of the propagation velocities for these mode types are given in Table 1.

If an ultrasonic wave is normally incident on a boundary between two media of different densities ρ and velocities V, the fractions of the wave amplitude which are reflected (R) and transmitted (T) are:

$$R = \frac{\rho_2 V_2 - \rho_1 V_1}{\rho_2 V_2 + \rho_1 V_1}$$

$$T = \frac{2\rho_2 V_2}{\rho_2 V_2 + \rho_1 V_1}.$$

For oblique incidence the reflection and transmission at a boundary are more complicated since the incident wave is transformed into reflected and refracted shear and longitudinal waves. *Snell's Law* governs the angles of *reflec-*

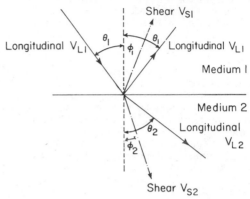

FIG. 1

tion and *refraction* for both types of waves. It states that

$$\frac{\sin \theta}{V} = \text{Constant}$$

where θ is the angle the beam makes with a normal to the intervening surface and V is the sound velocity. Therefore, in Fig. 1,

$$\text{Constant} = \frac{\sin \theta_1}{V_{L1}} = \frac{\sin \theta_2}{V_{L2}} = \frac{\sin \phi_2}{V_{S2}} = \frac{\sin \phi_1}{V_{S1}}$$

Transducers In general an electromechanical transducer is a device to convert electrical energy into mechanical vibrations which are then directed into the specimen. The transducer also commonly reconverts received mechanical oscillations back into electrical signals for amplification and recognition. Transducers can be *piezoelectric, ferroelectric, magnetostrictive,* or *optical* in nature.

Until recently, the only peizoelectric transducer in general use was *quartz*, but now several synthetic ceramic materials are employed. The application of a voltage across a piezoelectric crystal causes it to deform with an amplitude of deformation which is generally a simple function of the voltage. Reversal of the voltage causes reversal of the mechanical strain.

Transducer crystals are normally cut to vibrate at a resonant frequency, the thickness being one-half the acoustic wavelength. A bond between the crystal transducer and the speci-

TABLE 1. APPROXIMATE SOUND VELOCITIES

	In Water (m/sec)	In Steel (m/sec)
Longitudinal	1.5×10^3	6×10^3
Transverse	Cannot be supported	3×10^3
Rayleigh	Cannot be supported	3×10^3

men matches the acoustic impedance and carries the acoustic power into the specimen. Such a device can produce a wave motion in the material at the resonant and harmonic frequencies of the transducer. Backing layers may be affixed to the rear surface of the transducer and can be designed to perform a number of functions: transformation of the resonant transducer into a broadband transducer with center freuqency near the original resonance, reflection of power forward into the crystal and specimen, or simplification of the form of the signals measured in material testing applications.

A new transducer, of particular importance in research applications, is the *laser*. A high power pulsed laser can be used for the generation of ultrasonic pulses and a *laser interferometer* for the direct measurement of displacement amplitudes of ultrasonic waves. An obvious advantage of the laser is that it requires no mechanical contact with the specimen.

Ultrasonic Testing Most ultrasonic test equipment employs pulses of high-frequency sound (>1 MHz), the pulse width being adjustable between 0.1 and 1.0 msec and the repetition rate between 60 and 1000 pulses per second. Figure 2 shows a typical block diagram of ultrasonic testing equipment.

The synchronizer triggers a pulse in the generation circuit which is converted into an acoustic pulse by the transducer crystal. The synchronizer also starts the sweep circuit of the cathode ray tube (CRT) and the marker circuit, the latter making marker pips along the time base. Separate transducer crystals may be used for transmission and reflection, or one crystal may be made to carry out both functions. The echo from the back face of the specimen block arrives at twice the transmission time after the initial pulse. The marker pips can then be calibrated in terms of depth. A fault will show up as an extra echo whose depth into the specimen can be read from the marker pip which coincides with its position on the screen of the CRT.

The transducer is coupled into the specimen

under test by an oil or adhesive. In high-speed testing or in cases where rough surfaces are encountered, the specimen may be immersed in a liquid or a stream of liquid passed between the transducer and the specimen in order to give sonic coupling.

Examples of Ultrasonic Testing. Solid objects of thickness about 1 cm or greater may be tested for inhomogeneities. Special techniques also exist for sheet testing. Heavy forgings, crank shafts, rails, concrete structures and a host of other manufactured objects are now regularly scanned for porosity or internal faults. The frequencies employed are normally about 1–15 MHz for metal objects, about 0.5 MHz for many plastics, and 0.1 MHz for concrete.

Ultrasonic Scanning can be used to record the spatial variation of the reflection or transmission properties of an object. In earlier usage, echoes were recorded on the CRT which was circularly scanned (like the P.P.I. radar system). Nowadays automatic scanning and *data processing* are commonly used to obtain recognizable ultrasonic images of internal structure. Typical applications are the imaging of a fetus, tumor, or gallstone in medical applications; the imaging of the microstructure of solids using the acoustic microscope; the characterization of internal flaws in *nondestructive evaluation* applications. Typical measurements of these types are carried out at relatively high ultrasonic frequencies to achieve maximum resolution. In particular the *acoustic microscope* involves measurement of the attenuation of 100 MHz ultrasonic waves with a resulting resolution ~10 μm. In *sonar* underwater detection, on the other hand, frequencies must be low because of the high attenuation in water (60 dB/km at 100 kHz), with consequent reduction in resolution by many orders of magnitude compared to the acoustic microscope. Sea noises, fish, and underwater refraction by layers at different temperatures introduce difficulties of interpretation into the echo patterns. Depth recording is now accomplished by the use of many sending and receiving transducers coupled to a computing system which interprets differences of arrival times in terms of variation of path length and ultimately of depth. Continuous records of the ocean bed or the depth of surface ice can be made.

Examples of the Use of Ultrasonic Power. Longitudinal ultrasonic waves can be produced with sufficient intensity to cause pronounced physical changes in materials of industrial importance. In many cases, standing waves are set up in a containing vessel with resulting periodic variation of vibration amplitude. In almost all cases ceramic transducers are used. Emulsifying, mixing, dispersing, degassing, and cleaning applications generally employ unfocused acoustic radiation. In welding, soldering, drilling, machining, and the neurosurgical controlled damage of brain tissue, on the other hand, a focused acoustic beam is always used.

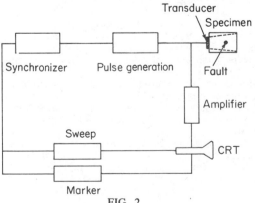

FIG. 2

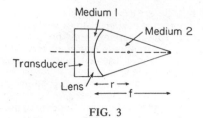

FIG. 3

Focusing may be done by the use of either an acoustic horn or by an acoustic lens. Figure 3 shows the principle of a simple plano-concave lens whose radius of curvature, r is given by

$$r = f\left(\frac{n-1}{n}\right)$$

where n, the index of refraction, is the ratio of velocities in the lens and adjacent medium:

$$n = \frac{V_1}{V_2}$$

Fundamental Research in Physics by Ultrasonics *Acoustic Velocity*. The acoustic velocity in a material is fundamentally related to the binding forces between the atoms or molecules. Measurement of the velocity is made by a pulse technique or by vibrating a specimen of known dimensions in one of its fundamental modes of oscillation and recording the frequency.

Acoustic Attenuation. Attenuation measurements are made by recording the decay in oscillation amplitude with time of a solid material set to ring in one of its natural modes or from the decrease in amplitude of successive echoes passing through the material. The attenuation of material is a measure of energy losses and may be used to study a variety of fundamental loss mechanisms.

Under the influence of *thermal activation*, impurity elements or defects may switch positions in a crystal lattice. This they do in a definite *relaxation time*, which is related exponentially to the temperature and to the difference in energy of the defect in its two positions. Now if the switching of position of the defect also causes a difference in dimension of the specimen, a maximum in attenuation is found when the period of an applied acoustic stress just matches the thermal relaxation time. A whole spectrum of relaxation attenuation maxima has been found in solids. Each is related to a definite internal process, each with its characteristic energy difference. Perhaps the best known of these is the relaxation maximum given by the diffusion of carbon atoms in iron from which the activation energy for carbon diffusion can be calculated (Snoek peaks).

Dislocations are line defects in crystalline material the movement of which is involved in all cases of plastic deformation. Acoustic waves vibrate the dislocations and, in general, the more freely the dislocation moves, the more the acoustic beam is attenuated. Attenuation measurements thus give information on how well the dislocation is locked in position in the solid, i.e., on the strength of the material. Pinning down of the dislocations may be effected by alloying or by radiation with high-energy particles which cause internal damage. Measurement of acoustic attenuation is therefore encountered in such studies.

Electrons in metals also attenuate acoustic waves if the frequencies are sufficiently high (>5 MHz) and the temperature sufficiently low (<10 K) so that the acoustic wavelength is comparable with the mean free path of the electrons. Attenuation measurements give information on relaxation times of electrons, and hence their energies. Such measurements can be used to determine the dimensions and shape of the *fermi surface* in metals and are intimately related to the phenomena of electrical and thermal conductivity. The onset of *superconductivity* in some metals and alloys at low temperatures is characterized by a fall to zero electrical resistance and also by a sharp fall in high-frequency acoustic attentuation.

In nonmetallic crystalline materials, thermal vibrations or *phonons* interact with one another and limit heat conductivity. Ultrasonic waves are phonons of known frequency and polarization, both being mechanical vibrational modes of the material. Hence, measurement of the attenuation of ultrasonic waves can be used to study phonon-phonon interaction and consequently the mechanisms of thermal conductivity. Such a study is usually carried out in materials below 100 K using ultrasonic waves at extremely high frequencies (>1000 MHz). Such information may lead to the elucidation of the spectrum of thermal vibrations in all materials above the absolute zero of temperature.

Acoustic Emission. Acoustic emission is a new technique which is evolving from ultrasonics. In acoustic emission only a detecting transducer is used, the information being generated by gross internal changes in the specimen material. Acoustic emission is the transient elastic wave generated by the rapid release of energy within a material under stress. Detection of the acoustic emission is usually accomplished using an ultrasonic transducer. During the last 25 years, several investigators have identified and studied the many sources of acoustic emission which include slip, dislocation breakaway and pinning, mechanical phase transitions, stress corrosion, and fatigue crack growth.

The stress waves emitted by the release of energy from defects can span a very wide frequency range, and indeed can include all frequencies up to thermoelastic waves. In the cases of the "cracking" of timber prior to failure, and the "cry" of tin undergoing bending, elastic energy is emitted in the audio range, and is audible to the human ear, while acoustic emission from crack advance is detectable up to several megahertz. Acoustic emission activity is often observed to be irreversible, in the sense that during the reloading of a material, acoustic emissions are not generated until the stress level exceeds its previous maximum value. This phenomenon is known as the *Kaiser effect*.

While acoustic emission is expected at much higher frequencies, the range over which most measurements are currently made is 0.1–1.0 MHz. This frequency range is chosen since it is high enough to avoid most environmental noise, and low enough to avoid excessive acoustic attenuation in most materials.

Two types of acoustic emissions are observed in ductile materials. *Continuous emissions* are attributed to the low-level, relatively steady noise which results from slow, undirectional loading of an unflawed specimen. Dislocation motion has been cited as a probable cause of this type of noise. *Burst emissions* are transients which appear only occasionally, are usually of much higher amplitude than the continuous type of emissions, and have been attributed to sudden gross structural changes such as micro-cracking. During fatigue failure, both continuous and burst emissions occur. The continuous emissions increase in amplitude prior to the occurrence of the high amplitude burst emissions.

T. S. HUTCHISON
S. L. MCBRIDE

References

Drouillard, T., "Acoustic Emission: A Bibliography with Abstracts," New York, Plenum Data Corporation, 1979. A complete collection of abstracts for the field of acoustic emission.

Mason, W. P., and Thurston, R. N., "Physical Acoustics," Vols. I–XV, Academic Press, New York. Definitive series on theory and techniques of ultrasonics.

Truell, R., Elbaum, C., and Chick, B. B., "Ultrasonic Methods in Solid State Physics," New York, Academic Press, 1969. Treats research techniques.

Cross-references: CONDUCTIVITY; ELECTRICITY; FERMI SURFACE; FERROICITY, FERROELECTRICITY, AND FERROELASTICITY; HEAT TRANSFER; MAGNETOSTRICTION; OSCILLOSCOPE; PHONONS; REFLECTION; REFRACTION; RELAXATION; SOLID STATE PHYSICS; SUPERCONDUCTIVITY; VIBRATION; WAVE MOTION.

ULTRAVIOLET RADIATION

Ultraviolet radiation comprises the region of the electromagnetic spectrum extending from the violet end of the visible, wavelength 4000Å, to the beginning of X-RAYS, arbitrarily taken as 100Å, a span of more than five octaves. The unit of wavelength generally used for the ultraviolet is $Å = 10^{-8}$ cm; it is named after A. J. Ångström, Swedish pioneer in SPECTROSCOPY, who made in 1868 the first accurate measurements of the wavelengths of spectral lines. Although still occasionally heard, the term ultraviolet rays has become obsolete. Ultraviolet light, however, is an expression which is usually acceptable even though ultraviolet cannot ordinarily be seen by the eye.

The ultraviolet is subdivided into several parts: The near ultraviolet, 4000 to 3000Å, present in sunlight, producing important biological effects, but not detectable by the eye; the middle ultraviolet, 3000 to 2000Å, called by biologists the far ultraviolet, not present in sunlight as it reaches the earth's surface, but well transmitted through air; the long range 2000 to 100Å known as the extreme ultraviolet, abbreviated XUV, since it connects the ultraviolet and x-rays. The XUV is known also as the vacuum ultraviolet, VUV, because it is not transmitted through air, and is sometimes called the far ultraviolet. The portion of the XUV from 2000 to 1350Å is known as the Schumann region, after its discoverer.

The boundary, 100Å, between XUV and x-rays is arbitrary; it is often preferred to make the distinction on the basis of the method of production and analysis. Radiation is called x-rays if produced by the classical x-ray tube in which bombardment by electrons removes inner-shell electrons from atoms of the target material, with radiation emitted when outer electrons fall back; x-rays are also generated by the BREMSSTRAHLUNG process when fast electrons are suddenly decelerated. In sparks, arcs, and electrical discharges through gases, the classical sources of ultraviolet, it is the outer electrons of either neutral atoms or ions that are excited; the radiation occurs when the outer electrons are recaptured and fall back to their ground states (see ELECTRICAL DISCHARGES IN GASES).

Ultraviolet radiation was discovered by Ritter in 1801; he found that silver chloride was blackened, as it is by visible light, if placed beyond the violet end of the sun's spectrum, where nothing can be seen. Stokes, in 1862, using a prism of quartz rather than glass, observed to a short-wavelength limit of 1830Å radiation produced by a spark discharge between aluminum electrodes by using a fluorescent plate detector. The break-through further into the ultraviolet was made between 1885 and 1903 by Victor Schumann, an instrument maker and machine shop owner of Leipzig. Schumann realized that there are three reasons why ultraviolet of shorter wavelengths had not been detected: (1) air is opaque, (2) quartz prisms and lenses do not transmit, (3) ordinary photographic emulsions are not sensitive, because of absorption by gelatin. He overcame these difficulties by constructing the first vacuum spectrograph by using optics of crystal fluorite instead of quartz, and by making photographic plates with almost no gelatin, now known as Schumann plates. Theodore Lyman of Harvard University, soon passed 1300Å, the limit reached by Schumann, by constructing a vacuum spectrograph with a reflection-type concave diffraction grating instead of the fluorite prism and lenses. He discovered the Lyman series of hydrogen, the most fundamental spectral series of the simplest and most abundant element, with first line, Lyman-alpha, at 1215.67Å and series limit at 911.7Å. Further progress reaching 140Å was made by R. A. Millikan and his students at the California Institute of Technology with the aid of a "hot" spark in vacuum. Finally Osgood, in 1927, closed the gap to x-rays by combining the hot spark and a grating used at grazing incidence; in the same year, Dauvillier reached 121Å in the XUV from the x-ray side, using an x-ray tube with a spectrometer utilizing a crystal of large lattice constant, made from a fatty acid.

Ultraviolet radiation is emitted by almost all light sources, to some extent. In general, the higher the temperature or the more energetic the excitation, the shorter are the wavelengths. Tungsten lamps in quartz envelopes radiate in the ultraviolet in accordance with Planck's law, slightly modified by the emissivity function of tungsten. Because of its high temperature, 3800K, the crater of an open carbon arc is an excellent source of ultraviolet extending to the air cutoff. Electrical discharges through gases produce intense ultraviolet emission, mainly in lines and bands. The most widely used is the quartz mercury arc; when the Hg pressure is allowed to rise to several atmospheres, the intensity becomes great, and the spectrum is a quasi continuum. For the shortest wavelengths, still higher temperatures are required, as produced by discharging a large capacitor at some 50 kV between metal electrodes in vacuum. This violent discharge vaporizes atoms from the electrodes, then strips off as many as 10 or 15 outer electrons. The emission line radiation emitted when these highly ionized atoms recapture electrons extends to very short wavelengths. A similar short-XUV "spark" can also be produced by focusing a high-power laser beam on a metal surface. Another source producing highly ionized atoms and emission lines in the XUV is the magnetically compressed plasma such as produced by the devices known as zeta and theta pinch, which reaches a temperature of half a million degrees. A relatively new source of XUV and x-rays is the synchrotron, in which electrons are accelerated to energies of the order of one billion electron volts while constrained to a circular orbit by a magnetic field. The electromagnetic radiation produced by the great centripetal acceleration is intense and is plane-polarized. Most important, its spectrum is a smooth, broad continuum which is far more useful for many purposes than any other XUV continuum. The position of the broad intensity peak depends on the electron energy and the magnetic field and can be placed at any wavelength between approximately 1000 Å and 1 Å by selecting the proper value of electron energy within the range several hundred MeV and a few GeV. The type of synchrotron known as the storage ring is particularly useful because of the high degree of control of the XUV beam that emerges. These sources offer powerful new technologies for the study of the fine structure of materials, particularly when samples are available only in minute quantities. Techniques are now available for producing the radiation in the form of very short pulses of extremely high intensity, which are especially useful in the study of transient reactions in matter. Another new source of radiation is the laser, with which extremely high values of intensity can be produced. The laser's spectrum is strictly monochromatic, exactly the opposite of the perfect continuum of the synchrotron. Progress is being made toward the development of lasers for the UV, and to a lesser degree for the XUV and x-rays. Applications, as yet, are few. For great XUV intensity, however, no man-made source has equaled the atomic bomb.

Solids, liquids, and gases usually transmit well in the near ultraviolet but always become opaque somewhere in the middle or extreme ultraviolet. Among solids, both crystal and fused quartz transmit to a short wavelength limit between 2000 and 1500Å, depending on purity; CaF_2 (fluorite), to 1230Å; MgF_2 to 1140Å; and LiF, the most transparent known solid, to 1050Å. These materials are invaluable for constructing lenses and prisms for ultraviolet instruments.

Solids in the form of very thin films transmit to some extent throughout the XUV, but the thickness must be less than a few thousand Ångstroms. Certain thin films are useful as optical filters; Al, for example transmits from 837Å corresponding to its plasmon frequency, to its x-ray L-edge at 170Å, and to a small extent as far as 150Å. Films of plastics, such as collodion of 300Å thickness, transmit fairly well below 500Å; they are useful as windows for low-pressure gas retention.

Gases vary greatly in their absorption characteristics. Molecular oxygen causes air to become opaque below about 1850Å, because of absorption in the Schumann-Runge band system, followed by continuous absorption from 1750 to 1290Å, and strong irregular absorption to shorter wavelengths. Molecular nitrogen, however, is relatively transparent all the way to 1000Å. Hydrogen absorbs in the Lyman series lines, and in an ionization continuum beyond the series limit, 911.7Å. Of all the gases, helium is the most transparent; absorption first takes place in the resonance lines, the longest lying at 584Å, and in a continuum beyond the series limit, 504Å. More complicated molecular gases, such as CO_2, NO, and N_2O are rather opaque throughout most of the XUV. Water vapor is much like O_2 with absorption commencing below 1850Å.

Reflection occurs for ultraviolet, just as for visible radiation. In general, the reflectance becomes less, as the wavelength decreases. Aluminum is the best reflector over much of the long-wavelength region; when properly prepared, by rapid evaporation of the pure material in an excellent vacuum, the reflectance is greater than 90 per cent to 2000Å, and can be maintained to 80 per cent at 1200Å by overcoating with a thin layer of MgF_2 to prevent growth of Al_2O_3. Below 1000Å, platinum is best with a reflectance of about 20 per cent at 600Å, but only about 4 per cent at 300Å.

Ultraviolet can be detected in a variety of ways, making use of effects produced when it is absorbed by matter, e.g., fluorescence with re-radiation of longer wavelengths; chemical reactions in solids; dissociation of gas molecules, with ensuing reactions; ionization of gases; emission of photoelectrons from surfaces of solids. In general, the shorter the wavelength, the more energetic is the reaction. This is in accordance with Einstein's law, $E = h\nu$, giving the energy of a photon in terms of Planck's constant, h, and its frequency ν (velocity of light ÷ wavelength). Thus, ultraviolet photons range up to about 10 000 times more energetic than visible photons, and effects produced by them are much easier to observe.

The simplest way to detect and measure ultraviolet radiation is by making use of the fluorescence process, converting the ultraviolet into visible radiation which can be seen, or into near ultraviolet which is easily photographed or measured with conventional photomultipliers. Calcium tungstate, for example, is an excellent converter for the middle ultraviolet. Materials much used for the extreme ultraviolet are oil and sodium salicylate. The latter is especially valuable because its quantum efficiency of fluorescence is high and is nearly independent of wavelength. An ordinary photomultiplier, with its glass window coated with a layer of sodium salicylate becomes a sensitive radiometer for use throughout the entire ultraviolet.

Ordinary photographic emulsions containing gelatin as a binder are useful only to about 2500Å; they can be sensitized easily by overcoating with oil or sodium salicylate. Eastman Kodak spectroscopic-type plates and films are available with ultraviolet sensitization, produced by overcoating with a fluorescent lacquer. Nearly gelatin-free Schumann-type emulsions combine greater sensitivity and higher resolving power than fluorescence-sensitized emulsions; they are available as Eastman Kodak 104 (formerly called SWR), the higher-speed 101, and Kodak Pathé SC5 and 7. All types must be handled so that the delicate emulsion surface is never touched.

Ultraviolet radiation is easily detected and measured directly by means of photomultipliers if they are equipped with ultraviolet transmitting windows or are used without an envelope in a vacuum. Almost all materials emit photoelectrons to some extent when ultraviolet-irradiated, but in applications, the problem is usually to devise a photocathode surface that has extremely high efficiency in a certain spectral region and is insensitive in others. When constructed with ultraviolet-transmitting envelopes, the various visible-sensitive photocathodes respond well throughout the ultraviolet. For applications in the presence of sunlight, however, it is often necessary to use a "solar blind" surface, having negligible sensitivity longward of 2900Å; one of the best surfaces is RbTe, with quantum efficiency of $< 10^{-3}$ at $\lambda > 3000$Å and ≈ 0.1 at $\lambda < 2600$Å.

For use at wavelengths shorter than 1300Å the best photocathodes are simple metal surfaces, such as tungsten, used directly in vacuum. Most metals exhibit a strong internal photoelectric effect at $\lambda < 1500$Å, which reaches a high value of about 15 per cent at 1000Å. This high work function results in low sensitivity to long-wavelength stray light, and low noise. By coating the metal cathode with LiF, MgF_2, or other compounds, the spectral response curve can be greatly modified and the long wavelength cut-off displaced, to adapt the detector for special applications. Metal photocathodes are available in electrostatically focused photomultipliers with photocathode and dynodes of Be-Cu or stainless steel, and magnetically focused strip-photomultipliers with tungsten photocathodes. The sensitivity is little changed by repeated exposure to air.

Ionization chambers and Geiger counters form another useful class of detectors of XUV radiation. Knowledge of the ionization efficiency of the gas makes it possible to use them for measurement of absolute energy. One of the most useful is filled with NO, responding from the ionization limit, 1350Å, to the transmission limit of the LiF window, 1050Å. To shorter wavelengths, it is possible to use Geiger counters without a window, by maintaining a slight positive gas pressure inside the tube. When filled with a rare gas, they count all incident photons of wavelengths shorter than the ionization limit of the gas, since the gas ionization efficiency is 100 per cent.

Ultraviolet radiation can, of course, be detected with a thermocouple, by direct conversion of its energy into heat. Because of low sensitivity, this fundamental absolute method of energy measurement is resorted to only when no other method is available, for example, in order to establish that the ionization efficiency of the rare gases is 100 per cent for radiation below their series limits.

Mankind's principal source of energy, the sun, emits strongly throughout the ultraviolet, but only the near ultraviolet reaches the surface of the earth. Wavelengths shorter than 2900Å are absorbed by a layer of ozone (O_3) with center at an altitude of about fifteen miles.

The principal action of solar ultraviolet is to produce sunburn, or erythema, but it is only the shortest wavelengths penetrating the ozone layer which have this action. Since the effective band is centered at 2967Å and extends to about 3100Å, sunlight becomes rapidly more effective in burning the skin when the subject goes to a high altitude and when the sun lies high in the sky. Excessive exposure is known to be a cause of skin cancer.

The normal eye does not sense the sun's ultraviolet, although it does have a small sensitivity below 4000Å. Young eyes transmit more than old, and do, indeed, detect ultraviolet from 4000Å to about 3130Å as a faint, bluish sensation, but sharp images are not formed without special corrective lenses, on account of the chromatic aberration of the eye's optical system. To shorter wavelengths, ultraviolet is absorbed by the cornea and causes fluorescence, which is seen as a general haze. Excessive exposure to wavelengths short of 3000Å, however, causes conjunctivitis, a painful burn of the cornea. For this reason, it is extremely important to wear glass goggles, when in the presence of intense sources of middle ultraviolet radiation. Similarly, on snowfields and glaciers, goggles must be worn to prevent the form of conjunctivitis known as snow blindness.

A new and unfortunate action of solar ultraviolet is the production of Los Angeles-type smog, containing molecules which irritate the eye and causes damage to plants. The photochemical processes are complicated and are not as yet completely understood. It is well established, however, that the principal atmospheric contaminants involved are nitrogen oxide, nitrogen dioxide, and various organic molecules present in gasoline engine exhausts. The initial process appears to be the absorption of solar ultraviolet by NO_2, which produces O and NO. The principal reactions, however, are the photolysis of mixtures of nitrogen oxides and hydrocarbons in air, caused by absorption of solar ultraviolet. The products of the reactions are ozone, aldehydes, acrolein, acetone, and peroxyacyl nitrates and nitrites (PAN); among these, formaldehyde, acrolein, and PAN are specific eye irritants. One principal phytotoxicant is ozone, which produces a mottling or bleaching of the upper surfaces of leaves; various others cause a bronzing or glazing of the underneath surfaces of leaves. Obviously, solar ultraviolet cannot be eliminated; photochemical smog relief must come from the chemist.

The principal present-day application of ultraviolet radiation is in increasing the efficiency of conversion of electrical energy into light. The fluorescent lamp utilizes a coating of a crystalline substance, such as manganese-activated zinc silicate, on the inside wall of a mercury arc lamp, to convert the middle- and near-ultraviolet radiation from the mercury vapor column into visible light and thus add to the visible-line spectrum emission of the mercury. The efficiency of fluorescent lamps may reach values 2.5 times greater than those of commonly used incandescent tungsten lamps.

Another widespread use of ultraviolet is in "black lighting," largely for the theater. By introducing an ultraviolet-transmitting, visible-opaque filter over a carbon arc projector, an intense ultraviolet beam can be projected onto the stage. There it causes different materials to glow brilliantly, with color determined by the particular dye molecule.

As a technical industrial tool, ultraviolet spectroscopic analysis has become of extreme importance. In the production of steel, for example, a sample can be analyzed in minutes, by introducing it into a source electrode, and analyzing the radiation with a multichannel spectrometer. Different exit slits select the strongest and most sensitive emission lines of the various elements present; with photomultipliers, their intensities are measured and converted at once to give the composition of the steel.

Among various biological and medical applications of middle-ultraviolet radiation, perhaps the most important are its uses to kill bacteria and fungi in hospitals and especially in operating rooms to eliminate the hepatitis virus from blood plasma, to keep foods sterile, and to treat skin diseases. Rickets and certain other diseases, can be cured by exposure of the body to ultraviolet, which produces vitamin D. Similarly, vitamin D is produced in milk by ultraviolet irradiation.

The middle- and extreme-ultraviolet radiation from the sun, although not able to reach the earth's surface, nevertheless affects man and his activities through its powerful influence on the upper atmosphere. First knowledge of the sun's ultraviolet spectrum was obtained in 1946, when a spectrograph was flown in a V-2 rocket by the U.S. Naval Research Laboratory and the solar spectrum was recorded from 3000 to 2200Å. In the thirty-six years since this event, great progress has been made in studying, from rockets and orbiting vehicles, the true solar spectrum, the reactions produced in the atmospheric gases when the sun's short wavelengths are absorbed, and the physical processes in the solar atmosphere giving rise to these radiations. Grouped together broadly as a space science, they comprise several fields, such as aeronomy, solar physics, solar-terrestrial relationships, and ionospheric research. In the years to come, it is certain that orbiting observatories will monitor the ultraviolet and x-ray emissions from the sun. An important advance in this direction was the solar experimentation carried out by instruments in the Apollo Telescope Mount (ATM) of NASA's Skylab. Operated on a daily basis by the astronauts, these spectrographs and solar imaging devices produced from May 1973 to

February 1974 a wealth of information on the XUV and x-ray emissions from the sun that is not likely to be surpassed for many years. Eventually, orbiting observatories will monitor routinely these solar emissions, complementing the work of the great ground-based observatories that study in visible light and in the near UV, the phenomena taking place in the solar atmosphere. The next major step is now being made from NASA's Space Shuttle, by experiments designed to record with high accuracy the spectrum of the sun's radiation in the UV and XUV. This work is planned to cover at least one entire solar cycle in order to learn more about the variability of the sun's short wavelength radiations and the connections with changes produced in the earth's atmosphere.

As affecting our present-day way of life, perhaps the most important action of solar short-ultraviolet and soft x-rays is in producing the several ionospheric layers; acting as mirrors, they reflect radio waves and so make possible radio communication over great distances. Far more important for mankind, however, is the solar radiation in the Schumann region, 2000 to 1300Å. It is upon this radiation that the human race relies for survival; absorption of these wavelengths by molecular oxygen in the high atmosphere gives rise to atomic oxygen, which then reacts with molecular oxygen to form ozone. It is this permanent layer of ozone which protects all forms of terrestrial life from the lethal effects of the sun's middle-ultraviolet radiation.

RICHARD TOUSEY

References

1. Green, Alex S., Ed., "The Middle Ultraviolet: Its Science and Technology," New York, John Wiley & Sons, Inc., 1966.
2. Samson, J. A. R., "Techniques of Vacuum Ultraviolet Spectroscopy," New York, John Wiley & Sons, 1967.
3. Winick, Herman, and Doniach, S., Eds., "Synchrotron Radiation Research," New York, Plenum Press, 1980.
4. Waynant, Ronald W., and Elton, Raymond C., "Review of Short Wavelength Laser Research," *Proc. I.E.E.E.*, 64, 1059–1092 (1976).
5. Eddy, John A., Guest Editor, "A Collection of Articles describing Optical Instrumentation on the NASA Skylab Spacecraft," *Applied Optics*, 16, 823–1008 (1977).

Cross-references: BREMSSTRAHLUNG, ELECTRICAL DISCHARGES IN GASES, PHOTOELECTRICITY, PLASMAS, SPECTROSCOPY, SYNCHROTRON, X-RAYS.

V

VACUUM TECHNIQUES

The term "vacuum," which strictly implies the unrealizable ideal of a space entirely devoid of matter, is used in vacuum technique to denote gas pressure below the normal atmospheric pressure of 760 torr = 1 bar (1 torr = 1 mm of mercury = 133 pascal; 1 bar = 1000 mbar, where 1 mbar = 1 millibar). The degree of quality of the vacuum attained is indicated by the total pressure of the residual gases in the vessel which is pumped. Table 1 shows the accepted terminology in denoting degrees of vacuum together with the pressure range concerned, the calculated molecular density (from the equation $p = nkT$ where p is the pressure, n is the molecular density, i.e., number of molecules per cubic centimeter, k is Boltzmann's constant, and T is the absolute temperature taken to be 293 K or 20°C) and the mean free path λ from the approximate equation for air: $\lambda = 5/p$ cm, where p is the pressure in millitorr.

Vacuum Pumps There are several types of vacuum pumps. The two most widely used are the mechanical rotary oil-sealed pump and the vapor pump. The former provides a medium vacuum and works relative to the atmosphere; the vapor pump, on the other hand, provides a high or very high vacuum and operates relative to a medium vacuum provided by a rotary pump, referred to as a backing pump in this connection. Thus, the most widely used high-vacuum system able to establish an ultimate pressure of about 10^{-6} torr or below consists of a vapor pump backed by a rotary pump.

Four or five patterns of rotary oil-sealed pump exist, but they have in common the fact that the volume between a rotor (or rotating plunger) and a stator is divided into two-crescent-shaped sections which are isolated from one another as regards the passage of gas. Further, they are furnished with an intake port and a discharge outlet valve to the atmosphere. On revolution of the rotor (speeds of 450 to models with direct motor drive of speeds of 1500 rpm are used) gas is swept from the intake port, compressed, and discharged to the atmosphere via the one-way outlet valve. The mechanism is immersed in a low-vapor-pressure oil for sealing and lubrication in a small pump; larger units have a separate oil reservoir and feed device. A spring-loaded vane type of rotary oil-sealed pump is shown in Fig. 1(a). A single-stage pump of this kind provides an ultimate pressure of about 10^{-2} torr; a two-stage one with two units in cascade will give an ultimate of about 10^{-4} torr. Rotary pumps with speeds from 20 to 20 000 liters/min are commercially available, the smallest being driven by an $\frac{1}{8}$-hp motor, the largest requiring a 40-hp motor.

These pumps handle permanent gases efficiently. Condensable vapors, e.g., water vapor, are not satisfactorily pumped because they may liquefy during the compression part of the rotation. To prevent this, gas ballast is a common provision whereby air from the atmosphere is admitted to the pump through a simple, adjustable screw valve to the region between the rotor and stator just before the discharge outlet valve. The amount of extra air admitted is readily adjusted to provide a compressed gas-vapor mixture which opens the discharge valve before vapor condensation occurs. Gas ballasting will clearly increase significantly the ultimate

TABLE 1. DEGREES OF VACUUM AND PRESSURE RANGES

Degree or Quality of Vacuum	Pressure Range (torr*)	Molecular Density, n (molecules/cm^3)	Mean Free Path, λ (cm)
Coarse or rough vacuum	760–1	2.69×10^{19}–3.5×10^{16}	6.6×10^{-6}–5×10^{-3}
Medium vacuum	1–10^{-3}	3.5×10^{16}–3.5×10^{13}	5×10^{-3}–5
High vacuum	10^{-3}–10^{-7}	3.5×10^{13}–3.5×10^{9}	5–5×10^{4}
Very high vacuum	10^{-7}–10^{-9}	3.5×10^{9}–3.5×10^{7}	5×10^{4}–5×10^{6}
Ultrahigh vacuum	$<10^{-9}$	$<3.5 \times 10^{7}$	$>5 \times 10^{6}$

*1 torr = 133 pascal = 1.32 mbar.

pressure provided by the pump, but this is not important since the gas-ballast valve can be closed after initial pumping has removed most of the water vapor.

Vapor pumps are of two main types: vapor diffusion pumps and vapor ejector pumps. Both employ vapor (of either mercury or a low-vapor-pressure oil) issuing from a nozzle as a means of driving gas in the direction from the intake port to the discharge outlet which is maintained at a medium vacuum by a backing rotary pump. In the diffusion pump [a two-stage design utilizing oil as the pump fluid is shown in Fig. 1(b)] the vapor issuing from the top, first-stage nozzle is directed downward towards the backing region. Gas molecules from the intake port diffuse into the streaming vapor. The directed oil molecules collide with the gas molecules to give them velocity components toward the backing region. A large pressure gradient is thereby established in the pump so that the intake pressure may be over 100 000 times less than the backing pressure. The intake pressure may therefore be 10^{-6} torr or lower with a backing pressure of 10^{-1} torr.

In the diffusion pump, the vapor stream is not essentially influenced by the gas pumped. In the vapor ejector pump, however, the vapor stream is enabled by a higher boiler pressure to be denser and of greater speed with a higher intake pressure, so that the gas is entrained by the high-speed vapor. Viscous drag and turbulent mixing now carry the gas at initially supersonic speeds down a pump housing of diminishing cross section. The ejector pump is designed to operate with a maximum pumping speed at an intake pressure of 10^{-1} to 10^{-3} torr and with a backing pressure of 0.5 to 1 torr or more. The diffusion pump, on the other hand, is designed to have a fairly constant speed from 10^{-3} torr down to an ultimate 10^{-6} torr or much lower in a modern, bakeable stainless steel system.

An important mechanical pump which operates in the same pressure region as the oil ejector is the Roots pump, capable of very great speeds and requiring backing by a rotary oil-sealed pump.

A vacuum system consisting of a diffusion pump and a backing pump, together with baffles, cold traps and isolation valves, is shown in Fig. 1(c). The cold trap is essential if a mercury vapor diffusion pump is used and is best filled with liquid nitrogen ($-196°C$); otherwise, the system will be exposed to the mercury vapor pressure, which is 10^{-3} torr at 18°C.

Ultrahigh-vacuum systems with stainless steel traps and metal sealing gaskets, and bakeable (except for the pumps) for several hours to 300°C, may be constructed on the lines of that shown in Fig. 1(c) to provide an ultimate pressure of 10^{-10} to 10^{-11} torr.

Other vacuum pumps include the sorption type based on the high gas take-up of charcoal or molecular sieve material at liquid nitrogen temperatures. Sorption pumps may be used in place of rotary pumps, with a desirable freedom from rotary pump oil vapor, especially in systems where the amount of gas to be handled is limited.

The chief rivals to the vapor diffusion pump at present are the getter-ion pump of the Penning discharge type with electrodes of titanium, sometimes called the sputter-ion pump, and the turbomolecular pump. The principle of operation of the former is illustrated by Fig. 2, where an egg-box type anode is situated between plane cathodes. The anode-cathode operating potential difference is of the order of 2–10 kV, and the magnetic flux density is about 3,000 gauss (0.3 tesla). The chief pumping action with active gases such as nitrogen and oxygen is to the anode, which receives deposited titanium (which has very high gas affinity) sputtered from the cathodes under the action of the positive ion bombardment. Some gas, especially inert gas like argon and the hydrogen gas present (which is certainly not inert) is pumped to the cathodes.

A typical multicell pump of moderate size of this type has a pumping speed of about 250 liters/sec. Much larger pumps with speeds of up to 5000 liters/sec are commercially available, as are small single-cell units with speeds of some 2 liters/sec.

The sputter-ion pumps provide a vapor-free system giving a so-called dry vacuum, and they are often incorporated in plant with molecular sieve sorption as the auxiliary pump. For medium-size laboratory plant able to provide ultrahigh vacuum they are most attractive. Probably their chief disadvantage is that the life of the pump is only about 40 hours at 10^{-3} torr, but this increases inversely with the pressure, so that it is 40 000 hours at 10^{-6} torr. They are therefore strong rivals to the diffusion pump for plant where moderate amounts of gas are handled in the lower pressure ranges. The turbomolecular pump (usually abbreviated TMP) has probably now become more important than the titanium sputter-ion pump as a rival to the diffusion pump, but it is more costly than either.

Introduced in 1958 by Becker as an important advance on the older molecular drag pumps originated by Gaede in 1913, Holweck's version in 1923, and that of Siegbahn in 1944, the TMP operates relative to a backing pressure of about 10^{-1} torr and provides an inlet pressure of 10^{-8} torr and down to 5×10^{-10} torr and below, provided that suitable bake-out is practiced.

In the turbomolecular pump (Fig. 3) the rotor consists of a special arrangement of separated, slotted disks mounted on a shaft. These disks are interleaved between similar stationary slotted vanes attached to the cylindrical housing. The rotor is driven at speeds of 16,000 rpm or greater to achieve movement of the gas from the inlet to the discharge outlet by virtue of a

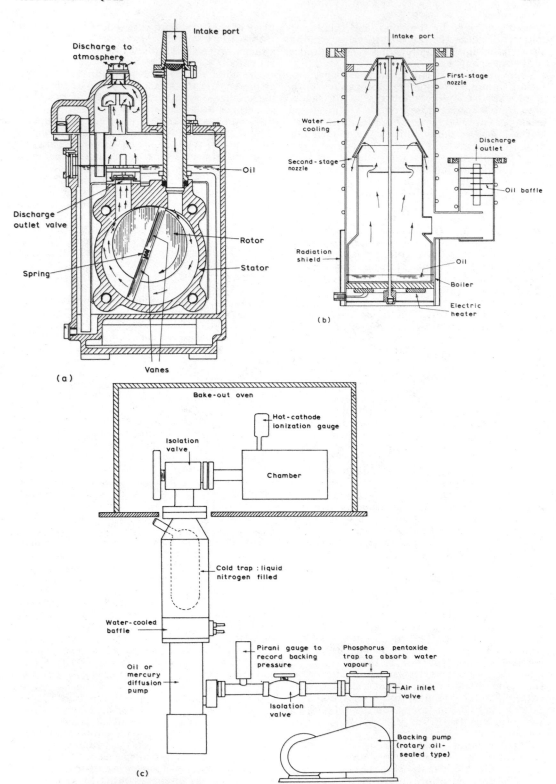

FIG. 1. (a) A spring-loaded vane type of rotary oil-sealed pump. (b) A two-stage or diffusion pump. (c) A diffusion pump-rotary backing pump vacuum system.

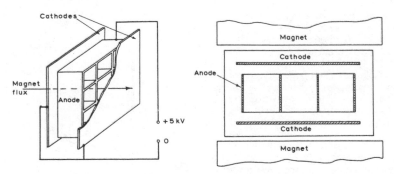

FIG. 2. A sputter-ion pump.

complex entrainment and drag of the gas within the wedge-shaped channels between the angled slots in the adjacent rotor and stator disks. As back-streaming is virtually absent (there is no back-streaming as in a vapor pump, only that of a modest amount of lubrication on the shielded axle bearings), turbomolecular pumps are increasingly being used in systems demanding the production of fluid-free high and ultra-high vacuum.

In addition, models with vertically mounted housings have been introduced instead of the horizontal ones such as that of the original Becker-type pumps. This possibility of vertical mounting enables the TMP to be used more readily in a vacuum system in place of the usu-

ally vertically mounted oil diffusion pump. Examples of the use of TMPs in general are in particle accelerators, where very low pressures are needed and they are used in conjunction with sputter-ion pumps after the initial pumping by conventional mechanical rotary gas-ballast pumps; and in modern electron microscopes and other specialized instruments used in surface science studies, especially in cases where high-speed pumping with freedom from oil contamination at pressures well below 10^{-6} torr is demanded.

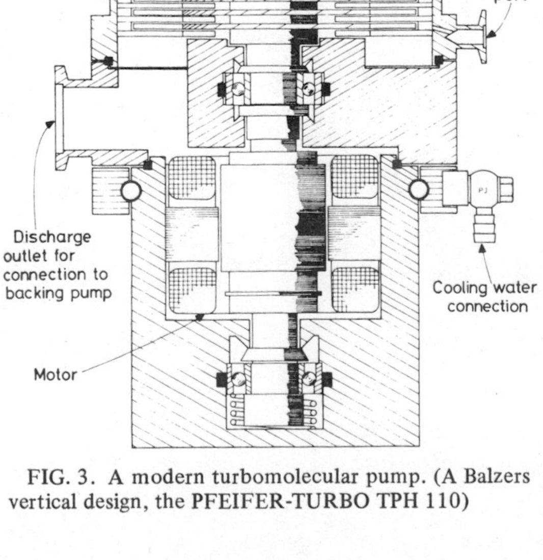

Titanium sublimation pumps (getter pumps) in a range of sizes are frequently used. Typically, the titanium is sublimed from an electrically heated tungsten or tantalum alloy filament loaded with titanium wire to become deposited on the interior wall of the surrounding casing, which may be air-cooled, water-cooled, or refrigerated with liquid nitrogen. The sorption of the active gases by the titanium provides high-speed pumping. Apart from their use as auxiliary pumping devices (particularly in the ultrahigh-vacuum region) frequently adopted practice is to use such pumps in conjunction with a sorption pump and a sputter-ion pump or with a turbomolecular pump backed by a mechanical rotary pump. Indeed, the latter system is able to provide readily ultrahigh vacuum at pressures of 10^{-11} torr or even lower.

Cryogenic pumping is presently receiving much attention. Here, basically, the provision is, within an initially evacuated system (at a total pressure of, say, 10^{-3} torr) a surface which is at such a low temperature that gas impinging on it is condensed. Apart from the fact that this enables an ideal vacuum pump to be produced in locations (such as within the vacuum chamber itself) where pumping is really needed, the use of cryogenic pumping has recently received a considerable boost because of the introduction of mechanical refrigerators (such as the Philips cryogenerator K20) able to produce temperatures as low as 20 K.

If the surface is maintained at the even lower temperature of liquid helium (4 K)—20 K is actually the same as the boiling point of liquid hydrogen—all other gases then have insignificantly low vapor pressures and molecules of

FIG. 3. A modern turbomolecular pump. (A Balzers vertical design, the PFEIFER-TURBO TPH 110)

these gases (except for helium itself and neon) impinging on the surface would remain there. A pumping speed for nitrogen of nearly 12 liters sec^{-1} cm^{-2} of cooled surface is hence theoretically possible. Liquid helium, together with molecular sieve and other sorbent surfaces as well as mechanical cryogenerators, are all being actively investigated with the possibility of providing very high pumping speeds (10^6 liters/sec is not out of question) in space simulators and possible other plant. Some experienced vacuum technologists consider that the future of vacuum pumping methods is going to depend on the developments in cryogenic techniques. As so often happens, however, the question of economy may well be decisive here.

Vacuum Gauges A considerable problem in vacuum technique is that there is only one straightforward gauge able to measure low gas pressures absolutely in the sense that its calibration is independent of the nature of the gas and can be directly referred to millimeters of mercury: the McLeod gauge (Fig. 4). This gauge is clumsy, contains mercury, and is easily broken. It is subject to important errors. It is best not used on a vacuum system, unless the system is intended for absolute low-pressure calibration purposes, and is used by a technologist who knows what he is doing.

More recently there has grown up considerable interest in the capacitance manometer (Fig. 5) (the original model is due to Alpert, Matland, and McCoubrey, as long ago as 1951). This provides a gauge the calibration of which is independent of the nature of the gas; it must, however, be calibrated for absolute pressure

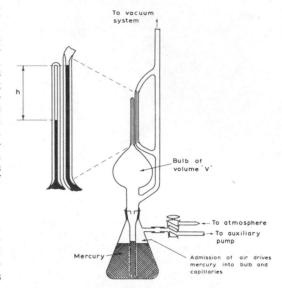

FIG. 4. The McLeod gauge.

measurement against a U-tube manometer or a McLeod gauge. The chief manufacturers of modern capacitance manometers (variously called the Baratron or the Barocel, depending on the manufacturer) have been MKS Instrument Inc. in the USA, who provide an instrument with a digital display of the pressure.

Capacitance manometers also lend themselves very well to the design of pressure-

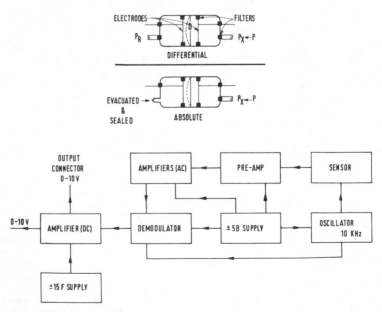

FIG. 5. The capacitance manometer.

controlled switches, in which the output is switched when the gas pressure exceeds a certain amount. This is because they depend essentially on the deflection of a diaphragm by gas pressure. They have also proved to be useful in the design of control systems based on gas flow ratios.

There are several alternative gauges for low-pressure gas measurement, but none of them is an absolute gauge and their calibrations all depend on the nature of the gas. Of the many possibilities, the thermal conductivity guages of the Pirani and thermocouple types are useful within the pressure range from 10^{-4} to 10 torr; these operate by virtue of the dependence of the thermal conductivity of a gas on the pressure at low pressures. Two types of ionization gauge have been used at pressures below 10^{-3} torr. The first of these is the Penning cold-cathode gauge, with a range from 5×10^{-3} to 10^{-7} torr which has been extended in the inverted magnetron type (Redhead gauge) to 10^{-11} torr or below. The second is the widely used hot-cathode ionization gauge (of which the most popular pattern is the Bayard-Alpert gauge, Fig. 6). The Bayard-Alpert gauge has become almost indispensable as a measuring instrument on high and ultrahigh vacuum systems; it has a range from 10^{-3} to 5×10^{-11} torr.

A Bayard-Alpert gauge will typically have a sensitivity for nitrogen of about 12 torr^{-1}; the positive ion current will be 12 μA with an electron current of 1 mA at a nitrogen pressure of 10^{-3} torr, and will decrease in direct proportion to the pressure until at about 10^{-10} torr (or less) the arrival of electrons at the positive

grid causes the production of x-rays. These x-rays are able to release further electrons from the ion collector, and the release of these negatively charged—electrons is indistinguishable from the arrival of pressure-dependent positive ions at the ion collector; i.e., the so-called "x-ray effect" sets the low pressure limit of the hot-cathode ionization gauge. This limit is at about 10^{-8} torr for conventional hot-cathode ionization gauges and about 10^{-10} torr for the Bayard-Alpert type.

In the measurement of gas pressure the determination of the partial pressures of the constituent gases is often as important as a knowledge of the total pressure. Gas analyzers for this purpose are based on the mass spectrometer of the magnetic deflection type, on the radio-frequency mass spectrometer, and on the quadrupole mass spectrometer. These gas analyzers also play an important part in leak detection techniques.

Recent developments concerned with gauges capable of giving an electrical output in the determination of low gas pressures in vacuum systems have been devoted to providing a digital readout in place of the usual analog one, to the link-up of the gauge output to a digital computer so as to provide a continuous record of the total and partial pressures prevailing in the vacuum system (a method which is particularly useful for partial pressure cases, as in the use of the quadrupole mass spectrometer), and to the control and automation of vacuum systems. The last possibility—providing an automatically operated vacuum plant—becomes more difficult as the total pressure in the plant is made lower because of the problems associated with the satisfactory operation of gas-metering and isolation vacuum values, especially when they are only very slightly open.

The applications of vacuum techniques are numerous. They include the coating of a substrate in a vacuum with metallic and insulating films in the production of optical mirrors, electrical resistors, and capacitors; microelectronics (in the manufacture of integrated circuits); and the production of semiconductor active devices by the methods of ion implantation. There should also be mentioned such specialties as antireflection and enhanced reflection coatings on glass, interference filters, and sorption and chemically active layers. The application of vacuum technology to the electron tube industry has been on the wane because with the advent of semiconductor electronics the only electron tube still in wide use is the cathode-ray tube—especially in domestic television. There, the manufacture and re-conditioning especially of color television receiver tubes has involved vacuum technology, but even cathode ray tubes may become obsolete with advances in opto-electronic devices.

Further important fields of activity which involve vacuum technology are electron beam welding (developments are likely to play a

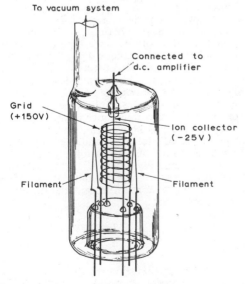

To vacuum system

Connected to d.c. amplifier

Grid (+150V)

Ion collector (−25V)

Filament

Filament

FIG. 6. The Bayard-Alpert hot-cathode ionization gauge.

major part in the automobile manufacturing industry), vacuum drying and freeze-drying, distillation and molecular distillation, vacuum metallurgy (including metal degassing), and space simulation. In some quarters there are also developments in the areas of rough vacuum use, probably due to the great current interest in oil exploration under the seas.

J. YARWOOD
K. J. CLOSE

References

O'Hanlon, J. F., "A User's Guide to Vacuum Technology," New York, John Wiley & Sons, Inc., 1980.

Wutz, M., Adam, H., and Walcher, W., "Theorie und Praxis der Vakuumtechnik," Brunswick, Federal Republic of Germany, F. Vieweg & Sons, 1982.

Holland, L., Steckelmacher, W., and Yarwood, J., "Vacuum Manual," London, E. & F. N. Spon Ltd., 1974.

Cross-references: DIFFUSION IN LIQUIDS, GAS LAWS, IONIZATION, KINETIC THEORY.

VAPOR PRESSURE AND EVAPORATION

Vapor Pressure Vapor pressure is the term applied to the driving force behind the apparently universal tendency for liquids and solids to disperse into the gaseous phase. All known liquids and solids possess this fundamental property, although in some cases it is too minute to be measurable. A typical liquid will exert a vapor pressure which is constant and reproducible. This pressure is dependent only upon the temperature of the system, and increases with increasing temperatures.

The molecular theory explains the phenomenon of vapor pressure through molecular activity. The molecules of a liquid are in rapid motion, even though they are in contact with each other. This motion or activity increases with temperature. At the vapor-liquid interface, this motion results in diffusion of some molecules from the liquid into the vapor. The attraction between molecules is strong, and some of the molecules dispersed into the vapor return to the liquid. The net number of molecules escaping produces the vapor pressure. For all practical purposes, this vapor pressure can be assumed constant whether the system is at equilibrium or not, due to the extremely high rate of molecular diffusion at the interface of the two phases.

In solids, the attractive forces of the molecules are so dominant that each is more or less frozen in place. Some diffusion does occur, however, as evidenced by the evaporation of ice, the odor of moth balls, and the slow diffusion or alloy formation of some metals kept in intimate contact. This vapor pressure increases with temperature, but is also a function of the molecular arrangement of the solid. As some solids such as sulfur are heated and molecular rearrangements take place, forming another allotrope of the same element, the vapor pressure changes sharply as this rearrangement occurs.

Vapor pressure can only be exhibited when the molecular activity is at a low enough level to permit continuous contact of the molecules and thus formation of a liquid. The maximum temperature at which this is possible is a fundamental property and is called the *critical temperature*. Above this temperature the material cannot be compressed to form a liquid, and only one phase results. This temperature is $374.0°C$ ($705.4°F$) for water and $-240.0°C$ ($-399.8°F$) for hydrogen.

The fundamental relationship between temperature and vapor pressure can be derived from thermodynamic laws. With certain limiting assumptions, the Clausius-Clapeyron equation is most often applied:

$$\frac{dp}{dT} = \frac{qP}{RT^2}$$

$\frac{dp}{dt}$ = slope of vapor pressure vs temperature curve at the point in question in cm mercury per °C

q = heat of vaporization in calories per gram mol

P = pressure in cm of mercury

R = gas constant in calories per °C per gram mol

T = temperature in Kelvin

Because of these limiting assumptions, the integrated form of this equation is used in practice primarily as a guide to develop methods of correlating and plotting vapor pressure data.

The vapor pressure of a solution containing a nonvolatile substance (e.g., salt in water) is lower than that of the pure liquid. This phenomenon can be explained by interference with the liquid molecular activity by the dissolved substances. The relationship between this vapor pressure depression and the concentration of the dissolved substance is valid for most substances at low concentrations. It was found to be dependent on the relative numbers of molecules of the solute and the solvent, and allowed accurate determinations of molecular weights of unknown solutes. If the Clausius-Clapeyron equation given above is combined with the above concentration relationship, it can be shown that:

$$\Delta T = \frac{RT^2}{q} \cdot C$$

where ΔT is the elevation of the boiling point. and C is the mol ratio of solute to solvent. This defines the effect of any solute on the vapor pressure exhibited by any solvent of latent heat q.

In the same manner, the vapor pressure of one component of a solution of two liquids has a different relationship with temperature than if it were pure. For many liquid mixtures, such as most hydrocarbon mixtures, the vapor pressures of the components vary directly from that exhibited in the pure form as their molar concentration in the solution. This relationship is known as Raoult's law:

$$\text{Partial pressure} = P_0 x$$

where x is the molar concentration of the component in the liquid and P_0 is the vapor pressure of the pure component at the same temperature as the mixture. A mixture following this rule is called an *ideal* solution, and its total volume is the sum of its components' volume.

If the gas phase above the liquid is also "ideal," the partial pressure of a component in this phase is equal to the total system pressure times the mole fraction of the component in the gas phase. This is called Dalton's law:

$$\text{Partial pressure} = P_t y$$

Combining these two formulas, it can be seen that:

$$\frac{y}{x} = \frac{P_0}{P_t} = K$$

for any particular temperature. This relationship can largely define many very complex liquid mixtures if the pressures used in the correlation are corrected by experimental data for deviation from the ideal.

A different relationship results if two liquids are relatively immiscible in each other. Molecular interference is minimal, and the total pressure exerted is equal to the sum of that of the individual pure components.

The fundamental property of vapor pressure is thus dependent on temperature and composition of the material considered. These known and reproducible relationships have great technical application.

Vapor pressure relations can be used to determine heats of solution, heats of sublimation, and heats of fusion. Problems dealing with the solution of gases in liquids and adsorption of gases by solids are best handled by vapor pressure concepts. In dealing with solutions of miscible liquids, the most simple and useful relationship involves plotting the mole fraction y of one component in the vapor against x, the mole fraction of the component in the liquid. The ratio of y/x is called the phase equilibrium constant K and is used for definition of bubble points and dew points of simple and complex hydrocarbon mixtures over temperature and pressure ranges to near the critical.

Some equations of state also define vapor pressure relationships. One of the most recognized is that developed by Benedict, Webb, and Rubin due to its ability to predict P-V-T properties in the two-phase region and to describe behavior of the superheated vapor. See *J. Chem. Phys.*, **8**, 334 (1940); ibid., **10**, 747 (1942); and *Chem. Eng. Progr.*, **47** (1951). The utility of this Benedict, Webb, and Rubin (BWR) equation has been expanded greatly since it was first proposed in 1940. It is now an indispensable tool in process design work. References for existing compounds, for utilization of experimental data, for mixtures, and for extending the BWR application where normal procedures cannot be applied are listed below:

(1) Existing compounds: Cooper, H. W., and Goldfrank, J. C., *Hydrocarbon Processing*, **46**, 141 (1967); Orye, R. V., *I & EC Process Design and Devel.*, **8**, 579 (1969).

(2) Improvement in agreement between experimental data and prediction from the BWR for *pure* components: (a) Improvements in methods of evaluating constants from experimental data: Selleck, F. T., Opfell, J. B., and Sage, B. H., *Ind. Eng. Chem.*, **45**, 1350, (1953); Eubank, P. T., and Fort, B. F., *Can. J. Chem. Eng.*, **47**, 177 (1969); Cox, K. W., Bond J. L., Kwok, Y. C., and Starling, K. E., *Ind. Eng. Chem. Fundam.*, **10**, 245 (1971). (b) Adjustment of constants as a function of temperature: Kaufmann, T. G., *I & E.C., Process Design and Devel.*, **7**, 115 (1968); Motard, R. L., and Organick, E. I., *A.I.Ch.E. Journal*, **6**, 39 (1960). (c) Addition of new constants: Starling, K. E., Natural Gas Processors Association Enthalpy Project Progress Report, Sept. 30, 1968 (Development of the BWRS); Nishiumi, H., and Saito, S., *J. Chem. Eng. Japan*, **8**, 356 (1975).

(3) Improvement in agreement for mixtures by including interaction parameters: Bishnoi, P. R., and Robinson, D. B., *Can. J. Chem. Eng.*, **50**, 506 (1972); Gugnoni, R. J., Eldridge, J. W., Okay, V. C., and Lee, T. J., *A.I.Ch.E. Journal*, **20**, 357 (1974).

(4) Extending the application of the BWR to systems where the experimental data is insufficient for normal procedures: Edmister, W. C., Vairogs, J., and Klekers, A. J., *A.I.Ch.E. Journal*, **14**, 479 (1968); Kaufmann, T. G., *I & E.C. Process Design and Devel.*, **7**, 115 (1968).

Evaporation The above effects of mixtures and of solutes on vapor pressure of a component are the very reasons why continuous generation of vapor (or *evaporation*) is the major tool of most process separations.

These effects permit us to separate salts from solutions, to separate liquid components from mixtures, and to use the energy relationships in process control of all kinds.

When molecules of a liquid do leave the surface and become vapor, they do so by overcoming the rather large attractive forces existing when they were in the liquid state. These forces were large since the molecules were in very close proximity in the liquid. Overcoming these attractive forces requires *energy*, heat energy, this is named the *"latent heat"* or "heat of vaporization" of the fluid. In general, this is in terms of heat units per weight unit of material, such as Btu per pound, or calories per gram. In magnitude, latent heat will decrease as the liquid temperature increases or as the kinetic energy of the molecules increases. At the critical temperature, there is no latent heat—the molecules are in such an excited state that formation of a liquid is not possible.

For continuous evaporation, a continuous supply of energy is required. An available utility such as steam is a typical source.

Depending on the process, the evaporation is done:

(1) In equipment named "evaporators" such as the popular LTV of forced or natural circulation design used in acid concentration, salt production, sugar solution concentration and others.

(2) In distillation towers in a stepwise manner, tray to tray, resulting in a slight change in composition at each tray until required terminal conditions of overhead and bottom composition are reached.

(3) In reactors of various designs where one or more components are driven off, frequently by the heat of reaction.

(4) In "cooling towers" where the desired effect is not the separation or concentration of components but the use of the latent heat of evaporating water to remove unwanted process heat.

(5) In all steam generating boilers.

(6) In any process step where a liquid/vapor phase change occurs.

The concentration and energy relationships discussed before apply.

Evaporation is thus the most widely used tool of nature and of industry. For this reason, an intimate knowledge of the theory of heat transfer, of the large quantity of experimental data available, and of the more sophisticated methods of utilizing equations of state to correlate these data, are all essential to understand and properly develop almost any industrial process.

DOUGLAS L. ALLEN

Cross-references: GAS LAWS, HEAT TRANSFER, KINETIC THEORY, LIQUID STATE, MOLECULES AND MOLECULAR STRUCTURE, SOLID-STATE PHYSICS, STATES OF MATTER, SURFACE PHYSICS.

VECTOR ANALYSIS

Introduction The concept of vectors and vector analysis was formulated in the last part of the 19th Century by Gibbs and others as an alternative to Hamilton's quaternions.[4] These concepts form an invaluable tool in the formulation and analysis of a wide range of physical problems, for two reasons. First, vector notation provides a compact and convenient mathematical description of many complex physical situations. Second, if the mathematical model of a physical situation is expressed in terms of vector equations, this model will be independent of any coordinate system or frame of reference. The generalization and abstraction of elementary vector concepts leads to such diverse mathematical structures as tensor analysis and the theory of linear vector spaces, which are the basic mathematical foundations of much of modern theoretical physics.

Vector Algebra[2,7,8,10,12,14] Although it is usually appropriate to begin a discussion of mathematical entities by defining the quantity under discussion, this is not the case for vectors. There are several reasons why it is not feasible to give an elementary definition of a vector. Chief of these is the fact that the definition is very dependent on the context in which it is used. Second, for the vectors discussed here, any initial definition would have to be greatly modified as the discussion proceeds. The quantities considered here are undefined geometrical entities called *vectors* (denoted by **boldface** letters) which can be represented by directed line segments in the two- or three-dimensional spaces of ordinary experience.

In the development of an algebra for geometrical vectors, it is necessary to introduce an additional set of quantities known as *scalars*. These scalars are elements of what is technically known as a commutative field. It is sufficient here to let the set of scalars be the real numbers. In a two- or three-dimensional space which satisfies the postulates of Euclidean geometry, the representation of geometric vectors by directed line segments endows them with the properties of length and direction. The length or magnitude of a vector is the length of its representative line segment, and its direction is represented by the direction of the line segment.

The preliminary algebraic concepts required are the idea of equality of two vectors, the addition of vectors, and the multiplication of a vector by a scalar. Two vectors **A** and **B** are said to be *equal* if their representative line segments are parallel and of the same length. Additional restrictions may be required on the equality of vectors when they represent physical quantities. This definition of equality implies that a given vector is unaltered, i.e., is invariant, when displaced parallel to itself. *Addition* of two vectors **A** and **B** is defined in terms of geometrical construction with their representative line segments. (See Fig. 1.) Bring the tail of the line segment **B** into coincidence with the head of the line segment **A** by means of a parallel displacement. Construct the directed line segment from the tail of **A** to the head of **B**. This line segment **C** represents the vector sum of the vec-

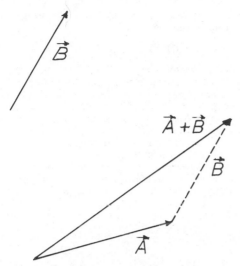

FIG. 1. The addition of vectors.

tors **A** and **B**,

$$C = A + B.$$

The *product of the scalar α and the vector* **A**, written α**A**, is represented by the line segment parallel (anti-parallel if $\alpha < 0$) to that representing **A** of length $|\alpha|\,\|\mathbf{A}\|$, where $\|\mathbf{A}\|$ is the magnitude of **A**.

These definitions of the sum of two vectors and the scalar multiple of a vector have the following properties:

(1) For every pair of vectors **A** and **B**, their sum **A** + **B** is a vector. This property is known as *closure under addition*.

(2) Addition is *commutative*, that is

$$A + B = B + A.$$

(3) For every three vectors **A**, **B**, and **C**,

$$A + (B + C) = (A + B) + C.$$

This property is known as *associativity of addition*.

(4) There is a vector **0**, known as the *null vector*, such that

$$A + 0 = A$$

for all vectors **A**.

(5) Corresponding to each vector **A**, there is a *negative* vector (−**A**) such that

$$A + (-A) = 0.$$

The negative vector (−**A**) is represented by a line segment of the same length as that representing **A** and anti-parallel to **A**.

(6) For every vector **A** and every scalar α, there is a unique vector **B** = α**A**, i.e., *the set of all vectors is closed under scalar multiplication*.

In particular, there is a scalar 1 such that 1**A** = **A** for all vectors **A**.

(7) *Scalar multiplication is cummutative and associative*, that is, for every vector **A** and every pair of scalars α and β,

$$(\alpha + \beta)\mathbf{A} = \alpha\mathbf{A} + \beta\mathbf{A}$$

$$\alpha(\beta\mathbf{A}) = (\alpha\beta)\mathbf{A} = \beta(\alpha\mathbf{A}).$$

(8) *Scalar multiplication distributes over addition*. Given two vectors **A** and **B** and a scalar α,

$$\alpha(A + B) = \alpha A + \alpha B.$$

Used in an abstract setting, these properties form part of the definition of an abstract linear vector space.

There are two additional operations, which are essentially algebraic, that can be defined for every pair of vectors **A** and **B**. The first of these is the *inner* or *dot product*, which associates a scalar with each pair of vectors **A** and **B**. This unique scalar is denoted **A** · **B**, and is defined in the following way: Let **A** and **B** have the same initial point, and let θ denote the angle between their representative line segments, measured from **A** to **B** as shown in Fig. 2. Then,

$$A \cdot B = \|A\|\,\|B\|\,\cos\theta.$$

It can be shown, although the proof is by no means trivial, that the inner product satisfies the following identities:

(9)   **A** · **B** = **B** · **A**.
(10) **A** · (**B** + **C**) = **A** · **B** + **A** · **C**.
(11)   α(**A** · **B**) = (α**A**) · **B** = **A** · (α**B**).
(12)   **A** · **A** ≥ 0, with the equality holding if and only if **A** = **0**.

The second product, known as the *cross product*, associates with each pair of vectors **A**

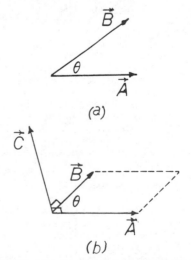

FIG. 2. The dot (a) and cross (b) products.

and **B** a vectorlike object denoted by

$$C = A \times B$$

which is defined as follows: Let **A** and **B** have a common initial point, and let the angle θ between them be measured from **A** to **B**. Then **C** is defined to have a magnitude

$$\|C\| = \|A \times B\| = \|A\| \, \|B\| \sin \theta$$

and a direction normal to the plane of **A** and **B** in the direction of advance of a right-hand screw as **A** is rotated into **B**. This construction is also illustrated in Fig. 2. The arbitrary choice of direction for the cross product prevents it from having all of the properties required of vectors. Quantities of this type are called *pseudo-* or *axial vectors*. In most applications the distinction between axial vectors and true vectors is unimportant. For any three vectors **A**, **B**, **C**, and every scalar α, the following identities are satisfied:

(13) $\qquad A \times B = - B \times A.$
(14) $\quad A \times (B + C) = A \times B + A \times C.$
(15) $\quad (\alpha A) \times B = A \times (\alpha B) = \alpha A \times B.$
(16) $\; A \times (B \times C) = (A \cdot C) B - (A \cdot B) C.$

There are a number of important consequences of the identity (16). One of the most important of these is the Jacobi identity:

$$A \times (B \times C) + B \times (C \times A)$$

$$+ C \times (A \times B) = 0.$$

The dot and cross products have a number of elementary applications to the three-dimensional geometry of lines and planes. Some of these are:[17]

(1) Two lines L_1 and L_2 are parallel (antiparallel) if and only if $L_1 \times L_2 = 0$, where L_1 and L_2 are vectors whose representative line segments lie on the given lines.

(2) Two lines L_1 and L_2 are perpendicular if and only if $L_1 \cdot L_2 = 0$.

(3) Relative to an arbitrary origin O, the vector **R** is described by the directed line segment from O to the point P. Hence, the point P is specified by the vector **R**, and there is a one-to-one correspondence between the points in space and the position vectors **R**. The equation of a line through the point P_0 parallel to a given line L can be written as $(R - R_0) \times L = 0$, where **R** is any other point on the line.

(4) The equation of a plane through the point P_0 perpendicular to a given vector **N** can be written as $(R - R_0) \cdot N = 0$, where **R** is the position vector of any other point in the given plane.

There is an additional product of three vectors which finds frequent application. If **A**, **B**, and **C** are three noncoplanar vectors, the scalar $A \cdot (B \times C)$ is equal to the volume of the parallelepiped defined by the three vectors. The mixed *triple product* satisfies the following commutation relations:

$$A \cdot (B \times C) = B \cdot (C \times A) = C \cdot (A \times B).$$

Vector Fields[1,2,7,8,10,11,12,13,14] In order to have a differential and integral calculus for vector quantities, it is necessary to introduce the concept of a *vector field*. A vector field associates with each point in some region of space a unique vector $\mathbf{F}(P)$. Although it is not necessary, it is convenient to describe the points in the given region and the associated vectors in terms of some coordinate system. A particularly simple choice of coordinates is the set of rectangular cartesian coordinates, with arbitrary origin, as shown in Fig. 3.

Before a description of the vector field $\mathbf{F}(P)$ can be obtained, a representation of the algebraic operations on vectors in terms of cartesian coordinates must be given. Introduce the triplet of unit vectors (vectors with unit magnitude) $\{i, j, k\}$ along the X-, Y-, and Z-coordinate axes, respectively, so that

$$i \times j = k, \quad j \times k = i, \quad k \times i = j.$$

This construction defines a *right-handed coordinate system*. In terms of the set of unit vectors $\{i, j, k\}$, a given vector **F** has the representation

$$F = F_1 i + F_2 j + F_3 k$$

where

$$F_1 = F \cdot i, \quad F_2 = F \cdot j, \quad F_3 = F \cdot k$$

are known as the *cartesian components of* **F**. In terms of the cartesian components, the various algebraic operations have the following representations:

$$(A + B) = (A_1 + B_1) i + (A_2 + B_2) j$$

$$+ (A_3 + B_3) k$$

$$\alpha A = \alpha A_1 i + \alpha A_2 j + \alpha A_3 k$$

$$A \cdot B = A_1 B_1 + A_2 B_2 + A_3 B_3$$

$$A \times B = (A_2 B_3 - A_3 B_2) i + (A_3 B_1 - A_1 B_3) j$$

$$+ (A_1 B_2 - A_2 B_1) k.$$

A given point P in space can be represented in terms of its cartesian coordinates, or its position vector $R = xi + yj + zk$. Then, a vector field is described by specifying the cartesian components of **F** as functions of the cartesian coordinates of P, i.e.,

$$F(P) = F_1(x, y, z) i + F_2(x, y, z) j$$

$$+ F_3(x, y, z) k.$$

Since a given vector is uniquely described by its cartesian components, the vector field $\mathbf{F}(P)$ is

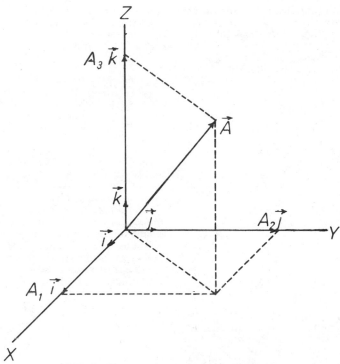

FIG. 3. The cartesian components of a vector.

said to be *differentiable if and only if each of* its cartesian components is a differentiable function of the coordinates.

If **F(R)** is a differentiable vector field, and $\phi(x, y, z)$ is a differentiable scalar function of the coordinates, the three basic differential operations involving vectors have the following cartesian representations:

$$\text{grad } \phi = \frac{\partial \phi}{\partial x} \mathbf{i} + \frac{\partial \phi}{\partial y} \mathbf{j} + \frac{\partial \phi}{\partial z} \mathbf{k}$$

$$\text{div } \mathbf{F} = \frac{\partial F_1}{\partial x} + \frac{\partial F_2}{\partial y} + \frac{\partial F_3}{\partial z}$$

$$\text{curl } \mathbf{F} = \left[\frac{\partial F_3}{\partial y} - \frac{\partial F_2}{\partial z} \right] \mathbf{i} + \left[\frac{\partial F_1}{\partial z} - \frac{\partial F_3}{\partial x} \right] \mathbf{j}$$

$$+ \left[\frac{\partial F_2}{\partial x} - \frac{\partial F_1}{\partial y} \right] \mathbf{k}.$$

These vector derivatives are known as the *gradient* of a scalar and the *divergence* and *curl* of a vector, respectively. The forms of these derivatives suggest the symbolic representations grad $\phi = \nabla \phi$, div $\mathbf{F} = \nabla \cdot \mathbf{F}$, curl $\mathbf{F} = \nabla \mathbf{X} \mathbf{F}$, where ∇ (del) is the vector differential operator

$$\nabla = \mathbf{i} \frac{\partial}{\partial x} + \mathbf{j} \frac{\partial}{\partial y} + \mathbf{k} \frac{\partial}{\partial z}.$$

The cartesian representations are inadequate as definitions of the derivative operations, since they are coordinate dependent. They do, however, provide a convenient method of computation.

Two important results can be obtained from the cartesian representations of the divergence and curl: (1) $\nabla \mathbf{X} \mathbf{F} = 0$ if and only if $\mathbf{F} = \nabla \phi$ for some scalar function ϕ; and (2) $\nabla \cdot \mathbf{F} = 0$ if and only if $\mathbf{F} = \nabla \mathbf{X} \mathbf{A}$ for some vector field **A**. The quantities ϕ and **A** are known as the *scalar* and *vector potentials* for the field **F**, respectively. It can be shown that any vector field **F** is determined by its scalar and vector potentials through the relation

$$\mathbf{F} = \nabla \phi + \nabla \mathbf{X} \mathbf{A}.$$

This result which is important in electromagnetic theory, is known as *Helmholtz' theorem*.

Symmetry considerations frequently make it desirable to describe the points of space in terms of coordinate systems which are not cartesian. In these cases, the descriptions of vectors and the derivative operations are somewhat more complicated. The form of the derivative operators with respect to these curvilinear coordinate systems must be derived from coordinate-free definitions of the gradient, divergence, and curl. Suppose that $\phi(P)$ and $\mathbf{F}(P)$ are differentiable scalar and vector fields respectively. Enclose any point P in the domain of definition by a small closed surface ΔS, and let $\mathbf{n}$ denote a unit vector at each point on ΔS which points out of the enclosed volume and is normal to ΔS (unit outward normal). If ΔV represents the volume enclosed by ΔS, the gradient, divergence, and curl are defined by the limiting processes

$$(\nabla \phi)(P) = \lim_{\Delta V \to 0} \frac{\iint_{\Delta S} \phi \, \mathbf{n} \, dA}{\Delta V}$$

$$(\nabla \cdot \mathbf{F})(P) = \lim_{\Delta V \to 0} \frac{\iint_{\Delta S} \mathbf{F} \cdot \mathbf{n} \, dA}{\Delta V}$$

$$(\nabla \times \mathbf{F})(P) = \lim_{\Delta V \to 0} \frac{\iint_{\Delta S} \mathbf{n} \times \mathbf{F} \, dA}{\Delta V}$$

if the limits exist and are independent of the way in which ΔV tends to zero. General discussions of curvilinear coordinates and explicit representations in the more common curvilinear coordinate system are given in Refs. 1, 7, 11, and 13.

In three dimensions, there are essentially three types of integrals involving vector fields. These are: (1) Integrals along some space curve C between two points P and Q; (2) integrals over an open surface S, bounded by some closed curve C; and (3) integrals over a volume enclosed by a closed surface S. The first of these integrals is motivated by the physical concept of the work done by a variable force $\mathbf{F}$ acting along the curve C. If $\mathbf{R}$ denotes the position vector of the points on $\mathbf{C}$, and s is the distance from P to R measured along C, then $d\mathbf{R}/ds$ is the vector tangent to C at $\mathbf{R}$. The scalar $\mathbf{F} \cdot (d\mathbf{R}/ds)$ is the component of the vector field $\mathbf{F}$ tangent to the curve C at each point on C. The integral

$$\int_C \mathbf{F} \cdot (d\mathbf{R}/ds) \, ds = \int_C \mathbf{F} \cdot d\mathbf{R}$$

is called the *line integral* of $\mathbf{F}$ from P to Q along the curve C. In general, the value of the integral depends not only on the end points P and Q, but also on the particular curve C connecting these points. If the integral is independent of the curve, the vector field $\mathbf{F}$ is called *conservative*. It is clear that the line integral of a conservative vector field vanishes whenever the curve C is closed. A necessary and sufficient condition for $\mathbf{F}$ to be conservative is that $\nabla \times \mathbf{F} = 0$.

Integrals involving the curl of a vector field over an open surface can frequently be trans-formed into line integrals over the bounding curve through the use of *Stokes's theorem*. Let S be an open surface with unit positive normal $\mathbf{n}$ which is bounded by the simple closed curve C. Then,

$$\iint_S (\nabla \times \mathbf{F}) \cdot \mathbf{n} \, dA = \oint_C \mathbf{F} \cdot d\mathbf{R}.$$

This result, which can be proved from the general definition of the curl, has an important role in the theory of differential forms.[5] In an elementary setting, Stokes' theorem is an extension to three dimensions of the two-dimensional Green's theorem.[17]

There is also a relation between the integral of the divergence of a vector field over a closed volume and the surface integral of the component of the vector field normal to the surface enclosing the volume. This relation, which follows immediately from the definition of the divergence, is known as the *divergence theorem*. Symbolically, the theorem can be stated as

$$\iiint_V \nabla \cdot \mathbf{F} \, dv = \iint_S \mathbf{F} \cdot \mathbf{n} \, dA$$

where S is any closed surface with unit outward normal $\mathbf{n}$ which encloses the volume V.

A combination of the divergence and gradient operations, known as the *Laplacian*, occurs in many of the partial differential equations of mathematical physics. The list of such equations included Poisson's equation, the heat equation, the diffusion equation, the wave equation and Schrödinger's equation. If ϕ is a twice differentiable scalar function, the Laplacian of ϕ is defined as $\nabla^2 \phi = \nabla \cdot \nabla \phi$. In terms of a cartesian coordinate system (x, y, z), the Laplacian of ϕ has the explicit representation

$$\nabla^2 \phi = \frac{\partial^2 \phi}{\partial x^2} + \frac{\partial^2 \phi}{\partial y^2} + \frac{\partial^2 \phi}{\partial z^2}.$$

The definition of the Laplacian of a vector field $\mathbf{F}(\mathbf{R})$ is somewhat more complicated. The cartesian components of the combination

$$\nabla \times (\nabla \times \mathbf{F}) - \nabla(\nabla \cdot \mathbf{F})$$

satisfy the partial differential equation

$$[\nabla \times (\nabla \times \mathbf{F}) - \nabla(\nabla \cdot \mathbf{F})]_i$$

$$= \frac{\partial^2 F_i}{\partial x^2} + \frac{\partial^2 F_i}{\partial y^2} + \frac{\partial^2 F_i}{\partial z^2} = \nabla^2 F_i.$$

Hence, the Laplacian of the vector field $\mathbf{F}$ is generally defined by

$$\nabla^2 \mathbf{F} = \nabla \times (\nabla \times \mathbf{F}) - \nabla(\nabla \cdot \mathbf{F}).$$

Vector Mechanics[3,6,7,16] The elementary use of vectors in mechanics arises from the fact that many mechanical quantities such as displacement, velocity, acceleration, and force require both a magnitude and direction for their complete specification. It was known before the introduction of vector concepts that these mechanical quantities combine according to a parallelogram law.

Consider a particle of mass m whose position is space at any given time t is specified by the cartesian coordinates $[x(t), y(t), z(t)]$ which are twice differentiable functions of t. The instantaneous position of the particle is alternatively described by the position vector

$$\mathbf{R}(t) = x(t)\mathbf{i} + y(t)\mathbf{j} + z(t)\mathbf{k}.$$

Velocity and acceleration vectors are defined as the first and second time derivatives of the position vector:

$$\mathbf{v}(t) = \frac{d\mathbf{R}}{dt} = \frac{dx}{dt}\mathbf{i} + \frac{dy}{dt}\mathbf{j} + \frac{dz}{dt}\mathbf{k}$$

$$\mathbf{a}(t) = \frac{d^2\mathbf{R}}{dt^2} = \frac{d^2x}{dt^2}\mathbf{i} + \frac{d^2y}{dt^2}\mathbf{j} + \frac{d^2z}{dt^2}\mathbf{k}.$$

The specification of the acceleration vector and specific values of the velocity and position vectors at $t = 0$ gives a complete kinematical description of the motion of the particle.

The dynamical description of the particle involves the relation between the particle's motion and its interaction with its environment. The interaction of the particle and its environment is described by a force vector $\mathbf{F}$, which is usually a function of the position of the particle, i.e. the environment gives rise to a force field which interacts with the particle. The vector sum $\mathbf{F} = \sum \mathbf{F}_i$ of all of the forces acting on a given particle is related to the motion of the particle through Newton's second law, which has the vector form

$$\mathbf{F} = d\mathbf{p}/dt$$

where the linear momentum $\mathbf{p}$ is the product of the (scalar) particle mass with its linear velocity $\mathbf{v}$. This relation gives a dynamical description of the motion of a simple mass particle.

In mechanics, it is not always possible to consider "free" vectors (vectors which are translationally invariant). As an example of this, consider a particle of mass m which moves under the action of a force $\mathbf{F}$, but which is constrained to remain a fixed distance from a given axis. This may be accomplished by fixing the particle at one end of a massless rod which is free to rotate about an axis through the other end. In this case, the rod exerts an unknown force on the particle which maintains the fixed distance. The only possible motion of the particle is a rotation about the axis, and we wish to find a description of this motion. It is convenient to define an *angular velocity vector* ω which is along the axis of rotation such that

if $\mathbf{R}$ is the instantaneous position vector of the particle relative to the axis, the linear velocity of the particle is given by

$$\mathbf{v} = \omega \times \mathbf{R}.$$

The time derivative of the angular velocity is defined to be the *angular acceleration* α. Differentiation with respect to time of the relation between the linear and angular velocities results in

$$\mathbf{a} = \alpha \times \mathbf{R} + \omega \times (\omega \times \mathbf{R}).$$

The second term in the equation for the linear acceleration is called the *centripetal acceleration*, and results from the force on the particle exerted by the rod. An analog of Newton's second law for the rotational motion can be obtained by first defining an angular momentum for the particle. If the particle has linear momentum $\mathbf{p}$, the angular momentum is defined to be $\boldsymbol{\ell} = \mathbf{R} \times \mathbf{p}$. Similarly, the torque acting on the particle is defined as $\tau = \mathbf{R} \times \mathbf{F}$. It is clear in this case, that the vectors $\mathbf{p}$ and $\mathbf{F}$ are not free, but must be attached to the particle (the concepts of free and fixed vectors are discussed in detail in Ref. 8). The analog of Newton's second law is

$$\tau = d\boldsymbol{\ell}/dt.$$

Detailed discussions of rotational motion and the motion of rigid bodies in vector terms is given in Ref. 16.

Many other mechanical concepts are most easily expressed in vector terms. Some of these are concepts such as the work done by a variable force moving along a given curve C. This work is given by the line integral

$$W = \int_C \mathbf{F} \cdot d\mathbf{r}.$$

If the work integral is independent of the path, the force $\mathbf{F}$ is conservative, and hence can be derived as the gradient of a scalar potential function. In particular, $\mathbf{F}$ is defined as

$$\mathbf{F} = -\nabla\phi.$$

Then, the scalar ϕ defines the potential energy of a particle in the given force field. Similarly, on the basis of Newton's second law and the work integral, the kinetic energy of a particle is defined as

$$T = \frac{\|\mathbf{p}\|^2}{2m}.$$

Electromagnetic Theory[7,9,15]   The study of the electromagnetic field employs the full power of vector notation and the analysis of vector quantities. The fundamental laws governing the electromagnetic field are expressed by the set of four Maxwell equations, a set of constituitive relations, and an equation of continuity which expresses the conservation of electric charge. The set of Maxwell equations is:

$$\nabla \times \mathbf{E} = \frac{\partial \mathbf{B}}{\partial t}$$

$$\nabla \cdot \mathbf{D} = \rho$$

$$\nabla \times \mathbf{H} = \mathbf{J} + \frac{\partial \mathbf{D}}{\partial t}$$

$$\nabla \cdot \mathbf{B} = 0.$$

In these equations, $\mathbf{E}$ is known as the *electric field intensity*, $\mathbf{D}$ the *electric displacement*, $\mathbf{H}$ the magnetic field intensity, $\mathbf{B}$ the *magnetic induction*, $\mathbf{J}$ the *electric current density*, and ρ the *electric charge density*. The electric field intensity and the electric displacement are related by the constituitive equation

$$\mathbf{D} = \epsilon \mathbf{E}$$

where ϵ is the *electric permittivity* and characterizes the electrical properties of the medium where the field exists. Similarly, the magnetic intensity and the magnetic induction are related by $\mathbf{B} = \mu \mathbf{H}$, where the magnetic permeability μ characterizes the magnetic properties of the medium. The final equation defining the electromagnetic field is the *equation of continuity:*

$$\nabla \cdot \mathbf{J} + \frac{\partial \rho}{\partial t} = 0.$$

Maxwell's equations can be derived from the experimental laws of Coulomb, Ampère, Biot-Savart, and Faraday.[7,9]

At a surface separating two different media which are characterized by $\{\epsilon_1, \mu_1\}$ and $\{\epsilon_2, \mu_2\}$ respectively, the four vector fields satisfy the boundary conditions

$$(\mathbf{B}_2 - \mathbf{B}_1) \cdot \mathbf{n} = 0$$

$$(\mathbf{D}_2 - \mathbf{D}_1) \cdot \mathbf{n} = \omega$$

$$\mathbf{n} \times (\mathbf{E}_2 - \mathbf{E}_1) = 0$$

$$\mathbf{n} \times (\mathbf{H}_2 - \mathbf{H}_1) = \mathbf{K}$$

where ω and $\mathbf{K}$ are a surface charge density and a surface current density, respectively. The unit normal $\mathbf{n}$ points from region 1 into region 2.

Much of modern electromagnetic theory consists of finding solutions of the set of Maxwell equations for given current and charge densities and specified boundaries between media. Details of such a program are given in Refs. 9 and 15.

FRED A. HINCHEY

References

The literature on vectors and their application to physical problems is so extensive that it is impossible to give a complete bibliography. The references listed are a representative sample of some of the more recent literature.

1. Arfken, G., "Mathematical Methods for Physicists," 2nd Ed., New York, Academic Press, 1970.
2. Chisholm, J. S., "Vectors in Three-Dimensional Space," New York, Cambridge Univ. Press, 1978.
3. Corben, H. C., and Stehle, P., "Classical Mechanics," 2nd Ed., New York, John Wiley & Sons, 1974.
4. Crowe, M. J., "A History of Vector Analysis," Notre Dame, Ind., Univ. Notre Dame Press, 1967.
5. Flanders, H., "Differential Forms with Applications to the Physical Sciences," New York, Academic Press, 1963.
6. Goldstein, H., "Classical Mechanics," Reading, Mass., Addison-Wesley, 1980.
7. Hinchey, F. A., "Vectors and Tensors for Engineers and Scientists," New York, Halsted Press, 1976.
8. Hoffman, B., "About Vectors," Englewood Cliffs, N. J., Prentice-Hall, 1966.
9. Jackson, J. D., "Electromagnetic Theory," 2nd Ed., New York, John Wiley & Sons, 1975.
10. Kemmer, N., "Vector Analysis," New York, Cambridge Univ. Press, 1977.
11. Kyrala, A., "Theoretical Physics: Applications of Vectors, Matrices and Quaternions," Philadelphia, W. B. Waunders Co., 1967.
12. Lass, H., "Vector and Tensor Analysis," New York, McGraw-Hill Book Co., 1950.
13. Morse, P. M., and Feshbach, H., "Methods of Theoretical Physics," New York, McGraw-Hill Book Co., 1953.
14. Shercliff, J. A., "Vector Fields," New York, Cambridge Univ. Press, 1977.
15. Stratton, J. A., "Electromagnetic Theory," New York, McGraw-Hill Book Co., 1941.
16. Synge, J. L., and Grifith, B., "Principles of Mechanics," New York, McGraw-Hill Book Co., 1959.
17. Thomas, G. B., Jr., and Finney, R. L., "Calculus and Analytic Geometry," 5th Ed. Reading, Mass., Addison-Wesley, 1979.

Cross-references: DYNAMICS, ELASTICITY, ELECTROMAGNETIC THEORY, FIELD THEORY, HYDRODYNAMICS, MECHANICS, RELATIVITY, STATICS, TENSORS.

VELOCITY OF LIGHT

Every elementary electric charge occupies a small volume. The charge is surrounded by a radially directed electric field, the strength of which decreases by the square of distance. By any motion with a velocity $v > 0$ of the charge relative to some fixed point, the distance and/or direction is changed and, thereby, the (vector) value of the field strength at the point. Here the field change occurs a while after the moment of charge motion. Otherwise expressed, the field change propagates with a finite velocity, usually labeled c. Moreover, during the short time for the action from a *certain part* of the charge volume to pass the rest of it, the volume moves a small distance proportional to v/c. This very small displacement creates the magnetic field always associated with an electric field change. In fact, magnetism is due to a (usually very small) part of the electric field.

Any change in the electromagnetic field propagates with the very high velocity of $c = 3 \times 10^{10}$ cm/sec, called the velocity of light. For, if the charge motion happens to be an oscillation of frequency v between 4×10^{14} and 8×10^{14} cps, we note the corresponding field variations as visible light. According to definition it is

$$c = v \cdot \lambda \qquad (1)$$

where λ is the wavelength in vacuum. The movements of the elementary charges are mostly oscillatory. By means of the field, after the delay due to c, they act on surrounding charges, and in that way all events in the atomistic world depend on the value of c. Therefore, the knowledge of c has turned out to be of extreme importance to our modern civilization.

The wave velocity according to Eq. (1) has a constant value, independent of all movements and, strictly speaking, independent of the medium where the propagation takes place. In a transparent body, in the intermediate space the action among the elementary charges of the atoms disperses with the velocity c. By inertia, the oscillating of the charges due to the active field is somewhat delayed as compared to the field. Now, the oscillating charge itself is a radiating source and its own delayed field interferes with the original one, creating a sum-wave, the velocity v of which is $v < c$. Exceptionally, in cases of resonance and absorption $v > c$. Our observation of light propagation in a substantial body relates to that slower interference wave. For the body, e.g., the medium of atmospheric air, we obtain corresponding to Eq. (1):

$$v = v \cdot \lambda_a \qquad (2)$$

From plain optical geometry we know that the refractive index of a medium is

$$n = \frac{c}{v} = \frac{\lambda}{\lambda_a} \qquad (3)$$

The technique for determining the wavelengths of light has progressed very far, and one can easily obtain an accurate value of n. For visible light, n depends on the wavelength used according to

$$n = A + \frac{B}{\lambda^2} + \frac{C}{\lambda^4} \qquad (4)$$

where A, B and C are positive constants. Close to resonance, the formula is not valid.

Direct determinations of the velocity of light, usually performed in air, are all based on the measurement of the time for a light pulse to cover a known distance. Such a pulse means an increase, followed by a decrease, of the amplitude of the light vibrations. What we observe is energy exchange associated with the amplitude changes, and as a result, we obtain the propagation velocity of the light energy. A change of amplitude is, however, equivalent to an interference among a series of adjacent wavelengths, since that change is created by just such an interference. The light pulse, therefore, consists of a whole group of adjacent wavelengths, interfering with each other. Interference is a sum-product. If the participating waves have different velocities, one will find by simple addition of two sine oscillations, that the group formed has a velocity different from those of the waves creating the group. On the surface of a calm sea, we observe a group of waves moving forward. The waves are created at the back of the group, travel through it, and disappear at the front. The difference between wave and group velocity is directly proportional to the wavelength and to the dependence of the wave velocity on the wavelength. Thus, calling the group velocity u, we obtain the difference:

$$v - u = \lambda_a \cdot \frac{dv}{d\lambda_a} \qquad (5)$$

Analogously to Eq. (3), we introduce a "group index"

$$n_g = \frac{c}{u} \qquad (6)$$

If, as in the case of air, $dv/d\lambda_a$ is a small quantity, we get, after some recalculation,

$$n_g = n - \lambda \cdot \frac{dn}{d\lambda} \qquad (7)$$

From Eq. (4) it is evident that $dn/d\lambda$ is a negative quantity, i.e., $n_g > n$. In vacuum, all velocities are equal: $u = v = c$.

The group velocity refers to the energy transport. Thus, in a medium there are the original waves of vacuum velocity c. By interference with waves from local charge oscillators, they create a wave system characteristic of the medium being considered and with the velocity $v < c$. By external energy action, that system is divided into an increasing number of adjacent waves, again interfering, of "second order," thereby forming groups of velocity $u < v$ (if normally $dn/d\lambda < 0$). Application of Eq. (7) to Eq. (4) yields:

$$n_g = A + \frac{3B}{\lambda^2} + \frac{5C}{\lambda^4} \qquad (8)$$

In the case of visible light or $\lambda = 0.4 - 0.8\mu$ ($\mu = 0.001$ mm) and in dry air of $0°C$, 760 mm Hg, $A = 1.000287619$, $B = 16.204 \times 10^{-7}$, $C = 0.1391 \times 10^{-7}$ (based on values derived by Edlén). Inserting n_{0g} corresponding to the λ used, we get

$$n_g = 1 + \frac{n_{0g} - 1}{1 + \alpha t} \cdot \frac{p}{760} - 0.55 \times 10^{-7} \cdot \frac{e}{1 + \alpha t}$$

$$(9)$$

where t is expressed in degrees Celsius, p in millimeters Hg, e in millimeters Hg of humidity, and $\alpha = 1/273$. For visible light, the error of Eq. (9) is less than 5×10^{-8}.

There is an experimental check of Eq. (8) for glass and calcite. Their refraction is 1000 times that of air. In calcite the group velocity ellipsoid was situated inside the ordinary one (Bergstrand, 1954). A more accurate control is proposed by Danielmeyer and Weber (Danielmeyer, 1971).

For micro- and radar waves with $\lambda > 700\mu$ the influence of λ in Eq. (8) is insignificant. Increasing influence of humidity, however, makes

$$n = 1 + \left[\frac{103.64}{T} (p - e) + \right.$$

$$\left. + \frac{86.26}{T} \left(1 + \frac{5748}{T} \right) e \right] \times 10^{-6} \qquad (10)$$

where $T = t + 273.16$. By using 103.64 in place of 103.49, as recommended by the International Association of Geodesy, the influence of a CO_2-term is partly compensated for. The uncertainty of Eq. (10) is about 5×10^{-7}.

There are a variety of methods for the determination of c. Among the direct ones, is the measurement of the travel time for a light pulse to cover a known distance. Usually there is a continuous series of pulses; i.e., the intensity is varied with a definite and known frequency. At every moment, there is a definite state of phase of the variation period. After reflection, the light returns, and one makes a phase comparison between emitted and received intensity. Thus, the intensity variation is used as a clock for the measurement of the running time. If the phase comparison is based on having the maximum of emitted light intensity coincide with the maximum of received intensity, the running time evidently is an even multiple of the period (=time) of a complete intensity variation cycle. Usually only one definite multiple number yields reasonable val-

ues, and there is no need for a special determination of the multiple.

The indirect methods are of much more varied character.

Determination of c Galileo was the first to try to show a finite light velocity. In his direct method, he used lanterns which could be screened off rapidly. The time elapsed on a distance of a few kilometers was, of course, too small to be observed visually.

In August 1676 at Paris, the Danish astronomer Ole Römer determined the revolution period of a satellite of Jupiter by its eclipse times into the planet's shadow. He used the result, 42 hours, to calculate the point of time for an eclipse occurring in November. It really occurred 10 minutes later. Römer explained this by the longer time for the light to reach the earth, now at a considerably greater distance from Jupiter. From the dimensions of the earth's orbit Römer defined c to be

$$c = 214\ 300 \text{ km/sec}$$

A modern value according to the same method is $299\ 840 \pm 60$ km/sec.

In 1725 the Englishman Bradley discovered astronomic aberration. Due to the velocity v of the earth in her orbit, the apparent direction to the stars normal to v is changed by v/c. Bradley could detect and measure the angle v/c because of its alternate sign for opposite parts of the earth orbit. Knowing v Bradley computed

$$c = 295\ 000 \text{ km/sec}$$

A modern value is $299\ 857 \pm 120$ km/sec.

In 1849 the Frenchman Fizeau took the next step. By projection of the image of a point source on the edge of a rapidly revolving cogged wheel, he got the required light pulse series. A blink or pulse passing through a cog interspace traveled the distance of 8633 meters to a reflector and back again along the optically identical path. Once more, now in the opposite direction, the pulse reached the cogged edge. For a suitable rotational speed the next cog interspace had now moved into position to let through the pulse; a case of "coincident maxs" (maxima). The time was computed from the revolution speed and the number of cogs. Fizeau's value was:

$$c = 315\ 300 \pm 500 \text{ km/sec}$$

In 1850, after an idea of Arago, using a point light source, Fizeau's compatriot Foucault directed a beam of constant intensity to a rapidly revolving plane mirror. During the time for the light to cover the distance to a second fixed mirror and back the revolving mirror had turned through a small angle, and the image of the light source was slightly displaced from its position when the mirror was stationary. Foucault's value was:

$$c = 298\ 000 \pm 500 \text{ km/sec}$$

In 1856 the German scientists Kohlrausch and Weber achieved a very important indirect determination. The magnetic field strength depends on v/c, the charge velocity v creating the magnetic field. Therefore, based on the same unit of force, the ratio between the unit of charge in electric and in magnetic measuring systems becomes c. Kohlrausch and Weber measured a charge (1) by its attraction to a second equal charge and (2) by the current produced at the passage of the charge through a galvanometer which showed a deflection due to the magnetic force. From (1)/(2) = c, they got:

$$c = 310\ 800 \text{ km/sec}$$

Rosa and Dorsey in 1906 measuring a capacity obtained 299 784 ± 30 km/sec.

In 1891 the Frenchman Blondlot transmitted Hertz waves along two straight and parallel wires, a Lecher guide system. After reflection, the waves created a standing wave system. Observing the nodal points Blondlot determined the wavelength. The frequency was known, and using Eq. (1) he obtained:

$$295\ 000 < c < 305\ 000 \text{ km/sec}$$

At this time, however, one was very uncertain as to the ratio between the velocities in vacuum and along the wires. Gutton solved this problem in 1911 by aid of the Kerr cell, making possible light variations of a frequency equal to that applied to the Lecher system. In a Kerr cell, the light passes between two condenser plates, lowered into nitrobenzene. Due to the directive action of the electric field between the plates on the oblong dipole molecules, the fluid acquires optical double-refracting properties. As soon as the field ceases, the molecules recover their random directions through the influence of their thermal movements. Placing the cell between an optic polarizer and a normally oriented analyzer, the light intensity leaving the combination, within certain limits, is directly proportional to the voltage applied to the condenser. The voltage may be that from a high-frequency radio transmitter.

Later on, Gutton's compatriot Mercier derived a theoretical formula for the ratio between light velocity and the velocities of guided waves. Using a valve oscillator for the high-frequency voltage on a Lecher system and applying his formula to the result, he obtained, in 1921,

$$c = 299\ 782 \pm 30 \text{ km/sec}$$

In the 1870's, Newcomb had introduced a revolving, reflecting multi-surface prism in place of Foucault's mirror. Michelson improved the system further, first in 1879 and finally at Mt. Wilson in 1926, his most accurate measurement. During the time required for the light to cover the distance of 35 km and back, the prism turned the next of its 12 surfaces in place for reflection. Only for exactly correct rotational speed was the position of the image of the suddenly illuminated distant reflector in the field of sight independent of the direction of rotation. The comparison of phase may be said, here too, to have occurred at "coincident maxs," and the high accuracy was due to the great displacement of the image for a turn of the revolving prism. Michelson's value was

$$c = 299\ 796 \pm 4 \text{ km/sec}$$

He overlooked a group correction of +2 km/sec. (Michelson, 1927)

Michelson, Pease and Pearson planned a determination in vacuum. It was completed by Pease and Pearson after Michelson's death in 1931. By multiple reflection in a 1-mile evacuated pipe, the light path was 10 miles. Due to the pipe, the geometry of the measuring device was considerably less symmetrical than in the case of the Mt. Wilson determination. The experiment was much talked about and the result was considered of high quality (Michelson, 1935),

$$c = 299\ 774 \pm 11 \text{ km/sec}$$

In 1928 the German scientists Karolus and Mittelstaedt performed the first direct determination using a Kerr cell. Their result was

$$c = 299\ 778 \pm 20 \text{ km/sec}$$

Here too, the group correction seems to have been overlooked.

Now Michelson's 1926 result was regarded as doubtful, and in 1940 Anderson, an American, tried to arrive at a decision. He used a Kerr cell. For phase comparison a photomultiplier received light from two nearby mirrors, the second of them at a somewhat greater distance in order to obtain intensity maxima from the one mirror coincident with minima from the second. The alternating photo-current then was equal to zero. Thereupon, the second mirror was moved away some 170 meters, and the distance was adjusted anew to yield a photo-current equal to zero. After the known mirror displacement, the light's travel time had increased an even multiple of the intensity period, known from the frequency. Anderson's value:

$$c = 299\ 776 \pm 14 \text{ km/sec}$$

Thereby the Michelson 1926 value was ruled out. Anderson carefully considered the group correction. Possibly, Anderson on occasions of his apparatus yielding values very different from the evacuated pipe result, might have felt particularly called upon to search for error sources and in that way unconsciously influenced the result. Or it might have been pure chance. In any case, his value superseded the pipe result (Anderson, 1941).

Anderson's value was used for radar systems. Thereby a microwave transmitter emits short pulses (10^{-6} second) which after distant reflection and return are received again. On the screen of a cathode ray tube the rapid sweep of the ray is marked by emitted and received pulses. The difference of mark positions and the sweep speed yields the pulse travel time. By knowing c the distance is obtained. "Oboe" (0.1-meter wave) and "Shoran" (1-meter) applied to known distances gave low values. Was the c-value used too low?

In 1947 Jones and Conford used Oboe in a manner opposite to its usual operation for the determination of c over known distances and got

$$c = 299\ 782 \pm 25\ \text{km/sec}$$

In the case of Shoran the main pulse-giving radar station is in an aircraft traversing the known straight distance between two points. In order to get strong echoes, there are radar slave stations at the points. Immediately after receiving the pulses, the slave stations reemit them on a slightly different wavelength. By that the main station receiver is not disturbed by its own transmitter. The distances to the two slave stations are continuously registered, and the shortest distance and also that used are obtained at the moment of transversing. A typical distance is 300 km. By careful treatment of Shoran results, Aslakson in 1949 obtained a velocity value

$$c = 299\ 792.3 \pm 2.3\ \text{km/sec}$$

In 1951 he used "Hiran," by which errors from varying signal strengths are avoided, and obtained

$$c = 299\ 794.2 \pm 1\ \text{km/sec}$$

(Aslakson, 1951).

Meanwhile Essen attacked the problem in a quite new manner by his indirect determination using a cavity resonator. Microwave guides may be circular tubes of sufficient diameter to let the waves through. If such a tube, whose length is an even multiple of half the wavelength, is closed by plane reflecting covers, the oscillating energy supplied remains in the cavity as a standing wave. This state only occurs in case of an exact, definite resonance frequency. In practice, there are several adjacent resonance frequencies depending on remaining oscillation states, e.g., including different multiples over a diameter or over several symmetrically oriented diameters. Sarbacher and Edson have derived the exact mathematical expression for all these resonance frequencies as functions of the tube length, tube diameter and the inside wave velocity. Essen and Gordon-Smith very carefully measured length, diameter and frequency of the evacuated cavity used. By calculations, they found the vacuum value c. Their result, in 1947, was:

$$c = 299\ 792.2 \pm 4.5\ \text{km/sec}$$

In 1949 Essen repeated the determination using a cavity whose length could be varied by an inserted movable plunger. By the dependence of the results on the cavity length, systematic errors could be rejected. The final value was

$$c = 299\ 792.5 \pm 1\ \text{km/sec}$$

Applying current practices, the error limits of 1 km/sec may be reduced perhaps by half (Essen, 1956).

At this time, Bergstrand performed direct experiments using visible light. The alternating voltage, supplied by a 10-M/Hz radio transmitter, coincidently fed a Kerr cell and a photo tube. In place of "coincident maxs" the phase comparison was carried out in moments of light pulses reaching half of maximum intensity. Then the photocurrent was strongly dependent on the distance to the reflector. By low-frequency phase shift of the emitted light intensity modulation (interchange of maximum and minimum), there were periodically two photocurrents, and by exactly chosen distance, they balanced each other to 0-current on the control instrument. In this way, by known displacement of the reflector through several successive such 0-points, the wavelength of intensity variation, i.e., the group length L_g, could be determined. The group velocity was obtained from $u = vL_g$, reduced to c by Eqs. (6) and (9). Besides some preliminary results from 1947–48, the 1950 value was:

$$c = 299\ 793.1 \pm 0.3\ \text{km/sec}$$

(Bergstrand, 1957).

Under the name of "Geodimeter," Bergstrand's instrument is used all over the world for measuring distances with high accuracy, c being considered as known. The error in 30 km need not exceed 5 cm. When the geodimeter was used to check the rocket-camera positions around Cape Kennedy, the accuracy turned out to be still better.

Using the geodimeter on known distances, of course, one can obtain c. The experience from recent, improved models now shows that the just related value of 1950 is slightly too high. In 1953 Mackenzie on the Ridge Way and Caithness base lines obtained

$$c = 299\ 792.3 \pm 0.5\ \text{km/sec}$$

Kolibayev performed measurements on many different base lines in the USSR during the years 1958 to 1963, and obtained the value

$$c = 299\ 792.6 \pm 0.1\ \text{km/sec}$$

Likewise, with a geodimeter, Grosse in Germany reported a result in 1967

$$c = 299\ 792.5 \pm 0.1\ \text{km/sec}$$

For the most recent geodimeters, the AGA Ltd. has applied a modulation system more suited for digital display. Also, by introducing a laser as the light source, full daylight measurements are possible.

In several countries nowadays, different forms of electro-optic distance-measuring instruments are manufactured. Intended for the same purpose are instruments using microwave (1 to 10 cm) radiation sources, such as Wadley's Tellurometer of 1957. By this method, a moving wave system is created between two equal instruments, one on each end of the distance, and having slightly different measuring frequencies (10.000 and 10.001 M/Hz). The phase comparison is done by the low-beat frequency of 1000 Hz.

Due to the marked influence of air humidity (Eq. 10) and occasionally occurring side reflections from ground-bound objects, the accuracy of these instruments is somewhat reduced. The results respecting c agree with those of the geodimeter.

One of the most accurate c determinations was that by Froome. In 1950 he started preliminary measurements and in 1958 he used a klystron transmitter delivering harmonics of 0.4 cm wavelength. By guiding pipes the microwave emission was divided into two equal parts. The two pipes terminated in transmitting horns supplied with lenses to form the front surface of the emitted wave as plane as possible. The pipes were bent 180° each in order to get the horns directed opposite to each other. The distance between the horns was 6 to 14 meters. Between the horns there was created a standing wave system, where the nodal points were observed by aid of a movable receiving detector. The displacement of the detector was measured interferometrically. In this way the microwave length was determined with great accuracy. By carrying through the determination for different horn distances, systematic errors were rejected. These errors may depend on disturbing reflections or on the wave fronts not being sufficiently plane or of known shape. The frequency was controlled by aid of an atomic clock. The result of 1958 was

$$c = 299\ 792.5 \pm 0.1\ \text{km/sec}$$

In 1967 Simkin, Lukin, Sikora, and Strelenski used a system very similar to that of Froome. The result was

$$c = 299\ 792.56 \pm 0.11\ \text{km/sec},$$

strongly supporting the Froome value and in excellent accordance with the latest geodimeter results (Simkin, 1967).

A further step towards shorter waves was made by V. Daneu et al. using a laser for the far infrared at $\lambda = 0.1$ mm. In a Michelson vacuum interferometer of 4 m path difference, the laser wave is compared to visible light of accurately known wavelength. The laser frequency is directly determined by comparison with high harmonics of a quartz oscillator. The preliminary results are in concordance with the present accepted value of c (V. Daneu, 1969).

A proposal by Mockler and Brittin is very interesting in two respects:

(1) The use of gamma rays.

(2) Light travels only once over the distance (as Roemer!).

First I refer to the article in this book on the MÖSSBAUER EFFECT. The vertical gable surface A of a quartz rod is directed towards the 3 m distant gable B of an identical rod. Both rods are vibrating longitudinally by stimulation from a common quartz-controlled 10^{10}-Hz oscillator. On A is a film of ^{57}Co, including ^{57}Fe decaying to ground state and thereby emitting gamma rays, marked by the Doppler-variation due to the vibrational movements of A. B is covered with ^{57}Fe, absorbing the emission by excitation from the ground state. Due to the very narrow common bandwidth of emission and absorption, 4 MHz of 4×10^{18} Hz, one will observe absorption at B only when its vibratory movement is in phase with the Doppler variation of the radiation hitting the surface. By then, there is an even number of "Doppler waves" over the known distance from A to B, yielding the "wave length." Again we use Eq. (1) to obtain c. Important for a good result is a strong quartz vibration (Mockler, 1961).

There is an indirect method which does not give such high accuracy: The exactly known frequency of an atomic clock is the ground rotational molecular frequency with quantum number =1 of some diatomic gas. From the quantum spectral formula of the rotational line series, one obtains the frequency at high quantum numbers of visible or infrared light, where the wavelength is known with great accuracy. The connection between wavelength and frequency yields c, and in 1955 Rank, Bennet and Bennet obtained the value 299 791.9 ± 2.2 km/sec.

Finally, E. Richard Cohen in this Encyclopedia quotes a numerical value based on what he considers to be the most accurate and consistent measurements available as of June 1982. The velocity of light was determined by measuring the wavelength and frequency of a methane-stabilized helium-neon laser and taking the product of these two quantities. He quotes a value of 299 792 458 m · s⁻¹ or 299 792.458 km/sec (see CONSTANTS, FUNDAMENTAL).

ERIK BERGSTRAND

References

Brillouin, Léon, "Wave Propagation and Group Velocity," New York, Academic Press, 1960.
Michelson, A., *Astrophys. J.*, 65 (1927).
Michelson, A., *Astrophys. J.*, 82 (1935).
Anderson, W. C., *J. Opt. Soc. Am*, 31 (1941).
Essen, L., *Proc. Roy. Soc. London Ser. A*, 204 (1950).

Froome, K. D., *Proc. Roy. Soc. London Ser. A*, 109 (1958).

Aslakson, C. I., *Trans. Am. Geophys. Union*, 32 (1951).

Dorsey, *Trans. Am. Phil. Soc.*, 34, Pt. 1 (1944).

Bergstrand, E., in Flügge, S., Ed., "Encyclopedia of Physics," Vol. 24, Berlin, Springer, 1957.

Bergstrand, E., *K. Sv. Vet. Akad: Arkiv för Fysik*, 8 (45), Stockholm (1954).

Simkin, G. S., Lukin, I. V., Sikora, S. V., and Strelenskii, V. E., 1967, *Izmeritel. Tekhn.*, 8, 92 (1967). (Translation: *Meas. Tech.*, 1967, 1018).

Danielmeyer, H. G., and Weber, H. P., *Phys. Rev. A*, 3 (5), 1708 (May 1971).

Daneu, V., Hocker, L. O., Javan, A., Ramachandra Rao, D., and Szoke, A., *Phys. Letters*, 29a (6), 2 (June 1969).

Mockler, R. C., Brittin, W. E., Nat. Bur. Stand., Rpt. Nr. 6762, Boulder, Colorado, 1961.

Froome, K. D., and Essen, L., "The Velocity of Light and Radio Waves," London and New York, Academic Press, 1969.

Cross-references: ATOMIC CLOCKS; CONSTANTS, FUNDAMENTAL; ELECTROMAGNETIC THEORY; INTERFERENCE AND INTERFEROMETRY; KERR EFFECTS; MÖSSBAUER EFFECT; OPTICS, GEOMETRICAL; PROPAGATION OF ELECTROMAGNETIC WAVES; RADAR; WAVEGUIDES; WAVE MOTION.

VIBRATION

There seems to be a reasonable consensus of opinion that, in physics, vibration is a periodic motion—a repeating to-and-fro motion—of a solid body or of the particles of an elastic body. Some examples will illustrate and clarify this definition. However, let us first point out that about us are an endless variety of vibrating systems, including the vibrations of vocal cords, piano and violin strings, bearings, loudspeaker cones, pneumatic drills, organ pipes, bells, woodwinds, and even structures. An historical example of the adverse effects of vibration was the Tacoma Narrows bridge disaster in 1940, in which the amplitude of vibration became so great that the bridge was entirely destroyed.

As a simple example of a vibrating system, let us consider what happens when a body of mass m fastened to a helical spring is moved slightly away from its equilibrium position and let go, assuming that there is no friction and that the mass of the spring is negligible compared with that of the attached body. The force on the body is directly proportional to the displacement x and oppositely directed, so the equation of motion is

$$m\ddot{x} = -kx, \qquad (1)$$

where k, the stiffness of the spring, is a positive constant and $\ddot{x}$ is the second derivative of the displacement with respect to time. The solution

of Eq. (1), as may be shown by substitution, is

$$x = A \cos \omega t + B \sin \omega t,$$

where A and B are constants, and ω, the so-called *angular frequency*, is $(k/m)^{1/2}$. If we agree to start counting time from the instant when $x = x_0$, and $v = 0$, the solution reduces to

$$x = x_0 \cos \omega t.$$

This sinusoidal motion is called *simple harmonic motion*. Its period T, the time for one complete vibration, is $2\pi (m/k)^{1/2}$, and its frequency f, is $1/T$, or

$$f = (1/2\pi) (k/m)^{1/2}. \qquad (2)$$

It is important to note that the frequency is independent of the amplitude x_0, and depends only on the mass of the attached body and on k.

Another simple system, somewhat different from the first, is the *simple pendulum*, consisting ideally of a point mass at one end of a massless string, the other end of the string being fastened to a rigid support. Using the idea that the torque is equal to the product of the moment of inertia and the angular acceleration, one gets for the equation of motion:

$$mL^2 \ddot{\theta} = -mgL \sin \theta, \qquad (3)$$

where L is the length of the string, g the acceleration of gravity, and θ the displacement angle from the vertical. For θ small enough to enable $\sin \theta$ to be replaced by θ, Eq. (3) has the same form as Eq. (1). Hence, if the boundary or initial conditions are $\theta = \theta_0$ and $\dot{\theta} = 0$ at $t = 0$, the solution is

$$\theta = \theta_0 \cos (g/L)^{1/2} t.$$

For this case the frequency is

$$f = (1/2\pi) (g/L)^{1/2}. \qquad (4)$$

In spite of the similarity between this example and the first one, one should note that here the frequency is independent of the mass; dependent on the value of g, which varies from place to place; and dependent on the amplitude—since it is valid only for small values of θ_0. None of these statements is true for a mass vibrating at the end of a spring.

Another—and somewhat more complicated—example of vibratory motion is that of the *transverse vibrations of a stretched string* of constant linear density ρ_L. The string, of length L, is fastened between two fixed supports, and is displaced so little that any displacement y is very small compared with L. Let us consider a small segment of the string of length dx under tension T. The equation of motion is

$$\rho_L \, dx \, \frac{\partial^2 y}{\partial t^2} = T\left(\frac{\partial y}{\partial x}\right)_{x + dx} - T\left(\frac{\partial y}{\partial x}\right)_x,$$

where x is the distance measured along the string. This reduces to

$$\frac{\partial^2 y}{\partial t^2} = \frac{T}{\rho_L} \frac{\partial^2 y}{\partial x^2}.$$

There are several ways of solving this equation. One is to assume that y may be written as the product of a function of x alone and a function of t alone, i.e., $y = X(x)\,\tau(t)$. This leads to solutions

$$X = C\cos\left(\omega x/c + D\right) \quad \text{and} \quad \tau = A\cos\left(\omega t + B\right),$$

if we let $c^2 = T/\rho_L$, and $\omega = 2\pi f$. By applying the boundary conditions, namely, that for all values of the time $y = 0$ at the fixed ends of the string (i.e., at $x = 0$ and at $x = L$), we find that $D = \frac{1}{2}\pi$ and $\omega L/c = n\pi$. Hence the possible frequencies with which the string can vibrate are given by

$$f_n = \frac{nc}{2L} = \frac{n}{2L}\left(\frac{T}{\rho_L}\right)^{1/2}, \quad n = 1, 2, 3, \ldots.$$

$$(5)$$

In contrast to the first two illustrations, where the system could vibrate with only one frequency, this time we have an infinite number of possible frequencies, each an integral multiple of the lowest. The lowest possible frequency is called the *fundamental* or *first harmonic*, the second lowest frequency the *second harmonic*, etc. The general solution is

$$y = \sum_{n=1}^{\infty} \sin\frac{n\pi x}{L}\left\{E_n\cos\frac{n\pi ct}{L} + G_n\sin\frac{n\pi ct}{L}\right\}.$$

Thus, knowing the initial position of the string, one can calculate its position at any future time. It turns out, for example, that if the string is pulled out at a point half-way between its ends and let go, only the odd harmonics are present in the motion.

In the previous example we had a system where there were an infinite number of possible frequencies of vibration, but all those higher than the lowest possible frequency were integral multiples of the lowest frequency. Let us look briefly at a system which vibrates somewhat differently. Let us consider the transverse vibra-

tions of a bar clamped at one end and free at the other—such a bar is often referred to as a *clamped-free bar*. The equation of motion is:

$$\frac{\partial^2 y}{\partial t^2} = -\kappa^2 c^2 \frac{\partial^4 y}{\partial x^4},$$

where $c^2 = Y/\rho$, Y being Young's modulus and ρ the density of the bar. κ is a constant depending on the shape of the bar; for example, for a bar of circular cross section, it is equal to one-half the radius. For such a bar the solution involves both trigonometric and hyperbolic functions. On applying the boundary conditions one finds that the allowed frequencies above the fundamental are not related to one another by a simple relationship. To demonstrate this, let us look at the following Table 1, where the ratio of the six lowest possible frequencies to the fundamental are listed both for the previous example and for this one. Higher frequencies which are not integral multiples of the fundamental are known as *overtones*, whereas those which are integral multiples of the fundamental are referred to both as overtones and as *harmonics*. Thus we see from Table 1 that in the case of the string the second lowest frequency is both an overtone and a harmonic, whereas the second lowest frequency in the case of the bar is an overtone but not a harmonic.

Up to now we have been discussing free vibrations in which the system is displaced and then let go. We know that due to friction, etc., the system will eventually come to rest. This process is known as *damping*. There is another kind of vibration, *forced vibration*, in which the system is driven by an external force—frequently a sinusoidal force—and made to vibrate at frequencies which may either be the same as or different from those at which it vibrates as a free system. These vibrations last as long as the force is applied. Perhaps the most interesting case of forced vibrations is that where the system is driven at a frequency close to that at which it vibrates as a free system. This is known as *resonance*. An example frequently cited is the situation where a person demolishes a goblet by singing loudly a note of just the correct frequency, a frequency near the resonant or "natural" frequency of vibration of the goblet. A sharp or narrow resonance may be an advantage, as in the case of tuning a radio receiver, or

TABLE 1. RATIO OF FREQUENCIES TO THE FUNDAMENTAL

	Lowest	2nd	3rd	4th	5th	6th
Transverse vibrations of a string fastened at both ends	1	2	3	4	5	6
Transverse vibrations of a clamped-free bar	1	6.267	17.55	34.39	56.85	84.93

a disadvantage, as in the case of a loudspeaker cone or cabinet.

FREDERICK E. WHITE

References

Barton, E. H., "A Textbook of Sound," London, Macmillan and Co., Ltd., 1922.

Kinsler, L. E., Frey, A. R., Coppens, A. B., and Sanders, J. V., "Fundamentals of Acoustics," Third Edition, New York, John Wiley & Sons, 1982.

Magrab, E. B., "Vibrations of Elastic Structural Members," Rockville, MD, Sijthoff & Noordhoff, 1979.

Rayleigh (Lord), "The Theory of Sound," First American Edition, New York, Dover Publication, 1945.

Stumpf, F. B., "Analytical Acoustics," Ann Arbor, MI, Ann Arbor Science Publishers, Inc., 1980.

Cross-references: ARCHITECTURAL ACOUSTICS, DYNAMICS, ELASTICITY, HEARING, MUSICAL SOUND, PHYSICAL ACOUSTICS, RESONANCE, SEISMOLOGY, WAVE MOTION.

VISCOELASTICITY

Viscoelasticity is a material property possessed by solids and liquids which, when deformed, exhibit both viscous and elastic behavior through simultaneous dissipation and storage of mechanical energy. The *material constants* linking stress and strain in the theory of elasticity become *time-dependent material functions* in the constitutive equations of viscoelastic theory. At sufficiently small (theoretically infinitesimal) strains the behavior of viscoelastic materials is well described by the *linear theory of viscoelasticity* epitomized by the celebrated *Boltzmann superposition principle*. Expressed in its simplest form

$$\sigma(t) = \int_0^t Q(t - u)\epsilon(u) \, du$$

$$\epsilon(t) = \int_0^t U(t - u)\sigma(u) \, du$$

it states that the stress (or strain) at time t under an arbitrary strain (or stress) history is a linear superposition of all strains (or stresses) applied at previous times u multiplied by the values of a weighting function $Q(t)$ [or $U(t)$] corresponding to the time intervals $t - u$ which have elapsed since imposition of the respective strains (or stresses).

Put another way (see *operator equation* below) the Boltzmann superposition principle expresses the fact that the material behavior can be described by linear differential equations with constant coefficients and time as variable.

Considerable simplification of the viscoelastic relations is achieved by mapping them from the real t axis into the complex s plane through the Laplace transformation. The resulting transforms in s may be manipulated algebraically and then inverted to regain the time-dependent form. Thus the above convolution integrals become

$$\sigma(s) = Q(s)\epsilon(s) \qquad \epsilon(s) = U(s)\sigma(s)$$

and $Q(t)$ and $U(t)$ can be seen to be the material functions representing the response to a unit impulse of strain or stress. The unit step response functions, the material functions under constant strain ϵ_0 and constant stress σ_0, are the *relaxation modulus* and *creep compliance*, obtained as

$$G(t) = \sigma(t)/\epsilon_0 = \mathcal{L}^{-1} Q(s)/s$$

$$J(t) = \epsilon(t)/\sigma_0 = \mathcal{L}^{-1} U(s)/s$$

where $\mathcal{L}^{-1}$ denotes inversion of the transform. For sinusoidal steady state strain $\epsilon(\omega)$ and stress $\sigma(\omega)$, there result the *complex modulus* and *complex compliance*

$$G^*(\omega) = \sigma(\omega)/\epsilon(\omega) = [Q(s)]_{s=i\omega}$$

$$J^*(\omega) = \epsilon(\omega)/\sigma(\omega) = [U(s)]_{s=i\omega}$$

whose real and imaginary parts are the *storage modulus* $G'(\omega)$, *storage compliance* $J'(\omega)$, *loss modulus* $G''(\omega)$, and *loss compliance* $J''(\omega)$. Their ratio is the loss tangent

$$\tan \delta \, (\omega) = G''(\omega)/G'(\omega) = J''(\omega)/J'(\omega)$$

Any of the above functions (or any other that may be derived, e.g., for constant rate of strain), if known over a sufficiently extended time scale (or, equivalently, over a sufficient range of frequencies), provides complete information on the viscoelastic properties of a homogeneous isotropic material. As formulated above, the equations refer to deformation in simple shear. When dealing with combined stresses, as in viscoelastic stress analysis, the tensorial character of stress and strain must, of course, be taken into account. The equations for other types of deformation, e.g., dilatation or uniaxal extension, are analogous to those for shear. The relations between the frequency-dependent viscoelastic functions in shear, extension, and dilatation are the same as those between the elastic constants, while for the time-dependent functions these relations are valid between the s-multiplied Laplace transforms. Thus the relaxation modulus in extension $E(t)$ is related to $G(t)$ through the *time-dependent Poisson's ratio* $v(t)$ by

$$E(s) = 2G(s) [1 + sv(s)]$$

and for $v(t) = 1/2$, $E(t) = 3G(t)$. Data obtained in shear and extension on incompressible bodies are thus readily combined. Moreover, if a viscoelastic function can be formulated analytically, it can be converted into any of the others, allowing combination of measurements made under different stress or strain histories. Use is made of this to extend the experimentally acces-

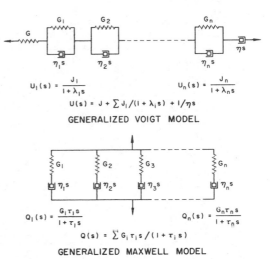

$$U_1(s) = \frac{J_1}{1 + \lambda_1 s} \qquad\qquad U_n(s) = \frac{J_n}{1 + \lambda_n s}$$

$$U(s) = J + \sum J_i /(1 + \lambda_i s) + 1/\eta s$$

GENERALIZED VOIGT MODEL

$$Q_1(s) = \frac{G_1 \tau_1 s}{1 + \tau_1 s} \qquad\qquad Q_n(s) = \frac{G_n \tau_n s}{1 + \tau_n s}$$

$$Q(s) = \sum G_i \tau_i s /(1 + \tau_i s)$$

GENERALIZED MAXWELL MODEL

FIG. 1

sible time scale. Thus at short times, dynamic (frequency-dependent) measurements are more convenient than transient (time-dependent) measurements, and *vice versa* for long times.

Another important extension of the time scale is available through *time-temperature superposition*. An increase in temperature generally shortens the time necessary for the molecular rearrangement processes responsible for viscoelastic behavior. If all such processes are affected by temperature in the same way, the material is *thermorheologically simple*, and a change in temperature simply shifts the viscoelastic functions along the logarithmic time or frequency axis. For polymers (the typical viscoelastic materials) above the *glass transition temperature* (at which main chain motion effectively ceases), the shift factor a_T is given by an equation of the form (WLF equation)

$$\ln a_T = - \frac{c_1(T - T_0)}{c_2 + T - T_0}$$

where T_0 is a suitably chosen reference temperature, and c_1 and c_2 are constants. Measurements made at different temperatures can thus be combined to yield a single *master curve*.

Measurements made under different stress or strain histories are often combined for experimental reasons and the linear theory of viscoelasticity furnishes a number of methods to this end. Interconversion of the viscoelastic functions is faciliated by the introduction of *spectral distribution functions*. The spectral functions may be derived conveniently by rewriting the Boltzmann superposition principle, given above as an integral operator equation, in the form of the *differential operator equation*:

$$[\sum p_m D^m]\, \sigma(t) = \sum [q_n D^n]\, \epsilon(t)$$

where $D^r = d^r/dt^r$. The Laplace transformation leads to

$$Q(s) = 1/U(s) = \sum q_n s^n / \sum p_m s^m$$

a definition of $Q(s)$ and $U(s)$ which is readily linked with the useful representation of viscoelastic behavior by *models* consisting of series-parallel combinations of springs (elastic or storage elements) and dashpots (viscous or dissipative elements). A parallel combination (Voigt model) is characterized by a *retardance* $U(s) = 1/(G + \eta s) = J/(1 + \lambda s)$, where $\lambda = \eta/G$ is the *retardation time*, η signifies viscosity, and $J = 1/G$. A series combination (Maxwell model) is characterized by a *relaxance* $Q(s) = 1/(J + 1/\eta s) = G\tau s/(1 + \tau s)$ where $\tau = \eta/G$ is the *relaxation time*.

Through the *combination rules*: "relaxances add in parallel, retardances add in series," $Q(s)$ and $U(s)$ are readily derived from a given more complex model as illustrated below.

These models are *conjugate*, i.e., with suitable choices for the parameters, they describe the

same viscoelastic material. Representation of viscoelastic behavior by a (small) finite number of elements is often inadequate. Molecular theories of polymer behavior lead to models with an infinite number of parameters, characterized by a *discrete distribution* (or *line spectrum*) *of relaxation or retardation times*. These distributions are normally so closely spaced that they cannot be resolved experimentally. One can therefore define *continuous distribution functions*

$$Q(s) = \int_0^\infty G(\tau)\, \frac{\tau s\, d\tau}{1 + \tau s} = \int_{-\infty}^\infty H(\tau)\, \frac{\tau s\, d \ln \tau}{1 + \tau s}$$

$$U(s) = J + \int_0^\infty J(\lambda)\, \frac{d\lambda}{1 + \lambda s} + 1/\eta s =$$

$$J + \int_{-\infty}^\infty L(\lambda)\, \frac{d \ln \lambda}{1 + \lambda s} + 1/\eta s$$

where $G(\tau)$ and $J(\lambda)$ are distributions of relaxation and retardation times (more properly: modulus and compliance densities on time) respectively, and $H(\tau) = \tau G(\tau)$ and $L(\lambda) = \lambda J(\lambda)$ are their counterparts on the more convenient logarithmic time scale. The spectral functions may be derived from experimentally determined curves of the viscoelastic functions through any of several approximation methods.

The various viscoelastic functions are illustrated above on the example of an uncross-linked amorphous polyvinyl acetate with a molecular weight of about 300 000. Data obtained in extension and shear at different temperatures were combined and interconverted.

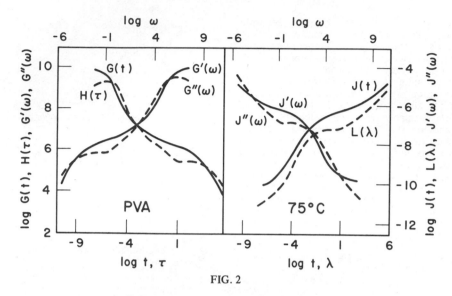

FIG. 2

The functions are grouped to display their qualitative symmetries.

Singling out the relaxation modulus for a broad interpretation of the viscoelastic behavior of an uncross-linked polymer, one sees that at short times (in the *glassy region*), the modulus approaches an asymptotic value, the *glass modulus*. It then drops through the glass-to-rubber transition region and levels out somewhat in the *plateau region* reflecting the effect of molecular entanglements. At still longer times, it decays to zero through the *terminal region*. For cross-linked networks this region is absent (if there is no chemical degradation). In this case, the flow term $1/\eta s$ is missing from $U(s)$ above; an additive constant, the *equilibrium modulus* G_e, characterizing the (level) *rubbery region*, appears in $Q(s)$.

Current efforts in the field of viscoelasticity are directed chiefly towards the development of nonlinear (large deformation) theory, theories of multiphase materials, anisotropic and inhomogeneous (semicrystalline and filled) systems, the solution of viscoelastic boundary problems (viscoelastic stress analysis), and the explanation of viscoelastic behavior on the molecular level (molecular theories).

N. W. TSCHOEGL

References

Alfrey, T., and Gurnee, E. F., "Dynamics of Viscoelastic Behavior," in Eirich, F. R., Ed., "Rheology," Vol. 1, New York, Academic Press, 1956.

Bueche, F., "Physical Properties of Polymers," New York, Interscience Publishers, 1962.

Christensen, R. M., "Theory of Viscoelasticity—An Introduction," New York, Academic Press, 1971.

Ferry, J. D., "Viscoelastic Properties of Polymers," New York, John Wiley & Sons, 1961.

Leaderman, H., "Viscoelasticity Phenomena in Amorphous High Polymeric Systems," in Eirich, F. R., Ed., "Rheology," Vol. 2, New York, Academic Press, 1958.

Nielsen, L. E., "Mechanical Properties of Polymers," New York, Van Nostrand Reinhold, 1962.

Staverman, A. J., and Schwarzl, F., "Linear Deformation Behavior of High Polymers," in Stuart, H. A., Ed., "Die Physik der Hochpolymeren," Vol. 4, Berlin, Springer, 1956.

Tobolsky, A. V., "Properties and Structure of Polymers," New York, John Wiley & Sons 1960.

Cross-references: ELASTICITY, MECHANICAL PROPERTIES OF SOLIDS, POLYMER PHYSICS, RHEOLOGY, VISCOSITY.

VISCOSITY

Materials in general show two broad kinds of behavior with regard to their reaction when subject to an applied force. In one type, they deform until a position of equilibrium is reached, when no further change of shape takes place. Many solids show this kind of behavior. Alternatively, there is no permanent resistance to change of shape, and continuous deformation takes place for as long as the force is applied. The material is said to flow. Substances possessing this property are called fluids and may generally be thought of as either liquids or gases. The class also includes some materials, such as glass and pitch, which at room temperature exhibit many of the characteristics normally associated with solids. When flow takes place in a fluid, it is opposed by internal friction arising from the cohesion of the molecules. This internal friction is the property of the fluid known as *viscosity*. It is clearly of great importance in many contexts; we may think, for example, of the flow of blood in the body, the flow of oil in pipe lines, the working of molten glass, and the process of lubrication. It may be used in the discrimination between streamlined and turbulent flow. It is also a useful property in providing information about the structure of complex organic molecules.

Definitions and Units The formal definition of viscosity arises from the concept put forward by Newton that under conditions of parallel flow, the shearing stress is proportional to the velocity gradient. If the force acting on each of two planes of area A parallel to each other, moving parallel to each other with a relative velocity V, and separated by a perpendicular distance X, be denoted by F, the shearing stress is F/A and the velocity gradient, which will be linear for a true liquid, is V/X.

Thus

$$\frac{F}{A} = \eta \cdot \frac{V}{X}$$

The constant η is known as the viscosity coefficient, dynamic viscosity, or viscosity of the liquid. The unit, expressed in dyne second per square centimeter, is known as the "poise," after Poiseuille who worked on viscosity in the early part of the nineteenth century. For practical purposes it is often more convenient to use a smaller unit, known as the centipoise, equal to a hundredth of a *poise*. It is useful to note that water at room temperature has a viscosity of approximately 1 centipoise. In many applications of viscometry, it is useful to note another quantity, called the kinematic viscosity, which is obtained by dividing the dynamic viscosity by the density. It is frequently denoted by the Greek letter ν, and the unit in which it is expressed is the "stokes," after Sir George Stokes, another pioneer worker in viscometry. In practical viscometry, viscosity is sometimes expressed in seconds; this is a measure of the time for a prescribed quantity of liquid to flow through a tube or aperture of defined dimensions.

In considering many problems of fluid motion, an important non-dimensional quantity is vd/ν, where v is the velocity of the fluid, ν its kinematic viscosity, and d a linear dimension, such as the diameter of a tube. This quantity is known as Reynolds' number; when it is less than about 1000, flow is generally streamlined, whereas at values over 1000 it is turbulent.

Values for the viscosity of a number of common substances are given in Table 1.

Methods of Measurement In measuring viscosity in the absolute sense, it is necessary to be able to use the basic relationship between shearing stress and velocity gradient in order to derive a relationship between the measurable quantities involved in flow in an apparatus of a particular shape. For example, it can be shown that the volume Q of a liquid of viscosity η

TABLE 1. COEFFICIENT OF VISCOSITY IN POISES (APPROXIMATE VALUES)

Water at 20°C	0.010
Mercury at 20°C	0.015
Carbon tetrachloride at 20°C	0.009
Olive oil at 20°C	0.84
Glycerine at 20°C	8.3
Golden Syrup at 20°C	1000
Glass at melting temperature	10^3
Glass at working temperature	10^7
Glass at annealing temperature	10^{13}
Air at 0°C	0.00017
Carbon dioxide at 0°C	0.00014
Steam at 100°C	0.00013
Pitch at 15°C	10^{10}

flowing per unit time through a tube of length l and internal radius a, when the pressure difference between the ends is P, is given by

$$Q = \frac{\pi P a^4}{8 l \eta}$$

Capillary viscometers enable all the variables in this equation to be measured and the viscosity to be determined.

Similarly, a viscometer can be constructed in which one cylinder rotates inside another with the liquid under test in the annular space between the two. Measurement of the angular velocity and the applied torque, together with a knowledge of the dimensions of the apparatus, enables the viscosity to be calculated.

Another method involves the measurement of the rate of fall of a sphere in a column of liquid. If the liquid has viscosity η and density ρ, and the sphere has radius r and density σ, then the velocity of fall is

$$v = \frac{2gr^2(\sigma - \rho)}{9\eta}$$

a relationship commonly known as Stokes' law.

There are also a number of empirical methods that can be used to obtain comparative values, the instruments being calibrated by using liquids of known viscosity.

Variation with Temperature The viscosity of liquids decreases rapidly with increasing temperature. For example, the viscosity of molten glass may be halved by raising the temperature 30°C. On the other hand, the viscosity of gases increases with temperature. This feature is of practical importance in two senses. In all methods of measurement strict temperature control is necessary, and when results are quoted, it is essential to state the temperature to which they refer.

The variation of viscosity with temperature is of great significance in the problem of lubrication, where oils may have to operate over widely different conditions. So-called viscostatic oils have a low temperature coefficient and can thus be used over a wide temperature range. As an indication of this property, the oil industry uses an empirical number known as the "viscosity index"; the higher the viscosity index, the less is the variation of viscosity with temperature.

Molecular Weight Determinations The viscosity of organic materials in solution was suggested by Staudinger as a useful index of their molecular structure, since the flow properties would be influenced by the size and shape of the molecules. For example, some high polymers give appreciable increases of viscosity even at low concentrations on account of the effect of randomly coiled long-chain molecules. Generally speaking, the higher the molecular weight, the greater is the increase in viscosity for a given weight in the solution. The values of molecular weight obtained by this method are not absolute, but they depend on the establishment of empirical relations by measurements on substances of known molecular weight in a given series (see MOLECULAR WEIGHT).

Non-Newtonian Systems A large number of industrially important materials do not obey the simple Newtonian relationship between shearing stress and shearing rate. In some cases the viscosity varies with the shearing rate, and in others, it varies with time when the shearing rate is constant. An important group of such materials is known as thixotropic substances—they become thinner when stirred. Paint, suspensions of clay in water, and many other substances behave in this way, and the property has considerable industrial significance. In the case of paint, for example, the low viscosity when brushed in a thin film makes for easy application, whereas the high viscosity under low shearing stresses enables the film to be retained on vertical surfaces. Some substances thicken up on stirring, and there are a number of other classes of behavior (e.g. VISCOELASTICITY) that constitute departure from Newtonian laws. The study of such anomalous systems is an important branch of modern RHEOLOGY.

A. DINSDALE

References

Newman, F. H., and Searle, V. H. L., "The General Properties of Matter," London, Ed. Arnold, 1962.

Dinsdale, A., and Moore, F., "Viscosity and Its Measurement," New York, Van Nostrand Reinhold, 1963.

Perry, J. H., "Chemical Engineers Handbook," 4th Ed., New York, McGraw-Hill Book Co., 1963.

Flory, P. J., "Principles of Polymer Chemistry," New York, Cornell University Press, 1963.

Scott-Blair, G. W., "A Survey of General and Applied Rheology," London, Pitman, 1949.

Eirich, F. R., Ed., "Rheology, Theory and Applications," Three volumes, New York, Academic Press, 1958.

Cross-references: ELASTICITY, FLUID DYNAMICS, MOLECULAR WEIGHT, RHEOLOGY, VISCOELASTICITY.

VISION AND THE EYE

The meaningful experience of vision requires light, the eye, and a conscious observer (animal or human) having an intact *visual system*. In its gross aspects, the visual system includes the eyes, the *extraocular muscles* which control eye position in the *bony orbit* (eye socket), the optic and other nerves that connect the eyes to the brain, and those several areas of the brain that are in neural communication with the eyes. This summary will stress the informational aspects of human vision; it should be realized however that no visual system could function without its *protective mechanisms* (tears and eyelids, especially) or if its normal metabolism (mediated through the vascular supply of eye and brain) were seriously interfered with.

The visual system is particularly well adapted for the rapid and precise extraction of spatial information from a more-or-less remote external world; it does this by analyzing, in ways that are as yet imperfectly understood, the continuously changing patterns of radiant flux impinging upon the surfaces of the eyes. Much of this light is reflected from objects which must be discriminated, recognized, attended to, and/ or avoided in the environment; this ability transcends enormous variations in intensity, quality, and geometry of illumination as well as vantage point of the observer. A block diagram of the visual system is given in Fig. 1.

Although image formation in the eye is importantly involved, the analogy between eye and photographic camera has been badly overworked and tends to create the erroneous impression that little else is needed to explain how we see. Image formation is greatly complicated by the movement of the eyes within the head, and of both eyes and head relative to the external sea of radiant energy. Such visual input is ordinarily sampled by discrete momentary pauses of the eyes called *fixations*, interrupted by very rapid ballistic motions known as *saccades* which bring the eyes from one fixation position to the next. Smooth movements of the eyes can occur when an object having a predictable motion is available to be followed.

Each eye controls many important functions within one mobile housing: it is a device to form an image upon a vast array of light sensitive *photoreceptors*, but it also contains systems to dissect, encode, and transmit information derived therefrom. A cross section of the human eye is shown in Fig. 2. The primary refracting surface is the *cornea*, a complex yet transparent structure which admits light through the anterior part of the outer surface of the eye. The *iris* contains muscles which alter the size of the entrance port of the eye, the *pupil*. The *crystalline lens* has a variable shape, under the indirect control of the *ciliary* muscle. Since it has a refractive index higher than the surrounding media, it gives the eye a variable focal length, allowing *accommodation* to objects at varying distances from the eye. The iris muscles and the ciliary muscle, known collectively as the *intra-*

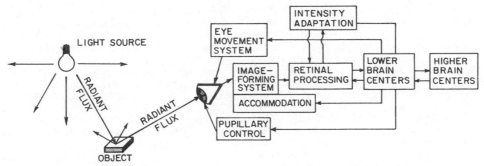

FIG. 1. Block diagram of the functional components of the visual system.

ocular muscles of the eye, are controlled by impulses having their origins in separate but interacting centers in the brain stem. These brain centers also receive nerve impulses from the eye. These loops, and those involving the extraocular musculature and thus eye position, have some of the properties of nonlinear servosystems, and have been actively investigated as such.

Much of the remainder of the eye is filled with fluids and materials under pressure, which help the eye maintain its shape. The *aqueous humor*—thin, watery, and continuously being replaced—fills the *anterior* chamber between cornea and lens. The *vitreous humor*—thinly jellylike and of very low metabolism—fills the majority of the eye's volume. The image produced through these structures is formed upon the *retina* at the back of the eye. The retinal image is very small, because the eye itself is small and has a short posterior focal length of about 19 to 23 mm, depending upon accommodative state. The retinal image has a point-

spread function on the order of two to three minutes of arc, corresponding to about 10μ on the retina for ideal conditions. These conditions include a 2 to 3 mm pupil, monochromatic light, optimal accommodation and a normal, young, and healthy eye. This quality approaches, but is somewhat worse, than that produced by diffraction-limited imagery in an ideal optical system. The retinal image is always in motion: even during the best efforts at steady fixation, there exists an irreducible tremor of the eye whose high-frequency components are in the 20 to 30 seconds-of-arc range, with larger drifting and saccadic movements up to 5 minutes of arc. It is possible to eliminate this residual motion by various optical techniques; such stabilization usually results in a total loss of vision, providing an elegant demonstration that the visual system responds primarily to *changes* in light patterns, rather than to steady states. Electrophysiological evidence from animals amply confirms this.

The *retina* is a thin structure of extreme com-

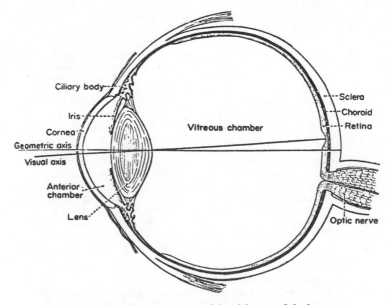

FIG. 2. Horizontal cross section of the right eye of the human.

plexity. It is considered embryologically to be a displaced part of the brain, and it is of clinical importance as the only part of the central nervous system that can be directly observed in the intact living subject. The receptors, the *rods* and *cones*, line the back surface of the retina, in immediate contact with dark layers (including the pigment epithelium and choroid) which help to nourish the receptors and to prevent multiple reflection of light. There are about 125 000 000 receptors in each human eye, of which only about 5 per cent are cones. The cones are however of an importance disproportionate to their relative number: in particular there is a small central bouquet of about 2000 of them, located in a rod-free depression of the retina known as the *fovea centralis* where they are packed together into a hexagonal array having a density of about 150 000/mm^2; these are capable of dissecting the finest details of the optimal retinal image. This process is aided by the lateral displacement of other retinal structures through which light must pass to reach the cone receptors. Moreover, this is the area of the retina where images have the highest attention value and which "projects" to a disproportionately huge area of the visual brain; the extraocular muscles move the eye more or less automatically, in the act of fixation, to put objects of interest into this region where their details can be most critically appreciated, while the accommodative mechanism alters the shape of the lens to produce the sharpest possible image in this region. The cones, including those in the fovea, function only at high luminance levels (approximately, above .01 candela/m^2), below which they are functionally blind and the rods take over. Thus the retina contains two systems intermixed: (a) the cone system (photopic), good for high-acuity vision, which also mediates all color vision; (b) the achromatic rod system, which has relatively poor spatial resolving power, but very high sensitivity.

The rods and cones are synaptically connected to the *bipolar cells*, which in turn relate to the *ganglion cells*, whose axons constitute the optic nerve fibers. There are also rich horizontal connections among the receptors, among the bipolar cells, and among the ganglion cells. In addition, there is a high degree of convergence: the 125 000 000 receptors ultimately feed into only 1 000 000 nerve fibers of the flexible optic nerve, which therefore constitutes the principal "bottleneck" of information flow in the visual system. The convergence ratio for the fovea is about 1:1, helping to preserve the high-detail vision of this region, while in the peripheral retina this ratio is many thousands to one, leading to high sensitivity at the sacrifice of resolving power.

The pathways from retina to brain are by no means independent, including those emerging from the central fovea. The horizontal interconnections are utilized to allow inhibitory processes to sharpen the "neural image" by a process of border enhancement, but much more complicated preprocessing of information occurs also. It is abundantly clear that the brain does not receive a replica of what is on the retina, although a spatial isomorphism between retina and brain does exist; rather, the messages sent to the brain tend to carry information that is already processed in complex ways to make efficient use of the limited communications pathways between eye and brain in an adaptively significant manner.

Horizontal cells form synaptic connections with receptors and bipolar cells, and by connecting with one another they integrate signals over long retinal distances. They probably play an important role in adaptation by providing inhibitory feedback to reduce the responses of receptors whose signals otherwise would saturate. In addition, horizontal cells are involved in the generation of opponent signals that relate to sums and differences of receptor outputs, and which are important both for spatial and color vision. *Amacrine cells* lie deeper in the retina, where they interconnect with bipolar and ganglion cells, as well as with other amacrines.

Cells in the visual system exist in a very wide variety of specialized types. This is true both within and between levels of visual processing, including the cells of the retina, the *lateral geniculate nucleus* (LGN) of the thalamus (to which the optic nerve fibers project) and the *visual cortex* of the brain, which receives its input from the LGN. Many other regions of the brain are also concerned with vision; most, but not all, receive their input from the LGN.

All of the cells of the retina except the ganglion cells (and to a limited extent the amacrines) exhibit graded potentials which carry information by means of an amplitude code. Most cells of the visual system, including the ganglion cells of the retina, show the classical all-or-none behavior according to which information is encoded in terms of impulse frequency for transmission over relatively long distances.

Visual cells have been extensively studied using microelectrodes, and the visual pathways have been traced using an ever expanding variety of techniques. At all levels of the visual system, the responsiveness of a cell relates to a restricted region of the visual field, but not to just a point. A ubiquitous feature of a visually responsive cell is its *receptive field*. This is defined as the region of visual space to which it responds. The reaction of a cell within such a region is not uniform. For example, ganglion cells exhibit a concentric center-surround organization: the output of the cells is increased by delivering stimuli to the center of its receptive field, but is inhibited by light in the surrounding region (or vice versa). In the visual cortex, cells respond instead to more complex stimuli, such as bars, lines, or gratings, depending upon such variables as orientation, movement, and length. Thus individual cells differ markedly in their responsiveness to important stimulus properties related to time, space, and color. Despite rapidly expanding knowledge of how the visual system analyzes the visual environment, as reflected in the activity of its constituent cellular components, almost nothing is known about the mechanisms of synthesis that presumably must underlie our integrated sensory experiences.

Because the two eyes are located in slightly different places in space, a disparity of the two retinal images results. Rather than to produce a blurred or confused picture, this *retinal disparity* results in the appearance of *stereoscopic depth*. Such depth judgments are remarkably precise, consistent with the findings that all but the smallest eye movements and accommodative adjustments are highly correlated between the two eyes, and that neural units in the visual brain are precisely connected, by way of intermediate synapses, to optically corresponding areas.

The normal eye exhibits a large amount of chromatic aberration which is not normally noticeable. There are at least two reasons for this: (a) the cone receptors exhibit a directional sensitivity which reduces the visual effectiveness of light entering the marginal zones of the pupil; (b) the visual system has a remarkable capacity to adapt to systematic distortions of almost any kind which do not carry useful information from the external world. For example, observers learn with practice to compensate for the effects of gross visual displacement caused by prisms placed before the eyes, and are no longer able to see the chromatic fringes produced by such prisms. Removal of the prisms produces reappearance of chromatic fringes and an apparent displacement in the opposite direction. The explanation of such effects is not simple: in this example, the adaptation to displacement is probably kinesthetic rather than visual, but the adaptation to fringes is almost certainly confined to the visual system. Related to this are many entoptic phenomena that are seldom preceived: (a) the shadows of the retinal blood vessels, which are in front of the receptors; (b) the blind spot in the visual field, caused by the receptor-free optic disc, large enough to contain 200 images of the moon; (c) "floaters," usually shadows of debris in the vitreous humor, clearly visible if attended to against bright, uniform surfaces such as the sky; (d) fleeting specks of light probably caused by the movements of corpuscles within the retinal blood vessels; (e) Maxwell's spot, probably corresponding to the region of the macular pigment, and many others.

The initial nonoptical event in the visual process is the absorption of single light quanta by single molecules of visual photopigment, of which millions are located in each rod or cone. Under ideal conditions, as few as a half-dozen of these elemental events within fairly broad bounds of time and area, are sufficient to lead to a visual sensation. The visual photopigment contained in the rods is *rhodopsin*, having a peak sensitivity at about 505 nm. It has been much studied and is found in most animals including man. Absorption of light by rhodopsin probably produces graded potentials at the receptors that trigger all-or-none nerve impulses by the time the ganglion cells are activated, if not before. The exact mechanisms whereby light absorption gives rise to receptor potentials (and these to nerve impulses), although under active investigation, cannot be said yet to be satisfactorily understood.

Color vision depends upon the existence of three classes of visual photopigment, all different from rhodopsin, housed in different proportions in different classes of cone receptors. When two fields of light that are physically different look exactly alike in color (*metameric matches*), it is probable that the rate at which light is being absorbed in the three classes of cone photopigment is the same from both fields, although this is not yet definitely established. The perception of color clearly involves the higher levels of the visual system as well.

Another important property of the eye is its adaptation to intensity by means of which the eye changes its gain and other characteristics, enabling it to respond discriminatively over a stimulus intensity range of about ten billion to one. At one time, it was felt that the bleaching of photopigments was primarily involved in this process; recent evidence indicates that this plays only a minor role and the true mechanisms are numerous and include changes in the organizational properties of retinal networks.

ROBERT M. BOYNTON

References

Brindley, G. S., "Physiology of the Retina and the Visual Pathway," Second Edition, Baltimore, Williams and Wilkins Co., 1970.

Boynton, R. M., "Human Color Vision," New York, Holt, Rinehart, and Winston, 1979.

Carterette, E. C., and Friedman, M. P., Eds., "Handbook of Perception," Vol. 5, "Seeing," New York, Academic Press, 1975.

Davson, H., Ed., "The Eye," Second Edition, New York, Academic Press, 1976.

Ditchburn, R. W., "Eye Movements and Visual Perception," Oxford, Clarendon Press, 1973.

Graham, C. H., "Vision and Visual Perception," New York, John Wiley & Sons, 1965.

Helmholtz, H. von, "Physiological Optics," translated by J. P. C. Southall, Optical Society of America, 1924; New York, Dover Publications, Inc., 1962.

Jameson, D., Ed., "Handbook of Sensory Physiology," Vol. VII/4, "Visual Psychophysics," Berlin and New York, Springer-Verlag, 1972.

LeGrand, Y., "Light, Colour, and Vision," Second Edition, translated by R. W. G. Hunt, J. W. T. Walsh, and F. R. W. Hunt, London, Chapman and Hall, Ltd., 1968.

LeGrand, Y., and El Hage, S. G., "Physiological Optics," Berlin and New York, Springer-Verlag, 1980.

Pirenne, M. H., "Vision and the Eye," London, Chapman & Hall, 1948.

Rodieck, R. W., "The Vertebrate Retina," San Francisco, Freeman, 1973.

Uttal, W. R., "A Taxonomy of Visual Processes," Hillsdale, NJ, Lawrence Erlbaum Associates, 1981.

Wolff, E., "The Anatomy of the Eye and Orbit," Philadelphia, Blakiston Co., 1948.

Wyszecki, G., and Stiles, W. S., "Color Science," Second Edition, New York, John Wiley & Sons, 1982.

Cross-references: COLOR; LENS; LIGHT, LUMINANCE; OPTICS, GEOMETRICAL, OPTICS, PHYSICAL; PHOTOGRAPHY; PHOTON; REFLECTION; REFRACTION.

VITREOUS STATE

The vitreous state, or the glassy state, is a special metastable condition in which a substance can exist. A material in the vitreous state may be termed a noncrystalline solid or a rigid liquid. The attainment of the vitreous state is illustrated by the volume-temperature relationship of an ideal system in Fig. 1. When a liquid is cooled slowly, crystallization will usually occur when the temperature reaches the melting point T_m as described by the path a-b. In the case of a complex liquid, crystallization will similarly commence when the temperature reaches the liquidus. Crystallization is generally considered to take place by a two-step mechanism, namely, nucleation and then crystal growth. In the ab-

sence of crystallization, a liquid can be supercooled below T_m, and the cooling path is now described by a-c. Since the fluidity of a liquid generally exhibits an exponential dependence on temperature, progressive undercooling along a-c will be accompanied by a rapid increase of viscosity. If crystallization has still not occurred when point d is reached, the viscosity of the liquid will have reached 10^{13} poises. (By comparison, the viscosity of glycerol at room temperature is only 10 poises.) For a particular cooling rate, say 5 deg/min, the cooling curve will now follow d-e. At any temperature below that corresponding to the point d, the material is said to be in the vitreous state. It now has the rigidity of the corresponding crystalline solid, but its structure is still devoid of long-range order similar to that of the parent liquid. A substance in the vitreous state is called a glass.

At the point d, the viscosity of the supercooled liquid is 10^{13} poises. At such high viscosity, molecular motions are retarded since the relaxation times are of the order of minutes and hours. The time taken for experimental measurements may now be actually less than the time needed to attain internal equilibrium. Thus for a particular cooling rate, an inflection will occur at d. For a slower cooling rate, d-e will be displaced to the first dashed line below it. For an even slower cooling rate, the inflection will occur at point f; the cooling curve will now be described by the lowest dashed curve. Similar behavior is shown by the heat content-temperature relationship of the system. The inflection points d and f, and hence the specific volume of a glass at the lower temperatures, are dependent on the cooling rate. The temperature region over which this inflection occurs is termed the glass-transition temperature T_g. Although

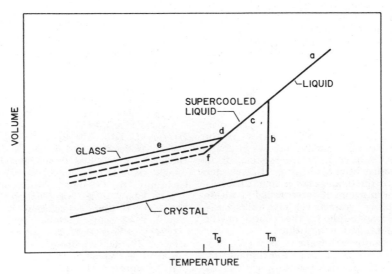

FIG. 1. Volume-temperature relationship of an ideal glass-forming system.

the supercooled liquid at above T_g is metastable with respect to the crystalline phase, it is in internal thermodynamic equilibrium. A glass, however, is generally considered not to be in internal thermodynamic equilibrium.[1]

Many substances are easily rendered into the vitreous state by supercooling the melt. These include organic liquids such as toluene and the alcohols, polymeric materials, fused salts such as $Ca(NO_3)_2$–KNO_3, ZrF_4–BaF_2, and many silicates, borates, phosphates, and their mixtures.[2] In general the ease of glass formation via the liquid is dependent on the crystallization kinetic constants as well as the cooling rate. Although the crystallization kinetic constants of metallic systems, for instance, are not favorable for glass formation, the vitreous state is attainable by very rapid cooling. Many metallic glasses are now known. The condensation of vapor at low temperatures can also yield noncrystalline solid phases and so can precipitation from solutions. Little is known about the relationship between such phases and the more common glassy phase obtained from the cooling of the melt. Present theoretical interests are centered on the nature of the glass transition at T_g. Two important questions are: (a) Is the glass-transition region an iso-free volume state for all substances? (b) Is there a theoretical lower temperature limit to T_g? No unambiguous answers to these questions are at present available.

JOHN D. MACKENZIE

References

Kauzmann, W., "The Nature of the Glassy State and the Behavior of Liquids at Low Temperatures," *Chem. Rev.* 43, 219 (1948).

Mackenzie, J. D., "Modern Aspects of the Vitreous State," Vol. 1, London, Butterworths, 1960.

Cross-references: AMORPHOUS METALS, CONDENSATION, CRYSTALLIZATION, CRYSTALLOGRAPHY, ELASTICITY, RHEOLOGY, VISCOELASTICITY, VISCOSTY.

VOLCANOLOGY

Volcanology is the science of volcanoes. Volcanologists, however, are generally geologists, geophysicists, or geochemists who specialize in studying volcanoes. With their roots in the Earth's mantle, volcanoes often explode their products into the stratosphere. This diversity of pheonomena requires the combined studies of many disciplines, physics being one of the more important with regard to the structure and processes of volcanoes.

Volcanic Activity An active volcano is defined as one that has a historic record of eruption, i.e., emission of lava flows or solid volcanic fragments. There are 529 active volcanoes known on Earth, and about 50 of these erupt during any one year. An active volcano can be either dormant or erupting. A volcano with no historic record of eruption may be considered potentially active if its record of prehistoric eruptions in geologically recent time implies that it has a moderate to high probability of future eruptions. An extinct volcano is one with very low probability of future eruptions. Life spans of volcanoes vary from 1 to 10,000,000 years and dormant periods from days to millions of years; thus it is difficult to distinguish potentially active from extinct volcanoes.

Types of Volcanoes Volcanoes have been classified by location, topographic form, general character of their eruptions, rock type, and tectonic setting. Most active volcanoes occur around the rim of the Pacific Ocean Basin. These often erupt explosively and build steep conical peaks known as *strato-volcanoes*, which are mixtures of lava and explosive debris (pyroclastic fragments). Hawaiian volcanoes, in contrast, are built mainly by lava flows poured out in gently sloping forms called *shield volcanoes*. Pacific rim volcanoes have rock types ranging from basalts ($\sim$50% SiO_2) through andesites ($\sim$55–60% SiO_2) and dacites ($\sim$60–65% SiO_2) to rhyolites ($\sim$65–75% SiO_2), whereas Hawaiian volcanoes are mainly basalts. As SiO_2 content increases, Fe and Mg generally decrease, and Na and K increase. The amount of dissolved gases in a magma (subsurface molten rock) and the viscosity of magma generally increase with SiO_2 content, and these both increase the potential for explosive eruptions.

Rift volcanoes occur along the separating edges of diverging tectonic plates. (See Fig. 1.) Iceland and the Mid-Atlantic Ridge are of rift volcanic origin. The products are generally basaltic lavas and most are nonexplosive submarine eruptions in water depths too great to allow explosive boiling of the dissolved magmatic gases or the contact water. The bedrock of the world's ocean basins, 70% of the Earth's area, is formed by the progressive rift volcanism which heals the spreading gap between the separating plates.

Subduction volcanoes occur on the overriding edges of converging plates, about 100–200 km from the oceanic trenches which mark the actual plate boundaries. (See Fig. 2.) These volcanoes form important components of island arcs like Japan and coastal mountain chains like the Andes. The rock name *andesite* comes from its common occurrence in the Andes Mountains. Subduction volcanoes, which comprise about 80% of the world's known volcanoes, typically erupt explosively and produce a variety of rock types. Thick viscous lava flows do occur, but most of their products are exploded pyroclastic fragments which range from fine volcanic ash to large blocks of pumice (rapidly cooled frothy volcanic glass).

The eruptions of Mount St. Helens (MSH) in

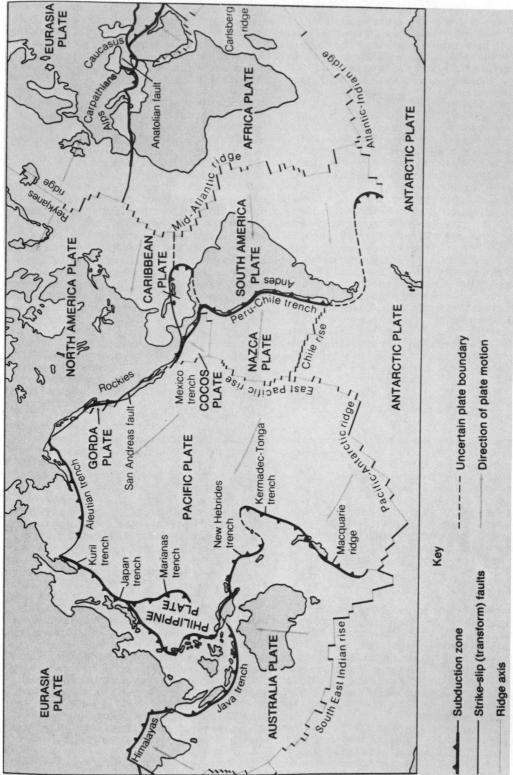

FIG. 1. Rigid plates of the Earth's surface are slowly moving horizontally away from and toward one another. The arrows shown assume that the Africa plate is not moving. Plates separate along the crests of mid-ocean ridges, slide past each other along strike-slip faults, and converge at subduction zones. (After J. F. Dewey, "Plate Tectonics," copyright © 1972 by Scientific American, Inc. All rights reserved.)

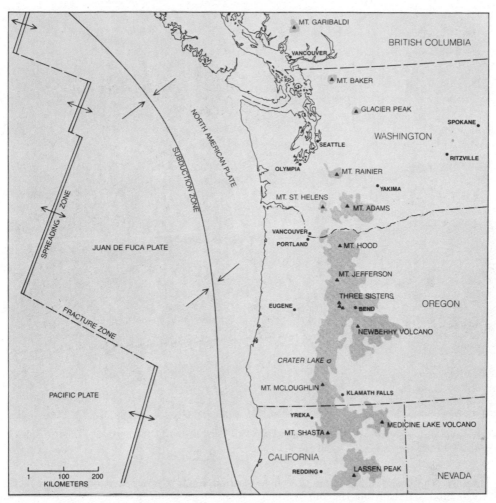

FIG. 2. Tectonic map shows the relation between the North American tectonic plate and the Juan de Fuca and Pacific plates to the west. At the subduction zone the Juan de Fuca plate is plunging under the North American plate, giving rise to the volcanoes of the Cascade Range (small black triangles). Shaded areas are volcanic deposits less than two million years old. (Data from U.S. Geological Survey. Copyright © 1981 by Scientific American, Inc. All rights reserved.)

northwestern United States provide a good case study of explosive volcanism. (See Figs. 3–7.) MSH was dormant from 1857 until March 20, 1980, when a magnitude 4 earthquake at shallow depth beneath MSH heralded the start of a major earthquake swarm that continued for 60 days. The eruption began on March 27 with a small steam explosion and formation of a crater in the ice-covered summit. These small explosions continued intermittently during April and early May, dusting the snow-clad peak with old rock particles but no new volcanic fragments. During this period of earthquakes and small explosions, a major topographic bulge was forming high on the mountain's north flank. Repeated aerial photographs and direct measure-

ments showed that an area about 1.5 by 2 km was moving outward at about 2 m per day. The volume increase of the bulge was close to 1,800,000 m^3 per day, and the bulge and shallow earthquakes were interpreted as being caused by the shallow injection of magma beneath MSH. On May 18 at 8:32 AM local time a magnitude 5.1 earthquake triggered a 2.3 km^3 avalanche of the over-steepened bulge. This suddenly released the pressure on the shallow magma intrusion and surrounding heated groundwater, and generated a massive explosion that was directed northward by the collapse of the north flank of the cone. This lateral blast was a fluidized mixture of hot expanding gases, mostly steam with some car-

FIG. 3. Eruption of Mount St. Helens on May 18, 1980. View is toward the northeast. (Photograph by the U.S. Geological Survey.)

bon dioxide (CO_2) and sulfur dioxide (SO_2), and fragmental debris from the old volcanic edifice as well as hot fragments of the newly intruded magma. The ratio of solids to gases in the mixture was high enough that much of the hot expanding blast cloud surged along the ground surface. The blast cloud had temperatures from 100 to 300°C; traveling at velocities from 100 to 325 m/sec it devastated 550 km² of forested mountain terrain. The mechanical energy of the blast was derived from explosive boiling, and was estimated at 3×10^{16} joules. The huge avalanche that preceded and accompanied the blast was probably partly fluidized by expanding pore gases, and accelerated rapidly to velocities as high as 80 m/sec. Some lobes swept into Spirit Lake and over a ridge 360 m high, but the bulk of the debris funneled down a valley, forming a hummocky deposit 22 km long, 1–2 km wide, and up to 195 m deep. The gravitational energy of the avalanche was about 5×10^{16} joules.

Following the initial avalanche and lateral blast, which lasted only a few minutes, the eruption produced a vertical jetting eruption lasting about 9 hours that generated a high ash cloud of volcanic gases (mostly steam), fractured silicate crystals and fine volcanic glass shards. Fallout from this ash cloud caused major nuisance, but little long-term damage, along a downwind path across central and eastern Washington, northern Idaho, and western Montana. The high jetting eruption from the boiling of the gases in the newly exposed magma caused progressively deeper coring of the crater until temporary exhaustion of the heat energy and slumping of the crater walls finally terminated the eruption. During this period of jetting eruptions, intermittent pyroclasic flows of fluidized mixtures of hot but solid volcanic rock fragments and gases spilled from the crater and down the north slopes of MSH onto the earlier avalanche and lateral blast deposits.

Mud flows, mixtures of water and the newly erupted volcanic and avalanche debris, flooded valleys well beyond the limits of the area devastated by the avalanche and lateral blast, and added greatly to the overall destruction. The count of dead and missing persons was 60, but it could have been 100 times worse if the

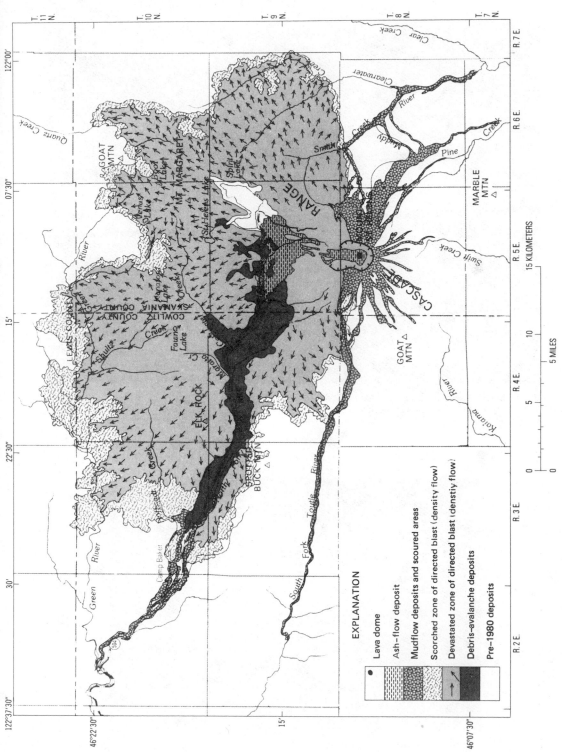

FIG. 4. Index map of Mount St. Helens area showing deposits and features of the 1980 eruptions. (Map by the U.S. Geological Survey.)

FIG. 5. Lava flows of aa (left) and pahoehoe both issued from Kilauea Volcano in Hawaii in 1974. The dark, rough aa is ~3–4 meters thick; it was covering the smoother, glistening pahoehoe when the flow stopped. (Photograph by Barbara Decker.)

FIG. 6. Mount St. Helens in June, 1970. The elevation of the summit, seen here from the north-northeast, was 2950 meters, rising from a base with an elevation of about 1000 meters. (Photograph by Robert Decker.)

FIG. 7. Mount St. Helens in July, 1980. The crater, seen here from the north, is 2 kilometers across. The elevation of the rim of the crater is between 2400 and 2550 meters, that of the floor of the crater between 1800 and 1900 meters. Pyroclastic flows cover much of the foreground. (Photograph by Ray Foster, Sandia Laboratories.)

general area north of the volcano had not been under restricted access. The total energy of the eruptions was largely contained in the thermal energy of the new magma involved (about 0.4 km^3 at 900–1000°C) and is estimated to be 1.7×10^{18} joules. Fortunately, it is typical of volcanic explosions that only a small fraction of thermal energy is converted to mechanical energy. Eruptions on the scale of the May 18, 1980, event at MSH occur somewhere in the world about every 20 years. Several much smaller eruptions at MSH involving ash explosions, pyroclastic flows, and extrusions of a viscous lava dome in the new crater have occurred since the major explosion.

Hot-spot volcanoes occurring within tectonic plates comprise only 5% of the world's active volcanoes. They are considered to be caused by some sort of melting anomalies in the Earth's upper mantle. The magma formed at hot spots is less dense than the overlying rocks, and forces and melts its way to the surface through the overlying lithospheric plate. As the plate slowly moves over the hot spot, a chain of volcanoes is formed that becomes progressively older in the direction of plate motion. The Island of Hawaii with its active volcanoes Mauna Loa and Kilauea

lies over a hot spot, and the other Hawaiian Islands which once formed over the same hot spot have moved away to the northwest. The Island of Kauai which has volcanic rocks about 5 million years old is 500 km northwest of the Island of Hawaii, giving an average plate velocity of 10 cm/year. Yellowstone National Park is another volcanic area that is apparently caused by a mantle hot spot. Hot-spot volcanoes in oceanic settings generally produce basaltic lava flows, but in continental settings hot-spot volcanoes produce varied rock types from basalts to rhyolites, and are often explosive. A major explosive eruption in the Yellowstone region about 600,000 years ago expelled 1000 km^3 of pyroclastic flows. This eruption was 2000 to 3000 times larger than the 1980 eruption of Mount St. Helens.

Volcanic products are classified by their chemistry, mineralogy, texture, and structure, or some combination of these characteristics. A basic goal in studying prehistoric volcanic products is to identify their origin as lava flows, air-fall deposits, pyroclastic flows, or mud flows. Air-fall deposits generally are well sorted and layered, whereas pyroclastic flows and mudflows are more poorly sorted and massive in appear-

TABLE 1. FORM OF VOLCANIC PRODUCTS.

Form	Name	Characteristics (dimensions)
Gas	Fume	
Liquid	Lavas	
	Aa	rough, blocky surface
	Pahoehoe	smooth to ropy surface
Solid	Airfall fragments	
	dust	$<\frac{1}{16}$ mm
	ash	$\frac{1}{16}$–2 mm
	cinders	2–64 mm
	blocks	>64 mm solid
	bombs	>64 mm plastic
	Pyroclastic flows	hot fluidized flows
	Mudflows	flows fluidized by rainfall, melting ice and snow, or ejected crater lakes

ance, like concrete. Table 1 shows the forms of some volcanic products and their textural and structural characteristics.

Volcanic landforms are quite variable. Stratovolcanoes and shield volcanoes have already been mentioned. Either of these volcanoes may have massive collapse basins on their summits, called *calderas*, which are larger than craters formed by volcanic vents. *Cinder cones* are angle-of-repose piles of loose volcanic fragments around vents, and *volcanic plateaus* are thick or multiple layers of low-viscosity pyroclastic flows or lava flows. Lava domes formed from high-viscosity magma build large steep-sided masses directly over their vents. Some volcanic vents are long lived and build complex volcanoes composed of several volcanic forms.

Roots of Volcanoes The deep structure and dynamics of volcanoes are of fundamental importance to the understanding of volcanic processes, but most data on these topics are indirect. Melting temperatures of rock in the 1200°C range occur at depths of 50 to 100 km beneath the Earth's surface. The source of heat is from radionuclides such as U^{238}, thorium^{232}, and potassium^{40}, or from other sources such as viscous dissipation from shear stresses in the Earth's mantle. Upward mass transfer of plastic material in the mantle could also cause melting from pressure reduction. Once enough molten rock has formed at depth it makes its way upward from bouyant forces by diapiric penetration or through fractures toward the surface. In volcanoes like Kilauea in Hawaii there is good evidence that magma accumulates in shallow reservoirs 2–6 km beneath the surface. Continuing accumulation in this shallow magma reservoir uplifts the ground surface and eventually causes fracturing upward or horizontally outward from the chamber. If the injection of magma into newly formed fractures reaches the surface, there is an eruption. Sometimes, particularly in lateral injections into the volcano's flanks, the magma stays beneath the surface and causes growth of the volcano by widening and uplift rather than by surface eruption. These shallow injections of magma are accompanied by earthquake swarms caused by progressive rock failure along the crack tip, and by rapidly changing stress gradients adjacent to the magma-filled fracture (*dike*). Geophysical monitoring of the local seismicity and ground-surface deformation on active volcanoes is a topic of active research, and much is being learned about the dynamics of magma movement beneath volcanoes.

Origin of Atmosphere and Hydrosphere The eruptions of volcanoes on Earth over geologic time have contributed significantly to the formation of the Earth's hydrosphere and atmosphere. The common gases from volcanoes contain oxygen, hydrogen, nitrogen, chlorine, and sulfur. Except for sulfur, their emissions from current volcanism can account for much of Earth's water and air if integrated over geologic time. Sulfur appears to be fifty times more abundant in volcanic gases than needed in this accounting. Probably it is recycled back to the mantle by precipitation as FeS_2 (pyrites) and subducted in oceanic trench sediments.

Geothermal Power and Mineral Deposits The heat energy contained in the upper 10 km of rock beneath the United States is estimated to be 3.3×10^{25} joules; however, in only a few places can drill holes into steam or hot water reservoirs make power extraction economically possible. Nearly all of these places are areas of geologically recent volcanism where the shallow emplacement of magma into or near subsurface water reservoirs has occurred. Normal heat flux from the Earth's interior is 0.06 watts per m^2 of the surface. In contrast, some geothermal steam wells can produce 6 megawatts of thermal power, equal to the normal heat flux from 100 km² of land surface. The key to this difference is the mass transfer of heat from deeper levels to near surface by volcanism, conductive and convective heat transfer from shallow magma bodies into subsurface water reservoirs, and finally mass transfer of heat by hot water or steam from drill holes. At The Geysers geothermal field north of San Francisco, superheated steam is directly tapped from porous underground reservoirs; but in most geothermal fields, water at or near its subsurface boiling temperature—about 300°C at 1 km depth—is allowed to flash into a steam and water mixture at reduced pressures in the bore hole. This two-phase fluid is separated at the well head, and the steam phase is used to drive a turbine. Exploration and development of geothermal-power fields is a young but rapidly expanding industry.

Many mineral deposits, including gold and silver, have been precipitated by the natural circulation of hot subsurface-water systems during geologic history. In most cases, shallow emplacements of magma provided the heat source for these natural distilleries, and in some cases, the valuable elements were themselves expelled from cooling magma as the more common rock-forming minerals crystallized and concentrated the excluded elements. In terms of geologic time, yesterday's geothermal fields are some of today's mineral deposits.

Other Research Topics There is some evidence that stratospheric aerosols, particularly sulfuric acid droplets, from violent volcanic explosions can lower the Earth's average surface temperature. Joint efforts by meteorologists and volcanologists are currently seeking to understand this effect. Forecasting volcanic eruptions is also a topic of active research. Local seismicity, ground-surface deformation, gas emissions, and electrical and magnetic fields often show anomalous behavior preceding eruptions at time scales of months to minutes. Finally, it is becoming evident that volcanism is a pervasive planetary process. Ancient volcanism on the moon, Mars, Mercury, and Venus, and active sulfur volcanism on Io, indicate that volcanic phenomena are not unique to our Earth.

ROBERT W. DECKER
BARBARA B. DECKER

References

Williams, Howell, and McBirney, A. R., "Volcanology," Freeman, Cooper and Co., 1979.

Decker, Robert, and Decker, Barbara, "Volcanoes," San Francisco, W. H. Freeman and Co., 1981.

Lipman, Peter W., and Mullineaux, Donal R. (Eds.), "The 1980 Eruptions of Mount St. Helens, Washington," U.S. Geological Survey Professional Paper No. 1250, 1982.

Decker, Robert, and Decker, Barbara, "The Eruptions of Mount St. Helens," *Scientific American* **244**(3), 68–80 (March 1981).

Simkin, T., Siebert, L., McClelland, L., Bridge, D., Newhall, C., and Latter, J., "Volcanoes of the World," Washington, D.C., Smithsonian Institution, 1981.

Cross-references: FALLOUT, GEODESY, GEOPHYSICS, HEAT TRANSFER, SEISMOLOGY.

W

WAVE MECHANICS

Introduction It has been known since the early part of the present century that systems of atomic dimensions do not obey the laws of classical mechanics, as formulated by Newton. A new mechanics has had to be constructed, and this is called *quantum mechanics*.

There were initially two apparently different formulations of quantum mechanics, one called MATRIX MECHANICS and the other *wave mechanics*. It was ultimately shown that the two formulations are completely equivalent and complicated quantum mechanical problems are nowadays usually solved by a combination of both. However, there are still many applications of quantum mechanics which can be treated by purely wave mechanical methods, and as these generally make a more direct appeal to physical intuition, it is perhaps worth while to give an account of wave mechanics quite separate from matrix mechanics.*

Waves and Particles The work of Planck, in 1900, on thermal radiation, and of Einstein, in 1905, on the photoelectric effect, suggested that light must, for some purposes, be regarded as consisting of particles, called *light quanta* or *photons*. The energy of a photon is given by

$$E = h\nu \qquad (1)$$

and its momentum by

$$p = h/\lambda \qquad (2)$$

where ν is the frequency and λ the wavelength of the light, and h is Planck's constant, which has the value 6.624×10^{-27} erg sec. On the other hand, diffraction phenomena, for example, can only be explained on the assumption that light consists of waves, so that neither the corpuscular nor the wave theory is completely satisfactory. In fact, it must simply be accepted that light can behave *either* as particles *or* as waves. The nature of a given experiment will emphasize one aspect or the other, and the relations between the two aspects are given by Eq. (1) and (2).

In 1924, de Broglie went further and suggested

*According to Dirac (1964), the validity of wave mechanics does not extend to quantum field theory, but it remains perfectly adequate in the domain of atomic and molecular physics.

that *any* moving particle, with mass m and speed v, will in some experiments display wave-like properties, the wavelength being given by Eq. (2), with $p = mv$. This purely theoretical suggestion received experimental confirmation in 1927, when Davisson and Germer observed the diffraction of a beam of electrons by a crystal of nickel, and similar results have since been obtained with beams of other kinds of particles, including atoms and molecules.

De Broglie's theory was the beginning of wave mechanics, but its further development was due to Schrödinger, who showed, in 1926, how the theory could be used to account for the existence of stationary states of atoms.

In 1911, Rutherford had proposed that an atom consists of a small nucleus surrounded by a planetary system of electrons. The electrons are continually accelerated; hence, according to classical electromagnetic theory, there should be a continual emission of radiation, accompanied by a diminution in the size of the electronic orbits and consequent increase in frequency. The existence of line spectra, that is, the emission of radiation in a discrete series of frequencies, and indeed the stability of matter itself, proves that this is not the case, and in 1913 Bohr suggested that an atom can exist in any one of a set of so-called *stationary states*, each with a definite energy, and with a finite energy difference between one and the next. Radiation is emitted only when an atom passes from a stationary state with energy E_1 to a state of lower energy E_2, and the frequency ν of the radiation is given by

$$h\nu = E_1 - E_2 \qquad (3)$$

Bohr attempted to graft this idea on to the classical picture of electrons describing planetary orbits about the nucleus by postulating that only certain orbits were permissible. This theory successfully explained the line spectrum of hydrogen and had several other partial successes, but it was soon superseded by the more revolutionary but also more versatile quantum mechanics.

The Schrödinger Equation According to de Broglie's theory a freely moving particle should be represented by a wave of the form

$$\Psi(x,t) = A \exp\left[2\pi i\left(\frac{x}{\lambda} - \nu t\right)\right]$$

where A is a constant. Using Eq. (1) and (2), this becomes

$$\Psi(x,t) = A \exp[i(px - Et)/\hbar]$$

$$= \psi(x) \exp(-iEt/\hbar) \qquad (4)$$

where $\hbar = h/2\pi$ and

$$\psi(x) = A \exp(ipx/\hbar) \qquad (5)$$

is the time-independent part of the wave (we shall deal with the time-dependent factor later). Differentiation shows that

$$\frac{d^2\psi}{dx^2} = -\frac{p^2}{\hbar^2}\psi \qquad (6)$$

However, the energy of the particle is

$$E = \frac{p^2}{2m} + V$$

the first term on the right-hand side being the kinetic energy and the second term the potential energy, which is constant if the particle is moving under no forces. Equation (6) thus becomes

$$\frac{d^2\psi}{dx^2} + \frac{2m}{\hbar^2}(E - V)\psi = 0 \qquad (7)$$

Although this equation has been plausibly derived only for a free particle, Schrödinger assumed that it also applies to a particle moving under a force, so that the potential energy V is a function of x. It is known as the *Schrödinger equation* or the *wave equation* for a particle moving in one dimension, and the function $\psi(x)$ is called the *wave function*. We will consider the interpretation of ψ below, but meanwhile let us note that ψ is to be *single-valued*, *continuous*, *smooth* [except at infinities of $V(x)$] and *finite everywhere*. Also, although $\psi = 0$ is always a solution of Eq. (7), it is not permitted as a wave function.

Particle in a One-dimensional Box That the Schrödinger equation does lead to discrete energy values, at least when classical mechanics would predict a limited range of motion or a *bound* particle, can easily be seen by considering a particle moving in one dimension between infinite potential barriers—this is known as a one-dimensional box.

Suppose that the potential energy is zero from $x = 0$ to $x = L$, and infinite everywhere else. In the region of zero potential energy, Eq. (7) becomes

$$\frac{d^2\psi}{dx^2} + \frac{2mE}{\hbar^2}\psi = 0 \qquad (8)$$

with the general solution

$$\psi = A \cos\sqrt{\frac{2mE}{\hbar^2}}x + B \sin\sqrt{\frac{2mE}{\hbar^2}}x \qquad (9)$$

The condition of finiteness on ψ (which we will not consider in detail) demands that it vanish in the region of infinite potential energy and because of the condition of continuity, we must then have $\psi(0) = \psi(L) = 0$. The only solutions in the region of zero potential energy satisfying these boundary conditions are of the form

$$\psi_n = B_n \sin\frac{n\pi x}{L}, \qquad n = 1, 2, 3, \cdots \qquad (10)$$

corresponding to the energy values E_n given by

$$\sqrt{\frac{2mE_n}{\hbar^2}} = \frac{n\pi}{L}$$

or

$$E_n = \frac{n^2\pi^2\hbar^2}{2mL^2} \qquad (11)$$

($n = 0$ gives $\psi \equiv 0$, which is not allowed).

The energy E can thus have any one of an infinite set of discrete values, corresponding to the integral values of n, but no other value. These values are called the *eigenvalues* of the Schrödinger equation or *energy levels* of the system, and the corresponding wave functions are called the *eigenfunctions* of the equation and represent the stationary states of the system.

Operators The Hamiltonian $H(x,p)$ of the one-dimensional system we have been considering is simply the energy expressed in terms of the momentum p and coordinate x, and the *equation of energy* is

$$H(x,p) \equiv \frac{p^2}{2m} + V(x) = E \qquad (12)$$

If we formally convert this into an *operator* equation by letting

$$p = \frac{\hbar}{i}\frac{d}{dx} \qquad (13)$$

(the coordinate x becomes the operator "multiply by x", but this is trivial), and allow both sides to operate on a function $\psi(x)$, we obtain

$$-\frac{\hbar^2}{2m}\frac{d^2\psi}{dx^2} + V\psi = E\psi \qquad (14)$$

which, when rearranged, is seen to be the Schrödinger equation [Eq. (7)].

The representation of dynamical variables by operators is fundamental to quantum mechanics, but the choice of operators made above is not unique. It may be confirmed by differentiation that

$$\frac{\hbar}{i}\frac{d}{dx}(x\psi) - x\frac{\hbar}{i}\frac{d\psi}{dx} = \frac{\hbar}{i}\psi$$

and using Eq. (13), this gives the operator equation

$$px - xp = \frac{\hbar}{i} \qquad (15)$$

The operator on the left is called the *commutator* of p and x, and is generally written $[p, x]$. Equation (15) is the basic equation of matrix mechanics and existed before Schrödinger's equation, p and x then being represented by MATRICES. This shows in an elementary way the close connection between wave mechanics and matrix mechanics.

Particles in Three Dimensions Equation (13) gives the clue to the extension of Schrödinger's equation to systems of one or more particles moving in three dimensions. We first write down the Hamiltonian of the system and transform each Cartesian component of momentum into a differential operator like Eq. (13) with respect to its corresponding (canonically conjugate) coordinate. For example, for a single particle moving in three dimensions, the Hamiltonian is

$$H(x, y, z, p_x, p_y, p_z) = \frac{1}{2m}(p_x{}^2 + p_y{}^2 + p_z{}^2)$$
$$+ V(x, y, z)$$

where p_x, p_y, p_z are the Cartesian components of the momentum, and the equation of energy is

$$H(x, y, z, p_x, p_y, p_z) = E \qquad (16)$$

If we transform the latter into an operator equation by writing

$$p_x = \frac{\hbar}{i}\frac{\partial}{\partial x}, \quad p_y = \frac{\hbar}{i}\frac{\partial}{\partial y}, \quad p_z = \frac{\hbar}{i}\frac{\partial}{\partial z}$$
$$(17)$$

(partial derivatives are now required as several variables are present), and operate upon a function $\psi(x, y, z)$, we obtain the equation

$$-\frac{\hbar^2}{2m}\left(\frac{\partial^2 \psi}{\partial x^2} + \frac{\partial^2 \psi}{\partial y^2} + \frac{\partial^2 \psi}{\partial z^2}\right)$$
$$+ V(x, y, z)\psi = E\psi \qquad (18)$$

which is the Schrödinger equation of the system. The essential correctness of this equation has been amply demonstrated by many successful applications (this is, of course, the *only* justification of any of the postulates of quantum mechanics). In particular, in the case of the hydrogen atom, where there is a single electron moving under the Coulomb attraction of a single proton (which may as a good approximation be considered to be at rest), the equation can be solved analytically, and the energy levels so found, together with the Bohr frequency rule [Eq. (3)], correctly give the line spectrum of hydrogen.

The extension to many particles is straightforward; in the Hamiltonian of the system, the substitutions of Eq. (17) are made for the momentum components of each particle of the system, and the equation of energy is thus transformed into the Schrödinger equation. Again the equation has had many successful applications, an early one being the calculation of the energy of the normal helium atom by Hylleraas in 1930, which the earlier Bohr theory failed to do. Unfortunately, owing to the interaction of the electrons, the equation cannot be solved analytically for many-electron systems and resort must be made to approximate methods of solution which have become extremely complicated in recent years [see e.g., Raimes (1972)].

Interpretation of the Wave Function Let us consider the Schrödinger equation for a single particle moving in three dimensions. This is a linear equation, so that if ψ is a solution, so also is $C\psi$, where C is any constant. This means that (in most cases) we can choose ψ so that

$$\int_{-\infty}^{\infty}\int_{-\infty}^{\infty}\int_{-\infty}^{\infty} |\psi(x, y, z)|^2 dx\, dy\, dz = 1$$
$$(19)$$

The function ψ is generally complex and $|\psi|$ is its *modulus* ($|\psi|^2 = \psi^*\psi$, where ψ^* is the complex conjugate of ψ). If it satisfies Eq. (19), the function ψ is said to be *normalized*.

The interpretation of ψ demands that it be normalized. If this is not possible, owing to the divergence of the integral in Eq. (19), as is the case with a free particle, then normalization has to be affected artificially by enclosing the system in a large box.

It was proposed by Born, in 1926, that if the particle is in a stationary state, with normalized wave function ψ, then $|\psi|^2$ may be interpreted as a *probability density*, such that

$$|\psi(x, y, z)|^2 dx\, dy\, dz$$

is the *probability* of finding the particle in the small volume element $dx\, dy\, dz$ at the point (x, y, z).

It is clear from this why normalization is necessary—Eq. (19) expresses the fact that the probability of finding the particle *somewhere* is unity.

An important extension of this interpretation, which we will not justify in detail, relates to the *average* or *expectation value* $\langle f \rangle$ of a dynamical variable f (such as momentum) for a system in a state ψ, i.e., the average of a large number of measurements of f made on a system in this state. It is found that

$$\langle f \rangle = \int_{-\infty}^{\infty}\int_{-\infty}^{\infty}\int_{-\infty}^{\infty} \psi^* f_{op} \psi\, dx\, dy\, dz$$
$$(20)$$

where f_{op} is the quantum mechanical operator representing f. Here we have restricted ourselves to a single-particle system, but the same applies to systems of any number of particles, so long as the integral is a multiple integral taken throughout the full range of all the variables appearing in ψ.

The interpretation of ψ, or rather of $|\psi|^2$ (ψ itself has no physical significance), which we have presented above, gives rise to a fundamental and important difference between classical mechanics and quantum mechanics. According to the latter, there is a finite probability of finding a particle in regions of space where its presence would be forbidden by classical mechanics— regions where its kinetic energy would be negative.

Let us take as an example the linear harmonic oscillator. Suppose a particle of mass m is moving along the x axis under the action of a force $-kx$ (k is positive) directed towards the origin. Classically, the particle executes simple harmonic motion with frequency $\omega/2\pi$, where $\omega = \sqrt{k/m}$. It is easy to see that the potential energy is $V(x) = \frac{1}{2} m\omega^2 x^2$, if the zero of potential energy is taken to be at $x = 0$, so that the Schrödinger equation [Eq. (7)] becomes

$$\frac{d^2 \psi}{dx^2} + \frac{2m}{\hbar^2} \left(E - \frac{1}{2} m\omega^2 x^2 \right) \psi = 0 \quad (21)$$

It may be verified by substitution that the function

$$\psi_0 = \exp(-m\omega x^2/2\hbar) \quad (22)$$

satisfies this equation, provided E has the value

$$E_0 = \frac{1}{2}\hbar\omega \quad (23)$$

This is, in fact, the *ground state* of the oscillator, that is to say, E_0 is the lowest energy level.

Now according to classical mechanics, the particle, having energy E_0, would be confined to a region of the x axis whose limits are given by setting $V(x) = E_0$, that is, $x = \pm \sqrt{\hbar/m\omega}$. However, ψ_0 tends to zero asymptotically as x tends to $\pm\infty$, so that, according to the interpretation of $|\psi|^2$, there is a finite probability of finding the particle at very large distances from the origin. This is known as the *tunnel effect* and has an important application in the theory of radioactive decay.

The foregoing example demonstrates another point in which quantum mechanics differs from classical mechanics. According to the latter, the oscillator could have zero energy (when the particle is at rest at the origin), but this is not permitted in quantum mechanics—the *lowest* energy the oscillator can have is $\frac{1}{2}\hbar\omega$, which is called its *zero-point energy*.

Time-Dependence So far we have only considered the stationary states of a system and, furthermore, only of a *conservative* system, i.e., one whose energy is constant. However, if a sys-

tem is subject to a disturbance which varies with time, for example, due to the passage of a charged particle or a light wave through it, there is a probability that the system will in a certain time make a transition from its initial stationary state to some other stationary state. In order to calculate such *transition probabilities*, it is necessary to consider explicitly the time dependence of wave functions, which we have so far neglected, and this entails the use of a modified Schrödinger equation.

Differentiation of the function $\Psi(x, t)$, Eq. (4) gives

$$E\Psi = i\hbar \frac{\partial \Psi}{\partial t} \quad (24)$$

which suggests that the energy E should in general be represented by the operator

$$E = i\hbar \frac{\partial}{\partial t} \quad (25)$$

If we convert the equation of energy [Eq. (12)] for a one-dimensional system into an operator, using Eqs. (13) and (25), and if we allow both sides to operate upon a function $\Psi(x, t)$, we obtain

$$-\frac{\hbar^2}{2m} \frac{\partial^2 \Psi}{\partial x^2} + V\Psi = i\hbar \frac{\partial \Psi}{\partial t} \quad (26)$$

[the derivative in Eq. (13) has been replaced by a partial derivative as we now have two variables, x and t]. This is known as the *time-dependent Schrödinger equation*; it applies even when the potential energy function V depends explicitly upon the time. If more than one dimension or several particles are involved, it is only necessary to use the appropriate quantum mechanical Hamiltonian operator on the left-hand side. It should be emphasized that this equation has *not* been rigorously derived, but only plausibly suggested. No such derivation exists, either for Eq. (26) or for Eq. (7)—their justification lies entirely in their successful application.

If we are, in fact, dealing with a conservative system, so that V does not depend explicitly upon t, it is easy to verify by substitution (and indeed this is obvious from the way in which the equation was derived) that Eq. (26) has solutions of the form

$$\Psi(x, t) = \psi(x) \exp(-iEt/\hbar) \quad (27)$$

where $\psi(x)$ is an eigenfunction of Eq. (7) and E is its corresponding eigenvalue. The function $\psi(x)$ is the time-independent wave function of a stationary state, and $\Psi(x, t)$ is the time-dependent wave function. Since it is not ψ but $|\psi|^2$ which has physical significance, and Eq. (27) shows that $|\Psi|^2 = |\psi|^2$, it is clear that so long as we are dealing with the stationary states of a conservative system, the time dependence of the wave function is of no importance.

Conclusion The foregoing account has been confined to the elementary ideas of wave mechanics, due to de Broglie and Schrödinger. No mention has been made, for example, of the relativistic wave equation, due to Dirac, or of the important concept of electron spin. Further information on wave mechanics, its applications, and its relationship with the rest of quantum mechanics can be obtained from the article on MATRIX MECHANICS in this book and from a vast number of books of which those listed in the references are a very small sample.

S. RAIMES

References

Born, M., "Atomic Physics," Eighth edition, London, Blackie and Son Limited, 1969.

Dirac, P. A. M., *Nature*, **203**, 115 (1964).

Merzbacher, E., "Quantum Mechanics," Second edition, New York, John Wiley & Sons, 1970.

Mott, N. F., and Sneddon, I. N., "Wave Mechanics and Its Applications," London, Oxford University Press, 1948; reprinted, New York, Dover, 1963.

Raimes, S., "The Wave Mechanics of Electrons in Metals," Amsterdam, North-Holland Pub. Co., 1961.

Raimes, S., "Many-Electron Theory," Amsterdam, North-Holland Pub. Co., 1972.

Cross-references: MATHEMATICAL PRINCIPLES OF QUANTUM MECHANICS, MATRICES, PHOTOELECTRICITY, PHOTON, PHYSICAL CHEMISTRY, QUANTUM THEORY, SCHRÖDINGER EQUATION.

WAVE MOTION

Wave motion can be said to be the most common and the most important type of motion that we know. It is through wave motion that sounds come to our ears, light to our eyes, electromagnetic waves to our radios and television sets, and tidal waves and earthquakes to our cities. Wave motion can be defined as that mechanism by which energy is transported from a source to a distant receiver without the transfer of matter between the two points.

Waves can be classified according to the manner of their production, namely, a vibrating material object or, in the case of electromagnetic waves, sources such as electrical oscillations in an aerial. The wind blowing across water causes surface waves; a piezoelectric quartz crystal vibrating under an applied electric field generates underwater wave motion. Or, waves could be classified according to the medium in which they travel. The most useful classification, however, involves the direction of motion of the particles of a medium (or of an electric or magnetic field in the case of electromagnetic waves) relative to the direction in which the energy of the wave is itself propagated. Such a classification is useful because

wave motions falling into the same class according to the selected criterion will have other similar properties.

Wave motion can be most easily understood if one considers first, as an example, wave motion in a horizontal, stretched string and then, by analogy, other types of wave motion. If one end of such a string is moved up and down, a rhythmic disturbance travels along the string. Each particle of the string moves up and down, while, at the same time, the wave motion moves along the length of the string. It is the state of the particles that advances, the medium as a whole returning to its initial condition after the disturbance has passed. Such a wave motion, one in which the vibratory motion of the medium is at right angles, or essentially at right angles, to the direction of propagation, is called transverse. Surface waves on liquids are transverse; so also are electromagnetic waves (x-rays, visible light, radio waves, and so forth), but here, since electromagnetic waves can travel in a vacuum, we must think of the electric and magnetic fields associated with such waves as changing in intensity in a direction at right angles to the direction of propagation.

Another type of wave motion, one termed longitudinal or compressional, can occur only in material media and has the particles of the medium moving forward and backward along the direction of propagation of the wave. Compressional waves are exemplified by sound waves in air, in which a volume in the path of the wave is alternately compressed and rarefied. These variations in pressure are very small. Even for the loudest sounds that an ear can tolerate, the pressure variations are of the order of 280 dyne/cm² (above and below atmospheric pressure of about 1 000 000 dyne/cm²).

Yet another type of wave motion is the torsional wave, which can take place only in solids. Less frequently seen, this type can be demonstrated by a long helical spring supported on a flat surface. As one end of the spring is given a quick, momentary twist about the axis of the spring, a pulse travels down the spring.

Whatever the type of wave, certain useful definitions can be set forth and general statements made. Phase describes the relative position and direction of movement of a particle in its periodic motion as it participates in wave motion, or the relative intensity at a point of the electric field accompanying electromagnetic waves. Frequency is the number of complete vibrations performed per unit of time by a particle (or field) through which a wave passes. Period is the time required for one complete vibration of a particle participating in wave motion. Wavelength is the distance between any two points that are in phase on successive waves or pulses. The velocity of a wave is the product of the frequency and the wavelength. All waves except electromagnetic waves require a medium for their propagation.

Wave motions may vary in the energy they

transport per unit of time. This property depends on the amplitude. The amplitude of a wave in a string, to take again an example, is the maximum displacement experienced by the particles of the string as they move from their equilibrium positions. The intensity of the wave is the power (energy per second) passing through a square centimeter perpendicular to the wave front, and it is related to the square of the amplitude. In the case of sound, intensity is related to loudness.

Two or more waves crossing one another's paths will not cause any change in the direction, frequency, or intensity of any of them. The displacement effects of two or more waves of the same kind passing through a medium are additive at any point; and at any moment, the displacement at a point is the vector sum of the separate displacements caused by the separate waves. Two transverse wave motions passing at right angles through a point in a medium cause a particle at that point to perform a path called a Lissajous figure, whose form depends upon the amplitudes and frequencies of the two waves and upon their phase relationship.

Standing (stationary) waves are produced by combining two similar wave trains moving in opposite directions. Not themselves waves, they are patterns of vibration that simulate waves standing still. An example is exhibited by a string one end of which is fastened rigidly and the other vibrated transversely at a constant frequency. Waves traveling down the string from the source meet waves reflected from the fixed end. If the tension in the string is adjusted properly, the string can be made to display nodes, points where the string does not move transversely because at those points the two waves cancel the effects of one another. Between the regularly spaced nodes are found the antinodes or loops where the two waves reinforce one another.

Wave motion also experiences the phenomena of absorption, reflection, refraction, interference, diffraction, beats, resonance, and polarization. However, longitudinal waves do not exhibit the property of polarization.

Complex waves can be analyzed into sets of simple waves according to the principles of Fourier analysis, where by "simple waves" are meant waves whose variations of displacement with time can be represented by sine curves.

Compressional waves require about 5 seconds to travel a mile in air, 1 second to travel a mile in water, and 1/3 second to travel a mile in iron. Though varying in speed from material to material, low-frequency compressional waves travel with the same speed in a particular medium, i.e., they do not exhibit dispersion. Small variations in speed sometimes found at high frequencies are due to relaxation phenomena.

Compressional waves in a fluid have a speed v that depends only on the density ρ and the adiabatic bulk modulus β of the medium according to the relation

$$v = \sqrt{\frac{\beta}{\rho}}$$

In solids, the speed of compressional waves is given by the relation

$$v = \sqrt{\frac{Y}{\rho}}$$

where Y is Young's Modulus. Thus it may be seen that a study of the propagation of waves in a medium gives important information about the medium.

Transverse waves on the surfaces of liquids do not travel with a fixed speed dependent only on the properties of the liquid. Their speed depends upon their wavelength and amplitude, the depth of the liquid, and whether the surface is confined, as in a canal. Surface tension waves on the surface of water have wavelengths less than 1.7 cm, while gravity waves on the surface of water have wavelengths greater than 1.7 cm. Ripples on water often move only 30 cm/sec. They have higher velocity as their wavelength becomes smaller. In contrast, for example, ocean waves measuring 244 m (800 feet) from crest to crest have been found to travel 20 m/sec (45 mph).

Transverse waves in strings (or wires) travel with a speed that depends only on the tension in the string T and the mass per unit length of the string m, according to the formula

$$v = \sqrt{T/m}$$

Electromagnetic waves of all frequencies travel with the same speed in a vacuum (2.9979×10^{10} cm/sec). In a particular medium, however, different frequencies (colors in the case of visible light) travel with different speeds. The speeds at particular frequencies also vary with the media.

The observed wavelength of a wave motion, whether it be longitudinal or transverse, depends on whether the source and the receiver are moving relative to one another, a phenomenon known as the Doppler effect. In the case of electromagnetic waves, the Doppler effect depends only on the relative velocity; in the case of a compressional wave, as for instance, sound, the magnitude of the effect depends not only on the relative velocity of the source and receiver but upon which is in motion with respect to the transmitting medium. In both cases the wavelength of the wave is increased if the source and receiver move away from one another and decreased if they move toward one another.

In some uses of wave motion, care must be taken to differentiate between "phase" and "group" velocity. The group velocity of a wave is the velocity usually observed, and the energy in a wave is transmitted with the group velocity. For example, measurements of the speed of light wherein the time for "chopped" pulses to travel a known distance is determined, result in values of the group velocity. On the other hand, again for example, if one carefully observes the expanding group of ripples when a stone is dropped into water, the group travels with one velocity, the group velocity, while a particular wave crest will advance through the group to the outer leading edge and exhibit the phase velocity. The difference in the two velocities depends on the wavelength in the material medium and on the dispersion, i.e., on the change in phase velocity with wavelength. In a vacuum—interstellar space for instance—the two velocities are the same for light, whatever the color.

ROBERT T. LAGEMANN

References

Alanso, M., and Finn, E. J., "Fundamental University Physics," Vol. II, "Fields and Waves," Reading, Mass., Addison-Wesley Press, 1967.

Towne, D. H., "Wave Phenomena," Reading, Mass., Addison-Wesley Press, 1967.

Strong, J., "Concepts of Classical Optics," San Francisco, W. H. Freeman and Co., 1957.

Temkin, Samuel, "Elements of Acoustics," New York, John Wiley, 1981.

Elmore, William C., and Heald, Mark H., "Physics of Waves," New York, McGraw-Hill, 1969.

Clay, Clarence S., and Medwin, Herman, "Acoustical Oceanography," New York, John Wiley, 1977.

Kock, Winston E., "Sound Waves and Light Waves: The Fundamentals of Wave Motion," Garden City, N.Y., Doubleday, 1965.

Cross-references: ACOUSTICS, DOPPLER EFFECT, DYNAMICS, ELECTROMAGNETIC THEORY, HEARING, INTERFERENCE AND INTERFEROMETRY, LIGHT, MICROWAVE TRANSMISSION, MUSICAL SOUND, OCEAN ACOUSTICS, PHYSICAL ACOUSTICS, REFLECTION, REFRACTION, RESONANCE, SEISMOLOGY, SONAR, ULTRASONICS.

WEAK INTERACTIONS

The weak interactions present a fascinating aspect of the problem of elementary particles. They appear at first to be totally unrelated to the other interactions that we know; nonetheless there have been discovered striking regularities among them, which have led to major advances in our knowledge of the laws of nature. We present here a brief history of the subject, which has been marked by many surprises.

*An average β-decay neutrino must pass through a thickness of matter of the order of 10^{20} g/cm^2 before suffering an interaction.

To avoid the difficulties found in understanding processes of β-decay (see RADIOACTIVITY), in 1931 Pauli suggested that the emission of a β-particle from a nucleus is always accompanied by that of a spin-$\frac{1}{2}$ neutral particle of small mass, which takes account of energy and angular momentum conservation in β-decay. Calorimetric measurements of the average energy release in a β-decay transition agree quite well with the value calculated from direct measurement of the β-decay spectrum, indicating that the hypothetical particles, which were called *neutrinos*, must interact extremely weakly with matter, since they escape from the apparatus without giving up any measurable energy. A quantitative theory of β-decay, incorporating the neutrino hypothesis (see NEUTRINO), was formulated in 1934 by Fermi, who wrote down the simplest relativistic expressions that would describe the basic β-transition: the transformation of a neutron into a proton accompanied by the creation of an electron and a neutrino. This theory has been strikingly successful in describing all aspects of β-decay—with one qualification to be described later—and is the one employed to this day. It could be anticipated from the weakness of the neutrino's interaction with matter that the coupling constant characteristic of β-decay would be small; it is in fact extremely small.* Adopting a simile of Lord Rutherford, one may say that from the point of view of the nucleus, β-decay practically never happens! To judge the scale, we note that lifetimes for β-decay are commonly several minutes, whereas γ-transitions of similar energy occur in nanoseconds or less.

The success of Fermi's theory in accounting for the shapes of β-spectra and the dependence of the lifetimes on the available energy release could be regarded as indirect evidence for the existence of the neutrino; there were also other indications, such as the occurrence of the predicted nuclear recoil after processes of orbital electron capture, a phenomenon predicted by the theory. Nonetheless, there was considerable satisfaction when the existence of the neutrino was directly demonstrated by Cowan and Reines in 1955 (for Reines' discussion, see NEUTRINO). They used the intense neutrino flux arising from free neutron decays near a reactor to induce inverse β-decay, i.e., the conversion of a proton into a neutron with the emission of a positron, at roughly the expected rate. A similar experiment by Davis gave a negative results. In Davis' case, however, the reaction sought was the inverse of a β^+-decay transition, although the neutrinos available to him were the same as those used by Cowan and Reines, i.e., neutrinos arising from the β^--decay of neutrons. The nonoccurrence of the Davis reactions demonstrates that the neutrinos emitted in β^--decays cannot be identical with the neutrinos associated with β^+-decay. This conclusion is supported by the absence of neutrinoless double β-decay, a process in which the neutrino from the β-decay of one nucleon could be reabsorbed within the same nucleus to induce the β-transition of a second nucleon. Davis subsequently enlarged his apparatus (using 600 tons of dry-cleaning fluid, C_2Cl_4!) to obtain the extreme sensitivity at which he should expect to see reactions induced by neutrinos emitted together with β^+-particles in fusion reactions in the interior of the sun. The rate of events which he finds[4] is about one-third of that expected from current solar-model calculations. Whether this reflects our ignorance of the solar interior or a failure in our understanding of weak interactions is not certain at the moment.

Universal Fermi Interaction The μ-mesons, or muons (see ELEMENTARY PARTICLES), which were originally identified with Yukawa's mesons, were found to interact only very weakly with matter. In fact, it was noted that their absorption could be regarded as a process exactly like electron capture, i.e., by an interaction just like that of β-decay, with even the same coupling constant! The decay of a muon into an electron and two neutral particles, presumably neutrinos, was also accounted for by assuming a Fermi coupling of these four particles, again with a coupling constant of the same magnitude. This led to the hypothesis of a *universal Fermi interaction* between fermions, and after the discovery of "strange" particles (see STRONG INTERACTIONS), the natural extension of this hypothesis to include hyperons led to a qualitative understanding of all strange-particle decays. The relatively meager data available on hyperon β-decays could be accounted for quite adequately by the Fermi theory, provided one used coupling constants somewhat smaller than for nuclear β-decay. This can be explained in terms of a universal Fermi interaction by modifying the definition of universality, in a way suggested by Cabibbo.

After the discovery of parity nonconservation in 1957, it was verified that the detailed space-time forms of the μ-decay interaction and the β-decay interaction are also very similar, lending further support to the idea of a universal Fermi interaction. Experiments showed that the form of the Fermi interaction is that of a current-current coupling, just as the electromagnetic interaction between two objects may be regarded as the interaction between their currents. It was therefore suggested that the Fermi Interaction may arise as a result of the interaction between currents which generate a vector field, similar to the electromagnetic field, which mediates the weak interaction. The currents appearing in the Fermi interaction induce a transfer of electric charge, therefore the intermediary W-field must carry electric charge. To account for the short range of the Fermi interaction, the quanta of this W-field must be very massive, which would account for the fact that these W-particles have not been seen so far. The selection rules

for weak interactions could be understood by assigning suitable properties to the generating currents. Gell-Mann's suggestion that these currents form a certain algebra codified these selection rules and also led to a mathematically precise definition of universality. It also led to the remarkable Adler-Weisberger relation between vector and axial-vector β-decay constants.

Fermi's theory predicts that the cross section for neutrino interactions increases quadratically with neutrino energy. For the reaction

$$\nu + n \rightarrow \begin{pmatrix} e^- \\ \mu^- \end{pmatrix} + p,$$

this growth is not expected to continue after the neutrino wavelength becomes smaller than nucleon dimensions, but the cross section at the corresponding energy, about 1 GeV, is already sufficient for experiments using high-energy accelerations. One of the first results of such experiments was to demonstrate the striking fact that the neutrino associated with muon capture, which is the one predominantly produced in meson decays, is unable to produce electrons, proving that it is physically distinct from the neutrino of β-decay. It now appears that there is at least one other neutrino associated with the heavy lepton τ found in e^+e^- annihilation. The *total* neutrino cross section continues to increase as the square of the neutrino energy, as measured in the neutrino-nucleon c.m. system, to the highest neutrino energies studied thus far, about 400 GeV at the Fermi National Accelerator Laboratory. This supports the idea of "pointlike" constituents (*partons*) in the nucleon, previously suggested by Feynman to explain the pointlike cross sections found in the inelastic scattering of high-energy electrons. All observations are consistent with the hypothesis of current-current interactions; the currents are simply described as those carried by quarks and leptons (see QUARKS and ELEMENTARY PARTICLES).

The similarity of form of the weak and electromagnetic interactions is a strong hint that the two are related. The principal difficulty with this notion was that while the carrier of electromagnetic interactions, the photon, has zero mass, the postulated W-meson must be very massive. Subsequently, it was realized that this extreme difference of masses can be understood in a gauge theory of weak and electromagnetic interactions if one is dealing with a situation of *spontaneous symmetry violation*, through a mechanism discovered by Higgs. A unified gauge theory of weak and electromagnetic interactions incorporating these ideas was proposed by Weinberg and also published by Salam. This scheme also possesses the advantage that calculations of higher-order effects of weak interactions, which previously yielded meaningless infinities, can be made finite in the same sense as in quantum electrodynamics. Earlier,

Glashow had proposed a specific model identical in form to the Weinberg-Salam theory without, however, explaining why the intermediary quanta of the weak interactions should be massive. A particular feature of this theory is that, in addition to the weak interactions mediated by charged $W^{\pm}$ bosons, there should be *neutral current* interactions transmitted by massive neutral Z bosons. The neutral current interactions of neutrinos were discovered in 1973, those of electrons in 1978, and of electrons with muons in 1981.

A difficulty in accepting the idea of weak neutral currents had been the absence of such interactions in strange particle decays. The problem was brilliantly solved by Glashow, Iliopoulos, and Maiani, who simultaneously explained the absence of $\Delta S = 2$ interactions by postulating the existence of a fourth, *charmed* quark which would be paired in weak interactions with the quark combination orthogonal to the one corresponding to Cabibbo's choice. Incorporation of this idea allows the Weinberg-Salam theory to be applied to hadrons as well as to leptons, and is now generally considered to provide the basic description of weak interactions. Glashow et al. were also able to deduce that charmed particles could not be too much more massive than "ordinary" hadrons, and the discovery of charmed particles as predicted, in 1974–76, must be regarded as one of the signal triumphs of theoretical physics. The intermediate bosons $W^{\pm}$ and Z^0 are yet to be detected. Their observation would be important discoveries in themselves. If they are found to have their predicted masses—slightly over 80 GeV/c^2 and 90 GeV/c^2, respectively—the confirmation of the Weinberg-Salam theory would be virtuall complete.*

The success achieved by the hypothesis that weak and electromagnetic interactions are related, naturally encouraged speculation about further unification of the forces of nature. A gauge theory called quantum chromodynamics, similar to the one used to describe electroweak interactions, is found to give an excellent qualitative account of strong iterations, although exact calculations cannot yet be performed for this theory. This led to the idea that both strong and electroweak interactions may be described by a gauge theory involving a larger symmetry group which encompasses both. The apparent difference in strength between the two classes of interactions must again be attributed to spontaneous symmetry breaking, which is expected to occur at a mass scale much higher than 100 GeV/c^2. Common to most such theories of grand unification is the idea that quarks and leptons may transform into each other just as one kind of lepton or quark transforms into another in weak interactions. If this is so, protons would no longer be absolutely stable as previously believed, but decay into leptons and radiation. The high mass scale of strong-electroweak symmetry breaking would assure that the X bosons which mediate such decays are extremely massive, causing the effective proton decay interaction to be exceedingly weak, consistent with the present experimental lower limit of 10^{30} years for the proton lifetime. Experiments to search for such extremely rare proton decays are currently in progress in many parts of the world.

As aspect of the weak interactions which we have not emphasized thus far is that they possess fewer space-time symmetries than the strong interactions. Until Lee and Yang pointed out the possibility in 1956, it was not generally suspected that any interaction could distinguish between right and left. It was then found that weak interactions disciminate between them very strongly; in the case of β-decay and related processes, this preference for a certain handedness could be attributed to the previously unknown fact that neutrinos are handed objects, a concept which can be given a relativistically invariant meaning only for massless particles. For other weak processes, the failure of reflection invariance remains totally unexplained; a new explanation is required even for neutrinos, which certain grand unified theories allow to have a small mass.

An answer to the question of how a left-handed coordinate system could be preferred over a right-handed one was offered by Landau's principle of symmetry under combined inversion CP. According to this principle, the apparent spatial asymmetry in a given experiment is predicted to be exactly the mirror image of that to be expected in an experiment performed with the corresponding antiparticles. This prediction has been verified in many experiments, and for several years all observations were in agreement with this hypothesis, until the discovery of K_L decays in 1964 proved that combined inversion CP could not be an exact symmetry. A consequence of the CPT theorem (see ANTIPARTICLES) is that, if CP-invariance fails, T-invariance must also fail; although this has not been directly observed in any phenomenon, there is indirect evidence from neutral K-meson decays that this is so. Many speculations have been advanced regarding the nature of the CP noninvariant interaction. All existing information is consistent with the hypothesis that CP noninvariance is confined to interactions much weaker than the usual weak interactions, which have observable consequences only in neutral K-meson decays, which are the only phenomena where CP noninvariance has been conclusively established. It is pos-

---

*Editor's note: The bosons W^+, W^-, and Z^0 were observed at CERN in 1938. Masses as determined experimentally matched those obtained from theory suprisingly well. See *Physics Today* **37**, S-30 (Jan. 1984).

sible that CP noninvariance may be associated with the existence of a third pair of quarks, composed of the b quark and its hypothetical partner. In that case, there is no chance of seeing possibly larger CP noninvariant effects outside the neutral K-meson system.

There is still much to learn about weak interactions.

P. K. KABIR

References

1. Fermi, E., "Elementary Particles," New Haven, Yale Univ. Press, 1951.
2. Marshak, R. E. Riazuddin, and Ryan, C. P., "The Theory of Weak Interactions in Particle Physics," New York, Wiley-Interscience, 1969.
3. Okun, L., "Leptons and Quarks," Amsterdam, North Holland, 1981.
4. Cleveland, B. T., Davis, R., and Rowley, J. K., in "Weak Interactions as Probes of Unification," (G. B. Collins et al., Eds.), AIP Conf. Proc. No. 72, p. 322, 1981.

Cross-references: ANTIPARTICLES, CONSERVATION LAWS AND SYMMETRY, CURRENT ALGEBRA, ELECTRON, ELECTROWEAK THEORY, ELEMENTARY PARTICLES, GAUGE THEORIES, NEUTRINO, NEUTRON, PROTON, RADIOACTIVITY, STRONG INTERACTIONS.

WORK, POWER, AND ENERGY

In the strict physical sense, work is performed only when a force is exerted on a body while the body moves at the same time in such a way that the force has a component in the direction of motion. In the simplest case where a constant force is applied in the same direction as the motion, it may be stated that work equals force multiplied by distance, or

$$W = Fs.$$

An example could be the raising of mass m from elevation h_1 to elevation h_2 in a constant gravitational field where the acceleration of gravity is g. The work performed would be

$$W = mg(h_2 - h_1).$$

To generalize for more complex situations, the amount of work done during motion from point "a" to point "b" can be expressed by:

$$W = \int_a^b F \cos \theta \, ds$$

where F is the total force exerted and θ is the angle between the direction of F and the direction of the elemental displacement ds. In the cgs system the unit of work is the dyne-centimeter

or erg, in the mks system the newton-meter or joule, and in the English system the unit of work is the foot-pound.

In rotational motion, the definition just given can be exactly applied, but it is often convenient to express the force as a torque and the motion as an angular displacement. The work done will be:

$$W = \int_a^b \tau \cos \theta \, d\omega$$

where in this case θ is always the angle between the torque τ expressed as a vector quantity and the elemental angular motion $d\omega$, also expressed as a vector. The units of work performed in angular motion will, of course, be the same as in the case of linear motion. Notice that the definition of work involves no time element.

Power is defined as the rate at which work is performed. The average power accomplished by an agent during a given period of time is equal to the total work performed by the agent during the period, divided by the length of the time interval. The instantaneous power can be expressed simply as

$$P = \frac{dW}{dt}.$$

In the cgs system, power has the units of ergs per second; in the mks system, units of joules per second (or watts), and in the English system, units of foot-pounds per second. A common engineering unit is the horsepower, defined as 550 foot-pounds per second, or 33 000 foot-pounds per minute.

Energy may be defined as the capacity for performing work. This definition may be better understood when stated as: the energy is that which diminishes when work is done by an amount equal to the work so done. The units of energy are identical with the units of work, previously given.

Energy can exist in a variety of forms, some more recognizable as being capable of performing work than others. Forms in which the energy is not dependent upon mechanical motion are generally referred to as forms of potential energy. The most common example in this category is gravitational potential energy. A body near the earth's surface undergoes a change in potential energy when it is changed in elevation, the amount being equal to the product of the weight of the body and the change in elevation.

Potential energy may also be stored in an elastic body such as a spring or a container of compressed gas. It may exist in the form of chemical potential energy, as measured by the amount of energy made available when given substances react chemically. Potential energy also exists in the nucleii of atoms and can be released by certain nuclear rearrangements.

Kinetic energy is the energy associated with mechanical motion of bodies. It is quantitatively equal to $\frac{1}{2}mv^2$ where m is the mass of a body moving with velocity v. In the case of rotational motion, the kinetic energy is more easily calculated using the expression $\frac{1}{2}I\omega^2$, where I is the moment of inertia of the body about its axis of rotation and ω is the angular velocity. Kinetic energy, like all forms of energy, is a scalar quantity. In a system made up of an assembly of particles, such as a given volume of gas, the total kinetic energy is equal to the sum of the kinetic energies of all the molecules contained in the volume. Calculation of the energy of such systems is very successfully treated theoretically on the basis of statistical averages.

Within a given system, energy may be transformed back and forth from one form to another, without changing the total energy in the system. A simple example is the pendulum, in which the energy is periodically converted from gravitational potential energy to kinetic energy and then back to gravitational potential energy. A similar situation, but on a submicroscopic scale, occurs in solid materials where the atoms are vibrating under the effect of interatomic rather than gravitational forces. As the temperature of a solid increases, the energy associated with the vibration of the atoms increases.

The example just given illustrates how, on a macroscopic scale, heat can be considered a form of energy. Regardless of the material involved, any amount of heat absorbed or released may be quantitatively expressed as an amount of energy. A gram-calorie of heat is equivalent to 4.19 joules, and in the English system a British thermal unit (Btu) is equivalent to 778 foot-pounds.

Potential energy is also present in electric and magnetic fields. The energy available in a region of electric field is equal to $E^2/8\pi$ per unit volume, where E is the electric field strength. Within a given volume, the total energy represented by the electric field is the integral of $E^2/8\pi$ over the volume. Similarly, the energy represented by a magnetic field may be independently calculated by integrating $H^2/8\pi$ over any given volume, where H represents the magnetic field strength. In the case of an electrically charged capacitor, the total energy in the electric field, and hence in the capacitor, can be shown to be $\frac{1}{2}CV^2$. Here C is the capacitance and V the electric potential to which the capacitor is charged. Similarly the total energy in the magnetic field associated with an inductor carrying an electric current is $\frac{1}{2}LI^2$, where L is the inductance and I the current.

Electromagnetic radiation is a combination of rapidly alternating electric and magnetic fields. Energy is associated with these fields and is exchanged between the electric and magnetic forms. This energy in a quantum of electromagnetic radiation, such as light or gamma radiation, can be represented in different ways, but is commonly expressed as $E = h\nu$. Here h is Planck's Constant and ν is the frequency of the radiation.

For particulate radiation or any very rapidly moving mass, the expression previously given for the kinetic energy, $\frac{1}{2}mv^2$, is not accurate when the velocity approaches that of the velocity of light. The theory of relativity requires a correction be made, and the exact kinetic energy, T, may be calculated in terms of mass, m_0, of the body measured when at rest, and the speed of light in vacuum, c, as follows:

$$T = m_0 c^2 \left[\left(1 - \frac{v^2}{c^2} \right)^{-1/2} - 1 \right]$$

Notice that this formula may also be written:

$$T = (m - m_0) c^2$$

where m is the variable quantity $m_0 \left(1 - \dfrac{v^2}{c^2} \right)^{-1/2}$.

This quantity represents the mass of the body, reducing to m_0 when v is zero, and approaching infinity as v approaches the speed of light.

This example illustrates another result of the theory of relativity, namely, the equivalence of mass and energy. Rewriting the last equation,

$$m = m_0 + \frac{T}{c^2}$$

The mass is seen to increase linearly with the kinetic energy of the body, the proportionality factor being c^2. Indeed, even the rest mass, m_0, represents an amount of energy equal to $m_0 c^2$. The total energy of a body of mass, m, can be generally given as:

$$E = mc^2 \text{ or } E = m_0 c^2 + T$$

In dealing with radiation, whether particulate or electromagnetic, it is customary to express energy in terms of electron volts. An electron volt is equal to the amount of work done when an electron moves through an electric field produced by a potential difference of one volt. One electron volt is equivalent to 1.60×10^{-12} erg. When charged particles such as electrons or protons are given kinetic energy by an accelerator, their kinetic energy is stated in terms of electron volts (eV) or million electron volts (MeV). In addition, any such particle will have a rest mass which can also be specified as an energy. For an electron,

$$m_0 = 9.11 \times 10^{-28} \text{ gram}$$

which is equivalent to 8.18×10^{-7} erg or 0.515 MeV.

A basic principle of physics is known as conservation of energy. This principle requires that within any closed system, the total energy must remain a constant. Energy can be changed from one form to another; but the total, so long as no energy is added to or lost from the system, must be constant. In the case of the swinging pendulum, decreases in kinetic energy reappear as increases in potential energy and vice versa. Eventually, of course, the pendulum will stop due to the effect of frictional forces. At that time, all of the kinetic and gravitational potential energy will have been converted to heat.

In another example involving a radioactive atom, the total energy represented by the atom and the emitted radiation must be constant. If a gamma ray is emitted, the rest mass of the atom will be decreased by an amount equivalent to the sum of the energy of the gamma ray and the recoil kinetic energy of the atom, which will be very small. If a beta ray is emitted, the rest mass of the atom will be decreased by an amount equivalent to the sum of the rest mass of the emitted electron, the kinetic energy of the electron, and the recoil kinetic energy of the atom.

WILLIAM E. PARKINS

Cross-references: CONSERVATION LAWS AND SYMMETRY; DYNAMICS; ENERGY LEVELS; ENERGY STORAGE, ELECTROCHEMICAL; ENERGY STORAGE, THERMAL-MECHANICAL; IMPULSE AND MOMENTUM; MECHANICS; RELATIVITY; ROTATION–CURVILINEAR MOTION; STATICS.

X

X-RAY DIFFRACTION

X-rays are electromagnetic radiations, and, like visible light, they can be diffracted (see OPTICS, PHYSICAL and DIFFRACTION). If a diffraction grating is used, the situation is as shown in Fig. 1. The points, B, B', B'', ··· along OX represent the lines of the grating seen end-on, and ABC, A'B'C', ··· are typical incident and scattered rays of a parallel incident beam. If D' is the foot of the perpendicular dropped from B on to A'B' and D is the foot of the perpendicular from B' to BC, the extra distance traveled by the ray A'B'C' is clearly D'B'−DB. This path difference can be expressed in terms of the grating space a, the angle of incidence ϕ_1, and the angle of scattering ϕ_2. From the triangle BB'D', D'B' = $a \cos \phi_1$, and from the triangle BB'D, DB = $-a \cos \phi_2$. If there is to be appreciable intensity diffracted in the direction BC, the path difference must be an integral multiple of the wavelength λ, say $h\lambda$. Then

$$h\lambda = a(\cos \phi_1 + \cos \phi_2) \qquad (1)$$

In order to obtain any appreciable scattering of x-rays from a grating it is necessary in practice to use very small angles of incidence, so that total external reflexion occurs. (The refractive index of matter for x-rays is very slightly less than unity, so that under suitable conditions they exhibit total external reflexion, whereas light exhibits total internal reflexion.) The points B, B', ··· thus correspond to the centers of the smooth reflecting portions of the grating. From measurements of the angles of incidence and diffraction under such conditions, the absolute values of the x-ray wavelengths have been derived.

A geometrical representation of Eq. (1) is helpful in extending the theory of diffraction by a grating (an object with a one-dimensional variation of scattering power) to diffraction by a crystal (an object with a three-dimensional variation of scattering power). Equation (1) can be rewritten:

$$\frac{\cos \phi_1}{\lambda} + \frac{\cos \phi_2}{\lambda} = \frac{h}{a} \qquad (2)$$

In Fig. 2, OP is a line of length $1/\lambda$ drawn parallel to BA, so that the angle XOP = ϕ_1, and PQ is a line of the same length drawn parallel to BC, so that it makes an angle ϕ_2 with OX. The projection of OP on OX, OP', is clearly of length $\cos \phi_1/\lambda$, and the projection of PQ, P'Q', is of length $\cos \phi_2/\lambda$. If Eq. (2) is satisfied, the length of OQ' = OP' + P'Q' must be h/a. Clearly, fixing h is not sufficient to fix the angles ϕ_1 and ϕ_2. If we imagine that the lines XO, OP and PQ are jointed together at O and P, Q is free to slide up and down the line GQ'H within wide limits without affecting the equality of Eq. (2), and so giving considerable freedom to the directions of incident and diffracted rays having a constant path difference $h\lambda$. Different values of h lead to different lines GH, G'H', ··· spaced at integral multiples of the distance $1/a$ along OX. All possible mutual relationships between the incident and strongly diffracted rays are represented

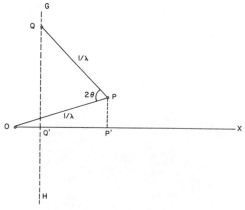

FIG. 2. Geometrical representation of the condition for strong reflection.

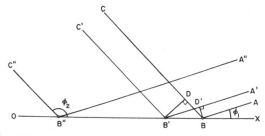

FIG. 1. Diffraction by a linear grating or line of scattering centres.

by those portions of the set of parallel lines of spacing $1/a$ that lie within a circle with radius $2/\lambda$ and center O.

The exterior angle between AB and BC is generally denoted by 2θ. From Fig. 1 it is clear that

$$2\theta = \phi_2 - \phi_1 \qquad (3)$$

and from Fig. 2

$$OQ = OP \sin\theta + PQ \sin\theta$$

$$= (2 \sin\theta)/\lambda \qquad (4)$$

With a little reinterpretation, Figs. 1 and 2 can be used to derive the condition for strong diffraction by a crystal. A crystal is a three-dimensionally periodic object, with a certain arrangement of atoms repeating indefinitely at intervals of a in one direction, b in another, and c in a third. The intervals a, b, c are not generally equal to each other, nor necessarily at right angles to each other. They define a parallelepiped called the *unit cell* of the crystal. For determining the geometrical conditions for the existence of a strong diffracted beam, the manner in which the atoms are arranged in the cell is not important, and for the moment we may think of them as forming a single scattering center at the corner of the cell. In Fig. 1, the points B, B′, $\cdots$ now represent a row of scattering centers parallel to a, and ABC, A′B′C′, $\cdots$ represent a sequence of incident and diffracted rays. The difference is that for a ruled grating one can expect constructive interference between the various scattered rays only if OX, AB, BC, A′B′, B′C′, $\cdots$ are coplanar, whereas an arrangement of atoms can scatter out of the plane of OX and AB, and hence BC need not lie in this plane. The condition for reinforcement, however, remains

$$D'B' - DB = h\lambda \qquad (5)$$

and its geometrical representation in Fig. 2 is changed only in that Q does not necessarily lie in the plane of OX and OP. With this relaxation, the condition for reinforcement is fulfilled if Q lies anywhere on a plane perpendicular to a and distant h/a from O. Different values of h lead to a set of parallel planes, and as far as the repetition of scattering centers in the direction of a is concerned, reinforcement will occur if Q lies anywhere on any of the planes within a sphere with center O and radius $2/\lambda$.

It is, however, necessary to satisfy similar conditions for the repetition of scattering centers in the directions of b and c. These conditions are exactly analogous: reinforcement for repetition along b will occur only if Q lies in one of a set of planes that are perpendicular to b and spaced $1/b$ apart, and reinforcement for repetition along c will occur only if Q lies in one of a set of planes perpendicular to c and spaced $1/c$ apart. There can be a strong beam diffracted by the crystal only if Q satisfies these three condi-

tions simultaneously. Two sets of planes intersect in a set of parallel lines, and the third set of planes will intersect these lines in a lattice of points. This is called the lattice reciprocal to the crystal lattice, or the reciprocal lattice for short. A strong diffracted beam can occur, then, only if Q coincides with one of the points of the reciprocal lattice, and only those points within a sphere of radius $2/\lambda$ are possible, since the maximum length of OQ is $2/\lambda$. This sphere is called the *limiting sphere*.

The repeat distances in the reciprocal lattice are usually called a^*, b^*, c^*. They have the same general directions as a, b, c, but are actually parallel to them only in those crystal systems that have their axes at right angles. There are a number of "reciprocal" relations between the two sets of axes, some obvious from the way in which the reciprocal lattice has been constructed, others obtainable only by analysis. For further details the article on CRYSTALLOGRAPHY should be consulted.

For a given wavelength λ, a given direction of incidence PO, and a fixed position of the crystal, the loci of possible positions of Q is a sphere with center P and radius $1/\lambda$. This sphere is called the *sphere of reflexion*, since strong diffraction can take place only if one of the points of the reciprocal lattice lies in or passes through its surface. With all parameters fixed, this is an unlikely coincidence, and in the practical study of x-ray diffraction, provision must be made for variation of either λ (the Laue method), or the orientation of the crystal (most types of single-crystal and crystal-powder cameras and diffractometers), or the direction of the incident x-rays (some special-purpose techniques). Space does not permit discussion of the practical details.

In the preceding treatment diffraction has been considered from the viewpoint of adding contributions scattered by centers arranged in a regular lattice. It can also be considered as reflexion from sets of parallel planes of atoms, the rays reflected by successive planes in a set being added. It can be shown that the connection between the two views is that (1) the line OQ joining the origin of the reciprocal lattice to the point hkl is perpendicular to the corresponding set of reflecting planes, and (2) its length is the reciprocal of their spacing d. From the second viewpoint, then, Eq. (4) may be rewritten

$$\lambda = 2d \sin\theta \qquad (6)$$

a relation known as *Bragg's law*.

The arrangement of atoms within the unit cell influences the intensity of the various orders of diffraction hkl, just as the shape of the ruling influences the intensities of the different orders produced by an optical diffraction grating. It is easy to see that x-rays diffracted by an atom at the position xa, yb, zc within the unit cell travel a shorter distance than those diffracted at the hypothetical scattering center at the origin, and that the corresponding phase difference is $2\pi(hx + ky + lz)$. If the cell contains n atoms whose scattering factors (the ratio of the actual scattered amplitude to the amplitude scattered by a free electron under the same conditions) are $f_1, f_2, \cdots, f_n$, the total amplitude scattered by the cell will thus have an in-phase component of

$$A = f_1 \cos 2\pi(hx_1 + ky_1 + lz_1) + \cdots$$
$$+ f_n \cos 2\pi(hx_n + ky_n + lz_n) \quad (7)$$

and an in-quadrature component of

$$B = f_1 \sin 2\pi(hx_1 + ky_1 + lz_1) + \cdots$$
$$+ f_n \sin 2\pi(hx_n + ky_n + lz_n) \quad (8)$$

the scattering by an atom at the origin of the cell being taken as the reference phase. The total intensity of scattering is thus proportional to

$$F^2 = A^2 + B^2 \quad (9)$$

The quantity F is called the *structure factor*; it is sometimes convenient to regard it as a complex quantity $F = A + iB$. It can vary in magnitude from zero to $f_1 + f_2 + \cdots + f_n$, with a root-mean-square value of $\sqrt{(f_1{}^2 + \cdots + f_n{}^2)}$. The total intensity of the order hkl is proportional to NF^2, where N is the number of unit cells in the crystal. The scatter of the actual values of the intensity above and below the mean value depends on the symmetry of the atomic arrangement. Although the relationship is statistical, a study of the scatter can often provide evidence about the *point group* of the crystal when other methods are not available. From Eq. (7) through (9) it is clear that for any hkl the actual intensity of diffraction depends on the atomic positions $x_1 y_1 z_1 \cdots x_n y_n z_n$. Measurement of the intensities of a sufficient number of diffraction maxima thus makes it possible to infer the atomic positions. Unfortunately there is no single procedure that can be guaranteed to lead from the measured intensities to the atomic coordinates, but there are many that are successful in appropriate cases: trial-and-error, isotypic (isomorphous) substitution, so-called "direct" methods. Direct methods make heavy demands on computing facilities, as they require statistical manipulation of a great body of data, but they succeed almost routinely for crystals whose unit cells contain up to about a hundred atoms (plus hydrogen, if the substance is organic).

Up to this point there has been a tacit assumption that the scattered intensity is a negligible fraction of that incident on the crystal. This assumption is justified for crystals that warrant the paradoxical description "ideally imperfect." In practice this covers most of the fine powders used in powder diffractometry and powder photography (see final paragraph below) and most of the reflections from organic crystals used in single-crystal investigations, but there are exceptions for inorganic crystals and for the stronger reflections from organic crystals. The observed effect is that the total intensity of the reflection is *less* than that calculated from (7)-(9) above, because the incident beam is weakened by the intensity diffracted out of it, and thus the parts of the crystal far from the source have less opportunity to diffract than parts close to the source. This effect is known as *secondary extinction*, and can be allowed for on a semiempirical basis for most crystals used in structure determination.

The treatment of x-ray diffraction outlined above is known as *kinematical theory*, and is adequate for most problems. It does not apply, however, for crystals that are approaching genuine perfection in atomic arrangement. Until recently such perfection was found only in rather rare mineral specimens, but it is now possible to grow practically perfect crystals of silicon and other substances in commercial quantities. In such crystals the incident beam is almost perfectly reflected once, the once-reflected beam is reflected back into parallelism with the incident beam, the twice-reflected beam is reflected back into parallelism with the once-reflected, and so on indefinitely. The parallel beams interfere with one another, and the mathematics, known as *dynamical theory*, becomes difficult. The most obvious result is that if the x-ray beam is incident at the correct angle it is almost totally reflected (there is a small reduction associated with absorption), but if the angle of incidence is wrong by more than a few seconds of arc the intensity of reflection is practically zero. One thus has the peculiar situation that a perfect crystal reflects perfectly while it is reflecting, but the total amount that it reflects when it is rotated through the reflecting position is very much less than that reflected by an ideally imperfect crystal of the same material. There are other peculiar phenomena associated with perfection, one of the most spectacular being transmission of the incident beam through a thickness of crystal that would completely absorb it if the angle of incidence were slightly altered.

At the opposite end of the scale there are crystals that are really imperfect, not merely ideally. The observed effect is that the intensity of reflection is appreciable over quite a range of Bragg angles; kinematical theory suffices to describe the phenomenon. In most cases [there are some exceptions, particularly for case (2) below], the maximum intensity of diffraction occurs when the point Q, defined above, coincides with one of the points hkl of the reciprocal lattice. As the point Q moves away from its ideal position, the intensity of diffraction does not drop instantaneously to zero, but falls off gradually in a manner depending on the size and perfection of the crystal. The study of the manner in which the falling off occurs is a fascinating exercise both experimentally and theoretically, but space permits only the quotation of a few qualitative results. The main varieties of imperfection considered are (1) the crystal is too small to give sharp diffraction maxima, (2) there are "mistakes" in the arrangement of atoms in the sequence of unit cells, and (3) the crystal is twisted, bent, or affected by dislocations. In case (1) the regions round each reciprocal-lattice point are identical except for variations in F, and extend in any direction to a distance that is roughly the reciprocal of the thickness of the crystal in the corresponding direction. In case (2) the size and shape of the regions may vary in a complex manner with hkl, but they show no general increase in size with increasing distance from the origin of reciprocal space. In case (3) the size of the regions shows a general increase in direct proportion to their distance from the origin of reciprocal space, as well as varying with hkl.

The intensity of many diffraction maxima may be recorded simultaneously on a photographic film with a suitable emulsion, and later measured with a densitometer, or individual diffraction maxima may be investigated one by one with a counter diffractometer. In spite of the complexity of the mechanism the latter is becoming the preferred method; various methods of automation, including on-line computer control, have removed the tedium associated with early diffractometric measurements. For determination of the atomic arrangement only the total ("integrated") intensity of each maximum is required, but for investigations of crystal imperfection much more detailed information about the variation of intensity with position in reciprocal space is needed. Often the material exhibiting imperfections is not available in crystals large enough to be examined individually, and the specimen has to take the form of an aggregate of many small crystallites in a more or less random orientation. In such cases measurement of the three-dimensional variation of intensity in reciprocal space is not possible, but by the use of a powder diffractometer the one-dimensional variation of intensity with Bragg angle [θ in Eq. (6) above] can be measured with considerable precision, and there is an extensive theory for the interpretation of such "line profiles."

A. J. C. WILSON

References

Amelinckx, S., Gevers, R. and van Landuyt, J., Eds., "Diffraction and Imaging Techniques in Material Science," Amsterdam, North-Holland, 1978. See especially the article by A. R. Lang, pp. 623–714.

Azároff, L. V., Ed., "X-ray Diffraction," New York, McGraw-Hill, 1974.

Giacovazzo, C., "Direct Methods in Crystallography," New York, Academic Press, 1980.

Guinier, A., "X-ray Diffraction in Crystals, Imperfect Crystals, and Amorphous Bodies," San Francisco, Freeman, 1963.

Hart, M., "Bragg Reflection X-ray Optics," *Reports on Progress in Physics* 34, 435–490 (1971).

Hosemann, R., and Bagchi, S. N., "Direct Analysis of Diffraction by Matter," Amsterdam, North Holland, 1962.

James, R. W., "The Optical Principles of the Diffraction of X-rays," London, Bell, 1948.

Ladd, M. F. C. and Palmer, R. A., "Structure Determination by X-ray Crystallography," New York, Plenum Press, 1977.

Richardson, M. F., Ramaseshan, S. and Wilson, A. J. C., Eds., "Crystallographic Statistics: Progress and Problems," Bangalore, Indian Academy of Science, 1982.

Wilson, A. J. C., "Elements of X-ray Crystallography," Reading, Mass., Addison-Wesley, 1970.

Wilson, A. J. C., "Mathematical Theory of X-ray Powder Diffractometry," Eindhoven, Philips Technical Library, 1963.

Zachariasen, W. H., "Theory of X-ray Diffraction in Crystals," New York, John Wiley & Sons, 1945.

Cross-references: CRYSTALLOGRAPHY; DIFFRACTION BY MATTER AND DIFFRACTION GRATINGS; ELECTRON DIFFRACTION; NEUTRON DIFFRACTION; OPTICS, PHYSICAL; PARACRYSTALS; X-RAYS.

X-RAYS

The portion of the electromagnetic spectrum known as x-rays has been recognized for nearly a century, yet radiation in this area continues to find new applications in many fields of science and technology. It seems unlikely that Wilhelm Conrad Roentgen could possibly have foreseen the impact that these rays would have when he discovered them in 1895 at the University of Wurzburg. He received the first Nobel prize (1901) in physics in recognition of the discovery. For seventeen years the exact nature of x-rays was obscure. They certainly deserved their designation of x- (or unknown) rays. In 1912, M. von Laue and his associates showed that x-rays could be diffracted from crystals which simulated three-dimensional gratings. This historic experiment not only proved the wavelike nature of x-rays, but also showed the regular three-dimensional arrangement of atoms in a crystal lattice. Two new fields of science were thus launched, namely, x-ray spectroscopy and x-ray crystallography. The crystal spectrometer was developed by W. L. Bragg; then Bragg, H. G. J. Moseley, M. Siegbahn, and others began the laborious task of cataloging lists of characteristic x-ray wavelengths. It was Moseley who recognized the simplicity and periodicity of the K series, for instance, whose main four lines varied step by step in going from one element to the next in the periodic table. The power of the technique is illustrated by the fact that spectral gaps were noted which indicated that certain elements were missing from the then known elemental scheme. In each case the missing elements were found and their x-ray spectra fell into the predicted places.

X-rays may be produced by a variety of methods, but the general method employed by Roentgen, that is, using electrons, continues to be the most popular. Roentgen operated his discharge tube at high voltage and with sufficient gas pressure to allow the cathode rays (electrons) to travel through the tube and collide with the walls to produce x-rays. Today, most x-ray tubes use a hot filament with electrons produced by the Edison effect. The fast-moving electrons convert their energy to an x-ray photon according to the law,

$$h\nu = E_1 - E_2$$

where E_1 and E_2 are the initial and final energies of the electron, h is Planck's constant, and ν is the frequency of the x-ray photon created. The radiation thus produced consists of a continuum of radiations, corresponding to the many possible energy losses that the electrons can undergo on striking the target of the x-ray tube. Such a continuum is represented in Fig. 1. When the electron loses all of its energy in a single collision, then $E_2 = 0$ and the x-ray photon of the highest energy is produced. This energy corresponds to the shortest possible wavelength or λ_{min}. Upon introduction of appropriate constants, λ_{min} in angstroms may be approximated by

$$\lambda_{min} = \frac{12{,}400}{E \text{ (volts)}}$$

where E is the x-ray tube voltage. Ten angstroms equal one nanometer. Therefore, if an x-ray tube is operated at 40 kV, then the shortest possible wavelength λ_{min} will be about 0.3 Å. The peak of the "white" or continuous radiation will be about $\frac{3}{2}$ times λ_{min}.

Superimposed on the continuum is a simple line and band spectrum characteristic of each element such as that shown in Fig. 1. These spectral features are produced by ejection of an electron in an inner energy level followed by the filling of this vacancy by an electron from a higher energy state. The energy of the x-ray quantum is the difference between the energies of the two levels involved. The x-ray series are named according to the final energy state involved in the transition with the familiar notations K, L, M, N, etc., as devised by Barkla in 1905 to emphasize the analogy with optical spectral series. It is not easy to place exact limits on the position of x-rays in the electromagnetic spectrum, since they overlap with γ-rays on the short-wavelength end and with vacuum ultraviolet radiation on the long-wavelength end. Generally, however, they are classified as ultrahard (<0.1 Å), hard (0.1–1.0 Å), soft (1–10 Å) and ultrasoft (>10 Å).

As the wavelength of x-rays is increased (or as they become "softer"), they are as a rule more easily absorbed by all materials. The decrease in intensity of a beam of x-rays passing through matter follows the law

$$I = I_0 e^{-\mu x}$$

where I_0 is the incident intensity, I the reduced intensity after a thickness x (cm) of homogeneous matter has been transversed, and μ the linear absorption or attenuation coefficient for the materials traversed.

In addition to this general attenuation in matter, there occur sharp absorption discontinuities in the x-ray spectrum corresponding to the energy necessary to just eject an electron from a given level, as shown in Fig. 2, where the structure near the K edge is shown for metallic iron. The large discontinuity suggests the use of metal foils to act as filters for various purposes. The edge is designated as K, L_I, L_{II}, L_{III}, etc., which indicates where the initial vacancy has occurred. The emission lines of a given element fall at longer wavelengths compared to the corresponding absorption edge. The absorption edge position and the fine

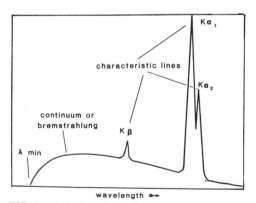

FIG. 1. Typical x-ray spectrum, showing continuum and characteristic radiation.

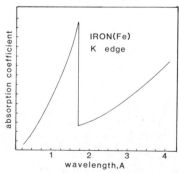

FIG. 2. Plot of absorption coefficient of iron against wavelength, showing characteristic K edge at 1.74 A.

structure on the edge depend strongly on the state of chemical combination of the emitting element, and detailed study of edge structure provides a powerful tool for the determination of chemical bonding.

The major uses of x-rays are in the areas of x-ray emission, x-ray diffraction, and x-ray absorption (to include interaction, such as ionization, with matter). X-ray emission analysis is a precision technique for elemental analysis which uses a high-intensity beam of x-rays from a primary source to excite characteristic x-rays in a sample. These x-rays are dispersed by a crystal spectrometer and detected by a scintillation or proportional detector which measures the angle at which the x-rays are diffracted by the analyzing crystal. The familiar Bragg equation

$$n\lambda = 2d \sin \theta$$

is used to deduce the wavelength λ of the secondary x-rays from the sample. θ is the angle that the x-ray beam makes with the crystal and is measured by the spectrometer, d is the lattice or interplanar spacing of the single-crystal analyzer, and n is the order of diffraction. This technique, commonly called *x-ray fluorescence*, is generally applicable to qualitative and quantitative analysis of all elements higher than atomic number 11 in the periodic table. With special instrumentation, the technique can be extended to low-atomic-number elements such as carbon and oxygen.

An extraordinary development in x-ray emission analysis has occurred with the development of the *electron microbeam probe* which collimates a beam of electrons to produce characteristic x-rays from a total sample volume of about one cubic micrometer. In addition to elemental analysis, the method provides a map of elemental distribution on a highly magnified image of the specimen. Scanning spectrometers and nondispersive Si (Li) detectors on scanning electron microscopes allow elemental distribution mapping in addition to the extremely high-resolution scanning pictures which are obtained from low-energy secondary electron emission from the sample surface.

An important new development in x-ray emission spectroscopy is the use of the fine features on x-ray lines and bands to deduce chemical bonding and the effects of chemical combination. Another significant use of monochromatic characteristic x-rays is to produce photoelectrons in the newly emerging technique known as ESCA (electron spectroscopy for chemical analysis), pioneered by Kai Siegbahn at Uppsala. The development of this technique, which is so closely related to x-ray spectroscopy, recently earned Siegbahn a Nobel prize in physics. This honor was especially fitting because his father was also honored for early pioneering work in x-ray spectroscopy.

X-ray diffraction is a widely used technique for characterization of solid crystalline materials in single-crystal analysis, powder diffraction analysis, and texture or topographical analysis. The same Bragg law mentioned earlier for x-ray emission analysis also applies to diffraction analysis, since it is a diffraction technique which is used to analyze or disperse the x-rays in the x-ray emission scheme. In x-ray diffraction, however, the wavelength λ is known, the angle θ (or 2θ) is measured, and the d spacing is calculated. Each crystalline material produces a diffraction pattern unlike that of any other material, and this pattern identifies each material unambiguously. An extensive file of diffraction patterns is provided by the American Society for Testing and Materials, which permits quick identification of crystalline materials. The reader is referred to the articles in this volume on X-RAY DIFFRACTION and DIFFRACTION BY MATTER AND DIFFRACTION GRATINGS for further details of the theory of diffraction phenomena, and to CRYSTALLOGRAPHY for applications to single-crystal analysis.

Perhaps the most familiar use of x-rays concerns the absorption (attenuation) and interaction with matter of an x-ray beam, since these phenomena provide the basis for radiography and uses in medical technology. *Radiography* is the registration on film or a fluorescent screen (fluoroscopy) of the differential absorption of a beam passing through a heterogeneous sample. This technique is used in medical diagnosis and to determine the soundness and perfection of manufactured products such as castings and forgings. An important development has been in cineradiography, where extremely high fluences of x-rays are pulsed to give sequential radiographs of transient phenomena lasting 10^{-9} sec or less. Microradiographic techniques are used to obtain magnifications of $100\times$ to $200\times$ of radiographs resulting from passage of monochromatic x-rays through a sample which is in direct contact with the recording film. Magnification is then realized by photographic enlargement. An improvement over contact microradiography is point projection radiography, where a fine point source of x-rays registers the sharp absorption discontinuities from a sample which is placed at varying distances from the film to achieve different sharp magnifications of the image. A further use of the absorption phenomenon is in absorptiometry, where measurement is made of the attenuation of a beam passing through a sample to determine, for instance, density, porosity, and coating thickness. Chemical analysis is also achieved by measuring characteristic absorption edges with a crystal spectrometer. Fine structure on these edges may also be used in certain cases to deduce bonding and chemical combination.

The areas of radiation chemistry, physics, biology, genetics, therapy, and protection all depend on the effects and changes produced when x-ray quanta are absorbed in matter. X-rays are, of course, ionizing radiations, and induce chemical reactions by formation of intermediate species, free radicals, and ions. These reactions may be used profitably for dosimetry, to produce mutations, and to irradiate selectively cancerous tissue, among many of the possible uses.

Because of space limitations, this review only sketchily refers to some of the numerous uses of x-rays. For further information it is suggested that the reader consult the references.

WILLIAM L. BAUN

References

Compton, A. H., and Allison, S. K., "X-Rays in Theory and Experiment," 2nd Edition, New York, Van Nostrand Reinhold, 1935.

Azaroff, Leonid, "X-Ray Diffraction," Vol. I, and "X-Ray Spectroscopy," Vol. II, New York, McGraw-Hill Book Co., Inc., 1974.

Kaelble, E. F., "Handbook of X-Rays," New York, McGraw-Hill Book Co., Inc., 1974.

Clark, G. L., "The Encyclopedia of X-Rays and Gamma Rays," New York, Van Nostrand Reinhold, 1963.

Flugge, S., "Encyclopedia of Physics, X-Rays," Vol. XXX, Berlin, Springer-Verlag, 1957.

Clark, G. L., "Applied X-Rays," 4th Edition, New York, McGraw-Hill Book Co., Inc., 1955.

Cross-references: ATOMIC SPECTRA, BIOMEDICAL INSTRUMENTATION, BIOPHYSICS, BREMS-STRAHLUNG, COMPTON EFFECT, CRYSTALLOG-RAPHY, DIFFRACTION BY MATTER AND DIF-FRACTION GRATINGS, ELECTRON, MEDICAL PHYSICS, X-RAY DIFFRACTION.

ZEEMAN AND STARK EFFECTS

Introduction The Zeeman and Stark Effects refer, respectively, to effects of external magnetic and external electric fields on the structure of atoms or molecules. In most cases the effects are observed through the modification of spectral features such as strength, polarization, width and position of emission or absorption lines. The Zeeman effect is named after Pieter Zeeman, who in 1896 observed a magnetic-field-induced broadening of the D emission lines in sodium vapor. (Zeeman's observed broadening was later found to be a small splitting.) The Stark effect is named after Johannes Stark, who in 1913 observed the electric-field-induced splitting of the Balmer transitions in atomic hydrogen.

Both the Stark and Zeeman effects have received extensive study since their discoveries. During the period of the development of the early quantum theories the effects played an important role by helping to stimulate the development of new ideas and computational techniques. In modern times the effects have

enjoyed continued interest due to newly discovered aspects of the intermediate and strong field problems. To a degree, much of the recent interest is also connected with the technological developments of tunable lasers and digital computers.

This article will examine the effects of external magnetic and electric fields on hydrogen and hydrogenlike atoms. The discussion has been limited to such atoms for the sake of simplicity but not at the cost of completeness. All the essential features of external field effects are brought out by the discussion of this simple example. It will be shown that the Zeeman and Stark effects are traceable to the interaction between the magnetic and electric dipole moments of atoms and the external fields. Dipole moments in atoms occur naturally or are induced by external fields.

Zeeman Effect *Introduction.* The Zeeman effect is the result of the interaction between the magnetic moment of the atom and the external magnetic field. The leading term to the interaction energy is,

$$W = -\boldsymbol{\mu} \cdot \mathbf{B}, \qquad (1)$$

where $\boldsymbol{\mu}$ is the magnetic dipole moment and $\mathbf{B}$ is the magnetic field. Magnetic dipole moments in atoms arise either from internal currents due to electron orbital motion, or from the natural (intrinsic) magnetism associated with the electron and the atomic nucleus. In the following discussions three well known cases of the Zeeman effect are considered: the normal effect, the anomalous effect, and the diamagnetic effect. (The terms "normal" and "anomalous" were coined before the Zeeman effect was well understood and unfortunately are misleading. The normal effect is actually a special case of the anomalous effect.) The magnetic moment associated with each case has a unique origin. The normal effect is caused by the interaction with the orbital magnetic moment. The anomalous effect is caused by the interaction with the combined orbital and intrinsic magnetic moments. The diamagnetic effect is caused by the interaction with the field-induced magnetic moment.

Normal Zeeman Effect. Figure 1 displays the simplest example of the normal Zeeman effect, that is, the effect on the s–p transition. (Spectroscopic notation is used to label quantum states. The letters $s, p, d, f, \ldots$ designate the orbital angular momentum, l, corresponding to values of $0, \hbar, 2\hbar, 3\hbar, \ldots, 2\pi\hbar$ is Planck's constant.) The p level splits into three equally spaced sublevels under the influence of the external field. The central p sublevel and the s level do not shift in response to the field. The energy spacings between the p sublevels increase linearly with the applied field. (An observation that applies to all normal-effect transitions is that the magnitude of the rate of separation between adjacent sublevels is a constant that is the same for all members of a Rydberg series. Thus, the

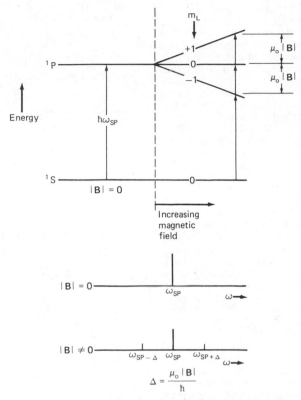

FIG. 1. Normal Zeeman effect for ^1S-^1P transition. Above is energy level diagram; below is spectrum with and without an external field.

appearance of the effect does not depend upon the principal quantum numbers of the states in question. This last observation is Preston's rule.) The spectrum, which is due to transitions between the s and p states, appears as a triplet as indicated in the lower portion of Figure 1. (The normal-effect triplet for the s-p transition is named after Lorentz, who described the effect in terms of his theory of electrons. Lorentz's explanation helped to support the notion of orbiting electrons in atoms.)

As indicated in the introduction, the normal effect is the result of the interaction between the magnetic moment due to the orbiting electron and the external field. From electromagnetic theory, the magnetic moment due to an electron circulating in a closed loop and having angular momentum l is,

$$\mu_l = \frac{-e}{2m_e C}\, l, \qquad (2)$$

where $-e$ is the electron charge, m_e is the electron mass, and c is the speed of light. (Unless otherwise noted all quantities are expressed in cgs units.) The interaction energy follows from

Eq. (1),

$$W = \frac{e}{2m_e C}\, l \cdot B. \qquad (3)$$

According to quantum mechanics the projection of l about an axis, here the magnetic field axis, assumes discrete values $m_l\hbar$, where m_l is an integer that ranges between $+l$ and $-l$. The expression for the interaction energy is, thus,

$$W = m_l \mu_0\, |B|, \qquad (4)$$

where μ_0 is the Bohr magneton constant defined by,

$$\mu_0 = \frac{e\hbar}{2m_e c} \simeq 1.4 \times 10^6 \text{ Hz/Gauss.}$$

The triplet spectrum in Figure 1 is thus completely described given that m_l for the p state has values $+1$, 0, and -1, while m_l for the s state has the value 0.

It happens that the normal effect does not actually occur for ordinary one-electron atoms because of the complication of electron spin.

The discussion above is thus strictly correct only if one assumes that the electron has no intrinsic magnetic moment of its own. The triplet does occur, however, in certain many-electron atoms (e.g., helium) where because of pairing of spins, the effects of intrinsic moments can be ignored.

Anomalous Zeeman Effect. Figure 2 diagrams examples of the anomalous Zeeman effect on a number of different atomic transitions. (One-electron atoms fall into the doublet category.) It is evident that the structure of the anomalous effect is more complicated than that of the normal effect. In addition, the field dependence is also more complicated. In weak fields, the levels separate linearly with the strength of the applied field as they did for the normal effect; however, the rates of separation of levels are different than those observed for the normal effect. In moderate fields, the levels shift in a complicated way that is not easily described by any simple power law. In strong fields, the levels again shift in proportion to the field strength but this time with rates of separation that are the same as those of the normal effect.

The anomalous effect was a mystery until 1925 when S. Goudsmit and G. Uhlenbeck introduced the concept of electron spin. Electron spin was conceived to explain why the fine structure separation of the ^2P levels of alkali metal atoms were so much larger than the corresponding levels of hydrogen. Goudsmit and Uhlenbeck suggested that electrons have both an intrinsic angular momentum and an intrinsic magnetic moment. Following this assumption, the fine structure splitting was shown to be the result of the magnetic interaction between the intrinsic magnetic moment of the electron and an internal magnetic field produced by the electron's orbital motion. This effect is referred to as the spin-orbit interaction. (Note that the spin-orbit effect in many-electron atoms is much more difficult to model because one must consider the magnetic couplings of all of the electrons with each other. These include, for example, the interaction between the magnetic field produced by the orbit of one electron with the magnetic moment of another electron. The problem is sufficiently complex that even sophisticated modern techniques have failed to reliably calculate the absolute magnitude and, in certain extreme cases, the sign of the spin-orbit interaction for the bulk of the atoms in the periodic table.)

Given the postulate of spin to explain fine structure, it became evident that spin could also be used to explain the anomalous effect. The only way that it was possible to reconcile both effects, however, was to assume that the maximum projection of the intrinsic angular momentum of an electron was $\hbar/2$ and that the intrinsic electron magnetic moment was,

$$\boldsymbol{\mu}_s = g_e \, \frac{-e}{2m_e C} \, \boldsymbol{s} \qquad (5)$$

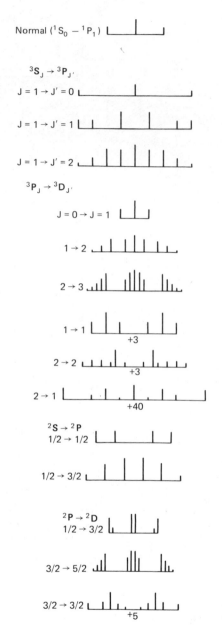

FIG. 2. Anomalous Zeeman spectra for numerous transitions. The normal Zeeman spectrum is shown for comparison. (After Condon and Shortley, 1971.)

where $\boldsymbol{s}$ is the electron spin angular momentum and g_e is an arbitrary factor added to preserve the form of the classical relationship between magnetic moment and angular momentum [Eq. (2)]. Early agreement with experiment was obtained by guessing that $g_e = 2$. (The Dirac equation, which combines quantum mechanics and relativity, justifies the ersatz of Eq. (5) and predicts that g_e is exactly two. It is now known that g_e has a value that is a fraction of a percent larger than two. One of the important successes of modern quantum electrodynamic field theory, which treats both particles and fields quantum mechanically, is the correct prediction of this constant.)

With the introduction of electron spin it is straightforward, although nontrivial, to explain shifts associated with the anomalous effect. The energy of the quantum state of a one-electron atom follows from the expression for the interaction energy between a magnetic moment and an external field,

$$W = - (\boldsymbol{\mu}_l + \boldsymbol{\mu}_s) \cdot B + a l \cdot s. \qquad (6)$$

The additional term here, $a l \cdot s$, represents the spin-orbit interaction. In the absence of external fields the spin and orbital angular momenta are coupled by the spin-orbit term such that the conserved quantity is j, which is defined as the sum of l and s, that is,

$$j \equiv l + s. \qquad (7)$$

The complete characterization of the anomalous effect is complicated because the external field interaction affects the internal coupling of l and s. In light of this, only the weak and strong field cases, which are the two cases most easily described, are to be considered. These two cases represent the extreme situations where l and s are either completely coupled or completely decoupled.

In weak fields, where the interaction terms in Eq. (6) due to the external field are small compared with the spin-orbit term, l and s are completely coupled. In this case the dominant contribution of the magnetic moment is due to the portion of the moment that is parallel to j. (Since $g_e \neq 1$, the magnetic moment does not point along j. As a consequence of the coupling, however, the contribution of the portion not parallel to j is generally negligible.) Therefore the net effective magnetic moment is

$$\boldsymbol{\mu}_j \simeq g_j \frac{-e}{2m_e C} j, \qquad (8)$$

where the expression for g_j,

$$g_j = 1 + \frac{j(j+1) + s(s+1) + l(l+1)}{2j(j+1)}, \qquad (9)$$

is derived from angular momentum coupling relations assuming $g_e = 2$. Here j, s, and l are the maximum projections of j, s, and l in units of $\hbar$. Eq. (9) was obtained first empirically by Landé in 1921 based on studies of the experimental data of Back. (Landé's observations extended the earlier rule of Runge, which expressed the fact that the weak-field anomalous effect splittings are related to the normal effect splittings by a rational fraction.) With the above

value for $\boldsymbol{\mu}_j$, the interaction energy for the weak-field anomalous effect according to Eq. (1) is,

$$W \simeq m_j g_j \mu_0 |B|, \qquad (10)$$

where m_j is an integer or half-integer ranging from $+j$ to $-j$ in unit steps. (Besides explaining this aspect of the anomalous effect, the comparison between the quantum mechanical derivation of g_j and the empirical results of Landé served to show that the norms (lengths) of the vectors l, s and j were $\sqrt{l(l+1)}\,\hbar, \sqrt{s(s+1)}\,\hbar$, and $\sqrt{j(j+1)}\,\hbar$ rather than $l\hbar$, $s\hbar$, and $j\hbar$ as predicted by the Bohr model.)

In strong fields, where the interaction terms in Eq. (6) due to the external field are large compared with the spin-orbit term, l and s are completely decoupled. The decoupling of l and s by the field is referred to as the *Paschen–Back effect*. As a result of the Paschen–Back decoupling, l and s are separately conserved so that the interactions associated with the orbital and spin moments contribute separately. Thus the expression for the strong-field anomalous effect is,

$$W \simeq (m_l + g_e m_s) \mu_0 |B|, \qquad (11)$$

where m_s is an integer or half-integer that ranges from $+s$ to $-s$ in unit steps.

The weak and strong field cases are best summarized by means of an example. In Figure 3 the anomalous Zeeman effect of a ^2P level is shown. In zero field, where l and s are coupled, there are two groupings of sublevels which correspond to states having $j = \frac{3}{2}$ and $j = \frac{1}{2}$. In weak fields, all of the j groupings split up according to their m_j values. The relative slopes of the levels are governed by the values of $g_j \mu_0$, which for $j = \frac{3}{2}$ is $4\mu_0/3$, and for $j = \frac{1}{2}$ is $2\mu_0/3$. In strong fields there are five groupings of sublevels which separate with relative slopes of μ_0. In intermediate fields the low and high field states connect smoothly with one another.

Up to this point the effect of the intrinsic magnetism of the atomic nucleus has been ignored. This is not a major omission, since nuclear moments are typically a few thousand times smaller than electronic moments. (This is due primarily to the differences in mass between nucleons and electrons.) The intrinsic nuclear moment enters into the anomalous effect in a way that is completely analogous to that in which the intrinsic electronic moment did. Here, instead of l and s coupling to form j, we have j and the nuclear spin angular momentum i coupling to form f. The coupling is the result of the interaction between the magnetic moment of the nucleus and the magnetic field produced both by the orbital motion and the intrinsic magnetism of the electrons. The so-called hyperfine splitting associated with this interaction leads to an energy term proportional to $i \cdot j$.

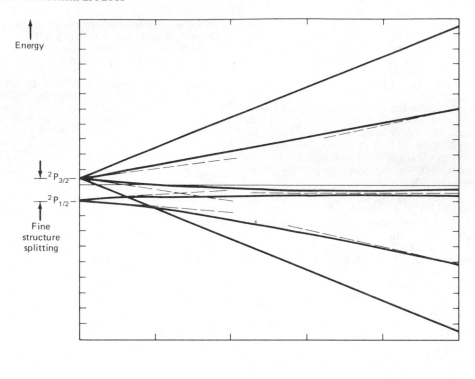

FIG. 3. The energy level diagram for ^2P level, showing anomalous Zeeman effect in weak to strong external fields. (After Condon and Shortley, 1971.)

As with the case involving electron spin, there are two field regimes in which nuclear spin effects are easily described, that is, the weak field regime and the strong field regime. In weak fields a Landé-like g-factor is introduced, g_f, and the levels split up according to values $g_f m_f$. Likewise, in strong fields i and j are decoupled such that the interactions associated with the nuclear moment and the electron moment contribute separately and the levels shift according to their values of $g_j m_j$ and $g_i m_i$. The strong field decoupling is the Back–Goudsmit effect, which is the analog of the Paschen–Back effect.

The complete picture of the anomalous effect now is that as the field increases from zero the contributions to the magnetic moment change from the case where l, s, and i, are completely coupled, to that where l and s are coupled but not i, to the final case where l, s, and i are completely decoupled.

Diamagnetic Effect. Even when all the angular momenta are zero so that there is no orbital or intrinsic magnetic moment, there is still a shift due to external fields. The shift is quadratic in its dependence on field strength and corresponds to an increase in internal energy. The effect is due to an induced moment and the increased energy indicates that the moment opposes the field. The opposition to the field means that this effect is inherently diamagnetic.

The diamagnetic shift can be understood simply as the result of the Larmor precession of an electron orbit when a magnetic field is applied. Here we consider the effect on an electron in a circular orbit. Larmor precession of the frame of reference due to the external field ($\omega_L = eB/2\ m_e C$) leads to a change in the orbital velocity of the electron when viewed in the laboratory frame. The changed orbital velocity results in a change in the net magnetic moment of an amount

$$\Delta \mu = -\frac{e^2 A}{4\pi m_e c^2}\ \mathbf{B}. \qquad (12)$$

where A is the orbital area projected on the plane perpendicular to $\mathbf{B}$. It is a bit surprising that the change in the moment is not dependent on the direction of electron circulation. (This fact is actually connected with Lenz's law of electromagnetism which expresses the tendency of electrical conductors to set up countercurrents that produce fields which oppose an externally applied magnetic field.) The energy shift due to diamagnetism follows from Eq. (12) by integrating the differential form of Eq (1) to give,

$$W = \frac{1}{2} \frac{e^2 A}{4\pi m_e c^2} |\mathbf{B}|^2. \quad (13)$$

The magnitude of the interaction energy associated with the diamagnetic effect can be estimated for an arbitrary state of principal quantum number n, by recognizing that the orbital area for a typical atom is $\pi(n^2 a_0)^2$, where a_0 is the Bohr radius, 0.5×10^{-8} cm. Thus according to Eq. (13) a frequency shift of $1.3 \times 10^{-4} n^4 |B|^2$ Hz is expected. (For 100 kG the $n = 1$ level is expected to shift 1.4 MHz. For the ground ^1S level of helium the observed shift is 2.4 MHz. Given that there are two $n = 1$ electrons in helium, 2.8 MHz might be expected. The agreement here is remarkably good.) The n^4 scaling of this effect means that for states of high n, the diamagnetic shift can be significant, especially in view of the fact that the Zeeman shifts due to the orbital and intrinsic moments are independent of n. The large relative size of the diamagnetic shift is one of the reasons that modern laser spectroscopists have studied this effect in the context of highly excited atoms.

Current Status. In recent years the Zeeman effect has received renewed interest due to the development of new experimental and theoretical techniques. One aspect of the problem that has received most of the recent attention is the case where the external fields cannot be considered as weak in comparison to the internal Coulomb field. This situation has been realized in the laboratory through the use of intense magnetic fields and high states of atomic excitation. It also occurs naturally in the atmospheres of certain collapsed stars for low states of excitation.

In a classic series of experiments on photoabsorption to the continuum (i.e., photoionization) by R. Garten and F. Thompkins, the transition from weak fields, where the electron is bound to the atom, to strong fields, where the electron is bound to the field, has been observed. The experimental evidence for the changeover from one regime to the other has been seen through the emergence of the so-called quasi-Landau resonances in the energy region near the threshold for photoionization. (A free electron in a magnetic field is bound in the plane perpendicular to the field axis by the Lorentz force. This binding gives rise to cyclotron orbits. Quantized cyclotron orbits are the basis for the discrete Landau levels.)

Considerable attention has recently been focused on measurements of the diamagnetic effect for highly excited atoms. One of the potentially important findings observed by Kleppner and co-workers are sharp crossings of sublevels. Such sharp crossings suggest the possibility of the existence of an unidentified symmetry of the magnetic field problem.

Spurred on by experiments such as the ones

mentioned above, theorists have been making marked progress towards developing a general theory of magnetic effects in fields of arbitrary strength. The problem, however, is still far from being solved.

Stark Effect *Introduction.* The Stark effect is due to the interaction between the electric moment of the atom and the external electric field. The leading term to the interaction energy is,

$$W = -\mathbf{p} \cdot \mathbf{E}, \quad (14)$$

where p is the electric dipole moment and E is the external field. Electric dipole moments in atoms arise as a consequence of the way that charge is distributed within atoms. In the following discussions two aspects of the Stark effect are to be considered: the linear effect and the quadratic effect. It will be shown that the linear effect is due to a dipole moment that arises from a naturally occurring nonsymmetric distribution of electron charge, while the quadratic effect is due to a dipole moment that is induced by the external field. For simplicity the effects of fine and hyperfine structure will be ignored.

Linear Stark Effect. Figure 4 diagrams the simplest case of the linear effect, that is, the transition from the ground level ($n = 1$) to the first excited level ($n = 2$) of atomic hydrogen. The excited level splits into three equally spaced sublevels under the influence of the external field. The spacings between sublevels increase linearly with the applied field. The central sublevels of $n = 2$ and the ground level do not shift in response to the field. (In the general case, an arbitrary n level splits into $2n - 1$ sublevels that separate in the field at a rate that scales with n. Thus the higher n states are more sensitive to the external field.)

As indicated above, the linear effect is the result of the interaction between the electric dipole moment due to the distribution of charge within the atom and the external field. The dipole moment can be understood by considering an eccentric elliptical orbit as shown in Figure 5. Here the nucleus is at one focus and the electron follows a Kepler orbit that sweeps out equal areas in equal times. It is evident that the further the electron is from the nucleus the slower it moves, and as a result the electron's position averaged over an orbit is not centered on the nucleus. The separation between positive and negative charge centers here necessarily leads to a dipole moment.

The size of the naturally occurring dipole moment corresponding to a highly eccentric orbit can be estimated, given that the mean radius of an atom in a state of principal quantum number n is on the order of $n^2 a_0$, where a_0 is the Bohr radius. Thus the dipole moment, which is the product of the charge and the charge separation, is $en^2 a_0$. This crude estimate is actually comparable with the observed moment

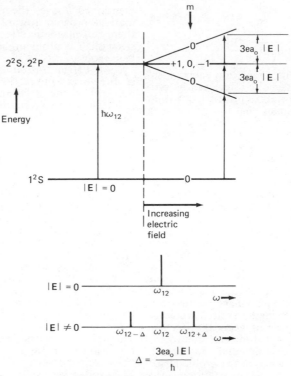

FIG. 4. Linear Stark effect for $n = 1$ to $n = 2$ transition. Above is energy level diagram; below is spectrum with and without external field.

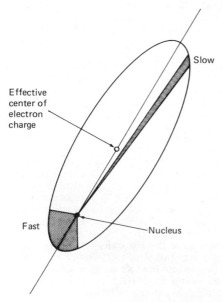

FIG. 5. Elliptical electron orbit. According to Kepler's law the electron is moving quickly when it is near the nucleus and slowly when it is far from the nucleus.

of the extreme-most Stark level, $3en(n-1)a_0/2$. The general result according to nonrelativistic quantum theory is that the linear shift is

$$W = \frac{3}{2}\, enka_0, \qquad (15)$$

where k is the electric quantum number, which ranges in value from $+(n - |m| - 1)$ to $-(n - |m| - 1)$ for all possible values of m, which ranges from $+(n-1)$ to $-(n-1)$.

The quantum mechanical predictions of charge distribution that correspond to the states described by Eq. (15) are most interesting. The charge distributions reveal parabolic symmetry. This symmetry is evident in Figure 6 from maps of the charge distributions corresponding to the eight $m = 0$ sublevels of the $n = 8$ level in atomic hydrogen. The ridges of charge are intersecting parabolas that open along and against the electric field axis. (This is most easily noted for the central states, $k = \pm 1$, which have the smallest dipole moments and correspond to nearly circular orbits.) The parabolic nature of the Stark problem is reflected in its usual analytical treatment, where the three-dimensional Schrödinger equation is separated into three one-dimensional equations in a parabolic coordinate system. (Incidentally, by comparison the magnetic field

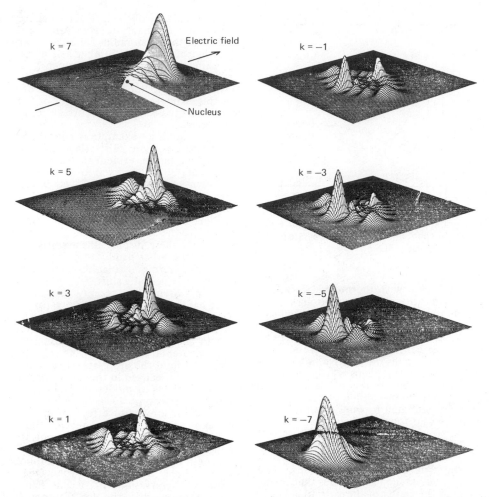

FIG. 6. Charge distribution for the eight $m_L = 0$ sublevels of the $n = 8$ level of hydrogen. The $k = +7$ and $k = -7$ sublevels are the extreme-most Stark states which display the largest dipole moments. The nucleus is in the center of the mesh. (Adapted from Kleppner, Littman, and Zimmerman, 1981.)

problem is more difficult to model analytically than the electric field problem because no similar separation of variables has ever been accomplished.)

Up to this point the discussion has concerned the linear effect for atomic hydrogen. More complex atoms also display linear effects, however, they can only do so when two or more states of different angular momentum have the same energy in zero field. Because of the high level of degeneracy in hydrogen this condition is met for all excited levels. In complex atoms, this condition is usually met only for the high angular momentum states. Isolated angular momentum states of a well defined parity cannot display linear shifts because of their inversion symmetry. (This is because the charge distribution must be the same when r is relaced by

$-r$ so that the center of electron charge for these states must coincide with the nucleus.)

The linear Stark effect has the distinction of being the first problem to which the well known perturbation method was applied. (Perturbation theory allows for the approximation of a solution of one problem in terms of an expansion of the solutions of another.) In this first application, the wavefunction for an atom in an electric field was expressed in terms of an expansion of wavefunctions for an atom in the absence of external fields. The linear shift was given by the first order term in the expansion for the energy.

Quadratic Stark Effect. For states with no linear shift, such as the ground state of hydrogen, there is still a shift due to external fields. The shift is quadratic in its dependence on field strength.

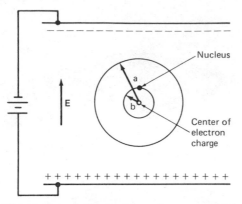

FIG. 7. The idealized effect of an external field on a uniform spherical charge distribution of radius *a*. The electron cloud and the nucleus separate by distance *b*. (After Purcell, 1965.)

As was the case with the quadratic shift in magnetic fields, the quadratic shift in electric fields is due to an induced moment. The induced moment can be understood by considering the external field effect on a uniform sphere of negative charge centered on a positive nucleus. The external field pulls the positive and negative charges in opposite directions which sets up an opposing force due to the displacement of charge centers. The magnitude of the displacement can be estimated by imposing the condition that the two forces be balanced. Figure 7 depicts the situation here. The charge outside of the inner sphere of radius b does not contribute to the net attractive force because the field inside a uniform spherical shell is zero. (This is the same phenomenon that causes the gravitational field at the center of the earth to be zero.) Thus, only the charge inside the inner sphere contributes. The amount of charge inside is $e(b^3/a^3)$, so that the force balancing condition is

$$e\,|\mathbf{E}| = \left(e\,\frac{b^3}{a^3}\right)\frac{e}{b^2}\,, \qquad (16)$$

or alternatively,

$$|\Delta\mathbf{p}| \equiv eb = a^3\,|\mathbf{E}|. \qquad (17)$$

Thus the strength of the induced dipole moment is roughly proportional to the field strength times the volume of the charge cloud. (The constant of proportionality is known as the *scalar polarizability*.) Given this expression for the induced moment, the interaction energy is obtained by integrating the differential form of Eq. (14), which gives

$$W = \tfrac{1}{2}\,a^3\,|\mathbf{E}|^2. \qquad (18)$$

The magnitude of the interaction energy associated with the quadratic effect can be estimated given that the mean radius for an arbitrary state of principal quantum number n is $n^2 a_0$. Thus, according to Eq. (18), a frequency shift of $11.2 n^6\,|E|^2$ Hz is expected, where $|E|$ is in units of stat-Volt/cm. For the ground state of hydrogen this estimate is low by a factor of 5, but for highly excited states it is correct to within a few percent.

At this point an example is appropriate. The linear and quadratic shifts that have been discussed thus far are evident in the laser spectroscopic data of sodium excited states near $n = 15$, as shown in Figure 8. The levels which enter from above and below here correspond to states from neighboring $n = 14$ and $n = 16$ levels. The grouping of states near the $15d$ level corresponds to the $n = 15$ high-angular-momentum states, and these display linear shifts. The $16p$ level, on the other hand, is far from any other states of differing angular momentum, and so it displays a quadratic shift.

Current Status. In recent years the Stark effect has received renewed interest due to the development of new experimental and theoretical techniques. Much of the recent interest has been in the situation where the external fields are comparable or stronger than the internal fields that hold the atom together. This condition has been realized in the laboratory through the use of fields of moderate strength and high states of excitation.

One of the interesting effects of intense fields on atoms that has received recent study is field-induced ionization of the neutral atom. Field ionization is interesting because the purely quantum mechanical effect of tunneling is known to play an important role. One of the significant observations of recent field ionization studies is that the ionization threshold for virtually all complex atoms occurs near the threshold energy that one would predict classically.

Another topic that has received recent attention is the lack of rigorous convergence of the Stark perturbation expansion. The convergence difficulty is connected to the fact that, in the presence of an electric field, bound states do not strictly exist because of a finite probability that any given state will spontaneously ionize via tunneling. (This problem is most important in strong fields.) Using a technique developed by Leibnitz and a digital computer, H. Silverstone and co-workers have calculated the perturbation expansion to 150th order and demonstrated explicitly the nature of the divergence. They also have shown how the convergence can be improved using the technique of Padé approximants.

Yet another topic that has drawn much attention is the observation of oscillations in the photoionization spectrum of an atom in an electric field. Here R. Freeman and co-workers have

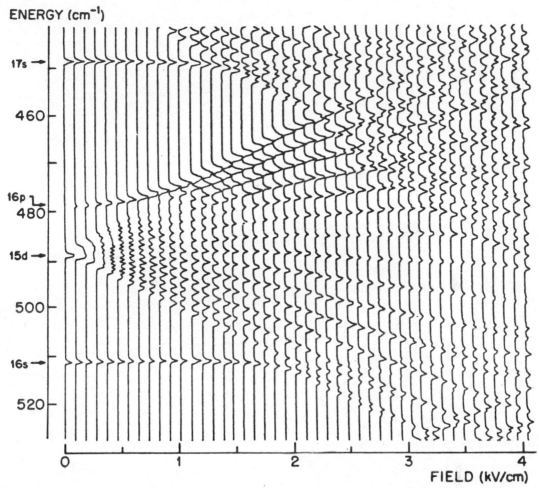

FIG. 8. Spectroscopic study of Stark effect on excited levels of sodium near $n = 15$. This is a composite map of data accumulated at many different values of applied field. [From M. G. Littman, M. L. Zimmerman, T. W. Ducas, R. R. Freeman, and D. Kleppner, *Phys. Rev. Lett.* **36**, 788 (1976).]

studied what appears to be the electrical analog of the quasi-Landau levels discussed previously (see the section of this article on Zeeman Effect—Current Status).

In spite of the fact that the Stark effect is an old topic, recent activity has shown that the area is indeed still vital.

MICHAEL G. LITTMAN

References

Bethe, H. A., and Salpeter, E. W., "Quantum Mechanics of One and Two Electron Atoms," Berlin, Springer-Verlag, 1957.

Ramsey, N. F., "Molecular Beams," Oxford, Clarendon Press, 1969.

Purcell, E. M., "Electricity and Magnetism," Berkeley Physics Course, Vol. 2, New York, McGraw-Hill, 1965.

Condon, E. V., and Shortley, G. H., "The Theory of Atomic Spectra, Cambridge, U.K., Cambridge Univ. Press, 1970.

Kleppner, D., Littman, M. G., and Zimmerman, M.L., "Highly Excited Atoms," *Sci. Amer.* **244**, 130 May (1981).

Cross-references: ATOMIC SPECTRA, DIPOLE MOMENTS, ELECTROMAGNETIC THEORY, FARADAY EFFECT, KERR EFFECTS, POLARIZED LIGHT, QUANTUM ELECTRODYNAMICS, QUANTUM THEORY, SPECTROSCOPY.

INDEX

Boldface numbers designate more important passages, and small capital letters are used for titles of articles in the book.